B K Ridley
April 1993

Handbook on Semiconductors

Volume 1

Handbook on Semiconductors

Completely revised and enlarged edition

Series editor

T. S. MOSS

Malvern, UK

Editor of volume 1 – P.T. Landsberg
Editor of volume 2 – M. Balkanski
Editor of volume 3 – S. Mahajan
Editor of volume 4 – C. Hilsum

NORTH-HOLLAND
AMSTERDAM · NEW YORK · LONDON · TOKYO

Volume 1

Basic Properties of Semiconductors

Volume editor

P. T. LANDSBERG

Faculty of Mathematical Studies
University of Southampton
Southampton SO9 5NH, UK

1992

NORTH-HOLLAND
AMSTERDAM · NEW YORK · LONDON · TOKYO

ELSEVIER SCIENCE PUBLISHERS B.V.
Sara Burgerhartstraat 25
P.O. Box 211, 1000 AE Amsterdam, the Netherlands

ISBN: 0 444 88855 1
0 444 89388 1 (Series)

This book is printed on acid-free paper

Printed in the Netherlands

General Preface

It is now ten years since publication of the first edition of the Handbook. During this time there have been tremendous advances and the subject continues to be of major scientific and commercial importance.

As before, the Handbook is in four volumes, each having a separate editor who is an internationally recognised authority in his field.

Many of the chapters necessarily have the same titles, but they have all been updated and extensively rewritten. In addition, there are many chapters with new titles covering areas of current scientific or technological interest. For example, there are chapters on preparation and properties of surfaces and low-dimensional structures, and various systems such as electron–hole liquids, heterojunctions and polymers. Superlattices are treated extensively. New phenomena include quantum Hall effect, localization, optical bistability, quantum confinement, non-linear effects and effects of strain or deformation.

There are several chapters on fabrication techniques and detailed properties of silicon – including five chapters on specific impurities – in view of its ever-increasing use in integrated circuits.

With its up-to-date treatment the new edition of the Handbook is an invaluable source of information for all workers in both semiconductor physics and device development.

T. S. Moss,
Malvern, UK

General Preface to the First Edition

The Handbook provides, in a set of four Volumes, comprehensive coverage of the whole field of semiconductor knowledge. Each Volume has a separate editor who is an internationally recognised leader in his field.

Volume 1 covers the basic aspects of crystal structure and energy bands of semiconductors. The second major area is that of transport properties, covering statistics, Boltzmann equation, scattering phenomena, effects of magnetic and electric fields, hot electron behaviour, quantum and polaronic transport, etc.

Volume 2 covers the optical properties of solids, both intrinsic and extrinsic, and includes the effects of temperature, pressure, electric field and doping. Excitons, phonons, polarons, etc., and free carrier effects are dealt with comprehensively. The second major topic in Volume 2 is that of opto-electronic effects, including internal and external photo-effects and generation of recombination radiation (spontaneous and stimulated) in light emitters and semiconductor lasers.

Volume 3 is concerned with the methods of preparation of semiconductors including crystal growth, purification and doping or implanting techniques. Characterisation, which is one of the most important aspects of materials technology, is given thorough treatment. Amorphous, magnetic and organic semiconductors are also considered.

Volume 4 is devoted to the physics of the many kinds of semiconductor devices which are now available, including the wide range of different diode types, i.e., rectifiers, mixers, varactors, transferred electron oscillators, tunnel and avalanche diodes, etc.; bipolar transistors and integrated circuits, and various field effect devices such as MOS and CCDs. Opto-electronic devices, such as photoconductive detectors, photo-diodes, solar cells, light-emitting diodes and lasers are covered.

The comprehensive and up-to-date treatment makes the Handbook an invaluable reference source for all workers in semiconductor physics or device development, in University, Government or industry, in the fields of electronics, optics, aerospace or computing. However, the treatment is sufficiently basic for the Handbook to be well suited to undergraduate studies, in either physics or electronic engineering.

T. S. Moss
Malvern, UK

The seven updated chapters of the present revised edition. (All other chapters are completely new.)

First edition			Present edition		
Ch.	Author(s)	Title	Ch.	Author(s)	Title
4B	M. L. Cohen and J. R. Chelikowsky	Pseudopotentials for semiconductors	3	J. R. Chelikowsky and M. L. Cohen	Ab initio pseudopotentials and the structural properties of semiconductors
4C	J. C. Phillips	Chemical models of energy bands	2	J. C. Phillips	Chemical models of energy bands
6	J. M. Baranowski, M. Grynberg and S. Porowski	Impurities in semiconductors: experimental	5	J. M. Baranowski and M. Grynberg	Impurities in semiconductors: experimental
7	P. T. Landsberg	Semiconductor statistics	6	P. T. Landsberg and O. Engström	Semiconductor statistics
8	L. M. Roth	Dynamics of electrons in semiconductors in electric and magnetic fields	10	L. M. Roth	Dynamics and classical transport of carriers in semiconductors
11A	D. K. Ferry	Fundamental aspects of hot electron phenomena	18	D. K. Ferry	Hot-electron transport phenomena
11C	J. R. Barker	Fundamental aspects of quantum transport	19	J. R. Barker	Fundamental aspects of quantum transport theory

Preface to Volume 1

In the ten years since Volume I of this Handbook was published the centers of interest in the basic physics of semiconductors have shifted. For example, amorphous and glassy materials, which were mentioned only incidentally in the original Volume I, are now areas which have their own conferences devoted to them, and we do not deal with them in this volume. Crystal structures and phonons were given a good treatment by R. Zallen, S. S. Mitra and N. E. Massa in the first edition and we do not deal with these topics either, but refer the reader to the first edition.

Volume I was called *Band Theory and Transport Properties* in the first edition. But the subject has broadened to such an extent that *Basic Properties* seemed to be a more suitable title. Some key topics have been updated and modernised by the authors of the 1982 edition (see the accompanying table).

The remaining chapters, and therefore the bulk of this work, are devoted to important "new" topics which give this volume an almost encyclopaedic form. But in each case at least a pedagogical introduction will help the novice in these areas. I will not introduce the authors here individually. Suffice is to say that they are sufficiently well known as experts that they are in no need of a special recommendation by a mere editor.

My job is done. The thrill of unexpected pitfalls and of last minute problems has gone. But a last and pleasant duty remains: it is to thank the contributors and publishers for their help in preparing what I feel is a very fine survey of basic ideas and developments in semiconductor physics.

P. T. Landsberg
University of Southampton
Southampton, UK

Contents of Volume 1

General Preface to the Revised Edition v
General Preface to the First Edition vii
Preface to Volume 1 ix
Contents of Volume 1 xi
Contributors to Volume 1 xiii

1. M. Rasolt
Contemporary topics in band theory 1
2. J. C. Phillips
Chemical models of energy bands 47
3. J. R. Chelikowsky and M. L. Cohen
Ab initio pseudopotentials and the structural properties of semiconductors 59
4. M. Lannoo
Deep and shallow impurities in semiconductors: Theoretical 113
5. J. M. Baranowski and M. Grynberg
Impurities in semiconductors: Experimental 161
6. P. T. Landsberg and O. Engström
Semiconductor statistics 197
7. L. J. Brillson
Surfaces and interfaces: Atomic-scale structure, band bending and band offsets 281
8. E. Schöll
Nonlinear dynamics, phase transitions and chaos in semiconductors . . 419
9. A. A. Rogachev
Electron–hole liquid in semiconductors 449
10. L. M. Roth
Dynamics and classical transport of carriers in semiconductors . . . 489
11. E. M. Conwell and H. A. Mizes
Conjugated polymer semiconductors: An introduction 583
12. P. J. Price
Electron tunneling in semiconductors 627
13. B. K. Ridley
Quantum confinement and scattering processes 665
14. B. L. Gallagher and P. N. Butcher
Classical transport and thermoelectric effects in low-dimensional and mesoscopic semiconductor structures 721

15. P. Voisin and G. Bastard
Coherence in III–V semiconductor superlattices 817
16. S. E. Ulloa, A. MacKinnon, E. Castaño and G. Kirczenow
From ballistic transport to localization 863
17. T. Chakraborty
The quantum Hall effect 977
18. D. K. Ferry
Hot-electron transport phenomena 1039
19. J. R. Barker
Fundamental aspects of quantum transport theory 1079

Author index 1129
List of main abbreviations used 1187
Subject index 1189

Contributors to Volume 1

J. M. Baranowski, Institute of Experimental Physics, University of Warsaw, Warsaw, Poland.

J. R. Barker, Nanoelectronics Research Centre, Department of Electronics and Electrical Engineering, University of Glasgow, Glasgow G12 8QQ, United Kingdom.

G. Bastard, Laboratoire de Physique de la Matière Condensée de l'Ecole Normale Supérieure, 24 rue Lhomond, F-75005 Paris, France.

L. J. Brillson, Xerox Webster Research Center for Technology, Webster, NY 14580, USA.

P. N. Butcher, Physics Department, University of Warwick, Coventry, CV4 7AL, United Kingdom.

E. Castaño, Departamento de Física, Universidad Autónoma Metropolitana-Iztapalapa, Apartado Postal 55-534, 09340 México, DF, México.

T. Chakraborty, Institute for Microstructural Sciences, National Research Council, Montreal Road, M-50, Ottawa, Ont. K1A 0R6, Canada.

J. R. Chelikowsky, Department of Chemical Engineering and Materials Science, University of Minnesota, Minneapolis, MN 55455, USA.

M. L. Cohen, Department of Physics, University of California, and Materials Sciences Division, Lawrence Berkeley Laboratory, Berkeley, CA 94720, USA.

E. M. Conwell, Xerox Webster Research Center, Webster, NY 14580, USA.

O. Engström, Department of Solid State Electronics, Chalmers University of Technology, S-41296 Göteborg, Sweden.

D. K. Ferry, Department of Electrical Engineering, Arizona State University, Tempe, AZ 85287-5706, USA.

B. L. Gallagher, Physics Department, University of Nottingham, Nottingham, NG7 2RD, United Kingdom.

M. Grynberg, Institute of Experimental Physics, University of Warsaw, Warsaw, Poland.

G. Kirczenow, Department of Physics, Simon Fraser University, Burnaby, BC V5A 1S6, Canada.

P. T. Landsberg, Faculty of Mathematical Studies, University of Southampton, Highfield, Southampton SO9 5NH, United Kingdom.

M. Lannoo, Laboratoire des Surfaces et Interfaces, Institut Supérieur d'Electronique du Nord, F-59046 Lille Cedex, France.

A. MacKinnon, Blackett Laboratory, Imperial College, London SW7 2BZ, United Kingdom.

H. A. Mizes, Xerox Webster Research Center, Webster, NY 14580, USA.

J. C. Phillips, AT&T Bell Laboratories, 600 Mountain Avenue, Murray Hill, NJ 079420-2070, USA.

P. J. Price, IBM Research Division, T. J. Watson Research Center, Yorktown Heights, NY 10598, USA.

M. Rasolt, Solid State Division, Oak Ridge National Laboratory, Oak Ridge, TN 37831-6032, USA.

B. K. Ridley, Physics Department, University of Essex, Colchester, United Kingdom.

A. A. Rogachev, A. F. Ioffe Physico-Technical Institute, Academy of Sciences, 194021 St. Petersburg, Russia.

L. M. Roth, Physics Department, State University of New York, Albany, NY, USA.

E. Schöll, Institut für Theoretische Physik, Technische Universität Berlin, Hardenbergstrasse 36, W-1000 Berlin 12, Germany.

S. E. Ulloa, Department of Physics and Astronomy, Condensed Matter and Surface Sciences Program, Ohio University, Athens, OH-45701-2979, USA.

P. Voisin, Laboratoire de Physique de la Matière Condensée de l'Ecole Normale Supérieure, 24 rue Lhomond, F-75005 Paris, France.

CHAPTER 1

Contemporary Topics in Band Theory

MARK RASOLT

Solid State Division
Oak Ridge National Laboratory
Oak Ridge, TN 37831-6032, USA

Handbook on Semiconductors
Completely Revised Edition
Edited by T.S. Moss
Volume 1, edited by P.T. Landsberg

Contents

1. Introduction 3
2. The two types of band structures 5
   2.1. Empirical pseudopotential method 5
   2.2. Ab initio pseudopotential method 12
   2.3. The self-energy formulation of band theory 15
3. Band gap discontinuities 22
   3.1. Weak external potential 23
      3.1.1. No interaction between the particles 23
      3.1.2. Interacting systems 24
   3.2. Expansion of $E[n]$ in powers of the nonuniform density deviation 25
   3.3. General derivation of the band gap discontinuities 30
4. Continuous internal symmetries in the band structure of donor-doped multivalley semiconductors 33
   4.1. Continuous symmetries within the effective-mass Hamiltonian 34
   4.2. The energy of $\hat{H}$ in the RPA 35
   4.3. Effective Hamiltonian for the valley electrons 36
      4.3.1. Frequency-dependent corrections 39
      4.3.2. Higher-order corrections from G_c 40
   4.4. The effective-mass Hamiltonian 41
References 43

1. Introduction

When we dynamically probe metals, semimetals, semiconductors, insulators (or for that matter any collection of electrons and ions) we are measuring the response of an excited many-body system to an external field. When we measure their static properties we are studying the thermodynamic properties of a *many-body* system. In both cases, when the nuclei are placed periodically (we will here exclusively consider only periodic systems and furthermore only systems where the electron density is commensurate with this periodicity) some form of band theory is generally the initial starting point. It is then very important to understand what such a band theory represents. For example, how does it differ when applied to the response of the system versus the thermodynamics of the system?

Until relatively recently (around the mid seventies) band theory (or more precisely band structure calculation) did not stress these differences. Initially, the single-electron potential, which generates the band structure, was treated as an empirical input (see § 2). Computational advances have made such calculations more fundamental by the ab initio construction of these potentials self-consistently in terms of the electron density (see § 2). Still the precise meaning of such calculations were not stressed. It is the purpose of this opening chapter to reexamine these subtle issues, with particular attention on their implication to the physics of semiconductors. There exist many articles on the numerical solution of this one-particle problem and this aspect will not be stressed here. Rather, this article will focus more on the many-body aspects that enter the band structure of semiconductors. We feel that further progress in understanding some finer points of semiconductor physics will depend on greater appreciation and efforts in this direction.

Two primary areas of such many-body effects, on band theory, have drawn considerable recent attention: (a) the size of the semiconducting band gap (Perdew et al. 1982, Perdew and Levy 1983, Sham and Schlüter 1983, Kleinman 1984, Sham 1985, Sham and Schlüter 1985, Lannoo et al. 1985, Godby et al. 1986, Gunnarson and Schönhammer 1986, Kleinman 1986, Perdew 1985, 1986, Rasolt 1987a, Norman and Koelling 1984, Ng 1989, Perdew 1990, Krieger and Li 1990, Schlüter and Sham 1990), and (b) the presence of continuous symmetries in indirect gap, multivalley, *n*-doped semiconductors (Rasolt 1983a, Rasolt et al. 1985, Rasolt and MacDonald 1986, Rasolt et al. 1986, Rasolt 1987b, Tešanović and Halperin 1987, Sachdev 1987, Rasolt 1990) (like Ge or Si). The difference between a metal and a semiconductor is precisely the presence of such a gap between the ground state and higher excited states of the full *many-body* Hamiltonian, as well as its single-particle simplification (i.e., band structure). In addition, of course, this gap then provides for conduction band doping which leads to the wide scope of semiconducting properties. Understanding this gap is therefore very important. As we shall see (see § 3) understanding the many-body aspect of the gap alone already captures much of the subtleties alluded to above. More specifically, we shall see that the effective band structure potential as given by the density functional theory (Hohenberg and Kohn 1964, Kohn and Sham 1965, Sham and Kohn 1966), suffers serious deficiencies in predicting the correct band gap; the so-called band gap discontinuities (Perdew et al. 1982, Perdew and Levy 1983, Sham and Schlüter 1983, Kleinman 1984, Sham 1985, Sham and Schlüter 1985, Lannoo et al. 1985, Godby et al. 1986, Gunnarson and Schönhammer 1986, Kleinman 1986, Perdew 1985, 1986, Rasolt 1987a, Norman and Koelling 1984, Ng 1989, Perdew 1990, Krieger and Li 1990, Schlüter and Sham 1990). We, in fact, will argue that the energy density functional in general (in systems with a gap) acquire important new modification over traditional expectations (Rasolt 1987a). Our discussion should leave no doubt that the description of the response to an external field and the description of thermodynamics requires two different effective band structure potentials.

Concerning point (b), i.e., the presence of continuous symmetries. In fig. 1 we show the band structure of Si along several symmetry directions. The shaded area represents the addition of the doped conduction electron states. These states possess the

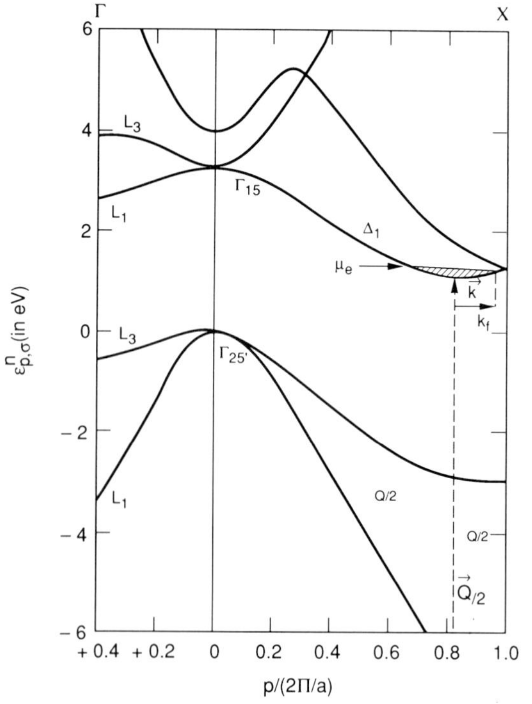

Fig. 1. The band structure of silicon along the X(100) direction and along L(111) direction. The pockets of donor-doped electrons are centered at crystal momentum $\boldsymbol{p} = \frac{1}{2}\boldsymbol{Q}$. The shaded area represents these valley electrons which in § 4 we will assign an isospin quantum number.

usual point group operations plus time reversal plus the continuous SU(2) spin group (if spin–orbit interaction is neglected). However, it has been observed recently (Rasolt 1983a, Rasolt et al. 1985, Rasolt and MacDonald 1986, Rasolt et al. 1986) that, in fact, to a *very high* degree of accuracy the valley electrons (shaded areas) related by time reversal symmetry (e.g., in Si the shaded areas along X and −X) possess internal SU(2) isospin rotation. More precisely, if we label one valley as isospin component one and the second valley as isospin component two (or equivalently spin-up, spin-down) we can rotate the two valleys just like two spins. This leads to a host of new many-body effects in the band structure of, e.g., Si, heretofore ignored (Rasolt 1990). We will cover (in § 4) in some detail this new aspect of the band structure of multivalley semiconductors and prove the existence of such almost perfect continuous internal symmetry in a band structure which in the strict sense has only discrete symmetries.

Our opening chapter should then bring the reader up to date on two relatively new and important issues of band theory of semiconductors. Since the band gap and the effect of carrier doping define and delineate semiconductors from other systems (like metals) they implicitly form the basis for many of the other properties of semiconductors to be discussed in the following chapters.

2. The two types of band structures

2.1. Empirical pseudopotential method

The full many-body Hamiltonian for a collection of ions and electrons is given by

$$\hat{H} = -\frac{\hbar^2}{2m}\sum_{i=1}^{N} \nabla_i^2 + \sum_{i=1}^{N} V(\boldsymbol{r}_i) + \tfrac{1}{2}\sum_{i\neq j}^{N} v(\boldsymbol{r}_i - \boldsymbol{r}_j), \tag{2.1}$$

where $V(\boldsymbol{r}_i)$ is the potential corresponding to some fixed periodic arrangement of the nuclear charges (here we ignore classical and quantum lattice displacement). Also in eq. (2.1) $v(\boldsymbol{r}_i - \boldsymbol{r}_j)$ is the interparticle Coulomb interaction $e^2/|\boldsymbol{r}_i - \boldsymbol{r}_j|$ and the indices i, j run over the number of electrons N.

In terms of the second-quantized field operators $\psi_\sigma^\dagger(\boldsymbol{r})$ and $\psi_\sigma(\boldsymbol{r})$, where $\psi_\sigma^\dagger(\boldsymbol{r})$ creates an electron of spin σ at site $\boldsymbol{r}$, $\hat{H}$ can be also written as (Abrikosov et al. 1963)

$$\begin{aligned}\hat{H} = &\sum_{\sigma_1}\int \mathrm{d}^3r\, \psi_{\sigma_1}^\dagger(\boldsymbol{r})\left(-\frac{\hbar^2}{2m}\nabla^2 + V(\boldsymbol{r})\right)\psi_{\sigma_1}(\boldsymbol{r})\\ &+\tfrac{1}{2}\sum_{\sigma_1,\sigma_2}\int \mathrm{d}^3r_1 \int \mathrm{d}^3r_2\, \psi_{\sigma_1}^\dagger(\boldsymbol{r}_1)\psi_{\sigma_2}^\dagger(\boldsymbol{r}_2)v(\boldsymbol{r}_1 - \boldsymbol{r}_2)\psi_{\sigma_2}(\boldsymbol{r}_2)\psi r_{\sigma_1}(\boldsymbol{r}_1).\end{aligned} \tag{2.2}$$

Traditionally the band structure of eq. (2.2) meant approximating eq. (2.2) by the quadratic form of $\psi_\sigma(\boldsymbol{r})$, $\psi_\sigma^\dagger(\boldsymbol{r})$ in the presence of some effective local external periodic potential $V_{\text{eff}}(\boldsymbol{r}, \varphi_{\text{G}})$, i.e., a "single-particle-like" effective Hamiltonian $\hat{H}_0(\varphi_{\text{G}})$ (the dependence of $\hat{H}_0$ on the ground state $|\varphi_{\text{G}}\rangle$ illustrates the self-consistent nature of $\hat{H}_0$). $V_{\text{eff}}(\boldsymbol{r}, \varphi_{\text{G}})$ has the same periodicity as $V(\boldsymbol{r})$ (eq. (2.1)); i.e., $V_{\text{eff}}(\boldsymbol{r} + \boldsymbol{R}_n) = V_{\text{eff}}(\boldsymbol{r})$

with $\boldsymbol{R}_n$ the lattice vector. The meaning of this $V_{\text{eff}}(\boldsymbol{r}, \varphi_{\text{G}})$ and whether it should in fact be local both in the position $\boldsymbol{r}$ and energy of the higher excited states of $\hat{H}$ (eq. (2.1)) will occupy much of this opening chapter of this book. For the moment, however, let us take the traditional view and treat $V_{\text{eff}}(\boldsymbol{r}, \varphi_{\text{G}})$ empirically.

We can then expand the field operators in terms of a specific set of Bloch states $\varphi^n_{\boldsymbol{p}}(\boldsymbol{r})$, i.e.,

$$\psi^{\dagger}_{\sigma}(\boldsymbol{r}) = \sum_{\boldsymbol{p},n} C^{\dagger}_{n\boldsymbol{p}\sigma}\varphi^{n*}_{\boldsymbol{p}\sigma}(\boldsymbol{r}) \tag{2.3}$$

such that

$$[\hat{H}_0(\varphi_{\text{G}}), C^{\dagger}_{n,\boldsymbol{p},\sigma}] = \varepsilon^n_{\boldsymbol{p},\sigma} C^{\dagger}_{n\boldsymbol{p}\sigma} \tag{2.4}$$

and $\varepsilon^n_{\boldsymbol{p}\sigma}$ is the dispersion, or band structure illustrated in, e.g., fig. 1. Here $\boldsymbol{p}$ is the crystal momentum of the electron and n the band index. Equations (2.3) and (2.4) are equivalent to the more familiar solution of the single-electron band structure given by

$$\left(\frac{-\hbar^2}{2m}\nabla^2 + V_{\text{eff}}(\boldsymbol{r}, \varphi_{\text{G}})\right)\varphi^n_{\boldsymbol{p}\sigma}(\boldsymbol{r}) = \varepsilon^n_{\boldsymbol{p}\sigma}\varphi^n_{\boldsymbol{p}\sigma}(\boldsymbol{r}). \tag{2.5}$$

As it stands, $V_{\text{eff}}(\boldsymbol{r}, \varphi_{\text{G}})$ represents some kind of self-consistent potential for all of the electrons (deep core states as well as the valence and conduction bands). Because of the orthogonality of the valence electrons to the core states, $V_{\text{eff}}(\boldsymbol{r}, \varphi_{\text{G}})$ can be mapped to a much weaker potential, a pseudopotential (Philips and Kleinman 1959, Cohen and Heine 1961, Austin et al. 1962, Ashcroft 1966, Heine and Abarenkov 1964, Animalu and Heine 1965, Anderson 1963) $V_{\boldsymbol{p}}(\boldsymbol{r})$ restricted, however, only to the valence and conduction electrons. This is very important, particularly when we treat the band structure of eq. (2.5) empirically. The point is that V_{eff} has a very strong $-Z(r)e^2/r$ potential ($r \to 0$, $Z(\boldsymbol{r}) = Z \equiv$ nuclear charge $= 14$ for Si) at small r. It, however, is largely ignored by the valence or conduction electrons due to the orthogonality to the core states, which leads to the exclusion of these states from this core region. This effect is then reproduced by the weaker $V_{\text{p}}(\boldsymbol{r})$. From an empirical point of view, the fact that $V_{\text{p}}(\boldsymbol{r})$ is relatively weak (more precisely smooth in $\boldsymbol{r}$) means that it can be expanded in only a few plane waves, i.e.,

$$V_{\text{p}}(\boldsymbol{r}) = \sum_{\boldsymbol{G}} V_{\text{p}}(\boldsymbol{G})F(\boldsymbol{G})\exp(\mathrm{i}\boldsymbol{G}\cdot\boldsymbol{r}), \tag{2.6}$$

where

$$F(\boldsymbol{G}) = \frac{1}{N^*}\sum_{\boldsymbol{\tau}} \exp(-\mathrm{i}\boldsymbol{G}\cdot\boldsymbol{\tau}) \tag{2.7}$$

when all the N^* atoms in the unit cell are the same species, $\boldsymbol{\tau}$ are the position vectors of these atoms in the unit cell, and $\boldsymbol{G}$ the reciprocal vectors. In this empirical pseudopotential method (EPM) $V_{\text{p}}(\boldsymbol{G})$ are the empirical parameters which are fitted to experiment (see below). These experimental measurements directly or indirectly reflect either the states $\varphi^n_{p\sigma}(\boldsymbol{r})$ or the corresponding eigenstates $\varepsilon^n_{p\sigma}$. Using the Bloch symmetry of $\varphi^n_{p}(\boldsymbol{r})$, i.e.,

$$\varphi^n_{\boldsymbol{p}}(\boldsymbol{r}) = \frac{1}{\sqrt{\Omega}} \exp(\mathrm{i}\boldsymbol{p}\cdot\boldsymbol{r}) \sum_{\boldsymbol{G}} A^n_{\boldsymbol{Gp}} \exp(\mathrm{i}\boldsymbol{G}\cdot\boldsymbol{r}) \tag{2.8}$$

(Ω is the volume of the system), eq. (2.5) for the valence and conduction states (with $V_{\text{eff}} \rightarrow V_{\text{p}}$) becomes a matrix equation for the expansion coefficients $A^n_{\boldsymbol{G}}$ and the eigenvalues (i.e., band structure) $\varepsilon^n_{\boldsymbol{p}}$. The entries of the matrix are the empirical $V_{\text{p}}(\boldsymbol{G})$ and the crystal momentum $\boldsymbol{p}$ of the Bloch state, i.e.,

$$\sum_{\boldsymbol{G}'} (H_{\boldsymbol{G},\boldsymbol{G}'} - \varepsilon^n_{\boldsymbol{p}} \delta_{\boldsymbol{G},\boldsymbol{G}'}) A^n_{\boldsymbol{G}',\boldsymbol{p}} = 0, \tag{2.9a}$$

where

$$H_{\boldsymbol{G},\boldsymbol{G}'} = \frac{\hbar^2}{2m} |\boldsymbol{p}+\boldsymbol{G}|^2 \delta_{\boldsymbol{G},\boldsymbol{G}'} + V_{\text{eff}}(\boldsymbol{G}-\boldsymbol{G}'). \tag{2.9b}$$

For the zinc blendes of the III–V and II–VI semiconductors there are minor changes in eq. (2.7) to incorporate the difference in the potentials of the two species; otherwise the empirical fit of eq. (2.9) to experiments proceeds in the same way (see below).

We could directly treat the $V_{\text{eff}}(\boldsymbol{G})$ as fitting parameters. Operationally, however, the way it is done (Brust et al. 1962, Cohen and Bergstresser 1966, Allen and Cohen 1969) is to first approximate the ionic potential $V(\boldsymbol{r})$ (in eq. (2.1)) by a parameterized spatial form of $V_{\text{p}}(\boldsymbol{r})$ (see fig. 2). The area in the core region of $V_{\text{p}}(\boldsymbol{r})$ and somewhat beyond is represented by some parameterized square well (or a Gaussian well, etc.) which outside the core region hooks up continuously to the Coulomb tail. The Fourier transform of $V_{\text{p}}(\boldsymbol{r})$ is then a function of these parameters, which parameterize

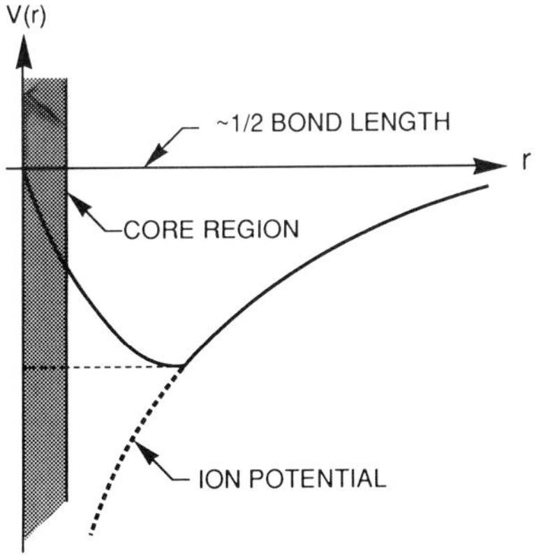

Fig. 2. A schematic plot of a spatial form for a parameterized pseudopotential $V_{\text{p}}(\boldsymbol{r})$; taken from Cohen and Chelikowsky (1988).

the square well (or the Gaussian well, etc.) and which are then used to fit the experiments via eq. (2.9).

One last technical point. On general grounds $V_p(\boldsymbol{r})$ can be shown to be nonlocal (i.e., $V_p(\boldsymbol{r}) \to V_p(\boldsymbol{r}, \boldsymbol{r}')$) when decomposed into its angular momentum components (Philips and Kleinman 1959, Cohen and Heine 1961, Austin et al. 1962, Ashcroft 1966, Heine and Abarenkov 1964, Animalu and Heine 1965, Rasolt and Taylor 1972, 1973a). A more precise fit to experiment is achieved by parameterizing $V_p(\boldsymbol{r})$ with different potential $V_{pl}(\boldsymbol{r})$ for each angular momentum l, i.e.,

$$V_p(\boldsymbol{r}, \boldsymbol{r}') = \sum_{l=0}^{\infty} V_{pl}(r) P_l(\cos\theta) P_l^*(\cos\theta') \frac{\delta_{r,r'}}{r'^2}, \tag{2.10}$$

where $P_l(x)$ are the Legendre polynomials. The Fourier transform of eq. (2.10) gives a $\boldsymbol{p}$-dependent $V_{\rm eff}(\boldsymbol{G})$; i.e., $V_{\rm eff}(\boldsymbol{G}) \to V_{\rm eff}(\boldsymbol{p}, \boldsymbol{G}, \boldsymbol{G}')$ because

$$\begin{aligned} \langle \boldsymbol{p}+\boldsymbol{G} | V_{pl} | \boldsymbol{p}+\boldsymbol{G}' \rangle = {} & \frac{4\pi}{\Omega_c}(2l+1) P_l(\cos\theta_{\boldsymbol{p}+\boldsymbol{G},\boldsymbol{P}+\boldsymbol{G}'} \\ & \times \int_0^{\infty} \mathrm{d}r\, r^2 V_{pl}(r) j_l(|\boldsymbol{p}+\boldsymbol{G}|r) j_l(|\boldsymbol{p}+\boldsymbol{G}'|r) \end{aligned} \tag{2.11}$$

(Ω_c is the unit cell volume). This dependence on the crystal momentum $\boldsymbol{p}$, which is absent for the full $V(\boldsymbol{r})$ or $V_{\rm eff}(\boldsymbol{r})$ and which is a consequence of the nonlocality of $V_p(\boldsymbol{r}, \boldsymbol{r}')$ in coordinate space, could be very important in this EPM. It might in fact (perhaps unintentionally) incorporate important nonlocal structure *intrinsic* to the *self-energy* of the electron (Sham and Kohn 1966, Kane 1972, Rasolt and Vosko 1974a,b, Wang and Rasolt 1977, MacDonald 1980, Wang and Pickett 1983, Pickett and Wang 1984, Hybersten and Louie 1985, 1986, 1988, von der Linden and Horsch 1988, Hott 1991, Gygi and Baldereschi 1989, Hanke and Sham 1989, Steiner et al. 1991) which we will argue below is the only correct way to think of band theory appropriate to the dynamic response of a many-body system. (The close resemblance of eq. (2.10) to this intrinsic structure of the self-energy is best seen in Wang and Rasolt (1977)). Finally, in the EPM these bare fitting parameters $V_p(\boldsymbol{G})$ are linearly screened by the dielectric response $\varepsilon(\boldsymbol{G})$ to give $V_{\rm eff}(\boldsymbol{G}) = V_p(\boldsymbol{G})/\varepsilon(\boldsymbol{G})$. For the nonlocal $V_p(\boldsymbol{r}, \boldsymbol{r}')$ nonlocal screening can be accounted for (Rasolt and Taylor 1975).

We are now set to solve eq. (2.9) for the eigenvalues $\varepsilon_{\boldsymbol{p}}^n$ and eigenfunctions $A_{\boldsymbol{G}',\boldsymbol{p}}^n$ in terms of $V_{\rm eff}(\boldsymbol{G}-\boldsymbol{G}')$. Use these to compare with experiment, adjust the parameters of $V_p(\boldsymbol{r})$ (or $V_{pl}(\boldsymbol{r})$) until a satisfactory agreement with experiment is achieved. Now come the tough questions: to what experiment should we fit? and what is the fundamental implication of such a fit? Should we fit to the dynamical response (e.g., optical properties) or should we fit to the thermodynamic properties (e.g., compressibility)? We will discuss these important issues shortly. For the moment we only remark that both of these fits are sensible as long as we *do not* use one set of parameters (e.g., from dynamics) to explain the thermodynamic properties and vice versa.

The parameters of $V_p(\boldsymbol{r})$ are usually extracted from the optical absorption. The optical absorption is described by the current–current response function; more precisely by the real part of the conductance given by

$$\operatorname{Re}\sigma(\omega)=\operatorname{Re}\left(\frac{\mathrm{i}}{3\Omega}\int_{-\infty}^{0}\mathrm{d}t\,\langle\varphi_{\mathrm{G}}[\boldsymbol{j}(0)\cdot\boldsymbol{j}(t)]|\varphi_{\mathrm{G}}\rangle\frac{\exp(-\mathrm{i}\omega t)}{\mathrm{i}\omega}-\frac{ne^2}{\mathrm{i}\omega m}\right), \tag{2.12}$$

where $\boldsymbol{j}(t)$ is the average paramagnetic current; i.e.

$$\begin{aligned}\boldsymbol{j}(t)&=\int\mathrm{d}^3r\,\boldsymbol{j}(\boldsymbol{r},t)\\&=\frac{\hbar}{2m_i}\int\mathrm{d}^3r\sum_{\sigma}[\psi_{\sigma}^{\dagger}(\boldsymbol{r},t)\nabla\psi_{\sigma}(\boldsymbol{r},t)-\psi_{\sigma}(\boldsymbol{r},t)\nabla\psi_{\sigma}^{\dagger}(\boldsymbol{r},t)],\end{aligned} \tag{2.13}$$

where ψ_σ are the field operators of eq. (2.2). Equation (2.12) is for zero temperature, which all of our discussions will be restricted to; at finite T we need an ensemble average of eq. (2.12). From eqs. (2.12) and (2.13), it is not difficult to see that $\sigma(\omega)$ is closely related to the two-particle Green function in the electron–hole channel. In fig. 3 we give some low-order graphs of $\sigma(\omega)$. Figure 3a is zeroth order and fig. 3b is first order in the electron–electron interaction; i.e., $v(\boldsymbol{r})$ of eq. (2.1) or (2.2). Equation (2.12) is exact; it is *exactly* what we measure in linear optical absorption; it is *this quantity* we must use to extract the fitting parameters of eq. (2.9). Now then what does eq. (2.9) have to do with eq. (2.12)? Let us take a look at fig. 3a. The double solid lines are single-particle Green functions given by a Dyson equation for a nonlocal self-energy (incidentally, $V_{\mathrm{eff}}(\boldsymbol{p},\boldsymbol{G})$ would have to have at least in addition an explicit

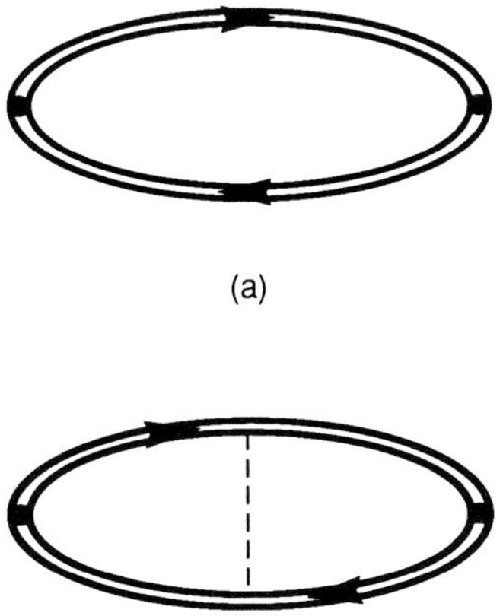

(a)

(b)

Fig. 3. (a) Zeroth-order, current–current response function. The double-solid arrowed lines are the full crystalline interacting single-particle Green function. It satisfies the Dyson equation; eq. (2.20) in the text. (b) First-order, current–current response function in the electron–electron interaction given by the dashed line.

energy dependence (Rasolt and Taylor 1973a, Kane 1972, Rasolt and Vosko 1974a,b, Wang and Rasolt 1977, MacDonald 1980, Wang and Pickett 1983, Pickett and Wang 1984, Hybersten and Louie 1985, 1986, 1988, von der Linden and Horsch 1988, Hott 1991, Gygi and Baldereschi 1989, Hanke and Sham 1989, Steiner et al. 1991)). Then these solid lines (i.e., Green function) are given in terms of eqs. (2.8) and (2.9) by

$$G(\boldsymbol{r}, \boldsymbol{r}', \omega) = \sum_{\boldsymbol{p},n} \frac{\varphi_{\boldsymbol{p}}^{*n}(\boldsymbol{r})\varphi_{\boldsymbol{p}}^{n}(\boldsymbol{r}')}{\omega - \varepsilon_{\boldsymbol{p}}^{n} - \mathrm{i}\delta\,\mathrm{sgn}(\varepsilon_{\boldsymbol{p}}^{n} - \mu)}, \tag{2.14}$$

where μ is the chemical potential. (To be perfectly rigorous we should recognize that the exact $G(\boldsymbol{r}, \boldsymbol{r}', \omega)$ can never look like that; e.g., in addition to a single pole at $\varepsilon_{\boldsymbol{p}}^{n}$ it *must* in general have a continuous branch cut for the lifetime of such quasiparticle states $\varepsilon_{\boldsymbol{p}}^{n}$. In the fit of the EPM the hope is that such corrections are small.)

Using eq. (2.14) in fig. 3a (i.e., eqs. (2.12) and (2.13)) we can express $\sigma(\omega)$ in terms of the $\varphi_{\boldsymbol{p}}^{n}$ and $\varepsilon_{\boldsymbol{p}}^{n}$, fit to experiment, return to eq. (2.9) and adjust V_{eff}, generate new $\varphi_{\boldsymbol{p}}^{n}$ and $\varepsilon_{\boldsymbol{p}}^{n}$, return to eq. (2.12) until a satisfactory fit is achieved.

Figure 3a directly reflects the band structure. Figure 3b (and higher-order graphs) is no longer directly a "single-particle" property. It corresponds to the interaction between the quasiparticles. These graphs are low-order electron–hole scattering on the way to form, perhaps, an exciton. Such graphs will also modify the valence to conduction band absorption (above the exciton lines, see fig. 4). In the EPM fits we assume that such corrections are small (they probably are). We can then use these results to predict such excitonic levels, and more. (For the reader more familiar with metal physics (i.e., Fermi liquid) fig. 3b shows strong similarity to the Fermi liquid parameters for the interaction of two quasiparticles close to the Fermi energy (Abrikosov et al. 1963, Nozières 1964).)

In figs. 1, 5, 6, and 7 we illustrate the level of success of the EPM in predicting the dynamic response. Such results are presented in great detail in Cohen and Chelikowsky (1988) and will not be stressed here. We should only note that the agreement with optical response is quite impressive in view of the small number of

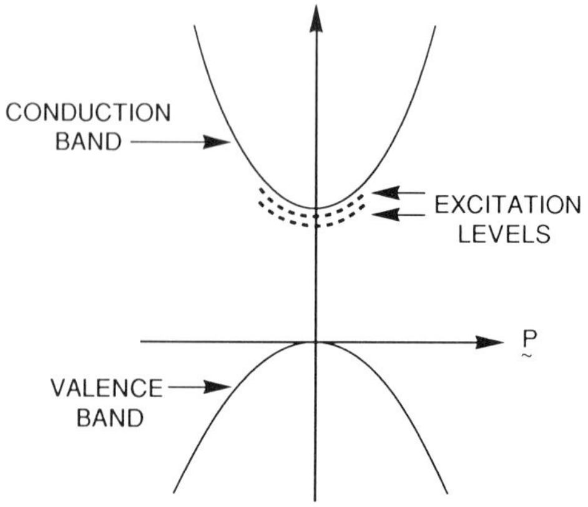

Fig. 4. Schematic band structure of a direct-gap semiconductor illustrating the exciton levels.

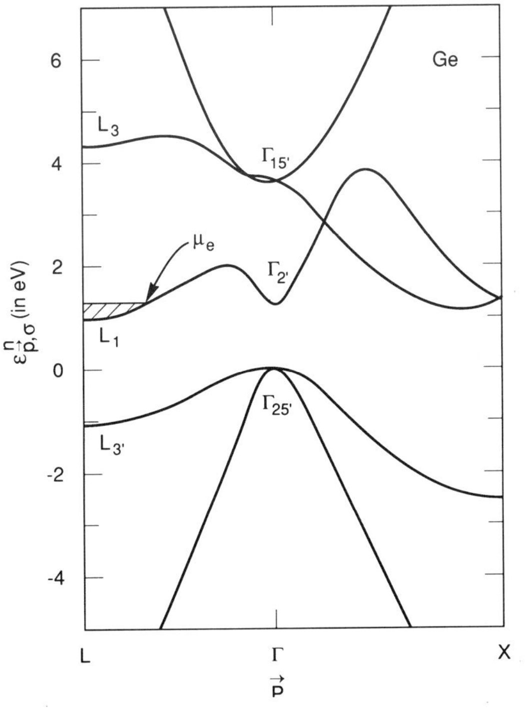

Fig. 5. The same as fig. 1 for germanium. The donor electron pockets are now at the zone boundary L.

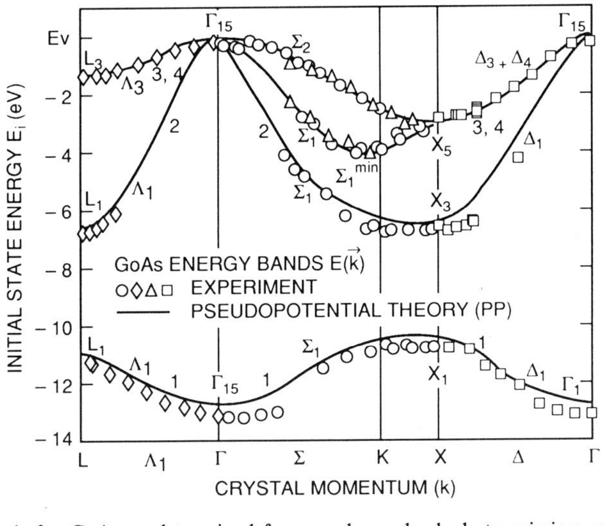

Fig. 6. Valence bands for GaAs as determined from angle-resolved photoemission and from the EPM (see eq. (2.9) of text) taken from Cohen and Chelikowsky (1988).

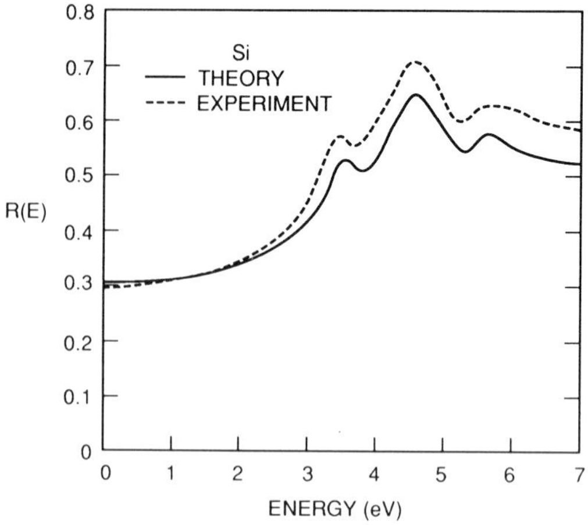

Fig. 7. Reflectivity spectrum of silicon. Taken from Cohen and Chelikowsky (1988).

the fitting parameters $V_{\text{eff}}(\boldsymbol{p}, \boldsymbol{G})$; in general only three fitting parameters per specie are necessary.

When we try to use these parameters to predict the thermodynamics of these systems like crystal structure, compressibility, lattice constants, lattice vibrations, solid–solid phase transformation, etc., the results are much less impressive. The reason for an upshot is that the $\varphi^n_{\boldsymbol{p}}$ and $\varepsilon^n_{\boldsymbol{p}}$ of eq. (2.9), when extracted from eq. (2.12), *cannot* be used for ground state properties in the framework of one particle-like structure (see below). While this was recognized quite early (Sham and Kohn 1966) its subtle implications were not stressed until relatively recently (Rasolt and Vosko 1974a,b, Wang and Rasolt 1977, MacDonald 1980, Wang and Pickett 1983, Pickett and Wang 1984, Hybersten and Louie 1985, 1986, 1988, von der Linden and Horsch 1988, Hott 1991, Gygi and Baldereschi 1989, Hanke and Sham 1989, Steiner et al. 1991, Mearns 1988, Schönhammer and Gunnarson 1988) (see § 3). To calculate this second class of properties, we need to generate a second type of band structure in the framework of the density functional theory (DFT). In practice this is done using the self-consistent ab initio pseudopotentials.

2.2. *Ab initio pseudopotential method*

Hohenberg and Kohn (1964) have proved two basic *ground state* properties of a nonuniform interacting many Fermion system, which were later extended to finite T by Mermin (1965) (we will consider only $T = 0$). The first is that the ground state energy $E = \langle \varphi_G | \hat{H} | \varphi_G \rangle$ ($\hat{H}$ given in eqs. (2.1) or (2.2)) is uniquely given by the density $n(\boldsymbol{r})$; i.e., $E = E[n]$ (perhaps more basic the $|\varphi_G\rangle$ is given uniquely by $n(\boldsymbol{r})$ from which $E = E[n]$). The second is that if we write

$$E[n] \equiv E_V = F[n] + \int \mathrm{d}^3 r\, V(\boldsymbol{r}) n(\boldsymbol{r}) \tag{2.15}$$

then the absolute minimum of E_V, when the external potential $V(\boldsymbol{r})$ is fixed, is given by this density $n(\boldsymbol{r})$. These two properties comprise the so-called density functional theory (DFT).

Let us already here (in preparation for our discussion of band gap discontinuities (see § 3)) restate the first property more carefully. What Hohenberg and Kohn (1964) have actually shown is that $E = E_N[n]$, i.e., it is a functional of $n(\boldsymbol{r})$ as well as a function of the total particle number N. Incidentally, since in the thermodynamic limit E should depend on intensive variables, then $E_N[n] = E[n_0, n]$ where n_0 can be taken as the average density of $n(\boldsymbol{r})$.

As we shall see, this additional dependence on n_0, which was almost entirely overlooked when approximate structures of $E[n]$ were calculated or suggested, is at the heart of the band gap discontinuity (see § 3).

Kohn and Sham (1965) went a step further and rigorously mapped $E[n]$ to a self-consistent single-particle equations, the so-called Kohn and Sham equations. Very briefly, Kohn and Sham (1965) split the internal energy $F[n]$ into

$$F[n] = T_s[n] + E_H[n] + E_{xc}[n], \tag{2.16}$$

where E_H is the Hartree energy, E_{xc} is the exchange and correlation energy and T_s is the kinetic energy of a *noninteracting* nonuniform fermions with the same density $n(\boldsymbol{r})$. They then write $T_s[n]$ in terms of auxiliary functions $\varphi_i(\boldsymbol{r})$ as

$$T_s[n] = \int d^3r \sum_{i=1}^{N} \left(\frac{-\hbar^2}{2m} \varphi_i^*(\boldsymbol{r}) \nabla^2 \varphi_i(\boldsymbol{r}) \right), \tag{2.17a}$$

with

$$n(\boldsymbol{r}) = \sum_{i=1}^{N} \varphi_i^*(\boldsymbol{r}) \varphi_i(\boldsymbol{r}) \tag{2.17b}$$

and minimize eq. (2.15). The set of equations they get, specialized again to a periodic system, are

$$\left[-\frac{\hbar^2}{2m} \nabla^2 + V_{\text{eff}}(\boldsymbol{r}) \right] \varphi_{\boldsymbol{p}}^n(\boldsymbol{r}) = \varepsilon_{\boldsymbol{p}}^n \varphi_{\boldsymbol{p}}^n(\boldsymbol{r}), \tag{2.18a}$$

where

$$V_{\text{eff}}(\boldsymbol{r}) = V(\boldsymbol{r}) + e^2 \int d^3r' \, n(\boldsymbol{r}')/|\boldsymbol{r} - \boldsymbol{r}'| + v_{xc}(n(\boldsymbol{r})), \tag{2.18b}$$

with

$$v_{xc}(n(\boldsymbol{r})) = \delta E_{xc}(n(\boldsymbol{r}))/\delta n(\boldsymbol{r}). \tag{2.18c}$$

Working out eqs. (2.15)–(2.18), eq. (2.15) can be rewritten as

$$\begin{aligned} E_V = \sum_{\boldsymbol{p},n} \varepsilon_{\boldsymbol{p}}^n - \tfrac{1}{2} e^2 \int d^3r \int d^3r' \, n(\boldsymbol{r}) n(\boldsymbol{r}')/|\boldsymbol{r} - \boldsymbol{r}'| \\ + E_{xc}[n] - \int d^3r \, v_{xc}(n(\boldsymbol{r})) n(\boldsymbol{r}). \end{aligned} \tag{2.19}$$

The "eigenvalues" ε_p^n of eq. (2.18) form again a band structure just like eq. (2.5) or equivalently eq. (2.9). (In fact, unlike eq. (2.9) in eq. (2.18) $V_{\text{eff}}(\boldsymbol{r})$ (or $v_{\text{xc}}(\boldsymbol{r})$) is rigorously local.) But here the ε_p^n are, at least rigorously, nothing more than Lagrange multipliers. They, e.g., play a role in setting the bounds of the sum over N from the lowest $\varepsilon_{p_1}^n$ to the highest N at $\varepsilon_{p_N}^n$.

Equation (2.19) gives the ground state energy rigorously in terms of this "band structure" ε_p^n and the density $n(\boldsymbol{r})$. Even though this second type of "band structure" is strictly appropriate to the thermodynamics, the issue of its relevance to some detail description of quasiparticle structure remained in contention until relatively recently. For example, it is not at all obvious that if we knew the exact form of $V_{\text{eff}}(\boldsymbol{r})$ (or more precisely $v_{\text{xc}}(\boldsymbol{r})$) the ε_p^n would not give the Fermi surface of metals correctly.

Initial attempts at such questions (Rasolt and Vosko 1974a,b, Wang and Rasolt 1977, Rasolt et al. 1975) showed that the Fermi surface of metals given by the self-energy formulation (see below) led to significant differences from the ε_p^n of eq. (2.18). However, the approximation used for the self-energy (Sham and Kohn 1966, Rasolt and Vosko 1974a) were uncontrolled and only recently it has been finally shown rigorously (Mearns 1988) that the two-band structures, even on the Fermi surface, are indeed not the same. (We should mention in passing that there now does exist a current density functional theory which must, in principle, reproduce the de Haas–van Alphen oscillations exactly (Vignale and Rasolt 1987, 1988, Rasolt and Vignale 1990).) In semiconductors, where the excitations from the valence to the conduction band span a large energy range (leaving for the moment the band gap issue aside), there is no doubt that these ε_p^n are inappropriate and the two band structures should be kept separate.

Equations (2.18) and (2.19) have been applied to the ground state properties of semiconductors with impressive success (Starkloff and Joannopoulos 1977, Zunger and Cohen 1979, Hamann et al. 1979, Kerker 1980, Cohen 1982, Vanderbilt 1987; see chapter 3 of this volume). Here is a brief description of the actual calculational procedure. Just as in the EPM the full $V(\boldsymbol{r})$ in eq. (2.18b) is mapped onto a weaker pseudopotential appropriate only for the valence and conduction electrons (Starkloff and Joannopoulos 1977, Zunger and Cohen 1979, Hamann et al. 1979, Kerker 1980). However, here this potential is *not* treated as a fitting parameter but is constructed ab initio (Starkloff and Joannopoulos 1977, Zunger and Cohen 1979, Hamann et al. 1979, Kerker 1980). One variance of such a pseudopotential is to use a weak pseudopotential in the core region but to insure that the atomic wavefunctions are reproduced very accurately outside the core region (see fig. 8). Also the pseudowavefunction is appropriately normalized. The radius of the pseudopotential, where the true wavefunction and the pseudowavefunction start deviating, is taken inside the first maximum of the valence state (see fig. 8). Such ab initio pseudopotentials (Starkloff and Joannopoulos 1977, Zunger and Cohen 1979, Hamann et al. 1979, Kerker 1980, Yin and Cohen 1982, Louie et al. 1982) $V_{\text{p}}(\boldsymbol{r})$ also reproduce the higher atomic eigenvalues and eigenfunctions very accurately. Incidentally, the insistance on the equality of the pseudodensity and full density (which is similar to the wavefunction) outside the core was the ingredient for the "ab initio" pair potentials in simple metals as proposed much earlier (Rasolt and Taylor 1975, 1973b, Dagens et al. 1975)

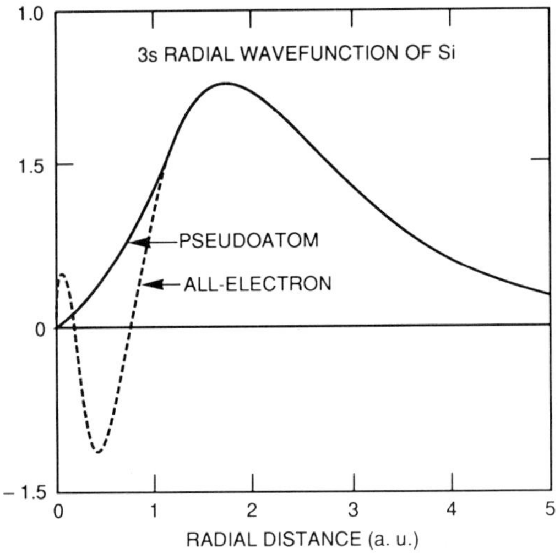

Fig. 8. The full electron and pseudoelectron radial wavefunction for the 3s electron in silicon; taken from Cohen and Chelikowsky (1988).

(although on a more modest scale). This $V_p(\boldsymbol{r})$ then replaces $V(\boldsymbol{r})$ in eq. (2.18). Equation (2.18) is generally solved in plane wave basis and the $v_{xc}(n(\boldsymbol{r}))$ is approximated in the local density approximation (LDA) (Cohen 1982). Because $V_{eff}(\boldsymbol{r})$ depends on the density eq. (2.18) has to be iterated to self-consistency. But unlike the iteration of the EPM to self-consistency with the optical measurements (described previously) here no such fitting is involved. As already mentioned, the resulting agreement with ground state properties is very impressive (Starkloff and Joannopoulos 1977, Zunger and Cohen 1979, Hamann et al. 1979, Kerker 1980, Cohen 1982, Vanderbilt 1987).

When we discussed the fit to optical measurements, in the context of the EPM, we made close connection of the fitting parameters to the self-energy formulation of the band structure; we turn to this next.

2.3. *The self-energy formulation of band theory*

In § 2.1 and § 2.2 we separated the two classes of properties of a many-body system to dynamic response and thermodynamics and presented two *different* "band" theories which underline their appropriate theoretical treatment. There is, however, one theoretical framework which incorporates both; it is a full many-body calculation. In fig. 9 we present some low-order graphs for the self-energy $\Sigma_{xc}(\boldsymbol{r}, \boldsymbol{r}', \varepsilon)$. It enters *rigorously* the single-particle line in the so-called "skeleton" expansion of the current–current correlation function (or equivalently conductivity) of fig. 3. This Green function satisfies the Dyson equation

$$\left[-\left(\frac{\hbar^2}{2m}\nabla^2+\varepsilon\right)+V(\boldsymbol{r})+V_H(\boldsymbol{r})\right]G(\boldsymbol{r},\boldsymbol{r}',\varepsilon)+\int d^3r''\,\Sigma_{xc}(\boldsymbol{r},\boldsymbol{r}'',\varepsilon)G(\boldsymbol{r}'',\boldsymbol{r}',\varepsilon) = -\delta(\boldsymbol{r}-\boldsymbol{r}'). \tag{2.20}$$

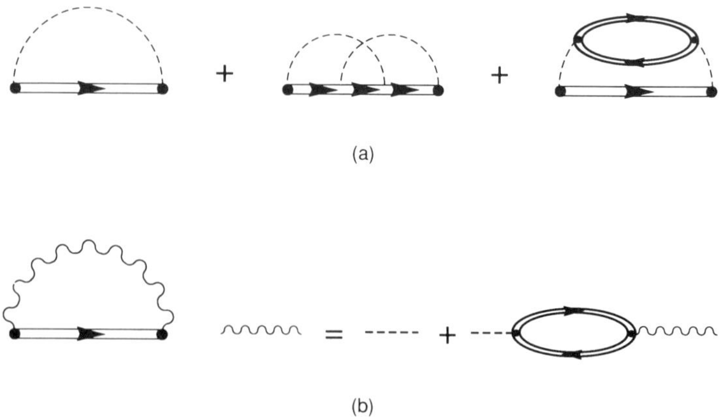

Fig. 9. (a) Low-order graphs for the self-energy Σ_{xc} in terms of the self-consistent Green function (double arrowed lines) given by eq. (2.20) in the text. (b) The self-energy Σ_{xc} in the self-consistent random phase approximation (RPA); i.e., the GW approximation.

Note that from fig. 9 the exact solution of eq. (2.20) is a self-consistent problem, since the arrowed lines in the skeleton expansion are again the full $G(\boldsymbol{r}, \boldsymbol{r}', \varepsilon)$. If we could solve this equation for G exactly to all orders in $v(\boldsymbol{r}_1 - \boldsymbol{r}_2)$ (the dashed lines in fig. 9) and then insert these G's in fig. 3 (again to all orders of v) we would rigorously get the linear optical response; this is of course technically not possible. (We will discuss the state of the art of this shortly.)

When $V(\boldsymbol{r})$ is periodic, general considerations suggest that

$$G_{n,n'}(\boldsymbol{p}, \varepsilon) \equiv \int \mathrm{d}^3 r \int \mathrm{d}^3 r' \; \varphi_{\boldsymbol{p}}^{*n}(\boldsymbol{r}) G(\boldsymbol{r}, \boldsymbol{r}', \varepsilon) \varphi_{\boldsymbol{p}}^{n'}(\boldsymbol{r}') \tag{2.21}$$

is strongly peaked in ε around $\varepsilon_{\boldsymbol{p}}^n \delta_{n,n'}$.

Ignoring the spread of these quasiparticle peaks (this could actually be a serious approximation, particularly for $E \gg E_{\mathrm{g}} \equiv$ the semiconductor gap) we can write $G(\boldsymbol{r}, \boldsymbol{r}', \varepsilon)$ in the form of eq. (2.14). Using eq. (2.14), eq. (2.20) can then be rewritten as

$$\left[-\frac{\hbar^2}{2m}\nabla^2 + V(\boldsymbol{r}) + V_{\mathrm{H}}(\boldsymbol{r})\right]\varphi_{\boldsymbol{p}}^n(\boldsymbol{r}) + \int \mathrm{d}^3 r' \, \Sigma_{\mathrm{xc}}(\boldsymbol{r}, \boldsymbol{r}', \varepsilon_{\boldsymbol{p}}^n)\varphi_{\boldsymbol{p}}^n(\boldsymbol{r}') = \varepsilon_{\boldsymbol{p}}^n \varphi_{\boldsymbol{p}}^n(\boldsymbol{r}). \tag{2.22}$$

(We should add that at the peaks of eq. (2.14) (or (2.21)) along the real axis there are changes in the quasiparticle amplitude (usually labeled Z and referred to as quasiparticle renormalization amplitudes (Nozières 1964)), which have not been included in eq. (2.21) (or eq. (2.14)). This would not change the $\varepsilon_{\boldsymbol{p}}^n$ in eq. (2.22) but will play a role, e.g., in calculating optical conductivity $\sigma(\omega)$.) These $\varphi_{\boldsymbol{p}}^n(\boldsymbol{r})$ and $\varepsilon_{\boldsymbol{p}}^n$, as we already alluded to in § 2.1, must therefore be closely related to eq. (2.9), i.e., the appropriate band theory for the dynamic response. (We should, however, mention that an extension of the DFT to a time-dependent external potential (i.e., $V(\boldsymbol{r}, t)$ in eq. (2.18b)) does exist (Runge and Gross 1984, Gross and Kohn 1985, 1990). It therefore does

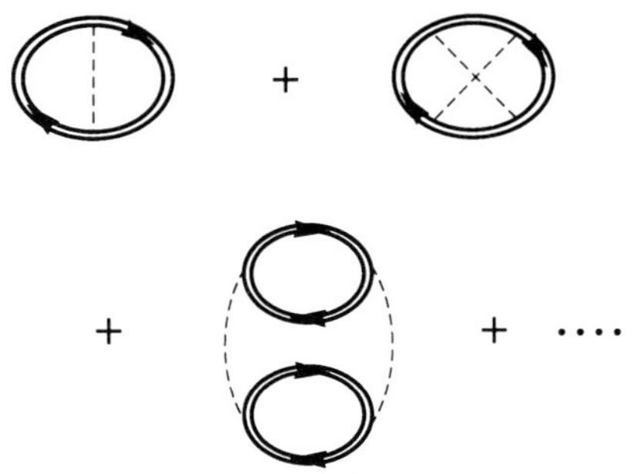

Fig. 10. Low-order skeleton expansion (to second order in the particle–particle interaction) related to the ground state energy.

provide a framework for calculating $\sigma(\omega)$ correctly within a "noninteracting" particle response; we will say a bit more about it below.)

The thermodynamics (i.e., the ground state energy E) is given by the linked cluster expansion of the skeleton graphs; some low order terms are listed in fig. 10. We will use this many-body formulation (for the energy E) in conjunction with the *equivalent* DFT method to discuss the band gap discontinuity (see § 3). Here we just add that the DFT has been, by far, the most widely applied method to ground state properties due to its much simpler structure and universality. The many-body calculation of E (in large systems like bulk Si) is generally too difficult to implement beyond the Hartree–Fock approximation.

Equation (2.20), of course, cannot be solved exactly. We next turn to an up to date review of the actual implimentation of eq. (2.20). One approximation employed by all calculations is eq. (2.14) which leads at once to eq. (2.22). The first to suggest the implimentation of eq. (2.22) for the calculation of quasiparticle excitations $\varepsilon_{\boldsymbol{p}}^n$ was Hedin (Hedin 1965a,b, Hedin and Lundqvist 1969). First he restricted the graphs of $\Sigma_{\mathrm{xc}}(\boldsymbol{r}, \boldsymbol{r}', \varepsilon)$ (fig. 9) to the random phase approximation (RPA), the so-called GW approximation. Again, for a few exceptions (e.g., Minnhagen (1974) who analyzed the effect of some vertex corrections Γ in the uniform electron gas) to our knowledge all calculations start in this GW spirit (Hedin 1965a,b, Hedin and Lundqvist 1969). We list the set of equations which define GW:

$$\Sigma_{\mathrm{xc}}(\boldsymbol{r}, \boldsymbol{r}', \varepsilon) = \frac{1}{2\pi \mathrm{i}} \int \mathrm{d}\omega' \exp(\mathrm{i}0^{+}\omega') G(\boldsymbol{r}, \boldsymbol{r}', \varepsilon - \omega') W(\boldsymbol{r}, \boldsymbol{r}', \omega'), \tag{2.23a}$$

where the screened interaction W (see fig. 9) is

$$W(\boldsymbol{r}, \boldsymbol{r}', \omega) = \frac{1}{\Omega} \int \mathrm{d}^3 r'' \, \varepsilon^{-1}(\boldsymbol{r}, \boldsymbol{r}'', \omega) v(\boldsymbol{r}'' - \boldsymbol{r}') \tag{2.23b}$$

and the dielectric response $\varepsilon(\boldsymbol{r}, \boldsymbol{r}', \omega)$ is

$$\begin{aligned} \varepsilon(\boldsymbol{r}, \boldsymbol{r}', \omega) = \delta(\boldsymbol{r} - \boldsymbol{r}') + \frac{1}{2\pi \mathrm{i}} \int \mathrm{d}^3 \boldsymbol{r}'' \, v(\boldsymbol{r} - \boldsymbol{r}'') \\ \times \int \mathrm{d}\omega' \exp(\mathrm{i}0^{+}\omega') G(\boldsymbol{r}'', \boldsymbol{r}', \omega - \omega') G(\boldsymbol{r}', \boldsymbol{r}'', \omega'). \end{aligned} \tag{2.23c}$$

Equations (2.14), (2.22), and (2.23) close the self-consistent loop; to our knowledge, no one as yet solved this self-consistent loop exactly for a periodic potential (e.g., Si). (Note again that eq. (2.23), as a consequence of the approximation in eq. (2.14), is not even the self-consistent RPA; i.e., the self-energy $\Sigma_{xc}(\boldsymbol{r}, \boldsymbol{r}', \omega)$ does not appear in eq. (2.14).) The most difficult part of eq. (2.23) to implement is the construction of W; eq. (2.23b). It requires the inversion of a nonlocal dynamical matrix which itself is a complicated function of $\varphi_{\boldsymbol{p}}^{n}$ and $\varepsilon_{\boldsymbol{p}}^{n}$ (eq. (2.23)). The first simplification (suggested by Hedin) is to ignore the ω dependence in W. Equation (2.23a) then takes the form (Hedin 1965a,b, Hedin and Lundqvist 1969)

$$\Sigma_{xc}(\boldsymbol{r}, \boldsymbol{r}', \varepsilon) \approx \tfrac{1}{2}\delta(\boldsymbol{r}-\boldsymbol{r}')[W(\boldsymbol{r}, \boldsymbol{r}', 0) - v(\boldsymbol{r}-\boldsymbol{r}')] - W(\boldsymbol{r}, \boldsymbol{r}', 0) \sum_{\substack{\boldsymbol{p},n \\ \text{occupied}}} \varphi_{\boldsymbol{p}}^{n}(\boldsymbol{r})\varphi_{\boldsymbol{p}}^{n}(\boldsymbol{r}'), \tag{2.24}$$

the so-called screened exchange. One of the first to apply this form to semiconductors (to Si) was Kane (1972). (He also tested the effect of the frequency dependence in W and found it to be small.) Equation (2.24) still requires the matrix inversion of $\varepsilon^{-1}(\boldsymbol{r}, \boldsymbol{r}', 0)$. Kane (1972) ignored the off-diagonal terms (the local field effects), he used an approximate diagonal form for $\varepsilon(\boldsymbol{q})$ (Kane 1972) and the quasipartical amplitudes $\varphi_{\boldsymbol{p}}^{n}(\boldsymbol{r})$ were not calculated self-consistently. Nevertheless, his results showed significant corrections to the LDA; an increase in the gap energy over the LDA (in agreement with experiment) (see fig. 11). The self-energy formulation, although numerically very approximate at that stage, was heading in the right direction. (We should also note that Kane (1972) made the important observation that the self-energy formulation (at least in screened exchange) would always lead to an increasing gap over its LDA version.)

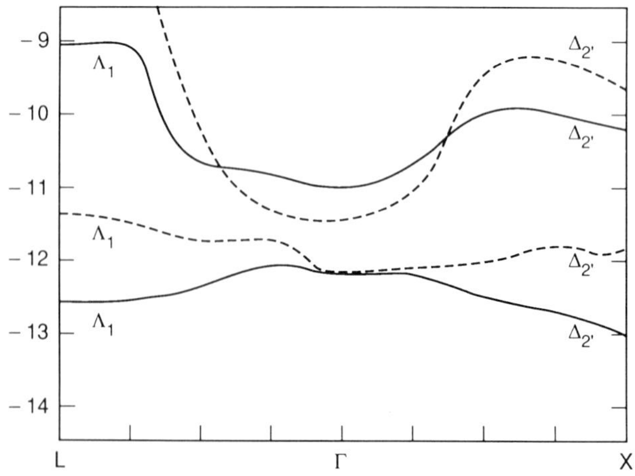

Fig. 11. Band structure of Si. Solid lines are the screened exchange self-energy for the Δ_2' and Λ_1 bands of silicon. Dashed lines are the results of a Slater approximation to the screened exchange $\alpha n^{1/3}$ with $\alpha = 0.88$; taken from Kane (1972).

Another approach for constructing the nonlocal self-energy $\Sigma_{xc}(\boldsymbol{r}, \boldsymbol{r}', \varepsilon)$ was suggested by Sham and Kohn (1966) and slightly modified by Hedin and Lundqvist (1969). It was then further generalized to include nonuniform corrections by Rasolt and Vosko (1974b). The idea of Sham and Kohn (1966) was to again exploit the uniqueness of $|\varphi_G\rangle$ as a functional of $n(\boldsymbol{r})$ (Hohenberg and Kohn 1964). Since the single-particle Green function $G(\boldsymbol{r}, \boldsymbol{r}', t)$ is given by (Abrikosov et al. 1963)

$$G(\boldsymbol{r}, \boldsymbol{r}', t) = -\mathrm{i}\langle\varphi_G|\mathscr{T}\psi(\boldsymbol{r}, t)\psi^\dagger(\boldsymbol{r}', 0)|\varphi_G\rangle \tag{2.25}$$

(where $\mathscr{T}$ is the time ordering operator) it *must* be given uniquely by $n(\boldsymbol{r})$; so must $G(\boldsymbol{r}, \boldsymbol{r}', \omega)$ and from eq. (2.20) so must $\Sigma_{xc}(\boldsymbol{r}, \boldsymbol{r}', \omega)$. The approximation for $\Sigma_{xc}(\boldsymbol{r}, \boldsymbol{r}', \omega)$ proceeds very much as in the LDA; i.e., we take the uniform density n_0 in the uniform electron gas form of the self-energy $\Sigma_h(\boldsymbol{r} - \boldsymbol{r}', \varepsilon, n_0)$ and replace it by $n(\boldsymbol{r})$. The form Sham and Kohn (1966) suggest is

$$\Sigma_{xc}(\boldsymbol{r}, \boldsymbol{r}', \varepsilon) \approx \Sigma_h(\boldsymbol{r} - \boldsymbol{r}', \varepsilon - \mu + \mu_h(n(\boldsymbol{r}_0)); n(\boldsymbol{r}_0)), \tag{2.26}$$

where μ is the chemical potential, $\mu_h(\boldsymbol{r}_0)$ is the chemical potential of the homogeneous electron gas and $\boldsymbol{r}_0 = \frac{1}{2}(\boldsymbol{r} + \boldsymbol{r}')$. Σ_h is now calculated in the RPA (i.e., the uniform electron gas version of eq. (2.23)); it contains, of course, both the energy dependence of the quasiparticle ε and the density variation (i.e., $n(\boldsymbol{r}_0)$). This form was first applied to metals by Rasolt and Vosko (1974a,b).

One can further reduce eq. (2.26) to a "WKB" form (Sham and Kohn 1966) by writing

$$\varphi_{\boldsymbol{p}}^n(\boldsymbol{r}) \approx A(\boldsymbol{r}) \exp[\mathrm{i}\boldsymbol{p}(\boldsymbol{r}) \cdot \boldsymbol{r}] \tag{2.27}$$

and then neglecting the $\boldsymbol{r}$ dependence of $A(\boldsymbol{r})$ and $\boldsymbol{p}(\boldsymbol{r})$ (when Σ_h of eqs. (2.20b), (2.22), and (2.26) operates on $|\varphi_{\boldsymbol{p}}^n\rangle$), we get (Sham and Kohn 1966)

$$\Sigma_h|\varphi_{\boldsymbol{p}}^n\rangle = \Sigma_h(\boldsymbol{p}(\boldsymbol{r}), \varepsilon - \mu + \mu_h(n(\boldsymbol{r})); n(\boldsymbol{r}))\varphi_{\boldsymbol{p}}^n(\boldsymbol{r}). \tag{2.28}$$

To solve for $\boldsymbol{p}(\boldsymbol{r})$ Sham and Kohn (1966) suggested the approximation $\mu = V(\boldsymbol{r}) + V_H(\boldsymbol{r}) + \mu_h(n(\boldsymbol{r}))$. Using eqs. (2.22), (2.27), and (2.28) we find, for states on the Fermi level that (Sham and Kohn 1966)

$$\Sigma_h|\varphi_{\boldsymbol{p}}^n\rangle = \Sigma_h(\boldsymbol{p}(\boldsymbol{r}), \mu_h(n(\boldsymbol{r})), n(\boldsymbol{r}))\varphi_{\boldsymbol{p}}^n(\boldsymbol{r}), \tag{2.29}$$

with $|\boldsymbol{p}(\boldsymbol{r})| = (3\pi^2 n(\boldsymbol{r}))^{1/3}$. (Incidentally, this is identical to v_{xc} in eq. (2.18) in the LDA; it makes a very *loose* connection between quasiparticle structure and DFT "eigenvalues".) This WKB approximation was applied to the band structure of Si by Wang and Pickett (1983). The modification to Si was to introduce a more appropriate diagonal form (diagonal in $\boldsymbol{q}$) for the dielectric constant ε of eq. (2.23) and for G in eq. (2.23a) (by using a dielectric constant which reflects the gap of the semiconductor (Levine and Louie 1982).) Their results for Si show significant improvement over their LDA (Wang and Pickett 1983), but the validity of all these uncontrolled approximations is hard to judge.

Serious advances over the initial work of Kane (1972) using eqs. (2.14) and (2.23) have been made by Hybersten and Louie (1985, 1986, 1988), Godby et al. (1986), von der Linden and Horsch (1988), and Hott (1991). The primary improvement is in the calculation of $\varepsilon^{-1}(\boldsymbol{r}, \boldsymbol{r}', \omega)$. In the work of Hybersten and Louie (1985, 1986, 1988) and Hott (1991) this dielectric constant was calculated in the following way:

(a) The nonlocal structure (i.e., local field effects) were calculated using the inverse of the static dielectric function $\varepsilon^{-1}_{\boldsymbol{G},\boldsymbol{G}'}(q, \omega = 0)$ in the LDA. (Since the static dielectric function is a ground state property it is given exactly in the DFT (see, however, our comment about the dynamic dielectric function below); but, of course, it is not given exactly in the LDA).

(b) To extend the dielectric matrix to finite frequency they proposed a generalized plasmon–pole model since Im $\varepsilon^{-1}_{\boldsymbol{G},\boldsymbol{G}'}(q, \omega)$ is generally a peaked function in ω, i.e.,

$$\operatorname{Im} \varepsilon^{-1}_{\boldsymbol{G},\boldsymbol{G}'}(\boldsymbol{q}, \omega) \approx A_{\boldsymbol{G},\boldsymbol{G}'}(\boldsymbol{q})[\delta(\omega - \tilde{\omega}_{\boldsymbol{G}\boldsymbol{G}'}(\boldsymbol{q})) - \delta(\omega + \tilde{\omega}_{\boldsymbol{G},\boldsymbol{G}'}(\boldsymbol{q}))]. \tag{2.30}$$

The $A_{\boldsymbol{G},\boldsymbol{G}'}(\boldsymbol{q})$ and $\tilde{\omega}_{\boldsymbol{G},\boldsymbol{G}'}(\boldsymbol{q})$ are given by the Kramers–Kronig relation which relates them to the static response, i.e.,

$$\varepsilon^{-1}_{\boldsymbol{G},\boldsymbol{G}'}(\boldsymbol{q}, 0) = \delta_{\boldsymbol{G},\boldsymbol{G}'} + \frac{2}{\pi} P \int_0^\infty \mathrm{d}\omega\, \omega^{-1} \operatorname{Im} \varepsilon^{-1}_{\boldsymbol{G},\boldsymbol{G}'}(\boldsymbol{q}, \omega) \tag{2.31}$$

and the Johnson sum rule (Johnson 1974)

$$\int_0^\infty \mathrm{d}\omega\, \omega \operatorname{Im} \varepsilon^{-1}_{\boldsymbol{G},\boldsymbol{G}'}(\boldsymbol{q}, \omega) = -\frac{\pi}{2}\omega_{\mathrm{p}}^2 \frac{n(\boldsymbol{G}-\boldsymbol{G}')}{n(0)} \frac{(\boldsymbol{q}+\boldsymbol{G})\cdot(\boldsymbol{q}+\boldsymbol{G}')}{|\boldsymbol{q}+\boldsymbol{G}|^2}, \tag{2.32}$$

where $n(\boldsymbol{G})$ is the Fourier component of the crystalline electron density and ω_{p} the plasma frequency.

We wish to make one final observation about the dynamical response (like $\varepsilon(\boldsymbol{r}, \boldsymbol{r}', \omega)$). It is often stated that such a dynamical response is not a ground state property and, therefore, outside the validity of the DFT. However, the dielectric constant $\varepsilon(\boldsymbol{r}, \boldsymbol{r}', t)$ is directly related to the time-dependent density–density response function $\chi(\boldsymbol{r}, \boldsymbol{r}', t)$, i.e.,

$$\chi(\boldsymbol{r}, \boldsymbol{r}', t) = -\mathrm{i}\langle \varphi_{\mathrm{G}} | \mathscr{T} \rho(\boldsymbol{r}, t) \rho(\boldsymbol{r}', 0) | \varphi_{\mathrm{G}} \rangle. \tag{2.33}$$

Just like the self-energy or single-particle Green function (see eq. (2.25)), it is then given uniquely by the density, i.e., $\chi(\boldsymbol{r}, \boldsymbol{r}', \omega) = \chi(\boldsymbol{r}, \boldsymbol{r}', \omega, n(\boldsymbol{r}))$. This is also true for the current–current response (eq. (2.12)) or for that matter any higher-order response. This is in a way related to the time-dependent DFT of Rünge and Gross (1984) already mentioned before. It is then possible in principle, to approach optical properties within this spirit. However, in practice, the uncertainty in the time-dependent functionals, and in particular vis-à-vis band gaps (see next §) will pose a very difficult challenge.

This $\varepsilon^{-1}_{\boldsymbol{G},\boldsymbol{G}'}(\boldsymbol{q}, \omega)$ is applied to eq. (2.23) without any further approximation (Hybersten and Louie 1985, 1986, 1988, Hott 1991). In both cases, the wavefunctions $\varphi^n_{\boldsymbol{p}}(\boldsymbol{r})$, which enter the $G(\boldsymbol{r}, \boldsymbol{r}', \varepsilon - \omega')$ of eq. (2.23a), are not calculated self-consistently using eq. (2.22). In Hybersten and Louie (1985, 1986, 1988) $\varphi^n_{\boldsymbol{p}}(\boldsymbol{r})$ from the LDA are used, in Hott (1991) the EPM $\varphi^n_{\boldsymbol{p}}(\boldsymbol{r})$ are used.

In table 1 we list their results. It is clear that there are significant differences; the later calculations of Hott (1991) are in considerably worse agreement with experi-

Table 1
Values for the band gaps of Si, Ge, and diamond in eV by Hybersten and Louie (1986) and by Hott (1991) as compared with experiments.

References	Si	Ge	Diamond
Hybersten and Louie (1986)	1.29	0.75	5.6
Hott (1991)	1.21	0.59	7.16
Experiment	1.17	0.86	5.48

ments. In figs. 12 and 13 we present the detailed self-energy band structure of Si and Ge (Hott 1991).

In summary, there is compelling evidence that the self-energy band structure calculations in the GW approximation capture much of the experimentally observed optical measurements of semiconductors (to what level of accuracy is still uncertain). Nevertheless, even low-order RPA (and even the non-self-consistent version) has not been solved exactly. Further effort on improving these calculations will help clear up some of these uncertainties. There is even stronger evidence that the DFT band structure is inappropriate for describing these optical properties; but what about the band gap? A band gap (ignoring quasiparticle–hole interactions which lead to

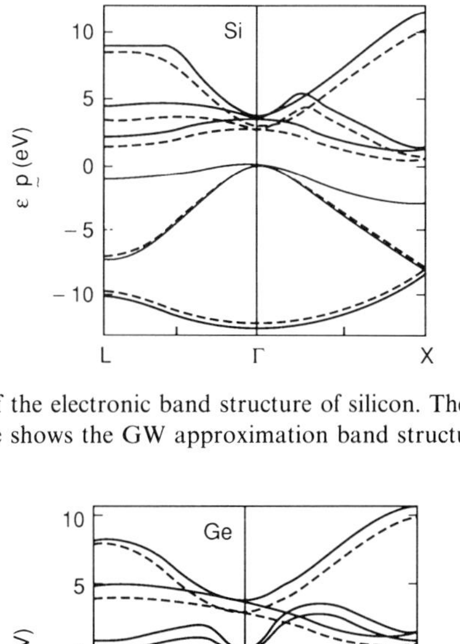

Fig. 12. A recent calculation of the electronic band structure of silicon. The dashed line shows the LDA band structure. The solid line shows the GW approximation band structure; taken from Hott (1991).

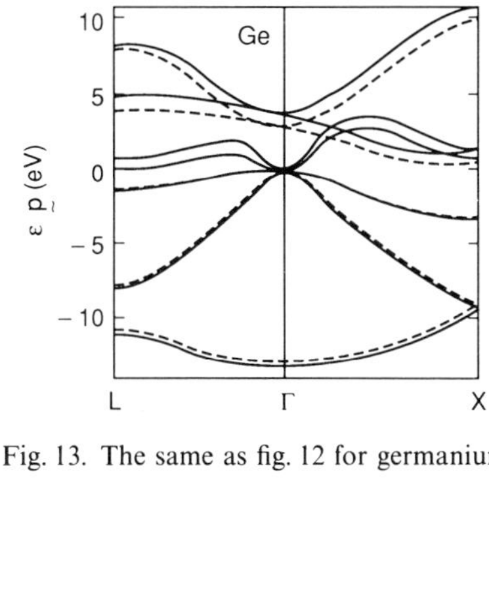

Fig. 13. The same as fig. 12 for germanium.

relatively very small corrections) is a ground state property (see below); so what is wrong with the DFT?

3. Band gap discontinuities

What makes a semiconductor is, of course, the gap between the ground state and the next higher excited state. Predicting the correct band gap has, therefore, played a central role in the band theory of semiconductors (see previous §). This will become even more important, e.g., when we need to understand the salient features in the calculation, which will change the gap when we form various heterojunction superlattices. Recent advances in pico- and femtosecond spectroscopy (Liu et al. 1982, von der Linde and Fabricius 1982, Lompre et al. 1982, Shank et al. 1983, Bigot et al. 1990, Rasolt and Kurz 1985, Rasolt 1986, Rasolt et al. 1987, Rasolt and Perrot 1988a,b) and in heavy doping (Sernelius 1986, 1988) and the effect of the corresponding carriers on the gap are other examples. The self-energy quasiparticle band structure takes care of it (see § 2), but there remains a fundamental issue to be resolved.

If we assume that in a neutral excitation the particle and hole do not interact (this is certainly not always true; excitons are just one counterexample; see § 2) then the gap energy E_g is given by (Perdew et al. 1982, Sham and Schlüter 1983)

$$E_g = E[n_{N+1}(N+1)] - 2E[n_N(N)] + E[n_{N-1}(N-1)], \tag{3.1}$$

where $n_M(N)$ is the density obtained by summing M orbitals in the Kohn–Sham (KS) equations of the interacting N-electron ground state (see eqs. (2.15)–(2.19)). (Note that for $E_V[n]$ the external potential $V(r)$ is fixed. Here, however, we will consider the *full* functional $E[n]$ of only the density.) All the energies in eq. (3.1) are rigorously calculable, in principle, in the FDT. The question then is whether the poor agreement (Perdew et al. 1982, Perdew and Levy 1983, Sham and Schlüter 1983, Kleinman 1984, Sham 1985, Sham and Schlüter 1985, Lannoo et al. 1985, Godby et al. 1986, Wang and Pickett 1983, Pickett and Wang 1984, Hybersten and Louie 1985, 1986, 1988, von der Linden and Horsch 1988, Hott 1991) in E_g originates in the lack of precise knowledge for the functionals $E[n_M(N)]$ or in discontinuities of $E[n_M(N)]$ which arise when we add or subtract a particle below or above the gap as suggested first by Perdew and Levy (1983) and Sham and Schlüter (1985). Here we start with a slightly different emphasis (Rasolt 1987a) and analyze the implication of a semiconducting gap on the structure of $E[n]$ at a fixed number N. We show that along with the discontinuities in $E[n_N(N)]$ as a function of particle number N (Perdew et al. 1982, Sham and Schlüter 1983), $E[n_N(N)]$, for a fixed number of particles, must be nonanalytic in the density deviation $n_N(\mathbf{r})$ from the uniform density n_0; the two are intimately related. Our discussion should shed some further insight on the assumed structure of $E[n]$ which was taken for granted even for ground state calculations.

3.1. Weak external potential

We start by considering a weak one-dimensional external potential $V(x)$ with only one reciprocal-lattice vector component G_x, i.e.,

$$V(x) = 2V(G_x)\cos(G_x x). \tag{3.2}$$

We take the $V(G_x)$ to be real and negative. For such a system the wavefunctions above and below the gap, at reciprocal vector $k_x = \frac{1}{2}G_x$, are, respectively,

$$\begin{aligned} \psi_c(x) &= \left(\frac{2}{L}\right)^{1/2} \sin \tfrac{1}{2}G_x x, \\ \psi_v(x) &= \left(\frac{2}{L}\right)^{1/2} \cos \tfrac{1}{2}G_x x. \end{aligned} \tag{3.3}$$

Here L is the length of the system. The corresponding densities are

$$\begin{aligned} \rho_c(x) &= \frac{2}{L}\sin^2 \tfrac{1}{2}G_x x, \\ \rho_v(x) &= \frac{2}{L}\cos^2 \tfrac{1}{2}G_x x. \end{aligned} \tag{3.4}$$

If we choose exactly the right number of electrons N then we have a legitimate model for a direct-gap semiconductor at $k_x = \frac{1}{2}G_x$.

3.1.1. No interaction between the particles

The Schrödinger equation in this case is

$$\left(-\frac{\hbar^2}{2m}\frac{\partial^2}{\partial x^2} + V(x)\right)\psi_{k_x}(x) = \varepsilon_{k_x}\psi_{k_x}(x). \tag{3.5}$$

Using eq. (3.3), we get at once a gap energy at $k_x = \frac{1}{2}G_x$ of

$$E_g = -2V(G_x); \tag{3.6}$$

this is rigorous. Let us next take the DFT approach. For N' particles and $M = N'$

$$E[n_{N'}(N)] = E_0(N') - \tfrac{1}{2}\int \mathrm{d}x \int \mathrm{d}x'\, \delta n_{N'}(N', x)\delta n_{N'}(N', x')K(x - x', N'), \tag{3.7}$$

where $E_0(N')$ is the kinetic energy for a uniform electron gas of N' particles, $\delta n_N(x) = n_N(x) - n_0$ and

$$K(x - x', N') \equiv \int \frac{\mathrm{d}q}{2\pi}\chi_0^{-1}(q, N')\exp[\mathrm{i}q(x - x')], \tag{3.8}$$

with $\chi_0(q, N')$ the uniform noninteracting susceptibility (Ziman 1964).

There is no interaction here between particles and therefore

$$n_{N+1}(N+1, x) = n_N(N, x) + \rho_c(x) \tag{3.9}$$

and

$$n_{N-1}(N-1, x) = n_N(N, x) - \rho_v(x),$$

and from linear response

$$\delta n_{N'}(N', q) = -V(q)\chi_0(N', q). \tag{3.10}$$

Using eqs. (3.1), (3.4), and (3.7)–(3.9) we find

$$E_g + 2V(G_x) = [E_0(N+1) + E_0(N-1) - 2E_0(N)]. \tag{3.11}$$

Now the E_0 are energies for the uniform noninteracting electron gas. These cannot be discontinuous to order $1/N$ in the particle number since such discontinuities would introduce gaps for excitations around the Fermi surface of a metal (for example). We therefore get the right energy gap for this noninteracting model of a small-gap semiconductor. As we show next, this result *does not* extend to the interacting case. In fact, while the noninteracting $E[n(N)]$ is continuous across the gap, as a function of particle number N, its form is different than eq. (3.7) even in this small-$V(G_x)$ limit. We need to *emphasize* that the discontinuity (or nonanaliticity) we are talking about is in the *internal structure* of $E[n]$, *not* in the density change from a valence state (i.e., ρ_v) to a conduction state (i.e., ρ_c). The internal structure of eq. (3.7) is obviously analytic in n_0 and n. The gap in eq. (3.11) ($E_g = -2V(G_x)$) arises strictly from the discontinuity of $\rho_c - \rho_v$; this is no big deal.

3.1.2. Interacting systems

There are only two changes in eqs. (3.7) (3.11) when interparticle interactions are included. First, obviously $\chi_0(N', q)$ is replaced by the interacting $\chi(N', q)$. Second, the density $n_N(N, x)$ in eq. (3.9) changes when $\rho_c(x)$ is added ($\rho_c(x)$ in this weak limit is independent of the potential and therefore remains the same as in eq. (3.4)). $\delta n_N(N, G_x)$ is now given by

$$\delta n_c(G_x) = -[V(G_x) + v(G_x)\rho_c(G_x)]\chi(N+1, G_x) \tag{3.12}$$

and

$$\delta n_v(G_x) = -[V(G_x) - v(G_x)\rho_v(G_x)]\chi(N-1, G_x).$$

This is the conventional definition for χ but the response is both to $V(G_x)$ and the additional particle $\rho(G_x)$. Turning again to eqs. (3.1), (3.4), (3.7), and (3.8) we get for the gap

$$E_g = -2V(G_x)[1 - v(G_x)\chi(G_x)]. \tag{3.13}$$

This result which follows rigorously from eq. (3.7) cannot be right. The correct E_g must follow from the Dyson equation (see § 2)

$$\left(-\frac{\hbar^2}{2m}\frac{\partial^2}{\partial x^2} + V(x) + V_H(x)\right)\psi_{k_x}(x) + \int dx'\, \Sigma_{xc}(x, x', \varepsilon_{k_x})\psi_{k_x}(x') = \varepsilon_{k_x}\psi_{k_x}(x). \tag{3.14}$$

As an example, consider the Hartree–Fock approximation described in fig. 14. The gap E_g is then given by

$$E_g = -2V(G_x)[1 - v(G_x)\chi(G_x)] - 2V(G_x)\Lambda(G_x - k_x, k_x), \quad k_x = \tfrac{1}{2}G_x, \tag{3.15a}$$

where

$$\Lambda(G_x - k_x, k_x) = \int dx \int dx' \exp[i(G_x - k_x)x] \exp(-ik_x x')\Lambda(x, x'), \tag{3.15b}$$

and the last term in eq. (3.15a) is *absent* in eq. (3.13). Incidentally, eq. (3.13) is not even the DFT gap which this model would predict from eqs. (2.18) and (2.19).

The only assumption we made in reaching these contradictions is that $E[n_N(N)]$ has a power expansion in $\delta n_N(x)$. As we will show, this is *not so* for a semiconductor. The key point is that electron states k_x next to the zone boundary are strongly modified no matter how small the external potential $V(G_x)$ is. We first, however, turn to the issue of the existence of such an expansion in δn. This, as we shall see, is at the heart of $v_{xc}(n(\boldsymbol{r}))$ (eq. (2.18c)); it is not generally true and it is the reason the DFT fails to give the correct band gaps.

3.2. Expansion of $E[n]$ in powers of the nonuniform density deviation

The following construction of the LDA of $E(n)$ has been largely taken for granted: calculate $E(n_0)$ for the uniform electron gas and replace n_0 by the local density $n(\boldsymbol{r}) = n_0 + \delta n(\boldsymbol{r})$. However, there is a lot more to it than that.

As we shall see next, this procedure, which is incorrect for systems with a gap, is at the root of the contradictions (just discussed) and the gap discontinuities. We, therefore, examine this procedure in some detail. We will also gain some important insights to $E[n]$ which have largely been left unappreciated (see, however, Geldart and Rasolt (1987)).

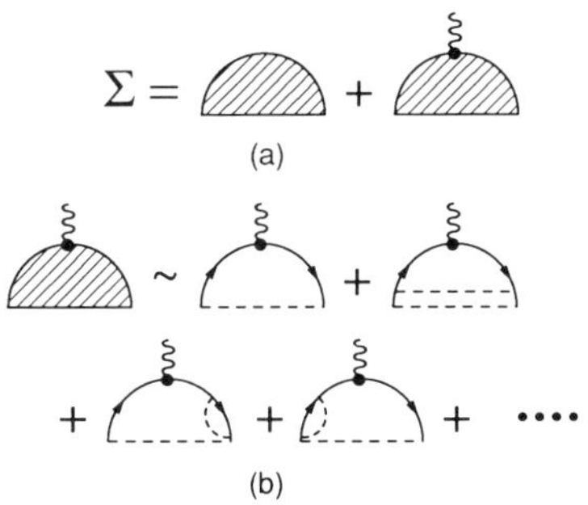

Fig. 14. (a) Diagrammatic expansion of the self-energy, to first order in the external potential $V(x)$. The cross and wiggly line corresponds to $V(x)$ screened by the interacting dielectric function and the shaded semicircle to the vertex function $\Lambda(x, x')$. (b) Lowest-order Hartree–Fock contributions to $\Lambda(x, x')$. Dashed lines are the interparticle interactions and arrowed lines the electron propagators.

Let us again consider eq. (3.7) for the exchange and correlation this time and now in three dimensions, i.e.,

$$E_{xc}[n] = E_{xc}(n_0) + \tfrac{1}{2}\int d^3r \int d^3r'\, \delta n(\boldsymbol{r}) K_{xc}(\boldsymbol{r}-\boldsymbol{r}', n_0)\delta n(\boldsymbol{r}'). \tag{3.16}$$

In Fourier space

$$E_{xc}[n] = E_{xc}(n_0) + \frac{1}{2\Omega}\sum_{\boldsymbol{q}} |\delta n(\boldsymbol{q})|^2 K_{xc}(\boldsymbol{q}, n_0). \tag{3.17}$$

Assuming $\delta n(\boldsymbol{r})$ to be slowly varying

$$E_{xc}[n] \approx E_{xc}(n_0) + \frac{1}{2\Omega}\sum_{\boldsymbol{q}} |\delta n(\boldsymbol{q})|^2 [K_{xc}(0, n_0) + \tfrac{1}{2}q^2 K''_{xc}(0, n_0) + \cdots], \tag{3.18a}$$

$$E_{xc}[n] = E_{xc}(n_0) + \int d^3r [\tfrac{1}{2}K_{xc}(0, n_0)|\delta n(\boldsymbol{r})|^2 + \tfrac{1}{4}K''_{xc}(0, n_0)|\nabla n(\boldsymbol{r})|^2 + \cdots], \tag{3.18b}$$

where the primes are derivatives with respect to $\boldsymbol{q}$. Notice the explicit dependence of $E[n]$ on $\delta n(\boldsymbol{r})$ and n_0 (or equivalently N) *separately*. We need to consider two more technical details: (i) in eqs. (3.16)–(3.18) the nonuniform electron gas is exactly neutralized by a positive background. (ii) Accounting for this positive background the K_{xc} is given by the irreducible (Abrikosov et al. 1963, Geldart and Vosko 1966, Geldart and Taylor 1970a,b) response function $\Pi(\boldsymbol{q})$, i.e.,

$$K_{xc}(\boldsymbol{q}, n_0) = [\Pi^{-1}(\boldsymbol{q}) - \Pi_0^{-1}(\boldsymbol{q})], \tag{3.19}$$

where the $\Pi(\boldsymbol{q})$ is related to the static dielectric function of the uniform electron gas by

$$\varepsilon(\boldsymbol{q}) = 1 + \frac{4\pi^2}{q^2}\Pi(\boldsymbol{q}). \tag{3.20}$$

Now, here is the key point. By virtue of $\partial E/\partial N = \mu$ and the Ward–Pitayevski identity (Ward 1950) leading to

$$\partial\mu/\partial n_0 = 1/\Pi(0), \tag{3.21}$$

the first two terms on the right-hand side of eq. (3.18b) are recognized as the first two terms in the expansion of the LDA in powers of δn; i.e.,

$$E_{xc}^{LDA}[n] = \int d^3r\, A_{xc}(n(\boldsymbol{r})), \tag{3.22}$$

where $A_{xc}(n(\boldsymbol{r}))$ is the exchange correlation energy per unit volume of a uniform system of the local density $n(\boldsymbol{r}) = n_0 + \delta n(\boldsymbol{r})$. Expansion of eq. (3.22) to second order in $\delta n(\boldsymbol{r})$ yields the above-mentioned terms of eq. (3.18b).

The third term on the right-hand side of eq. (3.18b) goes beyond viewing the system as quasi-uniform and yields the lowest-order gradient correction to the LDA. We must emphasize that while A_{xc} is given fully by the density dependence of the energy of a uniform system, $K''_{xc}(q=0)$ is not obtainable from this knowledge of the energy of a uniform system but requires information concerning the wavenumber dependence of the density fluctuation correlation function of the uniform system. Higher-order terms in powers of δn could be included beyond the second-order perturbation expansion in V. This would introduce higher-order response functions, involving triplet, quadruplet, etc., correlation functions of the interacting uniform system depending on a correspondingly increased number of independent wavevectors.

$$E_{xc}[n] = E_{xc}(n_0) + \Omega \sum_{n=2}^{\infty} \frac{1}{n!} \mathrm{K}_n(\boldsymbol{q}_1, \boldsymbol{q}_2, \ldots)(\delta n(\boldsymbol{q}_n))^n. \tag{3.23}$$

If the simple structure associated with gradient corrections is desired, we could expand all of these new terms in powers of their various wavevector dependences to obtain contributions involving

$$\begin{aligned} \int \mathrm{d}^3 r [&B_{xc}(n_0)(\nabla n(\boldsymbol{r}))^2 + C_{xc}(n_0)(\nabla^2 n(\boldsymbol{r}))^2 + D_{xc}(n_0)(\nabla^2 n(\boldsymbol{r}))(\nabla n(\boldsymbol{r}))^2 \\ &+ E_{xc}(n_0)(\nabla n(\boldsymbol{r}))^4 + \cdots]. \end{aligned} \tag{3.24}$$

Note that the last three terms are of second, third, and fourth order, respectively, in V just by counting factors of n. There will also be terms involving zero, two, four, etc., spatial derivatives each with various factors of $n(\boldsymbol{r})$. When all results are collected using the Ward–Pitayevski identities for the higher K_n (see Brovman and Kagan (1969) and Rasolt (1983)), i.e.,

$$\lim_{\boldsymbol{q} \to 0} K_{n+1}(\boldsymbol{q}_1, \cdots, \boldsymbol{q}_n, \boldsymbol{q}, n_0) = \frac{\partial K_n}{\partial n_0}(\boldsymbol{q}_1, \cdots, \boldsymbol{q}_n, n_0), \tag{3.25}$$

the net effect is that all the gradient coefficients $A_{xc}(n_0)$, $B_{xc}(n_0)$, etc., and their various derivatives with respect to n_0 will enter with *precisely* the right series of powers of $\delta n(\boldsymbol{r})$ so that each gradient coefficient is evaluated at the local density $n(\boldsymbol{r}) = n_0 + \delta n(\boldsymbol{r})$. Since $\nabla \delta n(\boldsymbol{r}) = \nabla n(\boldsymbol{r})$ is trivially true, it follows that this particular partial resummation of the perturbation expansion leads to (Geldart and Rasolt 1987)

$$E_{xc}[n] = E_{xc}^{\mathrm{LDA}}[n] + E_{xc}^{\mathrm{g}}[n], \tag{3.26}$$

where the LDA is indeed given by eq. (3.22) *to all* orders in the perturbation expansion and the gradient corrections are similarly given by (Rasolt and Geldart 1975, 1976)

$$\begin{aligned} E^{\mathrm{g}}[n] = \int \mathrm{d}^3 r [&B_{xc}(n(\boldsymbol{r}))(\nabla n(\boldsymbol{r}))^2 + C_{xc}(n(\boldsymbol{r}))(\nabla^2 n(\boldsymbol{r}))^2 \\ &+ D_{xc}(n(\boldsymbol{r}))(\nabla^2 n(\boldsymbol{r}))(\nabla n(\boldsymbol{r}))^2 + E_{xc}(n(\boldsymbol{r}))(\nabla n(\boldsymbol{r}))^4 + \cdots]. \end{aligned} \tag{3.27}$$

Let us summarize this very important result. If the expansion of eq. (3.23) for the exchange correlation density functional of the ground state energy exists then the initial expansion of eq. (3.18b), which is an explicit function of n_0 and a functional of $\delta n(\boldsymbol{r})$, becomes a functional of *only* $n(\boldsymbol{r}) \equiv n_0 + \delta n(\boldsymbol{r})$. Under these circumstances (which, as we will see below, are highly accurate in metals but are incorrect for semiconductors), no discontinuities can occur in $E[n]$. The gap is then given exactly by the DFT (i.e., by eqs. (2.18) and (2.19)), here is why.

If the dependence in V_{eff} of eq. (2.18) is only on the full density $n(\boldsymbol{r})$ then this gap is given exactly by $\varepsilon_g = \varepsilon_c - \varepsilon_v$; where ε_v is the highest $\varepsilon^n_{\boldsymbol{p}_N}$ in the valence band and ε_c is the next occupied $\varepsilon^{n+1}_{\boldsymbol{p}_{N+1}}$, i.e., the lowest conduction band state. *Proof*: In eq. (2.19) the additional electron adds a density $\rho_c(\boldsymbol{r})$ (to $n(\boldsymbol{r})$), an additional $\varepsilon^{n+1}_{\boldsymbol{p}_{N+1}} \equiv \varepsilon_c$, and changes the $\Sigma_{\boldsymbol{p}} \varepsilon^n_{\boldsymbol{p}}$ to $\Sigma_{\boldsymbol{p}} \varepsilon'^n_{\boldsymbol{p}}$. The change in this sum can be calculated exactly to the required order in $1/N$ using first-order perturbation theory, i.e.,

$$\begin{aligned} \sum_{\boldsymbol{p}} \varepsilon'^n_{\boldsymbol{p}} &= \sum_{\boldsymbol{p}} \left\langle \varphi^n_{\boldsymbol{p}} \left| \frac{\delta V_{\text{eff}}(n(\boldsymbol{r}))}{\delta n} \rho_c(\boldsymbol{r}) \right| \varphi^n_{\boldsymbol{p}} \right\rangle + \sum_{\boldsymbol{p}} \varepsilon^n_{\boldsymbol{p}} \\ &= \int \mathrm{d}^3 r \, \frac{\delta V_{\text{eff}}(n(\boldsymbol{r}))}{\delta n} \rho_c(\boldsymbol{r}) n(\boldsymbol{r}) + \sum_{\boldsymbol{p}} \varepsilon^n_{\boldsymbol{p}}. \end{aligned} \tag{3.28}$$

Using eq. (3.28) in (2.19) along with eq. (2.18b) we get at once that

$$E_g^{\text{DFT}} \equiv \varepsilon_g \equiv E_V[n + \rho_c] - E_V[n] = \varepsilon_c - \varepsilon_v. \tag{3.29}$$

Suppose, however, that the expansion of eq. (3.23) did not exist. Then $E(n)$ becomes an explicit functional of n and a functional of n_0; i.e., $E[n_0, n]$. Similar calculation using now eq. (3.1) leads to a gap given by

$$\begin{aligned} E_g = \varepsilon_g + \lim_{n_+ \to n_0, n_- \to n_0} \int \frac{\mathrm{d}^3 r}{\Omega} \Bigg\{ & \left[\frac{\delta E_{\text{xc}}}{\delta n_0^+} (n_0^+, n(\boldsymbol{r})) - v_{\text{xc}}(n_0, n(\boldsymbol{r})) \right] \\ & - \left[\frac{\delta E_{\text{xc}}}{\delta n_0} (n_0^-, n(\boldsymbol{r})) - v_{\text{xc}}(n_0, n(\boldsymbol{r})) \right] \Bigg\}. \end{aligned} \tag{3.30}$$

The limits of n_+ and n_- are from above and below n_0 and account for possible discontinuities (and nonanalyticities) in $E[n_0, n]$. The last term in eq. (3.30) corrects the DFT ε_g. This correction to the gap ε_g is the "band gap discontinuity." Now how does E or E_{xc} become nonanalytic in $n(\boldsymbol{r})$ (or equivalently $\delta n(\boldsymbol{r})$). To see this, we return to the functional of the noninteracting electron, one-dimensional model (see previous subsection) where the density and E are given exactly (Kleinman 1986), i.e.,

$$n(G_x) = \frac{2V(G_x)}{\pi G_x} \ln \left[\frac{|\frac{1}{2}\tilde{k}_F G_x| + [\frac{1}{4}\tilde{k}_F^2 G_x^2 + V^2(G_x)]^{1/2}}{\frac{1}{4}G_x^2 + [\frac{1}{16}G_x^4 + V^2(G_x)]^{1/2}} \right] \tag{3.31}$$

and

$$\begin{aligned} E[n] = \frac{1}{\pi} \Bigg[& \tfrac{1}{3}(\tilde{k}_F + \tfrac{1}{2}G_x)^3 + \frac{2V^2(G_x)}{G_x} \ln \left(\frac{|\frac{1}{2}\tilde{k}_F G_x| + [\frac{1}{4}\tilde{k}_F^2 G_x^2 + V^2(G_x)]^{1/2}}{\frac{1}{4}G_x^2 + [\frac{1}{16}G_x^4 + V^2(G_x)]^{1/2}} \right) \\ & + |\tilde{k}_F| [\tfrac{1}{4}\tilde{k}_F^2 G_x^2 + V^2(G_x)]^{1/2} - \tfrac{1}{2}\tilde{k}_F^2 G_x \\ & - \tfrac{1}{2}G_x \{ [\tfrac{1}{16}G_x^4 + V^2(G_x)]^{1/2} - \tfrac{1}{4}G_x^2 \} \Bigg], \end{aligned} \tag{3.32}$$

where this time we set $m = 1$ and where $\tilde{k}_F = (\frac{1}{2}\pi)n_0 - \frac{1}{2}G_x$ and n_0 is the uniform density (when $\tilde{k}_F = 0$ the system is a semiconductor). Clearly from eqs. (3.31) and (3.32), $E[n_N(N)]$ has no expansion in small $n(G_x)$ for a semiconductor and is entirely different from the assumed form of eq. (3.23) or (3.24). Again, the point is that electron states k_x next to the zone boundary are strongly modified no matter how small the external potential $V(G_x)$ is. (Note, in view of this discussion we expect $E(n)$ in a Mott insulator to behave differently.) While our model applies to zone boundary direct-gap semiconductors similar nonanalytic behaviour will be part of the band structure of any type of semiconductor.

To demonstrate such explicit nonanalyticity for E_{xc} is more difficult but such explicit dependences on n_0 are clearly there. We can, however, readily suggest an approximate form for the last two terms in eq. (3.30) from eq. (2.22) (the exact form will be discussed in § 3.3). If we add and subtract $v_{xc}(n_0, n(\boldsymbol{r}))$ in eq. (2.22), we get

$$-\frac{\hbar^2}{2m}\nabla^2\varphi_{\boldsymbol{p}}^n(\boldsymbol{r}) + [V(\boldsymbol{r}) + V_H(\boldsymbol{r}) + v_{xc}(n_0, n(\boldsymbol{r}))]\varphi_{\boldsymbol{p}}^n(\boldsymbol{r}) + \int d^3r'\,[\Sigma_{xc}(\boldsymbol{r}, \boldsymbol{r}', \varepsilon_{\boldsymbol{p}}^n) - v_{xc}(n_0, n(\boldsymbol{r}))\delta(\boldsymbol{r}-\boldsymbol{r}')]\varphi_{\boldsymbol{p}}^n(\boldsymbol{r}') = \varepsilon_{\boldsymbol{p}}^n\varphi_{\boldsymbol{p}}^n(\boldsymbol{r}). \tag{3.33}$$

Now, the full gap E_g is given by the $\varepsilon_{\boldsymbol{p}}^n$ of eq. (3.33). ε_g is given by the solution of the first two terms. If the full quasiparticle $\varphi_{\boldsymbol{p}}^n$ and the DFT $\varphi_{\boldsymbol{p}}^n$ are the same (they are not), then from eqn. (3.33) and (3.30) we get at once that the band gap discontinuity Δ_{xc} (Perdew et al. 1982, Sham and Schlüter 1983) is given by

$$\begin{aligned}\Delta_{xc} &\equiv \frac{1}{\Omega}\int d^3r\left\{\left[\frac{\delta E_{xc}}{\delta n_+}(n_+, n(\boldsymbol{r})) - v_{xc}(n_0, n(\boldsymbol{r}))\right] - \left[\frac{\delta E_{xc}}{\delta n_-}(n_-, n(\boldsymbol{r})) - v_{xc}(n_-, n(\boldsymbol{r}))\right]\right\}\\ &= \int d^3r\int d^3r'\left\{\psi_c^*(\boldsymbol{r})[\Sigma_{xc}(\boldsymbol{r}, \boldsymbol{r}', \varepsilon_c) - v_{xc}(n_0, n(\boldsymbol{r}))\delta(\boldsymbol{r}-\boldsymbol{r}')]\psi_c(\boldsymbol{r}') - \psi_v^*(\boldsymbol{r})[\Sigma_{xc}(\boldsymbol{r}, \boldsymbol{r}', \varepsilon_v) - v_{xc}(n_0 n(\boldsymbol{r}))\delta(\boldsymbol{r}-\boldsymbol{r}')]\psi_v(\boldsymbol{r}')\right\}\\ &= \int d^3r\int d^3r'\,\psi_c^*(\boldsymbol{r})[\Sigma_{xc}(\boldsymbol{r}, \boldsymbol{r}', \varepsilon_c) - v_{xc}(n_0, n(\boldsymbol{r}))\delta(\boldsymbol{r}-\boldsymbol{r}')]\psi_c(\boldsymbol{r}').\end{aligned} \tag{3.34}$$

The last equality again assumes that the DFT $\varphi_{\boldsymbol{p}}^n$ is the same as the quasiparticle $\varphi_{\boldsymbol{p}}^n$. Then, since the state at the top of the valence level is at the Fermi energy

$$\int d^3r\,d^3r'\,\psi_v^*(\boldsymbol{r})\Sigma_{xc}(\boldsymbol{r}, \boldsymbol{r}', \varepsilon_v)\psi_v(\boldsymbol{r}') = \int d^3r\,\psi_v^*(\boldsymbol{r})v_{xc}(n_0, n(\boldsymbol{r}))\psi_v(\boldsymbol{r}). \tag{3.35}$$

We end with several observations. The discontinuities in $E[n_0, n]$, as a function of n_0, are particularly important to the band structure and specifically to the relatively few states next to the gap. The nonanalyticity in $n(\boldsymbol{r})$ plays a role as well in the total ground state energy *at fixed* n_0. The extent that traditional forms like the LDA and gradient corrections (Rasolt and Geldart 1975, Geldart and Rasolt 1976) (see eq. (3.27)) apply, depends on the radius of convergence for the expansion of, e.g., eqs. (3.31) and (3.32). The scale is obviously $V(G_x)/(\tilde{k}_F^2/2m) < 1$. The suitability of applying $E[n]$ (derived from an electron gas, where the spectrum of excitations is continuous; with no gaps) has been challenged previously (Geldart and Rasolt 1987, 1976, Ma and Brueckner 1968, Rasolt and Geldart 1986, Mintmire and White 1987). We believe it shares some similarities with the above discussion. Concerning finite temperature effects, all of our discussion applies rigorously to $T = 0$. The sharp discontinuity across the gap is formally correct only at $T = 0$. However, when optical measurements are made, the current–current response is measured, and clearly the effect of such corrections will be seen at finite T as well. The finite-T corrections to "E_g" (E_g is no longer a true gap at finite T) will depend on $k_B T/E_g$.

3.3. General derivation of the band gap discontinuities

To express the band gap discontinuities explicitly in terms of a universal function of the uniform density n_0 and functional of the full density $n(\boldsymbol{r})$ (or equivalently, the density deviation $\delta n(\boldsymbol{r})$) (see eq. (3.32)) would be truly in the spirit of the DFT. This turns out to be very difficult. In practice such discontinuities are calculated using the last relation of eq. (3.34) which explicitly contains the particle orbitals ψ_c and ψ_v and the nonlocal self-energy $\Sigma_{xc}(\boldsymbol{r}, \boldsymbol{r}', \varepsilon)$. Although, in principle, $\Sigma_{xc}(\boldsymbol{r}, \boldsymbol{r}', \varepsilon)$, ψ_c and ψ_v are all implicit functionals of $n(\boldsymbol{r})$ and n_0 (see § 2) operationally these terms are calculated, from start, for each external $V(\boldsymbol{r})$ (§ 2) and are not treated as universal functionals.

The last term of eq. (3.34) follows from the assumption (in eq. (3.35)) that the quasiparticle orbital $\varphi_{\boldsymbol{p}}^n$ is equal to the DFT orbital. Numerical studies (Hybersten and Louie 1985, 1986) indicate that this assumption is quite accurate; but, of course, not exact. We, therefore, want to conclude this § 3 with a rigorous derivation of Δ_{xc}, entirely expressed in terms of DFT orbitals. This was done for semiconductors by Sham (1985); as we see below, his final expression for Δ_{xc} is very similar to eq. (3.34).

According to the second term, on the right-hand side of this equation, we need a workable form for E_{xc} when the number of particles *changes*. Since we want to express E_{xc} entirely with DFT orbitals, it is not surprising that our unperturbed Hamiltonian $\hat{H}_0$ (eq. (2.4)) should be given by eq. (2.18a). So, we add and subtract from eq. (2.2) the term $V_H(n(\boldsymbol{r})) + v_{xc}(n_0, n(\boldsymbol{r}))$ and rewrite (Sham 1985) eq. (2.2) as

$$\hat{H}(\lambda) = \hat{H}_0 + \lambda \hat{H}_1, \tag{3.36a}$$

where

$$\hat{H}_0 = \hat{T} + \hat{V}_{eff} - \mu \hat{N} \tag{3.36b}$$

and

$$\hat{H}_1 = \hat{U} - \hat{V}_{eff} + \hat{V}, \tag{3.36c}$$

where

$$\hat{T} = \sum_{\sigma} \int \mathrm{d}^3 r\, \psi_{\sigma}^{\dagger}(\boldsymbol{r}) \left(\frac{-\hbar^2}{2m} \nabla^2 \right) \psi_{\sigma}(\boldsymbol{r}), \tag{3.36d}$$

$$\hat{V}_{\mathrm{eff}} = \sum_{\sigma} \int \mathrm{d}^3 r\, \psi_{\sigma}^{\dagger}(\boldsymbol{r}) V_{\mathrm{eff}}(n_0, n(\boldsymbol{r})) \psi_{\sigma}(\boldsymbol{r}) \tag{3.36e}$$

(with $V_{\mathrm{eff}}(\boldsymbol{r})$ given in eq. (2.18b)),

$$\hat{V} = \sum_{\sigma} \int \mathrm{d}^3 r\, \psi_{\sigma}^{\dagger}(\boldsymbol{r}) V(\boldsymbol{r}) \psi_{\sigma}(\boldsymbol{r}) \tag{3.36f}$$

and $\hat{U}$ is given by the last term on the right-hand side of eq. (2.2). The density $n(\boldsymbol{r})$ entering the $V_{\mathrm{eff}}(n_0, n(\boldsymbol{r}))$ of eq. (3.36e) is taken as the corresponding ground state density of eq. (3.36a) with $\lambda = 1$.

First, to derive the $v_{\mathrm{xc}}(n(\boldsymbol{r}))$ of eq. (3.35) in terms of the DFT orbitals (or equivalently Green function) we use the self-consistency condition on the full Green function (Sham 1985, Rasolt and Vosko 1974b) (eq. (2.25)), i.e.,

$$\mathrm{i} \lim_{t \to -0} G(\boldsymbol{r}, \boldsymbol{r}, t) = n(\boldsymbol{r}) \tag{3.37a}$$

and the Green function of $\hat{H}_0$

$$\mathrm{i} \lim_{t \to -0} G_0(\boldsymbol{r}, \boldsymbol{r}, t) = n(\boldsymbol{r}) \tag{3.37b}$$

(eq. (3.37b) follows from eq. (2.17b)). Using next the Dyson equation (eq. (2.20)) rewritten slightly differently; i.e.,

$$G(\boldsymbol{r}, \boldsymbol{r}', t) = G_0(\boldsymbol{r}, \boldsymbol{r}', t) + \int \mathrm{d}^3 r_1\, \mathrm{d}t_1 \int \mathrm{d}^3 r_2\, \mathrm{d}t_2\, G_0(\boldsymbol{r}, \boldsymbol{r}_1, t - t_1) \times \Sigma(\boldsymbol{r}_1, \boldsymbol{r}_2, t_1 - t_2) G(\boldsymbol{r}_2, \boldsymbol{r}', t_2), \tag{3.38a}$$

or in matrix notation

$$G = G_0 + G_0 \Sigma G \tag{3.38b}$$

along with eq. (3.37) gives for the diagonal matrix elements

$$\mathrm{i} \lim_{t \to -0} G_0 \Sigma G = 0. \tag{3.39}$$

Because of the form of $\hat{H}_1$ the self-energy Σ (fig. 9) contains additional insertions from $-\hat{V}_{\rm eff}+\hat{V}$ so that

$$\Sigma(\boldsymbol{r},\boldsymbol{r}',t)=\Sigma_{\rm xc}(\boldsymbol{r},\boldsymbol{r}',t)-v_{\rm xc}(n_0,n(\boldsymbol{r}))\delta(\boldsymbol{r}-\boldsymbol{r}')\delta(t) \tag{3.40}$$

(where, e.g., $\Sigma_{\rm xc}$ in the RPA is given in fig. 9), and from eqs. (3.39) and (3.40)

$$\begin{aligned}&\lim_{t\to-0}\int \mathrm{d}^3r_1\int \mathrm{d}t_1\, G_0(\boldsymbol{r},\boldsymbol{r}_1,t-t_1)G(\boldsymbol{r}_1,\boldsymbol{r},t_1)v_{\rm xc}(n_0 n(\boldsymbol{r}))\\ &=\lim_{t\to-0}\int \mathrm{d}^3r_1\,\mathrm{d}t_1\int \mathrm{d}^3r_2\,\mathrm{d}t_2\, G_0(\boldsymbol{r},\boldsymbol{r}_1,t-t_1)\Sigma_{\rm xc}(\boldsymbol{r}_1,\boldsymbol{r}_2,t_1-t_2)G(\boldsymbol{r}_2,\boldsymbol{r},t_2).\end{aligned} \tag{3.41}$$

Equation (3.41) gives the full form of the DFT $v_{\rm xc}(n_0,n(\boldsymbol{r}))$ in terms of the DFT orbitals (Sham 1985); obviously in this form it is only an implicit function of n_0 and functional of $n(\boldsymbol{r})$.

This $v_{\rm xc}(n_0,n(\boldsymbol{r}))$ is a nonanalytic functional of $n(\boldsymbol{r})$ (see previous subsection) but continuous in the change of particle number n_0. To get at the discontinuities we need to next calculate $\delta E_{\rm xc}/\delta n_+$ and $\delta E_{\rm xc}/\delta n_-$ in eqn. (3.30) or (3.34).

The total energy E_V can be written as a coupling constant integral (Harris and Jones 1974), i.e.,

$$E_V=E_0+\int_0^1\frac{\mathrm{d}\lambda}{\lambda}\langle\varphi_{\rm G}(\lambda)|\lambda\hat{H}_1|\varphi_{\rm G}(\lambda)\rangle, \tag{3.42a}$$

where

$$E_0=\langle\varphi_{\rm G}^0|\hat{H}_0|\varphi_{\rm G}^0\rangle=T_{\rm s}[n_0,n]+\int \mathrm{d}^3r\, V_{\rm eff}(n_0,n(\boldsymbol{r}))n(\boldsymbol{r}), \tag{3.42b}$$

$|\varphi_{\rm G}(\lambda)\rangle$ is the ground state as a function of λ and $|\varphi_{\rm G}^0\rangle$ is the ground state of $\hat{H}_0$. What Sham (1985) does next is to rewrite this coupling constant integral in terms of the full Green function and self-energy using the equation of motion of the field operators $\psi(\boldsymbol{r},t)$ and $\psi^\dagger(\boldsymbol{r},t)$, i.e.,

$$\frac{\hbar}{\mathrm{i}}\frac{\partial\psi^\dagger(\boldsymbol{r},t)}{\partial t}=[\hat{H},\psi^\dagger(\boldsymbol{r},t)]. \tag{3.43a}$$

By multiplying both sides by $\psi(\boldsymbol{r}',0)$ and using eq. (2.25) we can get the equation of motion for $G_\lambda(\boldsymbol{r},\boldsymbol{r}',t)$, i.e.,

$$\frac{\partial}{\partial t}G_\lambda(\boldsymbol{r},\boldsymbol{r}',t)=\langle\varphi_{\rm G}(\lambda)|\mathscr{T}\psi(\boldsymbol{r}',0)[\hat{H},\psi^\dagger(\boldsymbol{r},t)]|\varphi_{\rm G}(\lambda)\rangle. \tag{3.43b}$$

Working out the commutator we find a relation for $\langle\varphi_{\rm G}(\lambda)|\lambda\hat{H}_1|\varphi_{\rm G}(\lambda)\rangle$ in terms of G_λ and $\Sigma_{\rm xc}^\lambda$ from which (Sham 1985)

$$\begin{aligned}E_V=E_0&+\int_0^1\frac{\mathrm{d}\lambda}{2\mathrm{i}\lambda}\int \mathrm{d}^3r\,\mathrm{d}t\int \mathrm{d}^3r_1\,\mathrm{d}t_1\\ &\times[\Sigma_{\rm xc}^\lambda(\boldsymbol{r},\boldsymbol{r}_1,t-t_1)-\lambda V_{\rm H}(n(\boldsymbol{r}))\delta(\boldsymbol{r}-\boldsymbol{r}_1)\delta(t-t_1)]G_\lambda(\boldsymbol{r},\boldsymbol{r}_1,t-t_1).\end{aligned} \tag{3.44}$$

The integral over λ can be performed by recognizing various relations between the derivatives of the "skeleton diagrams" of E_V (fig. 10) and of Σ_{xc} and G. The calculation has been carried out in great detail by Luttinger and Ward (1960), and more recently by Sham (1985) and will not be repeated here. The final expression for $E_{xc}[n_0, n(\boldsymbol{r})]$ is (Sham 1985)

$$E_{xc}[n_0, n(\boldsymbol{r})] = \mathrm{i}T_r[ln(1 - \Sigma G_0) + \Sigma G] - \mathrm{i}Y'_{xc}, \tag{3.45}$$

where Y'_{xc} are the skeleton diagrams of fig. 10 (without E_H). Again we note that implicitly all the G_0's, which make up Σ and G are functionals of $n(\boldsymbol{r})$. However, explicitly the structure of $E_{xc}[n_0, n(\boldsymbol{r})]$ in eq. (3.45) is calculated in terms of $V(\boldsymbol{r})$ and is, therefore, not universal.

Equation (3.45) can now be used to calculate the $(\delta E_{xc}/\delta n_+)$ and $(\delta E_{xc}/\delta n_-)$ in eq. (3.34). Upon the addition of an extra electron at the bottom of the conduction band (n_+)

$$\delta G_0 = 2\pi \mathrm{i}\delta(\omega - \varepsilon_c)\varphi_c(\boldsymbol{r})\varphi_c^*(\boldsymbol{r}') \tag{3.46a}$$

and upon the addition of an extra electron at the top of the valence band (n_-)

$$\delta G_0 = 2\pi \mathrm{i}\delta(\omega - \varepsilon_v)\varphi_v(\boldsymbol{r})\varphi_v^*(\boldsymbol{r}'). \tag{3.46b}$$

Calculating the functional change in eq. (3.45) for these two cases gives the result for $\delta E_{xc}/\delta n_+$ and $\delta E_{xc}/\delta n_-$. (Actually, eq. (3.46) represents both a change in n_+ (and n_-) as well as $\delta n(\boldsymbol{r})$. However, again $(\delta E_{xc}/\delta n)$ is nonanalytic in $n(\boldsymbol{r})$ but continuous in n_0.) Sham and Schlüter (1985) found the following form for Δ_{xc}

$$\Delta_{xc} = \int \mathrm{d}^3r \int \mathrm{d}^3r' \, [\varphi_c^*(\boldsymbol{r})\tilde{\Sigma}(\boldsymbol{r}, \boldsymbol{r}', \varepsilon_c)\varphi_c(\boldsymbol{r}') - \varphi_v^*(\boldsymbol{r})\tilde{\Sigma}(\boldsymbol{r}, \boldsymbol{r}', \varepsilon_v)\varphi_v(\boldsymbol{r}')], \tag{3.47a}$$

where in matrix notation

$$\tilde{\Sigma} = (\Sigma_{xc} - v_{xc})(1 + G_0\tilde{\Sigma}). \tag{3.47b}$$

This exact result for Δ_{xc}, entirely in terms of DFT orbitals, is indeed very similar to eq. (3.34).

Equation (3.47) provides an alternative procedure to the full self-energy band structure calculation discussed in § 2. However, since eq. (3.47) requires as well the form of the self-energy, its only benefit is that it does not require the calculation of the quasiparticle state itself; i.e., φ_p^n of eq. (3.33).

The issue of band-gap discontinuities is a relatively recent aspect in the band theory of semiconductors. Another recent development in the band structure of semiconductors is the recognition of continuous symmetries between the pockets of electrons in multivalley semiconductors like Si (Rasolt 1990). We turn to this in the following § 4.

4. Continuous internal symmetries in the band structure of donor-doped multivalley semiconductors

Silicon, when donor doped, contains six pockets of electrons in the first Brillouin zone. Groups of two pockets, along 100, 010, and 001 are related by inversion symmetry (see figs. 1 and 15). Of course, these pockets of electrons are connected as well by the various other point-symmetry operations. What has been so far largely overlooked is that to a *very high degree of accuracy* the two pockets related by inversion symmetry satisfy the continuous SU(2) symmetry Lie group.

In this § 4 we derive these internal symmetries along with the *very small* symmetry breaking terms (which destroy this perfect continuous symmetry) and discuss some of its consequences to the properties of donor doped semiconductors.

4.1. Continuous symmetries within the effective-mass Hamiltonian

Let us look at the band structure of fig. 1 and concentrate on the small pockets of electrons in the shaded region when Si is n doped. Of course, there are six equivalent such $\Lambda \to X$ directions in Si and the electron pockets have the configuration of fig. 15. From fig. 15 we see that in silicon only groups of two valleys are related by time-reversal symmetry. Then we can capture much of the physics (see next subsection for corrections to this) with the following effective-mass Hamiltonian $\tilde{H}$ for each group of the two valleys

$$\begin{aligned}\tilde{H} = -\tfrac{1}{2}\hbar^2 \sum_{\tau_1=1}^{2} \sum_{\sigma_1=1}^{2} \int \mathrm{d}^d r\, \psi^\dagger_{\tau_1\sigma_1}(\boldsymbol{r})(m^{-1})_{\alpha\beta} \frac{\partial}{\partial r_\alpha}\frac{\partial}{\partial r_\beta} \psi_{\tau_1\sigma_1}(\boldsymbol{r}) \\ + \tfrac{1}{2} \sum_{\tau_1,\tau_2}^{2} \sum_{\sigma_1,\sigma_1}^{2} \int \mathrm{d}^d r_1 \int \mathrm{d}^d r_2\, \psi^\dagger_{\tau_1,\sigma_1}(\boldsymbol{r}_1)\psi^\dagger_{\tau_2,\sigma_2}(\boldsymbol{r}_2)\tilde{v}(\boldsymbol{r}_1 - \boldsymbol{r}_2)\psi_{\tau_2,\sigma_2}(\boldsymbol{r}_2)\psi_{\tau_1,\sigma_1}(\boldsymbol{r}_1),\end{aligned} \tag{4.1}$$

where $(m^{-1})_{\alpha\beta}$ is the effective mass,

$$\tfrac{1}{2}(m^{-1})_{\alpha\beta} = \frac{\partial}{\partial k_\alpha}\frac{\partial}{\partial k_\beta} \varepsilon^n_{\boldsymbol{k}\sigma}\Big|_{k=0}, \tag{4.2}$$

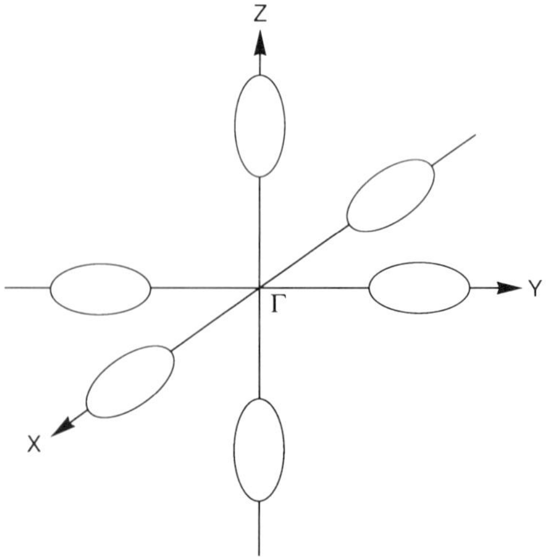

Fig. 15. The Fermi energy surface of the six equivalent valleys of electrons in silicon along the 100, 010, and 001 directions.

with α and β the Cartesian coordinates in d dimensions. The $\psi_{\tau\sigma}$ are the field operators of the $\tau = 2$ identical valleys with two electron spins $\sigma = 1, 2$, $\tilde{v}(\boldsymbol{r}_1 - \boldsymbol{r}_2)$ is the interparticle Coulomb repulsion screened by the dielectric constant ε (see next subsection)

$$\tilde{v}(\boldsymbol{r}_1 - \boldsymbol{r}_2) = \frac{e^2}{\varepsilon} \frac{1}{|\boldsymbol{r}_1 - \boldsymbol{r}_2|}. \tag{4.3}$$

Unlike the previous wave vector $\boldsymbol{p}$, the wave vector $\boldsymbol{k}$ is measured from the center of each valley (see figs. 1 and 5). $\tilde{H}$ is clearly invariant under SU(2) symmetry of the valley index τ (we refer to it as isospin space) and SU(2) of the spin index σ. $\tilde{H}$ is SU(2) $\times$ SU(2) invariant; in other words $\tilde{H}$ is invariant under the transformation in isospin space of any 2×2 unitary matrix U which transforms the $\psi \to \psi'$ according to

$$\psi'_{\tau_1} = \sum_{\tau_2 = 1}^{2} U_{\tau_1, \tau_2} \psi_{\tau_2}, \tag{4.4a}$$

or, equivalently,

$$\psi'_{\tau = 1} = a^* \psi_{\tau = 1} + b^* \psi_{\tau = 2}, \tag{4.4b}$$

$$\psi'_{\tau = 2} = b \psi_{\tau = 1} - a \psi_{\tau = 2}, \tag{4.4c}$$

with $|a|^2 + |b|^2 = 1$ and the same in spin space.

We will discuss a variety of phenomena associated with these internal symmetries shortly. Before we do so, however, we must justify the effective-mass Hamiltonian $\tilde{H}$(eq. (4.1)). Looking back at figs. 1 and 5 it is indeed surprising that in this rigid looking band structure the electron pockets can exhibit such continuous symmetries, *to a high degree of accuracy*. We must then set, with great care, the energy scales for which these continuous symmetries break down. This we do in the next subsections. We will confine our discussion to the random-phase approximation.

4.2. *The energy of $\hat{H}$ in the RPA*

Let us return to eq. (2.2) and add and subtract to it a self-consistent Hartree potential.

$$\hat{V}_{\mathrm{H}} = \sum_{\sigma_1} \int \mathrm{d}^d r \int \mathrm{d}^d r' \; v(\boldsymbol{r} - \boldsymbol{r}') n(\boldsymbol{r}') \psi^{\dagger}_{\sigma_1}(\boldsymbol{r}) \psi_{\sigma_1}(\boldsymbol{r}), \tag{4.5}$$

where

$$n(\boldsymbol{r}) = \sum_{\sigma} \langle \varphi_{\mathrm{G}} | \psi^{\dagger}_{\sigma}(\boldsymbol{r}) \psi_{\sigma}(\boldsymbol{r}) | \varphi_{\mathrm{G}} \rangle. \tag{4.6}$$

The self-consistent one-particle Hamiltonian of eq. (3.36b) is

$$\hat{H}_0 = \sum_{\sigma_1} \int \mathrm{d}^d r \; \psi^{\dagger}_{\sigma_1}(\boldsymbol{r}) \left(-\frac{\hbar^2}{2m} \nabla^2 + V(\boldsymbol{r}) \right) \psi_{\sigma_1}(\boldsymbol{r}) + \hat{V}_{\mathrm{H}}. \tag{4.7}$$

(We note that other choices could certainly have been made for $\hat{H}_0$, like the density-functional theory one-particle Hamiltonian of § 3.) We analyze the structure of the energy $E = \langle \varphi_G | \hat{H} | \varphi_G \rangle$ using the Feynman graphical representation. In fig. 16a we show the ring diagrams leading to the RPA interaction energy. We also show the Hartree contribution in fig. 16b. (Note that any additional terms like fig. 16c get cancelled by the counterterm of eq. (4.5) to reduce to figs. 16a and 16b.) The dashed lines are the Coulomb interactions $e^2/|\boldsymbol{r}_1 - \boldsymbol{r}_2|$. The solid arrowed lines are the single-particle propagators of $\hat{H}_0$, i.e.,

$$G_0(\boldsymbol{r}, \boldsymbol{r}', \omega) = \sum_{\boldsymbol{p},n} \frac{\varphi^{n*}_{\boldsymbol{p}\sigma}(\boldsymbol{r})\varphi^{n}_{\boldsymbol{p}\sigma}(\boldsymbol{r}')}{\omega - \varepsilon^n_{\boldsymbol{p}\sigma} + \mathrm{i}\delta\, \mathrm{sgn}(\varepsilon^n_{\boldsymbol{p}\sigma} - \mu)}, \tag{4.8}$$

4.3. Effective Hamiltonian for the valley electrons

The discussion so far treated both the many valence electrons and the *very few* valley electrons on an equal footing. We want to take advantage of the small number of valley electrons and separate them from the rest (see Rasolt (1990)). To do that, we manipulate the RPA Feynman graphs of fig. 16. Let us first examine the set of fig. 16a.

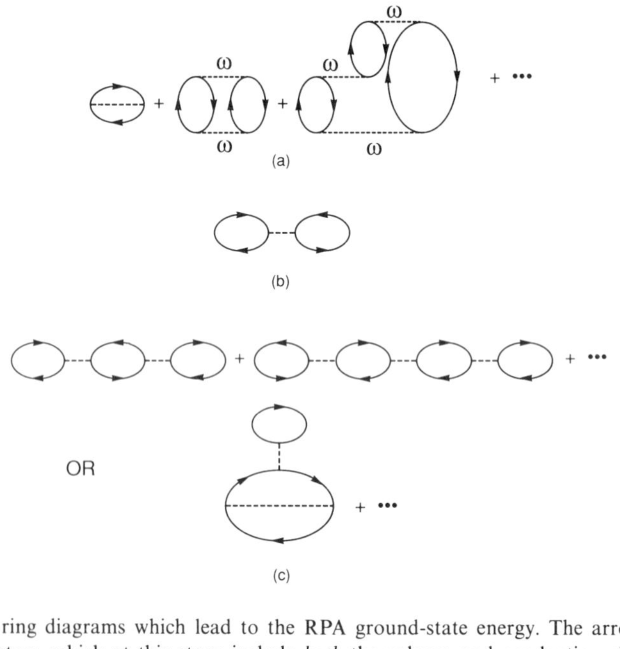

Fig. 16. (a) The ring diagrams which lead to the RPA ground-state energy. The arrowed lines are the electron propagators, which at this stage include *both* the valence and conduction electron states. The dashed line is the unscreened Coulomb interparticle interaction. The frequency ω is placed there as a reminder of the dynamic fluctuations entering the RPA ground-state calculation. (b) The Hartree electrostatic energy. (c) Example of self-energy insertions in the single-particle Green function, which renormalize the arrowed lines and reduce back to a and b through the counterterm in eq. (4.5).

We decompose the full G_0 of eq. (4.8) to a valence part G_v and a conduction part G_c, where

$$G_c = \sum_{k,\tau} \frac{\varphi_{k\tau}^{c*}(\boldsymbol{r})\varphi_{k\tau}^{c}(\boldsymbol{r}')}{\omega - \varepsilon_{p\sigma}^{\tau} + i\delta \operatorname{sgn}(\varepsilon_{p\sigma}^{\tau} - \mu)}, \tag{4.9}$$

(the sum over $\boldsymbol{k}$ is strictly inside the valleys) and

$$G_v = \sum_{p,n} \frac{\varphi_{p}^{n*}(\boldsymbol{r})\varphi_{p}^{n}(\boldsymbol{r}')}{\omega - \varepsilon_{p\sigma}^{n} + i\delta \operatorname{sgn}(\varepsilon_{p\sigma}^{n} - \mu)}, \tag{4.10}$$

where the sum over $\boldsymbol{p}$ and n are within all the rest.

We consider the contributions of fig. 16a to second order in G_c; these are shown in fig. 17. Now, we are not concerned with absolute energies, certainly not the very big energy which is independent of G_c. Actually we are not even concerned about energies within the valleys which are independent under isospin rotation of eq. (4.4). For example, from the Bloch symmetry of $\varphi_p^n(\boldsymbol{r})$ we can show that fig. 17a is independent of isospin rotation. To see this we write the Feynman graphs of fig. 17 in the form

$$E_a = \int d^d r \int d^d r' \int \frac{dy}{2\pi} F(\boldsymbol{r}, \boldsymbol{r}', y) G_c(\boldsymbol{r}, \boldsymbol{r}', y), \tag{4.11}$$

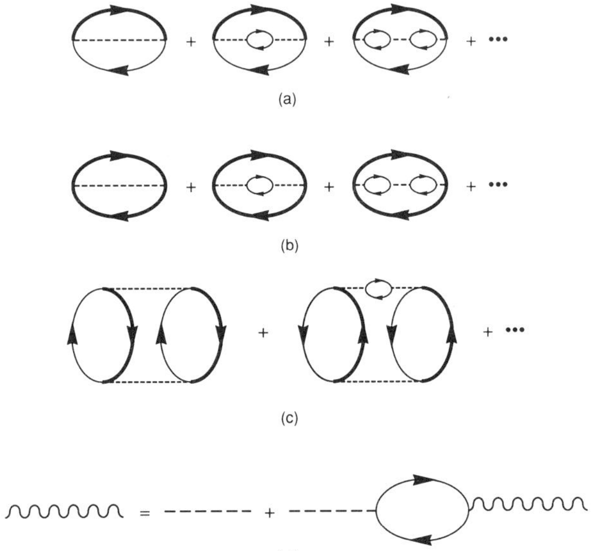

Fig. 17. Example of Feynman graphs which evolve from fig. 16a when the electron propagators are separated into a conduction band G_c and valence band G_v. (a) An example of a single conduction-band Green function (double solid line) interacting with the valence electrons. (b) Conduction electrons (i.e., valley electrons) to second order in G_c interacting via the screened interparticle interaction, screened by the valence electron *only* (see fig. 17d). (c) Other examples of conduction–conduction electron interaction to second order in G_c. (d) The screened interaction $V(\boldsymbol{r}, \boldsymbol{r}', y)$; see text.

where to avoid working with imaginary contributions of F we rotated ω to the imaginary axis, i.e., $\omega = \mathrm{i}y$. The function F represents all the contributions multiplying the double solid arrowed lines in fig. 17a. Now $F(\boldsymbol{r}, \boldsymbol{r}', y)$ is made up entirely of Bloch states and has therefore Bloch symmetry, i.e.,

$$F(\boldsymbol{r}, \boldsymbol{r}', y) = F(\boldsymbol{r} + \boldsymbol{R}, \boldsymbol{r}' + \boldsymbol{R}, y), \tag{4.12}$$

where $\boldsymbol{R}$ is a lattice vector. It then follows at once that F in Fourier space can be written as

$$F(\boldsymbol{r}, \boldsymbol{r}', y) = \sum_{\boldsymbol{G},\boldsymbol{G}'} \int \frac{\mathrm{d}^d q}{(2\pi)^d} F(\boldsymbol{q}, y)_{\boldsymbol{G},\boldsymbol{G}'} \exp[\mathrm{i}(\boldsymbol{q} + \boldsymbol{G}) \cdot \boldsymbol{r}] \exp[-\mathrm{i}(\boldsymbol{q} + \boldsymbol{G}') \cdot \boldsymbol{r}'], \tag{4.13}$$

where $\boldsymbol{G}$ is a reciprocal lattice vector. Now we perform the isospin rotation of eq. (4.4) on the valley electrons. This translates to writing G_{c} of eq. (4.8) as

$$G_{\mathrm{c}} = \sum_{\boldsymbol{k}} \frac{[a^* \varphi^*_{\boldsymbol{k},\tau=1}(\boldsymbol{r}) + b^* \varphi_{\boldsymbol{k},\tau=2}(\boldsymbol{r})][a\varphi_{\boldsymbol{k},\tau=1}(\boldsymbol{r}') + b\varphi_{\boldsymbol{k},\tau=2}(\boldsymbol{r}')]}{\omega - \varepsilon^{\mathrm{c}}_{\boldsymbol{k},\sigma,1} + \mathrm{i}\delta \operatorname{sgn}(\varepsilon^{\mathrm{c}}_{\boldsymbol{k},\sigma,1} - \mu)}. \tag{4.14}$$

Turning back to eq. (4.11), we need to show that terms like

$$E'_a = \int \mathrm{d}^d r \int \mathrm{d}^d r' \int \frac{\mathrm{d}y}{(2\pi)} F(\boldsymbol{r}, \boldsymbol{r}', y) b a^* \varphi^*_{\boldsymbol{k},\tau=1}(\boldsymbol{r}) \varphi_{\boldsymbol{k},\tau=2}(\boldsymbol{r}') \tag{4.15}$$

vanish.

This follows from the following: $\varphi_{\tau=1,\boldsymbol{k}}$ is the Bloch state from valley 1 and momentum $\boldsymbol{k}$ (measured from the center of the valley) and $\varphi_{\tau=2,\boldsymbol{k}}$ is the Bloch state from valley 2. From the Bloch symmetry we can write that

$$\varphi_{\tau=1,\boldsymbol{k}}(\boldsymbol{r}) = \exp[\mathrm{i}(\boldsymbol{k} - \boldsymbol{Q}/2) \cdot \boldsymbol{r}]\, u_{1,\boldsymbol{k}}(\boldsymbol{r}) \tag{4.16}$$

(where $u_{1,\boldsymbol{k}}(\boldsymbol{r})$ has the translational periodicity of the lattice) and that

$$\varphi_{\tau=2,\boldsymbol{k}}(\boldsymbol{r}) = \exp[\mathrm{i}(\boldsymbol{k} + \boldsymbol{Q}/2) \cdot \boldsymbol{r}]\, u_{2,\boldsymbol{k}}(\boldsymbol{r}), \tag{4.17}$$

$\frac{1}{2}\boldsymbol{Q}$ is the center of the valley (see fig. 1). From eqs. (4.16), (4.17), and (4.13) we get

$$E'_a = \int \mathrm{d}^d q\, F'(\boldsymbol{q})_{\boldsymbol{G},\boldsymbol{G}'} \delta(\boldsymbol{q} + \boldsymbol{k} + \boldsymbol{G} - \tfrac{1}{2}\boldsymbol{Q}) \delta(\boldsymbol{q} + \boldsymbol{k} + \boldsymbol{G}' + \tfrac{1}{2}\boldsymbol{Q}) \tag{4.18}$$

and eq. (4.15) vanishes. In short the contribution of fig. 17a is isospin invariant. The same applies to the quadratic contributions, in G_{c}, of fig. 17c. (Note that the effective Hamiltonian $\hat{H}_{\mathrm{eff}}$ and the effective mass Hamiltonian $\hat{H}$ are not to be confused as the same; they are two distinct Hamiltonians.)

The contribution from fig. 17b can be summed into the form

$$E_{\mathrm{RPA}} = \int \mathrm{d}^d r \int \mathrm{d}^d r' \int \frac{\mathrm{d}y}{2\pi} V(\boldsymbol{r}, \boldsymbol{r}', y) G_{\mathrm{c}}(\boldsymbol{r}, \boldsymbol{r}', y) G_{\mathrm{c}}(\boldsymbol{r}', \boldsymbol{r}, y), \tag{4.19}$$

where the $V(\boldsymbol{r}, \boldsymbol{r}', y)$ is the sum of the ring diagrams of fig. 17d. It is important to note that $V(\boldsymbol{r}, \boldsymbol{r}', y)$ is given entirely (to second order in G_{c}) by the valence electrons alone. It, therefore, can be considered as an external dynamically screened interaction between two isospin (or valley) electrons only. It also, of course, has the periodicity of eq. (4.13).

The sum of these ring diagrams is recognized in yet another way as the dynamical screening of the bare Coulomb interaction $v(\boldsymbol{r}-\boldsymbol{r}')$, i.e.,

$$V(\boldsymbol{r},\boldsymbol{r}',y)=\varepsilon^{-1}(\boldsymbol{r},\boldsymbol{r}',y)v(\boldsymbol{r}-\boldsymbol{r}'), \tag{4.20}$$

where $\varepsilon(\boldsymbol{r},\boldsymbol{r}',y)$ is the dielectric function for the valence electrons only and again has the periodicity of eq. (4.13).

The Hartree contribution of fig. 17b can also be opened up in terms of G_c and G_v. The result to quadratic order in G_c is

$$E_H=\tfrac{1}{2}\int \mathrm{d}^d r \int \mathrm{d}^d r' \,[V(\boldsymbol{r},\boldsymbol{r}',y=0)]\rho_c(\boldsymbol{r})\rho_c(\boldsymbol{r}'), \tag{4.21}$$

where

$$\rho_c(\boldsymbol{r})=2\int\frac{\mathrm{d}y}{2\pi}G_c(\boldsymbol{r},\boldsymbol{r}',y). \tag{4.22}$$

If we ignore the frequency dependence in eq. (4.20) then we can write an effective Hamiltonian, equivalent to quadratic order in G_c, as the noninteracting $\hat{H}_0$ of eq. (4.7), plus the two interacting contributions of eqs. (4.19) and (4.21), i.e.,

$$\hat{H}_{\mathrm{eff}}=\sum_{\boldsymbol{k},\sigma,\tau}\varepsilon_{\boldsymbol{k},\sigma,\tau}C^{\dagger}_{\tau,\sigma,\boldsymbol{k}}C_{\tau,\sigma,\boldsymbol{k}} + \tfrac{1}{2}\int \mathrm{d}^d r\int \mathrm{d}^d r' \sum_{\substack{\tau_1,\tau_2,\tau_3,\tau_4\\ \sigma_1,\sigma_2}} V(\boldsymbol{r},\boldsymbol{r}',y=0)\psi^{\dagger}_{\tau_1\sigma_1}(\boldsymbol{r})\psi^{\dagger}_{\tau_2\sigma_2}(\boldsymbol{r}')\psi_{\tau_3\sigma_2}(\boldsymbol{r}')\psi_{\tau_4\sigma_1}(\boldsymbol{r}) \tag{4.23}$$

where the field operators of the valleys are

$$\psi^{\dagger}_{\tau\sigma}=\sum_{\boldsymbol{k}}C^{\dagger}_{\tau,\sigma,\boldsymbol{k}}\varphi^{*}_{\tau\boldsymbol{k}}(\boldsymbol{r}). \tag{4.24}$$

This is then the final form for the effective Hamiltonian within the valleys. What have we sidestepped in deriving eq. (4.23)? The effect of the frequency dependence in eq. (4.19). The effect of higher orders in G_c. These, as we now show, are much smaller corrections to eq. (4.23).

4.3.1. *Frequency-dependent corrections*

We return to eq. (4.19). The frequency dependence in $V(\boldsymbol{r},\boldsymbol{r}',y)$ will modify the static results in eq. (4.23). To *estimate* the amount we consider the frequency dependence of the valence electron response. A rough treatment of the Lindhard dielectric function for these electrons introduces corrections at small frequencies ω of order ω^2/E_g^2 (where E_g is roughly the gap). As a characteristic frequency of the valley electrons we take $\omega \approx \hbar^2 k_F^2/2m$ (we could have taken the plasma frequency), and $E_g \approx \hbar^2 Q^2/2m$. The correction to eq. (4.19) we estimate as

$$\Delta E_{\mathrm{RPA}} \approx \left(\frac{k_F}{Q}\right)^4 \int \mathrm{d}^d r \int \mathrm{d}^d r' \; V'(\boldsymbol{r}, \boldsymbol{r}', y=0) \int \frac{\mathrm{d}y}{2\pi} G_c(\boldsymbol{r}, \boldsymbol{r}', y) G_c(\boldsymbol{r}', \boldsymbol{r}, y). \tag{4.25}$$

4.3.2. Higher-order corrections from G_c

In fig. 18a we illustrate corrections from higher orders in G_c in the RPA. In fig. 18b we rewrite the same contributions in terms of the screened interaction $V(\boldsymbol{r}, \boldsymbol{r}', y=0)$ (screened by the valley electrons). Therefore, a many-body calculation, within each valley, will include such higher-order corrections. At very small momentum $\boldsymbol{q}$ transfers (see fig. 18) such corrections are very important. Higher corrections from G_c are not then necessarily small due to the $1/q^2$ behavior of the Fourier transform of $V(\boldsymbol{r}, \boldsymbol{r}', y = 0)$. This, of course, will lead to important screening within the valley electrons. In any event, a many-body calculation using $\hat{H}_{\mathrm{eff}}$ includes these terms. For isospin

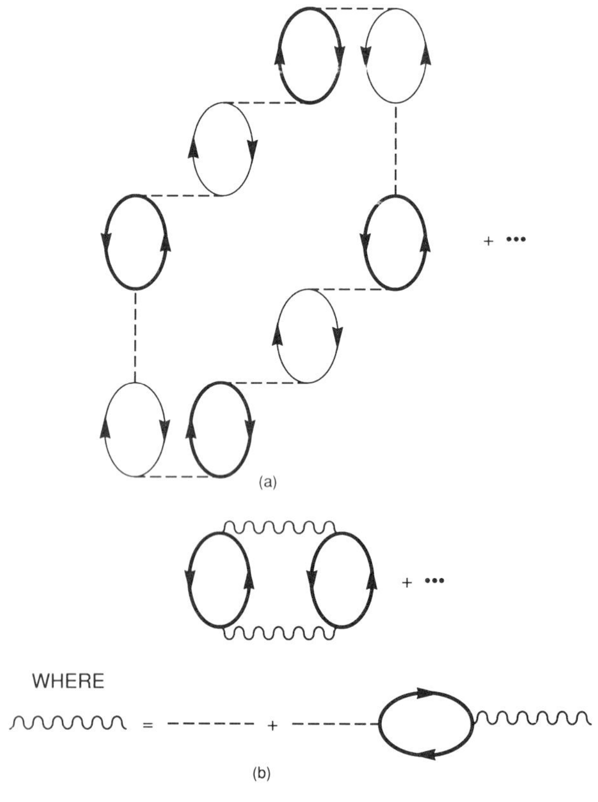

Fig. 18. (a) Examples of higher-order in G_c; in this case fourth order. (b) The interaction between two conduction electrons, which was previously screened by the valence electrons only (i.e., eq. (4.20)) will now be, in addition, screened by the conduction electrons among themselves.

rotation, where $\boldsymbol{q} \approx \boldsymbol{Q}$, correction from higher powers of G_c will be small; of the order of $(k_F/Q)^d$ (d is the dimension).

Finally, we remark on another derivation of $\hat{H}_{\text{eff}}$ (eq. (4.23)) which starts by considering the valley electrons as external particles. If we fix two particles at position $\boldsymbol{r}_1$ and $\boldsymbol{r}_2$ in the lattice, then their pair interaction energy, in linear response, is given by

$$E = V(\boldsymbol{r}_1, \boldsymbol{r}_2, y=0), \tag{4.26}$$

with $V(\boldsymbol{r}, \boldsymbol{r}', y=0)$ given in eq. (4.20). From this, $\hat{H}_{\text{eff}}$ of eq. (4.23) follows at once. But electrons in the valleys are not external charges, they share statistics with the valence electrons and they are not fixed. Equation (4.26) does not provide the kind of estimate for the corrections to $\hat{H}_{\text{eff}}$ discussed above.

4.4. *The effective-mass Hamiltonian*

So where does the SU(2)-invariant effective-mass Hamiltonian $\hat{H}$, of eq. (4.1), reside in $\hat{H}_{\text{eff}}$ of eq. (4.23)? and where are the symmetry-breaking (SB) corrections to $\hat{H}_{\text{eff}}$?

We can get access to these terms by turning first to an uncorrelated product ground state, i.e.,

$$|\varphi_G\rangle = \prod_{\substack{\tau=1 \\ \sigma=1,2}} C^{\dagger}_{\tau\sigma\boldsymbol{k}}|0\rangle, \tag{4.27}$$

where $|0\rangle$ is the vacuum, $C^{\dagger}_{\tau\sigma\boldsymbol{k}}$ are the creation operators of states with isospin τ, spin σ, and momentum $\boldsymbol{k}$ measured from the center of each valley (see fig. 1). Now apply the isospin rotation of eq. (4.4) to $|\varphi_G\rangle$ to give

$$|\varphi'_G\rangle = \prod_{\substack{\sigma=1,2 \\ |\boldsymbol{k}|=0\to k_F}} (a^* C^{\dagger}_{\tau=1,\sigma,\boldsymbol{k}} + b^* C^{\dagger}_{\tau=2,\sigma,\boldsymbol{k}})|0\rangle. \tag{4.28}$$

(We ignore the SU(2) rotations in spin space; spin rotations are totally uninteresting here since they leave $\hat{H}_{\text{eff}}$ invariant.) So we want to study the expectation of $\langle\varphi'_G|\hat{H}_{\text{eff}}|\varphi'_G\rangle$. $\hat{H}_{\text{eff}}$ in eq. (4.23) is written in a mixed representation. We write it next entirely in terms of the $C^{\dagger}_{\tau,\boldsymbol{k}}$, i.e.,

$$\hat{H}_{\text{eff}} = \sum_{\boldsymbol{k},\tau} \varepsilon_{\boldsymbol{k},\tau} C^{\dagger}_{\tau,\boldsymbol{k}} C_{\tau,\boldsymbol{k}} + \tfrac{1}{2} \sum_{\substack{\boldsymbol{k}_1,\boldsymbol{k}_2,\boldsymbol{k}_3,\boldsymbol{k}_4 \\ \tau_1,\tau_2,\tau_3,\tau_4}} F_{\substack{\boldsymbol{k}_1,\boldsymbol{k}_2,\boldsymbol{k}_3,\boldsymbol{k}_4 \\ \tau_1,\tau_2,\tau_3,\tau_4}} C^{\dagger}_{\boldsymbol{k}_1,\tau_1} C^{\dagger}_{\boldsymbol{k}_2,\tau_2} C_{\boldsymbol{k}_3,\tau_3} C_{\boldsymbol{k}_4,\tau_4}, \tag{4.29}$$

where

$$F_{\substack{\boldsymbol{k}_1,\boldsymbol{k}_2,\boldsymbol{k}_3,\boldsymbol{k}_4 \\ \tau_1,\tau_2,\tau_3,\tau_4}} = \int \mathrm{d}^d r_1 \int \mathrm{d}^d r_2 \; V(\boldsymbol{r}_1, \boldsymbol{r}_2, y=0) \varphi^*_{\tau_1,\boldsymbol{k}_1}(\boldsymbol{r}_1) \varphi^*_{\tau_2,\boldsymbol{k}_2}(\boldsymbol{r}_2) \varphi_{\tau_3,\boldsymbol{k}_3}(\boldsymbol{r}_2) \varphi_{\tau_4,\boldsymbol{k}_4}(\boldsymbol{r}_1). \tag{4.30}$$

Now it is not difficult to see that the quartic interaction E_{quar} can be written as

$$\begin{aligned} E_{\text{quar}} &= \tfrac{1}{2} \sum_{\substack{k_1,k_2,k_3,k_4 \\ \tau_1,\tau_2,\tau_3,\tau_4}} F_{\substack{k_1,k_2,k_3,k_4 \\ \tau_1,\tau_2,\tau_3,\tau_4}} \langle \varphi'_G | C^{\dagger}_{k_1,\tau_1} C^{\dagger}_{k_2,\tau_2} C_{k_3,\tau_3} C_{k_4,\tau_4} | \varphi'_G \rangle \\ &= \tfrac{1}{2} \sum_{\substack{k_1,k_2,k_3,k_4 \\ \tau_1,\tau_2,\tau_3,\tau_4}} F_{\substack{k_1,k_2,k_3,k_4 \\ \tau_1,\tau_2,\tau_3,\tau_4}} \{ \langle \varphi'_G | C^{\dagger}_{k_1,\tau_1} C_{k_4,\tau_4} | \varphi'_G \rangle \langle \varphi'_G | C^{\dagger}_{k_2,\tau_2} C_{k_3,\tau_3} | \varphi'_G \rangle \\ &\quad - \langle \varphi'_G | C^{\dagger}_{k_1,\tau_1} C_{k_3,\tau_3} | \varphi'_G \rangle \langle \varphi'_G | C^{\dagger}_{k_2,\tau_2} C_{k_4,\tau_4} | \varphi'_G \rangle \}. \end{aligned} \tag{4.31}$$

So the quartic interactions can be written as products of two quadratic expectation values (for such a rotated $|\varphi'_G\rangle$). The quadratic expectation values can be readily evaluated to give

$$\begin{aligned} \langle \varphi'_G | c^{+}_{k_1\tau_1} c_{k_2\tau_2} | \varphi'_G \rangle = \delta_{k_2,k} \delta_{k_1,k} (a^* a \delta_{\tau_{1,1}} \delta_{\tau_{2,1}} + b^* b \delta_{\tau_{1,2}} \delta_{\tau_{2,2}} \\ + a^* b \delta_{\tau_{1,1}} \delta_{\tau_{2,2}} + b^* a \delta_{\tau_{1,2}} \delta_{\tau_{2,1}}) \end{aligned} \tag{4.32}$$

and eq. (4.31) can be then calculated; the calculation is presented in great detail in Rasolt (1990). The final result for E is (Rasolt 1990)

$$E = \frac{1}{2} \frac{1}{\Omega} \sum_{kk'}^{k_F} \left((a^2 + b^2)^2 \frac{v(\boldsymbol{k} - \boldsymbol{k}')}{\varepsilon} \right) + E_{\text{SB}}, \tag{4.33}$$

where $a^2 + b^2 = 1$ and $v(\boldsymbol{k} - \boldsymbol{k}') = 4\pi e^2 / |\boldsymbol{k}' - \boldsymbol{k}|^2$, in $d = 3$. The first term on the right-hand side is, of course, invariant under isospin rotation and corresponds precisely to the isospin exchange contribution from the effective-mass Hamiltonian $\hat{H}$ of eq. (4.1). E_{SB} is the symmetry breaking correction to eq. (4.33) (or equivalently to eq. (4.1)), which we are after. There are many kinds of E_{SB} terms (see Rasolt (1990)). For example, one such contribution is

$$E_{\text{SB}} = 2a^* a b^* b \sum_{\boldsymbol{k},\boldsymbol{k}'}^{k_F} \left(F_{\substack{k',k,k,k' \\ 2,1,2,1}} - F_{\substack{k',k,k',k \\ 2,1,2,1}} \right). \tag{4.34a}$$

Using eq. (4.30) and expanding in powers of $\boldsymbol{k}$ and $\boldsymbol{k}'$ we get that (Rasolt 1990)

$$E_{\text{SB}} \approx \frac{1}{\Omega} \frac{e^2}{\varepsilon |\boldsymbol{Q}|^4} \sum_{\boldsymbol{k},\boldsymbol{k}'}^{k_F} k^2. \tag{4.34b}$$

All other E_{SB} terms scale the same (Rasolt 1990). So, the ratio R of eq. (4.34b) to eq. (4.33) is then

$$R \approx |k_F / Q|^{d+1}.$$

Since the size of the valleys (i.e., k_F) is *much* smaller than their separation, we clearly see that the symmetry breaking corrections to $\hat{H}$ are indeed very small.

Having proved the existence of this continuous symmetry in the band structure of multivalley semiconductors the analogy between isospin and spin space is immediate. For example, if the free energy predicts broken symmetry configuration in spin space (e.g., a ferromagnet) the corresponding SU(2) symmetry then predicts the existence of low-lying Goldstone modes (i.e., spin waves). Similarly, if the free energy predicts a preference for more electrons in one of the valleys (i.e., isospin polarization) then the corresponding SU(2) symmetry predicts another low-lying Goldstone mode which we refer to as valley waves (see Rasolt (1990)).

The implication of such low lying modes to various properties of donor-doped semiconductors in two and three dimensions in the presence and absence of an external magnetic field has been studied by Rasolt and co-workers (Rasolt 1983a, Rasolt et al. 1985, Rasolt and MacDonald 1986, Rasolt et al. 1986, Rasolt 1987b, 1990). Various other broken-symmetry states analogous to spin and charge-density waves have been considered by Tešanović and Halperin (1987). The effect of the isospin components on Fermi liquid properties were also studied recently by Sachdev (1987). We believe that isospin space in multivalley semiconductors carries a wide scope of possibilities, both experimentally as well as theoretically, which have just begun to be exploited (Rasolt 1990).

In conclusion, this first chapter should bring to the reader two current and important issues of semiconductor band theory. Since band theory is the first step in studying almost all of the rich phenomenas of semiconductors, it should set the way for the various presentations in the following chapters.

Acknowledgments

This research was sponsored by the Division of Materials Sciences, U.S. Department of Energy under Contract No. DE-AC05-84OR21400 with Martin Marietta Energy Systems, Inc. I wish to acknowledge very useful discussions with D.J.W. Geldart and J.P. Perdew, and also the help of V. Hendrix in the preparation of this manuscript.

References

Abrikosov, A.A., L.P. Gorkov and I.E. Dzyaloshinski, 1963, Methods of Quantum Field Theory in Statistical Physics (Prentice-Hall, Englewood Cliffs, NJ).

Allen, P.B., and M.L. Cohen, 1969, Phys. Rev. **187**, 525.

Anderson, P.W., 1963, Concepts in Solids (Benjamin, New York).

Animalu, A.D.E., and V. Heine, 1965, Philos. Mag. **12**, 1249.

Ashcroft, N.W., 1966, Phys. Lett. **23**, 48.

Austin, B.J., V. Heine and L.J. Sham, 1962, Phys. Rev. **127**, 276.

Bigot, J.Y., M.T. Portella, R.W. Schoenlein, J.E. Cunningham and C.V. Shank, 1990, Phys. Rev. Lett. **65**, 3429.

Brovman, E.G., and Y. Kagan, 1969, Zh. Eksp. & Teor. Fiz. **57**, 1329 [Sov. Phys.-JETP **30**, 721 (1970)].

Brust, D., J.C. Philips and F. Bassani, 1962, Phys. Rev. Lett. **9**, 94.

Cohen, M.H., and V. Heine, 1961, Phys. Rev. **122**, 1821.

Cohen, M.L., 1982, Phys. Scr. **T1**, 5.

Cohen, M.L., and T.K. Bergstresser, 1966, Phys. Rev. **141**, 789.

Cohen, M.L., and J.R. Chelikowsky, 1988, Electronic Structure and Optical Properties of Semiconductors (Springer, Berlin).

Dagens, L., M. Rasolt and R. Taylor, 1975, Phys. Rev. B **11**, 2726.

Geldart, D.J.W., and M. Rasolt, 1976, Phys. Rev. B **13**, 1477.
Geldart, D.J.W., and M. Rasolt, 1987, Single-Particle Density in Physics and Chemistry, p. 151.
Geldart, D.J.W., and R. Taylor, 1970a, Can. J. Phys. **48**, 155.
Geldart, D.J.W., and R. Taylor, 1970b, Can. J. Phys. **48**, 167.
Geldart, D.J.W., and S.H. Vosko, 1966, Can. J. Phys. **44**, 2137.
Godby, R.W., M. Schlüter and L.J. Sham, 1986, Phys. Rev. Lett. **56**, 2415.
Gross, E.K.U., and W. Kohn, 1985, Phys. Rev. Lett. **55**, 2850.
Gross, E.K.U., and W. Kohn, 1990, Adv. Quantum Chem. **21**, 255.
Gunnarson, O., and K. Schönhammer, 1986, Phys. Rev. Lett. **56**, 1968.
Gygi, F., and A. Baldereschi, 1989, Phys. Rev. Lett. **62**, 2160.
Hamann, D.R., M. Schlüter and C. Chiang, 1979, Phys. Rev. Lett. **43**, 1494.
Hanke, W., and L.J. Sham, 1989, Solid State Commun. **71**, 211.
Harris, J., and R.O. Jones, 1974, J. Phys. F **4**, 1170.
Hedin, L., 1965a, Ark. Fys. **19**, 231.
Hedin, L., 1965b, Phys. Rev. A **139**, 796.
Hedin, L., and B.I. Lundqvist, 1971, J. Phys. C **4**, 2064.
Hedin, L., and S. Lundqvist, 1969, Solid State Phys. **23**, 1.
Heine, V., and I. Abarenkov, 1964, Philos. Mag. **9**, 451.
Hohenberg, P., and W. Kohn, 1964, Phys. Rev. B **136**, 864.
Hott, R., 1991, Phys. Rev. B **44**, 1057.
Hybersten, M.S., and S.G. Louie, 1985, Phys. Rev. Lett. **55**, 1418.
Hybersten, M.S., and S.G. Louie, 1986, Phys. Rev. B **34**, 5390.
Hybersten, M.S., and S.G. Louie, 1988, Phys. Rev. B **37**, 2733.
Johnson, D.L., 1974, Phys. Rev. B **9**, 4475.
Kane, E.O., 1972, Phys. Rev. B **5**, 1493.
Kerker, G.P., 1980, J. Phys. C **13**, L189.
Kleinman, L., 1984, Phys. Rev. B **30**, 2223.
Kleinman, L., 1986, Phys. Rev. B **33**, 7299.
Kohn, W., and L.J. Sham, 1965, Phys. Rev. **140**, A1133.
Krieger, J.B., and Y. Li, 1990, Phys. Lett. A **146**, 256.
Lannoo, M., M. Schlüter and L.J. Sham, 1985, Phys. Rev. B **32**, 3890.
Levine, Z.H., and S.G. Louie, 1982, Phys. Rev. B **25**, 6310.
Liu, J.M., H. Kurz and N. Bloembergen, 1982, Appl. Phys. Lett. **41**, 643.
Lompre, L.A., J.M. Liu, H. Kurz and N. Bloembergen, 1982, Appl. Phys. Lett. **43**, 168.
Louie, S.G., S. Froyen and M.L. Cohen, 1982, Phys. Rev. B **26**, 1738.
Luttinger, J.M., and J.C. Ward, 1960, Phys. Rev. **118**, 1417.
Ma, S.K., and K.A. Brueckner, 1968, Phys. Rev. **165**, 18.
MacDonald, A.H., 1980, J. Phys. F **10**, 1737.
Mearns, D., 1988, Phys. Rev. B **38**, 5906.
Mermin, D., 1965, Phys. Rev. **137**, A1441.
Minnhagen, P., 1974, J. Phys. C **7**, 3013.
Mintmire, J.W., and C.T. White, 1987, Phys. Rev. B **35**, 4180.
Ng, T.K., 1989, Phys. Rev. B **39**, 1369.
Norman, M.R., and D.D. Koelling, 1984, Phys. Rev. B **30**, 5530.
Nozières, P., 1964, Theory of Interacting Fermi Systems (Benjamin, New York).
Perdew, J.P., 1985, Density Functional Methods in Physics, eds R.M. Dreizler and J. de Providencia (Plenum, New York).
Perdew, J.P., 1986, Int. J. Quantum Chem.: Quantum Chem. Symp. **19**, 497–523.
Perdew, J.P., 1990, Adv. Quantum Chem. **21**, 113.
Perdew, J.P., and M. Levy, 1983, Phys. Rev. Lett. **51**, 1884.
Perdew, J.P., R.G. Parr, M. Levy and J.L. Baldluz, Jr, 1982, Phys. Rev. Lett. **49**, 1691.
Philips, J.C., and L. Kleinman, 1959, Phys. Rev. **116**, 287.
Pickett, W.E., and C.S. Wang, 1984, Phys. Rev. B **30**, 4719.
Rasolt, M., 1983a, Phys. Rev. Lett. **50**, 778.

Rasolt, M., 1983b, Phys. Rev. B **27**, 5653.
Rasolt, M., 1986, Phys. Rev. B **33**, 1166.
Rasolt, M., 1987a, Phys. Rev. B **36**, 5041.
Rasolt, M., 1987b, Phys. Rev. Lett. **58**, 1482.
Rasolt, M., 1990, Solid State Phys. **43**, 93.
Rasolt, M., and D.J.W. Geldart, 1975, Phys. Rev. Lett. **35**, 1234.
Rasolt, M., and D.J.W. Geldart, 1986, Phys. Rev. B **34**, 1325.
Rasolt, M., and H. Kurz, 1985, Phys. Rev. Lett. **54**, 722.
Rasolt, M., and A.H. MacDonald, 1986, Phys. Rev. B **34**, 5530.
Rasolt, M., and F. Perrot, 1988a, Phys. Rev. B **37**, 2630.
Rasolt, M., and F. Perrot, 1988b, Phys. Rev. B **38**, 4130.
Rasolt, M., and R. Taylor, 1972, J. Phys. F **2**, 270.
Rasolt, M., and R. Taylor, 1973a, J. Phys. F **3**, 67.
Rasolt, M., and R. Taylor, 1973b, J. Phys. F **3**, 1678.
Rasolt, M., and R. Taylor, 1975, Phys. Rev. B **11**, 2717.
Rasolt, M., and G. Vignale, 1990, Phys. Rev. Lett. **65**, 1498.
Rasolt, M., and S.H. Vosko, 1974a, Phys. Rev. Lett. **32**, 297.
Rasolt, M., and S.H. Vosko, 1974b, Phys. Rev. B **10**, 4195.
Rasolt, M., S.B. Nickerson and S.H. Vosko, 1975, Solid State Commun. **16**, 827.
Rasolt, M., F. Perrot and A.H. MacDonald, 1985, Phys. Rev. Lett. **55**, 433.
Rasolt, M., B.I. Halperin and D. Vanderbilt, 1986, Phys. Rev. Lett. **57**, 126.
Rasolt, M., A.M. Malvezzi and H. Kurz, 1987, Appl. Phys. Lett. **51**, 2208.
Runge, E., and E.K.U. Gross, 1984, Phys. Rev. Lett. **52**, 997.
Sachdev, S., 1987, Phys. Rev. Lett. **58**, 2590.
Schlüter, M., and L.J. Sham, 1990, Adv. Quantum Chem. **21**, 97.
Schönhammer, K., and O. Gunnarson, 1988, Phys. Rev. B **37**, 3128.
Sernelius, B.E., 1986, Phys. Rev. B **34**, 5610.
Sernelius, B.E., 1988, Phys. Rev. B **37**, 10244.
Sham, L.J., 1985, Phys. Rev. B **32**, 3876.
Sham, L.J., and W. Kohn, 1966, Phys. Rev. **145**, 561.
Sham, L.J., and M. Schlüter, 1983, Phys. Rev. Lett. **51**, 1888.
Sham, L.J., and M. Schlüter, 1985, Phys. Rev. B **32**, 3883.
Shank, C.V., R. Yen and C. Hirlimann, 1983, Phys. Rev. Lett. **50**, 454.
Starkloff, T., and J.D. Joannopoulos, 1977, Phys. Rev. B **16**, 5212.
Steiner, M.M., R.C. Albers, D.J. Scalapino and L.J. Sham, 1991, Phys. Rev. B **43**, 1637.
Tešanović, Z., and B.I. Halperin, 1987, Phys. Rev. B **36**, 4888.
Vanderbilt, D., 1987, Phys. Rev. Lett. **26**, 1456.
Vignale, G., and M. Rasolt, 1987, Phys. Rev. Lett. **59**, 2360.
Vignale, G., and M. Rasolt, 1988, Phys. Rev. B **37**, 10685.
von der Linde, D., and N. Fabricius, 1982, Appl. Phys. Lett. **41**, 991.
von der Linden, W., and P. Horsch, 1988, Phys. Rev. B **37**, 8351.
Wang, C.S., and W.E. Pickett, 1983, Phys. Rev. Lett. **51**, 597.
Wang, J.S.Y., and M. Rasolt, 1977, Phys. Rev. B **15**, 3714.
Ward, J., 1950, Phys. Rev. **78**, 182.
Yin, M.T., and M.L. Cohen, 1982, Phys. Rev. B **25**, 7403.
Ziman, J.M., 1964, Principles of the Theory of Solids (Cambridge Univ. Press, Cambridge).
Zunger, A., and M.L. Cohen, 1979, Phys. Rev. B **20**, 4082.

CHAPTER 2

Chemical Models of Energy Bands

J. C. PHILLIPS

Bell Laboratories
Murray Hill, New Jersey 07974
USA

Handbook on Semiconductors
Completely Revised Edition
Edited by T.S. Moss
Volume 1, edited by P.T. Landsberg

Contents

1. Chemical models of energy bands 49
2. Physical properties of semiconductor materials 49
3. Dielectric model of semiconductors 50
4. Departures from molecular orbital theory 51
5. Basic formulae 51
6. Chemical trends in band edges 52
7. Alloys and interfaces 55
8. Summary of chemical trends 57

References 57

1. Chemical models of energy bands

Elsewhere in this volume the theoretical techniques which are used to define and to calculate energy bands in perfect crystals are outlined in principle, as are the many different experimental tools which have yielded precise and extensive information on the band structures of many semiconductors. In the early days it was suspected that the energy bands of different semiconductors would exhibit systematic trends (e.g., the bands of InAs and GaSb should be intermediate between those of GaAs and InSb), and recent data are indeed consistent with chemical interpolations based on the periodic table. The problem is to construct a simplified model which contains these trends in the most compact and accurate way.

There are many situations in which the translational periodicity of the perfect crystal is absent, e.g., alloys, surfaces, interfaces (such as Schottky metal–semiconductor barriers or semiconductor–insulator passivation layers), shallow and deep impurity levels, vacancies, interstitials, dislocations, etc. In all these situations the atomic positions are not known, and in general it is not feasible to determine them by means of the quantum-mechanical methods used to treat the energy bands of perfect crystals. However, it is often possible to idealize these more complicated geometries and to analyze the trends in their characteristic properties in terms of the chemical parameters appropriate to the much more accurately known energy bands of perfect crystals.

What are these chemical parameters? A phrase that is notable for its absence from the discussion so far is *chemical bond*. For historical reasons and because of the predominance of elements from the first period in molecular chemistry most scientists are predisposed to think of chemical bonds in terms of overlapping atomic orbitals. However, in semiconductors with predominantly tetrahedral coordination configurations chemical trends are much better understood in terms of the redistribution of bond charges, not the rehybridization (change in amplitude coefficients) of individual s and p atomic orbitals. The bond charges respond directly to the crystal pseudopotential in the bonding region between two atoms, and are hence describable in terms of the chemical parameters of this potential as well as the effective number of valence electrons which can respond to it.

2. Physical properties of semiconductor materials

Before we leave this subject, because so many scientists have the atomistic model so deeply ingrained in their outlook, we examine in more detail the reasons why the atomic orbital model is not so useful for semiconductors (except possibly for mnemonic reasons, e.g., in ascribing tetrahedral coordination to sp^3 hybridization) as it is for molecules. Atomistic models of the Hückel type have been enormously successful in discussions of substituted hydrocarbon molecules; why cannot the same methods be applied by rote to semiconductors?

The answer to this question lies in the many physical properties which make semiconductors a characteristic family of materials, separate and quite distinct from hydrocarbon molecules. Semiconductors lie between insulators (often formed from elements of the first period, e.g., oxides) and metals (usually formed from heavier simple elements such as Pb, or transition elements where the d electrons are important). Many semiconductors contain simple metallic elements from the Cu, Ag or Au periods (such as Ge, Sb or Hg) and in these periods the d core electrons often contribute to the chemical trends. The success of the Hückel theory rests primarily on its ability to exclude d electrons from its atomic orbital basis set because for elements from the first period the $2p \rightarrow 3d$ promotion energy is large. Because of their borderline position near metals we cannot treat most semiconductor crystals by the simplistic concepts which have worked so well for hydrocarbon molecules.

What about the three materials of greatest importance to semiconductor technology, Si, SiO_{2-x}, and Si_3N_{4-y} $(0 \leqslant x, y \leqslant 1)$? If these were the only compounds of interest to semiconductor physicists (i.e., if we were to exclude all light-emitting materials as well as all materials whose technological potential has not yet been developed), then we could do a passable job with the Hückel model. Silicon itself is quite stable and has few structural properties of a metallic type, and the addition of first-period elements such as N or O makes the materials less metallic and more molecular. However, the chemical model developed below treats Si and all semiconductor compounds not containing first-period elements with an accuracy $\sim 1\%$. Its worst errors ($\sim 5\%$) arise in materials such as ZnO or CdO which contain heavy metallic cations and first-period anions (very large electronegativity differences). Even in these worst cases the errors are much smaller than the typical errors ($\sim 15\%$) of atomistic models of semiconductors. In general, chemical trends in complicated (imperfect crystalline) geometries cannot be predicted with confidence from a chemical model of bulk energy bands unless that model is accurate to 5% or better.

3. Dielectric model of semiconductors

The quantum chemical model of semiconductors which has proved remarkably successful is the dielectric model (Phillips 1970, 1973), so called because the central features of the model are based on the fundamental optical spectra of the crystals and especially on the static electronic dielectric constant ε_0. It concerns all (~ 50) binary octet semiconductors such as Si, Ge, GaAs, ZnS, ...; the general chemical formula for these compounds is $A^N B^{8-N}$. As the compounds are predominantly tetrahedrally coordinated with cubic or nearly cubic crystal structures, it is sufficient for many chemical purposes to idealize the energy bands $E(k)$ as isotropic in extended reciprocal space (Penn model), and (at least initially) to treat the Jones zone (which contains all the valence electrons in a nearly free electron model) as spherical.

In the dielectric model of semiconductors, as in the Hückel model of hydrocarbons, we recognize that one kind of chemical bonding interaction dominates. For the Hückel model this is π bonding between C or other molecular framework atoms; for $A^N B^{8-N}$ semiconductors it is the interaction between oppositely directed, strongly overlapping sp^3 hybrids on nearest neighbors. In both cases these interactions can be decomposed into their covalent and ionic components, denoted by β and $\Delta\alpha$ in the Hückel theory and by E_h and C in the dielectric theory.

4. Departures from molecular orbital theory

The parallel between the Hückel theory and the dielectric theory, which has been perfect to this point, breaks down when we ask how many valence electrons are involved in the chemical bonds. In hydrocarbons we have one π orbital per planar framework atom, and in $A^N B^{8-N}$ semiconductors it would seem obvious that we have four valence electrons per atom on the average. This is true for elements from the C and Si periods, but for the Ge, Sn and Pb periods the effective number of valence electrons per atom is actually larger than four; in atomic orbital language the hybridization configurations have become sp^3d^z with $z \sim 1$–2. These d-hybridized configurations are more metallic than the sp^3 configurations characteristic of tetrahedral bonding of elements from the first and second periods; the effect is to enlarge the dimensionality of the basis function space, to weaken the tetrahedral directional character of the chemical bonds, and to reduce the magnitude of the bond charges.

5. Basic formulae

Covalent, ionic and metallic interactions are included quantitatively in the dielectric theory through the basic formula for the dielectric constant ε_0 in the Penn model,

$$\varepsilon_0 = 1 + D(\hbar\omega_p)^2(1 - E_g/4E_f)E_g^{-2}. \tag{1}$$

Here E_g is the Penn energy gap between bonding and antibonding energy levels, corresponding to the energy gap between the highest occupied valence band and the lowest unoccupied conduction band, averaged over the Jones zone surface. The width of the valence bands in a free electron model is denoted by E_f, and

$$\omega_p^2 = 4\pi Ne^2/m \tag{2}$$

in the free-electron plasma frequency, with N equal to eight electrons per diatomic unit cell volume. The effect of d-hybridization in an $A^N B^{8-N}$ compound is described by

$$D = (1 + d_A)(1 + d_B), \tag{3}$$

where d_A and d_B are constants which depend only on the periods to which atoms A and B belong. For A or B from the first or second periods, d_A or $d_B = 0$, as the case may be. The values of d_α for the third, fourth or fifth periods are determined experimentally from optical sum rules similar to the f-sum rule for atomic or molecular spectra (Van Vechten 1969).

The energy gap E_g contains both covalent and ionic contributions; these are separated in a symmetrical fashion by the relation

$$E_g^2 = E_h^2 + C^2. \tag{4}$$

Qualitatively speaking, if we choose our origin of coordinates halfway between atoms A and B in the unit cell, and separate the pseudopotential into its even part V_e and its odd part V_o, then we have the correspondence

$$E_h^2 \leftrightarrow \langle V_e^2 \rangle, \qquad C^2 \leftrightarrow \langle V_o^2 \rangle, \tag{5}$$

$$\langle V_e V_o \rangle = 0, \tag{6}$$

where the brackets represent an average over all the interactions which rearrange the valence pseudocharge from the constant value corresponding to $V_e = V_o = 0$. In chemical theories we are concerned with charge densities (not wavefunction amplitudes, which contribute only interference effects which are of little chemical significance). The simple algebra of eqs. (1)–(6) can be used to organize all physical properties which depend on isotropic atoms and isotropic bond charges. More refined models are needed to explain effects associated with axial bond charge asymmetries, such as the wurtzite crystal structure with its hexagonal symmetry, which will not be discussed here.

Explicit formulae for E_h and C for a given compound are obtained from

$$\log E_h = \text{const.} - 2.5 \log d, \tag{7}$$

where d is the bond length, and the const. is chosen to fit $\varepsilon_0 = 11.9$ in Si, $E_h(\text{Si}) = 4.8$ eV. Similarly for the ionic potential we have (in eV)

$$C = 1.5\left(\frac{Z_A}{r_A} - \frac{Z_B}{r_B}\right) e^{-ik_s r}, \tag{8}$$

where k_s is the Thomas–Fermi screening length and $2r = r_A + r_B$, with $r_{A,B}$ being the appropriate covalent radii. The prefactor in eq. (8) is a geometrical constant; substitution of eqs. (7) and (8) in eq. (1) and comparison with experiment shows that this number is indeed constant to within 7% (rms).

The constitutive parameters and the dielectric constants of $A^N B^{8-N}$ compounds have been summarized by Van Vechten (1969) and are listed in table 1. Included in the table are the static electronic dielectric constant (denoted by ε_∞) and the true static dielectric constant (including lattice polarizability) denoted by ε_0.

6. Chemical trends in band edges

By construction the dielectric theory provides a systematic account of chemical trends in the average dielectric properties of crystalline energy bands. Individual band edges (the top of the valence band, the lowest conduction band edges as $k = \Gamma$, X and L) follow the average trend of E_g, but with a large D-dependent correction

Table 1

Electronic and dielectric parameters of $A^N B^{8-N}$ tetrahedrally coordinated semiconductors. The Fermi width E_F of the valence electrons, regarded as a free-electron gas, is a good approximation to the overall width of the valence band, while $\hbar\omega_p$ (from eq. (2)) is another characteristic property of the valence electrons, their natural frequency of oscillations as a free-electron gas relative to a uniform positive background charge slab. The static and optical dielectric constants are also given where known: the predicted values from eqs. (1)–(8) are listed (with *) where experimental values are not available. Various parameters of the dielectric theory are listed in the remaining columns.

Substance	E_F (eV)	$\hbar\omega_p$ (eV)	ε_0	ε_∞	$\langle E_g \rangle$ (eV)	E_h (eV)	C (eV)	f_i (%)
C	28.9	31.2	5.66	5.66	13.5	13.5	0	0
Si	12.5	16.6	12.0	12.0	4.77	4.77	0	0
Ge	11.5	15.6	16.0	16.0	4.31	4.31	0	0
α-Sn	8.72	12.7	24	24	3.06	3.06	0	0
SiC	19.4	23.2	9.8	6.6	9.12	8.27	3.85	17.7
BN	28.1	30.6	7.1	4.5	15.2	13.1	7.71	25.7
BP	17.8	21.7		8.2*	7.66	7.44	1.84	5.8
BAs	16.1	20.1		10.2*	6.56	6.64	1.07	2.6
AlN*	19.3	23.0	9.1	4.8	11.0	8.17	7.32	44.6
AlP	23.4	16.5		7.6	6.03	4.72	3.76	38.8
AlAs	11.4	15.5		8.2	5.81	4.38	3.82	43.2
AlSb	9.77	13.8	14.4	10.2	4.68	3.52	3.08	43.3
GaN*	18.2	22.1		5.4	10.3	7.65	6.94	45.2
GaP	12.4	16.5	11.1	9.1	5.76	4.72	3.30	32.8
GaAs	11.5	15.6	13.0	10.9	5.20	4.32	2.90	31.0
GaSb	9.82	13.9	15.7	14.4	4.12	3.55	2.10	26.0
InN	14.9	18.9		6.3*	8.36	5.93	5.89	49.6
InP	10.7	14.8	12.4	9.6	5.16	3.93	3.35	42.1
InAs	10.1	14.2	14.6	12.3	4.58	3.67	2.74	35.9
InSb	8.76	12.7	17.8	15.7	3.76	3.08	2.15	32.7
BeO	25.3	28.3	7.4	3.0	18.8	11.5	14.7	62.0
BeS	15.6	19.7		7.1*	7.47	6.31	3.99	28.5
BeSe	14.3	18.4		8.5*	6.57	5.65	3.36	26.1
BeTe	12.0	16.1		11.6*	4.98	4.54	2.05	16.9
MgTe	9.04	13.0		8.3*	4.80	3.20	3.58	55.6
ZnO*	17.6	21.5	8.2	3.7	12.5	7.33	10.1	65.5
ZnS	12.6	16.7	8.3	5.1	7.85	4.82	6.20	62.3
ZnSe	11.4	15.6	8.1	6.0	6.98	4.29	5.51	62.3
ZnTe	9.91	14.0	10.1	7.4	5.66	3.59	4.38	59.9
CdS*	10.8	14.9	9.7	5.3	7.01	3.97	5.78	67.9
CdSe*	9.94	14.0	9.7	6.0	6.42	3.60	5.31	68.4
CdTe*	8.75	12.7	10.2	7.2	5.40	3.08	4.43	67.5
HgS*	10.8	14.9	23	4.9*	8.20	3.76	7.30	79.0
HgSe	9.91	14.0	25.6	7.8*	6.06	3.43	5.00	68.0
HgTe	8.75	12.6		6.6*	4.95	2.92	4.00	65.2
CuF	20.3	23.9		2.5*	18.1	8.73	15.8	76.6
CuCl	12.6	16.7	9	3.7	9.60	4.83	8.30	74.6
CuBr	11.1	15.2	8.0	4.4	7.35	4.14	6.90	73.5
CuI	10.1	14.1		5.5	6.61	3.66	5.50	69.2
AgI	8.77	12.8		4.9	6.48	3.09	5.70	77.3

(Γ), a moderate D-dependence (L) or negligible D-dependence (X). Explicit formulae for these band-edge trends, as well as extensive tables of band-edge energies, have been given by Van Vechten in his classic papers (Van Vechten 1969).

Of much greater consequence is the power of the theory to predict chemical trends in structural properties; success in this area suggests that E_h, C, D and, to a lesser extent, $\hbar\omega_p$ can be used to describe bonding in imperfect geometries. By defining the fractional bond ionicity f_i through

$$f_i = C^2/E_g^2, \tag{9}$$

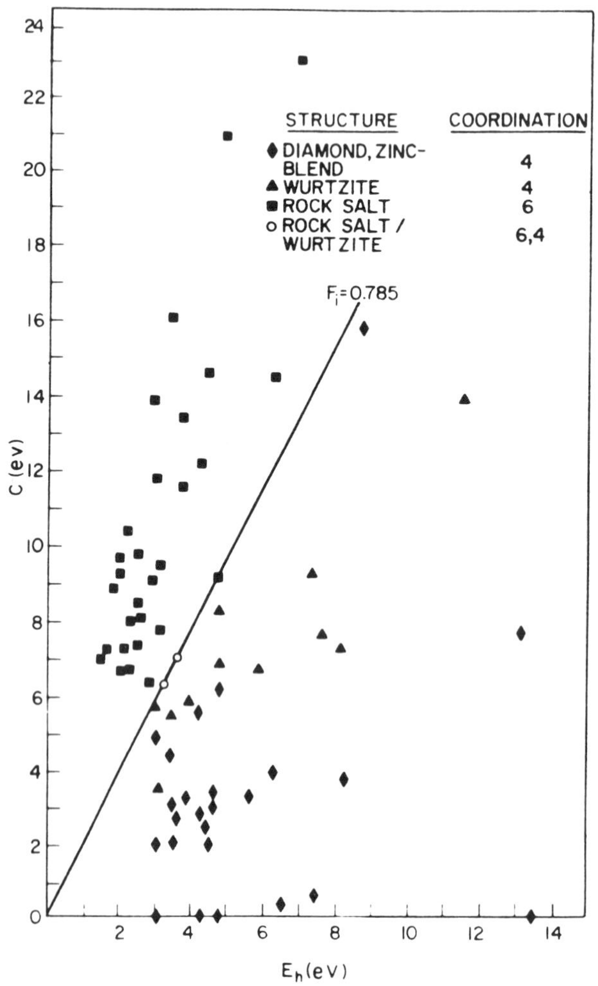

Fig. 1. A Cartesian population map of $A^N B^{8-N}$ compounds in (E_h, C) space, where E_h and C are the covalent and ionic energy gaps defined by the dielectric relations (eqs. (1)–(8) of text). Covalent structures (four-fold coordinated) are separated from ionic structures (six-fold coordinated) by the line $C^2/E_g^2 = 0.785 = F_i$, the critical ionicity.

we discover (Phillips 1970) that many integrated properties (such as cohesive energies, heats of formation, and bond-bending force constants) are linear functions of f_i. Thermochemical parameters (defect energies, melting or other phase transformation curves as a function of P and T) have also been treated with the dielectric theory by Van Vechten (1975). As illustrations of the accuracy and depth of the dielectric formulation the structural transition from four-fold to six-fold (NaCl) coordination configurations of $A^N B^{8-N}$ compounds is shown in fig. 1. The difference in Gibbs free energy $\Delta G = G(6) - G(4)$ between the two phases, measured in units of the heat of atomization (vaporization) is shown in fig. 2. The ratio of noncentral (covalent) directional force strength β to central force strength α is shown in fig. 3.

7. Alloys and interfaces

Chemical models of energy bands provide information on the local atomic environments of elements in isovalent alloys. Recent work has shown that the tetrahedral

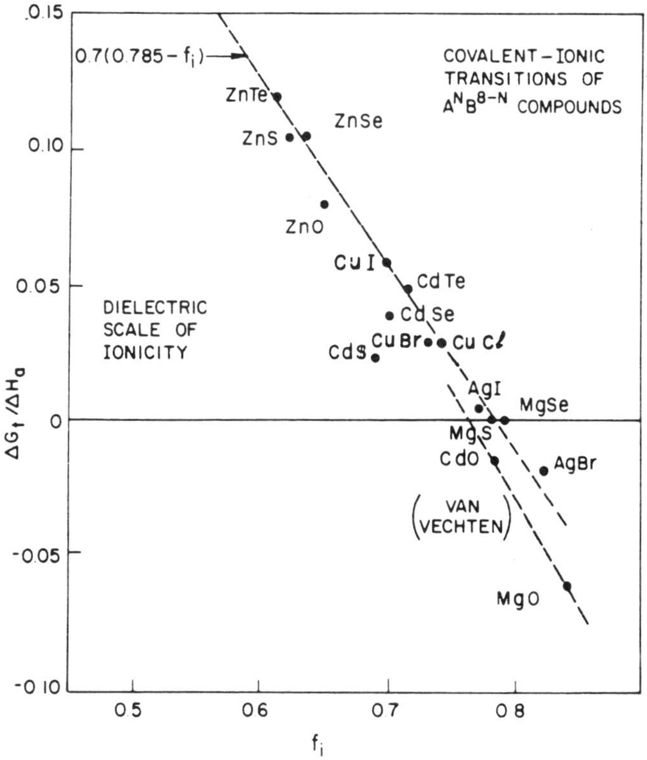

Fig. 2. The difference in Gibbs free energy $\Delta G = G(6) - G(4)$, normalized by the heat of atomization, is proportional to $x = f_i - F_i$ for small x. Here F_i is the critical ionicity. The values of ΔG are determined, e.g., from measurements of transition pressures and volume changes.

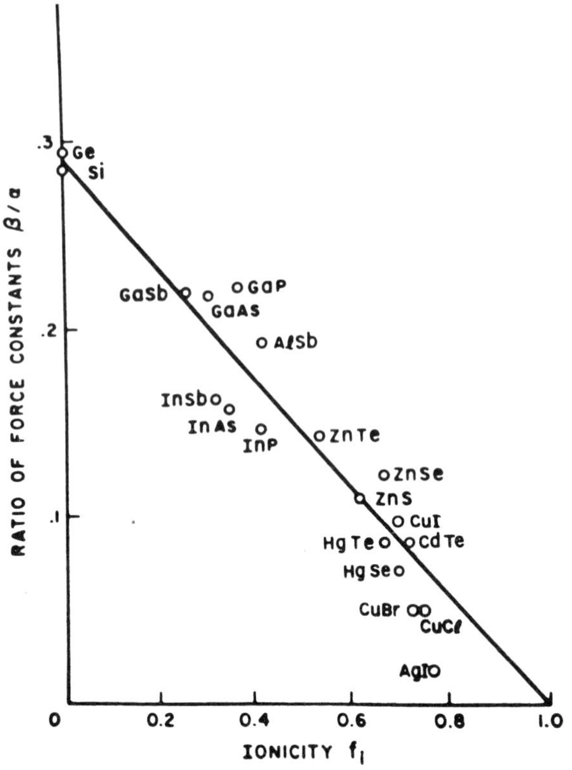

Fig. 3. The ratio of directional (three-body, bond-bending) to central (two-body, bond stretching) interatomic forces for tetrahedrally coordinated crystals as a function of ionicity. Note that although β/α extrapolates to 0 as $f_i \to 1$, there actually is a softening for $f_i \gtrsim 0.7$ as the critical ionicity $F_i = 0.785$ is approached. The lattice instability is heralded by a softening of zone-boundary phonons.

environment is chemically distorted (Mikkelsen and Boyce 1982). Similar distortions are present as optical band-edge bowing and can be related to mixing enthalpy (Phillips 1973), but the complete relationships between phase diagrams and local atomic order can be obtained only by extensive statistical analysis (Wei and Zunger 1990).

Valence band-edge offsets are important in quantum-well heterostructures which lie at the heart of diode laser technology. Chemical trends in these offsets are now well understood in the context of sophisticated self-consistent pseudopotential calculations (Van de Walle and Martin 1987). These offsets, when small, are additive. When large, they are not (Bylander and Kleinman 1989), because of heat-of-formation nonlinearities similar to those responsible for band-edge bowing and tetrahedral environmental distortion in bulk alloys (above).

8. Summary of chemical trends

The dielectric model has been successful in analyzing, unifying and predicting many chemical and physical properties of perfect and imperfect $A^N B^{8-N}$ semiconductors. With growing sophistication in semiconductor materials science other applications of this simple model should become possible. A successful example is the study by Littlewood and Heine (1979) of the structural properties of the PbS ($A^N B^{10-N}$) family.

References

Bylander, D.M., and L. Kleinman, 1989, Phys. Rev. B **39**, 5116.
Littlewood, P.B., and V. Heine, 1979, J. Phys. C **12**, 4431.
Mikkelsen, J.C., and J.B. Boyce, 1982, Phys. Rev. Lett. **49**, 1412.
Phillips, J.C., 1970, Rev. Mod. Phys. **42**, 317.
Phillips, J.C., 1973, Bonds and Bands in Semiconductors (Academic Press, New York) p. 208 ff.
Shay, J.L., S. Wagner and J.C. Phillips, 1976, Appl. Phys. Lett. **28**, 31.
Van de Walle, C.G., and R.M. Martin, 1987, Phys. Rev. B **35**, 8154.
Van Vechten, J.A., 1969, Phys. Rev. **182**, 891; **187**, 1007.
Van Vechten, J.A., 1975, J. Electrochem. **122**, 423.
Wei, S.-H., and A. Zunger, 1990, Appl. Phys. Lett. **56**, 662.

CHAPTER 3

Ab initio Pseudopotentials and the Structural Properties of Semiconductors

JAMES R. CHELIKOWSKY

Department of Chemical Engineering
and Materials Sciences
University of Minnesota
Minneapolis, MN 55455, USA

MARVIN L. COHEN

Department of Physics
University of California, and
Materials Sciences Division
Lawrence Berkeley Laboratory
Berkeley, CA 94720, USA

Handbook on Semiconductors
Completely Revised Edition
Edited by T.S. Moss
Volume 1, edited by P.T. Landsberg

Contents

1. Introduction . . . 61
   1.1. History . . . 61
   1.2. Background and definitions . . . 63
      1.2.1. Defining a pseudopotential . . . 63
      1.2.2. Self-consistent and ab initio pseudopotentials . . . 65
2. Pseudopotential methods . . . 68
   2.1. Empirical pseudopotential method (EPM) . . . 68
   2.2. Local density pseudopotentials . . . 71
   2.3. Pseudopotentials without the local density approximation . . . 80
3. Application to structure . . . 81
   3.1. Total energies . . . 81
   3.2. Bulk structures . . . 83
   3.3. Surface structures . . . 89
4. Applications to vibrational properties . . . 90
5. Other applications . . . 96
   5.1. Structural trends . . . 96
   5.2. Bulk moduli . . . 98
   5.3. Electron–phonon interactions and superconductivity . . . 102
   5.4. Molecular dynamics using ab initio pseudopotentials . . . 105
6. Conclusions . . . 106

References . . . 107

1. Introduction

1.1. History

A long-time goal of condensed matter physics has been the determination of properties of materials using only information about the constituent atoms. Implicit in the development of a scheme of this kind is the potential to predict the existence and properties of new solids not previously realized in nature, and the possibility of developing methods for producing solids with useful properties. At the least, success in this area can serve as proof of the validity and predictive power of the quantum theory of solids.

Achievements of this kind are now possible for some classes of solids. The first successes came in the early 1980's and this area is developing at a rapid rate. The major reasons for the almost 60 year hiatus between the development of quantum theory, and its direct application to structural properties of solids, are that a workable model for a real solid needed to be developed, and significant computational power was needed. The pseudopotential model of a solid led the way in providing a workable model, and modern computers provided the computational resources. Here we discuss the major developments in this area. Because tetrahedrally coordinated semiconductors have been the prototype materials in this area, they will be the focus of this review.

An essential step in the creation of a model of a solid capable of the precision needed for structural studies was the development of a scheme which could produce accurate information about the electronic structure of the materials studied. The empirical pseudopotential method (EPM) (Cohen and Chelikowsky 1982, 1988, Cohen and Bergstresser 1966, Chelikowsky and Cohen 1976, Bassani and Celli 1961, Brust et al. 1962, Cohen and Heine 1970, Fermi 1934) provided such a scheme through the use of optical spectra to determine the potentials governing the behavior of valence electrons in solids. With a few fit Fourier coefficients of the potential, extremely accurate electronic band structures can be computed. Although this approach does rely on empirical data, the form of the potential and the underlying scheme are rooted in pseudopotential theory (Cohen and Chelikowsky 1988, Cohen and Heine 1970).

Because the pseudopotential model describes a solid as a sea of valence electrons moving in a periodic background of cores (composed of nuclei and inert core electrons), some of the complexities of all-electron calculations are avoided. A group IV solid such as C with 6 electrons is treated in a similar fashion to Ge with 32 electrons since both have 4 valence electrons. In addition, the focus of the calculation is only on the accuracy of the valence electron wavefunction in the spatial region away from the core. The nodeless pseudowavefunction is taken to be identical to the appropriate all-electron wavefunction in the regions of interest for solid-state effects. For the core region, the wavefunction is extrapolated back to the origin in a manner consistent with the normalization condition. Various schemes have been developed to construct potentials which yield wavefunctions of this kind and these will be discussed later. It is interesting to note that a similar construction (fig. 1) was introduced by Fermi (Fermi 1934, Cohen 1984a,b) to account for the shift in the wavefunctions of high lying stakes of alkali atoms subject to perturbations from foreign atoms. In this remarkable paper, Fermi introduced the conceptual basis for both the pseudopotential and the scattering length.

Direct applications of the EPM to solids came in the mid-1960's and 1970's and implementation of ab initio approaches began in the 1980's. Important research on the physics of the pseudopotential itself was done in the late 1950's and 1960's. The derivation (Phillips and Kleinman 1959) of a repulsive potential arising from the orthogonalization of valence electrons to core states led to the concept of a "Pauli force", which kept valence electrons out of the core region, and to the view that there is a cancellation (Cohen and Heine 1961) of the attractive core potential and the repulsive potential arising from the orthogonality condition. Various forms and properties of the pseudopotentials were devised (Austin et al. 1962, Harrison 1966), and the underlying physical mechanisms giving rise to the properties of pseudopotentials were elucidated.

Much of the current work in this area is centered on the construction of useful potentials derivable with only the atomic number as input. Applications of these ab initio potentials are therefore considered to be free of prejudice or data related to the solid under consideration. For structural properties, total energies of candidate crystal structures are compared as a function of volume to determine the lowest energy configuration. For calculations of phonon spectra and electron–phonon

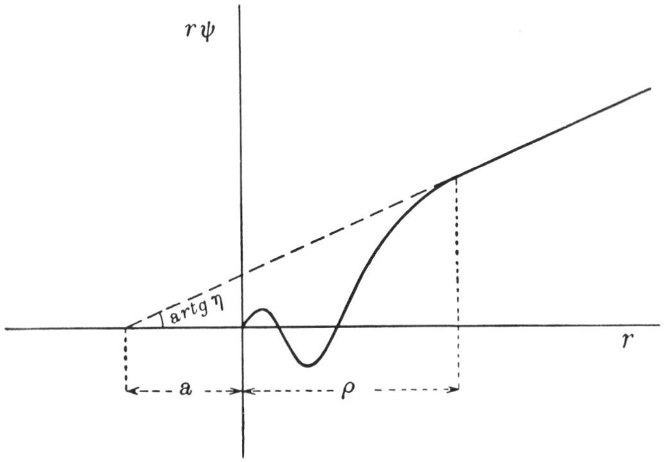

Fig. 1. Fermi's approximation to the position dependence of the electronic wavefunction for determining a pseudopotential (Fermi 1934).

interactions, the ionic mass is the only additional input, hence the scheme remains free of input about the solid-state except for the use of a finite number of candidate crystal structures.

The results of these studies have been highly satisfying (Cohen 1986, 1989). After presenting more details about the background of this area, the techniques and applications will be described.

1.2. Background and definitions

The transcription of the electronic structure problem to a "one-electron" Hamiltonian intimately depends on our ability to construct an accurate potential. There are several avenues open to us for such a construction. We could try to figure out a solution without experimental input. This procedure has several advantages. An ab initio potential is not in danger of being "biased" by preconceptions of the worker, for example, assuming a covalent bonding configuration when it might be more appropriate to assume a metallic configuration. Also, experimental data might not exist, or it might be inaccurate. The case of a surface is a good example. The potential at a surface might be very different than that in the bulk, but experimental data characteristic of the surface potential might be unavailable. Conversely, experiment can play a major role in defining a potential. For example, when reliable optical measurements of solids became routine, such measurements were used as input to define potentials which could be used to generate energy bands (Cohen and Chelikowsky 1988).

1.2.1. Defining a pseudopotential

Consider a simple system consisting of a perfect elemental crystal. We might hope to write the potential for this crystal as follows:

$$V(\boldsymbol{r}) = \sum_{\boldsymbol{R},\tau} V_a(\boldsymbol{r} - \boldsymbol{R} - \boldsymbol{\tau}), \tag{1}$$

where $\boldsymbol{R}$ is a lattice vector, $\boldsymbol{\tau}$ is a basis vector and $V_a(\boldsymbol{r})$ is a potential which we associate with each atom. It is possible to expand this potential in the reciprocal lattice as

$$V(\boldsymbol{r}) = \sum_{\boldsymbol{G}} V_a(\boldsymbol{G}) S(\boldsymbol{G}) \exp(\mathrm{i}\boldsymbol{G} \cdot \boldsymbol{r}), \tag{2}$$

where the structure factor is given by

$$S(\boldsymbol{G}) = \frac{1}{N_a} \sum_{s=1}^{N_a} \exp(-\mathrm{i}\boldsymbol{G} \cdot \boldsymbol{\tau}_s). \tag{3}$$

$V_a(\boldsymbol{G})$ is the form factor, or Fourier transform, for an atomic potential, V_a, and N_a is the number of basis atoms within the unit cell. The structure factor contains information on the "geometry" and the form factors contain information on the "electronic interactions." If the sum in eq. (2) were to converge rapidly, one would have a very convenient form for the potential. Unfortunately, the "true" crystalline potential does not converge rapidly. The potential near the nucleus varies rapidly and, in general, the number of the form factors required for an accurate solution for the electronic structure of the solid prohibits a sensible description of the potential via eq. (2). However, if we replace the "all-electron" potential by a weaker potential which reproduces only the properties of the valence state electrons, it is possible to achieve a dramatic improvement in convergence. This is the fundamental underpinning of the pseudopotential concept.

There are a number of ways of justifying the transformation from an "all-electron" potential ,which reproduces both core and valence electron state properties to a pseudopotential which reproduces only the valence properties. One of the earliest justification comes from the work of Phillips and Kleinman (Phillips and Kleinman 1959, Kleinman and Phillips 1960, Cohen and Heine 1961). Let us write the wavefunction of the solid as being composed of two parts:

$$\psi = \varphi + \sum_t b_t \varphi_t, \tag{4}$$

where the sum is over the core states φ_t, b_t are constants, and φ is taken to be a smooth, slowly varying function. This form of the wavefunction was first proposed by Herring (Herring 1940) where φ is taken as a plane wave, $\exp(i\boldsymbol{k}\cdot\boldsymbol{r})$, and $b_t = -\langle\varphi_t|\exp(\boldsymbol{k}\cdot\boldsymbol{r})\rangle$. This choice corresponds to the orthogonalized plane wave (OPW) method. OPW's are an appropriate form for the solid state wavefunction. Near the ion core the OPW looks "core like" and in the interstitial regions, where the potential varies slowly, the OPW appears "plane wave like." An OPW also satisfies the requirement of the valence states being orthogonal to the core states.

Substituting eq. (4) into $H\psi = E\psi$, and taking $b_t = -\langle\varphi_t|\varphi\rangle$ which ensures orthogonality of ψ to the core states, we obtain

$$H\varphi - \sum_t \langle\varphi_t|\varphi\rangle E_t\varphi_t = E\varphi - E\sum_t \langle\varphi_t|\varphi\rangle\varphi_t, \tag{5}$$

which can be rearranged to yield

$$H\varphi + \sum_t (E - E_t)\varphi_t\langle\varphi_t|\varphi\rangle = E\varphi. \tag{6}$$

This form can be written as

$$(H + V_{\mathrm{R}})\varphi = E\varphi, \tag{7}$$

where

$$V_{\mathrm{R}}\varphi = \sum_t (E - E_t)\varphi_t\langle\varphi_t|\varphi\rangle.$$

V_{R} acts like a short-ranged, non-Hermitian repulsive potential. Equation (6) is a new wave equation for φ, the pseudowavefunction. While φ does not correspond to the real wavefunction, it is important to recognize that the energy eigenvalue is not a "pseudo" energy, but the "true" energy corresponding to the "true" wavefunction ψ.

We can write eq. (6) as

$$H_{\mathrm{p}}\varphi = \left(\frac{-\hbar^2\nabla^2}{2m} + V_{\mathrm{p}}\right)\varphi = E\varphi, \tag{8}$$

where $H_{\mathrm{p}} = (p^2/2m) + V_{\mathrm{p}}$ and $V_{\mathrm{p}} = V + V_{\mathrm{R}}$ is the pseudopotential. The point is that V_{R} acts as a repulsive potential which effectively cancels the strong part of the all-electron potential within the core region. While V_{p} is weak, the potential, in principle, is not only state dependent, or angular momentum dependent, it is also energy dependent as indicated in eq. (7). In many cases, the energy and state dependence of this potential is weak, and we can replace the real potential by a simple pseudopotential which is a function of position alone.

If we assume that the "all-electron" potential can be replaced by a weak pseudopotential, then only a few form factors might be required to reproduce the crystalline potential. The empirical pseudopotential method (EPM) is based on the existence of such a rapidly convergent "Fourier" series for a potential. If only a few form factors are relevant and the experimental data are sufficient, it is often possible to fix the form factors on the basis of experiment. The EPM has been highly successful for describing the electronic structure of semiconductors and simple metals. For semiconductors, this method relies primarily on optical data. If the optical spectra of a solid is known, then the pseudopotential can be adjusted to reproduce the spectra. This technique remains one of the most effective means to analyze the optical spectra of solids. Using ab initio methods for the optical spectra of solids is possible, but such methods tend to be computationally intensive. With respect to simple metals, it is possible to consider other experimental data such as de Haas–van Alphen measurements which probe the Fermi-surface, to determine the form factors (Cohen and Heine 1970).

Provided the screening does not change too much, one can transfer a pseudopotential from one solid to another. One might extract an indium potential and an antimony potential from fitting the optical properties of indium antimonide. Using the elemental potentials, one might calculate the Fermi-surfaces of the individual metals. In the case of indium, the superconducting properties have also been determined successfully by such a procedure (Allen and Cohen 1969). If the screening change is significant, the transferability of the form factors is suspect. In this situation, the part of the pseudopotential corresponding to the ion core–valence electron interaction is transferable, but the "total" potential is not.

While the pseudopotential concept has met its greatest success in terms of applications to the solid state, there is nothing in the Phillips–Kleinman cancellation theorem which restricts the pseudopotential concept to solids. We mention a few representative references which indicate the use of pseudopotentials in molecular or atomic physics: Kahn and Goddard (1972), Melius and Goddard (1974), Bardsley (1974), Christiansen et al. (1979), Kerker et al. (1979), Schlüter et al. (1979), Sakai and Huzinaga (1982), Chelikowsky and Chou (1988a).

1.2.2. Self-consistent and ab initio pseudopotentials

A deficiency of the empirical pseudopotential method is that the potential is "biased" by the experimental data to which it was fit. If, e.g., we fit the optical properties of

GaSb and extract a Ga potential, we must be very careful in trying to use this Ga potential in another crystal. For example, this potential might work fine in GaAs which has a similar bonding configuration when compared to GaSb, but the potential may fail to describe GaN which is considerably more ionic than GaAs. Another problem can occur at a surface. In GaAs, each Ga atom is surrounded by four As atoms. For the cleavage plane, the (110) surface, the Ga atom is bonded to three As atoms. Given the coordination change, the surface Ga atom cannot be expected to retain a "bulk-like" screening potential. There is no reason to be confident that the Ga pseudopotential extracted from the crystalline environment will be very accurate at the surface.

We can consider the form factors to be composed of two interactions: the "ion core–valence electron" interaction and the "valence–valence" interaction. A fundamental postulate of the pseudopotential method is that the ion core is chemically inert. We assume that this part of the potential can be transferred with little loss in accuracy. The key issue is to determine the screening potential, i.e., the interaction between valence electrons. A usual procedure is to construct a *self-consistent potential.* The wavefunctions obtained in an electronic structure calculation can be used to construct a new screening potential which in turn can be used with the ion core potential to compute a new total potential and new wavefunctions. When no changes occur in a feedback loop of this kind, the solution is considered to be self-consistent.

A *one-electron* Hamiltonian can be written as

$$H = \frac{p^2}{2m} + V_{\text{ion}} + V_{\text{scr}}. \tag{9}$$

The ion-core pseudopotential, V_{ion}, can be taken as a linear superposition of spherical potentials. Determining the ionic potential can be accomplished by resorting to atomic structure calculations as will be outlined in detail in § 2.3. The screening potential is a separate subject and a topic which is intimately related to the study of *many-body* physics. A common approach is to divide the screening potential into two parts. One part arises from the Poisson equation and corresponds to an electrostatic potential

$$\nabla^2 V_{\text{H}} = -4\pi e^2 \rho(\boldsymbol{r}), \tag{10}$$

where V_{H} is called the Hartree potential, or the Coulomb potential, and $\rho(\boldsymbol{r})$ is the *valence* electron density. The second part of the screening potential, the exchange-correlation part of the potential, V_{xc}, is purely quantum mechanical in nature and arises from the Pauli principle. A common approximation for this part of the potential arises from the local density approximation, i.e., the potential depends only on the charge density at the point of interest, $V_{\text{xc}}(\boldsymbol{r}) = V_{\text{xc}}[\rho(\boldsymbol{r})]$. Perhaps the simplest form of this potential comes from the work of Slater (Slater 1951):

$$V_{\text{xc}}(\boldsymbol{r}) = -3\alpha(\tfrac{3}{8})^{1/3} e^2 [\rho(\boldsymbol{r})]^{1/3}. \tag{11}$$

This form depends on a parameter, α, and was employed in early self-consistent pseudopotential calculations. The screening potential is formed from the sum of V_{H} and V_{xc}. The procedure for generating a self-consistent potential is given in fig. 2.

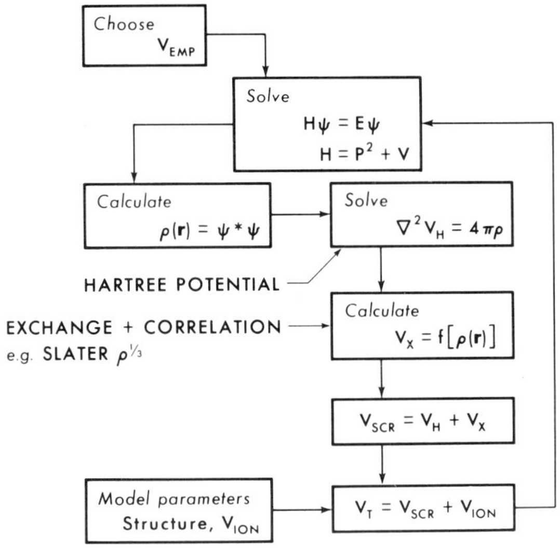

Fig. 2. Block diagram illustrating the procedure for obtaining a self-consistent field potential.

In the empirical pseudopotential method, the *total* potential, $V_T = V_{ion} + V_H + V_{xc}$, is fixed by fitting the form factors of this potential to experiment. In a self-consistent field calculation, the total potential is determined by screening the ion-core potential, V_{ion}. The construction of ion-core pseudopotentials has become an active area of electronic structure theory. Methods for constructing such potentials have centered on *ab initio* or *first-principles* pseudopotentials; i.e., the informational base on which these potentials are based does not involve any experimental input. The primary application of these potentials involves total energy or structural calculations as opposed to spectroscopic calculations (e.g., Yin and Cohen 1982a–d, Cohen 1984a,b). The reason for this emphasis is that the local density approximation is a *ground state* theory and is formally not applicable to excited state properties.

If we were to find a procedure from which we knew the total potential and the charge density of interest, we could write the ionic core potential as

$$V_{ion} = V_T - V_H[\rho(r)] - V_{xc}[\rho(r)]. \tag{12}$$

Given the wavefunctions and the eigenvalues, one can invert the Schrödinger equation and extract the total potential. For example, in the simplest picture where we have just one atomic state of interest

$$V_T = \frac{\hbar^2 \nabla^2 \psi}{2m\psi} + E. \tag{13}$$

This equation is well-behaved provided that ψ is nodeless. Normally, this will be the case for a pseudopotential wavefunction. One additional complication is that the potential is state dependent or angular momentum dependent, e.g., see eq. (7). For an s-state, ψ_s, we will get a $V_{T,s}$; for a p-state we will obtain $V_{T,p}$ and so on. In general, one can decompose the pseudopotential into a sum over angular momentum components: $V_T = V_s P_s + V_p P_p + V_d P_d + V_f P_f + \cdots$, where P_l projects out the lth angular momentum component.

Usually, one extracts the energy levels of interest by performing an atomic structure calculation starting from the all-electron potential, i.e., $V_{\text{all-electron}} = -Ze^2/r$, where Z is the atomic number. These calculations also yield the "all-electron" wavefunctions. It is possible to use the "all-electron" energy levels and wavefunctions with suitable modifications to construct ab initio ion-core potentials. These types of potentials are strongly dependent on the nature of the "many-body" physics which enters via V_{xc}. One can approach this problem using various levels of sophistication. A very simple approach is to consider "Thomas–Fermi" screening which might be only appropriate for simple metals (e.g., Chelikowsky 1980). At the other end of the spectrum, one might consider quantum Monte Carlo approaches (e.g., Kalos and Whitlock 1986, Fahy et al. 1988) and devise ion core potentials within this framework. It is important to recognize that the pseudopotential cannot be "better" than the worst approximation involved in its construction. Usually, the "weakest" approximations in the construction of the pseudopotential center on the "many-body" simplification as opposed to the separation into core and valence regimes of the all-electron potential.

2. Pseudopotential methods

In this section, we discuss specific approaches to constructing pseudopotentials. We will concentrate on ab initio methods which do not require experimental input. Most of these approaches are based on using the local density approximation for determining the screening potential. However, we will also touch briefly on new approaches and methods which do not rely on this approximation, or the one-electron approximation. In order to put these ab initio methods in perspective, we begin with a brief review of the empirical pseudopotential method. The EPM was the first method to describe accurately the optical properties of solids and stands as one of the landmark achievements in the study of optical properties of solids (Cohen and Chelikowsky 1988).

2.1. Empirical pseudopotential method (EPM)

From our earlier discussion, we noted that one could express the crystalline potential as an expansion in reciprocal lattice vectors. If the real potential is replaced by a pseudopotential, then the expansion in reciprocal lattice vectors may converge rapidly enough so that only a few form factors are required for an accurate description of the potential. The EPM is most successful for elements for which the pseudopotential can be expressed as a simple local potential, i.e., not energy dependent, or state dependent. This requirement is often met by simple metals, and metalloids, but is not met by first-row elements, or complex metals (either transition metals or the rare earths). In the case of first-row elements, no p-states exist within the ion core and no cancellation occurs for the valence p-states (as can be seen from the Phillips–

Kleinman cancellation theorem). The absence of cancellation results in p-states which are more localized than can be accounted for by simple, local pseudopotentials. Likewise, no cancellation occurs for the 3d, or 4f states, but the problem is more complex here than a simple cancellation. Localized states such as d- or f-states require a large number of plane waves in a plane wave basis and often involve potentials which cannot be expanded in terms of a rapidly convergent series.

For illustration purposes, we consider the elemental semiconductors silicon and germanium. These materials occur in the diamond crystal structure. The diamond lattice is composed of two interpenetrating face centered cubic lattices which are displaced from one another by $(\frac{1}{4}, \frac{1}{4}, \frac{1}{4})a$, where a is the lattice constant. If the origin is taken halfway between the two fcc lattices, then the basis vectors are given by $\boldsymbol{\tau} = \pm(\frac{1}{8}, \frac{1}{8}, \frac{1}{8})a$. The structure factor given by eq. (3) becomes $S(\boldsymbol{G}) = \cos(\boldsymbol{G} \cdot \boldsymbol{\tau})$. If one is interested in spectroscopy, then the $\boldsymbol{G} = 0$ term can be omitted since this term corresponds to the average crystalline potential and uniformly shifts all the energy levels. For cohesive energies, this term can be very important and cannot be arbitrarily set. The next several $\boldsymbol{G}$ vectors correspond to $\boldsymbol{G} = (2\pi/a)(1, 1, 1)$, $(2\pi/a)(2, 0, 0)$, $(2\pi/a)(2, 2, 0)$, $(2\pi/a)(3, 1, 1)$ and $(2\pi/a)(2, 2, 2)$, with sign permutations. It is convenient to categorize these vectors as $G^2 = 3, 4, 8, 11$, and 12 in units of $(2\pi/a)^2$. The form factors depend only on the magnitude of $\boldsymbol{G}$, provided that $V_a = V_a(|\boldsymbol{r}|)$ as is often the case. We note that $S(\boldsymbol{G}) = 0$ for $G^2 = 4$ and 12. Scattered waves corresponding to these vectors destructively interfere. Of the first six reciprocal lattice vectors, only the form factors corresponding to $G^2 = 3, 8$, and 11 are required to specify the crystalline potential. Provided we can terminate the series at this point, we need only three numbers, i.e., $V_a(G)$ for $G^2 = 3, 8$, and 11 to determine the pseudopotential.

In figs. 3 and 4, we illustrate the essential features of an empirical pseudopotential. In fig. 3, the all-electron potential and the pseudopotential are illustrated schematically in real space. The cancellation is strong only in the core region and at a distance

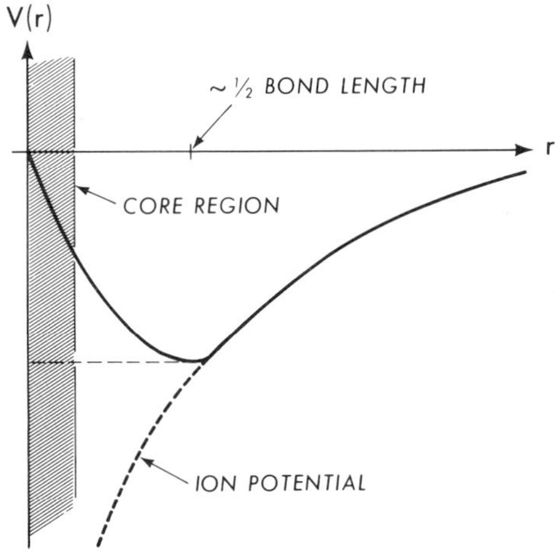

Fig. 3. Schematic plot of a pseudopotential in real space.

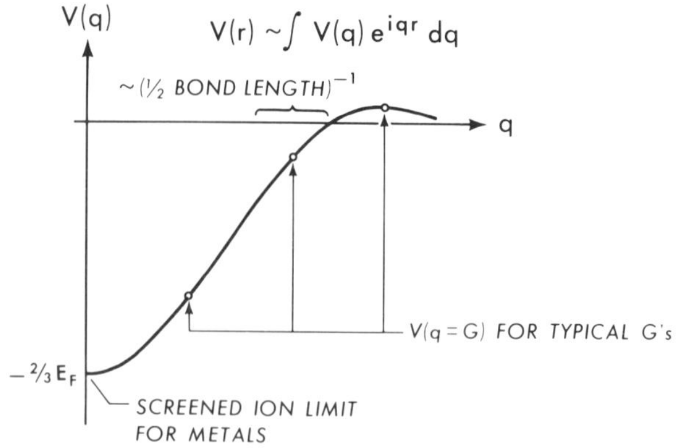

Fig. 4. Schematic plot of a pseudopotential in reciprocal space. The "typical G's" illustrated correspond to $G^2 = 3$, 8 and 11 for a diamond structure where G^2 is in units of $(2\pi/a)^2$.

of roughly $\frac{1}{2}$ the bond length the two potentials converge. In fig. 4, the pseudopotential is illustrated in reciprocal space. The size of the core region determines where the pseudopotential becomes positive. The values of the reciprocal lattice vectors and the $V_a(G)$'s necessary to determine the crystalline potential are also illustrated. In the case of silicon or germanium, one can relate these to $V_a(3)$, $V_a(8)$ and $V_a(11)$.

Given the rapidly convergent series for the crystalline potential, it is often convenient to write a plane wave basis for the wavefunctions:

$$\psi_{n,\boldsymbol{k}}(\boldsymbol{r}) = \sum_{\boldsymbol{G}} a_{n,\boldsymbol{k}}(\boldsymbol{G}) \exp[\mathrm{i}(\boldsymbol{k}+\boldsymbol{G})\cdot\boldsymbol{r}], \tag{14}$$

where n is the band index. The expansion coefficients, $a_{n,\boldsymbol{k}}(\boldsymbol{G})$, can be determined by solving a secular equation of the form:

$$\sum_{\boldsymbol{G}'} (H_{\boldsymbol{G},\boldsymbol{G}'}(\boldsymbol{k}) - E\delta_{\boldsymbol{G},\boldsymbol{G}'}) a_{n,\boldsymbol{k}}(\boldsymbol{G}') = 0, \tag{15}$$

where

$$H_{\boldsymbol{G},\boldsymbol{G}'}(\boldsymbol{k}) = \hbar^2 \frac{(\boldsymbol{k}+\boldsymbol{G})^2}{2m} \delta_{\boldsymbol{G},\boldsymbol{G}'} + V_a(\boldsymbol{G}-\boldsymbol{G}')S(\boldsymbol{G}-\boldsymbol{G}'). \tag{16}$$

The simplicity of a plane wave expansion makes this basis highly attractive for computational purposes. Often plane wave expansions are used in situations which require a large number of plane waves to achieve accurately converged solutions of the secular equation. The computational complexities which arise from the use of large numbers of plane waves are offset by the simplicity of the plane wave matrix elements in eq. (16). Numerical techniques have been developed which allow secular equations involving up to $\sim 10^5$ plane waves to be solved (Martins and Cohen 1988a,b). However, for the case of silicon or germanium, 50–100 plane waves are usually sufficient for accurate EPM calculations.

An EPM calculation proceeds as indicated in fig. 5. First, a set of $V(\boldsymbol{G})$'s are chosen. A simple model potential may be used as a starting point. Often these models are fixed by examining atomic term values (e.g., Heine and Abarenkov 1965, Ashcroft 1966, Cohen and Heine 1970). One might examine a Si^{+3} ion, i.e., one electron moving in the potential of the ion core. Using the measured term values for this system, one can fix a model potential for Si^{+4}. This potential is then screened for use in crystalline silicon, e.g., using a Thomas–Fermi dielectric function. Next, the structure is included via $S(\boldsymbol{G})$, and the Schrödinger equation is solved as indicated in eqs. (14)–(16). From the solution of the Schrödinger equation, we obtain energy bands and wavefunctions. Using $E_n(\boldsymbol{k})$ and $\psi_{n,\boldsymbol{k}}$, we can solve for the response functions such as the reflectivity, $R(E)$, and the density of states, $D(E)$. The procedures for determining these functions have been discussed elsewhere (Cohen and Chelikowsky 1982, 1988). These functions are compared with experiment, and the $V(\boldsymbol{G})$'s can be altered if good agreement is not achieved. This process is repeated until satisfactory agreement between experiment and theory is obtained. Only three potential parameters, $V_a(3)$, $V_a(8)$ and $V_a(11)$, are required. Using optical properties and photoemission measurements, the fitting procedure is clearly overdetermined. Moreover, the starting potential from model potentials is usually a reasonable approximation and the changes made are a matter of fine-tuning. It is impressive that the EPM works so well considering the limited input. For semiconductors such as silicon or germanium, the optical gaps can be reproduced within ~ 0.1 eV and the main photoemission features within ~ 0.5 eV.

2.2. *Local density pseudopotentials*

A significant advance in the study of electronic structure was the development of the local density approximation (LDA) for exchange and correlation (Slater 1951,

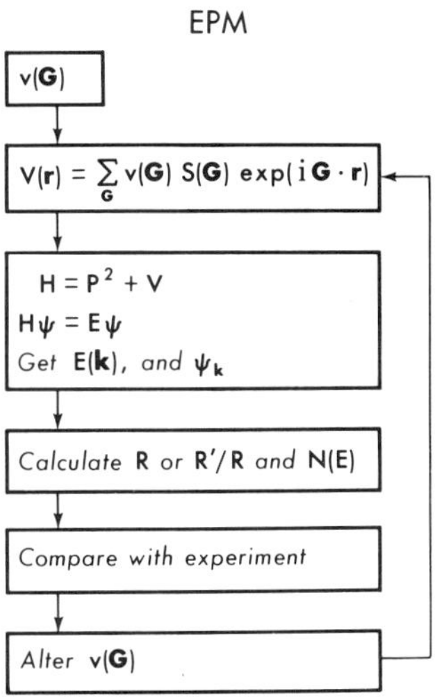

Fig. 5. Bloch diagram illustrating the procedure for obtaining an empirical pseudopotential.

Hohenberg and Kohn 1964, Kohn and Sham 1965). This approximation allows one to construct self-consistent field pseudopotentials for solids and to consider problems which are frequently outside the scope of current quantum chemistry techniques. As discussed in our introductory remarks, the LDA is appropriate for determining ground state properties of solids such as crystal structure energies, compressibilities, phonon spectra, elastic constants, and so forth.

We consider an atomic structure calculation within the LDA framework. For example, in the case of silicon the Schrödinger equation can be solved for the eigenvalues and wavefunctions. Knowing the valence wavefunctions, i.e., $3s^2$ and $3p^2$, and corresponding eigenvalues, the pseudowavefunctions can be constructed. We can invert the Schrödinger equation and find the total pseudopotential, and then "unscreen" the potential via eq. (12) and determine the ion-core pseudopotential. This ion-core potential, which arises from tightly bound core electrons and the nuclear charge, is not expected to change from one environment to another. It should be transferable from the atom to a molecular state to a solid state. The issue of this "transferability" is one which must be addressed according to the system of interest. The immediate issue here is how to define pseudowavefunctions which can be used to define the corresponding pseudopotential.

Suppose we insist that the pseudowavefunction be identical to the all-electron wavefunction outside of the core region, e.g., $\varphi_{3s}(r) = \psi_{3s}(r)$ for $r > r_c$ where r_c defines the size of the core. This will guarantee that the pseudowavefunction will possess identical properties in the region away from the ion core. For $r < r_c$, we alter the all-electron wavefunction as we wish the wavefunction in this region to be nodeless. One other criterion is mandated. Namely, the integral of the pseudocharge density within the core should be equal to the integral of the all-electron charge density. Without this condition, the pseudowavefunction differs by a scaling factor from the all-electron wavefunction, i.e., $\varphi_{3s}(r) = C\psi_{3s}(r)$ for $r > r_c$ where C differs from unity. Since we expect the bonding in a solid to be highly dependent on the tails of the valence wavefunctions, it is imperative that the normalized pseudowavefunction be identical to the all-electron wavefunctions for $r > r_c$. The criterion by which one insures $C = 1$ is called "norm conserving" (Hamann et al. 1979, Bachelet et al. 1982). Some of the earliest ab initio potentials did not incorporate this constraint (Topiol et al. 1977, Zunger and Cohen 1978, 1979, Zunger 1979). Such pseudopotentials are useful in defining indices which characterize chemical trends. These potentials are not used for total-energy computations as they tend to be "hard core" pseudopotentials; i.e., they diverge near the origin as $1/r^2$.

Kerker (1980) proposed a straightforward method for constructing a local density pseudopotential. He suggested that the pseudowavefunctions have the following form:

$$\varphi_l(r) = r^l \exp p(r), \quad r < r_c, \tag{17}$$

where $p(r) = -\alpha r^4 - \beta r^3 - \gamma r^2 - \delta$ and

$$\varphi_l(r) = \psi_l(r), \quad r > r_c.$$

This form of the pseudowavefunction for φ assures us that the function will be nodeless. Kerker proposed criteria for fixing the parameters (α, β, γ, and δ). One criterion is that the wavefunction be norm conserving. Other criteria include: (a) The all-electron and pseudowavefunctions have the same valence eigenvalue. (b) The pseudowavefunction be nodeless and be identical to the all-electron wavefunction for $r > r_c$. (c) The pseudowavefunction must be continuous as well as the first and second derivatives of the wavefunction at r_c. An example of an atomic pseudowavefunction for Si is given in fig. 6 where it is compared to an all-electron wavefunction.

Other early local density pseudopotentials include those proposed by Hamann et al. (1979), Bachelet et al. (1982) and Greenside and Schlüter (1983). These pseudopotentials were constructed from a different perspective. The all-electron potential was calculated for the free atom. This potential was multiplied by a smooth, short-range cut-off function which removes the strongly attractive and singular part of the potential. The cut-off function is adjusted numerically to yield eigenvalues equal to the all-electron valence eigenvalues and nodeless pseudowavefunctions. These wavefunctions are nodeless and they converge to the all-electron wavefunctions outside the core region. Again, the pseudocharge within the core is constrained to be equal to the all-electron value. Another interesting approach was taken by Starkloff and Joannopoulos (1979). They proposed a simple local potential, i.e., an ion-core pseudopotential which is not state dependent or angular momentum dependent. They did not require the pseudowavefunction to be nodeless. This method is advantageous in that the potentials are simple functions of r. The disadvantage of this method is that the potentials tend to be strong and the pseudowavefunctions require many plane waves to converge.

An additional advantage of the norm-conserving potential concerns the logarithmic

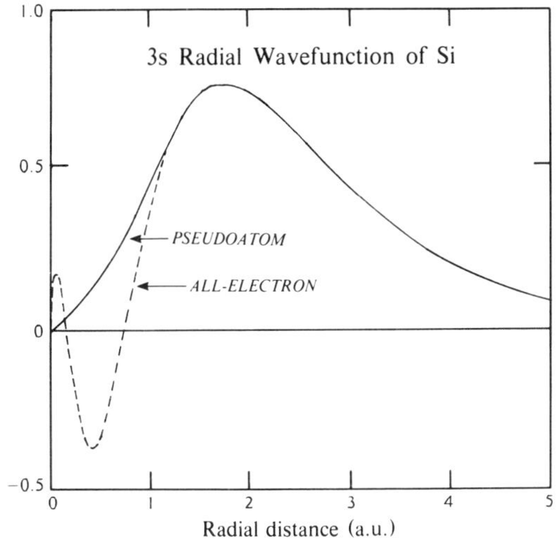

Fig. 6. All-electron and pseudowavefunction for the 3s radial wavefunction of a silicon atom.

derivative of the pseudowavefunction (Bachelet et al. 1982). An identity exists:

$$-2\pi\left((r\varphi)^2 \frac{\mathrm{d}^2 \ln \varphi}{\mathrm{d}\varepsilon\, \mathrm{d}r}\right)_R = 4\pi \int_0^R \varphi^2 r^2\, \mathrm{d}r = Q_{\mathrm{core}}(R). \tag{18}$$

The energy derivative of the logarithmic derivative of the pseudowavefunction is fixed by the amount of charge within a radius, R. The radial derivative of φ is related to the scattering phase shift from elementary quantum mechanics. For a norm-conserving pseudopotential, the scattering phase shift at $R = r_\mathrm{c}$ and at the eigenvalue of interest is identical to the all-electron case as $Q_{\mathrm{core,AE}}(r_\mathrm{c}) = Q_{\mathrm{core,pseudo}}(r_\mathrm{c})$. The scattering properties of the pseudopotential and the all-electron potential have the same energy variation to first order when transferred to other systems. Conditions for matching higher derivatives have been carried out (Shirley et al. 1989); however, the additional accuracy obtained in terms of the energy derivatives is usually small.

As indicated, there is some flexibility in constructing pseudopotentials. While virtually all the local density pseudopotentials impose the condition that $\varphi_l(r) = \psi_l(r)$ for $r > r_\mathrm{c}$, the construction for $r < r_\mathrm{c}$ is not unique. The non-uniqueness of the pseudowavefunction was recognized early in its inception. For example, within the Phillips–Kleinman formulation, one can always add a function, f, to the pseudowavefunctions without altering the pseudopotential provided f is orthogonal to the core states. Consider the matrix element in eq. (7): $\langle \varphi | \varphi_t \rangle$. If one changes φ to $\varphi + f$, then $\langle \varphi + f | \varphi_t \rangle = \langle \varphi | \varphi_t \rangle + \langle f | \varphi_t \rangle = \langle \varphi | \varphi_t \rangle$, if $\langle f | \varphi_t \rangle = 0$. Nothing is changed in the Phillips–Kleinman pseudopotential by this addition.

The non-uniqueness of the pseudopotential can be exploited to optimize the convergence of the pseudopotentials for the basis of interest. Much recent effort has been made to construct "soft" pseudopotentials. By "soft", one means a "rapidly" convergent calculation using plane waves as a basis. Typically, soft potentials are characterized by a "large" core. However, as the core becomes larger, the "goodness" of the pseudowavefunction can be compromised as the transferability of the pseudopotential becomes more limited.

We note that for some cases, such as copper, it may be possible to choose a large core radius without sacrificing transferability (Chelikowsky and Chou 1988b, Bar-Yam et al. 1989). In the case of copper, the maximum of the 3d wavefunction occurs near the nucleus relative to the nearest-neighbor separation in copper. In this situation, it is possible to choose the core size to be large. The maximum of the 3d wavefunction may not be accurately represented via a pseudopotential construction using such a large core, but the tails of the wavefunction in the "bonding" region may still be well represented. In fig. 7, we illustrate the situation for copper. The corresponding pseudopotential was used to calculate the structural properties of copper. The results agree well with both all-electron calculations and with other pseudopotential constructions.

Several schemes have been developed to generate "soft" pseudopotentials for species which extend effectively the core radius while preserving "transferability". The primary motivation for such schemes is to reduce the number of plane waves

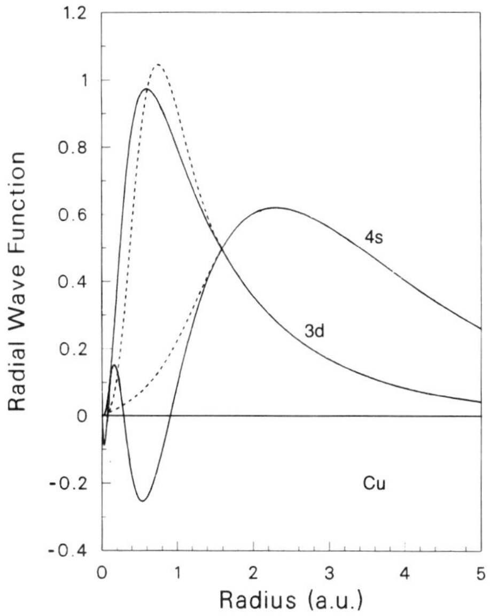

Fig. 7. All electron and pseudowavefunctions for the 3d and 4s states in atomic copper. The all-electron states are solid lines and the pseudowavefunctions are shown as dashed. The Wigner–Seitz radius in crystalline copper is about 2.7 au. Note how the 3d pseudowavefunction maximum does not correspond to the maximum of the all-electron wavefunction, but the two wavefunctions agree near and beyond about 1.5 au.

in the basis, but the use of "soft" potentials also facilitates the implementation of other bases such as Gaussians. The use of Gaussian bases often requires the pseudopotential also to be expanded in Gaussians (Chelikowsky and Louie 1984). This expansion is simplified for "smooth" potentials. One of the earliest discussions of such issues is from Vanderbilt (1985). A common measure of pseudopotential "softness" is to examine the behavior of the potential in reciprocal space. For example, a "hard-core" pseudopotential decays as $1/q$. This rate of decay is worse than using the bare Coulomb potential which scales as $1/q^2$. The Kerker pseudopotential does no better than the Coulomb potential as the Kerker pseudopotential has a discontinuity in its third derivative at the origin and at the cut-off radius. This gives rise to slope discontinuities in the potential at the same locations and a slow $1/q^2$ decay of the potential. One should examine each case, as the error introduced by truncation of such a potential in reciprocal space may still be acceptable in terms of yielding accurate wavefunctions and energies. Hamann–Schlüter–Chiang potentials often converge better than the Kerker potentials in that they contain no such discontinuities.

One issue which remains unresolved is the "best" criterion by which to construct an "optimal" pseudopotential. Clearly, an optimal pseudopotential is one which minimizes the number of plane waves required to achieve the desired goal, i.e., it yields a converged total energy yet does not sacrifice transferability. The task remains on what criteria to use to insure that the pseudopotential is optimal in this sense (Vanderbilt 1985, 1990). Often one requires the Fourier components of the potential to converge rapidly. It is not clear that this criterion is the best one. For example, it is possible that a "deep" pseudopotential can be "smoother", i.e., converge faster in reciprocal space than a "weak" potential. However, a "deep" pseudopotential can result in localized wavefunctions which may not be easily expressed with a limited number of plane waves. One suggestion has been to examine the convergence of the pseudoatomic wavefunction in reciprocal space (Rappe et al. 1990). This criterion is useful in judging the localization of the wavefunctions.

One straightforward approach to optimizing a pseudopotential is to build additional constraints into the polynomial given in eq. (17). For example, suppose we write.

$$p(r) = c_0 + \sum c_n r^n. \tag{19}$$

In Kerker's scheme, $n \leqslant 4$. However, there is no compelling reason for demanding that the series terminates at this particular point. If we extend the expansion, we may impose additional constraints. For example, one might try to constrain the reciprocal space expansion of the pseudowavefunction so that beyond some momentum cut-off, q_c, the wavefunction vanishes (Rappe et al. 1990).

Another approach has been suggested by Troullier and Martins (1991). They write eq. (19) as

$$p(r) = c_0 + c_2 r^2 + c_4 r^4 + c_6 r^6 + c_8 r^8 + c_{10} r^{10} + c_{12} r^{12}. \tag{20}$$

As usual, they constrained the coefficients to be norm conserving. In addition, they demanded continuity of the pseudowavefunctions and the first four derivatives at r_c. The final constraint was to demand zero curvature of the pseudopotential at the origin. This latter condition can be justified on "intuitive grounds". If the value of the pseudopotential is "very large" at the origin either positive or negative, the potential must show some significant variation to converge to $V_{p,\mathrm{ion}}(r \gg r_c) \to -Z_{\mathrm{val}}(e^2/r)$. By demanding that the second derivative vanish, the variation of $V_{p,\mathrm{ion}}(r < r_c)$ is minimized.

In fig. 8, we illustrate ionic pseudopotentials for several approaches for the case of carbon. The variety of approaches belies the fact that all of these potentials will produce similar electronic structures if a converged basis is employed.

Several other issues concerning pseudopotentials which have been optimized for use with plane waves have been discussed in the literature. For example, one complication involves the matrix elements of a nonlocal potential. If we consider an ion-core pseudopotential which replicates, the lth angular momentum state, we may write

$$V^{\mathrm{ion}}_{p,l}(r) = V_{T,l}(r) - V_H(r) - V_{xc}(r). \tag{21}$$

This is similar to the potential in eq. (12) except here we may have more than one

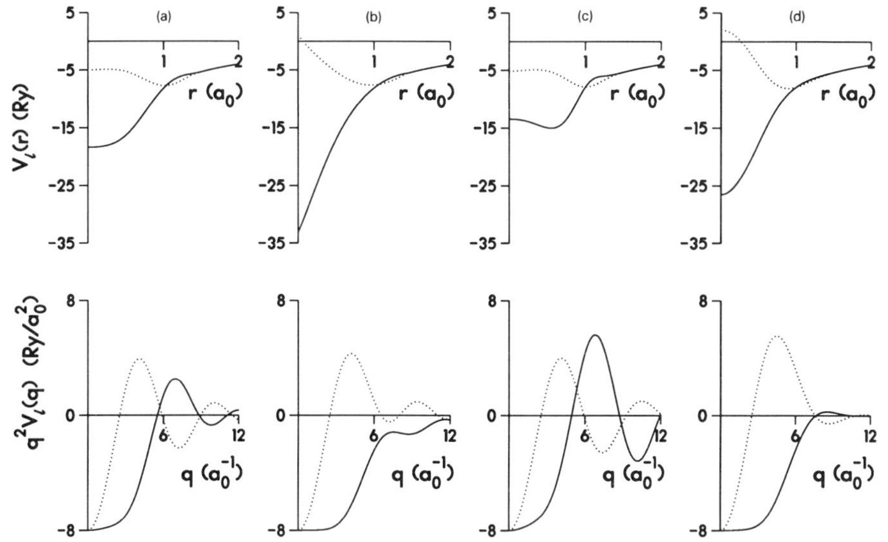

Fig. 8. Ionic pseudopotentials for carbon generated by four different methods in real and reciprocal space. The dotted and dashed lines correspond to the s and p pseudopotentials, respectively. (a) Troullier and Martins (1991). (b) Kerker (1980). (c) Hamann et al. (1979). (d) Vanderbilt (1985).

angular momentum component. It is often convenient to write the ionic core potential as $V_{\mathrm{p}}^{\mathrm{ion}}(r)$ as

$$V_{\mathrm{p}}^{\mathrm{ion}}(r) = V_{\mathrm{p,local}}^{\mathrm{ion}}(r) + \sum_{l} [V_{\mathrm{p},l}^{\mathrm{ion}}(r) - V_{\mathrm{p,local}}^{\mathrm{ion}}(r)] P_l, \tag{22}$$

where we have selected one l-component as the "local" reference ion pseudopotential and we project out corrections to this potential which are l-dependent. Usually, the sum in eq. (22) is restricted to a few angular momentum terms. For example, we might choose to consider the $l = 2$ component as the reference potential and project out $l = 0$ and $l = 1$ potential corrections. Such a projection would imply that for components of the wavefunction with $l > 2$ the ion-core pseudopotential would be identical. For tetrahedral semiconductors, this is not a serious issue as only s-, p-, and d-orbital characters are important.

The angular momentum dependence of the pseudopotential is reflected in a more complex matrix element than eq. (16). One may write eqs. (15) and (16) as

$$\sum_{G'} (H_{G,G'}(k) - E\delta_{G,G'}) a_{n,k}(G') = 0, \tag{23}$$

where

$$H_{G,G'}(k) = \hbar^2 \frac{(k + G)^2}{2m} \delta_{G,G'} + V_{\mathrm{a}}(k + G, k + G') S(G - G'). \tag{24}$$

The matrix elements for $V_{\mathrm{a}}(k + G, k + G')$ can be written as a local part as in eq. (16)

which includes the screening terms, i.e., $V_H(\boldsymbol{G}-\boldsymbol{G}')$ and $V_{xc}(\boldsymbol{G}-\boldsymbol{G}')$, and a nonlocal contribution. The matrix elements for the nonlocal part are given by

$$V_a(\boldsymbol{k}+\boldsymbol{G}, \boldsymbol{k}+\boldsymbol{G}') = \frac{2l+1}{4\pi\Omega_a} P_l(\cos\gamma) \int \Delta V_l(r)\, j_l(|\boldsymbol{k}+\boldsymbol{G}|r)\, j_l(|\boldsymbol{k}+\boldsymbol{G}'|r) r^2\, \mathrm{d}r, \tag{25}$$

where $\Delta V_l(r) = V^{\text{ion}}_{\text{p},l}(r) - V^{\text{ion}}_{\text{p,local}}(r)$ is the nonlocal term, Ω_a is the atomic volume, $P_l(\cos\gamma)$ is a Legendre polynomial with $\cos\gamma = (\boldsymbol{k}+\boldsymbol{G})\cdot(\boldsymbol{k}+\boldsymbol{G}')/|\boldsymbol{k}+\boldsymbol{G}||\boldsymbol{k}+\boldsymbol{G}'|$, and j_l is a spherical Bessel function. This matrix element presents some computational difficulties as the function involves a two-dimensional function of $\boldsymbol{k}+\boldsymbol{G}$ and $\boldsymbol{k}+\boldsymbol{G}'$. If several thousand $\boldsymbol{G}$-vectors are involved, the two-dimensional character of the matrix element can be quite undesirable. Kleinman and Bylander (1982), proposed that the matrix elements be modified to

$$\begin{aligned} V_a^{\text{KB}}(\boldsymbol{k}+\boldsymbol{G}, \boldsymbol{k}+\boldsymbol{G}') = & \left[\int \Phi_l(r)\Delta V_l(r)\, j_l(|\boldsymbol{k}+\boldsymbol{G}|r) r^2\, \mathrm{d}r\right] \\ & \times \left[\int \Phi_l(r)\Delta V_l(r)\, j_l(|\boldsymbol{k}+\boldsymbol{G}'|r) r^2\, \mathrm{d}r\right] \\ & \times \left[4\pi\Omega \int \Phi_l(r)\Delta V_l(r)\Phi_l(r) r^2\, \mathrm{d}r\right]^{-1} \sum_m Y^*_{lm}(\hat{\boldsymbol{l}}_{\boldsymbol{k}+\boldsymbol{G}}) Y_{lm}(\hat{\boldsymbol{l}}_{\boldsymbol{k}+\boldsymbol{G}'}), \end{aligned} \tag{26}$$

where $\Phi_l(r)$ is the atomic reference pseudowavefunction, and $Y^*_{lm}(\hat{\boldsymbol{l}}_{\boldsymbol{k}+\boldsymbol{G}})$ is a spherical harmonic with $\hat{\boldsymbol{l}}_{\boldsymbol{k}+\boldsymbol{G}} = (\boldsymbol{k}+\boldsymbol{G})/|\boldsymbol{k}+\boldsymbol{G}|$. This form has the advantage of generating a separable matrix element. However, some care must be used in its implementation. For example, it is possible to generate "ghost states". These states occur below the nodeless state of the atom and violate the Wronskian theorem. This theorem states that the atomic wavefunctions are ordered energetically so that the energy increases with the number of nodes. Such a pseudopotential clearly is not chemically sensible. Fortunately, well-defined criteria have been presented for avoiding such problems (Bylander and Kleinman 1990). For example, $\Delta V_l(r)$ should be made as small and short-ranged as possible. One can choose a different reference potential to achieve this. In cases where the potential components are quite different, the positive $\Delta V_l(r)$ component should be made small. A theorem has also been proved which provides a test for the existence of ghost states (Gonze et al. 1990).

Another approach has recently been proposed by Vanderbilt (1990) for constructing soft pseudopotentials. In this approach, "soft pseudopotentials" are constructed within a "generalized eigenvalue problem". The norm conservation constraint is relaxed, and the charge density within the ion core is not explicitly considered as part of the pseudowavefunction. The relaxation of the norm conservation constraint allows one to consider a much larger core radius and a softer potential. The method is complicated by the addition of a nonlocal overlap operator and formally resembles the original Phillips–Kleinman pseudopotential.

One issue which is relevant for pseudopotential constructions, regardless of whether the potential is intended for use with a plane wave basis or not, concerns the issue of unbound, or weakly bound, atomic states. If an atom does not bind a state of interest, then the atomic wavefunction corresponding to this state is clearly not localized, or normalizable. Nonetheless, the pseudopotential corresponding to this state might be of some interest, i.e., in the crystal such diverging wavefunctions are "captured" by the potentials of neighboring atoms (Hamann 1989). For example, Ba has a strong f-component resonance. However, these f-states are not bound for neutral Ba. In order to bind such states, one must consider highly ionized atomic states which result in very strong pseudopotentials. Sometimes these potentials are so strong as to be useless for plane wave bases, or so far removed from any chemical environment of interest that their transferability may be suspect. Hamann (1989) has suggested a method for handling such cases by integrating out the Schrödinger equation to a large distance, e.g., $2.5r_c$. The corresponding "terminated" wavefunction is then used to generate a pseudopotential for the component of interest.

One other issue merits mention: the construction of relativistic pseudopotentials (Kleinman 1980, Bachelet and Schlüter 1982). For most tetrahedral semiconductors involving light atoms, these effects are small and are manifested in small spin–orbit splittings. For example, in silicon the spin–orbit splitting of the upper valence band is less than 0.05 eV. However, for semiconductors involving heavy atoms such as InSb, the spin–orbit splitting can be quite large, e.g., $\sim$1 eV. In the lead salts, relativistic effects can be large both in terms of scalar corrections to the Hamiltonian, and in terms of the spin–orbit splitting.

In the EPM, spin–orbit effects are handled via a single parameter which can be adjusted to match the spin–orbit splittings of the valence band (Cohen and Chelikowsky 1988). In terms of ab initio pseudopotentials, the issue is more complex, and few calculations have been performed which include spin–orbit splittings. However, scalar terms are easily incorporated into relativistic potentials. For total energy calculations, the inclusion of scalar terms are not as crucial as might first appear. Relativistic effects are strongest within the ion core and do not tend to be altered strongly by the chemical environment. Thus, when one considers an energy difference, e.g., between the energy of an isolated atom and the energy per atom in the solid state, relativistic corrections will often be "subtracted out".

One important issue, which is closely related to including relativistic or spin–orbit effects in pseudopotentials, concerns the issue of core–valence charge interactions. We have abused local density theory in this sense. When we write $V_{xc} = V_{xc}[\rho_{val}(\boldsymbol{r})]$, we have not considered the possibility of core–valence interactions (Louie et al. 1982). We should write $V_{xc} = V_{xc}[\rho_T(\boldsymbol{r})]$, where $\rho_T(\boldsymbol{r}) = \rho_{core}(\boldsymbol{r}) + \rho_{val}(\boldsymbol{r})$ is the total charge density. For example, we could choose to define an ion core pseudopotential by rewriting eq. (12) as

$$V_{ion} = V_T - V_H[\rho_{val}(\boldsymbol{r})] - V_{xc}[\rho_{core}(\boldsymbol{r}) + \rho_{val}(\boldsymbol{r})]. \tag{27}$$

We need not be concerned with the Hartree term since linear superposition holds for electrostatics. However, linear superposition does not hold for the exchange–correlation term. Even if we assume that the core charge does not change, V_{xc} remains dependent on the nature of the overlap term. The issue of core–valence charge overlap is most important for magnetic systems, but in the case of non-magnetic solids like the tetrahedral semiconductors, it is usually assumed that this correction is not of crucial importance.

2.3. *Pseudopotentials without the local density approximation*

As we have noted in our previous discussions, the separation of core and valence properties is an independent approximation from the form of the one electron approximation for the electron–electron interactions. For example, one could use the solutions of the all-electron atom and ask what the form of the ion-core pseudopotential would be if a Hartree–Fock approximation were made. However, this approach would not be paticularly helpful for solid state semiconductor systems since the Hartree–Fock approximation is known generally to yield poor descriptions of the energy band structure.

Recent efforts have concentrated on approaches to this many-body problem which in principle are "exact". An example is the Green function Monte Carlo method (e.g., Kalos and Whitlock 1986). It would be advantageous in this method to remove the core electrons from consideration as the complexity of the problem grows rapidly with the number of electrons. However, a severe limitation of this approach is that nonlocal pseudopotentials such as those generated within the local density approximation are incompatible with certain technical limitations of this approach.

To outline the nature of this approach, one would like to keep the Green function,

$$G(\boldsymbol{R}, \boldsymbol{R}'; \Delta t) = \Psi(\boldsymbol{R})\Psi(\boldsymbol{R}')^{-1}\langle \boldsymbol{R}|\exp[-\Delta t(H - E_{\mathrm{T}})]|\boldsymbol{R}'\rangle, \tag{28}$$

nonnegative so that it can be sampled by a probability distribution. In eq. (28), Ψ is a trial wavefunction, R is a set of $3N$ electron coordinates, H is the Hamiltonian, E_{T} is a trial energy, and Δt is a small imaginary time step. The Green function is used to project out the ground-state component of a trial wavefunction. For fermion systems, the Green function has a *nodal* structure. One can assume an approximate solution and impose a fixed nodal structure. Within this framework, one can always deal with a nonnegative Green function. This fixed node formalism imposes conditions on the Hamiltonian: one of the conditions restricts the nature of the ion core pseudopotential in terms of its nonlocality. Unfortunately, purely local pseudopotentials are of limited accuracy. It has been suggested that one subsume the effects of the nonlocal pseudopotential by allowing the electronic mass to be "position" dependent (Bachelet et al. 1989). Let us write for the pseudo atom:

$$H_{\mathrm{p}} = -\boldsymbol{\nabla} a(r)\boldsymbol{\nabla} + b(r)\frac{\boldsymbol{L}^2}{2r^2} + V_{\mathrm{ion}}(r) \tag{29}$$

where $a(r)$ and $b(r)$ are radial functions. The function $a(r)$ can be considered as an electron mass which depends on position. This form is, in general, unacceptable. For some atoms, e.g., transition metal atoms, conditions which insure a nonnegative Green function cannot be obtained. It is required that $a(r) > -1$ and that $a(r) + b(r) > -1$. Nonetheless, this approach illustrates how one might use a different approach in constructing a "pseudo-Hamiltonian".

Another example of the use of pseudopotentials without the LDA concerns the issue of how ab initio pseudopotentials might be constructed to address spectroscopic questions. Most charge density methods, e.g., LDA, are ground-state methods based on the Hohenberg–Kohn–Sham theorem. As pointed out earlier, such methods are strictly speaking not appropriate for extracting "excited state" properties such as optical gaps, dielectric functions and so on. One might ask whether it is possible to construct a pseudopotential whose role is to reproduce the optical spectra of a solid. Empirically, this is clearly possible as evidenced by the great success of the EPM. Modifications of the local density approach have great appeal as the energy bands which result from this procedure are often a good "first" approximation to the EPM bands. The main flaw is that the computed band gaps are too small. Presumably if a local density prescription could be formulated for this application, then a corresponding pseudopotential could be developed.

Another approach (Hybertsen and Louie 1985, 1986, Zhu and Louie 1991) to this problem comes from the quasiparticle picture. The eigenvalues of a quasiparticle are obtained within a many-body Green function technique. Within this technique the inclusion of local fields in the screening and dynamical correlation effects is crucial. It is not apparent how to include such effects within a "pseudopotential" context. Indeed, such a potential would not be a "pseudopotential" in the sense of separating core and valence states, but a "many-body" pseudopotential which might include electron–electron interaction implicitly.

3. Application to structure

3.1. *Total energies*

As stated earlier, the general approach (Cohen 1982) developed for the determination of crystal structure and structural properties is to search for the lowest total energy structure from a number of candidate structures. The ordering of the structural energy for the various structures can change as a function of volume or density and this allows for the investigation of pressure-induced structural phase transitions. For the cases considered here, a plane wave basis set is used. This choice has the advantage of being free from bias toward a specific crystal structure. Angular dependences of the charge density such as that found in covalent systems are represented well by plane waves and this approach can describe the charge distributions for various crystal structures on an equal footing. One disadvantage of the plane wave basis is the large number of plane waves required to simulate localized configurations. In these cases, it is prudent to use localized orbitals and successful schemes (Chelikowsky and Louie 1984) based on localized orbitals are available. Combined plane wave and localized orbitals, the so-called "mixed basis approach" (Louie et al. 1979) is another powerful tool for calculating electronic and structural properties.

Because the output of standard electronic band structure calculations yield wavefunctions, charge densities, and other data expressed in terms of Fourier components in momentum space, it is useful to perform total energy calculations using a momentum-space formalism (Ihm et al. 1979). The total energy can be written as

$$E_{\text{tot}} = E_{\text{kin}} + E'_{\text{ec}} + E'_{\text{H}} + E_{\text{xc}} + E'_{\text{cc}}, \tag{30}$$

representing the electronic kinetic energy, the electron–core interaction energy, the Hartree electron–electron Coulomb energy, the electronic energy arising from exchange and correlation, and the Ewald core–core Coulomb energy. The cancellation of the divergent terms and the charge neutrality condition are accounted for through the use of renormalized terms (Ihm et al. 1979, Ihm and Cohen 1979) indicated by primes in eq. (30).

The momentum-space expressions for the terms in eq. (30) are as follows:

$$E_{\text{kin}} = \frac{1}{N}\sum_{i,\boldsymbol{G}} n_i |\psi_i(\boldsymbol{k}_i + \boldsymbol{G})|^2 \frac{\hbar^2 |\boldsymbol{k}_i + \boldsymbol{G}|^2}{2m}, \tag{31}$$

$$E'_{\text{ec}} = \frac{1}{N}\sum_{i,\boldsymbol{G},\boldsymbol{G}'} n_i \psi_i^*(\boldsymbol{k}_i + \boldsymbol{G})\psi_i(\boldsymbol{k}_i + \boldsymbol{G}') \times \left(V_{\text{ps}}(\boldsymbol{k}_i + \boldsymbol{G}, \boldsymbol{k}_i + \boldsymbol{G}') + \delta_{\boldsymbol{G}\boldsymbol{G}'} \frac{1}{\Omega_{\text{c}}} \int^{\infty} \frac{Ze^2}{r}\,\mathrm{d}\boldsymbol{r} \right), \tag{32}$$

$$E'_{\text{H}} = \tfrac{1}{2}\Omega_{\text{c}} \sum_{\boldsymbol{G}\neq 0} |\rho(\boldsymbol{G})|^2 \frac{4\pi e^2}{|\boldsymbol{G}|^2}, \tag{33}$$

$$E_{\text{xc}} = \tfrac{1}{2}\Omega_{\text{c}} \sum_{\boldsymbol{G}} \rho^*(\boldsymbol{G})\varepsilon_{\text{xc}}(\boldsymbol{G}), \tag{34}$$

$$E'_{\text{cc}} = \tfrac{1}{2}\sum_{s,s'} Z_s Z_{s'} e^2 \left\{ \frac{4\pi}{\Omega_{\text{c}}} \sum_{\boldsymbol{G}\neq 0} \left[\frac{1}{|\boldsymbol{G}|^2} \cos[\boldsymbol{G}\cdot(\boldsymbol{\tau}_s - \boldsymbol{\tau}_{s'})] \exp\left(-\frac{|\boldsymbol{G}|^2}{4\eta^2} \right) \right] - \frac{\pi}{\eta^2 \Omega_{\text{c}}} + \sum_{\boldsymbol{l}}{}' \left[\frac{\operatorname{erfc}(\eta x)}{x} \right]_{x = |\boldsymbol{l} + \boldsymbol{\tau}_s - \boldsymbol{\tau}_{s'}|} - \frac{2\eta}{\sqrt{\pi}} \delta_{ss'} \right\}. \tag{35}$$

Each term represents the energy per cell; N is the total number of cells, Ω_{c} is the cell volume, $\boldsymbol{l}$ is a lattice vector, $\boldsymbol{G}$ is a reciprocal lattice vector, Z is the total core charge which is a sum of the individual core charges Z_s, $\boldsymbol{\tau}_s$ is the basis vector, n_i is the occupation number, $\boldsymbol{k}_i$ and ψ_i represent the crystal momentum and wavefunction of state i, and $\rho(\boldsymbol{G})$ and $V_{\text{ps}}(\boldsymbol{k}_i + \boldsymbol{G}, \boldsymbol{k}_i + \boldsymbol{G}')$ represent the Fourier components of the charge density and pseudopotential. In eq. (35) the prime in the $\boldsymbol{l}$ summation excludes the $\boldsymbol{l} = 0$ term when $\boldsymbol{\tau}_s = \boldsymbol{\tau}_{s'}$ and η is a parameter controlling the convergency of the Ewald summation (Ihm and Cohen 1980a).

The total energy can be written in terms of the energy eigenvalues obtained from a solution of the secular equation solved for the energy bands

$$E_{\text{tot}} = \frac{1}{N}\sum_i n_i \varepsilon_i - E'_{\text{H}} + \Delta E_{\text{xc}} + E'_{\text{cc}}, \tag{36}$$

where

$$\Delta E_{xc} = \Omega_c \sum_{G} \rho^*(G)[\varepsilon_{xc}(G) - V_{xc}(G)] \tag{37}$$

and

$$V_{xc} = \frac{d(\rho\varepsilon_{xc})}{d\rho}. \tag{38}$$

For the calculations described here, we use the local density approximation where the functional

$$E_{xc}[\rho] = \int \rho(r)\varepsilon_{xc}[\rho(r)]\, dr \tag{39}$$

and $\varepsilon_{xc}[\rho(r)]$ is a function of $\rho(r)$.

3.2. Bulk structures

Once the pseudopotential is constructed, the band structure calculations can be done and E_{tot} can be evaluated according to the prescription given above. The first applications (Ihm and Cohen 1980b, Yin and Cohen 1980, 1982a) used Si and Ge as prototype materials. The total energy as a function of volume $E_{tot}(V)$ was calculated for several structures: fcc, bcc, hcp, cubic diamond (CD), hexagonal diamond (HD), β-tin, and simple cubic (SC). For anisotropic systems the c/a ratios and other internal parameters can be optimized for each volume. Depending on the system studied the range of volumes considered is varied. For example, for Si, the range examined, is from about one-half of the volume of the CD phase to an upper limit which is about 20% larger than the CD volume. The size of the basis set, the $\boldsymbol{k}$-point sampling, and self-consistency requirements are imposed to give an E_{tot} which is converged to about 10^{-5} Ry. For structures which yield metallic systems, the $\boldsymbol{k}$-point sampling is sometimes critical because the variations in the Fermi surface make is difficult to assure that occupied states are being sampled.

Once the $E_{tot}(V)$ is computed for a finite number of points, typically 15–20, these points are then fit using a least-squares procedure to an equation of state such as the Birch equation (Birch 1952) or the Murnaghan equation (Murnaghan 1944) given below

$$E_{tot}(V) = \frac{B_0 V}{B_0'}\left(\frac{(V_0/V)^{B_0'}}{B_0' - 1} + 1\right) + \text{const.}, \tag{40}$$

where B_0 and B_0' are the bulk modulus and its pressure derivative at the equilibrium volume V_0. Using eq. (40), the fit of the calculated points yields the minimum energy E_{min}, the equilibrium lattice constants and the bulk modulus. Representative calculated and measured (Donohue 1974, Brewer 1977, McSkimin and Andreatch 1963) values for the lattice constant and bulk modulus results are given in table 1 for Si and Ge in the diamond structure. The results are impressive when one considers that

Table 1
Comparison of calculated and measured static properties of Si and Ge (Yin and Cohen 1982a).

	Lattice constant (Å)	Cohesive energy (eV/atom)	Bulk modulus (Mbar)
Si			
Calculation	5.451	4.84	0.98
Experiment	5.429[a]	4.63[b]	0.99[c]
Ge			
Calculation	5.655	4.26	0.73
Experiment	5.652[a]	3.85[b]	0.77[c]

[a] Donohue (1974).
[b] Brewer (1977).
[c] McSkimin and Andreatch (1963).

this is an ab initio calculation requiring only the atomic number and crystal structure as input. Generally this method yields lattice constants and bulk moduli to an accuracy of about 1% and 5%, respectively.

The cohesive energy can also be calculated by comparing the total energy at the equilibrium lattice constant, E_m, with the energy of the isolated atoms. Spin polarization effects (von Barth and Hedin 1972, Günnarsson et al. 1974) and zero-point vibrational energy need to be considered. Early results (Yin and Cohen 1982a) computed within the LDA gave good estimates of the cohesive energies. Excellent results have been obtained when the LDA is replaced by quantum Monte Carlo calculations (Fahy et al. 1988) of E_{xc}.

In the study published in 1982, Yin and Cohen (Yin and Cohen 1980, 1982a) applied the total energy pseudopotential method to Si and Ge, and provided a fairly complete set of data. Since that time various parts of these calculations have been done by others, and the experimental data has been updated. Here we begin by giving the Yin–Cohen 1982 calculational results to provide a consistent background in this area. The experimental data listed in tables 1, 2 and 3 are therefore of the same vintage. Some discussion of newer calculational and experimental results are included after the discussion of the Yin–Cohen 1982 study.

The total energy curves of Si and Ge as a function of volume for the seven crystal structures discussed earlier are given in figs. 9 and 10. Table 2 contains E_{min}, $\Delta E_{min} = E_{min} - E_{min}^{CD}$, and the corresponding volume V_{min} for each crystal phase. For both Si and Ge, the CD phase is lowest in energy in agreement with experiment. At small volumes other structural phases are lower in energy, hence pressure-induced solid–solid structural phase transitions are expected. These transitions occur when the Gibbs free energy

$$G = E_{tot} + PV - TS, \tag{41}$$

for each of the two phases involved is the same. For $T = 0$, the pressure-induced phase transformation can be represented by the common tangent line between the $E_{tot}(V)$ curves of the two phases considered. The negative of the slope of the common

Table 2
Volumes at the minimum structure energies (V_{min}, normalized to measured free volume), the minimum energies (E_{min}), and $\Delta E_{min}(\equiv E_{min} - E_{min}^{diamond})$ for the seven plausible structures of Si and Ge (Yin and Cohen 1982a).

	Diamond	Hexagonal diamond	β-tin	sc	bcc	hcp	fcc
Si							
V_{min}	1.012	1.015	0.773	0.808	0.736	0.723	0.733
E_{min} (Ry)	−7.9086	−7.9074	−7.889	−7.883	−7.870	−7.868	−7.867
ΔE_{min} (eV)	0	0.016	0.27	0.35	0.53	0.55	0.57
Ge							
V_{min}	1.003	1.003	0.802	0.839	0.795	0.805	0.816
E_{min} (Ry)	−7.8885	−7.8874	−7.870	−7.866	−7.856	−7.855	−7.854
ΔE_{min} (eV)	0	0.015	0.25	0.31	0.44	0.45	0.46

Table 3
Comparison of the calculated and measured transition volumes ($V_t^{d,\beta}$) of the diamond and β phases, their ratios (V_t^{β}/V_t^{d}), and the transition pressures (P_t) for Si and Ge (Yin and Cohen 1982a). Volumes are normalized to the measured zero-pressure volumes.

	V_t^d	V_t^β	V_t^β/V_t^d	P_t (kbar)
Si				
Calculation	0.928	0.718	0.774	99
Experiment	0.918[a]	0.710[a]	0.773[a]	125[b]
Deviation	1.1%	1.1%	0.1%	−20%
Ge				
Calculation	0.895	0.728	0.813	96
Experiment	0.875[a]	0.694[a]	0.793[a]	100[c]
Deviation	2.3%	4.9%	2.5%	−4%

[a] Jamieson (1963).
[b] Piermarini and Block (1975), Weinstein and Piermarini (1975).
[c] Asaumi and Minomura (1978).

tangent line is the transition pressure P_t. Examples are given in figs. 9 and 10 for the transitions from CD to the β-tin phase for Si and Ge. As hydrostatic pressure is applied the path indicated in the figures, $1 \rightarrow 2 \rightarrow 3 \rightarrow 4$, illustrates the change from CD at 1 to β-tin at 4, with the transformation occurring along the $2 \rightarrow 3$ segment of the path where both structures exist. The transition pressures P_t and the initial and final transition volumes, V_t^d and V_t^b, are given in table 3.

The calculational results and comparisons above have been examined using other theoretical techniques such as the linear muffin tin orbital (LMTO) method (McMahan et al. 1981). More detailed studies and investigation of other structures using the same approach as that used by Yin and Cohen (Yin and Cohen 1980, 1982a) have been done (Chang and Cohen 1984, 1985a, 1986a, Needs and Martin 1984, Vohra et al. 1986, Liu et al. 1988, Corkill et al. 1991) for Si, Ge, and Sn. Of particular interest is the successful prediction of high-pressure structural phases such

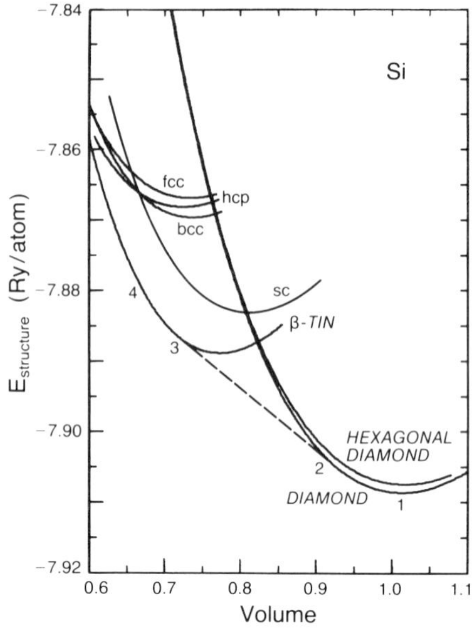

Fig. 9. Total-energy curves of the seven phases of Si as a function of the atomic volume normalized to Ω_{expt}. The dashed line is the common tangent of the energy curves for the diamond phase and the β-tin phase ($c/a = 0.552$) (Yin and Cohen 1982a).

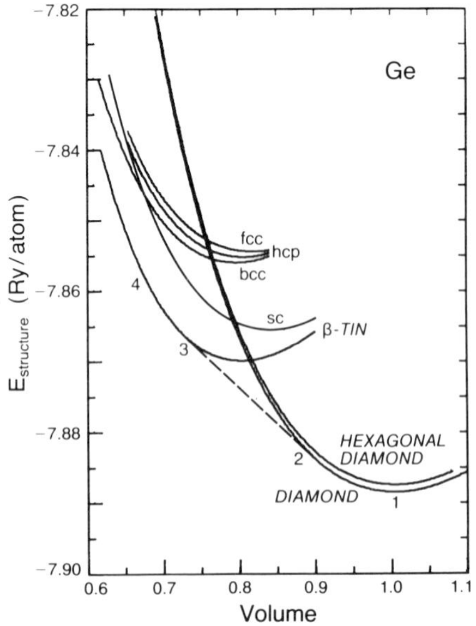

Fig. 10. Total-energy curves of the seven phases of Ge as a function of the atomic volume normalized to Ω_{expt}. The dashed line is the common tangent of the energy curves for the diamond phase and the β-tin phase ($c/a = 0.551$) (Yin and Cohen 1982a).

as hexagonal forms (Chang and Cohen 1984, Vohra et al. 1986) of Si and Ge and fcc Si (Liu et al. 1988). These high-pressure phases are metallic and predicted to be superconducting. Two of the predictions of superconductivity have been verified (Chang et al. 1985) while the others remain untested. It is interesting to note when updating the earlier calculations (Yin and Cohen 1980, 1982a) and experimental data, that the unusually larger error for P_t for the CD to β-Sn structural transition in Si has been greatly reduced by more recent experimental measurements (Olijnyk et al. 1984, Hu and Spain 1984).

The calculations for the group-IV semiconductors serve as prototypes for the application of the total-energy pseudopotential method to study structural properties of solids. The first extensions to other solids were made for the III-V semiconductors GaAs, AlAs, GaP, and AlP (Froyen and Cohen 1983a,b). In a more recent study (Zhang and Cohen 1987) these were examined again and AlSb, GaSb, InP, InAs, and InSb were added. The current status for these calculations is illustrated in figs. 11 and 12. It is likely that changes will occur as more precise calculations are done. Other studies on III–V zincblendes include BN, BP, and BAs (Wentzcovitch et al. 1986, 1987). Although these studies generally give good results for the lattice constants

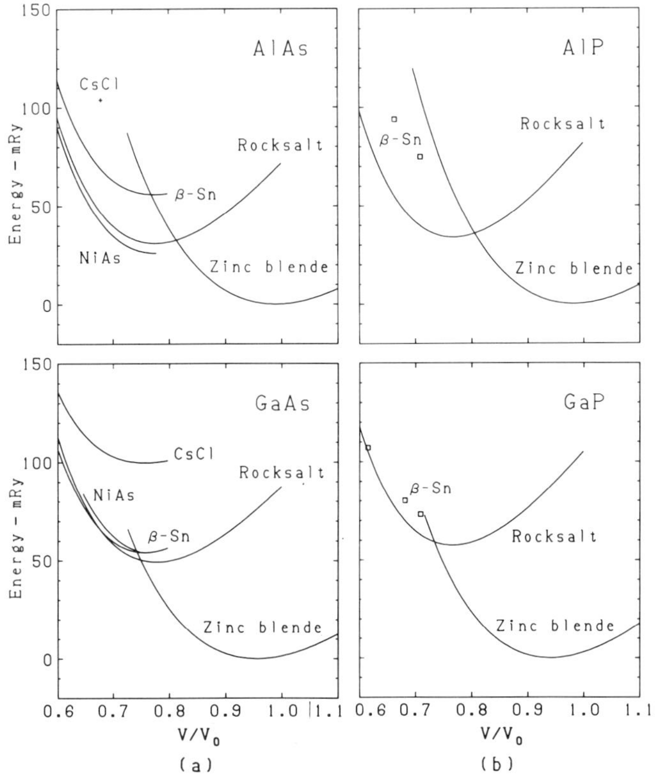

Fig. 11. Total-energy curves versus volume for some representative structures of AlAs, AlP, GaAs, and GaP (Froyen and Cohen 1983a,b).

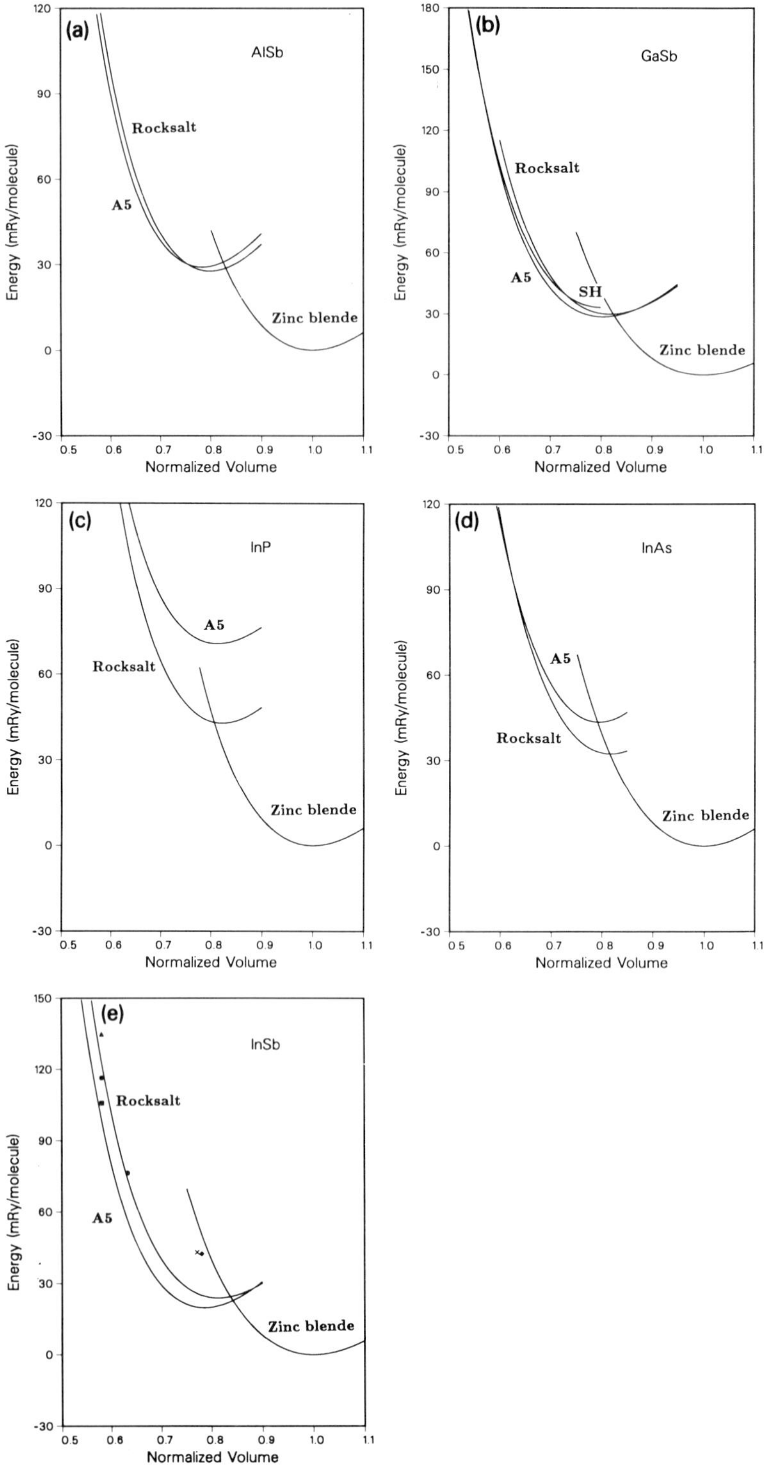
(a)
AlSb
Rocksalt
A5
Zinc blende
(b)
GaSb
Rocksalt
A5
SH
Zinc blende
(c)
InP
A5
Rocksalt
Zinc blende
(d)
InAs
A5
Rocksalt
Zinc blende
(e)
InSb
Rocksalt
A5
Zinc blende
Energy (mRy/molecule)
Normalized Volume

and bulk moduli, the agreement between the calculated and measured properties of the high-pressure phases and for phase transition pressures and volumes is not always satisfactory. As an example of the latter, the calculated transition pressures for the Al compounds are consistently lower by around 50% than the measured values (Zhang and Cohen 1987). In contrast, the calculated lattice constants are within 0.4%, 0.3%, and 0.3% of the measured values for AlP, AlAs, and AlSb, respectively (Zhang and Cohen 1987). At this point it is unclear why the transition pressures of the Al compounds are underestimated.

Although the II–VI zincblendes have not been studied as extensively as the III–V ones, some calculations exist (Engel and Needs 1990). Zincblende SiC (Chang and Cohen 1987) has also been examined, and this IV–IV semiconductor has many properties in common with the III–V zincblende systems. In addition to calculations of structural parameters such as lattice constants, elastic constants and transition pressures, the electronic charge density is also evaluated in most of the studies of the ground-state properties. The general trends observed (Cohen and Chelikowsky 1988) in calculations based on the EPM are reproduced. The group-IV materials have homopolar symmetric bonds whereas there is an asymmetric charge distribution for the III–V and II–VI ones with a build-up of charge closer to the anion in almost all cases. The exception occurs for the boron compounds (Wentzcovitch et al. 1987).

For the series BN, BP, and BAs, the distribution of valence electron charge density varies in an interesting manner. Since there are no p-states in the core for both B and N, the valence electrons with p-character are not repelled from the core region and exhibit a peak in their charge density which is close to the position of an atomic p-state. This is the origin of the double-hump structure in the charge density of diamond compared to a single-hump distribution for Si or Ge. The result is that in the case of BN, since both B and N are without core p-states, the charge concentrates near N as expected for a III–V compound. In contrast, for BP the p-like attraction of B competes with the group-V potential and this results in a compensation and a covalent-like charge distribution. Hence, BN should have a fairly large ionicity and BP will be almost totally covalent. The trend continues for BAs where there is a reversal of the usual III-V zincblende signature with somewhat more charge localized near the B than the As. It may be reasonable to refer to this compound as AsB, as has been done on occasion.

3.3. Surface structures

The application of the total energy pseudopotential method to the study of surface structures began at about the same time (Ihm and Cohen 1979) as the application to bulk solids. In most of the studies the approach is similar to that used for the bulk since a supercell is constructed in which the "surface" atoms are moved to simulate surface reconstructions. The total energies for the different configurations

---

Fig. 12. Total-energy curves versus volume for some representative structures of AlSb, GaSb, InP, InAs, and InSb (Zhang and Cohen 1987).

are then compared. Once a low-energy structure is obtained, the various lattice constants involved are varied until an energy minimum is reached. The limitations on this approach for surfaces are similar to those discussed for the determination of bulk structures. One simplifying aspect is that in some cases only the first-layer atoms are moved so that one is restricted to a two-dimensional rather than a three-dimensional space. However, for many reconstructions it is useful to know the positions of atoms several layers into the crystal, and these calculations are limited by the computational power available.

Another method which is particularly useful for studies of surface structure is the approach (Yin and Cohen 1981) based on the evaluation of Hellman–Feynman forces on each of the atoms being examined. By varying the atomic positions to continually reduce the Hellman–Feynman forces, a minimum in the total energy is found. This approach is often helpful for choosing good candidate structures for more detailed evaluation using the total-energy approach.

Early studies focused on Si(100) (Ihm et al. 1980, Yin and Cohen 1981) and Si(111) (Northrup et al. 1981, Pandey 1981, Northrup and Cohen 1982) where theoretical models such as the Chadi asymmetric dimer model for Si(100) and the Pandey π-bonded chain for Si(111) were tested. The calculations were usually restricted to smaller unit cells caused by (2×1) reconstructions. Extensions to adatom geometries (Northrup and Cohen 1984), surface adsorption (Zhang et al. 1985b), Schottky barriers (Zhang et al. 1985a), and heterostructures (Zhang et al. 1986) were made and provided useful information. Often a comparison with experiment of the resulting electronic band structures for different geometries was used for deciding between likely structures. In the earlier studies, because a local density approximation was used, the calculated excitation spectrum was not definitive since the band-gap energies were underestimated. In recent calculations (Northrup et al. 1991) quasiparticle spectra have been obtained from first-principles and direct comparison with experiment is possible.

Much of the recent experimental research in this area is focused on the use of scanning tunneling microscopy, low-energy electron diffraction, and transmission high-energy electron diffraction. These techniques have allowed the solution of long-standing surface structure problems such as the Si(111)-7×7 surface reconstruction (Takayanagi et al. 1985). Combining scanning tunneling microscopy techniques, photoemission spectroscopy, and total energy pseudopotential calculations (figs. 13 and 14) has been fruitful (Biegelsen et al. 1990). However, many of the reconstructed surface geometries are too complex for the ab inito approach. In these cases, a significant amount of the theoretical work has been done by Chadi (see, e.g., Chadi 1989) using tight-binding techniques.

4. Applications to vibrational properties

An important application of the total-energy pseudopotential approach has been its use in calculations of vibrational properties of solids. Again, the input needed for the calculations is minimal. In contrast to phenomenologic force constant models

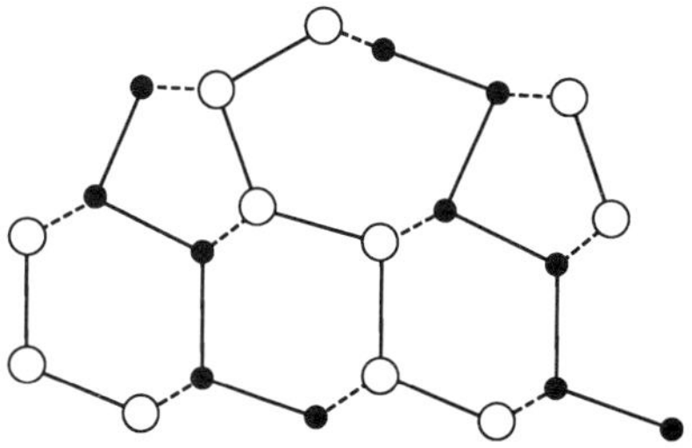

Fig. 13. Schematic side view of the π-bonded chain model. The two top surface atoms form the π-bonded chains. The solid and open circles denote atoms in different (110) planes (Northrup et al. 1991).

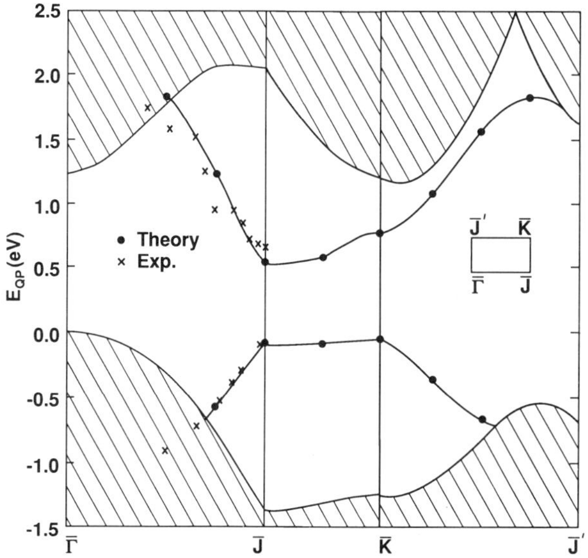

Fig. 14. Quasiparticle surface-state band structure (Northrup et al. 1991) for Si(111)2 × 1 compared to photoemission (Uhrberg et al. 1982) and inverse photoemission (Perfetti et al. 1987, Cricenti et al. 1990) experiments.

which often require 15 parameters to achieve reasonable fits to phonon dispersion curves, the total-energy approach uses only the masses and atomic numbers of the constituent atoms.

If we again rely on Si and Ge as prototype solids, then the method can be illustrated as a straightforward extension (Yin and Cohen 1980, 1982b) of the earlier discussion of the application of the total-energy pseudopotential approach to the study of bulk and surface structure. To calculate the phonon dispersion curve $\omega(\boldsymbol{q})$ along a specific direction, a supercell is chosen, and the atomic cores are displaced to simulate a particular phonon. This is the "frozen-phonon" approach (Heine and Weaire 1970, Chadi and Martin 1976, Wendel and Martin 1979, Yin and Cohen 1980). The change in the total energy arising from the distortion depends on the type of distortion assumed and the amplitude. Since a specific phonon wave vector $\boldsymbol{q}$ is chosen, this method is limited to calculating $\omega(\boldsymbol{q})$ at specific points in the Brillouin zone. The phonon frequencies can also be evaluated by computing the Hellman–Feynman forces on the displaced atoms (Ihm et al. 1981).

The above approach is simplest for computing $\omega(\boldsymbol{q})$ at symmetry points. Calculations at nonsymmetry points are possible but require larger unit cells. Standard calculations usually limit the computations to three or four $\boldsymbol{q}$-points along a symmetry direction. Another related approach was developed (Yin and Cohen 1982c) to compute $\omega(\boldsymbol{q})$ by calculating the force constants between atomic layers. This calculation yields $\omega(\boldsymbol{q})$ along the direction normal to the atomic layers. Both approaches give results in excellent agreement with experiment.

As an example of a frozen-phonon supercell calculation, we briefly examine the calculations of the phonon properties for the Γ and X points of the Brillouin zone for Si and Ge. The phonon polarizations for Γ and X can be determined using group theory (fig. 15), and the primitive cell for the distorted lattice contains two atoms for the LTO(Γ) mode and four atoms for the phonon modes at X. The phonon distorted lattice has inversion symmetry which simplifies the calculation.

For a given amplitude u, the change in the total energy arising from the distortion is $\Delta E_{\text{tot}}(u)$. The force constant k can be obtained from a second derivative of the $E_{\text{tot}}(u)$ or a first derivative of the force $F(u)$,

$$k = \left.\frac{\partial^2 E_{\text{tot}}}{\partial u^2}\right|_{u=0} \cong \frac{2\Delta E_{\text{tot}}(u)}{u^2} \tag{42}$$

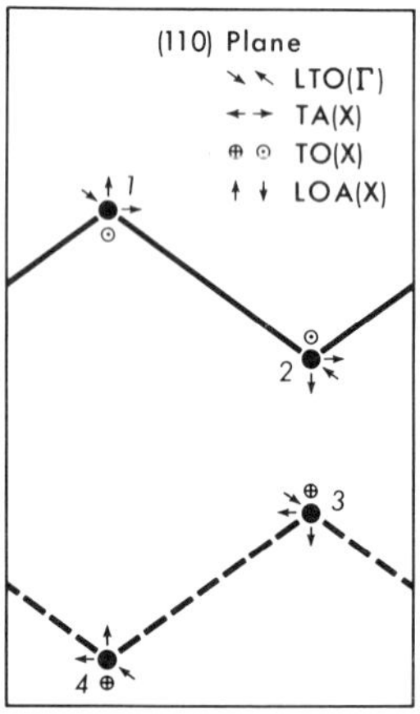

Fig. 15. Phonon polarization at Γ and X $(0, 0, 2\pi/a)$ for the diamond structure. Atoms are numbered and denoted by black dots. The solid lines denote an atomic chain in a (110) plane, and the dashed lines denote the projection of an atomic chain a distance $\frac{1}{4}\sqrt{2}a$ away from that plane where a is the lattice constant (Yin and Cohen 1982b).

or

$$k = \left.\frac{-\partial F}{\partial u}\right|_{u=0} \cong -\frac{F(u)}{u}. \tag{43}$$

The phonon frequency

$$\omega = 2\pi f = \sqrt{k/M}, \tag{44}$$

where M is the atomic mass. The calculation of the total energies or forces is done typically for five different amplitudes ranging from 0.01 to 0.1 Å. The total energies are fit by a quadratic function to about 1%. For some phonon mode calculations such as the LTO(Γ) modes in Si and Ge higher-order fits such as the third-order term are needed. These fits allow the evaluation of anharmonic terms. Table 4 contains the results. An impressive achievement of these calculations is the good agreement with experiment obtained for the TA(X) modes in Si and Ge. For empirical force constant model calculations, many-neighbor parameters are needed to obtain the correct values for these modes because of the important role played by long-ranged forces.

The use of eq. (42) or eq. (43) involves different approaches, but the results for the total-energy approach or for the force-constant approach are essentially the same. These are listed in table 4. Some extra information can be obtained from the force calculation. For example, for the LOA(X) mode, the forces on atoms 2 and 3 differ for finite amplitude (fig. 15). This arises from the effects of anharmonicity and illustrates that it is more difficult to compress a bond than to stretch it. For the phonon calculation, the average of the two force constants are used, and this quantity does not vary to within 0.5% with amplitude.

If the interlayer force constants are computed, then the full dispersion can be calculated as discussed earlier. This gives the value of the sound velocity. For example,

Table 4

Comparison of calculated phonon frequencies (in Thz) of Si and Ge (Yin and Cohen 1982b) at Γ and X with experiment (Dolling 1963a,b, Nilson and Nelin 1971, 1972) (f_{expt}). The values for f_{E}(f_{F}) are obtained from energy (force) calculations. The deviations from experimental values are given in parentheses.

	LTO (Γ)	LOA (X)	TO (X)	TA (X)
Si				
f_{E}	15.16 (−2%)	12.16 (−1%)	13.48 (−3%)	4.45 (−1%)
f_{F}	15.14 (−3%)	11.98 (−3%)	13.51 (−3%)	4.37 (−3%)
f_{expt}	15.53	12.32	13.90	4.49
Ge				
f_{E}	8.90 (−2%)	7.01 (−3%)	7.75 (−6%)	2.44 (2%)
f_{F}	8.89 (−3%)	6.96 (−3%)	7.78 (−6%)	2.45 (2%)
f_{expt}	9.12	7.21	8.26	2.40

the $q \to 0$ TA velocity $v_{TA}[110]$ propagating along the [110] direction with $[1\bar{1}0]$ polarization is associated with the shear modulus $C_{11} - C_{12}$;

$$v_{TA}[110] = \left(\frac{C_{11} - C_{12}}{2\rho_M}\right)^{1/2}, \tag{45}$$

where ρ_M is the mass density. The calculated results for the velocity and shear modulus are in good agreement with experiment as shown in table 5.

Another property of the vibrational spectrum which can be computed is the mode-Grüneisen parameters γ where

$$\gamma = -\mathrm{d}\ln\omega/\mathrm{d}\ln\Omega \cong -\Delta\ln\omega/\Delta\ln\Omega, \tag{46}$$

where Ω is the volume. The phonon frequencies are calculated at Γ and X for Si and Ge for variations in the lattice constants from 1% to 3%. The results given in table 6 reveal the excellent results obtained. It is particularly satisfying that the negative Grüneisen constant for the TA(X) mode is obtained since this is a difficult result to obtain using empirical methods. The results of the ab initio calculations can also be

Table 5
Comparison of calculated values (Yin and Cohen 1982b) of C_{11}–C_{12} and the TA velocity along [110] (with $[1\bar{1}0]$ polarization) of Si and Ge with experiment (McSkimin and Andreatch 1963).

	C_{11}–C_{12} (Mbar)	$v_{TA[110]}$ (10^5 cm/s)
Si		
Theory	1.07	4.79
Experiment	1.027	4.693
Ge		
Theory	0.74	2.64
Experimental	0.819	2.770

Table 6
Comparison of calculated (Yin and Cohen 1982b) mode-Grüneisen parameters for phonons at Γ and X of Si and Ge with experiment.

	LTO (Γ)	LOA (X)	TO (X)	TA (X)
Si				
Calculated	0.9	1.3	0.9	−1.5
Experimental	0.98	1.5	0.9	−1.4
Ge				
Calculated	0.9	1.4	1.0	−1.5
Experimental	1.12 ± 0.02[b] 0.88 ± 0.08[c]			

[a]Weinstein and Piermarini (1975).
[b]Buchenauer et al. (1971).
[c]Asaumi and Minomura (1978).

used to explore the origins of the negative sign on this effect by examining the various contributions to the total energy. In this case, the electronic contribution dominates in stabilizing the system. As the volume decreases, the electronic distribution becomes more metallic in nature and less dominant compared to the Ewald contribution which tends to lower the frequencies. The result is to reduce TA(X) and produce a negative Grüneisen parameter for TA(X).

When interlayer force constants are calculated, the resulting phonon spectrum is in good agreement with experiment. For example, for Si the Γ to X curve was computed (Yin and Cohen 1982c) using the calculated interlayer force constants and the frozen phonon results at Γ and X. The results are close to the experimentally determined points (fig. 16).

As discussed earlier, the frozen-phonon total-energy approach can be used to extract nonharmonic contributions to the phonon couplings. By examining the $E_{tot}(u)$ curves terms beyond the quadratic dependencies can be obtained. For example, a detailed study was done (Vanderbilt et al. 1986) of third- and fourth-order anharmonic coupling constants for optical phonons in Ci, Si, and Ge. These calculations were used to determine bare phonon–phonon scattering amplitudes. By including virtual processes, renormalized multiphonon vertices and phonon self-energies were calcu-

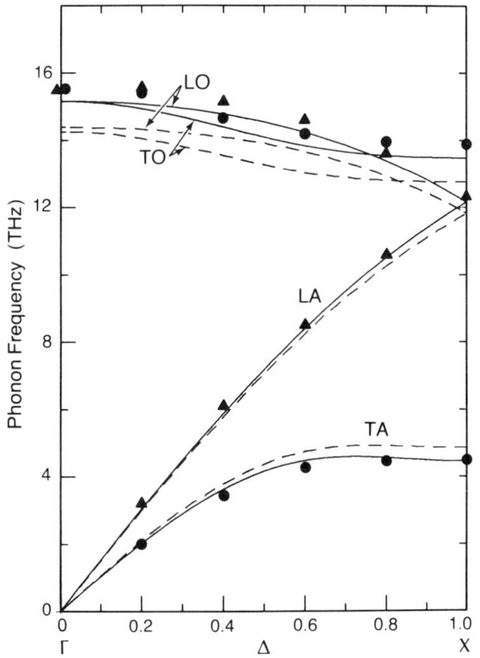

Fig. 16. Phonon frequencies from Γ to X in the [001] direction (Yin and Cohen 1982c). The dashed lines are calculated from the computed force constants. The solid lines are calculated using the computed third-nearest-layer force constants and the frozen-phonon results at Γ and X. The experimental points (Dolling 1963a,b, Nilsson and Nelin 1972, Sinha 1973) are denoted by dots for transverse modes and triangles for longitudinal modes.

lated. One specific application was the study of the renormalized four-phonon vertices. These were found to be negative which is the wrong sign for allowing the formation of a proposed two-phonon bound state.

Although Si and Ge were used as prototypes, other semiconductors such as the zincblendes (Kunc and Martin 1982, Chang and Cohen 1985b) have been studied with comparable success.

5. Other applications

5.1. Structural trends

Ab initio pseudopotential methods provide us with a general prescription for answering questions involving phase stability, and structural properties of a given system. However, a central goal of solid state physics, and materials science, has been the development of atomistic properties which will predict "global" trends in "all possible" compounds. For example, suppose one would like to know all possible solids which have structures similar to the high-temperature superconductors. Achieving such a goal is a formidable task. If we consider the ensemble of all elemental combinations, the number of possibilities is so great and the distinctions in energy so fine as to preclude the development of a deductive quantum mechanical solution from "first principles".

One of the most promising trends in this area is the construction of quantum based chemical indices which are based in part on pseudopotentials. Such chemical indices can be used to "screen" a number of possible structures and decide which ones would be most energetically favorable. Traditionally, these indices have been based on atomic properties such as atomic "size" and "electronegativity". Recently, pseudopotentials have been used to construct more accurate scales. The first such scales were based on empirical potentials and consisted of "orbital radii" (Bloch and Simons 1972, Simons and Bloch 1973, St. John and Bloch 1974, Chelikowsky and Phillips 1978). Orbital radii are angularly dependent radii which can be extracted from "hard-core" pseudopotentials.

Consider a model ion-core pseudopotential such as

$$V_l(r) = -Z_{\text{val}}\frac{e^2}{r} + \frac{B_l}{r^2}, \tag{47}$$

where Z_{val} is the number of valence electrons. B_l is a parameter fixed by spectroscopic data which reproduces the term value of interest, i.e., $l = 0, 1, 2, \ldots$, for the $Z_{\text{val}} - 1$ ionized species. In the case of silicon, we would examine the spectroscopic term values of one valence electron moving in the potential of the silicon ion core, i.e., the term values for Si^{3+}. By construction, this model pseudopotential is guaranteed to reproduce the energy levels for the species of interest (Bloch and Simons 1972, Simons and Bloch 1973). The particular form of the potential yields classical crossing points corresponding to $V_l(r_l) = 0$. These crossing points are used to determine the orbital radii, r_l. Physically such radii can be considered as an orbitally dependent measure of the "core size". Moreover, changes in these radii should be related to differences in the orbital cancellation of the all-electron potential, e.g., in the Phillips–Kleinman pseudopotential picture. Since the ion cores should not be sensitive to the chemical environment, the orbital radii should be transferable from one compound to another. An example of such a potential is given in fig. 17. Because of the divergence of the potential near the origin, the potential is called a "hard-core" pseudopotential.

Ab initio pseudopotentials can also be used to define orbital radii. As mentioned in § 2.2, some of the earliest ab initio pseudopotentials consisted of hard-core potentials (Zunger and Cohen 1978, 1979). It is possible to define orbital radii for these potentials by examining their turning points. Ab initio pseudopotentials have a number of advantages compared to empirical potentials in defining orbital radii. Spectroscopic data is not available for highly ionized species, e.g., Ni^{9+}, or Cl^{6+}. Ab initio pseudopotentials can easily be used to construct radii for such species. Also, ab initio orbital radii can be constructed for hypothetical situations using different definitions of core versus valence electrons, e.g., Cu^{+} and Cu^{10+}.

Ab initio orbital radii have been used to determine structural trends for compounds which include the tetrahedral semiconductor family. Consider the family of crystals $A^N B^{8-N}$ where A and B do not include the transition metals or rare earths and where N is the number of valence electrons of species A. These "octet" compounds correspond to crystals such as NaCl, GaAs, CaO, InP, and so on. Seventy or so members of this family exist; their structures correspond to six crystal types – diamond, graphite, zincblende, wurtzite, rocksalt and cesium chloride. Predicting the

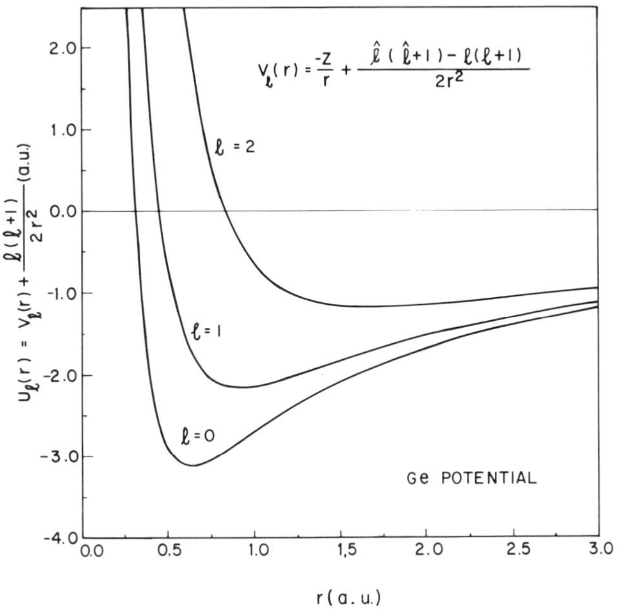

Fig. 17. Example of a hard-core ionic pseudopotential. The potential illustrated is for Ge^{3+}. Values for the orbital radii, r_s, r_p and r_d, can be obtained by the classical turning points.

most stable structure for a given octet AB compound is a nontrivial task. For example, wurtzite and zincblende structures differ only in the third-nearest neighbor, and the energy differences between these structures can be less than ~ 0.01 eV/atom.

The goal of a chemical scaling approach is to systematize structural trends so that similar structural types will group together in a "chemical index" space. Suppose one knows that a given AB compound is a wurtzite structure, then a chemically "similar" compound as defined by a chemical index should also possess the wurtzite structure. Defining what one means by "chemical" similarity is the *key* issue in defining a chemical index.

Given three orbital radii, r_s, r_p, and r_d, per atom, there are a number of possible combinations one could use to extract a chemical index. Two measures which have worked well for this group of compounds are "r_π" and "r_σ" which are defined as $r_\pi(AB) = [r_p(A) - r_s(A)] + [r_p(B) - r_s(B)]$ and $r_\sigma(AB) = [r_p(A) + r_s(A)] - [r_p(B) + r_s(B)]$. One can interpret $r_\sigma(AB)$ as a "size" difference between A and B. $r_\pi(AB)$ is more difficult to interpret. If $r_p(A) \approx r_s(A)$ then hybrid orbitals of s and p orbitals might form more readily. (One may recall that the orbital radii are defined by the spectroscopic term values so $r_p(A) \approx r_s(A)$ implies term energy levels are such that $E_s \approx E_p$. It should be more easy to mix s and p if $E_s \approx E_p$ than if $E_s \ll E_p$.) In fig. 18, we illustrate a structural plot from Zunger and Cohen (1979) in (r_σ, r_π) space in which each known compound is indicated by a symbol corresponding to its structure. This plot results in one of the best "groupings" to date. In particular, note how the wurtzite structures group together and are different from the placement of the zincblende structures. Previous plots using chemical indices based on a dielectric theory (Phillips 1973) were able to achieve a separation between six-fold coordinated structures (rocksalt) and four-fold structures (diamond, zincblende, and wurtzite), but not between the four-fold structures.

Recently, this scheme has been updated and new indices such as electronegativity scales and the average number of valence electrons have been added as other coordinates (Villars and Calvert 1985, Villars 1986). An example of this approach is the work of Villars et al. (1986) in which they predicted successfully a number of candidates for quasicrystals.

5.2. Bulk moduli

In the description of the total-energy pseudopotential approach, a fundamental parameter used in the equation of state is the bulk modulus at zero pressure B_0. This quantity is determined from the $E_{tot}(V)$ curve by fitting a few selected points. The ab initio calculations of B_0 are usually in agreement with experiment to within about 6%. Because hardness usually scales with B_0, this is an important physical quantity to obtain theoretically. For complex structures, one or two hours of supercomputer time is required for a calculation of this kind.

Another approach to obtain B_0 is to use an empirical theory which requires some experimental input. Although this approach (Cohen 1985a) is not a first-principles method, it has the advantage that a computer is not required and it allows an

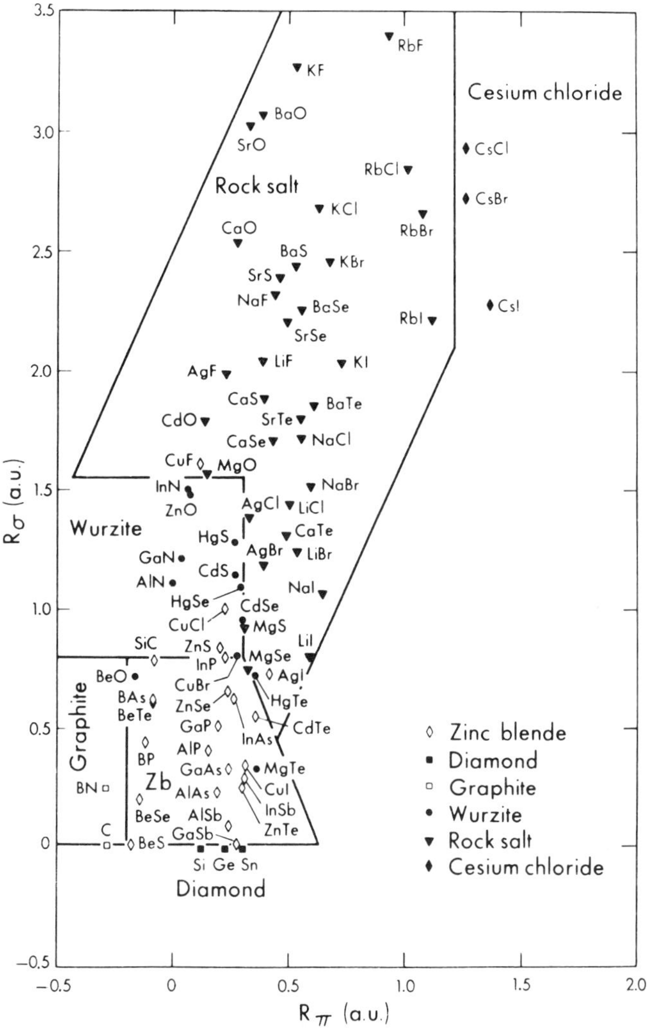

Fig. 18. Structural plot for the octet binary compounds. The indices have been obtained from ab initio pseudopotentials. Note how the different structural types are grouped within this chemical index space.

analysis of trends in bulk moduli and their dependence on other physical properties such as bond lengths. The development of an empirical theory to be used in this way could serve as a model for other empirical approaches. For example, the ab initio theory can be used to compute properties for specific materials. Analyses of these calculations allow the determination of the critical components of the theory necessary to obtain the results. At this point, it may be possible to use input of various kinds to compute the dominant contributions without resorting to a full first-principles calculation. Below we describe the empirical approach in some detail with the aim that it might serve as a prototype for other empirical extensions of the ab initio calculations.

The empirical theory (Cohen 1985a) for B_0 is based on the use of the Phillips average homopolar gap E_h (Phillips 1973). In the same manner as for metals where B_0 scales like the Fermi energy divided by the volume per atom, it can be argued that for semiconductors E_h plays the role of the Fermi energy and the bond volume scales like a constant divided by the bond length d. This yields the relation

$$B_0 = 45.6 E_h d^{-1}, \tag{48}$$

where B_0 is in GPa, E_h is in eV, and d is in Å. Since E_h scales as $d^{-2.5}$ (Phillips 1973), then

$$B_0 = 1761 d^{-3.5}. \tag{49}$$

To include the effects of ionicity an empirical parameter λ was introduced. The nonionic group-IV materials (C, Si, Ge, Sn) have $\lambda = 0$, while the III–V and II–V zincblends have $\lambda = 1$ and 2, respectively. This modification yields

$$B_0 = (1971 - 220\lambda) d^{-3.5}. \tag{50}$$

Although eq. (49) gives a good fit for most of the valence 8 four-fold coordinated semiconductors, eq. (50) is more accurate as it accounts for the weakening of the covalent bond through charge transfer.

Another modification allows the use of this approach for materials where the coordination number $N_c \neq 4$,

$$B_0 = \frac{\langle N_c \rangle}{4}(1971 - 220\lambda) d^{-3.5}, \tag{51}$$

where $\langle N_c \rangle$ is the average coordination number for the compound.

Returning to the valence 8, $N_c = 4$ systems, the results for B_0 are given in table 7. Generally, the comparison between theory and experiment is within a few percent. A few "bad actors" exist such as Ge. For BN, experimental extrapolations yielded $B_0 > 400$ GPa. However, recent measurements (Knittle et al. 1989) yield values in good agreement with the theoretical predictions. At this point, the calculated values for BP and BAs stand as predictions. They do agree with the values obtained from ab initio calculations.

As an example of the use of the empirical approach for predicting trends and new materials, it is interesting to examine the upper limits on B_0 for materials in this class. From eq. (51), we see that low compressibility solids are expected to have low ionicity and short bond lengths. For example, diamond, which has the highest measured B_0 at this time, has a short bond length of 1.54 Å, and the bond is completely ionic with $\lambda = 0$. However, it should be stressed that eq. (51) has no apparent limit for B_0 in the range of currently measured values. Hence, there is no obvious barrier limiting B_0 to a maximum equal to that of diamond at 443 GPa. Shorter bonds can occur even in the case of carbon where sp^2 bonding can contribute. An obvious example is the sp^2 bonds in graphite which are shorter than the sp^3 bonds of diamond. These bonds are in the layers and are stiff for compressions along

Table 7
Comparison of calculated bulk moduli from eq. (3) with measured values. Almost all of the experimental data are taken from Landolt–Bornstein (1982). A few of the quoted values are from references to the original experimental work given in the theoretical papers referred to in this article.

Material	d (Å)	Experimental B_0 (GPa)	Calculated B_0 (GPa)
IV			
C	1.54	443	435
Si	2.35	99	99
Ge	2.45	77	85
Sn	2.75	53	57
SiC	1.88	224	213
III–V			
BN	1.57	369	367
BP	1.96	–	166
BAs	2.07	–	138
AlP	2.36	86.0	86.7
AlAs	2.43	77.0	78.3
AlSb	2.66	58.2	57.0
GaP	2.36	88.7	86.7
GaAs	2.45	74.8	76.1
GaSb	2.65	57.0	57.8
InP	2.54	71.0	67.0
InAs	2.61	60.0	61.0
InSb	2.81	47.4	47.1
II–VI			
ZnS	2.34	77.1	78.1
ZnSe	2.45	62.4	66.5
ZnTe	2.64	51.0	51.2
CdS	2.52	62.0	60.3
CdSe	2.62	53.0	52.6
CdTe	2.81	42.4	41.2
HgSe	2.63	50.0	51.9
HgTe	2.78	42.3	42.7

the layer. Recently, an all-sp^2 three-dimensional structure was considered (Liu et al. 1991). Although this structure would have a relatively large B_0, it was found that it is metastable and would convert to diamond unless otherwise constrained.

Another promising approach is to explore compounds having CN bonds. Using conventional atomic radii, this bond should be in the range of 1.47–1.49 Å which is shorter than that of diamond, and λ should be between 0 and 1. For a zincblende tetrahedral configuration where $N_c = 4$, antibonding states would be occupied, and hence, this structure is unlikely to form. An alternative structure is β-C_3N_4 which is C–N in the β-Si_3N_4 structure. For this structure, $\langle N_c \rangle = 3.43$. Studies of both β-Si_3N_4 and β-C_3N_4 (Liu and Cohen 1990) using the total-energy pseudopotential approach yield calculated bulk moduli which are close to the estimates made using eq. (51). For β-Si_3N_4, the values are in agreement with the measurements. For

β-C_3N_4, the calculated value is $B_0 = 427$ GPa which is comparable with that of diamond.

The calculations for β-C_3N_4 are indicative of the use of empirical approaches. Even though these schemes involve some experimental data, they are often flexible and easy to use. By examining classes of materials via an empirical approach, it may be possible to find good candidates to study in more detail using the ab initio approach. In turn, as mentioned above, the ab initio calculations are often useful for developing similar methods since it is possible to find out the influence of various parameters and contributions. Hopefully, in the case of B_0, it will be possible to use the theoretical results to predict new materials in addition to β-C_3N_4 which have ultralow compressibilities.

5.3. Electron–phonon interactions and superconductivity

Ever since the development of the BCS theory of superconductivity (Bardeen et al. 1957), there have been attempts to calculate the parameters needed to compute the superconducting transition temperature T_c. In particular, the electron–phonon coupling parameter λ is needed with high precision because T_c depends on λ through an exponential function. For $\lambda < 1.25$, T_c can be evaluated using the McMillan equation (McMillan 1968),

$$T_c = \frac{\Theta_D}{1.45} \exp\left(\frac{-1.04(1+\lambda)}{\lambda - \mu^* - 0.62\lambda\mu^*} \right), \tag{52}$$

where Θ_D is the Debye temperature and μ^* is the repulsive Coulomb parameter. Because it is difficult to calculate μ^* from first-principles, it is usually treated as a parameter. Using tunneling measurements, estimates of $\mu^* \sim 0.1$ for sp metals are common. Sometimes refinements are made by scaling the parameter by the density of states at the Fermi energy $N(E_F)$.

Early estimates for λ were uncertain and, hence, gave fairly large variations in T_c. Few predictions of new superconductors were possible because of the uncertainties connected with the calculations of the normal state properties. One exception (Cohen 1969) was the evaluation of the parameters for doped semiconductors since the electronic structure and electron–phonon parameters of the host undoped systems were known. This brought the study of semiconductors into the realm of superconductivity. Although the successful prediction of superconductivity in degenerate semiconductors was also a success of BCS theory, it did rely on experimental input about the normal state. In contrast, the recent successful prediction of superconductivity (Chang et al. 1985) in highly compressed Si used a first-principles calculation of the crystal structure, electronic structure, phonon spectrum, and electron–phonon coupling constants.

The central parameter in the calculation of T_c is λ which is an average of the coupling parameters λ_{qv} for coupling an electron to a phonon of mode index v and wave vector q,

$$\lambda = \frac{1}{\Omega_{BZ}} \int \sum_{v} \lambda_{qv} \, d^3q, \tag{53}$$

where Ω_{BZ} is the Brillouin zone volume. The parameter $\lambda_{q\nu}$ can be expressed in terms of the phonon line width $\gamma_{q\nu}$ and the phonon frequency $\omega_{q\nu}$

$$\lambda_{q\nu} = \frac{\gamma_{q\nu}}{\pi N(E_F)\hbar\omega_{q\nu}^2}. \tag{54}$$

It can also be expressed in terms of the electron–phonon matrix element $g(n\boldsymbol{k}, n'\boldsymbol{k}', \boldsymbol{q}\nu)$

$$\lambda_{q\nu} = 2N(E_F)\frac{\langle\langle |g(n\boldsymbol{k}, n'\boldsymbol{k}', \boldsymbol{q}\nu)|^2\rangle\rangle}{\hbar\omega_{q\nu}}, \tag{55}$$

where

$$g(n\boldsymbol{k}, n'\boldsymbol{k}', \boldsymbol{q}\nu) = \sum_{\tau}\left(\frac{\hbar\Omega_{BZ}}{2M_\tau\omega_{q\nu}}\right)^{1/2}\delta(\boldsymbol{k}-\boldsymbol{k}'-\boldsymbol{q})\left\langle \psi^0_{n\boldsymbol{k}}\left|\hat{\boldsymbol{\varepsilon}}^\tau_{q\nu}\cdot\frac{\delta V}{\delta \boldsymbol{R}_\tau}\right|\psi^0_{n'\boldsymbol{k}'}\right\rangle. \tag{56}$$

The expression $\langle\langle |g|^2\rangle\rangle$ represents a Fermi surface average of g, M_τ is the atomic mass for the atom labeled by τ, $\hat{\boldsymbol{\varepsilon}}^\tau_{q\nu}$ is the polarization vector, and $\psi^0_{n\boldsymbol{k}}$ and $\psi^0_{n'\boldsymbol{k}'}$ are the Bloch states for band indices n and n' and states $\boldsymbol{k}$ and $\boldsymbol{k}'$ for the undistorted crystal.

If we now use a similar formalism to that introduced for the frozen-phonon calculation, then

$$\sum_{\nu}\hat{\boldsymbol{\varepsilon}}^\tau_{q\nu}\cdot\frac{\delta V}{\delta \boldsymbol{R}_t} = \frac{V_{q\nu}-V_0}{u_{q\nu}}, \tag{57}$$

where $V_{q\nu}$ and V_0 are the distorted and undistorted potentials, respectively. Their difference caused by the presence of a frozen phonon is divided by an average value of the displacement $u_{q\nu}$.

The phonon frequencies are calculated in the manner described earlier and because a supercell approach is used, $\lambda_{q\nu}$ are determined for only a finite number of q points in the Brillouin zone. Hence, the average obtained using eq. (53) is often only a rough approximation for λ.

The first applications of the above formalism were studies of Si at high pressures. Because of the covalent-like charge density computed for simple-hexagonal (sh) and hcp Si, it was felt (Cohen 1985b) that these materials might be superconductors. A detailed prediction and subsequent experimental confirmation (Chang et al. 1985) added considerable support to the validity of the calculations of λ and more generally to the total-energy pseudopotential method. Here was a case where the electronic structure, vibrational structure, and superconductivity and their pressure dependence (Cohen et al. 1985) (figs. 19 and 20) were all calculated from first-principles. In fact, in one case (hcp) even the existence of the material was predicted.

Other applications include the calculation of the pressure dependence of λ and T_c (Dacorogna et al. 1985), analysis of λ and T_c for the β-Sn phases of Si, Ge, and Sn (Chang and Cohen 1986b), and the prediction of superconductivity in sh Ge (Martins and Cohen 1988a,b). Although λ and T_c were not computed for high-pressure forms of GaAs, it was shown that contrary to the earlier conclusions, GaAsII and GaAsIII are metallic (Zhang and Cohen 1989). This theoretical work motivated an experimen-

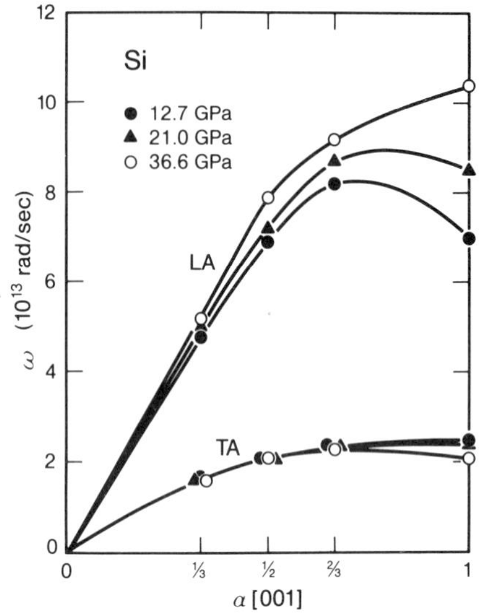

Fig. 19. Phonon frequencies along the [001] axis at different pressures for the sh phase of Si (Cohen et al. 1985). Lines are used as a guide.

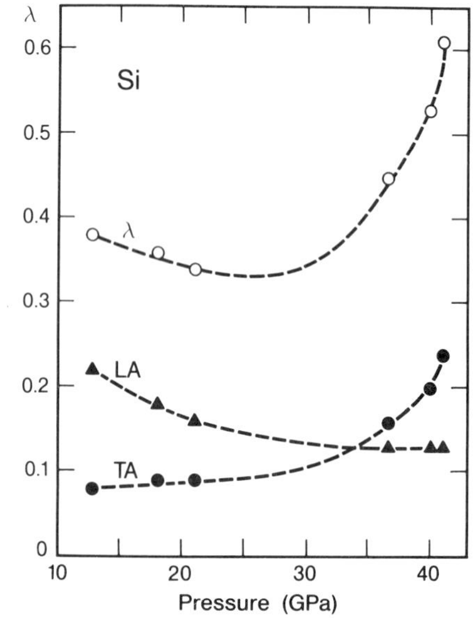

Fig. 20. Averaged electron–phonon coupling constants (λ) for the LA and TA modes and their sum as a function of pressure for the sh phase of Si (Cohen et al. 1985). Lines are used as a guide.

tal study (Zhang et al. 1989) which demonstrated that both high-pressure orthorhombic phases (GaAsII and GaAsIII) are metallic and superconducting.

5.4. Molecular dynamics using ab inito pseudopotentials

Pseudopotential methods have greatly enhanced our ability to perform electronic structure calculations, but for the most part, these methods have been applied to static systems. For example, in § 3.1. we concentrated on the total energy of a given structure at zero temperature. Recently, pseudopotential methods have been extended to finite-temperature computer simulations, i.e., molecular dynamical simulations.

Traditionally, molecular dynamics simulations have relied on interatomic forces which have been determined from empirical interatomic forces (e.g., Ciccotti et al. 1987). Such approaches work well for systems in which pairwise forces dominate, e.g., van der Waals gases, but for covalent systems where many-body or angular forces play an important role, e.g., silicon, these approaches have been less successful. Another drawback of such approaches is that they do not include a description of the electronic properties of the system of interest. If one could implement a scheme where the forces between particles were fully quantum mechanical, then many limitations of traditional molecular dynamics would be eliminated. Until recently, it was widely believed that such a goal was unattainable. Owing to computational constraints, quantum mechanical calculations of interatomic forces for molecular dynamics seemed beyond the capacity of even the largest supercomputers.

However, in 1985 Car and Parrinello proposed a molecular dynamics approach which for the first time was able to incorporate quantum mechanical forces. Their method was based on combining molecular dynamics with ab initio pseudopotentials constructed within the local density approximation. Within this approach they were able to (a) compute the ground state electronic properties of complex, i.e., large and/or disordered, systems and (b) perform ab initio molecular dynamics simulations where their only assumptions are the validity of classical mechanics to describe the motion of the ion cores and the Born–Oppenheimer approximation to separate the nuclear and electronic coordinates (Car and Parrinello 1985).

The starting point of the Car–Parrinello molecular dynamics approach is to introduce a classical Lagrangian of the form

$$L = \sum_i \tfrac{1}{2}\mu \int \mathrm{d}^3 r \left|\frac{\mathrm{d}\psi_i}{\mathrm{d}t}\right|^2 + \sum_j \tfrac{1}{2} M_j \left(\frac{\mathrm{d}\boldsymbol{R}_j}{\mathrm{d}t}\right)^2 - E[\{\psi_i\}, \boldsymbol{R}_j] + \sum_{i,j} \Lambda_{ij}\left(\int \mathrm{d}^3 r |\psi_i|^2 - \delta_{ij}\right). \tag{58}$$

The first term in eq. (58) is a classical kinetic energy term associated with the electronic states, ψ_i, and where μ is a parameter with dimension mass times length squared. M_j and $\boldsymbol{R}_j$ correspond to the mass and position of the ion cores. $E[\{\psi_i\}, \boldsymbol{R}_j]$ is the Hohenberg–Kohn energy functional which contains the electronic kinetic energy, the ion-core Coulomb repulsion, and the effective electronic potential energy (which includes contributions from the ionic pseudopotential, the Hartree potential and the exchange-correlation potential). The Λ_{ij} are Lagrange multipliers which impose orthonomality constraints between the occupied orbitals, ψ_i. This Lagrangian yields the following equations of motion for $\{\psi_i\}$ and $\{\boldsymbol{R}_j\}$:

$$\mu \frac{\mathrm{d}^2 \psi_i}{\mathrm{d}t^2} = -\frac{\delta E}{\delta \psi_i^*} + \sum_j \Lambda_{ij} \psi_j, \tag{59}$$

$$M_j \frac{\mathrm{d}^2 R_j}{\mathrm{d}t^2} = -\boldsymbol{\nabla}_{\boldsymbol{R}} E. \tag{60}$$

The ion-core dynamics as given by eq. (60) have real physical meaning and correspond to traditional molecular dynamics if E were given by a sum over interatomic potentials. The equation of motion for the $\{\psi_i\}$ is fictitious and is to be considered as a "tool to perform the dynamical simulated annealing". As noted by Car and Parrinello: By variation of the velocities, the temperature of the system can be slowly reduced and for $T \rightarrow 0$ the equilibrium state of minimal E is reached. At equilibrium, $\mathrm{d}^2\psi_i/\mathrm{d}t^2 = 0$, the eigenvalues of the Λ matrix in eq. (39) coincide with the occupied eigenvalues determined from the Kohn–Sham equation. Only when these conditions are satisfied does this Langrangian describe a real physical system whose representative point in configuration space lies on the Born–Oppenheimer surface. The coupling between the electronic and the classical degrees of freedom is controlled by adjusting the μ parameter. When external constraints are present, e.g., a strain, constant volume, or the like, the Langrangian in eq. (60) can be suitably modified.

The electronic states in the Car–Parrinello method are usually expressed in terms of plane waves. This has also motivated the use of "soft" ab initio pseudopotentials as discussed in § 2.2. The number of degrees of freedom which must be considered in the integration of eq. (59) are reduced if fewer plane waves can be considered. Although the Car–Parrinello method is clearly a great advance, ab initio molecular dynamics remains a computationally intensive method. Applications of the approach include liquid semiconductors and metals, clusters and disordered solids. Examples can be found in Galli et al. (1990), Allan and Teter (1987), Car and Parrinello (1988), Ballone and Galli (1989), Qian et al. (1990), Li et al. (1990), and Jones and Hohl (1990).

6. Conclusions

We have presented a review of an area which is still in a state of flux. There is still considerable activity both in the basic science of pseudopotentials and in their applications. New studies of solids, liquids, molecules, and clusters using pseudopotentials are being reported continuously. The approaches used in these studies differ, but the underlying concepts and many of the techniques have a common basis. It is the goal of this review to explore the properties of pseudopotentials and to use the area of structural properties of semiconductors to illustrate some standard techniques which were developed to apply them to problems of fundamental and practical interest.

It is inevitable that considerable evolution will occur in this area and that both scientific advances and developments in computer hardware and software will modify many of the details presented. However, if the area of empirical pseudopotentials and, in particular, the EPM can serve as a model, it is interesting to note that the EPM and the applications of this method done in the 1960's and 1970's are still "current". The pseudopotential area evolved by adding new subareas of the kind discussed in this review. Perhaps this trend will continue, and the research described here can be regarded as another component of the pseudopotential saga. Others are sure to follow.

Acknowledgements

JRC would like to acknowledge support for this work by the US Department of Energy of the Office of Basic Energy Sciences (Division of Materials Research) under Grant No. DE-FG02-89ER45391. He would also like to acknowledge computational support from the Minnesota Supercomputer Institute. MLC acknowledges support by National Science Foundation Grant No. DMR-8818404 and by the Director, Office of Energy Research, Office of Basic Energy Sciences, Materials Sciences Division of the US Department of Energy under Contract No. DE-AC03-76SF00098, and by the J.S. Guggenheim Foundation.

References

Allan, D.C., and M. Teter, 1987, Phys. Rev. Lett. **59**, 1136.
Allen, P.B., and M.L. Cohen, 1969, Phys. Rev. **187**, 525.
Andreoni, W., A. Baldereschi, N.O. Lipari and F. Meloni, 1981, Solid State Commun. **37**, 837.
Asaumi, K., and S. Minomura, 1978, J. Phys. Soc. Jpn. **45**, 1061.
Ashcroft, N.W., 1966, Phys. Lett. **23**, 48.
Austin, B.J., V. Heine and L.J. Sham, 1962, Phys. Rev. **127**, 276.
Bachelet, G.B., and M. Schlüter, 1982, Phys. Rev. B **25**, 2103.
Bachelet, G.B., D.R. Hamann and M. Schlüter, 1982, Phys. Rev. B **26**, 4199.
Bachelet, G.B., D.M. Ceperley and M.G.B. Chiochetti, 1989, Phys. Rev. Lett. **62**, 2088.
Ballone, P., and G. Galli, 1989, Phys. Rev. B **40**, 8563.
Bar-Yam, Y., S.T. Pantelides and J.D. Joannopoulos, 1989, Phys. Rev. B **39**, 3396.
Bardeen, J., L.N. Cooper and J.R. Schrieffer, 1957, Phys. Rev. **106**, 162.
Bardsley, J.N., 1974, Case Stud. At. Phys. **4**, 299.
Bassani, F., and V. Celli, 1961, J. Phys. & Chem. Solids **20**, 64.
Biegelsen, D.K., R.D. Bringans, J.E. Northrup and L.-E. Swartz, 1990, Phys. Rev. Lett. **65**, 452.
Birch, F., 1952, J. Geophys. Res. **57**, 227.
Bloch, A.N., and G. Simons, 1972, J. Am. Chem. Soc. **94**, 8611.
Brewer, L., 1977, Lawrence Berkeley Laboratory Report, LBL-3720.
Brust, D., J.C. Phillips and F. Bassani, 1962, Phys. Rev. Lett. **9**, 94.
Buchenauer, C.J., F. Cerdeira and M. Cardona, 1971, in: Proc. 2nd Int. Conf. on Light Scattering in Solids, ed. M. Balkanski (Flammarion, Paris) p. 280.
Bylander, D.M., and L. Kleinman, 1990, Phys. Rev. B **41**, 907.

Car, R., and M. Parrinello, 1985, Phys. Rev. Lett. **55**, 2471.
Car, R., and M. Parrinello, 1988, Phys. Rev. Lett. **60**, 204.
Chadi, D.J., 1989, Ultramicroscopy **31**, 1.
Chadi, D.J., and R.M. Martin, 1976, Solid State Commun. **19**, 643.
Chang, K.J., and M.L. Cohen, 1984, Phys. Rev. B **30**, 5376.
Chang, K.J., and M.L. Cohen, 1985a, Phys. Rev. B **31**, 7819.
Chang, K.J., and M.L. Cohen, 1985b, First principles study of the lattice dynamical behavior of AlAs, in: Proc. 17th Int. Conf. on the Physics of Semiconductors, eds D.J. Chadi and W.A. Harrison (Springer, Berlin) p. 1151.
Chang, K.J., and M.L. Cohen, 1986a, Phys. Rev. B **34**, 8581.
Chang, K.J., and M.L. Cohen, 1986b, Phys. Rev. B **34**, 4552.
Chang, K.J., and M.L. Cohen, 1987, Phys. Rev. B **35**, 8196, Erratum: 1991, Phys. Rev. B **43**, 12060.
Chang, K.J., M.M. Dacorogna, M.L. Cohen, J.M. Mignot, G. Chouteau and G. Martinez, 1985, Phys. Rev. Lett. **54**, 2375.
Chelikowsky, J.R., 1980, Phys. Rev. B **21**, 3074.
Chelikowsky, J.R., and M.Y. Chou, 1988a, Phys. Rev. B **37**, 6504.
Chelikowsky, J.R., and M.Y. Chou, 1988b, Phys. Rev. B **38**, 7966.
Chelikowsky, J.R., and M.L. Cohen, 1976, Phys. Rev. B **14**, 556. Erratum: 1984, Phys. Rev. B **30**, 4828.
Chelikowsky, J.R., and S.G. Louie, 1984, Phys. Rev. B **29**, 3470.
Chelikowsky, J.R., and J.C. Phillips, 1978, Phys. Rev. B **17**, 2453.
Christiansen, P.A., Y.S. Lee and K.S. Pitzer, 1979, J. Chem. Phys. **71**, 4445.
Ciccotti, G., D. Frenkel and I.R. McDonald, 1987, Simulation of Liquids and Solids (North-Holland, Amsterdam).
Cohen, M.H., and V. Heine, 1961, Phys. Rev. **122**, 1821.
Cohen, M.L., 1969, Superconductivity in low-carrier-density systems: Degenerate semiconductors, in: Superconductivity, ed. R.D. Parks (Marcel Dekker, New York) p. 615.
Cohen, M.L., 1982, Phys. Scr. **T1**, 5.
Cohen, M.L., 1984a, Am. J. Phys. **52**, 695.
Cohen, M.L., 1984b, Ann. Rev. Mat. Sci. **14**, 119.
Cohen, M.L., 1985a, Phys. Rev. B **32**, 7988.
Cohen, M.L., 1985b, Closing Address, in: Proc. 17th Int. Conf. on the Physics of Semiconductors, eds D.J. Chadi and W.A. Harrison (Springer, Berlin) p. 1151.
Cohen, M.L., 1986, Science **234**, 549.
Cohen, M.L., 1989, Nature **338**, 291.
Cohen, M.L., and T.K. Bergstresser, 1966, Phys. Rev. **141**, 789.
Cohen, M.L., and J.R. Chelikowsky, 1982, Pseudopotentials for semiconductors, in: Handbook on Semiconductors, Vol. 1, ed. W. Paul (North-Holland, Amsterdam) p. 219.
Cohen, M.L., and J.R. Chelikowsky, 1988, Electronic Structure and Optical Properties of Semiconductors (Springer, Berlin).
Cohen, M.L., and V. Heine, 1970, The fitting of pseudopotentials to experimental data and their subsequent application, in: Solid State Physics, Vol. 24, eds H. Ehrenreich, F. Seitz and D. Turnbull (Academic Press, New York) p. 37.
Cohen, M.L., K.J. Chang and M.M. Dacorogna, 1985, Physica B **135**, 229.
Corkill, J.L., A. Garcia and M.L. Cohen, 1991, Phys. Rev. B **43**, 9251.
Cricenti, A., S. Selci, K.O. Magnussen and B. Reihl, 1990, Phys. Rev. B **41**, 12908.
Dacorogna, M.M., K.J. Chang and M.L. Cohen, 1985, Phys. Rev. B **32**, 1853.
Dolling, G., 1963, in: Inelastic Scattering of Neutrons in Solids and Liquids, Vol. II (IAEA, Vienna) pp. 37, 249.
Donohue, J., 1974, The Structure of Elements (Wiley, New York).
Engel, G.E., and R.J. Needs, 1990, Phys. Rev. B **41**, 7876.
Fahy, S., X.W. Wang and S.G. Louie, 1988, Phys. Rev. Lett. **61**, 1631.
Fermi, E., 1934, Nuovo Cimento **11**, 157.
Froyen, S., and M.L. Cohen, 1983a, Phys. Rev. B **28**, 3258.

Froyen, S., and M.L. Cohen, 1983b, Static and structural properties of III–V zinc-blende semiconductors, in: Proc. 16th Int. Conf. on the Physics of Semiconductors, Part I (North-Holland, Amsterdam) p. 561.
Galli, G., R.M. Martin, R. Car and M. Parrinello, 1990, Phys. Rev. B **42**, 7470.
Gonze, X., J.P. Vigneron and J.-P. Michenaud, 1989, J. Phys.: Condensed Matter **1**, 525.
Gonze, X., P. Käckell and M. Scheffler, 1990, Phys. Rev. B **41**, 12264.
Greenside, H.S., and M. Schlüter, 1983, Phys. Rev. B **28**, 535.
Günnarsson, O., B.I. Lundqvist and J.W. Wilkins, 1974, Phys. Rev. B **10**, 1319.
Hamann, D.R., 1989, Phys. Rev. B **40**, 2980.
Hamann, D.R., M. Schlüter and C. Chiang, 1979, Phys. Rev. Lett. **43**, 1494.
Harrison, W.A., 1966, Pseudopotentials in the Theory of Metals (Benjamin, New York).
Heine, V., and I. Abarenkov, 1965, Philos. Mag. **9**, 451.
Heine, V., and D. Weaire, 1970, Solid State Phys. **24**, 249.
Herring, C., 1940, Phys. Rev. **57**, 1169.
Hohenberg, P., and W. Kohn, 1964, Phys. Rev. B **136**, 864.
Hu, J.Z., and I.L. Spain, 1984, Solid State Commun. **51**, 263.
Hybertsen, M., and S.G. Louie, 1985, Phys. Rev. Lett. **55**, 1418.
Hybertsen, M., and S.G. Louie, 1986, Phys. Rev. B **34**, 5390.
Ihm, J., and M.L. Cohen, 1979, Solid State Commun. **29**, 711.
Ihm, J., and M.L. Cohen, 1980a, Phys. Rev. B **21**, 3754.
Ihm, J., and M.L. Cohen, 1980b, Phys. Rev. B **21**, 1527. Erratum: 1980, Phys. Rev. B **22**, 2135.
Ihm, J., A. Zunger and M.L. Cohen, 1979, J. Phys. C **12**, 4409, Erratum: 1980, J. Phys. C **30**, 3095.
Ihm, J., M.L. Cohen and D.J. Chadi, 1980, Phys. Rev. B **21**, 4592.
Ihm, J., M.T. Yin and M.L. Cohen, 1981, Solid State Commun. **37**, 4911.
Jamieson, J.C., 1963, Science **139**, 762.
Jones, R.O., and D. Hohl, 1990, J. Am. Chem. Soc. **112**, 2590.
Kahn, L.R., and W.A. Goddard III, 1972, J. Chem. Phys. **56**, 2685.
Kalos, M.H., and P.A. Whitlock, 1986, Monte Carlo Methods, Vol. 1 (Wiley, New York).
Kerker, G.P., 1980, J. Phys. C **13**, L189.
Kerker, G.P., A. Zunger, M.L. Cohen and M. Schlüter, 1979, Solid State Commun. **32**, 309.
Kleinman, L., 1980, Phys. Rev. B **21**, 2630.
Kleinman, L., and D.M. Bylander, 1982, Phys. Rev. Lett. **48**, 1425.
Kleinman, L., and J.C. Phillips, 1960, Phys. Rev. **118**, 1153.
Knittle, E., R.M. Wentzcovitch, R. Jeanloz and M.L. Cohen, 1989, Nature **337**, 349.
Kohn, W., and L.J. Sham, 1965, Phys. Rev. A **140**, 1133.
Kunc, K., and R.M. Martin, 1982, Phys. Rev. Lett. **48**, 406.
Li, X.P., P.B. Allen, R. Car, M. Parrinello and J.Q. Broughton, 1990, Phys. Rev. B **41**, 3260.
Liu, A.Y., and M.L. Cohen, 1990, Phys. Rev. B **41**, 10727.
Liu, A.Y., K.J. Chang and M.L. Cohen, 1988, Phys. Rev. B **37**, 6344.
Liu, A.Y., M.L. Cohen, K.C. Hass and M.A. Tamor, 1991, Phys. Rev. B **43**, 6742.
Louie, S.G., K.M. Ho and M.L. Cohen, 1979, Phys. Rev. B **19**, 1774.
Louie, S.G., S. Froyen and M.L. Cohen, 1982, Phys. Rev. B **26**, 1738.
Madelung, O., M. Scholz and H. Weiss, eds, 1982, Semiconductors, Vol. 17, Landolt-Börnstein: Numerical Data and Functional Relationships in Science and Technology, New Series, ed. K.-H. Hellenge (Springer, Berlin).
Martins, J.L., and M.L. Cohen, 1988a, Phys. Rev. B **37**, 3304.
Martins, J.L., and M.L. Cohen, 1988b, Phys. Rev. B **37**, 6134.
McMahan, A.K., M.T. Yin and M.L. Cohen, 1981, Phys. Rev. B **24**, 7210.
McMillan, W.G., 1968, Phys. Rev. **167**, 331.
McSkimin, H.J., and P. Andreatch Jr, 1963, J. Appl. Phys. **34**, 651.
Melius, C.F., and W.A. Goddard III, 1974, Phys. Rev. A **10**, 1528.
Murnaghan, F.D., 1944, Proc. Nat. Acad. Sci. U.S.A. **30**, 244.
Needs, R.J., and R.M. Martin, 1984, Phys. Rev. B **30**, 5390.
Nilsson, G., and G. Nelin, 1971, Phys. Rev. B **3**, 364.
Nilsson, G., and G. Nelin, 1972, Phys. Rev. B **6**, 3777.

Northrup, J.E., and M.L. Cohen, 1982, Phys. Rev. Lett. **49**, 1349.
Northrup, J.E., and M.L. Cohen, 1984, Phys. Rev. B **29**, 1966.
Northrup, J.E., J. Ihm and M.L. Cohen, 1981, Phys. Rev. Lett. **47**, 1910.
Northrup, J.E., M.S. Hybertsen and S.G. Louie, 1991, Phys. Rev. Lett. **66**, 500.
Olijnyk, H., S.K. Sikka and W.B. Holzapfel, 1984, Phys. Lett. A **103**, 137.
Pandey, K.C., 1981, Phys. Rev. Lett. **47**, 1913.
Paul, W., 1982, Band Theory and Transport Properties (North-Holland, Amsterdam).
Perfetti, P., J.M. Nicholls and B. Reihl, 1987, Phys. Rev. B **36**, 6160.
Phillips, J.C., 1973, Bonds and Bands in Semiconductors (Academic Press, New York).
Phillips, J.C., and L. Kleinman, 1959, Phys. Rev. **116**, 287.
Pick, R.M., M.L. Cohen and R.M. Martin, 1970, Phys. Rev. B **1**, 910.
Piermarini, G.J., and S. Block, 1975, Rev. Sci. Instrum. **46**, 973.
Qian, G.X., M. Weinert, G.W. Fernando and J.W. Davenport, 1990, Phys. Rev. Lett. **64**, 1146.
Rappe, A.M., K.M. Rabe, E. Kaxiras and J.D. Joannopoulos, 1990, Phys. Rev. B **43**, 1227.
Sakai, Y., and S. Huzinaga, 1982, J. Chem. Phys. **76**, 2537, 2552.
Schlüter, M., A. Zunger, G.P. Kerker, K.M. Ho and M.L. Cohen, 1979, Phys. Rev. Lett. **42**, 540.
Sham, L.J., 1969, Phys. Rev. **188**, 1431.
Shirley, E.L., D.C. Allan, R.M. Martin and J.D. Joannopoulos, 1989, Phys. Rev. B **40**, 3652.
Simons, G., and A.N. Bloch, 1973, Phys. Rev. B **7**, 2754.
Sinha, S.K., 1973, CRC Crit. Rev. Solid State Sci. **3**, 273.
Slater, J.C., 1951, Phys. Rev. **81**, 385.
St. John, J., and A.N. Bloch, 1974, Phys. Rev. Lett. **33**, 1095.
Starkloff, T., and J.D. Joannopoulos, 1979, Phys. Rev. B **19**, 1077.
Stich, I., R. Car, M. Parrinello and S. Baroni, 1989, Phys. Rev. B **39**, 4997.
Takayanagi, K., Y. Tanishiro, M. Takahashi and S. Takahashi, 1985, J. Vac. Sci. & Technol. A **3**, 1502.
Topiol, S., A. Zunger and M.A. Ratner, 1977, Chem. Phys. Lett. **49**, 367.
Topp, W.C., and J.J. Hopfield, 1973, Phys. Rev. B **7**, 1295.
Troullier, N., and J.L. Martins, 1991, Phys. Rev. B **43**, 1993.
Uhrberg, R.I.G., G.V. Hansson, J.M. Nicholls and S.A. Flodström, 1982, Phys. Rev. Lett. **48**, 1032.
Vanderbilt, D., 1985, Phys. Rev. B **32**, 8412.
Vanderbilt, D., 1990, Phys. Rev. B **41**, 7892.
Vanderbilt, D., S.G. Louie and M.L. Cohen, 1986, Phys. Rev. B **33**, 8740.
Villars, P., 1986, J. Less-Common Metals **119**, 175.
Villars, P., and L.D. Calvert, 1985, Pearson's Handbook of Crystallographic Data for Intermetallic Phases (ASM, Metals Park, OH, USA).
Villars, P., J.C. Phillips and H.S. Chen, 1986, Phys. Rev. Lett. **57**, 3349.
Vohra, Y.K., K.E. Brister, S. Desgreniers, A.L. Ruoff, K.J. Chang and M.L. Cohen, 1986, Phys. Rev. Lett. **56**, 1944.
von Barth, U., and L. Hedin, 1972, J. Phys. C **5**, 1629.
Weinstein, B.A., and G.J. Piermarini, 1975, Phys. Rev. B **12**, 1172.
Wendel, H., and R.M. Martin, 1979, Phys. Rev. B **19**, 5251.
Wentzcovitch, R.M., K.J. Chang and M.L. Cohen, 1986, Phys. Rev. B **34**, 1071.
Wentzcovitch, R.M., M.L. Cohen and P.K. Lam, 1987, Phys. Rev. B **36**, 6058.
Yin, M.T., and M.L. Cohen, 1980, Phys. Rev. Lett. **45**, 1004.
Yin, M.T., and M.L. Cohen, 1981, Phys. Rev. B **24**, 2303.
Yin, M.T., and M.L. Cohen, 1982a, Phys. Rev. B **26**, 5668.
Yin, M.T., and M.L. Cohen, 1982b, Phys. Rev. B **26**, 3259.
Yin, M.T., and M.L. Cohen, 1982c, Phys. Rev. B **25**, 4317.
Yin, M.T., and M.L. Cohen, 1982d, Phys. Rev. B **25**, 7403.
Zhang, S.B., and M.L. Cohen, 1987, Phys. Rev. B **35**, 7604.
Zhang, S.B., and M.L. Cohen, 1989, Phys. Rev. B **39**, 1450.
Zhang, S.B., M.L. Cohen and S.G. Louie, 1985a, Phys. Rev. B **32**, 3955.
Zhang, S.B., M.L. Cohen and J.E. Northrup, 1985b, Surf. Sci. **157**, L303.

Zhang, S.B., M.L. Cohen and S.G. Louie, 1986, Phys. Rev. B **34**, 768.
Zhang, S.B., D. Erskine, M.L. Cohen and P.Y. Yu, 1989, Solid State Commun. **71**, 369.
Zhu, X., and S.G. Louie, 1991, Phys. Rev. B **43**, 14142.
Zunger, A., 1979, J. Vac. Sci. & Technol. **16**, 1337.
Zunger, A., and M.L. Cohen, 1978, Phys. Rev. B **18**, 5449.
Zunger, A., and M.L. Cohen, 1979, Phys. Rev. B **20**, 4082.

CHAPTER 4

Deep and Shallow Impurities in Semiconductors: Theoretical

M. LANNOO*

Laboratoire des Surfaces et Interfaces
(Unité Associée au Centre National de la Recherche Scientifique)
Institut Supérieur d'Electronique du Nord
59046 Lille Cedex, France

*Work partly done at AT&T Bell Laboratories, 600 Mountain Avenue, Murray Hill, NJ 07974-2070, USA.

Handbook on Semiconductors
Completely Revised Edition
Edited by T.S. Moss
Volume 1, edited by P.T. Landsberg

Contents

1. Introduction . . . 115
2. The basic theoretical methods for deep defects . . . 115
   2.1. Local density . . . 115
   2.2. Empirical tight-binding theory . . . 117
   2.3. Defect molecule models . . . 119
3. Shallow states, effective-mass theory (EMT) . . . 120
   3.1. Derivation of EMT for a single band . . . 121
   3.2. Extension to more complex cases . . . 123
4. Central-cell correction and shallow–deep instability . . . 125
   4.1. Central-cell correction . . . 125
   4.2. Effect of the electron–lattice interaction . . . 128
   4.3. The coexistence of deep and shallow states . . . 129
5. Transition-metal and rare-earth impurities . . . 132
   5.1. Transition-metal impurities . . . 132
      5.1.1. Renormalized molecular model of transition-metal impurities . . . 133
      5.1.2. The justification of empirical laws . . . 135
      5.1.3. Optical cross sections . . . 137
   5.2. Rare-earth (RE) impurities . . . 138
6. The DX center and EL2 . . . 144
   6.1. The DX center . . . 144
      6.1.1. The two extreme theoretical descriptions of donors (no atomic relaxation) . . . 144
      6.1.2. The importance of lattice relaxation or distortion . . . 147
      6.1.3. Magnetic resonance on the paramagnetic state . . . 149
   6.2. EL2 . . . 150
      6.2.1. Physical properties of EL2 . . . 151
      6.2.2. Recent developments . . . 152

References . . . 156

1. Introduction

Impurities in semiconductors are very interesting systems since they give rise to a broad variety of behaviors. Historically this began with shallow impurities, donors and acceptors, with binding energies of the electron or hole in the 10 meV range and slowly decaying wave functions. The picture became different with deeper impurities like double donors, for instance, where other theoretical techniques must be used. This difference becomes still more striking for transition-metal impurities and rare-earth impurities where electron–electron interactions are likely to play an important role. Finally even more complex effects have been observed in some famous cases like EL2 and DX centers for which the electron–lattice interaction seems to play a key role. The importance of EL2 centers is related to the fact that they are native defects, which tend to give GaAs a semi-insulating behavior. On the other hand, the DX centers are almost certainly a simple substitutional impurity (e.g., Si_{Ga} in GaAlAs), which has the peculiarity of giving rise to a deep donor level instead of a shallow one (the name DX comes from the previous belief that it was a complex involving a donor D and another defect X).

The aim of the present chapter is to show how these different types of behaviors can be analyzed in a coherent way. We first review the basic theoretical methods in the strongly localized or the shallow limit. We discuss the shallow–deep instability and the effects associated with the coexistence of deep and shallow states. We then consider the particular case of transition-metal and rare-earth impurities where the short-range potential plays the major role. We end up with EL2 and DX, and show that these centers have a mixed (deep + shallow) behavior, which is partly at the origin of their great complexity.

This chapter is not intended to be an extensive review of the literature on this rich subject but tries to give a comprehensive account of the subject as at present understood together with some recent developments.

2. The basic theoretical methods for deep defects

We briefly recall now the essential features of the local-density and empirical tight-binding theories, which are the two most popular methods for treating deep defects. We also give a short discussion of their accuracy.

2.1. Local density

The method is an extension of the Thomas–Fermi approximation based on the Hohenberg and Kohn (1964) theorem showing that the ground-state properties of an electron system are determined by a knowledge of its electron density $\rho(\boldsymbol{r})$. The total energy of the interacting electron system can then be written as

$$E(\rho) = T[\rho] + \frac{1}{2}\int \frac{\rho(\boldsymbol{r})\rho(\boldsymbol{r}')\,\mathrm{d}\boldsymbol{r}\,\mathrm{d}\boldsymbol{r}'}{|\boldsymbol{r}-\boldsymbol{r}'|} + \int V_{\mathrm{ext}}(\boldsymbol{r})\rho(\boldsymbol{r})\,\mathrm{d}\boldsymbol{r} + E_{\mathrm{xc}}[\rho], \tag{1}$$

where T represents the kinetic energy, the second term gives the electrostatic inter-electronic repulsion, V_{ext} is the potential due to the nuclei and E_{xc} is the exchange–correlation energy. A variational solution of the problem (Kohn and Sham 1965) allows one to derive a set of one-particle Schrödinger equations of the form

$$\left(t + V_{\mathrm{ext}}(\boldsymbol{r}) + \int \frac{\rho(\boldsymbol{r}')\,\mathrm{d}\boldsymbol{r}'}{|\boldsymbol{r}-\boldsymbol{r}'|} + V_{\mathrm{xc}}(\boldsymbol{r})\right)\psi_k(\boldsymbol{r}) = \varepsilon_k(\boldsymbol{r})\psi_k(\boldsymbol{r}), \tag{2}$$

with

$$\rho(\boldsymbol{r}) = \sum_{\text{occupied}\,k} |\psi_k(\boldsymbol{r})|^2, \tag{3}$$

and

$$V_{\mathrm{xc}}(\boldsymbol{r}) = \frac{\delta E_{\mathrm{xc}}[\rho(\boldsymbol{r})]}{\delta\rho(\boldsymbol{r})}. \tag{4}$$

Equations (2)–(4) have to be solved in a self-consistent way. The formulation is exact but the problem is that E_{xc} and V_{xc} are not known in general. The local-density approximation (LDA) consists in assuming that, locally, the relation between E_{xc} and $\rho(\boldsymbol{r})$ is the same as that of the free-electron gas of identical density. The intrinsic error associated with the LDA varies from system to system. From a large number of data (Schlüter and Sham 1982, Dahl and Avery 1984) one can say that the LDA describes correctly the energy variations near the minimum configurations, that is the calculated equilibrium geometries and vibrational frequencies are usually excellent (within a few per cent of experiment). On the other hand, the binding energies are usually too large: by 0.2 to 1 or 2 eV for diatomic molecules and 0.2 to 1 eV for Ge, Si and diamond.

It is tempting to use the differences between the eigenvalues of eq. (2) as particle excitation energies. This is not justified in general as shown by the fact that the predicted values of the energy gap are found to be substantially lower than the experimental ones (Hamann 1979) (0.6 eV for Si, about 0 eV for Ge, instead of 1.2 and 0.7 eV, respectively). Recent work has shown that this originates from a discontinuity in V_{xc} across the gap (Sham and Schlüter 1983, Perdew and Levy 1983) and successful improvements have been achieved with the use of the GW approximation (Hybertsen and Louie 1985, Lannoo et al. 1985, Godby et al. 1986) (GW is a schematic notation to denote a first-order expansion of the electron self-energy in terms of the Green's function G and the screened electron–electron interaction W). However, this increases enormously the complexity of the calculation and, for this reason, has not yet been applied to defect problems.

The common practice has thus been to correct this deficiency in an empirical way by shifting the conduction band rigidly to adjust the gap (the "scissor" operator

(Baraff and Schlüter 1981)). It is not easy to estimate error bars but they might be as large as 0.5 eV for migration barriers, about 1 eV for formation energies, and about 0.2, 0.3 eV for the absolute position of deep levels in the gap (Schlüter 1985). Errors might also arise from numerical uncertainties but these can, in principle, be remedied more easily. The practical techniques used for numerical defect calculations are the supercell method (Bar-Yam and Joannopoulos 1984) and the Green's function method (Car et al. 1984, Baraff and Schlüter 1984).

2.2. *Empirical tight-binding theory*

The central assumption here is that the wave function ψ can be expressed as a linear combination of atomic orbitals

$$\psi = \sum_{i,\alpha} c_{i,\alpha}\varphi_{i,\alpha}, \tag{5}$$

where $\varphi_{i,\alpha}$ is the αth free-atom orbital of the ith atom. The whole set of $\varphi_{i,\alpha}$ is complete and even overcomplete. However, for practical reasons one has to truncate this expansion. In tight binding (TB) this is usually restricted to a "minimal basis set", which only includes free-atom states belonging to the valence shell of interest (e.g., one s and three p states for zinc blende semiconductors). It is a description that provides the most appealing physical picture, allowing a clear understanding of the formation of molecular states or bands from the atomic limit.

With the truncated expansion (5) energy levels are obtained from

$$\det|\mathbf{H} - E\mathbf{S}| = 0, \tag{6}$$

where $\mathbf{H}$ is the Hamiltonian matrix and $\mathbf{S}$ the overlap matrix of elements

$$S_{i\alpha,j\beta} = \langle\varphi_{i,\alpha}|\varphi_{j,\beta}\rangle. \tag{7}$$

Strictly speaking the tight-binding approximation is defined as the use of a minimal basis set plus the neglect of interatomic overlaps. This amounts to replacing $\mathbf{S}$ by the unit matrix $\mathbf{I}$, i.e., one has to diagonalize the $\mathbf{H}$ matrix only. In the empirical tight-binding (ETB) method the matrix elements of $\mathbf{H}$ are truncated to first, second or at most third nearest neighbors and they are determined from a fit to the bulk band structure. Systematic fits exist which usually exhibit a d^{-2} dependence with respect to the bond length d for interactions between the s and p atomic states (Harrison 1980, Hjalmarson et al. 1980). In the applications to the defect problem one usually assumes that the interatomic matrix elements remain unmodified while the diagonal terms can be determined in a more or less self-consistent way (Pêcheur et al. 1981).

The advantages of the ETB method lie in its simplicity and in the fact that in most cases it provides qualitatively correct pictures. Difficulties are due to the fact that fitting bulk band structures does not lead to a unique answer. The consequence of this is a dispersion of the TB results for point defects. This is clearly shown in tables 1 and 2 for vacancies in Si and GaAs, respectively. The overall dispersion of the results is of the order of 0.6 eV. Such a large value is partly due to the fact that some of the

Table 1
Position of the A_1 and T_2 silicon vacancy levels (in eV) with respect to the top of the valence band.

A_1	T_2	Method	Reference
	0.27	ETB	Bernholc and Pantelides (1978)
−0.88	0.12	ETB	Kauffer et al. (1976, 1977)
	0.25	ETB	Talwar and Ting (1982)
−0.4	0.75	ETB 3 centers	Papaconstantopoulos and Economou (1980)
−1.03	0.77	ETB modified	Pêcheur et al. (1981)
	0.5	LDA	Louie et al. (1976)
−1.1	0.76	LDA	Bernholc et al. (1978, 1980)
	0.7	LDA	Baraff and Schlüter (1979)

Table 2
Position of the A_1 and T_2 Ga and As vacancy levels (in eV) with respect to the top of the valence band (OT denotes techniques other than ETB or LDA).

A_1, V_{Ga}	T_2, V_{Ga}	A_1, V_{As}	T_2, V_{As}	Method	Reference
−0.6	0.03	0.71	1.47	ETB	Bernholc and Pantelides (1978)
−0.4	0.4	0.05	1.1	ETB	Daw and Smith (1979)
		1.3	1.51	ETB	Das Sarma and Madhukar (1981)
	0.05		0.95	ETB	Talwar and Ting (1982)
−0.92	−0.11	0.32	0.95	ETB	Van der Rest and Pêcheur (1983)
VB	<0.1	0.6	1.45	ETB S*	Buisson et al. (1982)
−1.06	0.03	−0.81	1.40	ETB modified	Van der Rest and Pêcheur (1983)
−1	0.06	−0.8	1.08	LDA	Bachelet et al. (1981)
−0.05	0.15		1.75	OT	Jaros and Brand (1976)
0.2	0.3	0.12	1.25	OT	Li'in and Masterov (1976)
	0.55	0.86	1.33	OT	Fazzio et al. (1979)
	0.06		CB	OT	Srivasta (1980)

models include refinements beyond ETB or contain some inconsistencies. One can also note that the splitting of the A_1 and T_2 vacancy levels is consistently lower in ETB than in LDA. Strong criticisms have been raised against ETB (Krieger and Laufer 1981, Singh et al. 1982) especially concerning its prediction of chemical trends of sp impurities in semiconductors and the conclusion that there is a pinning energy at the vacancy level (Hjalmarson et al. 1980). However, it was shown by Lannoo (1984) that such criticisms are not well founded and that pinning should exist but with some possible deviations of the order of 0.1 or 0.2 eV. A striking illustration of this is provided by a self-consistent ETB calculation (Delerue et al. 1989a) of 3d transition-metal impurities in Si, which provides results of the same quality as the corresponding LDA calculation (Zunger and Lindefelt 1983). A justification of the different steps leading to TB is presented by Lannoo (1984) showing some possibilities of improvement but at the expense of simplicity.

2.3. *Defect molecule models*

In this section we show that defect molecular models can be used not only as qualitative tools to describe the physics in a simple way but also as quantitative tools. As will be demonstrated later they are also ideal for justifying empirical rules.

The defect molecular approach was used a long time ago for vacancy systems (Coulson and Kearsley 1957). In the absence of the Green's function techniques introduced later (Lannoo and Lenglart 1968) in this context, the only available method was to perform a cluster calculation (the point defect being at its center) with the techniques of quantum chemistry. A full calculation of this kind was performed for the vacancy in diamond (Coulson and Kearsley 1957) using the defect molecule of fig. 1, including the four sp^3 hybrids pointing towards the vacancy. This calculation even included multiplet splitting effects and was extended to the determination of Jahn–Teller distortions (Friedel et al. 1967). Later such cluster calculations have been applied to larger clusters either in the extended Hückel approximation (Messmer and Watkins 1971) or in the self-consistent $X\alpha$ scattered-wave approximation (Watkins and Messmer 1974).

In addition to their use as more or less quantitative tools, the defect molecule models play a much more important role as qualitative guides for building a coherent picture from the experimental data. Perhaps the most famous example of this is provided by the vacancy in silicon where the defect molecule of fig. 1 was used to obtain a complete description of the EPR data (Watkins 1963, 1964). The power of the model is that it provides information about the level structure (the four sp^3 orbitals of fig. 1 couple symmetrically to give an A_1 and a T_2 symmetry state, the last one being in the gap). It also describes formally the different possibilities of

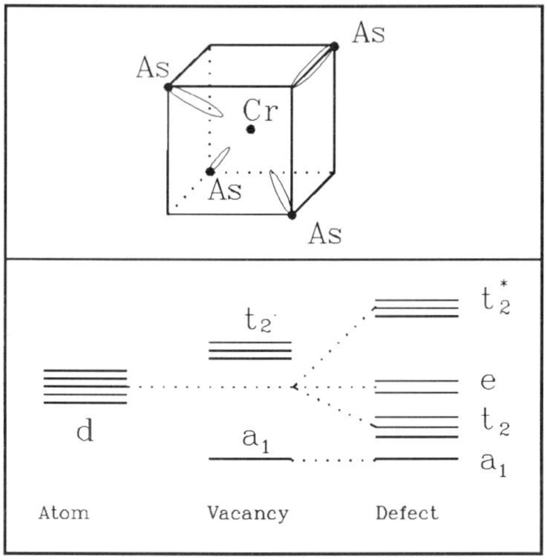

Fig. 1. Defect molecule showing the four dangling bonds of a vacancy (V_{Ga} in GaAs in the present case) and a substitutional impurity (taken to be Cr). The lower part corresponds to the level structure of a transition-metal impurity in the defect molecule model described in § 5.1.1.

distortion when the charge state (i.e., level occupation) is varied. It can even be used to understand the sign and magnitude of the shift in the **g** tensor and hyperfine structure of the defect (Watkins 1963, 1964), both quantities being obtained from electron paramagnetic resonance measurements (the **g** tensor characterizes the Zeeman splitting).

The power of defect molecule models seems quite general. The properties of deep levels due to s–p substitutional impurities are easily understood. One simply replaces the vacancy in fig. 1 by an impurity atom. The coupling between the atomic and vacancy states gives rise to two A_1 and two T_2 states (bonding and antibonding combination). For a donor-like impurity (i.e., an attractive potential) this set of levels is lowered with respect to the bulk situation. In such a case the A_1 antibonding level can be lowered in the gap. This is found for the As_{Ga} antisite in GaAs for instance, where the model also explains the magnitude of the observed hyperfine interaction (Mauger et al. 1987a,b). Such a picture can even be extended to transition-metal impurities, as will be discussed later.

3. Shallow states, effective-mass theory (EMT)

Deep defects or impurities represent one extreme case with a rapidly decaying wave function that only experiences the short-range part of the potential. At the other extreme one finds the shallow states encountered in the cases of single donors and acceptors. Here the wave function still decays exponentially, but quite slowly so that it extends over a large number of unit cells.

The physics can be understood fairly easily on the basis of a simple hydrogenic model. Indeed, considering for instance the case of single donors like P in Si, four of the five electrons of the phosphorus atoms saturate the P–Si bonds. The situation of the extra electron can be viewed as if it was in the field of the extranuclear charge $+e$. However, this occurs in a medium of dielectric constant ε, the origin of the energies being the bottom of the conduction band E_c where electrons are free with an effective mass m_n. This yields one hydrogenic equation with e^2 replaced by e^2/ε and m by m_n. One finds hydrogenic states of energies

$$E_n = E_c - \frac{m_n e^4}{2\hbar^2 \varepsilon^2} \frac{1}{n^2} \tag{8}$$

the ground-state wave function being

$$\begin{aligned} \psi &= A \exp(-r/a^*) \\ a^* &= \hbar^2 \varepsilon / m_n e^2. \end{aligned} \tag{9}$$

Here n is the principal quantum number and a^* the effective Bohr radius. This simple model is well known and has had much success. We now proceed to its justification and extensions.

3.1. Derivation of EMT for a single band

Let us then begin with the simplest case of a crystal whose electronic structure can be described in terms of a single energy band, the solution of the perfect-crystal Schrödinger equation:

$$H_0\varphi_{\boldsymbol{k}}(\boldsymbol{r}) = \varepsilon(\boldsymbol{k})\varphi_{\boldsymbol{k}}(\boldsymbol{r}). \tag{10}$$

If a perturbative potential $V(\boldsymbol{r})$ is applied to the system, we can describe a solution of the perturbed system $\psi(\boldsymbol{r})$ as a linear combination of the perfect-crystal eigenstates ($\boldsymbol{k}$ belongs to the first Brillouin zone)

$$\psi(\boldsymbol{r}) = \sum_{\boldsymbol{k}} a(\boldsymbol{k})\varphi_{\boldsymbol{k}}(\boldsymbol{r}) \tag{11}$$

and obtain the unknown coefficients by projecting the new Schrödinger equation

$$(H_0 + V)\psi(\boldsymbol{r}) = E\psi(\boldsymbol{r}) \tag{12}$$

onto the basis states $\varphi_{\boldsymbol{k}}(\boldsymbol{r})$. This immediately leads to the set of linear equations

$$\varepsilon(\boldsymbol{k})a(\boldsymbol{k}) + \sum_{\boldsymbol{k}'} \langle\varphi_{\boldsymbol{k}}|\nabla|\varphi_{\boldsymbol{k}'}\rangle a(\boldsymbol{k}') = Ea(\boldsymbol{k}). \tag{13}$$

At this stage we need to simplify the matrix elements of V, otherwise it is impossible to go further, except numerically. We use the fact that the $\varphi_{\boldsymbol{k}}(\boldsymbol{r})$ are Bloch functions

$$\varphi_{\boldsymbol{k}}(\boldsymbol{r}) = \mathrm{e}^{\mathrm{i}\boldsymbol{k}\cdot\boldsymbol{r}}u_{\boldsymbol{k}}(\boldsymbol{r}) \tag{14}$$

and express the matrix element of the potential as

$$\langle\varphi_{\boldsymbol{k}}|V|\varphi_{\boldsymbol{k}'}\rangle = \int V(\boldsymbol{r})\mathrm{e}^{\mathrm{i}(\boldsymbol{k}'-\boldsymbol{k})\cdot\boldsymbol{r}}u_{\boldsymbol{k}}^*(\boldsymbol{r})u_{\boldsymbol{k}'}(\boldsymbol{r})\,\mathrm{d}^3\boldsymbol{r}. \tag{15}$$

In view of the Bloch theorem the product $u_{\boldsymbol{k}}^*u_{\boldsymbol{k}'}$ is a periodic function of $\boldsymbol{r}$ and we can expand it in a Fourier series

$$u_{\boldsymbol{k}}^*(\boldsymbol{r})u_{\boldsymbol{k}'}(\boldsymbol{r}) = \sum_{\boldsymbol{G}} C_{\boldsymbol{k}\boldsymbol{k}'}(\boldsymbol{G})\mathrm{e}^{\mathrm{i}\boldsymbol{G}\cdot\boldsymbol{r}}, \tag{16}$$

where $\boldsymbol{G}$ are the reciprocal lattice vectors. The matrix elements (15) can thus be expressed exactly as

$$\langle\varphi_{\boldsymbol{k}}|V|\varphi_{\boldsymbol{k}'}\rangle = \sum_{\boldsymbol{G}} C_{\boldsymbol{k}\boldsymbol{k}'}(\boldsymbol{G})V(\boldsymbol{k}'+\boldsymbol{G}-\boldsymbol{k}), \tag{17}$$

where $V(\boldsymbol{k})$ is the Fourier transform of $V(\boldsymbol{r})$. At this level we must make some assumptions about $V(\boldsymbol{r})$. The first one is that it varies slowly in space (i.e., over distances large compared to the size of the unit cell). This means that its Fourier transform decreases very quickly with the modulus of the wave vector, i.e., that one can neglect terms with $\boldsymbol{G}\neq 0$ in (17) and also that only terms with $\boldsymbol{k}'\sim\boldsymbol{k}$ will contribute effectively.

We now make the second central assumption of EMT, i.e., we look for solutions whose energy E is close to a band extremum $\boldsymbol{k}_0$. If this is so, only states with $\boldsymbol{k}\simeq\boldsymbol{k}_0$

will have a value of $\boldsymbol{k}$ sensibly different from zero in eq. (13). This means that one can rewrite (13), using (17), in the approximate form:

$$\varepsilon(\boldsymbol{k})a(\boldsymbol{k}) + \sum_{\boldsymbol{k}'} C_{\boldsymbol{k}_0,\boldsymbol{k}_0}(0)V(\boldsymbol{k}' - \boldsymbol{k})a(\boldsymbol{k}') = 0. \tag{18}$$

However, $C_{\boldsymbol{k}_0,\boldsymbol{k}_0}(0)$ has the important property that it is given by the following integral over the crystal volume:

$$\begin{aligned} C_{\boldsymbol{k}_0,\boldsymbol{k}_0}(0) &= \int u^*_{\boldsymbol{k}_0}(\boldsymbol{r})u_{\boldsymbol{k}_0}(\boldsymbol{r})\,\mathrm{d}v \\ &= \int \varphi^*_{\boldsymbol{k}_0}(\boldsymbol{r})\varphi_{\boldsymbol{k}_0}(\boldsymbol{r})\,\mathrm{d}v = 1, \end{aligned} \tag{19}$$

if we have chosen to normalize the $\varphi_{\boldsymbol{k}}(\boldsymbol{r})$ over the unit-cell volume. The final form of the EMT equation is thus

$$\varepsilon(\boldsymbol{k})a(\boldsymbol{k}) + \sum_{\boldsymbol{k}'} V(\boldsymbol{k}' - \boldsymbol{k})a(\boldsymbol{k}') = Ea(\boldsymbol{k}), \tag{20}$$

where the sum over $\boldsymbol{k}'$ extends over the first Brillouin zone. As the contributions to (20) essentially come from $\boldsymbol{k} \sim \boldsymbol{k}' \sim \boldsymbol{k}_0$ one can extend this sum to the whole space with only small error. It is then interesting to derive from this a real-space equation by Fourier transforming this equation. To perform this we must take into account the fact that the function $a(\boldsymbol{k})$ is strongly peaked near $\boldsymbol{k}_0$. We thus introduce the following Fourier transform, which is named the envelope function and is defined as

$$F(\boldsymbol{r}) = \sum_{\boldsymbol{k}} a(\boldsymbol{k}) \exp[\mathrm{i}(\boldsymbol{k} - \boldsymbol{k}_0)\cdot\boldsymbol{r}] \tag{21}$$

such that $F(\boldsymbol{r})$ varies slowly in space when $a(\boldsymbol{k})$ only takes important values in the vicinity of $\boldsymbol{k} = \boldsymbol{k}_0$. To obtain an equation for $F(\boldsymbol{r})$ we multiply eq. (20) by $\exp[\mathrm{i}(\boldsymbol{k} - \boldsymbol{k}_0)\cdot\boldsymbol{r}]$ and sum over $\boldsymbol{k}$, assuming that one makes a negligible error in the potential term by extending the summation over $\boldsymbol{k}$ to the whole space. This leads to the real-space equation

$$[\varepsilon(\boldsymbol{k}_0 - \mathrm{i}\nabla_{\boldsymbol{r}}) + V(\boldsymbol{r})]F(\boldsymbol{r}) = EF(\boldsymbol{r}). \tag{22}$$

This is a differential equation in which the operator $\boldsymbol{k}_0 - \mathrm{i}\nabla_{\boldsymbol{r}}$ has been substituted for $\boldsymbol{k}$ in the dispersion relation $\varepsilon(\boldsymbol{k})$. As we have seen the function $F(\boldsymbol{r})$ is likely to vary slowly with $\boldsymbol{r}$ (or $a(\boldsymbol{k}) \neq 0$ only for $\boldsymbol{k} \sim \boldsymbol{k}_0$) so that one can expand $\varepsilon(\boldsymbol{k})$ to second order in the neighborhood of $\boldsymbol{k} \sim \boldsymbol{k}_0$. Calling $\boldsymbol{\alpha}$ the principal axes of this expansion we have

$$\varepsilon(\boldsymbol{k}) \sim \varepsilon(\boldsymbol{k}_0) + \sum_{\alpha} \frac{\hbar^2}{2m_\alpha}(k_\alpha - k_{0\alpha})^2, \tag{23}$$

which defines the effective masses m_α along the direction α. This enables one to rewrite eq. (22) as

$$\left(-\sum_\alpha \frac{\hbar^2}{2m_\alpha}\frac{\partial^2}{\partial x_\alpha^2} + V(\mathbf{r})\right)F(\mathbf{r}) = [E - \varepsilon(\mathbf{k}_0)]F(\mathbf{r}), \tag{24}$$

which represents the usual form of the EMT equation as derived by many authors (Kohn 1957, Luttinger and Kohn 1955, Landsberg 1969, Bassani et al. 1974, Pantelides 1978). Its application to the simple donor for which $V(r) = -e^2/\varepsilon r$ in the isotropic case (where all $m_\alpha = m_n$) readily leads to eqs. (8) and (9).

It is interesting to examine the meaning of the function $F(\mathbf{r})$. For this we start from the expansion (11) of $\psi(\mathbf{r})$, express the $\varphi_{\mathbf{k}}(\mathbf{r})$ as in (14) and factorize $\exp(i\mathbf{k}_0 \cdot \mathbf{r})$. This gives

$$\psi(\mathbf{r}) = \exp(i\mathbf{k}_0 \cdot \mathbf{r}) \sum_{\mathbf{k}} a(\mathbf{k}) \exp[i(\mathbf{k} - \mathbf{k}_0)\cdot\mathbf{r}]u_{\mathbf{k}}(\mathbf{r}). \tag{25}$$

As $a(\mathbf{k})$ is peaked near $\mathbf{k}_0$ we approximate $u_{\mathbf{k}}(\mathbf{r})$ by its value at $\mathbf{k}_0$, which leads us directly to

$$\psi(\mathbf{r}) \simeq F(\mathbf{r})\varphi_{\mathbf{k}_0}(\mathbf{r}). \tag{26}$$

This means that $F(\mathbf{r})$ can be rewritten as the product of the Bloch function (which varies over a length typically of the order of the interatomic distances) times a slowly varying "envelope function". The advantage of EMT is that one obtains $F(\mathbf{r})$ directly from a Schrödinger-like equation involving the effective masses.

All the assumptions that have been used in the derivation of EMT (i.e. slowly varying potential, extension of the sum over $\mathbf{k}'$ in eq. (20) over all space, expansions of $\varepsilon(\mathbf{k})$ to second order, neglect of contributions from higher bands) can be shown (Kohn 1957, Luttinger and Kohn 1955, Landsberg 1969, Bassani et al. 1974, Pantelides 1978) to correspond to the condition $a^* \gg a$ where a is the size of the crystal unit cell. In other words the wave function must be spread over a large number of unit cells. EMT will thus work better for excited states.

3.2. Extension to more complex cases

Three other situations can be encountered in zincblende semiconductors:

(i) the conduction band can have several equivalent minima,
(ii) the valence band has a degenerate maximum,
(iii) there can be resonant states.

We discuss briefly how EMT can be extended to these cases.

Let us first consider the case of several equivalent minima for the conduction band (e.g., Si with six minima along the (100) direction, Ge with four at the L points of the Brillouin zone). The idea is to write the donor electron wave function under the form

$$\psi(\mathbf{r}) = \sum_j \alpha_j \psi_j(\mathbf{r}) \tag{27}$$

$\psi_j(\mathbf{r})$ being a wave function like (11) derived from the jth minimum in $\mathbf{k}$ space. One must then write the Hamiltonian matrix in the basis of the $\psi_j(\boldsymbol{r})$. The non-diagonal terms (intervalley coupling) are likely to be small since the $\psi_j(\boldsymbol{r})$ correspond to distant minima in the Brillouin zone. In the spirit of perturbation theory we first neglect this intervalley coupling. Each minimum can thus be treated independently leading to the simple EMT eq. (24) in which, due to the axial symmetry, one usually has a longitudinal (m_ℓ) and a transverse (m_t) effective mass. Such an equation has cylindrical symmetry and its solutions have been tabulated by (Faulkner 1969). An easy order of magnitude estimate can be obtained by using the average mass

$$\frac{1}{m_n} = \frac{1}{3}\left(\frac{1}{m_\ell} + \frac{2}{m_t}\right). \tag{28}$$

Since they are equivalent the g minima, when treated independently, give rise to states with degeneracy g. Introduction of the intervalley coupling will split this degeneracy. Symmetry enables one to classify the resulting states and their degeneracy. One obtains:

- Si (degeneracy 6): $A_1(1\times)$, $E(2\times)$, $T_2(3\times)$,
- Ge (degeneracy 4): $A_1(1\times)$, $T_2(3\times)$,
- some III–V at the X point: $T_2(3\times)$ or $A_1(1\times)$ and $E(2\times)$.

The case of a degenerate extremum like the top of the valence band is also quite complex and requires special attention. The wave function of the acceptor state $\psi(\boldsymbol{r})$ must now be written as a combination of the Bloch states $\varphi_{n\boldsymbol{k}}(\boldsymbol{r})$ belonging to the different energy branches $\varepsilon_n(\boldsymbol{k})$. One then proceeds to the same simplifications as in the conventional EMT of § 3.1. As the terms $c_{n\boldsymbol{k},n'\boldsymbol{k}'}(\boldsymbol{G}=0)$ are approximated by $c_{n\boldsymbol{o},n'\boldsymbol{o}}(0) = \delta_{nn'}$, each band can be treated independently and one gets

$$\varepsilon_n(\boldsymbol{k})a_n(\boldsymbol{k}) + \sum_{\boldsymbol{k}'} V(\boldsymbol{k}'-\boldsymbol{k})a_n(\boldsymbol{k}') = Ea_n(\boldsymbol{k}) \tag{29}$$

identical to eq. (20) for each band. The problem, however, is that it is not possible to expand the $\varepsilon_n(\boldsymbol{k})$ to second order in $\boldsymbol{k}$ to obtain hydrogenic-like equations. In fact, as shown by Kane (1956), the quantities which can be expanded to second order in $\boldsymbol{k}$ are the elements $h_{ij}(\boldsymbol{k})$ of a 3×3 matrix (6×6 when spin–orbit coupling is included (Kane 1956)) whose eigenvalues are the $\varepsilon_n(\boldsymbol{k})$, known as the $\boldsymbol{k}\cdot\boldsymbol{p}$ matrix. Transforming back eq. (29) into the basis defining $h_{ij}(\boldsymbol{k})$ one obtains

$$\sum_j h_{ij}(\boldsymbol{k})a_j(\boldsymbol{k}) + \sum_{\boldsymbol{k}'} V(\boldsymbol{k}'-\boldsymbol{k})a_i(\boldsymbol{k}) = Ea_i(\boldsymbol{k}), \tag{30}$$

which can now be transformed to real space, giving

$$\sum_j h_{ij}(-\mathrm{i}\nabla_r)F_j(\boldsymbol{r}) + V(\boldsymbol{r})F_i(\boldsymbol{r}) = EF_i(\boldsymbol{r}). \tag{31}$$

If the h_{ij} are expanded to second order one gets a set of coupled differential equations for the $F_i(\boldsymbol{r})$.

A final case of some interest corresponds to resonant hydrogenic states, derived from secondary extrema (e.g., X and L for GaAs) and falling in the continuum of other band states (e.g., Γ in GaAs). If the coupling is weak the resonance can be sharp. Its width is given by the Fermi golden rule

$$\Delta E = \pi \overline{|V_{\ell k}|^2} n(E_l), \tag{32}$$

where E_ℓ denotes the energy of the hydrogenic state before coupling, $V_{\ell k}$ denotes the coupling with band states k and $n(E_\ell)$ is the density of continuum states at energy E_ℓ. One can obtain a simple upper bound for ΔE by replacing the functions by their envelope parts $e^{i\boldsymbol{k}\cdot\boldsymbol{r}}$ and $(1/\sqrt{\pi a^{*3}})\exp(-r/a^*)$ (this ignores the overlap of the periodic parts of the Bloch functions). Noting that the Fourier transform of $(e^2/\varepsilon r)\exp(-r/a^*)$ is equal to

$$\frac{e^2}{\varepsilon}\left(\frac{1}{q^2+\dfrac{1}{a^{*2}}}\right)$$

and that $n(E_\ell)$ is $(1/4\pi^2)k_\ell^3/E_\ell$ one obtains directly that

$$\Delta E = 4\frac{e^4}{\varepsilon^2 a^{*3}}\frac{1}{|q^2+1/a^{*2}|^2}\frac{k_\ell^3}{E_\ell}, \tag{33}$$

where q is the distance, in reciprocal space, between the secondary and absolute minima. Noting that $e^2/2\varepsilon a^*$ is the hydrogenic binding energy E_{H} and ignoring $1/a^* \ll 1/q$ we get, as an upper bound

$$\Delta E = \frac{16\,k_\ell^3\,E_{\mathrm{H}}^2}{a^*\,q^4\,E_\ell}. \tag{34}$$

A reasonable estimate can be made from the following values $q \sim 1/a$, $k_\ell \sim 1/5a$, $a^*/a \sim 40$, $E_{\mathrm{H}} \sim 13.6$ meV, $E_\ell \sim 200$ meV. This leads to $\Delta E \lessapprox 10^{-5}$ to 10^{-6} eV, showing that these resonances are likely to be fairly sharp.

4. Central-cell correction and shallow–deep instability

Even for shallow impurities the ground-state properties are not accurately predicted by EMT. This is due to the so-called central-cell correction which will first be discussed. When its magnitude is large it can lead to a shallow–deep instability, which can be enhanced by the electron–phonon interaction, as will be shown using Toyozawa's model. Finally if the ground state becomes really deep it can no longer be treated by EMT and we show in the last part of this section how deep states and shallow states can coexist.

4.1. Central-cell correction

Obviously the conditions of applicability of EMT are not always satisfied. In particular the condition $a^* \gg a$ is more difficult to realize for the 1s hydrogenic ground state than for the excited states. Clear evidence for this is provided by donors in silicon for which the experimental ground-state binding energies reported in table 3 differ considerably (by more than a factor of two) from the theoretical prediction of 24 meV given in table 3. This difference, called the "chemical shift", depends on the nature of the impurity and is thus due, at least partly, to the "central-cell correction" to the impurity potential. This correction contains the local differences between the true perturbative potential and its long-range Coulomb form $-e^2/\varepsilon r$ and is expected to become very small outside the impurity cell. Another effect is due to the fact that one has a situation of equivalent extrema discussed previously which, as shown later, is equivalent to a strengthening of the central-cell correction. Finally, in many cases, the double donors, for instance, the central-cell correction can be so strong that one gets a deep impurity state in the gap. In such cases EMT is no longer valid and, as we shall see for the case of Se in Ge, one can get peculiar effects.

A very good illustration of the existence of the chemical shift is provided by single donors in GaAs. These can be located either at the cation (Si, Ge, Sn) or at the anion (S, Se, Te) site. The conduction band lies at the Γ point and the effective mass is isotropic and very small ($m^* = 0.067$) so that this represents a case where EMT should be at its best. Indeed the Bohr radius given by eq. (9) with $\varepsilon = 12.5$ turns out to be of order 100 Å, much larger than the unit-cell radius. With these values the 1s binding energy is predicted to be 5.7 meV. The experimental values, whose range is given in table 3, are detailed by Skromme et al. (1985) and are close to the theoretical one but exhibit a significant chemical shift which, surprisingly, has an extremum for Ge. As discussed by Delerue and Lannoo (1992a) the corresponding central-cell correction originates partly from an intra-atomic impurity contribution but also partly from a significant size effect.

Another well documented case corresponds to single donors in silicon. We have seen in table 3 that the chemical shift is much more substantial but the problem is complicated now by the existence of $g = 6$ equivalent extrema, lying along the (100) axes. As shown in detail by Luttinger and Kohn (1955) and Kohn (1957), the ground state has A_1 symmetry, which means that if one uses an expansion like (27) for the

Table 3

Calculated and experimental values of the binding energy (meV) of s states in various semiconductors. In the case of diamond, the impurity considered is of the acceptor type (boron) and because there is a large uncertainty in the value of m^* observed for holes, we have arbitrarily taken an average value. For the other elements we have considered substitutional donor impurities (taking for m^* the effective mass of the electrons). The chemical shift corresponds to the difference between the experimental and theoretical values of the binding energy, which is significant mainly for E_{1s}.

Material	m_l/m	m_t/m	m^*/m	ε	Calculated		Experimental	
					E_{1s}	E_{2s}	E_{1s}	E_{2s}
Diamond	–	–	0.7	5.7	290	73	370	–
Silicon	0.98	0.19	0.26	12	24	6	43.1 –53.7	–
Germanium	1.60	0.08	0.12	16	6.6	1.6	10.3 –14.2	–
Gallium arsenide	0.066	0.066	0.066	12.5	5.74	1.43	5.89–6.08	1.44
Indium phosphide	0.080	0.080	0.080	12.6	6.85	1.71	7.28	–

wave function, the α_i are automatically known to be equal by symmetry. Thus the ground-state wave function takes the form

$$\psi_{A_1}(\boldsymbol{r}) = \frac{1}{\sqrt{6}} \sum_{j=1}^{6} \psi_j(\boldsymbol{r}), \tag{35}$$

where the ψ_j are the wave functions for the independent minima. However, as discussed in the basic paper by Kohn (1957), the use of the $\psi_j(\boldsymbol{r})$ given by simple EMT does not give a proper value of the spin density at the impurity nucleus, obtained from the hyperfine constants in electron spin resonance (ESR). The inclusion of central cell effects however can at the same time improve the ground state binding energy and the spin density (Kohn 1957, Luttinger and Kohn 1955).

In many cases the central cell correction is so strong that the EMT ground state can become deep, its wave function being thus localized in the vicinity of the impurity site. The proper theoretical description of this shallow–deep transition lies of course beyond the range of applicability of EMT. Nevertheless, some attempts have been made (Reiss 1956, Chaudhuri and Coon 1984) to give a qualitative description of this effect within the framework of EMT. In these approaches the potential is taken to have its long-range Coulomb form $-e^2/\varepsilon r$ outside a sphere of radius R of the order of the atomic radius. Within the sphere it is taken to be closer to an atomic or unscreened potential. In such work EMT is used in both regions with the bulk effective mass for $r > R$ and $m^* = 1$ for $r < R$. One then has to solve different Schrödinger equations in the two regions and match them. The parameter that can be varied is R and the result for the binding energy exhibits a somewhat sharp transition from the hydrogenic value to the deep value. This method, however, provides at best an illustration and a more quantitative theory should incorporate a more correct description of deep states based on the methods discussed before. We shall come back to such a theory in § 4.3.

To conclude this section let us investigate within EMT the relationship between the strength of the central-cell correction and the existence of several equivalent minima. This has been pointed out in (Resca and Resta 1979, 1980) and to derive their results we use the wave function (27) as a trial function in a variational treatment and calculate the corresponding expectation value of the total Hamiltonian

$$E(\psi) = \sum_{j,j'} \alpha_j^* \alpha_{j'} \langle \psi_j | H_0 + V | \psi_{j'} \rangle. \tag{36}$$

In this expression the contribution due to H_0 is diagonal and each term $\langle \psi_j | H_0 | \psi_j \rangle$ can be determined by the arguments of EMT and reduced to $\langle F_j | T_j | F_j \rangle$ where T_j is the kinetic energy term of eq. (24). For the potential matrix elements one can replace the $\psi_j(\boldsymbol{r})$ by their approximate form (26) so that $E(\psi)$ can be written as

$$E(\psi) = \sum_j |\alpha_j|^2 \langle F_j | T_j | F_j \rangle + \sum_{jj'} \alpha_j^* \alpha_{j'} \langle F_j | \varphi_{kj}^* V \varphi_{kj'} | F_{j'} \rangle. \tag{37}$$

It is now the envelope function F_j that plays the role of the trial function. If, as did Resca and Resta (1979, 1980), we make the approximation that all the $F_j(\boldsymbol{r})$ are independent of j and take the common form $F(\boldsymbol{r})$, then $E(\psi)$ simplifies to

$$E(\psi) = \left\langle F \,|\, T + \left|\sum_j \alpha_j \varphi_{\boldsymbol{k}j}\right|^2 V \,|\, F \right\rangle \tag{38}$$

In this expression $T = \Sigma_j |\alpha_j|^2 T_j$ corresponds to the isotropic kinetic energy operator with the average effective mass defined by (28). The interesting point with (38) is that it takes the conventional simple EMT form for a single extremum but with a renormalized perturbative potential given by the second term. The correction factor has been calculated in Resca and Resta (1979, 1980) for the A_1 state in Si (with $\alpha_j = 1/\sqrt{6}$). It takes a form such that the strength of the potential is increased by a factor of six within a radius r_0; outside r_0 one recovers the normal $V(\boldsymbol{r})$. This corresponds to an increase in the central-cell correction whose origin comes from interference effects between the Bloch waves in (38).

4.2. *Effect of the electron–lattice interaction*

We look here at the possible influence of the electron–lattice interaction on the shallow–deep instability. For this we make use of a model derived by Toyozawa (1983), which is a continuum description, in the EMT spirit. Let us call $\psi(\boldsymbol{r})$ the envelope function of the donor electron and discuss the 1s or A_1 symmetry case for which we take a trial function

$$\psi(\boldsymbol{r}) = \frac{1}{\sqrt{\pi a^3}} \exp\left(-\frac{r}{a}\right), \tag{39}$$

where a is the variational parameter (in the original paper ψ was taken as a Gaussian but this does not affect the qualitative conclusions). It is well known that the expectation values of the kinetic and long-range Coulomb energies for such a state are, respectively, given by $\hbar^2/2m^*a^2$ and $e^2/\varepsilon a$. Minimization of the sum of these with respect to a would thus produce directly the hydrogenic energy of the 1s term.

To go beyond this one needs to incorporate two more terms:

(i) the central-cell effect consisting of a local perturbative potential having roughly the atomic size,
(ii) an electron lattice coupling term.

The expectation value of the first term is easy to derive, at least in the limit where a is much larger than the atomic size, since it will be equal to $|\psi(\boldsymbol{r}=0)|^2$ times the average short-range potential. It will thus behave as $-V_0(a_0/a)^3$ where a_0 is of the order of the lattice constant. Finally the coupling to lattice modes can be analyzed simply by considering that the true electron wave function is

$$\psi = \sum_i \psi(\boldsymbol{R}_i) W_i \tag{40}$$

in the Wannier representation (Lannoo and Bourgoin 1981) (this is strictly valid for a $\boldsymbol{k}=0$ minimum in a one-band case but can be extended to other situations). Each Wannier function here is an antibonding state whose energy is sensitive to a change in bond length $\Delta(\boldsymbol{R}_i)$. The contribution of this to the expectation value is thus given by

$$C \sum_i |\psi(\boldsymbol{R}_i)|^2 \Delta(\boldsymbol{R}_i), \tag{41}$$

where C is a constant. To this one must add an elastic-energy term

$$\frac{1}{2} k_r \sum_i \Delta^2(\boldsymbol{R}_i)$$

and minimize with respect to each $\Delta(\boldsymbol{R}_i)$, leading to a relaxation energy

$$-\frac{C^2}{2k_r} \sum_i [\psi(\boldsymbol{R}_i)]^4. \tag{42}$$

In the continuum limit one can replace the sum by an integral, which again leads to a term that scales like $1/a^3$. Adding all contributions to the total energy one thus obtains

$$E = \frac{\hbar^2}{2m^* a^2} - \frac{e^2}{\varepsilon a} - K\left(\frac{a_0}{a}\right)^3. \tag{43}$$

Keeping in mind the condition that a should not become smaller than the lattice constant a_0 one can define a parameter $\lambda = a_0/a$, which should vary between zero and one and express E in terms of λ

$$E = E_0(\lambda^2 - g_\ell \lambda - g_s \lambda^3). \tag{44}$$

This expression is cast into the same form as in Toyozawa's (1983) paper, g_ℓ and g_s describing the strength of the long- and short-range effects. The interesting conclusions are the following.

(i) If $3g_s g_\ell > 1$ E is a continuously decreasing function of λ, which takes its lowest value for $\lambda = 1$ describing a deep state.
(ii) If $3g_s g_\ell < 1$ two cases can occur.
  (a) $3g_s + g_\ell > 2$ and $3g_s > 1$: E first decreases, reaches a minimum, increases up to a maximum and then decreases again until $\lambda = 1$, which thus represents another minimum. This situation shows metastable behavior.
  (b) If conditions (a) are not satisfied E has only one minimum in the interval.

These conclusions are interesting in the sense that they show that even EMT theory can predict metastable behavior by lattice relaxation. This is particularly interesting for the DX center, which we discuss in the last section and which is generally thought to involve a substantial electron–lattice interaction leading to metastable behavior.

4.3. The coexistence of deep and shallow states

The theoretical understanding of donors in semiconductors is still an open problem as evidenced by the cases of double donors (like EL2) and even single donors (like

DX). As we have seen, in most known cases the hydrogenic picture seems to be of quantitative value for the excited states. However, it is likely to fail for the ground state as soon as the so-called central-cell correction becomes important, leading to a shallow–deep transition. Among others, one basic question is to what extent the existence of a deep ground state modifies the nature of the excited shallow states. Up to now, for practical reasons, no first-principles calculation has been able to deal with both types of states simultaneously. We try here to perform a general analysis of this problem and illustrate its conclusions by a discussion of P and Se in Ge.

As introduced before, the dual nature of deep donors is usually treated from two extreme points of view.

(i) The localized description. This one is valid for deep states, which have a compact wave function. It is based on the use of a short-range potential U_{SR}, truncated in real space. This allows us to make use of Green's functions or supercell techniques, either in the local density (LDA) approximation or in the tight-binding approximation. As discussed in § 2.3 the net result of such treatments for donors is to provide two characteristic states, a non-degenerate one A_1(ab) and a triply degenerate one T_2(ab). Both are essentially built from antibonding combinations of the sp^3 orbitals belonging to the impurity and its nearest neighbors (Lannoo and Bourgoin 1981). For double donors like the As antisite in GaAs A_1(ab) is deep in the gap while T_2(ab) is resonant in the conduction band (Bachelet et al. 1983). For single donors the A_1(ab)–T_2(ab) system is shifted upwards, A_1(ab) falling in the neighborhood of the bottom of the conduction band (Yamaguchi et al. 1990, Foulon et al. 1990).

(ii) The effective-mass approximation. This applies in cases of a smooth long-range potential U_{LR} usually taken to be equal to $-Ze^2/\varepsilon r$. Each conduction-band minimum, when treated independently, gives rise to a series of hydrogenic states. The case of N equivalent minima thus leads to hydrogenic states with N-fold degeneracy, which is lifted by intervalley coupling, mainly due to the central-cell correction. For the L conduction band, one has $N = 4$, and the splitting of each ns hydrogenic state gives rise to a non-degenerate $A_1(ns)$ and a triply degenerate $T_2(ns)$ state. The normal ordering of levels is $A_1(ns) < T_2(ns)$ in view of the attractive nature of the potential.

Let us now discuss what should be the characteristic features of an "exact" calculation of the donor states. The full potential seen by the donor electron reduces to U_{SR} in the vicinity of the defect and to U_{LR} in the outer region. This has the following consequences: (i) any deep state (i.e., having a compact wave function) only experiences U_{SR} and is correctly predicted by a localized description ignoring U_{LR}. For the deep donors considered here this only occurs for the A_1(ab) state and not for the T_2(ab) state that is resonant within the conduction band. (ii) Once the deep A_1(ab) state is determined the treatment of the shallow states should then proceed by first orthogonalizing the hydrogenic trial wave functions to the deep-state wave function. This is equivalent to replacing the true potential by a pseudopotential (see Cohen et al. (1970) for a review in bulk crystals) whose short-range attractive part U_{SR} is suppressed leading now to a small repulsive central-cell correction. Such a correction concerns only the *ns* hydrogenic states, which are the only ones to have a non-vanishing amplitude at the impurity site and thus to experience the central-cell correction. The net result is that all hydrogenic *ns* states still exist and are given by conventional effective-mass theory. However, for equivalent minima the fact that the pseudopotential now leads to a small repulsive central-cell correction should lead to an inverted splitting. For the L band this corresponds to $A_1(ns) > T_2(ns)$. The same result can be viewed in another way: when the central-cell potential becomes strongly attractive $A_1(1s)$ becomes deep and, by mixing with other bands, becomes $A_1(ab)$; $A_1(2s)$ is lowered from its original 2s hydrogenic values, tends towards $A_1(1s)$ by getting mostly 1s character, and so on .... Of course the final result is that $A_1(1s)$ lies between the 2s and 1s EMT values as deduced from the orthogonalization argument.

These conclusions are fully confirmed by the detailed experimental results obtained for donors in Ge (Grimmeiss et al. 1988) for which the lowest conduction band corresponds to the four equivalent L minima. Figure 2 reproduces the results of Grimmeiss et al. (1988) with slight changes in notation. The part on the left gives the results of effective-mass theory when the L minima are treated independently (they are obtained from the calculation by Faulkner (1969) since the L minima of Ge are strongly anisotropic with $m_\ell = 1.58 m_t = 0.08$). They are in quantitative agreement with experiment for the single donor P with a small $A_1(1s)$–$T_2(1s)$ splitting of about 3 meV, these two levels being in normal order. The situation becomes

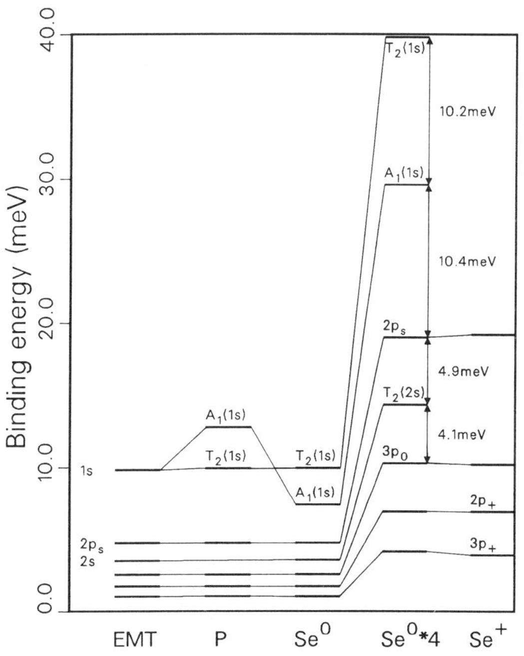

Fig. 2. The binding energies of shallow impurity states in Ge showing the results from EMT calculations and the experimental results for P, Se^0 and Se^+.

different for the double donor Se for which the A_1 state becomes deep, in which case it cannot be described correctly by EMT since it will have contributions from the different bands. The lowest two excited states of Se^0 at 9.95 and 7.4 meV were noted, respectively, as $T_2(1s)$ and $A_1(2s)$ in Grimmeiss et al. (1988). However, they are close to the EMT value of 9.81 meV and, according to our discussion, can be better viewed as $T_2(1s)$ and $A_1(1s)$ in inverted order, their splitting of 2.5 meV being small. For Se^+ the A_1 ground state is still deeper but the observed excited states also correspond to EMT at the condition of scaling the binding energies by the appropriate factor $Z^2 = 4$.

We show in the last section that this type of analysis applies to EL2, which is related to the double-donor antisite As_{Ga}.

5. Transition-metal and rare-earth impurities

5.1. Transition metal (TM) impurities

Transition-metal (TM) impurities create deep levels in most of the covalent and ionic semiconductors. With experiments like deep-level transient spectroscopy (DLTS) and electron paramagnetic resonances (EPR), several aspects of the behavior of 3d impurities are now well known. First, although free TM atoms show ionization energies larger than 10 eV, several charge states can be localized in the band gap of the semiconductor, indicating a strong interaction between the defect and the medium. However, the multiplet spectra of 3d impurities are quite similar to those of the free atoms and therefore have been described by the well-known Tanabe–Sugano diagrams (Sugano et al. 1970). We can add to this apparent contradictory behavior that the total spin of the EPR-detected defects seems to obey Hund's rules, like the free atom, but ENDOR data show an important spin delocalization.

The main features of these impurities are now better understood theoretically. Calculations of the electronic structure using cluster or Green's function approaches and based on the local-density or tight-binding approximations (see, for instance, Zunger and Lindefelt 1983, Hemstreet 1977, Hemstreet and Dimmock 1979, Deleo et al. 1981, Fazzio and Leite 1980, Pêcheur and Toussaint 1983, Vogl and Baranowski 1985, Picoli et al. 1984, Katayama-Yoshida and Zunger 1985, Beeler et al. 1985, Zunger 1986) are able to give a clear idea of the electronic configuration of these defects but the corresponding levels are obtained with an accuracy not better than 0.3 eV (Lannoo 1984). The effects left out like multiplet splitting (Zunger 1986) lead to a great difficulty in the interpretation of experimental data. In spite of this progress many properties of the TM impurities in III–V or in II–VI semiconductors remain unsatisfactorily explained, as is pointed out in reviews like (Clerjaud 1985).

The aim of the present section is to discuss the trends in several physical properties on the basis of an extremely simple and pedagogical molecular model, introduced by Picoli et al. (1984), which proves as successful as for the vacancy in silicon. It provides a good representation of the results of a self-consistent tight-binding Green's function calculation. These results are comparable to those of the LD calculations performed for Si and directly confirm the validity of the molecular description.

5.1.1. Renormalized molecular model of transition-metal impurities

We briefly recall here the molecular model of TM substitutional impurities (Picoli et al. 1984) in which the defect states result from the interaction of the transition-metal atom "d" states with the four dangling bonds of the Ga vacancy (see fig. 1). The "d" states, using symmetry considerations, are divided into two classes: e-like orbitals, which remain uncoupled leading to very localized states in a more complete scheme; t_2-like orbitals, which couple with the corresponding t_2 states of the vacancy creating t_2 bonding and t_2^* antibonding levels whose energies are given by the diagonalization of the following Hamiltonian matrix:

$$\begin{pmatrix} E_d & V \\ V & E_v \end{pmatrix} \tag{45}$$

where E_d and E_v are, respectively, the d and dangling-bond state energies and V is their coupling. The eigenvalues ε_{t_2} and $\varepsilon_{t_2^*}$ are

$$\varepsilon_{t_2}^{t_2^*} = \bar{\varepsilon} \pm (\delta^2 + V^2)^{1/2} \tag{46}$$

with $\bar{\varepsilon} = (E_d + E_v)/2$ and $\delta = (E_d - E_v)/2$.

The eigenstates are

$$|t_2\rangle = \alpha|t_{2d}) + \beta|t_{2v}), \tag{47a}$$

$$|t_2^*\rangle = \beta|t_{2d}) - \alpha|t_{2v}), \tag{47b}$$

$$\alpha^2 = \frac{1}{2}\left(1 - \frac{\delta}{(\delta^2 + V^2)^{1/2}}\right), \tag{48a}$$

with

$$\beta^2 = \frac{1}{2}\left(1 + \frac{\delta}{(\delta^2 + V^2)^{1/2}}\right). \tag{48b}$$

Writing $n_d = n_e + 6\alpha^2 + n_A\beta^2$ $n_v = 6\beta^2 + n_A\alpha^2$ for an electronic configuration $t_2^6 e^{n_e} t_2^{*n_A}$, the self-consistency can be simply achieved by imposing a linear dependence of the diagonal terms of the Hamiltonian with respect to the electronic population:

$$E_d = E_{d0} + U_d(n_d - n_{d0}), \tag{49a}$$

$$E_v = E_{v0} + U_v(n_v - n_{v0}), \tag{49b}$$

where n_d and n_v are, respectively, the electronic populations on the d orbitals and on dangling-bond hybrids, U_d is the average Coulomb energy for d electrons, U_v is the effective Coulomb energy for the t_2 state of the vacancy (U_v is usually very small compared to U_d (Picoli et al. 1984)). E_{d0} and E_{v0} are, respectively, the d-orbital energy for the TM ion with n_{d0} d electrons and the energy of the t_2 state of the vacancy occupied by n_{v0} electrons (see Delerue et al. (1989a) for details). This set of equations leads to the following one (if $\delta_0 = (E_{d0} - E_{v0})/2$):

$$\delta = \frac{\delta_0 + (U_d/2)(n_d - n_{d0} + 3) + (U_d - U_v)n_A/4 - U_v(3 - n_{v0})}{1 + \dfrac{(U_d + U_v)(6 - n_A)}{4(\delta^2 + V^2)^{1/2}}}, \tag{50}$$

which can be solved by iteration.

We shall take a value of 8 eV for the Coulomb parameter U_d (Picoli et al. 1984), but, as long as it is much larger than V, the exact value of this parameter is found to be unimportant for the accuracy of the results. The coupling parameter V is, in the simple defect molecule model, deduced from the Harrison's empirical rules (Harrison 1980).

This molecular approach might seem sufficient to explain the major physical features of the defect but inadequate to give quantitative values because of the effects left out of account. The main problem is the small size of the cluster used. However, Picoli et al. (1984) derived a renormalized defect molecule model that includes a modification of the previous parameters to take into account the rest of the crystal. This leads to a reduction of the coupling parameter V by the delocalization factor γ:

$$V = \gamma V_0, \tag{51}$$

where V_0 is the value calculated directly from Harrison's rule (Harrison 1980) and the reduction factor γ is due to the fact that the t_2 level of the vacancy is not totally localized on the four dangling bonds. Thus the molecular model takes into account most of the effects of the medium on the impurity molecule. In these conditions the defect molecule model becomes more quantitative as demonstrated in fig. 3, which

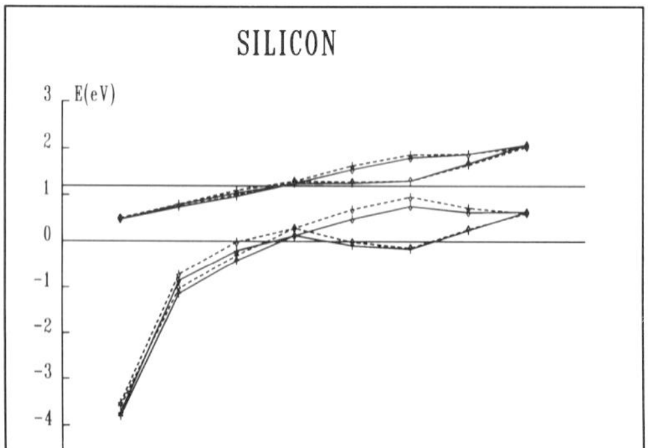

Fig. 3. Comparison between the results of the full Green's function calculation (full lines) and the molecular model (dashed lines).

compares its predictions to those of a full tight-binding Green's functions calculation (Delerue et al. 1989a).

5.1.2. The justification of empirical laws

One of the most interesting features of defect molecule models is that they can be used to derive or justify empirical laws. This will be illustrated by the case of transition-metal impurities for which there exist several empirical laws.

5.1.2.1. Alignment of TM levels. As shown in fig. 4 it is possible to align the experimental TM energy levels (at least for the 3d series) in different semiconductors by a single shift of their band structures (Langer and Heinrich 1984, Delerue et al. 1988, Langer et al. 1988). This property holds with surprising accuracy, of the order of 0.1 eV. A justification of this finding can be directly provided by the simplest version of the defect molecule model, as shown by Tersoff and Harrison (1987). For this one considers the self-consistency condition (50) and takes the limit of $U_d \to \infty$ in this expression. One then obtains an expression in which δ (which measures the difference in energy between the d state and the vacancy state) is fixed only by V and the filling of the d band, i.e., is practically independent of δ_0, its non-self-consistent value. This means that the impurity levels are pinned on the dangling-bond state at a distance that only depends on the nature of the impurity (with a weak dependence on the semiconductor due to the parameter V). This provides the basic explanation for the alignment of TM levels, an explanation of which has been confirmed by recent Green's functions calculations (Delerue et al. 1989a).

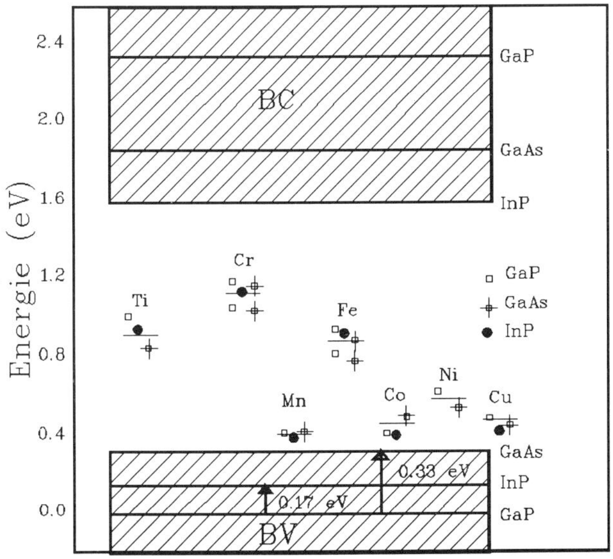

Fig. 4. Alignment of TM levels for different III–V semiconductors.

This alignment has led some authors to speculate that the corresponding shift in the band structures provides a natural value for the band offsets at heterojunctions (Langer and Heinrich 1984). This empirical approach works surprisingly well and can also be justified theoretically (Delerue et al. 1988, Langer et al. 1988).

5.1.2.2. Two other empirical laws. Recently, two new empirical laws for the crystal-field splitting Δ and the ionization energy E_{I} of transition-metal ions in semiconductors have been justified theoretically (Liro et al. 1987). The first one concerns the product $E_{\mathrm{I}}\Delta$, which turns out to be proportional to d^{-7} where d is the interatomic distance, at least for a given impurity in different semiconductors. Figure 5 shows that it can be justified by the Green's function results of Delerue et al. (1989a) but can also be understood directly on the basis of the molecular model. We have seen before that, in the molecular model, each t_2-like d orbital can be treated separately. Then the wave function ψ can be written as

$$\psi = a_{\mathrm{d}}\varphi_{\mathrm{d}} + \sum_{\alpha} a_{\alpha}\varphi_{\alpha}, \tag{52}$$

where φ_{d} is one of the d orbitals and the φ_{α} belong to the remaining crystal. It is

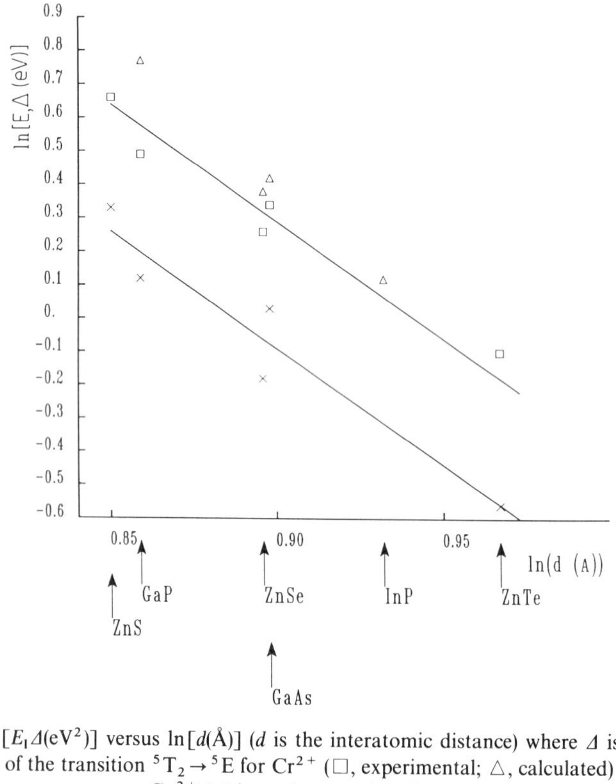

Fig. 5. Plot of $\ln[E_{\mathrm{I}}\Delta(\mathrm{eV}^2)]$ versus $\ln[d(\text{Å})]$ (d is the interatomic distance) where Δ is the energy of the zero-phonon line of the transition $^5T_2 \rightarrow {}^5E$ for Cr^{2+} (□, experimental; △, calculated) and $^4A_2 \rightarrow {}^4T_2$ for Co^{2+} (×) in various semiconductors.

easy to show that the eigenstates of the Hamiltonian follow the equation:

$$E = E_{\mathrm{d}} + \sum_{\alpha} \frac{|V_{\mathrm{d}\alpha}|^2}{E - E_{\alpha}}, \tag{53}$$

where E_α is the energy of the state φ_α and $V_{\mathrm{d}\alpha}$ is the coupling term between φ_{d} and φ_α. As the neighbors of the TM ion are the anions and the t_2-like combinations of the anion states mainly belong to the top of the valence band E_{v}, we can approximately replace E_α by E_{v} (this corresponds to another manner of building a molecular model), which leads to

$$(E - E_{\mathrm{d}})(E - E_{\mathrm{v}}) = \sum_{\alpha} |V_{\mathrm{d}\alpha}|^2. \tag{54}$$

As the crystal-field splitting is equal to $\Delta = E - E_{\mathrm{d}}$ and the ionization energy from the valence band to $E_{\mathrm{I}} = E - E_{\mathrm{v}}$, we obtain finally

$$E_{\mathrm{I}}\Delta = \sum_{\alpha} |V_{\mathrm{d}\alpha}|^2. \tag{55}$$

Appling Harrison's empirical rules (Harrison 1980), in which $V_{\mathrm{d}\alpha}$ varies as $d^{-7/2}$, $E_{\mathrm{I}}\Delta$ will vary as d^{-7}. Therefore, Harrison's laws receive a direct experimental confirmation.

The second empirical law described by Liro et al. (1987) shows that the crystal-field splitting varies linearly with the bulk Phillips' ionicity (Phillips 1970). Figure 6 makes evident the validity of this law for $Cr^{2+}(^5T_2 \to {}^5E)$ and $Co^{2+}(^4A_2 \to {}^4T_2)$ internal transitions. In the same figure the corresponding calculated crystal-field splittings show the same linear dependence (with a constant shift that is due to multielectron effects). A simple analytical formula can be obtained for this law again on the basis of the molecular model in the approximation of a strong screening, that is where the TM ion is assumed to be neutral. In this extreme case, the crystal-field splitting is fixed entirely by the bulk ionicity with a quasi-linear dependence. This empirical law still confirms the importance of screening effects. We again see that the molecular model gives the tendencies and provides the correct physical interpretation.

5.1.3. Optical cross sections

A possible approach to the calculation of the optical cross sections previously applied to the case of the vacancy in silicon (Petit et al. 1986) is based on a tight-binding Green's function formalism for the electronic contribution (i.e., without an electron–lattice interaction). This formalism has also been applied to the case of the TM substitutional impurities in semiconductors. A complete study of the optical cross sections of the 3d series in InP has been performed. Detailed results are given by Delerue et al. (1989b). This approach allows us to calculate directly the absolute value of the optical cross section versus frequency which can be compared with the experimental one. We shall see that the comparison between theory and experiment can definitely provide support to a model for the electronic configuration of a defect.

The shape of the experimental optical cross sections contains information about the transitions from the defect levels (fundamental or excited) to the bands or about

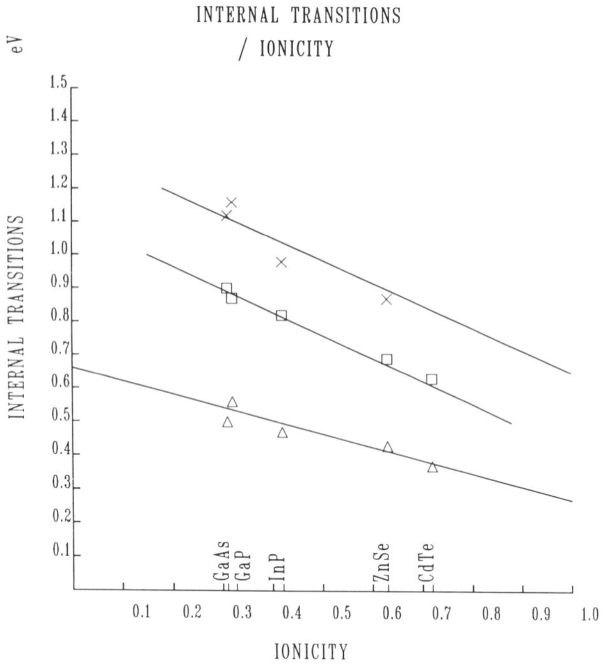

Fig. 6. Relationship between the effective crystal-field splitting and the ionicity for $Cr^{2+}(^5T_2 \rightarrow {}^5E)$ (□, experimental; ×, calculated) and for $Co^{2+}(^5A_2 \rightarrow {}^4T_2)$ (△ observed).

the presence of internal transitions. The analysis of the absolute values can be related to the nature of the wave functions of the defect (Martinez 1986). The direct comparisons in fig. 7 between theoretical and experimental spectra is also a test for the validity of a given model. Let us, for instance, in fig. 7 discuss the optical cross section spectra for Ti(0/+) in InP. The σ_n spectra (transition of an electron from the defect to the conduction band) shows an important peak centered around 0.62 eV. This is not observed on the theoretical spectra for a transition to the conduction band. This peak is in fact correlated with an internal transition (here $^2E \rightarrow {}^2T_2$) in the final state: the addition of this internal transition leads to a very good agreement between theory and experiment. The optical threshold at about 1.3 eV is associated with the transition to the bottom of a conduction band (L_1). The weak absolute value for $\sigma_n(10^{-17}\ cm^2)$ is evidence of the involvement of the e level as predicted by the calculations (Delerue et al. 1989b, Martinez 1986). Such an analysis has been pursued for the entire 3d series (Delerue et al. 1989b) allowing us to show, for instance, that the observed optical cross section for Co(−/0) is not consistent with an electron on a "e" orbital involved in the transition.

5.2. *Rare-earth (RE) impurities*

Rare-earth (RE) impurities in semiconductors have a fairly compact 4f open shell, which retains its atomic characteristics in the crystal. In semiconductors this can

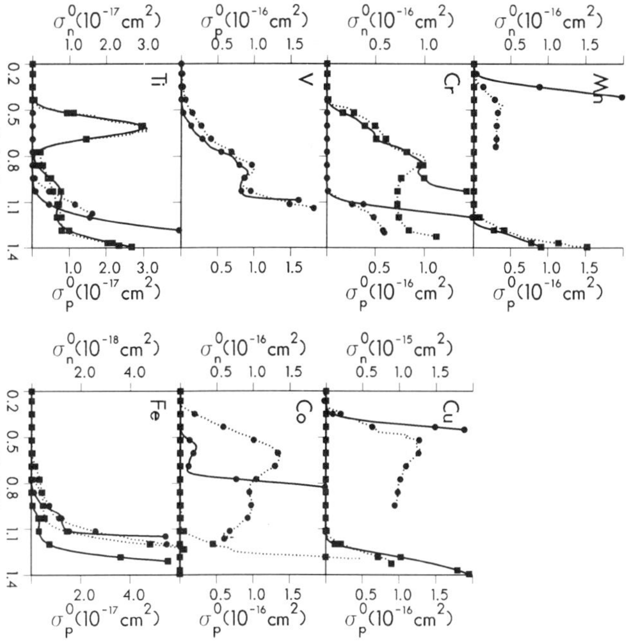

Fig. 7. Absolute photoionization cross-section spectra σ_n^0 and σ_p^0 for transition-metal impurities in InP (the theoretical spectra are shown by the solid lines, the experimental ones by dotted lines, (●) σ_p^0 and (■) σ_n^0).

result in useful magnetic or optical properties, leading for instance to the possibility of application as light-emitting devices (Ennen and Schneider 1984). Most studies have concentrated on the study of RE impurities in II–VI materials (Boyn 1988), but there exists some information about covalent (Tang et al. 1989, Ennen et al. 1987), III–V (Ennen and Schneider 1984, Ennen et al. 1987, Lambert et al. 1988, 1992, Pomrenke et al. 1986, Wagner et al. 1986, Kasatkin et al. 1981, Ennen et al. 1985, Aszodi et al. 1985, Kröber and Hangleiter 1988, Klein 1988) and II–VI semiconductors (Boyn 1988, Title 1964, 1967, Godlewski and Hommel 1986, Przybylinska et al. 1987). RE elements are mostly found in a 3+ oxidation state except in II–VI compounds where the 2+ state is also sometimes observed. Although this fact is consistent with a "core-like" behavior of the 4f shell it requires a consistent detailed explanation. We shall thus present here a theoretical description that can account in a simple physical manner for the trends of RE impurities in this family of materials. It is based on a self-consistent Green's function treatment in a tight-binding framework. A major advantage of this model is that is relates the properties of RE impurities not only to those of the free atoms but also to the situation in RE metals (Herbst et al. 1972, 1976, 1978).

Most of the free RE atoms along the lanthanide series, from Ce to Yb have a $4f^n 6s^2$ configuration, except for Ce and Gd, which have, respectively, $4f5d6s^2$ and $4f^7 5d6s^2$ configurations (Martin et al. 1974). The 4f shell is very contracted on the atom and this contraction increases with the atomic number. This is so effective that already for Nd($4f^4 6s^2$) the 4f function has its maximum inside the $5s^2 5p^6$ closed shells of the xenon structure (Wybourne 1965). Contrary to the 4f orbitals, the 6s orbitals are fairly extended (the 6s maximum for Gd lies at 1.8 Å from the nucleus compared with 0.3 Å for the 4f orbitals, which can thus be treated as core orbitals).

This difference in the size of the orbitals forms the basis for the understanding of the properties of RE metals (Herbst et al. 1972, 1976, 1978) and mixed-valence compounds (Falicov 1981). In the metals the 6s states give rise to a broad band so that the atomic configuration in the solid corresponds to an almost half-filled s band, i.e., is close to $6s^1$. It is thus reasonable to consider that the electronic structure in the metal essentially derives from the following three atomic configurations: $4f^{n+1}5d^1 6s^1$, $4f^n 5d^2 6s^1$ and $4f^{n-1}5d^3 6s^1$, corresponding, respectively, to two, three and four valence electrons per atom. Such an atomic-like point of view has been developed successfully for the RE metals by Herbst et al. (1972, 1976, 1978). In such cases there is experimental evidence that the lifetime of the 4f states (which are resonant) is of the order of 10^{-12} s, indicating a very small broadening ($\sim$1 meV) consistent with the fact that the 4f states can be considered as core states.

RE impurities in semiconductors can be discussed in exactly the same spirit, allowing a connection to be made with the behavior in metals. Again the RE impurity is described from the atomic configurations $4f^{n+3-m}5d^{m-1}6s^1$, where m is the number of valence electrons (we include, as before, the possibilities $m = 2, 3, 4$), the total number of electrons being $n + 3$ when the atom is neutral. With such a notation the oxidation state of the RE atom in a mixed-valence compound or as an impurity in a semiconductor is simply equal to $+m$ since, in the conventional definition of the oxidation state, all valence electrons are assumed to be captured by the other more electronegative atoms.

The calculation is based on the frozen 4f shell approximation, that is we consider the 4f states as core states that do not interact with the atomic s and p states of their neighbors in the crystal. The 4f electron population (i.e., $n + 3 - m$) is thus fixed at an integer value. This implies that a full calculation of the electronic structure including the 5d, 6s shells should be performed for each value $m = 2, 3, 4$ of interest, to select the value leading to the lowest energy. Such a calculation can be simplified by considering that the 6s state is so extended that its local density of states will be spread almost equally between the valence and conduction bands, corresponding to a $6s^1$ configuration as discussed above for metals. The problem is then reduced to the coupling of the 5d states with the neighboring s and p states of the crystal atoms. We here only consider the case of substitutional RE atoms on the cation site.

From the above discussion, the problem, in the frozen 4f shell approximation, becomes identical to the one of transition-metal impurities discussed above. We thus apply exactly the same procedure here, assuming that the energy E_d of the d state in the solid depends linearly on the excess d population, as given by eq. (49). Similar to Delerue et al. (1989a) we take E_{d0} to be the free-atom value calculated with the Hartree–Fock–Slater method of Herman and Skillman (1963) for the $5d^{m-1}6s^1$ configuration. We take $U_{dd} \simeq 8$ eV as in Delerue et al. (1989a) but the results are insensitive to this when it is large enough. We also have to determine the tight-binding matrix elements coupling the impurity d states with their neighbors. To get an idea of these we choose to scale the matrix elements used by Delerue et al. (1989a) for the 3d impurities by use of the Wolfsberg–Helmholtz formula (Wolfsberg and Helmholtz 1952) in which the intensity of the coupling is taken to be proportional to the average atomic energy and to the overlap integral, which we calculate. As the 5d orbitals are more extended we find that the strength of the coupling is increased by a factor of about 2.5. To perform the calculation we finally have to position the d level, E_d, with respect to the semiconductor band structure. Indeed, this level is calculated in the empirical tight-binding approximation described by Delerue (1989a) where all the parameters are determined except the absolute position of the average band structure. To determine this last quantity we simply deduce the average diagonal term of the bulk Hamiltonian (i.e., the average sp^3 energy) from the same free-atom Herman and Skillman (1963) procedure as for E_{d0}. Clearly this has the advantage of allowing for some cancellation of errors, which is probably the reason why this procedure has been so successful for transition-metal impurities.

All the parameters being specified we then determine the electronic states from a Green's function procedure, as for TM impurities. Again, the characteristic states are, in order of increasing energies: a triply degenerate t_2 bonding state, a doubly degenerate e state and finally a triply degenerate t_2^* antibonding state. However, contrary to TM impurities we obtain no gap states: the t_2 states fall within the valence band while e and t_2^* are located in the conduction band. The reasons for this are twofold, E_d being higher and the coupling strength being much larger than for transition-metal impurities. In all cases the electronic structure thus corresponds to the configuration $t_2^6 e^0 t_2^{*0}$. This means that the charge state of the impurity is completely defined by the number $n+3-m$ of f electrons of the rare-earth impurity and by the valence N of the cation element of the crystal. One easily finds that the net impurity charge q, in units of e, is given by

$$q = m - N. \tag{56}$$

This represents one simple and quite general condition for determining the stable charge state of rare-earth impurities. The full determination of q thus requires the knowledge of m, that is of the number $n+3-m$ electrons belonging to the frozen 4f shell in the stable state. One then needs to calculate the change in the total impurity energy when m is varied by one unit. The important quantity to define is

$$\varepsilon_{4f}(m-\tfrac{1}{2}) = E(m-1) - E(m), \tag{57}$$

where $E(m)$ is the total energy of the impurity with m valence electrons. With this definition $\varepsilon_{4f}(m-\frac{1}{2})$ is the 4f level in the solid whose relative position with respect to the Fermi level determines the number of 4f electrons or, in other words, the number of impurity valence electrons m. To achieve such a calculation we first write that ε_{4f} depends linearly upon the change δn_d of the number of 5d electrons from the neutral free atom to the solid

$$\varepsilon_{4f} = \varepsilon_{4f0} + U_{fd}\delta n_d. \tag{58}$$

We determine U_{fd} from the free-atom calculations (Herman and Skillman 1963) by a procedure described by Delerue and Lannoo (1992b). $\varepsilon_{4\mathrm{f}0}$ is the 4f level in the neutral free atom, which we obtain from the following expression:

$$\varepsilon_{4\mathrm{f}0}(m-\tfrac{1}{2})=\varepsilon_{5\mathrm{d}0}(m-\tfrac{1}{2})+\Delta E_{\mathrm{df}}(m,m-1). \qquad (59)$$

Here $\varepsilon_{5\mathrm{d}0}$ is the neutral free-atom 5d level which, to be coherent with what has been done before, is obtained via the Herman–Skillman technique in the intermediate state $4\mathrm{f}^{n+3.5-m}5\mathrm{d}^{m-1.5}6\mathrm{s}^1$. As discussed before this procedure should correctly position $\varepsilon_{5\mathrm{d}0}$ with respect to the band structure. On the other hand, ΔE_{df} is the energy needed, in the neutral free atom, to promote one electron from the d shell to the f shell. It is very sensitive to correlation effects and, for this reason, we have deduced it from the detailed work by Herbst et al. (1972) using either spin-unrestricted Hartree–Fock theory or experimental values to estimate the amount of correlation energy.

Figure 8 represents our results for $\varepsilon_{4\mathrm{f}}(2.5)$ and $\varepsilon_{4\mathrm{f}}(3.5)$ for the whole series of rare-earth elements at the substitutional site in Si, InP and CdTe (at the cation site in the last two cases). We see that a general behavior emerges corresponding to $\varepsilon_{4\mathrm{f}}(3.5)$ in the valence band and $\varepsilon_{4\mathrm{f}}(2.5)$ in the conduction band. This means that these impurities should all have $m=3$ valence electrons, corresponding to the so-called 3+ oxidation state. However, this result for the upper level $\varepsilon_{4\mathrm{f}}(2.5)$ is sensitive to the inclusion of correlation effects since this level can cross the forbidden gap near the center and the end of the series. This effect increases with the ionicity, i.e., when going from Si to InP and CdTe.

These general trends are confirmed by experimental observations.

(a) In Si, Er is observed in the 3+ state but it seems that most of the impurities are on interstitial sites (Tang et al. 1989, Ennen et al. 1987).

(b) In III–V materials one observes Er^{3+} in InP (Lambert et al. 1992, Pomrenke et al. 1986), GaAs (Ennen and Schneider 1984, Ennen et al. 1987, Pomrenke et al. 1986) and GaP (Ennen and Schneider 1984), Nd^{3+} in GaAs (Ennen and Schneider 1984) and GaP (Ennen and Schneider 1984, Wagner et al. 1986), Tm^{3+} in GaAs (Ennen and Schneider 1984), and Pr^{3+} in GaP (Kasatkin et al. 1981). An interesting case is Yb, which seems to be trivalent in GaAs and GaP (Ennen et al. 1985) as well as in n- or p-type InP (Lambert et al. 1988, Ennen et al. 1985, Aszodi et al. 1985, Kröber and Hangleiter 1988, Klein 1988). Our results only agree with this conclusion when correlation effects are included (dashed curve of fig. 8).

(c) In II–VI semiconductors there is much information for CdTe (Boyn 1988, Title 1967): Nd, Gd, Er and Yb are found in the 3+ state, Eu always in the 2+ state even in other compounds and Tm also in the 2+ state but the atomic site is uncertain. These trends are coherent with our results including correlation (i.e. with $\varepsilon_{4\mathrm{f}}(2.5)$ given by the dotted line on fig. 8) except for Yb which we predict to be Yb^{2+}. One should also note that, for ZnS, there is some evidence (Godlewski and Hommel 1986, Przybylinska et al. 1987) that the $\varepsilon_{4\mathrm{f}}(2.5)$ level lies in the forbidden gap for Eu and Yb, in correspondence with the general trend exhibited by fig. 8, where this level is found to fall within the gap at the middle and the end of the series.

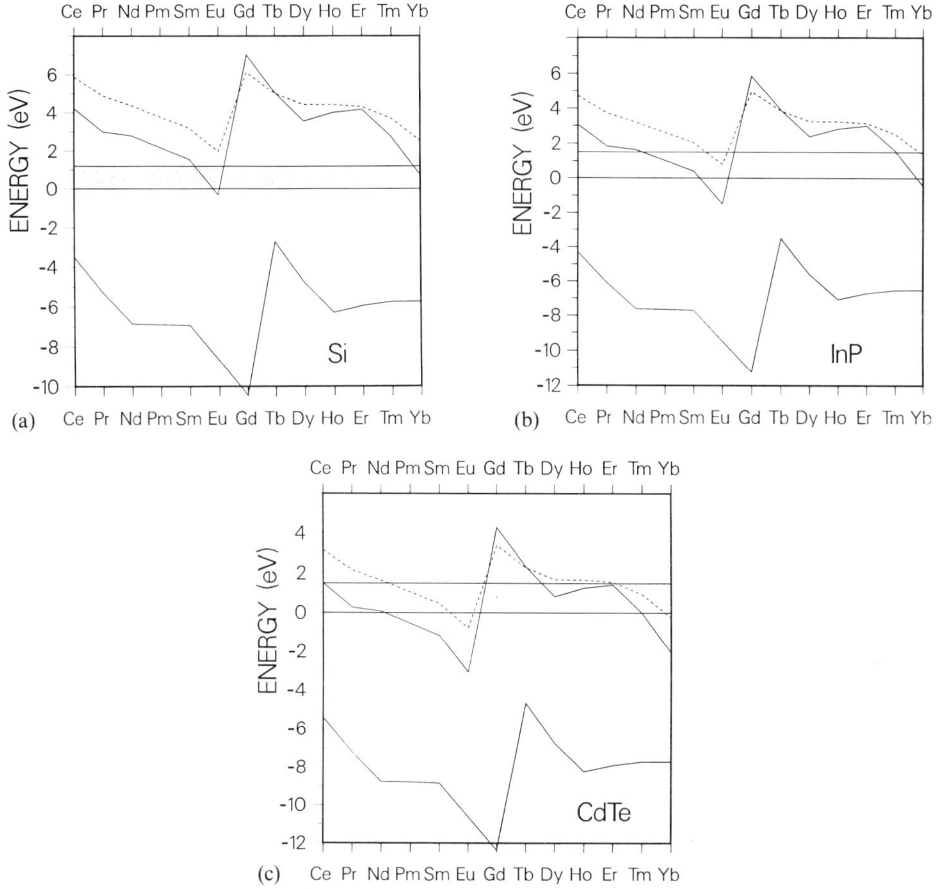

Fig. 8. (a) 4f levels for substitutional rare-earth impurities in silicon. These levels are positioned with respect to the band gap (horizontal lines), the top of the valence band being at the origin of the energies. The lowest level and the highest level plotted with a continuous line are, respectively, the $\varepsilon_{4f}(3.5)$ and $\varepsilon_{4f}(2.5)$ levels calculated with the Hartree–Fock data of Herbst et al. (1972) for rare-earth atoms. The level plotted with a dotted line is the $\varepsilon_{4f}(2.5)$ level calculated with the experimental data of the same reference to account for the correlation energy. (b) Same as (a) for InP. (c) Same as (a) for CdTe.

It is interesting to compare these results to those of other theoretical studies. The only other existing calculation in semiconductors concerns Yb in InP (Hemstreet 1986) and is based on the Xα scattered-wave cluster method. The main conclusion of this study was that some of the 4f states are substantially delocalized, in contradiction with experimental evidence coming from extremely sharp internal transitions. This is probably due to the fact that, in this work, the radius of the muffin-tin spheres was arbitrarily modified to adjust the small cluster gap to the experimental one, resulting in an overestimate of the interatomic coupling and thus of the delocalization of the 4f states.

6. The DX center and EL2

We now consider the two impurities that have led to much controversy in recent years.

6.1. *The DX center*

The DX center in $Al_xGa_{1-x}As$ is a very peculiar and interesting defect. There is strong evidence that it corresponds to an isolated single donor either on the cation (e.g., Si_{Ga}) or the anion (e.g., Te_{As}) site. From this point of view its normal behavior would be to give rise to effective-mass states. These should be localized but fairly shallow for the lowest Γ conduction band and resonant with a lifetime of order 10^{-8}, 10^{-9} s for the L and X relative minima (Kaplan 1963, Bassani et al. 1969, Altarelli and Iadonisi 1971, Bir 1971). Instead of this what is observed is a relatively deep level, for $x > 0.22$, which tends to follow the L band when x is varied but also when a hydrostatic pressure is applied (for recent reviews see Bourgoin 1990, Mooney 1990). This center leads to persistent photoconductivity and seems to exhibit a dual nature since, in addition to the DX states, one can observe four states labeled D_1, D_2, D_3 and D_4 (Henning et al. 1988, 1989), which follow perfectly the predictions of EMT. For $x < 0.22$ the DX state becomes resonant in the Γ band but application of a hydrostatic pressure enables one to localize it in the gap. Furthermore, after removal of the pressure, the electron remains trapped on the DX level, indicating vanishingly small overlap of this state with the Γ band, i.e., an almost infinite lifetime for the resonant level.

6.1.1. *The two extreme theoretical descriptions of donors (no atomic relaxation)*

As we have seen the dual nature of donors is usually treated from two different points of view, a localized description or EMT theory. We discuss both of them and try to synthesize their predictions.

6.1.1.1. Localized description. This is in principle valid for deep states whose spatial extension is fairly limited so that only the short-range part U_{SR} of the defect potential has to be known accurately. This allows use to be made of Green's functions or supercell techniques, either in the local-density (LDA) or in the tight-binding (TB) approximations.

To understand the physical nature of the deep donor states let us recall the simplest TB description, that is the defect molecule model of § 2.3, which we illustrate for the Si_{Ga} case. For this we first remove the Ga atom, creating a vacancy V_{Ga}. We then replace it by the Si impurity. In this view the s and p atomic states (respectively, of A_1 and T_2 symmetry) interact with the A_1 and T_2 symmetry combinations of the V_{Ga} dangling-bond states. Each pair of states of the same symmetry will lead to one bonding and one antibonding combination. This leads to the level structure of fig. 9. For attractive donor potentials the bonding A_1(b) and T_2(b) levels fall in the valence band and are filled. Among the antibonding combinations A_1(ab) can possibly be

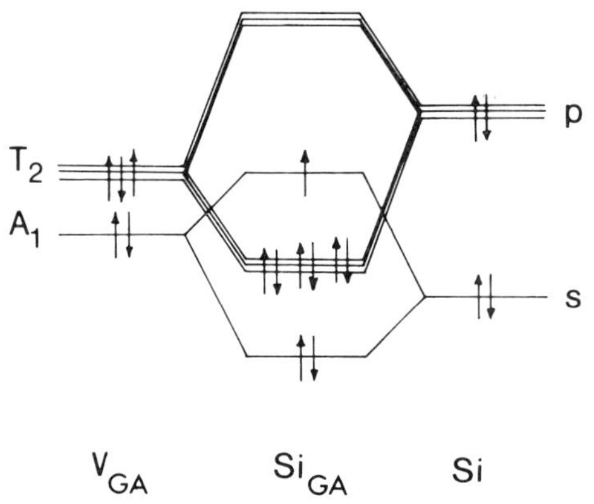

Fig. 9. The molecular model of Si_{Ga} showing the interactions of the s and p states of Si, and the A_1 and T_2 dangling-bond states of the Ga vacancy.

lowered into the gap while T_2(ab) will be resonant in the conduction band. The single donor will be neutral when A_1(ab) is occupied by one electron.

This simple picture is fully confirmed by numerical calculations. Let us begin with the results of a charge-dependent TB calculation (Foulon et al. 1990). Considering first the arsenic antisite As_{Ga} as a test one finds the level A_1(ab) at $E_c - 0.2$ eV in the neutral-charge state and $E_c - 0.63$ eV in the positive state, indicating the existence of a deep donor state in agreement with other calculations. The case of the single donor Si in GaAs or AlAs is more difficult since A_1(ab) falls in the vicinity of E_c and the method is not accurate enough. Anyway, from the results of Foulon et al. (1990), one can draw some general conclusions:

(i) A_1(ab) is probably resonant for Si_{Ga} in GaAs but certainly within the gap for Si_{Al} in AlAs,
(ii) the $A_1(\text{ab}) \rightarrow T_2(\text{ab})$ distance varies in energy from about 0.7 eV in GaAs to about 1 eV in AlAs,
(iii) the effective Coulomb interaction U is of order 0.4–0.5 eV.

Let us now come to the results of the first-principles LDA calculations. For the unrelaxed situation Chadi and Chang (1988a, 1989) simply indicate that A_1(ab) is resonant in the Γ band for GaAs. This is confirmed by the more recent work of Yamaguchi et al. (1990) whose results can be summarized as follows:

(i) the A_1(ab) state is at $E_c + 0.17$ eV for GaAs but can be stabilized within the gap by a hydrostatic pressure (fig. 10),
(ii) the $A_1(\text{ab}) \rightarrow T_2(\text{ab})$ transition lies between 0.8 and 0.9 eV,
(iii) A_1(ab) depends on the local environment in $Ga_{1-x}Al_xAs$ and can be found in the gap for $x > 0.3$. These results confirm the TB trends and will be discussed later.

6.1.1.2. EMT calculations. These are opposite in spirit since EMT applies to cases of a smooth long-range potential U_{LR} usually taken as $-Ze^2/\varepsilon r$. Each conduction-

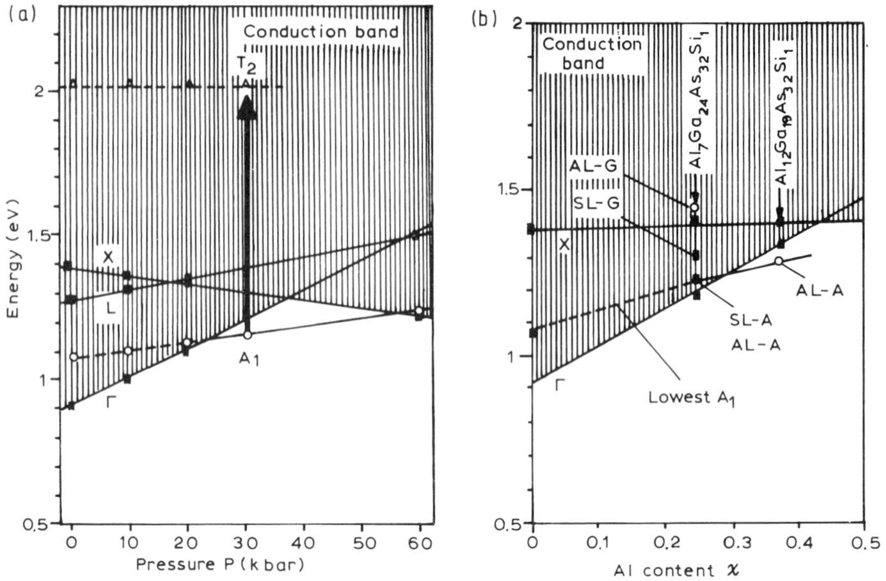

Fig. 10. Results from the LDA calculation of Yamaguchi et al. (1990): (a) versus hydrostatic pressure, (b) versus composition.

band minimum, when treated independently, gives rise to a series of hydrogenic states. Those connected with the lowest conduction band will be truly localized, the others will form resonant states with a lifetime of order 10^{-8}–10^{-9} s. Numerical values of the ground-state binding energy E_B in this independent-minimum approximation are, for GaAs:

- Γ minimum: $E_{B\Gamma} = 5.7$ meV, since $m_\Gamma = 0.065$.
- L minimum: one has anisotropy with $m_\ell = 1.9$ and $m_t = 0.075$. From the numerical calculations of Faulkner (1969) one obtains $E_{BL} = 14$ meV.
- X minimum: here $m_\ell = 1.9$ and $m_t = 0.19$ so that Faulkner (1969) gives $E_{BX} \sim 30$ meV.

The case of N equivalent minima ($N = 4$ for L, $N = 3$ for X) leads to hydrogenic states with N-fold degeneracy that is lifted by intervalley coupling, mainly due to the central-cell correction. An interesting point is that this will not occur for the X minima when considering group-IV donors on the Ga site. In this case, due to symmetry, the X minima will remain uncoupled leading to a triply degenerate T_2(X) state (Morgan 1986). This will not be true of the L minima ($N = 4$) for which the intervalley coupling gives, for the 1s states, an A_{1L}(1s)–T_{2L}(1s) splitting.

Bourgoin and Mauger (1988) have argued that the central-cell correction is strong enough for the A_{1L}(1s) to become moderately deep while retaining its pure L character. They identify this level with the DX state, which would explain why it follows the L band when x is varied. In their model, at the limit of applicability of EMT, the DX level is expected to be weakly coupled to the lattice.

6.1.2. The importance of lattice relaxation or distortion

One basic characteristic of the DX center in $Al_xGa_{1-x}As$ is its optical absorption centered at about 1 eV (Bourgoin 1990). This value, combined with the small thermal ionization energy which varies with x (from 0 to 200 meV) leads to a large Franck–Condon shift, which has been taken as evidence for a large lattice relaxation (LLR). The LLR model is also consistent with other features of the DX center like persistent photoconductivity and the existence of the barrier for capture.

The occurrence of such an important electron–lattice coupling is not expected for single donors, where one usually gets fairly delocalized hydrogenic states. Nevertheless, we have seen in § 4.2 that the EMT description of Toyozawa could lead to substantial effects in the case where a shallow–deep instability occurs. However, this model only concerns 1s states while the situation for DX is certainly more complex.

A model of the DX center has been proposed by Morgan (1986) which can be viewed as an extension of Toyozawa's model. In this the T_{2L}(1s) state, being triply degenerate, is considered to be subject to a Jahn–Teller splitting. The most natural motion to consider first is a motion of the impurity itself, forming a lattice mode of T_2 symmetry. It is known that such a case leads to (100) or (111) distortions. Morgan argues that the second category of displacements is favored, the displacement occurring in a (111) antibonding direction. Such a problem could be treated in a similar way as Toyozawa's model in the context of EMT. Again, if a shallow–deep instability occurs, then EMT fails and a localized description is needed.

The lack of electron paramagnetic (EPR) observations for the DX ground state has led to the proposal that DX should behave as a negative-U system. This was first discussed on the basis of Toyozawa's model (Kachaturyan et al. 1989). Strong support for this possibility was then provided by the LDA calculation of Chadi and Chang (1988a, 1989). From an 18-atom supercell calculation they found that DX has a metastable configuration corresponding to the large atomic displacement pictured in fig. 11. For the Si_{Ga} impurity this situation can be viewed as a distorted V_{Ga}–Si_I pair in which the Si atom forms three bonds and has two lone-pair electrons, leaving one As dangling bond. For the Si_{Ga} case in GaAs the energy of this metastable state is only slightly unfavorable (by about 0.22 eV). Later on similar calculations were

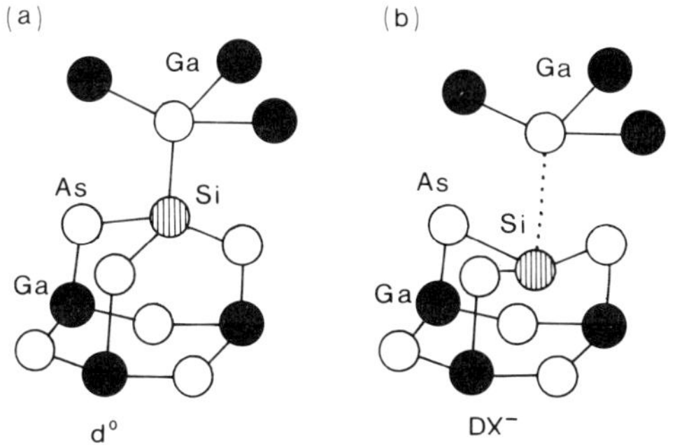

Fig. 11. The stable configuration DX^- of Chadi and Chang (Chadi and Chang 1988a, 1989): (a) normal Si_{Ga} configuration, (b) distorted configuration.

performed for GaAs under pressure and for the $Ga_{1-x}Al_xAs$ compounds as a function of x (Zhang and Chadi 1990). The results reproduced in fig. 12a,b are in striking agreement with experiment. They thus seem to provide extremely strong support in favor of DX^- (i.e. this strongly distorted configuration) being the observed DX ground state. However, let us see now why the situation is perhaps not so simple.

One of the main arguments in favor of the large lattice relaxation (LLR) model is the difference between the optical ionization energy (centered on 1 eV), and the thermal ionization energy ($\lesssim$200 meV). In a simple configuration coordinate diagram this implies a large Franck–Condon shift ($\sim$0.9 eV), which would thus support the strongly distorted DX state. However, we have seen before that the TB theories as well as the LDA calculation of Yamaguchi show that it is probable that the A_1(ab) state of the neutral donor lies in the gap (see fig. 10) for some range of x-values or

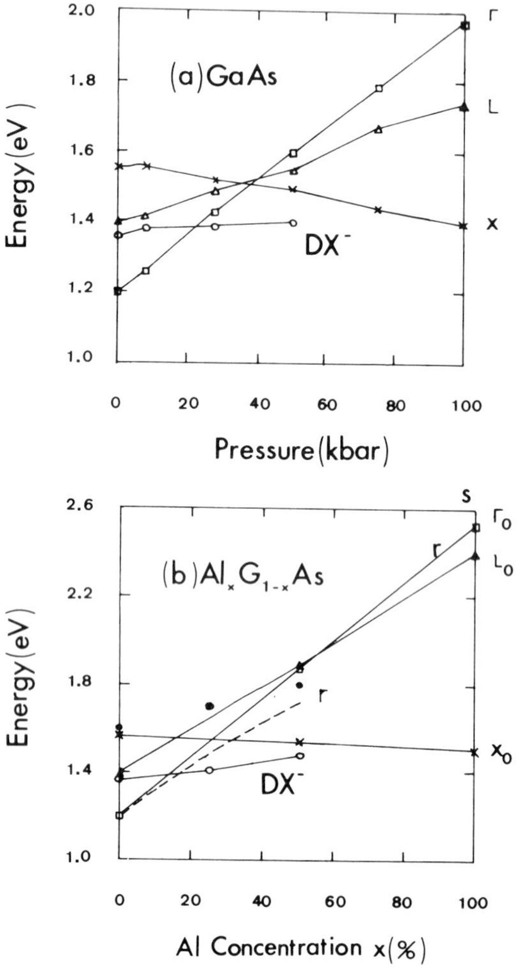

Fig. 12. The calculated pressure (a) and concentration dependences (b) of the DX level ε(+, −).

hydrostatic pressures. From such a state the dominant optical transition would be towards the T_2(ab) resonant state, located about 1 eV above. This corresponds directly to experiment, so that there is no need to invoke a LLR, at least for this property. However, other properties like the almost infinite lifetime of the resonant DX state in GaAs or the lack of EPR signal for the ground state have yet to be explained in a small-lattice-relaxation (SLR) model.

The fact that the DX state practically follows the L band has been taken by Bourgoin and Mauger (1988) as evidence that DX corresponds to the A_{1L}(1s) state. However, considering the large binding energy of DX with respect to the L band (~200 meV) this state is necessarily deep, so that EMT is no longer valid and this state should be given by the localized descriptions, that is should be A_1(ab).

Thus the relevant information must come from such descriptions, in particular from the first-principles LDA calculations. However, the two of these we have discussed lead to opposite conclusions. First, as we have seen, the main result of Chadi and Chang (1988a, 1989) is that DX corresponds to a strongly distorted DX state. On the other hand, Yamaguchi et al. (1990) find that this distortion only leads to a secondary minimum for the energy, 0.65 eV higher than the absolute one that retains T_d symmetry and corresponds to the A_1(ab) state with no negative-U properties. Furthermore, as shown in fig. 10, A_1(ab) exhibits the same behavior versus pressure and composition as the experimentally observed DX state. The comparison between these calculations is thus not decisive at all and one can wonder what the differences are between the two treatments. The following points emerge:

(i) Both calculations have different forms for the exchange–correlation potential V_{xc}.
(ii) Yamaguchi et al. consider large supercells and include directly pairs of Si atoms in the supercell.

It is hard to deduce which of the two predictions is the more accurate. The difference might well be due to the sensitivity to V_{xc} but could also be attributed to technical inaccuracies. At least one must have in mind that the drawback to LDA treatments is the band gap problem related to the local form of V_{xc}: with the "best" available V_{xc} the predicted gaps for Si and Ge are 0.6 and 0 eV compared to the experimental values 1.2 and 0.7 eV. This could be reflected in substantial inaccuracies concerning the energies of deep defects. Thus the most reasonable position is to consider that there are two likely possibilities for the DX ground state: the distorted DX^- configuration and the undistorted D^0 A_1(ab) state. Both of them will then have to be compared with the entire body of experimental information. At this point we should mention a third LDA calculation by Dabrowski et al. (1990) finding that a hydrostatic pressure can stabilize both states, with DX^- being the more stable situation.

6.1.3. Magnetic resonance on the paramagnetic state

There have been a lot of recent results on the paramagnetic state D_0 of the donors. For clarity we restrict ourselves to the group-IV donors Si and Sn, which seem to exhibit fairly different types of behavior. Published studies on Si_{Ga} have been performed for indirect-gap materials ($x > 0.4$) by ODMR or conventional EPR (Kennedy and Glaser 1990, von Bardeleben 1990) mainly for epitaxial layers $Ga_{1-x}Al_xAs$/GaAs. The spectrum is anisotropic and corresponds to near-tetragonal symmetry. It has been explained as resulting from the $T_2(X)$ state derived from the X minimum. However, in bulk material this should give rise to an isotropic spectrum as confirmed by experiment. The anisotropy in epitaxial layers is due to the lattice mismatch between the layer and the substrate which, although small (maximum value 0.18%), leads to a strained layer, with symmetry reduced to tetragonal. The X minima are no longer equivalent and one can show (Glaser et al. 1989, Kaufman et al. 1990) that X_z is raised in energy while X_x and X_y are lowered. The splitting should be linear with x with a maximum value of about 15 meV so that X_z can be assumed to be depopulated. The spectra can only be interpreted by assuming that the coupling between the equivalent minima X_x and X_y is completely quenched. This is thought to be due to random strains, which split these minima by an amount larger than their coupling due to the spin–orbit interaction. Finally, the anisotropy decreases with x, which could indicate an increased interaction between minima.

On the other hand, studies of Sn_{Ga} have been performed in direct-gap materials. They result in an isotropic line observed either by EPR or by MCDA and ODEPR (von Bardeleben et al. 1989, 1991, Fockele et al. 1990, 1992). The major feature of such studies is that one observes a strong hyperfine interaction whose value indicates an unpaired "s" electron density at the Sn nucleus of order 20%. This is typical of the deep A_1(ab) state. This result contradicts the Si_{Ga} case for which the $T_2(X)$ state was observed.

How can we try to reconcile these apparently different results? A possible answer comes from fig. 10 where the lowest A_1(ab) level found by Yamaguchi et al. (1990) is seen to cross the Γ band at $x \sim 0.3$. If we extrapolate it a crossing with the X level would occur at $x \sim 0.5$. This would explain perfectly the Sn_{Ga} results in this interval for which the stable D^0 state would be A_1(ab). On the other hand, for $x > 0.5$ the $T_2(X)$ level would be the lowest, explaining the Si_{Ga} results. However, a problem would arise in the overlap region where the Sn and Si situations are the same (the predicted A_1(ab) position are identical (Yamaguchi et al. 1990)).

In such cases one would expect to have a superposition of the two situations with approximate weighting factors. This might well correspond to the experimental situation but this still needs to be clarified. However, the main conclusion that emerges from this analysis is that, for D_0, the deep A_1(ab) antibonding state is likely to appear in the gap in the intermediate composition range. This is confirmed by the more detailed and careful discussion given by von Bardeleben et al. (1991).

As a general conclusion of this section it is clear that the electronic structure of DX is not yet clearly understood. The DX^- strongly distorted model is the most widely accepted. We have seen, however, that it is not yet firmly established and that the electronic structure of the photoexcited D^0 state is beginning to be elucidated.

6.2. EL2

EL2 has been one of the most studied defects in GaAs (Martin and Makram-Ebeid 1986). However, at the present time there are still two competing models, i.e., the isolated antisite As_{Ga} or a complex As_{Ga}–X. We first review some basic experimental information, then present arguments in favor of one or the other of the experimental models and finally discuss some recent developments that illustrate nicely the general theory of donor impurities.

6.2.1. Physical properties of EL2

We list here established features of EL2 that will serve as the basic elements for our discussion. First, EL2 is a donor (Mircea et al. 1976) characterized by a midgap donor level E_1 whose location has been obtained from DLTS experiments to be at 0.75 ± 0.02 eV below the conduction band. A double donor state has also been detected at $E_2 = 0.54$ eV above the valence-band edge and ascribed to EL2 (Osaka et al. 1986).

The temperature dependence of the electron capture cross-section σ_n of the E_1 level gives an idea of the Franck–Condon shift (Vincent et al. 1982) $S\hbar\omega = 140 \pm 10$ meV, confirmed by the more precise value deduced from the electric field dependence of the emission rate (Makram-Ebeid and Lannoo 1982). Such a value indicates a moderate, but non-negligible, electron–phonon interaction of the stable state of EL2. This level picture has been completed by a detailed analysis of the photoionization cross-sections σ_n^0 and σ_p^0, leading (Chantre et al. 1981) to a value of $S\hbar\omega$ of the same order of magnitude as the one above.

The most characteristic feature of EL2 is its quenching behavior, which is observed through various techniques. For instance, in photocapacitance studies, the sample is cooled in the dark at $T < 140$ K under conditions where EL2 is filled (the so-called O state). It is then illuminated under a reverse bias; for $h\nu$ between 0.9 and 1.35 eV the photocapacitance first increases rapidly and then decreases to a value of the order of the original one (Vincent et al. 1982). This indicates a transition from the stable state O to a new state (usually called O*), O and O* having the same charge state. Indeed, since this process can occur in the space charge region of a junction, a transfer of charge from EL2 to another nearby defect should be ruled out. The O* state is metastable since the stable state can be regenerated (Mitonneau and Mircea 1979) either thermally at a rate r:

$$r(S^{-1}) \simeq 10^{11} \exp(-0.3/kT)\,(\mathrm{eV}) \tag{60}$$

or under electron injection conditions with a rate proportional to the density of the injected electrons. The cross-section $\sigma^*(h\nu)$ for the optically induced transition from $O \rightarrow O^*$ is peaked at 1.1 eV, an energy at which $\sigma_n^0 \sim \sigma_p^0 \sim 10\sigma^*$ (Vincent et al. 1982) (of the order of 10^{-16} cm^{-2}). A comparison between the optical and the photoionization spectrum shows that such an optical transition is related to an intracentre optical absorption band (Kamińska et al. 1983) similar to σ^* $(h\nu)$ and whose zero-phonon line corresponds to an $A_1 \rightarrow T_2$ transition characteristic of an undistorted T_d centre.

In the last few years two microscopic models of EL2 have attracted attention. After early work (Kamińska et al. 1983, Weber et al. 1982) concluding with an identification of the isolated antisite, von Bardeleben et al. (1985) proposed the As_{Ga}–As_i pair model on the basis of electron-spin resonance and DLTS studies. This complex defect model was further supported by the results of electron–nuclear double-resonance experiments (Meyer et al. 1985). Theoretically it was even shown that the model could account for the metastable state in a split configuration (Delerue et al. 1987) and that it was capable of explaining many of the experimental results (Baraff et al. 1988). However, its basic weakness as it stands results from the fact that there seems to be no binding energy associated with the As_{Ga}–As_i pair.

This situation changed recently when two independent calculations (Dabrowski and Scheffler 1987, Chadi and Chang 1988b) came to the conclusion that the isolated arsenic antisite As_{Ga} has a metastable configuration, which can be identified as V_{Ga}–As_i. This is of the same nature as what has been proposed for the D^- configuration of DX. From the chemical point of view it corresponds to trivalent As with three electrons engaged in covalent bonds and two "s" lone-pair electrons. This model has the great advantage of being simple. It also gives a nice interpretation of some of the experimental data and especially the fact that the stress splitting of the zero-phonon line observed in optical absorption (Kamińska et al. 1985) seems a strong indication that the ground state has T_d and not a lower symmetry. However, the problem that remains to be solved is the apparent contradiction between this T_d symmetry of As_{Ga} and the other indications given above the EL2 has a lower symmetry consistent with a complex. To discuss this discrepancy further we now give details of two recent developments.

6.2.2. Recent developments

6.2.2.1. Optical absorption. Let us first consider the results obtained in optical absorption. The dominant feature is a broad band centered on 1.2 eV while a very weak zero-phonon line is observed at 1.04 eV. The splitting of this line under uniaxial stress was shown to be consistent with a transition from an A_1 ground state to a triply degenerate T_2 excited state (Kamińska et al. 1985, Davies 1990) in T_d symmetry. As we have seen this is a strong argument in favor of the isolated As_{Ga} level. However, the detailed interpretation of the splittings requires the existence of an A_1 excited state just above the T_2 one. This can be explained by the possibility of a dynamic Jahn–Teller effect in the T_2 excited state, which leads quite naturally to a T_2 vibronic state followed by A_1 (Davies 1990). The only weakness of the whole interpretation is that, for the moment, it does not give a satisfactory interpretation of the replica of the zero-phonon line and their intensities (Kamińska et al. 1985).

Recently von Bardeleben (1989) has discussed the possibility that the 1.04 eV line corresponds to the transition between the deep donor state A_1 and the 1s hydrogenic states derived from the L minima. Let us examine this possibility for the isolated As_{Ga} state as described by Lannoo (1991). This is a double donor so that, with respect to the L conduction band, the situation should be comparable to the case of Se in Ge discussed in § 4.3 (even the effective masses $m_\ell = 1.9$, $m_t = 0.075$ are of comparable magnitude). It must then be characterized by a deep A_1(ab) state (as observed) and an inverted splitting of the two lower states T_2(1s)–A_1(1s) derived from the L band. If the zero-phonon line is due to the A_1(ab)–T_2(1s) transition its stress splitting is entirely determined by the splitting of the T_2(1s)–A_1(1s) system under stress. From symmetry requirements the perturbation matrix with respect to the stress tensor components s_{ij} can be written as

$$\begin{pmatrix} \Delta + a's & ds_{yz} & ds_{zx} & ds_{xy} \\ ds_{yz} & as + (b/2)(\sqrt{3}s_e - s_t) & cs_{xy} & cs_{zx} \\ ds_{zx} & cs_{xy} & as - (b/2)(\sqrt{3}s_e + s_t) & cs_{yz} \\ ds_{xy} & cs_{zx} & cs_{yz} & as + bs_t \end{pmatrix}, \qquad (61)$$

where the basis states, correspond respectively, to the A_1 level and the x, y, z components of T_2. Here $s = s_{xx} + s_{zz} + s_{yy} + s_t = 2s_{zz} - s_{xx} - s_{yy}$, $s_e = 3^{1/2}(s_{xx} - s_{yy})$ and $\Delta > 0$ is the difference in energy between the $A_1(1s)$ and $T_2(1s)$ levels at zero stress. This matrix takes exactly the same form as the one used by Davies (1990). It can be further simplified if we assume that the modifications in intervalley coupling and binding energy of each independent one-valley 1s state are negligible. In that case, $a = a'$, $b = 0$ and $c = d$, which leads to the matrix used by Kamińska et al. (1985) (this can be seen by writing the perturbation matrix (61) in the basis of the one-valley 1s states). Then $d(=c)$ simply describes the shift of the (111) minima under stress. This can be determined from the fact that $\frac{4}{3}ds$ is the splitting of the L minima in the perfect crystal for stress in the (111) direction. Using the value measured by Mirlin et al. (1987), one gets $d = -77$ meV GPa^{-1}.

Kamińska et al. (1985) and Davies (1990) attributed the inverted order of the A_1 and T_2 levels to a dynamic Jahn–Teller effect and their parameters (Δ and d) were treated as adjustable quantities, while here they can be fully determined from independent arguments. As shown in both papers a perfect fit can be achieved with the matrix (61) for the following set of parameters: $\Delta = 7.5$ meV, $d = 76$ meV GPa^{-1} (from Kaminska et al. 1985) and $\Delta = 10$ meV, $d = 90$ meV GPa^{-1} (from Davies 1990) taking fig. 2 with $d = 90$ meV GPa^{-1}). The value we predict for d from the splitting of the L minima in the bulk falls exactly in this range. Furthermore, the value of the parameter Δ lies in the range 7.5–10 meV. In our model this is the difference in energy between the $A_1(1s)$ and $T_2(1s)$ states, which should be close to what is obtained for Se in Ge. Looking at fig. 2, this splitting is 2.55 meV for Se^0 and, when assuming that it scales by a factor of four, becomes 10.2 meV for Se^+. The first value is much too small but the second value has the correct order of magnitude implying that the observed charge state in optical absorption is $EL2^+$. $EL2^+$ cannot be As_{Ga}^+ since this one should be paramagnetic and visible in EPR. This means that $EL2^+$, as observed, must be a complex defect As_{Ga}^0–X^+, where X^+ could be As_i (von Bardeleben et al. 1985, Meyer et al. 1985). The presence of X^+ should not be seen in the final $T_2(1s)$ states because these hydrogenic states are extremely shallow and extended in space. The validity of this picture is strengthened by the predicted positions of the excited states. As obtained from fig. 2 (for Se^+ in Ge) the intervals in energy between these states are respectively 10.2, 10.4, 4.9, and 4.1 meV. Apart from the slight corrections necessary to account for the difference between As_{Ga} in GaAs and Se in Ge this explains exactly the first four peaks in the absorption spectrum – the so-called replica – for EL2 (the transition $A_1(ab)$–$A_1(2s)$, which is normally forbidden, may become allowed if the initial state for optical absorption is distorted by the presence of X^+). This picture also explains the excited levels for other antisite-related defects (Spaeth et al. 1990). The stress-splitting pattern for these defects should be similar to the one observed for EL2 and should be interpretable with eq. (61) (but not necessarily with the simple assumption that the binding energy and the intervalley coupling are stress invariant, i.e., $a = a'$, $b = 0$ and $c = d$).

6.2.2.2. Magnetic circular dichroism. Magnetic circular dichroism experiments give information about both optical and magnetic properties of the defects. The MCDA $S(h\nu)$ with respect to the optical excitation energy $h\nu$ is defined as the difference in optical absorption α for right (+) and left (−) circularly polarized light, the sample being in a static magnetic field (the axis of the polarization is the axis of the magnetic field) (Kaufmann and Windscheif 1988)

$$S(h\nu) \approx \alpha^+(h\nu) - \alpha^-(h\nu). \tag{62}$$

The absorption is simply related to the optical cross sections σ of the deep impurity ($\alpha \propto \sigma$). Petit et al. have derived a Green's function formalism of the optical cross-section (Petit et al. 1986). The advantage of this formalism is that it takes into account the full band structure and a calculated wave function. It also introduces the modifications undergone by the initial Bloch states in the vicinity of the impurity. We have extended this formalism to the calculation of the MCDA. From Petit et al. (1986) the optical cross-section is expressed as

$$\sigma(h\nu) \propto -\frac{1}{h\nu}\,\mathrm{Im}\,\langle \mathrm{a}_1 | pG(E_{\mathrm{a}_1} \pm h\nu)p | \mathrm{a}_1 \rangle, \tag{63}$$

where Im denotes the imaginary part, the sign +(−) corresponds to a transition involving the conduction (valence band), a_1 is the wave function of the deep A_1(ab) state of energy E_{a_1} and G is the Green's function. With circularly polarized light, p is given by $p_x + (-)\mathrm{i}p_y$ for $\sigma^{+(-)}$, where p_x and p_y are the momentum operators. After some simple algebra, we obtain for the MCDA

$$S(h\nu) \propto -\frac{1}{h\nu}\,\mathrm{Re}\langle \mathrm{a}_1 | p_y G(E_{\mathrm{a}_1} \pm h\nu)p_x - p_x G(E_{\mathrm{a}_1} \pm h\nu)p_y | \mathrm{a}_1 \rangle, \tag{64}$$

where Re stands for the real part. Equation (64) is completely general. The Green's function must be evaluated taking into account the spin–orbit coupling, which often leads to lengthy calculations. For simplicity we assume that $T = 0°$ K, i.e., that only the A_1(ab) state with a spin $m_s = -\frac{1}{2}$ is populated. The lattice coupling is also neglected. We develop our calculations for $\mathrm{As}_{\mathrm{Ga}}$ in GaAs in a tight-binding scheme with two s and six p orbitals per atom including spin. We retain interactions between the first and second neighbors only, and use the parameters of Talwar and Ting (1982) slightly modified to include the spin–orbit coupling. The spin–orbit parameters for As and Ga atoms are fitted to account for the experimental valence-band splitting ($\Gamma_8 - \Gamma_7 = 0.35$ eV). Here we consider that the impurity potential matrix has terms different from zero only on the defect and its four neighbors. The $\mathrm{As}_{\mathrm{Ga}}$ defect is created by replacing the s and p energies of a Ga atom by those of an As atom. The potential on the central atom and the first neighbors is adjusted to obtain the local neutrality of charge, which is usually a good approximation to the true potential. The ionization levels are calculated using half-integer occupation numbers (Slater 1960). The matrix elements of the momentum operator are calculated as by Petit et al. (1986).

For the undistorted As_{Ga} antisite we obtain the level $+/0$ at $E_{vb} + 1.29$ eV and $+/++$ at $E_{vb} + 0.90$ eV. The A_1(ab)–T_2(ab) splitting is 0.9 eV and the s density on the As_{Ga} atom is about 16%, in agreement with other calculations (Bachelet et al. 1983) and experiment (Meyer et al. 1985). As the ground state for the MCDA must be paramagnetic (Kaufmann and Windscheif 1988) – here the positive charge state –, we consider the transition from the $++/+$ $(+/0)$ levels to the conduction (valence) band. The results are plotted on fig. 13. We first look at the transition to the valence band (dashed line). The MCDA spectrum always keeps the same sign, unlike the results of Kaufmann and Windscheif (1988). We explain this difference by the simplified p-band structure used by these authors, which does not properly include the mixing between bands. Here we also use the realistic and calculated antibonding wave function. The transition to the conduction band (full line) is characterized by a strong transition to the T_2(ab) level, which is resonant in the conduction band. The T_2(ab) state is split into $J = \frac{3}{2}$ and $J = \frac{1}{2}$ states by spin–orbit coupling. This leads to a positive peak for the transition to the $J = \frac{1}{2}$ states and a negative one to the $J = \frac{3}{2}$ states. The contributions of the L and X bands ($h\nu > 1.3$ eV) is small in comparison. In fig. 13 we also note that the transition to the valence band gives a small MCDA spectrum compared to the transition to the T_2(ab) states. Therefore we conclude that the experimental MCDA spectra for As antisite-related defects must be characterized by strong A_1(ab)–T_2(ab) transitions. These transitions are strongly allowed because the A_1(ab) and T_2(ab) states are both antibonding and their wave functions are localized in the same region of space as already pointed out by Dabrowski and Scheffler (1988). This supports the analysis of the MCDA of Meyer et al. (1984).

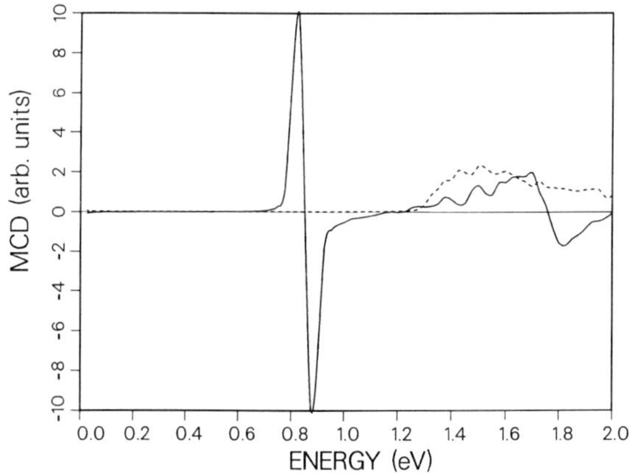

Fig. 13. Calculated MCD spectrum for the isolated arsenic antisite As_{Ga}.

As already stated our calculated MCDA spectrum has one positive peak followed by one negative peak (a simple derivative-like structure). These two peaks are sharp because the T_2(ab) states are weakly coupled to the conduction band. The difference in energy between the peak maxima is typical of the spin–orbit splitting, which is calculated to be about 0.09 eV. The effect of the electron–lattice coupling will be to broaden the transitions. A strong phonon coupling with non-symmetrical vibrational modes could also cause the $J=\frac{3}{2}$ and $J=\frac{1}{2}$ states to repel each other, as is actually observed for the F center in CsBr and CsCl (Henry et al. 1965). Very recently the isolated arsenic antisite defect has been identified by optically detected magnetic resonance (Krambrock et al. 1992). Its MCDA spectrum has a simple derivative-like structure as given by our calculation (but with very broad bands implying a strong phonon coupling). In contrast the MCDA spectrum of EL2 is characterized by one positive band followed by two negative bands (Spaeth et al. 1990, Meyer et al. 1984). This difference could be ascribed to a strong distortion in the T_2(ab) excited states of the isolated antisite (Henry et al. 1965) but this must be rejected due to the observation of the isolated antisite with a simpler MCDA structure (Krambrock et al. 1992). A spatial symmetry of the defect lower than T_d caused by the presence of another defect (e.g., As_i) would split the $J=\frac{3}{2}$ states into two Kramers doublets (some mixing with the $J=\frac{1}{2}$ state would also occur). In the MCDA spectrum this might result in the splitting of the negative peak into two peaks, as is actually observed (Meyer et al. 1984). Such a possibility for the MCDA spectrum of EL2 is under study.

In conclusion, these recent developments show that the only experimental argument against the As_{Ga}–As_i pair model for EL2 was based on the optical absorption, which was believed to reflect a defect with full T_d symmetry. The analysis of the optical absorption of EL2 shows that the stress splitting as well as the so-called replica can be fully understood from the transition of the deep state to the hydrogenic states of the L band. The MCDA spectrum of the isolated arsenic antisite has a simple derivative-like structure, unlike the situation observed for EL2. All these results imply or support the possibility that EL2 is a complex defect of the form As^0_{Ga}–X^+.

References

Altarelli, M., and G. Iadonisi, 1971, Nuovo Cimento B **5**, 21.

Aszodi, G., J. Weber, Ch. Vihlein, L. Pu-Lin, H. Ennen, U. Kaufman, J. Schneider and J. Windscheif, 1985, Phys. Rev. B **31**, 7767.

Bachelet, G.B., G.A. Baraff and M. Schlüter, 1981, Phys. Rev. B **24**, 943.

Bachelet, G.B., M. Schlüter and G.A. Baraff, 1983, Phys. Rev. B **27**, 2545.

Bar-Yam, Y., and J.D. Joannopoulos, 1984, Phys. Rev. Lett. **52**, 1129.

Baraff, G.A., and M. Schlüter, 1979, Phys. Rev. B **19**, 4965.

Baraff, G.A., and M. Schlüter, 1981, Inst. Phys. Conf. Ser. **59**, 287.

Baraff, G.A., and M. Schlüter, 1984, Phys. Rev. B **30**, 1853.

Baraff, G.A., M. Lannoo and M. Schlüter, 1988, Phys. Rev. B **38**, 6003.

Bassani, F., G. Iadonisi and B. Preziosi, 1969, Phys. Rev. **186**, 735.

Bassani, F., G. Iadonisi and B. Preziosi, 1974, Rep. Prog. Phys. **37**, 1099.

Beeler, F., O.K. Andersen and M. Scheffler, 1985, Phys. Rev. Lett. **55**, 1498.
Bernholc, J., and S.T. Pantelides, 1978, Phys. Rev. B **18**, 1780.
Bernholc, J., N.O. Lipari and S.T. Pantelides, 1978, Phys. Rev. Lett. **41**, 895.
Bernholc, J., N.O. Lipari and S.T. Pantelides, 1980, Phys. Rev. B **21**, 3545.
Bir, G.L., 1971, Sov. Phys.-Solid State **13**, 371.
Bourgoin, J.C., ed., 1990, Physics of DX Centers in GaAs Alloys, Solid State Phenomena, Vol. 10 (Sci. Tech. Publications, Vaduz, Liechtenstein).
Bourgoin, J.C., and A. Mauger, 1988, Appl. Phys. Lett. **53**, 749.
Boyn, R., 1988, Phys. Status Solidi **148**, 11.
Buisson, J.P., R.E. Allen and J.D. Dow, 1982, J. Phys. (France) **43**, 181.
Car, R., P.J. Kelly, A. Oshiyama and S.T. Pantelides, 1984, Phys. Rev. Lett. **52**, 1814.
Chadi, D.J., and K.J. Chang, 1988a, Phys. Rev. Lett. **61**, 873.
Chadi, D.J., and K.J. Chang, 1988b, Phys. Rev. Lett. **60**, 2187.
Chadi, D.J., and K.J. Chang, 1989, Phys. Rev. B **39**, 10063.
Chantre, A., G. Vincent and D. Bois, 1981, Phys. Rev. B **23**, 5335.
Chaudhuri, S., and D. Coon, 1984, Phys. Rev. B **30**, 3338.
Clerjaud, B., 1985, J. Phys. C **18**, 3615.
Cohen, M.L., V. Heine and D. Weaire, 1970, Solid State Physics, Vol. 24, eds H. Ehrenreich, F. Seitz and D. Turnbull (Academic Press, New York).
Coulson, C.A., and M.J. Kearsley, 1957, Proc. R. Soc. London A **241**, 433.
Dabrowski, J., and M. Scheffler, 1987, Phys. Rev. Lett. **36**, 1332.
Dabrowski, J., and M. Scheffler, 1988, Phys. Rev. Lett. **60**, 2183.
Dabrowski, J., M. Scheffler and R. Strehlov, 1990, in: The Physics of Semiconductors, Vol. 1, eds E.M. Anastassakis and J.D. Joannopoulos (World Scientific, Singapore) p. 489.
Dahl, J.P., and J. Avery, eds, 1984, Local Density Approximations in Quantum Chemistry and Solid State Physics (Plenum, New York).
Das Sarma, S., and A. Madhukar, 1981, Phys. Rev. B **24**, 2051.
Davies, G., 1990, Phys. Rev. B **41**, 12303.
Daw, M.S., and D.L. Smith, 1979, Phys. Rev. B **20**, 5150.
Deleo, G.G., G.D. Watkins and W.B. Fowler, 1981, Phys. Rev. B **23**, 1851.
Delerue, C., and M. Lannoo, 1992a, to be published.
Delerue, C., and M. Lannoo, 1992b, Phys. Rev. Lett. **67**, 3006.
Delerue, C., M. Lannoo, D. Stievenard, H.J. von Bardeleben and J.C. Bourgoin, 1987, Phys. Rev. Lett. **59**, 2875.
Delerue, C., M. Lannoo and J.M. Langer, 1988, Phys. Rev. Lett. **61**, 199.
Delerue, C., M. Lannoo and G. Allan, 1989a, Phys. Rev. B **39**, 1669.
Delerue, C., M. Lannoo, G. Bremond, G. Guillot and A. Nouailhat, 1989b, Europhys. Lett. **9**, 373.
Ennen, H., and J. Schneider, 1984, Proc. 13th Int. Conf. on Defects in Semiconductors, eds L.C. Kimerling and J.M. Parsey Jr (The Metallurgical Society of AIME, Warrendale, PA) p. 115.
Ennen, H., G. Pomrenke and A. Axmann, 1985, J. Appl. Phys. **57**, 2182.
Ennen, H., J. Wagner, H.D. Müller and R.S. Smith, 1987, J. Appl. Phys. **61**, 4877.
Falicov, L.M., ed., 1981, Valence Fluctuations in Solids (North-Holland, Amsterdam).
Faulkner, R.A., 1969, Phys. Rev. **184**, 713.
Fazzio, A., and J.R. Leite, 1980, Phys. Rev. B **21**, 4710.
Fazzio, A., J.R. Leite and M.L. De Siqueira, 1979, J. Phys. C **12**, 3469.
Fockele, M., J.M. Spaeth and P. Gibart, 1990, in: The Physics of Semiconductors, Vol. 1, eds E.M. Anastassakis and J.D. Joannopoulos (World Scientific, Singapore) p. 517.
Fockele, M., J.-M. Spaeth, H. Overhof and P. Gibart, 1992, to be published.
Foulon, Y., M. Lannoo and G. Allan, 1990, in: Physics of DX Centers in GaAs Alloys, Solid State Phenomena, Vol. 10, ed. J.C. Bourgoin (Sci. Tech. Publications, Vaduz, Liechtenstein) p. 195.
Friedel, J., M. Lannoo and G. Leman, 1967, Phys. Rev. **164**, 1056.
Glaser, E., T.A. Kennedy, R.S. Sillmon and M.G. Spencer, 1989, Phys. Rev. B **40**, 3447.
Godby, R.W., M. Schlüter and L.J. Sham, 1986, Phys. Rev. Lett. **56**, 2415.
Godlewski, M., and D. Hommel, 1986, Phys. Status Solidi A **95**, 261.

Grimmeiss, H.G., L. Montelius and K. Larsson, 1988, Phys. Rev. B **37**, 6916.
Hamann, D.R., 1979, Phys. Rev. Lett. **42**, 662.
Harrison, W.A., 1980, Electronic Structure and the Properties of Solids, The Physics of the Chemical Bond (Freeman, New York).
Hemstreet, L.A., 1977, Phys. Rev. B **15**, 834.
Hemstreet, L.A., 1986, in: Material Science Forum, Vols. 10–12, ed. H.J. von Bardeleben (Trans. Tech. Publ., Switzerland) p. 85.
Hemstreet, L.A., and J.O. Dimmock, 1979, Phys. Rev. B **20**, 1527.
Henning, J.C.M., J.P.M. Ansems and T.E.C. Brandsma, 1988, Trans. Tech. Publ., Special Volume.
Henning, J.C.M., E.A. Montie and J.P.M. Ansems, 1989, in: Materials Science Forum, Vols. 38–41 (Trans. Tech. Publ., Switzerland) p. 1085.
Henry, C.H., S.E. Schnatterly and C.B. Slichter, 1965, Phys. Rev. A **137**, 583.
Herbst, J.F., D.N. Lowy and R.E. Watson, 1972, Phys. Rev. B **6**, 1913.
Herbst, J.F., R.E. Watson and J.W. Wilkins, 1976, Phys. Rev. B **13**, 1439.
Herbst, J.F., R.E. Watson and J.W. Wilkins, 1978, Phys. Rev. B **17**, 3089.
Herman, F., and S. Skillman, 1963, Atomic Structure Calculations (Prentice Hall, New York).
Hjalmarson, H.P., P. Vogl and J.D. Dow, 1980, Phys. Rev. Lett. **44**, 810.
Hohenberg, P., and W. Kohn, 1964, Phys. Rev. B **136**, 864.
Hybertsen, M., and S.G. Louie, 1985, Phys. Rev. Lett. **55**, 1418.
Jaros, M., and S. Brand, 1976, Phys. Rev. B **14**, 4494.
Kachaturyan, K., E.R. Weber and M. Kamińska, 1989, in: Materials Science Forum, Vols. 38–41 (Trans. Tech. Publ., Switzerland) p. 1067.
Kaminska, M., M. Skowronski, J. Lagowski, J.M. Parsey and H.C. Gatos, 1983, Appl. Phys. Lett. **43**, 302.
Kamińska, M., M. Skowronski and V. Kuszko, 1985, Phys. Rev. Lett. **55**, 2204.
Kane, E.O., 1956, J. Phys. & Chem. Solids **1**, 82.
Kaplan, H., 1963, J. Phys. & Chem. Solids **24**, 1593.
Kasatkin, V.A., F.P. Kesamanly and B.E. Samorukov, 1981, Sov. Phys.-Semicond. **15**, 352.
Katayama-Yoshida, H., and A. Zunger, 1985, Phys. Rev. B **31**, 7877.
Kauffer, E., P. Pecheur and M. Gerl, 1976, J. Phys. C **9**, 2319.
Kauffer, E., P. Pecheur and M. Gerl, 1977, Phys. Rev. B **15**, 4107.
Kaufman, E., W. Wilkening, P.M. Mooney and T.F. Kuech, 1990, Phys. Rev. B **41**, 10206.
Kaufmann, U., and J. Windscheif, 1988, Phys. Rev. B **38**, 10060.
Kennedy, A., and E. Glaser, 1990, in: Physics of DX Centers in GaAs Alloys, Solid State Phenomena, Vol. 10, ed. J.C. Bourgoin (Sci. Tech. Publications, Vaduz, Liechtenstein) p. 53.
Klein, P.B., 1988, Solid State Commun. **65**, 1097.
Kohn, W., 1957, in: Solid State Physics, Vol. 5, eds F. Seitz and D. Turnbull (Academic Press, New York) p. 258.
Kohn, W., and L.J. Sham, 1965, Phys. Rev. A **140**, 1133.
Krambrock, K., J.-M. Spaeth, C. Delerue, G. Allan and M. Lannoo, 1992, Phys. Rev. B **45**.
Krieger, J.B., and P.M. Laufer, 1981, Phys. Rev. B **23**, 4063.
Kröber, W., and A. Hangleiter, 1988, Appl. Phys. Lett. **52**, 114.
Lambert, B., Y. Toudic, G. Grandpierre, A. Rupert and A. Le Corre, 1988, Electron. Lett. **24**, 1446.
Lambert, B., A. Le Corre, Y. Toudic, C. Lhomer, G. Grandpierre and M. Gauneau, 1992, to be published.
Landsberg, P.T., 1969, in: Solid State Theory: Methods and Applications (Wiley, London) p. 138.
Langer, J.M., and H. Heinrich, 1984, Phys. Rev. Lett. **45**, 671.
Langer, J.M., C. Delerue, M. Lannoo and H. Heinrich, 1988, Phys. Rev. B **38**, 7723.
Lannoo, M., 1984, J. Phys. C **17**, 3137.
Lannoo, M., 1991, in: Materials Science Forum, Vol. 83–87, eds G. Davies, G. de Leo and M. Stavola (Trans Tech. Publications, Switzerland) p. 865.
Lannoo, M., and J.C. Bourgoin, 1981, Point Defects in Semiconductors, Vol. 22, Theoretical Aspects, Springer Series in Solid State Science (Springer, Berlin).
Lannoo, M., and P. Lenglart, 1968, J. Phys. & Chem. Solids **30**, 2409.
Lannoo, M., M. Schlüter and L.J. Sham, 1985, Phys. Rev. B **32**, 3890.

Li'in, and V.F. Masterov, 1976, Sov. Phys.-Semicond. **10**, 496.
Liro, Z., C. Delerue and M. Lannoo, 1987, Phys. Rev. B **36**, 17.
Louie, S.G., M. Schlüter, J.R. Chelikowsky and M.L. Cohen, 1976, Phys. Rev. B **13**, 1654.
Luttinger, J.M., and W. Kohn, 1955, Phys. Rev. **97**, 869.
Makram-Ebeid, S., and M. Lannoo, 1982, Phys. Rev. B **25**, 6406.
Martin, G.M., and S. Makram-Ebeid, 1986, in: Deep Centers in Semiconductors, Vol. 6, ed. S.T. Pantelides (Gordon and Breach, New York).
Martin, W.C., L. Hagan, J. Reader and J. Sugar, 1974, J. Phys. Chem. Ref. Data **3**, 775.
Martinez, G., 1986, Proc. 14th Int. Conf. on Defects in Semiconductors, Materials Science Forum, Vols. 10–12, ed. H.J. von Bardeleben (Trans. Tech. Publications Ltd., Switzerland) p. 603.
Mauger, A., H.J. von Bardeleben, J.C. Bourgoin, F. Lannoo and M. Lannoo, 1987a, Europhys. Lett. **4**, 1151.
Mauger, A., H.J. von Bardeleben, J.C. Bourgoin and M. Lannoo, 1987b, Phys. Rev. B **36**, 5982.
Messmer, R.P., and G.D. Watkins, 1971, in: Radiation Damage in Semiconductors, eds J.W. Corbett and G.D. Watkins (Gordon and Breach, New York) p. 23.
Meyer, B.K., J.-M. Spaeth and M. Scheffler, 1984, Phys. Rev. Lett. **52**, 851.
Meyer, B.K., D.M. Hofmann, J.R. Niklas and J.-M. Spaeth, 1985, Phys. Rev. Lett. **47**, 970.
Mircea, A., A. Mitonneau, L. Hollan and A. Briere, 1976, Appl. Phys. **11**, 153.
Mirlin, D.N., V.F. Sapega, I.Ya. Karlik and R. Katilius, 1987, Solid State Commun. **61**, 799.
Mitonneau, A., and A. Mircea, 1979, Solid State Commun. **30**, 157.
Mooney, P.M., 1990, J. Appl. Phys. **67**, 3, R1.
Morgan, T., 1986, Phys. Rev. B **34**, 2664.
Osaka, J., H. Okamoto and K. Kobayashi, 1986, in: Semi-Insulating III–V Materials (Ohmsha) p. 421.
Pantelides, S.T., 1978, Rev. Mod. Phys. **50**, 797.
Papaconstantopoulos, D.A., and E.N. Economou, 1980, Phys. Rev. B **6**, 2903.
Pêcheur, P., and G. Toussaint, 1983, Proc. 12th Int. Conf. on Defects in Semiconductors, ed. C.A.J. Ammerlaan (North-Holland, Amsterdam) p. 112.
Pecheur, P., G. Toussaint and M. Lannoo, 1981, Inst. Phys. Conf. Ser. **59**, 147.
Perdew, J., and M. Levy, 1983, Phys. Rev. Lett. **51**, 1884.
Petit, J., G. Allan and M. Lannoo, 1986, Phys. Rev. B **33**, 8595.
Phillips, J.C., 1970, Rev. Mod. Phys. **42**, 317.
Picoli, G., A. Chomette and M. Lannoo, 1984, Phys. Rev. B **30**, 7138.
Pomrenke, G., H. Ennen and W. Haydl, 1986, J. Appl. Phys. **59**, 601.
Przybylinska, H., M. Godlewski and A. Stapor, 1987, Proc. XVI School on the Physics of Semicond. Compounds, Jaszowiec (Polish Scientific Publishing, Krakow).
Reiss, H., 1956, J. Chem. Phys. **25**, 681.
Resca, L., and R. Resta, 1979, Solid State Commun. **29**, 275.
Resca, L., and R. Resta, 1980, Phys. Rev. Lett. **44**, 1340.
Schlüter, M., 1985, The Role of Theory in Defect Physics, in: Mater. Res. Soc. Symp. Proc. Vol. 46.
Schlüter, M., and L.J. Sham, 1982, Phys. Today **2**, 36.
Sham, L.J., and M. Schlüter, 1983, Phys. Rev. Lett. **51**, 1888.
Singh, V.A., U. Lindefelt and A. Zunger, 1982, Phys. Rev. B **25**, 2781.
Skromme, B.J., S.S. Bose, B. Lee, T.S. Low, T.R. Lepkowsky, R.Y. DeJoule, G.E. Stillman and J.C.M. Hwang, 1985, J. Appl. Phys. **58**, 4685.
Slater, J.C., 1960, Quantum Theory of Atomic Structure (McGraw-Hill, New York).
Spaeth, J.-M., K. Krambrock and D.M. Hofmann, 1990, in: Proc. 20th Int. Conf. on the Physics of Semiconductors, eds E.M. Anastassakis and J.D. Joannopoulos (World Scientific, Singapore) p. 441.
Srivasta, G.P., 1980, Phys. Status Solidi **93**, 761.
Sugano, S., Y. Tanabe and H. Kaminura, 1970, Multiplets of Transition-Metal Ions in Crystals (Academic Press, New York).
Talwar, D.N., and C.S. Ting, 1982, Phys. Rev. B **25**, 2660.
Tang, Y.S., K.C. Heasman, W.P. Gillin and B.J. Sealy, 1989, Appl. Phys. Lett. **55**, 432.
Tersoff, J., and W.A. Harrison, 1987, Phys. Rev. Lett. **58**, 2367.
Title, R.S., 1964, Phys. Rev. A **133**, 198.

Title, R.S., 1967, Physics and Chemistry of II–VI Compounds, eds M. Aven and J.S. Prener (North-Holland, Amsterdam) ch. 6, p. 265.
Toyozawa, Y., 1983, Physica B **116**, 7.
Van der Rest, J., and P. Pêcheur, 1983, Physica B **116**, 121.
Vincent, G., D. Bois and A. Chantre, 1982, J. Appl. Phys. **53**, 3643.
Vogl, P., and J. Baranowski, 1985, 17th Int. Conf. on the Physics of Semiconductors, eds J.D. Chadi and W.A. Harrison (Springer, Heidelberg) p. 623.
von Bardeleben, H.J., 1989, Phys. Rev. B **40**, 12546.
von Bardeleben, H.J., 1990, in: Physics of DX centers in GaAs alloys, Solid State Phenomena, Vol. 10, ed. J.C. Bourgoin (Sci. Tech. Publications, Vaduz, Liechtenstein) p. 181.
von Bardeleben, H.J., D. Stievenard, J.C. Bourgoin and A. Huber, 1985, Appl. Phys. Lett. **47**, 970.
von Bardeleben, H.J., J.C. Bourgoin, P. Basmaji and P. Gibart, 1989, Phys. Rev. B **40**, 5892.
von Bardeleben, H.J., J.C. Bourgoin, C. Delerue and M. Lannoo, 1991, Phys. Rev. B **44**, 9060.
Wagner, J., H. Ennen and H.D. Müller, 1986, J. Appl. Phys. **59**, 1202.
Watkins, G.D., 1963, J. Phys. Soc. Jpn. **18**(II), 22.
Watkins, G.D., 1964, Effet des Rayonnements sur les Semiconducteurs (Dunod, Paris) p. 97.
Watkins, G.D., and R.P. Messmer, 1974, Phys. Rev. Lett. **32**, 1244.
Weber, E.R., H. Ennen, U. Kaufmann, J. Windscheif, J. Schneider and T. Wosinski, 1982, J. Appl. Phys. **53**, 6140.
Wolfsberg, M., and L. Helmholtz, 1952, J. Chem. Phys. **20**, 837.
Wybourne, B.G., 1965, Spectroscopic Properties of Rare Earths (Wiley Interscience, New York).
Yamaguchi, E., K. Shiraishi and T. Ohno, 1990, in: The Physics of Semiconductors, Vol. 1, eds E.M. Anastassakis and J.D. Joannopoulos (World Scientific, Singapore) p. 501.
Zhang, S.B., and D.J. Chadi, 1990, Phys. Rev. B **42**, 7174.
Zunger, A., 1986, Solid State Phys. **39**, 275.
Zunger, A., and U. Lindefelt, 1983, Phys. Rev. B **27**, 1191.

CHAPTER 5

Impurities in Semiconductors: Experimental

J. M. BARANOWSKI and M. GRYNBERG

Institute of Experimental Physics
University of Warsaw
Warsaw, Poland

Handbook on Semiconductors
Completely Revised Edition
Edited by T.S. Moss
Volume 1, edited by P.T. Landsberg

Contents

1. General 163
2. Absorption 163
3. Photoconductivity 173
4. Luminescence 177
5. Raman scattering 182
6. Electron paramagnetic resonance 182
7. Electrical measurements 186
   7.1. Classical transport 186
   7.2. Capacitance transient spectroscopy 188
8. Summary 191
References 192

1. General

The standard experimental methods of investigation of shallow and deep impurity and defect states in semiconductors and the most significant experimental results are presented in this chapter. The main stress is put on those experiments that supply information about the identification of the nature of the center, the determination of the energy position of the ground and excited states of the impurity, the impurity wavefunction, and lattice relaxation around the impurity.

This information may be obtained by applying the most common experimental techniques such as absorption, photoconductivity, photocapacitance, luminescence, deep-level transient spectroscopy, Raman scattering, electron paramagnetic resonance and transport measurements. The experimental data presented and discussed are given mainly for the most typical semiconductors such as Si, Ge, A_3–B_5 and A_2–B_6 compounds.

2. Absorption

Absorption of light is one of the most successful methods for the investigation of impurities in semiconductors. There are two areas in which absorption has been very fruitful. These are ground state to excited state optical transitions for shallow and deep impurities, and photoionization transitions. Ground state to excited state optical transitions as well as photoionization transitions provide information about the energies and the symmetry wavefunctions of shallow and deep impurities. The investigation of local modes may provide information about the local symmetry of the neighborhood around an impurity or a defect.

Shallow donors and acceptors in semiconductors have been studied in much more detail than any other impurities. These impurities are described by an effective-mass equation analogous to the Schrödinger equation for a hydrogen atom and are often called hydrogen-like impurities. The effective-mass theory explains the shallow impurities in terms of hydrogen-like system immersed in a medium of relative dielectric constant ε_r.

The typical relative dielectric constant of a semiconductor has a value of greater than 10, which leads to a reduction of binding energy E_B by more than two orders of magnitude. The effective mass, which is generally smaller than the free-electron mass, reduces the binding energy E_B. The binding energies of hydrogenic donors and acceptors in GaAs, Si and Ge range from a few meV to about 100 meV. Optical transitions in hydrogenic centers therefore take place in the far-infrared region of the electromagnetic spectrum (Ramdas and Rodriguez 1981).

At present Fourier transform spectroscopy has been one of the most important tools in the study of shallow impurities in semiconductors. At low temperatures, in crystals containing a relatively low concentration of impurities (Bohr radius smaller than the interimpurity distance) the ground state to excited state absorption is observed as a series of very sharp lines, in some cases as narrow as a few μeV (0.01 cm^{-1}). The shallow impurities in Si and Ge have been studied most extensively. This is because in Si and Ge, due to pure covalent bonds, there is an absence of the fundamental lattice absorption and the material is transparent in the far-infrared region. A typical example of intra-impurity absorption of phosphorus in Ge is shown in fig. 1 (Reuszer and Fisher 1964).

For donors in such semiconductors as Si and Ge the hydrogen-atom-like description is more complicated – the effective mass is anisotropic because the band extremes lie away from the center of the Brillouin zone. Donors in Si have a sixfold set of 1s states due to the location of the six conduction-band minima along the [100] crystal momentum space axes (k-space). The states are grouped into a 1s singlet (A_1), a 1s doublet (E) and a 1s triplet (T_2). The singlet 1s (A_1) is most sensitive to the short-range potential at the core of the impurity and is typically the lowest-lying state – called the ground state. The 1s (E) and 1s (T_2) states are less sensitive to the impurity core potential and lie near the energy predicted by the effective-mass theory. The situation for donors in Ge is different because the four conduction-band minima are located along the [111] axes in k-space leading to 1s (A_1) and 1s (T_2) sets of states. The particular band structures of Si and Ge also cause the p-states to split into a p_0 singlet state (magnetic quantum number $m = 0$) and a p_+ and p_- doublet state ($m = \pm 1$).

As is seen in fig. 1, at 10 K only transitions from the lowest 1s (A_1) state are observed; however, at about 18 K the population of the 1s (T_2) state is sufficient to observe the E_3 line, which corresponds to the 1s (T_2)–$2p_0$ transition. Direct transitions between the split sublevels of the 1s state are forbidden by symmetry. However, the nonsphericity of the impurity potential can mix the 1s ground state with the

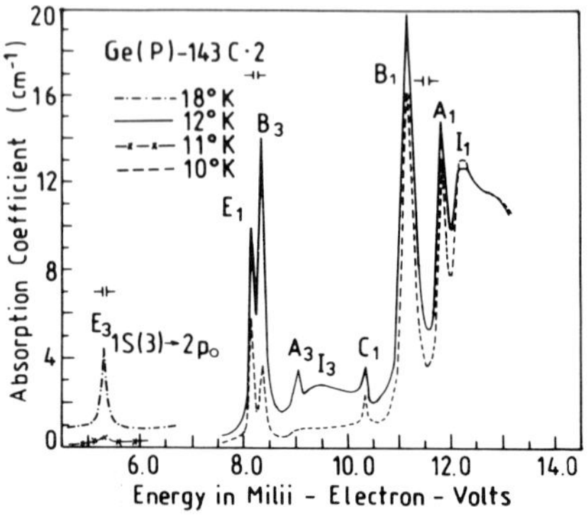

Fig. 1. The excitation spectrum of phosphorus impurities in germanium for different sample temperatures. $N_D = 8 \times 10^{14}$ cm^{-3} (after Reuszer and Fisher 1964).

higher p-state, and this allowed observation of the weak 1s (A_1)–1s (T_2) transition (Kobayashi and Norita 1977).

From these experiments, the ground- and excited-state energies can be evaluated. The summarized results for donor states in Ge are presented in fig. 2 and compared with effective-mass theory. It is seen that the position of the ground state depends on the impurity. The deviation of the energy of the ground state from the prediction of the effective-mass theory is called the chemical shift.

A hydrogen-like potential leads to the formation of impurity states not only below the lowest conduction-band minima, but also below the minima of the higher conduction bands (resonance states) (Paul 1968). An optical transition between the sulphur donor ground state and a resonance state in GaP:S has been observed in the absorption experiment by Gołdys et al. (1987) and is shown in fig. 3.

From an analysis of the result presented in fig. 3 it was possible to obtain the binding energy of the resonance state as well as the interaction of the resonance state with the lowest conduction-band continuum.

Shallow acceptors differ in several significant ways from donors. They are situated near the top of the valence band, which in diamond and zinc blende semiconductors is located at the center of the Brillouin zone. There are two bands, the heavy-hole and the light-hole band, which are degenerate at $k = 0$. The twofold band degeneracy together with the spin orientation lead to acceptor states of the Γ_8 symmetry. An

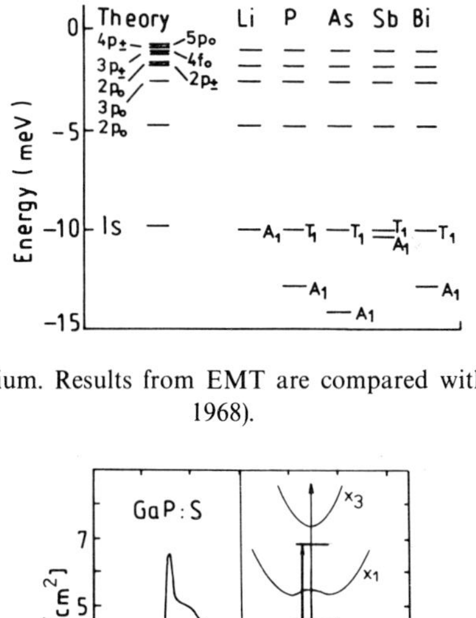

Fig. 2. Donor levels in germanium. Results from EMT are compared with experiment (after Faulkner 1968).

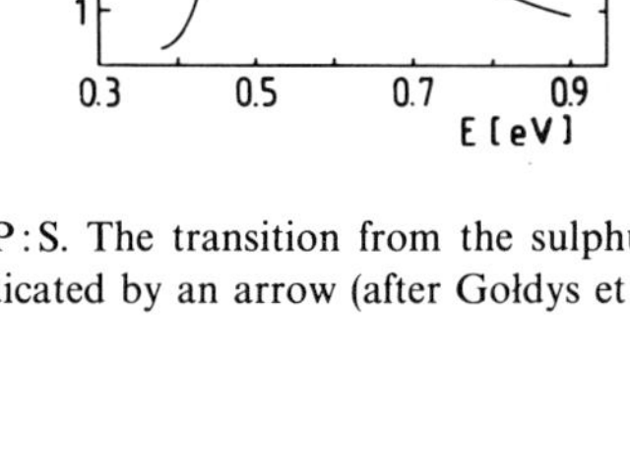

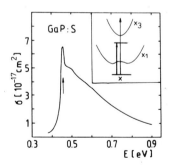

Fig. 3. Optical absorption for GaP:S. The transition from the sulphur ground state to the resonance state is indicated by an arrow (after Gołdys et al. 1987).

additional band lies below the top of heavy-hole and light-hole bands, separated by an energy difference called the spin–orbit splitting energy. This split-off band has an influence on the acceptor state in the case of Si because the split-off energy of 42.8 meV is of the same order of magnitude as the shallow-acceptor binding energies. In Ge the split-off energy is 300 meV and the influence of this band on the acceptor state is negligible. The optical spectra connected with shallow acceptors have been observed in Ge by Ahlburn and Ramdas (1968) and in Si by Onton et al. (1967). Figure 4 shows the characteristic absorption spectrum of boron acceptors in silicon. Two series of optical transitions, $P_{3/2}$ and $P_{1/2}$, are observed, which result from the relatively small spin–orbit splitting of the valence band. The Γ_8 symmetry of the ground state has been determined by Onton et al. (1967) using investigations of optical spectra under uniaxial stress.

In fig. 5 the results for different acceptors in Ge are summarized. As in the case of donors, the ground state of acceptors suffers from a clear chemical shift. On the other hand, all the non-s-like excited states of shallow impurities are insensitive to

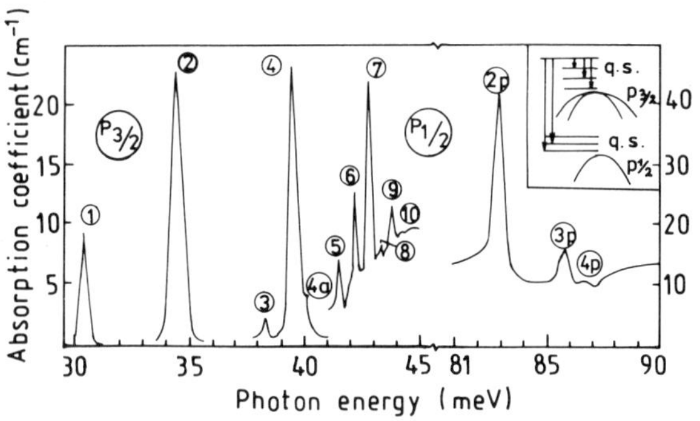

Fig. 4. Excitation spectrum of boron impurities in silicon, showing the two series of level associated with the $J = 3/2$ and $J = 1/2$ valence bands (after Onton et al. 1967).

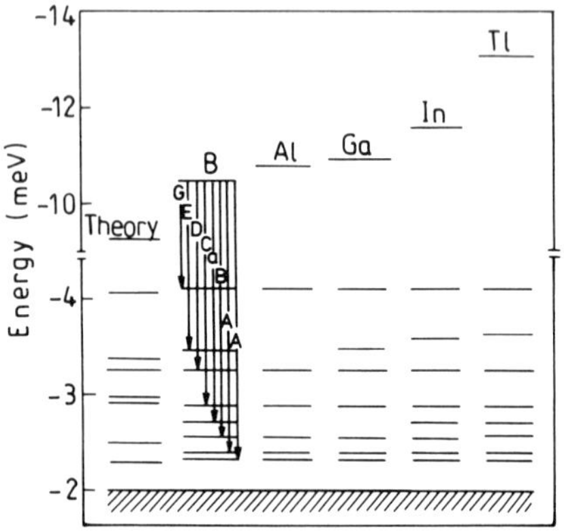

Fig. 5. Acceptor levels in germanium. Results from EMT are compared with experiment (after Jones and Fisher 1969).

the central-cell potential. The insensitivity is well understood on the basis of the nature of the p- or d-wavefunctions, which vanish at the impurity core. It also means that the central-cell potential is indeed localized, being limited to less than a lattice constant.

In A_3–B_5 and A_2–B_6 semiconductor compounds the inter-impurity transition of shallow impurities overlaps the strong lattice vibrational absorption. Only in a few cases was it possible to obtain results of a quality comparable with those for silicon and germanium (e.g., in AlSb, Jones and Fisher (1969)).

For compound semiconductors with a small effective mass or a high dielectric constant, the impurity Bohr radius is so large that even in the purest sample the impurity wavefunctions overlap and create an impurity band that merges with the band states. Application of a high external magnetic field for such cases leads to "shrinkage" of the impurity wavefunction and the creation of discrete states. This effect is called "magnetic freeze-out".

When the cyclotron energy $\hbar\omega_c$ is greater than the ionization energy of the impurity ($\hbar\omega_c/2R^* = \gamma \gg 1$ where $R^* \sim m^*/\varepsilon^2$, which takes place for n-InSb even for a weak external magnetic field, it is possible to observe the series of impurity states that are bound to each Landau sublevel (Kaplan 1969, Kuchar et al. 1977)). Optical transitions between the ground state and the excited states have been observed in extremely pure n-type InSb ($N_D - N_A = 5 \times 10^{13}\ \text{cm}^{-3}$) at pumped liquid helium temperatures in magnetic fields of 2–10 T. The results obtained (for energy, selection rules) are in good agreement with the effective-mass theory (EMT) calculation. Only some small correction to the ground-state energy is necessary. The experiment has been done by Kuchar et al. (1977) with known impurities, which have been obtained by nuclear transmutations. Comparing their result with the theoretical calculations for the excited-state energy as a function of the magnetic field, the magnetic field dependence of the impurity binding energy was obtained.

In zero-gap semiconductors like α-Sn, or HgTe, the impurity state is a resonant state. In the absence of an external magnetic field only conductivity measurements yield any information about the density of states connected with the impurity. In the quantized magnetic field the optical transitions from occupied impurity states to Landau subbands have been observed (Tuchendler et al. 1973). These transitions have been observed in transmission, using a far-infrared molecular laser or backward tube as a light source and a sweeping external magnetic field. From a linear extrapolation of the transition energy as a function of the magnetic field the ionization energy at zero magnetic field has been evaluated. For acceptor centers in HgTe two energies, 0.7 and 2.2 meV, have been obtained. Similar observations on acceptors in mixed crystals (HgTe–CdTe) show an increase of the ionization energy with a decrease of the Γ_6–Γ_8 gap (in alloys with a zero-gap structure). In zero-gap materials there is no observation of a discrete donor state, probably owing to the large half-width of the donor resonance state being degenerate with the high density of states in the valence band.

For shallow impurities in non-zero gap semiconductors the response to a magnetic field appears to be well described by effective-mass theory. An example of this is the

Zeeman effect for donors in GaAs, where very good agreement was found between experiment and a variational calculation (Stillman et al. 1971).

In highly compensated n-type crystals the donor states are completely ionized. Using "interband light excitation" one can "pump" electrons from deep states in the gap into the conduction band, from where they can subsequently drop onto shallow donor states. This gives rise to the possibility of observing intra-impurity transitions in transmission or magnetophotoconductivity experiments.

Each donor "feels" a different electric field which comes from the unscreened ionized impurities (no free carriers for screening). This leads to a substantial change of the magnetophotoconductivity, broadening of the 1s–2p$^+$ line, and the appearance of a broad structure at low magnetic fields. The results obtained for semi-insulating GaAs (Karpierz et al. 1990) are shown in fig. 6.

A comparison of the results obtained for n-GaAs and semi-insulating GaAs with a similar concentration of donors shows the influence of a fluctuating potential in strongly compensated materials.

A neutral donor can bind an additional electron, forming the D$^-$ center (in analogy with the H$^-$ ion (Chandrashekar 1944)).

The small binding energy $E^{B}_{D^-} = 0.055R^*$ makes the observation of these states extremely difficult. In the presence of a high magnetic field, due to wavefunction localization, the binding energy increases and photoionization transitions (D$^-$ ground state → Landau levels) can be observed. The results for epitaxially grown n-GaAs are shown in fig. 7 (Najda 1989).

The D$^-$ center in bulk crystals can be observed in the best-quality samples only, and "interband optical pumping" (making the concentration of D$^-$ centers high enough for observation) is needed.

For shallow donors in quantum wells (QW) the ground state energy depends on the donor position (Bastard 1981, Jarosik et al. 1985). For a donor located in the center of the QW, the ground-state energy is much greater (depending on the QW width, conduction-band offset and donor Bohr radius) than for a donor located at the interface or in the barrier. However, the energies of the excited states are

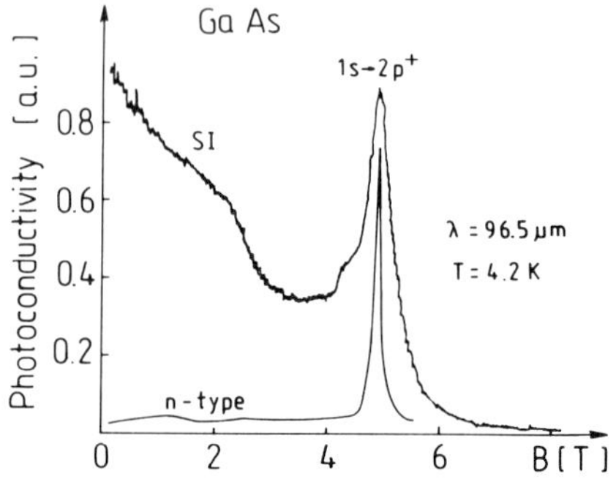

Fig. 6. Photoconductivity as a function of external magnetic field for n-type and SI (excited with 0.86 μm light) GaAs. Both samples have donor concentrations $N_D \sim 10^{16}$ cm^{-3} (after Karpierz et al. 1990).

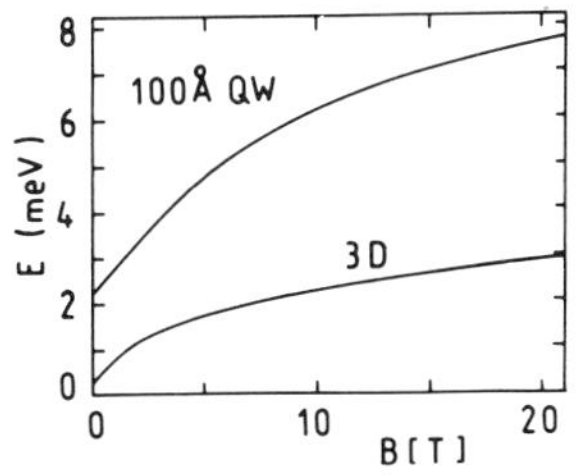

Fig. 7. D^- binding energy as a function of external magnetic field in bulk GaAs: 3D and 100 Å GaAs/GaAlAs MQW doped in the center of each QW and each barrier (after Huant et al. 1990).

practically independent of position, so that the $1s \rightarrow 2p^+$ transition energy is different for donors located at different distances from the QW centers (Huant et al. 1990).

Figure 8 shows the magnetotransmission results obtained on a multiple quantum well (MQW) of GaAs/GaAlAs with a well width of 100 Å and a barrier width of 200 Å, for a different level of doping. The $1s \rightarrow 2p^+$ transition energy difference for donors located in the center of the well and the center of the barrier is visible.

Structure B (fig. 8) is related to the photoionization of a D^- center formed in the center of the QW. Electrons from donors in the barrier fall into the QW and form D^- states in the QW.

The formation of the D^- states is facilitated because the number of electrons in the QW is greater than the number of donors. The formation of D^- centers in a QW is therefore more probable than in a bulk crystal, and the structure "B" can be observed without "interband optical pumping". Due to localization in the QW the D^- binding energy is much greater than in a bulk crystal. Localization caused by a high magnetic field strongly increases the D^- center binding energy; see fig. 7.

In contrast to hydrogenic-like impurities much less is known about the excited states of deep impurities. However, absorption measurements have been used to study the excited states of certain classes of deep impurities such as transition-metal and rare-earth impurities. A great deal of work has been done to study the excited states of the transition-metal impurities in II–VI and III–V compounds. These

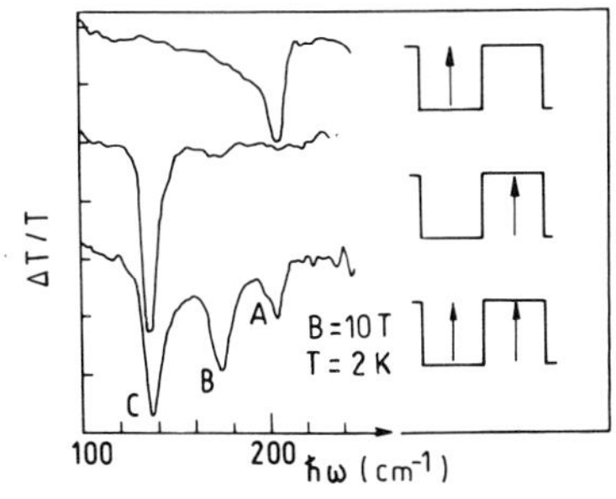

Fig. 8. Transmission as a function of energy of a GaAs/GaAlAs MQW. A and C: $1s \rightarrow 2p^+$ transitions of donors in a barrier and donors in a QW center, respectively. B: photoionization of a D^- center in the QW (after Huant et al. 1990).

impurities are well localized and because of the strong interactions between the 3d electrons they cannot be described using effective-mass theory. Early examples of the optical investigations are the works of Pappalardo and Dietz (1961), Weakliem (1962), Baranowski et al. (1967), Vallin et al. (1970), Wray and Allen (1971), Langer and Baranowski (1971). Optical transitions to the excited states of the $3d^n$ shells can identify the configuration and the site symmetry. Optical methods are complementary to EPR; some configurations are more easily seen by one method than by the other. It is generally found that the optical spectra of divalent transition-metal impurities in II–VI and III–V compounds are reasonably well described by crystal-field theory. A crystal-field approach uses some empirical parameters such as the Racah parameters B and C (Griffith 1964) describing the interaction between d electrons, and also the crystal-field parameter Δ connected with additional splitting of the d-states in solids by a crystal field arising from the nearest neighbors. The study of the fine structure of the absorption bands revealed that they are due to phonon and spin–orbit interactions, which are often coupled via the Jahn–Teller effect. The importance of the Jahn–Teller coupling has been recognized in the work of Ham and Slack (1971), Kaufmann and Koidl (1973), Koidl et al. (1973) and Uba and Baranowski (1978).

Piezoabsorption is an important technique for determining the symmetry of the defect. A good example of the application of this technique is the investigation of the uniaxial-stress splitting of the zero-phonon line arising from the optical transition to the excited state of the EL2 defect in GaAs. The EL2 defect has been identified with the midgap level in undoped GaAs and has been found to exhibit unusual properties; when exposed to light a metastable state which is electronically and optically inactive is obtained. This was shown for the first time by Bois and Vincent (1977) and by Martin (1981). The defect can be returned from the metastable to the equilibrium state by thermal treatment. The EL2 internal optical transition, which transfers the defect into the metastable state, possesses a characteristic zero-phonon line, which was found by Kamińska et al. (1983). Investigation of the zero-phonon line under uniaxial stress identified the presence of the A_1–T_2 optical transition and therefore of the tetrahedral symmetry of the EL2 defect in the normal state, as has been shown by Kamińska et al. (1985).

Absorption of light can cause the photoionization of the defect. In the photoionization process either the initial or the final state is not localized near the defect. This process can be observed not only in absorption but also using other experimental techniques such as photoconductivity measurements in bulk materials or photocapacitance and photocurrent measurements on p–n junctions or Schottky barriers.

The simplest information that can be obtained from the absorption measurements is the position of the impurity level within the forbidden energy gap of the semiconductor. For example, a study of the acceptor–conduction-band transition in InSb using the wavelength modulation technique revealed the ionization energy and splitting of the impurity level under uniaxial stress (Sharan and Heasell 1970).

However, optical methods can provide much more information than the energy of the impurity level. Due to the great sensitivity of the optical techniques used the spectral distribution of the photoionization cross section $\sigma(\hbar\omega) = \alpha(\hbar\omega)/N$ (where

$\alpha(\hbar\omega)$ is the absorption coefficient connected with photoionization transitions, and N the concentration of impurity centers) can be accurately measured over several orders of magnitudes. This type of measurement can provide information about the impurity wavefunctions and the lattice relaxation as well.

There are two ways of approaching the problem of photoionization transitions; one that considers an optical transition as purely electronic and the other that takes into account lattice relaxation effects as well. The different approaches to the electronic photoionization transitions are based on the impurity potential. For shallow impurities described by a Coulombic potential and the effective-mass approximation the quantum defect model proposed by Bebb (1969) has been used. The quantum defect method assumes:

(1) an isotropic, nondegenerate and parabolic energy band,

(2) an impurity initial state described by effective-mass theory being associated with one of the energy bands.

The expression for the absorption cross section given by Bebb (1969) uses as a fitting parameter the effective quantum number $\nu \equiv (R_0/E_B)^{1/2}$, where R_0 is the effective Rydgerg constant and E_B is the experimentally determined defect binding energy. For hydrogenic impurities ν tends to one and $\sigma(\hbar\omega)$ is in agreement with the calculations of Eagles (1960). The opposite extreme, ν close to zero, is appropriate for a potential for which there is no long-range Coulomb tail. The photoionization cross section (or absorption) in this limit is equivalent to that given by Lucovsky (1965). As the potential becomes less Coulombic the shape of the photoionization spectrum spreads out and its peak moves to higher energies. The quantum defect method interpolates between the two extremes.

Some shallow acceptors such as Al, In, Ga in Si and Hg in Ge have been described with the help of the quantum defect model proposed by Bebb and Chapman (1967). However, in the interpretation attributed to photoionization transitions of deep centers the Lucovsky formula has often been used. If the impurity potential is so localized that it can be approximated by a delta function we obtain the following formula for the optical cross section:

$$\sigma(\hbar\omega) = \frac{1}{n}\left(\frac{\varepsilon_{\text{eff}}}{\varepsilon_0}\right)^2 \frac{16\pi e^2 h}{3m^* c} \frac{E_1^{1/2}(\hbar\omega - E_1)^{3/2}}{(\hbar\omega)^3}, \tag{1}$$

where E_1 is the ionization energy, $\hbar\omega$ the photon energy, m^* the effective mass, and n the index of refraction. The ratio $\varepsilon_{\text{eff}}/\varepsilon_0$ is an effective-field correction, which takes account of the ratio between the actual local electric field at the impurity relative to the average macroscopic field of the photon and is usually approximated by $\frac{1}{3}(n^2+2)$, as was shown by Dexter (1958). It was found that the infrared photoionization absorption present in Au- and Ag-doped GaAs (Queisser 1971) is well described by the Lucovsky formula. Other examples are transition-metal acceptors such as Mn ($E_1 = 0.1$ eV, Queisser (1971), Chapman and Hutchinson (1967)) Co ($E_1 = 0.14$ eV, Baranowski et al. (1972a)), Cu ($E_1 = 0.157$ eV, Queisser (1971)).

The optical ionization energies of these impurities are in agreement with thermal ones obtained from the Hall effect measurements performed by Brown and Blakemore (1972) and Haisty and Cronin (1964), which indicates that the lattice relaxation effects are negligible. In the case of GaAs : Mn a sharp structure is present below the onset of the ionization continuum, as is seen in fig. 9. These sharp lines have been interpreted as being due to transitions from the ground state to hydrogenic-like excited states of the acceptor. These excited states indicate that in addition to the localized part of the impurity potential there is also a long-range Coulombic part. Remarkably the delta function limit appears to give a better description of the photoionization absorption. It has also been found by Messenger and Blakemore (1971) that In in Si is better described by the delta function potential than by the quantum defect model.

The Lucovsky formula can be obtained by assuming that an impurity wavefunction can be expressed by a combination of wavefunctions of the band to which transitions are taking place, as was shown by Kopylov and Pikhtin (1974). The square of the matrix element of the optical transitions in such a case is proportional to the energy in the band, which leads to a spectral dependence of the absorption coefficient

$$\alpha(\hbar\omega) \sim (\hbar\omega - E_\mathrm{I})^{3/2}/(\hbar\omega)^3$$

which is characteristic for "parity-forbidden" optical transitions.

In the case when an impurity wavefunction is composed of wavefunctions of opposite parity to those of the band to which optical transitions are taking place, the matrix element for optical transitions is independent of energy. The spectral dependence of the photoionization absorption reflects the density of states in the band; $\alpha(\hbar\omega) \sim (\hbar\omega - E_\mathrm{I})^{1/2}/(\hbar\omega)^3$ is characteristic of "allowed" optical transitions. In fact, when a localized impurity wavefunction is composed of the wavefunctions of several bands the spectral dependence of the photoionization absorption can be intermediate between the "forbidden" and "allowed" limits, as has been shown by Jaros (1975). Oxygen in GaP is an example of an impurity whose wavefunction consists of contributions from several bands, as has been shown by Jaros and Ross (1974) and Jaros (1975).

For the case of localized impurities it was pointed out by Jaros (1977) that it is important to consider the influence of lattice relaxation in the process of the photoionization transition. The equilibrium position of the lattice around the defect can be different for the initial and final states. For a strong impurity–lattice coupling it is customary to picture such an effect in a configuration coordinate diagram. The

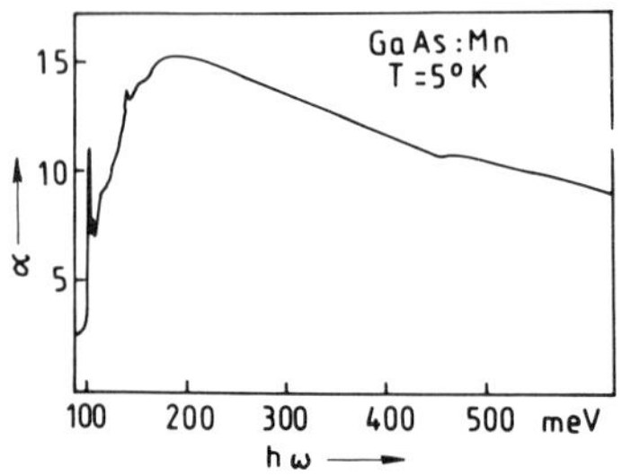

Fig. 9. Absorption due to transitions from the valence band to the acceptor level in GaAs : Mn (after Queisser 1971).

difference between the optical E_{opt} and thermal E_{th} ionization energies reflects the magnitude of the so-called Frank–Condon shift.

The study of photoionization absorption in GaP:Cu (Kopylov and Pithkin 1974) ZnS:Cr and ZnSe:Cr (Kamińska et al. 1978) has revealed the importance of lattice relaxation effects. The absorption due to the transition of an electron from the valence band into the $Cr^{2+}(d^4)$ charge state, resulting in the creation of a $Cr^{+}(d^5)$ state (identified by photoinduced EPR) in ZnSe is shown in fig. 10. The energy positions of the $Cr^{+}(d^5)$ charge state have been found to be 2.41 and 1.93 eV above the top of the valence band with Frank–Condon energies of 0.37 and 0.33 eV for ZnS and ZnSe, respectively. The value obtained for the lattice relaxation energy is much larger than the Jahn–Teller energy established in the $Cr(d^4)$ ground state, which is about 0.05 eV (Vallin et al. 1970, Vallin and Watkins 1974, Nygren et al. 1972). Because the ground state of $Cr^{+}(d^5)$ cannot suffer a Jahn–Teller distortion the lattice relaxation that takes place in the process of changing the charge state of the impurity has to be predominantly due to a totally symmetric distortion.

3. Photoconductivity

Photoconductivity measurements have been widely applied in the study of shallow and deep impurities. Information about the excited states of impurities, ionization energies, and emission and capture rates can be obtained from the photoconductivity measurements.

In the study of the excited states of shallow impurities the photoconductivity technique is more sensitive than the far-infrared absorption measurements. It was found that the photocurrent can be detected even when a strong background absorp-

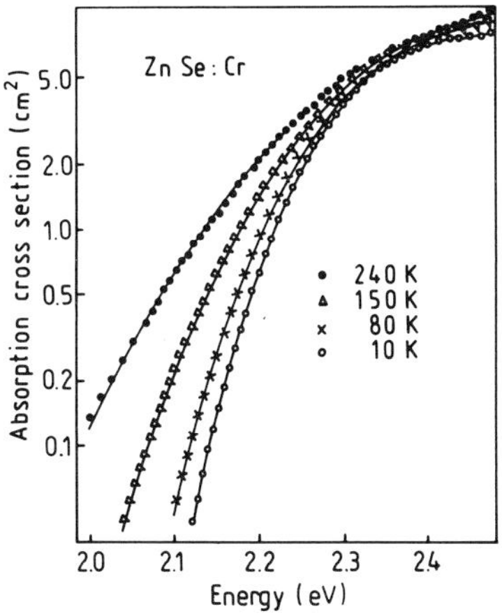

Fig. 10. Photoionization absorption cross section spectra connected with creation of the $Cr^{+}(d^5)$ charge state in ZnSe. Points are experimental and solid lines are calculated according to Kamińska et al. (1978), eq. (2).

tion is present. Therefore it was possible to follow the 1s→2p transition in III–V and II–VI compounds as far as the frequency of TO phonons, where an impurity study by absorption measurements is not possible. In addition, due to the high sensitivity of the photoconductivity measurements very pure samples could be studied. The line width could be reduced by as much as an order of magnitude in some cases compared to ordinary absorption measurements.

The detection of transitions between discrete states of an impurity by photoconductivity is possible using photothermal ionization as was shown by Lifshitz et al. (1968). The impurity being optically excited can be ionized due to the absorption of phonons or the impact ionization mechanism. A typical photoconductivity spectrum measured for GaAs containing three residual donors is shown in fig. 11 (Stillman et al. 1972). In addition to transitions from the 1s ground state it was also possible to study the transitions from the 2p state when the temperature was sufficient to populate it. The optical excitations from the 2p state enable one to find higher s- and d-states for which electric dipole transitions from the ground 1s state are forbidden by parity considerations. Using this technique, it was found that for s-like excited states the central-cell corrections are much less important than for the ground 1s state. The investigation of donors in GaAs done by Stradling (1976) showed, for example, that the differences in energy between the 2p, 2s states and the 3d state can be described on the basis of effective-mass theory with an accuracy of better than 1%. The results for donors in the much more polar material CdTe showed deviations by as much as 5%, which in part can be explained by polaron pinning and by a polaron-induced Lamb shift between the 2s and 2p states. It was also found by Stradling (1976) that for shallow donors in CdTe the parity-forbidden 1s→2s transition observed in the photocurrent is surprisingly strong when compared to the 1s→2p line.

Photoconductivity studies in the presence of an external magnetic field reveal the Zeeman splitting of the impurity levels. Examples of such investigations are the study of donors in compound semiconductors by Stradling (1976) and in Ge by Gershenzon

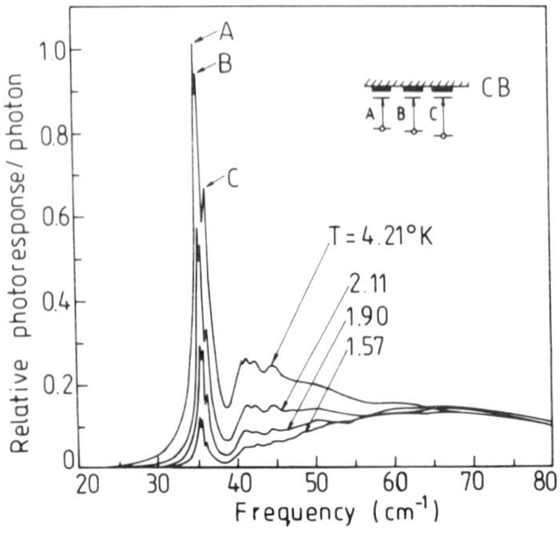

Fig. 11. Excited-state photoconductivity in GaAs. The inset shows the energy level diagram for the three different residual donors (after Stillman et al. 1972).

et al. (1973) and Nisida and Muro (1975). Two different techniques can be used in such studies: measurements in a constant magnetic field using a spectrophotometer (or Fourier interferometer) and measurements that use a molecular laser (or backward tube) as a monochromatic light source and a sweeping external magnetic field. Due to the power of the light source, the latter technique is more convenient for the study of transitions from the excited states or to investigate the extremely low impurity concentration. Such investigation of As and Sb donors in Ge by Nisida and Muro (1975) revealed interesting differences between these two impurities. It was found that despite the fact that the impurities have a similar scheme of excited levels the line widths for As are three times broader than for Sb. This implies that the lifetimes of the excited states are different for the two impurities.

The excited states of some deep impurities can also be studied by photoconductivity measurements. It was found that the excited levels of transition-metal impurities connected with states of the $3d^n$ shell may lie above the bottom of the conduction band and be resonant states. This was found, for example, in the study of the $Ti(d^2)$ impurity in CdS by Boyn et al. (1970), in CdTe by Baranowski et al. (1972b), and in ZnSe and CdSe by Kocot and Baranowski (1973, 1977). The internal optical transitions within the $3d^2$ shell of the Ti^{2+} impurity have been reflected in the photocurrent, which was interpreted as a manifestation of the autoionization of the impurity. There is strong experimental evidence that the same situation is present in $Cr(d^4)$ in GaAs (see, e.g., Stocker and Schmidt 1976).

Another type of photoconductivity measurement is connected with a study of the photoionization spectrum of the defect. These measurements provide information about the ionization energy, the impurity wavefunction, and the optical emission and capture rates. However, in the vast majority of cases the photoconductivity spectra have been normalized per incident photon. In such a case the photocurrent is a complex function of the optical emission and capture rates and information about these parameters cannot be obtained by a straightforward analysis. The threshold of such a normalized photocurrent can provide only approximate information about the ionization energy of an impurity (e.g., the increase of the sensitivity of the detection system can move the threshold to a lower energy). The reliable value for the position of the impurity level can be obtained from a properly interpreted spectral distribution of the photoionization cross section.

A technique for measuring the spectral distribution of the photoionization cross sections by photoconductivity measurements has been pioneered by Grimmeiss and Ledebo (1975). In this method the steady-state photocurrent is kept constant at different photon energies $h\nu$ by changing the intensity $I(h\nu)$ of the exciting light. A constant photocurrent is connected with a constant concentration of optically excited free carriers if changes of the carrier mobility are neglected. This implies that the occupancy of the impurity level is unchanged and the photoionization cross section $\sigma(h\nu)$ is obtained in a simple form:

$$\sigma(h\nu) = \text{const.}/I(h\nu). \tag{2}$$

Hence the spectral distribution of the photoionization cross section is obtained by plotting the inverse of the light intensity at a constant value of the photocurrent against the photon energy. It was shown by Grimmeiss and Ledebo (1975) that, in the case when more than one level is present, the photoionization cross sections are superimposed on each other in a plot of the inverse of light intensity. The above method employed by Grimmeiss and Ledebo (1975) in the study of GaAs:O leads to the photoionization spectrum shown in fig. 12. Interpretation of the photoionization cross section according to Lucovsky's formula permits the determination of three energy levels in GaAs:O. The spectral distribution of the photoionization cross section for GaP:O has also been obtained by Grimmeiss et al. (1974) using the constant-photocurrent technique. Due to the high photosensitivity of GaP:O it was possible to measure $\sigma(h\nu)$ over several orders of magnitude up to very small values of $\sigma(h\nu)$, which are difficult to obtain by other techniques. It should be pointed out that photoconductivity measurements, for example, in n-type material, can be used to investigate the spectral distribution of the photoionization cross section σ_n for electrons and σ_p for holes. Applying a measuring technique that uses two light sources allows, in some cases, the spectral distribution of the photoionization cross section σ_p to be obtained for the excitation of holes from the impurity energy levels to the valence band. First, illumination with light of a high intensity at a constant photon energy produces a photocurrent by two-step excitation, and holes created in the valence band can be captured by a hole trap. Additional simultaneous illumination $I(h\nu)$ in the infrared region is able to release holes from the hole trap to the valence band leading to some quenching of the photocurrent. If the intensity $I(h\nu)$ of this additional illumination is changed with photon energy in such a way that the amount of quenching is kept constant then the value of σ_p will follow the relation $\sigma_p = \text{const.}/I(h\nu)$. This technique permitted the determination of the spectral distribution of σ_p for the hole trap at 0.48 eV above the valence band in GaAs:O (Grimmeiss and Ledebo 1975). The advantage of a constant-photocurrent method is that this technique does not suffer from long decay times due to charge transfer between different levels and is not affected by the fact that the photocurrent has different intensity dependences at different photon energies.

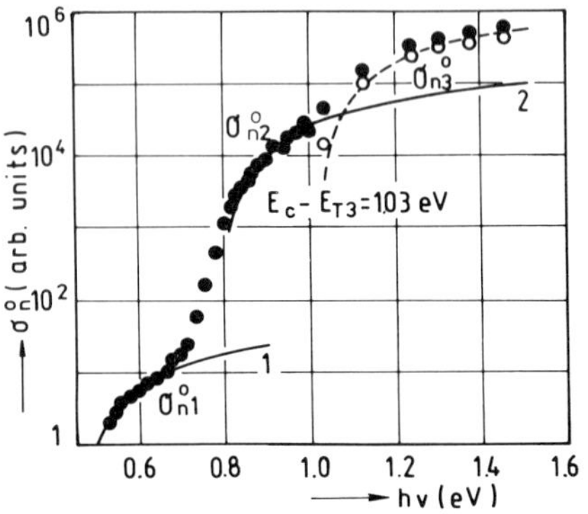

Fig. 12. Photoionization cross section for the excitation of electrons from the deep impurity levels to the conduction band versus photon energy. Solid curves have been calculated (after Grimmeiss and Ledebo 1974).

In addition to the bulk material techniques for detection and characterization of impurities, several techniques using p–n junctions and Schottky barriers have been developed. These techniques are especially useful for a study of deep impurities.

The photoionization cross section of impurities can be obtained from measurements of the photocurrent through a p–n junction. These can be measured by the transient techniques proposed by Sah et al. (1970) and Björklund and Grimmeiss (1970), which involve measurements of the decay time constant and the initial and final photocurrent when the sample is illuminated with monochromatic light. Another technique proposed by Ejder and Grimmeiss (1974) involves measurement of steady-state photocurrent using two light sources. A constant occupancy of the impurity levels is achieved using a high-intensity monochromatic light source with a properly chosen photon energy. The measurements of steady-state photocurrent due to additional monochromatic light can establish the photoionization cross section.

Some compound semiconductors exhibit a persistent impurity photoconductivity at low temperatures. This effect is characterized by a large photocurrent that persists for hours, or even days, after the optical excitation has been removed. It can usually be quenched only by heating the sample above some characteristic temperature. Such effects have been observed for III–V compounds by Bois and Vincent (1977) in GaAs, by Lang and Logan (1975, 1977) in $Al_xGa_{1-x}As$, by Crawford et al. (1968) in $GaAs_{1-x}P_x$, and for II–VI compounds by Lorenz et al. (1964), Iselev et al. (1972) and Dmowski et al. (1977a,b) in CdTe, by Burkey et al. (1967) in $Cd_{1-x}Zn_xTe$, and by Wright and Mooradian (1968) in CdS. A striking property of the centers responsible for permitting photoconductivity is an unusually large Stokes shift and an extremely small electron capture cross section at low temperatures. For example, the unidentified donor defect in n-type $Al_xGa_{1-x}As$ has a thermal depth $E_t \sim 0.1$ eV and an optimal depth $E_0 \sim 1.3$ eV, thus the Frank–Condon energy is 1.2 eV. The electron capture cross section at $T < 77$ K is very small ($<10^{-30}$ cm^2) (Lang and Logan 1977).

Persistent photoconductivity is the consequence of a strong impurity/lattice coupling and can be explained in a configuration-coordinate diagram. The equilibrium position of the lattice around the ionized impurity is different than for the impurity with a captured electron. Therefore an energy barrier exists between the two charge states. When the height of the barrier is large the tunneling probability is so small that a metastable photocurrent can persist for a long time.

4. Luminescence

Luminescence can be produced by optical excitation or by the injection of minority carriers, for example, across a p–n junction under an applied voltage (electroluminescence). It is usually associated with the localized states of impurities or bound excitons rather than a direct interband transition. The sharp luminescence lines obtained at low temperatures yield well defined energy assignments. In addition, the extreme sensitivity of the luminescence technique allows one to detect a very low concentration of impurities. There are, however, some shortcomings of the luminescence method that should be pointed out. In general no quantitative information about the impurity – such as its optical cross section – can be detected from the intensity of the luminescence lines. This is because the magnitude of the luminescence is in most cases limited by the unknown nonradiative recombination process. The luminescence intensity differs from sample to sample and often changes from place to place in the crystal investigated. It is known that charged defects and dislocations quench the luminescence in semiconductors. Microscopic studies of crystals that have been etched to reveal the dislocations have shown dark regions of little or no luminescence, a few micrometers wide around each dislocation, in GaP and other semiconductors (Heinke 1974, Queisser 1976, Kishino et al. 1976, Iwanomoto and Kasani 1976). It has been shown by Tasker and Stoneham (1977) that internal electric fields caused by the presence of charged dislocations and point defects are capable of preventing the binding of a carrier in a shallow state of an impurity and thus preventing luminescence from this center. In some cases quantitative information about the impurity can be obtained. For impurity centers showing radiative recombination, the photoionization cross section can be determined from photoluminescence excitation (PLE) and quenching (PLQ) spectra, as has been shown by Grimmeiss and Monemar (1973), Ejder and Grimmeiss (1974), and Monemar and Samuelson (1976a,b). Such experiments have so far been carried out only for a few centers in II–V compounds.

A study of bound excitons can give several types of information about the impurity centers in semiconductors. A bound exciton can be described as a free exciton localized by weak interaction with the impurity or defect. The bound excitons are usually observed in luminescence, and the emission lines are narrower than the lines corresponding to transitions in which free carriers are involved. The detailed identification of the charge state and type of the defect to which the exciton is bound can be found from Zeeman and uniaxial-stress experiments. Exciton binding to neutral donors and acceptors has been observed in many systems and shows systematic trends known as Haynes rule. The ratios of excitonic binding energy E_{ex} to the ionization energy of a donor E_D or of an acceptor E_A to which the exciton is bound, measured for a variety of defects in different crystals, are essentially the same in all cases. It was shown by Halstead and Aven (1965) that for acceptor states $E_{ex}/E_A = 0.1$, and for donor states $E_{ex}/E_D = 0.2$. However, the simple direct proportionality fails for several III–V compounds like GaP (Dean et al. 1971a,b, Dean 1973), GaAs (White et al. 1973, Ashen et al. 1975), and InP (White et al. 1972a,b). However, in any case, the Haynes rule still provides a very useful aid to the identification of centers.

A luminescence study of bound excitons led to the discovery of a new kind of impurity state in semiconductors known as an isoelectronic trap. Isoelectronic impurities are defined as substitutional impurities (usually on anion sites) that have the same valence-electron structure as the atom they replace. Thus, isoelectronic impurities are neutral impurities and they do not lead to doping of the semiconductor. However, some of the isoelectronic impurities produce discrete levels within the forbidden energy gap. This is because free carriers can be trapped by their short-range interaction with the defect. Three charge states of the isoelectronic impurities are important. They are the empty neutral center, the center with one trapped carrier, and the center with a bound exciton. The short-range potential around the isoelectronic impurity need only be attractive to one of the two carriers. Thus, if one electron can be bound, a hole can be attracted by the Coulomb interaction to form a bound exciton. If a hole is bound, the same Coulomb interaction can trap an electron.

In III–V compounds the most important cases have been N and Bi substituting for the anions in GaP (Thomas and Hopfield 1966, Faulkner 1968), in InP (Dean et al. 1971a,b), and in the alloys $In_{1-x}Ga_xP$ (Scifres et al. 1972) or $GaAs_{1-x}P_x$ (Scifres et al. 1971). Examples in II–VI compounds are an electron trap in ZnTe:O (Hopfield et al. 1966) and a hole trap in CdS:Te (Aten and Hoenstra 1964). Careful luminescence measurements have shown that the zero-phonon lines for excitons bound to isoelectronic impurities in ZnTe:O, GaP:N and GaP:Bi form a doublet. The emission spectrum of GaP:N (after Czaja 1971) is shown in fig. 13. The higher energetic line A is due to a recombination transition of an exciton bound to an isoelectronic impurity from the state with total angular momentum $J = 1$ to the crystal ground state with $J = 0$. The $J = 1$ state is due to the antiparallel alignment of the electron spin of 1/2 with the hole angular momentum of 3/2. The line B is connected with a $J = 2$ state arising from the parallel orientation of the electron spin and the hole angular momentum. The Zeeman splittings of the zero-phonon lines A and B have revealed information about the angular momenta.

An experimental tool with which to distinguish trapped electrons and holes can be obtained from the study of the structure of phonon wings in emission and absorption spectra. The Hopfield–Thomas model (Hopfield et al. 1966) predicts that trapped carriers are mainly responsible for the observed electron–phonon coupling. Considerations of the symmetry of the matrix elements that describe the phonon-assisted recombination of a bound exciton determines that only phonons of symmetry Γ_1 couple linearly to the trapped electrons and phonons of symmetries Γ_1 and Γ_{12}, and trapped holes of symmetry Γ_{15}. Such a study, for example, led to the conclusion that Bi in GaP acts as a hole trap. A hole trap in GaP:Bi has also been proved to exist by the observation of pair spectra between donors and the bound hole (Dean et al. 1969).

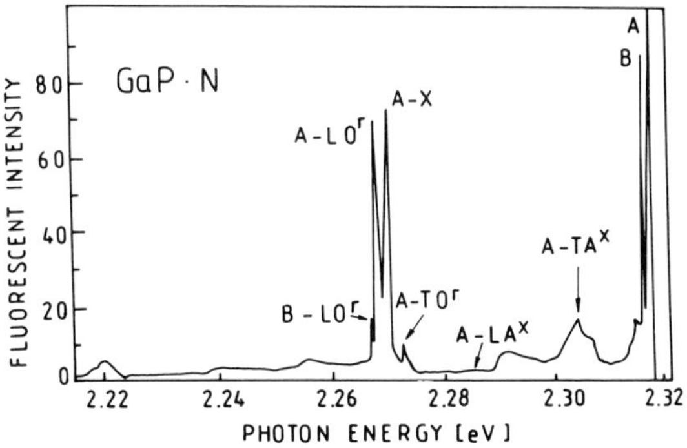

Fig. 13. Emission spectrum of GaP:N (after Czaja 1971).

The nitrogen impurity in GaP and $GaAs_{1-x}P_x$ and $In_{1-x}Ga_xP$ alloys introduces a shallow electron trap. This isoelectronic impurity has been very extensively studied because of its importance in LED. An investigative review of GaP:N has been given by Dean (1977).

An understanding of the electronic properties of several shallow electrically active impurities in GaP have largely been obtained from the study of pair spectra. The pair spectra observed in luminescence result from the recombination of an electron bound to a donor with a hole bound to an acceptor. When an electron bound to a donor recombines with a hole on an acceptor at a distance R from the donor, the Coulomb energy of interaction between the two ions lowers the energy of the final state by $e^2/\varepsilon R$. Therefore for each crystallographically inequivalent value of R (allowed by the lattice structure) a distinct emission line appears. The luminescence spectrum consists of several sharp lines whose energies are directly related to the sum of the donor E_D and the acceptor E_A ionization energies through the following simple relationship (Thomas et al. 1964):

$$h\nu = E_g - E_D + E_A + e^2/\varepsilon R + f(R). \tag{3}$$

The correction term $f(R)$ is significant only for relatively close pairs. The many donor and acceptor ionization energies in GaP collected by Dean (1977) were derived from the energy sums for donor–acceptor pair spectra. Independent knowledge of the donor ionization energies was obtained by Onton (1969) and Onton and Taylor (1970) from infrared photoionization spectra of donors and were confirmed through electrical transport measurements.

For semiconductors with a direct gap and a low electron effective mass it is often difficult to resolve discrete donor–acceptor pairs. For example, in GaAs and InP only the unresolved distant donor–acceptor pairs have been observed in the edge luminescence. It was also possible to observe the luminescence from free electrons to bound holes by a great increase in the optical pumping rate. These distant donor–acceptor pairs and free-electron–bound-hole luminescence bands have been used by Ashen et al. (1975) to distinguish and identify many shallow acceptors in GaAs.

In addition to the determination of the relative positions of the donor and acceptor levels the pair spectra offer a convenient means for mapping the strain fields around the donors. The strain fields generated by the donors can lift the orbital degeneracy of holes bound to nearby acceptors. The splitting of the acceptor level by the strain field of a donor depends on the magnitude and direction of the vector $\boldsymbol{R}$ from the donor to the acceptor (it also depends on the deformation potentials for acceptors along the $\langle 100 \rangle$ and $\langle 111 \rangle$ directions). Therefore the strain field around the donor can be mapped by measuring the splitting of the acceptor levels. It was found by Morgan (1972) that the strain fields associated with O, S, Se and Te donors in GaP are distributed over a volume of a crystal extending more than 20 Å from the impurity.

The luminescence measurements have also been used in the study of excited states of shallow and deep impurities.

The luminescence connected with transitions between excited states of hydrogenic-like impurities can be observed in the spectral region of 50–400 μm. This far-infrared radiation can be excited by electrical field pulses, which cause impact ionization of shallow impurities. Such luminescence has been observed for Sb, As, and P donors in Ge and for an unidentified shallow impurity in epitaxially grown n-GaAs by Thomas and Fan (1974). A typical emission spectrum for Ge:As for different rates of impact ionization is shown in fig. 14. The sharp peaks in the spectrum are attributed to the electron transition from the excited p states to the ground 1s state. The infrared emission experiment in the presence of a magnetic field was performed in InSb by Gornik (1972).

The binding energy of shallow impurity states in a QW depends strongly on the location of the impurity in the QW (Bastard 1981). In photoluminescence experiments in QW and MQW the electron transitions between donors and valence-band states (Stepniewski et al. 1989) and between conduction-band states and acceptors (Miller et al. 1982) were observed. For GaAs/GaAlAs QW with different donor doping geometries the photoluminescence experiments confirm the more precise results from intra-impurity magnetoabsorption in the far infrared, and show clearly the position dependence of the donor binding energy. For acceptors, photoluminescence seems to be a still more powerful and precise method in such investigations. The distribution of oxygen and carbon (uncontrolled acceptors) inside a QW was obtained from photoluminescence (Miller et al. 1982a,b).

The migration of some impurities (donors and acceptors) in the direction of crystal growth towards the interface was observed. A possible mechanism of such migration has recently been discussed (the role of the electric fields created by the charged surface during growth seems to be decisive).

The luminescence caused by well localized transition-metal impurities in II–VI compounds has been widely investigated. Examples of such work are the study of luminescence in near-infrared and visible light in V^{2+} by Le Manh Hoang and Baranowski (1977), in Cr^{2+} by Grebe et al. (1976), in Mn^{2+} by Busse et al. (1976), in Co^{2+} by Busse et al. (1970), Radliński (1978) and Dubenskii et al. (1965), and in Cu^{2+} by Broser et al. (1965). The observed luminescences are due to optical trans-

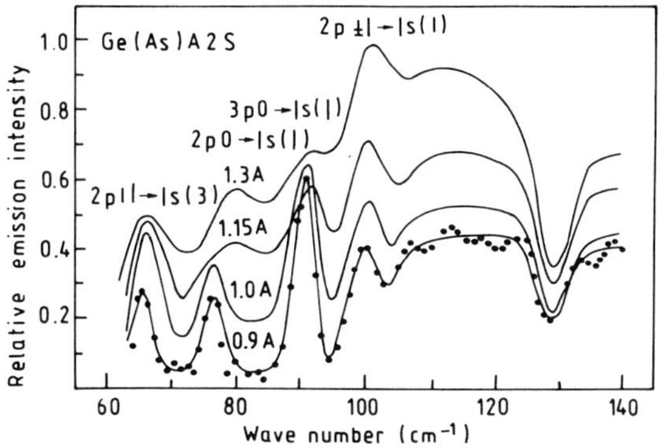

Fig. 14. Emission spectrum of the far-infrared luminescence of As impurities in germanium for different excitations (after Thomas and Fan 1974).

itions between the excited and ground states of the 3d shell of the cation substitutional transition-metal ion.

5. Raman scattering

In Raman scattering different elementary excitations of solids such as phonons, vibrational local modes, plasmons, and electronic transitions can be seen as Stokes or anit-Stokes lines. Normally one uses a laser operating in the visible region to cause the excitations. The intensity of the scattered light increases strongly when the energy of the excited light is in resonance with the electronic transitions in the crystal. This is known as the resonance Raman scattering (Barker and Loudon 1972). Electrons on donor atoms or holes on acceptor atoms scatter light very efficiently while transitions between bound states occur.

Measurements of the position, selection rules, and intensities of the scattered light from donors and acceptors can, in principle, yield information about the impurity wavefunction and the energy states. The first experiments of this type were done by Henry et al. (1966) on GaP containing Zn and Hg acceptors. They observed five lines whose energies corresponded to the transitions from the ground to the excited states of the acceptors. A more detailed study of the acceptors in GaP was performed by Manchon and Dean (1970). They found that the interimpurity transitions $1S_{p3/2} \rightarrow 1S_{p1/2}$ have a greater scattering intensity. The first donor study by light scattering was performed in Si:P by Wright and Mooradian (1968). The energy of the observed transitions agreed well with the energy of the $1S(1) \rightarrow 1S(2)$ valley–orbit transitions. They showed that these valley–orbit transitions should be the strongest donor transitions observed in Raman scattering. Similar transitions in GaP:S and GaP:Te were reported by Manchon and Dean (1970). Colwell and Klein (1972), and Dean and Hartman (1972) observed donor lines in Raman scattering in SiC (6H polytype). In CdS:Cl, where the minimum of the conduction band lies in the center of the Brillouin zone, there is no valley–orbit splitting. Henry and Nassau (1970) observed the strongest Raman scattering line, corresponding to the $1S \rightarrow 2S$ and $1S \rightarrow 2P_2$ transitions. This effect has been investigated in the resonance Raman regime. Some of the more intense interimpurity transitions observed in Raman scattering are not observed in absorption measurements because they are parity-forbidden transitions. Therefore Raman scattering is a very useful and complementary tool to other methods in the investigation of impurity states.

6. Electron paramagnetic resonance

Electron spin resonance (EPR) is one of the most powerful methods in the experimental physics of defects. The information that can be obtained from EPR includes the identification of the site of the defect and its symmetry and the wavefunction amplitudes at the various nuclei. An EPR study involves measuring microwave transitions, in the presence of a magnetic field, within a small group of energy levels. In a formal way it is possible to regard these ground-state levels (or metastable excited levels) as an isolated set whose properties can be described without reference to other levels. This method of description uses the so-called spin Hamiltonian and it is fully described, for example, by Abragam and Bleaney (1970). It is assumed that these levels can be represented by the effective spin S by equating the ground-state multiplicity to $2S+1$. Each level is therefore characterized by a quantum number M_S. The advantage of this method is that it permits the experimental results to be expressed simply in terms of a small number of parameters.

The simplest form of spin Hamiltonian is for $S<2$,

$$H_S = \sum_{ij} (\beta H_i g_{ij} S_j + I_i A_{ij} S_j + S_i D_{ij} S_j + I_i P_{ij} I_j). \tag{4}$$

The terms on the right-hand side of eq. (4) are the Zeeman energy, the hyperfine interaction, the zero-field splitting, and the quadrupole interaction, respectively. S is an electronic effective spin, I is a nuclear spin, H is a magnetic field and β is the Bohr magneton.

Anisotropy of the $\boldsymbol{g}$-tensor is directly related to the symmetry of the defect. For cubic symmetry the $\boldsymbol{g}$-tensor is isotropic. In the presence of an axial field in the crystal, uniaxial stress or a local deviation from cubic symmetry, due for example to Jahn–Teller distortion, the $\boldsymbol{g}$-tensor becomes anisotropic. The deviation of the g-factor due to spin–orbit coupling from a free-electron-spin value of $g=2.0023$ can be used as a guide to the identification of electron and hole centers. The g-shift equal to $g_{ii}-2.0023$, is negative for an electron and positive for a hole center. Information about the defect symmetry can also be obtained from zero-field splitting. The D_{ij} tensor of zero-field splitting vanishes in cubic symmetry and its form indicates the deviation of the environment from the cubic type. Important information about the amplitude of the impurity electronic wavefunction at the nuclei near a defect can be obtained from the observed hyperfine structure resulting from the coupling between the electron and nuclear spins. Detailed spin resonance and electron–nuclear double-resonance (ENDOR) measurements, which permit the determination of the hyperfine interaction between donor electron and lattice nuclei, are available for donors in Si (Hale and Mieher 1969, 1971). It was found that the amplitude of the impurity wavefunction on the impurity itself differs appreciably in magnitude from that expected in the effective-mass approximation. This directly indicates the necessity of a central-cell correction.

A large amount of detailed information about deep defects in crystals has been obtained using the EPR technique. Our understanding of vacancy centers in silicon is based almost entirely on the spin resonance work of Watkins (1963, 1968, 1975).

The vacancies in Si are striking in that they reorient and diffuse at low temperatures. The activation energies for reorientation of the vacancy in which the symmetry axis of the defect changes without any motion of the vacancy itself are rather low: 0.013 eV (V^+), 0.23 eV (V^0), 0.072 eV (V^{2-}). For the V^- center the unpaired electron wavefunction switches from one pair of atoms to another with an activation energy of 0.008 eV. The vacancy motion has been analyzed for V^0 and V^{2-}. The results also show the charge state dependence of the activation energy, 0.33 eV for V^0, and

0.18 eV for V^{2-}. A large number of vacancy impurity pairs in Si have been identified by EPR. The trapping of a vacancy by interstitial oxygen (Watkins and Corbett 1961), by the group-V atoms: P (Watkins and Corbett 1964), As and Sb (Elkin and Watkins 1968) can be given here as examples. The vacancy aggregates in Si have also been identified by EPR. The simplest vacancy aggregate, the divacancy, has been extensively studied in its single positive and negative $s=\frac{1}{2}$ charge state by Watkins and Corbett (1965), Corbett and Watkins (1965) and by Ammerlaan and Watkins (1972). More complicated vacancy aggregates such as the trivacancy, the tetravacancy and the pentavacancy have been identified using the spin resonance technique by Lee and Corbett (1973, 1974).

In II–VI compounds the chalcogen and metal vacancies have been identified. In ZnS an EPR center has been attributed by Schneider et al. (1965) to a single electron trapped by an isolated sulphur vacancy. The observed EPR spectrum was isotropic, and the resolved hyperfine interactions reveal that the electron wavefunction is spread equally over the four zinc nearest neighbors without a Jahn–Teller distortion. On the other hand, the metal vacancy observed in its single negative charge state in ZnSe, ZnS, ZnO and CdS tends to exhibit a large trigonal Jahn–Teller distortion (Watkins 1975). Estimates based on the effect of external stress on the EPR spectra indicate Jahn–Teller energies of the order of 1 eV. The metal vacancies in II–VI compounds are less mobile than those in Si. Activation energies for the migration of the zinc vacancy of 1.26 and 1.04 eV have been found. It has also been found that zinc vacancies in ZnS can create pairs with such impurities as chlorine, bromine, aluminum and gallium (Schneider et al. 1965).

A large number of EPR studies have been done on transition-metal impurities in semiconductors. In almost all II–VI compounds the divalent Cr^{2+}, Mn^{2+} and Co^{2+} ions have been identified. From the hyperfine structure the substitutional cation sites of these impurities have been established. As the transition metals with the electron configuration $3d^n4s^2$ are on substitutional cation sites in the II–VI compounds, two outer 4s electrons are promoted to bonding states giving a $3d^n$ configuration. An electron may be added or removed from the d shell giving $3d^{n-1}$ or $3d^{n+1}$, the trivalent or monovalent impurity charge state. These different charge states have been found for several transition-metal impurities in II–VI compounds. For example, the $Fe^{3+}(d^5)$ charge state has been found almost in all II–VI host lattices. The other examples are $Ni^{3+}(d^7)$ and $Ni^{+}(d^9)$ in ZnS (De Wit et al. 1964, Schulz and Wepfer 1972, Holton et al. 1964), ZnSe (Watts 1969), and CdS (Schulz and Wepfer 1972, Morigaki 1964) or $Fe^{+}(d^7)$ in ZnS (Holton et al. 1964). The monovalent and trivalent charge states can usually be detected in photosensitive EPR. The appearance of a new EPR signal after illumination of the sample with monochromatic light can be used for the identification of photoionization transitions and can help to determine the energy position of the particular charge state. The threshold energies in excitation spectra of the photoinduced EPR of the $Cr^{+}(d^5)$ have been found in ZnS (Fair et al. 1966), ZnSe (Suto et al. 1967), CdTe (Cieplak et al. 1975) and ZnTe (Suto and Aoki 1967). The isotropic $Cr^{+}(d^5)$ EPR was in this case created by the transition of an electron from the valence band into the $Cr^{2+}(d^4)$ charge state. In contrast to the isotropic $Cr^{+}(d^5)$ EPR spectrum the $Cr^{2+}(d^4)$ one revealed a purely tetragonal distortion (Vallin and Watkins 1974). The application of uniaxial stress was found to produce a substantial preferential alignment of the defects, indicating that the distortion is of Jahn–Teller origin. Thus the presence of a static Jahn–Teller distortion of the ground state of $Cr^{2+}(d^4)$ in II–VI compounds has been unambiguously determined in EPR measurements, and the Jahn–Teller energy has been found to be about 0.05 eV (Vallin and Watkins 1974).

EPR studies of impurities in III–V compounds are not as numerous as for II–VI compounds. The most extensively studied impurity was Cr in GaAs because of the importance of this material for semi-insulating substrates. The existence of three charge states, $Cr^{+}(d^5)$, $Cr^{2+}(d^4)$ and $Cr^{3+}(d^3)$ was identified using EPR by Kaufman and Schneider (1976) and Krebs and Strauss (1977a,b). It was found that it is possible to convert the charge state of a given Cr ion from 3+ or 2+ to 1+ with light of energy $h\nu$ smaller than the band gap of GaAs. The EPR spectrum connected with $Cr^{+}(d^5)$ reflects the undistorted tetrahedral symmetry of the gallium substitutional impurity. On the other hand, the reduction of the symmetry to D_{2d} and C_{2v} in the case of $Cr^{2+}(d^4)$ and $Cr(d^5)$, respectively, was observed, indicating the presence of a static Jahn–Teller distortion. Thus the EPR studies of GaAs:Cr not only proved the existence of different charge states of the Cr impurity but also showed that in an optical transition in which a change of the charge state of the Cr occurs the lattice distortion takes place.

In high magnetic fields, EPR on deep or shallow impurities can be studied using non-cavity methods. In a high field the Zeeman splitting of the ground state is greater than the thermal energy at liquid helium temperatures (LHeT), so that only the lowest state is populated. This leads to the spin polarization of the impurity system. For crystals with a finite free-electron concentration at LHeT, the free electrons also occupy the lowest possible state (the quantum limit). The two "polarized" systems, an impurity and free electrons, can interact via exchange (s–d, or p–d) interactions. In the EPR regime an impurity absorbs a photon, then relaxes by interacting with a conduction electron.

Such a spin "flip–flop" interaction contributes to the scattering of free electrons, thus increasing the resistivity. The results of resistivity measurements in the EPR regime for HgTe:Mn are shown in fig. 15 (Wittlin et al. 1980). The mechanism is effective for conduction electrons with wavefunctions having a mixed character, with both spin–up and spin–down parts (e.g., due to the spin–orbit interaction). A similar effect was observed in CdSe:Mn (Wittlin et al. 1980). For crystals without free electrons, another noncavity method – optical absorption – was used for high-magnetic-field EPR. In Cr-doped SI-GaAs, the Zeeman splitting of Cr^{2+} ions was observed (Wagner and White 1979). A more detailed study was done on GaP:S (Muller et al. 1989). In this optical method both electric dipole and magnetic dipole transitions can take place. However, even a small spin–orbit interaction causes "spin mixing" and leads to finite transition probabilities for electric dipole transitions, which become the dominant ones.

In crystals with a finite amount of compensation, hopping absorption in the far infrared (background absorption) leads to the "opening of an absorption window" in the "vicinity" of the EPR resonance, see fig. 16. The non-cavity EPR method in

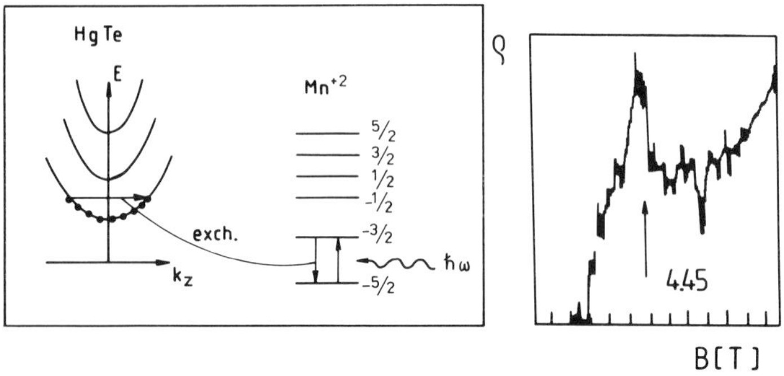

Fig. 15. Spin "flip–flop" mechanism in the EPR regime and resistivities plotted against external magnetic field B in HgTe:Mn at 4.2 K for an Mn concentration of about 3×10^{19} cm^{-3}, and $\hbar\omega = 0.5$ meV.

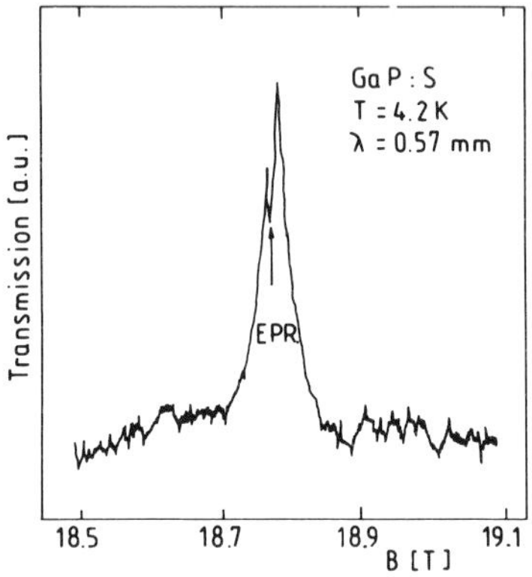

Fig. 16. Transmission plotted against external magnetic field. (Note the increase of transmission in the vicinity of the EPR resonance.) (After Muller et al. 1989.)

high magnetic fields due to a large Zeeman splitting is also very useful in studies of complex defects, e.g., thermal donors in silicon (Gregorkiewicz et al. 1991).

An EPR transition causes a change in the impurity system magnetization, and gives an important contribution to the total crystal magnetization. The relaxation of excited electrons occurs mainly due to spin–lattice interactions. The kinetics of this interaction can be directly observed via magnetization measurements under pulse excitation in the EPR regime. Experiments on heavily doped CdTe:Mn (Strutz et al. 1990) provide direct information about the spin relaxation in a wide range of magnetic fields and temperatures.

7. Electrical measurements

7.1. Classical transport

The Hall effect and conductivity measurements are the most common methods for obtaining the thermal ionization energies E_t of impurities in semiconductors. The

Hall effect gives direct information about the concentration of free electrons n or holes p. For the case where only one type of carrier is present $n = \pm r/Re$, where R is the Hall coefficient and r the ratio of the Hall mobility to the drift mobility. Measurements of the Hall coefficient R are usually very straightforward. The experimental determination of r is more difficult but fortunately its value never departs strongly from unity. The ionization energies of impurities are deduced from an analysis of the temperature dependence of the concentration $n(T)$. For the simplest case when n is determined by the ionization of only one impurity, the concentration n depends exponentially on temperature. The plot of $\ln n$ as a function of $1/T$ gives a straight line whose slope determines the ionization energy. If there is no compensation the slope is $E_t/2kT$. For a partially compensated semiconductor the slope is E_t/kT. If the electron gas is even partially degenerate, or if the free carriers come from more than one type of impurity, and also if the intrinsic excitations have to be taken into account, the equation of neutrals (see chapter 6 of this volume), which connects the concentration of free carriers with concentrations of ionized impurities, cannot be solved analytically. The ionization energies can therefore be derived only by numerical calculations. In such cases analysis requires additional information about the density of states in the band, the concentrations of impurities and the degeneracy of the impurity levels. For "multiellipsoid" semiconductors like Ge and Si splitting of the ground state of the impurity has to be considered (Koenig et al. 1962). For some impurities both the ground and excited states can influence the concentration of free carriers. This is especially important when levels connected with inequivalent conduction bands are close in energy, as for Se in GaSb (Paul 1968, Hoo and Becker 1976). Impurities and impurity complexes that are characterized by several charge states or several inequivalent lattice configurations require very careful analysis of the statistical model (Landsberg 1956, Blakemore 1962, Milnes 1973, Porowski et al. 1974, see also chapter 6 of this volume).

The values of the ionization energies of impurities E_t deduced from the temperature dependence of the free-carrier concentration are often substantially smaller than those obtained from optical measurements E_0. This is due to the fact that the lattice distortion and lattice energy in the vicinity of an impurity is a function of the level of occupation of the impurity state and therefore a Stokes shift is observed. For the so-called "hydrogenic" type of impurities the lattice distortion is usually very small. However, even for this type the difference between E_t and E_0 can be substantial. For instance, Koenig et al. (1962) found that E_t for As in germanium is about 10% smaller than E_0. Similar effects have been observed by Stillman et al. (1969) for donors in GaAs and by Casey et al. (1971) for GaP. However, some authors suggest that the observed difference between E_t and E_0 is the consequence of disregarding the conduction in the impurity band due to the overlap of the wavefunctions of the excited states or the different concentrations of impurities at which E_t and E_0 are measured.

Impurity–lattice coupling sometimes becomes dramatically stronger for deep impurities where the contribution of well localized impurity states to the bonds between the impurity and the nearest neighbors can be quite large. It was observed that in some cases the difference between E_t and E_0 can exceed 1 eV (Dmowski et al. 1977a,

Lang and Logan 1977). The changes of the electronic state of the impurities cause such strong lattice distortions, or even lattice rearrangements, that at low temperatures the potential barriers between different lattice configurations are high enough that the impurities cannot reach thermal equilibrium and metastable states are observed. This also causes effects such as a very slow relaxation of the conductivity and Hall constant, an abrupt change of ionization energies of impurities at a certain critical temperature and persistent photoconductivity below this critical temperature (Vul et al. 1970b, Iselev et al. 1972, Porowski et al. 1974, Burkey et al. 1967, Dmowski et al. 1977a). Hall and conductivity measurements have been widely used in the determination of the dependence of ionization energies of impurities on hydrostatic pressure (Paul 1963, 1968, Pitt 1977). Most of the pressure coefficients of localized levels in semiconductors fall within the range of pressure coefficients of the band states. This was interpreted in terms of wavefunctions of the localized level being a superposition of state functions of the valence and conduction bands. The pressure dependence of the energy of the shallow levels is almost the same as the energy of the nearest band. The observed small changes of ionization energy can be explained within the hydrogenic model due to the change of the effective mass and dielectric constant (Holland and Paul 1962, Paul 1963). However, it was shown by Paul (1968) and by Hoo and Becker (1976) that some deep levels also have pressure coefficients that are almost equal to those of the conduction-band minima. This was explained as a result of the wavefunction of the impurity level being described predominantly by the Bloch functions from a single minimum. Hall and conductivity measurements have been very useful in a study of resonant impurity states connected with the subsidiary minimum (Paul 1968). This was especially the case for zinc blende semiconductors with the resonant levels degenerate with the Γ band. Application of pressure increases the energy of the Γ minimum faster than for the resonant level. As a result one can observe that the resonant energy decreases with pressure and above a certain pressure the resonant state becomes a true bound state. For higher pressures its ionization energy increases and a very strong decrease of conductivity due to deionization of the level is observed. Such an effect was obtained for the first time by Foyt et al. (1966) in CdTe. The impurity levels associated with the subsidiary minimum have been studied extensively for GaSb doped with S, Se and Te (Kosicki et al. 1968, Paul 1968, Pitt 1969, Vul et al. 1970a, Hoo and Becker 1976, Dmowski et al. 1977b).

More difficult than the Hall effect are photo-Hall measurements. The photo-Hall measurements enable one to separate the hole and electron contributions when illumination with light of an energy smaller than the band gap is used. An example of the photo-Hall effect are the measurements of GaAs:Cr done by Lin and Bube (1976) and Look (1977).

7.2. *Capacitance transient spectroscopy*

The study of the electrical properties of semiconductors by capacitance measurements is now a well established technique in the experimental investigation of impurities. This technique is especially useful for highly conductive semiconductors. Employing such materials in the form of reverse-biased p–n junction structures or Schottky diodes made of conducting n- or p-type material it is possible to carry out capacitance measurements. Such measurements are made in a region of the semiconductor that is essentially free of mobile carriers. Therefore such measurements allow the study of localized electronic states in semiconductors. For p–n junctions and Schottky diodes the application of a reverse bias creates a depletion layer. The width W of the depletion (or space charge) layer is given by

$$W = \left(\frac{2\varepsilon(V_{bi} + V)}{eN} \right)^{1/2}.$$

Here ε is the dielectric constant of the depleted semiconductor material, V_{bi} is the built-in bias voltage, e is the charge of the electron, and N is the density of the ionized doping impurities in the depleted semiconductor. The above equation indicates that the reverse-biased p–n junction or Schottky barrier essentially forms a parallel-plate capacitor from a layer of dielectric of thickness W and dielectric constant ε between two conducting electrodes. The capacitance of this variable voltage capacitor is given by

$$c = \varepsilon A / W,$$

where A is the area of the junction.

It is clear from both equations that a plot of $1/c^2$ against the applied voltage V produces a straight line. The slope of this line gives the impurity concentration N in the semiconductor. A variety of capacitance and photocapacitance techniques has been used for a study of the concentration of impurities and the spectral dependence of the optical cross section of the impurities. One of the first studies was done for the deep oxygen donor in GaP (Kukimoto et al. 1973, Henry and Lang 1977), and provided reliable measurements of the concentration, optical cross section and binding energies for a one- and two-electron states of the impurity. A double-source differential photocapacitance technique removed a difficulty connected with a long capacitance response time (White et al. 1976). A different group of techniques is connected with the study of the capacitance transients under reverse-bias voltage step conditions (Sah et al. 1970, Lang 1974). The important feature of this method is that the trapping takes place outside the depletion layer and therefore the results are not affected by the strong electric field at the junction.

A review of all the experimental techniques for measuring the photocurrent and capacitance with p–n junctions, Schottky barrier contacts and with bulk materials has been given by Grimmeiss (1977). The junction photocurrent and photocapacitance techniques are among the most powerful tools for measuring the energy positions of impurities, both thermal E_t and optical E_0. The thermal ionization energies E_t are derived from the temperature dependence of the thermal emission cross sections (Sah and Forbes 1971, Sah et al. 1972, Zohta 1972, Sah et al. 1970). For the case of impurities with strong impurity–lattice couplings both the thermal-emission cross section and the thermal-capture cross section are thermally activated and thus the value of E_t is given as a difference between the activation energies of the emission and capture processes (Lang and Logan 1975). Optical ionization energies E_0 are obtained by comparing the measured spectral dependences of photoionization with theory (Rosier and Sah 1971, Grimmeiss and Monemar 1973, Olofsson 1974). However, for several cases the agreement of the measured spectrum of cross sections with Lucovsky's model is rather poor (Braun and Grimmeiss 1974, Herman and Sah 1973, White et al. 1976).

The fact that the capacitance connected with a depletion layer is voltage dependent led to the development of a very useful technique called DLTS (deep-level transient spectroscopy). If the junction is reversed biased with a fixed bias voltage, a depletion layer, and a corresponding capacitance, results. The junction is then cooled down to a sufficiently low temperature that the thermal-emission rate, for example, e_n for electrons, on the trap is negligible. The bias is then reduced to a substantially lower value by applying a filling pulse. During the filling pulse the traps in the corresponding region of the depletion layer are filled up with electrons. Following this the bias is returned to its original value, thereby leaving electrons trapped in the depletion layer. The width of the depletion layers is now larger than it would have been without the trapping process, since some of the positive charge of the ionized donors is now cancelled by the negative charge of the trapped electrons. As a result the junction capacitance has been decreased. Warming up the sample, while continuously monitoring the depletion layer capacitance, will lead to a temperature at which the thermal-emission rate $e_n \sim \exp(-E_T/kT)$ is no longer negligible (where E_T is the depth of the trap in respect to the bottom of the conduction band). Electrons thermally promoted to the conduction band are immediately swept out of the depletion layer by the electric field, resulting in a corresponding increase of the capacitance of the junction. This process will continue until all the charged traps in the depletion layer are emptied.

A very convenient method of measuring emission times has been introduced by Lang (1974). This DLTS method (introduced above) makes use of the exponential temperature dependence of e_n. As the temperature is swept through (hence the term "spectroscopy" in the name of the method) the emission will vary rapidly as schematically shown in fig. 17. In the DLTS, discrimination between the various emission rates is done by applying the "rate-window" method in which the capacitance transient at two different times, t_1 and t_2 is measured. As shown in fig. 17, the

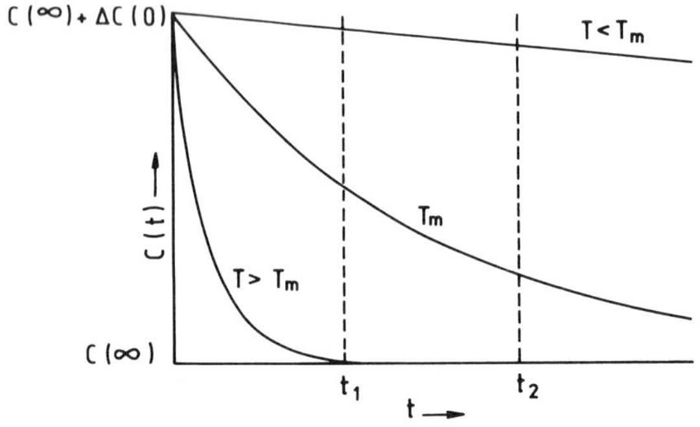

Fig. 17. Capacitance transients at various temperatures. At T_m, the signal $S \equiv C(t_1) - C(t_2)$ for a maximum.

quantity $S = C(T_1) - C(t_2)$ will be close to zero at low temperatures ($T \ll T_m$) and high temperatures ($T \gg T_m$), but will go through a maximum at the temperature T_m, which will satisfy the following condition

$$e_n(T_m) = \frac{\ln(t_2/t_1)}{t_2 - t_1}.$$

For variants of these procedures, see Landsberg and Shaban (1987).

In other words, observation of the repetitive capacitance transient through a rate window will lead to a peak in the temperature versus output plot. Such a plot is called a DLTS spectrum.

The dependence of e_n on T can be found by varying T_m, which in turn is accomplished by changing t_1 and t_2. In practice, it is common to vary t_1 and t_2 while the ratio of t_2/t_1 is kept constant. For each temperature at which a DLTS peak occurs an emission rate $e_n(T_m)$ can be calculated and the values of $e_n(T_m)/T_m^2$ may be plotted on a logarithmic scale versus T_m giving an Arrhenius plot. The slope of such a plot directly gives the trap energy. Originally the rate window technique was implemented by a double-boxcar integrator by Lang (1974). In addition to the boxcar technique there are several other ways to establish a rate window, such as the lock-in method proposed by Kimerling (1976) and the exponential correlation method developed by Miller et al. (1975). All three of these techniques are critically discussed by Miller et al. (1977).

A large number of traps have been found using the DLTS technique. One can find examples of the experimental results for electron and hole traps in GaAs in the paper by Martin et al. (1977) and by Mitonneau et al. (1977), respectively. In addition to the normal DLTS method a light source can be included in the system. By this means it is possible to excite minority carriers, and thus observe hole capture and emission. This method is called optical DLTS (ODLTS).

8. Summary

The present state of knowledge about impurities in semiconductors has been mostly obtained using the experimental methods that have been mentioned above. The understanding of the properties of shallow donors and acceptors in semiconductors is now reasonably good. This is due to a more extensive study of these impurities than any other defect systems. The interest which surrounded the shallow impurity states is that these systems are fundamental in solid state electronics. In addition, the theoretical description of the shallow-impurity states is relatively simple because of the validity of the effective-mass approximation.

On the other hand, the present state of knowledge about deep impurity centers is not so good. The progress in understanding deep impurities from the experimental side has often been slowed down by difficulties in the idenitification of the centers. The presence of localized impurities is closely related to the presence of lattice defects, vacancies and dislocations because of the tendency of one to form complexes with the other. Therefore it is often not clear to what extent the isolated impurity center is responsible for the investigated properties. The experimental techniques are slowly moving into the area of chemical identity of the defects. The other major difficulty connected with deep impurities is that a theory is lacking that could put the numerous scattered experimental data into one consistent framework.

References

Abragam, A., and B. Bleaney, 1970, Electron Paramagnetic Resonance (Clarendon Press, London).
Ahlburn, B.T., and A.K. Ramdas, 1968, Phys. Rev. **167**, 717.
Ammerlaan, C.A.J., and G.D. Watkins, 1972, Phys. Rev. B **5**, 3988.
Ashen, D.J., P.J. Dean, P.D. Green, D.T.J. Hurle, J.B. Mullin and A.M. White, 1975, J. Phys. & Chem. Solids **36**, 1041.
Aten, A.C., and J.J. Hoenstra, 1964, Phys. Lett. **11**, 97.
Baranowski, J.M., J.W. Allen and G.L. Pearson, 1967, Phys. Rev. **160**, 627.
Baranowski, J.M., M. Grynberg and E.M. Magerramov, 1972a, Phys. Status Solidi b **50**, 433.
Baranowski, J.M., J.M. Langer and S. Stefanowa, 1972b, in: Proc. 11th Int. Conf. on the Physics of Semiconductors (PWN, Warsaw) p. 1001.
Barker Jr, A.S., and R. Loudon, 1972, Rev. Mod. Phys. **44**, 18.
Bastard, G., 1981, Phys. Rev. B **24**, 4714.
Bebb, H.B., 1969, Phys. Rev. **185**, 1116.
Bebb, H.B., and R.A. Chapman, 1967, J. Phys. & Chem. Solids **28**, 2087.
Björklund, G., and H.G. Grimmeiss, 1970, Phys. Status Solidi **42**, K1.
Blakemore, J.S., 1962, Semiconductor Statistics (Pergamon Press, New York).
Bois, D., and G. Vincent, 1977, J. Phys. F **38**, L351.
Boyn, R., J. Dziesiaty and D. Wruck, 1970, Phys. Status Solidi b **42**, K197.
Braun, S., and H.G. Grimmeiss, 1974, J. Appl. Phys. **45**, 2658.
Broser, J., M. Maier and J.H. Schulz, 1965, Phys. Rev. **140**, A2135.
Brown, W.J., and J.S. Blakemore, 1972, J. Appl. Phys. **43**, 2242.
Burkey, B.C., R.P. Khosla, J.R. Fisher and D.L. Lesee, 1967, J. Appl. Phys. **47**, 1095.
Busse, J.W., H.E. Gummlich, D. Maier-Hosch, E. Neuman and H.J. Schulz, 1970, J. Lumin. **1/2**, 66.
Busse, J.W., H.E. Gummlich and B. Meissner, 1976, J. Lumin. **12/13**, 693.
Casey Jr, H.C., F. Ermanis, L.C. Luther and L.R. Dawson, 1971, J. Appl. Phys. **42**, 2130.
Chandrasekhar, S., 1944, Astrophys. J. **102**, 176.
Chapman, R.A., and W.G. Hutchinson, 1967, Phys. Rev. Lett. **18**, 2873.
Cieplak, M.Z., M. Godlewski and J.M. Baranowski, 1975, Phys. Status Solidi b **70**, 323.
Colwell, P.J., and M.V. Klein, 1972, Phys. Rev. B **6**, 498.
Corbett, J.W., and G.D. Watkins, 1965, Phys. Rev. **138**, A555.
Crawford, M.G., G.E. Stillman, J.A. Rossi and N. Holonyak Jr, 1968, Phys. Rev. **168**, 867.
Czaja, W., 1971, Festkörperprobleme **XI**, 65.
De Wit, M., T.L. Estle, W.C. Holton and J. Schneider, 1964, Bull. Am. Phys. Soc. **9**, 2491.
Dean, P.J., 1973, in: Proc. Leningrad Conf., p. 538.
Dean, P.J., 1977, Electroluminescence, Topics in Applied Physics, Vol.17 (Springer, Berlin) p. 63.
Dean, P.J., and R.L. Hartman, 1972, Phys. Rev. B **5**, 4911.
Dean, P.J., J.D. Cuthbert and R.T. Lynch, 1969, Phys. Rev. **179**, 754.
Dean, P.J., R.A. Faulkner, S. Kimura and M. Ilegems, 1971a, Phys. Rev. B **4**, 1926.
Dean, P.J., A.M. White, E.M. Williams and M.G. Astles, 1971b, Solid State Commun. **9**, 1955.
Dexter, R.L., 1958, in: Solid State Physics, Vol. 6, eds F. Seitz and D. Turnbull (Academic Press, New York) p. 353.
Dmowski, L., M. Baj, A. Iller and S. Porowski, 1977a, in: Proc. Int. Conf. on High Pressure and Low Temperature Physics, Cleveland, Ohio, 1977 (Plenum Press, New York) p. 575.

Dmowski, L., M. Baj and S. Porowski, 1977b, in: Proc. Int. Conf. on High Pressure and Low Temperature Physics, Cleveland Ohio 1977 (Plenum Press, New York) p. 505.
Dubenskii, K.K., Ya.E. Karris, A. Ryskin, P.P. Feofilov and G.I. Khilko, 1965, Opt. Spectrosc. **19**, 353.
Eagles, D.M., 1960, J. Phys. & Chem. Solids **16**, 76.
Ejder, E., and H.G. Grimmeiss, 1974, Appl. Phys. **5**, 275.
Elkin, E.L., and G.D. Watkins, 1968, Phys. Rev. **174**, 881.
Fair, H.D., R.D. Ewing and F.E. Williams, 1966, Phys. Rev. **144**, 298.
Faulkner, R.A., 1968, Phys. Rev. **175**, 991.
Foyt, A.G., R.E. Halstead and W. Paul, 1966, Phys. Rev. Lett. **16**, 55.
Gershenzon, E.M., G.N. Goltsman and N.G. Pitsina, 1973, Sov. Phys.-JETP **37**, 299.
Godlewski, M., 1985, Phys. Status Solidi a **90**, 11.
Gołdys, E., P. Galtier and G. Martinez, 1987, in: Proc. 18th Int. Conf. on the Physics of Semiconductors, ed. O. Engström (World Scientific, Singapore) p. 963.
Gornik, E., 1972, Phys. Rev. Lett. **29**, 595.
Grebe, G., G. Roussos and H.J. Schulz, 1976, J. Lumin. **12/13**, 701.
Gregorkiewicz, T., M.H.P.Th. Bekman, C.A.J. Ammerlaan, W. Knap, L.C. Brunel and G. Martinez, 1991, Phys. Rev., in press.
Griffith, J.S., 1964, The Theory of Transition Metal Ions (University Press, Cambridge).
Grimmeiss, H.G., 1977, Ann. Rev. Mat. Sci. **7**, 2155.
Grimmeiss, H.G., and L.A. Ledebo, 1975, J. Appl. Phys. **46**, 2155.
Grimmeiss, H.G., and B. Monemar, 1973, Phys. Status Solidi a **19**, 505.
Grimmeiss, H.G., L.A. Ledebo, C. Ovren and T.N. Morgan, 1974, in: Proc. 12th Int. Conf. on the Physics of Semiconductors (B.G. Taubner, Stuttgart) p. 386.
Haisty, R.W., and G.R. Cronin, 1964, in: Proc. Int. Conf. on Semiconductors, Paris (Dunod, Paris) p. 1161.
Hale, E.B., and R.L. Mieher, 1969, Phys. Rev. **184**, 739.
Hale, E.B., and R.L. Mieher, 1971, Phys. Rev. B **3**, 1955.
Halstead, R.E., and M. Aven, 1965, Phys. Rev. Lett. **14**, C4.
Ham, F.S., and G.A. Slack, 1971, Phys. Rev. B **4**, 777.
Heinke, W., 1974, in: Proc. Conf. on Lattice Defects in Semiconductors, ed. F. Huntley (Institute of Physics, London) p. 380.
Henry, C.H., and D.V. Lang, 1977, Phys. Rev. B **15**, 989.
Henry, C.H., and K. Nassau, 1970, Phys. Rev. B **2**, 997.
Henry, C.H., J.J. Hopfield and L.C. Luther, 1966, Phys. Rev. **17**, 1178.
Herman III, J.M., and C.T. Sah, 1973, J. Appl. Phys. **44**, 1259.
Holland, M.G., and W. Paul, 1962, Phys. Rev. **128**, 30.
Holton, W.C., J. Schneider and T.L. Estle, 1964, Phys. Rev. **133**, A1638.
Hoo, K., and W.M. Becker, 1976, Phys. Rev. B **14**, 5372.
Hopfield, J.J., D.G. Thomas and R.T. Lynch, 1966, Phys. Rev. Lett. **17**, 312.
Huant, S., S.P. Najda, W. Knap, G. Martinez, B. Etienne, C.J.G.M. Langerak, J. Singelton, R.A.J. Thomeer, G. Mai, E.M. Peeters and J.T. Devreese, 1990, in: Proc. 20th Int. Conf. on the Physics of Semiconductors, eds E.M. Anastasskis and J.D. Joannopoulos (World Scientific, Singapore) p. 1369.
Iselev, G.W., J.A. Kafalas, A.J. Strauss, H.P. McMillan and R.H. Bube, 1972, Solid State Commun. **10**, 619.
Iwanomoto, M., and A. Kasani, 1976, Appl. Phys. Lett. **28**, 591.
Jaros, M., 1975, J. Phys. C **8**, 2455.
Jaros, M., and S.F. Ross, 1974, in: Proc. Twelfth Int. Conf. on the Physics of Semiconductors (B.G. Taubner, Stuttgart) p. 401.
Jarosik, N.C., B.D. McCombe, B.V. Shanabrook, J. Comas, J. Ralston and G. Wicks, 1985, Phys. Rev. Lett. **54**, 1283.
Jones, R.L., and P. Fisher, 1969, J. Phys. & Chem. Solids **26**, 1125.
Kamińska, M., J.M. Baranowski and M. Godlewski, 1978, in: Proc. 14th Int. Conf. on the Physics of Semiconductors, Edinburgh (Institute of Physics, Bristol, London) p. 303.
Kamińska, M., M. Skowroński, J. Lagowski, J.M. Parsey and H.C. Gatos, 1983, Appl. Phys. Lett. **43**, 302.

Kamińska, M., M. Skowroński and W. Kuszko, 1985, Phys. Rev. Lett. **55**, 2204.
Kaplan, R., 1969, Phys. Rev. **181**, 1154.
Karpierz, K., M.L. Sadowski and M. Grynberg, 1990, Proc. 20th Int. Conf. on the Physics of Semiconductors, eds E.M. Anastasskis and J.D. Joannopoulos (World Scientific, Singapore) p. 609.
Kaufmann, U.G., and P. Koidl, 1974, J. Phys. C **7**, 791.
Kaufmann, U.G., and J. Schneider, 1976, Solid State Commun. **30**, 143.
Kimerling, L.C., 1976, IEEE Trans. Nucl. Sci. **NS-23**.
Kishino, S., N. Chinone, H. Nakashima and R. Ito, 1976, Appl. Phys. Lett. **29**, 488.
Kobayashi, M., and S. Norita, 1977, J. Phys. Soc. Jpn. **43**, 1455.
Kocot, K., and J.M. Baranowski, 1973, Phys. Status Solidi b **59**, K11.
Kocot, K., and J.M. Baranowski, 1977, Phys. Status Solidi b **81**, 629.
Koenig, S.H., R.D. Brown and W. Schillinger, 1962, Phys. Rev. **128**, 1668.
Koidl, P., O.F. Schirmer and U.G. Kaufmann, 1973, Phys. Rev. B **8**, 4926.
Kopylov, A.A., and A.N. Pithkin, 1974, Phys. & Technol. Semicond. **11**, 867.
Kosicki, B.B., A. Jayaraman and W. Paul, 1968, Phys. Rev. **172**, 764.
Krebs, J.J., and G.H. Stauss, 1977a, Phys. Rev. B **15**, 17.
Krebs, J.J., and G.H. Stauss, 1977b, Phys. Rev. B **16**, 971.
Kuchar, F., E. Fantner and G. Bauer, 1977, J. Phys. C **10**, 3577.
Kukimoto, H., C.H. Henry and F.R. Merrit, 1973, Phys. Rev. B **7**, 2486.
Landsberg, P.T., 1956, Proc. Phys. Soc. B **69**, 1056.
Landsberg, P.T., and E.H. Shaban, 1987, J. Appl. Phys. **61**, 5055.
Lang, D.V., 1974, J. Appl. Phys. **45**, 3023.
Lang, D.V., and R.A. Logan, 1975, J. Electron. Mater. **5**, 1053.
Lang, D.V., and R.A. Logan, 1977, Phys. Rev. Lett. **39**, 635.
Langer, J.M., and J.M. Baranowski, 1971, Phys. Status Solidi b **44**, 155.
Le Manh Hoang, and J.M. Baranowski, 1977, Phys. Status Solidi b **84**, 361.
Lee, Y.H., and J.W. Corbett, 1973, Phys. Rev. B **8**, 2810.
Lee, Y.H., and J.W. Corbett, 1974, Phys. Rev. B **9**, 4352.
Lifshitz, T.M., N.P. Lichtman and V.J. Sidorow, 1968, in: Proc. 9th Int. Conf. on the Physics of Semiconductors, Moscow (Nauka, Leningrad) p. 1081.
Lin, A.L., and R.H. Bube, 1976, J. Appl. Phys. **47**, 1859.
Look, D.C., 1977, Solid State Commun. **24**, 825.
Lorenz, M.R., B. Segall and H.H. Woodbury, 1964, Phys. Rev. **136**, A751.
Lucovsky, G., 1965, Solid State Commun. **3**, 299.
Manchon, D.D., and P.J. Dean, 1970, in: Proc. 10th Int. Conf. on the Physics of Semiconductors, eds S.P. Veller, J.C. Hensl and F. Stern (US Atomic Energy Commission, Washington, DC) p. 760.
Martin, G.M., 1981, J. Appl. Phys. **39**, 747.
Martin, G.M., A. Mitonneau and A. Mircea, 1977, Electron. Lett. **13**, 141.
Messenger, R.A., and J.S. Blakemore, 1971, Phys. Rev. B **4**, 1874.
Miller, G.L., J.V. Ramirez and D.A.H. Robinson, 1975, J. Appl. Phys. **46**, 2638.
Miller, G.L., D.V. Lang and L.C. Kimerling, 1977, Ann. Rev. Mater. Sci. 377.
Miller, R.C., A.C. Gossard, W.T. Tsang and O. Munteanu, 1982a, Phys. Rev. B **25**, 3871.
Miller, R.C., W.T. Tsang and O. Munteanu, 1982b, Appl. Phys. Lett. **41**, 374.
Milnes, A.G., 1973, Deep Impurities in Semiconductors (Wiley, New York).
Mitonneau, A., G.M. Martin and A. Mircea, 1977, Electron. Lett. **13**, 66.
Monemar, B., and B. Samuelson, 1976a, J. Lumin. **12/13**, 507.
Monemar, B., and B. Samuelson, 1976b, in: Proc. 13th Int. Conf. on the Physics of Semiconductors, Rome (Topografia Marres, Rome) p. 639.
Morgan, T.N., 1972, in: Proc. 11th Int. Conf. on the Physics of Semiconductors (Polish Scientific Publishers, Warsaw) p. 989.
Morigaki, K., 1964, J. Phys. Soc. Jpn. **19**, 1485.
Muller, F., L.C. Brunel, M. Grynberg, J. Blinowski and G. Martinez, 1989, Europhys. Lett. **8**, 291.
Najda, S.P., 1989, Semicond. Sci. & Technol. **4**, 439.
Nishida, Y., and K. Muro, 1975, Suppl. Progr. Theor. Phys. **57**, 77.

Nygren, B., J.T. Vallin and G.A. Slack, 1972, Solid State Commun. **11**, 35.
Olofsson, G., 1974, Phys. Status Solidi a **22**, 175.
Onton, A., 1969, Phys. Rev. **186**, 786.
Onton, A., and R.C. Taylor, 1970, Phys. Rev. B **1**, 2587.
Onton, A., P. Fisher and A.K. Ramdas, 1967, Phys. Rev. **163**, 686.
Pappalardo, R., and R.E. Dietz, 1961, Phys. Rev. **123**, 1188.
Paul, W., 1963, Solids under Pressure (McGraw-Hill, New York).
Paul, W., 1968, in: Proc. 9th Int. Conf. on the Physics of Semiconductors, Moscow (Nauka, Leningrad) p. 16.
Pitt, G.D., 1969, High Temp., High Pressure **1**, 118.
Pitt, G.D., 1977, Contemporary Phys., 1977, Vol. 18, p. 137.
Porowski, S., M. Kończykowski and J. Chroboczek, 1974, Phys. Status Solidi a **63**, 291.
Queisser, H.J., 1971, Festkörp. **XI**, 45.
Queisser, H.J., 1976, J. Appl. Phys. **10**, 275.
Radliński, A.P., 1978, Phys. Status Solidi b **86**, 41.
Ramdas, A.K., and S. Rodriguez, 1981, Rep. Prog. Phys. **44**, 1297.
Reuszer, J.K., and P. Fisher, 1964, Phys. Rev. **135**, A1125.
Rosier, L.L., and C.T. Sah, 1971, J. Appl. Phys. **42**, 4000.
Sah, C.T., and L. Forbes, 1971, Solid State Electron. **14**, 182.
Sah, C.T., L. Forbes, L.L. Rosier and A.F. Tasch, 1970, Solid State Electron. **13**, 759.
Sah, C.T., W.W. Chan, H.S. Fu and J.W. Walker, 1972, Appl. Phys. Lett. **20**, 193.
Schneider, J., A. Räuber, B. Dischler, T.L. Estle and W.C. Holton, 1965, J. Chem. Phys. **42**, 1839.
Schulz, M., and G.G. Wepfer, 1972, Solid State Commun. **10**, 405.
Scifres, D.R., N. Holonyak Jr, C.B. Duke, G.G. Kleiman, A.B. Kunz, N.C. Crawford, W.D. Groves and A.H. Herzog, 1971, Phys. Rev. Lett. **27**, 191.
Scifres, D.R., H.M. Macksey, N. Holonyak Jr, R.D. Dupuis, G.W. Zack, C.B. Duke, G.G. Kleiman and A.B. Kunz, 1972, Phys. Rev. B **5**, 2206.
Sharan, R., and E.L. Heasell, 1970, J. Phys. & Chem. Solids **31**, 541.
Stepniewski, R., S. Huant, G. Martinez and B. Etienne, 1989, Phys. Rev. B **40**, 9772.
Stillman, G.E., C.M. Wolfe and J.O. Dimmock, 1969, Solid State Commun. **7**, 921.
Stillman, G.E., D.M. Larsen, C.M. Wolfe and R.C. Brandt, 1971, Solid State Commun. **9**, 2245.
Stillman, G.E., C.M. Wolfe and D.M. Korn, 1972, in: Proc. 11th Int. Conf. on the Physics of Semiconductors (Polish Scientific Publishers, Warsaw) p. 863.
Stocker, H.J., and M. Schmidt, 1976, in: Proc. 13th Int. Conf. on the Physics of Semiconductors, Rome, ed. F.G. Fumi (Tipografia Marres, Rome) p. 611.
Stoneham, A.M., 1975, Theory of Defects in Solids (Clarendon Press, Oxford).
Stradling, R.A., 1976, in: Proc. Int. Conf. on The Application of High Magnetic Fields in Semiconductors Physics, Würzburg, p. 345.
Strutz, T., A.M. Witowski and P. Wyder, 1990, in: Proc. 20th Int. Conf. on the Physics of Semiconductors, eds E.M. Anastasskis and J.D. Joannopoulos (World Scientific, Singapore) p. 1811.
Suto, K., and M. Aoki, 1967, J. Phys. Soc. Jpn. **22**, 149.
Suto, K., M. Aoki, M. Nakada and S. Ibuki, 1967, J. Phys. Soc. Jpn. **22**, 1121.
Tasker, P.W., and A.M. Stoneham, 1977, J. Phys. Solid State Phys. C **10**, 5131.
Thomas, D.G., and J.J. Hopfield, 1966, Phys. Rev. **150**, 680.
Thomas, D.G., M. Gershonzon and F.A. Trumbore, 1964, Phys. Rev. **133**, A269.
Thomas, S.R., and H.Y. Fan, 1974, Phys. Rev. B **9**, 4295.
Tuchendler, J., M. Grynberg, Y. Couder, H. Thome and R. LeToullec, 1973, Phys. Rev. B **8**, 3884.
Uba, S., and J.M. Baranowski, 1978, Phys. Rev. B **17**, 69.
Vallin, J.T., and G.D. Watkins, 1974, Phys. Rev. B **9**, 2051.
Vallin, J.T., G.A. Slack, S. Roberts and A.E. Hughes, 1970, Phys. Rev. B **2**, 4313.
Vul, A.Ya., G.L. Bir and Yu.V. Shmartsev, 1970a, Fiz. Tekh. Poluprovodn. **4**, 2331.
Vul, A.Ya., L.V. Golubev, L.V. Shronova and Yu.V. Shmartsev, 1970b, Fiz. Tekhn. Poluprovodn. **4**, 234.
Wagner, R.J., and A.M. White, 1979, Solid State Commun. **32**, 399.
Watkins, G.D., 1963, J. Phys. Soc. Jpn. **11**, 22.

Watkins, G.D., 1968, in: Radiation Effects in Semiconductors, ed. F. Vook (Plenum Press, New York) p. 67.
Watkins, G.D., 1975, in: Defects and Their Structure in Nonmetallic Solids, eds B. Henderson and A.E. Hughes (Plenum Press, New York) p. 203.
Watkins, G.D., and J.W. Corbett, 1961, Phys. Rev. **121**, 1001.
Watkins, G.D., and J.W. Corbett, 1964, Phys. Rev. **134**, A1359.
Watkins, G.D., and J.W. Corbett, 1965, Phys. Rev. **138**, A543.
Watts, R.K., 1969, Phys. Rev. **188**, 569.
Weakliem, D.H., 1962, J. Chem. Phys. **36**, 2117.
White, A.M., P.J. Dean, K.M. Fairhurst, W. Bardsley, E.W. Williams and B. Day, 1972a, Solid State Commun. **11**, 1093.
White, A.M., P.J. Dean, L.L. Taylor and R.C. Clarke, 1972b, J. Phys. C5 **L110**, 1727.
White, A.M., D.J. Ashen, J.B. Mullin, M. Webb, B. Day and P.D. Green, 1973, J. Phys. **66**, L243.
White, A.M., P.J. Dean and P. Porteous, 1976, J. Appl. Phys. **47**, 3230.
Wittlin, A., M. Grynberg, W. Knap, J. Kossut and Z. Wilamowski, 1980, J. Phys. Soc. Jpn. 49 (Suppl. **A**) 635.
Wray, E.M., and W. Allen, 1971, J. Phys. C **4**, 512.
Wright, G.B., and A. Mooradian, 1968, Bull. Am. Phys. Soc. **13**, 479.
Zohta, Y., 1972, J. Appl. Phys. **43**, 1713.

CHAPTER 6

Semiconductor Statistics

P. T. LANDSBERG

Faculty of Mathematical Studies
University of Southampton
Southampton SO9 5NH, UK

O. ENGSTRÖM

Department of Solid State Electronics
Chalmers University of Technology
S-412 96 Göteborg, Sweden

Handbook on Semiconductors
Completely Revised Edition
Edited by T.S. Moss
Volume 1, edited by P.T. Landsberg

Contents

Introduction . . . 200
1. Equilibrium statistics . . . 200
1.1. Probability distributions . . . 200
1.2. Particles in bands . . . 202
1.3. Particles in traps . . . 204
1.4. The entropy of a particle system with vibronic states . . . 209
1.5. Negative-*U* centres . . . 211
1.6. Quasi-Fermi levels . . . 213
2. Recombination statistics . . . 215
2.1. Basic assumptions . . . 215
2.2. The recombination rate . . . 216
2.3. Band–band transitions . . . 219
2.4. Band–trap transitions . . . 219
2.5. Generalised Shockley–Read–Hall statistics . . . 221
2.6. Decay laws for band–band recombination . . . 223
2.7. Decay laws for band–trap recombination . . . 225
2.8. Detailed balance and its relation to measurements . . . 228
3. Radiative processes . . . 231
3.1. Radiative recombination . . . 231
3.2. Non-equilibrium kinetics for coupled photons and electrons in two-level systems . . . 234
3.3. Photon recycling . . . 238
4. Other non-radiative processes . . . 239
4.1. Truncated cascade recombination at traps with excited electronic states . . . 239
4.2. Multi-phonon processes: The configurational coordinate model . . . 243
4.3. Multi-phonon processes: Thermal capture and emission . . . 246
4.4. Metastability of defect states . . . 249
5. Continuous trap distributions . . . 251
5.1. Formal development . . . 251
5.2. Meyer–Neldel rule . . . 251
5.3. A tractable Gaussian-like trap distribution . . . 252
5.4. Exponential band tails . . . 253
5.5. Urbach tails . . . 257
6. Surface and interface effects . . . 258
6.1. Surface recombination . . . 258
6.2. Insulator–semiconductor interfaces . . . 261

6.3. Grain boundaries and recombination 264
7. Lattice defects and solubility 267
7.1. Statistics of simple lattice defects 267
7.2. A model for saturation solubilities of impurities 273
7.3. The connection between doping and the solubility of defects 274
References 277

Introduction

As observed in the 1982 version of this chapter (Landsberg 1982), semiconductor statistics is basic to much of the work discussed in this volume. The basic ingredients of this topic are therefore widely known. This has enabled us to produce a new chapter by omitting some more elementary topics (e.g., classical ideal gas theory, quantities referred to the intrinsic level), as well as some more esoteric subjects (e.g., exciton statistics, recurrence relations). This has left us with a slimmed down chapter which we hope will be useful also to experimentalists. The old chapter as well as the recent book by one of us (Landsberg 1992) are, of course, still available for some of the details which we did not take over into this chapter.

Among the new topics introduced here, we should emphasize that we struggled to explain clearly and honestly the use of Gibbs free energies (as against just "energies") for the energy levels of traps. We also updated our discussion by including, e.g., considerations which are needed for the interpretation of experiments in connection with the vibrational properties of traps. Further, we have included something about negative-U centres, more than before about continuous trap distributions and grain boundaries, etc.

It is hoped that in this way the chapter will fulfill a useful purpose.

1. Equilibrium statistics

1.1. Probability distributions

As a starting point for all our considerations on charge carrier statistics in semiconductors we will consider the carriers as indistinguishable particles belonging to a *grand canonical ensemble* with the electrochemical potential μ. It is shown in statistical mechanics that for an equilibrium system of given volume v, and temperature T, the probability of finding it with n particles and energy E_i is

$$P(v, \mu, T; n, E_i) = \Xi^{-1} \exp[(\mu n - E_i)/kT]. \tag{1.1.1}$$

Here the normalising factor is the *grand partition function*

$$\Xi = \sum_{n,i} [\exp(\mu n - E_i)/kT]. \tag{1.1.2}$$

These are general results for identical particles with general interaction. Even more general formulae are possible. However, we shall consider a system of particles which interact only when they collide (i.e., they have point interactions). One can consider the different quantum states $j = 1, 2, \ldots,$ in which they can be, and label the number in state j by n_j and the energy of the *single-particle* state e_j while E_i denotes the energies of the system. One can simply sum over all permitted integers $n_1, n_2, \ldots$, etc., instead of summing over all n and all i, since this procedure covers all possible states of the system if the particles involved are indistinguishable. Thus, eq. (1.1.2) becomes

$$\Xi = \sum_{n_1=0} \sum_{n_2=0} \cdots \exp\left[\frac{\mu(n_1+n_2+\cdots)-(e_1 n_1 + e_2 n_2 + \cdots)}{kT}\right]$$

$$= \sum_{n_1=0} \sum_{n_2=0} \cdots [t_1^{n_1} t_2^{n_2} \cdots],$$

$$\Xi = \prod_{j=1}\left[\sum_{n_j=0} t_j^{n_j}\right], \qquad t_j = \exp\left(\frac{\mu - e_j}{kT}\right), \quad j = 1, 2, \ldots. \tag{1.1.3}$$

The upper limits in the summations are either 1, in which case a quantum state is either empty or full, or infinity, then a quantum state can accommodate any number of particles. The first case (fermions) applies to electrons, the second case (bosons) applies to, e.g., photons.

Returning to eq. (1.1.1), using eq. (1.1.3), the probability can be written in a different way

$$P(v, \mu, T; n_1, n_2, \ldots) = \frac{t_1^{n_1} t_2^{n_2} \cdots}{\left(\sum_{n_1=0} t_1^{n_1}\right)\left(\sum_{n_2=0} t_2^{n_2}\right)\cdots}$$

$$= \left[\frac{t_1^{n_1}}{\sum_{n_1=0} t_1^{n_1}}\right]\left[\frac{t_2^{n_2}}{\sum_{n_2=0} t_2^{n_2}}\right]\cdots. \tag{1.1.4}$$

This gives the probability that an equilibrium system of particles with point interactions and of given volume, electrochemical potential and temperature is found in a state such that n_1 particles are in the first single-particle quantum state, n_2, in the second, etc., where t_j is given by eq. (1.1.3). The mean occupation number for fermions in the quantum state k is

$$\bar{n}_k = \sum_{n_1} \sum_{n_2} \cdots \left\{ n_k \left[\frac{t_1^{n_1}}{\sum_{n_1=0} t_1^{n_1}}\right]\left[\frac{t_2^{n_2}}{\sum_{n_2=0} t_2^{n_2}}\right]\cdots\left[\frac{t_k^{n_k}}{\sum_{n_k=0} t_k^{n_k}}\right]\cdots\right\}$$

$$= \sum_{n_k=0}^{1} n_k \frac{t_k^{n_k}}{\sum_{l=0}^{1} t_k^l} = \frac{0 t_k^0 + 1 t_k^1}{t_k^0 + t_k^1} = \frac{t_k}{1+t_k}. \tag{1.1.5}$$

Hence

$$\bar{n}_k = \frac{1}{\exp[(e_k - \mu)/kT] + 1}. \tag{1.1.6}$$

For bosons we sum to infinity and find similarly

$$\bar{n}_k = \frac{1 + t_k + 2t_k^2 + \cdots}{1 + t_k + t_k^2 + \cdots} = \frac{t_k/(1-t_k)^2}{1/(1-t_k)} \tag{1.1.7}$$

since the denominator is a geometrical progression and the numerator is t_k times the derivative with respect to t_k of the denominator. Hence

$$\bar{n}_k = \frac{1}{\exp[(e_k - \mu)/kT] + a}, \tag{1.1.8}$$

where $a = 1$ for fermions, $a = -1$ for bosons, a can be neglected for large negative μ. These are the famous Fermi–Dirac, Bose–Einstein and Maxwell–Boltzmann distribution functions. The electrochemical potential μ is more usually referred to as the Fermi level, which is the term we use here.

For electrons and holes in semiconductors one requires the first and the third distribution functions. The second one is needed when interactions with excitons, photons or phonons are considered. The above argument, apart from being based on the grand canonical ensemble through eq. (1.1.1), uses only the assumption of identical particles with point interaction. Alternative derivations have to rely on spurious assumptions of a mathematical nature. The method of the most probable distribution relies on the existence of large numbers so that the Stirling approximation can be used. Also, integer variables must be replaced by continuous variables so that differentiations can be performed. The method of steepest descent, used in connection with the canonical or microcanonical ensemble, also requires the neglect of certain terms. Thus, all these methods are mathematically less direct. If one wants to keep the total number of particles strictly constant, one has to use the canonical ensemble, which gives somewhat different results. One can compare the grand canonical ensemble and the canonical mean occupation numbers *exactly* for a few simple cases. For the lowest state of a two-state, one-particle system the discrepancy is 17% (fermions) and 9% (bosons) (Landsberg and Harshman 1988). It is unlikely to be higher in other cases when it is, however, hard to obtain exact results for the canonical ensemble.

1.2. Particles in bands

For the conduction band of a semiconductor, assumed to extend to infinite energy, the mean number of electrons in thermal equilibrium at temperature T and Fermi level μ is

$$\bar{n} = \int_{E_c}^{\infty} P(E)\, N(E)\, \mathrm{d}E, \tag{1.2.1}$$

where E_c is the energy at the bottom of the conduction band and $P(E)$ is the occupation probability of a quantum state of energy E. This drops exponentially for large energies, by eq. (1.1.6). Hence, the finite extent of a band is well simulated by the limit $E = \infty$ in eq. (1.2.1). The density of states function (number per unit volume per unit energy range) can be expressed, by using the effective-mass approximation, as

$$N(E) = A(E - E_c)^{1/2}, \tag{1.2.2}$$

where

$$A = \frac{1}{2\pi^2}\left(\frac{2m_c}{h^2}\right)^{3/2}, \tag{1.2.3}$$

and m_c is the carrier effective mass. Combining eqs. (1.1.6), (1.2.1) and (1.2.2), we get

$$\begin{aligned}\bar{n} &= A\int_{E_c}^{\infty}\frac{(E - E_c)^{1/2}\,\mathrm{d}E}{1 + \exp[(E - \mu)/kT]}\\ &= A(kT)^{3/2}\int_0^{\infty}\frac{x^{1/2}\,\mathrm{d}x}{1 + \exp[x - (\mu - E_c)/kT]} = N_c F_{1/2}[(\mu - E_c)/kT],\end{aligned} \tag{1.2.4}$$

where N_c is an effective electron concentration in the conduction band given by

$$N_c \equiv A(kT)^{3/2}\sqrt{\pi}/2 = 2(2\pi m_c kT/h^2)^{3/2} \tag{1.2.5}$$

and

$$F_s(a) \equiv \frac{1}{\Gamma(s+1)}\int_0^{\infty}\frac{x^s\,\mathrm{d}x}{1 + \exp(x - a)}. \tag{1.2.6}$$

Different approximations to eq. (1.2.4) can be used, depending on the value of the Fermi level μ compared to the conduction band edge E_c. When the difference $E_c - \mu$ is large enough such that the exponential function dominates the denominator in eq. (1.2.4), one obtains the non-degenerate, or classical, approximation

$$\bar{n} = N_c \exp[(\mu - E_c)/kT]. \tag{1.2.7}$$

If the integral in eq. (1.2.4) has to be taken into account one speaks of Fermi degeneracy. (This has to be distinguished from the degeneracy when several linearly independent wave functions belong to the same eigenvalue.) For highly (Fermi) degenerate material, when the Fermi level has a value well above the conduction band edge, a rough approximation is to replace the Fermi function by unity for energies up to μ and by zero for energies beyond μ. Then

$$\bar{n} = A\tfrac{2}{3}(\mu - E_c)^{3/2}. \tag{1.2.8}$$

The accurate result and the classical and degenerate approximation are illustrated in fig. 1.1.

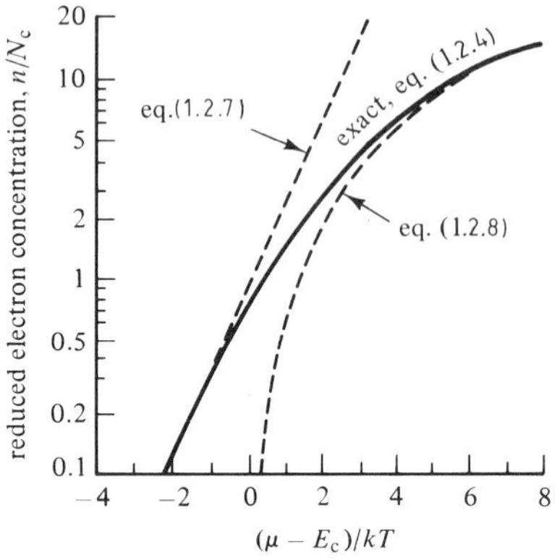

Fig. 1.1. The concentration of electrons in a conduction band.

For the valence band, holes are present at a quantum state with energy E and a probability of

$$1 - \frac{1}{1+\exp[(E-\mu)/kT]} = \frac{1}{1+\exp[(\mu-E)/kT]}. \tag{1.2.9}$$

In analogy with eq. (1.2.4), by assuming that the valence band extends from energy $-\infty$ to the valence band edge E_v, we find for the hole concentration p in the valence band

$$\bar{p} = B\int_{-\infty}^{E_c} \frac{(E_v - E)^{1/2}\,dE}{1+\exp[(\mu-E)/kT]} = N_v F_{1/2}[(E_v-\mu)/kT], \tag{1.2.10}$$

where

$$N_v = 2(2\pi m_v kT/h^2)^{3/2}. \tag{1.2.11}$$

Here m_v is the effective mass of holes in the valence band. For the non-degenerate case this gives

$$\bar{p} = N_v \exp[(E_v-\mu)/kT]. \tag{1.2.12}$$

In later sections we shall regard the bar for the average n and p to be understood.

1.3. Particles in traps

We now seek to derive the occupation probabilities of localised states acting as electron and hole traps in a semiconductor. As the trap exchanges particles with the rest of the solid, we start by writing the grand canonical partition function given in eq. (1.1.1) as

$$\Xi = \sum_{r=0}^{M} \lambda^r Z_r, \quad \lambda \equiv \exp(\mu/kT). \tag{1.3.1}$$

Here r is the number of particles captured by the trap and M is the maximum number of captured particles. The factor Z_r in eq. (1.3.1) is the canonical partition function for the trap with r captured electrons and reads

$$Z_r = \sum_l \exp[-E(l, r)/kT]. \tag{1.3.2}$$

The summation in eq. (1.3.2) is taken over all possible electronic and vibronic quantum states, l. The concentration ν_r of centres with r captured particles in the energy states $E(l, r)$ is obtained from the partition function in eq. (1.3.1) as

$$\nu_r = P(r)N = \frac{\lambda^r Z_r}{\sum_{s=0}^{M} \lambda^s Z_s} N, \tag{1.3.3}$$

where N is the total concentration of centres. In experimental studies of trap systems, the energies observed are thermodynamic quantities, averaged, either over an ensemble of traps or over time, when single-trap phenomena are observed (Ralls et al. 1984, Kirton and Uren 1986, Farmer et al 1987, Andersson et al. 1990). The energies measured in solid state physics are not normally energy eigenvalues of Hamiltonians. To take this into account leads to some delicate questions of statistical mechanics (Landsberg 1958, Engström and Alm 1983) to which we now turn.

The preceding main results, eqs. (1.2.4), (1.2.10) and (1.3.3), depend on the grand canonical ensemble, for which the volume is prescribed. The basis of the argument has, therefore, to be changed to a constant-pressure ensemble if one wants to have results which apply to the many experiments performed at constant pressure. It is fortunate that the result of this reformulation is to keep the above formulae intact, except that the energy eigenvalues E have to be replaced by the appropriate Gibbs free energies.

A rough procedure is to specify the pressure, temperature and the number of particles, N. A state of the system then arises by a selection of an energy $E_i(v)$ and (given $i = 1, 2, \ldots$) a volume $v_{ij} (j = 1, 2, \ldots)$. Of course, the volume is a continuous variable and not discrete, but this notational problem is easily overcome (Almbladh and Rees 1981). The probability of finding a state of the system is then (by Lagrangian multiplier method or otherwise)

$$P_i(T, p, N; E_i; v_{ij}) = Q^{-1} \exp(-h_{i,j}), \tag{1.3.4}$$

where the reduced "enthalpy" of the state (Lloyd and O'Dwyer 1963)

$$H_{i,j} \equiv E_i + pv_{i,j}, \qquad h_{i,j} \equiv H_{i,j}/kT,$$

has been introduced and Q is the constant-pressure ensemble partition function which normalises P_i. Its thermodynamic significance is found from the entropy S $(\equiv -k\Sigma_i P_i \ln P_i)$

$$TS = kT \sum_{N,i,j} \left(\frac{H_{i,j}}{kT} + \ln Q \right) P_i = H + kT \ln Q, \tag{1.3.5}$$

where

$$H \equiv \langle H_{i,j} \rangle.$$

This implies

$$kT \ln Q = -G(T, p, N), \tag{1.3.6}$$

where G is the Gibbs free energy. Hence, eq. (1.3.4) can be written

$$P_i(T, p, N; E_i, v_{i,j}) = \exp\left(\frac{G(T, p, N) - H_{i,j}}{kT}\right).$$

This is hardly surprising as the exponent is, as usual for ensembles, of the form $-S/k$.

From the normalising property of Q one has

$$Q = \exp\left[-\frac{G(T, p, N)}{kT}\right] = \sum_{i,j} \exp(-h_{i,j}).$$

Note that eq. (1.3.6) is analogous to the result

$$kT \ln Z = -F(T, v, N)$$

for the canonical ensemble where F is the Helmholtz free energy. (Thus, confusion would result if one were to use the symbol Z for Q in eq. (1.3.6): one would then ask for the G to be changed to F!). If one restricts the (i, j) sums in some way (indicated by a prime) a partial Gibbs free energy may be regarded as corresponding to it. For example, if one restricts the occupation of the single-particle energy level l to unity,

$$\begin{aligned} G(T, p, N; n_l = 1) &= -kT \ln\left[\sum_{i,j}{}' \exp(-h_{i,j}(n_l = 1))\right] \\ &= -kT \ln\left[\sum_{i,j}{}' \exp(-H_{i,j}(n_l = 1)/kT)\right]. \end{aligned} \tag{1.3.7}$$

Here we have kept the occupation of the single-particle energy level, l, at unity. Thus, the mean occupation number of level l is in simplified notation

$$\langle n_l \rangle = \frac{\sum' \exp(-h_1)}{\sum' \exp(-h_0) + \sum' \exp(-h_1)}.$$

It leads to a function which depends on the ratio of restricted sums, but the result is physically more transparent if one replaces these by appropriate partial Gibbs free energies.

For the partial G divided by kT we shall write $g(1, N-1)$ and $g(0, N)$ to correspond to h_1 and h_0, respectively, meaning that $N-1$ or N particles are in levels other than l in the two cases. Then

$$\begin{aligned} \langle n_l \rangle^{-1} &= 1 + \exp[g(1, N-1) - g(0, N)] \\ &= 1 + \exp\{g(1, N) - g(0, N) - [g(1, N) - g(1, N-1)]\} \\ &= 1 + \exp(\eta_l - \gamma). \end{aligned} \tag{1.3.8}$$

Thus, $kT\eta_l$ is the Gibbs free energy for adding an electron to level l at given T and

p. Also,

$$\mu = kT\gamma = G(n_l, N) - G(n_l, N-1) \quad (n_l = 0 \text{ or } 1), \tag{1.3.9}$$

is the Gibbs free energy for adding an electron to a level other than l. Finally, $(\eta_l - \gamma)kT$ is the Gibbs free energy for adding an electron to level l from some other level. In this way the energy in the usual Fermi–Dirac formula is interpreted as a Gibbs free energy when the pressure is kept constant.

For impurity centres, in a notation following Almbladh and Rees (1981), let the state of one of the centres be characterised by a (multiple) label α, let r be the number of electrons in the centre considered, let N be the number of electrons in states other than those of the impurity considered and let β specify remaining states. Then, from

$$\exp[-g(T, p, N)] = \sum_{i,j} \exp(-h_{i,j}), \tag{1.3.10}$$

the probability that a centre has captured r electrons is

$$\begin{aligned} P(T, p, N; r) &= \frac{\sum_{\alpha,\beta} \exp[-h(\alpha, r, \beta, N-r)]}{\sum_{\alpha',\beta'} \sum_{r'=0}^{M} \exp[-h(\alpha', r', \beta', N-r')]} \\ &= \frac{\exp[-g(r, N-r)]}{\sum_{r'=0}^{M} \exp[-g(r', N-r')]} \\ &= \frac{\exp[-g(r, N) - \gamma r]}{\sum_{r'=0}^{M} \exp[-g(r', N) - \gamma r']}, \end{aligned} \tag{1.3.11}$$

where the partial Gibbs free energies have absorbed the α-sums and the β-sums, and we have used eq. (1.3.9) in the form

$$r\gamma = g(r, N) - g(r, N-r).$$

Therefore, one finds a new version of eq. (1.3.3),

$$P(T, p, N; r) = \frac{\lambda^r Q_r}{\sum_{r'=0}^{M} \lambda^{r'} Q_{r'}}, \tag{1.3.12}$$

where, in full analogy with eq. (1.3.6),

$$Q_r(T, p, N) \equiv \exp\left[-\frac{G(r, N)}{kT}\right]. \tag{1.3.13}$$

This completes the transcription from energies to Gibbs free energies. Note that Q_r is the partition function for the constant-pressure ensemble. The constant-pressure ensemble discussed here is sometimes called the Gibbs ensemble. Other constant-pressure ensembles are known, but are not needed in the present context.

We now write for the concentration of centres, v_r, which have captured r particles

$$v_r = P(r)N = \frac{\exp\left(\dfrac{r\mu - G(r)}{kT}\right)}{\sum_{s=0}^{M} \exp\left(\dfrac{s\mu - G(s)}{kT}\right)} N, \tag{1.3.14}$$

where N is the concentration of centres. For the one-particle trap, the probability of capturing one particle becomes

$$v_1 = \frac{N \exp\left(\dfrac{\mu - G(1)}{kT}\right)}{\exp\left(-\dfrac{G(0)}{kT}\right) + \exp\left(\dfrac{\mu - G(1)}{kT}\right)} = \frac{N}{1 + \exp\left(\dfrac{G(\frac{1}{2}) - \mu}{kT}\right)}. \tag{1.3.15}$$

The energy difference $G(\frac{1}{2}) = G(1) - G(0)$ is the Gibbs free energy needed to take the captured particle from the trap and deposit it at infinity. We shall use more generally

$$\begin{aligned} &G(r - \tfrac{1}{2}) \equiv G(r) - G(r-1), \\ &S(r - \tfrac{1}{2}) \equiv S(r) - S(r-1), \\ &H(r - \tfrac{1}{2}) \equiv H(r) - H(r-1) \equiv G(r - \tfrac{1}{2}) + TS(r - \tfrac{1}{2}). \end{aligned} \tag{1.3.16}$$

Using the expression in eq. (1.3.16), v_1 can be written as

$$v_1 = \frac{N}{1 + \exp\left(\dfrac{G(\frac{1}{2}) - \mu}{kT}\right)} = \frac{N}{1 + \dfrac{1}{X(\frac{1}{2})} \exp\left(\dfrac{H(\frac{1}{2}) - \mu}{kT}\right)}, \tag{1.3.17}$$

where

$$X(\tfrac{1}{2}) = \exp\left[\frac{S(\frac{1}{2})}{k}\right] \tag{1.3.18}$$

is an entropy factor containing the influence of electronic degeneracy and vibronic entropy.

For a one-particle trap without the influence of atomic vibrations and with the electronic degeneracies g_0 and g_1, eq. (1.3.17) reduces to

$$v_1 = \frac{N}{1 + \dfrac{g_0}{g_1} \exp\left(\dfrac{E(\frac{1}{2}) - \mu}{kT}\right)}, \tag{1.3.19}$$

where $E(\frac{1}{2}) = E(1) - E(0)$ is the difference between eigenvalues. We have also used

$$X(\tfrac{1}{2}) = \exp\left[\frac{S(1) - S(0)}{k}\right] = \exp[\ln g_1 - \ln g_0] = g_1/g_0.$$

General expressions for the average number of captured particles N_e can be obtained from eq. (1.3.3) as (the partition function Z_i may be interpreted as Q_i's in the sense of eq. (1.3.6))

$$\frac{N_e}{N} = \frac{\sum_{j=0}^{M} j\lambda^j Z_j}{\sum_{r=0}^{M} \lambda^r Z_r} = \frac{M}{1 + \dfrac{\sum_{j=0}^{M-1} (M-j)\lambda^j Z_j}{\sum_{r=1}^{M} r\lambda^r Z_r}}, \tag{1.3.20}$$

which leads to the following expressions for N_e,

$$\frac{N_e}{N} = \frac{1}{1 + Z_0/\lambda Z_1} \qquad (M = 1), \tag{1.3.21}$$

$$\frac{N_e}{N} = \frac{2}{1 + \dfrac{2Z_0 + \lambda Z_1}{\lambda Z_1 + 2\lambda^2 Z_2}} \qquad (M = 2), \tag{1.3.22}$$

$$\frac{N_e}{N} = \frac{3}{1 + \dfrac{3Z_0 + 2\lambda Z_1 + \lambda^2 Z_2}{\lambda Z_1 + 2\lambda^2 Z_2 + 3\lambda^3 Z_3}} \qquad (M = 3). \tag{1.3.23}$$

These results have been reviewed with additional details elsewhere (Landsberg 1992).

1.4. The entropy of a particle system with vibronic states

The capture and emission of charge carriers at lattice defects or impurities are in many cases related to local phonon interactions (Dexter 1958, Henry and Lang 1977, Morgan 1983). The total number of eigenstates of such a system, therefore, is not determined only by the electronic potential set by long-range Coulomb forces and the defect core. Also because the vibrational properties of the defect centre differ from the rest of the lattice, the local vibrations give rise to energy states, which play a role in the statistical treatment. In the simplest case, the "vibronic states" can be assumed to belong to a w-dimensional ($w = 1, 2, 3, \ldots$) harmonic oscillator representing the vibrations of the trap. Another simplification, which can be used to gain understanding in relation to earlier common treatments of this problem, is to assume that the system consists of two independent systems: the trapped electron and the vibrating trap centre. In this case, the total eigenenergy becomes the sum of the eigenenergies of the electronic potential and the eigenenergies of the atomic oscillator. As the energies of the whole system are spread among the electronic degenerate and excited states and among the vibronic states, the electronic and vibronic parts both contribute to the entropy of the trap. In order to get a feeling for the magnitudes of these distributions, we calculate the entropies for the special model case, described above.

The energies of the system depend on the quantum numbers of the trapped electron and the w quantum numbers of the oscillator. The total energies are then simply the sums of the component energies. The canonical partition function is therefore a product $Z_r Z_a$ of the electronic and the atomic partition function with (e.g. Landsberg 1971)

$$Z_a = \left[\frac{\exp(-x(r))}{1-\exp(-2x(r))}\right]^w, \quad x(r) \equiv h\nu(r)/2kT. \tag{1.4.1}$$

The oscillator frequency can depend on the charge state, r, of the centre. The entropy $S(r)$ is the sum of the entropies of the components, i.e., of

$$S_e(r) = k\left[\ln Z_r + T\left(\frac{\partial \ln Z_r}{\partial T}\right)_V\right] \tag{1.4.2}$$

and

$$S_a(r) = k\left\{\frac{2x(r)}{\exp[2x(r)]-1} - \ln[1-\exp[-2x(r)]\right\} w, \tag{1.4.3}$$

where we have used

$$S = -\left(\frac{\partial F}{\partial T}\right)_V = \left(\frac{\partial(kT\ln Z)}{\partial T}\right)_V. \tag{1.4.4}$$

If Z_r is approximated as $g_r \exp[-E(r)/kT]$, then $S_e(r)$ is $k \ln g_r$. Hence, for a one-electron trap, the change in the electronic part of the entropy connected with the emission or capture of an electron, $S_e(\frac{1}{2})$, is (as already utilised in eq. (1.3.19))

$$S_e(\tfrac{1}{2}) = S_e(1) - S_e(0) = k \ln \frac{g_1}{g_0}. \tag{1.4.5}$$

For a trap with $g_1 = 4$, and $g_0 = 1$, this gives $S_e(\frac{1}{2}) = 1.39k$. The vibronic contribution, $S_a(r)$ varies with the separation between vibronic energy states, $h\nu$. The atomic entropy is plotted as a function of $h\nu/kT$ in fig. 1.2. If, on the release of a trapped electron, the energy separation $h\nu$ changes from $3kT$ to $0.1kT$, the entropy contribution S_a changes about $3k$ as seen in the figure. This gives a total entropy change of

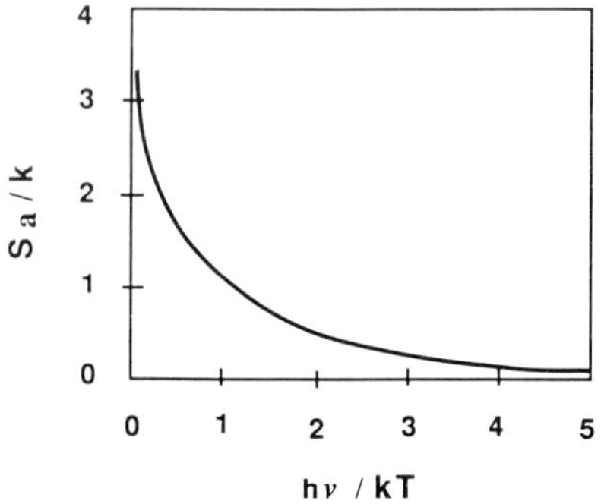

Fig. 1.2. The entropy S_a of eq. (1.4.3) in units of k as a function of the difference in vibronic energy levels in units of kT of a one-dimensional harmonic oscillator.

$4.39k$ which gives a value of 80.6 to the entropy factor $X(\frac{1}{2})$ in eq. (1.3.18). Therefore, the atomic contribution to $S(r-\frac{1}{2})$ can be considerable.

It is worth observing that the occupation probability given by eq. (1.3.19), which is also often found in the literature, is applicable only when the vibronic part $S_a(r-\frac{1}{2})$ of the entropy can be neglected. This is to be expected for traps, where the electronic wave function of the captured particle is extended over many lattice distances so that the binding properties of the defect centre are unaffected by the transition (Van Vechten and Thurmond 1976). Shallow impurities may fulfill this condition. In the opposite and more common case, eq. (1.3.17) must be used for a one-electron trap and eq. (1.3.14) in a more general case (Brooks 1955, Engström and Alm 1983).

1.5 Negative-U centres

This phenomenon applies to centres capable of trapping 0, 1 or 2 electrons ($M=2$ in our notation). If one neglects their mutual interaction and assumes that no other defects are present, then the *number* of trapped electrons, Θ, is in our usual notation

$$\Theta(\mu, T)=\frac{\lambda Z_1+2\lambda^2 Z_2}{Z_0+\lambda Z_1+\lambda^2 Z_2}. \tag{1.5.1}$$

Given Θ, Z_0, Z_1, and Z_2, one can regard this as an equation for λ. The solution of the resulting quadratic equation is

$$\lambda=x[(1+y^2)^{1/2}-1] \quad (\Theta\neq 1), \tag{1.5.2}$$

where

$$x\equiv\frac{1-\Theta}{2(2-\Theta)}\frac{Z_1}{Z_2}, \qquad y^2\equiv\frac{4\Theta(2-\Theta)}{(1-\Theta)^2}\frac{Z_0 Z_2}{Z_1^2} \quad (\Theta\neq 1).$$

If $\Theta=1$, then clearly

$$\lambda=(Z_0/Z_2)^{1/2}=xy \quad (\Theta=1). \tag{1.5.3}$$

For a negative-U centre, the correlation energy, U, between the two electrons is negative. Assuming a centre without vibrational properties and putting the energy scale such that $E(0)=0$, the correlation energy U is defined in connection with Z_2,

$$Z_0=1, \qquad Z_1=g_1\exp\left(-\frac{E(1)}{kT}\right), \qquad Z_2=g_2\exp\left(-\frac{E(2)}{kT}\right), \tag{1.5.4}$$

with

$$E(2)=2E(1)+U.$$

In this way, one obtains a generalization of the negative-U centre formalism developed by Adler and Yoffa (1976) and Hoffman (1980) to the partition function formalism. This possibility was noted by Landsberg (1992, p. 60). It allows for excited states of the centres in their various charge states, which is lost if one uses eq. (1.5.4).

Adler and Yoffa note that for negative-U (when $y^2 \gg 1$)

$$\mu = kT \ln \lambda = E(1) + \tfrac{1}{2}U - \tfrac{1}{2}\ln\left(\frac{2}{\Theta} - 1\right). \tag{1.5.5}$$

Thus, whereas the Fermi level, μ, rises from $E(1)$ to $E(1)+U$ as Θ is increased in the normal case ($U>0$), approximation (1.5.5) shows that for $U<0$ it does not vary much with Θ (for $0.2 \leqslant \Theta \leqslant 1.8$). A surprising result is that at $T=0$ the Fermi level can lie below the dominantly occupied states. A similar analysis was performed by Hoffman (1980) based on the doping level of the semiconductor.

The occupation probabilities are readily obtained in a general case from the occupation probabilities $P(r) = \nu_r/N$ as given by eq. (1.3.14). Putting zero energy at $G(0)$ and letting $G(2) = 2G(1) + U$ in analogy with the treatment above, in fig. 1.3 the occupation probabilities are shown for a centre with $U = +10kT$ and a centre with $U = -10kT$. For the negative-U centre (fig. 1.3b), one sees that when the Fermi level is at $-5kT$, the probability is higher that the centre has zero electrons than one electron in spite of the fact that the Fermi level is below the energy level $G(1)$ (equal to zero in fig. 1.3b). This contradicts our normal idea that occupied states tend to

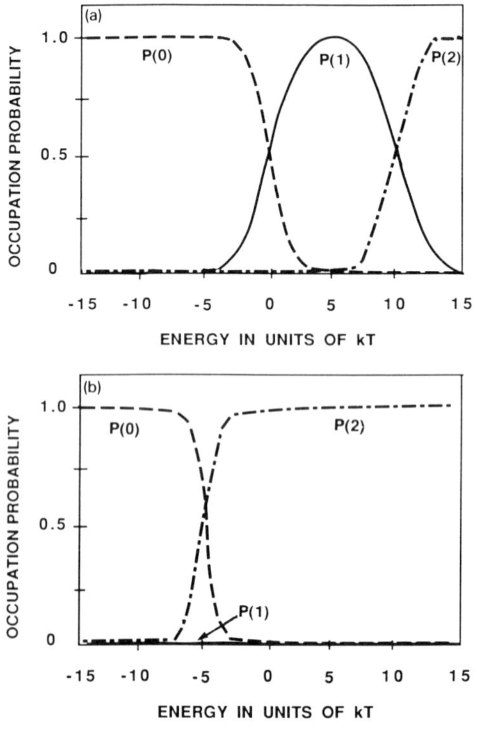

Fig. 1.3. The occupational probabilities $P(0)$, $P(1)$ and $P(2)$ for: (a) a positive-U centre with $U = +10kT$; (b) for a negative-U centre with $U = -10kT$. The energy scales denote the difference between the Fermi-level position and the energy $G(1)$ in units of kT.

lie below the Fermi level and are more numerous than empty centres, which tend to lie above the Fermi level.

1.6. Quasi-Fermi levels

The concentrations of electrons and holes in the energy bands of a semiconductor may exceed the values at thermal equilibrium deduced above. This may be caused by perturbations like irradiation or current injection. As discussed further in the next section, we will partition the available electron states into one group for the conduction band, one for the valence band and one group for each different set of traps. The transitions between states within the groups are assumed to be much more rapid than they are between groups. Then the particles occupying states of the same group can be considered as being in thermal equilibrium among themselves, while particles of different groups are not. This leads us to define a *quasi-Fermi level* for each group replacing the common Fermi level relevant for the whole system at thermal equilibrium. Excess electrons in the conduction band will, however, tend to drop back into holes in the valence band in a *recombination process.* The energy liberated may be given to the lattice, to other charge carriers, or it may be emitted as radiation. In this way, a steady state can be set up in which the number of carriers recombine at the same rate as they are generated by the external source, leaving the concentrations of electrons and holes constant in time.

One would expect the rate of recombination to be proportional to the concentration of electrons, n, and holes, p, since this governs the probability of an electron "meeting" a hole with the possibility of a recombination act. The product np at a non-equilibrium situation can be expressed, in analogy with eqs. (1.2.4) and (1.2.10), as

$$np = N_c N_v F_{1/2}(\gamma_e - \eta_c)\, F_{1/2}(\eta_v - \gamma_h), \tag{1.6.1}$$

which becomes, in the case of non-degeneracy,

$$np = N_c N_v \exp(\gamma_e - \eta_c) \exp(\eta_v - \gamma_h),$$

where $kT\gamma_e$ and $kT\gamma_h$ are the quasi-Fermi levels for electrons in the conduction band and holes in the valence band, respectively, while $kT\eta_c = E_c$ and $kT\eta_v = E_v$. We have replaced the chemical potential by two distinct ones: $\mu_e = kT\gamma_e$ for electrons in the conduction band and $\mu_h = kT\gamma_h$ for holes in the valence band. This takes account of the fact that the carriers in each band are assumed to be in equilibrium with each other, while the electrons and the holes are not in equilibrium with each other. Complete equilibrium would occur only if $\mu_e = \mu_h = \mu$. The quasi-Fermi levels μ_e and μ_h provide a means of describing recombination away from thermal equilibrium.

The existence of quasi-Fermi levels leads to the following generalisation of eq. (1.1.8). The probability of occupation, p_I, and of a vacancy, q_I, of a state, I, belonging to a group of states, i, with a quasi-Fermi level γ_i are, respectively,

$$p_I = \frac{1}{1 + \exp(\eta_I - \gamma_i)} \tag{1.6.2}$$

and

$$q_I = \frac{\exp(\eta_I - \gamma_i)}{1 + \exp(\eta_I - \gamma_i)}. \tag{1.6.3}$$

Here the energy of a state I is E_I and $\eta_I = E_I/kT$. It follows that

$$\frac{p_I}{q_I} = \exp(\gamma_i - \eta_I). \tag{1.6.4}$$

Consider now the recombination of a hole (h) and an electron (e) to yield an electron in the valence band (e_{vb}) as a chemical reaction

$$e + h \rightleftarrows e_{vb}. \tag{1.6.5}$$

The mass action law would lead one to expect a concentration-independent quantity np/n_{vb} where n_{vb} is the number of valence-band electrons. This latter quantity is, however, so great that it is essentially constant when electron and hole concentrations vary, provided we assume $n, p \ll n_{vb}$. For a non-degenerate semiconductor, we find from eq. (1.6.1)

$$np = N_c N_v \exp(\gamma_e - \gamma_h - \eta_g) = n_i^2 \exp(\gamma_e - \gamma_h), \tag{1.6.6}$$

where $\eta_g = E_g/kT$, E_g is the energy band gap and n_i is the intrinsic carrier concentration.

The recombination process may also take place by transferring an electron from the conduction band to an empty impurity energy level within the band gap or from such an impurity trap level to a hole in the valence band. In the former case, the chemical reaction between an electron in the conduction band (e) and an empty trap, i.e., a trapped hole (h_t) can be expressed

$$e + h_t \rightleftarrows e_t, \tag{1.6.7}$$

where e_t represents a trapped electron. The probability for the trap to capture either zero or one electron can be expressed similar to eq. (1.3.12) as

$$P(r) = \frac{\exp(r\gamma_r)Q_r}{\sum_{s=0}^{M} \exp(s\gamma_r)Q_s} \tag{1.6.8}$$

and

$$P(r-1) = \frac{\exp[(r-1)\gamma_{r-1})]}{\exp(r\gamma_r)} \frac{Q_{r-1}}{Q_r} P(r), \tag{1.6.9}$$

where $kT\gamma_r$ is the quasi-Fermi level of the trap with r captured electrons. Defining

$$\gamma_{r-\frac{1}{2}} \equiv r\gamma_r - (r-1)\gamma_{r-1}, \tag{1.6.10}$$

we can express eq. (1.6.9) as

$$P(r-1) = \exp(-\gamma_{r-\frac{1}{2}}) \frac{Q_{r-1}}{Q_r} P(r). \tag{1.6.11}$$

Some useful reaction kinetics for impurity centres, electrons and holes can, of course, be formulated but will not be discussed here (see the first edition of this Handbook; also Landsberg 1992).

2. Recombination statistics

2.1. Basic assumptions

In principle, the recombination problem in semiconductors is greatly complicated by the mutual interactions of the electrons. However, a simplified picture is succesful. In this the electron interactions are taken into account as perturbations leading to changes of the states of the electron pairs. These can be pictured as transitions due to electron collisions and they can be described within the framework of the single-particle states found in the unperturbed problem. In this context, we take into account only two-electron interactions, also known as Auger processes. In addition, electrons interact with the radiation and lattice fields and emit or absorb photons and phonons. These electron–boson interactions are represented by the transitions of single electrons in an energy band scheme.

The recombination problem is still complex because of the many states available to electrons in bands and traps. A key simplification arises from the fact that it is often possible, as already mentioned in § 1.6, to talk about small numbers of groups of quantum states: I labels quantum states in group i, J labels quantum states in group j, etc. Within each group it is supposed that the transitions are much more rapid than they are between groups. In a sense, therefore, electrons in each group approach thermal equilibrium among themselves much more rapidly than between the groups. This motivates one to assign quasi-Fermi levels $kT\gamma_i$, $kT\gamma_j$, etc., for each different group. With this assumption, recombination problems can be discussed by neglecting transitions confined to one group, thereby greatly reducing the number of transition types to be considered. For an M-electron centre, we then assume that as the number of captured carriers, r, increases, the free-energy levels $G(r)$ increase until a maximum value $G(M)$ is approached for the maximum number, M, of captured carriers.

In summary, we make three important assumptions on which the analysis below depends:

(i) Among all possible electron interactions, only two-electron processes (Auger processes) are taken into account.
(ii) The quasi-Fermi level concept is applicable.
(iii) There exists a maximum number, M, for the electrons which can be captured by the electron traps and the corresponding energy levels increase with increasing r

2.2. The recombination rate

Two main paths for recombination between electrons and holes can be distinguished: band-to-band and band-trap. The energy released by the recombination process can be absorbed by bosons or by electrons. This gives a number of possible recombination paths as described in fig. 2.1.

We consider three groups of energy states: those belonging to the conduction band, the valence band and the electron traps. To obtain the general expression for the net recombination rate we proceed as follows. The transition probability S_{IJ} for a single-electron transition from state I to state J depends on states I and J only. For this transition to occur, state I should be "occupied" and state J should be "vacant". The general expression for the average rate of the transition from I to J takes the form $p_I S_{IJ} q_J$, where p_I is the probability of state I being occupied and q_J is the probability of state J being vacant. The reverse process is the transition of an electron from state J to state I. State J will have to be occupied and state I to be vacant. Thus, the rate of this reverse process is $p_J S_{JI} q_I$. The net recombination rate per unit volume of the process $I \to J$ can be then written as

$$I \to J = p_I S_{IJ} q_J - p_J S_{JI} q_I = p_I S_{IJ} q_J \left(1 - \frac{p_J q_I S_{JI}}{p_I q_J S_{IJ}}\right). \tag{2.2.1}$$

By the principle of detailed balance, eq. (2.2.1) equals zero at thermal equilibrium. By virtue of this relation it can also be written as

$$p_I S_{IJ} q_I (1 - X_{IJ}),$$

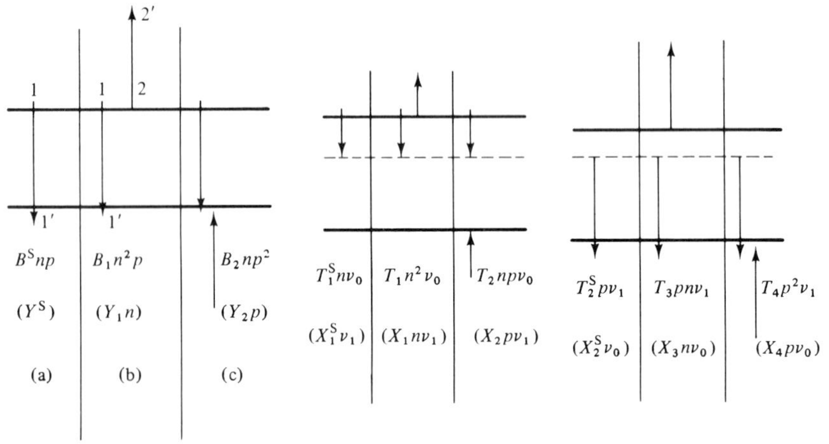

Fig. 2.1. Definition of recombination coefficients. Transition rates per unit volume are stated with each process, and in brackets for the reverse process. B_1, B_2, T_1 to T_4 refer to Auger processes; B^S, T_1^S, T_2^S refer to single-electron recombination; Y^S, X_1^S, X_2^S refer to carrier generation processes; Y_1, Y_2, X_1 to X_4 refer to impact ionisation processes. In this article arrows indicate transitions made by electrons.

where

$$X_{IJ} = \frac{S_{JI} q_I p_J}{S_{IJ} p_I q_J} = \frac{S_{JI}/S_{IJ}}{(S_{JI}/S_{IJ})_0} \frac{q_I p_{I0}}{q_{I0} p_I} \frac{p_J q_{J0}}{p_{J0} q_J}, \tag{2.2.2}$$

and subscript 0 denotes equilibrium quantities. For recombination by a single mechanism (e.g. radiative recombination or capture by traps, etc.) the overall rate of recombination will be the sum of expressions like eq. (2.2.1) over allowed states I in some group i and over allowed states J in some group j. The experimentally observed rate of recombination will be the sum of the overall rate for the different mechanisms; here we assume that the different mechanisms do not interfere with each other.

We illustrate the evaluation of X_{IJ} for three cases. When I and J refer to states in bands, p_I is the probability that state I is occupied and q_I is the probability that state I is vacant. Clearly $p_I + q_I = 1$, and using eq. (1.6.4), we have

$$\frac{q_I p_{I0}}{q_{I0} p_I} = \exp(\gamma_0 - \gamma_i), \tag{2.2.3}$$

$$X_{IJ} = \frac{S_{JI}/S_{IJ}}{(S_{JI}/S_{IJ})_0} \exp(\gamma_j - \gamma_i). \tag{2.2.4}$$

When a trap is involved matters are rather different. Because of the postulated interactions among the electrons in a centre it is not possible to talk about the *same* level being occupied or vacant. Consequently, identification of forward and reverse processes in terms of levels becomes impossible. Instead, it is sufficient to deal with a centre, say an r-electron centre, as a whole; we then need the probability that a given centre is an r-electron centre. For the reverse transition where the capture of an electron converts it back again into an r-electron centre, the capturing centre must be an $(r-1)$-electron centre. Thus, q_I is now replaced by the probability that the centre is an $(r-1)$ electron centre.

Consider the capture of an electron into an $(r-1)$-centre. Then, for a typical transition (as shown in fig. 2.2), eq. (2.2.2) applies. The first factor is formally

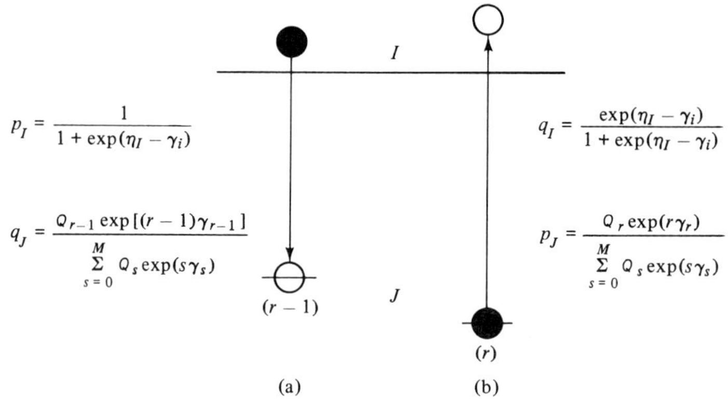

Fig. 2.2. Examples of factors p and q for a single-electron transition into traps (a), and their inverse (b).

unchanged, the second factor is given by eq. (2.2.3) and the third factor is, with eq. (1.3.14),

$$p_r p_{r-1\,0}/p_{r\,0} p_{r-1} = \exp[r\gamma_r + (r-1)\gamma_0 - r\gamma_0 - (r-1)\gamma_{r-1}]$$
$$= \exp[r\gamma_r - (r-1)\gamma_{r-1} - \gamma_0].$$

Using the reduced quasi-Fermi level (1.6.10) for an $(r-1 \to r)$ electron transition

$$r\gamma_r - (r-1)\gamma_{r-1} \equiv \gamma_{r-1/2}, \tag{2.2.5}$$

one finds in the case of electron capture by an $(r-1)$-electron centre

$$X_{IJ} = \frac{S_{JI}/S_{IJ}}{(S_{JI}/S_{IJ})_0} \exp(\gamma_{r-1/2} - \gamma_i) \quad (i = \mathrm{e} \text{ or } \mathrm{h}), \tag{2.2.6}$$

which again is of the form of eq. (2.2.4).

For single-boson transitions, eqs. (2.2.1) and (2.2.2) are still valid. One has to note that for fermions the occupation probability of a quantum state is exactly the same as the mean number of particles in that state. It is the latter interpretation that is more basic, and it holds also for bosons. In this case, one has

$$p_I = \frac{1}{\exp(\eta_I - \gamma_I) - 1}, \quad q_I = 1 + p_I \quad \text{(bosons)}, \tag{2.2.7}$$

in contrast to

$$p_I = \frac{1}{\exp(\eta_I - \gamma_I) + 1}, \quad q_I = 1 - p_I \quad \text{(fermions)}. \tag{2.2.8}$$

Equation (2.2.7) expresses the increased attraction a highly occupied boson state has for *additional* bosons. However, in all cases

$$q_I/p_I = \exp(\eta_I - \gamma_I). \tag{2.2.9}$$

If we assume that the ratio S_{IJ}/S_{JI} is independent on the excitation, we have at thermal equilibrium

$$p_{I0} S_{IJ} q_{J0} = p_{J0} S_{JI} q_{I0}. \tag{2.2.10}$$

Using eq. (1.6.4) in eq. (2.2.1), we obtain

$$\frac{S_{IJ}}{S_{JI}} = \frac{\exp(\gamma_0 - \eta_J)}{\exp(\gamma_0 - \eta_I)} = \exp(\eta_I - \eta_J) \tag{2.2.11}$$

and

$$\frac{p_J q_I}{p_I q_J} = \exp(\gamma_J - \gamma_I)\exp(\eta_I - \eta_J). \tag{2.2.12}$$

Using eqs. (2.2.11) and (2.2.12) in eq. (2.2.1) now gives for the net recombination rate per unit volume from state I to state J (when the S_{IJ} are excitation-independent) the key result

$$I \to J = p_I S_{IJ} q_J [1 - \exp(\gamma_J - \gamma_I)]. \tag{2.2.13}$$

2.3. Band–band transitions

The net recombination rate per unit volume, u_{cv}, for transitions between the conduction band the the valence band is expressed by the sum of eq. (2.2.13) taken over all states in the bands,

$$u_{cv} = \sum_{C,V} p_C S_{CV} q_V [1 - \exp(\gamma_h - \gamma_e)], \tag{2.3.1}$$

where γ_e is the reduced quasi-Fermi level for electrons in the conduction band and γ_h is the quasi-Fermi level for holes in the valence band. Since the total concentration of electrons, n, and holes, p, can be written as

$$\sum_C p_C = nV \quad \text{and} \quad \sum_V q_V = pV, \tag{2.3.2}$$

where V is the volume of the semiconductor, a transition coefficient, B, for the band–band processes can be defined by averaging over the states in the bands as

$$B = \left(\sum_{C,V} p_C S_{CV} q_V \right) \Big/ V^{-2} \sum_C p_C \sum_V q_V . \tag{2.3.3}$$

By such definitions, we introduce three band–band recombination coefficients, B^s, B_1, and B_2 (see fig. 2.1),

$$\sum_{C,V} p_C S_{CV} q_V = \begin{cases} B^s np & \text{(electron–boson interaction)} \\ B_1 n^2 p & \text{(electron–electron interaction)} \\ B_2 np^2 & \text{(hole–hole interaction).} \end{cases} \tag{2.3.4}$$

The total rate, F (in $cm^3\ s^{-1}$), for band-to-band recombination can be considered as the sum of the three possibilities given by eq. (2.3.4),

$$F = B^s + B_1 n + B_2 p, \tag{2.3.5}$$

and for the recombination rate per unit volume

$$u_{cv} = Fnp[1 - \exp(\gamma_h - \gamma_e)]. \tag{2.3.6}$$

For a non-degenerate semiconductor, we find by using eq. (1.6.6)

$$u_{cv} = F(np - n_i^2). \tag{2.3.7}$$

2.4. Band–trap transitions

The net rate per unit volume, $u_{r-\frac{1}{2}}$, for an electron in the conduction band recombining with a hole which occupies a trap with probability q_{r-1} such that the trap goes from $r-1$ to r captured electrons, can be expressed, analogously to eq. (2.3.1), as

$$u_{c\,r-\frac{1}{2}} = \left(\sum_C p_C S_{CT} \right) q_{r-1} [1 - \exp(\gamma_{r-\frac{1}{2}} - \gamma_e)]. \tag{2.4.1}$$

As for the band-to-band recombination processes, we now introduce averaged recombination coefficients T^s, T_1 and T_2 (see fig. 2.1):

$$\left(\sum_C p_C S_{CV}\right) q_{r-1} = \begin{cases} T_1^s n v_{r-1} & \text{(electron–boson interaction)} \\ T_1 n^2 v_{r-1} & \text{(electron–electron interaction)} \\ T_2 p n v_{r-1} & \text{(electron–hole interaction).} \end{cases} \tag{2.4.2}$$

The total recombination coefficient, $G(r-\frac{1}{2})$ for band–trap recombination is given by the sum of all the interaction possibilities above as

$$G(r-\tfrac{1}{2}) = T_1^s(r-\tfrac{1}{2}) + T_1(r-\tfrac{1}{2})n + T_2(r-\tfrac{1}{2})p \tag{2.4.3}$$

and, from eqs. (2.4.1) and (2.4.3),

$$u_{c\,r-\frac{1}{2}} = G(r-\tfrac{1}{2}) n v_{r-1}[1-\exp(\gamma_{r-\frac{1}{2}}-\gamma_e)]. \tag{2.4.4}$$

Using eq. (1.6.11) in eq. (2.4.4), we obtain

$$u_{c\,r-\frac{1}{2}} = G(r-\tfrac{1}{2}) n v_{r-1} - G(r-\tfrac{1}{2}) n \exp(-\gamma_e)\frac{Q_{r-1}}{Q_r} v_r. \tag{2.4.5}$$

The capture and emission coefficients c_n and e_n often used in experimental literature can be identified as

$$c_n(r-\tfrac{1}{2}) = G(r-\tfrac{1}{2}), \tag{2.4.6}$$

$$e_n(r-\tfrac{1}{2}) = G(r-\tfrac{1}{2}) n \exp(-\gamma_e)\frac{Q_{r-1}}{Q_r}. \tag{2.4.7}$$

In this notation, eq. (2.4.5) reads

$$u_{c\,r-\frac{1}{2}} = c_n(r-\tfrac{1}{2}) n v_{r-1} - e_n(r-\tfrac{1}{2}) v_r. \tag{2.4.8}$$

For a one-electron trap ($M = 1$) we have

$$P(0)N_N = v_0 \equiv p_t, \qquad P(1)N_t = v_1 \equiv n_t, \qquad p_t + n_t = N, \tag{2.4.9}$$

so that the recombination rate per unit volume, $u_{c\frac{1}{2}}$, for the one-electron trap is the familiar expression

$$u_{c\frac{1}{2}} = c_n(\tfrac{1}{2}) n p_t - e_n(\tfrac{1}{2}) n_t, \tag{2.4.10}$$

as often seen in the experimental literature (see, e.g., Sah et al. 1970).

The emission rate, $e_n(r-\frac{1}{2})$, given by eq. (2.4.7) is valid without any restriction on the concentration of free electrons in the conduction band. For a non-degenerate case, when

$$n \exp(-\gamma_e) = N_c \exp(-\eta_c), \tag{2.4.11}$$

we recognize the common expression for the emission rate (see Landsberg 1986, eq. (13))

$$e_n(r-\tfrac{1}{2}) = c_n(r-\tfrac{1}{2})N_c \exp\left(-\frac{E_c - G(r-\frac{1}{2})}{kT}\right). \tag{2.4.12}$$

Recombination of an electron, captured in a trap, with a hole in the valence band can be expressed by the recombination coefficients T_2^s, T_3 and T_4 in the same way as in eq. (2.4.2) (see fig. 2.1),

$$p_r\left(\sum_V S_{TV} q_V\right) = \begin{cases} T_2^s p v_r & \text{(hole–boson interaction)} \\ T_3 n p v_r & \text{(hole–electron interaction)} \\ T_4 p^2 v_r & \text{(hole–hole interaction)} \end{cases} \tag{2.4.13}$$

The total hole recombination rate, $H(r-\frac{1}{2})$, is the sum of the rates given by the coefficients in eq. (2.4.13),

$$H(r-\tfrac{1}{2}) = T_2^s(r-\tfrac{1}{2}) + T_3(r-\tfrac{1}{2})n + T_4(r-\tfrac{1}{2})p, \tag{2.4.14}$$

Again, identifying the capture and emission rates, c_p and e_n, respectively, we obtain

$$c_p(r-\tfrac{1}{2}) = H(r-\tfrac{1}{2}), \tag{2.4.15}$$

$$e_p(r-\tfrac{1}{2}) = H(r-\tfrac{1}{2})p \exp(\gamma_h)\frac{Q_r}{Q_{r-1}}. \tag{2.4.16}$$

In this notation, the recombination rate $u_{r-\frac{1}{2}\,v}$ reads

$$u_{r-\frac{1}{2}\,v} = c_p(r-\tfrac{1}{2})pv_r - e_p(r-\tfrac{1}{2})v_{r-1}. \tag{2.4.17}$$

For a one-electron trap,

$$u_{\frac{1}{2}v} = c_p(\tfrac{1}{2})pn_t - e_p(\tfrac{1}{2})p_t, \tag{2.4.18}$$

and for the non-degenerate case, where

$$p \exp(\gamma_h) = N_v \exp(\eta_v), \tag{2.4.19}$$

we find the familiar expression for the emission rate for holes, similar to eq. (2.4.12),

$$e_p(r-\tfrac{1}{2}) = c_p(r-\tfrac{1}{2})N_v \exp\left(-\frac{G(r-\frac{1}{2}) - E_v}{kT}\right). \tag{2.4.20}$$

2.5. Generalised Shockley–Read–Hall statistics

In many cases, especially for semiconductors with indirect band gaps, the lifetime of charge carriers in the bands are almost completely determined by band-to-trap recombination processes. After excitation and creation of excess charge carriers, a steady state is reached, where the recombination rates from the conduction band to the trap states and from the trap to the valence band states are equal,

$$u_{c\,r-\frac{1}{2}} = u_{r-\frac{1}{2}\,v}, \tag{2.5.1}$$

We define an *electron lifetime*, τ_n, and a *hole lifetime*, τ_p, by

$$\tau_n(r-\tfrac{1}{2}) \equiv \frac{1}{c_n[v_r + v_{r-1}]}, \qquad \tau_p(r-\tfrac{1}{2}) \equiv \frac{1}{c_p[v_r + v_{r-1}]}, \tag{2.5.2}$$

and free carrier concentrations $n(r-\frac{1}{2})$ and $p(r-\frac{1}{2})$ as

$$n(r-\tfrac{1}{2}) \equiv n\exp(-\gamma_e)\frac{Q_{r-1}}{Q_r}, \qquad p(r-\tfrac{1}{2}) \equiv p\exp(\gamma_h)\frac{Q_r}{Q_{r-1}}. \tag{2.5.3}$$

These concentrations have already been encountered in eqs. (2.4.7) and (2.4.16). Then using eqs. (2.4.8), (2.4.17), (2.5.2) and (2.5.3) in eq. (2.5.1) yields the steady-state recombination rate per unit volume expressed by the charge carrier lifetimes, as

$$u_{c\,r-\frac{1}{2}} = \frac{np[1-\exp(\gamma_h-\gamma_e)]}{\tau_n(r-\frac{1}{2})[p+p(r-\frac{1}{2})] + \tau_p(r-\frac{1}{2})[n+n(r-\frac{1}{2})]}, \tag{2.5.4}$$

or, by the capture rates, as

$$u_{c\,r-\frac{1}{2}} = \frac{np[1-\exp(\gamma_h-\gamma_e)]c_p(r-\frac{1}{2})c_n(r-\frac{1}{2})}{c_p(r-\frac{1}{2})[p+p(r-\frac{1}{2})] + c_n(r-\frac{1}{2})[n+n(r-\frac{1}{2})]}[v_r + v_{r-1}]. \tag{2.5.5}$$

The total recombination rate u, including band-to-band transitions and all possible recombination paths over the trap capable of capturing M electrons, is given by

$$u = u_{cv} + \sum_{r=0}^{M} u_{c\,r-\frac{1}{2}}. \tag{2.5.6}$$

Equations (2.5.5) and (2.5.6) are valid without any restrictions on the concentrations of free charge carriers or the trap multiplicity, M (for a review, see Landsberg 1992). For a one-electron trap in a non-degenerate semiconductor, eq. (2.5.3) changes to

$$n(\tfrac{1}{2}) = N_c\exp\left(-\frac{E_c - G(\frac{1}{2})}{kT}\right) \equiv n_1, \qquad p(\tfrac{1}{2}) = N_v\exp\left(-\frac{G(\frac{1}{2}) - E_v}{kT}\right) \equiv p_1. \tag{2.5.7}$$

For this special case, the concentrations $n(r-\frac{1}{2})$ and $p(r-\frac{1}{2})$ become equal to the free-carrier concentrations n_1 and p_1 introduced by Shockley and Read (1952). Also, under the same conditions, we find

$$\tau_n(\tfrac{1}{2}) = \frac{1}{c_n(\frac{1}{2})N} \equiv \tau_{n0}, \qquad \tau_p(\tfrac{1}{2}) = \frac{1}{c_p(\frac{1}{2})N} \equiv \tau_{p0}, \tag{2.5.8}$$

where τ_{n0} and τ_{p0} are the electron and hole lifetimes as defined by Shockley and Read (1952). With eqs. (2.5.7) and (2.5.8) put into eq. (2.5.5), we find the Shockley–Read expression for the recombination rate, valid for a non-degenerate semiconductor,

$$u_{c\frac{1}{2}} = \frac{(np - n_i^2)c_n(\frac{1}{2})c_p(\frac{1}{2})}{c_p(\frac{1}{2})(p+p_1)+c_n(\frac{1}{2})(n+n_1)} N_t = \frac{(np-n_i^2)}{\tau_{n0}(p+p_1)+\tau_{p0}(n+n_1)}. \tag{2.5.9}$$

It should be observed that this expression still allows for electron–electron *and* electron–boson interactions in the recombination mechanism depending on the meanings attached to c_n and c_p.

2.6. Decay laws for band–band recombination

The decay of excess carrier concentrations as a function of time is an important material parameter determining the switching properties of bipolar semiconductor devices. For example, in the case of high-power devices, in order to minimize losses, one is interested to keep a long decay time at high charge carrier injection levels when the device is conducting current. At low injection levels, on the other hand, just before the device is switched off, one is interested to have as small a decay time as possible in order to empty the last part of excess carriers from the active device regions and thereby increase the switching speed. The decay times are directly determined by the recombination process and depend on the charge carrier lifetime. Therefore, not only are different processes expected for different semiconductor materials, but also, within the same semiconductor, different mechanisms may dominate the lifetime depending on the concentration of excess charge carriers. For very high concentrations, the Auger-recombination rates determined by the factors B_1, B_2, T_1, T_2, T_3 and T_4 are expected to dominate. At lower concentrations, the electron–boson band–band factor B^s becomes strong in direct band gap semiconductors like gallium arsenide. For silicon, with an indirect band gap, two-step processes across trap energy levels are more probable at lower injection. In the latter case, the lifetime can be controlled through the concentration of trap centres responsible for the two-step recombination process. The band–band processes are determined by intrinsic properties of the semiconductor crystal. Especially, at higher charge carrier concentrations, the Auger recombination mechanism often sets an upper limit of lifetime available in a practical situation.

For the band–band recombination rate u_{cv} per unit volume, we find from eq. (2.3.7)

$$u_{cv} = (B^s + B_1 n + B_2 p)(np - n_0 p_0), \tag{2.6.1}$$

where n_0 and p_0 are equilibrium concentrations. Writing $n = n_0 + \delta n$, $p = p_0 + \delta p$ and assuming that the excess carrier concentrations δn and δp are much larger than the concentrations of charged states in the energy band gap of the semiconductor, one finds by neutrality arguments

$$\delta n \simeq \delta p. \tag{2.6.2}$$

For two-step processes one can use the expressions given in eqs. (2.5.4), (2.5.5) and (2.5.9) to obtain the corresponding decay times for recombination at traps.

For a case where the Auger terms in eq. (2.6.1) can be neglected compared with the electron–boson direct recombination term, we find, by using $u_{cv} = -\mathrm{d}(\delta n)/\mathrm{d}t$,

$$\frac{\mathrm{d}(\delta n)}{\mathrm{d}t} = -\,B^{s}(n_0 + p_0 + \delta n)\delta n, \tag{2.6.3}$$

which has the solution

$$\delta n(t) = \frac{(n_0 + p_0)\,\delta n(0)\exp\left(-\dfrac{t}{\tau_\infty}\right)}{n_0 + p_0 + \delta n(0)\left[1 - \exp\left(-\dfrac{t}{\tau_\infty}\right)\right]}, \tag{2.6.4}$$

where

$$\tau_\infty^{-1} \equiv B^{s}(n_0 + p_0). \tag{2.6.5}$$

For large t such that δn becomes much smaller than the equilibrium concentrations, $\delta n \ll n_0, p_0$, the exponential term in eq. (2.6.4) can be neglected and δn approaches

$$\delta n(t) = \delta n(0)\exp\left(-\frac{t}{\tau_\infty}\right). \tag{2.6.6}$$

For small t, with $\delta n \gg n_0, p_0$, eq. (2.6.3) is solved by

$$\delta n(t) = \delta n(0)\exp\left(-\frac{t}{\tau_0}\right), \quad \tau_0 \equiv \frac{1}{B^{s}(n_0 + p_0 + \delta n(0))}. \tag{2.6.7}$$

The decay of excess carriers due to single-particle processes from band to band, therefore, has an initially faster decay which is broken into a more slowly decaying exponential function for long time values.

From the point of view of application, two main cases are of particular interest concerning Auger-recombination processes. First, for very high charge carrier injection levels, Auger recombination may take over the dominating role in determining charge carrier lifetimes. Second, in highly doped regions, e.g., emitter junctions of devices, this process may dominate the recombination rates even at low injection levels. As the charge carrier lifetime in an emitter region determines the injection efficiency of bipolar devices, the recombination processes in these cases are of considerable technical importance.

We consider first the case of high injection into a moderately doped material such that $\delta n \gg n_0, p_0$. Then, we may assume that the two Auger capture rates, $B_1 n$ and $B_2 p$ dominate in eq. (2.6.1), and we write

$$\frac{\mathrm{d}(\delta n)}{\mathrm{d}t} = -\,(B_1 + B_2)\,\delta n^3. \tag{2.6.8}$$

This differential equation has the solution

$$\delta n(t) = \frac{\delta n(0)}{[2(B_1 + B_2)\,\delta n(0)^2 t + 1]^{1/2}}. \tag{2.6.9}$$

In the second case, we consider an n^{++} emitter where $n_0 \gg p_0$ and $\delta n \ll n_0$. Then, the capture rate $B_1 n$ dominates in the first bracket of eq. (2.6.1) while the second bracket can be written as $n_0\,\delta n$ and the differential equation reads

$$\frac{d(\delta n)}{dt} = -B_1 n_0\,\delta n, \tag{2.6.10}$$

which has the solution

$$\delta n(t) = \delta n(0)\exp(-B_1 n_0 t). \tag{2.6.11}$$

For a p^{++} emitter, the corresponding expression,

$$\delta n(t) = \delta n(0)\exp(-B_2 p_0 t), \tag{2.6.12}$$

is found in a similar way.

2.7. Decay laws for band–trap recombination

For the simplest case, trap recombination through a one-electron ($M = 1$) centre, we find from the Hall–Shockley–Read expression in eq. (2.5.9) by taking $-d(\delta n)/dt = u_{c-\frac{1}{2}}$

$$-\frac{d(\delta n)}{dt} = \frac{(n_0 + p_0 + \delta n)\delta n}{(\tau_{p0} + \tau_{n0})[(n_0 + p_0)\alpha + \delta n]}, \tag{2.7.1}$$

where

$$\alpha = \frac{\tau_{p0}(n_0 + n_1) + \tau_{n0}(p_0 + p_1)}{(\tau_{p0} + \tau_{n0})(n_0 + p_0)}. \tag{2.7.2}$$

Four different cases can be distinguished depending on doping type and the relative positions of the equilibrium Fermi level and the trap energy $G(\frac{1}{2})$ as shown in fig. 2.3.

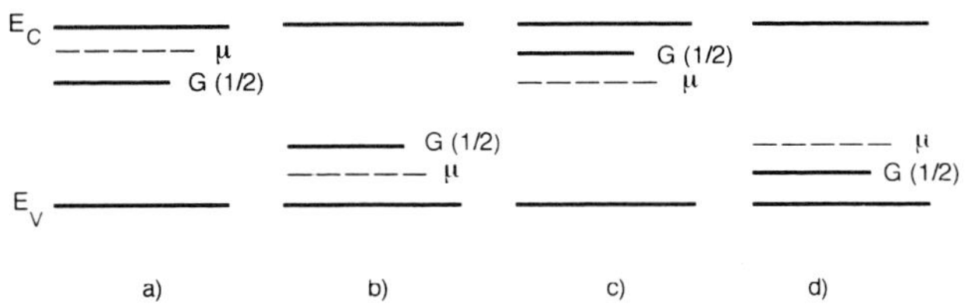

Fig. 2.3. Four different cases for the position of the energy level $G(\frac{1}{2})$ in relation to the Fermi level for n-type (a and c) and p-type (b and d) material. Corresponding charge carrier lifetimes are given in table 2.1.

The general solution of the decay (2.7.1) is (Landsberg 1982, eq. (20.8))

$$\delta n(t) = \begin{cases} \left(\dfrac{n_0 + p_0 + \delta n}{n_0 + p_0 + \delta n(0)}\right)^{1-(1/\alpha)} \delta n(0) \exp\left[-\dfrac{t}{\alpha(\tau_{n0} + \tau_{p0})}\right] & (\alpha \neq 1), \\ \delta n(0) \exp\left[-\dfrac{t}{(\tau_{n0} + \tau_{p0})}\right] & (\alpha = 1). \end{cases} \tag{2.7.3}$$

The special cases summarised in table 2.1 are readily obtained from this result. It shows that one finds again exponential decays for short and long times:

$$\frac{\delta n(t)}{\delta n(0)} = \exp\left(-\frac{t}{\alpha(\tau_{n0} + \tau_{p0})}\right) \times \begin{cases} 1 & \text{(short times)} \\ \left[\dfrac{n_0 + p_0}{(n_0 + p_0 + \delta n_0)}\right]^{1-(1/\alpha)} & \text{(long times).} \end{cases}$$

Note that other non-exponentials are quite possible, e.g., of the form (Evans and Landsberg 1965)

$$\frac{\delta n(t)}{\delta n(0)} = \left(1 + \frac{t}{\tau}\right)^{-m}.$$

From table 2.1, one sees that for an n-type semiconductor with the Fermi level well above the free-energy level of the trap, the charge carrier lifetime at low injection is given by the lifetime for holes, τ_{p0} (see eq. (2.5.8)). Correspondingly, for a p-type semiconductor with the Fermi level well below the trap energy level, the low-injection lifetime is given by the lifetime of electrons. This is so because the total recombination time is the sum of the two time intervals needed for the carrier to go from the conduction band to the trap and from the trap to the valence band. The bottleneck in this process is the time it needs to interact with the minority carriers because of their low concentration compared with the concentration of majority carriers.

In cases (c) and (d), where the Fermi level is below the trap level in n-type and above the trap level in p-type material, the occupation of the trap by majority carriers becomes smaller due to re-emission into the conduction and valence band, respectively. This increases the lifetime by the factors n_1/n_0 and p_1/p_0, respectively, because of the increased average time it takes for the trap to capture a majority carrier. In both cases these factors are larger than one.

Table 2.1

Case	τ	Comment
(a) $n_0 \gg n_1, p_0, p_1$	τ_{p0}	Minority carrier lifetime, n-type material
(b) $p_0 \gg n_1, n_0, p_1$	τ_{n0}	Minority carrier lifetime, p-type material
(c) $n_1 \gg n_0 \gg p_0, p_1$	$(n_1/n_0)\tau_{p0}$	As (a) with smaller trapped-electron population
(d) $p_1 \gg p_0 \gg n_0, n_1$	$(p_1/p_0)\tau_{p0}$	As (b) with smaller trapped-hole population
(e) $\delta n \gg n_0, p_0$	$\tau_{n0} + \tau_{p0}$	High injection

Of course, if one has an exponential decay, say $\delta n(t) = \delta n(0)\exp(-t/\tau(r-\frac{1}{2}))$, then the usual lifetime relation,

$$u_{c\,r-\frac{1}{2}} = \frac{\delta n}{\tau(r-\frac{1}{2})}, \tag{2.7.4}$$

follows. This lifetime represents the average transition time from the conduction band to the valence band via the energy level $G(r-\frac{1}{2})$. For a non-degenerate semiconductor, $u_{c\,r-\frac{1}{2}}$ is given by eq. (2.5.5) as

$$u_{c\,r-\frac{1}{2}} = \frac{(np-n_i^2)c_p(r-\frac{1}{2})\,c_n(r-\frac{1}{2})}{c_p(r-\frac{1}{2})[p+p(r-\frac{1}{2})]+c_n(r-\frac{1}{2})[n+n(r-\frac{1}{2})]}(v_r+v_{r-1}). \tag{2.7.5}$$

Taking into account all possible transitions via an M-electron centre, and no others, we find for the total transition time τ, which is then also the lifetime of the free carriers if the centre dominates the recombination process,

$$\tau = \delta n\left(\sum_{s=0}^{M} u_{c\,s-\frac{1}{2}}\right)^{-1}. \tag{2.7.6}$$

Normally some level, $G(r-\frac{1}{2})$ say, will lie closest to and below the Fermi level, e.g., the $G(\frac{1}{2})$ level in fig. 2.3. The levels below $G(r-\frac{1}{2})$ will then be effectively absent and the $s=r$ term is liable to dominate in eq. (2.7.6).

Thus, for an M-electron centre, where the energy levels $G(s-\frac{1}{2})$ are separated by at least a number of kT, only the uppermost filled energy level contributes to the low-injection lifetime. For an n-type semiconductor one finds

$$\tau = \frac{1}{c_p(r-\frac{1}{2})N}[\approx \tau_p(r-\frac{1}{2})], \tag{2.7.7}$$

and, in a similar way for a p-type semiconductor

$$\tau = \frac{1}{c_n(r-\frac{1}{2})N}[\approx \tau_n(r-\frac{1}{2})]. \tag{2.7.8}$$

these expressions are formally, similar to those obtained for a one-electron centre given in table 2.1 and approximate the lifetime of eq. (2.5.2).

The high-injection lifetime determined by an M-electron centre becomes more complicated in the general case because, when $\delta n \approx n \approx p \gg n_0, p_0$, the occupation probabilities $P(s)$ of the different energy levels $G(s-\frac{1}{2})$ now depend on the relations between all possible capture constants $c_n(s-\frac{1}{2})$ and $c_p(s-\frac{1}{2})$. For the r-electron centre the recombination rate per unit volume in the high-injection case is obtained from eqs. (2.5.4) and (2.5.5) as

$$u_{c\,r-\frac{1}{2}} = \frac{\delta n\, c_p(r-\frac{1}{2})c_n(r-\frac{1}{2})(v_r+v_{r-1})}{c_p(r-\frac{1}{2})+c_n(r-\frac{1}{2})} = \frac{\delta n}{\tau_n(r-\frac{1}{2})+\tau_p(r-\frac{1}{2})}. \tag{2.7.9}$$

The high-injection condition, $\delta n \approx n \approx p \gg n_0, p_0$, means that the thermal emission can be neglected, which, in turn, implies that the recombination rate from the conduction band to a certain energy level at $G(r-\frac{1}{2})$ is equal to the recombination rate for transferring an electron from that level to the valence band,

$$c_n(r-\tfrac{1}{2})nv_{r-1} = c_p(r-\tfrac{1}{2})pv_r. \tag{2.7.10}$$

This gives a condition for the relation between occupation probabilities as

$$\frac{v_{r-1}}{v_r} = \frac{c_p(r-\frac{1}{2})}{c_n(r-\frac{1}{2})}. \tag{2.7.11}$$

Equation (2.7.6) together with eqs. (2.7.9) and (2.7.11) and the obvious requirement

$$\sum_{s=0}^{M} v_s = N, \tag{2.7.12}$$

gives a chance of obtaining the high-injection lifetimes for different M values. For $M = 1$, one obtains

$$\tau_H = \frac{c_n(\frac{1}{2}) + c_p(\frac{1}{2})}{c_n(\frac{1}{2})c_p(\frac{1}{2})}\frac{1}{N} \tag{2.7.13}$$

and for $M = 2$,

$$\tau_H = \frac{c_n(\frac{1}{2})(c_p(\frac{3}{2}) + c_n(\frac{3}{2})) + c_p(\frac{1}{2})c_p(\frac{3}{2})}{c_n(\frac{1}{2})\,c_p(\frac{3}{2})\,(c_p(\frac{1}{2}) + c_n(\frac{3}{2}))}\frac{1}{N}. \tag{2.7.14}$$

As mentioned in § 2.6, a common requirement for high-power bipolar semiconductor devices is that the ratio between the high- and the low-injection lifetimes, τ_H/τ_L be large. The low-injection lifetime for an $M = 2$ centre in n-type material is obtained from eq. (2.7.7) as

$$\tau_L = \frac{1}{c_p(\frac{3}{2})N}. \tag{2.7.15}$$

Together with eq. (2.7.14) we then obtain

$$\frac{\tau_H}{\tau_L} = \frac{c_n(\frac{1}{2})[c_p(\frac{3}{2}) + c_n(\frac{3}{2})] + c_p(\frac{1}{2})c_p(\frac{3}{2})}{c_n(\frac{1}{2})(c_p(\frac{1}{2}) + c_n(\frac{3}{2}))}. \tag{2.7.16}$$

We see from eq. (2.7.16) that a centre with the properties

$$c_p(\tfrac{3}{2}) \gg c_p(\tfrac{1}{2}) \gg c_n(\tfrac{1}{2}), c_n(\tfrac{3}{2})$$

would give τ_H/τ_L larger than 1 for the $M = 2$ recombination centre. For a recombination centre with $M = 1$, which is less common in silicon, the corresponding requirement for n-type material is only $c_p(\frac{1}{2}) \gg c_n(\frac{1}{2})$.

2.8. *Detailed balance and its relation to measurements*

The concept of detailed balance has been used in earlier sections as a condition for over-all energy balance at thermal equilibrium for the capture and emission of carriers in a trap. More generally it means that, at thermal equilibrium, every different type of interaction giving rise to a rate of energy exchange is balanced by the reverse rate. For electron capture and emission at semiconductor traps, this means for example that a multi-phonon recombination is balanced by the corresponding multi-phonon excitation for every energy level pair of the system. Similarly, radiative recombination processes are balanced by optical absorption processes, Auger recombination by impact ionisation and so on.

This idea was used by van Roosbroeck and Shockley (1954) to calculate the radiative lifetime for free charge carriers in band-to-band recombination processes from optical absorption data. Similar reasoning was given by Blakemore (1967) and, in an additional development, by Bebb (1972) for the relation between quantities describing radiative capture and emission at impurities. Also, for the balance between Auger recombination and impact ionization the corresponding expressions have been deduced by Landsberg (1972) and compared with the radiative case.

As a consequence of detailed balance, emission rates can be expressed in a similar way, as was done in § 2.4, for every energy level in the conduction band and for every type of interaction process, e.g. those involved in eqs. (2.4.6) and (2.4.7). This has some implications for the interpretation of measured capture and emission data. Suppose, for example, that the thermal emission rate e_n and the capture rate c_n are measured by two independent measurement techniques in the same sample. The most common technique for the measurement of emission rates is a technique based on space-charge transients (Sah et al. 1970, Lang 1974, Schroder 1990, Andersson and Engström 1990), when the measurement takes place in the space-charge region of a p–n junction and no free-charge carriers exist in the bands. For the capture rate, a number of techniques can be used. One common method was developed by Lang (1974) and involves the capture of majority carriers into traps, which were emptied before the capture process. The capture conditions during such a measurement differ from a case where also minority carriers are present. It was recently demonstrated by Hangleiter (1987) that such a case gives the additional possibility of carrier recombination by exciton processes. These methods would give different values of the capture rate depending on which recombination mechanism dominates under given conditions. Therefore, when using equations like eq. (2.4.12) care must be taken so that the emission rates and the capture rates used in the expression are determined by the same interaction process.

In recent experimental works on recombination at impurities in silicon (Kleverman et al. 1985, Hangleiter 1987), some new emission and capture mechanisms have been observed which we will use as examples.

By utilising the radiative band–band recombination in silicon as a probe to measure free-carrier lifetimes upon the creation of excess carrier concentrations by YAG laser pulses, Hangleiter (1987) found evidence for a mechanism where the free excitons among the excess carriers constitute the precursors for the capture process. As illustrated in fig. 2.4a, the electron of the exciton may be captured by the trap and the energy may be released by exciting the hole down among the states of the valence band. A similar mechanism was proposed for the hole capture shown in the right-hand diagram of fig. 2.4a. The reverse of these processes are shown in fig. 2.4b. In order to create an emission event for the electron, a hole is needed deep in the valence band such that an electron close to the valence-band edge can recombine

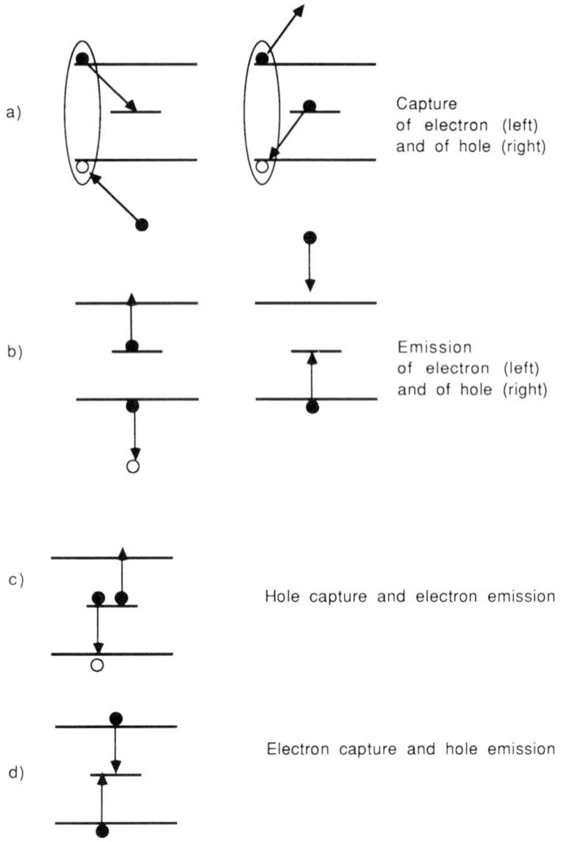

Fig. 2.4. (a) Capture by a trap of electrons and holes from a free exciton and (b) the corresponding emission processes (Hangleiter 1987). (c) and (d): A possible recombination–emission mechanism at a double donor (Kleverman et al. 1985, Landsberg 1992). Configurations just before the transitions are shown.

and give its energy to the captured electron. In an analogous manner, an electron is needed high up in the conduction band to give rise to the emission of a hole. As long as the measurement of thermal emission rates takes place in a p–n junction, one would rarely expect too high concentrations of free carriers in the bands. Some carriers may pass through the reverse-biased space-charge region by leakage, but in most cases to a very small extent, and certainly with the concentrations not corresponding to those in a material at thermal equilibrium. As the emission processes shown in fig. 2.4b rely on the existence of free-charge carriers in the bands, we can conclude that capture rates connected with the corresponding recombination mechanism cannot be used in expressions like eq. (2.4.12) when the emission rates are measured in a space charge region.

Another mechanism, experimentally observed by Kleverman et al. (1985) and earlier discussed by a number of authors (Bess 1958, Sheinkman 1965, Neumark 1972) requires a double-donor or a double-acceptor centre and was found for the two double donors selenium and sulfur in silicon. For further details, see Landsberg (1992, p. 433). When the trap is filled by two electrons as shown in fig. 2.4c, they may interact such that one electron is captured by a hole in the valence band, while the other electron absorbs the energy thus created and goes to the conduction band. This process involves the emission of two electrons; or equivalently, the simultaneous emission of an electron and the capture of a hole. A similar process may be thought of, where the electron is captured and the hole is emitted, as shown in fig. 2.4d. These two processes are clearly detailed-balance counterparts. In both cases, free carriers are needed, and one would therefore suspect that they are rarely found in a reverse-biased p–n junction. Hence, as in the Hangleiter case above, capture rates connected with this process would not satisfy eq. (2.4.12); when the emission rate is measured in a space-charge region.

Then, we are left with the question: which capture rates *are* going into eq. (2.4.12)? As no or very small concentrations of free carriers exist in the space-charge region where the thermal emission rates are measured, one would not expect any kind of electron-related process to occur. Therefore, the most probable emission processes for these situations are those created by phonons and photons. This favours the use of single-carrier capture experiments of the type suggested by Lang (1974). However, not even for one of the most investigated recombination centres, gold in silicon, a consistent picture has evolved (Engström and Grimmeiss 1974, 1975, Brotherton and Bicknell 1978, Lang et al. 1980, Ledebo and Wang 1983). Capture rate data collected by Hangleiter (1987) for various recombination centres in silicon indicate that the results found depend on the measurement method. What kind of capture rate is to be used in eq. (2.4.12) when the emission rate is measured in a space-charge region, still seems to be an open question. This problem is, therefore, an important subject for more detailed experimental studies.

3. Radiative processes

3.1 Radiative recombination

For radiative transitions, the rate per unit volume is given by eq. (2.2.1) and if $E = h\nu = E_I - E_J > 0$ then, putting dimensions in brackets,

$$S_{IJ} = B_{IJ} N_\nu + B_{IJ}, \qquad S_{IJ} = B_{IJ} N_\nu \quad (\mathrm{L}^{-3}\,\mathrm{T}^{-1}), \tag{3.1.1}$$

where N_ν denotes the occupation number of a mode of frequency ν, and

$$B_{IJ} = \frac{4e^2 n_c E}{Vm^2 h^2 c^3} |M_{IJ}|_{A\nu} \quad (\mathrm{L}^{-3}\,\mathrm{T}^{-1}). \tag{3.1.2}$$

Here V is the volume of the material, m is the normal electron mass, n_c is the refractive index, and the momentum matrix element (dimensionally also a momentum) has been introduced, averaged over the directions of photon propagation vectors and over polarizations (Adams and Landsberg 1969). The terms in eqs. (3.1.1) represent, in turn, induced and spontaneous emission and absorption. It follows that

$$X_{IJ} = \frac{\text{reverse rate}}{\text{forward rate}} = \frac{N_\nu}{N_\nu + 1} \frac{q_I p_J}{p_I q_J} = \frac{N_\nu}{N_\nu + 1} \exp[(\eta_I - \gamma_i) - (\eta_J - \gamma_j)], \tag{3.1.3}$$

where band states have been assumed. Hence

$$X_{IJ} = \frac{N_\nu}{N_\nu + 1} \exp\left(\frac{h\nu}{kT} - \gamma_i + \gamma_j\right), \quad \text{i.e.,} \quad \frac{S_{JI}/S_{IJ}}{(S_{JI}/S_{IJ})_0} = \frac{N_\nu}{N_\nu + 1} \exp\left(\frac{h\nu}{kT}\right). \tag{3.1.4}$$

Comparison of eq. (3.1.4) with eq. (2.2.4) shows the effect of excitation-dependence in radiative processes. However, for a Planck equilibrium distribution

$$N_\nu = N_{\nu 0} \equiv \left[\exp\frac{h\nu}{kT} - 1\right]^{-1}, \quad \text{i.e.} \quad \frac{S_{JI}/S_{IJ}}{(S_{JI}/S_{IJ})_0} = 1. \tag{3.1.5}$$

Solving eq. (3.1.4) for N_ν, one obtains a generalised distribution for photons in a non-equilibrium steady state with electron–hole pairs, namely,

$$N_\nu = \left[X_{IJ}^{-1} \exp\left[\frac{h\nu}{kT} - (\gamma_i - \gamma_j)\right] - 1\right]^{-1}.$$

The possible imbalance of forward and reverse rates and the separation between quasi-Fermi levels thus imposes a departure from the normal Planck distribution. The appearance of a non-zero chemical potential for photons has been noted by Ruppel and Würfel (1980) who have $X_{IJ} = 1$. The above result is a further generalisation, which will, however, not be pursued here (Landsberg 1981).

It is usual to subtract the absorption from the stimulated rate to find, with an obvious notation, $u_{IJ}^{\text{st}} = B_{IJ} N_\nu p_I q_J$, etc.,

$$u_{IJ}^{\text{st}} - u_{IJ}^{ab} = [1 - \exp(E/kT + \gamma_j - \gamma_i)] N_\nu u_{IJ}^{\text{sp}}. \tag{3.1.6}$$

On summing over all states I and J in the appropriate groups i and j of states,

$$(u_{ij}^{\text{st}} - u_{ij}^{ab})/N_\nu u_{ij}^{\text{sp}} = 1 - \exp(E/kT + \gamma_j - \gamma_i).$$

This is positive, and stimulated emission dominates the absorption ("negative absorption") if, and only if (Bernard and Duraffourg 1961),

$$\mu_i - \mu_j > h\nu. \tag{3.1.7}$$

The condition for *population inversion*,

$$\frac{1}{1 + \exp(\eta_I - \gamma_i)} > \frac{1}{1 + \exp(\eta_J - \gamma_j)},$$

is identical to eq. (3.1.7). If $E = h\nu = E_g$ is the forbidden energy gap of a semiconductor, condition (3.1.7) for lasing in the form $\mu_i - \mu_j > E_g$ implies that at least one band must be degenerate. Laser action in indirect-gap materials was also considered early in the 1960s (Basov et al. 1961, Benoit à la Guillaume and Tric 1961, Mashkevich and Vinetskii 1966), including direct-gap laser action in indirect materials (Adams and Landsberg 1968), and related general questions (Frova 1973).

The emitted photon can, in an indirect material, have an energy $h\nu < E_g$ because of the energy possibly absorbed by a phonon. These complex processes with phonon and/or additional electron cooperation have therefore been studied. For example, an additional electron can gain some of the energy which would normally go to a photon (Grinberg et al. 1966).

To discuss this situation (Landsberg 1967), let the states which lose or gain an electron as a result of this radiative transition be labelled respectively by l (= 1, 3, 5, ..., say, for definiteness), and g (= 2, 4, 6, ...). The expression for the transition rate between pairs of levels then involves the probabilities $p_l q_g$ of finding a state l in a condition to lose an electron, and other states ready to be converted into states g by capturing an electron. The electron energies are E_l and E_g. Similarly phonons of energies ε_a ($a = 1, 3, 5, \ldots$) may be absorbed and phonons of modes e and energy ε_e ($e = 2, 4, 6, \ldots$) may be emitted. For negative absorption we need $R > 1$, with

$$R = \frac{\text{rate of stimulated emission of photons of energy } h\nu}{\text{rate of absorption of photons } h\nu \text{ by the inverse process}}. \tag{3.1.8}$$

Each rate is a product of occupation numbers and matrix elements. The latter cancel from R. The phonon contribution to the ratio of products of their occupation numbers looks like the factor $N_\nu/(N_\nu + 1)$ in eq. (3.1.3) and is

$$\prod_{a,e} \frac{q_a p_e}{p_a q_e} = \prod_a \frac{n_a + 1}{n_a} \prod_e \frac{n_e}{n_e + 1} = \exp\left[\frac{1}{kT}\left(\sum_a \varepsilon_a - \sum_e \varepsilon_e\right)\right]. \tag{3.1.9}$$

The contribution from the electron states to R is of the form X_{IJ} in eq. (2.2.4). For excitation-independent S's it is

$$\prod_{l,g} \frac{p_l q_g}{q_l p_g} = \prod_l \frac{p_l}{q_l} \prod_l \frac{q_g}{p_g} = \exp\left[\sum_l (\gamma_l - \eta_l) - \sum_g (\gamma_g - \eta_g)\right]. \tag{3.1.10}$$

Equation (3.1.9) follows from eqs. (2.2.7) and (2.2.9) (with $\gamma_I = 0$), and eq. (3.1.10) follows from eq. (2.2.9) provided the electron states are in bands. Energy conservation yields

$$h\nu + \sum_a \varepsilon_a - \sum_e \varepsilon_e - \sum_l E_l + \sum_g E_g = 0. \tag{3.1.11}$$

Substitution of eqs. (3.1.9)–(3.1.11) in eq. (3.1.8) yields

$$R = \exp\left(\sum_l \gamma_l - \sum_g \gamma_g - h\nu/kT\right).$$

Thus, for negative absorption, eq. (3.1.7) generalises to the statement that the quasi-Gibbs free energy must decrease by an amount

$$\sum \mu_l - \sum \mu_g > h\nu. \tag{3.1.12}$$

This result holds also if the electronic states are in impurity levels.

For a review of some of the optoelectronic aspects of this work, see Casey and Stern (1976), Thompson (1980) and Penzkofer (1988).

3.2. *Non-equilibrium kinetics for coupled photon and electrons in two-level systems*

One need not remain near equilibrium. Rate equations have proved to be an important tool to analyse non-equilibrium situations not only in chemical kinetics but also in the theory of semiconductor lasers (Carrol 1985) and other problems of quantum electronics which involve the interaction between photons and electrons. These investigations have led to a better understanding of non-equilibrium radiation which is sometimes characterised by a change in the distribution function from Planck-type (chemical potential $\mu = 0$) to Bose-type ($\mu \neq 0$) (Chel'tzov 1983, Würfel 1982, Landsberg 1981, Schöll and Landsberg 1983).

Consider a pair of energy levels E_I and E_J, where $E_I - E_J = h\nu > 0$. Let the number of available quantum states at these levels be N_I and N_J and their occupation probability be p_I and p_J, respectively. Assume the total number of electrons,

$$N_e = p_I N_I + p_J N_J, \tag{3.2.1}$$

to be conserved. The relative population of these levels may change because of the following processes:

(1) Radiative transitions (absorption and spontaneous and stimulated emission) with a transition probability B per unit time.
(2) Non-radiative excitation with a transition probability per unit time $C = C_1 + P$, where P denotes external pumping and C_1 stands for all other (e.g. thermal) processes.
(3) Non-radiative recombination with a transition probability per unit time D. We put $E = B + D$ for the total downward-transition probability per unit time (excluding stimulated emission).

Assume that the transition rates of processes (1)–(3) are of the form

$$u^{\mathrm{rad}} = B(N+1)\alpha - BN\beta, \qquad u^{\mathrm{exc}} = C\beta. \qquad u^{\mathrm{rec}} = D\alpha, \tag{3.2.2}$$

where N is the number of photons of frequency ν, and

$$\alpha = N_I p_I N_J(1 - p_J), \qquad \beta = N_J p_J N_I(1 - p_I) \tag{3.2.3}$$

represent the probability factors for the transitions. In addition, the photon-dissipation rate (e.g., by other absorption processes, scattering and cavity loss) is assumed to have the form

$$u^{\mathrm{diss}} = \kappa(N - N_0) \quad (\kappa \geqslant 0), \tag{3.2.4}$$

where κ is a constant and N_0 is the photon number in thermal equilibrium. We denote the number of electrons in the upper level by $n = N_I p_I$; physically meaningful solutions must clearly satisfy

$$0 \leqslant n \leqslant N_e < N_I + N_J, \quad N \geqslant N_0.$$

The rate of change of photons and excited electrons is

$$\dot{N} = u^{\mathrm{rad}} - u^{\mathrm{diss}}, \tag{3.2.5}$$

$$\dot{n} = -u^{\mathrm{rad}} - u^{\mathrm{rec}} + u^{\mathrm{exc}}. \tag{3.2.6}$$

Using eqs. (3.2.1)–(3.2.4), one can express the non-linear rate equations (3.2.5) and (3.2.6) in terms of the independent variables N and n only. One way of doing this is to introduce the following characteristic values of n,

$$n_{\text{th}} \equiv \frac{N_e N_I}{N_I + N_J}, \qquad n_\kappa \equiv \frac{\kappa}{B(N_I + N_J)}. \tag{3.2.7}$$

The first can be shown to be characteristic of laser threshold, and n_{th} and $n_{\text{th}} + n_\kappa$ are significant as values of n at which the photon number N can diverge (see fig. 3.1). In addition, the parameters

$$\alpha_I \equiv \frac{N_I + N_e}{N_I + N_J}, \qquad \alpha_J \equiv \frac{N_j - N_e}{N_I + N_J}, \qquad \alpha_I + \alpha_J = 1 \tag{3.2.8}$$

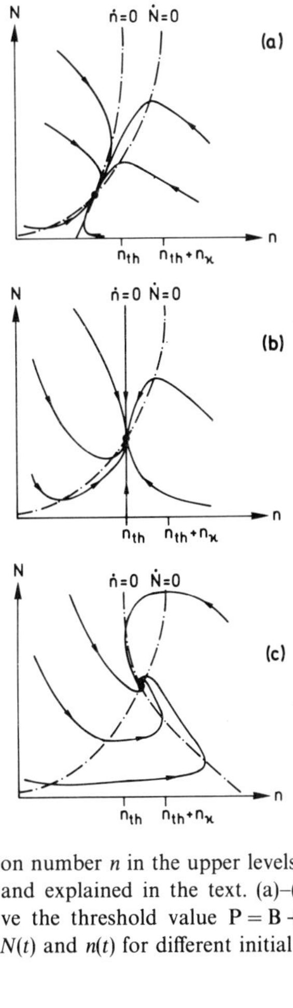

Fig. 3.1. Phase portraits of the electron number n in the upper levels and photon number N, as given by eqs. (3.2.9) and (3.2.10) for $N_J > N_e$ and explained in the text. (a)–(c) correspond to different pumping rates: (a) below, (b) at, and (c) above the threshold value $P = B + D - C_1$. The thick lines represent trajectories $N(t)$ and $n(t)$ for different initial conditions.

are useful. Hence, eqs. (3.2.5) and (3.2.6) become

$$\dot{N} = B(N_I + N_J)\left[\frac{n^2}{N_I + N_J} + \alpha_J n + n_\kappa N_0 + (n - n_{\text{th}} - n_\kappa)N\right], \tag{3.2.9}$$

$$\dot{n} = -(N_I + N_J)\left[\frac{E - C}{N_I + N_J}n^2 + (\alpha_I C + \alpha_J E)n - Cn_{\text{th}} + B(n - n_{\text{th}})N\right]. \tag{3.2.10}$$

This system of non-linear differential equations contains all information about the steady-state and the time-dependent solutions. It is most conveniently discussed in terms of phase portraits for the flows $n(t)$ and $N(t)$ (fig. 3.1).

The steady state of the electron–photon system is stable and is given by the intersection of the two curves $\dot{N} = 0$ and $\dot{n} = 0$ (null isoclines), which are, respectively,

$$N = \left(\frac{n^2}{N_I + N_J} + \alpha_J n + n_\kappa N_0\right)/(n_{\text{th}} + n_\kappa - n), \tag{3.2.11}$$

$$N = \left[\frac{E - C}{N_I + N_J}n^2 + (\alpha_I C + \alpha_J E)n - Cn_{\text{th}}\right]/B(n_{\text{th}} - n). \tag{3.2.12}$$

The curve [obtained from eq. (3.2.11)] depends on the pumping rate only indirectly through n but not explicitly through $C\,(= C_1 + P)$, as does that of eq. (3.2.12). In principle, one can trace the dependence of N and n on P from these equations. This rather tedious algebraic task is not carried out here.

The question of the path in the (N, n) plane by which an arbitrary initial state approaches the steady state is best answered by showing phase portraits. These are constructed by finding the sign of $\dot{N}$ on either side of the curve $\dot{N} = 0$ and similarly for the $\dot{n} = 0$ curve. This, together with the observation that a phase portrait must cross the $\dot{N} = 0$ line horizontally and the $\dot{n} = 0$ line vertically, yields the required curves. Three different situations are depicted in the phase portraits of fig. 3.1 for $N_J > N_e$, as given by eqs. (3.2.9) and (3.2.10). The $\dot{N} = 0$ curve is the same in each phase portrait but the curve $\dot{n} = 0$ changes as one passes with increasing pumping rate from fig. 3.1a to fig. 3.1b to fig. 3.1c.

Results (3.2.11) and (3.2.12) must, of course, hold in thermodynamic equilibrium, although this is not immediately evident. A proof is given by Schöll and Landsberg (1983) utilizing the detailed balance result

$$D = C_1 \exp(h\nu/kT). \tag{3.2.13}$$

It is of interest to illuminate the above results from a different angle. To this end we determine the photon distribution *without* eliminating the probability factors α and β in order to effect a comparison with the standard Bose distribution. We may do this either by putting $\dot{N} = 0$ in eq. (3.2.5), without making any steady-state assumption regarding the number of electrons n in the upper levels I in the first place, or by putting $\dot{n} = 0$ in eq. (3.2.6) without making any assumption about the photon number in the first place. Since often the electrons decay fast to a steady state, case $\dot{n} = 0$ seems to be more relevant than case $\dot{N} = 0$, since in the latter case the electrons need not then be in a steady state. We assume that the energy levels E_I and E_J

belong to groups of states i and j, which are characterised by quasi-Fermi levels $kT\gamma_i$ and $kT\gamma_j$, respectively. An example for groups i and j is provided by the conduction and the valence band in a semiconductor. Hence, eqs. (3.2.3) give

$$\beta/\alpha = \exp(\eta_g - \gamma), \quad \gamma = \gamma_i - \gamma_j. \tag{3.2.14}$$

A relation between n and γ is obtainable by introducing eq. (3.2.1) and $n = N_I p_I$ into eq. (3.2.3). This gives

$$\frac{\beta}{\alpha} = \frac{n^2/(N_I + N_J) - \alpha_I n + n_{\text{th}}}{n^2/(N_I + N_J) - \alpha_I n + n}. \tag{3.2.15}$$

These results show that

$$\begin{aligned} &\gamma < \eta_g \Leftrightarrow n < n_{\text{th}} \Leftrightarrow \beta > \alpha, \\ &\gamma > \eta_g \Leftrightarrow n > n_{\text{th}} \Leftrightarrow \alpha < \beta. \end{aligned} \tag{3.2.16}$$

Since the threshold for laser action in semiconductors in the absence of dissipation is $\gamma = \eta_g$ (see eq. (3.2.7) above), we see that $n = n_{\text{th}}$ for laser threshold, giving the promised interpretation of n_{th}. It occurs when $\alpha = \beta$ or in the special case when $N_I = N_J$, i.e. when the two groups of levels are equally populated.

In the steady state, γ is a function of the pumping rate P through n. If $P = 0$, the electronic system is in equilibrium and $\gamma = 0$. If $P > 0$, γ gives a measure of the departure from equilibrium.

The results of either $\dot{N} = 0$ or $\dot{n} = 0$ can be put in the form

$$N = [f^{-1} \exp(\eta_g - \gamma) - 1]^{-1}, \tag{3.2.17}$$

here γ has the meaning of an effective chemical potential divided by kT, arising from the coupling of the photons to a non-equilibrium electronic system. γ is constant for a given group of states i and j, whereas E_I and E_J, and thus $h\nu$, are allowed to vary over their appropriate groups. One finds that

$$f = \frac{1 + \kappa N_0/(B\alpha)}{1 + \kappa(1 + N_0)/(B\beta)} \qquad (\dot{N} = 0), \tag{3.2.18}$$

or

$$f = \frac{1 + (D/B)[e^{\gamma} - (C/D)\, e^{\eta_g}]e^{-\gamma}}{1 + (D/B)[e^{\gamma} - (C/D)\, e^{\eta_g}]e^{-\eta_g}} \quad (\dot{n} = 0). \tag{3.2.19}$$

The usefulness of using the factor f is limited by the fact that f can depend on the energy gap $E_g = h\nu$ (e.g., through α, β or γ). A merit of the factor f in eq. (3.2.17) is that it measures departures from certain simple conditions (when it is unity).

For $f = 1$, one requires either the equilibrium electron distribution ($\beta/\alpha = \exp \eta_g$) in eq. (3.2.18) or no photon dissipation ($\kappa = 0$). In the case of eq. (3.2.19), one requires either electron equilibrium ($D = C_1 \exp \eta_g$, $P = 0 = \gamma$) or $D/C = \exp(\eta_g - \gamma)$. The latter condition arises from steady-state electron and photon distributions without dissipation, whence $C\beta = D\alpha$ by eqs. (3.2.5) and (3.2.6).

Lastly, consider the full steady state. We use this term in the technical sense, $\dot{N} = 0$ *and* $\dot{n} = 0$. The phase portraits of fig. 3.1 show intersections of curves $\dot{n} = 0$ and

$\dot{N}=0$, and it is also possible to discuss this analytically, except that the algebra becomes rather involved.

One can also approach the above kinetics of coupled electrons and photons via a master equation, and by solving this equation one can discuss both the time-dependent and the steady-state photon statistics. This involves generating function techniques and will not be discussed here (Bădescu 1991).

3.3. *Photon recycling*

A photon produced by radiative recombination may be absorbed in the material and the electron and hole thus produced diffuse for a certain distance, followed by another act of radiative recombination. This repeated radiative recombination is called photon recycling. It gives rise to an effective externally observed radiative lifetime which is larger than what would be expected in the absence of recycling. It furthermore lengthens the electron or hole diffusion coefficients. This matter first attracted attention in 1957 (Moss 1957, Landsberg 1957, Dumke 1957) and some experimental evidence for its occurrence is known (see, e.g., Bensaid et al. 1989, Keyes et al. 1990).

In more sophisticated theories one calculates the photon recycling rate per unit volume in the form of an electron–hole pair volume generation rate (Kuriyama et al. 1977)

$$g_{\mathrm{pr}}(z)=\frac{1}{2}\int_0^{\infty}\mathrm{d}(h\nu)\int_0^{d}\mathrm{d}z'\,\frac{\eta\,\Delta n(z')}{\tau}\,\gamma(h\nu)\,\alpha(h\nu)\,[I_1(l)+I_2(l')]. \tag{3.3.1}$$

Using cylindrical coordinates (r,θ,z) the $I_1(l)$-term is due to direct propagation from the point (r,θ,z') of photon emission to the point $(0,0,z)$ where the function $g_{\mathrm{pr}}(z)$ is to be calculated. Then l is the distance $z'-z$ and the actual path length involved in $(l^2+r^2)^{1/2}$. The $I_2(l')$-term is due to photons reflected at the plane where the appropriate active layer terminates. Here, l' is $z'+z$. The optical path length between the same two points as before, but allowing for the extra distance due to reflection, is $(l'^2+r^2)^{1/2}$.

The remaining terms in eq. (3.3.1) will now be explained. The internal quantum efficiency η gives the fraction of electron–hole pairs which decay radiatively and the volume density of the photon emission rate is, therefore, $\eta\,\Delta n(z')/\tau$, where Δn is the carrier density at (r,θ,z'), assumed independent of r and θ, and τ is the minority carrier lifetime. This quantity has to be multiplied by the spontaneous emission spectrum $\gamma(h\nu)$ per particle so as to yield the energy spectrum of emission. The function γ is normalised such that

$$\int\gamma(h\nu)\,\mathrm{d}(h\nu)=1.$$

Lastly, multiplying by the absorption coefficient $\alpha(h\nu)$ yields the incremental volume generation rate of electron–hole pairs at $(0,0,z)$ due to the radiation. Integration over the whole active layer of thickness d and over all energies gives the total volume rate.

The quantities $I_j(l)$ $(j = 1, 2)$, are integrals

$$I_j(l) = \int_0^\infty \frac{r}{l^2 + r^2} R_j(\theta) \exp[-\alpha(l^2 + r^2)^{1/2}]\, \mathrm{d}r,$$

where $R_1(\theta) = 1$ (no reflection) and $R_2(\theta)$ is the reflection coefficient at $z = 0$. Reflection at $z = d$ has been neglected. The factor $[4\pi(l^2 + r^2)^{1/2}]^{-1}$ is due to the spread of a spherical wave of radius $(l^2 + r^2)^{1/2}$; it is multiplied by an element of area $2\pi r\, \mathrm{d}r$.

The phenomenon of photon recycling can be important for solar cells, and by taking it into consideration the agreement between theory and experiment can be improved. More information can be found in the literature (Asbeck 1977, Von Roos 1983).

4. Other non-radiative processes

Through Auger effects in earlier sections, we have already attended to important non-radiative transitions. The experimental consequences of this and other effects are widely studied. In this section some additional non-radiative mechanisms are discussed.

4.1. Truncated cascade recombination at traps with excited electronic states

A model based on the assumption of a ladder of electronic excited states, where the charge carrier is stepping down in energy by emitting one phonon at a time in a "phonon cascade" was introduced by Lax (1960) and was further developed by others (Ascarelli and Rodriguez 1961, Hamann and McWorther 1964, Lal and Landsberg 1965, Smith and Landsberg 1966, Brown 1966, Brown and Rodriquez 1967, Abakumov et al. 1991; reviewed by Stoneham 1975). This kind of modelling explains the existence of very large capture cross sections for impurities and defects with deep lying energy levels and their temperature dependence, which for many traps in covalent semiconductors follows a T^{-n}-law $(1 \leqslant n \leqslant 4)$. One problem with this phonon cascade theory is that for deep impurities, e.g. in silicon, there is no evidence for the existence of a ladder of excited states reaching all the way from the band edge to the ground state. Even if a hydrogen-like potential is assumed, a number of phonons must be emitted simultaneously during the last step from the lowest excited state to the ground state. An alternative two-step model was suggested (Gibb et al. 1977) where a carrier is first captured into an excited state followed by a subsequent transition to the ground state, in competition with thermal emission back to the band. In a capture experiment, which normally does not distinguish between the states of excitation of a trapped carrier, the measured cross section of the trap would be determined mainly by the first excited state. The transient behaviour of centres with one excited state was discussed by Rees et al. (1980) and in more detail by Dhariwal and Landsberg (1989) and for the influence of an electric field by Landsberg and Dhariwal (1989). In the latter studies, it was demonstrated that a trap with a set of excited states, can be treated as approximately a system where all the excited states are lumped together into one effective excited state giving a "truncated cascade recombination" (TCR) process. Ladders of levels due to clusters of impurities have also been invoked (Belyaev and Zvyagin 1991).

In the TCR description, an electron in the ground state can be emitted to the conduction band either directly with the emission rate e_{ng} or via the excited state. The second process takes place in two steps; first with a rate $(t'_{\text{n}})^{-1}$ from the ground state to the excited state followed by a thermal emission rate e_{ne} from the excited state to the band, as demonstrated in fig. 4.1

The total time needed for the second process can be expressed as

$$t_t = t'_{\text{n}} + \frac{t'_{\text{n}} + t_{\text{n}}}{t_{\text{n}}} \frac{1}{e_{\text{ne}}}. \tag{4.1.1}$$

The factor $(t'_{\text{n}} + t_{\text{n}})/t_{\text{n}}$ is motivated by the assumption that the particle spends an average time t_{n} in the excited state before going to the ground state and an average time t'_{n} in the ground state before being thermally transferred to the excited state. Of the total cycle time $t'_{\text{n}} + t_{\text{n}}$ spent in the two states, only a portion $t_{\text{n}}/(t'_{\text{n}} + t_{\text{n}})$ is spent in the excited state. As the time needed to be emitted to the conduction band from the excited state is $1/e_{\text{ne}}$ under the condition that the excited state is occupied by the particle, the time for the second step is $1/e_{\text{ne}}$ divided by the portion of the cycle time spent in the excited state.

The "effective" emission rate $e_{\text{n eff}}$ given by the two possible emission paths from the ground state is obtained as the sum

$$e_{\text{n eff}} = e_{\text{ng}} + t_{\text{t}}^{-1} = e_{\text{ng}} + \left(t'_{\text{n}} + \frac{t'_{\text{n}} + t_{\text{n}}}{t_{\text{n}}} \frac{1}{e_{\text{ne}}} \right)^{-1}. \tag{4.1.2}$$

This is the emission rate which would be found as the inverse time constant of a current or a capacitance transient obtained in an experiment using junction space-

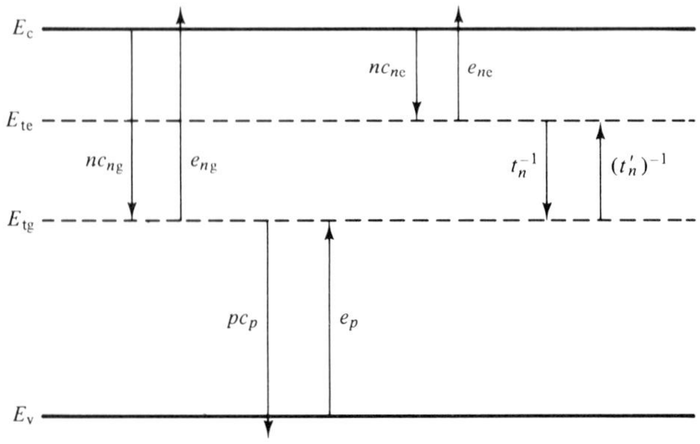

Fig. 4.1. Transition rates (*not* per unit volume) for the truncated cascade recombination.

charge techniques, where the space-charge region of a back-biased p–n junction is utilised to create a region depleted from free charge carriers in the bands (Sah et al. 1970, Lang 1974). Under these conditions, $e_{\mathrm{n\,eff}}$ is the time constant for the exponential decay of occupied traps as can be obtained from the general expressions deduced by Dhariwal and Landsberg (1989). (One has to use their result (3.15) and their table 2 with the assumption $n = p = c_{\mathrm{p}} = e_{\mathrm{p}} = c_{\mathrm{ng}} = 0$.)

At thermal equilibrium, the condition for detailed balance between the two trap states requires

$$t_{\mathrm{n}}^{-1} P_{\mathrm{e}}(1) = t_{\mathrm{n}}'^{-1} P_{\mathrm{g}}(1), \tag{4.1.3}$$

where the occupation probabilities $P_{\mathrm{g}}(1)$ for the ground state and $P_{\mathrm{e}}(1)$ for the excited state are given by eq. (1.3.12) with $Q_1 = Q_{\mathrm{e}1} + Q_{\mathrm{g}1}$. This gives

$$(t_{\mathrm{n}}')^{-1} = t_{\mathrm{n}}^{-1} \frac{Q_{\mathrm{e}1}}{Q_{\mathrm{g}1}}. \tag{4.1.4}$$

Here, $Q_{\mathrm{g}1}$ and $Q_{\mathrm{e}1}$ are the partition functions for the ground state and for the excited state, respectively (see eq. (1.3.13)).

Detailed balance for the traffic between the excited state and the conduction band implies

$$e_{\mathrm{ne}} P_{\mathrm{e}}(1) = n c_{\mathrm{ne}} P(0), \tag{4.1.5}$$

where c_{ne} is the rate for capturing electrons from the conduction band into the excited state and $P(0)$, given by eq. (1.3.12), is the probability that the trap is empty ($r = 0$). This gives

$$e_{\mathrm{ne}} = c_{\mathrm{ne}} n \frac{Q_0}{Q_{\mathrm{e}1} \exp \dfrac{\mu}{kT}}. \tag{4.1.6}$$

Combining eq. (4.1.1) with eqs. (4.1.4) and (4.1.6), we obtain

$$t_t^{-1} = \frac{t_{\mathrm{n}}^{-1}}{e_{\mathrm{ne}} + \dfrac{1 + t_{\mathrm{n}}/t_{\mathrm{n}}'}{t_{\mathrm{n}}}} c_{\mathrm{ne}} n \frac{Q_0}{Q_{\mathrm{g}1} \exp \dfrac{\mu}{kT}}. \tag{4.1.7}$$

The ratios of the partition functions in eqs. (4.1.4), (4.1.6) and (4.1.7) can be written in analogy with eq. (1.3.13) as

$$\frac{Q_{\mathrm{e}1}}{Q_{\mathrm{g}1}} = \exp\left[\frac{G_{\mathrm{g}}(1) - G_{\mathrm{e}}(1)}{kT}\right], \tag{4.1.8}$$

$$\frac{Q_0}{Q_{\mathrm{e}1}} = \exp\left[\frac{G_{\mathrm{e}}(1) - G(0)}{kT}\right] = \exp \frac{G_{\mathrm{e}}(\frac{1}{2})}{kT}, \tag{4.1.9}$$

$$\frac{Q_0}{Q_{\mathrm{g}1}} = \exp\left[\frac{G_{\mathrm{g}}(1) - G(0)}{kT}\right] = \exp \frac{G_{\mathrm{g}}(\frac{1}{2})}{kT}. \tag{4.1.10}$$

If $G_{\mathrm{e}}(1) - G_{\mathrm{g}}(1)$ is larger than several kT, we see from eqs. (4.1.8) and (4.1.4) that

$t'_n \gg t_n$, and eq. (4.1.7) can be approximated by

$$t_t^{-1} = P_e c_{ne} N_c \exp\left(-\frac{E_c - G_g(\frac{1}{2})}{kT}\right), \tag{4.1.11}$$

where

$$P_e \equiv \frac{1}{1 + t_n e_{ne}}. \tag{4.1.12}$$

It is interesting to observe that the sticking probability that an electron captured into the excited state will reach the ground state before it is emitted to the conduction band, i.e. the ratio $t_n^{-1}/(t_n^{-1} + e_{ne})$, is equal to the factor P_e defined by eq. (4.1.12). This shows that the TCR approach is consistent with the cascade theory (Lax 1960), simplified to have one effective excited level.

The emission rate e_{ng} for the direct transition from the ground state to the conduction band is similarly obtained by detailed-balance reasoning as

$$e_{ng} = c_{ng} N_c \exp\left(-\frac{E_c - G_g(\frac{1}{2})}{kT}\right), \tag{4.1.13}$$

and, thus, the effective emission rate given by eq. (4.1.2) is

$$e_{n\,eff} = (c_{ng} + P_e c_{ne}) N_c \exp\left(-\frac{E_c - G_g(\frac{1}{2})}{kT}\right). \tag{4.1.14}$$

The sum of the capture rate for the ground state c_{ng} and the capture rate for the excited state c_{ne} weighted by the factor P_e can be considered as an effective capture rate, $c_{n\,eff}$, for the trap,

$$c_{n\,eff} = c_{ng} + P_e c_{ne}, \tag{4.1.15}$$

When the trap is positioned in an electric field, the free-energy difference $E_c - G_g(\frac{1}{2})$ is lowered by the Poole–Frenkel effect (Hartke 1968). This effect of increasing the exponential factor in eq. (4.1.14) is counteracted by a corresponding increase in e_{ne} [eqs. (4.1.6) and (4.1.9)] which decreases P_e in eq. (4.1.14). It has been found that the influence by an electric field on deep traps with excited states is smaller than expected by the three-dimensional treatment of the Poole–Frenkel effect by Hartke (Rosier and Sah 1971). This can be understood as a decrease of the effective capture rate, c_{eff} as given by eq. (4.1.15) and a corresponding counteracting effect on the increase of $e_{n\,eff}$ (Dhariwal and Landsberg 1989).

The capture cross section, σ_n, for electrons is defined as

$$\sigma_n = \frac{c_n}{v_{th}}, \tag{4.1.16}$$

where v_{th} is the average thermal velocity of electrons in the conduction band. In a common method for the determination of this quantity (Lang 1974, Schroder 1990), the space-charge region of a back-biased p–n junction is decreased by a step voltage. Traps which are empty in the space-charge region are then filled by the electrons in the conduction band of the collapsed space-charge region. The time constant, τ_c, of this process is determined by the product of the free-electron concentration, n, and the capture rate, which for the present case means that

$$\tau_c = \frac{1}{nc_{n\,eff}} = [n(c_{ng} + c_{ne})/(1 + e_{ne}t_n)]^{-1}. \tag{4.1.17}$$

This gives a capture cross section which decreases with increasing temperature as observed for a number of deep centres in covalent semiconductors. Equation (4.1.17), therefore, can be viewed as an approximation to a cascade theory (Dhariwal and Landsberg 1989).

4.2. Multi-phonon processes: The configurational coordinate model

While cascade recombination theory, described in § 4.1, involves the consecutive transition of carriers from higher to lower excited electron states, there is still a possibility for capture directly into a deep electronic ground state by emission of phonons. In such a description, not only the electronic system is considered, but, in addition, the influence of local atomic vibrations on the energy eigen values of the trap is taken into account. The subject, known as multi-phonon recombination, involves a great number of subtle problems and has generated an extensive literature (see, e.g., Dexter 1958, Markham 1959, Stoneham 1981, Morgan 1983). As a useful concept to describe these complex phenomena in a simplified manner, the configurational coordinate model has been introduced. In its simplest representation one assumes, as we did earlier in § 1.4, that the eigenenergies for the total trap system, E, can be expressed as the sum of the electronic eigenenergies, E_e, and the eigenvalues, E_a, of the vibrating recombination centre:

$$E = E_e + E_a. \tag{4.2.1}$$

This is an oversimplification, but serves as a useful starting point for the understanding of these processes. Equation (4.2.1) shows that in addition to the eigenvalues of the Schrödinger equation for the bare electronic system, we also have to take into account the vibrational energies of the trap to which the captured charge carrier belongs. This is done by considering the elastic energies involved in the vibration of the trap about a lattice coordinate (configuration coordinate) obtained as a transformation from the displacement coordinates. If the vibrating centre is considered as a harmonic oscillator, its potential energy can be assumed parabolic as a function of the deviation of this lattice coordinate from its equilibrium point. Its eigenenergy values become then equally spaced.

Considering a trap, capable of capturing one electron ($M = 1$) without excited electron states but with the atomic eigenvalues given by a harmonic oscillator, the total energy of the system can be represented by two parabolas on a lattice coordinate as shown in fig. 4.2. The lower curve, labelled T (for "trap") in the figure, represents the trap energy, E, when the electron is captured. The upper curve, C (for "conduction band"), defines the eigenenergies of the system when the electron is excited to the conduction band and the trap is empty. Accordingly, the energy difference between

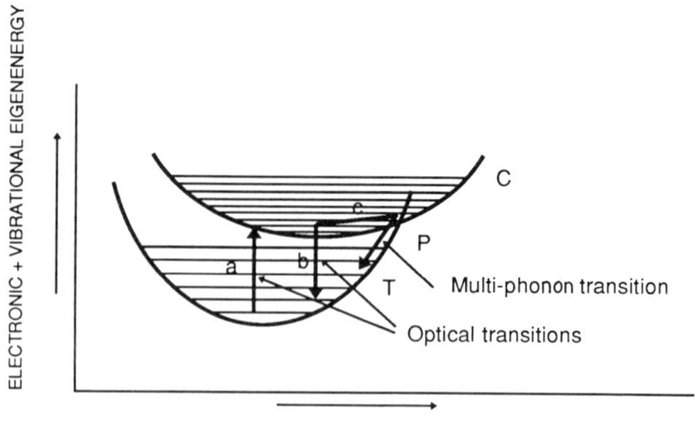

Fig. 4.2. The transition of an electron between a trap and the conduction band when the vibrational energy of the trap is taken into account. The curves represent the vibrational energies and are separated in energy by the electronic energy needed to take the electron from the trap to the conduction band. The lower curve (T) represents the electronic plus the vibrational energy of the trap when the electron is captured into it. The upper curve (C) gives the corresponding energy when the trap is empty and the electron is present in the conduction band. The separation in lattice coordinate depends on the change in the symmetry of atomic binding force when an electron is transferred.

the minima of the two oscillator potentials is determined by the electronic energy difference between the conduction-band edge and the ground state of the trap centre. When the electron is emitted from the trap into the conduction band, two main lattice structural effects may occur. For a trap, where the captured electron is very localised and thus may contribute considerably to the bonding with neighbouring lattice atoms, a large decrease in binding force is expected (Van Vechten and Thurmond 1976). This is followed by a decrease in the curvature of the potential curve and thus in vibrational frequency of the centre. As a second consequence, a smaller separation between the vibronic states occurs. In addition, for a non-symmetric change in binding force, the equilibrium point of the oscillator may change upon electron emission. In this model, an electron transition between the trap and the conduction band is accompanied by a translation of the curve of the atomic potential and by a change in its curvature, giving rise to a change in the density of vibronic states and the occurrence of an intersection point, *P*, between the two curves (fig. 4.2).

The probability of transferring the system from T to C depends not only on the matrix elements for electron interaction with exciting particles but also on the overlap in lattice coordinate between the wavefunctions of the two oscillators. As the eigenfunctions of a harmonic oscillator have their main contributions near the vibrational turning points, the most probable transitions take place between points along the two potential curves. For optical transitions it can be assumed that the transition time is much smaller than the periodic time of atomic vibrations. In this scheme, therefore, a transition where the captured electron is excited by a photon to the conduction band can be represented by a vertical change in the configurational coordinate diagram from the ground state of T to the turning point of C which can be reached without changing lattice coordinate as shown in fig. 4.2 (transition a). In this situation, the electron is in the conduction band while the trap is empty, but in an excited vibronic state due to the difference in equilibrium points between T and C. Thus, it will relax down to the ground state at the equilibrium point of C. When the electron in the conduction band is captured into the trap by emitting a photon, the ground state of C has to couple with one of the vibronic states of T. This process, represented by transition b in fig. 4.2, may require a smaller change in electronic energy than the electronic energy needed for the emission process. Thus the photon energy emitted during the recombination process may be considerably smaller than the photon energy required for the excitation process. Phenomena of this kind, known as "Franck–Condon shift" have been observed in a number of materials (Dexter 1958).

At the intersection point P in fig. 4.2, where a strong overlap exists between the vibronic wave functions of the two oscillators, a transition may take place without the exchange of energy. When the system is at thermal equilibrium and the trap has captured an electron, the trap centres are distributed among the vibronic states of T. As there exists a certain probability to find some of the oscillators T of this distribution at the intersection point P, the captured electrons can be delivered to the conduction band and these oscillators are transferred to C. The opposite process may also take place when one of the empty trap oscillators C, reaches P. Then, the electron can be transferred from the conduction band to the electronic state of the trap, whereupon the filled trap oscillator, T, relaxes down along the vibronic states emitting a phonon at every step. As these thermal emission and capture processes take place by raising and lowering the trap oscillator among its vibronic states through the capture and emission of phonons, they have been called "multi-phonon processes".

Three important properties are expected from the vibronic system:

(1) The transitions are associated with a change in the density of vibronic states of the oscillator and thus with a change of its entropy.
(2) The capture cross section for charge carriers from the bands to the trap may be thermally activated with an activation energy determined by the position of the intersection point P.
(3) The optically obtained ionisation energies may differ considerably from energies measured in a thermal experiment.

Phenomena demonstrating different aspects of (1)–(3) have been observed, e.g., in AlGaAs (Lang and Logan 1977, Lang et al. 1979), in GaAs and GaP (Henry and Lang 1977) and, with weaker effect, in silicon (Kleverman et al. 1985, Chantre and Bois 1985). Also for interface traps at the SiO_2/Si interface similar results have been obtained (Kirton and Uren 1986, Andersson and Engström 1989, Engström and Grimmeiss 1989, Cobden et al. 1990, Ricksand and Engström 1991), as further discussed in § 6.

4.3. Multi-phonon processes: Thermal capture and emission

The charge carrier traffic between the trap and the conduction band at thermal equilibrium is subject to detailed balance by

$$nc_{\mathrm{n}} P(0) = e_{\mathrm{n}} P(1). \tag{4.3.1}$$

The expressions developed in § 1.3 for a one-electron trap ($M = 1$) with electronic and vibronic states can be applied here and we find by using eqs. (1.2.7) and (1.3.12) in eq. (4.3.1)

$$e_{\mathrm{n}} = c_{\mathrm{n}} N_{\mathrm{c}} \exp\left[-\frac{E_{\mathrm{c}} - (G(1) - G(0))}{kT} \right] = c_{\mathrm{n}} N_{\mathrm{c}} \exp\left[-\frac{E_{\mathrm{c}} - G(\frac{1}{2})}{kT} \right]. \tag{4.3.2}$$

This is a special case of eq. (2.4.12) and we will see now how it works for a multi-phonon process. First, observe that the exponent of eq. (4.3.2) contains the Gibbs free energy $G(\frac{1}{2})$ of the trap together with the energy E_{c} which we often consider as the eigenenergy of the conduction-band edge. However, the meaning of E_{c} depends upon the meaning of N_{c}. In silicon, for instance, the most commonly used values for N_{c} (and N_{v}) are obtained by using the experimental density of states effective mass for electrons (and holes) (see, e.g., Barber 1967). As these values are obtained by assuming and taking into account a temperature dependence of $E_{\mathrm{c}} - E_{\mathrm{v}}$, the meaning of E_{c} in eq. (6.3) is a free energy. Thus, for convenience, we define

$$\Delta G_{\mathrm{n}} \equiv E_{\mathrm{c}} - G(\tfrac{1}{2}), \qquad \Delta S_{\mathrm{n}} \equiv S_{\mathrm{c}} - S(\tfrac{1}{2}), \qquad \Delta H_{\mathrm{n}} \equiv \Delta G_{\mathrm{n}} + T\Delta S_{\mathrm{n}}. \tag{4.3.3}$$

Also, observe that in the configuration coordinate model the capture process may be connected with a thermal activation when the empty-trap oscillator C reaches the intersection point P. Selecting a set of vibronic states, i, close to the intersection point and assuming that the capture process from each of these states are given by a capture rate $c_I (I \in i)$, we find an average capture rate, c_n, by weighting c_I with the probabilities to find the oscillator in the different states i:

$$c_{\mathrm{n}} = \frac{\sum_I c_I \exp\left(-\frac{Ihv}{kT} \right)}{Z_j}, \tag{4.3.4}$$

where hv is the energy separation of the vibronic states as earlier introduced in § 1.3 and Z_j is the canonical partition function of the oscillator C,

$$Z_j = \sum_J \exp\left(-\frac{Jhv}{kT} \right) = \exp\left(-\frac{F_j}{kT} \right). \tag{4.3.5}$$

Here, the summation is taken over the set j of all vibronic states of C and F_j is the Helmholtz free energy of C. The numerator of eq. (4.3.4) can be expressed by using a mean value, c_0, of the capture rates c_I and a free energy, F_i, of the states belonging to i, as

$$\sum_I c_I \exp\left(-\frac{Ihv}{kT} \right) = c_0 \sum_I \exp\left(-\frac{Ihv}{kT} \right) = c_0 \exp\left(-\frac{F_i}{kT} \right). \tag{4.3.6}$$

Combining eqs. (4.3.4)–(4.3.6), we find

$$c_{\mathrm{n}} = c_0 \exp\left(-\frac{F_i - F_j}{kT}\right). \tag{4.3.7}$$

As the state set i is a subset of the total set, j, of oscillator states, the corresponding entropy is lower than that of the total oscillator and we can define, in analogy with the definitions in eq. (4.3.3)

$$\Delta F_{\mathrm{B}} = F_i - F_j, \qquad \Delta S_{\mathrm{B}} = S_i - S_j, \qquad \Delta U_{\mathrm{B}} = \Delta F_{\mathrm{B}} + T\Delta S_{\mathrm{B}}, \tag{4.3.8}$$

where the suffix B refers to the transition between parabolas T and C (transition at P in fig. 4.2). Now using eqs. (4.3.3) and (4.3.8) in eq. (4.3.2) we find an expression for the thermal emission rate valid for multi-phonon processes

$$e_{\mathrm{n}} = c_{\mathrm{B}} \exp\left(-\frac{\Delta U_{\mathrm{B}}}{kT}\right) X_{\mathrm{n}} N_{\mathrm{c}} \exp\left(-\frac{\Delta H_{\mathrm{n}}}{kT}\right), \tag{4.3.9}$$

where

$$c_{\mathrm{B}} = c_0 \exp\left(\frac{\Delta S_j}{k}\right) \tag{4.3.10}$$

and, in analogy with eq. (1.3.18),

$$X_{\mathrm{n}} = \exp\left(\frac{\Delta S_{\mathrm{n}}}{k}\right). \tag{4.3.11}$$

It is important to keep in mind that when a measurement is performed, in general, a collection of traps are studied. Energy quantities extracted from measured results, therefore, are to be considered as average values taken over the whole ensemble and to be interpreted in the thermodynamic language mentioned above. Individual interface traps have also been studied (Ralls et al. 1984, Kirton and Uren 1986, Farmer et al. 1987, Andersson et al. 1990) and the results have been interpreted in a multi-phonon scheme. In these latter cases, however, the averaging must be taken over time and, according to the *ergodic hypothesis* formulated in statistical mechanics, such a procedure gives normally the same values as when ensemble averaging is used.

Obviously, in an experimental situation we have to deal with a number of different energy scales, depending on the measurement method. When the thermal emission rate is measured as a function of temperature by space-charge transient methods (Sah et al. 1970, Lang 1974) and the temperature dependence of the pre-exponential factors c_{B} and X_{n} are taken into account, the result can be represented in an Arrhenius plot, the slope of which represents an activation energy of the system. As seen from eq. (4.3.9) such a procedure gives a slope yielding the sum $\Delta U_{\mathrm{B}} + \Delta \mathrm{H}_{\mathrm{n}}$. This is the total change in enthalpy of the system when the charge carrier is transferred between the trap and the conduction band. As seen in eq. (1.3.5) the enthalpy is the thermal average energy taken across the different eigenenergies. Also, when independent measurements of the capture rate are made, the slope in an Arrhenius plot of the result is determined by the energy ΔU_{B}. Both these energy quantities must be distinguished from the changes in energy eigenvalues in fig. 4.2 and cannot easily be represented in the configuration coordinate diagram, which has the sum of the eigenenergies as given by eq. (4.2.1) on the ordinate axis. This must be kept in mind when theoretically calculated data are compared with measurements (Engström 1989). As one example we may mention capture cross sections calculated from theoretical considerations of multi-phonon transitions (Henry and Lang 1977, Morgan 1983). The quantities obtained in these cases are best identified as the average values c_0, defined above in eq. (4.3.6). From an Arrhenius plot of the capture rate measured by common methods (Lang 1974), however, one would obtain the quantity c_B, defined by eq. (4.3.10), by extrapolating the activation curve to infinite temperature. These two quantities differ by the factor $\exp(\Delta S_B/k)$. This relation needs to be taken into account when comparing theoretical and experimental values of capture rates. Other interesting cases arise when studying interface states as will be discussed in § 6.2.

Up to this point we have been dealing only with electron transitions between the trap and the conduction band. Expressions corresponding to eqs. (4.3.1)–(4.3.11) can be formulated for hole emission from the trap to the valence band, by changing subscript n for p and using N_v for the valence band instead of N_c. Therefore, we limit ourselves here to a brief discussion of the changes in entropy for a two-step transition from the valence band to the conduction band. Figure 4.3 shows the configuration coordinate diagram, now with the valence band included.

The curve V in fig. 4.3 thus represents the energy of the system when the electron belongs to the valence band and the trap is empty. As the trap is empty for both C and V, these two potentials have the same curvature and are separated by an energy

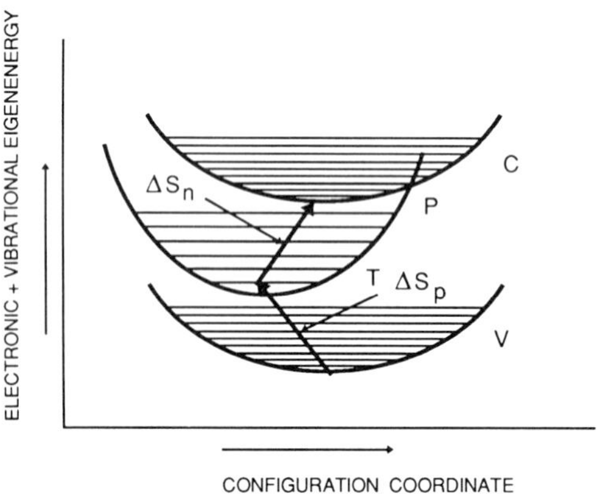

Fig. 4.3. Illustration of the entropy changes connected with a two-step transition of an electron from the valence band to the conduction band via a vibrating trap. The curves represent the sums of electronic and vibrational eigenenergies when the electron is in the valence band and the trap is empty (V), when the electron is in the trap (T) and, finally, when the electron is in the conduction band the the trap is empty (C). When the system goes from V to T, the trap entropy changes from $S(0)$ to $S(1)$. On going from T to C the trap entropy changes back from $S(1)$ to $S(0)$. The total entropy change, therefore, depends only on the entropy changes S_v and S_c of the bands (see text).

corresponding to the energy band gap of the semiconductor. The band gap can also be represented on a Gibbs free energy scale (Thurmond 1975) and its temperature dependence can be associated with an entropy change ΔS_{cv} when the electron is excited from the valence band to the conduction band. Partitioning this entropy into one contribution, S_v, for the valence band and another, S_c, for the conduction band, we find for the total entropy change in the transition of an electron from the valence band to the trap,

$$\Delta S_p = S(0) - S(1) - S_v = -S(\tfrac{1}{2}) - S_v. \tag{4.3.12}$$

Using the corresponding expression in eq. (4.3.3) for ΔS_n, we find for the sum

$$\Delta S_n + \Delta S_p = \Delta S_{cv}, \tag{4.3.13}$$

which means that the sum of the temperature dependences of the free energies ΔG_n and ΔG_p is equal to the temperature dependence of the semiconductor band gap.

4.4. Metastability of defect states

For a defect where the atomic relaxation following an electronic transition is large enough, the probability for optical transitions can become very small, leaving multi-phonon recombination as the only possibility. In the example demonstrated in fig. 4.4, an electron excited to the configuration coordinate curve A^+ can be transferred to B^0 only when the trap system receives thermal energy high enough to pass the energy barrier between the two potential wells. For low temperatures, the defect can be "frozen" in the state A^+. This kind of state is called *meta-stable.* The so-called DX-centre in AlGaAs (Lang and Logan 1977, Lang et al. 1979) has been described by such a scheme. There exists also a possibility for the defect to change its atomic configuration without changing its charge state. This means that the same charge

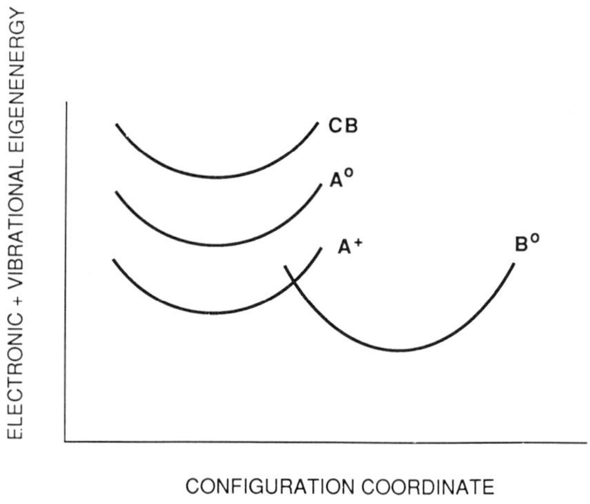

Fig. 4.4 Configuration coordinate diagram of a metastable state.

state can be represented by two or more different potential curves in the configuration coordinate diagram. Such a behaviour has been found for a radiation damage defect in silicon (Chantre and Kimerling 1986), Fe–Al (Chantre and Bois 1985) and the C–C (Song et al. 1988) pairs in silicon. This phenomenon is often referred to as multistability or bistability depending on the number of existing configurations (Chantre 1989).

We will take the bistable system of "thermal donors" (TD) in silicon (Chantre 1989) as an example to demonstrate the usefulness of the treatment developed earlier in this section. In CZ silicon, TD can be introduced by a heat treatment at about 425°C for 30 min or more. For such short annealing times, smaller TD clusters are generated with three charge states: *0, 1+ and 2+*, thus representing double-donor centres. Further, the neutral state is bistable with two different ground levels A^0 and B^0, respectively, depending on the atomic configuration. The ground state can be represented in a level scheme as shown in fig. 4.4. The transition from one of the bistable states may take place from A^+ by adding one electron and the choice for the final state between A^0 and B^0 depends on temperature.

The reaction rate R_{A-B} for a transition from A^+ to B^0 depends on the number, n, of free electrons in the conduction band and the probability $P(1)$ to find traps in the state A^+, i.e. traps which have captured one electron,

$$R_{A^+B^0} = c_n(\tfrac{1}{2})nP(1). \tag{4.4.1}$$

Using eq. (1.3.14) in eq. (4.4.1) gives

$$\begin{aligned} R_{A^+B^0} &= c_n(\tfrac{1}{2})n\frac{1}{1+\exp\left(\dfrac{G(\frac{1}{2})-\mu}{kT}\right)+\exp\left(-\dfrac{G(\frac{3}{2})-\mu}{kT}\right)} \\ &\approx c_n(\tfrac{1}{2})n\exp\left(-\frac{G(\frac{1}{2})-\mu}{kT}\right). \end{aligned} \tag{4.4.2}$$

The approximation in eq. (4.4.2) can be used if the Fermi level is positioned several kT below the energy level $G(\frac{3}{2})$ for the doubly charged state A^{2+}. Using the notation from § 4.3, we obtain

$$R_{A^+B^0} = \frac{n^2}{N_c}c_B\frac{1}{X_n}\exp\left(\frac{\Delta H_n - \Delta U_B}{kT}\right). \tag{4.4.3}$$

The reaction rate for the reverse process B^0–A^+ is determined by the thermal emission rate of electrons from B^0 to the conduction band, which is given by

$$R_{B^0A^+} = c_B N_c X_n \exp\left(-\frac{\Delta H_n + \Delta U_B}{kT}\right). \tag{4.4.4}$$

By comparing eq. (4.4.3) and eq. (4.4.4), one expects to find a smaller pre-exponential factor and a lower activation energy for the transition A^+–B^0. This was found to be the case for TD in silicon (Chantre 1989).

5. Continuous trap distributions

In this section we cannot give a full survey of the work on continuous trap distributions as the field is very extensive. Instead some typical examples will be given.

5.1. Formal development

Let $N(\eta)\,\mathrm{d}\eta$ ($\eta \equiv E/kT$) be the number of trapping states per unit volume in an energy range. Then the mean concentration of electrons in these states is by a Taylor expansion

$$\int_{\eta_\mathrm{v}}^{\eta_\mathrm{c}} \frac{N(\eta)\,\mathrm{d}\eta}{1+\exp(\eta-\gamma)} = \int_0^{\gamma} N(\eta)\,\mathrm{d}\eta + \frac{\pi^2}{6}\left(\frac{\mathrm{d}N}{\mathrm{d}\eta}\right)_{\eta=\gamma} + \frac{7\pi^4}{360}\left(\frac{\mathrm{d}^3 N}{\mathrm{d}\eta^3}\right)_{\eta=\gamma} + \cdots. \tag{5.1.1}$$

This series can be continued and holds for any density function, provided the Fermi level is at several kT from the band edges in the forbidden gap. It is difficult to handle in generality. As a special case, a density function decreasing exponentially from a value N_0 at the conduction band edge,

$$N(\eta) = N_0 \exp[-a(\eta_\mathrm{c}-\eta)] \quad (a>0), \tag{5.1.2}$$

is sometimes of interest. If $n=p=0$ at $T=0$, i.e. all these states are occupied at $T=0$, the Fermi level equation for electrical neutrality is at a general temperature

$$n-p = N_0 \exp(-a\eta_\mathrm{c}) \int_{\eta_\mathrm{v}}^{\eta_\mathrm{c}} \frac{\exp(a\eta)\exp(\eta-\gamma)}{1+\exp(\eta-\gamma)}\,\mathrm{d}\eta \to \frac{1}{a} N_0. \tag{5.1.3}$$

The last form holds only if the Fermi level lies well within the gap and the temperature is low enough i.e. if

$$\exp(\eta_\mathrm{v}-\gamma) \ll 1 \ll \exp(\eta_\mathrm{c}-\gamma)$$

[as explained in the first edition (Landsberg 1982, p. 424)]. These considerations have been used in a general way for the gap states in amorphous semiconductors. In a-Si:H they are often believed to be due to dangling bonds which can be in three different states of charge with energy levels which can normally be considered as doping-independent. There are many theoretical and experimental discussions of gap states (Street et al. 1983, Kocka et al. 1987) and this research is quite active.

5.2. Meyer–Neldel rule

Continuous distributions of energy levels in the forbidden energy gap have been assumed to explain the so-called Meyer–Neldel rule for sintered oxide semiconductors having a resistivity in excess of 100 Ω cm and an impurity activation energy in excess of 0.06 eV. The rule states that the conductivity satisfies an Arrhenius-type law with an activation-energy-dependent pre-exponential factor,

$$\sigma(E,T) = C(E)\exp(-E/kT), \quad C(E) = \exp(A+BE). \tag{5.2.1}$$

It was found by Meyer in 1933 and it is sometimes called the *thermodynamic compensation law*. But it also occurs in organic semiconductors (Gutman and Lyons 1967). A brief survey of the theories was given by Landsberg (1958) where it was shown to what extent the theories assumed unusual relationships between impurity concentrations and mobility or activation energy. More recently, other ideas have been proposed (Mialhe et al. 1988, Vainus 1991) but these are not as yet generally accepted (Wintle 1988). The occurrence of the law has also been conjectured in solar cells (Coutts and Pearsall 1984, Goradia and Weizer 1984). See also Peled and Schein (1991) and Chen and Huang (1991).

The flavour of these theories may be given as follows. Consider a non-degenerate semiconductor with N_D donors of one kind. Then the Fermi level equation is in the simplest case (no holes, degeneracy factors absorbed in the energy level E_D)

$$N_D = N_c \exp(\gamma - \eta_c) + N_D/[1 + \exp(\eta_D - \gamma)]. \tag{5.2.2}$$

Its solution is with $\varepsilon \equiv E_c - E_D$ (for single ionisation of the donors)

$$E_c - E_F = \tfrac{1}{2}\varepsilon - \tfrac{1}{2}kT \ln(N_D/N_c). \tag{5.2.3}$$

Hence, the conductivity σ is given by ($\boldsymbol{v}$ = mobility)

$$\ln \sigma = \ln(en\boldsymbol{v}) = \ln[e\boldsymbol{v}(N_D N_c)^{1/2}] - \varepsilon/2kT. \tag{5.2.4}$$

To these standard considerations we now add the assumption of dominant neutral-impurity scattering, i.e. $\boldsymbol{v} = L/N_D$, where L is a constant and most impurities are regarded as neutral. A further assumption expresses the fact that the activation energy ε of the donors decreases as N_D increases by the relation

$$N_D = N_0 \exp(-\lambda\varepsilon), \tag{5.2.5}$$

where N_0 and λ are positive constants. Hence

$$\ln \sigma = \ln[eL(N_c/N_0)^{1/2}] + \tfrac{1}{2}\lambda\varepsilon - \frac{1}{2kT}\varepsilon, \tag{5.2.6}$$

which is the Meyer–Neldel rule. The mobility in this approach has the form

$$\boldsymbol{v} = \boldsymbol{v}_0 \exp(\lambda\varepsilon), \quad \boldsymbol{v}_0 = L/N_0. \tag{5.2.7}$$

Thus, $\boldsymbol{v}_0$ (*not* a vector) is independent of impurity concentration and activation energy and the mobility increases exponentially with the activation energy. It is these consequences that would have to be tested by experiments. No recent study of this matter is known to us.

5.3. A tractable Gaussian-like trap distribution

A convenient distribution which has been used for theories of space-charge limited currents with mobility and trap density both position-dependent (de Blasi et al. 1983), and has been applied successfully to GaS_xSe_{1-x} (Micocci et al. 1985), is the following

(Mathur and Dahiya 1974)

$$N_{\mathrm{MD}}(E)=4N_0\frac{\exp[(E-E_t)/kT_t]}{\{1+\exp[(E-E_t)/kT_t]\}^2}. \tag{5.3.1}$$

This gives for the total concentration of traps (if $E_v\to-\infty$, $E_c\to\infty$)

$$N_{t,\mathrm{MD}}=\int_{E_v}^{E_c}N_{\mathrm{MD}}(E)\,\mathrm{d}E=4N_0kT_t, \tag{5.3.2}$$

N_0 is a constant number of states per unit volume per unit energy and T_t is a parameter such that

$T_t\to0$: $N(E)\to$ one energy level $E=E_t$, $T_t\to\infty$: $N(E)\to$ constant distribution.

Note also that

$$N_{\mathrm{MD}}(E_t)=N_0.$$

One can compare $N_{\mathrm{MD}}(E)$ with a Gaussian distribution

$$N_g(E)=N_0\exp[-\sigma(E-E_t)^2], \tag{5.3.3}$$

which gives for the total concentration

$$N_{t,g}=N_0(\pi/\sigma)^{1/2}. \tag{5.3.4}$$

If one compares $N_{\mathrm{MD}}(E)$ and $N_g(E)$ for the same total concentration of traps, then

$$\sigma=\frac{\pi}{16k^2T_t^2}. \tag{5.3.5}$$

Figure 5.1 holds for $kT_t=0.01$ eV (and, therefore, $\sigma=1963$) and shows that $N_{\mathrm{MD}}(E)$ can be arranged to have many properties of $N_g(E)$. Both distributions are certainly symmetrical about E_t.

So far integrals from $-\infty$ to ∞ have been considered. It is upon integration between finite limits that $N_{\mathrm{MD}}(E)$ is more convenient. One finds

$$N_{t,\mathrm{MD}}=4N_0kT_t\left[\frac{1}{1+\exp(\eta_v-\eta_t)}-\frac{1}{1+\exp(\eta_c-\eta_t)}\right], \tag{5.3.6}$$

where $\eta_{c,v}\equiv E_{c,v}/kT_t$. This is more convenient than the result

$$N_{t,g}=\tfrac{1}{2}\left(\frac{\pi}{\sigma}\right)^{1/2}N_0\{\mathrm{erf}[\sigma^{1/2}(E_c-E_t)]-\mathrm{erf}[\sigma^{1/2}(E_t-E_v)]\}. \tag{5.3.7}$$

That is why the distribution N_{MD} has found several uses in the past.

5.4. Exponential band tails

In the last three subsections various approaches to band tails will be discussed. In many cases they are due to disorder in space or in time, or both. Disorder in space can be due to heavy doping of semiconductors. This is employed to advantage in

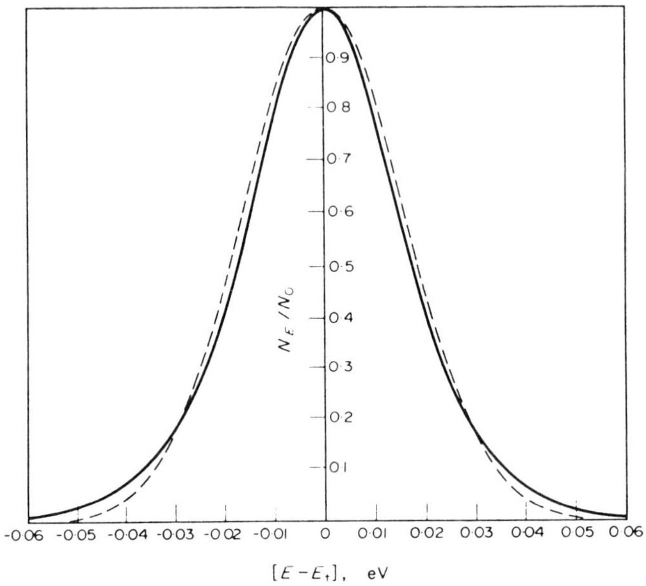

Fig. 5.1. Comparison of $N_{MD}(E)$ (full line) with $N_g(E)$ (dashed line) for $kT_t = 0.01\,\mathrm{eV}$ and $\sigma = 1963$.

the emitter part of some transistors, in semiconductor lasers, in thermoelectric devices, in tunnel diodes and other devices. In amorphous materials this disorder is also expected due to the many dangling bonds. Disorder in time is due to the thermal vibration of ions so that the configuration felt by an electron changes in time and this effect increases with temperature. For a review, see Shklovskii and Efros (1984). Amorphous materials, as is well known, have an increasing number of applications, e.g. in thin-film transistors and solar cells. They, too, exhibit band tails due to dangling bonds. For a review of amorphous semiconductors, see Pankove (1984). Band tails can also occur in pure materials but this will not be discussed here. For a review, see Landsberg (1992, pp. 220 and 388).

We note next that the notion of exponential band tails can be used in semiconductor injection laser theory owing to the high concentration of shallow impurities often required in these devices. One can then replace the $E^{1/2}$ density of states by two such distributions, one for each band (fig. 5.2). Each is characterised by its value at the band edge and a spreading parameter E_{0c} or E_{0v} whose magnitude can be of the order 36 meV (Popov 1967),

$$D_c(E) = D_c(E_c)\exp\left(\frac{E - E_c}{E_{0c}}\right),$$
$$D_v(E) = D_v(E_v)\exp\left(\frac{E_v - E}{E_{0v}}\right). \qquad (5.4.1)$$

This yields for the concentration of shallow donors and acceptors, if the integration

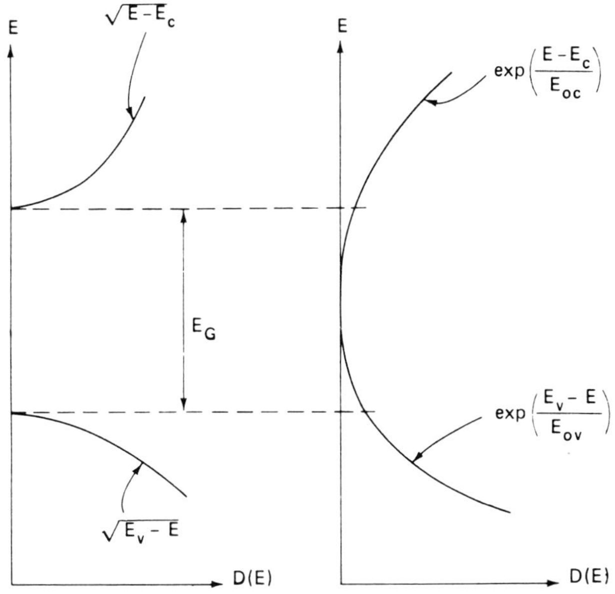

Fig. 5.2. Exponential band tails compared with the parabolic density of states.

is cut off at appropriate energies B_c and B_v,

$$N_D = \int_{-\infty}^{B_c} D_c(E)\,dE = D_c(E_c)\,E_{0c}\exp\left(\frac{B_c - E_c}{E_{0c}}\right),$$
$$N_A = \int_{B_v}^{\infty} D_v(E)\,dE = D_v(E_v)\,E_{0v}\exp\left(\frac{E_v - B_v}{E_{0v}}\right). \tag{5.4.2}$$

These represent an energy width multiplied by a density of states per unit energy range at the band edge. For the electron and hole concentration, one finds

$$n = D_c(E_c)kT\exp\left(\gamma_e - \frac{E_c}{E_{0c}}\right)\int_{-\infty}^{B_c/kT-\gamma_c}\frac{\exp(a_c x)}{1+e^x}\,dx,$$
$$p = D_v(E_c)kT\exp\left(\frac{E_v}{E_{0v}} - \gamma_v\right)\int_{-\infty}^{\gamma_v - B_v/kT}\frac{\exp(a_v x)}{1+e^x}\,dx, \tag{5.4.3}$$

where the γ's are reduced chemical potentials and

$$a_c \equiv kT/E_{0c}, \qquad a_v \equiv kT/E_{0v}. \tag{5.4.4}$$

It is desirable to attempt some simplifications in order to use expressions (5.4.1) in integrals of, e.g., stimulated emission rates (3.1.6) over all relevant states I, J (Adams 1969, Kressel 1973). This can be achieved if the contributions from high x in eq. (5.4.3) have a low weight. This, in turn, requires that the quantities (5.4.4) are less than unity. In that case the upper limits in eq. (5.4.3) can be replaced by ∞ and the formula

$$\int_{-\infty}^{\infty} \frac{\exp(ax)}{1+\mathrm{e}^x}\,\mathrm{d}x = \pi \operatorname{cosec}(a\pi) \quad (0 < a < 1), \tag{5.4.5}$$

can be used in eq. (5.4.3). In addition one can often assume $B_c \approx E_c$ and $B_v \approx E_v$, so that the exponential factor disappears in eq. (5.4.2). The theory lends itself for the discussion of threshold currents, light output and I–V characteristics of lasers.

The exponential tail model has also been used in the analyses of Fabry–Pérot amplifiers using multimode rate equations (Mukai and Yamamoto 1982) and in discussions of the energy gap in impure Si (Van Cong and Brunet 1985).

Instead of taking a phenomenological point of view of band tails, for heavy doping one can derive them from more microscopic considerations. To calculate the details of the band tails due to impurities, one may (in a simple model) consider N randomly distributed impurities located at R_i $(i = 1, \ldots, N)$ in a volume V. Let the electron impurity ion potential be $v(\boldsymbol{r} - \boldsymbol{R}_i)$, so that the one-electron potential is

$$\Phi(\boldsymbol{r}) = \sum_{i=1}^{N} v(\boldsymbol{r} - \boldsymbol{R}_i).$$

It fluctuates in space with an average

$$\Phi_0 \equiv \langle \Phi(\boldsymbol{r}) \rangle = \int \cdots \int \frac{\mathrm{d}\boldsymbol{R}_1}{V} \cdots \frac{\mathrm{d}\boldsymbol{R}_N}{V}\, \delta\left[\Phi - \sum_{i=1}^{N} v(\boldsymbol{r} - \boldsymbol{R}_i)\right]. \tag{5.4.6}$$

Also, the mean square value of $\Phi(\boldsymbol{r})$ is

$$\zeta \equiv \langle \Phi(\boldsymbol{r})^2 \rangle.$$

One then finds that the probability $P(\Phi)$ of finding a potential Φ in the range $(\Phi, \Phi + \mathrm{d}\Phi)$ is given approximately by a Gaussian distribution about the mean Φ_0 (Kane 1963, Sa-Yakanit and Glyde 1988, Landsberg 1992, p. 385). The potential Φ is called a Gaussian random potential. This presumes that the impurities are equally likely to be in any volume element.

The density of states for electrons is

$$N(E) = C E^{1/2} \boldsymbol{H}(E), \tag{5.4.7}$$

where $\boldsymbol{H}(E)$ is the Heaviside function. The density of states averaged over the potential Φ is

$$N_{\mathrm{av}}(E) \equiv \int_{-\infty}^{\infty} N(E - \Phi(r))\, P(\Phi)\, \mathrm{d}\Phi. \tag{5.4.8}$$

We have here treated the electron as free, i.e. with the same kinetic energy everywhere, neglecting the kinetic energy of localisation which may be important for the deeper potential wells. The result is

$$N_{\mathrm{av}}(E) = \frac{C}{(2\pi\zeta)^{1/2}} \int_{-\infty}^{E} (E - \Phi)^{1/2} \exp\left[-\frac{1}{2}\frac{(\Phi - \Phi_0)^2}{\zeta} \right] \mathrm{d}\Phi. \tag{5.4.9}$$

Substituting the appropriate expression for C,

$$N_{\mathrm{av}}(E) = \frac{m_{\mathrm{c}}^{3/2}\zeta^{1/4}}{2\pi^2\hbar^3} \exp\left[-\frac{(E - \Phi_0)^2}{4\zeta} \right] D_{-3/2}\left[-\frac{E - \Phi_0}{\zeta^{1/2}} \right], \tag{5.4.10}$$

where

$$D_{-p}(z) \equiv \frac{\exp(-\frac{1}{4}z^2)}{\Gamma(p)} \int_0^{\infty} x^{p-1} \exp(-zx - \tfrac{1}{2}x^2)\, \mathrm{d}x \tag{5.4.11}$$

is the parabolic cylinder function. This is essentially Kane's conclusion and gives us another continuous trap distribution function. The result may also be written as $[z = (E - \Phi_0)/(2\zeta)^{1/2}]$

$$N_{\mathrm{av}}(E) = \frac{m_{\mathrm{c}}^{3/2}\zeta^{1/4}}{2\pi^2\hbar^3} \exp(-\tfrac{1}{2}z^2) D_{-3/2}(-2^{1/2}z), \tag{5.4.12}$$

and is given in fig. 5.3 which illustrates the low-energy tail.

Various other and more sophisticated calculations have been performed, most notably by Halperin and Lax (1966) and Sritrakool et al. (1986). However, the underlying physical idea is that outlined here.

5.5. Urbach tails

Urbach tails are named after a paper by Urbach in 1953 and refer to the exponential rise in the optical absorption coefficient, α, as the photon energy $h\nu$ is increased. A simple semiconductor model would suggest a sudden onset of absorption from zero,

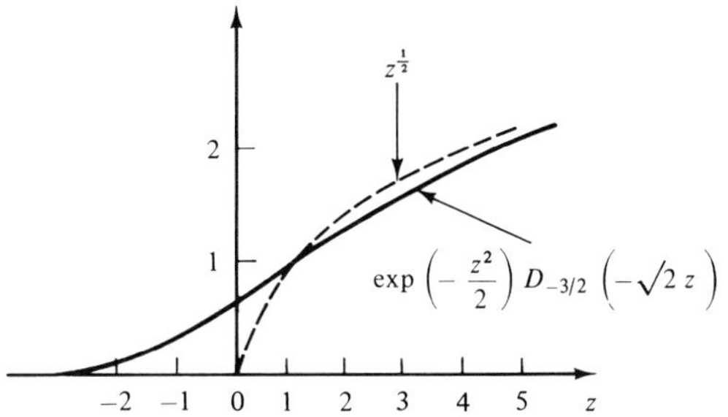

Fig. 5.3. A low-energy tail in the density of states of a disordered system (Sa-Yakanit and Glyde 1988).

when the photon energy is equal to the energy gap. In fact, the rise in absorption is often much gentler and the ln α versus $h\nu$ curve becomes flatter as doping is increased. This applies, e.g., to GaAs at 5 K (Pankove 1965). There is a great deal of information also about amorphous materials and fig. 5.4 is taken from a review of the optical absorption edge of amorphous Si (Cody 1984).

Such phenomena can be summarised in the formula

$$\alpha(E) = \alpha_0 \exp\left(\frac{E - E_1}{E_0}\right) \quad (E_0 \sim 0.05 \text{ eV}) \tag{5.5.1}$$

for the Urbach tail, where α_0, E_0 and E_1 are constants inferred from the experiment. If the matrix element for the transition is only weakly dependent on the energy, one can infer an exponential law for the density of states in the gap. Two tails are to be expected, one for the conduction band and one for the valence band, the latter being usually more pronounced. For a theory of the Urbach tail in terms of the density of states, see Sa-Yakanit and Glyde (1988).

6. Surface and interface effects

6.1. Surface recombination

As is well known, the mean concentration of current carriers multiplied by their mean velocity yields a number flux. If this is multiplied by the charge of these carriers

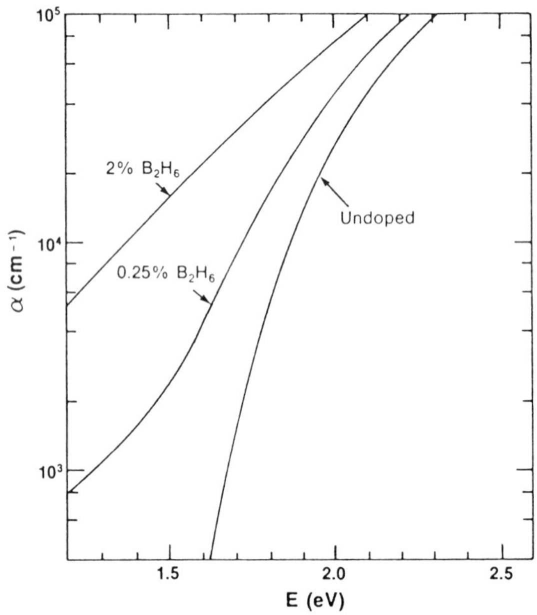

Fig. 5.4. The absorption edge of heavily boron-doped a-Si:H_x. The indicated concentrations refer to the gas phase.

one has a current density, $j = env$, say. Thus, a surface recombination current density may also be conceived to be a product of charge, recombining carrier density and a velocity. The latter is the surface recombination velocity, s. It is often defined by an equation of the type $j = es\,\delta n$ where δn is an appropriate excess carrier density, taken at the inner edge of the surface depletion region.

If one takes the recombination rate expression (per unit volume) into traps, typically eq. (2.4.4),

$$u_{c\,r-\frac{1}{2}} = G(r-\tfrac{1}{2})[1-\exp(\gamma_{r-\frac{1}{2}}-\gamma_e)]nv_{r-1}, \tag{6.1.1}$$

then the last factor has dimension L^{-3}. For surface recombination the same argument applies, only v_{r-1} has to be interpreted not as a volume density, but as an areal density of traps. This makes $u_{c\,r-\frac{1}{2}}$ a recombination rate per unit area. Thus the whole Shockley–Read–Hall theory of §§ 2.4–2.6 applies formally unchanged.

In an alternative exposition, one may write an equation like eq. (2.7.5) (for $r = 1$)

$$u = \frac{s_n s_p(np - n_i^2)}{s_n(n+n_1)+s_p(p+p_1)}, \tag{6.1.2}$$

where $s_n = c_n N_t$, $s_p = c_p N_t$ are inverse lifetimes. However, in the case of surface recombination, N_t is the number of traps per unit area, so that s_n and s_p become surface recombination velocities.

The value of s depends on the space-charge region within the surface. The usual assumption that s is insensitive to the excess carrier concentration δn at the inner edge of the space charge layer is not valid, and one often has to accept that s itself depends on δn. There has been a minor controversy on this matter (Rees 1985, de Visschere 1986, Dhariwal and Mehrotra 1988).

If one has a distribution of trapping levels, say $N(E_t)\,dE_t$ per unit surface area in a range $(E_t, E_t + dE_t)$, one can simply use eq. (2.7.5) with N_t (volume density) replaced by $N(E_t)\,dE_t$, the trap density in a small energy range,

$$du = \frac{c_n c_p(np - n_i^2)}{c_n(n+n_1)+c_p(p+p_1)} N(E_t)\,dE_t. \tag{6.1.3}$$

One can integrate this expression over the whole spectrum of trap states. We shall illustrate this result by graphs without going into the mathematical details. For these, the original paper should be consulted (Landsberg and Browne 1988). We shall study the ratio, R, of the recombination rates (u_1) with traps concentrated at one energy E_t to the rates (u_{sp}) when the *same* number of traps are spread over a spectrum of traps throughout the band gap.

Figure 6.1 assumes a constant density of states throughout the whole energy gap, this constant being the same for each curve. One finds that the maximum value of R occurs at an energy

$$E_{t\,max} = E_i - \tfrac{1}{2}kT\ln\frac{c_n}{c_p}, \tag{6.1.4}$$

where E_i is the intrinsic Fermi level. $E_{t\,max}$ is the most recombination-efficient level. It is clear from this that R_{max} is always greater than unity, since the spectrum will

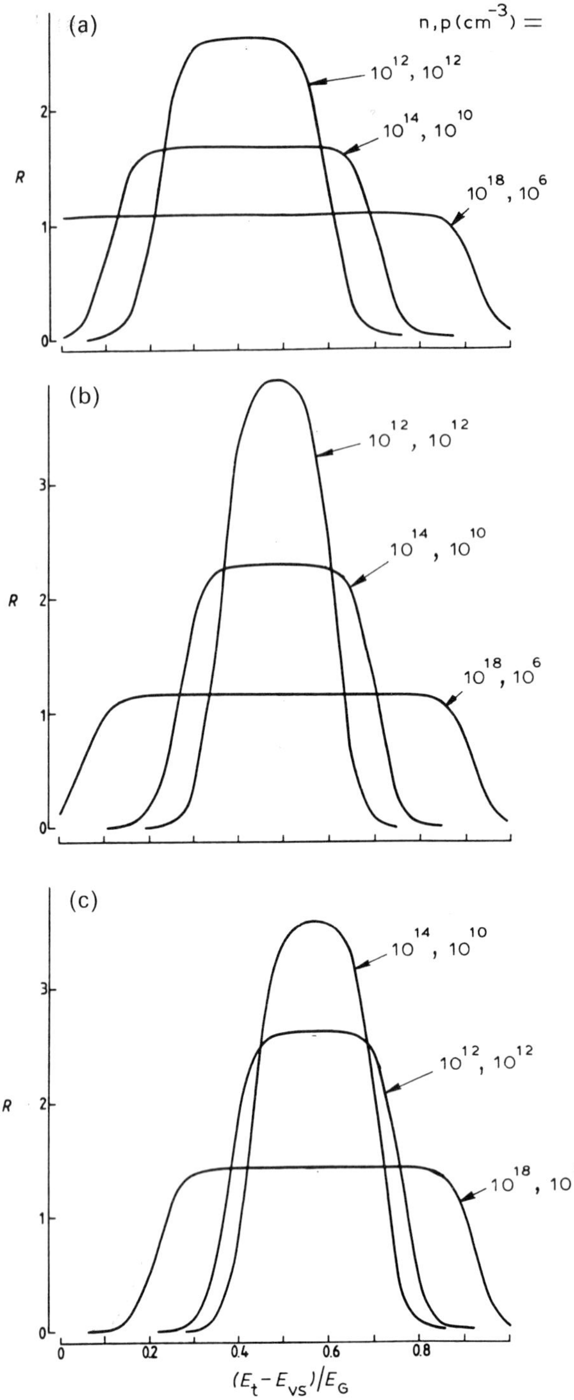

Fig. 6.1. The dependence of $R \equiv u_1/u_{sp}$ on the energy E_t of the single level. A constant density of states per unit area per unit energy range throughout the energy gap, with this constant the same for each curve, is assumed. The total number of traps cancels out of the equations and need not be specified. Carrier concentration (in $cm^3\ s^{-1}$): (a) $c_n = 10^{-8}$; $c_p = 10^{-11}$; (b) $c_n = 10^{-8}$; $c_p = 10^{-8}$; (c) $c_n = 10^{-11}$; $c_p = 10^{-8}$.

include some energies away from $E_{t\,max}$ that are not so recombination-efficient, whereas all of the N_t states placed at the single energy level $E_{t\,max}$ are recombination efficient. Hence, u_1 is greater than u_{sp} and so R is greater than unity. Similarly, if all of the N_t surface states are placed at a single energy level away from $E_{t\,max}$ (i.e., close to either of the band edges), then this single level will be recombination-inefficient, whereas the spectrum will include some levels which are close to $E_{t\,max}$ which may be much more recombination-efficient. Hence R may be considerably less than unity when the single energy level E_t is close to E_{cs} or E_{vs}.

An interesting feature of figs. 6.1a,b and c is that when the surface is strongly n-type (i.e. $n = 10^{18}\ \mathrm{cm}^{-3}$), R_{max} is closer to unity than for the cases where the surface is less strongly n-type, and also the plateau region is more extensive. One can also see from the figure that the ratio R may drop considerably below unity if the single energy level is located close to either of the band edges and that it can rise to just below 4 if the single energy level is located at $E_{t\,max}$, for the parameter values considered. There will, however, *always* be at least one energy level such that if all the N_t surface states (or bulk states) are located at that energy, the ratio, R, will be unity.

It emerges from these calculations that the more strongly n-type (or p-type) the surface (or bulk) becomes, the better the approximation of assuming a single energy level becomes, since R_{max} approaches unity, and the variation of R with E_t is less strong. Then as long as one replaces the spectrum with a single energy level, not to close to the band edges, the replacement of a uniform spectrum by a single level is reasonable.

6.2. Insulator–semiconductor interfaces

Interfaces between insulator layers and crystalline semiconductors are of considerable technical interest and, especially a great deal of attention has been paid to the SiO_2/Si system due to its application in metal-oxide silicon field-effect transistors (MOS-FET) (see Nicollian and Brews 1982). For the experimental investigations of this interface, very often metal-oxide semiconductors (MOS) capacitors are used. These structures allow an additional freedom to the choice of measured energies compared to what was discussed in § 2. This is due to the possibility of changing the Fermi level, μ, at the interface by changing the voltage applied to the capacitor. The band diagram of a MOS structure is shown in fig. 6.2.

Due to the isolating properties of the oxide, the metal is not in thermal equilibrium with the semiconductor when a voltage V_G is applied across the device. The electron states, on the other hand, which appear at the interface between the semiconductor and the oxide, are in thermal equilibrium with the bulk of the semiconductor. By changing the gate voltage, therefore, the Fermi-level can be swept across the different energy levels of interface states. Assuming that the interface states are single-electron traps, their occupation probability, $P(1)$, is determined by the position of the Fermi

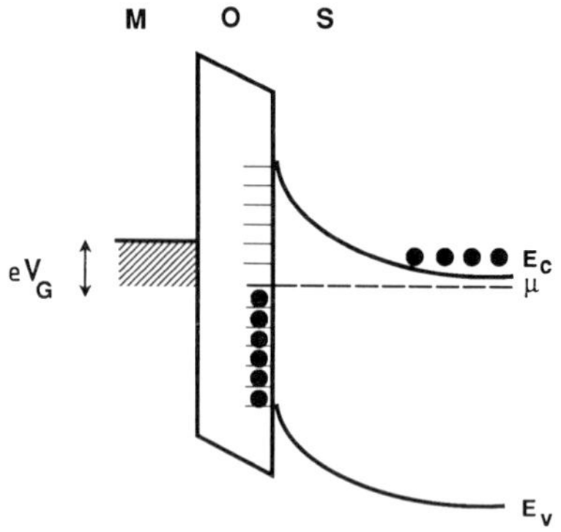

Fig. 6.2. Energy diagram of a MOS-structure with a voltage $-V_g$ applied to the gate metal. The Fermi level position at the oxide–semiconductor interface can be varied and used as a probe to select states at different energies.

level relative to the free energy of the trap as found earlier, e.g. in eq. (1.3.19),

$$P(1) = \frac{1}{1 + \exp\left(\dfrac{G(\frac{1}{2}) - \mu}{kT}\right)}. \tag{6.2.1}$$

The energy difference $G(\frac{1}{2}) - \mu$ depends on the surface potential Ψ_s which can be controlled by the voltage. This means that the Gibbs free energies of the different interface traps can be probed by the Fermi level at the interface. From measurements of the capacitance at low and high frequencies as a function of the voltage, the concentrations of interface states can be obtained at different energy positions in the band gap (Grovc ct al. 1965, Kuhn 1970). Within the energy resolution limits, which are about $\pm 1.7kT$ (Engström and Alm 1983), the interface states as a function of energy often exhibit a continuous distribution for the SiO_2/Si system as shown in fig. 6.3.

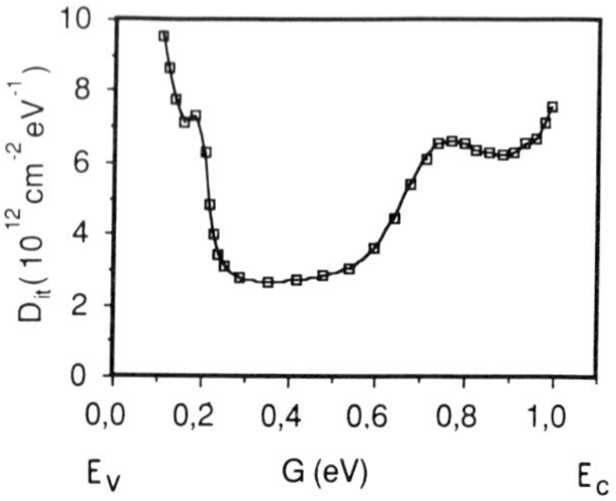

Fig. 6.3. The density of interface states as a function of the Gibbs free energy in the band gap of silicon at the interface between silicon dioxide and silicon. The interface traps were created by irradiation from a ^{60}Co-source (Andersson and Engström 1989).

It is obvious from eq. (6.2.1) that as long as the Fermi level is used as a probe to select the energy positions of the traps, the energy scale obtained from such a measurement is based on the Gibbs free energy.

Among the different methods available to investigate capture data of MOS interface states, deep level transient spectroscopy (DLTS) offers the greatest flexibility. The original method, for bulk traps (Lang 1974), has been further developed for MOS systems by Yamasaki et al. (1979) and by Johnson (1982). The method is based on the measurement of capacitance transients occurring when the occupation of interface states is changed as a result of a voltage step applied across the structure and the energy position of the interface states relative to the Fermi level is changed. Two main problems arise for this case which are not present when conventional DLTS is used on p–n junctions:

(1) Since the interface states have a continuous energy distribution, the DLTS spectra become rather unstructured, which gives rise to problems when interpreting the experimental data.
(2) When the capacitance changes at a constant applied voltage V_G, the whole change is due to a change in the capacitance of the space-charge region in the semiconductor. This means that the surface potential changes during the measurement, which in turn gives varying concentrations of charge carriers in the bands at the interface during the transient.

The recombination rate at an interface state (denoted by the suffix "it") with the energy position at G, expressed as the time derivative of the captured carrier concentration n_{it} is (Ricksand and Engström 1991)

$$\frac{dn_{it}(G, t)}{dt} = D_{it} c_n(G)\, n_s(t) - [c_n(G)\, n_s(t) + e_n(G)] n_{it}(G, t), \qquad (6.2.2)$$

where D_{it} is the density of interface states and n_s is the concentration of electrons in the conduction band at the interface. One sees that even if one discrete energy level were present, instead of a continuous distribution, another problem occurs because the differential equation is non-linear. This gives rise to non-exponential transients and thus to complications for the extraction of the capture and remission rates, e_n and c_n.

Problem (1) can be solved by selecting a certain energy level by a constant voltage and using a very small, superimposed step voltage to change the surface potential. In such a case, the capacitance transient evolves from the change in occupation of traps within a narrow energy region which may be treated as one discrete energy level. By changing the quiescent voltage and thus selecting different free-energy positions for the Fermi level, the energy distribution of interface states are "discretised" on a free-energy scale (Johnson 1982, Wang 1979).

Problem (2) can be solved in two different ways. One method is to keep the capacitance constant during the measurement by a feedback loop controlling the quiescent voltage and taking the change in this latter quantity as a measure of the trap occupation transient (Johnson 1982). A second possibility is to solve eq. (6.2.2) numerically and extract the statistical quantities from simulations (Goguenheim et al. 1991, Ricksand and Engström 1991).

The capture properties of MOS interface traps introduced during sample preparation have been investigated by different methods by a number of authors (Deuling et al. 1972, Schulz and Johnson 1978, Kamieniecki 1979, Kamieniecki et al. 1981, Kreigler et al. 1979). Even if these experimental data were not interpreted in a detailed statistical language (Engström and Grimmeiss 1989), one can conclude that the results are not in agreement as between the different authors. This may be due to differences between the samples investigated. A better definition of sample properties can be obtained by γ-irradiation of the MOS structure, thus creating interface states with characteristic features as shown in fig. 6.3. Such structures were investigated by DLTS and capture rates were extracted from fitting to simulated results. It was found that the capture cross sections of interface traps in these structures were thermally activated with activation energies of about 0.15 eV (Ricksand and Engström 1991). Other data from optical measurements on the same type of traps exhibited a discrepancy between optical and thermal ionization energies of about 0.2 eV (Andersson and Engström 1989). Together, these experiments strongly suggest that the capture properties of interface traps, introduced by irradiation at silicon dioxide–silicon interfaces are dependent on the vibrational properties of the trap system, following the scheme described in § 4.3.

6.3. Grain boundaries and recombination

An important application of the model of a continuous distribution of surface states is the electrical properties of grains. These are determined to a large extent by the grain boundary barrier height, which is strongly affected by the charged interface states and the compensating space charge regions on each side of the grain. This, in turn, is due to the recombination traffic which is here assumed to be in a steady state. In this sense the barrier height depends on recombination. A similar behaviour has been found for "thermally bonded" silicon interfaces (Bengtsson and Engström 1989, Bengtsson et al. 1992). An early model of grain boundaries due to Taylor et al. (1952) is here replaced by a more modern version. For a summary of relevant work up to 1979, see Kazmerski (1980).

Figure 6.4 shows a steady-state energy diagram for a model p-type grain due to acceptors at energy E_A. Surface recombination takes place through N_{DS} donor states and N_{AS} acceptor states at energy E_{DS} at the grain surface $x = 0$. From fig. 6.4 one sees that

$$\eta_A(x) = \eta_A(w) - e\varphi(x)/kT. \qquad (6.3.1)$$

The assumptions to be made are as follows:

(1) The surface of the grain is taken as a flat surface between identical grains. This makes a one-dimensional treatment appropriate.

(2) The recombination rates in the bulk and at the surface are regarded at not too different, so that one may assume parallel and flat quasi-Fermi levels of known

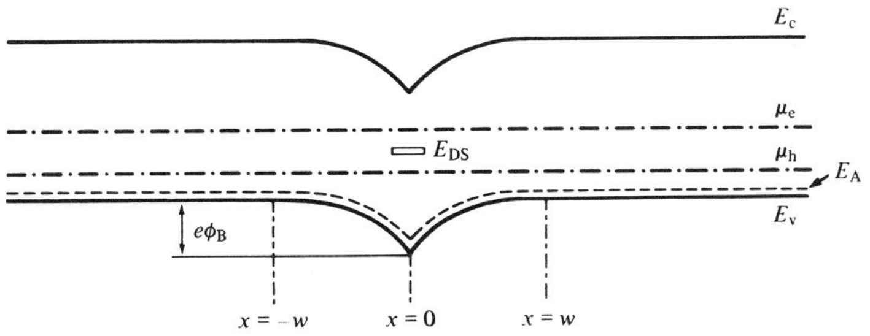

Fig. 6.4. Model of a grain boundary at $x = 0$ in a non-equilibrium steady state. E_{DS} denotes the level of a surface donor. For a surface acceptor it has to be relabelled E_{AS}.

separation ($\mu_e - \mu_h$). Thus bulk recombination needs to be introduced explicitly; its rate can be worked out from $\gamma_e - \gamma_h$ once appropriate recombination centre densities and trapping cross sections are settled.

(3) The carriers in the grains are non-degenerate.

(4) They recombine at the surface and in the bulk by the SRH process.

(5) The bulk recombination centres contribute only negligibly to the space charge when compared to the effect of the dopants, either N_D^+ or N_A^-.

The departure from equilibrium in the model is imposed by taking various specific separations between the quasi-Fermi levels as given. This separation can be established by injection or optical excitation which has to overcome recombination in the bulk of the grain. In this way of looking at the problem it is, therefore, not necessary to introduce bulk recombination explicitly. The band edges follow also the law (6.3.1). The carrier contribution $p - n$ to the space-charge density $\rho(x) = q(N_{D^+} - N_A^- + p - n)$ will here be taken into account as well.

For the mathematical details of such a theory, we refer to the literature. The general case is discussed (with references) by Landsberg (1992, eq. (2.7.16)). The formula for the case which arises when the forbidden gap has, at the surface, constant (i.e. energy-independent) numbers per unit area per unit energy range of donors (N_{DS}) and acceptors (N_{AS}) is considered by Landsberg and Abrahams (1984).

Consider first the effect of donor doping on the equilibrium barrier height $e\Phi_B$ in polycrystalline silicon:

(i) experimentally (points in fig. 6.5),

(ii) theoretically assuming a single level of surface acceptors at $E_{AS} - E_{vS} = 0.34$ eV, $N_{As} = 6 \times 10^{11}$ cm^{-2}, $kT = 0.026$ eV, $c_n(\frac{1}{2}) = 10^{-7}$ cm^3 s^{-1}, $c_p(\frac{1}{2}) = 10^{-9}$ cm^3 s^{-1}, $n_i = 4.67 \times 10^9$ cm^{-3}, $E_g = 1.11$ eV, $N_c = 2.8 \times 10^{19}$ cm^{-3}, $N_v = 10^{19}$ cm^{-3}, $\varepsilon = 11.8$; and

(iii) theoretically using the same values for a constant density of surface states in the forbidden gap given by $6 \times 10^{11}/1.1 = 5.45 \times 10^{11}$ cm^2 eV^{-1}. The experiments are due to Seager and Castner (1978).

To understand curve (a) qualitatively, regard the potentials at $x = 0$ as fixed. The Fermi level is raised by increasing the bulk donor concentration N_D, thus creating

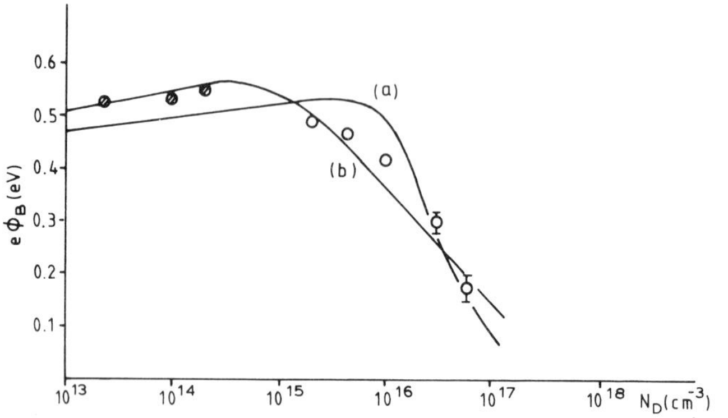

Fig. 6.5. Barrier heights at the boundary of polysilicon grains. Solid circles: from 270 K, resistivity data. Open circles: from potential profile measurements. These experimental points are compared with theory (see text): (a) single level of surface acceptors, (b) constant spectrum of surface acceptors.

more negatively charged (occupied) surface acceptors. This raises the bands with respect to the potentials at $x=0$ and, rigidly with them, the surface acceptor levels. Thus, one may think of these states as being pulled up in energy by a faster moving Fermi level μ_{eq} as N_{D} increases. Thus, $\Phi_{\text{B eq}}$ rises. A maximum is reached when $\mu_0 \approx E_{\text{As}}$. As μ_0 rises further, $\Phi_{\text{B eq}}$ declines. The reason is that all surface acceptors are now charged and the charge per unit area in the surface (Q_{s}) remains now constant as N_{D} rises further. However, the barrier thickness (w) declines as N_{D} increases and this leads to a lowering of $\Phi_{\text{B eq}}$. To understand these last remarks one may adopt (only temporarily!) the depletion approximation which yields after one and two integrations of the Poisson equation

$$Q_{\text{s}} \sim N_{\text{D}} w, \qquad \Phi_{\text{B eq}} \sim N_{\text{D}} w^2 \sim Q_{\text{s}}^2 / N_{\text{D}}. \tag{6.3.2}$$

Curve (b) may be understood similarly. The fit in both cases is reasonable.

Next, fig. 6.6 gives the non-equilibrium, steady-state barrier height as a function of illumination in n-type CdS. The crosses are experimental data (Wu and Bube 1974). The curve is the theoretical curve assuming a constant surface donor distribution of $8.5 \times 10^{11}\ \text{cm}^{-2}\ \text{eV}^{-1}$. Other data used are $N_{\text{D}} = 1.1 \times 10^{17}\ \text{cm}^{-3}$, $kT = 0.026\ \text{eV}$, $c_{\text{n}}(\frac{1}{2}) = c_{\text{p}}(\frac{1}{2}) = 2.5 \times 10^{-8}\ \text{cm}^3\ \text{s}^{-1}$, $n_{\text{i}} = 2.8 \times 10^{-2}\ \text{cm}^{-3}$, $E_{\text{g}} = 2.42\ \text{eV}$, $N_{\text{c}} = 2 \times 10^{18}\ \text{cm}^{-3}$, $N_{\text{v}} = 10^{19}\ \text{cm}^{-3}$ and $\varepsilon = 10$. This fit is again reasonable. Here the drop in Φ_{B} to half its dark value is due to the loss of charge on the interface due to illumination. The mechanism is this: the occupation of the surface acceptors increases owing to the rise in the electron quasi-Fermi level μ_{e}, rendering more of them neutral. The decrease in Q_{s} leads to a decrease in w and Φ_{B} by eq. (6.3.2).

Further applications of the grain boundary model have been made, but will not be given here. It has also been confirmed that the adoption of the depletion approximation can give high or low values of Φ_{B}, depending on the circumstances, with an error of the order up to 10%.

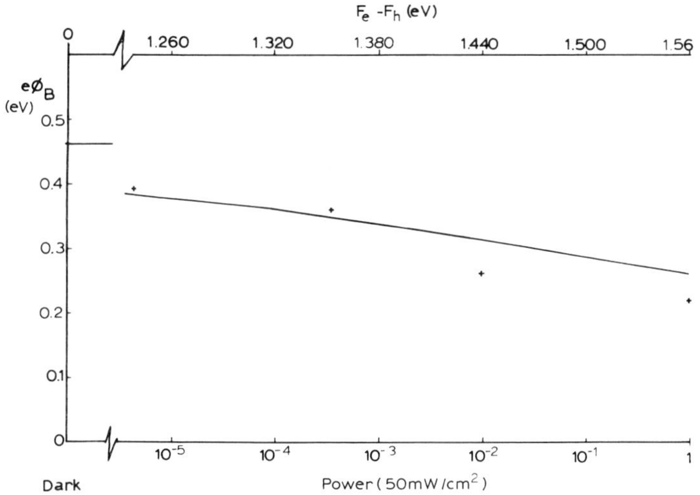

Fig. 6.6. The steady-state barrier height in CdS (n-type) as a function of illumination. The separation of the quasi-Fermi levels which corresponds to the illumination and some experimental points are also shown.

7. Lattice defects and solubility

7.1. Statistics of single lattice defects

The arguments for electron traps in § 1.3 can be used similarly for defects. For example, if L is a lattice site, then for a system at temperature T,

$$\Xi_L = Z_{L0} + \lambda_L Z_{L1}, \qquad \frac{Z_{L1}}{Z_{L0}} \equiv \exp\left(-\frac{E_L}{kT}\right), \qquad \lambda_L \equiv \exp\frac{\mu_L}{kT}. \tag{7.1.1}$$

Here Ξ_L is the general partition function for a lattice site, Z_{L0} is the canonical partition function for a vacant lattice site and Z_{L1} for a lattice site occupied by the lattice atom. If the solid is monatomic, there is only one type of lattice site and μ_L is the appropriate chemical potential. Let N_L and n_L be the number of available and occupied sites respectively. Then, the reaction giving the equilibrium between occupied and vacant lattice sites leads to a reaction constant

$$\frac{n_L}{N_L - n_L} = \frac{\lambda_L Z_{L1}}{Z_{L0}} = \exp\frac{\mu_L - E_L}{kT}. \tag{7.1.2}$$

Similarly for a defect site D

$$\frac{n_D}{N_D - n_D} = \frac{\lambda_D Z_{D1}}{Z_{D0}} = \exp\frac{\mu_D - E_D}{kT}. \tag{7.13}$$

It then follows that in equilibrium ($\mu_L = \mu_D$)

$$\frac{(N_L - n_L)n_D}{n_L(N_D - n_D)} = \exp\left(-\frac{E_D - E_L}{kT}\right). \tag{7.1.4}$$

Thus, $E_D - E_L$ is the energy required to populate the defect site, starting with the ideal lattice. This assumes that only one type of defect is present, and at low enough concentration, so that defect interaction can be neglected.

For Frenkel defects the number of unoccupied lattice sites is actually equal to the number of interstitials, which are the defects in this case. Denoting the number of Frenkel defects by n_F and the total number of interstitial sites by N_I,

$$N_L - n_L = n_F, \qquad N_D = N_I, \qquad n_D = n_I = n_F. \tag{7.1.5}$$

Hence, eq. (7.1.4) becomes

$$\frac{n_F^2}{(N_L - n_F)(N_I - n_F)} = \exp\left(-\frac{W_F}{kT}\right). \tag{7.1.6}$$

Here, $W_F = E_I - E_L$ is the energy needed to create a Frenkel defect starting with the perfect lattice. It corresponds to the "reaction"

$$L^{\bullet} + I^0 \rightleftarrows L^0 + I^{\bullet},$$

where $^{\bullet}$ means that the site is occupied and 0 means the site is vacant. If $n_F \ll N_L, N_I$, i.e. for a small number of Frenkel defects, eq. (7.1.6) yields

$$n_F = (N_L N_I)^{1/2} \exp\left(-\frac{W_F}{2kT}\right). \tag{7.1.7}$$

The activation energy for defect formation can then be obtained from an Arrhenius-type plot. The treatment given avoids combinatorics, Stirling's approximation and the normal complications of most probable distributions. Equation (7.1.7) gives a standard formula with W_F lying usually between 10–100 kcal/mol. Note that, following Landsberg (1982, p. 393) we have again applied the grand canonical ensemble with a gain in simplificity.

For a Schottky type defect in a material CA, a cation C and an anion A move to the surface leaving a vacancy of each in the bulk material. Thus n_F is replaced by n_s in eq. (7.1.6) and, instead of the number of interstitial sites N_I initially available for occupation, one has a number of surface sites, say N_σ. For the move of the cations and for the anions one has identical left-hand sides of eq. (7.1.6), so that

$$\frac{n_s^4}{(N_L - n_s)^2(N_\sigma - n_s)^2} = \exp\left(-\frac{2W_s}{kT}\right),$$

it follows that

$$\frac{n_s^2}{(N_L - n_s)(N_\sigma - n_s)} = \exp\left(-\frac{W_s}{kT}\right). \tag{7.1.8}$$

For small concentrations an Arrhenius-type law is found again

$$n_s = (N_L N_\sigma)^{1/2} \exp\left(-\frac{W_s}{2kT}\right).$$

If more than one monlayer is deposited on the surface, new considerations come into play as N_σ has to be more carefully defined. Indeed, one normally treats the surface sites to be in unlimited supply. In that case one merely needs eq. (7.1.3) for cations and also for anions, with each lattice site a potential defect site, i.e.,

$$n_D \to n_s, \qquad N_D \to N_L.$$

Hence,.

$$n_s = (N_L - n_s) \exp\left(-\frac{W'_s}{2kT}\right), \tag{7.1.9}$$

where W'_s and W_s are somewhat different, and eq. (7.1.9) is a more widely quoted formula.

The situation is more complicated if one has Schottky *and* Frenkel defects. In that case some additional notation is useful. We have $i = +$ or $-$ for positive or negative ion occupancy and α for a lattice, interstitial or surface site, i.e., $\alpha = \mathrm{L}$, I or σ. One then has eight *unknowns*:

$$n^i_\alpha \quad (i = +, -; \alpha = \mathrm{L, I}, \sigma); \qquad \lambda^{(i)} Z^{(i)}.$$

One can regard as *known* the total number $N^{(i)}_\alpha$ of possible (α, i) sites as well as the reduced energy level $\eta^{(i)}_\alpha \equiv E^{(i)}_\alpha / kT$ for a site of type α with i-occupancy. One then has eq. (7.1.2) or (7.1.3) for each of the six $n^{(i)}_\alpha$:

$$\lambda^{(i)} Z^{(i)} = \frac{n^{(i)}_\alpha}{N^{(i)}_\alpha - n^{(i)}_\alpha} \exp(\eta^{(i)}_\alpha). \tag{7.1.10}$$

In addition, the number of lattice vacancies of type i is at each temperature equal to the number of interstitial and surface sites occupied, giving a seventh and eighth equation:

$$N^{(i)}_\mathrm{L} - n^{(i)}_\mathrm{L} = n^{(i)}_\mathrm{I} + n^{(i)}_\sigma \quad (i = +, -). \tag{7.1.11}$$

These can be solved for our eight unknowns.

The first step is to use eq. (7.1.10) with $\alpha = \mathrm{I}$ and σ with the same i-value so that $\lambda^{(i)} Z^{(i)}$ cancels out. This gives

$$n^{(i)}_\sigma = \frac{N^{(i)}_\sigma n^{(i)}_\mathrm{I} \exp(\eta^{(i)}_\mathrm{I})}{N^{(i)}_\mathrm{I} \exp(\eta^{(i)}_\sigma) - n^{(i)}_\sigma \exp(\eta^{(i)}_\sigma) + n^{(i)}_\mathrm{I} \exp(\eta^{(i)}_\mathrm{I})}. \tag{7.1.12}$$

By working out also $n^{(i)}_\mathrm{I} + n^{(i)}_\sigma$, and then using eq. (7.1.10) with $\alpha = L$, and also using eq. (7.1.11), one eventually arrives at a cubic equation for $n^{(i)}_\mathrm{I}$. It is

$$\sum_{j=0}^{3} h_j \left(\frac{n^{(i)}_\mathrm{I}}{N^{(i)}_\mathrm{L}}\right)^j = 0, \tag{7.1.13}$$

where

$$h_3 \equiv (e^{\eta'^{(i)}_\sigma} - e^{\eta'^{(i)}_I})(1 - e^{\eta'^{(i)}_I})e^{2\eta^{(i)}_L},$$

$$h_2 \equiv e^{2\eta^{(i)}_L}\left[e^{\eta'^{(i)}_I} - e^{\eta'^{(i)}_\sigma} + \frac{N^{(i)}_\sigma}{N^{(i)}_L}e^{2\eta'^{(i)}_I} + \frac{N^{(i)}_I}{N^{(i)}_L}e^{(\eta'^{(i)}_\sigma + \eta'^{(i)}_I)} - 2\frac{N^{(i)}_I}{N^{(i)}_L}e^{\eta'^{(i)}_\sigma} + \left(\frac{N^{(i)}_I}{N^{(i)}_L} - \frac{N^{(i)}_\sigma}{N^{(i)}_L}\right)e^{\eta'^{(i)}_I}\right],$$

$$h_1 \equiv e^{2\eta^{(i)}_L}\frac{N^{(i)}_I}{N^{(i)}_L}\left[\left(2 + \frac{N^{(i)}_I}{N^{(i)}_L}\right)e^{\eta'^{(i)}_\sigma} + \left(\frac{N^{(i)}_\sigma}{N^{(i)}_L} - 1\right)e^{\eta'^{(i)}_I}\right],$$

$$h_0 \equiv -e^{2\eta^{(i)}_L}\left(\frac{N^{(i)}_I}{N^{(i)}_L}\right)^2 e^{\eta'^{(i)}_\sigma}.$$

Here $\eta'_\alpha \equiv (E_\alpha - E_L)/kT$ ($\alpha = I, \sigma$). Once $n^{(i)}_I$ is determined, $n^{(i)}_\sigma$ can be found from eq. (7.1.12) and hence $n^{(i)}_L$ can be obtained from eq. (7.1.11).

On ignoring surface sites one should regain our theory for Frenkel defects. However, as N_σ bears a definite ratio to N_L, this limit should not be attempted by letting $N_\sigma \to 0$. Instead it can be reached by the indefinite raising of E_σ, i.e., by letting $\eta'_\sigma \to \infty$. One then finds the relation

$$-(A-B)^2 + B(A-B)^2 + B^2(A-B)\exp(\eta'^{(i)}_I) = 0 \tag{7.1.14}$$

from eq. (7.1.13), where

$$A \equiv \frac{N^{(i)}_I}{N^{(i)}_L}, \qquad B \equiv \frac{n^{(i)}_I}{N^{(i)}_L}.$$

This is for $i = +$ easily shown to be equivalent to eq. (7.1.6) and so checks that the present theory represents a generalisation to a mixture of Frenkel and Schottky defects.

These theoretical results (Landsberg and Canagaratna 1984) are in need of practical tests. Further theoretical developments, by using two-site partition functions, have been given, but the whole matter needs to be explored further.

From the extensive literature, note the discussion of impurity atoms in a stress field of a dislocation which involves giving the fraction of neutral donors a Fermi-type distribution:

$$\frac{n^x_D}{n^+_D + n^x_D} = \frac{1}{1 + n^+_D/n^x_D} = \frac{1}{1 + Z_0/\lambda Z_1} = \frac{1}{1 + \exp(\eta_D - \gamma)}$$

(Beshers 1958, Johnson 1981). Note also a small controversy concerning the chemical potential of an impurity-vacancy pair (Howard and Lidiard 1964, Yoshida 1976, 1980). An earlier interesting discussion (Chatterjea and Hauser 1976) on the statistics of doubly ionised and neutral nearest-neighbour donor–donor impurity pairs also exists, but has not been developed further. The generation–recombination of vacancies and interstitials has, in addition, been shown to lead to distributions of defects which suggests that defects tend to bunch together in space (Vinetskii 1983). Non-uniform distributions of vacancies are induced by temperature gradients which tend to concentrate them at the colder end. This phenomenon has also been discussed from a theoretical point of view (Stark 1980).

We finally consider the classical theory of pairing which goes back to Bjerrum in 1926 and Fuoss in 1934 and is expounded by Reiss et al. (1956). One considers a solution containing equal concentrations (N) of ions of opposite sign. In equilibrium, each negative ion, say A, has a nearest neighbour ion (assumed to be positive) a distance x from it. Let it have probability $g(x)\,\mathrm{d}x$ to lie in a spherical shell of volume $4\pi x^2\,\mathrm{d}x$ and centred on A. The probability that this nearest neighbour lies in a shell of volume $4\pi r^2\,\mathrm{d}r$ is then given by the integral equation

$$g(r) = \left[1 - \int_a^r g(x)\,\mathrm{d}x\right] 4\pi r^2 c(r). \tag{7.1.15}$$

This arises as the product of two probabilities. The first is the chance that the sphere of volume $\frac{4}{3}\pi r^3$ is empty, a being the closest distance between two opposite ions. The second factor is the probability of finding a positive ion in the shell of volume $4\pi r^2\,\mathrm{d}r$. This is $4\pi r^2 c(r)\,\mathrm{d}r$, where $c(r)$ is the concentration of positive ions at a distance r from the negative one, which is increased above N by electrostatic attraction. The solution of eq. (7.1.15) (which is verified by substituting eq. (7.1.16) into it) is

$$g(r) = 4\pi r^2 c(r) \exp\left[-4\pi \int_a^r x^2 c(x)\,\mathrm{d}x\right]. \tag{7.1.16}$$

Each ion is surrounded by a cloud of opposite charge and this screening suggests that each ion interacts only with its neighbour, i.e. with potential energy $-q^2/\varepsilon x$, where ε is the dielectric constant. Hence, using $c(\infty) = N$,

$$c(x) = N \exp\left(\frac{q^2}{\varepsilon x k T}\right). \tag{7.1.17}$$

This gives the curve shown in fig. 7.1, where the assumption of a distance of closest approach is seen to suggest repulsive forces for $r < b$. One may regard all nearest neighbours at $r < b$ as ion pairs and the rest of the neighbours of A as unpaired.

One normally needs $g(r)$ only for $r < b$ in which case the exponential function in eq. (7.1.16) can be replaced by unity, so that

$$g(r) \approx 4\pi r^2 N \exp\left(\frac{q^2}{\varepsilon r k T}\right). \tag{7.1.18}$$

This result yields an estimate for the reaction constant

$$K = p/(N - p)^2, \tag{7.1.19}$$

which determines the equilibrium between a concentration p of paired ions and two sets of unpaired ions, each of concentration $N - p$. For the neighbourhood of infinite

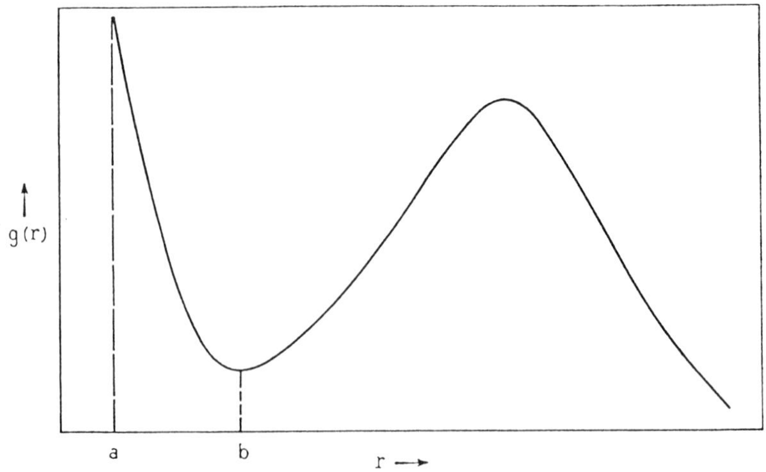

Fig. 7.1. Schematic distribution of neighbours in an assembly of particles when forces of interaction are present. Repulsive forces are reflected in the appearance of a distance, a, of closest approach of two particles, attractive forces by the maximum.

dilution ($p \ll N$)

$$K \approx \frac{p}{N^2} = \frac{\int_a^b g(r)\,\mathrm{d}r}{N} = 4\pi \int_a^b r^2 \exp\left(\frac{q^2}{\varepsilon r k T}\right) \mathrm{d}r. \tag{7.1.20}$$

We have used the fact that the fraction p/N of paired ions is the fraction of nearest neighbours lying at $r \leqslant b$. In this way, the pair concentration can be estimated in a remarkably elegant manner, using $b \sim q^2/2\varepsilon kT$, without the need to use a rough estimate for p in eq. (7.1.19).

If, in the generalisation of eq. (7.1.19), one has concentrations N_D and N_A of donors and acceptors, some of which are paired,

$$K = \frac{p}{(N_\mathrm{D} - p)(N_\mathrm{A} - p)}.$$

In that case, p can be calculated from

$$\frac{p}{N_\mathrm{D}} = \tfrac{1}{2}\left[1 + \frac{1}{KN_\mathrm{D}} + \frac{N_\mathrm{A}}{N_\mathrm{D}}\right] - \left[\tfrac{1}{4}\left(1 + \frac{1}{KN_\mathrm{D}} + \frac{N_\mathrm{A}}{N_\mathrm{D}}\right)^2 - \frac{N_\mathrm{A}}{N_\mathrm{D}}\right]^{1/2},$$

but now only if the reaction constant K is known.

The coexistence in Si of self-interstitials and vacancies in thermal equilibrium has been demonstrated by using dopant diffusion and the interstitial–substitutional diffusion of Pt in Si and of Cu in Ge (Frank and Stolwijk 1987). These and a host of other studies in the diffusion area makes the sort of calculation outlined here of interest. A short list of notable topics is the effect of p–n junction fields on diffusion profiles (Anthony 1982), diffusion of co-implanted Si and Ge in GaAs (Deal and Robinson 1990) and anomalous diffusion effects (Hildebrand 1982, Jung and Marschall 1988).

7.2. A model for saturation solubilities of impurities

The solubility of a neutral impurity in a semiconductor crystal in thermal equilibrium is solely a function of temperature. If the impurity atoms may exist in different charge states, their solubility is still determined by the concentration of those which are neutral (Shockley and Moll 1960). We consider a case where a specific impurity is neutral when it has captured s electrons. In a region where the Fermi level has a position such that the occupation probability for capture of s electrons by the impurity is $P(s)$ and the solubility for neutral atoms corresponds to a concentration ν_s, the total concentration, N, of atoms present depends on the relation

$$P(s)N = \nu_s. \tag{7.2.1}$$

$P(s)$ is given by eq. (1.3.12) and we find that the concentration of atoms which can be dissolved in the material depends on the Fermi level position as

$$N = \nu_s \frac{\sum_{r=0}^{M} [\exp(r\mu/kT)Q_r]}{\exp(s\mu/kT)Q_s}. \tag{7.2.2}$$

Thus the solubility of an impurity acting as a charge carrier trap, may be largely enhanced in a material where the Fermi level has a position such that the ratio is low between the concentration of neutral and charged traps. It is found that ν_s can be determined experimentally and so furnishes an important base line for the other concentrations which can be obtained from eq. (7.2.1). For an impurity with $M = 3$ and $s = 1$, the ratios between the concentration ν_0, ν_2 and ν_3 of charged impurities and ν_1 of the neutral impurities are given by eq. (7.2.2) as

$$\begin{aligned}
\ln\frac{\nu_0}{\nu_1} &= \ln\frac{Q_0}{Q_1} - \frac{\mu}{kT} = \frac{G(\frac{1}{2})}{kT} - \frac{\mu}{kT},\\
\ln\frac{\nu_2}{\nu_1} &= \ln\frac{Q_2}{Q_1} + \frac{\mu}{kT} = -\frac{G(\frac{3}{2})}{kT} + \frac{\mu}{kT},\\
\ln\frac{\nu_3}{\nu_1} &= \ln\frac{Q_3}{Q_1} + \frac{2\mu}{kT} = -\frac{G(\frac{3}{2}) + G(\frac{1}{2})}{kT} + \frac{2\mu}{kT}.
\end{aligned} \tag{7.2.3}$$

Figure 7.2 demonstrates $\ln \nu_r/\nu_1$ as a function of μ/kT for the impurity described by eq. (7.2.3). Note that if μ/kT is varied at constant temperature the lines drawn are exactly straight within the model. (If T is allowed to vary, then the ratio of the partition functions becomes temperature-dependent and one ceases to have straight lines.) The variation of μ can be assumed to be due to variable dopings by some other majority impurity. One sees that this affects the take-up by the material of the

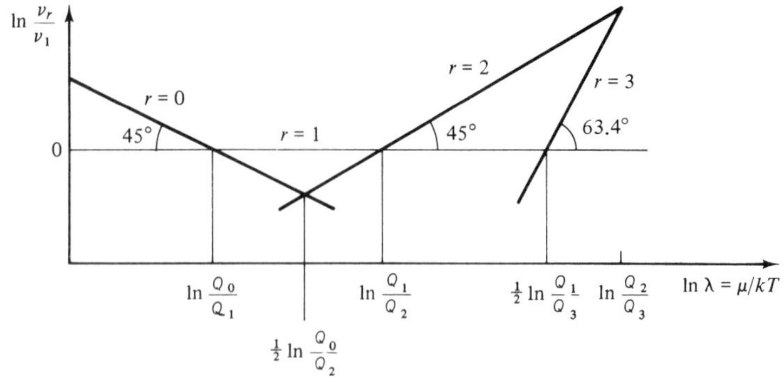

Fig. 7.2. Concentration of r-electron centres for $r = 0, 1, 2, 3$ as a function of Fermi level divided by kT. Saturation conditions and thermal equilibrium at a fixed temperature have been assumed.

multi-charge impurity under investigation, even if the source of this impurity is maintained at constant strength.

The concentrations ν_r can be determined, e.g., by radioactive tracer technique under different background doping conditions and the corresponding Fermi level positions can be determined by Hall or resistivity measurements. Then, from eq. (7.2.3), the Q_r/Q_s ratios can be determined from measurements of ν_r/ν_s obtained for different impurity diffusion temperatures.

The free energies $G(r - \frac{1}{2})$ in eq. (1.3.16) have a temperature dependence given by

$$G(r - \tfrac{1}{2}) = H(r - \tfrac{1}{2}) - TS(r - \tfrac{1}{2}), \tag{7.2.4}$$

as explained in § 1.3. Thus, plotting the experimentally obtained values of $\ln Q_r/Q_s$ as a function of $1/T$, one would obtain a straight line with a slope determined by the enthalpy values of the trap and with an intercept given by the entropies as seen by eqs. (7.2.3) and (7.2.4). This method has been used to estimate electronic degeneracies by partitioning the entropy into an electronic part equal to $k \ln(g_r/g_s)$ and an estimated atomic part consistent with eqs. (1.4.2) and (1.4.3) (Landsberg 1977).

If one is not interested in separating enthalpies (or energy levels) and entropies (or degeneracies), these can be lumped together in the partition function and eq. (7.2.2) gives the uptake, N, of a minority impurity in terms of the concentration ν_s of the neutral impurity *at a given temperature*. It will be assumed that ν_s is not affected by the heavy majority doping, which is assumed to exceed $100N$. A number of different possibilities arise depending on the charge multiplicity, M, as shown in table 7.1. Case 4, which applies to Au in Si, yields a curve of N against $\lambda = \exp \mu/kT$ which has a minimum when $\lambda^2 = Q_0/Q_2$ and $\nu_0 = \nu_2$. More details are given by Landsberg (1992, p. 64).

7.3. The connection between doping and the solubility of defects

The qualitative effect of doping on the solubility of defects can be seen as follows. Suppose that we have a donor defect. Then, acceptor doping lowers the Fermi level

Table 7.1
Saturation solubility of minority impurity as a function of $\lambda = \exp(\mu/kT)$ fixed by heavy primary doping.

Case number	M	s (number of electrons on neutral centre)	Equation for N/v_s
1	1	0	$1+\frac{\lambda Q_1}{Q_0}$
2	1	1	$\frac{Q_0}{Q_1}\lambda^{-1}+1$
3	2	0	$1+\frac{Q_1}{Q_0}\lambda+\frac{Q_2}{Q_0}\lambda^2$
4	2	1	$\frac{Q_0}{Q_1}\lambda^{-1}+1+\frac{Q_2}{Q_1}\lambda$
5	2	2	$\frac{Q_0}{Q_2}\lambda^{-2}+\frac{Q_1}{Q_2}\lambda^{-1}+1$

and increases the N_d^+ concentration and hence N_d. Conversely, donor doping brings down N_d^+ and N_d. For an acceptor defect donor doping raises the N_d^- concentration and hence N_d. Some curves are shown in fig. 7.3. Figure 7.4 gives in addition the expected lifetime dependence.

Details of the mathematical arguments to support this heuristic discussion are given by Landsberg (1992, pp. 95 and 127). Briefly, the normal semiconductor statistics for traps is used twice. First at the freezing-in temperature T_f of the acceptor-

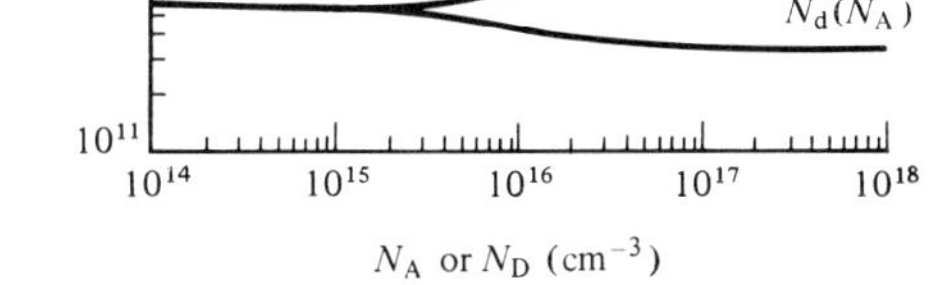

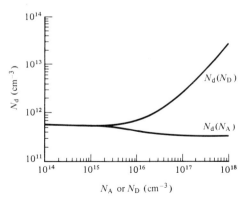

Fig. 7.3. Residual defect density (acceptor type) as a function of impurity concentration in n-type [N_d (N_D)] and p-type [N_d (N_A)] non-degenerate Si. For donor-type defects the top curve would be N_d (N_A) and the bottom curve would be N_d (N_D) (schematic).

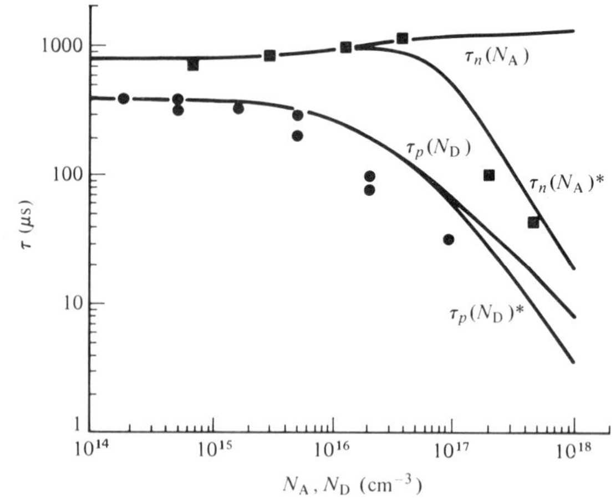

Fig. 7.4. Best room-temperature minority carrier lifetimes in non-degenerate Si at $T = 300$ K as a function of residual defect density (acceptor type). The effect of including the band–band Auger effect is shown as τ_p^* and τ_n^* (Landsberg and Kousik, 1984). For donor-type defects the top curve would be τ_p (N_D) and the bottom curve τ_n (N_A) (schematic). Experimental points are also shown.

type defects, where

$$N_d^x = 5 \times 10^{22} \exp\left(-\frac{E_a}{kT_f}\right) \quad (\text{cm}^{-3}). \tag{7.3.1}$$

For silicon, the concentration of host lattice sites is about 5×10^{22} cm^{-3}. The Fermi level equation,

$$p - n + N_D^+ = N_d^-,$$

can then be reduced to a cubic equation for N_d^-. Its solution yields N_d (N_A) in fig. 7.3.

Next we study the concept of the perfect quench, which implies infinitely rapid cooling from the quenching temperature T_q, preserving the total defect density N_d down to the lower measurement temperature T. Then one applies the normal semiconductor statistics for a second time, namely at the temperature T using the previously calculated value of N_d. Using the Shockley–Read mechanism, one arrives at τ_n (N_A) in fig. 7.4.

The best experimental lifetimes, when fitted to the theoretical curve, can then furnish an identification of key parameters. In the case of Si one finds

$$2T_1^s = T_2^s = 5 \times 10^{-9}\ \text{cm}^{-3}\ \text{s}^{-1} \qquad E_a = 1.375\ \text{eV}, \qquad T_f = 620\ \text{K}$$
$$\text{Position of defect level} = 45\ \text{meV above midgap.} \tag{7.3.2}$$

The cooling from T_q tends to lower N_d^x in eq. (7.3.1) so that $T_f < T_q$. However, this relation is not clear and we lose no generality by assuming $T_f \approx T_q$. It is possible that the defect defined by eq. (7.3.2) is a self-interstitial or a cluster of self-interstitials as found in swirl and dislocation-free float-zone grown Si. This is the attribution made by Jastrzebsky and Zanzucci (1981) of the defect they found at $E_v + 0.56$ eV with capture cross sections $\sigma_p^s = 2\sigma_n^s = 10^{-14}$ cm^2.

A diagram similar to fig. 7.3 has been obtained for AlGaAs, with

$$T_f = 823 \text{ K}, \qquad E_a = 1.39 \text{ eV}, \qquad \text{Defect level at } E_v + 63 \text{ meV}.$$

In this connection it has been shown (Krispin 1989) that dopant-induced solubility can be studied to advantage to lead to more insight into the nature of native defects in compound semiconductors.

References

Abakumov, V.N., V.I. Perel' and I.N. Yassievich, 1991, Nonradiative Recombination in Semiconductors (North-Holland, Amsterdam).

Adams, M.J., 1969, Solid-State Electron. **12**, 661.

Adams, M.J., and P.T. Landsberg, 1968, Int. Conf. Physics of Semiconductors, Moscow (Akademiya Nauk, Leningrad) p. 619.

Adams, M.J., and P.T. Landsberg, 1969, The Theory of the Injection Laser, in: Gallium Arsenide Injection Lasers, ed. C.H. Gooch (Wiley, London).

Adler, D., and E.J. Yoffa, 1976, Phys. Rev. Lett. **36**, 1197.

Almbladh, C.-O., and G.J. Rees, 1981, J. Phys. C **14**, 4575.

Andersson, G.I., and O. Engström, 1990, J. Appl. Phys. **67**, 3500.

Andersson, M.O., and O. Engström, 1989, Appl. Surf. Sci. **39**, 289.

Andersson, M.O., Z. Xiao, S. Norrman and O. Engström, 1990, Phys. Rev. B **41**, 9836.

Anthony, P.J., 1982, Solid-State Electron. **25**, 1003.

Asbeck, P., 1977, J. Appl. Phys. **48**, 820.

Ascarelli, G., and S. Rodriguez, 1961, Phys. Rev. **124**, 1321; Phys. Rev. **127**, 167.

Badescu, V., 1991, J. Phys.: Condens. Matter **3**, 6509.

Barber, H.D., 1967, Solid-State Electron. **10**, 1039.

Basov, N.G., O.N. Krokhin and Y.M. Popov, 1961, Sov. Phys. JETP **12**, 1033; **13**, 845.

Bebb, H.B., 1972, Phys. Rev. B **5**, 4201.

Belyaev, A.D., and I.P. Zvyagin, 1991, Sov. Phys. Semiconductors **25**, 19.

Bengtsson, S., and O. Engström, 1989, J. Appl. Phys. **66**, 1231.

Bengtsson, S., G.I. Andersson, M.O. Andersson and O. Engström, 1992, J. Appl. Phys. **72**, 144.

Benoit à la Guillaume, C., and C. Tric, 1961, J. Phys. Rad. Paris **22**, 834.

Bensaid, B., F. Raymond, M. Leroux, C. Vèrié and B. Fofana, 1989, J. Appl. Phys. **66**, 5542.

Bernard, M.G.A., and B. Duraffourg, 1961, Phys. Stat. Sol. **1**, 699.

Beshers, D.N., 1958, Acta Metall. **6**, 521.

Bess, L., 1958, Phys. Rev. **111**, 129.

Blakemore, J.S., 1967, Phys. Rev. **163**, 809.

Brooks, H., 1955, Adv. Electron. & Electron Phys. **7**, 117.

Brotherton, S.R., and J. Bicknell, 1978, J. Appl. Phys. **49**, 667.

Brown, R.A., 1966, Phys. Rev. **148**, 974.

Brown, R.A., and S. Rodriguez, 1967, Phys. Rev. **153**, 890.

Carrol, J.E., 1985, Rate Equations in Semiconductor Electronics (Cambridge University Press, Cambridge).

Casey Jr, H.C., and F. Stern, 1976, J. Appl. Phys. **47**, 631.

Chantre, A., 1989, Appl. Phys. A **48**, 3.

Chantre, A., and D. Bois, 1985, Phys. Rev. B **31**, 7979.

Chantre, A., and L.C. Kimerling, 1986, Appl. Phys. Lett. **48**, 1000.

Chatterjea, A., and J.R. Hauser, 1976, J. Phys. & Chem. Solids **37**, 1031.

Chel'tzov, V.F., 1983, J. Phys. C **16**, 3615.

Chen, Y.F., and S.F. Huang, 1991, Phys. Rev. B **44**, 13775.

Cobden, D.H., M.J. Uren and M.J. Kirton, 1990, Appl. Phys. Lett. **56**, 1245.
Cody, G.D., 1984, in: Semiconductors and Semimetals, Vol. 21B, ed. J.I. Pankove (Academic Press, New York) p. 11.
Coutts, T.J., and N.M. Pearsall, 1984, Appl. Phys. Lett. **44**, 134.
de Blasi, C., G. Micocci, A. Rizzo and A. Tepore, 1983, Solid-State Electron. **26**, 1095.
de Visschere, P., 1986, Solid-State Electron. **29**, 1161.
Deal, M.D., and H.G. Robinson, 1990, Solid-State Electron. **33**, 665.
Deuling, H., E. Klausmann and A. Goetzberger, 1972, Solid-State Electron. **15**, 559.
Dexter, D.L., 1958, in: Solid State Physics, eds F. Seitz and D. Turnbull, Vol. 6 (Academic, New York) p. 355.
Dhariwal, S.R., and P.T. Landsberg, 1989, J. Phys.: Condens. Matter **1**, 569.
Dhariwal, S.R., and D.R. Mehrotra, 1988, Solid-State Electron. **31**, 1355.
Dumke, W.P., 1957, Phys. Rev. **105**, 139.
Engström, O., 1989, Appl. Phys. Lett. **55**, 47.
Engström, O., and A. Alm, 1983, J. Appl. Phys. **54**, 5240.
Engström, O., and H.G. Grimmeiss, 1974, Appl. Phys. Lett. **25**, 413.
Engström, O., and H.G. Grimmeiss, 1975, J. Appl. Phys. **46**, 831.
Engström, O., and H.G. Grimmeiss, 1989, Semicond. Sci. & Technol. **4**, 1106.
Evans, D.A., and P.T. Landsberg, 1965, Solid-State Electron **26**, 315.
Farmer, K.R., C.T. Rogers and R.A. Buhrman, 1987, Phys. Rev. Lett. **58**, 2255.
Frank, W., and N.A. Stolwijk, 1987, Mater. Sci. Forum **15–18**, 369.
Frova, A., ed., 1973, The Physics and Technology of Semiconductor Light Emitters and Detectors, Pugnochiuso Symposium, 1972 (North-Holland, Amsterdam). See particularly the Panel Discussion on pp. 548–552. Also printed in J. Luminescence, Vol. 7 (1973).
Gibb, R.P., G.J. Rees, B.W. Thomas, B.L.H. Wilson, B. Hamilton, D.R. Wight and N.F. Mott, 1977, Philos. Mag. **36**, 1021.
Goguenheim, D., D. Vuillaume, G. Vincent and N.M. Johnson, 1991, J. Appl. Phys. **68**, 1104.
Goradia, C., and V.G. Weizer, 1984, Appl. Phys. Lett. **45**, 1298.
Grinberg, A.A., A.A. Rogachev and S.M. Ryvkin, 1966, Sov. Phys. Solid State **7**, 1774.
Grove, A.S., B.E. Deal, E.H. Snow and C.T. Sah, 1965, Solid-State Electron. **8**, 145.
Gutman, F., and L.E. Lyons, 1967, in: Organic Semiconductors (Wiley, New York) p. 429.
Halperin, B.I., and M. Lax, 1966, Phys. Rev. **148**, 772.
Hamann, D.R., and A.L. McWorther, 1964, Phys. Rev. **134**, A250.
Hangleiter, A., 1987, Phys. Rev. B **17**, 9149.
Hartke, J.L., 1968, J. Appl. Phys. **39**, 4871.
Henry, C.H., and D.V. Lang, 1977, Phys. Rev. B **15**, 989.
Hildebrand, O., 1982, Phys. Status Solidi (a) **72**, 575.
Hoffmann, H.J., 1980, Phys. Rev. Lett. **45**, 1733.
Howard, R.E., and A.B. Lidiard, 1964, Rep. Prog. Phys. **27**, 161.
Jastrzebski, L., and P. Zanzucci, 1981, in: Semiconductor Silicon, ed. H.R. Huff (Electrochemical Society, Princeton, NJ) p. 138.
Johnson, N.M., 1982, J. Vac. Sci. & Technol. **21**, 303.
Johnson, R.A., 1981, Phys. Rev. B **24**, 7383.
Jung, H., and P. Marshall, 1988, Jpn. J. Appl. Phys. **27**, L 2112.
Kamieniecki, E., 1979, Appl. Phys. Lett. **35**, 807.
Kamieniecki, E., N. Gomma, A. Kloc and R. Nitecki, 1981, J. Vac. Sci. & Technol. **18**, 883.
Kane, E.O., 1963, Phys. Rev. **131**, 79.
Kazmerski, L.L., 1980, Polycrystalline and Amorphous Thin Films and Devices (Academic Press, New York) ch. 3.
Keyes, B.M., D.J. Dunlavy, R.K. Ahrenkiel, S.E. Asher, L.D. Partain, D.D. Liu and M.S. Kuryla, 1990, J. Vac. Sci. & Technol. A **8**, 2004.
Kirton, M.J., and M.J. Uren, 1986, Appl. Phys. Lett. **48**, 1270.
Kleverman, M., H.G. Grimmeiss, A. Litwin and E. Janzén, 1985, Phys. Rev. B**31**, 3659.

Kocka, J., M. Vanececk and F. Schauer, 1987, J. Non-Cryst. Solids **97/98**, 715. (This volume contains the Proceedings of the 12th Int. Conference on Amorphous and Liquid Semiconductors, including at least 75 pages on gap states).
Kressel, H., 1973, in: Photonics, eds M. Balkanski and P. Lallemand (Gauthier-Villars, Paris).
Kriegler, R.J., T.F. Devenyi, K.D. Chik and J. Shappir, 1979, J. Appl. Phys. **50**, 398.
Krispin, P., 1989, J. Appl. Phys. **65**, 3470.
Kuhn, M., 1970, Solid-State Electron. **13**, 873.
Kuriyama, T., T. Kamiya and H. Yanai, 1977, Jpn. J. Appl. Phys. **16**, 465.
Lal, P., and P.T. Landsberg, 1965, Phys. Rev. **140**, A 46.
Landsberg, P.T., 1957, Proc. Phys. Soc. B **70**, 1175.
Landsberg, P.T., 1958, in: Semiconductors and Phosphors (Vieweg, Braunschweig) p. 52.
Landsberg, P.T., 1967, Phys. Status Solidi **19**, 777.
Landsberg, P.T., 1971, Problems in Thermodynamics and Statistical Physics (Pion, London) problem 3.9.
Landsberg, P.T., 1972, Proc. R. Soc. London A **331**, 103.
Landsberg, P.T., 1977, J. Phys. D **10**, 2467.
Landsberg, P.T., 1981, J. Phys. C **14**, L 1025.
Landsberg, P.T., 1982, in: Handbook on Semiconductors, Vol. 1, ed. W. Paul (North-Holland, Amsterdam).
Landsberg, P.T., 1986, J. Appl. Phys. **60**, 2189.
Landsberg, P.T., 1992, Recombination in Semiconductors (Cambridge University Press, Cambridge).
Landsberg, P.T., and M.S. Abrahams, 1984, Proc. IEEE 17th Photovoltaic Specialists Conference, Orlando, Florida, p. 597, eq. (1); J. Appl. Phys. **55**, 4284.
Landsberg, P.T., and D.C. Browne, 1988, Semicond. Sci. & Technol. **3**, 193.
Landsberg, P.T., and S.G. Canagaratna, 1984, Phys. Status Solidi (b) **126**, 141.
Landsberg, P.T., and S.R. Dhariwal, 1989, Phys. Rev. **39**, 91.
Landsberg, P.T., and P. Harshman, 1988, J. Stat. Phys. **53**, 475.
Landsberg, P.T., and G.S. Kousik, 1984, J. Appl. Phys. **56**, 1696.
Lang, D.V., 1974, J. Appl. Phys. **45**, 3023.
Lang, D.V., and R.A. Logan, 1977, Phys. Rev. Lett. **39**, 635.
Lang, D.V., R.A. Logan and M. Jaros, 1979, Phys. Rev. B **19**, 1015.
Lang, D.V., H.G. Grimmeiss, E. Meijer and M. Jaros, 1980, Phys. Rev. B **22**, 3917.
Lax, M., 1960, Phys. Rev. **119**, 1502.
Lebedo, L.Å., and Z.G. Wang, 1983, Appl. Phys. Lett. **42**, 680.
Lloyd, P., and J.J. O'Dwyer, 1963, Mol. Phys. **6** 573.
Markham, J.J., 1959, Rev. Mod. Phys. **31**, 956.
Mashkevich, V.S., and V.L. Vinetskii, 1966, Sov. Phys. Solid State **7**, 1605.
Mathur, V.K., and R.P. Dahiya, 1974, Solid-State Electron. **17**, 61.
Mialhe, P., J.P. Charles and A. Khoury, 1988, J. Phys. D **21**, 383.
Micocci, G., A. Rizzo and A. Tepore, 1985, J. Appl. Phys. **58**, 1274.
Morgan, T.N., 1983, Phys. Rev. B **28**, 7141.
Moss, T.S., 1957, Proc. Phys. Soc. B **70**, 247.
Mukai, T., and Y. Yamamoto, 1982, IEEE J. Quantum Electron. **QE-18**, 564.
Neumark, G.F., 1972, Phys. Rev. B **7**, 3802.
Nicollian, E.H., and J.R. Brews, 1982, MOS (Metal Oxide Semiconductor) Physics and Technology (Wiley, New York, 1982).
Pankove, J.I., 1965, Phys. Rev. **140**, A 2059.
Pankove, J.I., ed., 1984, Vols. 21A–D of Semiconductors and Semimetals, eds R.K. Willardson and A.C. Beer (Academic Press, Orlando, FL).
Peled, A., and L.B. Schein, 1991, Phys. Scr. **44**, 304.
Penzkofer, A., 1988, Progr. Quantum Electron. **12**, 291.
Popov, Y.M., 1967, Appl. Opt. **6**, 1818.
Ralls, K.S., W.J. Skocpol, L.D. Jackel, R.E. Howard, L.A. Fetter, R.W. Epworth and D.M. Tennant, 1984, Phys. Rev. Lett. **52**, 228.
Rees, G.J., 1985, Solid-State Electron. **28**, 517.
Rees, G.J., H.G. Grimmeiss, E. Janzén and B. Skarstam, 1980, J. Phys. C **13**, 6157.

Reiss, H., C.S. Fuller and F.J. Morin, 1956, Bell Syst. Tech. J. **35**, 535.
Ricksand, A., and O. Engström, 1991, J. Appl. Phys. **70**, 6915, 6927.
Rosier, L.L., and C.T. Sah, 1971, Solid-State Electron. **14**, 41.
Ruppel, W., and P. Würfel, 1980, IEEE Trans. Electron. Dev. **ED-27**, 877.
Sa-Yakanit, V., and H.R. Glyde, 1988, in: The Path Integral Method with Applications, eds A. Ranfagni, V. Sa-Yakanit and L.S. Schulman (World Scientific, Singapore).
Sah, C.T., L. Forbes, L.L. Rosier and A.F. Tasch Jr., 1970, Solid-State Electron. **13**, 759.
Schöll, E., and P.T. Landsberg, 1983, J. Opt. Soc. Am. **73**, 1197.
Schroder, D.K., 1990, Semiconductor Materials and Device Characterison (Wiley, New York).
Schulz, M., and N.M. Johnson, 1978, Solid State Commun. **25**, 481.
Seager, C.H., and T.G. Castner, 1978, J. Appl. Phys. **49**, 3879.
Sheinkman, M.K., 1965, Sov. Phys. Solid State **7**, 18.
Shklovskii, B.I., and A.L. Efros, 1984, in: Electronic Properties of Doped Semiconductors (Springer, Berlin).
Shockley, W., and J.L. Moll, 1960, Phys. Rev. **119**, 1480.
Shockley, W., and W.T. Read, 1952, Phys. Rev. **87**, 835.
Smith, E.F., and P.T. Landsberg, 1966, J. Phys. & Chem. Solids **27**, 1727.
Song, L.W., B.W. Benson and G.D. Watkins, 1986, Phys. Rev. B **33**, 1452.
Song, L.W., X.D. Zhan, B.W. Benson and G.D. Watkins, 1988, Phys. Rev. Lett. **60**, 460.
Sritrakool, W., V.Sa-Yakanit and H.R. Glyde, 1986, Phys. Rev. B **33**, 1199.
Stark, J.P., 1980, Phys. Rev. B **21**, 556.
Stoneham, A.M., 1975, Theory of Defects in Solids (Clarendon Press, Oxford).
Stoneham, A.M., 1981, Rep. Prog. Phys. **44**, 1251.
Street, R.A., J. Zesch and M.J. Thompson, 1983, Appl. Phys. Lett. **43**, 672.
Taylor, W.E., N.H. Odell and H.Y. Fan, 1952, Phys. Rev. **88**, 867.
Thompson, G.H.B., 1980, Physics of Semiconductor Laser Devices (Wiley, Chichester).
Thurmond, C.D., 1975, J. Electrochem. Soc. **122**, 1133.
Vainus, B., 1991, J. Phys.: Condens. Matter **3**, 3941.
van Cong, H., and S. Brunet, 1985, Solid-State Electron. **28**, 587.
van Roosbroeck, W., and W. Shockley, 1954, Phys. Rev. **94**, 1558.
van Vechten, J.A., and C.D. Thurmond, 1976, Phys. Rev. B **14**, 3539.
Vinetskii, V.L., 1983, Sov. Phys. Solid State **25**, 665.
von Roos, O., 1983, J. Appl. Phys. **54**, 1390.
Wang, K.L., 1979, IEEE Trans. Electron. Dev. **ED-26**, 819.
Wintle, H.J., 1988, J. Phys. D **21**, 1472.
Wu, C.-h., and R.H. Bube, 1974, J. Appl. Phys. **45**, 648.
Würfel, P., 1982, J. Phys. C **15**, 3967.
Yamasaki, K., M. Yoshida and T. Sugano, 1979, Jpn. J. Appl. Phys. **18**, 113.
Yoshida, M., 1976, Jpn. J. Appl. Phys. **15**, 2261.
Yoshida, M., 1980, Jpn. J. Appl. Phys. **19**, 1897.

CHAPTER 7

Surfaces and Interfaces: Atomic-Scale Structure, Band Bending and Band Offsets

L. J. BRILLSON

Xerox Webster Research Center
Webster, NY 14580
USA

Handbook on Semiconductors
Completely Revised Edition
Edited by T.S. Moss
Volume 1, edited by P.T. Landsberg

Contents

1. Introduction . . . 284
2. Semiconductor surfaces . . . 285
 2.1. Geometric structure . . . 286
  2.1.1. Characterization . . . 286
  2.1.2. Relaxation and reconstruction . . . 291
  2.1.3. Domains, steps, and defects . . . 300
 2.2. Chemical structure . . . 305
  2.2.1. Crystal growth . . . 306
  2.2.2. Kinetics of growth, evaporation, and diffusion . . . 307
  2.2.3. Etching . . . 309
 2.3. Electronic structure . . . 310
  2.3.1. Characterization and theory . . . 310
  2.3.2. Localized states and surface band structure . . . 314
  2.3.3. Reconstruction dependence . . . 318
3. Adsorbate–semiconductor and metal–semiconductor interfaces . . . 320
 3.1. Geometric structure . . . 321
  3.1.1. Adsorbate bonding . . . 321
  3.1.2. Epitaxical overlayers . . . 324
 3.2. Chemical structure . . . 327
  3.2.1. Overlayer growth . . . 328
  3.2.2. Diffusion, chemical reaction, and phase formation . . . 330
  3.2.3. Atomic-scale control . . . 337
 3.3. Electronic structure . . . 340
  3.3.1. Schottky barrier formation . . . 341
  3.3.2. Interface states . . . 345
  3.3.3. Interface state characterization . . . 361
  3.3.4. Atomic-scale control . . . 365
4. Semiconductor–semiconductor interfaces . . . 374
 4.1. Geometric structure . . . 376
  4.1.1. Epitaxical growth . . . 376
  4.1.2. Dislocations and strain . . . 378
 4.2. Chemical structure . . . 379
  4.2.1. Interdiffusion and chemical reaction . . . 379
  4.2.2. Template structures . . . 381

4.3. Electronic structure 382
4.3.1. Heterojunction band offsets 383
4.3.2. Band offset characterization 385
4.3.3. Interface dipoles 390
4.3.4. Atomic-scale control 397
5. Synopsis 399
References 401

1. Introduction

Semiconductor surfaces and interfaces have geometric, chemical, and electronic properties that differ substantially from their bulk counterparts. Extensive research of these microscopic properties over the past few decades has established their major contribution to macroscopic measurements of semiconductor electronic and optical properties. A comprehensive treatment of the huge array of observations for all semiconductor surfaces and interfaces would require several volumes. This chapter focuses primarily on the atomic-scale properties of surfaces and interfaces and, in particular, on the interrelationships between geometric, chemical, and electronic structures. Even with this restriction, space allows presentation of only the salient properties and physical processes involved. The main emphasis of this chapter will be on the movement of atoms away from their regular bulk lattice positions at semiconductor interfaces and the strong effect this movement has on their localized and extended electronic states. In turn, such microscopic changes manifest themselves on a macroscopic device level and indeed provide opportunities to tailor electronic properties via atomic-scale techniques.

The electronic properties of semiconductor surfaces and interfaces offer considerable evidence of their dependence on geometric and chemical features. For the semiconductor–vacuum interface, atomic rearrangements can occur that create or remove localized states inside the forbidden band gap. For metal–semiconductor or semiconductor–semiconductor interfaces, chemical reaction and diffusion is commonly observed, substantially changing the conventional electronic band structure used in describing transport and optoelectronic phenomena. A major consequence is that the interface electronic properties depend sensitively upon the particular surface preparation before contact, the nature and strength of chemical interaction between particular constituents, and the specifics of any subsequent processing. In order to derive measureable electronic properties for such systems, one must first know where the atoms and their charge moments are. In fact, it is possible to obtain extended, lattice-ordered, unstrained, and abrupt interfaces with which to obtain well-defined electronic properties. For epitaxial semiconductor (and even some metal) overlayers, it is possible to extract such atom and charge information in order to calculate band- and localized-state properties. Here again, the specifics of interface formation can play a major role in determining the particular geometric and chemical structures of the epitaxial interface. The next section of this chapter contains a description of the bonding changes of semiconductor surface atoms and their associated electronic features. This description includes the major techniques available to characterize the surface geometric, chemical, and electronic structures on an atomic scale. The key aspects of relaxation and reconstruction appear in the description of the geometric structure, along with the additional complexity introduced by domains, steps, and defects. Crystal growth, the dynamics of surface growth, diffusion, and evaporation, as well as etching, provide the context for the discussion of clean-surface chemical structure. For the electronic structure, this section includes an overview of the associated experimental and theoretical techniques available for analysis, the characteristic surface state and band structure features observed at clean surfaces, and the interrelationship between the electronic and the geometric structures.

Section 3 addresses both adsorbate– and metal–semiconductor interfaces, introducing a variety of microscopic chemical phenomena that can alter local atomic bonding and contribute to spatially-extended electronic interfaces. Included are the systematics of chemical reaction and diffusion on a microscopic scale and their relation to the observed Schottky barrier heights. The variety of Schottky barrier models used to account for measured semiconductor band bending appears here along with key observations in support of each. Of greater practical significance are the variety of atomic-scale approaches now demonstrated to alter and, in some cases, control Schottky barrier heights on a macroscopic scale. Not only do these phenomena provide new device strategies, but they serve to differentiate between many of the competing models for interface charge transfer and band bending. Metal–semiconductor interfaces yield the most diverse and detailed information relating atomic-scale geometric and chemical structures to electronic properties. Accordingly, a substantial portion of this chapter is devoted to the presentation of these features.

Section 4 addresses the analogous geometric, chemical, and electronic properties of semiconductor–semiconductor interfaces. Geometric considerations center on the lattice matching across the interface and the resultant quality of semiconductor film growth. Chemical properties are restricted to the cross-bonding and dipole formation that depend sensitively on the specifics of growth and processing. Particular emphasis is placed on heterojunction band offsets – the alignment of the different semiconductor band structures at their interface, and their relation to atomic-scale structural features. Appearing here are a variety of band offset measurement techniques and alternative theoretical models for their prediction. Finally, this section includes recent approaches to altering the heterojunction band offset by atomic-scale modifications of the interface.

Finally, § 5 summarizes the major points of each subsection and highlights new directions in characterizing and controlling semiconductor surface and interface properties.

2. Semiconductor surfaces

Semiconductor surfaces exhibit a rich diversity of phenomena, which provide insight into the physics, chemistry, electronics, and materials science of condensed matter on an atomic scale. The properties of semiconductor surfaces affect numerous macroscopic features and are critical in applications ranging from microelectronics to solar cells to catalysis. The clean semiconductor surface also exhibits a base set of geometric, chemical, and electronic properties with which to understand the adlayer, metal, and semiconductor interfaces.

2.1. Geometric structure

The geometric structure of semiconductor surfaces offers a natural introduction to the different properties of the semiconductor–vacuum interface compared with the bulk. This section provides an overview of (i) the characterization tools available for structural analysis, (ii) the relaxation and reconstruction features that alter the bulk-lattice bonding near the semiconductor surface, and (iii) the deviations from regular planar geometry due to domains, steps, and defects.

2.1.1. Characterization

Numerous experimental and theoretical techniques have evolved over the past few decades to characterize and understand semiconductor surfaces and interfaces on an atomic scale. Driving these developments has been the need to understand the numerous physical phenomena uncovered as well as the technological value of the resultant properties for microelectronic and optoelectronic device applications. For the experimentalist, the advent of ultrahigh-vacuum (10^{-10} Torr) technology provided the means to prepare and preserve surfaces and interfaces long enough (hours) for detailed characterization. Table 1 lists ultrahigh-vacuum electronic and optical techniques for characterizing semiconductor surfaces and interfaces. Detailed descriptions of many or most of this family of techniques are available in a wide array of review articles, monographs, and books. See, for example, Czanderna 1975, Brillson 1982a,b, Zangwill 1988, Van Hove et al. 1989, Schroeder 1990. These techniques involve various combinations of incoming and outgoing electrons, ions, and photons as well as electrostatic techniques. In general, a combination of such techniques is desirable in order to obtain a relatively unambiguous physical interpretation of the surface or interface properties measured. Electron spectroscopy techniques provide an extremely high surface sensitivity by virtue of the short mean-free path of electrons with kinetic energies in the 50–100 eV range. As shown in fig. 1, the inelastic collison scattering length λ_e, and therefore the electron scattering length, exhibits a minimum near 50–100 eV, such that only electrons within 4–6 Å of the surface can escape into vacuum without energy loss (cf. Seah and Dench 1979). This inelastic mean-free path increases significantly at much higher or lower energies ($E_k \lesssim 30$ eV); λ_e can vary significantly from material to material, especially for semiconductors whose band gaps reduce the probability of scattering by electron–hole-pair creation. For higher electron kinetic energies ($E_k \gtrsim 100$ eV), λ_e exhibits an $E^{1/2}$ dependence. Therefore, spectroscopy techniques involving electrons leaving a semiconductor surface can possess high surface sensitivity if the exiting electron's energy falls in the λ_e minimum. Figure 2 illustrates schematic energy diagrams of the optical and electronic excitation processes for four such techniques: UV photoelectron spectroscopy (UPS), X-ray photoelectron spectroscopy (XPS), Auger electron spectroscopy (AES), and low-energy electron-loss spectroscopy (LELS). UPS samples electrons excited by photons

Table 1
Ultrahigh-vacuum electronic and optical techniques for characterizing semiconductor surfaces and interfaces and the associated properties they measure.

Technique	Information
Auger electron spectroscopy (AES)	Surface chemical composition, depth distribution
X-ray photoemission spectroscopy (XPS)	Surface chemical compositon and bonding
UV photoemission spectroscopy	Fermi level with respect to band edges, work function, valence-band states
Soft X-ray photoemission spectroscopy (SXPS)	Surface chemical composition and bonding, Fermi level with respect to band edges, valence-band states
Constant initial (CIS) and final (CFS) state spectroscopies	Empty states above Fermi level
Angle-resolved photoemission spectroscopy (ARPES)	Atomic bonding symmetry, Brillouin zone dispersion
Surface extended X-ray absorption fine structure (SEXAFS)	Local surface bonding coordination
Inverse photoemission spectroscopy	Unoccupied surface state and conduction-band states
Laser-excited photoemission spectroscopy (LAPS)	Band gap states
Low-energy electron (LEED) and positron (LEPD) diffraction	Surface atomic geometry
X-ray diffraction	Bulk atomic geometry
Total external X-ray diffraction (TEXRD)	Interface lattice structure, interface strain
Low-energy electron-loss spectroscopy (LELS)	Interface reactions, electronic and atomic excitations
Surface photovoltage spectroscopy (SPS)	Band gap states, work function, band bending
Infrared absorption spectroscopy (IR)	Band gap states, atomic bonding and coordination
Cathodoluminescence spectroscopy (CLS)	Surface states within band gap, buried interface states, new compound band gap energies
Photoluminescence spectroscopy	Surface chemical compounds, states within band gap
Surface reflectance spectroscopy (SRS)	Surface dielectric response
Ellipsometry	Surface or interface dielectric response
Surface photoconductivity spectroscopy	States within band gap
Raman scattering spectroscopy	Interface compounds and bonding, band bending
Rutherford backscattering spectroscopy (RBS)	Surface atomic geometry, depth distribution
Secondary ion mass spectroscopy (SIMS)	Interface chemical composition, depth distribution
He beam scattering	Energy transfer dynamics, surface charge density
Scanning tunneling microscopy (STM)	Surface atomic geometry, surface morphology, filled- and empty-state geometries
Atomic force microscopy (AFM)	Surface electrostatic forces, magnetic polarization
Scanning tunneling spectroscopy (STS)	Band gap states, heterojunction band offsets
Ballistic electron energy microscopy (BEEM)	Barrier heights, heterojunction band offsets, barrier height lateral inhomogeneity
Field ion microscopy (FIM)	Surface atomic motion, atomic geometry
High-resolution transmission electron microscopy (HRTEM)	Interface lattice structure
Low-energy electron microscopy (LEEM)	Surface morphology, diffusion, phase transformations, grain boundary motion
Electron paramagnetic resonance (EPR)	Unpaired electron spins

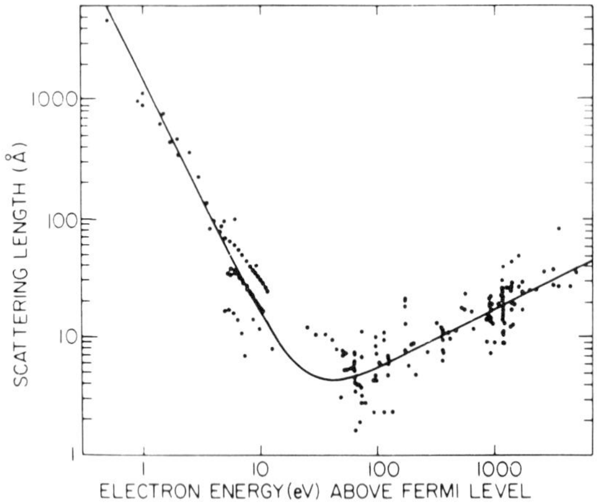

Fig. 1. Inelastic collision scattering length λ_e in angstroms of an electron in a solid as a function of energy above the Fermi level. The solid curve represents an empirical least-squares fit over the entire energy range for a range of elements. The pronounced minimum near 50–100 eV affords a high surface sensitivity for a variety of electron excitation techniques (after Seah and Dench 1979).

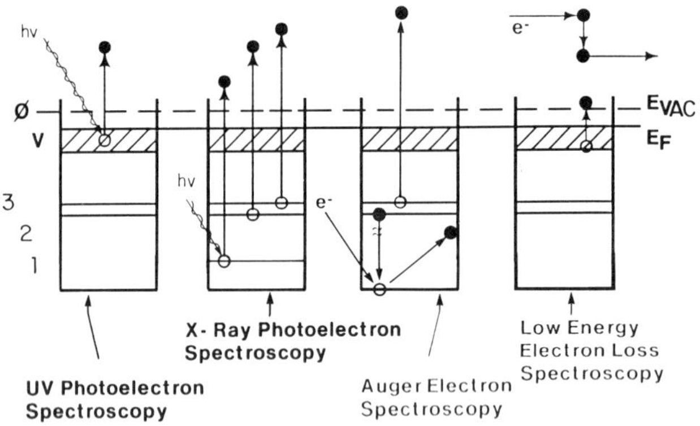

Fig. 2. Schematic diagrams of the optical and electronic excitation processes corresponding to: (a) UV photoemission, (b) X-ray photoemission, (c) Auger electron spectroscopy, and (d) low-energy electron-loss spectroscopy. In each case electron energies near the escape depth minimum shown in fig. 1 result in a high surface sensitivity.

from high-lying valence band states, XPS from these as well as deeper, core level, states. Here

$$E_k = h\nu - E_B - (E_{vac} - E_F), \tag{2.1}$$

where $h\nu$ is the incident-photon energy and E_B is the binding energy relative to E_F, the Fermi level. E_{vac} is the vacuum level, which the electron must exceed in order to escape from the solid. Likewise, XPS involves photoexcitation of an electron from a core hole deeper in energy, using photons of much higher energy. For AES, an incident electron rather than a photon creates a core hole and a second electron from a higher energy level fills the core hole, exciting a third electron to leave the solid. The kinetic energy of this ejected electron is

$$E^{123}_{\text{Auger}} = E_1(Z) - E_2(Z) - E_3(Z + \Delta) - E_{\text{vac}}, \tag{2.2}$$

where E_1, E_2, and E_3 are energy levels in the solid, Z is the atom's nuclear charge and Δ accounts for the doubly ionized state of the parent atom. For LELS, excitation within the solid causes an energy loss to an incident electron of an amount characteristic of the excitation. For XPS and ELS, one can adjust the incoming electron's excitation energy to move the exiting electron energy into the λ_e minimum. For AES, selection of transitions E^{123}_{Auger} with energies in the 100–1000 eV regime can provide a moderately high surface sensitivity as well. In addition to illustrating the principle of surface sensitivity, these spectroscopies represent the most common surface-science tools for characterizing surface composition (AES, XPS), energy band structure (UPS, XPS), and surface vibrational and electronic excitations (LELS).

Diffraction techniques such as low-energy electron diffraction (LEED) (fig. 3a) and reflection high-energy electron diffraction (RHEED) (fig. 3b) also achieve surface sensitivity by the λ_e dependence, whereas microscopy methods require highly localized probes either normal or parallel to the interface plane. Electrostatic techniques rely on the high sensitivity of the surface space charge to the relatively low densities of the surface or interface charge. Electron- or photon-generated electron–hole-pair recombination within the surface space-charge region can also provide interface-

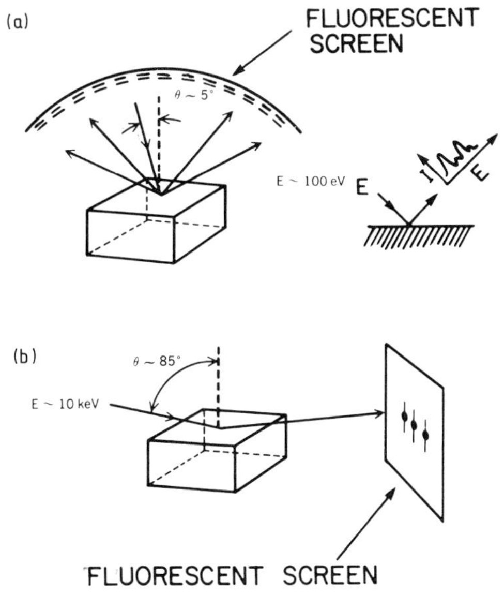

Fig. 3. Schematic geometry of (a) the low-energy electron diffraction (LEED) and (b) reflection high-energy electron diffraction (RHEED) techniques. The inset in (a) describes the dependence of the LEED spot intensity on the diffracted-electron energy (after Duke and Park 1972).

sensitive luminescence features. Each of these approaches is discussed in more detail in the appropriate subsections to follow.

Within the past decade, scanning tunneling microscopy (STM) has also become a standard tool for characterizing semiconductor surface geometry. The STM technique can provide a real-space surface image with atomic resolution by positioning a sharp metallic tip only a few angstroms from the surface under investigation and monitoring the tunneling current between tip and substrate as a function of lateral position. This tunneling current results from the finite probe and specimen wavefunction extension on an atomic scale; see fig. 4a. In the usual experimental configuration pictured in fig. 4b, the tunneling current in this two-terminal device provides a feedback signal suitable for controlling the tip spacing and charting the surface atomic morphology. Alternatively, one may hold the tip–substrate potential at a constant voltage to obtain images of filled and unfilled states across the surface. Indeed, it is necessary to interpret such maps with care since the tunneling current reflects both the surface atomic position and the projection of the surface atomic wavefunctions into the vacuum (Demuth et al. 1988). Scans of tunneling current versus tip–surface voltage can provide the distribution of filled and empty states near the band gap energies (cf. Feenstra 1990). STM has demonstrated a resolution of about 2 Å in the plane of the substrate and about 0.1 Å vertically. The STM technique is not limited to the solid–vacuum interface, unlike electron spectroscopy techniques, although vacuum conditions yield the highest spatial resolution. The well-defined geometrical features provided by STM along with the capability to determine the local electronic structure of these same features have made STM perhaps the leading new tool in elucidating the surface structure.

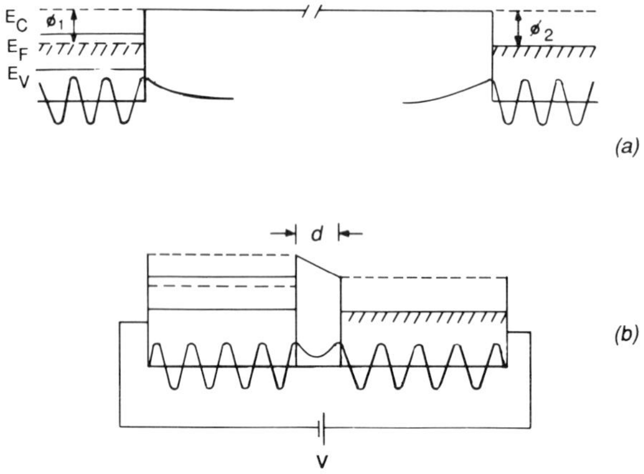

Fig. 4. Schematic energy diagram of the wavefunction tunneling that underlies the scanning tunneling microscope (STM) technique. Here the finite extent of the filled-state tails between the semiconductor and metal (probe tip) permit a finite current flow. Typical STM maps involve scanning a sharp probe tip across a surface at constant current or at a constant distance d, using a feedback technique for real-time compensation. Depending on the bias, the technique probes either filled or unfilled states (adapted from Zangwill 1988).

While LEED has historically provided the vast majority of surface structure data, supplemented to a considerable extent in recent years by STM, a number of other diffraction techniques have provided information. These include: UHV high-resolution transmission electron diffraction (HRTEM), glancing angle X-ray diffraction, and low-energy positron diffraction (LEPD). In addition, ion beam techniques such as He atom scattering and ion beam scattering are especially useful in measuring displacements parallel to the surface. However, the tabulations presented here are largely the result of LEED and STM studies.

2.1.2. Relaxation and reconstruction

Atoms at ordered surfaces change their positions in two ways to reflect the discontinuity of the crystal lattice. First, smoothing of the surface electron distribution alters the electrostatic equilibrium of the ion cores, resulting in a surface relaxation that modifies the bond angles and lengths of the surface atoms to lower the overall free energy. Generally, the in-plane size and shape of the unit cell structure remains the same as the bulk structure. The ionic character of the chemical bond is a major factor in determining the bulk crystalline structure of a solid (Phillips 1970). For semiconductors, the directional nature of the chemical bonds between atoms favors the tetrahedral coordination of the zinc blende and wurzite structures (Phillips 1970, Zangwill 1988). Prepared under different conditions, the surfaces of such lattice structures can assume a variety of complex atomic bonding arrangements, which modify the size and shape of the unit cell. Such reconstructions depend sensitively on the surface stoichiometry, orientation, and thermal/ambient processing history. In addition, multilayer atom displacements can occur as part of the overall minimization of the elastic and electronic energy.

The LEED technique provides a sensitive method of determining the dimensions of the surface unit cell and thereby any changes due to relaxation or reconstruction. Each layer of the crystal comprises a regular, two-dimensional array which acts as a diffraction grating. The crystal can be viewed as a set of such geometrically equivalent layers parallel to a given surface. Stacked together, these layers form a grating array, which reflects a low-energy beam of electrons into a set of diffracted beams. The resultant diffraction pattern can be imaged on a fluorescent screen, as represented in fig. 3a and pictured in fig. 5. A system of retarding and accelerating grids between the crystal and the screen act as a filter to pass and display electrons of a selected energy. A diffraction pattern also results when high-energy electrons (~ 10 keV) are reflected at glancing incidence from a crystal surface (fig. 3b). This configuration is used for determining the surface order during molecular-beam epitaxial growth of semiconductor heterojunctions.

The spatial distribution of the diffracted beams relative to the incident beam depends on the translational symmetry of the atoms in the outermost crystal layer. Conservation of momentum parallel to the surface requires that

$$\boldsymbol{k}_{\mathrm{i},\parallel} = \boldsymbol{k}_{\mathrm{f},\parallel} + \boldsymbol{g}, \tag{2.3}$$

where $\boldsymbol{k}_{\mathrm{i},\parallel}$ and $\boldsymbol{k}_{\mathrm{f},\parallel}$ are the components parallel to the surface of the incident and scattered wave vectors, respectively, and $\boldsymbol{g}$ is a vector of the two-dimensional recipro-

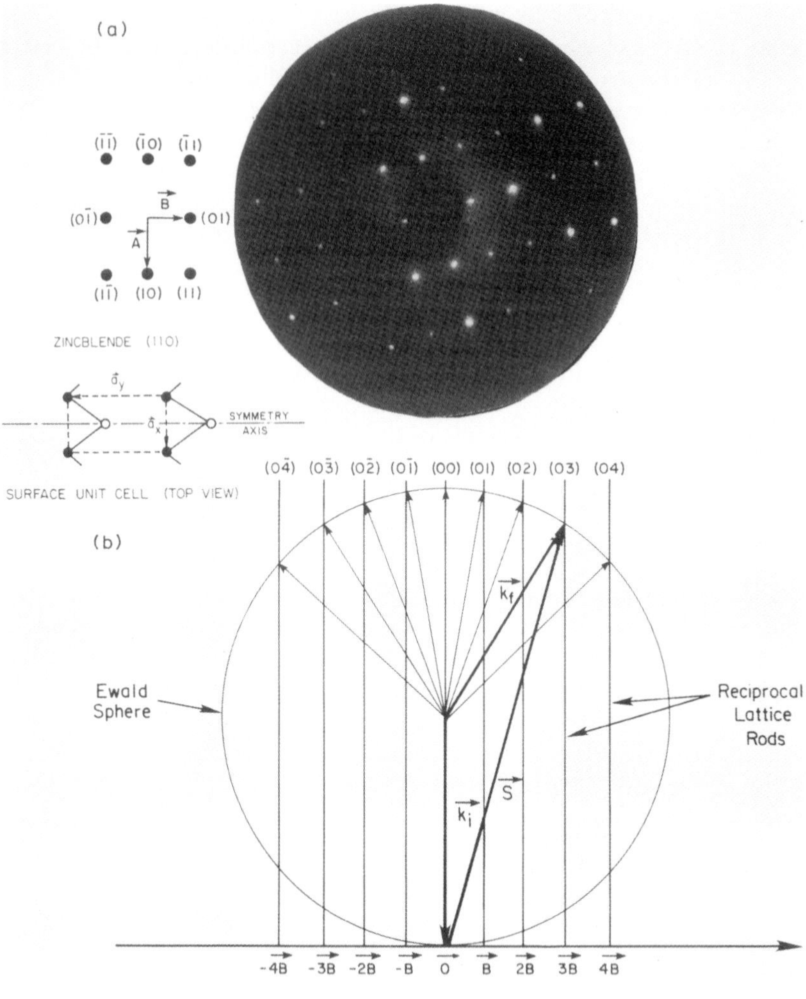

Fig. 5. Correspondence of the elastic LEED pattern to the two-dimensional atomic positions within the surface unit cell and their reciprocal lattice rods. (a) Photograph of the cleaned GaAs(110) normal-incidence LEED pattern at 150 eV electron primary energy. The beam indexing is given on the left-hand side of the picture along with the zinc blende (110) unit cell. (b) Construction of the Ewald sphere showing the rods of the 2D reciprocal lattice, the incident wave vector $\boldsymbol{k}_i$, the diffracted wave vector $\boldsymbol{k}_f$ corresponding to the (03) beam and the momentum transfer $\boldsymbol{S}$. $\boldsymbol{A}$ and $\boldsymbol{B}$ are reciprocal lattice vectors (after Kahn 1983).

cal lattice (Bravais lattice) associated with the surface plane (Pendry 1974). Here

$$\boldsymbol{k}_{i,\parallel} = (2mE/\hbar^2)^{1/2} \sin \Theta \tag{2.4}$$

for electrons of energy E and momentum $\boldsymbol{k}$ on a planar surface at an angle Θ relative to its exterior normal. The reciprocal lattice vector $\boldsymbol{g}$ can be expressed in terms of the reciprocal lattice's primitive vectors, $\boldsymbol{A}$ and $\boldsymbol{B}$, as

$$\boldsymbol{g} = h\boldsymbol{A} + k\boldsymbol{B}, \tag{2.5}$$

where h and k are integers. This expression combined with eqs. (2.3) and (2.4) relates the angles of the diffracted beams to the incident beam energy and the dimensions of the surface unit cells. In this way, one can extract the dimensions of the surface unit mesh from the energy dependence of diffracted beam angles along symmetry directions (Pendry 1974). Figure 5 illustrates a simple LEED pattern, its correspondence to the unit cell of the semiconductor surface (here, GaAs(110)), and to the reciprocal lattice rods $\boldsymbol{A}$ and $\boldsymbol{B}$, defined by the unit cell lattice parameters a_x and a_y (Kahn 1983). The Ewald sphere provides a geometric representation of momentum conservation in terms of the incident and final electron wavevectors and the reciprocal lattice rods. The rod intersections with the Ewald sphere (Henzler 1977) determine the direction of each diffracted beam, correspondingly termed (hk) and shown above the surface unit cell in fig. 5. Thus the diffracted spot pattern is a sensitive indicator of the surface atomic geometry.

In order to determine the three-dimensional unit cell structure within each layer and the geometry of these layers with respect to each other, one must analyze the diffracted-beam intensity as a function of energy. Figure 3a illustrates schematically the energy dependence of a particular diffracted beam, which can be measured from the fluorescent image intensity. Such an analysis is necessary because the presence of the surface destroys the translational invariance of the crystal normal to the surface: i.e., the unit cell structure cannot be obtained from the geometry of the diffracted beams alone. Further, extension to low-energy electrons of the theory of X-ray diffraction is not appropriate for describing the diffracted-beam intensities, since the electrons exhibit multiple scattering. The same strong electron–solid interaction responsible for the λ_e minimum in fig. 1 and the high surface sensitivity of LEED also result in large cross sections for elastic electron–ion-core scattering, necessitating the use of multiple scattering or "dynamical" models to interpret the energy (and temperature) dependence of LEED spot intensities. Several reviews are available of such models (Van Hove and Tong 1979, Duke 1974, Pendry 1974) and their application to the determination of the atomic structures of semiconductor surfaces (Kahn 1983, Somorjai and Van Hove 1979, MacLaren et al. 1987). In recent years, low-energy positron diffraction has proved to be an effective structural probe as well (Canter et al. 1990).

As a representative example of surface relaxation, fig. 6 illustrates the atomic geometry of the (110) surfaces of compound semiconductors with the zinc blende structure and a set of independent structural variables that characterize them. The top-layer tilt angle, ω_1, is defined by $\omega_1 = \tan^{-1}[(\Delta_{1,\perp}/(a_y - \Delta_{1,y})]$,where $\Delta_{1,y}$ is a vector shear between anion and cation along the y-axis and $\Delta_{1,\perp}$ is a normal vector shear as shown. For GaAs, a variety of surface structure analysis techniques have confirmed and refined this structure, where an approximately bond-length-conserving rotation (ω_1 in side view) of the top-layer Ga and As atoms causes a buckling of the surface zig-zag chains (top view). For a three-dimensional representation, see MacLaren et al. (1987). Tables of reference for these and other structures can be found in Van Hove et al. (1989). Surface structure analyses of the top-layer chain buckling for other III–V compound semiconductors suggest that surface bond lengths are conserved and $\omega_1 = 29 \pm 3°$ for all materials studied to date (Duke 1988) – with

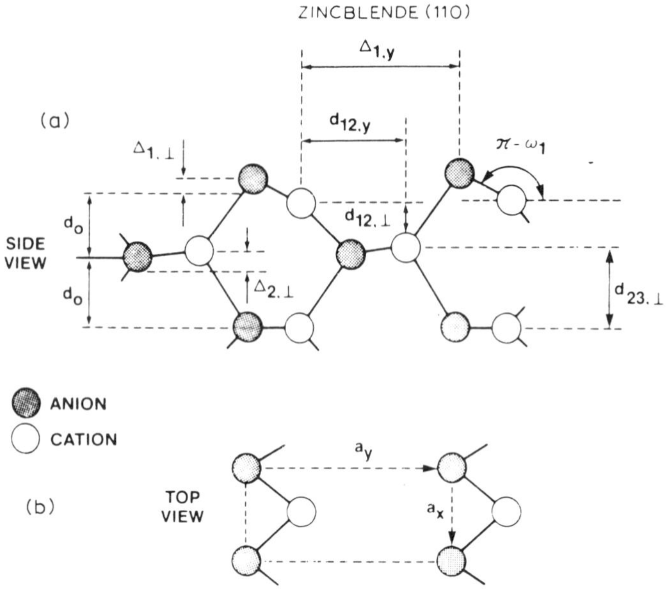

Fig. 6. Schematic representation of the surface relaxation for (110) surfaces of compound semiconductors with the zinc blende structure, along with the independent structural variables that characterize the atomic geometries. The top layer tilt angle, ω_1, is defined by $\omega_1 \equiv \tan^{-1}[\Delta_{1,\perp}/(a_y - \Delta_{1,y})]$ (from Duke 1988).

only a small, analysis-dependent variation with lattice dimensions if the y-axis values are allowed to vary (Duke 1983). This bond rotation is regarded as the primary characteristic of the atomic geometries of the (110) surfaces of all compound semiconductors with the zinc blende structure, with the anion rotating outward to a prismatic coordination and the cation rotating inward to a nearly planar coordination (Duke 1988). The driving force for this relaxation is a lowering of a dangling bond on the column-V element and a raising of a dangling bond on the column-III element, which minimizes the surface energy. In addition to surface state lowering, such relaxation can induce a filling of group-V, and an emptying of group-III, dangling bonds on polar surfaces (Applebaum et al. 1976, Chadi 1978, Harrison 1979, Biegelsen et al. 1990a). The atomic structure illustrated in fig. 6 appears to be universal for (110) cleavage surfaces of all zinc blende III–V and II–VI materials (Duke and Wang 1988). Furthermore, cleavage $(10\bar{1}1)$ and $(11\bar{2}0)$ surfaces of the wurzite II–VI compound semiconductors exhibit (i) a similar surface state lowering with relaxation, (ii) a stronger relaxation dependence on the bulk lattice constant, and (iii) characteristic surface-bonding surface states near the top of the valence band (Duke and Wang 1988). Overall, relaxations at these extensively-studied compound semiconductor surfaces exhibit regular and characteristic features, indicating that surface bond energy minimization depends primarily on the surface geometric structure rather than the detailed electronic structure (Duke 1988).

Both compound and elemental semiconductors provide illustrative cases of surface reconstruction. A common designation for a reconstructed surface is (Wood 1964)

$$M(hkl) - \alpha[(a_s/a) \times (b_s/b)]\zeta - S,$$

where $M(hkl)$ refers to the (hkl) face of a substrate M, α specifies the primitive (p) or centered (c) unit meshes of the composite system, and $[(a_s/a) \times (b_s/b)]$ denotes the ratios of the sides of the unit overlayer mesh vectors to those of the substrate mesh. In the case of an adsorbate, S is the overlayer element and ζ denotes the angle between the overlayer and substrate unit mesh vectors $\boldsymbol{a}_s$ and $\boldsymbol{a}$. For example, the cleaved, reconstructed Si(111) surface is termed (2 × 1) because its surface mesh is doubled along one direction, halving the distance between the diffraction spots along the same direction of this otherwise square array. A much more complicated reconstruction of this face is illustrated by the (7 × 7) pattern shown in fig. 7a. For this annealed (111) surface of Si, the brightest spots are "normal" diffraction and the

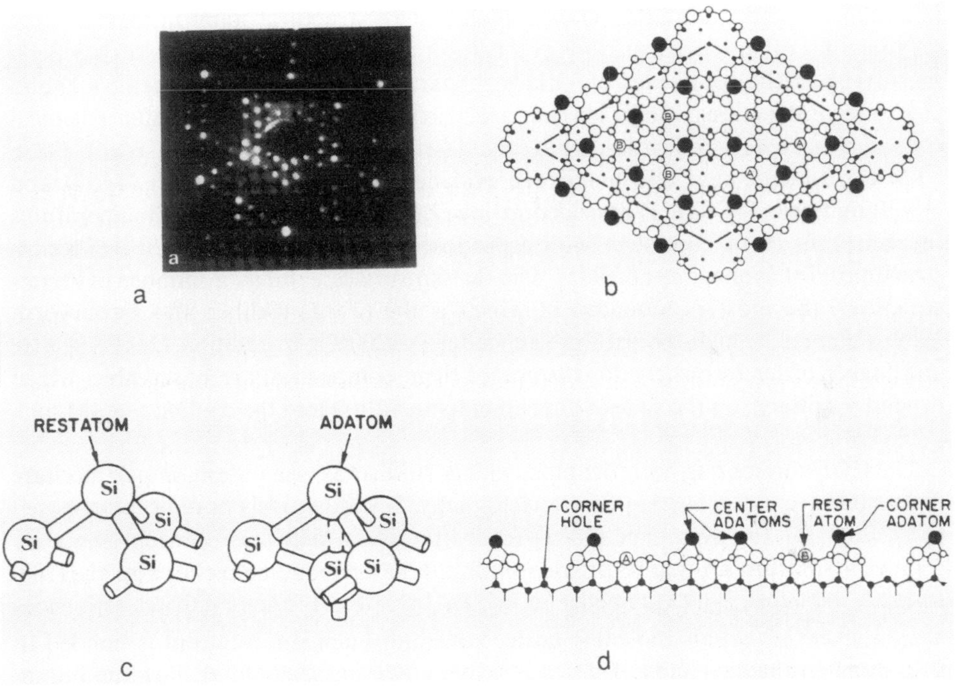

Fig. 7. Schematic representation of the surface reconstruction for the clean Si(111) (7 × 7) surface. (a) Characteristic LEED diffraction pattern (after Lander and Morrison 1964). (b) Top view of the dimer-adatom-stacking fault (DAS) model for Si(111)-(7 × 7) (Takayanagi et al. 1985). (c) Cross-sectional side view, taken along the long diagonal of the unit cell (after Golovchenko 1986), and showing the positions of the corner hole, center adatoms, rest atoms, and corner adatoms. The (7 × 7) unit cell containing the stacking fault is outlined. Circles of decreasing size indicate atoms of increasing distance from the surface. The large solid circles indicate the 12 adatoms. The circles A and B denote rest atoms in the faulted and unfaulted halves, respectively. (d) Ball and stick representation of the structure and bonding at rest atom and adatom sites.

other spots are one-seventh-order beams due to the surface structure. Figures 7b and c illustrate the dimer-adatom-stacking fault (DAS) proposed (Takayanagi et al. 1985) for the Si(111) (7 × 7) surface on the basis of transmission electron diffraction (TED) measurements. Here the reconstruction involves three double layers of atoms, with adatoms (large shaded circles) and two underlying double layers (large and small open, large and small filled circles). The cross-sectional view indicates a stacking fault within one part of the unit cell (left) producing a wurzite-type stacking of the outermost two double layers. The faulted and unfaulted top double layers are separated by a triangular network of partial dislocations with broken bonds partially healed by the formation of dimers, as well as holes at the crossing points of the dislocations, which expose the second double layer (Schlüter 1988). Results of STM (Binnig et al. 1983), LEED (Tong et al. 1990), transmission electron diffraction (Petroff and Wilson 1983), and ion channeling (Bennett et al. 1983) are consistent with this picture. A major implication of this currently accepted and long sought-after reconstruction is that extended, coherent features of the surface topology can play a role in minimizing the strain and the overall energy for structural stabilization.

Numerous other reconstructions of the Si surface have been reported which depend sensitively on the conditions of surface preparation. In addition to the (2 × 1) and (7 × 7) reconstructions, the Si(111) surface also exhibits a (1 × 1) type after chemical etching, adsorbate stabilization, or specific laser annealing (Zehner and White 1982). STM experiments have also uncovered evidence for domains of (5 × 5), (7 × 7), and (9 × 9) unit cells for Si(111) annealed at lower (~600°C versus 750°C) temperatures (Becker et al. 1986), consistent with a proposed family of $(2n-1) \times (2n-1)$ reconstructions (Takayanagi et al. 1985). The Si(100) surface exhibits a number of reconstructions, the most pronounced of which is the (2 × 1) (Schlier and Farnsworth 1959), obtained by high-temperature annealing (~1200°C). In addition, (2 × 2), (4 × 2), and higher order reconstructions (some of them composites) are observable, which depend sensitively on the surface annealing temperature and the cooling rate (Mönch et al. 1981).

The (100) surface has been the most widely studied Si orientation, in part because of its extensive use in current device technology. Several models have been proposed to account for the dominant (2 × 1) reconstruction which illustrate representative geometric structures. Figure 8 provides four alternative bonding pictures for the (100) surface: the ideal bulk geometry in fig. 8a involves a square lattice with sides $a_C/\sqrt{2}$ where a_C equals the cubic lattice constant. Each surface atom is bonded to two atoms in the next subsurface layer, leaving two unsaturated or dangling bonds. The relatively open structure of this surface permits several possibilities to account for the (2 × 1) reconstruction (cf. Schlüter 1988). These include (fig. 8b) a surface dimer model, where the top-layer atoms dehybridize to saturate or empty partially-filled dangling bonds (Schlier and Farnsworth 1959, Levine 1973, Chadi 1979a,b); (fig. 8c) a missing row model (Schlier and Farnsworth 1959, Poppendieck et al. 1978, Harrison 1976); and (fig. 8d) a conjugated-chain model (Seiwatz 1963) as well as several related modifications with more delocalized surface bonding (Northrup 1985, Pandey 1985). Recent STM observations of long dimer rows on this surface now

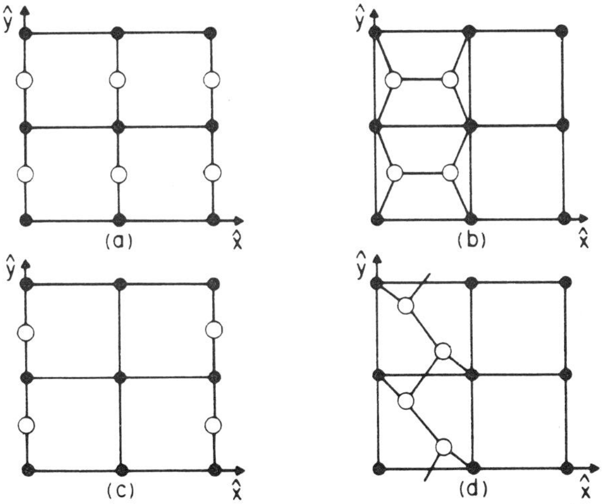

Fig. 8. Alternative top views of the reconstructed Si(100) outermost atomic layers according to: (a) the ideal bulk geometry; (b) the surface dimer model; (c) the missing row model; and (d) the conjugated-chain model. Open circles denote top-layer atoms, filled circles denote second-layer atoms (after Schlüter 1988).

confirm the dimer model (Tromp et al. 1985, Van der Veen 1985, Hamers et al. 1986a), both buckled (asymmetric) as well as unbuckled, in support of previous ion scattering. (Aono et al. 1982) and LEED (Yang et al. 1983, Holland et al. 1984) measurements. Further details can be extracted from electronic observations, which favor an asymmetric dimer reconstruction; see § 2.3. Conversely, the π-bonded chain model, involving a zig-zag pattern of connected dimers along the dimer row direction (Pandey 1981), appears to provide the best description of the cleaved Si(111) (2×1) surface (Mårtensson et al. 1985, Selloni and Del Sole 1986). As with the surface relaxations discussed earlier, the minimization of dangling bonds, taken with the overall strain, drives all these surface reconstructions.

Reconstructions of the III–V compound semiconductor surfaces emphasize the importance of specific growth conditions in determining the surface geometric structure. Figure 9 illustrates a map of GaAs(100) reconstructions measured by RHEED as a function of conditions during molecular-beam epitaxy (MBE) growth (Däweritz and Hey 1990). This surface is used more widely than any other in crystal growth and heteroepitaxy. Depending upon the As_4-to-Ga vapor flux ratio (beam equivalent pressures (BEP)) and substrate temperature, the growth surface exhibits a multiplicity of reconstructions as well as facetting, droplet formation and other surface degradations. Additional reconstructions result with successive annealings at different temperatures.

Reconstruction also depends on changes in the surface stoichiometry. Figure 10 shows the wide range in surface compositions as measured by SXPS Ga and As core level intensities from GaAs(100) surfaces characterized by LEED (Mönch 1985). The particular reconstructions are obtained by annealing in arsenic or vacuum in the 500–600°C temperature range, then cooling to room temperature. Additional, transi-

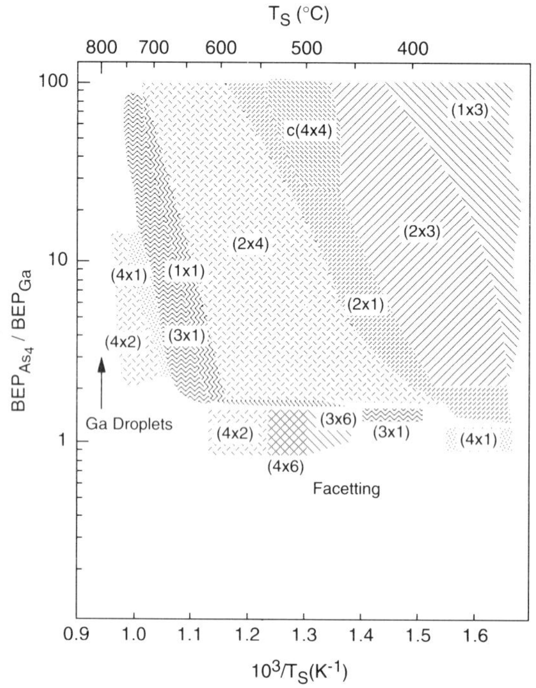

Fig. 9. GaAs(100) reconstructions measured by RHEED as a function of As_4-to-Ga vapor flux ratio (beam equivalent pressure (BEP)) and substrate temperature during MBE growth (after Däweritz and Hey 1990).

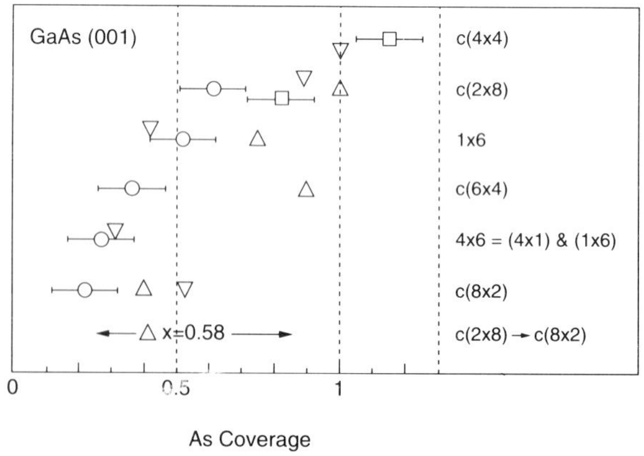

Fig. 10. Fractional surface coverage of As on GaAs(001) surfaces measured by AES and photoelectron spectroscopies as a function of various LEED reconstructions. The data points (Δ from van Bommel et al. (1978), ○ from Drathen et al. (1978), □ from Massies et al. (1980a), ▽ from Bachrach et al. (1981), and △ × from Arthur (1974)) illustrate a correlation between atomic structure and chemical composition (after Mönch 1985).

tional, reconstructions can be observed at high temperatures, for example, a (4 × 4) structure. The sequence of compositions has been well established, although the difficulty in obtaining the $c(8 \times 2)$ reconstruction (Bachrach et al. 1981) and its As-driven transitions to other reconstructions (van Bommel et al. 1978, Neave and Joyce 1978) has introduced some uncertainty with respect to its relative position (Mönch 1985). Conversely, fig. 10 indicates that reconstructions are observable over a range of surface stoichiometries. Such phenomena suggest that a particular reconstruction may in fact be either a composite of reconstructions or a structure involving more complexity than a regular extension of primitive cells.

The GaAs(001) (2 × 4)–$c(2 \times 8)$ reconstruction offers an instructive example of a geometric structure determined by the extended array of primitive cells. STM measurements provide a visualization of this reconstruction in fig. 11 (Pashley et al. 1988). The primary feature of this array is the unit cell consisting of three top-layer dimer pairs of As and a missing dimer. Starting with the unreconstructed arsenic-rich surface (fig. 11a), the As pairs dimerize along the [$\bar{1}$10] direction, giving rise to a 2× periodicity along [$\bar{1}$10]. The missing dimer results in a 4× periodicity along [110]. The unit cells can arrange themselves in domains, typically of about 100 Å across, with either in-phase or antiphase relations to each other. In-phase domain boundaries produce the (2 × 4) structure, antiphase boundaries the (2 × 8) structure. Thus the arrangement of basic (2 × 4) units determines the reconstruction. This

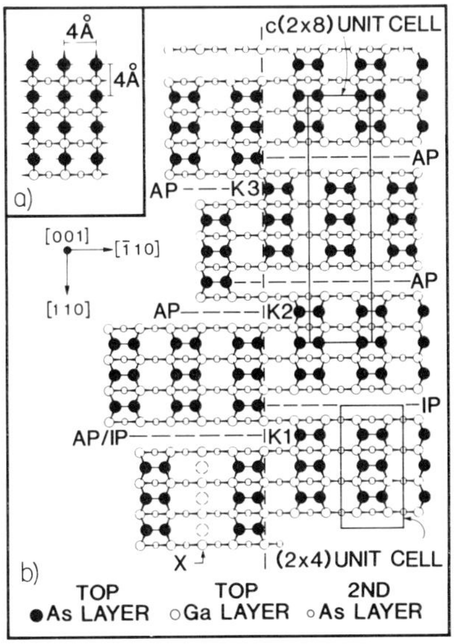

Fig. 11. Geometric structure for the GaAs(001) (2 × 4)–(2 × 8) reconstruction indicated by STM: (a) structure of the unreconstructed GaAs(100) arsenic-rich surface; (b) the missing dimer model for the GaAs(100) (2 × 4) surface producing (2 × 4) or (2 × 8) structures, depending on the in-phase (IP) or antiphase (AP) nature of the missing dimer boundary, respectively. Different boundary kinks – k_1, k_2 and k_3 – can arise, depending on the interaction of IP and AP domain boundaries. Disorder in the As pairing is denoted by X (after Pashley et al. 1988).

observation confirms the prediction of missing dimer arrays from a tight-binding-based total-energy calculation (Chadi 1987). In addition, several different kink features are observable, depending on the domain boundaries, as is disorder in the As pairing. This geometric structure also offers a natural explanation for the range of stoichiometry observable with the same LEED reconstruction.

Space does not permit a full description of the reconstructions for less well-studied surfaces of GaAs. However, it can be noted that several other features are believed to play a role in these structures, including adsorbed trimers (Biegelsen et al. 1990a,b), vacancies, and facets (McRae 1966). A wide variety of reconstructions and related preparation conditions for specific orientations of different semiconductors appear in table 2. Common reconstruction features occur for the same crystal orientation within the same family of III–V compounds. Additional features are evident for II–VI and other families of compound semiconductors (see Kahn (1983) and references therein), although much fewer data are available. Overall, this subsection serves to emphasize the wide variety of features that can contribute to the atomic geometry of the perfect, intrinsic semiconductor surface. In turn, these features depend on the energetics and strain associated with the detailed atomic coordinations, a topic to be addressed later in this section.

2.1.3. Domains, steps, and defects

In addition to the detailed bonding changes associated with relaxation and reconstruction, extrinsic features of the semiconductor surface introduce further complexities. For clean surfaces, these extrinsic features include multiple domains, steps, facets, and point defects. In addition to their possible effects on the electronic structure, such morphological features and related features on a macroscopic scale play a dominant role in crystal growth, epitaxy, and etching.

Domains with different geometric types of order can coexist on a scale of only tens of nanometers or less, as already illustrated in fig. 11. Depending on the domain size relative to the coherence length of an incident probe beam, typically several hundred angstroms, LEED provides a method for the systematic analysis of such domains in terms of spot broadening/splitting and a rise in background (Henzler 1982, 1985, Lagally 1982). Low-energy electron microscopy (LEEM) provides an effective means of observing the growth of domains, particularly at elevated temperatures (Bauer 1990), and underscores the importance of impurities, elastic strain, and dislocations, as well as other parameters. The LEEM technique is especially useful in understanding semiconductor growth kinetics.

Steps also break the translational invariance of the surface. Controlled misorientation of a crystal surface away from a low-index plane can provide a regular array of steps with an average step spacing defined by the misorientation angle and the step height. Such steps are desirable in crystal growth since they serve as nucleation sites that increase the rate of film growth. Figure 12 illustrates step arrays for a vicinal GaAs(001) surface cut in two, perpendicular vicinal directions (Pashley et al. 1991). For these GaAs(001)2° → [111]A and B surfaces, the STM image exhibits step spacings consistent with those expected from the vicinal angle. Each step has a single-layer height, but with terrace width variations from 50 to 120 Å.

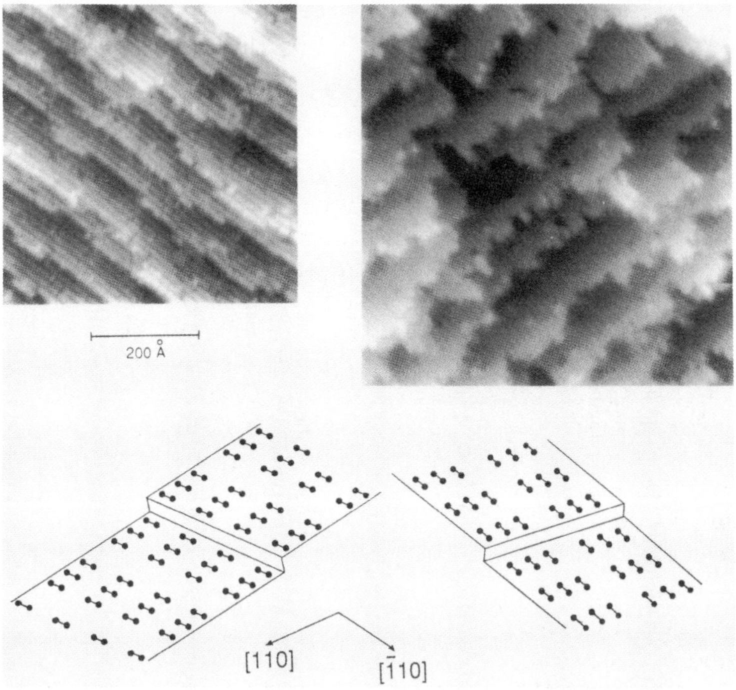

Fig. 12. Different in-step raggedness with surface orientation as indicated by STM filled-state images of steps on GaAs(001) misoriented 2° towards (a) the (111)A (Ga) surface and (b) the (111)B (As) surface. The corresponding schematic diagrams of the step edges with respect to the dimer direction appear below. For the GaAs(001)2° → [111]A, steps run parallel to the dimer rows and are relatively straight. For the GaAs(001)2° → [111]B, steps run perpendicular to the dimer rows and exhibit high kink densities (after Pashley et al. 1991).

In addition to disruption of the lateral invariance and introduction of dangling bonds, the steps introduce kink sites whose density depends on the misorientation direction. Figure 12 illustrates steps (a) along the dimer direction, which are relatively smooth, and steps (b) perpendicular to the dimer direction, which exhibit a high kink density. As shown, these kinks occur in units of 16 Å, i.e., in dimensions of the (2 × 4) unit cells. Higher spatial resolution STM images confirm the identity of these features. Since the (2 × 4) unit cell contains the correct number of electrons to fill the As dangling bonds and empty the Ga dangling bonds, this structure and extensions of such island structures along the 2x direction are stable. Such structures grow out from steps in the [110] misorientation and result in the ragged B-type step edges pictured in (b). In contrast, A-type edges are relatively straight, a result attributed to island cell structures growing out from A-type step edges having an insufficient number of electrons to fill all the As dangling bonds, and being thermodynamically less favorable to grow out from the steps (Pashley 1989). Hence local bonding associated with the reconstruction appears to be responsible for the contrast in the extended morphological features.

Table 2
Preparation techniques and corresponding reconstructions of the polar faces of various binary compounds (adapted from Kahn (1983) with permission).

Crystal	Bulk Structure	Surface	Preparation	ELEED pattern	Comments: structure	Selected refs.*
ZnO	Wurzite	(0001)Zn	Cleaving	1×1	Highly stepped, facets	[1, 2]
			ArB + anneal < 600°C	1×1	Top Zn layer contracted by 0.2 ± 0.1 Å	[3–5]
			ArB + 600 < anneal < 850°C	2×2; $\sqrt{3} \times \sqrt{3}$–30°	Patches with different reconstructions	[4]
		(0001)O	Cleaving	1×1	Highly stepped, facets	[4]
			ArB + anneal < 850°C	1×1	Relaxation unknown	[4]
CdS	Wurzite	(0001)Cd	ArB + anneal ⩽ 300°C	1×1	Sixfold symmetry; relaxation unknown	[6]
		(0001)S	ArB + anneal ⩽ 300°C	1×1 + weak $\sqrt{3}$–30°	Some evidence of facets	[6]
GaAs	Zinc blende	(100)	MBE, $T = 550$–660°C	$c(8 \times 2)$	Ga dimers	[7–9]
				4×6	Coexisting (2×6) and $c(8 \times 2)$ domains	[7–9]
				2×6	2 As dimer/4 missing As dimer cells	[7–9]
				"1×6"	(2×6) unit cell containing 2 As–As dimers	[9, 35]
				2×4	3 As dimer/1 missing As dimer cells; 2/2, 1/3 As dimer/missing As dimer	[9, 10, 30, 32]
				$c(2 \times 8)$	Antiphase array of (2×4) cells	[9, 10]
				$c(4 \times 8)$	Triple As dimer adatoms	[9, 31, 33]
				1×1	As saturated	[11, 12]
				2×1	$\frac{1}{2}$ monolayer of Ga vacancies	[10]
		(111)Ga	ArB + anneal 450°C	2×2		[11, 12, 13]
			MBE, $T = 377$°C	2×2	As stabilized, regular array of Ga vacancies	[13–15]
			ArB + anneal 550°C	2×2	$\frac{1}{4}$ missing Ga	[10]
		(111)As	ArB + anneal 450°C	3×3	(110) facets at $T \simeq 600$°C	[12]
			MBE, $T = 377$°C	2×2	Trimers on ideal As plane	[14, 16]
			MBE-grown + anneal 300°C	2×2	Unit cell: 7/4 As/Ga, adsorbed As trimer	[16]
			MBE-grown + anneal 500°C	$(\sqrt{3} \times \sqrt{3})$R30°	Intermediate between $\sqrt{19} \times \sqrt{3} \times$	[34]
			MBE, $T = 500$°C	$\sqrt{19} \times \sqrt{19}$–23.4°	Ga stabilized	[14]
			MBE-grown + anneal 300°C + anneal 500°C	$\sqrt{19} \times \sqrt{19}$–23.4°	As-capped hexagonal ring	[16]
GaP	Zinc blende	(111)Ga	ArB + anneal 550°C	2×2	$\frac{1}{4}$ missing Ga, nearly coplanar Ga–P bi-layer	[17]

InP	Zinc blende	(100)	ArB + anneal 330°C	4 × 1	Diffuse streaks	[18]
			Prolonged annealing	4 × 2		
GaSb	Zinc blende	(111)Ga	ArB + anneal 400°C	2 × 2		[12]
		(111)Sb	ArB + anneal 400°C	3 × 3	Facets for $T \simeq 600°C$	[12]
InSb	Zinc blende	(111)In	ArB + anneal	2 × 2		[11]
		(111)Sb	ArB + anneal	3 × 3		[11]
				2 × 2	Possible contamination	[19]
InAs	Zinc blende	(111)In	ArB + anneal	2 × 2		[20]
		(111)As	ArB + anneal	3 × 3	Facets form before superstructure	[20]
CdTe	Zinc blende	(111)Cd	ArB, no anneal	1 × 1	No comments on disorder	[21]
CoO	Cubic	(111)	Oxidation of Co(0001)	1 × 1	Top O layer contracted by 15%	[22]
MgO	Cubic	(111)	Chemical etch	1 × 1	(100) facets	[23]
NaO_2	CaF_2	(111)	ArB + anneal	1 × 1	Termination between 2 Na layers, bulk structure	[24]
TiO_2	Rutile	(100)	ArB + anneal + high T (1100°C)		(110) facets with (2 × 1) reconstruction (114) facets	[25]
MoS_2	Molybdenite	(0001)	Cleaving	1 × 1	S–Mo layer spacing contracted 5%, first Van der Waals gap contracted 3%	[26, 27]
$NbSe_2$	Molybdenite	(0001)	Cleaving	1 × 1	Se–Nb layer spacing contracted 0.6%, first Van der Waals gap contracted 1.5%	[26, 27]
TiS_2	CdI_2	(0001)	Cleaving	1 × 1	S–Ti layer spacing contracted 5%, first Van der Waals gap contracted 5%	[28]
$TiSe_2$	CdI_2	(0001)	Cleaving	1 × 1	Se–Ti layer spacing expanded 5%, first Van der Waals gap contracted 5%	[28]
LaB_6	CaB_6	(111)	ArB + anneal 1600°C	1 × 1	Relaxed, La atoms move towards the surface	[29]
		(110)	ArB + anneal 1600°C	$c(2 \times 2)$		[29]

*References: [1] Kohl et al. (1974); [2] Van Hove and Leysen (1972); [3] Duke and Lubinsky (1975); [4] Chang and Mark (1974); [5] Lubinsky et al. (1976); [6] Chang and Mark (1975); [7] Bachrach et al. (1981); [8] Drathen et al. (1978); [9] Biegelsen et al. (1990b); [10] Pashley et al. (1988); [11] Guichar et al. (1979); [12] MacRae (1966); [13] Haberern and Pashley (1990); [14] Cho (1970); [15] Tong et al. (1984); [16] Biegelsen et al. (1990a); [17] Xu et al. (1985); [18] Moisson and Bensoussan (1982); [19] Haneman (1960); [20] Grant and Haas (1970); [21] Solzbach and Richter (1976); [22] Ignatiev et al. (1977); [23] Heinrich (1976); [24] Anderson et al. (1977); [25] Firment (1982); [26] Mrstik et al. (1977); [27] Van Hove et al. (1977); [28] Van Hove and Tong (1979); [29] Nishitani et al. (1980); [30] Larsen et al. (1982); [31] Biegelsen et al. (1990c); [32] Larsen et al. (1983); [33] Sauvage-Simkin et al. (1989); [34] Jacobi et al. (1979).

There has been considerable effort devoted to describing the geometric structure of the vicinal Si surface. For the Si(001) surface, surface reconstruction also results in the formation of dimers. Due to the symmetry of the underlying diamond lattice, the dimer rows are perpendicular to each other on terraces separated by an odd number of monoatomic steps (Swartentruber et al. (1990) and references therein). For surfaces cut toward the [110] direction, each terrace has a dimer reconstruction running perpendicular to those of the previous step. Furthermore, adjacent steps are inequivalent, e.g., smooth (S_A) versus rough (S_B), analogous to the GaAs(001) case just described. Here a step of type S_A has an energy ε_{S_A}, connects a higher 2×1 terrace with a lower 1×2 terrace, and has dimer rows parallel to S_A. A step of type S_B has energy ε_{S_B}, connects these terraces in reverse order, and has dimer rows perpendicular to S_B. Kinks in one type of step are made up of dimer segments of the other type of step. For higher misorientation angles, terraces exhibit the same (2×1) or (1×2) periodicity but with a double-layer step height. The transition between the two morphologies is believed to occur at $\theta \gtrsim 4°$. The transition between these two surfaces has been predicted theoretically ($\theta_C \sim 1.2$–$2.5°$) based on minimization of step plus strain relaxation energies (Alerhand et al. 1990) as well as the entropy associated with the single/double-step distributions at thermal equilibrium (Kariotis et al. 1991).

It is possible to extract thermodynamic values for the step and kink energies from distributions of the kink separations and kink lengths (Swartentruber et al. 1990). Starting from the functional forms for a simple lattice, $E(n) = n\varepsilon + C$, and independent kinks of length n atoms with a Boltzmann probability,

$$N(n) \propto \exp[-E(n)/kT],$$

one obtains $\varepsilon_{S_A} = 0.028 \pm 0.002$ eV/atom, $\varepsilon_{S_B} = 0.09 \pm 0.01$ eV/atom and $C = 0.08 \pm 0.02$ eV, comparable to calculations of $\varepsilon_{S_A} = 0.01$ and $\varepsilon_{S_B} = 0.015$ eV/atom (Chadi 1987). A comprehensive overview of recent developments in the thermodynamics of surface morphology appears in Williams and Bartelt (1991), as well as in Blakely and Umbach (1988) and Bales and Zangwill (1990).

Overall, the free energy associated with the surface steps is due to both kinks and to the long-range strain fields associated with step wandering (effectively, a repulsive potential due to a reduction in entropy). Such strain effects have already been recognized to play a role in the Si(111) (7×7) reconstruction (Alerhand et al. 1988). As a result, the details of surface preparation (e.g., sample cleaning history) are important in the subsequent formation of metastable structures. The short-range (compared with terrace widths) kink energetics govern the step fluctuations while the longer-range strain fields control the extended average positions of the step segments on a scale comparable with the terrace widths. The latter determines the distribution of the terrace widths. Furthermore, the (7×7) reconstruction correlates strongly across the single steps of slowly cooled surfaces so that the widths of terraces come in multiples of the distance between adjacent rows of corner holes (Wang et al. 1990). Thus, reconstruction plays a role in forming the step configuration.

Atomic bonding influences the extrinsic surface structure of other Si orientations as well. For the dimerized Si(001) surfaces, the (2×1) reconstructions link between terraces via dimer bonds, where the bond lengths remain constant but the bond angles change (Wieranga et al. 1987). For the cleaved Si(111) (2 × 1) surface, the π-bonded nature of the step is evident (Feenstra and Stroscio 1987) and the stepped surface exhibits domains of either triangularly shaped terraces or long, narrow terraces with parallel [11$\underline{2}$] oriented steps, depending on the nature of the cleave (Tokumoto et al. 1990). Recently, Tung and Schrey (1989) have noted a one-to-one correspondence between silicide interface dislocations and Si(111) surface steps for epitaxial silicide layers grown at room temperature. The change in crystal symmetry at the interface produces a phase difference in the regular sequence of oriented lattice bonds across the steps, resulting in a dislocation. Again, both surface bonding and strain contribute to the extended features of the geometric lattice.

In addition to steps, the surface may be facetted, i.e., it may incorporate macroscopic, nonuniform morphologies with new crystallographic planes exposed. Impurities or adsorbates can alter the step/facet structure by either the kinetic or thermodynamic mechanisms. Examples include carbon contamination (Yang and Williams 1990) and As adsorption (Ohno and Willians (1990) and references therein), both of which cause macroscopic facetting with very low coverages (e.g., 0.3% C, 0.2 ML As).

Finally, semiconductor surfaces provide evidence for a number of point defects. These advacancies (Lapujoulade 1990) can be viewed as the analogue of surface adsorbates. STM observation of such surface defects include anion and cation vacancies for GaAs (Cox et al. 1990, Stroscio 1988), anion vacancies for InSb, and adjacent anion and cation vacancies (Schottky defects) (Whitman et al. 1991). Ga vacancy point defects appear to be negatively charged. Cox et al. found terraces with different step directions depending on the quality of the cleave and the cleavage direction. Edge atoms at such monoatomic steps displayed high enough densities ($>10^{13}$ cm^{-2}) to have significant electrostatic effects, as noted earlier for badly cleaved surface areas (Huijser and van Laar 1975). STM and high-resolution photoluminescence from surfaces of the ternary compound $CuInSe_2$ has yielded evidence for several different defect types – Cu vacancies, Cu at In sites, and In at Cu sites (Aboie-Elfotouh et al. 1991) as well as specific energies associated with these defect sites. These and other extrinsic features reduce the lattice geometric order, adding features to the semiconductor surface which, as shown in the following sections, generally exhibit electrical activity at the interface.

2.2. Chemical structure

Strictly speaking, the chemical structure of the clean semiconductor surface involves either the addition to, movement on, or removal from the outer layers of atoms of the same constituents as the bulk crystal. This subsection deals with these topics in terms of crystal growth, the dynamics of growth, diffusion, and evaporation, as well as the role of surface structure and bonding in etching phenomena.

2.2.1. Crystal growth

Among the wide variety of methods developed for the growth of semiconductor crystals (Carruthers and Witt 1975, Casey and Panish 1978b), epitaxial techniques performed under vacuum conditions have provided the greatest opportunity for analyzing the structural and energetic parameters that govern atom addition to and incorporation within the crystal lattice. As a result of the vacuum environment, a host of surface science techniques are usable, particularly the electron diffraction and chemical characterization tools described in § 2.1.1. In the case of molecular-beam epitaxy (MBE), neutral atom beams evaporated from crucibles impinge on a heated, crystalline substrate and interact to form ordered overlayers (Arthur 1968, Cho and Arthur (1975) and references therein). The most studied of such MBE-grown systems has been the GaAs(001) surface. Typically, monoatomic Ga and As tetramers or dimers evaporate from Knudsen cells and are deposited on a clean GaAs substrate. The latter is prepared by etching, to eliminate surface contaminants and to produce a protective, volatile oxide, which is thermally desorbed prior to growth. Additional Knudsen sources supply dopant elements, which can be controlled to produce free carrier concentrations in the 10^{14}–10^{19} cm^{-3} range. Maintenance of a UHV environment during growth provides the low contamination levels characteristic of high-quality semiconductor material.

The adsorption, dissociation, and desorption of the impinging species can be summarized in a mass balance equation for the surface concentration of Ga atoms (Θ) during growth (Joyce et al. 1988)

$$d\Theta/dt = J_{Ga} - nJ_{As_n}S_{As_n} + 2R_{As_2} + D_{Ga} - D_{As} \tag{2.6}$$

where J_{Ga} and J_{As_n} ($n = 2$ or 4) are the incident fluxes of Ga and As_n, respectively, S_{As_n} is the As_n sticking coefficient, R_{As_2} is the dissociation/description rate of As_2 molecules from the substrate, and D_{Ga} and D_{As} are the diffusion rates of Ga and As atoms, respectively, from the bulk to the surface. Over a range of substrate temperatures (e.g., 450–650°C for GaAs), all of the group-III element adsorbs onto the surface (Arthur 1968). The resident lifetime of Ga is much longer than that of As, which desorbs rapidly unless adsorbed Ga is present (e.g., R_{As_2} is composition and temperature dependent). Upon bonding to Ga, S_{As_n} increases then decreases again when the excess Ga is consumed. Thus a 1:1 ratio of Ga to As can be maintained, provided that $J_{As_n} > J_{Ga}$ (Panish 1980). Depending on the annealing temperature, duration, quench rate, crystal constituents may diffuse out from the bulk to the surface, producing surface accumulation and subsurface depletion of species (Massies et al. 1980a). As already noted in fig. 10 and elsewhere (Ranke and Jacobi 1981), reconstructions are based on stoichiometry, but particular surface compositions do not necessarily specify a particular structure.

It is possible to determine the symmetry of the growing crystal surface from observation of the RHEED diffraction pattern in two orthogonal directions or from a single backscattered LEED pattern. The RHEED analysis procedure involves monitoring the change in diffraction streaks due to surface disorder over the course of crystal growth (Joyce et al. 1984). Any lack of perfect lattice ordering restricts the average size of the ordered regions, thereby broadening the reciprocal lattice rod and increasing its area of intersection with the Ewald sphere. The resultant broadened or elongated rods are observable in the diffraction pattern. From the integral or half-integral position of the streaks, it is also possible to extract the periodicity of the surface disorder (Joyce et al. 1988). STM data for growth structures are available for surfaces prepared by quenching after homoepitaxial growth (annealing to obtain the Ga-rich phases) and imaging *in situ* at room temperatures (Biegelsen et al. 1990b). Furthermore, soft X-ray photoemission spectroscopy (SXPS) measurements provide evidence of multiple bonding sites for each of the surface elements (Le Lay et al. 1991). Ellipsometric measurement of the surface dielectric properties has also been achieved for the epitaxial surface during growth (Aspnes and Studna 1985, Aspnes et al. 1989). Modulated-beam mass spectrometry of the kinetic features of adsorption and desorption provides the surface lifetimes, the energies of the binding states, the sticking coefficients, the thermal accommodation coefficient of molecules interacting with the surface and the orders of reaction. From such measurements it is possible to derive detailed growth models that account for the features of the geometrical structure observed (Joyce et al. 1988).

The variation of RHEED intensities during crystal growth also provides information about the nature of crystalline layer formation (Neave et al. 1983, Van Hove et al. 1983). Figure 13 illustrates schematically the oscillatory behavior of the specular RHEED spot intensity usually associated with layer-by-layer two-dimensional growth. As depicted in the thin-film growth model, the onset of growth results in the formation of two-dimensional centers, distributed randomly across the surface. For a single-scattering process with the electron wavelength much shorter than the crystal step height (~0.1 Å versus 2.83 Å for GaAs), diffuse scattering increases and specular scattering decreases correspondingly. Such diffuse scattering reaches a maximum at half-layer coverage and a minimum at full monolayer coverage. Hence the oscillations have a period corresponding to monolayer-by-monolayer growth. The monotonic decrease of peak specular amplitude (Joyce et al. 1988) relates to the completeness of each layer prior to growth of the next layer. Kinetic models of growth use such RHEED behavior to distinguish between alternative processes and rates of atom adsorption, diffusion, and lattice incorporation (Madhukar and Ghaisas (1988) and references therein, Cohen et al. (1990)). Complicating this approach is the need to take multiple-scattering (dynamical) effects into account (Kawamura et al. 1984). Nevertheless, *in situ* diffraction analysis of MBE growth affords considerable information about the atomic layer features associated with crystal growth.

2.2.2. *Kinetics of growth, evaporation, and diffusion*

Regular RHEED oscillations provide a monitor of both the growth and the evaporation, provided they occur in a layer-by-layer fashion. For evaporation rates measured as a function of temperature, one can derive an activation energy of sublimation (4.7 eV for GaAs) (Kawai et al. 1985, Van Hove and Cohen 1983). RHEED oscillations also probe the surface diffusion parameters inherent in the growth process. Here vicinal surfaces are used whose steps act as sinks for the growth of deposited atoms. For a terrace width W and a diffusion length L of the rate-controlling species (e.g.,

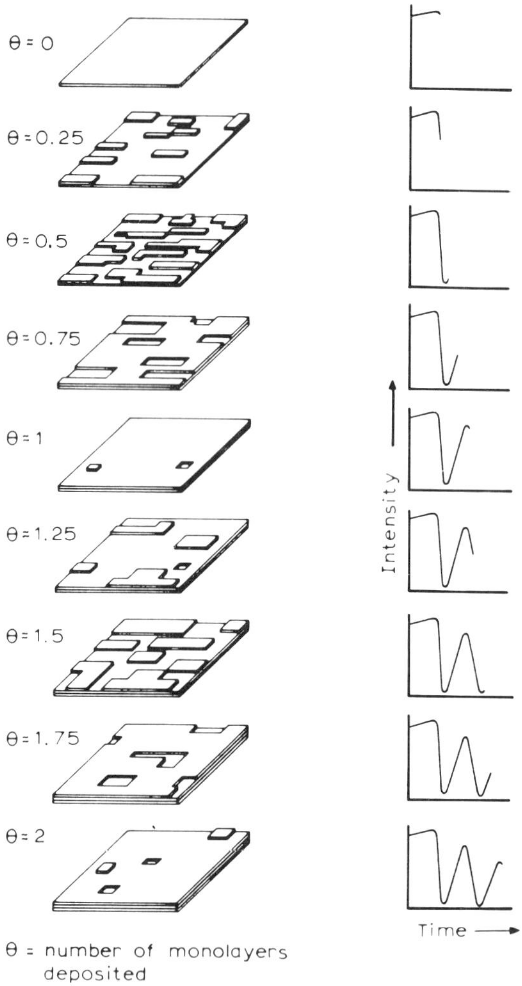

Fig. 13. Schematic representation of the RHEED intensity behavior associated with the laminar growth of epitaxial monolayers. Diffuse scattering is a maximum at half-layer coverage and a minimum at full monolayer coverage, leading to the oscillations shown (after Joyce et al. 1988).

Ga for GaAs), no growth occurs on the terrace as long as $L > W$. For a high enough incident flux or low enough temperatures $L < W$, growth occurs on the terraces and the RHEED oscillations from beams directed along terrace edges begin to show disorder (Neave et al. 1985). For example, Ga on GaAs(001) (2×4) exhibits an activation energy $E_D \approx 1.3$ eV and a diffusion coefficient $D_0 \approx 10^{-5}$ cm^2 s^{-1}. Such parameters determine the terrace widths under actual growth conditions and are critical in establishing the atomic-scale roughness of epitaxial interfaces.

Finally, RHEED oscillation intensities provide rates of surface smoothing from the increased specular spot intensities, which are evident with interrupted growth.

The RHEED intensities recover with both fast and slow exponential time constants. Values of the fast time constant as a function of temperature give an apparent activation energy for the smoothing process of about 2.3 eV (Lewis et al. 1985), considerably higher than the activation energy for surface migration already discussed and somewhat higher than the cohesive energy of GaAs (~1.7 eV) (Joyce et al. 1988). These dynamics suggest that a bond breaking mechanism is involved. Thus the energies associated with dissociation and surface diffusion are of central importance in determining the smoothness of multiple quantum wells, superlattice structures, as well as simple heterojunctions under a given set of growth conditions. In turn, roughness on this scale leads to geometric structure features that are associated with a variety of electronic features.

2.2.3. Etching

The removal of atoms by etching provides yet another aspect of the role surface structure and bonding play in the macroscopic features of the semiconductor surface. Numerous etching processes exist, including wet and dry (reactive ion and plasma) etching, as well as thermal, sputtering, electrolytic, ion bombardment and etching in melts. Examples of the latter include gratings, mesas, highly anisotropic holes and grooves, and selectively patterned heterostructures. Dry (i.e., plasma-assisted or reactive ion) etching performed in a low-pressure gaseous discharge provides increased control, selectivity, and definition for semiconductor etching (Mogab 1983). Both reactive ion and plasma etching techniques depend on chemical reactions to produce volatile compounds. However, the activated-ion process is sufficiently energetic to disrupt the local bonding geometry.

Only wet etching processes are addressed here since they illustrate particularly well the role of atomic-scale properties in macroscopic surface features. Wet etching is useful not only in surface preparation but also as a probe for surface defects, and the distribution of impurities, as well as forming specific geometric structures for devices (Gatos and Lavine 1965, Kern and Deckert 1978, Heimann 1982). The two primary aspects are the macroscopic rate of etching and the resultant microstructure of the surface. The geometric structure features that affect wet etching are: the surface orientation, defects, and impurities. The difference in chemical reactivity between surfaces with different orientations are not usually observed on a macroscopic scale, due to the small differences in their surface free energies (Gatos and Lavine (1965) and references therein) and the lack of atomically flat faces in "real" macroscopic surfaces. Such differences are evident, however, in relatively slow reactions, in which the geometric structure on an atomic scale introduces kinetic limitations to the process. Such orientation effects may be amplified by impurity adsorption, which may be specific to the particular surface (Gatos and Lavine 1960a). The ability to donate unshared electrons at dangling bond sites (Gatos and Lavine 1960b) provides a basis for understanding the etch rate differences between different crystallographic planes, as well as edge dislocations and other lattice non-uniformities. Hence for III–V compounds, surfaces terminating with group-V atoms (B{111} surfaces) exhibit more reactivity (e.g., higher etch rate) than those terminating with group III-atoms

(A{111} surfaces) (Gatos et al. 1960). Similar arguments can be applied to surfaces with a mixed composition. Likewise, the increased reactivity observed for edge dislocations (and the creation and propagation of etch pits) is related to the chemical bonding and structure of the atoms along the dislocation line. Here the relative reactivity of the dislocation relative to the surface plane determines the etch pit growth. Molecular kinetic theories of etching describe the etch kinetics in terms of the rate of step formation (Heimann 1982) and movement. Extensive tables of chemical etch treatments for III–V, II–VI, and several other families of semiconductors are available (Gatos and Lavine 1965, Kern and Deckert 1978, Aspnes and Studna 1981).

Impurities can alter the reactivity of dislocations, especially when segregated to dislocation sites. Changes in etch patterns result in etching, which depend on the reactivity of the dislocation relative to the impurity. In turn, the chemical activity of the impurity may depend upon its adsorption site. Doping (Winters and Haarer 1987) and light (Morrison 1977, Houle 1989) affect the etch rates – suggesting that free carriers as well as the bonding structure contribute to the surface chemical processes.

Overall, the growth, diffusion, and etching phenomena described in this section each provide evidence for the role of the atomic bonding geometry on macroscopic properties.

2.3. Electronic structure

The discontinuity in lattice potential at a semiconductor surface alters the electronic band structure and produces localized charge states. This section provides

(i) an overview of the experimental and theoretical techniques available for characterizing the surface electronic structure,

(ii) an introduction to localized states and altered band structure, and

(iii) representative examples of the surface state and band structure features observed at clean semiconductor surfaces, which illustrate the dependence of the electronic structure on the detailed geometrical structure of the surface.

2.3.1. Characterization and theory

A number of characterization and theoretical techniques are available with which to probe the electronic and electrostatic features of semiconductor surfaces on a monolayer scale. These are representative of the methods used to describe the electronic structure of metal–semiconductor and semiconductor–semiconductor interfaces as well. Chief among the experimental techniques has been photoelectron spectroscopy (PES) and its variants, which provide a measure of the surface density of states, its variation across the Brillouin zone, and the Fermi level position within the fundamental band gap. This technique has yielded much of the data to be discussed in the following sections. Figures 1 and 2 plus eq. (2.1) provide the basis for the high surface sensitivity and the energetics of the photoemission process. Figure 14 illustrates in greater detail the generation of photoelectrons from the semiconductor valence band

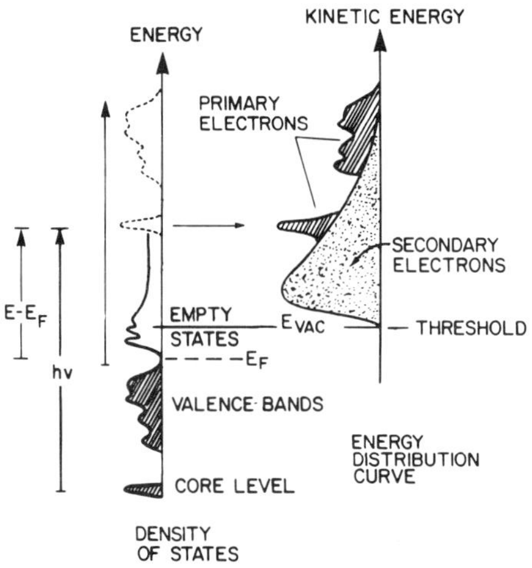

Fig. 14. Schematic energy diagram and excited-electron transitions associated with photoelectron spectroscopy. Incident photons of energy $h\nu$ excite photoelectrons from occupied valence bands and core levels. The energy distribution curve (EDC) consists of "primary" electrons above the vacuum level, which escape without energy loss (dark shading), and "secondary" electrons, which have scattered inelastically before emission (light shading). The energy $E - E_F$ governs the inelastic mean-free path and escape depth of electrons at a given E within the EDC (after Eastman and Nathan 1975). The EDC shifts rigidly in energy with band bending as the surface E_F moves with respect to the band structure.

and core levels. The energy distribution curve (EDC) shown consists of "primary" electrons above the vacuum level, which escape without energy loss, and "secondary" electrons, which have scattered inelastically. The EDC measurement involves collecting the photoemitted electrons and analyzing their energies at fixed incident $h\nu$ and monochromator resolution ΔE. For electrons with $E - E_F > 20$ eV, the EDC features correspond to the one-electron density of states in the solid, shifted upward by $h\nu$. For electrons with lower kinetic energies, selection rules involving conservation of crystal momentum can be important and these EDC features reflect the joint density of filled and empty states (after Grobman et al. 1975). The energy $E - E_F$ governs the inelastic mean-free path and escape depth of photoelectrons at a given energy within the EDC. Advanced light sources such as a synchrotron radiation storage ring (cf. Margaritondo 1988) provide soft X-rays tunable over a large energy range, which can maximize or vary the surface sensitivity. A particularly valuable aspect of PES is its simultaneous detection of both surface chemical composition and bonding environment, as measured by core level feature lineshapes and intensities.

Angle-resolved photoemission spectroscopy (ARPES) is a technique for utilizing surface photoemission to derive symmetry information about atomic bonding within the surface plane. Here one collects electrons emitted into a small solid angle as well as a narrow energy window. From these parameters, one can in principle determine the component of momentum parallel to the surface. During the emission process, only momentum parallel to the surface is conserved (Kane 1964), yielding

$$k_{\parallel} = (\sin\theta)(2mE_k)^{1/2}\hbar, \tag{2.7}$$

where $k_{\parallel}$ is the momentum along a direction in the three-dimensional Brillouin zone that is parallel to the surface, E_k is the kinetic energy, and θ is the angle of photoemitted electrons with respect to the surface normal. Angle-resolved density of states features can thus be analyzed along particular directions of the Brillouin zone rather than in the entire volume. The ARPES technique is particularly effective for layered materials (Smith et al. 1976) and surface states (Knapp and Lapeyre 1976), for which $k_{\perp}$ and $k_{\parallel}$ are decoupled. In general, detailed lineshape analyses are required to identify and analyze the surface state features. ARPES is valuable in establishing the symmetry of not only the substrate, but the adsorption sites as well. Polarization-dependent experiments made possible with soft X-ray photoemission spectroscopy (SXPS) provide additional information that can be used to determine adsorbate bonding sites (Rowe et al. 1977) and molecular orientation (Smith et al. 1976) on ordered surfaces.

Figure 14 also shows that the EDC can shift rigidly in energy as the surface E_F moves with respect to the band structure. With a fixed E_F position in the bulk semiconductor, this change in the surface value of E_F corresponds to a surface band bending. Photoemission techniques can monitor this E_F position with respect to the band edges for the clean surface as well as with successive overlayers of new material, which is a widely used method for measuring barrier heights and assessing E_F movement due to localized states (Brillson (1982a) and references therein). However, surface photovoltage effects due to the incident illumination can significantly alter the band bending as well (Hecht (1990) and references therein). One obtains surface band bending values from photoemission energy shifts only if the surface conductivity is sufficient to conduct away the photogenerated charge. See also § 3.3.4.

Related techniques include: photoelectric yield to measure the valence band edge with respect to the vacuum (Allen and Gobeli 1962), constant-initial-state spectroscopy (CIS) (Gudat and Kunz 1972) and constant-final-state spectroscopy (CFS) (Eastman and Freeouf 1974), the latter pair of which probe empty densities of states. Other electron spectroscopies that probe electronic features include: inverse photoemission spectroscopy (IPS) (Dose 1985, Himpsel 1990), which uses incident electrons to measure unoccupied densities of states and their dispersion, low-energy electron-loss spectroscopy (LEELS) (Raether 1978), which provides a measure of the unoccupied states, plasmon excitations and surface dielectric constant, as well as cathodoluminescence spectroscopy (CLS), which yields a measure of the surface and interface states within the band gap (Brillson and Viturro 1988, Yacobi and Holt 1990). Optical spectroscopies can also provide information about the surface and interface – despite their typical excitation depth of hundreds or thousands of angstroms – if the photoexcitation results in electrostatic changes near the surface or if the surface spectral features are amplified relative to the bulk via multiple surface scattering or modulation techniques. Electrostatic techniques such as surface photovoltage spectroscopy (SPS) measure the change in band bending with photostimulated population or depopulation of gap states (Gatos and Lagowski 1973, Brillson 1982a). Absorption and emission spectroscopies include infrared absorption, surface reflectance, ellipsometry, surface photoconductivity, photostimulated capacitance, Raman scattering spectroscopy, and luminescence spectroscopy. Each technique yields spectral features, a portion of which can be isolated to provide surface electronic information (Brillson 1982a). Finally, STS (scanning tunneling spectroscopy) provides a measure of both the filled and empty surface density of states, which can be spatially localized to atomic dimensions (Feenstra (1990) and references therein). Combinations of these techniques provide a considerable amount of information about the electronic properties of semiconductor surfaces, metal–semiconductor interfaces, semiconductor–semiconductor junctions, as well as their interdependence with associated interface chemical and geometric properties.

Theoretical calculations of electronic structure depend on the atomic structure and the correlation between electrons at the surface or interface. As a result, such calculations require either (a) a well established surface atomic geometry or (b) a self-consistency with both the electronic and the structural data. Until recently, well-defined surface or interface geometries have been the exception in surface and interface studies. Hence, theoretical approaches to characterize the atomic structure of semiconductor surfaces and interfaces fall into one of three categories:

(i) prediction of an interface's lowest energy configuration of bonding and structure based on a number of calculational schemes,

(ii) calculation and statistical comparison of LEED data to provide "best-fit" models, and

(iii) calculations of spectroscopic features which can be compared with the spectral features of PES, LEELS or other spectroscopy techniques described above in order to deduce structural models (Brillson 1982a, Duke 1988). Reviews of theoretical techniques for the semiconductor electronic structure are available (Lieski 1984, Pollmann et al. 1985, Dow et al. 1984, Bertoni 1990, LaFemina 1991).

Among the various theoretical approaches and computational methods, it is appropriate to distinguish between models of ground-state properties and models of excited-state (e.g., excitation spectra) properties (Duke 1988). Models designed to calculate one class of properties may not be appropriate for the other. An example of a ground-state model is the calculation of ground-state energy in terms of independent structural parameters, such as those shown in fig. 6, and the minimization of this energy to extract the surface atomic geometry. Here, the electronic structure of the surface is involved insofar as the charge density and surface state features contribute to the ground-state energy. Calculations based on LEED observations fall into this ground state category as well. Calculations of spectroscopic features to derive the electronic structure require the use of excitation spectra obtained from the experimental techniques described in this section. One typical approach involves comparison of the observed and calculated features for the surface state energies and their dispersion across the Brillouin zone (e.g., the eigenvalue spectrum). Yet another involves layer-by-layer calculations of occupied valence-band densities of states.

For Si surfaces, a wide variety of approaches have succeeded in accounting for many surface features observed by PES. These include: self-consistent calculations of the potential and charge density (Applebaum and Hamann 1973, 1975, Schlüter et al. 1975a,b), the empirical tight-binding method (Pandey and Phillips 1974, Ciraci and Batra 1976), and surface energy minimization (Chadi 1978, 1979c). For relaxed

(110) compound semiconductor surfaces, three methods used to evaluate the electronic structure are the pseudopotential (Chelikowsky et al. 1976, Manghi et al. 1982, Srivastara et al. 1983, Zunger 1980), the empirical tight-binding method (Mailhiot et al. 1984a, Swarts et al. 1980, Chadi 1979c, 1978, Pollmann et al. 1985), and ab initio quantum chemical calculations (Swarts et al. 1981, Chang and Goddard 1984). Empirical tight-binding calculations, which include electron–electron interactions (Northrup et al. 1991), have also been extended to both ground-state and excited electronic spectra. Electronic structure calculations for Si surfaces do not provide results of a consistency in detail with atomic structure measurements comparable to that obtained for the zinc blende compound (110) surfaces (Duke 1988). For Si(100) (2 × 1) surfaces, the lowest energy symmetric dimers (Batra 1990, Craig and Smith 1989, Payne et al. 1989) are nearly degenerate in energy with asymmetric, buckled dimers (Chadi 1979a) and require antiferromagnetic order between the spins of the electrons on the two atoms of the surface (Artacho and Yndurain 1989) in order to obtain qualitative agreement with photoemission results (Uhrberg and Hansson 1991). For Si(111) (2 × 1) surfaces, the π-bonded chain model (Pandey 1981) is inconsistent with LEED and ion scattering results. Similarly, an exact characterization of the atomic coordinates for Si(111) (7 × 7) surfaces is not yet known (Schlüter 1988). Thus, while many different approaches, models, and techniques are available for addressing the surface electronic structure, the theoretical calculations provide only an incomplete description of these well-studied Si surfaces.

2.3.2. Localized states and surface band structure

The presence of the semiconductor surface leads to two-dimensional (e.g., localized) states and, consequently, a surface band structure with distinctive differences from the bulk. Localized states are of major importance since they are responsible for the band bending and electronic barriers at the semiconductor interface. The trapped charge associated with localized states falls into one of four categories:

(i) intrinsic surface states present at the semiconductor–vacuum interface associated with the discontinuity in the lattice potential,
(ii) surface states produced after contact with a metal and associated with wavefunction tailing into the band gap of the semiconductor,
(iii) extrinsic surface states associated with imperfections of the semiconductor interface (e.g., impurities or defects common to both semiconductor surface and bulk), and
(iv) interface-specific states due to interface chemical reactions, interdiffusion, or adsorbate-specific local chemical bonding. Each is discussed in this chapter. Intrinsic surface states are considered first.

The discontinuity of the lattice potential results in dangling atomic bonds which, with or without reconstruction, produce localized states near the surface. Surface localization of such states within the forbidden band gap is easily understood from a consideration of the associated wavefunction, which decays exponentially into the vacuum and into the bulk. States outside of the forbidden gap, termed resonances, also contribute to the surface density of states.

Electronic structure calculations of semiconductor surfaces indicate a substantial influence of the surface atomic structure. The termination of the semiconductor lattice at the surface produces both dangling bonds projecting into the vacuum as well as changes in back bonding extending into the bulk. These charge distributions depend sensitively on the local coordination of atoms at the surface (i.e. the formation of dimers, trimers, multilayer bond configurations) as well as morphological features such as steps and long-range strain. Associated with these altered charge distributions are multiple energy bands, both inside and outside the forbidden band gap. The number of such bands within the band gap depends on the crystal orientation and the number of dangling bonds per unit cell (after Zangwill 1988). While the full set of surface states and resonances depends on the overall charge distribution in the top semiconductor layers, it is found that changes in the coordination of the top layer alter substantially the occupancy of these surface state bands. Thus a change in atomic coordination (i.e. pairing of dangling bonds) can change a "metallic" surface, which includes a partially filled surface state band, to a "semiconducting" surface, which includes fully occupied and unoccupied surface state bands (Bertoni 1990, LaFemina 1991). While a unique correspondence does not exist in general between these surface bonds and bands, such effects serve to emphasize the first-order effect of local atomic bonding on intrinsic surface state properties.

The experimental and theoretical results for Si intrinsic surface states show that:

(i) intrinsic surface states exist near or within the Si band gap with spectral features that vary from surface to surface and that correspond to experimental observations with varying degrees of success;

(ii) the density of intrinsic surface states depends sensitively on the detailed surface reconstruction of the Si surface;

(iii) complete characterization of the surface state eigenvalue spectrum for Si surfaces requires firmer determinations of their respective surface atomic geometries plus a detailed understanding of the effects that Coulomb interactions have on deviations from one-electron models of surface electronic structure.

Research on compound semiconductors shows that intrinsic surface states exist within the valence band of both cleaved (110) and non-cleaved surfaces and within the band gap of non-cleaved surfaces (Bertoni 1990, LaFemina 1991). Early experimental and theoretical studies suggested that intrinsic surface states existed within the band gap of cleaved semiconductor surfaces as well, but it is now generally accepted that the cleavage surfaces of most III–V as well as the few II–VI compound semiconductors studied have no significant intrinsic states within the forbidden gap (Brillson 1982a, Himpsel (1990) and references therein). Initial observations of such band gap states have been reinterpreted in light of (a) energy corrections due to surface core–hole relaxation and (b) electrically active defect (e.g., cleavage steps) phenomena (Gudat and Eastman 1976, Huijser and van Laar 1975). Calculations based on unreconstructed III–V cleavage surfaces reflected empty or both empty and filled states within the band gap. Refined to include modifications in surface atomic positions, i.e., relaxation and reconstruction, such calculations were then able to account for the absence of states within the semiconductor band gap (cf. Brillson 1982a and references therein).

The results for the GaAs(110) surface best illustrated the sensitivity of surface states to atomic structure since both experimental spectra (Knapp et al. 1978, Williams et al. 1978, Huijser et al. 1978, Knapp and Lapeyre 1976) and a surface atomic geometry (see fig. 6) were available. Calculations that take atomic relaxation into account predict intrinsic states out of the band gap, thereby reinforcing both structural and surface state analyses. Figure 15 illustrates a representative example of such a calculation for $\omega_1 = 27.3°$. The calculations of surface state dispersions (solid lines) for relaxed GaAs(110) surfaces (from a tight-binding model (Beres et al. 1983)) are compared with the measured dispersions from angle-resolved photoemission measurements of Williams et al. (1978) (closed circles). Similar results are obtained by other theoretical approaches. The shaded areas indicate the allowed energy bands of GaAs projected along the indicated symmetry directions. This calculation reveals eleven surface states, the character of which vary somewhat with the planar wave vector ***k***. States with energies only a few electron volts below E_V (above E_C) are derived primarily from anions (cations). States at the forbidden gap correspond to dangling bond states, while a state just below the forbidden gap corresponds to a back-bond state. The energy region between conduction (E_C) and valence (E_V) bands is of particular significance: here the band gap is free of intrinsic surface states. Qualitatively similar results are obtained for other III–V compounds, although GaP results indicate some state tailing into the band gap from the conduction band (Beres et al. 1983) because of the indirect nature of the band gap.

Table 3 provides the energy positions of unoccupied intrinsic surface states observed for a variety of elemental and compound semiconductor surfaces (Himpsel (1990) and references therein). In contrast to the elemental semiconductors, unoccu-

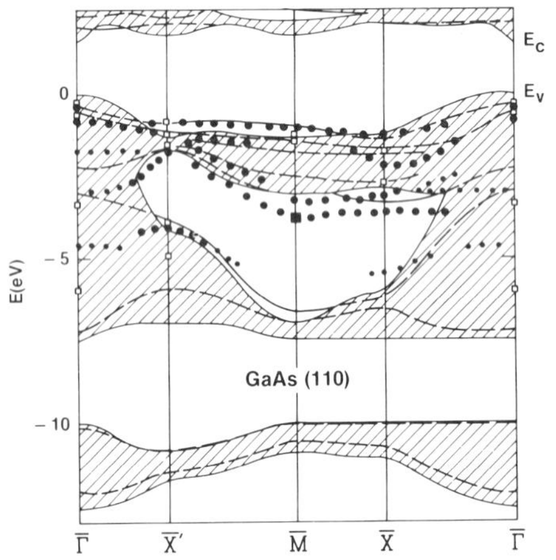

Fig. 15. Calculated surface state dispersions (solid lines) derived from a tight-binding model (Beres et al. 1983) compared with measured dispersions from the angle-resolved photoemission measurements of Williams et al. (1978) (closed circles) and Huijser et al. (1978) (open squares). Shaded areas indicate the projected bulk allowed energy bands of GaAs (adapted by Duke 1988). The relaxation sweeps the surface state bands out of the semiconductor band gap.

Table 3
Energies relative to the valence-band maximum of unoccupied surface states for different semiconductors and surface reconstructions, as determined by inverse photoemission spectroscopy (after Himpsel 1990).

Surface	Critical point	Energy (eV)	Ref.*
Si(111) (2 × 1)	$\bar{\Gamma}$	1.3	[1]
	$\bar{J}$	0.65	
	$\bar{J}'$	1.4	
Si(111) (7 × 7)	$\bar{\Gamma}$	1.2	[2]
Si(100) (2 × 1)	$\bar{\Gamma}$	0.7	[3]
Ge(111) (2 × 1)	$\bar{\Gamma}$	1.5	[4]
	$\bar{J}$	0.15	
Ge(111) *c*(2 × 8)	$\bar{\Gamma}$	1.0	[5]
GaP(110) (1 × 1)	$\bar{\Gamma}$	2.4	[6]
GaAs(110) (1 × 1)	$\bar{\Gamma}$	2.1	[7]
	$\bar{X}$	1.7	
	$\bar{X}'$	2.0	
GaSb(110) (1 × 1)	$\bar{\Gamma}'$	2.1	[8, 9]
InP(110) (1 × 1)	$\bar{\Gamma}'$	2.7	[10]
InAs(110) (1 × 1)	$\bar{\Gamma}'$	1.9	[10]
InSb(110) (1 × 1)	$\bar{\Gamma}'$	1.9	[10]
CdS(11$\bar{2}$0) (1 × 1)	$\bar{\Gamma}'$	3.8	[11]
	$\bar{\Gamma}'$	5.8	[11]
CdSe(11$\bar{2}$0) (1 × 1)	$\bar{\Gamma}'$	—	[11]
CdTe(110) (1 × 1)	$\bar{\Gamma}'$	2.9	[11]

*References: [1] Perfetti et al. (1987); [2] Kaxiras et al. (1990); [3] Himpsel and Fauster (1984); [4] Nicholls and Reihl (1989); [5] Drube et al. (1988); [6] Himpsel (1990); [7] Straub et al. (1985); [8] Carstensen et al. (1990); [9] R. Manzke et al. (1987); [10] Drube et al. (1987); [11] Magnusson et al. (1978).

pied surface states for all the compound semiconductors studied lie outside the forbidden band gap.

Non-cleavage surfaces of compound semiconductors can exhibit states within the band gap. However, such surfaces have until recently been prepared only by polishing, Ar^+ ion bombardment and/or vapor phase epitaxy on oriented substrates. Such treatments produce different surface structures, depending on the specific process conditions – see, for instance, fig. 9 for different growth conditions – and with the exception of epitaxial growth, they can produce damage to the semiconductor lattice many layers below the surface. EELS measurements performed on polar faces of III–V compounds such as GaAs(111)A (As rich), GaAs(111)B (Ga rich) and GaP(111)P (P rich) reveal intrinsic surface states in the band gap (Ludeke and Esaki 1974, Ludeke and Koma 1975, Ranke and Jacobi 1973), but these spectra are complicated by excitonic effects (Chadi and Cohen 1975a,b).

The absence of intrinsic surface states within the forbidden gap of cleaved compound semiconductors is significant since the metal deposited on such surfaces results in E_F stabilization within less than a few monolayers – indicative of a local electronic mechanism. The fact that intrinsic surface states for these materials play no direct role in E_F stabilization demonstrates that extrinsic and/or metal-induced features play a dominant role in Schottky barrier formation for many semiconductors.

2.3.3. Reconstruction dependence

Calculations and measurements of electronic excitation spectra reveal a sensitive dependence on the surface atomic geometry. In addition to the relaxation effects already described, surface reconstruction produces significant changes in the localized charge states and the surface electronic band structure. An illustrative case is the influence of buckling in the dimer reconstruction of Si(100) (2 × 1). This dimerization, already described in § 2.1.2, can yield a (2 × 1) symmetry with either asymmetric or symmetric dimers as pictured in fig. 16, upper and lower right, respectively. A surface structure calculation based on an energy minimization procedure (Chadi 1979a) serves to illustrate the dependence of the total energy and dimer charge transfer for such variant reconstructions. In the structural calculation of these geometries, Coulombic contributions are taken into account, which tend to reduce somewhat the asymmetry of the dimer and the magnitude of charge transfer. Subsurface relaxation and dipole–dipole interactions also modify such results. Tight-binding calculations based on these geometries yield the corresponding surface state dispersions shown in fig. 16, upper and lower left. Here the occupied and empty surface states

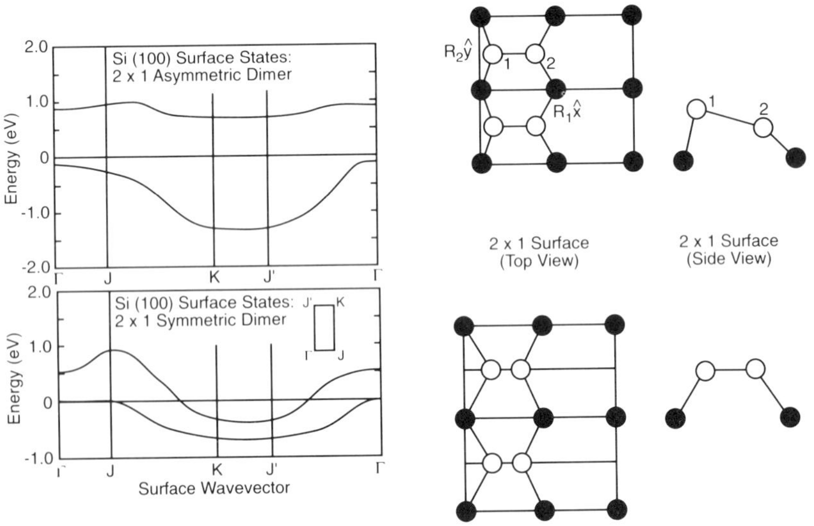

Fig. 16. Comparison of surface state dispersions calculated for asymmetric (top) and symmetric (bottom) Si(100) (2 × 1) dimer reconstructions, showing filled and empty bands near the valence-band maxima (after Chadi 1979a). The corresponding dimer geometries (side views not to scale) appear to the right of each band diagram (adapted from Chadi 1979b). The partially filled and empty energy bands of the symmetric dimer indicate a metallic character, in contrast to the semiconducting character of the asymmetric dimer.

near E_V show a metallic behavior for the symmetric dimer (partially-filled and empty energy bands), in contrast to the semiconducting character (filled states below E_V separated by an energy gap from empty states above) of the asymmetric dimer. ARPES measurements of Himpsel and Eastman (1979) and Uhrberg and Hansson (1991) confirm the semiconducting nature of this surface and the details of the energy dispersion in fig. 16 (top). More recent ab initio calculations (Payne et al. 1989, Craig and Smith 1989, Batra 1990) indicate that symmetric dimers are the lowest energy configuration but require antiferromagnetic local spin correlation to obtain a semiconducting surface energy gap (Artacho and Yndurain 1989). STM images provide evidence for both buckled and unbuckled dimers (Tromp 1986, Hamers et al. 1986a). Recent low-temperature STM measurements suggest that the buckled dimer is energetically more stable (Wolkow et al. 1991).

A different type of reconstruction best accounts for the spectroscopic measurements of Si(111) (2×1). The π-bonded chain model illustrated in fig. 8d (Pandey 1981) represents a topology whose zig-zag chains provide substantially lower total energies per surface atom than buckled geometries (Pandey 1982, Northrup and Cohen 1982, Nielsen et al. 1983) and which provides good agreement between the calculated surface state dispersions for Si(111) and the experimental photoemission data (Uhrberg et al. 1982, Mårtennson et al. 1985). This surface state dispersion (not shown) exhibits a strong dispersion along the chain direction and flat bands in the perpendicular direction (J–K). A strong anisotropy in the optical absorption and reflectivity predicted (Selloni and Del Sole 1986) and observed (Chiaradia et al. 1984, Olmstead and Amer 1984) between these two directions lends further support to this model. Nevertheless, details within this reconstruction such as buckling, tilting, or dimerization are not fully resolved because such structural models cannot be confirmed independently by LEED and ion scattering (Schlüter 1988).

Along the same lines, experiments on Si(111) (7×7) reconstructions show filled- and empty-state features (Hamers et al. 1986b, Wolkow and Avouris 1988) that agree with the photoemission and inverse photoemission densities of states (Himpsel and Fauster 1984). These measurements reveal spectroscopic features associated with specific atomic sites within the unit cell. Depending on the "adatom" or "rest" atom nature of the site (see figs. 7b and 7c), these spectra exhibit empty- and filled-state features, respectively, which reflect the charge transfer occurring between sites (Himpsel 1990). Here differences in the electronic features for different atomic geometries arise not only on the same surface but even within the same unit cell.

Reconstruction-dependent electronic features are also evident for III–V compound semiconductor surfaces. ARPES measurements of the surface state dispersion for the GaAs(100) (2×4) or $c(2 \times 8)$ reconstruction (Larsen et al. 1982) are in strong contrast to similar measurements for the GaAs(100) $c(4 \times 4)$ reconstruction (Larsen et al. 1983). In both cases, ARPES results indicate the symmetry of surface states to be (2×1) rather than the symmetries measured by RHEED or LEED (photoemission is more sensitive than LEED to structure within the unit cell compared with the long-range order seen by LEED). However, both the energies and dispersions differ markedly between the two reconstructions (Bringans and Bachrach 1990). By the same token, the Ga-rich GaAs(111) (2×2) (Bringans and Bachrach 1984) and As-rich GaAs($\bar{1}\bar{1}\bar{1}$) (2 × 2) (Bringans and Bachrach 1984, Jacobi et al. 1979) show strongly different surface state dispersions and local symmetries (e.g., (2 × 2) for GaAs($\bar{1}\bar{1}\bar{1}$) and (1 × 1) for GaAs(111), despite the same LEED symmetry). In addition, different reconstructions on (001) GaAs as determined by RHEED (depending on As flux and substrate temperature) exhibit distinct differences in the visible–near-ultraviolet dielectric response as measured by spectroellipsometry (Aspnes et al. 1990). These differences are representative of different optical transitions between Ga and As dimers and lone-pair states. Indeed, comparison between optical features and tight-binding calculations provides an identification of features associated with one or the other dimer. It is therefore evident that geometric and electronic features may be related on a scale as small as the unit cell.

We see that many advanced techniques are available with which to probe the geometric, chemical, and electronic structure properties of clean semiconductor surfaces, and that such a structure differs in numerous aspects from that of the bulk semiconductor, and that a strong relationship exists between the atomic-scale geometry, the growth and processing used to prepare such surface configurations, and the local surface state and electronic band structures. The enhanced ability to identify such features and the relations between the spectroscopic and unit-cell features with new growth, processing, and characterization techniques will enable even more refined determinations of the structural properties and their extrinsic modification.

3. Adsorbate–semiconductor and metal–semiconductor interfaces

The previous section contains an overview of the characterization methods, the salient physical phenomena, and the chief principles involved in understanding the properties of semiconductor surfaces. In this section many of the same methods, phenomena, and principles are operative but the addition of overlayer atoms and films introduces a new level of complexity. Section 3 combines features of the adsorbed atom with those of the multilayer film in discussing the geometric chemical and electronic structures, although the two are sufficiently rich in detail to otherwise warrant separate, expanded treatments. As in § 2, the sensitivity of the geometric structure to chemical processing and of the electronic structure to the geometric/chemical structure is a unifying theme of this section. Both disordered and epitaxial overlayers comprise this discussion, the former representing the bulk of observations to date, the latter a rapidly growing subset whose properties lend themselves to more refined analyses. While many of the geometric, chemical, and electronic phenomena presented here are representative of semiconductor interfaces in general, the focus of this section is on metal–semiconductor interfaces, the phenomenon of Schottky barrier formation, and the interface band bending, the latter of which has both fundamental and technological significance. Here, the ability to prepare metal–semiconductor interfaces in new ways appears to shed new light on the role of atomic-scale phenomena in forming macroscopic electronic properties.

3.1. Geometric structure

Although adsorption and growth of atomic species on semiconductor surfaces is in general polycrystalline or amorphous, a considerable body of information exists for ordered adlayers and overlayers. These geometric features offer an additional insight into the chemical and electronic properties of semiconductor surface sites as well as the kinetics and thermodynamics of heteroepitaxial growth. This subsection gives an overview of such adlayers and multilayer structures with an emphasis on (a) the site specificity and induced changes in reconstruction associated with adsorbates and (b) the properties of epitaxical (versus "epitaxial", see Donnay (1986)) overlayer growth.

3.1.1. Adsorbate bonding

The last few years have witnessed a rapid development in our knowledge of chemisorbed species on semiconductor surfaces and their effect on the surface geometric structure. A variety of forms of adsorbate bonding can occur on semiconductor surfaces. These can be distinguished by (a) the adsorbate site specificity, (b) the effect of adsorption on the surface reconstruction, and (c) the degree of chemical reactivity resulting in a new species. The last of these is discussed in § 3.2. Consistent with the structural studies for the clean semiconductor surfaces (cf. § 2.2), the Si and GaAs surfaces have been the subject of most adsorbate bonding studies. Indeed, there have been STM studies of over two dozen elements on Si alone in the past few years.

Adsorbates on Si exhibit several characteristic features relative to the clean-surfce geometry. For Si on Si(111) (e.g., homoepitaxy), Si atoms adsorb with a marked preference for orientations that nearly align the (7 × 7) unit cells of the substrate with the growing island (Kohler et al. 1989). These island growths form domains at whose boundaries enhanced nucleation sites for impinging atoms occur (Hamers et al. 1989). Such domains exhibit a marked anisotropy due to the higher energy of type B versus type A (see § 2.1.3) step edges (Chadi 1987), with evidence for both kinetic (Mo et al. 1989) and thermodynamic (Alerhand et al. 1990) contributions. Within the Si(111) (7 × 7) unit cell, adsorbates display a preference for particular chemisorption sites. Figures 7a and b represent the geometric structure for a number of inequivalent unit-cell sites – center adatoms, corner adatoms, and rest atoms. In turn, fig. 7d illustrates the local bonding and geometry for rest and adatoms. Si and Ge are believed to adsorb onto such adatom sites, also termed T_4 sites in fig. 17, for both Si (in a $c(2 \times 2)$ reconstruction) and Ge (in a $c(2 \times 8)$ reconstruction), respectively. Reactions of H_2O, NH_3, phosphine (PH_3), and disilane (Si_2H_6) indicate that rest atoms are more reactive than adatoms and, among adatoms, center sites appear to be more reactive than corner sites (Avouris and Lyo 1990). These results emphasize the role of local geometry in overlayer bonding.

Of the many adsorbates on Si studied by STM, the chemisorption features of metallic adsorbates have particular significance in the context of Schottky barrier formation. The column-III metals Al (Hamers and Demuth 1988), Ga (Nicholls et al. 1987) and In (Nogami et al. 1988) on Si(111) adsorb substitutionally on T_4 adatom

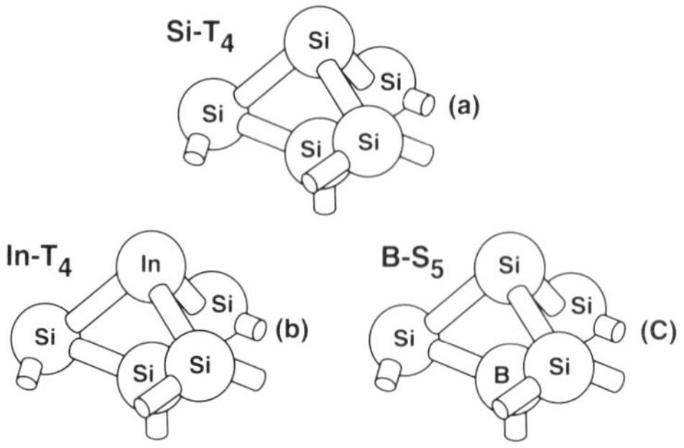

Fig. 17. Ball and stick representation of local structure and bonding for adatoms on Si(111) in (a) a Si T_4 site, (b) an In T_4 site, and (c) a B S_5 site (adapted from Avouris and Lyo 1990).

sites. Figure 18 illustrates the STM image for 0.03 ML nominal In coverage on Si(111) (7×7). In adatoms replacing Si in the reconstruction appear as brighter spots (Nogami et al. 1988). At these low densities, the replacement preserves the (7×7) structure, with a tendency of In atoms to occupy positions along edges of the unit cell as opposed to adjacent to corner holes. Agglomeration of In at higher densities around atomic steps can exhibit local areas of $(\sqrt{3} \times \sqrt{3})$ reconstruction and other reconstructions can arise at higher coverages.

In contrast to other column-III metals, B is thought to adsorb by substitution directly below an Si adatom, as shown in fig. 17c (Bedrossian et al. 1989, Lyo et al. 1989). This subsurface S_5 site exhibits a total energy below that of a T_4 site (Avouris and Lyo 1990). Of added significance is the fact that chemisorbed B alters drastically the surface chemistry of NH_3 on Si (7×7), as well as charge transfer between the alkali metal K and the Si substrate (Ma et al. 1990). The alkali metal Cs on Si exhibits periodic, one-dimensional chain-like structures on Si and GaAs (Stroscio et al. 1988, Wong et al. 1989), although the coverage-dependent morphology for other alkali metals is not settled (Hashizume et al. 1990, Badt et al. 1990).

The transition metals Fe, Co, Ni, and Cu form silicides that exhibit two (1×1) epitaxical variants rotated by 180° plus Si adatoms and trimers (Kubby et al. 1991). Pd (Koehler et al. 1988), Ag (Tosch and Neddermeyer 1988), and Li (Hasagawa et al. 1990) appear to chemisorb preferentially on the faulted half of the Si(111) (7×7) unit cell. In retrospect, one can rationalize all of these preferential chemisorption phenomena in terms of the relative energetics of adatom, dimer, back-bonding, and fault formation.

A more limited set of STM results exist for GaAs adsorption. Of particular note, Au on GaAs(110) appears to bond preferentially to Ga sites (Feenstra 1989) and Si adsorbs on multiple, inequivalent sites on GaAs(100) (Bringans and Bachrach 1990). Characteristic valence-band spectra indicate the chemisorption of Cl to anion sites on GaAs(110), characteristic of many III–V compound semiconductors (Margaritondo et al. 1979). Sb on GaAs(110) provides an example of epitaxially-constrained adsorbate bonding with Sb in zig-zag chains approximately where top-layer anions and cations would have been (Duke et al. (1982), Mårtensson and

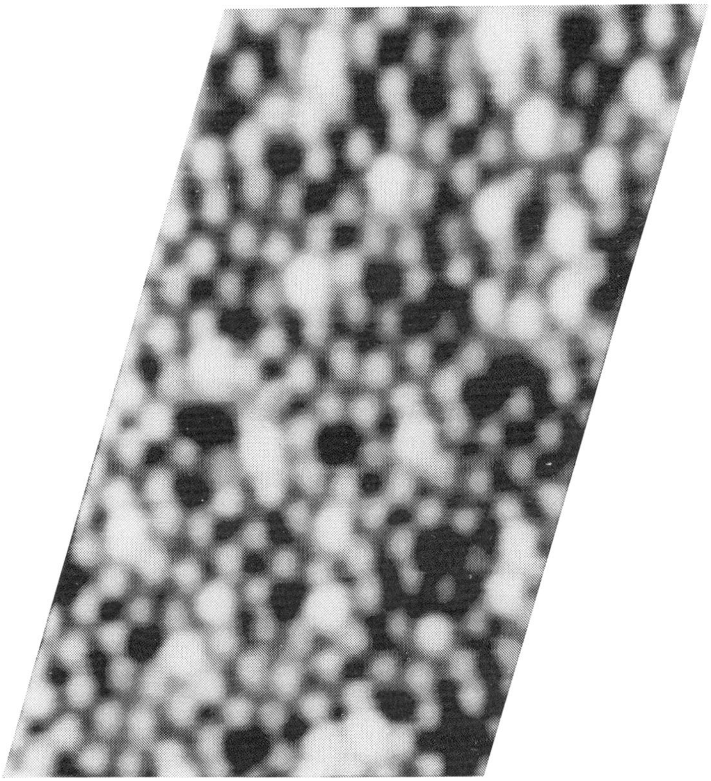

Fig. 18. Scanning tunneling microscopy image of the Si(111) (7 × 7) surface with 0.03 monolayers nominal In coverage, illustrating the preferential adsorption sites within the unit cell. Imaged area is about 75 × 120 Å^2. In adatoms replacing Si appear as brighter spots (after Nogami et al. 1988). The contrast between In and Si is due to differences in the tunneling densities of states.

Feenstra (1989) and references therein, Ford et al. (1990) and references therein). Here, the saturated surface valence eliminates the surface relaxation. Sm on GaAs(110) exhibits similar zig-zag chains with an analogous weak interaction at low coverage (0.01 ML) but stronger, chemically-disruptive reactions at intermediate and higher coverages (Trafas et al. (1990) and references therein).

Al on GaAs(110) provides an example of staged adsorption, reaction, and adsorbate incorporation into the semiconductor lattice. At very low coverages, Al exhibits mobile chemisorption, presumably in twofold sites (Ihm and Joannopoulos 1982, Daniels et al. 1982). Clustering occurs at higher coverages (Zunger 1983, Skeath et al. 1983), which is believed to overcome the activation barrier for bond breaking and Al–Ga place exchange. With further coverage and annealing, Al replaces Ga in second- and third-layer sites, which finally forms a diffusion barrier to further interdiffusion (Duke et al. 1981, Kahn et al. 1981).

Atomic adsorption can induce changes in semiconductor surface reconstruction, and such changes can occur at submonolayer coverage. For example, H on Si(100) quenches the (2 × 1) reconstruction and, depending upon H exposure and annealing temperature, gives rise to monohydride (2 × 1), disordered (3 × 1), then dihydride (1 × 1) reconstructions (Boland 1990). As on Si(100) rotates the (2 × 1) reconstruction by breaking Si dimers and forming As dimers (Bringans and Bachrach 1990). As on Si(111) converts the (7 × 7) to a (1 × 1) reconstruction, where Si and As atoms are all fully coordinated and with the fifth As valence electron forming a doubly occupied surface orbital. Similarly As converts the Ge(111) (2 × 8) surface to (1 × 1) (Bringans and Bachrach 1990). Cl on Si(111) exhibits an analogous type of behavior, as does F (which can also etch past the top layer (Bringans and Bachrach 1990). The subsurface B substitution converts the Si(111) (7 × 7) surface to a $(\sqrt{3} \times \sqrt{3})R30°$ structure. Somorjai and Van Hove (1979) have provided a general review of adsorbate-induced restructuring of surfaces. See Bringans and Bachrach (1990) for recent surface chemical bonding studies for GaAs and Si. Finally, Si–Si(111) homoepitaxy preserves the top layer reconstruction, requiring unreconstruction and subsequent reconstruction of a new top layer for growth (Kohler et al. 1989). This suggests that low barriers exist for altering reconstructions. Overall the new reconstructions induced by chemisorbed species reflect the sensitivity of bonding structure and associated charge transfer to the specific details of interface preparation.

3.1.2. Epitaxical overlayers

Although the vast majority of metal and other overlayer interfaces with crystalline semiconductors are polycrystalline, the subset of epitaxically-ordered metal/semiconductor interfaces provides the clearest opportunities to establish the relationships between the two-dimensional interface geometric and electronic structures. For elemental semiconductors, there are few examples of abrupt, elemental metal epilayers. The most notable case is Pb on Si(111) (Estrup and Morrison 1964, Heslinga et al. 1990, Le Lay et al. 1988), which grows in parallel epitaxy with $(111)Pb_{\parallel}(111)Si$ and $[1\bar{1}0]Pb_{\parallel}[1\bar{1}0]Si$. Prepared at room temperature, this interface displays a commensurate Pb/Si(111) (7 × 7) structure in parallel orientation with the substrate. After high-temperature annealing or growth, an incommensurate Si(111) $(\sqrt{3} \times \sqrt{3})R30°$-Pb structure appears, which differs only in the structure of the first layer of Pb and Si atoms at the interface and with identical bulk structures in both cases (Grey et al. 1989).

Metal silicides on Si provide several examples of epitaxically-ordered metal–semiconductor interfaces. Epitaxical silicides on Si include: Ni, Ti, V, Cr, Fe, Co, Zr, Nb, Mo, Pd, Ta, W, and Pt. Their properties are tabulated elsewhere (Chen et al. 1986). An especially important example of an expitaxically ordered metal/Si interface is the $NiSi_2$/Si(111) junction. $NiSi_2$ has a cubic CaF_2-type structure with a lattice constant only 0.44% smaller than that of Si ($a_{0,Si} = 5.4307$ Å at room temperature). Furthermore, $NiSi_2$ on Si(111) has two sevenfold coordinated epitaxical variants, rotated by 180°, which exhibit significant differences in electrical behavior (Tung et al. 1983). Figure 19 illustrates the two epitaxical variants of the $NiSi_2$/Si(111) interface. For "type A" $NiSi_2$/Si(111), the silicide keeps the same orientation as the substrate, whereas the "type B" silicide is rotated. Exclusive growth of either

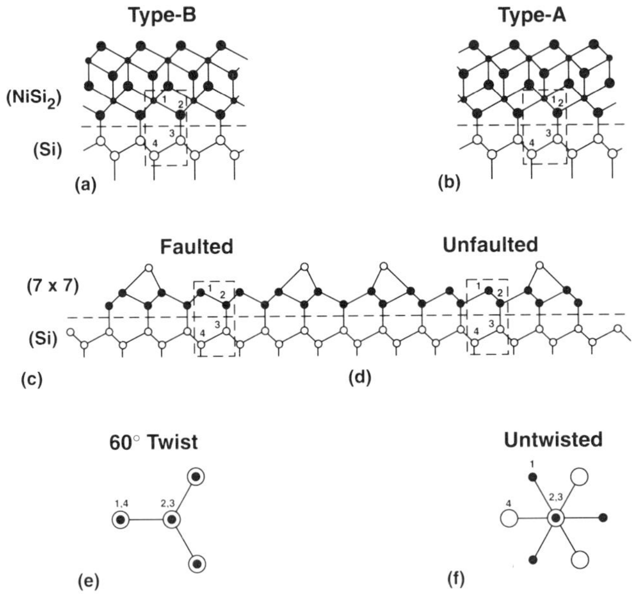

Fig. 19. Ball and stick models for (a) type-A and (b) type-B epitaxial $NiSi_2$/Si interfaces and, for comparison, the (c) unfaulted and (d) faulted sides of a Si(111) (7 × 7) unit cell, all viewed in the ⟨110⟩ direction. For type-A Ni_2Si/Si(111), the silicide retains the same orientation as the substrate, whereas the type-B silicide is rotated 180° about the surface normal. The dashed lines separate surface/interface atoms (1,2) from the substrate atoms (3,4). Dashed boxes indicate interface units for both systems. Top views of these interface units appear in (e) and (f). There is a one-to-one correspondence in mapping silicide interface units to the 7 × 7 interface units. In both cases, the two variants are related by a simple 60° twist along the ⟨111⟩ direction between atoms 2 and 3 (after Yeh 1989).

untwinned type-A epitaxy or twinned type-B epitaxy is possible, depending on the recipe for surface preparation, template structure, and annealing conditions (Tung et al. 1983). Figure 19 also shows for comparison ball and stick models for the unfaulted (c) and faulted (d) sides of a Si(111) (7 × 7) unit cell viewed along the same ⟨110⟩ direction (Yeh 1989). One can relate the interface units (shown as the dashed boxes) of the A- and B-type variants in the same way as the faulted and unfaulted parts of the (7 × 7) unit cell, i.e., by a simple 60° twist along the ⟨111⟩ direction between atoms 2 and 3 as illustrated in figs. 19e and f. The electronic significance of this distinction between different types of faults is discussed in § 3.3.

Several elemental metals exhibit epitaxical growth on compound semiconductors. Most notable are Al (Ludeke et al. 1973) and Ag (Ludeke 1984, Massies et al. 1982) on GaAs(001). Both exhibit multiple variants, depending on surface termination and growth temperature (Sands et al. 1990). Epitaxical Fe on GaAs(110) and (100) (Prinz and Krebs 1981) is distinguished by its ferromagnetic properties and good lattice match (to within 1.2%). A list of epitaxical metals on GaAs appears in table 4 (Sands et al. 1990b). In addition to a process-dependent epitaxy, chemical interdiffusion and/or reaction is evident for most systems, as it is for most if not all metals on GaAs at elevated temperatures (Palmstrøm and Morgan 1985).

Table 4
Epitaxial elemental metal films grown on UHV-prepared GaAs surfaces (after Sands et al. 1990b).

Metal	Melting point	Epitaxy on GaAs	Stability and morphology	Refs.*
Al	660°C	Multiple variants on {100}	Stable on GaAs, Al–Ga exchange on GaAs, reactive in As vapor, 3D growth, prone to agglomeration	[1–8]
Ag	961°C	Single unrotated variant possible for growth above 200°C on {100}	Stable on GaAs in closed system, can "etch" GaAs in an open system. 3D Volmer–Weber growth, prone to agglomeration	[1, 9–13]
Fe	1536°C	Unrotated b.c.c. Fe on {100} and {110} GaAs	Not stable on GaAs, forms arsenides or Fe–Ga–As phases during deposition, annealing and exposure to As vapor, smooth films on {110} GaAs after coalescence of 3D nuclei	[14–19]
Au	1064°C	Single variant on {100}, {110} and {111} possible, often twinned	Similar to Ag	[20–22]
Co	1495°C	Unrotated b.c.c. Co on {100} and {110} GaAs, similar to Fe	Similar to Fe, Co–Ga–As phase formed during annealing	[23–25]
In	156°C	Axes of pseudocubic In unit cell parallel to cube axes of GaAs on {100} GaAs	Stable on InAs but dissolution and regrowth of (In,Ga)As on GaAs substrates, 3D growth	[26, 27]

*References: [1] Ludeke (1984); [2] Cho and Dernier (1978); [3] Missous et al. (1986a); [4] Svensson et al. (1983); [5] Tadayon et al. (1988); [6] Okamoto et al. (1982); [7] Chambers (1989); [8] Oh et al. (1990); [9] Massies et al. (1982); [10] Farrow et al. (1989); [11] Panish (1967); [12] Pugh and Williams (1986); [13] Lilienthal-Weber (1988); [14] Ruckman et al. (1987); [15] Waldrop and Grant (1979); [16] Prinz and Krebs (1981); [17] Krebs et al. (1987); [18] Chambers et al. (1986); [19] Harris et al. (1987); [20] Takeda et al. (1974); [21] Snyman et al. (1977); [22] Leung et al. (1983); [23] Prinz (1985); [24] Xu et al. (1987); [25] Lind et al. (1988); [26] Ding et al. (1986); [27] Savage and Lagally (1986).

Metallic alloys provide greater flexibility in achieving lattice-matched interfaces to III–V compound semiconductors. Many transition-metal (TM)–group III-metal intermetallic phases crystallize in the CsCl structure, an ordered derivative of the body-centered cubic structure. The TM–III intermetallic phases have lattice parameters approximately one half those of many III–V compound semiconductors. In terms of growth, their advantages include a wide range of compositional homogeneity and abrupt junctions (due to the presence of the group-III constituent). As with the silicides, epitaxical variants are possible for TM–III/III–V interfaces that are associated with the rotational orientation relationship. Similarly, their preparation is highly dependent on growth temperature, initial surface reconstruction and the composition of the III–V surface (Sands et al. 1990b). Rare-earth lanthanides, actinides plus Sc and Y form monopnictides and monochalcogenides with a NaCl structure that exhibits a large range of lattice parameters as well. For the monopnictides of Sc, Y and the rare-earth elements (all termed RE–V here), the lattice parameter range and chemistry are quite suitable for epitaxial and stable growth of III–V heterostructures as well. Figure 20 illustrates the range of lattice parameters for TM–III and RE–V compounds in comparison with those of several III–V compounds (Sands et al. 1990b). These intermetallic alloys provide much greater flexibility than elemental metals in achieving lattice-matched conditions with most common III–V semiconductors. In addition, these alloys provide a high thermal stability due to their phase equilibria (Palmstrøm et al. 1990) as well as morphological stability (low film agglomeration) due to their high melting points. By careful deposition of template structures and control of growth temperatures, it is possible to achieve buried III–V/metal/III–V epitaxy (Palmstrøm et al. 1990b, Sands et al. 1990a).

These epitaxical metal–semiconductor systems provide examples of surface structures that determine the growth of macroscopic overlayers. As in § 3.3, they also offer ideal interface structures for probing the role of chemical bonding in forming local electronic structure, in particular the interface states responsible for Schottky barrier formation.

3.2. *Chemical structure*

Ultrahigh vacuum surface science techniques show in general that the deposition of metals on semiconductor surfaces leads to interface-specific chemical phenomena

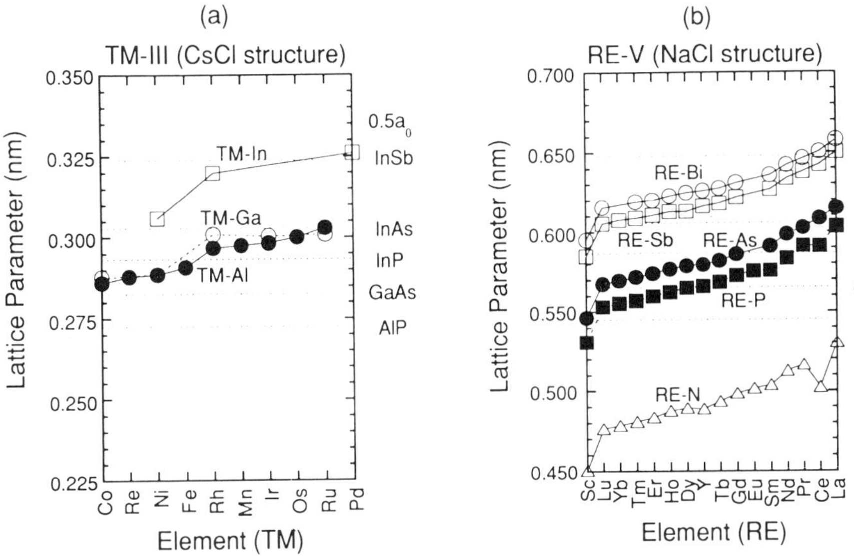

Fig. 20. Lattice parameters of several III–V compound semiconductors in comparison with the lattice parameters of (a) transition-metal TM–III phases with the CsCl (B2) structure and (b) the rare-earth RE–V compounds with the NaCl (B1) structure. Depending on the TM or RE element, it is possible to achieve lattice matching between the overlayer phase and the semiconductor substrate (after Sands et al. 1990b).

(Brillson 1982a, and references therein). These phenomena may alter the composition and bonding near the interface and lead to new chemical phases with properties different from those of the semiconductor and overlayer constituents. Depending on the nature of the local bonding interactions, overlayer growth can proceed in several ways – controlled by the specific kinetics and thermodynamics. The local bond strength and interface processing will also determine the nature and extent of the chemical reaction and interdiffusion, which exhibit a systematic behavior on a microscopic scale. As a result, several approaches are now available to control the interface chemistry via atomic-scale techniques. Both the interface chemical phenomena and their atomic-scale control are of significance since they can be associated with new electronic properties. The correlation of interface chemical and electronic structure discussed in § 3.3 reveals the significance of interface-specific phenomena on their macroscopic junction properties.

3.2.1. Overlayer growth

Overlayer growth occurs in several ways, depending upon the specific details of the chemical bonding, atomic structure, and processing at the interface. On a monolayer scale, gas phase adsorption on semiconductors has been the subject of considerable research, centered in large part on the role of local chemical bonding in catalytic processes and in the formation of electronic device structures (Brillson and Margaritondo (1988), Froitzheim (1988) and references therein). On a scale of microns, thin-film condensation, nucleation, and growth represents an area that has been extensively explored as well (Poate et al. (1978), Mayer and Lau (1990) and references therein). Intermediate between these regimes, the growth of adatoms into thin films depends on surface mobility, adatom–adatom and adatom–substrate bonding energies, substrate surface defects and their density, as well as the temperature and method of deposition (Weaver 1988, Weaver and Waddill 1991).

When atoms are incident on a surface, they can form bonds with the substrate by chemisorption or physisorption, as described in section 3.1.1. Without strong chemical interactions between adsorbate and substrate, overlayer growth may proceed in several fashions:

(i) layer by layer, with surface kinetics analogous to those described in § 2.2.1,
(ii) islanding, e.g., with three-dimensional growth commencing immediately following nucleation, and
(iii) a combination of (i) and (ii) in which a full monolayer grows before island growth begins.

Laterally uniform growth depends on the mobilities of adsorbates on the surface, which in turn depend on both kinetic and thermodynamic factors. Thermodynamic bond strengths are a surrogate for mobilities and are directly involved in island formation only in the high-mobility limit. In this kinetic limit, growth uniformity involves the strength of adatom–substrate versus adatom–adatom bonding, as well as the epitaxical relationship between adsorbate and substrate. Epitaxical growth facilitates lateral growth as oppposed to island growth and may be commensurate (e.g., the overlayer and substrate meshes are related by a simple mathematical transformation (Bauer 1982) or incommensurate (e.g., the overlayer and substrate are not rationally related).

Figures 21a,b illustrate the various non-interacting configurations for (a) isolated adatoms and (b) groupings of adatoms, respectively, depending upon their epitaxial

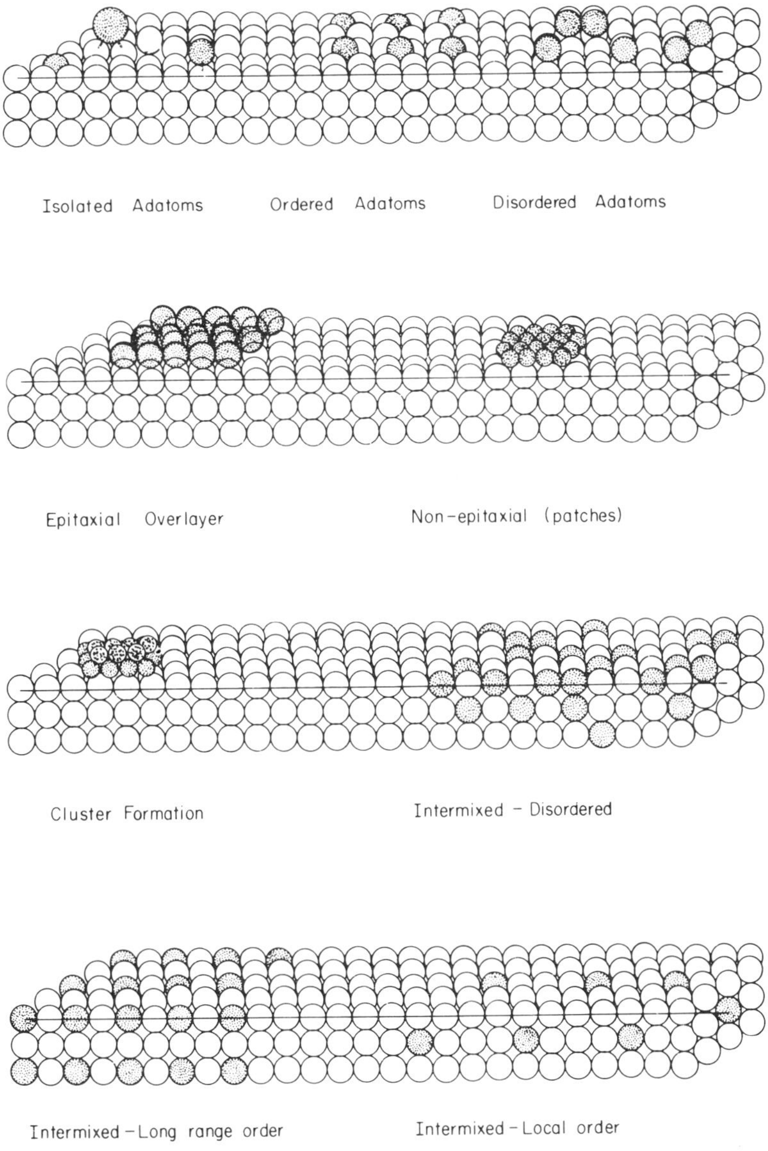

Fig. 21. Alternative pathways of interface evolution for atoms deposited on a surface. From top to bottom, the panels illustrate the atomic distribution and bonding with increasing deposition and/or time elapsed, starting with individual atoms, atomic groupings, subsurface distributions, and finally interface compounds or solutions (after Weaver 1988).

nature. Low adatom surface mobility due to strong adatom–substrate bonding (or high densities of adatom–defect bonding sites) favor more uniform lateral growth. On the other hand, a strong adatom–adatom bonding (i.e., a high heat of fusion or condensation) favors cluster formation, as shown in fig. 21c, which can provide the energy required to break substrate bonds and initiate chemical reactions, diffusion and other defect formation (Zunger 1983). Figure 21d illustrates the penetration of overlayer species into the substrate either in an ordered or a disordered fashion. A disordered solid solution is more likely when the chemical interactions of constituent atoms are nearly equivalent between different atoms and between like atoms entropy is the main driving force for diffusion. Conversely, a chemical phase with more long-range order and well-defined boundaries is expected when the energy gained by bonding between dissimilar atoms dominates the interaction (Weaver 1986). Segregation of substrate species to the growing film surface is also not an uncommon feature in this growth regime.

Energy processing by a variety of techniques can further modify the growth behavior. Here, energy processing refers to the use of heat, light, electrons, ions, or other energy sources to alter the interface chemistry. Examples include thermal and rapid thermal annealing, laser annealing, UV irradiation, and ion bombardment. See also § 3.2.3.

Activity in metal–semiconductor interface research over the past decade has focused on characterizing these interface atomic distributions and bonding as a function of the physical and chemical parameters and, in particular, on relating such chemical features to the interface electronic properties. The fully developed interface structure is in general not abrupt for most metal–semiconductor and other interfaces, even at room temperature. Instead, such interfaces can involve subsurface regions of interdiffusion and chemical reaction, as well as overlayer phases with dielectric properties unlike those of the chemical constituents (Brillson 1982a,b).

3.2.2. *Diffusion, chemical reactions, and phase formation*

Metal–semiconductor interfaces exhibit a systematic chemical behavior on a scale of both atomic layers and microns. Furthermore, many of these chemical features are characteristic of particular classes of semiconductors. This subsection provides examples of representative types of metal–semiconductor behavior to illustrate the general principles of interface chemistry, as well as distinctive aspects of different semiconductor classes which correlate with their electronic properties.

Photoemission spectroscopy, ion backscattering and other surface science techniques are used to characterize the detailed evolution of interface chemical phases. High-resolution SXPS provides an illustrative case of chemical structural analysis at a metal–semiconductor interface, showing the presence of several characteristic atomic-scale features. Figure 22a illustrates Si 2p core level spectra obtained as a function of increasing Ce deposition on Si(111). As indicated by the deconvoluted spectra, three distinct phases contribute to the evolving lineshape (Grioni et al. 1985). These are identified with (1) the Si substrate, (2) a reacted Ce–Si species, and (3) surface-segregated Si. Figure 22b provides core level intensities for each of these

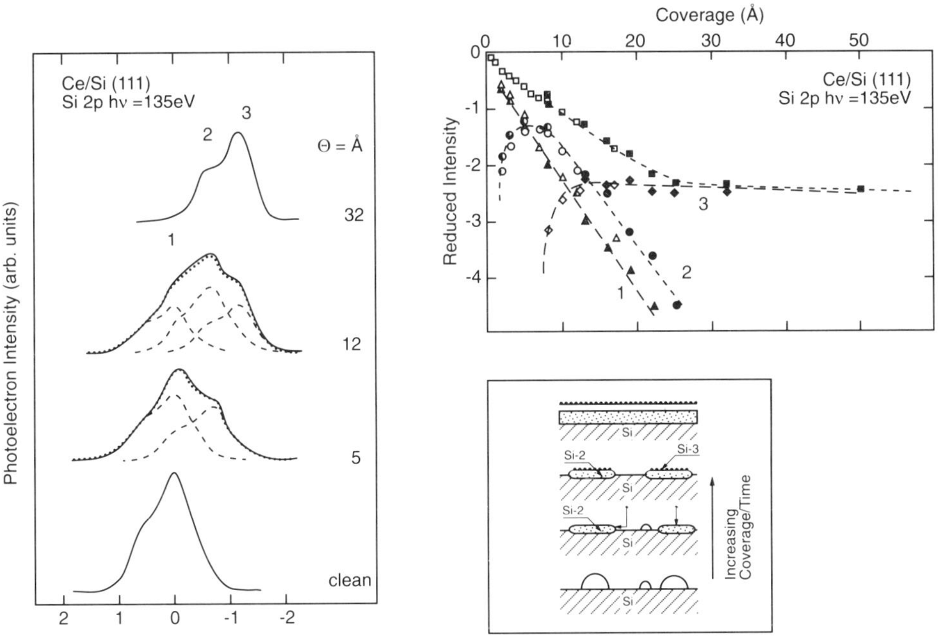

Fig. 22. Evolution of multiphase metal–semiconductor interface with increasing metal deposition: (a) SXPS core level lineshapes, (b) peak intensity attenuations, and (c) schematic model of the heterogeneous interface for Ce on Si(111). Decomposition of evolving Si 2p core level spectra taken at $h\nu = 135$ eV with increasing Ce deposition in (a) reveals the presence of three spin–orbit-split components attributed to clean Si (Si-1), a reacted phase (Si-2), and surface-segregated Si (Si-3). Attenuation of each component in (b) indicates interface localization of reacted component (after Grioni et al. 1985).

components as a function of Ce coverage. Appropriately normalized to the deposited metal emission, these intensities reveal (i) exponential attenuation of the substrate beneath the deposited film, (ii) the presence of a reacted phase localized to within 10–20 Å of the interface, and (iii) the presence of dissociated Si species. Figure 22c represents a schematic model for the heterogeneous growth of several phases simultaneously. The arrows point to regions that can exhibit different core level lineshapes. This model indicates that Ce–Si interface growth proceeds by cluster formation, reaction between the clusters and the underlying substrate, transport of impinging atoms along the surface to the cluster periphery, segregation of substrate species to the free overlayer surface, and ultimately, the formation of a continuous reacted overlayer (Van Loenen et al. 1984, Butera et al. 1986). Such a model of island growth and coalescence accounts for SXPS observations of Ce/Si(111), Ce/Ge(111) and V/Ge(111) (Butera et al. 1986).

Surface segregation depends on the cohesive energy and relative atomic sizes of the substrate and overlayer species (Lin et al. (1987), Weaver et al. (1990) and references therein), as well as the substrate atom solubility and diffusion rate within the overlayer. As illustrated in fig. 21d, substrate atoms may form a solid solution within the overlayer. Assuming that thermodynamic concepts are applicable to these ultrathin films, a Lever rule (Rhines 1956, Mayer and Lau 1990) may be applied to determine the relative amounts of the competing phases in the mixed-phase regime, which depends on the composition of each phase and the overall mole fractions of the constituents (Weaver 1986). Analogous surface behavior diagrams based on matrices of chemical reactions have also been proposed to describe multicomponent phase formation (Davis 1986).

For compound semiconductor–metal interfaces, interface chemistry is complicated by the increased number of substrate components, their combinations with the overlayer and their preferential outdiffusion. Thus, for the family of III–V compound semiconductors, both metal–anion and metal–cation phases form whose distribution and extent depend on the kinetics and thermodynamics (Brillson et al. 1983). Outdiffusion of semiconductor constituents into the metal overlayer is observed, even near room temperature, which is inversely proportional to the heat of semiconductor compound formation (Brillson 1982a, Grioni et al. 1986).

Semiconductor outdiffusion also depends on the interface reactivity (Brillson et al. 1981b), with metal–anion reaction products dominating the interface structure by providing diffusion barriers. Semiconductor anions dissociated from the semiconductor lattice can be "chemically trapped" by the metal overlayer, depending upon the strength of the metal–anion bonding. A measure of such metal reactivity with the semiconductor is the interface heat of reaction ΔH_R, expressed as (Brillson 1978a)

$$\Delta H_R = (1/x)[H_F(CA) - H_F(M_xA)] \tag{3.1}$$

for the reaction

$$M + (1/x)CA \rightarrow (1/x)[M_xA + C], \tag{3.2}$$

where ΔH_R is the difference in heat of formation (Wagman et al. 1968–1971, Mills 1974, Kubaschewski and Alcock 1979) between H_F for the compound semiconductor CA and the most stable metal–anion product M_xA, normalized per metal atom. Surface science measurements of interface chemical reactions confirm the existence of a critical heat of reaction dividing reactive and unreactive regimes for metals on II–VI (Brillson 1978a, Davis et al. 1985, 1988), III–V (Williams et al. 1978), III–VI (Hughes et al. 1982), II–V (Wyeth and Catalano 1980) and IV–VI (Baars et al. 1978) compound semiconductors. This expression can be generalized to include metal–cation alloying (McGilp 1984, McGilp and McGovern 1985) in terms of the heat of solution $\Delta H_{sol}(C; M)$ calculated (Miedema et al. 1980, Niessen et al. 1983) for cation C in metal M. In the limit of infinite dilution ($x \ll y$) for the alloy A_xM_y,

$$\Delta H_R = (1/r)\Delta H_F(M_rA_s) - (s/qr)\Delta H_F(C_pA_q) + (ps/qr)\Delta H_{sol}(C; M) \tag{3.3}$$

for semiconductor C_pA_q reacting with a metal M to form a compound M_rA_s. Inclusion of metal–cation alloying provides better agreement with SXPS observations of interface chemical reactivity for metal/III–V, III–VI, and II–VI interfaces where ΔH_{sol} makes a significant or dominant contribution to ΔH_R (McGilp and Montgomery 1989, McGilp and McLean 1988). Measures of metal–semiconductor reactivity such as ΔH_R are significant in that they correlate not only with interface chemistry but also in many cases with electronic properties as well.

Chemical trapping effects lead to outdiffusion of semiconductor constituent cations and anions whose stoichiometry (e.g., the ratio of cation to anion concentration) in the metal overlayer can be reversed, depending on the bonding at the interface (Brillson et al. 1981a). Figure 23 illustrates such a chemical trapping for atomic interlayers of Al at the Au/GaAs(110) interface. Here the formation of Al–As bonds results in an increased barrier to further As outdiffusion. As shown, the thickness of the interlayer controls the outdiffusion such that only a few atomic interlayers or less are capable of changing the outdiffusion stoichiometry by an order of magnitude. Even larger effects of interlayers on outdiffusion are evident with metals of different metal–anion reactivity (Brillson et al. 1980a). Such chemical phenomena can reverse the stoichiometry at the buried interface (Shapira and Brillson 1983), with electrical consequences to be discussed in § 3.3.

In contrast to the chemical behavior of the III–V compounds, metals in II–VI compound semiconductors display outdiffusion of semiconductor constituents whose anion/cation stoichiometry does not in general reverse (Brillson et al. 1983). In addition, strong metal–anion bonding produces extended anion outdiffusion and cation accumulation at the interface (Brucker and Brillson 1981). A notable exception is Yb on CdTe(110), where strong Yb–Te bonding results in a diffusion barrier to Te outdiffusion, little Cd outdiffusion, and relatively abrupt junctions (Shaw et al. 1988).

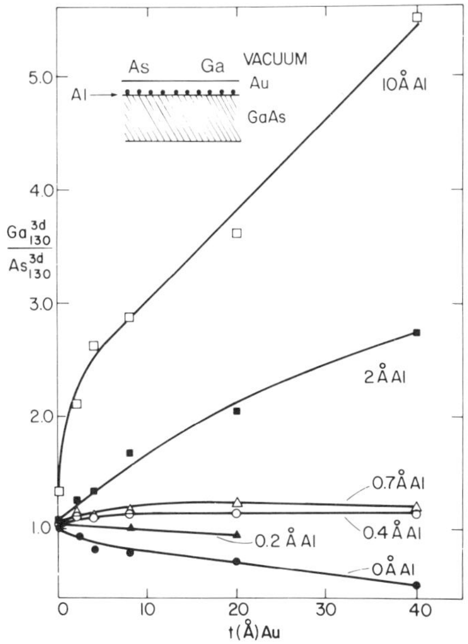

Fig. 23. Atomic-scale control of metal–semiconductor interdiffusion, as illustrated by SXPS core level ratios of Ga to As at $h\nu = 130$ eV for the Au/Al/GaAs interface versus Al interlayer thickness. The inset shows the interlayer configuration schematically. The thickness of the interlayer controls the outdiffusion of Ga and As into Au and has a significant effect on their relative abundance in the overlayer (after Brillson et al. 1980a).

General trends in chemical diffusion have not been established for other classes of compound semiconductors.

In addition to the variation in outdiffusion of dissociated semiconductor constituents, their reaction within the metal overlayer introduces additional complexity. For metals on III–V compounds such as InP and GaAs, SXPS studies reveal a wide range of chemical interactions, which depend on the particular metal (Kendelewicz et al. 1985a,b, delGuidice et al. 1986, Ruckman et al. 1986, Grioni et al. 1986, Ludeke 1986), including metal–anion bonding with different stoichiometries, metal–cation alloying as well as metal island formation. These features are evident at room temperature and, to a lesser extent, even at temperatures substantially below. As with elemental semiconductor–metal interfaces, elevated temperatures and thicker metallization for compound semiconductors permit the growth and evolution of additional chemical phases.

Metal interfaces with ternary compound semiconductors exhibit pronounced chemical interactions which reflect the relative strengths of the constituent bonds. Metals on HgCdTe promote disruption of the semiconductor lattice and preferential removal of the Hg constituent – indicative of weaker Hg–Te versus Cd–Te bond strength (Davis et al. 1983). Al–HgCdTe interfaces exhibit features of a microscopic AlCdTe compound. Majority carrier inversion may occur with metal-induced changes in crystal composition (Davis et al. 1984). Room temperature deposition of Ag on cleaved (110) HgCdTe produces Ag diffusion by 10^2–10^3 Å into the bulk lattice, displacing Hg away from the interface region (Friedman et al. 1986). These changes in local chemistry may cause electrical changes – in the case of Ag, either by the metal acting as a p-type dopant or by the creation of Hg vacancies. Preferential replacement of semiconductor constituents is evident for metals on $In_xGa_{1-x}As(100)$ $(0 \leqslant x \leqslant 1)$. (These noncleavage surfaces are prepared by As "capping" following growth and thermal "decapping" prior to metallization). Thus deposited Al replaces In in the (InGa)As lattice and forms a surface layer comprised of only Al, Ga, and As (Brillson 1986). This preferential replacement may arise due to the lower In- versus Ga–As bond strength or the larger mismatch in the InAs versus AlAs and GaAs lattice constants. Significantly, alloys of reduced In content ($x = 0.25$ versus 0.75) lead to proportionally less In outdiffusion – presumably due to reduced vacancy diffusion pathways or to the correspondingly smaller (In,Ga)As lattice spacing, which restrict metal indiffusion. As with binary compounds, metals of significantly different interface reactivity (Brillson 1978a) such as Au and Al on InGaAs exhibit qualitatively different local bonding and stoichiometry of outdiffusion. Furthermore, the stoichiometry of anion versus cation outdiffusion exhibit large and monotonic changes as the alloy composition varies from InAs to GaAs. These compositional changes provide a basis for interpreting the Schottky barrier changes that are observed for these semiconductors. Indeed, for such systems it is possible to relate the chemical features observed on an atomic scale to electronic changes observed spectroscopically and on a macroscopic scale (Brillson 1982a, Brillson and Viturro 1988, Brillson and Margaritondo 1988). Nevertheless, there exist no microscopic models that describe this chemical/electronic behavior comprehensively.

With increased metal coverage and temperature, reactions, diffusion and new phase formation occur on a scale of microns (Poate et al. 1978, Murarka 1983, Sands 1989). Much of our knowledge of the buried interfaces at thick metal–semiconductor junctions stems from Rutherford backscattering spectrometry (RBS) and transmission electron microscopy (TEM) studies of Si and III–V compound semiconductors. RBS measurements of metal–Si junctions show that interface formation occurs well below the lowest liquid-phase eutectic temperature and, in most cases, as temperatures of about one third to one half the melting point (in degrees K) (Mayer and Tu 1971). In general, a sequence of phases can evolve that depends on the temperature, metal thickness, type of metal, crystal structure, and impurities present. Depending on which element is consumed first, the interface system moves toward equilibrium by formation of compounds richer in the remaining element (Ottaviani 1981, Canali et al. 1979). As a representative example, fig. 24 illustrates schematically the sequence of phases formed in the SiNi film and the Si-crystal–Nb-film systems. In both cases, Ni_2Si forms initially at the Ni–Si interface. With annealing at increasing temperatures, a variety of phases appear, which depend on both the kinetics and the thermodynamics. As one of the elements is consumed the system is driven to equilibrium by the formation of compounds richer in the remaining element (Ottaviani 1979). Such compounds are known from the equilbrium phase diagrams of the two constituents (Hansen 1958).

In addition to the phase changes observed on a micron scale, secondary phases may be present in microscopic amounts, which are undetectable by RBS. In the Ni–

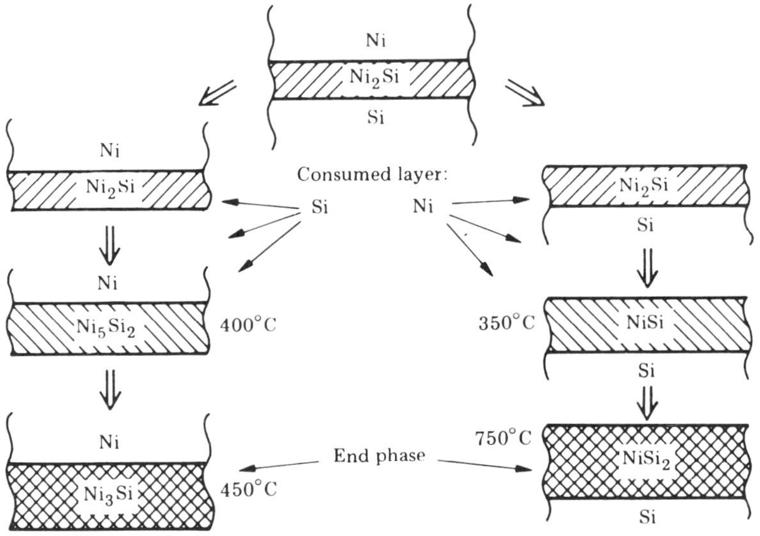

Fig. 24. Schematic diagram showing sequence of metallurgical phases formed in Si–Ni film and silicon crystal–Ni film systems. Initially Ni_2Si is formed. The indicated temperatures are rough guides of the phase formation for films several thousand angstroms thick and for an annealing time of 1 hour (after Ottaviani 1979). As one of the elements is consumed, the system is driven to equilibrium by formation of compounds richer in the remaining element.

Ni_2Si–Si case pictured in fig. 24, XPS reveals an Ni-rich silicide transition region, while the Ni_2Si–Si interface exhibits a corresponding Si-rich silicide region (Grunthaner et al. 1980). Such transition regions are only 10–20 Å in width. For this system, as well as Ni monolayers on Si, the transition region involves Si atoms displaced from lattice sites and Ni atoms diffusing to tetrahedral interstitial sites in Si (Cheung et al. 1981). Analogous behavior occurs for other metals. For epitaxical silicides such as Pd_2Si on Si(111), TEM lattice images reveal structurally sharp interfaces with a Si-rich silicide, misfit dislocations and atomic steps within several angstroms of the interface (Schmid et al. 1981).

A general scheme proposed for metal–Si interface growth (Braicovitch 1988) consists of

(i) an incubation stage in which the deposited metal reduces barriers to reaction and/or intermixing via rebonding, dielectric screening (Hiraki 1983), or other mechanisms,
(ii) an initial interaction involving lateral diffusion, indiffusion, lattice defect formation, clustering, and/or inhomogeneous adlayer growth,
(iii) a thicker coverage regime in which the details of these atomic-scale features influence the formation of various combinations of reacted/intermixed, homogeneous/inhomogeneous, and abrupt/extended features.

Thus, phase formation at metal–Si junctions demonstrates the influence of the local bond strength and of the atomic structure on the macroscopic composition and morphology.

Bulk-like metal films on III–V compounds exhibit a rich variety of chemical behaviors, analogous to the phase formation of their elemental counterparts. For metals on binary semiconductors, the three-component system will evolve during annealing to a combination of phases predicted by the appropriate isothermal section of the equilbrium phase diagram for the constituents (Beyers et al. 1987, Sands 1989, Williams et al. 1986). Such phases include not only compounds but also elemental clusters of these constituents (i.e. As clusters at GaAs interfaces (Woodall and Freeouf 1981, Lilienthal-Weber et al. 1986)). This assumes that the system is bounded such that species are not added or lost – a condition not applicable under many process conditions. Single-phase regions are separated from one another by two-phase regions consisting of "bundles" of tie lines (Sands et al. 1990b). For system compositions within a two-phase region, the composition of the two phases is given by the end points of the appropriate tie line, with the relative amounts of each phase being determined by the Lever rule.

The application of equilibrium bulk thermodynamics to metal–compound-semiconductor reactions has led to the design of thermally-stable, as-deposited metallizations. Deposition and prolonged annealing of a reactive elemental metal film on a compound semiconductor substrate leads to the formation of new, thermally-stable phases. In contrast, codeposition of the thermally-stable phase results in a thermally-stable contact that does not react with the semiconductor substrate (Sands et al. 1990b). Such thermally-stable phases, along with the epitaxial lattice properties and the high metal melting points required to prevent the agglomeration mentioned earlier, are basic requirements for the growth of abrupt, multilayer metal–semiconductor heterostructures. Compound overlayers on semiconductors may also exhibit metastability against chemical reaction as long as the compound phases formed by the initial overlayer–substrate reaction contain no low-melting-point phases to facilitate further interactions (Murakami et al. 1989). Microscopically abrupt semiconductor junctions with compound overlayers are also achievable via solid-phase regrowth (Marshall et al. 1989). While these concepts have been demonstrated for bulk-like metal films on III–V compounds, the presence of surfaces, interfaces and the associated strain are expected to alter the phase equilibria. Nevertheless, metal films as thin as about 100 Å have exhibited reactions consistent with the phase diagrams determined from bulk diffusion couples (Lin and Chang 1989).

Overall, these results indicate that the complexities of reaction, diffusion, and phase formation exhibit systematic features which are defined both by the kinetic and thermodynamic parameters of the macroscopic metal–semiconductor couples as well as the detailed bonding at their microscopic interface.

3.2.3. Atomic-scale control

Numerous atomic-scale techniques have been developed over the past decade for controlling the interface chemical structure. As semiconductor device dimensions shrink to ever smaller dimensions, such techniques acquire added significance in containing the extent of active volume altered by the interface chemistry. Furthermore, changes in local chemistry have electronic consequences, opening new avenues to control the interface electronic structure. Here the techniques developed to date are grouped in terms of: atomic-scale interlayer passivation, wet chemical surface passivation, modification via energy processing, and passivation via semiconductor growth.

Atomic-scale interlayer passivation may employ a wide variety of chemical species and physical mechanisms to modify the interface chemical structure. Figure 23 illustrates the order-of-magnitude change in outdiffusion stoichiometry for a reactive metal interlayer only a few monolayers thick at a compound semiconductor–metal interface (Brillson et al. 1980b). Here different thicknesses of the same metal display a monotonic increase in "chemical trapping" of outdiffusing anion with decreasing amounts of reactive species. Conversely, an interlayer 10 Å thick of different metals can alter the anion/cation stoichiometry observed at a free overlayer surface by over two orders of magnitude, depending on the metal–anion reactivity (Brillson et al. 1980a). Figure 25 illustrates the different metal- and reactivity-dependent P/In outdiffusion stoichiometry for single metals on InP. Reactive metals ($\Delta H_R \leqslant 0$ in the inset) such as Al, Ti, and Ni attenuate anion outdiffusion strongly, whereas metals such as Ag, Pd, and Cu with weak metal–anion bonding exhibit anion-rich outdiffusion. Also consistent with the ΔH_R correlation but not shown is the cation-rich outdiffusion reported for Cr and Mn and the anion-rich outdiffusion for Ag (Kendelewicz et al. 1985a). Sputter-depth profiling of such interfaces confirms the "chemical trapping" of anions at the intimate metal–semiconductor junction and an inverse relationship between cation/anion stoichiometry at the buried interface unlike that at the free overlayer surface (Shapira and Brillson 1983, Shapira et al. 1984).

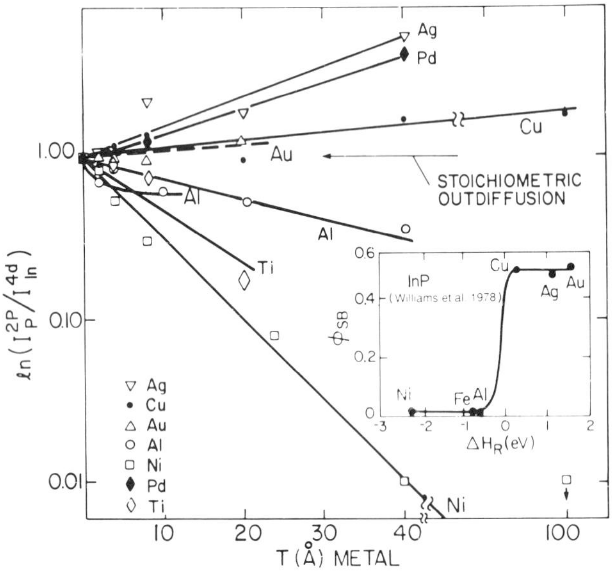

Fig. 25. SXPS ratio of surface anion/cation core level intensities I_P^{2p}/I_{In}^{4d} versus metal coverage on InP(110) relative to the UHV-cleaved surface ratio for Ag, Pd, Cu, Au, Al, Ti, and Ni. The inset contains a barrier height Φ_{SB} versus heat of reaction ΔH_R plot (after Williams et al. 1978) and emphasizes the correlation between Φ_{SB} and the stoichiometry of outdiffusion (after Brillson et al. 1981).

Hence interface chemical bonding affects the stoichiometry at buried metal–semiconductor interfaces, where such changes can have electrical effects. These are discussed in § 3.3. For metal interlayers at these compound semiconductor interfaces, the interface heat of reaction appears to be a semiquantitative thermodynamic tool with which to predict the interface chemical composition.

Metal interlayers at Si–metal interfaces both promote and reduce intermixing, depending on the interlayer coverage. This contrasting behavior relates to the different morphology or microscopic arrangement of Si atoms at the semiconductor surface (Franciosi et al. 1983). Similarly, interlayers can have a "catalytic" effect to enhance intermixing at Al–Si contacts (Brillson 1986) or, for alkali metals, to promote oxidation (Franciosi et al. 1987).

Nonmetallic interlayers may also influence the interface chemistry. Thus, an amorphous P layer (Olego et al. 1984) may limit outdiffusion by passivating the grain boundary and other pathways for outdiffusion. Silicide compounds can also serve to inhibit intermixing, as observed for monolayer thicknesses of $CoSi_2$ at Au–Si junctions (Xu et al. 1987). Combinations of metal overlayers and the chalcogens S, Se, and Te produce reacted and diffused species on III–V compound surfaces, which can alter the interface chemistry (Waldrop 1985a,b). The formation of a strongly-reacted interface compound can provide a barrier against semiconductor outdiffusion, a means to remove the excess anion or cation species from the interface, and/or a means to create an interfacial layer with new dielectric properties.

Gas adsorption can produce substantial changes in chemical structure at compound semiconductor surfaces. Examples include: H_2S adsorption on GaAs, which decreases the As-to-Ga surface concentration (Massies et al. 1980b, 1981); H and Cl adsorption on InP, which disrupts the semiconductor surface, leaving patches of In (Mönch 1983, Troost et al. 1987), oxidation and vacuum annealing of GaAs to remove As and produce a Ga_2O_3-covered surface (Grant et al. 1981); and air exposure of InGaAs, which causes a pronounced decrease in anion/cation outdiffusion with subsequent Au metallization (Brillson et al. 1986). The most inert interlayers used to promote changes in interface chemistry have been monolayers of the noble gas Xe. Xe condensed on a semiconductor surface at about 50 K acts as a buffer for subsequent deposition and clustering of metals, which is associated with semiconductor disruption (Weaver and Waddill (1991) and references therein). Upon Xe desorption at higher temperatures, an intimate metal–semiconductor contact forms with substantially reduced substrate disruption. Thus exposure of a semiconductor surface to specific gas ambients prior to metallization can have major effects on the chemical structure of the final metal–semiconductor junction.

Wet chemical treatments can modify the semiconductor interface chemistry in several ways. Photoelectrochemical washing of GaAs(100) surfaces in distilled water under band gap illumination results in dissolution of As and As oxides (Offsey et al. 1986). Inorganic sulphides such as Li_2S $(NH_4)_2S$, and $Na_2S \cdot 9H_2O$ are believed to etch As away and form stable Ga chalcogenides (Yablonovitch et al. 1987b, Sandroff et al. 1987). Preferential etchants can also serve to release epitaxical semiconductor layers from their growth substrates, thereby permitting "lift-off" and transfer to new substrates (Yablonovitch et al. 1987, 1990). In all three cases, wet chemistry removes active bonding sites in order to passivate the surface.

Energy processing techniques represent a different approach to interface chemistry modification. Perhaps the simplest in principle is high-temperature annealing. Brief (1 min) high-temperature (1250°C) annealing of (100)Si leads to clean, atomically-ordered (1 × 1) surfaces. These surfaces display two-to-three orders-of-magnitude less dissociation and interdiffusion (hundreds of angstroms versus tens of microns) when metallized with Al and annealed under conventional process conditions (400°C for 30 min) (Brillson et al. 1984). This type of behavior is attributed to the removal of surface disorder. Rapid thermal annealing (RTA) can promote chemical reactions preferentially. Using RTA one can produce preferential reactions at a Si–SiO_2 self-aligned gate structure such that Ti silicides form at the Ti–Si interface before silicides can form at the Ti–SiO_2 interface on the same surface (Brillson et al. 1985). In contrast to clean surface annealing, ion bombardment serves to promote surface disruption, which is reflected as enhanced interdiffusion after subsequent metallization (cf. Brillson et al. 1984). Ion bombardment is also useful in promoting the growth of novel deposited growth structures, due to the additional energy available for atomic diffusion and bond breaking (Greene et al. 1988). Finally, pulsed laser annealing is useful in preparing highly localized chemical reactions at metal–semiconductor interfaces (Poate and Mayer 1982). Pulsed-laser annealing can promote reactions at temperatures above the semiconductor melting point and at the same time restrict interdiffusion to only a few hundred angstroms (Richter et al. 1984, Richter and

Brillson 1986). Furthermore, the step-wise nature of the pulsed annealing leads to the production of intermediate chemical reaction products (Shaw et al. 1988a). The directed feature of the laser annealing permits lateral patterning of such reactions across the semiconductor surface. Thus energy processing can alter the interface chemistry either by the generation of a surface or interlayer, which alters the subsequent reactions/diffusion or by enhancement of such reactions/diffusion at the buried metal–semiconductor junction.

Epitaxical semiconductor growth represents the latest approach to control of interface chemical structure. Epitaxical (100) III–V surfaces capped with As following MBE growth, then thermally decapped prior to interface formation may exhibit less interface reaction with a reactive metal than does a UHV-cleaved (110) surface (Chang et al. 1990). For GaAs(100), the high-temperature anneal required for complete decapping may also introduce surface and subsurface As depletion (Massies et al. 1981). Such differences are most apparent at low temperatures (e.g., 50–100 K), where the interface chemical activity is otherwise reduced. Vicinal III–V compound semiconductor surfaces exhibit an enhanced interface chemistry. The extent of the Al–GaAs(100) exchange reaction at 90 K depends on the GaAs(100) misorientation direction (Chang et al. 1990) and misorientation angle (Chang et al. 1991), scaling with the density of active sites introduced by vicinal step edges. Epitaxical metal alloys form abrupt, thermally-stable interfaces with III–V compound semiconductors, as described in § 3.1.2. Overlayer epitaxy serves to inhibit the semiconductor outdiffusion otherwise possible at overlayer grain boundaries. Depending on the template structure formed on the III–V compound, an epitaxical metal alloy can grow with different crystal orientations (Palmstrøm et al. 1989). The growth temperature may also affect the crystallographic variant formed (Sands et al. (1990b) and references therein). In addition, such epitaxial metal–semiconductor interfaces provide a means to separate the crystallographic strain from the interface chemical composition, and the geometric orientation (Palmstrøm et al. 1990a).

The multiplicity of techniques described here all have the capability to alter and control the interface chemical structure on an atomic scale. As will be seen in § 3.3, many of these atomic-scale modifications can result in electronic changes, which manifest themselves on a macroscopic scale.

Overall, the chemical structure of adsorbate–semiconductor, and in particular metal–semiconductor, interfaces is rich and diverse. Local chemical bonding determines the mode of film formation and interface growth. Specific features of the metal–semiconductor bonding also govern the type and extent of the reaction and diffusion at the metal film–substrate contact. Furthermore, these atomic-scale features are useful as a means of preserving the junction integrity or of engineering new interface chemical structures.

3.3. Electronic structure

The electronic structure of the semiconductor–overlayer interface centers on the nature of the charge transfer and localization between the two media and the resultant band bending within the semiconductor surface space charge region. The vast majority of the research done to date pertains to the metal–semiconductor interface, both in terms of understanding the fundamental physical mechanisms involved as well as in developing electrical contacts for device applications. Accordingly, the focus of this subsection is the electronic structure of metal–semiconductor interfaces, categorized in terms of:

(a) the phenomenon of Schottky barrier formation,
(b) the physical nature of interface states responsible for semiconductor band bending, which can account for the observed barrier heights, and
(c) the atomic-scale methods now available to control the junction electronic properties. Many of these electronic properties can be correlated with chemical and/or geometric features described in §§ 3.1 and 3.2, to reveal the role of surface atomic properties in contact rectification.

3.3.1. Schottky barrier formation

The properties of junctions and barriers on a macroscopic scale have been reviewed in an earlier volume of the *Handbook on Semiconductors* series (Shaw 1981). See also Rhoderick and Williams (1988). The electronic properties of the metal–semiconductor interface depend upon the fundamental nature of charge transfer and localization at the microscopic heterointerface. Conventional techniques for measuring barrier heights macroscopically include current–voltage (I/V) methods for gauging charge transport across the interface, capacitance–voltage (C/V) methods for determining the height and width of the barrier region, and internal photoemission spectroscopy (I/P) for identifying optical transitions from the Fermi level to the majority-carrier band edge. With the advent of UHV surface science techniques, these physical mechanisms may now be addressed in greater detail. New techniques, such as soft X-ray photoemission, surface photovoltage, cathodoluminescence, photoluminescence, Raman and scanning tunneling charge spectroscopies, are now available that characterize band bending and localized states on an atomic scale. This subsection aims to complement the macroscopic, phenomenological approach (Sze 1981). Here we examine the role of localized charge in Schottky barrier formation and its relation to conventional charge exchange between metal and semiconductor.

When a metal contacts a semiconductor, charge exchange and band bending occur. Schottky barrier formation depends strictly on the potential difference between E_F in the metal and the majority-carrier band edge of the semiconductor. Figure 26 illustrates the band bending scheme for a metal at the surface of an n-type semiconductor (Schottky 1939). For the case of a metal with a high work function shown in the upper panel, electrons flow from the semiconductor to the metal after contact, depleting a characteristic surface region in the semiconductor of electrons. With the two Fermi levels E_F^M and E_F^{SC} aligned, a double layer forms with a voltage drop of qV_B equal to the contact potential difference between the metal and the interior of the semiconductor. The double layer consists of a surface space charge region, typically 10^{-4}–10^{-6} cm thick, and an induced charge on the metal surface. The n-type depletion region pictured is a layer of high resistance. Thus a voltage applied

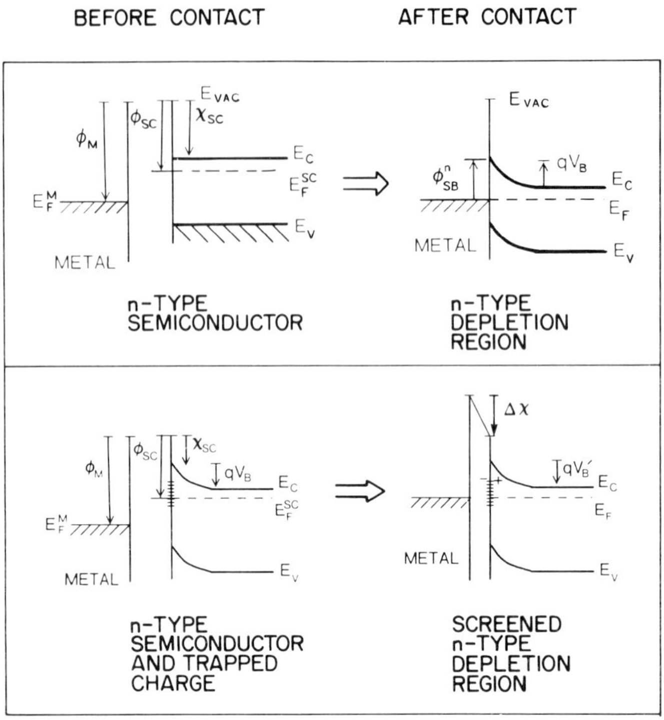

Fig. 26. Schematic diagram of band bending before and after metal–semiconductor contact for interfaces with (upper) or without (lower) trapped charge states. Trapped charge which screens the contact potential difference between metal and semiconductor may exist on the semiconductor surface before contact or may be created upon metallization.

to this junction will fall mostly across the surface space charge region. In the simple Schottky–Mott–Davidov model (Schottky 1939, Mott 1939, Davidov 1938, 1939), the barrier height Φ_{SB} depends on the difference in thermionic work function and is expected to be

$$\Phi_{SB} = \Phi_{M} - X_{SC}, \tag{3.4}$$

where Φ_M and X_{SC} are the metal work function and semiconductor electron affinity, respectively, defined with respect to the vacuum level E_{VAC}. According to this model, the potential Φ within the semiconductor satisfies Poisson's equation

$$\nabla^2 \Phi(x) = -4\pi\rho(x)/\varepsilon_s, \tag{3.5}$$

where ρ is the charge density in the surface space charge region of width w, x denotes the coordinate axis normal to the metal–semiconductor interface, and ε_s is the static dielectric constant of the semiconductor. This relation defines a parabolic band bending region for the abrupt metal–semiconductor junction. Furthermore, by assuming $\rho = qN$, the bulk concentration of ionized impurities within the surface space charge region, one obtains a depletion layer width of

$$w = [q\varepsilon_s(\Phi - \Phi_0)/2\pi N]^{1/2}, \tag{3.6}$$

where Φ_0 is the semiconductor potential for $x > w$. For a metal with a very low work function, band bending with the opposite sign can produce majority-carrier concentrations higher at the surface than in the bulk (accumulation region), resulting in no barrier to majority carrier transport across the interface. Metals on p-type semiconductors exhibit an analogous behavior with the opposite sign of band bending.

For most actual metal–semiconductor interfaces, eq. (3.4) does not accurately describe the dependence of the Schottky barrier heights on the metal work function. Until the last decade, barrier height measurements have indicated only a weak dependence of the semiconductor height on the particular metal contact. In the case of intimate metal–Si interfaces, metals with work functions ranging over 2 eV produce only a 0.5 eV variation in barrier height (Sze 1981). Likewise, common metals on compound semiconductors yield only a few tenths of an electron volt variation in band bending. Until recently, metal–GaAs junctions exhibited an E_F movement restricted to a range of only 0.2–0.25 eV (Palmstrøm and Morgan 1985). Table 5 provides Schottky barrier heights for metals on clean GaAs(110) and (100) surfaces (Newman et al. 1985). An analogous type of behavior is reported for I/V and C/V measurements on clean, cleaved InP(110) surfaces as well as chemically-etched surfaces of both semiconductors (Rhoderick and Williams 1988). Thus, for the vast majority of metal–semiconductor interfaces, eq. (3.4) does not in general describe the dependence of band bending on metal work function.

The insensitivity of semiconductor barrier heights to different metals can be explained by the presence of interface dipoles due to the additional, localized charge (Bardeen 1947). The lower portion of fig. 26 illustrates the formation of this dipole either before or after contact. If the existence of such states and dipoles reduces the movement of E_F within the semiconductor band gap with charge transfer – due to work function differences or an applied electrical bias – then the value of E_F is said to be "pinned" as opposed to merely stabilized at a given energy. Such states can be due to a variety of factors, as discussed in § 3.3.2.

The rectification process involves not just one but several dipoles. Figure 27a shows schematically the potential distributions for isolated metal and semiconductor surfaces. The work function of the metal Φ_M is comprised of two parts, an internal potential S_M and a surface dipole V_M. Correspondingly, the semiconductor electron affinity X_{SC} consists of the internal potential S_s, a surface dipole V_s, and ζ_n, the bulk electron E_F relative to the bottom of the conduction band. The surface dipoles are associated with electronic tunneling into the vacuum and are required to confine electrons within the solid at the surface. Figure 27b illustrates the potential distribution for the metal and semiconductor in contact. The Fermi level is constant across the junction so that the potential drop across the interface must equal the difference in internal potentials. The surface dipoles are replaced by an interface dipole V_i, and the potential drop is shared between V_i and the band bending V_b within the surface space charge region (depletion width λ) of the semiconductor. The relative magnitudes

Table 5
Schottky barrier heights (eV) for various metals on clean n-GaAs(110) and (100) surfaces (adapted from Rhoderick and Williams 1988).

Metal	(110)				(100): Waldrop 1984a)		Other	
	Newman et al. (1985)		Mead and Spitzer (1964)					
	I/*V*	*C*/*V*	*C*/*V*	*I*/*P*	*I*/*V*	*C*/*V*	*I*/*V*	*C*/*V*
Cu	0.87	0.94–1.08	0.83–0.90	0.82	0.96	0.96	0.85[b]	0.85[b]
Pd	0.85	0.88			0.91	0.93		
Ag	0.85–0.90	0.95–0.99	0.90–0.95	0.88	0.90	0.89	0.82[c]	1.03[c]
Au	0.92	0.99–1.05	0.93–0.98	0.90	0.89	0.87	0.88[c]	0.98[c]
Al	0.80–0.85	0.84–0.93	0.78–0.92	0.80	0.85	0.84	0.85[a], 0.78[b]	0.87[a], 0.78[b]
Ti					0.83	0.83		
Mn	0.72	0.75			0.81	0.89		
Pb					0.80	0.91		
Bi					0.77	0.79		
Ni	0.77	0.82			0.77	0.91		
Cr	0.67	0.72			0.77	0.81		
Fe					0.72	0.75		
Mg					0.62	0.66		
Pt			0.90–0.98	0.86				
Be			0.82	0.81				
Sn	0.77	0.82	0.68–0.74					
Ba			0.94					
Co					0.76	0.86		
Sb							0.73[b]	0.74[b]

[a](110): Svensson and Andersson (1985).
[b](100): Missous et al. (1986b).
[c](110): Ismail et al. (1984).

of V_i and V_b depend on the amount of charge localized in the interface dipole relative to the charge transferred into or out of the surface space charge region. From the relations illustrated in fig. 27, one obtains

$$V_b = (\Phi_M - X_{SC} + \zeta_n) + (V_s - V_M - V_i), \tag{3.7}$$

and the measured (n-type) barrier height Φ_B from the Fermi level to the conduction band edge $(X_{SC} + \zeta_n)$ is thus the sum of two terms, the classical expression involving the difference between the two observables Φ_M and X_{SC}, and a term involving the difference in dipoles (Duke and Mailhiot 1985). The p-type semiconductor expression is analogous. In practice, the dipole terms are not directly observable – one derives their magnitudes experimentally only from their effects on the observables Φ_M, X_{SC} and V_B.

Equation (3.7) shows that band bending and barrier height will vary as $\Phi_M - X_{SC}$ if the charges associated with the metal and semiconductor surfaces do not change when the interface forms – i.e., the dipole difference term is zero. This corresponds

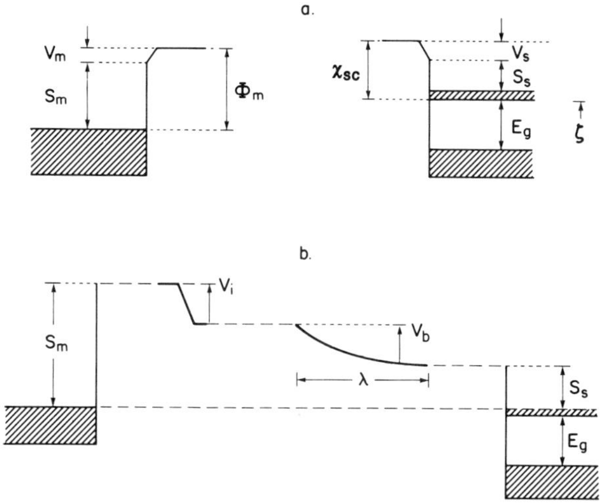

Fig. 27. Potential distribution for (a) separated metal and semiconductor surfaces and (b) the metal–semiconductor interface. The observable work function Φ_m (electron affinity X_{SC}) consists of a calculated internal potential S_m (S_s) plus a surface dipole V_m (V_s). The local interface dipole V_i plus the dipole V_b of the surface space charge region account for the difference in internal potentials when the metal and semiconductor are joined (after Duke and Mailhiot 1985).

to the simple band bending model described by eq. (3.4). However, for most interfaces studied previously,

$$\Phi_B \neq [\Phi_M - X_{SC}] \tag{3.8}$$

as shown by a weak dependence of Φ_B on Φ_M for different metals on a given semiconductor. For example, the barrier variations for metals on GaP, GaAs, Si, and CdS prepared under high-vacuum conditions in the 1960s showed slopes $d\Phi_B/d\Phi_M$ of only 0.27, 0.07, 0.27, and 0.38, respectively, instead of 1.0 (Kurtin et al. 1969, Sze 1981). See also table 5 (Newman et al. 1985) and fig. 33. This weak dependence implies $V_s - V_M - V_i \neq 0$ or, in other words, a screening of the potential difference between metal and semiconductor by additional charge trapped at the junction. These macroscopic electrical measurements demonstrate the general nature of the localized interface charge.

3.3.2. Interface states

Of the four types of localized states introduced in § 2.3, only intrinsic surface states of the semiconductor arising from the discontinuity in lattice potential and bonding at the vacuum–solid interface have been discussed so far. This subsection addresses interface-specific states, namely:

(i) localized, metal-induced gap states due to wave function tunneling from the metal into the semiconductor band gap,

(ii) extrinsic states due to contamination or lattice imperfections, and

(iii) metal-induced extrinsic features due to localized atomic bonding, interdiffusion, or chemical reaction.

Gap states associated with wavefunction tunneling are also considered intrinsic since they depend only on the bulk properties of the metal and semiconductor. Similar to intrinsic surface states (fig. 28a), the wavefunction tailing into the semiconductor band gap shown in fig. 28b constitutes a localized charge and an atomic-scale dipole which offsets part of the potential difference between metal and semiconductor. For simple metals on Si, such tails should extend on the order of 10 Å or less into the semiconductor (Heine 1965). Substitution of a featureless "jellium" of continuum states for the metal's electronic structure (de-emphasizing localized chemical bonding and any other structural properties) has afforded calculations that test the barrier sensitivity to different metal work functions for semiconductors across a wide spectrum (Schlüter (1982) and references therein). Such calculations display a weak dependence of Φ_{SB} on Φ_M for high dielectric constants of covalent semiconductors such as GaAs and Si and much less potential screening for ionic semiconductors such as CdS and SiO_2, consistent with experimental observations (Mead 1966, Kurtin et al. 1969).

An alternative approach to such tunneling charge redistribution involves the existence of a charge neutrality level in the semiconductor band gap, which minimizes the influence of the metal work function on the interface E_F position within the semiconductor band gap (Tejedor et al. 1977, Tersoff 1984). Here a new density of states due to the metal wavefunction is compensated for by a decrease in the semiconductor valence and conduction band density of states. A charge neutrality

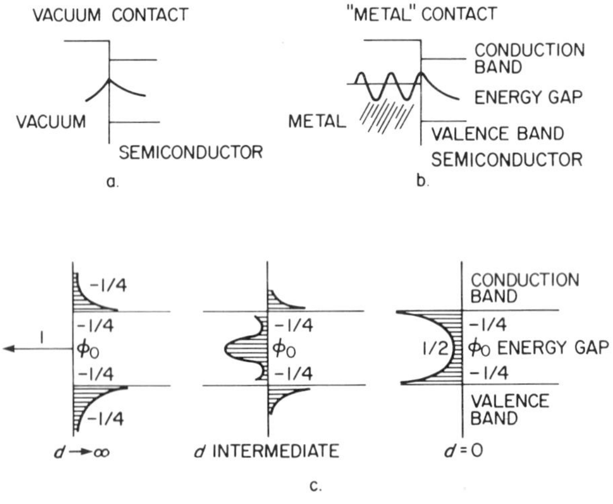

Fig. 28. Localized wavefunction tunneling: (a) into both vacuum and the semiconductor band gap (shown in (b)) at a metal–semiconductor interface. The exponentially decaying behaviour leads to the continuous density of interface states shown in (c) for a one-dimensional model of a covalent semiconductor–metal interface, where Φ_0 defines the charge neutrality energy of the semiconductor and d is the metal–semiconductor separation (after Tejedor et al. 1977).

level defines the energy below which the interface and valence-band densities of states compensate each other locally. Such a compensation follows from a sum rule which states that, at any site, the integrated local density of states, both filled and empty, is a constant (Applebaum and Hamann 1974, Kallin and Halperin 1984). Thus, the introduction of new gap states by changing the potential or boundary conditions alters the occupancy of states in the local conduction and valence bands according to its conduction- or valence-band character (Tersoff 1987). The effective mid-gap point between conduction- and valence-band character therefore determines an E_F position and local charge redistribution (dipole formation) required to maintain charge neutrality across the interface. Figure 28c illustrates the initial conduction- and valence-band densities without a metal, the formation of discrete gap states at intermediate distances, and the formation of a continuum of states at the intimate contact. Here the transfer of charge to or from the metal sets up a local dipole that tends to restore equilibrium of the metal E_F and charge neutrality levels and to render the barrier height insensitive to metal work function. The induced dipole potential D is given by (Tejedor et al. 1977, Mönch 1986):

$$D = (E_F - \Phi_0)/S, \tag{3.9}$$

where Φ_0 is the charge neutrality level and S is a dimensionless parameter that can be related to semiconductor dielectric properties. For $S \sim 0.1$–0.3 and $E_F - \Phi_0$ differing by 0.1 eV, D can be on the order of an electron volt. Metal wavefunction tunneling is calculated to produce large local dipole charge densities of about 10^{14}–10^{15} cm^{-2} eV^{-1}.

The calculated magnitude of this charge transfer and dipole depends sensitively on the assumed boundary conditions for charge penetration (Mailhot and Duke 1986), leading to conflicting assessments of the dipole magnitude and metal dependence (Harrison 1985). Factoring in local atomic bonding yields variations in charge neutrality level, depending on the metal–anion versus metal–cation bonding (Flores et al. 1989). Experimentally, a monotonic dependence of E_F "pinning" on the electronegativity of adsorbates on Si (Schmid 1985) and GaAs (Mönch 1989) is observed, as presented in fig. 29. As shown in fig. 29a, both transition-metal silicides and noninteracting metals on Si exhibit a near-linear dependence on the electronegativity. Figure 29b illustrates the variation in barrier height position versus electronegativity for a wide variety of metals measured by the J–V technique (Waldrop 1984a,b, 1985a,c, Ludwig and Heymann 1986) shown in comparison with the positions of E_F for Cs, S, and Cl adsorbates. The solid line corresponds to the Φ_B variation based on the changes in charge transfer due to the electronegativity difference. These Miedema electronegativity values derive from a semiempirical fit with thermodynamic solubility data (Miedema et al. 1980) and related to the Pauling (1960) electronegativity (Mönch 1989). Deviations from this line are attributed to extrinsic charge sites, i.e., defects, to be discussed. Here potential screening by metal-induced gap states serves to reduce the electrostatic effects of the local interface structure and may define a range of stabilization energies that depend on the relative density of extrinsic versus intrinsic gap states.

Additional possible causes for intrinsic E_F pinning include:

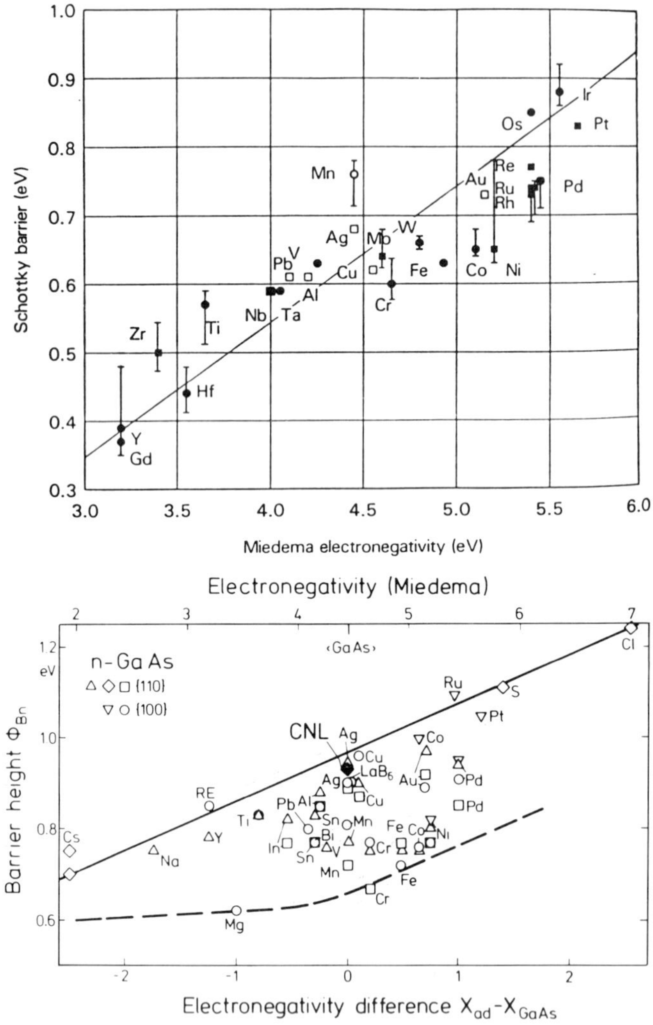

Fig. 29. Barrier height, (a) as a function of Miedema electronegativity for transition metal silicides and nonreacting metals on n-type Si (after Schmid (1985) and references therein), and (b) as a function of electronegativity difference between the adatoms and the substrate with $X_{sub} = 4.45$ metal–GaAs contacts and for adsorbates on GaAs(110) cleaned surfaces. Data points in (a) correspond to silicides (■ and ●) and nonreacting metals (□). The solid line in (b) indicates the Φ_B variation due to tunneling charge transfer as a function of $X_{ad} - X_{GaAs}$ and no defects. The dashed line indicates the E_F position with the largest density of defects compatible with the available data at $E_{CBM} - 0.65$ eV. The deviations in Φ_B from the solid line indicate high densities of interface defects (after Mönch 1988a,b, 1989).

(a) a mechanism for shifting electrostatically the dangling bond hybrid orbitals in conduction and valence bands (Harrison 1976) to keep E_F pinned at a constant gap energy,

(b) a "negative U" model of negative electron correlation between the interface electrons due to atomic disorder and electron localization (cf. Schlüter 1982),

(c) a narrowing of the semiconductor band gap at the interface – due to a reduction of exchange – correlation contributions to the band gap (Inkson 1974), and
(d) combinations of metal-induced gap states and defects.

Tests of all these metal-induced gap state models require atomically-abrupt metal–semiconductor interfaces, i.e., without additional chemical phases with their own dielectric properties. However, only a few such interfaces have been studied to date, as will be discussed in § 3.3.3, and these indicate barrier heights that depend on the interface atomic structure.

Extrinsic interface states due to lattice imperfections or contaminants can be further subdivided according to their surface or bulk origins. Surface chemisorbed species are known to produce new electronic states in the semiconductor band gap (Brattain and Bardeen 1953, Goodwin and Mark 1972). SXPS studies of E_F movements at low temperature, where clustering and/or new chemical bonding is retarded, exhibit "overshoot" phenomena believed to be due to charge exchange involving adsorbate donor or acceptor states (Stiles et al. 1987, Stiles and Kahn 1988, Cao et al. 1987). Figure 30 illustrates the E_F dependence on metal coverage and its relation to the proposed adsorbate level. In the case of a mid-gap donor, the effect of additional positive charge at the surface (fig. 30a) is to increase band bending and raise E_F closer to the donor level energy. Taken as a measure of the donor position, the "overshoot" can be related to the properties of the isolated adatom – e.g., the first ionization potential. Figure 30b shows the linear dependence for an aray of adsorbate-induced donors (○) and acceptors (●) on GaAs(110) (Kahn et al. 1989). This is analogous to adsorption on ionic surfaces (Baidyaroy et al. 1972) and in accordance with models of defect-modified tunneling (Mönch 1988b). In practice, it is possible to infer bonding configurations for such surface adsorbates from electronic spectra as altered by chemisorption (Margaritondo et al. 1979, Mele and (Joannopoulos 1987). Strongly bonded adsorbates also alter the surface electronic properties for more ionic semiconductors (Baidyaroy and Mark 1972, Spicer et al. 1976, Brillson 1976).

Chemical contamination on a monolayer scale also causes substantial changes in E_F stabilization at interfaces. Clean surfaces with and without exposure to the atmosphere exhibit qualitatively different E_F movements and band bending with subsequent metallization (Brillson et al. 1986). In the case of Au on InGaAs, such exposure to air alters both the premetallized band bending and the final Schottky barrier height, with the Φ_B values of air-exposed interfaces agreeing with the diode measurements of interfaces prepared under low-vacuum conditions (Kajiyama et al. 1973). Residual contamination has pronounced effects on the MBE growth and the J–V characteristics of epitaxical Al–GaAs(100) junctions (Missous et al. 1986b). Likewise, air exposure significantly degrades the stability of metal/GaAs(110) diode barriers with annealing (Newman et al. 1988).

Numerous structural imperfections exist which can introduce new states into the semiconductor band gap. These include exposed lattice steps due to cleavage or misorientation as well as near-surface point defects. Cleavage steps are well-known sources of electronic states in the semiconductor band gap (van Laar and Scheer 1967), with densities of states sufficiently high (e.g., of the order of $10^{14}\ \mathrm{cm}^{-2}$ or

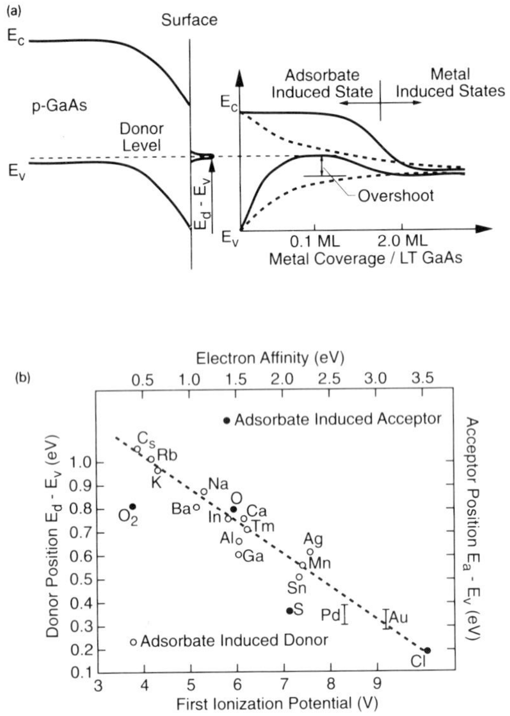

Fig. 30. Characteristic E_F movement (a) as a function of submonolayer and monolayer metal coverages on GaAs(110) at room temperature (dashed line) and low temperature (solid line). The adsorbate-induced donor level shown at the left corresponds with the initial maximum E_F energy for p-type GaAs. At higher coverages, the E_F convergence is attributed to metal-induced states. In (b), the adsorbate-induced donor (acceptor) level position extracted from the initial E_F movement on low-temperature p-type (n-type) GaAs appears as a function of the adatom's first ionization energy (electron affinity) (after Kahn et al. 1989).

higher) for E_F "pinning" to occur (Huijser and van Laar 1975) at poorly-cleaved surfaces. Such electrically-active sites also manifest themselves in terms of higher chemical activity, e.g., metal–compound-semiconductor exchange reactions as mentioned in § 3.2.3 (Chang et al. 1990a,b) and oxidation of Si (Ibach et al. 1973). Indeed, for intentionally stepped, vicinal surfaces of GaAs(100), a quantitative correlation exists between the density of interface states deep in the semiconductor band gap and the density of exposed, active chemical sites (Chang et al. 1991b). Point defects are also associated with the surface and interface charge. These include interstitials and vacancies introduced during surface cleaning, i.e., Ar^+ bombardment (Brillson 1975), and Schottky point defects resident at cleaved surfaces (Whitman et al. 1990). Native defects also occur at interfaces and bulk semiconductors exposed to high-energy electrons and ions (Aukermann 1973, Walukiewicz 1988a). For GaAs and possibly other compound semiconductors, the energy levels of such defects correspond rather well with the E_F "pinning" energies for the metal/UHV-cleaved semiconductor interfaces.

Foreign atoms within the semiconductor lattice can introduce deep levels, which can alter the surface space charge region (see, e.g., Milnes 1973, Halsted et al. 1965). Such impurities may be associated with the semiconductor growth process (i.e. C in GaAs), nonstoichiometry of the growth phase (i.e. excess As in GaAs), or as a result of indiffusion during metallization and processing (i.e. Au diffusion into GaAs) (Hiesinger 1976). As with III–V compound semiconductors (Milnes 1973), metal impurities and their complexes with native defects in II–VI compounds can introduce a multiplicity of deep levels that span the band gap (Halstead et al. 1965). In general, noble and near-noble metals such as Au, Ag, Cu, and In tend not to react strongly with semiconductors and to diffuse rapidly at relatively low temperatures (Boltaks 1963, Shaw 1973). Such diffusion is particularly pronounced for the more ionic semiconductors, where relatively large lattice constants and vacancy concentrations facilitate impurity movement.

Native defects within the semiconductor bulk lattice introduce a wide array of energy levels in semiconductors (Mirceau and Bois 1979). Orders-of-magnitude differences in deep-level emission from GaAs grown by different methods is observable. Similar variations in deep-level emission are evident for CdTe and other II–VI compound semiconductors. For both GaAs and CdTe, the deep-level luminescence energy correlates closely with the energies of E_F stabilization measured spectroscopically (Shaw et al. 1989). In addition, such imperfections may be segregated to the semiconductor surface or interface (Bartels et al. 1983) and their distribution may be sensitive to thermal processing (Makram-Ebeid et al. 1982). Finally, the metallization process itself may introduce deep-level segregation (Yahata and Nakajima 1984) with concentrations sufficiently high to produce measurable electrostatic effects at the interface.

Extended surface or bulk imperfections are electrically active as well. These include misfit dislocations, antiphase domain boundaries, and combinations of native defects produced during growth and/or processing. Arrays of misfit dislocations exhibit electrostatic band bending consistent with line charge densities and depletion regions extending radially outward (Woodall et al. 1983). As misfit density increases, depletion regions can overlap, thereby changing the average barrier height across the interface plane. Dislocation loops/dislocations introduce deep levels in trigonal Se (Brillson and Griffiths 1978). The interfacial stress introduced by metallization can also lead to deep levels, that is via silicide formation over Si/SiO_2 interfaces (Reuters et al. 1990) or plastic deformation of GaAs (Hoffman et al. 1990).

All these extrinsic features point to the significance of bulk and surface crystal quality in Schottky barrier formation. The deviations from ideal-crystal properties on an atomic or larger scale contribute electrically-active sites, which compete with interface-specific phenomena in controlling the contact charge transfer and band bending.

Considerable evidence exists for the role of interface-specific phenomena in forming localized states which dominate the Schottky barrier formation. These interface-specific, extrinsic states include: localized states due to metal–semiconductor bonding and interdiffusion, defect formation, and new phase formation. All three types are sensitive to the kinetics and thermodynamics of the chemical interactions at the intimate metal–semiconductor junction. Phenomenological evidence for localized states due to metal–semiconductor bonding extends across many interface systems. A linear correlation exists between the barrier heights of transition-metal silicide/Si junctions and heats of formation (enthalpy changes) of the transition-metal silicide (Andrews and Phillips 1975), which can be interpreted as the strength of chemical bond charge transfer and bond hybridization between metal and Si atoms. The barrier heights for transition metals on Si also exhibit a linear dependence on a measured thermodynamic property of the interface, namely, the eutectic temperature for the transition-metal silicide/Si system (Ottaviani et al. 1980). Figure 31 illustrates the increase in Φ_B with decreasing eutectic temperature, which extends over nearly half the Si band gap. The particular eutectic shown is that closest to the metal side for silicides whose growth is dominated by metal diffusion and vice versa for interfaces with growth determined by Si diffusion. See, e.g., § 3.2.2. The correlation in fig. 31 is thus based on a measured interfacial thermodynamic parameter and suggests that band bending is related to the bond strength of the interfacial layer.

The chemical dependence of the interface charge transfer is evident from photoemission measurements of metal chemisorption on semiconductor surfaces. For Si interfaces, these results emphasize the metal-specific nature of interface bonding (Rubloff (1983) and references therein). Indeed a correlation exists between metal d-band occupancy and the heats of silicide formation (Hara and Ohdomari 1988). Similarly the near-linear dependence of Φ_B on electronegativity shown in fig. 29 is evidence that Φ_B depends on local bond charge transfer (Schmid 1985, Mönch 1988a, 1989).

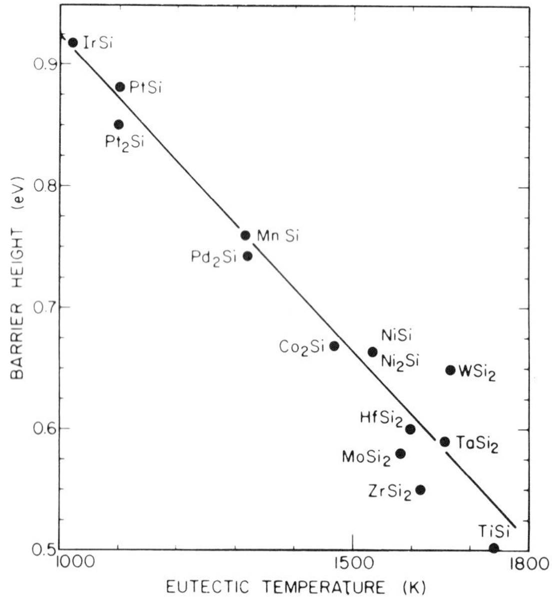

Fig. 31. Barrier heights of transition-metal silicides measured by conventional electrical methods plotted versus lowest melting eutectic temperature for the transition-metal silicide/silicon system (after Ottaviani et al. 1980).

A strong correlation between barrier heights and interface thermodynamics is also evident for compound semiconductors. Figure 32 illustrates a qualitative transition in Schottky barrier heights for metals on compound semiconductors, which depends on the interface heat of reaction ΔH_R according to eq. (3.1) (Brillson 1978a). As shown, this transition occurs at the chemical boundary between reactive and unreactive interfaces, $\Delta H_R \sim 0$ for semiconductors whose ionicity spans the range from ionic to covalent. Similar barrier dependences on ΔH_R are evident for InP (Williams et al. 1978, Hughes et al. 1982, Hökelek and Robinson 1983), Zn_3P_2 (Wyeth and Catalano 1980) and PbTe (Baars et al. 1978). This apparent relationship between Φ_B and chemical reactivity suggests that chemical stability is a factor in the barrier dependence on different metals. Figure 33 displays the variation of this Φ_B dependence

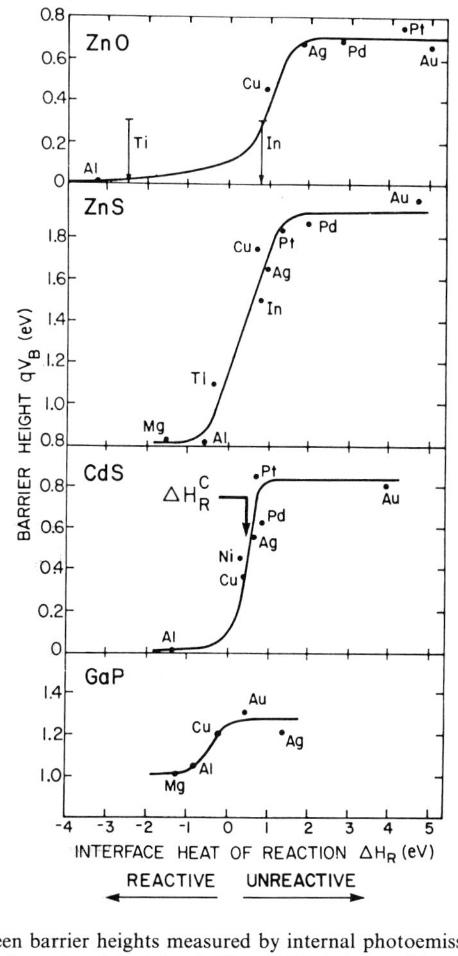

Fig. 32. Correlation between barrier heights measured by internal photoemission (after Mead 1966) as a function of interface heat of reaction ΔH_R (after Brillson 1978a). A transition from low to high Φ_B occurs at ΔH_R^C, observed spectroscopically. Qualitatively similar behavior occurs for a wide variety of semiconductors, regardless of ionicity.

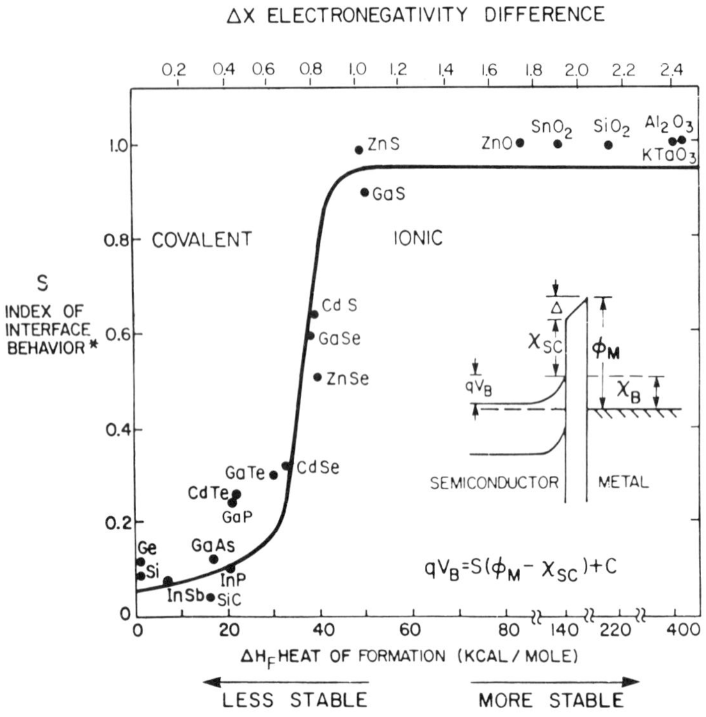

Fig. 33. Transition of interface behavior S as described in the inset as a function of electronegativity difference (upper scale) (after Kurtin et al. 1969), or, equivalently, the chemical heat of formation (lower scale) (after Brillson 1978a) for a wide variety of semiconductors.

on Φ_M in the terms of an index of interface behavior S as defined in the inset. The same set of S-values are plotted against both the electronegativity difference ΔX (Kurtin et al. 1969) and chemical heat of formation ΔH_F (Brillson 1978a). The S-values may increase above unity (Schlüter 1978). In addition to showing that E_F "pinning" is common to the more covalent semiconductors, fig. 33 also indicates that semiconductors with the lowest chemical stability exhibit the least Φ_B dependence on different metals, whereas semiconductors with high chemical stability exhibit the highest (Brillson 1978b, 1982a).

The chemical dependence of Φ_B for compound semiconductors can be related to the stoichiometry of anion/cation outdiffusion. The inset in fig. 23 illustrates a Φ_B versus ΔH_R plot analogous to fig. 32. Significantly, metals in fig. 23 producing anion-rich outdiffusion lead to high values of Φ_B, whereas, without exception, metals exhibiting "chemical trapping" and cation-rich outdiffusion produce low n-type Φ_Bs (Brillson et al. 1980b). This link between outdiffusion stoichiometry and high or low E_F position suggests that compound semiconductor outdiffusion can produce electrically-active sites with different deep-level energies and that these deep levels contribute to stabilizing the interface Fermi level. Here the chemical bond strengths can alter the band bending by an indirect mechanism.

Geometric effects at metal–semiconductor interfaces support the role of local bonding in Schottky barrier formation. In an extensive review of Si/transition-metal interface compounds Calandra et al. (1984) emphasized the importance of the detailed atomic positions of metal atoms at the semiconductor interface in determining the density of states in the band gap region. As shown in fig. 34, barrier measurements of epitaxial $NiSi_2$–Ni interfaces (Tung 1984) exhibit significant Φ_B differences for two epitaxical interfaces of $NiSi_2$/Si – 0.65 eV for type-A epitaxy and 0.78 eV for type-B epitaxy. The subtle differences in bonding between these two eptiaxical variants were discussed in association with fig. 19. While such differences can be complicated by differences in structural perfection rather than epitaxy (Liehr et al. 1985), the $NiSi_2$/Si differences have been confirmed experimentally (Hauenstein et al. 1986, Tung et al. 1986, Ospelt et al. 1988) and justified theoretically on the basis of tunneling charge

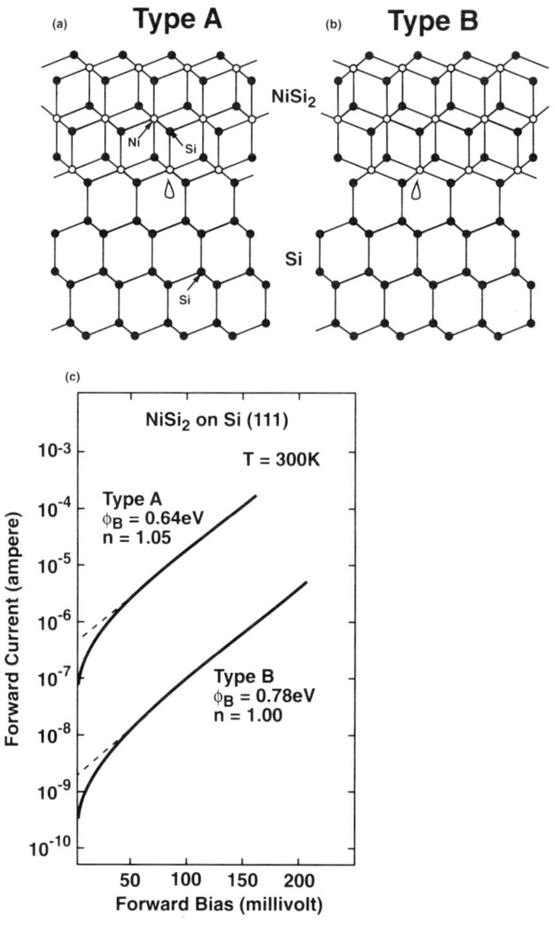

Fig. 34. Interface lattice geometry of (a) type-A and (b) type-B $NiSi_2$ on Si(111) surfaces. Current–voltage characteristics (c) for diodes where rectifying contacts are type-A and type-B $NiSi_2$ on Si(111) exhibit ideality factors near unity and significant Schottky barrier differences (after Tung 1984).

transfer (Fujitani and Asano 1990, Das et al. 1989). Analogous Φ_B differences have been reported for Pb/Si(111) interfaces differing only in the structure of the first interface layer of Pb and Si atoms (Heslinga et al. 1990). For both Ni–Si and Pb–Si junctions, the abrupt metal–semiconductor interfaces are free of extended chemical reactions, diffusional mixing, and lattice disorder. Instead, variations in crystal orientation can produce Φ_B differences of about 0.1–0.2 eV. Since the bulk lattice constants of each diode are identical, only local bonding and interface electronic structure can contribute to this Φ_B variation. Indeed valence-band photoemission spectra indicate marked differences in the occupied densities of states for the two Pb–Si phases (Le Lay et al. 1989). For these epitaxical junctions the interface states apparently depend on the interface geometric structure. Overall, there appear to be several strong correlations between the atomic bond strength and the geometry with band bending at metal–semiconductor interfaces.

Defects created by adsorbate chemisorption on semiconductor surfaces may also be a significant factor in Schottky barrier formation (Mark et al. 1975, 1977, Wieder 1978, Lindau et al. 1978, Williams et al. 1979, Spicer et al. 1980a). Here the energy released by chemisorption induces bond breaking and surface defects with deep-level densities sufficiently high to "pin" E_F. A basis for proposing the surface bond breaking mechanism is the rapid stabilization of E_F for GaAs with only submonolayer chemisorption and a similar behavior for both metals and oxygen (Spicer et al. 1980a). Besides the modifications in surface bonding induced by atom condensation, cluster formation can release sufficient energy to overcome the activation barriers for breaking bonds within the semiconductor surface (Zunger 1981). Evidence for the role of clusters in forming charge sites can be inferred from the different barriers produced by metals with and without an Xe interlayer, e.g., for clusters formed on the semiconductor directly versus prior to metal–semiconductor contact (Weaver and Waddill 1991). The adsorbate-independent nature of the E_F "pinning" level energies suggests that characteristic defects form for a given semiconductor (Spicer et al. 1980a).

Theoretical activity has focused on the deep levels of vacancies (Daw and Smith 1979, 1980, Srivastava 1979) and antisite defects (Allen and Dow 1981, 1982, Dow and Allen 1982) at or just below the semiconductor surface and their trends with particular metal or semiconductor compositions. Early interface defect models incorporated surface anion and cation vacancies (Wieder 1978, Spicer et al. 1980a). Subsequent calculations show that antisite defects have lower activation energies than simple defects in III–V compounds (Van Vechten 1975, 1980). At present, the limitations in predicting absolute energies with sufficient precision renders the comparison of theoretical and experimental energy values insufficient to rule out particular defects unequivocally.

Known deep-level energies provide an alternative approach to the identification of interface defects. The correspondence between energy levels derived from E_F pinning measurements and those observed for the deep level EL2 provide a basis for As_{Ga} antisite defects as the electrically-active sites at metal/GaAs(110) interfaces (Spicer et al. 1988). As shown in fig. 35, such an identification involves the formation of EL2 donor defects in the near-surface region with two donor levels – here at 0.75 and 0.5 eV. Furthermore, the crystal must have additional minority acceptor levels

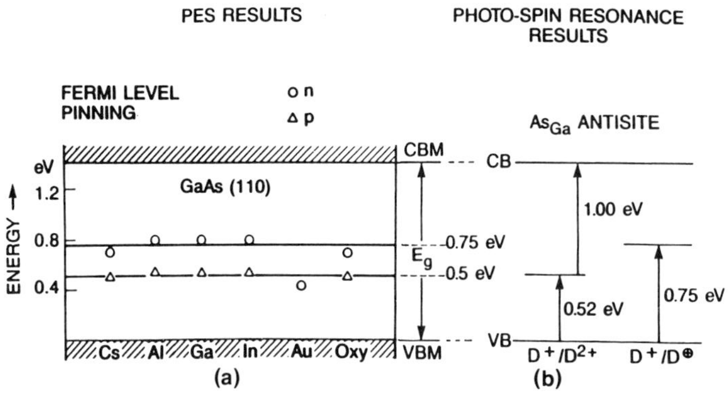

Fig. 35. Photoemission spectroscopy (after Spicer et al. 1980a) and photoinduced electron spin resonance measurements (after Weber et al. 1982) of defect levels in GaAs and their relation to E_F pinning. Pinning positions for n-type (○) and p-type (△) GaAs appearing for the indicated elements in the room temperature 1.42 eV GaAs band gap indicate defect energies at 0.65 and 0.5 eV, respectively. The photospin resonance results indicate nearly identical energies of the As_{Ga} antisite levels above the valence-band maximum at 8 K ($E_g = 1.52$ eV) (after Spicer et al. 1988).

below 0.5 eV to account for pinning at mid-gap in n-type GaAs – for example, a Ga_{As} antisite defect (Fornari et al. 1989). In this defect model the E_F position depends to some extent on the interface chemistry and, in particular, on the anion/cation stoichiometry, since the relative densities of donor and acceptor sites are dependent on composition. Thus the n- and p-type stabilization in the 0.5 to 0.75 eV energy range shown requires As_{Ga} antisite concentrations close to twice those of the Ga_{As} antisites. In turn, this requires generally As-rich interfaces – consistent with the growth of melt-grown cleavage material from the As-rich end of the GaAs bulk phase diagram. Less As-rich interfaces then lead to "pinning" at the 0.5 eV level or below. Hence the particular Ga/As stoichiometry resulting from metal–semiconductor interactions determines the ratio of different antisite defects and thereby the E_F position in the range between the defect levels. The formation of additional defect complexes could shift E_F outside this narrow range of energies. Other stoichiometry-dependent defect models incorporate additional deep-level defects (Nishizawa et al. 1990) and semiconductor doping (Walukiewicz 1987, 1988b). Studies of GaAs E_F versus surface reconstructions with annealing (Svensson et al. 1984, Cheng et al. 1984), Ga films on GaAs (McLean and Williams 1987) and LaB_6 Schottky barrier behavior with annealing (Yokotsuka et al. 1987) support a variation of E_F position with GaAs interface stoichiometry. Thus antisite defects offer a plausible explanation for the mid-gap E_F stabilization and its variation with interface chemistry observed at many metal/GaAs interfaces.

An alternative interpretation of GaAs Schottky barrier formation involves interface impurity levels forming within the semiconductor band gap which become delocalized through interaction with states of the metal overlayer (Ludeke et al. 1989). Here the effective charge per defect level depends self-consistently on the metal impurity energy as well as the energy shift and broadening due to the delocalization. However, evidence for such impurity-induced levels have to date been controversial (McLean et al. 1988).

Defects near semiconductor surfaces may also compete with other mechanisms during the initial stages of Schottky barrier formation. For example, the "overshoot" phenomenon pictured in fig. 30 can be interpreted as the influence of work function (Cao et al. 1987) or tunneling states (Mönch 1988b) prior to the creation of large densities of defects at higher metal coverages. As shown, the slopes with metal electronegativity in fig. 30 suggest a significant but secondary role for defects at Si–metal junctions, but a more dominant effect at GaAs contacts. Overall a variety of defect and defect plus tunneling models appear capable of accounting for the narrow range of Schottky barrier heights reported for most melt-grown GaAs–metal interfaces and their dependence on the particular metal used in terms of electronegativity or stoichiometry effects.

New alloy phases at the metal–semiconductor interface provide yet another basis for charge transfer and band bending. As described in § 3.2.2, chemical interactions between metal and semiconductor can lead to interfacial phases with work functions or other dielectric properties different from the starting materials. For metals on Si, a correlation between the alloy work function based on a common metal–Si_4 stoichiometry and the geometric mean of the constituent properties for the alloy work function suggests the presence and dielectric effect of such an interphase (Freeouf 1980). Indeed TEM lattice imaging and photoemission measurements provide evidence for such a Si-rich stoichiometry at the otherwise abrupt junction (Schmid et al. 1981). STM measurements of metals on Si may provide additional insight into the actual interfacial structure of such junctions.

For interfaces with III–V compound semiconductors, a variety of techniques reveal the presence of excess anion species segregated to the intimate metal–semiconductor interface (Brillson (1982a) and references therein, Brillson et al. (1981a), Lilienthal-Weber et al. (1986)). Hence, the work function of different metals on, for example, GaAs can be that of As and the E_F appears "pinned" at a common mid-gap energy defined by the classical charge transfer between the interfacial phase and the semiconductor (Freeouf and Woodall 1981, Woodall and Freeouf 1981). Figure 36 presents E_F positions on an absolute energy scale with respect to the valence and conduction bands of different III–V and pseudobinary alloy compound semiconductors. The relatively constant interface Fermi level for relatively unreactive (Au) contacts to these materials all lie close to the dashed line $E_{VAC} - E_F = 5.0$, corresponding to work functions of 4.8–5.1 eV, for example, close to the classical work functions for As and P. Significantly, these Φ_Bs for diodes prepared under low-vacuum conditions are in close agreement with the E_F positions determined for metals on bulk cleaved III–V crystals as well. Thus the effective work function can account for a wide range of contact measurements on different semiconductors under different conditions. The interface states associated with As-rich phases may also be considered chemisorption states at submonolayer coverages, with observable mid-gap energies (Viturro et al. 1988b), evolving into metallic edges for As clusters, precipitates or macroscopic phases. As with various defect models, one can interpret the resultant Φ_B changes in terms of changing interface anion compositions. Furthermore, such interfacial phases

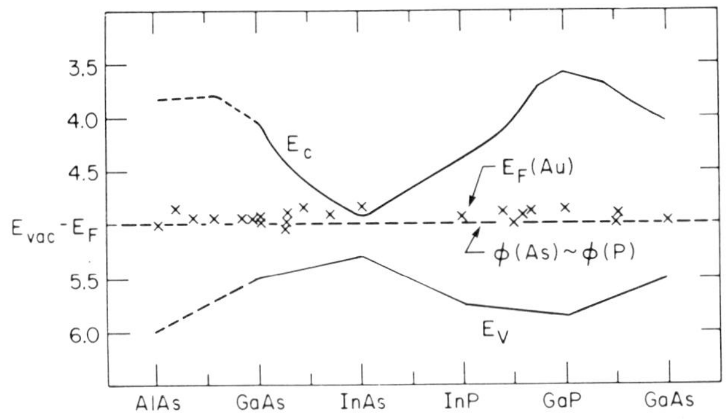

Fig. 36. Interface Fermi level positions for relatively unreactive (mostly) Au contacts to various III–V compounds and alloys. On an absolute scale, the Fermi level "pinning" energies lie within the band gaps at the anion work function energy. This correlation suggests the presence of an anion-rich interphase that dominates the Schottky-like charge transfer (after Freeouf and Woodall 1981).

are directly observable by microscopic techniques, allowing for straightforward experimental verification.

New interphase properties may result from lattice disorder, which also gives rise to localized states. Such lattice disorder can be associated with fluctuations in bond angle and length as well as the presence of stress and irregularity at the boundary, resulting in a distribution of bonding and antibonding states with energies around a characteristic hybrid orbital energy (Hasegawa and Ohno 1986, Hasegawa et al. 1987). Such disorder introduces dangling bonds with a characteristic continuum of energies in the band gap and a minimum density near an empirical hybrid energy, which can account for insulator–GaAs and metal–GaAs "pinning".

Independent of a particular physical mechanism, it is possible to gauge the effect of localized charge states on the semiconductor band bending. The presence of acceptors and donors located near the semiconductor interface leads to interface dipoles V_i that alter the overall band bending V_b according to eq. (3.7). The charge state of such electrically-active sites depends upon the E_F position relative to the energy levels of these states in the band gap. The total interface charge depends upon which donor and/or acceptor sites of this type are present, their densities, and their relative ordering in energy. The relation between V_b and Φ_M will exhibit a distinctive dependence on the detailed energetics and positions of the localized states. Figure 37 illustrates the V_b dependence on Φ_M for various interface charge-center densities of three-charge-state (e.g., positive, neutral, negative) centers as n-type or p-type semiconductor surfaces (Duke and Mailhiot 1985). Depending upon the exchange–correlation interaction potential U for such centers, acceptor energies lie above or below the donor energies. Here the center levels are at $E_d = E_g/3$ and $E_a = 2E_g/3$ for $U > 0$ and $E_a = E_g/3$ and $E_d = 2E_g/3$ for $U < 0$. They are situated an arbitrary distance of 10 Å below the surface. Figures 37a and b show that deviations from Schottky-like behavior (eq. (3.4)) occur outside the energy range between donor and acceptor states. Figures 37c and d are indicative of the E_F level "pinning" behavior addressed by the

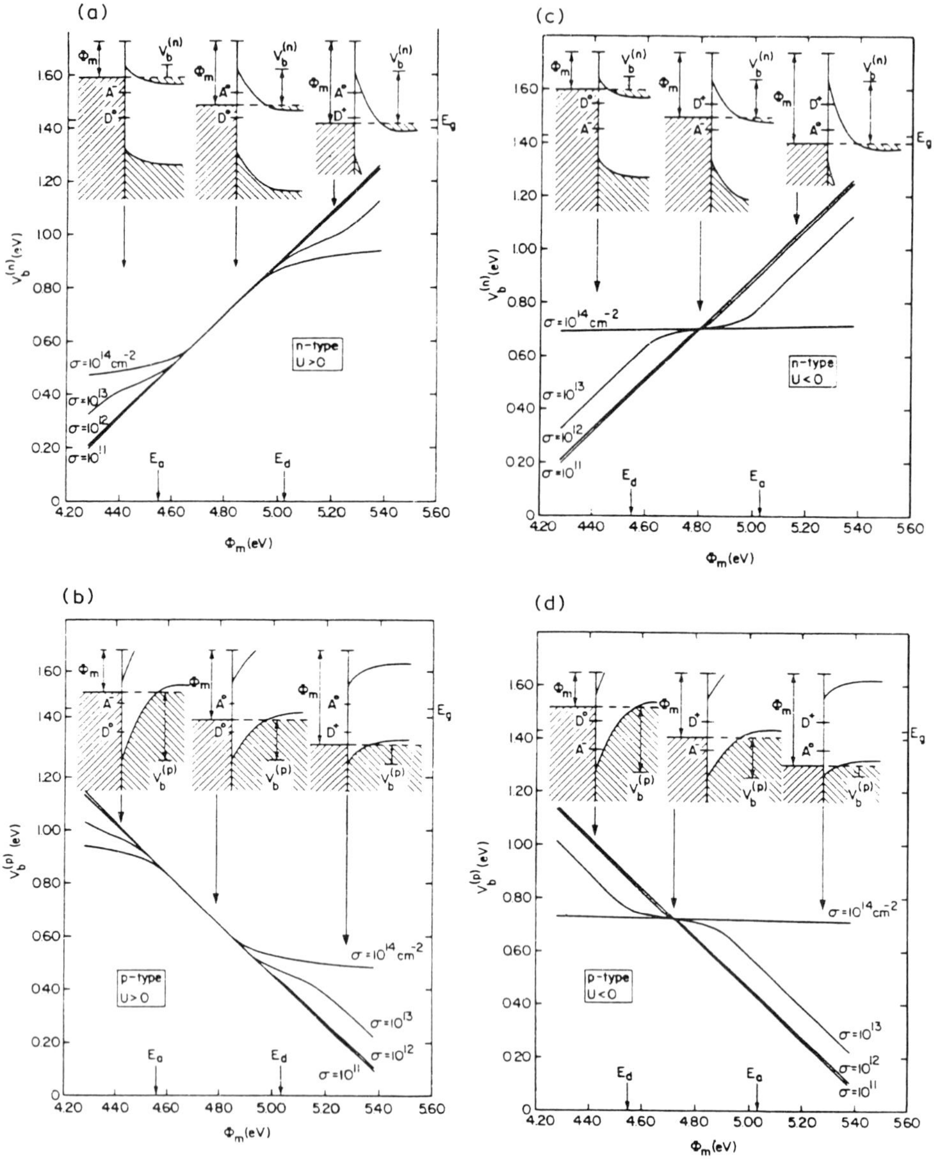

Fig. 37. Barrier height as a function of metal work function calculated for various interface charge-center densities in the case of a three-charge-state center with positive U (acceptor above donor in energy) for n-type (a) and p-type (b) GaAs with $E_d = E_g/3$ and $E_a = 2E_g/3$. For negative U on n-type (c) and p-type (d) GaAs, $E_a = E_g/3$ and $E_d = 2E_g/3$. Charge centers are situated 10 Å below the semiconductor surface. The qualitatively different functional dependences serve to differentiate among various charge-center combinations (after Duke and Mailhiot 1985).

numerous physical models described above, where V_b assumes a constant value over a wide range of Φ_M. Here the neutral state is unstable with respect to the charge donor and acceptor states, and electrons are exchanged in pairs between D^+ and A^- at E_F positions between the two. For surface densities of negative-U centers on the order of 10^{14} cm^{-2}, charge exchange between the metal and the semiconductor goes almost entirely into these centers, resulting in almost no change in the barrier height. Thus fig. 37 represents a generic, self-consistent method of gauging the effect that combinations of charge centers have on the V_b versus Φ_M dependence.

Overall, a diverse assortment of interface state mechanisms can create localized interface states. However, because each of these can account for the relatively narrow range of energies reported in strongly "pinned" junctions, i.e., metal/GaAs and several other interfaces, it has not yet been possible to identify a unique mechanism responsible for non-ideal Schottky barrier behavior. Such an identification requires both additional characterization of the interface states and a more detailed description of the E_F behavior during the interface formation. Several experimental techniques now make possible characterization of interface state energies, relative densities, and charge character. Atomic-scale processing techniques which widen the range of E_F stabilization for a given semiconductor surface also enable finer distinctions to be made between various physical mechanisms. The latter are discussed in § 3.3.4.

3.3.3. Interface state characterization

Experimental observations of interface states have been a challenging goal for several decades. Conventional techniques for measuring the presence of interface states are based on transport or electrostatic gauges of band bending across the semiconductor surface space charge region. While these have provided a basis for interface state characterization since the 1960s, they are at best indirect probes since they measure the effect of localized charge on the extended features of the junction – the change in energy band position and charge density within the surface space charge region. Furthermore, they are subject to a number of complications, including barrier tunneling, new dielectric layers, and recombination or trapping at states near the interface and within the band gap (Shaw 1981).

Compared with conventional methods, surface science techniques permit a more detailed characterization of localized states but are hampered by their relative insensitivity to low charge densities. For example, states with densities of 10^{10}–10^{12} cm^{-2} can produce significant electrostatic effects, whereas the limit of detection for electron emission techniques is typically 0.01 monolayer equivalent or about 10^{13} cm^{-2}. Techniques with a high surface sensitivity have a concomitant inability to probe the electronic and chemical properties more than a few monolayers below the surface. Nevertheless, a number of techniques have yielded significant information about the interface state properties including photoemission, electrostatic, and buried interface spectroscopies.

Photoemission, inverse photoemission, and their variant spectroscopies have provided measurements of intrinsic surface state energies outside and, to a limited extent, within the band gap, as described in § 2.3.2. However, localized metal–semiconductor interface states have been difficult to distinguish from simple metallization of the overlayer. Two-photon laser-pumped photoemission spectroscopy has provided evidence for states induced by metal overlayers (Haight and Bokor 1986). Electrostatic techniques provide a higher sensitivity to surface state concentrations but lack the atomic-scale, chemical information of the electron emission approaches. Surface photovoltage spectroscopy (SPS) represents perhaps the simplest of such techniques, involving only a vibrating Kelvin probe in a feedback circuit to provide a contact potential difference spectrum of the photoexcited semiconductor surface (Gatos and Lagowski 1973, Brillson 1975). Because the sensitivity is typically on the order of a millivolt and surface charge densities of 10^{12}–10^{13} cm^{-2} can produce barrier heights of a volt, SPS can detect densities of as little as 10^{9}–10^{10} cm^{-2}. While this technique is capable of identifying the conduction- or valence-band nature of gap-state optical transitions – a feature lacking in luminescence spectroscopies, its spectral sensitivity diminishes sharply with overlayer coverage as well.

Among the techniques capable of measuring the electronic properties of interface states on a monolayer scale, STM has significant advantages in its ability to provide laterally-specific information on an atomic scale (Binning et al. 1983) as well as both filled- and empty-state information. STM can detect the presence of filled or unfilled states from the dependence of the conductivity on the tip-to-specimen voltage. Figure 38 illustrates such suitably normalized spectra for Sb atoms and clusters on GaAs(110) surfaces (Feenstra and Mårtensson 1988). The top panel illustrates a GaAs(110) surface with an approximately 0.7 monolayer equivalent Sb forming a 1×1 ordered terrace on the semiconductor substrate (dark portion of the image) and the corresponding normalized conductivity versus voltage spectra (bottom) for edges ((a) and (b)) and terraces ((c) and (d)) of the Sb-covered GaAs. Here the Sb induces two pronounced filled-state features associated with dangling bonds on two inequivalent Sb atoms in the overlayer (Bertoni et al. 1983, Mailhiot et al. 1984a) as well as enhanced empty-state features above the band gap in energy. Such valence- and conduction-band changes with coverage and ordering are also evident from photoemission (Hansson et al. 1981, Bringans and Bachrach 1990, Himpsel 1990). For Sb clusters on GaAs, different band bending shifts of the voltage scale are evident, which indicate stronger band bending on compared with off the clusters. The cluster and terrace edges exhibit peak features not evident in the corresponding terrace spectrum and which bracket the "pinned" Fermi level. These edge states are believed to arise from dangling bonds localized near the edges of Sb terraces and to depend on the extent of incomplete bonding at the interface (Feenstra and Mårtensson 1988), although the chemical nature of such a bond perturbation is not yet known. Hence, fig. 38 illustrates the existence of discrete states associated with particular atomic-scale features whose energies can account for the E_F stabilization.

At higher metal coverages, STM and other surface-sensitive techniques are less effective at probing these interface states. For such buried interfaces, cathodoluminescence (CLS) and photoluminescence (PL) spectroscopies can measure the further evolution of the band gap states. Figure 39 represents the detection of interface states for metals on InP cleaved (110) surfaces (Viturro et al. 1986). The clean surface

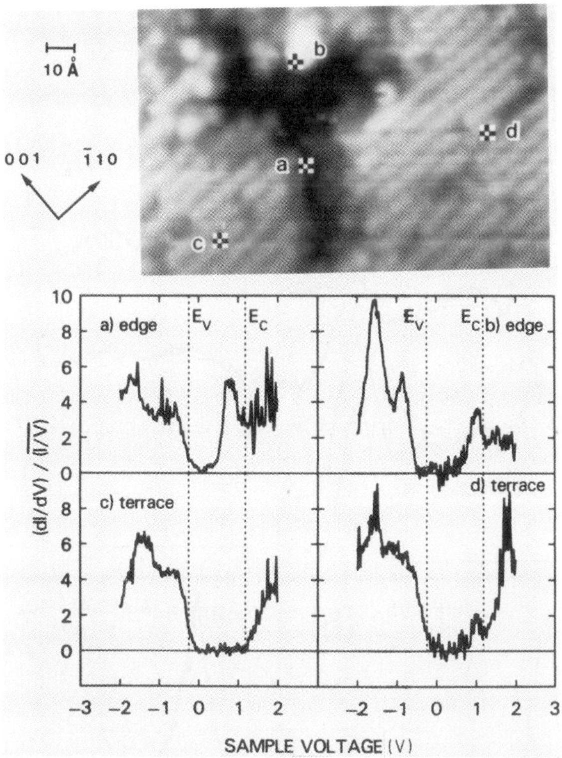

Fig. 38. Scanning tunneling microscopy and the corresponding normalized conductivity versus voltage spectra for the 0.7 monolayer equivalent on GaAs(110). Locations at edges of terraces (a) and clusters (b) as well as on top of terraces (c) and (d) label positions at which spectra were acquired. Ordered Sb overlayer regions and clusters give rise to new states outside the band gap. Cluster edges (a) and (b) provide evidence for new states within the gap region (Feenstra and Mårtensson 1988).

exhibits no gap features, only band-to-band recombination. With deposition of monolayer deposits of metal, discrete luminescence peak features appear whose energies and intensities evolve with metal coverage. Different metals produce features with different energies, a different evolution with coverage, and different attentuation of the band-bending-dependent band gap emission. The observed energies are consistent with the Schottky barrier heights measured by electrical and photoemission techniques, and their evolution is consistent with SXPS observations of E_F movements and chemical bond changes at characteristic multilayer coverages. Thus while such luminescence techniques are at present only semiquantitative in assessing the gap state densities, they nevertheless demonstrate that interface states below a multilayer metal film continue to change in energy and density and can account for E_F stabilization behavior.

Additional characterization of buried interface electronic structure is available with ballistic electron energy microscopy (BEEM), which permits spatial mapping of the Schottky barrier height across the metallized surface (Kaiser and Bell 1988). As

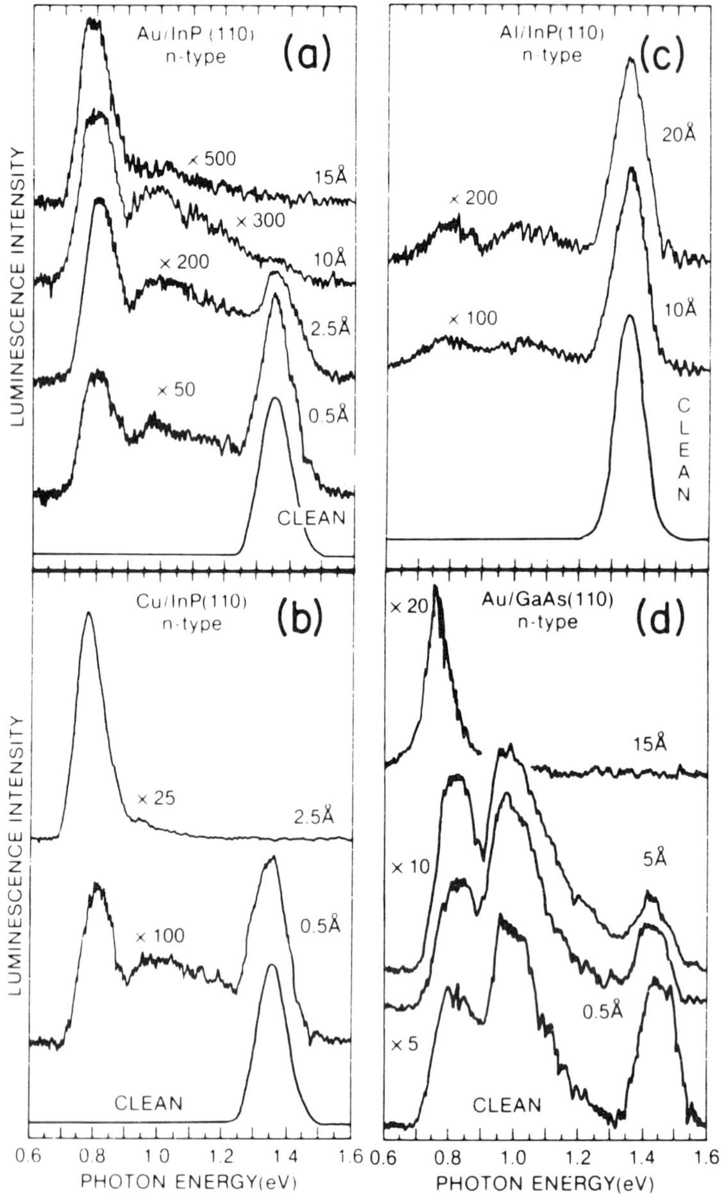

Fig. 39. Direct observation of interface states by low-energy cathodoluminescence spectroscopy for (a) Au, (b) Cu, (c) Al, and (d) Pd on UHV-cleaned InP(110), illustrating detection of optical transitions involving discrete states within the band gap which depend on the particular metal and which evolve with multilayer coverages (after Viturro et al. 1986).

a variant of the STM technique pictured in fig. 4, BEEM involves the onset in electron (hole) injection from the STM tip into the semiconductor conduction (valence) band and the variations in transport band edge with band bending (Bell et al. 1990). Thus it is possible not only to gauge Schottky barrier heights at buried interfaces but also to monitor their variations across an interface on a near-atomic scale. Such measurements exhibit considerable barrier heterogeneity, independent of morphology, for the several junctions studied to date (Kaiser and Bell 1988, Fowell et al. 1990, 1991).

These and other techniques yield considerable information about the physical nature of interface states during the initial stages of Schottky barrier formation. The properties of such interface states appear to be sensitive to the chemical and structural properties of the contact, consistent with the properties of semiconductor surfaces already described in § 2. Such sensitivity argues against models of interface states with little or no dependence on the chemical and geometric features of the intimate junction. Further discrimination between physical mechanisms is possible for interfaces with chemically and structurally-induced changes in barrier heights.

3.3.4. Atomic-scale control

Recent advances in fabricating and characterizing metal–semiconductor contacts have led to a variety of new techniques for controlling the contact electronic properties and to a further refinement of the atomic-scale mechanisms of interface state formation. Analogously to § 3.2.2, the numerous atomic-scale techniques available for modifying the electronic structure can be categorized as: atomic-scale interlayer passivation, wet-chemical passivation and passivation via semiconductor growth. Traditional methods of contact formation to most semiconductors involve contacts with appropriate processing aimed at forming tunnel diodes, localized-state-assisted contacts, or graded composition, reduced barrier junctions (Palmstrøm and Morgan 1985). Figure 40 illustrates schematic energy band diagrams for metals on GaAs, which represent such metallurgical approaches to Schottky barrier control (after Murakami et al. 1989, Sands et al. 1990b). Barriers lower than the conventional thermionic emission barrier of 0.8 eV pictured in the top panel are obtainable via tunneling through a highly doped, narrow surface depletion region (i.e. field emission barrier), via transport through a graded composition region with a lower barrier height and/or a lower resistance, or via transport through a reduced barrier, heterojunction contact. Conversely, graded composition can lead to increased barrier heights (bottom panel) via enlarged band gap regions. Such approaches typically require elevated temperatures and advanced metallurgical techniques to control the junction chemistry, as described in § 3.2.2.

An atomic-scale approach to controlling the interface states and semiconductor band bending is through interlayers of various elements – reactive metals, unreactive adsorbates, and even other semiconductors. Concomitant with the chemical changes described in § 3.2.3, reactive metal layers of only nanometer thickness produce major changes in electronic barrier characteristics. Such reactive layers at metal/III–V compound semiconductor interfaces produce Φ_B changes of tenths of an electron

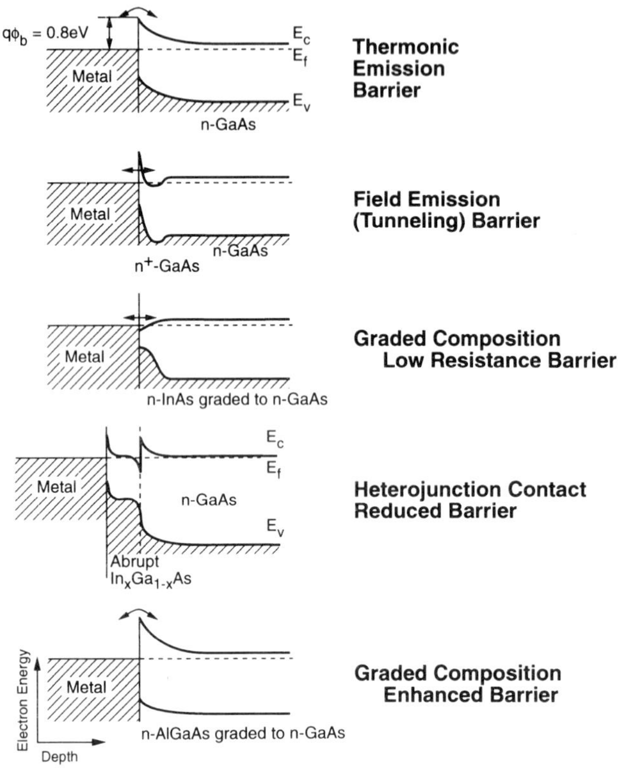

Fig. 40. Schematic band diagram of metal–semiconductor contacts illustrating the metallurgical approaches to barrier control. The top panel shows the 0.8 eV thermionic emission barrier conventionally observed for metals on GaAs. The accompanying panels illustrate decreased barrier heights via tunneling, graded composition, and heterojunction compound interfaces as well as increased barriers (bottom panel) resulting from larger band gap interface compounds (after Murakami et al. 1989, Sands et al. 1990b).

volt (Brillson et al. 1980a, 1981a,b). Yb, Sm, and Sm–Al alloy interlayers at Al/GaAs(001) MBE-grown surfaces also produce nearly 0.2 eV C–V and J–V Φ_B changes on n-type and p-type GaAs (Hirose et al. 1989). Al interlayers at Au–CdS diodes produce even more dramatic Φ_B changes. Figure 41 shows the J–V characteristics for Au diodes on UHV-cleaved CdS($10\bar{1}0$) surfaces as a function of Al interlayer thickness. The inset shows a cross-sectional schematic diagram of the interlayer structure. The barrier characteristics change from rectifying to Ohmic over an interlayer thickness of only 2 Å Al (Brucker and Brillson 1981). Here the reactive metal acts to retard Cd outdiffusion, resulting in a highly n-type surface structure and an effective Φ_B reduction due to tunneling. Similar phenomena occur for Al interlayers at Au–CdTe contacts (Williams and Patterson 1982).

The changes in the Φ_B properties with reactive metal interlayers may be accompanied by changes in the deep-level densities within the near-surface region. Photoluminescence spectroscopy of Au–CdTe(110) interfaces with a Yb interlayer exhibits

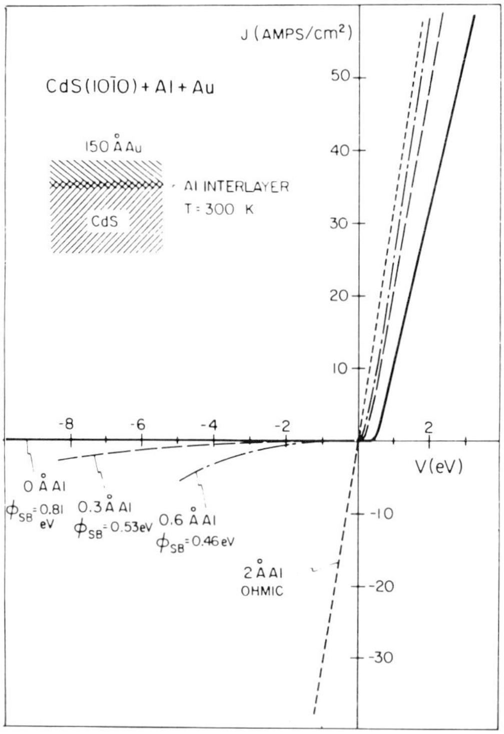

Fig. 41. Atomic-scale control of macroscopic device characteristics, as illustrated by the J–V characteristics of Au–CdS(1010) Schottky diodes versus Al interlayer thickness. The inset shows a cross-sectional schematic diagram of the interlayer structure (after Brucker and Brillson 1981).

lower deep-level emission associated with reduced Te outdiffusion (Shaw et al. 1988b). Furthermore, E_F appears to stabilize at the deep-level energies (Shaw et al. 1988a,b, 1989). Such correlations provide additional evidence for deep-level formation due to metal–semiconductor interdiffusion.

Less reactive adsorbates can introduce controlled changes to the electronic structure of metal–semiconductor contacts as well. Molecular adsorption of H_2S on clean, LEED-ordered InP introduces disorder and a Φ_B reduction from 0.5 to less than 0.3 eV at Au–InP(110) junctions (Montgomery et al. 1981). Surface doping by S atoms to produce shallow donor levels may account for this type of behavior. Adsorption of Xe monolayers on GaAs (Waddill et al. 1989), InP (Vitomirov et al. 1990) and ZnSe (Vos et al. 1990) provides a means of reducing the metal–semiconductor chemical interaction associated with deposition and clustering by either physical separation and dissipation of condensation/fusion energies. See § 3.2.3. Upon thermal desorption of Xe, the resultant metal cluster comes into semiconductor contact with minimal surface disruption due to the cluster formation per se. The resultant junctions exhibit significantly altered band bending with respect to the directly metallized surfaces. Thus for Xe-mediated metal clusters on p-type GaAs(110), metal-specific E_F stabilization energies range from $E_V + 0.37$ to $E_V + 0.62$ eV versus the $E_V + 0.5$ eV barrier conventionally seen (e.g., fig. 35). Similar cluster deposition on n-type GaAs(110) centers E_F stabilization at $E_C - 0.32 \pm 0.1$ eV, in contrast to the $E_C - 0.7$ eV position typically reported. Xe-mediated metallization also results in substantial Φ_B changes from unmediated values for InP and ZnSe junctions. A similar reduction in interface chemistry may account for the systematic dependence of Φ_B on deposition temperature and the enhanced barrier range (0.5 to 0.96 eV) for metals on cooled InP(100) surfaces (Shi et al. 1991).

Semiconductor interlayers have provided the most effective means for controlling Schottky barrier heights. Introduction of the chalcogenes S, Se, and Te, as well as Si or Ge interlayers, provide a wide range of E_F stabilization energies for GaAs (Waldrop 1985a,b, Waldrop and Grant 1986, 1987, 1988, Grant and Waldrop 1987). For nanometer thicknesses of S, Se, or Te, a representative set of metals produces a range of Φ_B values of over 0.6 eV. Metals that react strongly with the chalcogen produce the smallest n-type Φ_B and metals that react weakly produce the largest (in agreement with fig. 32) (Waldrop and Grant 1987, 1988). Combinations of Au, Ge, and Ni layers on LEC GaAs(100) lead to an n-type Φ_B variation from about 0.25 to 0.9 eV (Waldrop and Grant 1987). Here different metallurgical phases form in contact with GaAs, which may alter the Ge diffusion into the GaAs and/or produce a low Ge/GaAs barrier. Figure 42 illustrates an even wider range of E_F stabilization energies ranging from $E_V + 0.33$ to $E_V + 1.2$ eV (corresponding to $\Phi_B = 1.1$ to 0.23 eV, respectively) for doped Ge(As), Ge(P), Si(As), and Si(Ga) with particular growth treatments indicated. Dashed lines indicate conduction and valence bands determined from heterojunction band offset experiments (Waldrop et al. 1983, List et al. 1986). The clustering of E_F positions near the derived Ge and Si band edge energies suggests that the semiconductor work function dominates the E_F stabilization. As-rich GaAs epilayers grown on GaAs(100) at 200°C are also reported to yield low (0.1–0.2 eV) band bending (Look et al. 1990). Similarly, Al/GaAs interfaces with Si interlayers prepared under various Ga and As background conditions yield Φ_B values ranging from 0.3 to 1.04 eV (Costa et al. 1991). Each of these results suggests that E_F is unpinned at the interlayer–semiconductor interface and that the specific interface preparation conditions have large systematic effects on the interface band bending.

Wet-chemical treatments are another effective means of passivating semiconductor surfaces and reducing the interface state densities. Until a few years ago, semiconductor passivation research centered on electrochemical solution treatments to reduce the surface recombination velocity (SRV) in various device applications. However, even photochemical washing of GaAs with above-band gap light and using only H_2O reduces SRV and the interface state density by orders of magnitude (Offsey et al. 1986). Inorganic sulphides produce even larger SRV reductions as derived from the transient surface conductivity response probed via r.f. induction (Yablonovitch and Gmitter 1988). For unpassivated Si, GaAs, and InGaAs surfaces, the recombination velocity is about 10^6–10^7 cm s^{-1}, as shown in fig. 43. Na_2S, NaOH, and semiconductor epilayers cause an SRV reduction of over six orders of magnitude. For these sulphides, the interface state decrease is attributed to the formation of Ga chalcogenides, which rehybridizes dangling bond states out of the semiconductor band gap

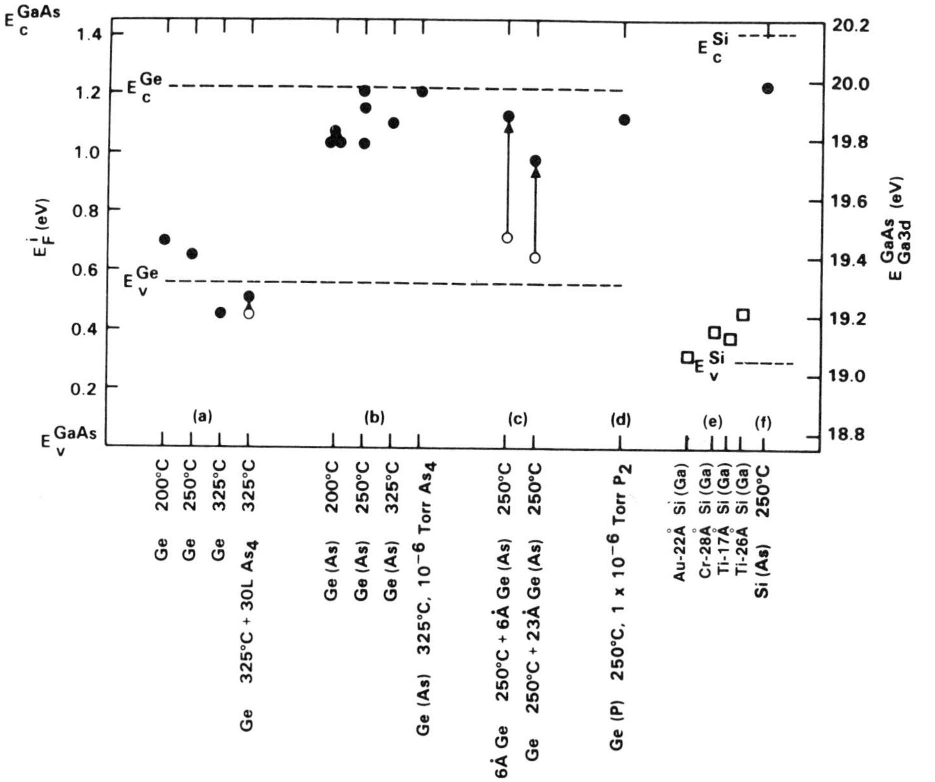

Fig. 42. Wide energy range of Fermi level positions within the GaAs(100) band gap for a variety of (doped) Si and Ge interlayers and metal overlayers. Unless otherwise noted, overlayer thicknesses are about 10 Å and the depositions intended to incorporate As or P in the overlayer were performed in a 10^{-7} Torr background pressure of As_4 or P_2: (a) Ge overlayers deposited in vacuum; (b) Ge overlayers deposited in As_4; (c) Ge overlayers initially deposited in vacuum, but completed with deposition in As_4; (d) Ge overlayer deposited in P_2; (e) Si overlayer deposited with a Ga monolayer in about a 2×10^{-6} Torr H_2 background pressure; (f) Si overlayer deposited in As_4. Dashed lines indicate the Ge and Si band edges derived from measured band offsets. E_F positions appear to range across the Ge and Si band energies (after Grant and Waldrop 1987, Waldrop and Grant 1988).

(i.e. Nelson et al. 1980). For H_2O treatments, the removal of water-soluble As oxides is proposed to account for the apparent surface state reduction (Offsey et al. 1986). Thermal oxides and hydrogen treatments have similar effects on the SRV. Here HF solutions with the pH adjusted to eliminate surface roughness produce microscopically smooth surfaces with a homogeneous monohydride passivation layer and SRV values as low as less than $1\ cm\ s^{-1}$ (Higashi et al. 1990). Thus the removal of electrically-active species or the formation of stable overlayers with the surface constituents passivates interface states against both E_F "pinning" and high surface recombination velocity.

Semiconductor growth provides a rich variety of atomic-scale approaches to modify and control Schottky barrier heights. More specifically, elements of crystal

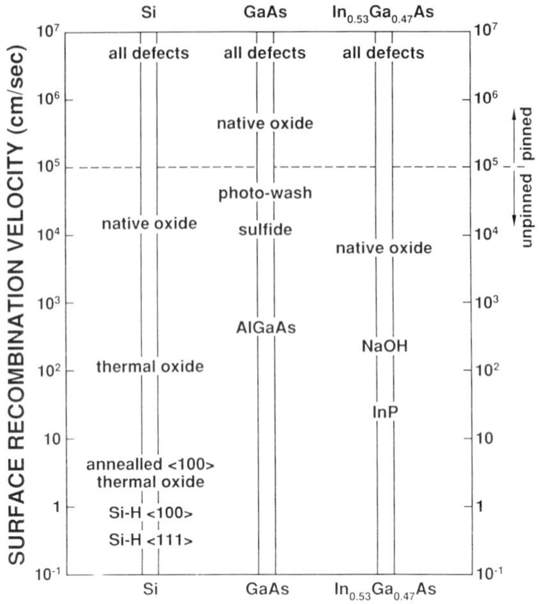

Fig. 43. Surface recombination velocites for Si, GaAs, and $In_{0.53}Ga_{0.47}As$ epitaxial layers on GaAs(100) for various surface treatments. Low SRV values correspond to reduced interface state densities and unpinned surface Fermi levels (after Yablonovitch and Gmitter (1988) and references therein).

growth, surface misorientation, and epitaxical interface modification have all provided evidence for a significant electronic dependence on extrinsic, structural properties.

Different techniques of crystal growth can result in variations of stoichiometry, point defect density, dislocation density and even elemental precipitates. Chemical interactions magnify these artifacts, most of which can introduce new electronic features. Comparison of E_F stabilization at metal–GaAs interfaces provides an example of the different electronic features associated with melt-grown (110) versus MBE-grown (100) semiconductor surfaces. As represented in fig. 44, these two types of metal/GaAs junctions exhibit qualitative differences: a narrow range of E_F stabilization with submonolayer metal deposits and different n-type versus p-type energies for the same metal on melt-grown GaAs(110) (upper) (Spicer et al. 1980b), versus a wide range of E_F stabilization with multilayer metal deposits and matching n- and p-type energies for the same metal on As-decapped MBE-grown GaAs(100) (lower) surfaces (Brillson et al. 1988). Corresponding differences are evident in deep-level emission for these bulk materials and their metallized surfaces. Both bulk and surface PL measurements reveal intense, mid-gap emission for melt-grown GaAs, as opposed to orders-of-magnitude lower emission shifted away from mid-gap for MBE-grown semiconductor material (Shaw et al. 1989, Viturro et al. 1986). These electrical differences may arise either from

(i) the difference in deep-level densities and energies,
(ii) the difference in orientation, and/or

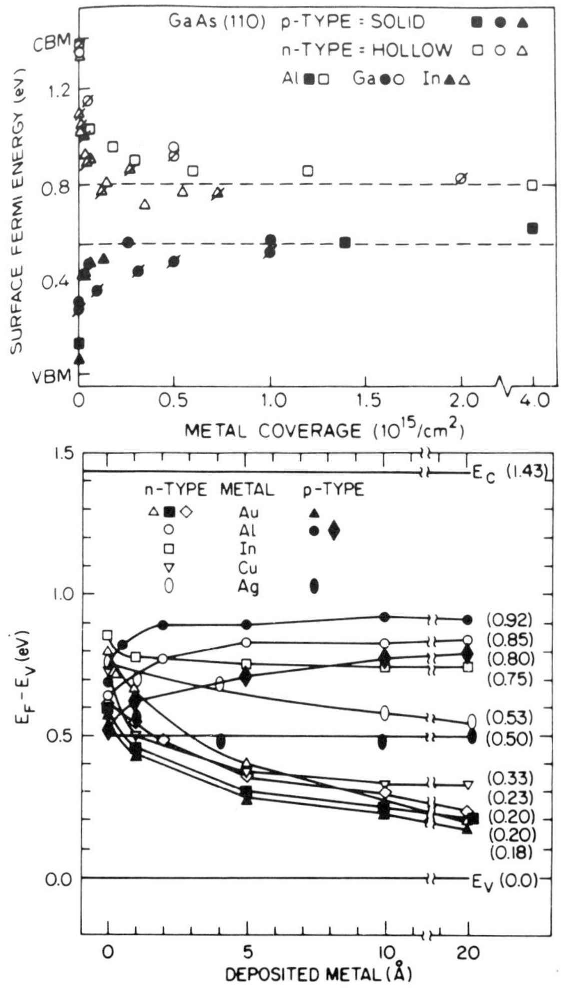

Fig. 44. Qualitative difference in metal-induced E_F experiments within the GaAs band gap measured via SXPS. Metals on UHV-cleaned melt-grown GaAs(110) (upper) exhibit a submonolayer E_F movement to a narrow mid-gap range of energies with different n- and p-type values (after Spicer et al. 1980b). Metals on As-decapped, MBE-grown GaAs(100) (lower) show E_F movement over several monolayers to a wide range of energies with matching n- and p-type values (after Brillson et al. 1988).

(iii) chemical changes associated with the As-decapping technique used to prepare clean surfaces.

Metals on II–VI compound semiconductors provide evidence for (i) (Shaw et al. 1988b, 1989), the influence of surface reconstruction on electronic structure described in § 2.3.3 provides a basis for (ii), and (iii) may involve surface As depletion (Makram-Ebeid et al. 1982) and/or surface smoothing in reducing localized charge states.

At lower temperatures an even wider range of metal/GaAs(100) barriers is evident, which can be plotted to show near-ideal behavior for all but the most reactive metals

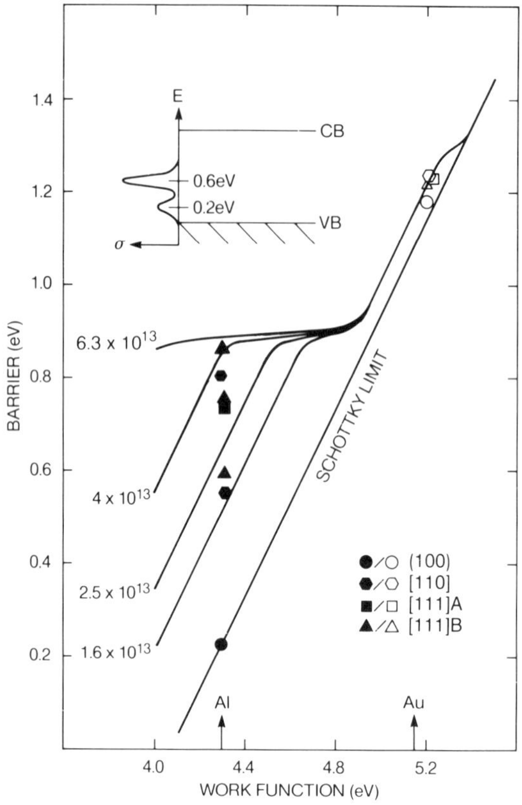

Fig. 45. Self-consistent analysis of barrier heights versus metal work function (a) and interface state densities versus active structural site densities (b) for Al and Au on vicinal GaAs(100) surfaces. Barrier heights increase monotonically with misorientation angle and direction. The family of density curves for the 0.6 eV and (constant density) 0.2 eV states pictured in the insets yield the interface states shown in (b), which display a near one-to-one dependence on the active structural site density (after Chang et al. 1991b).

(Viturro et al. 1988). Macroscopic internal photoemission measurements confirm the wide Φ_B range for Al, Cu, and Au on GaAs(100) (Chang et al. 1990a) and also show the chemically unstable nature of the reactive Al/GaAs contacts (Chang et al. 1992). Similar wide Φ_B ranges are reported for metals on GaP(110) (Chiaradia et al. 1987, Brillson et al. 1987). However, there is as yet no consensus on E_F positions for low-Φ_M metals, the reported values of which vary considerably (Alonso et al. 1990, Miyano et al. 1990, Ludeke et al. 1990). Such discrepancies may arise in part due to surface photovoltage effects (Hecht (1990) and references therein) at submetallic coverages (but which can be eliminated in the metallic overlayer regime) as well as the reactive metal sensitivity to specific preparation and measurement conditions. As a result, the SXPS measurements of E_F positions for layers on metal layers on semiconductors are subject to considerable interpretation at present, especially at low temperatures.

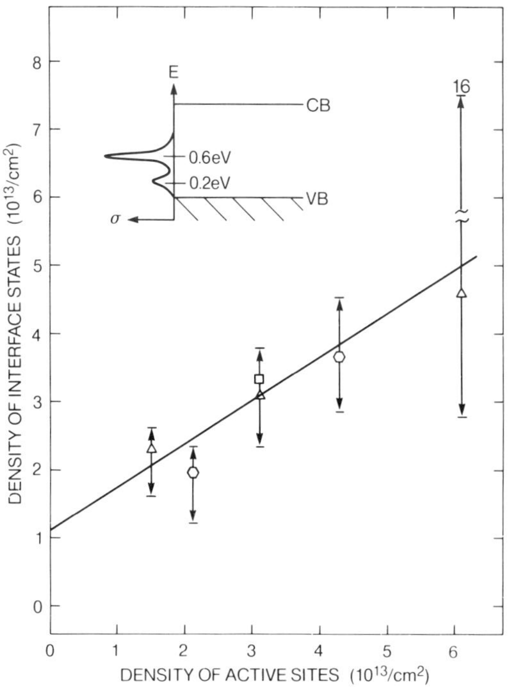

Fig. 45 *continued*

A wide range of E_F stabilization energies provides a test bed for investigating additional electrically-active features of metal–semiconductor interfaces. One such factor related to surface structure and growth is the presence of active sites associated with vicinal surface step edges. Such misoriented surfaces exhibit additional electrically-active sites in conjunction with the higher chemical activity mentioned in § 3.2.3. Figure 45 illustrates a self-consistent analysis analogous to fig. 37 (Duke and Mailhiot 1985), relating these electrical and structural sites. Figure 45a illustrates the values of Φ_B extracted from SXPS measurements of rigid core level shifts at 90–100 K for metals on As-decapped vicinal GaAs(100) (Chang et al. 1990b). Each data point corresponds to a different misorientation direction and/or angle from the GaAs(100) surface. The data exhibit Φ_Bs that increase with the density of active step (exposed As) sites for the same metal (Al) on GaAs. The relatively constant Au Φ_Bs serve to underscore the metal-specific nature of this behavior. The family of curves pictured in fig. 45a corresponds to different interface state densities of a level at $E_V + 0.6$ eV and a (constant) background density of states at $E_V + 0.2$ eV. Transitions involving such states are evident in CLS spectra for vicinal surfaces with Al (but not Au) overlayers. Analogous CLS transitions involving states at 0.8 and 0.2 eV are observable for metals on oriented GaAs(100), which account for the Φ_B versus Φ_M behavior according to a similar analysis. Figure 45b shows the density of interface charge states extracted from fig. 45a plotted for the same misorientations as a function of chemically-active edge sites (Chang et al. 1991b). The slope indicates about two thirds of an electron per active structural site (according to an idealized ball and stick picture of the vicinal steps), extrapolating down to a low ($1 \times 10^{13}\ \mathrm{cm}^{-2}$) density at the well-oriented surface. STM measurements of such vicinal surfaces (Pashley et al. 1991) reveal densities of kink sites at such vicinal surfaces that are sufficiently high to account for these electrical effects. Thus discrete deep levels calculated self-consistently and observed spectroscopically correlate with the density of associated structural imperfections on an atomic-scale.

Epitaxical growth of binary (semi-)metal alloys on compound semiconductors provides additional control of the interface chemical structure (see § 3.2.3) and Schottky barrier formation. Here it is possible to separate out the electrical effects of strain from those of misorientation and lattice mismatch. For $Sc_xEr_{1-x}As$ ($0 \leqslant x \leqslant 1$) epitaxy on GaAs grown at about 400°C, diodes formed with different composition and lattice mismatch exhibit only a weak Φ_B dependence (<0.1 eV). This relatively small Φ_B variation shown in fig. 46a contrasts with the Φ_B changes due to annealing (a) and a misorientation angle from [100] (b) (Palmstrøm et al. 1990a). Indeed fig. 46b exhibits a Φ_B variation of more than 0.4 eV for this epitaxical metal–semiconductor interface with changing orientation and annealing of the GaAs(100) growth surface. Interface structures are expected to be substantially different for different orientations, including different amounts of {100} and {111} ledges and steps for {h11}-type, $h > 1$, surfaces and different stacking sequences between {100} and {111} $Sc_xEr_{1-x}As$ ($0 \leqslant x \leqslant 1$). These results are analogous to those for $NiSi_2$/Si and $PbSi_2$/Si interfaces already discussed in § 3.3.2, showing an epitaxical dependence on lattice ordering. For Ni/Al overlayers on GaAs, the Φ_Bs change by 0.2–0.3 eV with deposit thickness as Ni/Al stoichiometry and surface geometric order change (Chambers and Loebs 1990) – related in part to the formation of an AlAs layer, Ni indiffusion, and the properties of GaAs grown at low (250°C) temperature. AlAs interlayers intentionally grown at such epitaxical junctions produce 0.1–0.2 eV Φ_B increases in accordance with the bottom panel of fig. 40 (Sands et al. 1990b). All these Φ_B measurements may be macroscopic averages, and even large Φ_B variations may exist within microscopic domains at a given junction (Tung 1991). Thus it now appears possible to vary Φ_B considerably even with epitaxical metal–semiconductor junctions.

Overall, many atomic-scale techniques provide substantial Φ_B control via different physical mechanisms. Furthermore, the correlation of geometric, chemical, and electrical properties for the metal–semiconductor interface provide considerable evidence for the role of localized bonding and composition in determining the electronic structure of metal–semiconductor contacts.

4. Semiconductor–semiconductor interfaces

Semiconductor heterojunction interfaces offer perhaps the most refined approach to designing new electronic materials and devices. Even more so than semiconductor–

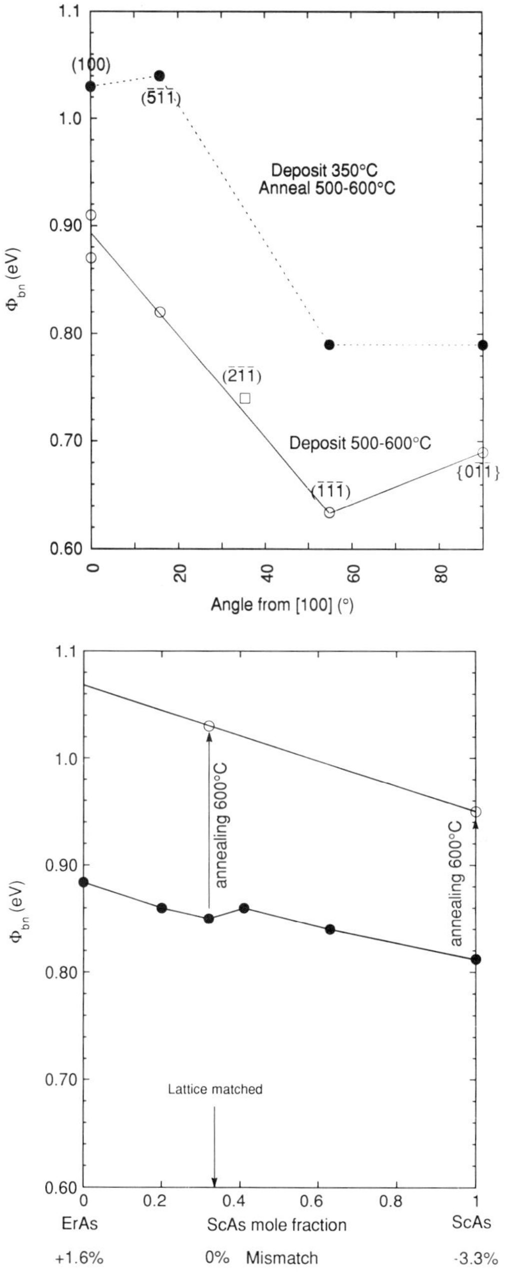

Fig. 46. Schottky barrier heights for $Sc_xEr_{1-x}As$ metal alloys on GaAs(100) as a function of (a) lattice mismatch and annealing and (b) interface misorientation and annealing. The weak dependence on lattice mismatch is in contrast to the strong Φ_B dependence on misorientation and annealing (after Palmstrøm et al. 1990a).

metal contacts, the interface properties dominate the characteristics of the full semiconductor–semiconductor structure. Furthermore, many of the physical phenomena described for semiconductor surfaces and metal interfaces are present at semiconductor heterojunctions (and homojunctions) as well. However, the characterization of interface geometric, chemical, and electronic properties of semiconductor–semiconductor junctions has been less extensive than that for semiconductor surfaces and metal–semiconductor interfaces, in large part because of:

(a) the need to monitor properties during the process of interface formation, and
(b) the requirements of a UHV epitaxial growth and analysis system to perform such measurements.

This subsection presents the salient features of semiconductor heterojunction geometric, chemical, and electronic structures. A primary focus is the heterojunction band offset, which is considered to be the single most important feature of the structure. The band structures at semiconductor–semiconductor junctions can line up in several different ways. For atomically-abrupt interfaces, fig. 47 illustrates the different types of energy-band line-ups, namely, (a) straddling, with one band gap nested within the other, (b) staggered, with only a partial overlap, and (c) broken gap, with no overlap between the two (Kroemer 1985). Here the heterojunction band offsets are the conduction- and valence-band discontinuities, termed ΔE_C and ΔE_V, respectively. As in previous subsections, the specific geometric and chemical structures of the semiconductor–semiconductor interface on an atomic scale contribute directly to the heterojunction's macroscopic electronic properties.

4.1. Geometric structure

The geometric features of semiconductor–semiconductor interfaces are related primarily to the quality of heteroepitaxial growth. The two main aspects are the constraints imposed by the bulk-crystal lattices and the crystalline imperfections, which result as heterojunctions deviate from perfect epitaxy conditions.

4.1.1. Epitaxical growth

The lattice parameters of semiconductor crystals are characteristic functions of composition, temperature, and pressure. See, for example, previous *Handbook on*

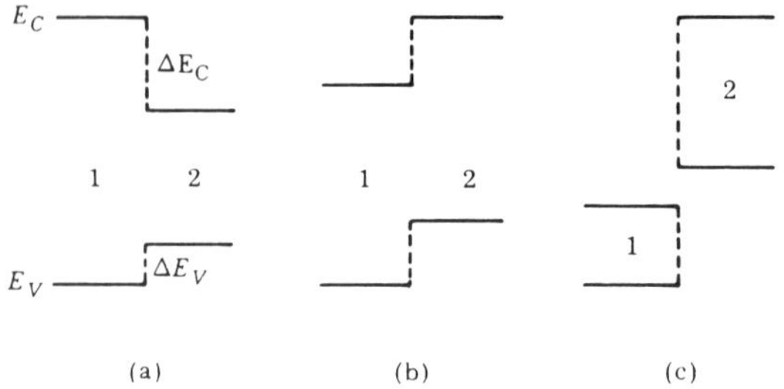

Fig. 47. Representative examples of heterojunction band offsets illustrating the different types of band-edge alignment: (a) straddling or nested; (b) staggered; (c) broken-gap (after Kroemer 1985).

Semiconductor chapters by Zallen (1982), Cohen and Chadi (1980), Martinez (1980), and Chang (1990). Near-ideal lattice match between semiconductor bulk lattices at room temperature is desirable to promote defect-free continuation of crystalline growth and to minimize subsequent generation of imperfections by strain built into the final lattice structure. Figure 48 presents lattice constants and energy band gaps for common III–V compound semiconductors as well as for Si and Ge (Cho 1985). This plot serves to illustrate semiconductor compounds which are obtainable with a given lattice constant for epitaxial growth and a given energy gap for particular optoelectronic applications. The solid lines between binary compounds denote ternary alloys with direct band gaps; dashed lines indicate alloys with indirect band gaps. Lines within areas bounded by more than two compounds denote the corresponding properties for quaternary alloy compounds. Of particular interest to surface and interface researchers have been the heterojunctions involving the GaAs/AlAs ternary alloys, as well as the Ge with GaAs, both of which exhibit relatively close lattice matching (lattice mismatch $f = \Delta a/a \leqslant 0/5\%$). More refined selection of lattice-matched semiconductors is available from lattice constant versus composition graphs for many of these ternary and quaternary alloy systems (Casey and Panish 1978a).

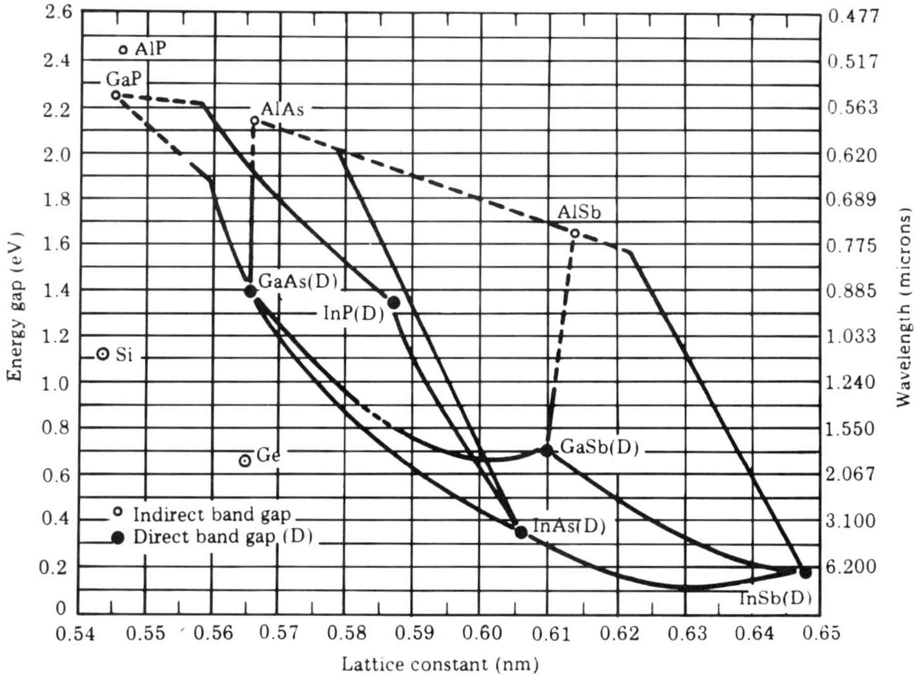

Fig. 48. Lattice constants and corresponding energy gaps for the common III–V compounds plus Ge and Si, which can serve as a guide to lattice-matched heterojunction growth. The lines between binary compounds provide data for the corresponding ternary alloys, with solid lines denoting direct band gap and dashed lines denoting indirect band gap compounds (after Cho 1985). Lines within the boundaries established by multiple compounds provide the lattice matching conditions for quaternary alloys.

Thus considerable information is available for the selection of lattice-matched semiconductors with a particular set of individual band gap properties.

Multilayer semiconductor structures termed superlattices represent a major alternative to the constraints imposed by lattice matching. By alternating subcritical-thickness layers of semiconductor films, one can obtain strained-layer superlattices (SLS) (Matthews and Blakeslee 1974, Osbourn et al. 1987) free of high dislocation densities. These SLS structures provide a wide continuous range of band gap properties with only a secondary dependence on the crystal lattice constant. The strain generated for SLSs depends on a given set of lattice mismatch, thickness and growth orientation. Indeed, the presence of strain-induced piezoelectric fields along specific orientations can lead to substantial modification of the electronic structure and optoelectronic properties of these materials (Mailhiot and Smith 1986, 1987, Smith and Mailhiot 1987). Because this strain contribution to the interface electronic bands can be large, it may dominate the overall optoelectronic response features of the macroscopic structure. In turn, these optoelectronic properties will depend on the layer widths, the layer compositions, and the dependence of the heterojunction strain on these parameters.

4.1.2. Dislocations and strain

As described in § 4.1.1, deviations from perfect lattice matching conditions result in strain within the epitaxial semiconductor growth layer. Below a critical epitaxical layer thickness h_C, the mismatch between lattices is taken up by strain in the growing film. Above this critical thickness, it is energetically more favorable to accommodate the lattice mismatch by the generation of misfit dislocations. The thickness-dependent balance between the energies of strain accommodation versus dislocation generation for single heterojunctions result in a detailed, material-dependent expression for the critical thickness (Matthews and Blakeslee 1974):

$$h_C = b(1 - \mu \cos^2 \Theta) \ln(h_C + 1)/2\pi f(1 + \mu) \cos \Phi, \tag{4.1}$$

where b is a Burgers vector, μ is the Poisson ratio, Θ the angle between the dislocation line and b, and Φ is the angle between the slip direction and that direction in the film plane which is perpendicular to the line of intersection of the slip plane and the interface (after Mayer and Lau 1990). In fig. 49 this relation is represented as a plot of critical thickness versus lattice mismatch. Furthermore, the dashed line defines a boundary line below which pseudomorphic (dislocation-free) growth occurs and above which dislocations and stacking faults are generated to relax the strain in the epitaxical film. For a relatively large lattice mismatch (above $\sim$1.5%), pseudomorphic two-dimensional growth is possible only for one or two monolayers. Thus in the thickness regime below the dashed line, subcritical films are likely to exhibit three-dimensional growth (Mayer and Lau 1990). Recent studies of Si–Ge heteroepitaxy suggests that growth interruption may mitigate the generation of misfit dislocations above this critical thickness (Eaglesham et al. 1990). As already described in § 3.3.2, the production of dislocations, stacking faults, grain boundaries, and other such extended lattice imperfections can have electrical activity (i.e. Woodall et al. 1983).

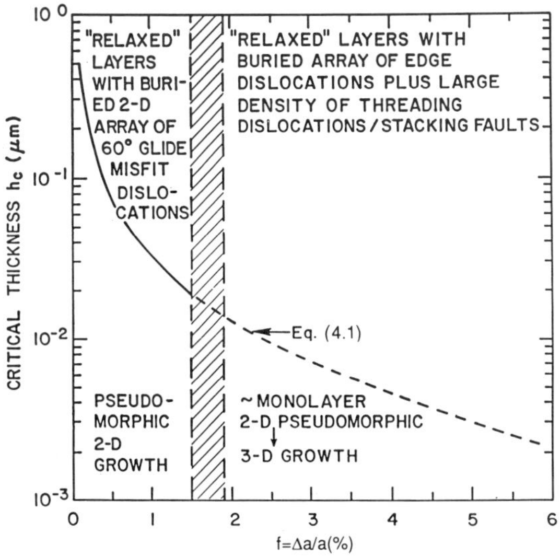

Fig. 49. Critical thickness h_C versus lattice mismatch $f = \Delta a/a(\%)$ for semiconductor overlayer growth, where the curve (after Matthews and Blakeslee 1974) displays the critical thickness for generation of misfit dislocations. For f smaller than about 1.5% pseudomorphic two-dimensional growth occurs in layers thinner than h_C. For f larger than 1.5% (dashed line), pseudomorphic two-dimensional growth continues up to only a few monolayers followed by three-dimensional growth. For thicknesses above h_C, dislocations and stacking faults form to relax the strain in the epitaxial layer (after Woodall et al. 1988).

For both single and multiple heterojunction semiconductor interfaces, the specific nature of bonding and composition at the microscopic junction controls the quality of semiconductor growth and the nature of extended lattice imperfections. The following subsections present the implications of such geometric features for chemical and electronic structure at the semiconductor–semiconductor interface.

4.2. Chemical structure

The chemical structure of semiconductor–semiconductor interfaces centers on the interactions that can occur between constituents of the two lattices which lead to new chemical phases and/or to variations in epitaxical growth. These two aspects of (a) interdiffusion and chemical reaction, and (b) template interfacial structures are both consistent with the key role of local bonding and interface preparation on the resultant interface properties.

4.2.1. Interdiffusion and chemical reaction

Analogous to the chemical structure of metal–semiconductor interfaces discussed in § 3.2, several chemical phenomena can act to render the semiconductor heterojunction less than atomically abrupt. In principle, lattice-matched heterojunctions provide atomically-abrupt and ordered interfaces that are free from all the complications of the imperfections common in Schottky barrier structures. Nevertheless, interdiffusion and chemical reaction are observed in numerous semiconductor classes of heterostructures, particularly at temperatures elevated above those required for epitaxical growth.

The Ge–Si interface exhibits chemical features illustrative of the thickness and temperature dependences of interdiffusion. While monolayer adsorption leads to ordered overlayers (Patel et al. 1985, Durbin et al. 1986, Dev et al. 1986), thicker interfaces exhibit significant interdiffusion for temperatures above about 200°C, as well as islanding and surface roughening at even higher temperatures (Hoeven et al. 1988). At 350°C, an intermediate Ge–Si alloy layer is evident (Chen et al. 1984). For Ge/Si superlattices, X-ray diffraction techniques provide measurements of activation energies for such interdiffusion, i.e., 3.1 ± 0.2 eV in the temperature range 640–780°C. For Ge-rich GeSi alloy interfaces with Si, Ge segregation occurs into Si. As with metal–semiconductor interfaces, this segregation can be suppressed by (Ga) interlayers (Zalm et al. 1989).

Diffusion effects are also evident for III–V compounds. Under MBE growth conditions, an abrupt interface of Ge on GaAs is formed at 320–360°C (Bauer and Sang 1983). However, further increases in substrate temperature result in an interdiffused junction. At temperatures of 650°C, for example, Ge diffuses fractions of microns into GaAs with concentrations of 10^{17}–10^{18} cm^{-3} (Sarma et al. 1984). Ge diffusion into III–V compound semiconductors is of particular importance since Ge acts as a p-type dopant, and even low concentrations (e.g., p.p.m.) alter the interface electronic properties significantly. Thus at III–V growth temperatures in the 450–650°C range, such diffusion may easily exceed the bulk crystal doping over depths comparable to the device dimensions. With further interdiffusion, solid-phase regrowth and production of a degenerately-doped GaAs layer is possible (Marshall et al. 1989).

For AlAs–GaAs, AlGaAs and GaAsP–GaAs superlattices, interfaces grown by organometallic vapor phase deposition (OMCVD) at 700–750°C exhibit secondary-ion mass spectrometry (SIMS) features characteristic of atomically abrupt interfaces (Coleman et al. 1986). Interdiffusion of Ga and Al in $Ga_xAl_{1-x}As$ ($0 \leqslant x \leqslant 1$) pseudobinary alloys shows that such III–V interdiffusion is composition dependent, decreasing in the 850–1100°C temperature range with increasing Al composition (Chang and Koma 1976). In addition, intermixing of semiconductor superlattices increases dramatically with the presence of dopant impurities such as Zn (Laidig et al. 1981), which alter the defect statistics controlling diffusion through the crystal lattice (Van Vechten 1982).

In addition to interdiffusion, evidence exists for multilayer compound formation at semiconductor–semiconductor interfaces. For example, TEM and in situ XPS measurements indicate the formation of a zinc blende Ga_2Se_3 interfacial layer at the ZnSe–GaAs interface (Li et al. 1990, Qiu et al. 1990). Similarly, XPS measurements of CdTe deposited under epitaxial conditions ($T \sim 150$–300°C) reveal the formation of In–Te layers whose thickness depends upon the substrate temperature (Mackay et al. 1986). Here it is possible to suppress such In telluride formation by application of a sufficiently high Cd overpressure during growth (Zahn et al. 1988). Both examples provide evidence that III–VI compounds form readily at II–VI/III–V interfaces. The degree to which such interface compounds form will depend upon the rate of semiconductor outdiffusion from the substrate and within the semiconductor overlayer.

The presence of chemical interactions at semiconductor heterojunctions can easily change the properties expected for the otherwise abrupt junction. While epitaxical growth of such structures requires elevated temperatures, the minimum temperatures for growth are desirable in order to achieve heterostructures with the least degree of chemical degradation.

4.2.2. Template structures

The structure and quality of semiconductor epitaxical growth is sensitive to the geometric and chemical structure of the heterointerface. Several types of template structures reflect this sensitivity, including:

(i) buffer layers to bridge the difference in lattice constants between the two semiconductors,
(ii) monolayer films for substrate bond passivation,
(iii) specific crystal orientations to optimize or modify growth,
(iv) monolayers films to maintain the uniform overlayer morphology, and
(v) specific atomic structure and stoichiometry to control interface dipole fields.

Buffer layers of changing composition and lattice constant provide a means to achieve epitaxical growth between materials with otherwise signficant lattice mismatch. Such buffer layers are typically fractions of a micron thick and may involve either (a) pseudobinary alloys of the two semiconductor constituents or (b) combinations of new materials with intermediate lattice properties. An example of the former is $Ga_xIn_{1-x}As$ $(0 \leqslant x \leqslant 1)$, varying compositions of which bridge the 7% lattice mismatch between GaAs and InAs (i.e. Woodall et al. 1981). An example of the latter are layers of BaF_2 on CaF_2, which minimize the 9% lattice mismatch between PbTe and Si (Maissen et al. 1988).

Monolayer template structures can promote the preferential growth of different epitaxical variants, as already mentioned in § 3.1.2 for $NiSi_2$ growth on Si (Tung 1984) and NiAl growth on GaAs (Sands et al. 1990b). Monolayer adsorbates have also proved effective in promoting heteroepitaxial growth between two semiconductors, as shown by the dependence of growth quality on surface stoichiometry and reconstruction. For GaAs on Si, STEM and SXPS measurements indicate that the lattice mismatch, the transition between covalent and partially-ionic bonding, and the sublattice selection all influence the mode of GaAs overlayer growth (Bringans and Bachrach 1990). Here SXPS measurements show a predominantly Si–As bonding at epitaxial GaAs on Si(111) interfaces, suggestive of a Ga–As double layer forming on top of the As-terminated Si. However, polar bonding considerations render this configuration energetically unstable (Harrison et al. 1978). In fact, total-energy calculations reveal that an interface consisting of three quarters of a layer of Si atoms above one As layer and one Ga layer yields a stable structure (Northrup 1988), and ARPES results indicative of a Ga–Si exchange reaction support this atomic geometry

(Northrup et al. 1988). Additional complexity is evident for GaAs on Si(100) involving mixed compositions between islands and Si on GaAs(100) involving mixed Si–As interfacial layers and dimerization (Bringans and Bachrach 1990).

The quality and nature of heterojunction growth depends on the substrate crystal orientation as well. It is argued that consistently superior morphology of GaAs(110) growth on Ge is due to the absence of surface reconstructions at (110) interfaces (Kroemer et al. 1980). Such reconstructions are required to reduce the high local electrostatic fields associated with interface dipoles on non-(110) surfaces (Harrison et al. 1978). These high fields can disrupt heteroepitaxial growth and lead to rough interfaces with poor electrical properties. Surface misorientation provides additional possibilities for the growth of quantum structures. By controlling the nucleation and growth kinetics with epitaxy on intentionally misoriented surfaces, one can obtain laterally periodic changes in a surface atomic composition and so-called "tilted" superlattices (Petroff et al. 1984).

Monolayer adsorbates can passivate the bonding not only at the intimate semiconductor–semiconductor interface but also within the growing epitaxical film. An As monolayer deposited on Si(001) under an As overpressure facilitates dislocation-free growth of epitaxical Ge films (LeGoues et al. 1989). Here As suppresses the tendency of Ge to form islands, the edges of which can promote misfit dislocations reaching down to the substrate (Woodall et al. 1988). Instead a small-size, uniformly distributed, unconnected set of novel defects form in order to relieve the strain in the growing film. As a result, it is possible to grow epitaxical films free of misfit dislocations above the critical thickness indicated in fig. 48.

The dependence of epitaxical growth on atomic structure and composition is also reflected in the heteroepitaxial growth of II–VI materials on III–V substrates. As with the choice of orientation, large electrostatic fields can be avoided by

(i) graded or mixed interfaces containing equal numbers of III–VI and II–V bonds (i.e. Ga–Se and Zn–As bonds for ZnSe on GaAs) or
(ii) by specific planar reconstructions that achieve the same purpose (i.e. by equating the number of Ga- and As-based interface bonds in the case of GaAs substrates) (Farrell et al. 1991). Such reconstructions require careful control of substate temperature and flux ratio.

Thus a diverse set of template structures are available to modify the epitaxical semiconductor–semiconductor growth, ranging in thickness from fractions of a micron down to single atomic layers. These template structures are representative of the effects that atomic-scale interface features have on the macroscopic semiconductor properties.

4.3. Electronic structure

The electronic properties of semiconductor–semiconductor interfaces provide the basis for a wealth of physical phenomena and device applications. These engineered semiconductor materials lead to novel band structures with features such as quantum confinement, ballistic transport, and unusually high carrier mobilities. The applications of such features extend to high-speed transistors, photodetectors, and lasers. The heterojunction features, which make possible these physical phenomena and device applications, center on the semiconductor band offset and, to a lesser extent, the semiconductor band bending within each of the space charge regions. In turn, these properties are associated with interface dipoles and localized states which, as already described in § 2 and § 3, both involve the specific geometric and chemical properties of the junction on an atomic scale.

This subsection provides:

(i) a general description of semiconductor band offsets and band bending,
(ii) the techniques employed to characterize energy band discontinuities,
(iii) the alternative interface dipole descriptions developed to account for these measurements, and
(iv) the atomic-scale techniques not available to modify or control heterojunction band offsets.

4.3.1. Heterojunction band offsets

The heterojunctions are characterized by several parameters. Figure 50 provides a schematic energy band diagram of a straddling semiconductor energy band line-up. In addition, band bending contributes to the variation of E_C and E_V across the heterojunction. In this case semiconductor A has an n-type (upward) band bending due to ionized impurities within the depletion region while semiconductor B has a p-type (downward) band bending due to charge accumulated in an inversion region. The single-headed arrows denote the sign convention for the ΔE_C and ΔE_V discontinuities: for the A–B heterojunction shown, ΔE_C is positive when the conduction-band edge of semiconductor A is above that of B; ΔE_V is positive when the valence band of A is below that of B. Thus in fig. 50, heterojunction A–B has positive ΔE_C and ΔE_V band offsets. Unless otherwise stated E_g(A) is larger than E_g(B) so that $\Delta E_g > 0$

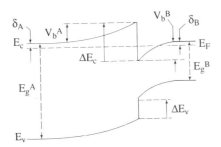

Fig. 50. Schematic energy band diagram of a semiconductor heterojunction, showing conduction-band minimum E_C and valence-band maximum E_V as functions of position across the interface between two n-type semiconductors with different band gaps, E_g(A) and E_g(B). S_A and S_B are the corresponding E_F separations from the majority-carrier band edges. Single-headed arrows denote the positive-sign convention for the valence-band discontinuity ΔE_V and the conduction-band discontinuity ΔE_C. Band bending in semiconductor A results from ionized impurities in the depletion region; band bending in B from charge accumulated in an inversion region.

where

$$\Delta E_g = E_g(\mathrm{A}) - E_g(\mathrm{B}) \tag{4.2}$$

and

$$\Delta E_C + \Delta E_V = \Delta E_g. \tag{4.3}$$

The vacuum levels for the two semiconductors are not in general aligned after contact. Instead they are shifted by a voltage called the built-in potential, V_{bi}, which depends on the difference in semiconductor surface and interface dipoles, and is equal to the sum of individual band bendings

$$V_{bi} = V_b(\mathrm{A}) + V_b(\mathrm{B}). \tag{4.4}$$

The application of Poisson's relation provides the band edge energies as a function of position, in analogy to eq. (3.6). Complete calculation of the heterojunction band diagrams, explicitly taking into account the built-in voltage, Fermi level positions, semiconductor dopings and effective masses appears in Casey and Panish (1978b).

Lattice-matched junctions with low interfacial strain provide the optimal cases for calculating or estimating these band discontinuities. Lattice-mismatched heterojunctions induce interfacial strain and deviations from perfect crystallinity that complicate any such calculations. Non-abrupt changes in composition produce graded band structure features whose spatial variation can be calculated by superimposing band offsets for the spatially-dependent composition onto the potential across the band bending regions and the built-in potential.

As in the case of the metal–semiconductor interface, the band alignment involves several dipoles. Analogous to fig. 27, these include not only the voltage drops across the two semiconductor band bending regions but also the surface dipoles for the two semiconductors and the interface dipole between them. Again equating the potential drop across the interface to the difference in internal potentials, but now including two rather than just one surface space charge region, one obtains (Mailhiot and Duke 1986)

$$\Delta E_C = (X_B - X_A) + (V_i - V_B - V_A), \tag{4.5}$$

where X_B and X_A are the measured electron affinities of the semiconductors B and A, respectively, V_B and V_A the corresponding surface dipoles, and V_i the interface dipole contribution. Thus the conduction-band discontinuity is just the difference in semiconductor electron affinities plus a term involving the difference in the dipoles. Again, the dipoles are not observable – one derives their magnitudes experimentally from their effects on the observables X_B, X_A, $V_b(\mathrm{A})$ and $V_b(\mathrm{B})$. For the case when the dipole difference term is zero, eq. (4.5) reduces to the electron affinity rule (Anderson 1962). This classical relation has not provided accurate predictions of band offsets, as addressed in § 4.3.3. The failure of the electron affinity rule implies that localized interface dipoles make a significant contribution to the electronic structure of semiconductor–semiconductor interfaces.

4.3.2. Band offset characterization

The macroscopic and microscopic techniques used to characterize the electronic structure at metal–semiconductor interfaces (i.e. § 3.3.3) have also been effective at probing the heterojunction band offsets. Macroscopic techniques for gauging the semiconductor–semiconductor band structure center on transport and optical measurements. Among transport methods, C–V measurements can extract the built-in potential V_{bi}, assuming a conventional square-root dependence for C on $V_{bi} - V$. Combining V_{bi} with δ_A and δ_B from the known bulk dopings, one can obtain a measure of ΔE_C and ΔE_V. However, impurity gradients and near-interface charge can alter such straightforward analyses (Forrest 1987). Similarly, these and additional factors such as carrier recombination, tunneling, and shunt currents introduce serious limitations on the interpretation of J–V measurements (Forrest 1987). The most reliable subset of such capacitance techniques are measurements of interfacial charge at the heterojunction interface with known doping profiles (Kroemer et al. 1980a, Wang et al. 1985). Another class of transport techniques involves measurements of carrier activation over a band discontinuity. These include variations of deep-level transient spectroscopy (DLTS) (Lang 1987) and internal photoemission (IP) spectroscopy (Heiblum et al. 1985, Haase et al. 1987). Although requiring special material structures to counter the experimental limitations, the latter appears to provide the most direct approach of the transport methods.

Optical techniques include absorption, luminescence, and light scattering spectroscopies involving the band features at the interface. Optical absorption measures transitions between sub-bands within heterojunction quantum wells defined by the band edge discontinuities (i.e. Duggan 1987). Optical features related to states on either side of a heterojunction yield characteristic direct transitions as a function of hydrostatic pressure (Wolford et al. 1987) and indirect transitions as a function of electric field for staggered alignment heterostructures (Wilson 1988). Light scattering also provides a measure of the band bending, charge densities, and energy levels associated with quantum well structures (Pinczuk 1984, Arbstreiter 1985). However, a quantitative interpretation of all these optical measurements requires a self-consistent calculation of the quantum confined states and the associated band offsets. Furthermore, while these optical techniques possess high energy precision, they require the fabrication of high-quality multilayer structures, whose features are dependent on thickness (Smith and Mailhiot 1987), and the results are restricted to geometrically abrupt, chemically uniform, and electronically defect-free interfaces. Unless the interface material structures are constrained in this way, macroscopic transport and optical methods provide, in general, less than direct measures of the band offsets.

Photoemission methods provide a means to gauge the band discontinuities during the monolayer-staged formation of the semiconductor–semiconductor junction. Such techniques monitor the relative energy displacement in the densities of states and thereby the band edges for the two semiconductors. Figure 51 illustrates the XPS method for determining the band edge discontinuities from the core level and valence-band energies (Grant et al. 1987). Analogous to fig. 14, XPS extracts the core level binding energies $E_{CL}(A) - E_V(A)$ and $E_{CL}(B) - E_V(B)$ with respect to edge features of

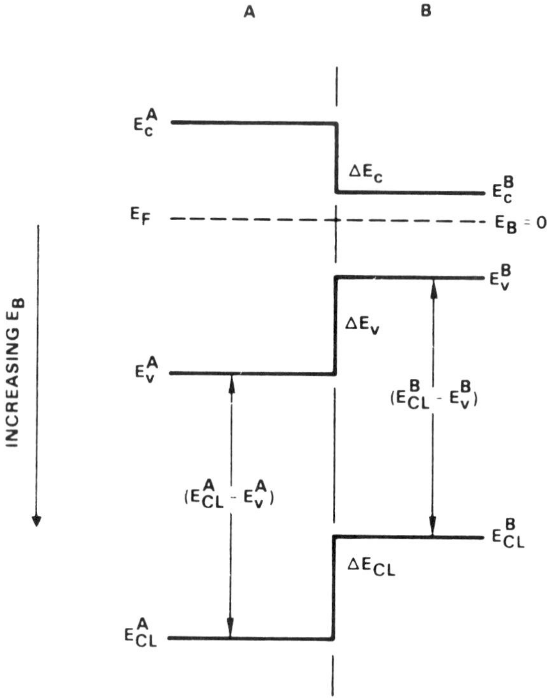

Fig. 51. Schematic energy band diagram for a semiconductor heterojunction illustrating the determination of valence-band offset from XPS measurements of core level binding and valence-band edge energies (after Grant et al. 1987).

the corresponding valence bands. From direct measurements of $E_{CL}(A) - E_V(A)$ and $E_{CL}(B) - E_V(B)$, and the difference in core level binding energies ΔE_{CL}, one may extract the valence-band discontinuity ΔE_V. Here, a well-defined determination of ΔE_V requires a lineshape fit of the XPS features to a theoretical valence-band density of states (Kraut et al. 1983). Likewise, the deconvolution of core level lineshapes can compensate for the effects of interface bonding on spectral energies E_{CL}. An alternative photoemission approach involves direct comparison of the valence-band edge features. Figure 52 illustrates the measurement of a double valence-band edge feature induced by deposition of a new semiconductor on a semiconductor substrate (Margaritondo and Perfetti 1987). This approach yields a direct measurement of ΔE_V in cases where a well-defined double-edge feature is present. In cases where such a double edge is indistinct, only an indirect measurement is possible from the shift of the composite features leading edge, taking into account the effect of band bending (Margaritondo and Perfetti 1987). Photovoltaic effects present in such photoemission measurements serve to reduce such band bending and in principle simplify the analysis of band offsets (Yu et al. 1990). The precision of these photoemission techniques is limited by the electron spectroscopy resolution (~0.1–0.2 eV) and is inferior to that of the optical techniques (< 0.01 eV) described here. However, signifi-

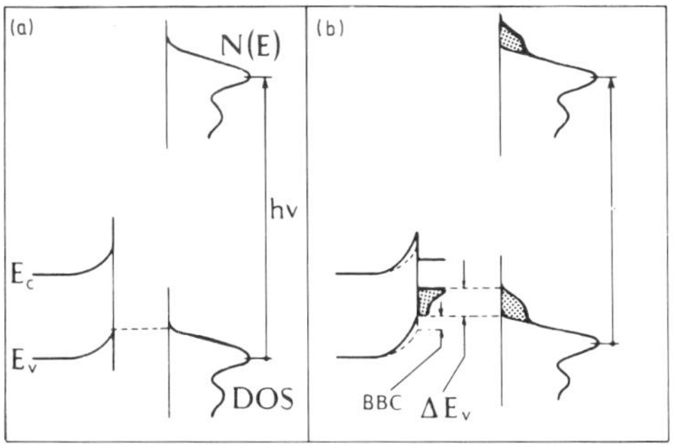

Fig. 52. Schematic energy band and density-of-states diagram for photoemission measurement of the valence-band discontinuity ΔE_V. Deposition of new semiconductor monolayers over the clean surface of a different semiconductor changes the measured valence-band edge from a single-edge structure (A) to a double-edge structure (B), which yields ΔE_V directly. For cases where two such edges are indistinct, measurements of the leading edge of the composite features provide corrections due to band bending (BBC) and an indirect gauge of ΔE_V (after Margaritondo and Perfetti 1987).

cantly better accuracies are achievable from photoemission energy shifts (Grant et al. 1987).

Finally, scanning tunneling microscopy techniques have now provided new information about heterojunction band offsets. Analogous to BEEM measurements of Schottky barrier height, STM (Albrektsen et al. 1990) and an inverse photoemission variant (Abraham et al. 1990) provide a nanometer-localized probe of heterojunction band discontinuities. Here the tunneling tip is scanned across the cross-sectioned interface to yield relative voltage shifts in current–voltage spectra or the onset of luminescence, respectively. Despite complications due to surface preparation and the effects of band bending, such localized techniques represent a powerful new tool for probing heterointerfaces.

Overall, the accuracy and reliability of all these band offset measurements can differ greatly, not only between experimental techniques but even for measurements by the same method (Capasso and Margaritondo 1987). Thus while the precision of many techniques is tens of meV, the discrepancies between such measurements can be an order of magnitude higher. Such effects result not only from variations in the experimental procedure and interface structure, but also from the uncertainties associated with the models linking the physical measurements to the band offsets. Overall, Margaritondo and Perfetti (1987) estimate the typical accuracy of ΔE_V and ΔE_C measurements to ± 0.1 eV.

Table 6 lists the heterojunction band offsets for a variety of elemental and compound semiconductor interface combinations (after Margaritondo and Perfetti (1987)). Where multiple measurements of a particular heterojunction are available, table 6 lists the average value and the experimental technique(s) employed. The signs of ΔE_V and ΔE_C follow the convention illustrated in fig. 50, namely that ΔE_V of the heterojunction A–B is positive if the valence-band edge of B is above that of A and that ΔE_C is positive if the conduction band of B is below that of A. The prefixes a and c refer to amorphous and crystalline materials, respectively. These techniques

Table 6

Heterojunction valence and conduction-band discontinuities extracted from photoemission (PH), capacitance–voltage (C/V), current–voltage (I/V), internal photoemission (IP), photoluminescence spectroscopy (PL), or other techniques (OT), with sign convention as shown in fig. 50 (adapted from Margaritondo and Perfetti (1987) with permission). Average values are italicized.

Interface	ΔE_V	ΔE_C	$\Delta E_C/\Delta E_G$ (%)	Experimental method	Ref.*	Comments
Si–Ge	*0.28*	*0.16*		PH, C/V	[1]	Average value: a-Si or c-Si on Ge(111), a-Ge or c-Ge on Si(111)
AlAs–Ge	*0.86*	*0.67*		PH	[1]	Average value: a-Ge or c-Ge on AlAs, AlAs on Ge(100)
AlAs–GaAs	*0.34*	*0.51*	*60*	PH, OT	[1]	Average value: GaAs on AlAs(100) and (110), AlAs on GaAs(100) and (110)
$Al_xGa_{1-x}As$–GaAs			*59*	PH, C/V, I/V, IP, OT	[1]	Average value: $x =$ 0.09–0.7
$Al_xIn_{1-x}As$–InP	0.49	−0.52		PL	[2]	$x = 0.48$, staggered gaps
AlSb–GaSb	0.4	0.5		OT	[3]	Superlattice
GaAs–Si	0.05	0.19		PH	[4]	a-Si on GaAs(110)
GaAs–Ge	*0.49*	*0.19*		PH, I/V, OT	[1]	Average value: a-Ge and c-Ge on GaAs(110), c-Ge on GaAs(100), c-Ge on GaAs(111)Ga, (100)Ga, (100)As, (111)As
GaAs–InAs	0.17	−0.09		PH	[5]	GaAs on InAs(100), staggered gaps
GaP–Ge	0.80	0.77		PH	[4]	a-Ge deposited on GaP(110)
GaP–Si	0.80	0.33		PH	[6]	a-Si or c-Si deposited on GaP(110)
GaSb–Ge	0.20	−0.20		PH	[4]	a-Ge on GaSb(110), staggered gaps
GaSb–Si	0.05	−0.49		PH	[4]	a-Si on GaSb(110), staggered gaps
InAs–Ge	0.33	0.27		PH	[4]	a-Ge on InAs(110)
InAs–Si	0.15	0.01		PH	[4]	a-Si on InAs(110)
InP–$In_xGa_{1-x}As$	0.37	0.23	*38*	C/V	[7]	$x = 0.53$
$In_xAl_{1-x}As$–$In_yGa_{1-y}As$	*0.36*	*0.36*	*51*	C/V, IP, PL, I/V, OT	[1]	$x = 0.52$, $y = 0.53$
InP–(In,Ga),(As,P)			*57*	C/V, PL	[1]	Different quarternary compositions, lattice matched with InP, with $E_G = 0.70$–1.20 eV
InP–Ge	0.64	−0.04		PH	[4]	a-Ge on InP(110)
InP–Si	0.57	−0.41		PH	[4]	a-Si on InP(110), staggered gaps
$In_xGa_{1-x}P$–GaAs	0.08	0.59	88	C/V	[8]	$x = 0.48$
InSb–Ge	0.0	−0.50		PH	[4]	a-Ge on InSb(110)

Table 6 (*continued*)

Interface	ΔE_V	ΔE_C	$\Delta E_C/\Delta E_G$ (%)	Experimental method	Ref.*	Comments
InSb–Si	0.0	−0.94		PH	[4]	a-Si on InSb(110)
CdS–Ge	1.75	0.00		PH	[4]	a-Ge on cleaved CdS
CdS-InP	0.77	0.30		PH	[9]	c-CdS on InP(110)
CdS–Si	1.55	−0.24		PH	[4]	a-Si on cleaved CdS, staggered gaps
CdSe–Ge	1.30	−0.23		PH	[4]	a-Ge on cleaved CdSe, staggered gaps
CdSe–Si	1.20	−0.57		PH	[4]	a-Si on cleaved CdSe, staggered gaps
CdTe–GaAs	0.21	−0.13		PH	[10]	c-CdT on cleaved GaAs(110)
CdTe–Ge	0.85	−0.08		PH	[4]	a-Ge on cleaved CdTe
CdTe–Si	0.75	−0.42		PH	[4]	a-Si on cleaved CdTe
CdTe–α-Sn	1.1	0.26		PH	[11]	α-Sn on CdTe(111)
CdTe–HgTe	*0.13*	*1.16*		PH, IR, OT	[1]	Average value
ZnS–Cu_2S	1.4	1.35		*I/V*	[12]	
ZnSe–Ge	*1.40*	*0.51*		PH, *I/V*	[1]	Average value: a-Ge and c-Ge on cleaved ZnSe, c-ZnSe on GaAs(110)
ZnSe–Si	1.25	0.22		PH	[4]	a-Si on cleaved ZnSe
ZnSe–GaAs	*1.03*	*0.20*		PH	[1]	Average value: annealed c-ZnSe on GaAs(110)
ZnSe–GaAs	1.25	0.05		PH	[13]	c-ZnSe on GaAs(100)
ZnTe–Ge	0.95	0.64		PH	[4]	a-Ge on cleaved ZnTe
ZnTe–Si	0.85	0.30		PH	[4]	a-Si on cleaved ZnTe
PbTe–Ge	0.35	−0.42		PH	[14]	a-Ge on PbTe(100), staggered gaps
GaSe–Ge	0.83	0.55		PH	[15]	a-Ge on cleaved GaSe
GaSe–Si	0.74	0.20		PH	[15]	a-Si on cleaved GaSe
CuBr–GaAs	0.85	0.74		PH	[16]	
CuBr–Ge	0.7	1.6		PH	[16]	
$CuInSe_2$–Ge	0.48	−0.25		PH	[17]	a-Ge on fractured $CuInSe_2$, staggered gaps
$CuInSe_2$–Si	0.00	−0.21		PH	[17]	a-Si on fractured $CuInSe_2$
$CuGaSe_2$–Ge	0.62	−0.33		PH	[17]	a-Ge on fractured $CuGaSe_2$, staggered gaps
$ZnSnP_2$–GaAs	0.13	−0.03		OT	[18]	Staggered gaps
ZnTe–GaSb	0.34	1.25		PH	[19]	c-ZnTe on GaSb(110)

*References: [1] Margaritondo and Perfetti (1987); [2] Caine et al. (1984); [3] Tejedor et al. (1985); [4] Katnani and Margaritondo (1983); [5] Kowalczyk et al. (1982); [6] Perfetti et al. (1984); [7] Forrest et al. (1984); [8] Hsieh et al. (1984); [9] Wilke et al. (1990); [10] Yu et al. (1990); [11] Takatani and Chung (1985); [12] Gorbik et al. (1980); [13] Colbow et al. (1991); [14] Cerrina et al. (1983); [15] Daniels et al. (1985); [16] Waldrop et al. (1985); [17] Turowski et al. (1985); [18] Patten et al. (1985); [19] Yu et al. (1991).

form the basis for evaluating the nature of the interface dipole contributions to the heterojunction band offsets.

4.3.3. Interface dipoles

The heterojunction band offsets do not exhibit a simple dependence on the difference in semiconductor electron affinities

$$\Delta E_C = (X_B - X_A) = \Delta X. \tag{4.6}$$

A plot of ΔE_C measured by core level photoemission spectroscopy as a function of electron affinity differences ΔX (i.e. Milnes and Feucht 1972) yields no systematic trend, as shown in fig. 53a (Bauer et al. 1983). The failure of this "electron affinity rule" (Anderson 1962) implies that the collection of dipole terms in eq. (4.5) makes a significant contribution to the band offset. As with the Schottky model (eq. (3.4)), these dipole contributions can be related to specific features of the microscopic interface.

A number of physical mechanisms have provided bases for calculating semiconductor band edge discontinuities and interface dipoles. These include: tunneling approaches involving charge neutrality levels analogous to eq. (3.9) (Flores and Tejedor 1979, Tersoff 1984), local-bond approaches incorporating tight-binding (Harrison 1977, 1985) or pseudopotential calculations (Frensley and Kroemer 1977) of electrostatic potentials at the interatomic junction, an effective work function model with a modified eq. (4.6) to take account of new chemical phases at the intimate junction (Freeouf and Woodall 1981, 1986), an empirical deep-level scheme relating the band edges to transition-metal levels tied to an average dangling bond energy (or the vacuum level) (Langer and Heinrich 1985, Zunger 1985), plus additional approaches based on the bulk semiconductor properties and involving the dielectric electronegativity (Van Vechten 1985), self-consistent first-principles calculations of semiconductor superlattice band structure (Christensen et al. 1990), and analytic calculations depending on empirical values of the bulk optical dielectric constant and the lattice constants (Jaros 1988). Arguably the most successful have been the local bond and tunneling approaches, as represented in figs. 53b and 54. In contrast to the electron affinity rule, ΔE_V offsets derived from band edge values calculated via Harrison's atomic-orbital method display a strong correlation with the experimental values (after Kroemer (1985)). For the covalent semiconductors represented, the tight-binding valence-band maximum can be represented as (Harrison 1977, Chadi and Cohen 1975c)

$$E_V = \tfrac{1}{2}(\varepsilon_a + \varepsilon_c) + [\tfrac{1}{4}(\varepsilon_a - \varepsilon_c)^2 + (4E_{xx})^2]^{1/2}, \tag{4.7}$$

where ε_a and ε_c are the cation and anion atomic energies, and the matrix element E_{xx}, given by

$$E_{xx} \approx -1.28\hbar/md^2, \tag{4.8}$$

depends on the nearest anion–cation distance d and free-electron mass m. In fig. 54 the results of this atomic-orbital method and of Tersoff's mid-gap energy approach are compared for an expanded set of heterojunction experiments. In fig. 54a calculated

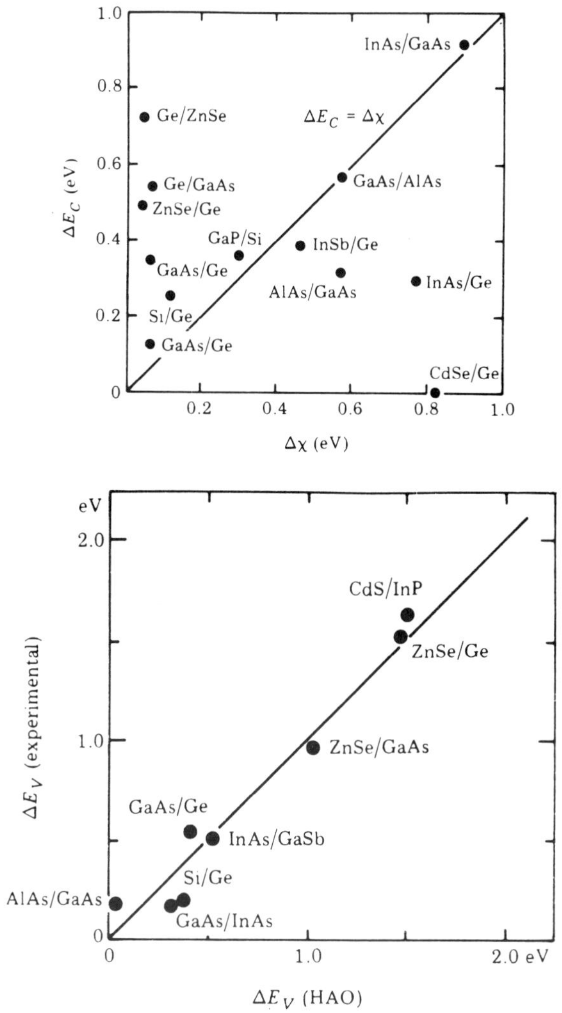

Fig. 53. Parameterization of band edge discontinuities in terms of (a) electron affinity differences ΔX and (b) atomic-like orbital theory. In (a) ΔE_C values measured by core-level photoelectron spectroscopy (with the substrate written first, the overlayer second) exhibit no systematic trend for an electron affinity rule $\Delta E_C = \Delta X$ (after Bauer et al. 1983). In (b), experimental ΔE_V values plotted versus ΔE_V offsets derived from band edge values calculated via the atomic-orbital method of Harrison (1977) (after Kroemer 1985). The improved systematic behavior in (b) reflects the incorporation of local potential differences at the atomic interface.

ΔE_V values have been corrected for lattice mismatch by substitution of an average lattice constant in eq. (4.8) (Katnani 1987). Paasch and von Faber (1991) have obtained an improved agreement between theory and experiment over fig. 54b, based in part on an explicit coupling between the band bending and the interface dipole. Both figs. 54a and b display comparable agreement with experiment, despite concep-

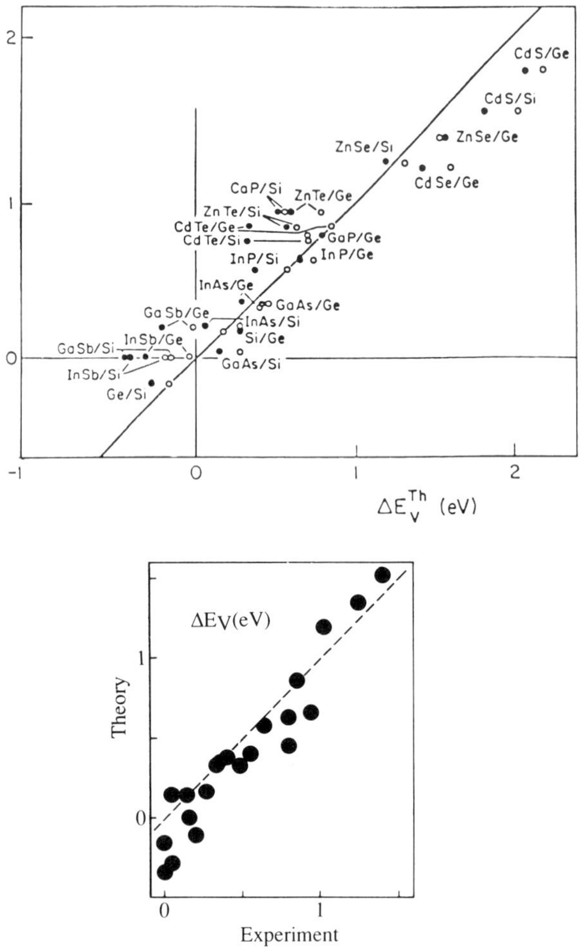

Fig. 54. Comparison of experimental ΔE_V values plotted versus ΔE_V: (a) calculated via the Harrison (1977) atomic-orbital method and (b) via the Tersoff (1984) mid-gap energy approach for an expanded set of heterojunctions (Margaritondo and Perfetti 1987). The theoretical values in (a) are correlated for interface lattice mismatch (after Katnani 1987). Both calculations yield agreement with experiment to within an accuracy of about 0.2 eV, despite conceptually different physical bases.

tually different physical bases. The mid-gap approach offers relative simplicity in terms of matching interface-independent values for different semiconductors. In addition, this model yields Φ_B values for Au on different semiconductors that are in good agreement with observations (Tersoff 1984). On the other hand, bond orbital approaches permit an explicit consideration of interface-specific geometric and chemical features. Despite the systematic correlations exhibited for both cases in fig. 54, limits to the prediction of heterojunction band discontinuities appear to be well outside the tolerances necessary for use in device design and fabrication – which are an order of magnitude more precise.

Measurements of heterojunction band offsets with variations in interface growth and processing have revealed additional features of semiconductor–semiconductor interface dipoles. These are related to the variation of band discontinuity with growth sequence, surface orientation, surface reconstruction, and interface chemical structure. The dependence of band offset on growth sequence can be expressed as the degree of commutativity and transitivity, and thereby the linearity of electronic structure variations. Measured band offsets appear to satisfy the linearity condition

$$\Delta E_{\mathrm{V}}(\mathrm{A–B}) + \Delta E_{\mathrm{V}}(\mathrm{B–A}) = 0 \tag{4.9}$$

and the transitivity condition

$$\Delta E_{\mathrm{V}}(\mathrm{A–B}) + \Delta E_{\mathrm{V}}(\mathrm{B–C}) + \Delta E_{\mathrm{V}}(\mathrm{C–A}) = 0. \tag{4.10}$$

Tests of linearity indicate that eqs. (4.9) and (4.10) are satisfied in varying degrees from limits below the experimental resolution to of the order of 0.2 eV (Margaritondo and Perfetti 1987). Following Grant et al. (1987) the sequence of growth in forming the junction can alter the heterojunction band offsets as a result of the interface chemical reactivity. As an example, linearity is not preserved for the Ge–CuBr(110) system, which reacts strongly for Ge deposited on CuBr but grows epitaxically for CuBr on Be (Waldrop et al. 1985). Similarly, ΔE_{V} for epitaxical Ge/ZnSe versus ZnSe/Ge junctions differs by 0.23 eV and ΔE_{V} for epitaxical Ge/GaAs versus GaAs/Ge junctions differs by 0.2 eV, consistent with the formation of interface dipoles. Furthermore, the GaAs/Ge(110) interface is unstable with time and annealing, varying by 0.2–0.3 eV (Grant et al. 1985, 1986). These latter growth sequence effects may be related to the antiphase disorder, i.e., the site allocation for compound semiconductor growth on an elemental semiconductor (Kroemer 1983). Such nonlinearities may arise because atom transfer across a heterojunction interface involving elements from different columns of the periodic table can produce interface dipoles (Harrison et al. 1978). For isocolumnar heterojunctions such as AlAs–GaAs(110), growth sequence effects are measurable but not large (~0.13 eV), considering the differences in surface morphology and reaction, which may depend on the particular type of surface preparation.

ΔE_{V} values exhibit a well-defined but relatively small dependence on the crystallographic orientation. Thus Ge on GaAs shows a variation of about 0.2 eV for Ge on (111), (100), (110) and (111) orientations of GaAs. These ΔE_{V} variations with crystallographic orientation appear in table 7 (Grant et al. 1987). Particular care is required in analyzing such XPS results since the spectral linewidths are about 1.1–1.2 eV, compared with ΔE_{V} variations of 0.2 eV. However, measurement reproducibility was to within ± 0.01 eV. The ΔE_{V} variations vary systematically according to the surface polarity (Grant and Harrison 1988), although theory predicts variations only half as large as are measured, based on a thermal distribution of energetically favored bonds. Such small variations are expected on the basis of charge redistribution under the otherwise high fields set up over atomic distances. Thus the stability of a charge accumulation involving a polar surface produces a charge rearrangement, which leads to a neutral transition layer. The orientation dependence of ΔE_{V} is also evident in pseudomorphically grown $\mathrm{Si/Si_{1-x}Ge_x}$ heterostructures on different Si substrates,

Table 7
Crystallographic orientation dependence of heterojunction valence-band offsets for Ge/GaAs interfaces. The Ge epitaxial layer thickness, Ge 3d and Ga 3d photoelectron linewidths, Ge 3d–Ga 3d core-level binding energy differences, the average variation in ΔE_V relative to the (110) interface, and the ΔE_V value for eight different Ge–GaAs interfaces (after Grant et al. 1987).

Substrate surface	Ge layer thickness (Å)	Γ(Ga 3d) (eV)	Γ(Ge 3d) (eV)	$(E^{GaAs}_{Ga\,3d} - E^{Ge}_{Ga\,3d})^b$	$\delta(\Delta E_V)_{AVE}$ (eV)	$(\Delta E_V)^b$ (eV)
(111)Ga (2 × 2)	13	1.17 ± 0.02	1.25 ± 0.01	−10.27	~ −0.085	0.50
	20	1.22 ± 0.02	1.26 ± 0.01	−10.31		0.46
(100)Ga *c*(8 × 2)	22	1.19 ± 0.02	1.25 ± 0.01	−10.22	−0.015	0.55
(110) (1 × 1)	14	1.13 ± 0.01	1.29 ± 0.01	−10.20	0	0.57
	17	1.16 ± 0.01	1.27 ± 0.01	−10.21		0.56
(100)As	14	1.15 ± 0.02	1.25 ± 0.01	−10.17	+0.035	0.60
(111)As (1 × 1)	13	1.21 ± 0.01	1.32 ± 0.01	−10.11	+0.10	0.66
	18	1.22 ± 0.01	1.28 ± 0.01	−10.10		0.67

[a] Uncertainty is ± 0.01 eV.
[b] Uncertainty is ± 0.04 eV.

which vary by 0.07 eV for Si(110), Si(100), and Si(111) due to strain-induced splitting of the band edges (Ni and Hansson 1990).

Surface reconstruction effects on heterojunction band offsets can arise due to differences in local geometric or chemical structure. However, the results for Ge/GaAs(100) heterojunctions demonstrate that surface reconstruction and stoichiometry have a large effect on E_F position and band bending but not on band offset values. Figure 55 show the E_F movement for Ge on different reconstructions of GaAs(100) prepared under UHV conditions and on cleaved GaAs(110). As in fig. 44, there is a qualitative difference in E_F movement with coverage. The E_F position stabilizes rapidly to a characteristic mid-gap value for GaAs(110) with submonolayer coverage, whereas E_F stabilizes over several monolayers to a wide range of energy positions for GaAs(100) (Chiaradia et al. 1984b). Figure 56 illustrates the final E_F positions for GaAs(100) for the three reconstructions indicated, exhibiting a greater than 0.3 eV variation in band bending but a ΔE_V constant to within ± 0.05 eV. Thus a major conclusion from these results is that differences in surface reconstruction do not affect the band discontinuities significantly, but they do alter the band bending on both sides of the semiconductor–semiconductor interface. Hence the charge rearrangement does not introduce a new dipole but rather a change in interface charge density at the atomic interface. Thus the independence of E_F and ΔE_V implies that the physical origins of the heterojunction band offsets and E_F stabilization are different at semiconductor–semiconductor interfaces (Chiaradia et al. 1984b). Similar measurements on As-doped Ge deposited on GaAs(100) also show different E_F positions with the same E_V, indicating in addition that As composition can contribute to E_F stabilization, both as an adsorbate and as a dopant (Chiaradia et al. 1984b).

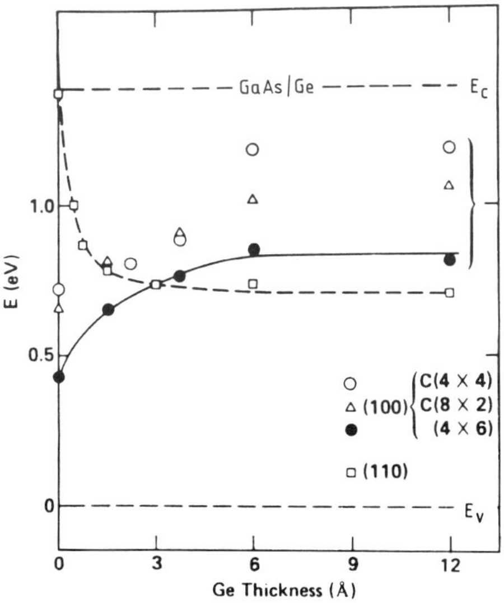

Fig. 55. Fermi level movement as a function of Ge deposition of MBE-grown GaAs(100) and melt-grown GaAs(110) surfaces as measured by SXPS core level movements. The Fermi level at the cleaned GaAs(110)–Ge interface exhibits rapid movement with submonolayer coverage to a characteristic mid-gap energy. In contrast E_F at the MBE-grown GaAs(100)–Ge interface exhibits more gradual movement to final positions, which vary substantially with the indicated surface reconstructions (after Chiaradia et al. 1984b).

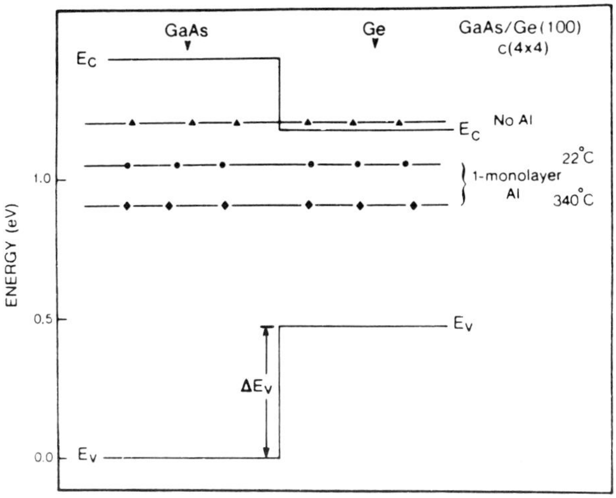

Fig. 56. Schematic energy band diagram of the GaAs(100)–Ge heterojunction showing the Fermi level position for a 10 Å Ge overlayer on the GaAs(100) surface with different reconstructions. The ΔE_V discontinuity remains unchanged despite a strong dependence of E_F on reconstruction and stoichiometry, demonstrating the independence from band bending and band offset (after Chiaradia et al. 1984b).

Surface reconstruction and stoichiometry can have a pronounced effect on the interface state energies and densities. The interface state densities of ZnSe–GaAs heterostructures range over at least two orders of magnitude, depending upon the RHEED reconstruction pattern of GaAs prior to epitaxical ZnSe growth (Qiu et al. 1990). Figure 57 illustrates the interface state energies and densities as extracted from a $C-V$ analysis for several well-defined reconstructions. The densities decrease in order of decreasing surface As concentration. The states shown here are consistent with those at low energy ($\leqslant 0.2$ eV) and 0.6 eV above E_V shown in fig. 45. Furthermore, the reduction of states with As depletion is consistent with the thermal processing steps required to produce the barriers shown in fig. 44. In fig. 57 this reduction is attributed to the formation of a GaAs interfacial compound, as discussed in § 4.2.1. The decreased density is shown by nearly ideal $C-V$ characteristics and the movement of E_F across the entire band gap. The improved electrical properties at ZnSe/GaAs heterojunctions are also confirmed by band bending measurements showing that E_F is not pinned at a fixed energy but instead follows the doping concentration-dependent E_F position in ZnSe (Olego and Cammack 1990). Deposition of Al interlayers also leads to changes in interface band bending but not ΔE_V. Here the presence of Al monolayers alters the chemical structure and growth kinetics of epitaxical GaAs/Ge heterojunctions. E_F variations are related to Al control of As diffusion out of the GaAs into Ge (Margaritondo et al. 1980) and the subsequent As doping of Ge (Katnani et al. 1985).

Overall, these interface contributions to heterojunction band offsets and band bending show that interface preparation can significantly alter the interface E_F position within the band gap at semiconductor–semiconductor junctions. While localized states at such interfaces are sensitive to the interface geometric and chemical structure, the heterojunction band offsets do not vary correspondingly. The introduc-

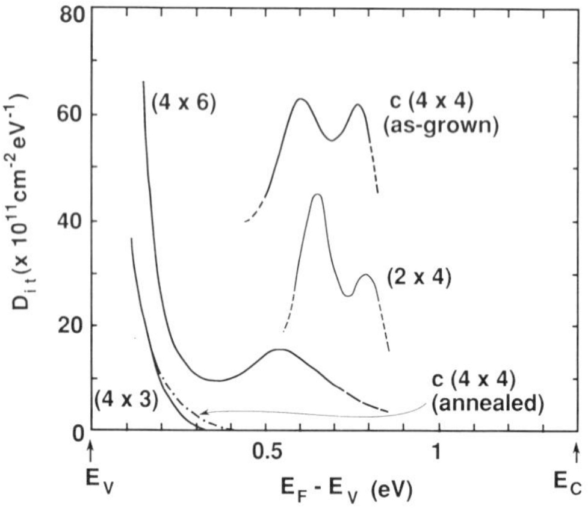

Fig. 57. Density of interface states calculated from the capacitance–voltage characteristics for a variety of GaAs epilayer surface reconstructions prior to ZnSe epitaxial growth. Interface state densities decrease with decreasing As concentration, emphasizing the sensitivity of interface state densities to local atomic bonding (after Qiu et al. 1990).

tion of new chemical species does not as a matter of course alter the band edge discontinuities (Christensen and Brey 1988). Indeed the heterojunction band offsets can vary, but only with the introduction of new interface dipoles. These interface dipoles are most evident among the different growth processes. Different surface orientations have larger effects on the band offsets than different surface reconstructions or growth sequences, unless major chemical or morphological effects are involved.

4.3.4. Atomic-scale control

Well-defined atomic-scale procedures are required in order to introduce significant dipole changes at semiconductor–semiconductor interfaces. Two such approaches have proved effective in creating sizeable and controllable changes in band offset:

(i) spatially-separated dopant layers and

(ii) chemical interlayers which exchange charge with both semiconductors.

Doping of ionized donor and acceptor charges in near-monolayer sheets ("delta" doping) by MBE techniques within a few monolayers of and on opposite sides of the interface introduces high electric field gradients across the junction and alters the effective barriers to charge transport. Figure 58 illustrates such effects on the band discontinuities ΔE_C and ΔE_V for a doping interface dipole at an intrinsic heterojunction (Capasso et al. 1985). Here, the interface dipole $\Delta\Phi$ produces triangular conduction band edge features which, in the limit of separations only a few monolayers thick, raise the bottom of the energy sub-band in the quantum well of the lower gap semiconductor, decreasing the thermal activation barrier for electrons by

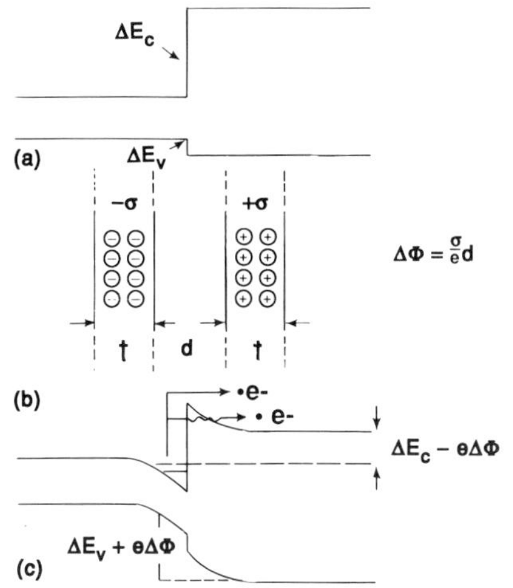

Fig. 58. Schematic energy band diagram of (a) an intrinsic heterojunction, (b) the introduction of a doping interface dipole with a sheet charge density $\pm\sigma$ of layer thickness t and separation d, and (c) the resultant band diagram incorporating an interface dipole $\Delta\Phi = \sigma d/\varepsilon$. In the limit of separation of only a few atomic layers, thermally-activated electrons can tunnel through the triangular barrier, which is effectively lowered to $\Delta E_C - e\Delta\Phi$ (after Capasso et al. 1985).

about $e\Delta\Phi/2$. On the high band gap side, the triangular barrier is transparent to tunneling, further reducing ΔE_C to $\Delta E_C - e\Delta\Phi$. For AlGaAs/GaAs pin diodes, such dipoles cause a 0.1 eV decrease in ΔE_C. Reversal of the donor and acceptor sheets can in principle increase ΔE_C and decrease ΔE_V. Shen et al. (1990) have introduced changes in ΔE_C for InAs/GaAs heterojunctions using a delta doping approach as well.

Chemical interlayers at heterojunctions (Margaritondo et al. 1980, Niles et al. 1985) have provided even greater control of the band edge discontinuities. For example, the introduction of thin, ordered Si layers within the interface region of GaAs–AlAs heterostructures introduces dipoles that alter ΔE_V continuously over the range 0.02–0.78 eV as a function of interlayer thickness and growth sequence of the semiconductors (Sorba et al. 1991). RHEED and XPS measurements indicate that the Si atoms remain at the interface, exhibit long-range order, and produce a dipole of magnitude 0.38 eV. Figure 59 displays the opposite sign in ΔE_V for the two growth sequences and the continuous changes with Si interlayer thicknesses from 0–2 monolayers.

Conversely, a compound semiconductor interlayer at an elemental semiconductor homojunction also produces sizeable band offset changes. One monolayer of GaAs at a Ge–Ge(111) homojunction leads to a 0.35–0.45 eV offset with the Ge valence-band edge lower on the As side of the junction (McKinley et al. 1991). Consistent with a dipole effect, ΔE_V changes with opposite sign result from an opposite growth sequence of the double Ga–As layer. Self-consistent calculations at such interlayer structures (Muñoz et al. 1990, Peressi et al. 1991) indicate stable atomic configurations with interlayer atoms occupying sites on consecutive lattice planes. Because of their valence differences with respect to the semiconductor matrix, these interlayer atoms would exhibit a charge transfer of −1 electron per atom in the Si monolayer that

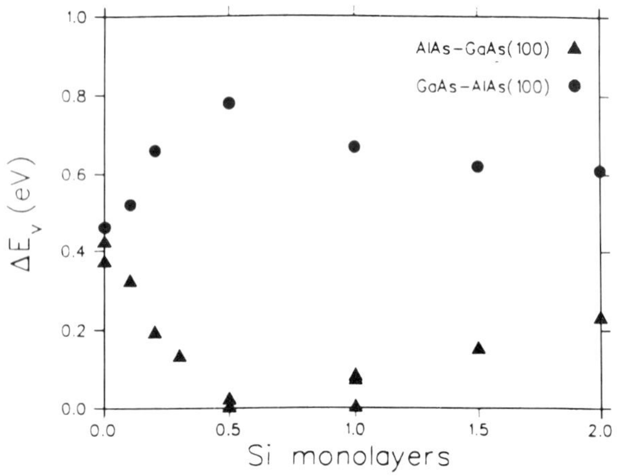

Fig. 59. Tunable valence-band offset ΔE_V for AlAs–GaAs(100) (△) and AlAs(100) (●) heterostructures as a function of ordered Si interlayer thickness. The Si layer is grown on an As-stabilized surface in both cases. Control of the Si coverage, growth sequence, and As flux alters ΔE_V continuously in the range 0.02–0.78 eV around the intrinsic $\Delta E_V = 0.40 \pm 0.07$ eV position (after Sorba et al. 1991).

replaces Ga in the structure and a charge transfer of +1 electron per atom in the Si monolayer replacing the As. A similar charge transfer and dipole are expected for the Ge–GaAs–Ge system prior to dielectric screening. Here the dipole sign is consistent with similar charge transfer arguments, whereas the screened-dipole magnitude is reduced (from 0.97 to 0.4 eV), possibly due to interface strain and interface roughness. Table 8 lists a variety of heterojunction interlayer structures with their corresponding interface dipoles. The large voltage shifts evident for numerous semiconductor–interlayer–semiconductor structures offer exciting new possibilities for altering the interface band structure using controlled atomic geometry and bonding.

Overall, the heterojunction features described in this subsection provide numerous examples of the influence that local chemical bonding and atomic composition have on lattice growth, band bending, and band edge discontinuities. Furthermore, the separation of the band bending and band offset dependences on interface growth and processing reveal the high degree of atomic-scale control required to engineer localized charge states and interface dipoles.

5. Synopsis

The aim of this chapter has been to provide an overview of semiconductor surfaces and interfaces and the interdependence of their geometric, chemical, and electronic structure. These relationships are evident on three levels of aggregation:

(i) on the atomic scale in terms of new local bonding and reconstruction;

Table 8
Interface dipoles measured for semiconductor–interlayer–semiconductor interface structures.

Semiconductor junction	Interlayer	ΔE_V (eV)	Refs.*
CdS($10\underline{1}0$)–Ge	Al	0.2	[1]
CdS($10\underline{1}0$)–Ge	Au	0.15–0.2	[1]
ZnSe(110)–Ge	Au	−0.25 to 0.08–0.26	[1]
ZnSe(110)–Ge	Al	0.3	[1]
GaP(110)–Si	Au	0.6	[1]
GaP(110)–Si	Al	0	[1]
GaAs(100)–Ge	Al	0	[2]
GaAs(100)–AlGaAs	Donor/ acceptor sheets	−0.1	[3]
GaAs(100)–AlAs	Si	0.38	[4]
GaAs(100)–GaAs	Si	−0.38	[4]
Ge(111)–Ge	Ga–As	0.35–0.45	[5]
Ge(111)–Ge	Al–As	0.4	[6]
Si(111)–Si	Ga–P	0.4	[6]
Si(111)–Si	Al–P	0.4	[6]

*References: [1] Niles et al. (1988); [2] Katnani et al. (1985); [3] Capasso et al. (1985); [4] Sorba et al. (1991); [5] McKinley et al. (1991); [6] Marsi et al. (1991).

(ii) on the nanometer scale in terms of new chemical phases, clusters, and interdiffusion;
(iii) on the micron scale in terms of macroscopic interfacial compound/phase formation and long-range, electrostatic fields generated by dipoles and charges near an interface.

For semiconductor surfaces, surface science techniques have shown how surface geometric structure differs in complex ways from that of the bulk lattice. New methods of growing and characterizing semiconductor structures have provided detailed information about the thermodynamics and kinetics associated with crystal lattice formation. Electronic and electrostatic techniques have shown that the surface band structure depends sensitively on the specific details of atomic reconstruction and that localized states form across the semiconductor band gap whose properties can be related to the atomic bonding features.

For overlayer–semiconductor interfaces, adsorption can induce surface geometric changes and provide a means to vary the epitaxial overlayer growth. Overlayer growth proceeds via several routes that depend upon local chemical interactions. Chemical reactions, diffusion, and new phase formation can occur which depend sensitively on local atomic bonding as well as extrinsic processing. Furthermore, such interface reactions and diffusion are controllable on an atomic scale. The electronic properties of metal–semiconductor interfaces center on the localized interface states and band bending, which are influenced by numerous extrinsic phenomena. Such localized states are directly observable near semiconductor surfaces and at buried interfaces. The associated Schottky barriers are controllable by a variety of atomic-scale techniques.

For semiconductor–semiconductor interfaces, the geometric structure involves primarily the epitaxical growth defined by the crystal lattice parameters and the imperfections generated outside the limits established for high-quality growth. As with metal–semiconductor interfaces, diffusion and reaction can occur on an atomic scale, which depend on surface bonding and stoichiometry. Furthermore, template structures provide a monolayer-scale technique with which to control the nature of epitaxy and growth. Measurements of the heterojunction and homojunction band offsets have provided an array of systematic results consistent with several disparate physical bases. Again, control of these electronic properties is now possible with atomic-scale techniques.

The strong relation between the atomic-scale structure and the electronic properties at semiconductor surfaces and interfaces provides a fertile ground for new research. The benefits are both fundamental and technological: an understanding (and control) of the semiconductor band structure and localized states on an atomic scale and a basis on which to produce the geometric, chemical, and electronic structures for devices on the smallest scale possible.

Acknowledgments

The author wishes to acknowledge valuable conversations with Drs. David Biegelsen, Jim Chadi, Esther Conwell, Robert Dandrea, Randy Feenstra, Ron Grant, Antoine

Kahn, Joel Kubby, Giorgio Margaritondo, Chris Palmstrøm, and Jerry Woodall during the course of writing this chapter. Special thanks are due to Dr. Charles Duke for a critical reading of the manuscript and to Ms. Marlene Elliott for her skillful office management during this period. I am indebted to the Office of Naval Research (George Wright) for the support which enabled much of the author's work included here. Most of all, I thank Janice, Lindsay, and Erica Brillson for their great patience and understanding.

References

Aboie-Elfotouh, F.A., L.L. Kazmerski, H.R. Moutinho, J.M. Wissel, R.G. Dhere, A.J. Nelson and A.M. Bakry, 1991, J. Vac. Sci. & Technol. **A9**, 554.
Abraham, D., L.A. Veider, Ch. Schönenberg, H.P. Meier, D.J. Arent and S.F. Alvarado, 1990, Appl. Phys. Lett. **56**, 1564.
Abstreiter, G., 1985, in: Molecular Beam Epitaxy and Heterostructures, eds L.L. Chang and K. Ploog (Martinus Nijhoff, The Hague).
Albrektsen, O., D.J. Arent, H.P. Meier and H.W.M. Salemink, 1990, Appl. Phys. Lett. **57**, 31.
Aldao, C.M., I.M. Vitomirov, F. Xu and J.H. Weaver, 1988, Phys. Rev. B **37**, 6019.
Alerhand, O.L., D. Vanderbilt, R.D. Meade and J.D. Joannopoulos, 1988, Phys. Rev. Lett. **61**, 1973.
Alerhand, O.L., A.N. Berker, J.D. Joannopoulos, D. Vanderbilt, R.J. Hamers and J.E. Demuth, 1990, Phys. Rev. Lett. **64**, 2406.
Allen, F.G., and G.W. Gobeli, 1962, Phys. Rev. **127**, 150.
Allen, R.E., and J.D. Dow, 1981, J. Vac. Sci. & Technol. **19**, 383.
Allen, R.E., and J.D. Dow, 1982, Phys. Rev. **B25**, 1423.
Alonso, M., R. Cimino, Ch. Maierhofer, Th. Chassé, W. Braun and K. Horn, 1990, J. Vac. Sci. & Technol. B **8**, 955.
Anderson, R.L., 1962, Solid-State Electron. **5**, 341.
Anderson, S., T.B. Pendry and P.M. Echenique, 1977, Surf. Sci. **65**, 539.
Andrews, J.M., and J.C. Phillips, 1975, Phys. Rev. Lett. **35**, 56; CRC Rev. Solid State Sci. **5**, 405.
Aono, M., Y. Hou, C. Oshima and Y. Ishizawa, 1982, Phys. Rev. Lett. **49**, 567.
Applebaum, J.A., and D.R. Hamann, 1973, Phys. Rev. Lett. **31**, 106.
Applebaum, J.A., and D.R. Hamann, 1974, Phys. Rev. B **10**, 4973.
Applebaum, J.A., and D.R. Hamann, 1975, Phys. Rev. B **12**, 1410.
Applebaum, J.A., G.A. Baraff and D.R. Hamann, 1976, Phys. Rev. B **14**, 1623.
Artacho, E., and F. Yndurain, 1989, Phys. Rev. Lett. **62**, 2491.
Arthur, J.R., 1968, J. Appl. Phys. **39**, 4032.
Arthur, J.R., 1974, Surf. Sci. **43**, 449.
Aspnes, D.E., and A.A. Studna, 1981, Appl. Phys. Lett. **39**, 316.
Aspnes, D.E., and A.A. Studna, 1985, Phys. Rev. Lett. **54**, 1956; J. Vac. Sci. & Technol. B **3**, 1498.
Aspnes, D.E., A.A. Studna, L.T. Florez, Y.C. Chang, J.P. Harbison, M.K. Kelly and H.H. Farrell, 1989, J. Vac. Sci. & Technol. **B7**, 901.
Aspnes, D.E., Y.C. Chang, A.A. Studna, L.T. Florez, H.H. Farrell and J.P. Harbison, 1990, Phys. Rev. Lett. **64**, 192.
Aukerman, L.W., 1973, in: Semiconductors and Semimetals, Vol. 4, eds R. K. Williardson and A. C. Beer (Academic Press, New York) p. 343.
Avouris, Ph., and In-Whan Lyo, 1990, in: Chemistry and Physics of Solid Surfaces, Vol. 8, eds R. Vanselow and R. Howe (Springer, Berlin) p. 371.
Baars, J., D. Bassett and M. Schulz, 1978, Phys. Status Solidi a **49**, 483.
Bachrach, R.Z., R.S. Bauer, P. Chiaradia and G.V. Hansson, 1981, J. Vac. Sci. & Technol. **18**, 797.
Badt, D., A. Brodde, St. Tosch and H. Neddermeyer, 1990, J. Vac. Sci. & Technol. A **8**, 251.
Baidyaroy, S., and P. Mark, 1972, Surf. Sci. **30**, 53.

Baidyaroy, S., W.R. Bottoms and P. Mark, 1972, Surf. Sci. **29**, 165.
Bales, G.S., and A. Zangwill, 1990, Phys. Rev. B **41**, 5500.
Bardeen, J., 1947, Phys. Rev. **71**, 717.
Bartels, F., H.J. Clemens and W. Mönch, 1983, Physica B **117/118**, 801; J. Vac. Sci. & Technol. B **1**, 149.
Batra, I.P., 1990, Phys. Rev. B **41**, 5048.
Bauer, E., 1982, Appl. Surf. Sci. **11/12**, 479.
Bauer, E., 1990, in: Chemistry and Physics of Solid Surfaces, Vol. 8, eds R. Vanselow and R. Howie (Springer, Berlin) p. 267.
Bauer, R.S., and H.W. Sang Jr, 1983, Surf. Sci. **132**, 479.
Bauer, R.S., P. Zürcher and H.W. Sang Jr, 1983, Appl. Phys. Lett. **43**, 663.
Becker, R.S., J.A. Golovchenko, G.S. Higashi and B.S. Swartzentruber, 1986, Phys. Rev. Lett. **57**, 1020.
Bedrossian, P., R.D. Meade, K. Mortensen, D.M. Chen, J.A. Golovchenko and D. Vanderbilt, 1989, Phys. Rev. Lett. **63**, 1257.
Bell, L.D., M.H. Hecht, W.J. Kaiser and L.C. Davis, 1990, Phys. Rev. Lett. **64**, 2679.
Bennett, P.A., L.C. Feldman, Y. Kuk, E.G. McRae and J.E. Rowe, 1983, Phys. Rev. B **28**, 3656.
Beres, R.P., R.E. Allen and J.D. Dow, 1983, Solid State Commun. **45**, 13.
Bertoni, C.M., 1990, in: Interactions of Atoms and Molecules with Solid Surfaces, eds V. Bortolani, N. H. March and M. P. Tosi (Plenum, New York) pp. 155–200.
Bertoni, C.M., C. Calandra, F. Manghi and E. Molinari, 1983, Phys. Rev. B **27**, 1251.
Beyers, R.K., B. Kim and R. Sinclair, 1987, J. Appl. Phys. **61**, 2195.
Biegelsen, D.K., R.D. Bringans, J.E. Northrup and L.-E. Swartz, 1990a, Phys. Rev. Lett. **65**, 452.
Biegelsen, D.K., R.D. Bringans, J.E. Northrup and L.-E. Swartz, 1990b, Phys. Rev. B **41**, 5701.
Biegelsen, D.K., R.D. Bringans and L.-E. Swartz, 1990c, in: Surface and Interface Analysis of Microelectronic Processing and Growth, eds L.J. Brillson and F.H. Pollak, SPIE Proc. **1186**, 136–143.
Binnig, G., H. Rohrer, Ch. Gerber and E. Weibel, 1983, Phys. Rev. Lett. **50**, 120.
Blakely, J.M., and C.P. Umbach, 1988, in: Diffusion at Interfaces: Microscopic Concepts, eds M. Grunze, H. Kreuzer and J.J. Weimer (Springer, Berlin) pp. 102–110.
Boland, J.J., 1990, Phys. Rev. Lett. **65**, 3325.
Boltaks, B.I., 1963, Diffusion in Semiconductors (Infosearch, London).
Braicovitch, L., 1988, in: Surface Properties of Electronic Materials, eds D.A. King and D.P. Woodruff, The Chemical Physics and Heterogenous Catalysis of Semiconductor Surfaces (Elsevier, Amsterdam) pp. 235–269.
Brattain, W.H., and J. Bardeen, 1953, Bell Syst. Tech. J. **82**, 1.
Brillson, L.J., 1975, J. Vac. Sci. & Technol. **12**, 76.
Brillson, L.J., 1976, J. Vac. Sci. & Technol. **13**, 225.
Brillson, L.J., 1978a, Phys. Rev. Lett. **40**, 260.
Brillson, L.J., 1978b, Phys. Rev. **18**, 2431.
Brillson, L.J., 1982a, Surf. Sci. Rep. **2**, 123.
Brillson, L.J., 1982b, Thin Solid Films **89**, 461.
Brillson, L.J., 1986, Surf. Sci. **168**, 260.
Brillson, L.J., and C.H. Griffiths, 1978, J. Vac. Sci. & Technol. **15**, 529.
Brillson, L.J., and G. Margaritondo, 1988, in: Surface Properties of Electronic Materials, eds D.A. King and D.P. Woodruff, The Chemical Physics and Heterogeneous Catalysis of Solid Surfaces, Vol. 5 (Elsevier, Amsterdam) pp. 119–181.
Brillson, L.J., and R.E. Viturro, 1988, Scanning Microscopy **2**, 789 and references therein.
Brillson, L.J., C.F. Brucker, G. Margaritondo, J. Slowik and N.G. Stoffel, 1980a, J. Phys. Soc. Jpn. **49**, 1089.
Brillson, L.J., G. Margaritondo and N.G. Stoffel, 1980b, Phys. Rev. Lett. **44**, 667.
Brillson, L.J., C.F. Brucker, A.D. Katnani, N.G. Stoffel and G. Margaritondo, 1981a, Appl. Phys. Lett. **38**, 784.
Brillson, L.J., C.F. Brucker, A.D. Katnani, N.G. Stoffel and G. Margaritondo, 1981b, Phys. Rev. Lett. **46**, 838.
Brillson, L.J., C.F. Brucker, N.G. Stoffel, A.D. Katnani, R. Daniels and G. Margaritondo, 1983, Surf. Sci. **132**, 212.

Brillson, L.J., M.L. Slade, A.D. Katnani, M. Kelly and G. Margaritondo, 1984, Appl. Phys. Lett. **44**, 110.
Brillson, L.J., M.L. Slade, H.W. Richter, H. Vander Plas and R.T. Falks, 1985, Appl. Phys. Lett. **47**, 1080.
Brillson, L.J., M.L. Slade, R.E. Viturro, M.J. Kelly, N. Tache, G. Margaritondo, J.M. Woodall, P.D. Kirchner and S.L. Wright, 1986, Appl. Phys. Lett. **48**, 1458; J. Vac. Sci. & Technol. B **4**, 919.
Brillson, L.J., R.E. Viturro, M.L. Slade, P. Chiaradia, D.G. Kilday, M.J. Kelly and G. Margaritondo, 1987, Appl. Phys. Lett. **50**, 1379.
Brillson, L.J., R.E. Viturro, J.L. Shaw, C. Mailhiot, N. Tache, J.T. McKinley, G. Margaritondo, J.M. Woodall, P.D. Kirchner, G.D. Pettit and S.L. Wright, 1988, J. Vac. Sci. & Technol. B **6**, 1263.
Bringans, R.D., and R.Z. Bachrach, 1984, Phys. Rev. Lett. **53**, 1954.
Bringans, R.D., and R.Z. Bachrach, 1990, in: Synchrotron Radiation Research: Advances in Surface Science, ed. R.Z. Bachrach (Plenum, New York) ch. 12.
Bringans, R.D., M.A. Olmstead, R.I.G. Uhrberg and R.Z. Bachrach, 1987, Appl. Phys. Lett. **51**, 523; Phys. Rev. B **36**, 9569.
Brucker, C.F., and L.J. Brillson, 1981, Appl. Phys. Lett. **39**, 67.
Butera, R.A., M. DelGuidice and J.H. Weaver, 1986, Phys. Rev. B **33**, 5345.
Caine, E.J., S. Subbanna, H. Kroemer, J.L. Merz and A.Y. Cho, 1984, Appl. Phys. Lett. **45**, 1123.
Calandra, C., O. Bisi and G. Ottaviani, 1984, Surf. Sci. Rep. **4**, 271.
Canali, C., G. Majri and G. Celotti, 1979, J. Appl. Phys. **59**, 255.
Canter, K.F., C.B. Duke and A.P. Mills Jr, 1990, in: Chemistry and Physics of Solid Surfaces, Vol. 8, eds R. Vanselow and R. Howe, Springer Series in Surface Sciences **22**, 183.
Cao, R.K., K. Miyano, T. Kendelewicz, K.K. Chin, I. Lindau and W.E. Spicer, 1987, J. Vac. Sci. & Technol. B **5**, 998.
Capasso, F., and G. Margaritondo, eds, 1987, Heterojunction Band Discontinuities: Physics and Device Applications (North-Holland, Amsterdam).
Capasso, F., A.Y. Cho, K. Mohammed and P.W. Foy, 1985, Appl. Phys. Lett. **46**, 664.
Carruthers, J.R., and A.F. Witt, 1975, in: Crystal Growth and Characterization, eds R. Veda and J.B. Mullin (North-Holland, Amsterdam).
Carstensen, H., R. Claessen, R. Manzke and M. Skibowski, 1990, Phys. Rev. B **41**, 9880.
Casey Jr, H.C., and M.B. Panish, 1978a, Heterostructure Lasers, Part B: Materials and Operating Characteristics (Academic Press, New York) ch. 6.
Casey Jr, H.C., and M.B. Panish, 1978b, Heterojunction Lasers, Part A: Fundamental Principles; Part B: Materials and Operating Characteristics (Academic Press, New York).
Cerrina, F., R.R. Daniels and V. Fano, 1983, Appl. Phys. Lett. **43**, 182.
Chabal, Y.J., 1986, Surf. Sci. **168**, 594 and references therein.
Chadi, D.J., 1978, Phys. Rev. Lett. **41**, 1062.
Chadi, D.J., 1979a, Phys. Rev. Lett. **43**, 43.
Chadi, D.J., 1979b, J. Vac. Sci. & Technol. **16**, 1290.
Chadi, D.J., 1979c, Phys. Rev. B **19**, 2074.
Chadi, D.J., 1987, J. Vac. Sci. & Technol. A **5**, 1691.
Chadi, D.J., and M.L. Cohen, 1975a, Solid State Commun. **16**, 691.
Chadi, D.J., and M.L. Cohen, 1975b, Phys. Rev. B **11**, 732.
Chadi, D.J., and M.L. Cohen, 1975c, Phys. Status Solidi b **68**, 405.
Chambers, S.A., 1989, Phys. Rev. B **39**, 12664.
Chambers, S.A., and V.A. Loebs, 1990, J. Vac. Sci. & Technol. B **8**, 724.
Chambers, S.A., F. Xu, H.-W. Chen, I.M. Vitomirov, S.B. Anderson and J.H. Weaver, 1986, Phys. Rev. B **34**, 6605.
Chang, L.L., 1990, in: Materials, Properties and Preparation, ed. C. Hilsum, Handbook on Semiconductors, Vol. 3 (North-Holland, Amsterdam) ch. 9.
Chang, L.L., and A. Koma, 1976, Appl. Phys. Lett. **29**, 138.
Chang, R., and W.A. Goddard III, 1984, Surf. Sci. **144**, 311.
Chang, S., R.E. Viturro and L.J. Brillson, 1990a, J. Vac. Sci. & Technol. A **8**, 3803.
Chang, S., L.J. Brillson, Y.J. Kime, D.S. Rioux, G.D. Pettit and J.M. Woodall, 1990b, Phys. Rev. Lett. **64**, 2551.

Chang, S., L.J. Brillson, Y.J. Kime, D.S. Rioux, G.D. Pettit and J.M. Woodall, 1991a, J. Vac. Sci. & Technol. A **9**, 902.
Chang, S., I.M. Vitomirov, L.J. Brillson, D.S. Rioux, P.D. Kirchner, G.D. Pettit and J.M. Woodall, 1991b, Phys. Rev. B **44**, 1391.
Chang, S., A. Raisanen, L.J. Brillson, J.L. Shaw, P.D. Kirchner, G.D. Pettit and J.M. Woodall, 1992, J. Vac. Sci. & Technol., to be published.
Chang, S.C., and P. Mark, 1974, Surf. Sci. **46**, 293.
Chang, S.C., and P. Mark, 1975, J. Vac. Sci. & Technol. **12**, 629.
Chelikowsky, J.R., S.G. Louie and M.L. Cohen, 1976, Phys. Rev. B **14**, 4724.
Chen, L.J., H.C. Cheng and W.T. Lin, 1986, in: Thin Films – Interfaces and Phenomena, eds R.J. Nemanich, P.S. Ho and S.S. Lau, MRL Symposia Proceedings **54**, 245.
Chen, P., D. Bolmont and C.A. Sébenne, 1984, Thin Solid Films **111**, 367.
Chen, S.J., K.L. Wang, R.C. Bowman Jr and P.M. Adams, 1989, Appl. Phys. Lett. **54**, 1253.
Cheng, H., X-J. Zhang and A.G. Milnes, 1984, Solid State Electron. **27**, 1117.
Cheung, N.W., P.J. Grunthaner and F.J. Grunthaner, 1981, J. Vac. Sci. & Technol. **18**, 917.
Chiaradia, P., A. Cricenti, S. Selci and G. Chiarotti, 1984a, Phys. Rev. Lett. **52**, 1145.
Chiaradia, P., A.D. Katnani, H.W. Sang Jr and R.S. Bauer, 1984b, Phys. Rev. Lett. **52**, 1246.
Chiradia, P., R.E. Viturro, M.L. Slade, L.J. Brillson, D.G. Kilday, M.J. Kelly, N. Tache and G. Margaritondo, 1987, J. Vac. Sci. & Technol. B **5**, 1075.
Cho, A.Y., 1970, J. Appl. Phys. **41**, 2780.
Cho, A.Y., 1985, in: Molecular Beam Epitaxy and Heterostructures, eds L.L. Chang and K. Ploog, NATO ASI Series E87 (Martinus Nijhoff, Dordrecht).
Cho, A.Y., and J.R. Arthur, 1975, Progr. Solid State Chem. **10**, 157.
Cho, A.Y., and P.D. Dernier, 1978, J. Appl. Phys. **49**, 3328.
Christensen, N.E., and L. Brey, 1988, Phys. Rev. B **38**, 8185.
Christensen, N.E., I. Gorczyca, O.B. Christensen, U. Schmid and M. Cardona, 1990, J. Crystal Growth **101**, 318.
Ciraci, S., and I.P. Batra, 1976, Solid State Commun. **18**, 1149.
Cohen, M.L., and D.J. Chadi, 1980, in: Optical Properties of Solids, Vol. 2, ed. M. Balkanski (North-Holland, Amsterdam) ch. 4B.
Cohen, P.I., G.S. Petrich, A. Dabiran and P.R. Pukite, 1990, in: Kinetics of Ordering and Growth at Surfaces, ed. M.G. Lagally (Plenum, New York) p. 225.
Colbow, K.M., Y. Gao, T. Tiedje, J.R. Dahn and W. Eberhardt, 1991, J. Vac. Sci. & Technol. A **9**, 2614.
Coleman, J.J., G. Costrini, S.J. Jeng and C.M. Wayman, 1986, J. Appl. Phys. **59**, 428.
Costa, J.C., F. Williamson, T.J. Miller, K. Beyzavi, M.I. Nathan, D.S.L. Mui, S. Strite and H. Morkoc, 1991, Appl. Phys. Lett. **58**, 382.
Cox, G., K.H. Graf, Szynka, U. Poppe and K. Urban, 1990, Vacuum **41**, 591.
Craig, B.I., and P.V. Smith, 1989, Surf. Sci. **218**, 569.
Czanderna, A.W., 1975, Methods of Surface Analysis (Elsevier, New York).
Daniels, R.R., A.D. Katnani, T.-X. Zhao, G. Margaritondo and A. Zunger, 1982, Phys. Rev. Lett. **49**, 895.
Daniels, R.R., G. Margaritondo, C. Quaresima, P. Perfetti and F.J. Levy, 1985, J. Vac. Sci. & Technol. A **3**, 479.
Das, G.P., P. Blöchl, O.K. Andersen, N.E. Christensen and O. Gunnarson, 1989, Phys. Rev. Lett. **63**, 1168.
Davidov, B., 1938, J. Tech. Phys. (USSR) **5**, 87.
Davidov, B., 1939, Soviet J. Phys. **1**, 167.
Davis, G.D., 1986, Surf. Int. Analysis **9**, 421.
Davis, G.D., N.E. Byer, R.R. Daniels and G. Margaritondo, 1983, J. Vac. Sci. & Technol. A **1**, 1726.
Davis, G.D., W.A. Beck, N.E. Byer, R.R. Daniels and G. Margaritondo, 1984, J. Vac. Sci. & Technol. A **2**, 546.
Davis, G.D., N.E. Byer, R.A. Riedel and G. Margaritondo, 1985, J. Appl. Phys. **57**, 1915.
Davis, G.D., W.A. Beck, M.K. Kelly, D.G. Kilday, Y.-M. Mo, N. Tache and G. Margaritondo, 1988, Phys. Rev. B **38**, 9694.
Daw, M.S., and D.L. Smith, 1979, Phys. Rev. B **20**, 5150.

Daw, M.S., and D.L. Smith, 1980, Appl. Phys. Lett. **8**, 690.
Däweritz, L., and R. Hey, 1990, Surf. Sci. **236**, 15.
del Guidice, M., M. Grioni, J.J. Joyce, M.W. Ruckman, S.A. Chambers and J.H. Weaver, 1986, Surf. Sci. **168**, 308.
Demuth, J.E., R.J. Hamers and R.M. Tromp, 1988, in: Solvay Conference on Surface Science, Springer Series in Surface Sciences, Vol. 14, ed. F. W. DeWette (Springer, Berlin) pp. 236–245.
Dev, B.N., G. Materlik, F. Grey, R.L. Johnson and M. Clausnitzer, 1986, Phys. Rev. Lett. **57**, 3058.
Ding, J., J. Washburn, T. Sands and V.G. Keramidas, 1986, Appl. Phys. Lett. **49**, 818.
Donnay, J.D.H., 1986, J. Mater. Res. **1**, vii.
Dose, V., 1985, Surf. Sci. Rep. **5**, 337.
Dow, J.D., and R.E. Allen, 1982, J. Vac. Sci. & Technol. **20**, 659.
Dow, J.D., R.E. Allen and O. Sankey, 1984, in: Chemistry and Physics of Solid Surfaces, Vol. V, eds R. Vanselow and R. Howe (Springer, New York) pp. 483 500.
Drathen, P., W. Ranke and K. Jacobi, 1978, Surf. Sci. **77**, L162.
Drube, W., D. Straub and F.J. Himpsel, 1987, Phys. Rev. B **35**, 5563.
Drube, W., R. Ludeke and F.J. Himpsel, 1988, in: Proc. 19th Int. Conf. on the Physics of Semiconductors, Warsaw 1988, ed. W. Zawadzki (Institute of Physics, Polish Academy of Sciences, Warsaw) p. 637.
Duggan, G., 1987, in: Heterojunction Band Discontinuities: Physics and Device Applications, eds F. Capasso and G. Margaritondo (North-Holland, Amsterdam) ch. 5.
Duke, C.B., 1974, Adv. Chem. Phys. **27**, 1.
Duke, C.B., 1983, J. Vac. Sci. & Technol. B **1**, 732.
Duke, C.B., 1988, in: Surface Properties of Electronic Materials, eds D.A. King and D.P. Woodruff, The Chemical Physics of Solid Surfaces and Heterogeneous Catalysis, Vol. 5 (Elsevier, Amsterdam, 1988) pp. 69–188.
Duke, C.B., and J. LaFemina, 1991, J. Vac. Sci.& Technol., in press.
Duke, C.B., and A.R. Lubinsky, 1975, Surf. Sci. **50**, 605.
Duke, C.B., and C. Mailhiot, 1985, J. Vac. Sci. & Technol. B **3**, 1170.
Duke, C.B., and R.L. Park, 1972, Phys. Today **25**, 24.
Duke, C.B., and Y.-R. Wang, 1988, J. Vac. Sci. & Technol. A **6**, 692.
Duke, C.B., A. Paton, R.J. Meyer, L.J. Brillson, A. Kahn, D. Kanani, T. Carelli, J.L. Yey, G. Margaritondo and A.D. Katnani, 1981, Phys. Rev. Lett. **46**, 440.
Duke, C.B., A. Paton, W.K. Ford, A. Kahn and J. Carelli, 1982, Phys. Rev. B **26**, 803.
Durbin, S.M., L.E. Berman, B.W. Batterman and J.M. Blakely, 1986, Phys. Rev. Lett. **56**, 236.
Eaglesham, D.J., H.-H. Gossman and M. Cerullo, 1990, Phys. Rev. Lett. **65**, 1227.
Eastman, D.E., and J.L. Freeouf, 1974, Phys. Rev. Lett. **33**, 1601.
Eastman, D.E., and M.I. Nathan, 1975, Phys. Today **28**, 44.
Estrup, P.J., and J. Morrison, 1964, Surf. Sci. **2**, 465.
Farrell, H.H., M.C. Tamargo and J.L. de Miguel, 1991, Appl. Phys. Lett. **58**, 355.
Farrow, R.F.C., V.S. Speriosu, S.S.P. Parkin, C. Chien, J.C. Bravman, R.N. Marks, P.D. Kirchner, G.A. Prinz and B.T. Zonker, 1989, in: Stresses and Mechanical Properties, eds J.C. Bravman, W.D. Nix, D.M. Barnett and D.A. Smith, Materials Research Society Symposium Proceedings (MRS, Pittsburgh, PA) Thin films **130**, p. 281.
Feenstra, R.M., 1989, Phys. Rev. Lett. **63**, 1412.
Feenstra, R.M., 1990, in: Interaction of Atoms and Molecules with Solid Surfaces, eds V. Bortolani, N.H. March and M.P. Tosi (Plenum, New York) p. 357.
Feenstra, R.M., and P. Mårtensson, 1988, Phys. Rev. Lett. **61**, 447.
Feenstra, R.M., and J.A. Stroscio, 1987, Phys. Rev. Lett. **59**, 2173.
Firment, L.E., 1982, Surf. Sci. **116**, 205.
Flores, F., and C. Tejedor, 1979, J. Phys. C **12**, 731.
Flores, F., A.H. Muroz and J.C. Duran, 1989, Appl. Surf. Sci. **41/42**, 144.
Ford, W.K., T. Guo, D.L. Lessor and C.B. Duke, 1990, Phys. Rev. B **42**, 8952.
Fornari, R., E. Gombia and R. Mosca, 1989, J. Electron. Mat. **18**, 151.
Forrest, S.R., 1987, in: Heterojunction Band Discontinuities: Physics and Device Applications, eds F. Capasso and G. Margaritondo (North-Holland, Amsterdam) ch. 8.

Forrest, S.R., P.-H. Schmidt, R.B. Wilson and M.L. Kaplan, 1984, Appl. Phys. Lett. **45**, 1199.
Fowell, A.E., R.H. Williams, B.E. Richardson and T.-H. Shen, 1990, Semicond. Sci. & Technol. **5**, 348.
Fowell, A.E., R.H. Williams, B.E. Richardson, A.A. Cafolla, D.I. Westwood and D.A. Woolf, 1991, J. Vac. Sci. & Technol. B **9**, 581.
Franciosi, A., J.H. Weaver and D.G. O'Neill, 1983, Phys. Rev. B **28**, 4889.
Franciosi, A., P. Soukassian, P. Philip, S. Chang, A. Wall, A. Raisanen and N. Troullier, 1987, Phys. Rev. B **35**, 910.
Freeouf, J.L., 1980, Solid State Commun. **33**, 1059.
Freeouf, J.L., and J.M. Woodall, 1981, Appl. Phys. Lett. **39**, 727.
Freeouf, J.L., and J.M. Woodall, 1986, Surf. Sci. **168**, 518.
Frensley, W.R., and H. Kroemer, 1977, Phys. Rev. B **16**, 2642.
Friedman, D.J., G.P. Carey, C.K. Shih, I. Lindau, W.E. Spicer and T.A. Wilson, 1986, Appl. Phys. Lett. **48**, 44.
Froitzheim, H., 1988, in: Surface Properties of Electronic Materials, eds D.A. King and D.P. Woodruff, The Chemical Physics and Heterogeneous Catalysis of Solid Surfaces, Vol. 5 (Elsevier, Amsterdam) pp. 183–233.
Fujitani, H., and S. Asano, 1990, Phys. Rev. B **42**, 696.
Gatos, H.C., and J. Lagowski, 1973, J. Vac. Sci. & Technol. **10**, 130.
Gatos, H.C., and M.C. Lavine, 1960, J. Appl. Phys. **31**, 743.
Gatos, H.C., and M.C. Lavine, 1965, in: Progress in Semiconductors, Vol. 9, eds A.F. Gibson and R.E. Burgess (Temple, London) pp. 1–45.
Gatos, H.C., P.L. Moody and M.C. Lavine, 1960, J. Appl. Phys. **31**, 212.
Golovchenko, J.A., 1986, Science **232**, 48.
Goodwin, T.A., and P. Mark, 1972, in: Progress in Surface Science, Vol. 1 (Pergamon, New York) p. 1.
Gorbik, P.P., V.N. Komashchenko and G.A. Fedorus, 1980, Sov. Phys. Semicond. **14**, 753.
Grant, J.T., and T.W. Haas, 1971, Surf. Sci. **26**, 669.
Grant, R.W., and W.A. Harrison, 1988, J. Vac. Sci. & Technol. B **6**, 1295.
Grant, R.W., and J.R. Waldrop, 1987, J. Vac. Sci. & Technol. B **5**, 1015.
Grant, R.W., J.R. Waldrop, S.P. Kowalczyk and E.A. Kraut, 1981, J. Vac. Sci. & Technol. **19**, 477.
Grant, R.W., J.R. Waldrop, S.P. Kowalczyk and E.A. Kraut, 1985, J. Vac. Sci. & Technol. B **3**, 1295.
Grant, R.W., J.R. Waldrop, S.P. Kowalczyk and E.A. Kraut, 1986, Surf. Sci. **168**, 498.
Grant, R.W., E.A. Kraut, J.R. Waldrop and S.P. Kowalczyk, 1987, in: Heterojunction Band Discontinuities: Physics and Device Applications, eds F. Capasso and G. Margaritondo (North-Holland, Amsterdam) ch. 4.
Greene, J.E., S.A. Barrett, J.-E. Sundgren and A. Rockett, 1988, in: Ion Beam Assisted Film Growth, ed. T. Itoh (Elsevier, Amsterdam) ch. 5.
Grey, F., R. Feidenhans'l, M. Nielsen and R.L. Johnson, 1989, J. Phys. (Paris) **50**, 7181.
Grioni, M., M. del Guidice, J.J. Joyce and J.H. Weaver, 1985, J. Vac. Sci. & Technol. A **3**, 907.
Grioni, M., J.J. Joyce and J.H. Weaver, 1986, J. Vac. Sci. & Technol. A **4**, 965.
Grobman, W.D., D.E. Eastman and J.L. Freeouf, 1975, Phys. Rev. **12**, 4405.
Grunthaner, P.J., F.J. Grunthaner and J.W. Mayer, 1980, J. Vac. Sci. & Technol. **17**, 924.
Gudat, W., and D.E. Eastman, 1976, J. Vac. Sci. & Technol. **13**, 831.
Gudat, W., and C. Kunz, 1972, Phys. Rev. Lett. **29**, 169.
Guichar, G.M., C.A. Sébenne and C.D. Thuault, 1979, J. Vac. Sci. & Technol. **16**, 1212.
Haase, M.A., S.C. Smith, J.J. Coleman and G.E. Stillman, 1987, Appl. Phys. Lett. **50**, 404.
Haberern, K.W., and M.D. Pashley, 1990, Phys. Rev. B **41**, 3226.
Haight, R., and J. Bokor, 1986, Phys. Rev. Lett. **57**, 1548.
Halstead, R.E., M. Aven and H.D. Coghill, 1965, J. Electrochem. Soc. **112**, 177.
Hamers, R.J., and J.E. Demuth, 1988, Phys. Rev. Lett. **60**, 2527.
Hamers, R.J., R.M. Tromp and J.E. Demuth, 1986a, Phys. Rev. **34**, 5343.
Hamers, R.J., R.M. Tromp and J.E. Demuth, 1986b, Phys. Rev. Lett. **56**, 1972.
Hamers, R.J., V. Kohler and J.E. Demuth, 1989, Ultramicroscopy **31**, 10.
Haneman, D., 1960, J. Phys. & Chem. Solids **14**, 162.
Hansen, M., 1958, Constitution of Binary Alloys (McGraw-Hill, New York).

Hansson, G.V., R.Z. Bachrach, R.S. Bauer and P. Chiaradia, 1981, Phys. Rev. Lett. **46**, 1033.
Hara, S., and I. Ohdomari, 1988, Phys. Rev. B **38**, 7554.
Harris, I.R., N.A. Smith, B. Cockayne and W.R. MacEwan, 1987, J. Cryst. Growth **82**, 450.
Harrison, W.A., 1976, Phys. Rev. Lett. **37**, 312.
Harrison, W.A., 1977, J. Vac. Sci. & Technol. **14**, 1016.
Harrison, W.A., 1979, J. Vac. Sci. & Technol. **16**, 1492.
Harrison, W.A., 1985, J. Vac. Sci. & Technol. B **3**, 1231.
Harrison, W.A., E.A. Kraut, J.R. Waldrop and R.W. Grant, 1978, Phys. Rev. B **18**, 4402.
Hasagawa, J., L. He, H. Ohno, T. Sawada, T. Haga, Y. Abe and H. Takahashi, 1987, J. Vac. Sci. & Technol. B **5**, 1097.
Hasagawa, Y., I. Kamiya, T. Hashizume and T. Sakurai, 1990, J. Vac. Sci. & Technol. A **8**, 238.
Hasegawa, H., and H. Ohno, 1986, J. Vac. Sci. & Technol. B **4**, 1130.
Hashimoto, K., Y. Hirakawa and T. Ikoma, 1990, Appl. Phys. Lett. **57**, 2555.
Hashizume, T., Y. Hasegawa, I. Kamiya, T. Ide, I. Sumita, S. Hyodo, T. Sakurai, H. Tochihara, M. Kubota and Y. Murata, 1990, J. Vac. Sci. & Technol. A **8**, 233.
Hauenstein, R.J., T.E. Schlesinger, T.C. McGill, B.D. Hunt and L.J. Schowalter, 1986, J. Vac. Sci. & Technol. A **4**, 860.
Hecht, M.H., 1990, Phys. Rev. B **41**, 7918.
Heiblum, M., M.I. Nathan and M. Eizenberg, 1985, Appl. Phys. Lett. **47**, 503.
Heimann, R.B., 1982, in: Crystals, Vol. 8, ed. H.C. Freyhardt (Springer, Berlin) pp. 175–224.
Heine, V., 1965, Phys. Rev. **138**, A1689.
Heinrich, V.E., 1976, Surf. Sci. **57**, 385.
Henzler, M., 1977, in: Electron Spectroscopy for Surface Analysis, ed. H. Ibach (Springer, Berlin) pp. 117–150.
Henzler, M., 1982, Appl. Surf. Sci. **11/12**, 450.
Henzler, M., 1985, in: The Structure of Surfaces, Springer Series in Surface Sciences, Vol. 2, eds M.A. Van Hove and R. Howe (Springer, New York) 351.
Heslinga, D.R., H.H. Weitering, D.P. vander Werf, T.M. Klapwijk and T. Hibma, 1990, Phys. Rev. Lett. **64**, 1589.
Hiesinger, P., 1976, Phys. Status Solidi a **33**, k39.
Higashi, G.S., Y.J. Chabal, G.W. Trucks and K. Raghavachari, 1990, Appl. Phys. Lett. **56**, 656.
Himpsel, F.J., 1990, Surf. Sci. Rep. **12**, 1.
Himpsel, F.J., and D.E. Eastman, 1979, J. Vac. Sci. & Technol. **16**, 1297.
Himpsel, F.J., and Th. Fauster, 1984, J. Vac. Sci. & Technol. A **2**, 815.
Hiraki, A., 1983, Surf. Sci. Rep. **3**, 355.
Hirose, K., H. Tsuda and T. Mizutani, 1989, Appl. Surf. Sci. **41/42**, 174.
Hoeven, J.J., J. Aarts and P.K. Larsen, 1988, J. Vac. Sci. & Technol. A **7**, 5.
Hoffmann, D.M., B.K. Meyer, J.-M. Spaeth, M. Wattenbach, J. Krüger, C. Kisielowski-Kemmerich and H. Alexander, 1990, J. Appl. Phys. **68**, 3381.
Hökelek, E., and G.Y. Robinson, 1983, J. Appl. Phys. **54**, 5199.
Holland, B.W., C.B. Duke and A. Paton, 1984, Surf. Sci. **140**, 1269.
Houle, F.A., 1989, Phys. Rev. B **39**, 10120.
Hsieh, S., J. Hsieh, E.A. Patten and C.M. Wolfe, 1984, Appl. Phys. Lett. **45**, 1125.
Hughes, G.J., J.T. McKinley, R.H. Williams and I.T. McGovern, 1982, J. Phys. C **15**, L159.
Huijser, A., and J. van Laar, 1975, Surf. Sci. **52**, 202.
Huijser, A., J. van Laar and T.L. Van Rooy, 1978, Phys. Lett. A **65**, 337.
Ibach, H., K. Horn, R. Dorn and H. Lüth, 1973, Surf. Sci. **38**, 433.
Ignatiev, A., B.W. Lee and M.A. Van Hove, 1977, in: Proc. 7th Int. Vacuum Congr. and 3rd Int. Conf. on Solid Surfaces, Vienna, eds R. Dobrozemsky, F. Rüdenauer, F.B. Viehböck and A. Breth (F. Berger und Söhne, Vienna) p. 1733.
Ihm, J., and J.D. Joannopoulos, 1982, Phys. Rev. B **26**, 4429.
Inkson, J.C., 1974, J. Vac. Sci. & Technol. **11**, 943.
Ismail, A., J.M. Palau and L. Lassabetere, 1984, Rev. Phys. Appl. **19**, 205.
Ivanov, I., and J. Pollmann, 1981, Phys. Rev. B **24**, 7275.

Jacobi, K., C.B. Mushwitz and W. Ranke, 1979, Surf. Sci. **82**, 270.
Jaros, M., 1988, Phys. Rev. B **37**, 7112.
Joyce, B.A., J.H. Neave, P.J. Dobson and P.K. Larsen, 1984, Phys. Rev. B **29**, 814.
Joyce, B.A., P.J. Dobson and P.K. Larsen, 1988, in: Surface Properties of Electronic Materials, eds D.A. King and D.P. Woodruff, The Chemical Physics of Solid Surfaces and Heterogeneous Catalysis, Vol. 5 (Elsevier, Amsterdam) pp. 271–307.
Kahn, A., 1983, Surf. Sci. Rep. **3**, 193.
Kahn, A., A.D. Katnani, J. Carelli, J.L. Yeh, C.B. Duke, R.J. Meyer, A. Paton and L.J. Brillson, 1981, J. Vac. Sci. & Technol. **18**, 792.
Kahn, A., K. Stiles, D. Mao, S.F. Horng, K. Young, J. McKinley, D.G. Kilday and G. Margaritondo, 1989, in: Metallization and Metal–Semiconductor Interfaces, NATO ASI Series B, Vol. 195, ed. I.P. Batra (Plenum, New York) p. 163.
Kaiser, W.J., and L.D. Bell, 1988, Phys. Rev. Lett. **60**, 1406.
Kajiyama, K., Y. Mizushima and S. Sakata, 1973, Appl. Phys. Lett. **23**, 458.
Kallin, C., and B.I. Halperin, 1984, Phys. Rev. B **29**, 2175.
Kane, E.O., 1964, Phys. Rev. Lett. **12**, 97.
Kariotis, R., M.B. Webb and M.G. Lagally, 1991, Phys. Rev. Lett., in press.
Katnani, A.D., 1987, in: Heterojunction Band Discontinuities: Physics and Device Applications, eds. F. Capasso and G. Margaritondo (North-Holland, Amsterdam) ch. 3.
Katnani, A.D., and G. Margaritondo, 1983, Phys. Rev. B **28**, 1944.
Katnani, A.D., P. Chiaradia, H.W. Sang Jr and R.S. Bauer, 1984, J. Vac. Sci. & Technol. B **2**, 471.
Katnani, A.D., P. Chiaradia, Y. Cho, P. Mahowald, P. Pianetta and R.S. Bauer, 1985, Phys. Rev. **32**, 4071.
Kawai, N.J., T. Kohjima, F. Sato, T. Sakamoto, T. Nakagawa and K. Ohta, 1985, in: 12th Int. Symp. on GaAs and Related Compounds, Karuizawa, Japan, Institute of Physics Conference Series, Vol. 79 (Adam Hilger, Bristol).
Kawamura, T., P.A. Maksym and T. Iijima, 1984, Surf. Sci. **148**, L671.
Kaxiras, E., K.C. Pandey, F.J. Himpsel and R.M. Tromp, 1990, Phys. Rev. B **41**, 1261.
Kendelewicz, T., N. Newman, R.S. List, I. Lindau and W.E. Spicer, 1985, Phys. Rev. B **3**, 1206.
Kendelewicz, T., M.D. Williams, W.G. Petro, I. Lindau and W.E. Spicer, 1985a, Phys. Rev. B **3**, 6503.
Kendelewicz, T., N. Newman, R.S. List, I. Lindau and W.E. Spicer, 1985b, J. Vac. Sci. & Technol. B **3**, 1206.
Kern, W., and C.A. Deckert, 1978, in: Thin Film Processes, eds J.L. Vossen and W. Kern (Academic Press, New York) pp. 432–498.
Knapp, J.A., and G.J. Lapeyre, 1976, J. Vac. Sci. & Technol. **13**, 757.
Knapp, J.A., D.E. Eastman, K.C. Pandey and F. Patella, 1978, J. Vac. Sci. & Technol. **15**, 1252.
Koehler, V.K., J.E. Demuth and R.J. Hamers, 1988, Phys. Rev. Lett. **60**, 2499.
Kohl, D., M. Henzler and G. Heilan, 1974, Surf. Sci. **41**, 403.
Kohler, V., J.E. Demuth and R.J. Hamers, 1989, J. Vac. Sci. & Technol. A **7**, 2860.
Kowalczyk, S.P., W.J. Schaffer, E.A. Kraut and R.W. Grant, 1982, J. Vac. Sci. & Technol. **20**, 705.
Kraut, E.A., R.W. Grant, J.R. Waldrop and S.P. Kowalczyk, 1983, Phys. Rev. B **28**, 1965.
Krebs, J.J., B.T. Jonker and G.A. Prinz, 1987, J. Appl. Phys. **61**, 2596.
Kroemer, H., 1983, Surf. Sci. **132**, 543.
Kroemer, H., 1985, in: Molecular Beam Epitaxy and Heterostructures, eds L.L. Chang and K. Ploog, NATO ASI Series E, Vol. 87 (Martinus Nijhoff, Dordrecht) p. 331.
Kroemer, H., J. Polasko and S.J. Wright, 1980a, Appl. Phys. Lett. **36**, 763.
Kroemer, H., W.-Y. Chien, J.S. Harris and D.D. Edwall, 1980b, Appl. Phys. Lett. **36**, 295.
Kubaschewski, O., and C.B. Alcock, 1979, Metallurgical Thermochemistry (Pergamon, New York).
Kubby, J.A., Y.R. Wang and W.J. Greene, 1991, Phys. Rev. Lett. **68**, 329.
Kurtin, S., T.C. McGill and C.A. Mead, 1969, Phys. Rev. Lett. **22**, 1433.
LaFemina, J., 1991, Surf. Sci. Rep., in press.
Lagally, M.G., 1982, in: Chemistry and Physics of Solid Surfaces IV, eds R. Vanselow and R. Howe (Springer, Berlin) p. 281.

Laidig, W.D., N. Holonyak, M.D. Camras, K. Hess, J.J. Coleman, P.D. Dapkus and J. Bardeen, 1981, Appl. Phys. Lett. **38**, 776.
Lander, J.J., and J. Morrison, 1964, Surf. Sci. **2**, 553.
Lang, D.V., 1987, in: Heterojunction Band Discontinuities: Physics and Device Applications, eds F. Capasso and G. Margaritondo (North-Holland, Amsterdam) ch. 9.
Langer, J.M., and H. Heinrich, 1985, Phys. Rev. Lett. **55**, 1414.
Lapujoulade, J., 1990, in: Interactions of Atoms and Molecules with Solid Surfaces, eds V. Bortolani, N.H. March and M.P. Tosi (Plenum, New York) pp. 381–405.
Larsen, P.K., J.F. van der Veen, A. Mazur, J. Pollman, J.H. Neave and B.A. Joyce, 1982, Phys. Rev. B **26**, 3222.
Larsen, P.K., J.H. Neave, J.F. van der Veen, P.J. Dobson and B.A. Joyce, 1983, Phys. Rev. B **27**, 4966.
LeGoues, F.K., M. Copel and R.M. Tromp, 1989, Phys. Rev. Lett. **63**, 1826.
Le Lay, G., J. Peretti and M. Hanbücken, 1988, Surf. Sci. **204**, 57.
Le Lay, G., K. Hricovini and J. Bonnet, 1989, Appl. Surf. Sci. **41/42**, 25
Le Lay, G., D. Mao, A. Kahn, Y. Hwu and G. Margaritondo, 1991, Phys. Rev. B **43**, 14301.
Lent, C.S., and P.I. Cohen, 1984, Surf. Sci. **139**, 121.
Leung, S., A.G. Milnes and D.D.L. Chung, 1983, in: Interfaces and Contacts, eds R. Ludeke and K. Rose, Materials Research Society Symposium Proceedings, Vol. 18 (MRS, Pittsburgh, PA) p. 109.
Levine, J.D., 1973, Surf. Sci. **34**, 90.
Lewis, B.F., F.J. Grunthaner, A. Madhukar, T.C. Lee and R. Fernandez, 1985, J. Vac. Sci. & Technol. B **3**, 1317.
Li, D., J.M. Gonsalves, N. Otsuka, J. Qiu, M. Kobayashi and R.L. Gunshor, 1990, Appl. Phys. Lett. **57**, 449.
Liehr, M., P.E. Schmid, F.K. LeGoues and P.S. Ho, 1985, Phys. Rev. Lett. **54**, 2139.
Lieski, N.P., 1984, J. Phys. & Chem. Solids **45**, 821.
Lilienthal-Weber, Z., R. Gronsky, T. Washburn, N. Newman, W.E. Spicer and E.R. Weber, 1986, J. Vac. Sci. & Technol. B **4**, 912.
Lilienthal-Weber, Z., A. Miret-Goutier, N. Newman, C. Jou, W.E. Spicer, J. Washburn and E.R. Weber, 1988, in: Epitaxy of Semiconductor Layered Structures, eds R.T. Tung, L.R. Dawson and R.L. Gunshor, Materials Research Society Symposium Proceedings, Vol. 102 (MRS, Pittsburgh, PA) p. 241.
Lin, J.C., and Y.A. Chang, 1989, in: Chemistry and Defects in Semiconductor Heterostructures, eds M. Kawabe, T.D. Sands, E.R. Weber and R.S. Williams, Materials Research Society Symposium Proceedings, Vol. 148 (MRS, Pittsburgh, PA) p. 3.
Lin, Zhangda, F. Xu and J.H. Weaver, 1987, Phys. Rev. B **36**, 5777.
Lind, D.M., Y.U. Idzerda, G.A. Prinz, B.T. Jonker and J.J. Krebs, 1988, J. Vac. Sci. & Technol. A **6**, 819.
Lindau, I., P.W. Chye, C.M. Garner, P. Pianetta, C.Y. Su and W.E. Spicer, 1978, J. Vac. Sci. & Technol. **15**, 1337.
List, R.S., P.H. Mahowald, J. Woicik and W.E. Spicer, 1986, J. Vac. Sci. & Technol. A **4**, 1391.
Look, D.C., C.E. Stutz and K.R. Evans, 1990, Appl. Phys. Lett. **57**, 2570.
Lubinsky, A.R., C.B. Duke, S.C. Chang, B.W. Lee and P. Mark, 1976, J. Vac. Sci. & Technol. **13**, 189.
Ludeke, R., 1984, J. Vac. Sci. & Technol. B **2**, 400 and references therein.
Ludeke, R., 1986, Surf. Sci. **168**, 291.
Ludeke, R., and L. Esaki, 1974, Phys. Rev. Lett. **33**, 653.
Ludeke, R., and A. Koma, 1975, CRC Crit. Rev. Solid State Sci. **5**, 259.
Ludeke, R., L.L. Chang and L. Esaki, 1973, Appl. Phys. Lett. **23**, 201.
Ludeke, R., A. Taleb-Ibrahimi and G. Jezequel, 1989, Appl. Surf. Sci. **41/42**, 1 51.
Ludeke, R., A.B. McLean and A. Taleb-Ibrahimi, 1990, Phys. Rev. B **42**, 2982.
Ludwig, M., and G. Heymann, 1986, J. Vac. Sci. & Technol. B **4**, 485.
Lyo, In-Whan, E. Kaxiras and Ph. Avouris, 1989, Phys. Rev. Lett. **63**, 1261.
Ma, Y., J.E. Rowe, E.E. Chaban, C.T. Chen, R.L. Headrick, G.M. Meigs, S. Modesti and F. Sette, 1990, Phys. Rev. Lett. **65**, 2173.
Mackay, K.J., P.B. Allen, W. Herrenden-Harker, R.H. Williams, G.M. Williams and C.R. Whitehouse, 1986, Appl. Phys. Lett. **49**, 354.

MacLaren, J.M., J.B. Pendry, P.J. Rous, D.K. Saldin, G.A. Somorjai, M.A. Van Hove and D.P. Vvedensky, 1987, Surface Crystallography Information Service: A Handbook of Surface Structures (Reidel, Dordrecht) 352pp.
MacRae, A.U., 1966, Surf. Sci. **4**, 247.
Madhukar, A., and S.V. Ghaisas, 1988, CRC Crit. Rev. Solid State & Mater. Sci. **14**, 2.
Magnusson, K.O., U.O. Karlsson, D. Straub, S.A. Flodström and F.J. Himpsel, 1978, Phys. Rev. B **36**, 6566.
Mailhiot, C., and C.B. Duke, 1986, Phys. Rev. B **33**, 1118.
Mailhiot, C., and D.L. Smith, 1986, J. Vac. Sci. & Technol. B **4**, 996.
Mailhiot, C., and D.L. Smith, 1987, Phys. Rev. B **35**, 1242.
Mailhiot, C., C.B. Duke and D.J. Chadi, 1984a, Phys. Rev. Lett. **53**, 2114.
Mailhiot, C., C.B. Duke and Y.C. Chang, 1984b, Phys. Rev. B **30**, 1109.
Mailhiot, C., C.B. Duke and D.J. Chadi, 1985, Surf. Sci. **149**, 366.
Maissen, C., J. Mašek, H. Zogg and S. Blunier, 1988, Appl. Phys. Lett. **53**, 1608.
Makram-Ebeid, S., D. Gautard, P. DeVillard and G.M. Martin, 1982, Appl. Phys. Lett. **40**, 161.
Manghi, F., E. Molinari, C.M. Bertoni and C. Calandra, 1982, J. Phys. C **15**, 1099.
Manzke, R., H.P. Barnscheidt, C. Janowitz and M. Skibowski, 1987, Phys. Rev. Lett. **58**, 610.
Margaritondo, G., 1984, Phys. Rev. B **30**, 4533.
Margaritondo, G., 1988, Introduction to Synchrotron Radiation (Oxford, New York).
Margaritondo, G., and P. Perfetti, 1987, in: Heterojunction Band Discontinuities: Physics and Device Applications, eds F. Capasso and G. Margaritondo (North Holland, Amsterdam) ch. 2.
Margaritondo, G., J.E. Rowe, C.M. Bertoni, C. Calandra and G. Manghi, 1979, Phys. Rev. B **20**, 1538.
Margaritondo, G., N.G. Stoffel, A.D. Katnani and L.J. Brillson, 1980, Appl. Phys. Lett. **37**, 917.
Margaritondo, G., A.D. Katnani, N.G. Stoffel and T.-X. Zhao, 1982, Solid State Commun. **43**, 163.
Mark, P., S.C. Chang, W.F. Creighton and B.W. Lee, 1975, CRC Critical Rev. Solid State Sci. **5**, 189.
Mark, P., E. So and M. Bonn, 1977, J. Vac. Sci. & Technol. **14**, 865.
Marshall, E.D., S.S. Lau, C.J. Palmstrøm, T. Sands, C.L. Schwartz, S.A. Schwarz, J.P. Harbison and L.T. Florez, 1989, in: Materials Research Society Symposium Proceedings, Vol. 148 (MRS, Pittsburgh, PA) p. 163.
Marsi, M., S. LaRosa, Y. Hwu, F. Gozzo, C. Coluzza, A. Baldereschi, G. Margaritondo, S. Baroni and R. Resta, 1991, unpublished.
Mårtensson, P., and R.M. Feenstra, 1989, Phys. Rev. B **39**, 7744.
Mårtensson, P., A. Cricenti and G.V. Hansson, 1985, Phys. Rev. **32**, 6959.
Martinez, G., 1980, in: Optical Properties of Solids, Vol. 2, ed. M. Balkanski (North-Holland, Amsterdam) ch. 4C.
Massies, J., P. Etienne, F. Dezaly and N.T. Linh, 1980a, Surf. Sci. **99**, 121.
Massies, J., F. Dezaly and N.T. Linh, 1980b, J. Vac. Sci. & Technol. **17**, 1134.
Massies, J., J. Chaplar, M. Laviron and N.T. Linh, 1981, Appl. Phys. Lett. **38**, 6693.
Massies, J., P. Delescluse, P. Etienne and N.T. Linh, 1982, Thin Solid Films **90**, 112.
Matthews, J.W., and A.E. Blakeslee, 1974, J. Cryst. Growth **27**, 118.
Mayer, J.W., and S.S. Lau, 1990, Electronic Materials Science: For Integrated Circuits in Si and GaAs (MacMillan, New York).
Mayer, J.W., and K.N. Tu, 1971, J. Vac. Sci. & Technol. **11**, 86.
McGilp, J.F., 1984, J. Phys. C **17**, 2249.
McGilp, J.F., and I.T. McGovern, 1985, J. Vac. Sci. & Technol. B **3**, 1641.
McGilp, J.F., and A.B. McLean, 1988, J. Phys. C **21**, 807.
McKinley, J.T., and G. Margaritondo, 1992, in: Contacts to Semiconductor Devices, ed. L.J. Brillson (Noyes, Park Ridge, NJ).
McKinley, J.T., Y. Hwu, B.E.C. Koltenbah, G. Margaritondo, S. Baroni and R. Resta, 1991, J. Vac. Sci. & Technol. A **9**, 917.
McLean, A.B., and R.H. Williams, 1987, Semicond. Sci. & Technol. **21**, 654.
McLean, A.B., R.H. Williams and J.F. McGilp, 1988, J. Vac. Sci. & Technol. B **6**, 1252.
McRae, A.V., 1966, Surf. Sci. **4**, 247.
McRae, E.G., and C.W. Caldwell, 1981, Phys. Rev. Lett. **46**, 1632.

Mead, C.A., 1966, Solid State Electron. **9**, 1023.
Mead, C.A., and W.G. Spitzer, 1964, Phys. Rev. **134**, A713.
Mele, G., and J.D. Joannopoulos, 1987, J. Vac. Sci. & Technol. **15**, 1287.
Miedema, A.R., P.F. de Châtel and F.R. de Boer, 1980, Physica B **100**, 1.
Mills, K.C., 1974, Thermodynamic Data for Inorganic Sulphides, Selenides and Tellurides (Butterworths, London).
Milnes, A.G., 1973, Deep Impurities in Semiconductors (Wiley–Interscience, New York).
Milnes, A.G., and D.L. Feucht, 1972, Heterojunctions and Metal–Semiconductor Junctions (Academic Press, New York).
Mircea, A., and D. Bois, 1979, in: Gallium Arsenide and Related Compounds, Inst. Phys. Conf. Ser. **46**, 82.
Missous, M., E.H. Rhoderick and K.E. Singer, 1984, Electron. Lett. **22**, 241.
Missous, M., E.H. Rhoderick and K.E. Singer, 1986a, J. Appl. Phys. **59**, 3189.
Missous, M., E.H. Rhoderick and K.E. Singer, 1986b, J. Appl. Phys. **60**, 2439.
Miyano, K., R.K. Cao, T. Kendelewicz, A.K. Wahi, I. Lindau and W.E. Spicer, 1990, Phys. Rev. B **41**, 1076.
Mo, Y.-M., B.S. Swartzentruber, Y. Kariotis, M.B. Webb and M.G. Lagally, 1989, Phys. Rev. Lett. **63**, 2393.
Mogab, C.J., 1983, in: VLSI Technology, ed. S. M. Sze (McGraw-Hill, New York) ch. 8.
Moisson, J.M., and M. Bensoussan, 1982, J. Vac. Sci. & Technol. **21**, 315.
Mönch, W., 1983, Thin Solid Films **104**, 285.
Mönch, W., 1985, in: Molecular Beam Epitaxy and Heterostructures, NATO ASI Series, Vol. 87 (Martinus Nijhoff, Dordrecht) p. 331.
Mönch, W., 1986, in: Festkörperprobleme – Advances in Solid State Physics, Vol. 26, ed. P. Grosse (Vieweg, Braunschweig) p. 67.
Mönch, W., 1988a, Phys. Rev. B **37**, 7129.
Mönch, W., 1988b, J. Vac. Sci. & Technol. B **6**, 1270.
Mönch, W., 1989, Appl. Surf. Sci. **41/42**, 128.
Mönch, W., 1990, Rep. Prog. Phys. **53**, 221.
Mönch, W., P. Koke and S. Krueger, 1981, J. Vac. Sci. & Technol. **19**, 313.
Montgomery, V., R.H. Williams and G.P. Srivastava, 1981, J. Phys. C **14**, L191.
Morrison, S.R., 1977, The Chemical Physics of Surfaces (Plenum, New York).
Mott, N.F., 1939, Proc. Roy. Soc. (London) A **171**, 27.
Mrstik, B.J., R. Kaplan, T.L. Reinecke, M.A. Van Hove and S.Y. Tong, 1977, Phys. Rev. B **15**, 897.
Muñoz, A., N. Chetty and R.M. Martin, 1990, Phys. Rev. B **41**, 2976.
Murakami, M., H.J. Kim, Y.-C. Shih, W.H. Price and C.C. Parks, 1989, Appl. Surf. Sci. **41/42**, 195.
Murarka, S.P., 1983, Silicides for VLSI Applications (Academic Press, New York).
Narayan, J., S. Sharan, A.R. Srivastava and A.S. Nandedkar, 1988, Mater. Sci. Eng. B **1**, 105.
Neave, J.H., and B.A. Joyce, 1978, J. Cryst. Growth **44**, 387.
Neave, J.H., B.A. Joyce, P.J. Dobson and N. Norton, 1983, Appl. Phys. A **31**, 1.
Neave, J.H., P.J. Dobson, B.A. Joyce and J. Zhang, 1985, Appl. Phys. Lett. **47**, 100.
Nelson, R.J., J.S. Williams, H.J. Leamy, B.J. Miller, H.C. Casey Jr, B.A. Parkinson and A. Heller, 1980, Appl. Phys. Lett. **38**, 76.
Newman, N., K.K. Chin, W.G. Petro, T. Kendelewicz, M.D. Williams, C.E. McCants and W.E. Spicer, 1985, J. Vac. Sci. & Technol. A **3**, 996.
Newman, N., Z. Lilienthal-Weber, E.R. Weber, J. Washburn and W.E. Spicer, 1988, Appl. Phys. Lett. **53**, 145.
Ni, W.-X., and G.V. Hansson, 1990, Phys. Rev. B **42**, 3030.
Nicholls, J.M., and B. Reihl, 1989, Surf. Sci. **218**, 237.
Nicholls, J.M., B. Reihl and J.E. Northrup, 1987, Phys. Rev. B **35**, 4137.
Nielsen, O.H., R.M. Martin, D.J. Chadi and K. Kunc, 1983, J. Vac. Sci. & Technol. B **1**, 714.
Niessen, A.K., F.R. de Boer, R. Boom, P.F. de Châtel, W.C.M. Mattens and A.R. Miedema, 1983, CALPHAD **7**, 51.

Niles, D.W., G. Margaritondo, P. Perfetti, C. Quaresima and M. Capozzi, 1985, Appl. Phys. Lett. **47**, 1092.
Niles, D.W., M. Tang, J.T. McKinley, R. Zanoni and G. Margaritondo, 1988, Phys. Rev. B **38**, 10949.
Nishitani, R., M. Aono, T. Tanaka, C. Oshima, S. Kawai, H. Iwasaki and S. Nakamura, 1980, Surf. Sci. **93**, 535.
Nishizawa, J., Y. Oyama and K. Dezaki, 1990, J. Appl. Phys. **67**, 1884.
Nogami, J., Park Sang-il and C.F. Quate, 1987, Phys. Rev. B **36**, 6221.
Nogami, J., Park Sang-il and C.F. Quate, 1988, J. Vac. Sci. & Technol. B **6**, 1479.
Northrup, J.E., 1985, Phys. Rev. Lett. **54**, 815.
Northrup, J.E., 1988, Phys. Rev. B **37**, 8513.
Northrup, J.E., and M.L. Cohen, 1982, Phys. Rev. Lett. **49**, 1349.
Northrup, J.E., R.D. Bringans, R.I.G. Uhrberg, M.A. Olmstead and R.Z. Bachrach, 1988, Phys. Rev. Lett. **61**, 2967.
Northrup, J.E., M.S. Hybertson and S.G. Louie, 1991, Phys. Rev. Lett. **66**, 500.
Offsey, S.D., J.M. Woodall, A.C. Warren, P.D. Kirchner, T.I. Chappell and G.D. Pettit, 1986, Appl. Phys. Lett. **48**, 475.
Oh, J.E., P.K. Bhattacharya, J. Singh, W. Dos Passos, R. Clarke, N. Mestres, R. Merlin, K.H. Chang and R. Gibala, 1990, Surf. Sci. **228**, 16.
Ohno, T.R., and E.D. Williams, 1990, J. Vac. Sci. & Technol. B **8**, 874.
Okamoto, K., C.E.C. Wood, L. Rathbun and L.F. Eastman, 1982, J. Appl. Phys. **53**, 1532.
Olego, D.J., and D. Cammack, 1990, J. Cryst. Growth **101**, 546.
Olego, D.J., R. Schachter and J.A. Baumann, 1984, Appl. Phys. Lett. **45**, 1127.
Olmstead, M.A., and N.M. Amer, 1984, Phys. Rev. Lett. **52**, 1148.
Osbourn, G.C., P.L. Gourley II, J. Fritz, R.M. Biefeld, L.R. Dawson and T.E. Zipperian, 1987, in: Semiconductors and Semimetals – Applications of Multiquantum Wells, Selective Dopings, and Superlattices, ed. R. Dingle (Academic Press, New York) Vol. 24.
Ospelt, M., J. Henz, L. Flepp and H. van Kanel, 1988, Appl. Phys. Lett. **52**, 227.
Ottaviani, G., 1979, J. Vac. Sci. & Technol. **16**, 1112.
Ottaviani, G., 1981, J. Vac. Sci. & Technol. **18**, 924.
Ottaviani, G., K.N. Tu and J.W. Mayer, 1980, Phys. Rev. Lett. **44**, 284.
Paasch, G., and E. von Faber, 1991, Progr. Surf. Sci. **35**, 19.
Palmstrøm, C.J., and D.V. Morgan, 1985, in: Gallium Arsenide: Materials, Devices and Circuits, eds M.G. Howes and D.V. Morgan (Wiley, New York) p. 195.
Palmstrøm, C.J., B.-O. Fimland, T. Sands, K.C. Garrison and R.A. Bartynski, 1989, J. Appl. Phys. **65**, 4753.
Palmstrøm, C.J., T.L. Cheeks, H.L. Gilchrist, J.G. Zhu, C.B. Carter and R.E. Nahory, 1990a, in: Electronic, Optical and Device Properties of Layered Structures, eds J.R. Hayes, M.S. Hyberston and E.R. Weber (Materials Research Society, Pittsburgh, PA) p. 63.
Palmstrøm, C.J., J.P. Harbison, T. Sands, R. Ramesh, T.G. Finstad, S. Mounier, J.G. Zhu, C.B. Carter, L.T. Florez and V.G. Keramidas, 1990b, Mater. Res. Soc. Symp. Proc. **198**, 153.
Pandey, K.C., 1981, Phys. Rev. Lett. **47**, 1913.
Pandey, K.C., 1982, Phys. Rev. Lett. **49**, 233.
Pandey, K.C., 1985, in: Proc. 17th Int. Conf. Phys. Semiconductors, eds D.J. Chadi and W.A. Harrison (Springer, New York) p. 55.
Pandey, K.C., and J.C. Phillips, 1974, Phys. Rev. Lett. **32**, 1433.
Panish, J.B., 1980, Science **208**, 916.
Panish, M.B., 1967, J. Electrochem. Soc. **114**, 516.
Pashley, M.D., 1989, Phys. Rev. **40**, 10481.
Pashley, M.D., K.W. Haberern, W. Friday, J.M. Woodall and P.D. Kirchner, 1988, Phys. Rev. Lett. **60**, 2176.
Pashley, M.D., K.W. Haberern and J.W. Gaines, 1991, Appl. Phys. Lett. **58**, 406.
Patel, J.R., J.A. Golovchenko, J.C. Bean and R.J. Morris, 1985, Phys. Rev. B **31**, 6884.
Patten, E.A., G.D. Davis, S.J. Hsieh and C.M. Wolfe, 1985, IEEE Electron Device Lett. **EDL-6**, 60.
Pauling, L.N., 1960, The Nature of the Chemical Bond (Cornell University, Ithaca).

Payne, M.C., N. Roberts, R.J. Needs, M. Needels and J.D. Joannopoulos, 1989, Surf. Sci. **211/212**, 1.
Pendry, J.B., 1974, Low Energy Electron Diffraction (Academic Press, London).
Peressi, M., S. Baroni, R. Resta and A. Baldereschi, 1991, Phys. Rev. B **43**, 7347.
Perfetti, P., F. Patella, F. Sette, C. Quaresima, C. Capasso, A. Savoia and G. Margaritondo, 1984, Phys. Rev. B **30**, 4533.
Perfetti, P., J.M. Nicholls and B. Reihl, 1987, Phys. Rev. B **36**, 6160.
Perfetti, P., F. Patella, F. Sette, C. Quaresima, C. Capasso, A. Savoia, W.G. Wilke, R. Seedorf and K. Horn, 1990, J. Cryst. Growth **101**, 620.
Petroff, P.M., and R.J. Wilson, 1983, Phys. Rev. Lett. **51**, 199.
Petroff, P.M., A.C. Gossard and W. Wiegmann, 1984, Appl. Phys. Lett. **45**, 620.
Phillips, J.C., 1970, Rev. Mod. Phys. **42**, 317.
Pinczuk, A., 1984, J. Phys. Colloq. (France) C **5**, 477.
Poate, J.M., and J.W. Mayer, 1982, Laser Annealing of Semiconductors (Academic Press, New York).
Poate, J.M., K.N. Tu and J.W. Mayer, 1978, Thin Films – Interdiffusion and Reactions (Wiley–Interscience, New York).
Pollmann, J., P. Krüger, A. Mazur and G. Wolfgarten, 1985, Surf. Sci. **153**, 977.
Poppendieck, T.D., T.C. Ngoc and M.B. Webb, 1978, Surf. Sci. **75**, 287.
Prinz, G.A., 1985, Phys. Rev. Lett. **54**, 1051.
Prinz, G.A., and J.J. Krebs, 1981, Appl. Phys. Lett. **39**, 397.
Pugh, J.H., and R.S. Williams, 1986, J. Mater. Res. **1**, 343.
Qiu, J., Q-D. Qian, R.L. Funshor, M. Kobayashi, D.R. Menke, D. Li and N. Otsuka, 1990, Appl. Phys. Lett. **56**, 1272.
Raether, H., 1978, J. Appl. Phys. 17, Suppl. **17**(2), 227 and references therein.
Ranke, W., and K. Jacobi, 1973, Solid State Commun. **16**, 691.
Ranke, W., and K. Jacobi, 1981, Prog. Surf. Sci. **10**, 1.
Reuters, P.J., M. Offenberg and P. Balk, 1990, Appl. Phys. Lett. **56**, 1903.
Rhines, F.N., 1956, Phase Diagrams in Metallurgy (McGraw-Hill, New York).
Rhoderick, E.H., and R.H. Williams, 1988, Metal–Semiconductor Contacts, 2nd Ed., in: Monographs in Electrical and Electronic Engineering, eds P. Hammond and R.L. Grimsdale (Clarendon Press, Oxford).
Richter, H.W., and L.J. Brillson, 1986, J. Appl. Phys. **60**, 1994.
Richter, H.W., L.J. Brillson, M.K. Kelly, R.R. Daniels and G. Margaritondo, 1984, J. Vac. Sci. & Technol. B **2**, 591.
Rowe, J.E., G. Margaritondo and S.B. Christman, 1977, Phys. Rev. B **16**, 1581.
Rubloff, G.W., 1983, Surf. Sci. **132**, 268.
Ruckman, M.W., J.J. Joyce and J.H. Weaver, 1986, Phys. Rev. B **33**, 2191.
Ruckman, M.W., J.J. Joyce and J.H. Weaver, 1987, Phys. Rev. B **33**, 7029.
Samorjai, G.A., and M.A. Van Hove, 1979, Adsorbed Monolayers on Solid Surfaces, Structure and Bonding Series, Vol. 38 (Springer, Berlin) 146pp.
Samorjai, G.A., and M.A. Van Hove, 1989, Progr. Surf. Sci. **30**, 201.
Sandroff, C.J., R.N. Nottenburg, J.-C. Bischoff and R. Bhat, 1987, Appl. Phys. Lett. **51**, 33.
Sands, T., 1989, Mater. Sci. Eng. B **1**, 289.
Sands, T., J.P. Harbison, N. Tabatabaie, W.K. Chan, H.L. Gilchrist, T.L. Cheeks, L.T. Florez and V.G. Keramidas, 1990a, Surf. Sci. **228**, 1.
Sands, T., C.J. Palmstrøm, J.P. Harbison, V.G. Keramidas, N. Tabatabaie, T.L. Cheeks, R. Ramesh and Y. Silberberg, 1990b, Mat. Sci. Rep. **5**, 99.
Sarma, K., R. Dalby, K. Rose, O. Aina, W. Katz and N. Lewis, 1984, J. Appl. Phys. **56**, 2703.
Sauvage-Simkin, M., R. Pinchaux, J. Massies, P. Calverie, N. Jedrecy, J.E. Bonnet and I.L. Robinson, 1989, Phys. Rev. Lett. **62**, 563.
Savage, D.E., and M.G. Lagally, 1986, J. Vac. Sci. & Technol. B **4**, 943.
Schlier, R.E., and H.E. Farnsworth, 1959, J. Chem. Phys. **30**, 917.
Schlüter, M., 1978, Phys. Rev. B **17**, 5044.
Schlüter, M., 1982, Thin Solid Films **93**, 3.
Schlüter, M., 1988, in: Surface Properties of Electronic Materials, eds D.A. King and D.P. Woodruff, The Chemical Physics of Solid Surfaces and Heterogeneous Catalysis, Vol. 5 (Elsevier, Amsterdam, 1988) pp. 37–68.

Schlüter, M., J.R. Chelikowsky, S.G. Louie and M.L. Cohen, 1975a, Phys. Rev. Lett. **34**, 1385.
Schlüter, M., J.R. Chelikowsky, S.G. Louie and M.L. Cohen, 1975b, Phys. Rev. B **12**, 4200.
Schmid, P.E., 1985, Helv. Phys. Acta **58**, 371.
Schmid, P.E., P.S. Ho, H. Föll and G.W. Rubloff, 1981, J. Vac. Sci. & Technol. **18**, 937.
Schottky, W., 1939, Z. Phys. **133**, 367.
Schroeder, D.K., 1990, Semiconductor Material and Device Characterization (Wiley Interscience, New York).
Seah, M.P., and W.A. Dench, 1979, Surf. & Interface Anal. **1**, 2.
Seiwatz, R., 1963, Surf. Sci. **2**, 473.
Selloni, A., and R. Del Sole, 1986, Surf. Sci. **168**, 35.
Shapira, Y., and L.J. Brillson, 1983, J. Vac. Sci. & Technol. B **1**, 618.
Shapira, Y., L.J. Brillson, A.D. Katnani and G. Margaritondo, 1984, Phys. Rev. B **30**, 4586.
Shaw, D., 1973, Atomic Diffusion in Semiconductors (Plenum, New York).
Shaw, J.L., R.E. Viturro, L.J. Brillson, D.G. Kilday, M.K. Kelly and G. Margaritondo, 1988a, J. Electron. Mater. **17**, 149.
Shaw, J.L., R.E. Viturro, L.J. Brillson and D. LaGraffe, 1988b, Appl. Phys. Lett. **53**, 1723.
Shaw, J.L., R.E. Viturro, L.J. Brillson and D. LaGraffe, 1989, J. Electron. Mater. **18**, 59.
Shaw, M.P., 1981, in: Handbook on Semiconductors, Vol. 4, ed. C. Hilsum (North-Holland, Amsterdam) pp. 1–90.
Shen, T.-H., M. Elliott, R.H. Williams and D.I. Westwood, 1991, Appl. Phys. Lett. **58**, 842.
Shi, Z.Q., R.L. Wallace and W.A. Anderson, 1991, Appl. Phys. Lett. **59**, 446.
Shockley, W., and G.L. Pearson, 1948, Phys. Rev. **74**, 232.
Skeath, P., C.Y. Su, I. Lindau and W.E. Spicer, 1980, J. Vac. Sci. & Technol. **17**, 874.
Skeath, P., I. Lindau, C.Y. Su and W.E. Spicer, 1983, Phys. Rev. B **28**, 7051.
Smith, D.L., and C. Mailhiot, 1987, Phys. Rev. Lett. **58**, 1264.
Smith, N.V., M.M. Traum, J.A. Knapp, J. Anderson and G.J. Lapeyre, 1976, Phys. Rev. B **13**, 4462.
Smith, R.J., J. Anderson and G.J. Lapeyre, 1976, Phys. Rev. Lett. **37**, 1081.
Snyman, L.W., J.S. Vermaak and F.D. Auret, 1977, J. Cryst. Growth **42**, 132.
Solzbach, U., and H.J. Richter, 1976, Surf. Sci. **97**, 385.
Somorjai, G.A., and M.A. Van Hove, 1979, Structure and Bonding, Vol. 38 (Springer, Berlin) 146pp.
Sorba, L., G. Bratina, G. Ceccone, A. Antonini, J.F. Walker, M. Micovic and A. Franciosi, 1991, Phys. Rev. B **43**, 2450.
Spicer, W.E., I. Lindau, P.E. Gregory, C.M. Garner, P. Pianetta and P.W. Chye, 1976, J. Vac. Sci. & Technol. **13**, 780.
Spicer, W.E., I. Lindau, P. Skeath and C.Y. Su, 1980a, Phys. Rev. Lett. **44**, 420.
Spicer, W.E., I. Lindau, P. Skeath and C.Y. Su, 1980b, J. Vac. Sci. & Technol. **17**, 1019.
Spicer, W.E., Z. Lilienthal-Weber, E.R. Weber, N. Newman, T. Kendelewicz, R.K. Cao, C. McCants, P. Mahowald, K. Miyano and I. Lindau, 1988, J. Vac. Sci. & Technol. B **6**, 1245.
Srivastava, G.P., 1979, Phys. Status Solidi b **93**, 761.
Srivastava, G.P., I. Singh, V. Montgomery and R.H. Williams, 1983, J. Phys. C **16**, 3627.
Stiles, K., and A. Kahn, 1988, Phys. Rev. Lett. **60**, 440.
Stiles, K., A. Kahn, D.G. Kilday and G. Margaritondo, 1987, J. Vac. Sci. & Technol. B **5**, 987.
Straub, D., M. Skibowski and F.J. Himpsel, 1985, Phys. Rev. B **32**, 5237.
Stroscio, J.A., R.M. Feenstra, D.M. Newns and A.P. Fein, 1988, J. Vac. Sci. & Technol. **A6**, 499.
Svensson, S.P., and T.G. Andersson, 1985, J. Vac. Sci. & Technol. B **3**, 760.
Svensson, S.P., G. Landgren and T.G. Andersson, 1983, J. Appl. Phys. **54**, 4474.
Svensson, S.P., J. Kanski, T.G. Andersson and P.O. Nilsson, 1984, J. Vac. Sci. & Technol. B **2**, 235.
Swarts, C.A., W.A. Goddard III and T.C. McGill, 1980, Surf. Sci. **110**, 400.
Swartzentruber, B.S., Y.-M. Mo, R. Kariotis, M.G. Lagally and M.B. Webb, 1990, Phys. Rev. Lett. **65**, 1913.
Sze, S.M., 1981, Physics of Semiconductor Devices, 2nd Ed. (Wiley–Interscience, New York) ch. 5.
Tadayon, B., S. Tadayon, M.G. Spencer, G.L. Harris, L. Rathbun, J.T. Bradshaw, W.J. Schaffer, P.W. Tasker and L.F. Eastman, 1988, Appl. Phys. Lett. **53**, 2664.
Takatani, S., and Y.W. Chung, 1985, Phys. Rev. B **31**, 2290.

Takayanagi, K., Y. Tanishiro, M. Takahashi and S. Takahashi, 1985, J. Vac. Sci. & Technol. A **3**, 1502.
Takeda, K., T. Hanawa and T. Shimojo, 1974, Jpn. J. Appl. Phys., Suppl. 2, Part 1, p. 589.
Tejedor, C., F. Flores and S.G. Louie, 1977, J. Phys. C **10**, 2163.
Tejedor, C., J.M. Calleja, F. Meseguer, E.E. Mendez, C.-A. Chang and L. Esaki, 1985, in: Proc. Int. Conf. on the Physics of Semiconductors, San Francisco, CA, eds J.A. Chadi and W.A. Harrison (Springer, Berlin) p. 559.
Tersoff, J., 1984, Phys. Rev. Lett. **52**, 465; Phys. Rev. B **30**, 4875.
Tersoff, J., 1985, Phys. Rev. B **32**, 6968.
Tersoff, J., 1987, in: Heterojunction Band Discontinuities: Physics and Device Applications, eds F. Capasso and G. Margaritondo (North-Holland, Amsterdam) ch. 1.
Tokumoto, H.S., Wakiyama, K. Miki and S. Okayama, 1990, Appl. Phys. Lett. **56**, 743.
Tong, S.Y., W.N. Mei and G. Xu, 1984, J. Vac. Sci. & Technol. B **2**, 393.
Tong, S.Y., H. Huang and C.M. Wei, 1990, in: Chemistry and Physics of Solid Surfaces VIII, eds R. Vanselow and R. Howe, Springer Series in Surface Sciences, Vol. 22 (Springer, New York) p. 395.
Tosch, St., and H. Neddermeyer, 1988, Phys. Rev. Lett. **61**, 349.
Trafas, B.M., I.M. Vitomirov, C.M. Aldao, Y. Gao, F. Xu, J.H. Weaver and D.L. Partin, 1989, Phys. Rev. B **39**, 3265.
Trafas, B.M., D.M. Hill, R.L. Siefert and J.H. Weaver, 1990, Phys. Rev. B **42**, 3231.
Tromp, R.M., R.J. Hamers and J.E. Demuth, 1985, Phys. Rev. Lett. **55**, 1303.
Tromp, R.M., R.J. Hamers and J.E. Demuth, 1986, Phys. Rev. **34**, 1388.
Troost, D., L. Koendens, L.-Y. Fan and W. Mönch, 1987, J. Vac. Sci. & Technol. B **5**, 1119.
Tung, R.T., 1984, Phys. Rev. Lett. **52**, 461.
Tung, R.T., 1991, Appl. Phys. Lett. **58**, 2821.
Tung, R.T., and F. Schrey, 1989, Phys. Rev. Lett. **63**, 1277.
Tung, R.T., J.M. Gibson and J.M. Poate, 1983, Phys. Rev. Lett. **50**, 429.
Tung, R.T., K.K. Ng, J.M. Gibson and F.J. Levy, 1986, Phys. Rev. B **33**, 7077.
Turowski, M., G. Margaritondo, M.K. Kelly and R.D. Tomlinson, 1985, Phys. Rev. B **31**, 1022.
Uhrberg, R.I.G., and G.V. Hansson, 1991, Crit. Rev. Solid State Mater. Sci. **17**, 133.
Uhrberg, R.I.G., G.V. Hansson, J.M. Nicholls and S.A. Flodstrom, 1982, Phys. Rev. Lett. **48**, 1032.
Van Bommel, A.J., J.E. Crombeen and T.G.J. van Dirschot, 1978, Surf. Sci. **72**, 95.
Van der Veen, J.F., 1985, Surf. Sci. Rep. **5**, 199.
Van Hove, H., and R. Leysen, 1972, Phys. Status Solidi a **9**, 361.
Van Hove, J.M., and P.I. Cohen, 1983, Appl. Phys. Lett. **47**, 726.
Van Hove, J.M., C.S. Lent, P.R. Pukite and P.I. Cohen, 1983, J. Vac. Sci. & Technol. B **1**, 741.
Van Hove, M.A., and S.Y. Tong, 1979, Surface Crystallography by LEED (Springer, Berlin).
Van Hove, M.A., and D.P. Vvedensky, 1987, Surface Crystallography Information Service: A Handbook of Surface Structures (Reidel, Dordrecht) 352pp.
Van Hove, M.A., S.Y. Tong and M.H. Elconin, 1977, Surf. Sci. **64**, 85.
Van Hove, M.A., S.-W. Wang, D.F. Ogletree and G.A. Somorjai, 1989, Adv. Quant. Chem. **20**, 1.
Van Laar, J., and J.J. Scheer, 1967, Surf. Sci. **8**, 342.
Van Loenen, E.J., J.W.M. Frenken and J.F. Van der Veen, 1984, Appl. Phys. Lett. **45**, 42.
Van Vechten, J.A., 1975, J. Electrochem. Soc. **122**, 423.
Van Vechten, J.A., 1980, in: Materials, Properties and Preparation, ed. S.P. Keller, Handbook on Semiconductors, Vol. 3 (North-Holland, Amsterdam) p. 1.
Van Vechten, J.A., 1982, J. Appl. Phys. **53**, 7082.
Van Vechten, J.A., 1985, J. Vac. Sci. & Technol. B **3**, 1240.
Vitomirov, I.M., C.M. Aldao, G.D. Waddill, C. Capasso and J.H. Weaver, 1990, Phys. Rev. B **41**, 8465.
Viturro, R.E., M.L. Slade and L.J. Brillson, 1986, Phys. Rev. Lett. **57**, 487.
Viturro, R.E., J.L. Shaw, C. Mailhiot, L.J. Brillson, N. Tache, J.T. McKinley, G. Margaritondo, J.M. Woodall, P.D. Kirchner, G.D. Pettit and S.J. Wright, 1988a, Appl. Phys. Lett. **52**, 2052.
Viturro, R.E., J.L. Shaw, L.J. Brillson, J.M. Woodall, P.D. Kirchner, G.D. Pettit and S.L. Wright, 1988b, J. Vac. Sci. & Technol. B **6**, 1397.
Vos, M., C.M. Aldao, D.J.W. Aaustuen and J.H. Weaver, 1990, Phys. Rev. B **41**, 991.
Waddill, G.D., I.M. Vitomirov, C.M. Aldao and J.H. Weaver, 1989, Phys. Rev. Lett. **62**, 1568.

Wagman, D.D., W.H. Evans, V.B. Parker, I. Halow, S.M. Bailey and R.H. Schumm, 1968–1971, Natl. Bur. Std. Technical Notes 270-3–270-7 (US Governmental Printing Office, Washington, DC).
Waldrop, J.R., 1984a, J. Vac. Sci. & Technol. B **2**, 445.
Waldrop, J.R., 1984b, Appl. Phys. Lett. **44**, 1002.
Waldrop, J.R., 1985a, Appl. Phys. Lett. **47**, 1301.
Waldrop, J.R., 1985b, J. Vac. Sci. & Technol. B **3**, 1197.
Waldrop, J.R., 1985c, Appl. Phys. Lett. **46**, 864.
Waldrop, J.R., and R.W. Grant, 1979, Appl. Phys. Lett. **34**, 630.
Waldrop, J.R., and R.W. Grant, 1986, J. Vac. Sci. & Technol. B **5**, 1015.
Waldrop, J.R., and R.W. Grant, 1987, Appl. Phys. Lett. **50**, 250.
Waldrop, J.R., and R.W. Grant, 1988, Appl. Phys. Lett. **52**, 1794.
Waldrop, J.R., E.A. Kraut, S.P. Kowalczyk and R.W. Grant, 1983, Surf. Sci. **132**, 513.
Waldrop, J.R., R.W. Grant, S.P. Kowalczyk and E.A. Kraut, 1985, J. Vac. Sci. & Technol. A **3**, 835.
Walukiewicz, W., 1987, J. Vac. Sci. & Technol. B **5**, 1062.
Walukiewicz, W., 1988a, Phys. Rev. B **37**, 4760.
Walukiewicz, W., 1988b, J. Vac. Sci. & Technol. B **6**, 1257.
Wang, W.I., E.E. Mendez and F. Stern, 1985, Appl. Phys. Lett. **45**, 639.
Wang, X.S., J.L. Goldberg, N.C. Bartelt, T.L. Einstein and E.D. Williams, 1990, Phys. Rev. Lett. **65**, 2430.
Weaver, J.H., 1986, Phys. Today **1**, 24.
Weaver, J.H., 1988, in: Analytical Techniques for Thin Films, eds K.N. Tu and R. Rosenberg, Treatise on Materials Science and Technology (Academic Press, New York) pp. 15–53.
Weaver, J.H., and G.D. Waddill, 1991, Nature **251**, 1444 and references therein.
Weaver, J.H., Z. Lin and F. Xu, 1990, in: Surface Segregation and Related Phenomena, eds P.A. Kowben and A. Miller (CRC Press, New York) ch. 10.
Weber, E.R., H. Ennen, V. Kaufmann, J. Windscheif, J. Schneider and T. Wasinki, 1982, J. Appl. Phys. **53**, 6140.
Whitman, L.J., J.A. Stroscio, R.A. Dragoset and R.J. Celotta, 1990, Phys. Rev. B **42**, 7228.
Whitman, L.J., J.A. Stroscio, R.A. Dragoset and R.J. Celotta, 1991, J. Vac. Sci. & Technol., in press.
Wieder, H.H., 1978, J. Vac. Sci. & Technol. **15**, 1478.
Wieranga, P.E., J.A. Kubby and J.E. Griffith, 1987, Phys. Rev. Lett. **59**, 2169.
Wilke, W.G., R. Seedorf and K. Horn, 1990, J. Cryst. Growth **101**, 620.
Williams, E.D., and N.C. Bartelt, 1991, Science **251**, 393.
Williams, G.P., R.J. Smith and G.J. Lapeyre, 1978, J. Vac. Sci. & Technol. **15**, 1249.
Williams, R.H., and M.H. Patterson, 1982, Appl. Phys. Lett. **40**, 484.
Williams, R.H., V. Montgomery and R.R. Varma, 1978, J. Phys. C **11**, L735.
Williams, R.H., R.R. Varma and V. Montgomery, 1979, J. Vac. Sci. & Technol. **16**, 1418.
Williams, R.S., J.R. Lince, T.C. Tsai and J.H. Pugh, 1986, in: Thin Films – Interfaces and Phenomena, eds R.J. Nemanich, P.S. Ho and S.S. Lau, Materials Research Society Symposium Proceedings, Vol. 54 (MRS, Pittsburgh, PA) p. 335.
Wilson, B.A., 1988, IEEE J. Quantum Electron. **QE-24**, 1763.
Winters, H.F., and D. Haarer, 1987, Phys. Rev. B **36**, 6613.
Wolford, D.J., T.F. Kuech and M. Jaros, 1987, in: Heterojunction Band Offsets: Physics and Device Applications, eds F. Capasso and G. Margaritondo (North-Holland, Amsterdam) ch. 6.
Wolkow, R., 1992, Phys. Rev. Lett. **68**, 2636.
Wolkow, R., and Ph. Avouris, 1988, Phys. Rev. Lett. **60**, 1049.
Wong, T.M., D. Heskett, N.J. Dinardo and E.W. Plummer, 1989, Surf. Sci. **208**, L1.
Wood, E.A., 1964, J. Appl. Phys. **35**, 1306.
Woodall, J.M., and J.L. Freeouf, 1981, J. Vac. Sci. & Technol. **19**, 794.
Woodall, J.M., J.L. Freeouf, G.D. Pettit, T.N. Jackson and P. Kirchner, 1981, J. Vac. Sci. & Technol. **19**, 626.
Woodall, J.M., G.D. Pettit, T.N. Jackson, C. Lanza, K.L. Kavanaugh and J.W. Mayer, 1983, Phys. Rev. Lett. **51**, 1783.
Woodall, J.M., P.D. Kirchner, D.L. Rogers, M. Chisholm and J.J. Rosenberg, 1988, Mat. Res. Soc. Symp. Proc. **126**, p. 3.

Wyeth, N.C., and A. Catalano, 1980, J. Appl. Phys. **51**, 2286.
Xu, F., G.W.Y. Hu, M.W. Puga, S.Y. Tong, J.L. Yeh, S.R. Wang and B.W. Lee, 1985, Phys. Rev. B **32**, 8473.
Xu, F., C.M. Aldao, I.M. Vitomirov and J.H. Weaver, 1987a, Appl. Phys. Lett. **51**, 1946.
Xu, F., J.J. Joyce, M.W. Ruckman, H.-W. Chen, F. Boscherini, D.M. Hill, S.A. Chambers and J.H. Weaver, 1987b, Phys. Rev. B **35**, 2375.
Yablonovitch, E., and T.J. Gmitter, 1988, Proc. Electrochem. Soc. **88–20**, 207.
Yablonovitch, E., C.J. Sandroff, R. Bhat and T.J. Gmitter, 1987a, Appl. Phys. Lett. **51**, 439.
Yablonovitch, E., T.J. Gmitter, J.P. Harbison and R. Bhat, 1987b, Appl. Phys. Lett. **57**, 2222.
Yablonovitch, E., D.M. Hwang, T.J. Gmitter, L.T. Florez and J.P. Harbison, 1990, Appl. Phys. Lett. **56**, 2419.
Yacobi, B.G., and D.B. Holt, 1990, Cathodoluminescence Microscopy of Inorganic Solids (Plenum, New York).
Yahata, A., and M. Nakajima, 1984, Jpn. J. Appl. Phys. **23**, L313.
Yang, W.S., F. Jona and P.M. Marcus, 1983, Phys. Rev. B **28**, 2049.
Yang, Y-N., and E.D. Williams, 1990, J. Vac. Sci. & Technol. A **8**, 2481.
Yeh, J.-J., 1989, Appl. Phys. Lett. **55**, 1241.
Yokotsuka, T., J. Narusawa, Y. Uchida and H. Nakashima, 1987, Appl. Phys. Lett. **50**, 591.
Yu, E.T., M.C. Phillips, J.O. McCaldin and T.C. McGill, 1991, J. Vac. Sci. & Technol. B **9**, 2233.
Yu, X., A. Raisanen, G. Haugstad, G. Ceccone, N. Troullier and A. Franciosi, 1990, Phys. Rev. B **42**, 1872.
Zahn, D.T.R., R.H. Williams, D.T. Golding, J.H. Dinan, K.J. Mackay, J. Geurts and W. Richter, 1988, Appl. Phys. Lett. **53**, 2409.
Zallen, R., 1990, in: Band Theory and Transport Properties, ed. W. Paul, Handbook on Semiconductors, Vol. 1 (North-Holland, Amsterdam) ch. 1.
Zalm, P.C., G.F.A. van derWalle, D.J. Gravesteijn and A.A. van Gorkum, 1989, Appl. Phys. Lett. **55**, 2520.
Zangwill, A., 1988, Physics at Surfaces (Cambridge University Press, Cambridge).
Zehner, D.M., and C.W. White, 1982, in: Laser Annealing of Semiconductors, eds J.M. Poate and J.W. Mayer (Academic Press, New York) p. 282.
Zunger, A., 1980, Phys. Rev. B **22**, 959.
Zunger, A., 1981, Phys. Rev. B **24**, 4372.
Zunger, A., 1983, Thin Solid Films **104**, 301.
Zunger, A., 1985, Ann. Rev. Mater. Sci. **15**, 411.

CHAPTER 8

Nonlinear Dynamics, Phase Transitions and Chaos in Semiconductors

ECKEHARD SCHÖLL

Institut für Theoretische Physik
Technische Universität Berlin
Hardenbergstr. 36, W-1000 Berlin 12, Germany

Handbook on Semiconductors
Completely Revised Edition
Edited by T.S. Moss
Volume 1, edited by P.T. Landsberg

Contents

1. Semiconductors as nonlinear dynamic systems far from thermodynamic equilibrium . . . 421
   1.1. Introduction . . . 421
   1.2. Dynamic systems . . . 422
   1.3. Nonequilibrium transport . . . 425
2. Generation–recombination induced nonequilibrium phase transitions . . . 428
3. Oscillatory instabilities and deterministic chaos . . . 431
   3.1. Survey of physical mechanisms . . . 431
   3.2. Routes to chaos . . . 435
4. Self-organized spatial structures . . . 439
   4.1. Current filaments and field domains . . . 439
   4.2. Nucleation and growth of filaments . . . 440
   4.3. Breathing filaments . . . 443
References . . . 445

1. Semiconductors as nonlinear dynamic systems far from thermodynamic equilibrium

1.1. Introduction

Semiconductors are complex many-body systems whose physical, e.g., electrical or optical, properties are governed by a variety of nonlinear dynamic processes. An important class of such processes which influences, in particular, the transport properties, but also optical phenomena, are generation and recombination (g–r) processes of nonequilibrium charge carriers. These are generally described by rate equations for the change in time of the carrier concentrations, which are nonlinear functions of these concentrations (see chapter 6 of this volume) and of the electric field. Another class of nonlinearities is given by nonlinear scattering processes of hot carriers which may lead to a field-dependent mobility, and thus to a current density which is a nonlinear function of the local electric field (see chapter 18 of this volume). Although this has been known for a long time, the view of a semiconductor as a nonlinear dynamic system is a fairly recent development. Such nonlinear dynamic systems can exhibit a variety of complex behaviour such as bifurcations, phase transitions, spatio-temporal pattern formation, self-sustained oscillations, and deterministic chaos (Guckenheimer and Holmes 1983, Thompson and Stewart 1986, Schuster 1987).

Semiconductors are *dissipative* dynamic systems, i.e., a steady state can only be maintained by a continuous flux of energy, and possibly matter, through them. Mathematically, this is described by the feature that volume elements in a suitable space of dynamic variables – the *phase space* – shrink with increasing time. In the language of thermodynamics, this represents an *open* system which is driven by external fluxes and forces so far from thermodynamic equilibrium that linear dynamic laws no longer hold. Due to the driving forces and the inherent nonlinearities of these systems, they may spontaneously evolve into a state of highly ordered spatial or temporal structures, so-called dissipative structures. Unlike an isolated, closed system, which after a perturbation always returns to a thermal equilibrium state characterized by maximum entropy, an open dissipative nonlinear system may exhibit a process of self-organization, in which the entropy is locally decreased. Such processes usually involve qualitative changes in the state of the system, similar to phase transitions. Nonequilibrium phase transitions have been noted in a great number of very different dissipative systems occurring in physics, chemistry, biology, and even social sciences (Haken 1983a,b, Nicolis and Prigogine 1977, Ebeling and Feistel 1986), but the observed phenomena are similar. Famous examples are the laser, the Bénard and Taylor instabilities in hydrodynamics, and chemical reaction systems. In the field of semiconductor physics, nonlinear charge transport processes (Schöll 1987) or nonlinear optical effects (Haug 1988) may give rise to nonequilibrium phase transitions. In the first case they are manifest as electrical instabilities like current runaway, threshold switching between a nonconducting and a conducting state, spontaneous oscillations of the current or voltage, nucleation and growth of current filaments or high-field domains, if sufficiently high electric or magnetic fields, injected currents, or optical irradiation are applied.

The analogy of an overheating instability of the electron gas in a semiconductor with an equilibrium phase transition was first pointed out by Volkov and Kogan (1969); Pytte and Thomas (1969) drew this analogy in the case of the Gunn instability of the electron drift velocity. Generation–recombination induced nonequilibrium phase transitions in semiconductors were first noted by Landsberg and Pimpale (1976). Impact ionization of electrons or holes by hot carriers across the bandgap or from localized levels was recognized as the main autocatalytic process which is necessary for nonequilibrium phase transitions (Schöll 1979). Low-temperature impurity breakdown and threshold switching were subsequently explained as g–r induced phase transitions (Landsberg et al. 1978, Schöll and Landsberg 1979, Robbins et al. 1981). Current filamentation was recognized as a process of self-organization in a system far from equilibrium, and treated theoretically in the framework of a nonequilibrium phase transition (Schöll 1981, 1982a,b, 1983). The interest in this field was greatly enhanced by the experimental discovery of deterministic chaos in semiconductors (Aoki et al. 1981, Teitsworth et al. 1983). The introduction of concepts and methods of nonlinear dynamics subsequently stimulated a large amount of experimental and theoretical work on chaos and spatio-temporal self-organized pattern formation in semiconductors. While a fairly comprehensive account of the developments up to 1987 has been given by Schöll (1987), the field has grown so much by now that a complete survey can no longer be presented. There exists a number of more recent reviews covering different aspects of the latest advances, e.g., on theoretical approaches to nonlinear and chaotic dynamics of generation–recombination processes (Schöll 1989a), on a general dynamic systems approach to nonlinear semiconductor transport (Schöll 1989b), on the derivation of various levels of description for macroscopic nonlinear behaviour from hot-electron transport theory (Schöll 1989c), on a microscopic theory of hot-carrier recombination and ionization processes (Reggiani and Mitin 1989), on nonlinear waves (Kerner and Osipov 1989), on experimental aspects (Brandl and Prettl 1990, Aoki 1992, Peinke et al. 1992), and on semiconductor device applications (Shaw et al. 1992). Recent collections of review papers and pedagogical introductions have been edited by Abe (1989) and Thomas (1992), respectively. Physical aspects of current instabilities have also been treated in a number of older monographs (e.g., Glicksman 1971, Bonch-Bruevich et al. 1975, Asche et al. 1979, Požela 1981). While the present article is intended to provide a survey of the main concepts and ideas of the field, the above references may serve as a more detailed guide to the original literature.

1.2. Dynamic systems

Generally, a dynamic system is specified by (i) a set of dynamic variables $\boldsymbol{q} \equiv (q_1, \dots, q_N) \in \Gamma$ which depend upon time t and spatial coordinates $\boldsymbol{r}$, and form the *phase space* Γ, (ii) a law determining the temporal evolution $\boldsymbol{q}(t)$, i.e. the phase trajectories or orbits, for given initial conditions $\boldsymbol{q}_0$, and (iii) a set of *control parameters* $\boldsymbol{k} \equiv (k_1, ..., k_M)$ by which the dynamic law can be changed. In a semiconductor the dynamic variables may represent, for instance, the local electric field $\mathscr{E}$, the carrier densities in the conduction and valence band, and in impurity levels, the mean carrier energy E and momentum $\boldsymbol{p}$, or the radius of a current filament. The control parameters are given by external parameters like applied currents, electric or magnetic fields, or optical irradiation intensities.

The modelling of a spatially extended nonlinear dynamic system, such as a semiconductor, may be cast into one of the following common forms, depending on whether space, time, and the dynamic variables are chosen as continuous or discrete:

(i) A set of partial differential equations
(space, time, and dynamic variables all continuous)
(ii) A set of ordinary differential equations
(space discrete, time and dynamic variables continuous)
(iii) Iterated maps
(space and time discrete, dynamic variables continuous)
(iv) Cellular automata
(space, time, and dynamic variables all discrete)

Semiconductor transport theory and g–r kinetics are usually formulated in terms of (i) or (ii), e.g., semiclassical balance equations or coupled rate equations, and the existing models for nonlinear dynamics in semiconductors are mostly of these types. One of the few exceptions is a model for the onset of chaos in semiconductors which was derived in the form of an iterated map (iii), starting from a stochastic Chapman–Kolmogoroff equation (Landsberg et al. 1988). Work generated towards modelling current filaments and high-field transport in terms of a cellular automaton (iv) also exists (Rieger and Vogl 1989, Kometer et al. 1992a,b).

Nonequilibrium phase transitions show up as bifurcations of a dynamic system. In the following we shall briefly review some basic aspects of stability and bifurcation, confining attention to the spatially homogeneous case (ii). The dynamics is then given by

$$\dot{\boldsymbol{q}} = \boldsymbol{F}(\boldsymbol{q}, \boldsymbol{k}) \tag{1.1}$$

where $\boldsymbol{F} \equiv (F_1, ..., F_N)$ is a set of nonlinear functions of $\boldsymbol{q}$. For every phase point $\boldsymbol{q}$, except for the *fixed points* $\boldsymbol{q}^*$ in which $\boldsymbol{F}(\boldsymbol{q}^*, \boldsymbol{k}) = 0$, the differential equations (1.1) give a unique direction of the phase flow $\boldsymbol{q}(t)$. Thus one can construct a phase portrait of the trajectories, similar to the streamlines of a fluid. The fixed points represent time-independent steady states, which are in general states far from thermodynamic equilibrium. Any solution $\boldsymbol{q} = \boldsymbol{u}_{\boldsymbol{q}_0}(t)$, which may be regarded as the path of a particle in $\boldsymbol{q}$-space is uniquely determined by its initial value $\boldsymbol{q}_0$ at the initial time t_0. Since the initial values are normally subject to perturbations, the question of interest becomes: what happens to the path of the particle if the initial condition is slightly different from the initial choice? A trajectory is said to be *locally stable* if other trajectories, obtained from initial conditions, different from, but close to $\boldsymbol{q}_0$, remain close to the original trajectory for all later times. It is *locally asymptotically stable* if for all trajectories $\boldsymbol{v}_{\boldsymbol{q}_0'}$ close to $\boldsymbol{u}$ whose phase points fulfill the stability criterion, the condition

$$|\boldsymbol{u}_{\boldsymbol{q}_0}(t) - \boldsymbol{v}_{\boldsymbol{q}_0'}(t)| \rightarrow 0 \quad \text{as} \quad t \rightarrow \infty$$

also holds. A bounded, indecomposable, invariant and locally asymptotically stable subset of the phase space is called an *attractor*. An attractor is *globally stable* if its basin of attraction includes the whole phase space. Note that attractors can only occur in *dissipative* dynamic systems, where phase volumes contract. These definitions include the special case that $\boldsymbol{u}$ consists of a single fixed point: a point attractor. If the phase space is at least 2-dimensional, periodic attractors (limit cycles) are also possible. Three dynamic degrees of freedom are required for a qualitatively new type of attractor: a *strange* or *chaotic* attractor. It consists of an attracting, bounded set of phase points with the property that trajectories which are initially very close to each other on the attractor, are separated exponentially as time goes on. Intuitively, this implies a sensitive dependence upon initial conditions, and an unstable motion *within* the attractor. It has in general a noninteger "fractal" dimension. For any bounded set of points in $\mathbb{R}^p$, representing the strange attractor, its dimension d is defined as

$$d = \lim_{\varepsilon \rightarrow 0} [\log N(\varepsilon)/\log(1/\varepsilon)], \tag{1.2}$$

where $N(\varepsilon)$ is the minimum number of p-dimensional cubes of side length ε needed to cover the set. Thus $N(\varepsilon) \sim \varepsilon^{-d}$ for $\varepsilon \rightarrow 0$, which gives $d = 0$, 1, and 2 for the trivial cases of a point, a line, and a surface, as expected.

Bifurcation is a phenomenon peculiar to *nonlinear* dynamic systems and is closely related to the loss of stability. It describes the branching of solutions (steady state, oscillatory, or spatially nonuniform solutions) as a control parameter k is varied. When different solution branches intersect or coalesce, they usually change their stability character. The simplest bifurcation is generated by the coalescence of two (saddle-node bifurcation or transcritical bifurcation) or three (pitchfork bifurcation) fixed points. This type of bifurcation is also called a *soft-mode instability* because the corresponding linear mode softens, i.e., the regression of small fluctuations from the steady state slows down critically. In a *Hopf bifurcation* a periodic attractor (stable limit cycle) is generated from a stable fixed point which turns unstable. At the bifurcation, the oscillation amplitude tends to zero, while the frequency remains nonzero (*hard-mode instability*). If a stable and an unstable limit cycle coalesce and disappear by *condensation of paths*, both the amplitude and the frequency are nonzero at the bifurcation. Finally, limit cycles may also bifurcate from a *homoclinic orbit* (i.e., a trajectory beginning and ending at the same saddle fixed point) with zero frequency and nonzero amplitude. Further bifurcations may lead to more complicated, quasiperiodic or chaotic oscillations following some universal scenarios (Schuster 1987).

1.3. Nonequilibrium transport

In order to specify the dynamic system (1.1), a physical model of semiconductor change transport and nonequilibrium g–r kinetics must be applied. This will in general involve the Maxwell equations coupled with an appropriate set of nonlinear transport equations. A systematic procedure is to derive a set of hydrodynamic balance equations for slow macroscopic observables like the carrier density $n(\boldsymbol{r}, t)$, the mean momentum per carrier $\boldsymbol{p}(\boldsymbol{r}, t)$, and the mean energy per carrier $E(\boldsymbol{r}, t)$. This can be done, e.g., by a moment expansion of the Boltzmann transport equation for the semiclassical carrier distribution function $f(\boldsymbol{r}, \boldsymbol{k}, t)$, where $\hbar\boldsymbol{k}$ is the crystal momentum:

$$\frac{\partial f}{\partial t} + \boldsymbol{v}_{\mathrm{g}} \cdot \nabla_{\mathrm{r}} f + q\hbar^{-1}\mathscr{E} \cdot \nabla_{\mathrm{k}} f = \left(\frac{\partial f}{\partial t}\right)_{\mathrm{coll}}. \tag{1.3}$$

Here $\boldsymbol{v}_{\mathrm{g}}$ is the group velocity, $\mathscr{E}$ is the electric field, and the subscript coll denotes the collision integral, which includes all dissipative processes like phonon and impurity scattering, carrier–carrier scattering and impact ionization. The carrier charge is $q = \pm e$ for holes or electrons, respectively.

With the definition of an ensemble average

$$\langle A \rangle \equiv n^{-1} \int A f(\boldsymbol{r}, \boldsymbol{k}, t) z \, \mathrm{d}^3 k \tag{1.4}$$

and $n \equiv \int f(\boldsymbol{r}, \boldsymbol{k}, t) z \, \mathrm{d}^3 k$, $z = 2\,(2\pi)^{-3}$ (density of states in $\boldsymbol{k}$-space) we obtain

$$\boldsymbol{p}(\boldsymbol{r}, t) = \langle \hbar \boldsymbol{k} \rangle, \tag{1.5}$$

$$\boldsymbol{v}(\boldsymbol{r}, t) = \langle \boldsymbol{v}_{\mathrm{g}} \rangle, \tag{1.6}$$

$$E(\boldsymbol{r}, t) = \langle E(\boldsymbol{k}) \rangle, \tag{1.7}$$

where $E(\boldsymbol{k})$ is the band structure. For a nondegenerate, isotropic, parabolic band $E(\boldsymbol{k}) = \hbar^2 k^2/(2m^*)$ with an effective mass m^*,

$$p(\boldsymbol{r}, t) = m^* \boldsymbol{v} \tag{1.8}$$

$$E(\boldsymbol{r}, t) = \tfrac{1}{2} m^* \langle \boldsymbol{v}_{\mathrm{g}}^2 \rangle = \tfrac{1}{2} m^* \boldsymbol{v}^2 + \tfrac{3}{2} k_{\mathrm{B}} T_{\mathrm{e}}. \tag{1.9}$$

Equation (1.9) defines an electron temperature T_{e}, and k_{B} is the Boltzmann constant. The mean energy is thus composed of a convective and a thermal contribution. Nonequilibrium transport will in general involve thermal and chemical nonequilibrium, i.e., T_{e} will be different from the lattice temperature T_{L}, and the distribution of carriers into different bands and localized levels will not be fixed by a common Fermi level (see chapter 6 of this volume), respectively.

Multiplying eq. (1.3) with appropriate powers of $\boldsymbol{k}$ and integrating over the first Brillouin zone, we obtain a closed system of moment equations:

$$\dot{n} + \nabla \cdot (n\boldsymbol{v}) = \varphi_0(n, E), \tag{1.10}$$

$$\dot{\boldsymbol{p}} + (\boldsymbol{v} \cdot \nabla)\boldsymbol{p} + \frac{1}{n}\nabla(n k_{\mathrm{B}} T_{\mathrm{e}}) - q\mathscr{E} = -\frac{\boldsymbol{p}\varphi_0}{n} - \boldsymbol{p}/\tau_{\mathrm{m}}(E), \tag{1.11}$$

$$\dot{E} + (\boldsymbol{v} \cdot \boldsymbol{\nabla})E + \frac{1}{n}\boldsymbol{\nabla}(nk_B T_e \boldsymbol{v}) - \frac{\kappa}{n}\Delta T_e - q\boldsymbol{v} \cdot \boldsymbol{\mathscr{E}} = -\frac{E\varphi_0}{n} - \frac{(E - E_0)}{\tau_e(E)}, \tag{1.12}$$

if the following assumptions (Baccarani et al. 1986) are used:

(i) The carrier temperature is a scalar:

$$\langle (v_{gi} - \langle v_{gi} \rangle)(v_{gj} - \langle v_{gj} \rangle) \rangle = \frac{3}{m^*} k_B T_e \delta_{ij}, \tag{1.13}$$

i.e., the momentum-flux tensor reduces to the scalar electron pressure $nk_B T_e$.

(ii) The energy-flow density is

$$\tfrac{1}{2} m^* n \langle v_g^2 \boldsymbol{v}_g \rangle = n\boldsymbol{v}E + n\boldsymbol{v}k_B T_e + \boldsymbol{j}_Q, \tag{1.14}$$

where the heat flux is approximated phenomenologically by $\boldsymbol{j}_Q = -\kappa \boldsymbol{\nabla} T_e$ with thermal conductivity κ.

(iii) The collision integrals are approximately evaluated (Schöll and Quade 1987, Quade et al. 1992) yielding the generation–recombination rate including impact ionization,

$$\varphi_0(n, E) = \int \left(\frac{\partial f}{\partial t}\right)_{\text{coll}} z \, d^3 k, \tag{1.15}$$

the momentum relaxation rate,

$$-n\frac{\boldsymbol{p}}{\tau_m(E)} = \int \hbar \boldsymbol{k} \left(\frac{\partial f}{\partial t}\right)_{\text{coll}} z \, d^3 k, \tag{1.16}$$

and the energy relaxation rate,

$$-n\frac{E - E_0}{\tau_e(E)} = \int \frac{\hbar^2 k^2}{2m^*} \left(\frac{\partial f}{\partial t}\right)_{\text{coll}} z \, d^3 k, \tag{1.17}$$

with mean energy-dependent momentum and energy relaxation times τ_m and τ_e, respectively, and $E_0 \equiv \frac{3}{2} k_B T_L$. This is equivalent to a self-consistent parametrization of the nonequilibrium distribution function by n, $\boldsymbol{p}$, and E. Equations (1.10)–(1.12) with (1.8) and (1.9) constitute a closed set of nonlinear partial differential equations for n, $\boldsymbol{p}$, and E. Generalizations to nonparabolic bands have also been considered (Thoma et al. 1991). All information about the microscopic physical processes is contained in the functions $\varphi_0(n, E)$, $\tau_m(E)$, $\tau_e(E)$, and κ. In particular, impact ionization contributes to the g–r rate *and* to the energy relaxation rate. The latter may be approximated by

$$n\frac{E - E_0}{\tau_e} = n\frac{E - E_0}{\tau'_e} + E_{th}\varphi_{ii}(n, E) \tag{1.18}$$

where τ'_e includes phonon scattering, and φ_{ii} and E_{th} are the impact ionization rate and threshold energy, respectively. The impact ionization rate of impurities, e.g., is given by

$$\varphi_{ii}(n, E) = X_1(E) n n_t \tag{1.19}$$

where n_t is the density of trapped carriers, and $X_1(E)$ is the impact-ionization coefficient which depends on the mean energy in a thresholdlike way. The impact-ionization and capture coefficients involving shallow impurity levels can be calculated from analytical approximations of the distribution function or by Monte Carlo techniques (Reggiani and Mitin 1989).

The transport equations (1.10)–(1.12) are coupled to the local electric field $\mathscr{E}$ by the Maxwell equations

$$\varepsilon_0 \varepsilon_s \nabla \cdot \mathscr{E} = \rho, \tag{1.20}$$

with the local charge density $\rho = e(N_D - N_A) + q(n + n_t)$ (N_D and N_A are donor and acceptor densities, n_t is the concentration of trapped electrons, ε_0 and ε_s are absolute and relative permittivity, respectively), and

$$\nabla \times \boldsymbol{H} = \varepsilon_0 \varepsilon_s \dot{\mathscr{E}} + \boldsymbol{j} \equiv \boldsymbol{j}_0, \tag{1.21}$$

with conduction current density $\boldsymbol{j} = qn\boldsymbol{v}$ and magnetic field $\boldsymbol{H}$. We shall confine attention to the important case that $\boldsymbol{H}$ is not a dynamic variable, but can be regarded as an external control parameter. From eq. (1.21) it follows by applying ∇ that the total external current density $\boldsymbol{j}_0$ satisfies $\nabla \cdot \boldsymbol{j}_0 = 0$, and may be regarded as an additional control parameter, composed of displacement and conduction currents:

$$\varepsilon_0 \varepsilon_s \dot{\mathscr{E}} = \boldsymbol{j}_0 - \boldsymbol{j}. \tag{1.22}$$

The time-scale on which the variables $n, \boldsymbol{p}, E, \mathscr{E}$ occurring in eqs. (1.10)–(1.12) and (1.22) change is determined by the g–r lifetime, the momentum relaxation time, the energy relaxation time, and the Maxwell dielectric relaxation time, respectively. Since macroscopic dynamics is dominated by the slow variables, it depends on the relations between these time-scales which of the above variables may be eliminated adiabatically from the dynamics (van Kampen 1985).

Momentum relaxation often occurs faster than other processes, so that $\boldsymbol{p}$ can be eliminated adiabatically from eq. (1.11) by setting $\mathrm{d}\boldsymbol{p}/\mathrm{d}t \equiv \dot{\boldsymbol{p}} + (\boldsymbol{v} \cdot \nabla)\boldsymbol{p} = 0$. Introducing the mobility

$$\mu(E) \equiv \frac{e}{m^*} \tau'_m(E), \qquad \frac{1}{\tau'_m} \equiv \frac{1}{\tau_m} + \frac{\varphi_0}{n}, \tag{1.23}$$

the drift-diffusion equation

$$\boldsymbol{j} = qn\boldsymbol{v} = en\mu(E)\mathscr{E} - qD(E)\nabla n - \frac{q}{e}\mu(E) n k_B \nabla T_e \tag{1.24}$$

is recovered if the diffusion coefficient D is defined via the Einstein relation $eD = \mu k_B T_e$.

In conclusion, the nonlinear dynamic system which describes semiconductors far from equilibrium is given by the g–r rate equation (1.10), the energy balance equation (1.12), and the current density equation (1.22) with (1.24). Additional g–r rate equations may have to be taken into account for the charge carriers in the valence band and in localized impurity levels. Simplifications may arise if additional fast variables can be eliminated, or if spatial gradients may be neglected.

2. Generation–recombination induced nonequilibrium phase transitions

In a nonequilibrium phase transition the dynamic variables (*order parameters*) change in a singular way at certain threshold values of the control parameters. Following Ehrenfest's classification, a phase transition is of nth order if the $(n-1)$th derivative of the order parameter with respect to the control parameter changes discontinuously, while all lower derivatives are continuous. Contrary to the equilibrium case, where the state variables and control parameters are well defined as extensive observables and adjoint intensive parameters, the choice of the variables and control parameters in nonequilibrium systems is not a priori uniquely determined. In the simplest case of spatially homogeneous, time independent, nonlinear charge transport the current density

$$j_0 = en\mu\mathscr{E} \tag{2.1}$$

is a possible, but not compulsory choice of the order parameter, and the electric field $\mathscr{E}$ may be regarded as a control parameter. The steady state current-density-field characteristic (2.1) corresponds to the homogeneous fixed point of the dynamic system (1.10)–(1.12), (1.22) where n and

$$E = E_0 + e\tau_e\mu\mathscr{E}^2 \tag{2.2}$$

are determined by the g–r kinetics and the energy balance, respectively.

The phase transition analogies include the similarity of the static current–field characteristic with the p–V diagram of a van der Waals gas in case of a first-order transition, the spontaneous symmetry breaking of the order parameter in case of a second-order transition, critical and tricritical points, critical slowing down of relaxation times, the build-up of long-range spatial correlations, spatial coexistence of phases described by "equal-areas rules" similar to Maxwell's rule, the growth of long-wavelength fluctuations, spinodal decomposition, and nucleation of current filaments similar to droplet nucleation phenomena (Schöll 1987). Some of these predictions have been verified experimentally for semiconductors in the regime of low-temperature impurity breakdown (Weispfenning et al. 1985, Röhricht et al. 1988).

The nonlinearity of the $j_0(\mathscr{E})$ characteristic (2.1) can be due to a field dependence of n (g–r instability), μ (drift instability) or τ_e (overheating instability). In the first case, on which we shall focus in this section, the steady carrier concentration n represents an appropriate order parameter of the phase transition. For nondegenerate statistics the g–r rates are polynomials of the involved carrier concentrations and obey the "mass-action law" of chemical reaction kinetics (cf. chapter 6). The first models for g–r induced phase transitions (Landsberg and Pimpale 1976, Schöll and Landsberg 1979) were based upon this similarity of semiconductor rate equations to Schlögl's (1972) chemical reaction models for nonequilibrium phase transitions, and upon their analogy with the van der Waals equation for equilibrium phase transitions. A necessary ingredient for an instability is impact ionization of electrons or holes across the bandgap or from localized levels, since this is the only "autocatalytic" (positive feedback) g–r process, i.e., it increases the carrier density at a rate proportional to the concentration of carriers which are already in the respective band. In the absence of external electric fields and optical irradiation the g–r coefficients of each process and its reverse are reciprocally linked by the principle of detailed balance. As a result, at or close to thermodynamic equilibrium, there exists a unique stable steady state, and instabilities cannot occur even though the g–r rates are strongly nonlinear functions of the carrier densities (Schöll 1987, § 1.2.2). It is a subtle point that for a g–r instability it is not sufficient that the carrier *densities* deviate strongly from their thermodynamic equilibrium values, but rather the *g–r coefficients* must meet additional conditions driving them far from their equilibrium values. This applies, in particular, to the impact ionization coefficient which is a threshold-like function of the applied electric field. Here we shall confine attention to unipolar conduction with one type of carriers (say electrons) and neglect the minority carriers (holes). We assume that the donors are partially compensated by fully ionized acceptors of concentration $N_A < N_D$ such that $N_D^* \equiv N_D - N_A$ is the effective donor concentration. Extensions to ambipolar conduction have been discussed elsewhere (Schöll 1987). The g–r rate of the conduction electrons is then given by

$$\varphi_0 = -T_1^S n p_t - T_1 n^2 p_t + X_1^S n_t + X_1 n n_t, \tag{2.3}$$

where n_t, $p_t = N_D - n_t$ denote the densities of neutral and ionized donors, respectively, and the g–r coefficients T_1^S, T_1, X_1^S, X_1 denote capture, Auger recombination, ionization, and impact ionization of donors, respectively (for the notation see chapter 6). In a spatially homogeneous state the local neutrality condition $N_D^* - n - n_t = 0$ holds by eq. (1.20), and n_t, p_t can be eliminated from the g–r rate φ_0. The steady-state concentration n is then given by the polynomial rate equation

$$\begin{aligned}\varphi_0(n, \mathscr{E}) &= X_1^S N_D^* + n(X_1 N_D^* - T_1^S N_A - X_1^S) - n^2(X_1 + T_1^S + T_1 N_A) - T_1 n^3 \\ &= 0,\end{aligned} \tag{2.4}$$

which depends upon $\mathscr{E}$ through the g–r coefficients, in particular through the impact ionization coefficient $X_1(\mathscr{E})$ which is a threshold-like function of $\mathscr{E}$. For $X_1^S > 0$ there is always a unique steady-state solution $n > 0$. For $X_1^S = 0$, i.e., for negligible thermal and optical ionization, eq. (2.4) admits the solution $n = 0$, and an additional solution $n > 0$ if

$$X_1 > X_{1c} \equiv T_1^S N_A / N_D^*. \tag{2.5}$$

The steady state $n^* = 0$ is asymptotically stable only for $X_1 \leqslant X_{1c}$, and $n^* > 0$ is asymptotically stable for $X_1 > X_{1c}$. This can easily be seen by linearizing eq. (2.4) around the steady state for small fluctuations $\delta n \equiv n - n^*$ which gives $\delta \dot{n} = \lambda\, \delta n$ with an eigenvalue $\lambda < 0$ in case of asymptotic stability. (For a complete analysis of all linear modes of the dynamic system (1.10), (1.21) see Schöll (1987).) Regarding $\mathscr{E}$ as a control parameter which is related monotonically with X_1, one finds that $n^*(\mathscr{E})$ is continuous and monotonic, but has a discontinuous derivative $\mathrm{d}n^*/\mathrm{d}\,\mathscr{E}$ at the critical field $\mathscr{E}_c$ defined by $X_1(\mathscr{E}) = X_{1c}$. This is a second-order nonequilibrium phase transition from a nonconducting ($n^* = 0$) to a conducting state ($n^* > 0$), corresponding to a transcritical bifurcation. Physically, this describes low-temperature impurity breakdown at a critical threshold field, occurring in a monotonic (if $\mu(\mathscr{E})$ is monotonic!) $j(\mathscr{E})$ characteristic, i.e., *without* negative differential conductivity (NDC).

In general, electrons bound at shallow donors possess a whole spectrum of energy levels, similar to the energy spectrum of a hydrogen atom, and the g–r process between the ground state and these excited states must also be taken into account. The simplest class of models involves the ground state and the first excited state only. This is appropriate for situations where higher excited states are thermally ionized but the first excited states is not, or where a group of excited levels can be lumped into a single "effective" level. Omitting Auger processes, which become important at large carrier densities only, we obtain the g–r rate of the conduction electrons

$$\varphi_0 = -T_1^S np_t + X_1^S n_2 + X_1 nn_1 + X_1^* nn_2 \tag{2.6}$$

and of the electrons in the donor ground state

$$\varphi_1 = T^* n_2 - X^* n_1 - X_1 nn_1, \tag{2.7}$$

where the g–r coefficients are defined in fig. 1, and n_1, n_2 denote the carrier densities in the donor ground and first excited state, respectively. The homogeneous steady states are now given by $\varphi_0 = \varphi_1 = 0$ with the local neutrality condition

$$n_2 = N_D^* - n - n_1 \tag{2.8}$$

and $p_t = N_A + n$.

For a range of applied fields $\mathscr{E}$ there are three positive solutions, two of which are stable, leading to an S-shaped $n(\mathscr{E})$ relation and a $j_0(\mathscr{E})$ characteristic with a region of negative differential conductivity (SNDC). This model is representative for a large class of g–r mechanisms involving M ($M = 2, 3, \ldots$) localized levels which, in the steady state, give a nonmonotonic ρ versus n relation due to the depletion of the impurity ground state by impact ionization (Schöll 1987, § 2.1.3). With a single localized level and nondegenerate statistics (mass-action g–r kinetics) SNDC cannot occur. Threshold switching from the high to the low resistivity branch of the $j_0(\mathscr{E})$ characteristic occurs when $\mathscr{E}$ exceeds a threshold field $\mathscr{E}_{th}$. With decreasing field $\mathscr{E}$ the system switches back to the high-resistivity ("off") state at a lower holding field

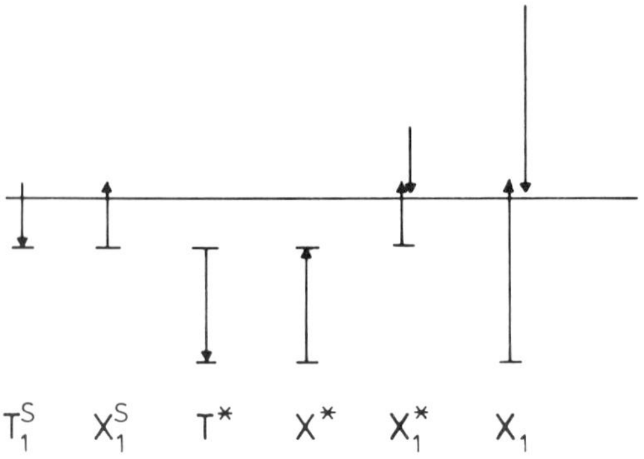

Fig. 1. Generation–recombination processes involving the conduction band, the donor ground state, and its first excited state.

$\mathscr{E}_h < \mathscr{E}_{th}$. Thus the typical hysteresis behaviour observed in threshold switching is produced. It presents a discontinuous, i.e., first-order nonequilibrium phase transition (Schöll 1982a). Second-order phase transitions are also possible, if the generation coefficients X^* or X_1^S vanishes.

Moreover, there exists a *critical point*: The $j_0(\mathscr{E})$ characteristic is S-shaped only below a critical value of the (thermal or optical) generation coefficients X^* and X_1^S which are thus the analog of the temperature in the van der Waals transition, whereas j_0 and $\mathscr{E}$ correspond to volume and pressure, respectively (fig. 2). Mean-field critical and even tricritical scaling exponents are obtained at the critical point (Schöll 1987). It can be shown by linearizing the dynamic equations around the steady state that small fluctuations decay proportionally to $\exp(-|\lambda|t)$ where $\lambda \to 0$ at the critical point. This is associated with critical slowing down of the relaxation time $1/|\lambda|$.

3. Oscillatory instabilities and deterministic chaos

3.1. Survey of physical mechanisms

A dynamic system of the type (1.10)–(1.12) and (1.22) can also give rise to the bifurcation of time-dependent oscillatory solutions (periodic or chaotic attractors). Physically, these will show up as self-sustained oscillations of the current or the voltage. Such spontaneous oscillations under DC bias conditions are ubiquitous in a variety of semiconductors. They may result from the coupling of a negative differential conductivity element with reactive circuit components like capacitors and inductors, or from transit-time oscillations due to the motion of high-field domains

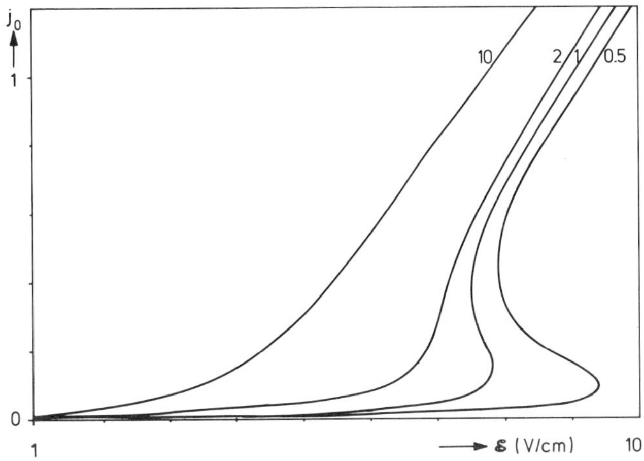

Fig. 2. Current density j_0 versus electric field $\mathscr{E}$, calculated for the g–r model of fig. 1 with $T_1^S = 10^{-6}\,\text{cm}^3\,\text{s}^{-1}$, $T^* = X^* = 10^6\,\text{s}^{-1}$, $X_1^* = 10^{-6}\exp(-1.5\,\text{V cm}^{-1}/\mathscr{E})\,\text{cm}^3\,\text{s}^{-1}$, $X_1 = 5\times10^{-8}\exp(-6\,\text{V cm}^{-1}/\mathscr{E})\,\text{cm}^3\,\text{s}^{-1}$, $N_D^* = 10^{15}\,\text{cm}^{-3}$, $N_A/N_D^* = 0.5$, $\mu = \text{const}$. The parameter of the different $j_0(\mathscr{E})$ curves is X_1^S in μs^{-1}. Similar plots are obtained for fixed X_1^S and variable X^* (Schöll 1982a).

as, e.g., in the Gunn effect, or from plasma instabilities like helicon waves in parallel electric and magnetic fields, or from intrinsic oscillatory instabilities, e.g., due to nonlinear g–r processes. Chaotic electrical oscillations were first reported by different groups in high-purity n-GaAs (Aoki et al. 1981, Maracas et al. 1985, Brandl et al. 1987, Knap et al. 1988), p-Ge (Teitsworth et al. 1983, Held et al. 1984, Peinke et al. 1985, Bumelienė et al. 1985, Spinnewyn et al. 1989), n-InSb (Seiler et al. 1985, Fuji et al. 1989), n-Si (Yamada et al. 1988), in n-GaInAs (Godlewski et al. 1992), and in heterostructures like GaAs MESFETS (Yano et al. 1992), or GaAs/AlGaAs layers (periodic oscillations: Belyantsev et al. 1986, Kolodzey et al. 1988, Vickers et al. 1989, Balkan and Ridley 1989, Hendriks et al. 1991) under a variety of experimental conditions, ranging from 4 K to room temperature and including weak infrared or visible illumination as well as complete shielding against external light, and in some cases parallel or transverse magnetic fields. In the majority of these experiments the oscillations occurred in the regime of impurity impact ionization, indicating that g–r processes between localized states and extended band states played a crucial role. The experiments can be divided into two classes: (i) driven chaos, which is induced by a periodically modulated bias voltage, (ii) self-generated chaos, which is observed under DC conditions and is widely independent of external circuit conditions.

Intrinsic, self-generated oscillatory instabilities may be due to different physical mechanisms. A number of models have been proposed and analysed. Within the framework of semiconductor transport theory, as outlined in § 1.3, these may be classified according to the slow dynamic variables which they involve.

(i) *Carrier density and electric field.* This refers to the case of a high-purity "relaxation semiconductor", i.e., the dielectric relaxation of the field $\mathscr{E}$ occurs on a slow time-scale. The mean energy E is a fast variable and can be expressed as a function of the field by the energy balance (2.2). In the regime of impurity impact ionization, dielectric relaxation oscillations displaying a period-doubling route to chaos have been found in spatially homogeneous models, either driven by an AC field (Teitsworth and Westervelt 1984, Aoki et al. 1986) or self-generated (Schöll 1985, 1986b, Pyragas et al. 1987, Hüpper and Schöll 1991). Spatio-temporal dielectric relaxation oscillations in terms of breathing current filaments have also been demonstrated (Schöll and Drasdo 1990).

(ii) *Carrier density, mean energy, and electric field.* Self-generated oscillations involving these dynamic variables may be induced by a combination of dielectric relaxation, impact ionization, and optical-phonon emission in the post-breakdown regime of p-Ge (Hüpper et al. 1989) or by real-space transfer of electrons, energy transfer, and space-charge dynamics in modulation-doped GaAs/AlGaAs heterostructures (Schöll and Aoki 1991). Real-space transfer was suggested as a mechanism for NNDC by Gribnikov (1973) and Hess et al. (1979). An oscillatory instability has also been found for *vertical* transport in a heterostructure-hot-electron-diode model using the mean energy and the fields in the two heterolayers as dynamic variables (Wacker and Schöll 1991). The spatially homogeneous model of Hüpper et al. (1989) uses mean-energy-dependent momentum and energy relaxation times $\tau_m(E)$, $\tau_e(E)$

obtained from a Monte Carlo simulation rather than phenomenological functions, and is thus founded at a more microscopic level. The dramatic decrease of the mobility and the energy relaxation time when the threshold energy for emission of optical phonons is exceeded, leads to strong nonlinearities and to NNDC in the post-breakdown regime. In the preceding positive differential conductivity regime, a subcritical Hopf bifurcation generates an unstable limit cycle which annihilates with a stable limit cycle by condensation of paths. The stable limit cycle undergoes a global bifurcation from a homoclinic orbit with critical slowing-down of the oscillation frequency. Hysteresis between oscillatory and stationary states is also found. Both phenomena, hysteresis and global bifurcations, have been observed in p-Ge at low-temperature impurity breakdown by Peinke et al. (1989).

(iii) *Carrier density and mean energy.* If the electric field relaxes fast, eq. (1.22) with eq. (1.24) reduces to the static current density versus field relation. Self-generated energy relaxation oscillations described by eqs. (1.10) and (1.12) may arise as a result of nonlinear energy loss through impact ionization (Schöll 1987, 1989a). If two or more oscillating cells are coupled by energy exchange (via phonons), this may serve as a simple model for the excitation and interaction of different spatio-temporal degrees of freedom. In a two-cell model, resonance transitions between linearly correlated oscillations (Schöll et al. 1989), mode-locking structures obeying the Farey tree ordering and quasi-periodicity (Naber and Schöll 1990a) have been found, in accordance with experiments (Rau et al. 1987). The bit-number cumulants of the invariant density of the underlying dynamic system have been shown to represent a characteristic measure for the build-up of internal correlations between different localized oscillation centers (Naber and Schöll 1990b). In particular, the bit-number variance may be viewed as a generalized specific heat in the framework of the thermodynamic formalism of dynamic systems (Schlögl and Schöll 1988).

Spontaneous energy relaxation oscillations can also be generated in an ambipolar carrier plasma by optical generation of electron–hole pairs, and band–band single-electron and Auger recombination. Models involving a common carrier temperature (Degtyarenko et al. 1974) or separate electron and hole temperatures (Bonch-Bruevich and Le Vu Ky 1984) have been studied, and the possibility of periodic or chaotic oscillations, respectively, has been pointed out.

The different oscillation mechanisms may be discussed from a unified viewpoint (Schöll 1989b) by analyzing the bifurcations of the underlying dynamic system. To this purpose we restrict ourselves in this section to the simplest, spatially homogeneous form, although extensions to include current filaments are possible. More complex spatio-temporal structures will be discussed in § 4. If the semiconductor is operated in a resistive external circuit described by the Kirchoff law $U_0 = IR_L + V$, where U_0 is the applied bias voltage, R_L is the load resistance, I and V are the sample current and voltage, respectively, eq. (1.22) with eq. (1.24) can be cast into the form

$$\dot{\mathscr{E}} = \{J_0 - [\sigma_L + \sigma(\boldsymbol{x}, \mathscr{E})]\mathscr{E}\}/\varepsilon_0 \varepsilon_s, \tag{3.1}$$

where $\sigma \equiv ne\mu$ is the conductivity which depends on the field $\mathscr{E}$ and additional transport variables $\boldsymbol{x} \equiv (x_1, x_2, \ldots, x_N)$, such as the density n and the mean energy E, and

$$J_0 \equiv U_0/R_L A \quad \text{and} \quad \sigma_L \equiv L/R_L A$$

are control parameters, A is the cross-section of the current flow, and L is the contact distance.

The static differential conductivity is given by

$$\sigma_{\text{diff}} = \frac{\mathrm{d}}{\mathrm{d}\mathscr{E}}[\sigma(\boldsymbol{x}(\mathscr{E}^*), \mathscr{E}^*)\mathscr{E}^*], \tag{3.2}$$

where * denotes the steady-state values obtained from $\dot{\boldsymbol{x}} = 0$, $\dot{\mathscr{E}} = 0$ in eqs. (1.10)–(1.12) and (3.1). The stability of the steady state with respect to small fluctuations $(\delta\boldsymbol{x}, \delta\mathscr{E}) \sim \exp \lambda t$ follows from linearizing (3.1) and the additional transport equations for $\dot{\boldsymbol{x}}$ around $(x^*, \mathscr{E}^*)$, which yields a secular equation of degree $N+1$ for the eigenvalues $\lambda_1, \ldots, \lambda_{N+1}$.

The simplest oscillatory instability is the Hopf bifurcation of a limit cycle. It occurs when a pair of complex conjugate eigenvalues crosses the imaginary axis. Here we shall not consider the general case (Schöll 1989b), but restrict ourselves to $N = 1$. The condition of a Hopf bifurcation can then be stated as

$$\lambda_1 + \lambda_2 = 0, \quad \lambda_1 \lambda_2 > 0. \tag{3.3}$$

One can prove the general relation

$$\lambda_1 \lambda_2 = -\tilde{\lambda}\lambda_{\mathrm{M}}^{\text{diff}}, \tag{3.4}$$

where $\tilde{\lambda} \equiv (\partial\varphi_0/\partial n)^*$ is an effective inverse g–r lifetime, and

$$\lambda_{\mathrm{M}}^{\text{diff}} \equiv (\sigma_L + \sigma_{\text{diff}})/\varepsilon_0 \varepsilon_s \tag{3.5}$$

is an effective inverse dielectric relaxation time.

Hence, eq. (3.3) becomes

$$\begin{aligned} &\tilde{\lambda} = \tilde{\nu}, \\ &\tilde{\lambda}(\sigma_L + \sigma_{\text{diff}}) < 0, \end{aligned} \tag{3.6}$$

where

$$\tilde{\nu} \equiv \left(\sigma_L + \frac{\partial}{\partial\mathscr{E}}[\sigma(\boldsymbol{x}, \mathscr{E})\mathscr{E}]\right)\Big/ \varepsilon_0 \varepsilon_s.$$

For $\sigma_L = 0$, $\tilde{\nu}$ is proportional to the differential mobility $\mathrm{d}v/\mathrm{d}\mathscr{E}$. Equation (3.6) can be satisfied by either

(i) $\sigma_L + \sigma_{\text{diff}} < 0 \quad \text{and} \quad \tilde{\lambda} = \tilde{\nu} > 0$

or

(ii) $\sigma_L + \sigma_{\text{diff}} > 0 \quad \text{and} \quad \tilde{\lambda} = \tilde{\nu} < 0.$

This demonstrates that oscillatory instabilities are possible for either *negative* differential conductivity (and *positive* differential mobility, in case of $\sigma_L = 0$), or *positive* differential conductivity (and *negative* differential mobility, for $\sigma_L = 0$). With $N > 1$, more complex situations are possible, but $\sigma_{diff} < 0$ and $\sigma_{diff} > 0$ may still occur (Schöll 1989b). Note that a Hopf bifurcation is not the only way to create self-generated oscillations; global bifurcations are another possibility (Thompson and Stewart 1986, Hüpper et al. 1989, Döttling and Schöll 1992), which has also been found experimentally (Peinke et al. 1989).

3.2. Routes to chaos

The onset of chaos is usually preceded by other bifurcations, like those discussed above. For example, with increasing value of the control parameter, the dynamic system may first exhibit a Hopf bifurcation of a limit cycle, and then, at still higher values of the control parameter, a transition to chaos. These transitions are in general universal scenarios which occur in very different physical systems (Schuster 1987). The most important scenarios are the period-doubling (or Feigenbaum) route, the intermittency (or Pomeau–Manneville) route, and the quasi-periodic (or Ruelle–Takens–Newhouse) route. A number of semiconductor experiments obeying these scenarios have been reviewed by Schöll (1987). It is remarkable that all three scenarios can be observed in semiconductors, in the regime of low-temperature impurity breakdown, by only slightly changing the bias current, the magnetic field, or the lattice temperature (e.g., Aoki 1992, Peinke et al. 1992).

A simple model for impact ionization induced self-generated voltage oscillations displaying a period-doubling route to chaos under a time-independent bias J_0 was first proposed on the basis of the two-level g–r mechanism shown in fig. 1 (Schöll 1985, 1986b). It involves the carrier densities n, n_1 and the electric field $\mathscr{E}$ as slow dynamic variables, and is given by

$$\begin{aligned} \dot{n} &= \varphi_0(n, n_1, \mathscr{E}), \\ \dot{n}_1 &= \varphi_1(n, n_1, \mathscr{E}), \\ \dot{\mathscr{E}} &= (J_0 - ne\mu\mathscr{E})/\varepsilon_0\varepsilon_s, \end{aligned} \tag{3.7}$$

with the g–r rates (eqs. (2.6), (2.7)) where the local neutrality condition (2.8) is used to eliminate n_2. The phase space of the model consists of $(\mathscr{E}, n, n_1)$, and thus has the minimum dimension which is necessary to allow for chaos under time-independent bias. An explicit condition for a Hopf bifurcation of a limit cycle can be derived by linearizing the system (3.7), in analogy with (3.3)–(3.6), but for $N = 2$. A necessary, but not sufficient, condition is that the dielectric relaxation time $1/\lambda_M^{diff}$ is larger than the impact ionization time $1/\tilde{\lambda}_1$, which singles out relaxation semiconductors, and that NDC occurs.

For typical numerical parameters, the system (3.7) shows a sequence of period-doubling bifurcation leading to chaos (fig. 3). Here $\mathscr{E}_0$, defined by the NDC branch of the S-shaped static current-density-versus-field characteristic $J_0 = en(\mathscr{E}_0)\mu(\mathscr{E}_0)\mathscr{E}_0$ has been chosen as a control parameter. As $\mathscr{E}_0$ is increased, a limit cycle of period T is generated by a Hopf bifurcation. Upon further increase, oscillations of period

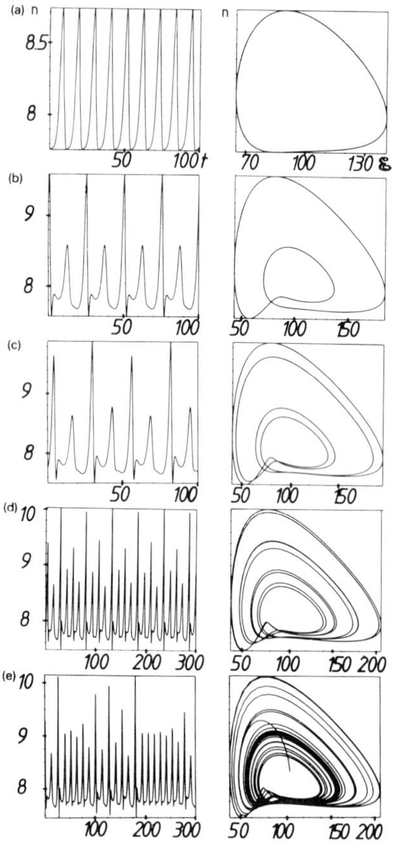

Fig. 3. Period-doubling route to chaos, exhibited by the g–r model of fig. 1, coupled with dielectric relaxation. The carrier density n in units of 10^{-3} N_D^*, versus time t in arbitrary units (left column), and phase portraits of n versus $\mathscr{E}$ in arbitrary units (right column) is shown for increasing values of the control parameter $\mathscr{E}_0$. (a) 102, (b) 105, (c) 105.3, (d) 105.42, (e) 105.5. The numerical parameters are as in fig. 2 except for $X_1^S = X^* = 10^4\ s^{-1}$, $N_A/N_D^* = 0.3$, $\mu(\mathscr{E}) = (\arctan 0.3\mathscr{E})/0.3\mathscr{E}$ (after Schöll 1986b).

2T (b), 4T (c), 8T (d), and chaotic oscillations (e) are successively displayed. Figure 3e shows the chaotic attractor beyond the accumulation point of the period-doublings. It is characterized by aperiodic but bounded spiraltype motion in phase space, which depends sensitively upon the initial conditions. A bifurcation diagram can be obtained by plotting the local maxima $\mathscr{E}_n$ of the field $\mathscr{E}(t)$ versus the control parameter $\mathscr{E}_0$. It displays a period-doubling cascade, followed by chaotic bands and periodic windows as $\mathscr{E}_0$ is increased. Successive local maxima can be used to construct the *first return map* f: $\mathscr{E}_n \rightarrow \mathscr{E}_{n+1}$. Its graph of $\mathscr{E}_{n+1}$ versus $\mathscr{E}_n$ resembles a parabola and belongs to Feigenbaum's (1978) universality class of one-dimensional non-invertible maps with quadratic extremum. These maps are found in many nonlinear systems including semiconductors, irrespective of their different physical nature and are generic for the period-doubling route to chaos in the sense that they yield the same scaling behaviour of the sequence of bifurcations. The simplest example is the logistic map $f(x) = rx(1-x)$ which becomes chaotic for $r > 3.5699\ldots$ (Schuster 1987).

In the case of eq. (3.7) the physical mechanism of the oscillatory instability may be visualized as follows: injected carriers are trapped in donors (or, in case of p-type material, acceptors), building up a space charge field. When this field reaches a certain threshold, impact ionization of the trapped carriers sets in, increasing the free carrier density, which results in enhanced dielectric relaxation of the field. Subsequently, the field drops below the impact ionization threshold, and the carriers are retrapped which completes the oscillation cycle. Thus an autocatalytic, i.e., explosive step provided by impact ionization, is coupled to a restoring force, i.e., a negative feedback, provided by delayed dielectric relaxation of the field.

In many semiconductor experiments, self-generated oscillations are only observed, or essentially modified, under the simultaneous application of a static magnetic field (e.g., Song et al. 1989, Spinnewyn et al. 1989, Aoki et al. 1990, Peinke et al. 1989, Brandl et al. 1990). A model which combines impact ionization of hot carriers with an extension of the classical Hall effect into the dynamic regime was recently proposed as an explanation of transverse-magnetic-field-controlled oscillations and chaos (Hüpper and Schöll 1991). The basic new idea is to consider both the applied electric drift field $\mathscr{E}_x$ and the induced Hall field $\mathscr{E}_z$ as dynamic variables whose time dependence is governed by dielectric relaxation, while the magnetic induction B is considered as an external control parameter. Further dynamic degrees of freedom may be provided by the carrier densities, which are coupled to the electric fields by strongly nonlinear interactions as a result of impact ionization of shallow impurities, or by other transport variables.

In the simplest model, for moderate magnetic fields, spatial imhomogeneities are neglected. Choosing the coordinate sytem such that the drift current j is is the x-direction, and B is in the y-direction, the following dynamic field equations are obtained:

$$\begin{aligned} \varepsilon_0 \varepsilon_s \dot{\mathscr{E}}_x &= J_0 - en\mu_B \mathscr{E}_x + qn\mu_B \mu B \mathscr{E}_z, \\ \varepsilon_0 \varepsilon_s \dot{\mathscr{E}}_z &= -en\mu_B \mathscr{E}_z - qn\mu_B B \mathscr{E}_x, \end{aligned} \tag{3.8}$$

where $\mu_B \equiv \mu/(1+\mu^2 B^2)$ incorporates the magnetoresistance. The steady state of eq. (3.8) (indicated by *) reproduces the classical static Hall effect $\mathscr{E}_z^* = -\mu B \mathscr{E}_x^* q/e$, $J_0 = en^* \mu \mathscr{E}_x^*$. The dynamics of the free and bound carriers is governed by the respective nonlinear g–r rates φ_0, φ_i, as outlined in § 2. Analytical conditions for the onset of oscillatory instabilities via a Hopf bifurcation can again be derived by a linear stability analysis (Hüpper and Schöll 1991). Self-generated limit-cycle oscillations and period-doubling routes to chaos occur in regimes of either positive or negative differential conductivity (Schöll et al. 1992, Hüpper et al. 1992). (B, J_0) or $(B, \mathscr{E}_x^*)$ may be regarded as a set of control parameters. Figure 4a shows an example of a regime of oscillatory instability for a single-level g–r model using the rate equation

$$\dot{n} = \varphi_0(n, \mathscr{E}) \tag{3.9}$$

with φ_0 as given by eq. (2.3) with $T_1 = 0$, and $\mathscr{E} \equiv (\mathscr{E}_x^2 + \mathscr{E}_z^2)^{1/2}$. By following different paths in the $(B, \mathscr{E}_x^*)$ plane, complex dynamic behaviour is found. At fixed current

($\triangleq \mathscr{E}_x^*$) and *decreasing* B, or at fixed B and *increasing* $\mathscr{E}_x^*$, period-doubling scenarios occur.

Further, the model shows typical Pomeau–Manneville scenarios to chaos via *intermittency* (Schöll et al. 1992). This is illustrated by a sequence of voltage oscillations $\mathscr{E}_x(t)$ for increasing values of B in figs. 4b–d. Intermittent behaviour is characterized by long regular (laminar) phases interrupted by short irregular bursts. The number of chaotic bursts increases with increasing B, until finally the signal becomes fully chaotic (fig. 4d). With *decreasing* B the mean laminar length $\langle l \rangle$ increases, and diverges at B_c according to a critical scaling law $\sim (B - B_c)^{-1/2}$ which is in agreement with universal predictions of type-I intermittency. This behaviour has been verified experimentally in p-Ge in the regime of low-temperature impurity breakdown (Richter et al. 1991). In general, other scaling laws of the mean laminar length $\langle l \rangle$, and of the probability distribution $P(l)$ are associated with type-II and type-III intermittency (Schuster 1987).

In the quasi-periodic route to chaos the Hopf bifurcation of a limit cycle (angular frequency ω_1) is followed by a second Hopf bifurcation which introduces a second, incommensurate frequency ω_2. It is convenient to parametrize the phase flow by the two angular coordinates $\varphi_1 = \omega_1 t$ and $\varphi_2 = \omega_2 t$. The doubly periodic motion can thus be visualized as the motion on a two-dimensional torus. For incommensurate ω_1 and ω_2 the motion is quasi-periodic and fills up the whole torus. After a subsequent Hopf bifurcation (frequency ω_3) the three-dimensional "torus" becomes unstable in favour of motion on a strange attractor. The quasi-periodic transition also exhibits universal properties, which can be investigated by studying simple one-dimensional

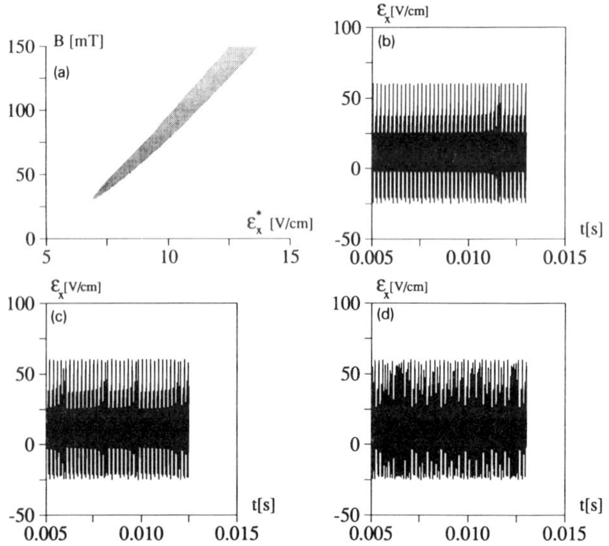

Fig. 4. (a) Regime of oscillatory instability (shaded) in the (B, $\mathscr{E}_x^*$) control parameter plane calculated for p-Ge at 4 K. (b)–(d) Calculated time series $\mathscr{E}_x(t)$ showing type-I intermittency for $\mathscr{E}_x^* = 12.66$ V/cm with (b) $B = 130.0467$ mT, (c) $B = 130.05$ mT, (d) $B = 130.09$ mT (after Schöll et al. 1992).

non-invertible maps (e.g., the *circle map*, Schuster 1987). There exist finite regions in the control parameter space (*Arnold tongues*) in which the motion on the two-dimensional torus is mode-locked, i.e., the frequency ratio $\omega_1/\omega_2 = p/q$ is rational. The resulting hierarchy of Arnold tongues can be conveniently represented by a *Farey tree* which orders all rationals $0 \leqslant p/q \leqslant 1$ according to their increasing denominators. Such behaviour can be reproduced by coupled cell models of nonlinear semiconductor oscillators (Naber and Schöll 1990a). The quasi-periodic transition to chaos has been observed, e.g., in n-GaAs in the regime of low temperature impurity breakdown (Brandl et al. 1990). Typically, it involves the excitation of additional spatio-temporal degrees of freedom, and thus is not reproduced by spatially homogeneous models of the kind described.

4. Self-organized spatial structures

4.1. Current filaments and field domains

The onset of chaos can often be understood in terms of simple, spatially homogeneous models as outlined in § 3. However, when such a system is driven deeper into the nonlinear regime, spatial and spatio-temporal degrees of freedom are excited which may eventually lead to fully developed chaotic and turbulent behaviour.

In semiconductors, spatially homogeneous states of negative differential conductivity are in general unstable against spatio-temporal fluctuations which may give rise to self-organized pattern formation. If the $j_0(\mathscr{E})$ characteristic is N-shaped (NNDC), such as in the Gunn effect, layered field inhomogeneities (high-field domains) can be formed. If the $j_0(\mathscr{E})$ characteristic is S-shaped (SNDC), such as in threshold switching, the current flow can become inhomogeneous over its cross section (current filaments). These spatial structures may be static or time-dependent. In the latter case current oscillations can arise due to domains moving in the direction of the current flow, or filaments "breathing" transversally to the current flow, respectively (Schöll 1987).

At this point a word of warning is indicated. First, NDC does not always imply instability of the steady state, and vice versa. For example, SNDC states can be stabilized (and experimentally observed!) by a heavily loaded circuit, and on the other hand, the Hopf bifurcation of a limit cycle oscillation can occur on a $j(\mathscr{E})$ characteristic with positive differential conductivity. Second, there is no one-to-one correspondence between SNDC and filaments, or between NNDC and domains. Finally, it is important to distinguish between the local $j(\mathscr{E})$ characteristic and the global $I(V)$ characteristic which is determined by the total current

$$I = \int_A \boldsymbol{j} \cdot \mathrm{d}\boldsymbol{f}$$

and the total voltage drop

$$V = \int_0^L \mathscr{E}(z)\, \mathrm{d}z.$$

Only for spatially homogeneous states, the $j(\mathscr{E})$ and the $I(V)$ characteristics are identical, up to rescaling.

Moving domains due to the Gunn instability (Shaw et al. 1979, 1992) or a recombination instability (Bonch-Bruevich et al. 1975) have been extensively investigated experimentally and theoretically in the 1960s and 1970s. Current filaments have recently attracted much interest in connection with SNDC and low-temperature impurity breakdown in n-GaAs (Schöll 1982b, Aoki and Yamamoto 1983, Mayer et al. 1988, Brandl et al. 1989) and p-Ge (Peinke 1992), and with p-i-n diodes at room temperature (Jäger et al. 1986, Purwins et al. 1987, Symanczyk et al. 1991, Jäger and Symanczyk 1992). They have been directly observed by a number of experimental, spatially resolved techniques ranging from scanning electron microscopy, optical scanning, IR absorption measurements to potential probe techniques.

Domains and filaments can be theoretically described as special solutions of the semiconductor transport equations (1.10)–(1.12) and (1.22), subject to appropriate boundary conditions. A linear mode analysis around the spatially homogeneous steady state for small space- and time-dependent fluctuations of the electromagnetic field and the carrier densities yields conditions for the onset of filamentary or domain-type instabilities. This can be put into the general form of a linear response $\delta \boldsymbol{j} = \sigma(\lambda)\, \delta \boldsymbol{\mathscr{E}}$ between the field fluctuation $\delta \boldsymbol{\mathscr{E}}$ and the current density fluctuation $\delta \boldsymbol{j}$, where the dynamic differential conductivity tensor $\sigma(\lambda)$ contains all eigenmodes λ resulting from drift, diffusion, and g–r processes (Schöll 1987). Unstable modes can be shown to lead to the bifurcation of stationary current filaments or travelling domains.

The fully developed spatial structures must be calculated from the *nonlinear* transport equations. While a full numerical solution of these equations provides the most detailed information about the spatial profiles, it does not give immediate physical insight. Therefore there is need for approximations. In particular, it is desirable to derive simple analytical relations which provide some relevant information about the domains and filaments, like the peak field and the propagation velocity of the domains, or the filament radius as a function of the material parameters and the applied bias, without explicitly solving the differential equations for the profiles. This is the purpose of *equal areas rules* which may be visualized geometrically in analogy with the Maxwell rule in thermodynamics which describe the coexistence of a gas and a liquid phase with a plane boundary (Schöll and Landsberg 1988). Mathematically, equal areas rules are definite first integrals of a dynamic system, connecting control parameters with certain boundary values of the profiles. A simple equal areas rule for travelling Gunn domains associated with NNDC was derived by Butcher (1965). Conditions for plane or cylindrical current filaments associated with SNDC were developed in the form of equal areas rules for both unipolar and ambipolar g–r mechanisms by Schöll (1982b, 1986a). Extensions to time-dependent filaments have also been given (Schöll and Drasdo 1990, Kunz and Schöll 1992, Wacker and Schöll 1992).

4.2. Nucleation and growth of filaments

By way of example, current filaments due to nonlinear g–r processes in unipolar n-type materials will be discussed within the following approximations:

(i) The mean energy and momentum are eliminated adiabatically as fast variables.

(ii) There is cylindrical symmetry, and $\mathscr{E} = \mathscr{E}_{\|} + \mathscr{E}_{\perp}$, where $\|$ and $\perp$ denote components parallel and perpendicular to the applied voltage.

(iii) Longitudinal inhomogeneities are neglected. Assuming planar Ohmic contacts gives $n = n(R, t)$, $\mathscr{E}_{\|} = \mathscr{E}_{\|}(t)$, $\mathscr{E}_{\perp} = \mathscr{E}_{\perp}(R, t)$, where R is the radial coordinate.

(iv) The occupation of the different impurity levels relaxes on a faster time-scale than $\mathscr{E}$ and n, such that the concentration of trapped carriers distributed over the ground state and the excited states of the impurities can be expressed as $n_t(n, \mathscr{E})$ in eq. (1.20) through the steady-state g–r rate equations.

(v) $\| \mathscr{E}_{\|} | \gg | \mathscr{E}_{\perp} \|$.

Equations (1.10), (1.20) and (1.22) then reduce to

$$\dot{n} = D\left(n'' + \frac{n'}{R}\right) + \mu(\mathscr{E}_{\|})[\mathscr{E}_{\perp} n' + n\rho(n, \mathscr{E}_{\|})/\varepsilon_0 \varepsilon_s], \tag{4.1}$$

$$\varepsilon_0 \varepsilon_s \dot{\mathscr{E}}_{\perp} = -en\mu(\mathscr{E}_{\|})\mathscr{E}_{\perp} - eDn', \tag{4.2}$$

$$\varepsilon_0 \varepsilon_s \dot{\mathscr{E}}_{\|} = j_0 - en\mu(\mathscr{E}_{\|})\mathscr{E}_{\|}, \tag{4.3}$$

$$\varepsilon_0 \varepsilon_s \left(\mathscr{E}'_{\perp} + \frac{\mathscr{E}_{\perp}}{R}\right) = \rho(n, \mathscr{E}_{\|}). \tag{4.4}$$

The prime denotes the derivative with respect to the radial coordinate. The spatially homogeneous steady states are given by

$$\rho(n, \mathscr{E}_{\|}) = 0, \qquad \mathscr{E}_{\perp} = 0. \tag{4.5}$$

The essential nonlinearity is contained in $\rho(n, \mathscr{E}_{\|})$ due to the g–r processes implicit in $n_t(n, \mathscr{E}_{\|})$. Impact ionization from at least two impurity levels may induce a nonmonotonic dependence of ρ on n in some range of applied fields $\mathscr{E}_0$, resulting in bistability of the uniform steady states (low conductivity n_1, high conductivity n_3), and in S-shaped negative differential conductivity, as discussed in § 2. The time-independent, spatially nonuniform solutions $n^*(R)$ and $\mathscr{E}^*(R)$ satisfy the equations

$$\mathscr{E}^{*\prime}_{\perp} + \frac{\mathscr{E}^*_{\perp}}{R} = \rho(n^*, \mathscr{E}^*_{\|})/\varepsilon_0 \varepsilon_s \tag{4.4$'$}$$

$$\mu n^* \mathscr{E}^*_{\perp} + D n^{*\prime} = 0. \tag{4.6}$$

They include filamentary solutions fulfilling the boundary conditions $n^*(0) \approx n_3$, $n^*(\infty) = n_1$, and $\mathscr{E}^*_{\perp}(0) = 0$ (fig. 5a). The filament has a radius $R^*_0(\mathscr{E}_0)$, $\mathscr{E}_0 \equiv \mathscr{E}^*_{\|}$, and a thin wall (transition layer) of thickness $\Delta R \ll R^*$.

Combining eqs. (4.4′) and (4.6) gives

$$\left(\frac{n^{*\prime}}{n}\right)' + \frac{n^{*\prime}}{Rn} + \frac{\mu}{\varepsilon_0 \varepsilon_s D} \rho(n^*, \mathscr{E}_0) = 0. \tag{4.7}$$

Equation (4.7) an be integrated over $\int_0^\infty (n^{*\prime}/n^*)\, \mathrm{d}R$ yielding

$$\Sigma + \frac{\mu}{\varepsilon_0 \varepsilon_s D} \int_{n_3}^{n_1} \rho(n^*, \mathscr{E}_0) \frac{\mathrm{d}n^*}{\mathrm{d}n} = 0, \tag{4.8}$$

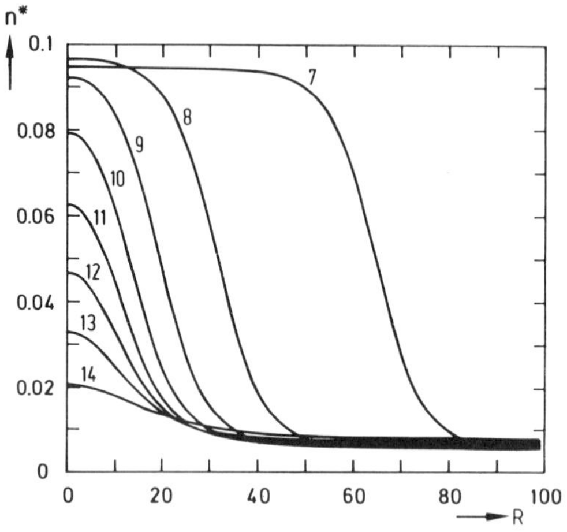

Fig. 5. Stationary radial electron density profile $n^*(R)$ of a current filament, calculated for the g–r model of fig. 1 with parameters of fig. 2, and $X_1^S = 5 \times 10^5\ \mathrm{s}^{-1}$. The curves 7–14 correspond to increasing values of the control parameter $\mathscr{E}_0$ from 7.9 to 9.3 V/cm (n in units of N_D^*, R in units of 10^{-6} cm) (after Schöll 1986a).

where

$$\Sigma \equiv \int_0^\infty \frac{1}{R}\left(\frac{n^{*\prime}}{n^*}\right)^2 \mathrm{d}R \approx \frac{1}{R_0^*}\int_0^\infty \left(\frac{n^{*\prime}}{n^*}\right)^2 \mathrm{d}R \tag{4.9}$$

can be evaluated self-consistently for $R_0^* \gg \Delta R$ (Schöll 1986a); it is analogous to a surface tension term in thermodynamics. Equations (4.8) and (4.9) determine $R_0^*(\mathscr{E}_0)$ as a monotonically decreasing function of the applied field $\mathscr{E}_0$. There is a value $\mathscr{E}_0 = \mathscr{E}_{c0}$, defined by

$$\int_{n_1}^{n_3} \rho(n^*, \mathscr{E}_{co}) \frac{\mathrm{d}n^*}{n^*} = 0, \tag{4.10}$$

for which $R_0^* \to \infty$, i.e., a high (n_3) and a low (n_1) conductivity phase coexist with a plane interphase. Equation (4.10) requires that the two hatched areas in fig. 6 are equal, and is thus analogous to the Maxwell rule for equilibrium phase transitions. The filamentary state $R_0^*(\mathscr{E}_0)$ leads to an additional, monotonically decreasing branch in the stationary current-voltage characteristic. A detailed analysis of the stability of filaments and of the influence of finite lateral boundaries shows that different nonequilibrium phase transitions associated with the nucleation (at a minimum radius) and growth of current filaments can be induced (Schöll 1987). These phenomena are analogous to droplet formation in a first-order gas–liquid phase transition. The formation of stable stationary filaments can be understood as a result of nonlocal circuit interactions in analogy with Ostwald ripening in a van der Waals gas

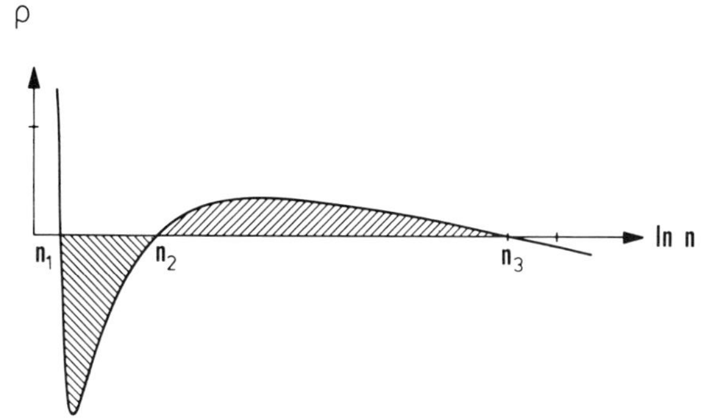

Fig. 6. Equal areas rule for two current density phases with a plane interface (schematic).

(Schimansky-Geier et al. 1991). For a load resistance $R_L \neq 0$, $\mathscr{E}_{\|}$ in eq. (4.1) is given by $J_0 = (\sigma_L + e\bar{n}\mu)\mathscr{E}_{\|}$, where the laterally averaged carrier density $\bar{n} = \int n(R, t)\, df/A$ introduces a nonlocal interaction in eq. (4.1). As the filament radius R increases, $\bar{n}$ and the current through the sample increase. Thus, the voltage drop across the load resistor increases, and the sample voltage $\mathscr{E}_{\|} L$ decreases. This results in an increase of R_0^* to a stable, finite value. If several current filaments are coupled via this nonlocal circuit interaction, the initially largest filament will eventually survive. Note that bistability between homogeneous and filamentary states may occur in appropriate ranges of the lateral dimensions even without SNDC, and the threshold and holding fields of the switching transitions can be shifted by adjusting the boundary conditions (e.g., by optical surface generation).

4.3. *Breathing filaments*

Stationary current filaments may become unstable under certain conditions, resulting in nonlinear spatio-temporal dynamics. "Breathing" current filaments were suggested as a possible mechanism for self-generated current oscillations (Schöll 1987, Schöll and Drasdo 1990). Recent space- and time-resolved experimental investigations have indeed demonstrated that in a certain regime of the S-shaped current–voltage characteristic, spontaneous nonlinear oscillations arise which are localized at the filamentary boundaries, and which represent a breathing motion of the filament walls (Brandl et al. 1989, Rau et al. 1991). Other nonlinear oscillation modes are associated with rocking on travelling current filaments (Niedernostheide et al. 1992).

Breathing filaments are characterized by a rigid shift of the filament wall with negligible changes in the profile. In order to model such dynamics, we identify as the most relevant dynamic variables the radius of the current filament $R_0(t)$, and the longitudinal field $\mathscr{E}_{\|}(t)$. For stationary filaments $R_0 = R_0^*$ holds.

We use the ansatz

$$n(R, t) = n^*(R - R_0(t) + R_0^*). \tag{4.11}$$

The spatial degrees of freedom may be projected out in eqs. (4.1) and (4.2) by multiplying with a suitable weight function G and integrating over $\int_0^\infty \mathrm{d}R$. The choice of G is guided by the requirement that it should be sharply peaked at the filament wall in order to project out the *relevant* contribution of the profile, and moreover G should allow for analytical evaluation of some of the integrals occurring in eq. (4.1) without explicit use of $n^*(R)$, as in the static case (4.8).

With $G = n^{*\prime}(R - R_0 + R_0^*)$ we obtain from eq. (4.1) approximately:

$$\dot{R}_0(t) = \frac{k_B T_L}{e} \mu(\mathscr{E}_\parallel) \left[\frac{1}{R_0^*(\mathscr{E}_\parallel)} - \frac{1}{R_0} \right]. \tag{4.12}$$

Equation (4.12) holds only if the filament radius is larger than the width of the filament wall; it has to be modified for R_0^*, $R_0 \approx \Delta R$ (Schöll and Drasdo 1990).

Integrating eq. (4.3) over the cross-section A of the current flow and connecting a load R_L as in eq. (3.1) gives

$$\dot{\mathscr{E}}_\parallel = [J_0 - (\sigma_L + e\bar{n}(R_0)\mu(\mathscr{E}_\parallel))\mathscr{E}_\parallel]/\varepsilon_0 \varepsilon_s, \tag{4.13}$$

with

$$\bar{n}(R_0) = \frac{1}{A} \int_A n(R, t)\, \mathrm{d}f \approx n_1 + (n_3 - n_1)\pi R_0^2 / A. \tag{4.14}$$

Equations (4.12) and (4.13) represent a nonlinear dynamic system, which may be regarded as an expansion of the partial differential equations (4.1) and (4.3) in terms of the "breathing" mode.

Self-generated oscillations induced by periodically breathing current filaments correspond to limit cycles in the $(R_0, \mathscr{E})$ phase space. The condition for a Hopf bifurcation of such a limit cycle can be obtained analytically by linearizing eqs. (4.12) and (4.13) around the steady state $R_0 = R_0^*$, $\mathscr{E}_\parallel = \mathscr{E}_0$. It is a special case of the general conditions discussed in § 3.1. A Hopf bifurcation occurs for sufficiently large load resistance R_L and negative differential conductance with numerical parameters for p-Ge and n-GaAs (Schöll and Drasdo 1990).

The oscillatory mechanism is based upon two features:

(i) An instability of the filament radius $R_0(t)$, such that R_0 tends to increase further if R_0 is above a critical radius $R_0^*(\mathscr{E})$, and to decrease for $R_0 < R_0^*(\mathscr{E}_\parallel)$. Thus R_0^* is analogous to the critical droplet radius (*critical nucleus*) in equilibrium or nonequilibrium phase transitions. The microscopic mechanism in our case is impact ionization of impurities, but this enters only implicitly into the dynamics through the function $R_0^*(\mathscr{E}_\parallel)$. Any other autocatalytic mechanism which yields a *decreasing* function $R_0^*(\mathscr{E}_\parallel)$ will furnish similar results.

(ii) A restoring force, which is provided by dielectric relaxation of the longitudinal electric field $\mathscr{E}_\parallel$. It is essentially controlled by the average carrier density $\bar{n}(R_0)$ which forces $\mathscr{E}_\parallel$ to decrease with increasing R_0.

The physical mechanism underlying the breathing oscillations is thus similar to the spatially homogeneous model for dielectric relaxations oscillations (eq. (3.7)). The two models are simple approximations of two different modes of dielectric relaxation oscillations: breathing or bulk-dominated, respectively. Both types of oscillations are predicted to occur on the falling branch of the (filamentary or uniform, respectively) current–voltage characteristic, generated by Hopf bifurcations with decreasing applied current. This agrees well with the experiments by Rau et al. (1991) who found in p-Ge at 4 K – with increasing current – two different types of oscillations in different regimes of the falling current–voltage characteristic, associated with large-amplitude circuit-limited (CLO = bulk dominated) modes and small-amplitude structure-limited (SLO = breathing) modes with lower frequency, followed by a regime of stable stationary current filaments. Going through the breakdown regime with increasing current, a variety of complex nonlinear spatio-temporal dynamics was observed: First, below breakdown, a random firing mode of current filaments occurs. The distribution of time intervals between individual impact ionization breakdown events appears to obey the universal power laws of self-organized criticality (Bak et al. 1987). Next, as the single breakdown events become so frequent that they start to overlap, periodic circuit-limited oscillations appear. Finally, at higher currents, in the regime of breathing oscillations, transitions to chaos, and chaos–chaos transitions characterized by a concomitant increase of the fractal dimension and excitation of further dynamic spatio-temporal degrees of freedom may also be induced. Here, the coupling of different localized oscillations centers gives rise to fully developed spatio-temporal chaos.

References

Abe, Y., ed., 1989, Special Issue on Nonlinear and Chaotic Transport in Semiconductors, Appl. Phys. A **48**, 93.

Aoki, K., 1992, Semicond. Sci. & Technol. B **7**, 474.

Aoki, K., and K. Yamamoto, 1983, Phys. Lett. A **98**, 72.

Aoki, K., T. Kobayashi and K. Yamamoto, 1981, J. Phys. Colloq. (Paris) C **-7**, 51.

Aoki, K., N. Mugibayashi and K. Yamamoto, 1986, Phys. Scr. **T14**, 7.

Aoki, K., Y. Kawase, K. Yamamoto and N. Mugibayashi, 1990, J. Phys. Soc. Jpn. **59**, 20.

Asche, M., Z.S. Gribnikov, V.V. Mitin and O.G. Sarbey, 1979, Goryachie Elektrony v Mnogodolinikh Poluprovodnikakh (Hot Electrons in Many-Valley Semiconductors) (Naukova Dumka, Kiev).

Baccarani, G., M. Rudan, R. Guerrieri and P. Ciampolini, 1986, in: Advances in CAD for VLSI, Vol. 1, ed. W. Engl (North-Holland, Amsterdam) p. 107.

Bak, P., C.L. Tang and K. Wiesenfeld, 1987, Phys. Rev. Lett. **59**, 381.

Balkan, N., and B.K. Ridley, 1989, Superlattices Microstructure **5**, 539; Semicond. Sci. & Technol. **3**, 507.

Belyantsev, A.M., A.A. Ignatov, V.I. Piskarev, M.A. Sinitsyn, V.I. Shashkin, B.S. Yavich and M.L. Yakovlev, 1986, JETP Lett. **43**, 437.

Bonch-Bruevich, V.L., and Le Vu Ky, 1984, Phys. Status Solidi B **124**, 111.

Bonch-Bruevich, V.L., I.P. Zvyagin and A.G. Mironov, 1975, Domain Electrical Instabilities in Semiconductors (Consultant Bureau, New York).

Brandl, A., and W. Prettl, 1990, in: Festkörperprobleme, ed. U. Rössler (Vieweg, Braunschweig) p. 371.

Brandl, A., T. Geisel and W. Prettl, 1987, Europhys. Lett. **3**, 401.

Brandl, A., M. Völcker and W. Prettl, 1989, Appl. Phys. Lett. **55**, 238.

Brandl, A., W. Kröninger, W. Prettl and G. Obermair, 1990, Phys. Rev. Lett. **64**, 212.

Bumelienė, S.B., J. Požela, K.A. Pyragas and A.V. Tamaševičius, 1985, Physica B **134**, 293.

Butcher, P.N., 1965, Phys. Lett. **19**, 546.

Degtyarenko, N.N., V.F. Elesin and V.A. Furmanov, 1974, Sov. Phys. Semicond. **7**, 1147.

Döttling, R., and E. Schöll, 1992, Phys. Rev. B **45**, 1935.

Ebeling, W., and R. Feistel, 1986, Physik der Selbstorganisation und Evolution (Akademie-Verlag, Berlin).
Feigenbaum, M.J., 1978, J. Stat. Phys. **19**, 25.
Fuji, K., T. Ohyama and E. Otsuka, 1989, Appl. Phys. A **48**, 189.
Glicksman, M., 1971, Solid State Physics: Advances in Research and Applications, Vol. 26 (Academic Press, New York).
Godlewski, M., K. Fronc, M. Gajewska, W.M. Chen and B. Monemar, 1992, Semicond. Sci. & Technol. B **7**, 483.
Gribnikov, Z.S., 1973, Sov. Phys.-Semicond. **6**, 1204.
Guckenheimer, J., and P. Holmes, 1983, Nonlinear Oscillations, Dynamical Systems and Bifurcations of Vector Fields (Springer, Berlin).
Haken, H., 1983a, Synergetics, An Introduction, 3rd Ed. (Springer, Berlin).
Haken, H., 1983b, Advanced Synergetics (Springer, Berlin).
Haug, H., ed., 1988, Optical Nonlinearities and Instabilities in Semiconductors (Academic Press, New York).
Held, G.A., C.D. Jeffries and E.E. Haller, 1984, Phys. Rev. Lett. **52**, 1037.
Hendriks, P., E.A.E. Zwaal, J.G.A. Dubois, F.A.P. Blom and J.H. Wolter, 1991, J. Appl. Phys. **69**, 302.
Hess, K., H. Morkoc, H. Shichijo and B.G. Streetman, 1979, Appl. Phys. Lett. **35**, 469.
Hüpper, G., and E. Schöll, 1991, Phys. Rev. Lett. **66**, 2372.
Hüpper, G., E. Schöll and L. Reggiani, 1989, Solid State Electron. **32**, 1787.
Hüpper, G., A. Rein and E. Schöll, 1992, to be published.
Jäger, D., and R. Symanczyk, 1992, in: Nonlinear Dynamics in Solids, ed. H. Thomas (Springer, Berlin) p. 68.
Jäger, D., H. Baumann and R. Symanczyk, 1986, Phys. Lett. A **117**, 141.
Kerner, B.S., and V.V. Osipov, 1989, Sov. Phys.-Usp. **32**, 101.
Knap, W., M. Jezewski, J. Lusakowski and W. Kuszko, 1988, Solid State Electron. **31**, 813.
Kolodzey, J., J. Laskar, T.K. Higman, M.A. Emanuel, J.J. Coleman and K. Hess, 1988, IEEE Electron Device Lett. **EDL-9**, 272.
Kometer, K., G. Zandler and P. Vogl, 1992a, Semicond. Sci. & Technol. B **7**, 559.
Kometer, K., G. Zandler and P. Vogl, 1992b, Phys. Rev. B, to be published.
Kunz, R., and E. Schöll, 1992, Mod. Phys. Lett. B, to be published.
Landsberg, P.T., and A. Pimpale, 1976, J. Phys. C **9**, 1243.
Landsberg, P.T., D.J. Robbins and E. Schöll, 1978, Phys. Status Solidi A **50**, 423.
Landsberg, P.T., E. Schöll and P. Shukla, 1988, Physica D **30**, 235.
Maracas, G.N., W. Porod, D.A. Johnson and D.K. Ferry, 1985, Physica B **134**, 276.
Mayer, K.M., J. Parisi and R.P. Huebener, 1988, Z. Phys. B **71**, 171.
Naber, H., and E. Schöll, 1990a, Z. Phys. B **78**, 301.
Naber, H., and E. Schöll, 1990b, Z. Phys. B **78**, 305.
Nicolis, G., and I. Prigogine, 1977, Self-Organization in Non-Equilibrium Systems (Wiley, New York).
Niedernostheide, F.J., M. Arps, R. Dohmen, H. Willebrand and H.G. Purwins, 1992, Phys. Status Solidi b, to be published.
Peinke, J., 1992, in: Nonlinear Dynamics in Solids, ed. H. Thomas (Springer, Berlin) p. 51.
Peinke, J., A. Mühlbach, R.P. Huebener and J. Parisi, 1985, Phys. Lett. A **108**, 407.
Peinke, J., U. Rau, W. Clauss, R. Richter and J. Parisi, 1989, Europhys. Lett. **9**, 743.
Peinke, J., J. Parisi, O.E. Roessler and R. Stoop, 1992, Encounter with Chaos (Springer, Berlin).
Požela, J., 1981, Plasma and Current Instabilities in Semiconductors (Pergamon, Oxford).
Purwins, H.G., G. Klempt and J. Berkemeier, 1987, Festkörperprobleme **27**, 27.
Pyragas, K.A., J. Požela, A. Tamaševičius and J. Ulbikas, 1987, Sov. Phys.-Semicond. **21**, 335.
Pytte, E., and H. Thomas, 1969, Phys. Rev. **179**, 431.
Quade, W., E. Schöll and M. Rudan, 1992, to be published.
Rau, U., J. Peinke, J. Parisi, R.P. Huebener and E. Schöll, 1987, Phys. Lett. **124**, 335.
Rau, U., W. Clauss, A. Kittel, M. Lehr, M. Bayerbach, J. Parisi, J. Peinke and R.P. Huebener, 1991, Phys. Rev. B **43**, 2255.
Reggiani, L., and V.V. Mitin, 1989, Riv. Nuovo Cimento **11**(12), 1.

Richter, R., U. Rau, A. Kittel, G. Heinz, J. Peinke, J. Parisi and R.P. Huebener, 1991, Z. Naturforsch. A **46**, 1012.
Rieger, M., and P. Vogl, 1989, Solid State Electron. **32**, 1399.
Robbins, D.J., P.T. Landsberg and E. Schöll, 1981, Phys. Status Solidi A **65**, 353.
Röhricht, B., R.P. Huebener, J. Parisi and M. Weise, 1988, Phys. Rev. Lett. **61**, 2600.
Schimansky-Geier, L., Ch. Zülicke and E. Schöll, 1991, Z. Phys. B **84**, 433.
Schlögl, F., 1972, Z. Phys. **253**, 147.
Schlögl, F., and E. Schöll, 1988, Z. Phys. B **71**, 231.
Schöll, E., 1979, Proc. R. Soc. A **365**, 511.
Schöll, E., 1981, J. Phys. Colloq. (Paris), C **-7**, 57.
Schöll, E., 1982a, Z. Phys. B **46**, 23.
Schöll, E., 1982b, Z. Phys. B **48**, 153.
Schöll, E., 1983, Z. Phys. B **52**, 321.
Schöll, E., 1985, Physica B **134**, 271.
Schöll, E., 1986a, Solid State Electron. **29**, 687.
Schöll, E., 1986b, Phys. Rev. B **34**, 1395.
Schöll, E., 1987, Nonequilibrium Phase Transitions in Semiconductors (Springer, Berlin).
Schöll, E., 1989a, Appl. Phys. A **48**, 95.
Schöll, E., 1989b, Phys. Scr. **T29**, 152.
Schöll, E., 1989c, Solid State Electron. **32**, 1129.
Schöll, E., and K. Aoki, 1991, Appl. Phys. Lett. **58**, 1277.
Schöll, E., and D. Drasdo, 1990, Z. Phys. B **81**, 183.
Schöll, E., and P.T. Landsberg, 1979, Proc. R. Soc. London A **365**, 495.
Schöll, E., and P.T. Landsberg, 1988, Z. Phys. B **72**, 515.
Schöll, E., and W. Quade, 1987, J. Phys. C **20**, L861.
Schöll, E., H. Naber, J. Parisi, B. Röhricht, J. Peinke and S. Uba, 1989, Z. Naturforsch. A **44**, 1139.
Schöll, E., G. Hüpper and A. Rein, 1992, Semicond. Sci. & Technol. B **7**, 480.
Schuster, H.G., 1987, Deterministic Chaos, 2nd Ed. (Physik-Verlag, Weinheim).
Seiler, D.G., C.L. Littler, R.J. Justice and P.W. Milonni, 1985, Phys. Lett. A **108**, 462.
Shaw, M.P., H.L. Grubin and P.R. Solomon, 1979, The Gunn–Hilsum Effect (Academic Press, New York).
Shaw, M.P., V.V. Mitin, E. Schöll and H.L. Grubin, 1992, The Physics of Instabilities in Solid State Electron Devices (Plenum, New York).
Song, X.N., D.G. Seiler and M.R. Loloee, 1989, Appl. Phys. A **48**, 137.
Spinnewyn, J., H. Strauven and O.B. Verbeke, 1989, Z. Phys. B **75**, 159.
Symanczyk, R., D. Jäger and E. Schöll, 1991, Appl. Phys. Lett. **59**, 105.
Teitsworth, S.W., and R.M. Westervelt, 1984, Phys. Rev. Lett. **53**, 2587.
Teitsworth, S.W., R.M. Westervelt and E.E. Haller, 1983, Phys. Rev. Lett. **51**, 825.
Thoma, R., A. Emunds, B. Meinerzhagen, H.J. Peifer and W.L. Engl, 1991, IEEE Trans. Electron. Devices **ED-38**, 1343.
Thomas, H., ed., 1992, Nonlinear Dynamics in Solids (Springer, Berlin).
Thompson, J.M.T., and H.B. Stewart, 1986, Nonlinear Dynamics and Chaos (Wiley, New York).
van Kampen, N.G., 1985, Phys. Rep. **124**, 69.
Vickers, A.J., A. Straw and J.S. Roberts, 1989, Semicond. Sci. & Technol. **4**, 743.
Volkov, A.F., and Sh. M. Kogan, 1969, Sov. Phys.-Usp. **11**, 881.
Wacker, A., and E. Schöll, 1991, Appl. Phys. Lett. **59**, 1702.
Wacker, A., and E. Schöll, 1992, to be published.
Weispfenning, M., I. Hoeser, W. Böhm, W. Prettl and E. Schöll, 1985, Phys. Rev. Lett. **55**, 754.
Yamada, K., N. Takara, H. Imada, N. Miura and C. Hamaguchi, 1988, Solid State Electron. **31**, 809.
Yano, H., N. Goto and Y. Ohno, 1992, Semicond. Sci. & Technol. B **7**, 491.

CHAPTER 9

Electron–Hole Liquids in Semiconductors

A. A. ROGACHEV

A.F. Ioffe Physico-Technical Institute
Academy of Sciences
194021, St. Petersburg, Russia

Handbook on Semiconductors
Completely Revised Edition
Edited by T.S. Moss
Volume 1, edited by P.T. Landsberg

Contents

1. Introduction 451
2. The electron–hole liquid binding energy 457
3. Condensation of excitons into electron–hole drops 459
4. Electrical conductivity of electron–hole drops 471
5. Phase diagram of electron–hole liquids in germanium 473
6. Multiexciton complexes in semiconductors 475
7. Many electron–hole effects at the semiconductor surface 479

References 484

1. Introduction

In the more than twenty years of intensive investigations of the properties of condensed excitons various reviews devoted to this subject have been published (Jeffries 1975, Thomas 1976, Voos and Benoit à la Guillaume 1975, Rice 1976, 1977, 1978, Hensel et al. 1977, Jeffries and Keldysh 1983). Therefore, to write a new review in such a competitive situation a very good reason is required. In the present review the author will attempt to discuss how the different valleys, both in conduction and valence bands, that have different, statistically independent, electrons or holes, determine the properties of the electron–hole liquid. Thus, a new quantum number may be introduced for semiconductors, namely, the number of a valley. A discussion of this problem can be found in the literature (Keldysh 1983, Rice 1976).

In the real world there are no statistically independent electrons, positrons or other particles. For this reason exciton molecules and exciton liquids might be fundamentally different from hydrogen (or alkali metals) and molecular liquids. Up to 1968 there existed only theoretical predictions that at increasing exciton density in a semiconductor with single-valley conduction and valence bands the following development of the system state may be expected (Keldysh 1989, 1972): excitons → biexcitons → dielectric condensate → Bose condensate.

At present there is no definite answer to the question whether the phase of the condensed excitons in a single-valley semiconductor is metallic or dielectric. The only statement which can be derived from the available theory is that the dielectric state is more probable. (The energy difference between the dielectric and the metallic state does not exceed a few percent of the exciton binding energy.) Undoubtedly, very convincing arguments in favour of the dielectric state can be given. Keldysh and Kopaev (1964) showed that for the case of equal effective masses of electrons and holes the dielectric state persists at high electron–hole pair concentrations. Beni and Rice (1978) showed that in polar semiconductors some stabilization of the metallic state takes place. But the question whether this is a metallic or dielectric phase is still open.

In the case of a large number of valleys, the many-exciton system changes its state with increasing exciton density in the following way: exciton gas → exciton molecules (i.e., biexcitons and multiexciton complexes) → metal liquid.

As experiments show, this scheme can be applied well to germanium and silicon, and theory predicts with a good accuracy the exciton binding energy in the liquid state and the density of the electron–hole (e–h) liquid. The energy scheme of an e–h drop is shown in fig. 1. Below we shall regard the terms "exciton liquid" and "e–h liquid" as synonymous.

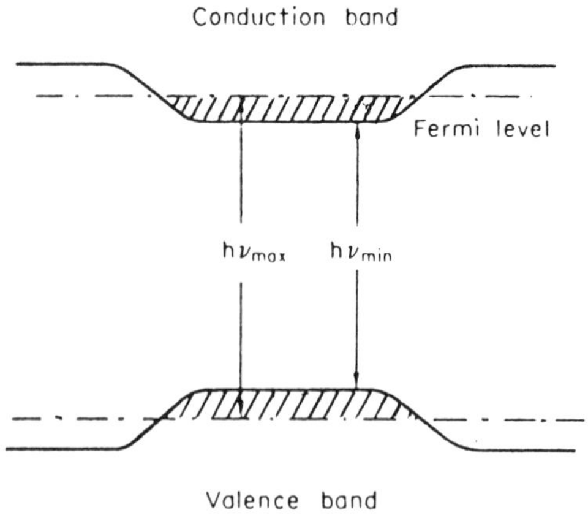

Fig. 1. Scheme of an e–h drop. The electron and hole quasi-Fermi levels are in the forbidden gap, making the high-density state energetically stable.

The effective-mass approximation is usually applied to describe the electron–hole interaction in semiconductors. In this approximation the Wannier–Mott exciton is treated as a result of the Coulomb interaction of an electron and a hole ($e^2/\varepsilon r$, where ε is the dielectric constant). The motion of the interacting electron and hole is described by the reduced mass $m = m_e m_h / m_e + m_h$ and by the exciton binding energy $E_x = e^4 m / 2h^2 \varepsilon^2$. This makes an exciton a hydrogen-like structure, so that it is possible to proceed further by supposing that excitons can form molecules (Lampert 1958, Moskalenko 1958) and also a dielectric liquid (Agranovich and Toshich 1967). However, the excitons cannot form a solid because of large zero-point oscillations. Therefore, the excitons form a liquid at low temperature, and that is why they exhibit Bose condensation. There is no transport of mass by the exciton liquid, but superfluidity may show up in its energy, its dielectric constant, its magnetic properties and in other effects (Keldysh 1972). Here we use a general statement that for any small attraction of molecules the condensation can still occur. The problem seems to be simple only at first sight. As the result of condensation the exciton density increases and a dielectric–metal transition can take place. In such a case any property of the electrons and holes (which are both fermions) are important. With increasing exciton density there are two important transitions in the many-exciton system. They are the gas–liquid and dielectric–metal transitions. And what is more, Keldysh and Kopaev (1964) have shown theoretically that at low temperature ($T = 0$) the high-density e–h gas is a dielectric, i.e., the occupied and free states are separated by a gap. The nature of this gap is similar to that of Cooper pairs described in the BCS superconductivity theory. We know from the BCS theory that a weak mutual attraction of electrons results from their interaction with phonons, and that this creates bound electron pairs formed from electrons located in the vicinity of the Fermi surface. In the dense electron–hole gas the formation of the Cooper pairs is possible due to this mutual attraction of electrons and holes. This attraction is weak because of the strong screening of the Coulomb interaction. These Cooper pairs form a Bose condensate, which is a dielectric, but not a superconductor, since the pairs have no charge. Thus, the system is a dielectric, both at low and high densities of electrons and holes.

If the isoenergetic surfaces of electrons and holes are not similar, the binding energy of an electron–hole pair appears to be many times smaller than that in the case of spherical isoenergetic surfaces (Kozlov and Maksimov 1965). This is the result of the fact that the integral used to determine the electron–hole binding energy converts from a surface integral to a contour integral. Keldysh and Kozlov (1968) have shown for the case when the Fermi energy is approximately equal to the exciton binding energy that a difference between the anisotropy of the effective masses of electrons and holes has not a very strong effect on the magnitude of the electron–hole binding energy. This is due to the substantial "thickness" of the integration contour line.

Thus, by the time of the experimental detection of the metallization of excitons, when their concentration is high enough (Asnin et al. 1967, Asnin and Rogachev 1968, Rogachev 1968), only theoretical considerations concerning the Bose condensation and the transition into the dielectric liquid existed.

At the IXth International Conference on Semiconductor Physics, held in Moscow in 1968, Keldysh predicted, in a discussion of a report about the metallization of excitons (Rogachev 1968), that the formation of a metallic liquid at an increase of the exciton gas density is possible. He pointed out in his closing remark (Keldysh 1968) that this idea was not checked experimentally. Then he said, that

> "On the other hand the results of Rogachev, who has observed the metallization of excitons in germanium at sufficiently high concentrations, are very interesting. In some way excitons in semiconductors constitute an autonomous system, a peculiar atomic gas. Accordingly, they must behave under increasing concentration just as any other gas behaves under increasing pressure: at some critical concentration which depends on temperature, it makes a transition into a 'liquid' state, i.e., into a state in which they are all bound together and the distance between particles is of the order of the exciton radius. One can make more or less convincing arguments, based on the smallness of the exciton mass, to the effect that during the 'liquefaction' the excitons will behave not like hydrogen which forms a molecular liquid, but more like alkali metal atoms, i.e., they form a metallic kind of liquid in which each electron is not rigidly bound to its hole, but moves more or less freely. The presence of such mobile metallic regions within a semiconductor undoubtedly would show up clearly in many of its properties. It is not excluded that it is just in this liquid phase that some properties of the type of Bose condensation could manifest themselves".

It is difficult to explain what makes excitons to be similar to the alkali metal atoms, rather than to hydrogen. The difference between hydrogen atoms and alkali metal atoms is primarily that the alkali metal atoms have deep electron shells loosening-up the last s-shell. For this reason excitons are more similar to hydrogen or positronium, which have no inner shells. Hypothetical metallic hydrogen has also a higher binding energy than any alkali metal. The first and most important step towards understanding of the high binding energy of the e–h liquid in germanium has been made by Keldysh and co-workers (1969), who noted that in multivalley semiconductors the metallic state has a higher binding energy than in single-valley semiconductors. Soon after this Brinkmann and Rice (1973), Cambescot and Nozières (1972) and Inoue and Hanamura (1973) confirmed this observation theoretically. Both the binding energy of excitons in the liquid phase and density of the liquid phase appeared to be in a good agreement with the experimental data available that time.

The exciton binding energy of the metallic exciton phase in germanium and silicon appeared to be approximately equal to half the exciton binding energy E_x, while in a biexciton it is estimated theoretically to be about $0.1E_x$, or even less. The semi-empirical Mott criterion $rn^{1/3} \approx 0.2$ requires that the binding energy for the dielectric (molecular) state must not differ strongly from that for the metallic state, and, correspondingly, distances between atoms in these states have to be approximately equal. The metallic exciton phase in germanium and silicon seems to violate this rule.

From 1968 till 1973 the author of this review was wrong in accepting the idea of Haynes (1966) that the recombination radiation is due to biexcitons, and that transition into the metallic state occurs at higher density (Rogachev 1974). An experimental argument in favour of this point of view was that in the thin (less than 10 μm) samples the dielectric–metal transition had a threshold density, but depended weakly on temperature. Criticism of this statement was published by Rice and Brinkmann (1973).

The true reason why the exciton condensation in thin and thick samples occurs differently, will be explained in the following part of the review. One can suppose that the dielectric–metal transition in the exciton system is a transition from multi-exciton complexes to a metallic liquid. The exciton binding energy in complexes is close to that in drops. However, being in essence the smallest possible drops, the complexes have a lower binding energy than the condensate drops, for the same reason why small drops have always a smaller binding energy than large ones. Generally speaking, the Mott criterion is not an absolute law of nature; however, it would be interesting to find a deviation from it. It is clear that in the case of the multiexciton complexes–exciton drops transition the Mott criterion is fulfilled.

To realize multiexciton complexes the existence of a multivalley structure is required both in the conduction and in the valence bands. The conduction-band structure in indirect semiconductors, such as germanium and silicon, allows one to have eight or twelve electrons with an s-type wavefunction. The valence band has both heavy- and light-hole bands. Rice (1977) has made the erroneous assumption that only heavy holes participate in the formation of the valence-band structure of the complex, and light holes do not, due to their large kinetic energy. This mistake was repeated in the works of Shore and Pfeiffer (1978). Actually, this assumption is equivalent to the statement that the nondiagonal matrix elements determining the Coulomb interaction V are equal to zero, i.e., $\langle \frac{1}{2}|V|\frac{3}{2}\rangle = 0$. These matrix elements are equal, as regards order of absolute magnitude, to the diagonal matrix elements of the form $\langle \frac{1}{2}|V|\frac{1}{2}\rangle$ and $\langle \frac{3}{2}|V|\frac{3}{2}\rangle$.Thus, the states with spins $\frac{1}{2}$ and $\frac{3}{2}$ make an equal contribution to the valence band of the complex. For this reason a complex may have four holes, envelope wavefunctions of which have no nodes (s-envelope wavefunctions). Experiment has shown that the multiexciton complex may have six holes, four of which occupy the Γ_8 and two Γ_7^+ states (Altukhov et al. 1980a). The electron shell is never full. Such a structure enables one to consider the multiexciton complexes as nucleation centers in the process of condensation of the electron–hole drops.

Let us consider a model of the multiexciton complex in which the number of electrons and holes with an s-type wavefunction is not limited. This means that we suppose the presence of a large number of statistically independent electrons and holes (i.e., many valleys). Actually, the number of independent valleys cannot be very large because of the overlap of the regions occupied by the valleys in k-space. The multiexciton complex wavefunction is shown schematically in fig. 2.

Since the wavefunctions of electrons and holes are identical, the charge density is equal to zero ($\rho = 0$), and, correspondingly, the potential energy calculated in the Hartree approximation vanishes. The kinetic energy E_k in the case of hydrogen-like functions is equal to $\hbar^2/2mr^2$. The stability of the exciton molecule is determined in this approximation only by the correlation and kinetic energies. The exchange energy is absent, since electrons and holes from different valleys have different quantum numbers (a spin and the valley number). The correlation energy per pair consisting of one electron and one hole is equal to $e^2/\varepsilon r_s$ (where $r_s = \sqrt{\varepsilon E_k/6\pi e^2 n}$ and $n = \gamma(\beta/\pi r^3)$, where β stands for a number of equal to about 10 and allows for a particle density decrease with distance from the center). The ground-state energy for an electron–hole pair is equal to

$$-E_g = \frac{e^2}{\varepsilon\sqrt{r_h r/\gamma}} - \frac{\hbar^2}{mr^2}, \tag{1}$$

as will be justified at the beginning of § 2.

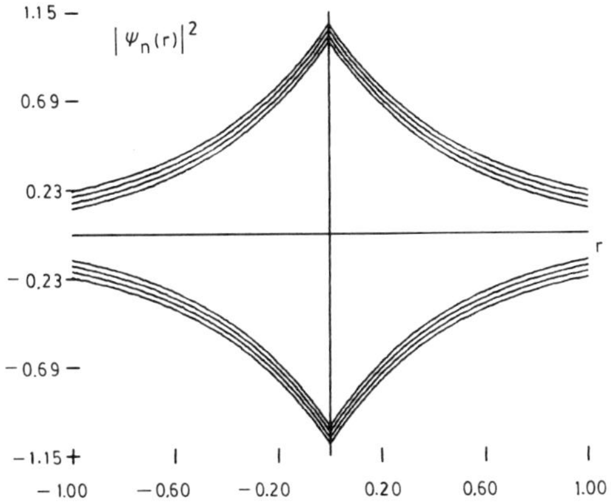

Fig. 2. Wavefunction of multiexciton complex. As electrons and holes are independent, $(\Sigma_{n=1}^{n=2\gamma}\psi_n(r))^2 = \Sigma_{n=1}^{n=2\gamma}\psi_n^2(r)$. So it is possible to depict the wavefunction of electrons as negative and the hole wavefunction as positive. In the Hartree approximation $\rho(r) = 0$. The potential energy is a correlation energy which may be treated as due to screening. The $\psi_n(r)$ functions are moved upward for clarity. γ is the number of valleys.

In this expression numerical coefficients of order unity are omitted. The complex size r is determined from the condition $dE_g/dr = 0$, and, correspondingly, $-E_g \approx \gamma^{2/3}$ Ry and $r \approx \gamma^{-1/3} r_h$, where $\mathrm{Ry} = \frac{1}{4}e^4 m/\varepsilon\hbar^2$. Basically, this calculation is similar to the calculation of Wand and Kittel (1978a,b) for complexes with heavy holes. If we take these complexes for our calculations as analogues of the Wigner–Seitz cells, used in the case of metals, we find that in the metallic exciton liquid the ground-state energy is proportional to $\gamma^{2/3}$ and close to the energy of the complex. Thus, multiexciton complexes, rather than biexcitons, are involved in the metallic liquid. This means that the Mott criterion is valid also for the metallic exciton liquid.

However, the question, as to the dielectric ground state in a single-valley semiconductor is still as open. The arguments in favour of the dielectric state in the single-valley case presented by Keldysh and co-workers (Keldysh 1972, Keldysh and Kopaev 1964) should be taken into account. The metallic liquid was experimentally found in such single-valley semiconductors, e.g., in CdS. The consideration that the binding energy in the metallic exciton liquid is higher in polar semiconductors (Keldysh and Silin 1975, Beni and Rice 1978) does not provide a final answer in favour of the metallic state, since the exciton binding energy also becomes higher at the cost of the "chattering" effect (Kane 1978). The exciton energy appears to be higher than the donor energy.

There also exist surface phenomena related to the many-electron effects. If there are no states on the semiconductor surface, and no electric field is applied, electrons, holes and excitons are pushed away from the surface. If an electric field is applied to the semiconductor surface, it seems natural that it attracts carriers of one sign and repels carriers of opposite sign, since due to screening the field strength decreases, but does not change its sign. This answer is, however, not correct. An experiment and a subsequent calculation (see § 7) of the quantum screening have shown that, actually, something paradoxical takes place: both electrons and holes are attracted to the surface. This results in the formation of two layers of carriers of opposite sign. It should be noted that the quasiclassical approximation does not show this effect, giving an attraction of carriers of only one sign and repulsion of carriers of the opposite sign. In the Hartree approximation the attraction of the carriers of both signs exists as an "overscreening" phenomenon. It turns out that for a very large range of electric field strengths the overscreening is realized in the following way: more carriers than are needed for screening are attracted to the surface, an electric field of the opposite direction being formed in the bulk. This field is screened by the carriers of the opposite sign (Altukhov et al. 1985) or by the field of donors or acceptors, if there are no injected e–h pairs. In the Hartree approximation the second layer of carriers also overscreens this new field, and so on.

Thus, when an electric field is applied to the sample, the potential in the vicinity of the surface oscillates with the distance from the surface, and it vanishes in the bulk (Altukhov et al. 1985, Monakhov and Rogachev 1988). If the exchange-correlation corrections are taken into account in the calculations, it appears that the first oscillations are enhanced and the number of nodes sharply decreases. It is likely that, even under favourable experimental conditions, not more than two potential nodes can be observed. The electron–hole pair binding energy on the surface has to be higher than the exciton binding energy, or, strictly speaking, it should be higher than the sum of the exciton formation energy and the exciton binding energy for the bulk electron–hole drops. The electron–hole pair binding energy on the surface as a function of the applied electric field strength has a maximum. Calculations in the Hartree approximation shows that this maximum value of the binding energy is proportional to the square of the number of valleys (γ^2). As in the case of the bulk electron–hole drops, the pair binding energy on the surface increases mainly at the expense of a decrease of the Fermi energy and an increase (in absolute value) of the correlation energy. The capture of e–h pairs by the surface has the nature of a transition from a three-dimensional gas to a two-dimensional gas. The condensation of the three-dimensional gas to a two-dimensional liquid has been observed in a semiconductor, the surface of which was covered with an oxide layer and contained a large number of deep impurity centers (Asnin et al. 1987a,b). Let us suppose that electrons and holes come close to such a surface, and electrons are captured by the surface centers more strongly than holes. Then, the negative charge of the surface centers attracts holes, and holes, in turn, create an electron layer; this increases the negative charge of the surface centers, increasing the number of holes on the surface, and so on. As a result of growth of the Fermi energy of holes on the surface, the capture of them is weakened, and, hence, the electron flow towards the surface decreases. Further growth of the semiconductor excitation results in an increase of the size of surface drops or of their amount. This condensation is, thus, the result of the appearance of a recombination instability.

2. The electron–hole liquid binding energy

Since the electric field of electrons and holes in the condensed phase disappears at a distance of the order of the spacing, r, between particles, the potential energy is equal as regards order of magnitude to $-e^2/\varepsilon r$. This is a "gain" in the binding energy. A "loss" results mainly from an increase in the kinetic energy with decreasing spacing between particles. The ground-state energy is expressed by

$$E_{\mathrm{g}} = E_{\mathrm{k}} + E_{\mathrm{ex}} + E_{\mathrm{cor}}, \tag{2}$$

where E_{k} and E_{cor} are kinetic and correlation energies, respectively. The first two terms comprise the ground-state energy in the Hartree–Fock approximation ($E_{\mathrm{HF}} = E_{\mathrm{k}} + E_{\mathrm{ex}}$). The difference between the true ground-state energy value and E_{HF} is the correlation energy E_{cor}. At a large enough e–h pair density, E_{HF} makes a significant contribution to the ground-state energy, while E_{cor} can be considered as a small correction. In the case of multivalley semiconductors, the exchange energy E_{ex} becomes a small correction, since at a large number of valleys (γ) the interaction of particles in one valley appears to be strongly screened, and the energy E_{cor} becomes a significant contribution to the ground-state energy E_{g}. This statement justifies formula (1).

Since the Hartree–Fock approximation has proved to be a good basis for a number of calculations, it is of interest to consider which properties of the e–h system can be derived from it. For the simplest case, when both the conduction and the valence band consist of one valley with an isotropic effective mass, E_{HF} is expressed in the following way:

$$E_{HF} = \frac{3}{10}\frac{\hbar^2}{\mu}(3\pi^2 n)^{2/3} - \frac{3}{2}\frac{e^2(3\pi^2 n)^{1/3}}{\pi\varepsilon}, \tag{3}$$

where ε is the dielectric constant, n is the electron–hole pair concentration, $\mu^{-1} = m_e^{-1} + m_h^{-1}$, with m_e and m_h the effective masses of an electron and a hole, respectively.

Expressing the energy in atomic units $E_x = e^2/2\varepsilon r_h$ and putting $r_s = (3/4\pi n)^{1/3} r_h^{-1}$ (where $r_h = \varepsilon\hbar^2/e^2\mu$ is the exciton Bohr radius), we can write eq. (3) in the following way:

$$E_{HF} = \frac{2.21}{r_s^2} - \frac{1.832}{r_s}. \tag{4}$$

Let us consider now a semiconductor with γ valleys in the conduction band and with the effective mass of holes several times larger than that of electrons. If the valence band had also γ valleys, E_{HF} would still be determined by expression (3), in which n is understood as a particle concentration in one valley. This result has a simple physical sense: in the Hartree–Fock approximation electrons in one valley "do not feel" the presence of electrons (or holes) in another valley. If one takes a semiconductor with $m_e \ll m_h$ and a large number of valleys ($\gamma > 12$) in the conduction band, one can in principle obtain a negative binding energy of excitons in the e–h liquid. Of course, the correlation energy, which is not into taken account in this case, cannot be considered as small, since the screening by electrons having a small kinetic energy due to the large number of valleys is rather strong. We make the approximation that the sum $E_{ex} + E_{cor}$ depends weakly on the number of valleys and, on the other hand, that the kinetic energy E_k depends strongly on this factor. Then, for the case of γ valleys in both the conduction and the valence (heavy holes) bands we obtain, instead of eq. (4), for an e–h drop

$$E_{HF} = \frac{2.21}{\gamma^{2/3} r_s^2} - \frac{1.832}{r_s}. \tag{5}$$

For $dE_{HF}/dr_s = 0$ this gives $E_{HF} = -0.38\gamma^{2/3}$. Thus, the exciton binding energy in drops, according to this simple theory, is numerically higher than unity for $\gamma > 4$.

E_{HF} for germanium and silicon has been calculated by Rice and Brinkmann (1973) with allowance being made for the anisotropic electron effective mass and for the presence of bands of both light and heavy holes as

$$\begin{aligned} E_{HF} &= \frac{0.468}{r_s^2} - \frac{1.136}{r_s} \quad \text{(Ge)}, \\ E_{HF} &= \frac{0.717}{r_s^2} - \frac{1.157}{r_s} \quad \text{(Si)}, \end{aligned} \tag{6}$$

where $r_h = \varepsilon\hbar^2/e^2\mu$, $\mu^{-1} = m_{0e}^{-1} + m_{0h}^{-1}$, $m_{0e}^{-1} = \frac{1}{3}(m_{e\parallel}^{-1} + m_{e\perp}^{-1})$ and $m_{0h} = 0.075 m_0$ (Ge) and $0.234m_0$ (Si). Minimum values of E_{HF} obtained from $dE_{HF}/dr_s = 0$ are equal to -0.689 and -0.467 for Ge and Si, respectively.

Calculation of the correlation energy was carried out by the random phase approximation (RPA) method, taking into account the Nozières–Pines (Combescot and Nozières 1972) or Hubbard (Brinkmann and Rice 1973) corrections. These corrections appeared to be comparatively strong because Ge and Si have many valleys. Essentially a greater role is played by the anisotropy of the electron effective mass (Combescot and Nozières 1972), or its neglect (Inoue and Hanamura 1973). Another method is based on the solution of the self-consistent correlation equations (Bhattacharya et al. 1974). The agreement of the calculations and the experimental data is wholly satisfactory. The results of calculations of the enhancement function $g_{eh}(r)$ by the RPA method and the method of the self-consistent equations appeared to be essentially different. The latter gave a value several times higher than obtained by the first method. This is not surprising, since the RPA does not take into account the multiple scattering of electrons by holes. The value of $g_{eh}(0)$, which is proportional to the radiative recombination intensity, obtained from measurements of the radiation lifetime in germanium, appeared to be 5.5, i.e., higher than the value given by the theory (Westervelt et al. 1974a,b). For a discussion of the electron–hole correlation function $g_{eh}(r)$ see the review by Singwi and Tosi (1981).

3. Condensation of excitons into electron–hole drops

The process of the electron–hole drop formation, at least in such semiconductors as germanium and silicon, can be described in the following way. When a sample is illuminated by light, electrons and holes are generated in the conduction and the valence bands. They then lose their energy through emission of optical and acoustical phonons until they come to the equilibrium with the lattice. At liquid-helium temperatures this process takes approximately 10^{-10} s. Then (or during "cooling down") electrons and holes are bound into excitons. The formation cross section is equal to 10^{-10}–10^{-11} cm^{-2} and, correspondingly, the formation time is approximately equal to 10^{-8} s. Since the exciton formation time is much smaller than the exciton lifetime, one can consider that light creates excitons directly. For a drop to be in the steady state (i.e., for the drop radius to be constant in time) it is necessary that the carrier recombination rate in the drop be equal to the rate of the exciton capture by the drop. If the drop radius, R, is less than the exciton mean free path (Pokrovskii et al. 1970), one can write

$$\tfrac{4}{3}\pi R^3 \frac{N_0}{\tau_d} = 4\pi R^2 \bar{v}(n_{ex} - n_s), \tag{7}$$

where N_0 is the carrier density in a drop, $\bar{v} = (8kT/\pi m^*)^{1/2}$ is the thermal exciton velocity, $m^* = m_e + m_h$, $n_s = N_{ex} \exp[-E_b(R)/kT]$, $E_b(R)$ is the work function of the exciton in a drop, N_{ex} is the density of states for excitons and n_{ex} the density of excitons.

Since the surface energy is a part of the drop energy, E_b depends on the drop radius R,

$$E_b(R) = E_b - \frac{2\sigma}{RN_0}, \tag{8}$$

where E_b is the energy of binding an exciton in the e–h liquid, σ is the surface tension coefficient. A drop of radius R is in equilibrium with the exciton gas, if the exciton gas density is equal to

$$n_{ex} = 3\frac{RN_0}{\bar{v}\tau_d} + n_s. \tag{9}$$

This formula is not valid if R is larger than the exciton mean free path ($R \gg l_{ex}$). The value of l_{ex} is not known precisely, but can be estimated as equal to 5×10^{-4} cm at 2 K. In the case when $R \gg l_{ex}$ the capture rate of excitons is limited by their diffusion, and

$$n_{ex} = \frac{1}{3}\frac{N_0 R^2}{D\tau_d} + n_s, \tag{10}$$

where D is the exciton diffusion coefficient.

Equations (9) and (10) show the most essential difference between the exciton liquid and ordinary liquids. This difference results from the fact that the e–h pair lifetime is limited. For this reason an e–h drop can exist only when oversaturation takes place. The bigger the drop, the higher is the value of this oversaturation. The concentration of excitons as a function of the drop radius, $n_{ex}(R)$, has a minimum in equilibrium. At small R, the limitation of n_{ex}, as in the case of ordinary liquids, will be determined by the exciton evaporation from the surface of small drops.

At larger R, the exciton concentration n_{ex} is determined by the recombination rate of the e–h pairs in a drop. The exciton condensation kinetics was investigated by Keldysh (Bagaev et al. 1976a) and Silver (1975a,b). The minimum of n_{ex} is reached at $dn_{ex}/dR = 0$ or

$$\frac{1}{3}\frac{N_0}{\tau_d} - \frac{2\sigma\bar{v}N_{ex}}{kTN_0R_{min}^2}\exp\left[-\left(\frac{E_b}{kT} - \frac{2\sigma}{kTN_0R_{min}}\right)\right] = 0. \tag{11}$$

If drops are big enough, this equation can be rewritten in the form

$$R_{min} = \left(\frac{6\bar{v}\tau_d\sigma N_{ex}\exp(-E_b/kT)}{kTN_0^2}\right)^{1/2}. \tag{12}$$

It follows from eq. (11) that at $T = 2$ K, $R = 10^{-5}$ cm; and at $T = 4.2$ K, $R = 10^{-4}$ cm. These drop sizes could be stationary only in the absence of fluctuations of the generation–recombination process. Since the drops are not eternal, generation of new drops and, correspondingly, new embryos of drops are required. Drops can be destroyed not only by fluctuations of the exciton capture rate and the recombination of excitons in the drops, but also by the Brownian movement of drops from the sample region in which carriers are generated, towards the unilluminated regions. To increase the rate of the nucleation, it is necessary to raise the exciton concentration.

Thus, we have practically always two values of the drop radius, R_{min} and R_{max}, for which eq. (9) is valid. The drops with R_{min} are in equilibrium with the exciton gas mainly at the expense of exciton evaporation due to the small work function of excitons in small drops; the drops with R_{max} are in equilibrium with the exciton gas mainly at the expense of recombination of carriers in a drop. If the radiative excitation is increased, the value of n_{ex} rises, but immediately the process starts, which tends to decrease n_{ex} down to the previous value. Actually, if n_{ex} rises, then the number of embryos which grow up to big drops increases, and this decreases n_{ex}. Thus, if the excitation is intensified gradually enough, n_{ex} and, similarly R_{min} and R_{max}, remain practically unchanged, but the number of drops grows. The growth rate is proportional to $\exp(-\frac{4}{3}\pi\sigma R_{min}/kT)$, and since the value of this exponent is small the growth rate will depend heavily on R_{min}. The presence of condensation centers in the semiconductor is likely to result in a drastic lowering of the nucleation barrier. However, at present, all known experimental results can be explained by homogeneous condensation. The question on the heterogeneous condensation is still open.

A very interesting phenomenon relating to the condensation mechanism of excitons in germanium has been found by Lo et al. (1973). They have shown that in very pure germanium two thresholds occur in the excitation intensity. The first one corresponds to disappearance of the drop recombination radiation upon decreasing the excitation intensity. Drops in this case have a radius given by eq. (11). Destruction of the drops supports the equilibrium described by eq. (9). The second threshold takes place at strong excitation, i.e., upon increasing the exciton concentration needed to increase the number of growing embryos. Thus, the recombination radiation intensity dependence on excitation exhibits a hysteresis, i.e., the ascending and descending branches do not coincide. The theory of the formation of exciton drop embryos was developed by Westervelt (Westervelt 1976a,b, Westervelt et al. 1978a) on the basis of the Becker–Döring equation (Becker and Döring 1935). This theory is based on the following physical model. A drop containing ν excitons (EHD) is located in a homogeneous gas of free excitons (FE). The drop can be formed by the incorporation of an exciton into a drop with $\nu - 1$ excitons,

$$\mathrm{EHD}_\nu = \mathrm{EHD}_{\nu-1} + \mathrm{FE}, \tag{13}$$

or due to a drop with $\nu + 1$ excitons losing one exciton by evaporation or recombination,

$$\mathrm{EHD}_\nu = \mathrm{EHD}_{\nu+1} - \mathrm{FE}. \tag{14}$$

Then it was supposed that all drops with $\nu > 1$ are stable and can participate in a series of captures, recombinations and evaporations determined by reactions (13) and (14). Westervelt has calculated equilibrium values of the drop radii for both the ascending and descending branches of the hysteresis mentioned above in dependence on temperature (fig. 3). As is seen from fig. 3 the values of radii for both branches coincide at low temperatures. This is ascribed to destruction of drops due to fluctuations of the exciton recombination and evaporation. The drop size increases with temperature resulting in a sharp decrease of the drop destruction probability. The dashed line gives the radius for the ascending branch obtained by allowing for the

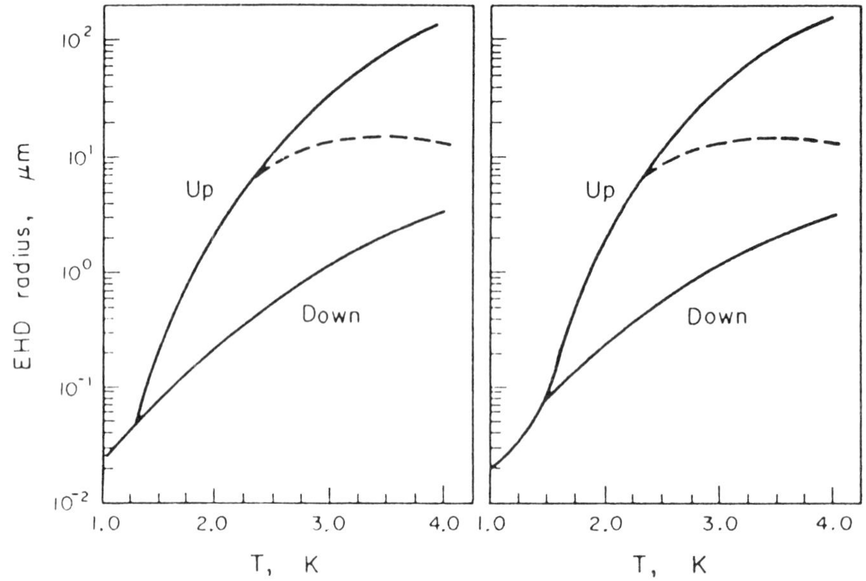

Fig. 3. Drop radius in germanium for the ascending and descending branches of the luminescence hysteresis curves for homogeneous (a) and inhomogeneous (b) nucleation, as calculated by Westervelt (1976a). The diffusion-limited radius is shown by the dashed curves.

limitation of the capture rate of excitons by their diffusion (eq. (10)). Westervelt has also made the assumption that as a result of the Brownian movement drops never leave the region illuminated with the exciting light. This is practically impossible, and one has to assume that drops are captured by impurity centers. Westervelt et al. (1978b) have demonstrated that drops remain motionless for more than 10^4 s. The drop diffusion coefficient estimated from the experiment was $D_p \leqslant 10^{-9}\ \mathrm{cm^2\ s^{-1}}$.

Such a small value of the diffusion coefficient is an indication that drops are captured by impurity centers. Etienne et al. (1978) have given experimental evidence that the hysteresis is associated with the immobility of a drop. They observed an unusual form of the hysteresis (fig. 4). It existed at medium temperatures and was absent at both low and high temperatures. The explanation of this appeared to be simple. At temperatures below the helium λ-point heat was removed through the sample surface, while at temperatures higher than the helium λ-point it dissipated via the copper sample holder. This generated a phonon stream from the sample which "blew away" the exciton drops captured by impurity centers.

The first measurements of drop sizes were managed in the experiments with Ge p–n junction (Asnin et al. 1970, 1972, Benoit à la Guillaume et al. 1971). The experimental idea is a very simple one. If a drop enters a strong electric field region, it is destroyed and this gives a big current pulse through the p–n junction. The charge carried by this current pulse yields the number of electron–hole pairs in a drop. Drops with sizes from 2 to 40 μm at $T = 2$ K have been detected. Small drops were registered at low excitation. Very big drops were observed on illuminating the

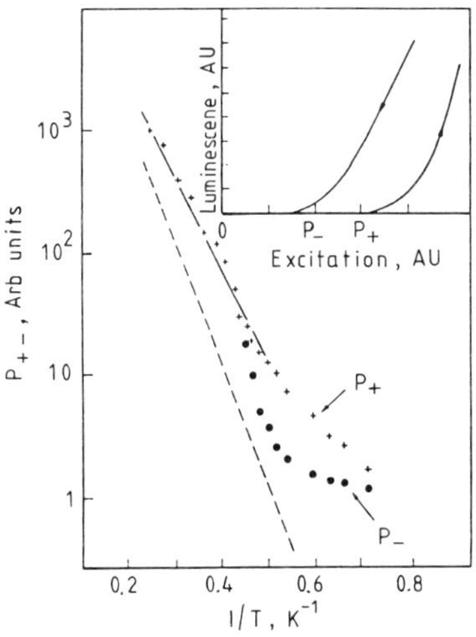

Fig. 4. Temperature dependence of the formation P_+ and breakup P_- excitation intensities of the hysteresis luminescence curve. The solid line corresponds to the slope at 16 K. The dashed line to that at 23 K. The experiment was made under conditions for which the flux of non-equilibrium phonons was created in the sample which had a temperature above the λ-point (Etienne et al. 1978).

sample by pulsed light (Asnin et al. 1972) (fig. 5). Measurements of drop sizes by means of this technique were carried out also by Benoit à la Guillaume et al. (1971) and Christensen and Hvam (1974). These experiments have shown that drops grow with both temperature and exciting light intensity. Determinations of drop sizes from light scattering measurements were made by Pokrovskii and Svistunova (1971). They

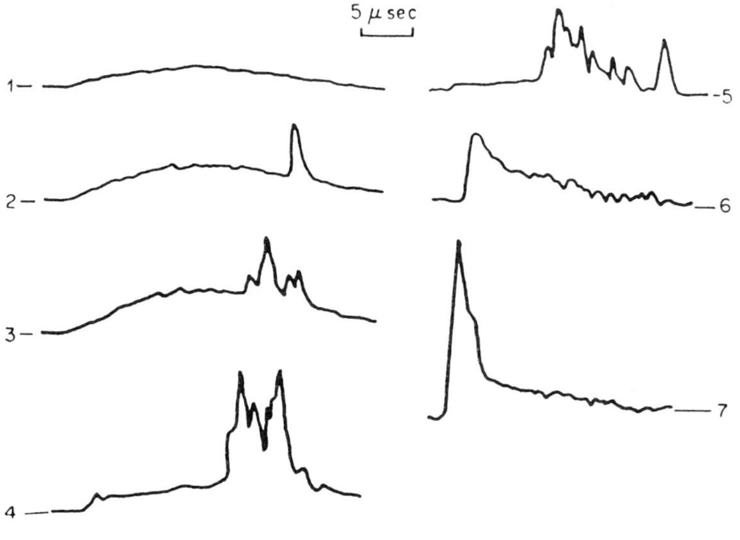

Fig. 5. Oscillograms of a p–n junction current against time. The excitation level increases as one passes from the arrows labelled 1 to 7 (Asnin et al. 1972).

have found drops at $T = 2$ K with a radius (in the vicinity of the illuminated surface) of $R = 7.6$ μm and at a distance of 1 mm from the surface, with $R = 3.4$ μm. Worlock et al. (1974) have obtained, at $T = 2.0$ K and higher excitations, drop radii of 4 μm and 2.5 μm. Voos et al. (1974) have estimated the drop sizes at $T = 2.0$ K at $R = 2 \pm 0.5$ μm. Bagaev et al. (1976b) have found that the drop size depends on both the temperature and the intensity, and also on the front steepness of the exciting light pulse: the steeper the leading edge of the light pulse, the smaller were the drops. This fact is in qualitative agreement with the Westervelt theory of drop nucleation (Westervelt 1976a,b). The drop radius was obtained from a shift of the blue part of the drop luminescence spectrum. Altukhov and Rogachev (1981) have found in silicon a blue shift of the e–h drop luminescence spectrum which was of the order of 1.5 meV from the moment when the drops show up in the spectrum up to rather high excitation level. They have estimated the radius of small drops as 200 Å. One should also take into account that the region between the exciton spectrum and that of e–h drops may be filled with the spectrum of multiexciton complexes. This problem will be discussed in § 6 of this review. The drop size growth on increasing the excitation level shows (eq. (9)) that the exciton concentration also rises.

The Westervelt theory (Westervelt 1976a,b) suggests that an increase of the exciton concentration by only 10% leads to a rise in the rate of the condensate nucleation by several orders of magnitude. Experiment has shown that the exciton concentration may exceed the threshold concentration of the e–h drop formation by more than an order of magnitude. The experimental facts discussed below indicate, however, that a drastic growth of condensation is not observed. Rogachev (1980) has shown that in order to explain this, at first sight, paradoxical situation one should take into account the capture of embryos by drops. A simple calculation illustrates this idea. At a high excitation level the exciton evaporation from drops cannot be accounted for. Let us also suppose that each exciton becomes, with a certain probability, an embryo which can capture other excitons. Thus, after a time t the exciton becomes an embryo with a radius R,

$$R = \frac{3}{4}\frac{n_{ex}\bar{v}}{N_0}(t + \Delta t). \tag{15}$$

where $\Delta t \equiv \frac{4}{3}(r_h N_0/\bar{v}n_{ex})$ is introduced to make the initial embryo radius equal to the Bohr radius, r_h, of the exciton. As the radius of the embryo grows, its thermal velocity decreases, resulting in a lowering of the probability of capture by big drops. Only those embryos that are never captured by drops in time t may grow into big drops, during which the embryos will grow up to such a size that one may regard them as at rest. Due to lifetime fluctuations some embryos can exist longer than the average lifetime. Then, an embryo converts without obstacles into a big drop, the radius of which is determined by formula (9). The drop size is limited by the e–h pair recombination rate inside the drop. An average number μ of captures of embryos during time t is equal to

$$\int_0^t \frac{dt}{\tau_c} = \int_0^t \frac{2\pi R^2 \bar{v} N_d}{\{(\pi/3)[\frac{3}{4}\bar{v}(n_{ex}/N_0)]^3 N_0\}^{1/2}(t + \Delta t)^{3/2}}\, dt = \mu, \tag{16}$$

where $N_d = 3G\tau_d/4\pi R^3 N_0$, G is the exciton generation rate and τ_d is the e–h pair lifetime in a drop.

The value of this integral does not depend much on the upper limit for large t, and one may consider that $t = \infty$. Then,

$$\mu = \left(\frac{3}{\pi}\right)^{1/2} \frac{G\bar{v}\tau_d^2}{r_h^{1/2} R^2 N_0^{3/2}}. \tag{17}$$

The probability that no capture will occur as a result of the fluctuation is proportional to $e^{-\mu}$. Assuming that the rate of the drop coalescenses due to the Brownian motion is equal to the drop formation rate, and omitting numerical constants of the order of unity, one obtains

$$R^2 = \frac{G\tau_d \bar{v}}{r_h^{1/2} N_0^{3/2}} \left[\ln\left(\frac{N_0^{5/2} R^{6.5}}{\bar{v} l_0 G \tau_d^2} \right) \right]^{-1}, \tag{18}$$

where $l_0 = \bar{v}\tau'_p$ and τ'_p is the time, during which a drop loses its momentum. For germanium τ'_p was calculated by Keldysh and Tikhodeev (1975). Using values of N_0 and τ_d from the Westervelt paper (Westervelt 1976b) one obtains for germanium $R = 2.2$ μm and for silicon $R = 0.1$ μm. In case of germanium $G\tau_d$ is taken as 2×10^{15} cm^{-3}, and for silicon as 10^{16} cm^{-3}.

The drop sizes in dependence on the excitation level were studied experimentally by Grossman et al. (1977) by means of the p–n junction technique involving giant pulses (Asnin et al. 1970, 1972). The results of their experiment are shown in fig. 6. The solid curve represents the calculation according to formula (18), in which $G\tau_d$ was used as a fitting parameter. It should be noted that the e–h pair concentration,

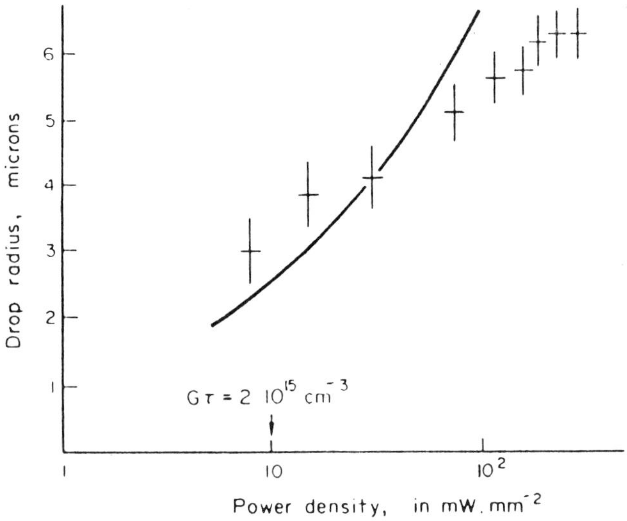

Fig. 6. Drop radius in germanium measured by means of a p–n junction current pulses technique by Grossman et al. (1977). The solid curve is based on eq. (18) and on the assumption that on exciton power of 10 mW/mm^2 corresponds to $G\tau = 2 \times 10^{15}$ cm^{-3} (Rogachev 1980).

taken as being equal to 2×10^{15} cm^{-3}, corresponds to this estimate when the e–h pair recombination rate is taken equal to the generation rate. Note also that in the experiments on the determination of drop sizes at large excitation levels the volume occupied by the drop cloud is expanded due to the phonon drag effect. The idea that drops can be dragged by the phonon wind was first proposed by Keldysh (Keldysh 1976, Bagaev et al. 1976c) who considered the effect of both long-wavelength and short-wavelength phonons. It was shown in a number of papers (Astemirov et al. (1976), Asnin et al. (1980)) that the main drag mechanism is the absorption by the e–h drops of the long-wavelength acoustical phonons with the wave vector $\boldsymbol{q} \leqslant 2\boldsymbol{k}_F$ ($\boldsymbol{k}_F$ is the Fermi vector in the e–h drops). It was experimentally proved that the e–h drops themselves may be the main source for such phonons (Asnin et al. 1977, 1978, Jochler and Worlock 1978, Asnin et al. 1981). Then, if one generates the e–h drop cloud with the help of a sharply focused light beam, the volume occupied by the e–h drops will be expanding with rising light intensity (fig. 7). Figure 8 presents the experimental result showing that dragging of drops takes place due to phonons generated by the drops themselves. To detect the e–h drops which pass some distance from the place where excitons were created, the p–n junction technique was used (Asnin et al. 1970, 1972). The time of flight of e–h drops over the distance from the generation point to the detection point (of the order of 1 mm) depended on the

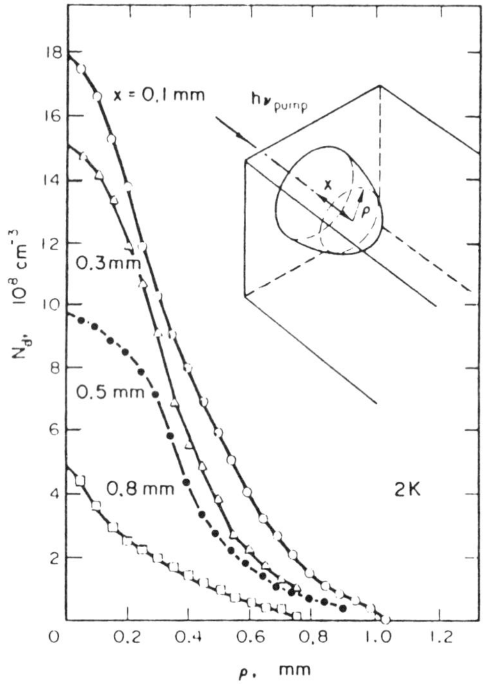

Fig. 7. Droplet concentration N_d in a hemispherical cloud in germanium determined by deconvolution of the absorption profile data. The distribution is shown as a function of the coordinate ρ measured from the beam axis with depth x as a parameter (Mattos et al. 1976).

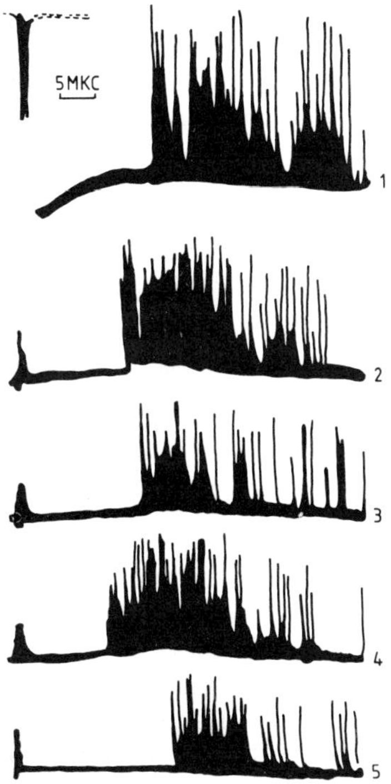

Fig. 8. Oscillograms of current pulses in a p–n junction which detects droplets moving along a magnetic field with strength (in T) equal to: (1) 0.8; (2) 1; (3) 1.25; (4) 1.7; (5) 2.3. Time mark is equal to 5×10^{-6} s. Each small pulse corresponds to one droplet of $R = (1–1.5) \times 10^{-4}$cm (Asnin et al. 1987b).

magnetic field. Oscillations in the time of flight occurred as the density of states are on the Fermi surface of the e–h liquid. The experiment has also demonstrated that the drops that fly away have approximately similar radii (of the order of 1 μm). Pokrovskii and Svistunova (1974) measured dimensions of the luminous spot in experiments on the e–h drop luminescence. As is seen from fig. 9, upon increasing the power by a factor of 40, the luminous spot diameter increases from 0.2 to 1 mm.

Embryo destruction plays an important role in the formation of the e–h liquid in thin samples. As a matter of fact, this was understood only in the last few years. Historically, the liquid was first found in a thin sample (Asnin and Rogachev 1969a,b). Condensation of e–h drops in thin samples appears to be very different from that in thick ones. Many researchers considered that "good" results could be obtained only with thick samples. It is easy to explain the reason for this. A luminescence line observed by Haynes in silicon (fig. 10) was ascribed by him to biexcitons. Two main circumstances caused Haynes to do so. First, the intensity of the new line grows as the exciton line intensity squared. Second, the line had a significant width, and for this reason it is clear that, when one exciton recombines, a part of the energy is simultaneously transferred to the exciton which is left behind, or the energy is used up in breaking a bond of the exciton left behind. The similar line in germanium

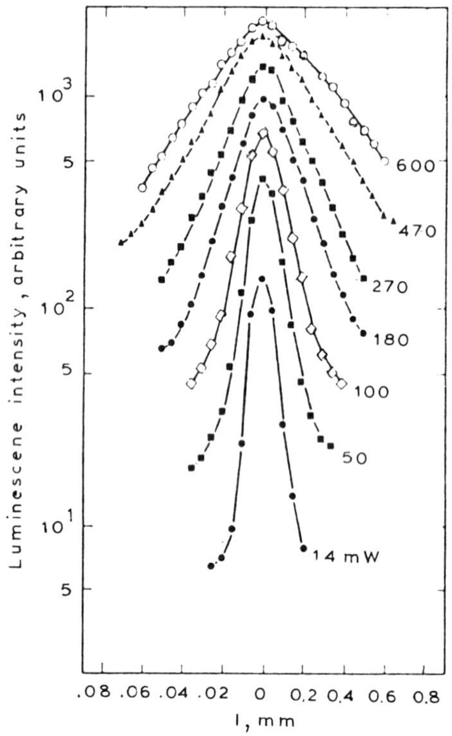

Fig. 9. Expansion of the e–h drop cloud in germanium with increasing excitation intensity (Pokrovskii and Svistunova 1974).

behaves in the same way with the only difference, as was pointed out by Pokrovskii and Svistunova (1969), that the new line intensity is proportional to the cubed, rather than the squared, exciton line intensity. It is clear that both square-law and cube-law dependences can be attributed to the exciton molecules, but cannot be attributed to the exciton liquid. Today it is well known that the new lines mentioned above belong to the liquid.

Although being hardly distinguished in the luminescence spectra, the exciton condensate and biexcitons (and multiexciton-complexes) differ considerably in other respects. The exciton condensate was observed first in studying the edge of the direct-transition modulation spectra in germanium at the Γ-point (Asnin and Rogachev 1969a,b, Asnin 1973). The experimental idea is a simple one: in places with high free-carrier concentration the absorption spectrum of the direct transitions near the exciton absorption line changes drastically, the latter vanishing due to screening of the Coulomb potential by free electrons and holes. At still higher carrier concentration the Burstein–Moss shift of the absorption edge is observed, resulting from the filling of the valence band edge. The threshold of this phenomenon is well known in absorption modulation spectra. However, the temperature dependence of the threshold was very weak: in raising the temperature from 2 to 4.2 K it changed by a factor

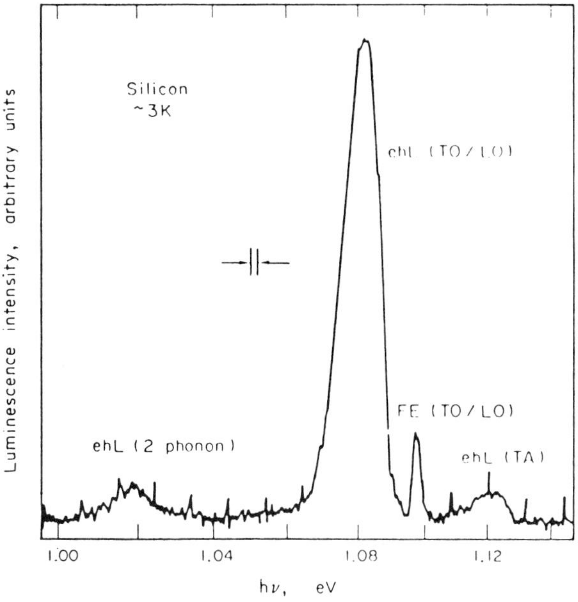

Fig. 10. Radiative recombination discovered by J.R. Haynes (1966).

of 2 (either an increase or a decrease). The threshold of the luminescence onset showed the same behaviour. A telling argument in favour of the e–h drops has been adduced in the work of Pokrovskii and Svistunova (1969), namely: the temperature dependence of the new luminescence line intensity is considerably stronger than would be expected for biexcitons (but not multiexcitons). Different is the physics of condensation in thick and thin samples (Asnin et al. 1983a,b). In thick samples (of the order of 1 mm) the e–h drop embryo has to "try" not to evaporate, while in thin samples it has to avoid striking the surface, where it may recombine.

The embryos which appeared on the surface are destroyed during a time $\tau_d = d/S$, where S is the surface recombination rate. Since in a well-etched sample $S \simeq 100$ cm/s and $\tau_d \simeq 10^{-7}$ s, while the exciton lifetime $\tau_{ex} = d/S$ for a 10^{-3} cm thickness sample is approximately equal to 10^{-5} s. Hence, one may consider that the embryos are destroyed on the surface instantaneously. In this consideration embryos are supposed to adhere to the surface. Let us denote density of drops formed by ν excitons as g_ν. Then, eqs. (15) and (16) may be written (Asnin et al. 1983b):

$$\frac{dg_\nu}{dt} = J_\nu - J_{\nu+1} + D_\nu \frac{d^2 g_\nu}{dx^2}, \tag{19}$$

$$J_\nu = g_{\nu-1} \beta n(\nu - 1)^{2/3} - g_\nu [\alpha_\nu \nu^{2/3} + (\nu/\tau_0)], \tag{20}$$

with boundary conditions $g_\nu(0) = 0$ and $g_\nu(d) = 0$, determining the surface recombination rate as equal to infinity. In eq. (20) $\beta = \pi r_0^2 \bar{v}$, $r_0 = 3/(4\pi n)^{1/3}$, $\alpha_\nu = \beta n_s \exp(8\pi r_0^2 \sigma / 3\nu^{1/3} kT)$, $n_s = N_c \exp(-\Phi/kT)$, Φ is the exciton work function in a drop and it is supposed to be independent of v. J_v is the conversion rate of an embryo with $v-1$ excitons into one with v excitons.

For the steady state and uniform exciton distribution eqs. (19) and (20) can be reduced to the following equation

$$g_{v-1}\beta n(v-1)^{3/2} - g_v\left[v^{2/3}(\beta n + \alpha) + \frac{v}{\tau_0} + \frac{\pi^2 D_0}{v d^2}\right] + g_{v+1}\left[\alpha_{v+1}(v+1)^{2/3} + \frac{v+1}{\tau_0}\right] = 0, \tag{21}$$

where $D_v = D_0 v^{-1}$ is chosen as the simplest dependence of D_v on v. Equation (18) differs from the equation used by Westervelt (Westervelt 1976a,b) only by the presence of the term $\pi^2 D/vd^2$.

Summing eq. (21) from some v to infinity, we obtain the following recurrence relation which is convenient for a numerical solution:

$$g_{v-1}\beta n(v-1)^{2/3} - g_v\left[v^{2/3}\alpha_v + \frac{v}{\tau_0}\right] = \sum_{j=v}^{\infty} \frac{\pi^2 D_0}{j d^2} g_v. \tag{22}$$

The results of calculations for germanium samples of different thicknesses are shown in fig. 11. The exciton diffusion coefficient was taken as $D_0 = kT\tau_p/m_{ex}$, where m_{ex} is the exciton mass. The value of τ_p was chosen as a fitting parameter, and is equal to 5×10^{-9} s at $T = 1.6$ K.

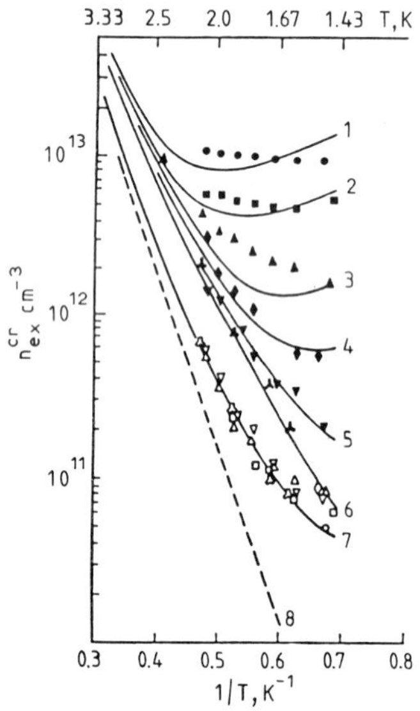

Fig. 11. Phase diagram of the exciton-gas–e–h liquid transition for the germanium samples of different thickness, d (in cm): (1) 0.003; (2) 0.005; (3) 0.0011; (4) 0.021; (5) 0.085; (6) 0.3; (7) thick samples; (8) saturated vapour (Asnin et al. 1987b).

For a thin sample ($d = 10^{-3}$ cm) with a donor concentration of 10^{15} cm^{-3} the phase diagram was the same as for a thick sample. It was assumed that τ_p depends on temperature in the way predicted by the Keldysh theory (Keldysh 1971). The exciton condensation phase diagram displays a sharp exponential dependence of the critical exciton concentration on the left-hand side of fig. 11, and in the lower temperature region one finds a weak temperature dependence. In the first region mentioned above the surface recombination acts mainly like other types of recombination. Thus the high-temperature part of the diagram is the same as discussed by Westervelt (1976a,b). At lower temperatures the surface recombination starts to play a more important role in the destruction of embryos, the temperature affecting in the main only the value of D_v. Correspondingly, the critical $n^{cr}{}_{ex}$ increases upon lowering of the temperature. It has been found experimentally that n^{cr}_{ex} tends to be independent of T (fig. 11), although in the first experiments on exciton condensation (Asnin and Rogachev 1969b) a weak growth of n^{cr}_{ex} was observed upon decreasing T. The theory predicts that, if one makes both embryos and drops immovable, the experimental data obtained on thin and thick samples must not differ substantially. An experiment actually proves this (fig. 12).

4. Electrical conductivity of electron–hole drops

As may be inferred from the discussion presented in the preceding sections, there exists a little hope that in germanium, silicon and other multivalley semiconductors the electron–hole gas will be dielectric. This is evident from the standpoint of the Keldysh and Kopaev (1964) theory in which it is shown that the e–h gas at low temperature has to be a dielectric when the semiconductor has only one valley in each of the

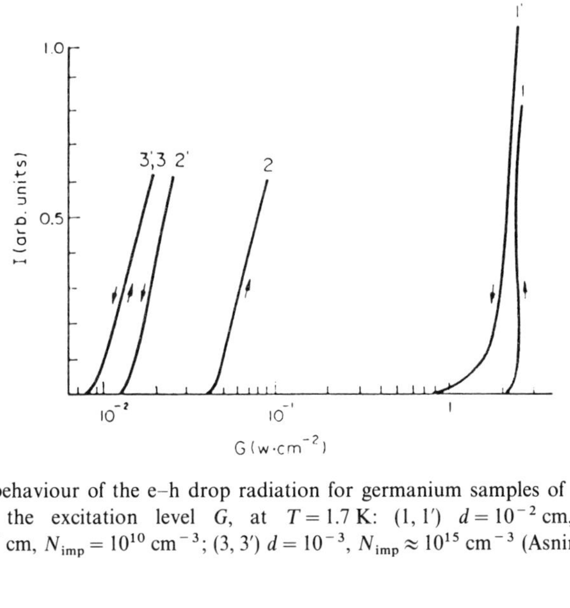

Fig. 12. Threshold behaviour of the e–h drop radiation for germanium samples of different thicknesses as a function of the excitation level G, at $T = 1.7$ K: (1, 1′) $d = 10^{-2}$ cm, $N_{imp} \approx 10^{10}$ cm^{-3}; (2, 2′) $d = 10^{-7}$ cm, $N_{imp} = 10^{10}$ cm^{-3}; (3, 3′) $d = 10^{-3}$, $N_{imp} \approx 10^{15}$ cm^{-3} (Asnin et al. 1983b).

allowed bands and the electron and hole effective masses are isotropic. It should be noted that as has been shown by Asnin and Rogachev (1968), Rogachev (1968), Asnin and Rogachev (1972), in germanium the exciton gas of sufficient density becomes a conductor, at least at $T > 1.7$ K. For experimental investigation of the dielectric–metal transition, semiconductors with a large carrier lifetime are required, since only then will the e–h gas temperature be close to the lattice temperature. At present, only two semiconductors are known that satisfy this condition. They are germanium and, in part, silicon (Rogachev 1968). However, these semiconductors are, unfortunately, multivalley ones and have anisotropic carrier effective masses. Thus the chances that the conclusions of the important work of Keldysh and Kopaev (1964) can be confirmed experimentally are, unfortunately, not very great. The main reason is that all semiconductors with isotropic carrier effective masses investigated up to now have direct transitions in the Γ-point. This means, as a rule, that the carrier lifetime is small.

Basic difficulties in investigating the e–h drop electrical conductivity are, first, creation of a significant e–h pair density without overheating the sample and, second, measurements of the created e–h pair concentrations. To decrease overheating the sample thickness is usually chosen as small as 10 μm (approximately). As far as the author knows, the problem of the e–h pair concentration measurements has never been solved satisfactorily. Asnin and Rogachev (1968), Rogachev (1968) and Asnin and Rogachev (1972) carried out such measurements in the following way. First, the sample photoconductivity was measured at room temperature. The sample was illuminated with pulsed light with a pulse duration much less than the carrier lifetime. In addition, it was supposed that at liquid-helium temperatures the light pulse creates as many carriers as it does at room temperature. Determination of the concentration at room temperature and not very high injection level presents no special problems, since the electron and hole mobilities in pure germanium are well known. After the e–h pair concentration in the exciton liquid was reliably measured by the optical method, it became clear that the earlier measurements underestimated the e–h density by approximately a factor of 5–10. Figure 13 shows the data in the form in which they were presented by Asnin and Rogachev (1972). Now it is well known that the magnitude of the carrier concentration is larger by a factor of 10. At pair concentrations less than $10^{16}\,\mathrm{cm}^{-3}$ (or, with correction, $< 10^{17}\,\mathrm{cm}^{-3}$) the conductivity is determined by the carriers that were the last to be bound into excitons. The mobility of these carriers drops with carrier concentration, but the conductivity does not depend significantly on it, being approximately equal to $0.1\,\Omega^{-1}\,\mathrm{cm}^{-1}$. In these samples the majority of n carriers was bound into the e–h drops, the rest being bound into excitons or remained free. At $n > 10^{16}\,\mathrm{cm}^{-3}$ a percolation over drops starts (Benoit à la Guillaume et al. 1972). Conductivity in this concentration region depends heavily on n, and fluctuates strongly in time. When the drops occupy almost the whole volume of the sample ($n = 2 \times 10^{16}\,\mathrm{cm}^{-3}$), the conductivity σ begins to depend weakly on n. σ may be approximated by $\sigma \approx n^3/T^2$. At $n = 2 \times 10^{17}\,\mathrm{cm}^{-3}$ and $T = 2$ K the mobility reaches $10^6\,\mathrm{cm}^2\,\mathrm{V}^{-1}\,\mathrm{s}^{-1}$. In the e–h plasma, when the e–h density is high, the magnitude of the carrier mobility is determined by the electron–hole scattering. Only those electrons and holes participate in the scattering that are

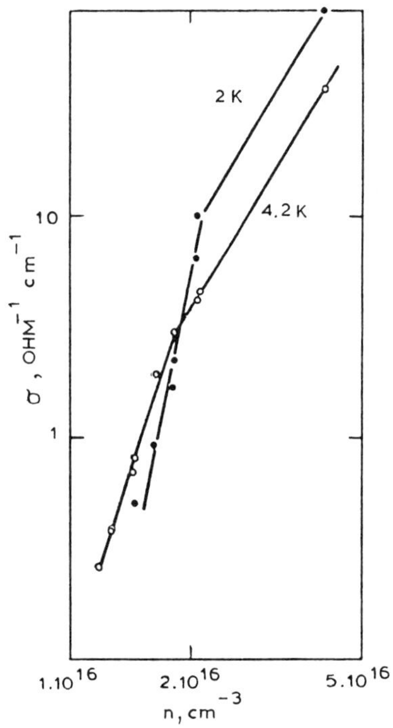

Fig. 13. Low-temperature photoconductivity of pure germanium as function of the e–h pair density (Asnin and Rogachev 1972).

located in the kT layer in the vicinity of the Fermi level (Barber 1937)

$$\sigma = e\mu n \approx \frac{E_{F,e}E_{F,h}n}{(kT)^2} \approx \frac{n^{7/3}}{T^2}, \tag{23}$$

where $E_{F,e}$ and $E_{F,h}$ are the Fermi energies of electrons and holes.

Thus, at least for $T > 1.7$ K and at high carrier concentration in germanium the dielectric gap has not been found.

The region of a low pair generation level has been studied in detail by Glicksman et al. (1976).

5. Phase diagram of electron–hole liquids in germanium

The exciton work function in the e–h liquid in germanium has a value higher than one half of the exciton binding energy. It exceeds the binding energy of the molecules of any known liquid. The high exciton binding energy in the liquid shows up in the liquid–gas phase diagram of the exciton system in germanium. If the phase diagram is plotted in dimensionless units (T/T_c and n/n_c, where T_c and n_c are critical values), one obtains a curve practically similar for all ordinary liquids but not for excitons.

Figure 14 presents a phase diagram for the e–h liquid in germanium (Thomas et al. 1978). The curve is plotted on the basis of the experimental luminescence data. As is seen in fig. 14, the experimental points differ essentially from the theoretical curve, the first being much broader. In the opinion of Thomas et al. (1978) this means that along with the spectral lines of radiation by excitons, e–h pairs, biexcitons and e–h drops, there is also present a line attributed to trions. A trion is a molecular biexciton ion constituted *either* of two electrons and one hole or of two holes and one electron. Trions, like exciton complexes not captured by impurity ions, have never been observed as a separate line in the spectrum probably due to the very large width of the spectral lines they give rise to. However, it is quite conceivable that such particles exist.

Let us consider now the phase diagram starting from low e–h pair concentrations. The Mott metal–dielectric transition criterion has the form $r_h q_D = 1$, where r_h is the Bohr radius of the exciton and q_D is the reciprocal screening radius. In the Fermi–Thomas approximation the Mott concentration is

$$n_M = \frac{3}{4\pi}\left(\frac{1}{9.9 r_h}\right)^3 = 1.7 \times 10^{14}\ \mathrm{cm}^{-3}. \quad (24)$$

This concentration is indicated by an arrow in fig. 14. The estimate (24) gives apparently, a very low value of n_M (Brinkmann and Rice 1973), since the influence of the kinetic energy is underestimated. Thomas et al. (1978) have proposed the following upper estimate of n_M. They took as n_M the concentration at which the

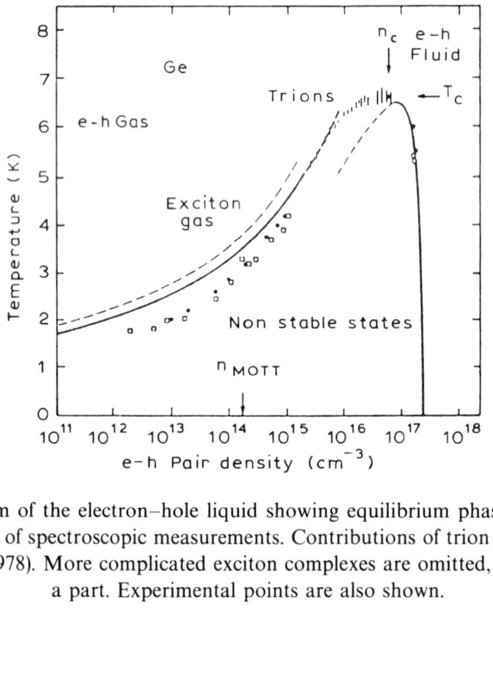

Fig. 14. A phase diagram of the electron–hole liquid showing equilibrium phase boundaries (solid line) calculated from the data of spectroscopic measurements. Contributions of trion and biexciton are shown dashed (Thomas et al. 1978). More complicated exciton complexes are omitted, but are expected to play a part. Experimental points are also shown.

exciton wavefunctions are destroyed,

$$n_{\mathrm{M}} = \frac{3}{4\pi}\left(\frac{1}{2r_{\mathrm{h}}}\right)^{3} = 2 \times 10^{16}\ \mathrm{cm}^{-3}. \tag{25}$$

This estimate, however, also does not reach the critical concentration $n_{\mathrm{c}} = (6 \pm 1) \times 10^{16}\ \mathrm{cm}^{-3}$. Thus, there exists a concentration region where excitons convert into a plasma, but drops still do not come into play. In alkali metals the phase boundary and the Mott boundary coincide. Broadening of the phase diagram with respect to the universal one (valid for many simple gases) was termed by Thomas et al. (1978) as the Mott distortion. The question of what is responsible for luminescence in the concentration and temperature regions which correspond to the Mott distortion, has not an agreed answer. Actually, at e–h pair concentrations higher than n_{M}, excitons and therefore trions and biexcitons are destroyed. One may assume that at these concentrations only plasma, metallic drops and the smallest drops (also called exciton complexes) exist. However, neither specific experiments nor comprehensive calculations have been carried out to confirm this assumption.

6. Multiexciton complexes in semiconductors

Multiexciton complexes are observed only in semiconductors with a complicated band structure and, for this reason, have no similarity with real atomic systems. Investigations of the exciton complexes help to solve the problem of how separate excitons and biexcitons convert into exciton drops. Active experimental studies were conducted on bound multiexciton complexes (BMEC) captured by donors and acceptors. Such complexes are present in the luminescence spectra as very narrow characteristic lines, and, what is more, they give several replicas, attributed to phonons and to scattering by impurity centers. Donors and acceptors entered into the composition of the BMEC have higher binding energy than that of excitons. This means that the binding energy of BMEC exceeds that of free multiexciton complexes, but, apparently, not by very much. Below we shall discuss some attempted experimental observations of such complexes.

Since BMEC in silicon with a number (m) of excitons larger than six have never been observed experimentally, $m = 6$ may be accepted as the upper limit of the complex stability in such crystals. Larger formations are, apparently, unstable. A new stability region appears with the formation of small drops. An answer to the question about the width of the instability region is of paramount importance for the embryo formation theory. The stability of small drops is determined by the surface energy and by the increase of the kinetic energy due to the restricted region in which the carriers move. An attempt to calculate the BMEC binding energy has been undertaken by Wünsche et al. (1978a,b) and Shore and Pfeiffer (1978), and calculations for small drops have been made by Rose et al. (1979). It has been shown that the complex appeared to be stable for the number of excitons not up to $m = 2$ (or 4 taking into account that the S-state contains, actually 4, rather than 2 holes), but up to $m = 10$ or, if for germanium $E_{\mathrm{ex}} + 1$ meV is considered as a sufficient binding energy, up to $m = 7$. A surprising result of these calculations is that drops of more than 10 excitons remain stable, showing some instability at $m = 40$ (for germanium). Drops consisting of more than 100 e–h pairs possess practically all the properties of a bulk e–h liquid (Rose et al. 1979). Another important result of these calculations is that they show that the interpretation of the line shift in terms of the surface tension coefficient σ is questionable, even for larger drops.

The multiexciton complexes have been experimentally discovered by Kaminskii and Pokrovskii (1970), Kaminskii et al. (1970) and Pokrovskii (1972) in silicon doped with boron and phosphorus. They found a series of new narrow luminescence lines located between the free-exciton and e–h drop lines. Sauer (1974) has shown that the dependence of each new line intensity on the excitation level and the luminescence kinetics is in correspondence with that expected for BMEC. The main experimental fact was that each new line appeared on the long-wavelength side of the spectra as the exciton intensity increased. The energy distance in the spectrum between a new line and a line which arose at a lower excitation level was correctly interpreted in the early works as the energy lost by the complex on extracting one exciton from it. However, a first paradox seemed to arise: the binding energy of each exciton goes up with increasing m in a complex. Doubts about the original assumption that the new lines belong to BMEC arose when it was found that all lines related to a different number m of excitons in a complex have an equal number of Zeeman splitting components (fig. 15). The position and the number of Zeeman splitting components in the spectra were the same as for the line due to one exciton bound on a neutral phosphorus atom. The only noticeable difference between the Zeeman spectra of different BMEC resided in distinct relative intensities of the components corresponding to different m. This was explained by a different degree of thermalization of the different Zeeman states, and this appeared to be an error. A correct explanation, given by Altukhov et al. (1977), was based on the shell model.

The first hole shell in germanium and silicon contains 4 holes whereas the first electron shell contains either 8 (Ge) and 12 (Si) electrons. Interaction of electrons with the central-cell potential splits the electron shell. The number of states in every subshell is determined by the impurity center symmetry (T_d). According to the symmetry of the central-cell field in silicon the electron subshell is split into three groups of states with symmetries Γ_1, Γ_3 and Γ_5 containing, respectively, 2, 4 and 6 states. The sequence of them is determined by the sign of the central-cell potential.

In the case of BMEC with a donor in its central cell the sequence is the same as for a neutral donor, i.e., $\Gamma_1, \Gamma_5, \Gamma_3$. Different transitions in BMEC based on a donor which are allowed in the framework of the Kirczenow shell model (Kirczenow 1977) are shown in fig. 16. Let us consider a BMEC with $m = 1$. Then two electrons completely fill the Γ_1 shell, and one hole is located in the Γ_8 shell. Only the Γ_8–Γ_1 transition is allowed from the ground state. The first excited state is the state in which one electron occupies the Γ_5 or Γ_3 shell. Since Γ_5 and Γ_3 states differ in energy, two luminescence lines occur, corresponding to the transitions Γ_8–Γ_5 (γ^1-line) and Γ_8–Γ_3 ($\gamma^{1'}$-line). To resolve these two lines a powerful spectrometer is required. The energy difference between the Γ_1–Γ_5 and Γ_1–Γ_3 levels depends on the chemical nature of the donor. In the case of a donor BMEC with 2 excitons ($m = 2$), one electron

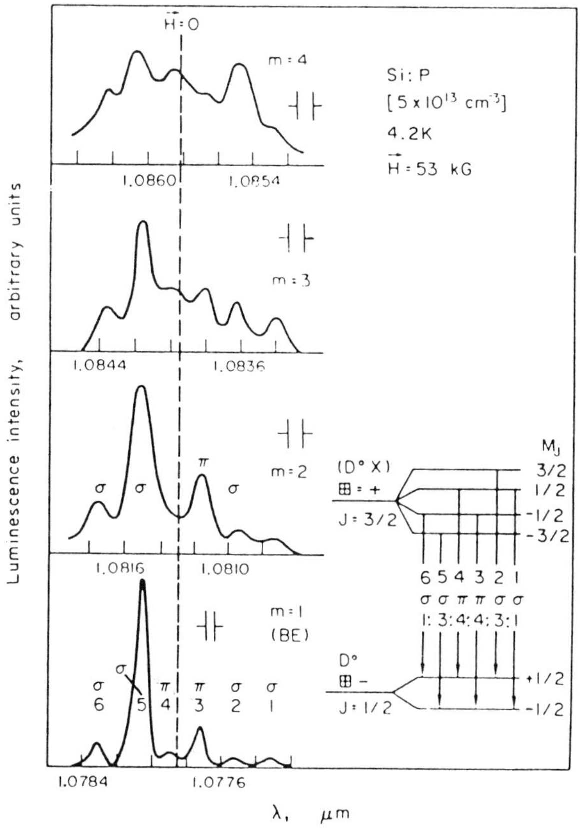

Fig. 15. Zeeman splitting of the phonon-less luminescence line of BMEC. The diagram shows transitions in the complex formed by one exciton bound to the neutral donor in silicon. σ is the circular polarization and π the linear polarization (Sauer and Weber 1976).

locates in the Γ_3 or Γ_5 shell. Further, we shall consider the energy spacing of the Γ_1, Γ_3 and Γ_5 shells to be small, as shown in fig. 16. The transition between Γ_8 and Γ_3, Γ_5 levels leaves the complex in the ground state. This line in the Kirczenow model is denoted by β_1. The Γ_8–Γ_1 transition leaves the complex in an excited state, since one electron will be in the Γ_1 and another in the Γ_3 or Γ_5 shells.

Filling of the hole shell takes place in a similar way up to $m = 4$ (or $m = 3$ in the case of the BMEC based on an acceptor), when the first hole shell becomes completely filled. Then the next shell, with a lower binding energy, has to be filled up. The Kirczenow model is virtually only a way of classifying the transitions. It does not answer, e.g., the question of whether or not the complex remains stable when the next hole shell is filled. Figure 17 shows luminescence spectra of the BMEC containing 6 excitons, two holes appearing to be in the shell following the Γ_8 shell. Thewalt (1977a,b) has noticed a general agreement between the Kirczenow model predictions and the experimental data for the BMEC based on phosphorus in silicon. The Γ_8–Γ_1

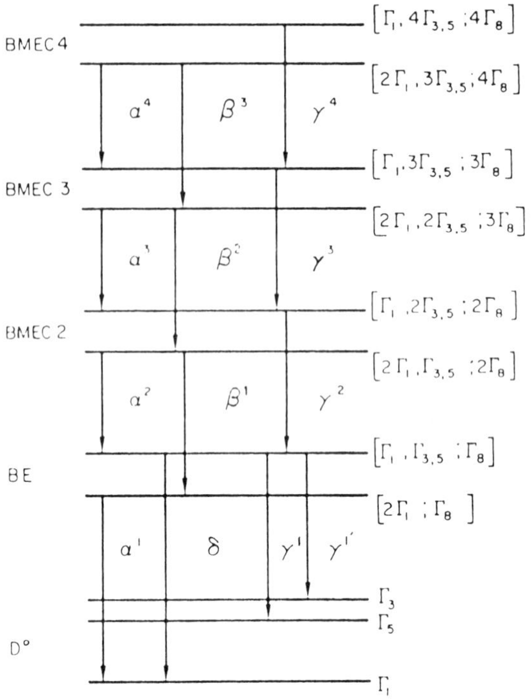

Fig. 16. Transitions in multiexciton complexes based on a substitutional donor in silicon for $m \leqslant 4$, D° is the neutral donor (Kirczenow 1977.)

transitions (α-lines) can be found in the phonon-less transitions, since the Γ_1 state has a maximum electron density in the central cell. For $m \geqslant 2$ these transitions leave the complex in an excited state. The Γ_8–$\Gamma_{3,5}$ transitions (β-lines) are possible at low temperature only for $m \geqslant 2$. Since the electron density in the $\Gamma_{3,5}$ state vanishes in the central cell, the β-lines may be observed only in the phonon replicas. Altukhov et al. (1980) have shown that the Γ_8 shell (4 states) is followed by the Γ_7 shell with 2 holes (fig. 17). Very little is known about the complexes not captured by donors and acceptors. Ashkinadze and Bel'kov (1989) have found that in a strong radio cm-wave field the intensity of the e–h drop line in silicon decreases only in the region of high energies. This is explained by the fact that the line ascribed to the drop luminescence in silicon is, actually, the superposition of multiexciton complex (MEC) and drop lines. The radiofrequency field warms up only free electrons and holes. The fast electrons destroy complexes and small drops and do not, in practice, affect big drops. Luminescence in the visible spectrum region associated with the simultaneous recombination of two excitons from a MEC was observed by Steele et al. (1987).

Difficulties in studying the MEC luminescence result from the large width of the spectrum line. Upon recombination a part of the energy is transferred to a mechanical acceleration of MEC. This is conceivable, since in silicon the radiative transitions are indirect ones.

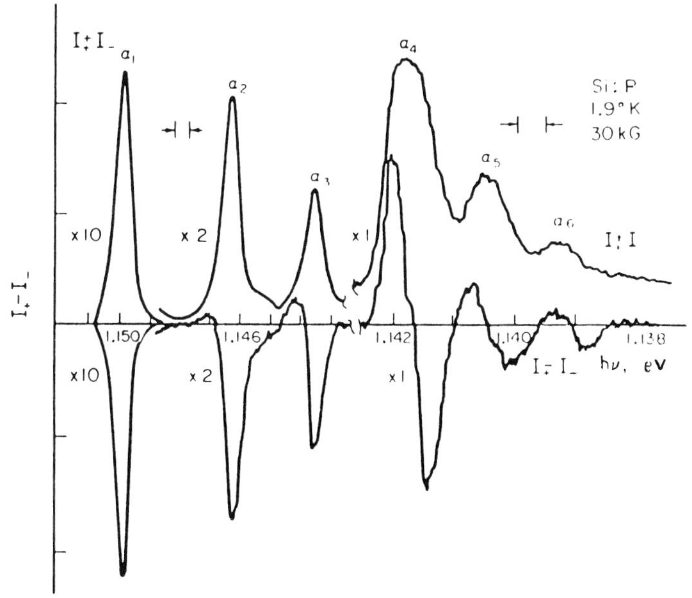

Fig. 17. Luminescence $(I_+ + I_-)$ and circular polarized luminescence $(I_+ - I_-)$ of phosphorus multi-exciton complexes in silicon. Zero-phonon α-line. $B = 3$ T, $T = 1.9$ K (Altukhov et al. 1980b).

7. Many electron–hole effects at the semiconductor surface

If an electric field is applied to a semiconductor surface, as was shown in § 1, electrons and holes participate in the screening of this field, and both electrons and holes are attracted by the surface. Let us consider this phenomenon in more detail. As a result of the field, in silicon new recombination lines have been found (Altukhov et al. 1983, 1984, Martelli 1985), the spectral positions of which depend on the strength of the electric field applied to the semiconductor surface. Upon increasing the electric field from low values to values as high as $\sim 10^7 \text{ V cm}^{-1}$ the long-wavelength boundary of these lines shifts towards low energies by roughly 5×10^{-2} eV. The width of these lines increases with electric field strength. Magneto-optical investigations have shown that the electrons and holes participating in the recombination are quasi-two-dimensional (Altukhov et al. 1984, Altukhov and Rogachev 1986). Therefore, in addition to the potential well for the carriers attracted by the field, also a potential well for the carriers of opposite sign is formed. Thus, a system is created in the semiconductor consisting of two spacially separated layers with opposite charges. For simplicity the layer closest to the surface will be considered as n-type, and the next as p-type. If one changes the direction of the field, the sign of carriers in the layers will naturally also change to the opposite one. The first layer is, in essence, a quasi-two-dimensional one, properties of which have been studied in detail (Ando et al. 1982). The existence of the second layer seems to be paradoxical, since it is constituted by the carriers repelled by the field from the semiconductor surface. If the carrier density in the first layer reaches 10^{13} cm^{-2}, the carrier density in the second layer is smaller by one or two orders of magnitude, depending on the optical excitation intensity. The carrier density in the first layer is mainly determined by the strength of the electric field applied to the surface. The addition to the carrier density of the first layer caused by illumination does not exceed 10%. Figure 18 shows the luminescence spectra at different electric field strengths. Both the distance between the layers and the thickness of the second layer may be estimated by the line shift towards higher energies with the intensity of the excitation. Both these values are of the order of 100 Å, which is several times larger than the thickness of the first layer. The small value of both the electron and the hole layer thicknesses leads to a strong overlap of the electron and hole wavefunctions and, consequently to a high intensity of the radiative recombination.

Experiments on the electrical conductivity (Altukhov et al. 1987a) enables one to obtain the carrier concentration in the second layer. In the region of relatively weak fields the e–h pairs created by light "adhere" to the surface in the form of excitons or small e–h drops. At higher fields the plasma state becomes the preferable one. The question why the electric field can attract electrons and holes to the surface, cannot be answered in the Fermi–Thomas approximation. The reason is that this approximation presupposes that the electron field strength drops monotonously to zero in the bulk of the sample, one type of carriers being attracted to the surface

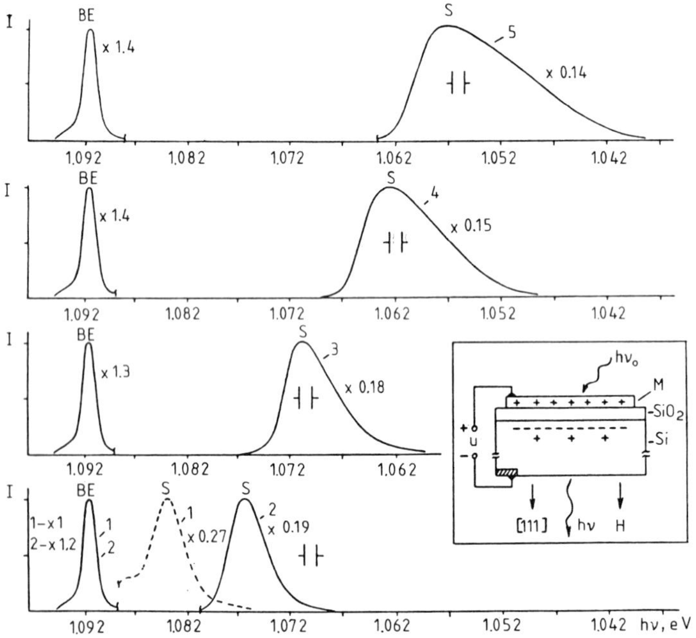

Fig. 18. Radiative recombination spectra of Si : P MOS-structure ($n = 2 \times 10^{15}$ cm^{-3}). The gate voltage (V) is: (1) 0; (2) 12.5; (3) 25; (4) 50; (5) 75. The excitation level is 0.2 W cm^{-2} of the green argon laser. BE denotes an exciton bound on a donor line. S denotes the surface line (Altukhov et al. 1984).

and another type of carriers moving in the opposite direction. The lowest approximation in which solution of the screening problems of two or more layers of carriers of opposite sign is probable, appears to be the Hartree approximation. However, for a correct description of the experimental data, consideration of the exchange and correlation corrections is required. The Hartree approximation does not allow correctly for the repulsive interaction between electrons and between holes. A part of this repulsion energy is compensated by the exchange energy.

A carrier from the second layer polarizes the first one, so that the charge induced by it may be approximately considered as the image of a carrier of the second layer in the first one (Altukhov et al. 1983, 1984). Such a potential forms a state with binding energy equal to $\frac{1}{32}e^4 m_e/r\varepsilon^2\hbar^2$, and for electrons and holes in silicon this is about 2–3 meV. This approximation is valid only for high carrier concentrations in the first layer. At lower concentrations, interaction of a hole with the electrons in the first layer ceases to have a collective character, and the main role is played by the interaction of pairs of carriers of opposite sign. This results in formation of surface excitons (Averkiev and Pikus 1988). The transition from a state in which surface excitons are formed to a state in which carriers in the second layer may participate in conductivity occurs rather sharply, when the first-layer carrier concentration attains some threshold values (Altukhov et al. 1987a).

In the quantum theory of screening (Ando et al. 1982) the electrochemical potential in the vicinity of the surface (fig. 19a) depends only on the electric field strength and carrier effective mass. It is characteristic of the theory that some electron work function Δ is involved which is a function only of the field of the surface. If one solves the problem in the Fermi–Thomas approximation, at $T=0$ $\Delta=0$. The main cause for $\Delta>0$ is that electrons occupy one or a small number of levels of the dimensional quantization. Since the probability to find electrons beyond the classical turning point for the ground states is approximately $\frac{1}{4}$, Δ is smaller than the depth of the self-consistent quantum well in the vicinity of the surface.

For the electric field in the semiconductor bulk to be equal to zero, the surface potential well has to contain $N_s=\varepsilon E/4\pi e$ electrons (E is the electrostatic field strength). This state is shown in fig. 19a, and is not necessarily an equilibrium state (it depends only on surface charge). The Fermi level (or the quasi-Fermi level) of

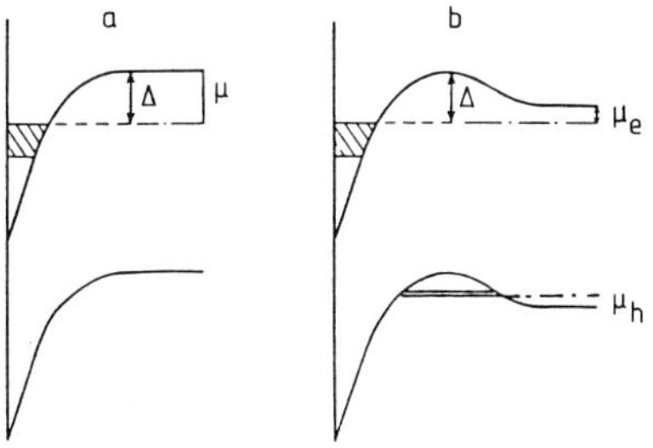

Fig. 19. The electrostatic potential near the surface: (a) $\mu=\Delta$, (b) $\mu<\Delta$. Applied field is the same in both cases, (b) is an action of illumination. It describes "overscreening"; i.e., the change of direction in the electric field.

electrons in the bulk (μ) is independently predetermined by temperature, doping, illumination, and so on. Let us assume that in the semiconductor bulk the Fermi level μ is smaller than Δ. Then, for establishing the diffusion–drift equilibrium of electrons some of them have to pass into the surface potential well. The resulting electric field is screened by donors and holes, and is opposite in direction to the external field (fig. 19b). The situation at hand corresponds to overscreening, i.e., to the case when the number of electrons in the surface potential well is larger than $\varepsilon E/4\pi e$. If the distance between the Fermi level and the conduction band bottom were larger than Δ, the number of electrons in the surface well would be less than $\varepsilon E/4\pi e$, and the remaining field would be screened by the Ohmic contact charge or by the acceptor charge (in the presence of compensation). In the same way overscreening takes place upon illuminating the sample at low enough temperatures. The value of Δ can exceed kT at liquid-helium temperature by a factor of 10–30. For the same reason the electrons generated by light are captured by the surface potential well; this produces an electric field in the opposite direction to the external one. This new field is screened by the holes which rush towards the surface. Further, the formation of a hole self-consistent well takes place, as in the case of electrons due to the external field. Since the quasi-Fermi level in the hole potential well may be further from the valence band top than it is in the bulk, overscreening occurs again. This is compensated by the formation of a new electron potential well, and so on. Thus, it follows from these qualitative considerations, which correspond to the Hartree approximation, that the electrostatic potential oscillates with distance. The oscillation amplitude attenuates, and the oscillation period increases (Altukhov et al. 1985). A numerical solution of the screening problem in the Hartree approximation by the e–h plasma, which confirms the qualitative considerations presented above, is given by Monakhov and Rogachev (1988). Figure 20 shows the potential in the vicinity of the surface and also the electron and hole wavefunction. Usually, only the case of one potential well for electrons and one for holes is observed experimentally. An observation of more than two self-consistent wells is only possible in multivalley semiconductors.

As in the case of the three-dimensional plasma, the two-dimensional plasma binding energy depends essentially on the number of valleys in the conduction and valence bands. This may be explained in the following way. The self-consistent well depth in the Hartree approximation is proportional to $E^{2/3}$, and the Fermi energy is proportional to E/γ (where γ is a number of valleys). Due to this $\Delta_{max} \approx \gamma^2$.

The e–h pair binding energy on the surface decreases with an increase of their concentration. That is why the e–h pair capture by the surface does not lead to condensation. It appears, however, that e–h pair condensation on the surface may occur also without an external electric field (Asnin et al. 1987a). Let us assume that the semiconductor (germanium) surface is covered by an oxide layer. Let us also assume that there exist many impurity centers in the dielectric forbidden gap and that electrons are captured more strongly than holes. Then, on illumination of the semiconductor a field of charged centers appears on the surface. In the places where the e–h pair concentration is greater, more electrons are captured by the impurity centers in the oxide and more holes come close to the surface, and together with them, as was discussed above, also electrons are attracted. The e–h pair concentration

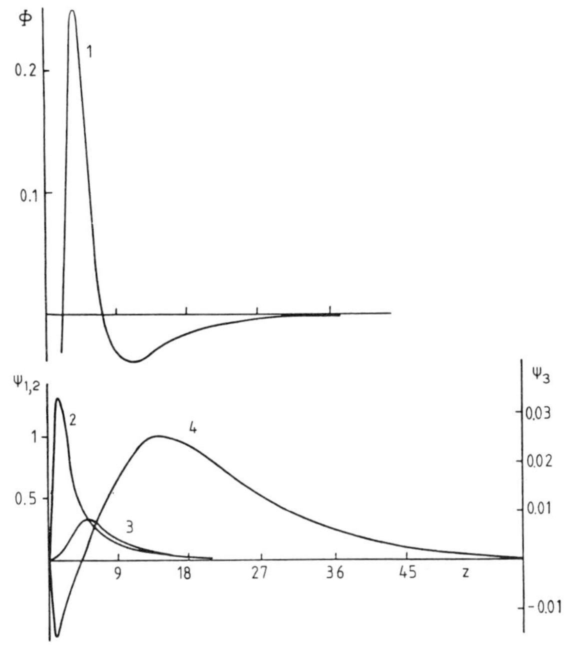

Fig. 20. Numerical solution of electrostatic potential near the surface (Monakhov and Rogachev 1988). (1) electrostatic potential; (2) electron wavefunction; (3) hole wavefunction; (4) second layer of electrons, scale $\times 30$.

in drops stops to grow, either due to overcharging of all impurity centers or due to the increase of the hole Fermi energy. From an analysis of the surface luminescence spectra it has been found that the hole concentration in surface e–h drops is $(2–3) \times 10^{12}\ cm^{-2}$ and that the Fermi energy reaches 10^{-2} eV. Similar problems of e–h drops on the surface were described by the group of Litovchenko (Litovchenko 1976, Litovchenko and Korbutyak 1981).

Acknowledgements

In the course of preparation of this review, the author enjoyed the opportunity to discuss many subjects with V.M. Asnin, N.S. Averkiev, A.M. Monakhov and G.E. Pikus. I am indebted for short but impressive discussions of the many-valley problem with L.V. Keldysh and Yu.V. Kopaev. For the translation of the review into English the generous help of G.V. Il'menkov was indispensible. Some important technical problems have been solved by Miss Ekaterina Rogacheva and Miss Natalia Sablina. The author would like to express his most profound gratitude to his wife for her understanding and patience during the work on this review, as well as for her active help in correction of the numerous mistakes.

References

Agranovich, V.M., and B.S. Toshich, 1967, Zh. Eksp. & Teor. Fiz. **53**, 149 [1968, Sov. Phys.-JETP **26**, 104].

Altukhov, P.D., and A.A. Rogachev, 1981, Fiz. Tverd. Tela **21**, 1956 [Sov. Phys.-Solid State **23**, 1142].

Altukhov, P.D., and A.A. Rogachev, 1986, Izv. AN USSR Ser. Fiz. **50**, 232.

Altukhov, P.D., K.N. El'tsov, G.E. Pikus and A.A. Rogachev, 1977, Pis'ma Zh. Eksp. & Teor. Fiz. **26**, 468 [JETP Lett. **26**, 337].

Altukhov, P.D., K.N. El'tsov, G.E. Pikus and A.A. Rogachev, 1980a, Fiz. Tverd. Tela **22**, 599 [Sov. Phys.-Solid State **22**, 350].

Altukhov, P.D., K.N. El'tsov, G.E. Pikus and A.A. Rogachev, 1980b, Fiz. Tverd. Tela **22**, 239 [Sov. Phys.-Solid State **22**, 140].

Altukhov, P.D., A.V. Ivanov, Yu.N. Lomasov and A.A. Rogachev, 1983, Pis'ma Zh. Eksp. & Teor. Fiz. **38**, 5 [JETP Lett. **38**, 4].

Altukhov, P.D., A.V. Ivanov, Yu.N. Lomasov and A.A. Rogachev, 1984, Pis'ma Zh. Eksp. & Teor. Fiz. **39**, 432 [JETP Lett. **39**, 524].

Altukhov, P.D., A.M. Monakhov, A.A. Rogachev and V.E. Khartsiev, 1985, Fiz. Tverd. Tela **27**, 576 [Sov. Phys.-Solid State **27**, 576].

Altukhov, P.D., A.A. Bakun, Yu.A. Kontsevoi, Yu.A. Kuznetsov, A.A. Rogachev, T.L. Romanova and G.P. Rubtsov, 1987a, Fiz. Tverd. Tela **29**, 2412 [Sov. Phys.-Solid State **29**, 1388].

Ando, T., A.B. Fowler and F. Stern, 1982, Rev. Mod. Phys. **53**(2).

Ashkinadze, B.M., and V.V. Bel'kov, 1989, Fiz. Tverd. Tela **31**, 167 [1981, Sov. Phys.-Solid State **31**, 1008].

Asnin, V.M., 1973, Fiz. Tverd. Tela **15**, 3298 [1974, Sov. Phys.-Solid State **15**, 2197].

Asnin, V.M., and A.A. Rogachev, 1968, Pis'ma Zh. Eksp. & Teor. Fiz. **7**, 464 [JETP Lett. **7**, 360].

Asnin, V.M., and A.A. Rogachev, 1969a, Pis'ma Zh. Eksp. & Teor. Fiz. **9**, 415 [JETP Lett. **9**, 248].

Asnin, V.M., and A.A. Rogachev, 1969b, Proc. 3rd Int. Conf. Photoconductivity (Stanford University Press, Stanford, CA) p. 13.

Asnin, V.M., and A.A. Rogachev, 1972, Pis'ma Zh. Eksp. & Teor. Fiz. **14**, 484 [JETP Lett. **14**, 338].

Asnin, V.M., A.A. Rogachev and S.M. Ryvkin, 1967, Fiz. Tekhn. Poluprovodn. **1**, 1740 [1968, Sov. Phys.-Semicond. **1**, 1447].

Asnin, V.M., A.A. Rogachev and N.I. Sablina, 1970, Pis'ma Zh. Eksp. & Teor. Fiz. **11**, 162 [JETP Lett. **11**, 49].

Asnin, V.M., A.A. Rogachev and N.I. Sablina, 1972, Fiz. Tverd. Tela **14**, 399 [Sov. Phys.-Solid State **14**, 332].

Asnin, V.M., A.A. Rogachev, N.I. Sablina and V.I. Stepanov, 1977, Fiz. Tverd. Tela **19**, 3150 [Sov. Phys.-Solid State **19**, 1844].

Asnin, V.M., N.I. Sablina and V.I. Stepanov, 1978, Pis'ma Zh. Eksp. & Teor. Fiz. **27**, 584 [JETP Lett. **27**, 551].

Asnin, V.M., N.I. Sablina and V.I. Stepanov, 1980, Fiz. Tverd. Tela **22**, 418 [Sov. Phys.-Solid State **22**, 244].

Asnin, V.M., A.A. Rogachev, N.I. Sablina and V.I. Stepanov, 1981, Fiz. Tverd. Tela **23**, 177 [Sov. Phys.-Solid State **23**, 99].

Asnin, V.M., V.V. Bel'kov, A.A. Rogachev, V.I. Stepanov and I.M. Fishman, 1983a, Zh. Eksp. & Teor. Fiz. **84**, 2129 [Sov. Phys.-JETP **84**, 1239].

Asnin, V.M., V.V. Bel'kov, A.A. Rogachev, V.I. Stepanov and I.M. Fishman, 1983b, Solid State Commun. **48**, 611.

Asnin, V.M., A.A. Rogachev, V.I. Stepanov and A.B. Churilov, 1987a, Pis'ma Zh. Eksp. & Teor. Fiz. **45**, 436 [JETP Lett. **45**, 558].

Asnin, V.M., A.A. Rogachev, V.I. Stepanov and A.B. Churilov, 1987b, Fiz. Tverd. Tela **29**, 1713 [Sov. Phys.-Solid State **29**, 985].

Astemirov, T.H., V.S. Bagaev, L.J. Paduchih and A.G. Poiarkov, 1976, Pis'ma Zh. Eksp. & Teor. Fiz. **24**, 225 [JETP Lett. **24**, 200].
Averkiev, N.S., and G.E. Pikus, 1988, Fiz. Tekhn. Poluprovodn. **21**, 1493 [Sov. Phys.-Semicond. **21**, 908].
Bagaev, V.S., T.I. Galkina, O.V. Gogolin and L.V. Keldysh, 1969, Zh. Eksp. & Teor. Fiz. Pis'ma Red. **10**, 309 [JETP Lett. **10**, 195].
Bagaev, V.S., N.V. Zamkovets, L.V. Keldysh, N.N. Sibeldin and V.A. Tsvetkov, 1976a, Zh. Eksp. & Teor. Fiz. **70**, 1501 [Sov. Phys.-JETP **43**, 783].
Bagaev, V.S., N.V. Zamkovets, L.V. Keldysh, N.N. Sibel'din and V.A. Tsvetkov, 1976b, Zh. Eksp. & Teor. Fiz. **70**, 150 [Sov. Phys.-JETP **43**, 783].
Bagaev, V.S., L.V. Keldysh, N.N. Sibel'din and V.A. Tsvetkov, 1976c, Zh. Eksp. & Teor. Fiz. **70**, 702 [Sov. Phys.-JETP **43**, 362].
Barber, W.G., 1937, Proc. R. Soc. London **154**, 383.
Becker, R.S., and W. Döring, 1935, Ann. Phys. **24**, 719.
Beni, G., and T.M. Rice, 1978, Phys. Rev. **18**, 768.
Benoit à la Guillaume, C., M. Voos, F. Salvan, J.M. Laurant and A. Bonnot, 1971, C.R. Acad. Sci., Ser. B **2**, 236.
Benoit à la Guillaume, C., M. Voos and F. Salvan, 1972, Phys. Rev. B **5**, 3079.
Bhattacharya, P.K., V. Massida, K.S. Singwi and P. Vashishta, 1974, Phys. Rev. B **10**, 5127.
Brinkmann, W.F., and T.M. Rice, 1973, Phys. Rev. **87**, 1508.
Christensen, O., and J.V. Hvam, 1974, Proc. 12th Int. Conf. on Physics of Semiconductors, Stuttgart, 1974 (Teubner, Stuttgart) p. 56.
Combescot, M., and P. Nozières, 1972, J. Phys. C **5**, 2369.
Etienne, B., J.M. Worlock and M. Voos, 1978, Solid State Electron. **21**, 1383.
Glicksman, M., M.N. Gurnee and J.R. Meyer, 1976, Proc. Auger Semin.: Physics of Highly Excited States, Tomakomai, 1975 (Springer, Berlin) p. 219.
Grossman, B., K.L. Shacklee and M. Voos, 1977, Solid State Commun. **23**, 271.
Haynes, J.R., 1966, Phys. Rev. Lett. **17**, 860.
Hensel, J.C., T.G. Phillips and G.A. Thomas, 1977, Solid State Phys. **32**, 87.
Inoue, M., and E. Hanamura, 1973, J. Phys. Soc. Jpn. **35**, 643.
Jeffries, C.D., 1975, Science **189**, 555.
Jeffries, C.D., and L.V. Keldysh, eds, 1983, Electron–Hole Droplets in Semiconductors (North-Holland, Amsterdam).
Jochler, J., and J.M. Worlock, 1978, Solid State Commun. **27**, 229.
Kaminskii, A.S., and Ya.E. Pokrovskii, 1970, Pis'ma Zh. Eksp. & Teor. Fiz. **11**, 381 [JETP Lett. **11**, 225].
Kaminskii, A.S., Ya.E. Pokrovskii and N.V. Alkeev, 1970, Zh. Eksp. & Teor. Fiz. **59**, 1937 [Sov. Phys.-JETP **32**, 1048].
Kane, E.O., 1978, Physics of Semiconductors, Proc. 14th Int. Conf. on the Physics of Semiconductors, Edinburgh, September 1978, Inst. Phys. Conf. Ser., Vol. 43 (The Institute of Physics, Bristol and London) p. 132.
Keldysh, L.V., 1968, Proc. IXth Int. Conf. on Physics of Semiconductors (Nauka, Moscow) p. 1303.
Keldysh, L.V., 1971, Eksitony v Poluprovodnikah (Nauka, Moscow) p. 5.
Keldysh, L.V., 1972, Problem of Theoretical Physics, devoted to the memory of I.E. Tamm, p. 433.
Keldysh, L.V., 1976, Pis'ma Zh. Eksp. & Teor. Fiz. **23**, 100 [JETP Lett. **23**, 86].
Keldysh, L.V., 1983, in: Electron–Hole Droplets in Semiconductors, eds C.D. Jeffries and L.V. Keldysh (North-Holland, Amsterdam) Introduction.
Keldysh, L.V., 1989, private communication.
Keldysh, L.V., and Yu.V. Kopaev, 1964, Fiz. Tverd. Tela **6**, 2791 [Sov. Phys.-Solid State **6**, 2219].
Keldysh, L.V., and A.N. Kozlov, 1968, Zh. Eksp. & Teor. Fiz. **54**, 978 [Sov. Phys.-JETP **27**, 521].
Keldysh, L.V., and A.P. Silin, 1975, Zh. Eksp. & Teor. Fiz. **69**, 1053 [Sov. Phys.-JETP **42**, 535].
Keldysh, L.V., and S.G. Tikhodeev, 1975, Zh. Eksp. & Teor. Fiz. Pis'ma Red. **21**, 582 [JETP Lett. **21**, 273].

Kirczenow, G., 1977, Can. J. Phys. **55**, 1787.
Kozlov, A.N., and L.A. Maksimov, 1965, Zh. Eksp. & Teor. Fiz. **48**, 1184 [Sov. Phys.-JETP **21**, 790].
Lampert, M.A., 1958, Phys. Rev. Lett. **1**, 450.
Litovchenko, V.G., 1976, Phys. Status Solidi B **77**, K71.
Litovchenko, V.G., and D.V. Korbutyak, 1981, Surf. Sci. **104**, L189.
Lo, T.K., B.J. Feldman and C.D. Jeffries, 1973, Phys. Rev. Lett. **31**, 1224.
Martelli, F., 1985, Solid State Commun. **55**, 905.
Mattos, J.V.C., K.L. Shacklee, M. Voos, T.C. Damen and J.M. Worlock, 1976, Phys. Rev. B **13**, 5603.
Monakhov, A.M., and A.A. Rogachev, 1988, Fiz. Tverd. Tela **30**, 1153 [Sov. Phys.-Solid State **30**, 1153].
Moskalenko, S.A., 1958, Opt. & Spektrosk. **5**, 147.
Pokrovskii, Ya.E., 1972, Phys. Status Solidi A **11**, 385.
Pokrovskii, Ya.E., and K.I. Svistunova, 1969, Pis'ma Zh. Eksp. & Teor. Fiz. **9**, 435 [JETP Lett. **9**, 261].
Pokrovskii, Ya.E., and K.I. Svistunova, 1971, Pis'ma Zh. Eksp. & Teor. Fiz. **13**, 287 [JETP Lett. **13**, 212].
Pokrovskii, Ya.E., and K.I. Svistunova, 1974, Proc. 12th Int. Conf. on Physics of Semiconductors, Stuttgart, 1974 (Teubner, Stuttgart) p. 71.
Pokrovskii, Ya.E., A.S. Kaminskii and K.I. Svistunova, 1970, Proc. 10th Int. Conf. on Physics of Semiconductors, Cambridge, MA (CONF - 70080) p. 504.
Rice, T.M., 1976, Contemp. Phys. **20**(3), 241.
Rice, T.M., 1977, Solid State Phys. **32**, 1.
Rice, T.M., 1978, in: The Metal Non-Metal Transition in Disordered Systems, eds L.R. Friedman and D.P. Tunspall (Scottish Universities Summer School in Physics Publication, Edinburgh) p. 369.
Rice, T.M., and W.F. Brinkmann, 1973, Comm. Solid State Phys. **5**, 151.
Rogachev, A.A., 1968, Proc. 9th Int. Conf. on the Physics of Semiconductors, Moscow, 1968 (Nauka, Moscow) p. 407.
Rogachev, A.A., 1974, Excitons at High Density, Springer Tracts in Modern Physics, Vol. 73 (Springer, Berlin) p. 127.
Rogachev, A.A., 1980, Prog. Quantum Electron. **6**, 141.
Rose, J.H., R.S. Pfeiffer, L.M. Sanders and H.B. Shore, 1979, Solid State Commun. **30**, 697.
Sauer, R., 1974, Proc. 12th Int. Conf. on Physics of Semiconductors, Stuttgart, 1974 (Teubner, Stuttgart) p. 42.
Sauer, R., and J. Weber, 1976, Phys. Rev. Lett. **36**, 48.
Shore, H.B., and R.S. Pfeiffer, 1978, Physics of Semiconductors, Proc. 14th Int. Conf. on Physics of Semiconductors, Edinburgh, September 1978, Inst. Phys. Conf. Ser., Vol. 43 (The Institute of Physics, Bristol and London) p. 627.
Silver, R.N., 1975a, Phys. Rev. B **11**, 1569.
Silver, R.N., 1975b, Phys. Rev. B **12**, 5689.
Singwi, K.S., and M.P. Tosi, 1981, Solid State Physics, Vol. 36 (Academic Press, New York) p. 177.
Steele, A.G., W.L. McMillan and M.L.W. Thewalt, 1987, Phys. Rev. Lett. **59**, 2899.
Thewalt, M.L.W., 1977a, Phys. Rev. Lett. **38**, 521.
Thewalt, M.L.W., 1977b, Can. J. Phys. **55**, 1463.
Thomas, G.A., 1976, Sci. Am. **6**, 28.
Thomas, G.A., G.B. Mock and M. Capizzi, 1978, Phys. Rev. B **18**, 4250.
Voos, M., and C. Benoit à la Guillaume, 1975, Optical Properties of Solid, New Developments (North-Holland, Amsterdam) p. 143.
Voos, M., K.L. Shacklee and J.M. Worlock, 1974, Phys. Rev. Lett. **33**, 61.
Wand, J.S., and C. Kittel, 1978a, Phys. Lett. A **42**, 189.
Wand, J.S., and C. Kittel, 1978b, Inst. Phys. Conf. Ser., Vol. 43 (The Institute of Physics, Bristol and London) p. 359.
Westervelt, R.M., 1976a, Phys. Status Solidi B **74**, 727.
Westervelt, R.M., 1976b, Phys. Status Solidi B **76**, 35.

Westervelt, R.M., T.K. Lo, J.L. Staehli and C.D. Jeffries, 1974a, Phys. Rev. Lett. **38**, 1054.
Westervelt, R.M., T.K. Lo, J.L. Staehli and C.D. Jeffries, 1974b, Phys. Rev. Lett. **32**, 1331.
Westervelt, R.M., J.L. Staehli and E.E. Haller, 1978a, Phys. Status Solidi B **50**, 557.
Westervelt, R.M., J.C. Culbertson and B.S. Black, 1978b, Physics of Semiconductors, in: Proc. 14th Conf. on Physics of Semiconductors, Edinburgh, September 1978, Inst. Phys. Conf. Ser., Vol. 43 (The Institute of Physics, Bristol and London).
Worlock, J.M., T.C. Damen, K.L. Shacklee and J.P. Gordon, 1974, Phys. Rev. Lett. **33**, 771.
Wünsche, H.-J., K. Henneberger and V.E. Khartsiev, 1978a, Phys. Status Solidi B **86**, 505.
Wünsche, H.-J., K. Henneberger and V.E. Khartsiev, 1978b, Physics of Semiconductors, Edingburgh, 1978, Inst. Phys. Conf. Ser., Vol. 43 (The Institute of Physics, Bristol and London) p. 615.

CHAPTER 10

Dynamics and Classical Transport of Carriers in Semiconductors

LAURA M. ROTH

Physics Department
State University of New York, Albany
USA

Handbook on Semiconductors
Completely Revised Edition
Edited by T.S. Moss
Volume 1, edited by P.T. Landsberg

Contents

1. Introduction . . . 492
2. Classical motion of carriers in external fields . . . 494
   2.1. Same results from energy band theory . . . 494
   2.2. Classical equation of motion . . . 495
   2.3. Energy bands of interest for semiconductors . . . 497
   2.4. Motion in a uniform magnetic field; cyclotron resonance . . . 503
   2.5. Motion in a steady electric field . . . 506
   2.6. Motion in combined electric and magnetic fields . . . 507
3. Quantum effects in energy bands due to external fields . . . 508
   3.1. Quantum mechanical representations for Bloch electrons in external fields . . . 508
   3.2. Semiclassical quantization of magnetic orbits . . . 510
   3.3. Beyond semiclassical quantization . . . 512
   3.4. The effective-mass approximation . . . 513
   3.5. Magnetic energy levels near band edges . . . 515
   3.6. Quantum states of band electrons in an electric field . . . 521
   3.7. Quantum states of band electrons in combined electric and magnetic fields . . . 524
4. Classical transport in semiconductors: introduction and elementary model . . . 525
   4.1. Particle kinetic theory for simple bands . . . 526
   4.2. Finite-frequency effects . . . 528
   4.3. Magnetic effects at high frequency . . . 529
5. The Boltzmann equation . . . 531
   5.1. Derivation of the Boltzmann equation . . . 531
   5.2. Linearized Boltzmann equation . . . 534
   5.3. The relaxation time: isotropic energy bands . . . 536
   5.4. Beyond the relaxation time approximation: variational principle . . . 538
   5.5. Ellipsoidal energy surfaces . . . 539
   5.6. Holes with warped energy surfaces . . . 541
   5.7. Inelastic scattering . . . 543
6. Electrical conductivity . . . 545
   6.1. Simple bands and classical statistics . . . 545
   6.2. Spherical but non-parabolic energy bands . . . 548
   6.3. Ellipsoidal energy surfaces . . . 549
   6.4. Warped hole bands . . . 550
7. Transport in electric and magnetic fields . . . 553
   7.1. Spherical bands . . . 555
   7.2. Several groups of carriers . . . 558

7.3. Ellipsoidal energy surfaces 559
7.4. Warped hole bands 563
8. Thermal gradients and fields 564
8.1. Zero magnetic field 565
8.2. Galvanomagnetic and thermomagnetic effects 567
9. Scattering mechanisms 568
9.1. Ionized-impurity scattering 569
9.2. Neutral-impurity scattering 570
9.3. Deformation potential acoustic-phonon scattering 571
9.4. Piezoelectric acoustic-phonon scattering 573
9.5. Deformation potential optical-phonon scattering 575
9.6. Polar optical-phonon scattering 575
9.7. Intervalley phonon scattering 576
9.8. Other scattering mechanisms 577
References 577

1. Introduction

In this chapter we shall give an introduction to the dynamics of electrons and holes in semiconductors in applied electric and magnetic fields, and the low-field classical transport of carriers in electric and magnetic fields and temperature gradients.

In discussing the dynamics of electrons we shall assume for the most part that the fields are uniform in space, although much of the theory applies to slowly varying fields. The case of impurities will be covered by Lannoo in chapter 4 and by Baranowski and Grynberg in chapter 5, and the special properties of carriers confined in quantum wells, inversion layers and superlattices will be covered by Ridley in chapter 13 and by Voisin and Bastard in chapter 15. The classical motion of carriers is based on the equation of motion for an electron in external fields. After a review of the energy bands of interest in semiconductors, which includes spherical and ellipsoidal conduction-band edges, and the complex valence-band edge, we shall discuss the classical motion of carriers in magnetic fields, electric fields, and combined fields.

While the classical approach gives a great deal of insight into the behavior of charged carriers in solids, it is also of interest to consider the quantization of carriers in magnetic and electric fields. We first discuss representations for Bloch electrons in external fields, and show how an effective Hamiltonian leads to the semiclassical quantization of levels in a magnetic field, including the spin splitting of the levels. We shall touch on the strange behavior of magnetic levels for arbitrary fields, and then concentrate primarily on energy states near band edges, which are of greatest interest in the case of semiconductors. The Luttinger–Kohn (1955) representation or the effective-mass approximation will be used to discuss the magnetic energy levels near both degenerate and non-degenerate band edges. For the case of a uniform field the band edge eigenstates appear to be continuous, but when states of an entire band are considered the levels may be quantized into the Stark ladder found by Wannier (1962), and tunneling can occur. For combined electric and magnetic fields there can be a crossover from closed to open magnetic orbits.

Turning to the transport problem, the approach in this chapter will be based on the classical motion of carriers. Quantum transport will be taken up by Barker in chapter 19. We shall first introduce the basic transport coefficients in electric and magnetic fields with the aid of a simple particle kinetic model, based on the classical equation of motion. The particle kinetic model gives insight into a wide variety of semiconductor experiments without explicitly including the full details of transport theory. This will be applied to study finite-frequency effects in electric and magnetic fields, including cyclotron resonance and magnetoplasma effects.

Then we shall turn to the more accurate Boltzmann equation, which is the primary tool currently used to study classical transport in electric and magnetic fields and thermal gradients. The Boltzmann equation can be considered as an equation of motion for the distribution function $f(\boldsymbol{r}, \hbar\boldsymbol{k}, t)$, the probability of finding a particle with crystal momentum $\hbar\boldsymbol{k}$ at a position $\boldsymbol{r}$. In a steady state the time rate of change of the distribution function due to applied fields, temperature gradients, and collisions, must vanish. The result is an integrodifferential equation whose main complication comes from the collision term. The various scattering mechanisms found to be important in semiconductors will be introduced, though a derivation of the transition probabilities is postponed until the last section.

We shall be concerned with the linearized Boltzmann equation, i.e., the special case of low electric fields. The hot-electron problem will be discussed by Ferry in chapter 18. For the linearized Boltzmann equation a relaxation time approximation is often made to simplify the collision term, and this is valid for spherically symmetric bands and collisions that are either elastic or "$\boldsymbol{k}$-randomizing", which means that a scattered particle is equally likely to emerge with diametrically opposite velocities. However, in real semiconductors it often happens that these conditions are not met. We will therefore give a brief introduction to the more general methods that have been developed to go beyond the relaxation time approximation. Even with elastic or $\boldsymbol{k}$-randomizing scattering there is anisotropy in both many-valley conduction bands and warped hole bands. For ellipsoids, the methods of Herring and Vogt (1956) and Samoilovich et al. (1961a–c) involve an expansion of the distribution function in spherical harmonics on the energy surface, and for zero magnetic field leads to a tensor relaxation time. For the hole bands the method of Lawaetz (1968) involves an expansion in cubic harmonics. In addition, even for spherical bands, polar optical scattering is neither randomizing nor elastic, and the more recent approaches for handling this involve matrix (Delves 1959) or iterative (Rode 1975) solutions of a difference equation.

After this introduction we shall discuss the electrical conductivity, Hall effect, and magnetoresistance for various energy bands of interest, as well as giving a brief discussion of thermopower and other thermal effects. We shall derive explicit results for the cases in which a relaxation time or a tensor relaxation time can be used.

In the final section we shall give a derivation of the transition probability for the various scattering mechanisms that have been found to be important for semiconductors.

This chapter is a shortened and updated version of my own chapter on electron dynamics, plus those of Arthur Smith on the particle kinetic model, Esther Conwell on the Boltzmann equation, and Wlodek Zawadzki on scattering mechanisms. I have relied heavily on these authors for material. I have covered the Boltzmann chapter fairly completely, but have not been able to present anything approaching the breadth or depth of the Zawadzki article, which will still have to remain a resource for the many details of transport theory that Zawadzki has studied in a wide variety of materials, including the two-band model for InSb and materials with inverted-band structures such as HeTe.

2. Classical motion of carriers in external fields

2.1. Some results from energy band theory

The energy states for an electron in a solid (Blount 1962a,b, Ziman 1960, Calloway 1974) are the Bloch states, whose wave functions are given in terms of the wave vector $\boldsymbol{k}$ and band index n by

$$b_{n\boldsymbol{k}}(\boldsymbol{r}) = \mathrm{e}^{\mathrm{i}\boldsymbol{k}\cdot\boldsymbol{r}} u_{n\boldsymbol{k}}(\boldsymbol{r}), \tag{2.1}$$

where $u_{n\boldsymbol{k}}(\boldsymbol{r})$ is periodic in space with the period of the crystalline lattice and $\boldsymbol{k}$ goes over N states in the Brillouin zone, where N is the number of unit cells in the crystal. The energy bands are solutions of the Schrödinger equation

$$[p^2/2m + V(\boldsymbol{r})]b_{n\boldsymbol{k}}(\boldsymbol{r}) = \mathscr{E}_n(\boldsymbol{k})b_{n\boldsymbol{k}}(\boldsymbol{r}), \tag{2.2}$$

where $\boldsymbol{p} = (\hbar/\mathrm{i})\boldsymbol{\nabla}$ and where $V(\boldsymbol{r})$ is the periodic potential. We can regard $V(\boldsymbol{r})$ as including relativistic effects such as the spin–orbit interaction, in which case the electron spin is included in the band index. The Bloch functions obey the following orthonormality relationship:

$$\int b^*_{n\boldsymbol{k}}(\boldsymbol{r})b_{n'\boldsymbol{k}'}(\boldsymbol{r})\,\mathrm{d}\boldsymbol{r} = \delta_{nn'}\delta_{\boldsymbol{k}\boldsymbol{k}'}, \tag{2.3}$$

where

$$\delta_{\boldsymbol{k}\boldsymbol{k}'} = \frac{1}{N}\sum_{\boldsymbol{R}} \mathrm{e}^{\mathrm{i}(\boldsymbol{k}-\boldsymbol{k}')\cdot\boldsymbol{R}} \tag{2.4}$$

is a crystalline (Kronecker) δ-function, with $\boldsymbol{R}$ going over all the lattice sites. The periodic functions $u_{n\boldsymbol{k}}$ for a given wave vector $\boldsymbol{k}$ are also orthonormal:

$$\int u^*_{n\boldsymbol{k}}(\boldsymbol{r})u_{n'\boldsymbol{k}}(\boldsymbol{r})\,\mathrm{d}\boldsymbol{r} = \delta_{nn'}. \tag{2.5}$$

In dealing with perturbing fields it is useful to write down several further results. The matrix elements of the momentum are diagonal in $\boldsymbol{k}$,

$$\langle n\boldsymbol{k}|\boldsymbol{p}|n'\boldsymbol{k}'\rangle = \delta_{\boldsymbol{k}\boldsymbol{k}'}\boldsymbol{p}_{nn'}, \tag{2.6}$$

and in particular m^{-1} times the diagonal matrix element, which is the expectation value of the velocity operator, is given by

$$\boldsymbol{v}_n(\boldsymbol{k}) = \frac{1}{m}\boldsymbol{p}_{nn} = \frac{1}{\hbar}\boldsymbol{\nabla}_{\boldsymbol{k}}\mathscr{E}_n(\boldsymbol{k}). \tag{2.7}$$

This is simply the group velocity of a wave packet centered at $\boldsymbol{k}$, as we shall discuss below. The quantity $\hbar\boldsymbol{k}$ is analogous to the classical momentum, and is referred to as the crystal momentum. Another result is the matrix element of the coordinate

$$\langle n\boldsymbol{k}|\boldsymbol{r}|n'\boldsymbol{k}'\rangle = \mathrm{i}\nabla_{\boldsymbol{k}}\delta_{\boldsymbol{kk}'}\delta_{nn'} + \delta_{\boldsymbol{kk}'}X_{nn'}(\boldsymbol{k}), \tag{2.8a}$$

$$X_{nn'}(\boldsymbol{k}) = \mathrm{i}\int u^*_{n\boldsymbol{k}}(\boldsymbol{r})\nabla_{\boldsymbol{k}}u_{n\boldsymbol{k}}(\boldsymbol{r})\,\mathrm{d}\boldsymbol{r}. \tag{2.8b}$$

The first term is singular and is the same as for free electrons, while the second term has interband matrix elements and is important in tunneling.

2.2. *Classical equation of motion*

We consider the classical motion of an electron in the presence of external electric and magnetic fields $\boldsymbol{E}$ and $\boldsymbol{B}$. An important result of band theory is that an electron wave packet centered at $\boldsymbol{r}$ and $\boldsymbol{k}$ in an energy band n moves according to the force equation

$$\hbar\frac{\mathrm{d}\boldsymbol{k}}{\mathrm{d}t} = \boldsymbol{F} = -e\boldsymbol{E} - \frac{e}{c}\boldsymbol{v}_n(\boldsymbol{k}) \times \boldsymbol{B}. \tag{2.9}$$

The proof for this equation, as well as the result for the group velocity, eq. (2.7), may be found in Callaway (1974), Roth (1982) and Landsberg (1992). The starting point is the Schrödinger equation for an electron of charge $-e$ moving in the crystalline potential plus external fields described by the scalar potential φ and the vector potential $\boldsymbol{A}$. The Hamiltonian operator is

$$\mathscr{H} = \frac{1}{2m}\left(\boldsymbol{p} + \frac{e}{c}\boldsymbol{A}\right)^2 + V(\boldsymbol{r}) - e\varphi(\boldsymbol{r}), \tag{2.10}$$

and the electric and magnetic fields are given by $\boldsymbol{E} = -\nabla\varphi$ and $\boldsymbol{B} = \nabla \times \boldsymbol{A}$.

We shall usually deal with uniform or slowly varying fields. For example, for uniform $\boldsymbol{B}$ in the symmetric gauge we have $\boldsymbol{A} = \frac{1}{2}\boldsymbol{B} \times \boldsymbol{r}$, and for uniform $\boldsymbol{E}$ we have $\varphi = -\boldsymbol{E}\cdot\boldsymbol{r}$. In the case of a uniform electric field and zero magnetic field the equation of motion, eq. (2.9), is actually exact, provided that $\boldsymbol{k}$ is interpreted as the average $\langle\boldsymbol{k}\rangle$ over a many-band wave packet. The magnetic case, however, involves the approximation of taking the group velocity at the center of the wave packet, and in general we are thinking of an electron in a particular energy band. Lifshitz and Kaganov (1959) outline the conditions for the validity of the classical approximations: the de Broglie wavelength must be much smaller than the characteristic length associated with a trajectory, the latter length must be much larger than a lattice spacing, and the magnetic splittings must be small compared to the electron energy.

It is interesting to notice (Lifshitz and Kaganov 1959, Zeiger and Pratt 1973, Lifshitz et al. 1973) that the classical motion of band electrons can be described by treating the electrons as particles with the classical Hamiltonian

$$\mathscr{H} = \mathscr{E}_n\left(\boldsymbol{p} + \frac{e}{c}\boldsymbol{A}(\boldsymbol{r})\right) - e\varphi(\boldsymbol{r}). \tag{2.11}$$

Then the equation of motion, eq. (2.9), is obtained by using Hamilton's equations for

$\boldsymbol{p}$ and $\boldsymbol{r}$, the canonical momentum and coordinate. Thus,

$$\dot{\boldsymbol{r}} = \nabla_{\boldsymbol{p}}\mathscr{H} = \nabla_{\boldsymbol{p}}\mathscr{E}_n\left(\boldsymbol{p} + \frac{e}{c}\boldsymbol{A}(\boldsymbol{r})\right), \tag{2.12}$$

$$\dot{\boldsymbol{p}} = -\nabla_{\boldsymbol{r}}\mathscr{H} = -(e/c)\nabla_{\boldsymbol{r}}\boldsymbol{A}\cdot\nabla_{\boldsymbol{p}}\mathscr{E}_n + e\nabla_{\boldsymbol{r}}\varphi(\boldsymbol{r}). \tag{2.13}$$

Using the vector identity, $\boldsymbol{C}\times(\nabla\times\boldsymbol{A}) = \nabla\boldsymbol{A}\cdot\boldsymbol{C} - \boldsymbol{C}\cdot\nabla\boldsymbol{A}$ we find

$$\dot{\boldsymbol{p}} + e\dot{\boldsymbol{A}}/c = -(e/c)\boldsymbol{v}_n\times\boldsymbol{B} - e\boldsymbol{E}, \tag{2.14}$$

which is the same as eq. (2.9) if $\hbar\boldsymbol{k}$ is identified as the kinetic momentum $\boldsymbol{p} + e\boldsymbol{A}/c$, rather than the canonical momentum $\boldsymbol{p}$. This is a compact way of describing the classical motion.

Equations (2.7) and (2.9) determine the classical motion both in real space and in $\boldsymbol{k}$-space. As an example, consider states close to a non-degenerate energy band edge (at which $\nabla_{\boldsymbol{k}}\mathscr{E}_n(\boldsymbol{k}) = 0$). The energy is given to second order in $\boldsymbol{k}$ by

$$\mathscr{E} = (\hbar^2/2)[k_x^2/m_x + k_y^2/m_y + k_z^2/m_z] = (\hbar^2/2)\boldsymbol{k}\cdot\boldsymbol{m}^{*-1}\cdot\boldsymbol{k}, \tag{2.15}$$

where

$$\boldsymbol{m}^{*-1} = \hat{\boldsymbol{i}}\hat{\boldsymbol{i}}/m_x + \hat{\boldsymbol{j}}\hat{\boldsymbol{j}}/m_y + \hat{\boldsymbol{k}}\hat{\boldsymbol{k}}/m_z \tag{2.16}$$

is the inverse effective mass tensor, which we have written in dyadic form using a principal-axis system. The origin of $\boldsymbol{k}$ is taken at the band edge, which need not be at the zone center. The velocity is given by

$$\boldsymbol{v} = \hbar^{-1}\nabla_{\boldsymbol{k}}\mathscr{E} = \boldsymbol{m}^{*-1}\cdot\hbar\boldsymbol{k}, \tag{2.17}$$

or we can say $\hbar\boldsymbol{k} = \boldsymbol{m}^*\cdot\boldsymbol{v}$, which gives a generalization of the usual relation between velocity and momentum. Actually it is possible to define an inverse effective-mass tensor to any order in $\boldsymbol{k}$ through eq. (2.17), by assuming a Taylor expansion of $\mathscr{E}(\boldsymbol{k})$, and $\boldsymbol{m}^*$ can be shown to be symmetric. This will be most useful for spherical bands (see eq. (2.25) below).

We suppose here that the band edge is a minimum or maximum of $\mathscr{E}_n(\boldsymbol{k})$, i.e., that the three effective-mass coefficients have the same sign. The constant-energy surfaces are ellipsoids. If the masses had varying signs we would have a saddle point. The simplest case occurs if $m_x = m_y = m_z = m^*$, i.e., a simple band edge such as occurs at the center of the Brillouin zone in GaAs and other III–V compounds. We find simply $\mathscr{E} = \hbar^2k^2/2m^*$, $\hbar\boldsymbol{k} = m^*\boldsymbol{v}$, and the force equation is

$$\boldsymbol{F} = m^*\mathrm{d}\boldsymbol{v}/\mathrm{d}t = -e\boldsymbol{E} - (e/c)\boldsymbol{v}\times\boldsymbol{B}, \tag{2.18}$$

so the electrons behave as though they have an effective mass of m^* and their normal charge $-e$. Near a band minimum m^* is positive, while for a maximum m^* is negative. In the latter case it is more interesting to discuss holes in a filled band rather than electrons at the top of an empty band. If an electron in state $\boldsymbol{k}$ is removed from a filled band, the total crystal momentum of the system of electrons becomes $\hbar\boldsymbol{k}' = -\hbar\boldsymbol{k}$, since the total crystal momentum of the filled energy band is zero. Using $m'^* = -m^* > 0$, the velocity is $\boldsymbol{v} = \hbar\boldsymbol{k}/m^* = \hbar\boldsymbol{k}'/m'^*$, i.e., it is unchanged on reversing the sign of both the mass and the crystal momentum. The force equation for a hole is then given by eq. (2.18) with positive mass and charge $+e$.

2.3. *Energy bands of interest for semiconductors*

In semiconductors the carriers of interest are near minima in the conduction band and maxima in the valence band. A simple model would have spherical parabolic electron and hole band edges at $\boldsymbol{k}=0$ (the Γ-point in the Brillouin zone). In actual materials the energy bands are usually more complex. While III–V compounds like GaAs have a spherical conduction band at Γ, non-parabolic effects are often important as in InSb. The conduction bands for Si and Ge have sets of ellipsoidal constant-energy surfaces away from the zone center. The valence bands of most semiconductors are degenerate with warped energy surfaces.

The wave functions as well as the energies for electrons near band edges can be described by the $\boldsymbol{k}\cdot\boldsymbol{p}$ perturbation method (Kane 1957, 1959, 1982). Consider a group of band edges at $\boldsymbol{k}=0$ that are either degenerate or closely spaced. The wave functions are given to lowest order in $\boldsymbol{k}$ by

$$b_{n\boldsymbol{k}}(\boldsymbol{r}) = \sum_i c_i \mathrm{e}^{\mathrm{i}\boldsymbol{k}\cdot\boldsymbol{r}} b_{i0}(\boldsymbol{r}), \tag{2.19}$$

where i goes over the group of bands. To second order in $\boldsymbol{k}$ the energy bands are determined by the set of coupled equations (see § 3.4)

$$\sum_j (\mathscr{H}_{ij} - \mathscr{E}\delta_{ij})c_j = 0, \tag{2.20}$$

where $\mathscr{H}$ is the effective-mass Hamiltonian

$$\mathscr{H}_{ij} = \left(\mathscr{E}_i + \frac{\hbar^2 k^2}{2m}\right)\delta_{ij} + \frac{\hbar\boldsymbol{k}\cdot\boldsymbol{p}_{ij}}{m} + \frac{1}{m^2}\sum_n{}' \frac{\hbar\boldsymbol{k}\cdot\boldsymbol{p}_{in}\boldsymbol{p}_{nj}\cdot\hbar\boldsymbol{k}}{\mathscr{E}-\mathscr{E}_n}. \tag{2.21}$$

Here n goes over all bands except the set of interest, as indicated by the prime on the summation, and $\mathscr{E}_i$ is the band edge energy of the ith band. The energy appearing in the denominator of the third term is usually replaced by the band edge energy to give the energy to lowest order in $\boldsymbol{k}$. This result is easily generalized to cover band edges away from the zone center by measuring $\boldsymbol{k}$ from the wave vector at the band edge.

For a non-degenerate band edge the diagonal momentum matrix element vanishes. The energy to second order in $\boldsymbol{k}$ is of the form of eq. (2.15), with the effective-mass tensor determined by the momentum matrix elements connecting nearby bands,

$$\boldsymbol{m}^{*-1} = \frac{\mathbf{1}}{m} + \frac{1}{m^2}\sum_n{}' \frac{\boldsymbol{p}_{in}\boldsymbol{p}_{ni} + \boldsymbol{p}_{ni}\boldsymbol{p}_{in}}{\mathscr{E}_i - \mathscr{E}_n}. \tag{2.22}$$

If the interaction with a close-by band dominates the effective-mass Hamiltonian, a two-band model can be used to describe deviations from a quadratic dependence on $\boldsymbol{k}$. For the isotropic case the electron energy relative to the conduction-band edge is

$$\mathscr{E}(1 + \mathscr{E}/\mathscr{E}_{\mathrm{g}}) = \hbar^2 k^2/2m_0^*, \tag{2.23}$$

where m_0^* is the effective mass at the band edge, and $\mathscr{E}_{\mathrm{g}}$ is the energy band gap. This can easily be solved for $\mathscr{E}(\boldsymbol{k})$. An example is InSb. For spherical, but non-parabolic, bands we can define an energy-dependent effective mass by writing

$$v_k = \hbar^{-1}\nabla_k \mathscr{E} = \hbar k/m^*(\mathscr{E}); \quad 1/m^* = (1/\hbar^2 k)\partial\mathscr{E}/\partial k. \tag{2.24}$$

For the two-band model we have

$$m^*(\mathscr{E}) = m_0^*(1 + 2\mathscr{E}/\mathscr{E}_g). \tag{2.25}$$

This is not the only way to define an effective mass. We could define an inverse effective-mass tensor as $\hbar^{-1}\nabla_k\nabla_k\mathscr{E}$, by generalizing Newton's second law $\boldsymbol{a} = \boldsymbol{F}/m$, in which case the effective mass would be a tensor even for spherical bands. The present definition corresponds to eq. (2.17), and gives a generalization of Newton's second law in the form $\boldsymbol{F} = \mathrm{d}(m^*\boldsymbol{v})/\mathrm{d}t$. It is useful for mobility calculations (Zawadzki 1982).

Figure 1 shows a portion of the band structure of Ge in the region of the energy gap, which shows typical behavior near the band edges. Some representative values of the parameters are given in table 1. Figure 2 gives constant-energy surfaces for several cases. For GaAs and InSb the conduction-band edge is spherical and at the Γ-point. The conduction-band edge for Ge is at the L-point (111). There are four equivalent valleys, and the energy surfaces are ellipsoids of revolution with the major axis along (111). In the principal-axis system the energy is

$$\mathscr{E} = (\hbar^2/2)[(k_x^2 + k_y^2)/m_t + k_z^2/m_\ell], \tag{2.26}$$

where m_ℓ and m_t are the longitudinal and transverse effective masses. The conduction-band edge in silicon has six equivalent valleys along Δ (100), which are displaced by 0.17 (2π/a) from the X-point, where there is a degeneracy. The energy surfaces are again ellipsoids of revolution, with the major axis along (100). Values of the effective

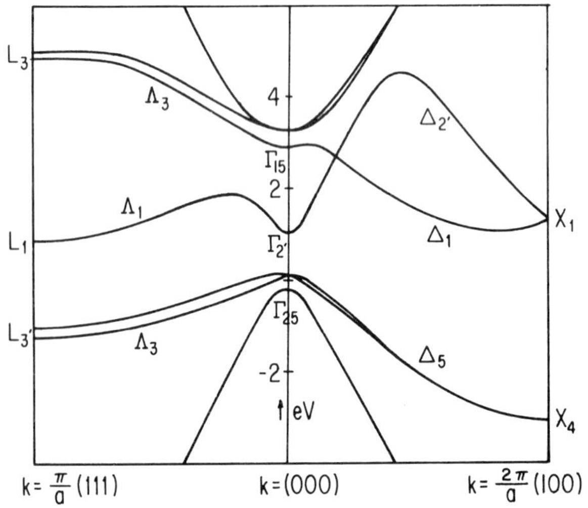

Fig. 1. Band structure of Ge obtained by the $\boldsymbol{k}\cdot\boldsymbol{p}$ method (after Cardona 1969).

Table 1
Energy band parameters for typical semiconductors[a].

Parameter	Ge	Si	InSb	GaAs
$\mathscr{E}_g$ (at Γ)	0.89	3.40	0.237	1.52
Δ (at Γ)	0.29	0.04	0.81	0.34
Conduction band:	L	Δ	Γ	Γ
m_t/m	0.0819[b]	0.192[c]		
m_ℓ/m	1.87[b]	0.90[c]		
$g_\perp$	1.92[d]	$g-0.0028$[e]		
$g_\parallel$	0.87[d]	$g-0.0040$[e]		
$m^*/m(\Gamma)$	0.038		0.014	0.067
$g(\Gamma)$	−2.86		−48.4	−0.06
Valence band:				
m_h	0.35	0.53	0.47	0.62
m_ℓ	0.043	0.16	0.015	0.074
m_{so}	0.092	0.24	0.107	0.15
A	13.38[f]	4.22	35.0	7.63
B	8.48[f]	0.79	31.4	4.86
C	13.1[f]	5.27	20.9	7.52
γ_1	13.35	4.22	35.08	7.65
γ_2	4.25	0.39	15.64	2.41
γ_3	5.69	1.44	16.91	3.28
κ	3.41	−0.26	14.76	1.72
q	0.07	0.01	0.15	0.04

[a] From the $\boldsymbol{k}\cdot\boldsymbol{p}$ calculation of Lawaetz (1971), and references cited therein, unless otherwise noted.
[b] Dexter et al. (1956).
[c] Rauch et al. (1960).
[d] Feher et al. (1959), Feher and Wilson (1960).
[e] Wilson and Feher (1961); $g = 2.0023$.
[f] Hensel and Suzuki (1970).

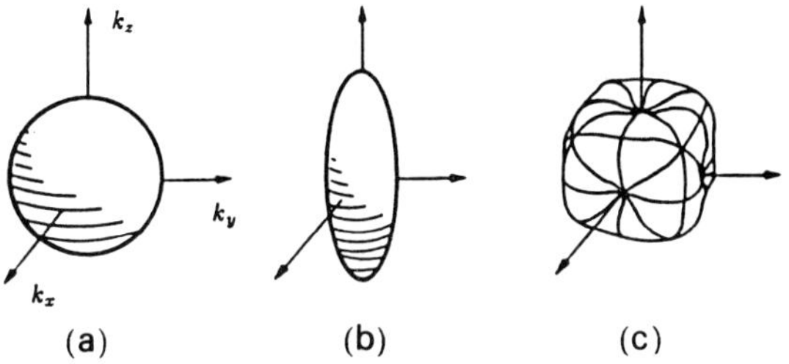

Fig. 2. Constant-energy surfaces for several models: (a) sphere; (b) ellipsoid; (c) warped hole surface (after Böer 1990).

masses for Si and Ge are included in table 1. Several III–V compounds have multiple-valley structures at the conduction-band edge (Stirn 1972). In GaP the conduction-band valleys are along Δ and have a "camel-back" structure near X (Böer 1990). This is because in the absence of inversion symmetry, there is a splitting of the conduction band at X. In GaAs and a number of III–V compounds, even though the band edge is at Γ, the X-point and L-point sets of valleys are close by in energy, so they can be populated under hot-electron conditions in strong electric fields (Gunn 1964).

The valence bands in diamond and zinc blende semiconductors are complex because there is a degeneracy at the zone center. The band edge in the absence of the spin–orbit interaction is Γ'_{25} (diamond) or Γ_{15} (zinc blende) and the wave functions are the p-like functions X, Y, Z, multiplied by the spin functions $\uparrow$ and $\downarrow$. The orbital part of the effective-mass Hamiltonian has the form

$$\mathscr{H} = \mathscr{E}'_{\mathrm{v}} + \hbar^2 k^2/2m + (\hbar^2/2m)\begin{pmatrix} Lk_x^2 + M(k_y^2+k_z^2) & Nk_xk_y & Nk_xk_z \\ Nk_xk_y & Lk_y^2 + M(k_x^2+k_z^2) & Nk_yk_z \\ Nk_xk_z & Nk_yk_z & Lk_z^2 + M(k_x^2+k_y^2) \end{pmatrix}. \tag{2.27}$$

The parameters L, M, and N are related to the interband sums in eq. (2.21) (Dresselhaus et al. 1955). The inclusion of the spin–orbit interaction term

$$\mathscr{H}_{\mathrm{so}} = \frac{\hbar}{4m^2c^2}\nabla V \times \boldsymbol{p} \cdot \sigma \tag{2.28}$$

where σ is the Pauli spin vector, gives a splitting of the band edge at $\boldsymbol{k}=0$ into a twofold degenerate set (the split-off band, Γ_7^+ or Γ_7) and a fourfold set (Γ_8^+ or Γ_8), separated by an energy Δ. The fourfold set is higher in energy. The spin–orbit interaction can be represented by the following additional term in the effective-mass Hamiltonian

$$\mathscr{H}_{\mathrm{so}} = \tfrac{1}{3}\Delta\begin{pmatrix} 0 & -\mathrm{i}\sigma_z & \mathrm{i}\sigma_y \\ \mathrm{i}\sigma_z & 0 & -\mathrm{i}\sigma_x \\ -\mathrm{i}\sigma_y & \mathrm{i}\sigma_x & 0 \end{pmatrix}. \tag{2.29}$$

Since the Pauli spin vectors are 2×2, when this term is added to eq. (2.27) (multiplied by the 2×2 unit matrix) it gives a 6×6 matrix that must be diagonalized to obtain the eigenvectors and energies. Because of Kramers theorem the states are twofold degenerate. If we neglect any $\boldsymbol{k}$-dependence introduced by the spin–orbit interaction, the effective-mass Hamiltonian obeys inversion symmetry, and the Kramers operator is $-\mathrm{i}\sigma_y J\mathscr{C}$, where J is the inversion operator, $\mathscr{C}$ takes the complex conjugate, and $-\mathrm{i}\sigma_y$ reverses the spins. The three bands that result are sketched in fig. 3.

The solutions near $\boldsymbol{k}=0$ can be obtained by exploiting the analogy with the splitting of a p level by spin–orbit interaction into J, M_J states with $J=3/2$ and $J=1/2$, for which the spin–orbit interaction is diagonal. The basis functions are linear

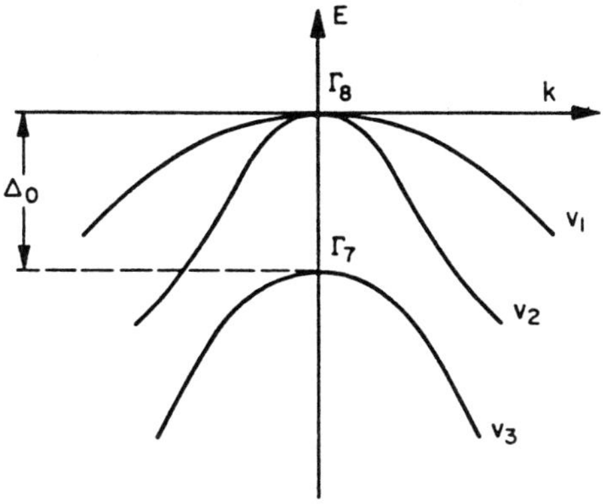

Fig. 3. Hole bands near the Γ-point (after Wiley 1975).

combinations:

$$\Gamma_8, J=3/2, \quad \frac{1}{\sqrt{2}}(X+\mathrm{i}Y)\uparrow, \quad \frac{1}{\sqrt{6}}(X+\mathrm{i}Y)\downarrow - \frac{2}{\sqrt{6}}Z\uparrow,$$
$$-\frac{1}{\sqrt{6}}(X-\mathrm{i}Y)\uparrow - \frac{2}{\sqrt{6}}Z\downarrow, \quad \frac{1}{\sqrt{2}}(X-\mathrm{i}Y)\downarrow; \tag{2.30}$$
$$\Gamma_7, J=1/2, \quad \frac{1}{\sqrt{3}}(X+\mathrm{i}Y)\downarrow + \frac{1}{\sqrt{3}}Z\uparrow, \quad -\frac{1}{\sqrt{3}}(X-\mathrm{i}Y)\uparrow + \frac{1}{\sqrt{3}}Z\downarrow.$$

The result for the energies of light, heavy and split-off band holes near $\boldsymbol{k}=0$ is (Dresselhaus et al. 1955, Kane 1982)

$$\mathscr{E}(\boldsymbol{k}) = \mathscr{E}_\mathrm{v} - (\hbar^2/2m)\{Ak^2 \pm [B^2k^4 + C^2(k_x^2k_y^2 + k_x^2k_z^2 + k_y^2k_z^2)]^{1/2}\},$$
$$\mathscr{E}(k) = \mathscr{E}_\mathrm{v} - \Delta - (\hbar^2/2m)Ak^2, \tag{2.31}$$

where $\mathscr{E}_\mathrm{v} = \mathscr{E}'_\mathrm{v} + \Delta/3$, where the upper sign is for light holes and the lower is for heavy holes, and where

$$A = -\tfrac{1}{3}(L+2M) - 1, \quad B = -\tfrac{1}{3}(L-M), \quad C^2 = \tfrac{1}{3}[N^2 - (L-M)^2]. \tag{2.32}$$

Another set of parameters often used is that of Luttinger (1956), introduced in § 3.5,

$$\gamma_1 = A, \ \gamma_2 = B/2, \quad \gamma_3 = \tfrac{1}{2}\sqrt{B^2 + C^2/3} = -N/6. \tag{2.33}$$

A spherical approximation is often made for the light and heavy holes by replacing the coefficient of C^2 by an angular average

$$\frac{m}{m_{\ell,\mathrm{h}}} \cong A \pm (B^2 + C^2/6)^{1/2}. \tag{2.34}$$

Actually, in most cases there is considerable warping of the bands, which is particularly important in the heavy-hole band.

For energies comparable to the spin–orbit splitting the bands become non-parabolic. Non-parabolic effects have been found to be important in considering transport in Ge and Si by Szmulowicz (1983), who found it necessary to use the complete 6×6 representation for the energies and wave functions.

In the case of narrow-gap semiconductors like InSb Kane (1982) showed that it is necessary to include the Γ_1 (Γ_2' for diamond) conduction-band states $S\uparrow$ and $S\downarrow$ in the set of band edges to be treated exactly. This adds an extra row and column to eq. (2.27). The row to lowest order is $[(\hbar^2/2m)A'k^2, \mathrm{i}Pk_x, \mathrm{i}Pk_y, \mathrm{i}Pk_z]$, where $\mathrm{i}P = \langle S|\hbar p_x/m|X\rangle$. The second-order terms L and N also now must omit the contribution from the conduction-band edge, giving

$$L' = L - (2m/\hbar^2)P^2/(\mathscr{E}_\mathrm{v} - \mathscr{E}_\mathrm{c}), \quad N' = N - (2m/\hbar^2)P^2/(\mathscr{E}_\mathrm{v} - \mathscr{E}_\mathrm{c}).$$

In the special case in which the only interband interaction that is important is between the s- and p-like bands Kane (1982) transformed the basis set to make the direction of quantization for the spin and angular momentum to be along $\boldsymbol{k}$. In terms of the states $\mathrm{i}S'\uparrow', (1/\sqrt{2})(X' + \mathrm{i}Y')\downarrow', Z'\uparrow'$, the $\boldsymbol{k \cdot p}$ matrix for one of the Kramers states is

$$\mathscr{H} = \hbar^2 k^2/2m + \begin{pmatrix} \mathscr{E}_\mathrm{c} & 0 & kP \\ 0 & \mathscr{E}_\mathrm{v} - 2\Delta/3 & \sqrt{2}\Delta/3 \\ kP & \sqrt{2}\Delta/3 & \mathscr{E}_\mathrm{v} - \Delta/3 \end{pmatrix}. \tag{2.35}$$

This gives a three-band approximation for the light-hole, electron, and split-off energies,

$$(\mathscr{E}' - \mathscr{E}_\mathrm{c})(\mathscr{E}' - \mathscr{E}_\mathrm{v})(\mathscr{E}' - \mathscr{E}_\mathrm{v} - \Delta) - k^2P^2(\mathscr{E}' - \mathscr{E}_\mathrm{v} + 2\Delta/3) = 0, \tag{2.36}$$

where $\mathscr{E}' = \mathscr{E} - \hbar^2k^2/2m$. In the case of InSb (fig. 4a), the spin–orbit splitting is much greater than the energy gap $\mathscr{E}_\mathrm{g} = \mathscr{E}_\mathrm{v} - \mathscr{E}_\mathrm{c}$, and the two-band model of eq. (2.23) suffices *with* $\hbar^2/2m_0^* = 2P^2/3$. The curvature of the heavy holes must be determined by the more distant band edges. According to Kane (1982)

$$\mathscr{E} = \mathscr{E}_\mathrm{v} + (\hbar^2/2m)[(1 + M)k^2 + (L - M - N)(k_x^2k_y^2 + k_x^2k_z^2 + k_y^2k_z^2)/k^2]. \tag{2.37}$$

Another interesting case is grey tin (Groves and Paul 1963) and HgTe, for which the band edges Γ_6 and Γ_8 are reversed ($\mathscr{E}_\mathrm{v} > \mathscr{E}_\mathrm{c}$), so that the degenerate band is now on top, and the energy gap is zero. This is sketched in fig. 4b.

In the case of III–V compounds inversion symmetry does not hold, which means that the Kramers degeneracy does not hold for every point in $\boldsymbol{k}$-space. The Kramers operator is now $-\mathrm{i}\sigma_y\mathscr{C}$, so the Kramers pair is $\boldsymbol{k}\uparrow$ and $-\boldsymbol{k}\downarrow$. There are small extra terms, which give inversion asymmetry splittings to the bands. The B-term of Kane (1982) adds to the extra row mentioned above the terms Bk_yk_z, Bk_xk_z, Bk_xk_y connecting S with X, Y, and Z. This gives a k^3 contribution to the conduction band, which leads to an inversion asymmetry splitting (Seiler and Becker 1967). In the presence of the spin–orbit interaction there are linear k terms in the effective-mass Hamiltonian (Dresselhaus 1955, Kane 1982). For the 4×4 degenerate band edge

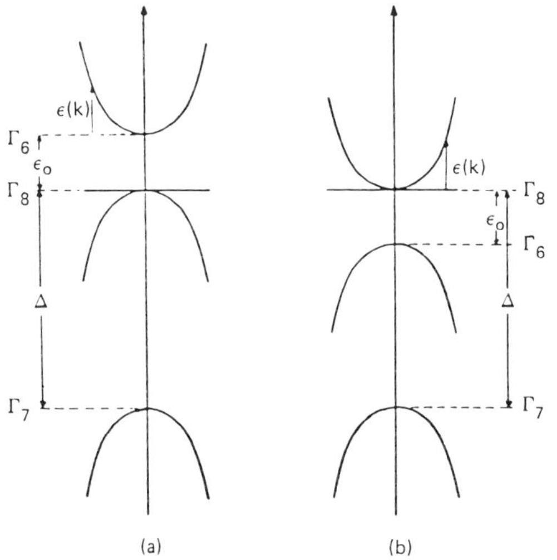

Fig. 4. (a) Band structure of InSb near the Γ-point. (b) Inverted band structure of HgTe after Zawadzki (1982).

this has the form

$$\mathscr{H} = -2C'/\sqrt{3}\pi\hbar[k_x\{J_x, J_y^2 - J_z^2\} + k_y\{J_y, J_z^2 - J_x^2\} + k_z\{J_z, J_x^2 - J_y^2\}], \tag{2.38}$$

when expressed in terms of the angular momentum operators for $J = \frac{3}{2}$ (see § 3.5).

2.4. *Motion in a uniform magnetic field; cyclotron resonance*

For a simple band edge the equation of motion of an electron or hole with charge $q = \pm e$ in a magnetic field is

$$\boldsymbol{a} = \frac{\hbar}{m^*}\frac{\mathrm{d}\boldsymbol{k}}{\mathrm{d}t} = \frac{q}{m^*c}\boldsymbol{v} \times \boldsymbol{B}. \tag{2.39}$$

It is well known that this corresponds to circular motion, and equating $\boldsymbol{a}$ to the centripetal acceleration $v^2/r = v\omega$ gives the cyclotron angular frequency $\omega_c = qB/m^*c$. This is the same as that for free electrons, except that the mass is replaced by the effective mass m^*.

Consider now the case of an ellipsoidal band edge. We can scale the coordinate system (Herring and Vogt 1956) to make the effective mass isotropic, and the problem is to see how the magnetic field scales. The equation of motion is

$$\frac{\mathrm{d}\boldsymbol{k}}{\mathrm{d}t} = \frac{q}{c}(\boldsymbol{m}^{*-1} \cdot \boldsymbol{k}) \times \boldsymbol{B}. \tag{2.40}$$

Let us write out a typical component in a coordinate system in which $\boldsymbol{m}^*$ is diagonal:

$$\frac{\mathrm{d}k_x}{\mathrm{d}t} = \frac{q}{c}\left(\frac{k_y B_z}{m_y} - \frac{k_z B_y}{m_z}\right).$$

This can be rewritten in the form,

$$\frac{\mathrm{d}}{\mathrm{d}t}\frac{k_x}{\sqrt{m_x}} = \frac{q}{c}\left(\frac{k_y}{\sqrt{m_y}}\frac{B_z}{\sqrt{m_x m_y}} - \frac{k_z}{\sqrt{m_z}}\frac{B_y}{\sqrt{m_x m_z}}\right).$$

If we use a coordinate system in which the momentum components are $k'_x = k_x\sqrt{m/m_x}$, etc., then the equation of motion has the isotropic form if $\boldsymbol{B}$ is replaced by $\bar{\boldsymbol{B}}$ such that $\bar{B}_x/m = B_x\sqrt{m_y m_z}$, etc. The cyclotron frequency is then

$$\omega_{\mathrm{c}} = \frac{q}{c}\left(\frac{B_x^2}{m_y m_z} + \frac{B_y^2}{m_z m_x} + \frac{B_z^2}{m_x m_y}\right)^{1/2}. \tag{2.41}$$

For the case of an ellipsoid of revolution, eq. (2.26), as in the conduction bands of Ge and Si, this becomes

$$\omega_{\mathrm{c}} = \frac{qB}{c}\left(\frac{\cos^2\theta}{m_{\mathrm{t}}^2} + \frac{\sin^2\theta}{m_\ell m_{\mathrm{t}}}\right)^{1/2}, \tag{2.42}$$

where θ is the angle between $\boldsymbol{B}$ and the axis of the ellipsoid (Dresselhaus et al. 1955, Dexter et al. 1956).

For the case of a general energy band the equation of motion is most easily considered in $\boldsymbol{k}$-space:

$$\frac{\mathrm{d}\boldsymbol{k}}{\mathrm{d}t} = \frac{q}{\hbar^2 c}\nabla_{\boldsymbol{k}}\mathscr{E} \times \boldsymbol{B}. \tag{2.43}$$

The velocity $\boldsymbol{v} = \nabla_{\boldsymbol{k}}\mathscr{E}/\hbar$ is a vector perpendicular to the energy surface, so that the motion is around the energy surface and in a plane perpendicular to the magnetic field as shown in fig. 5. Thus k_{B}, the component of $\boldsymbol{k}$ in the field direction, is fixed.

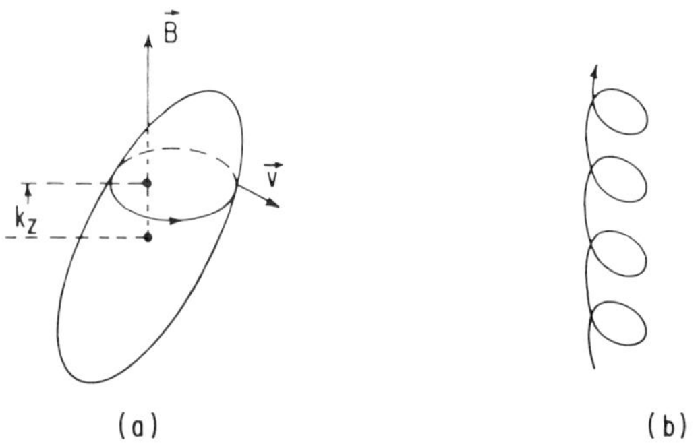

Fig. 5. Magnetic orbits: (a), in k-space; (b), in real space.

For the path element we have

$$\left(\frac{e}{\hbar c}|\boldsymbol{v} \times \boldsymbol{B}|\right)^{-1} \mathrm{d}k_{\mathrm{p}} = \mathrm{d}t. \tag{2.44}$$

This can be integrated, and the period of the motion corresponds to an integral around the entire path:

$$T = \frac{2\pi}{\omega_{\mathrm{c}}} = \frac{\hbar c}{e} \oint \frac{\mathrm{d}k_{\mathrm{p}}}{|\boldsymbol{v} \times \boldsymbol{B}|}. \tag{2.45}$$

This is the Shockley tube integral result (Shockley 1950) for the cyclotron period. We can readily show that it reduces to the previous result for ellipsoids.

The orbit in real space is found by noticing that

$$\frac{\mathrm{d}\boldsymbol{k}}{\mathrm{d}t} = \frac{q}{\hbar c}\left(\frac{\mathrm{d}\boldsymbol{r}}{\mathrm{d}t} \times \boldsymbol{B}\right).$$

Integrating this we find that

$$\boldsymbol{k}(t) = \boldsymbol{k}_0 + \frac{q}{\hbar c}[\boldsymbol{r}(t) - \boldsymbol{r}_0] \times \boldsymbol{B}. \tag{2.46}$$

The motion perpendicular to the field is then at right angles to that in $\boldsymbol{k}$-space, and the orbit has the same shape. The orbit radius is scaled over the $\boldsymbol{k}$-space orbit radius by the factor $\hbar c/eB = l^2$, where l is a parameter whose dimension is length.

The momentum $\hbar k_{\mathrm{B}}$ parallel to the field is fixed, and for a simple spherical band edge the velocity in the field direction $\hbar k_{\mathrm{B}}/m^*$ is constant, giving a helical motion. For other cases the $\boldsymbol{B}$-component of the velocity varies around the orbit (see fig. 5) so that the motion along the field is more complicated. For a closed orbit as in fig. 5 we can show that the average transverse velocity vanishes:

$$\oint \boldsymbol{v} \times \boldsymbol{B}\,\mathrm{d}t = (c/e) \oint \mathrm{d}\boldsymbol{k}_{\mathrm{p}} = 0,$$

so that the electron has in general a net motion along the direction of $\boldsymbol{B}$. For metals, in which the magnetic orbits of interest are on the Fermi surface, it is possible to have open orbits in which the electron traverses the whole zone. In this case there is a net transverse motion. The existence of open orbits accounts for the remarkable anisotropy in the magnetoresistance of metals (Chambers 1960, Callaway 1974).

An example of a complex energy surface is found in the valence bands of Ge and Si, for which the energies of light and heavy holes are given by eq. (2.31). The cyclotron frequency has been evaluated from eq. (2.45) and the result is (Dresselhaus et al. 1955) $\omega_{\mathrm{c}} = eB/m^*c$, where

$$\frac{m^*}{m} \cong \frac{1}{A \pm (B^2 + C^2/4)^{1/2}}\left(1 \pm \frac{C^2(1 - 3\cos^2\theta)^2}{64(B^2 + C^2/4)^{1/2}[A \pm (B^2 + C^2/4)^{1/2}]}\right) \tag{2.47}$$

Figure 6 shows a recorder tracing of cyclotron resonance for germanium showing both electron and hole cyclotron resonance. Note that harmonics of the cyclotron

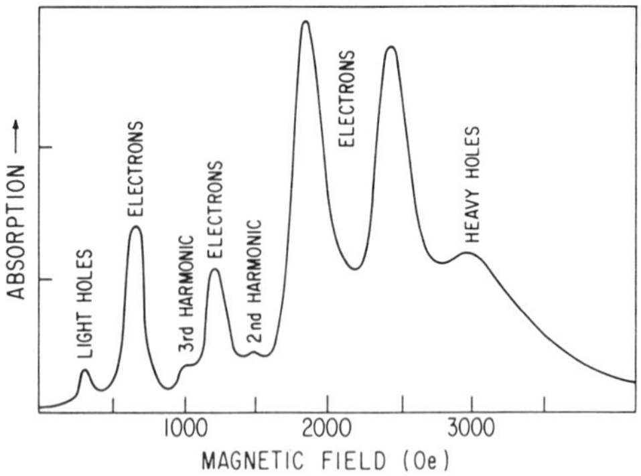

Fig. 6. Cyclotron resonance in Ge (after Dexter et al. 1956).

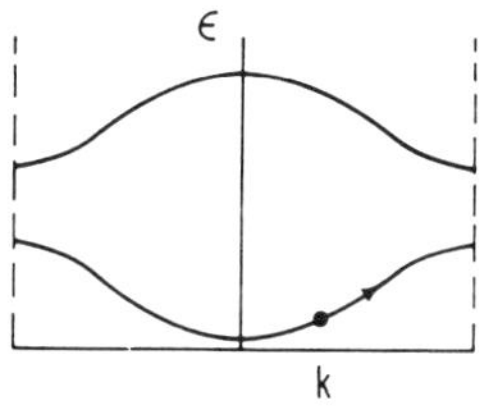

Fig. 7. The ***k***-motion of a wave packet in an electric field.

resonance frequency are seen, which is due to the warping of the heavy-hole orbits (Zeiger et al. 1957). We see that cyclotron resonance is useful for determining the energy band parameters. Further information can be found from studying the quantum effects due to the coupling of heavy- and light-hole bands, as will be discussed below.

2.5. Motion in a steady electric field

In a steady electric field $\mathrm{d}\boldsymbol{k}/\mathrm{d}t$ is constant and the electron is accelerated. As indicated in fig. 7 if the electron's motion is unimpeded (i.e. if there is no scattering) and if it remains in the same band, it traverses the Brillouin zone and reappears on the left. While the electron absorbs energy from the field in the first part of the motion in which $\boldsymbol{v}$ is parallel to the force, in the latter part of the motion near the top of the band the electron gives up energy to the field, so it ends up where it began. There seems to be no net acceleration. However, we have said that the force equation, $\mathrm{d}\boldsymbol{k}/\mathrm{d}t = (1/\hbar)q\boldsymbol{E}$, is exact if $\boldsymbol{k}$ is interpreted as the average wave vector for a wave packet that includes all the bands. The electron can tunnel to higher bands, and this is especially likely at band edges, so there is in general a net absorption of energy from the field. In practice there is also scattering by impurities, phonons and other carriers, which usually occurs long before the zone edges are reached.

Suppose, however, an electron does traverse the zone several times before either tunneling or being scattered. If the electric field is along an axis of the crystal as in fig. 7, the electron's motion is periodic in $\boldsymbol{k}$-space. If the lattice spacing in the field direction is b, the period of k is $2\pi/b$, so that the period of the motion is $T = 2\pi\hbar/eEb$ and the frequency is $\omega = eEb/\hbar$. In real space the motion is given by

$$\boldsymbol{r}(t) = \boldsymbol{r}_0 + \int \boldsymbol{v}(\boldsymbol{k}(t))\,\mathrm{d}t. \tag{2.48}$$

Since $\boldsymbol{k}$ and hence $\boldsymbol{v}$ are periodic in time, the motion in real space is also periodic.

2.6. Motion in combined electric and magnetic fields

We consider now motion in combined electric and magnetic fields. For the case in which $\boldsymbol{E}$ is along $\boldsymbol{B}$, there is a component of $\mathrm{d}\boldsymbol{k}/\mathrm{d}t$ in the $\boldsymbol{B}$-direction, so that the motion is no longer in a plane of constant k_z. There is a net acceleration of the electron in the z-direction.

A more interesting case is that of crossed electric and magnetic fields. If we have $\boldsymbol{B} \perp \boldsymbol{E}$, then eq. (2.9) can be rewritten by replacing $\boldsymbol{E}$ by $[(\boldsymbol{B}\times\boldsymbol{E})\times\boldsymbol{B}]/B^2$,

$$\frac{\mathrm{d}\boldsymbol{k}}{\mathrm{d}t} = \frac{q}{\hbar c}\left(\frac{\boldsymbol{B}\times\boldsymbol{E}}{B^2}c + \boldsymbol{v}\right)\times\boldsymbol{B}. \tag{2.49}$$

The electron now moves as it would on the energy surface $\mathscr{E}'(\boldsymbol{k}) = \text{const.}$, where

$$\mathscr{E}'(\boldsymbol{k}) = \mathscr{E}(\boldsymbol{k}) + (\hbar\boldsymbol{k}\cdot\boldsymbol{B}\times\boldsymbol{E}c)/B^2. \tag{2.50}$$

For a simple band this is

$$\mathscr{E}'(\boldsymbol{k}) = \frac{\hbar^2}{2m^*}\left(\boldsymbol{k} + \frac{\boldsymbol{B}\times\boldsymbol{E}}{B^2}\frac{m^*c}{\hbar}\right)^2 - \left(\frac{E}{B}\right)^2\frac{m^*c^2}{2}. \tag{2.51}$$

Thus the effective mass is unchanged but the orbit center (assuming $k_z = 0$) is at $\hbar\boldsymbol{k}_0 = -m^*c\boldsymbol{B}\times\boldsymbol{E}/B^2$, and the electron moves with an average velocity cE/B in the direction of $\boldsymbol{k}_0$. This is the well known cycloidal motion, and it is the origin of the Hall effect.

In the case of a general energy band and relatively small electric fields, the magnetic orbits are only slightly distorted, but we note that an orbit on a surface of constant $\mathscr{E}$ need not be closed if E is sufficiently large compared to B. Zawadzki and Lax (1966) have looked at the conduction band of InSb, for which the two-band model of eq. (2.23) gives

$$\mathscr{E} = \mathscr{E}_g/2 + (\mathscr{E}_g^2/4 + \hbar^2k^2\mathscr{E}_g/2m_0^*)^{1/2}. \tag{2.52}$$

For sufficiently large k, $\mathscr{E} \sim \hbar k(\mathscr{E}_g/2m_0^*)^{1/2}$ becomes linear in k, so that $\mathscr{E}'$ becomes

$$\mathscr{E}' \cong \hbar k[(\mathscr{E}_g/2m_0^*)^{1/2} + \hat{\boldsymbol{k}}\cdot\boldsymbol{B}\times\boldsymbol{E}c)/B^2]. \tag{2.53}$$

The orbit will be open if $Ec/B > (\mathscr{E}_g/2m_0^*)^{1/2}$, in which case the behavior is like that in an electric field. The present case is analogous to that of a relativistic electron, because the electron velocity cannot exceed a limiting value of $(\mathscr{E}_g/2m_0^*)^{1/2}$. We consider this quantum mechanically in § 3.7.

3. Quantum effects in energy bands due to external fields

It was conjectured by Peierls in 1933 that the energy levels for a band electron in a magnetic field could be found by treating the energy band function as a quantum mechanical Hamiltonian. Including the effects of external potentials such as electric fields and impurities, the suggestion is that the general effective-mass Hamiltonian

$$\mathscr{H} = \mathscr{E}_n(\boldsymbol{p} + e\boldsymbol{A}/c) - e\varphi(\boldsymbol{r}) \tag{3.1}$$

be used to obtain the quantum levels in the external fields. This is the quantum generalization of eq. (2.11). The results can be justified for sufficiently small electric and magnetic fields by using the appropriate representations. We shall give a brief discussion of the general case of band electrons in external fields, including the semiclassical quantization of magnetic levels and its extensions. We shall then turn to the case of states near energy band edges, and derive the effective-mass approximation using the representation of Luttinger and Kohn (1955). This is the most important case for semiconductors, and the method will be applied to determine the magnetic energy levels near the band edges. We shall then consider the motion of electrons in electric fields, including quantum levels and tunneling, and also motion in combined electric and magnetic fields.

3.1. *Quantum mechanical representations for Bloch electrons in external fields*

For the case of a steady electric field, the Bloch representation of eq. (2.1), often called the crystal momentum representation, is usually the most convenient:

$$\psi(\boldsymbol{r}) = \sum_{n\boldsymbol{k}} \mathrm{e}^{\mathrm{i}\boldsymbol{k}\cdot\boldsymbol{r}} u_{n\boldsymbol{k}}(\boldsymbol{r})\psi_{n\boldsymbol{k}}. \tag{3.2}$$

We shall make use of this representation in § 3.6. Another important representation is in terms of the Wannier functions $a_n(\boldsymbol{r} - \boldsymbol{R})$ (Wannier 1937, Callaway 1974), which have the properties

$$b_{n\boldsymbol{k}} = \frac{1}{\sqrt{N}} \sum_{n\boldsymbol{k}} \mathrm{e}^{\mathrm{i}\boldsymbol{k}\cdot\boldsymbol{R}} a_n(\boldsymbol{r} - \boldsymbol{R}), \tag{3.3}$$

$$\int a_n^*(\boldsymbol{r} - \boldsymbol{R}) a_n(\boldsymbol{r} - \boldsymbol{R}')\,\mathrm{d}\boldsymbol{r} = \delta_{nn'}\delta_{\boldsymbol{R}\boldsymbol{R}'}. \tag{3.4}$$

These functions are localized, and are the analogues of the atomic functions that occur in the tight-binding approximation.

When a magnetic field is present it is necessary to modify the representation because the lattice translations no longer leave the Hamiltonian invariant. The appropriate operation is a magnetic translation (Brown 1968), which includes gauge transforming the origin of the vector potential by a lattice vector. The Wannier representation is modified by including the Peierls phase factor (Luttinger 1951, Adams 1952). In particular, for the symmetric gauge $A = \frac{1}{2}B \times r$, we make the replacement

$$a_n(r - R) \rightarrow \exp[-\mathrm{i}(e/2\hbar c)B \times R \cdot r]a_n(r - R).$$

However, an entirely equivalent scheme is to modify the Bloch functions directly (Roth 1962). Consider the operator $P = p + eA/c$. For free electrons and plane wave states this corresponds to the k-space operator $\kappa = k + (e/c)A(\mathrm{i}\nabla_k)$. We express the wave function of the system in the following form:

$$\psi(r) = \sum_{nk} \mathrm{e}^{\mathrm{i}k \cdot r} u_{n\kappa}(r)\psi_n(k). \tag{3.5}$$

That is, we replace k in u_{nk} of eq. (3.2) by the operator κ, which operates on the function $\psi_n(k)$ on the right. We are assuming that either a Taylor or Fourier expansion of u_{nk} can be made.

The Schrödinger equation for the Hamiltonian of eq. (2.10) becomes in this modified Bloch representation (Roth 1962):

$$\sum_{n'} \int \mathrm{d}r\, u_{n\kappa}^{+}(r)\left(\frac{(p + \kappa)^2}{2m} + V(r) - e\varphi(\mathrm{i}\nabla_k) - \mathscr{E}\right) u_{n'\kappa}(r)\psi_{n'}(k) = 0. \tag{3.6}$$

Here $P + eA/c$ has been replaced by $p + \kappa$ and r by $\mathrm{i}\nabla_k$ in the Hamiltonian. Equation (3.6) can be compared to the zero-field result,

$$\int \mathrm{d}r\, u_{nk}^{+}(r)\left(\frac{(p + k)^2}{2m} + V(r)\right) u_{n'k}(r) = \mathscr{E}_n(k)\delta_{nn'}. \tag{3.7}$$

In writing eq. (3.6) we should note that the different components of κ do not commute. The commutation relations are the same as those for P and are given by eq. (3.11) below. Therefore, in defining $u_{n\kappa}$ we must define the order of factors of κ, and it is usual to take the completely symmetric order. Also, the modified Bloch functions are not orthogonal. (The same holds for the Wannier functions with the Peierls phase.) Equation (3.6) has the general form

$$\sum_{n'} [\mathscr{H}_{nn'}(\kappa, \mathrm{i}\nabla_k) - \mathscr{E}\mathscr{N}_{nn'}(\kappa)]\psi_{n'}(k) = 0, \tag{3.8}$$

and so defines an effective many-band Hamiltonian.

If we look to the lowest order in the magnetic field, we can neglect the non-commutativity of the κ-factors. In this case, from eq. (2.5) we have $\mathscr{N}_{nn'} \cong \delta_{nn'}$, and from eq. (3.7) the magnetic part of $\mathscr{H}_{nn'}$ is $\delta_{nn'}\mathscr{E}_n(\kappa)$. If we also neglect the commutator of $\varphi(\mathrm{i}\nabla_k)$ with $u(\kappa)$, we obtain

$$[\mathscr{E}_n(\kappa) - e\varphi(\mathrm{i}\nabla_k) - \mathscr{E}]\psi_{n'}(k) = 0. \tag{3.9}$$

For the uniform electric field the approximation to φ amounts to neglecting the second term of eq. (2.8). For other potentials, we are assuming that $\varphi(r)$ is slowly varying in space. Equation (3.9) is the k-space equivalent of eq. (3.1), and thus provides us a justification for treating $\mathscr{E}(\boldsymbol{P}) - e\varphi(\boldsymbol{r})$ as a quantum mechanical Hamiltonian.

Considering the magnetic problem, it is possible to go to higher order in B. By means of a suitable transformation, the off-diagonal parts of $\mathscr{H}_{nn'}$ can be eliminated and $\mathscr{N}_{nn'}$ can be reduced to $\delta_{nn'}$ to any order in the magnetic field, and this results in an effective one-band Hamiltonian (Roth 1962). The first-order term in B is an effective Zeeman interaction. In order to derive this we must include the spin–orbit interaction, eq. (2.28), as well as the Zeeman energy of the electron spin, $g\mu_B \boldsymbol{S}_0 \cdot \boldsymbol{B}$, where $g = 2.0023$ and μ_B is the Bohr magneton $e\hbar/2mc$. The first-order term in the effective Hamiltonian is then

$$\mathscr{H}_{1,nn'} = \frac{e\boldsymbol{B}}{4c}\left(\boldsymbol{X}_{nn'} \times (\boldsymbol{v}_n + \boldsymbol{v}_{n'}) + \frac{1}{m}\sum_{n''} (\boldsymbol{X}_{nn''} \times \boldsymbol{p}_{n''n'} - \boldsymbol{p}_{nn''} \times \boldsymbol{X}_{n''n'}) \right) + g\mu_B \boldsymbol{S}_{0nn'} \cdot \boldsymbol{B} \tag{3.10}$$

where $\boldsymbol{X}_{nn'}$ is given by eq. (2.8) and $\boldsymbol{v}_n = \nabla_{\boldsymbol{k}} \mathscr{E}_n/\hbar$ is the diagonal part of $\boldsymbol{p}$. Assuming a crystal with inversion symmetry, n and n' go over the Kramers degenerate pair at $\boldsymbol{k}$, and the Zeeman interaction can be put into the form of an effective $\boldsymbol{g}$-tensor. The matrix elements* in this expression are functions of $\boldsymbol{k} \to \boldsymbol{\kappa}$. The first part of eq. (3.10) has an obvious interpretation in terms of orbital angular momentum.

Similar approaches leading to an effective Hamiltonian have been developed by several authors (Wannier 1962, Wannier and Fredkin 1962, Blount 1962a,b, Zak 1972) for both the electric and magnetic cases. For the magnetic case such a procedure does not establish the existence of the effective-mass Hamiltonian for arbitrary magnetic fields. However, Brown (1964) has shown through group theoretical arguments that an effective Hamiltonian does exist for certain "rational" values of the magnetic field (see § 3.3). More recently Nenciu (1991) has shown that the effective Hamiltonian is valid for isolated bands.

3.2. Semiclassical quantization of magnetic orbits

The occurrence of periodicity in classical orbits is related by the correspondence principle to the quantization of the energy levels. We now apply the effective Hamiltonian to find the semiclassical quantization condition for the magnetic levels, which was first obtained by Onsager (1952). We assume a single non-degenerate band and for the present we ignore spin effects. As outlined above we treat the band energy as a quantum mechanical Hamiltonian, $\mathscr{E}(\boldsymbol{P}) = \mathscr{E}(\boldsymbol{p} + e\boldsymbol{A}/c)$, where $\boldsymbol{p} = (\hbar/\mathrm{i})\nabla$ and $\boldsymbol{r}$ are quantum mechanical operators. Taking $\boldsymbol{B}$ to be in the z-direction we find the commutation relations for the components of $\boldsymbol{P}$:

$$[P_x, P_y] = \frac{\hbar e}{\mathrm{i}c}\left(\frac{\partial A_y}{\partial x} - \frac{\partial A_x}{\partial y}\right) = -\mathrm{i}eB\hbar/c. \tag{3.11}$$

*In earlier work we replaced $\boldsymbol{p}_{nn'}$ by $\pi_{nn'} = p_{nn'} + [\boldsymbol{S} \times \nabla V(\boldsymbol{r})]_{nn'}/(2\pi mc^2)$, but the second term is very small and we shall neglect it here.

Comparing with $[p_x, p_y] = \hbar/\mathrm{i}$ we see that P_x and P_y can be regarded as a canonical coordinate and momentum pair, with $\hbar$ replaced by $\hbar eB/c$. Then from the Bohr–Sommerfeld quantization condition, $\int p\,\mathrm{d}q = (n+\gamma)2\pi\hbar$, where γ is a constant, we find the area in $\boldsymbol{k}$-space to be (Onsager 1952)

$$A = \int k_y\,\mathrm{d}k_x = (n+\gamma)2\pi eB/\hbar c. \tag{3.12}$$

Thus the area of the magnetic orbit in $\boldsymbol{k}$-space is quantized.

We can find the degeneracy of the quantum orbits by a phase space argument. A given area A in $\boldsymbol{k}$-space contains $L_xL_yA/4\pi^2$ states for a sample of dimensions L_x, L_y and L_z, so that the degeneracy of the magnetic levels in two dimensions is $L_xL_yeB/2\pi\hbar c$. In terms of the flux quantum $hc/e = \varphi_0$, the degeneracy is L_xL_yB/φ_0, or the number of flux quanta in the area of the crystal. The k_z degeneracy is described as usual by $(L_z/2\pi)\,\mathrm{d}k_z$ states in an interval $\mathrm{d}k_z$.

The quantum condition also gives the spacing between the levels, which should be $\hbar\omega_\mathrm{c}$, where ω_c is the cyclotron frequency. From eq. (3.12)

$$A(\mathscr{E}+\hbar\omega_\mathrm{c}) - A(\mathscr{E}) \cong \hbar\omega_\mathrm{c}\partial A/\partial\mathscr{E} = 2\pi eB/\hbar c, \tag{3.13}$$

but if $\mathrm{d}k_\perp$ is an increment of $\boldsymbol{k}$ normal to the orbit in the constant-k_B plane (see fig. 3),

$$(\hbar\omega_\mathrm{c})^{-1} = \frac{\hbar c}{2\pi eB}\frac{\partial A}{\partial\mathscr{E}} = \frac{\hbar c}{2\pi eB}\oint\frac{\mathrm{d}k_\perp}{\mathrm{d}\mathscr{E}}\,\mathrm{d}k_\mathrm{p} = \frac{c}{2\pi e}\oint\frac{\mathrm{d}k_\mathrm{p}}{|\boldsymbol{v}\times\boldsymbol{B}|}. \tag{3.14}$$

This is exactly the Shockley tube integral result for the cyclotron frequency.

To include the effects of spin we make use of the first-order term in the effective one-band Hamiltonian, eq. (3.10), which gives a Zeeman interaction. This can be expressed as

$$\mathscr{H} = \mathscr{E}(\boldsymbol{k}) + \mu_\mathrm{B}\boldsymbol{S}\cdot\boldsymbol{g}(\boldsymbol{k})\cdot\boldsymbol{B}, \tag{3.15}$$

where the $\boldsymbol{g}$-tensor depends in general on $\hbar\boldsymbol{k}$, which is actually the kinetic momentum $\boldsymbol{p} + e\boldsymbol{A}/c$. Let us first consider the classical motion, and include the equation of motion for the electron spin (Zeiger and Pratt 1973). To lowest order in B,

$$\frac{\mathrm{d}\boldsymbol{S}}{\mathrm{d}t} = -\hbar^{-1}\mu_\mathrm{B}\boldsymbol{S}\times[\boldsymbol{g}(\boldsymbol{k}(t))\cdot\boldsymbol{B}]. \tag{3.16}$$

On the right-hand side of this equation $\boldsymbol{k}(t)$ varies with time as the electron goes around the cyclotron orbit (see eqs. (2.43), (2.44)). In general $\boldsymbol{g}(\boldsymbol{k})$ varies along the orbit, so that the spin sees an effective magnetic field that varies periodically with time. This is a well defined problem that does not usually have a solution in closed form. However, if the variation is not too great, Zeiger and Pratt (1973) suggest replacing $\boldsymbol{g}(\boldsymbol{k})$ by its average around the cyclotron orbit:

$$\bar{\boldsymbol{g}}(\mathscr{E}, k_\mathrm{B}) = \oint\frac{\mathrm{d}k_\mathrm{p}}{|\boldsymbol{v}\times\boldsymbol{B}|}\,\boldsymbol{g}(\boldsymbol{k}) \Big/ \oint\frac{\mathrm{d}k_\mathrm{p}}{|\boldsymbol{v}\times\boldsymbol{B}|}. \tag{3.17}$$

In this case the motion can readily be shown to be a precession with the frequency

$$\omega_s = \hbar^{-1}\mu_B|\bar{\boldsymbol{g}} \cdot \boldsymbol{B}|. \tag{3.18}$$

When we look at the semiclassical quantization of the magnetic levels, we find that they have a spin splitting $\hbar\omega_s$. The quantization can also be described (Roth 1966) by having two values of γ in eq. (3.12).

In addition to cyclotron resonance, which can be described as transitions among the levels, the quantization of electron orbits in a magnetic field is responsible for a number of interesting effects. Magnetoabsorption at optical frequencies shows oscillations due to transitions between quantum levels, as do other magneto-optical effects (Mavroides 1972). The de Haas–van Alphen effect (de Haas and van Alphen 1930) and the Shubnikov–de Haas effect (Shubnikov and de Haas 1930) are oscillations in the susceptibility and magnetoresistance that occur for metals and degenerate semiconductors. Consider a fixed energy, i.e., the Fermi energy, and a certain value of k_B. Then a level sweeps through this energy whenever

$$\frac{1}{B} = \frac{2\pi e}{\hbar c}\frac{n+\gamma}{A(k_B, \mathscr{E})}. \tag{3.19}$$

Thus any property will have a periodicity in $1/B$. However, there will be various periods for different k_B, and it is only *extremal* areas that are actually seen. The theory of the de Haas–van Alphen effect for general energy bands was worked out by Lifshitz and Kosevich (1955) and a good exposition is found in Zeiger and Pratt (1973). The periodicity in $1/B$ gives information on the extremal areas of the Fermi surface, and has been a major tool in determining the Fermi surfaces of metals and semimetals. The Shubnikov–de Haas effect has been observed in a number of semiconductors (see Roth and Argyres 1966) for which of course the Fermi surface depends on the level of doping. Figure 8 shows the oscillatory magnetoresistance in HgSe as observed by Whitsett (1965).

3.3. Beyond semiclassical quantization

The effective Hamiltonian discussed above has been used to study the steady magnetic susceptibility (Wannier and Upadyaya 1964, Roth 1962). In order to calculate this, the semiclassical quantum conditions have to be extended to give γ in eq. (3.12) to first order in B, which involves calculating a second-order term in the effective Hamiltonian, as well as using the Zeeman term, which contributes to γ in zeroth order (Roth 1966). The above representation has also been used to calculate the interband Faraday rotation in semiconductors (Roth 1964, Ray 1967, Mavroides 1972).

Another way of extending the semiclassical picture is based on the WKB method (see Brown 1968). This has been used in analyzing magnetic breakdown effects (Cohen and Falicov 1961), which are of importance in de Haas–van Alphen experiments in metals, and originate when magnetic orbits are close to each other in $\boldsymbol{k}$-space, so that electrons can jump from one orbit to another.

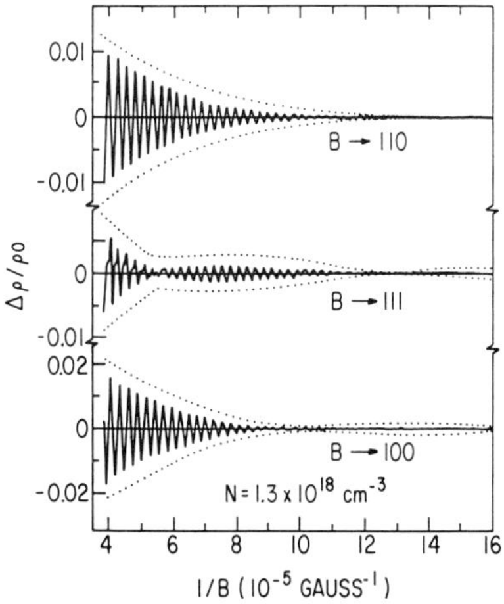

Fig. 8. The Shubnikov–de Haas effect in HgSe (after Whitsett 1965).

In seeking solutions for the effective Hamiltonian in a magnetic field of arbitrary strength, Brown (1968) has studied the translational properties of the wave functions in detail. The magnetic translation operations which leave the Hamiltonian invariant do not form a group, but rather a ray representation of the translation group. Brown found that eigenstates exist if $\boldsymbol{B}$ obeys a rationality condition, namely that $\boldsymbol{B}$ is along a primitive lattice vector (suitably defined) and that the magnetic flux through the base of the unit cell defined by the other two primitive lattice vectors is a rational fraction l/N of a magnetic flux quantum hc/e. The quantum levels exhibit banding, that is splitting of the degeneracy that depends sensitively on l and N, and this can change drastically for small changes in B.

Hofstadter (1976) has studied the solutions of the effective Hamiltonian for a simple two-dimensional tight-binding model, first investigated by Harper (1955). He found that the banding of the quantum levels had an overall width that varied rather uniformly with field, but there was fine structure that looked like a butterfly and exhibited fractal properties such as self-similarity. For fields not obeying the rationality condition, he found that the distribution of levels resembled Cantor sets.

These interesting and puzzling effects would be important only for extremely large magnetic fields. Brown estimates $N > 1000$ for the largest attainable fields. However, it is possible that banding effects might be observed for transverse magnetic fields in superlattices (Maan 1984).

3.4. The effective-mass approximation

In semiconductors the carriers of interest are primarily near the band edges, so it is of interest to find the quantum levels near both the non-degenerate and the degenerate band edges. The theory of Luttinger and Kohn (1955) is particularly well suited to this purpose, and in this section we outline this theory.

We consider once more an electron in a crystal subjected to an external field. For generality we shall impose an electromagnetic field described by vector and scalar potentials $A(r)$ and $\varphi(r)$, both assumed to vary slowly as a function of position. The Hamiltonian including external fields is given by eq. (2.10). If we consider a band edge at the center of the Brillouin zone, the Luttinger–Kohn representation uses, instead of Bloch functions, the functions

$$\chi_{nk} = \mathrm{e}^{\mathrm{i}\boldsymbol{k}\cdot\boldsymbol{r}} u_{n0}(\boldsymbol{r}), \tag{3.20}$$

where $\boldsymbol{k}$ goes over one Brillouin zone and where u_{n0} is the periodic function for $k = 0$. We can easily generalize the theory to take the origin at any point in $\boldsymbol{k}$-space. The representation leads in the absence of external fields to the $\boldsymbol{k}\cdot\boldsymbol{p}$ approximation discussed in § 2.3.

Actually, the many-band effective Hamiltonian is now the same as eq. (3.6), except that we replace u_{nk} by u_{n0}, which leads to a great simplification. However, for the present analysis it is interesting to work in real space rather than $\boldsymbol{k}$-space, and to write the wave function as

$$\varphi(\boldsymbol{r}) = \sum_n f_n(\boldsymbol{r}) u_{n0}(\boldsymbol{r}), \tag{3.21}$$

where the f_n are slowly varying "envelope functions" and the sum is over band edges at $\boldsymbol{k} = 0$. We can regard the f_n as having $\boldsymbol{k}$ Fourier components only within one Brillouin zone. Operating on ψ with $\mathscr{H}$, we find

$$\mathscr{H}\psi = \sum_n f_n(\boldsymbol{r})\mathscr{H}_0 u_{n0}(\boldsymbol{r}) + \sum_{n'} (1/m)\boldsymbol{P} f_{n'}(\boldsymbol{r})\cdot \boldsymbol{p} u_{n'0}(\boldsymbol{r}) + \sum_n [\boldsymbol{P}^2/2m - e\varphi(\boldsymbol{r})] f_n(\boldsymbol{r}) u_{n0}(\boldsymbol{r}). \tag{3.22}$$

Here $\boldsymbol{P} = \boldsymbol{p} + e\boldsymbol{A}/c$ operates on the envelope function, while $\boldsymbol{p}$ operates on the $\boldsymbol{k} = 0$ periodic function. Using

$$\boldsymbol{p} u_{n'0} = \sum_n u_{n0}(\boldsymbol{r}) \boldsymbol{p}_{nn'},$$

eq. (3.22) can be written as

$$\sum_n \left([\mathscr{E}_n - \mathscr{E} + \boldsymbol{P}^2/2m - e\varphi(\boldsymbol{r})] f_n(\boldsymbol{r}) + \sum_{n'} (1/m)\boldsymbol{P} f_{n'}(\boldsymbol{r})\cdot \boldsymbol{p}_{nn'} \right) u_{n0}(\boldsymbol{r}) = 0. \tag{3.23}$$

Here $\mathscr{E}_n$ is the energy of the nth band edge. We now argue that the f are slowly varying functions over a unit cell, while the u are rapidly varying. If we neglect the variation of f, its derivatives, and the potentials $\boldsymbol{A}$ and φ over a unit cell, then the coefficient of each of the u_{n0} must vanish. To show this we can multiply the equation by $u^*_{n''0}$ on the left and integrate over a unit cell. From orthogonality of the u, only the coefficient of $u_{n''0}$ will remain, and that equals the right-hand side, which is zero. This leads to the many-band set of equations for the f,

$$[\boldsymbol{P}^2/2m - e\varphi(\boldsymbol{r})] f_n(\boldsymbol{r}) + \sum_{n'} (1/m)\boldsymbol{p}_{nn'}\cdot \boldsymbol{P} f_{n'}(\boldsymbol{r}) = (\mathscr{E} - \mathscr{E}_n) f_n(\boldsymbol{r}). \tag{3.24}$$

Actually these equations give a solution of eq. (3.23) even without the "slowly varying" argument, so they may be regarded as exact. Equations (3.24) are the basic equations for the quantum states in the Luttinger–Kohn representation.

Since we are interested in states that are close to one or a group of band edges, we now wish to transform eq. (3.24) to eliminate the unwanted off-diagonal terms involving $\boldsymbol{p}_{nn'}$. As in the $\boldsymbol{k}\cdot\boldsymbol{p}$ approximation, most cases of interest can be included in the following scheme. We consider a set of band edges (i), which are close together in energy, and we assume that the f are largest for this set of states. For all other states we approximate $f_n(\boldsymbol{r})$ by

$$f_n(\boldsymbol{r}) \cong (\bar{\mathscr{E}}_i - \mathscr{E}_n)^{-1} \sum_i (1/m)\boldsymbol{p}_{ni}\cdot \boldsymbol{P} f_i(\boldsymbol{r}).$$

We have neglected the first term in eq. (3.14), which we assume is small for the $f_{n'}$, and we have approximated the energy $\mathscr{E}$ by the average band edge energy $\bar{\mathscr{E}}_i$ among the set of interest. Then we can eliminate the f from distant band edges to obtain the effective-mass equations for our set of band edges,

$$\mathscr{E} f_i(\boldsymbol{r}) = [\mathscr{E}_i + \boldsymbol{P}^2/2m - e\varphi(\boldsymbol{r})] f_i(\boldsymbol{r}) + \sum_j \left(\frac{1}{m}\boldsymbol{p}_{ij}\cdot\boldsymbol{P} + \sum_n{}' \frac{1}{m^2}\frac{\boldsymbol{p}_{in}\cdot\boldsymbol{P}\boldsymbol{p}_{nj}\cdot\boldsymbol{P}}{\bar{\mathscr{E}}_i - \mathscr{E}_n}\right) f_j(\boldsymbol{r}), \tag{3.25}$$

where the prime indicates that the set (i) is excluded from the sum. For most cases of interest the linear term in $\boldsymbol{P}$ vanishes.

The effective-mass equation can be applied to a variety of situations. We shall consider here only uniform electric and magnetic fields. We can also let φ correspond to an impurity potential, in which case we can calculate the shallow impurity levels, with or without additional fields. The perturbation can correspond as well to a quantum well, inversion layer or superlattice, which can confine or modify the motion of carriers in one dimension. These problems will be discussed in the other chapters.

3.5. Magnetic energy levels near band edges

We consider in this section the solutions of eq. (3.25) for the case of a magnetic field. For a band edge at energy $\mathscr{E}_0$ with arbitrary degeneracy, assuming that we can take $\boldsymbol{p}_{ij} = 0$, the envelope wave functions f_i are the solutions of the Schrödinger equation:

$$\sum_j \mathscr{H}_{ij} f_i(\boldsymbol{r}) = (\mathscr{E} - \mathscr{E}_0) f_i(\boldsymbol{r}), \tag{3.26}$$

with the effective-mass Hamiltonian

$$\mathscr{H}_{ij} = \frac{1}{2}\sum_{\alpha\beta} D_{ij}^{\alpha\beta} P_\alpha P_\beta + g\mu_{\mathrm{B}}(\boldsymbol{S}_0\cdot\boldsymbol{B})_{ij}, \tag{3.27}$$

where $\boldsymbol{P} = \boldsymbol{p} + e\boldsymbol{A}/c$, $\boldsymbol{S}_0$ is the electron spin, and

$$D_{ij}^{\alpha\beta} = \frac{1}{m}\delta_{ij}\delta_{\alpha\beta} + \frac{2}{m^2}\sum_n{}' \frac{p_{in}^{\alpha} p_{nj}^{\beta}}{\mathscr{E}_0 - \mathscr{E}_n}. \tag{3.28}$$

Since the components of P do not commute with each other, it is convenient to split $\mathscr{H}_{ij}$ into two parts $\mathscr{H}_{ij}^{(\mathrm{S})} + \mathscr{H}_{ij}^{(\mathrm{A})}$, which are symmetric and antisymmetric in the components of $\boldsymbol{P}$. For the symmetric part we have

$$\mathscr{H}_{ij}^{(S)} = \frac{1}{2}\sum_{\alpha\beta} \bar{D}_{ij}^{\alpha\beta}\{P_\alpha, P_\beta\}, \tag{3.29}$$

where $\bar{D}_{ij}^{\alpha\beta} = \frac{1}{2}(D_{ij}^{\alpha\beta} + D_{ij}^{\beta\alpha})$ is the symmetric part of $\boldsymbol{D}$, and $\{P_\alpha, P_\beta\}$ is the symmetric product of P_α and P_β. For the antisymmetric part we use eq. (3.11) to obtain

$$\mathscr{H}_{ij}^{(A)} = g\mu_B(\boldsymbol{S}_0 \cdot \boldsymbol{B})_{ij} + \sum_{\alpha\beta\gamma} \varepsilon_{\alpha\beta\gamma} D_{ij}^{\alpha\beta} \frac{\hbar e}{4ic} B_\gamma, \tag{3.30}$$

where $\varepsilon_{\alpha\beta\gamma}$ is the completely antisymmetric tensor. $\mathscr{H}_{ij}^{(A)}$ is thus proportional to the magnetic field. For zero magnetic field only the symmetric part of $\mathscr{H}$ survives and reduces to the $\boldsymbol{k}\cdot\boldsymbol{p}$ energy band equations. In this case we can take $f_j = c_j \exp(i\boldsymbol{k}\cdot\boldsymbol{r})$, and the energy bands are then obtained as eigenvalues of the equation

$$\sum_{j\alpha\beta} \frac{\hbar^2}{2} \bar{D}_{ij}^{\alpha\beta} k_\alpha k_\beta c_j = (\mathscr{E} - \mathscr{E}_0)c_i \tag{3.31}$$

as we have found in § 2.2.

Consider now a band edge with only the twofold Kramers degeneracy. In the absence of spin–orbit interaction the states are simply up- and down-spin states, while in the presence of spin–orbit interaction "up" and "down" refer to mixed-spin states. The effective-mass Hamiltonian has the general form

$$\mathscr{H} = \tfrac{1}{2}(\boldsymbol{p} + e\boldsymbol{A}/c)\cdot \boldsymbol{m}^{*-1}\cdot(\boldsymbol{p} + e\boldsymbol{A}/c) + \mu_B \boldsymbol{S}\cdot\boldsymbol{g}^*\cdot\boldsymbol{B}. \tag{3.32}$$

The effective-mass tensor is the symmetric part of the interband sum in eq. (3.28), i.e., $\boldsymbol{m}^* = \bar{\boldsymbol{D}}_{11}$ where 1 is the "up" state. $\boldsymbol{S}$ is an effective spin and $\boldsymbol{g}^*$ is the $\boldsymbol{g}$-tensor (Yafet 1963), a typical component of which is

$$g_{zz}^* = (2S_{0z})_{11} g + \frac{m}{2i}(D_{11}^{xy} - D_{11}^{yx}). \tag{3.33}$$

If the band edge has spherical symmetry, $\boldsymbol{m}^*$ and $\boldsymbol{g}^*$ will both be scalars. If we use the asymmetric gauge $\boldsymbol{A} = (0, Bx, 0)$, then the Schrödinger equation has the form

$$\frac{1}{2m^*}\left[p_x^2 + \left(p_y + \frac{eBx}{c}\right)^2 + p_z^2\right] f(\boldsymbol{r}) \pm \tfrac{1}{2}\hbar\omega_s f(\boldsymbol{r}) = \mathscr{E} f(\boldsymbol{r}), \tag{3.34}$$

where as usual $\boldsymbol{p} = (\hbar/i)\nabla$, and $\omega_s = g^*\mu_B B$. This is the same form as that for a free electron with mass m^* and g-factor g^* in a magnetic field. It is well known that this equation describes discrete Landau levels (Landau 1930) for the motion transverse to $\boldsymbol{B}$. The effective potential depends here only on x, and the orbital part of the wave function has the form

$$\psi_{n\boldsymbol{k}}(\boldsymbol{r}) = \frac{1}{\sqrt{L_y L_z}} \varphi_n(x - X_{\boldsymbol{k}}) \exp(ik_y y) \exp(ik_z z). \tag{3.35}$$

Here $\boldsymbol{k}$ is the two-dimensional wave vector associated with free motion in the y- and z-directions, and L_y, L_z are the corresponding normalization lengths. The φ are the normalized wave functions of a simple harmonic oscillator centered at the point $X_{\boldsymbol{k}} = -l^2 k_y$ where $l = (\hbar c/eB)^{1/2}$ is the classical radius of the lowest cyclotron orbit. Wave functions for other gauges are given by Johnson and Lippmann (1949).

The harmonic oscillator wave functions can be found by giving the $n = 0$ harmonic oscillator wave function as

$$\varphi_0(x) = (2l^2\pi)^{-1/4} \exp(-x^2/2l^2), \tag{3.36}$$

and then defining the raising and lowering operators as

$$a^\dagger = (l/2^{1/2}\hbar)(P_x + \mathrm{i}P_y), \quad a = (l/2^{1/2}\hbar)(P_x - \mathrm{i}P_y). \tag{3.37}$$

In the chosen gauge, $P_x = (\hbar/\mathrm{i})\partial/\partial x$ and $P_y = (\hbar/\mathrm{i})\partial/\partial y + eBx/c = \hbar(x - X_{\boldsymbol{k}})/l^2$. These operators have the following well-known properties,

$$a^\dagger_{\psi_{n\boldsymbol{k}}} = \sqrt{n+1}\,\psi_{n+1\boldsymbol{k}}, \quad a\psi_{n\boldsymbol{k}} = \sqrt{n}\,\psi_{n-1\boldsymbol{k}}. \tag{3.38}$$

We can therefore obtain the wave function $\psi_{n\boldsymbol{k}}$ by applying the raising operator n times to $\psi_{0\boldsymbol{k}}$.

The energy levels are given by

$$\mathscr{E}_n^{\pm} = (n + \tfrac{1}{2} \pm \tfrac{1}{2}\nu)\hbar\omega_c + \hbar^2 k_z^2/2m^*, \tag{3.39}$$

where $\nu = \omega_s/\omega_c$. The Landau levels have the spacing of the cyclotron frequency $\hbar\omega_c = \hbar eB/m^*c$, and in addition have a spin splitting. For free electrons, $\nu = 1$. The quantum levels have a degeneracy that is determined by requiring the orbit center to lie in the crystal, i.e., $-L_x/2 \leqslant X_{\boldsymbol{k}} \leqslant L_x/2$, and since the k_y values are spaced $2\pi/L_y$ units apart there are $L_x L_y/2\pi l^2$ of them for each n and k_z value. This result agrees with the phase space argument of § 3.3. The density of states is now a quasi-continuous function of energy (Roth and Argyres 1966).

An example of such a simple band edge is the $\boldsymbol{k} = 0$ conduction band of Ge, InSb and similar systems which have a p-like valence band split by the spin–orbit interaction, as described in § 2.2. If the only p matrix element of importance is that connecting the conduction band and the valence band, as in eqs. (2.35) and (2.36), the effective mass and g-factor are given by (Roth et al. 1959):

$$\frac{m}{m^*} = 1 + \frac{2mP^2}{\hbar^2 3}\left(\frac{2}{\mathscr{E}_g} + \frac{1}{\mathscr{E}_g + \Delta}\right), \tag{3.40}$$

$$g^* = 2\left[1 - \frac{2mP^2}{\hbar^2 3}\left(\frac{1}{\mathscr{E}_g} - \frac{1}{\mathscr{E}_g + \Delta}\right)\right]. \tag{3.41}$$

This leads to a simple relation between g^* and m^*. Thus the origins of the large g-shift (Bemski 1960) and of the small effective mass in InSb (see table 1) are very similar. Zawadzki (1973) has studied the spin properties of conduction electrons in a number of III–V compounds.

Non-parabolic effects in the conduction band can be included by generalizing Kane's three-band model of eqs. (2.35) and (2.36) to this case, which amounts to replacing "$\mathscr{E}_g$" by "$\mathscr{E} + \mathscr{E}_g$" in eqs. (3.40) and (3.41). Furthermore, if we assume that $\Delta \gg \mathscr{E}_g$, as in InSb, then a two-band model suffices (Zawadzki and Lax 1966), and, neglecting the first terms of eqs. (3.40) and (3.41), we can write for the energy levels in a magnetic field,

$$\mathscr{E}(1+\mathscr{E}/\mathscr{E}_g)=(n+\tfrac{1}{2}\pm\tfrac{1}{2}\nu_0)\hbar\omega_{c0}+\hbar^2k_z^2/2m_0^*, \tag{3.42}$$

where m_0^*, ν_0 and ω_{c0} refer to band edge quantities.

Consider now a more general anisotropic band edge for which eq. (3.32) applies. For the orbital part of the Hamiltonian, we can make a scale transformation of the momenta, just as in the classical case. Since $\boldsymbol{P}=\boldsymbol{p}+e\boldsymbol{A}/c$, this must also apply to the vector potential, and in order to preserve the commutation relations between $\boldsymbol{p}$ and $\boldsymbol{r}$, the coordinates must also be scaled. Assuming a principal-axis system for $\boldsymbol{m}^*$, we have for the x-component,

$$x'=\sqrt{m_x/m}x,\quad p'_x=\sqrt{m/m_x}p_x,\quad A'_x=\sqrt{m/m_x}A_x \tag{3.43}$$

and similarly for y and z. The Hamiltonian in the new coordinate system is

$$\mathscr{H}^{(\mathrm{S})}=(\boldsymbol{p}'+e\boldsymbol{A}'/c)^2/2m. \tag{3.44}$$

The effective magnetic field in the new coordinate system is

$$\bar{B}_z=\frac{\partial A'_y}{\partial x'}-\frac{\partial A'_x}{\partial y'}=\left(\frac{m^2}{m_xm_y}\right)^{1/2}B_z,\text{ etc.}, \tag{3.45}$$

which is the same as in the classical case. The wave functions are given by eq. (3.35) in the scaled system, and the energy levels are given by eq. (3.39), where the cyclotron frequency has the classical value, eq. (2.41), or for ellipsoids of revolution as in Ge and Si, eq. (2.42), and $k_z\to\boldsymbol{k}'\cdot\bar{\boldsymbol{B}}/\bar{B}$.

The spin part of the Hamiltonian is easily diagonalized, giving a spin splitting of $\hbar\omega_s=\mu_B|\boldsymbol{g}\cdot\boldsymbol{B}|$. For the conduction-band edges in Ge and Si, the $\boldsymbol{g}$-tensor has axial symmetry with components $g_\parallel$ and $g_\perp$. The spin splitting is

$$\hbar\omega_s=\mu_BB(g_\parallel^2\cos^2\theta+g_\perp^2\sin^2\theta)^{1/2}. \tag{3.46}$$

The g-factors are given in table 1. In evaluating the $\boldsymbol{g}$-tensor it is often sufficient to go to first order in $\mathscr{H}_{so}=2\boldsymbol{S}_0\cdot\boldsymbol{h}$, and this gives (Roth 1960),

$$\boldsymbol{g}=2\boldsymbol{I}+\frac{2}{im}\sum_{nn'}\frac{1}{\mathscr{E}_{in}\mathscr{E}_{in'}}(\boldsymbol{h}_{in}\boldsymbol{p}_{nn'}\times\boldsymbol{p}_{n'i}+\boldsymbol{h}_{nn'}\boldsymbol{p}_{in}\times\boldsymbol{p}_{n'i}+\boldsymbol{h}_{n'i}\boldsymbol{p}_{in'}\times\boldsymbol{p}_{nn'}), \tag{3.47}$$

where $\boldsymbol{I}$ is the unit dyadic. For the case of Ge, the large shift in $g_\parallel$ is due to the spin–orbit splitting of the nearby L_3 band edge, in analogy with eq. (3.41). The g-shifts in Si have been analyzed by Liu (1962), and these and other cases are reviewed by Yafet (1963).

If we consider states near the degenerate valence-band edge in semiconductors, described in § 2.2, the effective-mass equation, eq. (3.1) gives a set of coupled equations for the components f_i. Luttinger (1956) expressed the 4×4 effective-mass equation in terms of angular momentum matrices for $J=\frac{3}{2}$. He showed that the most general form for $\mathscr{H}$ is, from cubic symmetry,

$$\begin{aligned}\mathscr{H} = (1/m)[(\gamma_1 + 5\gamma_2/2)P^2/2 - \gamma_2(P_x^2J_x^2 + P_y^2J_y^2 + P_z^2J_z^2) - 2\gamma_3(\{P_x, P_y\}\{J_x, J_y\}\\ + \{P_y, P_z\}\{J_y, J_z\} + \{P_x, P_z\}\{J_x, J_z\})\\ - 2\mu_B\kappa \boldsymbol{J}\cdot\boldsymbol{B} - 2\mu_B q(J_x^3B_x + J_y^3B_y + J_z^3B_z)].\end{aligned} \tag{3.48}$$

For $B = 0$ and $\boldsymbol{P} = \hbar\boldsymbol{k}$ the terms involving γ_1, γ_2 and γ_3 provide a way of describing the effective-mass equations for the energies of the light- and heavy-hole bands. The Luttinger parameters are related to the other parameters by eq. (2.23), and we include them in table 1. As in the twofold-degenerate case, there are extra terms that are proportional to the magnetic field. The parameter κ can be related to antisymmetric terms in the $\boldsymbol{D}$-matrix for zero spin–orbit splitting. However, q is an additional parameter introduced by the spin–orbit interaction, which can be related to the spin–orbit splitting Δ_{15} of the Γ_{15} higher-conduction-band edge (Hensel and Suzuki 1969). Lawaetz (1971) has included q in a $\boldsymbol{k}\cdot\boldsymbol{p}$ calculation for a number of systems. (See Wiley 1975).

The solutions of the coupled equations resulting from the effective-mass Hamiltonian in eq. (3.48) have been studied by Luttinger (1956) and Goodman (1961). Assuming that the energy surfaces are nearly spherical, it is possible to make an expansion of the wave functions in terms of harmonic oscillator functions for the coordinates transverse to the magnetic field. In fig. 9 results for Ge are shown, calculated by Hensel and Suzuki (1970). The levels at $k_B = 0$ form four ladders, two for heavy and two for light holes. For n large, the level spacings correspond to the classical cyclotron frequency, together with a spin splitting, which for the isotropic case is $\kappa \pm \kappa$, in units of the free-electron cyclotron frequency. For small n the levels become unequally spaced, and cyclotron resonance experiments at low temperatures have yielded a number of extra lines from these quantum effects (Fletcher et al. 1955, Stickler et al. 1962). Hensel and Suzuki (1970) have also calculated the energy

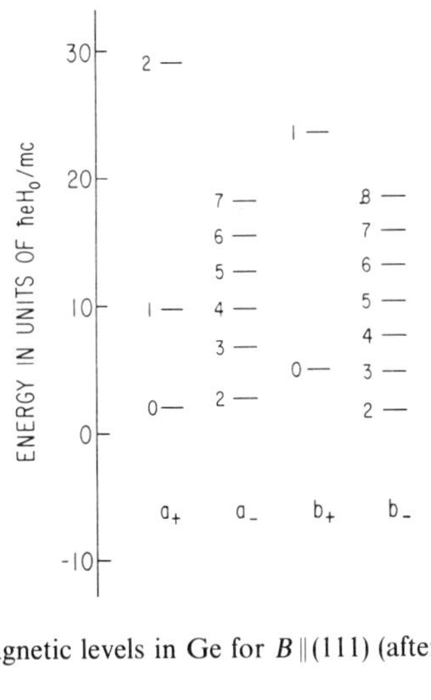

Fig. 9. Valence-band magnetic levels in Ge for $B \parallel (111)$ (after Hensel and Suzuki 1970).

spectrum for finite k_B. In order to determine the parameters to use for their fit, they carried out cyclotron and spin resonance experiments in a strained Ge crystal. The presence of strain splits the valence-band degeneracy and simplifies the level structure. They thereby obtained very good values of the parameters for Ge. In analyzing the quantum effects, they found that the cyclotron and spin resonance peaks correspond to critical points arising from maxima and minima of the energy separation between levels as a function of k_B. These do not always occur at $k_B = 0$.

An important probe of the band structure of semiconductors has been found in interband magneto-optical effects. The basic theory of interband magneto-absorption is given in Roth et al. (1959), and a review of magneto-optical effects in semiconductors is given by Mavroides (1972). The magneto-absorption for energies near the direct gap in Ge (Zwerdling et al. 1959), InSb (Pidgeon and Brown 1966), GaAs (Vrehen 1968) and other materials shows oscillations corresponding to transitions between Landau levels in the valence and conduction bands, modified by exciton effects. Figure 10 shows the level scheme and transitions for GaAs (Vrehen 1968), and fig. 11 shows the results of Pidgeon and Brown (1966) for InSb. In fitting the spectra, the

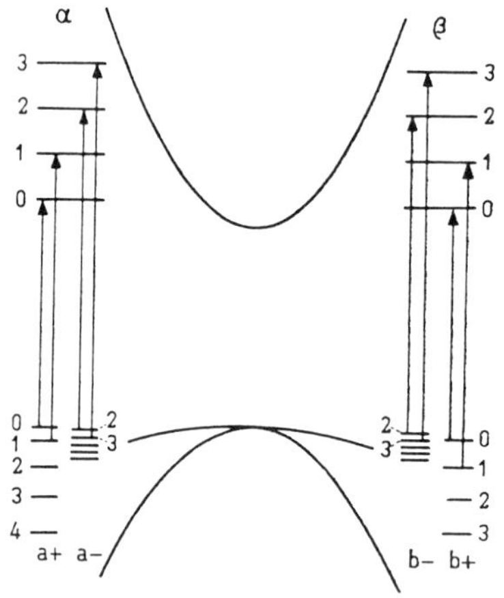

Fig. 10. Energy level scheme for the Landau levels in the valence and conduction bands in GaAs, showing some of the allowed optical transitions (after Vrehen 1968).

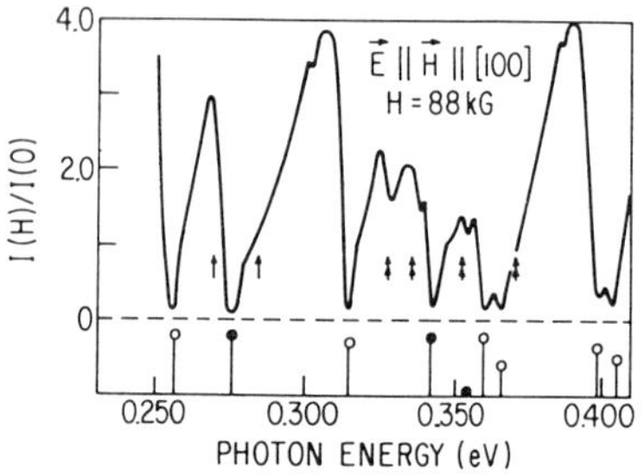

Fig. 11. Interband magnetoabsorption in InSb (after Pidgeon and Brown 1966).

level structure as described above is used, but it is necessary to include non-parabolic effects. Pidgeon and Brown solved eq. (3.1) with i and j going over all eight bands of fig. 4a. The inverted band structure of HgTe, shown in fig. 4b, was investigated by Pidgeon (1969). There are also interesting effects in alloys of CdTe and HgTe, for which the energy gap goes through zero (Weiler et al. 1977).

In the case of III–V compounds, which lack inversion symmetry, the assumption that $\boldsymbol{p}_{ij}=0$ in eq. (3.25) is no longer valid, and there are linear terms in the effective-mass Hamiltonian corresponding to eq. (2.38). Effects of this linear term have been seen in magnetoabsorption studies in InSb (Pidgeon and Groves 1969). When inversion asymmetry is combined with non-parabolicity a spin-dependent k^3 term occurs in the conduction band, which gives rise to an inversion asymmetry splitting there also. This effect accounts for the beats observed in the Shubnikov–de-Haas effect in HgSe (Whitsett 1965, Roth et al. 1967, Roth 1968) and in GaSb (Seiler and Becker 1967). The beats are in fact evident in the HgSe results presented in fig. 8.

3.6. *Quantum states of band electrons in an electric field*

We now consider the problem of a band electron in a uniform electric field described by the potential energy $e\boldsymbol{E}\cdot\boldsymbol{r}$. We shall first look at the electron states near a band edge. Considering the case of a simple non-degenerate band, and neglecting spin effects, the effective mass Schrödinger equation obeyed by the envelope function $f(\boldsymbol{r})$ is

$$(p^2/2m^* + eEx - \mathscr{E})f(\boldsymbol{r}) = 0. \tag{3.49}$$

The solution of this equation for energy $\mathscr{E}=\mathscr{E}_x+(\hbar^2/2m^*)(k_y^2+k_z^2)$ is given by (Landau and Lifshitz 1959):

$$f_{\mathscr{E},k}(\boldsymbol{r}) = \eta\frac{1}{\sqrt{L_xL_y}}\exp[\mathrm{i}(k_yy+k_zz)]\,\mathrm{Ai}\left(\frac{2m^*eE}{h^2}(x-\mathscr{E}_x/eE)\right). \tag{3.50}$$

Here Ai is the Airy function (Abromowitz and Stegun 1965):

$$\mathrm{Ai}(x) = \frac{1}{\pi}\int_0^\infty \cos(sx+s^3/3)\,\mathrm{d}s. \tag{3.51}$$

This function decreases exponentially for large positive x and oscillates with increasing phase for large negative x. There is also a solution Bi of eq. (3.49) – not an eigenfunction – that increases exponentially for $x \gg 0$. The eigenvalues $\mathscr{E}_x$ here are continuous. If ρ is the density of states and if the normalization factor η is chosen as

$$\eta = (2m^*eE/\hbar^2)^{1/3}(\pi eE\rho)^{-1/2} \tag{3.52}$$

then the normalization is given by (Reine et al. 1967):

$$\int f^*_{\mathscr{E},k}(\boldsymbol{r})f_{\mathscr{E},k}(\boldsymbol{r})\,\mathrm{d}\boldsymbol{r} = \delta(\mathscr{E}_x-\mathscr{E}'_x)/\rho. \tag{3.53}$$

It is not at first evident how to determine the density of states ρ. However, suppose that the electric field is along a (cubic) crystalline axis with spacing b. Then as suggested in fig. 12, the energy bands are tilted by the electric field, and the N_x states in the x-direction of the Brillouin zone are run through in a distance $N_x b$, i.e., the length of the crystal. This corresponds to an energy change $eEN_x b$, and so we can take $\rho = (eEb)^{-1}$.

While this approach is useful, we see from fig. 12 that the wave function should actually run through the entire band. This leads to quantization of the energy levels, and this must be dealt with by using states of the entire band. We therefore consider the problem using the Bloch representation of eq. (3.2) (Callaway 1974). The eigenvalue problem in this representation is obtained, using the matrix element of x, eq. (2.8), as

$$\mathscr{E}\psi_{n\boldsymbol{k}} = \mathscr{E}_n(\boldsymbol{k})\psi_{n\boldsymbol{k}} + \mathrm{i}eE\frac{\partial}{\partial k_x}\psi_{n\boldsymbol{k}} + \sum_{n'} eEX_{nn'}\psi_{n'\boldsymbol{k}}. \tag{3.54}$$

Let us neglect the interband matrix element $X_{nn'}$ for $n \neq n'$, and let $\bar{\mathscr{E}}_{n\boldsymbol{k}} = \mathscr{E}_{n\boldsymbol{k}} + eEX_{nn'}$. We then have

$$(\mathscr{E} - \bar{\mathscr{E}}_{n\boldsymbol{k}})\psi_{n\boldsymbol{k}} = \mathrm{i}eE\partial\psi_{n\boldsymbol{k}}/\partial k_x. \tag{3.55}$$

The solution of this equation is

$$\psi_{n\boldsymbol{k}} = \frac{1}{\sqrt{k_0}}\delta_{k_y k_{y0}}\delta_{k_z k_{z0}} \exp\left(-\frac{\mathrm{i}}{eE}\int_0^{k_x} (\mathscr{E} - \bar{\mathscr{E}}_{n\boldsymbol{k}'})\,\mathrm{d}k'_x \right). \tag{3.56}$$

We are again assuming that E is in a lattice direction, and we let $k_0 = 2\pi/b$ be the width of the Brillouin zone in that direction. We have normalized the wave function over the zone. Since the Bloch states are defined only over one zone, $\psi_{n\boldsymbol{k}}$ must be periodic in $\boldsymbol{k}$, which gives a condition on the energy,

$$\mathscr{E}_\nu = eEb\nu + \int_0^{k_0} \frac{\bar{\mathscr{E}}_{n\boldsymbol{k}}}{k_0}\,\mathrm{d}k_x, \tag{3.57}$$

where ν is an integer. Thus the band breaks up into a series of equally spaced energy levels called a Stark ladder or Wannier levels (Wannier 1962, Callaway 1974). The spacing eEb of the Wannier levels corresponds to the oscillation frequency $eBb/\hbar$ of

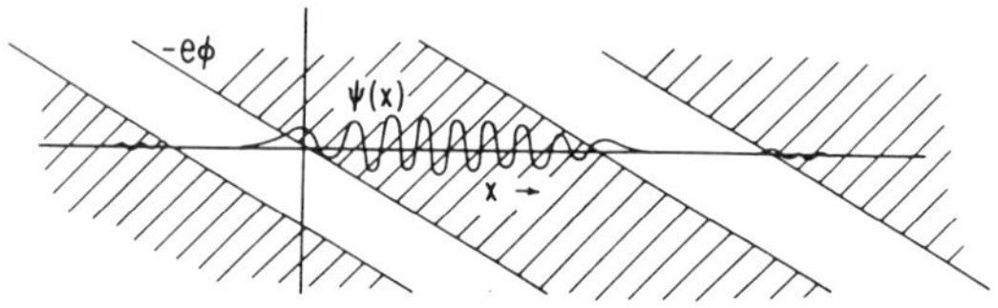

Fig. 12. Wave function for band electrons in an electric field.

the classical motion described in § 2.5. There has been some controversy over the existence of the Wannier levels (Zak 1968), because the derivation depends on ignoring the interband matrix element $X_{nn'}$. Certainly the existence of tunneling gives them a finite lifetime. However, Nenciu (1991) has shown that the levels exist as well-defined resonances. The Wannier levels have been observed by Koss and Lambert (1972) as steps in the electroabsorption in GaAs, and more recently by Mironov et al. (1985) as oscillations in the electro-optic effect in semi-insulating GaAs(Cr). The latter results show clearly the dependence of the level spacing on electric field, as can be seen from fig. 13.

The presence of the interband term gives rise to tunneling between bands, as indicated schematically in fig. 12. Let us calculate the rate of tunneling per unit volume between the Wannier levels ν and ν' in the bands n and n'. Using the Fermi golden rule, we have

$$W_{nn'} = \frac{2}{V}\frac{2\pi}{\hbar} \sum_{\nu\nu'\boldsymbol{k}_\perp} |\boldsymbol{M}_{n\nu n'\nu'}(\boldsymbol{k}_\perp)|^2 \, \delta(\mathscr{E}_{n\nu}(\boldsymbol{k}_\perp) - \mathscr{E}_{n'\nu'}(\boldsymbol{k}_\perp)). \tag{3.58}$$

The matrix element is

$$\begin{aligned} \boldsymbol{M}_{nn'}(\boldsymbol{k}_\perp) &= eE \int \psi_{n\nu}(\boldsymbol{k})^* X_{nn'}(\boldsymbol{k}) \psi_{n'\nu'}(\boldsymbol{k}) \, \mathrm{d}k_x \\ &= \frac{eE}{k_0} \int X_{nn'}(\boldsymbol{k}) \exp\left(\frac{\mathrm{i}}{eE} \int_0^{k_x} \bar{\mathscr{E}}_{nn'}(k'_x, \boldsymbol{k}_\perp) \, \mathrm{d}k'_x \right) \mathrm{d}k_x. \end{aligned} \tag{3.59}$$

Here $\bar{\mathscr{E}}_{nn'} = \bar{\mathscr{E}}_n - \bar{\mathscr{E}}_{n'}$, and the energy-conserving δ-function has eliminated the dependence on ν and ν'.

An approximate treatment by Kane (1959) replaces the summation over Wannier

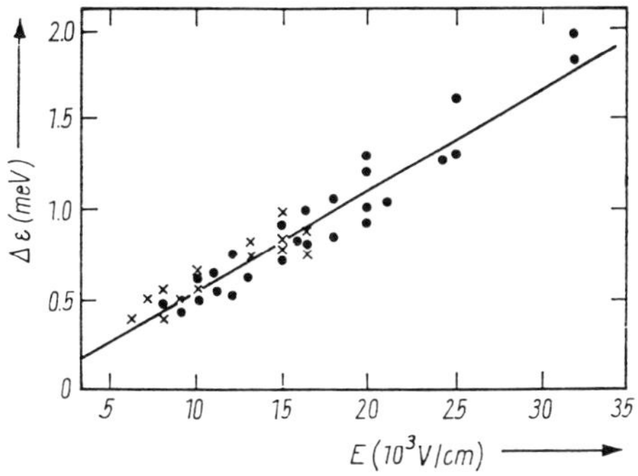

Fig. 13. Energy spacing of oscillations versus electric field for electro-optic effect in semi-insulating GaAs(Cr) (after Mironov et al. 1985).

levels by an integration, which is in fact over the density of states ρ of eq. (3.53),

$$W_{nn'} = \frac{k_0^2}{\pi\hbar eE} \int \frac{\mathrm{d}\boldsymbol{k}_\perp}{(2\pi)^2} |\boldsymbol{M}_{nn'}(\boldsymbol{k}_\perp)|^2. \tag{3.60}$$

Kane evaluates the tunneling rate for the two-band model of eq. (2.23). The result is

$$W = \frac{(eE)^2 m_r^{1/2}}{18\pi\hbar^2 \mathscr{E}_g^{1/2}} \exp\left(\frac{-\pi m_r^{1/2} \mathscr{E}_g^{3/2}}{2\hbar eE}\right). \tag{3.61}$$

Here $m_r = m_0^*/2$ is the reduced mass. Note that the tunneling rate goes to zero for zero field, but in a non-analytic way, so that this result is not obtainable from a simple perturbation treatment. A more precise theory, due to Argyres (1962), also includes oscillatory terms due to the Wannier quantization.

Tunneling transitions in the complex valence bands of semiconductors have been studied by Aleshkin and Romanov (1986) and Gorbovitskii (1988). In addition to Zener tunneling, it is possible to observe photon-assisted tunneling in the optical absorption in an electric field. This is the Franz–Keldysh effect (Franz 1958, Keldysh 1958, Tharmalingham 1963). In this case the matrix element is the optical dipole moment. A more extensive treatment of tunneling is given by Price in chapter 12.

3.7. Quantum states of band electrons in combined electric and magnetic fields

If we consider magnetic states near an energy band edge, we can see from eq. (3.39) that the motion along the magnetic field $\boldsymbol{B}$, assumed to be in the z-direction, is independent of the transverse motion. Thus a potential depending only on z modifies only the longitudinal motion. The most interesting case here is that of a quantum well or inversion layer, in which the electrons are confined, so that the levels are quantized in the z-direction. The above theory (as well as the theory including transverse electric fields considered below) holds for the motions in the x- and y-directions, but the levels are now completely quantized, and this leads to many interesting effects such as the quantum Hall effect, which are considered in other chapters.

Here we consider a uniform electric field, for which the effective-mass equation is, assuming simple bands and neglecting spin,

$$[(\boldsymbol{p} + e\boldsymbol{A}/c)^2/2m^* + e\boldsymbol{E}\cdot\boldsymbol{r}]f(\boldsymbol{r}) = \mathscr{E}f(\boldsymbol{r}). \tag{3.62}$$

As mentioned, the effect of the z-component of the electric field will be independent of B, and the motion in the z-direction is described by the Airy function. For the case of crossed fields, if we assume $\boldsymbol{E} \parallel x$ and use the asymmetric gauge $\boldsymbol{A} = (0, Bx, 0)$ as before, the effective-mass equation can be written as

$$\left\{\frac{1}{2m^*}\left[p_x^2 + \left(p_y + \frac{m^*Ec}{B} + \frac{eBx}{c}\right)^2 + p_z^2\right] - \frac{Ec}{B}\left(p_y + \frac{m^*Ec}{2B}\right)\right\} f(\boldsymbol{r}) = \mathscr{E}f(\boldsymbol{r}). \tag{3.63}$$

The wave functions are given by eq. (3.35) as in the case of the magnetic field only, except that now the origin is shifted so that

$$X_{\boldsymbol{k}} = -l^2(k_y + m^* Ec/\hbar B).$$

The energy has additional terms in it:

$$\mathscr{E}_{n\boldsymbol{k}} = (n + 1/2)\hbar\omega_c + \hbar^2 k_z^2/2m^* + eEX_{\boldsymbol{k}} + m(Ec/b)^2/2. \tag{3.64}$$

The third term here is the potential energy of the orbit center, and the last term is the kinetic energy due to a uniform velocity in the y-direction.

In the case of non-parabolic bands we have seen in § 2.6 that the classical motion can change from periodic-orbit to open-orbit behavior for large transverse electric fields. The effect can be studied quantum mechanically by using the two-band model of eq. (3.42), modified to include the electric field (Zawadzki and Lax 1966, Praddaude 1965, Zawadzki 1969). If we neglect spin and some small terms we have

$$(\mathscr{E} - eEx)\left(1 + \frac{\mathscr{E} - eEx}{\mathscr{E}_g}\right) f(\boldsymbol{r}) = \frac{1}{2m^*}\left[p_x^2 + \left(p_y + \frac{eBx}{c}\right)^2 + p_z^2\right] f(\boldsymbol{r}). \tag{3.65}$$

In this equation the effective harmonic oscillator potential is given by

$$V_{\text{ho}} = [e^2 B^2/(2m^* c^2) - E^2 e^2/\mathscr{E}_g](x - X_0)^2 + \text{const.}, \tag{3.66}$$

where X_0 is a constant. We see that the solutions will be magnetic-field-like for $E < B\sqrt{\mathscr{E}_g/2m^* c^2}$, but for larger values of E the levels will no longer be quantized.

A shift in the magnetoabsorption lines in crossed fields was predicted by Aronov (1963) and observed by Vrehen (1965) in germanium. The changeover from magneto-absorption to photon-assisted tunneling was observed by Reine et al. (1967) and calculated by Weiler et al. (1967) and Aronov and Picus (1967).

4. Classical transport in semiconductors: introduction and elementary model

If a weak electric field is imposed upon a semiconductor, with or without a magnetic field, a current flow is induced as a response (Ziman 1960, Beer 1963, Butcher 1986). The microscopic relationship between current density and field is

$$j_\alpha = \sum_\beta \sigma_{\alpha\beta}(\boldsymbol{B}) E_\beta, \tag{4.1}$$

where $\boldsymbol{\sigma}$ is the conductivity tensor, which depends upon the magnetic field. The conductivity components obey the Onsager relation $\sigma_{\alpha\beta}(\boldsymbol{B}) = \sigma_{\beta\alpha}(-\boldsymbol{B})$ (Callen 1948). Experimentally it is usually true that the current direction is determined instead of the field, for example, by having a long thin sample with leads at the ends. Therefore it is usually the resistivity tensor $\boldsymbol{\rho} = \boldsymbol{\sigma}^{-1}$ that is of interest:

$$E_\alpha = \sum_\beta \rho_{\alpha\beta} j_\beta. \tag{4.2}$$

(An exception is the Corbino disk geometry (see Seeger (1982) or Beer (1963)), in which the conductivity components are measured directly.) Suppose a magnetic field is imposed in the z-direction and a current flows in the x-direction. Then a Hall electric field appears in the y-direction. For isotropic systems the Hall constant is defined by $R = E_y/j_x B_z = \rho_{yx}/B_z$. For such systems E_y reverses on reversing the magnetic field, but when anisotropy is present there can be a component to ρ_{xy} even in B, and the Hall constant should be defined as (Beer 1963)

$$R = (\rho_{yx} - \rho_{xy})/2B_z. \tag{4.3}$$

The transverse magnetoresistance is given by $\Delta\rho_\perp/\rho_0 = (\rho_{xx} - \rho_0)/\rho_0$. If the current is in the z-direction the longitudinal magnetoresistance is $\Delta\rho_\parallel/\rho_0 = (\rho_{zz} - \rho_0)/\rho_0$. The even component of ρ_{xy} contributes to what is called the planar Hall effect (see Seeger 1982), which is actually an anisotropic magnetoresistance.

If the temperature gradients and heat currents are included there are additional transport coefficients. These will be discussed in § 8.

4.1. *Particle kinetic theory for simple bands*

The elementary theory of transport under electric and magnetic fields is based on the equation of motion for an electron in a simple band with effective mass m^* and charge q:

$$\mathrm{d}\boldsymbol{p}/\mathrm{d}t = qE + q\boldsymbol{v} \times \boldsymbol{B}/c - \boldsymbol{p}/\tau. \tag{4.4}$$

Here the momentum $\boldsymbol{p} = \hbar\boldsymbol{k}$ is related to the velocity by $\boldsymbol{p} = m^*\boldsymbol{v}$. In this simple particle kinetic model (Smith 1982), which is essentially the Drude model (Ashcroft and Mermin 1976), we include a phenomenological damping term involving the relaxation time τ. If we allow for several types of carriers, the particle current is given by

$$\boldsymbol{j} = \sum_i q_i n_i \boldsymbol{v}_i. \tag{4.5}$$

In the more exact Boltzmann equation treatment of § 5, we shall find that for simple bands a relaxation time approximation is often valid, provided that τ is allowed to depend on energy. A suitable average of the current over energy for a distribution of carriers is then necessary. It is possible to include an energy-dependent relaxation time as well as a tensor effective mass, and even an energy-dependent effective mass within the particle kinetic model. However, we shall postpone such complications until we consider the Boltzmann equation, while being mindful of the need for averages.

Considering first the case of steady electric and magnetic fields, we write the kinetic equation as

$$\boldsymbol{v} = (q\tau/m^*)\boldsymbol{E} + \omega_c\tau\boldsymbol{v} \times \hat{\boldsymbol{B}}, \tag{4.6}$$

where $\omega_c = qB/m^*c$ is the cyclotron frequency and $\hat{\boldsymbol{B}}$ is a unit vector in the direction of the magnetic field $\boldsymbol{B}$. By dotting and crossing $\hat{\boldsymbol{B}}$ into eq. (4.6) we can solve for $\boldsymbol{v}$:

$$\boldsymbol{v} = \frac{q\tau/m^*}{1 + (\omega_c\tau)^2}[\boldsymbol{E} - \omega_c\tau\hat{\boldsymbol{B}} \times \boldsymbol{E} + (\omega_c\tau)^2\hat{\boldsymbol{B}}\hat{\boldsymbol{B}}\cdot\boldsymbol{E}]. \tag{4.7}$$

Assuming a single group of carriers and a constant value of τ the current is $\boldsymbol{j} = nq\boldsymbol{v}$, and with eq. (4.7) we can define a conductivity tensor. If we assume that the magnetic field is in the z-direction, the conductivity tensor has the simple form,

$$\boldsymbol{\sigma} = \begin{pmatrix} \sigma_{xx} & \sigma_{xy} & 0 \\ -\sigma_{xy} & \sigma_{xx} & 0 \\ 0 & 0 & \sigma_{zz} \end{pmatrix}, \tag{4.8}$$

where

$$\sigma_{xx} = \frac{\sigma_0}{1+(\omega_c\tau)^2}, \quad \sigma_{xy} = \frac{\sigma_0 \omega_c \tau}{1+(\omega_c\tau)^2}, \quad \sigma_{zz} = \sigma_0, \tag{4.9}$$

and where $\sigma_0 = nq^2\tau/m^*$ is the classical result for the conductivity in the absence of a magnetic field. The conductivity transverse to the magnetic field depends on the field, but the longitudinal conductivity is unaffected by the field.

As mentioned earlier the important experimental quantity is the resistivity tensor, which we can obtain by taking the inverse of σ. However, it is easier to solve eq. (4.6) for $\boldsymbol{E}$

$$\boldsymbol{E} = \rho_0 \boldsymbol{j} + \frac{1}{nqc}\boldsymbol{B} \times \boldsymbol{j}, \tag{4.10}$$

where $\rho_0 = 1/\sigma_0 = m^*/nq^2\tau$ is the ordinary resistivity. Defining a resistivity tensor by analogy with σ, again assuming B to be in the z-direction, we have

$$\boldsymbol{\rho} = \begin{pmatrix} \rho_{xx} & \rho_{xy} & 0 \\ -\rho_{xy} & \rho_{xx} & 0 \\ 0 & 0 & \rho_{zz} \end{pmatrix}, \tag{4.11}$$

whcrc

$$\rho_{xx} = \rho_{yy} = \rho_{zz} = \rho_0; \quad \rho_{yx} = RB = -\rho_{xy}, \tag{4.12}$$

and where $R = 1/nqc$ is the classical result for the Hall constant, whose sign has the well known dependence on the sign of the carrier.

For a single group of carriers and constant τ the transverse and longitudinal magnetoresistance both vanish. The model is too simple! It is possible for an isotropic system to have a finite transverse magnetoresistance if τ depends on the energy or if there are more than one group of carriers. The different carriers conduct in parallel, so their contribution to the conductivity tensor must be added before taking the inverse. For simple bands the longitudinal magnetoresistance vanishes. For carriers in bands with more complex and anisotropic band structures, the classical theory gives a finite longitudinal magnetoresistance. Also, in the case of magnetic fields sufficiently large for quantization of the magnetic levels to be important, there is a finite longitudinal magnetoresistance for simple bands. We shall consider complex bands in the following sections using the Boltzmann equation, but first we apply the elementary model to the case of time-dependent electric fields.

4.2. Finite-frequency effects

We shall assume that the applied electric field and the current that flows in response to it have a sinusoidal time dependence, so they are given in the complex notation by $\boldsymbol{E}e^{i\omega t}$ and $\boldsymbol{j}e^{i\omega t}$, where $\omega > 0$. The physical quantities are the real parts of $\boldsymbol{E}$ and $\boldsymbol{j}$, and the average power absorbed per unit volume is $P = \frac{1}{2}\,\mathrm{Re}(\boldsymbol{E}\cdot\boldsymbol{j}^*)$, where $\boldsymbol{j}^*$ is the complex conjugate of $\boldsymbol{j}$. To include optical effects we write the wave equation as

$$\nabla^2 \boldsymbol{E} - \nabla\nabla\cdot\boldsymbol{E} = (4\pi i\omega/c^2)\boldsymbol{j} - (\kappa_0\omega^2/c^2)\boldsymbol{E}, \tag{4.13}$$

where κ_0 is the background dielectric constant. We shall assume that the electric field varies slowly enough in space that we can ignore the spatial dependence in calculating the electric properties, so we can simply use the conductivity tensor of eq. (4.1) as a constitutive relation in eq. (4.13). In the equation of motion, eq. (4.4), we must now include the explicit time dependence, but this is done simply by replacing $1/\tau$ by $1/\tau + i\omega$. The conductivity tensor is given by eq. (4.8) with $\tau \to \tau/(1 + i\omega\tau)$, and $\sigma_0 \to \sigma_0/(1 + i\omega\tau)$.

$$\sigma_{xx} = \frac{\sigma_0(1 + i\omega\tau)}{(1 + i\omega\tau)^2 + (\omega_c\tau)^2}, \quad \sigma_{xy} = \frac{\sigma_0\omega_c\tau}{(1 + i\omega\tau)^2 + (\omega_c\tau)^2}, \quad \sigma_{zz} = \sigma_0/(1 + i\omega\tau). \tag{4.14}$$

If there is no magnetic field, the conductivity is a scalar, $\sigma(\omega) = \sigma_0/(1 + i\omega\tau)$, and to describe the propagation of an electromagnetic wave, eq. (4.13), we replace $\boldsymbol{j}$ by $\sigma(\omega)\boldsymbol{E}$ on the right, which is equivalent to writing

$$\nabla^2 \boldsymbol{E} - \nabla\nabla\cdot\boldsymbol{E} = -\kappa(\omega^2/c^2)\boldsymbol{E}; \quad \kappa = \kappa_0 - 4\pi i\sigma(\omega)/\omega. \tag{4.15}$$

Here κ is a complex dielectric constant:

$$\kappa = \kappa' - i\kappa''; \quad \kappa' = \kappa_0'\left(1 - \frac{(\omega_p\tau)^2}{1 + (\omega\tau)^2}\right), \quad \kappa'' = \kappa_0'' + \kappa_0'\frac{\omega_p^2\tau/\omega}{1 + (\omega\tau)^2}, \tag{4.16}$$

where we have introduced the plasma frequency $\omega_p = 4\pi nq^2/m^*\kappa_0'$. For a wave travelling in the z-direction, the propagation can be described in terms of a complex index of refraction

$$\boldsymbol{E} = \boldsymbol{E}_0 e^{-i(n' - in'')\omega z/c} = \boldsymbol{E}_0 e^{-i(\beta - i\alpha/2)z}, \quad n'^2 - n''^2 = \kappa', \quad 2n'n'' = \kappa''. \tag{4.17}$$

Without a detailed solution we can see that when κ' is negative the attenuation part n'' of the refractive index is dominant, and the wave does not propagate, but is totally reflected. For large $\omega\tau$ the changeover from total reflection to propagation occurs at the plasma frequency ω_p.

Consider now the effect of including the magnetic field. It is often convenient to study the case of circularly polarized electric fields, transverse to $\boldsymbol{B}$, for which

$$\boldsymbol{E}(t) = E_0(\hat{\boldsymbol{i}}\cos\omega t \pm \hat{\boldsymbol{j}}\sin\omega t) = \mathrm{Re}(\boldsymbol{E}_\pm e^{i\omega t}); \quad \boldsymbol{E}_\pm = E_0(\hat{\boldsymbol{i}} \mp i\hat{\boldsymbol{j}}). \tag{4.18}$$

From the conductivity tensor, eq. (4.14), we find that the transverse current can be written as

$$\boldsymbol{j}_\pm = \sigma_\pm \boldsymbol{E}_\pm, \quad \sigma_\pm = \sigma_{xx} \mp \mathrm{i}\sigma_{xy} = \frac{\sigma_0}{1 + \mathrm{i}(\omega \pm \omega_\mathrm{c})\tau}. \tag{4.19}$$

To study cyclotron resonance (Lax and Mavroides 1960), we find the power absorbed per unit volume, $P = \frac{1}{2}\mathrm{Re}\, \boldsymbol{E}^* \cdot \boldsymbol{j} = \mathrm{Re}\, E_0^2 \sigma_\pm$, so that

$$P_\pm = \frac{\sigma_0 E_0^2}{1 + (\omega \pm \omega_\mathrm{c})^2 \tau^2}. \tag{4.20}$$

This has a maximum at $\mp$ the cyclotron frequency. For the positively rotating component the maximum is at $-\omega_\mathrm{c}$, which occurs only for negatively charged carriers. Similarly the negatively rotating component has a resonance only for positively charged carriers. Therefore, the use of circularly polarized fields makes it possible to determine the sign of the carrier. If the electric field is linearly polarized, e.g., in the x-direction, then the absorption is given by the sum of the two directions of polarization

$$P = (P_+ + P_-)/2 = \frac{1}{2} \frac{\sigma_0 E_0^2 [1 + (\omega^2 + \omega_\mathrm{c}^2)\tau^2]}{[1 + (\omega_\mathrm{c}^2 - \omega^2)\tau^2]^2 + 4\omega^2\tau^2}. \tag{4.21}$$

If $\omega_\mathrm{c}\tau \gg 1$, this expression has a maximum at $\omega \cong |\omega_\mathrm{c}|$, independent of the sign of the carriers. However, as $\omega_\mathrm{c}\tau$ decreases the resonance no longer occurs at the cyclotron frequency, and if $\omega_\mathrm{c}\tau \leqslant 1/\sqrt{3}$ there is no maximum at all.

4.3. Magnetic effects at high frequency

Consider now the effect of a magnetic field on a circularly polarized electromagnetic wave propagating in the z-direction, parallel to the magnetic field. In the wave equation the effective dielectric constant is

$$\kappa_\pm = \kappa_0 - 4\pi \mathrm{i} \sigma_\pm(\omega)/\omega, \tag{4.22}$$

with real and imaginary parts

$$\kappa'_\pm = \kappa'_0 \left(1 - \frac{\omega_\mathrm{p}^2 (\omega \pm \omega_\mathrm{c})\tau^2/\omega}{1 + (\omega \pm \omega_\mathrm{c})^2\tau^2}\right), \quad \kappa''_\pm = \kappa''_0 + \kappa'_0 \frac{\omega_\mathrm{p}^2 \tau/\omega}{1 + (\omega \pm \omega_\mathrm{c})^2 \tau^2}. \tag{4.23}$$

The magnetic field leads to a shift of the plasma edge, which is called the magnetoplasma effect (Palik and Wright 1967). For $\omega_\mathrm{p}\tau \gg 1$ and $\omega_\mathrm{p} \gg \omega_\mathrm{c}$, the shift is $\pm\frac{1}{2}\omega_\mathrm{c}$.

For propagation perpendicular to the magnetic field, the tensor conductivity must be used in eq. (4.13). There are two modes, one with $\boldsymbol{E}$ parallel to $\boldsymbol{B}$ which is unaffected by the magnetic field, and one with $\boldsymbol{E}$ perpendicular to $\boldsymbol{B}$, which for $\omega\tau \gg 1$ has an effective dielectric constant of

$$\kappa' = \kappa_0 \left(1 - \frac{\omega_\mathrm{p}^2(\omega_\mathrm{p}^2 - \omega^2)}{\omega^2(\omega_\mathrm{p}^2 + \omega_\mathrm{c}^2 - \omega^2)}\right). \tag{4.24}$$

This case has an interesting structure with two regions of propagation near the plasma frequency (Palik and Wright 1967).

Returning to the case of circular polarization, if we look at frequencies well above the plasma frequency, the refractive index is different for the two senses of circular polarization. If the attenuation is small ($\omega\tau \gg 1$), we have $n'_{\pm} \cong \sqrt{\kappa'_{\pm}}$, and the difference in propagation constant is, from eq. (4.17),

$$\beta_+ - \beta_- = (n_+ - n_-)\omega/c \cong (\kappa'_+ - \kappa'_-)\omega/2nc \cong \kappa'_0 \omega_p^2 \omega_c / n\omega^2 c. \tag{4.25}$$

This leads to the rotation of the plane of polarization of a plane wave, which is the Faraday rotation. To see this, a plane wave with electric vector at an angle θ with the x-axis is described by the superposition of two circularly polarized waves:

$$\boldsymbol{E} = \tfrac{1}{2}(\boldsymbol{E}_+ \mathrm{e}^{\mathrm{i}\theta} + \boldsymbol{E}_- \mathrm{e}^{-\mathrm{i}\theta}) = E_0(\hat{\boldsymbol{\imath}} \cos\theta + \hat{\boldsymbol{\jmath}} \sin\theta). \tag{4.26}$$

If the electric field is along x at $z = 0$, then at a distance L, $\boldsymbol{E}$ becomes

$$\begin{aligned} \boldsymbol{E} &= [\boldsymbol{E}_+(0)\exp(-\mathrm{i}\beta_+ L) + \boldsymbol{E}_-(0)\exp(-\mathrm{i}\beta_- L)]/2 \\ &= \exp(-\mathrm{i}\beta L)\{E_+(0)\exp(\mathrm{i}VL) \\ &\quad + E_-(0)\exp(\mathrm{i}VL)\}, \end{aligned} \tag{4.27}$$

where β is the average propagation constant, and V is the Verdet coefficient

$$V = -(\beta_+ - \beta_-)/2 = \kappa'_0 \omega_p^2 \omega_c / 2n\omega^2 c. \tag{4.28}$$

The polarization is therefore rotated by an angle $\theta = VL$, Faraday rotation in semiconductors is related to interband optical effects as well as free carrier effects (Piller 1972).

An interesting situation can arise at somewhat lower frequencies. In the absence of the magnetic field, if $\omega\tau \ll 1$, the waves will be rapidly attenuated, but if B is large enough so that $\omega_c\tau \gg 1$, it is possible that a circularly polarized wave can propagate with little attenuation. Such a wave is called a helicon wave (Bowers et al. 1961). In the limit described above,

$$\kappa'_{\pm} = \mp \kappa'_0 \omega_p^2/\omega\omega_c, \qquad \kappa''_{\pm} = \kappa'_0 \omega_p^2/\omega\omega_c^2 \tau. \tag{4.29}$$

For the polarization for which κ' is negative the wave will be rapidly attenuated, but for the other sense of polarization κ' is positive and the attenuation will be small. The propagation constant is $\beta = (n/c)\omega_p\sqrt{\omega/\omega_c}$. The wavelength of this unattenuated wave will be much shorter than the free-space wavelength, i.e., $\lambda/\lambda_0 \cong \sqrt{\omega\omega_c}/\omega_p$. Thus it is possible to have sample dimensions comparable to a wavelength even at relatively low frequencies and to observe cavity resonances as the magnetic field is varied.

5. The Boltzmann equation

5.1. Derivation of the Boltzmann equation

We assume a group of carriers described by the energy $\mathscr{E}(\boldsymbol{k})$. In the unperturbed case the carriers obey the Fermi distribution function

$$f_0(\mathscr{E}) = 1/\{\exp[(\mathscr{E} - \mathscr{E}_{\mathrm{F}})/kT] + 1\}, \tag{5.1}$$

where the Fermi level is related to the carrier density by

$$n = \int f_0(\mathscr{E})\,\mathrm{d}\boldsymbol{k}/4\pi^3. \tag{5.2}$$

For the case of classical statistics, the Fermi level is well below the band edge, and the distribution function is Maxwellian,

$$f_0(\mathscr{E}) \cong \exp[(\mathscr{E}_{\mathrm{F}} - \mathscr{E})/kT]. \tag{5.3}$$

This is valid for semiconductors when the carrier concentration is relatively low. For high doping the electron gas becomes degenerate and the Fermi distribution must be used. The distribution functions apply for holes as well as electrons with the hole energy taken as positive (reversed from the energy band diagram). We shall assume that all the energies are measured from the relevant band edge.

In the presence of external fields and temperature gradients, we expect the distribution function for the carriers to change to $f(\boldsymbol{r}, \boldsymbol{k})$, which depends on position as well as momentum. The density of carriers $n(\boldsymbol{r})$ is given by eq. (5.2) with f_0 replaced by f. To obtain the Boltzmann equation we examine how f changes in time under the influence of external fields, concentration and temperature gradients, and collisions of the carriers with impurities and phonons:

$$\frac{\mathrm{d}f}{\mathrm{d}t} = \left.\frac{\partial f}{\partial t}\right|_{\text{forces}} + \left.\frac{\partial f}{\partial t}\right|_{\text{diff}} + \left.\frac{\partial f}{\partial t}\right|_{\text{coll}}. \tag{5.4}$$

The change in f due to forces comes from the equation of motion, eq. (2.9),

$$\hbar\frac{\mathrm{d}\boldsymbol{k}}{\mathrm{d}t} = q\left(\boldsymbol{E} + \frac{\boldsymbol{v}(\boldsymbol{k}) \times \boldsymbol{B}}{c}\right), \tag{5.5}$$

where $\boldsymbol{v} = (1/\hbar)\nabla_{\boldsymbol{k}}\mathscr{E}$, and $q = \pm e$. This produces a drift of carriers in momentum space. Consider a point $\boldsymbol{k}$ in momentum space, and let the distribution function at time t be $f(\boldsymbol{r}, \boldsymbol{k})$. At a slightly later time $t + \Delta t$ the carriers at $\boldsymbol{k}$ will be those that were at $\boldsymbol{k} - \Delta t\,\mathrm{d}\boldsymbol{k}/\mathrm{d}t$ at time t, so the change in f is $f(\boldsymbol{r}, \boldsymbol{k} - \Delta t\,\mathrm{d}\boldsymbol{k}/\mathrm{d}t) - f(\boldsymbol{r}, \boldsymbol{k})$. Expanding f to first order in Δt and letting $\Delta t \to 0$, this gives

$$\left.\frac{\partial f}{\partial t}\right|_{\text{forces}} = -\frac{\mathrm{d}\boldsymbol{k}}{\mathrm{d}t}\cdot\nabla_{\boldsymbol{k}}f = -q\left(\boldsymbol{E} + \frac{\boldsymbol{v} \times \boldsymbol{B}}{c}\right)\cdot\frac{1}{\hbar}\nabla_{\boldsymbol{k}}f. \tag{5.6}$$

Next, if there is a spatial variation in the distribution function due to concentration and temperature gradients, this results in a diffusion of carriers in coordinate space. By an argument similar to that used above, the spatial motion $\mathrm{d}\boldsymbol{r}/\mathrm{d}t = \boldsymbol{v}$ of the carriers leads to a change in f:

$$\left.\frac{\partial f}{\partial t}\right|_{\text{diff}} = -\boldsymbol{v}\cdot\nabla_{\boldsymbol{r}} f. \tag{5.7}$$

Assuming steady external fields, the distribution function reaches a steady state condition such that the net change in f due to the above effects as well as collisions vanishes, i.e., $\mathrm{d}f/\mathrm{d}t = 0$. This gives

$$q\left(\boldsymbol{E} + \frac{\boldsymbol{v}\times\boldsymbol{B}}{c}\right)\cdot\frac{1}{\hbar}\nabla_{\boldsymbol{k}} f + \boldsymbol{v}\cdot\nabla_{\boldsymbol{r}} f = \left.\frac{\partial f}{\partial t}\right|_{\text{coll}}. \tag{5.8}$$

We can easily add a sinusoidal time dependence for the electric field as in the last section.

Let us look now at the collision term. It is assumed that there is a probability per unit time $W(\boldsymbol{k}\to\boldsymbol{k}')$ of collisions that take particles from $\boldsymbol{k}$ to $\boldsymbol{k}'$, and a probability per unit time $W(\boldsymbol{k}'\to\boldsymbol{k})$ that take particles from $\boldsymbol{k}'$ to $\boldsymbol{k}$. Taking account of the exclusion principle the net change in f is

$$\left.\frac{\partial f}{\partial t}\right|_{\text{coll}} = -\int \{W(\boldsymbol{k}\to\boldsymbol{k}')f(\boldsymbol{k})[1-f(\boldsymbol{k}')] - W(\boldsymbol{k}'\to\boldsymbol{k})f(\boldsymbol{k}')[1-f(\boldsymbol{k})]\}\,\mathrm{d}\boldsymbol{k}'. \tag{5.9}$$

We have not explicitly included scattering into other bands, but this can always be added when it is of importance, such as near degenerate band edges.

In the case of classical statistics the collision term is somewhat simpler

$$\left.\frac{\partial f}{\partial t}\right|_{\text{coll}} = -\int [W(\boldsymbol{k}\to\boldsymbol{k}')f(\boldsymbol{k}) - W(\boldsymbol{k}'\to\boldsymbol{k})f(\boldsymbol{k}')]\,\mathrm{d}\boldsymbol{k}'. \tag{5.10}$$

In table 2 are listed expressions for W for the principal scattering mechanisms that have been found to be important in semiconductors. These scattering mechanisms are described in § 9.

Equations (5.8) and (5.9) or (5.10) define an integrodifferential equation for the distribution function. In the absence of perturbing fields and temperature gradients the distribution function must be the Fermi function f_0. In this case the collision term must vanish, and the principle of detailed balance gives for all $\boldsymbol{k}$ and $\boldsymbol{k}'$ and all scattering mechanisms,

$$W(\boldsymbol{k}\to\boldsymbol{k}')f_0(\boldsymbol{k})[1-f_0(\boldsymbol{k}')] - W(\boldsymbol{k}'\to\boldsymbol{k})f_0(\boldsymbol{k}')[1-f_0(\boldsymbol{k})] = 0. \tag{5.11}$$

This holds if the phonons interacting with the electrons are in thermal equilibrium, as can be seen from the scattering mechanisms considered in § 9 and listed in table 2, for which $W(\boldsymbol{k}'\to\boldsymbol{k})/W(\boldsymbol{k}\to\boldsymbol{k}') = \exp[\beta(\mathscr{E}_{\boldsymbol{k}'} - \mathscr{E}_{\boldsymbol{k}})]$.

Once the Boltzmann equation has been solved, we must calculate the current for

Table 2
Scattering mechanisms. (For definitions, see § 9.)

Scattering mechanism	Potential $U(\boldsymbol{r})$	Intraband transition rate $W(\boldsymbol{k}\to\boldsymbol{k}')$
Ionized impurity	$-e^2 e^{-r/\lambda}/\kappa_0 r$	$\frac{N_i}{4\pi^2\hbar}\left(\frac{4\pi e^2/\kappa_0}{\lvert\boldsymbol{k}-\boldsymbol{k}'\rvert^2+1/\lambda^2}\right)^2 \times G(\boldsymbol{k},\boldsymbol{k}')\delta(\mathscr{E}_{\boldsymbol{k}'}-\mathscr{E}_{\boldsymbol{k}})$
Neutral impurity	Hydrogenic atom	$\frac{N_i\hbar^2 20a_0^*}{4\pi k m^{*2}}\delta(\mathscr{E}_{\boldsymbol{k}'}-\mathscr{E}_{\boldsymbol{k}})$
Acoustic deformation potential	$\sum_{\alpha\beta} D_{\alpha\beta}\frac{\partial u_\alpha}{\partial x_\beta}\sim E_1\nabla\cdot\boldsymbol{u}(\boldsymbol{r})$	$\frac{1}{4\pi^2\hbar}\sum_{\lambda,\mathrm{ac}}\frac{kT}{c_\lambda}\lvert\boldsymbol{D}(\boldsymbol{k},\boldsymbol{k}')\cdot\boldsymbol{e}_\lambda\rvert^2\delta(\mathscr{E}_{\boldsymbol{k}'}-\mathscr{E}_{\boldsymbol{k}})$ $\sim\frac{1}{4\pi^2\hbar}\frac{kTE_1^2}{c_\ell}\delta(\mathscr{E}_{\boldsymbol{k}'}-\mathscr{E}_{\boldsymbol{k}})$
Acoustic piezoelectric	$\nabla^2 U=-\frac{4\pi e}{\kappa_0}\sum_{\alpha\beta\gamma}h_{\alpha\beta\gamma}\frac{\partial^2 u_\gamma}{\partial x_\alpha\partial x_\beta}$	$G(\boldsymbol{k},\boldsymbol{k}')\frac{e^2kT}{\pi\hbar\kappa_0}\frac{P^2}{\lvert\boldsymbol{k}-\boldsymbol{k}'\rvert^2}\delta(\mathscr{E}_{\boldsymbol{k}'}-\mathscr{E}_{\boldsymbol{k}})$;
Optical deformation potential	$\boldsymbol{D}\cdot\boldsymbol{u}(\boldsymbol{r})$	$\frac{1}{4\pi^2\hbar}\sum_{\lambda,\mathrm{op}}\lvert\boldsymbol{D}(\boldsymbol{k},\boldsymbol{k}')\cdot\boldsymbol{e}_\lambda\rvert^2\frac{\hbar V}{2NM\omega_\lambda}$ $\times[N_\lambda(\mathscr{E}_{\boldsymbol{k}'}-\mathscr{E}_{\boldsymbol{k}}-\hbar\omega_\lambda)$ $+(N_\lambda+1)\delta(\mathscr{E}_{\boldsymbol{k}'}-\mathscr{E}_{\boldsymbol{k}}+\hbar\omega_\lambda)]$
Optical polar	$\nabla^2 U=-\frac{4\pi e^* e}{\Omega}\nabla\cdot\boldsymbol{u}(\boldsymbol{r})$	$G(\boldsymbol{k},\boldsymbol{k}')\frac{2e^*e}{\Omega M_r\omega_0}\frac{1}{\lvert\boldsymbol{k}-\boldsymbol{k}'\rvert^2}$ $\times[N_o\delta(\mathscr{E}_{\boldsymbol{k}'}-\mathscr{E}_{\boldsymbol{k}}-\hbar\omega_o)$ $+(N_o+1)\delta(\mathscr{E}_{\boldsymbol{k}'}-\mathscr{E}_{\boldsymbol{k}}+\hbar\omega_o)]$
Intervalley phonon	$\boldsymbol{D}\cdot\boldsymbol{u}(\boldsymbol{r})$	$W(1\boldsymbol{k}\to 2\boldsymbol{k}')=\frac{D_{12}^2}{8\pi^2\rho\omega_p}$ $\times[N_p\delta(\mathscr{E}_{1\boldsymbol{k}'}-\mathscr{E}_{2\boldsymbol{k}}-\hbar\omega_p)$ $+(N_p+1)\delta(\mathscr{E}_{2\boldsymbol{k}'}-\mathscr{E}_{1\boldsymbol{k}}+\hbar\omega_p)]$

the group of carriers

$$\boldsymbol{j}=\int f(\boldsymbol{r},\boldsymbol{k})q\boldsymbol{v}\,\mathrm{d}\boldsymbol{k}/4\pi^3 \tag{5.12}$$

and also the heat current. It is convenient to use the flux of energy relative to the Fermi energy, which is the chemical potential (Ziman 1960). We call this $\boldsymbol{w}^*$:

$$\boldsymbol{w}^*=\int f(\boldsymbol{r},\boldsymbol{k})(\mathscr{E}-\mathscr{E}_\mathrm{F})\boldsymbol{v}\,\mathrm{d}\boldsymbol{k}/4\pi^3. \tag{5.13}$$

5.2. Linearized Boltzmann equation

In this chapter we shall consider only the case of small electric fields and temperature gradients. Hot-electron effects will be taken up by Ferry in chapter 18. The magnetic field can be taken as arbitrary, but since we are not considering quantum effects, the results for high magnetic fields must be used with caution.

In the force term of eq. (5.8) we assume that the electric field $\boldsymbol{E}$ is small, and that deviations of f from f_0 are of first order in E and ∇T. Therefore, to first order, we replace f by f_0 in the coefficient of $\boldsymbol{E}$ and, writing

$$\hbar^{-1}\nabla_{\boldsymbol{k}} f_0 = \boldsymbol{v}\partial f_0/\partial\mathscr{E},$$

we have

$$\frac{\partial f_0}{\partial\mathscr{E}}\boldsymbol{v}\cdot q\boldsymbol{E} + \frac{q}{c}\boldsymbol{v}\times\boldsymbol{B}\cdot\frac{1}{\hbar}\nabla_{\boldsymbol{k}} f.$$

The term involving $\boldsymbol{B}$ is not approximated at this point. This term is actually of first order in $\boldsymbol{E}$ and ∇T. The contribution due to f_0 vanishes, because it is proportional to $\boldsymbol{v}\times\boldsymbol{B}\cdot\boldsymbol{v}$.

We can evaluate $\nabla_{\boldsymbol{r}} f$ by assuming a local equilibrium distribution function given by the Fermi function, with the Fermi level and temperature both dependent on position:

$$\nabla_{\boldsymbol{r}} f = \frac{\partial f_0}{\partial\mathscr{E}_{\mathrm{F}}}\nabla_{\boldsymbol{r}}\mathscr{E}_{\mathrm{F}} + \frac{\partial f_0}{\partial T}\nabla_{\boldsymbol{r}} T = \frac{\partial f_0}{\partial\mathscr{E}}\left(-\nabla_{\boldsymbol{r}}\mathscr{E}_{\mathrm{F}} - \frac{\mathscr{E}-\mathscr{E}_{\mathrm{F}}}{T}\nabla_{\boldsymbol{r}} T\right).$$

This gives

$$\frac{\partial f_0}{\partial\mathscr{E}}\boldsymbol{v}\cdot\left(q\boldsymbol{E}^* - \frac{\mathscr{E}-\mathscr{E}_{\mathrm{F}}}{T}\nabla_{\boldsymbol{r}} T\right) + \frac{q}{c}\boldsymbol{v}\times\boldsymbol{B}\cdot\frac{1}{\hbar}\nabla_{\boldsymbol{k}} f = \left.\frac{\partial f}{\partial t}\right|_{\mathrm{coll}}, \tag{5.14}$$

where

$$\boldsymbol{E}^* = \boldsymbol{E} - \nabla_{\boldsymbol{r}}\mathscr{E}_{\mathrm{F}}/q \tag{5.15}$$

is the negative gradient of the electrochemical potential. In addition to the applied electric field, the gradient of the Fermi level acts as an effective field. An immediate consequence of this is the close connection between the diffusion constant $D = -j/(qn^{-1}\,\mathrm{d}n/\mathrm{d}x)$ and the drift mobility $\mu = j/nqE$; clearly

$$D/\mu = (n/q)\,\mathrm{d}\mathscr{E}_{\mathrm{F}}/\mathrm{d}n. \tag{5.16}$$

The ratio is kT/q for classical statistics, and this is known as the Einstein relation.

We can simplify further by replacing the round brackets in eq. (5.14) by

$$q\boldsymbol{E}_1 = q\boldsymbol{E}^* - \frac{\mathscr{E}-\mathscr{E}_{\mathrm{F}}}{T}\nabla_{\boldsymbol{r}} T. \tag{5.17}$$

If there are no thermal gradients, $\boldsymbol{E}_1 = \boldsymbol{E}$, so we can treat that case first and then readily extend the result to include the thermal effects.

If we assume that

$$f = f_0 - \varphi(\boldsymbol{k})\frac{\partial f_0}{\partial \mathscr{E}} \tag{5.18}$$

and use the relationship

$$\frac{\partial f_0}{\partial \mathscr{E}} = -\frac{1}{kT} f_0(1 - f_0) \tag{5.19}$$

as well as the principle of detailed balance, eq. (5.11), then the collision term becomes

$$\left.\frac{\partial f}{\partial t}\right|_{\text{coll}} = -\int W(\boldsymbol{k} \to \boldsymbol{k}')\frac{f_0(\boldsymbol{k})[1 - f_0(\boldsymbol{k}')]}{kT}[\varphi(\boldsymbol{k}) - \varphi(\boldsymbol{k}')]\,\mathrm{d}\boldsymbol{k}'. \tag{5.20}$$

The magnetic term can be written in the form

$$\frac{q}{c}\boldsymbol{v} \times \boldsymbol{B} \cdot \frac{1}{\hbar}\nabla_{\boldsymbol{k}} f = \frac{q}{c}\boldsymbol{v} \times \boldsymbol{B} \cdot \frac{1}{\hbar}\nabla_{\boldsymbol{k}}\left(f_0 - \varphi(\boldsymbol{k})\frac{\partial f_0}{\partial \mathscr{E}}\right) = -\frac{\partial f_0}{\partial \mathscr{E}}\frac{q}{c}\boldsymbol{v} \times \boldsymbol{B} \cdot \frac{1}{\hbar}\nabla_{\boldsymbol{k}}\varphi,$$

since the gradients of f_0 and $\partial f_0/\partial \mathscr{E}$ are proportional to $\boldsymbol{v}$ and therefore give no contribution because of the cross product. The linearized Boltzmann equation for φ then becomes

$$\begin{aligned}\frac{\partial f_0}{\partial \mathscr{E}}\boldsymbol{v} \cdot q\boldsymbol{E}_1 - \frac{\partial f_0}{\partial \mathscr{E}}\frac{q}{c}\boldsymbol{v} \times \boldsymbol{B} \cdot \frac{1}{\hbar}\nabla_{\boldsymbol{k}}\varphi &= \left.\frac{\partial f}{\partial t}\right|_{\text{coll}} \\ &= -\int W(\boldsymbol{k} \to \boldsymbol{k}')\frac{f_0(\boldsymbol{k})[1 - f_0(\boldsymbol{k}')]}{kT}[\varphi(\boldsymbol{k}) - \varphi(\boldsymbol{k}')]\,\mathrm{d}\boldsymbol{k}'.\end{aligned} \tag{5.21}$$

If we put

$$\left.\frac{\partial f}{\partial t}\right|_{\text{coll}} = \frac{\partial f_0}{\partial \mathscr{E}} W_{\text{op}}\varphi, \tag{5.22}$$

then a convenient way to write the linearized Boltzmann equation is

$$\boldsymbol{v} \cdot q\boldsymbol{E}_1 - \frac{q}{c}\boldsymbol{v} \times \boldsymbol{B} \cdot \frac{1}{\hbar}\nabla_{\boldsymbol{k}}\varphi = W_{\text{op}}\varphi, \tag{5.23a}$$

$$W_{\text{op}}\varphi = \int W(\boldsymbol{k}, \boldsymbol{k}')[\varphi(\boldsymbol{k}) - \varphi(\boldsymbol{k}')]\,\mathrm{d}\boldsymbol{k}' \tag{5.23b}$$

where

$$W(\boldsymbol{k}, \boldsymbol{k}') = W(\boldsymbol{k} \to \boldsymbol{k}')\frac{1 - f_0(\boldsymbol{k}')}{1 - f_0(\boldsymbol{k})} = W(\boldsymbol{k}' \to \boldsymbol{k})\frac{f_0(\boldsymbol{k}')}{f_0(\boldsymbol{k})}. \tag{5.24}$$

5.3. The relaxation time: isotropic energy bands

A simplification that is often made is the relaxation time approximation, in which the collision term is approximated by

$$\left.\frac{\partial f}{\partial t}\right|_{\text{coll}} = -\frac{f - f_0}{\tau}. \tag{5.25}$$

For eq. (5.23), this implies $W_{\text{op}}\varphi = \varphi/\tau$. There are several conditions under which a relaxation time exists, which mainly apply to isotropic energy bands, in which $\mathscr{E}(\boldsymbol{k})$ depends only on $|\boldsymbol{k}| = k$.

(1) Elastic or inelastic scattering which is "$\boldsymbol{k}$-randomizing" (Herring and Vogt 1956). In this case, $W(\boldsymbol{k}' \to \boldsymbol{k})$ is independent of the sense of $\boldsymbol{k}'$, i.e., it is an even function of $\boldsymbol{k}'$ measured from an appropriate energy minimum. For isotropic systems this tends to randomize the velocities.

(2) Elastic scattering and isotropic energy bands. In addition, it is assumed that the transition probability $W(\boldsymbol{k} \to \boldsymbol{k}')$ depends only on the angle between $\boldsymbol{k}$ and $\boldsymbol{k}'$. We shall refer to this as isotropic scattering, which is not the same as $\boldsymbol{k}$-randomizing scattering. If the scattering is anisotropic in $\boldsymbol{k} - \boldsymbol{k}'$ an isotropic approximation is usually made.

Consider the first condition, $\boldsymbol{k}$-randomizing scattering. The collision term in eq. (5.23b) contains two terms, the first of which we can think of as involving scattering out of $\boldsymbol{k}$, and the second scattering in. Make the ansatz that the "in" scattering vanishes. Then there is a relaxation time from the "out" scattering

$$1/\tau_{\text{r}}(\mathscr{E}) = \int W(\boldsymbol{k}, \boldsymbol{k}')\, \mathrm{d}\boldsymbol{k}'. \tag{5.26}$$

Ignoring the magnetic field for the moment, we see that $\varphi(\boldsymbol{k})$ is proportional to the velocity $\boldsymbol{v}(\boldsymbol{k})$, which is an odd function of $\boldsymbol{k}$, measured from an energy minimum. Therefore $\varphi(\boldsymbol{k}')$ must be an odd function of $\boldsymbol{k}'$ and, since W is assumed to be an even function of $\boldsymbol{k}'$, the integral over $\boldsymbol{k}'$ of the "in" scattering vanishes. The magnetic field term is proportional to $\boldsymbol{v} \times \boldsymbol{\nabla}_{\boldsymbol{k}}\varphi$, which is also an odd function of $\boldsymbol{k}$, so the argument is unchanged when a magnetic field is included. The relaxation time depends on energy for isotropic energy bands. This result also holds for anisotropic bands, but in that case τ may be anisotropic.

To show that the case of elastic isotropic scattering for spherically symmetric energy bands leads to a relaxation time, we make the ansatz that φ is of the form

$$\varphi = \boldsymbol{u}(\mathscr{E}) \cdot \hbar\boldsymbol{k}, \tag{5.27}$$

that is, the distribution function is

$$f = f_0 - \boldsymbol{u}(\mathscr{E}) \cdot \hbar\boldsymbol{k}\,\partial f_0/\partial\mathscr{E}. \tag{5.28}$$

Here $\boldsymbol{u}$ is a fixed vector for a given energy, i.e., it is independent of the direction of $\boldsymbol{k}$. We shall see later that $\boldsymbol{u}$ is an energy-dependent drift velocity. For elastic scattering the collision term is now

$$W_{\mathrm{op}}\varphi = \int W(\boldsymbol{k}, \boldsymbol{k}')[\boldsymbol{u}(\mathscr{E}) \cdot \hbar(\boldsymbol{k} - \boldsymbol{k}')]\, \mathrm{d}\boldsymbol{k}', \tag{5.29}$$

where we have used the fact that $\mathscr{E}' = \mathscr{E}$. The term in square brackets is

$$u(\mathscr{E})\hbar k(\cos\theta_{\boldsymbol{k},\boldsymbol{u}} - \cos\theta_{\boldsymbol{k}'\boldsymbol{u}}).$$

In the angular part of the $\boldsymbol{k}'$-integral we can use spherical coordinates with the polar axis in the $\boldsymbol{k}$-direction. Expressing $\cos\theta_{\boldsymbol{k}',\boldsymbol{u}}$ in this system,

$$\cos\theta_{\boldsymbol{k}',\boldsymbol{u}} = \cos\theta_{\boldsymbol{k}',\boldsymbol{k}}\cos\theta_{\boldsymbol{u},\boldsymbol{k}} - \sin\theta_{\boldsymbol{k}',\boldsymbol{k}}\sin\theta_{\boldsymbol{u},\boldsymbol{k}}\cos(\varphi_{\boldsymbol{k}'} - \varphi_{\boldsymbol{u}}),$$

we see that the integral of $\cos(\varphi_{\boldsymbol{k}'} - \varphi_{\boldsymbol{u}})$ vanishes, because by assumption W only depends on $\theta_{\boldsymbol{k},\boldsymbol{k}'}$ and so does not depend upon $\varphi_{\boldsymbol{k}'}$. Therefore we can replace $\cos_{\boldsymbol{k}',\boldsymbol{u}}$ by $\cos_{\boldsymbol{k}',\boldsymbol{k}}\cos\theta_{\boldsymbol{u},\boldsymbol{k}'}$ and we find that eq. (5.25) holds with the relaxation time given by

$$1/\tau_{\mathrm{e}}(\mathscr{E}) = \int W(\boldsymbol{k}, \boldsymbol{k}')[1 - \cos\theta_{\boldsymbol{k}',\boldsymbol{k}})]\, \mathrm{d}\boldsymbol{k}'. \tag{5.30}$$

The relaxation time involved in the low-field transport effects is the relaxation time for the momentum. In high fields the relaxation of the energy of the electrons becomes important.

If there are several scattering mechanisms, the scattering rates and hence the inverse τ add:

$$1/\tau(\mathscr{E}) = \sum_i 1/\tau_i(\mathscr{E}). \tag{5.31}$$

For spherical bands, when there is a relaxation time, the Boltzmann equation, eq. (5.23) with φ given by eq. (5.27), becomes

$$\boldsymbol{v} \cdot q\boldsymbol{E}_1 - \frac{q}{c}\boldsymbol{v} \times \boldsymbol{B} \cdot \boldsymbol{u}(\mathscr{E}) = \boldsymbol{u}(\mathscr{E}) \cdot \hbar\boldsymbol{k}/\tau(\mathscr{E}). \tag{5.32}$$

The magnetic term can be rewritten as $\boldsymbol{v} \cdot (q/c)\boldsymbol{u}(\mathscr{E}) \times \boldsymbol{B}$. We see that $\boldsymbol{u}(\mathscr{E})$ has the appearance of a drift velocity in the force equation, by comparing this with the term in eq. (5.32) involving the electric field. In fact, since for spherical bands $\boldsymbol{v} = \hbar\boldsymbol{k}/m^*(\mathscr{E})$, where from eq. (2.24), $1/m^* = (1/\hbar^2 k)\partial\mathscr{E}/\partial k$, we can factor out the velocity, and this gives an equation for $\boldsymbol{u}$:

$$q\boldsymbol{E}_1 + \frac{q}{c}\boldsymbol{u}(\mathscr{E}) \times \boldsymbol{B} = m^*(\mathscr{E})\boldsymbol{u}(\mathscr{E})/\tau(\mathscr{E}). \tag{5.33}$$

If we interpret $\boldsymbol{u}$ as a drift velocity, then this looks exactly like the equation of motion in the particle kinetic model, except that we have added the thermal terms, and various quantities depend on energy. This shows that the particle kinetic model is actually valid for simple bands, provided that we average the drift velocity in an appropriate way. The correct average to use is described in § 6.

5.4. Beyond the relaxation time approximation: variational principle

There are several important cases in which the relaxation time approximation is not very good, and various methods have been used to deal with this. These methods are becoming increasingly sophisticated and powerful, and their use is important for understanding real semiconductors. One case in which the relaxation time approximation fails is that of anisotropic bands. For ellipsoidal energy surfaces we shall find in § 5.5 that a tensor relaxation time is a good approximation, but for the warped valence band edges a more general method involving expansion of the distribution function in cubic harmonics is appropriate, and we shall take this up in § 5.6. Another case is inelastic scattering that is not $\boldsymbol{k}$-randomizing, i.e., polar optical scattering. Fortunately, this seems to be important mainly for spherical energy surfaces, and we shall discuss methods for dealing with this case in § 5.7.

One important approach to solving the Boltzmann equation has been the Monte Carlo method, which has been especially powerful for studying hot-electron transport. We shall not discuss this here, but the method is thoroughly reviewed by Jacobini and Reggiani (1983). Other methods applied to the hot-electron problem include the expansion of the distribution function in Legendre polynomials and iterative solutions of the non-linear transport equations. All these methods are discussed in chapter 20.

We shall now discuss a time-honored method of going beyond the relaxation time approximation, namely the variational principle of Kohler (Kohler 1948, Ziman 1960, Nag 1980). It is convenient here to write the scattering operator as

$$L_{\mathrm{op}}\varphi(\boldsymbol{k}) = -\frac{\partial f_0}{\partial \mathscr{E}} W_{\mathrm{op}}\varphi(\boldsymbol{k}) = \int W(\boldsymbol{k}\to\boldsymbol{k}')\frac{f_0(\boldsymbol{k})[1-f_0(\boldsymbol{k}')]}{kT}[\varphi(\boldsymbol{k})-\varphi(\boldsymbol{k}')]\,\mathrm{d}\boldsymbol{k}'. \tag{5.34}$$

Because of the principle of detailed balance, L_{op} is self-adjoint:

$$\int \psi(\boldsymbol{k})L_{\mathrm{op}}\varphi(\boldsymbol{k})\,\mathrm{d}\boldsymbol{k} = \int \varphi(\boldsymbol{k})L_{\mathrm{op}}\psi(\boldsymbol{k})\,\mathrm{d}\boldsymbol{k} \tag{5.35}$$

and L_{op} is also positive definite, because it can easily be shown that

$$\int \varphi L_{\mathrm{op}}\varphi\,\mathrm{d}\boldsymbol{k} = \tfrac{1}{2}\int W(\boldsymbol{k}\to\boldsymbol{k}')\frac{f_0(\boldsymbol{k})[1-f_0(\boldsymbol{k}')]}{kT}[\varphi(\boldsymbol{k})-\varphi(\boldsymbol{k}')]^2\,\mathrm{d}\boldsymbol{k}\,\mathrm{d}\boldsymbol{k}' \geqslant 0. \tag{5.36}$$

If there is no magnetic field the Boltzmann equation has the form

$$L_{\mathrm{op}}\varphi(\boldsymbol{k}) = x(\boldsymbol{k}), \quad x(\boldsymbol{k}) = -\frac{\partial f_0}{\partial \mathscr{E}}\boldsymbol{v}\cdot\boldsymbol{E}_1. \tag{5.37}$$

The variational principle states that if φ is a function that obeys the condition

$$\int \varphi(\boldsymbol{k})L_{\mathrm{op}}\varphi(\boldsymbol{k})\,\mathrm{d}\boldsymbol{k} = \int \varphi(\boldsymbol{k})x(\boldsymbol{k})\,\mathrm{d}\boldsymbol{k} \tag{5.38}$$

then the functional

$$I = \int \varphi(\boldsymbol{k}) L_{\text{op}} \varphi(\boldsymbol{k}) \, \mathrm{d}\boldsymbol{k} \tag{5.39}$$

is a maximum when φ obeys the Boltzmann equation, eq. (5.37).

To give a short proof of the maximum principle, let us find the change in I if $\varphi \to \varphi + \delta\varphi$,

$$\Delta I = 2 \int \delta\varphi(\boldsymbol{k}) L_{\text{op}} \varphi(\boldsymbol{k}) \, \mathrm{d}\boldsymbol{k} + \int \delta\varphi(\boldsymbol{k}) L_{\text{op}} \delta\varphi(\boldsymbol{k}) \, \mathrm{d}\boldsymbol{k} = \int \delta\varphi(\boldsymbol{k}) x(\boldsymbol{k}) \, \mathrm{d}\boldsymbol{k}, \tag{5.40}$$

the last equality coming from the condition in eq. (5.38). If the Boltzmann equation is satisfied, that is if $L_{\text{op}}\varphi = x$, then the last two equalities give

$$\int \delta\varphi(\boldsymbol{k}) x(\boldsymbol{k}) \, \mathrm{d}\boldsymbol{k} = -\int \delta\varphi(\boldsymbol{k}) L_{\text{op}} \delta\varphi(\boldsymbol{k}) \, \mathrm{d}\boldsymbol{k} < 0, \tag{5.41}$$

but from eq. (5.40), this is simply equal to ΔI, which means that I is a maximum for the correct solution.

If φ is expanded in a set of linearly independent functions $y_i(\boldsymbol{k})$,

$$\varphi(\boldsymbol{k}) = \sum_i \varphi_i y_i(\boldsymbol{k}), \tag{5.42}$$

then the variational principle leads to an algebraic equation for the coefficients φ_i

$$\sum_j \left(\int y_i(\boldsymbol{k}) L_{\text{op}} y_j(\boldsymbol{k}) \, \mathrm{d}\boldsymbol{k} \right) \varphi_j = \int y_i(\boldsymbol{k}) x(\boldsymbol{k}) \, \mathrm{d}\boldsymbol{k}. \tag{5.43}$$

Furthermore, if the scattering operator only connects states with the same energy, which occurs if the scattering is either $\boldsymbol{k}$-randomizing or elastic, then the integrals over $\boldsymbol{k}$ can be replaced by integrals over a constant-energy surface. This justifies expanding the distribution function in a truncated set of surface harmonics, which is the method used in the case of ellipsoids and warped hole bands. An example for spherical bands is piezoelectric scattering, which is elastic but anisotropic. The trial function $q\tau(\mathscr{E})\boldsymbol{v}_x E$ gives

$$\frac{1}{\tau} = \int W(\boldsymbol{k} \to \boldsymbol{k}') \cos\theta_{\boldsymbol{k},x} (\cos\theta_{\boldsymbol{k},x} - \cos\theta_{\boldsymbol{k}',x}) \, \mathrm{d}\boldsymbol{k}' \, 3d\Omega_{\boldsymbol{k}}/4\pi. \tag{5.44}$$

This result was obtained by Zook (1964) by using the arguments of Herring and Vogt (1956).

If there is a finite magnetic field it is still possible to prove an extremum principle, but the functional for that case is no longer a maximum (Ziman 1960).

5.5. Ellipsoidal energy surfaces

In the case of carriers in bands with ellipsoidal energy surfaces the relaxation time approximation is not valid because of the anisotropy. However, a relaxation time tensor may be a good approximation. To study this case Herring and Vogt (1956) constructed a transformation to turn the ellipsoidal energy surfaces into spheres, which we have met before in § 2.4 and 3.5. If we make the change of variables

$$p_x = k_x\sqrt{m/m_x}, \quad p_y = k_y\sqrt{m/m_y}, \quad p_z = k_z\sqrt{m/m_z}, \tag{5.45}$$

then the energy becomes $\mathscr{E} = \hbar^2 p^2/2m$ and $\mathrm{d}\boldsymbol{k} = \mathrm{d}\boldsymbol{p}(m_x m_y m_z/m^3)^{1/2}$.

Actually the scattering, e.g., from ionized impurities, is isotropic in the original coordinates, as can be seen in table 2. However, when expressed in the new coordinates, the scattering rate is anisotropic. The scattering term has the general form

$$W_{\mathrm{op}}\varphi = \int W'(\boldsymbol{p},\boldsymbol{p}')[\varphi(\boldsymbol{p}) - \varphi(\boldsymbol{p}')]\,\mathrm{d}\boldsymbol{p}', \quad W'(\boldsymbol{p},\boldsymbol{p}') = W(\boldsymbol{k},\boldsymbol{k}')(m_x m_y m_z/m_0^3)^{1/2}, \tag{5.46}$$

with $W(\boldsymbol{k},\boldsymbol{k}')$ defined in eq. (5.24). We now expand the distribution function in spherical harmonics

$$\varphi(\boldsymbol{p}) = \sum \varphi_{lm}(p) Y_{lm}(\hat{\boldsymbol{p}}), \tag{5.47}$$

where the Y have the normalization,

$$\int Y^*_{lm}(\hat{\boldsymbol{p}}) Y_{l'm'}(\hat{\boldsymbol{p}})\,\mathrm{d}\Omega_p = \delta_{ll'}\delta_{mm'}. \tag{5.48}$$

If W is assumed to be either $\boldsymbol{k}$-randomizing or elastic, the collision term, eq. (5.46), involves $\varphi(\boldsymbol{p}')$ only for $p' = p$, and the collision operator can be expanded in spherical harmonics on the surface $p = \text{const.}$ (Herring and Vogt 1956)

$$W_{\mathrm{op}}\varphi = \sum W_{lml'm'}\varphi_{l'm'}(p) Y_{lm}(\hat{\boldsymbol{p}}), \tag{5.49}$$

where

$$W_{lml'm'} = \int Y^*_{lm}(\hat{\boldsymbol{p}}) W'(\boldsymbol{p},\boldsymbol{p}')[Y_{l'm'}(\hat{\boldsymbol{p}}) - Y_{l'm'}(\hat{\boldsymbol{p}}')]\,\mathrm{d}\boldsymbol{p}'\,\mathrm{d}\Omega_p. \tag{5.50}$$

The only contribution to the second term is from elastic scattering, for which W contains an energy-conserving δ-function.

The Boltzmann equation, eq. (5.23), in the new coordinates is

$$\boldsymbol{p}\cdot q\bar{\boldsymbol{E}}_1/m + \frac{q}{mc}\bar{\boldsymbol{B}}\cdot\boldsymbol{p}\times\nabla_{\boldsymbol{p}}\varphi = W_{\mathrm{op}}\varphi, \tag{5.51}$$

where $\bar{\boldsymbol{E}}_1$ and $\bar{\boldsymbol{B}}$ are transformed fields ($\boldsymbol{E}_1$ transforms like $\boldsymbol{k}$, and $\bar{\boldsymbol{B}}$ is given by eq. (3.45)). On the left-hand side the electric field term can be expanded in spherical harmonics involving only $l = 1$. The magnetic term can also be expanded in spherical harmonics, which only connect states with the same value of l. Thus a set of coupled equations for the φ is obtained, of the form:

$$\bar{\boldsymbol{E}}_1\cdot\boldsymbol{G}_{1m}\delta_{l1} + \bar{\boldsymbol{B}}\cdot\sum_{m'}\boldsymbol{H}_{lmlm'}\varphi_{lm'}(p) = \sum_{l'm'} W_{lml'm'}\varphi_{l'm'}(p), \tag{5.52}$$

where

$$G_{1m} = \int Y_{1m}^*(\hat{p})(qp/m)\, d\Omega_p, \tag{5.53}$$

$$H_{lmlm'} = \int Y_{lm}^*(\hat{p})(q/mc)\boldsymbol{p} \times \nabla_{\boldsymbol{p}} Y_{lm'}(\hat{p}')\, d\Omega_p. \tag{5.54}$$

Herring and Vogt (1956) truncated the equations at $l=1$. The $l=1$ spherical harmonics can be replaced by the vector components, p_x, p_y, and p_z, suitably normalized. The scattering operator is then a second-rank tensor, the inverse of a relaxation time tensor. In the principal-axis system a typical component is

$$W_{xx} = 1/\tau_x = \int W'(\boldsymbol{p},\boldsymbol{p}')\cos\theta_{\boldsymbol{p},x}(\cos\theta_{\boldsymbol{p},x} - \cos\theta_{\boldsymbol{p}',x})\, d\boldsymbol{p}'\, 3d\Omega_{\boldsymbol{p}}/4\pi. \tag{5.55}$$

In other words, if we write $\varphi = \boldsymbol{p} \cdot \boldsymbol{u}'(\mathscr{E}) = \boldsymbol{k} \cdot \boldsymbol{u}(\mathscr{E})$, then

$$W_{\rm op}\varphi \cong \boldsymbol{p} \cdot \boldsymbol{\tau}^{-1} \cdot \boldsymbol{u}' = \boldsymbol{k} \cdot \boldsymbol{\tau}^{-1} \cdot \boldsymbol{u}, \tag{5.56}$$

the last inequality indicating that the relaxation time tensor does not change on transforming back to $\boldsymbol{k}$.

Samoilovich et al. (1961a–c) have studied this system in the case of the conductivity $(B=0)$, using the above spherical harmonics expansion. For acoustic deformation potential scattering, they found that $l=1$ is sufficient, but for ionized impurity scattering they found it necessary to go to $l=3$ (only odd l-terms are involved). For ellipsoids of revolution the scattering operator only connects states with the same value of m, so that $W_{lml'm'} = W_{ll'}(m)\delta_{mm'}$. They take the inverse of $W_{ll'}(m)$, and the 11 matrix element still defines a relaxation time tensor for $B=0$, since only $l=1$ components of φ contribute to the current. They find longitudinal and transverse components $\tau_\ell = W_{11}^{-1}(0)$ and $\tau_t = W_{11}^{-1}(1)$.

Eagles and Edwards (1965) have shown that the tensor relaxation time is a good approximation for finite B. With the tensor relaxation time, the Boltzmann equation in the usual coordinates becomes

$$q\boldsymbol{E}_1 + (q/c)\boldsymbol{u}(\mathscr{E}) \times \boldsymbol{B} = \boldsymbol{m}^*\boldsymbol{\tau}^{-1} \cdot \boldsymbol{u}(\mathscr{E}), \tag{5.57}$$

where $\boldsymbol{m}^*$ is the effective-mass tensor, and as usual $\varphi = \hbar\boldsymbol{k} \cdot \boldsymbol{u}(\mathscr{E})$.

5.6. *Holes with warped energy surfaces*

The case of holes is more complex because the energy surfaces are warped, and there is more than one band. The Boltzmann equation becomes a set of coupled equations for the distribution functions for the heavy- and light-hole bands and possibly the split-off band. For the case of $\boldsymbol{k}$-randomizing or elastic collisions, the collision term couples states in the two or three bands that have the same energy. Lawaetz (1968a) has made a study of the effect of including symmetry for this problem. The collision operator $W_{\rm op}$ possesses the symmetry of the system, which for Si and Ge is cubic, and the same analysis could also be used for spherical symmetry and cylindrical symmetry, the latter being the case for a single valley of the conduction band. The perturbation in the distribution function φ transforms under rotation as the basis for some representation of the symmetry group of the system. For example, in the case of spherical symmetry in the linearized Boltzmann equation φ transforms as the $l=1$ spherical harmonic, and in the case of cubic symmetry we are particularly interested in the vectorial representation Γ_{15}, which contributes to the current. The scattering operator is then shown by Lawaetz only to connect φ with other terms belonging to the same irreducible representation of the symmetry group of the problem. This results in an enormous simplification and allows us to go beyond the relaxation time approximation.

More specifically, the linearized Boltzmann equation including multiband scattering can be written in the form

$$q\boldsymbol{E}_1 \cdot \boldsymbol{v}_n + q\boldsymbol{B} \cdot \boldsymbol{v}_n \times \frac{1}{\hbar} \boldsymbol{\nabla}_{\boldsymbol{k}} \varphi_n(\boldsymbol{k}) = \sum_{n'} \int W_{nn'}(\boldsymbol{k}, \boldsymbol{k}')[\varphi_n(\boldsymbol{k}) - \varphi_{n'}(\boldsymbol{k}')]\, \mathrm{d}\boldsymbol{k}'. \tag{5.58}$$

This is a generalization of eq. (5.23). We are assuming that $\varphi_{n'}(\boldsymbol{k}')$ corresponds to the same energy as $\varphi_n(\boldsymbol{k})$, which happens because the only inelastic collisions considered are assumed to be $\boldsymbol{k}$-randomizing, so they do not contribute to the "in" scattering term. For elastic collisions, W has an energy-conserving δ-function.

We now suppose that φ is expanded in some system of angular harmonics on the constant-energy surface,

$$\varphi_n(\mathscr{E}, \hat{\boldsymbol{k}}) = \sum_i \varphi_i^n(\mathscr{E}) Y_i(\hat{\boldsymbol{k}}), \tag{5.59}$$

where the Y_i are orthonormal,

$$\int Y_i^*(\hat{\boldsymbol{k}}) Y_j(\hat{\boldsymbol{k}})\, \mathrm{d}\Omega_{\boldsymbol{k}} = \delta_{ij}. \tag{5.60}$$

We emphasize that this expansion is on the constant-energy surface, and the variables are $\mathscr{E}$ and $\hat{\boldsymbol{k}}$, so that the magnitude of k depends on the energy and direction. The volume element for a given band n is

$$\mathrm{d}\boldsymbol{k} = \left(k^2 \frac{\partial k}{\partial \mathscr{E}}\right)_n \mathrm{d}\Omega_{\boldsymbol{k}}\, \mathrm{d}\mathscr{E} \equiv \Gamma_n(\mathscr{E}, \hat{\boldsymbol{k}})\, \mathrm{d}\Omega_{\boldsymbol{k}}\, \mathrm{d}\mathscr{E}. \tag{5.61}$$

Substituting the expansion into the Boltzmann equation, multiplying by Y_i^* and integrating over angles gives

$$\boldsymbol{E}_1 \cdot \boldsymbol{G}_i^n(\mathscr{E}) + \boldsymbol{B} \cdot \sum_j \boldsymbol{H}_{ij}^n(\mathscr{E}) \varphi_j^n(\mathscr{E}) = \sum_{jn'} W_{ij}^{nn'}(\mathscr{E}) \varphi_j^{n'}(\mathscr{E}), \tag{5.62}$$

where

$$\boldsymbol{G}_i^n(\mathscr{E}) = q \int Y_i^*(\hat{\boldsymbol{k}}) \boldsymbol{v}_n\, \mathrm{d}\Omega_{\boldsymbol{k}}, \tag{5.63}$$

$$H^n_{ij}(\mathscr{E}) = q \int Y^*_i(\hat{k}) \left(v_n \times \frac{1}{\hbar} \nabla_k \right) Y_j(\hat{k})\, \mathrm{d}\Omega_k, \tag{5.64}$$

$$\begin{aligned} W^{nn'}_{ij}(\mathscr{E}) = \delta_{nn'} \sum_{n''} \int Y^*_i(\hat{k}) W_{nn''}(k, k') Y_j(\hat{k}) \Gamma_{n''}(\mathscr{E}', \hat{k}')\, \mathrm{d}\mathscr{E}'\, \mathrm{d}\Omega_{k'}\, \mathrm{d}\Omega_k \\ - \int Y^*_i(\hat{k}) W_{nn'}(k, k') Y_j(\hat{k}') \Gamma_{n'}(\mathscr{E}, \hat{k}')\, \mathrm{d}\mathscr{E}'\, \mathrm{d}\Omega_{k'}\, \mathrm{d}\Omega_k. \end{aligned} \tag{5.65}$$

Equation (5.62) is to be solved for φ^n_i, and then the current is determined from eq. (5.12). In general the set of coupled equations must be truncated, and a certain authority to do so comes from the variational principle discussed in § 5.4.

The key point made by Lawaetz is that the scattering operator only connects those φ^n_i belonging to the same row of the same irreducible representation of the symmetry group of the problem. Considering cubic symmetry, G_i corresponds to the vectorial representation Γ_{15} of the cubic group, so if $B = 0$ this is the only representation needed. The natural expansion system is cubic harmonics (von der Lage and Bethe 1947). To introduce some notation, let $K^i_{L\mu}$ be the cubic harmonic for the ith irreducible representation, with μ labeling the row (which can be x, y, z, for example) and L the angular momentum from which K is formed. Note that there is more than one function belonging to each representation. The Boltzmann equation then has the form

$$E_1 \cdot G^{in}_{L\mu}(\mathscr{E}) + B \cdot \sum_{jL'\mu'} H^{ijn}_{L\mu L'\mu'}(\mathscr{E}) \varphi^{jn}_{L'\mu'}(\mathscr{E}) = \sum_{L'n'} W^{inn'}_{LL'}(\mathscr{E}) \varphi^{in'}_{L'\mu}(\mathscr{E}). \tag{5.66}$$

Clearly this will be much more difficult to work with than the relaxation time approximation, but the inversion of the matrices involved is straightforward.

We can see how this scheme gives a relaxation time in the case of a single spherical band and isotropic scattering. The irreducible representations are the spherical harmonics, and there is only one representation for each l, with the row corresponding to m. The scattering operator is diagonal in l and independent of m. For $l = 1$ we have the momentum relaxation time. The case of ellipsoids of revolution can also be included here. If we use spherical harmonics the representations with cylindrical symmetry are labeled by m, so the scattering matrix connects spherical harmonics with different l but the same m.

In the case of coupled valence bands if there is no anisotropy the equations break off at $l = 1$, and a relaxation time can be found for each band. For warped hole bands, the scattering mixes in cubic harmonics for odd l beyond $l = 1$, and the magnetic field mixes in representations other than Γ_{15}. These methods have been used by Lawaetz (1968b) and more recently by Szmulowicz (1983, 1986), to calculate transport coefficients in Si and Ge.

5.7. *Inelastic scattering*

For inelastic but isotropic scattering, in the case of spherical bands, we can still write $\varphi = u(\mathscr{E}) \cdot k$, in terms of an energy-dependent drift velocity, and the distribution function is given by eq. (5.28) (Nag 1980, Rode 1975). The collision term in the Boltzmann equation is

$$W_{op}\varphi = \int W(\boldsymbol{k}, \boldsymbol{k}')[\boldsymbol{u}(\mathscr{E})\cdot \hbar\boldsymbol{k} - \boldsymbol{u}(\mathscr{E}')\cdot \hbar\boldsymbol{k}']\, \mathrm{d}\boldsymbol{k}', \tag{5.67}$$

where $W(\boldsymbol{k}, \boldsymbol{k}')$ is given by eq. (5.24). Just as for the elastic case, in the $\boldsymbol{k}'$-angular integral we can take $\boldsymbol{k}$ as the polar axis and make the replacement $\cos\theta_{\boldsymbol{k}',\boldsymbol{u}} \rightarrow \cos\theta_{\boldsymbol{k},\boldsymbol{u}}\cos\theta_{\boldsymbol{k},\boldsymbol{k}'}$, so that

$$W_{op}\varphi = -m^*(\mathscr{E})\boldsymbol{v}\cdot \int W(\boldsymbol{k}, \boldsymbol{k}')[\boldsymbol{u}(\mathscr{E}) - \boldsymbol{u}(\mathscr{E}')\cos\theta_{\boldsymbol{k}',\boldsymbol{k}} k'/k]\, \mathrm{d}\boldsymbol{k}'. \tag{5.68}$$

Therefore the velocity can be factored on the two sides of eq. (5.23), and the Boltzmann equation becomes

$$(q/m^*)\boldsymbol{E}_1 + (q/m^*c)\boldsymbol{u}(\mathscr{E}) \times \boldsymbol{B} = \int W(\boldsymbol{k}, \boldsymbol{k}')[\boldsymbol{u}(\mathscr{E}) - \boldsymbol{u}(\mathscr{E}')\cos\theta_{\boldsymbol{k}',\boldsymbol{k}} k'/k]\, \mathrm{d}\boldsymbol{k}'. \tag{5.69}$$

For those processes that are described by a relaxation time the right-hand side gives $\tau_0^{-1}\boldsymbol{u}(\mathscr{E})$. For the rest let the kernels integrated over angle be W_{out} and W_{in}

$$\mathrm{RHS} = \left(\tau_0^{-1} + \int W_{out}(\mathscr{E}, \mathscr{E}')\, \mathrm{d}\mathscr{E}'\right)\boldsymbol{u}(\mathscr{E}) - \int W_{in}(\mathscr{E}, \mathscr{E}')\, \mathrm{d}\mathscr{E}'\, \boldsymbol{u}(\mathscr{E}'), \tag{5.70}$$

$$W_{out}(\mathscr{E}, \mathscr{E}') = \int W(\boldsymbol{k}, \boldsymbol{k}')k'^2 \frac{\partial k'}{\partial \mathscr{E}'}\, \mathrm{d}\Omega_{\boldsymbol{k}'}, \tag{5.71}$$

$$W_{in}(\mathscr{E}, \mathscr{E}') = \int W(\boldsymbol{k}, \boldsymbol{k}')\cos\theta_{\boldsymbol{k}',\boldsymbol{k}}(k'/k)k'^2 \frac{\partial k'}{\partial \mathscr{E}'}\, \mathrm{d}\Omega_{\boldsymbol{k}'}. \tag{5.72}$$

Consider the case of an electric field only, in the x-direction, i.e., the conductivity problem. For this case there is an integral equation in a single function u_x. This is also true for the case of thermal gradients, but for the case of a magnetic field in the z-direction, there are two coupled integral equations for the transverse components u_x and u_y.

In actual practice for polar optical-phonon scattering there is an energy-conserving δ-function such that $\mathscr{E}' = \mathscr{E} \pm \hbar\omega_o$, and the result is a difference equation:

$$\mathrm{RHS} = W_0\boldsymbol{u}(\mathscr{E}) + W_+\boldsymbol{u}(\mathscr{E} + \hbar\omega_o) + W_-\boldsymbol{u}(\mathscr{E} - \hbar\omega_o). \tag{5.73}$$

In order to solve these equations the variational method was used by Ehrenreich (1957) in his classic treatment of polar optical scattering in InSb. More recent treatments of polar optical scattering have depended upon the fact that eq. (5.69) with eq. (5.73) is a difference equation. A matrix method was developed by Delves (1959) and applied to the study of the transport properties of GaAs and GaP by Fletcher and Butcher (1972). An iterative method of solution was developed by Rode (1970). These methods are reviewed by Nag (1980).

6. Electrical conductivity

We consider now solutions of the linearized Boltzmann equation, eq. (5.23), in an electric field, when there are no magnetic fields or thermal gradients. To obtain the current we use eqs. (5.12) and (5.18) and note that the zero-order distribution function makes no contribution,

$$\boldsymbol{j} = \int q\boldsymbol{v}\left(-\frac{\partial f_0}{\partial \mathscr{E}}\right)\varphi(\boldsymbol{k})\,\mathrm{d}\boldsymbol{k}/4\pi^3. \tag{6.1}$$

In the relaxation time approximation, $W_{\mathrm{op}}\varphi = \varphi/\tau$, and the solution of the Boltzmann equation is $\varphi = q\tau\boldsymbol{v}\cdot\boldsymbol{E}$. The current is given by eq. (4.1) in terms of the conductivity tensor

$$\sigma_{\alpha\beta} = q^2\int \tau(\mathscr{E})v_\alpha v_\beta\left(-\frac{\partial f_0}{\partial \mathscr{E}}\right)\mathrm{d}\boldsymbol{k}/4\pi^3. \tag{6.2}$$

This applies to a general energy band. However, we have seen that the relaxation time approach is only valid for spherical bands.

6.1. Simple bands and classical statistics

We begin with the case of simple bands when there is a relaxation time. We can use eq. (6.2), but it is interesting to recall the solution for spherical bands $\varphi = \hbar\boldsymbol{k}\cdot\boldsymbol{u}(\mathscr{E})$, where from eq. (5.33) the energy-dependent drift velocity $\boldsymbol{u}$ is given by $[\tau(\mathscr{E})/m^*]q\boldsymbol{E}$. The current is given by

$$\boldsymbol{j} = \int q\boldsymbol{v}\left(-\frac{\partial f_0}{\partial \mathscr{E}}\right)\hbar\boldsymbol{k}\cdot\boldsymbol{u}(\mathscr{E})\,\mathrm{d}\boldsymbol{k}/4\pi^3. \tag{6.3}$$

For spherical energy surfaces, the integral simplifies, because $\boldsymbol{v} = \hbar\boldsymbol{k}/m^*$ and the averages over angle of $\overline{k_x k_y}$, etc. vanish, while the averages $\overline{k_x^2} = \overline{k_y^2} = \overline{k_z^2} = \frac{1}{3}k^2$, so

$$\boldsymbol{j} = \int \frac{\hbar k v}{3}\left(-\frac{\partial f_0}{\partial \mathscr{E}}\right)q\boldsymbol{u}(\mathscr{E})\,\mathrm{d}\boldsymbol{k}/4\pi^3. \tag{6.4}$$

For parabolic bands and classical statistics, we have $\partial f_0/\partial\mathscr{E} = -f_0/kT$ and $\hbar k v/3 = 2\mathscr{E}/3$, so that

$$\boldsymbol{j} = \int q\boldsymbol{u}(\mathscr{E})\frac{\mathscr{E}}{\frac{3}{2}kT}f_0\,\mathrm{d}\boldsymbol{k}/4\pi^3 = nq\langle\boldsymbol{u}(\mathscr{E})\rangle. \tag{6.5}$$

This defines a particular average over energy, since from the equipartition theorem, if $q\boldsymbol{u}$ were a constant the coefficient would just be n. In fact it is the relaxation time

τ that is averaged, since $\langle \boldsymbol{u} \rangle = \langle \tau \rangle q\boldsymbol{E}/m^*$, where

$$\langle \tau \rangle = \frac{\int \tau \mathscr{E} f_0 \, \mathrm{d}\boldsymbol{k}/4\pi^3}{\int \mathscr{E} f_0 \, \mathrm{d}\boldsymbol{k}/4\pi^3} = \frac{\int \tau \mathscr{E}^{3/2} \mathrm{e}^{-\mathscr{E}/kT} \mathrm{d}\mathscr{E}}{\int \mathscr{E}^{3/2} \mathrm{e}^{-\mathscr{E}/kT} \mathrm{d}\mathscr{E}}. \tag{6.6}$$

The last equality uses the fact that the density of states is proportional to $\mathscr{E}^{1/2}$.

The conductivity is

$$\sigma = ne^2 \langle \tau \rangle / m^*, \tag{6.7}$$

which is the same as the simple particle kinetic model, except for the average of τ. We can also write σ in terms of the mobility,

$$\sigma = ne\mu, \quad \mu = e\langle \tau \rangle / m^*. \tag{6.8}$$

The mobility depends on the interplay of the various scattering mechanisms. In table 3 we give results for the relaxation time for scattering mechanisms of interest in semiconductors, for the simple band model (Nag 1972, Zawadzki 1982). For isotropic elastic scattering (acoustic phonon deformation potential and impurity

Table 3
Relaxation times. (For definitions, see §9.)

Scattering mechanism	Scattering rate $1/\tau$	Power law for τ	Temperature dependence of τ
Ionized impurity	$\dfrac{\pi e^4 N_i \ln[1+(2k\lambda)^2]}{\sqrt{2m^*}\kappa_0^2 \mathscr{E}^{3/2}}$	$\mathscr{E}^{3/2}$	$T^{3/2}$
Neutral impurity	$\dfrac{N_n 20 a_0^*}{m^*}$	$\mathscr{E}^0$	T^0
Acoustic deformation potential	$\dfrac{\sqrt{2kTm^{*3/2}}E_1^2}{4\hbar^4 c_l}\mathscr{E}^{1/2}$	$\mathscr{E}^{-1/2}$	$T^{-3/2}$
Acoustic piezoelectric	$\dfrac{e^2\sqrt{2m^*}kT\langle P^2\rangle}{\hbar^2\kappa_0\mathscr{E}^{1/2}}$	$\mathscr{E}^{1/2}$	$T^{-1/2}$
Optical deformation potential	$\dfrac{(2m^*)^{3/2}D^2}{4\pi\hbar^3\rho\omega_o}[(n_o+1)(\mathscr{E}-\hbar\omega_o)^{1/2} + n_o(\mathscr{E}+\hbar\omega_o)^{1/2}]$	$[\mathscr{E}^{-1/2}]$	$T^{-1/2}/n_o$
Optical polar	$\dfrac{e^2\omega_o m^{*1/2}}{\sqrt{2}\hbar\mathscr{E}^{1/2}}\left(\dfrac{1}{\kappa_\infty}-\dfrac{1}{\kappa_0}\right)(2n_o+1)$	$[\mathscr{E}^{1/2}]$	$T^{1/2}/n_o$
Intervalley phonon	$\dfrac{(Z-1)(2m^*)^{3/2}D^2}{4\pi\hbar^3\rho\omega_p}[(n_p+1) \times (\mathscr{E}-\hbar\omega_p)^{1/2} + n_p(\mathscr{E}+\hbar\omega_p)^{1/2}]$	$[\mathscr{E}^{-1/2}]$	$T^{-1/2}/n_p$

scattering) τ is obtained from eq. (5.30). For piezoelectric scattering, which is important in polar semiconductors, eq. (5.44) is used. In the case of non-polar optical and intervalley scattering, which is $\boldsymbol{k}$-randomizing, τ is obtained from eq. (5.26). For polar optical scattering the relaxation time approximation is not valid except at high temperatures when $kT \gg \hbar\omega_o$, so we include only that result. We have discussed the methods of solution for this case in § 5.7.

For several scattering mechanisms τ has a power law dependence on energy:

$$\tau = \tau_1(\mathscr{E}/kT)^r. \tag{6.9}$$

If one such mechanism is dominant, then the average of τ is

$$\langle\tau\rangle = \tau_1(r+\tfrac{3}{2})!/\tfrac{3}{2}!. \tag{6.10}$$

In table 4 the numerical factors are given for several cases, along with similar averages which come up in describing the magnetic effects.

The temperature dependence of τ is given by τ_1, and this is included in table 3. In general there is a mixture of scattering mechanisms, and the scattering rates must be added as in eq. (5.31) before averaging over energy. The optical and intervalley phonon scattering rates depend on the phonon occupation number governed by the Bose factor. For high temperatures optical and intervalley phonons become important, for somewhat lower temperatures acoustic phonons dominate, while for the lowest temperatures impurity scattering is strongest, and the results depend on the concentration of the impurities.

This general picture has been borne out by many experiments. This is illustrated by the result of Fletcher and Butcher (1972) for GaAs, shown in fig. 14, in which the mobilities for the various single mechanisms are compared with the mobility for the combination of mechanisms. Since this is a polar semiconductor, the piezoelectric and polar optical scattering mechanisms are important in addition to the impurity scattering. Polar optical scattering is not described by a relaxation time, so their

Table 4
Classical averages for $\tau = \tau_1(\mathscr{E}/kT)^r$.

Average		$r=-1/2$	$r=1/2$	$r=3/2$
$\langle\tau\rangle$	$\tau_1\dfrac{(r+\frac{3}{2})!}{\frac{3}{2}!}$	$\dfrac{4}{3}\dfrac{\tau_1}{\sqrt{\pi}}$	$\dfrac{8}{3}\dfrac{\tau_1}{\sqrt{\pi}}$	$8\tau_1/\sqrt{\pi}$
$r_H = \langle\tau^2\rangle/\langle\tau\rangle^2$	$\dfrac{(2r+\frac{3}{2})!\frac{3}{2}!}{[(r+\frac{3}{2})!]^2}$	$3\pi/8 = 1.178$	1.104	1.933
$A = \langle\tau^3\rangle\langle\tau\rangle/\langle\tau^2\rangle^2$	$\dfrac{(3r+\frac{3}{2})!(r+\frac{3}{2})!}{[(2r+\frac{3}{2})!]^2}$	1.273	1.086	1.576
$\langle 1/\tau\rangle\langle\tau\rangle$	$\dfrac{(\frac{3}{2}-r)!(r+\frac{3}{2})!}{[\frac{3}{2}!]^2}$	$\dfrac{32}{9\pi} = 1.132$	1.132	$\dfrac{32}{3\pi} = 3.395$

$$\langle g(\mathscr{E}/kT)\rangle = \int g(x)x^{3/2}e^{-x}\,dx \bigg/ \int x^{3/2}e^{-x}\,dx$$

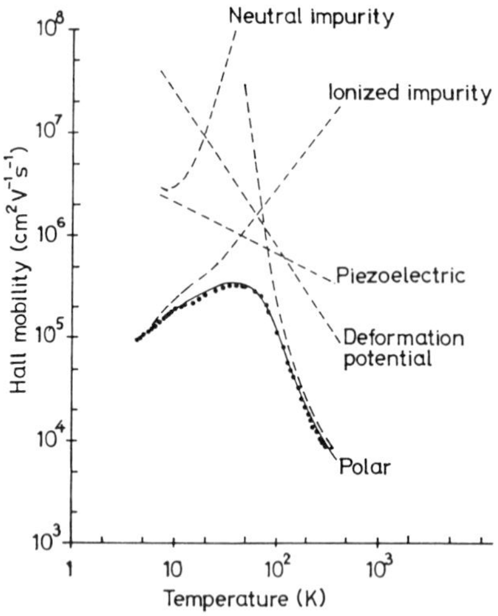

Fig. 14. Hall mobility of n-GaAs versus temperature, as measured by Stillman et al. (1970). Dashed lines are calculated from single scattering modes, solid line from mixed modes (after Fletcher and Butcher 1972).

calculation made use of the matrix method of Delves (1959) for solving the difference equation of § 5.7. The plot is actually for the Hall mobility, but the conductivity mobility has the same general behavior.

6.2. Spherical but non-parabolic energy bands

When there is a relaxation time, the solution of the Boltzmann equation described above still holds, except that the effective mass now depends on the energy, $m^{*-1} = (1/\hbar^2 k)\partial\mathscr{E}/\partial k$. Equation (6.4) is still valid, and defines the average of $q\boldsymbol{u}$ to be used for non-parabolic bands. If $q\boldsymbol{u}$ were a constant and factored out of eq. (6.4), then the integral would be

$$\int \frac{\hbar k v}{3}\left(-\frac{\partial f_0}{\partial \mathscr{E}}\right)\frac{\mathrm{d}\boldsymbol{k}}{4\pi^3} = \int\left(-\frac{\partial f_0}{\partial k}\right)\frac{k^3\,\mathrm{d}k}{3\pi^2} = \int \frac{k^2\,\mathrm{d}k}{\pi^2} f_0 = n, \tag{6.11}$$

where we have integrated by parts (Kolodsiejczak 1961, Zawadzki 1982). The conductivity σ becomes

$$\sigma = nq^2\langle\tau/m^*\rangle, \tag{6.12}$$

$$\langle\tau/m^*\rangle = \frac{1}{n}\int \frac{\tau(\mathscr{E})}{m^*(\mathscr{E})}\left(-\frac{\partial f_0}{\partial \mathscr{E}}\right)\frac{k^3(\mathscr{E})\,\mathrm{d}\mathscr{E}}{3\pi^2}. \tag{6.13}$$

We see that for non-parabolic bands the quantity to be averaged is τ/m^*. The mobility is now given by $\mu = q\langle\tau(\mathscr{E})/m^*(\mathscr{E})\rangle$.

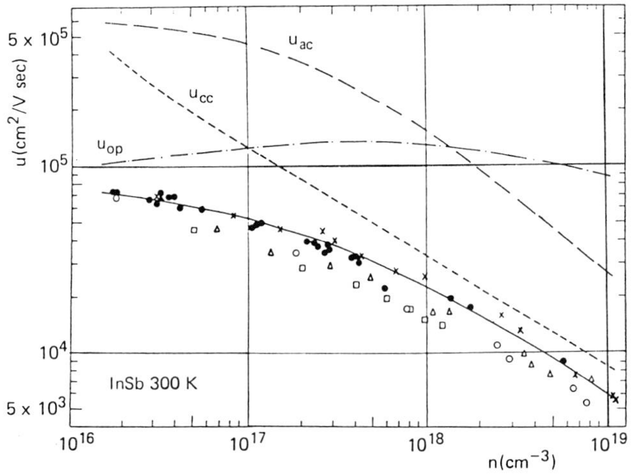

Fig. 15. Mobility in n-InSb at room temperature versus free-electron concentration, as measured by various authors. The dashed lines show the theoretical mobilities for charged center, polar optical, and acoustic scattering modes, the solid line from mixed modes (after Zawadzki 1982).

For spherical bands the relaxation time can fail in the case of polar optical scattering. The methods for solution of the difference equation discussed in § 5.7 are not restricted to parabolic bands.

Zawadzki (1982) has made extensive calculations for non-parabolic bands. An example is the result for the concentration dependence of the mobility of n-InSb shown in fig. 15. Screening of the optical modes was found to be important as well as non-parabolicity. Zawadzki gives a variational argument to show that the use of the relaxation time approximation at room temperature is equivalent to modifying e^*.

6.3. *Ellipsoidal energy surfaces*

For an ellipsoidal energy surface (eq. (2.15)), which we assume to be parabolic, the effective mass is a tensor, and we assume we have a tensor relaxation time. The first-order distribution function can still be written as $\varphi = \boldsymbol{u} \cdot \hbar \boldsymbol{k}$, and $\boldsymbol{u}$ is still an effective drift velocity, but instead of eq. (5.33) we use eq. (5.57),

$$qE_1 + \frac{q}{c}\boldsymbol{u}(\mathscr{E}) \times \boldsymbol{B} = \boldsymbol{A}^{-1} \cdot \boldsymbol{u}(\mathscr{E}), \tag{6.14}$$

where $\boldsymbol{A}$ is a tensor, expressed in dyadic form in the principal-axis system by

$$\boldsymbol{A} = (\tau_x/m_x)\hat{\boldsymbol{i}}\hat{\boldsymbol{i}} + (\tau_y/m_y)\hat{\boldsymbol{j}}\hat{\boldsymbol{j}} + (\tau_z/m_z)\hat{\boldsymbol{k}}\hat{\boldsymbol{k}}. \tag{6.15}$$

When there is only an electric field present $\boldsymbol{u}$ is given by

$$u_\alpha = (\tau_\alpha/m_\alpha)qE_\alpha, \quad \alpha = x, y, z. \tag{6.16}$$

Using $v_\alpha = k_\alpha/m_\alpha$ the conductivity tensor for an ellipsoid is

$$\sigma_{\alpha\beta} = q^2 \int \frac{\tau_\alpha(\mathscr{E})\,\hbar k_\alpha \hbar k_\beta}{m_\alpha \quad m_\beta}\left(-\frac{\partial f_0}{\partial \mathscr{E}}\right) \mathrm{d}\boldsymbol{k}/4\pi^3. \tag{6.17}$$

From symmetry considerations the conductivity components vanish unless $\alpha = \beta$. If we make use of the Herring–Vogt transformation, eq. (5.45), in evaluating the integral over $\boldsymbol{k}$, the factors $\hbar^2 k_\alpha^2/m_\alpha = \hbar^2 p_\alpha^2/m_0$ all give equal contributions $2\mathscr{E}/3$, because everything else depends on the energy, which is proportional to p^2. Then the conductivity and mobility tensors are given by

$$\sigma_{\alpha\beta} = \delta_{\alpha\beta} nq^2 \langle \tau_\alpha \rangle / m_\alpha, \quad \mu_{\alpha\beta} = \delta_{\alpha\beta} e \langle \tau_\alpha \rangle / m_\alpha, \tag{6.18}$$

where the average is the same as in eq. (6.6).

For a system of equivalent valleys such as are found in silicon and germanium, the conductivity must be summed, or the mobility averaged, over the valleys, and they are both scalars:

$$\mu = e\langle 2\tau_\mathrm{t}/m_\mathrm{t} + \tau_\ell/m_\ell \rangle /3 = e\langle (\tau_\ell/m_\ell)(2K+1)/3 \rangle, \tag{6.19}$$

where we assume ellipsoids of revolution, and introduce the anisotropy factor $K = m_\ell \tau_\mathrm{t}/m_\mathrm{t}\tau_\ell$. K is often assumed to be independent of energy.

If τ is isotropic (or for high frequency) we can define a mobility effective mass

$$m^{*-1} = (2/m_\mathrm{t} + 1/m_\ell)/3, \tag{6.20}$$

which differs from the density of states effective mass $(m_\mathrm{t}^2 m_\ell)^{1/3}$.

Rode has calculated the electron mobility of pure Ge and Si by using a spherical model for the carriers, and including acoustic phonons as well as intervalley scattering as discussed in § 9.7. The results are shown in fig. 16, and we see that the change in slope with the onset of intervalley scattering is reproduced. The deformation potentials E and D were used as adjustable parameters, and correspond to averages of the anisotropic quantities (Conwell 1967). For Ge, deformation potential optical-phonon scattering is also allowed, but it was not possible to distinguish it from intervalley scattering. For impurity scattering it is not a good approximation to use a spherical average. Dakhovskii (1963) applied the theory of Samoilovich et al. (1961a–c) to fit the results for n-type Si of Long and Myers (1959) including acoustic-phonon and ionized-impurity scattering. The results are shown in fig. 17.

6.4. *Warped hole bands*

In the case of light and heavy holes near the degenerate band edge at the Γ-point, the simplest theory would make the approximation of treating the bands as spherical and using a relaxation time. It has been pointed out (Lax and Mavroides 1955, Wiley 1975) that it is best to take τ as the same for both light and heavy holes, since the scattering is dominated by processes whose final state is in the heavy-hole band, which has the larger density of states.

The first attempt to include the warping of the bands, which is larger for the heavy holes, was by Lax and Mavroides (1955), who assumed a relaxation time and parabolic bands. The energy surfaces, eq. (2.29), can be written as

$$\mathscr{E} = \frac{\hbar^2 k^2}{2m^*} g(\theta, \varphi), \tag{6.21}$$

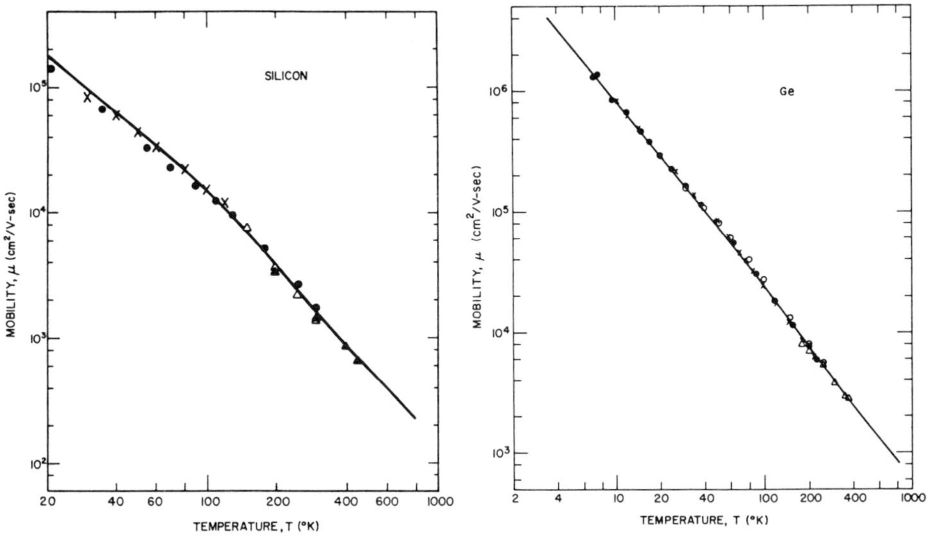

Fig. 16. Theoretical electron drift mobility in (a) pure Si and (b) pure Ge, including acoustic and intervalley scattering, compared with the Hall mobility from various experiments (after Rode 1972).

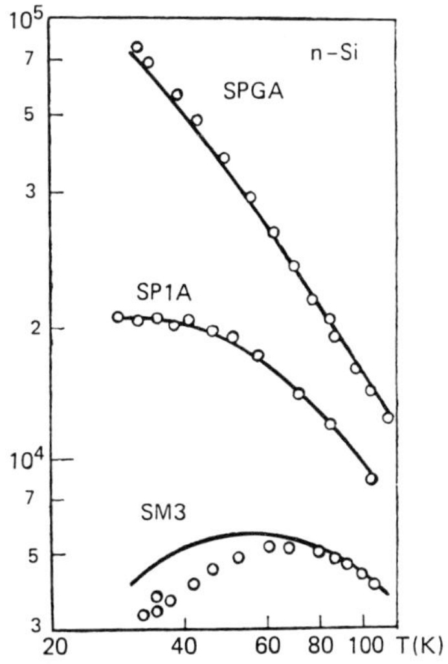

Fig. 17. Hall mobility of electrons in three n-Si samples versus temperature, as measured by Long and Myers (1959). Solid lines are theoretical, calculated for a mixture of ionized impurity and acoustic scattering (after Dakhovskii 1963).

where m^* is a spherical effective mass, $m^* = m_0/(A \pm B')$, where $B'^2 = B^2 + C^2/6$. Lax and Mavroides expanded the anisotropic factor g to first order in the anisotropy

$$g(\theta, \varphi) \cong \{1 - \Gamma'[(k_x^2 k_y^2 + k_y^2 k_z^2 + k_z^2 k_x^2)/k^4 - \tfrac{1}{6}]\}, \tag{6.22}$$

where $\Gamma' = \mp C^2/[2B'(A \pm B')]$. To give an idea of their analysis, let us think of the bands as a function of energy and angle, and use the density of states factors $\Gamma(\mathscr{E}, \hat{\boldsymbol{k}}) = k^2 \partial k/\partial \mathscr{E}$ of eq. (5.61) to write the number of holes p in a particular band as

$$p = \int \exp[(\mathscr{E}_F - \mathscr{E})/kT] \Gamma(\mathscr{E}, \hat{\boldsymbol{k}}) \, d\mathscr{E} \, d\Omega / 4\pi^3. \tag{6.23}$$

However, Γ can be written as a spherical part times as angular factor

$$\Gamma = \Gamma_s(\mathscr{E})/g(\theta, \varphi)^{3/2}, \quad \Gamma_s(\mathscr{E}) = \tfrac{1}{2}(2m^*/\hbar^2)^{3/2} \mathscr{E}^{1/2}, \tag{6.24}$$

then

$$p = \int \exp[(\mathscr{E}_F - \mathscr{E})/kT] \Gamma_s(\mathscr{E}) \frac{d\mathscr{E}}{\pi^2} \int \frac{1}{g(\theta, \varphi)^{3/2}} \frac{d\Omega}{4\pi}. \tag{6.25}$$

The angular integral is unity if the anisotropy is neglected, so all the anisotropy is lumped into the angular integral.

Similarly, the conductivity can be written using eq. (6.2) and invoking cubic symmetry,

$$\sigma = \int \exp[(\mathscr{E}_F - \mathscr{E})/kT] \frac{v_s^2}{3kT} \Gamma_s(\mathscr{E}) \frac{d\mathscr{E}}{\pi^2} \int (v^2/v_s^2) \frac{1}{g(\theta, \varphi)^{3/2}} \frac{d\Omega}{4\pi}, \tag{6.26}$$

where $v_s^2 = 2\mathscr{E}/m^*$. Again all the anisotropy is lumped into the angular integral. Thus the hole number and conductivity for each carrier are given by the spherical parabolic results multiplied by anisotropy factors. Lax and Mavroides evaluated the angular factors as expansions in powers of the anisotropy constant Γ'. The method works because the bands are assumed to be parabolic even though they are warped. The dependence of τ on energy can then be treated just as in spherical parabolic bands. Applications of this theory have been discussed by Stirn (1972) and by Wiley (1975).

However, the relaxation time approximation is not accurate for the valence band because of the anisotropy and interband coupling. We have discussed the more general method proposed by Lawaetz (1968a,b), who applied it to Ge. Extensive work has been done by Szmulowicz (1983, 1986) in applying this approach to p-type Ge and Si (see also Peterson et al. 1990). Szmulowicz's calculation takes into account warping and interband scattering. The 6×6 effective-mass equations are used for the wave functions and energies, which allows for non-parabolic effects and includes the split-off band. This is especially important in the case of Si. The Boltzmann equation is solved by expanding the distribution function in cubic harmonics up to $L = 7$. Results for the conductivity mobility of p-type Si, and the conductivity and Hall mobility of p-type Ge are shown in fig. 18, in the phonon-limited regime. The calculation included acoustic- and optical-phonon scattering, and the optical-phonon deformation constant d_o was used as an adjustable parameter. The work has been

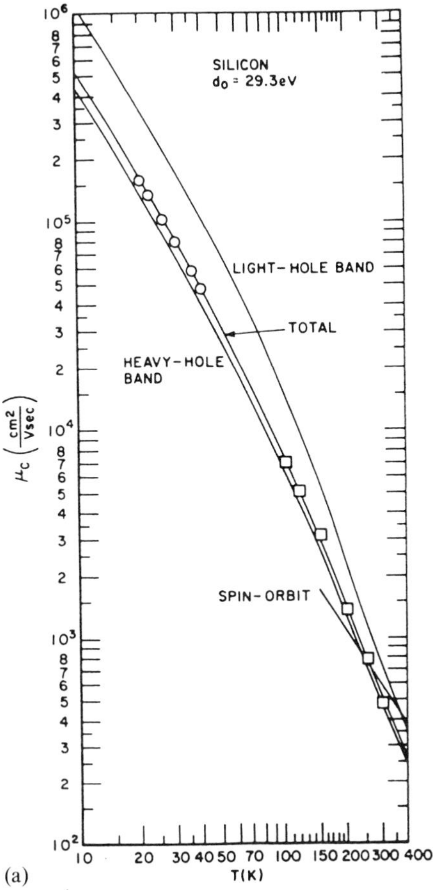

(a)

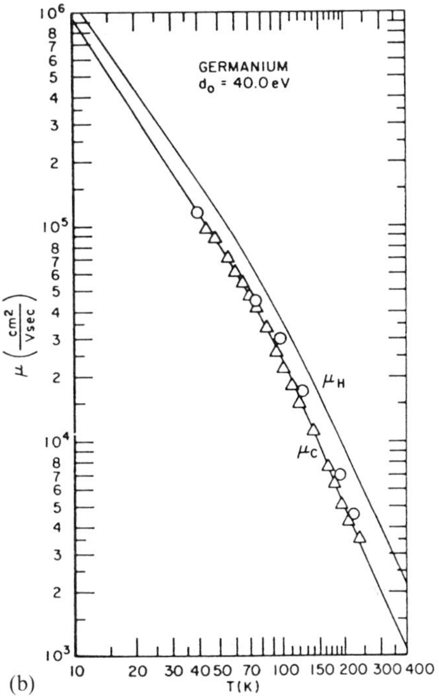

(b)

Fig. 18. (a) Total and partial conductivity mobilities for p-Si, calculated from mixed acoustic- and optical-phonon scattering, compared with experiment (after Szmulowicz 1983). (b) Conductivity and Hall mobilities for p-Ge, calculated from mixed acoustic- and optical-phonon scattering, compared with the experimental drift mobility (after Szmulowicz 1983).

extended to doped samples (Szmulowicz 1986, Peterson et al. 1990), including ionized-impurity scattering based on the Brooks–Herring theory and neutral-impurity scattering based on modification by Ralph (Ralph 1977) of the theory of Ergensoy (1950). Results for a Si sample are shown in fig. 19.

7. Transport in electric and magnetic fields

We now include a magnetic field in the Boltzmann equation, eq. (5.23), and consider solutions primarily for spherical and ellipsoidal bands, with a brief discussion of warped bands.

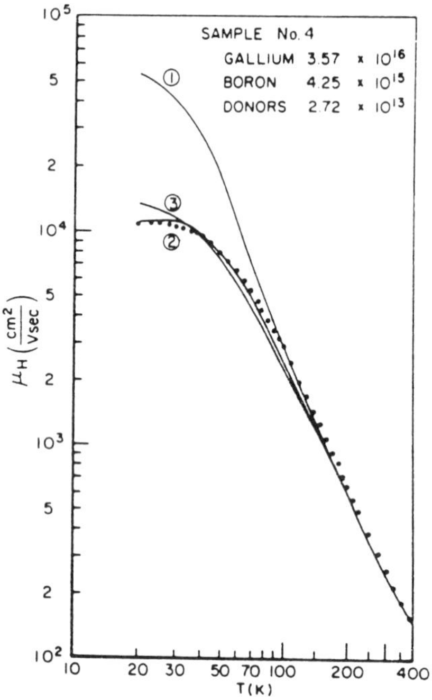

Fig. 19. Theoretical Hall mobility for a Ga-doped Si sample, calculated by including acoustic, optical, and ionized-impurity scattering (curve 1) plus neutral-impurity scattering according to Ralph (curve 2) or Erginsoy (curve 3), compared to experimental points (after Peterson et al. 1990).

It is important to note that within the relaxation time approximation there is an exact solution due to Chambers (1956), which comes from recalling the classical motion around the cyclotron orbit, described in § 2 (eq. 2.44)). The derivative in eq. (5.23) is actually along the cyclotron orbit, so the equation can be written as

$$\frac{\mathrm{d}\varphi}{\mathrm{d}s} + \frac{\varphi}{\tau} = \boldsymbol{v}_{\boldsymbol{k}}(s) \cdot q\boldsymbol{E}, \tag{7.1}$$

where $\mathrm{d}s = \hbar \mathrm{d}k_{\mathrm{p}}/|q\boldsymbol{v} \times \boldsymbol{B}/c|$ and s is the time of the classical motion along the cyclotron orbit (eq. 2.44). This is easily integrated formally

$$\varphi(s) = \int_{-\infty}^{s} \boldsymbol{v}_{\boldsymbol{k}}(s') \cdot q\boldsymbol{E}\mathrm{e}^{(s'-s)/\tau}, \tag{7.2}$$

and the current is obtained from eq. (6.1). This approach was used by Chambers (1956, 1960) to study the magnetoresistance of metals, and especially the effects of open orbits. In the case of semiconductors, McClure (1956) showed that eq. (7.1) can be solved by noticing that $\boldsymbol{v}_{\boldsymbol{k}}(s)$ is a periodic function of s, with the cyclotron period T. He found a solution as a Fourier expansion in s. This approach has been used for warped energy surfaces, but again we must be cautious since the use of the relaxation time approximation is a limitation.

7.1. Spherical bands

For spherical bands, assuming a relaxation time, we have found that the solution of the Boltzmann equation is obtained by solving eq. (5.33) for the drift velocity $\boldsymbol{u}(\mathscr{E})$. Just as in the case of the conductivity the current is found by averaging $\boldsymbol{u}$ over the energy

$$\boldsymbol{j} = nq\langle \boldsymbol{u}(\mathscr{E})\rangle, \tag{7.3}$$

where the average is given in eq. (6.6) for simple bands. However, eq. (5.33) is virtually the same as eq. (4.6) in the particle kinetic model, so we can take over the particle kinetic results and then average the drift velocity over energy. The solution for $\boldsymbol{u}$ is given by eq. (4.7), so the current is given by

$$\boldsymbol{j} = \left\langle \frac{nq^2\tau}{m^*} \frac{1}{1+(\omega_c\tau)^2} [\boldsymbol{E} - \omega_c\tau\hat{\boldsymbol{B}} \times \boldsymbol{E} + (\omega_c\tau)^2 \hat{\boldsymbol{B}}\hat{\boldsymbol{B}} \cdot \boldsymbol{E}] \right\rangle. \tag{7.4}$$

We can express this again in terms of the conductivity tensor, whose form is the same as eq. (4.8), but where the components are now

$$\sigma_{xx} = \left\langle \frac{nq^2\tau}{m^*} \frac{1}{1+(\omega_c\tau)^2} \right\rangle, \tag{7.5}$$

$$\sigma_{xy} = \left\langle \frac{nq^2\tau}{m^*} \frac{\omega_c\tau}{1+(\omega_c\tau)^2} \right\rangle, \tag{7.6}$$

$$\sigma_{zz} = \frac{nq^2\langle\tau\rangle}{m^*} = \sigma_0. \tag{7.7}$$

and where σ_0 is the zero-field conductivity.

To find the magnetoresistance and Hall constant, we need to obtain the resistivity tensor. After averaging the expressions in eqs. (7.5)–(7.7) over energy, and also possibly summing over various groups of carriers, we then invert the conductivity tensor. The resistivity tensor has the same form as eq. (4.11), with the components

$$\rho_{xx} = \frac{\sigma_{xx}}{\sigma_{xx}^2 + \sigma_{xy}^2} = \rho_{yy}, \tag{7.8}$$

$$\rho_{yx} = \frac{\sigma_{xy}}{\sigma_{xx}^2 + \sigma_{xy}^2} = -\rho_{xy}, \tag{7.9}$$

and

$$\rho_{zz} = \rho_0 = \sigma_0^{-1}. \tag{7.10}$$

The low-field Hall constant is obtained by expanding ρ_{yx} to first order in the magnetic field:

$$R = \frac{\rho_{yx}}{B} \cong \frac{\langle (nq^2\tau/m^*)(\omega_c\tau/B)\rangle}{\langle nq^2\tau/m^*\rangle^2} = \frac{1}{nqc} \frac{\langle\tau^2\rangle}{\langle\tau\rangle^2}. \tag{7.11}$$

The Hall constant is thus the classical value $1/nqc$ multiplied by a factor related to the energy dependence of the scattering. This is called the Hall factor r_H:

$$r_H = nqcR = \langle\tau^2\rangle/\langle\tau\rangle^2 \tag{7.12}$$

The Hall factor is important because the Hall mobility

$$\mu_H = |R|\sigma c = r_H\mu, \tag{7.13}$$

is an easier quantity to measure than the conductivity mobility μ, which requires measuring the drift velocity in an electric field. Thus if the Hall factor can be calculated accurately, the conductivity mobility can be obtained from the Hall mobility. If we consider a single scattering mechanism whose energy dependence is a power law ($\tau = \tau_1(\mathscr{E}/kT)^r$), the Hall factor is

$$r_H = (2r+\tfrac{3}{2})!\tfrac{3}{2}!/[(r+\tfrac{3}{2})!]^2. \tag{7.14}$$

Values for several scattering mechanisms are given in table 4. The Hall factor tends to be strongly dependent on temperature, and table 4 suggests that this is due to changes in the important scattering mechanism with temperature.

To obtain the high-field limit of the Hall constant we notice that in the conductivity tensor, $\sigma_{xx} = \sigma \sim 1/B^2$, while $\sigma_{xy} \sim 1/B$. Therefore

$$R = \frac{\rho_{xy}}{B} \cong \frac{1}{\sigma_{xy}B} \cong \left\langle \frac{nq^2\tau_B}{m^*}\frac{1}{\omega_c\tau}\right\rangle^{-1} = \frac{1}{nqc}, \tag{7.15}$$

which is just the simple result, and is independent of the scattering mechanism. We see that the Hall factor varies with magnetic field, and approaches one for high fields.

The Hall factor was calculated for GaAs by Fletcher and Butcher, and the results are shown in fig. 20, for the same sample as in fig. 14.

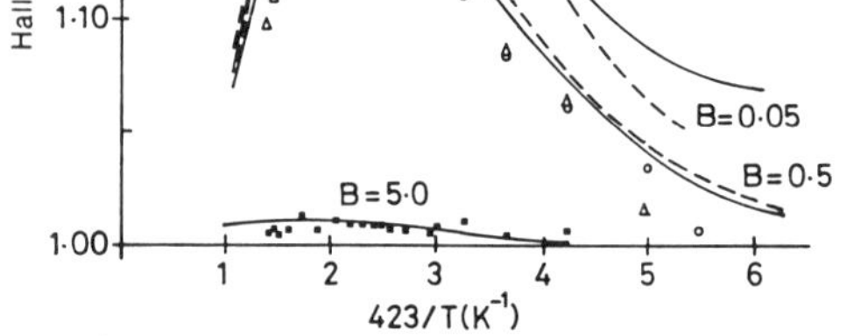

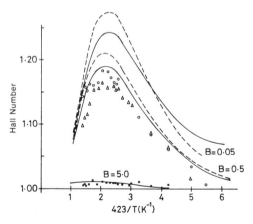

Fig. 20. Temperature dependence of the Hall factor for n-GaAs, calculated for several magnetic field strengths, compared with the experimental results of Stillman et al. (1970). Dashed curve, polar optical scattering, solid curve, all scattering modes (after Fletcher and Butcher 1972).

Considering now the magnetoresistance, we see that for spherical bands the longitudinal magnetoresistance vanishes. Longitudinal magnetoresistance comes into the classical model only when there is anisotropy. However, when quantum effects are important there is a finite longitudinal magnetoresistance for spherical bands (Roth and Argyres 1966). The transverse magnetoresistance is

$$\frac{\Delta\rho}{\rho} = \frac{\rho_{xx} - \rho_0}{\rho_0}. \tag{7.16}$$

For low fields we expand ρ_{xx} to second order in B:

$$\rho_{xx}^{-1} = \sigma_{xx} + \sigma_{xy}^2/\sigma_{xx} \cong \left\langle \frac{nq^2\tau}{m^*}(1 - \omega_c^2\tau^2) \right\rangle + \left\langle \frac{nq^2\tau}{m^*}\omega_c\tau \right\rangle^2 \bigg/ \left\langle \frac{nq^2\tau}{m^*} \right\rangle, \tag{7.17}$$

and therefore

$$\rho_{xx} = \rho_0(1 + \omega_c^2\langle\tau^3\rangle/\langle\tau\rangle - \omega_c^2\langle\tau^2\rangle^2/\langle\tau\rangle^2). \tag{7.18}$$

The transverse magnetoresistance can be written as

$$\frac{\Delta\rho}{\rho} = (\mu_H/c)^2 B^2(A - 1), \tag{7.19}$$

$$A = \langle\tau^3\rangle\langle\tau\rangle/\langle\tau^2\rangle^2 = r_3/r_H^2, \tag{7.20}$$

where $r_3 = \langle\tau^3\rangle/\langle\tau\rangle^3$. For $\tau = \tau_1(\mathscr{E}/kT)^r$, the result is

$$A = (3r + \tfrac{3}{2})!(r + \tfrac{3}{2})!/[(2r + \tfrac{3}{2})!]^2, \tag{7.21}$$

which is evaluated in table 4. For the case of constant τ, the transverse magnetoresistance vanishes for spherical parabolic bands, as we found from the particle kinetic model.

To find the high-field limit of the magnetoresistance we again note that $\sigma_{xx} \ll \sigma_{xy}$, so

$$\rho_{xx} \cong \sigma_{xx}/\sigma_{xy}^2 \cong \left\langle \frac{nq^2\tau}{m^*}\frac{1}{\omega_c^2\tau^2} \right\rangle \bigg/ \left\langle \frac{nq^2\tau}{m^*}\frac{1}{\omega_c\tau} \right\rangle^2 = \langle m^*/nq^2\tau\rangle, \tag{7.22}$$

so that

$$\Delta\rho/\rho_0 = \langle 1/\tau\rangle\langle\tau\rangle - 1. \tag{7.23}$$

The magnetoresistance saturates. The resistance at high fields is proportional to the scattering rate, and so is the conductivity. When quantum effects are important, the saturation does not occur (Roth and Argyres 1966).

For spherical, but non-parabolic, bands the results of this section can be generalized in the same way as the conductivity. The effective mass depends on the energy, so we must average the conductivity tensor over the energy in the sense of eq. (6.13), and we find that we always have the factor τ/m^*.

7.2. Several groups of carriers

If there are several groups of carriers, they conduct in parallel, so we must add the conductivity tensors. Therefore, for low magnetic fields,

$$\sigma_{xx} = \sum_i n_i e \mu_i, \tag{7.24}$$

$$\sigma_{xy} = \sum_i n_i q_i \langle (q\tau_i/m_i^*)^2/c \rangle B = \sum_i n_i q_i r_{\mathrm{H}i} \mu_i^2 B/c. \tag{7.25}$$

Assuming that the scattering mechanism is the same for the various carriers, the Hall constant is given by

$$R = \rho_{yx}/B = \sigma_{xy}/\sigma_{xx}^2 B = \frac{r_{\mathrm{H}}}{c} \frac{\sum_i n_i q_i \mu_i^2}{\left(\sum_i n_i e \mu_i\right)^2}. \tag{7.26}$$

For densities n and p of electrons and holes,

$$R = \frac{r_{\mathrm{H}}}{ec} \frac{-n\mu_{\mathrm{n}}^2 + p\mu_{\mathrm{p}}^2}{(n\mu_{\mathrm{n}} + p\mu_{\mathrm{p}})^2} = \frac{r_{\mathrm{H}}}{ec} \frac{p - nb^2}{(p + nb)^2}, \tag{7.27}$$

where b is the mobility ratio, $\mu_{\mathrm{n}}/\mu_{\mathrm{p}}$. Therefore the Hall constant will change sign as a function of carrier concentration at a point that depends on the mobility ratio (Beer 1963).

For high magnetic fields, the Hall constant for several types of carriers is

$$R \sim 1 \bigg/ \left(\sum_i n_i q_i c\right), \tag{7.28}$$

which gives $R \propto 1/(p - n)$ for the case of electrons and holes.

Considering the low-field magnetoresistance for several carriers, we have

$$\frac{\Delta\rho}{\rho B^2} = \frac{\sum_i n_i \mu_i^3 r_{3i}/c^2}{\sum_i n_i \mu_i} - \left(\frac{\sum_i n_i q_i \mu_i^2 r_{\mathrm{H}i}/c}{e \sum_i n_i \mu_i}\right)^2. \tag{7.29}$$

From this result we can see that the second term will tend to cancel for a combination of electrons and holes, giving a larger effect.

These results can be applied to p-type semiconductors. Here the heavy holes are more numerous ($p \propto m^{*3/2}$), but the light holes are more mobile. The light holes contribute strongly to the low-field Hall effect and the magnetoresistance. In addition, the Hall constant depends strongly on magnetic field, and this behavior can be understood qualitatively from the simple theory, although the details are affected by the anisotropy (Beer 1963).

7.3. Ellipsoidal energy surfaces

Let us look first at the case of a single ellipsoid, for which $\mathscr{E} = (\hbar^2/2)\boldsymbol{k} \cdot \boldsymbol{m}^{*-1} \cdot \boldsymbol{k}$ and $\boldsymbol{v} = \boldsymbol{m}^{*-1} \cdot \hbar \boldsymbol{k}$. The equation for the drift velocity $\boldsymbol{u}(\mathscr{E})$ is obtained from eq. (5.57), including the electric and magnetic fields as well as the tensor mass and the relaxation time:

$$q\boldsymbol{E} + \frac{q}{c}\boldsymbol{u}(\mathscr{E}) \times \boldsymbol{B} = \boldsymbol{A}^{-1} \cdot \boldsymbol{u}(\mathscr{E}), \tag{7.30}$$

where $\boldsymbol{A} = \boldsymbol{\tau}/\boldsymbol{m}^*$ is defined as in eq. (6.15). It is not difficult to show that the solution for $\boldsymbol{u}$ is of the form (Beer 1963)

$$\boldsymbol{u} = \frac{\boldsymbol{A} \cdot q\boldsymbol{E} + (\det \boldsymbol{A})(q^3/c^2)\boldsymbol{B}\boldsymbol{B} \cdot \boldsymbol{E} - (q^2/c)(\boldsymbol{L} \cdot \boldsymbol{B}) \times \boldsymbol{E}}{1 + \boldsymbol{B} \cdot \boldsymbol{L} \cdot \boldsymbol{B}(q/c)^2}, \tag{7.31}$$

where $\boldsymbol{L}$ is the tensor

$$\boldsymbol{L} = \boldsymbol{A}^{-1} \det \boldsymbol{A} = \begin{pmatrix} \tau_y\tau_z/m_y m_z & 0 & 0 \\ 0 & \tau_x\tau_z/m_x m_z & 0 \\ 0 & 0 & \tau_x\tau_y/m_x m_y \end{pmatrix}. \tag{7.32}$$

In the last term $(\boldsymbol{L} \cdot \boldsymbol{B}) \times \boldsymbol{E}$ can also be written as $\boldsymbol{A} \cdot [\boldsymbol{B} \times (\boldsymbol{A} \cdot \boldsymbol{E})]$ or as $\sqrt{\boldsymbol{A}} \cdot [(\sqrt{\boldsymbol{L}} \cdot \boldsymbol{B}) \times \sqrt{\boldsymbol{A}} \cdot \boldsymbol{E})]$, the last term being analogous to the Herring–Vogt transformation.

To obtain the conductivity tensor we must average $\boldsymbol{u}$, which depends on the τ, over energy, in the sense of eq. (6.6):

$$\boldsymbol{j} = \boldsymbol{\sigma} \cdot \boldsymbol{E} = \left\langle \frac{nq^2\boldsymbol{A} + n(q^4/c^2)(\det \boldsymbol{A})\boldsymbol{B}\boldsymbol{B}}{1 + \boldsymbol{B} \cdot \boldsymbol{L} \cdot \boldsymbol{B}q^2/c^2} \right\rangle \cdot \boldsymbol{E} - \left\langle \frac{n(q^3/c)(\boldsymbol{L} \cdot \boldsymbol{B})}{1 + \boldsymbol{B} \cdot \boldsymbol{L} \cdot \boldsymbol{B}q^2/c^2} \right\rangle \times \boldsymbol{E}. \tag{7.33}$$

For a system with a single ellipsoidal energy surface we invert this conductivity tensor to obtain the resistivity tensor, leading to the Hall effect and magnetoresistance. If the components of the relaxation time tensor are independent of energy, the magnetoresistance actually vanishes for a single ellipsoid, just as it does for a sphere.

For cubic semiconductors we are interested in sets of ellipsoids. We must now add the contributions of the various ellipsoids to the conductivity to obtain the total conductivity. For ellipsoids of revolution, we have

$$\boldsymbol{A}_i = (\tau_\ell/m_\ell)[\boldsymbol{1}K - (K-1)\hat{\boldsymbol{n}}_i\hat{\boldsymbol{n}}_i], \tag{7.34}$$

$$\boldsymbol{L}_i = (\tau_\ell/m_\ell)^2[\boldsymbol{1}K + K(K-1)\hat{\boldsymbol{n}}_i\hat{\boldsymbol{n}}_i], \tag{7.35}$$

where $K = \tau_t m_\ell/\tau_\ell m_t$ is the generalization of the mass ratio encountered already in the conductivity. Also $\hat{\boldsymbol{n}}_i$ is a unit vector along the axis of the ith ellipsoid. For the three pairs of ellipsoids in silicon the $\hat{\boldsymbol{n}}_i$ are along the cube axes, and for the four ellipsoids in Ge, the components of $\hat{\boldsymbol{n}}_i$ are $(1/\sqrt{3})(1, 1, 1)$, $(1/\sqrt{3})(1, -1, -1)$,

$(1/\sqrt{3})(-1, 1, -1)$ and $(1/\sqrt{3})(-1, -1, 1)$. Another possible direction is (110) for which the ellipsoids are not necessarily ellipsoids of revolution (Allgaier 1959).

To find the low-field Hall effect and magnetoresistance the conductivity is expanded in powers of the magnetic field. The general expression is

$$\sigma_{\alpha\beta}(\boldsymbol{B}) = \sigma_{\alpha\beta} + \sum_{\gamma} \sigma_{\alpha\beta\gamma} B_{\gamma} + \sum_{\gamma\delta} \sigma_{\alpha\beta\gamma\delta} B_{\gamma} B_{\delta} + \cdots. \tag{7.36}$$

For cubic crystals it is convenient to use the magnetoconductivity coefficients introduced by Seitz (1950)

$$\boldsymbol{j} = \sigma_0 \boldsymbol{E} + \alpha \boldsymbol{E} \times \boldsymbol{B} + \beta B^2 \boldsymbol{E} + \gamma \boldsymbol{BB} \cdot \boldsymbol{E} + \delta T \cdot E, \tag{7.37}$$

$$\boldsymbol{T} = (B_x^2 \hat{\boldsymbol{i}}\hat{\boldsymbol{i}} + B_y^2 \hat{\boldsymbol{j}}\hat{\boldsymbol{j}} + B_z^2 \hat{\boldsymbol{k}}\hat{\boldsymbol{k}}). \tag{7.38}$$

The last term reflects the cubic anisotropy, and δ vanishes for an isotropic system. We are primarily interested in the inverse of this relation, which in the form due to Seitz is (Pearson and Suhl 1951)

$$\boldsymbol{E} = \rho_0(\boldsymbol{j} + a\boldsymbol{j} \times \boldsymbol{B} + bB^2\boldsymbol{j} + c\boldsymbol{BB} \cdot \boldsymbol{j} + d\boldsymbol{T} \cdot \boldsymbol{j}). \tag{7.39}$$

The inverse Seitz coefficients are easily found to be

$$a = -\alpha\rho_0, \quad b = -(\beta + \rho_0\alpha^2)\rho_0, \quad c = (-\gamma + \rho_0\alpha^2)\rho_0, \quad d = -\rho_0\delta. \tag{7.40}$$

To find the Seitz coefficients we expand the current in powers of $\boldsymbol{B}$. The zero-order term is

$$\sigma_0 \boldsymbol{E} = \sum_i n_i q^2 \langle \boldsymbol{A}_i \rangle \cdot \boldsymbol{E} = nq^2 \langle (\tau_\ell/m_\ell)(2K+1)/3 \rangle \boldsymbol{E}, \tag{7.41}$$

which agrees with the previous result for the mobility, eq. (6.19). The first-order term is

$$\alpha \boldsymbol{E} \times \boldsymbol{B} = (q^3/c)\boldsymbol{E} \times \sum_i n_i \langle \boldsymbol{L}_i \cdot \boldsymbol{B} \rangle = (nq^3/c)\langle (\tau_\ell/m_\ell)^2 K(K+2)/3 \rangle \boldsymbol{E} \times \boldsymbol{B}, \tag{7.42}$$

which gives the Hall coefficient

$$R = \rho_0^2 \alpha = \frac{1}{nqc} \frac{\langle \tau_\ell^2 K(K+2)/3 \rangle}{\langle \tau_\ell (2K+1)/3 \rangle^2}. \tag{7.43}$$

This is independent of the orientation of the ellipsoids. If K is a constant, i.e., if τ_ℓ and τ_t both have the same energy dependence, the Hall factor is

$$r_{\mathrm{H}} = \frac{\langle \tau^2 \rangle}{\langle \tau \rangle^2} \frac{3K(K+2)}{(2K+1)^2}. \tag{7.44}$$

This is just the simple band result, depending on the energy dependence of τ, multiplied by an anisotropy factor depending on K.

The second-order term in $\boldsymbol{\sigma} \cdot \boldsymbol{E}$ is

$$\frac{q^4}{c^2} \sum_i n_i [\langle \det \boldsymbol{A}_i \rangle \boldsymbol{BB} \cdot \boldsymbol{E} - \langle (\boldsymbol{B} \cdot \boldsymbol{L}_i \cdot \boldsymbol{B}) \boldsymbol{A}_i \rangle \cdot \boldsymbol{E}]. \tag{7.45}$$

This gives different results for (100) and (111) multivalley models. The results for the inverse Seitz coefficients are given in table 5, for the case in which K is independent of energy. The low-field magnetoresistance is given by

$$\Delta\rho/\rho_0 B^2 = b + c(\boldsymbol{j}\cdot\boldsymbol{B})^2/j^2B^2 + d(j_x^2B_x^2 + j_y^2B_y^2 + j_z^2B_z^2)/j^2B^2. \tag{7.46}$$

By measuring the low-field magnetoresistance in various directions, the inverse Seitz coefficients can be determined. For example:

$$\begin{aligned} &M_{100} = b + c + d, \quad M_{110} = b + c + d/2, \\ &M_{110}^{1\bar{1}0} = b + \tfrac{1}{2}d, \quad M_{100}^{010} = M_{110}^{001} = b. \end{aligned} \tag{7.47}$$

The notation is $M_{\hat{j}}$ for the longitudinal magnetoresistance, and $M_{\hat{j}}^{\hat{B}}$, for the transverse magnetoresistance. The symmetry relations for the two models are shown in table 5 and compared with those for isotropic systems, for which $d = 0$ and $b + c = 0$, so there is no longitudinal magnetoresistance. In the multivalley conduction band, there is a finite longitudinal magnetoresistance, and the magnetoresistance is anisotropic.

At finite magnetic fields the magnetoresistance deviates from quadratic behavior, and for large fields it saturates and the Hall coefficient approaches $1/nqc$. This behavior was studied by Abeles and Meiboom (1954), Shibuya (1954), and Gold and Roth (1957).

To illustrate these results, the magnetoresistance of n-Si is shown in fig. 21 as measured by Pearson and Herring (1954) in samples with three different current directions. The longitudinal magnetoresistance vanishes for the 100 direction and is a maximum in the (111) direction. For Ge the behavior is different, and there is no direction in which the longitudinal magnetoresistance vanishes. Some values of the inverse Seitz coefficients for Ge and Si at 77 K are shown in table 6. Goldberg and Howard (1953) have investigated the symmetry relations for Ge. Stirn (1972) has reviewed how information on the symmetry of the magnetoresistance can be used to identify the conduction band structure of semiconductors.

The magnetoresistance measurements have been found to be a useful way to determine the anisotropy factor K, which depends on the anisotropy of the scattering

Table 5
Low-field magnetoresistance coefficients for multivalley band model.

Coefficient	Isotropic	(100) valleys	(111) valleys
b	$(\mu_{\mathrm{H}}c)^2(A-1)$	$(\mu_{\mathrm{H}}/c)^2[AF(K^2+K+1) \times K/3 - 1]$	$(\mu_{\mathrm{H}}/c)^2[AF(2K+1) \times K(K+2)/9 - 1]$
c	$-(\mu_{\mathrm{H}}/c)^2(A-1)$	$-(\mu_{\mathrm{H}}/c)^2(AFK^2-1)$	$-(\mu_{\mathrm{H}}/c)^2[AF(2K+1) \times K(K+2)/9 - 1]$
d	0	$-(\mu_{\mathrm{H}}/c)^2 AF(K-1)^2K/3$	$(\mu_{\mathrm{H}}/c)^2 AF2(K-1)^2K/9$
Symmetry relations	$d = 0, \quad b = -c$	$d = -(b+c) < 0$	$b = -c, \quad d > 0$
	$A = \langle\tau^3\rangle\langle\tau\rangle/\langle\tau^2\rangle^2,$	$F = 3(2K+1)/[K(K+2)]^2.$	$K = m_t\tau_\ell/m_\ell\tau_t$

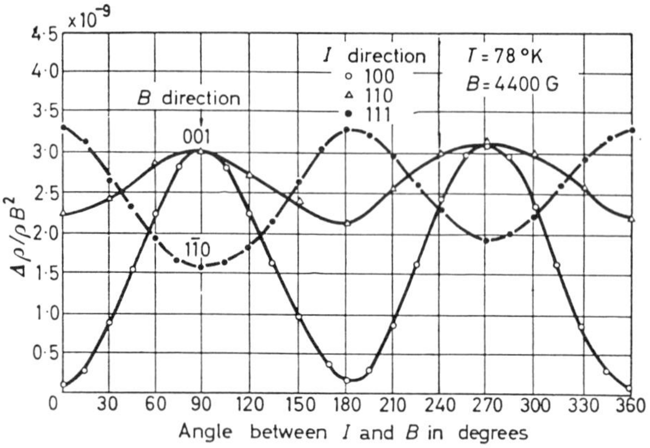

Fig. 21. Anisotropy of magnetoresistance in n-Si at 78 K, for three different current directions, as a function of the angle between the current and the magnetic field (after Pearson and Herring 1951).

Table 6
Inverse Seitz coefficients at 77 K. a in units of T^{-1}, b, c, d in units of T^{-2}.

Coefficient	n-Ge[a]	n-Si[b]	p-Ge[c]	Theory[d]	p-Si[e]	Theory[d]
a	3.7	2.0	1.94		0.95	
b	4.65	1.60	3.22	3.22	1.42	1.42
c	−4.83	0.72	−2.50	−2.54	−1.00	−0.64
d	13.7	−2.27	−1.04	−0.67	0.08	−0.77

[a] $a = \mu_H/c$ from Rode (1972); b, c, d from Goldberg and Howard (1958).
[b] a from Rode (1972); b, c, d from Pearson and Herring (1954).
[c] Pearson and Suhl (1951).
[d] Mavroides and Lax (1956). Constant τ results, scaled to match b.
[d] Long (1957).

as well as the mass. The factor A in table 5 can be eliminated by defining the quantity (Laff and Fan 1958, Long and Myers 1960)

$$W = \frac{d}{b + (\mu_H/c)^2} = \frac{2(K-1)^2}{(2K+1)(K+2)}, \quad (111) \text{ ellipsoids;}$$
$$= -\frac{(K-1)^2}{K^2 + K + 1}, \quad (100) \text{ ellipsoids.} \tag{7.48}$$

Laff and Fan (1958) studied K for Ge as a function of temperature. They found $K = 20$ at room temperature, i.e., close to the mass ratio, which is consistent with $\tau_\parallel/\tau_\perp$ being close to one for lattice scattering. On lowering the temperature, they found that K decreased, which is explained by the increase in ionized impurity scattering, for which $\tau_\parallel/\tau_\perp > 1$. [Eagles and Edwards (1965) find $\tau_\parallel/\tau_\perp \sim 11$.] In fig. 22 is shown the decrease in K with concentration for Ge at 77 K (Zawadzki 1982; see also Seeger 1982) due to increased ionized impurity scattering. At very low

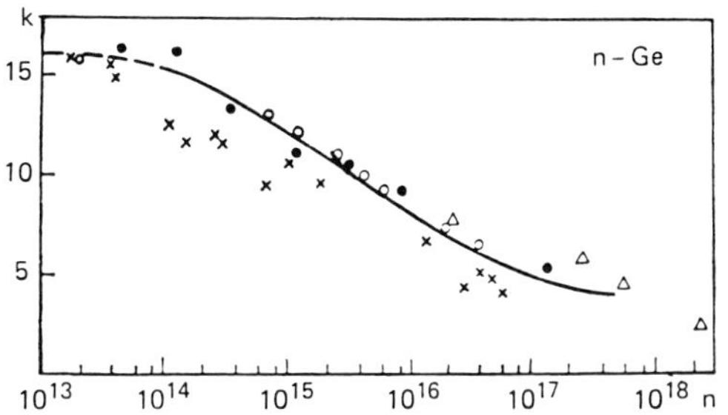

Fig. 22. Dependence of anisotropy factor in n-Ge at 78 K on carrier concentration, calculated from the theory of Samoilovich et al. (1961a–c) and compared with various experimental results after Zawadzki.

temperatures Laff and Fan found that K increased again, due to freeze-out and to neutral-impurity scattering.

Long and Myers (1958) and Neuringer and Little (1962) studied the change of K with temperature and impurity concentration in Si by using high-field as well as low-field magnetoresistance. For Si, acoustic phonon scattering gives $\tau_{\parallel}/\tau_{\perp} \sim 0.67$, while impurity scattering gives 4, and intervalley scattering 1. Dakhovskii (1963) and Dakhovskii and Mikha (1964) have calculated the behavior of K for Ge and Si respectively using the theory of Samoilovich et al. (1961a–c).

7.4. Warped hole bands

The magnetoresistance for holes in p-type semiconductors also exhibits anisotropy, which is primarily due to the warping of the heavy-hole band. Results for the inverse Seitz coefficients for the magnetoresistance of Ge and Si are included in table 6. In the theory of Lax and Mavroides (1955) and Mavroides and Lax (1957) the low-field Hall effect and magnetoresistance were obtained by a direct expansion of the Boltzmann equation, eq. (5.23), in the relaxation time approximation. The Hall factor was found to be multiplied by an anisotropy factor, and that factor as well as the low-field magnetoresistance coefficients were obtained as expansions in the anisotropy parameter Γ' of eq. (6.22). The application of the theory to semiconductors has been reviewed by Stirn (1972) and Seeger (1982), and the inverse Seitz coefficients for 77 K are included in table 6. One prediction is that the longitudinal magnetoresistance is small for $\boldsymbol{B}$ in the (100) direction. The theory worked well for Ge, as seen from the table, but was less good for Si, for which non-parabolic effects are important.

For higher fields, the series solution of the Boltzmann equation fails, and Beer and Willardson (1958) have used the Fourier series expansion method of McClure (1956) based on eq. (7.2) to fit the variation of the Hall factor with the magnetic field in Ge.

In the more recent theory of Lawaetz (1968a), discussed in §§ 5.6 and 6.3 the distribution function is also expanded in powers of B. This has been applied by Lawaetz (1968b) to study the Hall factor and magnetoresistance in Ge, and more recently by Szmulowicz (1983, 1986) and Peterson et al. (1990) to study the Hall

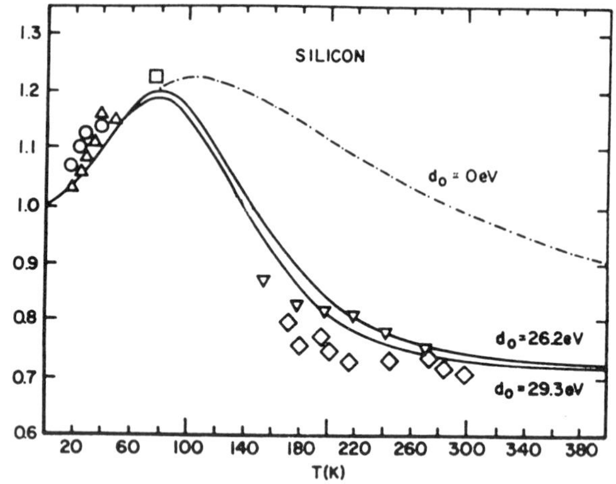

Fig. 23. Temperature dependence of the Hall factor for p-type Si, calculated for mixed acoustic- and optical-phonon scattering, for several choices of the optical deformation potential parameter d_0, compared with various experiments (after Szmulowicz 1983).

factor as a function of temperature. The Hall factor is very sensitive to the scattering mechanism, anisotropy, and non-parabolicity. Szmulowicz's results for Si are shown in fig. 23. The agreement for Ge was not quite as good (Szmulowicz 1983), but was an improvement over previous work.

8. Thermal gradients and fields

We now consider briefly the case in which there are thermal gradients and heat currents in addition to electric and magnetic fields. The driving term in the linearized Boltzmann equation, eqs. (5.21) or (5.23), is

$$qE_1 = qE^* - \frac{\mathscr{E} - \mathscr{E}_F}{kT} \nabla_r T. \tag{8.1}$$

where $E^* = E - \nabla_r \mathscr{E}_F / q$ is the negative gradient of the electrochemical potential. After the Boltzmann equation is solved, the electric current and the electronic contribution to the thermal current are given from eqs. (5.11) and (5.12) by

$$j = \int qv \left(-\frac{\partial f_0}{\partial \mathscr{E}} \right) \varphi(k)\, dk/4\pi^3, \tag{8.2}$$

$$w^* = \int (\mathscr{E} - \mathscr{E}_F) v \left(-\frac{\partial f_0}{\partial \mathscr{E}} \right) \varphi(k)\, dk/4\pi^3. \tag{8.3}$$

These are related to the electric field and thermal gradient by a set of transport coefficients

$$j_\alpha = \sum_\beta (\sigma_{\alpha\beta} E^*_\beta + L_{\alpha\beta} \nabla_\beta T), \tag{8.4}$$

$$w^*_\alpha = \sum_\beta (M_{\alpha\beta} E^*_\beta + N_{\alpha\beta} \nabla_\beta T). \tag{8.5}$$

The transport coefficients are in general tensors.

As in the case of the ordinary conductivity, in the usual experimental arrangement the direction of the current is given, so the transport equations are more conveniently rewritten in the form

$$E^*_\alpha = \sum_\beta (\rho_{\alpha\beta} j_\beta + \alpha_{\alpha\beta} \nabla_\beta T), \tag{8.6}$$

$$w^*_\alpha = \sum_\beta (\pi_{\alpha\beta} j_\beta - \kappa_{\alpha\beta} \nabla_\beta T), \tag{8.7}$$

where $\boldsymbol{\rho}$ is the resistivity, $\boldsymbol{\alpha}$ the thermoelectric power, $\boldsymbol{\pi}$ the Peltier coefficient and $\boldsymbol{\kappa}$ the thermal conductivity, all of which are tensors in general. They are related to the transport coefficients in eqs. (8.4) and (8.5) by

$$\begin{aligned} \boldsymbol{\rho} &= \boldsymbol{\sigma}^{-1}, \quad \boldsymbol{\alpha} = -\boldsymbol{\sigma}^{-1}\boldsymbol{L}, \\ \boldsymbol{\pi} &= \boldsymbol{M}\boldsymbol{\sigma}^{-1}, \quad \boldsymbol{\kappa} = \boldsymbol{M}\boldsymbol{\sigma}^{-1}\boldsymbol{L} - \boldsymbol{N}. \end{aligned} \tag{8.8}$$

The Onsager relations (Ziman 1960) take the form

$$\alpha_{\alpha\beta}(\boldsymbol{B}) = \pi_{\beta\alpha}(-\boldsymbol{B})/T, \quad \rho_{\alpha\beta}(\boldsymbol{B}) = \rho_{\beta\alpha}(-\boldsymbol{B}), \quad \kappa_{\alpha\beta}(\boldsymbol{B}) = \kappa_{\beta\alpha}(-\boldsymbol{B}) \tag{8.9}$$

The first of these was first obtained by Kelvin for the isotropic case and zero magnetic field, and is simply $\alpha = \pi/T$.

8.1. Zero magnetic field

We consider first conduction in the presence of electric fields and thermal gradients for isotropic energy bands. The theory for the conductivity in § 6 is easily generalized by replacing $\boldsymbol{E}$ by $\boldsymbol{E}_1$ given by eq. (5.17), in the Boltzmann equation, eq. (5.23). The transport coefficients are all scalars. In the case that a relaxation time holds, the transport coefficients are

$$\sigma = nq^2 \langle \tau/m^* \rangle, \tag{8.10}$$

$$L = -nq \langle (\tau/m^*)(\mathscr{E} - \mathscr{E}_\mathrm{F}) \rangle / T, \tag{8.11}$$

$$M = nq \langle (\tau/m^*)(\mathscr{E} - \mathscr{E}_\mathrm{F}) \rangle, \tag{8.12}$$

$$N = -n \langle (\tau/m^*)(\mathscr{E} - \mathscr{E}_\mathrm{F})^2 \rangle / T. \tag{8.13}$$

The average is given by eq. (6.13), or eq. (6.6) in the case of simple bands. We see that the Kelvin relation is satisfied, since $L = -M/T$. In the case of holes, we are measuring the energy and Fermi energy from the valence-band edge and they are reversed in sign from the band electron description. It is often more convenient to use the electron Fermi energy $\mathscr{E}_\mathrm{Fe} = -\mathscr{E}_\mathrm{g} - \mathscr{E}_\mathrm{Fh}$, measured from the conduction-band edge. The Fermi energy is of course the same for all carriers. For classical statistics both the hole and electron Fermi levels are negative.

The thermoelectric power α is given by

$$\alpha = -L/\sigma = \frac{\langle(\tau/m^*)(\mathscr{E} - \mathscr{E}_{\mathrm{F}})\rangle}{q\langle(\tau/m^*)\rangle T} = (k/q)(\delta - \mathscr{E}_{\mathrm{F}}/kT), \tag{8.14}$$

in terms of the dimensionless parameter $\delta = \langle(\mathscr{E}/kT)\tau/m^*\rangle / \langle\tau/m^*\rangle$. For a single carrier this is generally of the same sign as the carrier charge. For parabolic bands, classical statistics, and τ given by the power law of eq. (6.9), we have $\delta = r + \frac{5}{2}$. The result also holds for ellipsoidal energy surfaces.

If there are several carriers, the thermoelectric power is obtained by adding the L coefficients, and the result is

$$\alpha = \frac{\sum_i n_i \mu_i \alpha_i}{\sum_i n_i \mu_i}. \tag{8.15}$$

For electrons and holes this becomes (Ure 1972)

$$\alpha = \frac{k}{e} \frac{p[\delta_{\mathrm{p}} + (\mathscr{E}_{\mathrm{g}} + \mathscr{E}_{\mathrm{Fe}})/kT] - nb[\delta_{\mathrm{n}} - \mathscr{E}_{\mathrm{Fe}}]}{nb + p}, \tag{8.16}$$

so that α will change sign as a function of relative concentration.

In the case in which there is no relaxation time the methods applied to the conductivity hold equally well for the thermoelectric effects.

An additional effect which can be important in calculating the thermoelectric power is the phonon drag effect. We have implicitly assumed thus far that the phonons are in thermal equilibrium. However, when there is a temperature gradient, the phonon distribution is disturbed, and because of the electron–phonon interaction, the phonons disturb the electron distribution to produce an additional current and so an additional contribution to the thermoelectric power. This is the phonon drag effect, which was investigated by Herring (1954). Conversely, when the electrons are flowing the phonons are disturbed through the electron–phonon interaction, and this produces an additional contribution to the Peltier coefficient. The two are related by the Kelvin relation. Herring (1954) found for the phonon drag thermoelectric power

$$\alpha_{\mathrm{p}} = \frac{k}{e} f \frac{m v_{\mathrm{s}}^2 \tau_{\mathrm{p}}}{kT\langle\tau\rangle}, \tag{8.17}$$

where v_{s} is an average phonon velocity, τ_{p} an average phonon relaxation time, and f is the fraction of the electron collisions that are due to phonons. This can be especially important in high magnetic fields (Puri and Geballe 1966). Ure (1972) has reviewed thermoelectric effects in III–V compounds.

The thermal conductivity for spherical bands is

$$\kappa = LM/\sigma - N = -\frac{n\langle(\tau/m^*)(\mathscr{E} - \mathscr{E}_{\mathrm{F}})\rangle^2}{\langle(\tau/m^*)\rangle T} + \frac{n\langle(\tau/m^*)(\mathscr{E} - \mathscr{E}_{\mathrm{F}})^2\rangle}{T}. \tag{8.18}$$

We can again evaluate this for the parabolic classical case and a power law dependence of τ on energy, to give

$$\kappa = n(\tau_1/m^*)k^2 T(r+\tfrac{5}{2})!/\tfrac{3}{2}!, \tag{8.19}$$

and comparing this with the conductivity we see that the Wiederman–Franz relation is obeyed, $\kappa = L'\sigma T$, with the Lorentz number L' given by

$$L' = (r+\tfrac{5}{2})(k/q)^2. \tag{8.20}$$

This is not much different from the value of $L' = (\pi^2/3)(k/e)^2$ for degenerate statistics, which is characteristic of metals (Ziman 1960).

To obtain the total heat conductivity we must add the contribution from the phonons, which for most semiconductors is much larger than the electronic contribution.

For electrons and holes the thermal conductivity actually has a term proportional to the square of the energy gap (the difference between the electron and hole Fermi levels), given by the Lorentz number

$$L' = (k/e)^2\{r+\tfrac{5}{2}+[2(r+\tfrac{5}{2})+\mathscr{E}_g/kT]^2\sigma_n\sigma_p/\sigma^2\}. \tag{8.21}$$

This comes from pairs of electrons and holes produced at one end of the sample that go to the other end and recombine. This is still apparently small (Seeger 1982).

8.2. Galvanomagnetic and thermomagnetic effects

When there is a magnetic field, the thermoelectric power, Peltier coefficient and thermal conductivity become tensors, as well as the resistivity. For isotropic systems, all of these tensors have the same form as the resistivity tensor given by eq. (4.11). There are twelve components, which are reduced to nine independent ones by the Kelvin relations.

The effects described by these tensors are called galvanomagnetic and thermomagnetic effects. Galvanomagnetic effects occur when a conductor carrying an electric current is placed in a magnetic field, while thermomagnetic effects occur when there is no electric current, only a thermal current. A number of these effects have been described (Harman and Honig 1967, Conwell 1982, Böer 1990). For semiconductors, because the thermal conductivity is dominated by the phonons, not all of the possible effects are important. We have already discussed the isothermal galvanomagnetic effects, i.e., the Hall effect and the magnetoresistance. The most important thermomagnetic effects arise from the change of the thermoelectric tensor $\alpha_{\alpha\beta}$ in a magnetic field. Changes in the transverse and longitudinal thermoelectric power α_{xx} and α_{zz} are analogous to the magnetoresistance. A finite α_{xy} gives rise to an electric field transverse to the thermal gradient. This is called the (transverse) Nernst effect, or sometimes the Ettingshausen–Nernst effect. Some of these effects are reviewed by Zawadzki (1982) and Seeger (1982).

Other galvanomagnetic effects are the Ettingshausen effect, which is a temperature gradient that appears in a direction perpendicular to a current and a transversely applied magnetic field, and the (longitudinal) Nernst effect, which is the observation of a temperature gradient parallel to the current flow when there is a transverse magnetic field. In addition there are changes in the galvanomagnetic effects when the transverse boundary condition is adiabatic, which come about because the thermoelectric tensor has off-diagonal matrix elements.

Other thermomagnetic effects are the dependence of the thermal conductivity on the magnetic field, and the Righi–Leduc effect, which is the thermal analogue of the Hall effect: a temperature gradient that appears in a direction perpendicular to a thermal current and a transversely applied magnetic field. Again there are differences depending on whether the transverse boundary condition is adiabatic or isothermal.

9. Scattering mechanisms

We review briefly in this section the principle scattering mechanisms that have been found to be important for transport in semiconductors. For intrinsic semiconductors the interaction with acoustic and optical phonons dominates the scattering. At lower temperatures scattering by impurities, especially ionized impurities, becomes important, depending on the doping of the material. A very complete review of these effects is given by Gantmakher and Levinson (1987), and Zawadzki (1982) also gives many details. See also Butcher (1986).

It is simplest conceptually to begin with impurity scattering, which is elastic. If we use the Born approximation, according to the Fermi golden rule the transition rate from the Bloch state $n\boldsymbol{k}$ to the state $n'\boldsymbol{k}'$ due to a perturbation U is

$$T(n\boldsymbol{k} \to n'\boldsymbol{k}') = \frac{2\pi}{\hbar} |U_{n'\boldsymbol{k}'n\boldsymbol{k}}|^2 \delta(\mathscr{E}_{n'}(\boldsymbol{k}') - \mathscr{E}_n(\boldsymbol{k})). \tag{9.1}$$

To obtain the function W in eq. (5.9), the collision term in the Boltzmann equation, we must multiply T by the phase space density $(V/8\pi^3)$, where V is the volume of the crystal. We suppose that the electron or hole is scattered by a potential $U(r)$ that is slowly varying in space. The matrix element between Bloch functions is

$$U_{n'\boldsymbol{k}'n\boldsymbol{k}} = \int b^*_{n'\boldsymbol{k}'}(\boldsymbol{r})U(\boldsymbol{r})b_{n\boldsymbol{k}}(\boldsymbol{r})\,\mathrm{d}\boldsymbol{r} = \int \mathrm{e}^{\mathrm{i}(\boldsymbol{k}-\boldsymbol{k}')\cdot\boldsymbol{r}}\,U(\boldsymbol{r})u^*_{n'\boldsymbol{k}'}(\boldsymbol{r})u_{n\boldsymbol{k}}(\boldsymbol{r})\,\mathrm{d}\boldsymbol{r}, \tag{9.2}$$

where u is the periodic function defined by eq. (2.1). We assume that b is normalized over the whole crystal.

If U varies slowly, and the $u_{n\boldsymbol{k}}$ vary rapidly over a unit cell (see § 3.4), it is a good approximation to average the product of the $u_{n\boldsymbol{k}}$ over a unit cell, and since the $u_{n\boldsymbol{k}}$ are periodic, this is the same as averaging over the entire crystal of volume V,

$$\int u^*_{n'\boldsymbol{k}'}(\boldsymbol{r})u_{n\boldsymbol{k}}(\boldsymbol{r})\,\mathrm{d}\boldsymbol{r}/V = I_{nn'}(\boldsymbol{k},\boldsymbol{k}')/V. \tag{9.3}$$

Here I is called the overlap integral, and $I \cong 1$ for states close to a non-degenerate band edge where $n\boldsymbol{k} \cong n'\boldsymbol{k}'$. The matrix element then becomes

$$U_{n'\boldsymbol{k}'n\boldsymbol{k}} = \int \mathrm{e}^{\mathrm{i}(\boldsymbol{k}-\boldsymbol{k}')\cdot\boldsymbol{r}}\,U(\boldsymbol{r})\,\mathrm{d}\boldsymbol{r}I_{nn'}(\boldsymbol{k},\boldsymbol{k}')/V = U_{\boldsymbol{k}-\boldsymbol{k}'}I_{nn'}(\boldsymbol{k},\boldsymbol{k}')/V, \tag{9.4}$$

in terms of the Fourier transform $U_{k-k'}$ of the scattering potential. Suppose that there are $N_i V$ impurities distributed randomly over the volume V, so they scatter incoherently. The transition probability is then

$$W(nk \to n'k') = \frac{1}{8\pi^3} N_i \frac{2\pi}{\hbar} |U_{k-k'}|^2 G_{nn'}(k, k') \delta(\mathscr{E}_{n'}(k') - \mathscr{E}_n(k)), \tag{9.5}$$

where $G = |I^2|$. The scattering does not change the spin, but in the bands near the zone center in which the spin–orbit interaction is important, the spin states are mixed, so it is necessary (see Nag 1980) to sum over the final Kramers states σ' and average over the initial ones σ:

$$G_{nn'}(k, k') = \tfrac{1}{2} \sum_{\sigma\sigma'} |I_{n\sigma n'\sigma'}(k, k')|^2. \tag{9.6}$$

The overlap factor is important in the case of the coupled hole bands, where it gives an extra angular dependence to the scattering (Wiley 1975). For states close to the band edge, this is

$$\begin{aligned} G_{11} &= G_{22} = (1 + 3\cos^2\theta)/4, \\ G_{12} &= 3\sin^2\theta/4, \end{aligned} \tag{9.7}$$

where 1 and 2 are the heavy- and light-hole bands, and θ is the angle between $\boldsymbol{k}$ and $\boldsymbol{k}'$. G is also important for non-parabolic bands such as the conduction bands of III–IV compounds. For simple bands and $n = n'$, we have $G = 1$.

9.1. *Ionized-impurity scattering*

For ionized-impurity scattering, if we assume that electrons are scattered from impurities of charge e, the basic potential is $-e^2/\kappa_0 r$. (We revert to using e for the charge here as we shall later use q for a wave vector.) The Coulomb potential is screened by the low-frequency dielectric constant κ_0 of the crystal. In Dingle's theory (Dingle 1955), there is also screening by the conduction electrons. According to Poisson's equation for the potential $\varphi = -U/e$ away from the center

$$\nabla^2 \varphi = -4\pi\rho/\kappa_0 = 4\pi e \delta n/\kappa_0 \tag{9.8}$$

where δn is the change in electron density. δn is obtained by noticing that in the vicinity of the impurity the effective Fermi level becomes $\mathscr{E}_F + e\varphi$, so that to first order in φ, $\delta n = e\varphi \, dn/d\mathscr{E}_F$ and

$$\nabla^2 \varphi = \varphi/\lambda^2, \tag{9.9}$$

where

$$1/\lambda^2 = (4\pi e^2/\kappa_0) \, dn/d\mathscr{E}_F. \tag{9.10}$$

If we assume a single type of carrier, such as electrons, and classical statistics, the classical Debye result for the screening constant is obtained, from eqs. (5.2) and (5.3)

$$1/\lambda^2 = 4\pi e^2 n/\kappa_0 kT. \tag{9.11}$$

More generally

$$\partial n/\partial \mathscr{E}_{\mathrm{F}} = \int \frac{f_0(1-f_0)}{kT} \frac{\mathrm{d}\boldsymbol{k}^3}{4\pi}. \tag{9.12}$$

In the Brooks–Herring theory (Brooks 1955, Herring 1954), account is also taken of the screening due to impurities and holes (see also Falikov and Cuevas (1967) and Nag (1980)). For the compensated case, n is replaced by

$$n' = n + (n + n_{\mathrm{A}})[1 - (n + N_{\mathrm{A}})/N_{\mathrm{D}}]. \tag{9.13}$$

The solution of eq. (9.9) is the well known screened Coulomb potential

$$U(\boldsymbol{r}) = -e^2 \mathrm{e}^{-r/\lambda}/\kappa_0 r, \tag{9.14}$$

and the matrix element is

$$U_{\boldsymbol{k}-\boldsymbol{k}'} = -\frac{4\pi e^2/\kappa_0}{|\boldsymbol{k}-\boldsymbol{k}'|^2 + 1/\lambda^2}. \tag{9.15}$$

The Brooks–Herring transition rate for ionized-impurity scattering for a simple band is

$$W(\boldsymbol{k}\to\boldsymbol{k}') = \frac{1}{4\pi^2\hbar} N_{\mathrm{i}} \left| \frac{4\pi e^2/\kappa_0}{|\boldsymbol{k}-\boldsymbol{k}'|^2 + 1/\lambda^2} \right|^2 \delta(\mathscr{E}_{\boldsymbol{k}'} - \mathscr{E}_{\boldsymbol{k}}), \tag{9.16}$$

while for multiple and non-parabolic bands the overlap function must be included, so the transition rate is given by eqs. (9.5) and (9.15).

In the earlier approach of Conwell and Weisskopf (1950) a cut-off at small scattering angles was introduced, instead of screening. The results for the mobility are very similar (Debye and Conwell 1954).

9.2. Neutral-impurity scattering

At low temperatures neutral impurities may be important because the phonon scattering is weak, and because freeze-out diminishes the amount of ionized-impurity scattering and builds up the concentration of neutral impurities. In this case Erginsoy (1950) adapted the theory of electron scattering by atomic hydrogen due to Massey and Moiseiwitsch (1950), which included polarization and exchange effects, to semiconductors by introducing an effective mass and dielectric constant. The scattering is described by partial waves, and for low energies only the $l = 0$ phase shift is needed. Erginsoy found that the total cross section was well approximated by

$$Q_0 = 20a_0^*/k, \tag{9.17}$$

where $a_0^* = \kappa_0\hbar^2/m^*e^2$ is the effective Bohr radius. This calculation was for simple bands, and the result for the relaxation time is

$$1/\tau = (N_{\mathrm{n}}/V)20a_0^*/m^*. \tag{9.18}$$

This is independent of energy. Ralph (1977) has shown how to modify this theory so that it can be applied to the degenerate valence bands. Zawadzki (1982) has discussed deviations from the Erginsoy formula at low temperatures.

9.3. Deformation potential acoustic-phonon scattering

For the long-wavelength acoustic phonons involved in scattering carriers near a band edge, Bardeen and Shockley (1950) proposed the deformation potential interaction. If a strain is applied to a crystal the energy band edges are shifted. For a simple band the shift depends on the dilation, the fractional volume change. If there is a displacement $u(r)$ at each point, the dilation is $\nabla \cdot \boldsymbol{u}$. The interaction is then

$$U(r) = E_1 \nabla \cdot \boldsymbol{u}(\boldsymbol{r}), \tag{9.19}$$

where E_1 is the deformation potential constant. For a lattice wave the displacement depends on the position, and the local shift in the band edge is taken as a slowly varying potential.

In the more general approach of Picus and Bir (1959) the perturbation due to lattice strain has a rapidly varying component given by

$$U(\boldsymbol{r}) = \sum_{\alpha\beta} D_{\alpha\beta}\varepsilon_{\alpha\beta}, \tag{9.20}$$

where

$$\varepsilon_{\alpha\beta} = \tfrac{1}{2}(\partial u_\alpha/\partial x_\beta + \partial u_\beta/\partial x_\alpha) \tag{9.21}$$

is the strain tensor. $D_{\alpha\beta}$ is a deformation potential operator, which can have interband and intraband matrix elements. In the matrix element of U corresponding to eq. (9.2), the rapidly varying part D must be placed between the $u_{n\boldsymbol{k}}$, while the strain components are slowly varying and go into the Fourier transform factor.

For a simple band the band edge matrix element is simply the deformation potential constant, $\langle 0|D_{\alpha\beta}|0\rangle = \delta_{\alpha\beta}E_1$, so for states close to the band edge the effective potential is given by eq. (9.19). For more complex bands the band edge matrix elements are also related to band shifts and splittings under static strain. For an ellipsoid there are two deformation potential constants. In the principal-axis system, the effective potential is

$$U = E_1\Delta + E_2(\hat{\boldsymbol{n}}\cdot\boldsymbol{\varepsilon}\cdot\hat{\boldsymbol{n}} - \Delta/3), \tag{9.22}$$

where $\hat{\boldsymbol{n}}$ is a unit vector along the major axis. A uniaxial strain will shift the various valleys differentially.

For the case of the degenerate hole band edge, there are three deformation potential constants, which have been labeled by Picus and Bir (1959) as a, b, and d. We can write the 4×4 matrix in terms of the $j=\frac{3}{2}$ angular momentum matrices discussed in § 3 for the Luttinger Hamiltonian (Gantmakher and Levinson 1987)

$$U = (a+5b/4)\Delta - b(J_x^2\varepsilon_{xx} + J_y^2\varepsilon_{yy} + J_z^2\varepsilon_{zz}) - (d/2\sqrt{3})[(J_xJ_y + J_yJ_x)\varepsilon_{xy} + \text{c.p.}], \tag{9.23}$$

where c.p. stands for cyclic permutations. The parameter a gives the displacement of the band edge under dilation, while b and d give splittings of the fourfold band edge under uniaxial strain in the (110) and (111) directions, respectively. The parameters for the complete set of eight bands near the zone center are given by Zawadzki (1982).

Let us first work out the transition probability for the case of a simple band and the Bardeen–Shockley deformation potential of eq. (9.19). The expression for the displacement $\boldsymbol{u}$ for acoustic waves in a crystal is (Callaway 1974)

$$\boldsymbol{u}(\boldsymbol{r}) = \sum_{\lambda \boldsymbol{q}, \mathrm{ac}} \sqrt{\frac{\hbar}{2MN\omega_{\boldsymbol{q}\lambda}}} (\boldsymbol{e}_{\boldsymbol{q}\lambda} \mathrm{e}^{\mathrm{i}\boldsymbol{q}\cdot\boldsymbol{r}} a_{\boldsymbol{q}\lambda} + \boldsymbol{e}^*_{\boldsymbol{q}\lambda} \mathrm{e}^{-\mathrm{i}\boldsymbol{q}\cdot\boldsymbol{r}} a^\dagger_{\boldsymbol{q}\lambda}), \tag{9.24}$$

where M is the total mass in a unit cell, N is the number of unit cells, and $\boldsymbol{e}$ is a polarization vector for the acoustic mode λ. We are using a continuum approximation, which is valid for long wavelengths. Here $a^\dagger$ and a are creation and annihilation operators for the phonons, which operate on the phonon states according to

$$a_{\boldsymbol{q}\lambda}|N_{\boldsymbol{q}\lambda}\rangle = \sqrt{N_{\boldsymbol{q}\lambda}}|N_{\boldsymbol{q}\lambda} - 1\rangle, \quad a^\dagger_{\boldsymbol{q}\lambda}|N_{\boldsymbol{q}\lambda}\rangle = \sqrt{N_{\boldsymbol{q}\lambda} + 1}|N_{\boldsymbol{q}\lambda} + 1\rangle. \tag{9.25}$$

The scattering potential is now easily written down:

$$U(\boldsymbol{r}) = \mathrm{i}E_1 \sum_{\boldsymbol{q}, \lambda(\mathrm{ac})} \sqrt{\frac{\hbar}{2MN\omega_{\boldsymbol{q}\lambda}}} (\boldsymbol{q}\cdot\boldsymbol{e}_{\boldsymbol{q}\lambda} \mathrm{e}^{\mathrm{i}\boldsymbol{q}\cdot\boldsymbol{r}} a_{\boldsymbol{q}\lambda} - \boldsymbol{q}\cdot\boldsymbol{e}^*_{\boldsymbol{q}\lambda} \mathrm{e}^{-\mathrm{i}\boldsymbol{q}\cdot\boldsymbol{r}} a^\dagger_{\boldsymbol{q}\lambda}). \tag{9.26}$$

This expression is of the general form

$$U(\boldsymbol{r}) = \sum_{\boldsymbol{q}, \lambda} (U^+_{\boldsymbol{q}\lambda} \mathrm{e}^{\mathrm{i}\boldsymbol{q}\cdot\boldsymbol{r}} a_{\boldsymbol{q}\lambda} + U^-_{\boldsymbol{q}\lambda} \mathrm{e}^{-\mathrm{i}\boldsymbol{q}\cdot\boldsymbol{r}} a^\dagger_{\boldsymbol{q}\lambda}). \tag{9.27}$$

The matrix element for scattering should be taken between the initial and final states that include an occupation number description of the phonon states:

$$\langle\{N'\}\boldsymbol{k}|U|\{N\}\boldsymbol{k}\rangle = \sum_{\boldsymbol{q}, \lambda} (U^+_{\boldsymbol{q}\lambda} \delta_{\boldsymbol{k}', \boldsymbol{k}+\boldsymbol{q}} \sqrt{N_{\boldsymbol{q}\lambda}} \delta_{N'_{\boldsymbol{q}\lambda} N_{\boldsymbol{q}\lambda} - 1} + U^-_{\boldsymbol{q}\lambda} \delta_{\boldsymbol{k}', \boldsymbol{k}-\boldsymbol{q}} \sqrt{N_{\boldsymbol{q}\lambda} + 1} \delta_{N'_{\boldsymbol{q}\lambda} N_{\boldsymbol{q}\lambda} + 1}). \tag{9.28}$$

The first term corresponds to absorbing a phonon of wave vector $\boldsymbol{q}$. There is a momentum selection rule $\boldsymbol{k}' = \boldsymbol{k} + \boldsymbol{q}$, and there will be an energy selection rule $\mathscr{E}(\boldsymbol{k}') = \mathscr{E}(\boldsymbol{k}) + \hbar\omega_{\boldsymbol{q}\lambda}$. Similarly, the second term corresponds to the emission of a phonon of wave vector $\boldsymbol{q}$, with $\boldsymbol{k}' = \boldsymbol{k} - \boldsymbol{q}$ and $\mathscr{E}(\boldsymbol{k}') = \mathscr{E}(\boldsymbol{k}) - \hbar\omega_{\boldsymbol{q}\lambda}$.

The transition probability is found from eq. (9.1) by summing over final phonon states and averaging over initial phonon states:

$$T(\boldsymbol{k} \to \boldsymbol{k}') = \sum_{\lambda} \frac{2\pi}{\hbar} (|U^+_{\boldsymbol{q}\lambda}|^2 N_{\boldsymbol{q}\lambda} \delta(\mathscr{E}_{\boldsymbol{k}'} - \mathscr{E}_{\boldsymbol{k}} - \hbar\omega_{\boldsymbol{q}\lambda})|_{\boldsymbol{q}=\boldsymbol{k}'-\boldsymbol{k}} + |U^-_{\boldsymbol{q}\lambda}|^2 (N_{\boldsymbol{q}\lambda} + 1) \delta(\mathscr{E}_{\boldsymbol{k}'} - \mathscr{E}_{\boldsymbol{k}} + \hbar\omega_{\boldsymbol{q}\lambda})|_{-\boldsymbol{q}=\boldsymbol{k}'-\boldsymbol{k}}). \tag{9.29}$$

Here $N_{q\lambda}$ is the thermal average of the phonon occupation number, given by the well-known Bose factor

$$N_{q\lambda} = \frac{1}{e^{\hbar\omega_{q\lambda}/kT} - 1}. \tag{9.30}$$

In the case of acoustic modes the energy of the phonon is small compared to kT, so the phonon energy can be neglected in the δ-function, and the phonon occupation probabilities are given by the classical values:

$$N_{q\lambda} = N_{q\lambda} + 1 = kT/\hbar\omega_{q\lambda}. \tag{9.31}$$

Also, from eq. (9.25), only the longitudinal mode is involved. Substituting for $U^{+}_{k\lambda}$ from eq. (9.25) and setting $MN = \rho V$, where ρ is the density, gives

$$T(\boldsymbol{k} \to \boldsymbol{k}') = \frac{2\pi}{\hbar} \frac{2kT}{\hbar\omega_{q\ell}} \frac{E_1^2 \hbar q^2}{2\rho V \omega_{q\ell}} \delta(\mathscr{E}_{\boldsymbol{k}'} - \mathscr{E}_{\boldsymbol{k}}). \tag{9.32}$$

Then, using $\omega_{q\ell} = v_\ell q = \sqrt{c_\ell/\rho}\, q$, where c_ℓ is an elastic constant, the transition rate for the Boltzmann equation for simple bands becomes finally

$$W(\boldsymbol{k} \to \boldsymbol{k}') = \frac{1}{4\pi^2 \hbar} \frac{kTE_1^2}{c_\ell} \delta(\mathscr{E}_{\boldsymbol{k}'} - \mathscr{E}_{\boldsymbol{k}}). \tag{9.33}$$

For the general case we must use the deformation potential operator of eq. (9.20), and the result is

$$W(n\boldsymbol{k} \to n'\boldsymbol{k}') = \frac{1}{4\pi^2 \hbar} \sum_\lambda \frac{kT}{c_\lambda} |\hat{\boldsymbol{q}} \cdot \boldsymbol{D}_{nn'}(\boldsymbol{k}, \boldsymbol{k}') \cdot \boldsymbol{\varepsilon}_{q\lambda}|^2 \delta(\mathscr{E}_{n'}(\boldsymbol{k}') - \mathscr{E}_n(\boldsymbol{k})), \tag{9.34}$$

where

$$\boldsymbol{D}_{nn'}(\boldsymbol{k}, \boldsymbol{k}') = \int u^*_{n'\boldsymbol{k}'}(\boldsymbol{r}) \boldsymbol{D} u_{n\boldsymbol{k}}(\boldsymbol{r})\, \mathrm{d}\boldsymbol{r}, \tag{9.35}$$

in which we have used a dyadic notation and assumed that $\boldsymbol{D}$ is symmetric. For ellipsoidal valleys (see eq. (9.22)) and coupled valence bands, there is an interaction with transverse as well as with longitudinal phonons.

9.4. *Piezoelectric acoustic-phonon scattering*

In polar crystals the motion of the ions produces polarization fields that scatter the carriers. This *macrofield* is a slowly varying potential in contrast to the *microfield*, which is described by the deformation potential operator (see Zawadzki 1982, Gantmakher and Levinson 1987). In the case of polar crystals without a center of symmetry, an elastic strain can induce a polarization of the crystal. This is the piezoelectric effect, and it is described by a piezoelectric tensor h such that

$$P_\alpha = \sum_{\beta\gamma} h_{\alpha\beta\gamma} \varepsilon_{\beta\gamma} = \sum_{\beta\gamma} h_{\alpha\beta\gamma} \partial u_\beta / \partial x_\gamma. \tag{9.36}$$

The second form assumes that h is symmetric. For the zinc blende structure, there is a single non-vanishing independent component $h_{123}=h_{132}=h_{231}$, etc., or h_{14} in the system with six strain coefficients.

To find the interaction of an electron with this polarization, we note that if there is no free charge the divergence of the electric displacement vanishes,

$$\nabla\cdot\boldsymbol{D}=4\pi\nabla\cdot\boldsymbol{P}+\kappa_0\nabla\cdot\boldsymbol{E}=0, \tag{9.37}$$

where κ_0 is the zero-frequency dielectric constant. The electrostatic potential φ for the electric field then obeys Poisson's equation:

$$\nabla^2\varphi=-\nabla\cdot\boldsymbol{E}=4\pi\nabla\cdot\boldsymbol{P}/\kappa_0=\frac{4\pi}{\kappa_0}\sum_{\alpha\beta\gamma}h_{\alpha\beta\gamma}\frac{\partial}{\partial x_\alpha}\frac{\partial u_\beta}{\partial x_\gamma}. \tag{9.38}$$

The displacement $\boldsymbol{u}$ is again given by the plane wave of eq. (9.24), and the Laplacian of the plane wave just gives a factor $-q^2$. Therefore the interaction potential $U=-e\varphi$ is

$$U(r)=-e\sum_{\boldsymbol{q}\lambda(\mathrm{ac})}\sqrt{\frac{\hbar}{2MN\omega_{\boldsymbol{q}\lambda}}}(Q_{\boldsymbol{q}\lambda}\mathrm{e}^{\mathrm{i}\boldsymbol{q}\cdot\boldsymbol{r}}a_{\boldsymbol{q}\lambda}+Q^*_{\boldsymbol{q}\lambda}\mathrm{e}^{-\mathrm{i}\boldsymbol{q}\cdot\boldsymbol{r}}a^\dagger_{\boldsymbol{q}\lambda}), \tag{9.39}$$

where

$$Q_{\boldsymbol{q}\lambda}=\frac{4\pi}{\kappa_0}\sum_{\alpha\beta\gamma}\left(\frac{h_{\alpha\beta\gamma}q_\alpha q_\beta e_{\boldsymbol{q}\lambda,\gamma}}{q^2}\right). \tag{9.40}$$

This has the same form as eq. (9.27), and assuming again elastic scattering and classical statistics for acoustic phonons, the transition probability for a single band is

$$W(\boldsymbol{k}\to\boldsymbol{k}')=G(\boldsymbol{k},\boldsymbol{k}')\frac{e^2kT}{\pi\hbar\kappa_0}\frac{P^2}{|\boldsymbol{k}-\boldsymbol{k}'|^2}\delta(\mathscr{E}_{\boldsymbol{k}'}-\mathscr{E}_{\boldsymbol{k}}), \tag{9.41}$$

where P^2 is the dimensionless anisotropic function

$$P^2=\sum_\lambda\frac{\kappa_0}{4\pi c_\lambda}Q^2_{\boldsymbol{k}-\boldsymbol{k}'\lambda} \tag{9.42}$$

and c_λ is the elastic constant for the mode λ. As we have mentioned, the anisotropic function P^2 is to be averaged according to eq. (5.44) as shown by Zook (1964). Rode (1975) gives expressions* for this averaged P^2 ($\langle P^2\rangle\sim(4\pi/\kappa_0)\,\Sigma_\lambda\langle h^2/c_\lambda\rangle$).

It is also possible to include a screening factor (Zawadzki 1982), obtained from the same considerations as for ionized-impurity scattering

$$S(|\boldsymbol{k}-\boldsymbol{k}'|)=\left(\frac{\lambda^2|\boldsymbol{k}-\boldsymbol{k}'|^2}{1+\lambda^2|\boldsymbol{k}-\boldsymbol{k}'|^2}\right)^2. \tag{9.43}$$

*I believe Rode's expression should have the factor ε_0 in the denominator.

9.5. Deformation potential optical-phonon scattering

A long-wavelength optical phonon can induce shifts of band edges, because of the relative displacement of the two ions in a unit cell. We can write

$$U(\boldsymbol{r}) = \boldsymbol{D}\cdot\boldsymbol{u}(\boldsymbol{r}), \tag{9.44}$$

where $\boldsymbol{D}$ is a rapidly varying deformation potential operator, which has matrix elements at the band edges. Since it is a vector, we expect $\boldsymbol{D}$ to vanish for s bands at the zone center, but it does have matrix elements between p-like valence states, so the effect is important for the valence band, as has been argued by Wiley (1975). For the multivalley conduction bands, Harrison (1956) has shown that $\boldsymbol{D}$ vanishes for (100) valleys but may be finite for (111) valleys. For the valence bands there is a single deformation potential parameter d_0 (Gantmakher and Levinson 1987), and the interaction is

$$U(\boldsymbol{r}) = \frac{1}{2\sqrt{3}} d_0 [(J_x J_y + J_y J_x) u_z + \text{c.p.}], \tag{9.45}$$

where the J are the Luttinger matrices.

Long-wavelength optical phonons are described by the plane waves

$$\boldsymbol{u}(\boldsymbol{r}) = \sum_{\lambda \boldsymbol{q},\text{op}} \sqrt{\frac{\hbar}{2M_\text{r} N\omega_\lambda}} (\boldsymbol{e}_{\boldsymbol{q}\lambda} e^{i\boldsymbol{q}\cdot\boldsymbol{r}} a_{\boldsymbol{q}\lambda} + \boldsymbol{e}^*_{\boldsymbol{q}\lambda} e^{-i\boldsymbol{q}\cdot\boldsymbol{r}} a^\dagger_{\boldsymbol{q}\lambda}). \tag{9.46}$$

Here $\boldsymbol{u}$ is the relative displacement of the two atoms in a unit cell, and the mass is now the reduced mass M_r. The potential energy of interaction in this case is

$$U(\boldsymbol{r}) = \sum_{\boldsymbol{q},\lambda,\text{op}} \sqrt{\frac{\hbar}{2M_\text{r} N\omega_\lambda}} \boldsymbol{D}\cdot(\boldsymbol{e}_{\boldsymbol{q}\lambda} e^{i\boldsymbol{q}\cdot\boldsymbol{r}} a_{\boldsymbol{q}\lambda} + \boldsymbol{e}^*_{\boldsymbol{q}\lambda} e^{-i\boldsymbol{q}\cdot\boldsymbol{r}} a^\dagger_{\boldsymbol{q}\lambda}). \tag{9.47}$$

The transition rate for a homopolar crystal, which has only a single optical frequency ω_o, is

$$\begin{aligned} W(n\boldsymbol{k}\to n'\boldsymbol{k}') = {} & \frac{1}{4\pi^2\hbar} \sum_{\lambda,\text{op}} |\boldsymbol{D}_{nn'}(\boldsymbol{k},\boldsymbol{k}')\cdot\boldsymbol{e}_\lambda|^2 \frac{\hbar V}{2NM_\text{r}\omega_\text{o}} \\ & \times [N_\text{o}\delta(\mathscr{E}_{n'}(\boldsymbol{k}') - \mathscr{E}_n(\boldsymbol{k}) - \hbar\omega_\text{o}) + (N_\text{o}+1)\delta(\mathscr{E}_{n'}(\boldsymbol{k}') - \mathscr{E}_n(\boldsymbol{k}) + \hbar\omega_\text{o}]. \end{aligned} \tag{9.48}$$

Here $D_{nn'}$ is defined in analogy with eq. (9.35). The temperature dependence is dominated by the Bose factor, eq. (9.30).

9.6. Polar optical-phonon scattering

In polar crystals the longitudinal optical phonons carry a polarization, which scatters the electrons. This is a slowly varying potential, i.e., a macrofield. The theory of the polar optical modes is well known (Born and Huang 1954, see Callaway 1974), and the polarization can be described by $\boldsymbol{P} = \boldsymbol{u}e^*/\Omega$, where $\boldsymbol{u}$ is the relative displacement of the two ions in the unit cell, Ω is the volume of a unit cell, and e^* is an effective charge, given by (Callen 1949)

$$(e^*)^2 = \frac{\Omega M_r \omega_o^2}{4\pi}\left(\frac{1}{\kappa_0} - \frac{1}{\kappa_\infty}\right). \tag{9.49}$$

Here ω_o is the longitudinal optical frequency and κ_∞ is the high-frequency dielectric constant. The displacement for long-wavelength optical phonons is given by eq. (9.46). To find the Frohlich interaction of the electron with the polarization, we again use the condition that the divergence of the electric displacement vanishes. In this case $\nabla \cdot \boldsymbol{E} = -4\pi \nabla \cdot \boldsymbol{P}$, and using Poisson's equation the interaction with longitudinal waves is obtained as

$$U(\boldsymbol{r}) = \mathrm{i}\frac{4\pi e^* e}{\Omega} \sum_{\boldsymbol{q},\mathrm{LO}} \sqrt{\frac{\hbar}{2M_r N \omega_o q}}\frac{1}{-}(\mathrm{e}^{\mathrm{i}\boldsymbol{q}\cdot\boldsymbol{r}} a_{\boldsymbol{q}} - \mathrm{e}^{-\mathrm{i}\boldsymbol{q}\cdot\boldsymbol{r}} a_{\boldsymbol{q}}^\dagger). \tag{9.50}$$

The transition probability for a single band that may be non-parabolic, such as InSb, can be written as

$$W(\boldsymbol{k} \to \boldsymbol{k}') = G(\boldsymbol{k}, \boldsymbol{k}') \frac{2e^2 e^{*2}}{\Omega M_r \omega_0} \frac{1}{|\boldsymbol{k} - \boldsymbol{k}'|^2} \times [N_o \delta(\mathscr{E}_{\boldsymbol{k}'} - \mathscr{E}_{\boldsymbol{k}} - \hbar\omega_o) + (N_o + 1)\delta(\mathscr{E}_{\boldsymbol{k}'} - \mathscr{E}_{\boldsymbol{k}} + \hbar\omega_o)]. \tag{9.51}$$

Ehrenreich (1959) has included screening of polar optical scattering based on eq. (9.43). This interaction is inelastic and not $\boldsymbol{k}$-randomizing, since it depends on the scattering angle. Therefore, even for isotropic bands the relaxation time is not valid except for high temperatures, and the methods of § 5.7 must be used to solve the Boltzmann equation.

9.7. *Intervalley phonon scattering*

In the many-valley conduction bands of Ge and Si Herring (1955) showed that intervalley phonon scattering is important at high temperatures. Since the change in $\boldsymbol{k}$ between valleys is large, the energy of the phonons involved is large. The interaction depends again on the phonon displacement, and the transition rate between two valleys has the form

$$W(1\boldsymbol{k} \to 2\boldsymbol{k}') = \frac{D_{12}^2}{8\pi^2 \rho \omega_p}[N_p \delta(\mathscr{E}_{1\boldsymbol{k}'} - \mathscr{E}_{2\boldsymbol{k}} - \hbar\omega_p) + (N_p + 1)\delta(\mathscr{E}_{2\boldsymbol{k}'} - \mathscr{E}_{1\boldsymbol{k}} + \hbar\omega_p)], \tag{9.52}$$

where ω_p is the intervalley phonon energy and D is a deformation potential parameter defined in analogy with eq. (9.48).

Rode (1972) has made a study of intervalley scattering in Ge, Si and GaP. The four equivalent (111) valleys of Ge are located at the L-points in the Brillouin zone, such as $\boldsymbol{k} = (1, 1, 1)\pi/a$, with opposite $\boldsymbol{k}$ being equivalent. Phonons at the X-points, such as $\boldsymbol{k} = (1, 0, 0)2\pi/a$ are required for scattering between two such valleys. In Si the six valleys are located along (100), at points like $(0.83, 0, 0)2\pi/a$, and there are two types of intervalley phonons. The g-scattering is between valleys along the same axis. The phonon wave vector is $(0.34, 0, 0)2\pi/a$, and the scattering is an Umklapp process. The scattering between valleys along different axes is called f-scattering, and involves an Umklapp phonon of the form $(1, 0.17, 0.17)2\pi/a$, which is on the X-face of the Brillouin zone. In GaP, Rode took valleys at the X-point, so that X-point phonons are involved. Rode (1975) gives diagrams to show the relationship between the phonons and the valleys.

Intervalley phonon scattering is important at high electric fields in transferring carriers from one valley to another. An example is the Gunn effect (Gunn 1964).

9.8. Other scattering mechanisms

The scattering mechanisms we have outlined above are the principle ones of importance for semiconductors. Other mechanisms have been studied. Electron–electron scattering modifies the scattering from external perturbations by redistributing the electron energies. For ionized impurity scattering and classical statistics, the mobility is reduced by a factor of 0.57 (Appel 1961), while for acoustic-phonon scattering there is little effect. Electron scattering by heavy holes is similar to that from charged impurities. Scattering by the charge and strain fields of dislocations has been studied and compared with experiments in plastically deformed semiconductors. These are reviewed by Zawadzki (1982), and also Gantmakher and Levinson (1987) consider many other interesting effects.

References

Abeles, B., and S. Meiboom, 1954, Phys. Rev. **95**, 1385.

Abramowitz, M., and I.A. Stegun, 1965, Handbook of Mathematical Functions (Dover, New York) pp. 475–478.

Adams, E.N., 1952, Phys. Rev. **85**, 41.

Aleshkin, Yu.A., and Yu.A. Romanov, 1986, Fiz. Tekh. Poluprovodn. **20**, 281 [Sov. Phys. Semicond. **20**, 176].

Allgaier, R.S., 1959, Phys. Rev. **115**, 1185.

Appel, J., 1961, Phys. Rev. **122**, 1760.

Argyres, P.N., 1962, Phys. Rev. **126**, 1386.

Aronov, A.G., 1963, Sov. Phys. Solid State **5**, 402.

Aronov, A.G., and G.E. Picus, 1967, Sov. Phys. Solid State **51**, 281, 505.

Ashcroft, N.W., and N.D. Mermin, 1976, Solid State Phys. (Rinehart & Winston, New York).

Bardeen, J., and W. Shockley, 1950, Phys. Rev. **80**, 72.

Beer, A.C., 1963, in: Galvanomagnetic Effects in Semiconductors, Solid State Physics, Suppl. 4, eds. F. Seitz and D. Turnbull (Academic Press, New York).

Beer, A.C., and R.K. Willardson, 1958, Phys. Rev. **110**, 1286.

Bemski, G., 1960, Phys. Rev. Lett. **4**, 62.

Blount, E.I., 1962a, Phys. Rev. **126**, 1636.

Blount, E.I., 1962b, Solid State Phys. **13**, 305–373.

Böer, K.W., 1990, Survey of Semiconductor Physics (Van Nostrand Reinhold, New York).

Born, M., and K. Huang, 1954, Dynamical Theory of Crystal Lattices (Clarendon Press, Oxford).

Bowers, R., C. Legendy and F. Rose, 1961, Phys. Rev. Lett. **7**, 339.

Brooks, H., 1955, in: Advances in Electronics and Electron Physics, ed. L. Marton (Academic Press, New York) Vol. 7, p. 85.
Brown, E., 1964, Phys. Rev. **133**, A1038.
Brown, E., 1968, Solid State Phys. **22**, 313.
Butcher, P.N., 1986, The theory of electron transport in crystalline semiconductors, in: Crystalline Semiconducting Materials and Devices, eds P.N. Butcher, N.H. March and M.P. Tosi (Plenum Press, New York) p. 131.
Callaway, J., 1974, Quantum Theory of the Solid State (Academic Press, New York) chs. 4–6.
Callen, H.B., 1948, Phys. Rev. **73**, 1340.
Callen, H.B., 1949, Phys. Rev. **76**, 1394.
Cardona, M., 1969, in: Modulation Spectroscopy, Solid State Physics, Suppl. 11, eds. F. Seitz and D. Turnbull (Academic Press, New York).
Chambers, R.G., 1956, Proc. Roy. Soc. A **238**, 344.
Chambers, R.G., 1960, Magnetoresistance, in: The Fermi Surface, eds W.A. Harrison and M.B. Webb (Wiley, New York) pp. 100–124.
Cohen, M.H., and L.M. Falicov, 1961, Phys. Rev. Lett. **7**, 231.
Conwell, E.M., 1967, in: High Field Transport in Semiconductors, Solid State Physics, Suppl. 11, eds. F. Seitz and D. Turnbull (Academic Press, New York).
Conwell, E.M., 1982, Transport: the Boltzmann equation, in: Handbook of Semiconductors, Vol. 1, ed. W. Paul (North-Holland, Amsterdam).
Conwell, E.M., and V.F. Weisskopf, 1950, Phys. Rev. **1977**, 388.
Dakhovskii, I.V., 1963, Fiz. Tverd. Tela **5**, 2322 [1964, Sov. Phys.-Solid State **5**, 1695].
Dakhovskii, I.V., and E.F. Mikhai, 1964, Fiz. Tverd. Tela **6**, 3479 [1965, Sov. Phys.-Solid State **6**, 2785].
De Haas, W.J., and P.M. van Alphen, 1930, Leiden Comm. No. 212A.
Debye, P.P., and E.M. Conwell, 1954, Phys. Rev. **93**, 693.
Delves, R.T., 1959, Proc. Phys. Soc. London **73**, 572.
Dexter, R.N., H.J. Zeiger and B. Lax, 1956, Phys. Rev. **104**, 637.
Dingle, R.B., 1955, Philos. Mag. **46**, 831.
Dresselhaus, G., 1955, Phys. Rev. **100**, 580.
Dresselhaus, G., A.F. Kip and C. Kittel, 1955, Phys. Rev. **98**, 368.
Eagles, P.M., and D.M. Edwards, 1965, Phys. Rev. **138**, A1706.
Ehrenreich, H., 1957, J. Phys. & Chem. Solids **2**, 131.
Ehrenreich, H., 1959, J. Phys. & Chem. Solids **9**, 129.
Erginsoy, C., 1950, Phys. Rev. **79**, 1013.
Falicov, L.M., and M. Cuevas, 1967, Phys. Rev. **164**, 1025.
Feher, G., and D.K. Wilson, 1960, Bull. Am. Phys. Soc. **5**, 60.
Feher, G., D.K. Wilson and E.A. Gere, 1959, Phys. Rev. Lett. **3**, 25.
Fletcher, K., and P.N. Butcher, 1972, J. Phys. C **5**, 212.
Fletcher, R.J., W.A. Yager and F.R. Merrit, 1955, Phys. Rev. **100**, 747.
Franz, W., 1958, Z. Naturforsch. **13A**, 484.
Gantmakher, V.F., and Y.B. Levinson, 1987, Carrier Scattering in Metals and Semiconductors (North-Holland, Amsterdam).
Gold, L., and L.M. Roth, 1957, Phys. Rev. **107**, 358.
Goldberg, C., and W.E. Howard, 1953, Phys. Rev. **110**, 1035.
Goodman, R.R., 1961, Phys. Rev. **122**, 397.
Gorbovitskii, B.M., 1988, Fiz. Tekh. Poluprovodn. **22**, 1894 [Sov. Phys. Semicond. **22**, 1201].
Groves, S.H., and W. Paul, 1963, Phys. Rev. Lett. **11**, 505.
Gunn, J.B., 1964, IBM J. Res. Dev. **8**, 141.
Harman, T.C., and J.M. Honig, 1967, Thermoelectric and Thermomagnetic Effects and Applications (McGraw-Hill, New York).
Harper, P.G., 1955, Proc. Roy Soc. London A **68**, 874.
Harrison, W.A., 1956, Phys. Rev. **104**, 1281.
Hensel, J.C., and K. Suzuki, 1969, Phys. Rev. Lett. **22**, 838.

Hensel, J.C., and K. Suzuki, 1970, Quantum resonance spectroscopy in the valence bands of germanium, in: Proc. 10th Intern. Conf. on the Physics Semiconductors, Cambridge, eds S.P. Keller, J.C. Hensel and F. Stern (U.S.A.E.C.) pp. 541–551.
Herring, C., 1954, Phys. Rev. **96**, 1163.
Herring, C., 1955, Bell Syst. Tech. J. **34**, 237.
Herring, C., and E. Vogt, 1956, Phys. Rev. **101**, 944.
Hofstadter, D., 1976, Phys. Rev. B **14**, 2239.
Jacoboni, C., and L. Reggiani, 1983, Rev. Mod. Phys. **55**, 645.
Johnson, M.H., and B.A. Lippmann, 1949, Phys. Rev. **76**, 828.
Kane, E.O., 1956, J. Phys. & Chem. Solids **1**, 82.
Kane, E.O., 1957, J. Phys. & Chem. Solids **1**, 249.
Kane, E.O., 1959, J. Phys. & Chem. Solids **12**, 181.
Kane, E.O., 1982, Energy band theory, in: Handbook of Semiconductors, Vol. 1, ed. W. Paul (North-Holland, Amsterdam) ch. 4A.
Keldysh, L.V., 1958, Sov. Phys. JETP **7**, 788.
Kjeldaas, T., and W. Kohn, 1957, Phys. Rev. **105**, 806.
Kohler, M., 1948, Z. Phys. **124**, 772.
Kolodziejczak, J., 1961, Acta Phys. Polon. **20**, 379.
Koss, R.W., and L.M. Lambert, 1972, Phys. Rev. B **5**, 1979.
Laff, R.A., and H.Y. Fan, 1958, Phys. Rev. **112**, 317.
Landau, L., 1930, Z. Phys. **64**, 629.
Landau, L.D., and I.M. Lifshitz, 1959, Quantum Mechanics (Pergamon Press, New York) pp. 70–71, 491.
Landsberg, P.T., 1992, Recombination in Semiconductors (Cambridge University Press, Cambridge, UK) p. 529.
Lawaetz, P., 1968a, Phys. Rev. **166**, 763.
Lawaetz, P., 1968b, Phys. Rev. **174**, 867.
Lawaetz, P., 1971, Phys. Rev. **B4**, 3460.
Lax, B., and J.G. Mavroides, 1955, Phys. Rev. **100**, 1650.
Lifshitz, I.M., and M.I. Kaganov, 1959, Usp. Fiz. Nauk **69**, 419 [1960, Sov.-Phys. Usp. **2**, 831].
Lifshitz, I.M., and A.M. Kosevich, 1955, J. Exp. Theor. Phys. USSR **29**, 730 [1956, Sov. Phys. JETP **2**, 636].
Lifshitz, I.M., M.Ya. Asbel and M.I. Kaganov, 1973, Electron Theory of Metals (Consultants Bureau, New York).
Liu, S.H., 1962, Phys. Rev. **126**, 1317.
Long, D., 1957, Phys Rev. **107**, 672.
Long, D., and J. Myers, 1959, Phys. Rev. **115**, 1107.
Long, D., and J. Myers, 1960, Phys. Rev. **120**, 39.
Luttinger, J.M., 1951, Phys. Rev. **84**, 814.
Luttinger, J.M., 1956, Phys. Rev. **102**, 1030.
Luttinger, J.M., and W. Kohn, 1955, Phys. Rev. **97**, 869.
Maan, J.C., 1984, in: Two-Dimensional Systems, Heterostructures and Superlattices, eds G. Bauer, F. Kuchar and H. Heinrich (Springer, Berlin) p. 183.
Massey, H.S.W., and B.L. Moiseiwitsch, 1950, Phys. Rev. **78**, 180.
Mavroides, J.G., 1972, Magneto-optical properties, in: Optical Properties of Solids, ed. F. Abeles (North-Holland, Amsterdam) pp. 351–528.
Mavroides, J.G., and B. Lax, 1957, Phys. Rev. **107**, 1530; **108**, 1648.
McClure, J.W., 1956, Phys. Rev. **101**, 1642.
Mironov, A.G., V.A. Morozova and V.S. Vavilov, 1985, Phys. Status Solidi b **127**, 359.
Nag, B.R., 1972, Theory of Electrical Transport in Semiconductors (Pergamon, Oxford).
Nag, B.R., 1980, Electron Transport in Compound Semiconductors (Springer, Berlin).
Nenciu, G., 1991, Rev. Mod. Phys. **63**, 91.
Neuringer, L.J., and W.J. Little, 1962, Proc. Int. Conf. Physics of Semiconductors, Exeter (Institute of Physics, London) p. 614.
Onsager, L., 1952, Philos. Mag. **43**, 1006.

Palik, E.D., and G.B. Wright, 1967, Free carrier magneto-optical effects, in: Semiconductors and Semimetals, Vol. 3, eds. R.K. Willardson and A.C. Beer (Academic Press, New York) p. 421.
Pearson, G.L., and C. Herring, 1954, Physica **20**, 975.
Pearson, G.L., and H. Suhl, 1951, Phys. Rev. **83**, 768.
Peierls, R.E., 1933, Z. Phys. **80**, 763.
Peterson, T.L., F. Szmulowicz and P.M. Hemenger, 1990, J. Cryst. Growth **106**, 16.
Picus, G.E., and G.L. Bir, 1959, Sov. Phys. Solid State 1, **136**, 1502.
Pidgeon, C.R., 1969, Interband magneto-optics in small band gap semiconductors and semimetals, in: Electronic Structures in Solids, ed. E.D. Haidemenakis (Plenum, New York) pp. 47–67.
Pidgeon, C.R., and R.N. Brown, 1966, Phys. Rev. **146**, 575.
Pidgeon, C.R., and S.H. Groves, 1969, Phys. Rev. **186**, 824.
Piller, H., 1972, Faraday rotation, in: Semiconductors and Semimetals, Vol. 8, eds. R.K. Willardson and A.C. Beer (Academic Press, New York) p. 103.
Praddaude, H.C., 1965, Phys. Rev. **140**, 1292.
Puri, S.M., and T.H. Geballe, 1966, Thermomagnetic effects in the quantum region, in: Semiconductors and Semimetals, Vol. 1, eds R.K. Willardson and A.C. Beer (Academic Press, New York) p. 203.
Ralph, H.I., 1977, Phillips Res. Rept. **32**, 160.
Rauch, F.J., T.J. Stickler, H.J. Zeiger and G.S. Heller, 1960, Phys. Rev. Lett. **4**, 64.
Ray, B.K., 1967, Ph.D. Thesis (Tufts University).
Reine, M., Q.H.F. Vrehen and B. Lax, 1967, Phys. Rev. **163**, 726.
Rode, D.L., 1970, Phys. Rev. B **2**, 1012.
Rode, D.L., 1972, Phys. Status Solidi **53**, 245.
Rode, D.L., 1975, Low field electron transport, in: Semiconductors and Semimetals, Vol. 10, eds R.K. Willardson and A.C. Beer (Academic Press, New York) p. 1.
Roth, L.M., 1960, Phys. Rev. **118**, 1534.
Roth, L.M., 1962, J. Phys. & Chem. Solids **23**, 433.
Roth, L.M., 1964, Phys. Rev. **133**, A542.
Roth, L.M., 1966, Phys. Rev. **144**, 434.
Roth, L.M., 1968, Phys. Rev. **173**, 755.
Roth, L.M., 1982, Dynamics of electrons in semiconductors, in: Handbook of Semiconductors, Vol. 1, ed. W. Paul (North-Holland, Amsterdam) ch. 8.
Roth, L.M., and P.N. Argyres, 1966, Magnetic quantum effects, in: Semiconductors and Semimetals, Vol. 1, eds R.K. Willardson and A.C. Beer (Academic Press, New York) pp. 159–202.
Roth, L.M., B. Lax and S. Zwerdling, 1959, Phys. Rev. **114**, 90.
Roth, L.M., S.H. Groves and P.W. Wyatt, 1967, Phys. Rev. Lett. **19**, 576.
Samoilovich, A.G., I.Ya. Korenblit, I.V. Dakhovskii and V.D. Iskra, 1961a, Dokl. Akad. Nauk **139**, 355 [1962, Sov. Phys.-Dokl. **6**, 606].
Samoilovich, A.G., I.Ya. Korenblit, I.V. Dakhovskii and V.D. Iskra, 1961b, Fiz. Tverd. Tela **3**, 2939 [1962, Sov. Phys.-Solid State **3**, 2148].
Samoilovich, A.G., I.Ya. Korenblit, I.V. Dakhovskii and V.D. Iskra, 1961c, Fiz. Tverd. Tela **3**, 3285 [1962, Sov. Phys.-Solid State **3**, 2385].
Seeger, K., 1982, Semiconductor Physics (Springer, Berlin).
Seiler, D.G., and W.M. Becker, 1967, Phys. Lett. A **26**, 96.
Seitz, F., 1950, Phys. Rev. **79**, 372.
Shibuya, M., 1954, Phys. Rev. **95**, 1385.
Shockley, W., 1950, Phys. Rev. **79**, 191.
Shubnikov, L., and W.J. de Haas, 1930, Leiden Comm. No. 207D.
Smith, A.C., 1982, Transport using the particle kinetic model, in: Handbook of Semiconductors, Vol. 1, ed. W. Paul (North-Holland, Amsterdam) ch. 9.
Stickler, T.J., H.J. Zeiger and G.S. Heller, 1962, Phys. Rev. **127**, 1077.
Stillman, G.E., C.M. Wolfe and J.O. Dimmock, 1970, J. Phys. & Chem. Solids **31**, 1199.
Stirn, R.J., 1972, Band structure and galvanomagnetic effects in III–V compounds with indirect band gaps, in: Semiconductors and Semimetals, eds R.K. Willardson and A.C. Beer (Academic Press, New York) p. 1.

Szmulowicz, F., 1983, Phys. Rev. B **28**, 5943.
Szmulowicz, F., 1986, Phys. Rev. B **34**, 4031.
Tharmalingham, K., 1963, Phys. Rev. **130**, 2204.
Ure Jr, R.W., 1972, Thermoelectric effects in III–V compounds, in: Semiconductors and Semimetals, Vol. 8, eds R.K. Willardson and A.C. Beer (Academic Press, New York).
Von der Lage, F.C., and H. Bethe, 1947, Phys. Rev. **71**, 612.
Vrehen, Q.H.F., 1965, Phys. Rev. Lett. **14**, 558.
Vrehen, Q.H.F., 1968, J. Phys. & Chem. Solids **29**, 129.
Wannier, G.H., 1937, Phys. Rev. **52**, 191.
Wannier, G.H., 1962, Rev. Mod. Phys. **34**, 645.
Wannier, G.H., and D.R. Fredkin, 1962, Phys. Rev. **125**, 1910.
Wannier, G.H., and U.N. Upadyaya, 1964, Phys. Rev. **136**, A803.
Weiler, M.H., W. Zawadzki and B. Lax, 1967, Phys. Rev. **163**, 733.
Weiler, M.H., R.J. Aggarwal and B. Lax, 1977, Phys. Rev. B **16**, 3603.
Whitsett, C.R., 1965, Phys. Rev. **138**, A829.
Wiley, J.D., 1975, Mobility of holes in III–V compounds, in: Semiconductors and SemiMetals, Vol. 10, eds. R.K. Willardson and A.C. Beer (Academic Press, New York) p. 91.
Wilson, D.K., and G. Feher, 1961, Phys. Rev. **124**, 1068.
Yafet, Y., 1963, Solid State Phys. **14**, 1.
Zak, J., 1968, Phys. Rev. Lett. **20**, 1477.
Zak, J., 1972, Solid State Phys. **27**, 1.
Zawadzki, W., 1969, Bloch electrons in crossed electric and magnetic fields, in: Physics of Solids in Intense Magnetic Fields, ed. E.D. Haidemenakis (Plenum, New York) pp. 311–328.
Zawadzki, W., 1973, Spin properties of conduction electrons, in small-gap semiconductors, in: New Developments in Semiconductors, eds. P.R. Wallace, R. Harris and M.J. Zuckermann (Noordhoff, Leiden).
Zawadzki, W., 1982, Mechanisms of electron scattering in semiconductors, in: Handbook of Semiconductors, Vol. 1, ed. W. Paul (North-Holland, Amsterdam).
Zawadzki, W., and B. Lax, 1966, Phys. Rev. Lett. **16**, 1001.
Zeiger, H.J., and G.W. Pratt, 1973, Magnetic Interactions in Solids (Clarendon Press, Oxford) chs. 5 and 6.
Zeiger, H.J., B. Lax and R.N. Dexter, 1957, Phys. Rev. **105**, 495.
Ziman, J., 1960, Electrons and Phonons (Oxford University Press, Oxford).
Zook, J., 1964, Phys. Rev. **136A**, 869.
Zwerdling, D., B. Lax, L.M. Roth and K. Button, 1959, Phys. Rev. **114**, 80.

CHAPTER 11

Conjugated Polymer Semiconductors: An Introduction

E. M. CONWELL and H. A. MIZES

Xerox Webster Research Center
Webster, NY 14580, USA

Handbook on Semiconductors
Completely Revised Edition
Edited by T.S. Moss
Volume 1, edited by P.T. Landsberg

Contents

1. Introduction . . . 586
   1.1. Early history . . . 586
   1.2. Some differences from conventional semiconductors . . . 586
2. Preparation, crystal structure, and morphology . . . 590
   2.1. Synthesis . . . 590
   2.2. Mechanisms of doping . . . 591
   2.3. Structure of doped material . . . 592
3. Bonds and bands . . . 593
   3.1. σ and π bands . . . 593
   3.2. Band structure calculation for polymers . . . 596
4. Ground state of polyacetylene . . . 597
   4.1. Electron–phonon coupling . . . 597
   4.2. Calculation of the gap . . . 597
   4.3. Order parameter and continuum model . . . 599
5. Solitons . . . 599
   5.1. The order parameter . . . 599
   5.2. Soliton creation energy and mass . . . 600
   5.3. Experimental evidence for solitons . . . 602
6. Polarons and bipolarons . . . 603
7. Optical properties . . . 604
   7.1. Absorption of doped t-PA . . . 604
   7.2. Photoinduced absorption in t-PA . . . 605
   7.3. Absorption in NDGS polymers . . . 606
8. Soliton and polaron mobility . . . 607
   8.1. Experimental data – neutral soliton . . . 607
   8.2. Scattering of solitons by phonons . . . 608
   8.3. Trapping of solitons . . . 611
   8.4. Polaron mobility . . . 612
9. Photoconductivity . . . 614
   9.1. Long-time photoconductivity in t-PA . . . 614
   9.2. Short-time photoconductivity in t-PA . . . 615
   9.3. Photoconductivity in NDGS polymers . . . 617

10. Conductivity 617
10.1. Anisotropy of conductivity 617
10.2. Pristine and lightly doped samples 618
10.3. "Spinless conductivity" range 619
10.4. The "metallic" range 622
References 623

1. Introduction

1.1. Early history

A polymer is a substance consisting of macromolecules formed by linking together low-molecular-weight repeat units called monomers. Until the late 1970's polymers were considered insulators, of interest mainly for their mechanical properties. The first polymer to be made highly conducting by doping was polyacetylene. Polyacetylene, a chain of C–H's (in chemical notation $(CH)_x$) had been known to chemists for many years as a grey-brown powder. The discovery in H. Shirakawa's lab that it could be synthesized as a metallic-looking, although still quite insulating, thin film led to the suggestion that its conductivity might be increased by doping. Shortly thereafter it was demonstrated that this was indeed the case, with conductivities as high as $\sim 300\ \Omega^{-1}\ \mathrm{cm}^{-1}$ quickly attained. (For recent reviews see Heeger et al. 1988, Kanatzidis 1990, and references therein.) This demonstration attracted researchers from many disciplines and developments came rapidly. With improvements in the material, it was found that with doping the conductivity of polyacetylene could be varied from $10^{-9}\ \Omega^{-1}\ \mathrm{cm}^{-1}$ to $\sim 10^{5}\ \Omega^{-1}\ \mathrm{cm}^{-1}$, close to that of copper. Many other conjugated polymers were then successfully doped. Some of the most studied of these are listed in table 1. This table also shows the monomers of these polymers, the energy gaps, in general determined from optical absorption measurements, and the acronyms by which we will refer to them. We note that, except for polyacetylene, the conjugated polymers in general contain rings of C–H's or of C–H's and a different atom, such as nitrogen or sulfur, linked by short chains and possibly containing side groups. The significance of the structure of the monomers will be discussed further in § 3, as will the two modifications of polyacetylene. It is clear, however, that the polymers of table 1 are chains of quite simple repeat units. Whether the chains go into an amorphous solid ("spaghetti") or form a good crystal structure, the polymers are quasi-one-dimensional (1D) in their properties, in contrast to conventional semiconductors, which are 3D.

1.2. Some differences from conventional semiconductors

The quasi-1D nature of polymeric semiconductors is the source of some important differences from conventional semiconductors. The quasi-1D nature permits deformation, i.e., changes in atom positions and bonding, to occur with relative ease along the chain; such deformations are prevented in conventional semiconductors by the rigidity of the 3D binding. As will be seen later it is this deformability, in fact, that

Table 1

Polymer	Monomer	Bandgap (eV)
trans-polyacetylene (t-PA)	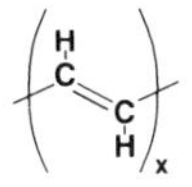	1.4
cis-polyacetylene (c-PA)	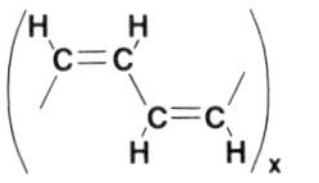	2.0
polypyrrole (PPy)	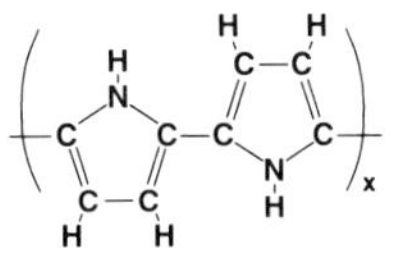	3.2
polythiophene (PT)	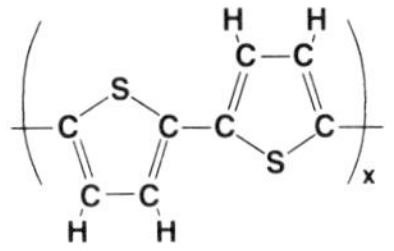	1.6
polyparaphenylene (PPP)	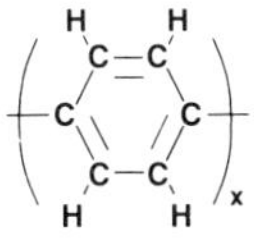	3.4
poly (phenylenevinylene) (PPV)	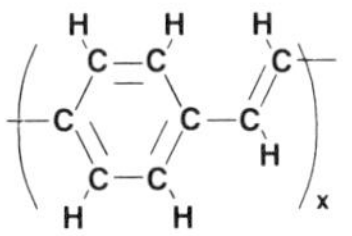	3.0
polyaniline (PANI) leucoemeraldine form	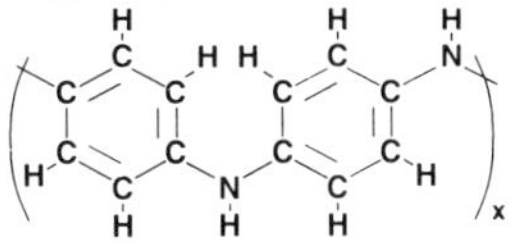	3.0

makes polyacetylene a semiconductor rather than a metal. In this section we consider the differences in behavior between 1D and 3D semiconductors that result when electrons are added or removed, and when impurities are introduced.

Consider first what happens when a single electron is added to the conduction band. Similar considerations hold for a hole added to the valence band. In the conventional semiconductor, if there is no empty level below the conduction band for the electron to drop into, it stays in the conduction band and can contribute to transport. In the conducting polymer, if there is no empty level below the conduction band, the electron will cause the chain it is on to deform around it, creating a characteristic pattern of bond deformation about 20 sites long called a polaron (see fig. 1a). In the deformation process a level is pulled out of the valence band with its two electrons and a level is pulled out of the conduction band. Two levels in the gap are created, the lower one filled with the two electrons brought up from the valence band, the upper containing the added electron (see fig. 1b). The stability of the polaron is due to the energy gained when the electron moves from the conduction

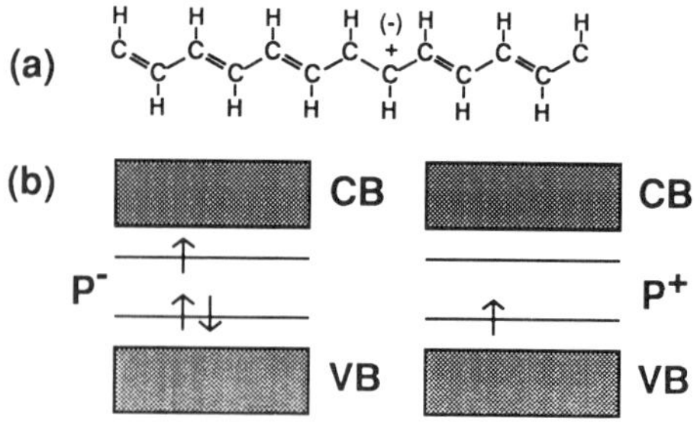

Fig. 1. (a) Schematic representation of bond distortion due to a polaron in t-PA. The actual distortion takes place over ~ 20 sites. The + (–) represents the hole (electron) added to the chain for the case of a positively (negatively) charged polaron. (b) The energy levels of a negatively charged polaron, P^-, at the left and of a positively charged polaron, P^+, at the right. CB stands for conduction band, VB for valence band.

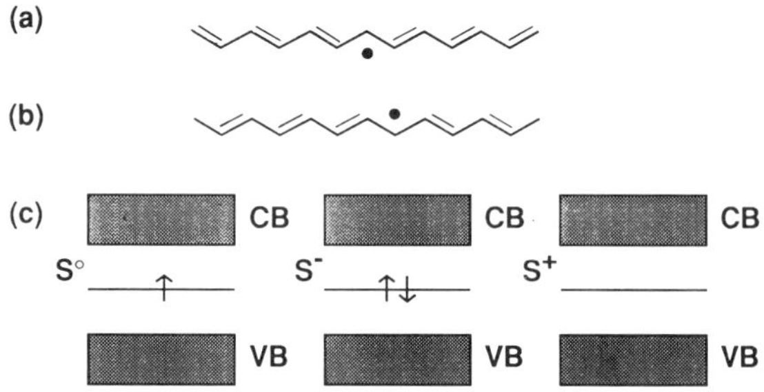

Fig. 2. Schematic representation of bond distortion due to a soliton or an antisoliton: (a) may be taken to represent a soliton, S; (b) an antisoliton, $\bar{S}$, or vice versa. The dot at the center represents the bound electron for a neutral soliton S°; (c) shows the filling of the energy level for neutral and charged solitons.

band into a lower level exceeding the elastic energy required to form this level. Although its energy levels are in the gap, the polaron can move freely on its own chain, its lattice distortion moving with it.

When an electron and a hole are created in a conventional semiconductor, e.g., by a photon, a stable exciton may result. If this is not the case, and there are no energy levels in the gap, the electron and hole will both move freely, contributing to transport, until they recombine. In a conducting polymer creation of an electron and hole may also result in an exciton. If a stable exciton is not created, however, the added electron and hole will create a pair of polarons, one positively charged, the other negatively charged, in the manner described above. If these polarons meet they can recombine. An exception to this behavior is trans-polyacetylene, in which solitons, a different type of excitation from polarons, are formed.

Trans-polyacetylene differs from all the other polymers in table 1 in that two bonding configurations can occur in the ground state. In one of these the double bond slants downward in going from left to right, as shown in the table, while in the other the double bond slants upward in going from left to right. These two structures are, of course, degenerate. In fact both of them can occur on the same chain, as indicated schematically in fig. 2. They are then separated by a "domain wall", with a characteristic bond distortion taking place over ~14 sites (rather than the two indicated in the figure) which is called a soliton. Solitons, in fact, occur in all t-PA chains with an odd number of C-H's because the chains are more stable with double bonds at both ends. These solitons are neutral; they occur without any charges being added to the chains. As seen in fig. 2, however, the fourth electron of the C at the center of the soliton, although still bound to the C nucleus, is neither bonding nor antibonding, and in first approximation, occupies a level at midgap. An electron added to the chain could also go into this level, giving rise to a negatively charged soliton, denoted S^-, as shown in fig. 2c. Alternatively, the original bound electron could be removed from the soliton making it positive, S^+, as also shown in fig. 2c. Both neutral and charged solitons are mobile, moving with their characteristic distortions. An interesting property of the solitons is that neutral solitons obey Fermi statistics while charged solitons obey Bose statistics, the reverse of typical charged particle behavior.

Returning to our original considerations, we note that addition of one electron to a t-PA chain without a neutral soliton cannot create a soliton, because it would require changing all the bonds to one side or the other, and it therefore creates a polaron excitation. By extension of this argument, addition of an electron and hole, or of two electrons or two holes, far removed from each other would create a pair of polarons. If these polarons were free to move throughout the chain, when they met they would be annihilated if the charges were unlike or, for like charges, transformed into a pair of solitons because solitons have lower energy. Finally, if two electrons or two holes were added to a t-PA chain close to each other two charged solitons would be created directly.

For most other known conducting polymers only one bonding configuration is favored for the ground state. We refer to them as non-degenerate ground state (NDGS) polymers. Addition of two electrons far removed to an NDGS polymer chain results in two separate P^-'s. If these two P^-'s met, they could form a single distortion called a bipolaron, BP^{--}, with two gap levels filled by four electrons. With the exception of t-PA, a BP^{--} is usually more stable than two P^-'s.

Conducting polymers also differ from conventional semiconductors in the way they accept impurities. In the conventional semiconductor an impurity atom differing in valence by unity from the semiconductor atoms is likely to go into the lattice substitutionally, with the extra electron or hole bound in a large hydrogenic orbit at low temperatures and usually free at room temperature. Some impurities go in interstitially, generally providing deep levels for electrons or holes. In the case of the polymers impurities do not go into the chain substitutionally (although some impurities may "attack" the chains chemically, destroying the polymer). The one-dimensionality allows impurities to line up in columns between the chains, even impurities that are large compared to the monomers. This process is the 1D analog of intercalation between graphite planes. Impurities that serve as dopants are ones that will give up an electron to the chain or accept an electron from the chain. Once on the chain, the electrons or holes go into soliton, polaron or bipolaron states as discussed earlier.

In conventional semiconductors the solubility for all impurities is quite small, generally no larger than 1 part in 10^4. The one-dimensionality of the polymers allows very large concentrations of impurity to go in between the chains, 20% dopant/monomer, e.g., being not unusual. When the amount of added impurity is small (by polymer standards), less than perhaps a few hundredths of a percent, the impurities go in randomly. At higher concentrations they tend to go in channels, or intercalated planes, with more or less regular spacing. Not surprisingly, where the impurity concentration goes above a few percent it may cause a reorganization of the lattice structure. Some examples of this will be given in § 2.

Given the large doping that is possible, it is also not surprising that a polymer semiconductor may become metallic. Trans-polyacetylene, polyparaphenylene, polypyrrole and poly(3 hexylthiophene) have been found to undergo an "insulator–metal" transition at high doping. The resulting metals are not simple, however, in that solitons in the first case and bipolarons in the other cases appear to be retained in the metallic state (Conwell and Mizes 1991). Discussion of the transition and the metallic state are beyond the scope of this article.

In what follows we will first discuss synthesis and doping of the polymeric semiconductors and what is known about structure. § 3 will be devoted to bonding and the resulting band structure. A more detailed description of the ground state of polyacetylene and of solitons, polarons and bipolarons will be given in §§ 4–6. This will be followed by sections on optical absorption, transport, photogeneration, photoconductivity, and finally conductivity.

2. Preparation, crystal structure, and morphology

2.1. Synthesis

Many different chemical methods have been used to synthesize conducting polymers (for a recent review, see Kanatzidis 1990). For example, $(CH)_x$ is frequently prepared by exposing acetylene gas to smooth surfaces prewetted by Ziegler–Natta catalyst. This technique is called the Shirakawa process. Although t-PA is the stable high-temperature form, this technique generally produces c-PA. The conversion to t-PA, or isomerization, is accomplished by either heating or doping. In what follows we deal almost entirely with t-PA. The films produced as described above are metallic in appearance, but microscopic examination shows them to consist of a tangled mass of thin fibrils 5 to 50 nm in diameter. The fibrils may fill only as little as one-third of the total volume, depending on synthesis conditions. They are 75 to 90% crystalline according to X-ray studies, the crystal structure being the herringbone structure shown in fig. 3. Even in the best polyacetylene samples, which are quite close to having full density, the coherence length deduced from X-rays is only 10 to 20 nm along the chains and less than 10 nm perpendicular to the chains (Djurado et al. 1989). The short coherence length is attributed to a series of small changes in unit cell parameters, or isolated chain defects, rather than abrupt changes such as grain boundaries (Pouget 1985).

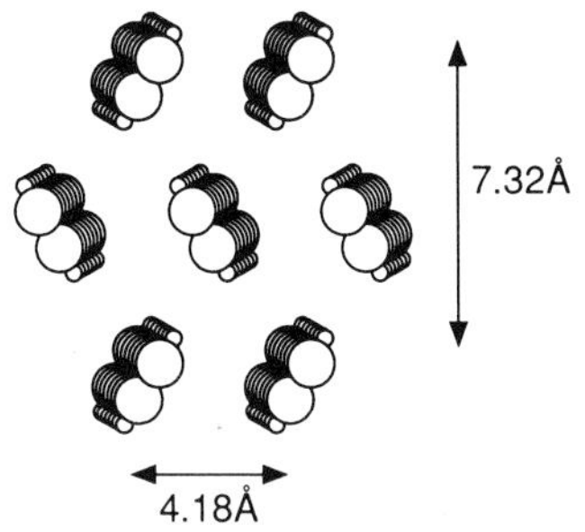

Fig. 3. Unit cell of undoped t-PA looking along the chain direction. The spacing between C–H's along the chain is 1.22 Å.

Another synthesis route, frequently used for t-PA and PPV, e.g., starts from a soluble precursor polymer which may be cast in a film form from solution. When this polymer is heated, a volatile fraction is lost, leaving the desired polymer. If the material is stretched while the transformation reaction is carried out, the resulting films are highly oriented, homogeneous and fully dense. Still another synthesis method, frequently used for polypyrrole, is electrochemical. Application of a potential between two electrodes in a cell containing the monomer in an appropriate solution results in the deposition of a film of the desired polymer on one of the electrodes. This method produces a doped film because counter ions from the solution are incorporated into the film as it grows. Films produced by this method may be amorphous or crystalline depending on growth conditions.

In films produced without stretching the overall orientation of the chains is usually random. The films may be oriented by subsequent stretching. Appropriately made films may be stretched by a factor as high as 15. The resulting strong orientation of the chains along the stretch direction gives rise to much higher conductivity and larger elastic constants along that direction than perpendicular to it. Conductivity anisotropies greater than a factor 200, and elastic constants approaching those of the strongest polymers, have been achieved by stretching t-PA (Akagi et al. 1989, Cao et al. 1991).

2.2. *Mechanisms of doping*

Doping is usually accomplished by exposure of the polymer to vapor or liquid containing the desired dopant, which then diffuses in, e.g., as in silicon. Doping can also be accomplished by using the undoped polymer film as an electrode in an electrochemical cell with the desired dopant in solution as ions. This allows precise control over the amount of doping. Dopant atoms or molecules are either highly electropositive or electronegative. Commonly used donors are alkali metal atoms. Iodine is a commonly used acceptor, the resulting ions being predominantly I_3^- although I_5^- is also seen at high dopant concentrations. Other commonly used acceptor ions are AsF_5^-, ClO_4^- and $FeCl_4^-$. Pristine PA is typically p-type with a conductivity σ of 10^{-5} to $10^{-6}\,\Omega^{-1}\,cm^{-1}$. Compensation with NH_3, presumably a donor, can decrease σ to as little as 10^{-9} to $10^{-10}\,\Omega^{-1}\,cm^{-1}$. Doping can bring the conductivity up to $10^5\,\Omega^{-1}\,cm^{-1}$, constituting a change by a factor 10^{14} or more.

2.3. Structure of doped material

As suggested earlier, for low dopant concentrations, up to perhaps a few tenths of a percent, the dopant ions are distributed randomly throughout the columns between the aligned polymer chains. For larger doping the polymer becomes two phase, one phase being dopant-rich with a regular arrangement of impurity ions in a reorganized lattice structure, while the other, dopant-poor, phase has the remaining dopant ions distributed randomly in the undoped polymer structure (Baughman et al. 1985, Pouget 1985). The detailed structure of the reorganized lattice depends on the concentration of the dopant ions and on their nature, particularly their size. Two general classes of structures can be identified: those in which columns are formed and those in which layers are formed. In the former case, the polymer strands rotate around their chain axes to form one-dimensional tunnels, each one of which is filled by a column of dopant ions. For example, for Na^+ doping up to ~6% the dopant-rich part of the lattice has a dopant concentration of 6.67%, with three t-PA chains arranged around a column of Na^+ ions (Baughman et al. 1983, 1985), as shown in fig. 4a. The spacing between Na^+ ions in this phase is $5a$, where a is the component along the chain of the average distance between C–H's. Because there are three t-PA chains for every ion column, the average distance along the chain between donated electrons is $15a$, thus the added electron or ion concentration of 6.67% cited earlier. With increasing doping, past the point where the sample would be filled with the 6.67% phase, new phases occur, in which the triangular structure is maintained but the spacing between ions is reduced. Na-doped PPV also displays a columnar structure with three PPV chains surrounding each ion column (Winokur et al. 1991). For K doping of t-PA the dopant-rich phases are tetragonal, as shown in fig. 4b, with columns of K^+ ions surrounded by four $(CH)_x$ chains (Baughman et al. 1983, 1985). K-doped PPP has a similar structure (Baughman et al. 1983, 1985). For large molecular dopants, such as AsF_6^-, ClO_4^- or iodine, the structure of the dopant-rich

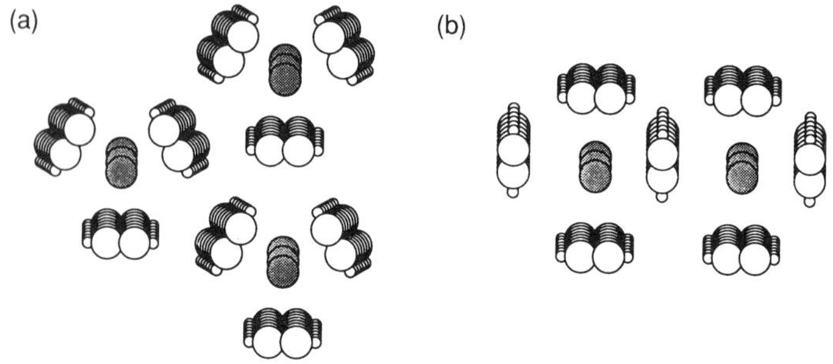

Fig. 4. (a) Structure of t-PA for Na^+ doping. (b) Structure of t-PA for K^+ doping.

regions is usually lamellar, with monolayers in which the host polymer chains are organized with their backbones parallel alternating with monolayers of dopant organized in columns. For not too high doping, in the dopant-containing monolayers dopant columns may alternate with polymer chains (Pouget 1985). Such structures have been studied particularly for iodine-doped t-PA (Murthy et al. 1988).

3. Bonds and bands

3.1. σ and π bands

Conjugated polymers are in general hydrocarbons with alternating double and single bonds. Unlike the most familiar semiconductors, where the atoms crystallize in the diamond or zincblende structure and have four nearest neighbors, the carbon atoms in conjugated polymers have only three nearest neighbors. The bonding configuration no longer consists of sp^3 hybrids, but has sp^2 hybrids oriented towards the three nearest neighbors, plus one p orbital, which we take as p_z, perpendicular to the sp^2 hybrids and thus perpendicular to the polymer backbone. These bonds are shown for a section of t-PA in fig. 5. The angle between any pair of the three sp^2 orbitals of an atom is close to 120°, which tends to constrain conducting polymers to a plane. In compounds containing rings, such as PPP and PT, stearic repulsion, i.e., the repulsion between nearby H atoms, may force the rings to rigidly rotate out of the plane.

Neighboring sp^2 hybrids pointing at each other couple strongly and form a bonding and an antibonding energy level with a separation of approximately 20 eV. The two carbon atoms participating in this bond each donate one electron, filling the bonding energy level. These electrons, called σ electrons, lie relatively deep in energy and contribute to the rigidity of the polymer, but not the electronic properties. There is a small coupling between neighboring sp^2 orbitals not oriented toward each other, giving the σ electrons a bandwidth, but this coupling is less than in conventional semiconductors.

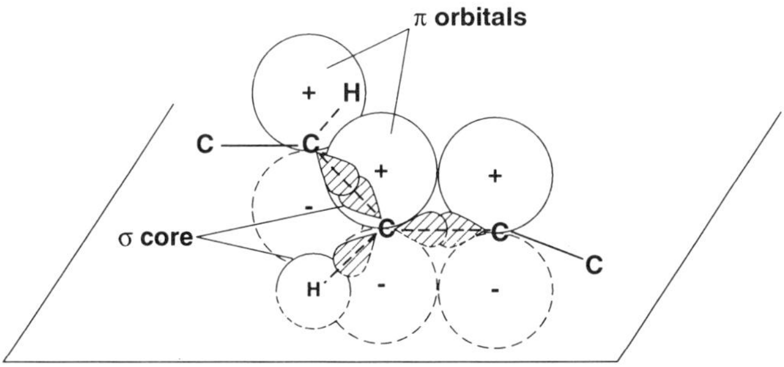

Fig. 5. $sp^2(\sigma)$ and $p_z(\pi)$ orbitals of carbon atoms in t-PA. The sp^2 orbitals are cross-hatched (after Salem 1966).

A carbon atom gives three of its four valence electrons to the σ band, leaving one electron to occupy the p_z orbital. There is a small coupling, on the order of 2.5 eV, between the p_z orbitals on neighboring carbon atoms. The band formed by the coupling between p_z orbitals is called the π band.

To understand the π band of t-PA, we begin by assuming the p_z orbitals are spaced uniformly on the chain, as shown schematically in fig. 6a. (The zigzag is neglected.) We assign the energy t to the coupling between neighboring p_z orbitals. On a simple tight-binding model

$$t = -\int \Phi_\pi^*(\boldsymbol{r}+\boldsymbol{a}) H \Phi_\pi(\boldsymbol{r})\, \mathrm{d}\boldsymbol{r}, \tag{3.1}$$

where $\Phi_\pi(\boldsymbol{r})$ and $\Phi_\pi(\boldsymbol{r}+\boldsymbol{a})$ are orbitals on nearest neighbors and H the Hamiltonian. t is called the resonance or transfer integral. Usually it is taken as a parameter adjusted to fit macroscopic properties. Because of the overlap a band forms, indicated by the dashed line in fig. 6c. The energy dispersion relation is

$$\varepsilon_k = -2t_0 \cos ka, \tag{3.2}$$

where t_0 is the value of t when the carbon atoms are equally spaced, k is the wave vector of an electron in this band and a is the spacing along the t-PA backbone between carbon atoms, 1.22 Å. Because each p_z orbital can accommodate two electrons, but each carbon atom donates only one electron to the band, this band is half-filled, and t-PA would be a metal.

However, the chain is dimerized, as shown in fig. 6b, the lattice constant being $2a$. The dimerization causes an energy gap to open at midband, as indicated by the solid

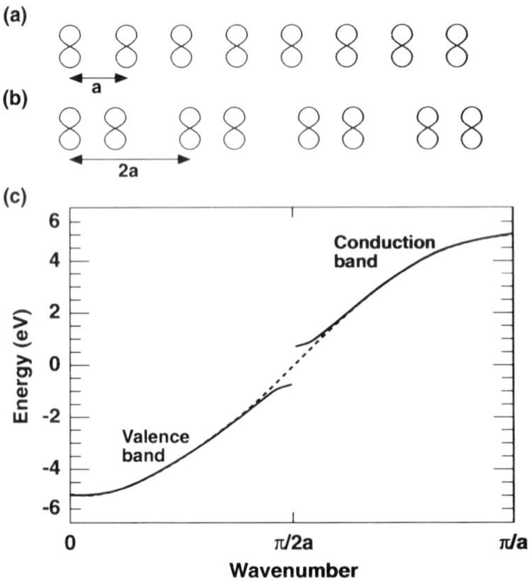

Fig. 6. p_z orbitals on undimerized chain (a), on dimerized chain (b). In (c) energy versus k for the π band is shown as dashed line for undimerized chain, solid line for dimerized chain.

line in fig. 6c. In chemical terminology the lower band is called the π band, the upper the π* band. In line with semiconductor terminology, however, the lower band is now usually called the valence band, the upper the conduction band. The reason for the dimerization, as shown by Peierls, is that the gain in electronic energy due to the filled states being lowered when a gap is formed is more than the energy required to distort the lattice. Experiments show that the difference in bond lengths is approximately 0.07 Å in trans-polyacetylene, much less than the 0.20 Å typical of single and double bonds in organic materials (Heeger et al. 1988). However, as has already been seen, it is useful as an intuitive picture to think of t-PA as having alternating single and double bonds.

Peierls' demonstration is easily generalized to show that a one-dimensional metal with any degree of band filling is unstable. Nevertheless, t-PA is the only conducting polymer that is not metallic because of Peierls' theorem. Other polymers contain larger monomers which result in a number of π bands with the highest occupied band being completely filled. For example, the monomer of PPP is a phenyl ring, which differs from benzene only by small changes in the bonds (Brédas et al. 1982). We may to a good approximation take the π energy levels of the monomer to be those of benzene (see, e.g., Salem 1966), shown in fig. 7. Each carbon atom donates 3 electrons to the σ levels and one electron to the π band. The π electrons fill the lower two levels in the figure, the upper of the two being doubly degenerate. When the phenyls are coupled to form PPP, the levels broaden into bands. However, there is no crossing of the individual levels, and the energy gap in these polymers is due to the inherent structure.

The monomers of PPy and PT have five π energy levels because there are only five atoms in the ring (see table 1). The nitrogen atom donates two electrons to the σ band, one electron to the hydrogen bond, and two electrons to the π band, while the sulfur atom donates two electrons to the σ band, two electrons to a lone pair, and two electrons to the π band. For both these materials, each monomer has two π electrons from the noncarbon atom and four π electrons from the carbon atoms.

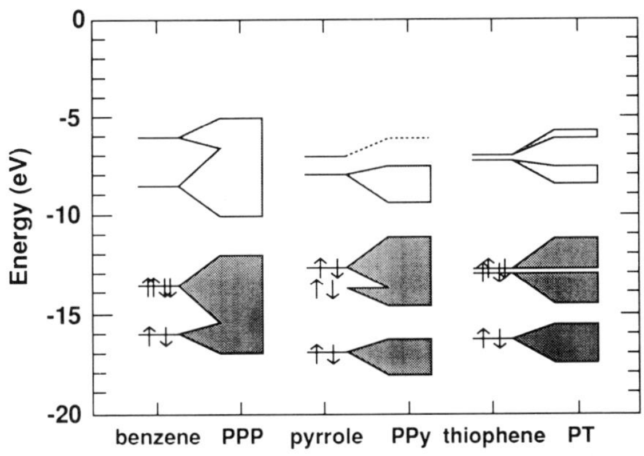

Fig. 7. π energy levels of monomer at left, of polymer at right, for PPP, PPy and PT. The π bands were calculated with a Hückel model. Light hatched areas indicate unoccupied bands.

The six electrons completely fill the lower three energy levels of the monomer, and these remain filled as the monomers are linked to form a band. Therefore, just as in PPP, the polymers PPy and PT are semiconductors due to their inherent structure.

3.2. Band structure calculations for polymers

Conjugated polymers that form three-dimensional crystalline solids are amenable to the usual band structure calculation techniques. However, the unit cells of most of these polymers are too large for the most accurate calculation techniques, and an ab initio calculation has been performed only on t-PA. The band structure calculated within the local density approximation (LDA) is shown in fig. 8. The π bands are indicated by arrows at the right. The fact that interchain spacing is much larger than intrachain spacing, leads to a rod-shaped Brillouin zone. The wave vector in the chain direction, which corresponds to the wave vector in a one-dimensional model of the polymer, lies along the long axis of the Brillouin zone (symmetry direction ΓB in fig. 8). The other symmetry axes correspond to electron wave vectors perpendicular to the chain. As expected, the dispersion of the conduction band perpendicular to the bands (0.53 eV according to the calculation of Vogl and Campbell 1990) is much less than its dispersion parallel to the bands (approximately 10 eV).

For other conjugated polymers, including PPP, PT and PPV, the band structure has been calculated in two steps. In the first step, the theoretical geometry of the polymer is optimized by performing an energy minimization on the largest possible oligomer using quantum chemical calculation methods (modified neglect of differential overlap or MNDO). In the second step, the electronic structure calculation is performed using a parametrized valence effective Hamiltonian (VEH) (Brédas et al. 1983). The VEH allows calculations for large structures because only one-electron integrals need to be evaluated and the atoms are frozen in place, thus eliminating the computationally intensive step of iterating for the optimal geometry.

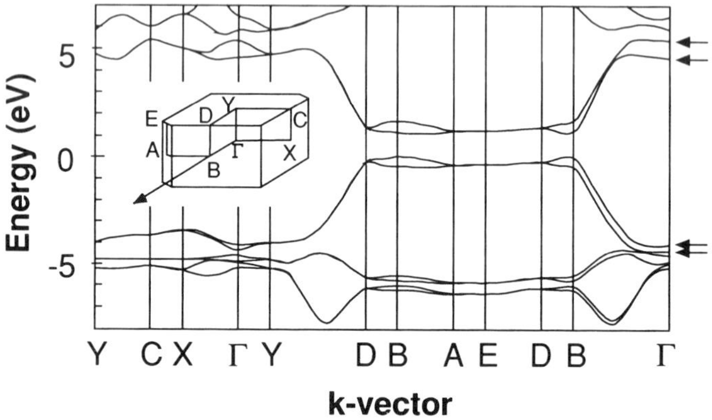

Fig. 8. Calculated electronic band structure of t-PA (after Vogl and Campbell 1990). The arrows indicate the π bands.

4. Ground state of polyacetylene

4.1. Electron–phonon coupling

To calculate the gap due to the Peierls distortion it is useful to take the view that the Peierls distortion results from the electron–phonon interaction. Because phonon energies are quite small compared to the Fermi energy E_F, there is a strong interaction of electrons with phonons that produces transitions from states with the Fermi wave vector k_F to $-k_F$ and vice versa. This strong interaction with $2k_F$ phonons produces a charge-density wave with displacement u_n for the atom at position x_n on the chain given by

$$u_n = u_0 \cos Qx_n, \quad Q \equiv 2k_F, \tag{4.1}$$

where u_0 is a constant.

The effect of introducing a new periodicity Q on the chain is, as is well known, to cause a mixing of states with wave vector k and those with $k \pm Q$. The amount of that mixing is proportional to the amplitude of the frozen-in-distortion u_0 and the strength of the electron–phonon coupling. The source of the coupling between electrons and phonons is the dependence of the transfer integral t on the spacing between molecules. Because the displacements are small compared to the spacing between atoms, we may write

$$t = t_0 + \alpha(u_n - u_{n+1}), \quad \alpha \equiv (\partial t/\partial u)_0, \tag{4.2}$$

α being a measure of the electron–phonon coupling. From eq. (4.2) we anticipate that the amount of mixing of states with k and $k \pm Q$ is proportional to α as well as u_0. As will be shown shortly the Peierls gap is, in fact, proportional to αu_0.

4.2. Calculation of the gap

To calculate the gap we introduce the Hamiltonian

$$H_{el} = -\sum_{n=1}^{N} [t_0 + \alpha(u_n - u_{n+1})][c_{n+1}^{\dagger} c_n + c_n^{\dagger} c_{n+1}] \tag{4.3}$$

where c_n^+ and c_n are the creation and annihilation operators for a π electron on the nth C atom and N the number of C atoms on the chain. Note that we neglect the chain zigzag. To find the effect of mixing states of different k, we transform the Hamiltonian (4.3) to the momentum representation and diagonalize it (see, e.g., Conwell 1987). This leads to the eigenvalues for the Peierls distorted state

$$E_k = \pm[(2t_0 \cos ka)^2 + (4\alpha u_0 \sin ka)^2]^{1/2}. \tag{4.4}$$

the + sign referring to the upper band, the − sign to the lower. The first term will be recognized, from eq. (3.2), as the energy of an electron with wave vector k in a uniform chain. Because the states strongly affected by the Peierls transition have $k \simeq k_F = \pi/2\alpha$, we set sin $k\alpha$ equal to unity in eq. (4.4), resulting in

$$E_k = \pm(\varepsilon_k^2 + \Delta_0^2)^{1/2}, \tag{4.5}$$

where

$$\Delta_0 = 4\alpha u_0. \tag{4.6}$$

The Peierls gap is then $8\alpha u_0$. Although this equation relates the Peierls gap to u_0, it does not yet determine the gap because the equilibrium value of u_0 is not yet known.

To determine the equilibrium value of the gap we must find the value of Δ_0 or u_0 that minimizes the total energy of the system, the sum of the electronic energy and the lattice distortion energy. The total electronic energy E_{el} for zero temperature is just the sum of eq. (4.5) over all the states in the lower band. For small dimerization, or small Δ_0/t_0, E_{el} is well approximated by (Heeger et al. 1988, Conwell 1987),

$$\frac{E_{el}}{N} = \frac{-4t_0}{\pi} - \frac{\Delta_0^2}{2\pi t_0}\left[\ln\left(\frac{8t_0}{\Delta_0}\right) - \frac{1}{2}\right]. \tag{4.7}$$

The energy of the lattice distortion may be obtained from

$$H_{lat} = \tfrac{1}{2}K \sum_{n=1}^{N} (u_{n+1} - u_n)^2, \tag{4.8}$$

where K is a stiffness constant, taken as that of the σ electrons. When the displacement of each site is $\pm u_0$, the sum over n gives the lattice energy

$$E_{lat}/N = 2Ku_0^2 = 2K\Delta_0^2/16\alpha^2. \tag{4.9}$$

Minimizing the total energy of the system, eq. (4.7) plus (4.9), with respect to Δ_0, we obtain

$$\Delta_0 = (8t_0/e)\exp(-1/2\lambda), \tag{4.10}$$

where λ, the dimensionless electron–phonon coupling constant, is given by

$$\lambda = 2\alpha^2/\pi K t_0. \tag{4.11}$$

Because the Hamiltonian we have been using is greatly simplified, neglecting, among other things, electron–electron interactions, it is usual to choose the parameters to fit experimental data. X-ray measurements give $u_0 = 0.04$ Å (Heeger et al. 1988). The size of the gap is somewhat uncertain because processes other than electron–hole creation may contribute to optical absorption near the band edge. We choose 1.4 eV as the gap. With Δ then 0.7 eV and $u_0 = 0.04$ Å, eq. (4.6) requires $\alpha = 4.2$ eV/Å. The usual values taken for t_0 and K are 2.5 eV and 21 eV/Å^2, respectively, leading to the electron–phonon coupling constant $\lambda = 0.19$.

4.3. Order parameter and continuum model

If we include the kinetic energy of the molecules, the total Hamiltonian becomes the SSH Hamiltonian (Su et al. 1980)

$$\begin{aligned} H = & -\sum_{n=1}^{N} [t_0 + \alpha(u_n - u_{n+1})](c_{n+1}^{\dagger} c_n + c_n^{\dagger} c_{n+1}) \\ & + \tfrac{1}{2} K \sum_n (u_n - u_{n+1})^2 + \tfrac{1}{2} M \sum_n \dot{u}_n^2, \end{aligned} \tag{4.12}$$

where M is the mass of a C–H. Despite its simplifications this Hamiltonian has been widely used and has led to many useful results, one of which was seen in the last section. It is not, however, useful for obtaining analytic results for such defects as solitons and polarons that involve distortions of the dimerization pattern. Fortunately, for these defects the dimerization pattern varies slowly on the scale of the lattice constant. In order to describe the defects quantitatively it is useful to define the order parameter

$$\tilde{u} = (-1)^n u_n. \tag{4.13}$$

Introducing the order parameter makes it possible to specify the local dimerization with a single number. Treating the order parameter as a continuous variable, Takayama et al. (1980), with some additional approximations, have derived a Hamiltonian that is the continuum limit of (4.12). From this Hamiltonian, equations of motion and a gap equation have been derived that are quite useful for obtaining analytic results. Although we will not derive these equations here, we will quote some of the analytic results. Because of somewhat different approximations, the expression for the half gap in the continuum model is

$$\Delta_0 = 4t_0 \exp(-1/2\lambda), \tag{4.14}$$

which differs from the results (4.10) obtained from the SSH Hamiltonian in the replacement of $8t_0/e$ by $4t_0$. For $\lambda = 0.19$ and $t_0 = 2.5$ eV eq. (4.14) leads to $\Delta_0 =$ 0.7 eV, consistent with our set of basic parameters, whereas Δ_0 obtained from eq. (4.10) is only 0.53 eV. We will therefore use eq. (4.14) rather than eq. (4.10).

5. Solitons

5.1. The order parameter

As discussed previously, a soliton is a domain wall between the two possible bond arrangements of t-PA. It can be described quantitatively using the SSH Hamiltonian or the continuum model. Numerical solution of the SSH Hamiltonian has been carried out for the case of the soliton (Su et al. 1980) and leads to the result stated earlier that the extent of the soliton is 14 sites. An analytic solution within the continuum model may be obtained by using the order parameter introduced in the last section. For the soliton shown in fig. 2a the first bond on the chain is a short one, obtained by moving the first ($n = 1$) C–H to the right, the second ($n = 2$) to the left, etc. Thus for the perfectly dimerized region to the left of the soliton $(\tilde{u}/u_0) = -1$. Because of the phase shift due to the soliton, for the perfectly dimerized region to the right of the soliton, $(\tilde{u}/u_0) = +1$. The fact that $(\tilde{u}/u_0) \rightarrow -1$ as $x \rightarrow -\infty$, and $(\tilde{u}/u_0) \rightarrow +1$ as $x \rightarrow +\infty$, suggests a tanh function for the order parameter of the soliton. In fact, the order parameter of the soliton obtained from the solution of the SSH Hamiltonian is well fitted by the function.

$$(\check{u}/u_0) = \pm \tanh[(x - x_0)/\xi], \tag{5.1}$$

where x_0 is the coordinate of the center of the soliton and 2ξ approximately its length, and the + sign holds for the soliton of fig. 2a, the − sign for that of fig. 2b. By using the order parameter (5.1) in the continuum model, it is found that (Takayama et al. 1980)

$$\xi = 2t_0 a/\Delta_0. \tag{5.2}$$

With the values of the parameters given earlier, ξ is found to be $7a$, in agreement with the SSH result. The order parameter for the soliton of fig. 2a is plotted versus distance in fig. 9a.

5.2. Soliton creation energy and mass

Within the approximations of the SSH Hamiltonian and the continuum model the energy E_s required to create a soliton is readily calculated. E_s is the sum of: (1) the energy Δ_0 required to raise an electron from the top of the valence band to the soliton level at midgap; (2) the energy of the lattice deformation and (3) the energy change of the valence electrons because the soliton causes a phase shift. The

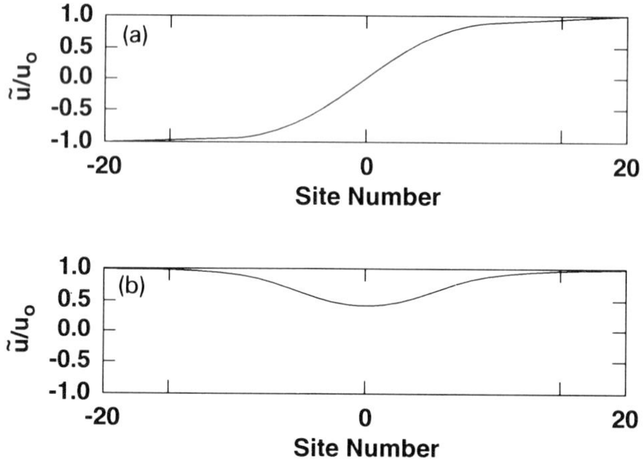

Fig. 9. (a) Order parameter versus distance of the soliton in fig. 2a from eq. (5.1). (b) Order parameter versus distance of the polaron from eq. (6.1). This order parameter is seen to be a superposition of the soliton order parameter $-\tilde{u}/u_0$, i.e., fig. 9a reversed right to left, and fig. 9a.

contribution (2) is readily calculated by turning the summation in eq. (4.8) into an integration and using $(u_{n+1} - u_n)^2 \simeq (2\tilde{u}_n)^2$, with $\tilde{u}_n$ taken from eq. (5.1) written in the continuum limit. The result of the calculation (Takayama et al. 1980) is

$$E_s = 2\Delta_0/\pi. \tag{5.3}$$

For the parameters quoted earlier, the soliton creation energy is found to be 0.45 eV.

The soliton, also called kink, that we have been discussing is one representative of a general class of solitons (Scott et al. 1973). An essential feature of this class of objects is the ability to move without change of shape. In the present case, the shape is determined by the set of chain distortions specified by $\tanh(x/\xi)$ to the accuracy of the continuum model. Thus we would expect that, at least for a speed v that the C–H groups can follow, i.e., $v \ll v_s$, the sound velocity, the displacements of the moving solitons would be described by

$$\tilde{u}(x, t) = u_0 \tanh[(x - vt)/\xi]. \tag{5.4}$$

Associated with the displacements of the C–H groups described by eq. (5.4) there is a kinetic energy

$$\mathrm{KE} = (M/2) \sum_{n=1}^{N} (\mathrm{d}u_n/\mathrm{d}t)^2 = (M/2a) \int_{-L/2}^{L/2} (\mathrm{d}u(x, t)/\mathrm{d}t)^2 \,\mathrm{d}x, \tag{5.5}$$

where M is the mass of the C–H group and L the length of the chain. With $\tilde{u}$ given by eq. (5.4) the result of the integration in eq. (5.5) (taking $L = \infty$) is

$$\mathrm{KE} = \frac{1}{2}\left(\frac{4Mu_0^2}{3a\xi}\right)v^2 \tag{5.6}$$

Thus the apparent mass of the soliton m_s is given by

$$m_s^0 = 4Mu_0^2/3a\xi. \tag{5.7}$$

The superscript ^{0} has been added to m_s in eq. (5.7) to emphasize that it is the mass in the limit $v \ll v_s$. With the parameters we have been using $m_s^0 \simeq 5m_e$, where m_e is the electron mass.

The motion of a soliton with large amounts of kinetic energy has been studied by numerical integration of the SSH Hamiltonian (4.12) for long chains of t-PA. It was found that for kinetic energies up to $\sim 0.1\Delta_0$ there is uniform soliton translation, well described by eqs. (5.4)–(5.6). Above $0.1\Delta_0$ an oscillatory tail develops and the soliton velocity does not continue to increase (Bishop et al. 1984). A limiting soliton velocity is found which, for the parameters given above for t-PA, is $c = 2.7v_s$ (Bishop et al. 1984, Guinea 1984). The existence of a limiting velocity indicates that the soliton mass is a function of v_s, approaching infinity as $v \to c$. Such "relativistic" behavior of mass with velocity is familiar for solitons obtained for other 1D systems, e.g., from sine-Gordon and Φ^4 field theories (Bishop and Schneider 1978).

Within the framework of the SSH Hamiltonian, where Coulomb effects are not explicitly included (although they are included implicitly by choosing parameter values that agree with experiment) the description above of solitons applies equally to neutral and charged solitons. A study incorporating Coulomb effects to the extent of the MNDO technique finds that there are indeed differences between neutral and charged solitons (Chance et al. 1986). Comparison is complicated by the fact that the "length" of the soliton is found to depend on the property being measured: the spatial extent of the departure of the bonds from perfect dimerization, or of the localized electron (hole) wavefunction or of the charge density associated with a charged soliton. The lengths associated with the latter two characteristics, which are the ones more accessible to measurements, are comparable to the length deduced from the SSH Hamiltonian and the continuum model.

Soliton defects can be added to a chain in pairs. The interaction of the two solitons splits the midgap level into two levels – "bonding" and "antibonding" between the two localized soliton wavefunctions. As more soliton pairs are added to the chain, a midgap soliton band forms, with the width of the band increasing as the average soliton spacing decreases.

5.3. *Experimental evidence for solitons*

Various experiments support the existence of solitons in trans-polyacetylene. Electron spin resonance measurements on undoped t-PA made by the Shirakawa process indicate the presence of about 1 spin per 1000 to 3000 atoms, the spin being characterized by the g-value of a π electron (Weinberger et al. 1980). Various types of magnetic resonance experiments indicate that these spins undergo 1D diffusion along the chains, as would be expected for solitons. It is now widely accepted that they are neutral solitons. (For a review, see Clarke and Scott 1986.) The good agreement of theory with experimental data for their diffusion rate, to be discussed in § 8, is further verification that the spins are due to S^0's. In view of the large creation energy for solitons, ~ 0.5 eV, it is clear that such a large concentration could not be thermally activated at room temperature and below. It has been suggested that the S^0's are created during the cis–trans isomerization and prevented from recombining by the presence of other defects, such as crosslinks.

As discussed earlier, charged solitons may be created either by doping or by irradiating t-PA with photons of suitable energy. The results of irradiation turn out to be quite complex and will be discussed in § 9. In the doping process, when an electron or hole is added to a chain where there is an S^0 present, it accepts the electron or hole to become an S^- or S^+, respectively. Because a charged soliton has no spin, as is clear from fig. 2c, doping should, at least initially, lead to a decrease in the number of spins. This has indeed been observed. (See, e.g., Moraes et al. 1985.) Still stronger evidence that doping, up to at least 5 or 6%, produces predominantly charged solitons, comes from the magnetic susceptibility observed for such doping. (See, e.g., Moraes et al. 1985.) After correction for the core susceptibility and the Curie susceptibility due to isolated defects, the remaining susceptibility, that due to the electron spins, is found to be very small. With the conductivity nevertheless quite high at those doping levels, the situation has been characterized as "spinless conductivity". Whether this characterization is deserved will be discussed in § 10. Further evidence for the existence of solitons in t-PA will be detailed in § 7.

6. Polarons and bipolarons

The SSH Hamiltonian and its continuum form predict not only the existence of soliton or kink defects but also polarons. Polarons may occur in low density doping, where solitons cannot form because they must form in pairs. They also occur when the electron and hole are created on separate chains. The chain distortion due to the polaron is shown schematically in fig. 1. The order parameter for the polaron distortion in the continuum model is given by (Campbell and Bishop 1981)

$$\frac{\tilde{u}}{u_0} = 1 - \frac{1}{\sqrt{2}}\left[\tanh\left(\frac{(x+x_0)}{\sqrt{2}\xi}\right) - \tanh\left(\frac{(x-x_0)}{\sqrt{2}\xi}\right)\right], \tag{6.1}$$

where ξ is the decay length for the soliton, given in eq. (5.2), and x_0 is determined by

$$x_0 = (\xi/\sqrt{2})\tanh^{-1}(1/\sqrt{2}). \tag{6.2}$$

The order parameter is shown in fig. 9b. It becomes smaller, i.e., the chain is less dimerized, in the middle of the polaron. The form of the order parameter is that of a closely spaced soliton–antisoliton pair, with one soliton charged and the other uncharged. With this in mind, one can understand why the polaron localizes two states in the gap. The two states due to the two soliton-like deformations, which would be at midgap for large x_0, couple significantly and split into an upper level and a lower level. The lower level is filled with the two electrons from the valence band. As shown in fig. 1b, a donated electron which causes the formation of a negatively charged polaron partially fills the higher midgap level, while a donated hole empties one of the electrons in the lower midgap state.

In NDGS polymers added charges also distort the chain, giving rise to a polaron with two gap levels, but because these polymers are more complicated, the chain distortion cannot be described simply by eq. (6.1). Calculations have shown that, in PPP, e.g., the polaron is characterized by the rings becoming slightly quinoid, i.e., the horizontal bonds shorten while the other bonds lengthen (Brédas et al. 1982). This is shown schematically for a P^+ in fig. 10. The bipolaron, which results from the combination of the two like-sign polarons close together has a similar structure (fig. 10b). Because the length of the bipolaron is about the same as that of the polaron, the strain involved in the bipolaron is greater. Also tending to destabilize the bipolaron is the repulsion of the two like charges. Nevertheless data on many polymers show that at high doping, where the polarons come close together, the

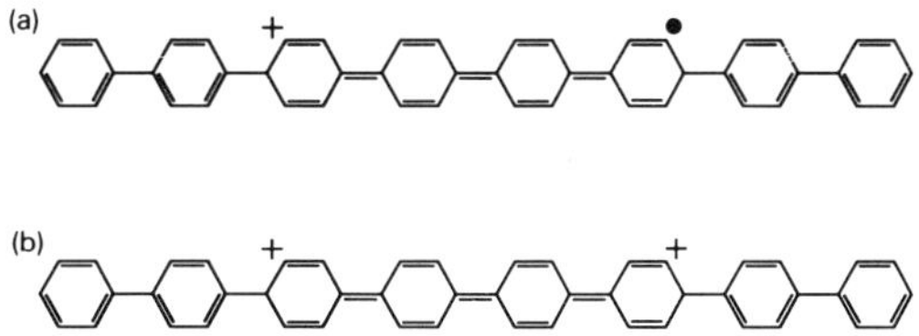

Fig. 10. Schematic drawings of (a) a polaron in PPP, (b) a bipolaron in PPP.

charges donated to the chains are housed in bipolaron states. The most telling data in this regard are optical absorption, to be discussed in the next section, and magnetic susceptibility. Just as in the case of t-PA, the spin susceptibility is found to be quite small, despite relatively high conductivity, to high doping levels for PPy, and PT, for example. This is unambiguous evidence for the donated charges being in bipolaron states because the bipolaron, with two charges, has no net spin. (For detailed discussion of the experimental evidence for bipolarons in doped polymers see Conwell and Mizes 1991.)

7. Optical properties

7.1. Absorption of doped t-PA

As discussed earlier, the formation of solitons, polarons, or bipolarons introduces states in the band gap of the conjugated polymers. The presence and energetics of these states can be directly observed via changes in the subbandgap optical absorption. According to fig. 2c, formation of a soliton should result in an optical transition at an energy of half the gap. Figure 11 shows a series of optical absorption spectra for t-PA electrochemically doped with ClO_4 as the doping percentage increases (Feldblum et al. 1982). The spectrum of undoped t-PA is shown by curve 1 and indicates a semiconductor with a band gap of approximately 1.4 eV. Doping the polymer causes a midgap absorption peak centered at ~0.7–0.8 eV to appear. This peak increases in height and broadens as the dopant concentration increases. The midgap absorption is found to be universal for a variety of both donor and acceptor dopants. The universality indicates that the midgap level is associated with the polymer chain, rather than electronic states on the dopant, and is thus a direct confirmation of the existence of solitons.

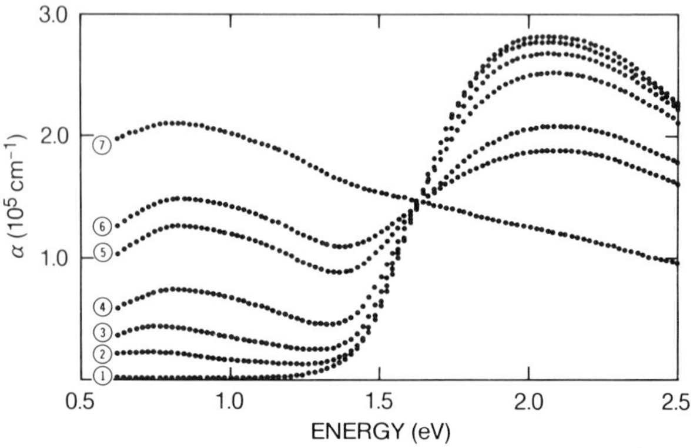

Fig. 11. Absorption spectra for doped t-PA, the doping increasing as the numbers do (after Feldblum et al. 1982).

The simple theory of § 1.2 predicts that both neutral and charged solitons form a level at midgap. Electron spin resonance (ESR) shows undoped t-PA has a concentration of neutral solitons that should significantly absorb light (Eckhardt and Chance 1983). However, photothermal deflection spectrometry, a technique to detect small absorption via the heating of the sample it causes, shows on an undoped sample two orders of magnitude less absorption at midgap than expected (Weinberger et al. 1984). One concludes from this measurement that the energy of the neutral soliton is shifted from midgap to a position so close to the conduction band edge that its absorption is difficult to detect. This shift is not unexpected, because the simple theory we presented neglects electron–electron interactions.

In addition to the midgap absorption two strong infrared absorption lines, at $\sim$900 and 1370 cm^{-1}, are found to occur upon light doping for any dopant and grow linearly in intensity with increasing doping. A third, weaker vibration associated with the doping is also observed, at 1270 cm^{-1} (for references and further discussion see Heeger et al. 1988). There are two possible origins for such infrared modes:

(1) The coupling of the π electrons to C-atom vibrations along the chain, which results in oscillations of the excess charge in the soliton level in response to the chain vibrations. This results in a much larger dipole moment along the chain than these C-atom vibrations would otherwise have, making them *i*nfra-*r*ed *a*ctive *v*ibrations, abbreviated IRAV.

(2) The oscillations of the charged soliton in the Coulomb field of the impurity ion to which it is bound (Su et al. 1980). These oscillations are called the "pinned mode". Subsequent work, to be discussed in § 9, indicated that the 900 cm^{-1} absorption represents the pinned mode. Indeed, it is found to vary somewhat in frequency and intensity with different dopants. The 1270 and 1370 cm^{-1} modes are due to the coupling with the C=C stretch and the C–C modes (Horovitz 1982, Vardeny et al. 1983).

7.2. Photoinduced absorption in t-PA

Irradiation of a conjugated polymer with photons of energy greater than the band gap creates electrons and holes, as in any semiconductor. In the polymers, however, as has been discussed at length, unless an exciton is created, each electron or hole causes the chain to relax around it, creating a characteristic excitation. The most direct evidence for this comes from studies of photoinduced light absorption, carried out in greatest detail for t-PA. This work was stimulated by dynamical calculations of Su and Schrieffer (1980), based on the SSH Hamiltonian, eq. (4.12), of the time evolution of the relaxation pattern on a t-PA chain after generation of a hole–electron pair. They found that the photoinjected pair evolves to an S–$\bar{S}$ pair in $\sim 10^{-13}$ s. Many experiments followed. Orenstein and Baker (1982) found, with experiments having microsecond time resolution, that irradiation with photons above gap energy produced two different transient optical absorption bands at energies less than the gap. The excitation associated with the higher energy absorption, at 1.35 eV, was found to be neutral, while that associated with the lower energy absorption, centered at ~0.43 eV, was charged (Orenstein et al. 1983). The origin of the neutral absorption is still uncertain, but the lower energy absorption has been clearly identified as due to charged solitons. One type of evidence for this identification is the observation, along with the 0.43 eV centered absorption, of two narrow infrared absorption lines close in frequency to the 1350 and 1270 cm^{-1} lines, the IRAV, observed in doped samples (Vardeny et al. 1983). The fact that the 900 cm^{-1} absorption line seen for doping induced solitons is not seen for the photogenerated solitons was taken as evidence that this line represents the pinned mode, resulting from the vibration of the soliton in the Coulomb well provided by the dopant ion (Vardeny et al. 1983). The Coulomb well provided by the dopants is also presumably the reason for the difference in the absorption band of solitons produced by doping, which is centered at ~0.7 eV, and that of the photogenerated solitons, centered at 0.43 eV. Additional evidence for the production of charged solitons was obtained by Flood et al. (1982) who showed that the photoinduced excitations remaining at microseconds and longer do not have spin.

7.3. Absorption in NDGS polymers

In NDGS polymers, the excitations produced by doping are polarons and bipolarons. The levels in the gap formed for a polaron are shown in fig. 1b. For a bipolaron there are also two levels in the gap, but these are both filled or both empty. As a result, although polarons have three subgap absorption lines (two absorptions have the same energy due to charge conjugation symmetry) bipolarons have only two. This fact has made it possible to distinguish between polarons and bipolarons. In PPy, e.g., although there are three absorption lines at low doping, one is found to disappear at high doping (Brédas et al. 1984). As discussed in the last section, the formation of bipolarons at high polaron density is expected, and has been seen in many other NDGS polymers as well.

The excitations that can be generated directly by light in NDGS polymers are polarons and excitons. The excitons in this case have two energy levels in the gap, just as the polaron does (see fig. 1b) but differ in having one electron in the upper and one electron in the lower state. They are sometimes called polaron excitons or polar excitons and may have larger binding energies than 3D semiconductor excitons because they include chain deformations. These excitons are expected to decay rapidly, either by emission of radiation or by non-radiative processes. Luminescence from these excitons has been observed in c-PA, PT, and PPV, to mention a few cases. In PPV luminescence has also been observed as a result of injecting carriers at contacts (Burroughes et al. 1990). This electroluminescence is in the visible and has been used in light-emitting diodes.

Despite the high probability of forming excitons that have a rapid geminate decay, there is evidence, particularly from photoconductivity measurements to be discussed in § 9, that some polarons are also created by photon absorption. As in t-PA these may occur due to the initial electron and hole being created on different chains. Also polarons may result from excitons that escape recombination, freeing their electrons and holes. Indirect evidence for the initial creation of polarons arises also from the finding of bipolarons in steady state experiments of various kinds, because bipolarons must result from the combination of like sign polarons. Steady-state experiments essentially observe the excitations microseconds or later after the onset of illumination. The induced steady-state photoabsorption for PT and PPV, e.g., shows two peaks rather than three, unambiguous evidence for bipolarons (Vardeny et al. 1986, Bradley and Friend 1989). The induced steady-state photoabsorption and the photo-induced IRAV (Moraes et al. 1984) are found to be quite similar to the absorption and IRAV of doped samples (Chung et al. 1983) indicating that the same species is dominant in both cases. Further evidence that this species is bipolarons is given by ESR measurements on the steady-state photogenerated excitations in PT (Vardeny et al. 1986). These measurements showed that only a relatively small number of spins, presumably polarons, are left at microseconds and later. After improving the structural order and increasing the molecular weight by annealing the PT samples at high temperature, Mo et al. (1985) found the photogenerated spins dropped below the detectable limit, whereas the photoinduced absorption signal remained the same. This indicates that polarons can exist at long times in low concentration in samples in which defects and imperfections are sufficiently numerous to restrict polaron mobility and thus inhibit the formation of bipolarons.

It may be noted that because the monomers are more complex than that of polyacetylene, there are more backbone vibrations to couple to and a larger number of IRAV than in polyacetylene. For a discussion of the IRAV in PT and references, see Heeger et al. (1988).

8. Soliton and polaron mobility

8.1. Experimental data – neutral soliton

Extensive data on the diffusion of the neutral soliton have been obtained by magnetic resonance techniques, specifically ESR, NMR and nuclear dynamic polarization. Deduction from the experimental results of the magnitude of the diffusion rate $D_{\|}$, and its variation with temperature, has, however, been fraught with controversy. A major source of the difficulty is that a fraction of the solitons are trapped, that fraction varying with temperature, and the trapping has different effects on different types of magnetic resonance measurements. As pointed out by Clarke and Scott (1986) the crux of the problem of measuring $D_{\|}$ lies in determining what fraction of the spins is trapped. This determination has been carried out with some care by two different experimental groups, whose results for $D_{\|}$ are shown in fig. 12. It is seen that at 300 K $D_{\|}$ is in the range 0.02 to 0.04 $cm^2\ s^{-1}$, considerably lower than is found for electrons in silicon, for example. Interestingly, a value in good agreement, 0.02 $cm^2\ s^{-1}$, was obtained optically from the length of time for the loss of polarization memory (Vardeny et al. 1982). $D_{\|}$ does not vary much with temperature T from 300 K down to 200 or 150 K and then decreases fairly rapidly with further decrease in T. The agreement between the two sets of magnetic resonance data is reasonably

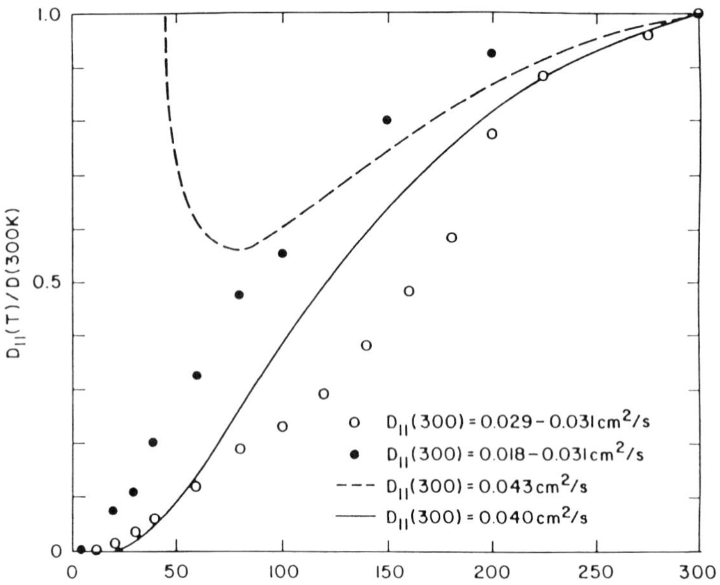

Fig. 12. Intrachain diffusion coefficient of neutral solitons in t-PA, normalized to 300 K values, versus temperature. Closed circles: data of Nechtschein et al. (1983); open circles: data of Mizoguchi et al. (1986); --- theoretical results for phonon scattering only; — theoretical results for phonon scattering plus barriers of 0.01 eV (after Jeyadev and Conwell 1987a).

good at the high temperature end but less good below ~200 K where, as will be seen, trapping becomes significant. The lack of agreement may be due to insufficient accuracy of the calculation of the number of trapped carriers or to genuine differences in trapping between the samples.

8.2. Scattering of solitons by phonons

One effect that is expected to be important in determining the S^0 diffusion rate is scattering by phonons. Because optical phonons in t-PA have very high energy, ~0.2 eV, the important scattering at 300 K and below must be due to the lower energy acoustic phonons. Even in undoped materials the S^0 density is not high, as noted earlier, and the solitons may be taken to have a Maxwell–Boltzmann distribution of velocities. In the absence of scattering the soliton has a constant wave vector $\boldsymbol{k} = m_s \boldsymbol{v}/\hbar$, where m_s is the soliton mass discussed in § 5. Because the soliton is an electron or hole dressed with lattice vibrations, it is reasonable to look for a treatment of soliton scattering by phonons similar to that of electron or hole scattering by phonons in a semiconductor. It was assumed that the internal state of the soliton is unchanged by the scattering, which only causes a change in wave vector from $\boldsymbol{k}$ to $\boldsymbol{k}'$.

For solitons treatment of scattering is simplified because they are constrained to move along the chain. The conditions of conservation of momentum and energy lead to the possible values of phonon wave vector $\boldsymbol{q}$ for emission or absorption. In the range of soliton energy where the mass is constant at m_s^0,

$$q = 2(|k| - k_s) = \frac{2m_s^0}{\hbar}(|v| - v_s) \quad \text{emission,} \tag{8.1}$$

where

$$k_s = m_s^0 v_s/\hbar. \tag{8.2}$$

It is seen from eq. (8.1) that phonon emission is not possible unless the soliton velocity is greater than the sound velocity. Solitons with $v < v_s$ may still be scattered by absorbing a phonon. The absorption process may take place either with the $\boldsymbol{q}$ vector of the phonon absorbed in the same direction as $\boldsymbol{k}$ or in the opposite direction to $\boldsymbol{k}$. Conservation of energy and momentum for these cases leads to

$$q = 2(k_s \pm |k|) = \frac{2m_s^0}{\hbar}(v_s \pm |v|) \quad \text{absorption,} \tag{8.3}$$

where the upper sign holds when the velocity after scattering is opposite in direction to the initial velocity, the lower sign when the velocities are in the same direction.

From eqs. (8.1) and (8.3) it is possible to evaluate the wavelength $\lambda_p = 2\pi/q$ of the phonons important in scattering. The average kinetic energy of a freely moving soliton at the low concentrations we are concerned with is $\frac{1}{2}k_B T$, k_B being Boltzmann's constant. Thus the average velocity $\langle v \rangle = (k_B T/m_s^0)^{1/2}$. Taking $m_s^0 = 5.45 m_e$ (Su et al. 1980) we find for $T = 100$ K, $\langle v \rangle = 1.67 \times 10^6$ cm/s. The velocity of sound in polyacetylene has been measured as 1.5×10^6 cm/s (Moses et al. 1980). Thus at 100 K, $\langle v \rangle$ is already larger than v_s. It is still sufficiently smaller than the limiting soliton velocity $c = 2.7 v_s = 4 \times 10^6$ cm/s discussed in § 5, however, to justify taking $m_s = m_s^0$ as was done above in calculating $\langle v \rangle$. With $\langle v \rangle > v_s$ the average soliton will relax by emission, which is more important than absorption, particularly at low temperatures (Jeyadev and Conwell 1987b). For $|v| = \langle v \rangle$ at 100 K, λ_p obtained from eq. (8.1) is equal to a few hundred lattice constants. To obtain λ_p for a temperature of 300 K it is necessary to take into account the increase in mass of the soliton with $\langle v \rangle$. The result is that the wavelength of the phonon involved in the scattering (emission) process is ~40 lattice constants (Jeyadev and Conwell 1987a). Thus the wavelengths of the acoustic phonons responsible for scattering in the range 100 K to 300 K which, as will be seen, is the range in which phonon scattering determines $D_{\parallel}$, are, at the least, several times the length of the soliton.

The fact that the important phonons for scattering have wavelengths many times the soliton length means that the scattering may be thought of as due to the compression or expansion of the soliton by the phonon or sound wave. The compression or expansion changes the energy E_s of the soliton by changing the lattice constant a. The energy change δE_s due to a strain $\delta a/a$ is given by

$$\delta E_s = (\partial E_s/\partial a)\delta a = (a\,\partial E_s/\partial a)(\delta a/a) \tag{8.4}$$

The quantity $(a\,\partial E_s/\partial a)$ is a deformation potential, to be denoted E_D. To evaluate E_D for this case we use $E_s = 2\Delta_0/\pi$ (eq. (5.3)). (The kinetic energy contribution to the soliton is negligible compared to the creation energy.) Taking the dependence of Δ_0 on a from eq. (4.14), where both t_0 and λ depend on a, we obtain (Jeyadev and Conwell 1987a)

$$E_{\mathrm{D}} = -(8\alpha a/\pi)(1 + 1/2\lambda)\,\mathrm{e}^{-1/2\lambda}. \tag{8.5}$$

With the parameter values we have been using, eq. (8.5) gives $E_{\mathrm{D}} = 3.3$ eV.

The deformation potential E_{D} is similar in spirit to the deformation potential introduced by Bardeen and Shockley (1950) for long-wavelength acoustic phonons in semiconductors. The importance of that deformation potential arises from the proof by Bardeen and Shockley that the matrix element for scattering of electrons from the state $\boldsymbol{k}$ to $\boldsymbol{k}'$ by one of these phonons is the matrix element between these states of $E_{\mathrm{D}}(\partial y(x)/\partial x)$, E_{D} being the shift in electron energy per unit strain and $\partial y(x)/\partial x$ the strain, $y(x)$ being the displacement at the point x due to the scattering phonon. A similar theorem may be proved for solitons scattered by long-wavelength phonons. The matrix element for scattering from $\boldsymbol{k}$ to $\boldsymbol{k}'$ is then

$$M_{k'k} = \langle k', N'(q) | E_{\mathrm{D}}(\partial y(x)/\partial x) | k, N(q) \rangle, \tag{8.6}$$

where $N(q)$ and $N'(q)$ are, respectively, the number of phonons with wavevector q before and after the collision. With $y(x)$ given by (Kittel 1976)

$$y(x) = \sum_q (\hbar/2MNv_{\mathrm{s}}q)(a_q\,\mathrm{e}^{\mathrm{i}qx} + a_q^\dagger\,\mathrm{e}^{-\mathrm{i}qx}), \tag{8.7}$$

where M is the mass of a CH and N the number of sites on the chain, the absolute square of the matrix element for the case $m_{\mathrm{s}} = m_{\mathrm{s}}^0$ takes the familiar form (Jeyadev and Conwell 1987c)

$$|M_{k'k}|^2 = (\hbar/2MNv_{\mathrm{s}})E_{\mathrm{D}}^2 q \begin{cases} N(q)\delta_{k',q\mp k} & \text{absorption,} \\ (N(q)+1)\delta_{k',k-q} & \text{emission.} \end{cases} \tag{8.8}$$

With eq. (8.8) and first order perturbation theory the scattering time τ_{tot} including emission and absorption processes is readily obtained. It is found that τ_{tot} is inversely proportional to the soliton mass, the square of the deformation potential and the total number of phonons with wave vectors that satisfy conservation of energy and momentum for the three different scattering processes (Jeyadev and Conwell 1987a). The diffusion coefficient is obtained from $\tau_{\mathrm{tot}}(k)$ by means of

$$D_{\parallel} = \langle v(k)^2 \tau_{\mathrm{tot}}(k) \rangle \tag{8.9}$$

where the average is taken over the velocity distribution of the solitons.

Above 100 K the soliton mass cannot be taken as m_{s}^0 and the “relativistic” dispersion relation referred to in § 5 must be used. This changes the values of q for emission and absorption (i.e., eqs. (8.1) and (8.3) are no longer valid) and the equation for conservation of energy. The resulting expressions are complicated (Jeyadev and Conwell 1987c) and must be evaluated numerically.

The resulting $D_{\parallel}$ versus T is shown as a dashed line in fig. 12. It is seen that from ~200 K to 300 K $D_{\parallel}$ is in good agreement with the measured values, perhaps better than could be expected because of the scatter of the experimental values and the uncertainty in some of the parameters involved in both the theoretical and experimental determinations. It is noteworthy that there are no arbitrary parameters involved in this fit. Reasonable agreement persists down to perhaps 100 K but below that the calculated $D_{\parallel}$ increases rapidly, indicating that phonon scattering is ineffective. That is not unexpected; at low temperatures the number of phonons excited is very low, making absorption events rare, and phonon emission is no longer allowed for most solitons because $\langle v \rangle < v_s$. Thus at low temperatures the scattering must be due primarily to defects or traps in the materials, as is usually the case for electrons in semiconductors.

8.3. Trapping of solitons

Because a soliton cannot, in principle, leave a chain (except for the improbable event of two nearby solitons jumping simultaneously) any defect on the chain presents a barrier to soliton motion. A barrier of height V_0 would essentially block the motion of solitons with energy less than V_0. To obtain the effect of a barrier on the diffusion rate, the lower limit of the integration (8.9) for $D_{\parallel}$ could be chosen as V_0 rather than zero. The solid line in fig. 12 shows the fit to the experimental data obtained by choosing $V_0 = 0.01$ eV. This value was chosen because of an earlier suggestion by Gibson et al. (1985), that remanent cis-polyacetylene linkages in trans-polyacetylene chains present a barrier to soliton motion that they estimated as 0.01 eV. Further, they demonstrated, by measurements of infrared absorption, that there is always a minimum of 5% of cis linkages in trans-polyacetylene obtained by the Shirakawa process. There is thus in principle a sufficient density of these barriers to trap all the neutral solitons. Because the decrease in the theoretical $D_{\parallel}$ with decreasing T is quite similar to the decreases found experimentally, and the experimental samples were made by the Shirakawa process, it does appear that cis inclusions are responsible for the trapping below ~150 K. More work, both theoretical and experimental, on better defined samples, remains to be done in this area but it appears that the basic processes determining the transport of neutral solitons in pristine polyacetylene are understood.

In principle the theory for phonon scattering given above should apply to charged solitons as well. The mobility $\mu_{\parallel}$ of charged solitons moving along a chain may then be very simply calculated from $D_{\parallel}$ obtained above by use of the Einstein relation

$$D = \mu k_B T / |e|, \qquad (8.10)$$

where e is the charge on the soliton. The result is shown in fig. 13. It should be emphasized that the mobility calculated is that for charged solitons in low concentration drifting along a chain in the absence of impurities or other pinning centers. It should be applicable to photogenerated charge solitons in pristine trans-polyacetylene samples before they are trapped. Note also that in the temperature range concerned the exact value of the maximum soliton velocity c has little effect on the value of $\mu_{\parallel}$ (Jeyadev and Conwell 1987a,b,c).

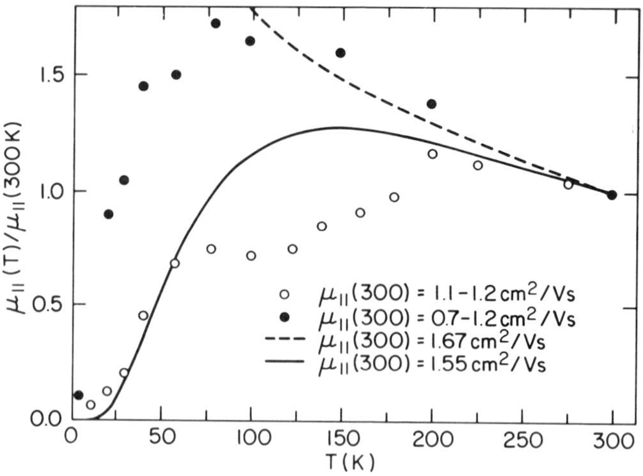

Fig. 13. Intrachain mobility of solitons in t-PA (normalized to 300 K value) calculated from $D_{\parallel} e/k_B T$, the $D_{\parallel}$ values taken from fig. 12. Closed circles: $\mu_{\parallel}$ from data of Nechtschein et al. (1983); open circles: $\mu_{\parallel}$ from data of Mizoguchi et al. (1986); dashed line, theoretical results for phonon scattering only; solid line, theoretical results for phonon scattering plus barriers of 0.01 eV (after Conwell and Jeyadev 1988).

8.4. *Polaron mobility*

The calculations described above for $D_{\parallel}$ and $\mu_{\parallel}$ of the solitons should also be valid for polarons because there too the wavelengths of the scattering phonons are at least several times the length of the polaron (Jeyadev and Conwell 1987a,c). The deformation potential E_D for the polarons is larger than that given in eq. (8.5) for the solitons by the factor $\sqrt{2}$ because the creation energy is larger by that factor. Thus E_D for polarons is $3.3\sqrt{2} = 4.7$eV. The maximum velocity c has not been calculated for polarons but is expected to be about the same as c for solitons. Calculations of $D_{\parallel}$ and $\mu_{\parallel}$ for polarons scattered by phonons, along the lines already described for solitons, lead to slightly larger values than for solitons because although E_D is larger, the low-energy mass of the polarons is smaller than that of the solitons by a factor greater than 5 (Jeyadev and Conwell 1987a,c). At 300 K $\mu_{\parallel} \simeq 2$ cm V^{-1} s^{-1}, about the same value as obtained for the soliton. The smaller mass also results in the effects of mass variation being larger over all, and starting at lower temperatures, for the polarons. Of course the actual $D_{\parallel}$ and $\mu_{\parallel}$ may be smaller than those calculated, particularly at low temperatures, due to scattering by defects. As in the case of solitons, the calculations apply only to polarons present in low concentration, drifting along a chain in the absence of impurities or other pinning or trapping centers. There is evidence, to be discussed in § 9, that photogenerated polarons in pristine samples satisfy these conditions for short times after their generation. Significantly, the value of $\mu_{\parallel}$ deduced from the measured picosecond photoconductivity on the basis of a rough estimate of the carrier concentration is ~2 cm^2 V^{-1} s^{-1} (Sinclair et al. 1986), in good agreement with the calculated values for $\mu_{\parallel}$ for soliton and polaron.

Unlike the soliton, the polaron can move between chains. This has been demonstrated experimentally in oriented t-PA samples by showing that there is photocurrent perpendicular to the chain direction in the picosecond regime (Walser et al. 1988) where, as will be discussed later, there is evidence that polarons exist. There are two possibilities for the interchain motion. One possibility is band motion, in which the lattice distortion moves simultaneously with the electron, as is the case for the intrachain soliton and polaron motion we have been discussing. The other possibility is polaron hopping between chains, with phonon assistance to compensate for a difference in distortion patterns. Both types of motion are predicted to occur for the small polaron found in ionic crystals, e.g., which by definition extends no more than one lattice constant in any direction (Holstein 1959). The polarons in polyacetylene and other conducting polymers, although satisfying this definition perpendicular to the chain direction, are quite extended along the chain direction, ~ 20 sites for polyacetylene, for example.

The critical quantity for determining whether the polaron motion transverse to the chains is bandlike, i.e., coherent, or hopping, is the transverse polaron bandwidth W_{tr}. Note that this bandwidth is smaller than the transverse electronic bandwidth, $4t_{\perp}$, where $t_{\perp}$ is the transfer integral (see eq. (3.1) for the direction transverse to the chain. There are two reasons for the smaller bandwidth: (1) the CHs also move in the transition, giving rise to a Franck–Condon factor, and (2) the valence electron wavefunctions are changed in the transition by the shift, mentioned earlier for the soliton, due to the presence of the polaron. As pointed out by Jeyadev and Schrieffer (1984), because the polaron motion requires simultaneous relaxation of two chains, $W_{tr} = 4t_{\perp}|\langle\psi_0|\psi_p\rangle|^2$, where ψ_p and ψ_0 are the wavefunctions for the chain with and without the polaron, respectively. In the adiabatic approximation ψ_p and ψ_0 are each products of the wavefunction of the CHs and that of the valence electrons. The factor in $|\langle\psi_0|\psi_p\rangle|^2$ from the C–H wavefunctions, the Franck–Condon factor for the transition, was found to have a value of 0.1 at $T = 0$ (Jeyadev and Schrieffer 1984). The factor from the valence electrons depends on chain length. Evaluation of this factor for a typical chain length of 70 sites gave $(0.2)^2$ (Su and Yu 1983). The value of the transverse bandwidth $4t_{\perp}$ varies with the energy value in the chain direction. Reasonably, for the case of a polaron, the transverse bandwidth should be taken at the top of the valence band or bottom of the conduction band. The transverse bandwidth at the top of the valence band obtained in band structure calculations for t-PA ranges from 0.1 eV (Grant and Batra 1983) to 0.5 eV (Vogl and Campbell 1990), with 0.3 eV apparently a reasonable average. Collecting the three factors we obtain $W_{tr} = 0.001$ eV for $T = 0$. The Franck–Condon factor is expected to decrease with increasing temperature due to the thermal vibrations. This decrease has been evaluated by Choi and Conwell (1991) and found to be small, only $\sim 10\%$ at 300 K.

Whether polaron transport perpendicular to the chains at a given temperature is bandlike or hopping is a function of the relative size of kT and W_{tr}. For $kT \gg W_{tr}$ band structure and therefore band motion are meaningless and hopping should predominate. For $kT \ll W_{tr}$, on the other hand, band motion should be dominant. These predictions are satisfied for the small polaron (Holstein 1959). For t-PA, with the calculated W_{tr} of 0.001 eV corresponding to kT at ~ 15 K, hopping should prevail at all but the lowest temperatures. In the band motion regime mobility should decrease with temperature due to phonon scattering. However, the actual temperature dependence in this regime is difficult to predict because it is likely that at the low temperatures concerned phonon scattering would be less important than some type of defect scattering or barriers. Hopping is a phonon-assisted process and hopping mobility increases with increasing temperature (Holstein 1959, Choi and Conwell 1991).

The experimental data of Walser et al. (1988, 1991) on transverse photocurrent of t-PA in the picosecond regime cover the range $\sim 150 < T < 300$ K. These data should therefore be characteristic of the hopping regime. Above 200 K the photocurrent is clearly increasing with increasing temperature. Below 200 K, although there is considerable scatter in the data, the photocurrent appears also to be increasing with T, although more slowly than above 300 K. The increase in photocurrent with T over the entire temperature range of the measurements is in agreement with the theoretical calculation of the bandwidth, which would lead to hopping conductivity over this range. It should be noted, however, that photocurrent is proportional to the product of carrier concentration and mobility, and carrier concentration could also be a function of temperature, in particular through dependence of carrier lifetime on temperature.

Calculations have not been done for polaron or bipolaron mobility in the NDGS polymers. It is expected, however, that techniques similar to those described for polyacetylene could be used, and that they would lead to qualitatively similar results.

9. Photoconductivity

9.1. Long-time photoconductivity in t-PA

Photoconductivity parallel to the chains was first observed in t-PA in experiments with a nsec time resolution (Etemad et al. 1981). It was found to be activated, with an activation energy of 0.2 eV (Etemad et al. 1981). Photoconductivity is found for photon energies less than 1 eV, well below the gap energy of $\sim$1.4 eV, suggesting the direct generation of solitons, i.e., without the initial creation of electron–hole pairs. The photocurrent is long lived, persisting for milliseconds. The decay of the photocurrent I with time t was found to be nonexponential. Rather it was found that

$$I(t) \propto t^{-\beta}, \quad \beta = 0.6, \tag{9.1}$$

independent of temperature and of the light intensity. As pointed out by Etemad et al. (1981), the power-law dependence of $I(t)$ is suggestive of dispersive transport (Scher and Montroll 1975), in which the hopping rate varies because of either variable distance or variable energy difference between pairs of sites. This characteristic of the transport, plus the existence of an activation energy, indicates that the charged solitons are not able to drift along the chains, but are trapped. Because solitons cannot hop between chains, as we have seen, the hopping process must consist of electrons hopping between solitons. Specifically, it has been suggested that an electron hops between a bound charged soliton and a moving S^0. This process will be discussed in § 10. Another interesting feature of the photoconductivity in t-PA was discovered in experiments on samples oriented by stretching. In such samples Dorsinville et al. (1985) found, surprisingly, that the photoinduced absorption and photoconductivity are a maximum when the polarization of the laser beam is perpendicular to the chains.

The various kinds of evidence cited above for the existence of charged solitons microseconds after the end of the exciting laser pulse were taken by many to be confirmation of the Su and Schrieffer (1980) calculation predicting that the electron–hole pairs become S–$\bar{S}$ pairs within a fraction of a picosecond. Orenstein et al. (1984), however, suggested that the observed long-lived charged solitons arise from conversion by polarons of the neutral solitons normally present in the Shirakawa process polyacetylene samples used. In this picture the original electron and hole are either created on different chains, or separate to different chains by hopping before they have become solitons. Lattice relaxation must then produce one positively charged and one negatively charged polaron. These diffuse along the chains as discussed in the last section. Upon meeting a neutral soliton, which provides a lower energy level, they transfer their charge and disintegrate into phonons. Alternatively, as a result of hopping or diffusion along a chain, a polaron might find another polaron on the same chain and the two would be transformed into a pair of charged solitons.

To decide between these two pictures of the charged soliton generation process, picosecond time-resolved measurements were made, after the exciting laser pulse, of the 0.43 eV photoinduced absorption band that is the signature of photogenerated charged solitons. It was found that there is a strong photogeneration of charged soliton pairs on a femtosecond time scale, in agreement with the Su and Schrieffer prediction. However, these charged solitons were found to decay on a time scale of <1 ps. A second group of charged solitons is subsequently created, peaking in number at ~ 40 ps after the laser pulse for 25 K ambient temperature (Rothberg et al. 1986). This latter group decays quite slowly, accounting for the charged solitons seen as long as milliseconds after the exciting pulse. The generation of the second group of solitons is reasonably attributed to polaron conversion either by collision with other polarons or with neutral solitons, with the delay time reflecting the polaron diffusion rate. This mechanism can explain the result of Dorsinville et al. (1985), cited earlier. Illumination perpendicular to the chains should enhance polaron creation, by increasing the probability of electron and hole creation on separate chains, and thus the ultimate number of charged solitons and the photoconductivity.

9.2. Short-time photoconductivity in t-PA

Photoconductivity has also been investigated for shorter times after the exciting pulse, with a time resolution of ~ 50 ps. The photoconductivity from 50 to ~ 300 ps was found to differ significantly from the long-time, i.e., μs to ms, photoconductivity described earlier. The short-time conductivity is orders of magnitude larger, has a characteristic decay time of ~ 300 ps and is about the same for temperatures from 10 to 300 K (Sinclair et al. 1986).

Given the very large magnitude and the fact that it occurs a very short time after the initiating pulse, the photoconductivity just described is reasonably interpreted as due to carriers drifting along the chains rather than hopping. In light of the work of Rothberg et al. (1986) the photoconductivity measured over the first 50 ps should include contributions from polarons and any solitons they give rise to, as well as from the originally photogenerated charged solitons. With increasing time after the pulse the originally photogenerated solitons and polarons disappear, the latter with considerable probability giving rise to solitons that should drift freely along the chains until either they find an oppositely charged soliton and recombine, or get trapped. The possibilities for trapping are larger for charged solitons than for neutrals; an ionized impurity, e.g., would be an effective trap for the former. Once trapped, these solitons can still contribute to photoconductivity by the electron-hopping mechanism described earlier. In the trapped condition their lifetime could be quite long. Recombination would require that an electron hop from an S^- to an S^+; with the soliton concentration quite low at this stage, some solitons would be unfavorably located for this particular hopping process to occur. The electron hopping process would result in recombination sooner or later, however, because, by interchanging the positions of charged and neutral solitons, it ultimately changes the relative positions of S^+'s and S^-'s.

The inference that the short-time photoconductivity is due to drifting charged solitons and polarons makes applicable the theory for the mobilities of these excitations developed in the last section. As noted there, an estimate of the number of photocarriers (Sinclair et al. 1986) led to the conclusion that their mobility must be $\sim 2\ cm^2\ V^{-1}\ s^{-1}$, in good agreement with the calculated value for either solitons or polarons. The theory can account also for another experimental finding, that of the short-time photoconductivity being approximately independent of temperature over a wide range, found experimentally to be 10 to 300 K. This finding led Sinclair et al. (1986) and others to suggest that the carriers are hot, i.e., have kinetic energies much greater than $k_B T$ due to the excess energies with which they are created. Indeed, allowing for the fact that the energy of the initiating photon in excess of the band gap energy $2\Delta_0$ must be dissipated by the electron and hole before soliton or polaron formation begins, solitons are created with an excess energy of $2\Delta_0$ minus the creation energy $4\Delta_0/\pi$, or 0.5 eV. According to the calculations of Bishop et al. (1984), for those solitons that separate, most of this energy goes into a well-localized neutral lattice excitation called a breather, halfway between the solitons. The remaining excess over thermal energy would be dissipated by acoustic phonon emission in a fraction of a ps (Jeyadev and Conwell 1987a,b,c). In fact, the rate of phonon emission is so high that the solitons would thermalize in less than 1 ps even if a large part of their excess energy were not lost to a breather. Polarons have a comparable rate of energy loss through acoustic phonon emission and, being created with less excess energy due to their larger creation energy, would thermalize even faster than solitons. Thus the experiments, with a resolution of ~ 50 ps could not be expected to detect hot solitons or polarons. The explanation for the observed temperature independence of photocurrent suggested by the theory is that the mobility is essentially independent of lattice temperature over a wide range. As seen in fig. 13, the variation of soliton mobility with T is small from 300 down to 50 K. Polaron mobility could well vary little to even lower temperatures. Finally, it has been questioned whether the sample is not heated well above 10 K by the laser pulse during the photoconductivity measurement (Conwell and Jeyadev 1988).

9.3. Photoconductivity in NDGS polymers

Photoconductivity in a number of these materials has been found to be quite similar to that just described for t-PA. In PPV, e.g., there is a steady-state or long-time photoconductivity and a short-time (ps to ns) photoconductivity several orders of magnitude larger (Bradley et al. 1988). The short-time photoconductivity also shows little T-dependence. The long-time photoconductivity is activated, with an activation energy in excess of 0.16 eV (Hörhold and Opfermann 1970). The mobility of the short-time photoconductivity, inferred from an estimate of number of carriers, is $0.08\ \mathrm{cm^2\ V^{-1}\ s^{-1}}$ (Bradley et al. 1988) while that for the long-time photoconductivity, measured by the time-of-flight technique, is only $2 \times 10^{-4}\ \mathrm{cm^2\ V^{-1}\ s^{-1}}$ (Takiguchi et al. 1987). The short-time mobility is presumably due to polaron drift along the chains, while the long-range mobility is undoubtedly due to hopping.

10. Conductivity

10.1. Anisotropy of conductivity

A plot of conductivity versus dopant concentration obtained recently for two t-PA samples doped with iodine is shown in fig. 14. Conductivities as high as those shown in the figure, over $10^5\ \Omega^{-1}\ \mathrm{cm}^{-1}$, are attained only along the chain direction in highly oriented, carefully prepared samples. These particular samples were oriented by stretching by a factor 5. The high conductivity $\sigma_{\parallel}$ parallel to the chains and a reflectivity of ~100% (Leising 1988) indicate that at high doping t-PA is metallic. Conductivity perpendicular to the chains, $\sigma_{\perp}$, is much smaller. For films heavily doped with I_3^-, Cao et al. (1991) found that $\sigma_{\parallel}/\sigma_{\perp}$ increased with stretching, reaching a value ~250 for a draw ratio of 15. This fact, plus the small optical absorption for light polarized perpendicular to the chains (Leising 1988), indicates that even at high doping t-PA is a semiconductor perpendicular to the chains. The ratio $\sigma_{\parallel}/\sigma_{\perp}$ is a function of doping and one may speculate that it approaches unity for very low doping where the conductivity is by hopping.

Conductivity versus doping has been measured for various dopants and some characteristic differences are found. For example, the steps in σ seen at ~6% and perhaps at 2–3% in fig. 14 are thought to indicate transitions to a new phase, i.e., a new lattice structure or a different packing of dopant chains in the same structure. These steps are seen more clearly in plots of σ versus time of doping (Tsukamoto et al. 1990). Because, as indicated earlier, different dopants give rise to different structures, the steps or discontinuities would in general occur at different concentrations for different dopants. Beyond this, there have been insufficient data taken

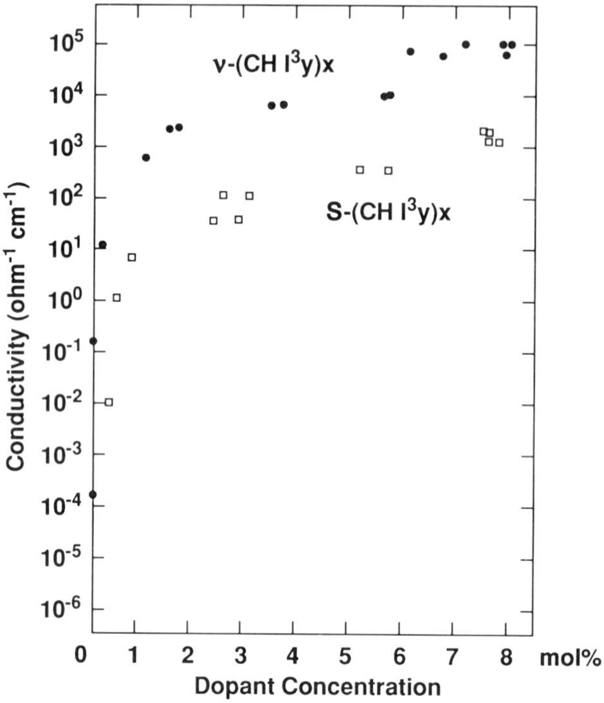

Fig. 14. Conductivity versus dopant concentration for two iodine doped polyacetylene samples. S denotes preparation by the Shirakawa method, ν preparation by the method of Tsukamoto et al. (after Tsukamoto et al. 1990).

on σ versus doping for samples stretched to the same extent to establish any significant differences between dopants. The highest values of $\sigma_{\|}$ have been obtained for p-type dopants, but n-type samples, e.g., K-doped samples, though doped as heavily, have not been stretched as much because of the difficulties of handling.

10.2. Pristine and lightly doped samples

Because the rates of change of conductivity with doping and the temperature dependence of σ vary as the doping changes, it is useful to consider the different doping ranges separately. For pristine or very lightly doped samples the only data reported are on unoriented samples. It is found that the conductivity increases quite rapidly with temperature. At very low doping levels electrons and holes contributed by donor or acceptor impurities are most likely to go into charged soliton levels in the gap, probably by converting neutral solitons. These solitons are bound by strong Coulomb attraction to the ions that gave rise to them and cannot contribute to σ directly. An electron or hole in a charged soliton level could not get to the conduction band or valence band, respectively, because these levels are much deeper than kT. Kivelson (1982) suggested a conduction mechanism involving mobile neutral solitons. To understand this mechanism, consider for specificity the case of donors and negatively charged solitons. An electron could not hop from an S^- to an S^0 at 300 K or below because that would require essentially the same energy as making the transition to the conduction band. If, however, the S^0 came close to a donor ion the hop could take place, with the S^0 becoming negatively charged and bound to the donor ion afterward, while the original S^- would become an S^0 and no longer be bound to its donor. The conductivity and thermoelectric power arising from this mechanism have been calculated by Kivelson (1982) using, for the most part, methods familiar in treatments of hopping transport. Among other difficulties with this calculation the matrix element for the electron–phonon coupling that makes the hopping possible is not known. Kivelson assumed the electron–phonon coupling constant g to vary with phonon energy $\hbar\omega$ as $(\hbar\omega)^x \exp(-\hbar\omega/E_0)$. x and E_0 were chosen to have g peak at an optical phonon energy, 0.15 eV. This leads to a large value of x and a DC conductivity $\sigma_{DC} \propto T^9$. Experimental data from ~100 to 300 K showed $\sigma_{DC} \propto T^{13.7}$ (Epstein et al. 1981a and Moses et al. 1981), which was considered to be in reasonable agreement with Kivelson's theory. Epstein and co-workers found also that the dependences on frequency and temperature of σ_{AC}, AC conductivity, for pristine t-PA are consistent with Kivelson's theory (Epstein 1981b). However, the data of Epstein and co-workers were also fitted by Emin and Ngai (1983), given σ_{DC} and one parameter, using general hopping theory for polarons strongly coupled to the lattice. It should also be noted that not all pristine t-PA samples behave precisely like those of Epstein et al. (1983) or Moses et al. (1981). Other authors have not found a good fit of σ_{DC} or σ_{AC} versus T with Kivelson's theory (Summerfield and Chroboczek 1985, Bott et al. 1986). Also there is reason to expect that there may not be a good supply of mobile neutral solitons in some samples at least. It can only be concluded that Kivelson's theory of transport, if correct, can apply only to some t-PA samples, not all. (For further discussion see, Conwell 1987.) For all the other conjugated polymers discussed, which do not have solitons, conduction in this low doping range should be by polaron hopping.

10.3. "Spinless conductivity" range

For doping from ~0.5 to ~5% in t-PA electron or hole concentration is considerable. However, as noted in § 5, the spin susceptibility χ_p is quite small in this doping range, indicating that the spins of the electrons and holes added to the chains are paired in charged solitons. This fact led to the suggestion that conduction in this doping range is due to moving solitons (Chung et al. 1984). No convincing theory has been advanced, however, as to how the solitons could overcome the very strong pinning to the donor or acceptor ions. Indeed, as pointed out by Epstein (1986), no shift in the soliton pinning frequency is observed even for large doping. Another serious difficulty for the idea of "spinless conductivity" arises from the two-phase nature of doped t-PA. In Na-doped t-PA, e.g., between ~0.4 and ~6% doping there is a dopant-rich ordered phase with 6.67% doping and a dopant-poor phase with ≦0.4% doping and the impurities random (Baughman et al. 1983, 1985). It is to be expected that there are different conduction mechanisms in these two phases. In the dopant-rich region solitons, with a length of ~14 sites, are spaced 15 sites apart, leaving them little room to move in any case. Overlap of electron wavefunctions in adjacent solitons is not large enough for the electrons to move freely from soliton to soliton within the bound region; if it were there would be high spin susceptibility, as is indeed seen at larger doping. A possible source of conduction in the dopant-rich region is electrons that have been thermally excited from the soliton band to the conduction band. Calculations lead to 10^{18} electrons/cm^3 in the conduction band for 6.67% doping (Conwell 1986). This number would not lead to higher spin susceptibility values than found experimentally. With the mobility of conduction electrons scattered by phonons in t-PA calculated to be $\sim 600\ cm^2\ V^{-1}\ s^{-1}$ (Menendez and Guinea 1983) an electron density of $10^{18}\ cm^{-3}$ could more than account for the measured σ. If the poorly doped regions are short enough so that electrons can travel from one dopant-rich region to another by tunneling, the poorly doped regions would not add resistance to the sample. For doping below ~5 or 6% this is apparently not the case because conductivity rises exponentially rather than linearly with doping. Thus at low doping the poorly doped regions must provide most of the resistance. As pointed out by Conwell and Jeyadev (1987), the difference in doping will result in a barrier, estimated to be 0.25 eV at 300 K, between adjacent regions of high and low doping. Current continuity requires that electrons go from dopant-rich regions into dopant-poor regions. This can be accomplished by electrons from the dopant-rich regions tunneling through the barrier into polaron levels or going over the barrier and then becoming polarons, or dropping into polaron levels.

A very different view of conductivity in this doping range has been presented by Epstein et al. (1983) and others. For iodine doped samples with I_3^- concentrations of 0.017, 0.033, 0.042 and 0.048 Epstein and co-workers found the temperature dependence of σ in the range 10 to 300 K to be well fitted by $T^{1/2} \exp[(T_0/T)^{1/4}]$. This is the T-dependence predicted for variable range hopping (VRH) within a constant density of states at the Fermi energy E_F (Mott 1969). VRH applies to a system with a large number of localized states occupying a range of energies which includes E_F. It was therefore assumed by Epstein and co-workers that E_F lies within the soliton band. For a well-ordered p-type sample all soliton levels would be empty and E_F would lie between the soliton band and the valence band. It is likely, however, that there is considerable disorder in iodine-doped polyacetylene because of the large size of the I_3^- ions, and disorder is expected to spread the soliton band. According to simple theory (Mott 1969),

$$T_0 = 16/k_B N_0 \xi^3, \tag{10.1}$$

where N_0 is the density of the states at E_F and ξ is a measure of the rate of decay of the wavefunction. ξ enters the calculation because it determines the overlap of the wavefunctions of the sites between which the electron jumps. Because the hopping is assumed to be 3D in this case, ξ was taken as $(\xi_{\parallel} \xi_{\perp}^2)^{1/3}$, where $\xi_{\parallel}$ is the extent of the soliton along the chain, given by eq. (5.2) and $\xi_{\perp}$ determines the decay rate of the soliton perpendicular to the chain. With $\xi_{\perp}$ taken from the decay rate of a p-type wavefunction perpendicular to the chain, this leads to $\xi = 3.6$ Å. In comparing experiment with theory, Epstein and co-workers obtained N_0 from the measured Pauli susceptibility χ_p (the susceptibility remaining after core and Curie contributions have been subtracted) using the free electron equation

$$\chi_p = 2\mu_B^2 N_0, \tag{10.2}$$

μ_B being the Bohr magneton. The resulting N_0 values were in the range 7–30 × 10^{20} states eV^{-1} cm^{-3}. Using the N_0 values and the measured T_0 values in eq. (10.1) led to $\xi = 5 \pm 2$ Å. This value for ξ compares resonably well with the 3.6 Å obtained from $(\xi_{\parallel}\xi_{\perp}^2)^{1/3}$. Calculating σ from Mott's expression (Mott 1969)

$$\sigma = 2(9\xi N_0/8\pi k_B T)^{1/2} \nu_{ph} \exp[-(T_0/T)^{1/4}], \tag{10.3}$$

with ν_{ph} an optical phonon frequency, Epstein and co-workers found σ to be too small by a factor 10^2 to 10^3. This had been found also for VRH in amorphous semiconductors (Mott and Davis 1979). Values of σ in good agreement with experiment could be obtained by replacing ν_{ph} with $\nu_{ph} \exp(2R/\xi)$, where R is the average hopping distance, according to a suggestion of Colson and Nagels (1980). It should be noted that different *T*-dependence for samples doped with iodine in this concentration range has been reported by others (see, Conwell 1987, for references and discussion). Also, it has been objected that the VRH theory, used by Epstein and co-workers, is not valid over a good part of the temperature range of their measurements (Ehinger et al. 1984). Specifically, for T_0 given by eq. (10.1) validity of the theory requires that $(T_0/2T)^{1/4} > 10$. For the two most heavily doped samples this condition is not satisfied above 20 K. Even for the lowest doped samples the condition for validity is not met above 95 K. Ehinger et al. (1984) have fitted σ versus T for iodine-doped t-PA samples with a more general hopping model based on the extended pair approximation. The density of states at E_F was not taken from eq. (10.2) on the grounds that it is not clear how well a free electron model works for electrons bound in soliton states, but was treated as a parameter. A good fit was obtained by Ehinger et al. (1984) to the experimental σ versus T for a density of states that varies considerably with energy around E_F.

The variation of σ with T was investigated for other dopants, also. For 6.5% AsF_5-doped t-PA σ was found also to be fitted well by $\exp[-(T_0/T)^{1/4}]$ (Audenaert 1984). However, T_0 was only ~2 × 10^4 K, which makes use of the VRH theory questionable. For 1.5% ClO_4^--doped t-PA it was also found that σ versus T could be fitted with $\exp[(-T_0/T)^{1/4}]$. However, χ_p is smaller in this case than for the iodine-doped samples and N_0 determined from eq. (10.2) led to σ calculated from eq. (10.3) being five orders of magnitude too small (Epstein et al. 1985). It was noted for ClO_4^- doping that χ_p varied with temperature and with doping, suggesting that the density of states is not constant with energy. The simplest energy dependence would be

$$N(E) = N_0 + (N_2/2)E^2. \tag{10.4}$$

This energy-dependence of the density of states leads to $\sigma \propto \exp[(-T_0/T)^{1/2}]$ (Hamilton 1972), which also fitted the observed σ versus T reasonably well in the range from ~70 to 300 K. Neglecting N_0 in eq. (10.4), Epstein and co-workers obtained N_2 from the measured χ_p and used this and the measured T_0 to obtain ξ. The value found for ξ was 14 Å, several times the value expected for a soliton. However, it appears to be in agreement with structural studies of ClO_4^- tetrahedra between layers of t-PA chains (Pouget et al. 1985). With the values of T_0, N_2 and ξ calculated as described, Epstein et al. (1985) found that the calculted σ was in reasonable agreement with experiment.

The VRH theory applied to iodine- and ClO_4^--doped samples by Epstein and co-workers would not work at all for Na-doped t-PA. In that case, χ_p is so low in this doping range that the density of states at E_F deduced from χ_p could not possibly support the measured conductivity. It is generally agreed that alkali-metal doped t-PA is more ordered than t-PA doped with I_3^-, AsF_5^-, ClO_4^-, etc., because the alkali-metal ions are smaller and fit better the vacancies in the t-PA lattice. In the discussion of the conductivity of Na-doped samples in the earlier part of this section, it was assumed that, as expected for well-ordered two-phase material, the soliton band is completely filled with electrons and E_F lies between that band and the conduction band. In the VRH calculations for iodine and ClO_4^- doping, on the other hand, the two-phase nature of the material was neglected. It was implicitly assumed that the hopping centers, the solitons, are randomly distributed. This assumption is in contradiction, however, with the large χ_p found for the iodine-doped samples. It is generally agreed that the large χ_p results from nonuniform doping, with the doping sufficiently high in some regions that the solitons overlap sufficiently to allow the electrons or holes to move freely along the chains, as in a metal. It has been, in fact, documented that the iodine tended to accumulate on the outside of the fibrils, with the inside remaining undoped (Epstein et al. 1981c). One can only conclude that the nature of the conduction mechanism in the doping range from a few tenths of a percent to ~5 or 6% has not been definitely established, and could in fact be different for different samples. It would be useful to revisit this doping range with the better oriented samples now available.

The general behavior of conductivity versus doping, temperature, etc., of other conjugated polymers is similar to that of polyacetylene. For a number of the other polymers, it is also found that χ_p is quite low until high doping concentrations are reached. The suggestion of spinless conductivity, in these other cases current carried by bipolarons, has also been made for these polymers. Similar objections can be made to the current being carried by bipolarons as were made in this section to the current being carried by solitons.

10.4. The "metallic" range

For doping beyond around 4 to 6%, χ_p is found to increase greatly with increasing doping, quite steeply in some samples. This is taken to be the signature of an insulator–metal transition, and indeed, as has been seen, $\sigma_{\parallel}$ may come close to the conductivity of copper. There has been considerable debate about the nature of the metallic state. One interesting feature is that the infrared absorption characteristic of solitons, the IRAV, continues to grow linearly to the highest doping, e.g., 18% K-doping (Tanner et al. 1989). This indicates that solitons persist in the metallic state, although they overlap sufficiently to allow the free motion of electrons required to achieve the high χ_p (Conwell et al. 1989). Despite the high conductivity and high χ_p, the temperature dependence of $\sigma_{\parallel}$ is not metallic. Although in the highest conductivity samples σ is generally T-independent at low temperatures, it typically increases with increasing T by ~30% or more up to 300 K. In a few samples σ is observed to peak at ~200 K and decrease slightly from 200 to 300 K. The increase of σ with T has been attributed to defect scattering being predominant (Kaiser and Graham 1990, Conwell and Mizes 1990).

The steep increase of χ_p at some high doping level has also been found for a number of NDGS polymers. More direct evidence for a transition to a metallic state has been the observation in photoemission experiments on highly doped PPy and a derivative of PT of states at the Fermi energy (Löglund et al. 1989, Bätz et al. 1990). Persistence of the IRAV and the characteristic absorption of bipolarons into the metallic state has led to the contention that bipolarons persist in the metallic state (Conwell and Mizes 1991).

References

Akagi, K., M. Suezaki, H. Shirakawa, H. Kyotani, M. Shimomura and Y. Tanabe, 1989, Synth. Met. **28**, D1.
Audenaert, M., 1984, Phys. Rev. B **30**, 4609.
Bardeen, J., and W. Shockley, 1950, Phys. Rev. **80**, 72.
Bätz, P., D. Schmeisser and W. Göpel, 1990, Solid State Commun. **74**, 461.
Baughman, R.H., N.S. Murthy, G.G. Miller, L.W. Shacklette and R.M. Metzger, 1983, J. Phys. (Paris) Colloq. **44**, C3-53.
Baughman, R.H., L.W. Shacklette, N.S. Murthy, G.G. Miller and R.L. Elsenbaumer, 1985, Mol. Cryst. & Liq. Cryst. **118**, 253.
Bishop, A.R., and T. Schneider, eds, 1978, Solitons and Condensed Matter Physics (Springer, New York).
Bishop, A.R., D.K. Campbell, P.S. Lomdahl, B. Horovitz and S.R. Phillpot, 1984a, Synth. Met. **9**, 223.
Bishop, A.R., D.K. Campbell, P.S. Lomdahl, B. Horovitz and S.R. Phillpot, 1984b, Phys. Rev. Lett. **52**, 671.
Bott, D.C., C.S. Brown, C.K. Chai, N.S. Walker, W.J. Feast, P.J.S. Foot, P.D. Calvert, N.C. Billingham and R.H. Friend, 1986, Synth. Met. **14**, 245.
Bradley, D.D.C., and R.H. Friend, 1989, J. Phys.: Condensed Matter **1**, 3671.
Bradley, D.D.C., Y.Q. Shen, H. Bleier and S. Roth, 1988, J. Phys. C **21**, L515.
Brédas, J.-L., B. Thémans and J.M. André, 1982, Phys. Rev. B **26**, 6000.
Brédas, J.-L., B. Thémans and J.M. André, 1983, J. Chem. Phys. **78**, 101.
Brédas, J.-L., J.C. Scott, K. Yakushi and G.B. Street, 1984, Phys. Rev. B **30**, 1023.
Burroughes, J.H., D.D.C. Bradley, A.R. Brown, R.N. Marks, K. Mackay, R.H. Friend, P.L. Burns and A.B. Holmes, 1990, Nature **347**, 539.
Campbell, D.K., and A.R. Bishop, 1981, Phys. Rev. B **24**, 4859.
Cao, Y., P. Smith and A.J. Heeger, 1991, Synth. Met. **41**, 181.
Chance, R.R., D.S. Boudreaux, J.-L. Brédas and R. Silbey, 1986, in: Handbook of Conducting Polymers, ed. T.A. Skotheim (Marcel Dekker, New York) p. 825.
Choi, H.-Y., and E.M. Conwell, 1991, Mol. Cryst. & Liq. Cryst. **194**, 23.
Chung, T.-C., J.H. Kaufman, A.J. Heeger and F. Wudl, 1983, Phys. Rev. B **30**, 702.
Chung, T.-C., F. Moraes, J.D. Flood and A.J. Heeger, 1984, Phys. Rev. B **29**, 2341.

Clarke, T.C., and J.C. Scott, 1986, in: Handbook of Conducting Polymers, ed. T.A. Skotheim (Marcel Dekker, New York) p. 1127.
Colson, R., and P. Nagels, 1980, J. Non-Cryst. Solids **35**, 129.
Conwell, E.M., 1986, Phys. Rev. B **33**, 2465.
Conwell, E.M., 1987, IEEE Trans. Electr. Insul. **EI-22**, 591.
Conwell, E.M., and S. Jeyadev, 1987, Synth. Met. **17**, 69.
Conwell, E.M., and S. Jeyadev, 1988, Mol. Cryst. & Liq. Cryst. **160**, 443.
Conwell, E.M., and H.A. Mizes, 1990, Synth. Met. **38**, 319.
Conwell, E.M., and H.A. Mizes, 1991, Phys. Rev. B **44**, 937.
Conwell, E.M., H.A. Mizes and S. Jeyadev, 1989, Phys. Rev. B **40**, 1630.
Djurado, D., J. Ma, N. Theophilou and J.E. Fischer, 1989, Synth. Met. **30**, 395.
Dorsinville, R., S. Krimchansky, R.R. Alfano, J.L. Birman, R. Tubino and G. Dellepiane, 1985, Solid State Commun. **56**, 857.
Eckhardt, H., and R.R. Chance, 1983, J. Chem. Phys. **79**, 5698.
Ehinger, K., S. Summerfield, W. Bauhofer and S. Roth, 1984, J. Phys. C **17**, 3753.
Emin, D., and K.L. Ngai, 1983, J. Phys. Colloq. (Paris) **44**, C3-471.
Epstein, A.J., 1986, in: Handbook of Conducting Polymers, ed. T. Skotheim (Marcel Dekker, New York) p. 1041.
Epstein, A.J., H. Rommelmann, M. Abkowitz and H.W. Gibson, 1981a, Mol. & Cryst. Liq. Cryst. **77**, 81.
Epstein, A.J., H. Rommelmann, M. Abkowitz and H.W. Gibson, 1981b, Phys. Rev. Lett. **47**, 1549.
Epstein, A.J., H. Rommelmann, M.A. Druy, A.J. Heeger and A.G. MacDiarmid, 1981c, Solid State Commun. **38**, 683.
Epstein, A.J., H. Rommelmann, R. Bigelow, H.W. Gibson, D.M. Hoffman and D.B. Tanner, 1983, Phys. Rev. Lett. **50**, 1866.
Epstein, A.J., R.W. Bigelow, H. Rommelmann, H.W. Gibson, R.J. Weagley and A. Feldblum, 1985, Mol. Cryst. & Liq. Cryst. **117**, 147.
Etemad, S., T. Mitani, M. Ozaki, T.-C. Chung, A.J. Heeger and A.G. MacDiarmid, 1981, Solid State Commun. **40**, 75.
Feldblum, A., J.H. Kaufman, S. Etemad, A.J. Heeger, T.-C. Chung and A.G. MacDiarmid, 1982, Phys. Rev. B **26**, 815.
Flood, J.D., E. Ehrenfreund, A.J. Heeger and A.G. MacDiarmid, 1982, Solid State Commun. **44**, 1055.
Gibson, H.W., R.J. Weagley, R.A. Mosher, S.B. Kaplan, W.M. Prest Jr and A.J. Epstein, 1985, Phys. Rev. B **31**, 2328.
Grant, P., and I.P. Batra, 1983, J. Phys. Colloq. (Paris) **44**, C3-437.
Guinea, F., 1984, Phys. Rev. B **30**, 1884.
Hamilton, E.M., 1972, Philos. Mag. **26**, 1043.
Heeger, A.J., S. Kivelson, J.R. Schrieffer and W.-P. Su, 1988, Rev. Mod. Phys. **60**, 781.
Holstein, T., 1959, Ann. Phys. (New York) **8**, 343.
Hörhold, H.H., and J. Opfermann, 1970, Makromol. Chem. **131**, 105.
Horovitz, B., 1982, Solid State Commun. **41**, 729.
Jeyadev, S., and E.M. Conwell, 1987a, Phys. Rev. Lett. **58**, 258; Phys. Rev. B **36**, 3284.
Jeyadev, S., and E.M. Conwell, 1987b, Phys. Rev. B **35**, 5917.
Jeyadev, S., and E.M. Conwell, 1987c, Phys. Rev. B **35**, 6253.
Jeyadev, S., and J.R. Schrieffer, 1984, Phys. Rev. B **30**, 3620.
Kaiser, A.B., and S.C. Graham, 1990, Synth. Met. **36**, 367.
Kanatzidis, M.G., 1990, Chem. & Eng. News, Dec. 3, 36.
Kittel, C., 1976, Introduction to Solid State Physics, 5th Ed. (Wiley, New York) p. 450.
Kivelson, S., 1982, Phys. Rev. B **25**, 3798.
Leising, G., 1988, Phys. Rev. B **38**, 10, 313.
Löglund, M., R. Lazzaroni, S. Stafström and W.R. Salaneck, 1989, Phys. Rev. Lett. **63**, 1841.
Menendez, C., and F. Guinea, 1983, Phys. Rev. B **28**, 2183.
Mizoguchi, K., K. Kume and H. Shirakawa, 1986, Abstracts Int. Conf. on Science and Technology of Synthetic Metals, Kyoto, p. 55.
Mo, Z., K.-B. Lee, Y.B. Moon, M. Kobayashi, A.J. Heeger and F. Wudl, 1985, Macromolecules **18**, 1972.

Moraes, F., H. Schaffer, M. Kobayashi, A.J. Heeger and F. Wudl, 1984, Phys. Rev. B **30**, 2948.
Moraes, F., J. Chen, T.-C. Chung and A.J. Heeger, 1985, Synth. Met. **11**, 271.
Moses, D., A. Denenstein, A. Pron, A.J. Heeger and A.G. MacDiarmid, 1980, Solid State Commun. **36**, 219.
Moses, D., J. Chen, A. Denenstein, M. Kaveh, T.-C. Chung, A.J. Heeger and A.G. MacDiarmid, 1981, Solid State Commun. **40**, 1007.
Mott, N.F., 1969, Philos. Mag. **19**, 835.
Mott, N.F., and E.A. Davis, 1979, Electronic Processes in Non-Crystalline Materials (Clarendon, Oxford).
Murthy, N.S., G.G. Miller and R.H. Baughman, 1988, J. Chem. Phys. **89**, 2523.
Nechtschein, M., F. Devreux, F. Genoud, M. Guglielmi and K. Holczer, 1983, Phys. Rev. B **27**, 61.
Orenstein, J., and G.L. Baker, 1982, Phys. Rev. Lett. **49**, 1043.
Orenstein, J., G.L. Baker and Z. Vardeny, 1983, J. Phys. Colloq. (Paris) **44**, C3-407.
Orenstein, J., Z. Vardeny, G.L. Baker, G. Eagle and S. Etemad, 1984, Phys. Rev. B **30**, 786.
Pouget, J.P., 1985, in: Electronic Properties of Polymers and Related Compounds, eds H. Kuzmany, M. Mehring and S. Roth (Springer, Berlin) p. 26.
Pouget, J.P., J.C. Pouxviel, P. Robin, R. Comes, D. Begin, D. Billaud, A. Feldblum, H.W. Gibson and A.J. Epstein, 1985, Mol. Cryst. & Liq. Cryst. **117**, 75.
Rothberg, L., T.M. Jedju, S. Etemad and G.L. Baker, 1986, Phys. Rev. Lett. **57**, 3229.
Salem, S., 1966, The Molecular Orbital Theory of Conjugated Systems (Benjamin, New York).
Scher, H., and E.W. Montroll, 1975, Phys. Rev. B **12**, 2455.
Scott, A.C., F.Y.F. Chu and D.W. McLaughlin, 1973, Proc. IEEE **61**, 1443.
Sinclair, M., D. Moses and A.J. Heeger, 1986, Solid State Commun. **59**, 343.
Su, W.-P., and J.R. Schrieffer, 1980, Proc. Nat. Acad. Sci. USA **77**, 5626.
Su, W.-P., J.R. Schrieffer and A.J. Heeger, 1980, Phys. Rev. B **22**, 2099; 1983, Erratum: Phys. Rev. B **28**, 1388.
Su, Z., and L. Yu, 1983, Phys. Rev. B **27**, 5199.
Summerfield, S., and J.A. Chroboczek, 1985, Solid State Commun. **53**, 129.
Takayama, M., Y.R. Lin-Liu and K. Maki, 1980, Phys. Rev. B **21**, 2388.
Takiguchi, T., D.H. Park, H. Ueno and K. Yoshino, 1987, Synth. Met. **17**, 657.
Tanner, D.B., G.L. Doll, A.M. Rao, P.C. Eklund, G.A. Arbuckle and A.G. MacDiarmid, 1989, Synth. Met. **28**, D141.
Tsukamoto, J., A. Takahashi and K. Kawasaki, 1990, Jpn. J. Appl. Phys. **29**, 125.
Vardeny, Z., J. Strait, D. Moses, T.-C. Chung and A.J. Heeger, 1982, Phys. Rev. Lett. **49**, 1657.
Vardeny, Z., J. Orenstein and G.L. Baker, 1983a, J. Phys. Colloq. (Paris) **44**, C3-325.
Vardeny, Z., J. Orenstein and G.L. Baker, 1983b, Phys. Rev. Lett. **50**, 2032.
Vardeny, Z., E. Ehrenfreund, O. Brafman, M. Nowak, H. Schaffer, A.J. Heeger and F. Wudl, 1986, Phys. Rev. Lett. **56**, 671.
Vogl, P., and D.K. Campbell, 1990, Phys. Rev. B **41**, 12797.
Walser, A.D., A. Seas, R. Dorsinville, R.R. Alfano and R. Tubino, 1988, Solid State Commun. **67**, 333.
Walser, A.D., R. Dorsinville, R.R. Alfano and R. Tubino, 1991, J. Appl. Phys. **69**, 1116.
Weinberger, B.R., E. Ehrenfreund, A. Pron, A.J. Heeger and A.G. MacDiarmid, 1980, J. Chem. Phys. **72**, 4749.
Weinberger, B.R., C.B. Roxlo, S. Etemad, G.L. Baker and J. Orenstein, 1984, Phys. Rev. Lett. **33**, 86.
Winokur, M.J., D. Chen and F.E. Karasz, 1991, Synth. Met. **41**, 341.

CHAPTER 12

Electron Tunneling in Semiconductors

PETER J. PRICE

IBM T.J. Watson Research Center
Yorktown Heights
New York 10598, USA

Handbook on Semiconductors
Completely Revised Edition
Edited by T.S. Moss
Volume 1, edited by P.T. Landsberg

Contents

1. Introduction 629
2. Tunneling in semiconductors 635
3. Diode structures 644
4. Diode conduction 648
5. Resonant tunneling 651
6. Additional topics 655
   6.1. Magnetic field effects 655
   6.2. Phonon-induced tunneling 656
   6.3. Photon-induced tunneling 656
   6.4. Effects of impurities and traps 657
   6.5. Image effects 658
   6.6. Low-dimensional structures 659
   6.7. Tunneling carrier injection 659
Appendix: the envelope representation 660
References 662

1. Introduction

Tunneling is a quantum process in which a particle penetrates into and traverses a barrier region within which its potential energy exceeds its initial (kinetic plus potential) energy. According to classical physics, conservation of energy would not permit this to occur. It is possible in quantum physics, where the particle's wave function, and hence its probability density, is spread out over both classically "allowed" and classically "forbidden" parts of the path. Wave functions penetrate from allowed regions into adjoining forbidden regions, for a limited distance, and this allows non-classical particle motions that can cross a sufficiently narrow barrier. Tunneling processes were conceived and investigated as early as 1927, soon after the inception of the wave-mechanical (Schrödinger equation) version of quantum physics. They included *field emission*, the escape of an electron from an atom that is in an externally applied electric field (Oppenheimer 1928a) and similarly escape from a metal into vacuum (Oppenheimer 1928b,c, Fowler and Nordheim 1928, Nordheim 1929); penetration of a surface barrier by an electron in *thermionic emission* (Nordheim 1928, 1929, Frenkel 1929); and *alpha decay* of an atomic nucleus, in which the alpha particle escapes by tunneling through a barrier potential at the periphery of the nucleus (Gamow 1928, Gurney and Condon 1929). Soon afterwards it was proposed that the current flow through "an electrical contact between two solid conducting bodies" involved electron tunneling through a potential barrier at the interface (Frenkel 1930), and that electron tunneling was the basis of rectification by semiconductor–metal junctions (Schottky 1931, Frenkel and Joffe 1932, Wilson 1932). Thus from the beginning atomic, nuclear, and solid-state physics all supplied instances of tunneling. The word "tunneling" does not occur in the original papers. It did not come into general use before 1959, although it meanwhile had appeared in a number of books and papers. Schottky (1931) was perhaps the first to use it in print: "... einen wellenmechanischen Tunneleffekt".

Among these phenomena, two types of tunneling process are to be distinguished:

(a) The particle is initially confined in a discrete localized state, and escapes to a virtually infinite "allowed" region, and thus into a continuum of itinerant states. This process is described by a particle state that varies in time – the probability of being still in the initial localized state decreases, at a rate which normally can be characterized by a decay lifetime.

(b) The particle tunnels through a barrier potential that separates two semi-infinite "allowed" regions, and thus passes from one continuum of itinerant states to the other. The process is then normally characterized by a dimensionless quantity, the probability for the incident particle to be transmitted through (rather than reflected from) the barrier – say $T(E)$, a function of the particle's energy E.

Tunneling of electrons in solids involves the relatively complex Bloch states, compared with the "de Broglie wave" solutions of the Schrödinger equation for a free particle. It is useful, however, to first discuss tunneling according to the free-particle one-dimensional Schrödinger equation

$$-(\hbar^2/2m)\,\mathrm{d}^2\psi/\mathrm{d}x^2 + U(x)\,\psi = E\psi, \tag{1.1}$$

so as to present a partial "anatomy of tunneling transmission". We consider situations where the potential energy $U(x)$ is $> E$ for a range $a < x < b$. It is convenient to also assume that U is constant outside a larger range $(x_\mathrm{L}, x_\mathrm{R})$, i.e. $U = U_\mathrm{L}$ for $x < x_\mathrm{L} \leqslant a$ and $U = U_\mathrm{R}$ for $x > x_\mathrm{R} \geqslant b$, as in fig. 1. The wave functions in these outer ranges may then be taken to be linear combinations,

$$\psi(x) = k^{-1/2}[A\exp(\mathrm{i}kx) + B\exp(-\mathrm{i}kx)] \tag{1.2}$$

with $A = A_\mathrm{L}$, $B = B_\mathrm{L}$ and $k = k_\mathrm{L}$ for $x < x_\mathrm{L}$, and similarly $A = A_\mathrm{R}$, etc., for $x > x_\mathrm{R}$, where $k(x)$ is the positive root of

$$(\hbar^2/2m)k^2 = E - U(x). \tag{1.3}$$

The tunneling probability, T, may be obtained from a solution of (1.1) over the whole range from $x < x_\mathrm{L}$ to $x > x_\mathrm{R}$, expressed in terms of the relation between the (A, B) coefficients of (1.2) on the two sides. It is equal to the value of $|A_\mathrm{R}/A_\mathrm{L}|^2$ for the solution having $B_\mathrm{R} = 0$. (The same value of T for given E is obtained with L and R interchanged and A and B interchanged in this prescription.) Within $(x_\mathrm{L}, x_\mathrm{R})$ but

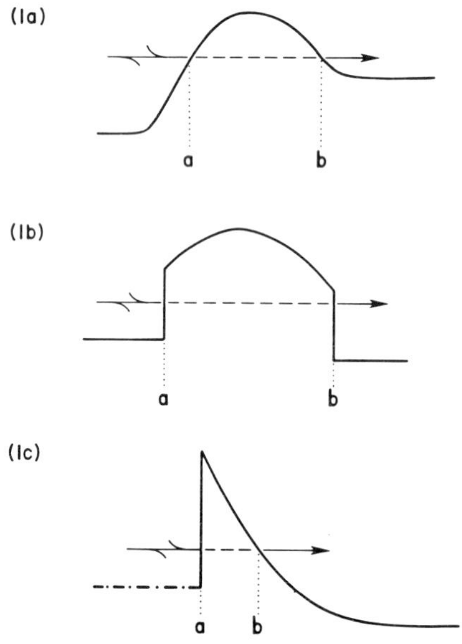

Fig. 1. Types of tunneling barrier.

outside (a, b) we may assume the approximate solutions

$$\psi(x) = p(x)\{A\exp[w(x)] + B\exp[-w(x)]\} \tag{1.4}$$

with

$$w(x) = \mathrm{i}\int^{x} k(x')\,\mathrm{d}x' \quad \text{and} \quad p(x) = k^{-1/2} \tag{1.5}$$

so long as $k(x)$ *varies slowly enough* (in particular, if $|\mathrm{d}k/\mathrm{d}x|$ is small compared to k^2). In the conditions of this *WKB approximation*, eq. (1.4) with $A = 0$ or $B = 0$ represents particle motion according to classical physics. The solution (1.4), (1.5) is also the clue to what applies in the range $a < x < b$, which is "forbidden" according to classical physics. Essentially the same mathematics gives (1.4) again but with new definitions

$$w(x) = \int^{x} \alpha(x')\,\mathrm{d}x' \quad \text{and} \quad p(x) = \alpha^{-1/2} \tag{1.6}$$

for $\alpha(x)$ the positive root of

$$(\hbar^2/2m)\alpha^2 = U(x) - E. \tag{1.7}$$

(Again, this particular form of solution is predicated on $|\mathrm{d}\alpha/\mathrm{d}x| \ll \alpha^2$.) The important part of the variation of the wave function in (a, b) is then given by the factors $\exp[\pm\int_a^x \alpha(x')\,\mathrm{d}x']$, and the tunneling probability T depends, substantially, on the exponent

$$\beta(E) \equiv \int_a^b \alpha(x)\,\mathrm{d}x. \tag{1.8}$$

In an allowed range, corrections to the WKB solution of (1.1) give a backward scattering effect. For a forbidden range where $U(x) > E$, however, the two *evanescent wave* solutions corresponding to the two terms of (1.4) should retain their separate increasing or decreasing character, even though modified numerically, as the Airy functions (see below) illustrate.

A sufficient approximation, when β is appropriately large, is

$$T = C\exp(-2\beta), \tag{1.9}$$

where the E dependence of C, the *prefactor*, normally is unimportant in view of the E dependence of β. The value of the prefactor C in the approximate result (1.9) depends on the particular situation at the points a, b where $U(x) - E$ changes sign (where "allowed" and "forbidden" meet). The WKB-like solutions $\psi(x)$ on each side have to be joined correctly as x increases through a and as x increases through b. Figure 1 illustrates the two kinds of situation that normally prevail: either there is a "classical turning point" where $U(x)$ is smoothly continuous, or $U(x)$ is discontinuous where $U - E$ changes sign. At a classical turning point, the mathematical details of the splicing of the wave functions are complicated, because the solutions (1.4) do not hold in its neighborhood (Kemble 1937, Landau and Lifshitz 1977, Merzbacher 1970). The result, however, is simple: with two classical turning points as in fig. 1a (and U elsewhere continuous),

$$C = 1. \tag{1.10}$$

On the other hand, with two discontinuities as in fig. 1b the solutions (1.4) still hold as $x \to a$ and as $x \to b$, but the result is not as simple:

$$C = K_a K_b, \tag{1.11}$$

where at each interface ($x = a, b$)

$$K = 4\alpha k/(\alpha^2 + k^2). \tag{1.12}$$

The disparity between (1.10) and (1.11) signifies that the solution of (1.1) on which the former is based becomes inapplicable when the variation of $U(x)$, in the neighborhood of a zero of $U - E$, becomes increasingly steep as it tends to a step form.

For the case of fig. 1a the general result (still within the WKB approximation) in place of the exponential form (1.9) is

$$T^{-1} = [\exp \beta + \tfrac{1}{4} \exp(-\beta)]^2 \tag{1.13}$$

For E greater than the maximum of $U(x)$, the WKB treatment gives the classical-physics result, $T = 1$. However, there will actually be some reflection, $T < 1$, corresponding to corrections to WKB due to the variation of $U(x)$ not being infinitely slow; and $T(E)$ in fact only approaches 1 as $E \to \infty$.

For the case of fig. 1b, the corresponding generalization (again, within the WKB approximation) of the exponential form (1.9) is

$$\frac{1}{T} = \frac{1}{T_0} + \frac{1}{C} 4 \sinh^2 \beta, \tag{1.14}$$

where C is still given by (1.11), (1.12) and

$$T_0 = \frac{4k_a \alpha_a k_b \alpha_b}{(k_a \alpha_b + k_b \alpha_a)^2}. \tag{1.15}$$

This formula for T is exact, as a consequence of (1.1), when $U(x)$ is piecewise constant, so that in the barrier $U = \text{constant} = U_0$ and $\alpha = \alpha_0$ $(= \alpha_a = \alpha_b)$, with $\beta = (b - a)\alpha_0$. Then, in particular, when $E = U_0$ (energy just at the top of the barrier)

$$4k_a k_b / T = (k_a + k_b)^2 + [k_a k_b (b - a)]^2. \tag{1.16}$$

For E greater than the maximum of $U(x)$ – the situation where transmission is classically allowed – the equivalent of (1.14) is conveniently obtained by substituting $\mathrm{i}\kappa_a$ for α_a and $\mathrm{i}\kappa_b$ for α_b in (1.12) and (1.15), and $-\sin^2 \tilde{\beta}$ in place of $\sinh^2 \beta$, where $\kappa(x)$ is the positive root of $(\hbar^2/2m)\kappa(x)^2 = E - U(x)$ and $\tilde{\beta} = \int_a^b \kappa(x)\, \mathrm{d}x$.

For illustration, fig. 2 shows T versus E calculated for the case of (1.14) shown in the inset. The height of the barrier is U_0 on the left (from $E = 0$ to $E = U_0$) and $2U_0$ on the right (from $E = U_0$ to $E = -U_0$), and the barrier width is W. The result, as a function of E/U_0, depends on the dimensionless parameter $\alpha_0 W = (W/\hbar)(2mU_0)^{1/2}$, which is here chosen to be equal to 4.0. The central part of the range $0 < E < U_0$ has $T(E)$ almost exponential like (1.9). Near $E = 0$, eq. (1.9) remains applicable but

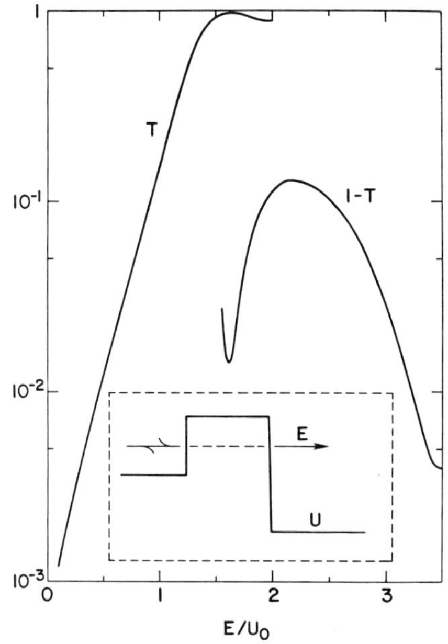

Fig. 2. Tunneling probability versus energy, for a rectangular potential profile.

the prefactor C varies as $E^{1/2}$. (Since $\mathrm{d}\ln T/\mathrm{d}\ln E \to \frac{1}{2} + (\alpha_0 W/U_0)E$ in this limit, T becomes proportional to $E^{1/2}$ when $E \ll U_0/2\alpha_0 W$.) For $E \lessapprox U_0$, the exponential form (1.9) is not applicable. The oscillations of $T(E)$ for $E > U_0$ are due, of course, to coherent reflections from the two steps in $U(x)$.

For the "Fowler–Nordheim" case of fig. 1c, the WKB result is

$$\frac{1}{T} = 1 + \frac{1}{C}[\exp 2\beta + \tfrac{1}{4}\exp(-2\beta)], \tag{1.17}$$

where in this case

$$C = K_a. \tag{1.18}$$

Thus for the three cases of fig. 1 the prefactors (1.10), (1.11) and (1.18) are heuristically related as products of two factors, one from each end of the forbidden range.

When $U(x)$ in the barrier varies linearly over the tunneling path,

$$U(x) = (b - x)F \tag{1.19}$$

(as for a constant electric field F/e), the exponent (1.6) in case (c) becomes $w(x) = \beta - \zeta(x)$ where

$$\zeta(x) = \int_x^b \alpha(x')\,\mathrm{d}x' = \frac{(2mF)^{1/2}}{\hbar}\frac{2}{3}(b - x)^{3/2} \tag{1.20}$$

and

$$\beta(E) = \zeta(a) = \frac{2}{3}\frac{(2m)^{1/2}[U(a)-E]^{3/2}}{\hbar}\frac{1}{F}. \tag{1.21}$$

As with piecewise-constant potentials $U(x)$, for linear variations (1.19) exact solutions of (1.1) are available. They are given by replacing $\alpha^{-1/2}\exp(-\zeta)$ by $2\pi^{1/2}\mathrm{Ai}(\zeta)$ and replacing $\alpha^{-1/2}\exp\zeta$ by $\pi^{1/2}\,\mathrm{Bi}(\zeta)$, where ζ is given by (1.20) and where $\mathrm{Ai}(z)$ and $\mathrm{Bi}(z)$ are the Airy functions, solutions of $\mathrm{d}^2\psi/\mathrm{d}z^2 = z\psi$ (Abramowitz and Stegun 1972).

One can have in practice a combination of cases (b) and (c), in which the barrier potential has two discontinuities but the variation $U(x)$ and the value of E are such that the forbidden range (a, b) extends from one discontinuity to a classical turning point between the discontinuities. Then reflections at the second discontinuity result in an oscillatory variation of $T(E)$ superposed on the monotonic trend. This situation has been analyzed, specifically for $U(x)$ in the barrier range varying linearly as in (1.19), by Gundlach (1966). One also may have, obviously, potentials such that there are two forbidden ranges separated by a finite "quantum-well" allowed range (or with multiple wells). This is the *resonant tunneling* situation (Bohm 1951, Merzbacher 1970), with its own special properties, which will be discussed in § 5.

The foregoing discussion of $T(E)$ refers to a stationary state, constant in time, although it concerns a transition: the particle was in the L half-space, and is now in the R half-space. A popular way of describing that time dependence is in terms of a hypothetical *wave packet* which combines quantum states coherently over some small energy range δE and thereby has the large amplitude of a coherent superposition, at a certain point x, for a time interval $\delta t \simeq \hbar/\delta E$. This coherence occurs at a time t that depends on x, such that the wave packet moves with velocity $\mathrm{d}x/\mathrm{d}t = (1/\hbar)\,\mathrm{d}E/\mathrm{d}k$. Similarly the passage of such a wave packet through the barrier system (MacColl 1932, Merzbacher 1970) normally takes a time $t_{\text{trans}} = \hbar\,\mathrm{d}(\arg(A_{\mathrm{R}}/A_{\mathrm{L}}))/\mathrm{d}E$. This t_{trans} is ordinarily of order of magnitude $\hbar$/(barrier energy), and thus in practice is short on the time scale of interest. More plausible discussions of time dependence would then not be expected to give a very different outcome. But the interaction of a single electron with collective interactions of an actual solid-state barrier system may depend on such time scales. On the other hand, for a particle escaping from a localized state by tunneling it is the lifetime, rather than a transmission probability, that is the primary physical quantity. The escape lifetime characterizes a non-stationary quantum state with a complex, rather than real, energy value (Gamow 1929, 1931), on which see § 5.

A phenomenon that also depends on evanescent waves, and so is connected with tunneling, is the coupling of localized electron states when their spatial separation is not large. This is also sometimes referred to as a tunneling effect. An instance is two quantum wells separated by a barrier. A symmetrical one-dimensional geometry is appropriate for discussion: then the lowest stationary state will have a symmetric wave function $\psi_+(x) = \psi_+(-x)$, and the next lowest will have an antisymmetric wave function $\psi_-(x) = -\psi_-(-x)$. The difference between their energy eigenvalues, $E_{\mathrm{s}} = E_- - E_+$, is a measure of the coupling effect. If the separation were large (the barrier thick enough) they would tend to double degeneracy and $E_s = 0$. Since the states are localized, it follows from (1.1) that this energy difference is given by

$$E_s \int_{-\infty}^{0} \psi_+ \psi_- \,\mathrm{d}x = \frac{\hbar^2}{2m}\left(\psi_- \frac{\mathrm{d}\psi_+}{\mathrm{d}x} - \psi_+ \frac{\mathrm{d}\psi_-}{\mathrm{d}x}\right)_{x=0}. \tag{1.22}$$

In the barrier range, ψ_+ and ψ_- must be the linear combinations $\Phi_+ \pm \Phi_-$ of the evanescent solutions of (1.1) satisfying $\mathrm{d}\Phi_\pm/\mathrm{d}x = \pm\alpha(x)\Phi_\pm$ (with the α value at $E = E_+$ for ψ_+ and at $E = E_-$ for ψ_-). From the exact equation (1.22), one may hence obtain

$$E_s \simeq 2(\hbar^2/2m)\,\alpha(0)\,\psi_+(0)^2. \tag{1.23}$$

Since, with the normalization, $\psi_+(0)$ is essentially proportional to $\exp(-\beta/2)$, where β is the integral (1.8) taken over the barrier width, eq. (1.23) shows that the splitting E_s is essentially proportional to $\exp(-\beta)$. For slightly different E values at which each well in the absence of the other well has an energy eigenstate, the $\Phi_\pm(x)$ functions can be extended as solutions of (1.1) on their "own sides" (x positive for Φ_+, negative for Φ_-), and in the other direction so that they are orthogonal. Then, for these basis orbitals, E_s is given approximately by (i.e. is $\simeq 2$ times the modulus of) the off-diagonal matrix element, $\langle \Phi_- | \mathscr{H}_{\mathrm{int}} | \Phi_+ \rangle$, of

$$\mathscr{H}_{\mathrm{int}} = \mathrm{i}\hbar \mathscr{F}(0), \tag{1.24}$$

where $\mathscr{F}(x_1) = -\mathrm{i}(\hbar/2m)[(\mathrm{d}/\mathrm{d}x)\delta(x - x_i) + \delta(x - x_1)(\mathrm{d}/\mathrm{d}x)]$ is the usual one-dimensional flux (i.e. current/charge) quantum operator, at x_1.

A solution of (1.1) in a finite interval is specified by the values of two convenient parameters, such as ψ and $\mathrm{d}\psi/\mathrm{d}x$ at some position x, or similarly the amplitude coefficients of two waves in opposite directions like the A and B of (1.2). Then the pairs of values at two points x_1 and x_2 are linearly related by a corresponding 2×2 matrix; and a property such as transmission probability is a function of the matrix elements. This transfer-matrix formalism is, obviously, not limited to a particular choice for the parameter pairs. One may multiply the matrices for convenient sub-intervals to obtain the overall matrix for a system of interest.

2. Tunneling in semiconductors

The discussion in § 1 applies to electrons in free space. The actual potentials and wave functions in a crystalline solid are quite different on a microscopic scale, but phenomena of interest have been described with some success by just replacing the free-electron mass m in eq. (1.1) by the applicable effective mass $\overline{m}$ (or, in a many-valley context, with the tensor generalization $(1/\overline{m})\nabla^2 \to \nabla \cdot (1/\overline{\boldsymbol{m}}) \cdot \nabla$). The resulting wave equation is then taken to apply not to the actual wave function $\psi(\boldsymbol{r})$ but to an *envelope function* $\overline{\psi}(\boldsymbol{r})$, and not with the complete potential $U(\boldsymbol{r})$, including the "crystal lattice potential" on an atomic scale $U_0(\boldsymbol{r})$, but containing only the additionally imposed potential, $W = U - U_0$, with both $\overline{\psi}(\boldsymbol{r})$ and $W(\boldsymbol{r})$ varying slowly on the scale of the crystal lattice. This treatment is valid and adequate in appropriate conditions, but it requires elaboration, which is outlined here and amplified in the appendix.

For a homogeneous crystal, when the electron state can be characterized by a single band energy function $\varepsilon(\boldsymbol{k})$, its *envelope representation* is given by (Kittel and Mitchell 1954)

$$\bar{\psi}(\boldsymbol{r}) = \sum_{\boldsymbol{k}} b(\boldsymbol{k}) \exp(\mathrm{i}\boldsymbol{k} \cdot \boldsymbol{r}),$$

where

$$\psi(\boldsymbol{r}) = \sum_{\boldsymbol{k}} b(\boldsymbol{k}) \phi(\boldsymbol{k}, \boldsymbol{r})$$

is summed over the Bloch wave functions $\phi(\boldsymbol{k}, \boldsymbol{r})$ with wave vectors $\boldsymbol{k}$. The envelope function $\bar{\psi}$ satisfies the wave equation (for a stationary state)

$$\varepsilon(-\mathrm{i}\boldsymbol{\nabla})\bar{\psi} + \iint \mathrm{d}^3 \boldsymbol{r}_1 \, \mathrm{d}^3 \boldsymbol{r}_2 \, K(\boldsymbol{r}; \boldsymbol{r}_1, \boldsymbol{r}_2) W(\boldsymbol{r}_1) \bar{\psi}(\boldsymbol{r}_2) = E\bar{\psi}(\boldsymbol{r}) \tag{2.1}$$

where the kernel $K(\,)$ localizes W and $\bar{\psi}$ in $\sim$ a lattice-constant distance around $\boldsymbol{r}$, and

$$\iint \mathrm{d}^3 \boldsymbol{r}_1 \, \mathrm{d}^3 \boldsymbol{r}_2 \, K(\boldsymbol{r}; \boldsymbol{r}_1, \boldsymbol{r}_2) = 1.$$

Thus if $\bar{\psi}$ and W are slowly varying, and if $\varepsilon(\boldsymbol{k}) = \varepsilon_0 + (\hbar^2/2\overline{m})k^2$, we obtain the above "effective-mass equation" for $\bar{\psi}$. (If $\varepsilon - \varepsilon_0$ is proportional to the square of $\boldsymbol{k} - \boldsymbol{k}_0$, with constants $\boldsymbol{k}_0$ and $\varepsilon_0 = \varepsilon(\boldsymbol{k}_0)$, then it is $\boldsymbol{k} - \boldsymbol{k}_0$ that is replaced by $-\mathrm{i}\boldsymbol{\nabla}$.) An extension of this formalism applies to continuously varying composition, as in a semiconductor alloy: a slowly varying band-edge energy contributes a term $\varepsilon_0(\boldsymbol{r})$ to $W(\boldsymbol{r})$ (and similarly $(1/\overline{m})\boldsymbol{\nabla}^2 \rightarrow \boldsymbol{\nabla} \cdot [1/\overline{m}(\boldsymbol{r})]\boldsymbol{\nabla}$ or $\boldsymbol{\nabla} \cdot [1/\overline{\boldsymbol{m}}(\boldsymbol{r})] \cdot \boldsymbol{\nabla}$); but effects of the alloy disorder are, of course, additional to this "virtual-crystal" model. One may note that the equivalent of the evanescent wave of § 1 is a Bloch (Bloch-like) state with imaginary or complex wave number (wave-vector component), belonging to a "forbidden-gap" energy range; but it is being represented here by a sum over the "allowed" Bloch states. The evaluation of (2.16) illustrates this relationship.

In one-dimensional terms, so long as the $\varepsilon(k)$ function is proportional to k^2, or α^2, the WKB solution (1.4) applies to $\bar{\psi}$, with $U(x)$ replaced by $W(x)$. In a non-parabolic energy range, the left-hand side of (1.3) is to be replaced by $\varepsilon(k)$, so that $\varepsilon(k) = E - W$, and similarly the left-hand side of (1.7) by the equivalent for α, while $p(x)$ is proportional to $|\mathrm{d}\varepsilon/\mathrm{d}k|^{-1/2}$ or $|\mathrm{d}\varepsilon/\mathrm{d}\alpha|^{-1/2}$ (or, in terms of (2.7), $|\mathrm{d}\varepsilon/\mathrm{d}q|^{-1/2}$). However, we now have a slow-variation condition both for $k(x)$ or $\alpha(x)$ and for $W(x)$.

Within this scheme, an offset energy – the difference between the two band-edge energies, ε_0, for two crystalline semiconductors joined epitaxially at an interface – appears as a step-function contribution to $W(\boldsymbol{r})$. Two such steps at parallel interfaces (such as plus and minus the same offset) provide a barrier layer, in particular a tunneling barrier like that of fig. 1b. However, we still have to prescribe boundary conditions for $\bar{\psi}$ at the epitaxial interface, where the underlying variation of composition is on an atomic, rather than macroscopic, scale. The free-particle conditions $\bar{\psi}_L = \bar{\psi}_R$ and $\varphi_L = \varphi_R$, where φ means the component of $\nabla\bar{\psi}$ in the normal direction and where "L", "R" label the two sides of the interface, cannot apply, because they do not satisfy continuity of current density. For the "parabolic" case $\varepsilon \sim k^2$, the carrier flux (current density/charge) is

$$I/e = (\hbar/2\bar{m})\,\mathrm{i}(\bar{\psi}\varphi^* - \bar{\psi}^*\varphi). \tag{2.2}$$

The boundary conditions

$$\bar{\psi}_L = \bar{\psi}_R, \quad (1/m_L)\varphi_L = (1/m_R)\varphi_R \tag{2.3}$$

(BenDaniel and Duke 1966) do satisfy continuity of current according to (2.2), and are commonly assumed to apply. (We shall omit the "bar" of $\bar{m}$ where the context makes it superfluous.) A more general form, still for "parabolic" energy functions, is

$$\begin{aligned} \bar{\psi}_L &= s_{11}\bar{\psi}_R + s_{12}\vartheta_R, \\ \vartheta_L &= s_{21}\bar{\psi}_R + s_{22}\vartheta_R, \end{aligned} \tag{2.4}$$

where $\vartheta \equiv (m/\bar{m})a\varphi \equiv (m/\bar{m})a\,\mathrm{d}\bar{\psi}/\mathrm{d}x$ (in terms of the free-electron mass m, the effective mass $\bar{m}$, and the lattice constant a) and where the constants s_{11}, etc., are real and satisfy $s_{11}s_{22} - s_{12}s_{21} = 1$. Akera et al. (1988, 1989) have calculated the four coefficients $[s]$ (their $[\tilde{t}]$) for the (Al,Ga)As system and other semiconductors of interest. For the former, with the Γ electrons (Al concentration below 0.4), they find that s_{12} and s_{21} are generally small while s_{11} and s_{22} are fairly close to 1, so that (2.3) is acceptable in this case. These boundary matching conditions still may need to be adapted to non-parabolic energy ranges.

For equivalent "valley" band edges, or other multiple bands overlapping in their energy ranges, the envelope function needs to be generalized to an array of functions

$$\bar{\psi}_n = \sum_{\boldsymbol{k}} b_n(\boldsymbol{k}) \exp(\mathrm{i}\boldsymbol{k}\cdot\boldsymbol{r}), \qquad n = 1, 2, \ldots, \tag{2.5}$$

corresponding to the Bloch representation

$$\psi = \sum_n \sum_{\boldsymbol{k}} b_n(\boldsymbol{k})\phi_n(\boldsymbol{k}, \boldsymbol{r}) \tag{2.6}$$

and (2.1) is generalized accordingly to a set of equations. Except for the situations to be considered in the ensuing paragraphs, the important coupling between these components (2.5) for different n values is at an interface, where (2.4) generalizes to include interband (intervalley) terms: these have been calculated, in particular, for interface mixing of the Γ and X functions in III–V semiconductors (Akera et al. 1988, 1989). An incident Bloch electron belonging to one band or band edge can thereby tunnel through a barrier and emerge in a Bloch state belonging to a different band or band edge. (With an ordered interface, the component of wave vector parallel to the interface will be the same for incident and transmitted waves. The pairing of the two normal-direction components of $\boldsymbol{k}$ is then determined by the equality of the two $\varepsilon(\boldsymbol{k})$ functions.)

Transfer between bands can also occur continuously over the spatial range of the tunneling process. The important instance in semiconductors is tunneling between valence band and conduction band, when there is an imposed electric field that makes their allowed energy ranges overlap, at a small spatial separation, as depicted in fig. 3. (It will also apply where there are degenerate bands having a common band-edge point, e.g., $\varepsilon_1(\boldsymbol{k}_0) = \varepsilon_2(\boldsymbol{k}_0)$, as with the valence bands of the diamond-type semiconductors, or having intersecting $\varepsilon(\boldsymbol{k})$ functions as with zincblende-type semiconductors. Valence-band degeneracy is an important complication of interband tunneling.) Zener (1934) proposed a mechanism of dielectric breakdown in which, as shown in fig. 3a, electrons in the filled states of the valence band tunnel to empty states of the conduction band, over a distance $d = \varepsilon_G/F$, where F is the applied field times electron charge (i.e. force on the electron due to the field) and ε_G is the energy gap. Essentially this process has been identified in reverse-biased p–n junctions (Chynoweth and McKay 1957), with a strong electric field over most of the junction width, as shown in fig. 3b. A similar situation occurs in an *Esaki diode* or *tunnel diode* (Esaki 1958), which is illustrated in fig. 3c. This is a p–n junction in which the two sides are very heavily doped, so that the electron and hole distributions are degenerate. The junction field is then very strong ($\sim 10^6$ V cm^{-1}), and electrons incident from either side of the short field range have an appreciable probability of tunneling through to the other band, in either direction. A net current (necessarily zero at zero bias) results, in the forward direction as well as the backward direction,

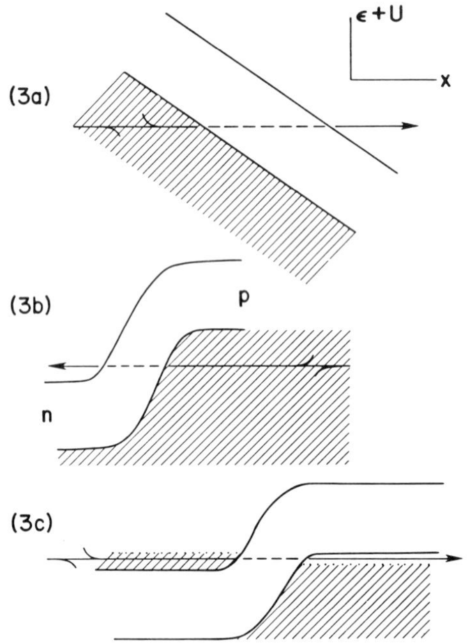

Fig. 3. Interband tunneling.

and has a characteristic dependence on forward bias, as will be discussed in §§ 3 and 4.

Beyond a turning point, x_0 where $E = \varepsilon_0 + W(x_0)$, the envelope function should be given by a WKB formula like (1.4) with (1.6) and (1.7), at distances $|x - x_0|$ large compared with $(\hbar^2/\overline{m}F)^{1/3}$, and by its non-parabolic generalization. In particular, we expect the probability of interband tunneling to be given (when suitably small) by a formula like (1.8), (1.9) with the appropriate $\alpha(x)$. As in fig. 1a, $\alpha(x)$ must increase from zero as x moves inwards from one of the turning points and decrease again to zero as it approaches the other turning point. In the present case, however, this happens while $E - W(x)$ increases from the valence-band edge (say, $\varepsilon = \varepsilon_{01}$) to the conduction-band edge ($\varepsilon = \varepsilon_{02}$). In terms of the complex wave vector

$$q = k + \mathrm{i}\alpha \tag{2.7}$$

the situation may be represented by fig. 4. As this suggests, the band energy functions $\varepsilon_n(k)$ in one dimension – and similarly for two and three dimensions – may be considered as belonging to a single analytic function $\varepsilon(q)$, with ε_n "sheets" meeting at branch points on the q plane.

For the case of constant F, eq. (1.8) becomes $\beta = (1/F)\int_1^2 \alpha(\varepsilon)\,\mathrm{d}\varepsilon$, which may be transformed here to

$$\beta = \frac{1}{F}\int_0^{\alpha_{12}} [\varepsilon_2(\alpha) - \varepsilon_1(\alpha)]\,\mathrm{d}\alpha, \tag{2.8}$$

where $\varepsilon_1(q)$ and $\varepsilon_2(q)$ meet as shown at $\varepsilon_{12} = \varepsilon(\alpha_{12})$, the maximum point of $\alpha(\varepsilon)$ – i.e. at a branch point of $\varepsilon(q)$. For the standard "two-band model" (Kane 1959), we have

$$\varepsilon_2 - \varepsilon_1 = \varepsilon_G[1 + (q/\alpha_{12})^2]^{1/2}, \tag{2.9}$$

where

$$(\hbar^2/\varepsilon_G)\,\alpha_{12}^2 = m_1 m_2/(m_1 + m_2) \equiv m_{12} \tag{2.10}$$

in terms of the two band masses m_1, m_2. Then, with $q = \mathrm{i}\alpha$, evaluating (2.8) gives

$$\beta = \frac{\pi}{4}\frac{1}{F}\varepsilon_G \alpha_{12}. \tag{2.11}$$

The basic theory (Franz 1956, Keldysh 1957, Kane 1959, Krieger 1966) is in terms of the situation of fig. 3a, with a uniformly constant F. In the Bloch representation (2.6), the equivalent of the set of envelope equations is

$$[\varepsilon_n(\boldsymbol{k}) - E - \mathrm{i}F\cdot(\partial/\partial\boldsymbol{k})]b_n(\boldsymbol{k}) = F\cdot\sum_l \boldsymbol{X}_{nl}(\boldsymbol{k})b_l(\boldsymbol{k}), \tag{2.12}$$

where

$$\boldsymbol{X}_{nl}(\boldsymbol{k}) = \mathrm{i}\int \mathrm{d}^3\boldsymbol{r}\; u_n^*(\boldsymbol{k},\boldsymbol{r})\,\partial u_l(\boldsymbol{k},\boldsymbol{r})/\partial\boldsymbol{k}. \tag{2.13}$$

Here $u(\boldsymbol{k},\boldsymbol{r}) = \phi(\boldsymbol{k},\boldsymbol{r})\exp(-\mathrm{i}\boldsymbol{k}\cdot\boldsymbol{r})$ is the periodic part of the Bloch function, normalized by $\delta_{ln} = \int \mathrm{d}^3\boldsymbol{r}\; u_l^* u_n$ integrated over the same domain as in (2.13). The diagonal terms

$(l=n)$ on the right-hand side of (2.12) may be discarded. When the off-diagonal terms are replaced by zero, the resulting decoupled equations have solutions

$$b_n(k) = \text{const.} \times \exp\left(\frac{1}{\mathrm{i}F}\int_0^k [\varepsilon_n(k') - E]\,\mathrm{d}k'\right), \tag{2.14}$$

where k is the component of $\boldsymbol{k}$ in the direction of $\boldsymbol{F}$ (the x direction) and for simplicity the lateral (y, z) components of $\boldsymbol{k}$ and $\boldsymbol{r}$ are disregarded. For each n, eq. (2.14) represents a state confined to one band, with a "Stark ladder" of eigenvalues $E = \varepsilon(n, \nu)$ spaced at an interval of $\varepsilon(n, \nu+1) - \varepsilon(n, \nu) = aF \equiv \Delta\varepsilon$, where a is the lattice constant. The off-diagonal terms $(l \neq n)$ on the right-hand side of (2.12) then couple states $|n, \nu\rangle$ to states $|l, \mu\rangle$, and so represent the interband tunneling process.

The role of interband matrix elements of the system Hamiltonian is supplied here, for the representation (2.6), by the quantities $M_{ln} = -\boldsymbol{F}\cdot\Sigma_{\boldsymbol{k}}\boldsymbol{X}_{ln}b_l^* b_n$. Correctly normalized for the present states, they become (dropping the ladder index)

$$M_{ln} = -(F/\kappa)\int_{-\kappa/2}^{\kappa/2} \mathrm{d}k\, X_{ln}(k) \exp\left(\frac{1}{\mathrm{i}F}\int_0^k [\varepsilon_n(k') - \varepsilon_l(k') + E_l - E_n]\,\mathrm{d}k'\right) \tag{2.15}$$

in one-dimensional terms, where $\kappa = 2\pi/a$ is the width of the Brillouin zone. We may boldly calculate an interband tunneling probability, T, by taking the interband *rate*, from one of the Stark ladder states, to be given by the "golden rule" as $(2\pi/\hbar)|M|^2$ times a density of states equal to $1/\Delta\varepsilon$ (Kane 1959), and by equating this to T times the "attempt frequency" for a wave packet cycling between band edges, equal to $\Delta\varepsilon/h$. Thus $(\Delta\varepsilon/h)T = (2\pi/\hbar)|M|^2(1/\Delta\varepsilon)$, and hence $T = (\kappa/F)^2|M|^2$, with initial and final energies E set equal. That is

$$T_{12} = \left|\int_{-\kappa/2}^{\kappa/2} \mathrm{d}k\, X_{12}(k) \exp\left(\frac{1}{\mathrm{i}F}\int_0^k [\varepsilon_2(k') - \varepsilon_1(k')]\,\mathrm{d}k'\right)\right|^2 \tag{2.16}$$

(Franz 1959, Argyres 1962) for a valence band "1" and a conduction band "2".

A version of this argument (Argyres 1962) evaluates the total tunneling current, per unit volume of solid, by also integrating over k_y and k_z. The golden-rule delta function in total energy difference then does not entail $E_1 = E_2$. However, this is not the same as evaluating T itself for a given incident-electron state (nor can it be applied to a strictly one-dimensional case). As Kane points out, for the situation of fig. 3a, T is applicable when there are scattering processes such that an electron approaching one band edge does not retain coherence with a preceding reflection from the opposing band edge of the band. Then the scattering rate $1/\tau$ is $\gg 2\,\Delta\varepsilon/h$ (twice the attempt frequency), and the associated energy broadening, $\hbar/\tau$, will be $\gg \Delta\varepsilon/\pi$. When $\hbar/\tau$ becomes $\gtrsim \Delta\varepsilon$, properties such as tunneling will no longer be based on the Stark-ladder $|n, \nu\rangle$ states. The golden-rule method can, however, be validly applied to the system of fig. 3c; but in order to calculate T it is necessary to evaluate the interband matrix element for that case, by finding its relationship to the matrix element (2.15) of the fig. 3a, "Zener case", as in the following paragraph.

For either band in (2.15), the envelope function given by (2.5) and (2.14), normalized by $\int|\bar{\psi}|^2\,\mathrm{d}x=1$, is

$$\bar{\psi}(x)=(2\pi\kappa)^{-1/2}\int_{-\kappa/2}^{\kappa/2}\mathrm{d}k\,\exp[\mathrm{i}\chi(x,k)],$$
$$\chi(x,k)=\int_{k_0}^{k}[x+(1/F)(E-\varepsilon(k'))]\,\mathrm{d}k'. \tag{2.17}$$

The WKB form results from approximating (2.17) by the method of stationary phase (Kane 1959):

$$\bar{\psi}(x)\simeq(2\pi\kappa)^{-1/2}[2\pi/\mathrm{i}\chi''(k_s)]^{1/2}\exp[\mathrm{i}\chi(k_s)], \tag{2.18}$$

where k_s is given by

$$\chi'(k_s)=0 \tag{2.19}$$

subject to $\operatorname{Im}\chi''(k_s)\geqslant 0$. Then (after some transformation of the exponent)

$$\bar{\psi}(x)\simeq\left(\frac{F}{\mathrm{i}\hbar\kappa\,v(k_s)}\right)^{1/2}\exp\left(\frac{\mathrm{i}}{F}\int_{\varepsilon_0}^{E+xF}k(\varepsilon)\,\mathrm{d}\varepsilon\right), \tag{2.20}$$

where $v(k)=(1/\hbar)\,\mathrm{d}\varepsilon/\mathrm{d}k$ gives the particle velocity. In the allowed range ($xF+E>\varepsilon_0$) there are two real values of $k_s(x)$, equal to $\pm k(\varepsilon=xF+E)$. By taking the WKB function (2.20) as extending into a zero-field range as in a p–n junction, we see that normalization in a macroscopic length L where $v_s\to v$ then requires multiplying the foregoing state vector (2.14) by $(\hbar\kappa v/2FL)^{1/2}$. For the case of fig. 3c there is one such factor on each side (L_1, L_2 and v_1, v_2). Since the density of states is multiplied by a factor $2FL_2/\hbar v_2\kappa$ relative to the case of fig. 3a while the attempt frequency is multiplied by $\hbar v_1\kappa/2FL_1$ (or similarly with "1" and "2" transposed), we finally obtain the same result (2.16) for the transmission probability. We shall make the assumption that the geometries can be combined in this way, and accordingly that (2.16) may be evaluated to give the applicable prefactor of eq. (1.9), while the exponent may be adapted to the continuously varying field of a p–n junction. The more satisfying formulation would be (Price and Radcliffe 1959, Shuey 1965) to solve the wave equation in this junction case, for states constrained to span the function space (Bloch states) of *one band*, and then treat the interband matrix elements of the complete Hamiltonian, connecting these solutions for the two isolated bands, as the tunneling perturbation.

Thus, calculation of tunneling probability may be based on evaluation of (2.16), with a constant F. To effect this, the integral over k is taken as an integral along a contour in the complex q space of (2.7), with the integrand an analytic continuation of the function of real k (Keldysh 1957, Kane 1959). (For $X_{nl}(k)$ given by (2.13), the continuation of $u(k)^*$ is not in general the complex conjugate of the continuation of $u(k)$, but it still provides the continuation $X_{nl}(q)$.) The meeting of two band $\varepsilon(q)$ functions on the q plane then provides a small (i.e. it needs to be small) neighborhood that contributes substantially all of the integral, as in the stationary-phase method. In the situation of fig. 4, that neighborhood is the vicinity of $q=q_{12}$, where $\varepsilon_1(q)=\varepsilon_2(q)$. A constant factor equal to $\exp(-2\beta)$, with β given by (2.8), may then be separated out of the integrand. What is left gives, for the prefactor of (1.9),

$$C=\left|\int_{-\kappa/2}^{\kappa/2} \mathrm{d}q\, X_{12}(q) \exp\left(\frac{1}{\mathrm{i}F}\int_{q_{12}}^{q} [\varepsilon_2(q')-\varepsilon_1(q')]\,\mathrm{d}q'\right)\right|^2. \tag{2.21}$$

In view of the nature of tunneling, it is reasonable that its rate depends mathematically on the relationship of the band energy functions for the complex wave vectors in the intervening energy gap, and thereby on the connection of the two wave functions in the classically excluded range of x.

From (2.9), we have $q_{12}=\pm\mathrm{i}\alpha_{12}$ and $\varepsilon_2-\varepsilon_1=(\varepsilon_G/\alpha_{12})(q-q_{12})^{1/2}$ with α_{12} given by (2.10). The branch-point singularity is in fact a general property of Bloch states; it occurs whenever two $\varepsilon(q)$ functions come together,

$$\varepsilon_n-\varepsilon_l \to \lambda_{nl}(q-q_{nl})^{1/2}, \tag{2.22}$$

and is then accompanied by a simple pole in X_{nl} given by

$$X_{nl}(q)\to \pm 1/4(q-q_{nl}) \tag{2.23}$$

(Keldysh 1957, Krieger 1966). On substituting (2.22) and (2.23) in (2.21), an integration on a path with the usual small half-circle under the branch point gives

$$C=(\pi/3)^2 \tag{2.24}$$

(Homilius and Franz 1954, Kane 1961, Krieger 1966). Since (2.24) is independent of F one may propose to use it in (1.9) with (1.8), for p–n junctions, interpreting it as the interband connecting coefficient for two band WKB functions.

In the treatment represented by eqs. (2.14)–(2.21), we have basis states that describe total reflection of the electron (or hole) from the tunneling structure; then a perturbation due to the complete system Hamiltonian $\mathscr{H}$ causes transitions between the "L"

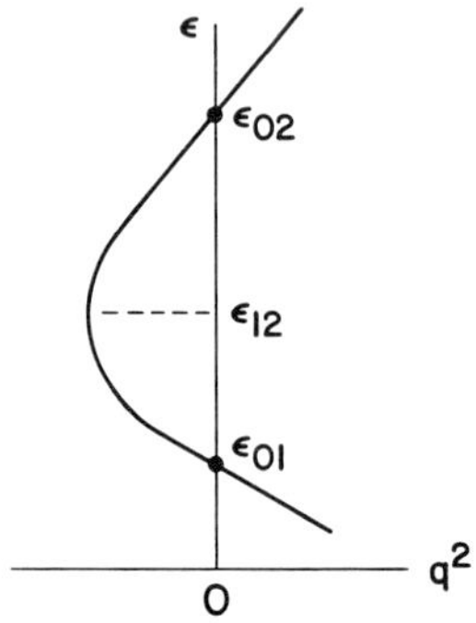

Fig. 4. Energy/wave-vector relationship for the two-band model.

(left-hand) set and the "R" (right-hand) set. This is often a convenient formulation, although obviously one loses the phase relation between the *A*- and *B*-coefficients of (1.2) for the two sides, if only the customary transition rates are calculated. For the situation of fig. 3, these basis sets are generated uniquely by band projection operators, but for situations such as are represented by fig. 1 the prescription of L and R sets is more artificial. A model (Duke 1969a, § 18) in which the L and R functions are associated with "local" Hamiltonians $\mathscr{H}_L$ and $\mathscr{H}_R$, and the intervening structure associated with a so-called *transfer Hamiltonian* $\mathscr{H}_T$ which operates on both of these sets of functions, is often employed for physically complex systems.

The discussion so far, for tunneling between just a pair of bands, is applicable for a unidimensional system, which may be represented by a "two-band model" as in eq. (2.9). But in the diamond and zincblende types of semiconductor, in particular, the valence band is degenerate at the band edge (at $\boldsymbol{k}_0 = 0$, the Γ point, apart from a small asymmetry in the case of the zincblendes), and near the band edge it involves coupled bands derived from three Γ-point orbitals having the symmetry of x, y and z. Spin–orbit coupling displaces a "split-off band" $\varepsilon_s(\boldsymbol{k})$ downward, leaving the "light hole" $\varepsilon_\ell(\boldsymbol{k})$ and the "heavy hole" $\varepsilon_h(\boldsymbol{k})$ which, for any given direction of $\boldsymbol{k}$, meet at a branch point (Dresselhaus et al. 1955). The theory of tunneling between this system and the conduction band (Krieger 1966) is much more complicated. Only the conclusions will be outlined here. (See also the penultimate paragraph of § 4.) Krieger shows, specifically for tunneling in a (100) direction, that

(a) in the four-band model of Kane (1957), ε_s and ε_ℓ meet at four additional branch points (e.g., for k_y and k_z real, $q_x = \pm k_{s\ell} \pm i\alpha_{s\ell}$), while $\varepsilon_h(\boldsymbol{k})$ does not meet these bands and does not meet the adjoining conduction band above them, $\varepsilon_c(\boldsymbol{k})$;
(b) to obtain the hole wave functions, in a constant field, one must solve a pair of equations like (2.12) with the coupling term from $X_{s\ell}$ retained;
(c) when the integrals generalizing (2.17) are evaluated correctly, one then obtains WKB functions analogous to (2.20);
(d) with correct treatment, in terms of this analytical structure, of (2.16) for $T_{\ell c}$ and T_{sc}, formulas like (1.9) are obtained with appropriate evaluations of (1.8);
(e) in these formulas, the prefactor is again $C = (\pi/3)^2$.

Although ε_h presumably is connected similarly to a higher band $\varepsilon(\boldsymbol{k})$, and incident heavy holes would tunnel to this band in a situation where that was energetically allowed, there is not the basis for tunneling in normal fashion, represented by (1.9), from the heavy-hole band to the Γ-symmetry (zone center) conduction band. While (2.16) still applies, it may be assumed to have a much smaller numerical value in this case.

In the diamond-like semiconductors Si and Ge, the lowest minimum of $\varepsilon(\boldsymbol{k})$ above the valence band is, of course, not at the zone center. It has been verified by computation (Jones 1966) that there is not a branch point in $\boldsymbol{q}$ space connecting this band-edge point with the Γ band edge of the valence band. It may be assumed that coherent propagation between them is negligible in an ideal structure, and interband tunneling in practice is due to effects of disorder, such as from impurities, and to inelastic tunneling processes in which a phonon is emitted or absorbed. This is often called *indirect tunneling*. The theory of eqs. (2.12)–(2.16) may be extended to calculate phonon-assisted tunneling rates (Keldysh 1958a, Tiemann and Fritzsche 1965) by replacing a sum over $\boldsymbol{k}$ in $b_2^*(\boldsymbol{k})\boldsymbol{F}\cdot\boldsymbol{X}_{21}(\boldsymbol{k})b_1(\boldsymbol{k})$ by sums over $\boldsymbol{k}_1$ and $\boldsymbol{k}_2$ in $b_2^*(\boldsymbol{k}_c)M_{12}(\sigma, \boldsymbol{k}_2 - \boldsymbol{k}_1)b_1(\boldsymbol{k}_1)$, where $M(\sigma, \boldsymbol{j})$ is a matrix element for interaction with phonons of wave vector $\boldsymbol{j}$ and branch σ, proportional to $N_{\sigma\boldsymbol{j}}^{1/2}$ (absorption) or $(N_{\sigma\boldsymbol{j}} + 1)^{1/2}$ (emission), where N is the phonon occupation. For an ideal parallel-plane structure, the initial and final states have discrete pairs of lateral components of $\boldsymbol{k}_1, \boldsymbol{k}_2$ for given lateral component of $\boldsymbol{j}$. Values of the latter are normally near the lateral component of $\Delta\boldsymbol{k}_0$, the difference between the two band-edge wave vectors $\boldsymbol{k}_0$. For the *normal* component of $\boldsymbol{j}$, however, there is no selection rule: only that the contributions to the phonon-induced tunneling process come substantially from a small range of values near the normal component of $\Delta\boldsymbol{k}_0$. (It is incorrect to invoke "momentum conservation" in this connection. The tunneling electrons are acted on by the force $\boldsymbol{F}$ as well as the phonon interaction field, so states incident on and transmitted from the tunneling region are not connected by their wave-vector difference being equal to $\pm\boldsymbol{j}$.) This tunneling can no longer be described in terms of unidimensional transmission probabilities, since a unique incident lateral wave vector is scattered incoherently into a continuum of transmitted lateral wave vectors.

3. Diode structures

There are several types of essentially parallel-plane structures in which conductors are weakly connected by a "forbidden" region that is transversed by tunneling electrons (or holes):

(a) The most primitive is the Schottky diode, formed by a metal contact to a semiconductor. If the offset (the relative alignment in energy of their Bloch bands) is such that the semiconductor Fermi level would be above the metal Fermi level, space charge induces a "band bending" such that in equilibrium the energy diagram resembles fig. 1c (with the semiconductor on the right). For the metal side negative, the tunneling probability increases with bias, because the barrier effectively becomes thinner. For the semiconductor side negative, the thermionic (Richardson) current increases with bias, because of the relative rise of the (quasi) Fermi level of the semiconductor.

(b) The tunneling barrier may be a thin layer of insulating solid between a metal and a semiconductor. In particular, metal-oxide–semiconductor (MOS) structures are used for field effect transistors (FETs): the barrier commonly is silicon oxide grown on silicon, the superposed metal is the gate, and the silicon contains the conducting channel. Then a natural question is the applicability of the formalism and results of § 2 to amorphous material – the monotonic evanescent wave functions, and even the conservation of lateral wave vector. An empirical result for silicon oxide barriers (Maserjian and Zamani 1982) is that a Fowler–Nordheim characteristic is obtained for current versus bias, presumably corresponding to (1.9) with (1.21), with

Gundlach oscillations superposed, implying coherent propagation and reflection between turning point and silicon interface.

(c) As already discussed, p–n junctions can behave as interband tunnel diodes. Esaki diodes have a characteristic dependence of current I on bias V (see § 4). There is a range of forward bias (the n side negative) in which the differential conductance dI/dV is negative. The reason is indicated in fig. 3c, with the dotted lines marking the quasi-Fermi levels. For elastic processes (and with some modification for inelastic ones), tunneling electrons go from occupied energies of the conduction band on the n side (hatch shaded) to unoccupied (unshaded) energies of the valence band on the p side. As the forward bias increases from zero, the number of states (measured by energy range) contributing to the tunneling increases and the current increases accordingly; but when eventually eV exceeds the sum of the two Fermi energies, this tunneling current must shut off. (The tunneling probability T decreases, meanwhile, with increasing forward bias.) When the forward bias increases further, the total current increases again, from non-tunneling contributions. It can also be augmented in the intermediate range, where $I(V)$ would otherwise have a minimum, by "excess current" due to a modified tunneling process (§ 6.4).

(d) Tunneling in which the electron passes through an epitaxial interface between two semiconductors was demonstrated for Ge/GaAs by Anderson (1960, 1962) and for Si/GaP by Zeidenbergs and Anderson (1967). In epitaxy, the two solids have the same crystal symmetry and near-enough the same lattice constant, enabling an ordered interface with continuity of the lattice to be realized. Figure 5a shows a potential profile for a p–n junction of Ge/GaAs. The difference of the forbidden gaps, about 0.8 eV, becomes the sum of the conduction-band and valence-band offsets. In the case illustrated, the wave function of the tunneling electron undergoes a transition from the Ge to the GaAs Bloch function, at the interface, as well as from conduction-band function to valence-band function. In an n–n or a p–p diode, the offset together with space charge can result in a barrier potential resembling that of fig. 1c (but with comparable potential gradients on the two sides of the interface).

These types of tunneling structure, and related properties, are extensively discussed in the monograph by Duke (1969a). Since that book was completed, there has occurred a burgeoning of *heterostructure* physics based on the fabrication technique of *molecular beam epitaxy* (MBE), as well as development of the vapor-growth chemistry used (Marinace 1960) to make Anderson's diodes. The semiconductor heterostructures are epitaxial combinations, in layers of thickness down to a few (even single) lattice constants. They have been made with a number of semiconductors and semiconductor alloys, especially with compounds of column-III and column-V elements and their ternary alloys, and more recently combining Ge and Si. The most developed and exploited category is combinations of GaAs, AlAs, and their alloy (Al,Ga)As. The alloy may be regarded as having the electronic properties of an "averaged" crystal plus disorder effects of the actual arrangement of Al and Ga atoms on the cation sites of the zincblende lattice. An important property is the crossover from the GaAs type of conduction band (Γ minimum lowest) to the AlAs type (X minimum lowest), at about $x = 0.4$ in $Ga_{1-x}Al_xAs$.

(e) A simple tunneling barrier may be essentially realized with an epitaxial heterostructure layer, and homopolar doping (n–n or p–p) on each side, as illustrated by fig. 5b. One still has the individuality of Bloch functions, of course. Wave functions belonging to different semiconductors differ on a lattice scale, apart from their envelopes $\bar{\psi}(x)$, and their continuity along the tunneling path entails a matching construction such as (2.4), or its extensions (Akera et al. 1989) to $\boldsymbol{k}$ pairs belonging to different regions of the Brillouin zone, but still having the same lateral value. A difference in the lateral $\boldsymbol{k}$ value is more drastic. For example, Solomon et al. (1986) investigated the $I(V)$ dependence of GaAs/(Al,Ga)As/GaAs n–n diodes in the Fowler–Nordheim regime of eqs. (1.17)–(1.21), in which the bias is large enough so that the second turning point of the tunneling path is within the barrier layer. For alloy barriers with Al concentration more than 0.5, they found that the effective mass for the tunneling exponent β corresponded to the component of the X-minimum mass tensor proper to the four lateral valleys in the Brillouin zone, rather than that of the two valleys aligned with the direction of the normal to the interface, although tunneling via the former from the Γ minimum would be precluded by selection rules for an ideal structure. They inferred that the predominant tunneling process entailed a lateral scattering. The theory of this phenomenon, ascribing the scattering to alloy disorder (Price 1988a), is an example of use of a transfer-Hamiltonian technique.

(f) For some epitaxial pairs, the offset is such that the valence band edge of one is above the conduction band edge of the other. For GaSb/InAs (Chang and Esaki 1980) the overlap exceeds 0.1 eV, and one can also pair the alloys (In,Ga)As and Ga(Sb,As). In "polytype" combinations like GaSb/AlSb/InAs (Esaki et al. 1981) the alignment is such that the AlSb layer acts as a tunneling barrier, for allowed energies in both GaSb and InAs. Then, for only light p and n doping of the GaSb and InAs

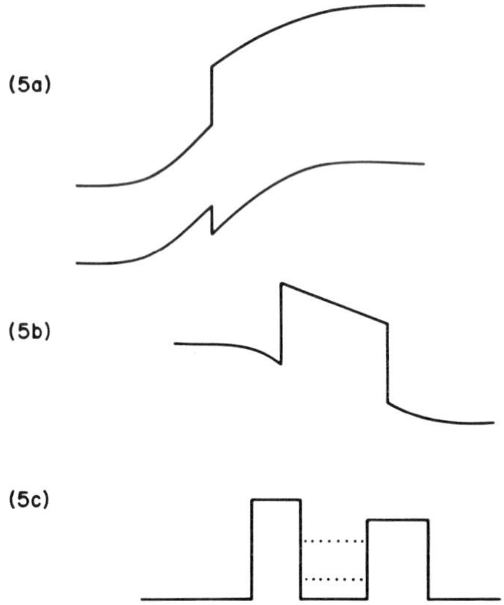

Fig. 5. Band-edge profiles in heterostructures.

respectively, one has a tunneling current with bias dependence similar to that of an Esaki diode (Luo et al. 1989, Chen et al. 1991). The lattice-scale component of the tunneling wave function will be that associated with the Γ point in both GaSb (valence band) and InAs (conduction band), while again in the AlSb the closest band edge in energy is that of the valence band at the Γ point (albeit the AlSb conduction band edge is at the X points); and so, at least if the layer is not too thin, one expects interface matching to be provided for by (2.4). (For a GaSb/InAs interface, Akera et al. find that s_{11} and s_{22} are small while s_{12} and s_{21} are relatively large, but they do not give corresponding results for the interfaces with AlSb.)

(g) One may have alternating epitaxial layers of two semiconductors having allowed and forbidden states, respectively, in the energy range of interest. Such structures have been investigated most extensively for the combination of GaAs and (Al,Ga)As, with the latter providing a forbidden range, and hence a tunneling barrier, for the conduction electrons of the former. With two such barriers, as in fig. 5c, they enclose a "well" layer that will have one or more quasi-levels, indicated by the dotted line. In the neighborhood of such a level, $T(E)$ has a resonance peak, to be discussed in § 5. Such structures have been investigated in the Al–Ga–As system (Chang et al. 1974); in the Si–Ge system (Schuberth et al. 1991); in the Hg–Cd–Te system (Reed et al. 1986); and in the "polytype" Al–Ga–As–Sb system (Yang et al. 1990, Chow et al. 1990, Longenbach et al. 1991), for which there is interband tunneling between the well layer and the exterior semiconductors. Because the transmission resonance requires only coherent interference between two (strong) reflections on the propagation path, the tunneling barriers may be replaced by strongly reflecting interfaces of a single "well" layer (Chow et al. 1988, Beresford et al. 1989).

(h) An extended sequence of "well" and "barrier" layers is a *superlattice*, which imitates natural crystal lattices, but with a larger "lattice constant" and hence a smaller zone, and with narrower bands of itinerant states. For energy ranges outside the latter, such an array can function as a tunneling barrier. For a virtually infinite array of identical wells and barriers, the superlattice wave vector is proportional to $\ln \lambda$, where λ is the factor by which the wave function changes between (the centers of) neighboring wells. Then

$$\lambda + \lambda^{-1} = 2\,\mathrm{Re}[1/t \exp(\mathrm{i}\chi)] \equiv 2G(E), \tag{3.1}$$

where $\chi(E)$ is equal to well width times the actual $k(E)$ in a well and $t(E)$ is the transmission amplitude of a single barrier ($T = |t|^2$). Within the small E range(s) where $-1 \leqslant G \leqslant +1$, this gives $|\lambda| = 1$ and hence the state is in a Bloch band. Otherwise, λ is real and $\neq 1$ and so the superlattice wave function is evanescent. For small barrier T values, predominantly G will be large and hence $\lambda \gg 1$. The itinerant states especially will be sensitive to applied fields on the scale of the Bloch band width divided by the interbarrier spacing. To describe its functioning as a composite barrier within a larger structure, analysis of the matching of the wave functions between inside and outside the finite superlattice is also needed.

For a barrier such as that of fig. 5b, one may apply the matching formula (2.4), when appropriate, to take account of the interfaces in the tunneling probability. In the simplest case, the barrier is symmetrical with the same enclosing semiconductor on each side. Then the tunneling prefactor, as in eq. (1.9), is found to be

$$C = \frac{16(a_1 a_2 k\alpha)^2}{[(a_1 \alpha s_{11} - s_{21})^2 + (ka_2)^2(a_1 \alpha s_{12} - s_{22})^2]^2}, \tag{3.2}$$

where the wave vector is $q = k$ outside and $q = \mathrm{i}\alpha$ inside the barrier, and where $a_1 = (m_1/m)a$ and $a_2 = (m_2/m)a$, with m_1 and m_2 the effective masses respectively inside and outside the barrier. The $[s]$ coefficients are defined so that the left-hand side of (2.4) refers to the barrier side of the interface. When $s_{ij} = \delta_{ij}$, then (3.2) reduces to the result for the case of eq. (2.3). In the opposite limit $s_{11} = s_{22} = 0$ and accordingly $s_{12}s_{21} = -1$, which would be a suitable approximation for GaSb/InAs (Akera et al. 1989), eq. (3.2) reduces to $C = 16/(\rho + 1/\rho)^2$ where $\rho = s_{12}^2 a_1 a_2 k\alpha$. Thus, C has an upper bound of 4 for both of these ideal limiting cases.

4. Diode conduction

The diode current $I(V)$ for a static bias V is conveniently calculated by assuming that electrons (holes) incident on the barrier structure from either side (L or R) have the thermal equilibrium distribution $f(E(\boldsymbol{k}))$ for that side, and are transmitted to the far side (and so contribute to the current) with probability T. (It is convenient here to denote band energy functions by $E(\boldsymbol{k})$, instead of the $\varepsilon(\boldsymbol{k})$ of § 2.) Then the number of electrons (holes) per unit time impinging from one side is (according to this model) equal to

$$2\left(\frac{\mathscr{A}}{(2\pi)^2}\right)\int \mathrm{d}^2\boldsymbol{k}'' \int_0 v' f \frac{\mathrm{d}k'}{2\pi},$$

where the normal and lateral components of $\boldsymbol{k}$ are, respectively, k' and $\boldsymbol{k}''$, and v' is similarly the normal component of the Bloch-state velocity $\boldsymbol{v}$, and the lateral area is $\mathscr{A}$. For an ideal parallel-plane structure, the lateral wave vector is conserved in elastic tunneling. Unless the semiconductor materials on the two sides are the same, however, their band energy functions $E_\mathrm{L}(\boldsymbol{k})$ and $E_\mathrm{R}(\boldsymbol{k})$ should be assumed to be different, and consequently for given energy, E, of the tunneling state the two (unsigned) k' values for the two incident directions in general will differ. That is,

$$\begin{aligned} &E_\mathrm{L}(\boldsymbol{k}_\mathrm{L}) = E_\mathrm{R}(\boldsymbol{k}_\mathrm{R}) = E, \\ &\boldsymbol{k}''_\mathrm{L} = \boldsymbol{k}''_\mathrm{R} \equiv \boldsymbol{k}'', \quad k'_\mathrm{L} \neq k'_\mathrm{R}. \end{aligned} \tag{4.1}$$

Accordingly it is convenient to take the "constants of tunneling" $\boldsymbol{k}''$ and E as the independent variables. The current from the L side to the R side is then

$$I_\mathrm{LR} = 2\frac{e}{h}\frac{\mathscr{A}}{(2\pi)^2}\int \mathrm{d}^2\boldsymbol{k}'' \int \mathrm{d}E\; T(\boldsymbol{k}'', E) f_\mathrm{L}(1 - f_\mathrm{R}), \tag{4.2}$$

where we have made use of the relation $\hbar\boldsymbol{v} = \partial E/\partial \boldsymbol{k}$. The net current $I = I_\mathrm{LR} - I_\mathrm{RL}$ is obtained by replacing $f_\mathrm{L}(1 - f_\mathrm{R})$ in the integrand of (4.2) by $f_\mathrm{L}(1 - f_\mathrm{R}) - f_\mathrm{R}(1 - f_\mathrm{L}) = f_\mathrm{L} - f_\mathrm{R}$. The difference, $\zeta_\mathrm{L} - \zeta_\mathrm{R}$, between the Fermi levels ζ of these Fermi functions

f_L and f_R may be equated to eV. (In practice the applied bias will be divided between the diode itself and a "screening range" on each side, but this complication need not be considered here.) Then the net current is

$$I = 2\frac{e}{h}\frac{\mathscr{A}}{(2\pi)^2}\int \mathrm{d}^2\boldsymbol{k}'' \int \mathrm{d}E\ T(\boldsymbol{k}'', E)[f(E) - f(E+eV)]. \tag{4.3}$$

In (4.3) as derived, the Fermi level ζ in $f(E)$ is to be equated with ζ_L, while the R side's band energy function $E_R(\boldsymbol{k})$ is displaced by the bias to $\tilde{E}_R(\boldsymbol{k}) = E_R(\boldsymbol{k}) - eV$. The lower limit of the E integral is then the greater of the minimum of E_L and the minimum of $\tilde{E}_R$, as functions of k', for each $\boldsymbol{k}''$ value. The ohmic conductance is obtained on replacing $f(E) - f(E+eV)$ in (4.3) by $-eV\,\mathrm{d}f/\mathrm{d}E$. Then for the low-temperature limit, with degenerate distributions on the two sides, it becomes

$$\left(\frac{\mathrm{d}I}{\mathrm{d}V}\right)_{V=0} = 2\frac{e^2}{h}\frac{\mathscr{A}}{(2\pi)^2}\int \mathrm{d}^2\boldsymbol{k}''\ T(\boldsymbol{k}'', \zeta). \tag{4.4}$$

When (1.9) applies and β is sufficiently large, we may expect to approximate (4.3) by expanding $\beta(\boldsymbol{k}'', E)$ on the constant-E surface about a minimum at $\boldsymbol{k}'' = \boldsymbol{k}_0(E)$:

$$\beta(\boldsymbol{k}'', E) \simeq \beta_0(E) + \Omega(E)|\boldsymbol{k}'' - \boldsymbol{k}_0(E)|^2 \tag{4.5}$$

(or the more general "ellipsoidal" form). On neglecting any $\boldsymbol{k}''$ dependence of the prefactor C, and approximating the resulting $\boldsymbol{k}''$ integral of (4.3) in the obvious way, this gives

$$I = 2\frac{e}{h}\frac{\mathscr{A}}{(2\pi)^2}\int \mathrm{d}E\ [f(E) - f(E+eV)]T(\boldsymbol{k}_0, E)\frac{\pi}{2\Omega}[1 - \exp(-\Omega\Xi)], \tag{4.6}$$

where $\Xi(E)$ is the lesser of the two upper limits of $|\boldsymbol{k}'' - \boldsymbol{k}_0|^2$, where $k' = 0$, for given E. (In the simple parabolic case, Ξ is the lesser of the two values of $(2\overline{m}/\hbar^2)|E - E_0|$, where E_0 is the band-edge value of E on either side.) In the same way, the integral in (4.4) would be evaluated as

$$T(\boldsymbol{k}_0(\zeta), \zeta)[\pi/2\Omega(\zeta)]\{1 - \exp[-\Omega(\zeta)\Xi(\zeta)]\}.$$

Separability of the energy function, within the barrier, leads to the form $\Omega = (\mathrm{d}\beta_0/\mathrm{d}E)\hbar^2/2m''$, where m'' is the applicable lateral effective mass. Specifically, the standard square barrier and parabolic energy function, with $\beta = \alpha(E')d$ for barrier width d, gives $\Omega(E) = d/2\alpha(E)$. The result is more complicated when the potential or band-edge energy of the barrier is not constant along the tunneling path – in the zero-bias state, or otherwise as a result of the bias potential V. The $T(\,)$ quantities in eqs. (4.2)–(4.6) will be, in general, functions of V. Detailed discussions of $I(V)$, for unipolar non-resonant diodes, are given in particular by Stratton (1962, 1969), by Duke (1969b) and by Gundlach (1973). The case of resonant tunneling (see § 5) is different because of the sharp peak in T versus energy, rather than an essentially monotonic dependence.

For the case of an Esaki diode, illustrated by fig. 3c, the E integral of eq. (4.3) has lower limit 0 and upper limit

$$\hat{E}(V) = E_{F1} + E_{F2} - eV, \tag{4.7}$$

where E_{F1} and E_{F2} are the Fermi energies relative to the band edge for holes on the p side and electrons on the n side. For this case we may take

$$\Xi = (2/\hbar^2)\min(m_1 E, m_2(\hat{E} - E)). \tag{4.8}$$

With increasing forward bias ($V > 0$) there is a current cut-off at $\hat{E} = 0$, i.e. at $V = (E_{F1} + E_{F2})/e$. A useful approximation (even if not accurate) is eq. (4.6) together with (a) assuming that F is constant over the tunneling path, and (b) taking both $T(0, E) = C\exp(-2\beta_0)$ and Ω as independent of E. Then (4.6) reduces to

$$\frac{I}{\mathscr{A}} = 2\frac{e}{h}\frac{C}{(2\pi)^2}\frac{\pi}{2\Omega}\exp(-2\beta_0)Z, \tag{4.9}$$

where

$$Z(V) = \int_0^{\hat{E}(V)} [f(E) - f(E + eV)]\{1 - \exp[-\Omega\Xi(E, V)]\}\,\mathrm{d}E. \tag{4.10}$$

The $I(V)$ dependence includes the V dependence of β_0 and Ω. On dropping the factor $1 - \exp(-\Omega\Xi)$, the integral in (4.10) can be evaluated exactly for the Fermi function; but it is useful to just consider the low-temperature limit, taking $f(E)$ as 1 for $E < \zeta$ and 0 for $E > \zeta$. Then

$$Z(V) = \min(eV, E_{F1}, E_{F2}, E_{F1} + E_{F2} - eV) \tag{4.11}$$

– a trapezoidal curve, for forward bias. However, omitting the final factor $1 - \exp(-\Omega\Xi)$ in the integrand, for the present case, obviously cannot be valid as the forward current cut-off is approached. The analysis for the low-temperature limit without this approximation is presented by Kane (1961).

In (4.9), with the Kane two-band model of eq. (2.9)–(2.11), we should substitute $C = (\pi/3)^2$,

$$\beta_0 = (\pi E_G/4F\hbar)(m_{12}E_G)^{1/2}, \tag{4.12}$$

and, extending the analysis in § 2 to three-dimensional wave vectors in the same way as Kane (1959),

$$\Omega\beta_0 = \left(\frac{\pi}{4}\frac{E_G}{F}\right)^2 \tag{4.13}$$

with E_G the gap energy and m_{12} the reduced mass as in eq. (2.10). Substituting $C = (\pi/3)^2$ and (4.13) in (4.9) gives

$$\frac{I}{\mathscr{A}} = \frac{2e}{h}\frac{2}{9\pi}\left(\frac{F}{E_G}\right)^2 \beta_0 \exp(-2\beta_0)Z, \tag{4.14}$$

proportional to $(1/\beta_0)\exp(-2\beta_0)Z$. Thus $I(V)/Z(V)$ should be expected to decrease with increasing forward bias, because β_0 – or more generally $\beta(0, E)$ – increases. For reverse bias (Chynoweth and McKay 1957) the analysis has $Z = -eV$ instead of the trapezoid, and the coefficient of Z in (4.14) would increase numerically with increasing $-V$.

For the four-band model, with tunneling in a (100) direction, Krieger (1966) obtained results corresponding to the foregoing for $\beta(\boldsymbol{k}'', E)$. In particular for tunneling between the light-hole band and the conduction band, eq. (4.13) holds unchanged, and so (4.14) is unchanged. In place of (4.12), however, the result is

$$\beta_0 = (\pi E_G/4F\hbar)(m_r E_G)^{1/2}\xi, \tag{4.15}$$

where

$$m_r = m_c m_\ell/(m_c + m_\ell) \tag{4.16}$$

is equivalent to the foregoing reduced mass m_{12}, in terms of the conduction-band effective mass m_c and light-hole (100)-direction effective mass m_ℓ, and where

$$\xi = \left(\frac{\Delta + \frac{1}{2}E_G}{\Delta + \frac{3}{4}E_G}\right)^{1/2}, \tag{4.17}$$

where $\Delta = \varepsilon_\ell l(0) - \varepsilon_s(0)$ is the zone-center "split-off energy". Thus β_0 is decreased relatively to the (4.12) value by the factor ξ. For $\Delta \gg E_G$ we have $\xi = 1$ and hence effectively the two-band result. The greatest difference from the latter is the factor $\xi = 0.8165$ for $\Delta \to 0$.

The foregoing treatment of the p–n Esaki diode still is idealized, beyond the explicit approximations made, in disregarding effects of impurity disorder even in a parallel-plane structure, in not considering the indirect-tunneling situation where the band edges are not at the same point in the Brillouin zone, and in not applying to tunneling "excess current". These aspects are discussed briefly in §§ 6.2 and 6.4. For the "type II" version of the tunnel diode (Luo et al. 1989, Chen et al. 1991) the variation of T in (4.3) of course will be more complicated than for Γ–Γ tunneling within one semiconductor, as above.

5. Resonant tunneling

A special case of interest is illustrated by fig. 5c. The essential feature is that an "allowed" range of the transmission path is enclosed by two coherent partial reflectors of the electron (hole) waves. Each of them is normally a tunneling barrier, although it can be instead an epitaxial interface between differing semiconductors. For the case of two "smooth" tunneling barriers, such as depicted in fig. 1a, the $T(E)$ function was calculated by Bohm (1951). Davis and Hosack (1963) proposed this as a model of a "thin-film triode" device. For the case of "square" barriers like that of fig. 1b, Tsu and Esaki (1973) investigated the double-barrier structure, and also the comparable multiwell superlattice. The first heterostructure implementation of double-barrier tunneling (Chang et al. 1974) had a single GaAs "quantum well" layer and doped exterior semiconductors of GaAs, and two barrier layers of $Ga_{1-x}Al_xAs$. (The Al concentration in the latter was such that the X minimum was lowest in energy, and presumably dominant in the tunneling wave functions. This complicates the wave-mechanical description, but the principle is not affected.)

The resonant transmission is sometimes referred to as a "Fabry–Pérot effect", by analogy with the optical interferometer (just as electromagnetic and optical wave effects were once explained in terms of the then more familiar fluid mechanics). Away from a resonance energy, E_0, the transmission probability is $\approx T_L T_R$ where $T_L(E)$ and $T_R(E)$ are the transmission probabilities of each barrier in the absence of the other. Near E_0, however,

$$T(E) = T_0/\{1 + [(E - E_0)/\Delta E]^2\} \equiv T_0 \Lambda(E) \tag{5.1}$$

in which ΔE is proportional to $T_L + T_R$ (and is characteristically small), while the maximum $T(E_0) = T_0$ depends on T_L/T_R and for $T_L = T_R$ is equal to 1. With T_L and $T_R \ll 1$, the Lorentzian form (5.1) applies over a range of $|E - E_0|$ many times the half-width ΔE. Thus the resonant transmission of the double-barrier system is quite different from ordinary tunneling transmission.

The electron (hole) is almost trapped at (normal-direction) energy $E \simeq E_0$. For the open system, this "quasi-level" is in place of an actual bound state and energy eigenvalue. The E_0 values are in fact close to what would be the energy levels if T_L and T_R were zero, with the same "well" boundaries. Consideration of these quasi-level states provides a general treatment of transmission resonances. For a Bloch-state pair on either side (L or R) of the structure, the wave function may be written as

$$\psi = v^{-1/2}[A\phi(k) + B\phi(-k)] \tag{5.2}$$

– conveniently presented here in one-dimensional terms – and the coefficients A, B on either side, at a given energy E, accordingly related by a 2×2 transfer matrix:

$$\begin{aligned} A_R &= c_{11} A_L + c_{12} B_L, \\ B_R &= c_{21} A_L + c_{22} B_F, \end{aligned} \tag{5.3}$$

(As with the coefficients in (2.4), we have det $[c] = 1$.) We now consider $c_{11}(E)$, etc., as analytic functions of complex E. A quasi-level state having outward waves only, $A_L = B_R = 0$, will be given by a zero of $c_{22}(E)$, occurring at a complex value $E = E_0 - i\hbar/2\tau$, with the state lifetime consequently equal to τ (Gamow 1929, Bahder et al. 1987). One may then show (Price 1988b) that, with τ assumed large, eq. (5.1) holds, for real E near E_0, with

$$\Delta E = \hbar/2\tau \tag{5.4}$$

– the Breit–Wigner relation – and with

$$T_0 = 4 I_L I_R/(I_L + I_R)^2, \tag{5.5}$$

where, for the quasi-level state, I_L and I_R are the outward decay currents, such as given by (2.2), on the two sides. For the case of a simple double-barrier structure, (5.5) may be written

$$T_0 = 4 T_L T_R/(T_L + T_R)^2 \tag{5.6}$$

and ΔE in (5.4) may be obtained from

$$1/\tau = \gamma(T_L + T_R), \tag{5.7}$$

where the "attempt frequency" γ is given by

$$1/\gamma = \hbar\, d(2\chi + \theta_L + \theta_R)/dE \tag{5.8}$$

at $E = E_0$, in terms of the phase elapse χ for a Bloch wave crossing the well and the reflection phases θ_L, θ_R of the two barriers. In quasi-classical terms, therefore, $1/\gamma$ is equal to twice the time to cross the well, $2\int dx/v$, plus the sum of the two reflection times $\hbar\, d\theta_L/dE$ and $\hbar\, d\theta_R/dE$, which is an heuristically convincing result. However, the system of (5.1)–(5.5) is more general, and equally applies to more complicated linear structures, such as (Ricco and Azbel 1984) superlattice-like arrays of wells and barriers which need not be ordered, exhibiting transmission peaks that are associated with localized quasi-level states.

Some time-like quantities are associated with resonant transmission (and, except for (a), apply to diodes more generally):

(a) The quasi-level decay lifetime, τ.

(b) Normally the quantity $t_{trans} = \hbar\, d\theta_{trans}/dE$, where θ_{trans} is the Bloch-wave phase elapse in transmission, gives the time required for a wave packet to cross the structure. For the system of (5.1)–(5.5), this t_{trans} is equal to $2\tau\Lambda(E)$, which becomes large (for E near E_0) when ΔE is typically small. However, as is shown by Merzbacher (1970), in the resonance range the incident wave packet is greatly distorted in transit; so t_{trans} has only a formal status here.

(c) The diode conductance does, however, have operational significance, and its frequency dependence characterizes the time scale of the resonance phenomenon. It has been shown analytically and by direct computation (Jacoboni and Price 1990) that, when the Fermi level is near E_0, the ohmic conductance varies over a frequency range $\sim 1/\tau$, as one might anticipate, although in a more complicated way than would be obvious. Computer simulation in terms of the Wigner function (also taking account of space charge effects) gives comparable results for differential conductance at non-zero bias (Frensley 1987, 1988, Kluksdahl et al. 1989).

In more general terms, the time-dependent properties of the system are given by quantities like the steady-state correlation function, $((A, s, B))$, for two variables A and B at times differing by the time interval s. In particular, A and B could be the complementary step functions $S_L(x)$ and $S_R(x)$ whose expectations are the probability for an electron to be on one side or on the other side of a diode, and $((S_L, s, S_R))$ would then give a measure of the time elapse for transit. Because $eVS(x)$ is essentially the perturbation for an applied bias V, apart from screening effects, $((S_L, s, S_R))$ gives the linear time-dependent charge-transfer response following a step-function $V(t)$, and its Fourier transform is simply related to the conductance versus frequency. In this sense, the frequency-dependent conductance gives a measure of the time-dependent behavior of a diode system. It can be written in terms of the current variable (charge times dS/dt) substituted for both the A and B quantities (Kubo 1957); and then its zero-frequency limit is as though (1.24) were the (fictitious) Hamiltonian perturbation giving, by the golden rule, the transition rate between two continuum sets of states belonging to the L and to the R sides of the diode.

(d) The "dwell time" (Büttiker 1983) is a measure of the concentration of the electron density inside the tunneling structure. It is defined by $t_{\mathrm{dwell\,S}} = l_S/v_S$ ($S = L$ or R) where the normalization length is

$$l_S = \int_{\text{inside}} |\psi_S|^2 \,\mathrm{d}x \tag{5.9}$$

for ψ_S the wave function due to an incident Bloch wave *normalized in unit length*, incident from side S = L or R, and where v_S is the incident velocity. It has been shown, at least for the double-barrier case (Gu and Gu 1989), that in the resonance range

$$t_{\mathrm{dwell\,S}} = \tau_S T(E), \tag{5.10}$$

where τ_L and τ_R are the partial decay times

$$\tau/\tau_S = I_S/(I_L + I_R). \tag{5.11}$$

Thus t_{dwell} is in general comparable with τ.

At thermal equilibrium with the Fermi level above the resonance energy, the contribution of the resonance to the effective number of electrons within the structure is given by these normalization lengths. In one-dimensional terms, it is (for each spin component)

$$\mathcal{N} = \int l_L \,\mathrm{d}k_L/2\pi + \int l_R \,\mathrm{d}k_R/2\pi = (1/h)(\tau_L + \tau_R) \int T \,\mathrm{d}E. \tag{5.12}$$

(It can be shown that, for these two incident Bloch waves together with their resulting reflected and transmitted waves, at the same E value, the two state vectors are orthogonal: $\int \psi_L^* \psi_R \,\mathrm{d}x = 1$.) But it follows from (5.1), (5.4), (5.5) and (5.11) that

$$\int T \,\mathrm{d}E = h/(\tau_L + \tau_R), \tag{5.13}$$

where again the integral is over the resonance range. Therefore $\mathcal{N} = 1$, which is just as if the quasi-level were an isolated true quantum level. In the three-dimensional diode, the potential due to this resonant space charge could result in a bistable current–voltage characteristic; but there are doubts about its role in bistabilities that have been observed.

The calculation of the diode current differs from § 4 because of the resonant energy dependence of T. Except when f varies significantly over the resonance range $\sim\Delta E$, we may replace the Lorentzian form (5.1) by a delta function of energy times (5.13). The ideal case has the resonance quantities (E_0, etc.) independent of $\boldsymbol{k}''$. If we may also assume separability of E' and E'', and a parabolic $\boldsymbol{k}''$ dependence of the latter so that $\mathrm{d}^2\boldsymbol{k}'' \to (2\pi m''/\hbar^2)\,\mathrm{d}E''$, then (4.3) becomes

$$\frac{I}{\mathscr{A}} = 2\frac{e}{h}\frac{m''}{\hbar}\frac{1}{\tau_L + \tau_R} eV \tag{5.14}$$

(so long as E_0 for $\boldsymbol{k}''=0$ is below the cathode Fermi level and above the cathode band edge). More generally, however, E_0 will depend on $\boldsymbol{k}''$, especially if the well and exterior semiconductors differ (Ohno et al. 1990); τ_L and τ_R (or equivalently T_0 and ΔE) will depend on $\boldsymbol{k}''$ especially because of a different lateral effective mass in the barriers. As V increases in the forward direction, E_0 will be lowered relative to the cathode band edge by some fraction of eV. The resonance contribution to the current will cut off when E_0 falls below the cathode band edge (and if E_0 is above the Fermi level at $V=0$, it will first "cut on" at a smaller V value). If, however, E_0 depends significantly on $\boldsymbol{k}''$, the cut-off (and cut-on) will not be sharp. The resonance quantities $T_0, \Delta E$ must also be expected to depend on V directly. If at zero bias the diode is symmetrical then by (5.5) or (5.6) T_0 will be equal to 1, and accordingly it will fall off from 1 for non-zero V. The value of the integral (5.13) – and hence the coefficient of V in (5.14) – will fall off because the increase of one τ_S will outweigh the decrease of the other. The result (5.14) is the same as, with equivalent assumptions about $\boldsymbol{k}''$ dependence, would be obtained (Weil and Vinter 1987) from the *sequential tunneling* model, in which the quasi-level is replaced by a distinct localized "well" state with independent tunneling transitions, through the L and R barriers, between it and the exteriors on the two sides.

6. Additional topics

6.1. Magnetic field effects

A uniform magnetic field $\boldsymbol{B}$ resolves the three-dimensional Bloch states of a semiconductor into a ladder (or ladders) of Landau levels, provided that the field strength is large enough for the level spacing to be large compared to $\hbar$ times scattering rates. Each level corresponds to a one-dimensional continuum of itinerant states, propagating in the field direction with wave number k and energy

$$\varepsilon_0 + (n+\tfrac{1}{2})(e\hbar/mc)B + (\hbar^2/2m)k^2$$

in the parabolic case. If the field direction is normal to the planes of a parallel-plane tunneling structure, then the tunneling process is not essentially changed. The total current is given by summation over levels (over n) in place of integration over the lateral wave vector $\boldsymbol{k}''$, and de Haas-like oscillations may occur in the $I(V)$ characteristic of p–n tunnel diodes (Roth et al. 1966) and unipolar tunneling structures (Hickmott 1985, Leadbeater et al. 1989, Mendez et al. 1991). But if $\boldsymbol{B}$ is parallel to the structure planes (or in an intermediate direction) then there is an essentially different and more complicated situation, with the "Larmor orbit" motion combining with the tunneling motion (Haering and Adams 1961, Hickmott 1991, Leadbeater et al. 1991, Schuberth et al. 1991). One might say that the hopping motion which provides the normal-direction current in the surrounding semiconductors has to include a tunneling transition between the magnetic edge states on either side of the structure.

6.2. Phonon-induced tunneling

Tunneling with absorption or emission of a phonon is a principal source of the current in an "indirect" tunnel diode, in which the valence- and conduction-band edges are located at different points in the Brillouin zone. Since the wave vectors $\boldsymbol{j}$ of the contributing phonons normally belong to a relatively small range about $\boldsymbol{j}_0$, the wave vector (or wave vectors) connecting these band edge points, the phonon energies $\hbar\omega(\boldsymbol{j})$ are close to a central value $\hbar\omega_0 = \hbar\omega(\boldsymbol{j}_0)$. Consequently, for such a tunnel diode at low temperature, a result of energy conservation together with the exclusion principle for degenerate cathode and anode electrons is that the current due to this phonon-induced process is appreciable only for $|eV| \gtrsim \hbar\omega_0$. For several phonon modes, there are corresponding multiple thresholds in $I(V)$ (Esaki and Miyahara 1960). Display of the experimental $\mathrm{d}I/\mathrm{d}V$, and $\mathrm{d}^2I/\mathrm{d}V^2$, versus V reveals these phonon energies (Chynoweth et al. 1962, Payne 1965), providing a sensitive phonon spectroscopy. The "zero-bias anomaly", which is often observed, typically a lower conductance $\mathrm{d}I/\mathrm{d}V$ over a range of some millivolts around $V=0$, is believed to be due at least in part to a progressive onset, with increasing bias, of current induced by absorption and emission of acoustic-mode phonons (Bennett et al. 1968).

For unipolar diode conduction through a single heterostructure barrier, plots of $\mathrm{d}^2I/\mathrm{d}V^2$ versus V similarly display features identified as energies of the principal phonons inducing tunneling (Hirakawa et al. 1990). In a double-barrier resonant tunneling diode, it is possible to observe a *phonon replica* feature, a secondary peak in $I(V)$ at a greater bias than the main current peak, attributed to tunneling with emission of an optical-mode phonon (Goldman et al. 1987). Wingreen et al. (1989) have developed a comprehensive theory in terms of the model in which the two sides of the structure and the well region have distinct states, with Hamiltonians $\mathscr{H}_\mathrm{L}$, $\mathscr{H}_\mathrm{R}$ and $\mathscr{H}_\mathrm{C}$ together with connecting terms $\mathscr{H}_{\mathrm{L-C}}$ and $\mathscr{H}_{\mathrm{C-R}}$, with addition of a phonon system represented by Hamiltonian $\mathscr{H}_\mathrm{p}$ and connecting terms $\mathscr{H}_{\mathrm{C-p}}$ for well state(s) and phonons. With the addition of a magnetic field normal to the structure planes, much sharper phonon-associated features appear (Yang et al. 1989).

6.3. Photon-induced tunneling

In the presence of a strong electric field, the conduction-band and valence-band wave functions penetrate into the forbidden gap of a semiconductor, and consequently interband (direct) transitions can be induced by photons (or by an optical-frequency coherent field) with an angular frequency ω less than $\varepsilon_\mathrm{G}/\hbar$ (Franz 1958, Keldysh 1958b). From the WKB dependence $\bar{\psi}(x) = \text{const.} \times \exp(-\zeta)$ given by (1.20), a rough estimate of this displacement is, in terms of photon energy displacement $\delta(\hbar\omega)$,

$$\delta_n(\hbar\omega) \approx (\hbar/2m_n)^{1/3} F^{2/3} \tag{6.1}$$

for each band. (For m_n equal to the free-electron mass and $F = 1 \times 10^6\ \mathrm{eV\ cm^{-1}}$, the value of (6.1) is 72 meV.) Detailed analysis of this *Franz–Keldysh effect* shows (Callaway 1964, Seraphin and Bottka 1965) that the absorption coefficient has an oscillatory dependence on $\hbar\omega$ near the absorption edge. In the case of indirect phonon-induced absorption, the edge is similarly shifted to lower photon energies (Chester and Fritsche 1965). Experimental data on absorption spectra in p–n junctions display these effects (Frova et al. 1966). In heterostructures, with an applied bias in the role of the field, the corresponding spectrum shifts have the character of the Stark effect (Miller et al. 1985); the subject is reviewed by Mendez and Agulló-Rueda (1989).

6.4. Effects of impurities and traps

The tunneling processes described in previous sections are those proper to an ideal parallel-plane structure. The lateral component $\boldsymbol{k}''$ of the wave vector is then conserved in an elastic coherent tunneling process, while in phonon-induced tunneling it changes by $\pm$ the lateral component $\boldsymbol{j}''$ of the phonon wave vector. A source of deviation from this geometry and selection rule is the doping elements that provide the electrons (holes) in the semiconductors bounding the diode, and which in a p–n tunnel diode give rise to the junction potential itself. The dopant atoms contribute short-range potential fields that can couple electron states with differing $\boldsymbol{k}''$, and they may also provide localized electron states that are the initial or final states of tunneling transitions. The same applies to defects such as radiation-induced vacancies. The disorder of atomic configurations in (Al,Ga)As and other semiconductor alloys has the same "symmetry breaking" effects on the $\boldsymbol{k}''$ selection rule.

In an "indirect" p–n diode, the observed current in the tunneling range of bias, other than that due to phonon-induced tunneling, is attributed mainly to tunneling induced by the impurity (and defect) fields. A measure of this contribution is given by the effective squared matrix element

$$|M|^2 = \sum_{\boldsymbol{R}} \left| \int \mathrm{d}^3\boldsymbol{r}\, \psi_{\mathrm{c}}^* \psi_{\mathrm{v}} \Phi(\boldsymbol{r} - \boldsymbol{R}) \right|^2, \tag{6.2}$$

where ψ_{c} and ψ_{v} are the wave functions, in the junction region, of stationary states constrained to be linear combinations of, respectively, conduction- and valence-band Bloch states, where $\Phi(\boldsymbol{r} - \boldsymbol{R})$ is the substitution potential of an impurity (or defect) at location $\boldsymbol{R}$, and the sum is over the impurities. In double-barrier resonant diodes, lateral scattering effects, especially in the well layer where the resonant wave functions have large amplitude, have the effect of broadening the transmission resonance and causing the transmitted wave due to an incident Bloch wave with given lateral wave vector to be distributed over a continuum of transmitted lateral wave vectors (Fertig et al. 1990, Leo and MacDonald 1991). Obviously, the calculation of the diode tunneling current becomes more elaborate than the treatment in § 4. In eq. (4.3), the integral over $\boldsymbol{k}''$ will be replaced by a double integral; and with inelastic processes the integral over E will be replaced in general by a double integral.

In p–n tunnel diodes, the forward-bias range between the cutoff of elastic interband tunneling current (for eV greater than the sum of the two Fermi energies) and the onset of appreciable thermal (minority-carrier) current, which typically should be a deep minimum, frequently is found to have a secondary current peak, of variable magnitude – the *excess current*. This is attributed (Yajima and Esaki 1958) to processes in which electrons tunnel from the n side to a localized state within the junction and locally within the forbidden gap, then make an inelastic transition (or cascade of transitions) locally to the valence band on the p side, or else make an inelastic transition on the n side from the conduction band to a localized state then tunnel out of it to the valence band on the p side. Observation of an increase of the excess current following irradiation known to generate defects supports this interpretation (Chynoweth et al. 1961). The unit process of elastic tunneling between a localized state – in fact a quasi-level, having a finite decay lifetime – and a band continuum should be essentially the phenomenon that was originally analyzed by Oppenheimer, but with the complication of involving combinations of Bloch states, and possibly phonon processes (Makram-Ebeid and Lannoo 1982).

With a localized state within a single heterostructure barrier, one can have resonant transmission over a range $\sim\hbar/$(escape lifetime) around the quasi-level energy. The geometry is more complicated than in the parallel-plane situation of § 5, but it can be shown that the theoretical behavior is essentially the same in terms of the applicable τ_L and τ_R (Combescot 1971, Wolf 1975 § 3, Kalmeyer and Laughlin 1987). Again, interaction with phonons contributes inelastic resonant tunneling processes (Glazman and Shekhter 1988).

6.5. *Image effects*

An electric charge near a conductor induces an equal and opposite screening charge in it. Ideally, a "point" charge Q in an insulator having dielectric constant K, at a distance x from the interface, induces a charge $-Q$ in the conductor, distributed so that its resulting field in the insulator is the same as if it were concentrated at a point an equal distance x beyond the interface. Then the screening space charge pulls the inducing charge towards the conductor with force $Q^2/4Kx^2$. This would suggest that, for an electron tunneling through a semiconductor barrier into a metal, there should be a potential energy $-Q^2/4Kx$, in addition to a band-edge offset. The degenerate carrier-filled regions on each side of an Esaki diode presumably provide a similar image force. The singularity at $x=0$ must in reality be rounded off within some distance x_0, which it is reasonable to associate with the de Broglie wavelength of the Bloch electrons of the conductor (Zheng and Ritchie 1991), as well as their screening length, with the image potential similarity extending into the conductor (Inkson 1973). (For an Esaki diode, this x_0 could be a substantial part of the interband tunneling path.) Another limitation of the effective image potential is in the time domain. The screening charge takes some time to move into position, and the tunneling event could occur in less time, and then presumably the image force would not apply. The time for the charge to move may be resolved into the frequencies of its excitation modes, here in particular the surface plasmons of a metal bounding a barrier, coupled dynamically to the tunneling electron (Rudberg and Jonson 1991). A substantial literature is reviewed up to 1990 in Echenique et al. (1991).

There is also a classical image force for a charge near an interface between a pair of dielectrics (such as semiconductors of a barrier heterostructure). The resulting potential on side "1" is equal to const. $+K_{12}Q^2/4x$ where $K_{12}=(K_1-K_2)/(K_1+K_2)$, the force on both sides being towards the medium with the greater K value. The consequent potential step can be finite only because this position dependence is rounded (Nordheim 1928) for small x, on some scale x_0 that can be expected to be $\sim$ lattice constants (Stern 1978, Sun and Gu 1990). If the dielectric polarization comes from displacement of ions (in a polar semiconductor), the associated time scale (Guéret 1992) should be $\gtrsim 1/$(optical-mode frequency). This would normally be greater than a tunneling time in a simple barrier, but less than the natural time scale for resonant tunneling. For the latter accordingly one may expect self-energy effects of coupling to phonons to be significant.

A similar consideration arises from the *Coulomb blockade* effect (Beenakker and van Houten 1991, § 17). If an electron effectively rests within a tunneling structure, while traversing it, then induced opposite charge in the bounding conductors is deemed to require an energy $e^2/2C$, with C the applicable capacity. (Thus the charge here is taken as spread over lateral dimensions, as opposed to the point charges in the foregoing with energy proportional to $1/x$.) It is inferred that for current to flow by this mode will entail a minimum bias equal to (energy)$/e=e/2C$. Such a threshold for conduction evidently has been realized in a resonant tunneling diode with small lateral dimensions (Su et al. 1992). Again, however, the tunneling electron should be expected to need a "residence time", at the location which terminates the capacitor field, greater than the applicable charging time of the capacity.

6.6. Low-dimensional structures

Parallel-plane diode structures as in §§ 3–5 may be fabricated by established processes. Techniques for producing structures constrained, on a nanometer scale, to lower dimensionality of carrier propagation are surveyed by Beenakker and van Houten (1991, §§ 2–3). Patterned gates overlaying a parallel two-dimensional electron gas have been used to impose barriers constraining the two-dimensional motion (Sivan et al. 1990); and in particular with a gate-induced pair of parallel linear barriers resonant tunneling may be observed where there is only one lateral degree of freedom (Chou et al. 1989, Ismail et al. 1989). The further step is to structures allowing carrier propagation in one dimension only, with a tunneling barrier or barriers interposed on this linear motion (Reed 1988, Smith et al. 1990). A narrowing of the channel will itself induce a local increase in the effective potential for one-dimensional propagation of the mode(s), and hence can provide a tunneling barrier. In practice, the lateral dimensions of "quantum wires" have accommodated a number of wave-function modes, with the quasi-linear tunneling motion consequently quite complicated (Bryant 1989).

6.7. Tunneling carrier injection

A tunneling barrier, especially in the Fowler–Nordheim configuration, may be used to inject a flux of electrons collimated to a narrow range in energy and direction, from a high density of electrons on the source side of the barrier into an "empty" region on the other side. From (1.12), (1.21), and (4.6) without the factor $1-\exp(-\Omega\Xi)$ (since concentration of the tunneling in the normal direction requires $\Omega\Xi$ to be large), we may infer that the flux after integrating over the lateral wave vector is substantially proportional to $f(E)\exp[-2\beta(E)]$. This must fall off as E increases towards the top of the barrier, so long as the Fermi level in the source is substantially below the top; and it must fall off as E decreases towards the source Fermi level, since it becomes just proportional to the exponential factor. The design objective is that this energy peak be narrow. The structure that receives the beam may be one that operates as a transistor device (Heiblum 1981), or it may be used to investigate inelastic scattering processes (Levi et al. 1985).

Appendix: the envelope representation

Equations (2.5) and (2.6) for a single band are connected by

$$\psi(\boldsymbol{r}) = \int \Gamma(\boldsymbol{r}, \boldsymbol{r}')\, \bar{\psi}(\boldsymbol{r}')\, \mathrm{d}^3 \boldsymbol{r}' \tag{A1}$$

and

$$\bar{\psi}(\boldsymbol{r}) = \int \Gamma(\boldsymbol{r}', \boldsymbol{r})\, \psi(\boldsymbol{r}')\, \mathrm{d}^3 \boldsymbol{r}', \tag{A2}$$

where

$$\Gamma(\boldsymbol{r}_1, \boldsymbol{r}_2) = \frac{1}{V} \sum_{\boldsymbol{k}} \phi(\boldsymbol{k}, \boldsymbol{r}_1) \exp(-\mathrm{i}\boldsymbol{k} \cdot \boldsymbol{r}_2) \tag{A3}$$

(real here on account of the Kramers degeneracy). The $\phi(\boldsymbol{k}, \boldsymbol{r})$ in (A3) are normalized in unit volume, like the $\exp(\mathrm{i}\boldsymbol{k}\cdot\boldsymbol{r})$, and V here is the normalization volume for ψ and $\bar{\psi}$, equal to NV_0 where N is the number of lattice cells, of volume V_0, in V. To obtain the envelope equivalent of the Schrödinger equation, we require $\bar{\mathscr{H}}$ such that

$$\langle \bar{\psi}_1 | \bar{\mathscr{H}} | \bar{\psi}_2 \rangle = \langle \psi_1 | \mathscr{H} | \psi_2 \rangle.$$

With $\mathscr{H} = \mathscr{H}_0 + W(\boldsymbol{r})$ and $\mathscr{H}_0 \phi(\boldsymbol{k}, \boldsymbol{r}) = \varepsilon(\boldsymbol{k}) \phi(\boldsymbol{k}, \boldsymbol{r})$, eqs. (A1)–(A3) give eq. (2.1) as $\bar{\mathscr{H}}\bar{\psi} = E\bar{\psi}$ with

$$K(\boldsymbol{r}; \boldsymbol{r}_1, \boldsymbol{r}_2) = \Gamma(\boldsymbol{r}_1, \boldsymbol{r})\, \Gamma(\boldsymbol{r}_1, \boldsymbol{r}_2). \tag{A4}$$

The kernels (A3) and (A4) may be written in terms of the Wannier function (Koster 1953) of the band,

$$w(\boldsymbol{r}) = (NV)^{-1/2} \sum_{\boldsymbol{k}} \phi(\boldsymbol{k}, \boldsymbol{r}) \tag{A5}$$

and its "empty-lattice" equivalent

$$w_0(\boldsymbol{r}) = (NV)^{-1/2} \sum_{\boldsymbol{k}} \exp(\mathrm{i}\boldsymbol{k}\cdot\boldsymbol{r}), \tag{A6}$$

where the sums here are over the Brillouin zone. Then

$$\Gamma(\boldsymbol{r}_1, \boldsymbol{r}_2) = \sum_{\boldsymbol{R}} w(\boldsymbol{r}_1 - \boldsymbol{R})\, w_0(\boldsymbol{r}_2 - \boldsymbol{R}), \tag{A7}$$

where the sum is over the lattice cells, given by the translation vectors $\boldsymbol{R}$. The localization property of $K(\boldsymbol{r}; \boldsymbol{r}_1, \boldsymbol{r}_2)$ is evident from this form. Similarly, eq. (A1) may be written as

$$\psi(\boldsymbol{r}) = V_0^{1/2} \sum_{\boldsymbol{R}} w(\boldsymbol{r} - \boldsymbol{R})\, \bar{\psi}(\boldsymbol{R}) \tag{A8}$$

and the envelope function may be interpolated by

$$\bar{\psi}(\boldsymbol{r}) = V_0^{1/2} \sum_{\boldsymbol{R}} w_0(\boldsymbol{r} - \boldsymbol{R})\, \bar{\psi}(\boldsymbol{R}). \tag{A9}$$

The coefficients $\eta(\boldsymbol{R}) \equiv V_0^{1/2}\bar{\psi}(\boldsymbol{R})$ of the $w(\boldsymbol{r} - \boldsymbol{R})$ in (A8) are in fact a representation of the state vector $|\psi\rangle$, and the linear-difference equation connecting the $\eta(\boldsymbol{R})$ (Koster and Slater 1954) is equivalent to the Schrödinger equation. Corresponding to the WKB solution of the latter is a WKB-like solution of the Koster–Slater equation, which can be applied to tunneling theory (Price and Radcliffe 1959).

A slowly varying position dependence of composition might be described by defining

$$\varepsilon(\boldsymbol{R}_1, \boldsymbol{R}_2) = \int w(\boldsymbol{r} - \boldsymbol{R}_1)\, \mathscr{H}_0\, w(\boldsymbol{r} - \boldsymbol{R}_2)\, \mathrm{d}^3\boldsymbol{r} \tag{A10}$$

using a homogeneous (not position dependent) Wannier function, but with this $\varepsilon(\boldsymbol{R}_1, \boldsymbol{R}_2)$ not depending only on $\boldsymbol{R}_1 - \boldsymbol{R}_2$, and by a kind of Wigner–Moyal transformation to

$$\tilde{\varepsilon}(\boldsymbol{r}, \boldsymbol{k}) \equiv V_0^{1/2} \sum_1 \sum_2 \varepsilon(\boldsymbol{R}_1, \boldsymbol{R}_2)\, w_0(\boldsymbol{r} - (\boldsymbol{R}_1 + \boldsymbol{R}_2)/2) \exp[\mathrm{i}(\boldsymbol{R}_1 - \boldsymbol{R}_2)\cdot\boldsymbol{k}]. \tag{A11}$$

The envelope-function Hamiltonian operator may then be taken as $\tilde{\varepsilon}(\boldsymbol{r}, -\mathrm{i}\nabla)$.

In the foregoing, the state vector was assumed to span one complete band of Bloch states, with the sums in (A5), etc., being over a Brillouin zone. But the formalism works as well if these sums are over parts of a zone ("valley" sectors) or one of several bands. The general form is a set of $\boldsymbol{k}$ sums, each associated with an envelope function and with a Wannier function; and the coupling between the envelope wave equations for these projected-out components $\bar{\psi}_n$ completes the scheme.

Acknowledgements

I thank my colleagues, and Professor Landsberg, for information and suggestions.

References

Of the works cited below, the book by Duke (1969a) is still the predominant source, with a wealth of theoretical and experimental material. The book chapter by Wolf (1975) gives an extensive account of aspects that are barely mentioned here. The book containing the chapter by Duke (1969b) includes a number of other useful contributions. The article "Semiconductor Tunneling", by D.C. Tsui, in the first volume (1982) of the first edition of this Handbook (which emphasizes "using tunneling to probe the physics of semiconductors", whereas the present account principally presents basic tunneling theory) remains a valuable source. The journal literature on semiconductor tunneling is by now enormous, and the citations below are a sparse selection, undoubtedly overlooking some important papers.

Abramowitz, M., and I.A. Stegun, 1972, Handbook of Mathematical Functions (Dover, New York).
Akera, H., S. Wakahara and T. Ando, 1988, Surf. Sci. **196**, 694–699.
Akera, H., S. Wakahara and T. Ando, 1989, Phys. Rev. B **40**, 11609–11618 and 11619–11633.
Anderson, R.L., 1960, IBM J. Res. Dev. **4**, 283–287.
Anderson, R.L., 1962, Solid-State Electron. **5**, 341–351.
Argyres, P.N., 1962, Phys. Rev. **126**, 1386–1393.
Bahder, T.B., C.A. Morrison and J.D. Bruno, 1987, Appl. Phys. Lett. **51**, 1089–1090.
Beenakker, C.W.J., and H. van Houten, 1991, Quantum transport in semiconductor nanostructures, in: Solid State Physics, Vol. 44, eds. H. Ehrenreich and D. Turnbull (Academic Press, New York).
BenDaniel, D.J., and C.B. Duke, 1966, Phys. Rev. **152**, 683–692.
Bennett, A.J., C.B. Duke and S.D. Silverstein, 1968, Phys. Rev. **176**, 969–992.
Beresford, R., L.F. Luo and W.I. Wang, 1989, Appl. Phys. Lett. **54**, 1899–1901.
Bohm, D., 1951, Quantum Theory (Prentice-Hall, Englewood Cliffs, NJ).
Bryant, G.W., 1989, Phys. Rev. B **39**, 3145–3152.
Büttiker, M., 1983, Phys. Rev. B **27**, 6178–6188.
Callaway, J., 1964, Phys. Rev. **134**, A998–A1000.
Chang,.L.L., and L. Esaki, 1980, Surf. Sci. **98**, 70–89.
Chang, L.L., L. Esaki and R. Tsu, 1974, Appl. Phys. Lett. **24**, 593–595.
Chen, J.F., L. Yang, M.C. Wu, S.N.G. Chu and A.Y. Cho, 1991, J. Cryst. Growth **111**, 659–663.
Chester, M., and L. Fritsche, 1965, Phys. Rev. **139**, A518–A525.
Chou, S.Y., D.R. Allee, R.F.W. Pease and J.S. Harris Jr, 1989, J. Appl. Phys. **55**, 176–178.
Chow, D.H., T.C. McGill, I.K. Sou, J.P. Faurie and C.W. Nieh, 1988, Appl. Phys. Lett. **52**, 54–56.
Chow, D.H., E.T. Yu, J.R. Söderström, D.Z.-Y. Ting and T.C. McGill, 1990, J. Appl. Phys. **68**, 3744–3746.
Chynoweth, A.G., and K.G. McKay, 1957, Phys. Rev. **106**, 418–426.
Chynoweth, A.G., W.L. Feldman and R.A. Logan, 1961, Phys. Rev. **121**, 684–694.
Chynoweth, A.G., R.A. Logan and D.E. Thomas, 1962, Phys. Rev. **125**, 877–881.
Combescot, R., 1971, J. Phys. C **4**, 2611–2622.
Davis, R.H., and H.H. Hosack, 1963, J. Appl. Phys. **34**, 864–866.
Dresselhaus, G., A.F. Kip and C. Kittel, 1955, Phys. Rev. **98**, 368–384.
Duke, C.B., 1969a, Tunneling in Solids (Academic Press, New York).
Duke, C.B., 1969b, in: Tunneling Phenomena in Solids, eds E. Burstein and S. Lundqvist (Plenum Press, New York) p. 31.
Echenique, P.M., F. Flores and R.H. Ritchie, 1991, Surf. Sci. **251/252**, 119–126.
Esaki, L., 1958, Phys. Rev. **109**, 603–604.
Esaki, L., and Y. Miyahara, 1960, Solid-State Electron. **1**, 13–21.
Esaki, L., L.L. Chang and E.E. Mendez, 1981, Jpn. J. Appl. Phys. **20**, L529–L532.
Fertig, H.A., S. He and S. Das Sarma, 1990, Phys. Rev. B **41**, 3596–3607.
Fowler, R.H., and L.W. Nordheim, 1928, Proc. Roy. Soc. A **119**, 173–181.

Franz, W., 1956, Dielektrische Durchschlag, in: Encylopedia of Physics (Springer, Berlin) pp. 155–263.
Franz, W., 1958, Z. Naturforsch. **13a**, 484–489.
Franz, W., 1959, Z. Naturforsch. **14a**, 415–418.
Frenkel, J., 1929, Einführung in die Wellenmechanik (Springer, Berlin).
Frenkel, J., 1930, Phys. Rev. **36**, 1604–1618.
Frenkel, J., and A. Joffe, 1932, Phys. Z. Sowjetunion **1**, 60–87.
Frensley, W.R., 1987, Phys. Rev. B **36**, 1570–1580.
Frensley, W.R., 1988, Superlatt. & Microstruct. **4**, 497–501.
Frova, A., P. Handler, F.A. Germano and D.E. Aspnes, 1966, Phys. Rev. **145**, 575–583.
Gamow, G., 1928, Z. Phys. **51**, 204–212.
Gamow, G., 1929, Z. Phys. **53**, 601–604.
Gamow, G., 1931, Constitution of Atomic Nuclei and Radioactivity (Oxford University Press, Oxford).
Glazman, L.I., and R.I. Shekhter, 1988, Zh. Eksp. & Teor. Fiz. **94**, 292–306 [Sov. Phys. JETP **67**, 163–170].
Goldman, V.J., D.C. Tsui and J.E. Cunningham, 1987, Phys. Rev. B **36**, 7635–7637.
Gu, L., and B.Y. Gu, 1989, Solid State Commun. **72**, 1215–1218.
Guéret, P., 1992, Ultramicroscopy, to be published.
Gundlach, K.H., 1966, Solid-State Electron. **9**, 949–957.
Gundlach, K.H., 1973, J. Appl. Phys. **44**, 5005–5010.
Gurney, R.W., and E.U. Condon, 1929, Phys. Rev. **33**, 127–140.
Haering, R.R., and E.N. Adams, 1961, J. Phys. & Chem. Solids **19**, 8–17.
Heiblum, M., 1981, Solid-State Electron. **24**, 343–366.
Hickmott, T.W., 1985, Phys. Rev. B **32**, 6531–6543.
Hickmott, T.W., 1991, Phys. Rev. B **44**, 12880–12890.
Hirakawa, K., H. Sakaki and T. Ikoma, 1990, Surf. Sci. **229**, 161–164.
Homclius, J., and W. Franz, 1954, Z. Naturf. **9a**, 5–14; 205–210.
Inkson, J.C., 1973, J. Phys. F **3**, 2143–2156.
Ismail, K., D.A. Antoniadis and H.I. Smith, 1989, Appl. Phys. Lett. **55**, 589–591.
Jacoboni, C., and P.J. Price, 1990, Solid State Commun. **75**, 193–196.
Jones, R.O., 1966, Proc. Phys. Soc. (London) **89**, 443–451.
Kalmeyer, V., and R.B. Laughlin, 1987, Phys. Rev. B **35**, 9805–9808.
Kane, E.O., 1957, J. Phys. & Chem. Solids **1**, 249–261.
Kane, E.O., 1959, J. Phys. & Chem. Solids **12**, 181–188.
Kane, E.O., 1961, J. Appl. Phys. **32**, 83–91.
Keldysh, L.V., 1957, Zh. Eksp. & Teor. Fiz. **33**, 994–1003 [1958, Sov. Phys. JETP **6**, 763–770].
Keldysh, L.V., 1958a, Zh. Eksp. & Teor. Fiz. **34**, 962–968 [Sov. Phys. JETP **6**, 665–669].
Keldysh, L.V., 1958b, Zh. Eksp. & Teor. Fiz. **34**, 1138–1141 [Sov. Phys. JETP **6**, 788–790].
Kemble, E.C., 1937, The Fundamental Principles of Quantum Mechanics (McGraw-Hill, New York).
Kittel, C., and A.H. Mitchell, 1954, Phys. Rev. **96**, 1488–1493.
Kluksdahl, N.C., A.M. Kriman, D.K. Ferry and C. Ringhofer, 1989, Phys. Rev. B **39**, 7720–7735.
Koster, G.F., 1953, Phys. Rev. **89**, 67–77.
Koster, G.F., and J.C. Slater, 1954, Phys. Rev. **95**, 1167–1176.
Krieger, J.B., 1966, Ann. Phys. **36**, 1–60.
Kubo, R., 1957, J. Phys. Soc. Jpn. **12**, 570–586.
Landau, L.D., and E.M. Lifshitz, 1977, Quantum Mechanics, Non-relativistic Theory (Pergamon Press, Oxford).
Leadbeater, M.L., E.S. Alves, L. Eaves, M. Henini, O.H. Hughes, A. Celeste, J.C. Portal, G. Hill and M.A. Pate, 1989, Phys. Rev. B **39**, 3438–3441.
Leadbeater, M.L., F.W. Sheard and L. Eaves, 1991, Semicond. Sci. & Technol. **6**, 1021–1024.
Leo, J., and A.H. MacDonald, 1991, Phys. Rev. B **43**, 9763–9771.
Levi, A.F.J., J.R. Hayes, P.M. Platzman and W. Wiegmann, 1985, Physica B **134**, 480–486.
Longenbach, K.F., L.F. Luo, S. Xin and W.I. Wang, 1991, J. Cryst. Growth **111**, 651–658.
Luo, L.F., R. Beresford and W.I. Wang, 1989, Appl. Phys. Lett. **55**, 2023–2025.
MacColl, L.A., 1932, Phys. Rev. **40**, 621–626.
Makram-Ebeid, S., and M. Lannoo, 1982, Phys. Rev. B **25**, 6406–6424.

Marinace, J.C., 1960, IBM J. Res. Dev. **4**, 280–282.
Maserjian, J., and N. Zamani, 1982, J. Appl. Phys. **53**, 559–567.
Mendez, E.E., and F. Agulló-Rueda, 1989, J. Lumin. **44**, 223–231.
Mendez, E.E., H. Ohno and L. Esaki, 1991, Phys. Rev. B **43**, 5196–5199.
Merzbacher, E., 1970, Quantum Mechanics (Wiley, New York).
Miller, D.A.B., D.S. Chemla, T.C. Damen, A.C. Gossard, W. Wiegmann, T.H. Wood and C.A. Burrus, 1985, Phys. Rev. B **32**, 1043–1060.
Nordheim, L.W., 1928, Proc. Roy. Soc. A **121**, 626–639.
Nordheim, L.W., 1929, Phys. Z. **30**, 177–196.
Ohno, H., E.E. Mendez and W.I. Wang, 1990, Appl. Phys. Lett. **56**, 1793–1795.
Oppenheimer, J.R., 1928a, Phys. Rev. **31**, 66–81.
Oppenheimer, J.R., 1928b, Proc. Nat. Acad. Sci. **14**, 363–365.
Oppenheimer, J.R., 1928c, Phys. Rev. **31**, 914.
Payne, R.T., 1965, Phys. Rev. **139**, 570–582.
Price, P.J., 1988a, Surf. Sci. **196**, 394–398.
Price, P.J., 1988b, Phys. Rev. B **38**, 1994–1998.
Price, P.J., and J.M. Radcliffe, 1959, IBM J. Res. Dev. **3**, 364–371.
Reed, M.A., 1988, Superlattices & Microstruct. **4**, 741–747.
Reed, M.A., R.J. Koestner and M.W. Goodwin, 1986, Appl. Phys. Lett. **49**, 1293–1295.
Ricco, B., and M.Ya. Azbel, 1984, Phys. Rev. B **29**, 4356–4363.
Roth, H., W. Bernard, W.D. Straub and J.E. Mulhern Jr, 1966, Phys. Rev. **145**, 667–674.
Rudberg, B.G.R., and M. Jonson, 1991, Phys. Rev. B **43**, 9358–9361.
Schottky, W., 1931, Phys. Z. **32**, 833–842.
Schuberth, G., G. Abstreiter, E. Gornik, F. Schäffler and J.F. Luy, 1991, Phys. Rev. B **43**, 2280–2284.
Seraphin, B.O., and N. Bottka, 1965, Phys. Rev. **139**, A560–A565.
Shuey, R.T., 1965, Phys. Rev. **137**, A1268–A1277.
Sivan, U., A. Palevski, M. Heiblum and C.P. Umbach, 1990, Solid-State Electron. **33**, 979–986.
Smith, C.G., M. Pepper, H. Ahmed, J.E.F. Frost, D.G. Hasko, R. Newbury, D.C. Peacock, D.A. Ritchie and G.A.C. Jones, 1990, Surf. Sci. **228**, 387–392.
Solomon, P.M., S.L. Wright and C. Lanza, 1986, Superlatt. & Microstruct. **2**, 521–525.
Stern, F., 1978, Phys. Rev. B **17**, 5009–5015.
Stratton, R., 1962, J. Phys. & Chem. Solids **23**, 1177–1190.
Stratton, R., 1969, in: Tunneling Phenomena in Solids, eds E. Burstein and S. Lundqvist (Plenum, New York) p. 105.
Su, B., V.J. Goldman and J.E. Cunningham, 1992, Science **255**, 313–315.
Sun, H., and S.-W. Gu, 1990, Phys. Rev. B **41**, 3145–3152.
Tiemann, J.J., and H. Fritzsche, 1965, Phys. Rev. **137**, A1910–A1913.
Tsu, R., and L. Esaki, 1973, Appl. Phys. Lett. **22**, 562–564.
Weil, T., and B. Vinter, 1987, Appl. Phys. Lett. **50**, 1281–1283.
Wilson, A.H., 1932, Proc. Roy. Soc. A **136**, 487–498.
Wingreen, N.S., K.W. Jacobsen and J.W. Wilkins, 1989, Phys. Rev. B **40**, 11834–11850.
Wolf, E.L., 1975, Nonsuperconducting electron tunneling spectroscopy, in: Solid State Physics, Vol. 30, eds H. Ehrenreich, F. Seitz and D. Turnbull (Academic Press, New York).
Yajima, T., and L. Esaki, 1958, J. Phys. Soc. Jpn. **13**, 1281–1287.
Yang, C.H., M.J. Yang and Y.C. Kao, 1989, Phys. Rev. B **40**, 6272–6276.
Yang, L., J.F. Chen and A.Y. Cho, 1990, J. Appl. Phys. **68**, 2997–3000.
Zeidenbergs, G., and R.L. Anderson, 1967, Solid-State Electron. **10**, 113–123.
Zener, C., 1934, Proc. Roy. Soc. A **145**, 523–529.
Zheng, X.-Y., and R.H. Ritchie, 1991, Phys. Rev. B **43**, 4002–4006.

CHAPTER 13

Quantum Confinement and Scattering Processes

B. K. RIDLEY

Physics Department
University of Essex
Colchester, UK

Handbook on Semiconductors
Completely Revised Edition
Edited by T.S. Moss
Volume 1, edited by P.T. Landsberg

Contents

1. Introduction . . . 667
2. Confinement of electrons . . . 668
 2.1. The envelope function . . . 668
 2.2. Non-parabolicity . . . 673
 2.3. Band-mixing . . . 674
3. Confinement of optical phonons . . . 678
 3.1. Microscopic models . . . 678
 3.2. Elastic-continuum models . . . 683
4. The electron–phonon interaction . . . 690
 4.1. General . . . 690
 4.2. Model rates . . . 692
 4.3. Well-width resonances . . . 697
 4.4. Hot phonons . . . 697
 4.5. Plasmon-LO coupling . . . 698
 4.6. Non-polar optical-phonon interaction . . . 699
 4.7. Acoustic-phonon interaction . . . 700
5. Other electron-scattering processes . . . 702
 5.1. Charged-impurity scattering . . . 702
 5.2. Carrier–carrier scattering . . . 706
 5.3. Interface-roughness scattering . . . 707
 5.4. Alloy scattering . . . 709
 5.5. Mobilities . . . 710
6. Phonon scattering . . . 710
 6.1. Phonon–phonon processes . . . 710
 6.2. Charged-impurity scattering . . . 712
 6.3. Alloy fluctuations and neutral impurities . . . 713
 6.4. Interface-roughness scattering . . . 714
References . . . 716

1. Introduction

The advent of high-quality, layered semiconductor material, following advances in crystal-growing techniques, has opened the way to an exploration of electrons and phonons in systems in which quantum confinement is the principle feature. Essentially, this development has extended the study of low-dimensionality in silicon FETS in the seventies to the III–V binary, ternary and quarternary compounds in the eighties, and the whole field is still extremely active in the nineties. In spite of over two decades of intense international research in this field there are still many outstanding problems to be tackled, and the present writer is very conscious that some of what is written here is far from being well-accepted, standard knowledge. A good deal is still being actively researched, particularly the topic of phonon-confinement, and the reader should be aware of this. The focus here will be on quasi-two-dimensionality, mostly associated with III–V quantum wells and superlattices, basically because single heterojunctions are well treated in the review by Ando et al. (1982), and because the 2D double heterojunction quantum well provides the simplest and most researched system. The paradigm system will be that consisting of GaAs/AlAs layers.

One of the notable features of the experimental study of electrons and phonons in 2D systems is that quantum confinement, for many good practical reasons, has usually been accompanied by high carrier concentrations – much higher than was usual in the study of bulk medium-to-high band-gap semiconductors. This association has complicated matters considerably, with hot-phonon and plasmon effects adding to, and usually obscuring, the effects of quantum confinement. The latter are most easily observed in optical experiments such as absorption or Raman spectroscopy where the carrier density is low, which in this context is of order 10^{10} cm^{-2}. Otherwise, the interpretation of experiments quite often requires numerical simulation, usually using Monte Carlo techniques. These latter have grown in sophistication in recent years, and they rely heavily on input concerning the electronic and vibrational bandstructures of the system they wish to describe, and input concerning the interactions which couple the electron/phonon populations and govern their dynamic properties.

Our emphasis here will be to describe briefly the basic quantum processes which flourish in 2D systems (fig. 1). We begin in § 2 with the quantum confinement of electrons, discussing the envelope-function method of describing energy subbands, and how non-parabolicity and band-mixing affect matters. How optical modes suffer confinement is a currently active research topic with a good deal of healthy controversy concerning the correct boundary conditions to be used in the envelope-function method. In § 3 we give an account of phonon confinement which emphasizes the

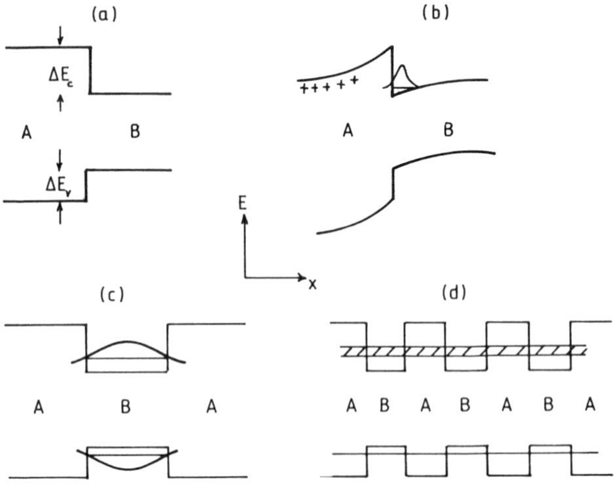

Fig. 1. Some examples of heterostructures: (a) a heterojunction, (b) a heterojunction with electrostatic quantum well, (c) a double-heterojunction quantum well, (d) a superlattice. (Energy, vertical; distance, horizontal.)

elastic properties of optical modes, this being the only approach which allows the confinement of optical phonons to be described in non-polar as well as polar material. § 4 describes the electron–phonon interaction in terms of an elastic-continuum model, covering intra- and intersubband rates, well-capture, well-width resonances, hot phonons, plasmon/LO coupling and other topics. Electron collisions not involving phonons, i.e., charged-impurity, carrier–carrier, interface roughness and alloy scattering, are covered in § 5. § 6 is unusual in that it discusses the annihilation and scattering of optical phonons, a topic virtually untouched in bulk-semiconductor research but necessary here in view of the prevalence of hot optical phonons produced in experiments on electronic transport and experiments involving high intensity optical excitation.

2. Confinement of electrons

2.1. *The envelope function*

The problem of calculating the eigenvalues of electrons in a square-well potential is treated in undergraduate texts (fig. 2). The boundary conditions are that the amplitude and gradient of the wavefunction be continuous, thereby ensuring the continuity of the probability and electron currents. These rules must apply in all cases. The difficulty of the problem of confinement in heterostructures is that the electron is already bound inside a solid. Its wavefunction is no longer a simple plane wave. In a periodic structure it is a Bloch wave given by

$$\psi(\boldsymbol{r}) = V^{-1/2} u_{\boldsymbol{k}}(\boldsymbol{r})\, e^{i\boldsymbol{k}\boldsymbol{r}}, \tag{2.1}$$

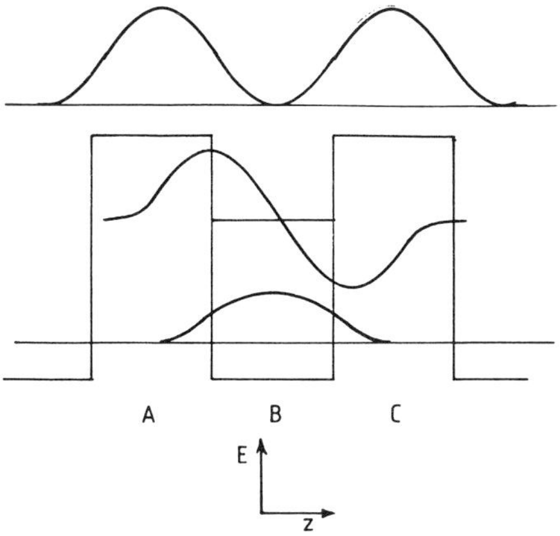

Fig. 2. Electron wavefunctions in a quantum well. The wavefunctions depicted are for the ground state, the first excited state and the second excited state, the latter being the lowest-energy continuum state.

where V is the normalization volume, and $u_{\boldsymbol{k}}(\boldsymbol{r})$ has the periodicity of the lattice and a characteristic symmetry at $\boldsymbol{k} = \boldsymbol{k}_0$, where $\boldsymbol{k}_0$ is the position in the Brillouin zone of the conduction-band edge. In a layered heterostructure which is lattice-matched, the wavefunction will still possess periodic components but there will also be reflections which will set up standing waves and, moreover, there may be more than one type of conduction valley involved. Under these circumstances it is necessary to follow the approach of Luttinger and Kohn (1955) and express the wavefunction in the form of an envelope function expansion:

$$\psi(\boldsymbol{r}) = \sum_n u_n(\boldsymbol{r}) F_n(\boldsymbol{r}), \tag{2.2}$$

where the $u_n(\boldsymbol{r})$ are a complete set of linearly independent periodic functions and the $F_n(\boldsymbol{r})$ are envelope functions expandable in plane waves with wave vectors restricted to the first Brillouin zone.

The simplest system to treat is one in which the lowest conduction valleys in the adjacent layers are closer to one another in energy than any other and are of the same symmetry and arise from closely similar electronic composition (e.g., GaAs/$Al_xGa_{1-x}As$ with small x). In this case the momentum matrix elements involving the periodic part of the wavefunction will be similar in magnitude, whence the effective mass in the two regions will not differ too markedly and the envelope function will be slowly varying. Ignoring non-parabolicity then leads one to postulate the one-dimensional effective-mass equation:

$$\frac{-\hbar^2}{2}\frac{\mathrm{d}}{\mathrm{d}z}\left(\frac{1}{m^*(z)}\frac{\mathrm{d}}{\mathrm{d}z}F(z)\right) + \hbar^2 k_{\parallel}^2/2m^* + E(z)F(z) = EF(z), \tag{2.3}$$

where

$$\frac{1}{m^*(z)} = \frac{1}{m} + \frac{2}{m^2}\sum_n \frac{|p_n|^2}{E(z) - E_n(z)}. \tag{2.4}$$

Here m is the free-electron mass and p_n is the momentum matrix element between band n and the lowest conduction band – the usual $\boldsymbol{k}\cdot\boldsymbol{p}$ result for the effective mass. We have taken the confinement to be along the z-direction only, and so have taken the envelope function to be of the form

$$F(\boldsymbol{r}) = F(z)\, e^{i\boldsymbol{k}_\parallel\cdot\boldsymbol{r}_\parallel}, \tag{2.5}$$

with $\boldsymbol{r}_\parallel$ the position vector and $\boldsymbol{k}_\parallel$ the wavevector in the plane of the layer. It follows that we must have

$$F(z) \quad \text{and} \quad \frac{1}{m^*(z)}\frac{dF(z)}{dz} \tag{2.6}$$

continuous everywhere, and in particular at the heterojunction.

These boundary conditions have been heavily exploited to calculate eigenvalues and eigenfunctions in many heterostructures. Comprehensive accounts can be found in the literature (Bastard and Brum 1986, Altarelli 1986, Bastard 1989). One solves the following equation for a single quantum well:

$$Z \equiv \frac{m_w^* k_B}{m_B^* k_w} = \begin{cases} +\,i\tan(k_w d_w/2) \\ -\,i\cot(k_w d_w/2), \end{cases} \tag{2.7}$$

with

$$\frac{\hbar^2 k_w^2}{2m_w^*} = E, \quad \frac{\hbar^2 k_B^2}{2m_B^*} = E - V_0, \tag{2.8}$$

where V_0 is the barrier height and E_w the energy measured from the bottom of the conduction band in the well; or the equation for a superlattice:

$$\cos KD = \cos k_w d_w \cos k_B d_B - \frac{1}{2}\left(Z + \frac{1}{Z}\right)\sin k_w d_w \sin k_B d_B, \tag{2.9}$$

where K is the superlattice wavevector and $D = d_w + d_B$ is the superlattice period. Figure 3 shows the subband structure for a $GaAs/Al_{0.3}Ga_{0.7}As$ superlattice. The energy of an electron in a subband consists of a part associated with its confined or modified motion in the z-direction plus its kinetic energy in the plane, viz.

$$E = E_n + \frac{\hbar^2 k_\parallel^2}{2m^*}, \tag{2.10}$$

where E_n is determined from eqs. (2.7) or (2.9).

The above empirical effective-mass formalism appears to work reasonably well in that its predictions are more or less in agreement with experiment. Lack of precise knowledge about band-offsets and the virtually inevitable imprecision associated with interface position, width and quality makes detailed tests of theory impracticable, which is unfortunate in view of the uncertainty which surrounds the validity of (eq. (2.3)) the boundary conditions in eq. (2.6). To circumvent this problem Burt (1988a, 1989) has proposed an exact formulation of the envelope-function method and has used it to provide a systematic derivation of the effective-mass equation (2.3)

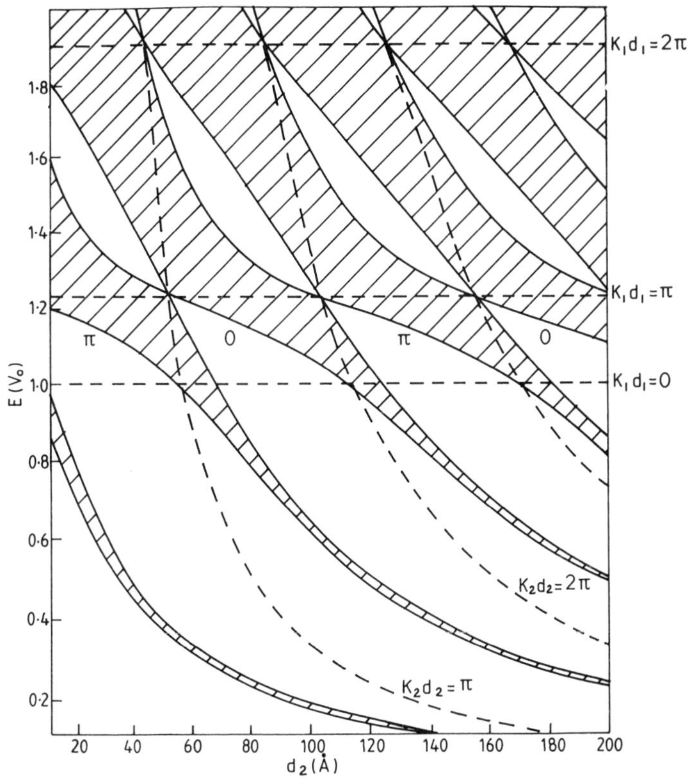

Fig. 3. Superlattice band structure for GaAs/$Al_{0.3}Ga_{0.7}$As system ($V_0 = 0.192$ eV) with $d_1 = 100$ Å barriers, as a function of well width d_2. The dashed lines labelled $K_2d_2 = n\pi$ depict the energies in an infinitely deep well.

for the case in which the effective-mass changes are small. The restriction to small effective-mass changes would appear to make the derivation inapplicable to the CdTe/HgTe system in which the mass reverses sign across the interface, going from light-hole like in CdTe to electron-like in HgTe, and leading to interface states (Bastard 1989). However, Burt (1992) has managed to rederive the effective-mass equation (2.3) under less restrictive conditions so that the prospects for justifying the application of eq. (2.3), a priori, to such systems are improving.

In Burt's formulation, $F(z)$ and its gradient are strictly continuous everywhere, and hence the boundary condition (2.6) at an abrupt interface can only be approximately true. As pointed out by Burt (1988b), the effective mass discontinuity in the slope appears as a discontinuity only on a mesoscopic scale for which his smoothly varying effective mass function is viewed as discontinuous. This is illustrated in fig. 4, in which the envelope function for the ground state of a one-dimensional superlattice is shown (Burt 1991). The envelope function and its gradient are continuous everywhere yet there is a rapid change in the gradient in the region of the abrupt interfaces corresponding to the change in effective mass. The kink in the envelope function

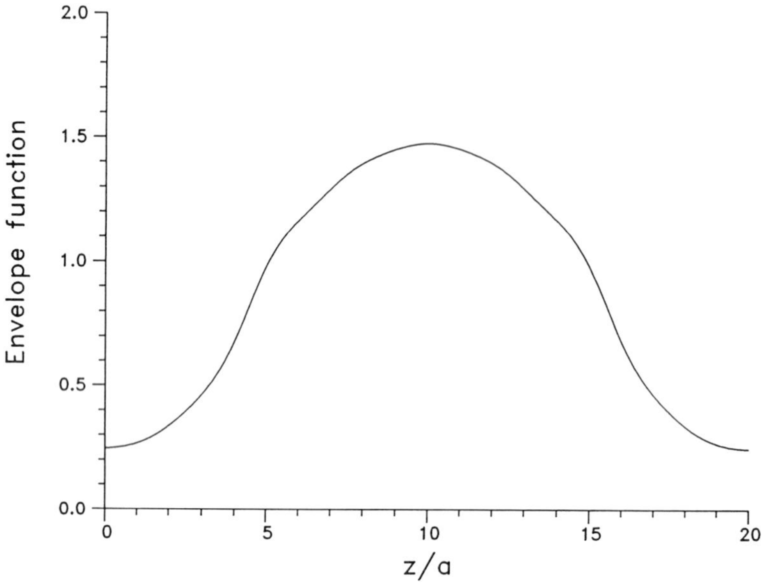

Fig. 4. Electron envelope function for a 10/10 superlattice ground state computed using crystal pseudopotentials. This shows that the envelope function is continuous on a microscopic scale, yet displays a discontinuous shape when viewed mesoscopically (Burt 1991).

introduced by using the boundary condition (2.6) is an approximation to the rapid change of gradient of the true envelope function. It should be emphasised that these comments on the boundary condition only apply to envelope functions defined by eq. (2.2) in which the same periodic basis functions, u_n, are used in both the wells and barriers. If different envelope function expansions, i.e., with different choices of u_n are used in the well and barrier layers, then different boundary conditions may apply.

All this said, the boundary condition (2.6) does lead to a curious prediction. If all other parameters are kept constant, the ground state energy of a square quantum well falls when the barrier mass is increased (Burt 1989). The reverse happens when the effective mass is omitted from the boundary condition. Intuitively one might expect the energy level to rise (the barrier is less penetrable and hence the kinetic energy due to increased confinement would increase) and this would appear to put a question mark over the boundary condition (2.6) and its derivation. However, calculations for the system described in fig. 4 show that the true eigenvalue does indeed fall as the microscopic barrier potential is modified so as to increase the effective mass but keep the barrier height the same (Burt 1992). Hence the boundary condition (2.6) is vindicated. The lowering in energy must come about because the fall in potential energy in shifting charge from the barrier region to the well region overcomes the increase in confinement energy.

2.2. *Non-parabolicity*

Effective-mass theory has been used to study the effect of non-parabolicity on subband energies. The conduction band in direct-gap III–V compounds has the form

$$E(\boldsymbol{k}) = \frac{\hbar^2 k^2}{2m_0^*} + \alpha_0 k^4 + \beta_0 (k_x^2 k_y^2 + k_y^2 k_z^2 + k_z^2 k_x^2) \pm \gamma_0 [k^2 (k_x^2 k_y^2 + k_y^2 k_z^2 + k_z^2 k_x^2) - 9 k_x^2 k_y^2 k_z^2]^{1/2}. \tag{2.11}$$

For confinement along the z-direction (the $\langle 001 \rangle$ direction) the subband minima are determined by setting $k_x = k_y = 0$, whence

$$E(k_z) = \frac{\hbar^2 k_z^2}{2m_0^*} + \alpha_0 k_z^4 \tag{2.12}$$

or, for weak non-parabolicity,

$$k_z^2 \approx \frac{2m_0^*}{\hbar^2} [E(1 + \alpha E)], \quad \alpha = -\left(\frac{2m}{\hbar^2}\right)\alpha_0. \tag{2.13}$$

(Note that α_0, like β_0 and γ_0, is negative.) This defines a "confinement mass", $m_z^*(E)$,

$$m_z^*(E) = m_0^*(1 + \alpha E). \tag{2.14}$$

A similar expression is defined for the mass in the barrier by replacing E by $E - V_0$. If $E - V_0$ is large, recourse must be made to the complex bandstructure of the barrier (Chang 1982). Non-parabolicity tends to increase the mass in the well and reduce the mass in the barrier. Increasing the mass in the well tends to lower the energy and reducing the mass in the barrier (as we have just discussed) increases the energy. Either may prevail. Introducing motion in the plane affects the energy dependence involving k_z and hence the boundary conditions which, in turn, modifies the energy. This has the effect of modifying the effective mass in the plane, leading to a mass enhancement 2–3 times larger than that for the confinement mass (Ekenberg 1989). Similar effects occur in valence-band states, as we will see.

It should be noted that some authors (e.g., Welch et al. 1984) use the "momentum mass", i.e., $\hbar k/v$, where v is the group velocity, in the matching condition of eq. (2.6). The basis for this is that the matching condition involves essentially currents, and therefore using an effective mass derivable from crystal momentum and group velocity is more appropriate than using the confinement mass of eq. (2.14). For weak non-parabolicity, the momentum mass is given by

$$m_p^* = m_0^*(1 + 2\alpha E), \tag{2.15}$$

which shows that it is twice as strongly affected by non-parabolicity than is the confinement mass. The increase of momentum mass with velocity turns out to be of the form familiar in special relativity (Landsberg 1986), viz.

$$m_p^* = \frac{m_0^2}{\sqrt{1 - v^2/v_0^2}}, \tag{2.16}$$

where v_0 is the limiting velocity given by $(2\alpha m_0^*)^{-1/2}$. But the straightforward solution of the Schrödinger equation shows that this is not the appropriate mass to use to determine bound states, though it more nearly describes the mass parallel to the layers than does the confinement mass.

The presence of non-parabolicity really entails more elaborate matching conditions than those of eq. (2.6) with eq. (2.14), but the latter are useful if errors of a few meV volts can be tolerated. For further discussion see the paper by Ekenberg (1989).

2.3. Band-mixing

When substantially different conduction bands are involved, such as the Γ and X valleys in the GaAs/AlAs system, the confined state becomes a mixture and we return to the basic envelope-function formalism of eq. (2.2). The gap between Γ states in GaAs and AlAs is not known precisely, but it is about 1 eV. With such an energy difference effective-mass theory is inapplicable and it is necessary to use the complex bandstructure of the bulk materials. Calculations of subband structure use either pseudo-potential or tight-binding techniques. The latter have been carried out by Shulman and Chang (1981, 1985) and by Ando and Akera (1989), and pseudo-potential calculations have been done by Jaros and co-workers (1984, 1985). These show that confinement of a Γ electron in GaAs is very largely due to the Γ barrier rather than to the Γ–X barrier. In other words, mixing of Γ and X states is small, though finite, and this mixing can be directly observed in experiments of resonant tunnelling in which tunnelling paths involve Γ–X transitions (Landheer et al. 1989).

In other structures such as GaAs/$Al_xGa_{1-x}As$ (x small), HgTe/CdTe and GaSb/InAs the envelope function method in the effective-mas approximation works quite well (Ando et al. 1989). Even when the adjacent lattices are mismatched a shift of position variable, following the technique of Pikus and Bir (1959), allows the envelope function method to be adapted to strained layers (Burt 1989). Besides tight-binding and pseudo-potential calculations there are a number of applications of the effective-mass method to the problem of quantum confinement of holes in strained-layer systems (e.g., Altarelli 1986, Marzin 1986, O'Reilly and Witchlow 1986, Andreani et al. 1987). Even without strain the problem of describing the quantum confinement of holes has the special feature of light- and heavy-hole degeneracy, in some cases exacerbated by the proximity of the split-off band. The presence of interfaces causes substantial mixing between light- and heavy-hole wavefunctions and the in-plane mass can show striking changes (fig. 5).

If all bands other than the light-hole (LH) and heavy-hole (HH) band are ignored

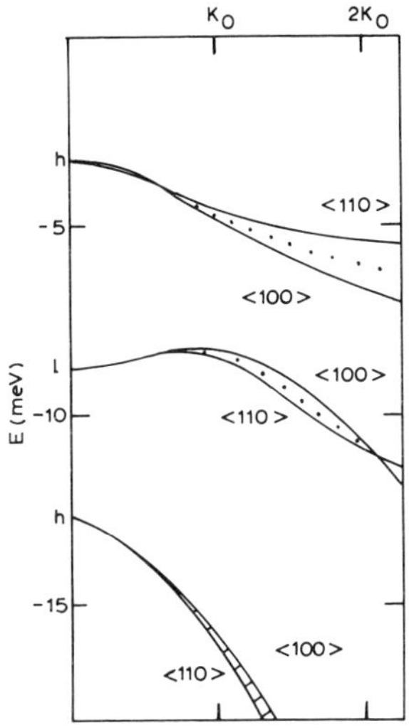

Fig. 5. Hole subband structure in a 140 Å GaAs/200 Å $Al_{0.21}Ga_{0.79}As$ superlattice in the plane perpendicular to the ⟨001⟩ growth axis. Solid lines: dispersion for $\boldsymbol{k}$ in ⟨100⟩ or ⟨110⟩ directions. Dotted lines: axial approximation ($\gamma_3 \approx \gamma_2 \approx (\gamma_2 + \gamma_3)/2$). The symbols h and l denote heavy and light and $k_0 = \pi/(340\ \text{Å}) = 9.24 \times 10^5\ \text{cm}^{-1}$ (Altarelli et al. 1985).

the dispersion in the presence of a biaxial strain is

$$E = E_s + \frac{\hbar^2}{2m}\{-\gamma_1 k^2 \pm [4\gamma_2^2 k^4 + 12(\gamma_3^2 - \gamma_2^2)(k_x^2 k_y^2 + k_y^2 k_z^2 + k_z^2 k_x^2) + 2\gamma_2(k_x^2 + k_y^2 - 2k_z^2)K^2 + K^4]^{1/2}\}, \tag{2.17}$$

where γ_1, γ_2, and γ_3 are the Luttinger parameters (Luttinger 1956), E_s is the shift in energy due to the dilatation, and K is the wavevector associated with splitting energy δ, viz.

$$K = \frac{2m\delta}{\hbar^2}, \quad E_s = 2a\frac{(c_{11} - c_{12})}{c_{11}}e_s, \quad \delta = b\frac{(c_{11} + 2c_{12})}{c_{11}}e_s, \quad e_s = \frac{\Delta a_0}{a_0}. \tag{2.18}$$

Here, c_{11} and c_{12} are the elastic constants, e_s is the strain defined as the fractional shift in lattice constant a_0, and a, b are valence band deformation potentials. There are four eigenfunctions (counting spin) each a linear combination of LH and HH wavefunctions. The total angular momentum of these p-orbitals is $\frac{3}{2}$, and it can be taken to be in the confinement direction. The HH states have spin $\pm\frac{3}{2}$, the LH states have spin $\pm\frac{1}{2}$. Suitable quantum-well wavefunctions for one of the doubly degenerate HH states and one of the doubly degenerate LH states are

$$H_1((\boldsymbol{k}_{11}, k_z)) = \begin{vmatrix} R_- \cos k_z z \\ -\mathrm{i}L^* \sin k_z z \\ -M^* \cos k_z z \\ 0 \end{vmatrix} \begin{vmatrix} |\frac{3}{2}, \frac{3}{2}\rangle \\ |\frac{3}{2}, \frac{1}{2}\rangle \\ |\frac{3}{2}, -\frac{1}{2}\rangle \\ |\frac{3}{2}, -\frac{3}{2}\rangle \end{vmatrix} \mathrm{e}^{\mathrm{i}\boldsymbol{k}_\parallel \cdot \boldsymbol{r}_\parallel}, \tag{2.19a}$$

$$L_1((\boldsymbol{k}_{11}, k_z)) = \begin{vmatrix} -L\cos k_z z \\ -\mathrm{i}R_+ \sin k_z z \\ 0 \\ -M^* \sin k_z z \end{vmatrix} \begin{vmatrix} |\frac{3}{2}, \frac{3}{2}\rangle \\ |\frac{3}{2}, \frac{1}{2}\rangle \\ |\frac{3}{2}, -\frac{1}{2}\rangle \\ |\frac{3}{2}, -\frac{3}{2}\rangle \end{vmatrix} \mathrm{e}^{\mathrm{i}\boldsymbol{k}_\parallel \cdot \boldsymbol{r}_\parallel}, \tag{2.19b}$$

where

$$R_\pm = P \pm Q + E - E_s, \quad P = \frac{\hbar^2}{2m}\gamma_1 k^2, \quad Q = \frac{\hbar^2}{2m}\gamma_2(k_x^2 + k_y^2 - 2k_z^2) + \delta,$$

$$L = -\mathrm{i}\frac{\hbar}{2m}3\sqrt{3}\gamma_3[(k_x - \mathrm{i}k_y)]k_z, \quad M = \frac{\hbar^2}{2m}[\gamma_2\sqrt{3}(k_x^2 - k_y^2) - \mathrm{i}\gamma_3 2\sqrt{3}k_x k_y]. \tag{2.19c}$$

The states are purely HH or LH only when $k_x = k_y = 0$. The energy is given by

$$E = E_s - \frac{\hbar^2}{2m}(\gamma_1 \pm 2\gamma_2)k_z^2 \pm \delta. \tag{2.20}$$

The upper sign gives the LH energy, the lower the HH energy. The states are split by an amount 2δ plus effects arising from confinement. The latter are determined in the usual way using the appropriate masses, viz.

$$m_H^* = m/(\gamma_1 - 2\gamma_2), \quad m_L^* = m/(\gamma_1 + 2\gamma_2). \tag{2.21}$$

Compressional stresses raise the HH band above the LH band, and vice versa (fig. 6).

Motion in the plane is anisotropic with an anomalously low mass for the HH band and an anomalously high mass for the LH band, a phenomenon referred to as mass reversal. In general, a detailed description of the subband structure and its intense non-parabolicities demands numerical computation. A simple analytical expression for the in-plane mass was derived for a strain-free system with infinitely high barriers by Nedorezov (1971). A similar expression including the effects of strain has been derived by Ridley (1990) in the spherical approximation (fig. 7), viz.

$$\left(\frac{m}{m_\parallel^*}\right)_H = (\gamma_1 - \eta_H\bar{\gamma}) + \frac{3(\gamma_1 - 2\bar{\gamma})(\cos k_L L + (-1)^{n+1})}{(1 - S_H)(1 - S_L)k_L L \sin k_L L}, \tag{2.22a}$$

$$\left(\frac{m}{m_\parallel^*}\right)_L = (\gamma_1 + \eta_L\bar{\gamma}) + \frac{3(\gamma_1 + 2\bar{\gamma})(\cos k_H L + (-1)^{n+1})}{(1 - S_H)(1 - S_L)k_H L \sin k_H L}, \tag{2.22b}$$

where

$$\bar{\gamma}^2 = \tfrac{1}{5}(2\gamma_2^2 + 3\gamma_3^2), \quad \eta_H = \frac{2 + S_H}{1 - S_H}, \quad \eta_L = \frac{2 + S_L}{1 - S_L}, \quad S_H = \frac{K^2}{2\bar{\gamma}k_H^2}, \quad S_L = \frac{K^2}{2\bar{\gamma}k_L^2}. \tag{2.22c}$$

The wave vectors k_H, k_L, are those defining equal energies for HH and LH bands at $k_x = k_y = 0$, thus

$$(\gamma_1 + 2\bar{\gamma})k_L^2 - K^2 = (\gamma_1 - 2\bar{\gamma})k_H^2 + K^2. \tag{2.23}$$

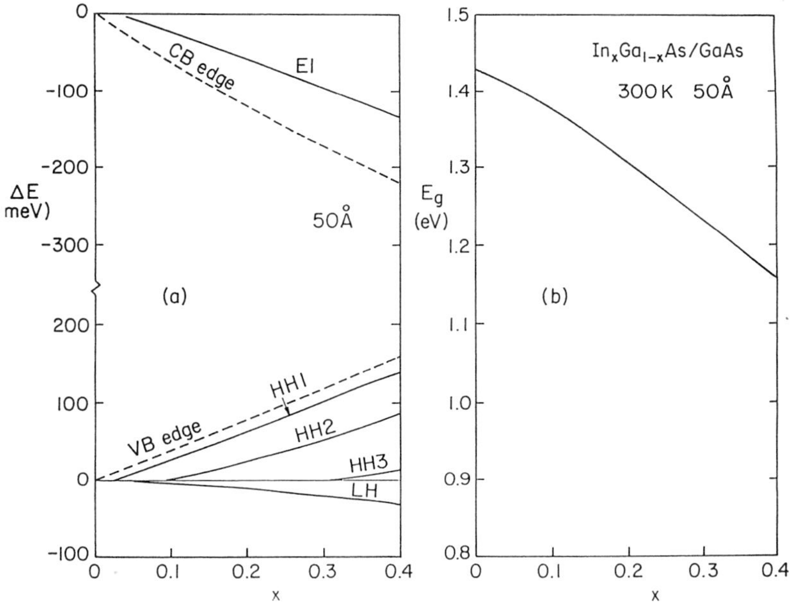

Fig. 6. Energy structure of a 50 Å well strained-layer system in $In_xGa_{1-x}As/GaAs$. (a) Subbands. (b) Energy gap (HH1 – E1). The effect of the split-off band has been ignored (Ridley 1990).

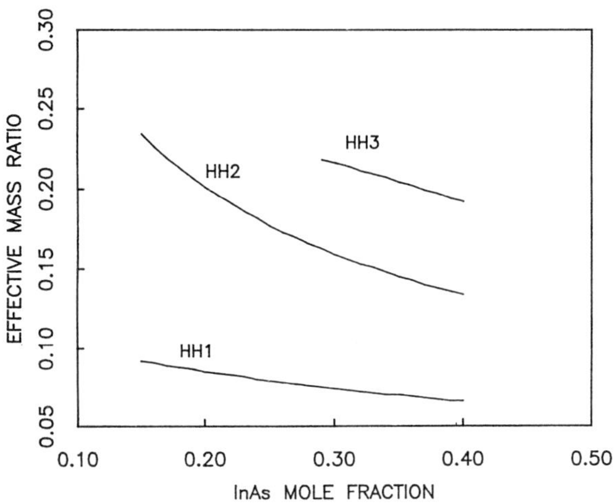

Fig. 7. In-plane effective masses in spherical approximation for a 50 Å well in $In_xGa_{1-x}As/GaAs$ (Ridley 1990).

In the HH subband labelled n (an integer), (e.g., eq. (2.22a)) $k_H = n\pi/L$ where L is the well width; in the LH subband (e.g., eq. (2.22b)), $k_L = n\pi/L$. Non-parabolocity factors near the zone centre have also been derived, allowing the bandstructure near the zone centre to be calculated (fig. 8). In spite of the simplifications introduced (infinitely high barriers, averaging out the anisotropy) these equations contain the essential physics of the result of confinement and yield magnitudes in reasonable agreement with experiment.

For tensional strains it can happen that $S_H = S_L = 1$, in which case the above equations become invalid. In this case, for small $k_{\parallel}$ ($k_{\parallel}^2 = k_x^2 + k_y^2$), the energy bands are described by

$$E = E_s + \frac{\hbar^2}{2m}[-\gamma_1(k_H^2 + k_{\parallel}^2) \pm 2k_H k_{\parallel}\bar{\gamma}\sqrt{3}] \tag{2.24}$$

corresponding to two bands degenerate at $k_{\parallel} = 0$ but with maxima shifted from the zone centre.

3. The confinement of optical phonons

3.1. Microscopic models

Optical phonons are confined whenever there is no overlap of the frequency bands in the adjacent materials. The existence of confined longitudinal optic (LO) and transverse optic (TO) modes has been amply confirmed by experiments on Raman scattering (see reviews by Klein 1986, Menendez 1989, Cardona 1989, 1990). On the other hand, the frequency bands of acoustic modes always overlap over a considerable range and this, in a superlattice, results in zone-folding (fig. 9) and the appearance of mini-gaps at the superlattice zone boundaries (Narayanamurti et al. 1979). The theory of acoustic modes in layered material has been given by Rytov (1956) and by Kelly (1985). In view of the problem of describing the confinement of optical modes it is worth remarking here that purely elastic boundary conditions were used to describe folded LA modes propagating along the superlattice axis, that is, u_z, the displacement, and the stress, $K\,du_z/dz$, where K is the appropriate elastic constant, were taken to be continuous. This leads to a dispersion relation identical in form to eq. (2.9) but with $Z = \rho_w v_w/\rho_B v_B$, where ρ is density and v is the LA velocity, and $\omega = v_w k_w = v_B k_B$. For arbitrary directions of propagation through the superlattice it is necessary to apply boundary conditions involving shear as well as dilatational stresses and this leads to mixing of LA and TA waves, as is well known (Lord Rayleigh 1889 and, e.g., Landau and Lifshitz 1986).

The confinement of optical phonons is most readily treated theoretically when there is no in-plane propagation for then linear-chain models of the lattice dynamics can be applied (Colvard et al. 1985, Molinari et al. 1986, Fasolino et al. 1990). The result for the AlAs/GaAs superlattice whose axis is along the $\langle 100 \rangle$ direction (fig. 10) can be expressed in terms of an envelope function for the optical displacement as

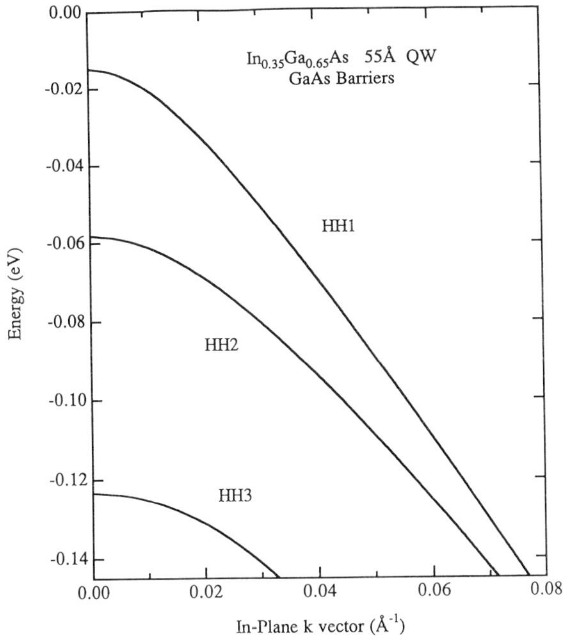

Fig. 8. Subband dispersion for a 55 Å quantum well in $In_{0.35}Ga_{0.65}As/GaAs$. (Spherical approximation including non-parabolicity) (Lester et al. 1991).

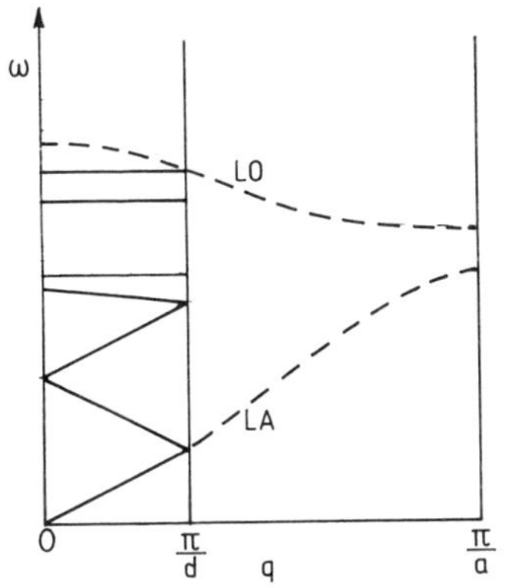

Fig. 9. Folding of acoustic modes and confinement of optical modes.

follows for the GaAs LO modes:

$$u_z = A \sin k_z z, \quad k_z = n\pi/L, \quad 0 \leqslant z \leqslant L, \tag{3.1}$$

where L is the width of the GaAs layer. A similar expression for the higher frequency AlAs modes can be written. In this system no overlap of the frequency bands occurs and the modes are rigidly confined. (Actually, some vibration always creeps across

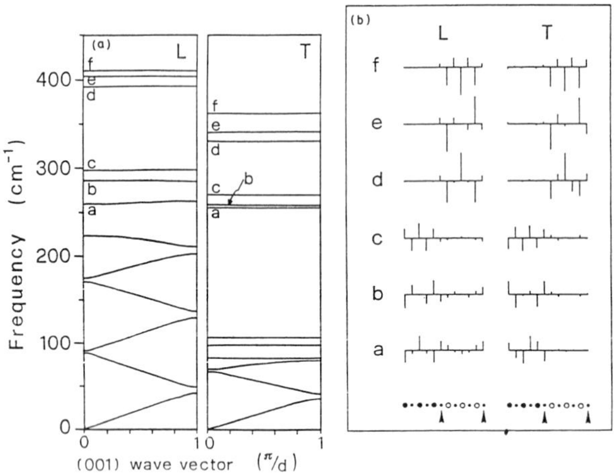

Fig. 10. (a) Longitudinal and transverse frequency dispersion along the growth direction ⟨001⟩ of a $(GaAs)_3(AlAs)_3$ superlattice. (b) Corresponding ionic displacement – modes are confined either to GaAs or to AlAs.
Key: ● As, · Ga, ○ Al (Molinari et al. 1986).

the interface and L should be replaced by $L+\delta L$, with δL of order of the atomic spacing, but for layers consisting of many unit cells this correction is negligible.) This quantization of k_z is reflected in a quantization of frequency through the dispersion relation which for long-wavelength modes can be expressed approximately as follows:

$$\omega_{\mathrm{L}}^2(k)=\omega_{\mathrm{L}}^2-v_{\mathrm{L}}^2k_z^2, \tag{3.2}$$

where v_{L} is a velocity intimately related to the velocity of LA modes, and this quantization has been observed in Raman experiments. The form of eqs. (3.1) and (3.2) is also applicable to TO modes with obvious redefinitions.

Associated with the LO displacement in GaAs (and in AlAs) is an electric field E_z and a scalar potential φ. In a polar isotropic dielectric medium the electric field and permittivity satisfy the relations

$$\boldsymbol{E}=-\left(\frac{e^*}{V_0\varepsilon_0}\right)\frac{\omega^2-\omega_{\mathrm{T}}^2(k)}{\omega_{\mathrm{L}}^2(k)-\omega_{\mathrm{T}}^2(k)}\boldsymbol{u},\qquad \varepsilon(\omega)=\varepsilon_\infty\frac{\omega^2-\omega_{\mathrm{L}}^2(k)}{\omega^2-\omega_{\mathrm{T}}^2(k)}, \tag{3.3}$$

where e^* is the effective ionic charge, V_0 is the volume of the primitive unit cell, $\omega_{\mathrm{L}}(k)$, $\omega_{\mathrm{T}}(k)$ are the LO, TO frequencies and ε_0, ε_∞ are the permittivities of free space and of the medium and high frequencies. Applying the conventional electromagnetic boundary conditions at the interfaces leads to the so-called dielectric-continuum (DC) model. Unfortunately this leads to mode patterns of opposite symmetry to those predicted by microscopic theory, and by elastic-continuum theory (fig. 11). For an LO mode $\varepsilon(\omega)=0$ and there is no electric displacement, hence no magnetic field, hence no electromagnetic energy, so there appears to be no physical necessity for electromagnetic boundary conditions. For a confined LO mode whose displacement

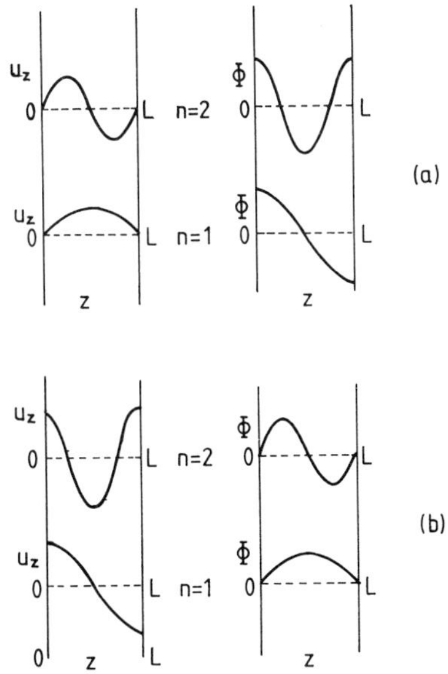

Fig. 11. Different symmetries of confined LO modes predicted by (a) hydrodynamic model, (b) dielectric continuum model. u_z is the optical displacement along the growth direction, Φ is the scalar potential.

is described by eq. (3.1) the field and associated potential are

$$E_z = \frac{-e^*}{V_0 \varepsilon_0} A \sin k_z z, \quad \varphi = \frac{e^*}{V_0 \varepsilon_0} \frac{A}{k_z} \cos k_z z, \quad k_z = n\pi/L, \tag{3.4}$$

thus, although the field vanishes at the interfaces, the potential does not. The question arises: "Should the potential be continuous or not?". Conventional electrostatics and the DC model demand that it must be continuous, in which case there exists a spatially independent oscillating potential in the AlAs. This solution is favoured by Bechstedt and Gerecke (1989), whose results are shown in fig. 12, who point out that such a potential would be undetectable in Raman experiments. Huang and Zhu (1988), on the other hand, introduce ad hoc components to "save the phenomenon" and make the potential vanish, viz.

$$\varphi = \frac{e^*}{V_0 \varepsilon_0} \frac{(-1)^{n/2}}{k_z} (\cos k_z z - 1), \quad n = 2, 4, \ldots \tag{3.5}$$

and a more complicated prescription for odd numbered modes which, moreover, necessitates the exclusion of the $n = 1$ solution. Their mode patterns are depicted in fig. 13. Both pairs of authors construct microscopic theories and obtain vibrational envelopes which describe their results all within the constraints imposed by the electromagnetic boundary conditions. This approach, among other things, maintains the basic symmetries of the DC model, which has relevance for the electron–phonon interaction. A quite different approach has been adopted by Ridley and Babiker

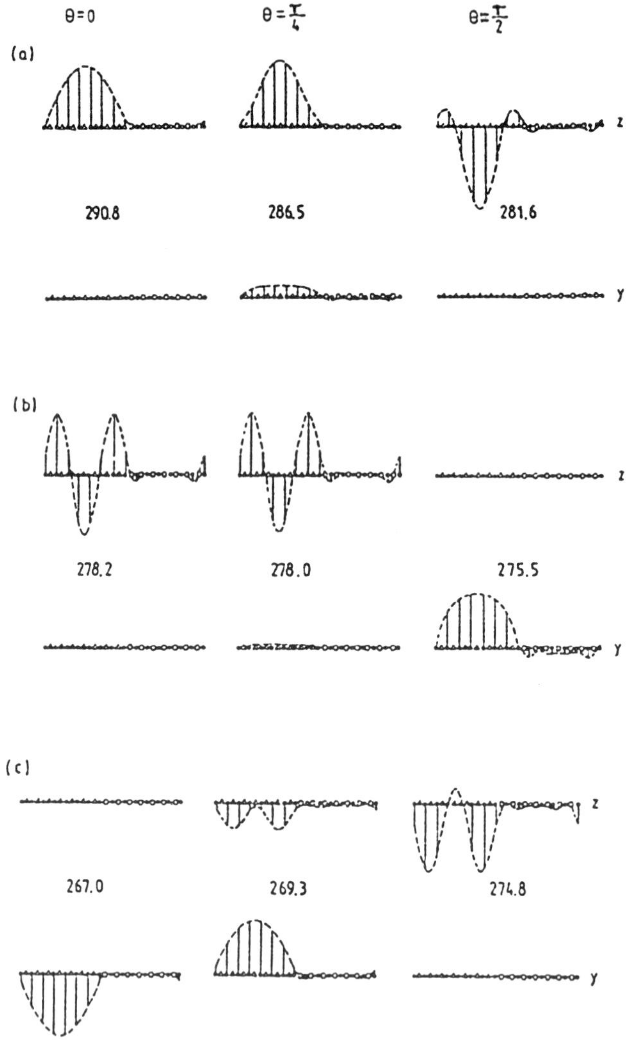

Fig. 12. Anion displacements in a $(GaAs)_7(AlAs)_7$ superlattice for symmetrical p-polarized GaAs modes. Vertical lines: microscopic theory. Dashed lines: envelope function theory. z: superlattice axis, y: in-plane direction. θ is the angle of propagation through the superlattice with respect to z (Gerecke and Bechstedt 1991).

(1991) who question the applicability of electromagnetic boundary conditions to a situation not involving electromagnetic energy. Their conclusion is that as far as an envelope-function representation is concerned the scalar potential is, indeed, discontinuous on the mesoscopic scale. In their view eq. (3.4) should be accepted as it stands. Fortunately these conflicts can be resolved by generalizing this hydrodynamic (HD) model as we now go on to discuss.

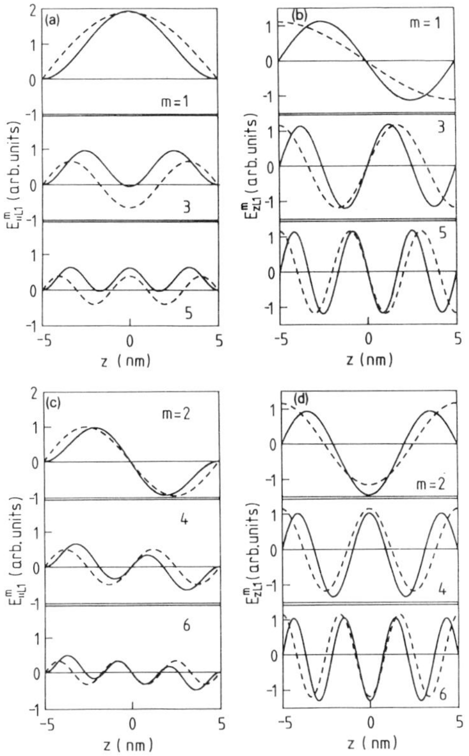

Fig. 13. Envelope functions of confined LO modes in a GaAs/AlAs superlattice. (a) In-plane field components. (b) Axial field components for symmetric LO modes. (c) and (d) ditto for asymmetrical LO modes. Solid lines: modified DC theory. Dashed lines: DC theory (Huang and Zhu 1988, Wendler 1990).

3.2. Elastic-continuum models

It is clear that mechanical boundary conditions are indispensable for describing the confinement of optical phonons, just as they are indispensable for describing the zone-folding of acoustic modes. This was appreciated first by Babiker (1986). Using the same boundary conditions as for acoustic modes led to eq. (2.9) with $Z = (v_L^2 \rho^2 k)_B / (v_L^2 \rho^2 k)_W$, and to complete agreement with linear-chain theory. Babiker extended his theory to incorporate in-plane propagation by assuming the continuum to be isotropic and by using hydrodynamic boundary conditions – the continuity of the normal component of the particle velocity (equivalent to the continuity of displacement) and the continuity of pressure – and obtained the usual form for the superlattice (eq. (2.9)), but now with

$$Z = \frac{(v_L \rho)_B^2 (k_B^2 + k_\parallel^2) k_w}{(v_L \rho)_w^2 (k_W^2 + k_\parallel^2) k_B}. \tag{3.6}$$

This theory could also describe interface LO modes as well as guided modes, and it is applicable to non-polar as well as to polar material. More recently, these boundary conditions have been used by Guillemot and Clerot (1990) to discuss LO modes in GaAs/AlGaAs structures. It is useful at this point to emphasize that the interface modes arising out of elastic continuum models are to be distinguished from the short-range interface modes which are predicted in lattice dynamical calculations as a result of a modification of force constants at the surface (e.g., Maradudin et al. 1971). Continuum models describe long-range interface modes, analogous to the acoustic Rayleigh wave. Elastic-continuum interface modes are, of course, also to be distinguished from interface polaritons, of which more later.

Regarding a solid as a fluid in which hydrodynamic boundary conditions prevail is not satisfactory. A fuller description invokes the continuity of all relevant elastic stress components including shear (Ridley 1991c). This approach starts from the dispersion relation for long wavelengths in three dimensions which can be written for a III–V compound as follows:

$$(\omega^2-\omega_T^2)\boldsymbol{u}(\boldsymbol{k})=[-H(\boldsymbol{k})+(\omega_L^2-\omega_T^2)I(\boldsymbol{k})]\boldsymbol{u}(\boldsymbol{k}) \tag{3.7a}$$

$$H(\boldsymbol{k})=\begin{vmatrix} Ak_x^2+B(k_y^2+k_z^2) & Ck_xk_y & Ck_xk_z \\ Ck_yk_x & Ak_y^2+B(k_z^2+k_x^2) & Ck_yk_z \\ Ck_zk_x & Ck_zk_y & Ak_z^2+B(k_x^2+k_y^2) \end{vmatrix} \tag{3.7b}$$

$$I(\boldsymbol{k})=\frac{1}{k^2}\begin{vmatrix} k_x^2 & k_xk_y & k_xk_z \\ k_yk_x & k_y^2 & k_yk_z \\ k_zk_x & k_zk_y & k_z^2 \end{vmatrix}, \tag{3.7c}$$

where A, B and C are related to (but are not identical to) the corresponding parameters in ordinary elasticity theory for the acoustic modes. Thus

$$A\sim v_{LA}^2=c_{11}/\rho,\quad B\sim v_{TA}^2=c_{44}/\rho,\quad C\sim(c_{12}+c_{44})/\rho, \tag{3.8}$$

where c_{11}, c_{12}, c_{44} are the elastic constants and v_{LA}, v_{TA} are the LA, TA velocities. The factors relating A, B and C to the acoustic parameters are all of order unity (as can be deduced from the work of Tsuchiya et al. 1989). For non-polar material we can put $\omega_L=\omega_T=\omega_0$. Optical strains may be defined as follows:

$$\begin{aligned} &S_1=-\frac{\partial u_x}{\partial x},\quad S_2=-\frac{\partial u_y}{\partial y},\quad S_3=-\frac{\partial u_z}{\partial z},\quad S_4=-\frac{1}{2}\left(\frac{\partial u_z}{\partial y}+\frac{\partial u_y}{\partial z}\right),\\ &S_5=-\frac{1}{2}\left(\frac{\partial u_x}{\partial z}+\frac{\partial u_z}{\partial x}\right),\quad S_6=-\frac{1}{2}\left(\frac{\partial u_y}{\partial x}+\frac{\partial u_x}{\partial y}\right), \end{aligned} \tag{3.9}$$

and stresses are then given by

$$\begin{aligned} &T_1=c_{11}^0S_1+c_{12}^0(S_2+S_3),\qquad T_2=c_{11}^0S_2+c_{12}^0(S_3+S_1),\\ &T_3=c_{11}^0S_3+c_{12}^0(S_1+S_2),\qquad T_4=2c_{44}^0S_4,\qquad T_5=2c_{44}^0S_5,\qquad T_6=2c_{44}^0S_6, \end{aligned} \tag{3.10}$$

where $c_{11}^{0}, c_{12}^{0}, c_{44}^{0}$ are the optical elastic constants derivable from the observed or calculated dispersion curves. In general the appropriate boundary conditions entail the continuity of energy flux and stress. Just as in the acoustic-wave case, the latter implies the mixing of LO and p-polarized TO modes when there is inplane propagation. (The term p-polarized refers to TO modes with displacements entirely in the plane of incidence, s-polarized refers to TO modes with displacement perpendicular to the plane of incidence.) An approach along similar lines is briefly mentioned in the paper of Tsuchiya et al. (1989), but no details are given. They claim that such an approach gives results much more in accord with lattice dynamics than does the dielectric-continuum model.

Taking the solid to be an isotropic elastic continuum allows the decomposition of modes into longitudinally and transversely polarized parts whatever the direction of propagation:

$$\begin{aligned} \boldsymbol{u} &= \boldsymbol{u}_{\mathrm{L}} + \boldsymbol{u}_{\mathrm{T}} \\ \nabla \times \boldsymbol{u}_{\mathrm{L}} &= 0, \quad \nabla \cdot \boldsymbol{u}_{\mathrm{T}} = 0. \end{aligned} \tag{3.11}$$

This simplification has been used (Ridley 1991c) to predict the existence of the optical-mode analogue of Rayleigh surface waves in a non-polar layer with stress-free surfaces and also of LO/TO hybrids in non-polar layers with infinitely rigid interfaces (Ridley 1991d). With hydrodynamic boundary conditions surface modes cannot exist (Constantinou and Ridley 1990) so their appearance on stress-free surfaces is a consequence of introducing shear stress into the theory.

The theory has been successfully applied to polar layers (Ridley 1991e) in which both electromagnetic and elastic boundary conditions can be satisfied by taking linear combinations of LO, TO and interface polaritons (IP), all at the same frequency. (Earlier attempts (Ridley 1991c,d) involved double hybridization and were only partially successful.) This scheme of triple hybridization yields mode patterns close to those obtained in microscopic theory (fig. 14) (although the latter has dealt only with superlattices, and here we are dealing with a single well) and so has resolved many of the conflicts in this field. It turns out that the HD model correctly predicts the asymptotic form when $k_x L \ll 1$, where k_x is the wavevector component in the plane and L is the well-width, and the DC model correctly predicts the asymptotic form when $k_x L$ is very large. Since an angle of propagation θ can be defined by $\tan^{-1}(k_x/k_z)$, where k_z is the wavevector component perpendicular to the plane, this variation constitutes a substantial angular dispersion of mode pattern. A double-hybrid model for polar material has recently been proposed by Zianni et al. (1992).

For the case of a single quantum well with infinitely rigid interfaces ($\boldsymbol{u} = 0$), roughly corresponding to the GaAs/AlAs system, the mode patterns for $k_x L \ll 1$ are given below. One type of LO-like hybrid is:

$$u_x = 2k_x A\, \mathrm{e}^{\mathrm{i}k_x x}\left(\cos k_{\mathrm{L}} z + \frac{\sinh k_0(z - L/2)}{\sinh k_0 L/2}\right), \tag{3.12a}$$

$$u_z = 2\mathrm{i}k_{\mathrm{L}} A\, \mathrm{e}^{\mathrm{i}k_x x} \sin k_{\mathrm{L}} z, \tag{3.12b}$$

$$k_{\mathrm{L}} L = n\pi, \quad n \text{ odd}. \tag{3.12c}$$

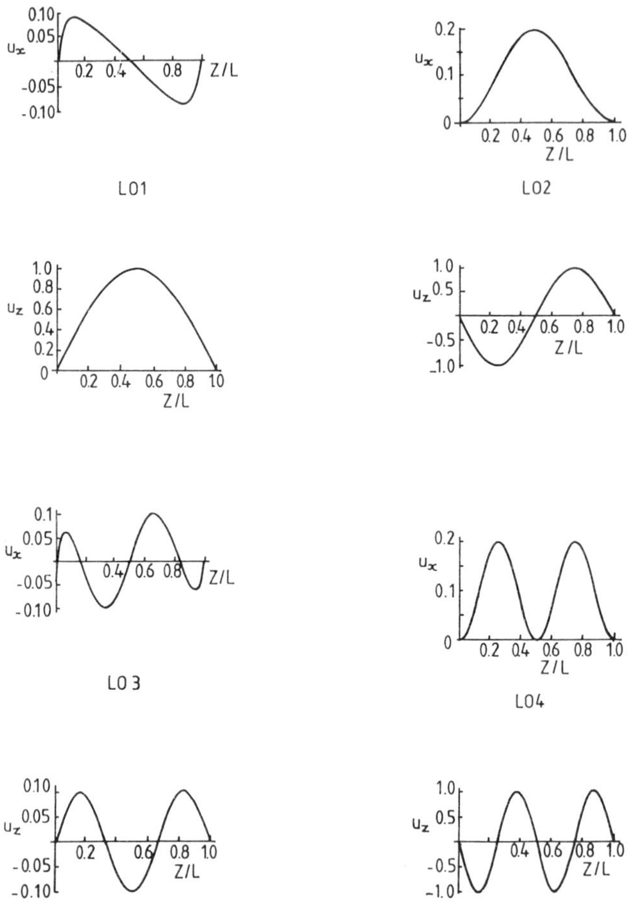

Fig. 14. LO hybrid modes in a quantum well of width 38 Å with rigid barriers for $k_x/k_L = 0.1$. (a) LO 1, (b) LO 2, (c) LO 3, (d) LO 4. (Note that these mode designations agree with those of fig. 12 but not with those of fig. 13 where $m = 1$ corresponds to our LO 2.)

This odd-order mode is really a triple hybrid. The IP components have negligible ionic displacement and, for brevity, are not shown, but they are essential for accounting correctly for the EM fields in the barrier. Because such fields exist these modes have been called "Coulomb modes" by Bechstedt and Gerecke (1989). In the above expressions k_L is the wavevector component along the z-direction (normal to the plane of the layers) of the LO component and k_0 is the corresponding parameter for the TO modes, which is purely imaginary and given by

$$k_0^2 = \frac{\omega_L^2 - \omega_T^2}{v_T^2}. \tag{3.13}$$

It is assumed that $k_0 L \gg 1$, an assumption which covers most cases. The even-order mode is also approximately a double hybrid, but this time the hybridization is principally that of the LO and IP modes, viz.

$$u_x = 2k_x A\, e^{ik_x x}\left(\cos k_L z - \frac{\cosh k_x(z - L/2)}{\cosh k_x L/2}\right), \tag{3.14a}$$

$$u_z = 2ik_L A\, e^{ik_x x} \sin k_L z, \tag{3.14b}$$

$$k_L L = n\pi, \quad n \text{ even}. \tag{3.14c}$$

The TO involvement is small provided the difference between LO and IP permittivities is small, which is a good approximation for the lower order modes. In this case there are no electric fields in the barrier, hence the term "mechanical" modes used by Bechstedt and Gerecke (1989). Note that for both odd- and even-order modes the LO component is that of HD theory. Some patterns are depicted in fig. 14.

The TO modes do not have electric fields and so can be hybridized solely with LO modes:

$$u_x = 2ik_T C\, e^{ik_x x} \sin k_T z, \tag{3.15a}$$

$$u_z = -2k_x C\, e^{ik_x x}(\cos k_T z - \cos k_L z), \tag{3.15b}$$

$$k_T L = n\pi, \quad k_L L = (2m + n)\pi, \tag{3.15c}$$

with n and m chosen so that TO and LO components have the same frequency. The latter condition also restricts k_x.

Increasing $k_x L$ leaves the even-order LO modes looking much the same, but this masks the fact that the LO component gradually changes over to the DC form. The latter metamorphosis also occurs for the odd-order modes but the overall change in mode shape is very noticeable (fig. 15). The most rapid change is for the $n = 1$ mode

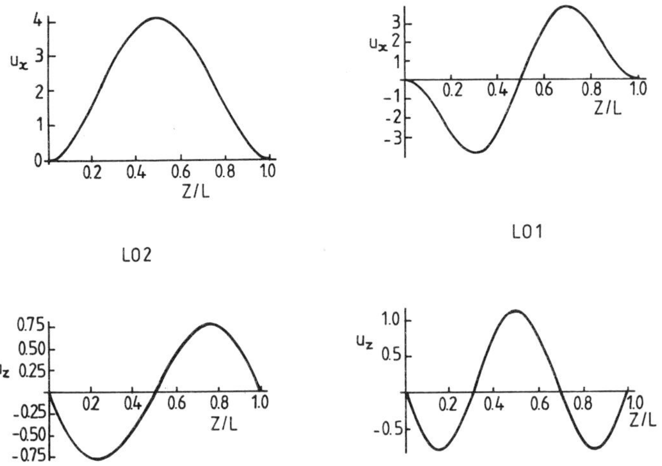

Fig. 15. LO hybrids and their angular variation. (a) LO 2, $k_x L = 3$, $k_L L = 1.67\pi$; (b) LO 3, $k_x L = 3$, $k_L L = 2.79\pi$.

which converts to a mode with $n=3$ characteristics. The complex dispersion is depicted in fig. 16. There are no interface hybrids in this case (rigid interfaces).

The LO and TO hybrids described above have their frequencies determined by their respective dispersion relations. The IP components are essentially EM waves obeying

$$\omega^2 = \frac{c^2 k^2}{\varepsilon_p(\omega)}, \tag{3.16}$$

where $k^2 = k_x^2 - k_p^2$, where k_p is the decay constant for the z-direction, which must always be close to k_x when k_x is far from the light-line if ω is to remain close to ω_L and ω_T. Surface polaritons were described by Fuchs and Kliewer (1965) who ignored mechanical boundary conditions and obtained for a single slab the dispersion relation

$$\tanh k_x L = -\frac{2r}{1+r^2}, \quad r = \frac{\varepsilon_B}{\varepsilon_p(\omega)}, \tag{3.17}$$

where ε_B is the barrier permittivity (assumed independent of frequency in the range considered). This relationship is for the so-called "unretarded" limit, i.e., for $k_x^2 \gg \omega^2 \varepsilon_\infty / c^2$, and it, and its corresponding expression for a superlattice, have been much used to describe IP modes in layered materials. Constraining IP modes to satisfy elastic as well as EM boundary conditions means that hybridization with LO modes occurs, and this has the effect of limiting the application of eq. (3.17) to only a few values of k_x in narrow layers (more values in wider layers). The usual IP dispersion is valid only if the tangential electric field of the LO mode vanishes at the interfaces. This restricts k_L ($k_L L = n\pi$) and hence the frequency, and it follows that eq. (3.17) can be satisfied for only certain values of k_x (fig. 16). At all other values of

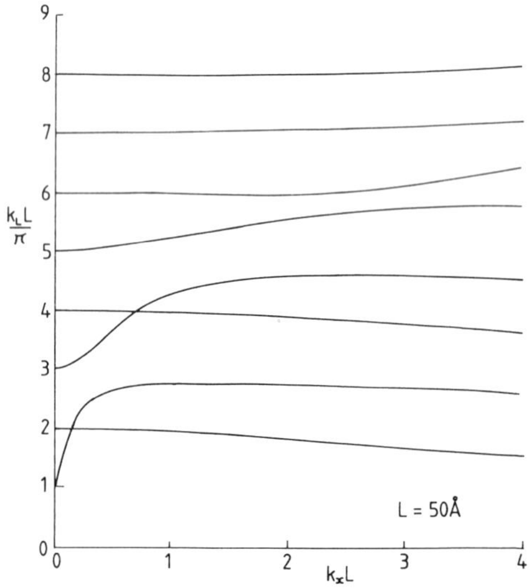

Fig. 16. Dispersion relation (50 Å GaAs well, AlAs barriers).

k_x the frequency of the IP mode is determined by the LO or TO mode with which it hybridizes. Basically it would appear that the Fuchs–Kliewer mode must play a heavily modified role in thin-layered material in which hybridization occurs because of this restriction. However, in thick layers surface polaritons could still be important. A discussion of this case has been given by Wendler and Haupt (1987). Another situation where IP modes may be relatively unaffected by hybridization is when the LO band of frequencies is too narrow to overlap the TO band – a situation which occurs in AlAs. In such a case, strong LO/IP hybridization cannot occur over the whole IP frequency range.

The triple hybrids of this elasto-electromagnetic continuum theory have unique properties. Besides having mixed polarization they have both scalar (φ) and vector ($\boldsymbol{A}$) potentials, the one donated by the LO component, the other by the IP component. It is clear that discontinuities of both these potentials occur at the interface in spite of the EM boundary condition being satisfied. The electric field $\boldsymbol{E}$ of the hybrid is

$$\boldsymbol{E} = -\nabla\varphi - \frac{\partial \boldsymbol{A}_{\rm p}}{\partial t}. \tag{3.18}$$

In the rigid barrier there is only a vector potential $A_{\rm B}$, so that continuity of E_x implies

$$A_{{\rm B}x} = -\frac{k_x}{\omega}\varphi + A_{{\rm p}x}. \tag{3.19}$$

The scalar potential vanishes discontinuously across the interface and $A_{{\rm p}x}$ jumps to $A_{{\rm B}x}$. These discontinuous changes, however, do not imply discontinuity of the energy of an electron in the layer at rest with respect to the wave. Thus the relevant Hamiltonian describing the interaction is $eA_{{\rm B}x}v_x$ in the barrier and $-e\varphi + eA_{{\rm p}x}v_x$ in the well, where v_x is the group velocity of an electron propagating along the layer. This argument breaks down when $k_x = 0$, but this never happens provided proper normalization of the EM mode within a finite cavity is carried out.

Electrons will interact with these hybrids via both types of potential. Holes, in addition, will interact with all three components via deformation potentials. The quantum involved will be neither a phonon nor a polariton, but a hybridon. In general the hybrid will possess both mechanical and EM energy, and it will be the continuity of the energy flux and the stability of the interface which will determine the general boundary conditions.

Finally it should be mentioned that any continuum theory of optical phonons is open to the criticism that it ignores the detailed ionic motion at the interface. Akera and Ando (1989) have compared the connection rules which emerge from the linear chain model with those of Babiker's hydrodynamic model. Although the latter correctly satisfies the continuity of energy flux, they apply only when the anion/cation mass rates are nearly the same across the interface. This is somewhat like the criterion for effective-mass theory to work in the analogous electron case. The Akera–Ando connection rules depend on where the interface is situated and on its nature. In practice interfaces are far from perfect and it is possible that a continuum approach leads to a more realistic picture. Another difficulty in relating an isotropic-continuum theory in detail to lattice dynamics is the neglect of anisotropy in the former. There is no intrinsic difficulty in incorporating anisotropy into a continuum theory but this has not been attempted. In any case, when it comes to calculating the interaction with electrons it is often convenient to perform spherical or axial averages.

Recently, optical phonons in superlattices have been extensively discussed by Enderlein et al. (1991).

4. The electron–phonon interaction

4.1. General

An electron in layered material has the possibility of interacting with several different types of phonons, viz.

(1) Guided LO/TO/IP hybrids in the well and in the barrier.
(2) Interface LO/TO/IP hybrids.
(3) Folded LA/TA modes.
(4) Interface LA/TA modes (Rayleigh waves).
(5) At high carrier densities, hybrids involving coupled LO/plasmon modes.

Symmetry restrictions mean that an electron in a Γ valley interacts with LO modes via the Fröhlich interaction, with IP modes via the $\boldsymbol{A}\cdot\boldsymbol{p}$ interaction and with LA modes via the deformation-potential interaction. Holes add a deformation coupling to LO, TO, TA and, presumably, IP modes. Electrons in L and X valleys have well-known characteristic couplings. Scattering in bulk material must distinguish between intravalley and intervalley processes, and in layered material it must further distinguish between intrasubband and intersubband processes and between those and well capture (fig. 17). This added complexity along with that introduced by mode mixing has meant that only limited progress has been made up to the time of writing in describing the electron–phonon interaction in quantum-confinement systems. In this section we will focus on some of the basic features of the process and briefly review the experimental evidence as we go along.

The scattering rate is usually assumed to be given adequately by first-order perturbation theory, viz.

$$W = \frac{2\pi}{\hbar}\int |\langle \mathrm{f}|H|\mathrm{i}\rangle|^2 \delta(E_\mathrm{f} - E_\mathrm{i})\, \mathrm{d}N_\mathrm{f} \tag{4.1}$$

where i, f refer to initial and final states. The matrix element for phonons can be expressed

$$|\langle \mathrm{f}|H|\mathrm{i}\rangle|^2 = \frac{\hbar}{2NM\omega} C^2(q) I^2(\boldsymbol{k}', \boldsymbol{k}) G^2(\boldsymbol{k}', \boldsymbol{k}, \boldsymbol{q})(n(\omega) = \tfrac{1}{2} \pm \tfrac{1}{2}), \tag{4.2a}$$

$$I(\boldsymbol{k}', \boldsymbol{k}) = \int_{\mathrm{cell}} u^*_{\boldsymbol{k}'}(\boldsymbol{r}) u_{\boldsymbol{k}}(\boldsymbol{r})\, \mathrm{d}\boldsymbol{r}, \tag{4.2b}$$

$$G(\boldsymbol{k}', \boldsymbol{k}, \boldsymbol{q}) = \int \psi^*(\boldsymbol{k}', \boldsymbol{r}) \Phi(\boldsymbol{q}, \boldsymbol{r}) \psi(\boldsymbol{k}, \boldsymbol{r})\, \mathrm{d}\boldsymbol{r}. \tag{4.2c}$$

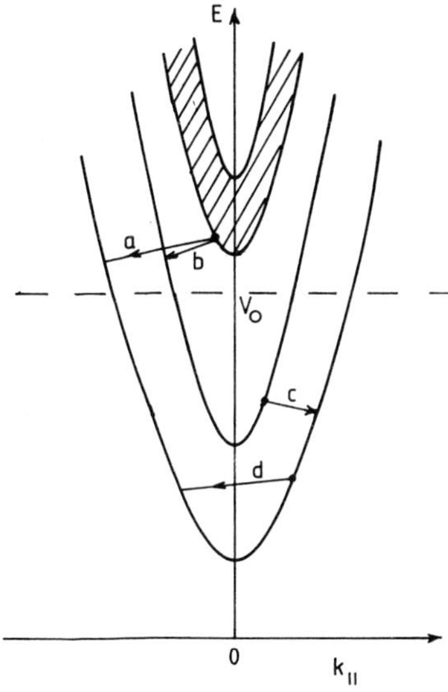

Fig. 17. Phonon-emission processes in a quantum well. States above V_0 form a (partial) continuum. Processes depicted are: (a) and (b) well-capture; (c) intersubband scattering; (d) intrasubband scattering.

Here M is the mass of each of the N oscillators (unit-cell mass for acoustic modes, reduced mass for optic modes). The frequency of the mode is ω, its Bose–Einstein occupancy is $n(\omega)$, its wavevector is $\boldsymbol{q}$ and its envelope function is $\Phi(\boldsymbol{q}, \boldsymbol{r})$. The upper sign is taken for phonon emission, the lower for phonon absorption. The initial electron state is characterized by wavevector $\boldsymbol{k}$ and envelope function $\psi(\boldsymbol{k}, \boldsymbol{r})$. The final state is notationally distinguished by the prime. Only one-phonon processes are allowed, and energy conservation implies

$$E_f - E_i = E(\boldsymbol{k}') - E(\boldsymbol{k}) \pm \hbar\omega, \tag{4.3}$$

where $\hbar\omega$ is the energy of the phonon. The interaction strength is $C^2(\boldsymbol{q})$ which for unscreened processes is:

$$\begin{aligned} C^2(q) &= e^2 \varphi_F^2(q) \quad &&\text{(Fröhlich)},\\ &= D_0^2 &&\text{(deformation-potential, optical)},\\ &= \Xi^2 q^2 &&\text{(deformation-potential, acoustic)}, \end{aligned} \tag{4.4}$$

where φ_F is the LO scalar potential per unit optical displacement and D_0, Ξ are the optical and acoustic deformation constants. The first overlap integral is for bulk material:

$$\begin{aligned} I^2(\boldsymbol{k}', \boldsymbol{k}) &= 1 &&\text{intra-parabolic conduction valley},\\ &= \tfrac{1}{4}(1 + 3\cos^2\theta_{\boldsymbol{k}}) &&\text{intra HH, or intra LH},\\ &= \tfrac{3}{4}\sin^2\theta_{\boldsymbol{k}} &&\text{inter HH and LH}, \end{aligned} \tag{4.5}$$

where θ_k is the angle between $\boldsymbol{k}$ and $\boldsymbol{k}'$. The expressions for holes are expected to be modified by quantum confinement. In 2D the second overlap integral is zero unless in-plane crystal momentum is conserved, thus

$$G(\boldsymbol{k}', \boldsymbol{k}, \boldsymbol{q}) = \delta_{\boldsymbol{k}'_{\parallel}, \boldsymbol{k}_{\parallel} \pm \boldsymbol{q}_{\parallel}} G(k'_z, k_z, q_z), \tag{4.6}$$

where

$$G(k'_z, k_z, q_z) = \int \psi^*(k'_z, z)\Phi(q_z, z)\psi(k_z, z)\, \mathrm{d}z. \tag{4.7}$$

Here is where the envelope functions discussed in the previous sections enter the problem. Because there is confinement in the z-direction there is no automatic conservation of momentum in that direction.

4.2. Model rates

At this point accurate calculations become numerical. In order to illustrate the basic feature of this 2D scattering we will adopt simplifying expressions for the electron and phonon envelope functions. For an infinitely deep quantum well with infinitely rigid walls we can put for electrons

$$\psi(k_z, z) = (2/L)^{1/2} \sin k_z z, \quad k_z L = m\pi. \tag{4.8}$$

The rates with bulk LO modes (Riddoch and Ridley 1983) are shown in fig. 18. For guided LO modes, neglecting the mixed-in TO/IP components,

$$\Phi(q_z, z) = \cos q_z z, \quad q_z L = m\pi. \tag{4.9}$$

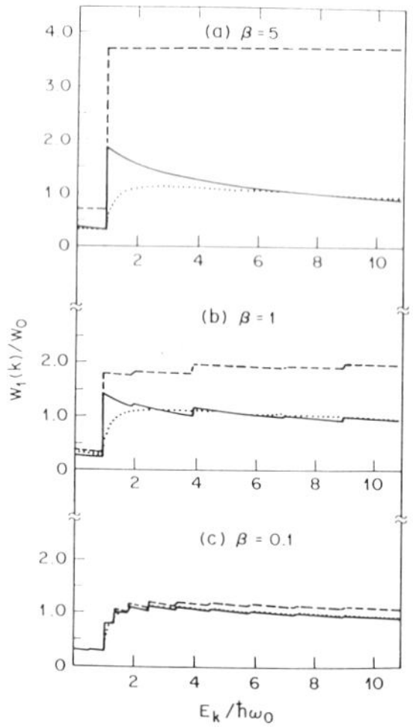

Fig. 18. Scattering rates for an electron in an infinitely deep quantum well by emission and absorption of bulk LO phonons via the Fröhlich interaction. $\beta = E_1/\hbar\omega_0$. For E_1 and W_0 see eq. (4.12d). Dashed lines: rates were momentum to be conserved. Dotted lines: bulk rates (Riddoch and Ridley 1983).

(Note that earlier work (e.g., Lassnig 1984, Wendler 1985) relying on the DC model, used $\sin q_z z$ instead of $\cos q_z z$ and obtain different selection rules.) With these envelope functions axial momentum is conserved, viz.

$$\begin{aligned} q_z &= k_z' \pm k_z, \quad k_z' \neq \text{intersubband} \\ &= 2k_z, \qquad k_z' = k_z \ \text{intrasubband} \end{aligned} \tag{4.10}$$

and $G^2(k_z', k_z, q_z) = \frac{1}{4}$ for intrasubband transitions the selection rule is $m = 2n$, and for intersubband transitions it is $m = n' \pm n$. The Fröhlich potential is given by

$$\Phi_F = -\frac{ie^*}{V_0 \varepsilon_0} \frac{1}{q_\parallel + q_z}, \tag{4.11}$$

thus the larger q_z is, the smaller the scattering rate. It follows for this interaction that intrasubband processes are weaker than scattering between adjacent subbands, and both rates diminish with decreasing well width. If interface LO modes are allowed q_z is effectively zero and the intrasubband rate turns out to be large (fig. 19).

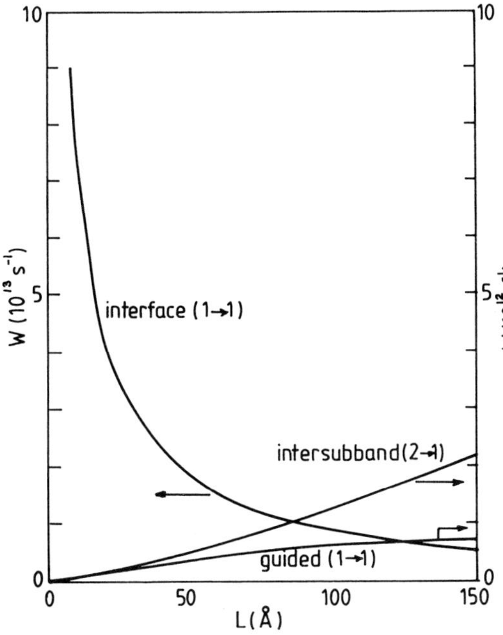

Fig. 19. Scattering rates for LO phonon emission in an infinitely deep GaAs well neglecting hybridization (Ridley 1989).

Explicit expressions are:

$$\text{Interface (if allowed) } W(1,1) = W_0\left(\frac{E_1}{\hbar\omega_L}\right), \tag{4.12a}$$

$$\text{guided modes} \begin{cases} W(1,\ 1) = \frac{1}{2}W_0\left(\frac{\hbar\omega_L}{E_1}\right)^{1/2} \dfrac{1}{4+(\hbar\omega_L/E_1)}, \\ \hbar^2 k^2/2m^* = \hbar\omega_L, \\ W(2,1) = \frac{1}{2}W_0\left(\frac{\hbar\omega_L}{E_1}\right)^{1/2}\left[\dfrac{1}{4-(\hbar\omega_L/E_1)} + \dfrac{1}{12-(\hbar\omega_L/E_1)}\right], \\ (\hbar^2 k^2/2m^*) = 0, \end{cases} \begin{matrix} \\ (4.12b) \\ \\ (4.12c) \end{matrix}$$

$$W_0 = \frac{e^2}{4\pi\hbar}\left(\frac{2m^*\omega_L}{\hbar}\right)^{1/2}\left(\frac{1}{\varepsilon_\infty}-\frac{1}{\varepsilon_s}\right), \quad E_1 = \frac{\hbar^2\pi^2}{2m^*L^2}. \tag{4.12d}$$

For GaAs, $W_0 = 7.7 \times 10^{12}\ s^{-1}$. The intersubband rate is for an electron at the bottom of subband 2 which implies a subband separation of $\hbar\omega_L$ or more. In our model the subband separation is just $3E_1$, and when the well width is such that $3E_1 = \hbar\omega_L$ the intersubband rate is maximum (fig. 20).

Apart from its neglect of mode mixing, this simple model is applicable to the AlAs/GaAs system. For a well width of 120 Å it gives $\tau_{12}\ (= W^{-1}(2,1)) = 650$ fs. A relaxation time of 600 fs was obtained by Grahn et al. (1990) in an AlAs/GaAs superlattice with 120 Å wells and 20 Å barriers, in perhaps serendipitous agreement. For comparison the intrasubband rate associated with the guided wave (eq. (4.12a)) predicts $\tau_{11} =$ 1.33 ps. This magnitude is roughly comparable with the time constant calculated by

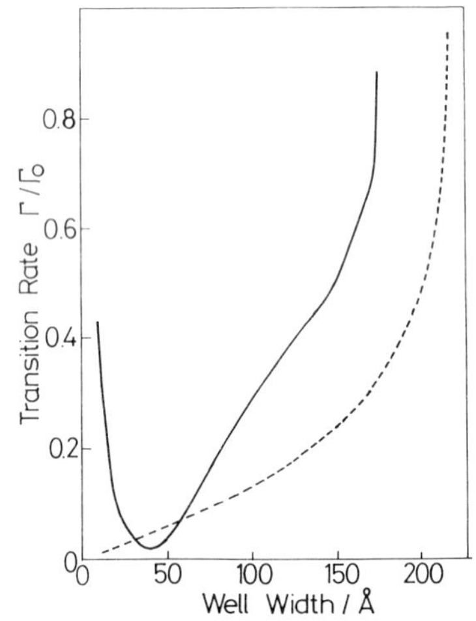

Fig. 20. Intersubband ($2 \to 1$) scattering rate for bulk LO phonon emission in a GaAs/AlGaAs superlattice (100 Å barriers) as a function of well width. Dashed line: infinitely deep well. Solid line: finite well in $GaAs/Al_{0.3}Ga_{0.7}As$ system. The resonance in wide wells occurs when $E_2 - E_1 = \hbar\omega_L$ ($\Gamma_0 \equiv W_0$) (Chamberlain and Babiker 1989).

Mori and Ando (1989) for interface polaritons in the same system and 120 Å well, although they assume a Fröhlich-like interaction rather than the correct electromagnetic $\boldsymbol{A}\cdot\boldsymbol{p}$ interaction. However, the results of Al-Dossary et al. (1991) suggest that though using the $\boldsymbol{A}\cdot\boldsymbol{p}$ interaction leads to smaller rates the difference is not too great. But both calculations assume the existence of pure IP modes which is certainly invalid for GaAs. Mori and Ando also use the DC model which leads to the intersubband rate being weaker than the intrasubband rate, in contradiction to eq. (4.12). It is clear that experiment has a role here in resolving this controversy. The effect of scattering by the AlAs/GaAs interface polaritons is shown in figs. 21 and 22. A comparison of the results of the Huang–Zhu and Babiker models was made by Rudlin and Reinecke (1990, 1991).

Since Babiker (1986) predicted the existence of pure LO interface modes it has been clear that their interaction with electrons could be strong. It was shown (Ridley 1989, Chamberlain et al. 1991) that this was indeed the case in a system such as GaAs/$Al_xGa_{1-x}As$ where x was not too large ($\leqslant 0.3$) (fig. 19). According to elasticity theory there will be mixed LO/TO/IP interface modes in general whose interaction with electrons for the GaAs/AlAs system is depicted in fig. 23. It seems safe to assume, however, that interface modes, if they are allowed, will dominate subband scattering in narrow wells. (Readers of the literature in this field should be warned that many authors use the term interface modes without qualification to mean IP modes. In general, both LO-like and IP-like interface modes are present and must be distinguished from one another.)

It should be noted that the interaction between an electron and the quantum of a hybrid (hybridon), whose strength is dipicted in fig. 23, incorporates the scattering strength of the IP component of the hybrid. Adding the contribution of the IP components of other modes increases the rates by roughly 20%.

Recently, rates for the GaAs/AlAs system have been computed by Rücker et al. (1991) based on a lattice-dynamics model which agree well with those of the HZ and

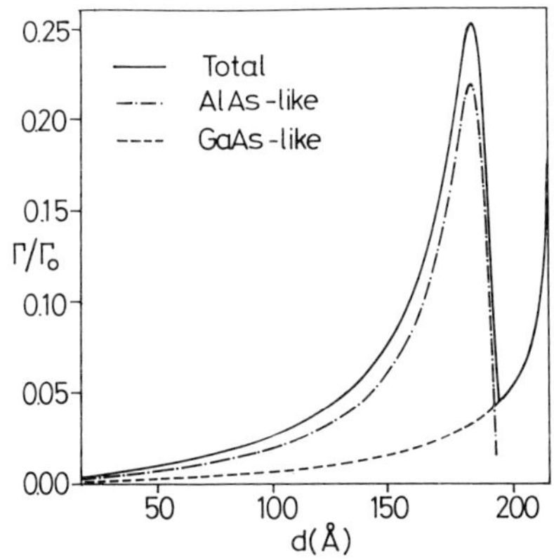

Fig. 21. Intersubband (2 → 1) scattering rate by emission of unhybridized interface polaritons via the $\boldsymbol{A}\cdot\boldsymbol{p}$ interaction in a GaAs/AlAs quantum well. Note the double resonance associated with the two frequencies ($\Gamma_0 \equiv W_0$) (Al-Dossary et al. 1991).

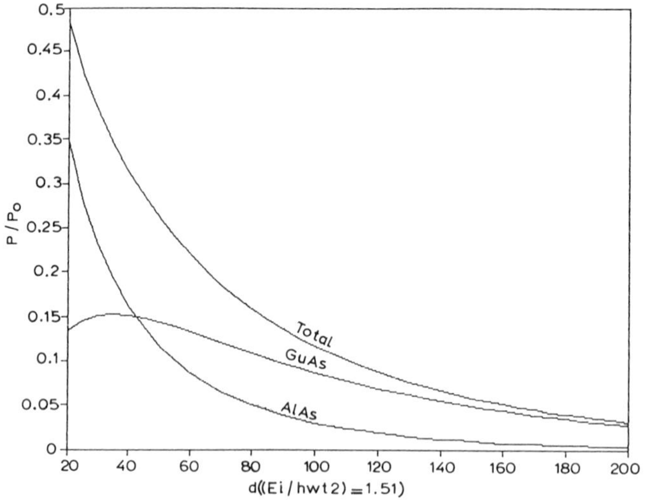

Fig. 22. Intrasubband scattering by emission of unhybridized interface polaritons via the $\boldsymbol{A} \cdot \boldsymbol{p}$ interaction in a GaAs/AlAs quantum well (Al-Dossary et al. 1991).

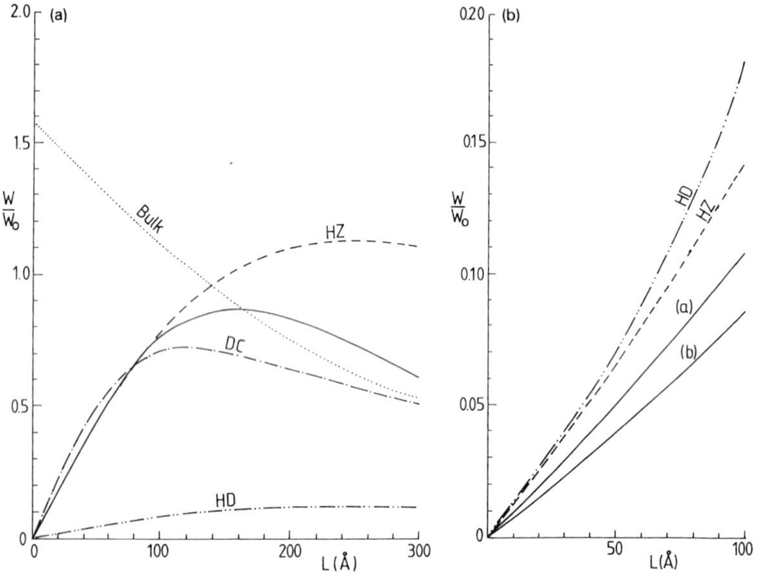

Fig. 23. Electron–hybridon scattering rates in GaAs/AlAs quantum well caused by GaAs modes. (a) Intrasubband rate. The continuous curve is that predicted by hybridon theory for an electron in the lowest subband with energy $E = \hbar\omega_L$ emitting a LO2 hybridon. Also shown are the rates predicted by bulk-phonon, DC, HD and HZ guided-mode models. (b) Intersubband rate. The continuous curves are those predicted by hybridon theory for an electron at the bottom of subband 2 scattering to subband 1 (a) $k_L L = 2.8\pi$ (b) $k_L L = 3\pi$. Also shown are the rates predicted by HD and HZ models (Ridley 1991e).

DC models, and by implication the hybrid model. It would thus appear that in the continuum theory of triple hybrids we have a viable instrument for calculating electron scattering rates in low-dimensional structures.

4.3. Well-width resonances

The condition which allows LO/TO interface modes to occur is sufficiently non-rigid interfaces, and this condition also allows leakage of the vibrational activity in the well across the interface to appear in the barrier in the form of an evanescent or propagation wave. In the GaAs/$Al_xGa_{1-x}As$ system with low x the optical-mode frequency bands overlap for frequencies at and below that of the GaAs-like LO mode in the alloy. Whenever the well width is such as to just confine a new mode in the well there is always a large evanescent component in the barrier, and this interacts strongly via the Fröhlich interaction with electrons, even those in the well because of the penetration of the wavefunction into the barrier. As a consequence, all rates are enhanced when this condition is satisfied. These well-width resonances show up particularly strongly in intersubband scattering and in well capture (Babiker et al. 1987, 1988, 1989) (figs. 24 and 25). In the latter case these resonances have the effect of keeping capture rates high as the well width diminishes, effectively countering the opposite trend predicted on the basis of purely electronic resonances, i.e., when the well width is just right to incorporate a new bound electron state (Brun and Bastard 1986, Kozyrev and Shik 1986). Well-width resonances, being only a few angstrom units wide, have not been observed to date, and it is unlikely that they will appear unless the interfaces are of exceptionally high quality. Smoothing out the resonances suggests that the capture time should be largely independent of well width and be between 1 and 2 ps in GaAs/$Al_{0.3}Ga_{0.7}As$. Experimental results scatter between 0.66 and 10 ps (Bimberg et al. 1985, Westland et al. 1988, Levine et al. 1988, Kuhn and Mahler 1989), and Devaud et al. (1988) report 0.3 ps for holes and less than 1 ps for electrons in the GaInAs/InP system. The mention of holes and the disparity of the electron/hole rates reminds us that in experiments involving both types of carrier, and even in others, space-charge effects may modify the capture process.

4.4. Hot phonons

Another factor which makes the interpretation of experiments anything but straightforward is the appearance of hot phonons, which are produced in numbers proportional to the electron volume density. Most experiments use areal electron densities in excess of 10^{11} cm^{-2}, and it is now well established that the lifetime of the LO optical phonon (about 4 ps at room temperature, rising to 8 ps at 10 K) is long enough to produce a bottle-neck in the flow of energy from the electrons to the thermal reservoir. Once the phonon occupation of the modes interacting most strongly with the electrons approaches or exceeds unity a significant amount of phonon reabsorption takes place, and this slows down all relaxation processes, which include well capture, intersubband scattering and intrasubband relaxation. Whether

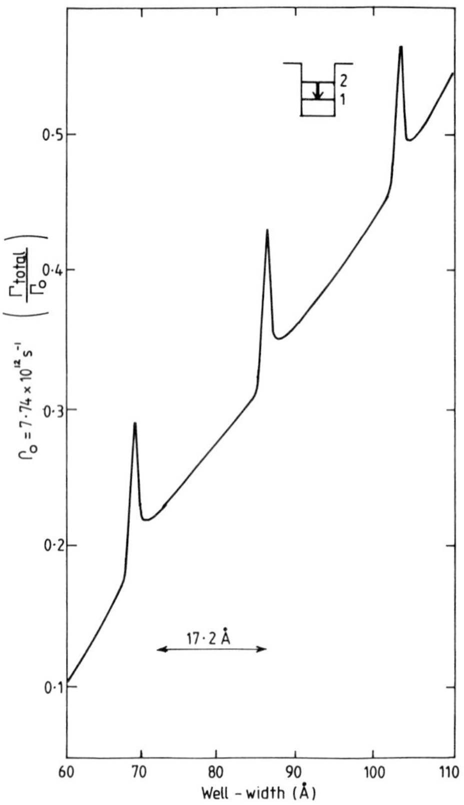

Fig. 24. Well-width resonances in intersubband scattering via emission of LO guided modes in a GaAs/$Al_{0.3}Ga_{0.7}As$ quantum well (numerical) (Babiker et al. 1987).

hot phonons produced in a transport experiment affect the momentum-relaxation time as well as the energy-relaxation time depends on how much they share the electron drift which, in turn, depends upon how rapidly they relax their own momentum. Phonon scattering thus becomes a highly relevant topic, and we will return to it in a later section. Hot phonons are discussed further in review articles by Shah (1986), Lyon (1986), Kocevar (1987), Esipov and Levinson (1987) and Ridley (1991a). Monte Carlo simulations have been carried out by Lugli et al. (1989).

4.5. *Plasmon LO coupling*

Another concomitant of high carrier densities is the plasmon and its coupling to the LO phonon. This coupling produces two modes each with mixed plasmon/phonon character and characteristic interaction with electrons. Plasmon coupling to confined modes has not been fully researched as yet, though low-dimensional plasmons have received considerable attention (e.g., Ando et al. 1982, Bechstedt and Enderlein 1985,

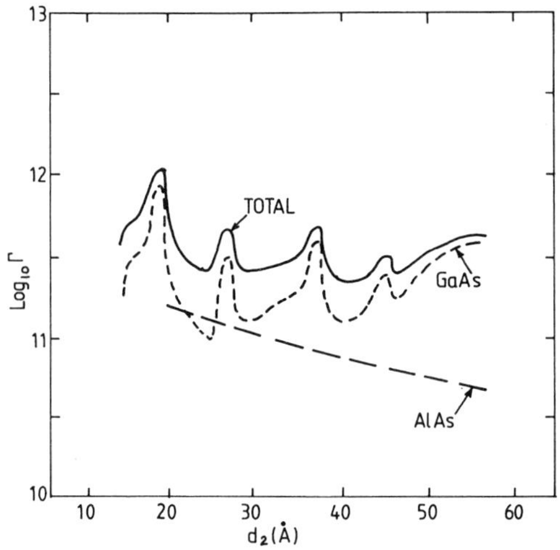

Fig. 25. Well-capture rates by emission of LO guided modes in GaAs and in the $Al_{0.3}Ga_{0.7}Al$ barriers of a GaAs/$Al_{0.3}Ga_{0.7}As$ superlattice with 100 Å barriers. The dashed curves indicated the contributions from GaAs-like modes in well and barrier and from AlAs-like modes in the barrier. (Babiker et al. 1989).

Wendler and Pechstedt 1986, 1987, Jain et al. 1988). In bulk material the effect is first to antiscreen the Fröhlich interaction, that is, to enhance it, but as the density of electrons rises ordinary screening takes over and eventually the electron–phonon interaction becomes negligible. The interaction with the plasmon-like mode always remains strong, but since energy and momentum transferred from the single particle to the collective oscillation gets rapidly returned to another single particle, the overall effect is to enhance the randomization of both dynamic quantities within the electron gas. Nevertheless this process is important for the thermalization of an electron injected into the gas at high energies (see § 5.2).

4.6. Non-polar optical-phonon interaction

In non-polar materials electrons may interact with optical modes via a deformation potential, in the way holes can in both polar and non-polar materials. The rate for emission is independent of wavevector and proportional to the density of states (fig. 26). In an infinitely deep well with unit overlap integral and bulk-like modes the intrasubband rate is

$$W_b = \frac{3D_0^2 m^*}{4\rho\hbar^2\omega_0 L}, \tag{4.13}$$

where D_0 is the deformation-potential constant, ρ is the density and ω_0 is an average LO/TO frequency (Ridley 1982). This rate increases as L diminishes. A full study of this interaction for holes in low-dimensional systems taking into account phonon confinement and mixing has yet to be presented.

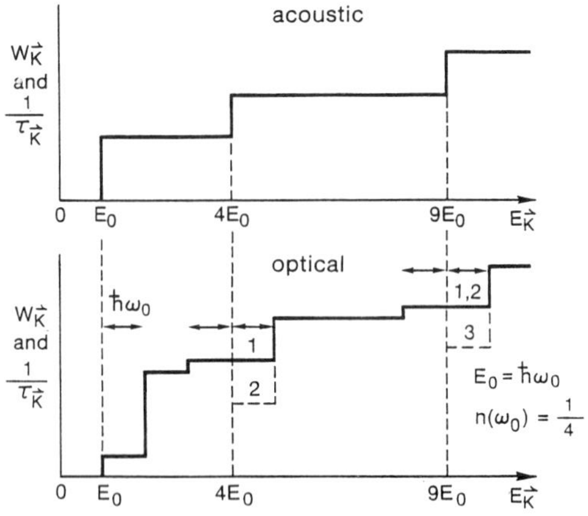

Fig. 26. Schematic scattering rates (equal to momentum-relaxation rates) for the deformation-potential interaction with bulk phonons. The rate follows the 2D density of states.

4.7. Acoustic-phonon interaction

At temperatures above 40 K in GaAs the interaction of the electron with acoustic modes in much weaker than that with LO modes. However, once electrons are made to drift faster than the velocity of the acoustic wave large acousto-electric effects can occur via the piezoelectric effect. The acoustic frequency most rapidly amplified rises with electron density, but so do the non-electronic losses. It turns out that for densities in the usual range of 10^{11} cm^{-2} and above, acousto-electric effects are likely to be inhibited (Balkan et al. 1989). The interaction with acoustic modes is thus most readily studied at temperatures below 40 K.

Treating the acoustic modes as essential bulk-like is likely to be a reasonable approximation. At temperatures high enough to ensure equipartition the rate is simply proportional to the density of states and given for electron intrasubband scattering by (Price 1981)

$$W_{AC} = \frac{3\Xi^2 m^* k_B T_L}{2\hbar^3 c_L L}, \tag{4.14}$$

where Ξ is the deformation potential, c_L is the appropriate LA elastic constant and T_L is the lattice temperature (fig. 26). For holes there is also an interaction with TA modes. In addition, both types of carrier interact with LA and TA modes via the piezoelectric interaction, though this becomes screened at high densities. Indeed, experiments are generally performed with degenerate electron populations exploiting the Shubnikov–de Haas effect to study the process of energy relaxation via the dependence of electron temperature on applied electric fields. At temperatures below 1 K equipartition fails for scattering across the Fermi surface and the Bloch–Grüneisen regime is entered in which the unscreened non-polar energy-relaxation rate reduces as $T_e^5 - T_L^5$ (T_e is electron temperature) and the momentum-relaxation rate as $(T_e^6 + T_L^6)/T_e$ (Ridley 1991a). The equivalent variations for piezoelectric scattering are $T_e^3 - T_L^3$ and $(T_e^4 + T_L^4)/T_e$ respectively. Screening increases the powers by the addition of 2 (Price 1982). Experiments report a deformation potential in the GaAs/AlGaAs system for electrons of 11 eV (Hirakawa and Sakaki 1986) and 7.5 eV (Daniels et al. 1989).

In the type of experiments covering the temperature range 2–10 K it is common to find the temperature dependence of the energy-relaxation rate quoted as a power law of the sort $T_e^n - T_L^n$, where T_e, T_L are the electron and lattice temperatures and n is an integer. The quoted range of n in the literature is between 2 and 7! Observation of exponents like $n = 7$ no doubt is associated with unwanted parallel conduction in the barrier layers, but another factor is that the temperature range spans the transition between Bloch–Grüneisen and high-temperature regimes, which is described only numerically. An approximate expression based on the theory of the 3D acoustic-phonon interaction was used successfully by Daniels et al. (1989), viz.

$$P = F(x)(C_{np} + C_p)(k_B T_e - k_B T_L), \tag{4.15}$$

where C_{np}, C_p are parameters related to the non-polar and polar interactions, viz.

$$C_{np} = \frac{3\Xi^2 m^{*2} k_F}{\pi \rho \hbar^3}, \tag{4.16}$$

$$C_p = \frac{3e^2 K_{av}^2 m^{*2} v_s}{2\pi \varepsilon_s \hbar^3 k_F} \tag{4.17}$$

and $F(x)$ is a factor which approaches unity at high electron temperatures, viz.

$$F(x) = \frac{\sinh(x_L - x_e)}{\sinh x_L \sinh x_e} \frac{\langle \hbar\omega \rangle}{2(k_B T_e - k_B T_L)}, \tag{4.18}$$

where $x_L = \langle \hbar\omega \rangle / 2k_B T_L$, $x_e = \langle \hbar\omega \rangle / 2k_b T_e$ and $\langle \hbar\omega \rangle = 2^{1/2} \hbar v_s k_F$, the latter being an average acoustic-phonon frequency. In the above, Ξ is the deformation potential, ρ is the density, v_s is an average phonon velocity, K_{av} is an average electromechanical coupling coefficient and k_F is the Fermi wavevector. Vickers (1991) has shown that eq. (4.15) is a good simulation of the correct numerical solution even in the 2D situation. Analytic expressions for the Bloch–Grüneisen regime have been given by Milsom and Butcher (1986) and by Ridley (1991a), but only numerical solutions exist for the high-temperature regime. As far as the author is aware, a full theory embodying the folding and mixing of LA and TA modes in layered material has not been given.

Between 10 K and about 40 K there exists a regime of energy loss describable neither by that due to acoustic modes nor by that expected for optical modes. Speculations have focussed on low-frequency plasma modes (Jain et al. 1988) or on "forbidden" two-phonon TA processes (e.g., Stradling and Wood 1970). The origin remains uncertain at the time of writing.

5. Other electron scattering processes

5.1. Charged-impurity scattering

The Coulombic interaction of electrons with their donor atoms (holes with their acceptor atoms) is as important in layered structures as it is in the bulk. An extra degree of freedom is present in layered structures in that the electrons and donors can be spatially separated in the technique usually known as modulation doping. Electrons thermally excited from donors grown into an AlGaAs barrier layer can be

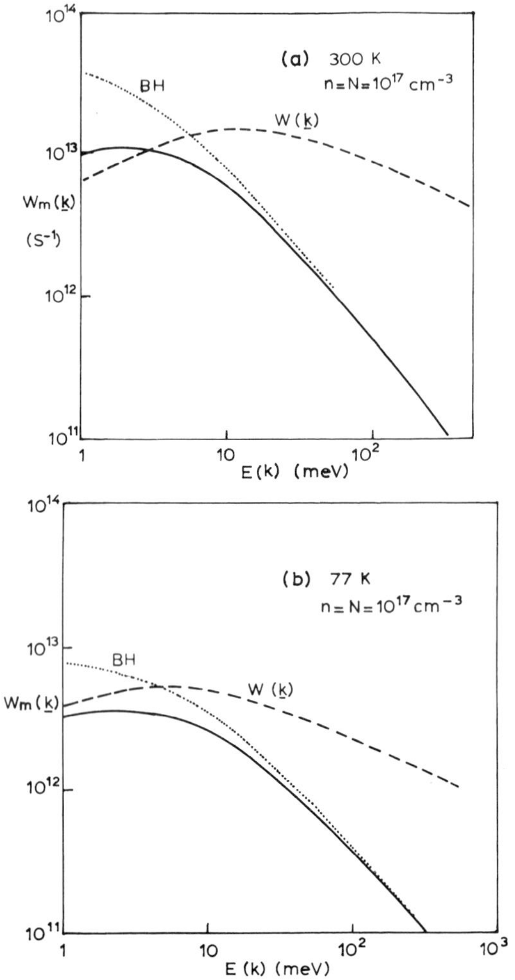

Fig. 27. Effect of statistical screening on the charged-impurity scattering $W(\boldsymbol{k})$ and momentum-relaxation $W_m(k)$ rates in bulk GaAs. Dotted line: Brooks–Herring model. Note that in the Brooks–Herring model $W(k)$ is infinite!

captured in the adjacent GaAs well and are, in consequence, only weakly scattered. This technique is responsible for the achievement of enormous electron mobilities ($\sim 10^7\ \mathrm{cm^2\ V^{-1}\ s^{-1}}$) at low temperatures, mobilities limited only by the interaction with acoustic phonons and other residual scattering mechanisms.

Calculating the scattering rate in low-dimensions cannot avoid the well-known problems associated essentially with the long-range nature of the Coulomb interaction. For local impurity scattering the problem is reduced by screening, at any rate for all but very low energy electrons, which imposes a cut-off to the interaction so that it becomes effectively a nearest-neighbour interaction. But for remote impurity scattering it is important not to overestimate the strength of the interaction. It is well known that in the bulk, adopting a random-phase approximation in the absence of screening leads to an infinite cross section, but that introducing screening via the Brooks–Herring formula leads to overestimates of the mobility in highly doped systems (Abram et al. 1978, Chattopadhay and Queisser 1981, Anderson et al. 1985). Thus there are two problems: (1) dealing with long-range effects, (2) screening.

In the present context the problem of long-range effects reduce to the question of how far remote scatterers can be considered to act independently. Answers to that question are (1) only fluctuations from the average count (Van Hall et al. 1988, Lassnig 1988, Schubert et al. 1989) and (2) the only interaction is with the nearest neighbour – so-called "statistical screening" (Ridley 1988). The latter idea has been shown to bridge the conceptual gap between the Brooks–Herring and Conwell–Weisskopf formulae (Ridley 1977). Figure 27 shows that statistical screening can reduce the scattering rate significantly.

Conventional screening takes the interaction to be static and long-range and uses a permittivity, $\varepsilon(\omega, q)$, with $\omega = 0$ and $q \to 0$, where ω is the angular frequency and q is the wavevector of the Fourier component of the potential. For local-impurity scattering at high densities the conventional screening function should be modified to include the effect of short-range encounters (Takimoto 1959, Ridley 1991b):

$$\varepsilon(0, q) = \varepsilon_s \left(1 + \frac{q_0^2}{q^2} F(q)\right)_{3\mathrm{D}} = \varepsilon_s \left(1 + \frac{q_0}{q} F(q)\right)_{2\mathrm{D}}, \tag{5.1}$$

where q_0 is the screening wavevector, and in bulk material $F(q) = 1$ for long-range interactions but $F(q) = \frac{1}{2}$ at high densities. In 2D it turns out that $F(q) = 1$ for all scattering events. In general, taking the frequency to be zero is also incorrect: it should be $\omega = \boldsymbol{v}_{\mathrm{cm}} \cdot \boldsymbol{q}$ where $\boldsymbol{v}_{\mathrm{cm}}$ is the velocity of the centre-of-mass of the two-body system (Meyer and Bartoli 1983). In general, we must use the full expression for the permittivity which in 2D systems at $T = 0$ K is given by (Stern 1967):

$$\varepsilon(\omega, q) = \varepsilon(\omega, q) + \mathrm{i}\varepsilon_2(\omega, q), \tag{5.2a}$$

$$\varepsilon_1(\omega, q) = \varepsilon_{\mathrm{L}}(\omega) + \frac{e^2 m^* n_s}{\hbar^2 k_{\mathrm{F}} q^2}\left[\frac{q}{k_{\mathrm{F}}} - c_+(a_+^2 - 1)^{1/2} - c_-(a_-^2 - 1)^{1/2}\right], \tag{5.2b}$$

$$\varepsilon_2(\omega, q) = \frac{e^2 m^* n_s}{\hbar k_{\mathrm{F}} q^2}[d_-(1 - a_-^2)^{1/2} - d_+(1 - a_+^2)^{1/2}], \tag{5.2c}$$

$$\varepsilon_{\mathrm{L}}(\omega) = \varepsilon_\infty \frac{\omega^2 - \omega_{\mathrm{L}}^2}{\omega^2 - \omega_{\mathrm{T}}^2}, \tag{5.2d}$$

$$c_{\pm} = \operatorname{sgn} a_{\pm} \quad \text{and} \quad d_{\pm} = 0 \quad \text{for} \quad |a_{\pm}| > 1, \tag{5.2e}$$

$$c_{\pm} = 0 \quad \text{and} \quad d_{\pm} = 1 \quad \text{for} \quad |a_{\pm}| < 1, \tag{5.2f}$$

$$a_{\pm} = \frac{q}{2k_{\mathrm{F}}} \pm \frac{\omega}{v_{\mathrm{F}} q}.$$

Here a subscript F refers to Fermi-level quantities. For static impurities (infinite mass) the centre-of-mass velocity is zero, and, further, wavevector changes cannot be greater than $2k_{\mathrm{F}}$. Thus $a_{\pm} < 1$, $\varepsilon_{\mathrm{L}}(\omega) = \varepsilon_{\mathrm{L}}(0) = \varepsilon_{\mathrm{s}}$, and $\varepsilon_2 = 0$, whence

$$\varepsilon_1(0, q) = \varepsilon_{\mathrm{s}}\left(1 + \frac{q_0}{q}\right), \quad q_0 = \frac{e^2 m^*}{2\pi\hbar^2 \varepsilon_{\mathrm{s}}}. \tag{5.3}$$

In 2D, therefore, $F(q) = 1$ for all scattering events.

Following Lee et al. (1983), we assume a ground-state wavefunction for the electron to be a delta function and obtain with statistical screening the momentum relaxation rates:

$$W_{\text{local}} = \frac{1}{2} W_0 \int_0^{\pi} \frac{\sin\alpha[1 - \exp(-2kL\sin\alpha)]\exp(-NaL|b|)}{[\sin\alpha + (q_0/2k)]^2}\,\mathrm{d}\alpha, \tag{5.4}$$

$$\begin{aligned} W_{\text{remote}} = {} & W_0 k \int_0^{\pi} \int_0^{z_{\max}} \sin\alpha \\ & \times \exp\left(-4k\sin\alpha + \left\{Na(b^2 + z^2)\left[\cos^{-1}\left(\frac{z_0}{(b^2 + z)^{1/2}}\right) - z_0 \frac{(b^2 + z^2 - z_0^2)^{1/2}}{b^2 + z^2}\right]\right\}\right) \\ & \times [\sin\alpha + (q_0/2k)]^{-2}\,\mathrm{d}z\,\mathrm{d}\alpha, \end{aligned} \tag{5.5}$$

where N is the density of impurities, a is the average spacing and b is the impact parameter given for local- and remote-impurity scattering by

$$b_{\text{local}} = -\frac{\pi R^2}{4L} \int_{\pi}^{\theta} \frac{1 - \exp[-2kL\sin(\theta/2)]}{\sin(\theta/2)[\sin(\theta/2) + (q_0/2k)]^2}\,\mathrm{d}\theta, \tag{5.6}$$

$$b_{\text{remote}} = -\frac{\pi R^2 k}{2} \int_{\pi}^{\theta} \frac{\exp[-4kz\sin(\theta/2)]}{[\sin(\theta/2) + (q_0/2k)]^2}\,\mathrm{d}\theta, \tag{5.7}$$

with

$$W_0 = 2\pi R^2 v N, \quad R = \frac{Ze^2}{4\pi\varepsilon_{\mathrm{s}} m^* v^2}. \tag{5.8}$$

For remote-impurity scattering the impurities are assumed to be uniformly spread between z_0 and $z_{\max}$ (2D gas at $z = 0$). The wavevector $\boldsymbol{k}$ is that of the electron in the 2D layer and v is the velocity. These expressions give the rates of momentum relaxation when the electron interacts only with the nearest scattering centre, its interaction with all others being time-averaged to zero (fig. 28).

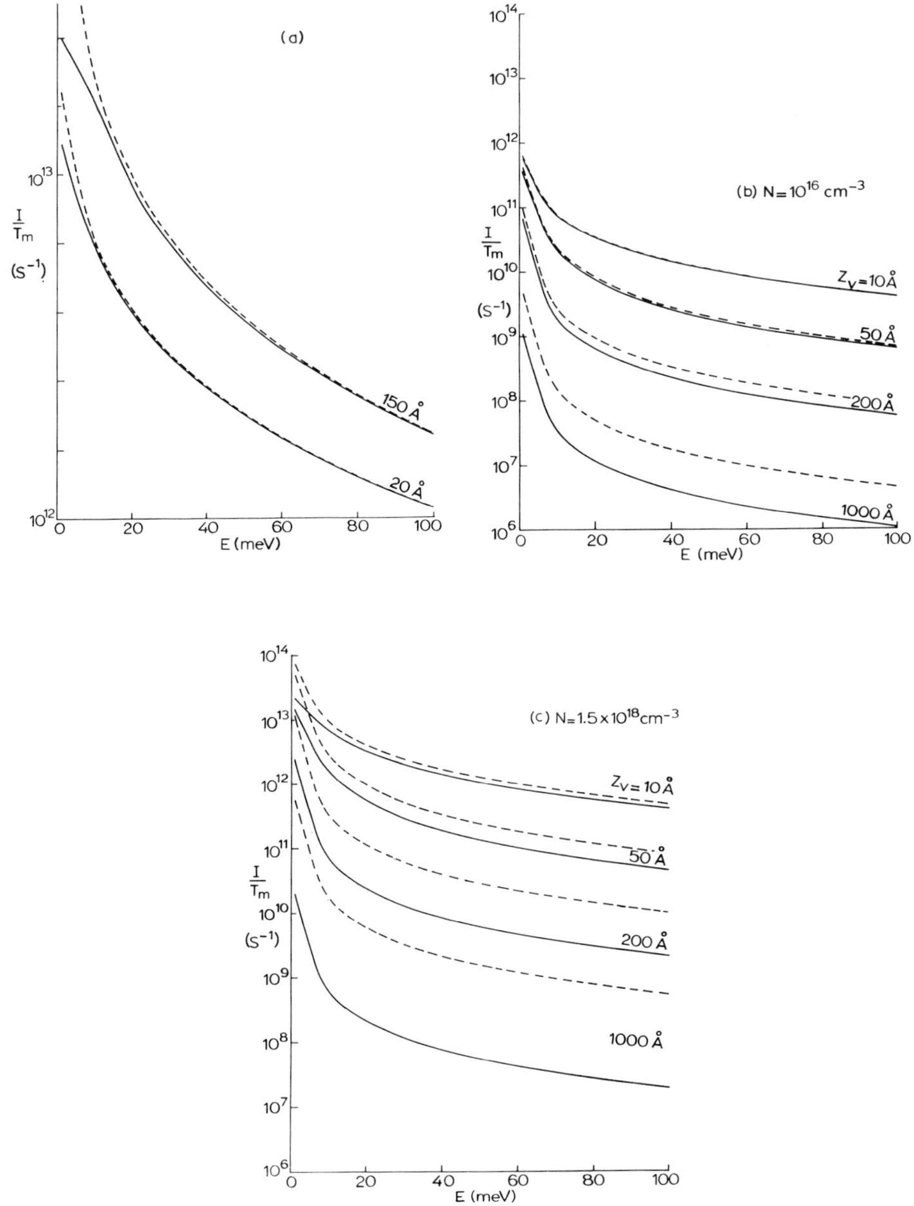

Fig. 28. Momentum-relaxation rate for statistically screened charged-impurity scattering in a GaAs quantum well. Solid curves: numerical. Dashed curves: without statistical screening. (a) Local impurities: $N = 10^{18}$ cm^{-3}, well widths 20 and 150 Å. Remote impurities: (b) $N = 10^{16}$ cm^{-3} (c) $N = 1.5 \times 10^{18}$ cm^{-3}. The space layer widths range from 10 to 1000 Å, and doping out to infinity (Ridley 1988).

5.2. Carrier–carrier scattering

In quantum-well lasers and in experiments involving photoexcited carriers the diffusion and drift of electrons and holes will be affected by their mutual Coulombic interaction. All the considerations of the previous section apply here, but two main differences stand out. One is that the calculation has to be carried out in a moving relative coordinate system. This means that the frequency dependence of screening has to be taken into account. The other is that not only is momentum exchanged but also energy. In bulk material the disparity of masses which exist between electrons and holes means that one can often ignore energy relaxation, but this is not the case in confined systems where the in-plane mass of holes can be very different from the bulk value. In view of these complications it is not surprising that little headway has been made to date on this topic. Even in bulk systems ignoring energy relaxation and refinements like statistical screening the expressions for momentum relaxation as derived by Meyer and Bartoli (1983) are too complicated for a useful analytic formula to be extracted. Discussions of energy relaxation resort to the expression for the 3D rate derived from plasma physics (Dreicer 1960), viz.

$$\frac{\mathrm{d}E_1}{\mathrm{d}t} = -\frac{e^4 n_2}{3.2^{1/2}\pi^{3/2}\varepsilon_s^2 m_1^* m_2^*}\frac{k_B(T_1 - T_2)}{(k_B T/m_1^*)^{3/2} + (k_B T_2/m_2^*)^{3/2}} \ln\left(\frac{3}{2}\frac{(4\pi\varepsilon_s k_B T_1)^{3/2}}{e^3(\pi n)^{1/2}}\right), \tag{5.9}$$

where subscripts 1, 2 denote the two populations and n is the screening density. Since holes cool more rapidly than electrons they provide a significant channel in which electrons can relax energy.

As in the case of electron–hole scattering discussions of electron–electron scattering resort to 3D formulations in which the frequency dependence of screening is ignored. Usually the interest is in the relaxation of a fast injected electron in which case the rate for relaxing momentum or energy is given by (Pines 1953)

$$W_1 = \frac{e^4 n_2}{8\pi\varepsilon_s^2 (2\bar{m}^*)^{1/2} E_{12}^{3/2}} L(E_{12}), \tag{5.10}$$

where n is the electron density, $\bar{m}^*$ is the reduced mass, E_{12} is given by

$$E_{12} = \frac{\hbar^2(\boldsymbol{k}_1 - \boldsymbol{k}_{cm})^2}{2m_1^*} + \frac{\hbar^2(\boldsymbol{k}_2 - \boldsymbol{k}_{cm})^2}{2m_2^*}, \tag{5.11}$$

$\boldsymbol{k}_{cm}$ is the wavevector of the centre-of-mass and $L(E_{12})$ is a logarithmic screening function (e.g., Ridley 1988b), and for a fast electron $E_{12} \sim \frac{1}{2}E_1$.

These expressions, applicable to bulk material, are no more than a guide to the situation in low dimensions. The complexity introduced by the quantum confinement of electrons is most effectively dealt with, though inevitably with a loss of generality, by Monte Carlo techniques (Goodnick and Lugli 1988, Ferry et al. 1988, Stanton et al. 1988). Screening, for example, is generally time-dependent in situations where hot carriers are injected, and scattering times range from 10 to 100 fs at densities around $5 \times 10^{11}\ \mathrm{cm}^{-2}$, with typically 12 meV of energy exchanged in each electron–

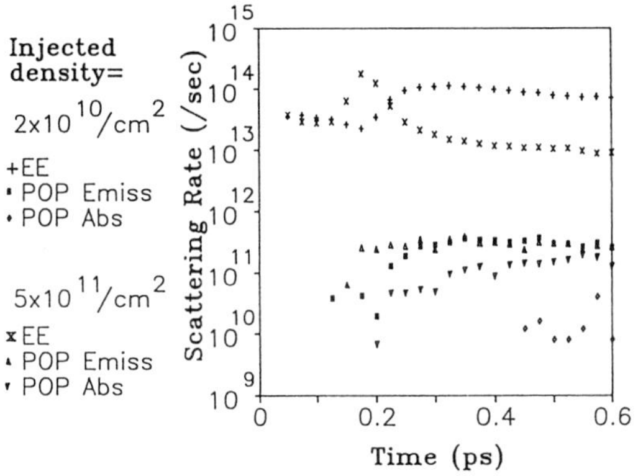

Fig. 29. Scattering rates during laser excitation at $t = 0.2$ ps. Upper pair of curves: electron–electron scattering, (+) $2 \times 10^{10}\ \mathrm{cm}^{-2}$, $\times$ $5 \times 10^{11}\ \mathrm{cm}^{-2}$. Lower curve: bulk LO phonon rates: (□, ◇) emission and absorption at $2 \times 10^{10}\ \mathrm{cm}^{-2}$; and (△, ▽) at $5 \times 10^{11}\ \mathrm{cm}^{-2}$ (Monte Carlo simulation by Goodnick and Lugli (1988)).

electron collision (fig. 29). Maxwellian distributions establish themselves in times of about 200 fs under these sircumstances.

Another interaction which randomizes momentum and energy within the electron gas is that between a single electron and a plasmon. A long-wave 3D-like plasmon in a superlattice has a frequency which makes the permittivity

$$\varepsilon(\omega) = \varepsilon_\infty \left(1 - \frac{\omega_p^2}{\omega^2}\right), \quad \omega_p^2 = \frac{ne^2}{\varepsilon_\infty m^*} \tag{5.12}$$

vanish. Here ω_p is the plasmon frequency and n is the volume density of electrons. The process is identical to the Fröhlich interaction (if plasmon damping is neglected) except that $\omega_L \to \omega_p$ and the static permittivity in the Fröhlich expression is taken as infinite, i.e.,

$$W_0 = \frac{e^2}{4\pi\hbar\varepsilon_0}\left(\frac{2m^*\omega_p}{\hbar}\right)^{1/2}. \tag{5.13}$$

In GaAs $\omega_p \approx \omega_L$ when $n = 5 \times 10^{17}\ \mathrm{cm}^{-2}$, whence $W_0 \approx 5 \times 10^{13}\ \mathrm{s}^{-1}$, suggesting collision times of the order of 20–100 fs, which translates to mean free paths for an electron with a velocity of $10^8\ \mathrm{cm\ s}^{-1}$ of 200–1000 Å. This compares with a figure of 650 Å for $n = 10^{17}\ \mathrm{cm}^{-3}$ calculated by Rorison and Herbert (1986) for combined single-particle (energy = 300 meV) and plasmon energy relaxation in bulk GaAs, and is consistent with the findings by Heiblum et al. (1985, 1986) and Levi et al. (1985) from experiments on ballistic electrons in hot-electron transistors.

5.3. *Interface-roughness scattering*

Although enormous mobilities ($10^7\ \mathrm{cm}^2\ \mathrm{V}^7\ \mathrm{s}^{-1}$) are reported for electrons in modulation-doped single heterojunctions of GaAs/$Al_xGa_{1-x}As$ ($x \leqslant 0.3$) at low temperatures, only modest mobilities appear to be achievable for electrons in quantum wells. This is usually attributed to the poor quality of the interface formed by growing GaAs on AlGaAs. Some degree of interface roughness is very difficult to avoid and its effect on scattering rate may well be the limiting factor in narrow wells.

The influence of interface roughness (IFR) on the energy of a 2D electron at a single heterojunction is difficult to work out because of its effect on the potential profile of the quantum well both directly and through the space-charge variations it induces (see, e.g., Ando et al. 1982). Here we concern ourselves with the simpler case of double-heterojunction quantum well, following initially the discussion of Sakaki et al. (1987).

A change $\Delta(\boldsymbol{r})$ in the well width ($\boldsymbol{r}$ is a position vector in the plane) produces a change in energy

$$\delta E(\boldsymbol{r}) = \frac{\mathrm{d}E}{\mathrm{d}L}\Delta(\boldsymbol{r}). \tag{5.14}$$

In the case of an infinitely deep well,

$$E = \frac{n\pi^2\hbar^2}{2m^*L^2}, \quad n \text{ integer}, \tag{5.15}$$

so the gradient is expressible analytically. Strictly, we need to know the exact spatial variation of $\Delta(\boldsymbol{r})$ in order to describe the perturbation on the electron's energy. Thus, if we can expand $\Delta(\boldsymbol{r})$ in a Fourier expansion the matrix element in the Born approximation is

$$|\langle \boldsymbol{k}'|\delta E(\boldsymbol{r})|\boldsymbol{k}\rangle|^2 = \left(\frac{\mathrm{d}E}{\mathrm{d}L}\right)^2 |\Delta(\boldsymbol{q})|^2, \quad \boldsymbol{q} = \boldsymbol{k}' - \boldsymbol{k}, \tag{5.16}$$

where $\boldsymbol{k}, \boldsymbol{k}'$ are the in-plane wavevectors of the electron, and we assume that the transition is intrasubband. Only the "power spectrum" of $\Delta(\boldsymbol{r})$ is required and it is usual to assume that the autocovariance function is isotropic and Gaussian:

$$\langle \Delta(\boldsymbol{r}')\Delta(\boldsymbol{r}' - \boldsymbol{r})\rangle = \varDelta^2\,\mathrm{e}^{-r^2/\Lambda^2}. \tag{5.17}$$

(But note that experimental evidence for Si/SiO_2 interfaces point to an exponential autocovariance function (Goodnick et al. 1985).) Thus $\varDelta$ is an average jog in the interface and λ, the correlation length, is a length of order of the range of spatial variation of the jog along the surface. It follows that

$$|\Delta(\boldsymbol{q})|^2 = \pi\varDelta^2\Lambda^2\,\mathrm{e}^{-q^2\Lambda^2/4}. \tag{5.18}$$

Including 2D screening the momentum-relaxation rate is then

$$W_{\mathrm{IFR}} = \frac{4(\mathrm{d}E/\mathrm{d}L)^2\varDelta^2\Lambda^2 m^*}{\hbar^3}\int_0^1 \frac{x^4\,\mathrm{e}^{-k^2\Lambda^2x^2}}{[x + (q_0/2k)]^2(1 - x^2)^{1/2}}\,\mathrm{d}x. \tag{5.19}$$

where q_0 is the usual screening vector. The rate tends to zero for both small and large electron wavevectors and exhibits a maximum determined by q_0 and λ (fig. 30). In an infinitely deep well the rate varies as L^{-6}.

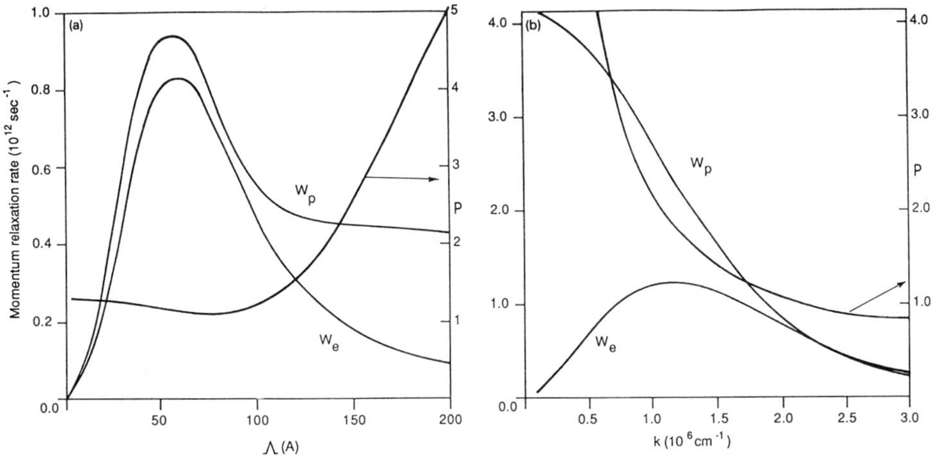

Fig. 30. Interface-roughness scattering against momentum-relaxation rates in a 50 Å GaAs infinitely deep well. W_e, W_p are electron and interface LO phonon rates, respectively, and $P = W_p/W_e$. (a) As a function of correlation length Λ with $\Delta = 1$ monolayer (ML) and $k = 2 \times 10^6\ \mathrm{cm}^{-1}$. (b) As a function of wavevector with $\Delta = 1$ ML and $\Lambda = 60$ Å (Gupta and Ridley 1991b).

Sakaki et al. (1987) found that for an GaAs-on-AlAs interface Δ was 3–5 Å and λ was 50–70 Å, and that the mobility variation with well width obeyed the L^{-6} law. X-ray analysis of the GaAs-to-AlAs interface shows interface widths to be about 4 monolayers (1 monolayer = 2.83 Å in GaAs) (Fewster 1988). This suggests Δ is about two monolayers and it is thought that $\Delta \approx 1$ ML represents a realistic limit. Using $\Delta = 1$ ML and $\lambda = 35$ Å for the GaAs-on-AlGaAs interface Dharssi and Butcher (1990) confirm that it is this interface which is mainly responsible for scattering electrons in wells of finite depth (250 meV) and width 30 Å.

5.4. *Alloy scattering*

Fluctuations around the average potential in an alloy $A_xB_{1-x}C$ defined by the virtual-crystal approximation causes scattering. We write the perturbing potential as follows:

$$V(\boldsymbol{r}) = \sum_{\boldsymbol{q}} V(\boldsymbol{q})\, e^{i\boldsymbol{q}\cdot\boldsymbol{r}}. \tag{5.20}$$

If V_A, V_B are potentials associated with cations A and B the virtual-crystal average is defined as

$$\bar{V}_0 = V_A x + V_B(1-x). \tag{5.21}$$

Where x varies, say to x', the deviation is

$$(V' - \bar{V}_0) = (V_A - V_B)(x' - x). \tag{5.22}$$

The rms deviation is that for a binomial distribution, hence,

$$|\langle V' - V_0 \rangle| = (V_A - V_B)\left[\frac{x(1-x)}{N_c}\right]^{1/2}, \tag{5.23}$$

where N_c is the number of cation sites.

The matrix element is then

$$\langle \boldsymbol{k}|H|\boldsymbol{k}\rangle = |V_A - V_B|\left[\frac{x(1-x)}{N_c}\right]^{1/2} \sum_{\boldsymbol{q}} \delta_{\boldsymbol{k}\pm\boldsymbol{q}-\boldsymbol{k}',0}\, G(k', q_z, k_z), \tag{5.24}$$

where, for scattering in an infinitely deep well,

$$G(k_z', q_z, k_z) = \frac{2}{L}\int_0^L \sin k_z' z\, e^{iq_z z} \sin k_z z\, dz. \tag{5.25}$$

The scattering rate between subbands n, m, in this case equal to the momentum-relaxation rate, is then

$$W_{n,m} = \frac{2\pi}{\hbar}(V_A - V_B)^2 x(1-x)V_0(1 + \tfrac{1}{2}\delta_{n,m})N(E), \tag{5.26}$$

where V_0 is the unit-cell volume and $N(E)$ is the subband density-of-states function for a given spin ($= m^*/2\pi\hbar^2 L$). (For a discussion of the overlap integral in eq. (5.25) see Price (1981) and Ridley (1982).)

In alloys the quantities V_A and V_B are not well known. The difference is taken to be the difference of pseudo-potentials, or of band-edge energies in the related binaries, or of electron affinities. A value of 0.5 eV has been used for GaInAs and 1.0 eV for AlInAs, but these values have to be regarded as stop-gaps in lieu of anything more firmly grounded.

Note that eq. (5.26) can also be used to estimate neutral-impurity scattering.

5.5. Mobilities

The mobility of electrons in low-dimensional structures arising from the above scattering mechanisms has been reviewed recently by Harris et al. (1989). Typical electron mobilities at toom temperature in GaAs heterostructures are a few thousand $cm^{-2}\,V^{-1}\,s^{-1}$ but can rise to $10^7\,cm^2\,V^{-1}\,s^{-1}$ at 1.5 K in modulation-doped structures (Foxon et al. 1989), but generally only to $10^5\,cm^{-2}\,V^{-1}\,s^{-1}$ in quantum wells.

6. Phonon-scattering

6.1. Phonon–phonon processes

As mentioned in § 4.4 the strength of hot-phonon effects is determined by phonon lifetime and scattering rate. These quantities for acoustic modes have received a good deal of attention in the past, principally because the associated processes are much more accessible to experiment than those for optical phonons. Although the area needs revisiting in the light of the zone-folding induced by superlattice structures we will limit our attention to optical phonons since much less attention has been given to scattering processes other than those involving electrons in which optical phonons are involved. What follows is based on the recent work of Ridley and Gupta (1991) and Gupta and Ridley (1991).

The lifetime of an optical phonon is determined by the anharmonicity of the lattice. The Hamiltonian for the lattice can be written to lowest order as follows:

$$H = \sum_{\boldsymbol{r},i,j} M_i^{1/2} \omega_i M_j^{1/2} \omega_j \boldsymbol{u}_i \cdot \boldsymbol{u}_j, \tag{6.1}$$

where $\boldsymbol{u}_i$ denotes the optical (or acoustic) displacement for the mode of vibration i and M_i is the corresponding mass of the oscillator. The sum is over all lattice sites and $\boldsymbol{u}_i$ can be written in second-quantization notation thus:

$$\boldsymbol{u}_i = \sum_{\boldsymbol{q}} \left(\frac{\hbar}{2NM_i\omega_i} \right)^{1/2} (\boldsymbol{e}_i a_{qi}\, \mathrm{e}^{\mathrm{i}\boldsymbol{q}\cdot\boldsymbol{r}} + \boldsymbol{e}_i^* a_{qi}^+\, \mathrm{e}^{-\mathrm{i}\boldsymbol{q}\cdot\boldsymbol{r}}), \tag{6.2}$$

where N is the number of unit cells in the cavity, $\boldsymbol{e}$ is a unit polarization vector, and a, a^+ are the usual annihilation and creative operators. A small perturbation can be regarded as changing the frequencies by an amount $\delta\omega_i$ so that

$$H = H_0 + H', \tag{6.3}$$

$$H_0 = \sum_{\boldsymbol{q}i} \hbar\omega_{\boldsymbol{q}i}(a_{\boldsymbol{q}i}^+ a_{\boldsymbol{q}i} + \tfrac{1}{2}), \tag{6.4}$$

$$H' = \sum_{\boldsymbol{r},i,j} M_i^{1/2} \omega_i M_j^{1/2} \omega_j \left(\frac{\delta\omega_i}{\omega_i} + \frac{\delta\omega_j}{\omega_j} \right) \boldsymbol{u}_i \cdot \boldsymbol{u}_j. \tag{6.5}$$

This form of the perturbation will be applied to a number of interactions in what follows.

The lifetime of the optical phonon can be calculated by taking the phonon to cause a frequency change in all other modes by virtue of the intrinsic anharmonicity of the lattice, viz.

$$\frac{\delta\omega}{\omega} = \boldsymbol{\Gamma} \cdot \boldsymbol{u}, \tag{6.6}$$

where we have introduced the quantity Γ, which is the optical-phonon analogue of the Grüneisen constant. For simplicity we assume an isotropic medium which allows us to replace $\boldsymbol{\Gamma} \cdot \boldsymbol{u}$ by Γu. Equation (6.5) becomes

$$H' = 2\Gamma \sum_{\boldsymbol{r},i,j} M\omega_i\omega_j \boldsymbol{u}_i \cdot \boldsymbol{u}_j u_k, \tag{6.7}$$

where we have $M_i = M_j = M$ in anticipation that modes i, j are acoustic modes and we use the subscript k to denote the optical mode. Energy and momentum conservation is satisfied when a long-wavelength LO phonon decays to two almost oppositely propagating LA modes such that $\omega_i = \omega_j = \omega_L/2$. The rate from first-order perturbation theory, with $M_k = M/2$, is

$$W = \frac{\Gamma^2 \hbar \omega_L^3 [2n(\omega_L/2) + 1]}{16\pi\rho v_L^3}, \tag{6.8}$$

which is the expression derived by Klemens (1966) if we take $\Gamma = \gamma\omega_L/v_L$, where γ is the Grüneisen constant and v_L is the velocity of LA modes, and ignore a factor $\frac{2}{3}$. The quantity $n(\omega_L/2)$ is the Bose–Einstein factor. For the case of GaAs experiment gives us a lifetime of about 8 ps at 10 K (Tsen et al. 1988, 1989) which means that Γ (otherwise to be calculated in an a priori theory) for GaAs is $1.7 \times 10^8\ \mathrm{cm}^{-1}$. This value leads to agreement with experiment at other temperatures (4 ps at 300 K, von der Linde et al. 1980; 7 ps at 77 K, Kash et al. 1985).

Higher-order phonon–phonon processes turn out to be much weaker than the three-phonon interaction which ends the life of an LO phonon. Thus, the lifetime is also the momentum-relaxation time for phonon–phonon processes. A calculation taking into account phonon confinement and mode mixing (which would affect the acoustic modes involved also) has not been carried out, but it is unlikely that this general conclusion will prove incorrect for layered material. Hence it is unlikely, if phonon–phonon processes are the only ones active, that hot phonons relax electron momentum appreciably.

An excellent review of phonon properties in semiconductors as observed in experiments on light scattering has been written recently by Tsang and Kash (1990).

6.2. Charged-impurity scattering

The Fröhlich interaction allows a novel interaction to take place between charged-impurity atoms and LO modes. A charged impurity polarizes the medium surrounding it, so that its effective charge is

$$e_{\mathrm{eff}} = \frac{Ze}{\kappa_s}, \tag{6.9}$$

where Z is the charge number and κ_s is the static dielectric constant. The LO mode interacts Coulombically with this charge and anharmonically with the surrounding polarization. Assuming that the internal dynamics of the impurity remain unaffected and that the impurity absorbs crystal momentum, we can calculate the scattering rate for the Fröhlich interaction in second order, i.e.

$$W = \frac{2\pi}{\hbar} \int \sum_n \left| \frac{\langle f | e_{\mathrm{eff}} \varphi | n \rangle \langle n | e_{\mathrm{eff}} \varphi | i \rangle}{E_i - E_n} \right|^2 \delta(E_f - E_i)\, dN_F \tag{6.10}$$

in usual notation. The result is

$$W = \frac{Z^4 e^4 (n(\omega_L) + 1)\omega_L N_I}{4\pi\kappa_s^4 \hbar^2 v_L^2 q^3} \left(\frac{1}{\varepsilon_\infty} - \frac{1}{\varepsilon_s} \right)^2. \tag{6.11}$$

We have assumed a dispersion relation for the LO mode to be of the quadratic form of eq. (3.2), which is why v_L, the velocity closely associated with the velocity of the LA mode, enters the above formula. The smaller v_L is the weaker the dispersion and the greater the number of available final states which fall within the energy uncertainly. For GaAs with $Z=1$ and $q=2\times10^6\ \mathrm{cm}^{-1}$, which is a typical wavevector involved in the electron–phonon interaction, the rate becomes $1.1\times10^7 N_I\ \mathrm{s}^{-1}$ with N_I in cm^{-3}. It matches the lifetime when $N_I=10^{18}\ \mathrm{cm}^{-3}$. As is usual for the Fröhlich interaction the rate is stronger for smaller wavevectors. Typical wavevectors involved in Raman scattering are $10^5\ \mathrm{cm}^{-1}$ or less, increasing the rate by a factor of 10^3 or more! Well-defined long-wavelength LO modes cannot exist in these circumstances.

The anharmonic interaction with the surrounding polarization actually leads to an infinite cross section as a consequence of the purely Coulombic nature of the polarization field. However, weighting for momentum relaxation reduces the rate far below that for the direct Fröhlich one.

Both interactions will be screened if free carriers are available, but the effect of screening has not been calculated. Once again, only bulk interactions have been considered to date.

6.3. *Alloy fluctuations and neutral impurities*

A foreign atom alters the phonon frequency by virtue of its different mass and different bond strength (different force constant). The effect is well known in the form of isotope scattering of acoustic waves. In the Born approximation eq. (6.5) with eq. (6.2) leads to

$$H'_{ij}=\frac{1}{N}\sum_{\boldsymbol{r}}\hbar\delta\omega\,\boldsymbol{e}_i\cdot\boldsymbol{e}_j\,\mathrm{e}^{\mathrm{i}(\boldsymbol{q}_i-\boldsymbol{q}_j)\cdot\boldsymbol{r}}a_ia_j^+, \tag{6.12}$$

which describes phonon i being annihilated and phonon j being created, with $\boldsymbol{r}$ such that zero of spatial coordinates coincides with the atom. When both modes have the same polarization $\boldsymbol{e}_i\cdot\boldsymbol{e}_j=\cos\theta$, where θ is the angle between $\boldsymbol{q}_i$ and $\boldsymbol{q}_j$.

The material will be regarded as an alloy of form $A_xB_{1-x}C$, and we can use the arguments of § 5.4 to write

$$\langle\delta\omega\rangle^2=(\omega_A-\omega_B)^2x(1-x). \tag{6.13}$$

where ω_A is the LO frequency for a particular mode when x approaches zero, ω_B for the same mode when x approaches unity in a real alloy, or for neutral impurities including isotopes

$$\omega_A-\omega_B\rightarrow\frac{\Delta M}{2M}\omega_L, \tag{6.14}$$

where ΔM is the mass difference. (Strictly another term $(\Delta F/2F)\omega_L$ should be added to denote the effect of differing force constants, but the contribution is usually small.) The rate in 3D is

$$W=\frac{(\omega_A-\omega_B)^2\omega_L qV_0x(1-x)(n(\omega_L)+1)}{3\pi v_L^2} \tag{6.15}$$

Table 1
Phonon momentum relaxation time associated with alloy scattering.

Material	Phonon mode	ω_L (meV)	$\Delta\omega_{AB}$ (meV)	τ(p)	
				3D[d]	2D (50 Å)
$Al_{0.3}Ga_{0.7}As$[a]	GaAs	34.8	5.27	30	8.0
	AlAs	47.4	5.54	20	5.3
$Al_{0.53}In_{0.47}As$[b]	InAs	29.4	0	∞	∞
	AlAs	45.5	8.7	6.6	1.8
$Ga_{0.52}In_{0.48}As$[c]	InAs	28.9	1.24	484	141
	GaAs	33.8	5.47	21	7.0

[a]Kim and Spitzer (1979).
[b]Emura et al. (1987).
[c]Lucovsky and Chen (1970).
[d]$q = 2.5 \times 10^6\ \mathrm{cm}^{-1}$.

and in 2D

$$W = \frac{(\omega_A - \omega_B)^2 \omega_L V_0 x(1-x)(n(\omega_L)+1)}{2v_L^2 L}. \tag{6.16}$$

Here V_0 is the volume of the unit cell. Rates for realistic impurity concentrations are negligible. Isotopes exist in relatively high concentrations but $\delta\omega$ is rather small and in the present context the isotopic scattering rate is usually negligible. In alloys the rate depends on the mode which can be AC-like or BC-like. Table 1 list some magnitudes. The InAs mode in AlInAs has the same frequency for all x and so does not suffer scattering by alloy fluctuations. Note that the 2D rate is q-independent and larger than the 3D rate by a factor $3\pi/(2qL)$ and many of the time-constants are of order of the GaAs phonon lifetime.

6.4. *Interface-roughness scattering*

Following the spirit of the approach in § 5.3 for guided phonon modes we obtain a frequency shift given by

$$\delta\omega = \frac{d\omega}{dq_z}\frac{dq_z}{dL}\Delta = \frac{n^2 v_L^2 \pi^2}{\omega_L L^3}\Delta, \tag{6.17}$$

with $q_z = n\pi/L$. Choosing a Gaussian autocovariance (for mathematical convenience) we obtain the rate

$$W = \frac{n^4 \pi^4 v_L^2 \Delta^2 \Lambda^2}{2\omega_L L^6} I, \tag{6.18a}$$

$$I = \int_0^{2\pi} (1-\cos\theta)\cos^2\theta \exp[-q_\parallel^2 \Lambda^2 \sin^2(\theta/2)]\, d\theta, \tag{6.18b}$$

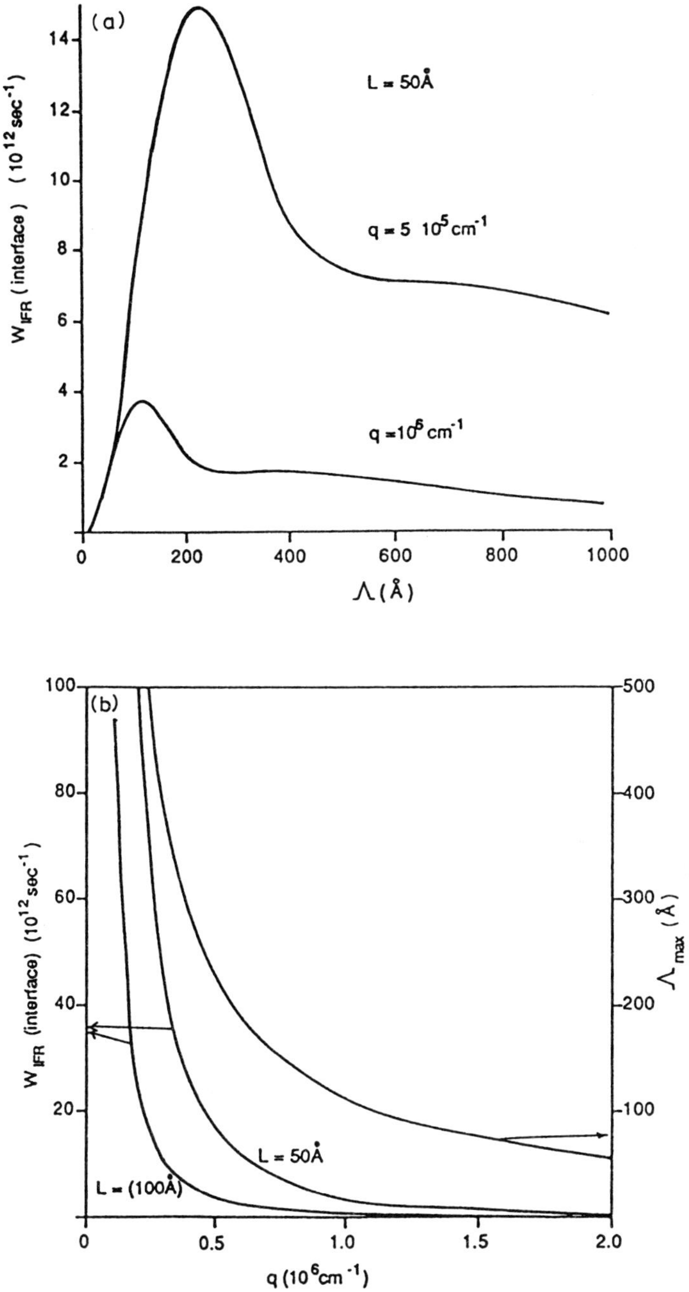

Fig. 31. Interface-roughness scattering of interface LO modes: momentum-relaxation rates in a 50 Å GaAs quantum well. (a) As a function of correlation length Λ for $q = 5 \times 10^5$ cm^{-1} and $q = 1 \times 10^6$ cm^{-1}. (b) As a function of phonon wavevector for $L = 50$ Å and $L = 100$ Å with Λ chosen to maximize the rate. Λ_{max} is depicted in the right-hand scale. For all cases $\Delta = 1$ ML (Gupta and Ridley 1991a).

where $q_{\parallel}$ is the wavevector in the plane. For realistic values of the parameters this rate turns out to be relatively small.

It might be expected, however, that interface modes will be more sensitive to interface quality. Instead of eq. (6.17), which is only weakly dependent on IFR because of the weak dispersion, we take $\delta\omega$ to be proportional to the difference in LO frequencies in the adjacent material. (Of course, if $\delta\omega$ is too large there will not be any interface modes at all, as pointed out in § 3.2.) Thus we take

$$\delta\omega = \delta\omega_{AB}\frac{\Delta}{L}, \tag{6.19}$$

whence

$$W = \frac{(\Delta\omega_{AB})^2\omega_L \Delta^2 \Lambda^2}{2v_L^2 L^2} I, \tag{6.20}$$

where I is given by eq. (6.18b). Figure 31 shows some features of this rate for the $GaAs/Al_{0.3}Ga_{0.7}As$ interface taking $\hbar\delta\omega_{AB} = 1.84$ meV and $\Delta = 1$ ML. Clearly, significant rates emerge especially for small wavevectors, just these, in fact, which the Fröhlich interaction favours and which are copiously present in the hot phonon spectrum. Unlike the case for electrons IFR scattering of phonons is not screened (see fig. 30).

Equation (6.20) has been used to explain the experimental results on hot electrons and hot phonons in quantum wells by Gupta et al. (1991). In these experiments there is good agreement found between the temperature–field curves measured at a fixed lattice temperature using photoluminescence techniques and those deduced from the variation of mobility with field and with lattice temperature. This agreement suggests that hot phonons must be randomized, and the above theory of IFR scattering helps to account for that.

Acknowledgements

The author is indebted to his colleagues at Essex for their assistance with many aspects of this work, and to Dr. M.G. Burr of British Telecom for making available his results on electron confinement.

References

Abram, R.A., G.J. Rees and B.L.H. Wilson, 1978, Adv. Phys. **27**, 799.
Akera, H., and T. Ando, 1989, Phys. Rev. B **40**, 2914.
Al-Dossary, O., M. Babiker and N.C. Constantinou, 1992, Phys. Rev. Lett., to be published.
Altarelli, M., 1986, in: Heterojunctions and Semiconductor Superlattices, eds G. Allan, G. Bastard, N. Boccara, M. Lannou and M. Voos (Springer, Berlin) p. 12.
Altarelli, M., U. Ekenberg and A. Fasolino, 1985, Phys. Rev. B **32**, 5738.
Anderson, D.A., N. Apsley, P. Davies and P.L. Giles, 1985, J. Appl. Phys. **58**, 3059.
Ando, T., and H. Akera, 1989, Phys. Rev. B **40**, 11619.

Ando, T., A.B. Fowler and F. Stern, 1982, Rev. Mod. Phys. **54**, 437.
Ando, T., S. Wakahara and H. Akera, 1989, Phys. Rev. B **40**, 11609.
Andreani, L.C., A. Pasquarello and F. Bassani, 1987, Phys. Rev. B **36**, 5887.
Babiker, M., 1986, J. Phys. C **19**, 683.
Babiker, M., M.P. Chamberlain and B.K. Ridley, 1987, Semicond. Sci. & Technol. **2**, 582.
Babiker, M., M.P. Chamberlain, A. Ghosal and B.K. Ridley, 1988, Surf. Sci. **196**, 422.
Babiker, M., A. Ghosal and B.K. Ridley, 1989, Superlatt. & Microstruct. **5**, 133.
Balkan, N., B.K. Ridley and J.S. Roberts, 1989, Superlatt. & Microstruct. **5**, 539.
Bastard, G., 1989, Wave Mechanics Applied to Semiconductor Heterostructures (Les Editions de Physique, Paris, France).
Bastard, G., and J.A. Brun, 1986, IEEE J. Quantum Electron. **QE-22**, 1625.
Bechstedt, F., and R. Enderlein, 1985, Phys. Status Solidi b **l29**, 349.
Bechstedt, F., and H. Gerecke, 1989, Phys. Status Solidi b **l56**, 151.
Bimberg, D., J. Christen, A. Steckenborn, G. Welmann and W. Schlapp, 1985, J. Lumin. **30**, 562.
Brun, J.A., and G. Bastard, 1986, Phys. Rev. B **33**, 1420.
Burt, M.G., 1988a, Semicond. Sci. & Technol. **3**, 739-753.
Burt, M.G., 1988b, Semicond. Sci. & Technol. **3**, 1224-1226.
Burt, M.G., 1989, in: Band Structure Engineering in Semiconductor Microstructures, eds R..A. Abram and M. Jaros, NATO ASI Series B (Plenum Press, New York) pp. 99-109.
Burt, M.G., 1991, private communication.
Burt, M.G., 1992, J. Phys.: Condensed Matter, Invited Paper, to appear.
Cardona, M., 1989, in: Spectroscopy of Semiconductor Microstructures, eds G. Fasol, A. Fasolino and P. Lugli (Plenum Press, New York) p. 1.
Cardona, M., 1990, Superlatt. Microstr. **7**, 183.
Chamberlain, M.P., and M. Babiker, 1989, Solid State Electron. **32**, 1675.
Chamberlain, M.P., M. Babiker and B.K. Ridley, 1991, Superlatt. & Microstruct. **9**, 227.
Chang, Y.-C., 1982, Phys. Rev. B **25**, 605.
Chattopadhay, D., and H.J. Queisser, 1981, Rev. Mod. Phys. **53**, 745.
Colvard, C., T.A. Grant, M.V. Klein, R. Fischer, H. Morkoç and A.C. Gossard, 1985, Phys. Rev. B **31**, 2080.
Constantinou, N.C., and B.K. Ridley, 1990, Phys. Rev. **41** 10622, 10627.
Daniels, M.E., B.K. Ridley and M. Emery, 1989, Solid State Electron. **32**, 1207.
Devaud, B., J. Shah, T.C. Damen and W.T Tsang, 1988, Appl. Phys. Lett. **52**, 1886.
Dharssi, I., and P.N. Butcher, 1990, J. Phys.: Condens. Matter **2**, 4629.
Dreicer, H., 1960, Phys. Rev. **117**, 343.
Ekenberg, U., 1989, Phys. Rev. B **40**, 7714.
Emura, S., T. Nakagawa, S. Gonda and S. Shimizu, 1987, J. Appl. Phys. **62**, 4632.
Enderlein, R., D. Suisky and J. Röseler, 1991, Phys. Status Solidi b **165**, 9.
Esipov, S.E., and Y.B. Levinson, 1987, Adv. Phys. **36**, 331.
Fasolino, A., E. Molinari and K. Kunc, 1990, Phys. Rev. B **41**, 8302.
Ferry, D.K., R.P. Joshi and M-J. Kann, 1988, Ultrafast Laser Probe Phenomena, SPIE Proc. **942**, 2.
Fewster, P.F., 1988, J. Appl. Cryst. **21**, 524.
Foxon, C.T., J.J. Harris, D. Hilton, J. Hewitt and C. Roberts, 1989, Semicond. Sci. & Technol. **4**, 582.
Fuchs, R., and K.L. Kliewer, 1965, Phys. Rev. A **140**, 2076.
Gerecke, H., and F. Bechstedt, 1991, Phys. Rev. B, to be published.
Goodnick, S.M., and P. Lugli, 1988, Solid State Electron. **31**, 463.
Goodnick, S.M., D.K. Ferry, C.W. Wilmsen, Z. Liliental, D. Fathy and Krivanek, 1985, Phys. Rev. B **32**, 8171.
Grahn, H.T., H. Scheider, W.W. Rühle, K. von Klitzing and K. Ploog, 1990, Phys. Rev. Lett. **64**, 2426.
Guillemot, C., and F. Clerot, 1990, Superlatt. & Microstruct. **8**, 263.
Gupta, R., and B.K. Ridley, 1991a, Physical Concepts of Materials for Novel Optoelectronic Devices, SPIE Proc. **1362**, 790.
Gupta, R., and B.K. Ridley, 1991b, to be published.
Gupta, R., N. Balkan, B.K. Ridley and M. Emery, 1991, SPIE **1362**, 798.

Harris, J.J., J.A. Pals and R. Woltzer, 1989, Rep. Prog. Phys. **52**, 1217.
Heiblum, M., M.I. Nathan, D.C. Thomas and C.M. Knoedler, 1985, Phys. Rev. Lett. **55**, 2200.
Heiblum, M., I.M. Anderson and C.M. Knoedler, 1986, Appl. Phys. Lett. **49**, 207.
Hirakawa, K., and H. Sakaki, 1986, Appl. Phys. Lett. **49**, 889.
Huang, K., and B. Zhu, 1988, Phys. Rev. B **38**, 2183, 13377.
Jain, J.K., R.A. Jalabert and S. Das Sarma, 1988, Phys. Rev. Lett. **60**, 353.
Jaros, M., and K.B. Wong, 1984, J. Phys. C **17**, L765.
Jaros, M., K.B. Wong and M.A. Gell, 1985, Phys. Rev. B **31**, 1205.
Kash, J.A., J.C. Tsang and J.M. Huam, 1985, Phys. Rev. Lett. **54**, 2151.
Kelly, M.J., 1985, J. Phys. C **18**, 5963.
Kim, O.K., and W.G. Spitzer, 1979, J. Appl. Phys. **50**, 4362.
Klein, M.V., 1986, IEEE J. Quantum Electron. **QE-22**, 1760.
Klemens, P.G., 1966, Phys. Rev. **148**, 845.
Kliewer, K.L., and R. Fuchs, 1966, Phys. Rev. **144**, 495.
Kocevar, P., 1987, Festkörperprobleme **27**, 197.
Kozyrev, S.V., and Y.A. Shik, 1986, Sov. Phys. Semicond. **19**, 1024.
Kuhn, T., and G. Mahler, 1989, Solid State Electron. **32**, 1857.
Landau, L.D., and E.M. Lifshitz, 1986, Theory of Elasticity, 3rd Ed. (Pergamon Press, Oxford).
Landheer, D., H.C. Liu, M. Buchanan and R. Stoner, 1989, Appl. Phys. Lett. **54**, 1784.
Landsberg, P.T., 1986, Phys. Rev. B **33**, 8321.
Lassnig, R., 1984, Phys. Rev. B **30**, 7132.
Lassnig, R., 1988, Solid State Commun. **65**, 765.
Lee, J., H.N. Spector and V.K. Arora, 1983, J. Appl. Phys. **54**, 6995.
Lester, L.F., S.D. Offsey, B.K. Ridley, W.J. Schaft, B.A. Foreman and L.F. Eastman, 1991, Appl. Phys. Lett. **59**, 1162.
Levi, A.F.J., J.R. Hayes, P.M. Platzman and W. Wiegmann, 1985, Physica B **134**, 480.
Levine, B.F., K.K. Choi, G.G. Bethea, J.F. Walker and R.J. Malik, 1988, Solid State Electron. **31**, 583.
Lucovsky, G., and M.F. Chen, 1970, Solid State Commun. **8**, 1397.
Lugli, P., P. Bordone, L. Reggiani, M. Rieger, P. Kocevar and S.M. Goodnick, 1989, Phys. Rev. B **39**, 7834.
Luttinger, J.M., 1956, Phys. Rev. **102**, 1030.
Luttinger, J.M., and W. Kohn, 1955, Phys. Rev. **97**, 869.
Lyon, S.A., 1986, J. Lumin. **35**, 121.
Maradudin, A.A., E.W. Montroll, G.H. Weiss and I.P. Ipotova, 1971, Theory of Lattice Dynamics in the Harmonic Approximation (Academic Press, New York).
Marzin, J.Y., 1986, in: Heterojunctions and Semiconductor Superlattices, eds G. Allan, G. Bastard, N. Boccara, M. Lannoo and M. Voos (Springer, Berlin) p. 161.
Menendez, J., 1989, J. Lumin. **44**, 285.
Meyer, J.R., and F.J. Bartoli, 1983, Phys. Rev. B **25**, 915.
Milsom, P.K., and P.N. Butcher, 1986, Semicond. Sci. & Technol. **1**, 58.
Molinari, E., A. Fasolino and K. Kunc, 1986, Superlatt. & Microstruct. **2**, 397.
Mori, N., and T. Ando, 1989, Phys. Rev. B **40**, 6175.
Narayanamurti, V., H.L. Störmer, M.A. Chin, A.C. Gossard and W. Wiegmann, 1979, Phys. Rev. Lett. **43**, 2012.
Nedozerov, S.S., 1971, Sov. Phys. Solid State **12**, 1814.
O'Reilly, E.P., and G.P. Witchlow, 1986, Phys. Rev. B **34**, 6030.
Pikus, G.E., and G.L. Bir, 1959, Sov. Phys. Solid State **1**, 1502.
Pines, D., 1953, Phys. Rev. **92**, 626.
Price, P.J., 1981, Ann. Phys. N.Y. **133**, 217.
Price, P.J., 1982, J. Appl. Phys. **53**, 6863.
Rayleigh, Lord, 1889, Proc. London Math. Soc. **20**, 225.
Riddoch, F.A., and B.K. Ridley, 1983, J. Phys. C **16**, 6971.
Ridley, B.K., 1977, J. Phys. C **10**, 1589.
Ridley, B.K., 1982, J. Phys. C **15**, 5899.

Ridley, B.K., 1988a, Semicond. Sci. & Technol. **3**, 111.
Ridley, B.K., 1988b, Quantum Processes in Semiconductors (Oxford University Press, Oxford).
Ridley, B.K., 1989, Phys. Rev. B **39**, 5282.
Ridley, B.K., 1990, J. Appl. Phys. **68**, 4667.
Ridley, B.K., 1991a, Rep. Prog. Phys. **54**, 169.
Ridley, B.K., 1991b, Solid State Electron. **34**, 111.
Ridley, B.K., 1991c, Phys. Rev. B **44**, 9002.
Ridley, B.K., 1991d, unpublished.
Ridley, B.K., 1991e, Compound Semiconductor Physics and Devices, SPIE Proc. **1675**, to be published.
Ridley, B.K., and M. Babiker, 1991, Phys. Rev. B **43**, 9096.
Ridley, B.K., and R. Gupta, 1991, Phys. Rev. B **43**, 4939.
Rorison, J.M., and D.C. Herbert, 1986, J. Phys. C **19**, 3991, 6357.
Rücker, H., E. Molinari and P. Lugli, 1991, Phys. Rev. B **44**, 3463.
Rudlin, S., and T.L. Reinecke, 1990, Phys. Rev. B **41**, 7713.
Rudlin, S., and T.L. Reinecke, 1991, Phys. Rev. B **43**, 9298.
Rytov, S.M., 1956, Sov. Phys. Acoustics **2**, 67.
Sakaki, H., T. Noda, K. Hirakawa, M. Tanaka and T. Matsusue, 1987, Appl. Phys. Lett. **51**, 1934.
Schubert, E.F., L.N. Pfeiffer, K.W. West and A. Izabelle, 1989, Appl. Phys. Lett. **54**, 1350.
Shah, J., 1986, IEEE J. Quantum Electron. **QE-22**, 1728.
Shulman, J.N., and Y.-C. Chang, 1981, Phys. Rev. B **24**, 4445.
Shulman, J.N., and Y.-C. Chang, 1985, Phys. Rev. B **31**, 2056.
Stanton, C.J., D.W. Bailey and K. Hess, 1988, IEEE J. Quantum Electron. **QE-24**, 1614.
Stern, F., 1967, Phys. Rev. **18**, 546.
Stradling, R.A., and R.A. Wood, 1970, J. Phys. C **3**, 2425.
Takimoto, N., 1959, J. Phys. Soc. Jpn. **14**, 1142.
Tsang, J.C., and J.A. Kash, 1990, in: Light Scattering in Solids VI, eds M. Cardona and G. Gunterodt (Springer, Berlin).
Tsen, K.J., S.Y. Tsen and H. Morkoc, 1988, SPIE **942**, 114.
Tsen, K.T., R.P. Joshi, D.K. Ferry and H. Morkoc, 1989, Phys. Rev. B **39**, 1446.
Tsuchiya, T., H. Akera and T. Ando, 1989, Phys. Rev. B **39**, 6025.
van Hall, P.J., T. Klaver and J.H. Wolter, 1988, Semicond. Sci. & Technol. **3**, 120.
Vickers, A.J., 1991, private communication.
von der Linde, D., J. Kuhle and H. Klingenberger, 1980, Phys. Rev. Lett. **44**, 1505.
Welch, D.F., G.W. Wicks and L.F. Eastman, 1984, J. Appl. Phys. **55**, 3176.
Wendler, L., 1985, Phys. Status Solidi b **129**, 513.
Wendler, L., 1990, private communication.
Wendler, L., and R. Haupt, 1987, Phys. Status Solidi b **141**, 493.
Wendler, L., and R. Pechstedt, 1986, Phys. Status Solidi b **138**, 197.
Wendler, L., and R. Pechstedt, 1987, Phys. Rev. B **35**, 5887.
Westland, D.J., J.F. Ryan, M.D. Scott, J.F. Davies and J.R. Riffar, 1988, Solid State Electron. **31**, 431.
Zhu, B., 1988, Phys. Rev. B **38**, 7694.
Zianni, X., P.N. Butcher and I. Dharssi, 1992, J. Phys.: Condens. Matter **4**, L77.

CHAPTER 14

Classical Transport and Thermoelectric Effects in Low-Dimensional and Mesoscopic Semiconductor Structures

B. L. GALLAGHER

Physics Department, University of Nottingham
Nottingham, NG7 2RD, UK

P. N. BUTCHER

Physics Department, University of Warwick
Coventry, CV4 7AL, UK

Handbook on Semiconductors
Completely Revised Edition
Edited by T.S. Moss
Volume 1, edited by P.T. Landsberg

Contents

1. Introduction . . . 723
2. Thermoelectric coefficients . . . 723
3. Diffusion thermopower . . . 725
   3.1. Simple models . . . 725
   3.2. Diffusion thermopower in the Boltzmann approach . . . 726
   3.3. Diffusion thermopower in 3D systems for specific scattering mechanisms . . . 728
   3.4. Diffusion thermopower for heterostructures . . . 728
   3.5. Diffusion thermopower for MOSFETs . . . 730
   3.6. Diffusion thermopower of 1D electron systems . . . 734
   3.7. Effects of sub-band structure on diffusion thermopower . . . 734
   3.8. Diffusion thermopower in superlattices . . . 743
   3.9. Thermopower in periodically modulated 2D conductors . . . 744
4. Phonon-drag thermopower . . . 745
   4.1. Simple models: the Seebeck and Peltier approach . . . 747
   4.2. The Boltzmann approach applied to heterostructures and MOSFETs . . . 750
5. Magnetothermopower . . . 762
   5.1. Diffusion magnetothermopower and "quantised" thermopower in heterostructures and MOSFETs . . . 762
   5.2. Phonon-drag magnetothermopower in heterostructures and MOSFETs . . . 770
   5.3. Effects of sub-band structure on phonon-drag thermopower . . . 778
6. Quantum interference effects in thermopower . . . 782
   6.1. Weak localisation in thermoelectric transport . . . 782
   6.2. Universal conductance fluctuations . . . 789
   6.3. Universal thermopower fluctuations . . . 791
7. Thermal and electrical transport in microstructures and edge states . . . 796
   7.1. The generalised Landauer–Büttiker formalism . . . 796
   7.2. Thermoelectric effects in quantum point contacts . . . 800
   7.3. Thermal conductivity and Peltier coefficient of a quantum point contact . . . 806
8. Strong localisation and metal–insulator transitions . . . 808

References . . . 813

1. Introduction

There now exists a massive amount of literature on the electrical transport properties of low-dimensional semiconductor structures. Technological advances have made possible the realisation of high-quality two-dimensional electron and hole gases and one-dimensional quantum point contacts. A large range of mesoscopic structures in which the dimensions of interest, though larger than the electron wavelength, are smaller than the mean free path or phase breaking length have also been extensively studied. Most of the experimental and theoretical work has concerned the conductivity tensors of such systems. However, a very large body of work has now accumulated on the corresponding thermoelectric properties. In this chapter we attempt to introduce this subject assuming no prior knowledge but go on to critically review the extensive research literature.

The thermoelectric properties have always played an important role in elucidating the transport and scattering mechanisms in conductors. In classical transport they are intimately related to the energy dependence of the conductivity. They thus provide an important test of any transport theory. In § 3 after a brief review of the simple theory it is shown that the diffusion themopower can shed much light on the nature of the scattering mechanisms limiting the mobility in low-dimensional structures. The thermoelectric coefficients are also shown to be very sensitive to sub-band structure in 2D and 1D conductors, superlattices, and other perodically modulated structures. In § 4 the additional phonon-drag thermopower is introduced and is shown to provide an important test of the theory of electron–phonon coupling and screening in MOSFETs and heterostructures. § 5 considers diffusion and phonon-drag in the presence of a magnetic field. In particular the current status of the long-standing prediction of a "quantised thermopower effect" in the quantum Hall regime is reviewed. § 6 moves on to recent observations of quantum interference effects in thermopower and universal thermopower fluctuations. In § 7 a generalisation of the Landauer–Büttiker formalism to include thermal and thermoelectric transport is presented. This is then used as a basis for understanding the very recently observed thermal and thermoelectric properties of quantum point contacts. For completeness the thermoelectric effects associated with metal–insulator transitions are briefly reviewed in § 8.

2. Thermoelectric coefficients

The most familiar of the thermoelectric effects is the Seebeck effect often loosely referred to simply as thermopower. Thermocouples work on the basis that in a circuit of the type shown in fig. 1 a temperature difference ΔT between junctions of different conductors A and B gives rise to a voltage difference ΔV. The Seebeck coefficient of the circuit is then defined by

$$S_{\mathrm{AB}} = \lim \left(\frac{\Delta V}{\Delta T} \right)_{\Delta T \to 0}. \tag{1}$$

S_{AB} is not a property of the junctions but of the bulk conductors and we can write $S_{\mathrm{AB}} = (S_{\mathrm{A}} - S_{\mathrm{B}})$ where S_{A} and S_{B} are the characteristic "absolute" thermopowers of the materials A and B.

Two related manifestations of the same material property are the Peltier and Thomson effects. Passing a current through the circuit in fig. 1 will lead to heat being given out at one of the junctions and absorbed at the other. The Peltier coefficient π_{AB} is then just the heat per unit positive charge. In other words the ratio of the rate of heat flow Q to the electrical current i, thus

$$\pi_{\mathrm{AB}} = Q/i, \tag{2}$$

where Q is taken as positive for heat produced. It may also be expressed as $\pi_{\mathrm{AB}} = \pi_{\mathrm{A}} - \pi_{\mathrm{B}}$. The Thomson effect may be observed in a single conductor. The Thomson coefficient σ_{T} is defined as the heat absorbed or produced when unit charge passes through unit temperature difference in the conductor.

Thermodynamical arguments (see Barnard (1972) for example) give the interrelation of these coefficients for any conductor as

$$S = \pi/T \tag{3}$$

and

$$S = \int_0^T \sigma_{\mathrm{T}}/T \, \mathrm{d}T. \tag{4}$$

Thus, as emphasised earlier, there is only one independent coefficient and one is free to measure or calculate whichever is the more convenient for a given material or theory. S is usually the easiest thing to measure but π is more straightforward to calculate.

Measurement of σ_{T} is very difficult but once this has been done for a single material the result can be used to provide the reference against which the coefficients of other materials can be measured. This has been done by Roberts (1981) for Pb. Alternatively, superconductors for which the coefficients are zero may be used.

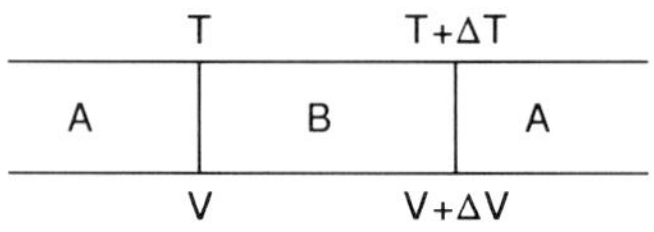

Fig. 1. Normal arrangement for the measurement of the difference in the Seebeck coefficient of two conductors A and B.

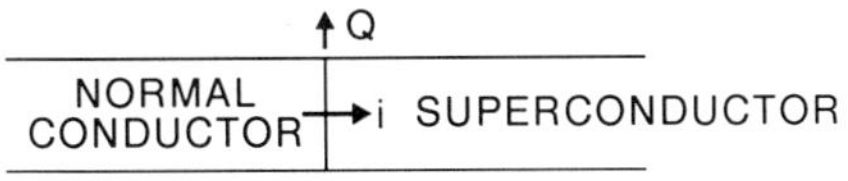

Fig. 2. Situation in which the Peltier coefficient of a normal conductor may be measured. The current through the junction, i, leads to the generation of heat at a rate Q.

The thermoelectric coefficients arise from two broadly distinct processes. We distinguish, for example, between the *diffusion thermopower* S_d which is due to the diffusive motion of the charge carriers in the temperature gradient and the *phonon-drag thermopower* S_g* which is due to the net momentum transfer from phonons to carriers. In the limit of weak electron–phonon coupling or low temperature these are formally distinct effects and one can write

$$S = S_d + S_g. \tag{5}$$

When the carrier diffusion coefficients are significantly modified by phonon scattering this distinction becomes less clear but it is usually still maintained.

3. Diffusion thermopower

3.1. Simple models

It is straightforward to write down an estimate of the magnitude of the diffusion thermopower. Consider the situation shown in fig. 2. Associated with the electrical current will be a heat current in the normal conductor but not the superconductor. This heat must be given out at the junction. For the non-degenerate case the thermal energy per carrier will be $\sim k_B T$ so from eq. (2)

$$\pi \sim \pm \frac{k_B T}{|e|}, \tag{6}$$

$$S \sim \pm \frac{k_B}{|e|}, \tag{7}$$

where the positive sign is for holes and the negative sign is for electrons. This suggests that while π is a measure of the ratio of the average thermal energy to the charge of the carriers, S is a measure of the ratio of the average entropy to the charge (see Dugdale (1977) for an elaboration of this idea). For the degenerate case the average thermal energy will be reduced by $\sim k_B T/\varepsilon_F$ where ε_F is the Fermi energy and so we expect

$$S \sim \pm \frac{k_B}{|e|}\left(\frac{k_B T}{\varepsilon_F}\right). \tag{8}$$

In the case of intrinsic semiconductors where the electrons have a significantly

*The subscript g is from the German gitter meaning lattice.

higher mobility than the holes the latter can be neglected. All the electrons will have a thermal energy $\sim(\varepsilon_c - \varepsilon_F)$, where ε_c is the conduction band edge and so

$$S \sim -\frac{(\varepsilon_c - \varepsilon_F)}{|e|T}. \tag{9}$$

Equation (8) agrees with the linear dependence upon T usually observed in metals in higher temperatures when the phonon drag is unimportant and the predicted magnitudes of a few μV/K at room temperature are also in reasonable agreement. In the intrinsic semiconductor case a T^{-1} dependence is predicted and for $(\varepsilon_c - \varepsilon_F) \sim 0.5$ eV one expects magnitudes of ~1 mV/K at room temperature. This is again in reasonable agreement with the behaviour observed for silicon and germanium.

It should be noted that eq. (9) applies to any activated conduction, when E_F lies in a gap in the density of states. This is a key point in the discussion of § 8. Since S tends to infinity as T tends to zero this might seem to be in conflict with the requirement that the entropy tends to zero. However, when this is considered carefully one finds that thermodynamics only requires that the product of S and the conductivity tends to zero (Tauc 1962).

3.2. Diffusion thermopower in the Boltzmann approach

Although the above simple estimates are helpful they are fundamentally flawed and are, for example, incapable of explaining the positive thermopowers observed in many metals. The problem lies in the averages which we have taken. In the case of the Peltier coefficient we require the heat current associated with the electrical current. Thus the appropriate average is that of the product of the thermal energy and the conductivity of each group of electrons. If we can ascribe a conductivity $\sigma(\varepsilon_i)$ to an electron in the energy level ε_i in a conductor with a chemical potential μ then we should write

$$S = \frac{-1}{|e|T} \frac{\sum_i (\varepsilon_i - \mu)\sigma(\varepsilon_i)}{\sum_i \sigma(\varepsilon_i)}. \tag{10}$$

The corresponding Boltzmann transport result is

$$S = \frac{-1}{|e|T} \frac{\int_0^\infty (\varepsilon - \mu)\sigma(\varepsilon)(\mathrm{d}f_0/\mathrm{d}\varepsilon)\,\mathrm{d}\varepsilon}{\int_0^\infty \sigma(\varepsilon)(\mathrm{d}f_0/\mathrm{d}\varepsilon)\,\mathrm{d}\varepsilon}, \tag{11}$$

where f_0 is the equilibrium Fermi–Dirac distribution. We see that the thermopower is closely related to the energy dependence of the conductivity. If $\sigma(\varepsilon)$ were constant S would be zero since $(\mathrm{d}f_0/\mathrm{d}\varepsilon)$ is symmetrical about μ. Physically this corresponds to the electrical current due to the net diffusion of electrons with energies greater than μ towards the cooler ends being exactly equal to that of the cooler electrons in the opposite direction. We also see that the sign of the thermopower does not simply follow from the charge of the carriers. A negative sign for electrons normally occurs only because $\sigma(\varepsilon)$ usually increases with ε. A positive sign is normal case for holes for the same reason.

For degenerate electrons a good approximation to eq. (11) may be obtained from the Taylor expansion of $\sigma(\varepsilon)$ about ε_F since $df_0/d\varepsilon$ falls rapidly to zero outside the range $\varepsilon_F \pm kT$. Writing

$$\sigma(\varepsilon) = \sigma(E_F) + \left.\frac{d\sigma}{d\varepsilon}\right|_{E_F} (\varepsilon - \varepsilon_F) + \left.\frac{1}{2}\frac{d^2\sigma}{d\varepsilon^2}\right|_{E_F} (\varepsilon - \varepsilon_F)^2 + \cdots, \tag{12}$$

only those terms anti-symmetric in ε will contribute and so, retaining these first three terms, eq. (11) becomes

$$S = \frac{-1}{|e|T}\frac{1}{\sigma}\left.\frac{d\sigma}{d\varepsilon}\right|_{E_F} \int_0^\infty (\varepsilon - \varepsilon_F)^2 \frac{df_0}{d\varepsilon}\, d\varepsilon. \tag{13}$$

The "Fermi integral" in eq. (13) has the value $(\pi kT)^2/3$ and so

$$S = \frac{-\pi^2}{3}\frac{k_B}{|e|}\frac{k_B T}{\sigma}\left.\frac{d\sigma}{d\varepsilon}\right|_{\varepsilon_F}. \tag{14}$$

This can be conveniently expressed as

$$S = \frac{-\pi^2}{3}\frac{k_B}{|e|}\frac{k_B T}{\varepsilon_F}\left.\frac{d\ln\sigma}{d\ln\varepsilon}\right|_{\varepsilon_F}, \tag{15}$$

which is generally referred to as the Mott formula. This can be more simply expressed if the conductivity is in terms of an energy dependent relaxation time $\tau(\varepsilon) \propto \varepsilon^p$ where p is a constant. Then, writing

$$\sigma = \frac{n(\varepsilon)e^2\tau(\varepsilon)}{m^*}, \tag{16}$$

where n is the electron number density and m^* the effective mass, eq. (15) gives

$$S = \frac{-\pi^2}{3}\frac{k_B}{|e|}\frac{k_B T}{\varepsilon_F}(\tfrac{3}{2} + p) \tag{17}$$

for the three-dimensional case with $n(\varepsilon) \propto \varepsilon^{2/3}$ and

$$S = \frac{-\pi^2}{3}\frac{k_B}{|e|}\frac{k_B T}{\varepsilon_F}(1 + p) \tag{18}$$

for two-dimensional electron systems with $n(\varepsilon) \propto \varepsilon$.

In the derivation of eq. (11) it has been implicitly assumed that the conductivity of a group of electrons is only dependent upon energy. If inelastic scattering, normally by phonons, significantly changes an electron's energy then this is no longer valid. In going from the more general expression (11) to the Mott formula (15) we have assumed that the conductivity is slowly varying on an energy scale kT. This is often a good approximation because the first correction term involves the third derivative of σ with respect to ε.

3.3. *Diffusion thermopower in 3D systems for specific scattering mechanisms*

In principle, then, the calculation of the thermoelectric coefficients is straightforward, but in practice the calculation of the required $\sigma(\varepsilon)$ can be difficult. However, in some circumstances it is easier to identify particular scattering mechanisms by their characteristic energy dependence than it is to calculate their absolute contribution to the total scattering rate.

Let us first look at some simple models. For simple hard-sphere scattering the electron mean free path is constant and we simply have $\tau \propto v_F^{-1} \propto \varepsilon^{-1/2}$ thus $p = -\frac{1}{2}$ in this case though the overall thermopower remains negative for electrons. Generally, we would expect τ to increase with ε for more realistic scattering potentials.

For the three-dimensional case Rutherford-type scattering from a Columbic potential gives a cross section $\propto \varepsilon^{-2}$ as does scattering from the "square-well potential" often used to model homovalent impurities and alloy scattering. Thus a p of 1.5 is predicted in both these cases. For the screened Columbic potential the situation is more complicated. In the weak and very strong screening limits we recover the unscreened and hard-sphere results. In the general case, however, p depends upon the scattering angle.

Thus a negative thermopower linear in T is the usual expectation with $p \sim -0.5$ for hard-sphere scattering and $p \sim 1.5$ for remote ionised impurities. Screening will generally reduce the latter value.

3.4. *Diffusion thermopower for heterostructures*

Detailed calculations of the diffusion thermopower for GaAs/AlGaAs heterostructures have been carried out by Kundu et al. (1987, 1988a) and by Karavolas and co-workers (1990, 1991). Kundu et al. use eq. (11) while Karavolas and co-workers employ the degenerate approximation (15) since they are only interested in the behaviour for $T \sim 1$ K. This should make no difference since $k_B T/E_F$ is small even up to 10 K and Kundu et al. only find departures from linearity for a very low surface electron density sample, $n_s = 1.7 \times 10^{15}$ m^2, above 10 K. Kundu et al. also consider deformation potential and piezoelectric scattering but not surprisingly find that these make no significant contribution to the diffusion thermopower even up to 15 K. Both Kundu et al. and Karavolas and co-workers use the Fang–Howard wavefunction (Ando et al. 1982).

Kundu et al. also use the expressions for remote ionised and background impurity scattering times quoted by these authors. Alloy scattering is also included when treating InGaAs heterostructure. Intersubband scattering is neglected which should be reasonable. The main weakness is that the energy dependence of the screening is not included. No spacer layer is included and the number density of background and remote impurities are chosen to obtain the correct mobility.

The main aim of the work of Kundu et al. was to try to account for the very different results obtained by different experimental groups on different samples. However, it may well be that some of this experimental work is in error (see later) and so we do not consider these comparisons directly but concentrate on the main theoretical predictions. From the graphical results given it is possible to compare the results with eq. (18) and hence obtain the predicted values of p. The values obtained lie in the range 1.3 to 1.6 for a wide range of electron, remote and local impurity densities. This is in surprisingly good agreement with our simple 3D estimate of $p \sim 1.5$. Kundu et al. also consider the effects of long-range potential fluctuations which might lead to variations of the density of states. They find the effect to be very small.

Karavolas and co-workers use similar expressions for the remote and background impurity scattering but retain the full energy dependent screening. They also consider surface roughness scattering. Otherwise the calculations are very similar. For heterostructures remote ionised impurity scattering is found to give $p \sim 1.5$ (see fig. 3a) with a weak dependence on electron density while background impurities give $p \sim 1.0$. The results for $n_s = 6.5 \times 10^{15}$ m^{-2} if surface roughness is neglected is $p \sim 1.3$ which is the same as that obtained by Kundu et al. for this electron density. The energy dependence of the screening thus does not have a marked effect in heterojunctions. We shall see that in MOSFETs it is crucial. A value of p of ~ 0 is obtained for the surface roughness contribution with a rather weak dependence upon the parameters characterising surface roughness magnitude and correlation length. The main conclusions are that the total resultant p is of order 1.4 (Kundu et al.) or 1.0 if surface roughness and energy dependent screening are included. One therefore expects

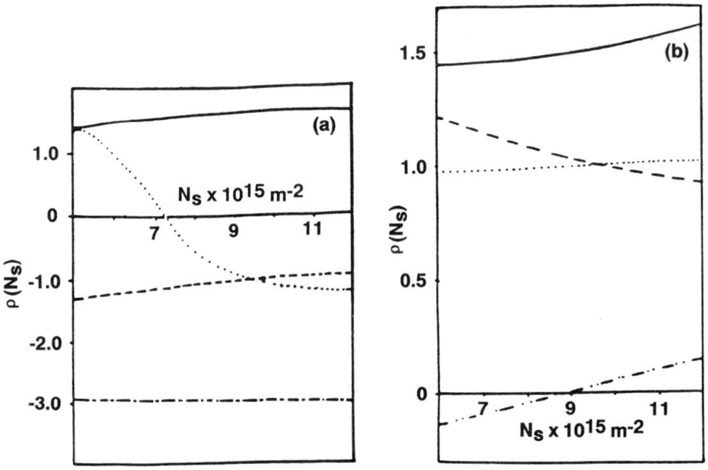

Fig. 3. (a) Calculated $p(n_s)$ (dotted line) due to interface roughness (dashed line), remote impurities (full line) and background impurities (chain line) for a Si MOSFET (Karavolas and Butcher 1991). (b) Calculated $p(n_s)$ (dotted line) due to contributions from interface roughness (chain line), remote impurities (full line) and background impurities (dashed line) for a GaAs heterostructure (Karavolas and Butcher 1991).

$S \sim 4\, T \mu V\, K^{-1}$ for a typical electron density of $3 \times 10^{15}\, m^{-2}$ with no significant deviation from linearity for $T < 10$ K.

As noted earlier, the experimental situation is somewhat confused. Obloh et al. (1984) obtained thermopowers of the predicted order for "low-mobility" samples ($\mu < 4\, m^2\, V^{-1}\, s^{-1}$) but larger values for a high-mobility sample ($\mu \sim 20\, m^2\, V^{-1}\, s^{-1}$). Nicholas and co-workers (Vuong et al. 1986, Brummel et al. 1986) found magnitudes less than one would expect and a surprising *decrease* in S with increasing T even down at 2 K for a range of samples with mobilities of $\mu \sim 2$–$50\, m^2\, V^{-1}\, s^{-1}$. Davidson et al. (1986) obtained thermopowers which appear to be of the right magnitude with a roughly linear T dependence below ~ 7 K followed by a surprising fall at higher T. In this work, however, the temperature gradient along the sample was not measured directly and the results are perhaps best viewed as qualitative.

More recent work descrbed in § 4.2 on a wide range of samples with mobilities ranging from ~ 4 to $\sim 80\, m^2\, V^{-1}\, s^{-1}$ has shown that the phonon-drag contribution dominates the thermopower down to ~ 0.6 K. The reason for this discrepancy is unclear. Ruf et al. (1988) speculate that phonon drag might be suppressed for low mobilities but as we attempt to show in § 4.2 there is no theoretical reason or experimental evidence for this. Phonon drag could be suppressed if the phonon mean free path was very short due to phonon scattering in heavily Cr doped substrates. In § 4.2 we show that this is an important effect but that it does not account for the discrepancy. This casts doubt on the early work and shows the need for more measurements on very low mobility heterostructures.

The prediction for heterostructure is best compared with the work of Ruf et al. (1988). As fig. 4 shows they observe a diffusion contribution for $T < 0.6$ K which goes over to a $\sim T^3$ dependence at higher T when phonon drag dominates. For a high-mobility sample with $\mu = 810\,000\, cm^2\, V^{-1}\, s^{-1}$ and $n_s = 3.3 \times 10^{15}\, m^{-2}$ a linear dependence with $S = 5.1\, T\, \mu V\, K^{-1}$ corresponding to $p = 1.2$ is observed. This is in excellent agreement with expectation if, as one would expect, remote ionised impurity scattering dominates. In a second low-mobility sample with $\mu = 100\,000\, cm^2\, V^{-1}\, s^{-1}$ a smaller thermopower and a much lower value of $p = 0.1$ is found. On the basis of the calculation of Karavolas and Butcher (1991) (fig. 3b) this suggests that surface roughness is much more important in this sample. This is consistent with the lower mobility.

3.5. *Diffusion thermopower for MOSFETs*

Zavaritskii and co-workers (Zavaritskii 1984) have studied the thermopower of Si MOSFETs grown on both the (100) and (111) surfaces. As fig. 5 shows they observe a roughly T^3 dependence at higher T with the emergence of a diffusion term below ~ 1 K. The calculated S_d indicated is obtained from eq. (18) with $p = 0$. The behaviour is therefore very similar to that observed in recent work on heterostructures despite the low mobility of order $1\, m^2\, V^{-1}\, s^{-1}$.

Karavolas and co-workers (1990, 1991) also apply the above approach to Si MOSFETs. The predicted behaviour is much more interesting since scattering by

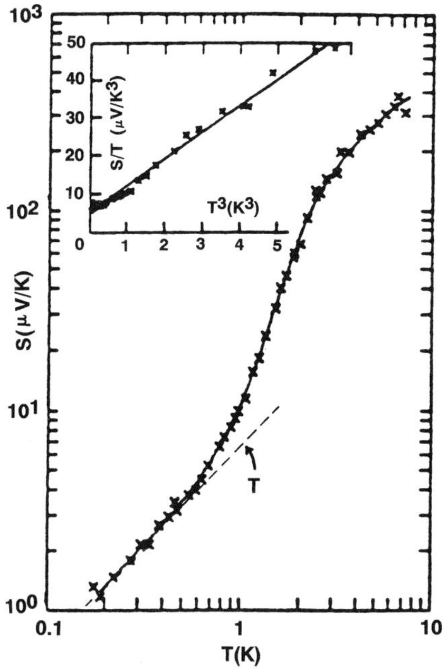

Fig. 4. Transition from diffusion to phonon drag thermopower in a GaAs/AlGaAs heterostructure ($n_s = 3.3 \times 10^{15}$ m^{-2}, $\mu = 81$ m^2/V s) (Ruf et al. 1988).

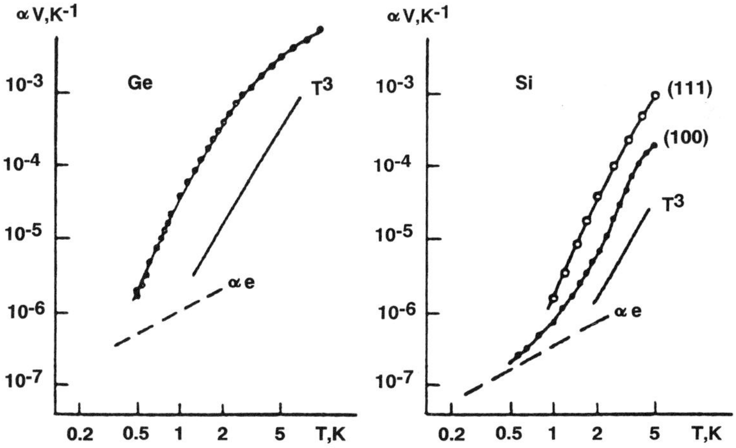

Fig. 5. Measured thermopower for Si MOSFETs on the (100) and (111) plane. The T^3 line is the estimated phonon drag and α_e the diffusion contribution calculated for $p = 0$ (Zavaritskii 1984).

remote ionized impurities and background impurities are found to give thermopowers of different signs. Furthermore, a change of sign as a function of electron density can occur. For their calculation a background impurity concentration of 1.7×10^{21} m^{-3} is used. This is a little larger than the known boron acceptor level of the samples studied by Gallagher et al. (1990a). The remote impurity concentration of 2×10^{22} m^{-3}, the surface roughness amplitude of 5 Å and correlation length of 22 Å are chosen to fit the observed electron density dependent mobility and thermopowers. As fig. 3b shows, the remote ionised impurity scattering is again found to give a weakly electron density dependent $p \sim 1.5$. The result for background impurities is, however, totally different with $p \sim -3.0$. This arises from the strong screening and the dominance of large-angle scattering from impurities in the conducting channel.

In the limit of small n_s the screening will be weak and scattering from the remote impurities will dominate. As n_s increases only those remote impurities within a range of $\sim k_F^{-1}$ will have significant cross section (Ando et al. 1982). For large n_s very few remote impurities contribute and background impurities and surface roughness dominate. The changeover will also occur in heterostructures, but since p is similar for remote and background impurities this has no observable effect. In MOSFETs increasing n_s can lead to a transition from $p \sim 1.5$ and a negative thermopower to $p < -1$ and a positive thermopower. It should be emphasised that the predicted change of sign disappears if the energy dependency of the screening is neglected. Interface roughness scattering gives mostly large-angle scattering in MOSFETs since the small correlation length potential is poorly screened. The resultant p is ~ -1. For heterostructures the smaller k_F^{-1} gives better screening and this along with the higher electron energies gives a smaller cross-section, more small-angle scattering and $p \sim 0$. Interface roughness dominates for large n_s in MOSFETs as the electrons are confined closer to the interface. This is well known and gives a decrease of mobility with increasing n_s despite the increasing screening (Ando et al. 1982). Thus at the largest n_s the value of p is close to -1.

These predictions are in excellent agreement with experiment. Gallagher et al. (1990a) find that the thermopower is dominated by the diffusion contribution for $T \sim 1$ K. For boron doping levels of $\sim 1.4 \times 10^{21}$ m^{-3} they observe a change from negative to positive thermopower as n_s is increased. As shown in fig. 6 the magnitude and form of $S(n_s)$ can be very well accounted for by the theory. Moreover, in recent work (Oxley et al. 1992) the thermopower has been measured for samples in which the boron doping level has been reduced by two orders of magnitude but which in other respects are identical to those used in the earlier work of Gallagher et al. (1990a). Remote impurity scattering will then dominate up to much larger n_s and the diffusion thermopower would be expected to remain negative. As fig. 7 shows, this is what happens. At higher temperatures the phonon-drag dominated behaviour is essentially unchanged, but at lower T we have a negative divergence ($\sim T^{-2}$) rather than a positive one. It is surprising that S remains negative up to very high n_s when surface roughness should dominate. This may be because of the increased screening. For $n_s = 4 \times 10^{16}$ m^{-2} the value of $k_F^{-1} \sim 28$ Å which is comparable with the correlation length of 22 Å.

We see that the overall behaviour in both heterojunctions and MOSFETs is in accordance with the theoretical predictions. Moreover, the thermopower has been shown to be an important means of identifying the relative importance of difference scattering mechanisms.

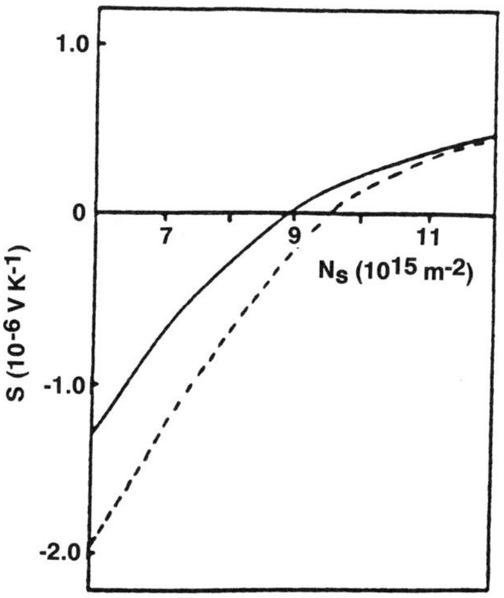

Fig. 6. Plot against electron density n_s of the calculated thermopower (dashed line) and the experimental data of Gallagher et al. (1990a) (full line) at 1.5 K (from Karavolas et al. (1990)).

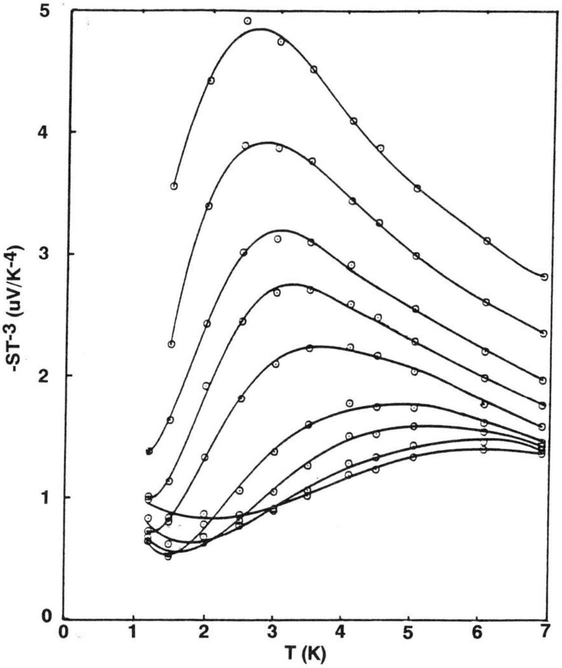

Fig. 7. Thermopower divided by the temperature cubed of a low boron doped Si MOSFET. Areal densities from uppermost to lowest curve are 7.08×10^{15}, 8.01×10^{15}, 9.35×10^{15}, 10.7×10^{15}, 13.4×10^{15}, 20.0×10^{15}, 26.7×10^{-2}, 34.7×10^{15}, 42.7×10^{15} m^{-2} (Oxley et al. 1992).

3.6. *Diffusion thermopower of 1D electron systems*

Kundu et al. (1988b, 1990) have carried out calculations of the low-field mobility and diffusion thermopower of a model of 1D conducting channel. They are not concerned with the problem of the influence of the sub-band structure on transport properties and consider all the electrons to be in the lowest sub-band at all temperatures. The main aim of their work is to examine how much the mobility of a 1D conductor might be enhanced by the restriction on allowed scattering angle. Scattering from background impurities, surface roughness, acoustic and optic phonons are considered and dynamical screening is included. As fig. 8 shows, the calculated mobilities are extremely large when screening is included. There are presumably rather later uncertainties in the values obtained because of the absence of real material parameters for, e.g., surface roughness and the problem of the singularity which is present in the polarisability in 1D. In contrast to the mobility the calculated diffusion thermopower shown in fig. 9 is rather unspectacular. It increases with decreasing electron density and saturates in the non-degenerate limit as would be expected. Phonon drag might also play a significant role (Kubakaddi et al. 1989). Unfortunately, there is no experimental data on such a 1D conductor against which to test these predictions.

3.7. *Effects of sub-band structure on diffusion thermopower*

In a confining potential quasi-2D electrons occupying only the lowest energy subband have a constant density of states of $m^*/\pi\hbar^2$ when spin but not valley degeneracy is

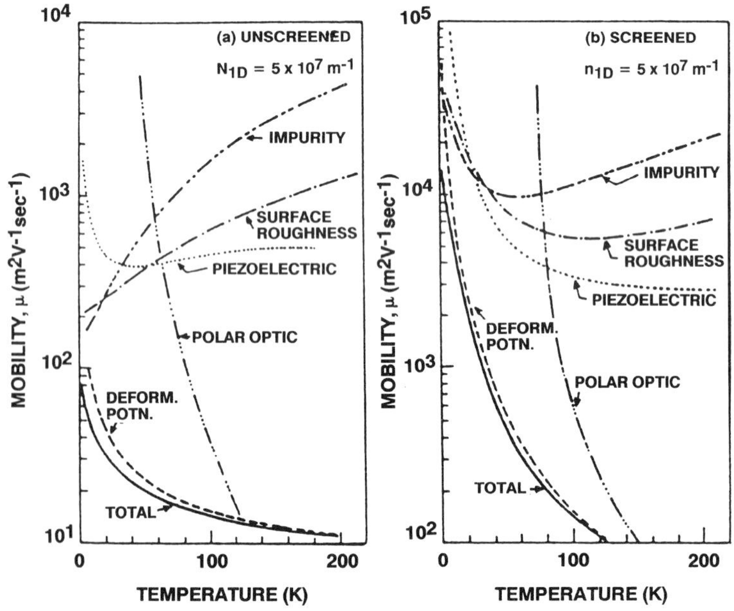

Fig. 8. The calculated mobility variation of a 1D electron gas in GaAs with temperature for 1D electron density $n_{1D} = 1.0 \times 10^8\ m^{-1}$ (a) unscreened, and (b) screened (Kundu et al. 1990).

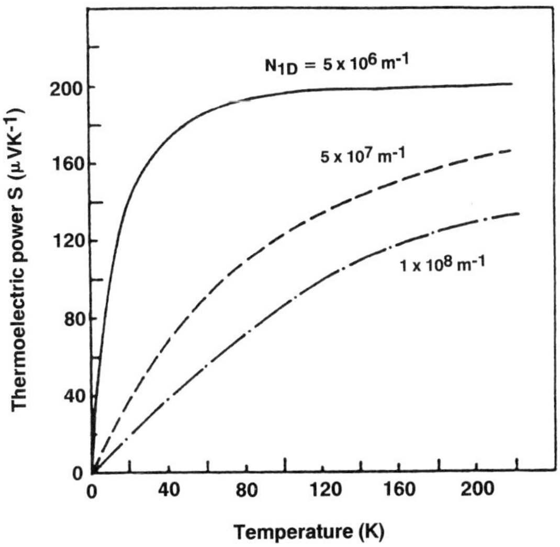

Fig. 9. Calculated thermopower of a 1D electron gas as a function of temperature (screened case) for three different values of 1D electron concentration (Kundu et al. 1990).

included. If there are a number of sub-bands of energy ε_n the density of states will have the staircase structure sketched in fig. 10a. In the corresponding ideal one-dimensional system the individual sub-bands have a density of states given by

$$N(\varepsilon) = \frac{1}{h}\left(\frac{2m^*}{\varepsilon - \varepsilon_n}\right)^{1/2} \tag{19}$$

and the total density of states is as indicated in fig. 10b. Thus in 2D we have discontinuities at the sub-band minima and in 1D square-root singularities. Such strong structure in the density of states would be expected to manifest itself in the transport properties. There have been a number of theoretical studies of this problem. The diffusion thermopower of such systems has received particular attention since it is determined by the energy dependence of the conductivity and is thus expected to show stronger structure than the conductivity itself. The 1D case is a useful model system for calculation but is extremely difficult to realise in practise. It is possible to define narrow 1D channels within 2D conductors electrostatically (Fowler et al. 1982, Thornton et al. 1986) or lithographically (Skocpol et al. 1987) but to our knowledge structures in which sub-band structure is observable have not been realised. In contrast, it is relatively simple to realise a device in which a small number of 2D sub-bands can be investigated by varying the electron density or sub-band separation.

For a given model confining potential it is relatively straightforward to calculate the sub-band structure. For example, Poole et al. (1985) calculate the density of states for a potential appropriate to the device shown in fig. 11a. In fig. 11c the calculated $N(\varepsilon)$ and Fermi energy is shown as a function of the thickness of the conducting channel. As the channel is narrowed a series of 2D sub-bands is progressively

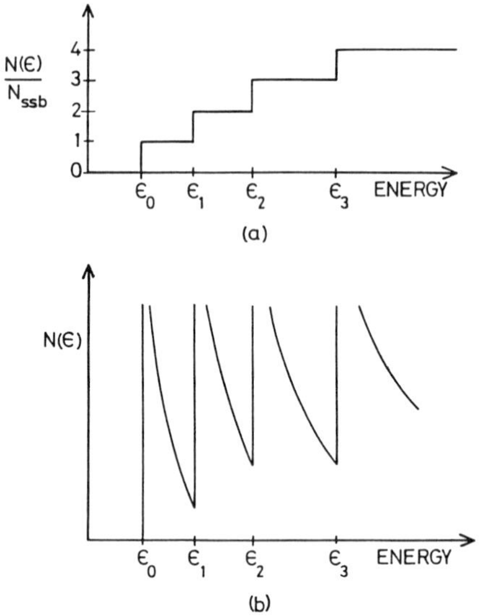

Fig. 10. Schematic diagrams of the density of states for (a) a 2DEG in units of the single sub-band density and (b) a 1DEG. The minima of the first four sub-bands are at $\varepsilon_0, \varepsilon_1, \varepsilon_2$ and ε_3.

depopulated. The calculated and measured Fermi energy are in good agreement indicating that the basic model is sound. The Boltzmann treatment of the zero-temperature electrical conductivity $\sigma(\varepsilon_F)$, as a function of Fermi energy, yields discontinuous behaviour of the type shown in fig. 12 (Ando et al. 1982, Sernelius et al. 1985). The smooth rise of $\sigma(\varepsilon_F)$ as ε_F moves between sub-band minima is due primarily to the increasing electron density. The discontinuous drop when ε_F enters a new sub-band is due to the sudden onset of scattering to the new sub-band from all the previous occupied ones.

On the basis of the Mott expression (15) we expect a negative diffusion thermopower of moderate size away from the discontinuities reflecting the positive value of $d\sigma/d\varepsilon$. A naive application of the Mott expression for ε_F moving through a sub-band minimum predicts a positive δ-function peak of area $k_B T$ in the thermopower. Since σ is not slowly varying on energy scales $k_B T$, we must actually use the integral form for the thermopower (eq. (11)). By doing so, Cantrell and Butcher (1985a,b) find that S is a sum of a negative contribution which is proportional to T and a positive contribution. They behave as is shown in fig. 13. The positive contribution is the thermally broadened version of the δ-function peak and has a width $\sim k_B T$. Since

Fig. 11. (a) Cross section of a variable conducting channel width GaAsFET. Broken lines show depletion region edges which locate the conducting away from the interfaces. (b) Variation of potential through the device (Poole et al. 1982). (c) Calculated sub-band positions and ε_F as a function of channel width for a donor density of $5 \times 10^{23}\ m^{-2}$ for the GaAsFET of Poole et al. (1982). The crosses are the measured values.

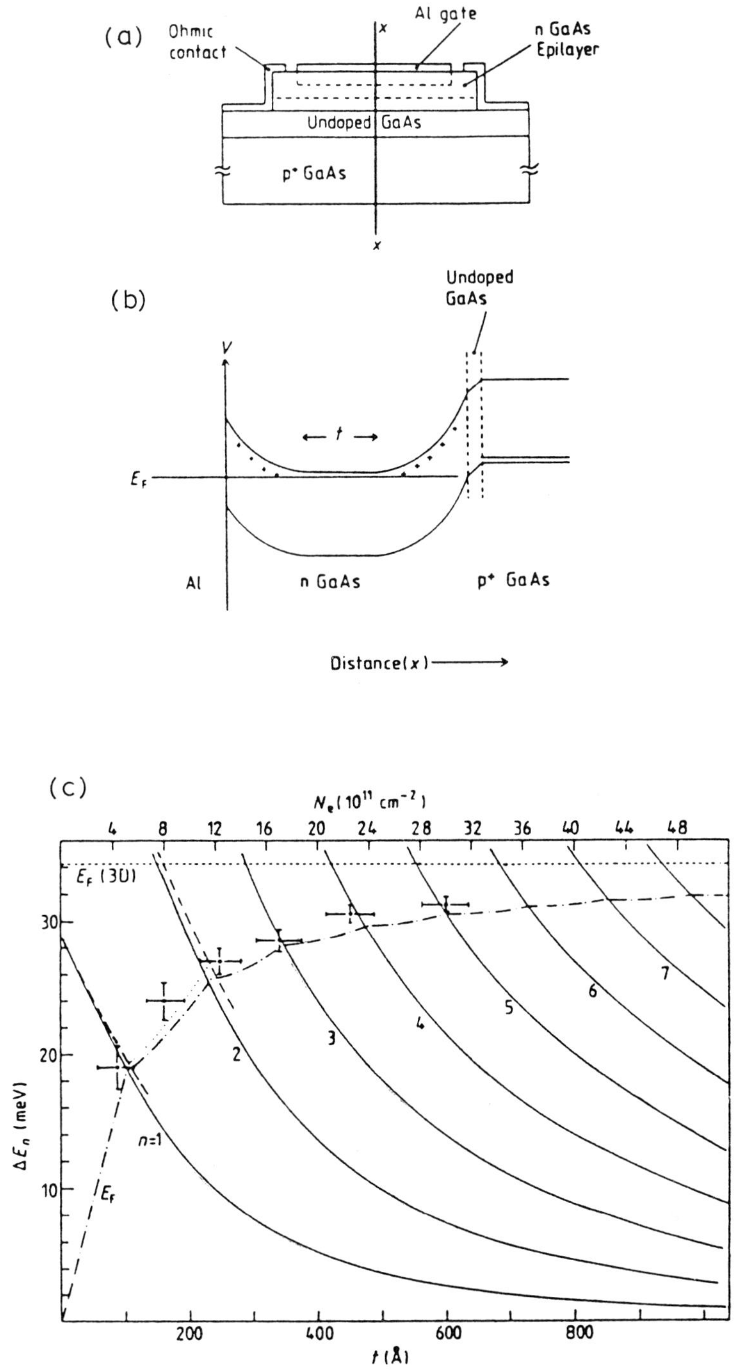
(a)
Ohmic contact
Al gate
n GaAs Epilayer
Undoped GaAs
p⁺ GaAs
x
x
(b)
Undoped GaAs
V
t
E_F
Al
n GaAs
p⁺ GaAs
Distance(x)
(c)
N_e (10^{11} cm^{-2})
4 8 12 16 20 24 28 32 36 40 44 48
E_F (3D)
30
20
10
ΔE_n (meV)
n=1
2
3
4
5
6
7
E_F
200
400
600
800
t (Å)

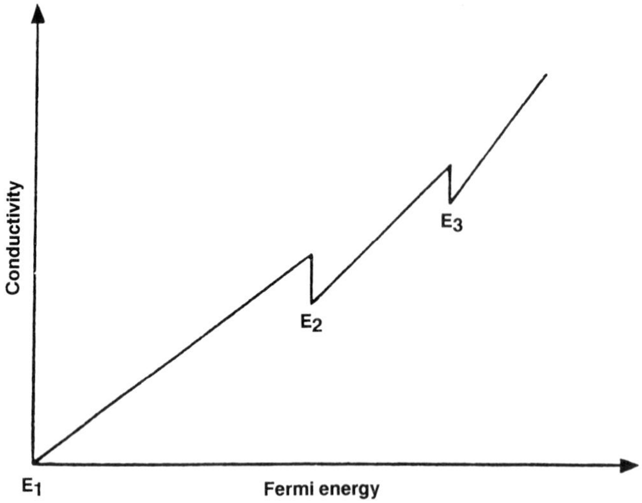

Fig. 12. Schematic plot of the zero-temperature conductivity as the Fermi energy moves through the energy minima of the first three sub-bands, ε_1, ε_2 and ε_3 (Cantrell and Butcher 1985a).

the area and width are both proportional to $k_B T$ the peak height remains constant and has a value of $\sim(k_B/e)2 \ln 2$.

In the discussion so far we have assumed that the levels are sharp. In fact they are broadened by the scattering potential and the ideal "staircase" structure of $N(\varepsilon)$ will be rounded off. This can be taken into account in an approximate way by use of the general Born approximation (Cantrell and Butcher 1985c,d). When this is done the predicted features in $\sigma(\varepsilon_F)$ and $S(\varepsilon_F)$ are less pronounced. A useful measure of the broadening of the states is the value of $k_F l$, where l is the electron mean free path and k_F is the Fermi wavevector. For $k_F l \gg 1$ the k-states are well defined while for $k_F l \sim 1$ the broadening in the states is comparable with k_F. In fig. 14 the calculated $\sigma(\varepsilon_F)$ for a fixed value of $k_F l$ are shown. We see that strong structure remains for $k_F l \sim 11$ but that it is already very weak for $k_F l \sim 4$. Experimental results for the conductivity of the electrostatically controlled device of Poole et al. (1985) which has $k_F l \sim 2$ shows no observable effects, which is consistent with these calculations. In high-mobility GaAs–AlGaAs heterostructures one can increase the carrier concentration by sub-band-gap illumination which excites extra electrons from the DX traps in the AlGaAs. The larger mobility of this system leads to the predicted structure being observable. An example of this from Fletcher et al. (1988c) is shown in fig. 15, and further data is presented in § 5.3.

Cantrell and Butcher (1985d) go on to calculate the corresponding thermopower including lifetime-broadening effects. Figures 16 and 17 show the results obtained for the cases $k_F l = 50$ (which is effectively the Boltzmann limit) and quite strongly disordered cases where $k_F l = 6$ and 3. The weak-scattering case shows the expected large positive-going peaks. In contrast to the conductivity the structure in S is still quite pronounced for $k_F l$ as small as 3 but the sign changes are eliminated. This

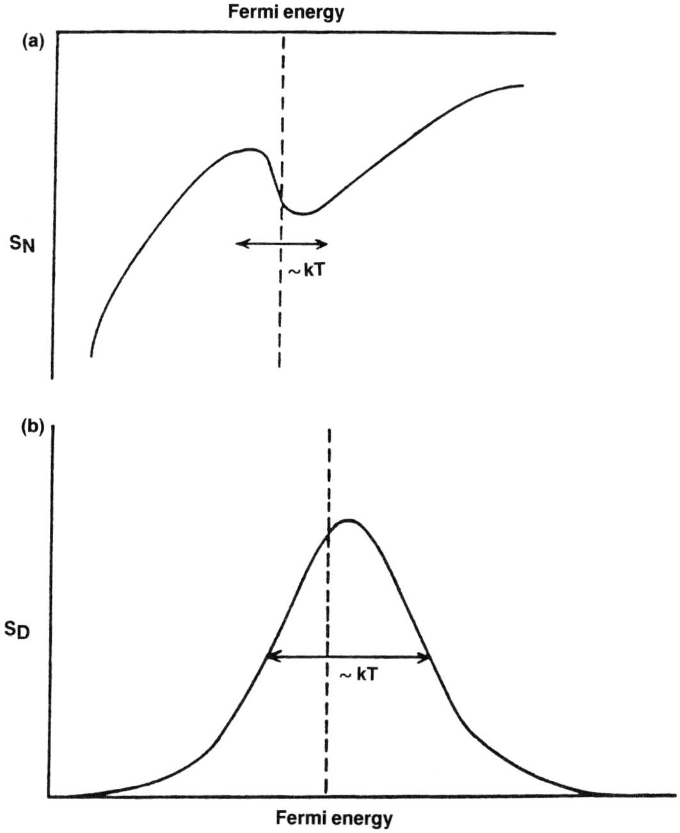

Fig. 13. Schematic plot of (a) the negative and (b) the positive contribution to the thermopower against Fermi energy in the neighbourhood of a sub-band minimum (Cantrell and Butcher 1985a).

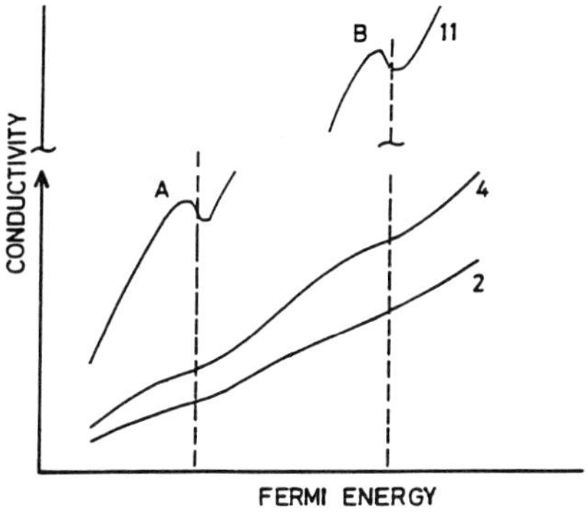

Fig. 14. The variation of the 2DEG conductivity with Fermi energy for the values $(k_F l)_{2D}$ given by the curves. The dashed lines indicate the positions of the second and third sub-band minima (Cantrell and Butcher 1985c).

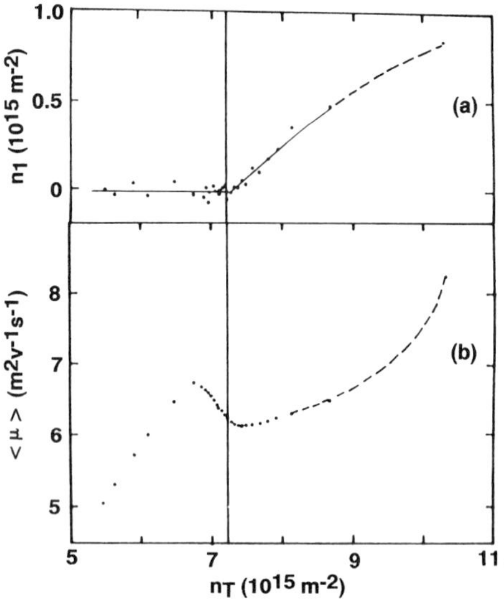

Fig. 15. (a) Second sub-band density n_1 versus total density n_T. (b) Average mobility versus n_T. The vertical line at $n_T = 7.2 \times 10^{15}$ m^{-2} denotes the first observation of second sub-band occupancy in ρ_{xx}. The lines through the data points are simply a guide to the eye (from Fletcher et al. 1988c).

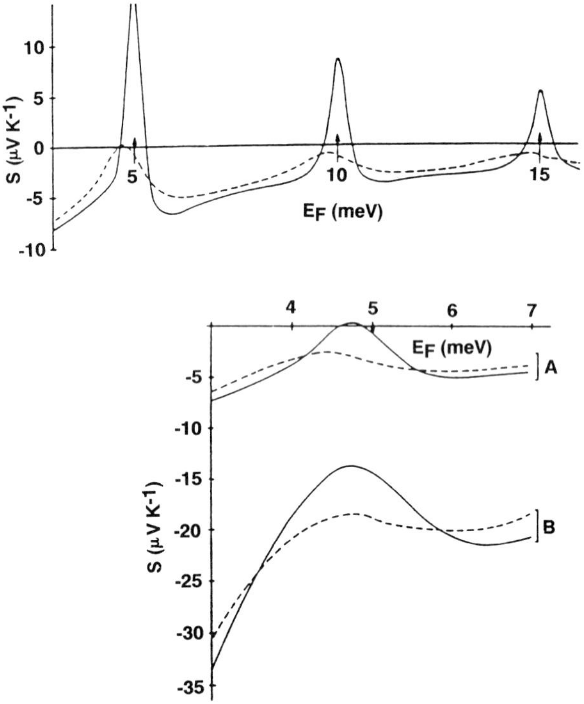

Fig. 16. Thermopower against Fermi energy (top) $(k_F l)_{2D} = 50$ (full curve) and 6 (broken curve); both at 1 K; (bottom) $(k_F l)_{2d} = 6$ (full curves) and 3 (broken curves) for $T =$ (A) 1 K and (B) 5 K (Cantrell and Butcher 1985c).

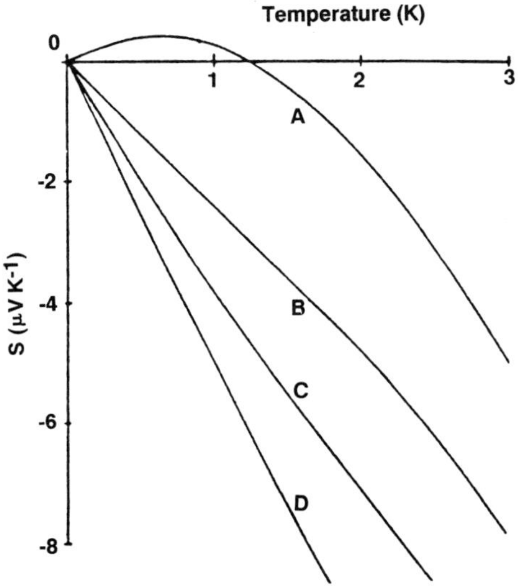

Fig. 17. Thermopower against temperature for $(k_F l)_{2D} = 6$ for values of ε_F of (a) 4.8 meV, (b) 5.2 meV, (c) 4 meV and (d) 6 meV, around a sub-band minimum at 5.0 meV (Cantrell and Butcher 1985c).

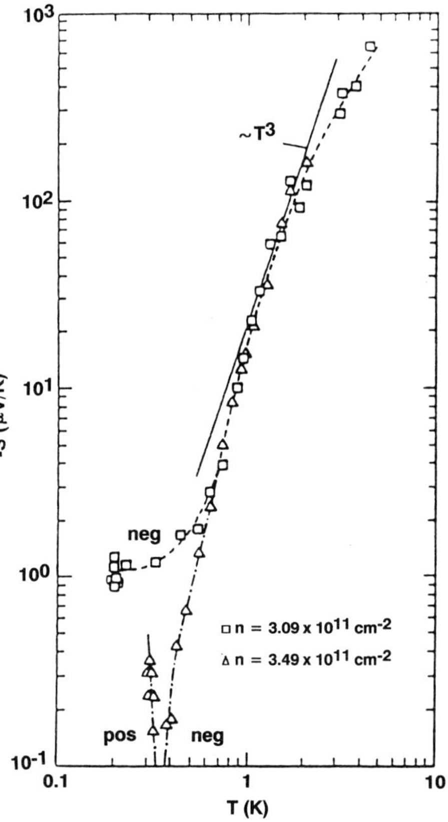

Fig. 18. Experimental plots of $-S$ against T for $n_s = 3.09$ and $3.49 \times 10^{15}\ \mathrm{m}^{-2}$. Phonon drag dominates above 1 K. S changes sign at 0.35 K in the latter case because ε_F is close to the bottom of the second sub-band (Ruf et al. 1989).

strong structure in $S(\varepsilon_F)$ will lead to large deviations from the usual linear dependence of the diffusion thermopower upon T as is shown in fig. 17 and can, for ε_F near a sub-band minimum, lead to $S(T)$ changing from negative to positive as T is reduced. Such quantum size effects have recently been observed in high-mobility heterostructures by Ruf et al. (1989). They again use persistent photoconductivity to vary electron density. As noted in § 3.4 they had previously shown that the diffusion thermopower was the dominant contribution in their samples below ~0.6 K. When the electron density was such that ε_F was close to the second sub-band minimum, however, the low-temperature behaviour is far from a simple T dependence and shows the predicted sign change (fig. 18). Plotting S/T against ε_F (fig. 19) shows the expected positive-going peak in $S(\varepsilon_F)$ when second sub-band occupation starts at ~12 meV. This seems to be in excellent agreement with theory though the temperature dependence of the peak is unexpectedly strong. The further sign change at higher ε_F is possible due to starting to occupy a further sub-band. Fletcher et al. (1991) have recently studied the effects of second sub-band occupation on the thermopower of high-mobility heterojunction at higher temperatures. In this case the effects, which are considered in § 5.3, are probably due to the phonon-drag thermopower.

Kearney and Butcher (1986, 1987) have calculated σ and S for the corresponding 1D case. The square-root singularities in the density of states leads to a vanishing $\sigma(\varepsilon_F)$ at the sub-band minima in the Boltzmann approach. The inclusion of lifetime broadening removes the singularities but the predicted structures in $\sigma(\varepsilon_F)$ and $S(\varepsilon_F)$, which are shown in figs. 20 and 21, are stronger than in the 2D case. As noted in

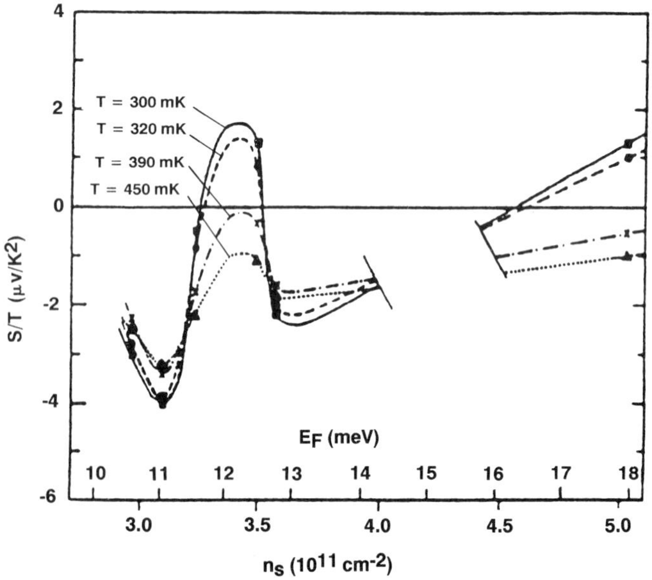

Fig. 19. Experimental plots of S/T against ε_F and n at four temperatures. The positive peak is due to the presence of the second sub-band minimum at 12.5 meV (Ruf et al. 1989).

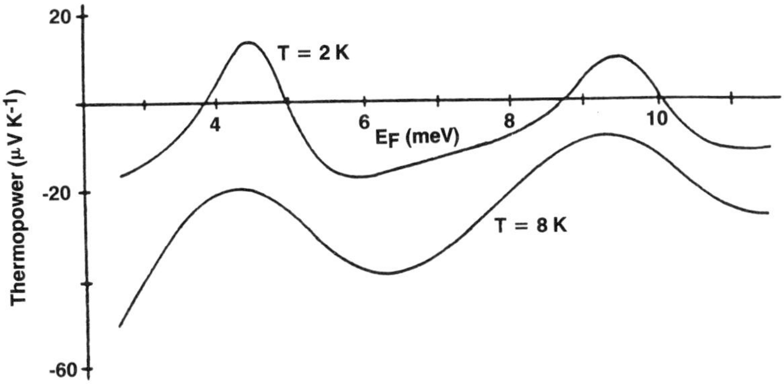

Fig. 20. The calculated electrical conductivity as the chemical potential moves through the first three 1D sub-bands for $k_F l \sim$ (a) 30, (b) 5, at a temperature of $T = 2$ K (Kearney and Butcher 1987).

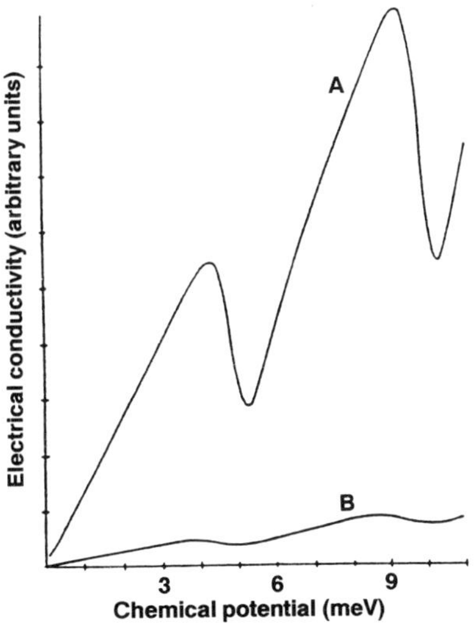

Fig. 21. The calculated thermopower for the $k_F l \sim 5$ case of fig. 20 (Kearney and Butcher 1987).

§ 3.6, however, a device in which such effects might be observed has not yet been realised.

3.8. Diffusion thermopower in superlattices

In principle, one can obtain an artificial mini-band structure in superlattices of alternating semiconductor materials with precise interfacial and dimensional control (Chang and Esaki 1980). The corresponding density of states will have a form similar to that of the 2D sub-bands as is shown in fig. 22. One therefore expects to see the same sort of structure in the thermopower with the usual negative thermopower away from the mini-band edges and the possibility of positive values close to them.

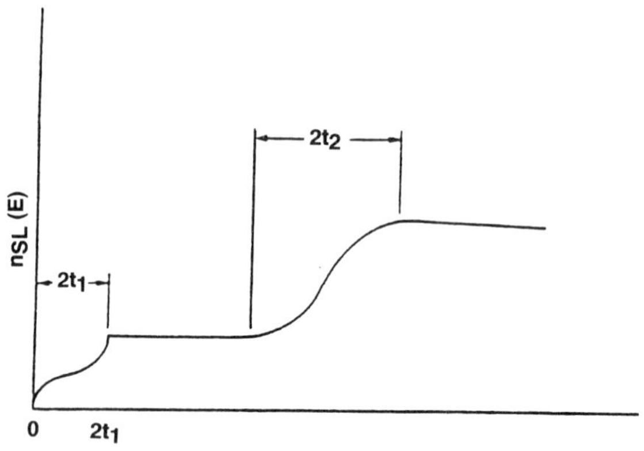

Fig. 22. Density of states distribution for a superlattice (Tao and Friedmann 1985).

Friedmann (1984) and Tao and Friedmann (1985) considered this problem using a simple model and the Mott formula (15). This precludes calculating S very close to the mini-band edges. The predicted thermopower shows the expected behaviour (fig. 23). Kubakaddi et al. (1991) calculate the corresponding phonon-drag thermopower. This shows similar sign changes due to U-processes. Their results indicated that this will generally be the dominant term. Unfortunately well-defined mini-bands have not been observed in practise except in novel ballistic devices (Kouwenhoven et al. 1990) and no thermopower measurements have been attempted.

3.9. *Thermopower in periodically modulated 2D conductors*

It has recently been found by Weiss et al. (1989) that if one applies a weak 1D or 2D periodic potential to a two-dimensional electron gas either by use of persistent

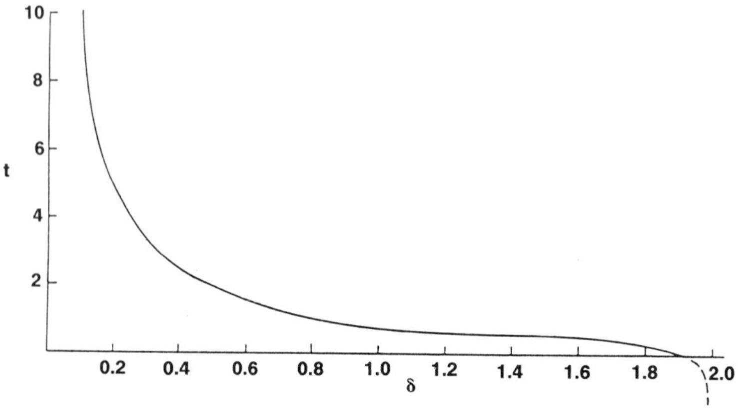

Fig. 23. Energy dependence of thermopower for a temperature gradient perpendicular to layers of a superlattice (Tao and Friedmann 1985).

photo-conductivity or a fine gate structure (see fig. 24) one can observe novel low-field magnetoresistance oscillations. These effects arise from the commensurability between the modulation period and the cyclotron orbit at the Fermi energy. With a 1D modulation the symmetry is broken and the transport tensors become asymmetric. Examples of these new oscillations are shown in fig. 25. Recently Peeters and Vasilopoulus (1990) have calculated the corresponding oscillatory structure expected in the thermopower and electronic thermal conductivity. The resistivities ρ_{xx} and ρ_{xy} shown in fig. 24 oscillate out of phase with ρ_{yy}. The corresponding thermopowers S_{xx}, S_{yy}, and S_{xy} are all predicted to oscillate in phase and to show the dramatic oscillatory structure shown in fig. 26. The predicted magnitudes are however of less than $\sim 1\ \mu V\ K^{-1}$. The measurement of such an effect would be an important test of the theory but would be very challenging experimentally.

4. Phonon-drag thermopower

In a thermopower experiment we establish a temperature gradient and measure the resultant voltage difference. The diffusive motion of the electrons in the temperature

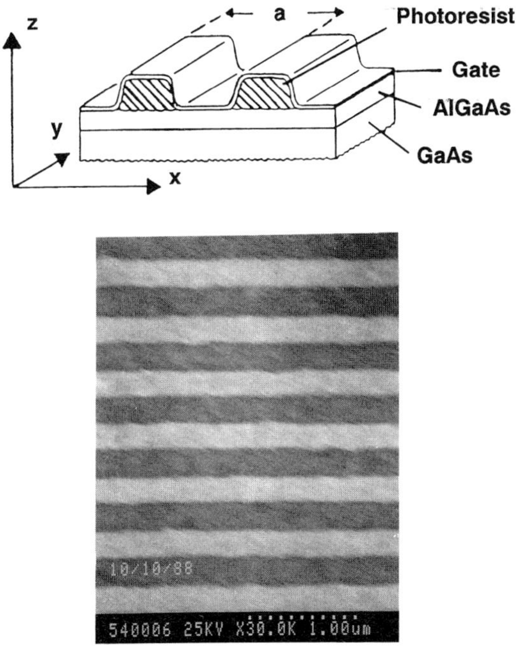

Fig. 24. A schematic diagram of the process used to fabricate one-dimensional superlattices and a scanning-electron microscope photograph of a typical device (Beton et al. 1990).

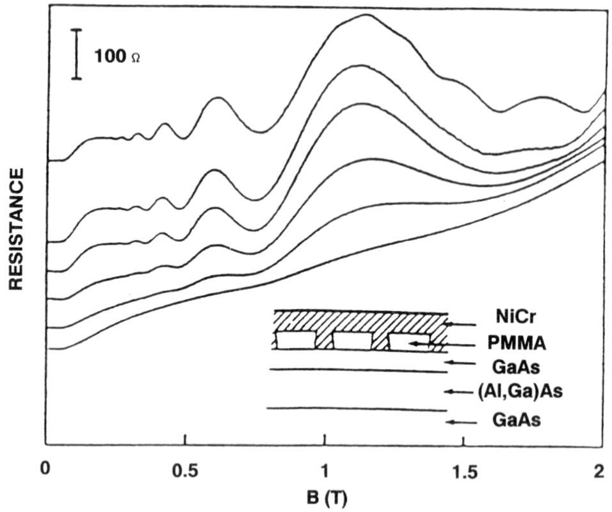

Fig. 25. Magnetoresistance at various temperatures of a 2DEG subjected to a periodic potential of period 0.15 μm. The curves are offset for clarity and correspond to temperatures 4.2 K (top), 6.9, 10, 15, 23 and 41 K (bottom). Inset: schematic representation of the gate fabrication (Beton et al. 1990).

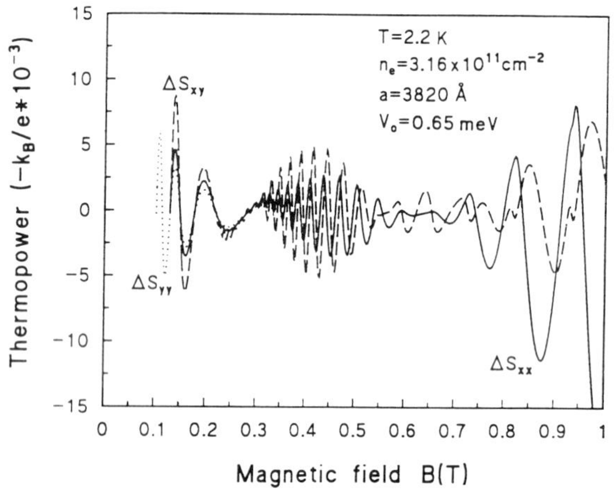

Fig. 26. Calculated correction to the thermopower due to a periodic 1D modulation potential (Peeters and Vasilopoulos 1990).

gradient gives rise to the diffusion thermopower which is an intrinsic property of the electron gas. However, there will also be a net phonon momentum associated with the heat flow down the specimen. Electron–phonon scattering will then lead to the transfer of a fraction of this momentum to the electrons and a resultant electric field.

This effect is called phonon-drag thermopower, S_g. The calculation of S_g is complicated since details of both the non-equilibrium electron and phonon distribution functions are required. Furthermore, as emphasised by Smith and Butcher (1990), while the simple models which can be constructed may be illuminating, care must be taken as they can also be misleading.

4.1. Simple models: the Seebeck and Peltier approach

The most useful of the simple approaches is the force balance argument for the Seebeck coefficient (Blatt et al. 1976, Guénault 1971). Consider first the 3D case. If the phonons in a conductor are treated as a gas of energy density $U(T)$ they will exert a pressure

$$P(T)=\tfrac{1}{3}U(T). \tag{20}$$

In a temperature gradient $\mathrm{d}T/\mathrm{d}x$ there is then a net force

$$F_x=\frac{-\mathrm{d}P}{\mathrm{d}x}=\frac{-1}{3}\frac{\mathrm{d}U}{\mathrm{d}T}\frac{\mathrm{d}T}{\mathrm{d}x}=\frac{-C_\mathrm{v}}{3}\frac{\mathrm{d}T}{\mathrm{d}x}. \tag{21}$$

We suppose that some fraction f_{pe} of this force is exerted on the carriers with density n. Thus f_{pe} is the fraction of momentum lost by the phonon to the carriers. Since $j=0$, the force balance condition for the carriers is

$$n|e|E_x=-f_{\mathrm{pe}}\frac{C_\mathrm{v}}{3}\frac{\mathrm{d}T}{\mathrm{d}x} \tag{22}$$

and we obtain

$$S_\mathrm{g}=-f_{\mathrm{pe}}\frac{C_\mathrm{v}}{n|e|}. \tag{23}$$

If the phonon momentum relaxation time excluding electron–phonon scattering is τ_p' and the phonon momentum relaxation time due only to scattering from carriers is τ_{pe} this equation can be written as

$$S_\mathrm{g}=\frac{-C_\mathrm{v}}{3n|e|}\frac{\tau_\mathrm{p}'}{\tau_\mathrm{p}'+\tau_{\mathrm{pe}}}. \tag{24}$$

This approximate form can be obtained in a more rigorous manner from the general theory (Smith and Butcher 1990).

At high temperature, τ_p' is limited by phonon–phonon scattering and $\tau_\mathrm{p}'\ll\tau_{\mathrm{pe}}$ so that

$$S_\mathrm{g}=\frac{-C_\mathrm{v}}{3n|e|}\frac{\tau_\mathrm{p}'}{\tau_{\mathrm{pe}}}. \tag{25}$$

This expression will also be correct for small n. At high temperature, $\tau_p' \propto T^{-1}$ and τ_{pe} is independent of T. Consequently since C_v is constant, S_g falls as T^{-1}. In the opposite limit of large electron density and strong electron–phonon scattering we obtain $S_g = -C_v/3ne$. In this limit the phonons and electrons drift with the same velocity.

For low T and low n, τ_p' will be limited by boundary scattering and eq. (24) becomes

$$S_g = \frac{-C_v}{3n|e|}\frac{\Lambda}{v_s\tau_{pe}}, \tag{26}$$

where Λ is the phonon mean free path (which is approximately the minimum specimen dimension) and v_S is the sound velocity. For the case of a quasi-2D conducting channel within a bulk specimen, n will usually be very low and eqs. (25) and (26) will apply. In this case the natural electron density is the surface density n_s. We must take care in deciding whether n should be replaced by n_s/δ (where δ is the channel width) or n_s/L_z (where L_z is the specimen width in the confinement direction). If we consider eq. (22) we see that for a specimen of volume V the total force on the right-hand side is obtained by multiplication by V. On the left-hand side, however, we must replace n by $n_s A$ where $AL_z = V$, but f_{pe} is still the fraction of the total phonon momentum delivered to the electron gas. The low dimensionality of the conducting channel is thereby already accounted for. That is, one would expect f_{pe} in this case to be smaller than if the electron density extended through the whole sample by a factor $\sim\delta/L_z$. Consequently,

$$S_g = \frac{-C_v L_z}{3n_s|e|}\frac{\Lambda}{v_s\tau_{pe}}. \tag{27}$$

If τ_{pe} is independent of n_s and only weakly temperature dependent then eq. (27) gives reasonable qualitative agreement with the experimental data. In heterojunctions and MOSFETs at helium temperatures, S is found to increase as $\sim T^3$ and to show the predicted n_s^{-1} and Λ dependencies (see § 4.2). It is important to emphasise that this behaviour is characteristic of the *degenerate* case. In a non-degenerate electron gas, τ_{pe}^{-1} is proportional to n_s and S_g is independent of n_s (Smith and Butcher 1990). It is difficult to predict the *magnitude* of S_g from eq. (26) since τ_{pe} must be estimated.

Inspection of eq. (26) suggests an alternative form since the lattice thermal conductivity, K, is given by

$$K = \tfrac{1}{3}C_{vS}\Lambda. \tag{28}$$

Hence

$$S = \frac{K}{nev_S^2}\tau_{pe}^{-1}, \tag{29}$$

and τ_{pe} can be obtained approximately from S/K. This expression can in fact be obtained directly by a simple argument. In a temperature gradient the heat flux is

$$\boldsymbol{Q} = -K\,\nabla T \tag{30}$$

and the net phonon momentum density $\boldsymbol{P}$ is Qv_S^{-2} (Herring 1954) so that

$$\boldsymbol{P} = \frac{K}{v_S^2}\nabla T \tag{31}$$

and the rate at which momentum is given to the electrons is just P/τ_{pe}. Consequently the force balance condition is

$$-n|e|\boldsymbol{E} = \frac{\boldsymbol{P}}{\tau_{pe}} = \frac{K}{v_S^2\tau_{pe}}\nabla T, \tag{32}$$

which gives eq. (29).

In the alternative Peltier approach one must consider an electric current dragging phonons which then bring additional thermal energy to the junctions. In the first theoretical treatment of phonon drag Herring (1954) develops a number of formulae including a simple model based on this approach which is further discussed by Nicholas (1985).

Consider an electric field $\boldsymbol{E}$ acting on the carriers, again we have a force $-nve\boldsymbol{E}$. The fraction of this force which acts on the phonon system is the same as the fraction of the momentum transferred to the phonons, i.e., f_{ep}. Now if the phonons have a momentum density $\boldsymbol{P}$ and momentum relaxation time τ_p the force balance equation is

$$f_{ep}n|e|\boldsymbol{E} = \frac{\boldsymbol{P}}{\tau_p}.$$

As noted above the phonon heat flux, Q, is $v_S^2 P$ and so we have

$$S = \frac{\pi}{T} = \frac{1}{T}\frac{\dot{Q}}{j} = \frac{v_S^2 n|e|\tau_p}{T} f_{ep}\frac{\boldsymbol{E}}{j} = \frac{v_S^2\tau_p f_{ep}}{\mu T}, \tag{33}$$

where μ is the carrier mobility. Equation (33) may be written as

$$S = \left(\frac{\Lambda\, v_S}{\mu T}\right) f_{ep} \tag{34}$$

or

$$S = \frac{\Lambda\, v_S}{\mu_p T}, \tag{35}$$

where μ_p is the phonon limited mobility. Now, while eqs. (34) and (35) should be reasonable estimates of S and entirely equivalent to those obtained in the Seebeck approach, their implications are unclear. The quantities f_{ep} and μ_p will generally be strong functions of T and dependent upon n. For metals at low temperature, $\mu_p \sim T^{-5}$ and so $S \sim T^4$ might be expected. Thus these formulae are of little use unless μ_p can be measured or independently calculated since the expected dependencies upon T and n_s are obscured.

Equation (35) can also be obtained by approximation to the full expression arising from the Boltzmann approach. However, it is found to be correct only in the non-degenerate limit or the limit of elastic scattering (Smith and Butcher 1990).

4.2. The Boltzmann approach applied to heterostructures and MOSFETs

The formal theory of phonon drag thermopower for a 2DEG is developed by Cantrell and Butcher (1987a,b). They considered the coupled electron and phonon Boltzmann equations for the distribution functions. On the assumption that the electron-phonon coupling is "weak" the equations are linearised and solved in a manner similar to the 3D calculation of Bailyn (1958, 1967). "Weak" in this context means that, for the electrons, the scattering rate from impurities is much greater than that due to phonon absorption and emission. For the phonons the scattering rate from boundaries and phonon–phonon interactions must be much greater than that from the electrons. Both of these conditions are easily met even in high mobility heterostructures. The specific model discussed in these papers is particularly simple. The two-dimensional system is considered to be a sheet of charge confined by an infinite square-well potential embedded within a bulk homogeneous sample of semiconductor. Thus the true electron wavefunctions and problems of interface modes and scattering are ignored. The electron–phonon coupling is considered to be due only to the unscreened deformation potential. It is worth noting that this general formalism yields not only the diffusion and phonon-drag thermopower but also the modification to the electron dispersion relationships due to the electron–phonon interaction. This renormalisation of the electron effective mass will generally be small and has been ignored. It should be noted, however, that such renormalisation would directly modify the thermopower (Howson and Gallagher 1988).

Before considering the subsequent improvements in the detail of the model and the comparisons with the experimental results it is useful to consider the general expectation for the dependencies upon temperature and electron density. From the simple model (§ 4.1) we have

$$S_g \propto \frac{\Lambda(T)}{n_A} C_v(T) f(n_s, T). \tag{36}$$

In the CB theory one also obtains a factor Λ/n_s on the assumption of constant phonon mean free path but the integrals over the phonon wavevectors cannot simply be decomposed into a measure of the total momentum in the phonon systems ($C_v(T)$) and the effectiveness of momentum transfer from the phonons to the electrons ($f(n_s, T)$). We can still, however, describe the general behaviour in such a language.

Consider firstly how the requirements of wavevector conservation in the interaction between the 3D phonons and 2D electrons will influence $f(n_s, T)$. For a 2D system with a very small spatial extent in the direction normal to the plane (the z-direction) the uncertainty, Δk_z, in the z-component of the electron wavevector will be very large and momentum conservation in this direction will be unimportant. For a finite spacial extent, δ, the upper limit on the q_z value of the phonons which can couple with the electrons will be

$$q_{z,\max} \sim \Delta k_z \sim 2\pi/\delta.$$

At a temperature T the dominant phonon wavevector is given by $q_{z,\mathrm{d}} \sim 3k_B T/\hbar v_S$

where v_S is the sound velocity. So for $q_{z,d} > q_{z,\max}$, i.e.,

$$T \geqslant \frac{hv_S}{3k_B\delta} = T_z, \tag{37}$$

we expect $f(n_A, T)$ to rapidly decrease. This is illustrated in fig. 27.

Wavevector conservation is strictly maintained in the xy-plane. This means that electrons can only interact with phonons of wavevector $\boldsymbol{q}$ in the xy-plane if $q \leqslant 2k_F$. In comparing this with the phonon wavevectors excited at a given temperature the appropriate "dominant phonon wavevector" does not correspond to the peak in the phonon density of states (as above) but the peak in the product of this and the wavevector, i.e., it is weighted by the momentum transferred. In this case $q_0 \sim 5k_B T/\hbar v_S$. Thus we again expect a rapid fall-off in $f(n_A, T)$ above a characteristic temperature

$$T_p \approx \left(\frac{n_s}{\pi}\right)^{1/2} \frac{hv_S}{5k_B}. \tag{38}$$

The dependence of S_g on the relative size of $2k_F$ and q_d is, however, much more

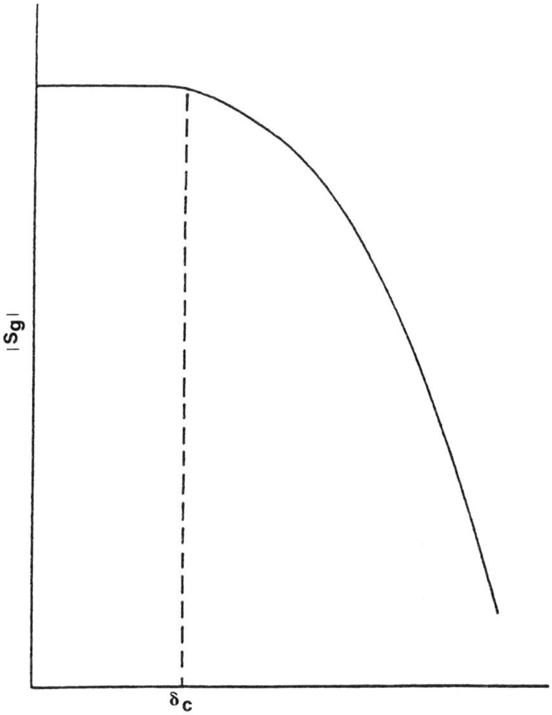

Fig. 27. A schematic plot of the phonon-drag thermopower against channel width for fixed temperature. The broken vertical line indicates the channel width $\delta \sim 3\pi hv/k_B T$ beyond which thermal phonons with $q > q_{z,\max}$ no longer contribute to S_g (Cantrell and Butcher 1987a).

dramatic than might be imagined from the discussion so far. This is because the electron–phonon scattering rate is also a strong function of $q/2k_F$. For small $q/2k_F$ the density of final available states into which the electrons can scatter is small. As q approaches $2k_F$, however, this increases dramatically, becoming divergent at $q = 2k_F$ in the absence of level broadening. This "Kohn anomaly" is much stronger in 2D than in 3D because of the restriction in phase space. The expected behaviour due to this is illustrated in fig. 28 where the contributions to the integrand occuring in the CB theory are shown. For T much less than T_p the scattering rate, and thus $f(n, T)$, is very small. As T increases $f(n, T)$ will increase until it reaches a maximum

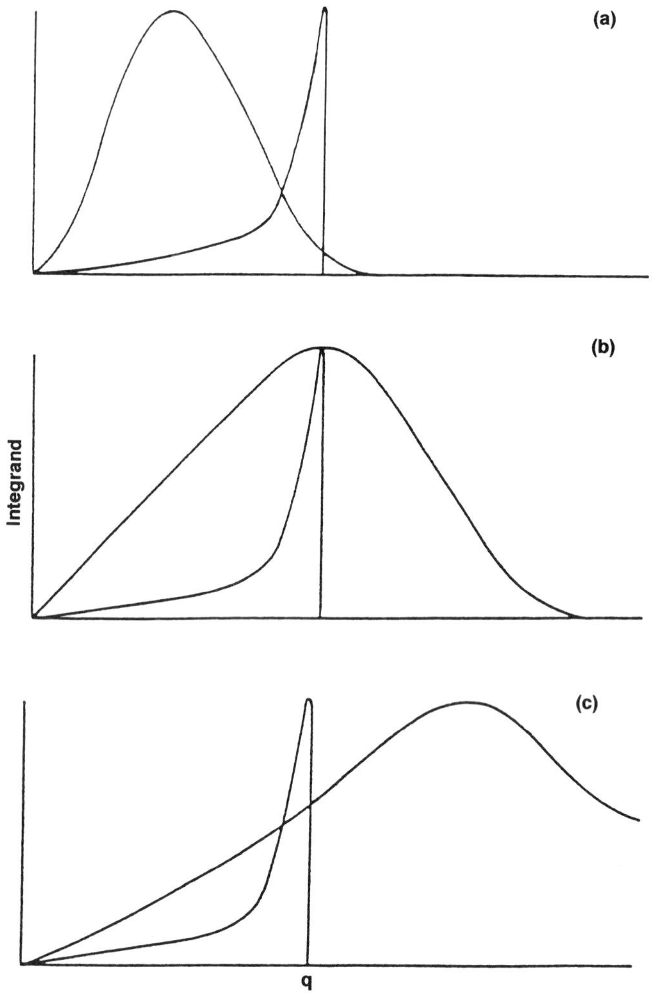

Fig. 28. Phonon distribution curve (smooth peak) and electron–phonon scattering rate (sharp peak) as a function of wavevector, q. (a) $T < T_p$, the temperature at which peak occurs in $|S_g|/T^3$; (b) $T = T_p$; (c) $T > T_p$ (Cantrell and Butcher 1987a).

for $T \sim T_p$. Above this temperature it will decrease as noted above. One therefore expects peaks in $f(n, T)$ when $2k_F \sim q_d$.

This behaviour has been seen in Si MOSFETs by Zavaritskii (1984) and Gallagher et al. (1987, 1990a). In this case $\delta \sim 80$ Å and $v_S \sim 6000\ \mathrm{ms}^{-1}$. This gives $T_z \sim 13$ K while for the accessables range of electron densities $T_p \sim 3$ K and so the peaks should be apparent and the cut-off in q_z of no importance for $T \leqslant 10$ K. When the measured thermopowers are plotted as a function of T (fig. 29) one sees that the behaviour appears close to T^3 showing that the phonon-drag contribution is dominant. This implies that the temperature dependence of $f(n_A, T)$ is relatively weak compared to that of $C_v(T)$ as expected. It is also observed that S is roughly inversed proportional to the number of "free" electrons for a given T (see § 8). The behaviour of $f(n_A, T)$ can, however, be revealed if S/T^3 or S/K, where K is the thermal conductivity, is plotted against T as is done in fig. 30 which shows the expected peaks. Furthermore

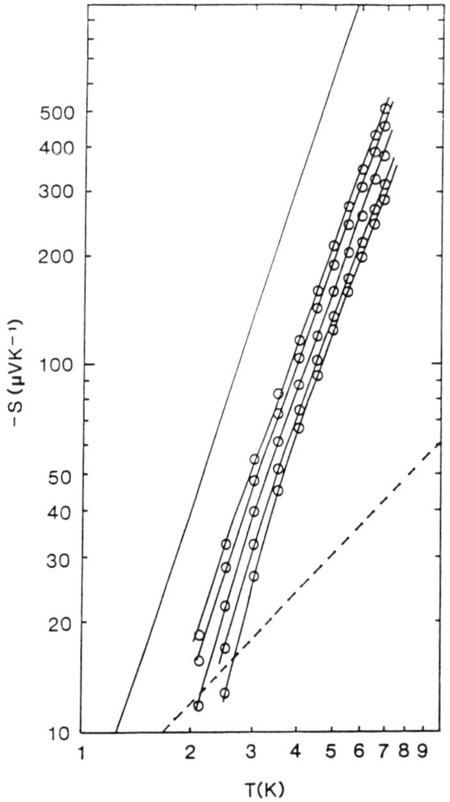

Fig. 29. Variation of thermopower with temperature for a MOSFET. Electron number densities uppermost line $4.90 \times 10^{15}\ \mathrm{cm}^{-2}$, second $5.47 \times 10^{15}\ \mathrm{m}^{-2}$, third $6.57 \times 10^{15}\ \mathrm{m}^{-2}$, fourth $8.35 \times 10^{15}\ \mathrm{m}^{-2}$, fifth $10.7 \times 10^{15}\ \mathrm{m}^{-2}$. The straight line is proportional to T^3. The curves through the points are a guide to the eye. The dashed line is the diffusion contribution given by the Mott formula with $p = 0.2$ for $n_s = 5 \times 10^{15}\ \mathrm{m}^{-2}$ (Gallagher et al. 1987).

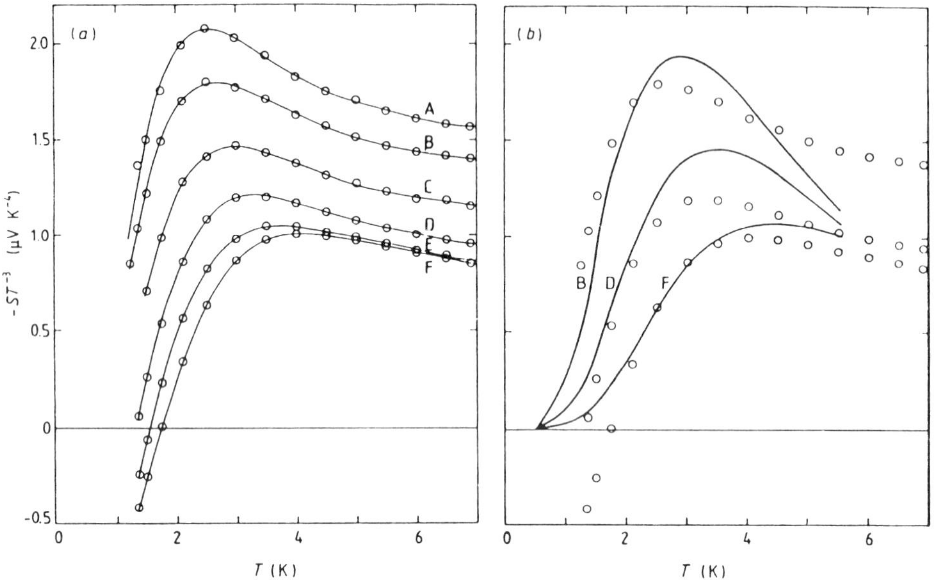

Fig. 30. (a) Thermopower divided by T^3 for electron number densities (A) $4.90 \times 10^{15}\ \mathrm{m}^{-2}$, (B) $5.47 \times 10^{15}\ \mathrm{m}^{-2}$, (C) $6.57 \times 10^{15}\ \mathrm{m}^{-2}$, (D) $8.35 \times 10^{15}\ \mathrm{m}^{-2}$, (E) $10.7 \times 10^{15}\ \mathrm{m}^{-2}$, (F) $14.1 \times 10^{15}\ \mathrm{m}^{-2}$. The curves through the points are a guide to the eye. (b) Data points and theory curves for (B) $5.47 \times 10^{15}\ \mathrm{m}^{-2}$, (D) $8.35 \times 10^{15}\ \mathrm{m}^{-2}$, (F) $14.1 \times 10^{15}\ \mathrm{m}^{-2}$ (Gallagher et al. 1990a).

when the peak positions are plotted against $n_A^{1/2}$ (fig. 31) they are found to be in remarkable good agreement with the simple prediction of eq. (38). The difference in slope for the (100) and (111) cases is due to the difference in valley degeneracy and enabled Zavaritskii to identify the (111) degeneracy as six. Syme et al. (1988, 1989) measured S/K in silicon on sapphire samples and found two peaks corresponding to the two different k_F values in the two occupied sub-bands (fig. 32). In fact, the peaks are considerably broadened by the fact that both longitudinal and transverse phonons with differing velocities can couple to the electrons.

For the case of GaAs the coupling is predominantly to longitudinal phonons and one expects a sharper peak. Such a peak has recently been observed by Oxley et al. (1992) as is shown in fig. 33. The dramatic fall on the high-temperature side is not, however, entirely due to in-plane momentum conservation. If one looks at the observed S_g found in a number of studies one sees that in this case the underlying T^3 dependence is not always apparent. In the data of Ruf et al. (fig. 4) one can see that $S \sim T^3$ in a rather narrow range around 1.5 K. This is because for the sample $T_p \sim 1.5$ K, so $f(n_A, T)$ is slowly varying in this range, and the underlying T^3 is apparent. For lower temperatures S_g falls rapidly and the linear diffusion terms come to dominate (see § 3.4). For higher temperatures we have $T > T_p$ and $f(n_s, T)$ falls rapidly. However, for a typical GaAs heterostructure sample $T_z \sim 6$ K and so for $T > 6$ K the fall-off in $f(n_s, T)$ is more dramatic and S_g increases much more slowly than T^3. As the temperature is further increased Fletcher et al. (1986) find that S_g

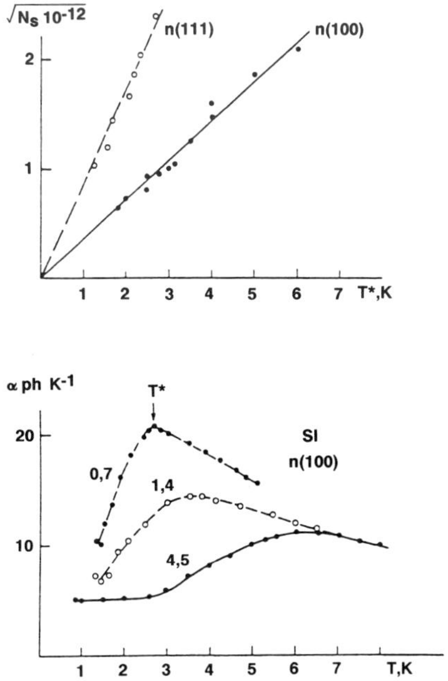

Fig. 31. (a) "Kohn anomaly" peaks observed in Si MOSFETs grown on the (100) and (111) surface. (b) Peak position as a function of $n_s^{1/2}$ (Zavaritskii 1984).

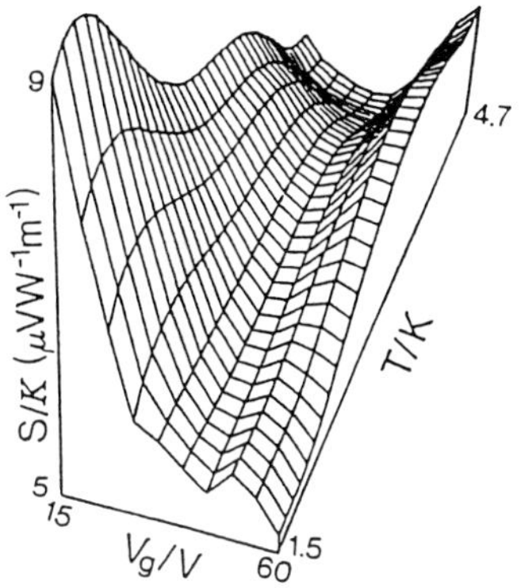

Fig. 32. Surface plot of the thermopower divided by thermal condivity, S/K, against temperature and gate voltage for a silicon-on-sapphire MOSFET (Syme et al. 1989).

reaches a maximum and then actually decreases with increasing T (fig. 34). The situation is further complicated by the fact that $\Lambda(T)$ decreases below the boundary scattering value at these higher temperatures due to phonon–phonon scattering. This is apparent in the measured thermal conductivities, $K(T)$, shown in fig. 35. This effect can be removed to a great extent by considering the temperature dependence of the

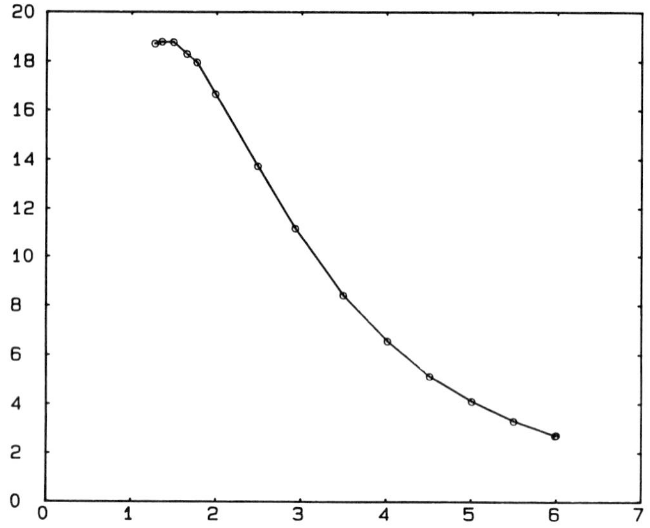

Fig. 33. "Kohn anomaly" peak for a GaAs heterostructure observed in a plot of S/T^3 (Oxley et al. 1992).

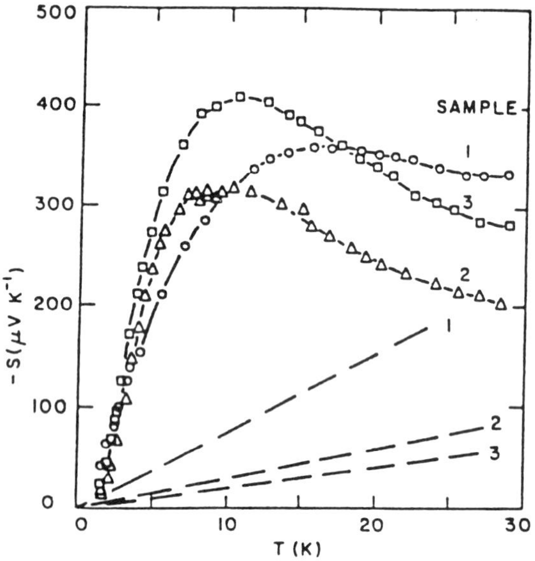

Fig. 34. Zero-field thermopower of three GaAs heterostructure samples. The straight lines through the origin are estimates of the diffusion contribution according to the Mott equation (Fletcher et al. 1986).

ratio S/K which in the simple picture is independent of Λ. This yields a much clearer "Kohn anomaly" peak (fig. 36). Similarly one can remove the dependence of $S(T)$ on $\Lambda(T)$ by normalising this by the factor $\Lambda(T)/\Lambda(T\rightarrow 0)$. Doing so yields a less dramatic departure from T^3 at higher temperatures (fig. 37).

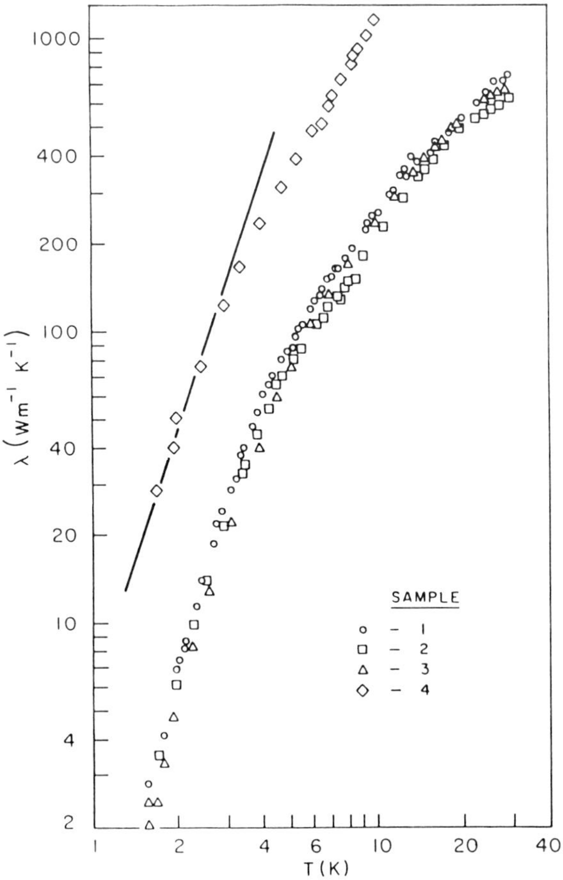

Fig. 35. The thermal conductivity λ for the samples of fig. 34 plus an additional sample. The line through sample 4 represents a T^3 variation as would be expected for boundary scattering at low temperature (Fletcher et al. 1986).

The Cantrell and Butcher (1987a,b) theory of phonon-drag thermopower gives good qualitative agreement with the observed thermopowers. To obtain good detailed agreement in form and magnitude, however, a number of refinements are required. For Si MOSFETs electrons in the anisotropic spheroidal bands are characterised by two effective masses and both longitudinal and transverse phonon modes couple to the electrons. In this case appropriate averages over the crystal directions of the deformation potentials for pure shear strain and pure dilatation must be made. When this is done Gallagher et al. (1987) obtain very good qualitative agreement with experiment. However, the predicted magnitudes are too large by a factor of ~ 35. This is principally due to the neglect of screening. Smith and Butcher (1989a) replace the artificial square well with the more appropriate Fang–Howard wavefunction, improve the choice of deformation potentials and sound velocities and the numerical integration near the singularities. This still leaves a discrepancy of ~ 15. They then

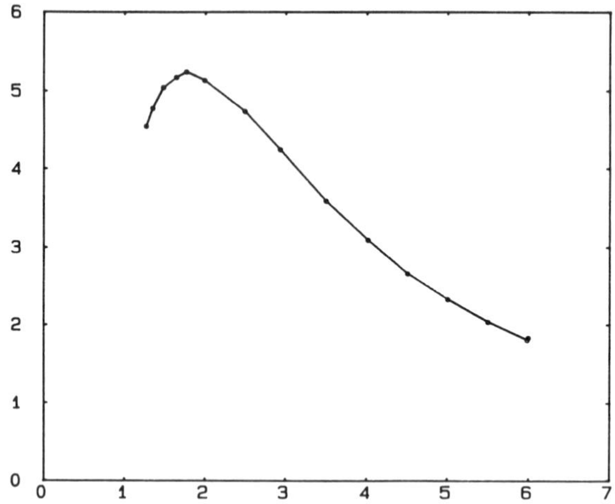

Fig. 36. "Kohn anomaly" peak for the same sample as fig. 33 but observed in S/K (Oxley et al. 1992).

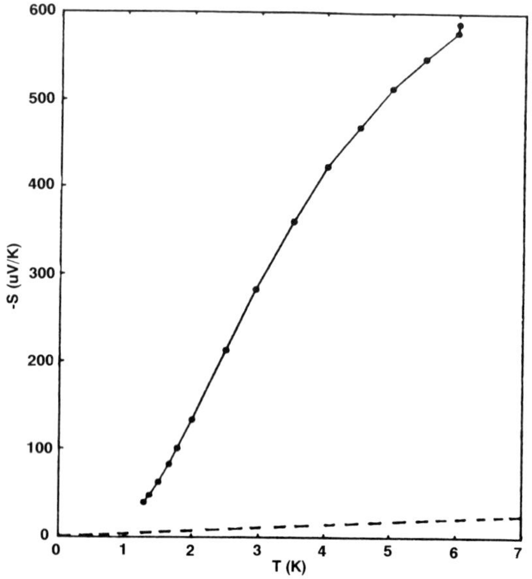

Fig. 37. Temperature dependence of the thermopower of a GaAs heterostructure and the "normalised" dependence scaled by $\Lambda(T)/\Lambda(T\to 0)$ (Oxley et al. 1992).

go on to consider the full multi-sub-band screening but conclude that the single sub-band approximation introduced no significant errors. The final theoretical predictions are then within $\sim$20% of the experimental values. In a further paper (Smith and Butcher 1989b) finite temperature screening is also considered but found to effect the results by only a few percent. Another small correction to the phonon-drag thermopower is found to arise from the energy dependence of the total electron scattering rate which is an interesting result because it is this which determines the diffusion thermopower. Gallagher et al. (1990a) compare the full theory including the temperature-dependent longitudinal and transverse phonons mean free paths with much improved experimental data. As is shown in fig. 30 the agreement in form and magnitude is excellent except below 2 K when a positive diffusion contribution emerges (see § 3.5) and at the highest temperatures when the theory predicts that S/T^3 should fall more rapidly than is observed. This level of agreement shows the crucial importance of screening in MOSFETs and indicates that the generally accepted values of the deformation potentials used ($\Xi_u = 9$ eV and $\Xi_d = -6$ eV) are probably quite accurate since it is their square which determines the predicted magnitude.

For GaAs based heterostructures there has been much controversy over the value of the deformation potential. Many studies have yielded values in the range 11–16 eV which is considerably larger than the values of 7–8 eV found in bulk GaAs. However, a recent study by Nolte et al. (1987) gave 9.3 eV (see Okuyama and Tokuda (1990) for detailed references). Phonon-drag thermopower might therefore throw light on this issue. Cantrell and Butcher (1987a,b) obtain reasonable agreement in form but magnitudes which are two to three times too large using a value of 8 eV and no screening. Lyo (1988) carried out the calculation using the Fang–Howard wavefunction and included screening. He also included piezoelectric coupling to both longitudinal and transverse acoustic phonons in addition to deformation potential coupling and showed this to be the dominant contribution below ~2 K. Reasonable agreement

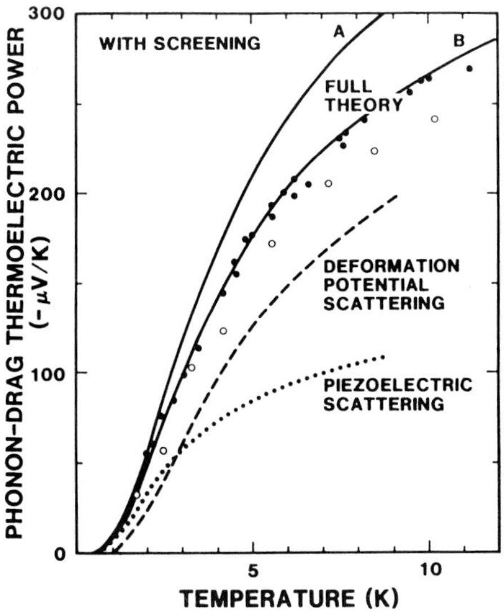

Fig. 38. The phonon-drag thermopower calculated in the presence of dielectric screening for phonon-mean free paths of 0.36 mm (curve A) and 0.30 mm (curve B) compared with experimental data (Lyo 1988).

with experimental data was obtained with a deformation potential of 9.3 eV and a constant phonon mean free path of 0.3 mm as is shown in fig. 38. However, Lyo (1988) uses an approximate form of the CB theory which leads to errors of order 50% (Smith and Butcher 1989b, Okuyama and Tokuda 1990). Furthermore, Lyo's statement that no adjustable parameters are used is misleading since the exact value of the deformation potential is controversial, and the true phonon mean free path is strongly temperature dependent. Smith and Butcher (1989b) evaluate S_g in the CB theory including piezoelectric and deformation potential coupling, using the Fang–Howard wavefunctions, temperature dependent screening and include other minor corrections. Comparison with three different sets of experimental data suggest a value for the deformation potential close to 16 eV. However, the uncertainties in the phonon mean free path introduce considerable uncertainty in this value. The comparison with the data of Ruf et al. (1988) is shown in fig. 39. Below ~6 K the agreement is excellent but the predicted S_g increases too slowly with T above this temperature as it does in the case of the Si MOSFETs considered earlier. Okuyama and Tokuda (1990) also apply the CB theory using the Fang–Howard wavefunction with deformation and piezoelectric scattering. They draw the unlikely conclusion that the deformation potential is *unscreened* and give it a value of ~8 eV. The agreement with experiment is very poor with the predicted S_g increasing much more rapidly than is observed experimentally and showing no tendency to saturate. The discrepancies are attributed to a temperature-dependent phonon mean free path.

Besides the detailed comparison between experimental results and theoretical predictions there is also a large body of direct experimental evidence of the dominance of phonon drag. Zavaritskii (1984) showed that the large thermopowers they observed in MOSFETs were consistent with phonon drag. Syme and Pepper (1989) found large thermopowers with a T^3 dependence in silicon on sapphire MOSFETs (fig. 40). These devices had mobilities as low as 0.05 $m^2\ V^{-1}\ s^{-1}$, indicating that phonon drag

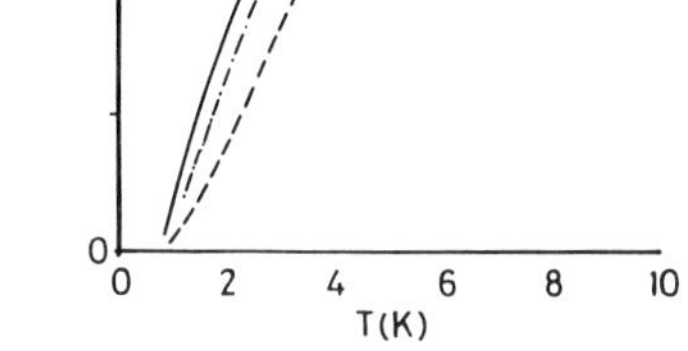

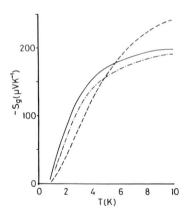

Fig. 39. The calculated thermopower compared with the data of Fletcher et al. (1986); the broken curve gives the experimental results, the chain curve is the calculated result with all corrections included and the full curve is the elastic result (see Smith and Butcher 1989b).

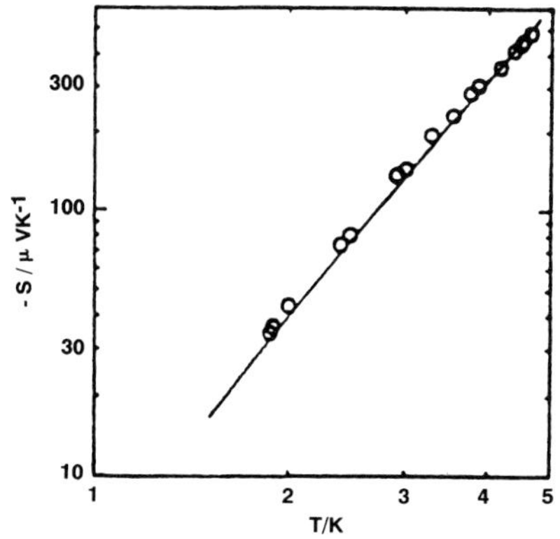

Fig. 40. The measured thermopower of a silicon-on-sapphire MOSFET. The circles are experimental points at an electron concentration of 3.7×10^{16} m^{-2}. The straight line is the fit $S = (-4.96\ \mu\text{V}\ K^{-4})T^3$ (Syme and Pepper 1989).

was not suppressed for low-mobility structures. Oxley et al. (1992) find that the phonon-drag thermopower was unrelated to the mobility of MOSFETs for devices with mobility values as low as 0.1 m^2 V^{-1} s^{-1}. Similarly, D'Iorio et al. (1988) varied the mobility of a heterostructure using the persistent photoconductivity effect in the range 20–80 m^2 V^{-1} s^{-1} and saw no indication of suppression of phonon drag at low mobility.

The most direct experimental demonstration of the dominance of phonon drag has come from the work of Fletcher et al. (1985) and D'Iorio et al. (1988). In the former paper samples with smaller phonon mean free paths due to phonon scattering in Cr-doped substrates were found to have smaller thermopowers than those with larger phonon mean free paths. In fact, the measured thermopowers and thermal conductivities were found to scale in the same way as would be expected. A more direct test of the phonon mean free path dependence of the thermopower was made by D'Iorio et al. (1988). They took a heterostructure specimen with a rough back surface and measured thermal conductivity and thermopower. They then polished this surface to optical quality and repeated the measurements. The polishing changes the phonon scattering from diffusive to specular and increases the phonon mean free path by almost a factor of two. The measured thermal conductivity shows the expected twofold increase and so does the thermopower. In this careful and definitive study they abraded the sample again to confirm that the transport properties returned to their original values, which they did.

D'Iorio et al. (1988) also looked at the electron density dependence of the thermopower in their persistent photoconductivity experiment. They found that S decreases with increasing n_s in the range 1.5 to 5 K. Thus the basic n_s^{-1} dependence seems, in this case, to be stronger than any dependence on n_s in $f(T, n_s)$.

5. Magnetothermopower

The thermopower, S, is a tensor in the presence of a magnetic field. For a 2D system in a perpendicular magnetic field B_z there are, however, only two independent components, $S_{xx} = S_{yy}$ and $S_{xy} = -S_{yx}$, the latter being the Nernst–Ettingshausen coefficient. In calculations the most convenient thermoelectric tensor to work with is M which determines the heat flux in an electric field, i.e., $\boldsymbol{U} = M\boldsymbol{E}$. The Peltier tensor is $M\rho$ where ρ is the resistivity tensor. Symmetry arguments show that for an isotropic 2DEG we have (Fromhold et al. 1992)

$$TS_{xx} = \rho_{xx} M_{xx} - \rho_{yx} M_{yx}, \tag{39}$$

and

$$TS_{yx} = \rho_{yx} M_{xx} + \rho_{xx} M_{yx}, \tag{40}$$

where all the tensor components are for the same orientation of B_z. In what follows we take $B_z > 0$. In the high-field limit the diagonal components of $\boldsymbol{\rho}$ and M are much smaller than the off-diagonal components tending to zero for weak disorder and so $S_{yx} \sim 0$ and

$$TS_{xx} \sim -\rho_{yx} M_{yx}. \tag{41}$$

5.1. *Diffusion magnetothermopower and 'quantised' thermopower in heterostructures and MOSFETs*

The general behaviour of the diffusive part of S_{xx} can be estimated by using the generalization of the Mott formula (Jonson and Girvin 1984, Oji 1984a),

$$M_{ij} = -\frac{\pi^2}{3}\frac{(k_B T)^2}{e}\frac{\partial \sigma_{ij}}{\partial \mu}. \tag{42}$$

As in our derivation of eq. (15) it is assumed that σ_{ij} is slowly varying on energy scales $\sim k_B T$. Between Landau levels the Hall conductivity σ_{yx} is constant and so we expect M_{yx} and thus S_{xx} to be zero. As μ moves through a Landau level, σ_{yx} moves between quantised values and its derivative and thus S_{xx} goes through a peak at around the centre of the Landau level. In this simple high-field picture then one expects S_{xx} to have the same general form as ρ_{xx} which is what is observed (see § 5.2). Equation (42) predicts that the peak heights will increase as T as is normal for diffusion thermopower. To observe such behaviour one would have to go to very low temperatures for two reasons. Firstly, the diffusion term must dominate the phonon-drag term. Secondly, we must have $k_B T$ much less than the Landau level width to ensure that σ_{yx} is slowly varying. Ruf et al. (1988) fail to see such behaviour down to 300 mK in a high-mobility heterostructure observing instead a much more complex behaviour as is shown in fig. 41.

In most experiments, however, we have $k_B T \sim \Gamma$, the Landau level width. The most interesting behaviour is predicted to occur in the limit $\hbar\omega_c \gg k_B T > \Gamma$ where ω_c is the cyclotron frequency. In this limit σ_{xx} varies rapidly within $k_B T$. We may

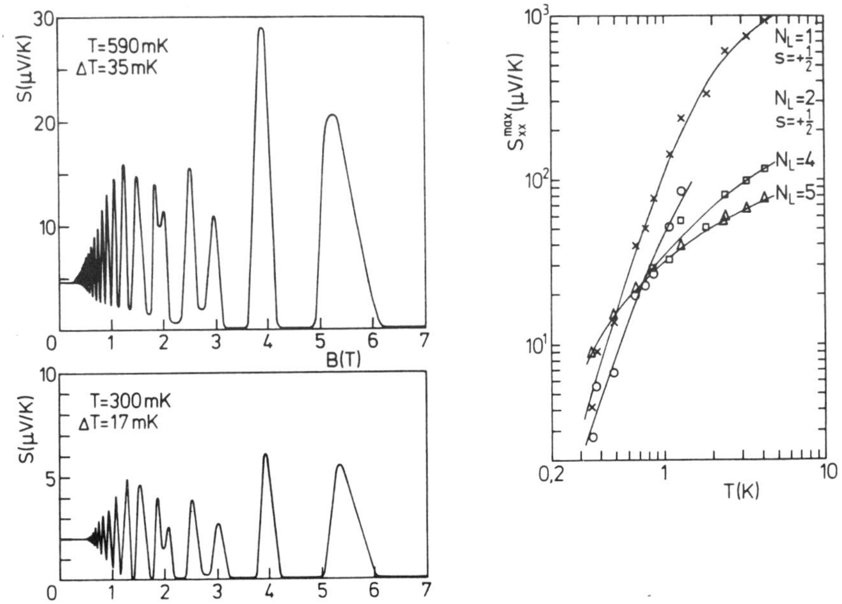

Fig. 41. (Left) Thermoelectric power oscillations observed in a GaAs heterostructure. (Right) Temperature dependence of S_{xx} maxima at different Landau levels (Ruf et al. 1988).

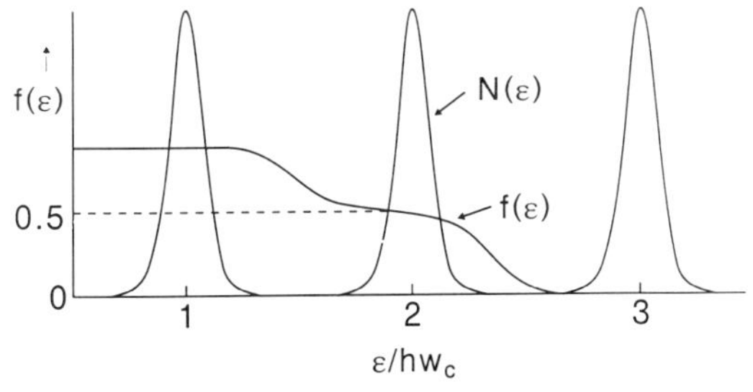

Fig. 42. Schematic diagram of the density of states $N(\varepsilon)$ and Fermi functions, $f(\varepsilon)$ for "quantised" thermopower.

then use the simple argument which estimates S_{xx} as the ratio of the entropy of the electrons to their charge. As emphasised in § 3.1 this argument can give an estimate of magnitude but normally fails because it takes no account of the energy dependence of the scattering. However, in the high-field weak scattering limit we need only consider the off-diagonal transport coefficients and these do not depend upon electron scattering in this limit. This extremely useful approach has been used by Zelenin et al. (1982), Zawadzki and Lassnig (1984) and Oji (1984a). Consider the case $\hbar\omega_c \gg kT > \Gamma$ illustrated in fig. 42. When the chemical potential μ lies within a range of localised states between Landau levels (a mobility gap) the average thermal energy and entropy of the carriers will be small and will tend exponentially to zero as the temperature is lowered. S_{xx} should then also be exponentially small. For the case when μ lies within the Landau level of index n we have the interesting situation that all levels below n are full and thus have zero associated entropy while within the nth level the electrons are non-degenerate. As μ passes through the level then the total entropy divided by the total number of carriers increases until μ is at the centre of the level. At this point half the available states will be filled and the probability of occupancy of any state is a half. The entropy then takes the maximum value of $k_B \ln 2$ per electron within the level and we obtain

$$S_{xx,\,\max} = \frac{k_B \ln 2}{e(n+\frac{1}{2})}. \tag{43}$$

This is the same result as that due to 2D sub-bands (§ 3.9). If spin splitting is resolved eq. (43) becomes,

$$S_{xx,\,\max} = \frac{k_B \ln 2}{e(2n+1\pm\frac{1}{2})} \tag{44}$$

for the lower energy $(-\frac{1}{2})$ and higher energy $(+\frac{1}{2})$ levels.

These results for the $\hbar\omega_c \gg kT \gg \Gamma$ limit have been shown to be correct using several different approaches. One important point common to all of them is that, to obtain the correct results, one must include conduction by edge states. This was first demonstrated for the 3D case by Obraztsov (1965) who showed that one must include the induced surface currents. Girvin and Jonson (1982) and Streda (1983, 1984) have shown that the currents carried by edge states in the 2D case are crucially important and their neglect leads to totally incorrect answers. Girvin and Jonson (1982) treat the case of a strip of 2D free electrons in a perpendicular high field and obtain

$$S_{xx} = -\frac{1}{eT}\,\frac{\sum\limits_n \int_{\varepsilon_n}^{\infty} (\varepsilon-\mu) f_0'(\varepsilon)\,d\varepsilon}{\sum\limits_n f_0(\varepsilon_n)}, \tag{45}$$

where the Landau level energies are $\varepsilon_n = (n+\frac{1}{2})\hbar\omega_c$, f_0 is the equilibrium Fermi function and f_0' is its derivative. This shows that, in this limit, S_{xx} is a universal function of $(kT/\hbar\omega_c)$ and $(\mu/\hbar\omega_c)$. S_{xy} is found to be zero. It is straightforward to show that this gives an exponential, activated behaviour between levels and a figure close to the "quantised" result (43) for $\mu = \varepsilon_n$. The full predicted behaviour is illustrated in fig. 43. However, the dependence of the denominator upon μ seems to imply that in fact the peak positions are moved slightly from ε_n and that one can have $S_{xx,\,\max}$ slightly larger than the values predicted by eq. (43). This has not been emphasised in the literature. Oji (1984a,b) obtains the same results as Jonson and Girvin (1982) using the arguments of Halperin (1982) for the quantization of the Hall conductivity in a Corbino geometry. Streda (1983) formulates the problem in the Kubo approach making use of the appropriate high-field relations for the transport coefficients (Smrcka and Streda 1977) and he also obtains eq. (45), as do Karyagin et al. (1988a). Lyo (1984) considers the case of coupled 2D layers appropriate to a

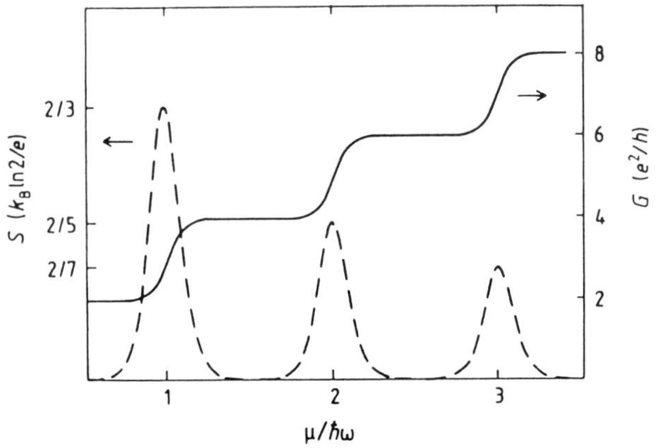

Fig. 43. Calculated thermopower, S_{xx}, (broken curve) and Hall resistance, R_{xy}, (full curve) as a function of the reduced chemical potential (Streda 1983).

superlattice, and obtains expressions giving the transition from the quantised behaviour to the normal 3D case as a function of interlayer bandwidth.

In a further series of papers weak disorder is included. Johnson and Girvin (1984), Streda (1984), Oji (1984a,b) and Oji and Streda (1985) all employ the Kubo formula and the self-consistent Born approximation (SCBA) for short-range scatters. This leads to semi-elliptical Landau levels of half width Γ. S_{xx} and S_{xy} (which is now non-zero) are now also dependent upon Γ and any "universal" behaviour is lost since the precise functional form will depend on the details of the scattering. As shown in fig. 44 as the strength of the disorder, and hence Γ/kT, is increased the peaks in S_{xx} are reduced and start to develop additional structure. At the same time peaks in S_{xy} grow. These oscillate about zero changing sign close to the centres of the Landau levels. The magnitude of these peaks appears to be independent of the Landau level index and so the magnitude of S_{xy} increases relative to S_{xx} with Landau level index. For $\Gamma \ll \hbar\omega_c$ both S_{xx} and S_{xy} are zero in the region of the Hall plateaux. However, as $\Gamma/\hbar\omega_c$ increases, S_{xy} increases much more rapidly than S_{xx} and so will be the more important contribution in this region. Zawadzki and Lassnig (1984) introduced disorder by using Gaussian Landau levels and do not insist that the levels are non-overlapping, i.e., $\hbar\omega_c \gg \Gamma$. They obtain the expected behaviour for $\hbar\omega_c \gg kT \gg \Gamma$ but also obtain a prediction of the behaviour down to low fields shown in fig. 45 with a value of $\Gamma = 0.5$ meV which is the right order for GaAs heterostructures. They also plot the size of the predicted maxima in S_{xx} relative to the "quantised" values (fig. 46). We see that, even for $\Gamma \sim kT$, the values are close to the "quantised" prediction. Futhermore, the very similar results of Jonson and Girvin (1984) show that this behaviour is not particularly sensitive to the model of the disorder used. The practical condition for observing the peaks of magnitude close to the "quantised" values is therefore $kT > \Gamma$ and not $kT \gg \Gamma$.

Confirmation of the "quantised" diffusion magnetothermopower predicted by eq. (43) was apparently found in one of the very first preliminary studies of the

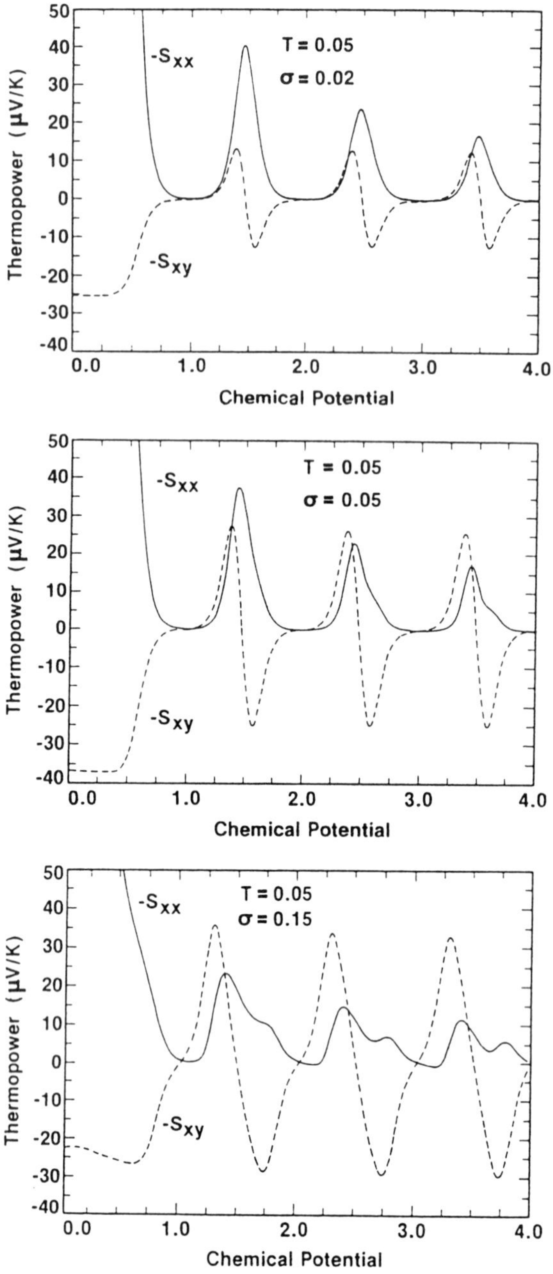

Fig. 44. The calculated effect of increasing disorder on S_{xx} and S_{xy}. Here, the units are such that $\hbar\omega_c = 1$, $kT = 0.05$ and Landau level width $= 0.08$ (a), 0.2 (b), 0.6 (c) (Jonson and Girvin 1984).

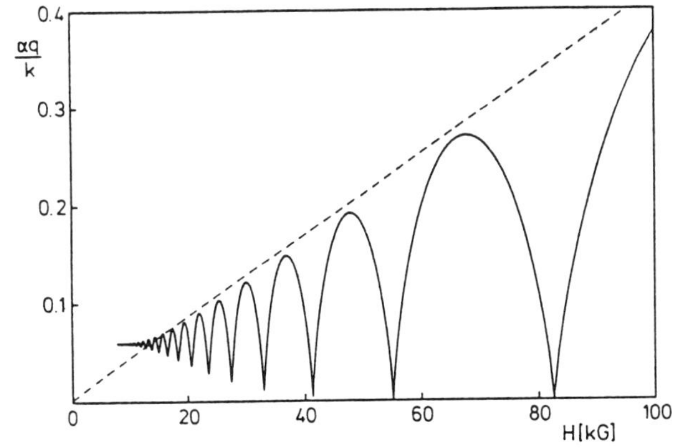

Fig. 45. Calculated diffusion magnetothermopower for a GaAs heterostructure at 6 K. The dashed line indicates values of $(-k_B/e)\ln 2/(n+\frac{1}{2})$ (Zawadski and Lassnig 1984).

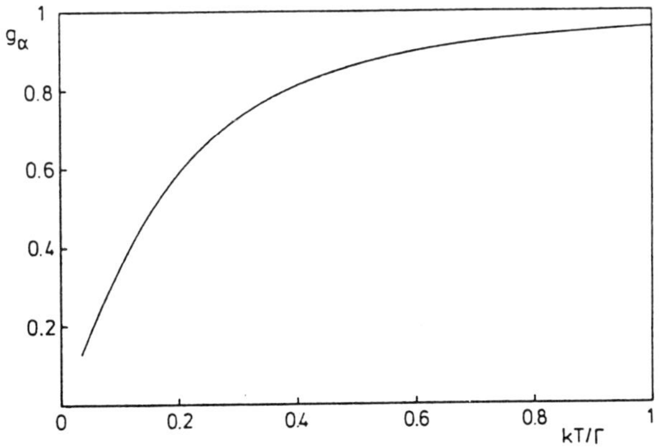

Fig. 46. Calculated peak height of diffusion magnetothermopower relative to free electron value as a function of kT/Γ in the range $\hbar\omega_c \gg kT$ and $\hbar\omega_c \gg \Gamma$ (Zawadski and Lassnig 1984).

thermopower of heterojunctions by Obloh et al. (1984). As is shown in fig. 47, the correct temperature independent maximum value of S_{xx} is obtained. This has never been observed in any of the many subsequent measurements. In a number of studies in which a small thermopower, consistent with the diffusion contribution, was subsequently observed, peak values less than those predicted were found (Obloh et al. 1984, Davidson et al. 1986). This was analyzed in terms of the finite Γ and values for Γ extracted. However, as noted in § 3.4 and § 4.2 some doubt has now been cast over this earlier work since all subsequent studies have found that phonon drag dominates the thermopower above ~1 K in GaAs heterostructures.

Ruf et al. (1988) observe the diffusion contribution in a heterostructure below ~0.6 K as can be seen from fig. 4. The corresponding magnetothermopower shown in fig. 41a has the expected form but the peak heights at the lowest temperature are

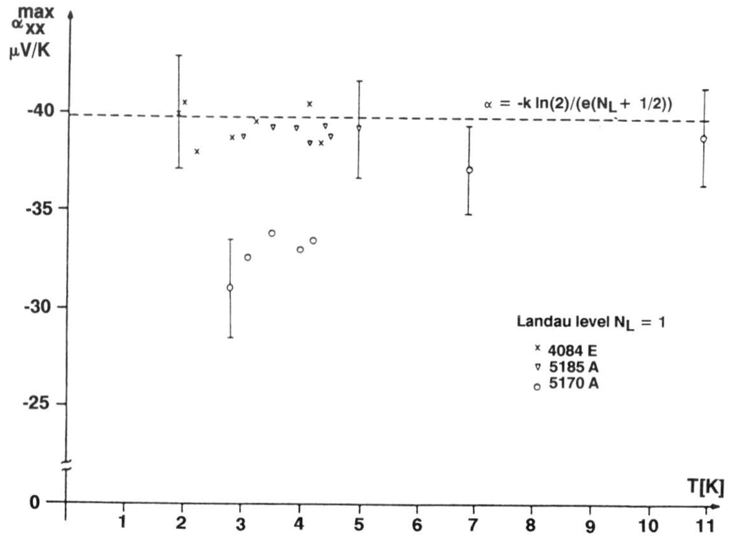

Fig. 47. Temperature dependence of the measured magnetothermopower peak height for three different samples of Landau quantum number $n_L = 1$ (Obloh et al. 1984).

smaller than the "quantised" values. This is presumably because at these low temperatures $k_B T < \Gamma$. As the temperature is raised one would then expect S_{xx} to increase linearly with T, as discussed above, until it reached the quantised value for $kT > \Gamma$. However, in practise the increasingly important phonon-drag contribution rapidly begins to dominate the thermopower and the peaks become much larger than the "quantised" values. The resulting peak height as a function of temperature is quite complex (fig. 41b) with the relative height of the different peaks changing dramatically.

Measured and calculated diffusion magnetothermopowers are compared by Oxley et al. (1990). They present accurate magnetothermopower results for a Si inversion layer. This system has the disadvantage of relatively low mobility and thus large Γ but the great advantage that the electron–phonon interaction is strongly screened. This means that the diffusion contribution dominates up to about 2 K in Si MOSFETs compared to $\sim$6 K in GaAs heterostructures. At low fields ($\omega_c \tau < 1$) S_{xx} is found to increase linearly with B. The overlap between the rather broad Landau levels leads to a non-oscillatory background and to S_{xx} only having zeros above 10 Tesla.

The diffusion magnetothermopowers are calculated by Oxley et al. (1990) by using the lowest-order cumulant approximation (Gerhardts 1975). In this approach the electron scattering rate only enters through the Landau level width. A Gaussian form is used for the density of states with a half width calculated from the self-consistent Born approximation to be consistent with the mobility. This should be a reasonable approximation because the scattering is predominantly short range in Si MOSFETs. The value obtained was $\Gamma = 0.6\sqrt{B}$ meV. The theoretical calculations show that at 1.47 K the oscillatory thermopower is almost entirely due to the diffusion contribution. The magnitudes of the oscillatory part of S_{xx} is found to be in excellent agreement with experiment but the non-oscillatory background is not very well reproduced. In our original paper (Oxley et al. 1990) there is unfortunately an important error. An electron density of 1.18×10^{16} m^{-2} rather than the correct value of 1.06×10^{16} m^{-2} was used which led to apparent phase differences between experiment and theory. In fig. 48 the magnetic field axis has been appropriately re-scaled to partially correct for this error. The theory used is rather crude and though agreement with the magnitude is reasonable the agreement in form is poor. Clearly a proper treatment which considers both bulk and edge state conduction is required.

Havlová and Smrcka (1986) have calculated the conductivity, thermopower and thermal conductivity of 2D electrons in the weak to intermediate field limit ($\omega_c \tau < 1$) using the Kubo formalism for weak disorder. They find, as expected, that the Mott rule (eq. (15)) is valid in both the high-field and low-field limits, since in both cases all the transport coefficients are essentially smooth on energy scales $k_B T$, but not for intermediate fields. The conductivity σ_{xx} obtained shows relatively simple cosine

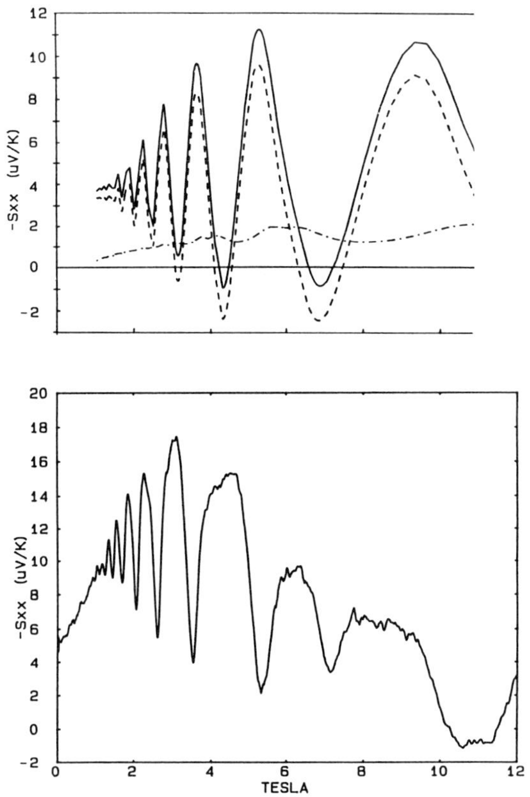

Fig. 48. Magnetothermopower for a Si MOSFET with $n_s = 11.8 \times 10^{15}$ m^{-2} at 1.47 K. (Top) theoretical calculation showing diffusion (dashed line) and drag (chained line) contributions to total (solid line) (Oxley et al. 1990). (Bottom) experimental data.

oscillations and the ρ_{xx} oscillations are in phase with those of the density of states. In contrast to this, S_{xx} is predicted to have cosine and sine components with different temperature dependences. This makes S_{xx} and ρ_{xx} in phase at higher temperature but almost 180° out of phase at low temperatures as is shown in fig. 49. At low fields one does not expect such phase differences for the phonon-drag contribution (see § 5.2). Fletcher (1983) studied these effects in bulk metals. He found excellent agreement between the measured and predicted oscillatory conductivity and thermopower and was able to conclude that the diffusion contribution was totally dominant at helium temperatures in the metals studied. D'Iorio et al. (1988) carried out a similar experimental test of the predictions of Havlová and Smrcka (1986) using GaAs heterostructures. Their high-accuracy measurements (fig. 50) show that the phase difference between the oscillations in S_{xx} and ρ_{xx} is $(0.01 \pm 0.11)\pi$ radians. The estimated uncertainty would seem to be the maximum possible and so one can conclude that S_{xx} and ρ_{xx} are in phase to high accuracy. This indicates that either Havlová and Smrcka are in error or the thermopower is dominated by phonon drag. D'Iorio et al. draw the latter conclusion which, as we try to show it, is almost certainly true. As can be seen in fig. 50 D'Iorio et al. also observe that S_{xx} is more strongly damped than ρ_{xx} at low field, which is not understood.

5.2. *Phonon-drag magnetothermopower in heterostructures and MOSFETs*

Since the large zero-field thermopowers of MOSFETs and heterostructures seem to be well accounted for by the predicted phonon-drag contribution it seems reasonable to assume that the observed large oscillatory magnetothermopower is also due to phonon drag. The general expectation concerning the qualitative behaviour is straightforward. In the disorder-free high-field limit ρ_{xx} and M_{xx} will tend to zero,

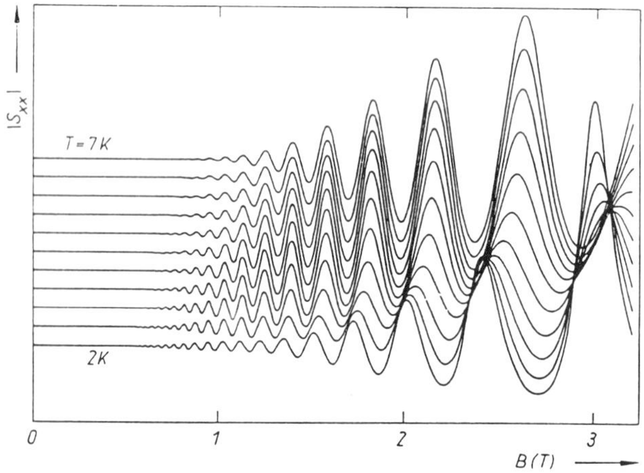

Fig. 49. Calculated effect of temperature upon phase of oscillations in S_{xx}. Temperature incremented in steps of 0.5 K. At 7 K S_{xx} essentially in phase with ρ_{xx} (Havlová and Smrcka 1986).

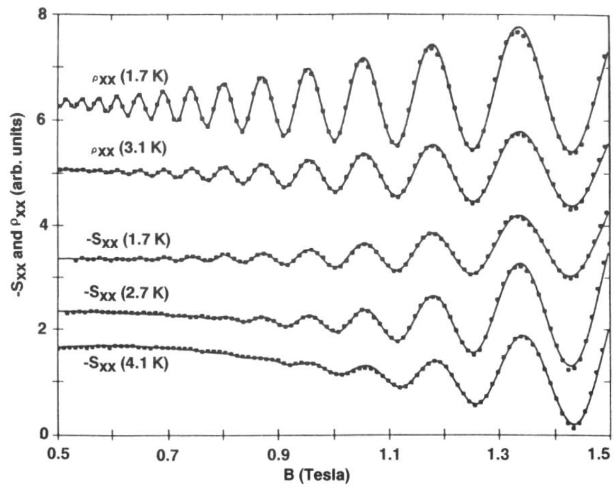

Fig. 50. Experimental results for the thermopower S_{xx} and resistivity ρ_{xx} of a GaAs heterostructure (Fletcher et al. 1988b).

as noted above, and so from eq. (41) S_{xy} will tend to zero. We might also expect the form of S_{xx} to reflect that of the density of states at E_{F}. So between Landau levels if E_{F} lies in a mobility gap we expect S_{xx} to tend exponentially to zero as T is reduced reflecting the number of available states an electron can scatter into on absorption of a phonon. Similarly, when E_{F} is close to the centre of a Landau level the number of available initial and final states is a maximum and S_{xx} should be a maximum. Thus S_{xx} is again expected to have a similar form to ρ_{xx}, however, no "universal" or "quantised" behaviour is expected since S_{xx} is no longer a property intrinsic to the 2D electron system but arises from the electron–phonon interaction. In this picture, however, we have ignored the effect of screening which, as we will show later, can dramatically modify these expectations.

Kubakaddi et al. (1989) set out the basic theory of phonon-drag magnetothermopower. Like the zero-field theory of Cantrell and Butcher (1987a,b) it considers the coupled electrons and phonon Boltzmann equations but the thermoelectric properties are calculated via the thermoelectric tensor $\boldsymbol{M}$ introduced at the beginning of § 4. In this first paper the weak-disorder high-field approximation is made in which $\rho_{xx} \sim 0$ and ρ_{yx} is given the classical value of $-B_z/n|e|$. S_{xx} then only depends upon M_{yx}. To find M_{yx} the Boltzmann equations are linearised and solved on the assumption of weak electron–phonon coupling. Comparison is made with the quantum limit magnetothermopower data of Fletcher et al. (1985) for a sample with $n_{\mathrm{A}} = 1.78 \times 10^{15}\,\mathrm{m}^{-2}$ and $\mu = 22.6\,\mathrm{m}^2\,\mathrm{V}^{-1}\,\mathrm{s}^{-1}$. The Fang–Howard wavefunction is assumed and Lorentzian Landau levels are used. For simplicity, screening is neglected and an appropriate unscreened deformation potential of 11.5 eV is used along with a piezoelectric constant of $1.2 \times 10^7\,\mathrm{V\,cm}^{-1}$. The phonon mean free path of 0.36 mm is obtained from the thermal conductivity data at 3 K. From the measured mobility one obtains a Landau level width of $0.1\sqrt{B}$ meV using the self-consistent Born approximation which, however, does not give reasonable agreement with the experimental results. Setting $\Gamma = 0.4\sqrt{B}$ meV, however, yields very good agreement as is shown in fig. 51. This larger value of Γ indicates that the dominant scattering mechanisms are long range and not short range as the SCBA formula assumes. This is to be expected for high-mobility heterostructures where the dominant scattering is from remote ionised impurities. The excellent agreement in form and magnitude suggests that the approximations made are reasonable.

Lyo (1989) calculates the oscillatory phonon-drag magnetothermopower as a function of field in a similar manner to Kubakaddi et al. (1989) but uses Gaussian Landau levels and includes static screening and inter Landau level scattering. The deformation potential is taken to be 9.3 eV. Both the phonon mean free path, Λ, and the Landau level width, Γ, are used as adjustable parameters. Furthermore, to reproduce the observed broad zeros in S_{xx} when E_F is between levels, a range of states in the wings of the Landau levels are considered to be localised. As figs. 52 and 53 show, very good agreement with the data of Fletcher et al. (1988a,b) is obtained using $\Lambda = 2.5$ mm and a constant width of $\Gamma = 1.6$ meV. This phonon mean free path is much larger than the specimen thickness but is in reasonable agreement with the value of ~2 mm inferred from the measured thermal conductivity. This is because the sample has been polished to produce specular reflections of phonons. The double maxima seen in the data would seem to be the expected spin splitting. Lyo, however, suggests that this could be due to screening which for small Γ leads to a minima at the Landau level centre as we discuss below.

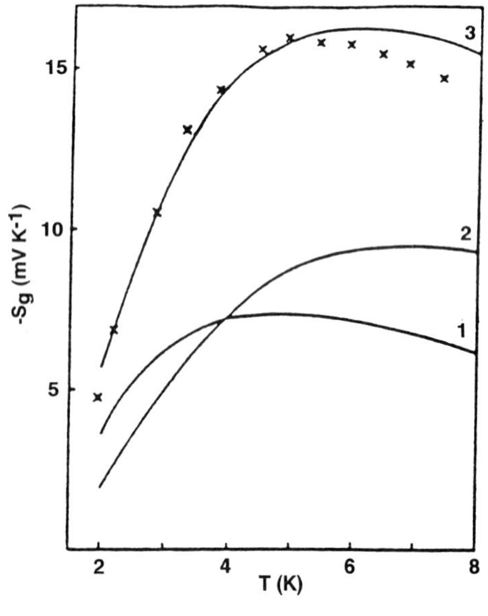

Fig. 51. Comparison of calculated thermopower for a GaAs/AlGaAs heterojunction with data of Fletcher et al. (1985) at 15 T. Curve 1: piezoelectric scattering alone; curve 2: deformation potential scattering alone; curve 3: 1 and 2 combined (Kubakaddi et al. 1989).

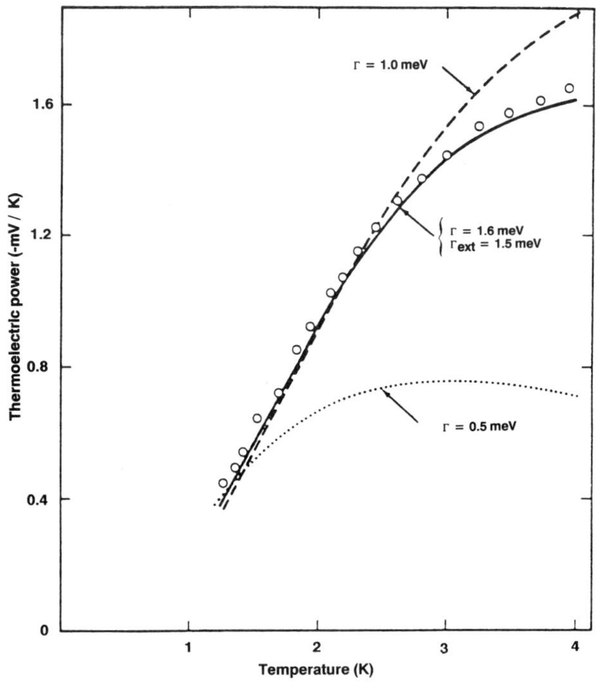

Fig. 52. The temperature dependence of the calculated thermopower with (solid curve) and without (dashed, dotted curves) localized states at $H = 2.85$ T. The data is from Fletcher et al. (1985) (Lyo 1989).

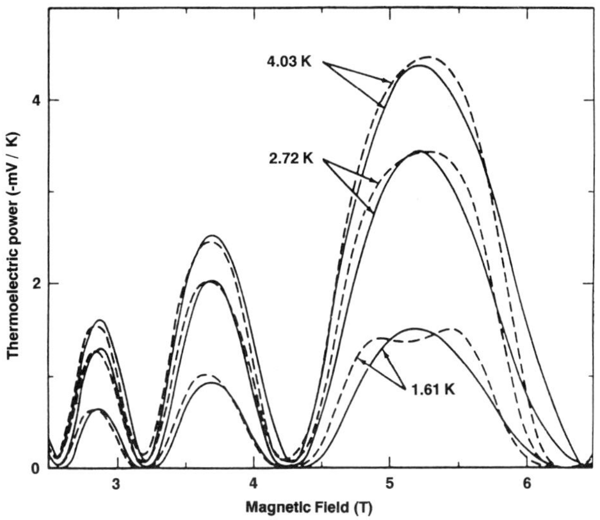

Fig. 53. Comparison of calculated thermopower (solid curves) with data of Fletcher et al. (1985) (Lyo 1989).

Oxley et al. (1990) apply the theory of Kubakaddi et al. and Lyo to the case of Si MOSFETs comparing its predictions with measured magnetothermopowers. Thomas–Fermi screening is used and scattering processes between the five Landau levels nearest E_F are included. This allows the application of the theory down to low fields at which $\hbar\omega_c \sim \Gamma$. For Si, short-range scattering dominates and so the SCBA, which gives $\Gamma = 0.6\sqrt{B}$, should be reasonably accurate. Phonon mean free paths are obtained from thermal conductivity data. The main achievement of this work is to demonstrate the changeover from diffusion to phonon-drag dominated magneto-thermopower at ~ 2 K. The magnitudes of S_{xx} and S_{xy} are reasonably well accounted for (see fig. 54) but agreement in the detailed form is rather poor. The main reason for this are the error in number density noted above and that in these relatively low mobility samples ($\mu \sim 1$ m^2 V^{-1} s^{-1}) the assumption that $\rho_{xy} \gg \rho_{xx}$ is poor.

Fromhold et al. (1992a,b) present accurate experimental results for S_{xx}, S_{yx}, ρ_{xx} and ρ_{yx} for a high-mobility (44 m^2 V^{-1} s^{-1}) heterostructure. These are then used to calculate experimental results for components M_{xx} and M_{yx}. It is these tensor components which are the direct outcome of the theoretical calculations. We shall show below that the theoretical predictions for M_{yx} are in reasonable accord with the data. However, the crude treatment of the response of the electron system used in the theory leads to the incorrect result $M_{xx} = 0$. The "theoretical" values of S_{xx} and S_{yx} discussed below are obtained from eqs. (39) and (40) by inserting the theoretical values of M_{yx} and *experimental* values of ρ_{xx}, ρ_{yx} and M_{xx} because no reliable formulae are available for any of these qualities in the regime under investigation. It is found that setting $M_{xx} = 0$ has little effect on S_{xx} but significantly worsens the agreement of the theoretical values of S_{yx} with the data. This is because the second term in eq. (39) is overwhelmingly dominant as emphasised by Fletcher et al. (1986).

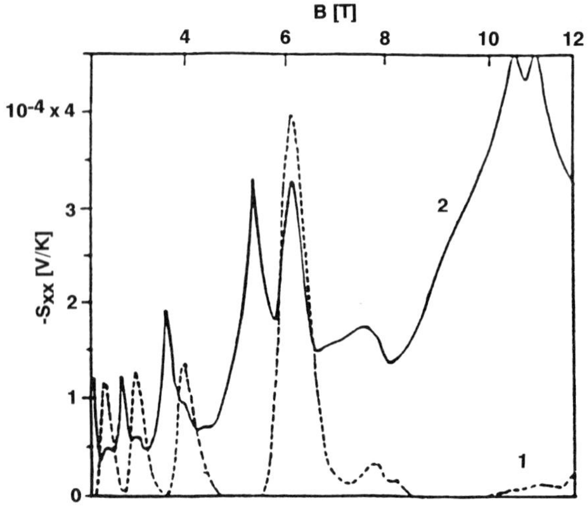

Fig. 54. Plots of $-S_{xx}$ at 1.28 K. (1) Experiment. (2) Theory when $\Gamma = 0.06B^{1/2}$. Note the appearance of the anti-phase peaks in the theory for such narrow Landau levels (Fromhold et al. 1992a).

These authors also suggested that the first term in eq. (40) is the more important. Fromhold et al. (1992a,b) find on the contrary that the second term is the more important although not overwhelmingly so. In particular, sign changes in M_{yx} do not produce sign changes in S_{yx}.

The model used and method of calculation are those of Kubakaddi et al. and Lyo. The temperature-dependent phonon mean free path is obtained directly from the measured thermal conductivity. Gaussian Landau level broadening is assumed and the only adjustable parameter in the calculation is the Landau level width. The SCBA gives $\Gamma \sim 0.1\sqrt{B}$ meV from the sample mobility. When such a value is used, however, the predicted peak positions in S_{xx} are totally wrong and, in fact, for $\Gamma < 0.06\sqrt{B}$ it is found that the predicted peaks are $\sim 180°$ out of phase with those in the density of states at the Fermi energy, $N(E_F)$, as is shown in fig. 54. This is because the screening increases with $N(E_F)$. For small enough Γ the screening close to the centre of a Landau level becomes so strong that it leads to a minimum rather than a maximum in S_{xx}. If localisation had been included the presence of a mobility gap between Landau levels would have precluded strictly anti-phase oscillations. Nevertheless, field-dependent screening would have a profound influence on the observed thermopower for very narrow Landau levels. In the present case good agreement in form and magnitude is obtained for $\Gamma = 0.5\sqrt{B}$ meV as is shown in figs. 55 and 56. Again this demonstrates the inadequacy of the SCBA indicating that the long-range scatterers which dominate in GaAs heterostructures are significantly broadening the Landau levels. Spin splitting is not included in the theory but is apparent in the experimental data at high fields.

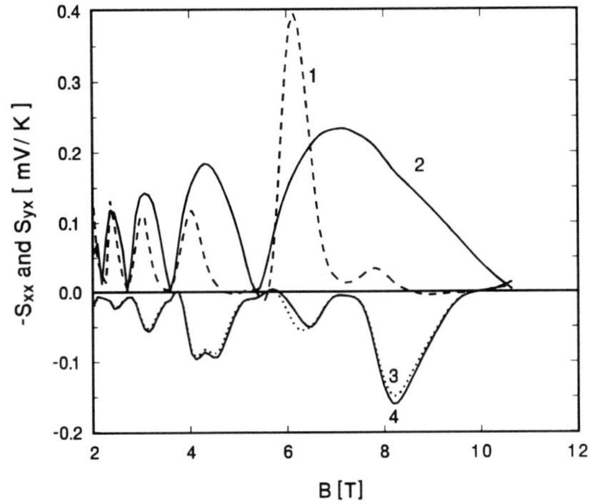

Fig. 55. Comparison of experimental and theoretical values of $-S_{xx}$ and S_{yx} for a sample with $n_s = 0.52 \times 10^{16}\ \mathrm{m}^{-2}$ at 1.28 K. Curve 1: experimental value of $-S_{xx}$. Curve 2: theoretical values of $-S_{xx}$. Curve 3: experimental values of S_{yx}. Curve 4: theoretical values of S_{yx}. For curves 2 and 4 the Landau level width is $0.5B^{1/2}$ meV and the phonon mean free path obtained from the thermal conductivity is 0.75 mm (Fromhold et al. 1992b).

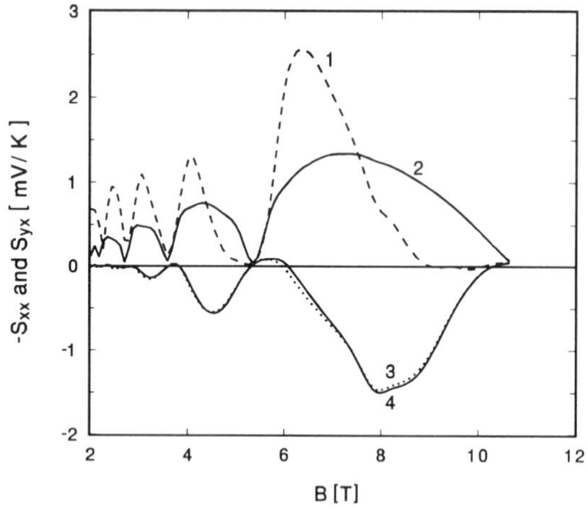

Fig. 56. Same comparison as in fig. 55 but at 5.0 K; the phonon mean free path has now fallen to 0.41 mm (Fromhold et al. 1991b).

The experimentally obtained M_{xx} and M_{yx} along with the calculated M_{yx} at 2.94 K are shown in figs. 57 and 58. The location and magnitudes of the maxima in $|M_{xy}|$ are fairly accurately predicted by the theory. What is most striking, however, is that contrary to the theoretical result $M_{xx} = 0$ the experimental values are little smaller than M_{xy} even in these high-mobility samples and shows quite complex oscillations.

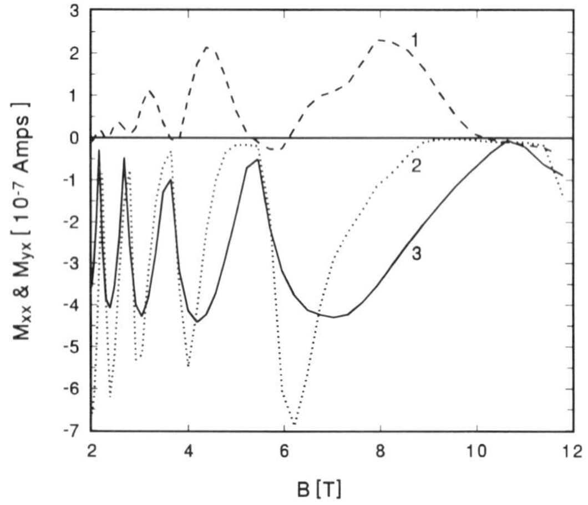

Fig. 57. Comparison of experimental and theoretical values of M_{yx} and M_{xx} for a sample with $n_s = 0.52 \times 10^{16}\ \mathrm{m}^{-2}$ at the intermediate temperature of 2.94 K. Curve 1: experimental values of M_{xx}. Curve 2: experimental values of M_{yx}. Curve 3: theoretical values of M_{yx}. For curve 3 $\Gamma = 0.5B^{1/2}$ and the phonon mean free path is 0.41 mm (Fromhold et al. 1992b).

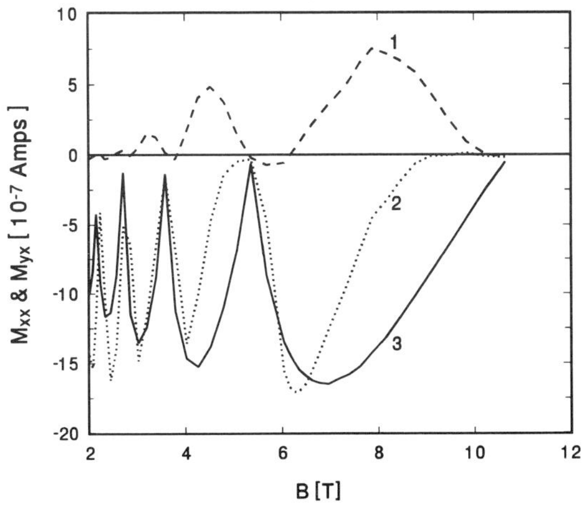

Fig. 58. Same comparison as fig. 57 but at 5.0 K (Fromhold et al. 1991b).

The final predictions for S_{xx} and S_{xy} are given in figs. 55 and 56. Similar good agreement is found at a range of temperatures between 1.275 and 5.005 K. Thus with a reasonable choice of only one adjustable parameter it has proved possible to reasonably account for the full complex behaviour of $S_{xx}(B, T)$ and $S_{xy}(B, T)$.

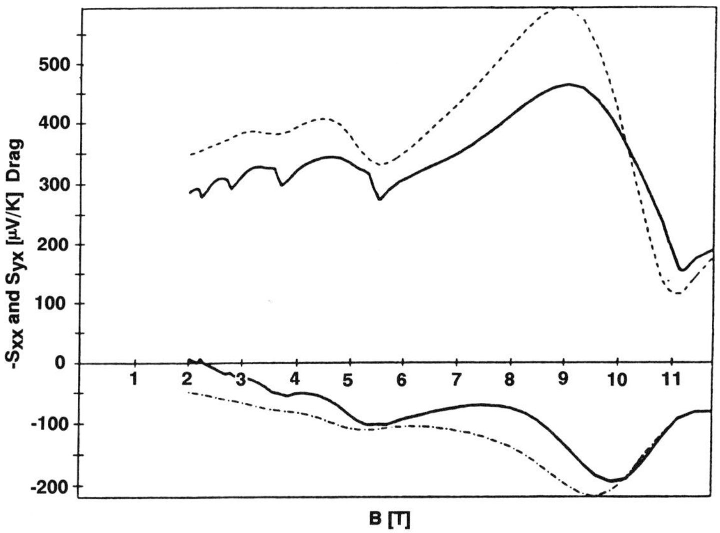

Fig. 59. Comparison of the measured (dashed lines) and calculated (solid lines) magnetothermopower of a Si MOSFET at 5.02 K. At these temperatures phonon drag is totally dominant (Qin et al. 1992).

M_{xx} is expected to be more important for the lower-mobility Si MOSFET samples and this is at least part of the reason for the poorer agreement obtained for such samples by Oxley et al. (1990). Qin et al. (1992) have recently applied the approach of Fromhold et al. (1992a,b) to Si MOSFET data. The agreement with data shown in fig. 59 is very good. M_{xx} has yet to be calculated and this remains an outstanding theoretical challenge in this area.

Large phonon-drag magnetothermopowers have been seen by a number of groups. Fletcher and co-workers have obtained good data for specimens with a range of electron densities and phonon mean free paths (see fig. 60). They have also been able to directly demonstrate the phonon-drag origin of the thermopower by the polishing experiment described in § 4.2. Results for the magnetothermopower of the same specimens with different phonon mean free paths are shown in fig. 61. In this case the increase is less than the factor of two expected. Fromhold et al. (1992a,b) analyse some of the data of Fletcher et al. (1986) and find a similar measure of agreement to that shown in figs. 55 and 56.

Karyagin et al. (1988a,b) use an alternative treatment of S_{xx} to interpret the data of Fletcher et al. (1986) for a GaAs/AlGaAs heterojunction in the quantum limit. Both the piezoelectric contribution to the electron–phonon interaction and the boundary contribution to phonon scattering are ignored. For a reasonable choice of material parameters the predicted thermopowers are an order of magnitude too small. Nevertheless, the alternative theoretical approach deserved closer examination as a possible avenue to the calculation of M_{xx}.

It should be noted, however, that there is a fundamental objection to all the theory so far advanced. This is that ρ_{xx} cannot be strictly defined at high field since R_{xx} becomes non-local (see § 7.1). For example, we find that higher B-field spin-split "ρ_{xx}" peaks show non-local behaviour and are suppressed at low currents just as is the corresponding peak S_{xx} shown in fig. 55. So a theory which considers bulk state *and* edge state conduction would seem to be required even though phonon absorption in edge states may well be suppressed.

5.3. *Effects of sub-band structure on phonon-drag thermopower*

Fletcher et al. (1991) have used persistent photoconductivity and magnetic depopulation to study the thermopower of a high-mobility heterostructure as ε_F is swept through the bottom of the second 2D sub-band. This sample shows a very large thermopower consistent with total dominance by phonon drag. As ε_F moves into the second sub-band the mobility drops (fig. 62) due to the increased scattering (see § 2.7); the corresponding thermopower increases. If the electron density is fixed and a B-field parallel to the 2D electron is steadily increased the additional confinement pushes the second sub-band to higher energy. Figure 63 shows that depopulation of the second sub-band occurs at progressively higher B-fields for higher electron densities and is accompanied by the expected increase in mobility; the corresponding thermopower now drops (fig. 64). The reason for this is very likely to be that suggested by Fletcher et al.; that the increased density of states when the second sub-band is

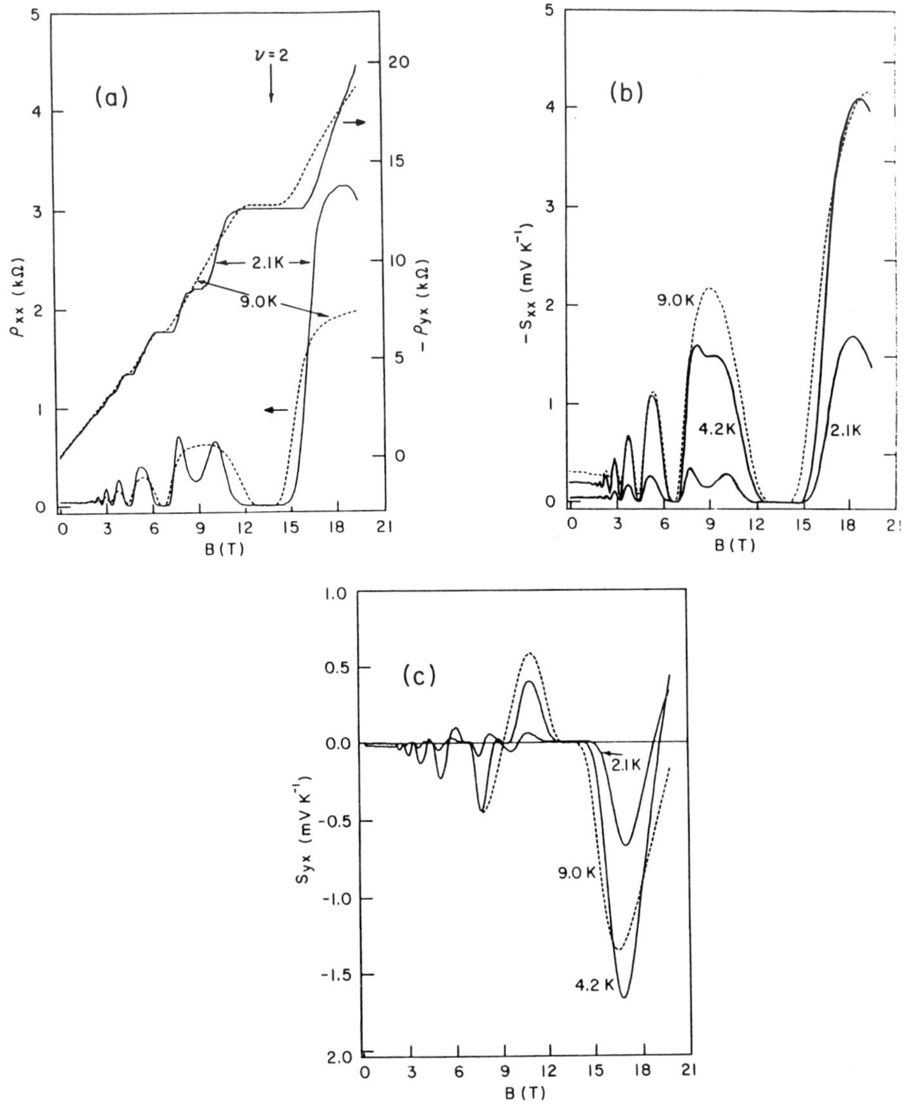

Fig. 60. Experimental data for a GaAs heterostructure: (a) ρ_{xx} and $-\rho_{yx}$, (b) $-S_{xx}$ and (c) S_{yx} (Fletcher et al. 1986).

occupied leads directly to an increase in electron–phonon scattering and hence an increase in phonon drag. However, they also see a more complicated behaviour at lower temperatures, when an additional negative peak becomes apparent. This is the opposite behaviour to that predicted for the diffusion thermopower. The origin of this peak is unclear but one possibility is that it is due to the low-q phonons coupling

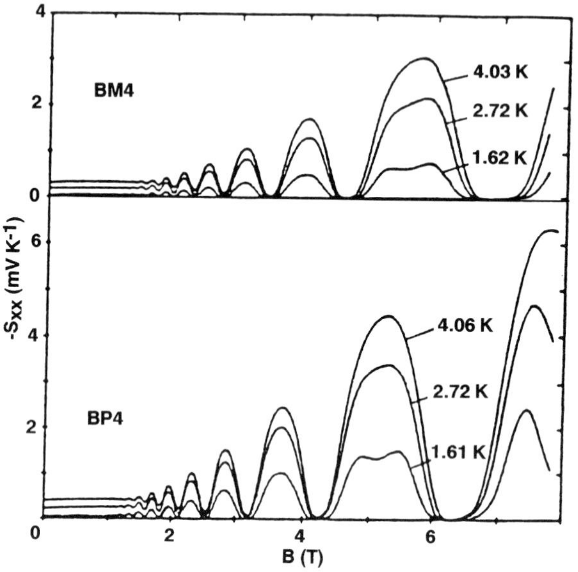

Fig. 61. The thermopower of a polished GaAs heterostructure sample (BP4) which has a phonon mean free path twice as large as the abraded (BM4) sample at three temperatures. The phase shift is due to a slightly different number of carriers in the two samples (D'Iorio et al. 1988).

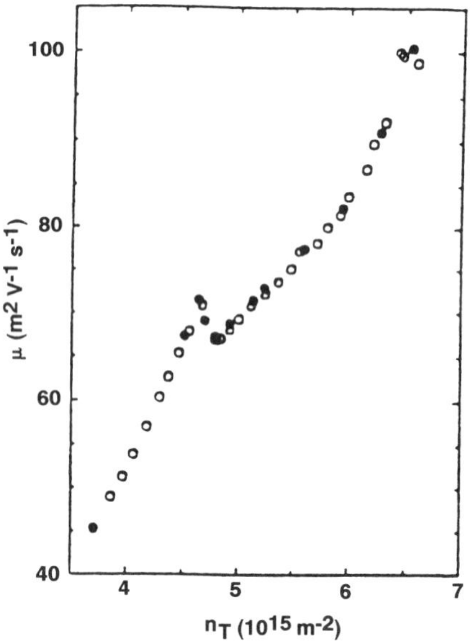

Fig. 62. The mobility μ as a function of total carrier density n_T for a GaAs heterostructure showing second sub-band occupation. The data were all taken in the temperature range 1.17–1.25 K. Measurements from two experimental runs are shown (Fletcher et al. 1991).

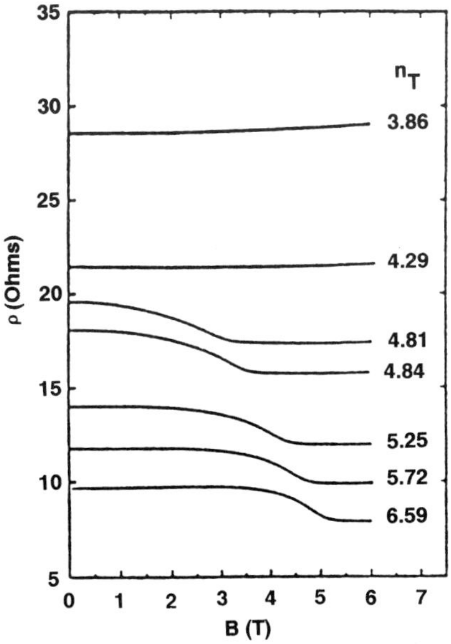

Fig. 63. The variation of the resistivity ρ with field B at various electron densities n_T (in units of 10^{15} m^{-2}). The magnetic depopulation of the second sub-band is clearly apparent (Fletcher et al. 1991).

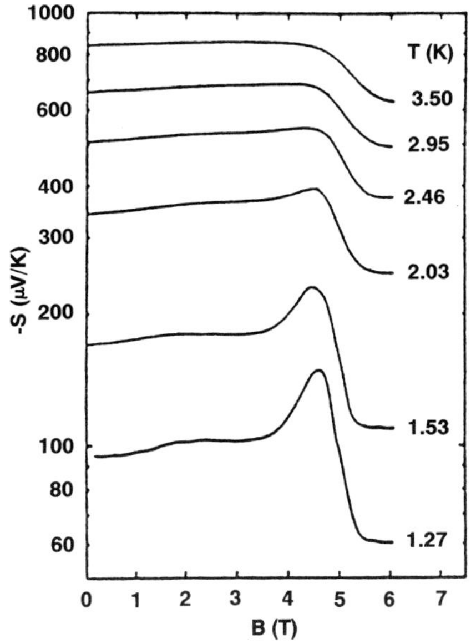

Fig. 64. The thermopower S as a function of field B at various temperatures for an electron density of $n_T = 6.50 \times 10^{15}$ m^{-2}. Note that at low temperature one has a peak before the drop in the magnitude of S because of the depopulation of the second sub-band (Fletcher et al. 1991).

very effectively to the second sub-band electrons which have a small Fermi circle radius k_F. That is, it could be related to the "Kohn" anaomaly discussed in § 4.2.

6. Quantum interference effects in thermopower

In the usual semiclassical theory of electron conduction the interference between scattered waves is neglected on the assumption that such effects average to zero. In recent years it has been realised that this is not necessarily the case and that such interference effects can be very important particularly in strongly disordered conductors. The most important of these effects are "weak localisation" and "universal conductance flutuations". Both of the corresponding thermoelectric effects have recently been observed for the first time.

6.1. Weak localisation in thermoelectric transport

A relatively simple picture of the weak localisation effect has been given by Bergmann (1984) (see also Lee and Ramakrishnan 1985). Consider the set of scattering centres shown in fig. 65. Let an electron wave scatter from site 1. Now the outgoing partial waves will be subsequently scattered from the other scattering centres. The trajectory indicated which takes us back to site 1 is one possible path for the multiply scattered electron partial wave. But the time-reversed, anticlockwise, path is also equally probable. Waves travelling these two paths arrive back at site 1 exactly in phase and with the same amplitude and constructively interfere. The important point about the type of paths we have considered is that the interference is *always* constructive for all such loops. This coherent effect gives an enhanced probability of backscattering for the electrons. This is only of importance if there is a reasonable probability of an electron traversing the path without undergoing a change of phase due to some inelastic scattering event so the condition for this effect to be apparent is that the

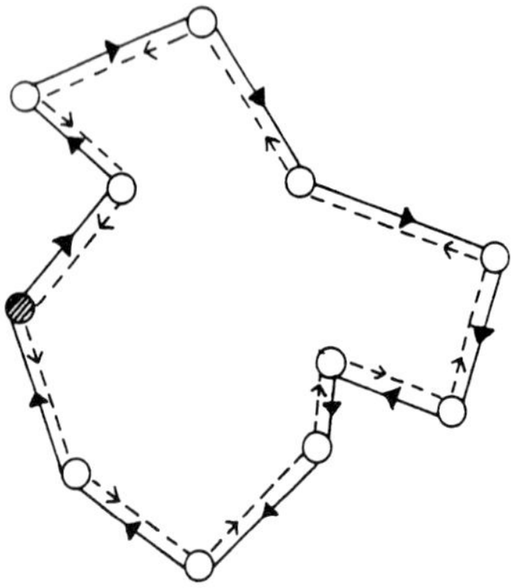

Fig. 65. Indirect interference.

electron phase coherence length is much longer than the elastic scattering length. For a system showing weak localisation we therefore expect a negative temperature coefficient of resistance. The weak localisation correction to the Boltzmann conductivity σ_B can be written in the form

$$\sigma = \sigma_B - \Delta\sigma_{WL}$$

where

$$\Delta\sigma_{WL} \sim \frac{e^2}{\pi h} \ln\left(\frac{\tau}{\tau_\varphi}\right) \quad \text{in 2D.} \tag{46}$$

Here τ_φ is the phase breaking time and τ the elastic scattering time. The transition in the behaviour from 3D to 2D to 1D has been treated by Cantrell and Butcher (1985e) and Kearney and Butcher (1988b). Such dependences are most easily seen in strongly disordered systems where τ is small and σ_B very weakly temperature dependent. This effect can dominate the temperature dependence for amorphous metals (Howson and Gallagher 1988) and low-mobility semiconductor structures. In fig. 66 the resulting temperature dependence for a low-mobility 2D electron gas in a Si MOSFET is shown. In this case τ_φ is dominated by electron scattering and is proportional to T^{-1} (Lee and Ramakrishnan 1985) so

$$\Delta\sigma_{WL} = \frac{e^2}{\pi h} \ln\left(\frac{T}{T_0}\right), \tag{47}$$

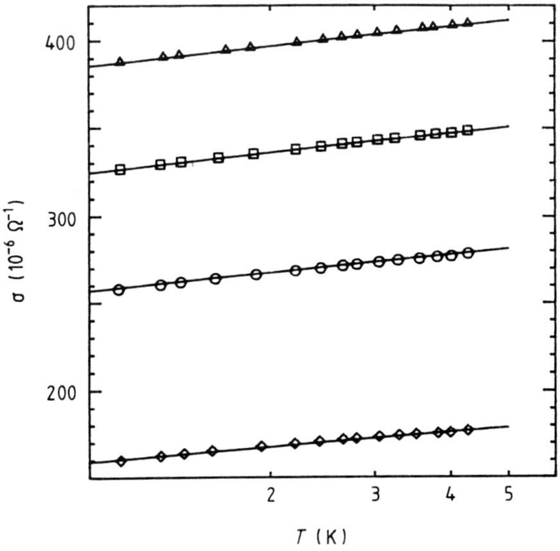

Fig. 66. The conductivity of silicon-on-sapphire MOSFETs plotted against temperature for different carrier concentrations. The full lines are the best fits to $\sigma = A + B \ln T$; the expected form for weak localisation (Syme and Pepper 1989).

where T_0 is a characteristic constant, in agreement with the observed logarithmic behaviour. Application of a magnetic field reduces $\Delta\sigma_{WL}$ since electrons traversing the paths of fig. 65 will pick up a phase change of opposite sign if they go clockwise or anticlockwise, i.e., the time reversal symmetry is broken. We therefore expect a negative magnetoresistance as well as a negative temperature coefficient. Such behaviour is shown in fig. 67 for a Si MOSFET. Spin–orbit coupling modifies weak localisation in a way which is well understood (Bergmann 1984). The situation is often complicated in real disordered systems where, because of the decreased electron diffusion coefficient, the electron–electron interaction is enhanced. This also gives a logarithmic temperature correction to the conductivity in 2D, and a positive magnetoresistance when the Zeeman energy, $g_L \mu_B B$, becomes comparable with the thermal energy $k_B T$ (g_L is the Landé g-factor and μ_B the Bohr magneton).

The influence of such corrections upon the thermopower has been theoretically established recently and observed experimentally in very low mobility MOSFETs. Applying the Mott formula (eq. (15)) to the weak localisation correction in 2D Kearney and Butcher (1988a,b) obtain, for $k_F l \gg 1$,

$$S \sim -\frac{\pi^2}{3}\frac{k_B}{|e|}\frac{k_B T}{E_F}[1-(k_F l)^{-1}\ln(T/T_0)], \tag{48}$$

where l is the elastic mean free path, and T_0 is a characteristic temperature. However, as in the case of magnetothermopower (§ 4) it is more illuminating to work in terms of the thermoelectric coefficient η which is defined by the general expression for the

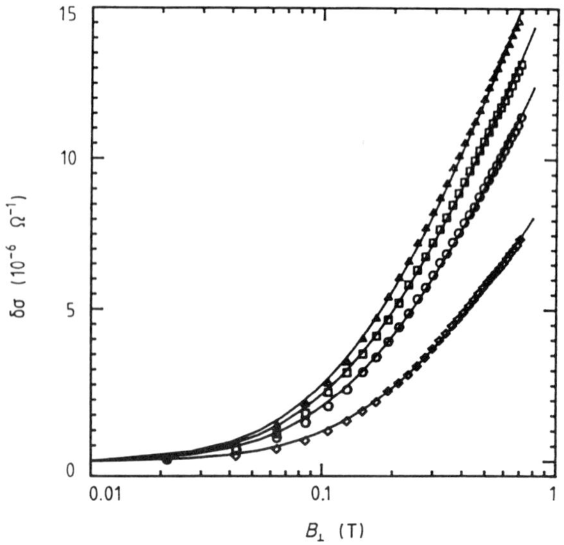

Fig. 67. The magnetoconductivity plotted against perpendicular magnetic field for the samples of fig. 66 at a temperature of 1.9 K. The full curves are fits to the weak localisation theory (Syme and Pepper 1989).

current density $\boldsymbol{j}$ due to an applied electric field $\boldsymbol{E}$ and temperature gradient $\boldsymbol{\nabla}T$,

$$\boldsymbol{j} = \sigma(\boldsymbol{E} + |e|^{-1}\Delta\mu) - \eta\boldsymbol{\nabla}T. \tag{49}$$

Note that $\eta = M/T$ where M is defined in § 4. Setting $\boldsymbol{j} = 0$ we have $S = \eta/\sigma$ and η can be written explicitly as a sum of diffusion, η_d, and phonon drag, η_g, contributions. So we can see that corrections to S arise directly from those to σ but also from η. In an early paper Ting et al. (1982) suggested that the correction to η was such that the diffusion thermopower remained unaffected by quantum interferences, i.e. $\Delta S_d / S_d \equiv \Delta\eta_d/\eta_d - \Delta\sigma/\sigma = 0$. This was later corrected by a number of authors (see Kearney et al. (1991) and references therein) who showed that such a cancellation did not occur. Kearney et al. (1991) and Castellani et al. (1988) obtained a Mott-type formula for η using diagrammatic perturbation theory which gives

$$\Delta\eta_d(B, T, \varepsilon_F) = -\frac{\pi^2}{3}\frac{k_B}{|e|}k_B T(1+\lambda)\left(\frac{d}{d\varepsilon}\Delta\sigma(B, T, \varepsilon)\right)_{\varepsilon=\varepsilon_F}. \tag{50}$$

The factor $(1+\lambda)$ is due to electron–phonon mass renormalisation (Howson and Gallagher 1988) and should be close to unity in semiconductor systems. There are therefore two distinct contributions to the thermopower correction. One arises directly from the correction to the conductivity the other from the energy dependence of η.

Syme and Pepper (1989), Syme et al. (1989) investigate weak localisation and electron–electron interaction corrections to the thermopower of silicon-on-sapphire MOSFETs. Due to the large lattice mismatch between these materials the structures have very low mobilities of order $0.05\ \mathrm{m^2\,V^{-1}\,s^{-1}}$.

The thermopower is measured using the resistance of the device itself to measure the temperature gradient. The observed thermopower is found to be dominated by phonon drag and increases as $\sim T^3$ (see fig. 40). Since, as they argue, one does not expect to see quantum corrections in η_g any observed correction to the measured thermopower should arise solely from $\Delta\sigma$ and

$$\frac{\Delta S}{S} \sim -\frac{\Delta\sigma}{\sigma}.$$

Because of the strong temperature dependence of S the characteristic temperature dependence of ΔS is unobservable but the magnetic field dependence may be observable. Weak localisation is found to dominate the magnetoresistance at low fields ($B < 1$ T) and excellent agreement with theory is obtained (fig. 67). The corresponding magnetothermopower is shown in fig. 68. As predicted $\Delta S/S$ falls as $\Delta\sigma/\sigma$ rises. In fig. 69 the ratio of the relative changes in S and σ are plotted. We see that it is very close to the expected value of -1 which is strong evidence that these are indeed weak localisation corrections to the thermopower.

Syme and Pepper also observe corrections due to electron–electron interactions. Since the magnetic field dependence of these corrections is a spin effect it can be observed with B parallel to the plane of the electrons. In contrast weak localisation

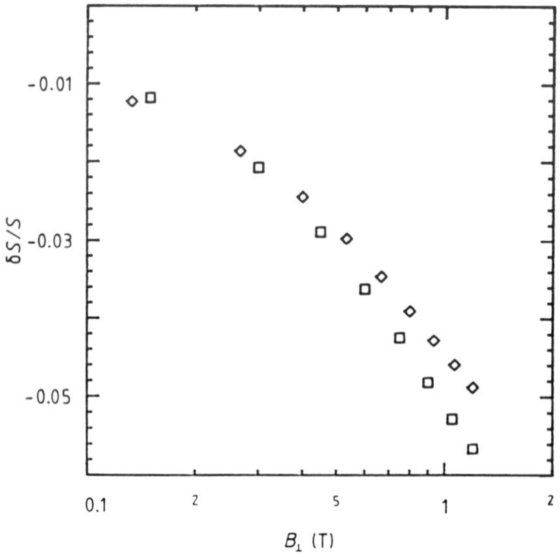

Fig. 68. The relative change in the thermopower plotted against perpendicular magnetic field for two of the samples of figs. 66 and 67 at a temperature of 1.9 K (Syme and Pepper 1989).

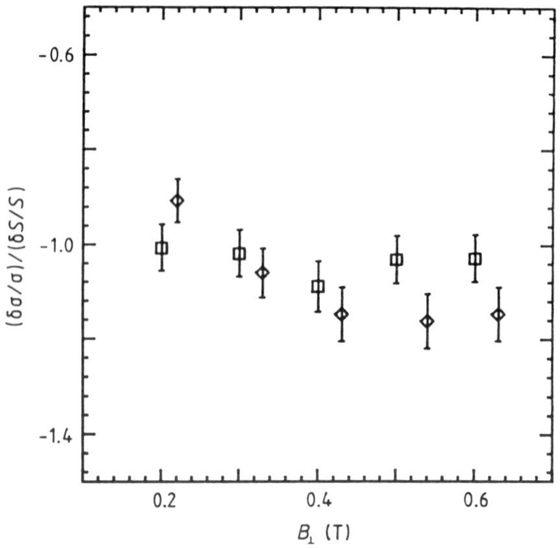

Fig. 69. The ratio of the relative changes in the thermopower and conductivity plotted against perpendicular magnetic field for the two samples of fig. 68 (Syme and Pepper 1989).

and classical magnetoresistance depend only upon the perpendicular component of B. Figure 70 shows that the negative high-field magnetoconductance in the parallel configuration is well accounted for by the electron–electron interaction theory. The corresponding thermopower given in fig. 71 now increases with B and the ratio of the relative changes in S and σ is again close to -1. Thus both types of quantum correction to the thermopower seem to be only manifestations of the corrections to the conductivity and not the thermoelectric coefficient η.

Kearney, Syme, and Pepper (1991) have recently presented evidence for weak localisation corrections to η itself for their silicon-on-sapphire devices. To do so they calculate $\eta(B)$ directly from the measured $S(B)$ and $\sigma(B)$. Figures 72 and 73 show the results obtained. The conductivity correction is well accounted for by the weak localisation theory although the value of a constant arising for intervalley scattering must be used as a fitting parameter. The observed $\Delta\eta(B)$ is then compared directly with eq. 50 by a procedure which involves obtaining the energy derivative of $\Delta\sigma$ from its variation with electron number density. The excellent agreement obtained indicates that $\Delta\eta(B)$ does indeed arise from weak localisation corrections.

It is important that further experimental and theoretical work is carried out on this problem since these results have far reaching implications. Firstly, the numerical differentiation used to evaluate eq. (50) relies upon the assumptions that one can equate the energy derivative evaluated at ε_F to that *with respect to* ε_F which is not obvious. Secondly, as emphasised in their earlier papers, the observed $\eta(B)$ is predominantly due to phonon drag. Gallagher et al. (1990b) showed the presence of quantum

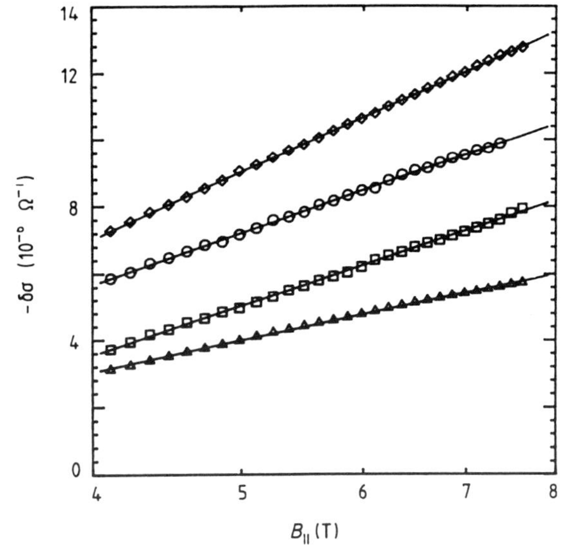

Fig. 70. The magnetoconductivity plotted against parallel magnetic field for the samples of fig. 66 at a temperature of 1.9 K. The full lines are fits to the electron–electron interaction theory (Syme and Pepper 1989).

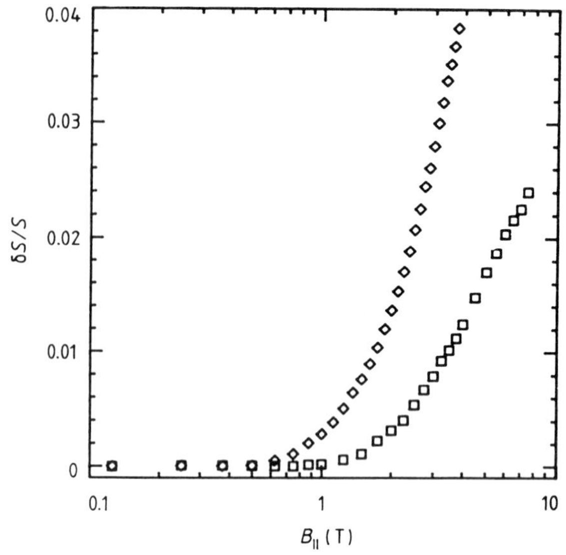

Fig. 71. The relative change in the thermopower plotted against parallel magnetic field for the two samples of fig. 68 at a temperature of 1.8 K (Syme and Pepper 1989).

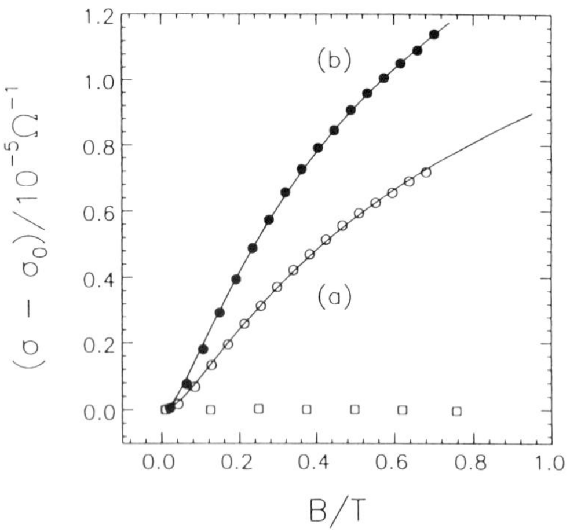

Fig. 72. The conductivity correction $\Delta\sigma$ plotted against applied magnetic field for an SOS MOSFET. The circles are experimental data for $B_{\perp}$, a temperature of 1.85 K, and electron concentrations of, curve (a) 2.10×10^{12} cm^{-2} ($k_F l = 2.1$) and, curve b, 3.18×10^{12} cm^{-2} ($k_F l = 3.3$). The curves are least-squares fits to the weak localisation theory. The squares are experimental data for $B_{\|}$, a temperature of 1.40 K, and an electron concentration of 2.10×10^{12} cm^{-2} (Kearney et al. 1991).

interference effects in the diffusion thermoelectric coefficient η_d (§ 6.2). What is being claimed by Kearney et al. (1991) is that because the weak localisation corrections are absent in phonon-drag coefficients one may still observe the correction to η_d. This clearly deserves further study.

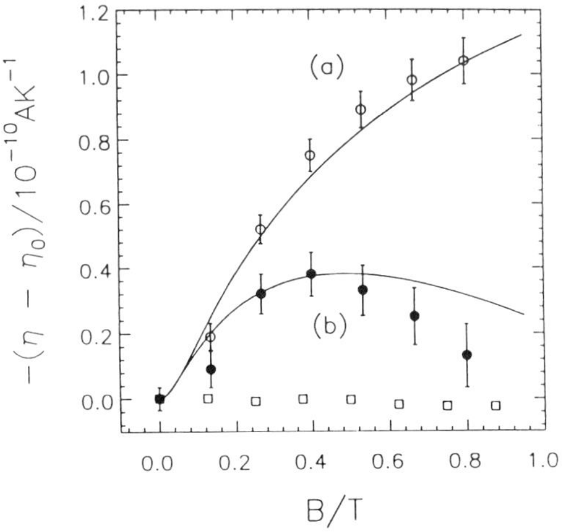

Fig. 73. The correction to the thermoelectric coefficient η plotted against applied magnetic field with the same symbols and parameters as in fig. 72. The curves are the theoretical predictions obtained from the conductivity data (Kearney et al. 1991).

6.2. *Universal conductance fluctuations*

Weak localisation is the result of interference between specific time-reversed trajectories and is observable in macroscopic samples. Random "direct" interference of the type shown in fig. 74 will also occur between partial waves travelling along any two paths. However, such partial waves have no specific phase relationship and in a macroscopic sample "self-averaging" will lead to the interference having no observable consequences. That is, such samples can be considered as comprising of an ensemble of subunits over which in a measurement we effectively take any ensemble average. This point of view has been the standard one in electrical transport until very recently. However, the rapidly diminishing size of electronic "microstructures" and "nanostructures" has now made it untenable in many cases at helium temperatures. Conductance measurements, e.g., often exhibit reproducible, sample-specific, aperiodic, fluctuations when a parameter like electron density or magnetic field strength is varied. We show examples of this in fig. 75, which is taken from a recent review of universal fluctuation by Stone (1988). Such fluctuations arise directly from the non-self-averaging of the random interference effects; the changes in electron

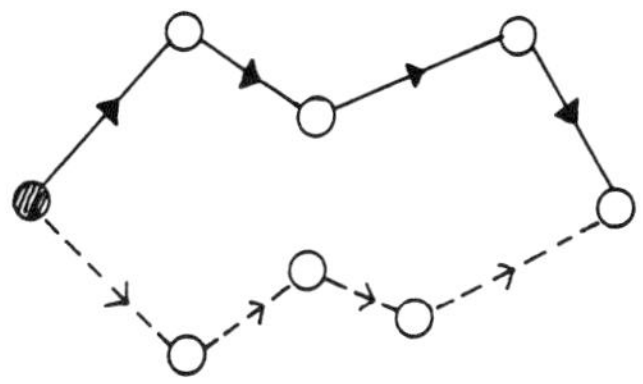

Fig. 74. Direct interference.

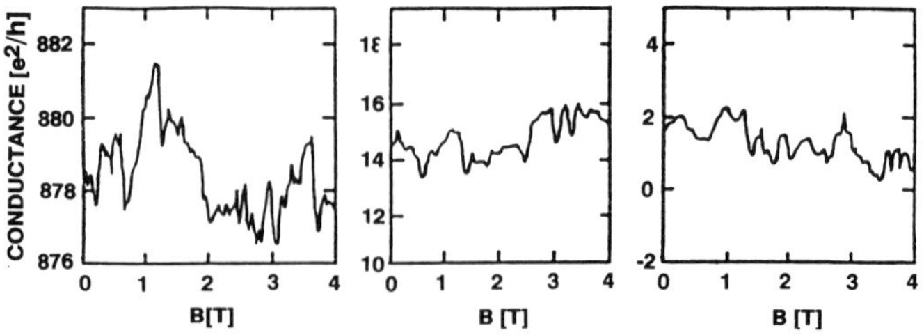

Fig. 75. Comparison of conductance fluctuations in units of e^2/h as B varies. Left-hand box: results for a 0.8 μm diameter gold ring (left), a quasi-1D Si MOSFET (middle), and an Anderson model (right) (after Stone 1988).

wavelength or phase cause the random fluctuations for a given specific distribution of static scatterers. That such fluctuations occur in very small systems is perhaps not surprising. What is surprising is that such effects can be observed in relatively large samples and that the rms deviation δG_{rms} of the conductance, G, takes on a "universal" value of order e^2/h for small enough samples. One might guess that the length scale beyond which interference effects start to average was of the order of the elastic scattering length, l. It is found, however, that the important length scale is that over which an electron retains phase coherence, L_φ, which can be very much longer than l. Furthermore, on length scales smaller than L_φ, quantum mechanical effects, which arise from the spatial correlation in the one-electron wavefunctions, always give rise to fluctuations of order e^2/h. This is illustrated in fig. 75. This is far from obvious and no simple picture has been put forward. For length scales larger than L_φ, however, classical self-averaging occurs. Thus for a sample of size L we have

$$\delta G_{\text{rms}} \sim e^2/h \left(\frac{L_\varphi}{L}\right)^{-(d-4)/2}, \tag{51}$$

where d is the dimensionality with respect to the length L_φ. For the 1D case (when only the length of the specimen is longer than L_φ) one has

$$\delta G_{\text{rms}} = \alpha_1 \frac{e^2}{h}\left(\frac{L_\varphi}{L}\right)^{2/3}. \tag{52}$$

As indicated by the subscript the prefactor is dependent on dimensionality. It is also dependent upon whether the average is over a range of ε_{F} or B (Lee et al. 1987). However, it is always very close to unity, hence the use of the term universal conductance fluctuations ("UCFs").

This effect was originally discovered in numerical calculations by Stone (1985), and confirmed analytically by Lee et al. (1987). These authors also verified their "ergodic hypothesis" that δG_{rms} is essentially independent of whether it is the distribution of scatters, B, or E_{F} which is varied. Their numerical calculations for all three cases are illustrated in fig. 76. This is an important result since it allows the result of the analytical theory, in which the distribution of scatters is varied, to be used to interpret experimental results which involve the variation of B or ε_{F}.

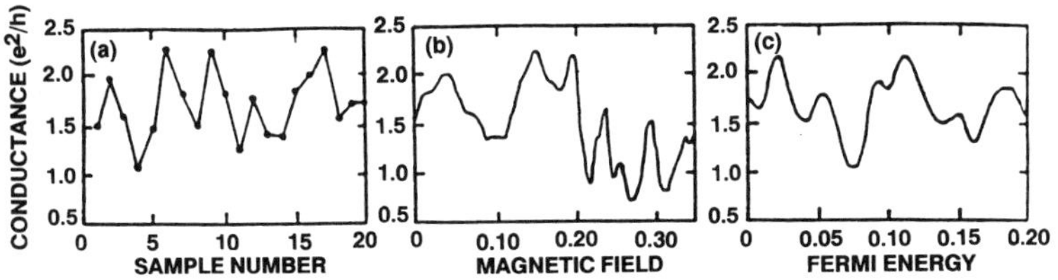

Fig. 76. Numerical calculations of the conductance fluctuations in units of e^2/h when the quantity varied is (a) the impurity distribution, (b) the magnetic field and (c) the Fermi energy (Lee et al. 1987).

In the generally accepted picture (Lee and Stone 1985), one important distinction between UCFs and weak localisation effects is that another length scale besides L_φ, may be important. This is the thermal length,

$$L_{\mathrm{T}} = \left(\frac{\hbar D}{k_{\mathrm{B}} T}\right)^{1/2}. \tag{53}$$

This arises from the fact that the thermal broadening of energy corresponds to a broadening in phase. Thus when $k_{\mathrm{B}} T$ becomes large compared to the energy broadening arising from inelastic scattering, h/τ_{in}, this can become the important energy scale and L_{T} the corresponding length scale. The problem is, however, a subtle one which depends upon dimensionality. In particular Lee et al. (1987) conclude that, whatever the relative size of L_{T} and L_{in}, the energy scale can never be much larger than $k_{\mathrm{B}} T$. Considering fig. 75 this energy scale will be just the typical spacing between peaks and valleys in the UCFs as ε_{F} is varied. More precisely, it can be identified with E_{c} the energy over which the fluctuations are correlated (Lee and Stone 1985). A correlation magnetic field, B_{c}, can be defined in a similar way. In this case the fluctuations become uncorrelated when the induced phase change around the largest loop is 2π. In 2D this corresponds to one flux quantum passing through an area L_φ^2, i.e.

$$B_{\mathrm{c}} = \frac{h/e}{L_\varphi^2}. \tag{54}$$

Fluctuations in conductance also manifest themselves in the Hall conductivity (Al'tshuler and Khmel'nitskii 1985) and as fluctuating non-Ohmic conduction as a function of applied voltage (Kaplan 1988), which gives rise to strong harmonic generation (Webb et al. 1988, de Vegvar et al. 1988), and rectification fluctuations (Galloway et al. 1990a). Recently universal thermopower fluctuations have also been observed (Gallagher et al. 1990b, Galloway et al. 1990b).

6.3. *Universal thermopower fluctuations*

A simple estimate of the expected diffusion thermopower fluctuations for a degenerate conductor can be obtained from the Mott formula (eq. (15)) written in terms of the conductance

$$S = \frac{-\pi^2}{3}\frac{k_B}{|e|}\frac{k_B T}{G}\left(\frac{dG}{dE}\right)_{E=E_F}. \tag{55}$$

Since quantum interference effects lead to a conductance which is a rapidly fluctuating function of E this immediately implies that the associated thermopower might fluctuate about zero with a large amplitude. An estimate of the amplitude may now be obtained by replacing the derivative in eq. (55) by $\delta G_{rms}/E_c$, where E_c is the correlation energy. We then have

$$\delta S_{rms} = -\beta\frac{\pi^2}{3}\frac{k_B}{|e|}\frac{k_B T}{E_c}\frac{\delta G_{rms}}{G}, \tag{56}$$

with $\beta \sim 1$. The most interesting case is then $E_c = k_B T$ since we then have the simple result

$$\delta S_{rms} = -\beta\frac{\pi^2}{3}\frac{k_B}{|e|}\frac{\delta G_{rms}}{G}. \tag{57}$$

This would appear to be a crude estimate since the Mott formula assumes that G is slowly varying on energy scales kT which is not true in this case. Furthermore, we have failed to distinguish between the fluctuations in S arising simply from G and those of the thermoelectric coefficient η. Esposito et al. (1987) obtained a general expression for δS_{rms}. Rewriting eq. (49) in the appropriate form we have the general linear response expression for the current

$$I = G\left(V - \frac{\Delta\mu}{e}\right) - \eta\,\Delta T, \tag{58}$$

where

$$\eta = -\frac{1}{eT}\int_0^\infty f'_0(\varepsilon - \mu)G(\varepsilon)\,d\varepsilon \tag{59}$$

and $S = \eta/G$.

Esposito et al. then obtained an expression for the rms amplitude of the fluctuations in η in terms of the correlation function for the conductance fluctuations

$$F(\varepsilon) = \langle G(\mu)G(\mu+\varepsilon) - \langle G(\mu)\rangle\langle G(\mu+\varepsilon)\rangle\rangle, \tag{60}$$

where the brackets indicate an ensemble average. They find

$$\delta\eta_{rms} = -\frac{1}{eT}\int_0^\infty\int_0^\infty f'(\varepsilon)f'(\varepsilon')(\varepsilon-\mu)(\varepsilon'-\mu)F(\varepsilon-\varepsilon')\,d\varepsilon\,d\varepsilon'. \tag{61}$$

This is a general formal expression requiring only that the scattering be predominantly elastic. Our simple expressions can now be recovered for the case where the relative fluctuations G are much smaller than those in η since then

$$\delta S_{rms} = \delta\eta_{rms}/G. \tag{62}$$

Equation (61) might seem to suggest that one cannot calculate the thermopower without knowledge of the detailed form of $F(\varepsilon)$. For the case $\mu \gg k_B T = E_c$, however, the integral is insensitive to the form of $F(\varepsilon)$ and the result (57) is recovered with $\beta = 0.17 \pm 0.01$ for a range of model forms for $F(\varepsilon)$ (Gallagher et al. 1990b). Lesovik and Kheml'nitskii (1988) also show that the fluctuations in η can be very much larger than the normal contribution and obtain the appropriate thermopower correlation function. Kearney (1988) obtains similar results using a model form of the conductance in a numerical calculation. Serota et al. (1988) go on to establish the validity of the result for multi-terminal devices.

The ratio of δS_{rms} to $(\delta G_{rms}/G)$ for $E_c = k_B T$ is thus expected to take a simple universal value, independent of temperature, size of specimen, or the degree of disorder. This remarkable result is implicit in the theoretical work refered to; where the complicated temperature and size dependence of δS_{rms} obtained in fact arises solely from those of $\delta G_{rms}/G$.

The fundamental problem in measuring the thermopower of samples showing UCFs is how to establish and measure a significant temperature difference along such micron-scale devices. The length scale over which an appreciable difference in lattice temperature can be established is the phonon mean path. Gallagher et al. (1990b) (see also Galloway et al. 1990b) found that in practise it is not possible to produce a temperature drop of more than $\sim$1 mK across a 10 μm sample in which case the thermoelectric voltage will be a few nanovolts which is of the same order as the Johnson noise at helium temperature. This led them to use a novel "hot electron thermopower" technique in which gradients are established in the *electron temperature*. Such gradients, which are easily established, have the same results as the conventional technique as long as electron–phonon scattering is much weaker than electron–impurity scattering. It also eliminates the phonon drag contribution. Gallagher et al. (1990b) study n^+ GaAs wires of cross section $\sim 0.4 \times 0.03$ μm in which $E_c = k_B T$ and the simple universal behaviour is expected. Their measurement arrangement is shown in fig. 77. An AC field is used to heat the electrons along the wire and the thermoelastic voltage, V_{AB}, is measured between the ends of the side arms. The electrons at C will be at a temperature $T_{E,max}$ and those at A and B close to the lattice temperature T_L since the distance AC = CB is about twice the electron energy relaxation length. For a homogeneous conductor the voltages V_{AC} and V_{CB} would cancel. The thermopower fluctuations in each arm are, however, uncorrelated and the symmetry is broken because the phase coherence length is $\sim$0.5 μm. Consequently, V_{AC} and V_{CB} simply add in a rms fashion.

The magnetoresistance obtained (fig. 78) shows weak localisation at low fields and relatively large conductance fluctuations which persist to the highest fields and coexist with the Shubnikov–de Haas oscillations. The large fluctuations about zero seen in the thermoelectric voltage are shown in fig. 79. They are found to be totally reproducible over a period of weeks and are essentially free of noise. To make an accurate comparison of rms amplitudes obtained with the theoretical predictions the form of $T_E(x)$ along the arms AC, CB is calculated. Finally $\delta V_{rms}/(\delta G_{rms}/G)$ was plotted against ΔT. The data for different T_L all fall close to the same line as expected and are in good agreement with the predicted near-linear dependence (fig. 80). The results are thus in agreement with theory in form, dependence on temperature, and

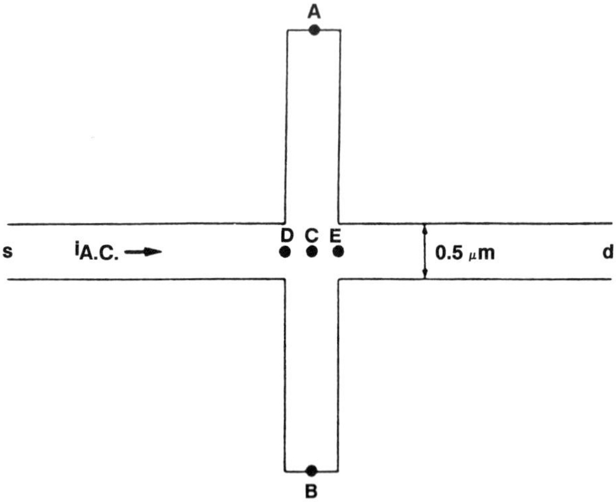

Fig. 77. Schematic of the wire geometry used to measure the "hot electron thermopower" on an n^+ GaAs wire. Length of AB is 9 μm (Gallagher et al. 1990b).

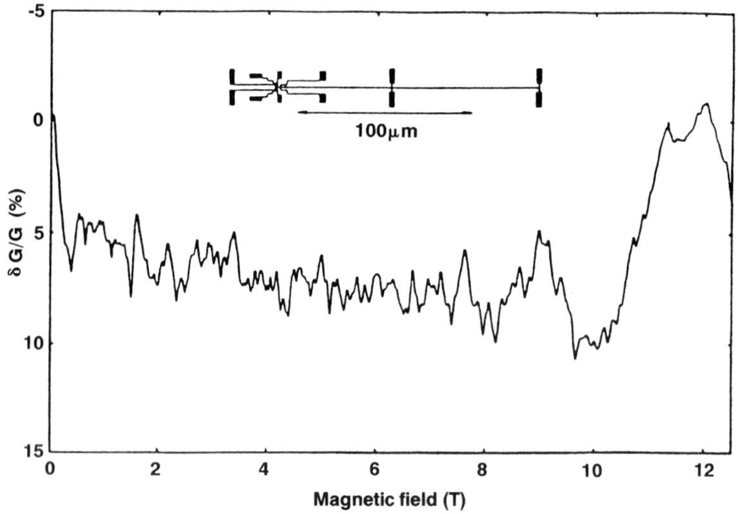

Fig. 78. Two-terminal resistance fluctuations in a 9 μm section of a very small cross section n^+ GaAs wire. Inset: the wire geometry. The resistance at zero field is 22 kΩ (Gallagher et al. 1990b).

temperature gradient. However, absolute magnitudes differ from theoretical predictions by a factor of about two. This is perhaps not surprising because of the relatively simple theory used and a full calculation of the expected absolute magnitude of the thermopower fluctuations would seem most worthwhile. Further thermopower fluctuation measurements might well yield much new insight into the basic physics of conductance fluctuation since they depend in a very direct way upon the energy

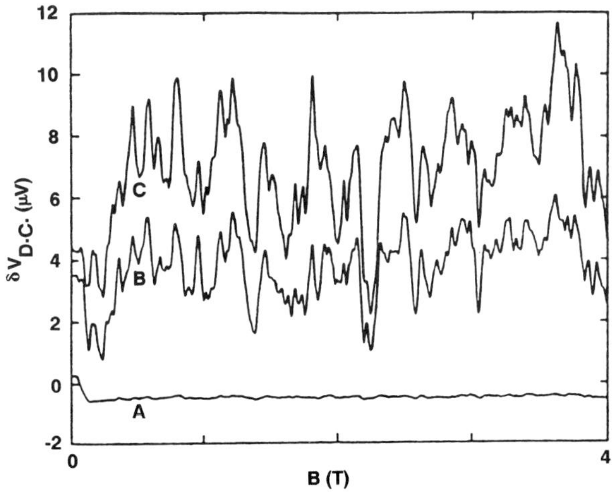

Fig. 79. dc thermoelectric voltage fluctuations due to the electron temperature gradient for traces (A) $\Delta T = 0$, (B) $\Delta T = 1$ K, and (C) $\Delta T = 2$ K. Trace B has been offset by 4 μV and C by 7 μV (Gallagher et al. 1990b).

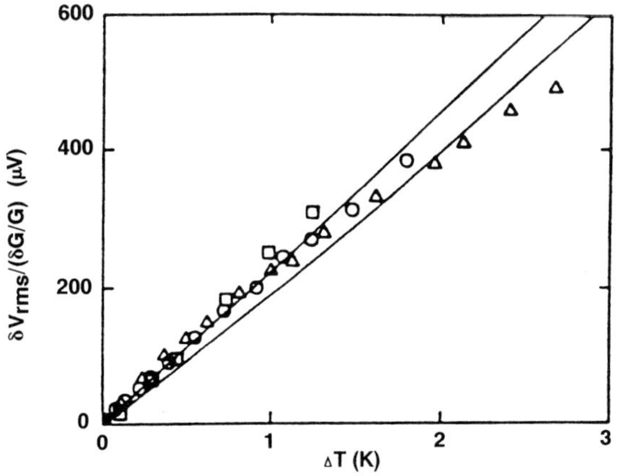

Fig. 80. rms dc thermoelectric voltage fluctuations as a function of ΔT compared with the theoretical predictions for the true temperature gradient. Upper line: predictions for a lattice temperature, T_L, of 4 K. Lower line: for $T_L = 1$ K. The predicted values have been increased by a factor of 2. Data points: $T_L = 1.08$ K (△), 2.00 K (○), and 4.00 K (□) (Gallagher et al. 1990b).

correlation function. Of particular interest is the conductivity/thermopower cross-correlation function (Lesovik and Kheml'nitskii 1988).

Gusev et al. (1990) claim to observe thermopower fluctuations by the normal approach of establishing a gradient in the lattice temperature. No indication of the temperature differences across the 1 μm sample is given and no comparison with theory is made. Using their values for the parameters in eq. (56) gives $\delta S_{rms} \sim 0.4\ \mu\mathrm{V\,K}^{-1}$ so one would expect signals of order a nanovolt for a temperature difference of a few mK. This shows that the microvolt signals observed must have some other origin.

7. Thermal and electrical transport in microstructures and edge states

7.1. The generalised Landauer–Büttiker formalism

In conductors which are comparable in size with or smaller than the electron elastic or phase coherence length the *conductance*, G, becomes non-local and one can no longer characterise the system by a local *conductivity*, σ. Similarly in the quantum Hall effect one has quantised *conductances* and even when the Hall conductance is *not* quantised the conductance is non-local and need not scale with device size in any simple way (McEwan et al. 1991). To deal with this situation Landauer (1957) proposed a general formula for the electrical conductance of a two-terminal electron system. This has been expressed by Büttiker (1988a,b, 1989) in a way which is convenient for a many-terminal system of the type shown in fig. 78. The Büttiker formalism directly relates the conductance matrix G to the electron scattering matrix S. The formulae that emerge shows that concepts of *conductivity* are naturally replaced by the *electron transmission probabilities* for the various parts of the electron paths. This new formalism is ideal for dealing with non-local conduction in ballistic conductors and in quantum point contacts in particular. It also helps a great deal in understanding phase coherent effects such as universal conductance fluctuations and Aharanov–Bohm oscillations. Furthermore its application to the quantum Hall effect provides profound insights into the reason for quantisation.

In this section we will briefly review the Landauer–Büttiker formalism for G and its extension to include the thermal and thermoelectric transport matrices (Sivan and Imry 1986, Butcher 1990). The general formalism shows how these matrices are the direct analogues of the local tensors describing thermopower, the Peltier effect and thermal conductivity in bulk solids and leads to predictions of a class of new effects. The case of quantum point contacts studied theoretically by Streda (1989) and experimentally by Molenkamp et al. (1990) will be considered in detail. Application of the formalism to the quantum Hall effect is then shown to shed new light on the question of "quantised thermopower".

We follow Büttiker (1988a) and suppose that free electrons with effective mass m^* enter a generalised microstructure like that of fig. 81 through ideal terminals in the form of long, straight electron waveguides. For simplicity we concentrate on the case when $\boldsymbol{B} = 0$ and quote results when $\boldsymbol{B} \neq 0$. The one-electron Hamiltonian in a terminal is then $H = \boldsymbol{p}^2/2m^* + V(x, y)$ where $\boldsymbol{p} = -i\hbar\nabla$ and $V(x, y)$ is the potential energy field confining the electrons. The normalised energy eigenfunctions in a length l of the terminal take the form

$$\psi_{\alpha k}(x, y, z) = l^{-1/2} \exp(ikz)\, \varphi_\alpha(x, y), \tag{63}$$

with energies

$$\varepsilon_{\alpha k} = E_\alpha + h^2 k^2/2m^*. \tag{64}$$

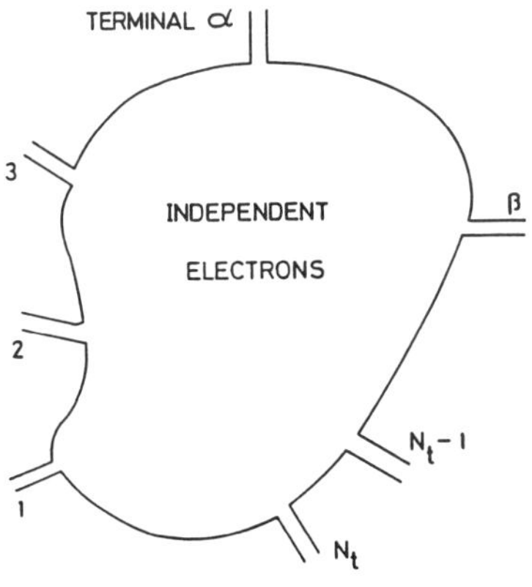

Fig. 81. Schematic diagram of an independent electron microstructure with N_t terminals labelled 1, 2, 3, ... α, ... β, ... $N_t - 1$, N_t (see text for details).

Here, z increases towards the microstructure and α labels the normalised transverse eigenfunctions $\varphi_\alpha(x, y)$ which are determined, together with the transverse energies E_α, by the 2D Schrodinger equation $(p_x^2 + p_z^2)\varphi_\alpha/2m^* = \varepsilon_\alpha \varphi_\alpha$.

It is assumed that all the scattering inside the microstructure is elastic so that we may confine our attention to electrons with a particular energy ε. Then eq. (64) shows that the channels (i.e., eigenfunctions) with $E_\alpha > \varepsilon$ are evanescent and decay to zero away from the microstructure. In the asymptotic regions of the terminals we are concerned only with the propagating channels for which $\varepsilon > E_\alpha$ and k is real. We may then quantise k by introducing periodic boundary conditions over the terminal length l. The mean longitudinal velocity is the group velocity $v_{\alpha k} = \hbar^{-1}\, \mathrm{d}\varepsilon_{\alpha k}/\mathrm{d}k = \hbar k/m^*$ and the density of states $N_{\alpha k}$ per unit energy range per unit length of the terminal is

$$N_{\alpha k} = |\pi\, \mathrm{d}\varepsilon_{\alpha k}/\mathrm{d}k|^{-1} = 2/h|v_{\alpha k}|, \tag{65}$$

where we have included a factor of 2 to allow for spin degeneracy.

Equation (64) determines $|k|$ for each propagating mode and we may identify an incident wave with $k = |k|$ and $v_{\alpha k} = \hbar|k|/m^*$ and a reflected wave for which both these quantities are negated. It is convenient to write $k_i = |k|$ and $k_r = -|k|$. Then the wavefunction in the asymptotic regions of the terminals has the form:

$$\psi = \sum_\alpha \left[a_\alpha (l/v_{\alpha k_i})^{1/2} \psi_{\alpha k_i} + b_\alpha (l/|v_{\alpha k_r}|)^{1/2} \psi_{\alpha k_r}\right]. \tag{66}$$

In this equation we have generalised the interpretation of α: it now labels a channel in *any* of the terminals and the sum ranges over the propagating channels in *all* the terminals. The square root factors have been introduced to give a convenient normalisation to the coefficients a_α and b_α of the incident and reflected waves. Their contributions to the longitudinal particle flux towards the microstructures are $|a_\alpha|^2$ and $-|b_\alpha|^2$, respectively.

The scattering matrix S determines the relation imposed by the microstructure between the coefficients $\boldsymbol{a} = \{a_\alpha\}$ of the incident waves and the coefficients $\boldsymbol{b} = \{b_\alpha\}$ of the reflected waves at a particular value ε of the electron energy. We write the relation in the form $\boldsymbol{b} = S\boldsymbol{a}$ where $\boldsymbol{a}$ and $\boldsymbol{b}$ are column matrices. Since particles are conserved we must have $|\boldsymbol{b}|^2 = |\boldsymbol{a}|^2$ for all $\boldsymbol{a}$. Hence S is unitary, i.e., $S^{-1} = S^\dagger$. Another important symmetry property follows from the time-reversal symmetry of Schrodinger's equation for the entire microstructure at energy ε: S is symmetrical. When $\boldsymbol{B} \neq 0$ the scattering matrix remains unitary but the "reciprocity symmetry" generalises to $S(-\boldsymbol{B}) = \tilde{S}(\boldsymbol{B})$.

The terminal transport relations involve real quantities. The scattering matrix enters into them through the real matrix T of transmission and reflection probabilities with elements $T_{\alpha\beta} = |S_{\alpha\beta}|^2$. We see that $T_{\alpha\beta}$ is the probability that an electron incident in channel β will appear in channel α, when $\alpha \neq \beta$, or will be reflected in channel β (when $\alpha = \beta$). The reciprocity symmetry of S obviously carries over to T, i.e., $T(-\boldsymbol{B}) = \tilde{T}(\boldsymbol{B})$. Moreover, the unitary character of S may be used to show that

$$\sum_\alpha T_{\alpha\beta} = \sum_\beta T_{\alpha\beta} = 1. \tag{67}$$

To calculate the electric current in channel α we suppose that the occupation probability $f_\alpha(\varepsilon)$ of the incident wave is given by the Fermi–Dirac function:

$$f_\alpha(\varepsilon) = \{\exp[(\varepsilon - \mu_\alpha)/k_\beta T_\alpha] + 1\}^{-1}, \tag{68}$$

where μ_α is the chemical potential and T_α is the temperature in the reservoir feeding the channel. The total charge flux towards the microstructure is

$$\begin{aligned} J_\alpha &= -e \int \mathrm{d}\varepsilon\, f_\alpha(lN_{\alpha k_\mathrm{i}})(v_{\alpha k_\mathrm{i}}/l) + e \sum_\beta \int \mathrm{d}\varepsilon\, f_\beta(lN_{\beta k_\mathrm{i}})(v_{\beta k_\mathrm{i}}) \\ &= -e^{-1} \sum_\beta \int \mathrm{d}\varepsilon\, f_\beta \Gamma_{\alpha\beta}, \end{aligned} \tag{69}$$

where we have used eq. (65) and the notation

$$\Gamma_{\alpha\beta} = \frac{2e^2}{h}(\delta_{\alpha\beta} - T_{\alpha\beta}). \tag{70}$$

To obtain the total heat flux Q_α flowing towards the microstructure we have to divide J_α by e and insert a factor $(\varepsilon - \mu_\alpha)$ in the integrand of eq. (69):

$$Q_\alpha = e^{-2} \sum_\beta \int \mathrm{d}\varepsilon_\beta\, \Gamma_{\alpha\beta}(\varepsilon - \mu_\alpha). \tag{71}$$

In these equations α and β always refer to propagating channels.

Equations (69) and (71) are non-linear transport equations. To linearise them we put $\mu_\alpha = \varepsilon_\mathrm{F} - eV_\alpha$ and $T_\alpha = T - \theta_\alpha$ where eV_α and θ_α are small perturbations of the chemical potential and temperature in channel α from equilbrium values ε_F and T which are common to all channels. Then

$$f_\alpha \simeq f_0 + f'_0\left(eV_\alpha - \frac{\varepsilon-\varepsilon_F}{T}\theta_\alpha\right), \tag{72}$$

where f_0 is given by eq. (68) with $\mu_\alpha = \varepsilon_F$ and $T_\alpha = T$ and f'_0 is the energy derivative of f_0. When eq. (72) is substituted into eqs. (69) and (71) we see that f_0 makes no contribution to J_α and Q_α because of eq. (67). Hence we obtain the linear transport equations

$$J_\alpha = \sum_\beta (G_{\alpha\beta}V_\beta + L_{\alpha\beta}\theta_\beta), \tag{73a}$$

$$Q_\alpha = \sum_\beta (M_{\alpha\beta}V_\beta + N_{\alpha\beta}\theta_\beta), \tag{73b}$$

where

$$G_{\alpha\beta} = -\int \mathrm{d}\varepsilon\, f'_0 \Gamma_{\alpha\beta} \simeq \Gamma_{\alpha\beta}, \tag{74a}$$

$$L_{\alpha\beta} = -\frac{1}{eT}\int \mathrm{d}\varepsilon\, f'_0 \Gamma_{\alpha\beta}(\varepsilon-\varepsilon_F) \simeq L_0 eT\Gamma'_{\alpha\beta}, \tag{74b}$$

$$M_{\alpha\beta} = -TL_{\alpha\beta} \sim -L_0 eT^2\Gamma'_{\alpha\beta} \tag{74c}$$

and

$$N_{\alpha\beta} = \frac{1}{e^2T}\int \mathrm{d}\varepsilon\, f'_0 \Gamma_{\alpha\beta}(\varepsilon-\varepsilon_F)^2 \sim -L_{0T}\Gamma_{\alpha\beta} \tag{74d}$$

with $L_0 = (\pi k_B/e)^2/3$ denoting the Lorenz number. In eq. (74) the first formula is exact. The second formula is the leading term in a Sommerfeld expansion at low temperatures in which $\Gamma_{\alpha\beta}$ and its energy derivative $\Gamma'_{\alpha\beta}$ are evaluated at $\varepsilon = \varepsilon_F$ (see the derivation of eq. (14)).

In a microstructure eq. (73) replace the local relations between fluxes and gradients which are normally used to describe the linear response of bulk solids. However, some care is needed in developing the analogy. The Greek subscripts in eq. (73) label propagating channels. We are actually interested in *terminals*. Each terminal may contain several propagating channels which are all fed from a common reservoir so that they have common values of V_β and θ_β. Moreover, only the total fluxes of charge and heat in each terminal are accessible to measurement. To allow for these facets of the microstructure problem we have only to reinterpret α and β as *terminal* labels and replace $\Gamma_{\alpha\beta}$ in eq. (73) by

$$\Gamma_{\alpha\beta} \to \sum_{\alpha'}\sum_{\beta'} \Gamma_{\alpha'\beta'}, \tag{75}$$

where the summation is over all propagating channels α' in terminal α and β' in terminal β. It is then convenient to write the terminal relations as matrix equations:

$$\boldsymbol{J} = G\boldsymbol{V} + L\boldsymbol{\theta} \tag{76a}$$

$$\boldsymbol{Q} = M\boldsymbol{V} + N\boldsymbol{\theta}. \tag{76b}$$

Equations (76) may be further simplified because eqs. (67) and (74) show that the rows and columns of all the transport matrices in eq. (76) sum to zero. The first property reflects the fact that $\boldsymbol{J}$ and $\boldsymbol{Q}$ are determined only by *differences* of the terminal voltages and temperatures. The second property reflects the fact that $\Sigma_\alpha J_\alpha = \Sigma_\alpha Q_\alpha = 0$ because of particle conservation. We are therefore free to choose one terminal as a reference (ground) terminal at which we set $V_\alpha = \theta_\alpha = 0$. Moreover, we have no need to calculate J_α and Q_α in the reference terminal because they are determined by the fluxes in all the other terminals.

To take these observations into account we remove, e.g., the last row and column from all the matrices in eq. (76). We leave this operation understood. The result is the "reduced form" of the linearised terminal transport equations. It has the advantage that G now has an inverse $R = G^{-1}$ so that we may rewrite the equations as

$$\boldsymbol{V} = R\boldsymbol{J} + S\boldsymbol{\theta} \tag{77a}$$

$$\boldsymbol{Q} = \Pi\boldsymbol{J} - \kappa\boldsymbol{\theta} \tag{77b}$$

where $S = -RL$, $\pi = MR$ and $\kappa = MRL - N$.

The reduced form of eq. (76) is the analogue of the way theoreticians conventionally write linear transport relations in bulk solids. The square matrix G is called the conductance matrix but L, M and N have no names. Equation (77) is the analogue of the way experimentalists conventionally write linear transport relations. The square matrix R is the resistivity matrix, and by analogy with bulk solid nomenclature, S, π and κ are conveniently called the thermopower, Peltier coefficient and thermal conductivity matrices, respectively. All these matrices have dimensions one less than the number of terminals. We see from eq. (70) that they all show reciprocity symmetry because Γ does, i.e., they are transposed when the magnetic field is reversed. This property of G and R leads to the familiar reciprocity theorem of electrical circuit theory: measured resistances are unchanged when ideal current sources and voltmeters are interchanged (Büttiker 1988) L and S, M and π and N and κ all have the same property, but its experimental consequences remain to be fully investigated in thermal and thermoelectric studies of microstructures (Butcher 1990).

7.2. *Thermoelectric effects in quantum point contacts*

In § 7.1 we defined a quantum point contact to be a two-terminal electronic microstructure in which every transmission coefficient $T_{\alpha'\beta'}$ is either 0 or 1. Then G is quantised in units of $2e^2/h$. The real problem is not how to *define* a quantum point contact, it is how to *make* one. Similar solutions to the fabrication problem were presented independently in 1988 by Van Wees et al. and Wharam et al.

We show the structure used by the Netherlands group in fig. 82. It consists of a 2DEG at a modulation-doped GaAs/AlGaAs heterojunction which is overlayed by

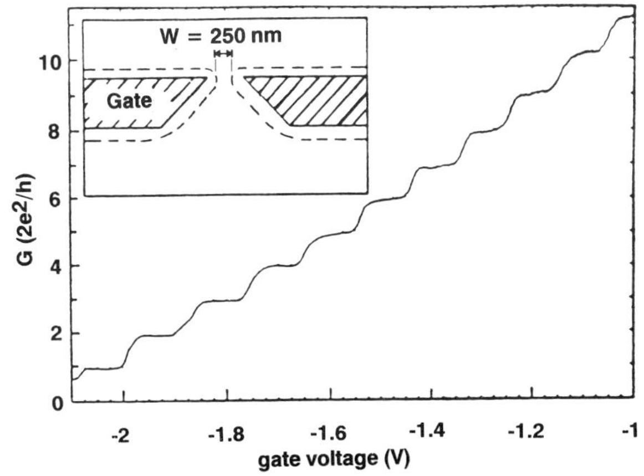

Fig. 82. Point contact conductance as a function of gate voltage at 0.6 K, demonstrating the conductance quantization in units of $2e^2/h$. The data are obtained from the two-terminal resistance after subtraction of a background resistance. The constriction width increases with increasing voltage on the gate (see inset) (van Wees et al. 1988).

a metal gate. When the gate bias V_g is negative, electrons are driven away from the region underneath the gate. The two terminals of the device are the wide 2DEGS on either side of the gate. These are so broad that their resistances are unquantised and are independent of V_g. However, to get from one terminal to the other, electrons must pass through the constriction produced by the gate bias. This becomes narrower and narrower as V_g becomes more negative. Consequently, fewer and fewer of the channels in the constriction can propagate at the Fermi level.

The constriction is the quantum point contact. We see from fig. 82 that its width $w < 0.25$ μm and length $L < 1.0$ μm. Both these lengths are much less than elastic mean free path in the 2DEG (8.5 μm) and the inelastic mean free path at the temperature of the experiment (0.6 K). Consequently, the electrons do not make a significant number of collisions with either impurities or phonons while traversing the construction. Moreover, provided w varies sufficiently slowly, they do not make a significant number of collisions with the edges of the constriction either. Instead, each of the channels which can propagate at the Fermi level in the input terminal either passes through the constriction without reflection or is cut off and reflected back completely. The sum in eq. (75) therefore reduces to the number of channels which can propagate through the constriction which decreases by one each time another channel is cut off by the progressive reduction of w. This is the origin of the staircase structure of the resistance which is shown in fig. 82. The quantisation in units of $2e^2/h$ is clearly exhibited. Similar results are reported by Wharam et al. (1988).

It can be objected that the assumption on which this interpretation of the data is based is possibly untenable. We have supposed that channels change adiabatically en route through the constriction and this might not be the case because of the sharp changes of width which are involved. However, Stone and Szafer (1989) have made detailed calculations of G on the basis of eq. (74a) for the two structures shown at the top of fig. 83a in which spin degeneracy is ignored. In spite of a 4:1 change of width between the terminals and the constriction, sharp quantisation is predicted.

In § 7.1 we developed the Landauer–Büttiker formalism to deal with the thermal conductance κ and the thermopower S of a multi-terminal microstructure. It is interesting to apply these ideas to a quantum point contact. Then κ and S are scalars. At low temperatures κ is related to G by the Wiedemann–Franz law $K = L_0 TG$ (where L_0 is the Lorenz number) which follows immediately from eq. (74a) and (74d). Moreover, the thermopower S is related to G by the Mott formula (15) with σ replaced by G. Consequently, in a two-terminal microstructure, κ has the same staircase structure as G but with a step height of $2L_0 Te^2/h$.

Let us now consider the thermopower of a point contact. Replacing $\sigma(\varepsilon)$ by the transmission coefficient $T(\varepsilon)$ in eq. (11) gives

$$S = \frac{-\Delta V}{\Delta T} = -\frac{\Delta\mu}{e\Delta T} = \frac{-1}{eT} \frac{\int_{-\infty}^{\infty} f_0'(\varepsilon)T(\varepsilon)(\varepsilon-\mu)\,\mathrm{d}\varepsilon}{\int_{-\infty}^{\infty} f_0(\varepsilon)T(\varepsilon)\,\mathrm{d}\varepsilon}. \tag{78}$$

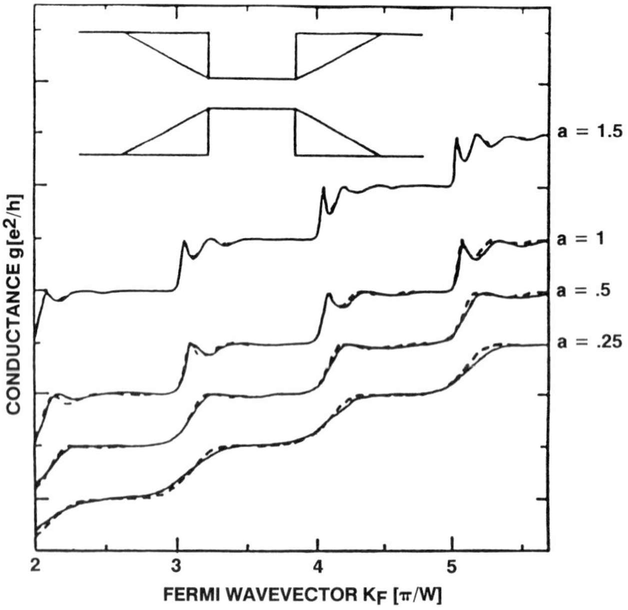

Fig. 83. Conductance of the 2DEG structures shown at the top of the figure plotted against Fermi wave number. The width of the terminals is four times the width of the constrictions. The quantity a is the ratio of the length of the transition (between the tapers) to the width of the constriction. The solid lines are exact numerical results for the abrupt transition. The dashed line, which is almost hidden by the top curve, is for the tapered transition. The dashed lines on the lower three curves show the results of an analytical approximation for the abrupt transition. The curves are offset vertically (Stone and Szafer 1989).

Consider now the idealised staircase-like structure for $T(\varepsilon)$ shown in fig. 84. For this case the contribution to the integrals of each sub-band is distinct and we can write

$$S = \frac{-1}{eT} \frac{\sum_{n=0}^{\infty} \int_{\varepsilon_n}^{+\infty} f'_0(\varepsilon)(\varepsilon-\mu)\,\mathrm{d}\varepsilon}{\sum_{n=0}^{\infty} f_0(\varepsilon_n)}, \tag{79}$$

where ε_n denotes the lowest energy of the nth sub-band. This is exactly the same expression as that obtained for the corresponding sum over the Landau levels (eq. (45)) in the case of the quantum Hall effect. For a real point contact device eq. (79) will still be valid provided that the sub-band separation, $\Delta\varepsilon$, is large compared to the energy interval over which $T(\varepsilon)$ changes by unity, Γ. For values of μ on the steps in the conductivity $T(\varepsilon)$ is constant and thus S is zero. For μ between steps S is finite and for μ midway between steps we have (Streda 1989)

$$S_{\max} = -\frac{k_B}{e}\frac{\ln 2}{(n+\frac{1}{2})} = -\frac{60}{(n+\frac{1}{2})}\ \mu\mathrm{V\,K^{-1}} \tag{80}$$

provided that $\Delta\varepsilon \gg k_B T > \Gamma$. The predicted behaviour (fig. 84) is exactly the same as that for 2D sub-bands (§ 3.7) and the quantum Hall effect (§ 5.1). The thermopower is thus independent of the origin of the quantisation of the energy spectrum and both magnetic and size quantisation yield the same result. The situation for the case of a point contact is, however, simpler as the conducting channel is *always* one dimensional and one does not need to assume $\sigma_{xy} \gg \sigma_{xx}$.

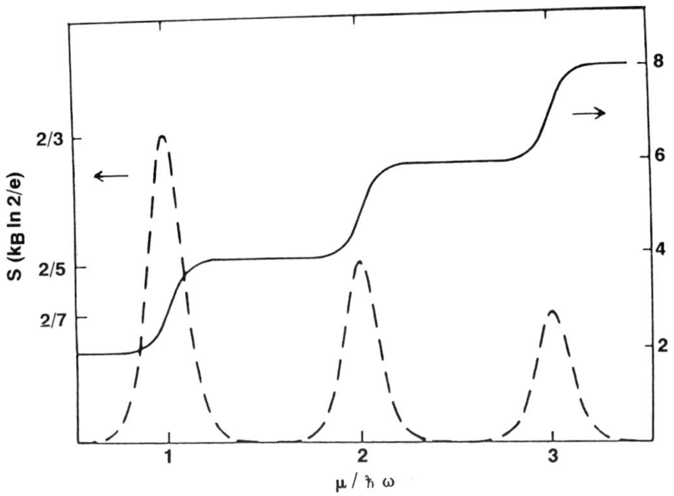

Fig. 84. The two-terminal conductance G (full curve) and the thermopower S (broken curve) of a channel in the form of the parabolic potential well as functions of the reduced chemical potential $\mu/\hbar\omega$ ($\hbar\omega$ denotes the sub-band spacing, $k_B T/\hbar\omega = 0.05$) (Streda 1989).

Experimentally the situation is also simpler because the electron–phonon interaction is, by definition, negligible for a ballistic channel and phonon drag will not be present. However, by the same token, it will be impossible to induce a finite temperature difference across such a microscopic channel by inducing a lattice temperature gradient in the microscopic substrate. The solution to this puzzle is that one must directly heat the *electrons* in one of the reservoirs. The hot-electron thermopower technique introduced in § 6.3 is therefore essential in the ballistic regime. This was employed by Molenkamp et al. (1990) using the devices shown schematically in the insert of fig. 85. A current is passed along the main channel to heat the electrons above the lattice temperature. The low electrical thermal conductivity of the point contacts then ensures that the electron temperature on the other side of the point contact is essentially that of the lattice. The transverse voltage V_{12} is then measured with the point contacts set to pass *different* numbers of 1D channels. The hotter region is in the *centre* of a thermoelectric circuit and a thermoelectric voltage is observed because of the lack of symmetry in the two arms of the circuit. This has the advantage of eliminating any bulk contributions since only the difference in the two arms, in this case the number of channels in the point contact, contributes.

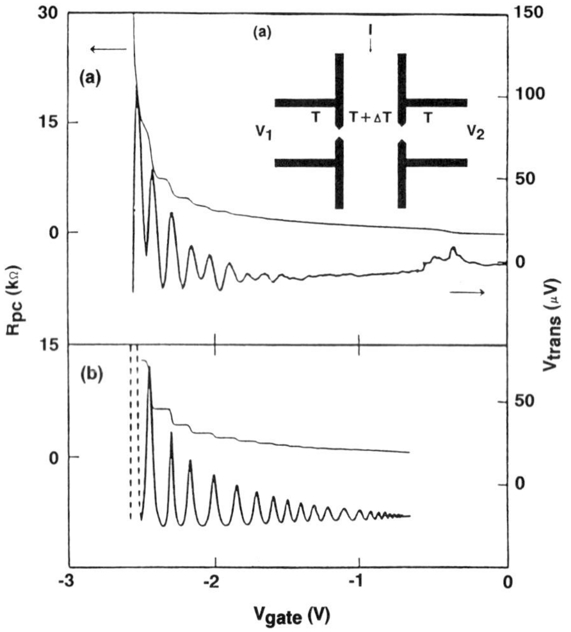

Fig. 85. (a) Experimental traces of the transverse voltages $V_{trans} = V_{12}$ (thick curve) and the point contact resistance, R_{pe} (thin curve) as a function of gate voltages V_{gate} at a lattice temperature of 1.65 K for $I =$ 5 μA. (b) Calculation of the transverse voltage (thick curve) with an electron temperature of 4 K and a lattice temperature of 1.65 K. The thin curve gives the dependence of R_{pc} on V_{gate} calculated at the lattice temperature (Molenkamp et al. 1990).

In the experiment of Molenkamp et al. the voltage on one gate is held fixed and this is considered to give the "reference thermopower" while that on the other is varied. The result is shown in fig. 85. As can be seen the observed thermoelectric voltage shows a series of peaks, when the resistance is between plates, which increase in magnitude as the number of 1D channels is successively decreased. This is just what is expected. Furthermore, when a moderate magnitude field is applied the states within the channel are further confined and the energy spacing of the resulting magneto-electric subbands increases with increasing field. This manifests itself in the steady reduction in the number of subbands for a given gate voltage. As fig. 86 shows, the corresponding thermoelectric voltage oscillations observed are again in agreement with expectation. The increase in magnitude of these voltage oscillations is attributed to the increased heating in the channel caused by the enhancement of the longitudinal resistance due to the Shubnikov–de Haas contribution.

The electron temperature is not measured by Molenkamp et al. but is inferred from the energy balance equation using an estimated energy relaxation time. This is only expected to give a rough estimate of ΔT. With $\tau_E \sim 10^{-10}$ s they estimate $\Delta T \sim 1$ K

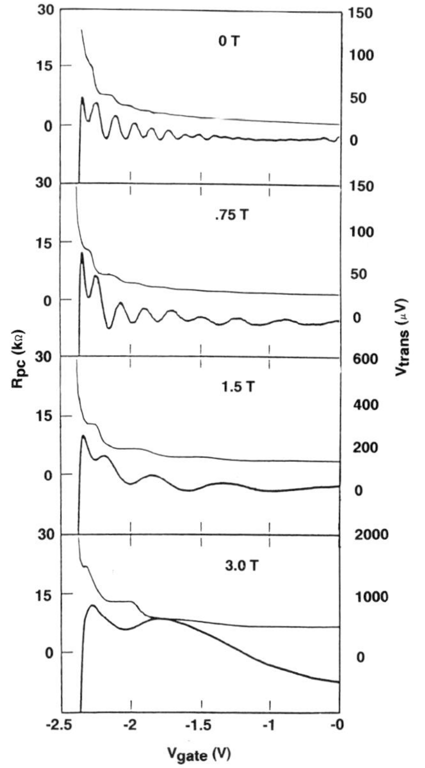

Fig. 86. The effects of magnetic depopulation of V_{trans} for the point contact device of fig. 85. Again, the thick curves give V_{trans} and the thin curves R_{pc} (Molenkamp et al. 1990).

for a current of 5 μA. When the calculation of the expected thermoelectric voltage is carried out assuming an idealised staircase transmission probability but not a small ΔT the results are in good agreement with observation (fig. 87). Unfortunately, such a calculation is not possible when $B \neq 0$. It is a pity that in this important piece of work ΔT was not determined directly since showing that the observed voltage was linear in ΔT would go a long way to firmly establishing its thermoelectric origin.

7.3. Thermal conductivity and Peltier coefficient of a quantum point contact

For real point contact devices the thermopower peaks will generally be smaller than the predicted "quantised" values of eq. (80) due to departure from the ideal staircase transmission probability. This is illustrated in fig. 87 where Van Houten et al. (1992) have calculated the thermopower for a reasonable potential. They also calculate the thermal conductance which still shows the step-like behaviour which follows from the Wiedemann–Franz law (§ 7.2).

Despite these deviations from ideal behaviour it is still possible to use the thermopower of point contacts as a crude microthermometer. Van Houten et al. (1992) and Molenkamp et al. (1992) have been able to use this technique to qualitatively verify the form of the thermal conductivity of a quantum point contact and obtain preliminary results for the Peltier coefficient.

The experimental arrangements for the thermal conductivity measurement is that of fig. 88. The current heats the electrons in the main channel as in the thermopower

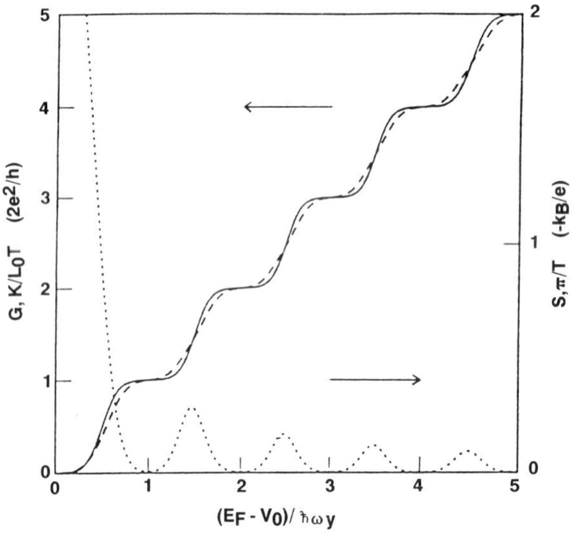

Fig. 87. Calculated conductance G (full curve), thermal conductance κ/L_0T (dashed curve), and the thermopower S and Peltier coefficient $\pi/T = S$ (same dotted curve) for a quantum point contact with a saddle-shaped potential, as a function of Fermi energy at 1 K. L_0 is the Lorentz number (van Houten et al. 1992).

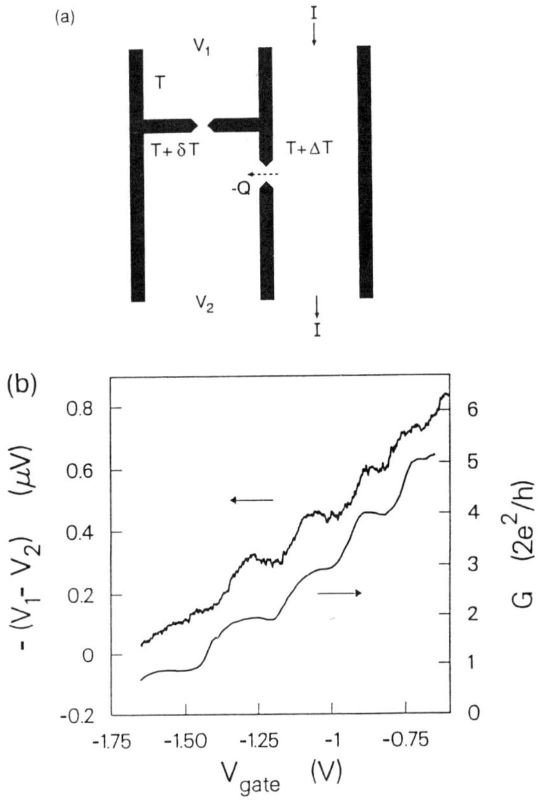

Fig. 88. (a) Schematic representation of the device used to demonstrate quantised steps in the thermal conductance of a quantum point contact, using another point contact as a miniature thermometer. The main channel is 0.4 μm wide. (b) Measured conductance and rms value of the second harmonic component of the voltage $V_1 - V_2$ as a function of the gate voltage defining the point contact in the main channel boundary, at a lattice temperature of 1.4 K and and ac current of rms amplitude 0.6 μA. The gates defining the other point contact were kept at -1.4 V, so that its conductance is $G = 1.5\ (2e^2/h)$ (van Houten et al. 1992).

measurement. Heat flow through the point contact now heats the electrons in region 2. If the heat capacity in the region is constant then the variation of δT with the voltage on the point contact through which the heat is flowing will be directly proportional to its thermal conductivity. This in turn is proportional to the thermoelectric voltage $(V_1 - V_2)$. As fig. 86 shows reasonable qualitative agreement with expectation is obtained. The expected magnitude of $(V_1 - V_2)$ can be obtained by assuming reasonable values for the thermopower of the point contacts and that the heat flowing through the point contact is deposited in an area of size roughly equal to the energy relaxation length squared. Van Houten et al. find that the observed magnitudes are consistent with expectation.

They go on to perform a yet more difficult experiment to try to observe the Peltier effect directly. The essential modification shown in fig. 89a is to introduce a point

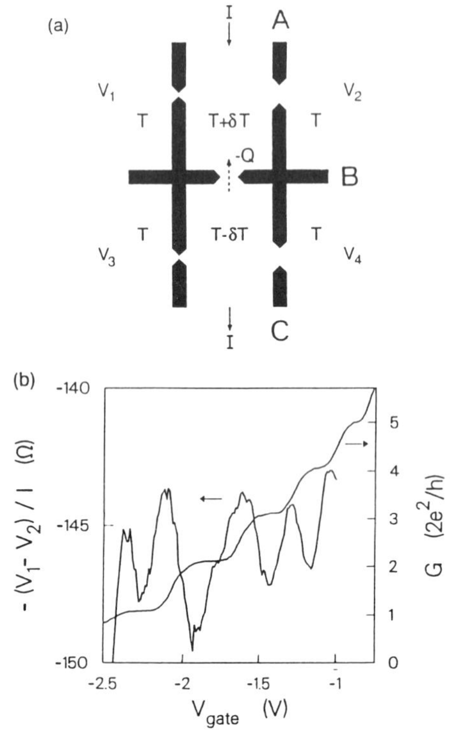

Fig. 89. (a) Schematic representation of the device used to demonstrate quantum oscillations in the Peltier coefficient of a quantum point contact. Arrows indicate direction of positive flow. The main channel is 4 μm wide, and the distance between the pairs of point contacts in its boundaries is 20 μm. (b) Measured conductance and thermovoltage $-(V_1 - V_2)$ divided by the current I as a function of the voltage on gate B, defining the point contact in the channel. The lattice temperature is 1.6 K and the current is about 0.1 μA near $G = 2e^2/h$. Gates defining point contacts 1 and 3 were adjusted so that their conductance was $G = 1.5\ (2e^2/h)$. Gates A and C were unconnected (van Houten et al. 1992).

contact into the main channel in which the current flows. Besides the Joule heating in the channel we will now have Peltier cooling on one side of the point contact and heating on the other. With fixed bias on the point contacts between 1 and 2 the voltage $(V_1 - V_2)$ should oscillate due to the oscillations of δT as a function of the voltage on the central point contact. One then expects thermopower like peaks at the steps in the conductance. The results obtained (fig. 89b) show evidence of this. In practice, the experiment is complicated by the variation of the Joule heating and a rather more complex experimental arrangement was employed to eliminate this effect.

8. Strong localisation and metal–insulator transitions

In severely disordered systems the electronic states can become strongly localised. This can lead to transport by hopping mechanisms. Hopping conduction in low-

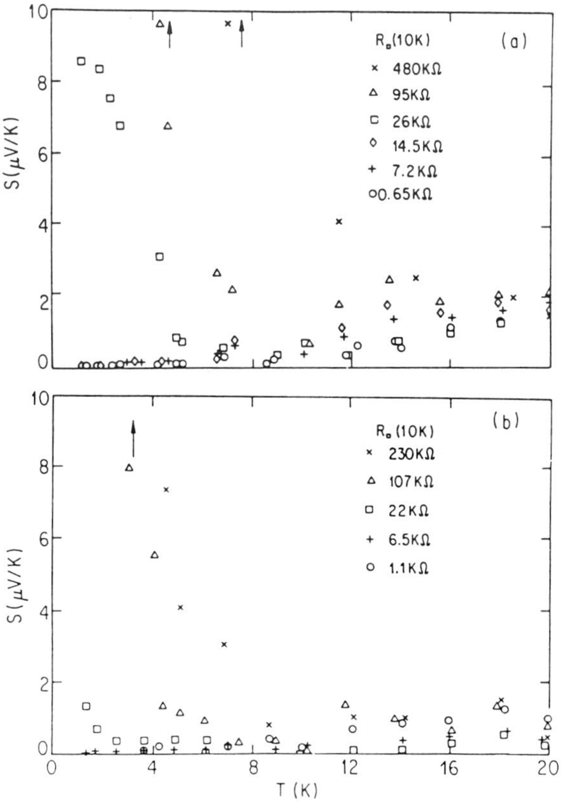

Fig. 90. Absolute thermopower as a function of temperature for (a) Pd and (b) Pd–Au thin films. Samples with $R_\square > 30\ \mathrm{k}\Omega/\square$ show the presence of an energy gap (Burns and Chaiken 1983).

dimensional structures has been reviewed by Butcher (1985), Bishop et al. (1985), Timp et al. (1986a,b), Fowler et al. (1988) and Butcher and McInnes (1988). One area of particular interest has been systems in which one can observe a transition from metallic to insulating behaviour either as a function of electron density or magnetic field. A key theoretical issue in strongly localised systems is whether one has a finite density of states at the Fermi energy or if, due to electron–electron interactions, a gap opens up at E_F.

Since thermopower depends upon the energy dependence of the conductivity it might well be able to shed light on the contentious issues involved. If a mobility edge, E_c, separates strongly and weakly localised states and a gap is not present then Sivan and Imry (1986) and Kearney and Butcher (1988b) argue that the thermopower should be given by

$$S = -\frac{k_B}{|e|}\left[A - B\left(\frac{E_F - E_c}{k_B T}\right)\right],$$

where A and B are constants. This implies that S will vary smoothly across the mobility edge and, in sharp contrast to the conductivity, it is not expected to show any structure for $E_F = E_c$ in the limit $T \to 0$. Furthermore at $E_F - E_c$ it will have taken on a universal value of $-Ak_B/|e|$.

Burns et al. (1981), Burns and Chaiken (1983) and Burns (1989) argue that thermoelectric measurements might be able to distinguish between metal–insulator transitions with and without a gap at E_F. The third law of thermodynamics requires that $\sigma S \to 0$ as $T \to 0$ so for the metallic case since $\sigma \neq 0$ at $T = 0$ it must always be the case that $S \to 0$ as $T \to 0$. On the insulating side of a transition $\sigma \to 0$ as $T \to 0$ so S may tend to zero, some finite value or diverge less rapidly than σ^{-1}. For the situation where there is no gap conduction on the insulating side as $T \to 0$ will be by hopping. In this case if electron–electron interactions are ignored, one expects $S \to 0$ as $T \to 0$. A system with an energy gap, ΔE, in which the conduction is activated will have the same general behaviour as an intrinsic semiconductor (eq. (9)), i.e., $S \sim -\Delta E/k_B T$. So the expectation is that $S \to \infty$ as $T \to 0$. Simple calculations for the situation where a Coulomb gap arises from the electron–electron interactions also suggest that S does not tend to zero in this case (Burns and Chaiken 1983, Mott 1987). Burns et al. (1981), and Burns and Chaiken (1983) study disordered two-dimensional Pd and Pd–Au films. They find that a metal–insulator transition occurs at a critical resistance of $\sim 30\ \mathrm{k}\Omega\ \square^{-1}$. The thermopower of these samples is dominated by the diffusion contribution. On the insulating side of the transition it is found to increase as T is reduced (fig. 90) indicating the presence of a gap at E_F. Burns (1989) carries out a similar study of the magnetic-field induced metal–insulator transition in degenerately doped n-type Ge. On the insulating side of the transition the thermopower is consistent with a T^{-1} dependence (fig. 91). This they take as evidence of the existence of a "soft" Coulomb gap. From these measurements S is clearly continuous through

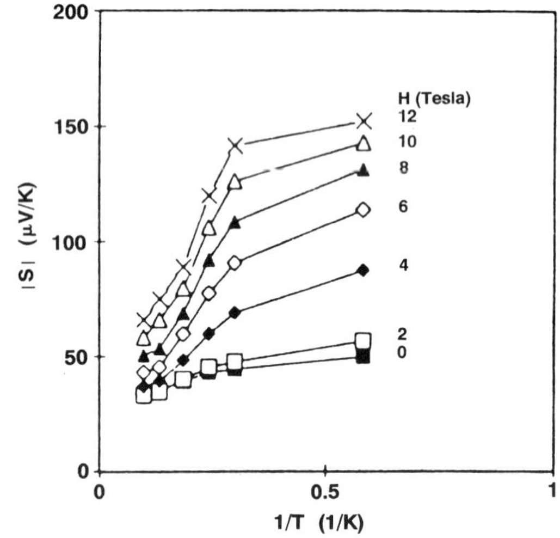

Fig. 91. Thermopower at constant magnetic field for a degenerately doped n-type Ge sample through a magnetically induced metal–insulator transition (Burns et al. 1989).

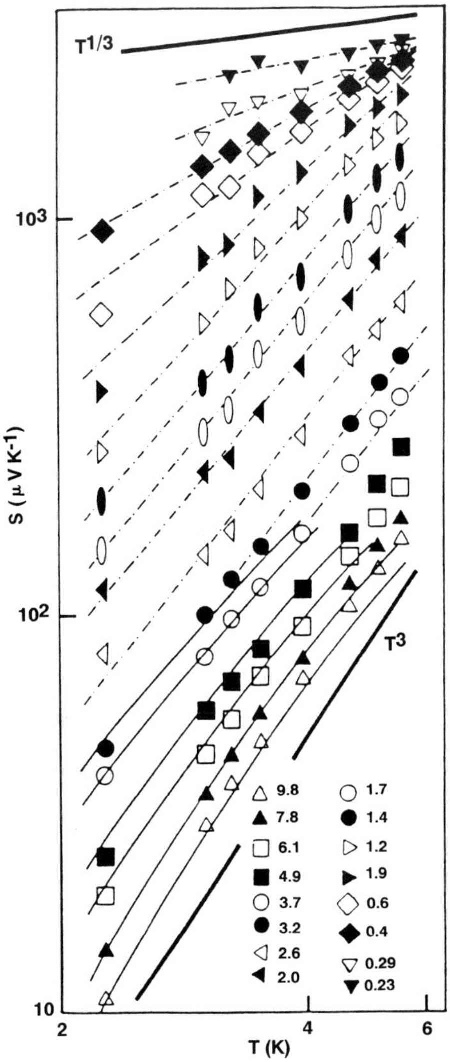

Fig. 92. The measured thermopowers of a Si MOSFET through the metal–insulator transition. The full curves are the calculated dependences at metallic densities. The chain curves are guides to the eye. The electron densities (in units of $10^{15}\ m^{-2}$) are given in the key. The density at which the transition occurs is $n_{MIT} = 1.2 \times 10^{15}\ m^{-2}$ (Gallagher et al. 1987).

the transition but this probably has little bearing on the theoretical issue which concerns the behaviour in the limit of $T \to 0$.

Gallagher et al. (1987) have measured the thermopower of a Si MOSFET through the metal–insulator transition as the electron areal density, n_S, is gradually lowered obtaining the results shown in figs. 92 and 93. The thermopower in this case is dominated by phonon drag. There have not to our knowledge been any theoretical

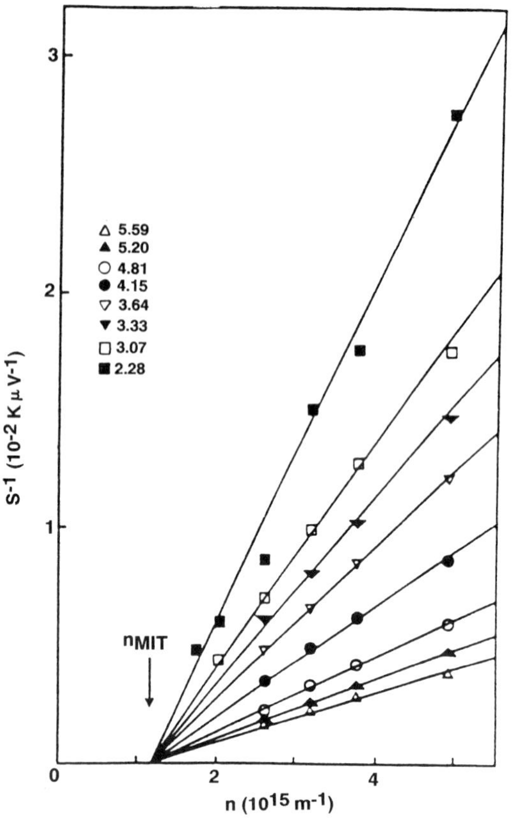

Fig. 93. The inverse thermopower as a function of electron density. The metal–insulator transition occurs at n_{MIT}. The temperatures (in K) are given in the key note that the values represented by full squares are reduced by a factor of 0.7 (Gallagher et al. 1987).

studies of this problem. Gallagher et al. suggest that both diffusion and drag thermopower should depend inversely upon the number of delocalised carriers and not the total number of carriers as is the case for the Hall coefficient. Evidence for this is given in fig. 93. They observe a transition from metallic to activated conduction for a density of $n_{MIT} \sim 1.2 \times 10^{15}\ m^{-2}$. They find for $n_s > n_{MIT}$ the $\sim T^3$ phonon-drag thermopower discussed in § 4.2. As the electron density is lowered the power of T falls, slowly at first and then for $n_s < n_{MIT}$ much more rapidly. For $(E_c - E_F)/kT \ll 1$ it is suggested that variable-range hopping becomes the dominant conduction mechanism. This should give a large diffusion thermopower with a $T^{1/3}$ dependence (Mott and Davis 1979) consistent with the values measured for the lowest electron densities. The thermopower varies smoothly through the transition suggesting that phonon-drag effects are present *through* the transition. Furthermore, for the lowest electron densities it appears that, as $T \rightarrow 0$, S is slowly decreasing or tending to a constant value. This suggests that the metal–insulator transition is due to a mobility edge rather than a gap at E_F. Timp et al. (1986a) reach the same conclusion for a two-dimensional impurity band.

Acknowledgements

We would like to acknowledge the many and varied contributions to this work of Richard Barraclough, Peter Beton, Dave Cantrell, Malcolm Carter, Mike Davies, Laurence Eaves, Mark Fromhold, Chris Gibbings, Trevor Galloway, Tony Hartland, Mohamed Henini, V. C. Karavolas, Mike Kearney, S. S. Kubakaddi, Peter Main, B. G. Mulimni, John Oxley, G. Qin, Mark Smith and Pavel Streda.

One of us (BLG) is particularly grateful to Dennis Greig and Mike Pepper for introducing him to thermoelectric effects and low-dimensional physics, respectively.

The financial support of the SERC is gratefully acknowledged.

Most importantly we would like to thank Alison Gilbert for all the excellent secretarial and editorial work without which this article would not have been possible.

References

Al'tshuler, B.L., and D.E. Khmel'nitskii, 1985, JETP Lett. **42**, 359.
Ando, T., A.B. Fowler and F. Stern, 1982, Rev. Mod. Phys. **54**, 437.
Bailyn, M., 1958, Phys. Rev. **112**, 1587.
Bailyn, M., 1967, Phys. Rev. **157**, 480.
Barnard, R.D., 1972, Thermoelectricity in Metals and Alloys (Taylor and Francis, London).
Bergmann, G., 1984, Phys. Reports **107**, 1.
Beton, P.H., P.C. Main, M. Davison, M. Dellow, R.P. Taylor, E.S. Alves, L. Eaves, S.P. Beaumont and C.D.W. Wilkinson, 1990, Phys. Rev. B **42**, 9689.
Bishop, D.J., D.C. Tsui and R.C. Dynes, 1985, in: Localisation, Interaction and Transport Phenomena, eds B. Kramer, G. Bermann and Y. Bruynseraeda (Publisher, Residence) pp. 31–37.
Blatt, F.J., P.A. Schroeder, C.L. Foiles and D. Greig, 1976, Thermoelectric Power of Metals (Plenum, New York).
Brummell, M.A., T.H.H. Vuong, R.J. Nicholas, J.C. Portal, M. Razeghi, K.Y. Chang and A.Y. Cho, 1986, Solid State Commun. **57**, 377.
Burns, M.J., 1989, Phys. Rev. B **40**, 5473.
Burns, M.J., and P.M. Chaiken, 1983, Phys. Rev. B **27**, 10, 5924.
Burns, M.J., W.C. McGinnis, R.W. Simon, G. Deutscher and P.M. Chaiken, 1981, Phys. Rev. Lett. **47**, 22, 1620.
Butcher, P.N., 1985, in: Amorphous Solids and the Liquid State, eds N.H. March, R.A. Street and M. Tosi (Plenum, New York) p. 311.
Butcher, P.N., 1990, J. Phys. C **2**, 4869.
Butcher, P.N., and J.A. McInnes, 1988, in: Physics and Technology of Submicron Structures, eds H. Heinreich, G. Bauer and F. Kuchar (Springer, Berlin) p. 218.
Büttiker, M., 1988a, Phys. Rev. B **38**, 9375.
Büttiker, M., 1988b, IBM J. Res. & Dev. **32**, 303.
Büttiker, M., 1989, Phys. Rev. Lett. **62**, 229.
Cantrell, D.G., and P.N. Butcher, 1985a, J. Phys. C **18**, L587.
Cantrell, D.G., and P.N. Butcher, 1985b, J. Phys. C **18**, 6627.
Cantrell, D.G., and P.N. Butcher, 1985c, J. Phys. C **18**, 5111.
Cantrell, D.G., and P.N. Butcher, 1985d, J. Phys. C **18**, 6639.

Cantrell, D.G., and P.N. Butcher, 1985e, J. Phys. C **18**, 6627.
Cantrell, D.G., and P.N. Butcher, 1987a, J. Phys. C **20**, 1985.
Cantrell, D.G., and P.N. Butcher, 1987b, J. Phys. C **20**, 1993.
Castellani, C., C. DiCastro, M. Grilli and G.S. Strinati, 1988, Phys. Rev. B **37**, 6663.
Chang, L.L., and L. Esaki, 1980, Surf. Sci. **98**, 70.
Davidson, J.S., E.D. Dahlberg, A.J. Valois and G.Y. Robinson, 1986, Phys. Rev. B **33**, 8238.
de Vegvar, P.G.N., G. Timp, P.M. Mankiewich, J.E. Cunningham, R.E. Behringer and R.E. Howard, 1988, Phys. Rev. B **38**, 4326.
D'Iorio, M., R. Stoner and R. Fletcher, 1988, Solid State Commun. **65**, 697.
Dugdale, 1977, The Electrical Properties of Metals and Alloys (Edward Arnold, London).
Esposito, F.P., B. Goodman and M. Ma, 1987, Phys. Rev. B **36**, 4507.
Fletcher, R., 1983, Phys. Rev. B **28**, 1721.
Fletcher, R., J.C. Maan and G. Weimann, 1985, Phys. Rev. B **32**, 8477.
Fletcher, R., J.C. Maan, K. Ploog and G. Weimann, 1986, Phys. Rev. B **33**, 7122.
Fletcher, R., M. D'Iorio, A.S. Sachrajda, R. Stoner, C.T. Foxon and J.J. Harris, 1988a, Phys. Rev. B **38**, 3137.
Fletcher, R., M. D'Iorio, W.T. Moore and R. Stoner, 1988b, J. Phys. C **21**, 2681.
Fletcher, R., E. Zaremba, M. D'Iorio, C.T. Foxon and J.J. Harris, 1988c, Phys. Rev. B **38**, 7866.
Fletcher, R., J.J. Harris and C.T. Foxton, 1991, Semicond. Sci. & Technol. **6**, 54.
Fowler, A.B., A. Harstein and R.A. Webb, 1982, Phys. Rev. Lett. **48**, 196.
Fowler, A.B., J.J. Wainer and R.A. Webb, 1988, IBM J. Res. & Dev. **32**, 372.
Friedmann, L., 1984, J. Phys. C **17**, 3999.
Fromhold, T.M., P.N. Butcher, G. Qin, B.G. Mulimani, J.P. Oxley and B.L. Gallagher, 1992a, Surf. Sci. **263**, 183.
Fromhold, T.M., P.N. Butcher, G. Qin, B.G. Mulimani, J.P. Oxley and B.L. Gallagher, 1992b, J. Phys. C, in press.
Gallagher, B.L., C.J. Gibbings, M. Pepper and D.G. Cantrell, 1987, Semicond. Sci. & Technol. **2**, 456.
Gallagher, B.L., J.P. Oxley, T. Galloway, M.J. Smith and P.N. Butcher, 1990a, J. Phys. C **2**, 755.
Gallagher, B.L., T. Galloway, P.H. Beton, J.P. Oxley, M. Carter, S.P. Beaumont, S. Thoms and C.D.W. Wilkinson, 1990b, Phys. Rev. Lett. **64**, 2058.
Galloway, T., B.L. Gallagher, P.H. Beton, J.P. Oxley, S.P. Beaumont, S. Thoms and C.D.W. Wilkinson, 1990a, J. Phys. C **2**, 5614.
Galloway, T., B.L. Gallagher, P.H. Beton, J.P. Oxley, M. Carter, S.P. Beaumont, S. Thoms and C.D.W. Wilkinson, 1990b, Surf. Sci. **229**, 326.
Gerhardts, R.R., 1975, Z. Phys. B **21**, 285.
Girvin, S.M., and M. Jonson, 1982, J. Phys. C **15**, L1147.
Guénault, A.M., 1971, J. Phys. F **1**, 373.
Gusev, G.M., Z.D. Kvon and A.G. Pogosov, 1990, JETP Lett. **51**, 3, 171.
Halperin, B.I., 1982, Phys. Rev. B **25**, 2185.
Havlova, H., and L. Smrcka, 1986, Phys. Stat. Sol. B **137**, 331.
Herring, C., 1954, Phys. Rev. **96**, 1163.
Howson, M.A., and B.L. Gallagher, 1988, Phys. Rep. **170**, 265.
Jonson, M., and S.M. Girvin, 1984, Phys. Rev. B **29**, 1939.
Kaplan, S.B., 1988, Surf. Sci. **196**, 93.
Karavolas, V.C., and P.N. Butcher, 1991, J. Phys. C **3**, 2597.
Karavolas, V.C., M.J. Smith, T.M. Fromhold, P.N. Butcher, B.G. Mulimani, B.L. Gallagher and J.P. Oxley, 1990, J. Phys. C **2**, 10401.
Karyagin, V.V., I.I. Lyapilin and V.V. Dyakin, 1988a, Phys. Stat. Sol. B **148**, 501.
Karyagin, V.V., I.I. Lyapilin and V.V. Dyakin, 1988b, Sov. Phys. Semicond. **22**, 954.
Kearney, M.J., 1988, Ph. D. Thesis, University of Warwick.
Kearney, M.J., and P.N. Butcher, 1986, C**19**, 5429.
Kearney, M.J., and P.N. Butcher, 1987, J. Phys. C **20**, 47.
Kearney, M.J., and P.N. Butcher, 1988a, J. Phys. C **21**, L265.
Kearney, M.J., and P.N. Butcher, 1988b, J. Phys. C **21**, 2539.

Kearney, M.J., R.T. Syme and M. Pepper, 1991, Phys. Rev. Lett. **66**, 1622.
Kouwenhoven, L.P., F.W.J. Hekking, B.J. van Wees, C.J.P.M. Harmans, C.E. Timmering and C.T. Foxon, 1990, Phys. Rev. Lett. **65**, 361.
Kubakaddi, S.S., and P.N. Butcher, 1989, J. Phys. C **1**, 3989.
Kubakaddi, S.S., P.N. Butcher and B.G. Mulimani, 1989, Phys. Rev. B **40**, 1377.
Kubakaddi, S.S., P.N. Butcher and B.G. Mulimani, 1991, J. Phys. C **3**, 5445.
Kundu, S., C.K. Sarkar and P.K. Basu, 1987, Surf. Sci. **196**, 700.
Kundu, S., C.K. Sarkar and P.K. Basu, 1988a, J. Appl. Phys. **61**, 5080.
Kundu, S., C.K. Sarkar and P.K. Basu, 1988b, Phys. Rev. B **38**, 5730.
Kundu, S., C.K. Sarkar and P.K. Basu, 1990, J. Appl. Phys. **68**, 1070.
Landauer, R., 1957, IBM J. Res. Dev. **1**, 223.
Lee, P.A., and T.V. Ramakrishnan, 1985, Rev. Mod. Phys. **57**, 287.
Lee, P.A., and A.D. Stone, 1985, Phys. Rev. Lett. **55**, 1622.
Lee, P.A., A.D. Stone and H. Fukuyama, 1987, Phys. Rev. B **35**, 1039.
Lesovik, G.B., and D.E. Khmel'nitskii, 1988, Sov. Phys. JETP **67**, 957.
Lyo, S.K., 1984, Phys. Rev. B **30**, 3257.
Lyo, S.K., 1988, Phys. Rev. B **38**, 6345.
Lyo, S.K., 1989, Phys. Rev. B **40**, 6458.
McEwan, P.L., A. Szafer, C.A. Richter, B.W. Alphenaar, J.K. Jain, A.D. Stone, R.G. Wheeler and R.N. Sacks, 1990, Phys. Rev. Lett. **64**, 2062.
Molenkamp, L.W., H. van Houten, C.W.J. Beenakker, R. Eppenga and C.T. Foxon, 1990, Phys. Rev. Lett. **65**, 1052.
Molenkamp, L.W., Th. Gravier, H. van Houten, O.J.A. Bujik, M.A.A. Mabesoone and C.T. Foxon, 1992, Phys. Rev. Lett. **68**, 3765.
Mott, N.F., 1987, Conduction in Non-Crystalline Materials (Oxford University Press, Oxford).
Mott, N.F., and E.A. Davies, 1979, Electronic Properties of Non-Crystalline Materials, 2nd Ed. (Oxford University Press, Oxford).
Nicholas, R.J., 1985, J. Phys. C **18**, L695.
Nolte, D.D., W. Walukiewicz and E.E. Haller, 1987, Phys. Rev. Lett. **59**, 501.
Obloh, H., K. von Klitzing and K. Ploog, 1984, Surf. Sci. **142**, 236.
Obraztsov, Y.N., 1965, Sov. Phys.-Solid State **7**, 455.
Oji, H., 1984a, J. Phys. C **17**, 3059.
Oji, H., 1984b, Phys. Rev. B **29**, 3148.
Oji, H., and P. Streda, 1985, Phys. Rev. B **31**, 7291.
Okuyama, Y., and N. Tokuda, 1990, Phys. Rev. **42**, 7078.
Oxley, J.P., B.L. Gallagher, T. Galloway, P.N. Butcher, V. Karavolas and T.M. Fromhold, 1990, Proc. 20th Int. Conf. Physics on Semiconductors, p. 853.
Oxley, J.P., B.L. Gallagher, T.M. Fromhold, G. Qin and P.N. Butcher, 1992, to be published.
Peeters, F.M., and P. Vasilopoulos, 1990, Phys. Rev. B **42**, 5899.
Poole, D.A., M. Pepper, K.F. Berggren, G. Hill and H.W. Myron, 1982, J. Phys. C **15**, L121.
Qin, G., T.M. Fromhold, P.N. Butcher, B.G. Mulimani, J.P. Oxley and B.L. Gallagher, 1992, J. Phys. C, in press.
Roberts, R.B., 1981, Philos. Mag. B **43**, 1125.
Ruf, C., H. Obloh, B. Junge, E. Gmelin, K. Ploog and G. Weimann, 1988, Phys. Rev. B **37**, 6377.
Ruf, C., M.A. Brummell, E. Gmelin and K. Ploog, 1989, Superlatt. & Microstr. **6**, 175.
Sernelius, B.E., K.F. Berggren, M. Tomak and C.M. Fadden, 1985, J. Phys. C **18**, 225.
Serota, R.A., M. Ma and B. Goodman, 1988, Phys. Rev. B **37**, 6540.
Sivan, U., and Y. Imry, 1986, Phys. Rev. B **33**, 551.
Skocpol, W.J., P.M. Mankiewich, R.E. Howard, L.D. Jackel, D.M. Tennant and A.D. Stone, 1987, Phys. Rev. Lett. **58**, 2347.
Smith, M.J., and P.N. Butcher, 1989a, J. Phys.: Condensed Matter **1**, 1260.
Smith, M.J., and P.N. Butcher, 1989b, J. Phys.: Condensed Matter **1**, 4859.
Smith, M.J., and P.N. Butcher, 1990, J. Phys.: Condensed Matter **2**, 2375.
Smrcka, L., and P. Streda, 1977, J. Phys. C **10**, 2153.

Stone, A.D., 1985, Phys. Rev. Lett. **54**, 2692.

Stone, A.D., 1988, in: Physics and Technology of Submicron Structures, eds H. Heinreich, G. Bauer and F. Kuchar (Springer, Berlin) p. 108.

Stone, A.D., and A. Szafer, 1989, Phys. Rev. Lett. **62**, 300.

Streda, P., 1983, J. Phys. C **16**, L369.

Streda, P., 1984, Phys. Stat. Sol. B **125**, 849.

Streda, P., 1989, J. Phys. C **1**, 1025.

Syme, R.T., and M. Pepper, 1989, J. Phys. C **1**, 2747.

Syme, R.T., M.J. Kelly and M. Pepper, 1988, J. Phys. **1**, 3375.

Syme, R.T., M. Pepper, A. Gundlach and A. Ruthvan, 1989, Superlatt. & Microstr. **5**, 103.

Tao, Z., and L. Friedmann, 1985, J. Phys. C **18**, L455.

Tauc, J., 1962, Photo- and Thermoelectric Effects in Semiconductors (Pergamon, New York).

Thornton, T.J., M. Pepper, H. Ahmed, D. Andrews and G.J. Davies, 1986, Phys. Rev. Lett. **56**, 1198.

Timp, G., A.B. Fowler, A. Harstein and P.N. Butcher, 1986a, Phys. Rev. B **33**, 1499.

Timp, G., A.B. Fowler, A. Harstein and P.N. Butcher, 1986b, Phys. Rev. B **34**, 8771.

Ting, C.S., A. Houghton and J.R. Senna, 1982, Phys. Rev. B **25**, 1439.

van Houten, H., L.W. Molenkamp, C.W.J. Beenakker and C.T. Foxon, 1992, Semicond. Sci. & Technol. B **7**, 215.

van Wees, B.J., H. van Houten, C.W.J. Beenakker, J.G. Williamson, L.P. Kouwenhoven, D. van der Marel and C.T. Foxon, 1988, Phys. Rev. Lett. **60**, 848.

Vuong, T.H.H., R.J. Nicholas, M.A. Brummell, J.C. Portal, F. Alexandre, J.M. Masson and T. Kerr, 1986, Solid State Commun. **57**, 381.

Webb, R.A., S. Washburn and C.P. Umbach, 1988, Phys. Rev. B **37**, 8455.

Weiss, D., K. von Klitzing, K. Ploog and G. Weimann, 1989, Europhys. Lett. **8**, 179.

Wharam, D.A., T.J. Thornton, R. Newbury, M. Pepper, H. Ahmed, J.E.F. Frost, D.J. Hasko, D.C. Peacock, D.A. Ritchie and G.A.C. Jones, 1988, J. Phys. C **21**, L209.

Zavaritskii, N.V., 1984, Physica B **126**, 396.

Zawadzki, W., and R. Lassnig, 1984, Surf. Sci. **142**, 225.

Zelenin, S.P., A.S. Kondrat'ev and A.E. Kuchma, 1982, Sov. Phys. Semicond. **16**, 3, 355.

CHAPTER 15

Coherence in III–V Semiconductor Superlattices

PAUL VOISIN and GERALD BASTARD

Laboratoire de Physique de la Matière Condensée de l'Ecole Normale Supérieure
24 rue Lhomond, F-75005 Paris, France

Handbook on Semiconductors
Completely Revised Edition
Edited by T.S. Moss
Volume 1, edited by P.T. Landsberg

Contents

1. Introduction . . . 819
2. Remarks on superlattice structural properties, materials and types . . . 820
3. Approaches to superlattice band structure . . . 826
4. Superlattice optical properties . . . 833
5. Excitons in superlattices . . . 837
6. Superlattices in strong magnetic fields . . . 840
7. Superlattices in an external electric field: Wannier quantization . . . 843
8. Electro-optical absorption in semiconductor superlattices . . . 848
9. Field-induced quantum boxes . . . 852
10. Diffusion and vertical transport in superlattices . . . 854
11. Concluding remarks . . . 855

References . . . 858

1. Introduction

At the end of the sixties, at a time when many physicists thought that fundamental semiconductor physics was "finished", a short paper by Leo Esaki and Ray Tsu, rejected by the Physical Review for being too speculative, appeared in the IBM Journal of Research and Development (Esaki and Tsu 1970). This was the starting point of the history of semiconductor heterostructures, which, twenty years after, represents about 50% of the effort in semiconductor physics worldwide. In this seminal paper, the authors speculated that a periodic modulation of the composition or doping of a semiconductor at a length scale shorter than the electron mean free path would result in a folding of the Brillouin zone into mini-zones showing strong energy dispersion effects leading to novel electronic properties, and in particular to original negative differential conductance mechanisms. As anticipated by the authors, the practical realization of such "superlattice" structures was a "formidable task", but "efforts directed to this end would open new areas of investigation in the field of semiconductor physics". Indeed, the initial development of sophisticated epitaxial growth techniques allowing sharp composition modulation was rather slow and convincing evidence of the quantum confinement of carriers into semiconductor quantum wells was not obtained before the mid-seventies. Early attempts to investigate the axial transport at that time (Esaki and Chang 1974) were somewhat disappointing, as they revealed a tendency to electric field domain formation making the observation of "miniband" transport impossible. On the contrary, at the same time, optical absorption experiments in multiple quantum well structures brought direct evidence of the quantum confinement in narrow quantum wells (Dingle et al. 1974) and then of the coupling of such quantum wells by the resonant tunnel effect through narrow barriers (Dingle et al. 1975, Dingle 1975). The richness of quantum well optical properties and in-plane transport properties of modulation-doped heterojunctions then led the field for several years, with an increasing number of researchers involved in the fabrication and the study of a variety of quantum wells. For about ten years, the original idea of superlattices in the sense of periodic structures with long-range phase coherence of electronic states was left out of the limelight, while the static and dynamic lattice properties of superlattices were thoroughly investigated. With the increasing quality of the samples, and the first signs of a saturation in the topic of quantum well physics, superlattices finally regained increasing interest in the last few years.

There already exist a number of recent reviews (Smith and Mailhiot 1990, Bastard et al. 1988, 1991) and even text books (Allan et al. 1986, Bastard 1988) dealing with the electronic structure and optical properties of quantum well structures, and we shall not attempt to cover such an extended subject here. Several important topics like doping superlattices (Döhler 1986, 1987, 1988), whose properties depend self-consistently on the carrier density, or quasi-3D plasmons in doped superlattices (Jain and Das Sarma 1988, and references therein), remain out of our scope. We shall concentrate on those particular properties which rely on the periodicity of the layer arrangement and phase coherence of electronic states in superlattices. In § 2, we review the various material systems and the resulting types of superlattices. Then we discuss in an elementary way the models which allow realistic calculations of superlattice band structure (§ 3), interband optical properties (§ 4) and excitonic properties (§ 5). § 6 is devoted to the effects of an in-plane or an axial magnetic field, and § 7 and § 8 concentrate on the effects of an axial electric field, or "Wannier–Stark" effects. The quasi-zero-dimensional states, or "field-induced quantum boxes" which result from the combination of Landau and Wannier quantizations are described in § 9. Transport properties are briefly discussed in § 10.

2. Remarks on superlattice structural properties, materials and types

We consider compositional superlattices (SL) which are periodic stacks of thin alternate layers of different semiconductors with identical crystalline structures (for III–V and most II–VI compounds, the zincblende structure) and similar lattice parameters. In principle, the SL unit cell could contain an arbitrary number of layer types or thicknesses (Esaki et al. 1981), but in practice only binary superlattices with two materials and layer thicknesses (say A, L_A and B, L_B) have been investigated in some detail. Thus, superlattices are highly anisotropic structures in which two spatial periods coexist: the host crystal period $a \approx 5.6$ Å, which is also the SL period in the directions of the layer plane, and the superperiod $d = L_A + L_B \approx 70$ Å which is the SL period along the growth axis. In fact, the perfect lattice matching is a highly restrictive condition, since no pair of binary materials have exactly the same lattice parameter. Fortunately, this condition is not necessary for the epitaxial growth of heterostructures with perfect crystalline integrity because some lattice mismatch can be accommodated by biaxially straining the materials (Osbourn 1982). In such a case, the layers have the same in-plane lattice constant resulting from the mechanical equilibrium of the whole structure, and slightly different lattice parameters along the growth axis. A representative figure for the lattice mismatch which can be accommodated by elastic deformations is 1%. In this range, the related modifications of the host electronic properties are adequately described by the deformation potential theory, and the strain-induced effects (bandgap change and valence band splitting, as illustrated in fig. 1) can compare quantitatively with the effect of the quantum confinement. The consideration of lattice mismatched materials has considerably enriched the physics of semiconductor heterostructures, both by multiplying the number of systems and by introducing specific electronic properties. Topical reviews of strained-layers materials have recently been published (Pearsall 1990, 1991).

The ideal tool for the structural analysis of superlattices is X-ray double diffraction (Segmuller et al. 1977, Speriosu 1981, Quillec et al. 1984, Picraux et al. 1991). Diffraction spectra (or "rocking curves") consist of a set of narrow peaks (since the

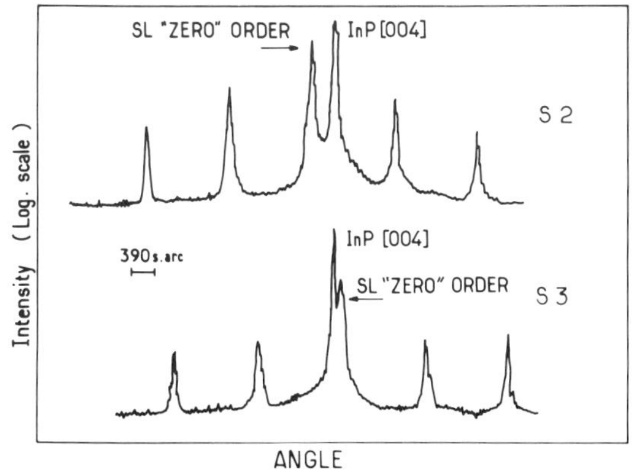

Fig. 1. X-ray double diffraction spectra (near the InP[004] substrate peak) of two InP–$Al_xIn_{1-x}As$ superlattices which have identical nominal parameters (40–100 Å) but actually have slightly different AlInAs composition $x = 0.506$ in S2 and $x = 0.465$ in S3, respectively, which is deduced immediately from the position of "zero" SL order with respect to the substrate peak.

structure is periodic), whose spacing is proportional to the reciprocal SL period. If $n_{A,B}$ and $a^z_{A,B}$ are the numbers of A,B host monolayers and the corresponding lattice parameters along the growth direction, Bragg's condition for the angular position θ_N of the diffraction peaks is

$$2d \sin \theta_N = N\lambda \quad \text{with} \quad d = n_A a^z_A + n_B a^z_B \quad (N = 1, 2, \ldots). \tag{1}$$

The calculation of the structure factor, which is proportional to the peak intensity, shows that the SL diffraction peaks can be divided in replicas of subsets centered close to the respective orders (K) of the host (or substrate) crystals. Using $n_{SL} = n_A + n_B$ for the total number of monolayers and introducing the average crystal parameter $\langle a^z \rangle = d/n_{SL}$, we get

$$2n_{SL}\langle a^z \rangle \sin \theta_{KM} = KM\lambda. \tag{2}$$

The $M = n_{SL}$ peak in replica K is usually called the "zeroth" order of SL diffraction. The position of this peak immediately gives $\langle a^z \rangle$, which straightforwardly measures the average composition if the in-plane lattice parameter is known. More information can be obtained from the quantitative fit of the intensities of the other "satellites", which are indexed by $i = M - n_{SL}$: the envelope of these intensities results from the interference of diffraction patterns associated with the A and B layers, which have a width proportional to the reciprocal layer thickness $1/L_{A,B}$, an integrated intensity proportional to $L_{A,B}$, and are centered on the angular position of the Kth order of bulks with lattice parameters $a^z_{A,B}$. These features are more easily seen in diffraction spectra from strained-layer superlattices where $a^z_A \neq a^z_B$. Figure 2 shows typical X-ray rocking curves, obtained in two $Al_xIn_{1-x}As$(100 Å)–InP(40 Å) SLs (S2 and S3) having a small difference (4%) in Al(In)As composition x (Lugagne-Delpon et al. 1991a,b). The spacing of the SL satellites measures the SL period which is the same for the two samples, while the position of the SL "zeroth" order relative to the

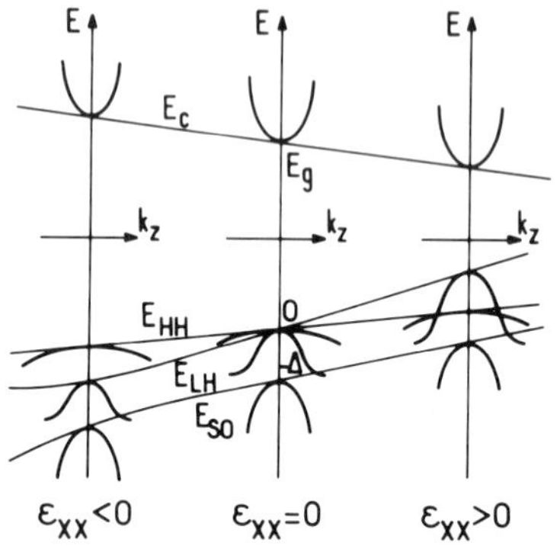

Fig. 2. Effect of biaxial compressive ($\varepsilon_{xx} < 0$) or tensile ($\varepsilon_{xx} > 0$) strain on the band structure of direct-gap semiconductors.

InP[004] substrate peak measures the deviation to the lattice-matched composition. Precise and non-destructive determination of the structural properties with X-ray diffraction is a considerable advantage of SLs over single quantum wells. Finally, let us stress that the periodicity of the layer arrangement tested by X-ray diffraction does not imply a strong coupling of the quantum wells in the sense of the electronic properties: a "superlattice" in the sense of structural properties may very well be a series of uncoupled quantum wells!

Superlattice structural properties can also be studied from the lattice dynamics point of view. While acoustical-like vibration modes basically experience the periodic modulation of sound velocity and photo-elastic constants, optical phonons are more sensitive to details of the interfaces. Raman spectroscopy of lattice dynamics of superlattices has been extensively studied both experimentally and theoretically, and is a remarkable characterization tool for high-quality materials. Reviews on this rich topic can be found in the literature (Jusserand and Paquet 1986, Fasolino and Molinari 1990, and references therein).

The first heterostructure system in which convincing results were obtained, by far the most studied and still the most popular, is GaAs–$Al_xGa_{1-x}As$. In this system, the lattice matching is almost perfect, and the bandgap difference ΔE_g (eV) = 1.34x is shared between the conduction and valence bands in such a way that, as illustrated in fig. 3a, the GaAs layers form a quantum potential well for both the electron and the hole states, which defines the "type I" band edge alignment. The consideration of fig. 3a is an invitation to imagine an alternative configuration, illustrated in fig. 3b, where the electrons are confined in the layers of one material (say A) while the holes are confined in the other layers (say B). This is the type II configuration which is observed in the almost unexplored Al(In)As–InP system (Caine et al. 1984, Aina et al. 1988, Lugagne-Delpon et al. 1991a,b). A somewhat pathological version of the type II configuration has been studied in the InAs–GaSb system (fig. 4a,b), where the offset of the conduction bands is so large (larger than GaSb bandgap) that the bottom of the InAs conduction band lies ≈150 meV below the top of the GaSb valence band (Sai-Halasz et al. 1977, 1978). This has a remarkable consequence: if

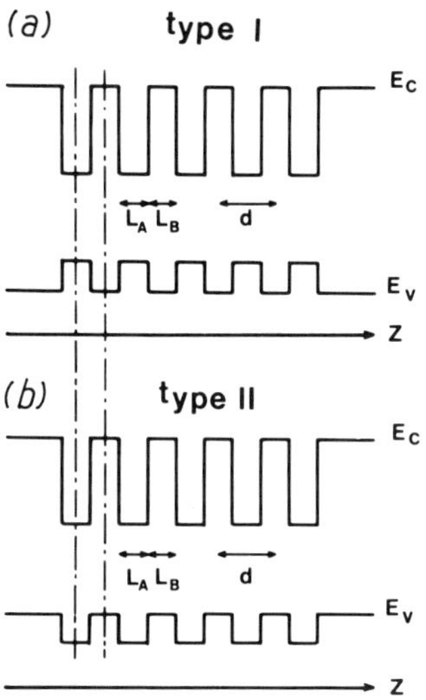

Fig. 3. Spatial modulation of the energies of conduction and valence band extrema in type I (a) and type II (b) superlattices. The two "Canonical" examples are the GaAs–Al_xGa_{1-x}As and InP–Al_xIn_{1-x}As systems, respectively.

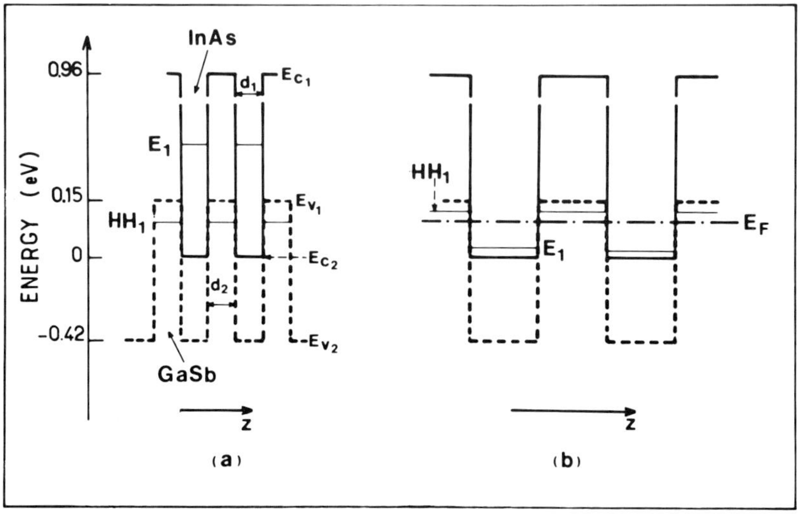

Fig. 4. Band-edge configuration and the "semiconductor to semimetal" transition in InAs–GaSb superlattices.

the sum of the confinement energies of the electron and hole in their respective quantum wells is large enough (narrow-layers regime, fig. 4a), the ground SL conduction state E1 is above the ground valence state H1, and one certainly deals with a regular semiconductor (Voisin et al. 1981, Chang et al. 1981b, Voisin, 1987). On the contrary, if the layers are thick enough (fig. 4b), E1 will lie below H1, which was first (and rather naturally) analyzed as a semimetallic situation (Guldner et al. 1980, Chang et al. 1981a, Voisin 1983). In fact, the calculation of the in-plane dispersion of the subbands (Altarelli 1983, Fasolino and Altarelli 1984) shows strong anticrossings, and the situation is much more complicated than the naive picture of a semiconductor-to-semimetal transition. Most of the experimental work on the InAs–GaSb system was made at an early stage of our understanding of the physics of superlattice band structure and before recent advances in epitaxial growth, and we believe that this system would deserve a second generation of studies.

The consideration of strained layers adds a few other band offset configurations, because the strain lifts the degeneracy of the heavy and light valence bands, as shown in fig. 5 (Voisin 1988, Marzin et al. 1990): different quantum well potentials actually act on heavy and light holes, and the sign of the splitting, which depends on the nature (tensile or compressive) of the strain, may change. As can be seen in fig. 5, in addition to configurations directly inherited from the type I and type II SLs, "mixed type" configurations can be obtained, in which the ground light-hole and heavy-hole states are confined in adjacent layers. Examples of such mixed type SLs are the $In_xGa_{1-x}As$–GaAs, $x \approx 0.15$ (Marzin et al. 1985) and $In_xGa_{1-x}Sb$–GaSb, $x \approx 0.15$ (Warburton et al. 1990) systems, where the layers of small-gap ternary alloy (which experience a biaxial compressive stress) confine the conduction and heavy-hole states while the light-hole states are confined in the larger-gap binary material.

The simple picture of electron and holes in simple square well shaped quantum wells relies on several assumptions which are not necessarily verified. For example, it is assumed that the A–B and B–A interfaces are equivalent, which is not experimentally proved, and presumably wrong in some particular cases like InAs–GaSb where quite different chemical bonds (Ga–As and In–Sb, respectively) are involved at the two interfaces (Capasso and Magaritondo 1987, Priester 1991, and references therein). It is also implicitly assumed that the atomic part of the host Bloch functions out of which the SL eigenstates are built are similar, or at least have the same symmetry. In some cases, conduction states associated with different minima in the host conduc-

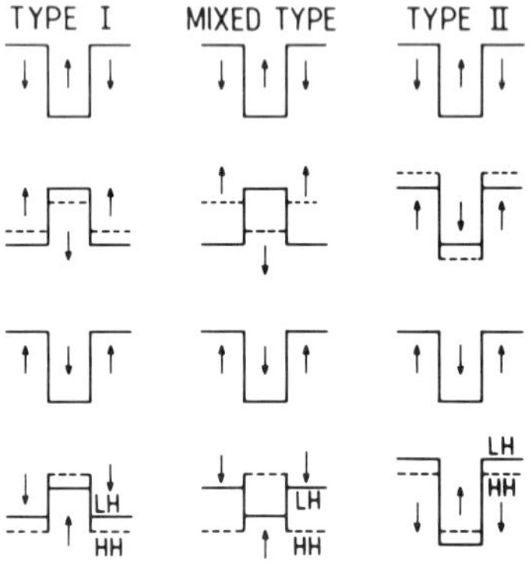

Fig. 5. The six possible band-edge configurations in strained-layer superlattices. The converging and diverging arrows indicate the sign of the strain (compressive or tensile, respectively) in each layer. Biaxial strain of the order of 1% correspond to valence band splittings of the order of 80 meV.

tion bands (Γ, X, L) come into play in the same energy range, as illustrated schematically in fig. 6. In such situations, one has to consider separately the SL states built out of the same valley in the A and B Brillouin zone (say, the states close to the Γ minima in A and B), which gives a set of Γ-like, X-like, etc., SL states. In general, these states are weakly coupled to each other by the SL potential, and, to a good approximation, can be considered as SL eigenstates. This situation has been studied in detail in the case of GaAs–AlAs superlattices where the Γ-like states confined in the GaAs layers and the X-like states confined in AlAs can cross if the layer thicknesses are small enough (Meynadier et al. 1988, Scalbert et al. 1989, Pulsford et al. 1990, Young et al. 1991). However, this weak-coupling approach could be inadequate in other circumstances, e.g., when the involved evanescent states in the barrier are very deep in the bandgap, as in the GaSb–AlSb system (Voisin et al. 1984b, Voisin 1987). A subtle effect of the strain has been predicted (Smith and Mailhiot 1987, Mailhiot and Smith 1988), and found (Laurich et al. 1989, Gammon et al. 1990, Shanabrook et al. 1990) in the case of strained-layer superlattices grown along the (111) direction: for this direction, the piezo-electric tensor is non-zero, and axial electric fields of opposite signs appear in the A and B layers, which adds a "sawtooth" component to the superlattice potential. Finally, the very concept of quantum well can break down in a system like HgTe–CdTe (fig. 7) where the conduction band has a prominent P-like character in HgTe while it has the normal S-like character in CdTe, and vice-versa for the light-hole bands. In this case, one has to match states (e.g., $P_{1/2}$-like states) which have a positive mass in one layer (HgTe) and a negative mass in the other (CdTe) (Schulman and McGill 1979, Bastard 1982). This unique situation allows the existence of SL states which are evanescent in both types of layers, with eigenfunctions peaking at the interfaces (Chang et al. 1985, Berroir 1988, Berroir et al. 1989). The HgTe–CdTe case is so original that it forms a new category of superlattices, the type III SL. All the heterostructures investigated so far fit into this three-fold classification.

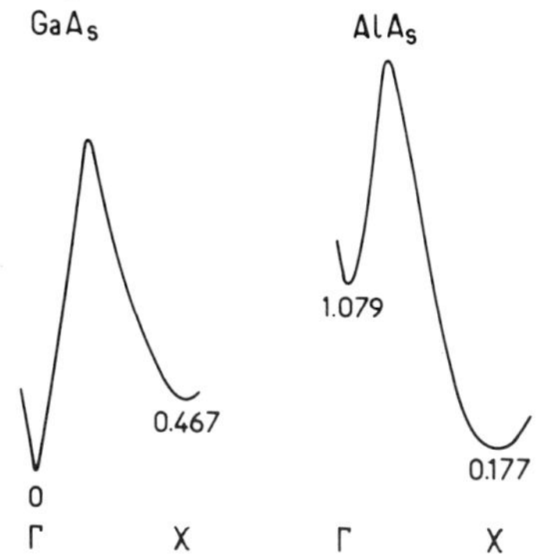

Fig. 6. Conduction bands of bulk GaAs and AlAs (schematic). Energies (in eV) of the involved minima, relative to the Γ_6 minimum of GaAs, are indicated.

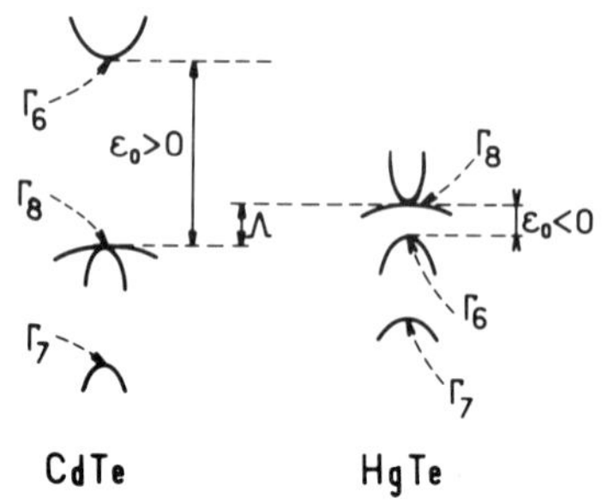

Fig. 7. The "inverted" band structure of HgTe and the "normal" band structure of CdTe. The now accepted value for the Γ_8 band offset is $\approx$400 meV.

3. Approaches to superlattice band structure

The physics of superlattices is that of a series of quantum wells which are coupled together by the resonant tunnel effect through the narrow potential barriers. However, this origin of the miniband structure is not always transparent in the various methods used in the theoretical calculations, which we review in this section.

Long before the idea of semiconductor superlattices was first put out, the motion of an electron in a square-wave potential of period d and amplitude V had been considered as an elementary quantum mechanics exercise (the Kronig–Penney problem). The wavefunction $f(z)$ inside each well (A) or barrier (B) region is the sum of incoming and reflected plane waves, and at each potential singularity the wavefunction and its derivative must be continuous. Noting that the wavefunction is periodic (Bloch theorem), the eigenstates are obtained as the solutions of

$$\cos(qd) = \cos(k_A L_A)\cos(k_B L_B) - \tfrac{1}{2}(\xi + 1/\xi)\sin(k_A L_A)\sin(k_B L_B), \tag{3}$$

where $\hbar^2 k_A^2/2m_0 = \varepsilon$, $\hbar^2 k_B^2/2m_0 = \varepsilon - V$ (m_0 is the rest mass of the electron), and $\xi = k_A/k_B$. For energies ε lower than the barrier height V, k_A is real and k_B is imaginary, while for energies above V, both wavevectors are real. In all cases, allowed states correspond to energy bands for which the modulus of the right-hand side of eq. (3) is smaller than unity. Eigenstates are parametrized by a band index and the wavevector q. It is noteworthy that expansion of eq. (3) for vanishing L_A, L_B and q gives the miniband edge $\varepsilon(q = 0) = VL_B/(L_A + L_B)$, which corresponds to the average potential. This is the "pseudo-alloy" behavior of short-period superlattices.

The problem of wavefunction matching conditions for heterostructures, when the electron has different effective masses (m_A, m_B) on both sides of the interface was also considered a long time ago (Ben-Daniel and Duke 1966): in this case, the continuity of the wavefunction and that of the current probability imply the continuity of $f(z)$ and $m^{-1}\,df/dz$. Using such continuity relations in the Kronig–Penney problem simply changes the quantity ξ in eq. (3) into $\xi = (k_A/m_A)/(k_B/m_B)$. Hence, eq. (3) gives the dispersion relations of superlattice minibands in an effective mass approach, as long as non-parabolicity effects can be neglected.

The need for taking non-parabolicity (in other words, the multi-band nature of the Hamiltonian) into account in narrow-gap heterostructures was the original motivation for the development of the framework of the "envelope function approximation" (White and Sham 1981, Bastard 1981, 1982, Altarelli 1983). In this effective mass approach, the microscopic details of the host band structure are neglected and dispersion relations near the Γ-point (as illustrated in fig. 8) are obtained using the $\boldsymbol{k} \cdot \boldsymbol{p}$ model: host material properties are thus summarized by a few constants such as the bandgap and spin–orbit energies, and the Kane matrix element (Kane 1957) between the conduction and valence Bloch functions at $k = 0$. The model is based on the assumption that the cell periodic part of the Bloch wavefunctions are the same in the two materials, which may seem very strong. In practice, the fact that Kane's matrix elements in the different III–V or II–VI semiconductors are remarkably similar is apparently sufficient to justify this assumption (Smith and Mailhiot 1986). Detailed descriptions of the envelope function formalism can be found in several review papers (Smith and Mailhiot 1990, Bastard et al. 1988, 1991) and textbooks (Allan et al. 1986, Bastard 1988). Here we only summarize the salient points of the method and the main results.

In the envelope function model, a superlattice eigenstate is a spinor written as

$$\psi(z, r_t) = \sum_i e^{ik_t r_t} f_i(z) u_i(z, r_t), \qquad (4)$$

where r_t and k_t are the in-plane coordinate and wavevector, $f_i(z)$ are slowly varying envelope functions and u_i the host Bloch wavefunction of band i at $k = 0$, assumed to be identical in the A and B materials. In practice, for Γ-related states, one retains only the ($\Gamma_6, \Gamma_8, \Gamma_7$) basis ($i = 1$ to 8) corresponding to the fundamental conduction band and heavy-hole, light-hole and split-off valence bands, while the influence of remote bands is treated as a perturbation. There is unfortunately no consensus on the notation for heterostructure coordinates: considering the superlattice as a uniaxial material, the ∥ and ⊥ (or l and t for *l*ongitudinal and *t*ransverse) subscripts should refer to the natural quantization axis, which is the growth axis z. For historical reasons, many authors have preferred the opposite notation where the ∥ subscript refers to the layer plane considered as a more "intuitive" reference, while the ⊥

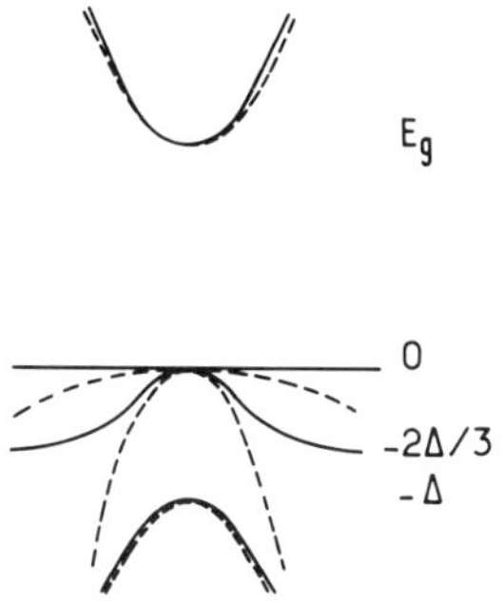

Fig. 8. The $\boldsymbol{k} \cdot \boldsymbol{p}$ description of the host band structure near the Γ point: Kane's dispersion relations (solid lines), corrected for the interaction with remote bands (dashed lines).

subscript means parallel to the growth axis. Here, we stick to the first type of notation which we find more satisfactory.

Letting the superlattice Hamiltonian act on ψ yields an effective Hamiltonian acting of the envelopes f_i. In each bulk material, this 8×8 Hamiltonian is the sum of a matrix H resulting from the exact treatment of the Hamiltonian within the restricted $(\Gamma_6, \Gamma_8, \Gamma_7)$ basis (Kane matrix), and a matrix δH resulting from a perturbative treatment of remote bands up to k^2 terms (Luttinger matrix). For a given k_t, and a given energy, the solution of the bulk problem gives a set of 8 axial wavevectors k_i, which may be real or imaginary. One then expands the f_i's inside each layer as linear combinations of bulk functions which are plane waves with real or imaginary wavevectors k_i:

$$f_j(z) = \sum_i \alpha_{ji}^{\mathrm{A,B}}(k_t)\exp(\mathrm{i}k_i^{\mathrm{A,B}}z) + \beta_{ji}^{\mathrm{A,B}}(k_t)\exp(-\mathrm{i}k_i^{\mathrm{A,B}}z). \tag{5}$$

The consideration of matching conditions at two consecutive interfaces finally provides 2×16 linear equations for the $\alpha_{ji}^{\mathrm{A,B}}$ and $\beta_{ji}^{\mathrm{A,B}}$ coefficients, and the eigenvalues are obtained as the zeros of a 32×32 determinant. 2×8 matching conditions are obtained by writing that the f_i's are continuous at the interfaces and 2×8 others are obtained by integrating the δH matrix across the interfaces and using again the continuity of the f_i's. In addition, the f_i's are Bloch waves with the superperiodicity d, which introduces a specific quantum number q, the superlattice wavevector along the growth axis (k_t is a good quantum number for each host layer and for the superlattice itself). This conceptually simple method, which has been particularly successful in dealing with the HgTe–CdTe case (Berroir 1988, Guldner et al. 1991), is, however, relatively involved from the computational point of view.

Fortunately, the formalism is considerably simplified if one sets $k_t = 0$, which in most cases is enough to evaluate desired quantities such as miniband energies. In this case, the heavy-hole and light-particle states decouple (Bastard 1981). For the heavy holes, we are left with a one-dimensional problem with parabolic dispersion relations inside each layer. One readily finds the modified Kronig–Penney dispersion relation, with $\xi = (k_A/M_A)/(k_B/M_B)$, where $M_{A,B}$ are the heavy hole effective masses in the A,B materials. The corresponding energy spectrum is illustrated in fig. 9 for the case of GaAs–AlGaAs superlattices. For periods larger than 50 Å, "bound" subbands (i.e., those located below the top of the barrier) are quite narrow, and the gaps between them relatively large, while "unbound" subbands are broad and separated by narrow gaps. In fact, in most systems and for SL periods of practical interest, heavy-hole bound subbands at $k_t = 0$ can be considered as almost dispersionless, while unbound subbands are reasonably approximated by an energy continuum. In fact, the first heavy-hole subbands at $k_t = 0$ are usually so narrow that the concept of extended states for the heavy holes is probably misleading: any potential fluctuation will produce a broadening comparable to or even larger than the miniband width, and will result in a strong localization of the eigenstates. Note, however, that the calculation of energy levels is not affected by this effect, since the energies of the narrow heavy-hole subbands are identical to the corresponding states in isolated quantum wells.

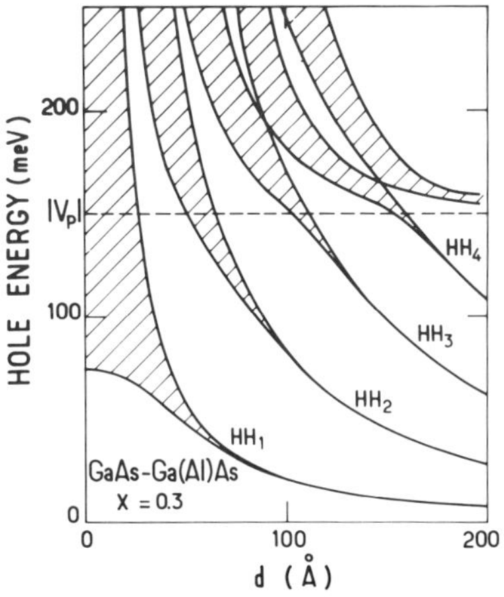

Fig. 9. Allowed heavy-hole bands (hatched areas) as a function of the superlattice period in GaAs–$Al_xGa_{1-x}As$ superlattices with equal well and barrier thicknesses. The valence band offset $|V_P|$ is equal to 150 meV for an alloy composition $x = 0.3$. The zero of hole energies is taken at the top of the GaAs Γ_8 valence band.

For the light-particle states, the convenient approximation is to neglect the influence of the δH matrix. This is usually legitimate for the conduction states whose dispersion in the hosts are governed by the interaction with the light-hole and split-off bands. The approximation is more questionable when dealing with the light-hole states in materials having a small spin–orbit coupling constant Δ. Indeed, the energy $-2\Delta/3$ below the top of the Γ_8 valence band is a singular point in Kane's dispersion relation, and there are no eigenenergies in the $-2\Delta/3$, $-\Delta$ energy range (fig. 8). Realistic light-hole dispersion relations in this energy range can be recovered only by the inclusion of the interaction with remote bands. Discarding δH and removing the decoupled heavy-hole terms leaves us with two sets of three coupled first-order differential equations, because positive and negative m_j's are not coupled. Projecting these coupled first-order equations onto the S subspace (i.e., the Γ_6 conduction band edge) finally gives a second-order equation formally written as an effective Schrödinger equation acting on the f_S envelope:

$$[-\mathrm{d}/\mathrm{d}z\,[\hbar^2/2\mu(\varepsilon, z)]\,\mathrm{d}/\mathrm{d}z + V_S(z)]f_S(z) = \varepsilon f_S(z), \tag{6}$$

where V_S is the SL potential associated with the Γ_6 extremum (we take $V_S(z) = 0$ in the A layers, and $V_S(z) = V_S$ in the B layers), and the energy-dependent and position-dependent effective mass $\mu(\varepsilon, z)$ is obtained from Kane's dispersion relation of the hosts:

$$\begin{aligned} \varepsilon &= \hbar^2 k_A^2/2\mu_A(\varepsilon) = [\hbar^2 k_A^2 |P|^2/3][2/(\varepsilon + \varepsilon_A) + 1/(\varepsilon + \varepsilon_A + \Delta_A)], \\ \varepsilon - V_S &= \hbar^2 k_B^2/2\mu_B(\varepsilon) = [\hbar^2 k_B^2 |P|^2/3][2/(\varepsilon - V_S + \varepsilon_B) + 1/(\varepsilon - V_S + \varepsilon_B + \Delta_B)], \end{aligned} \tag{7}$$

where $P = (-\mathrm{i}/m_0)\langle S|p_x|X\rangle$ is the Kane matrix element, $\varepsilon_{A,B}$ are the A,B material bandgaps and $\Delta_{A,B}$ the corresponding spin–orbit coupling constants. The matching condition for df_S/dz is the continuity of $1/\mu(\varepsilon, z)\, df_S/dz$ across the interface, in complete analogy with the heavy-hole case. This continuity relation thus appears as the generalization of the probability current conservation to the multiband situation. The SL light-particles dispersion relations are obtained following the same method as before, and have the same functional form as eq. (3), with the quantity ξ now equal to $\{k_A/\mu_A(\varepsilon)\}/\{k_B/\mu_B(\varepsilon)\}$. In this approximation, electron, light-hole and split-off minibands are independent solutions of the same equation, but, as already mentioned, the approximation is often questionable for the light-hole states. Figure 10 shows calculated conduction subbands in GaAs–AlGaAs superlattices, which are qualitatively similar to heavy-hole subbands, but quantitatively quite different: due to the small effective mass (which favors tunneling), conduction minibands are much wider than heavy-hole subbands. The concept of traveling Bloch waves, which usually is an unrealistic picture of the physical reality for heavy-hole states, becomes relevant for electrons in a superlattice with a period $d \leqslant 100$ Å where the ground conduction subband width is larger than 10 meV.

The effect of built-in strain can be included almost straightforwardly in this formalism by adding the strain Hamiltonian to the $\boldsymbol{k} \cdot \boldsymbol{p}$ Hamiltonian in the description of the host band structures (Marzin et al. 1990). This merely amounts to modifying the bandgaps and the dispersion relations accordingly.

This non-parabolic envelope function model has been extensively used to calculate electronic properties of semiconductor heterostructures, and is clearly the only way of handling properly the HgTe–CdTe and InAs–GaSb cases, where non-parabolicity and band mixings play a major role. Its necessity to interpret experimental data in large-gap materials is more questionable: numerical results for GaAs–AlGaAs superlattices obtained by replacing the functions $\mu_{A,B}(\varepsilon)$ by the band-edge effective mass

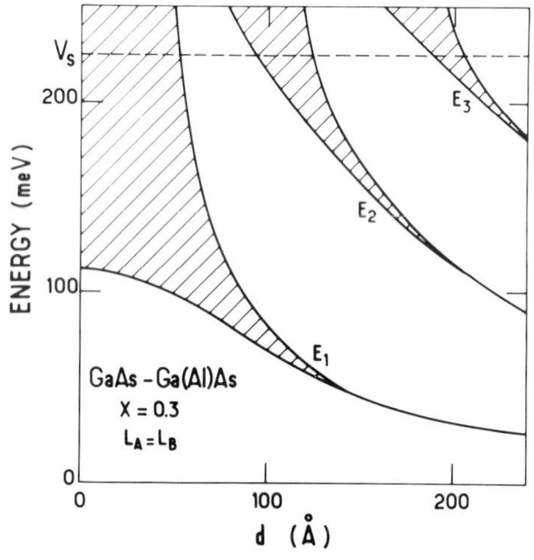

Fig. 10. Conduction minibands (hatched areas) as a function of the superlattice period in GaAs–$Al_{0.3}Ga_{0.7}As$ superlattices with equal well and barrier thicknesses. The conduction band offset V_S is equal to 224 meV for the alloy composition $x = 0.3$. The zero of energy is taken at the bottom of the GaAs Γ_6 conduction band.

of GaAs (parabolic approximation) are amazingly similar to the prediction of the non-parabolic model. This is in part due to the fact that the non-parabolicity in the well is somehow compensated by the non-parabolicity in the barrier.

The necessity of a multiband approach, however, appears immediately when considering the in-plane dispersion relations of the valence subbands: indeed, at $k_t \neq 0$, the heavy hole and light hole are strongly coupled by the off-diagonal terms of the δH matrix, and calculated dispersion relations using the "diagonal" approximation (i.e., neglecting the coupling) are quite different from the exact solutions, as illustrated in fig. 11. The complexity of the k_t dispersion relations of the valence subbands is a permanent and inherent difficulty of heterostructure physics. Although the effects are now well understood on experimental as well as on theoretical grounds, the computational effort to take them into account is most of the time out of proportion. Fortunately, rough approximations like the "diagonal approximation" are often enough to account for the main physical effects. The major exception is the interpretation of magneto-optical data which, in most cases, is even qualitatively impossible within oversimplified approaches (Ancilotto et al. 1987, Brum et al. 1988). A remarkably simple (but approximate) method for the calculation of these in-plane dispersion relations of the valence minibands consists in a projection of the Hamiltonian on the truncated basis formed by the $k_t = 0$ eigenstates (see, e.g., Bastard and Brum 1986). Hence, eigenstates at finite k_t are described as mixtures of states which have simple and general symmetry properties (thanks to the $k_t = 0$ decoupling of heavy and light particles). This allows one to give a conceptually simple and intuitive discussion of the related optical properties.

Finally, a generalization of the envelope function theory to situations (such as GaAs–AlAs) where different extrema of the host band structure have to be considered has been proposed in terms of transfer and interface matrices (Ando and Akera 1989). The spinor corresponding to the manifold band structure is propagated through each layer by the "transfer" matrices, and connected across the interfaces by the "interface" matrices. In this approach, the physics of band mixing is introduced (in

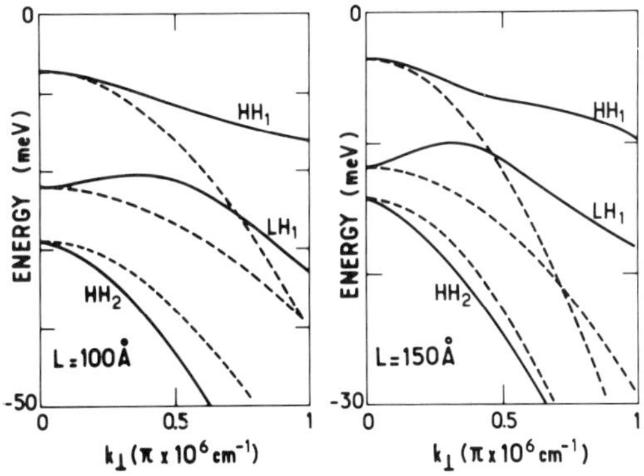

Fig. 11. In-plane dispersion relations in GaAs–AlGaAs quantum wells. Solid lines show the exact solutions, and the dashed lines are the nearly parabolic dispersion relations obtained in the diagonal approximation.

general, phenomenologically) by adding terms in the "interface matrix" (Nicholas et al. 1991).

At the opposite of the effective mass approach, which neglects the microscopic structure of the superlattice and accounts for energy states close enough to a host band extremum, atomic-scale tight-binding methods provide a framework well suited to the description of both the interface details and host band structure in the entire Brillouin zone. Such methods have indeed been employed successfully to calculate the band structure and the optical spectra of short-period superlattices (Schulman and Chang 1981, 1985, Chang and Schulman 1985). Pseudo-potential formalisms have also been applied to superlattices (Jaros et al. 1985, Ninno et al. 1985, Gell et al. 1986). However, for a reasonable degree of sophistication of the model, the complexity of the computation (which increases dramatically with the size of the SL unit cell) becomes considerable, and these methods are not convenient to handle the problems of external fields, excitonic effects, etc.

A conceptually simple method which permits a physical discussion of superlattice effects consists in a tight-binding approach applied at the scale of the superperiod, i.e., to the quantum well envelope functions themselves (Voisin et al. 1984a). Here, we consider effective-mass solutions of the isolated quantum well problem yielding a set of envelope functions $\varphi_j(k_t, z)\,e^{ik_t r_t}$, where j is a level index, and we construct a superlattice eigenstate in a simplified tight-binding scheme:

$$\psi_j(k_t, z)\,e^{ik_t r_t} = \sum_n \alpha_{nj}\varphi_j(k_t, z - nd)\,e^{ik_t r_t}. \tag{8}$$

In constructing such a wavefunction, we implicitly assume that the resulting minibands will be sufficiently narrow compared to the energy separation between the corresponding quantum well energy levels, which is correct only for the first bound subbands, and for large enough periods. If in addition we assume that QW functions centered in adjacent wells are orthogonal to each other ($\langle\varphi_i|\varphi_j\rangle = \delta_{ij}$), the tight-binding Hamiltonian is solved analytically and gives the superlattice dispersion relation

$$\varepsilon_j(q) = E_j - 2\lambda_j \cos qd, \tag{9}$$

where λ_j is the transfer integral between nearest neighbors,

$$\lambda_j = -\langle\varphi_j(z)|\,V_{SL}\,|\varphi_j(z - d)\rangle. \tag{10}$$

The superlattice potential V_{SL} in eq. (10) has to be defined as an attractive potential, $V_{SL} = 0$ in the barrier and $V_{SL} = -V$ in the quantum well. The corresponding SL eigenstates are of course traveling Bloch waves:

$$\psi_j(k_t, q, z)\,e^{ik_t r_t} = \sum_n e^{iqnd}\varphi_j(k_t, z - nd)\,e^{ik_t r_t}. \tag{11}$$

In spite of the large number of approximations, this simple model gives sensible quantitative results (in particular for the conduction bandwidth $\Delta E_1 = 4\lambda_1$) for the first conduction and valence minibands in a variety of superlattices. In addition, the physical origin of the superlattice minibands appears very clearly, since the quantity h/λ_1 can be interpreted as the time to tunnel from one well to the other. This tight-binding method is also well suited to the situation of finite-size superlattices, possibly in the presence of external fields (Bleuse et al. 1988a).

If more accuracy is needed, the problem of a finite-size superlattice is conveniently treated in the envelope function approach using numerical solution of the Schrödinger equation with a transfer-matrix method analogous to that of Ando and Akera (Bleuse et al. 1988a, Soucail et al. 1990a,b, 1991). Such numerical solutions are presently becoming quite popular because the method has a great flexibility and can be used to calculate the energy levels and eigenfunctions in all kinds of situations. The heterostructure, possibly including an external potential $W(z)$, is decomposed into a set of layers or sublayers where the potential is constant. Inside each layer an envelope function (e.g., f_S) is the sum of an incoming and a reflected plane wave. The envelope function and its derivative on both sides of an interface are connected using the rules of the envelope function formalism. Thus, at the beginning of each sublayer i, the values of the envelope function, $(f_S)_i$ and of its derivative $(df_S/dz)_i$ can be expressed as the product of $[(f_S)_{i-1}, (df_S/dz)_{i-1}]$ by a transfer matrix T_i. Starting with $f = 0$, $df/dz \neq 0$ at one edge of the heterostructure, one obtains the wavefunction at the other edge by multiplying the matrices T_i. Eigenenergies are those for which f_S vanishes at the extremity of the structure.

4. Superlattice optical properties

In the spirit of the envelope function method, the dipolar matrix element $M_{if} = \langle\psi_f|p|\psi_i\rangle$ for an optical transition between SL states i and f is a sum of terms:

$$M_{if} = \sum_{\nu\mu} \{\langle f_\nu^f | f_\mu^i\rangle\langle u_\nu|p|u_\mu\rangle + \langle u_\nu|u_\mu\rangle\langle f_\nu^f|p|f_\mu^i\rangle\}. \tag{12}$$

The first terms are relevant for interband transitions, and the second ones for intraband transitions (like transitions between conduction subbands of different indices). A complete calculation of the optical matrix elements can be achieved straightforwardly after a complete solution of the 32 × 32 envelope function Hamiltonian, but, fortunately, many results can be obtained with simple considerations and little algebra. Extended discussions of quantum well and superlattice optical properties can be found in the literature (Voisin 1986a, Bastard et al. 1988, Chang and Schulman 1985). Here, we discuss the selection rules in the case of interband transitions. If we neglect complexities associated with the multi-band approach, only one envelope function (that of the $|S\rangle$ state, hereafter f^e) comes into play for the conduction band, and only one (that of the relevant $|P\rangle$ state, hereafter f^h) for the valence band. The optical matrix element reduces to the product of the overlap of these envelopes and the bulk optical matrix element. Polarization selection rules (identical to those in a bulk material) are associated with the optical matrix element, and two types of selection rules with the overlap integral: a (k_t, q) wavevector conservation rule results from the periodicity of the envelope functions, and a parity selection rule comes from the symmetries of the wavefunctions in the SL period, as discussed below.

Binary superlattices are symmetric with respect to the planes P_A and P_B bisecting each A,B layer. At $q=0$ and $q=\pi/d$, the envelope functions are standing waves, and they have simultaneously a definite parity with respect to P_A and P_B. In other words, they are eigenfunctions of the reflections R_A (with respect to P_A) and R_B (with respect to P_B) with eigenvalues ± 1. As the product $R_B R_A$ is equal to the translation operator t_d by an SL period, and the envelope functions are eigenstates of t_d with eigenvalues e^{iqd}, the parity with respect to one of the plane (say P_A) must be the same at $q=0$ and $q=\pi/d$, while the parity with respect to the other (say P_B) must change. For a type I SL, both the conduction and valence envelopes are confined in the same layers (say A), and usually they will retain the same symmetry with respect to P_A at $q=0$ and $q=\pi/d$. This means that a transition which is parity-allowed at $q=0$ (e.g., both f^e and f^h are even with respect to P_A) will remain parity-allowed at $q=\pi/d$, as illustrated in fig. 12a, while a forbidden transition at $q=0$ (f^e and f^h having different parities with respect to P_A) will remain forbidden at $q=\pi/d$. The situation is qualitatively different in a type II SL because the conduction and valence envelopes are concentrated in adjacent layers, and usually retain their symmetry with respect to the center of the layer where they are confined. If a transition is allowed at $q=0$ (say that f^e and f^h are symmetric with respect to P_A, as in fig. 12b), it will become parity forbidden at $q=\pi/d$ because f^e remains even while f^h becomes odd with respect to P_A. Hence, parity selection rules in a type I SL are basically identical to those in isolated quantum wells: $M_{nm}(k_t, q)$ weakly depends on the SL wavevector q, is parity-allowed if $n-m$ is even and parity forbidden if $n-m$ is odd, and the

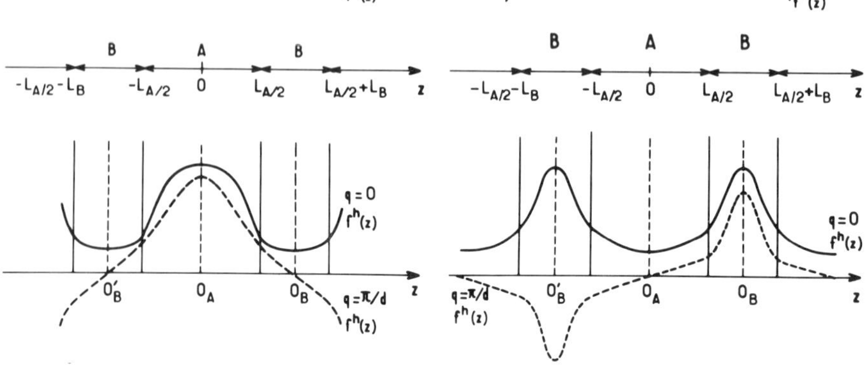

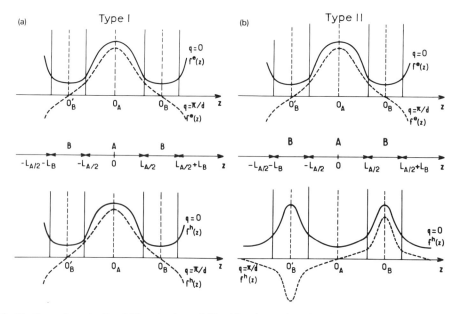

Fig. 12. Ground conduction (f^e) and valence (f^h) subbands envelope functions at $q=0$ and $q=\pi/d$ in a type I (a) and a type II (b) superlattice. The vertical dashed lines $O_{A,B}$ are the traces of the planes $P_{A,B}$ bisecting each A,B layer.

$n-m=0$ transitions are by far the most intense. Completely different conclusions are reached for type II SLs: $M_{nm}(k_t, q)$ strongly depends on q, since transitions with $n-m$ even are allowed at $q=0$ and become forbidden at the miniband edge, while transitions with $n=m$ odd are forbidden at $q=0$ and become allowed at $q=\pi/d$. Transitions with $n-m=0$ or $n-m\neq 0$ have equivalent strengths, always small compared to that of an $n-m=0$ transition in a type I SL. Using the tight-binding description of the envelope functions (eqs. (8)–(11)), one finds analytically the formula

$$|M_{nm}(q)|^2 = |\langle \varphi_n^{\mathrm{e}} | \varphi_m^{\mathrm{h}} \rangle|^2 [1 + (-1)^{n-m} \cos qd], \tag{13}$$

where the type II parity selection rules are readily observed. It is noteworthy that this type II selection rule relies on the phase coherence of only one of the particles involved, and it prevails as soon as the conduction miniband width is significantly larger than the level broadening. On the other hand, for narrow minibands the $q=0$ and $q=\pi/d$ edges are almost degenerate and therefore we get no selection rule at all on the subband index.

The shape of the band-to-band absorption reflects the three-dimensional character of superlattices. Using the tight-binding model and parabolic in-plane dispersion relations ("diagonal approximation"), we get the following formulas:

$$\begin{aligned}
&\alpha_{\mathrm{I}}(h\nu) = \beta \frac{\alpha_0}{\pi} \arccos(1-2\xi), && \text{if } 0 \leqslant \xi \leqslant 1,\\
&\alpha_{\mathrm{I}}(h\nu) = \beta\alpha_0, && \text{if } \xi > 1,\\
&\alpha_{\mathrm{II}}(h\nu) = 2\beta \frac{\alpha_0}{\pi} \{\sqrt{\xi(1-\xi)} + \arccos(1-2\xi)\}, && \text{if } 0 \leqslant \xi \leqslant 1,\\
&\alpha_{\mathrm{II}}(h\nu) = 2\beta\alpha_0, && \text{if } \xi > 1,
\end{aligned} \tag{14}$$

for type I and type II superlattices, respectively, where ξ is the reduced photon energy $\xi = (h\nu - E_{\mathrm{g}}^{\mathrm{SL}})/(\Delta E + \Delta H)$, $\beta = |\langle \varphi_n^{\mathrm{e}} | \varphi_m^{\mathrm{h}} \rangle|^2$ is the overlap of the relevant quantum well functions (β is close to unity for a type I SL, but can be much less in a type II SL), and α_0 is a constant of the order of $6\times 10^{-3}/d$ for a heavy hole to conduction transition:

$$\alpha_0 d = \frac{\pi e^2}{2n\varepsilon_0 ch} \frac{2m_0|P|^2}{h\nu} \frac{\mu_t}{m_0}, \tag{15}$$

where $P = (-\mathrm{i}/m_0)\langle \mathrm{S}|p_x|\mathrm{X}\rangle (2m_0P^2 \approx 23$ eV for all III–V semiconductors), n is the refractive index of the SL, and μ_t is the in-plane reduced mass of the electron–hole pair. The absorption profiles corresponding to the ground subbands in type I and type II SLs are shown in fig. 13. In the type I SL, the absorption spectrum is proportional to the joint density of states of the conduction and valence minibands, and shows a Van Hove singularity at $q=\pi/d$, while this singularity is washed out by the q-dependence of the wavefunction overlap in a type II SL. In situations of practical interest (SL period $\lesssim 100$ Å), there are only one (at most two) electron and light-hole bound subbands, but there may be more heavy-hole subbands. Only one (or two) transitions will be observed in a type I SL like GaAs–AlGaAs, but many

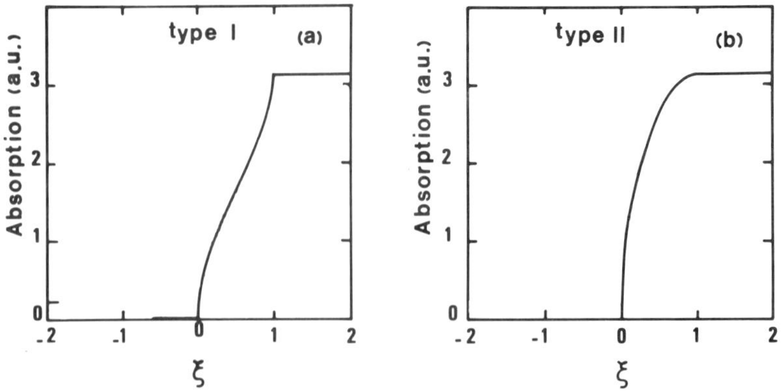

Fig. 13. Dimensionless absorption (αd) profiles for the ground subbands in a type I (a) and a type II (b) superlattice, as a function of the reduced photon energy $(h\nu - Eg^{SL})/(\Delta E + \Delta H)$. The absorptions are not on the same vertical scale: allowed transitions are much stronger in a type I SL than in a type II SL.

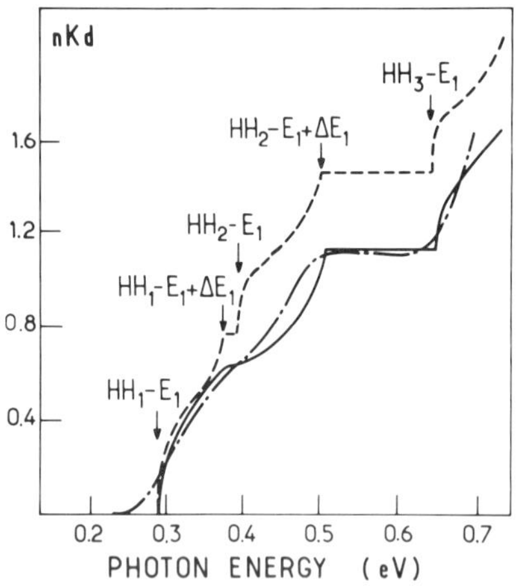

Fig. 14. Experimental (dash-dot-dash line) optical absorption spectrum of an InAs(27 Å)–GaSb(44 Å) superlattice from Chang et al. (1981b). The vertical axis is the total absorption, $N\alpha d$, where N is the number of periods. Joint density of states of the conduction and valence subbands (dashed line), arbitrarily scaled with respect to the experimental curve, and theoretical absorption (solid line), fitted to the experimental curve. The calculation uses thicknesses 27–40 Å. The experimental absorption is in fact twice the calculated one, which may be due to the Coulomb enhancement. $HH_i–E_1$ labels the transition from the heavy hole subband HH_i to the first conduction band E_1 at the miniband extremum, $q = 0$. $HH_i - E_1 + \Delta E_1$ corresponds to the same transition at the miniband edge $q = \pi/d$.

more will be seen in a type II SL like InP–AlInAs or InAs–GaSb. The experimental and calculated absorption spectra in a 27 Å–44 Å InAs–GaSb superlattice are shown in fig. 14 to illustrate this remark.

These selection rules are exact at $k_t = 0$, and they still hold in non-parabolic (narrow gap) materials. However, they are nothing more than a sensible guideline, because the mixing of different valence subbands at $k_t \neq 0$ can obviously relax them (Chang and Schulman 1985). Special configurations with a strong repulsion (hence a strong mixing) of the H2 and L1 subbands at small k_t have been designed and have experimentally proved this violation of the parity selection rule at finite k_t (Miller et al. 1985).

5. Excitons in superlattices

Excitonic effects modify considerably the optical spectra of III–V or II–VI bulk semiconductors, and they are even more pronounced in semiconductor quantum wells. Isotropic three-dimensional excitons in bulk materials are well described by a hydrogenic model, and both the bound states energies and oscillator strengths as well as the modification of the absorption by unbound pairs (Coulomb enhancement or Sommerfeld factor) can be obtained analytically (Lederman and Dow 1976). This would also hold for purely 2D excitons, but quantum well excitons are not purely 2D (since the electron and hole wavefunctions have a finite extent along the third direction, the growth axis z), and variational methods are needed to evaluate their binding energies and oscillator strengths. For narrow quantum wells where the subband separation is much larger than the exciton binding energy, separation of the z-motion (unperturbed by Coulomb interaction) and in-plane motion is an excellent approximation, which suggests the "natural" variational excitonic wavefunction (Bastard et al. 1982):

$$\psi_{\mathrm{exc}}(z^{\mathrm{e}}, z^{\mathrm{h}}, r_{\mathrm{t}}^{\mathrm{e}} - r_{\mathrm{t}}^{\mathrm{h}}) = \sqrt{2/\pi\lambda^2}\,\varphi^{\mathrm{e}}(z^{\mathrm{e}})\varphi^{\mathrm{h}}(z^{\mathrm{h}}) \exp[-(|r_{\mathrm{t}}^{\mathrm{e}} - r_{\mathrm{t}}^{\mathrm{h}}|)/\lambda]. \tag{16}$$

More sophisticated models (Sanders and Chang 1985, Broido and Sham 1986, Bauer and Ando 1988) have shown that this simple variational approach actually gives reasonably good results. Little attention has been paid to the Coulomb enhancement, which is often assumed to behave as in the purely 2D limit, i.e., it is believed to increase the band-to-band absorption by a factor $S(\varepsilon)$ equal to 2 at the absorption edge and decreasing slowly at higher energies (Lederman and Dow 1976):

$$S(\varepsilon) = 2/[1 + \exp(-2\pi\sqrt{\varepsilon/R^*})], \tag{17}$$

where R^* is the 3D effective Rydberg.

Excitons in superlattices correspond to a strongly anisotropic 3D situation. A two-parameter variational approach providing sensible results in both the 2D and the bulk limit was first carried out (Chomette et al. 1987) with the trial function:

$$\psi_{\mathrm{exc}}(z^{\mathrm{e}}, z^{\mathrm{h}}, r_{\mathrm{t}}^{\mathrm{e}} - r_{\mathrm{t}}^{\mathrm{h}}) \approx N \exp[-(|r_{\mathrm{t}}^{\mathrm{e}} - r_{\mathrm{t}}^{\mathrm{h}}|)/\lambda] \sum_q \exp(-q^2/2\beta^2) f_q^{\mathrm{e}}(z^{\mathrm{e}}) f_{-q}^{\mathrm{h}}(z^{\mathrm{h}}), \tag{18}$$

where N is a normalization constant and λ and β are the two variational parameters. Comparison of this theory with experimental data is shown in fig. 15 for the case of GaAs–AlGaAs SLs. SL excitons have a binding energy close to the bulk Rydberg up to SL periods of ≈ 70 Å (this regime corresponds to a conduction miniband width $\Delta E_1 \geqslant 40$ meV). The exciton binding energy has a sharp rise up to ≈ 9 meV near $d \approx 120$ Å, which corresponds to the regime of almost uncoupled quantum wells where the binding energy is considerably enhanced by the 2D character.

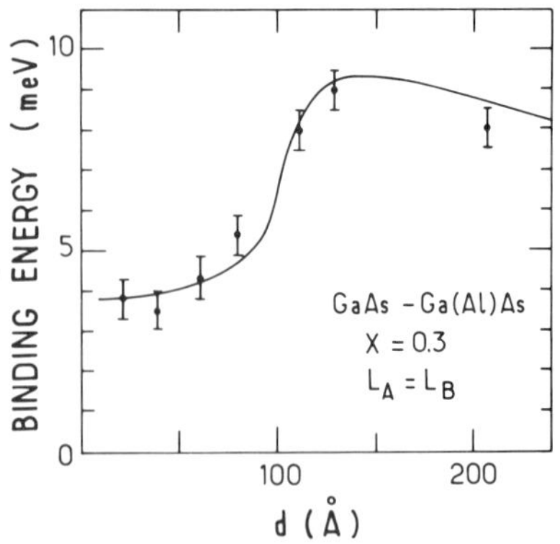

Fig. 15. Binding energies of excitons as a function of the SL period in GaAs–$Al_xGa_{1-x}As$ ($x = 0.3$) superlattices with equal well and barrier thicknesses, from Chomette et al. (1987).

Another approach to SL excitons consists in their interpretation as tight binding of "two-well" excitons (Whittaker 1990, Dignam and Sipe 1990a,b, 1991). Here, the Coulomb interaction between a hole quantum well function in the well "i" and an electron quantum well function in the well "j" gives a "two-well" exciton which is treated variationally. A tight binding of these excitons then gives an insight to the excitonic correction to the whole absorption miniband. The method is both elegant and suitable to study the effect of external fields, and gives sensible results for the exciton ground state. However, the excited states and continuum of these two-well excitons are not taken into account, and it is difficult to discuss how they will influence the absorption lineshape. The crucial point here is the shape of the absorption associated with the "saddle-point" exciton formed out of miniband states close to the $q = \pi/d$ miniband edge.

More generally, variational methods are effective and reliable in predicting ground state energies, but they are no help for excited states. A complete solution of the problem implies the diagonalization of the SL + Coulomb Hamiltonian in a large part of the SL Brillouin zone (Chu and Chang 1987, Chang and Chu 1991), which was achieved numerically in the case of GaAs–AlGaAs SLs. Impressive agreement was obtained between experimental excitation spectra and calculated absorption spectra (Deveaud et al. 1989), as shown in fig. 16. According to these results, SL absorption spectra differ so much from the band-to-band absorption that it is in general difficult to "measure" the miniband width in terms of the energy separation between a ground state and a saddle-point exciton. In most cases, no clear absorption structure associated with the Van Hove singularity at $q = \pi/d$ is predicted. In contrast, other experimental results show clear evidence of a miniband. For example, the data shown in fig. 17a correspond to a strained-layer $In_{0.15}Ga_{0.85}As$–GaAs SL (Soucail et al. 1990a) where the strain increases the heavy to light hole splitting, and a H1–E1 absorption band well separated from the L1–E1 absorption is clearly observed. The width of this absorption band agrees very well with the calculated miniband width displayed in fig. 17b. In InGaAs-based SLs, excitonic features are weaker (both

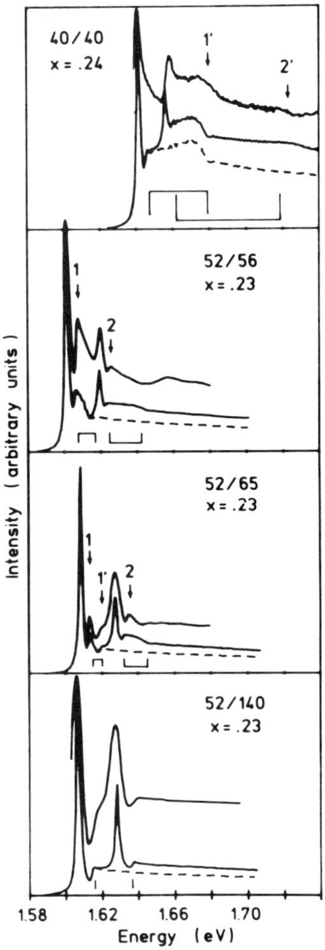

Fig. 16. Experimental photoluminescence excitation spectra (upper trace) and theoretical absorption spectra (lower trace) in GaAs–$Al_xGa_{1-x}As$ superlattices, from Deveaud et al. (1989). The dashed lines show the calculated absorption for the sole heavy-hole band. Layer thicknesses L_A/L_B are indicated in Å. Arrows 1 and 2 point to the onset of absorption continuum for heavy-hole and light-hole transitions, respectively, and arrows labelled 1′ and 2′ point to the absorption structures associated with the $q = \pi/d$ Van Hove singularity.

because the exciton binding energy is much smaller and the linewidths are larger than in GaAs–AlGaAs SLs), and absorption spectra are more like the band-to-band absorption, as shown in fig. 18 (Bleuse et al. 1988b, Ferreira et al. 1990).

The linewidth of exciton peaks in heterostructures is always governed by inhomogeneous broadening. An obvious culprit for such broadening is the unavoidable presence of layer thickness fluctuations, as, in the regime of narrow quantum wells, energy levels depend considerably on monolayer changes in the quantum well thickness. Somewhat surprisingly, exciton linewidths are narrower in short-period super-

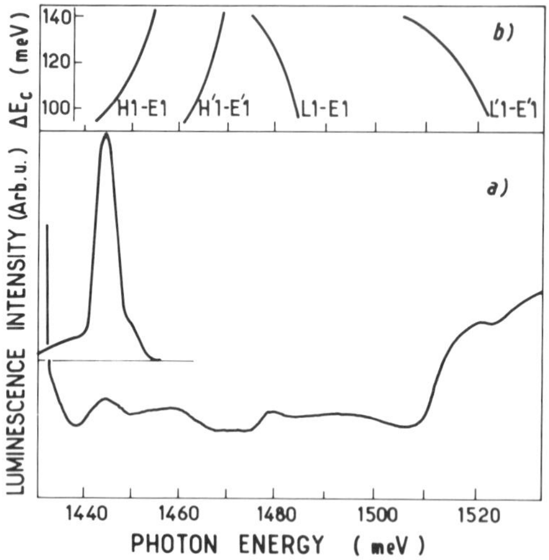

Fig. 17. Luminescence and excitation spectra of an $In_{0.15}Ga_{0.85}As$–GaAs strained-layer superlattice (a), and calculated transition energies (b). H1–E1 (L1–E1) is the onset of the ground heavy hole (light hole) to conduction transition, and H′1–E′1 (L′1–E′1) is the corresponding mini-zone edge transition. From Soucail et al. (1990a).

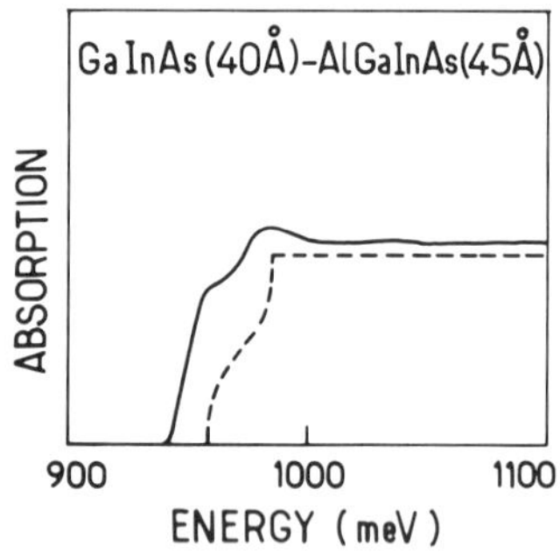

Fig. 18. Optical absorption spectrum (arbitrary units) in a $Ga_{0.47}In_{0.53}As$(40 Å)–$Al_{0.24}Ga_{0.24}In_{0.52}As$ (45 Å) superlattice (solid line), from Bleuse et al. (1988b). The dashed line shows schematically what band-to-band absorption should be.

lattices than they are in the corresponding isolated quantum wells (Jung et al. 1989). This is because the wavefunction of the bound state associated with a thicker quantum well is more extended in the SL than in the QW case, hence the perturbation by the localized defect is smaller: the wavefunction delocalization smoothes out potential fluctuations.

6. Superlattices in strong magnetic fields

Spectroscopies using the quantization of the electronic motion by an external magnetic field are a convenient and popular way to explore the band structure of a solid away from the band edge. In this respect, several aspects of semiconductor superlattices are remarkable: (i) the energy separation between Landau levels of the conduction or light hole subbands may be comparable to the subband width, which gives a unique opportunity to observe one-dimensional density of states; (ii) the situation with $B \parallel z$ quantizes the in-plane motion but leaves the axial motion unchanged (to the extent that these two motions can be decoupled), or at least free, while in the $B \perp z$ configuration, the superlattice and magnetic potentials interact; this configuration allows the observation of effects linked to the axial motion of carriers in an SL. Unfortunately, there are common and unavoidable drawbacks to all interband magneto-optical investigations, which are the complexity of the valence band physics and the difficulty of taking properly excitonic effects into account. As already mentioned, the envelope function theory provides a framework allowing complete calculations of magneto-optical spectra, and this was indeed done in several cases (Ancilotto et al. 1987, Brum et al. 1988). However, the computational effort is really considerable, and experimentalists are often left with hand-waving types of discussion.

$B \parallel z$ and $B \perp z$ interband magneto-absorption spectra in a very short period AlGaAs(11 Å)–GaAs(39 Å) superlattice are shown in fig. 19. (Belle et al. 1985). These spectra evidence an interesting qualitative effect: in the $B \parallel z$ configuration, sharp absorption peaks are observed in the whole energy range, while in the $B \perp z$ configuration, the spectra become featureless above 1.8 eV. The interpretation is the following: when the magnetic field quantizes the in-plane motion, Landau levels are formed in the conduction miniband. Over a very large energy range, they are reasonably approximated by

$$E_N(B, q) = E_1 + (N + \tfrac{1}{2})\hbar\omega_c - (\Delta E_1/2) \cos qd. \tag{19}$$

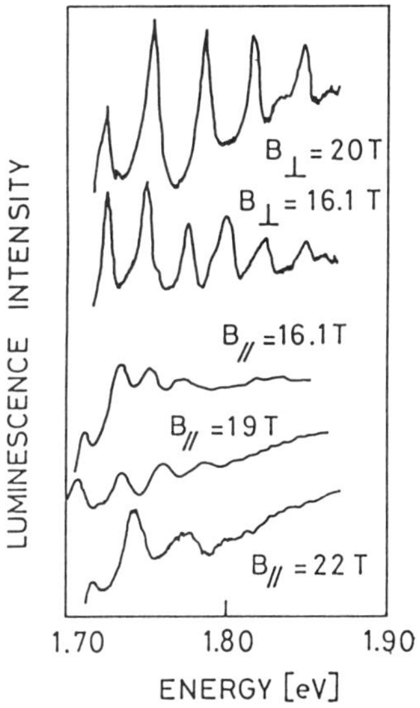

Fig. 19. Excitation spectra of a short-period GaAs(11 Å)–AlGaAs(39 Å) superlattice in the $B \parallel z$ (labelled $B_\perp$) and $B \perp z$ (labelled $B_\parallel$) configurations. From Maan (1987a,b).

In the case of strong non-parabolicity, the miniband width ΔE_1 and the cyclotron frequency $\hbar\omega_c = eB/m$ would depend on the energy, but this would not change the qualitative picture. Although the dispersion of valence band Landau levels is much more complex (which leads to complex fan charts), sharp interband transitions are observed, just as in single quantum wells. The situation is completely different when B is applied along the x-direction, perpendicular to the growth axis. With the appropriate gauge, the vector potential $A = (0, -Bz, 0)$ lies in the y-direction, and the Hamiltonian for the envelope function (in a parabolic approximation) becomes, after a few manipulations,

$$H_z = \frac{-\hbar^2}{2m}\frac{\partial^2}{\partial z^2} + \tfrac{1}{2}m\omega_c^2(z - z_0)^2 + V_{SL}(z). \tag{20}$$

The corresponding eigenstates are indexed by the wavevector k_x, the position of the center of the cyclotron orbit $z_0 = \lambda^2 k_y(\lambda^2 = \hbar/eB$ is the magnetic length), and a level index n. The z_0 dependence of the energy levels shown in fig. 20 (Maan 1987a,b) clearly explains qualitatively the effect observed in fig. 19: for energies within the first miniband (more generally, for energies within a zero-field miniband), Landau levels are flat (they have no dispersion with respect to z_0), and this will give rise to sharp transitions. Furthermore, using the tight-binding approximation, it can be proved analytically that these flat levels are evenly spaced. On the opposite, for energies outside a miniband, the Landau levels show a strong dispersion with respect to the position of the orbit center z_0. The corresponding transitions will be broadened, and the spectra become featureless. The observation of sharp transitions at low energies in this $B \perp z$ configuration is convincing evidence that electrons actually tunnel coherently through a multitude of barriers.

The data shown in fig. 19 correspond to a large miniband width $\Delta E_1 \approx 150$ meV, and the separation between Landau levels always remains small compared to ΔE_1. Absorption spectra for $B \parallel z$, $B = $ O and 20 T for the same GaInAs–AlGaInAs superlattice as in fig. 18 are shown in fig. 21. (Ferreira et al. 1990, Soucail 1990). At this

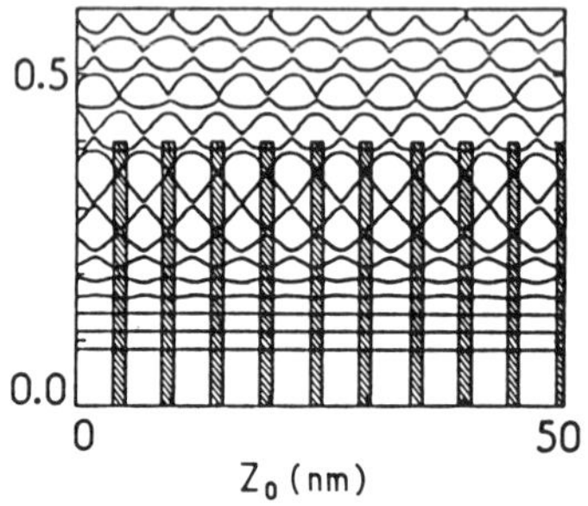

Fig. 20. Dependence of the energies of the SL Landau levels (in eV) on the z-position of the orbit center z_0, from Maan (1987). The hatched areas show the position and height of the potential barriers. For energies within the miniband, the Landau levels are almost dispersionless, which gives rise to sharp optical transitions. They become dispersive for energies outside the miniband, which explains the disappearance of the optical transitions above 1.8 eV in fig. 19.

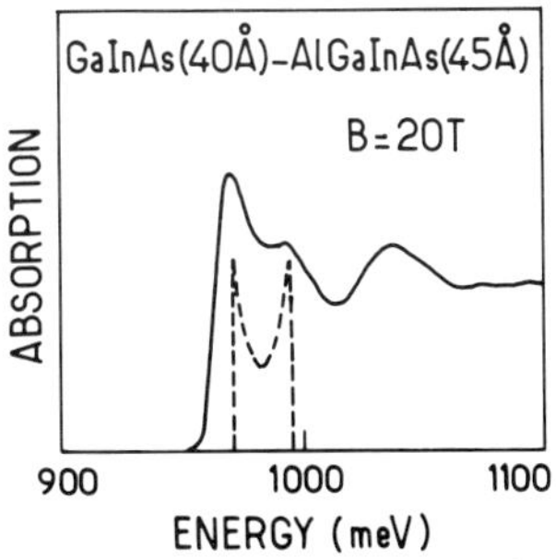

Fig. 21. Magneto-absorption spectrum at $B = 20$ T in the same InGaAs–AlGaInAs superlattice as in fig. 18 (solid line), from Ferreira et al. (1990). The dashed line shows schematically what band-to-band absorption should be.

field $\hbar\omega_c \approx 45$ meV is significantly larger than the miniband width $\Delta E_1 \approx 35$ meV, and the one-dimensional density of states corresponding to a Landau subband (eq. (19)) is clearly observed. The situation of normal and "saddle point" excitons are much more symmetrical in this large magnetic field regime, as the density of states peaks sharply at $q = 0$ and $q = \pi/d$, and the miniband width can be more reliably measured as the energy separation between the two first absorption peaks than at $B = 0$. This miniband is actually significantly narrower than expected from the calculation. Unfortunately, the identification of upper lying Landau subbands is not unambiguous, because valence band mixing effects come into play.

7. Superlattices in an external electric field: Wannier quantization

The nature of the eigenstates in a crystal in the presence of an external electric field F has remained controversial ever since the original contributions of G. Wannier in the early sixties (Wannier 1962, for a recent review, Nenciu 1991). The terms of the controversy are, roughly speaking, the following: in a simplified one-band tight-binding approach, it is found that the energy spectrum is a Stark ladder of discrete energy levels and the corresponding wavefunctions are localized. On the other hand, the field induces interband transitions and finally the electron should become unbound: thus the Wannier–Stark states should be an artefact of the one-band approximation (Zak 1968). In the effective mass approach, the energy spectrum remains continuous and the eigenfunctions are Airy functions which behave like plane waves with a linearly varying wavevector, which means that they remain delocalized. It is now generally accepted that Wannier's concept of field-induced quantization is correct with the restriction that the energy levels are narrow resonances rather than real bound states (Bentosela et al. 1982).

On the experimental side, however, there is no convincing evidence of Wannier quantization in bulk materials, and observed electro-optical properties are correctly described by the effective mass theory or Franz–Keldysh effect (Lederman and Dow 1976). The main reason for this is the following: the relevant parameter for the localization effect is the reduced field $\eta = eFd/\Delta$, where Δ is the band width and d the lattice period. In bulk materials where $\Delta \approx 1$ eV and $d \approx 6$ Å, η always remains smaller than 10^{-2}. This means that the high-field regime where Wannier quantization should dominate is not experimentally accessible. On the other hand, in superlattices, $d \approx 60$ Å and $\Delta \approx 50$ meV, and a novel high-field regime $\eta \geqslant 1$ is easily attained. The dominant effect here is the breaking of the resonance condition for the inter-well tunnel effect, which transforms the unbound Bloch waves into strongly localized states.

The simplest approach to the problem of superlattices in an axial electric field F is based on the tight-binding description of the envelope functions sketched at the end of § 3 (eqs. (8)–(11)). As proved by Wannier thirty years ago, the addition of a linearly varying potential eFz changes completely the solutions of the one-band tight-binding Hamiltonian. In the limit of an infinite system, the energy spectrum becomes a "Stark ladder":

$$\varepsilon_{\mathrm{i}\nu} = E_{\mathrm{i}} + \nu eFd, \tag{21}$$

where ν is an integer, and the associated wavefunction $\psi_{\mathrm{i}\nu}$ is strongly localized near the quantum well of index ν:

$$\psi_{\mathrm{i}\nu}(z) = \sum_n J_{n-\nu}(2\lambda_{\mathrm{i}}/eFd)\varphi_{\mathrm{i}}(z - nd), \tag{22}$$

where $J_p(x)$ is the Bessel function of index p, which decreases factorially with increasing p.

In fact, the miniband continuum (eq. (9)) and the Stark ladder (eq. (21)) spectra are asymptotic solutions valid in the limit of an infinite SL. To reconcile them, one has to consider a finite superlattice. Here, we take a finite superlattice with $(2N + 1)$ wells, in the presence of an electrostatic potential eFz. For convenience, we choose the origin at the center of the structure. At zero electric field, the tight-binding Hamiltonian with finite-size boundary conditions is exactly solvable, and gives $2N + 1$ discrete eigenstates forming a quasi-continuum of width ΔE_{c}, and obeying (Bleuse et al. 1988a)

$$\varepsilon_q = E_1 - 2\lambda_{\mathrm{c}} \cos qd, \quad \text{with} \quad qd = i\pi/2(N+1) \quad (1 \leqslant i \leqslant 2N+1), \tag{23}$$

where λ_{c} is the modulus of the transfer integral between nearest neighbors (which is equal to $\Delta E_{\mathrm{c}}/4$) and E_1 is the ground state of the isolated wells (eventually corrected by a negative shift term s_1 due to the effects of the other wells on the ground level of a given well). As a result of the finite size, the probability density of a given state is not evenly distributed over the structure like that of a traveling Bloch wave, but eigenfunctions are nevertheless extended through the entire structure, since the expansion coefficient c_{nq} (eq. (8)) are harmonic functions:

$$c_{nq} = \sqrt{1/(N+1)} \begin{cases} \sin nqd \\ \cos nqd \end{cases} \quad \text{with} \quad qd = \begin{cases} j\pi/(N+1) & (1 \leqslant j \leqslant N) \\ (j'+1/2)\pi/(N+1) & (0 \leqslant j' \leqslant N) \end{cases} \tag{24}$$

The simplicity of the tight-binding results (eqs. (9), (21)) owes much to the assumption that QW functions centered in adjacent wells are othogonal and the shift integral negligible, which is usually a satisfying first-order approximation. As can be seen in fig. 22, when increasing F, the spectrum evolves continuously from the quasi-continuum forming the miniband at $F=0$ to the discrete Stark ladder at large F. The difference between the exact spectrum and the Stark ladder is an edge effect which is more pronounced for the states localized close to the edge of the structure. For a large enough SL, a huge majority of the eigenstates are correctly described by eqs. (21) and (22), which can therefore be used reliably to calculate the optical properties, even at a relatively small electric field (see below). A sketch of the finite SL, the presence probability of a representative eigenstate (actually obtained from a numerical solution of the problem) and the strong localization of this eigenstate in a moderate electric field are illustrated in fig. 23.

As in bulk materials, however, the predictions of the one-band tight-binding model are doubtful because, in an infinite SL, Stark ladder levels associated with different subbands should come into resonance and couple to each other for some values of the field. In addition, the effect of the approximations in the tight-binding model have to be examined. To address these issues, we have used numerical solutions of the Schrödinger equation for the envelope functions (in the potential sketch in the inset of fig. 24, corresponding to a superlattice clad between 150 Å thick "buffer" layers), using the transfer matrix method (Soucail et al. 1991). This approach is limited to structures with a limited number of wells, and is somewhat inelegant, but it is fundamentally exact, and allows one to take into account details like the host non-parabolicity, the intra-well (Stark) effect of the field, etc. Here, we discuss the numerical results for the specific case of the first conduction subbands in a GaInAs(39 Å)–AlGaInAs(46 Å) SL, which actually corresponds to our reference sample. In this system, the barrier height is 250 meV. The first subband E1, extending

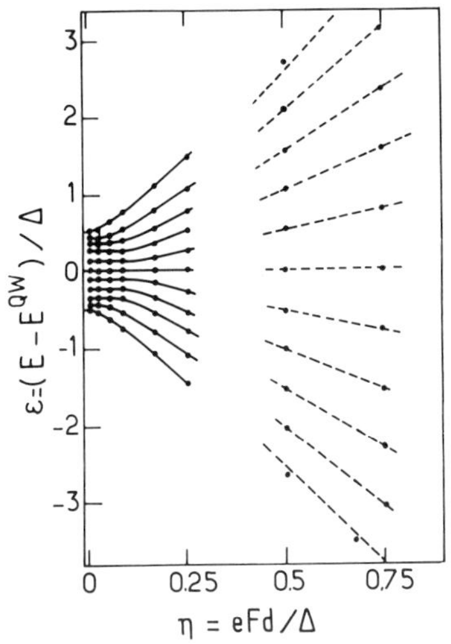

Fig. 22. Computed eigenenergies (dots) versus applied electric field in a 9-period superlattice, in the simplified tight-binding approach. The dashed lines show the Stark-ladder asymptotes.

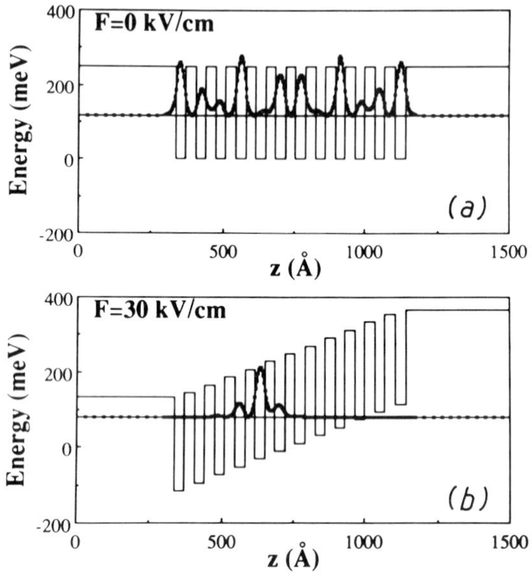

Fig. 23. Potential profile and square modulus of the wavefunction of a representative state at $F = 0$ (a) and $F = 30$ kV/cm (b) in a 12-period GaAs(35 Å)–AlGaAs(35 Å) superlattice, obtained from a numerical solution of Schrödinger equation for the envelope functions (from Bleuse et al. (1988a)).

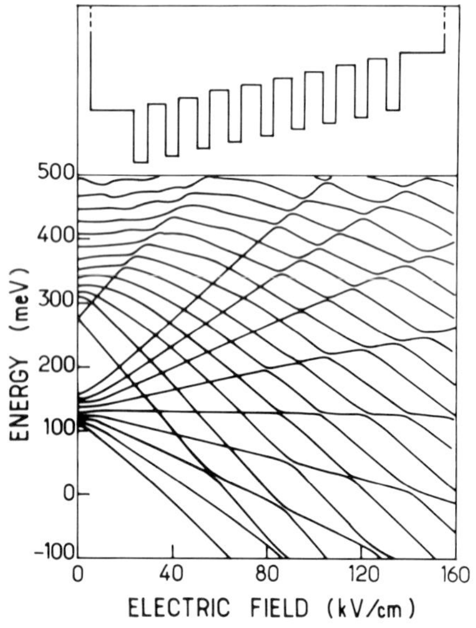

Fig. 24. Field dependence of the calculated conduction energy levels in the 9-well InGaAs(39 Å)–AlGaInAs(46 Å) superlattice structure schematized in the inset. The conduction band offset for this system is 250 meV (from Soucail et al. 1991).

from 109 to 157 meV, is far below the top of the barrier, while the excited subbands E2 and E3 are above the barrier. Two "edge states" associated with the buffer layers are obtained at 280 meV. Figure 24 shows the energy levels over broad energy and field ranges in the case of a 9-well superlattice, and fig. 25 shows the details of the low-field dependence of the first miniband for a 49-well SL.

At first sight, the levels of the E1 subband display a fan chart quite similar to fig. 22, showing a smooth transition from the "miniband" spectrum at $F = 0$ to the Stark ladder at large field ($F \geqslant 40$ kV/cm). Field-induced matrix elements do produce anticrossings of these Stark ladder levels with those arising from the upper-lying subbands. For each level, the size of the anticrossings increases monotonically with the field at which they occur, and reaches typically 4 meV at $F = 100$ kV/cm: anticrossings remain quite small compared to the spacing between the Stark ladder levels, $eFd = 85$ meV at the same field. It is a general result that these anticrossings become sizeable when eFd becomes comparable to the gap between the subbands. Incidentally, they have in the present case a negligible effect on the calculated optical absorption spectra.

More surprisingly, it can be observed in fig. 24 that the energy levels associated with the excited subbands E2 and E3 do not behave like the ground subband E1: no "Stark ladder" can be observed, at any field. Instead, we get a rather complex pattern of levels with oscillatory spacings. The reason for this behavior can easily be traced back to the fact that the gap between E2 and E3 is very small: these bands strongly interact, and cannot be individually described by a one-band tight-binding

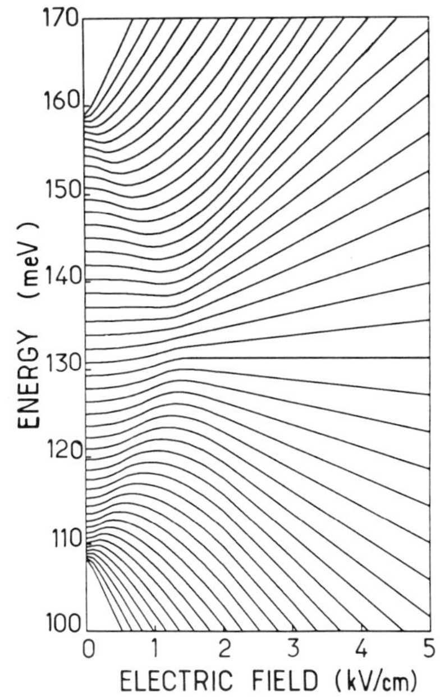

Fig. 25. Detail of the low-field behavior of the calculated energy levels in a 49-well InGaAs(39 Å)–AlGaInAs(46 Å) superlattice, illustrating the instability of the whole energy spectrum.

approach. This is a completely general feature of our numerical results that energy bands which are reasonably well decoupled from the others at $F=0$ (say they have a width small compared to the neighboring gaps) are well described by the Stark ladder at large F, while those which are strongly coupled to each other (say the gap between them is smaller than their width) do not obey Wannier quantization. This is of practical importance, since for SL parameters ensuring a significant bandwidth there is usually one and only one subband located below the potential barrier and weakly coupled to the others. Hence, the criterion for field-induced localization in band i is $eFd/\Delta_i \approx 1$, that for large anticrossings between Stark levels of bands i and j (or field-induced "delocalization", Schneider et al. 1990b) is $eFd/G_{ij} \approx 1$ (where G_{ij} is the gap between the bands i and j), and the criterion for the breakdown of Wannier quantization for bands i and j is $\Delta_i/G_{ij} \approx 1$.

A more careful examination of the low-field behavior of the E1 subband (fig. 25) reveals two differences with respect to fig. 22: (i) the band is not symmetrical, the spacing of the lower levels being significantly smaller than that of the upper levels and (ii) the central level, which is not exactly at the midband energy at $F=0$, is not flat: it first rises by a couple of meVs before it reaches a saturation value which is not at the midsubband energy either. In fact, independent of the SL size the energy of this central level at $F=0$ is almost exactly that of the isolated quantum well, and the saturation value is exactly the average of the band. The latter result was already derived in Wannier's original papers, as a simple consequence of the one-band approach. The "strange" low-field behavior is also observed on the other levels: the upper ones first go down before going up, and the lower ones first go up before coming down, which produces significant changes in the shape of the density of states. The "critical" field at which the central level becomes (rather abruptly) horizontal is nearly equal to Δ/Ned: it depends on the size of the structure, and becomes arbitrarily small if the structure is large enough, which means that the whole spectrum becomes unstable near $F=0$.

The physical origin of this effect has to be found among the various approximations of the tight-binding approach: details of the host band structure like non-parabolicity, interaction with upper-lying subbands, assumption of quasi-orthogonal quantum well functions, etc. We have checked that it is in fact entirely due to the last one, and can be completely recovered in a tight-binding approach taking into account the finite overlap of adjacent quantum well functions: such a model using overlap and transfer integrals obtained from envelope-function calculations of the isolated quantum well problem gives energy spectra almost identical to those of fig. 24, which proves that intersubband effects and non-parabolicity are usually negligible in actual superlattices at zero or low electric field.

8. Electro-optical absorption in semiconductor superlattices

Coming back to the simplified tight-binding method which gives sensible analytical results for both the energy spectrum and eigenfunctions, it is clear that the absorption between valence band and conduction band Stark ladders (in a type I SL) has to be a series of steps at the energies

$$h\nu_p = E_g^{QW} + peFd, \tag{25}$$

where p is a relative integer and E_g^{QW} the isolated quantum well bandgap. It turns out from the algebra of Bessel functions that the height α_p of these steps is proportional to the square of a simple Bessel function (Bleuse et al. 1988a):

$$\alpha_p = N\alpha_0 J_p^2 \frac{2(\lambda_c + \lambda_v)}{eFd} = N\alpha_0 J_p^2 \frac{1}{2\eta}, \tag{26}$$

where $\alpha_0 \approx 0.006$ is the absorption for an allowed heavy hole to conduction transition in a quantum well and η is now the reduced field $(\Delta_c + \Delta_v)/eFd$. This means that the plot of the absorption coefficient using the reduced photon energy $\varepsilon = (h\nu - E_g^{QW})/(\Delta_c + \Delta_v)$ and the reduced field η is universal. At large field, only the J_0^2 term survives, corresponding to electron and hole wavefunctions localized in the same well (vertical transition in real space), and the absorption is a single step at E_g^{QW}, like that of a series of uncoupled quantum wells. The transitions at $h\nu_p \neq 0$ correspond to electrons and holes whose wavefunctions are centered in QWs distant from pd, and we call them "oblique" transitions. Figure 26a compares the absorption profiles at $\eta = 0$ and $\eta = 1$, evidencing the $(\Delta_c + \Delta_v)/2$ blue shift of the effective absorption edge which was the intuitive starting point of this Wannier–Stark venture (Voisin

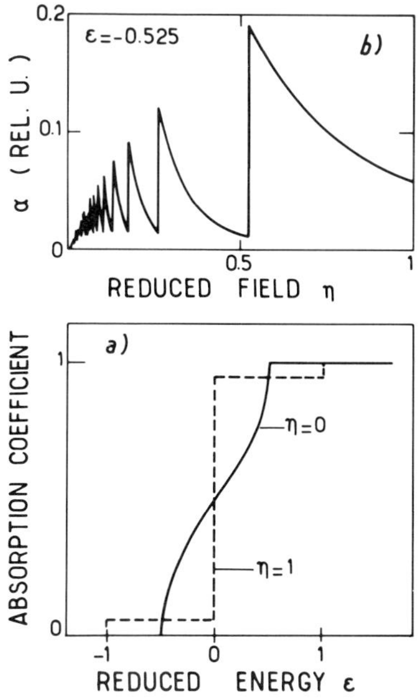

Fig. 26. Absorption profiles at zero and large field (a) and electro-optical oscillations periodic in F^{-1} (b), for a photon energy just below the SL bandgap.

1986b). Note that the strength of the absorption on the two-dimensional plateau does not depend on the field, which, algebraically, appears as a consequence of the Graf sum rule on the J_p^2. When the electric field is swept at a constant photon energy, the oblique transitions cross this constant energy and then fade away as the field increases, which gives rise to a novel type of electro-optical oscillations, as illustrated in fig. 26b. These oscillations are periodic in F^{-1}.

These effects have been observed, soon after being predicted, in a variety of superlattices and at room temperature as well (Mendez et al. 1988, Voisin et al. 1988, Bleuse et al. 1988b, Yan et al. 1989, Bar-Joseph et al. 1989, Agullo-Rueda et al. 1990, Soucail et al. 1990a, Bigan et al. 1990, Barrau et al. 1990, Schneider et al. 1990b). We show in fig. 27 transmission spectra obtained in an InGaAs–AlGaInAs SL forming the intrinsic part of a PIN diode. The lowest trace is the flat band spectrum (corresponding to a forward bias of 1 V) already shown in fig. 18, with a steep absorption edge (but no marked exciton feature) followed by a broad shoulder and the absorption plateau. As soon as the field is decreased from the value giving flat bands, the "vertical" transition appears and develops in the middle of the band, while the zone-center and zone-edge transitions progressively transform into oblique transitions which vanish as they split away from the vertical one. A remarkable (and general) feature of these data is the enhancement of the excitonic character of the absorption edge which contrasts with the decrease of excitonic features observed in both the 3D (bulk) and 2D (QW) limits. This paradoxal result is a simple consequence of the change in dimensionality undergone by the exciton with field-induced localization: as Δ is much larger than the Rydberg energy, the SL exciton is essentially three-

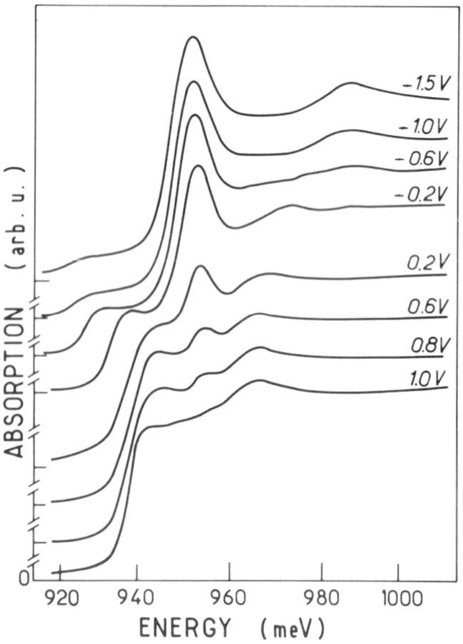

Fig. 27. Absorption spectra in a 50-period InGaAs(40 Å)–AlGaInAs(45 Å) SL. The lower trace shows the flat-band spectrum, and the upper one corresponds to an applied electric field of ≈50 kV/cm.

dimensional, while in the high-field limit, the electron and hole wavefunctions are localized within a single narrow quantum well; then the exciton becomes nearly two-dimensional, and shows increased binding energy and oscillator strength. While the low-field and high-field limits were easily predictable (Bleuse et al. 1988a,b), the effect of Coulomb interaction at arbitrary field is not so easy to handle. This problem was solved in recent papers (Dignam and Sipe 1990a,b, 1991, Whittaker 1990). It has been shown that Coulomb interaction is mostly responsible for the observed asymmetry between the oscillator strength of $\pm p$ oblique transitions at low field. Along the same line, it was proved experimentally that the binding energy of the "spatially indirect" excitons associated with the oblique transitions $p = \pm 1$ is quite significant, even in the high-field limit where they loose all oscillator strength: it saturates, typically, to half the binding energy of the "spatially direct" or vertical exciton (Brum and Agullo-Rueda, 1990, Soucail et al. 1990a).

Finally, it is noteworthy that the same physics of field-induced localization produces qualitatively different results in the case of type II SLs, where electron and hole wavefunctions localize in adjacent layers. In this case, the high-field limit is a double-step absorption at the energies $E_g^{QW} \pm \frac{1}{2}eFd$. The evolution of the absorption spectra with increasing electric fields for type I and type II superlattices are shown in fig. 28a,b (Bleuse 1988, Voisin 1990a). The type II behavior was actually observed experimentally in the case of short-period AlAs–GaAs SLs, where the electron ground

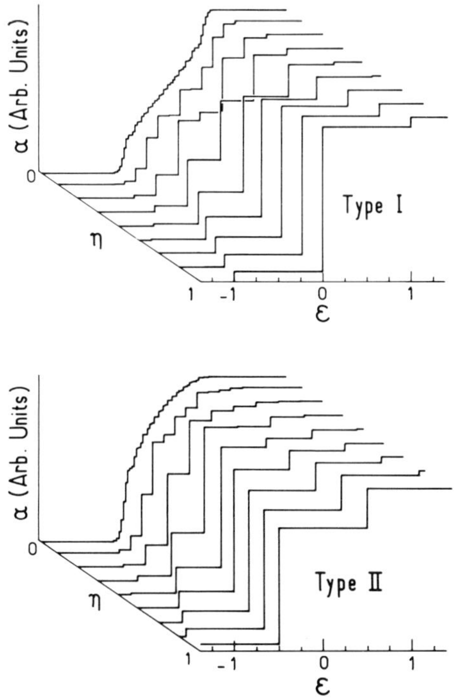

Fig. 28. Theoretical electro-absorption spectra of type I and type II superlattices.

state is related to X-valleys in AlAs while the holes are confined as usual in GaAs (Whittaker et al. 1990). The physics in this system is, however, more subtle, because of (i) the heavy mass of electrons in X-valleys (which increases the intra-well Stark effect), (ii) the quasi-indirect character of optical transitions in k-space, (iii) the problem of valley degeneracy, (iv) the finite coupling of these type II states with the normal type I states localized in GaAs. InP–AlInAs SLs are a more likely candidate for the observation of the type II behavior in a "canonical" situation.

An important aspect of these "Wannier–Stark" effects is their potential application to electro-optical modulators. A waveguide modulator using the redshift of oblique transitions and having a promising potential for telecommunication applications has recently been demonstrated (Bigan et al. 1990, 1992), as well as a high-contrast reflection modulator using the blue shift effect and leading to outstanding electro-optical bistability when used in a self-electro-optical effect device (SEED) configuration (Law et al. 1990, Olbright et al. 1991). Applications of the Wannier–Stark effects (Voisin 1990a,b) presently form a competitive and rapidly expanding field.

9. Field-induced quantum boxes

The combination of Wannier quantization of the axial motion with Landau quantization of the in-plane motion allows the reduction of the dimensionality of the electronic motion from 3D to 2D, 1D and 0D. The effect of parallel electric and magnetic fields (applied parallel to the growth axis) on the optical absorption of the same InGaAs–AlGaInAs SL is shown in fig. 29 (Ferreira et al. 1990). The absorption features become sharper and sharper as both fields increase, as a result of the singularization of the joint density of state. At $B=20$ T, the cyclotron energy $h\omega_c$ becomes larger than the band width Δ, and at $F=0$, as in fig. 21, the density of states of a 1D Landau subband (with sharp maxima at the zone center and zone edge) can be clearly observed. Wannier quantization of this Landau subband then occurs when the electric field is turned on (fig. 30). Unfortunately, these features can be observed on the ground Landau level only, as the mixing of the heavy and light valence minibands at finite B relaxes efficiently the optical selection rules and makes the interpretation of the magnetic fan charts particularly difficult. A remarkable feature is that the complexity itself of these fan charts is qualitatively modified by the electric field for the following reason: the in-plane dispersions of the valence subbands result from their mutual coupling via the $\boldsymbol{k}_t \cdot \boldsymbol{p}$ interaction, which scales with the energy separation between the bands at $k_t=0$. Due to the considerable width of the light-hole miniband, the H1–L1 splitting is much smaller in the SL ($F=0$) than in the isolated QW (high-field limit), therefore the $\boldsymbol{k}_t \cdot \boldsymbol{p}$ interaction of these bands is modified by the field, hence their in-plane dispersions (Ferreira and Bastard 1988).

An interesting aspect of the data in figs. 29 and 30 is the fact that the oblique transition $p=-1$ at large electric field is seen much more clearly at large magnetic field than at $B=0$. This is a magneto-exciton effect due to the difference between the vertical and oblique exciton binding energies: the latter, being less bound, reaches the high magnetic field regime sooner than the former, which is more resistive to the

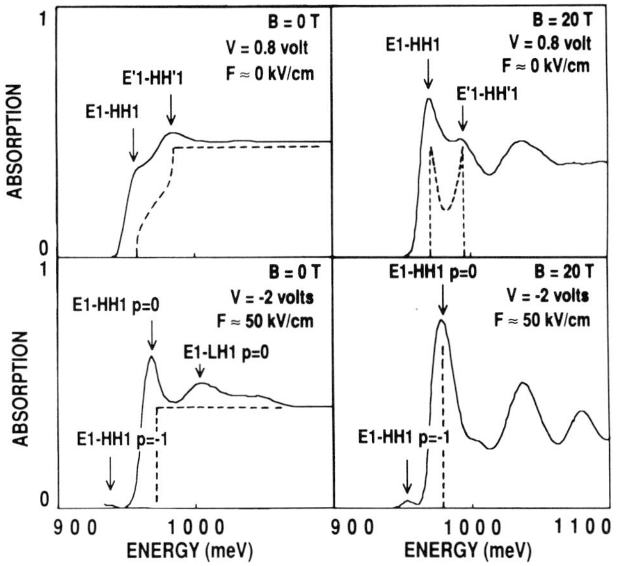

Fig. 29. Absorption spectra in the same InGaAs(40 Å)–AlGaInAs(45 Å) SL at various electric and magnetic fields, illustrating the formation of "field-induced quantum boxes". The dashed lines show schematically the shape of band-to-band absorption.

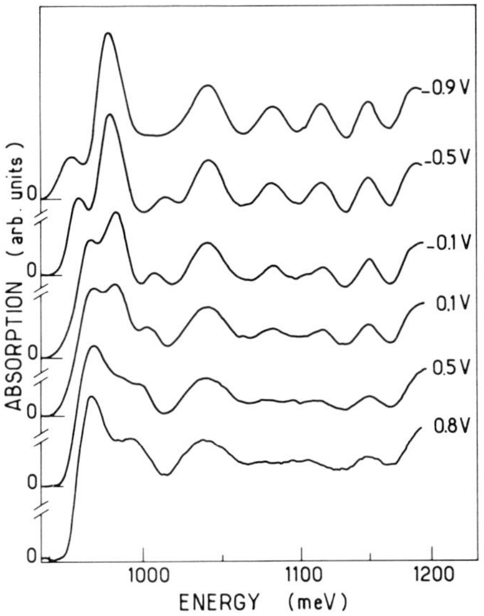

Fig. 30. Electric-field dependence of the spectra at $B = 20$ T, showing Wannier quantization of the ground Landau subband.

magnetic perturbation. The ratio of the binding energies $E_b(p=-1)/E_b(p=0)$ thus increases at first with the field, hence the ratio of the oscillator strengths. A simple variational approach of this problem accounts reasonably well for our observations (Ferreira et al. 1990).

A superlattice in parallel electric and magnetic fields thus appears as a three-dimensional array of quantum dots with flexible size and coupling (of course, this representation is not quite gauge-invariant!), and the investigation of vertical transport by elastic or inelastic hopping between these quasi-zero-dimensional states will certainly increase our understanding of the microscopic physics of transport phenomena at finite fields. Calculations of the scattering times (Ferreira and Bastard 1989a, Ferreira 1991), as illustrated in fig. 31, show huge oscillations (by several orders of magnitude!) associated with the commensurability of $nh\omega_c$, $peFd$ (where n and p are integers), and $h\omega_{LO}$. These oscillations have a simple physical origin, since they are essentially due to the huge variations of the density of final states for the scattering, but they have not yet been found experimentally.

10. Diffusion and vertical transport in superlattices

The extraordinary flexibility of epitaxial growth allows the design of heterostructures specially suited for the study of given properties. An example of this is the inclusion of an "enlarged" quantum well at a certain position inside a superlattice: the comparison of the luminescence intensities of this enlarged quantum well and of the SL itself brings rich information on the diffusion of carriers and/or excitons created or injected in the SL (Chomette et al. 1986), and time-resolved experiments (Deveaud et al. 1987, Schneider et al. 1990a) in such structures allow almost direct measurement of the

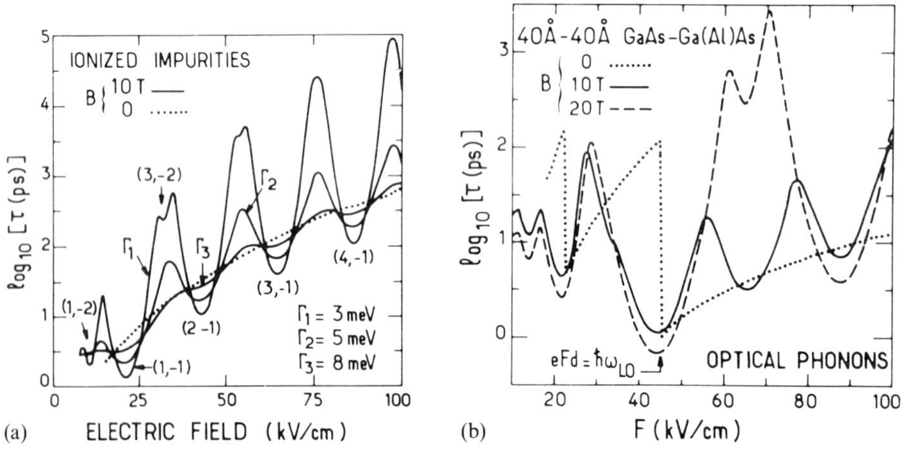

Fig. 31. Oscillations of the interwell elastic (a) and inelastic (b) hopping time for a superlattice in parallel electric and magnetic fields, from Ferreira (1991). Calculations are for a GaAs(40 Å)–AlGaAs(40 Å) superlattice. Γ is a phenomenological broadening of the quantum well (discrete) Landau levels. An areal density of 10^{10} impurities/cm^2 is used in (a).

carrier (or exciton) "time of flight". The corresponding physics is rather subtle, and a critical discussion of reported results is out of the scope of this review. We wish to stress two contradictory aspects of the exciton diffusion problem. Firstly, superlattice excitons are localized particles in the same sense as the heavy holes are localized: the very concept of excitonic Bloch wave is irrelevant. Indeed, the effective mass for the tunneling of an exciton is the sum of the electron and hole masses: excitons tunnel even less than a naked hole, they do not benefit from the miniband structure. One should not confuse exciton minibands, which are dispersionless and therefore unphysical, with the fact that different excitons (like ground-state and saddle-point excitons) are actually formed in the SL. Secondly (and conversely), the fact that heavy-hole subbands at $k_t = 0$ are dispersionless should not be considered too seriously. Indeed, at finite k_t, there is some admixture of the light-hole character into the "heavy-hole" spinor, and this produces a considerable acceleration of heavy hole tunneling time (Ferreira and Bastard 1989b), which, of course, will help excitons as well. A variety of situations have been investigated with the technique of the "enlarged quantum well", like SLs with intentional layer thickness disorder (Chomette et al. 1986) which are an interesting test for disorder-induced localization in quasi-one dimension (as illustrated in fig. 32), or SLs with quasi-electric fields produced by grading the Al content (Lambert et al. 1987) in AlGaAs quantum wells or barriers, etc. However, detailed theoretical interpretation of the measured diffusion coefficients (accounting for the above remarks) seem presently out of reach.

Along the same line, the layers surrounding the superlattice can be designed in such a way that there is no (or very strong) barrier for electron or hole injection, which corresponds to perfect Ohmic or rectifying contacts. However, the study of the axial or "vertical" transport (for which superlattices were invented) is still in its childhood. Many results, including the observation of negative differential conductance (Sibille et al. 1989, 1990a,b, and references therein) have been interpreted phenomenologically in terms of effective medium with field-dependent drift velocity. A few theoretical (Kazarinov and Suris 1972, Tsu and Döhler 1975, Döhler et al. 1975, Calecki et al. 1984) and experimental (e.g., Palmier et al. 1985, Beltram et al. 1990) works have attempted to take Wannier quantization at finite field into account (Wannier quantization in SLs was actually first considered in the context of vertical transport), but we still lack a complete microscopic description of the interplay between intra-well energy relaxation and inter-well hopping (Ferreira and Bastard 1989a, Ferreira 1991) associated with a satisfactory description of electron population. Like the growth of superlattices themselves, the understanding of vertical transport in superlattices "will be a formidable task, but effort directed to this end will open new areas of investigations in the field of semiconductor physics".

11. Concluding remarks

The recent developments of optical studies of superlattices has certainly established on a firm ground the existence of extended electronic states in these structures. More specifically, the Wannier–Stark effects show quite clearly that at zero-electric field,

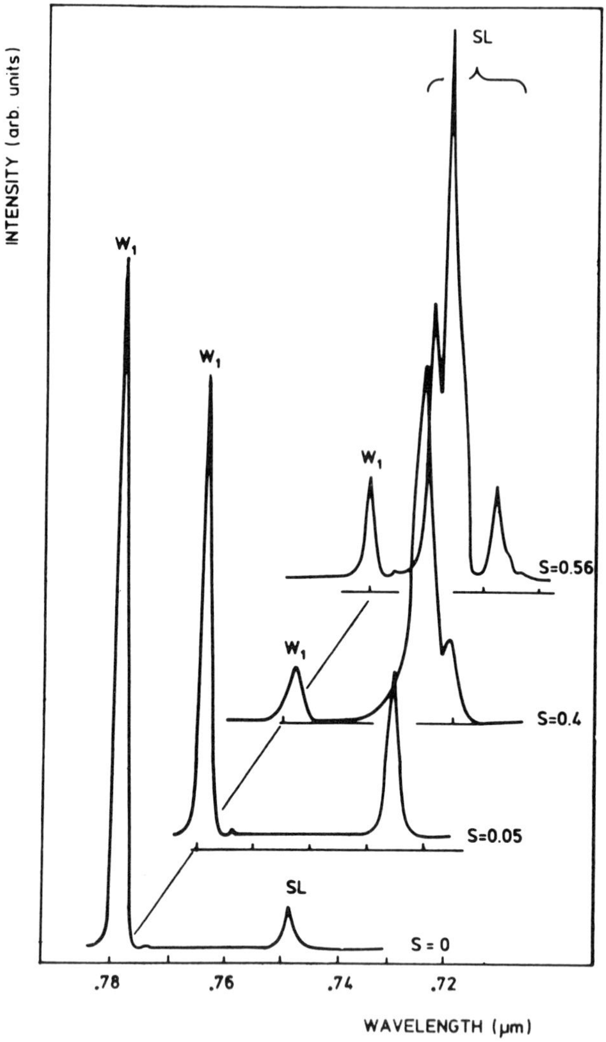

Fig. 32. Luminescence spectra of heterostructures comprising an intentionally disordered superlattice (SL) and an enlarged quantum well (W1). S (nm) is the amplitude of the intentional layer thickness fluctuations. At small disorder, there is a fast diffusion in the SL and the enlarged quantum well collects nearly every photo-created carrier, while at large disorder, carriers localize and recombine in the SL. From Chomette et al. (1986).

electrons do experience the superlattice potential, which lowers their ground state by half the miniband width ΔE (or, in the tight-binding scheme, twice the transfer integral λ). However, it is not yet clear that this "perfect" superlattice model, which works qualitatively, accounts quantitatively for all the observations. In particular, it is noteworthy that the observed bandwidth in the GaInAs–AlGaInAs superlattice which we have used as a reference throughout this review, $\Delta E \approx 30$ meV, is significantly smaller than the calculated value, $\Delta E_{th} \approx 48$ meV. Uncertainties on actual sample parameters are unlikely to account for such a discrepancy, which we have observed systematically on a series of equivalent samples. Results reported by Mendez et al. (1988) and Agullo-Rueda et al. (1990), on GaAs–AlGaAs SLs also indicate bandwidths smaller than predicted from nominal parameters. Conversely, data in InGaAs–GaAs strained-layer SLs (Soucail et al. 1990a) and in GaAs–AlAs SLs (Schneider et al. 1990b) are in agreement with theoretical estimates. Whether there is a fundamental effect here, for example associated with the alloy nature of the well and/or barrier layers, or with residual doping, or with interface roughness, is presently uncertain.

Along the same line, it was pointed out that the number of "oblique" transitions observed in electro-optical absorption spectra should somehow fix a lower limit to the coherence length of electronic states (Agullo-Rueda et al. 1990, Barrau et al. 1990), and coherence lengths larger than 15 periods have been claimed. This is not quite correct since what is observed is essentially the spatial extent of excitonic bound states along the *z*-direction, which has little to do with the coherence length of a free electron, as noticed by Dignam and Sipe (1990b, 1991). What actually limits the coherence length, and the consequences on both the energy levels and "vertical" transport properties at low electric field is a matter of current interest.

More generally, the role that the "alloy" nature of the layers might play for heterostructure energy levels is a virgin topic of investigations. It is well known that alloy disorder contributes to the "bowing" of alloy bandgap versus alloy composition, i.e., it affects significantly the properties of the bulk materials. It is certainly worth examining if alloy disorder can affect significantly the properties of heterostructures.

An important material parameter for all sort of opto-electronic applications (ranging from guided-wave devices such as lasers or modulators to "vertical micro-cavity" devices) is the refractive index. The usual approximation of an "average" refractive index is far too rough for the optimization of actual guided-wave devices, whose properties depend on relatively small layer-to-layer differences of the refractive index. Calculations and accurate measurements of SL (more generally, heterostructure) refractive index would certainly be a useful topic.

Twenty years after the seminal contribution of Esaki and Tsu, superlattices remain a lively topic, and many fundamental questions, such as the relevance of the original concept of "Bloch oscillations" (which is the semiclassical counterpart of the quantum mechanical concept of Wannier quantization) are still open. At the same time, the Wannier–Stark electro-optical effects are giving rise to a new generation of useful electro-optical devices, which will certainly sustain both the fundamental and device-oriented efforts in this field.

Acknowledgements

The authors are indebted to so many colleagues with whom they have collaborated for the last ten years that it is impossible to list them exhaustively. The work done at Ecole Normale Supérieure results from close collaboration with J.M. Berroir,

J. Bleuse, J.A. Brum, C. Delalande, R. Ferreira, Y. Guldner, B. Soucail, and J.P. Vieren. Special thanks are due to L.L. Chang, L. Esaki, and M. Voos who have initiated, suggested and continuously encouraged our efforts.

References

Agullo-Rueda, F., E.E. Mendez and J.M. Hong, 1989, Phys. Rev. B **40**, 1357.
Agullo-Rueda, F., E.E. Mendez, J.A. Brum and J.M. Hong, 1990, Surf. Sci. **228**, 80.
Aina, L., M. Mattingly and L. Stecker, 1988, Appl. Phys. Lett. **53**, 1620.
Allan, G., G. Bastard, N. Boccara, M. Lannoo and M. Voos, eds, 1986, Heterojunctions and Semiconductor Superlattices, Proc. 1985 Les Houches, Winterschool (Springer, Berlin).
Altarelli, M., 1983, Phys. Rev. B **28**, 842.
Altarelli, M., 1986, in: Heterojunctions and Semiconductor Superlattices, eds G. Allan, G. Bastard, N. Boccara, M. Lannoo and M. Voos (Springer, Berlin).
Ancilotto, F., A. Fasolino and J.C. Maan, 1987, Superlatt. & Microstr. **3**, 187.
Ando, T., and H. Akera, 1989, Phys. Rev. B **40**, 11604, 11619.
Bar-Joseph, I., K.W. Goossen, J.M. Kuo, R.F. Kopf, D.A.B. Miller and D.S. Chemla, 1990, Appl. Phys. Lett. **55**, 340.
Barrau, J., K. Khirouni, Do Xuan Than, T. Amand, M. Brousseau, F. Laruelle and B. Etienne, 1990, Solid State Commun. **74**, 147.
Bastard, G., 1981, Phys. Rev. B **24**, 5693.
Bastard, G., 1982, Phys. Rev. B **25**, 7584.
Bastard, G., 1988, Wavemechanics Applied to Semiconductor Heterostructures (Editions du Physique, Paris).
Bastard, G., and J.A. Brum, 1986, IEEE J. Quantum Electronics **QE-22**. 1625.
Bastard, G., E.E. Mendez, L.L. Chang and L. Esaki, 1982, Phys. Rev. B **26**, 1974.
Bastard, G., C. Delalande, Y. Guldner and P. Voisin, 1988, in: Advance in Electronics and Electron Physics, Vol. 72, ed. P. Hawkes (Academic Press, New York).
Bastard, G., J.A. Brum and R. Ferreira, 1991, Solid State Phys. **44**, 229.
Bauer, G.E.W., and T. Ando, 1988a, Phys. Rev. B **38**, 6015.
Bauer, G.E.W., and T. Ando, 1988b, Phys. Rev. B **37**, 3130.
Belle, G., J.C. Maan and G. Weimann, 1985, Solid State Commun. **56**, 65.
Beltram, F., F. Capasso, D.L. Sivco, A. Hutchinson, S.-N.G. Chu and A.Y. Cho, 1990, Phys. Rev. Lett. **64**, 3167.
Ben-Daniel, D.J., and C.B. Duke, 1966, Phys. Rev. **152**, 682.
Bentosela, F., V. Grecchi and F. Zironi, 1982, J. Phys. C **15**, 7119.
Berroir, J.M., 1988, Th.D. Université Paris 6 (unpublished).
Berroir, J.M., Y. Guldner, J.P. Vieren, M. Voos, X. Chu and J.P. Faurie, 1989, Phys. Rev. Lett. **62**, 2024.
Bigan, E., M. Allovon, M. Carré and P. Voisin, 1990, Appl. Phys. Lett. **57**, 327.
Bigan, E., M. Allovon, M. Carré, C. Braud, A. Carenco and P. Voisin, 1992, IEEE J. Quantum Electron. **QE-28**, 214.
Bleuse, J., 1988, Th.D. Université Paris 6 (unpublished).
Bleuse, J., G. Bastard and P. Voisin, 1988a, Phys. Rev. Lett. **60**, 220.
Bleuse, J., P. Voisin, M. Allovon and M. Quillec, 1988b, Appl. Phys. Lett. **53**, 2632.
Broido, D.A., and L.J. Sham, 1986, Phys. Rev. B **34**, 3917.
Brum, J.A., and F. Agullo-Rueda, 1990, Surf. Sci. **229**, 472.
Brum, J.A., P. Voisin, M. Voos, L.L. Chang and L. Esaki, 1988, Surf. Sci. **196**, 545.
Caine, E.J., S. Subbanna, H. Kroemer, J.L. Merz and A.Y. Cho, 1984, Appl. Phys. Lett. **45**, 1123.
Calecki, D., J.F. Palmier and A. Chomette, 1984, J. Phys. C **17**, 5017.
Capasso, F., and G. Magaritondo, eds, 1987, Heterojunction Band Discontinuities: Physics and Device Applications (North-Holland, Amsterdam).

Chang, L.L., N.J. Kawai, E.E. Mendez, C.-A. Chang and L. Esaki, 1981a, Appl. Phys. Lett. **38**, 30.
Chang, L.L., G.A. Sai-Halasz, L. Esaki and R.L. Aggarwal, 1981b, J. Vac. Sci. & Technol. **19**, 589.
Chang, Y.C., and H. Chu, 1991, in: Condensed Systems of Low Dimensionality, NATO-ASI Series, Vol. B 253, ed. J.L. Beeby (Plenum, New York) p. 217.
Chang, Y.C., and J.N. Schulman, 1985, Phys. Rev. B **31**, 2069.
Chang, Y.C., J.N. Schulman, G. Bastard, Y. Guldner and M. Voos, 1985, Phys. Rev. B **31**, 2557.
Chomette, A., B. Deveaud, A. Regreny and G. Bastard, 1986, Phys. Rev. Lett. **57**, 1464.
Chomette, A., B. Lambert, B. Deveaud, F. Clrot, A. Regreny and G. Bastard, 1987, Europhys. Lett. **4**, 461.
Chu, H., and Y.C. Chang, 1987, Phys. Rev. B **36**, 2946.
Chu, H., and Y.C. Chang, 1989, Phys. Rev. B **39**, 10861.
Deveaud, B., J. Shah, T.C. Damen, B. Lambert and A. Regreny, 1987, Phys. Rev. Lett. **58**, 2582.
Deveaud, B., A. Chomette, F. Clrot, A. Regreny, J.C. Maan, R. Romestain, G. Bastard, H. Chu and Y.C. Chang, 1989, Phys. Rev. B **40**, 5802.
Dignam, M.M., and J.E. Sipe, 1990a, Phys. Rev. B **41**, 2865.
Dignam, M.M., and J.E. Sipe, 1990b, Phys. Rev. Lett. **64**, 1797.
Dignam, M.M., and J.E. Sipe, 1991, Phys. Rev. B **43**, 4097.
Dingle, R., 1975, in: Festkörperprobleme, Vol. XV, ed. H.J. Queisser (Pergamon/Vieweg, Braunschweig) p. 21.
Dingle, R., W. Wiegmann and C.H. Henry, 1974, Phys. Rev. Lett. **33**, 827.
Dingle, R., A.C. Gossard and W. Wiegmann, 1975, Phys. Rev. Lett. **34**, 1327.
Döhler, G.H., 1986, IEEE J. Quantum Electron. **QE-22**, 1682.
Döhler, G.H., 1987, CRC Crit. Rev. Solid State Sci. **13**, 97.
Döhler, G.H., 1988, SPIE 943 (Bellingham) p. 129.
Döhler, G.H., R. Tsu and L. Esaki, 1975, Solid State Commun. **17**, 317.
Esaki, L., and L.L. Chang, 1974, Phys. Rev. Lett. **33**, 495.
Esaki, L., and R. Tsu, 1970, IBM J. Res. & Dev. **14**, 61.
Esaki, L., L.L. Chang and E.E. Mendez, 1981, Jpn. Appl. Phys. Lett. **20**, L529.
Fasolino, A., and M. Altarelli, 1984, Surf. Sci. **142**, 322.
Fasolino, A., and E. Molinari, 1990, Surf. Sci. **228**, 112.
Ferreira, R., 1991, Phys. Rev. B **43**, 9336.
Ferreira, R., and G. Bastard, 1988, Phys. Rev. B **38**, 8406.
Ferreira, R., and G. Bastard, 1989a, Phys. Rev. B **40**, 1074.
Ferreira, R., and G. Bastard, 1989b, Europhys. Lett. **10**, 279.
Ferreira, R., B. Soucail, P. Voisin and G. Bastard, 1990, Phys. Rev. B **42**, 11404.
Gammon, D., B.V. Shanabrook, J.C. Ryan and D.C. Katzer, 1990, Phys. Rev. B **41**, 12311.
Gell, M.A., K.B. Wong, D. Ninno and M. Jaros, 1986, J. Phys. C **19**, 3821.
Guldner, Y., J.P. Vieren, P. Voisin, M. Voos, L.L. Chang and L. Esaki, 1980, Phys. Rev. Lett. **45**, 1719.
Guldner, Y., J. Manass, J.P. Vieren, M. Voos and J.P. Faurie, 1991, in: Condensed Systems of Low Dimensionality, NATO-ASI Series, Vol. B 253, ed. J.L. Beeby (Plenum, New York) p. 97.
Jain, J.K., and S. Das Sarma, 1988, Surf. Sci. **196**, 466.
Jaros, M., K.B. Wong and M.A. Gell, 1985, Phys. Rev. B **31**, 1205.
Jung, P.S., J.M. Jacob, J.J. Song, Y.C. Chang and C.W. Tu, 1989, Phys. Rev. B **40**, 6454.
Jusserand, B., and D. Paquet, 1986, in: Heterojunctions and Semiconductor Superlattices, eds G. Allan, G. Bastard, N. Boccara, M. Lannoo and M. Voos (Springer, Berlin).
Kane, E.O., 1957, J. Phys. & Chem. Solids **1**, 249.
Kazarinov, R.F., and R.A. Suris, 1972, Fiz. Tekh. Poluprovodn. **6**, 148 [Sov. Phys.-Semicond. **6**, 120].
Lambert, B., A. Chomette, B. Deveaud and A. Regreny, 1987, Semicond. Sci. & Technol. **2**, 705.
Laurich, B.K., K. Elcess, C.G. Fonstad, J.G. Berry, C. Mailhiot and D.L. Smith, 1989, Phys. Rev. Lett. **62**, 649.
Law, K.K., R.H. Yan, J.L. Merz and L.A. Coldren, 1990a, Appl. Phys. Lett. **56**, 1886.
Law, K.K., R.H. Yan, J.L. Merz and L.A. Coldren, 1990b, Appl. Phys. Lett. **57**, 1345.
Lederman, F.L., and J.D. Dow, 1976, Phys. Rev. B **13**, 1633.

Lugagne-Delpon, E., P. Voisin, J.P. Vieren, M. Voos, J.P. André and J.N. Patillon, 1992a, Semicond. Sci & Technol. **7**, 524.
Lugagne-Delpon, E., P. Voisin, M. Voos, J.P. André and J.N. Patillon, 1992b, Surf. Sci. **267**, 479.
Luttinger, J.M., 1956, Phys. Rev. **102**, 1030.
Maan, J.C., 1987a, in: Physics and Applications of Quantum Wells and Superlattices, NATO-ASI series, Vol. 170, eds E.E. Mendez and K. von Klitzing (Plenum, New York) p. 347.
Maan, J.C., 1987b, in: Festkörperprobleme, Advances in Solid State Physics, Vol. 27 (Pergamon/Vieweg, Braunschweig) p. 137.
Mailhiot, C., and D.L. Smith, 1988, Phys. Rev. B **37**, 10415.
Marzin, J.Y., M.N. Charasse and B. Sermage, 1985, Phys. Rev. B **31**, 8298.
Marzin, J.Y., J.M. Grard, P. Voisin and J.A. Brum, 1990, in: Semiconductors and Semimetals, Vol. 32, ed. T.P. Pearsall (Academic Press, New York).
Mendez, E.E., F. Agullo-Rueda and J.M. Hong, 1988, Phys. Rev. Lett. **60**, 1946.
Meynadier, M.H., R.E. Nahory, J.M. Worlock, M.C. Tamargo, J.L. de Miguel and M.D. Sturge, 1988, Phys. Rev. Lett. **60**, 1138.
Miller, R.C., A.C. Gossard, G.D. Sanders, Y.C. Chang and J.N. Schulman, 1985, Phys. Rev. B **32**, 8452.
Nenciu, G., 1991, Rev. Mod. Phys. **44**, 229.
Nicholas, R.J., N.J. Pulsford, G. Duggan, C.T. Foxon, K.J. Moore, C. Roberts and K. Woodbridge, 1991, in: Condensed Systems of Low Dimensionality, NATO-ASI Series, Vol. B 253, ed. J.L. Beeby (Plenum, New York) p. 201.
Ninno, D., K.B. Wong, M.A. Gell and M. Jaros, 1985, Phys. Rev. B **32**, 2700.
Olbright, G.R., T.E. Zipperian, J. Klem and G.R. Hadley, 1991, J. Opt. Soc. Am. B **8**, 346.
Osbourn, G.C., 1982, J. Appl. Phys. **53**, 1586.
Palmier, J.F., H. Le Person, C. Minot, A. Chomette, A. Regreny and D. Calecki, 1985, Superlatt. & Microstr. **1**, 67.
Pearsal, T.P., ed., 1990/1991, Semiconductors and Semimetals, Vols. 32 and 33 (Academic Press, New York).
Picraux, S.T., B.L. Boyle and J.Y. Tsao, 1991, in: Semiconductors and Semimetals, Vol. 33, ed. T. Pearsall (Academic Press, New York) p. 139.
Priester, C., 1991, J. Phys. III (France) **1**, 481.
Pulsford, N.J., R.J. Nicholas, P. Dawson, K.J. Moore, G. Duggan and C.T. Foxon, 1990, Surf. Sci. **228**, 62.
Quillec, M., L. Goldstein, G. Le Roux, J. Burgeat and J. Primot, 1984, J. Appl. Phys. **55**, 2904.
Sai-Halasz, G.A., R. Tsu and L. Esaki, 1977, Appl. Phys. Lett. **30**, 651.
Sai-Halasz, G.A., L.L. Chang, J.M. Welter, C.-A. Chang and L. Esaki, 1978, Solid State Commun. **27**, 935.
Sanders, G.D., and Y.C. Chang, 1985a, Phys. Rev. B **31**, 6892.
Sanders, G.D., and Y.C. Chang, 1985b, Phys. Rev. B **32**, 5517.
Scalbert, D., J. Cernagora, C. Benoit a la Guillaume, M. Maaref, F.F. Charfi and R. Planel, 1989, Solid State Commun. **70**, 945.
Schneider, H., H.T. Grahn and K. von Klitzing, 1990a, Surf. Sci. **228**, 362.
Schneider, H., H.T. Grahn, K. von Klitzing and K. Ploog, 1990b, Phys. Rev. Lett. **65**, 2720.
Schulman, J.N., and Y.C. Chang, 1981, Phys. Rev. B **24**, 4445.
Schulman, J.N., and Y.C. Chang, 1985, Phys. Rev. B **31**, 2056.
Schulman, J.N., and T.C. McGill, 1979, Appl. Phys. Lett. **34**, 663.
Segmuller, A., P. Krisna and L. Esaki, 1977, J. Appl. Crystallogr. **10**, 1.
Shanabrook, B.V., D. Gammon, R. Beresford, W.I. Wang, R.P. Leavitt and D.A. Broido, 1990, Superlatt. & Microstr. **7**, 363.
Sibille, A., J.F. Palmier, F. Mollot, H. Wang and J.C. Hesnault, 1989, Phys. Rev. B **39**, 6272.
Sibille, A., J.F. Palmier, A. Celeste, J.C. Portal and F. Mollot, 1990a, Europhys. Lett. **13**, 279.
Sibille, A., J.F. Palmier, H. Wang and F. Mollot, 1990b, Phys. Rev. Lett. **64**, 52.
Smith, D.L., and C. Mailhiot, 1986, Phys. Rev. B **33**, 8345, 8360.
Smith, D.L., and C. Mailhiot, 1987, Phys. Rev. Lett. **58**, 1264.
Smith, D.L., and C. Mailhiot, 1990, Rev. Mod. Phys. **62**, 173.

Soucail, B., 1990, Th.D. Université Paris 6 (unpublished).
Soucail, B., N. Dupuis, R. Ferreira, P. Voisin, A.P. Roth, D. Morris, K. Gibb and C. Lacelle, 1990a, Phys. Rev. B **41**, 8568.
Soucail, B., P. Voisin, M. Voos, D. Rondi, J. Nagle and B. de Crémoux, 1990b, Semicond. Sci. & Technol. **5**, 918.
Soucail, B., R. Ferreira, G. Bastard and P. Voisin, 1991, Europhys. Lett. **15**, 857.
Speriosu, V.S., 1981, J. Appl. Phys. **56**, 1591.
Tsu, R., and G.H. Döhler, 1975, Phys. Rev. B **12**, 680.
Voisin, P., 1983, Th.D. d'Etat, Orsay (unpublished).
Voisin, P., 1986a, in: Heterojunctions and Semiconductor Superlattices, eds G. Allan et al. (Springer, Berlin).
Voisin, P., 1986b, French Patent 86 16576.
Voisin, P., 1987, in: Optical Properties of Narrow Gap Low-Dimensional Structures, NATO-ASI series, Vol. B 152, eds C.M. Sotomayor-Torres et al. (Plenum, New York) p. 85.
Voisin, P., 1988, SPIE, Vol. 861 (Bellingham) p. 88.
Voisin, P., 1990a, Superlatt. & Microstr. **8**, 323.
Voisin, P., 1990b, 48th Dev. Res. Conf., Santa Barbara, CA (unpublished).
Voisin, P., G. Bastard, C.E.T. Gonzalves da Silva, M. Voos, L.L. Chang and L. Esaki, 1981, Solid State Commun. **39**, 79.
Voisin, P., G. Bastard and M. Voos, 1984a, Phys. Rev. B **29**, 935.
Voisin, P., C. Delalande, M. Voos, L.L. Chang, A. Segmuller, C.-A. Chang and L. Esaki, 1984b, Phys. Rev. B **30**, 2276.
Voisin, P., J. Bleuse, C. Bouche, S. Gaillard, C. Alibert and A. Regreny, 1988, Phys. Rev. Lett. **61**, 1639.
Wannier, G.H., 1962, Rev. Mod. Phys. **34**, 645.
Warburton, R.J., G.M. Sundaram, R.J. Nicholas, S.K. Haywood, G.J. Rees, N.J. Mason and P.J. Walker, 1990, Surf. Sci. **228**, 270.
White, S., and L.J. Sham, 1981, Phys. Rev. Lett. **47**, 879.
Whittaker, D.M., 1990, Phys. Rev. B **41**, 3238.
Whittaker, D.M., M.S. Skolnick, G.W. Smith and C.R. Whitehouse, 1990, Phys. Rev. B **42**, 3591.
Yan, R.H., R.J. Simes, H. Ribot, L.A. Coldren and A.C. Gossard, 1989, Appl. Phys. Lett. **54**, 1549.
Young, J.F., S. Charbonneau and P. Coleridge, 1991, in: Condensed Systems of Low Dimensionality, NATO-ASI Series, Vol. B 253, ed. J.L. Beeby (Plenum, New York) p. 181.
Zak, J., 1968, Phys. Rev. Lett. **20**, 1477.

CHAPTER 16

From Ballistic Transport to Localization

S. E. ULLOA

Department of Physics and Astronomy and
Condensed Matter and Surface Sciences Program
Ohio University, Athens, Ohio 45701-2979, USA

A. MACKINNON

Blackett Laboratory, Imperial College
London SW7 2BZ, UK

E. CASTAÑO

Departamento de Física
Universidad Autónoma Metropolitana-Iztapalapa
Apartado Postal 55-534
09340 México, DF, México

G. KIRCZENOW

Department of Physics
Simon Fraser University
Burnaby, British Columbia, Canada V5A 1S6

Handbook on Semiconductors
Completely Revised Edition
Edited by T.S. Moss
Volume 1, edited by P.T. Landsberg

Contents

1. Introduction . . . 866
2. System fabrication . . . 868
   2.1. Nanostructures . . . 869
3. Ballistic transport in two-dimensional systems . . . 875
   3.1. Magnetic focusing and steering . . . 876
   3.2. Electron beam refraction . . . 879
4. Ballistic transport in nanostructures . . . 880
   4.1. Quantized conductance in narrow constrictions . . . 881
      4.1.1. Experimental results . . . 881
      4.1.2. Theoretical description . . . 883
      4.1.3. Adiabatic transport . . . 886
   4.2. Multiple nanoconstrictions . . . 888
      4.2.1. Constrictions in series . . . 888
      4.2.2. Constrictions in parallel . . . 889
   4.3. Impurity and geometrical resonances . . . 892
      4.3.1. Single-impurity effects . . . 892
      4.3.2. Multiple-impurity effects . . . 895
      4.3.3. Geometrical resonances . . . 896
   4.4. Non-linear transport in constrictions . . . 899
   4.5. Magnetic field effects . . . 902
      4.5.1. Edge states . . . 903
   4.6. Quantum wire junctions . . . 906
      4.6.1. Classical considerations . . . 909
      4.6.2. Quantum effects . . . 911
         4.6.2.1. Simple junctions . . . 913
         4.6.2.2. Quantum collimation and scrambling in rounded junctions . . . 915
         4.6.2.3. Quantum dots at quantum wire junctions . . . 919
   4.7. Quantum dot systems . . . 921
      4.7.1. Single dots . . . 922
      4.7.2. Quantum dot chains . . . 925
         4.7.2.1. Non-linear response . . . 928
      4.7.3. Two-dimensional arrays . . . 928
5. Brief history of localization . . . 932
6. Experiments on localization . . . 934
   6.1. Hopping transport in amorphous semiconductors . . . 934
   6.2. Weak localization . . . 936
   6.3. Conductance fluctuations . . . 938

7. Basic concepts 939
    7.1. Models of disorder 939
    7.2. Properties of the electronic spectrum 940
    7.3. Quantities of interest 941
8. One-dimensional localization 943
9. Weak localization and quantum interference 946
    9.1. Quantum interference corrections to the average conductivity 946
    9.2. Negative magnetoresistance 947
    9.3. Oscillations of magnetoresistance 948
10. The scaling approach 949
    10.1 Single-parameter scaling 949
    10.2. Perturbation theory 950
    10.3. Field-theoretical formulation 951
    10.4. Results for numerical scaling calculations 952
11. Localization in a magnetic field 953
    11.1. Perturbational approach 953
    11.2. Thouless number study 954
    11.3. The percolation limit 954
    11.4. The participation number 954
    11.5. Numerical scaling 955
    11.6. Comparison with experiment 955
12. Fluctuations 956
    12.1. The statistics of transport in 1D disordered systems 956
    12.2. Fluctuations in the metallic limit 957
    12.3. Fluctuations and one-parameter scaling 959
References 959

1. Introduction

Remarkable advances in the growth of semiconductor materials and their postprocessing have opened new areas of research in condensed-matter systems, a recurrent theme in this volume. This is especially true of electronic transport. It is now possible to produce systems with a wide range of impurity and defect concentrations, ranging from the ultraclean nearly ideal electronic systems, to the other extreme of disordered structures where impurity and defect scattering play an important role. It is also possible to modify systems to have effectively different dimensionalities, such that these effects can be studied and compared with theoretical developments. This interplay between theory and experiment, together with the flexibility in achieving different system characteristics, has produced a variety of interesting results of fundamental importance in the advancement of the field.

For example, one can now routinely achieve extremely high low-temperature mobilities, giving rise to mean free paths (and coherence lengths, at low temperatures) exceeding 50 μm. This regime of nearly macroscopic coherence is in stark contrast to the situation in commonly available semiconductor materials, and typically encountered in metals, where the mean free paths are closer to the 100 Å length scale. These two extremes, as well as a variety of intermediate situations, have received a great deal of attention, partly because the phenomena exhibited by these structures are numerous and quite interesting, and also for their possible practical applications. In this chapter we review many of the phenomena exhibited by systems in different regimes. We also discuss the concepts introduced to provide an interpretation for the many and diverse observations. Our present understanding of transport in semiconductor systems has benefitted in no small part from the interplay and parallel development of experiments and theory, and our discussions reflect that. We place emphasis on the relationships between theoretical and experimental work, as we believe that this will aid our understanding of the material and should facilitate future advances. A brief discussion on the materials techniques that have allowed these areas to flourish will be given in § 2.

The very large mean free paths in ultraclean systems, nearly four orders of magnitude larger than in metallic ones, allow electrons to experience very few collisions with either impurities or lattice vibrations during transport. As a consequence, the carriers are able to sample macroscopic distances of thousands of lattice constants in a free-particle or *ballistic* fashion. This truly ballistic regime has been achieved only in recent years, but has already yielded a number of very interesting results, as we discuss in the first part of this chapter. Section 3 will describe the recent ballistic transport experiments in mostly two-dimensional (2D) systems. This section will review some of the exciting new results and fascinating promises for new physics and device fabrication in these systems. Section 4 then discusses the effects of imposing additional barriers and constrictions on the electronic motion, so that carriers are confined to quasi-one- or zero-dimensional regions (*nanostructures*). The quantization of the conductance (van Wees et al. 1988a, Wharam et al. 1988a), the quenching of the Hall effect (Roukes et al. 1987, Timp et al. 1988a, Ford et al. 1988a), the appearance of nonlocal resistances (Timp et al. 1988a, Takagaki et al. 1988a), and transmission resonance effects in these systems (McEuen et al. 1990) will be reviewed. These phenomena represent some of the most impressive discoveries in this area.

On the other hand, disordered systems have been studied within the framework of *localization*, a concept first discussed by P.W. Anderson in 1958 in connection with quantum mechanical diffusion (Anderson 1958). Disordered systems are characterized by large electron backscattering produced by impurities, reflecting quantum mechanical interference effects. Numerous studies have given rise to notable advances in understanding many of the consequences of impurity backscattering. Many theoretical techniques, some rather sophisticated – scaling and field-theoretical methods, for example – have been used in this field. We focus our attention on these fascinating topics in § 5 and following below.

Reviews emphasizing various developments in transport have already appeared (Harris et al. 1989, Reed 1990, Beaumont and Sotomayor-Torres 1990, Chamberlain et al. 1990, Beenakker and van Houten 1991a). More technical details of many of the developments in these fields can be found in the proceedings of several periodic conferences, such as Proc. Int. Conf. on Modulated Semiconductor Structures 1–5 (MSS1–MSS5), EP2DS (Proc. IV–IX Int. Conf. on Electronic Properties of Two-Dimensional Systems), and nanostructure physics and fabrication (NPF) (Reed and Wiley 1989, 1991).

In the area of disordered systems there is, in addition to a number of textbooks on disordered systems which may be used for introductory reading (Mott and Davis 1971, Zallen 1983, Ziman 1979, Bonch-Bruevich et al. 1984, Shklovskii and Efros 1984, Lifshitz et al. 1988, Cusack 1987), a (still increasing) number of review articles on various aspects (Thouless 1974, Elliott et al. 1974, Kramer and Weaire 1979). The problem of localization in one-dimensional random systems has been treated in several reviews (Ishii 1973, Abrikosov and Ryzkhin 1978, Erdös and Herndon 1982, Gogolin 1982). The localization problem was considered in the review by Lee and Ramakrishnan (1985) with special emphasis on the metallic limit. Efetov (1983) treated the field-theoretical aspects using the supersymmetric formulation. Experiments in the regime of *weak localization* have been reviewed by Bergmann (1984). Chakravarty and Schmid (1986) treated the limit of weak localization from the wave mechanical point of view. Vollhardt and Wölfle (1991) described the perturbational aspects. There is also a number of introductory summaries (Vollhardt 1987, MacKinnon 1988).

Conference volumes devoted to the topic of localization are also available (Friedman and Tunstall 1978, Nagaoka and Fukuyama 1982, Nagaoka 1985, Kramer et al. 1985, Weller and Ziesche 1984, Finlayson 1986, Ando and Fukuyama 1987, Kramer and Schön 1990, Benedict and Chalker 1991).

2. System fabrication

As already mentioned, it is due to advances in micro- (and nano-) fabrication techniques on semiconductor materials that the field of electronic transport has developed to its present stage, especially in the case of ballistic transport. This section discusses briefly the experimental techniques that allow the fabrication of these systems, and presents some recent promising suggestions and pioneering results for new systems.*

A basic prerequisite for the fabrication of many of the experimental systems is the availability of the nearly ideal semiconductor materials grown with techniques such as molecular beam epitaxy (MBE) or organometallic chemical vapor deposition (OMCVD) (Chang and Giessen 1985). These techniques produce crystals by depositing, on a given substrate, single monolayers of material at a time, with remarkable control of both the atomic species and the growth rates. Indeed, by abruptly changing the species being deposited it is possible to create atomically flat interfaces between different materials, with extremely low impurity and defect concentrations (Chang and Giessen 1985).

Moreover, the field of ballistic electronic transport is based on the availability of two-dimensional electron gas (2DEG) systems. These systems are usually achieved in one of two ways (although inventiveness gives rise to innumerable possibilities for combinations and new suggestions, as has been the case):

(a) the use of a gate to produce an inversion or accumulation layer as in a MISFET (metal–insulator–semiconductor field effect transistor), or
(b) the use of band discontinuities to confine carriers to the interface region between two different semiconductors in a heterojunction.

These structures, with the addition of extra carriers by the introduction of dopant species at appropriate locations, produce electron or hole gases that exhibit an effective 2D dynamics.

The prototypical MISFET is a layered structure consisting of silicon (the semiconductor), its oxide (the insulator) and a metallic gate with the pairing electrode being the heavily doped substrate on which the silicon layer is deposited. Application of voltages to the gate bends the semiconductor and insulator bands and consequently controls the carrier concentration at the semiconductor–insulator interface. For small voltages, the carrier concentration near the interface is only partially depleted, forming the so-called "depletion layer". However, for stronger gate voltages, the carriers near the interface region are of the opposite kind to those in the rest of the structure and an "inversion layer" is formed. These carriers are confined (for strong enough electric fields across the interface) to the vicinity of a 2D plane near the

*In what follows we refer to electronic systems only for brevity, although many of the concepts are equally applicable to holes, if in a more complicated fashion due to the band degeneracies typical of the valence bands of these materials.

interface, their degrees of freedom in the perpendicular direction frozen to only a few discrete quantum mechanical levels. These structures represent some of the first realizations of low-dimensional systems, giving rise to a large number of very interesting results (Ando et al. 1982).

However, the existence of extremely clean and nearly perfect *semiconductor heterojunctions* has given rise to many exciting results in ballistic transport. Out of the many possible materials, the system of choice has been the interface between GaAs and the alloy compound $Al_xGa_{1-x}As$, due to the high quality achieved in their growth. By alloying GaAs with Al for small concentration values, $x \leqslant 0.3$, the direct band gap of the material is increased with concentration relative to that of the parent GaAs (Chang and Giessen 1985), allowing a variety of "band engineering" concepts to be implemented (Esaki 1986). In particular, by producing a nearly flat interface where the Al alloy concentration is changed suddenly, it is possible to create a large conduction-band discontinuity, which has an effect similar to that of a potential step on the electronic motion in the direction perpendicular to the interface. This effective potential step ($\approx 50\%$ of the band gap), together with the Coulomb attraction of ionized donor impurities – placed on the alloy side of the interface and relatively far (≈ 50–500 Å) from it – creates a potential well for the electrons that limits their motion in the direction perpendicular to the interface. Typical potentials induce a strong quantization of the motion in the perpendicular direction, producing a two-dimensional electron system.

It is in such 2D structures that many interesting developments in condensed-matter physics have appeared, such as the integer and fractional quantum Hall effects (see Prange and Girvin (1990) and chapter 17 in this volume), carrier tunneling studies (see chapter 12 in this volume), hot-electron phenomena (see chapter 18 in this volume), and the ballistic and metallic transport phenomena discussed below.

2.1. *Nanostructures*

In addition to the existence of clean 2D systems, ballistic electronic transport has benefitted a great deal from the ability to produce small structural features that further confine electrons. Quasi-one-dimensional structures were pioneered early in the 1980s, obtaining interesting signatures of quantum confinement of carriers by comparing conductance and magnetoresistance observations (Wheeler et al. 1982, Skocpol et al. 1982, Proc. IV–IV Int. Conf. on Electronic Properties of Two-Dimensional Systems (see EP2DS4, 1982)), with theoretical predictions for one-dimensional (1D) localization (see § 8). More direct evidence of carrier confinement was, however, limited by the impurity and boundary scattering present in these systems. Refinement of the techniques used to confine carriers to regions comparable to their Fermi wavelength (typically on the order of 500 Å in these systems) and the availability of high-mobility samples has been critical to the development of this field. In particular, the production of so-called *point contacts* (nearly-point-like regions connecting two 2D regions of electrons) has allowed the unequivocal identification of quantum confinement effects, as will be reviewed in the following sections (Thornton et al.

1986, Berggren et al. 1986, van Wees et al. 1988a, Wharam et al. 1988a, van Houten et al. 1990).

Several techniques are used to achieve carrier confinement (see Roukes et al. (1990a) for a recent review), the more common methods being electron-beam lithography (Wheeler et al. 1982, Thornton et al. 1986), ion-beam damage (Scherer et al. 1987, Hirayama et al. 1988, Hirayama and Saku 1989, Wieck and Ploog 1990, Nakata et al. 1991), and material etching (Skocpol et al. 1982, Kirtley et al. 1986, van Houten et al. 1986, Grambow et al. 1989). These techniques have been used to produce lateral barriers to electronic motion, on both the macroscopic and microscopic length scales. This has allowed the fabrication of an assortment of interesting electronic devices, further discussed below, such as finite-size barriers, "lenses", "nozzles", "collectors", and point contacts.

The technique using focused ion beams induces damage in well-controlled regions by implantation of high-energy ions ($\approx$ 100 keV; e.g., Ga, Si, and Be), often followed by sample annealing. The electron mobility is drastically reduced in the regions implanted with a high dose, producing effective barriers for electronic motion. This technique is capable of producing arbitrary implantation patterns, being able to create thin wires, and point contacts with various geometries. It is also possible to utilize a metal gate to provide an additional control of the carrier density (Nakata et al. 1991, Roukes et al. 1990a).

A nanofabrication technique that has received much attention is electron-beam (and optical) lithography. This makes use of highly focused and stable electronic beams from a high-resolution electron microscope (or laser light beams) to "write" a desired pattern onto the surface of a photosensitive material deposited on the semiconductor sample. After chemical "development" of the pattern, the structure can then be used in various ways to produce confinement of the 2D electrons a few atomic layers below.

One of the pioneering uses of lithographic techniques to produce narrow wires was introduced by Scherer et al. (1987). These authors produce a patterned geometry on photoresist material, deposited over a 2D electron gas system in GaA–AlGaAs, as a mask for further processing with ion bombardment. By exposing the unprotected areas of the surface of their samples to low-energy (100 eV) Ar ions, Scherer et al. deplete the exposed regions of electrons. The unexposed regions then remain as conducting wires with widths as small as 750 Å, or even less. Roukes et al. have used these devices to study the transport properties of ultrasmall systems (Roukes et al. 1987), as discussed in more detail below (§ 4.6).

In a different approach, various etching, deposition and "lift-off" steps can be used after development of the pattern to deposit a thin metallic layer (typically Ti and Au). This layer is used as a top gate, deposited directly onto the surface of the semiconductor, so that application of voltages to it produces a strong electrostatic potential extending into the region of the 2D electrons (see fig. 1). By appropriate design of these top gates it is possible to induce a variety of confinement potentials and "reflector" surfaces that control the electronic states. Careful control of the gate geometries has been used to provide electrostatic steering of electronic beams within the 2D layer, and even to produce extreme quantum confinement when the feature

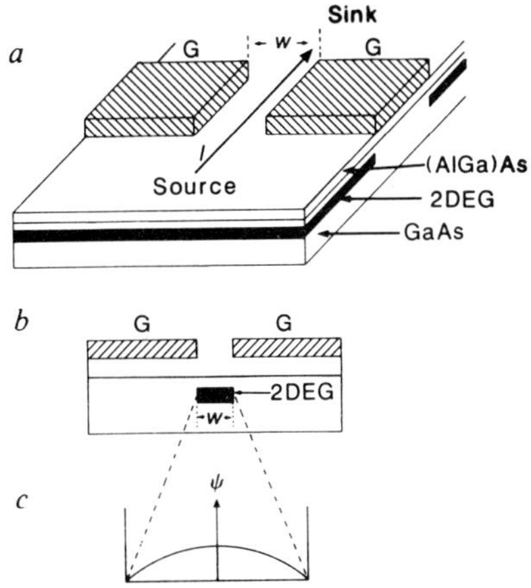

Fig. 1. (a) Schematic diagram of semiconductor heterostructure showing a 2DEG just below the interface between GaAs and (AlGa)As. A split metallic gate deposited on the surface is shown shaded (G). (b) Negatively biased gate depletes electrons underneath, defining a narrow channel of width w. Current I goes along channel from source to drain. (c) Electronic wavefunction of the lowest transverse mode ψ confined inside the channel. From Sheard and Eaves (1988).

sizes have appropriately small values. Indeed, it is now possible to produce features in the range of 5 to 10 nm, rapidly approaching the level of control possible in the growth direction of a few angstroms. Notice, however, that the size of the *electrostatically defined* confinement features in the 2D electron gas is affected by other parameters, such as the screening length, and not only by the geometrical sizes of the metallic layers used as gates. This imposes a natural limit on the small size and sharpness of the features one can achieve, which is many times larger than the one introduced by the resolution of the lithographic technique itself.

For example, using two narrow (1 or 2 μm) neighboring gates, separated by a small distance (≈ 0.25 μm), various groups have created *quantum point contacts* in semiconductors (van Wees et al. 1988a, Wharam et al. 1988a, van Houten et al. 1990). This "split-gate" technique, pioneered by Thornton et al. (1986) and Berggren et al. (1986), also offers the flexibility to vary the gate potential to achieve varying levels of electronic confinement. Since the metal gate can lie rather close to the electronic 2D interface (as close as 400 Å), without significantly degrading the carrier mobility, variations in the gate potential can produce a wide range of electronic confinement situations. Figure 1 shows a diagram of such an arrangement (from Sheard and Eaves (1988)). In this structure, when a sufficiently negative gate voltage is applied (typically -0.5 V), the electrons underneath and close to the gates are repelled, leaving only a narrow strip connecting the two large 2D electron gas regions. Further lowering of the gate voltage narrows the strip even more, producing a short *constriction*, which exhibits quasi-one-dimensional electron dynamics. As will be discussed later (§ 4), it is this extreme confinement that gives rise to the interesting quantization of the conductance and other effects.

Detailed calculations of the potentials for electronic motion in various situations are now possible. These calculations include the simultaneous, self-consistent solution of both the Schrödinger and Poisson equations with boundary conditions appropriate to each structure. Laux et al. (Laux and Stern 1986, Laux et al. 1988, Kumar et al. 1989, 1990), Davies (1988, 1989), Glazman and Larkin (1991), and Stopa (1992), have presented results for various split-gate, narrow-gate, and other confinement geometries in Si- and GaAs-based systems. These theoretical studies have provided valuable insights into the electronic properties of many nanostructures, and have been helpful in the design and interpretation of experiments. However, they have been based on Hartree or Thomas–Fermi treatments of the electron–electron interactions, and thus neglect exchange and correlation effects that are known to become increasingly important as the dimensionality is reduced. The possible importance of exchange effects for the electronic structure of quantum wires has been pointed out by Nakamura and Okiji (1991). Recently, Sun and Kirczenow (1992) have shown that in some models of quantum wires, the self-consistent Hartree approximation yields *qualitatively* incorrect ground states, and that a full treatment including exchange and correlation effects as in the density functional theory of Kohn and Sham (1965) is essential.

It is also possible to fabricate split-gate systems with smaller scale structures within the constriction, so as to induce additional potential barriers into the electronic path. Experimental realization of such theoretically proposed constriction geometries (Ulloa et al. 1990a,b, Castaño et al. 1990) has been marvelously achieved recently by Kouwenhoven et al. (1990a) and by Haug et al. (1990a). Figure 2 shows a micrograph of the device used by Haug et al. (1990a), and illustrates how the patterning of a "fingered" split gate induces a potential modulation producing interesting resonance effects (see § 4.7.2 below).

Other researchers have used narrow split-gate geometries but with somewhat larger characteristic openings, even 1 or 2 μm across, to produce "nozzles" which collimate an injected electronic current, very much as is done with light in optical systems. Collimated beams have been used to study focusing and steering of electrons by external magnetic and electrostatic fields (van Houten et al. 1989, Molenkamp et al. 1990, Spector et al. 1990a,b,c,d, 1991, Sivan et al. 1990), as well as lateral tunneling and transport of high-energy (hot) electrons (see chapter 18 in this volume, and Heiblum and Sivan (1990) and Ferry (1989) for reviews). Some of these interesting experiments will be further described in the next section. As an illustration of the structures used, fig. 3 shows a micrograph of an electrostatic "lens" fabricated by Spector et al. (1990a).

Another way to produce lateral carrier control in a 2D system is to etch away regions of material in a predetermined geometry. The techniques of shallow or deep mesa etching (typically using electron-beam lithography to design a protective mask with the desired geometry) have indeed produced a number of different structures. Prominent examples of these are the quantum wires and quantum dots used in recent experiments (Timp et al. 1987a, Grambow et al. 1989, 1990, Reed et al. 1988, 1990, Roukes et al. 1987). These techniques are highly developed, so that early problems with the large lateral carrier depletion, produced by the pinning of the Fermi level

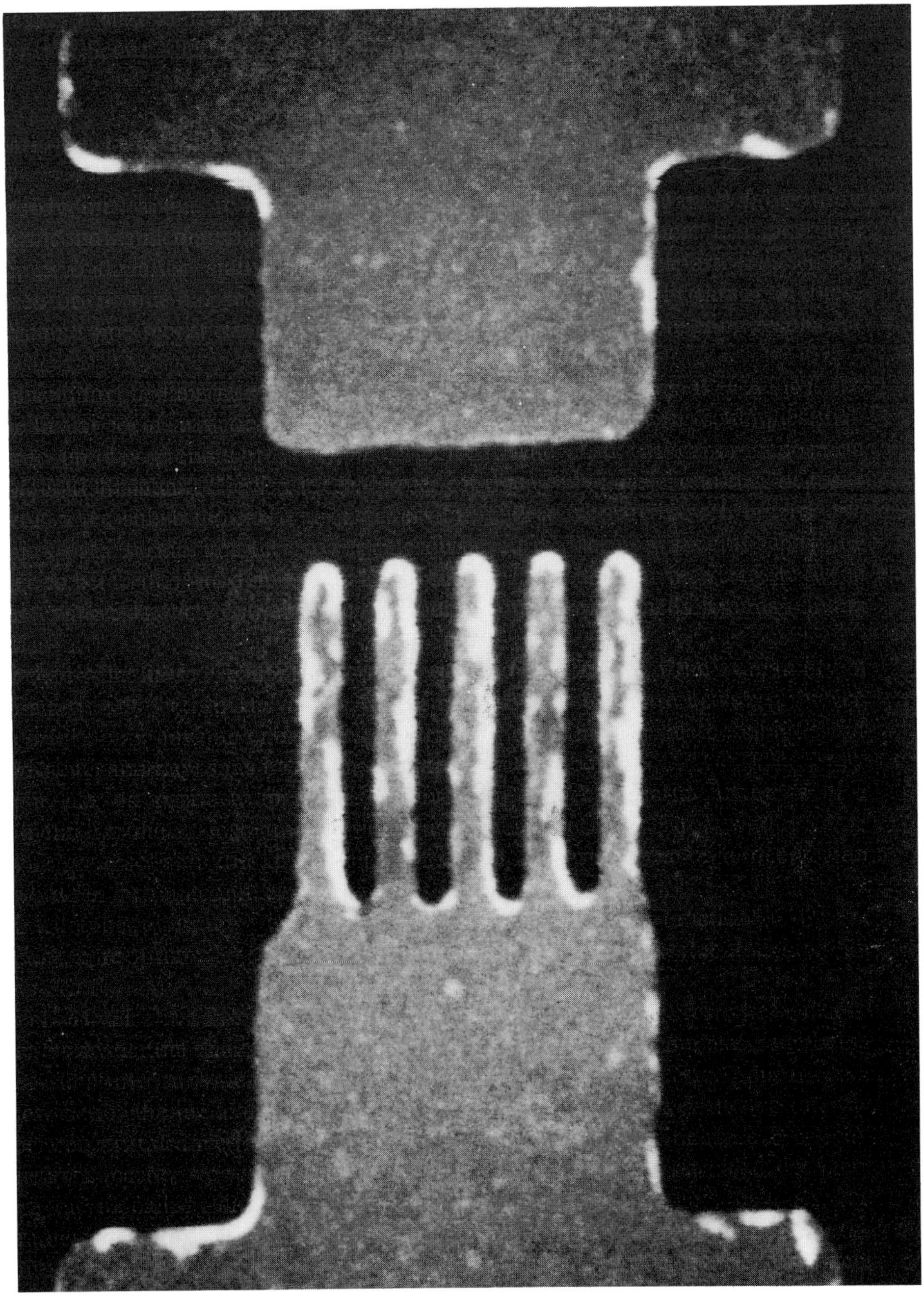

Fig. 2. Electron micrograph of the patterned gate used in the experiments of Haug et al. (1990a). The "fingers" on one side of the split gate are 50 nm wide (period of 100 nm). Potential modulations along the nanoconstriction have a strong effect on the conductance of the device.

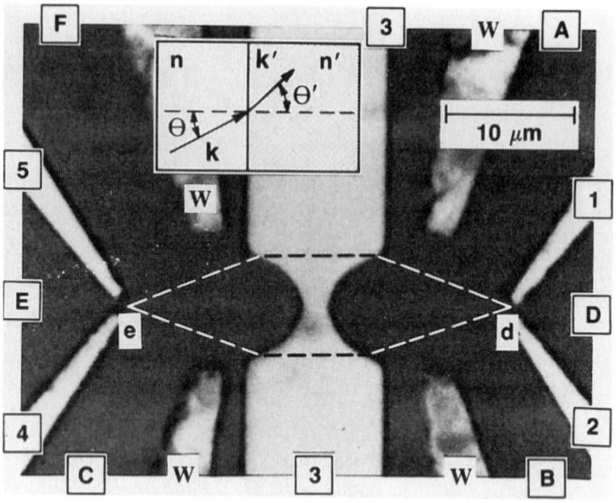

Fig. 3. Micrograph of sample used to demonstrate electrostatic focusing of electrons. Various ohmic contacts A–F, electron absorbers W, and emitter and detector orifices e and d are indicated. Inset shows the refraction of electrons at a boundary line between two regions of different electron densities, $n' < n$. From Spector et al. (1990a).

to defect states at the mesa surface, are now well understood and characterized (Randall et al. 1990). Indeed, fig. 4 shows an excellent illustration of this technique, where quantum dot columns are fabricated by etching a multi-heterojunction sample (Reed et al. 1990).

Before leaving this section on structure fabrication, we should call attention to a number of interesting recent proposals where the high controllability achieved in epitaxial growth techniques is used to produce sharp potential modulations in the electronic plane. Pioneering proposals have been put forth by Arakawa and Sakaki (1982) and some experimental realizations have been reported recently (Notzel et al. 1991, Christen et al. 1991), where growth on crystalline surfaces with different orientations produces quantum wires along V-shaped grooves. Also recently, the group of Petroff and co-workers (Tsuchiya et al. 1989) has reported the fabrication of "serpentine" superlattices where the alternate growth of heterojunction materials on a cleaved surface with large Miller indices produces narrow wires with wandering edges on a stepped surface. Finally, the first realization of a radically different scheme has been reported recently by Pfeiffer et al. (1990) and Störmer et al. (1991). There, a high-quality multilayer structure is cleaved on one of its edges and subjected to overgrowth on the lateral surface just exposed. The cleaving appears to introduce no significant damage, and the precursor multilayer structure determines an "atomically precise" potential modulation along the overgrowth interface. Similar results on laterally-cleaved multilayer systems have also been reported by Haug et al. (1992) and Chang et al. (1991).

A number of interesting alternate procedures for obtaining electronic systems with reduced dimensionalities have been proposed or tested, and new results can be expected. While a complete list is beyond the scope of this chapter, and the reader

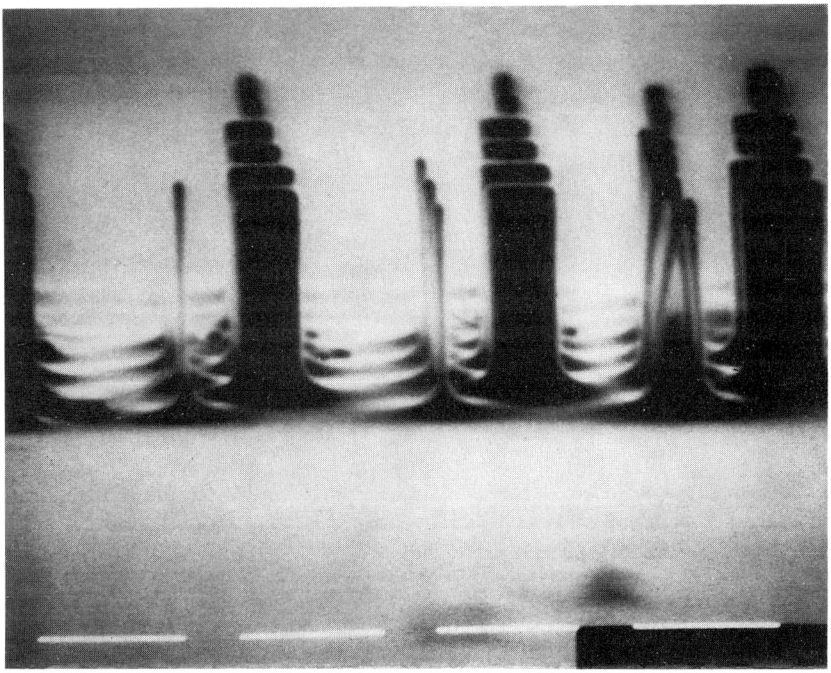

Fig. 4. Electron micrograph of an anisotropically etched heterostructure. The etching into the GaAs is over 1 μm deep. The narrowest quantum dot columns are about 300 Å wide. From Randall et al. (1990).

is referred to the cited reviews (Reed 1990, Beaumont and Sotomayor-Torres 1990, Chamberlain et al. 1990, Beenakker and van Houten 1991a, Proc. Int. Conf. on Modulated Semiconductor Structures 1–5 (MSS1–MSS5), Proc. IV–IX Int. Conf. on Electronic Properties of Two-Dimensional Systems, Reed and Wiley 1989, 1991), one could mention the use of intrinsic-strain effects by Kash et al. (1988, 1991), fabrication of lateral p–n junctions by Meirav et al. (1988), use of selective OMCVD growth to fabricate nanostructures (Yamada et al. 1989, Fukui et al. 1991, Takahashi et al. 1991), and the use of strong light beams by Prentiss et al. (1992) and Timp (1991) to influence the lateral growth of structures.

3. Ballistic transport in two-dimensional systems

Transport experiments at low temperatures on large systems with high mobilities, where the carriers are not restricted to move along nanometer structures but rather have a 2D character, have also yielded a number of interesting results recently. The technology has produced samples with low-temperature (≈ 1 K) mobilities reaching as high as several "million" ($\mu \approx 5\text{–}10 \times 10^6 \mathrm{cm}^2\,\mathrm{V}^{-1}\,\mathrm{s}^{-1}$), and with electronic densities of about $2 \times 10^{11}\,\mathrm{cm}^{-2}$. This is equivalent to Fermi wavelengths $\lambda_F \approx 560$ Å and elastic mean free paths $l_e = mv_F\mu/e \approx 50$ μm (where v_F is the Fermi velocity and m is

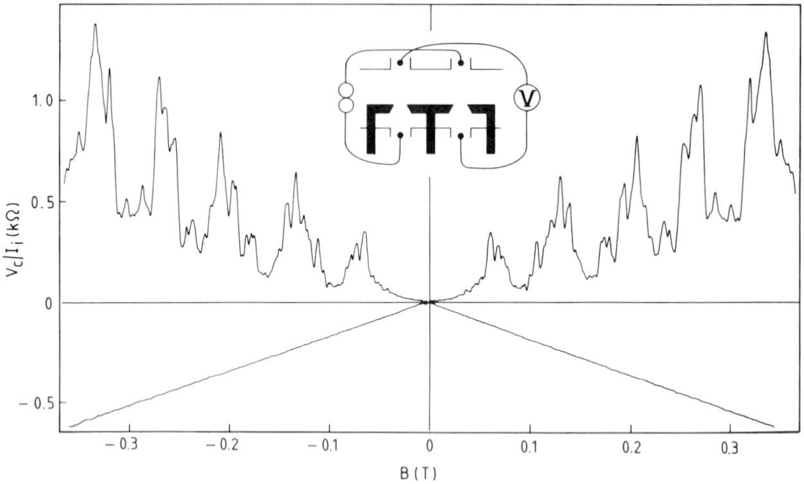

Fig. 5. Magnetic focusing peaks in generalized Hall resistance measurements at $T = 50$ mK. Current and voltage probes are shown in the inset. The two nearly symmetrical traces correspond to interchanged current and voltage leads, demonstrating the symmetry properties of the transport coefficients. Successive peaks correspond to higher numbers of electron reflections off the central region separating the constrictions (shown by thick lines in inset), as the cyclotron radius decreases with increasing magnetic field. From van Houten et al. (1989).

the electronic effective mass, $m = 0.07m_e$ in GaAs). With these strikingly large values of the elastic mean free path and at low temperatures – to quench phonon scattering events – experiments in the ballistic regime have been very successful. In the next two sections we discuss the "steering" and "focusing" of 2D electrons via magnetic and electric fields.

3.1. Magnetic focusing and steering

Early on, soon after the discovery of the quantization of the conductance in point contacts (see § 4.1), these devices were used as current injectors in experiments studying *magnetic focusing* effects. Work by van Houten et al. proved that ballistic transport over rather long distances was taking place in 2D systems (van Houten et al. 1988a, 1989). In their experiments, two nanoconstrictions, separated by 1 to 3 μm, were used as a current injector and collector pair in the presence of an external magnetic field (see inset in fig. 5). Figure 5 shows traces of the collector voltage versus magnetic field for a fixed split-gate voltage (fixing the constriction widths) (van Houten et al. 1989). The most striking feature of these curves is the set of nearly equidistant voltage peaks for increasing values of the magnetic field. Notice also the asymmetry in these curves, since there is absolutely no structure when the magnetic field is reversed. The overall envelope of this structure, observed for temperatures in the range of 0.05 to 7 K, can be explained classically as arising from the "steering" of the collimated electron beam by the external magnetic field. Indeed, for a magnetic field pointing into the plane of the inset in fig. 5, and for the current being injected by the point contact on the left, electrons would experience a Lorentz force forcing them to turn to their right. The peaks in fig. 5 then arise from multiple nearly-specular reflections by the central portion of the double-constriction arrangement.* The sample used to obtain this figure had a separation between openings of $L = 3$ μm, so that multiples of the orbit diameter for electrons at the Fermi energy occur for values of the magnetic field given by $B = 2\hbar k_F p/eL$, where $k_F = (2\pi n_s)^{1/2}$ is the Fermi wavenumber, $n_s = 3.5 \times 10^{11}$ cm^{-2} is the electron density, and $p - 1$ is the number of specular reflections needed to reach the right constriction (or current collector). From this relation, peaks in the collected current would be expected at multiples of $B_{focus} = 0.066$ T, in agreement with the observed structure. Notice that in this sample the low-temperature mobility is reported as $\mu = 0.9 \times 10^6$ cm^2 V^{-1} s^{-1}, giving an elastic mean free path of $l_e \approx 9$ μm. The close agreement with the expected values of B_{focus} shows that indeed electrons move ballistically over regions comparable to the elastic mean free path, and furthermore that the reflection off the potential barrier induced by the middle portion of the gate is nearly specular. The reader will also notice small-scale structure in the focusing peaks of fig. 5. This structure was interpreted as being due to quantum interference effects between the modes propagating in the system, especially in the limit of narrow constrictions, where the injector would be expected to behave as a spatially-coherent point source. The expected effects of impurity and phonon scattering events, which tend to degrade the more subtle quantum interference, are still not as well understood, however.

A Philips–Delft-University collaboration has also studied the effects of beam steering and collimation by quantum point contacts in more detail, using pairs of constrictions *across* a channel a few microns wide (see inset of fig. 6) (Beenakker and van Houten 1989a, Molenkamp et al. 1990). Having two constrictions opposite to each other, to be used as the injector/collector pair, allows the exploration of electronic motion without intervening collisions with the sample boundaries (other than those in the constriction itself). Application of magnetic fields perpendicular to the 2D electron gas steers the beam off the "line of sight", decreasing the voltage across the channel (see fig. 6). Notice that the curve is symmetric under field reversal, as expected from the Onsager relations (Büttiker 1986). Figure 6 also shows how a classical simulation of the current, which includes some flaring of the opening to account for the collimation of the electronic beam (Beenakker and van Houten 1989a), agrees well with the measurements. Here again, the fine structure in the experimental trace was attributed to quantum interference effects in combination with remnant impurity scattering. The group of Taylor et al. (1991a) at NRC has also reported detailed studies of the magnetic steering of electronic beams using two point contacts, and show good agreement with the transmission probabilities calculated in the classical trajectory method of Beenakker and van Houten (1989a). The results of Taylor et al. also provide an interesting demonstration of the reciprocity

*The degree of specularity of reflections has been discussed by Thornton et al. (1989) for samples produced by both ion-damage and split-gate techniques.

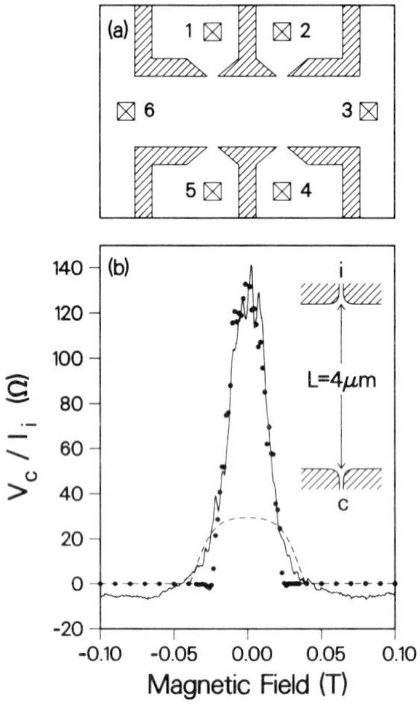

Fig. 6. (a) Schematic diagram of a device used to demonstrate magnetic focusing and collimation with opposing constrictions. Dashed lines indicate gates, the squares are ohmic contacts to the 2DEG. (b) Plots of generalized magnetoresistance $V_c/I_i = R_{16,53}$ plotted against magnetic field, where I_i is injected between contacts 1 and 6, and voltage V_c on contact 5 is measured with respect to 3. The solid line is the experimental curve. The black dots are the results of classical electron trajectory calculations for a smooth widening of the constriction opening. The dashed-line results assume openings with sharp corners (and therefore no collimation). From Molenkamp et al. (1990).

under magnetic-field reversal, in their purposely asymmetric arrangement (Büttiker 1986).

The importance of beam collimation by a narrow constriction is further illustrated below in the discussion of the measurements in constrictions in series and in other multiprobe geometries with even smaller sizes (see §§ 4.2 and 4.6). The concept of beam collimation has also been extended to the study of electronic emission by clean metal tips (Tekman et al. 1990, García et al. 1989, Sáenz et al. 1990).

All the experiments discussed above show that ballistic electrons in 2D systems can be deflected magnetically over distances of a few microns. On the other hand, the measurements of Spector et al. in samples with some of the highest mobilities reported to date, illustrate the ballistic transport of carriers over nearly macroscopic distances (≈ 100 μm) (Spector et al. 1990a,b,c,d). These systems are characterized by high mobilities and a relatively large separation between the top gate and the 2D interface (≈ 5000 Å), in addition to larger-scale gate features (1 or 2 μm gate openings, for example). As a consequence, the injector and collector constrictions are purposely not in the quantum regime (i.e. $\lambda_F \ll W$, where λ_F is the Fermi wavelength and W is the constriction width). In some of their first experiments, ballistic transport beyond 100 μm is reported in samples with a low-temperature mobility of about

$5 \times 10^6\ \mathrm{cm^2\,V^{-1}\,s^{-1}}$ (and elastic transport mean free path $l_e \approx 28\ \mu$m) (Spector et al. 1990d). Their multiprobe arrangement is used to demonstrate magnetic focusing features like those of van Houten et al. (1988a, 1989), but with injector–collector separations L of up to 64 μm. Moreover, a fit of the height of the focusing peaks to a decaying exponential function of the separation L allows the determination of an "effective" decay length of 15 μm, smaller than l_e, perhaps due to a different effective angular averaging of the impurity scattering events. Spector et al. also describe the effectiveness of using ohmic contacts as point sources and detectors instead of the split-gate constrictions, as well as using them as electron absorbers and randomizers (Spector et al. 1990c). Similarly, Hirayama et al. have reported measurements in large Hall-bar geometries where classical electron focusing effects over large distances are predominant (see § 4.6.1) (Hirayama et al. 1991a,b,c).

3.2. *Electron beam refraction*

One of the more tantalizing experiments in 2D transport is the demonstration of the "refraction" of ballistic electrons (Spector et al. 1990a,b, Sivan et al. 1990). Spector et al. (1990a) showed that electrons propagating through the region under the gates shown in fig. 3, experience a deflection similar to a refracted light beam. The role of relative refractive index in this configuration is played by the ratio of the electronic densities both outside and under the gate, so that electrons are deflected according to the relation

$$\frac{\sin\theta}{\sin\theta'} = \left(\frac{n_s'}{n_s}\right)^{1/2}, \tag{1}$$

where n_s (n_s') is the electronic density in the region surrounding (under) the gate, which in turn defines the local Fermi energy ($E_F \propto k_F^2 \propto n_s$, in a simple 2D gas). Equation (1) is obtained from conservation of momentum parallel to the interface separating the two density regions (Spector et al. 1990a). Notice that since lowering the voltage reduces n_s', the region under the gates has a relative index of refraction $(n_s'/n_s) \leqslant 1$, and a *converging* lens is actually concave, as shown in fig. 3. A sample experimental trace is shown in fig. 7, where the ratio of currents in the collector (I_d) and the emitter (I_e) is shown as a function of gate bias (or n_s'/n_s). For n_s' (denoted as n' on the top axis) approaching zero, the bias has depleted the carriers under the gate and the incoming beam is totally reflected, so that I_d is nearly zero. As the bias increases, some electrons traverse the "lens" region but miss the collector d since the focal point lies far in front of the constriction. As the focal point moves closer, the collected current reaches a maximum (at $(n_s'/n_s)^{1/2} \approx 0.87$), before decaying again as the focal point moves behind d. This behavior agrees well with ray-tracing calculations of the electronic paths, as shown by the dashed curve in fig. 7, although some discrepancies in the peak widths and backgrounds may be due to residual impurity scatterings. Sivan et al. (1990) have also demonstrated electron focusing using electrostatic lenses, and Spector et al. have further demonstrated the operation of a "prism" (Spector et al. 1990b).

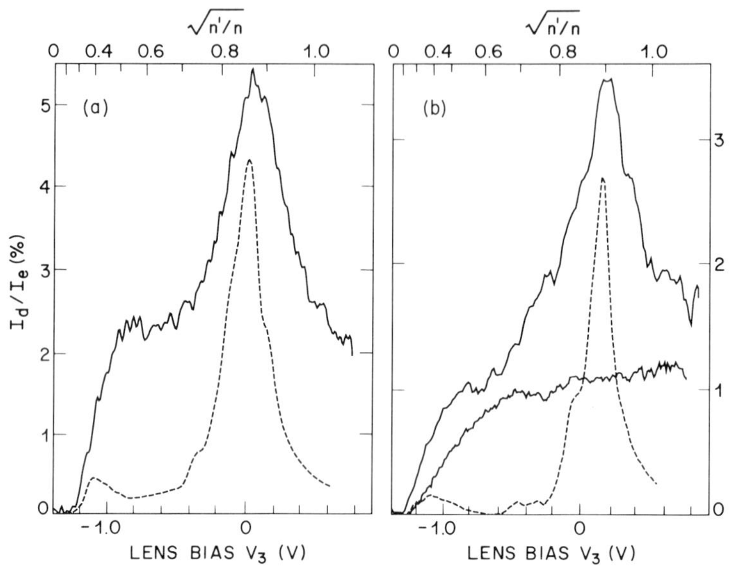

Fig. 7. Electrostatic focusing data, as shown by the upper solid traces, for a "lens" device such as that of fig. 3. Panels (a) and (b) are the results for different geometries. As the bias is changed, the density beneath the lens (denoted by n') changes and the fraction of electrons through the drain I_d/I_e varies non-monotonically. This fraction reaches a maximum when the focal point of the lens falls near the drain probe. From Spector et al. (1990a).

Other analogs of optical experiments performed with electrons are sure to follow soon. Smith et al. (1989c, 1990a) have demonstrated the operation of an electronic Fabry–Pérot interferometer, by constructing two reflectors separated by a distance of about 1 μm. Avishai et al. have analyzed this configuration theoretically (Avishai et al. 1990), and show general agreement with experiments. Spector et al. have also recently reported the penetration of mutually intersecting electronic beams with negligible interactions (Spector et al. 1991). While these results are so far limited to low currents (≈ 1 μA), they are interesting electronic analogs of the penetration of optical beams.

Other interesting transport experiments in 2D systems include "hot" electron (energies higher than the local Fermi energy) focusing and tunneling through barriers (Heiblum and Sivan 1990, Sivan et al. 1989a, Williamson et al. 1990a,b), where the possibilities of studying phonon scattering events (Heiblum and Sivan 1990), and determining the local potential drops in a conductor (Williamson et al. 1990a,b, Büttiker 1989), have sparked a lot of activity. A detailed description of hot-electron phenomena is also found in chapter 18 in this volume.

4. Ballistic transport in nanostructures

The experiments of the previous section give an idea of the high quality achieved in the fabrication of 2D electron gas systems. It is now the turn of microlithography.

The various phenomena discussed in this section highlight advances in the lithographic techniques that have produced nanostructures with very small characteristic length scales. The combination of various lithographic techniques and advanced sample fabrication has given rise to a number of very interesting results, discussed in what follows.

One of the more striking observations in this field, and one which received a great deal of attention (Sheard and Eaves 1988, Khurana 1988), is the quantization of conductance in narrow constrictions (van Wees et al. 1988a, Wharam et al. 1988a). Using split-gate geometries – first studied by Thornton et al. (1986) and Berggren et al. (1986) – van Wees et al. (1988a) and Wharam et al. (1988a) discovered independently that the conductance shows a series of plateaus at integer multiples of the conductance constant $2e^2/h$, as a function of the gate voltage. Using the geometry shown in fig. 1, they were able to study the electronic motion in the quantum regime (such that $\lambda_F \approx W$, where W is the width of the constriction and λ_F is the Fermi wavelength). As we shall see below, it is this extreme ballistic quantum confinement that forces the quantization of the conductance and gives rise to a number of other fascinating consequences. It is indeed the availability of narrow channels in the quantum regime that has allowed many of the observations discussed below. Our discussion is grouped according to the physical elements of the systems studied and the concepts needed in the description of the experiments.

4.1. Quantized conductance in narrow constrictions

4.1.1. Experimental results

The observations of van Wees et al. and Wharam et al. were performed in samples clearly in the ballistic regime of the unpatterned parent 2D electron gases (long mean free paths, $l_e \approx 10$ μm, and low temperatures, $T \leqslant 1$ K), and in narrow constrictions (or "point contacts") created with split gates producing quasi-one-dimensional channels with characteristic widths and lengths $\leqslant 3000$ Å $\ll l_e$. Experimental traces showing the quantization of the conductance as the gate bias is varied (which in turn changes the constriction width), are shown in figs. 8 and 9. Both figures show a plot of the resistance after subtracting a background resistance, associated with residual impurity scattering in the wide 2D regions, in series with the constriction. Figure 8 shows the constriction resistance, while fig. 9 shows the reciprocal of this quantity (after subtracting an additional resistance appearing at the theshold bias, $V_{gate} \approx -0.5$ V in both cases, the voltage at which the electrons are depleted from under the gates). Figure 8 also shows that a change in the Fermi energy induced by illuminating the sample changes the quantization index at a given gate bias.

The appearance of plateaus in figs. 8 and 9 is clearly reminiscent of the quantum Hall effect (Prange and Girvin 1990), although these observations are made in the *absence* of magnetic fields. However, the original observations and subsequent work have shown that the resistance values at the plateaus are quantized only to a much lower accuracy ($\leqslant 1\%$ (Timp et al. 1989a, Timp 1990)) than is typically achieved in the quantum Hall effect (which is better than $10^{-5}\%$ (Prange and Girvin 1990,

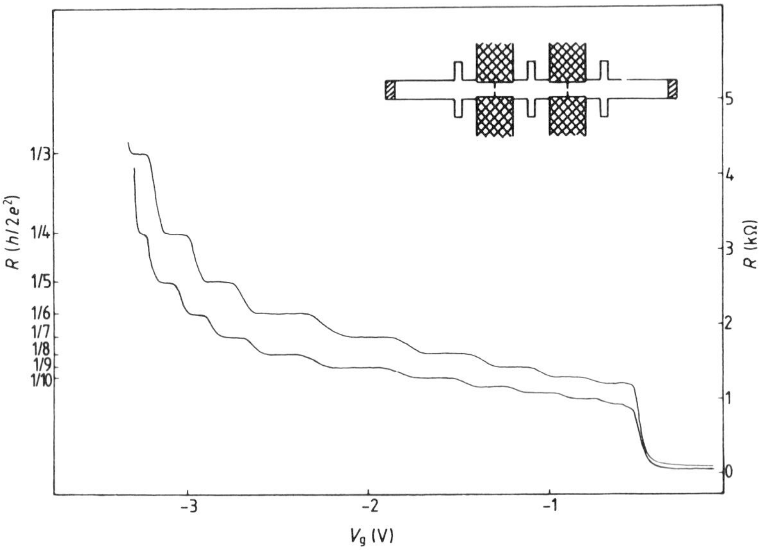

Fig. 8. Resistance of a single-nanoconstriction device at $T = 0.1$ K plotted against voltage V_g. The two curves have different carrier concentrations induced by sample illumination. Notice the quantized resistance values in units of $h/2e^2$. The inset shows a diagram of the device with two split gates, one of which is used in the experiment. The split gate is 0.5 μm wide and 0.4 μm long. From Wharam et al. (1988a).

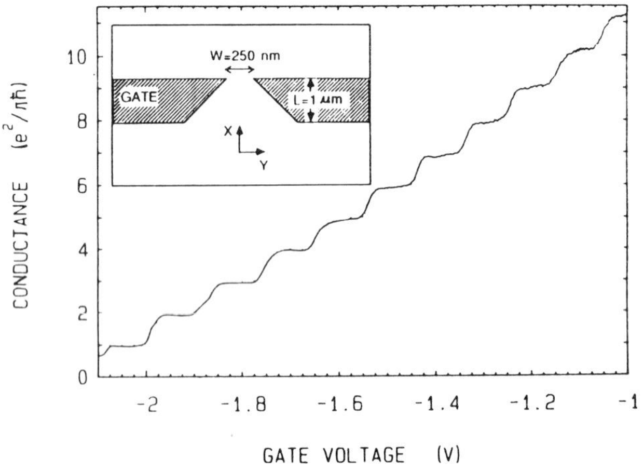

Fig. 9. Conductance of nanoconstriction created by split gate against gate voltage, at $T = 0.6$ K. The conductance (after subtraction of the lead resistance) shows clear plateaus quantized at multiples of $2e^2/h$. Inset: diagram of device with split gate shown as dashed region. From van Wees et al. (1988a).

Hartland et al. 1991, 1992, Taylor and Witt 1989)), and are sample dependent. It is also observed that the plateaus deteriorate gradually with increasing temperature, acquiring a non-zero slope, and that at moderately high values ($T \approx 5$ K) the conductance curves show a monotonic behavior with gate bias (van Wees et al. 1991, van

Houten et al. 1990). Similarly, at lower temperatures ($\approx$100 mK), the conductance curves also deteriorate, developing impurity-related fluctuations. They also often exhibit strong time-dependent quantum "telegraphic noise" due to movement of electrons in traps surrounding the constriction, which alters the effective constriction potential geometry (Timp et al. 1990a, Cobden et al. 1991, Li et al. 1990, Dekker et al. 1991, Ford et al. 1991, Taylor et al. 1991b). See § 4.3 for a discussion of these and other impurity effects.

4.1.2. Theoretical description

From the description of the experimental results above it is evident that the regime of ballistic transport is required to observe the quantization of the conductance. This suggests that a theoretical explanation would not require the consideration of impurity scattering events to understand this phenomenon. In this situation it is the backscattering of electrons at the constriction openings that is responsible for the observed resistance. A simple classical explanation in terms of the particle flux through the constriction would be expected to yield a current I proportional to both the source-to-drain voltage V_{sd} and the width of the constriction W, since increasing either one would give rise to a larger flux. Consequently, the conductance of the system would be given by $G = I/V_{sd} \propto W$, where the proportionality factors involve the Fermi wavenumber and natural constants. The resulting expression would be the 2D analog of the conductance of three-dimensional metallic point contacts discussed by Sharvin and co-workers (Sharvin 1965, Sharvin and Bogatina 1969) and van Houten et al. (1990). It is clear, however, that this simple linear dependence on constriction width (or the associated featureless gate voltage dependence) is not observed.

On the other hand, since the constriction width is comparable to the characteristic electronic wavelength ($\lambda_F \approx W$), it is to be expected that quantum mechanical considerations should play an important role. Indeed, the electronic states within the constriction would be characterized by a wavefunction vanishing (or nearly so, depending on the strength and type of the lateral confinement potentials) at the side walls of the constriction (see fig. 1c). The eigenstates would have a quasi-one-dimensional character, with a free-particle dispersion in the longitudinal direction (in the effective-mass approximation), superimposed on the discrete transverse energy levels E_p resulting from the lateral confinement. These electronic states are qualitatively similar to the propagating modes in electromagnetic waveguides, and are called "transverse modes" (or subbands). In the simplest approximation the current through the constriction is calculated by adding the contributions from all the propagating transverse modes (with the subband minimum below the Fermi level E_F), so that

$$I = \sum_{p=1}^{N} I_p = \sum_{p}^{N} e v_p \delta\sigma, \tag{2}$$

where N is the largest mode number $p \geqslant 1$ obeying $E_p \leqslant E_F$, v_p is the mode velocity, and $\delta\sigma$ is the traveling charge imbalance due to V_{sd}. The charge imbalance can be expressed in terms of the density of states g_p by $\delta\sigma = eV_{sd}g_p$. Furthermore, since the density of states for each 1D transverse mode is given simply by the slope of the dispersion relation, $1/2\pi\hbar v_p$, one obtains

$$G = \frac{I}{V_{sd}} = \frac{2e^2}{h} N, \tag{3}$$

where the factor of two arises from spin degeneracy. This simple expression would seem to explain the quantization: plateaus arise in the conductance curve every time that a new subband falls below the Fermi energy by either widening the constriction or by raising the Fermi energy at a fixed gate voltage.

Notice that this expression can be generalized to the case in which each mode is not transmitted totally, but with probability T_p (summed over exit channels), so that the resulting conductance is given by

$$G = \frac{2e^2}{h} \sum_{p=1}^{N} T_p. \tag{4}$$

In terms of the **t**-matrix, which denotes the probability amplitude of scattering mode q into mode p by the matrix element t_{pq}, this equation can also be written as

$$G = \frac{2e^2}{h} \operatorname{Tr} \mathbf{t}\mathbf{t}^{\dagger}, \tag{5}$$

where Tr denotes the trace of the matrix. This expression is the Fisher and Lee (1981) multichannel version of the Landauer formulas (Landauer 1970, 1989a), first obtained for 1D systems by Economou and Soukoulis (1981a,b) and interpreted in terms of contact resistances by Imry (1986a) and Büttiker (1986). These formulas have been further generalized by Büttiker (1986) to the multiprobe multichannel case, as discussed in § 4.6.2. The degree to which various Landauer formulas are consistent with the nonlocal character of the quantum mechanics of coherent electron states has been examined by Castaño and Kirczenow (1989).* Further use of this formalism in other situations has been discussed by van Wees (1991a) and Sols (1991).

Notice however, that the simple picture described by eq. (3) would appear to be valid only in the limit of long constriction channels, since it is then that only the propagating transverse modes would contribute to the current. The experimental systems typically have comparable lengths and widths, and the effect of evanescent states tunneling from one end of the constriction to the other would seem to suggest a deviation from the quantization. In other words, the process of injection of carriers from the 2D gas into and out of the constriction is not addressed by the simple argument above.

Soon after the reported observations, several groups developed detailed theoretical models addressing these and related questions. These papers solve the quantum mechanical problem of carrier injection into a constriction, typically modeled by either a hard-wall square-well potential or a harmonic oscillator lateral potential, by expanding the wavefunction into waveguide modes (Kirczenow 1988a, 1989a,b, Szafer and Stone 1989, Escapa and García 1989, García and Escapa 1989, Tekman and

*The various versions of the Landauer formulas have in the past been controversial; see for example Landauer (1989a,b) and Stone and Szafer (1988).

Ciraci 1989a, Avishai and Band 1989a), by various tight-binding techniques (Haanappel and van der Marel 1989, van der Marel and Haanappel 1989a), and even by analytic solutions of special geometries of the confinement potentials (Kawabata 1989, Levinson 1988, Matulis and Segzda 1989, Büttiker 1990a). Although the details of the solution of the Schrödinger equation vary, the conclusions are similar. Thermodynamic potential arguments and their relevance to conduction quantization have also been discussed by Widom and Tao (1988) and van der Marel (1991).

In particular, it was shown in these works that evanescent states tend to weakly erode the quantization in short constrictions, but even a constriction with an aspect ratio of unity (length ≈ width) exhibits clearly-developed plateaus (see fig. 10, curve b). These calculations also predict that at low temperatures the conductance curves can exhibit oscillations as a function of the gate bias (or Fermi energy), superimposed on the plateaus expected from the subband quantization arguments of eq. (3). These oscillations, corresponding to longitudinal resonances arising from partial reflections of the wavefunction at the constriction openings, are stronger for sharply defined constriction orifices (as the "impedance mismatch" is larger). Figure 10 shows these "organ pipe" resonances for constrictions with sharp corners and various lengths.

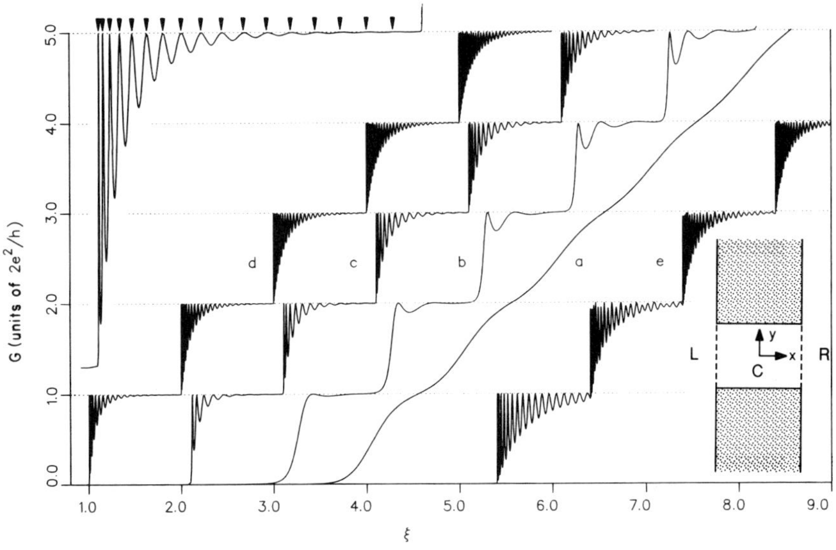

Fig. 10. Calculated conductance curves of short ballistic constriction with sharp openings. See inset lower right: shaded areas are regions inaccessible to electrons. Curves shown at $T = 0$ against $\xi = (\hat{E}_F - \hat{U})^{1/2}$, where $\hat{E}_F = E_F/\Delta$, $\hat{U} = U/\Delta$ is a potential step on entering the constriction, and $\Delta = \hbar^2\pi^2/2mW^2$ is the energy of lowest transverse level in the constriction. The Fermi level varies while keeping W, U, and l (the length of the constriction) fixed. Curves a–d are for $U = 0$ and $l/W = 0, 1, 5, 10$, respectively. Curve e is for $\hat{U} = 2.5$ and $l/W = 10$. Notice that the resonances become more numerous for longer channels. Top left inset: blow-up of first step in curve d. Markers at top indicate values of ξ such that $i\lambda_F/2 = l\gamma$, where i is an integer (= 1, 2, ...), and γ (= 1.044 in this case) is a fitting factor close to unity that accounts for end effects. From Kirczenow (1989b).

The resonances occur when an integral multiple of half the Fermi wavelength coincides with the length of the constriction (corrected for end effects by a factor close to unity), as indicated by the markers in the top left inset of fig. 10. It is evident that the resonances are more pronounced for a longer channel, as the effect of evanescent modes decreases, and are more numerous, as the resonance condition can be satisfied more times. Notice that flaring of the openings of the nanoconstriction, as well as the potential rise present inside some structures near pinch-off, tends to reduce the reflections at the channel ends, causing a deterioration of the resonance condition (Kirczenow 1988a, 1989a,b, Szafer and Stone 1989, Escapa and García 1989, Tekman and Ciraci 1989a, Shikin 1989). The effects of flaring or tapering of the constriction ends are discussed further in § 4.1.3.

Unequivocal experimental identification of these "organ pipe" oscillations has proved difficult, due to electrostatic rounding of the constriction orifices, and the effects of remnant impurity scattering. However, several tentative reports have appeared. Prominent among these are those of Hirayama et al. in Ga-implanted samples, in which it may be easier to produce sharp-constriction features (Hirayama et al. 1989a,b), and those of van Wees et al. (1991), Brown et al. (1989a), and Martín-Moreno and Smith (1989), in split-gate structures. Notice that finite-temperature broadening of the Fermi surface would tend to smooth out the fine structure shown in fig. 10, making it disappear when kT exceeds the energy width and separation of the resonances.

It is interesting to mention again the close analogy between the transport of electrons in the ballistic regime and the propagation of waves. Notice that the arguments for the conductance formula described above would be equally valid for the propagation of classical waves through narrow openings. This point has been discussed in detail by van Houten and Beenakker (1990). Indeed, realization of an optical analog has been reported recently by Montie et al. (1991). They observe that the total transmitted power of *diffuse light* (non-plane wave incidence) through a narrow flat slit of variable width increases in a sequence of plateaus as the width of the slit increases. It is natural to conclude that similar effects would be observed in the case of any wave propagation process through a constriction. Also, Beenakker and van Houten (1991b) have presented arguments for the observation of quantized critical Josephson currents through a channel shorter than the coherence length, as a *superconducting* analog of the quantized conductance phenomenon in normal constrictions.

As an example of a reverse analogy (where an optical device serves as the model for one with electrons) Eugster and del Alamo have proposed a quantum field-effect "directional coupler" (del Alamo and Eugster 1990, Eugster and del Alamo 1991). Using two electronic waveguides which come into close proximity over a certain length – such as to allow electron tunneling between them – they propose the electronic equivalent of an optical switch.

4.1.3. Adiabatic transport

The problem of carrier injection into the constriction has also been treated theoretically in the so-called *adiabatic* regime by several authors (Glazman et al. 1988, Imry

1989, Payne 1989, Yacoby and Imry 1990, Yosefin and Kaveh 1991, Büttiker 1990a). These authors consider the potential defining the constriction to be a smoothly-varying function of position along the constriction. This resembles the experimental situation in the region near the constriction opening, where electronic screening is expected to heal the sharp features in a split-gate or ion-damage structure, for example. Adiabaticity allows the definition of local well-defined transverse modes which evolve *without mixing* as the constriction width, $W(x)$, narrows from the ends (see fig. 11). The existence of these non-mixing local modes would clearly guarantee the quantization of the conductance, according to eq. (3), where N would be given by the number of transmitting modes *in the narrowest* region of the constriction, W_0. Departures from quantization due to mode-mixing corrections were shown to be exponentially small in the ratio of the radius of curvature of the constriction to its smallest width (Glazman et al. 1988, Yacoby and Imry 1990).

This idea of complete adiabatic transport through the constriction has also been recognized as an extreme idealization of the experimental situation, since a given constriction is actually defined over only a finite region of the otherwise infinite 2D gas (Payne 1989, Yacoby and Imry 1990). The adiabatic picture, on the other hand, would require the width of the constriction to change gradually over an *infinite* region. Glazman and Jonson have introduced the notion of local versus *global* adiabatic transport, pointing out the necessity of the application of magnetic fields to achieve a crossover into the latter (Glazman and Jonson 1989, 1990). Despite these considerations it is still widely believed that adiabatic transport is the fundamental explanation of the quantization of the conductance in nanoconstrictions. However, Castaño and Kirczenow (1992) have recently pointed out that this may not be the case. They consider the lack of global adiabatic transport in realistic structures, in

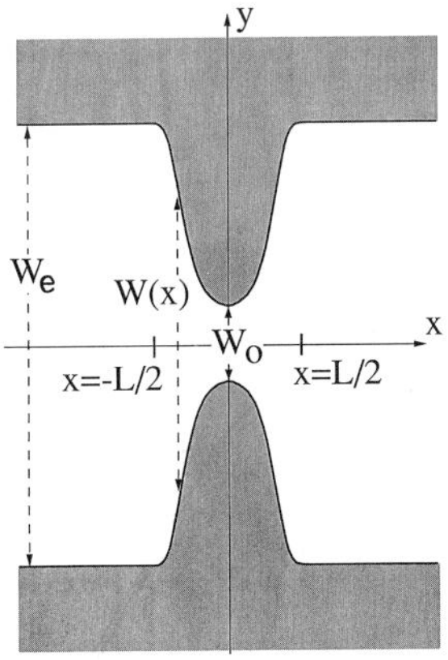

Fig. 11. Schematic diagram of constriction used in calculations of non-adiabatic transport. Shaded areas are inaccessible to electrons. Width of the constriction $W(x)$ varies smoothly, although not adiabatically, from W_0 in the narrowest part at $x=0$, to W_e, for $|x| \geqslant L/2$. Exact calculations of the conductance for the sample function, $W(x) = W_e W_0/[(W_e - W_0)\cos(x\pi/L)^4 + W_0]$, for $|x| \leqslant L/w$, as depicted in the figure, yield quantized plateaus only when mode mixing is allowed within the constriction. From Castaño and Kirczenow (1992).

the absence of magnetic fields, by including the variation in constriction width over a *finite* region of the system. Exact calculations of the conductance for a model width function, $W(x)$ in fig. 11, show that quantization of the conductance follows only if *non-adiabatic mode mixing* is allowed. This surprising result brings into question the conventional wisdom about adiabatics. While the results presented are for a specific width function, the conclusions are believed to be valid in general, and arising from the finite length of the constriction. More discussion about this interesting controversy is expected, especially in light of recent reports by Roukes and collaborators of strong experimental evidence of non-adiabaticity in constrictions with well-developed conductance plateaus (Roukes 1991).

4.2. *Multiple nanoconstrictions*

4.2.1. *Constrictions in series*

After the surprising observation of conductance quantization in a single constriction, as described above, it is natural to consider the behavior of multiple constrictions. Wharam et al. (1988b) were the first to explore the behavior of two successive ballistic constrictions "connected" in series. Other authors have also explored the conductance and magnetoconductance of similar arrangements of constrictions or point contacts (Beton et al. 1989, Main et al. 1989a,b, Kouwenhoven et al. 1989a, Hirayama and Saku 1990a,b, Staring et al. 1990, Yamada and Yamamoto 1991, Taylor et al. 1991a,b,c). These authors find an anomalous (non-ohmic) resistance addition rule: the resistance of the series connection is *smaller* than the sum of the components. The departure from the ohmic regime has been interpreted as being due to adiabatic transmission right through the whole structure formed by the pair of constrictions (Glazman and Jonson 1990). In the extreme fully-adiabatic situation, the resistance of the pair arrangement is given by the largest resistance of the two (Beenakker and van Houten 1989a), just as the resistance of a single constriction is determined by its width at the narrowest point. This result has also been explained in terms of the semiclassical collimation of electrons as they traverse each of the constrictions (Beenakker and van Houten 1989a). As an example of typical results, fig. 12d shows a diagram of the experimental system used by Kouwenhoven et al. (1989a), while figs. 12a–c show a comparison between the conductances of the individual elements A and B, and the conductance of the series connection (G_{ser}) for several values of the magnetic field.

The beam collimation mechanism considers electrons as they leave a smoothly shaped constriction opening, so that the transverse momentum decreases (as the lateral confinement width increases), producing an increase in the forward momentum and longitudinal kinetic energy (Beenakker and van Houten 1989a). Since most experimental constrictions have smoothly-varying openings, the collimation results in enhanced transmission through the second opening nearby, as long as the region joining the two constrictions does not disturb the momentum distribution (Kouwenhoven et al. 1989a). This produces an enhancement in the total conductance of the pair, relative to the ohmic addition of conductances. Notice that this argument

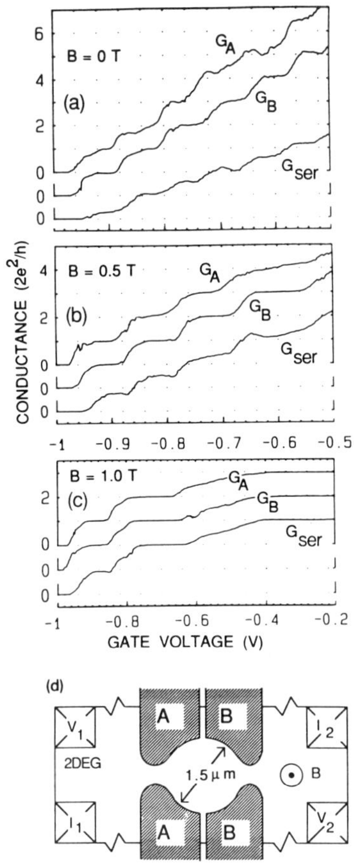

Fig. 12. Conductance curves plotted against gate voltages for a series combination of nanoconstrictions A and B. Panels (a)–(c) show a comparison of individual conductances, G_A and G_B, with the conductance of a complete device, G_{ser}. Panel (d) shows a schematic layout of device with dual constrictions. Labeled current and voltage contacts are attached to 2DEG regions. (a) At zero magnetic field, the two resistances add, so that $G_{ser}^{-1} \approx G_A^{-1} + G_B^{-1}$, and transport is ohmic. At higher fields adiabatic transport sets in, so that little or no scattering occurs between the two constrictions. $G_{ser} \approx G_A$, or G_B in this case. From Kouwenhoven et al. (1989a).

is entirely classical and provides qualitative agreement with the experimental results. However, a detailed analysis of different situations shows some discrepancies, perhaps resulting from diffraction effects (Kouwenhoven et al. 1989a, Beton et al. 1989) and, more importantly, non-adiabatic mode mixing (Castaño and Kirczenow 1992).

4.2.2. Constrictions in parallel

In the case of constrictions connected in parallel, a number of experimental results have been reported by Smith et al. (1989a), Hirayama and Saku (1990c,d), and

Simmons et al. (1991). Theoretical analyses have been presented by Avishai et al. (1989), Castaño and Kirczenow (1990a,b) and Ji and Berggren (1990).

Smith et al. (1989a) use a device with two split gates on a high-mobility heterostructure. Their results suggest that the two constrictions act cooperatively, thus forming a quantum coupled system if the interconstriction separation is smaller than the phase coherence length of the system. Using an energy minimization argument they explain their results in terms of a simultaneous depopulation of the subbands in each constriction. The locking of the plateau regions usually produces conductance jumps by *even* multiples of $2e^2/h$, despite the fact that the constrictions are not identical. Figure 13 shows some of these interesting results. By applying a magnetic field they have also been able to observe Aharonov–Bohm oscillations in the resistance, due to circulating edge states passing through both openings (see § 4.5.1). This is an elegant demonstration of the coherence of the electronic wavefield over the region of the constrictions. Similar results have been found by Simmons et al. (1991). This group developed a novel self-alignment split-gate technique that allows two *independently* tunable parallel quantum point contacts. Hirayama and Saku (1990c,d) have made systems with multiple parallel constrictions using focused-ion-beam techniques. They fabricate devices with up to 100 point contacts. In these experiments, as in the ones discussed above, it is observed that the conductances of the constrictions are additive.

The theoretical work concerned with parallel constrictions uses direct and exact (numerical) solutions of the Schrödinger equation to obtain reflection and transmission coefficients and the reservoir-to-reservoir conductance of the system (Avishai et al. 1989, Castaño and Kirczenow 1990a,b, Ji and Berggren 1990). A typical

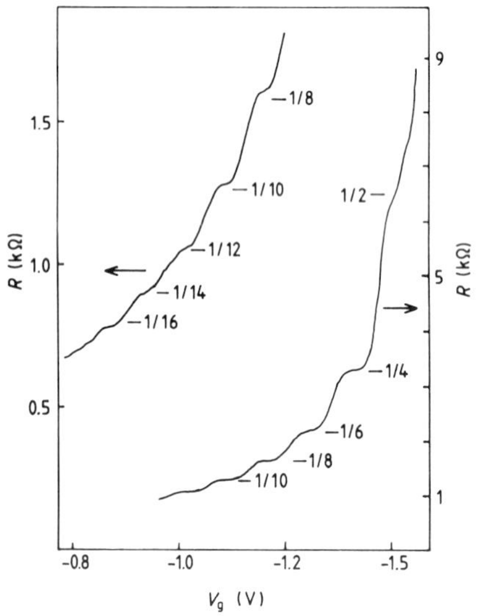

Fig. 13. Resistance versus gate voltage for parallel arrangement of two nanoconstrictions. Fractions of $h/2e^2$ are marked on the graph. Notice resistance jumps by even factors only. From Smith et al. (1989a).

schematic description of the system considered theoretically is shown in fig. 14. It is observed that the conductance of the system is nearly perfectly additive, so that $G \approx G_1 + G_2$, where G_i is the conductance of the ith constriction when the other is not present. As shown in fig. 15, the additivity breaks down, although only weakly, when the constrictions are contiguous and relatively short (note that the differences between G and $G_1 + G_2$ are small even in this case). This illustrates the independence of the two constrictions. The collimation helps to eliminate the effective communica-

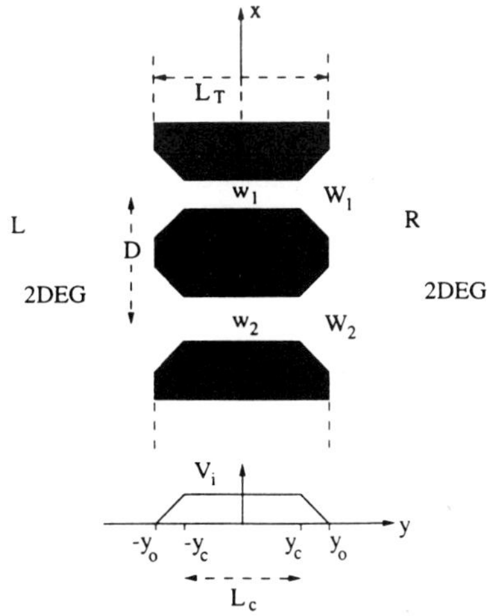

Fig. 14. Schematic drawing of parallel arrangement of constrictions. The lower trace indicates a potential inside each of the constrictions, simulating the situation in experimental devices. From Castaño and Kirczenow (1990b).

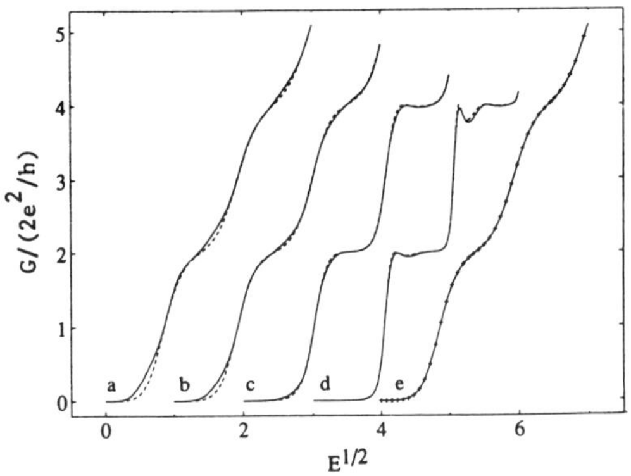

Fig. 15. Calculated conductance of parallel constrictions in fig. 14 versus Fermi energy $E_F^{1/2}$. $W_1 = W_2 = w_1 = w_2 = 1$, and $L_T = L_C$ in all curves. In curves a–d the solid line is G and the dashed line is $G_1 + G_2$, $D = 1$, and $L_T = 0.1, 0.2, 0.5$, and 1, respectively. In curve e, $G_1 + G_2$ is represented with crosses, $L_T = 0.1$, and $D = 2$. Both curves in each pair are displaced successively by one unit to the right. From Castaño and Kirczenow (1990b).

tion between the constrictions: for constrictions that are *not* contiguous the differences between G and $G_1 + G_2$ are negligible, as shown in fig. 15, curve e.

These results can be explained qualitatively (Castaño and Kirczenow 1990b) by noting that the reservoir-to-reservoir conductance is proportional to the total transmission, i.e.,

$$G = \frac{2e^2}{h} \sum_{mn} T_{mn}, \tag{6}$$

where T_{mn} is the probability of transmission form the nth transverse level in one of the reservoirs into the mth level of the other. Since the reservoirs considered are semi-infinite, their distribution of modes is continuous, and is limited only by the energy of the incident electrons. This means that G is equivalent to a total transmission probability averaged over a *continuous* distribution of incident wave vectors, i.e. an *effective* phase averaging is performed. This, together with the collimation along the constrictions, renders them effectively independent. However, if the distribution of carriers in the different modes in the reservoirs were lumped in any way, the phase averaging would not be as effective and the additivity of the constrictions could be affected.

4.3. *Impurity and geometrical resonances*

The systems used in ballistic transport studies are some of the cleanest structures ever grown. Nevertheless, some scattering due to random potential fluctuations, boundary roughness and impurities occurs in many samples, and studies of these have appeared in the literature (Roukes et al. 1990c). The study of these effects has included several interesting phenomena, such as transmission resonances induced by impurities (McEuen et al. 1990). Moreover, it can be said that these studies represent, in a sense, the middle ground between the ballistic transport discussed so far in these sections, and the diffusive transport discussed in the sections on localization (see § 5 et seq. below).* On the other hand, the analysis of constrictions with different geometries, especially those with quasi-localized states within them, has also emphasized the importance of resonances, and it is therefore closely related to the study of impurity effects.

4.3.1. *Single-impurity effects*

The effects of scattering by impurities or defects inside constrictions or quantum wires (McEuen et al. 1990, Timp et al. 1990a, Cobden et al. 1991) or at the boundaries of quantum wires (Thornton et al. 1989, Roukes et al. 1990c) have been observed and characterized in experimental systems. McEuen et al. (1990) reported a transmission resonance feature in the conductance of a split-gate device for large gate bias.

*A related "quasi-ballistic" regime has been defined by some authors (Beenakker and van Houten 1991a) as the situation when the mean free path is larger than the width of the narrow wire or constriction but smaller than its length.

At these voltages, the constriction was barely conducting, and the resonance feature resulted from electronic tunneling via an attractive impurity inside the constriction (see fig. 16). Identification of this feature was carried out through a systematic study of the magnetic field and asymmetrical gate bias dependence of the conductance, in a fashion similar to the analysis of Kopley et al. (1988) for a MOSFET system. Theoretical studies of the effects of single impurities in nanoconstrictions have been presented by Haanappel and van der Marel (1989), van der Marel and Haanappel (1989), Chu and Sorbello (1989), Bagwell (1990), Kumar and Bagwell (1991a), Tekman and Ciraci (1990a, 1991), Kriman et al. (1990), Joe and Ulloa (1990), and Levinson et al. (1991). The transmission through a state localized around the impurity (a quasi-bound state) is responsible for the appearance of a resonance feature in the conductance (or transmission coefficient of the constriction). This is similar to the situation in a double quantum barrier device where electrons are transmitted via a state localized in an intermediate region.* However, the different dimensionality of the electronic states in the quasi-one-dimensional conductor makes this effect appear directly in the conductance curve of the system. This is in contrast to the situation in higher dimensions where the resonance is convoluted with the density of states in the directions perpendicular to the tunneling (Bagwell et al. 1990). This point is also important in the tunneling through quantum dots discussed in § 4.7. The gate bias (or channel width) at which the resonance feature appears in the conductance curve is directly related to the strength of the impurity binding, as well as to the location of the impurity within the constriction (Tekman and Ciraci 1990a, Joe and Ulloa

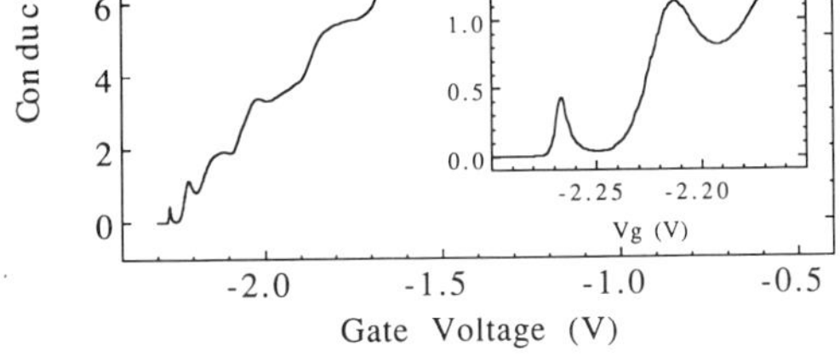

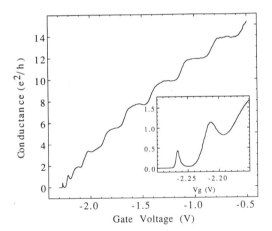

Fig. 16. Conductance of nanoconstriction against gate voltage at $T = 0.55$ K. Inset displays the low-conductance region with an isolated peak appearing before the onset of the first transverse mode. This transmission resonance feature is associated with an isolated impurity-ion state in the constriction. Studying the temperature and magnetic field dependence of the peak allows the determination of energy and approximate location of the impurity state. From McEuen et al. (1990).

*See chapter 12 on tunneling in this book.

1990). Figure 17 illustrates this point for a model *attractive* delta-function potential in the constriction.*

Faist et al. (1990) have observed conductance "dips" before the onset of each plateau in a split-gate geometry, although using Si-based devices (with somewhat smaller mean free paths). These features in the conductance curves (notice the dip in fig. 17 before the onset of the third plateau) were predicted by Chu and Sorbello (1989), and have been studied by several theoretical groups (Bagwell 1990, Masek et al. 1989). This sharp reduction in transmission appears as the quasi-bound state associated with each of the transverse subbands enhances the carrier localization at the corresponding energy.

Other authors have explored "random telegraph signals" (RTS) possibly introduced by single impurities changing their location or ionization state in quasi-ballistic systems. Timp et al. (1990a) have shown that RTS are strongly enhanced as a transverse subband is about to be depleted, since low-angle scattering is favored under those conditions, as described by Büttiker (1990b, 1991), who generalizes the work by Lesovik (1989) and Yurke and Kochanski (1990). These observations have been corroborated by Cobden et al. (1991) and reflect the insensitivity of the quantized resistance to smooth potential fluctuations of the device (if one remembers the arguments of adiabatic transport of each subband). Other measurements on the spontaneous resistance switching and noise signals in quantum point contacts have been reported by Li et al. (1990), Dekker et al. (1991), Ford et al. (1991), and Taylor et al. (1991b), who find similar results although strongly sample dependent. Further

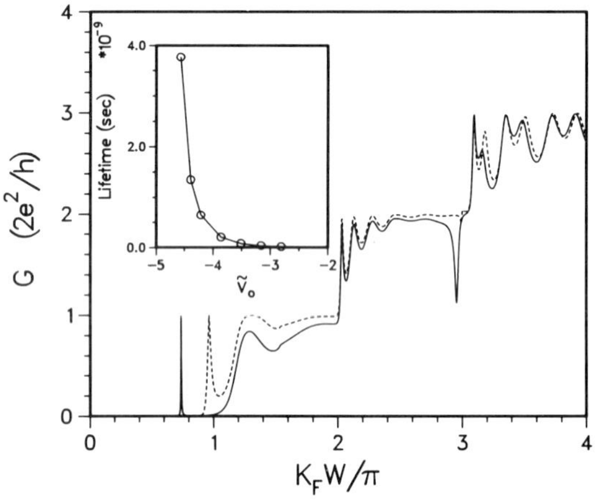

Fig. 17. Calculated conductance of attractive delta-function impurity inside a nanoconstriction. The solid line shows the curve for $\hat{V}_0 = V_0/E_1 = -3.5$, where V_0 is the strength of the delta potential and E_1 is the energy of the first transverse mode in the absence of an impurity. The dashed line is the trace for $\hat{V}_0 = -1.75$. Notice that the resonance becomes sharper and deeper (further from the onset of the first transverse mode at $k_F W/\pi = 1$) as the impurity becomes more attractive. Inset: resonance width against impurity strength. From Joe and Ulloa (1990).

*A repulsive potential mainly shifts the onset of the conductance plateaus without producing additional curve features (Joe and Ulloa 1990).

exploration of the kinetics (time dependence) of electronic transport in nanostructures promises interesting results.

4.3.2. Multiple-impurity effects

Another problem of interest, and closer to the regime of diffusive transport, is how possible large potential fluctuations in the structures affect the conductance of nanoconstrictions. Masek et al. (1989), Akera and Ando (1990a,b, 1991a,b), Ando and Akera (1990), He and Das Sarma (1989a,b), Nixon and Davies (1990), Nixon et al. (1991a,b), Laughton et al. (1991), and Kumar and Bagwell (1991b) have explored the theoretical consequences of multiple impurities in various approximations. Davies et al. use realistic screened ionized-impurity potentials to describe the background over which electrons move in narrow wires and constrictions. They find that the quantization of the conductance, for example, breaks down for constrictions that are an order of magnitude shorter than the mean free path. This would emphasize the importance of clean samples with well-defined boundaries to allow observation of ballistic phenomena. Figures 18 and 19 show characteristic results of their calculations, illustrating the breakdown of the ballistic regime. The authors comment on

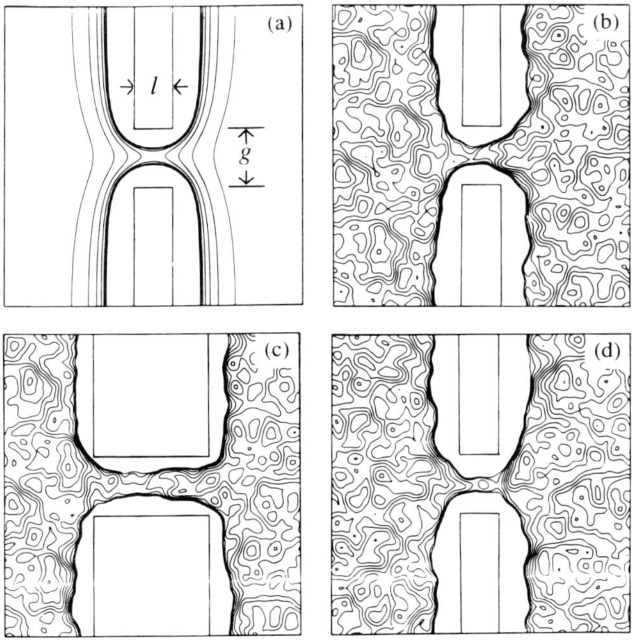

Fig. 18. Gate pattern and electron density profiles for nanoconstrictions with a split-gate separation $g = 0.3$ μm. Contours start from zero and are 4.2×10^{10} cm^{-2} apart. All cases correspond to two transverse modes occupied in the narrowest part of the constriction. Panel (a) shows the situation with no impurities, while (b)–(d) show various cases with donor impurity potentials with a correlation length of about 2 μm. In panel (c) the gate length $l = 0.6$ μm, while $l = 0.2$ μm in the others. Notice the potential well in the channel of (d), which produces transmission resonances. From Nixon et al. (1991a).

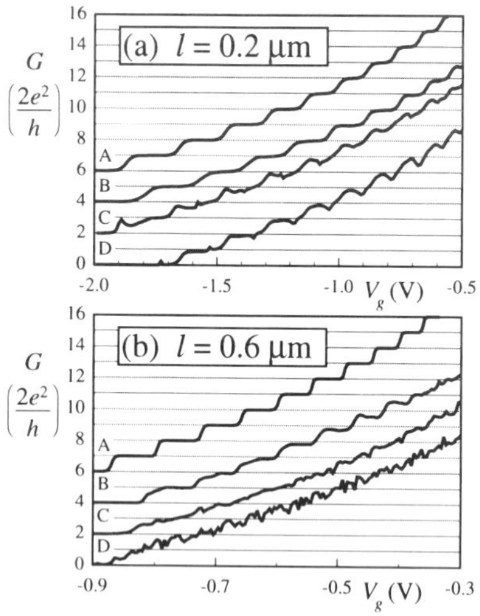

Fig. 19. Calculated conductance against gate voltage for gates with $l = 0.2$ and 0.6 μm. Curve A in each panel shows the conductance in the clean system, while other curves are the results for different impurity configurations. Curves C and D show peaks due to resonances in the constriction. Notice that the breakdown of quantization is more pronounced for the longer channel device of panel (b). Curves A, B, and D in (a) are the resulting conductance for situations in fig. 18a,b, and d, respectively. Curve B in (b) corresponds to fig. 18c. From Nixon et al. (1991a).

the need to reduce the fraction of ionized donors in their calculation to yield better agreement with experiments, as well as to introduce correlations between ionized centers which should tend to smooth out large potential fluctuations. This point has been analyzed in an interesting fashion by Efros et al. (1990).

4.3.3. Geometrical resonances

The importance of different geometries in the conductance of nanoconstrictions has been pointed out in several theoretical works. Peeters (1989a), Tekman and Ciraci (1989b, 1990b, 1991), Lent and Kirkner (1989), Lent and Sivaprakasam (1990), and Lent (1990a, 1991) have studied the transmission of a widening in a constriction, a "cavity", finding a number of interesting transmission features associated with states localized in that region. Weisshaar et al. have also reported theoretical calculations of this structure in the nonlinear regime (Weisshaar et al. 1990a, 1991a) (see § 4.4). An approximate experimental realization of cavities, "quantum dots", will be discussed in a different section below (§ 4.7).

Peeters (1989b) has also studied the magnetic field dependence of the binding energy of a bound state formed at the intersection of narrow quantum wires. These states have also been studied by various authors, such as Schult et al. (1989), Kirczenow (1990a), Büttiker (1988a, 1990c), Berggren and co-workers (Berggren and Ji 1991, Weisz and Berggren 1990, Rundquist et al. 1990), Ono (1992) and Paranjape (1991). A detailed discussion of the effects of resonant states on the transport through junctions of quantum wires is presented in § 4.6 below.

Sols et al. (1989a–c) studied the conductance properties of a constriction with an additional finite-length stub, as a proposal for a novel "quantum-modulated transistor". This structure has also been proposed independently by Datta (1989), and studied further (together with other related systems) by Berggren and Ji (1990), Avishai and Band (1990), and Price (1991). An additional gate is deposited to modulate the length of the stub, which in turn affects the conductance of the system due to the coherent interference of the various current paths. Figure 20 shows a diagram of the stub and fig. 21 an example of the rich conductance properties of this system. No experimental realization of this device has been reported to date. Other device-oriented proposals have been made, exploiting different characteristics of ballistic transport, such as the interference of coherent beams (Datta 1989), diffraction on exit through a short constriction (Kriman et al. 1989), or focusing produced by a barrier attached to a constriction (De Raedt et al. 1989). Analyses of various proposals in terms of their usefulness for devices have been presented by Landauer (1989c,d, 1990) and Subramaniam et al. (1990).

Lent (1990b), Lent and Sivaprakasam (1990), Sols and Macucci (1990), and Weisshaar et al. (1989, 1990a,b, 1991a,b) have also studied theoretically the transmission characteristics of bends (a narrow wire with a ninety-degree turn) and double bends (two narrow wires connected by either a rounded or straight perpendicular segment). These authors report a number of interesting shape-dependent transmission features, such as resonant effects arising from interference produced by partial electronic reflections from the sharp confinement potential in the case of double-bend structures (Weisshaar et al. 1989, 1990b, 1991b). A double-bend structure has

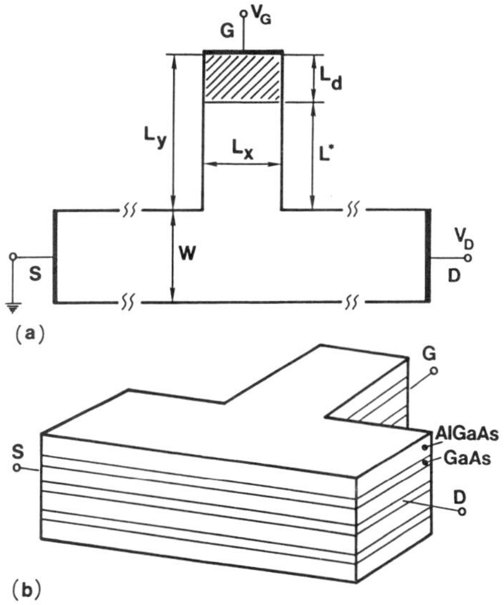

Fig. 20. Schematic diagram of the "quantum-modulated transistor" of Sols et al. The gate voltage V_G controls the effective length of the stub L^*, which regulates transmission from source S to drain D. From Sols et al. (1989b).

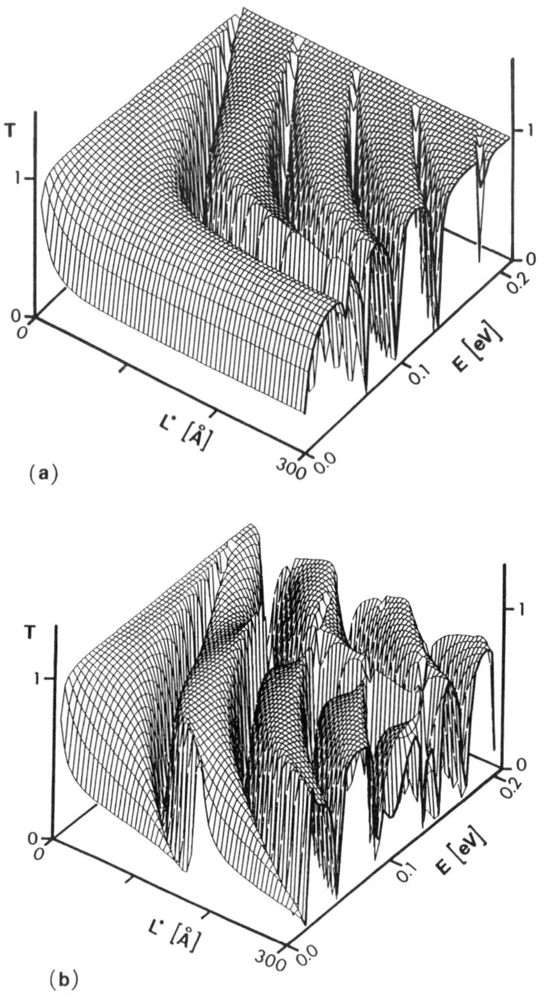

Fig. 21. Calculated transmission probability from S to D in structure of fig. 20, plotted against the stub length L^* and electron energy E. Here $W = 100$ Å, and (a) $L_x = 80$ Å, and (b) $L_x = 120$ Å. The transmission resonance features are drastically affected as the geometry of the device changes. From Sols et al. (1989b).

been constructed by Wu et al. (1991) using a device with a structured split gate. The low-temperature conductance exhibits a number of peaks, which the authors identify with the theoretical predictions of Weisshaar et al. (1989, 1990b, 1991b). The observed resonance features are also reminiscent of the impurity scattering events observed by McEuen et al. (1990), and although difficult to identify unequivocally, these results are promising for exploiting the geometrical resonances in these devices.

Other ballistic systems with complex geometries have also been considered by Pendry et al. (1991), and Joe and Ulloa (1991a).

4.4. Non-linear transport in constrictions

Interesting work has been reported on ballistic nanostructures in the high-field regime, where novel non-linear effects are expected due to the application of a finite voltage along narrow constrictions (Glazman and Khaetskii 1989a, Kouwenhoven et al. 1989b, Lent et al. 1989, Kelly et al. 1989, Kelly 1989, Brown et al. 1989b, Castaño and Kirczenow 1990c, Castaño et al. 1990, Weisshaar et al. 1990a, 1991a, Patel et al. 1990). A number of different systems have been considered theoretically, such as sharp and adiabatic constrictions, as well as constrictions with an added potential modulation along their length. The different, and sometimes contrasting, results emphasize the importance of different aspects of this interesting regime. Comparison with the experimental results for some of these structures has yielded interesting conclusions, highlighting the significance of various theoretical assumptions.

The application of a finite voltage V_0 produces a current proportional to the transmission coefficient, which is in general voltage dependent, averaged over an energy window of size equal to $|eV_0|$ if $E_F \geqslant |eV_0|$, but equal to E_F otherwise, so as to take into account all of the current-carrying states in the system. Then,

$$I(E_F, eV_0) = \frac{2e}{h} \int_{\nu(E_F - |eV_0|)}^{E_F} T(E, V_0)\, dE, \tag{7}$$

where $T(E, V_0)$ is the transmission coefficient along the constriction for a given energy E and a given applied voltage V_0; $\nu(x) = x$ if $x \geqslant 0$ and $\nu(x) = 0$ otherwise. It is clear that the effect of V_0 is twofold: it introduces an energy window of states contributing to the current, and modifies the transmission coefficient directly.

The first experimental study was reported by Kouwenhoven et al. (1989b) who used a split-gate device with large applied voltages. Their results are characterized by strikingly non-linear I–V curves, as shown in fig. 22. Notice also that the quantization breaks down for large enough applied voltages. These results were explained using a model that generalizes the linear-response situation: a phenomenological parameter m is introduced to take into account the fraction of the voltage drop at the *ends* of the constriction. It is also assumed that the potential energy is nearly flat around the center of the point contact. The value of this parameter is not universal, being either 0.5 or 0.2, depending on the plateau in which it is measured. This non-universality might be due to the fact, as the authors point out, that their model does not take into account the interdependence of the gate and the applied voltages, a problem that has not received a definite answer as yet.

Glazman and Khaetskii (1989a) used an adiabatic model to study theoretically the high-field regime of narrow ballistic constrictions. These authors predicted the occurrence of extra plateaus between those found in the linear-response regime. These extra plateaus, now known as "half-plateaus", appear together with the usual low-field quantized plateaus, although the latter have become smeared as a result of the applied voltage. Their model uses adiabatic injection and implicitly assumes that the

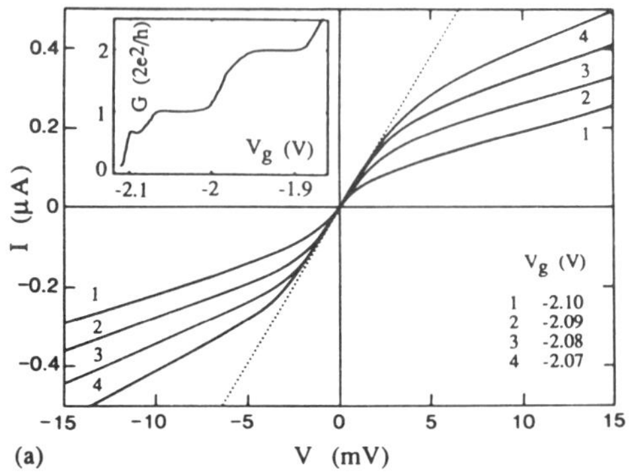

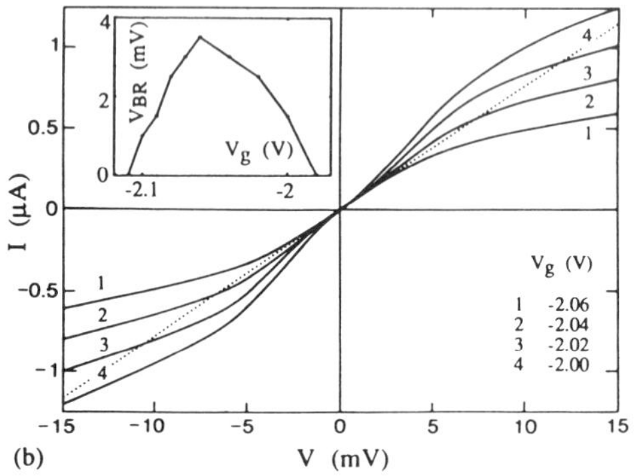

Fig. 22. Current–voltage characteristics at different gate voltages. Nanostructure shows quantized conductance at $2e^2/h$ for small values of V, as shown by the straight dashed line, and in the inset of (a). Inset in (b) shows breakdown voltage V_{BR} against gate voltage V_g. From Kouwenhoven et al. (1989b).

potential drop occurs symmetrically at the ends of the constriction, with a potential energy that is fairly flat around the center of the device. In their model, the window of integration is between $E_F + eV/2$ and $E_F - eV/2$, and the transmission coefficient is that of a parabolic barrier, which is used to represent the adiabatic constriction.

The experimental results of Patel et al. (1990) seem to confirm the predictions of Glazman and Khaetskii. Their device is a split-gate point contact that shows excellent quantized plateaus in the low-field regime. Some of their results are shown in figs. 23 and 24, where extra plateaus somewhat close to the "half-position" are clear. This occurs for a wide variety of physical parameters, including the application of a magnetic field perpendicular to the plane of their device. These authors ascribe the existence of half-plateaus to the presence of adiabatic transport in the manner considered by Glazman and Khaetskii (1989a).

Other theoretical studies do not obtain these half-plateaus (Castaño and Kirczenow 1990c, Lent et al. 1989). These theoretical works used very different models: the model constrictions are not adiabatic – the constrictions are sharply defined – and more importantly, the assumed potential distribution is linear along the constriction. Thus the results differ significantly from those of Glazman and

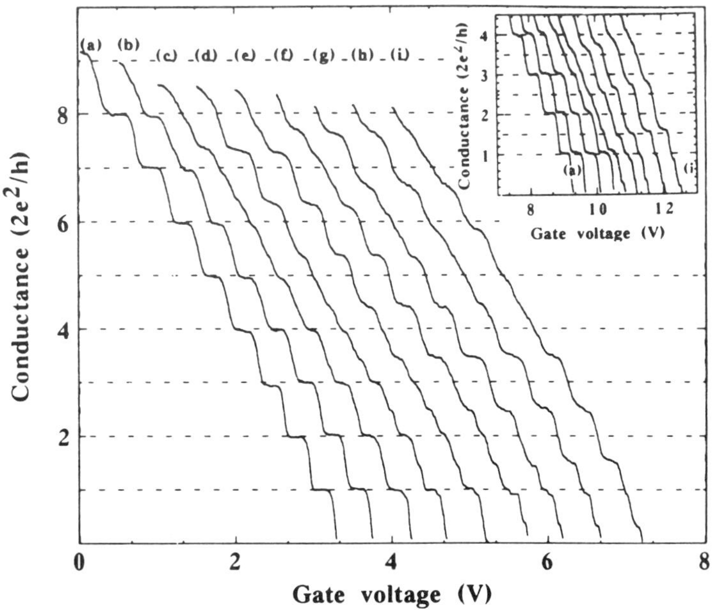

Fig. 23. Differential conductance against gate voltage for different d.c. bias voltages. Traces are offset successively by 0.2 V. The d.c. bias increases from zero in graph (a), to 4 mV in graph (i), in steps of 0.5 mV. Measurements are carried out at 0.5 T and 30 mK. The inset shows the same data at zero magnetic field. Notice the appearance of "half plateaus" as the d.c. bias increases. From Patel et al. (1990).

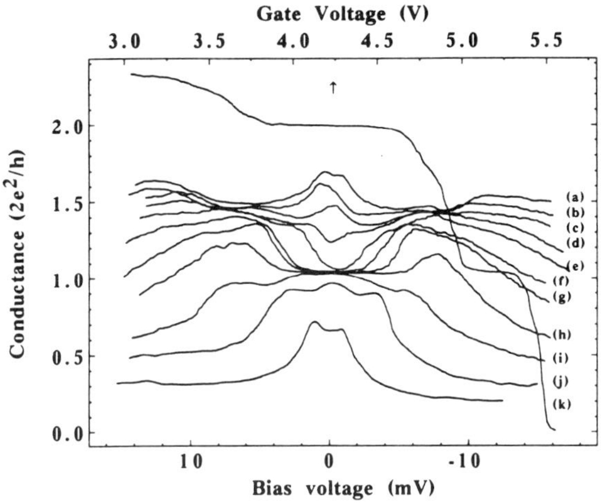

Fig. 24. Upper trace shows the conductance of the device plotted against the gate voltage (top axis). Other curves show the differential conductance at fixed gate voltages as the d.c. bias is swept. Trace (a) begins at a gate voltage of 4.90 V, to end with trace (k) at 5.40 V, in increments of 50 mV. Traces converge at integer plateaus for zero bias, while they converge at half-integer values for high biases. From Patel et al. (1990).

Khaetskii (1989a). The difference serves to illustrate that details of the geometry and potential profile can alter the non-linear transport properties in a significant manner.

The theoretical work of Lent et al. (1989) uses a wide–narrow–wide configuration and finite-element techniques to study the effects of high temperatures and high fields on electronic transport in constrictions. Their results show, among other things, that the behavior of a narrow constriction is qualitatively different from that of a potential barrier (in contrast to the situation occurring in the low-field regime), and that the current saturates for high voltages.

The work of Castaño and Kirczenow (1990c) considers a sharply defined narrow constriction connecting two semi-infinite 2DEGs. Their results show that non-linear effects influence the conductance in several different ways with respect to the zero applied voltage results: the transmission resonances are smoothed out, and the quantized plateaus are shifted and degraded. As we have mentioned, G does not develop half-plateaus in this case. For high voltages, the energy window reaches its maximum and the current saturates at values strongly dependent on the Fermi energy. If the applied voltage is increased significantly, the potential energy mismatch between the two 2DEGs becomes extremely strong. In this case, the energy window is very large; however, the potential mismatch produces strong quantum mechanical reflections that can overcome the increased energy window in a very effective way. In this case, the differential conductance of the system becomes negative. This extremely high applied-voltage regime has been studied in a series of theoretical and experimental papers by the Cavendish group (Kelly et al. 1989, Kelly 1989, Brown et al. 1989b).

4.5. *Magnetic field effects*

Application of magnetic fields perpendicular to the 2D electron gas layer has very interesting consequences in heterostructures, such as the focusing effect discussed above (§ 3.1). In the case of a nanoconstriction (or quantum point contact), in the regime of quantized conductance, the magnetic field has three main effects.

(1) Moderate fields ($B \leqslant 1$ T) tend to reduce the effects of impurities (*magnetic backscattering reduction* (Büttiker 1988b, van Houten et al. 1988b), as discussed below in detail), which in turn reduces fluctuations in the plateau regions appearing at low temperatures ($T \approx 50$ mK); in this sense, moderate magnetic fields would have a qualitatively similar effect to that of the moderate temperatures at which the first conductance quantization observations were optimized (van Wees et al. 1991, van Houten et al. 1990).

(2) Higher magnetic fields result in magnetic depopulation of the subbands (Wharam et al. 1988a, van Wees et al. 1988b, Berggren et al. 1986, Berggren and Newson 1986), which produces progressively wider plateaus (in bias voltage) with well-quantized values, and smoothly makes the transition to the quantum Hall regime in this system. See fig. 25 for an illustration of this type of behavior (van Wees et al. 1988b).

(3) Lastly, for the largest fields ($B \approx 10$ T), the appearance of additional plateaus at odd integer multiples of e^2/h is observed (see fig. 25) (van Wees et al. 1988b).

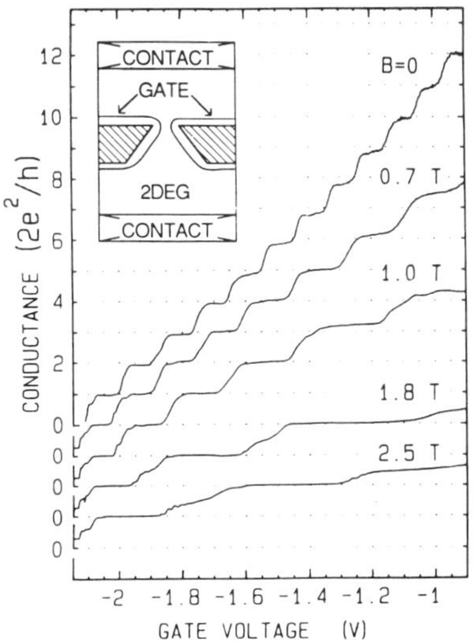

Fig. 25. Conductance of nanoconstriction against gate voltage for various values of magnetic field at 0.6 K. Inset shows a schematic diagram of the top view of the device. The thin contour lines indicate depletion under and around the gate. Notice that quantized plateaus become wider as the magnetic field increases, increasing the mode separation and producing *magnetic depopulation*. From van Wees et al. (1988b).

This spin splitting effect is, however, more clearly visible when a magnetic field is applied *parallel* to the 2D gas, as originally reported by Wharam et al. (see fig. 26).

Theoretical studies have been able to explain the robustness of the quantization to magnetic fields. The reason for this is that the conductance formula (eq. (3) in § 4.1.2) should be valid for any type of subband structure (Glazman and Khaetskii 1989b, Avishai and Band 1989b, Efetov 1989, Büttiker 1990a, Prêtre et al. 1991). These studies also explain more subtle dependences, such as the additional plateaus at higher fields (associated with spin splitting of the subbands) appearing first at higher conductance values (large N values in eq. (3)) (Glazman and Khaetskii 1989b).

Detailed comparison with experimental results, using realistic model potentials to describe the constrictions, has verified that *magnetic depopulation* of the bands is the reason for the shift and widening of the plateaus (van Wees et al. 1988b, Wharam et al. 1989a, Weisz and Berggren 1989). Indeed, as the magnetic field increases, the effective confinement potential shifts the energy of the subbands upwards, as they approach Landau levels in the limit of high fields, forcing an offset (shift in the plateaus) and larger gate bias for the successive population of each subband (which explains the widening of the plateaus).

4.5.1. Edge states

Apart from the effects of magnetic field on the conductance quantization, many other interesting observations have been reported. Examples of these are the suppression

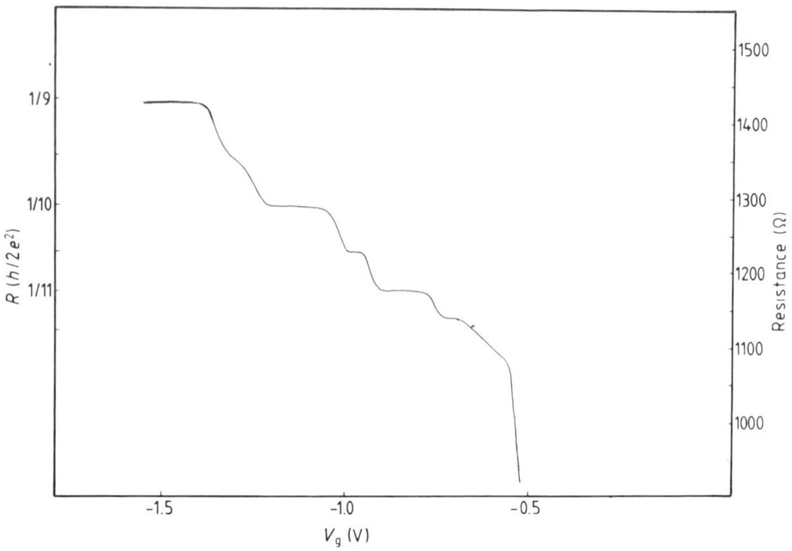

Fig. 26. Magnetoresistance of constriction versus gate voltage. A magnetic field is applied *parallel* to the 2DEG plane, $B = 13.6$ T, at $T \approx 0.1$ K. Spin splitting of energy levels gives rise to additional plateaus at $R = h/21e^2$, $h/23e^2$, and an ill-defined plateau at $h/19e^2$. From Wharam et al. (1988a).

of backscattering by a constriction in the presence of a magnetic field (van Houten et al. 1988b), a number of transport phenomena in multiprobe systems (discussed below in the section on quantum wire junctions, 4.6), and the observation of Aharonov–Bohm oscillations in various systems: a single nanoconstriction (point contact) (van Loosdrecht et al. 1988), small rings (Timp et al. 1987b, 1989b, Ford et al. 1988b, 1989a), and small cavities or "dots" (van Wees et al. 1989a, Brown et al. 1989c, Wharam et al. 1989b, Taylor et al. 1992, Ford et al. 1991) (see also §§ 4.6.2 and 4.7 on quantum dots).

A common concept entering the description of all of these observations is the appearance of *edge states* at high magnetic fields. Edge states are the quantum-mechanical analog of the classical skipping orbits that appear when a charged particle moves in the presence of a magnetic field in the neighborhood of a boundary. Edge states were introduced by Halperin to explain the quantum Hall effect in macroscopic 2D electron systems (Halperin 1982). They are the extended current-carrying states of such systems, in contrast to the bulk states that, except near the centers of Landau levels, are localized by disorder (Prange and Girvin 1990). The picture of edge states in macroscopic systems has now been extended by Streda et al. (1987), Jain and Kivelson (1988a,b) and Büttiker (1988b) in terms of the Landauer-type formulas of Economou and Soukoulis (1981a,b), Fisher and Lee (1981) and Büttiker (1986). However, in the context of narrow constrictions and wires, the strong lateral confinement suppresses the edge states, except at high magnetic fields (such that the characteristic cyclotron radius $l_c = \hbar k_F/eB \ll W$, where W is the width of the channel). A classical picture of this high-B regime (see fig. 27) would have the electrons traveling

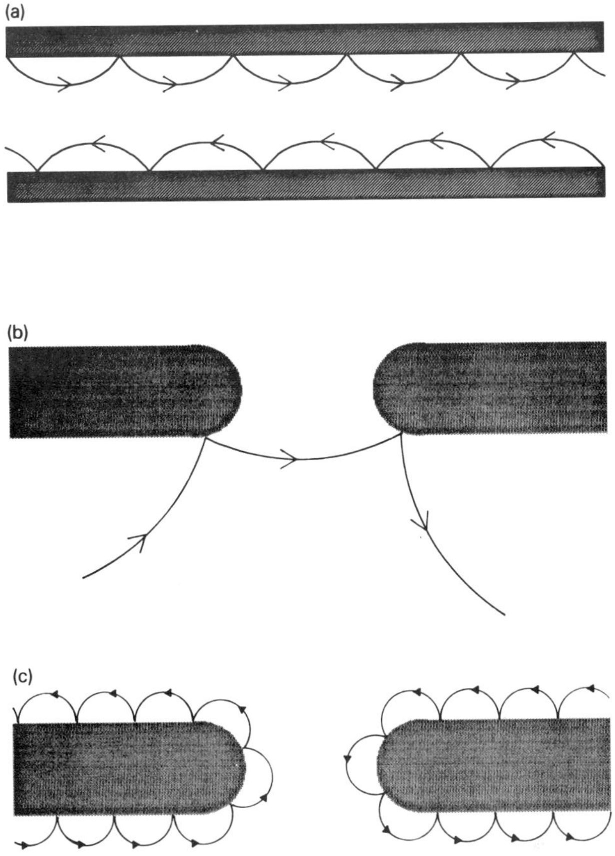

Fig. 27. Classical skipping orbits of electrons in magnetic fields and near potential barriers. Shaded regions are not accessible to electrons. (a) In a wire, orbits at opposite ends of sample move in opposite directions. (b) At low fields the cyclotron radius is larger than the opening and electrons are frequently deflected back. (c) At high fields a small cyclotron radius allows for electronic motion along equipotential lines, reducing backscattering. After van Houten et al. (1990).

along the constriction or wire pushed against one of the lateral walls by the Lorentz force, so that a skipping motion is set up with electrons moving in opposite directions on both sides of the channel (Beenakker et al. 1989). Quantum mechanically, this corresponds to wavefunctions with guiding centers close to either edge, and with a lateral extent given approximately by l_c. A detailed quantum mechanical description of edge states in narrow conductors has been given by several authors (Peeters 1988, Kirczenow 1988b, 1989e, Ravenhall et al. 1989, Schult et al. 1990). Also, van Wees (1991b) has studied the effect of a time-dependent magnetic flux on the electronic transport via edge states defined on a mesoscopic Corbino disk.

The suppression of backscattering by the constriction can be easily explained within the semiclassical skipping-orbit picture, as shown in fig. 27. For low magnetic fields, the geometry of the constriction dominates the backscattering since incident electrons are easily reflected back, as shown in fig. 27b. However, for higher field values, the cyclotron radius for electrons at the Fermi energy, l_c, is much smaller than the constriction width and the electrons are guided through the opening by the edge that defines it (van Houten et al. 1988b, Beenakker et al. 1989). In the language of quantum-mechanical edge states, the reduction in backscattering can be understood by considering that the edge states moving in opposite directions are separated by a distance comparable to the channel width (since the location of their guiding centers follows equipotential lines in an adiabatic scheme (Glazman and Jonson 1989)), and the matrix element connecting the two edge states becomes small as the field increases, due to an exponentially decreasing overlap. It is this reduction in backscattering probability that explains a decaying amplitude of the Aharonov–Bohm (AB) oscillations observed in small rings as the intensity of the magnetic field is increased (Timp et al. 1987b, 1989b, Ford et al. 1988b, 1989a). The presence of well-defined edge states would also explain AB oscillations in the singly-connected geometry of a quantum dot, as seen by van Wees et al. (1989a), Brown et al. (1989c) and Taylor et al. (1992). Figure 28 shows typical AB magnetoresistance oscillation curves for this geometry (Brown et al. 1989c). Furthermore, as discussed in the next section, it is the presence of edge states in narrow wire junctions that produces a host of interesting effects in magnetic fields.

We should mention one last point. At low magnetic fields the edge states in narrow channels overlap with each other. However, there is no backscattering in a truly-ballistic system. It is therefore interesting to realize that, in a curious way, ballistic systems behave very similarly, *whether the magnetic field is large enough to give rise to edge states or not.* This interesting "crossover" from the edge state to the fully-ballistic regime provides an alternative way to understand the quantization of the conductance in nanoconstrictions (point contacts) as the zero-field limit of the quantum Hall effect (Kirczenow 1988b).

4.6. Quantum wire junctions

A system that has played a major role in the development of the present understanding of ballistics in semiconductor nanostructures is the junction of intersecting narrow conductors or "quantum wires". This system is depicted schematically in fig. 29, where quantum wires labeled 1–4 come together at a junction, the shaded area. The quality of the best samples available to date has been such that at low temperatures the mean free path of the electrons in the quantum wires is shorter than the length of the quantum wires (see, e.g., Nixon and Davies 1990, Nixon et al. 1991a,b, Thornton et al. 1989, Roukes et al. 1990c, Akera and Ando 1989a,b, 1990a,b, Ando and Akera 1990, Ando 1990, Timp et al. 1989a, 1990b, Timp 1990, Takagaki et al. 1990a, Glazman and Jonson 1991) but larger than the size of the junction, as was found by Ford et al. (1989b, 1991), Ford (1991a,b), and Roukes et al. (1990b). Thus while the quantum wires themselves are not ballistic, their junctions with characteristic dimensions about 0.1 μm, may be viewed as such.

The experimental measurements of greatest interest have been those of the Hall resistance R_H and the "bend" resistance R_B of the junctions. Referring to fig. 29, the

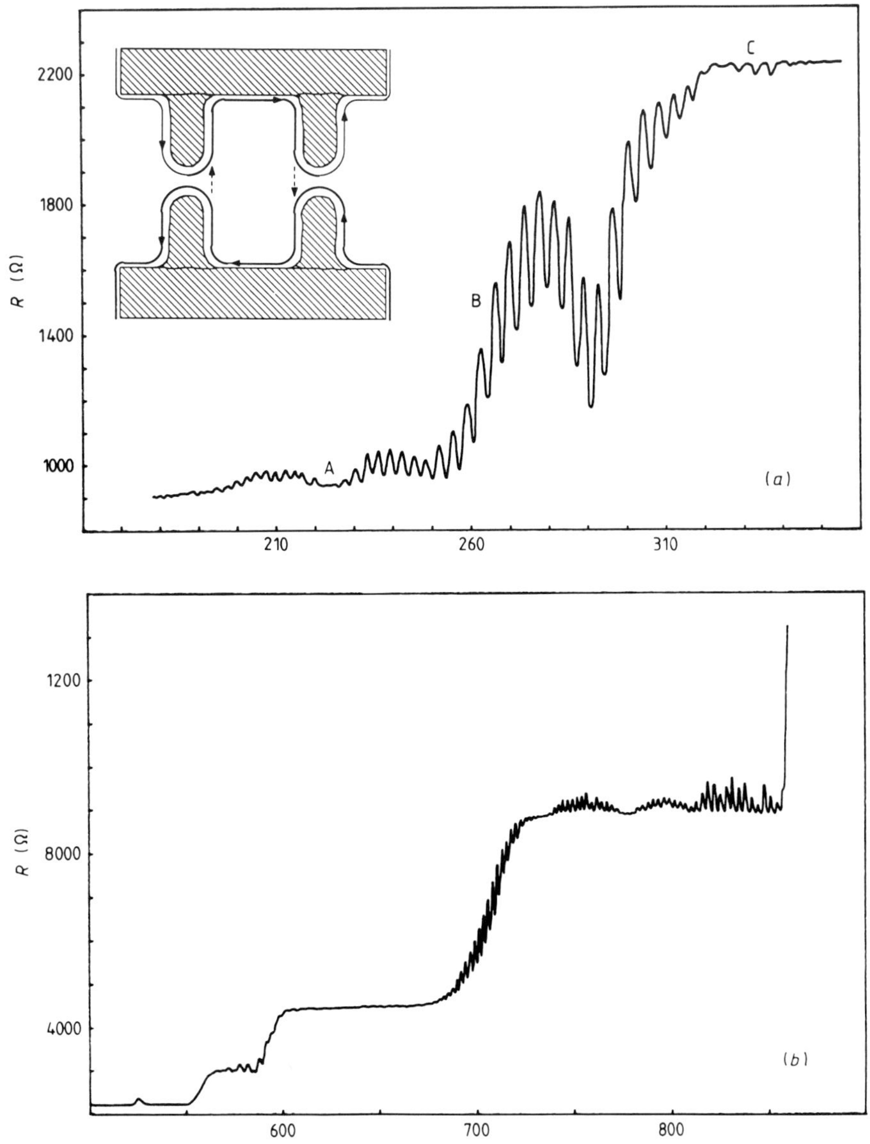

Fig. 28. Magnetoresistance of constriction versus gate voltage (in mV) in the region between the first and second plateaus. Trace shows Aharonov–Bohm oscillations superimposed on the resistance rise. Inset: dashed regions are inaccessible to electrons. Thin lines show relevant edge states, with broken lines indicating tunneling between right- and left-propagating states. A closed-orbit state is formed with a well defined area that can be changed with the gate voltage. From Brown et al. (1989c).

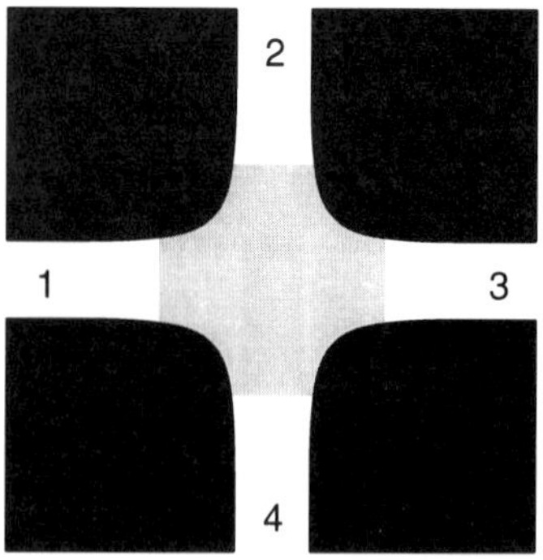

Fig. 29. Schematic representation of a junction (shaded) of four quantum wires. The black areas are depletion regions from which electrons are electrostatically excluded. Note the smooth geometry of the junction with rounded corners. As in the ballistic constrictions discussed earlier, this rounding is due to the electrostatic nature of the potential defining the quantum wires and their junction.

Hall resistance is defined as $R_H = V/I$, where V is the Hall voltage between the reservoirs attached to leads 2 and 4 through which no net current flows and I is the current flowing through leads 1 and 3 in a magnetic field oriented perpendicular to the plane of the cross. The bend resistance was defined by Takagaki et al. (1988a, 1989a) analogously, except that the current flows into the junction through lead 1 and out through lead 4 while the voltage is measured between the reservoirs attached to leads 2 and 3 and is adjusted so that these leads carry no net currents. Under suitable conditions, experimental measurements of these Hall and bend resistances are sensitive to the local behavior of the electrons in the junctions, and thus can be used to probe directly the physics of these small ballistic regions. Clearly for this to be the case the temperature must not be too high or the electron mean free path would be smaller than the size of the junction and the junction would not be ballistic. However, the temperature must also not be so low that the quantum coherence length is much larger than the junction size, because then coherent multiple scattering of electrons between the junction and the potential irregularities in the leads becomes important.

Usually measurements are performed on structures containing more than one junction. This can lead to nonlocal resistance effects because of electrons propagating from one junction to another without being totally randomized by scattering processes in between. This has been discussed by Timp et al. (1988a), Timp and Howard (1991), Timp (1990), Kirczenow (1989e), Roukes et al. (1990c), Takagaki et al. (1988b, 1989b, 1990a,b,c), Sukhorukov and Levinson (1990), Baranger (1990), and Schult et al. (1990).

Some early measurements on quantum wire junctions by Simmons et al. (1988) and Timp et al. (1987c) showed anomalies in their Hall resistance. However, it was the observation by Roukes et al. (1987) of the apparent disappearance or "quenching" of the Hall resistance of the junctions of their narrowest quantum wires at low magnetic fields that sparked much of the subsequent interest in these structures. The results of these measurements are shown in fig. 30. The suppression of the Hall

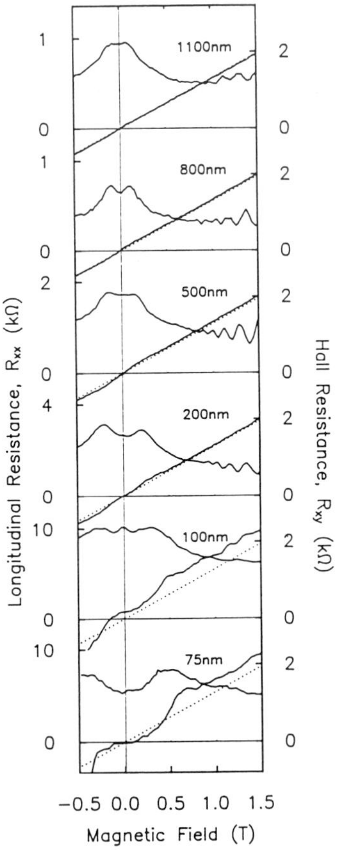

Fig. 30. The low magnetic field Hall resistance of junctions of quantum wires ranging in width from 1100 to 75 nm at $T = 4.2$ K. For the wider wires, the Hall resistance follows the linear magnetic field dependence (dotted) of macroscopic samples, but for the narrowest 75 nm wires it appears to vanish at low magnetic fields, before the magnetic field reaches zero. From Roukes et al. (1987).

voltage at low magnetic fields was subsequently confirmed by Timp et al. (1988a) and Ford et al. (1988a, 1989b). It presented an interesting puzzle since it suggested that the symmetry between the two Hall leads attached to a narrow current-carrying conductor that is broken by the application of a transverse magnetic field is somehow restored for very narrow conductors at low magnetic fields, but it was unclear what mechanism could be responsible for this.

4.6.1. Classical considerations

A reasonable intuitive explanation of the observed quenching of the Hall resistance was suggested by the experiments of Ford et al. (1989b) on junctions with modified geometries. The essential idea is illustrated in fig. 31 (Ford et al. 1989b). The quantum wires under consideration, while narrow, are wide enough to have several transverse

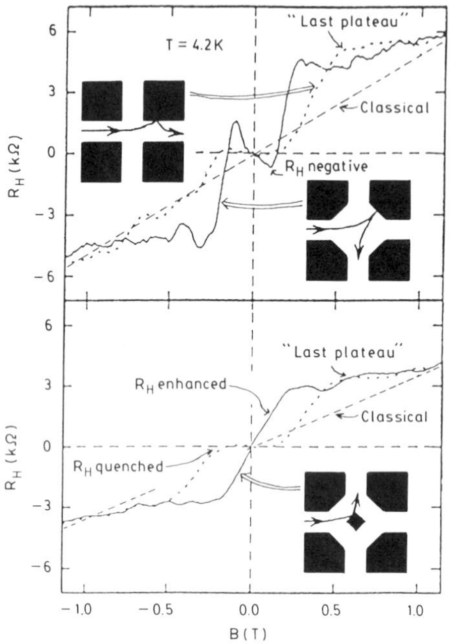

Fig. 31. Hall resistance of modified junctions. Black areas represent gates defining the junctions. Unmodified junctions (top left) give rise to quenched Hall resistances (dotted), while widened junctions (top right) show negative Hall resistance at low B due to electrons rebounding from the junction boundary. Widened junctions with a reflecting dot in the center show an enhanced Hall resistance at low B due to electrons rebounding from the dot. The semiclassical electron trajectories that give rise to these Hall resistances are shown. From Ford et al. (1989b) and Ford (1991a).

levels occupied by electrons, i.e. they are not single-mode waveguides. This means that qualitatively one may think of the electrons as following classical billiard-ball trajectories, bouncing off the edges of the depleted regions (the black areas of fig. 29). A dramatic consequence of this occurs in widened junctions defined by gates with truncated corners (fig. 31 top right inset). Here electrons that enter the junction from the left and are deflected by the magnetic field towards the top Hall lead, do not arrive in that Hall lead at low magnetic fields but are instead reflected into the opposite (bottom) Hall lead, giving rise to a negative Hall voltage and a negative Hall resistance. One can imagine that in a structure with square gates, where the junction geometry is smoothed only by the electrostatics as in fig. 29, this effect would be less pronounced and give rise instead to a quenching of the Hall resistance. Clearly the details of the junction geometry are very important.

This classical picture has been developed theoretically in considerable detail by several authors (Beenakker and van Houten 1989b, 1990, 1991a, Roukes and Alerhand 1990, Baranger 1990, Jalabert et al. 1990, Baranger et al. 1991) who performed computer simulations of the electron billiard-ball kinetics. Beenakker and van Houten (1989b, 1990) demonstrated qualitative agreement between such simulations and various features of the experimental data. They associated quenching of the Hall resistance with complex trajectories of electrons in junctions with smooth geometries, involving many successive collisions of an electron with the junction's wall. A weak magnetic field is not effective in deflecting electrons traveling on such "scrambled" trajectories into the Hall lead "favored" by the Lorentz force, resulting in the disappearance of the Hall resistance. A closer examination of these scrambled classical trajectories by Roukes and Alerhand (1990) and by Jalabert et al. (1990), Baranger et al. (1991) showed them to be an example of classical chaos. Beenakker and van Houten (1989b, 1990) were also able to reproduce the negative Hall resistance observed by Ford et al. (1989b), and obtained an anomalous Hall plateau at higher values of B qualitatively similar to those observed experimentally (Simmons et al. 1988, Timp et al. 1987c, Roukes et al. 1987, 1990b, Timp et al. 1988b, Ford et al. 1988c, 1989b, Chang 1989, Chang et al. 1988a). The classical picture could also account qualitatively for the behavior of the bend resistance of quantum wire junctions observed by Takagaki et al. (1988a, 1989a,c,d, 1990d), Timp et al. (1988a), Timp (1990) and Roukes et al. (1990a,b,c). It can also explain the magnetotransport coefficients observed by Main et al. (1990) in structures with more open geometries where the four quantum wires are replaced by four quadrants of a 2D electron gas separated by narrow gates.

A detailed comparison between the classical simulations of Beenakker and van Houten (1989b, 1990) and experiments on specially tailored quantum wire junctions was carried out by Roukes et al. (1990b). They found broad qualitative agreement, but large deviations from a number of universal scaling rules predicted by the classical scattering theory. They tentatively attributed these discrepancies to the influence of random scattering and/or quantum diffraction effects that were not considered by Beenakker and van Houten (1989b, 1990).

It is interesting to compare the above results with ballistic experiments on structures having a similar geometry but much larger dimensions (several microns), that have recently been carried out by Hirayama et al. (1991a–c). By comparison, these larger structures are in the extreme classical limit, so that interference effects can be ruled out, and the geometry is influenced little by electrostatic smoothing effects. In contrast to the small junctions, the Hall resistance of these systems does not display quenching or negative values but shows oscillations and a plateau. These features are due to classical electron focusing that is somewhat analogous to that studied earlier by van Houten et al. (1989) in a two-terminal geometry with quantum point contacts, and is reviewed in § 3.1.

4.6.2. Quantum effects

The classical phenomenology can account qualitatively for a number of phenomena observed in quantum wire junctions at low magnetic fields, provided that the temperature is high enough to smear out interference effects. However, as the magnetic field increases there is a crossover to the quantum Hall regime, which cannot be discussed classically (Chang et al. 1988b, Chang 1989, Roukes et al. 1990a, Ford et al. 1990, 1991, Timp 1990); even some evidence of a fractional quantum Hall effect has been reported (Timp et al. 1989). Also, since the number of transverse quantum levels populated in the quantum wires is not very large (typically it is in the range 3–10), it is clear that the junctions are inherently quantum systems, and a full quantum treatment of them is more appropriate. As still narrower quasi-one-dimensional conductors of high quality are developed this will be increasingly the case. Other related systems such as junctions of quantum wires constructed out of chains of quantum dots that are predicted to exhibit interesting Hall behavior (Shi and Kirczenow 1991), have no classical analog at all.

Most quantum theories of transport through junctions have been based on the scattering approach to conduction in low-dimensional systems, already discussed above for the simpler case of two-probe geometries. That is, following Landauer (1957, 1970, 1985), transport is treated as a scattering problem, but here the scattering center is the junction. For multilead junctions, the most convenient formulation is in terms of Büttiker's (1986) equations

$$I_\alpha = \left(N_\alpha \mu_\alpha - \sum_\beta T_{\alpha\beta} \mu_\beta \right) \frac{q_e}{h}. \tag{8}$$

These equations express the current I_α in each 1D lead α as the current $N_\alpha \mu_\alpha q_e/h$ injected directly into it from the reservoir at an electrochemical potential μ_α that is connected to it, summed algebraically with the currents $T_{\alpha\beta}\mu_\beta q_e/h$ injected into lead α from leads β via scattering processes occurring at the junction. Here $T_{\alpha\beta}$ is the probability (summed over subbands and spin) that an electron incident on the junction from lead β is scattered into lead α. $T_{\alpha\alpha}$ are reflection coefficients. N_α is the number of populated channels (subbands and spin) in lead α, and q_e is the electron charge. Equation (8) is the direct extension to the multilead multichannel case of the well-known two-probe 1D Landauer conductance formula $G = 2Te^2/h$ of Economou and Soukoulis (1981a,b) and Fisher and Lee (1981), as interpreted by Imry (1986a) and Büttiker (1986), where T is the transmission probability across the scatterer in a one-dimensional conductor.* Equation (8) can also be obtained from the formal arguments of Baranger and Stone (1989a), Feng (1990a–c), Sols (1992), Datta (1991), and Prêtre (1991). Recently Büttiker has extended the theory to include the effects of noise (Büttiker 1990b). Pastawski (1991b), Sols (1991) and van Wees (1991a), among others, have also considered other uses and extensions of this formalism.

The main task of the theoretical work on junctions has been to calculate the scattering coefficients $T_{\alpha\beta}$, since if they are known the Hall resistance and the other transport coefficients can be obtained by solving eq. (8). Strictly speaking, $T_{\alpha\beta}$ in eq. (8) is the probability of an electron being transmitted from reservoir β to reservoir α, including the effects of electron scattering by imperfections in the leads, but usually ideal leads have been assumed for simplicity. An experimental determination of all of the $T_{\alpha\beta}$ coefficients of quantum wire junctions, by current and voltage measurements and application of eq. (8), has recently been reported (Roukes 1991b).

*It should be noted that, as in the case of ballistic constrictions (see § 4.1.2 and Szafer and Stone (1989)), the considerations leading to Landauer's two-terminal conductance formula $G = (2e^2/h)T/(1-T)$ (Landauer 1970, 1989a) are not relevant here, since we are interested in measured voltages of the macroscopic electron reservoirs that are attached to the leads (Imry 1986a, Stone and Szafer 1988).

The first quantum theories of the Hall resistance of a junction of narrow conductors were those of Peeters (1988, 1989b), Akera and Ando (1989) and Devenyi and Imry (Imry 1989) who considered quantum wires with Hall leads weakly coupled to them through tunnel barriers. They found that such models do not yield a quenching of the Hall resistance. This result was later confirmed by Li and Thouless (1990). The question of whether the Hall effect can be meaningfully discussed at all in a ballistic quasi-one-dimensional conductor undisturbed by the presence of *any* Hall probes, was considered by Kirczenow (1988b). He concluded that a Hall resistance could be identified and measured non-invasively in such as system, and that it would display a quantum Hall effect directly related to the quantized conductance of ballistic constrictions (van Wees et al. 1988a, Wharam et al. 1988a), but that there would be no quenching of the Hall resistance at low magnetic fields (Kirczenow 1988b). Thus it is clear that the quenching of the Hall resistance is not an intrinsic property of narrow quantum conductors as such, or of quasi-one-dimensional conductors with weakly-coupled Hall probes connected to the quantum conductor by tunnel barriers. However, very recent experiments by Roukes and collaborators (Shepard et al. 1992) have shown that Hall probes weakly coupled to the conductor by narrow point contacts (not tunnel barriers) behave differently; the measured Hall resistance exhibiting both negative values and quenching at low B.

4.6.2.1. Simple junctions. Quantum theories of transport through simple model junctions with strongly coupled current and Hall leads were developed by Kirczenow (1989c–e), Ravenhall et al. (1989), Schult et al. (1990) and Akera and Ando (1990a,b). In these theories the junctions were of uniform conductors, without electrostatic rounding of the corners where the quantum wires meet. The Hall and current leads were treated on an equal footing, with no tunnel barriers separating any of the leads. Quantum wires with soft, parabolic confining potentials (Kirczenow 1989c–e) and with infinite square-well potentials (Ravenhall et al. 1989, Schult et al. 1990) were both studied, with similar results. Despite the simplicity of these models, they display a remarkable richness of interesting physics.

Some examples of the calculated Hall resistances as functions of magnetic field are shown in fig. 32. The Hall resistance shows obvious quantum Hall plateaus. This is interesting since the structure considered, being a junction of 1D ballistic conductors, is very different from the 2D electron gases with disorder in which the quantum Hall effect has usually been observed (Prange and Girvin 1990). The 2D quantum Hall effect has been described by Streda et al. (1987) and Jain and Kivelson (1988a,b) in terms of the Landauer formalism applied to 1D magnetic electron edge states, the quantization of the Hall resistance being due to the suppression of backscattering of the edge states in a strong magnetic field, as discussed by Büttiker (1988b). Although the electron modes of a narrow quantum wire are not true edge states, the quantum Hall plateaus in fig. 32 are also associated with the suppression of backscattering, in this case by the junction; all of the electrons incident on the junction from any lead being transmitted into the lead favored by the Lorentz force (Kirczenow 1989d).

There are also anomalous "last plateaus" visible at lower magnetic fields in fig. 32. These are not quantum Hall plateaus since their Hall resistances are not close to the

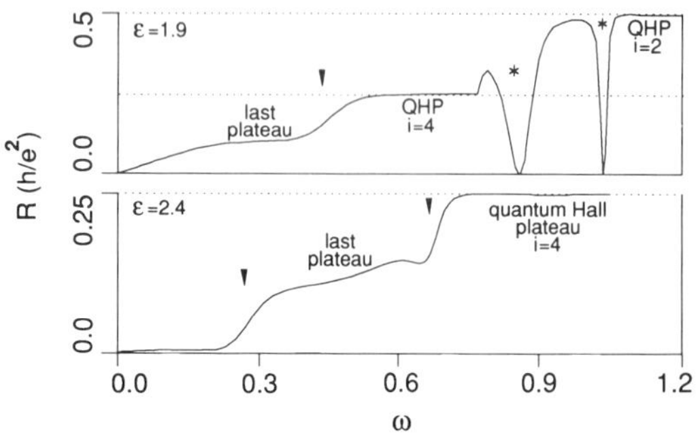

Fig. 32. Examples of the Hall resistance calculated for junctions of uniform quantum wires with parabolic confining potentials at constant Fermi energy plotted against magnetic field. $\omega = \omega_c/\omega_0$, where $\omega_c = eB/m$ is the cyclotron frequency, and ω_0 is the frequency of the harmonic confinement potential in the absence of magnetic fields. $\varepsilon = E_F/\hbar\omega_0$ is the normalized Fermi energy. Notice the quantum Hall plateaus, anomalous "last plateaus", the quench of the Hall resistance at low magnetic fields in the lower panel, and resonances indicated by asterisks and pointers. From Kirczenow (1989e).

canonical quantized values and the "plateaus" have a noticeable slope. In some cases the Hall resistance is quenched at low magnetic fields as in the lower panel of fig. 32. There are also sharp dips in the Hall resistance that are due to resonant states at the junction of the quantum wires. Such resonant states have also been studied by a variety of different techniques by Schult et al. (1989), Peeters (1989a,b), Kirczenow (1990a), Büttiker (1988a, 1990c), Paranjape (1991a), Berggren and Ji (1991), Weisz and Berggren (1990), Rundquist et al. (1990) and Ono (1992). Physically they can be viewed as being due to electrons temporarily trapped in the junction, because an electron's quantum kinetic energy can be reduced by moving it from a lead to the junction where the geometrical confinement is less stringent. A careful analysis shows (Kirczenow 1989c–e, Ravenhall et al. 1989, Schult et al. 1990) that such resonant states are responsible not only for the sharp structures marked with asterisks (*) in fig. 32 but also for the anomalous "last plateaus" and quenching of the Hall resistance that are shown.

The resonances are also predicted to have characteristic signatures in the bend resistance of junctions (Kirczenow 1989d, 1990a). *In addition* to these resonances, the bend resistance in these models displays extra structures when the Fermi level crosses the bottom of any of the 1D subbands of the quantum wires, as was predicted by Avishai and Band (1989c) and Kirczenow (1989d). Such band-edge crossing effects in the junction bend resistance have been confirmed experimentally by Behringer et al. (1991) and modeled in greater detail by Baranger and Stone (Behringer et al. 1991, Baranger 1990, Baranger and Stone 1990a,b). The calculations of Avishai and Band (1989c), Kirczenow (1989d) and Baranger and Stone (Behringer et al. 1991, Baranger 1990, Baranger and Stone 1990a,b) were all quantum mechanical. However, Roukes and his collaborators (Shepard et al. 1992) have recently pointed out that quantum phase coherence is not required for such band-edge bend resistance anomalies in realistic models of junctions, possibly explaining why they are seen even at relatively high temperatures, above 1 K (Behringer et al. 1991).

It should be noted that the quenching of the Hall resistance found in the models of junctions of uniform quantum wires (Kirczenow 1989c–e, Ravenhall et al. 1989, Schult et al. 1990) occurs only in a series of relatively narrow ranges of Fermi energy since it is associated with resonances, and it is only found for very narrow quantum wires with one to three transverse modes populated. It differs fundamentally from the quenching of the Hall resistance that has been observed experimentally to date. The experiments have been performed on wider quantum wires, and the quenching is usually observed over a single wide range of Fermi energy or gate voltage. This is referred to as "generic quenching" (Baranger and Stone 1989b, 1990a,b). Non-generic quenching of the Hall resistance occurring in two separate ranges of electron density has also been observed in some samples by Roukes et al. (1990b), but not in the regime in which the resonant quenching is predicted by the above models. It was pointed out by Baranger and Stone and co-workers (Baranger and Stone 1989b, 1990a,b, Baranger et al. 1991) that in the wider quasi-one-dimensional conductors that are currently available, generic quenching of the Hall resistance requires a rounded junction geometry.

The sharp Hall resonances shown at high magnetic fields in fig. 32 have been predicted by Kirczenow, using analytic arguments (Kirczenow 1990a), to occur also in rounded junction geometries, but their observation will most likely require true ballistic quantum wires, and preferably single-mode electron waveguides, neither of which are as yet available. At the present time, the only theory of the Hall effect in junctions of quantum wires with defects is that of Akera and Ando (1990a,b) who considered junctions with unrounded corners and used a Boltzmann approach to estimate the gross effects of the potential irregularities in the quantum wires. They found that the junction Hall resonances may survive in the presence of the quantum wire defects, however, this finding is open to question since the effects of coherent electron scattering between the defects and the junction, which should alter the Hall resonances, were not included in the calculations.

4.6.2.2. Quantum collimation and scrambling in rounded junctions. Baranger and Stone (1989b, 1990a,b) carried out quantum mechanical calculations of the Hall resistance of junctions of wide quasi-one-dimensional conductors with a smooth geometry at low magnetic fields, and found the Hall resistance to be quenched over a wide range of model parameters (generic quenching). Their results are shown in fig. 33. Their physical interpretation of these results (reviewed in depth recently by Baranger et al. (1991)) was as follows: as an electron in a particular transverse mode approaches the junction, the quantum wire broadens smoothly, and the transverse kinetic energy of the electron decreases because the electron is less confined in the transverse direction. Accordingly the kinetic energy and momentum of the electron in the forward direction increase, an adiabatic "collimation" effect. Thus, at low magnetic fields, the electron accelerates as it approaches the junction, and passes straight through the junction almost without noticing the Hall leads, so that the Hall voltage and Hall resistance are suppressed. This quenching mechanism is quite the opposite of the mechanism advocated by Beenakker and van Houten (1989b, 1990) in which the quenching occurs because an electron undergoes many collisions with the walls of the junction,

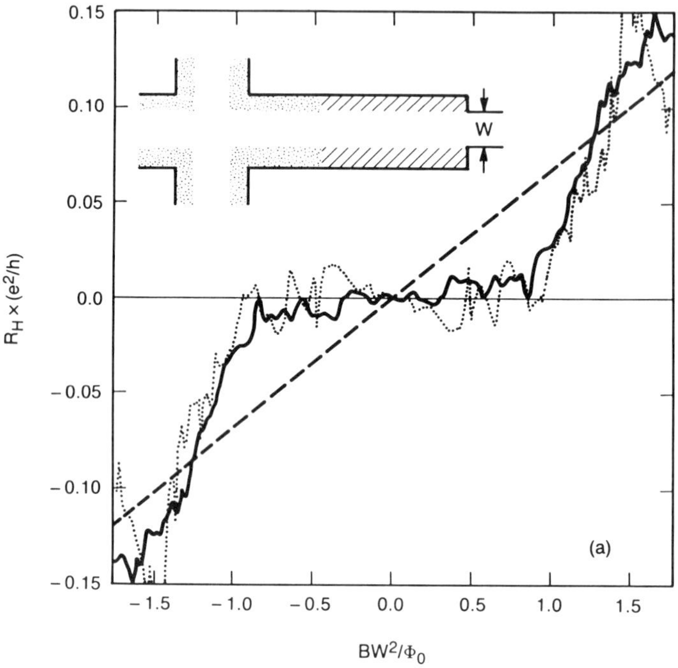

Fig. 33. Calculated Hall resistance against magnetic field for a smoothly tapered quantum wire junction. The shading in the inset indicates the region over which the effective widths of the quantum wires are graded. The lowest four transverse modes are populated with electrons in the leads. Dotted line, $T = 0$; solid line, $T = 3E_1$, where E_1 is the energy of the lowest transverse level; dashed line, Hall resistance of macroscopic diffusive 2D system. Notice roughness of R_H in the quench at low B. From Baranger and Stone (1989b).

"scrambling" its classical trajectory. Note, however, that both mechanisms require junctions with smooth geometries and both involve some degree of collimation of the electron beam incident on the junction.

Chang et al. (1989) found that quenching could be obtained by surrounding a wide junction by narrow constrictions, and interpreted this as a collimation effect. The importance for generic quenching of a smooth geometry that encourages collimation was confirmed experimentally by Roukes et al. (1990b). However, the identification by Baranger and Stone (1989b, 1990a,b) of simple transmission of electrons straight through the junction as the important quenching mechanism was not tested directly in the above experiments. Very recently, Greene et al. (1991) have reported measurements of the probability that an electron is transmitted straight through the junction as well as the probabilities of the electron being transmitted "around the bend" into either side lead. They found that all of these transmission probabilities are very similar in magnitude when the Hall resistance is quenched. This would support the classical scrambled-trajectory picture of Beenakker and van Houten (1989b, 1990), rather than the straight-through transmission advocated by Baranger and collabora-

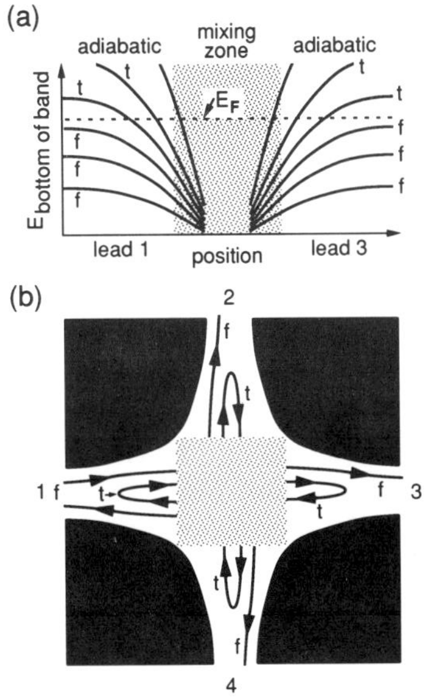

Fig. 34. Schematic representation of quasi-adiabatic (white) and non-adiabatic (shaded) regions of a junction with a smooth geometry. Mode mixing is strongest in the non-adiabatic central region of the junction. (a) Energies of the bottoms of the quasi-one-dimensional subbands of the quantum wires as a function of position in the junction. (b) Electrons at the Fermi energy that belong to modes labeled f are free to escape from the junction, while those belonging to modes labeled t are trapped in the junction because the energies of the bottoms of their subbands rise above the Fermi level, as depicted in (a). From Kirczenow (1990b).

tors (Baranger and Stone 1989b, 1990a,b, Baranger et al. 1991). However, it is clear that the final word in this debate has not yet been said.

The quantum analog of the classical scrambled trajectories of Beenakker and van Houten (1989b, 1990) has been studied theoretically by Kirczenow (1990b). He noted that *adiabatic* transmission of an electron straight through a junction in a single transverse mode of a quantum wire *cannot* occur because at the core of the junction these transverse modes do not exist. Thus while an electron can approach the junction quasi-adiabatically, non-adiabatic mode mixing *must* occur in the junction itself.* This is illustrated in fig. 34. An electron at the Fermi energy that enters the non-adiabatic junction core may be emitted from the core into any transverse mode of any lead provided that that mode is also at the Fermi energy. The bottom-of-the-band energies of some of those modes (labeled t) lie above the Fermi energy far from the junction, so that an electron emitted from the core into such a mode cannot escape from the junction but is reflected back into the junction core. That is, an electron entering the junction can become trapped there by non-adiabatic mode mixing and be reflected many times before being finally emitted from the junction

*This is somewhat analogous to the non-adiabatic mode mixing during propagation of electrons through ballistic constrictions that has recently been discussed by Glazman and Jonson (1990), Laughton et al. (1991) and Castaño and Kirczenow (1992).

in a “free” mode f whose bottom-of-the-band energy is below the Fermi level. This quantum analog of classical-trajectory scrambling in rounded junctions has been simulated numerically (Kirczenow 1990b). It was found that for junctions of narrow quantum wires with one or two populated 1D subbands, it gives rise at $T=0$ to very strong coherent interference effects and results in Hall resistances that are quenched or strongly positive or negative at low magnetic fields, the behavior depending very sensitively on the junction geometry and Fermi energy, as shown in fig. 35. However, for wider quantum wires similar to those studied by Baranger and Stone (1989b, 1990a,b), one finds relatively weak fluctuations in the Hall resistance (Kirczenow 1990b) similar to those visible in fig. 33. These weak quantum fluctuations due to coherent multiple reflections of trapped modes in large rounded junctions have recently been discussed in terms of “quantum chaos” by Jalabert et al. (1990) and Baranger et al. (1991). Other work analyzing these quantum fluctuations includes that of Takagaki and Ferry (1991) and Oakeshott and MacKinnon (1991).

The simple kT smearing of the Fermi surface that is reflected in the finite-temperature plot in fig. 33 is not very effective at smoothing out these quantum fluctuations in the Hall resistance. However, if phase breaking is introduced into the calculations, as was done by Kirczenow (1990b), the Hall resistance curves come to resemble the experimental data at He temperatures, which are fairly smooth. An interesting result was predicted in the presence of strong phase breaking (Kirczenow 1990b): even for junctions of very narrow quantum wires with only one or two subbands populated, there is a progression from positive to quenched to negative Hall resistances at low B as the Fermi level increases or the rounding of the junction becomes more pronounced. These trends are very similar to those observed experi-

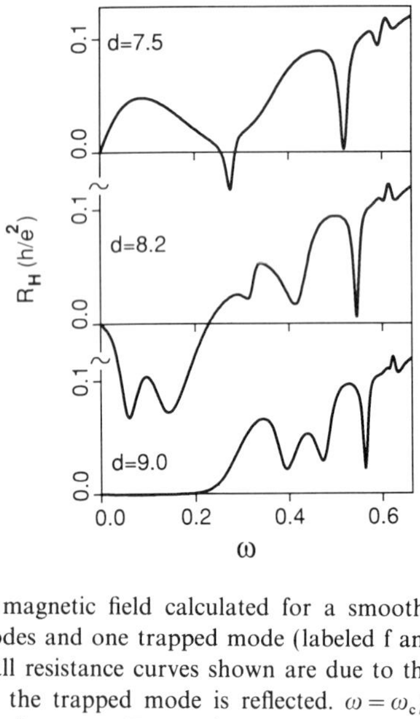

Fig. 35. Hall resistance against magnetic field calculated for a smoothly tapered junction of narrow quantum wires with two free modes and one trapped mode (labeled f and t, respectively, in fig. 34). The differences between the three Hall resistance curves shown are due to the different distances d from the center of the junction at which the trapped mode is reflected. $\omega=\omega_c/\omega_0$, where ω_c is the cyclotron frequency. From Kirczenow (1990b).

mentally in the junctions of wide quantum wires, where they can be understood qualitatively in classical terms (Roukes et al. 1990b, Ford et al. 1989b). In the case of the very narrow quantum wires discussed by Kirczenow (1990b), a classical description is entirely inappropriate, but the trends persist nonetheless, provided only that phase coherence is suppressed.

4.6.2.3. Quantum dots at quantum wire junctions. Quantum fluctuations in the Hall resistance of quantum wire circuits are observed at temperatures in the milliKelvin range (Williamson 1991), but it is normally very difficult to distinguish fluctuations that are associated with the junctions themselves from those that involve imperfections in the leads. In fact, if the quantum coherence length is large, these effects are inseparable, even in principle. For this reason no plausible experimental observations of either the resonances predicted to occur in rounded junctions at higher magnetic fields (Kirczenow 1990a), or of the quantum fluctuations due to multiple scattering of modes trapped in large junctions at low magnetic fields (Kirczenow 1990b, Jalabert et al. 1990, Baranger et al. 1991) have as yet been published. Another exotic quantum interference effect recently predicted by Avishai and Band (1991) to occur in a junction penetrated by a single narrow tube of magnetic flux will also be difficult to observe convincingly for similar reasons.

This does not mean, however, that the observation of coherent quantum interference effects occurring in junctions of quantum wires is impossible. Indeed, Ford et al. (1991) have recently observed such an interference effect in the Hall resistance of a modified junction in the form of a quantum dot connected to four quantum wires by narrow constrictions. Their results are shown in fig. 36. When the constrictions are narrowed close to pinch-off (fig. 36c), the Hall resistance exhibits four very strong minima extending down to zero, between $B = 0.5$ and 1.5 T, as well as finer irregular structure. These minima are unusually robust for a mesoscopic interference effect, remaining very strong even above 1 K, but finally smooth out into a plateau at $R_{\mathrm{H}} \approx h/4e^2$ for $T \approx 8$ K. This interesting type of behavior has been explained by Kirczenow and Castaño (1991) as a magnetic diffraction effect. In their model the main Hall resistance minima are the result of destructive quantum interference between the two outermost eigenmodes of the quantum dot as they travel between adjacent constrictions, for example, through the hatched region in fig. 37. The robustness of the minima to thermal smearing is due to the phase differences controlling the diffraction pattern being unusually insensitive to the electron energy. The experimental high-temperature plateau at $R_{\mathrm{H}} \approx h/4e^2$ is reproduced by the theory in the case of strong phase breaking. A similar physical picture involving interference between a pair of quantum dot modes as they propagate between constrictions may account for the strong peaks in the two-terminal magnetoresistance of quantum dots, observed recently by Taylor et al. (1992).

The theory of Kirczenow and Castaño (1991) requires mixing to occur between the quantum dot modes at the constrictions. Thus the modes of the quantum dot behave quite differently from the edge states of macroscopic 2D electron gases, which have been found to propagate over large distances without appreciable mixing in the experiments of Komiyama and Hirai (1989a,b), van Wees et al. (1989b) and

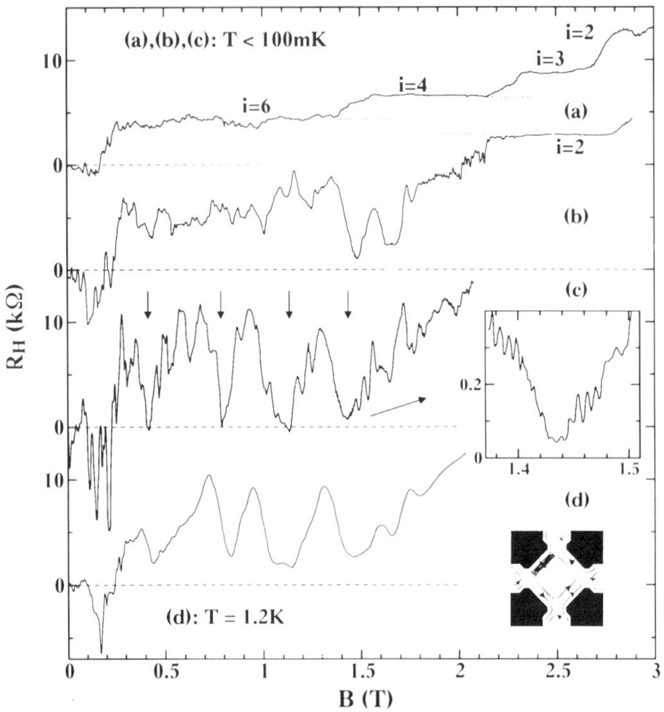

Fig. 36. Inset: the quantum dot is connected to four leads by relatively narrow constrictions. Traces (a)–(c) show the Hall resistance of the dot as the constrictions are progressively narrowed. While curves (a) and (b) show quantum Hall plateaus modulated by random fluctuations, curve (c), where the constrictions are narrowest, shows four strong Hall resistance minima going down to $R_H = 0$. Notice also quenching of R_H and negative R_H values at lower values of B in some curves. From Ford et al. (1991) and Ford (1991a).

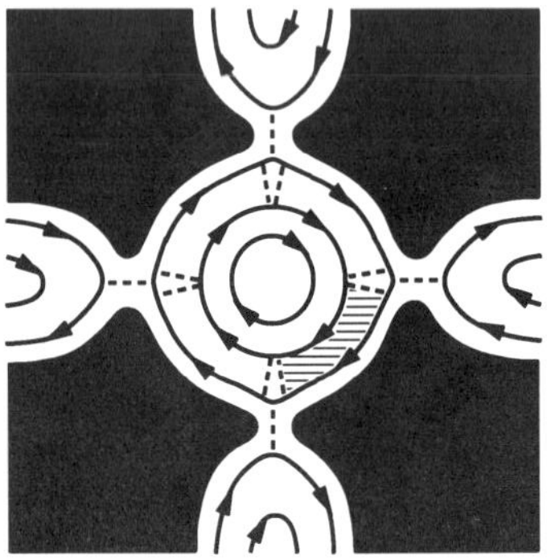

Fig. 37. Schematic drawing of quantum dot coupled to leads by constrictions. Interference resulting in Hall diffraction minima occurs across four regions, one of which is shown as being hatched. From Kirczenow and Castaño (1991).

Alphenaar et al. (1990). The reasons for this difference appear to be the confined geometry of the dot and the lower magnetic fields in the experiments of Ford et al. (1991), both of which are conducive to mode mixing in the presence of the strong disturbance of the dot potential by the constrictions. This interpretation is also consistent with the fact that the Aharonov–Bohm conductance oscillations observed earlier by van Wees et al. (1989a) in a much larger two-terminal quantum dot at higher magnetic fields could be adequately explained without considering mixing between different quantum dot modes.

Kirczenow and Castaño (1991) predict that the Hall diffraction minima should be present even if only two adjacent constrictions connecting the dot to the leads are close to pinch-off, while the other two are relatively wide. In this regime one would expect to observe the quantum interference without complications arising from Coulomb blockade effects (Averin and Likharev 1991), and this appears to be the situation in the initial experiments of Ford et al. (1991) where the effective widths of the individual constrictions could not be tuned independently, and thus were presumably different. More recently, in quantum dots with independently tunable constrictions, Ford (1991b) has observed the diffraction minima in the Hall resistance, but also Coulomb blockade oscillations in the two-probe conductance qualitatively similar to those observed in other devices by Meirav et al. (1990), Kouwenhoven et al. (1991a, 1992), Staring et al. (1991), and Williamson (1991). The interplay of such single-electron tunneling effects with the quantum interference involved in the magnetic diffraction phenomenon should be a fascinating topic for further study.

The following sections deal with arrays of quantum dots and with quantum dots produced by other techniques (not at wire junctions, but rather connected to only two probes): imposing barriers along single quantum wires, purposely changing the width of the wire over a certain region, or confining electrons by etching of the semiconductor structures.

4.7. Quantum dot systems

Since nanofabrication techniques have achieved a high degree of sophistication, a natural line to follow has been to progressively restrict the electronic motion within the solid. The extreme confinement achieved by various techniques is the creation of quasi-zero-dimensional or "quantum dot" systems. (One example of such a system, where two quantum wires intersect one another, has already been discussed in the preceding section.) In these systems, the electron is confined by electrostatic potentials or material boundaries in all three directions within regions comparable to its wavelength. It is clear that confinement of electrons to "boxes" or "dots" constructed with split-gate geometries and in short channels (with characteristic widths of several hundred to a few thousand angstroms) is enough to promote strong quasi-zero-dimensional effects. Indeed, numerous studies of various properties of quantum dots have already appeared in the literature.

The simplest and perhaps the most widely used model of an isolated quantum dot adopts the approximation of non-interacting electrons, confined in a 2D parabolic potential well. The single-electron energy spectra and wavefunctions for this model have been known for many years, both in the presence and absence of magnetic fields (Fock 1928, Darwin 1930). The qualitative validity of this very simple model has been supported by the fully 3D self-consistent Hartree calculations of Kumar et al. (1990), who considered a quantum dot defined by a gate on a textured GaAs–$Al_xGa_{1-x}As$ heterostructure. However, Maksym and Chakraborty (1990) and Merkt and collaborators (Merkt et al. 1991, Wagner et al. 1992) have pointed out that a more complete treatment of electron–electron interactions, albeit in a simpler two-dimensional model of the quantum dot, can lead to a qualitatively different energy level structure as the magnetic field increases. In their theory the total angular momentum of the interacting electrons in the quantum dot ground state exhibits "magic quantum numbers" and can differ from that predicted by the simple non-interacting electron picture. However, a different exactly-solvable many-electron quantum dot model studied by Johnson and Payne (1991) yields qualitatively different results, and the precise nature of the many-body quantum dot ground state is at present controversial (Johnson and Kirczenow 1992).

Explorations of the electronic states via capacitive measurements (see Smith III (1990) for a review) and of the collective electronic modes via infrared spectroscopy (see Que et al. (1991), Que and Kirczenow (1988), Brey et al. (1989), Dempsey et al. (1990), Chakraborty et al. (1991), and Kempa et al. (1991) for recent theoretical papers, and Merkt et al. (1989), Lorke (1992), and Heitmann et al. (1992) for brief overviews) are but a few examples of other work in this area. Transport studies of closely related systems, such as the grid potential modulations imposed on a 2D electron gas have also been discussed in the literature, although the characteristic sizes of these systems put them outside of the ballistic regime (Gerhardts et al. 1991, Weiss et al. 1991, Lorke et al. 1990, Washburn 1991). On the other hand, attaching current probes to individual dots or to several of them in chains allows studies of the electronic states via transport *through* quantum dots. It is these effects that are discussed in this section.

It is noteworthy that a similar degree of confinement in three directions is now possible with a combination of epitaxial and etching techniques, as described in § 2. "Vertical" quantum dots (where the electronic current is *perpendicular* to the semiconductor heterojunction, and clearly different from the laterally conducting electronic motion devices discussed so far) have interesting transport properties as well, as exemplified by the beautiful experiments of Reed et al. (1988), Tarucha et al. (1990), and Tewordt et al. (1990). The following section discusses tunneling through single dots of both the vertical and horizontal types, while the section after that (§ 4.7.2) covers chains of dots (see also the discussion for dots produced at the intersection of quantum wires in § 4.6.2).

4.7.1. Single dots

Reed et al. (1988) performed experiments on dots similar to those shown in fig. 4, where lateral confinement is achieved by etching the material, while confinement along the direction of electronic tunneling is provided by the conduction band discontinuities of the constituent semiconductors (GaAs–AlGaAs–InGaAs). In these systems, the confinement lengths are only a few hundred angstroms, so that indeed quasi-zero-dimensional confinement is achieved (in the absence of leads). Their experimental $I-V$ curves show a series of features that correlate well with the calculated resonances through the discrete states (see fig. 38) (Reed et al. 1988, Tarucha et al. 1990, Tewordt et al. 1990). Bryant has presented calculations taking into account the expected mixing of the transverse levels to explain the fine structure appearing in the experimental curves (Bryant 1989, 1990, 1991a). These features appear as the effective width of the dot changes rather suddenly in the various layers of the structure, due to differences in lateral depletion widths, producing "non-adiabatic" mode mixing. Tewordt et al. (1990) report evidence of inelastic scattering in the $I-V$ curve of vertical tunneling experiments, from which they conclude that the tunneling in their devices is sequential (see chapter 12 in this volume).

Similar in spirit to the vertical quantum dot, although in a 2D realization of the tunneling structure, the Cavendish (Smith et al. 1988, Wharam et al. 1989b, Brown et al. 1989c), Delft University–Philips (van Wees et al. 1989a, Kouwenhoven et al. 1989a), and NRC (Taylor et al. 1991c, 1992) groups have presented results for dots in split-gate geometries. Smith et al. (1988, 1989b, 1990b) have presented conductance measurements in a dot formed by shaping the gate in the form of a small nearly square box, so as to form a region of size about 3000 × 3000 Å^2, with finite barriers along the tunneling direction. Figure 39 shows a diagram of the gate used and a typical resistance trace for this device. Notice the appearance of strong resonance-like features associated with the transmission through the quasi-bound states residing in the dot (see discussion on geometrical resonances above, § 4.3.3). These features are robust to moderate temperatures ($T \leqslant 2$ K) and magnetic fields ($B \leqslant 0.3$ T), where the thermal or magnetic energy remains smaller than the binding energy of the localized state (about 1 meV at about 11 K). These characteristics are reproduced

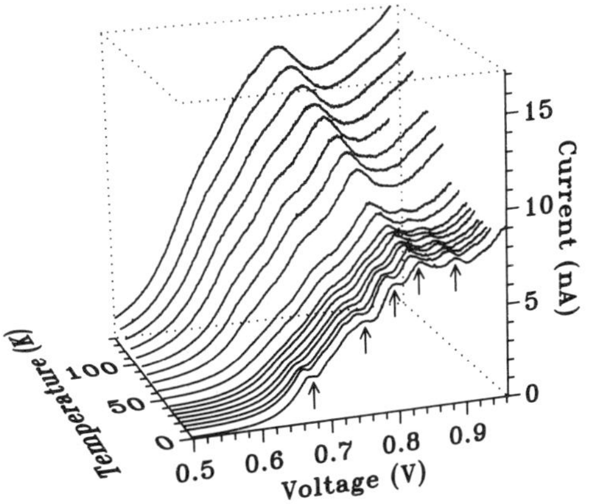

Fig. 38. Current–voltage characteristics of single "vertical" quantum dot structure as a function of temperature. Resonant tunneling through isolated quasi-bound states produces features in low-temperature traces, as indicated by the arrows. From Reed et al. (1988).

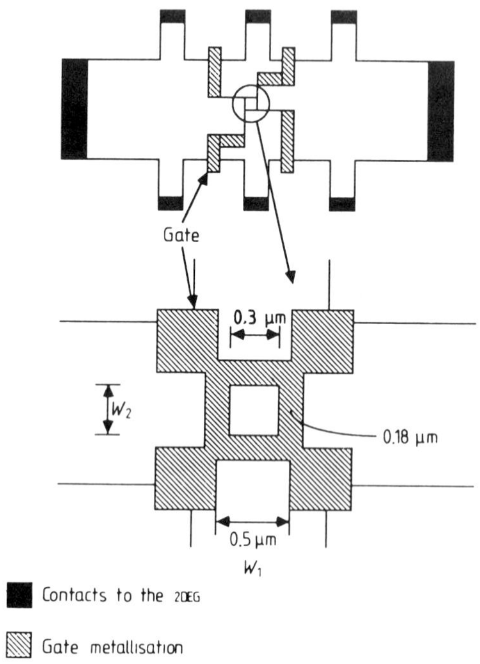

Fig. 39. (a) Schematic diagram of "horizontal" quantum dot gated device. (b) Resistance of device at 330 mK versus gate voltage. Scale changes by a factor Y for different traces. Peaks in the curve correspond to resonant tunneling through the quantum dot. Dashed horizontal segments indicate values of $R = h/2ie^2$, for different values of i. From Smith et al. (1988).

well by a calculation of the conductance in a model structure that simulates this device (Martín-Moreno and Smith 1989). Agreement is best in the high-conductance regime ($G \geqslant 2e^2/h$). The discrepancy in the weak tunneling (small-conductance) regime may be due to difficulties in a precise evaluation of the device parameters. On the other hand, discrepancies between model and experiments in the weak tunneling regime may be due to the effects of charge accumulation in the device. This interesting regime, in which a *Coulomb blockade* may be important in controlling the conductance of a device, has recently given rise to a lot of activity in this field. We refer the reader to some of the original experimental (Scott-Thomas et al. 1989, Field et al. 1990, Meirav et al. 1989, 1990, McEuen et al. 1991, Brown et al. 1990) and theoretical (Glazman and Shekhter 1989, van Houten and Beenakker 1989, Groshev et al. 1991, Meir et al. 1991) articles, as well as to some recent brief overviews of this rapidly developing field (van Houten 1992, Averin and Likharev 1991, Beenakker and van Houten 1991a).

Lateral quantum dots have also been studied in the regime of high magnetic fields (van Wees et al. 1989a, Brown et al. 1989c, Wharam et al. 1989b, Taylor et al. 1991c, 1992), where the Aharonov–Bohm effect in this singly-connected geometry has been observed thanks to the appearance of edge states, as discussed in the magnetic field effects section, § 4.5. Microscopic tight-binding calculations giving a possible alternative description of such Aharonov–Bohm resonances and also of the related persistent

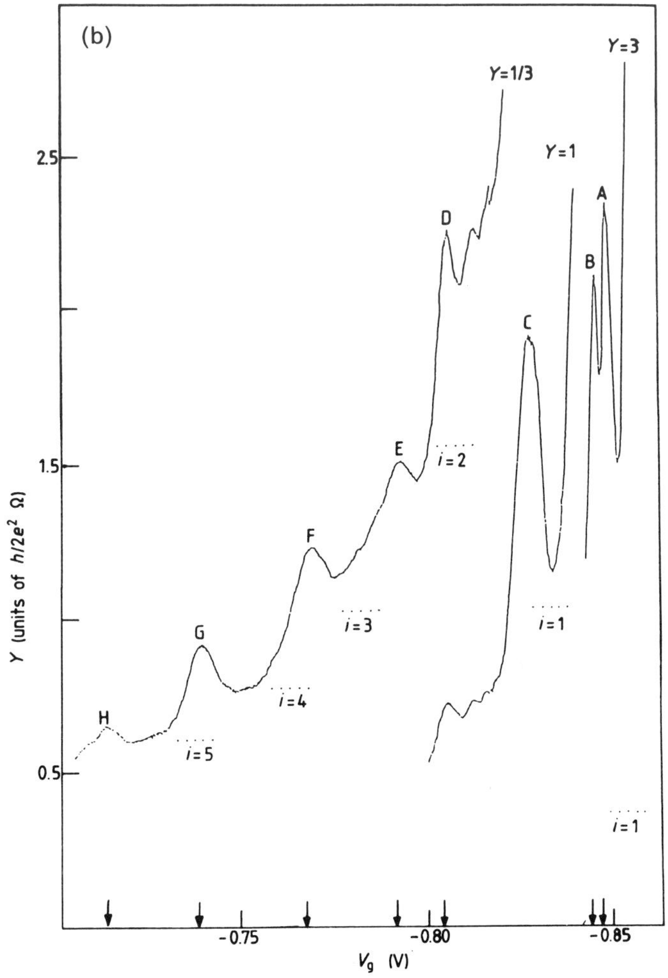

Fig. 39. *continued*

currents in a single quantum dot have been published by Sivan et al. (1989b) and Sivan and Imry (1988). Interesting Hall resonances in quantum dots with four side probes, created at the intersection of quantum wires, are also discussed in § 4.6.2.

4.7.2. *Quantum dot chains*

The simplest structure with coupled quantum dots is the double quantum well device of Reed et al. (1989), also described theoretically by Bryant (1990, 1991b). It is found theoretically that tunneling occurs via states localized mostly in one of the wells, or via a state that couples resonantly to both dots, producing fine structure in the I–V curve. That this structure is not observed in experiments is perhaps due to thermal averaging effects.

A different type of coupled quantum dot system is the structure proposed by Ulloa and co-workers (Ulloa et al. 1990a,b, Castaño et al. 1990), and realized in ingenious experiments by Kouwenhoven et al. (1990a,b, 1991b) and Haug et al. (1990a,b, 1992). Further calculations on similar structures have also been presented by Brum (1991a,b), Weisshaar et al. (1990b, 1991b), and de Aguiar and Wharam (1991). The structure is a chain of quantum dots connected in sequence so as to achieve a series of coherent quasi-bound states across the whole chain. These states are the precursors of the Bloch states of the quasi-one-dimensional *crystal*, in the limit of an infinite quantum dot chain. In a tight-binding scheme, they can be viewed as arising from the overlap of multiply-degenerate states isolated in each dot (for a chain of nearly identical dots), mixed by the finite transparency of the intervening barriers. If probes are attached to the structure, it is reasonable that these states will appear as transmission resonances in a conductance experiment. Figure 40 shows the predicted conductance curve for a device with four connected dots, clearly showing a series of four resonance peaks associated with each transverse subband.

Figure 41 shows an experimental conductance curve of Kouwenhoven et al. for a chain with fifteen quantum dots (Kouwenhoven et al. 1990a). The trace shown displays groups of 15 conductance peaks separated by miniband gaps in a conductance plateau. On the other hand, Haug et al. (1990a) used several devices to probe the transmission structure in chains of fewer dots. Figure 42 shows their experimental data for a chain with four dots (see fig. 2), which clearly exhibits features well-correlated with the designed gate structure. Similar structure for higher conductance values was also observed (Haug et al. 1992). Notice that both groups report the need to apply a magnetic field ($B \approx 2$ T) in order for the conductance features to be well

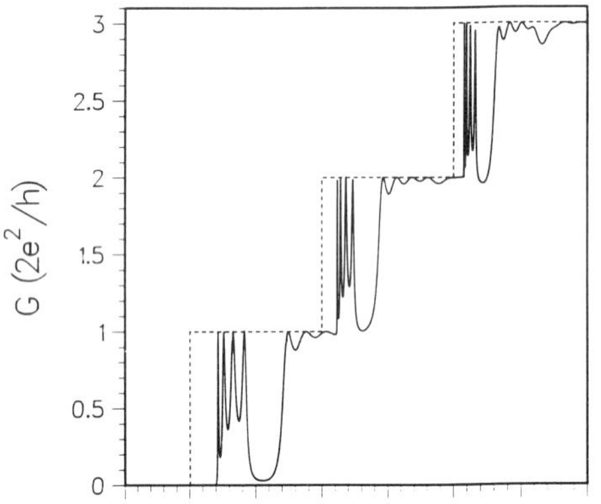

Fig. 40. Conductance of quantum dot chain with five barriers versus Fermi wavenumber (solid line), at $T=0$. Conductance steps of unmodulated ideal constriction are shown dashed. Potential modulation provided by barriers along the constriction produces strong resonances and upward shifts of the subband onsets. Conductance drops at $k_F W/\pi \approx 1.6$, 2.4, and 3.2 are identified with the first miniband gap of each transverse mode in the structure. The second gap of the third transverse mode is weakly visible at $k_F W/\pi \approx 3.7$. From Ulloa et al. (1990a).

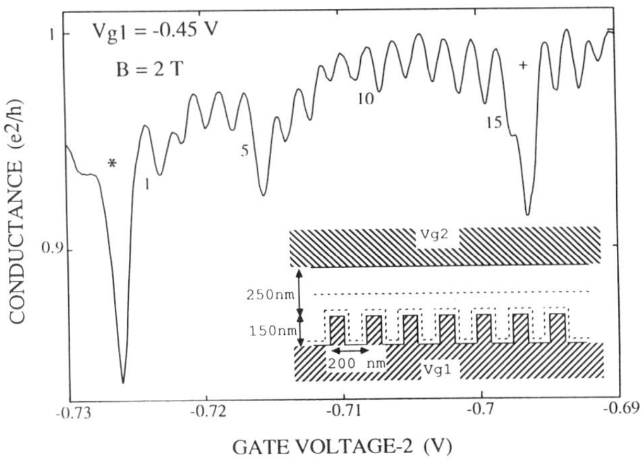

Fig. 41. Conductance of patterned device against gate voltage V_{g2}, at 2 T and 10 mK. Curve shows 15 oscillations, corresponding to the same number of quasi-localized states produced by 16 fingers in the gate (schematically shown in the inset). The upper depletion region (shown as a dashed line) is shifted down towards the fingers when V_{g2} is made more negative. Miniband gaps are labeled by * and +. From Kouwenhoven et al. (1990a).

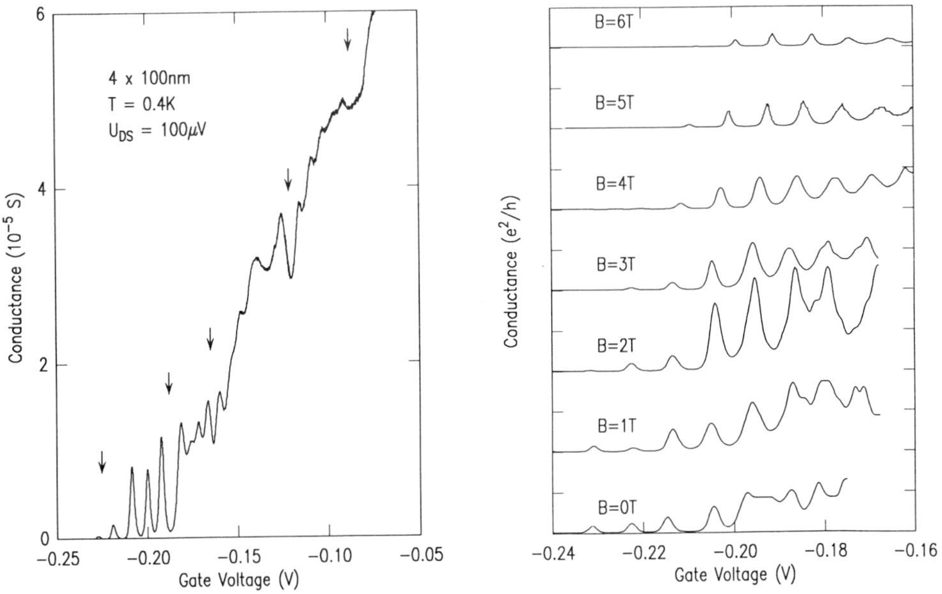

Fig. 42. (a) Conductance plotted against gate voltage for the device of fig. 2 with five barriers (fingers), at 0.4 K and zero magnetic field. Arrows indicate miniband gaps, as the number of resonances between any two of them is four. (b) Magnetoconductance plotted against gate voltage for device in (a) for several magnetic fields. Notice non-monotonic dependence of the resonance structure strength as a function of the field. From Haug et al. (1990a).

defined. These magnetic fields suppress impurity backscattering (see § 4.5), so that the scattering induced by the structured gate appears more prominently. Haug et al. report, however, that a further increase of the magnetic field decreases the transmission features (see fig. 42), due perhaps to the formation of edge states on both sides of the channel, which makes backscattering by the structure more difficult (see section on edge states, § 4.5.1) (Lent and Leng 1991a,b).

It should again be noted that ignoring Coulomb blockade effects in the problem of transmission through quantum dots may be warranted in the high-conductance regime ($G \approx e^2/h$). However, more weakly conducting regimes may involve those effects, although the details of the specific structure would be significant in determining their relative importance.

Another interesting structure proposed is a junction of four quantum-dot chains, in a geometry similar to the junctions of quantum wires discussed in § 4.6. Shi and Kirczenow (1991) show that, under suitable conditions, this array of crossed dot chains may exhibit a negative Hall resistance and a quenching of the Hall effect due to quantum interference. This is a generalization of the quantum interference phenomena found in theory (Kirczenow and Castaño 1991) and experiment (Ford et al. 1991) in single dots formed at the junctions of quantum wires.

4.7.2.1. Non-linear response. Castaño et al. (1990) have also studied the I–V characteristics of a quantum dot chain in the interesting non-linear regime. This is a conceptually important issue, since these systems may allow the observation of the elusive Bloch oscillations in solids, as well as the so-called Stark ladders. As described above, the conductance in the low-field regime is strongly modulated by the presence of minibands and minigaps associated with the superlattice, even for systems with only a few periods. This modulation is not lost in the high-field regime. In fact, the number of features in the differential conductance is doubled, due to the explicit twofold action of the applied voltage described in § 4.4. This is shown in fig. 43. As the applied voltage is increased, the states of the miniband associated with a given dot become misaligned, producing a strong reflection and a decreasing current until the next state falls in line. This process continues until the field shifts the minigap into resonance with the given state and the current decreases, as shown in fig. 44. This situation is repeated as the states in the next minibands and minigaps are shifted into alignment.

More work is needed in furthering our understanding of the non-linear response of quantum dot chains. No experiments have been reported on these systems in the non-linear regime to date. Both theoretical and experimental studies are needed to clarify the competing roles of electric-field localization (in a Stark ladder) and the band conduction in this quasi-one-dimensional crystal. Moreover, the predicted conductance drops could allow the fabrication of a versatile quantum mechanical ballistic oscillator in 2DEGs.

4.7.3. Two-dimensional arrays

The quantum mechanics of electrons in two-dimensional periodic potentials and transverse magnetic fields has fascinated theorists for many years. Some interesting

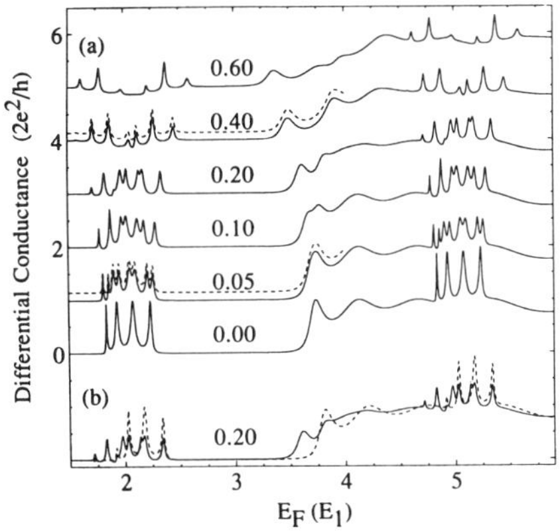

Fig. 43. (a) Differential conductance plotted against Fermi energy E_F (in units of E_1, the first transverse mode energy) for a quantum dot chain and various values of d.c. bias V_0 (see eq. (7)), as indicated. Solid lines are the results for adiabatic constrictions, dashed lines for sharp constrictions (offset upwards by 0.15 for clarity). The system depicted has five barriers (four resonances) in constriction. The differences introduced by injection (solid and dashed lines) are minimal, proving that the potential modulation in the constriction dominates the behavior. (b) The solid line is the differential conductance and the dashed line is the transmission coefficient for a static potential. The double-peak structure in the solid line comes from the V_0-dependence of the transmission coefficient. From Castaño et al. (1990).

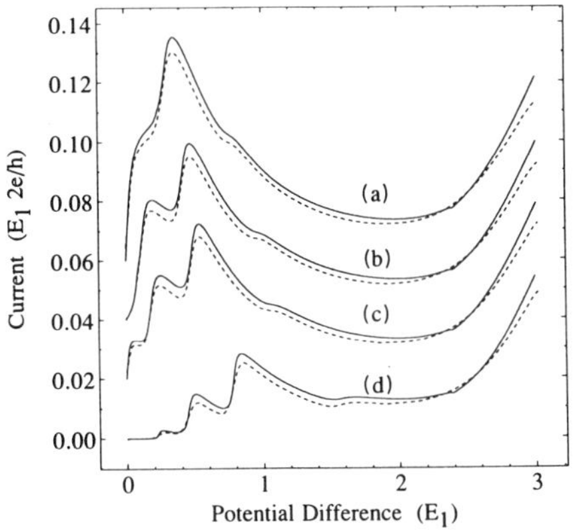

Fig. 44. Current–voltage characteristics for various Fermi energy values: 2.080, 2.000, 1.835, and 1.700 in curves (a)–(d), respectively. Solid lines are the results for adiabatic constrictions, dashed lines for sharp constrictions. Each pair is displaced by 0.02 from the preceding pair for clarity. Resonant tunneling through successive levels of the quantum dot chain produces the structure at low bias voltages. From Castaño et al. (1990).

predictions have been that the single-particle energy spectra of such systems should exhibit a fractal gap structure as a function of magnetic field (the Hofstadter "butterfly" spectrum) (Zilberman 1957a,b, Brown 1964, Azbel 1964, Rauh et al. 1974, Hofstadter 1978, Wannier 1978, Wannier et al. 1979, Schellnhuber and Obermair 1980) and that when the Fermi level lies in a spectral gap the Hall conductance should be quantized with positive or negative index (Thouless et al. 1982, Rammal et al. 1983, MacDonald 1984, Avron et al. 1983, Niu and Thouless 1984, Kohmoto 1985, Niu et al. 1985, Aoki 1985, 1992). That is, such systems are predicted to exhibit a novel type of quantum Hall effect that is due to their spatial periodicity, and fundamentally different from the integer and fractional quantum Hall effects that have been observed to date (Thouless 1990). The experimental observation of these effects is a long-standing goal of condensed-matter physics and materials science. The requirements are 2D spatial periodicity, quantum coherence, and that the magnetic flux threading a unit cell be of the order of a flux quantum $\Phi_0 = h/e$. For periodic semiconductor nanostructures with periods about 100 nm the magnetic flux requirement can be satisfied by easily realized magnetic fields of about 1 T. However, the fabrication of the required very high quality samples with a sufficient degree of spatial order remains a major challenge in this field. Some progress in this direction has been achieved recently with the fabrication of laterally conducting 2D arrays of quantum dots (Ismail et al. 1989, Puechner et al. 1990, Liu et al. 1990, Lorke et al. 1990), "antidots" (Fang and Stiles 1990, Ensslin and Petroff 1990, Kern et al. 1991, Weiss et al. 1991) and related systems (Alves et al. 1989, Smith et al. 1990c, Weiss et al. 1990, Gerhardts et al. 1991, Washburn 1991). However, these structures extend typically over regions with sizes larger than the electronic mean free path in the sample. Although some hints of the Hofstadter "butterfly" spectrum have been reported (Gerhardts et al. 1991, Gerhardts and Pfannkuche 1992), there have as yet been no definitive observations of this or of the associated quantum Hall effect.

Recently calculations of the magnetoconductance properties of finite arrays of coupled quantum dots have been reported by Kirczenow (1992a,b). Such arrays appear to be promising candidates for experiments directly probing the magnetic properties of electrons in 2D periodic potentials. One of the structures that was studied is shown in fig. 45. The current and voltage leads required for measurements were explicitly included in the model and the details of how they are connected to the array were found to be very important. The energy spectra of such arrays turn

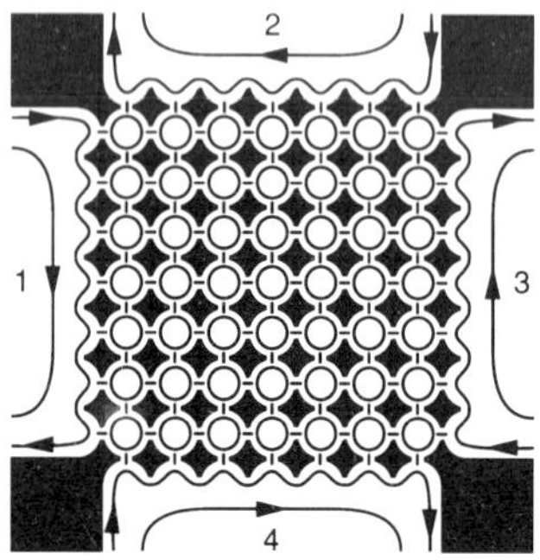

Fig. 45. Schematic representation of quantum dot array coupled to leads 1–4. The black areas are inaccessible to electrons. The white circular areas are the quantum dots. Circles represent the radial modes in each dot with dashes indicating tunneling among them. Two edge states are shown in each lead. From Kirczenow (1992a).

out to be quite intricate; the spectral features appear as transmission and Hall resonances when the leads are coupled to the array via individual quantum dots (Kirczenow 1992a). Bulk spectral gaps of *two* types were predicted (Kirczenow 1992a,b): gaps that contain edge states orbiting the array, and insulating spectral gaps (without edge states) that derive from the gaps between the energy levels of isolated quantum dots. An example of the calculated Hall resistance of a quantum dot array is shown in fig. 46. Note the oscillatory negative Hall resistance at lower values of β, which is due to an exotic edge state that propagates around the array in the direction *opposite* to that which a classical skipping orbit would take. If the Aharonov–Bohm-like oscillations in the Hall resistance are suppressed by suitably coupling the leads to the array, a *negative* quantum Hall plateau with $R_H = -h/2e^2$ emerges in this Hall resistance region (Kirczenow 1992b). In general, it is found that if the Fermi level lies in a bulk spectral gap of the array that contains edge states then, for a proper arrangement of leads, the array exhibits a quantum Hall effect. However, if the Fermi level lies in a bulk spectral gap of the array without edge states, then the Hall resistance is ill defined and no quantum Hall effect is possible.

Recently Johnson and Kirczenow (1992) have shown that in a *ballistic* quantum dot array coupled to *ideal* reservoirs the Hall conductance should be quantized in fractional as well as integer multiples of e^2/h, even in the absence of electron–electron interactions. A theoretical analysis of the likely effects of sample imperfections on these phenomena (Kirczenow 1992b) has shown that observation of the predicted negative quantum Hall plateaus and insulating spectral gaps will require high-quality arrays with the number of electrons per quantum dot varying across the array by less than about one half or one quarter, depending on the degree of spin polarization

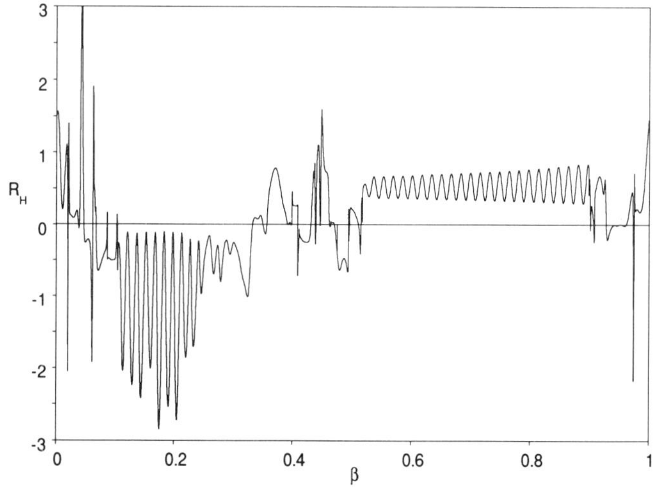

Fig. 46. Hall resistance for a 9 × 9 quantum dot array plotted against magnetic field ($\beta = Bd^2/\Phi_0$, where d is the period of the array and $\Phi_0 = h/e$ is the flux quantum). Negative oscillations at low β-values arise from an edge state propagating in the array in the opposite direction to the classical skipping orbits. From Kirczenow (1992b).

of the array. This is a stringent quality requirement and meeting it will test the limits of nanofabrication technology.

5. Brief history of localization

Perhaps the first paper in which the problem of localization was discussed in connection with quantum mechanical diffusion is that of Anderson (1958). He formulated the problem and gave a first quantitative estimate of the strength of the random potential which is necessary for the absence of diffusion in certain random lattices. The relevance of localization with regard to the transport properties of amorphous semiconductors was discussed by Mott (1968). He proposed the concept of the mobility edge, which separates energetically the localized from the extended states (fig. 47).

In 1970 Landauer pointed out that since the d.c. conductivity is vanishing in the localized regime at zero temperature, it is no longer a useful quantity for the description of transport through *finite* systems. Instead, the conductance must be considered. He proposed an alternative description of the conductance of 1D disordered systems in terms of their transmission properties. The Landauer relation (Landauer 1970) gives explicitly the scaling properties of the conductance as a function of the length of the system. There have been several generalizations to quasi-1D systems with many transmission channels (Fisher and Lee 1981, Economou and Soukoulis 1981a,b, Anderson et al. 1980, Langreth and Abrahams 1981, Thouless 1981, Anderson 1981, Büttiker et al. 1985) (see also § 4.1.2).

In order to describe the conductance of a hypercube of volume L^d, $g(L)$, its logarithmic derivative β was introduced,

$$\beta = \frac{\mathrm{d}\ln g}{\mathrm{d}\ln L}. \tag{9}$$

It was assumed that it depends only on the conductance itself, and not on energy,

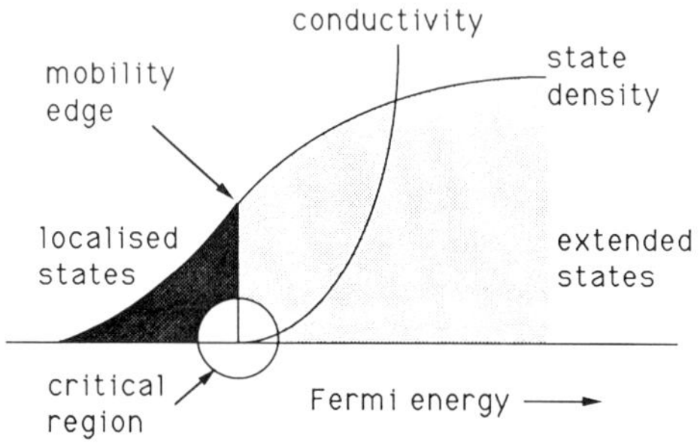

Fig. 47. The concept of the mobility edge. Electronic states below and above the mobility edge are localized and extended, respectively. If the Fermi energy lies in the region of the localized states, the system is insulating at $T = 0$. In the extended-states region it is metallic.

disorder, or L separately. The qualitative behavior of $\beta(g)$ was obtained by interpolating from the known asymptotic behavior at large and small conductance assuming that β is a continuous, monotonically increasing function (fig. 48).

If $\beta > 0$ the conductance increases with the size of the sample reflecting a metallic behavior. The metallic region may be characterized by the classical behavior, namely $\beta(g) = d - 2$, which may be obtained from the classical relation between the conductance and the conductivity. On the other hand, if $\beta < 0$, $g(L)$ decreases with L, eventually terminating in the localized regime where $\beta(g) = \ln g$. A fixed point is defined by $\beta(g_c) = 0$. It corresponds to a disorder-induced metal–insulator transition (MIT). One of the essential results of the one-parameter scaling theory is that such a MIT can only exist in 3D since this is the only dimensionality where β can have positive and negative values. In 1D and 2D $g(L)$ is always decreasing with L. Thus, the insulating regime is always eventually reached in the thermodynamic limit at the absolute zero of the temperature, for non-interacting electrons, and without magnetic scattering effects.

During the time when the scaling theory of localization was developed experimental techniques became available that made explicit tests of the theory possible. In the metallic regime the asymptotic (perturbation) theory for weak disorder (weak elastic scattering) predicted a disorder-induced logarithmic correction to the temperature dependence of the conductivity at very low temperature for 2D systems. The corresponding quantitative theory was formulated by Hikami et al. (1980) and by Altshuler et al. (1980), and was verified experimentally in a series of beautiful experiments done on very thin Mg films (Bergmann 1982a–c, 1984).

The numerical values for the localization length near the critical disorder in 3D (MacKinnon and Kramer 1981, 1983a) and for small disorder in 2D, which were obtained in these calculations, turned out to be macroscopically large. The question arises of how the wavefunctions behave before the asymptotic exponential decay sets

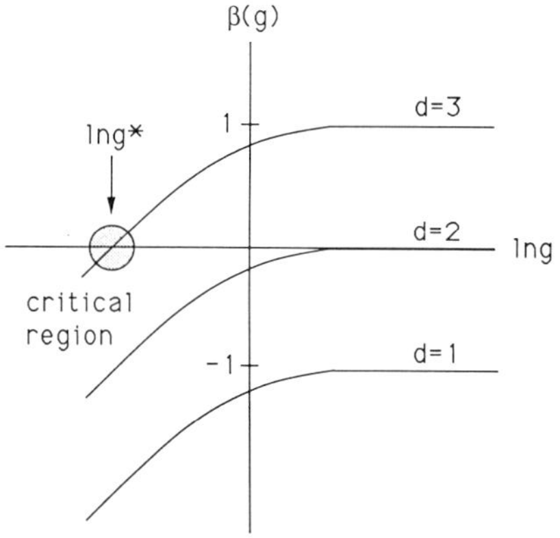

Fig. 48. The β-function for the zero-temperature conductance of a disordered system for dimensionality $d = 1, \ldots, 3$. $g(L)$ increases with increasing L if $\beta > 0$, but decreases for $\beta < 0$. $\beta = 0$ defines the critical point corresponding to the Anderson transition. It is only achievable in $d = 3$.

in. The idea of an inverse power law decrease as a function of the distance from some localization center was introduced by Kaveh and Mott (1981).

Power law localization has also been found recently in 1D disordered systems that are subjected to an electric field (Delyon et al. 1984, Cota et al. 1985, Leo and Movaghar 1988). However, such a behavior would be in severe disagreement with the one-parameter scaling theory. Presumably, further theoretical and numerical studies are necessary in order to clarify the issue.

6. Experiments on localization

A large number of phenomena exist that can be explained in terms of localization of quantum states. A *classical* example is Mott's celebrated $T^{-1/4}$ law for the low-temperature behavior of the conductivity of amorphous semiconductors (Mott and Davis 1971). The existence of quantum interference has been demonstrated in thin metallic films (Bergmann 1984). The metal–insulator transition has been investigated in a variety of systems including doped semiconductors and amorphous metal–non-metal mixtures (Thomas 1986, Katsumoto 1991), and magnetic-field-induced transitions (Biskupski and Briggs 1988, Chen et al. 1989). Conductance fluctuations as a result of disorder were observed in small systems at low temperature (Washburn and Webb 1986, Lee et al. 1987). The existence of the quantized Hall resistance in inversion layers has been interpreted as a direct manifestation of the existence of localized states in the presence of a magnetic field (Aoki and Ando 1981).

6.1. Hopping transport in amorphous semiconductors

Figure 49 shows the d.c. conductivity of amorphous silicon as a function of the temperature. It can be seen directly from the plot that at low enough temperatures (Beyer 1974)

$$\sigma_{\mathrm{d.c.}}(T) = \sigma_0 \exp(-K/T^{-1/4}), \tag{10}$$

over two orders of magnitude, with suitable constants σ_0 and K. At higher temperatures this behavior changes into an activated one.

The behavior of the conductivity can be understood when assuming that the transport is mediated by *phonon-assisted hopping processes between localized states*, that are energetically close to each other but with localization lengths that are small compared with the spatial distance between the centers of localization.

At low temperatures the hopping probability p between the states will be proportional to the overlap integral of the two wavefunctions, which depends exponentially on their spatial separation R, and a Boltzmann factor containing their mean energetic separation Δ,

$$p \propto \exp(-\alpha R - \beta\Delta). \tag{11}$$

Here α is proportional to the inverse of the diameter of the states and β is the inverse temperature. For small hopping distances R, necessary in order to obtain a large

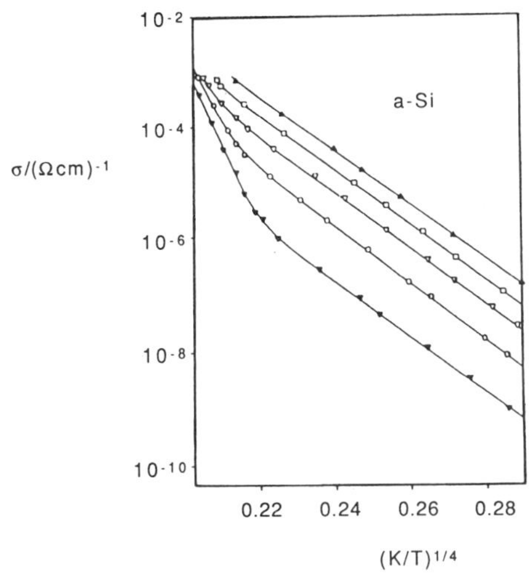

Fig. 49. Hopping transport in amorphous silicon. The logarithm of the d.c. conductivity depends on the temperature according to Mott's $T^{-1/4}$-law. The different curves are for different deposition temperatures (redrawn after Beyer (1974)).

hopping probability, the number of states that are available for a hopping process from a given state will be small on the average. Correspondingly, the mean energy separation of the states will be large. The Boltzmann factor will cause a reduction in the total hopping probability. On the other hand, when R is large, there are many states to which an electron can hop from a given site. Δ is small. The Boltzmann factor will enhance the total hopping probability. In order to maximize p one has to know how Δ depends on R.

Assuming that the localization centers are distributed homogeneously in space we estimate (for a d-dimensional system)

$$\Delta \propto [R^d n(E_F)]^{-1}. \tag{12}$$

where $n(E_F)$ is the density of states at the Fermi level. The distance R_{max} that maximizes p is then obtained by minimizing the exponent in eq. (11),

$$R_{max} = \left(\frac{d\beta}{\alpha n(E_F)}\right)^{1/(d+1)}. \tag{13}$$

Inserting this result into eq. (11), and noting that $p \propto \sigma$, yields the d-dimensional version of the $T^{-1/4}$-law.

More rigorous treatments of this remarkable idea for explaining the low-temperature transport properties of many amorphous materials which is, as many others, due to Mott, can be found in the literature (Efros and Shklovskii 1975, Gogolin 1982).

6.2. Weak localization

At low temperatures thin, weakly disordered metal films exhibit anomalies in the behavior of the electrical resistance that can only be understood when taking into account the quantum mechanical nature of the electrons. These anomalies are found in the temperature dependence as well as in the magnetic field dependence of the resistance. This is the regime of *almost classical* transport (large conductance, see fig. 50) in a 2D system. Since the quantum corrections to the conductivity can be interpreted as being due to an interference of the electron wavefunctions that favors backscattering, and since the localization length considerably exceeds all of the other relevant lengths in the system, this is called the regime of *weak localization.*

One of the most striking effects is seen in the temperature dependence. Many thin metallic films show a logarithmic increase of the resistance when the temperature is decreased. Many examples can be found in the review article by Bergmann (1984).

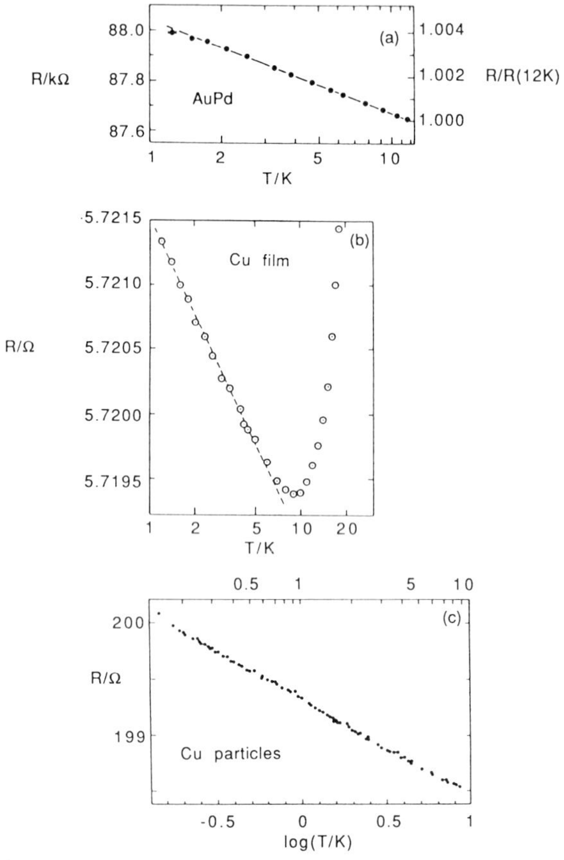

Fig. 50. (a) Logarithmic temperature dependence of the resistance of thin Au–Pd films (Dolan and Osheroff 1979), (b) thin Cu films (van den Dries et al. 1981) and (c) fine Cu particles (Kobayashi et al. 1980).

Results of this kind were first taken as confirming the scaling theory of localization in the asymptotic regime of weak localization where the β-function appeared to be given by

$$\beta(g) = \frac{\mathrm{d}\ln g}{\mathrm{d}\ln L} = -\frac{\text{const.}}{g} \tag{14}$$

in 2D. Integrating with respect to the system size L one obtains

$$g - g_0 = -\text{const.} \ln (L/L_0), \tag{15}$$

where g_0 and L_0 are constants of integration. For $L < L_0, g\,(>g_0)$ can be treated classically. For $L > L_0$ the conductance decreases logarithmically with the length L. In order to obtain the temperature dependence the geometrical system size has to be replaced by an effective system size L_i, the mean distance between successive inelastic scattering events suffered by the particle (Anderson et al. 1979, Thouless 1980, Altshuler and Aronov 1985). With the diffusion constant D and the (temperature-dependent) inelastic scattering time τ_i (phase coherence time) we have

$$L_\mathrm{i}^2 = D\tau_\mathrm{i}(T). \tag{16}$$

Taking $\tau_\mathrm{i} \propto T^{-p}$ at low temperatures one obtains the desired type of log T behavior.

However, some time later doubts were cast on this interpretation since, based on earlier work on the density of states (Altshuler and Aronov 1979, Altshuler et al. 1980), it was discovered that the Coulomb interaction between the electrons leads to the same temperature dependence in the presence of disorder, except that the prefactor is different (Fukuyama 1980, 1981). Thus, a tool was needed that could distinguish between localization and interaction effects.

The latter was provided by applying an external magnetic field (Uren et al. 1980). Localization effects lead to a negative magnetoresistance, whereas a Coulomb interaction gives a positive magnetoresistance.

The fact that the regime of weak localization in metallic transport in disordered metals can be described within the picture of the interference between quantum mechanical probability waves has been demonstrated explicitly by Sharvin and Sharvin (1981). When measuring the magnetoresistance of a Mg cylinder, which was about 1 μm in diameter and 1 cm long, they observed oscillations when varying the magnitude of an externally applied axial magnetic field. They were periodic with a period $\Delta\Phi = h/2e$, half the normal Aharonov–Bohm period (fig. 51). The results were in quantitative agreement with the theoretical predictions made a few months earlier by Altshuler et al. (1981) who applied the same technique that was used for the treatment of *weak localization* (see below) to describe the low-temperature magnetotransport properties of loops and cylinders.

Magnetoresistance oscillations have been observed by several other groups in cylindrical systems as well as in large 2D arrays of loops (Gijs et al. 1984a,b, Pannetier et al. 1984, Washburn and Webb 1986, Aronov and Sharvin 1987). Thus, the asymptotic regime of weak scattering is well understood physically and well confirmed experimentally.

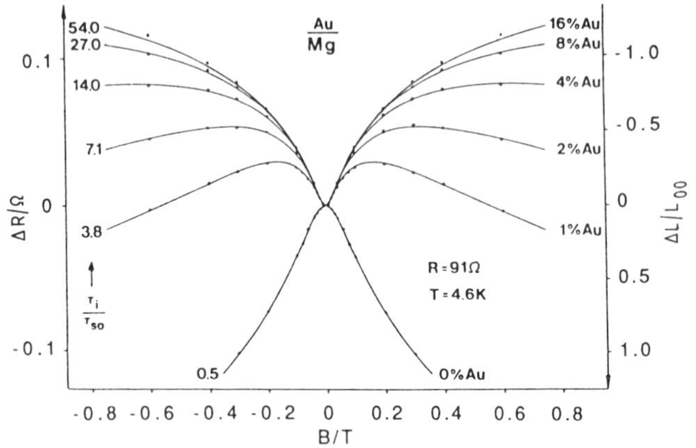

Fig. 51. The magnetoresistance ΔR of thin Mg films. The clean film shows a negative magnetoresistance indicating localization. When the film is covered with a small amount of gold atoms the magnetoresistance becomes positive due to increasing spin–orbit scattering. The right-hand scale shows the magnetoconductance ΔL. On the left, the ratio of the inelastic scattering time and the spin–orbit scattering time is indicated (redrawn after Bergmann (1984)).

6.3. Conductance fluctuations

A phenomenon that is closely related to localization is the reproducible fluctuations of the conductance in *mesoscopic* samples. These have geometric dimensions of the order of or less than the phase coherence length (inelastic scattering length), i.e. a few hundred angstroms when considering temperatures of the order of 1 K. Measuring the conductance or the resistance of thin wires as a function of the magnetic field in the case of metallic samples and of quasi-1D inversion layers in high-quality heterostructures one observes irregular but nevertheless (for a given sample) reproducible structures (Pichard and Sanquer 1990, Mailly and Sanquer 1991, Caro et al. 1991). Changing the gate voltage, i.e. the charge density in quasi-1D inversion layers in MOSFETs (fig. 52), similar random fluctuations are detected at low temperature (Fowler et al. 1982, 1988, Washburn and Webb 1986).

These conductance fluctuations can be understood when assuming that at low enough temperature the transport takes place coherently within single quantum states throughout the whole sample. If the Fermi energy is such that it accidentally coincides with the energy of a quantum state the transmission probability (and hence the conductance) is very high (almost one) as compared with the situation where the Fermi energy does not coincide with a quantum state energy. In this case the variance of the fluctuations is independent of the size of the sample and other parameters like the degree of disorder (universal conductance fluctuations, UCF). In this regime, which corresponds to that of weak localization since the states are essentially assumed to extend throughout the whole sample, one can use perturbational techniques for the determination of the magnitude of the fluctuations.

When the sample size is larger than the localization length, which is the case in the insulating regime, one expects a second type of fluctuations, the variance of which

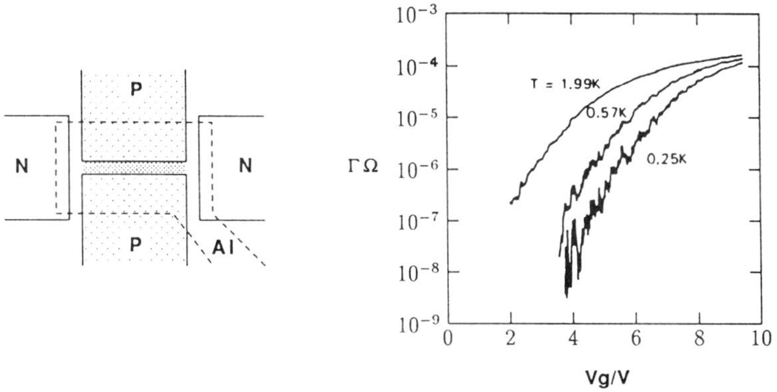

Fig. 52. The reproducible fluctuations of the conductance Γ as observed at low temperatures in a quasi-1D inversion layer channel when changing the voltage V_g applied to the Al gate on top of a Si MOSFET as shown schematically in the left-hand part (redrawn after Fowler et al. (1982)).

increases exponentially with the length of the sample. These are due to the non-ergodic nature of the localized phase. To our knowledge they have not yet been directly observed experimentally. They could, however, be connected with the above mentioned conductance fluctuations in the quasi-1D MOSFET channels.

7. Basic concepts

7.1. Models of disorder

The characterization of the properties of ideally ordered materials is comparatively easy. Due to the presence of the long-range order one has translational symmetry. The quantum objects (electrons, phonons, magnons, etc.) are of the Bloch type. As a consequence, they are freely itinerant and can move unrestrictedly throughout the whole system. In reality, however, there are no ideally ordered media. There are always distortions of the ideal order due to the presence of impurities, dislocations, vacancies, and other defects. As long as the concentration of these is small, one may still use the concepts developed for the translationally symmetric systems as a starting point for the understanding of the properties of the distorted systems. However, if the concentration of the distortions is large, it is necessary to leave translational symmetry, and to develop new methods.

Starting from the ideal crystal, models of disorder may be constructed in various ways. Models for glassy systems and amorphous semiconductors may be obtained by relaxing the lattice structure (structural disorder). A lattice with two or more different kinds of atoms distributed at random establishes the most simple model of an alloy (compositional disorder).

A simple model for a disordered system is provided by the Hamiltonian

$$H = \frac{\hat{\boldsymbol{p}}^2}{2m} + \sum_{j=1}^{N} V(\boldsymbol{r} - \boldsymbol{R}_j), \tag{17}$$

where $\hat{\boldsymbol{p}}$ is the momentum operator, m the effective mass of the particle, and V the potential energy of an atom at the site $\boldsymbol{R}_j$. The distribution of the atomic potentials in space may be described by a (normalized) probability density distribution function $P(\{\boldsymbol{R}_j\})$. This model has been used extensively in the theory of *weak localization* (Lee and Ramakrishnan 1985, Bergmann 1984, Vollhardt and Wölfle 1991, Vollhardt 1987).

The following model that is defined on a lattice is very commonly used:

$$H = \sum_{j\nu} \varepsilon_{j\nu} | j\nu\rangle\langle j\nu | + \sum_{j\nu,k\mu} V_{j\nu,k\mu} | j\nu\rangle\langle k\mu |. \tag{18}$$

Here, $\varepsilon_{j\nu}$ are the energies associated with the states labeled by ν at the sites j of the lattice, and the non-diagonal elements $V_{j\nu,k\mu}$ denote the hopping matrix elements between the states. The diagonal part of the Hamiltonian corresponds to the potential energy and the non-diagonal part to the kinetic energy in the continuous-space description (17). Disorder is introduced by taking the site energies and/or the hopping matrix elements at random and assuming some probability distribution function for them.

Although looking rather specialized at first glance, Hamiltonians of the type of eq. (18) have many applications in various areas of the physics of disordered systems. They may be used to describe the vibrational properties (Dean 1972) as well as the electronic properties of amorphous semiconductors (Kramer and Weaire 1979) and their alloys (Elliott et al. 1974).

7.2. Properties of the electronic spectrum

In the case of an ideal crystal a characteristic feature of the density of states is the occurrence of van Hove singularities that are due to long-range order. There are sharp band edges, for instance. A disordered system does not have any long-range order. There are no singularities in the density of states. This can be proven rigorously for the tight-binding model considered above, with the diagonal elements being given by a smooth distribution function (Wegner 1979). In particular, there are no sharp edges, but smooth band tails, instead. A number of theoretical approaches deal with their analytic form. The exponential behavior of the band tails,

$$n(E) = A \exp\{B[\pm(E_0^{\pm} - E)]^{-d/2}\}, \tag{19}$$

can easily be obtained for a random d-dimensional two-component alloy and for the Anderson model with a rectangular distribution of the site energies by using a famous argument that is due to Lifshitz (1965). $E_0^{\pm}$ are the true upper and lower (Lifshitz) bounds of the spectrum in these cases and A, B are model-specific constants.

At energies far away from the band edges one may expect that a weak random potential cannot localize the states to finite regions of space. Instead, the amplitude of the states, although fluctuating more or less randomly, will be nonzero essentially everywhere. Consequently, these states will be called extended. If this is the case, there must exist certain energies, denoted by E_c and E'_c, which separate the extended from the localized states. As we shall see later, localized states do not contribute to transport, even if they are situated energetically at the Fermi energy, whereas extended states do. Thus, E_c and E'_c are denoted as *mobility edges*. In general, the mobility edges depend on the disorder. If the latter is large enough they will merge into the center of the band. The system becomes an insulator.

Up to now, the only mathematically rigorous statement concerns the existence of extended states in 1D. For any finite amount of disorder *there are no extended states* in $d = 1$, a statement that was already made by Mott and Twose (1961). In 2D the present general belief is that there are also no extended states, even for an infinitesimally small amount of disorder.

In order to understand the physical mechanisms leading to localization of the quantum states in random potentials better it is instructive to consider first a classical particle moving in a random potential $V(x)$ in one dimension. For simplicity, we take the potential to be restricted to values smaller than E_0. Then one can decide by simple energy considerations whether or not the particle is *localized*. If the total energy E is smaller than E_0 the particle is confined to finite intervals, within the (accidental) potential wells. On the other hand, if $E > E_0$ then it can move along the whole space.

For a quantum mechanical particle it is more complicated to distinguish the character of the states. On the one hand, the potential barriers cannot absolutely confine the particle to a certain well because of tunneling. This may even lead to complete delocalization of a classically localized particle. On the other hand, for $E > E_0$ repeated scattering at the potential fluctuations may eventually lead to a superposition of destructively interfering waves in such a way that a classically *extended* particle may become localized. In addition to potential localization it is the competition between tunneling and interference that determines whether or not a state is localized.

An example for localization via quantum interference is the above mentioned one-dimensional localization. An example for delocalization via tunneling, although not related to disordered systems, are the Bloch states in a crystal, in particular those that correspond to the core states.

7.3. Quantities of interest

(a) The simplest quantity which may be considered is the spectral density of energy levels E_ν

$$n(E) = \frac{1}{\Omega} \sum_\nu \delta(E - E_\nu), \tag{20}$$

where Ω is the sample volume.

(b) As mentioned above, the localization properties of the states influence the transport properties of the system. Therefore, the theory of localization is essentially a theory of the transport properties, or the electrical conductivity σ. The latter is given by the Kubo formula,

$$\sigma(T, \omega) = \frac{\pi e^2}{m^2 \omega \Omega} \sum_{\alpha} \sum_{\beta \neq \alpha} |\langle \alpha | \hat{p} | \beta \rangle|^2 [f(E_\alpha) - f(E_\beta)] \delta(\mathrm{E}_\alpha - \mathrm{E}_\beta - \hbar\omega). \tag{21}$$

Here, $f(E_\alpha)$ is the Fermi function, $E_{\alpha,\beta}$ are the energy eigenvalues corresponding to the eigenstates denoted by $|\alpha\rangle$, $|\beta\rangle$, and T and ω are the temperature and the frequency of the electric field, respectively. $\hat{p}$ is the projection of the momentum operator onto the direction of the electric field.

A first restriction on the *classical* result is given by the Ioffe–Regel criterion (Ioffe and Regel 1960), which states that the mean free path l should be greater than the wavelength. At least, if l is of the order of or smaller than the wavelength, then a full quantum mechanical calculation of the conductivity is needed.

(c) As we have replaced the system to be considered by a statistical ensemble, by introducing a probability distribution in the preceding section, all of the physical quantities have to be configurationally averaged. However, this will yield physically meaningful results only if the quantity considered is *self-averaging* in the usual statistical sense, i.e. if

$$\langle A \rangle = \lim_{\Omega \to \infty} A(\Omega). \tag{22}$$

In connection with localization this property is not trivially fulfilled, even in the metallic limit. If a quantity is not self-averaging, then, in principle, one has to consider its probability distribution function (Lerner 1991), or, equivalently, all its moments.

(d) The *asymptotic behavior* of a localized state is usually described by the exponential decay length of its envelope, λ, the localization length, i.e.

$$\Psi(\boldsymbol{r}) = f(\boldsymbol{r}) \mathrm{e}^{-r/\lambda}, \tag{23}$$

where $f(\boldsymbol{r})$ is a randomly varying function. $\lambda \to \infty$ corresponds to an extended state. In practice this definition is not very useful, since its application would require the calculation of single eigenstates.

(e) In order to decide whether or not a state is localized it is often sufficient to consider the second moment of the probability density (Wegner 1980),

$$P^{-1} = \sum_{\boldsymbol{r}} |\Psi(\boldsymbol{r})|^4, \quad \|\Psi\| = 1. \tag{24}$$

This is the *inverse participation number*. It is a measure of the portion of the space where the amplitude of the wavefunction differs markedly from zero. It may also be considered as providing a measure for an average diameter R of the state via $R = P^{1/d}$. For plane waves one obtains that $P \sim L^d$, the volume of the system, and this diverges in the thermodynamic limit. Such a behavior may be considered as being representative of extended states.

(f) One may define the *fractal dimensionality* d^* of a state by (Aoki 1983, 1986, Kramer et al. 1988)

$$\lim_{L\to\infty} P = L^{d^*}. \tag{25}$$

For plane waves $d = d^*$. For a general extended state d^* may be different from the Euclidean dimensionality, i.e. $d^* \leqslant d$. For a localized state P is proportional to the volume in which the state has a nonvanishing amplitude. This volume tends to a constant in the thermodynamic limit. Thus, the fractal dimensionality vanishes in the case of localized states.

(g) Another possibility to investigate the localization properties of the states is the shift of the energy eigenvalues of a finite system due to small changes in the boundary conditions, as was proposed by Edwards and Thouless (1972). The average energy shift, δE, in second-order perturbation theory, is related to the conductivity by

$$\sigma L^{d-2} = \frac{e^2}{h}\frac{\delta E}{\Delta E} = \frac{e^2}{h} g(L), \tag{26}$$

where ΔE is the average energy spacing of the eigenvalues. $g(L)$ is called the Thouless number. The idea is that for localized states the mean energy shift will become very small for large system sizes such that $g(L)$ vanishes exponentially, whereas in the metallic regime the boundary conditions will always influence the energy levels, even for infinite systems. Note from eq. (26) that $g(L)$ is simply related to the conductance G by $G = (e^2/h)g$. Thus, g is often called the *dimensionless conductance* and will be used interchangeably with G in further discussions.

(h) For 1D and quasi-1D systems the localization length may be calculated from the limiting behavior of the products of random matrices (Mehta 1967, Ishii 1973, Derrida et al. 1987, Muttalib et al. 1987, Pichard 1990, 1991a,b). In essence this is equivalent to a calculation of the transmission coefficient T of the random medium. If T decays exponentially for large distances then the mean diffusion distance is finite. The localization length λ may now be defined by using the exponential decay of the transmission probability as

$$\frac{2}{\lambda} = -\lim_{|\boldsymbol{r}-\boldsymbol{r}'|\to\infty} \frac{\ln T(\boldsymbol{r}, \boldsymbol{r}'; E)}{|\boldsymbol{r}-\boldsymbol{r}'|}. \tag{27}$$

If the localization length is finite, the states near the Fermi energy are asymptotically localized, as defined in eq. (23).

8. One-dimensional localization

1D disordered systems play a key role in the understanding of the properties of solids since many features of the electronic states, and of the related transport properties, can be discussed rigorously. We concentrate in this section on the localization of the states, the behavior of the conductivity of the infinite system, the conductance of a finite system at the absolute zero of temperature, and the statistical properties of the transport quantities.

We consider the *tight-binding* Hamiltonian (18), lattice spacing $a \equiv 1$, with only nearest-neighbor hopping and one state per lattice site, i.e.

$$Ea_j = \varepsilon_j a_j + V(a_{j+1} + a_{j-1}), \tag{28}$$

V is the constant hopping matrix element, a_j, $a_{j\pm 1}$ are amplitudes of the wavefunctions associated with the lattice sites, and ε_j are the corresponding energies. Only the case of statistically independent site energies is considered here. The influence of statistical correlations has been discussed in the literature.

In 1D all the eigenstates of a random Hamiltonian are exponentially localized in the asymptotic sense. The inverse exponential decay length of the eigenstates may be calculated from

$$-2\gamma = \lim_{j\to\infty} \frac{1}{j} \ln(a_j^2 + a_{j+1}^2), \tag{29}$$

with $\{a_j\}$ the solutions of the time-independent Schrödinger equation. In general, it will depend on the particular realization of the random potential, $\gamma = \gamma(\varepsilon_1, \ldots, \varepsilon_L)$. However, the localization length (and its inverse) are self-averaging quantities. Their configurational averages agree with the most probable values in the thermodynamic limit $L \to \infty$.

Since γ is always finite in 1D, an infinitely long 1D disordered system cannot be transparent, i.e. its conductance becomes exponentially small as the length of the system is increased.

There is a relation between the spectral properties and the localization length, which was discovered by Herbert and Jones (1971):

$$\gamma(E) = \int_{-\infty}^{+\infty} \rho(x) \ln|E - x| \,\mathrm{d}x = \mathrm{Re} \int_{-\infty}^{E} \lim_{L\to\infty} \frac{1}{L} \sum_{j=1}^{L} G_{jj}(x) \,\mathrm{d}x, \tag{30}$$

where $\rho(E)$ is the density of states and G_{jj} are the diagonal elements of the Green's function, which may be evaluated in second-order perturbation theory (Thouless 1979). In the limit of small disorder one obtains ($|E| < 2$)

$$\gamma(E) = W^2/24(4V^2 - E^2), \tag{31}$$

so that $\gamma(0) = W^2/96V^2$ for small W, where W is the width of a rectangular distribution, $-\frac{1}{2}W < \varepsilon < \frac{1}{2}W$.

The result that the localization length of a 1D disordered system diverges at $W = 0$ as W^{-2} remains true also for the case of a spatially correlated potential.

That the d.c. conductivity vanishes in 1D disordered systems has been shown analytically by Kunz and Souillard (1980) and numerically for the one-dimensional Anderson model with a rectangular distribution of the site energies by MacKinnon (1980), Czycholl et al. (1981), and Thouless and Kirkpatrick (1981) by using the d.c.-Kubo formula at the absolute zero of temperature in a recursive procedure similar to the one described in connection with eq. (27) for the localization length.

Classically, the conductance $g(L)$ of a d-dimensional hypercube of volume L^d is related to the conductivity by

$$g(L) = \sigma L^{d-2}. \tag{32}$$

Since the conductivity vanishes in the localized regime, it is no longer a useful parameter when considering the transport properties of a macroscopic, but finite, sample. One should have a theory of the conductance (or, equivalently, of its inverse, the resistance) without referring to the conductivity.

The various *Landauer* formulas for the conductance or resistance of a 1D wire of length L have been discussed in § 4.1.2. For a purely 1D system (containing a single conduction channel) the resistance $R(L)$ may be expressed in the form

$$R(L) = (h/e^2)T(L)^{-1}. \tag{33}$$

A dimensionless resistance may be defined by $r(L) = R(L)e^2/h = T(L)^{-1}$. As the logarithm of $T(L)$ is statistically a well behaved quantity (it obeys the central-limit theorem) and its configurational average is asymptotically equal to $-\gamma L$, one may expect that the resistance is an exponentially increasing function of L, on the average. In addition, the probability distribution of r will not fulfill the central-limit theorem. The exponential increase of the resistance has been proven for a variety of models by several authors (Abrahams and Stephen 1980, Andereck and Abrahams 1980, Stone et al. 1981, Kree and Schmid 1981, Sak and Kramer 1981, Kirkman and Pendry 1984a,b, Mel'nikov 1981).

The dimensionless conductance, $g(L)$, is defined as the inverse of the dimensionless resistance, $r(L)$. The calculation of the corresponding configurational average is much more complicated than in the case of the resistance. It has been performed by Abrikosov and Ryzhkin (1978) in the limit of weak disorder. Kirkman and Pendry (1984b) have treated the general case. The result for weak disorder, and in the center of the band ($E = 0$), is

$$\langle g(L) \rangle \propto \sigma_2^{-3/2} \exp(-\tfrac{1}{4}\sigma_2 L) \tag{34}$$

in the limit of large L. Thus, $\gamma_g = \frac{1}{4}\sigma_2$ is the inverse of the *localization length*, which is characteristic for the exponential decrease of the average conductance.

In the weak-disorder limit there are relationships between the various localization lengths, namely

$$\ln \langle g \rangle = \tfrac{1}{4}\langle \ln g \rangle, \tag{35a}$$

$$\ln \langle r \rangle = 2\langle \ln r \rangle, \tag{35b}$$

$$\langle \ln r \rangle = -\langle \ln g \rangle. \tag{35c}$$

The fact that the three lengths do not agree with each other reflects the fact that the resistance, and the conductance are *not* self-averaging in 1D disordered systems (Sak and Kramer 1981).

In addition to the fundamental statistical effects in the d.c. transport properties one can expect interesting and novel statistical features in time- and frequency-dependent transport. First studies were done for 1D (Pendry et al. 1986, Mašek and Kramer 1988) and in the weakly disordered metallic regime (Fal'ko 1989).

9. Weak localization and quantum interference

The limit of a weak random potential was considered as rather unimportant for the localization problem until it was rediscovered (Abrahams et al. 1979, Gor'kov et al. 1979) that there exists a certain class of diagrams, the *maximally crossed diagrams* (Langer and Neal 1966), in the perturbation expansion of the conductivity that could be summed exactly for non-interacting time-reversal-invariant systems, and that give logarithmic corrections to the low-temperature low-frequency conductivity in 2D. The conclusion was that the zero-temperature zero-frequency conductivity of disordered systems without interactions and for zero magnetic field always vanishes for $d \leqslant 2$, and consequently that all quantum states must be localized. A considerable body of quantitative theoretical and experimental work followed this discovery (Lee and Ramakrishnan 1985, Bergmann 1984). One of the important points was that the diagrams could be interpreted physically as quantum interference processes that give rise to an enhanced backscattering. This point of view could be verified directly using a type of Aharonov–Bohm configuration but with normal metal cylinders instead of superconductors (Aronov and Sharvin 1987). We shall not repeat the formal theory here. It has already been treated very elaborately in many review papers (Lee and Ramakrishnan 1985, Altshuler and Aronov 1985, Fukuyama 1985, Kawabata 1985, Bergmann 1984). Instead, we shall stress the physical point of view of the quantum interference in deriving the main results.

9.1. Quantum interference corrections to the average conductivity

Let us consider the limit of a weak random potential $V(\boldsymbol{r})$. Then the Born series for the Green's functions may be used in order to evaluate eq. (27). One notes that the zero-temperature conductivity can be written as an infinite sum of terms of the form

$$T(\boldsymbol{r}, \boldsymbol{r}'; E) = \overline{\left| \sum_n A_n(\boldsymbol{r}, \boldsymbol{r}') \right|^2} = \sum_n \overline{|A_n|^2} + \sum_{n \neq m} \overline{A_n A_m^*}, \qquad (36)$$

where

$$A_n(\boldsymbol{r}, \boldsymbol{r}') = \sum_{\boldsymbol{r}_1} \dots \sum_{\boldsymbol{r}_n} G_0(\boldsymbol{r}, \boldsymbol{r}_1) V(\boldsymbol{r}_1) G_0(\boldsymbol{r}_1, \boldsymbol{r}_2) V(\boldsymbol{r}_2) \dots V(\boldsymbol{r}_n) G(\boldsymbol{r}_n, \boldsymbol{r}') \qquad (37)$$

is the probability amplitude for a transition from $\boldsymbol{r}$ to $\boldsymbol{r}'$. It may be interpreted as a superposition of all of the transition amplitudes corresponding to specific paths denoted by $\boldsymbol{r}_1 \dots \boldsymbol{r}_n$.

Due to the configurational average most of the terms in eq. (36) vanish except for those in which the sites $\boldsymbol{r}_1 \dots \boldsymbol{r}_n, \boldsymbol{r}'_1 \dots \boldsymbol{r}'_m$ are close to each other. For a white noise potential they must be at least pairwise equal. The first of the terms in eq. (36) corresponds to classical diffusion from $\boldsymbol{r}$ to $\boldsymbol{r}'$ whereas the second is due to interference between different paths. If $\boldsymbol{r} \neq \boldsymbol{r}'$ the interference term is always small compared to the classical diffusion term. For closed paths, $\boldsymbol{r} = \boldsymbol{r}'$, however, there is a class of interference terms that contribute to the total transmission probability with an amount that is equal to that of the corresponding diffusion term if the system is time-reversal invariant, i.e. does not contain any magnetic effects. They are characterized by the condition $\boldsymbol{r}_1 = \boldsymbol{r}'_m, \boldsymbol{r}_2 = \boldsymbol{r}'_{m-1}, \dots, \boldsymbol{r}_n = \boldsymbol{r}'_1$. The total probability is now twice the classical probability. Backscattering is thus strongly enhanced. Consequently, forward scattering must be decreased.

The corrections to the average conductivity can be calculated explicitly (Langer and Neal 1966, Abrahams et al. 1979, Gor'kov et al. 1979, Vollhardt and Wölfle 1980a,b, Hikami et al. 1980). The conductivity may be written as

$$\sigma = \frac{ne^2\tau}{m} - \frac{4e^2}{h}\frac{1}{L^d}\sum_Q \left(\frac{1}{Q^2}\right). \tag{38}$$

The first term is the classical result, σ_0 (n corresponds to the electron density, cf. eq. (21)). The second term is the quantum correction due to weak localization, $\delta\sigma \equiv \sigma - \sigma_0$. The summation over Q is restricted to a region between a lower cutoff $Q_{\min} \propto 1/L$ and an upper cutoff $Q_{\max} \propto 1/l$. Evaluation of the sum (after transformation into an integral by performing the thermodynamic limit) yields for the quantum correction

$$\delta\sigma = -\frac{2e^2}{h}\begin{cases}(L-l), & d=1,\\ \pi^{-1}\ln(L/l), & d=2,\\ \pi^{-2}(l^{-1}-L^{-1}), & d=3.\end{cases} \tag{39}$$

L is here an effective system size, which is, e.g., given by the temperature-dependent phase coherence time τ_φ according to the Thouless relation, $L = \sqrt{D\tau_\varphi}$ (Thouless 1979, 1980, Anderson et al. 1979, Altshuler and Aronov 1985), and l is the mean free path for elastic scattering. The logarithmic temperature behavior of the conductivity at low temperature (see § 6) follows immediately from eq. (39) when assuming $\tau_\varphi = T^{-p}$.

9.2. *Negative magnetoresistance*

The influence of a magnetic field on the transport can be studied using this approach by bearing in mind that the transition amplitudes acquire additional phase factors in the presence of a magnetic field,

$$A(\boldsymbol{r}, \boldsymbol{r}') = A^0(\boldsymbol{r}, \boldsymbol{r}') \exp\left(\frac{ie}{\hbar}\int_{\boldsymbol{r}}^{\boldsymbol{r}'} \boldsymbol{A}\cdot \mathrm{d}\boldsymbol{s}\right). \tag{40}$$

In the case of a closed loop the phase is $\varphi = \int \boldsymbol{A}\cdot\mathrm{d}\boldsymbol{s} = 2\pi\Phi/\Phi_0$, where Φ is the magnetic flux through the loop and Φ_0 the flux quantum h/e. The return probability for a given path is multiplied by

$$1 + \cos(4\pi\Phi/\Phi_0). \tag{41}$$

For small fluxes the cosine may be expanded. The return probability decreases with increasing magnetic field. Consequently, the transmission probability must increase, and this induces an increase of the conductivity. This is the mechanism for the negative magnetoresistance that has been observed for a long time but is explained only in the course of the development of the theory of weak localization by Kawabata (1985), Kühl (1980), and Altshuler et al. (1980).

The complete quantitative theory of the magnetoresistance of 2D systems also taking into account spin–orbit scattering was formulated by Hikami et al. (1980), and by Altshuler et al. (1980), and verified experimentally by Bergmann (1984).

9.3. *Oscillations of magnetoresistance*

Equation (41) shows an interesting feature. For a given path the contribution to the magnetoresistance oscillates as a function of the magnetic flux with a period that is given by $\Delta\Phi = h/2e = \frac{1}{2}\Phi_0$. If it were possible to select experimentally only paths whose areas are the same when projected onto a plane perpendicular to the direction of the magnetic field, these oscillations should be observable. This effect was predicted to occur in thin metallic cylinders, when placed in a magnetic field parallel to the cylinder axis, by Altshuler et al. (1981). Experimentally, the oscillations were first observed in thin-walled Mg and Li cylinders by Sharvin and Sharvin (1981) (see § 6). The experiment was repeated on Mg cylinders and Al cylinders by Gijs et al. (1984a,b).

Another possibility for selecting paths of a given area is to use small metallic rings (Webb et al. 1985a,b, Chandrasekar et al. 1985) and networks (Pannetier et al. 1984). Whereas the dominant period in the cylinder experiments is $\frac{1}{2}\Phi_0$, the ring and network experiments also show periods of Φ_0. This can easily be understood by evaluating according to eq. (40) the phase difference between two paths entering a ring structure on one side and leaving on the other, but following the left and the right branch, respectively.

The important point in all of these interference experiments is that the electrons must be able to behave coherently around the circumference of the cylinder or ring. Therefore, the mean distance $L_\varphi = \sqrt{D\tau_\varphi}$ between phase-breaking processes must be of the order of or larger than this circumference. This condition restricts the experiments to low temperatures and diameters of the order of 1 μm. As L_φ enters the theory as a parameter, the phase-breaking length and its temperature dependence may be determined by fitting to the experimental data.

Reviews of this work up to the present time have been given by Washburn and Webb (1986), by Aronov and Sharvin (1987), and van Haesendonck et al. (1991).

10. The scaling approach

10.1. Single-parameter scaling

Thouless (1974) introduced an argument which suggests that the conductance G_{2L} of a block of size $(2L)^d$ was related solely to the conductance G_L of the 2^d blocks that were combined to build the larger block. We consider here the most general form of such an argument, in order to derive some general conclusions and to arrive at some idea of the limitations.

We consider a set of properties of a system of size L^d, which we represent by a vector $\boldsymbol{a}_L$, whose elements are chosen to be dimensionless. It is often useful to think of L as an effective system size, such as the inelastic scattering length L_i or even the resolution with which an external observer could measure the system. We assume that the set $\boldsymbol{a}$ is complete in the sense that we can write

$$\boldsymbol{a}(bL) = \boldsymbol{F}(\boldsymbol{a}(L), b). \tag{42}$$

The set $\boldsymbol{a}$ may well have to be infinitely large. In fact, in order to describe the distribution of values required for a full description of a disordered system it must be so.

At this point we make the first approximation, namely that eq. (42) can be rewritten in the differential form

$$\frac{\mathrm{d}\boldsymbol{a}}{\mathrm{d}\ln L} = \boldsymbol{f}(\boldsymbol{a}). \tag{43}$$

Equations such as eq. (43) have been extensively studied in the context of non-linear dynamics, chaos, etc., from which much of the language is derived, as well as in more conventional phase transitions. The most important property of such equations is that for increasing L the vector $\boldsymbol{a}$ tends towards a simple subspace, often a line or even a point. Such subspaces or points are called *attractors* or *fixed points*, respectively. Asymptotically we can describe the behavior of the system completely in terms of the properties of the attractor. It is important to remember, however, that this is really a description of the behavior of almost infinite systems. It is not clear a priori whether any given real or numerical experiment is sufficiently close to the attractor that the deviations from it may safely be ignored.

For the moment we consider only the case where eq. (43) has a 1D attractor. A theory based on this property is often called a *one-parameter scaling theory*. The attractive line may be parametrized in several different ways. Conventionally the original suggestion of Thouless is followed and the single parameter used is the dimensionless conductance, $g = (h/e^2)G$, or more precisely its arithmetic mean, $\langle g \rangle$. In the following paragraphs we shall use the abbreviation g instead of $\langle g \rangle$.

The equation describing the flow along the line is (Abrahams et al. 1979)

$$\frac{\mathrm{d}\ln g}{\mathrm{d}\ln L} = \beta(\ln g). \tag{44}$$

This approach will only be valid as long as the effective system size L is larger than any other length scale associated with the irrelevant contributions. In particular the mean free path l constitutes a lower limit for L.

Once we have identified a single parameter several results follow from quite general considerations.

(a) The general solution of the 1D scaling equation (44) has the form

$$g = g(L/\xi). \tag{45}$$

Everything is defined in terms of a single length scale ξ. Note, however, that the function $g(x)$ may be multivalued. There may be several values of g corresponding to the same value of x.

(b) For very strong disorder, $g \ll 1$, we expect exponential localization, i.e.

$$g \sim \exp(-2L/\lambda), \quad \beta(\ln g) \sim \ln g. \tag{46}$$

By comparing eq. (45) with eq. (46) we identify ξ as the localization length λ.

(c) For very weak disorder we expect classical ohmic conductivity,

$$g \sim \sigma L^{d-2}, \quad \beta(\ln g) \sim d-2. \tag{47}$$

Again by comparing eq. (45) with eq. (47) we find that ξ is related to the conductivity σ by

$$\xi \sim \sigma^{-1/(d-2)}. \tag{48}$$

The β-function can now be sketched as shown in fig. 48. Note, in particular, that, for $\beta > 0$, g tends towards case (c), extended behavior, whereas, for $\beta < 0$ g tends towards localized states. $\beta = 0$ represents a fixed point. In 1D β is almost certainly always negative. The flow is always towards small g, i.e. localized states. In 3D β must cross zero. There is always a fixed point and a metal–insulator transition. 2D represents the marginal case. It is impossible to tell whether β crosses zero without further information.

By combining (a), (b) and (c) we obtain a relationship between the conductivity exponent s in

$$\sigma \sim |\tau - \tau^*|^s \tag{49}$$

and the localization length exponent ν in

$$\lambda \sim |\tau - \tau^*|^{-\nu} \tag{50}$$

of the form

$$s = (d-2)\nu. \tag{51}$$

This relation, originally derived by Wegner (1976) was earlier interpreted by many authors as a prediction of a minimum metallic conductivity; $s = 0$ for $d = 2$. Note, however, that this interpretation presupposes the existence of a fixed point.

10.2. Perturbation theory

As described in the previous section, it is possible to calculate corrections to these results using diagrammatic perturbation theory. These corrections are equivalent to taking into account enhanced backscattering. From the quantum corrections to the classical conductivity one easily derives the following form of the β-function for the conductance

$$\beta(\ln g) = (d-2) - b/g. \tag{52}$$

Abrahams et al. (1979) noted that this implies that β is always negative in 2D and hence that all states are localized even for infinitesimal disorder. The solution of eq. (52) for 2D can be written as

$$g = \begin{cases} g_0 - b\ln L, \\ -b\ln(L/\xi). \end{cases} \tag{53}$$

Clearly both forms are equivalent to that given for 2D in eq. (39). The weak logarithmic dependence is often termed *weak localization.* It was discussed in the previous section.

In 1D it is well known that all states are exponentially localized (see § 8). By combining this with the formula derived by Landauer (1970) connecting the conductance with the transmission coefficient, where $T = \exp(-\gamma L)$, it is possible to derive a complete expression for the β-function

$$\frac{\mathrm{d}\ln g}{\mathrm{d}\ln L} = -(1+g)\ln\left(\frac{1+g}{g}\right). \tag{54}$$

These perturbation theory results are expected to be valid as long as the effective system size L, usually the inelastic scattering length, is larger than the mean free path l. However, in 1D and 2D the correction to the classical Ohm's law behavior increases with L, so that eventually the theory must break down. We expect the length scales for this breakdown to be of the order of the localization length, which can be written as $\xi_{\mathrm{loc}}^{1\mathrm{D}} \approx \pi l$ and $\xi_{\mathrm{loc}}^{2\mathrm{D}} \approx l\exp(\pi k_{\mathrm{F}} l/2)$, which are perturbative estimates (Lee and Ramakrishnan 1985).

10.3. Field-theoretical formulation

Here we attempt only the briefest introduction to these ideas. For a full discussion we refer the reader to the review by Efetov (1983) and the lecture notes by Wegner (1979). Consider the integral form

$$\int_{-\infty}^{+\infty} \mathrm{D}\varphi^\dagger\, \mathrm{D}\varphi \exp\{\mathrm{i}[\varphi^\dagger(E+\mathrm{i}\eta-\varepsilon)\varphi]\} = \frac{\mathrm{i}\pi}{E+\mathrm{i}\eta-\varepsilon}. \tag{55}$$

Note that this is a 2D integral, because we integrate over φ and $\varphi^\dagger$ separately. Note, also, that the integral only converges when $\eta > 0$ because the real part of the exponent is $-\eta\varphi^\dagger\varphi$. Equation (55) can be generalized,

$$\int_{-\infty}^{+\infty} \prod_{ij}^{N} \mathrm{D}\varphi_i^\dagger\, \mathrm{D}\varphi_j \exp\{\mathrm{i}[\varphi_i^\dagger(E+\mathrm{i}\eta-H_{ij})\varphi_j]\} = \frac{(\mathrm{i}\pi)^N}{\det(E+\mathrm{i}\eta-H_{ij})}. \tag{56}$$

Using the same technique it is possible to show that the following relation for the matrix elements of the Green function holds:

$$\int_{-\infty}^{+\infty} \prod_{ij}^{N} \mathrm{D}\varphi_i^\dagger \, \mathrm{D}\varphi_j \, \varphi_m \varphi_n^\dagger \exp(\ldots) = G_{mn} \frac{-\mathrm{i}(\mathrm{i}\pi)^N}{\det[E + \mathrm{i}\eta - H_{ij}]}. \tag{57}$$

Note that this is nothing but the statistical average $\langle \varphi_m \varphi_n^\dagger \rangle$, where the integral in eq. (56) represents a thermodynamic partition function.

A calculation of the conductivity requires quantities like $G^+ G^-$, which cannot be calculated from the same partition function because the G^- part violates the condition for convergence of the integral. Instead one calculates $\langle \varphi_k^1 \varphi_l^{\dagger 1} \varphi_m^2 \varphi_n^{\dagger 2} \rangle$ using the exponent

$$\sum_{\alpha=1}^{n} \mathrm{i}[\varphi_{i\alpha}^{\dagger 1}(E + \mathrm{i}\eta - H_{ij})\varphi_{j\alpha}^1 - \varphi_{i\alpha}^{\dagger 2}(E - \mathrm{i}\eta - H_{ij})\varphi_{j\alpha}^2]. \tag{58}$$

According to the symmetry of eq. (58) the Hamiltonian may be classified into four *universality classes* (see table 1).

10.4. Results from numerical scaling calculations

Using the Anderson model with diagonal disorder MacKinnon and co-workers formulated a numerical scaling procedure for the average of the logarithm of the quantum mechanical transmission probability, log *T*, through *d*-dimensional strips and bars of finite cross section and essentially infinite lengths (MacKinnon 1980, 1985, MacKinnon and Kramer 1981, 1983a). The transmission probability corresponds to the conductance through the system.

The results may be summarized as follows.

(1) Within the statistical accuracy of the raw data (1%), in almost all of the cases, the one-parameter scaling function can be established. This constitutes a quantitative criterion for the validity of the scaling behavior.

(2) The method yields complete localization in 2D with an essential singularity at zero disorder, and an Anderson transition in 3D.

(3) Box and Gaussian distribution of the site energies yield the same scaling function. This is the explicit demonstration of the universality of scaling in the orthogonal class.

Table 1
Universality classes and symmetry groups

Case	Without spin	With spin
Time-reversal symmetry	Orthogonal	Symplectic
No time-reversal symmetry	Unitary	Unitary

11. Localization in a magnetic field

In order to understand the quantized Hall effect (von Klitzing et al. 1980, von Klitzing 1986), some knowledge of the basic localization features of the electronic states in 2D disordered systems is unavoidable (Aoki 1987).

Neglecting interactions, etc. one has to solve the Schrödinger equation for an electron (mass m, charge e) that moves under the influence of a homogeneous magnetic field $\boldsymbol{B}$ parallel to the z-axis (represented by a suitable vector potential $\boldsymbol{A}$) and a random potential $V(\boldsymbol{r})$,

$$\left(\frac{1}{2m}(\boldsymbol{p}-e\boldsymbol{A})^2+V(\boldsymbol{r})\right)\Psi(\boldsymbol{r})=E\Psi(\boldsymbol{r}). \tag{59}$$

The presence of a magnetic field introduces as an additional complication the non-trivial nature of the energy spectrum and the states in the limit of vanishing disorder. In the simplest 2D effective mass we have the degenerate spectrum of the discrete Landau levels and the corresponding Landau states.

The random potential is specified as described in § 7. The special case of a Gaussian white noise potential as well as the more general case of a spatially correlated potential has been used in perturbation theories (Ono 1982a,b, 1984, Hikami and Brézin 1986) and in the recently developed field-theoretical treatments (Pruisken 1984, Weidenmüller 1987). In the asymptotic limit where the correlation length of the potential $a \gg l_c$, where $l_c=(h/e|\boldsymbol{B}|)^{1/2}$ is the cyclotron length, the electronic problem is equivalent to a classical percolation problem (Tsukada 1976, Iordansky 1982, Kazarinov and Luryi 1982, Ono 1982a, Trugman 1983, Luryi and Kazarinov 1983). In this case one can use the results of percolation theory in order to discuss the localization properties of the states. However, as we shall see later, it is imperative to study the influence of quantum mechanical effects (tunneling and interference). One possibility is to put in these effects by hand, as has been done recently by Chalker and Coddington (1988) in a very interesting study. It would certainly be more satisfactory to have a model that is able to cover both of the limits $a/l_c \to 0$ (white noise limit) and $a/l_c \to \infty$ on an equal footing. A first attempt to develop such a model has been made by Huckestein and Kramer (1989).

11.1. Perturbational approach

The perturbational approach has been carried out most elaborately by Ono (1982a,b, 1983, 1984, 1985).

The explicit result for the critical behavior for the case of δ-function-like impurity potentials is

$$\lambda(E)=\text{const.}\exp[\alpha/(E-E_N)^2]. \tag{60}$$

The constant α is of the order of the square of the disorder-induced bandwidth and E_N is the energy of the unperturbed Landau level. There is an essential singularity of λ in the band center. Although there are a number of assumptions in this approach that are not easy to justify, this was the first hint obtained from theory that extended states might exist only at a single energy in each of the Landau bands in the quantum mechanical high-magnetic-field limit.

11.2. Thouless number study

One of the earliest numerical studies of localization was performed by Ando (1984). An inverse localization length $\gamma(E)$ was defined via the exponential behavior of $g(L)$,

$$g(E; L) = g_0 \exp[-\gamma(E)/L]. \tag{61}$$

The results for the case of short-range scatterers were consistent with a divergence of the localization length only at the centers of the Landau bands, but $\gamma(E) \propto |E - E_N|^{\nu}$ with $\nu < 2$ in contrast to the essential singularity obtained from the renormalized perturbation expansion.

11.3. The percolation limit

In the limit $|\boldsymbol{B}| \to \infty$ the quantum mechanical problem can be replaced by a classical percolation problem provided that the random potential has a finite correlation length. The basis of the percolation argument has been formulated by Tsukada (1976).

The question of whether or not an eigenstate is localized may be decided by investigating the percolation problem for the equipotential lines (Trugman 1983). An intuitive insight is obtained by considering the analogy with a hilly landscape (potential energy) that is gradually filled with water, the water level corresponding to the energy E. For very low water levels the water runs into the deep valleys. All shore lines (equipotential lines) are closed paths, i.e. the states are localized. The same is true for very high water levels. Only a few mountains are high enough to reach above the water level. The shore lines are again closed paths. It is clear that in between there must be one water level at which it is possible to travel either by boat or on foot from one side of the system to the other. This corresponds to an extended state. Percolation theory says that there is exactly one percolating path. The length t of a cluster (equipotential line) is given by

$$t = |E - E_0|^{-1/\sigma}, \tag{62}$$

with $\sigma = 36/91$. The diameter ξ of the area covered is

$$\xi = |E - E_0|^{-\nu}, \tag{63}$$

with $\nu = \frac{4}{3}$.

Equations (62) and (63) may be used to estimate the critical behavior of the states.

11.4. The participation number

The (inverse) participation number (ratio) has been studied for the random Landau model in different geometries using a variety of methods (Kramer et al. 1988, Hikami

1986, Hikami and Brézin 1986, Aoki 1983, 1986, Ando 1984, Kunz and Souillard 1982, Zirnbauer 1986, Ohtsuki and Ono 1989). The shape of the wavefunctions were studied numerically (Aoki 1983, 1986, Ono et al. 1989, 1991). The spatial properties of the states turned out to be rather peculiar showing a pronounced self-similar network structure. Fractal behavior has been inferred near the critical energy. The fractal dimension d^* of the states near the centers of the Landau bands was estimated from the behavior of the amplitudes of the wavefunction, $d^* = 1.57$ (Aoki 1986).

11.5. Numerical scaling

The scaling properties of the asymptotic exponential decay length of the modulus of the Green function with respect to energy, disorder, and the width of the system have been studied for both the lattice model (Schweitzer et al. 1984, MacKinnon et al. 1984, MacKinnon and Kramer 1983b, Kramer and MacKinnon 1986) and the random Landau model. In the latter δ-function potentials as well as Gaussian potentials have been considered (Aoki 1987, Ando and Aoki 1985a,b, Ando 1985, Huckestein and Kramer 1989).

The application of the scaling method to the random Landau model yielded the most precise and reliable information about the nature of the singularity in the center of the band. For the equivalent of the Gaussian white noise model the attempt to establish a one-parameter scaling relation was successful (Huckestein and Kramer 1990). This could be achieved due to the possibility of treating larger system sizes than before (Huckestein 1990). The critical exponent for the lowest Landau band was extracted with an accuracy of better than 2%, $\nu = 2.34 \pm 0.04$. Comparison of the data with those obtained from different models, including the quantum mechanically treated quasi-classical percolation limit (Chalker and Coddington 1988, Mil'nicov and Sokolov 1989), indicates that this result is independent of the microscopic details of the potential and is truly universal.

11.6. Comparison with experiment

The results of the scaling approach can be compared with the experimental data when identifying the size of the system with the temperature-dependent phase coherence length L_φ (see above). This idea (Pruisken 1984, Levine et al. 1984), although not being extremely well supported theoretically for systems in the localized regime, yields a striking agreement with the results of temperature- and frequency-dependent measurements of the transport properties in the quantized Hall regime (Ebert et al. 1982, Wei et al. 1988, Kuchar et al. 1990).

Since the scaling function depends only on the variable $L/\xi \sim E^\nu L$ near the center of the band the derivatives satisfy $\mathrm{d}^n f/\mathrm{d}E^n \propto L^{n/\nu}$ near $E = 0$. Near the band center $E \propto B - B^*$ for fixed particle density, where B^* is the magnetic field at half-integer filling. Letting $L \equiv L_\varphi T^{-p/2}$, it follows that if the nth-order derivative of the scaling function has extrema at B^* they must diverge according to $T^{-n\kappa}$, where $\kappa = p/2\nu$. Assuming further that the magnetoresistance and the Hall resistance depend on the same variable, by fitting to the measured temperature dependences, the values $\kappa = 0.42$ and $p = 2.0 \pm 0.2$ were obtained.

This result for p is at present not understood theoretically.

Experimentally the above value of κ was obtained for a variety of magnetoresistance peaks corresponding to an integer filling factor as well as a number of fractional fillings (Engel et al. 1990). This strongly suggests that the above scaling picture may also be valid in the regime of the fractionally quantized Hall effect.

The temperature dependence of the widths of the plateaus corresponding to integer quantization of the Hall resistance were determined (Kuchar et al. 1990). Noting that the phase coherence length introduces effective (temperature-dependent) mobility edges into the system the plateau widths as a function of the temperature can be calculated from the number of states between the two successive mobility edges belonging to two successive Landau bands (Huckestein and Kramer 1990).

12. Fluctuations

We have seen above that the average conductance and the inverse of the average resistance of 1D disordered systems do not agree with each other (35). They are not self-averaging in the sense of statistical physics. As in 1D the localization length is always finite, irrespective of the energy and disorder, it is tempting to conclude that non-self-averaging of transport is also an intrinsic property of the localized regime in 2D and 3D.

In the metallic limit the conductivity may be calculated from a configurational average of a quantum mechanical transition probability, as we have seen in § 7. Due to the configurational average, most of the interference terms do not contribute to the transition probabilities (36). If we consider the conductance of a given sample, however, the interference terms dominate the total transition probability in a random manner. As a consequence, microscopic changes of the random potential in an impure metal of finite size should result in large changes of the (coherent) conductance. Thus, even in metals it is by no means guaranteed that quantum coherent transport is self-averaging.

It is also not clear what the consequences of non-self-averaging for the nature of the Anderson transition would be. The study of the distribution functions of the physical quantities of interest is therefore of crucial importance for a thorough understanding of Anderson localization.

12.1. The statistics of transport in 1D disordered systems

The central-limit theorem for the localization length has been shown to be valid for the disordered harmonic chain by O'Connor (1975). Approximate treatments for the electronic problem have been given by Anderson et al. (1980) using a Landauer-type approach, and by Mel'nikov (1981) by estimating the distribution function of the resistance, and calculating from that the distribution function of its logarithm. The case of a Gaussian white noise potential has been treated by Abrikosov (1981) and by Kree and Schmid (1981).

The most important result is that the logarithm of the resistance (or of the conductance, alternatively) is a statistically well-behaved quantity, its limiting distribution function being a Gaussian with a finite variance. The corresponding relative average fluctuations can be calculated explicitly by using approximate methods (Kree and Schmid 1981, Mel'nikov 1981, Abrikosov 1981, Tankei and Takano 1986) and by numerical procedures (Andereck and Abrahams 1980, Sak and Kramer 1981, Kantor and Kapitulnik 1982). They *decrease* with an increasing length of the system, i.e.

$$\frac{\langle(\Delta \ln r)^2\rangle^{1/2}}{\langle \ln r\rangle} = \left(\frac{2}{\gamma N}\right)^{1/2}, \quad \gamma N \gg 1, \tag{64}$$

where the notation is as in § 9. This holds for small γ (Sak and Kramer 1981, Abrikosov 1981, Tankei and Takano 1986). In the limit of large γ, however, there seem to be deviations from this behavior (Sak and Kramer 1981, Tankei and Takano 1986). The reasons are not yet understood.

As the distribution of $\ln r$ is asymptotically well behaved, it is intuitively clear that the resistance as well as the conductance must have statistical distributions that yield asymptotically divergent fluctuations.

The second moment of the resistance grows more rapidly with the length of the system than the average resistance itself (Abrahams and Stephen 1980, Stone and Joannopoulos 1982). Similar conclusions can be drawn for the higher moments of the resistance, and for a large degree of disorder. A full account of the asymptotic behavior of all the moments of the resistance and the conductance has been given by Kirkman and Pendry (1984a,b), Abrikosov (1981), and Mel'nikov (1981) by using approximate methods.

These results imply that the root mean square (r.m.s.) fluctuations of the resistance grow exponentially with the length of the system. The resistance is not a self-averaging quantity. The physical reason for this behavior is the exponential increase of the resistance in the localized regime when increasing the size of the system. Small statistical fluctuations of the localization length in the exponent will thus cause exponentially large changes of the resistance. Similar statements also hold for the conductance.

Although there are important quantitative aspects to be resolved in the future there is a certain probability that the reproducible conductance fluctuations observed originally in quasi-1D confined inversion layers in Si-MOSFETs (Fowler et al. 1982) can be identified with the fluctuations induced by the strong localization described above.

12.2. Fluctuations in the metallic limit

The experimental and theoretical investigations of recent years revealed a most surprising feature of the transport properties of metallic systems that are defined by the condition that the mean free path l is much smaller than the system diameter L (Wheeler et al. 1982, Umbach et al. 1984, Skocpol et al. 1986, Licini et al. 1985, Washburn and Webb 1986, Webb 1988, Mailly and Sanquer 1991, Caro et al. 1991). At very low temperatures, when inelastic scattering processes are frozen out to such a degree that in a sample of finite size almost no phase randomization takes place, the sample-specific statistical fluctuations of the conductance as a function of the Fermi level, an applied magnetic field, or the configuration of the impurities were observed to be larger than expected. They were *well reproducible* for a given sample. For a slightly modified sample (for instance, by heating up and cooling down again) their behavior changed in its details (fig. 53). However, the root mean square deviation turned out to be approximately a constant,

$$\delta G = f = O(e^2/\hbar), \tag{65}$$

and universal within certain limits in the sense that it did not depend on the average conductance of the sample. There is a small dependence on its dimensionality and geometrical shape. There is also a striking dependence of the magnitude of the

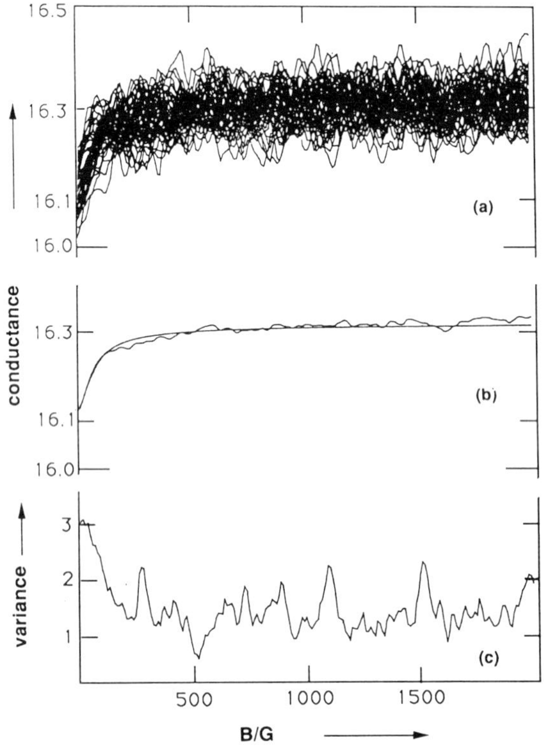

Fig. 53. (a) Reproducible fluctuations of the magnetoconductance in units of e^2/h at $T = 45$ mK of a SiGaAs wire after heating and cooling down again (46 cycles). (b) The mean value of the 46 curves and a 1D weak-localization fit ($L_\Phi = 3$ μm). (c) The variance of the 46 curves in units of $10^{-3}(e^2/h)^2$ (redrawn after Mailly and Sanquer (1991)).

fluctuations on the universality class of the system. In the presence of a magnetic field the amplitude is reduced by a factor of a half. All of these findings were well supported by numerical and analytical calculations (Stone 1985, Lee and Stone 1985, Altshuler 1985, Altshuler and Khmel'nitskii 1988, Imry 1986a,b, Muttalib et al. 1987, Serota et al. 1987, Zanon and Pichard 1988, Mello 1987, Mašek and Kramer 1989, Pichard and Sanquer 1990, Pichard 1991a,b). The conductivity is related to the conductance by the classical relation $G = \sigma L^{d-2}$, therefore the relative fluctuations vanish only according to $\delta G/G \propto L^{2-d}$ in contrast to what one would expect from classical statistical physics, $\delta G/G \propto L^{-d/2}$. This means that even in 3D metallic systems the zero-temperature conductance and resistance (and hence the conductivity and resistivity) are not self-averaging when the system is coherent.

12.3. Fluctuations and one-parameter scaling

From the results of the studies of both of the asymptotic limits we can conclude that a complete theory of Anderson localization must necessarily be a theory of the distribution functions of the relevant quantities, and not only the configurational averages. This viewpoint has been stressed during the last few years by an increasing number of researchers (Kravtsov and Lerner 1984, Altshuler et al. 1986, Shapiro 1986, 1987, Efetov 1980, 1987, 1988, Cohen et al. 1988, Chase and MacKinnon 1987, Schreiber and Kramer 1987, Altshuler et al. 1990, Lerner 1991). The theory is at present far from being complete.

References

Abrahams, E., and M. Stephen, 1980, J. Phys. C **13**, L377.

Abrahams, E., P.W. Anderson, D.C. Licciardello and T.V. Ramakrishnan, 1979, Phys. Rev. Lett. **42**, 673.

Abrikosov, A.A., 1981, Solid State Commun. **37**, 997.

Abrikosov, A.A., and I.A. Ryzkhin, 1978, Adv. Phys. **27**, 147.

Akera, H., and T. Ando, 1989, Phys. Rev. B **39**, 5508.

Akera, H., and T. Ando, 1990a, Phys. Rev. B **41**, 1964.

Akera, H., and T. Ando, 1990b, Surf. Sci. **229**, 268.

Akera, H., and T. Ando, 1991a, in: Proc. 20th Int. Conf. on the Physics of Semiconductors, Thessaloniki, Greece, 1990 (World Scientific, Singapore).

Akera, H., and T. Ando, 1991b, Phys. Rev. B **43**, 11676.

Alphenaar, B.W., P.L. McEuen and R.G. Wheeler, 1990, Phys. Rev. Lett. **64**, 677.

Altshuler, B.L., 1985, Sov. Phys.-JETP Lett. **41**, 530(R); 649(E).

Altshuler, B.L., and A.G. Aronov, 1979, Zh. Eksp. & Teor. Fiz. **77**, 2028 [Sov. Phys.-JETP **50**, 968].

Altshuler, B.L., and A.G. Aronov, 1985, in: Electron–Electron Interactions in Disordered Systems, eds A.L. Efros and M. Pollak (North-Holland, Amsterdam) p. 1.

Altshuler, B.L., and D.E. Khmel'nitskii, 1988, Sov. Phys.-JETP Lett. **42**, 359.

Altshuler, B.L., A.G. Aronov and P.A. Lee, 1980, Phys. Rev. Lett. **44**, 1288.

Altshuler, B.L., A.G. Aronov and B.Z. Spivak, 1981, Sov. Phys.-JETP Lett. **33**, 94.

Altshuler, B.L., V.E. Kravtsov and I.V. Lerner, 1986, Pis'ma Zh. Eksp. & Teor. Fiz. **43**, 342 [JETP Lett. **43**, 441].

Altshuler, B.L., V.E. Kravtsov and I.V. Lerner, 1990, in: Anderson Transition and Mesoscopic Fluctuations, eds B. Kramer and G. Schön, Physica A **167**, 1–314.

Alves, E.S., P.H. Benton, M. Henini, L. Eaves, P.C. Main, O.H. Hughes, G.A. Toombs, S.P. Beaumont and C.D.W. Wilkinson, 1989, J. Phys.: Condens. Matter **1**, 8257.
Andereck, B., and E. Abrahams, 1980, J. Phys. C **13**, L383.
Anderson, P.W., 1958, Phys. Rev. **109**, 1492.
Anderson, P.W., 1981, Phys. Rev. **23**, 4828.
Anderson, P.W., E. Abrahams and T.V. Ramakrishnan, 1979, Phys. Rev. Lett. **43**, 718.
Anderson, P.W., D.J. Thouless, E. Abrahams and D.S. Fisher, 1980, Phys. Rev. B **22**, 3519.
Ando, T., 1984, J. Phys. Soc. Jpn. **53**, 3101.
Ando, T., 1985, in: Anderson localizaiton, ed. Y. Nagaoka, Prog. Theor. Phys. **84**, Suppl. p. 69.
Ando, T., 1990, Phys. Rev. B **42**, 5626.
Ando, T., and H. Akera, 1990, in: Proc. Int. Conf. on Application of High Magnetic Fields in Semiconductor Physics, Würzburg.
Ando, T., and H. Aoki, 1985a, Phys. Rev. Lett. **54**, 832.
Ando, T., and H. Aoki, 1985b, J. Phys. Soc. Jpn. **54**, 2238.
Ando, T., and H. Fukuyama, eds, 1987, Anderson Localization, Springer Proc. in Physics, Vol. 28 (Springer, Berlin).
Ando, T., A.B. Fowler and F. Stern, 1982, Rev. Mod. Phys. **54**, 437.
Aoki, H., 1983, J. Phys. C **16**, 1205.
Aoki, H., 1985, Phys. Rev. Lett. **55**, 1136.
Aoki, H., 1986, Phys. Rev. B **33**, 7310.
Aoki, H., 1987, Rep. Prog. Phys. **50**, 655.
Aoki, H., 1992, Surf. Sci. **263**, 137.
Aoki, H., and T. Ando, 1981, Solid State Commun. **18**, 1079.
Arakawa, Y., and H. Sakaki, 1982, Appl. Phys. Lett. **40**, 939.
Aronov, A.G., and Yu.V. Sharvin, 1987, Rev. Mod. Phys. **59**, 755.
Averin, D.V., and K.K. Likharev, 1991, in: Mesoscopic Phenomena in Solids, eds B.L. Altshuler, P.A. Lee and R.A. Webb (North-Holland, Amsterdam).
Avishai, Y., and Y.B. Band, 1989a, Phys. Rev. B **40**, 12535.
Avishai, Y., and Y.B. Band, 1989b, Phys. Rev. B **40**, 3429.
Avishai, Y., and Y.B. Band, 1989c, Phys. Rev. Lett. **62**, 2527.
Avishai, Y., and Y.B. Band, 1990, Phys. Rev. B **41**, 3253.
Avishai, Y., and Y.B. Band, 1991, Phys. Rev. Lett. **66**, 1761.
Avishai, Y., M. Kaveh, S. Shatz and Y.B. Band, 1989, J. Phys.: Condens. Matter **1**, 6907.
Avishai, Y., M. Kaveh and Y.B. Band, 1990, Phys. Rev. B **42**, 5867.
Avron, J.E., R. Seiler and B. Simon, 1983, Phys. Rev. Lett. **51**, 51.
Azbel, M.Ya., 1964, Zh. Eksp. & Teor. Fiz. **46**, 929 [Sov. Phys.-JETP **19**, 634].
Bagwell, P.F., 1990, Phys. Rev. B **41**, 10354.
Bagwell, P.F., T.P.E. Broekaert, T.P. Orlando and C.G. Fonstad, 1990, J. Appl. Phys. **68**, 4634.
Baranger, H.U., 1990, Phys. Rev. B **42**, 11479.
Baranger, H.U., and A.D. Stone, 1989a, Phys. Rev. B **40**, 8169.
Baranger, H.U., and A.D. Stone, 1989b, Phys. Rev. Lett. **63**, 414.
Baranger, H.U., and A.D. Stone, 1990a, Surf. Sci. **229**, 212.
Baranger, H.U., and A.D. Stone, 1990b, Science and Engineering of 1- and 0-Dimensional Semiconductors, eds S.P. Beamont and C.M. Sotomayor-Torres (Plenum Press, London).
Baranger, H.U., D.P. DiVincenzo, R.A. Jalabert and A.D. Stone, 1991, Phys. Rev. B **44**, 10637.
Beaumont, S.P., and C.M. Sotomayor-Torres, eds, 1990, Science and Engineering of 1- and 0-Dimensional Semiconductors (Plenum Press, London).
Beenakker, C.W.J., and H. van Houten, 1989a, Phys. Rev. B **39**, 10445.
Beenakker, C.W.J., and H. van Houten, 1989b, Phys. Rev. Lett. **63**, 1857.
Beenakker, C.W.J., and H. van Houten, 1990, Electronic Properties of Multilayers and Low-Dimensional Semiconductor Structures, NATO-ASI Series, eds J.M. Chamberlain, L. Eaves and J.C. Portal (Plenum Press, London).
Beenakker, C.W.J., and H. van Houten, 1991a, Solid State Phys. **44**, 1.
Beenakker, C.W.J., and H. van Houten, 1991b, Phys. Rev. Lett. **66**, 3056.

Beenakker, C.W.J., H. van Houten and B.J. van Wees, 1989, Superlatt. & Microstr. **5**, 127.
Behringer, R.E., G. Timp, H.U. Baranger and J.E. Cunningham, 1991, Phys. Rev. Lett. **66**, 930.
Benedict, K.A., and J.T. Chalker, eds, 1991, Localisation 1990 (Institute of Physics, Bristol).
Berggren, K.F., and Z.-L. Ji, 1990, Superlatt. & Microstr. **8**, 59.
Berggren, K.F., and Z.-L. Ji, 1991, Phys. Rev. B **43**, 4760.
Berggren, K.F., and D.J. Newson, 1986, Semicond. Sci. & Technol. **1**, 327.
Berggren, K.F., T.J. Thornton, D.J. Newson and M. Pepper, 1986, Phys. Rev. Lett. **57**, 1769.
Bergmann, G., 1982a, Phys. Rev. Lett. **49**, 162.
Bergmann, G., 1982b, Phys. Rev. Lett. **48**, 1046.
Bergmann, G., 1982c, Phys. Rev. B **25**, 2937.
Bergmann, G., 1984, Phys. Rep. **107**, 1.
Beton, P.H., B.R. Snell, P.C. Main, A. Neves, J.R. Owers-Bradley, L. Eaves, M. Henini, O.H. Hughes, S.P. Beaumont and C.D.W. Wilkinson, 1989, J. Phys.: Condens. Matter **1**, 7505.
Beyer, W., 1974, Ph.D. Thesis, Universität Marburg.
Biskupski, G., and A. Briggs, 1988, J. Non-Cryst. Solids **87**, 683.
Bonch-Bruevich, V.L., R. Enderlein, B. Esser, R. Keiper, A.G. Mironov and I.P. Zvyagin, 1984, Elektronentheorie ungeordneter Halbleiter (VEB Deutscher Verlag der Wissenschaften, Berlin).
Brey, L., N.F. Johnson and B.I. Halperin, 1989, Phys. Rev. B **40**, 10647.
Brown, E., 1964, Phys. Rev. A **133**, 1038.
Brown, R.J., M.J. Kelly, R. Newbury, M. Pepper, B. Miller, H. Ahmed, D.G. Hasko, D.C. Peacock, D.A. Ritchie, J.E.F. Frost and G.A.C. Jones, 1989a, Solid State Electronics **32**, 1179.
Brown, R.J., M.J. Kelly, M. Pepper, H. Ahmed, D.G. Hasko, D.C. Peacock, J.E.F. Frost, D.A. Ritchie and G.A.C. Jones, 1989b, J. Phys.: Condens. : Matter **1**, 6285.
Brown, R.J., C.G. Smith, M. Pepper, M.J. Kelly, R. Newbury, H. Ahmed, D.G. Hasko, J.E.F. Frost, D.C. Peacock, D.A. Ritchie and G.A.C. Jones, 1989c, J. Phys.: Condens.: Matter **1**, 6291.
Brown, R.J., M. Pepper, H. Ahmed, D.G. Hasko, D.A. Ritchie, J.E.F. Frost, D.C. Peacock and G.A.C. Jones, 1990, J. Phys.: Condens. Matter **2**, 2105.
Brum, J.A., 1991a, Phys. Rev. B **43**, 12082.
Brum, J.A., 1991b, in: Proc. 20th Int. Conf. on Physics of Semiconductors, Thessaloniki, Greece, 1990 (World Scientific, Singapore).
Bryant, G.W., 1989, Phys. Rev. B **39**, 3145.
Bryant, G.W., 1990a, in: Science and Engineering of 1- and 0-Dimensional Semiconductors, eds S.P. Beamont and C.M. Sotomayor-Torres (Plenum Press, London).
Bryant, G.W., 1990b, Mater. Res. Soc. Ext. Abstr. **EA-26**, 107.
Bryant, G.W., 1991a, in: Int. Conf. on Nanostructure Physics and Fabrication, eds M. Reed and W.P. Wiley (Academic Press, New York).
Bryant, G.W., 1991b, Phys. Rev. B **44**, 3064.
Büttiker, M., 1986, Phys. Rev. Lett. **57**, 1761.
Büttiker, M., 1988a, Phys. Rev. B **38**, 12724.
Büttiker, M., 1988b, Phys. Rev. B **38**, 9375.
Büttiker, M., 1989, Phys. Rev. B **40**, 3409.
Büttiker, M., 1990a, Phys. Rev. B **41**, 7906.
Büttiker, M., 1990b, Phys. Rev. Lett. **65**, 2901.
Büttiker, M., 1990c, Semiconductors and Semimetals, ed. M.A. Reed (Academic Press, New York).
Büttiker, M., 1991, in: Proc. 20th Int. Conf. on the Physics of Semiconductors, Thessaloniki, Greece, 1990 (World Scientific, Singapore) p. 49.
Büttiker, M., Y. Imry, R. Landauer and S. Pinhas, 1985, Phys. Rev. B **31**, 6207.
Caro, J., J.A. Gao, A.H. Verbruggen, S. Radelaar and J. Midelhoek, 1991, in: Quantum Coherence in Mesoscopic Systems, NATO-ASI Series B, Vol. 24, ed. B. Kramer (Plenum, New York).
Castaño, E., and G. Kirczenow, 1989, Solid State Commun. **70**, 801.
Castaño, E., and G. Kirczenow, 1990a, Phys. Rev. B **41**, 5055.
Castaño, E., and G. Kirczenow, 1990b, Proc. Nanostructures and Microstructure Correlation with Physical Properties of Semiconductors, SPIE Proc. Series **1284**, 101.
Castaño, E., and G. Kirczenow, 1990c, Phys. Rev. B **41**, 3874.

Castaño, E., and G. Kirczenow, 1992, Phys. Rev. B **45**, 1514.
Castaño, E., G. Kirczenow and S.E. Ulloa, 1990, Phys. Rev. B **42**, 3753.
Chakraborty, T., V. Halonen and P. Pietiläinen, 1991, Phys. Rev. B **43**, 14289.
Chakravarty, S., and A. Schmid, 1986, Phys. Rep. **140**, 193.
Chalker, J.T., and P.D. Coddington, 1988, J. Phys. C **21**, 2665.
Chamberlain, J.M., L. Eaves and J.C. Portal, eds, 1990, Electronic Properties of Multilayers and Low-Dimensional Semiconductor Structures, NATO-ASI Series, Vol. 231 (Plenum Press, London).
Chandrasekar, V., M.J. Rooks, S.J. Wind and D.E. Prober, 1985, Phys. Rev. Lett. **55**, 1610.
Chang, A.M., 1989, in: Electronic Structure Properties of Semiconductors, ed. W. Schörter (VCH Verlagsgesellschaft, Weinheim).
Chang, A.M., G. Timp, R.E. Howard, R.E. Behringer, P.M. Mankiewich, J.E. Cunningham, T.Y. Chang and B. Chelluri, 1988a, Superlatt. & Microstr. **4**, 515.
Chang, A.M., G. Timp, T.Y. Chang, J.E. Cunningham, P.M. Mankiewich, R.E. Behringer and R.E. Howard, 1988b, Solid State Commun. **67**, 769.
Chang, A.M., T.Y. Chang and H.U. Baranger, 1989, Phys. Rev. Lett. **63**, 996.
Chang, L.L., and B.C. Giessen, eds, 1985, Synthetic Modulated Structures (Academic Press, New York).
Chang, S.S., S. Ando and T. Fukui, 1991, in: Proc. 5th Conf. on Modulated Semiconductor Structures. Nara, 1991.
Chase, K.S., and A. MacKinnon, 1987, J. Phys. C **20**, 6189.
Chen, C.Y., R.J. Birgeneau, D.R. Gabbe, H.P. Jenssen, M.A. Kastner, P.J. Picone, N.W. Preyer and T. Thio, 1989, Physica C **162**, 1031.
Christen, J., E. Kapon, E. Colas, D.M. Hwang, L.M. Schiavone, M. Grundmann and D. Bimberg, 1991, in: Proc. 5th Conf. on Modulated Semiconductor Structures, Nara, 1991.
Chu, C.S., and R.S. Sorbello, 1989, Phys. Rev. B **40**, 5941.
Cobden, D.H., N.K. Patel, M. Pepper, D.A. Ritchie, J.E.F. Frost and G.A.C. Jones, 1991, Phys. Rev. B **44**, 1938.
Cohen, A., Y. Roth and B. Shapiro, 1988, Phys. Rev. B **38**, 12125.
Cota, E., J.V. José and M.Ya. Azbel, 1985, Phys. Rev. B **32**, 6187.
Cusack, N.E., 1987, The Physics of Structurally Disordered Materials: An Introduction (Adam Hilger, Bristol).
Czycholl, G., B. Kramer and A. MacKinnon, 1981, Z. Phys. B **43**, 5.
Darwin, C.G., 1930, Proc. Cambridge Philos. Soc. **27**, 86.
Datta, S., 1989, Superlatt. & Microstr. **6**, 83.
Datta, S., 1991, in: Physics of Low-Dimensional Semiconductor Structures, eds P.N. Butcher, N.H. March and M.P. Tosi (Plenum, London).
Davies, J.H., 1988, Superlatt. & Microstr. **3**, 995.
Davies, J.H., 1989, in: Proc. Int. Conf. on Nanostructure Physics and Fabrication, eds M. Reed and W.P. Wiley (Academic Press, New York) p. 107.
de Aguiar, F.M., and D.A. Wharam, 1991, Phys. Rev. B **43**, 9984.
De Raedt, H., N. García and J.J. Sáenz, 1989, Phys. Rev. Lett. **63**, 2260.
Dean, P., 1972, Rev. Mod. Phys. **44**, 127.
Dekker, C., A.J. Scholten, F. Liefrink, R. Eppenga, H. van Houten and C.T. Foxon, 1991, Phys. Rev. Lett. **66**, 2148.
del Alamo, J.A., and C.C. Eugster, 1990, Appl. Phys. Lett. **56**, 78.
Delyon, F., B. Simon and B. Soulliard, 1984, Phys. Rev. Lett. **52**, 2187.
Dempsey, J., N.F. Johnson, L. Brey and B.I. Halperin, 1990, Phys. Rev. B **42**, 11708.
Derrida, B., K. Mecheri and J.-L. Pichard, 1987, J. Phys. (Paris) **48**, 733.
Dolan, G.J., and G.D. Osheroff, 1979, Phys. Rev. Lett. **43**, 721.
Ebert, G., K. von Klitzing, C. Probst and K. Ploog, 1982, Solid State Commun. **44**, 95.
Economou, E.N., and C.M. Soukoulis, 1981a, Phys. Rev. Lett. **46**, 618.
Economou, E.N., and C.M. Soukoulis, 1981b, Phys. Rev. Lett. **47**, 973.
Edwards, J.T., and D.J. Thouless, 1972, J. Phys. C **5**, 807.
Efetov, K.B., 1980, Zh. Eksp. & Teor. Fiz. **65**, 360 [Sov. Phys.-JETP **92**, 638].
Efetov, K.B., 1983, Adv. Phys. **32**, 53.

Efetov, K.B., 1987, Zh. Eksp. & Teor. Fiz. **66**, 634 [Sov. Phys.-JETP **93**, 1125].
Efetov, K.B., 1988, Zh. Eksp. & Teor. Fiz. **67**, 357 [Sov. Phys.-JETP **94**, 199].
Efetov, K.B., 1989, J. Phys.: Condens. Matter **1**, 5535.
Efros, A.L., and B.I. Shklovskii, 1975, J. Phys. C **8**, L49.
Efros, A.L., F.G. Pikus and G.G. Samsonidze, 1990, Phys. Rev. B **41**, 8295.
Elliott, R.J., J.A. Krumhansl and P.L. Leath, 1974, Rev. Mod. Phys. **46**, 465.
Engel, L., H.P. Wei, D.C. Tsui and M. Shayegan, 1990, Surf. Sci. **229**, 13.
Ensslin, K., and P.M. Petroff, 1990, Phys. Rev. B **41**, 12307.
Erdös, P., and R.C. Herndon, 1982, Adv. Phys. **31**, 65.
Esaki, L., 1986, IEEE J. Quantum Electron. **QE-22**, 1611.
Escapa, L., and N. García, 1989, J. Phys.: Condens. Matter **1**, 2125.
Eugster, C.C., and J.A. del Alamo, 1991, Phys. Rev. Lett. **67**, 3586.
Faist, J., P. Guret and H. Rothuizen, 1990, Phys. Rev. B **42**, 3217.
Fal'ko, V., 1989, Europhys. Lett. **8**, 785.
Fang, H., and P.J. Stiles, 1990, Phys. Rev. B **41**, 10171.
Feng, S., 1990a, Physica A **168**, 439.
Feng, S., 1990b, Phys. Lett. A **151**, 176.
Feng, S., 1990c, Phys. Lett. A **143**, 400.
Ferry, D.K., 1989, Acta Polytechnica Scandinavica, Electron. Eng. Ser. **64**, 271.
Field, S.B., M.A. Kastner, U. Meirav, J.H.F. Scott-Thomas, D.A. Antoniadis, H.I. Smith and S.J. Wind, 1990, Phys. Rev. B **42**, 3523.
Finlayson, D.M., ed., 1986, Localisation and Interactions in Disordered and Doped Semiconductors (SUSSP, Edinburgh).
Fisher, D.S., and P.A. Lee, 1981, Phys. Rev. B **23**, 6851.
Fock, V., 1928, Z. Phys. **47**, 446.
Ford, C.J.B., 1991a, in: Localisation 1990, eds K.A. Benedict and J.T. Chalker (Institute of Physics, Bristol) p. 85.
Ford, C.J.B., 1991b, Low and high field queching of the Hall effect and Coulomb blockade in ballistic junctions, to appear in Phys. Scr.
Ford, C.J.B., T.J. Thornton, R. Newbury, M. Pepper, H. Ahmed, D.C. Peacock, D.A. Ritchie, J.E.F. Frost and G.A.C. Jones, 1988a, Phys. Rev. B **38**, 8518.
Ford, C.J.B., T.J. Thornton, R. Newbury, M. Pepper, H. Ahmed, C.T. Foxon, J.J. Harris and C. Roberts, 1988b, J. Phys. C **21**, L325.
Ford, C.J.B., T.J. Thornton, R. Newbury, M. Pepper, H. Ahmed, D.C. Peacock, D.A. Ritchie, J.E.F. Frost and G.A.C. Jones, 1988c, Phys. Rev. B **38**, 8518.
Ford, C.J.B., T.J. Thornton, R. Newbury, M. Pepper, H. Ahmed, D.C. Peacock, D.A. Ritchie, J.E.F. Frost and G.A.C. Jones, 1989a, Appl. Phys. Lett. **54**, 21.
Ford, C.J.B., S. Washburn, M. Büttiker, C.M. Knoedler and J.M. Hong, 1989b, Phys. Rev. Lett. **62**, 2724.
Ford, C.J.B., S. Washburn, M. Büttiker, C.M. Knoedler and J.M. Hong, 1990, Surf. Sci. **229**, 298.
Ford, C.J.B., S. Washburn, R. Newbury, C.M. Knoedler and J.M. Hong, 1991, Phys. Rev. B **43**, 7339.
Fowler, A.B., A. Hartstein and R.A. Webb, 1982, Phys. Rev. Lett. **48**, 196.
Fowler, A.B., J.J. Wainer and R.A. Webb, 1988, IBM J. Res. & Dev. **32**, 372.
Friedman, L.R., and D.P. Tunstall, eds, 1978, The Metal–Non-Metal Transition in Disordered Solids (SUSSP, Edinburgh).
Fukui, T., S. Ando, Y. Tokura and T. Toriyama, 1991, Appl. Phys. Lett. **58**, 2018.
Fukuyama, H., 1980, J. Phys. Soc. Jpn. **48**, 2169.
Fukuyama, H., 1981, J. Phys. Soc. Jpn. **50**, 3407.
Fukuyama, H., 1985, in: Electron–Electron Interactions in Disordered Systems, eds A.L. Efros and M. Pollak (North-Holland, Amsterdam).
García, N., and L. Escapa, 1989, Appl. Phys. Lett. **54**, 1418.
García, N., J.J. Sáenz and H. de Raedt, 1989, J. Phys.: Condens. Matter **1**, 9931.
Gerhardts, R.R., and D. Pfannkuche, 1992, Surf. Sci. **263**, 324.
Gerhardts, R.R., D. Weiss and U. Wulf, 1991, Phys. Rev. B **43**, 5192.
Gijs, M., C. Van Haesendonck and Y. Bruynseraede, 1984a, Phys. Rev. Lett. **52**, 2069.

Gijs, M., C. Van Haesendonck and Y. Bruynseraede, 1984b, Phys. Rev. B **30**, 2964.
Glazman, L.I., and M. Jonson, 1989, J. Phys.: Condens. Matter **1**, 5547.
Glazman, L.I., and M. Jonson, 1990, Phys. Rev. B **41**, 10686.
Glazman, L.I., and M. Jonson, 1991, Phys. Rev. B **44**, 3810.
Glazman, L.I., and A.V. Khaetskii, 1989a, Europhys. Lett. **9**, 263.
Glazman, L.I., and A.V. Khaetskii, 1989b, J. Phys.: Condens. Matter **1**, 5005.
Glazman, L.I., and A.I. Larkin, 1991, Superlatt. & Microstr. **6**, 32.
Glazman, L.I., and R.I. Shekhter, 1989, J. Phys.: Condens. Matter **1**, 5811.
Glazman, L.I., G.B. Lesovik, D.E. Khmel'nitskii and R.I. Shekhter, 1988, JETP Lett. **48**, 238.
Gogolin, A.A., 1982, Phys. Rep. **86**, 2.
Gor'kov, L.P., A.I. Larkin and D.E. Khmel'nitskii, 1979, Pis'ma Zh. Eksp. & Theor. Fiz. **30**, 248 [Sov. Phys.-JETP Lett. **30**, 228].
Grambow, P., T. Demel, D. Heitmann, M. Kohl, R. Schüle and K. Ploog, 1989, Microelectron. Eng. **9**, 357.
Grambow, P., E. Vasiliadou, T. Demel, K. Kern, D. Heitmann and K. Ploog, 1990, Microelectron. Eng. **11**, 47.
Greene, S.K., M. Pepper, D.A. Wharam, D.C. Peacock, D.A. Ritchie, J.E.F. Frost, D.G. Hasko, H. Ahmed and G.A.C. Jones, 1991, J. Phys.: Condens. Matter **3**, 1961.
Groshev, A., T. Ivanov and V. Valtchinov, 1991, Phys. Rev. Lett. **66**, 1082.
Haanappel, E.G., and D. van der Marel, 1989, Phys. Rev. B **39**, 5484.
Halperin, B.I., 1982, Phys. Rev. B **25**, 2185.
Harris, J.J., J.A. Pals and R. Woltjer, 1989, Rep. Prog. Phys. **52**, 1217.
Hartland, A., K. Jones, J.M. Williams, B.L. Gallagher and T. Galloway, 1991, Phys. Rev. Lett. **66**, 969.
Hartland, A., K. Jones, J.M. Williams, B.L. Gallagher and T. Galloway, 1992, Surf. Sci. **263**, 112.
Haug, R.J., K.Y. Lee, T.P. Smith III and J.M. Hong, 1990a, in: Proc. 20th Int. Conf. on the Physics of Semiconductors, Thessaloniki, Greece (World Scientific, Singapore, 1991).
Haug, R.J., K.Y. Lee and J.M. Hong, 1990b, Mater. Res. Soc. Ext. Abstr. **EA-26**, 29.
Haug, R.J., H. Munekata and L.L. Chang, 1992a, Surf. Sci. **263**, 374.
Haug, R.J., J.M. Hong and K.Y. Lee, 1992b, Surf. Sci. **263**, 415.
He, S., and S. Das Sarma, 1989a, Solid State Electron. **32**, 1695.
He, S., and S. Das Sarma, 1989b, Phys. Rev. B **40**, 3379.
Heiblum, M., and U. Sivan, 1990, in: Hot Carriers in Semiconductor Microstructures: Physics and Applications, ed. J. Shah (Academic Press, New York).
Heitmann, D., K. Kern, T. Demel, P. Grambow, K. Ploog and Y.H. Zhang, 1992, in: Proc. 9th Int. Conf. on Electronic Properties of Two-Dimensional Systems, Surf. Sci., to appear.
Herbert, D.C., and R. Jones, 1971, J. Phys. C **4**, 1145.
Hikami, S., 1986, Prog. Theor. Phys. **76**, 1210.
Hikami, S., and E. Brézin, 1986, Surf. Sci. **170**, 262.
Hikami, S., A.I. Larkin and Y. Nagaoka, 1980, Prog. Theor. Phys. **63**, 707.
Hirayama, Y., and T. Saku, 1989, Appl. Phys. Lett. **54**, 2556.
Hirayama, Y., and T. Saku, 1990a, Solid State Commun. **73**, 113.
Hirayama, Y., and T. Saku, 1990b, Phys. Rev. B **41**, 2927.
Hirayama, Y., and T. Saku, 1990c, Jpn. J. Appl. Phys. **29**, L368.
Hirayama, Y., and T. Saku, 1990d, Phys. Rev. B **42**, 11408.
Hirayama, Y., S. Tarucha, Y. Suzuki and H. Okamoto, 1988, Phys. Rev. B **37**, 2774.
Hirayama, Y., T. Saku and Y. Horikoshi, 1989a, Phys. Rev. B **39**, 5535.
Hirayama, Y., T. Saku and Y. Horikoshi, 1989b, Jpn. J. Appl. Phys. **28**, L701.
Hirayama, Y., T. Saku, S. Tarucha and Y. Horikoshi, 1991a, Appl. Phys. Lett. **58**, 2672.
Hirayama, Y., S. Tarucha, T. Saku and Y. Horikoshi, 1991b, in: Symposium on Analogies in Optics and Microelectronics, Eindhoven, The Netherlands, May 1–3, 1991.
Hirayama, Y., S. Tarucha, T. Saku and Y. Horikoshi, 1991c, Phys. Rev. B **44**, 3440.
Hofstadter, D., 1978, Phys. Rev. B **14**, 2239.
Huckestein, B., 1990, in: Anderson Transition and Mesoscopic Fluctuations, eds B. Kramer and G. Schön, Physica A **167**, 1–314.

Huckestein, B., and B. Kramer, 1989, Solid State Commun. **71**, 445.

Huckestein, B., and B. Kramer, 1990, Phys. Rev. Lett. **64**, 1437.

Imry, Y., 1986a, in: Directions in Condensed Matter Physics, Vol. 1, eds G. Grinstein and G. Masenko (World Scientific, Singapore) p. 101.

Imry, Y., 1986b, Europhys. Lett. **1**, 249.

Imry, Y., 1989, Proc. Int. Conf. on Nanostructure Physics and Fabrication, eds M. Reed and W.P. Wiley (Academic Press, New York).

Ioffe, A.F., and A.R. Regel, 1960, Prog. Semicond. **4**, 237.

Iordansky, S.V., 1982, Solid State Commun. **43**, 1.

Ishii, K., 1973, Prog. Theor. Phys. Suppl. **53**, 77.

Ismail, K., T.P. Smith III, W.T. Masselink and H.I. Smith, 1989, Appl. Phys. Lett. **55**, 2766.

Jain, J.K., and S.A. Kivelson, 1988a, Phys. Rev. Lett. **60**, 1542.

Jain, J.K., and S.A. Kivelson, 1988b, Phys. Rev. B **37**, 4276.

Jalabert, R., H.U. Baranger and A.D. Stone, 1990, Phys. Rev. Lett. **65**, 2442.

Ji, Z.-L., and K.F. Berggren, 1990, Semicond. Sci. & Technol. **5**.

Joe, Y.S., and S.E. Ulloa, 1990, Mater. Res. Soc. Ext. Abstr. **EA-26**, 47.

Joe, Y.S., and S.E. Ulloa, 1991a, Ballistic transport anomalies in asymmetric quantum wire nanostructures, preprint.

Johnson, B.L., and G. Kirczenow, 1992a, What is the ground state of a quantum dot?, preprint.

Johnson, B.L., and G. Kirczenow, 1992b, Phys. Rev. Lett. **69**, 672.

Johnson, N.F., and M.C. Payne, 1991, Phys. Rev. Lett. **67**, 1157.

Kantor, Y., and A. Kapitulnik, 1982, Solid State Commun. **42**, 161.

Kash, K., J.M. Worlock, M.D. Sturge, P. Grabbe, J.P. Harbison, A. Scherer and P.S.D. Lin, 1988, Appl. Phys. Lett. **53**, 782.

Kash, K., B.P. Van der Gaag, D.D. Mahoney, A.S. Godz, L.T. Florez, J.P. Harbison and M.D. Sturge, 1991, Phys. Rev. Lett. **67**, 1326.

Katsumoto, S., 1991, in: Localisation 1990, eds K.A. Benedict and J.T. Chalker (Institute of Physics, Bristol).

Kaveh, M., and N.F. Mott, 1981, J. Phys. A **14**, 259.

Kawabata, A., 1985, in: Anderson Localization, ed. Y. Nagaoka, Prog. Theor Phys. **84**, Suppl., p. 16.

Kawabata, A., 1989, J. Phys. Soc. Jpn. **58**, 372.

Kazarinov, R.F., and S. Luryi, 1982, Phys. Rev. B **25**, 7626.

Kelly, M.J., 1989, J. Phys.: Condens. Matter **1**, 7643.

Kelly, M.J., R.J. Brown, C.G. Smith, D.A. Wharam, M. Pepper, H. Ahmed, D.G. Hasko, D.C. Peacock, J.E.F. Frost, R. Newbury, D.A. Ritchie and G.A.C. Jones, 1989, Electron. Lett. **25**, 992.

Kempa, K., D.A. Broido and P. Bakshi, 1991, Phys. Rev. B **43**, 9343.

Kern, K., D. Heitmann, P. Grambow, Y.H. Zhang and K. Ploog, 1991, Phys. Rev. Lett. **66**, 1618.

Khurana, A., 1988, Search and Discovery, in: Physics Today **41**, 21.

Kirczenow, G., 1988a, Solid State Commun. **68**, 715.

Kirczenow, G., 1988b, Phys. Rev. B **38**, 10958.

Kirczenow, G., 1989a, J. Phys.: Condens. Matter **1**, 305.

Kirczenow, G., 1989b, Phys. Rev. B **39**, 10452.

Kirczenow, G., 1989c, Phys. Rev. Lett. **62**, 1920.

Kirczenow, G., 1989d, Solid State Commun. **71**, 469.

Kirczenow, G., 1989e, Phys. Rev. Lett. **62**, 2993.

Kirczenow, G., 1990a, Phys. Rev. B **42**, 5357.

Kirczenow, G., 1990b, Solid State Commun. **74**, 1051.

Kirczenow, G., 1992a, Surf. Sci. **263**, 330.

Kirczenow, G., 1992b, Theory of two-dimensional quantum dot arrays in magnetic fields: electronic structure and lateral quantum transport, preprint.

Kirczenow, G., and E. Castaño, 1991, Phys. Rev. B **43**, 7343.

Kirkman, P.D., and J.B. Pendry, 1984a, J. Phys. C **17**, 4327.

Kirkman, P.D., and J.B. Pendry, 1984b, J. Phys. C **17**, 5707.

Kirtley, J.P., Z. Schlesinger, T.N. Theis, F.P. Milliken, S.L. Wright and L.F. Palmateer, 1986, Phys. Rev. B **34**, 5414.
Kobayashi, S., F. Komori, Y. Ootuka and W. Sasaki, 1980, J. Phys. Soc. Jpn. **49**, 1635.
Kohmoto, M., 1985, Ann. Phys. (NY) **160**, 343.
Kohn, W., and L.J. Sham, 1965, Phys. Rev A **140**, 1133.
Komiyama, S., and H. Hirai, 1989a, Phys. Rev. B **40**, 7767.
Komiyama, S., and H. Hirai, 1989b, Phys. Rev. B **40**, 12566.
Kopley, T.E., P.L. McEuen and R.G. Wheeler, 1988, Phys. Rev. Lett. **61**, 1654.
Kouwenhoven, L.P., B.J. van Wees, W. Kool, C.J.P.M. Harmans, A.A.M. Staring and C.T. Foxon, 1989a, Phys. Rev. B **40**, 8083.
Kouwenhoven, L.P., B.J. van Wees, C.J.P.M. Harmans, J.G. Williamson, H. van Houten, C.W.J. Beenakker, C.T. Foxon and J.J. Harris, 1989b, Phys. Rev. B **39**, 8040.
Kouwenhoven, L.P., F.W.J. Hekking, B.J. van Wees, C.J.P.M. Harmans, C.E. Timmering and C.T. Foxon, 1990a, Phys. Rev. Lett. **65**, 361.
Kouwenhoven, L.P., B.J. van Wees, F.W.J. Hekking, C.J.P.M. Harmans and C.E. Timmering, 1990b, in: Localization and Confinement of Electrons in Semiconductors, Springer Series in Solid State Sciences, Vol. 97, eds F. Kuchar, H. Heinrich and G. Bauer (Springer, Berlin).
Kouwenhoven, L.P., N.C. van der Vaart, A.T. Johnson, C.J.P.M. Harmans, J.G. Williamson, A.A.M. Staring and C.T. Foxon, 1991a, in: Proc. German Physical Society Meeting, Münster; Festkörperprobleme/Advances in Solid State Physics, Vol. 31.
Kouwenhoven, L.P., B.J. van Wees, B. van der Enden, C.J.P.M. Harmans and C.E. Timmering, 1991b, in: Proc. 20th Int. Conf. on Physics of Semiconductors, Thessaloniki, Greece, 1990 (World Scientific, Singapore).
Kouwenhoven, L.P., A.T. Johnson, N.C. van der Vaart, D.J. Maas, C.J.P.M. Harmans and C.T. Foxon, 1992, Surf. Sci. **263**, 405.
Kramer, B., and A. MacKinnon, 1986, in: Localization in Disordered Systems, Teubner texte zur Physik, Vol. 3, eds W. Weller and P. Ziesche (Teubner, Leipzig).
Kramer, B., and G. Schön, eds, 1990, Anderson Transition and Mesoscopic Fluctuations, Physica A **167**, 1–314.
Kramer, B., and D. Weaire, 1979, Topics in Applied Physics, Vol. 36 (Springer, Berlin) pp. 9–39.
Kramer, B., G. Bergmann and Y. Bruynseraede, eds, 1985, Localisation, Interaction, and Transport Phenomena, Springer Series Solid State Science, Vol. 61 (Springer, Berlin).
Kramer, B., Y. Ono and T. Ohtsuki, 1988, Surf. Sci. **196**, 127.
Kravtsov, V.E., and I.V. Lerner, 1984, Solid State Commun. **52**, 593.
Kree, R., and A. Schmid, 1981, Z. Phys. **42**, 297.
Kriman, A.M., G.H. Bernstein, B.S. Haukness and D.K. Ferry, 1989, Superlatt. & Microstr. **6**, 381.
Kriman, A.M., B.S. Haukness and D.K. Ferry, 1990, Proc. Nanostructures and Microstructure Correlation with Physical Properties of Semiconductors, SPIE Proc. Ser. **1284**, 82.
Kuchar, F., et al., 1990, in: Proc. ICALS, Greece.
Kühl, H., 1980, Ph.D. Thesis, Universität zu Köln.
Kumar, A., and P.F. Bagwell, 1991a, Phys. Rev. B **43**, 9012.
Kumar, A., and P.F. Bagwell, 1991b, Phys. Rev. B **44**, 1747.
Kumar, A., S.E. Laux and F. Stern, 1989, Appl. Phys. Lett. **54**, 1270.
Kumar, A., S.E. Laux and F. Stern, 1990, Phys. Rev. B **42**, 5166.
Kunz, H., and B. Souillard, 1980, Commun. Math. Phys. **78**, 201.
Kunz, H., and B. Souillard, 1982, J. Phys. (Paris) Lett. **43**, L39.
Landauer, R., 1957, IBM J. Res. & Dev. **1**, 223.
Landauer, R., 1970, Philos. Mag. **21**, 863.
Landauer, R., 1985, Localisation, Interaction, and Transport Phenomena, Springer Series Solid State Science, Vol. 61, eds B. Kramer, G. Bergmann and Y. Bruynseraede (Springer, Berlin).
Landauer, R., 1989a, J. Phys.: Condens. Matter **1**, 8099.
Landauer, R., 1989b, Physica D **38**, 226.
Landauer, R., 1989c, in: Proc. Int. Conf. on Nanostructure Physics and Fabrication, eds M. Reed and W.P. Wiley (Academic Press, New York) p. 17.

Landauer, R., 1989d, Physics Today, (October) **42**, 119.
Landauer, R., 1990, Physica A **168**, 75.
Langer, J.S., and T. Neal, 1966, Phys. Rev. Lett. **16**, 984.
Langreth, D.C., and E. Abrahams, 1981, Phys. Rev. B **31**, 2978.
Laughton, M.J., J.A. Nixon and J.H. Davies, 1991, Phys. Rev. B. **44**, 1150.
Laux, S.E., and F. Stern, 1986, Appl. Phys. Lett. **49**, 91.
Laux, S.E., D.J. Frank and F. Stern, 1988, Surf. Sci. **196**, 101.
Lee, P.A., and T.V. Ramakrishnan, 1985, Rev. Mod. Phys. **57**, 287.
Lee, P.A., and A.D. Stone, 1985, Phys. Rev. Lett. **55**, 1622.
Lee, P.A., A.D. Stone and H. Fukuyama, 1987, Phys. Rev. B **35**, 1039.
Lent, C.S., 1990a, Appl. Phys. Lett. **57**, 1678.
Lent, C.S., 1990b, Appl. Phys. Lett. **56**, 2554.
Lent, C.S., 1991, in: Computational Electronics, eds K. Hess, J.P. Leburton and U. Ravaioli (Kluwer, Boston) p. 259.
Lent, C.S., and D.J. Kirkner, 1989, in: Proc. Int. Conf. on Nanostrucutre Physics and Fabrication, eds M. Reed and W.P. Wiley (Academic Press, New York).
Lent, C.S., and M. Leng, 1991a, Appl. Phys. Lett. **58**, 1650.
Lent, C.S., and M. Leng, 1991b, J. Appl. Phys. **70**, 3157.
Lent, C.S., and S. Sivaprakasam, 1990, in: Proc. Nanostructures and Microstructure Correlation with Physical Properties of Semiconductors, SPIE Proc. Series **1284**, 31.
Lent, C.S., S. Sivaprakasam and D.J. Kirkner, 1989, Solid State Electron. **32**, 1137.
Leo, J.L., and B. Movaghar, 1988, Phys. Rev. B **38**, 8061.
Lerner, I.V., 1991, in: Quantum Coherence in Mesoscopic Systems, NATO-ASI Series B, Vol. 24, ed. B. Kramer (Plenum, New York).
Lesovik, G.B., 1989, JETP Lett. **49**, 592.
Levine, H., S.B. Libby and A.M.M. Pruisken, 1984, Nucl. Phys. B **240** (FS12), 30, 40, 57.
Levinson, I.B., 1988, Sov. Phys.-JETP Lett. **48**, 301.
Levinson, Y.B., M.I. Lubin and E.V. Shukhorukov, 1991, Pis'ma Zh. Eksp. & Teor. Fiz. **54**, 405.
Li, Q., and D.J. Thouless, 1990, Phys. Rev. Lett. **65**, 767.
Li, Y.P., D.C. Tsui, J.J. Heremans, J.A. Simmons and G.W. Weimann, 1990, Appl. Phys. Lett. **55**, 774.
Licini, J.C., D.J. Bishop, M.A. Kastner and J. Melngailis, 1985, Phys. Rev. Lett. **55**, 2987.
Lifshitz, I.M., 1965, Sov. Phys.-Usp. **7**, 549.
Lifshitz, I.M., S.A. Gredescul and L.A. Pastur, 1988, Introduction to the Theory of Disordered Systems (Wiley, New York).
Liu, C.T., K. Nakamura, D.C. Tsui, K. Ismail, D.A. Antoniadis and H.I. Smith, 1990, Surf. Sci. **228**, 527.
Lorke, A., 1992, Surf. Sci. **263**, 307.
Lorke, A., J.P. Kotthaus and K. Ploog, 1990, Phys. Rev. Lett. **64**, 2559.
Luryi, S., and R.F. Kazarinov, 1983, Phys. Rev. B **27**, 1386.
MacDonald, A.H., 1984, Phys. Rev. B **29**, 6563.
MacKinnon, A., 1980, J. Phys. C **13**, L1031.
MacKinnon, A., 1985, in: Localisation, Interaction, and Transport Phenomena, Springer Series in Solid State Science, Vol. 61, eds B. Kramer, G. Bergmann and Y. Bruynseraede (Springer, Berlin).
MacKinnon, A., 1988, Anderson Localization (VCH Verlagsgesellschaft, Weinheim, Germany) p. 71.
MacKinnon, A., and B. Kramer, 1981, Phys. Rev. Lett. **47**, 1546.
MacKinnon, A., and B. Kramer, 1983a, Z. Phys. **53**, 1.
MacKinnon, A., and B. Kramer, 1983b, in: High Magnetic Fields in Semiconductors, Lecture Notes in Physics, Vol. 177, ed. G. Landwehr (Springer, Berlin) p. 74.
MacKinnon, A., B. Kramer and L. Schweitzer, 1984, in: Proc. LITPIM Suppl., Vol. PTB-PG-1, eds B. Kramer and L. Schweitzer (PTB, Braunschweig) p. 314.
Mailly, D., and M. Sanquer, 1991, in: Quantum Coherence in Mesoscopic Systems, NATO-ASI Series B, Vol. 24, ed. B. Kramer (Plenum, New York).
Main, P.C., P.H. Beton, B.R. Snell, A.J.M. Neves, J.R. Owers-Bradley, L. Eaves, S.P. Beaumont and C.D.W. Wilkinson, 1989a, Phys. Rev. B **40**, 10033.

Main, P.C., P.H. Beton, B.R. Snell, A.J.M. Neves, J.R. Owers-Bradley, L. Eaves, M. Davies, M. Henini, O.H. Hughes, S.P. Beaumont and C.D.W. Wilkinson, 1989b, Solid State Electronics **32**, 1303.
Main, P.C., B.R. Davidson, P.H. Beton, L. Eaves, J.R. Owers-Bradley, A.J.M. Neves, S.P. Beaumont and C.D.W. Wilkinson, 1990, J. Phys.: Condens. Matter **2**, 6541.
Maksym, P.A., and T. Chakraborty, 1990, Phys. Rev. Lett. **65**, 108.
Martín-Moreno, L., and C.G. Smith, 1989, J. Phys.: Condens. Matter **1**, 5421.
Mašek, J., and B. Kramer, 1988, Solid State Commun. **68**, 611.
Mašek, J., and B. Kramer, 1989, Z. Phys. B **75**, 37.
Mašek, J., P. Lipavski and B. Kramer, 1989, J. Phys.: Condens. Matter **1**, 6395.
Matulis, A., and D. Segzda, 1989, J. Phys.: Condens. Matter **1**, 2289.
McEuen, P.L., B.W. Alphenaar, R.G. Wheeler and R.N. Sacks, 1990, Surf. Sci. **229**, 312.
McEuen, P.L., E.B. Foxman, U. Meirav, M.A. Kastner, Y. Meir, N.S. Wingreen and S.J. Wind, 1991, Phys. Rev. Lett. **66**, 1926.
Mehta, M.L., 1967, Random Matrices and the Statistical Theory of Energy Levels (Academic Press, New York).
Meir, Y., N.S. Wingreen and P.A. Lee, 1991, Phys. Rev. Lett. **66**, 3048.
Meirav, U., M. Heiblum and F. Stern, 1988, Appl. Phys. Lett. **52**, 1268.
Meirav, U., M.A. Kastner, M. Heiblum and S.J. Wind, 1989, Phys. Rev. B **40**, 5871.
Meirav, U., M.A. Kastner and S.J. Wind, 1990, Phys. Rev. Lett. **65**, 771.
Mello, P.A., 1987, Phys. Rev. B **35**, 1082.
Mel'nikov, V.I., 1981, Sov. Phys.-JETP Lett. **32**, 225.
Merkt, U., Ch. Sikorski and J. Almeier, 1989, in: Spectroscopy of Semiconductor Microstructures, eds G. Fasol, A. Fasolino and P. Lugli (Plenum, New York) p. 89.
Merkt, U., J. Huser and M. Wagner, 1991, Phys. Rev. B **43**, 7320.
Mil'nicov, G.V., and I.M. Sokolov, 1989, Sov. Phys.-JETP Lett. **48**, 536.
Molenkamp, L.W., A.A.M. Staring, C.W.J. Beenakker, R. Eppenga, C.E. Timmering, J.G. Williamson, C.J.P.M. Harmans and C.T. Foxon, 1990, Phys. Rev. B **41**, 1274.
Montie, E.A., E.C. Cosman, G.W. 't Hooft, M.B. van der Mark and C.W.J. Beenakker, 1991, Nature **350**, 594.
Mott, N.F., 1968, J. Non-Cryst. Solids **1**, 1.
Mott, N.F., and E.A. Davis, 1971, Electronic Processes in Non-Crystalline Materials (Clarendon Press, Oxford).
Mott, N.F., and W.D. Twose, 1961, Adv. Phys. **10**, 107.
Muttalib, K.A., J.-L. Pichard and A.D. Stone, 1987, Phys. Rev. Lett. **59**, 2475.
Nagaoka, Y., ed., 1985, Anderson Localization, Prog. Theor. Phys. **84**, Suppl.
Nagaoka, Y., and H. Fukuyama, eds, 1982, Anderson Localization, Springer Series Solid State Science, Vol. 39 (Springer, Berlin).
Nakamura, A., and A. Okiji, 1991, J. Phys. Soc. Jpn. **60**, 1873.
Nakata, S., Y. Hirayama, S. Tarucha and Y. Horikoshi, 1991, J. Appl. Phys. **69**, 3633.
Niu, Q., and D.J. Thouless, 1984, J. Phys. A **17**, 2453.
Niu, Q., D.J. Thouless and Y.-S. Wu, 1985, Phys. Rev. B **31**, 3372.
Nixon, J.A., and J.H. Davies, 1990, Phys. Rev. B **41**, 7929.
Nixon, J.A., J.H. Davies and H.U. Baranger, 1991a, Phys. Rev. B **43**, 12638.
Nixon, J.A., J.H. Davies and H.U. Baranger, 1991b, Superlatt. & Microstr. **9**, 187.
Notzel, R., L. Daweritz, N.N. Ledentsov and K. Ploog, 1991, in: Proc. 5th Conf. on Modulated Semiconductor Structures, Nara, 1991.
Oakeshott, R.B.S., and A. MacKinnon, 1991, in: Proc. Conf. on Nanostructure Physics and Fabrication, eds M. Reed and W.P. Wiley (Academic Press, New York).
O'Connor, A.J., 1975, Commun. Math. Phys. **5**, 63.
Ohtsuki, T., and Y. Ono, 1989, J. Phys. Soc. Jpn. **58**, 956.
Ono, Y., 1982a, J. Phys. Soc. Jpn. **51**, 237.
Ono, Y., 1982b, J. Phys. Soc. Jpn. **51**, 2055, 3544.
Ono, Y., 1983, J. Phys. Soc. Jpn. Suppl. 247, **52**, 2492.
Ono, Y., 1984, J. Phys. Soc. Jpn. **53**, 2342.

Ono, Y., 1985, in: Anderson Localization, ed. Y. Nagaoka, Prog. Theor Phys. **84**, Suppl.
Ono, Y., 1992, Surf. Sci. **263**, 129.
Ono, Y., T. Ohtsuki and B. Kramer, 1989, J. Phys. Soc. Jpn. **58**, 1705.
Ono, Y., T. Ohtsuki and B. Kramer, 1991, J. Phys. Soc. Jpn. **60**, 270.
Pannetier, B., J. Chaussy, R. Rammal and P. Gandit, 1984, Phys. Rev. Lett. **53**, 718.
Paranjape, V.V., 1991, J. Phys.: Condens. Matter **3**, 6751.
Pastawski, H.M., 1991b, Phys. Rev. B **44**, 6329.
Patel, N.K., L. Martín-Moreno, M. Pepper, R. Newbury, J.E.F. Frost, D.A. Ritchie, G.A.C. Jones, J.T.M.B. Janssen, J. Singleton and J.A.A.J. Perenboom, 1990, J. Phys.: Condens. Matter **2**, 7247.
Payne, M.C., 1989, J. Phys.: Condens. Matter **1**, 4939.
Peeters, F.M., 1988, Phys. Rev. Lett. **61**, 589.
Peeters, F.M., 1989a, Science and Engineering of 1- and 0-Dimensional Semiconductors, eds. S.P. Beaumont and C.M. Sotomayor-Torres (Plenum Press, London).
Peeters, F.M., 1989b, Superlatt. & Microstr. **6**, 217.
Pendry, J.B., P.D. Kirkman and E. Castaño, 1986, Phys. Rev. Lett. **57**, 2983.
Pendry, J.B., A.B. Prêtre, P.J. Rous and L. Martín-Moreno, 1991, Surf. Sci. **244**, 160.
Pfeiffer, L.N., K.W. West, H.L. Störmer, J.P. Eisenstein, K.W. Baldwin, D. Gershoni and J. Spector, 1990, Appl. Phys. Lett. **56**, 1697.
Pichard, J.-L., 1990, in: Anderson Transition and Mesoscopic Fluctuations, eds B. Kramer and G. Schön, Physica A **167**, 1–314.
Pichard, J.-L., 1991a, in: Quantum Coherence in Mesoscopic Systems, NATO-ASI Series B, Vol. 24, ed. B. Kramer (Plenum, New York).
Pichard, J.-L., 1991b, in: Localisation 1990, eds. K.A. Benedict and J.T. Chalker (Institute of Physics, Bristol).
Pichard, J.-L., and M. Sanquer, 1990, in: Anderson Transition and Mesoscopic Fluctuations, eds B. Kramer and G. Schön, Physica A **167**, 1–314.
Prange, R.E., and S.M. Girvin, eds, 1990, The Quantum Hall Effect, 2nd Ed. (Springer, New York).
Prentiss, M., G. Timp, N. Bigelow, R.E. Behringer and J.E. Cunningham, 1992, Appl. Phys. Lett. **60**, 1027.
Prêtre, A.B., 1991, J. Phys.: Condens. Matter **3**, 8037.
Prêtre, A.B., L. Martín-Moreno and J.B. Pendry, 1991, in: Proc. Int. Conf. on Nanostructure Physics and Fabrication, eds M. Reed and W.P. Wiley (Academic Press, New York).
Price, P.J., 1991, in: NPF, Electronic Processes in Non-Crystalline Materials (Clarendon Press, Oxford).
Proc. Int. Conf. on Electronic Properties of Two-Dimensional Systems IV to IX, appearing in various volumes of Surf. Sci. **113** (1982), **142** (1984), **170** (1986), **196** (1988), **229** (1990), and to appear (1992).
Proc. Int. Conf. on Modulated Semiconductor Structures, MSS1: Pasadena, 1982; MSS2: Kyoto, 1985; MSS3: Montpelier, 1987; MSS4: Ann Arbor, 1989 [Surf. Sci. **228** (1990)]; and MSS5: Nara, 1991.
Pruisken, A.M.M., 1984, Nucl. Phys. B **235** (FS11), 277.
Puechner, R.A., J. Ma, R. Mezenner, W.P. Liu, A.M. Kriman, G.N. Maracas, G. Bernstein, D.K. Ferry, P. Chu, H.H. Wieder and P. Newman, 1990, Surf. Sci. **228**, 520.
Que, W., and G. Kirczenow, 1988, Phys. Rev. B **38**, 3614.
Que, W., G. Kirczenow and E. Castaño, 1991, Phys. Rev. B **43**, 14079.
Rammal, R., G. Toulouse, M.T. Jackel and B.I. Halperin, 1983, Phys. Rev. B **27**, 5142.
Randall, J.N., M.A. Reed, J.H. Luscombe, G.F. Frazier, W.R. Frensley, A.C. Seabaugh, Y.C. Kao, T.M. Moore and R.J. Matyi, 1990, Proc. Nanostructures and Microsctructure Correlation with Physical Properties of Semiconductors, SPIE Proc. Ser. **1284**, 66.
Rauh, A., G.H. Wannier and G.M. Obermair, 1974, Phys. Status Solidi B **63**, 215.
Ravenhall, D.G., H.W. Wyld and R.L. Schult, 1989, Phys. Rev. Lett. **62**, 1780.
Reed, M.A., ed., 1990, Semiconductors and Semimetals (Academic Press, New York).
Reed, M.A., and W.P. Wiley, eds, 1989 and 1991, Proc. Int. Conf. on Nanostrucutre Physics and Fabrication (Academic Press, New York).
Reed, M.A., J.N. Randall, R.J. Aggarwal, R.J. Matyi, T.M. Moore and A.E. Wetsel, 1988, Phys. Rev. Lett. **60**, 535.
Reed, M.A., J.N. Randall, J.H. Luscombe, W.R. Frensley, R.J. Aggarwal, R.J. Matyi, T.M. Moore and A.E. Wetsel, 1989, Festkörperprobleme **29**, 267.

Reed, M.A., J.N. Randall and J.H. Luscombe, 1990, Nanotechnology **1**, 63.
Roukes, M.L., 1991, invited lecture presented at the 38th Nat. Symp. of the American Vacuum Soc., Seattle.
Roukes, M.L., and O.L. Alerhand, 1990, Phys. Rev. Lett. **65**, 1651.
Roukes, M.L., A. Scherer, S.J. Allen Jr, H.G. Craighead, R.M. Ruthen, E.D. Beebe and J.P. Harbison, 1987, Phys. Rev. Lett. **59**, 3011.
Roukes, M.L., T.J. Thornton, A. Scherer, J.A. Simmons, B.P. Van der Gaag and E.D. Beebe, 1990a, in: Science and Engineering of 1- and 0-Dimensional Semiconductors, eds J.M. Chamberlain, L. Eaves and J.C. Portal (Plenum Press, London).
Roukes, M.L., A. Scherer and B.P. Van der Gaag, 1990b, Phys. Rev. Lett. **64**, 1154.
Roukes, M.L., T.J. Thornton, A. Scherer and B.P. Van der Gaag, 1990c, in: Electronic Properties of Multilayers and Low-Dimensional Semiconductor Structures, NATO-ASI Series, eds J.M. Chamberlain, L. Eaves and J.C. Portal (Plenum Press, London).
Rundquist, J., Z.-L. Ji and K.F. Berggren, 1990, Mater. Res. Soc. Ext. Abstr. **EA-26**, 87.
Sáenz, J.J., N. García, V.T. Binh and H. de Raedt, 1990, in: Scanning Tunneling Microscopy and Related Methods, eds R.J. Behm, N. García and H. Rohrer (Kluwer, Amsterdam) pp. 409–441.
Sak, J., and B. Kramer, 1981, Phys. Rev. B **24**, 1761.
Schellnhuber, H.J., and G.M. Obermair, 1980, Phys. Rev. Lett. **45**, 276.
Scherer, A., M.L. Roukes, H.G. Craighead, R.M. Ruthen, E.D. Beebe and J.P. Harbison, 1987, Appl. Phys. Lett. **51**, 2133.
Schreiber, M., and B. Kramer, 1987, in: Anderson Localization, Springer Proc. in Physics, Vol. 28, eds T. Ando and H. Fukuyama (Springer, Berlin).
Schult, R.L., D.G. Ravenhall and H.W. Wyld, 1989, Phys. Rev. B **39**, 5476.
Schult, R.L., H.W. Wyld and D.G. Ravenhall, 1990, Phys. Rev. B **41**, 12760.
Schweitzer, L., B. Kramer and A. MacKinnon, 1984, J. Phys. C **17**, 4111.
Scott-Thomas, J.H.F., S.B. Field, M.A. Kastner, H.I. Smith and D.A. Antoniadis, 1989, Phys. Rev. Lett. **62**, 583.
Serota, R.A., R.K. Kalia and P.A. Lee, 1987, Phys. Rev. B **32**, 8441.
Shapiro, B., 1986, Phys. Rev. B **34**, 4394.
Shapiro, B., 1987, Philos. Mag. **56**, 1031.
Sharvin, Yu., and Yu.V. Sharvin, 1981, Sov. Phys.-JETP Lett. **34**, 272.
Sharvin, Yu.V., 1965, Sov. Phys.-JETP **21**, 655.
Sharvin, Yu.V., and N.I. Bogatina, 1969, Sov. Phys.-JETP **29**, 419.
Sheard, F.W., and L. Eaves, 1988, Nature **333**, 600.
Shepard, K.L., M.L. Roukes and B.P. van der Gaag, 1992, Phys. Rev. Lett. **68**, 2660.
Shi, H., and G. Kirczenow, 1991, J. Phys.: Condens. Matter **3**, 955.
Shikin, V.B., 1989, Sov. Phys.-JETP Lett. **50**, 167.
Shklovskii, B.I., and A.L. Efros, 1984, Electronic Properties of Doped Semiconductors, Springer Series Solid State Science, Vol. 45 (Springer, Berlin).
Simmons, J.A., D.C. Tsui and G. Weimann, 1988, Surf. Sci. **196**, 81.
Simmons, J.A., S.W. Hwang, D.C. Tsui and M. Shayegan, 1991, in: Proc. Int. Conf. on Nanostructure Physics and Fabrication, eds M. Reed and W.P. Wiley (Academic Press, New York).
Sivan, U., and Y. Imry, 1988, Phys. Rev. Lett. **61**, 1001.
Sivan, U., M. Heiblum and C.P. Umbach, 1989a, Phys. Rev. Lett. **63**, 992.
Sivan, U., Y. Imry and C. Harzstein, 1989b, Phys. Rev. B **39**, 1242.
Sivan, U., M. Heiblum, C.P. Umbach and H. Shtrikman, 1990, Phys. Rev. B **41**, 7937.
Skocpol, W.J., L.D. Jackel, E.L. Hu, R.E. Howard and L.A. Fetter, 1982, Phys. Rev. Lett. **49**, 951.
Skocpol, W.J., P.M. Mankiewich, R.E. Howard, L.D. Jackel, D.M. Tennant and A.D. Stone, 1986, Phys. Rev. Lett. **56**, 2865.
Smith, C.G., M. Pepper, H. Ahmed, J.E.F. Frost, D.G. Hasko, D.C. Peacock, D.A. Ritchie and G.A.C. Jones, 1988, J. Phys. C **21**, L893.
Smith, C.G., M. Pepper, R. Newbury, H. Ahmed, D.G. Hasko, D.C. Peacock, J.E.F. Frost, D.A. Ritchie, G.A.C. Jones and G. Hill, 1989a, J. Phys.: Condens. Matter **1**, 6763.

Smith, C.G., M. Pepper, H. Ahmed, J.E.F. Frost, D.G. Hasko, D.C. Peacock, D.A. Ritchie and G.A.C. Jones, 1989b, Superlatt. & Microstr. **5**, 599.
Smith, C.G., M. Pepper, H. Ahmed, J.E.F. Frost, D.G. Hasko, R. Newbury, D.C. Peacock, D.A. Ritchie and G.A.C. Jones, 1989c, J. Phys.: Condens. Matter **1**, 9035.
Smith, C.G., M. Pepper, H. Ahmed, J.E.F. Frost, D.G. Hasko, R. Newbury, D.C. Peacock, D.A. Ritchie and G.A.C. Jones, 1990a, Surf. Sci. **228**, 387.
Smith, C.G., M. Pepper, H. Ahmed, J.E.F. Frost, D.G. Hasko, D.C. Peacock, D.A. Ritchie and G.A.C. Jones, 1990b, Surf. Sci. **228**, 387.
Smith, C.G., M. Pepper, R. Newbury, H. Ahmed, D.G. Hasko, D.C. Peacock, J.E.F. Frost, D.A. Ritchie, G.A.C. Jones and G. Hill, 1990c, J. Phys.: Condens. Matter **2**, 3405.
Smith III, T.P., 1990, in: Proc. Nanostructures and Microstructure Correlation with Physical Properties of Semiconductors, SPIE Proc. Ser. **1284**, 12.
Sols, F., 1991, Phys. Rev. Lett. **67**, 2874.
Sols, F., 1992, Ann. Phys. (U.S.A.) **214**, 386.
Sols, F., and M. Macucci, 1990, Phys. Rev. B **41**, 11887.
Sols, F., M. Macucci, U. Ravaioli and K. Hess, 1989a, Appl. Phys. Lett. **54**, 350.
Sols, F., M. Macucci, U. Ravaioli and K. Hess, 1989b, J. Appl. Phys. **66**, 3892.
Sols, F., et al., 1989c, in: Proc. Int. Conf. on Nanostructure Physics and Fabrication, eds M. Reed and W.P. Wiley (Academic Press, New York).
Spector, J., H.L. Störmer, K.W. Baldwin, L.N. Pfeiffer and K.W. West, 1990a, Appl. Phys. Lett. **56**, 1290.
Spector, J., H.L. Störmer, K.W. Baldwin, L.N. Pfeiffer and K.W. West, 1990b, Appl. Phys. Lett. **56**, 2433.
Spector, J., H.L. Störmer, K.W. Baldwin, L.N. Pfeiffer and K.W. West, 1990c, Appl. Phys. Lett. **56**, 967.
Spector, J., H.L. Störmer, K.W. Baldwin, L.N. Pfeiffer and K.W. West, 1990d, Surf. Sci. **228**, 283.
Spector, J., H.L. Störmer, K.W. Baldwin, L.N. Pfeiffer and K.W. West, 1991, Appl. Phys. Lett. **58**, 263.
Staring, A.A.M., L.W. Molenkamp, C.W.J. Beenakker, L.P. Kouwenhoven and C.T. Foxon, 1990, Phys. Rev. B **41**, 846.
Staring, A.A.M., J.G. Williamson, H. van Houten, C.W.J. Beenakker, L.P. Kouwenhoven and C.T. Foxon, 1991, in: Symp. on Analogies in Optics and Microelectronics, Eindhoven, The Netherlands, May 1–3, 1991.
Stone, A.D., 1985, Phys. Rev. Lett. **54**, 2592.
Stone, A.D., and J.D. Joannopoulos, 1982, Phys. Rev. B **25**, 2500.
Stone, A.D., and A. Szafer, 1988, IBM J. Res. & Develop. **32**, 384.
Stone, A.D., J.D. Joannopoulos and D.J. Chadi, 1981, Phys. Rev. B **24**, 5583.
Stopa, M.P., 1992, Surf. Sci. **263**, 433.
Störmer, H.L., L.N. Pfeiffer, K.W. Baldwin, K.W. West and J. Spector, 1991, Appl. Phys. Lett. **58**, 726.
Streda, P., J. Kucera and A.H. MacDonald, 1987, Phys. Rev. Lett. **59**, 1973.
Subramaniam, S., S. Bandyopadhyay and W. Porod, 1990, Bull. Am. Phys. Soc. **35**, 493, paper I23 10.
Sukhorukov, E.V., and Y.B. Levinson, 1990, Sov. Phys.-JETP **70**, 782.
Sun, Y., and G. Kirczenow, 1992, Density functional theory of the electronic structure of a quantum wire, preprint.
Szafer, A., and A.D. Stone, 1989, Phys. Rev. Lett. **62**, 300.
Takagaki, Y., and D.K. Ferry, 1991, Phys. Rev. B **44**, 8399.
Takagaki, Y., K. Gamo, S. Namba, S. Ishida, S. Takaoka, K. Murase, K. Ishibashi and Y. Aoyagi, 1988a, Solid State Commun. **68**, 1051.
Takagaki, Y., K. Gamo, S. Namba, S. Ishida, S. Takaoka, K. Murase, K. Ishibashi and Y. Aoyagi, 1988b, Solid State Commun. **68**, 1051.
Takagaki, Y., K. Gamo, S. Namba, S. Takaoka, K. Murase and S. Ishida, 1989a, Solid State Commun. **71**, 809.
Takagaki, Y., K. Gamo, S. Namba, S. Takaoka, K. Murase, S. Ishida, K. Ishibashi and Y. Aoyagi, 1989b, Solid State Commun. **69**, 811.
Takagaki, Y., K. Gamo, S. Namba, S. Takaoka, K. Murase, S. Ishida, K. Ishibashi and Y. Aoyagi, 1989c, Solid State Commun. **69**, 811.
Takagaki, Y., F. Wakaya, S. Takaoka, K. Gamo, K. Murase and S. Namba, 1989d, Jpn. J. Appl. Phys. **28**, 2188.

Takagaki, Y., K. Gamo, S. Namba, S. Takaoka and K. Murase, 1990a, Solid State Commun. **75**, 873.

Takagaki, Y., K. Gamo, K. Murase and S. Namba, 1990b, Ext. Abstr. of the 22nd Int. Conf. on Solid State Devices and Materials, Sendai, p. 721.

Takagaki, Y., T. Sakamoto, K. Gamo, S. Namba, S. Takaoka and K. Murase, 1990c, in: 5th Int. Conf. on Physics of Electro-Optic Microstructures and Microdevices, Superlatt. & Microstruct. **8**, 97.

Takagaki, Y., Y. Kusumi, S. Takaoka, K. Gamo, K. Murase and S. Namba, 1990d, Jpn. J. Appl. Phys. **29**, 192.

Takahashi, T., Y. Arakawa and M. Nishioka, 1991, Appl. Phys. Lett. **58**, 2372.

Tankei, K., and F. Takano, 1986, J. Phys. Soc. Jpn. **55**, 3516.

Tarucha, S., Y. Hirayama, T. Saku and T. Kimura, 1990, Phys. Rev. B **41**, 5459.

Taylor, B.N., and T.J. Witt, 1989, Metrologia **26**, 47.

Taylor, R.P., A.S. Sachrajda, J.A. Adams, P. Zawadzki, P.T. Coleridge and P. Marshall, 1991a, in: Symp. Analogies in Optics and Microelectronics, Eindhoven, The Netherlands, May 1–3, 1991.

Taylor, R.P., S. Fortin, A.S. Sachrajda, J.A. Adams, M. Fallahi, M. Davies, P.T. Coleridge and P. Zawadski, 1991b, Low frequency noise in multiple quantum point contact systems, preprint.

Taylor, R.P., A.S. Sachrajda, J.A. Adams, C.R. Leavens, P. Zawadzki and P.T. Coleridge, 1991c, in: Proc. Int. Conf. on Nanostructure Physics and Fabrication, eds M. Reed and W.P. Wiley (Academic Press, New York).

Taylor, R.P., A.S. Sachrajda, J.A. Adams, P. Zawadski, P.T. Coleridge and M. Davies, 1992, Surf. Sci. **263**, 247.

Tekman, E., and S. Ciraci, 1989a, Phys. Rev. B **39**, 8772.

Tekman, E., and S. Ciraci, 1989b, Phys. Rev. B **40**, 8559.

Tekman, E., and S. Ciraci, 1990a, Phys. Rev. B **42**, 9098.

Tekman, E., and S. Ciraci, 1990b, in: Scanning Tunneling Microscopy and Related Methods, eds R.J. Behm, N. García and H. Rohrer (Kluwer, Amsterdam) p. 157.

Tekman, E., and S. Ciraci, 1991, Phys. Rev. B **43**, 7145.

Tekman, E., S. Ciraci and A. Baratoff, 1990, Phys. Rev. B **42**, 9221.

Tewordt, M., V.J. Law, M.J. Kelly, R. Newbury, M. Pepper, D.C. Peacock, J.E.F. Frost, D.A. Ritchie and G.A.C. Jones, 1990, J. Phys.: Condens. Matter **2**, 8969.

Thomas, G.A., 1986, in: Localisation and Interactions in Disordered and Doped Semiconductors, ed. D.M. Finlayson (SUSSP, Edinburgh).

Thornton, T.J., M. Pepper, H. Ahmed, D. Andrews and G.J. Davies, 1986, Phys. Rev. Lett. **56**, 1198.

Thornton, T.J., M.L. Roukes, A. Scherer and B.P. Van der Gaag, 1989, Phys. Rev. Lett. **63**, 2128.

Thouless, D.J., 1974, Phys. Rep. **13**, 93.

Thouless, D.J., 1979, in: Ill-Condensed Matter, eds G.D. Toulouse and R. Balian (North-Holland, Amsterdam) p. 1.

Thouless, D.J., 1980, Solid State Commun. **34**, 683.

Thouless, D.J., 1981, Phys. Rev. Lett. **47**, 92.

Thouless, D.J., 1990, in: The Quantum Hall Effect, 2nd Ed., eds R.E. Prange and S.M. Girvin (Springer, New York) ch. 4.

Thouless, D.J., and S. Kirkpatrick, 1981, J. Phys. C **14**, 235.

Thouless, D.J., M. Kohmoto, M.P. Nightingale and M. den Nijs, 1982, Phys. Rev. Lett. **49**, 405.

Timp, G., R.E. Behringer, J.E. Cunningham and R.E. Howard, 1989, Phys. Rev. Lett. **63**, 2268.

Timp, G., 1990, in: Semiconductors and Semimetals (Academic Press, New York).

Timp, G., 1991, in: Proc. Int. Conf. on Nanostructure Physics and Fabrication, eds M. Reed and W.P. Wiley (Academic Press, New York).

Timp, G., and R.E. Howard, 1991, in: Nanostructure Engineering, ed. R.F. Pease, Proc. IEEE **79**, 1188.

Timp, G., A.M. Chang, P.M. Mankiewich, R.E. Behringer, J.E. Cunningham, T.Y. Chang and R.E. Howard, 1987a, Phys. Rev. Lett. **59**, 732.

Timp, G., A.M. Chang, J.E. Cunningham, T.Y. Chang, P.M. Mankiewich, R.E. Behringer and R.E. Howard, 1987b, Phys. Rev. Lett. **58**, 2814.

Timp, G., A.M. Chang, P.M. Mankiewich, R.E. Behringer, J.E. Cunningham, T.Y. Chang and R.E. Howard, 1987c, Phys. Rev. Lett. **59**, 732.

Timp, G., H.U. Baranger, P. deVegvar, J.E. Cunningham, R.E. Howard, R.E. Behringer and P.M. Mankiewich, 1988a, Phys. Rev. Lett. **60**, 2081.
Timp, G., H.U. Baranger, P.G.N. de Vegvar, J.E. Cunningham, R.E. Howard, R.E. Behringer and P.M. Mankiewich, 1988b, Phys. Rev. Lett. **60**, 2081.
Timp, G., R.E. Behringer, S. Sampere, J.E. Cunningham and R.E. Howard, 1989a, in: Proc. Int. Conf. on Nanostructure Physics and Fabrication, eds M. Reed and W.P. Wiley (Academic Press, New York).
Timp, G., P.M. Mankiewich, P.G.N. de Vegvar, R.E. Behringer, J.E. Cunningham, R.E. Howard, H.U. Baranger and J.K. Jain, 1989b, Phys. Rev. B **39**, 6227.
Timp, G., R.E. Behringer and J.E. Cunningham, 1990a, Phys. Rev. B **42**, 9259.
Timp, G., R.E. Behringer and J.E. Cunningham, 1990b, Phys. Rev. B **42**, 9259.
Trugman, S.A., 1983, Phys. Rev. B **27**, 7539.
Tsuchiya, M., J.M. Gaines, R.H. Yan, R.J. Simes, P.O. Holtz, L.A. Coldren and P.M. Petroff, 1989, Phys. Rev. Lett. **62**, 466.
Tsukada, M., 1976, J. Phys. Soc. Jpn. **41**, 1466.
Ulloa, S.E., E. Castaño and G. Kirczenow, 1990a, Phys. Rev B **41**, 12350.
Ulloa, S.E., E. Castaño and G. Kirczenow, 1990b, Proc. Nanostructures and Microstructure Correlation with Physical Properties of Semiconductors, SPIE Proc. Ser. **1284**, 57.
Umbach, C.P., S. Washburn, R.B. Laibowitz and R.A. Webb, 1984, Phys. Rev. B **30**, 4048.
Uren, M.J., R.A. Davies and M. Pepper, 1980, J. Phys. C **13**, L985.
van den Dries, L., C. van Haesendonck, Y. Bruynseraede and G. Deutscher, 1981, Phys. Rev. Lett. **46**, 565.
van der Marel, D., 1991, Phys. Rev. B **43**, 3469.
van der Marel, D., and E.G. Haanappel, 1989, Phys. Rev. B **39**, 7811.
van Haesendonck, C., H. Vloeberghs and Y. Bruynseraede, 1991, in: Ref. Quantum Coherence in Mesoscopic Systems, NATO-ASI Series B, Vol. 24, ed. B. Kramer (Plenum, New York).
van Houten, H., 1992, Surf. Sci. **263**, 442.
van Houten, H., and C.W.J. Beenakker, 1989, Phys. Rev. Lett. **63**, 1893.
van Houten, H., and C.W.J. Beenakker, 1990, in: Analogies in Optics and Micro Electronics, eds W. van Haeringen and D. Lenstra (Kluwer, Amsterdam) p. 203.
van Houten, H., B.J. van Wees, M.G.J. Heijman and J.P. André, 1986, Appl. Phys. Lett. **49**, 1781.
van Houten, H., B.J. van Wees, J.E. Mooij, C.W.J. Beenakker, J.G. Williamson and C.T. Foxon, 1988a, Europhys. Lett. **5**, 721.
van Houten, H., C.W.J. Beenakker, P.H.M. van Loosdrecht, T.J. Thornton, H. Ahmed, M. Pepper, C.T. Foxon and J.J. Harris, 1988b, Phys. Rev. B **37**, 8534.
van Houten, H., C.W.J. Beenakker, J.G. Williamson, M.E.I. Broekaart, P.H.M. van Loosdrecht, B.J. van Wees, J.E. Mooij, C.T. Foxon and J.J. Harris, 1989, Phys. Rev. B **39**, 8556.
van Houten, H., C.W.J. Beenakker and B.J. van Wees, 1990, in: Semiconductors and Semimetals, ed. M.A. Reed (Academic Press, New York).
van Loosdrecht, P.H.M., C.W.J. Beenakker, H. van Houten, J.G. Williamson, B.J. van Wees, J.E. Mooij, C.T. Foxon and J.J. Harris, 1988, Phys. Rev. B **38**, 10162.
van Wees, B.J., 1991a, Phys. Rev. Lett. **66**, 2033.
van Wees, B.J., 1991b, Phys. Rev. Lett. **66**, 2033.
van Wees, B.J., H. van Houten, C.W.J. Beenakker, J.G. Williamson, L.P. Kouwenhoven, D. van der Marel and C.T. Foxon, 1988a, Phys. Rev. Lett. **60**, 848.
van Wees, B.J., L.P. Kouwenhoven, H. van Houten, C.W.J. Beenakker, J.E. Mooij, C.T. Foxon and J.J. Harris, 1988b, Phys. Rev. B **38**, 3625.
van Wees, B.J., L.P. Kouwenhoven, C.J.P.M. Harmans, J.G. Williamson, C.E. Timmering, M.E.I. Broekaart, C.T. Foxon and J.J. Harris, 1989a, Phys. Rev. Lett. **62**, 2523.
van Wees, B.J., E.M.M. Willems, L.P. Kouwenhoven, C.J.P.M. Harmans, J.G. Williamson, C.T. Foxon and J.J. Harris, 1989b, Phys. Rev. B **39**, 8066.
van Wees, B.J., L.P. Kouwenhoven, E.M.M. Willems, C.J.P.M. Harmans, J.E. Mooij, H. van Houten, C.W.J. Beenakker, J.G. Williamson and C.T. Foxon, 1991, Phys. Rev. B **43**, 12431.
Vollhardt, D., 1987, Festkörperprobleme; Advances in Solid State Physics, Vol. 27 (Vieweg, Braunschweig) p. 63.

Vollhardt, D., and P. Wölfle, 1980a, Phys. Rev. Lett. **45**, 842.

Vollhardt, D., and P. Wölfle, 1980b, Phys. Rev. B **22**, 4666.

Vollhardt, D., and P. Wölfle, 1991, Self-Consistent Theory of Anderson Localization, in: Electronic Phase Transitions, eds W. Hanke and Yu.A. Kopaev (North-Holland, Amsterdam).

von Klitzing, K., 1986, Rev. Mod. Phys. **58**, 519.

von Klitzing, K., G. Dorda and M. Pepper, 1980, Phys. Rev. Lett. **45**, 494.

Wagner, M., U. Merkt and A.V. Chaplik, 1992, Phys. Rev. B **45**, 1951.

Wannier, G.H., 1978, Phys. Status Solidi B **88**, 757.

Wannier, G.H., G.M. Obermair and R. Ray, 1979, Phys. Status Solidi B **93**, 337.

Washburn, S., 1991, Nature **353**, 119.

Washburn, S., and R.A. Webb, 1986, Adv. Phys. **35**, 375.

Webb, R.A., 1988, in: Proc. Mauterndorf.

Webb, R.A., S. Washburn, C.P. Umbach and R.B. Laibowitz, 1985a, Phys. Rev. Lett. **54**, 2696.

Webb, R.A., S. Washburn, C.P. Umbach and R.B. Laibowitz, 1985b, in: Localisation, Interaction, and Transport Phenomena, Springer Series in Solid State Science, Vol. 61, eds B. Kramer, G. Bergman and Y. Bruynseraede (Springer, Berlin).

Wegner, F.J., 1976, Z. Phys. B **25**, 327.

Wegner, F.J., 1979a, Phys. Rev. B **19**, 783.

Wegner, F.J., 1979b, in: Recent Advances in Statistical Mechanics, Brasov, p. 63.

Wegner, F.J., 1980, Z. Phys. B **36**, 209.

Wei, H.P., D.C. Tsui, M.A. Paalanen and A.M.M. Pruisken, 1988, Phys. Rev. Lett. **61**, 1294.

Weidenmüller, H.A., 1987, Nucl. Phys. B **290** (FS20), 87.

Weiss, D., K. von Klitzing, K. Ploog and G. Weimann, 1990, Surf. Sci. **229**, 88.

Weiss, D., M.L. Roukes, A. Menschig, P. Grambow, K. von Klitzing and G. Weimann, 1991, Phys. Rev. Lett. **66**, 2790.

Weisshaar, A., J. Lary, S.M. Goodnick and V.K. Tripathi, 1989, Appl. Phys. Lett. **55**, 2114.

Weisshaar, A., J. Lary, S.M. Goodnick and V.K. Tripathi, 1990a, Proc. NATO-ASI on Granular Electronics, Il Ciocco, Italy, July 1990.

Weisshaar, A., J. Lary, S.M. Goodnick and V.K. Tripathi, 1990b, SPIE Proc. Ser. **1284**, 45.

Weisshaar, A., J. Lary, S.M. Goodnick and V.K. Tripathi, 1991a, IEEE Electron. Dev. Lett. **12**, 2.

Weisshaar, A., J. Lary, S.M. Goodnick and V.K. Tripathi, 1991b, J. Appl. Phys. **70**, 355.

Weisz, J.F., and K.F. Berggren, 1989, Phys. Rev. B **40**, 1325.

Weisz, J.F., and K.F. Berggren, 1990, Phys. Rev. B **41**, 1687.

Weller, W., and P. Ziesche, eds, 1984, Localization in Disordered Systems, Teubner texte zur Physik, Vol. 3 (Teubner, Leipzig).

Wharam, D.A., T.J. Thornton, R. Newbury, M. Pepper, H. Ahmed, J.E.F. Frost, D.G. Hasko, D.C. Peacock, D.A. Ritchie and G.A.C. Jones, 1988a, J. Phys. C **21**, L209.

Wharam, D.A., M. Pepper, H. Ahmed, J.E.F. Frost, D.G. Hasko, D.C. Peacock, D.A. Ritchie and G.A.C. Jones, 1988b, J. Phys. C **21**, L887.

Wharam, D.A., U. Ekenberg, M. Pepper, D.G. Hasko, H. Ahmed, J.E.F. Frost, D.A. Ritchie, D.C. Peacock and G.A.C. Jones, 1989a, Phys. Rev. B **39**, 6283.

Wharam, D.A., M. Pepper, R. Newbury, H. Ahmed, D.G. Hasko, D.C. Peacock, J.E.F. Frost, D.A. Ritchie and G.A.C. Jones, 1989b, J. Phys.: Condens. Matter **1**, 3369.

Wheeler, R.G., K.K. Choi, A. Goel, R. Wisnieff and D.E. Prober, 1982, Phys. Rev. Lett. **49**, 1674.

Widom, A., and R. Tao, 1988, J. Phys. C **21**, L1061.

Wieck, A.D., and K. Ploog, 1990, Appl. Phys. Lett. **56**, 928.

Williamson, J.G., 1991, in: Proc. Conf. on Nanostructure Physics and Fabrication, eds M.A. Reed and W.P. Wiley (Academic Press, New York).

Williamson, J.G., H. van Houten, C.W.J. Beenakker, M.E.I. Broekaart, L.I.A. Spendeler, B.J. van Wees and C.T. Foxon, 1990a, Phys. Rev. B **41**, 1207.

Williamson, J.G., H. van Houten, C.W.J. Beenakker, M.E.I. Broekaart, L.I.A. Spendeler, B.J. van Wees and C.T. Foxon, 1990b, Surf. Sci. **229**, 303.

Wu, J.C., M.N. Wybourne, W. Yindeepol, A. Weisshaar and S.M. Goodnick, 1991, Appl. Phys. Lett. **59**, 102.

Yacoby, A., and Y. Imry, 1990, Phys. Rev. B **41**, 5341.
Yamada, S., and M. Yamamoto, 1991, Phys. Rev. B **43**, 9369.
Yamada, S., H. Asai, Y.K. Fukai and T. Fukui, 1989, Phys. Rev. B **39**, 11199.
Yosefin, M., and M. Kaveh, 1991, Phys. Rev. B **44**, 3355.
Yurke, B., and G.P. Kochanski, 1990, Phys. Rev. B **41**, 8184.
Zallen, R., 1983, The Physics of Amorphous Solids (Wiley, New York).
Zanon, N., and J.-L. Pichard, 1988, J. Phys. (Paris) **49**, 907.
Zilberman, G.E., 1957a, Zh. Eksp. & Teor. Fiz. **32**, 296 [Sov. Phys.-JETP **5**, 208].
Zilberman, G.E., 1957b, Zh. Eksp. & Teor. Fiz. **33**, 387 [Sov. Phys.-JETP **6**, 299].
Ziman, J.M., 1979, Models of Disorder (University Press, Cambridge).
Zirnbauer, M.R., 1986, Nucl. Phys. B **235** (FS15), 375.

CHAPTER 17

The Quantum Hall Effect

TAPASH CHAKRABORTY

Institute for Microstructural Sciences
National Research Council
Montreal Road, M-50
Ottawa, Canada K1A 0R6

Handbook on Semiconductors
Completely Revised Edition
Edited by T.S. Moss
Volume 1, edited by P.T. Landsberg

Contents

1. Introduction . . . 979
2. Two-dimensional electron gas . . . 979
   2.1. Systems with a two-dimensional electron gas . . . 980
      2.1.1. Inversion layers in Si-MOSFET structures . . . 980
      2.1.2. Electron layers in semiconductor heterostructures . . . 981
   2.2. Electrons in a strong magnetic field . . . 982
3. Integral quantum Hall effect . . . 985
   3.1. Experimental work . . . 985
      3.1.1. Classical Hall effect . . . 987
      3.1.2. Quantum mechanical approach . . . 988
   3.2. Integral quantization: theoretical work . . . 988
      3.2.1. Kubo formula approach . . . 989
      3.2.2. The gauge invariance approach . . . 991
      3.2.3. The topological invariance approach . . . 992
   3.3. Other developments . . . 993
      3.3.1. Electron localization in the quantum Hall regime . . . 993
      3.3.2. Renormalization group approach . . . 995
      3.3.3. Current-carrying edge states . . . 996
      3.3.4. Further topics . . . 998
4. Fractional quantum Hall effect . . . 999
   4.1. Ground state . . . 1001
      4.1.1. Theoretical work . . . 1002
      4.1.2. Recent developments . . . 1006
   4.2. Elementary excitations . . . 1008
      4.2.1. Quasiholes and quasiparticles . . . 1008
      4.2.2. Quasiparticle statistics . . . 1009
      4.2.3. Hierarchies of quasiparticles . . . 1011
      4.2.4. Energy gap . . . 1013
      4.2.5. Spin-reversed quasiparticles . . . 1016
      4.2.6. Quantization condition . . . 1020
   4.3. Other developments . . . 1021
      4.3.1. Collective modes . . . 1021
      4.3.2. Even-denominator filling fractions . . . 1024
      4.3.3. Multiple layer systems . . . 1027
      4.3.4. Nature of long-range order in the Laughlin state . . . 1028

References . . . 1032

1. Introduction

The quantization of the Hall effect discovered by von Klitzing et al. (1980) is a remarkable macroscopic quantum phenomenon which occurs in two-dimensional electron systems at low temperatures and strong perpendicular magnetic fields. Under these conditions, the Hall conductivity exhibits plateaus at integral multiples of e^2/h (a universal constant). The striking result is the accuracy of the quantization (better than one part in ten million) which is totally indifferent to impurities or geometric details of the two-dimensional system. Each plateau is accompanied by a deep minimum in the diagonal resistivity, indicating a dissipationless flow of current. In 1982, there was yet another surprise in this field. Working with much higher mobility samples, Tsui et al. (1982) discovered the fractional quantization of the Hall conductivity. The physical mechanism responsible for the integer quantum Hall effect (IQHE) and the fractional quantum Hall effect (FQHE) are quite different, despite the apparent similarity of the experimental results. In the former case, the role of the random impurity potential is quite decisive, while in the latter case electron–electron interaction plays a predominant role resulting in a unique collective phenomenon.

In the following sections, we shall briefly describe the theoretical and experimental developments in the QHE. It should be mentioned, however, that the QHE has been one of the most active fields of research in condensed matter physics for over a decade. It is, therefore, quite impossible to describe here all the details of the major developments. Our aim in this chapter is to touch upon the most significant theoretical and experimental work to construct a reasonably consistent picture of the QHE. For more details on the topics discussed, the reader is encouraged to read the original work cited here and some of the reviews available in the literature (Aoki 1987, Chakraborty and Pietiläinen 1988a, Hajdu and Landwehr 1985, Halperin 1983, MacDonald 1989, Morandi 1988, Prange and Girvin 1987, Rashba and Timofeev 1986).

2. Two-dimensional electron gas

The major impetus in the studies of the QHE is due to experimental realization of almost ideal two-dimensional electron systems. The electrons are dynamically two-dimensional because they are free to move in only two spatial dimensions. In the third dimension, they have quantized energy levels [in reality, the wave functions have a finite spatial extent in the third dimension (Ando et al. 1982)]. In the following, we provide a very brief discussion on the systems where the electron layers are created. For details see the reviews by Ando et al. (1982) and Störmer (1984).

2.1. Systems with a two-dimensional electron gas

Electron layers have been created in many different systems. Electrons on the surface of liquid helium provides an almost ideal two-dimensional system (Grimes 1978, Monarkha and Shikin 1982). They are trapped on the surface by a combination of an external field and an image potential. The electron concentration in this system is, however, very low (10^5–10^9 cm^{-2}) and the system behaves classically. The high-density electron systems where the QHE is usually observed are typically created in the metal-oxide–semiconductor field effect transistor (MOSFET) and in semiconductor heterojunctions.

2.1.1. Inversion layers in Si-MOSFET structures

A schematic picture of an n-channel Si-MOSFET is shown in fig. 1a. The system consists of a semiconductor (p-Si) which has a plane interface with a thin film of insulator (SiO_2), the opposite side of which carries a metal gate electrode. Application of a voltage (gate voltage V_G) between the gate and the Si/SiO_2 interface results in bending of the electron energy bands. For a strong enough electric field, as the bottom of the conduction band is pushed down below the Fermi energy E_F, electrons accumulate in a two-dimensional quasi-triangular potential well close to the interface (fig. 1b). As the width of the well is small (~ 50 Å), electron motion perpendicular to the interface is quantized but the electrons move freely parallel to the interface. In the plane, the energy spectrum is

$$\varepsilon_i(\boldsymbol{k}) = \varepsilon_i^0 + \frac{\hbar^2 k_{\|}^2}{2m^*},$$

where m^* is the effective mass of the electrons, $k_{\|}$ is the two-dimensional wave vector and ε_i^0 is the bottom of the corresponding subband. The system is called an inversion

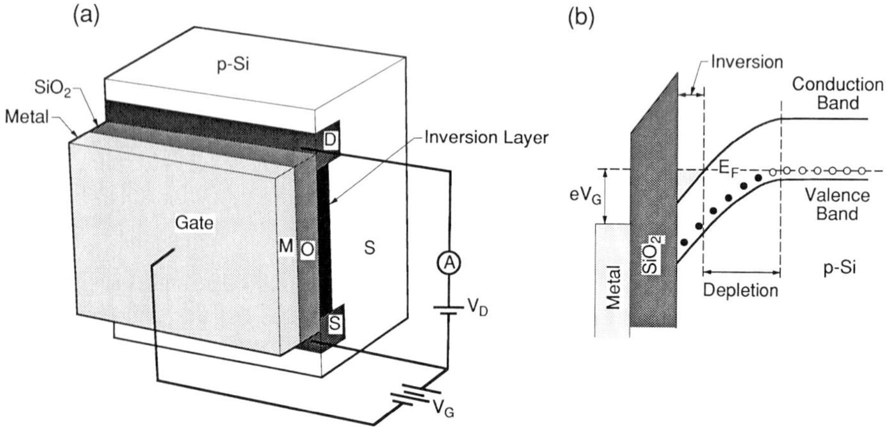

Fig. 1. (a) Schematic view of an Si-MOSFET and (b) energy level diagram.

layer because here the charge carriers are the electrons while the semiconductor is p-type.

At low temperatures ($kT \ll \Delta E$, the energy spacing) the electrons are trapped in the lowest subband and the system is purely two-dimensional. The MOSFET is quite useful in the present study because by varying the gate voltage the electron concentration can be varied within a wide range ($n_0 \sim 0$–10^{13} cm^{-2}).

2.1.2. Electron layers in semiconductor heterostructures

Two-dimensional electron layers are also created in semiconductor heterostructures at a nearly perfectly lattice-matched semiconductor/semiconductor interface. One such widely used system is the GaAs/$Al_xGa_{1-x}As$ ($0 < x \leq 1$) heterostructure. The lattice constants of GaAs and $Al_xGa_{1-x}As$ are almost the same so that the interface is nearly free from any disorder. The band gap of the alloy is wider than that of GaAs and it increases with the aluminum concentration x. Carriers in the neighborhood of the heterojunction transfer from the doped AlGaAs alloy across the interface to the low-lying band edge states of the narrow band gap material (GaAs). The electric field due to the charge transfer bends the energy bands as shown in fig. 2. A quasi-triangular potential well ($\sim$100 Å) formed in the GaAs traps the electrons as two-dimensional carriers.

The mobile carriers are spatially separated from their parent ionized impurities via modulation doping. This leads to very high carrier mobilities and, in fact, the FQHE was discovered in these high-mobility GaAs-heterostructures (Störmer 1984). However, unlike MOSFETs, the electron concentration in heterostructures can be varied only within a very narrow range. Carrier densities in these systems typically range from 1×10^{11} cm^{-2} to 1×10^{12} cm^{-2}.

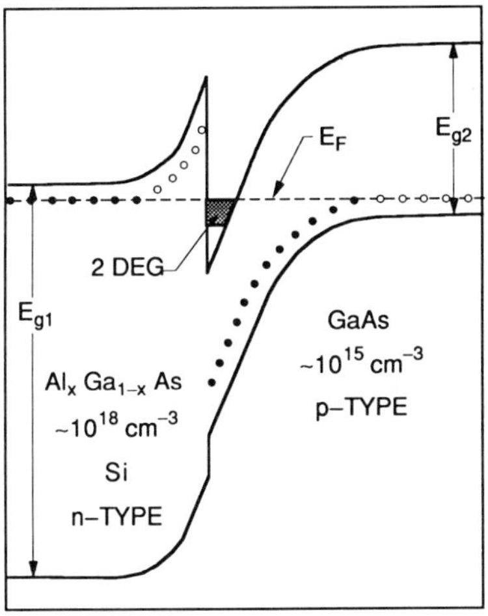

Fig. 2. Energy diagram at a GaAs-heterostructure interface.

2.2. Electrons in a strong magnetic field

Let us begin with the problem of a free electron (with effective mass m^*) in a uniform magnetic field B. The Hamiltonian is then written

$$\mathscr{H}_0 = (\Pi_x^2 + \Pi_y^2)/2m^*, \tag{1}$$

where $\boldsymbol{\Pi} = -\mathrm{i}\hbar\nabla + (e/c)\boldsymbol{A}$ is the kinetic momentum and the vector potential $\boldsymbol{A}$ is related to the magnetic field as $\boldsymbol{B} = \nabla \times \boldsymbol{A}$. Following Kubo et al. (1965), we introduce the center coordinates of the cyclotron motion (X, Y) as

$$X = x - \xi, \qquad Y = y - \eta, \tag{2}$$

where

$$\xi = (c/eB)\Pi_y, \qquad \eta = -(c/eB)\Pi_x, \tag{3}$$

are the relative coordinates. It can be easily seen that (ξ, η) represents a cyclotron motion with frequency

$$\omega_\mathrm{c} = \frac{eB}{m^*c} \quad \text{(cyclotron frequency)}. \tag{4}$$

Defining the magnetic length

$$l_0 \equiv \left(\frac{\hbar c}{eB}\right)^{1/2} \quad \text{(cyclotron radius)} \tag{5}$$

and from the commutation relation

$$[\xi, \eta] = -\mathrm{i}l_0^2,$$

it is clear that ξ and η are subject to an uncertainty of order l_0. The Hamiltonian (1) is now rewritten in terms of (ξ, η) as

$$\mathscr{H}_0 = \frac{\hbar\omega_\mathrm{c}}{2l_0^2}(\xi^2 + \eta^2), \tag{6}$$

whose eigenenergies are the discrete *Landau levels* (Fock 1928, Landau 1930)

$$E_n = (n + \tfrac{1}{2})\hbar\omega_\mathrm{c}, \quad n = 0, 1, 2, \ldots. \tag{7}$$

The Hamiltonian (6) does not contain (X, Y), which means that electrons in cyclotron motion with different center coordinates have the same energy. The center coordinates also follow the commutation rule, $[X, Y] = \mathrm{i}l_0^2$.

Choosing now the Landau gauge such that the vector potential $\boldsymbol{A}$ has only one nonvanishing component, say, $A_y = Bx$, the Hamiltonian is

$$\mathscr{H}_0 = \frac{1}{2m^*}\left[p_x^2 + \left(p_y + \frac{eB}{c}x\right)^2\right]. \tag{8}$$

The variables are easily separable and an eigenfunction is written in the form

$$\varphi = \mathrm{e}^{ik_y y}\,\chi(x), \tag{9}$$

where the usual identification is made, $p_y = -i\hbar\,\partial/\partial y \rightarrow \hbar k_y$. The function $\chi(x)$ is the eigenfunction of the time-independent Schrödinger equation

$$-\frac{\hbar^2}{2m^*}\chi'' + \tfrac{1}{2}m^*\omega_c^2(x-X)^2\chi = E\chi(x), \tag{10}$$

where $X = -k_y l_0^2$. The above equation is easily recognized as the Schrödinger equation corresponding to a *harmonic oscillator* of spring constant $\hbar\omega_c = \hbar^2/m^* l_0^2$, with equilibrium point at X.

The eigenfunction (ignoring the normalization factor) is now written

$$\varphi_{nX} = \mathrm{e}^{ik_y y}\exp[-(x-X)^2/2l_0^2]\,H_n[(x-X)/l_0] \tag{11}$$

with H_n the Hermite polynomial. The functions are extended in y and localized in x. The localization remains unaffected under a gauge transformation. When the system is confined in a rectangular cell with sides L_x and L_y, the degeneracy of each Landau level (number of allowed states) is, in fact, the number of allowed values of k_y, such that the center X lies between 0 and L_x. With use of periodic boundary conditions we get, $k_y = 2\pi n_y/L_y$, with n_y an integer. The allowed values of n_y are then determined by the condition

$$X = \frac{2\pi n_y}{L_y} l_0^2, \quad 0 < X < L_x. \tag{12}$$

The degeneracy N_s can then be expressed in terms of the magnetic length l_0 as

$$N_s = \frac{L_x L_y}{2\pi l_0^2}. \tag{13}$$

Equation (13) can also be reexpressed in terms of the magnetic flux $\boldsymbol{\Phi}$ and the flux quantum $\boldsymbol{\Phi}_0 = hc/e$ as

$$N_s = \frac{e}{hc}\boldsymbol{\Phi} = \frac{\boldsymbol{\Phi}}{\boldsymbol{\Phi}_0}. \tag{14}$$

The Landau-level degeneracy is thus the total number of flux quanta in the external magnetic field. One other important quantity is the dimensionless density of the electrons expressed as the *filling factor* of the Landau level

$$\nu = 2\pi l_0^2 n_0, \tag{15}$$

where n_0 is the electron density in the system.

Thus far, we have ignored the presence of any impurities in the system. In a more general case, the Hamiltonian is written as

$$\mathscr{H} = \mathscr{H}_0 + U(\boldsymbol{r}), \tag{16}$$

where $U(\boldsymbol{r})$ is the electron–impurity interaction. Following Kubo et al. (1965), the equation of motion for (X, Y) can be derived as

$$\begin{aligned} \dot{X} &= \frac{\mathrm{i}}{\hbar}[\mathscr{H}, X] = \frac{\mathrm{i}}{\hbar}[U, X] = \frac{l_0^2}{\hbar}\frac{\partial U}{\partial y}, \\ \dot{Y} &= \frac{\mathrm{i}}{\hbar}[\mathscr{H}, Y] = \frac{\mathrm{i}}{\hbar}[U, Y] = -\frac{l_0^2}{\hbar}\frac{\partial U}{\partial x}, \end{aligned} \tag{17}$$

where we have employed the commutation relation: $[x, \eta] = -[y, \xi] = \mathrm{i}l_0^2$. Due to the presence of the impurity potential, the degeneracy of the states with different (X, Y) is lifted and the Landau levels instead of being a series of δ-functions broaden into bands (fig. 3).

The density of states $D(E)$ of the two-dimensional electron system is an important quantity for the understanding of the QHE. Several authors have contributed theoretically and experimentally to our present understanding of the density of states (DOS) in the presence of a random potential (Ioffe and Larkin 1981, Wegner 1983, Brezin et al. 1984, Ando and Uemura 1974, Ando 1983a, 1984, Ando and Aoki 1985, Aoki and Ando 1985). The DOS can be determined by measuring thermodynamic quantities like magnetization (Eisenstein et al. 1985), electron heat capacity (Gornik et al. 1985) and quantum oscillations of the chemical potential (Pudalov et al. 1985), from magnetocapacitance measurements (Smith et al. 1986), from activated conduction (Stahl et al. 1985), from the temperature dependence of the slope of the Hall plateau (Wei et al. 1985a) or from an analysis of the shape of the Hall plateau (Pudalov and Semenchinsky 1985). A review of the experimental results is available in the literature (Gornik 1987). The results of all these studies can be stated as follows: the DOS between the $D(E)$ peaks is approximately constant and is a significant fraction of the value at zero magnetic field. The width of each peak is $\Gamma \sim B^{1/2}$. It is also established now that there are localized states in the tails of $D(E)$ and extended

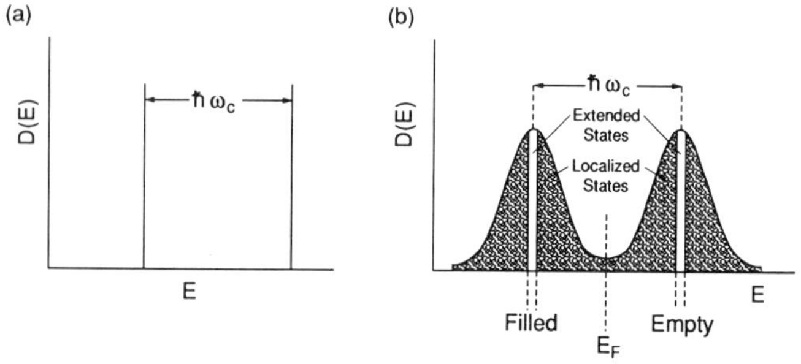

Fig. 3. Density of states versus energy. (a) impurity-free system, $U = 0$. (b) $U \neq 0$, with mobility edges and the localized regions.

states which carry the Hall current located in the region of the maximum of $D(E)$. The regions of localized states are known as the *mobility gaps* and their boundaries with the regions of extended states are called the *mobility edges*. For further details, see § 3.3.1.

3. Integral quantum Hall effect

Before the discovery of the IQHE by von Klitzing et al. (1980), there were experimental indications of the existence of such an effect. Plateau-like behavior was actually observed in ρ_{xy} (Englert and von Klitzing 1978) and in σ_{xy} (Wakabayashi and Kawaji 1980). Accurate quantization of the Hall plateau was not achieved in those experiments. The possibility of ρ_{xy} quantization was also considered theoretically by Ando et al. (1975).

3.1. Experimental work

In fig. 4, we present the experimental results of von Klitzing et al. (1980) and von Klitzing (1982, 1986) for a Si-MOSFET inversion layer in a magnetic field of $B = 19$ T. The diagonal resistance R_{xx} ($\approx \rho_{xx}$) and the Hall resistance R_{xy} ($\approx R_H$) are plotted as a function of gate voltage V_G ($\propto$ electron concentration). The diagonal

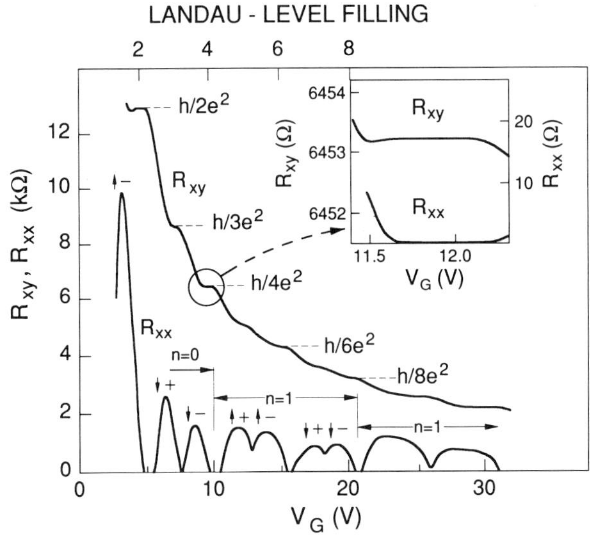

Fig. 4. The quantum Hall effect observed in Si(100) MOS inversion layer in a magnetic field of $B = 19$ T at temperature $T = 1.5$ K. The diagonal resistance and the Hall resistance are shown as a function of gate voltage ($\propto$ electron concentration). The oscillations in R_{xx} are labeled by the Landau-level index (n), spin ($\uparrow$, $\downarrow$) and the valley (+, −). The upper scale indicates the Landau-level filling described in the text, and the inset shows the details of a plateau for $B = 13.5$ T.

resistance is seen to vanish at different regions of V_G, indicating a current flow without any dissipation. In the same regions, R_{xy} develops *plateaus* with $R_{xy}=h/ne^2$ (n integer). The quantization condition on the plateaus are found to be obeyed with *extreme accuracy*. The experimental accuracy so far achieved is better than one part in 10^7 while resistivities as low as $\rho_{xx}<10^{-10}\Omega/\square$ have been established (Störmer 1984). Another interesting feature of the above findings is that the quantization condition of the conductivity is very insensitive to the details of the sample (geometry, amount of disorder, etc.). The quantized Hall resistance is more stable and more reproducible than any wire resistor and since 1 January 1990, the quantized Hall resistance has been used as an international reference resistor. The Hall resistance of a quantized plateau is expressed as $R_H=R_K/n$ and $R_K\approx 25\,813\,\Omega$ is the *von Klitzing constant* which appears to be a universal quantity (Taylor 1989, Quinn 1989).

The accurate quantization of the Hall resistance can be used to determine the value of the fine-structure constant α (von Klitzing et al. 1980). This is a quantity of fundamental importance in quantum electrodynamics and is related to the Hall resistance in the manner

$$\alpha=\tfrac{1}{2}\mu_0 c\frac{e^2}{h}=\tfrac{1}{2}\mu_0 c(R_K)^{-1},$$

where μ_0 is the permeability of the vacuum and c is the velocity of light.

After the initial experiment by von Klitzing et al., there were several other experiments demonstrating integral quantization in a variety of systems. Tsui and Gossard (1981) observed the effect in a GaAs-heterostructure. This system was also used by other authors (Paalanen et al. 1982, Ebert et al. 1982, von Klitzing 1986). Because of the small effective mass m^* of the electrons in GaAs [$m^*_{Si}/m^*_{GaAs}>3$], the Landau-level splitting is larger compared to that in Si and the high-quality of the interface results in a high mobility of the two-dimensional electrons (fig. 5). In addition, the IQHE has been observed in systems like InGaAs/InP (Nicholas et al. 1982, Guldner et al. 1982, 1986), HgTe/CdTe (Kirk et al. 1986) and two-dimensional electrons and holes formed in InAs/GaSb heterostructures (Mendez et al. 1985), in GaAs heterostructures (Störmer et al. 1983a) and in Si-MOSFETs (Gusev et al. 1984).

It is interesting to note that the quantum Hall effect presents a very special situation viz. the conductivity σ_{xx}, describing the current density along the electric field, and resistivity ρ_{xx}, defining the electric field strength along the current path, vanish simultaneously. In a two-dimensional system, in the presence of a magnetic field, the current density $\boldsymbol{j}$ is related to an electric field $\boldsymbol{E}$ by

$$\boldsymbol{j}=\hat{\sigma}\boldsymbol{E},\quad \boldsymbol{E}=\hat{\rho}\boldsymbol{j}, \tag{18}$$

where $\hat{\sigma}$ is the conductivity tensor and $\hat{\rho}=(\hat{\sigma})^{-1}$ is the resistivity tensor. They are defined as

$$\hat{\rho}=\begin{pmatrix}\sigma_{xx} & \sigma_{xy}\\ -\sigma_{xy} & \sigma_{xx}\end{pmatrix}^{-1}=\frac{1}{\sigma_{xx}^2+\sigma_{xy}^2}\begin{pmatrix}\sigma_{xx} & -\sigma_{xy}\\ \sigma_{xy} & \sigma_{xx}\end{pmatrix}, \tag{19}$$

where we have used the Onsager reciprocity relations (Kubo 1957), $\sigma_{xx}=\sigma_{yy}$ and $\sigma_{yx}=-\sigma_{xy}$. Similarly,

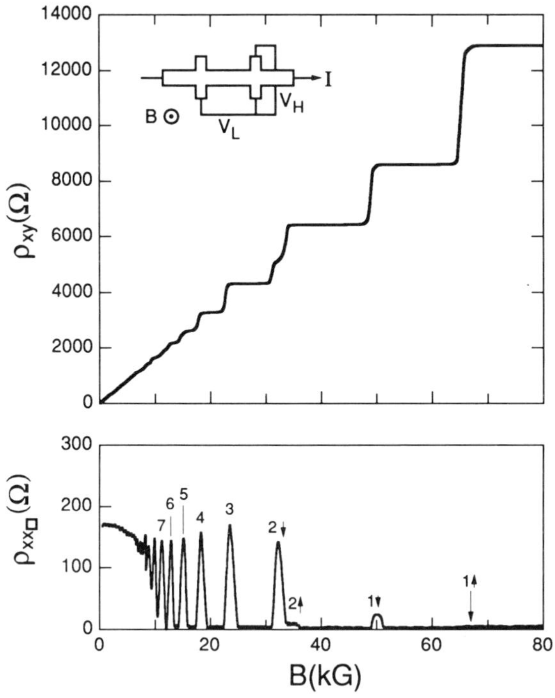

Fig. 5. The Hall resistivity ($\rho_{xy} = V_H/I$) and magnetoresistivity ($\rho_{xx} = V_L/I$) of a modulation-doped GaAs-heterostructure. The inset shows the sample configuration.

$$\hat{\sigma} = \frac{1}{\rho_{xx}^2 + \rho_{xy}^2} \begin{pmatrix} \rho_{xx} & -\rho_{xy} \\ \rho_{xy} & \rho_{xx} \end{pmatrix}. \tag{20}$$

Experimental results show that when a plateau appears, $\rho_{xx} = 0$, which means that, $\sigma_{xx} = \rho_{xx}/(\rho_{xx}^2 + \rho_{xy}^2) = 0$, provided that $\rho_{xy} \neq 0$, and $\sigma_{xy} = -\rho_{xy}/(\rho_{xx}^2 + \rho_{xy}^2) = -1/\rho_{xy}$. For a sample of length L in each dimension, the Hall resistance R_H is related to the Hall resistivity ρ_{xy} as $R_H = -\rho_{xy} L^{2-\delta}$, where δ is the dimensionality. In two dimensions, resistance and resistivity are, therefore, the same quantity. The dimension of R_H is usually expressed as $\Omega/\square$.

3.1.1. Classical Hall effect

The motion of an electron moving classically in crossed electric (E_x) and magnetic (B_z) fields is a superposition of a circular motion with frequency ω_c and a uniform drift perpendicular to E_x with a velocity $\boldsymbol{v} = (0, v_D)$, $v_D = cE_x/B$ (Störmer and Tsui 1983). The resulting orbit is a *trochoid* and is a consequence of the Lorentz force. The current density is then $\boldsymbol{j} = en_0\boldsymbol{v}$. From the discussion above, we can readily deduce that, $\sigma_{xx} = 0$ and $\sigma_{xy} = -n_0ec/B$. The classical Hall conductivity as a function of electron concentration does not show any quantization. In order to introduce scattering due to random potentials in the system, one adds a term $m^*\boldsymbol{v}/\tau$, where τ

is the scattering relaxation time, to the equation of motion. The resulting diagonal and Hall conductivities are

$$\sigma_{xx} = \frac{n_0 e^2 \tau}{m^*} \frac{1}{1+(\omega_c \tau)^2}; \qquad \sigma_{xy} = -\frac{n_0 ec}{B} \frac{(\omega_c \tau)^2}{1+(\omega_c \tau)^2} = -\frac{n_0 ec}{B} + \frac{\sigma_{xx}}{\omega_c \tau}. \tag{21}$$

The free-electron result is trivially recovered by considering the limit $\omega_c \tau \to \infty$. As we shall see below, this scattering-free situation will occur only in very special circumstances.

3.1.2. *Quantum mechanical approach*

The quantum mechanical results for the dynamics of an electron in a magnetic field are presented in § 2.2. In the case of an electron in crossed electric and magnetic fields, the eigenstates (7) and (9) can be written as (Brown 1968)

$$\begin{aligned} E_{nX} &= (n+\tfrac{1}{2})\hbar\omega_c + eE_x X - \tfrac{1}{2}m^* v_D^2, \\ \varphi_{nX} &= e^{ik_y y}\,\chi(x - X + v_D/\omega_c). \end{aligned} \tag{22}$$

These states carry a current in the y-direction. The contribution to the current from an occupied state is

$$j_y = -e\left\langle nX \left| \frac{1}{m^*}\left(p_y + \frac{eBx}{c}\right)\right| nX \right\rangle = \frac{c}{B}\frac{\partial E_{nX}}{\partial X} = ev_D \tag{23}$$

since, $\dot{\boldsymbol{r}} = (1/m^*)[\boldsymbol{p} - (e/c)\boldsymbol{A}]$. All the electronic states $|nX\rangle$ carry the same Hall current. The total current is then $j_y = n_0 ev_D$ and the Hall conductivity, as discussed above, is readily obtained as $\sigma_{xy} = -n_0 ec/B$. From the definition of the filling factor (15), we get the quantization as: $\sigma_{xy} = -\nu e^2/h$. According to the Kubo formula, the diagonal term of the conductivity tensor can be expressed entirely in terms of the states at the Fermi surface. The off-diagonal elements of $\hat{\sigma}$ are, on the other hand, determined by all the states below the Fermi level. Therefore, if the Fermi energy is located inside the mobility gap, $\sigma_{xx} = 0$ (and $\rho_{xx} = 0$) at $T = 0$. For $T \neq 0$, the diagonal conductivity is non-zero, but exponentially small. The non-zero contribution is due to activated excitation of the electrons to the extended states belonging to higher Landau levels, or due to variable-range hopping (Ono 1982). The Hall-conductivity σ_H is finite, even at $T = 0$, due to the contribution from extended states below E_F. As the variation of E_F (due to change in n_0 or B) within the mobility gap has no effect on the occupancy of the states (at $T \to 0$), $\sigma_H (= \sigma_{xy})$ remains constant. This accounts for the observation of the plateaus in σ_H as a function of n_0 or B. The accurate quantization of σ_H in the region of a plateau still remained to be explained and is the central issue of several theoretical studies to be discussed below.

3.2. *Integral quantization: theoretical work*

An explanation of some aspects of the IQHE is possible without introducing the localized states, by invoking the reservoir theory (Baraff and Tsui 1981). According to this theory, the ionized donors in the heterojunctions serve as a reservoir. The electrons tunnel across a potential barrier to the electron channel. When the density of states in the channel $D(E)$ is a sum of delta-functions, the thermodynamic equilibrium requires that ν is an integer in certain ranges of B. Then, $\sigma_H = \text{const.}$ and so is ν. The reservoir theory does not account for the exact quantization, cannot properly explain the Si-MOSFET system where $D(E)$ is governed by the gate potential and has not received any experimental support (Pudalov et al. 1984).

The theoretical approaches discussed below address the important aspects of the IQHE, viz., the high precision of the QHE and independence of the effects of the sample boundary or the impurities. Also, they shed light on the physical mechanism that is responsible for the occurrence of IQHE.

3.2.1. *Kubo formula approach*

The Kubo formula (Kubo 1957, Kubo et al. 1965) is a general expression for the current, regarded as a linear response to an external field. In this approach the Hall conductivity is written as

$$\sigma_{xy} = -\frac{n_0 ec}{B} + \Delta\sigma_{xy}, \tag{24}$$

where $\Delta\sigma_{xy}$ is given by the correlation function of the velocities of the center coordinates X and Y as (Aoki and Ando 1981, Ando 1982, 1983b, Aoki 1987)

$$\begin{aligned}\Delta\sigma_{xy} = \frac{e^2\hbar}{\mathrm{i}A}\sum_\alpha \Big\langle f(E_\alpha)\sum_\beta \Re(E_\alpha - E_\beta + \mathrm{i}0)^{-2} \\ \times [\langle\alpha|\dot{X}|\beta\rangle\langle\beta|\dot{Y}|\alpha\rangle - \langle\alpha|\dot{Y}|\beta\rangle\langle\beta|\dot{X}|\alpha\rangle]\Big\rangle,\end{aligned} \tag{25}$$

where $f(E)$ is the Fermi distribution function, $|\alpha\rangle$ is the eigenstate of the Hamiltonian (16) and A is the area of the system. If $|\alpha\rangle$ is a localized state, then, for any $|\beta\rangle$,

$$\langle\alpha|\dot{X}|\beta\rangle = (\mathrm{i}\hbar)^{-1}\langle\alpha|X|\beta\rangle(E_\alpha - E_\beta). \tag{26}$$

Using eqs. (25) and (26) and the relation $[X, Y] = \mathrm{i}l_0^2$, the contribution from the state $|\alpha\rangle$ to $\Delta\sigma_{xy}$ in eq. (25) is $\Delta\sigma_{xy}^\alpha = f(E_\alpha)ec/B$.

From this result, several important observations (at $T = 0$) can be made:

(a) As long as the Fermi level lies in the energy regime of the localized states, σ_{xy} is constant.

(b) If all the states below the Fermi level E_F are localized, $\sigma_{xy} = 0$ because $\Delta\sigma_{xy}$ exactly cancels* $-n_0 ec/B$. For the QHE to exist, at least one state per Landau level has to be extended. This shows that the presence of a magnetic field provides a situation different from that predicted by the scaling theory (Abrahams et al. 1979). According to that theory, particles moving in a two-dimensional random potential are always localized in the absence of a magnetic field and at $T = 0$.

*A similar result was also obtained by Usov and Ulinich (1982).

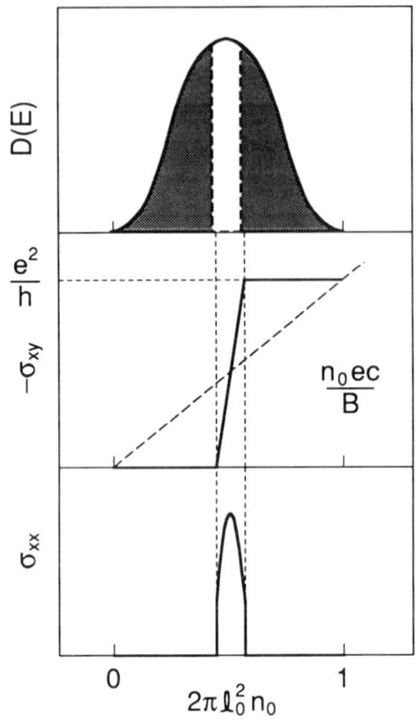

Fig. 6. The diagonal conductivity σ_{xx}, the Hall conductivity σ_{xy} and the density of states $D(E)$ as a function of the Landau-level filling factor.

In the case of a strong magnetic field and when E_F lies in a gap between the nth and $(n+1)$th Landau levels, i.e., electrons occupy states up to the nth Landau level such that $v = n$, Aoki and Ando (1981) showed that $\sigma_{xx} = 0$ and $\sigma_{xy} = -ne^2/h$. The situation is depicted in fig. 6 where, at the onset of the upper plateau, $\sigma_{xy} = -e^2/h$ which is what one expects when *all* the states (including localized states) would carry the Hall current.

Prange (1981) explained this apparent paradox by studying a model of free electrons interacting with a δ-function impurity. He concluded that the Hall current at the integral quantization is exactly the same as that for free electrons because the loss of Hall current due to the formation of one localized state is exactly compensated by an appropriate increase of the Hall current carried by the remaining extended states.

There are several other important contributions based on the Kubo formula which should also be mentioned. Thouless (1981) has shown that for the integer quantization, the Hall conductivity (as derived from the Kubo formula) is unaffected by a weak variation of the impurity potential. Streda (1982) reformulated the Kubo approach to write the Hall conductivity as

$$\sigma_H = \sigma_H^I + \sigma_H^{II}, \quad \sigma_H^{II}(E) = -ec \left. \frac{\partial N(E)}{\partial B} \right|_{E=E_F}. \tag{27}$$

Here $N(E)$ is the number of states with energy $\leqslant E$, and the first term σ_H^I depends on the material parameters and impurity potential $U(\boldsymbol{r})$. If, at $T=0$, the Fermi energy lies in a gap of the energy spectrum of the system, σ_{xx} and σ_H^I vanish. If that gap is the nth Landau gap and the degeneracy of the Landau band is the same as that for free-electron Landau levels $N(E_F)=neB/hc$ and $\sigma_H^{II} = -ne^2/h$. Relation (27) is thermodynamic.

3.2.2. *The gauge invariance approach*

The universal character of the QHE suggests that the effect is due to a fundamental principle. Laughlin (1981) proposed that the effect is due to the gauge invariance and the existence of a mobility gap. Following Laughlin, we consider a Gedankenexperiment involving the *measurement* of Hall conductivity in the geometry of fig. 7. A two-dimensional electron system is bent into the form of a ribbon. A magnetic field $\boldsymbol{B}$ pierces the ribbon everywhere normal to the surface and a voltage V (Hall voltage) is applied across the edges of the ribbon.

Let us now imagine passing a flux $\boldsymbol{\Phi}$ through a solenoid as shown. The current I is then expressed in terms of the total electronic energy of the system U as (Byers and Yang 1961)

$$I = c\frac{\partial U}{\partial \boldsymbol{\Phi}}. \tag{28}$$

Under a gauge transformation $\boldsymbol{A} \to \boldsymbol{A} + \delta A = \boldsymbol{A} + \delta\boldsymbol{\Phi}/L$, where L is the circumference of the loop, the wave function of the electron acquires a phase factor

$$\psi' \to \psi \exp\left(\mathrm{i}\frac{e}{\hbar c}\frac{\delta\boldsymbol{\Phi}}{L}y\right) = \psi \exp\left(2\pi\mathrm{i}\frac{\delta\boldsymbol{\Phi}}{\boldsymbol{\Phi}_0}\frac{y}{L}\right). \tag{29}$$

If the electron is localized (e.g., trapped by an impurity), the wave function vanishes outside a region of size smaller than L and will not respond to the flux. The energies of localized states are unchanged by the adiabatic process of varying $\boldsymbol{\Phi}$. On the other hand, if the electron is extended, such a transformation is not allowed unless $\delta\boldsymbol{\Phi}/\boldsymbol{\Phi}_0$ is an integer (the wave functions are required to be single-valued). We have seen earlier that the electrons in Landau levels have extended wave functions and will contribute to the current.

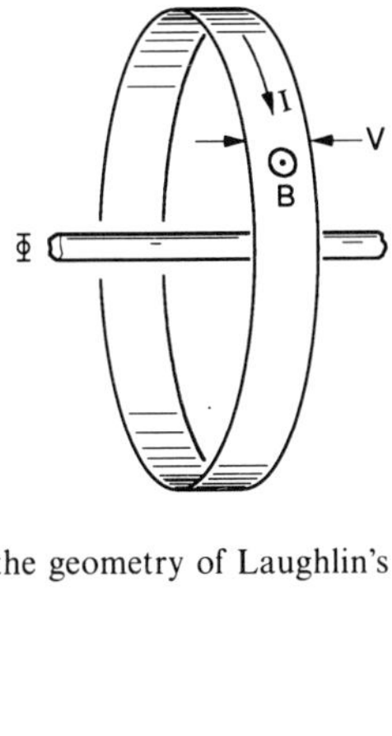

Fig. 7. Hall effect in the geometry of Laughlin's Gedankenexperiment.

Let us now imagine that the magnetic flux through the loop is increased adiabatically from zero. If the Fermi level lies in a mobility gap (i.e., $\sigma_{xx} = 0$ as discussed above) any localized states which may be present will not *see* the change. The electrons in the extended states will however, respond to the change until the flux reaches $\boldsymbol{\Phi}_0$, when each of these states must map identically onto themselves (the transformation is unitary). For free electrons, Laughlin (1981) showed that as the flux is increased adiabatically, each state moves to its neighboring state in the direction of the electric field, as in a shift register (Laughlin 1984a). When the flux increases by one quantum, the electron distribution must look exactly the same as before. The net result of the adiabatic process is a transfer of charge from one edge of the loop to the other. If n electrons are transferred during a change $\Delta\boldsymbol{\Phi}$ of one flux quantum, the net change in electronic energy is $\Delta U = -neV$, where V is the potential difference between the edges. Writing $\delta U/\delta\boldsymbol{\Phi} \to \Delta U/\Delta\boldsymbol{\Phi}$, we get $I = -c(neV/\Delta\boldsymbol{\Phi}) = -n(e^2/h)V$, and $\sigma_\mathrm{H} = -ne^2/h$. According to Laughlin's approach, the quantization is so accurate because it is based on two very general conditions: the gauge invariance of the system and the existence of a mobility gap.

Laughlin's work has been extended to the case of a dirty system by Giuliani et al. (1983) and also to study the gauge transformation for disordered systems numerical experiments have been performed (Aoki 1982, Aers and MacDonald 1984). Halperin (1982) extended the Laughlin approach to include the role of edge states which will be discussed in § 3.3.3.

3.2.3. The topological invariance approach

We have thus far, considered only the one-electron system and ignored the many-body interaction. The quantization condition in the presence of a many-body interaction was studied by Niu et al. (1985) (see also Avron and Seiler 1985, Aoki and Ando 1986a), who showed that the Hall conductance can be expressed in a topologically invariant form. In the following, we briefly discuss the work of Niu et al. (1985).

The Hall conductivity of a rectangular plane with sides L_1 and L_2 is written via the Kubo formula as (see, e.g., Morandi 1988)

$$\sigma_\mathrm{H} = \frac{\mathrm{i}e^2\hbar}{A}\sum_{n>0}\frac{(v_1)_{0n}(v_2)_{n0} - (v_2)_{0n}(v_1)_{n0}}{(E_0 - E_n)^2}, \tag{30}$$

where $A = L_1L_2$, $|0\rangle$ is the ground state and $|n\rangle$ corresponds to the excited states of the N_e-electron Hamiltonian, E_0 and E_n are the corresponding eigenenergies and the velocity operators v_1 and v_2 are defined as

$$\boldsymbol{v} = \sum_{i=1}^{N_\mathrm{e}}\frac{1}{m_i^*}\left(-\mathrm{i}\hbar\nabla_i + \frac{e}{c}\boldsymbol{A}_i\right), \quad \boldsymbol{v} = (v_1, v_2). \tag{31}$$

Considering the Landau gauge $\boldsymbol{A} = (0, Bx)$, the wave function ψ for the opposite edges are related by magnetic translations (Brown 1968)

$$\psi(x_i + L_1) = \mathrm{e}^{\mathrm{i}\alpha L_1}\,\mathrm{e}^{\mathrm{i}y_iL_1/l_0^2}\,\psi(x_i), \quad \psi(y_i + L_2) = \mathrm{e}^{\mathrm{i}\beta L_2}\psi(y_i), \tag{32}$$

where α and β are phase parameters. Making the unitary transformation

$$\varphi_n = \exp\left(-\mathrm{i}\alpha\sum_{i=1}^{N_\mathrm{e}} x_i\right)\exp\left(-\mathrm{i}\beta\sum_{i=1}^{N_\mathrm{e}} y_i\right)\psi_n,$$

one can write the Hall conductivity as (Niu et al. 1985)

$$\sigma_{\mathrm{H}} = \frac{\mathrm{i}e^2}{\hbar}\left(\left\langle\frac{\partial\varphi_0}{\partial\theta}\middle|\frac{\partial\varphi_0}{\partial\varphi}\right\rangle - \left\langle\frac{\partial\varphi_0}{\partial\varphi}\middle|\frac{\partial\varphi_0}{\partial\theta}\right\rangle\right), \tag{33}$$

where $\theta = \alpha L_1$ and $\varphi = \beta L_2$.

One major condition for quantization in this approach is that the Hall conductivity is a *local* response function, insensitive to the boundary condition. We can, therefore, average over all the phases $(0 \leqslant \theta < 2\pi, 0 < \varphi \leqslant 2\pi)$ that specify different boundary conditions

$$\sigma_{\mathrm{H}} = \bar{\sigma} = \frac{e^2}{h}\int_0^{2\pi} \mathrm{d}\theta \int_0^{2\pi} \frac{\mathrm{d}\varphi}{2\pi\mathrm{i}}\left(\left\langle\frac{\partial\varphi_0}{\partial\varphi}\middle|\frac{\partial\varphi_0}{\partial\theta}\right\rangle - \left\langle\frac{\partial\varphi_0}{\partial\theta}\middle|\frac{\partial\varphi_0}{\partial\varphi}\right\rangle\right). \tag{34}$$

This can be evaluated as

$$\sigma_{\mathrm{H}} = \frac{e^2}{h}\int_0^{2\pi} \mathrm{d}\theta \int_0^{2\pi} \frac{\mathrm{d}\varphi}{2\pi\mathrm{i}} \nabla \times \boldsymbol{A} = \frac{e^2}{h}\frac{1}{2\pi\mathrm{i}}\oint \mathrm{d}\boldsymbol{l}\cdot\boldsymbol{A}. \tag{35}$$

The vector $\boldsymbol{A}$ is defined by its components

$$A_\gamma = \frac{1}{2}\left(\left\langle\frac{\partial\varphi_0}{\partial\gamma}\middle|\varphi_0\right\rangle - \left\langle\varphi_0\middle|\frac{\partial\varphi_0}{\partial\gamma}\right\rangle\right), \quad \gamma = \theta, \varphi. \tag{36}$$

Now, if the ground state is nondegenerate and is separated from the excited states by a gap, the ground state can only change by a phase factor depending on θ and φ. For example, the ground state must go back to itself (up to an overall phase factor) as θ and φ change by 2π. Therefore, $\oint \mathrm{d}\boldsymbol{l}\cdot\boldsymbol{A} = 2\pi\mathrm{i} \times \text{(integer)}$ and $\sigma_{\mathrm{H}} = \text{integer} \times e^2/h$.

Integral (34) is a topological invariant and was originally obtained for noninteracting electrons in a two-dimensional periodic potential (Thouless et al. 1982). An excellent introduction to the connection between the IQHE and the topological idea can be found in the book by Morandi (1988). Similar arguments for the FQHE have also been put forward by Niu et al. (1985). A better treatment of the latter case was given, however, by Tao and Haldane (1986) and will be discussed briefly in § 4.2.6.

3.3. Other developments

3.3.1. Electron localization in the quantum Hall regime

We have seen above that the extended and localized states form an integral part of our understanding of the IQHE. Several authors have studied the role of various disorder potentials in the IQHE regime (Iordansky 1982, Kazarinov and Luryi 1982, Luryi and Kazarinov 1983, Trugman 1983, Joynt and Prange 1984, Apenko and Lozovik 1985).

For an external potential $U(\boldsymbol{r})$ which varies slowly over length scales of order l_0, and neglecting the inter-Landau-level transition, the motion of an electron reduces to a drift of the center of cyclotron motion along an equipotential line $U(\boldsymbol{r}) = \text{const}$. The state of an electron with energy E (measured from the center of the Landau band) is localized near the line $U(\boldsymbol{r}) = E$, and decreases exponentially from this line. The properties of the equipotential lines for a two-dimensional random potential are fairly well known from the classical percolation model (Zallen and Scher 1971). If the potential $U(\boldsymbol{r})$ is symmetric about $E = 0$ (e.g., for equal number of attractive and repulsive scatterers), the equipotentials with $E \neq 0$ are closed curves of finite length, and correspond to localized states. If we call the regions with $U(\boldsymbol{r}) > E$ land and those with $U(\boldsymbol{r}) < E$ water, we get a set of islands for $E > 0$ or a number of lakes for $E < 0$. In that case, we have at least one coast line at $E = 0$. In this picture, all the states in a two-dimensional system subjected to a strong magnetic field are localized, except for the state at the center of the Landau band, which is extended. In an external electric field, analyzing the delocalized drift trajectories, Iordansky (1982) and others (Kazarinov and Luryi 1982, Luryi and Kazarinov 1983, Trugman 1983) obtained the quantization for σ_{xy}.

The behavior of the localization length in strong magnetic fields has been a very active area of research. Using diagrammatic techniques, Ono (1982) was the first to suggest that only the center of the Landau level is extended while the other states are exponentially localized. Ando studied the electron localization by numerically diagonalizing the Hamiltonian matrix for a system with short-range δ-function (Ando 1983a) and long-range (Gaussian form) impurities (Ando 1984) which are randomly distributed. The localization length was studied by Ando via the Thouless-number method. The Thouless number $g(L)$ is defined as the ratio of the shifts ΔE of the individual energy levels due to a change in boundary conditions (from periodic to anti-periodic) to the level separation $[L^2 D(E)]^{-1}$, where $D(E)$ is the density of states per unit area. For localized states, one can determine the extent of the localized wave functions or the inverse localization length $\alpha(E)$, from

$$g(L) = g(0)\, \mathrm{e}^{-\alpha(E)L},$$

where L is the sample length (Licciardello and Thouless 1978). The conclusion was that the states are exponentially localized except in the vicinity of the center of each Landau band. The inverse localization length was found to decrease almost linearly with energy when the energy is far from the center of the Landau band, and then smoothly approaches zero with energy. Extremely large localization lengths in the center of the Landau band were also found in numerical studies by other groups (Schweitzer et al. 1984). Finite-size studies with sample size $\sim 50\,000$ times the cyclotron radius (Aoki and Ando 1985, 1986b, Ando and Aoki 1985) resulted in $\alpha(E) \propto |E - E_n|^s$ near the center E_n, of each Landau level with the critical exponent $s \lessgtr 2$ and $s \lessgtr 4$ for $n = 0$ and $n = 1$, respectively. From a perturbative calculation, Hikami (1986) estimated $s = 1.9 \pm 0.2$. The critical exponent from the classical percolation model is $\frac{4}{3}$ (Trugman 1983) and including quantum tunneling one gets $s = \frac{7}{3}$ (Mil'nikov and Sokolov 1988). A percolation model, including quantum tunneling and interference near the percolation threshold (Chalker and Coddington 1988) yielded a one-parameter scaling behavior and a critical exponent $s = 2.5 \pm 0.5$. Recent finite-size scaling studies in the lowest Landau band (Huckestein 1990, Huckestein and Kramer 1990) revealed a universal one-parameter scaling behavior and a critical exponent $s = 2.34 \pm 0.04$. For more recent results on the finite-size scaling, see the paper by Huo and Bhatt (1992).

3.3.2. Renormalization group approach

This approach is a result of the attempts to unify the weak localization in a two-dimensional electron system (Abrahams et al. 1979, Gorkov et al. 1979) and the quantization of the Hall conductivity. According to the two-parameter scaling theory of the IQHE (Pruisken 1984, 1985, Levine et al. 1983, Khmel'nitzkii 1983, 1984), σ_{xx} and σ_{xy} (in units of e^2/h) vary with a length scale L, given by the renormalization group equations

$$\frac{\mathrm{d}\sigma_{\eta\nu}}{\mathrm{d}\xi} = \beta_{\eta\nu}(\sigma_{xx}, \sigma_{xy}), \quad \xi = \ln L, \tag{37}$$

where $\beta_{\eta\nu}$ is a periodic function of σ_{xy} with a period of e^2/h. The resulting phase diagram is shown in fig. 8. The beginning of each flow line is selected at a point corresponding to a spatial scale $\xi = 0$, where $\hat{\sigma}$ can be estimated from the classical formula (21). With increasing system size, all flow lines merge into one of the fixed points at $(\sigma_{xx}, \sigma_{xy}) = (0, -ne^2/h)$ which are the *localization fixed points* and describe localized wave functions of the electrons near the Fermi energy E_{F}. In addition, the system also has unstable fixed points (denoted by $\otimes$) where the flow lines with $\sigma_{xy} = -(n + \frac{1}{2})e^2/h$ terminate. At these points, $\sigma_{xx} > 0$ and describes the singular behavior in the renormalized transport coefficients, corresponding microscopically to the occurrence of a diverging localization length. These *delocalization fixed points* are associated with the extended states at E_{F}. It should be noted that the explicit form of the β-function is not known, only its asymptote is calculated. An approximate

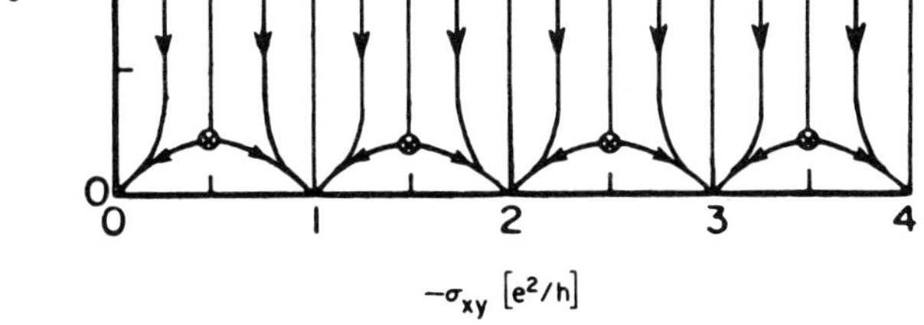

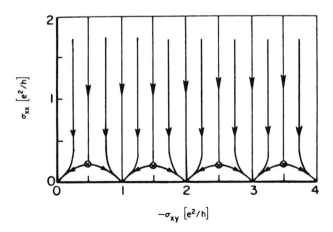

Fig. 8. Scaling diagram for the integral quantum Hall effect.

calculation of σ_{xx} at the unstable fixed points gives $\sigma_{xx} = (1.4/\pi)e^2/h$ for a white-noise random potential (Hikami 1984).

The predictions of the two-parameter scaling theory have been tested experimentally (Wei et al. 1985b, 1986). The study was carried out on the 2DES in InGaAs/InP heterostructures. It was assumed that the effective sample size is governed by inelastic scattering which can be varied via the temperature T, and a phase diagram was constructed (fig. 9) from the measurements of $\sigma_{xx}(T)$ and $\sigma_{xy}(T)$. In this figure, the dashed lines are from the temperature range 4.2–10 K, where lowering of T is accompanied by the usual enhancement of the Shubnikov–de Haas oscillations as a result of a change of the Fermi distribution. In the scaling region ($T \sim 0.5$–4.2 K) the results, presented as full lines, show a tendency (as $T \to 0$) to flow out toward the fixed points $(0, e^2/h)$, $(0, 2e^2/h)$, $(0, 3e^2/h)$ and $(0, 4e^2/h)$. Detailed studies by these authors indicated that there is a remarkable symmetry about the line $\sigma_{xy} = 1\frac{1}{2}$. The data are consistent with the existence of an intermediate coupling, delocalization critical point. A phase diagram for the FQHE based on the scaling hypothesis has also been proposed (Laughlin et al. 1985). The theories are still in a state where a fair amount of *guesswork* is required as input.

The two-parameter scaling theory of the IQHE has not received much support from other theoretical work. Finite-size studies of the diagonal and off-diagonal conductivities (Ando 1986) indicate that they are *not* independent but possess single flow lines dependent on the Landau-level index which contradicts the scaling theory results discussed above.

3.3.3. *Current-carrying edge states*

In calculating the Hall current in § 3.1.2, we assumed that the current is distributed over the entire surface of the two-dimensional layer. Some authors have pointed out

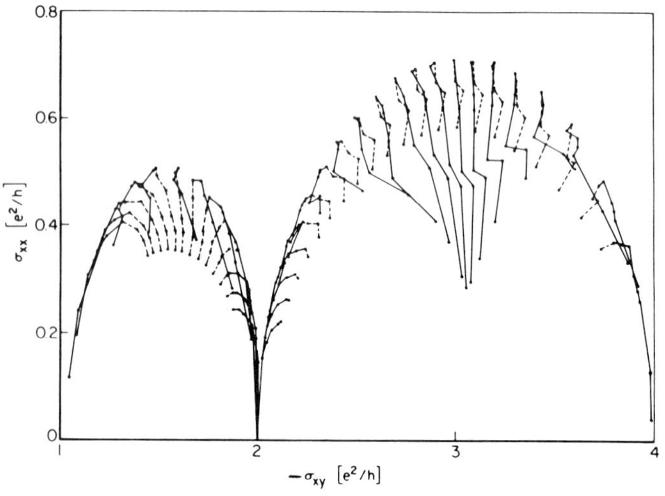

Fig. 9. Experimental results for the conductivity plotted as T-driven flow lines from $T = 10$ to 0.5 K. The dashed lines are from 10 to 4.2 K and the solid lines from 4.2 to 0.5 K.

that, in fact, the current flows along the one-dimensional edge of the layer (Halperin 1982, Thouless 1982, MacDonald and Streda 1984).

Let us consider the noninteracting electrons of § 2.2 but now confined in a potential

$$V(x) = 0, \qquad x \in (-\tfrac{1}{2}L_x, \tfrac{1}{2}L_x)$$
$$\quad = +\infty, \quad \text{otherwise.} \tag{38}$$

Far from the boundary inside the sample, $V(x) \equiv 0$ and the eigenstates are as derived above. However, near the boundary, the energy eigenvalues deviate from the Landau energy (7) and behave as in fig. 10, as a function of X. The total current carried by the nth branch of the energy spectrum when all the bulk states for that branch are occupied is given by (MacDonald and Streda 1984)

$$I_n = \frac{e}{h}(\mu^{\mathrm{R}} - \mu^{\mathrm{L}}) \tag{39}$$

where μ^{R} is the chemical potential at the right (positive-x) edge of the sample and μ^{L} is the chemical potential at the left edge.

In the equilibrium case, $\mu^{\mathrm{R}} = \mu^{\mathrm{L}}$, the bulk current which flows in the presence of an electric field (in which case the above result also holds) is exactly cancelled by the surface diamagnetic current. For noninteracting electrons all the current would flow at the edges and the quantized Hall current would be just the difference between the two edge currents. MacDonald and Streda (1984) also showed that these results are unaltered by the inclusion of the random potential in the Hamiltonian (Apenko and Lozovik 1985). The role of edge currents has also been studied by Smrcka (1984), who derived the Hall current from the Kubo formula, taking into account the edge states explicitly. He concluded that the IQHE is exclusively due to the edge currents. It has been proposed that the edge currents might be observed from the oscillations in the magnetic susceptibility (Azbel 1985). Several experiments exploring the current-carrying edge states have been reported in the literature (van Wees et al. 1989, Müller et al. 1990).

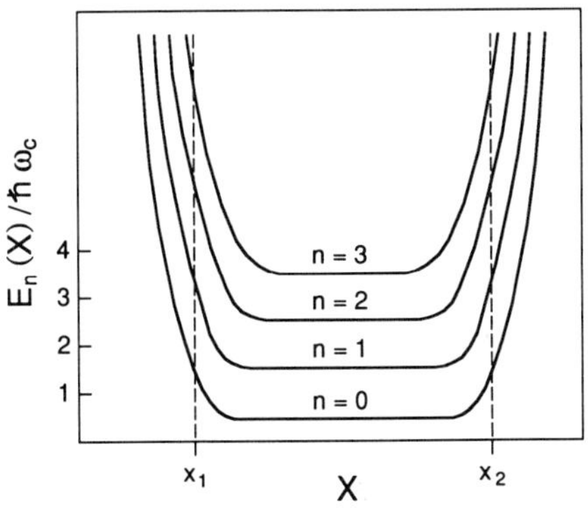

Fig. 10. Energy spectrum of a noninteracting electron in an infinite well confining potential.

3.3.4. Further topics

Finally, we should also mention the work on the IQHE based on the Landauer (1970) approach to transport theory in a two-dimensional system subjected to a strong perpendicular magnetic field. In this approach the conductivity is related to the transmission and reflection probabilities of electrons which are incident from the leads. Streda et al. (1987) studied the transport in the case where the current is incident from a disorder-free (ideal) lead on the left of the sample and is collected by an ideal lead on the right. They argued that the magnetoresistance R and the Hall resistance R_H satisfy the sum rule $(R + R_H)^{-1} = (e^2/h)\,\mathrm{Tr}(t^\dagger t)$, where t is the transmission matrix. Büttiker (1988) has proposed a theory of the IQHE in open multiprobe conductors based on the formation of the edge states and the suppression of backscattering (from one side of the sample to the other). Experiments on the transport through narrow channels (Haug et al. 1989, Washburn et al. 1988) have revealed interesting results but the literature on transport in narrow channels is too numerous to cover in this general review and the interested reader is referred to an excellent review by Beenakker and van Houten (1991).

The IQHE breaks down at a critical current density j_c (the corresponding critical Hall field $E_H = 60$ V/cm at a magnetic field $B = 5$ T) where the resistivity increases abruptly by orders of magnitude and the Hall plateau disappears (Ebert et al. 1983, Sakaki et al. 1984, Pudalov and Semenchinsky 1984a). A typical result is shown in fig. 11. At a current density of $j_x = j_c = 0.5$ A/m, ρ_{xx} at the center of the $\nu = 2$ plateau increases dramatically. A pronounced hysteresis is observed near the breakdown threshold.

In the literature several mechanisms are proposed as possible causes for the breakdown.

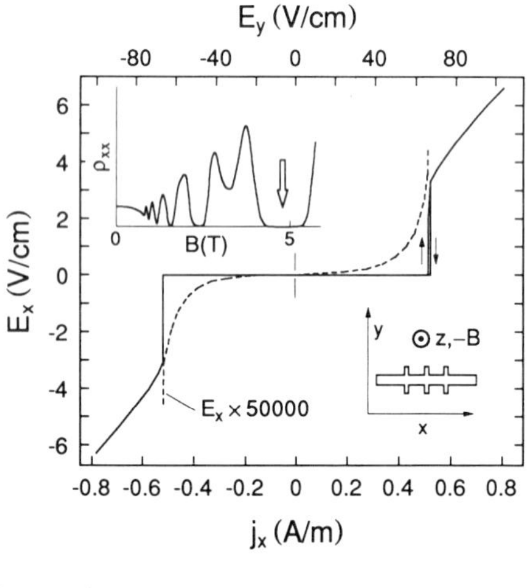

Fig. 11. Breakdown of the IQHE under quantum field effect conditions. The current–voltage characteristic of a GaAs-heterostructure at a filling factor $\nu = 2$. The device geometry and the $\rho_{xx}(B)$ curve are shown as insets.

(a) Transition between levels accompanied by the absorption or emission of phonons (Streda and von Klitzing 1984).

(b) Zener effect: tunneling of the carriers between occupied and empty Landau levels (Cage et al. 1983).

(c) Heating instability associated with the breakdown of the balance between the energy gained and released by electrons (Ebert et al. 1983, Komiyama et al. 1985).

4. Fractional quantum Hall effect

The FQHE was realized in the high-mobility 2DEG in $GaAs/Al_xGa_{1-x}As$ heterostructures (Tsui et al. 1982, Störmer et al. 1983b, Störmer 1984, Tsui and Störmer 1986) and in high mobility Si-MOSFETs (Pudalov and Semenchinsky 1984b). It is characterized by the fact that the Hall conductance has plateaus quantized to certain simple *fractions* ν of the unit e^2/h and at the same place, the longitudinal resistivity shows an almost dissipationless current flow. Here ν is a rational fraction with an odd denominator [in fact, an even denominator fraction has also been found to exhibit the FQHE (Clark et al. 1986, Willett et al. 1987)].

Figure 12 shows the effect for a few values of the filling factor ν. For $\nu > 1$, the characteristic features of the integral QHE are clearly visible. However, in the extreme quantum limit, i.e., for $\nu < 1$, and at low temperatures ($T = 0.48$ K), one observes a clear minimum in ρ_{xx} and a quantized Hall plateau at $\nu = \frac{1}{3}$. In later experiments

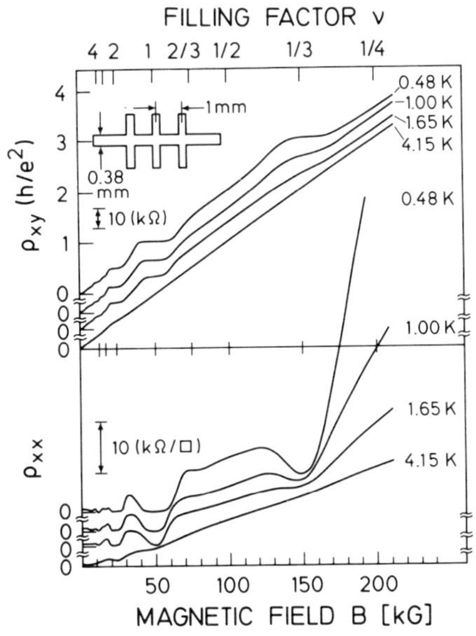

Fig. 12. The Hall resistivity ρ_{xy} and the diagonal resistivity ρ_{xx} as a function of magnetic field in a GaAs-heterostructure with electron density $n_0 = 1.23 \times 10^{11}$ cm^{-2}, where the FQHE was first discovered. The scale on the top shows the *filling factor*, defined in the text.

with higher mobility samples the plateau was found to be quantized to better than three parts in 10^5 and ρ_{xx} was lower than $0.1\Omega/\Box$. A weak structure around $\nu = \frac{2}{3}$ is also visible at the lowest temperatures. Subsequent experiments with higher mobility samples (Chang et al. 1984, Willett et al. 1987, Sajoto et al. 1990) revealed several other fractions which are displayed in fig. 13 and summarized below:

$$\nu = \tfrac{1}{3}, \tfrac{1}{5}, \tfrac{1}{7}^*, \tfrac{2}{5}, \tfrac{2}{7}, \tfrac{2}{9}^*, \tfrac{2}{11}^*, \tfrac{3}{7}, \tfrac{3}{11}^*, \tfrac{3}{13}^*, \tfrac{3}{17}^*, \tfrac{4}{9}, \tfrac{4}{11}^*, \tfrac{4}{13}^*, \tfrac{5}{11}, \tfrac{6}{13}^*, \tfrac{7}{15}^*$$

$$= \tfrac{2}{3}, \tfrac{4}{5}, \tfrac{3}{5}, \tfrac{5}{7}, \tfrac{4}{7}, \tfrac{5}{9}^*, \tfrac{6}{11}^*, \tfrac{7}{13}^*$$

$$= \tfrac{4}{3}, \tfrac{7}{5}, \tfrac{9}{7}, \tfrac{5}{3}, \tfrac{8}{5}, \tfrac{10}{7}, \tfrac{13}{9}^*, \tfrac{11}{7}, \tfrac{7}{3}^*, \tfrac{8}{3}^*, \tfrac{5}{2}.$$

The fractions with an asterisk (*) have as yet shown structures in ρ_{xx} only. The first row of fractions are simply $\nu = p/q$ $(2p < q)$. The second row contains the fractions $\nu = 1 - p/q$ and the last row contains the other fractions $\nu = 1 + p/q$, $1 + (1 - p/q)$, $2 + p/q$, etc. The arrangements of the fractions are made with the *electron–hole symmetry* in mind which will be described below.

The major characteristics of FQHE, viz., the appearance of plateaus in ρ_{xy} and minima in ρ_{xx}, are similar in nature to those of the IQHE. In spite of this apparent similarity the physical origin of the former effect is different from the latter as can be seen from the fact that:

(a) plateaus and minima appear at *fractional* filling factors where no structure is expected in the single-electron density of states, and

(b) the effect is observed only in samples of very high mobility.

This leads us naturally to believe that the electron–electron interaction plays a major role in the FQHE. The main theoretical step in explaining the FQHE, therefore, is to determine the properties of an interacting 2DEG with a neutralizing background

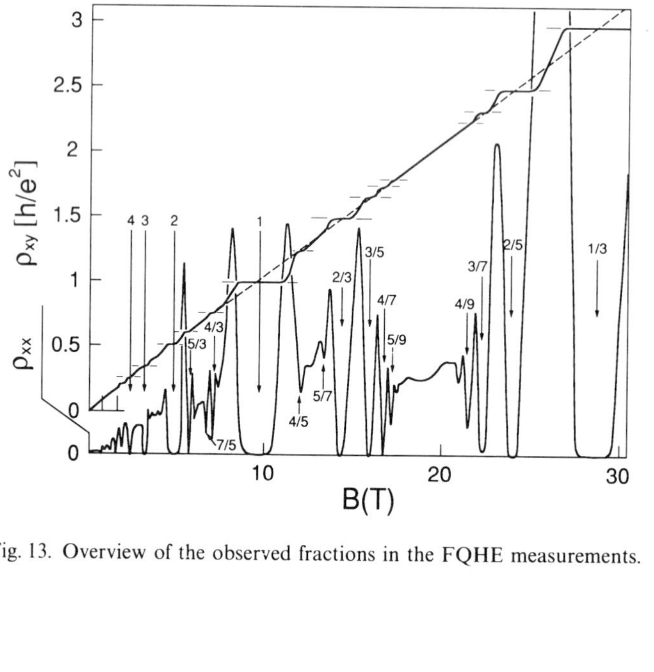

Fig. 13. Overview of the observed fractions in the FQHE measurements.

subjected to a strong perpendicular magnetic field such that only the lowest Landau level is partially filled.

In our attempt to explain the fractional Hall steps, we have to understand the exceptionally stable states of the electron system at particular rational values of ν. The pinning of the density at those values of ν would require that the energy versus density curve shows a cusp-type behavior. A cusp would imply a discontinuity of chemical potential, which would in turn, mean that the electron system is, in fact, *incompressible** at those stable states.

4.1. Ground state

As the fractional Hall steps are observable only in samples of very high mobility, impurity potentials are not expected to be very important in comparison with electron–electron interactions. The first step in the explanation of the FQHE, as stated above, would, therefore, be to study the properties of a system of two-dimensional interacting electrons in a uniform positive background with the magnetic field strength such that only the lowest Landau level is partially filled. One could then consider the effect of impurities as a perturbation. The unit of potential energy is $e^2/\varepsilon l_0$, which is taken to be the energy scale throughout. Here, ε is the background dielectric constant. For a magnetic field $B \gtrsim 10$ T, where the FQHE is generally observed, using the values $\varepsilon \approx 12.9$ and $m^* \approx 0.067 m_e$ which are appropriate for GaAs, it is easy to verify that $e^2/\varepsilon l_0 \lesssim \hbar\omega_c$, the cyclotron energy. The admixture of states in higher Landau levels can thus be safely ignored as a first approximation.

In the extreme quantum limit (no Landau-level mixing), the Hamiltonian describing the two-dimensional electron gas can be written as

$$\mathscr{H} = \tfrac{1}{2}\sum_q V(q)[\bar\rho(q)\,\bar\rho(-q) - \rho e^{-q^2 l_0^2}], \tag{40}$$

where $\bar\rho(q)$ is the projected density operator (discussed in § 4.3.1.) and ρ is the average density of the particles in the system. In the absence of kinetic energy, the ground state is expected to be a solid – which means that the Hamiltonian describes a set of interacting classical particles. The operators $\bar\rho(q)$ do not commute with one another. However, in the limit $B \to \infty$ we get (Fukuyama et al. 1979), $[\bar\rho_q, \bar\rho_{-q}] \to l_0$ and one obtains a (presumably triangular) solid. The earlier attempts to explain the FQHE were mostly centered on crystal-state calculations. Such calculations (Yoshioka and Lee 1983, Yoshioka et al. 1983) however, did not find any singularity at $\nu = \frac{1}{3}$. It is also difficult conceptually to understand how the crystal could carry electric current with no resistive loss, since the charge-density wave (CDW) would be pinned by impurities. Also, if the CDW is not allowed to move with a drift velocity $\boldsymbol{E} \times \boldsymbol{B}c/B^2$, the electron contribution to the Hall conductivity would vanish (Laughlin 1987).

The earliest numerical calculation of the ground state including Coulomb interactions, was by Yoshioka et al. (1983). They investigated the eigenstates of an electron system in a periodic rectangular geometry, by numerically diagonalizing the

* The compressibility $\kappa \sim [d^2E/d\nu^2]^{-1} \to 0$ at the point where the cusp appears.

Hamiltonian. Their results, as we shall discuss in the following section, revealed several interesting features; the most important result was, of course, that the ground state has a significantly lower energy than that of a Hartree–Fock (HF) Wigner crystal.

4.1.1. *Theoretical work*

In the lowest Landau level, the single-particle wave function is (except for the normalization factor),

$$\varphi = \exp[\mathrm{i}Xy/l_0^2 - (X-x)^2/2l_0^2], \tag{41}$$

where $X = k_y l_0^2$ is the center coordinate of the cyclotron motion. Following Yoshioka et al. (1983), let us now put a finite number of electrons in the cell and introduce interactions among them. Consider the situation where there are a few electrons in a rectangular cell of sides a and b, and a strong magnetic field perpendicular to the x–y plane. The electrons are considered to be in the lowest Landau level and are spin-polarized. Applying periodic boundary conditions in the y-direction, one obtains $k_y = X_j/l_0^2 = 2\pi j/b$ for an integer j. For the periodic boundary condition along the x-direction, let us write, for an integer m, $X_m = a$. Clearly, $ab = 2\pi l_0^2 m$, and from eq. (13), m is the Landau-level degeneracy N_s. For N_e electrons in the cell, the filling factor is easily calculated as $\nu = 2\pi l_0^2 N_e/ab = N_e/N_s$. The Hamiltonian for the finite-electron system, suitably modified for the periodic rectangular geometry, is diagonalized numerically to obtain the eigenstates. Details of the calculation can be found in the literature (Yoshioka et al. 1983, Yoshioka 1984a, Chakraborty and Pietiläinen 1988a).

In fig. 14, we present the numerical results of Yoshioka et al. (1983), for the ground-state energies per particle as a function of filling fraction in the lowest Landau level. There are several interesting features noticeable in the result. Let us first consider the case $\nu = \frac{1}{3}$. The ground-state energy per particle for four-, five- and six-electron systems are extremely insensitive to the system size. They are also very close to the infinite system result to be discussed below.

The ground-state energies tend to have downward *cusps* for $\nu = \frac{1}{3}$ and $\frac{2}{5}$. (A cusp is also visible at $\nu = \frac{1}{2}$ but the ground-state energies at this filling fraction strongly depend on the system size.) Cusps at the experimentally observed filling fractions are quite naturally required, in order to describe the incompressible fluid state first proposed by Laughlin.

In fig. 14, also presented are the energy of the crystal state for the $N_e = 4$ system obtained in the exact diagonalization (closed squares) and HF approximation (open squares) (Yoshioka et al. 1983). The long- and short-dashed lines show the energy of the electron and hole crystals resulting from the HF approximation for the infinite system. These crystal state results show a smooth behavior at $\nu = \frac{1}{3}$ and are not the lowest energy state.

In the lowest Landau level, there is electron–hole symmetry. The system with N_e electrons in N_s sites is thus equivalent to that with $(N_s - N_e)$ holes in N_s sites. When we choose the products of the single-electron eigenstates as a basis, the off-diagonal matrix elements for $\nu = N_e/N_s$ are the same as those for $\nu = (1 - N_e/N_s)$ filling

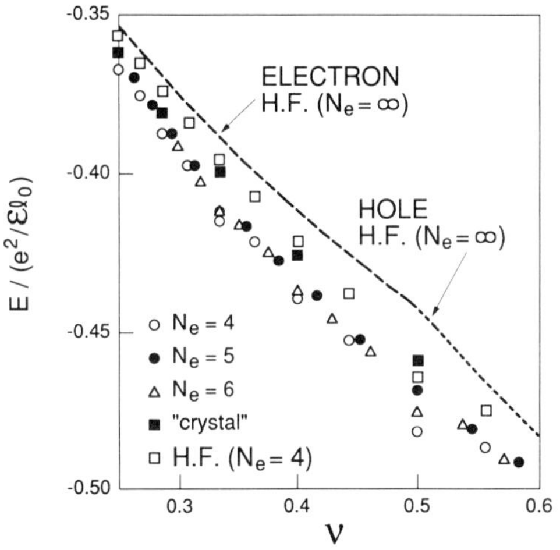

Fig. 14. Energies per particle for finite-electron systems in a periodic rectangular geometry, as a function of filling factor in the lowest Landau level. The long- and short-dashed lines are the energy of the electron and hole crystals within the HF approximation for the infinite system. Closed squares denote the crystal state energies for the $N_e = 4$ system. Open squares show the energy of the crystal state for the $N_e = 4$ system obtained in the HF approximation.

fractions for the same J values. The diagonal matrix elements differ only by a constant. The results for $\nu > \frac{1}{2}$, shown in fig. 14, are obtained by the above arguments.

While these numerical results demonstrate that the ground state is not crystalline, as Halperin (1983) pointed out, not much insight is gained about the ground-state wave function. The crucial step in obtaining a ground-state wave function for a translationally invariant liquid state, and the mechanism for stabilizing the system at particular densities, was made by Laughlin (1983a, 1984b, 1987), and is described below. It should be pointed out that similar finite-size calculations were also done by Haldane and Rezayi (1985) in a spherical geometry.

In discussing Laughlin's theory, we will closely follow the seminal paper by Laughlin (1983a) as well as papers by other authors (Halperin 1983, Chakraborty and Pietiläinen 1988a). Let us consider the electrons to be confined in the x–y plane and subjected to a magnetic field perpendicular to the plane. Considering the symmetric gauge vector potential, $A = \frac{1}{2}B(x\hat{y} - y\hat{x})$, it is convenient to regard the x–y plane as a complex plane. For the lowest Landau level, the single-particle wave functions are eigenfunctions of orbital angular momentum

$$\varphi_m(z) \equiv |m\rangle = \frac{1}{(2\pi l_0^2 2^m m!)^{1/2}} \left(\frac{z}{l_0}\right)^m \mathrm{e}^{-|z|^2/4l_0^2}, \tag{42}$$

with $z = x - \mathrm{i}y$ being the electron position. The many-electron system is described by a Hamiltonian of the form

$$\mathscr{H} = \sum_j \left[\frac{1}{2m_e}\left| -\mathrm{i}\hbar\nabla_j - \frac{e}{c}A_j\right|^2 + V(z_j)\right] + \sum_{j<k} \frac{e^2}{|z_j - z_k|}, \tag{43}$$

where z_j is the location of the jth electron and $V(z)$ is the potential generated by a uniform neutralizing background. It is easy to verify (Laughlin 1983b) that

$$\langle m|r^2|m\rangle = 2(m+1)l_0^2, \tag{44}$$

which means that the area covered by a single electron in state $|m\rangle$ moving in its cyclotron orbit is proportional to m. This result might be considered as an indication of the relation between the interelectron spacing and the angular momentum. From eq. (13), it is readily noticeable that the degeneracy of a Landau level N_s is the upper bound to the quantum number m. This is seen by requiring that, $\pi\langle r^2\rangle \leqslant A$, where A is the area of the system. One then obtains from the above relations, $m \leqslant N_s - 1$, where N_s is the Landau-level degeneracy. Therefore, the state space of an electron in the lowest Landau level is spanned by $1, z, z^2, \ldots, z^{N_s-1}$ times the exponential factor $e^{-|z|^2/4l_0^2}$.

The *Jastrow-type* many-electron wave function proposed by Laughlin for the $\nu = 1/m$ state is

$$\psi_m = \prod_{\substack{j,k=1\\ j<k}}^{N_e} (z_j - z_k)^m \prod_{j=1}^{N_e} e^{-|z_j|^2/4l_0^2}. \tag{45}$$

For m being an odd integer, this wave function obeys Fermi statistics. The wave function is entirely made up out of states in the lowest Landau level. It is also an eigenstate of the angular momentum with eigenvalue $M = \frac{1}{2}N_e(N_e - 1)m$. The total angular momentum M is the degree of the polynomial (conservation of angular momentum). In order to gain more insight about the wave function, let us expand the first product in powers of z_1, keeping all other coordinates fixed. The highest power would then be $m(N_e - 1)$, which must be equal to $N_s - 1$. For large N_e, we then obtain

$$m \cong \frac{A}{2\pi l_0^2 N_e} = \frac{1}{\nu}. \tag{46}$$

The parameter m is thus fixed by the density and, unlike in conventional Jastrow theory, we do not have a variational parameter in the trial wave function. For $m = 1$ (filled Landau level), the polynomial $\Pi_{j<k}(z_j - z_k)$ is the Vandermonde determinant of order N_e. As $N_e \to \infty$, the particle density in this state tends to $1/2\pi l_0^2$ (Bychkov et al. 1981).

For $m > 1$, the wave function vanishes as a high power of the two-electron separation, and thus tends to minimize the repulsive interaction energy. The probability distribution of the electron for ψ_m is given by

$$\begin{aligned} |\psi_m|^2 &= e^{-\mathscr{H}_m}, \\ \mathscr{H}_m &= -2m \sum_{j<k} \ln|z_j - z_k| + \sum_j |z_j|^2/2l_0^2. \end{aligned} \tag{47}$$

This is the Hamiltonian for a *charge neutral* two-dimensional classical plasma where the particles are interacting via a two-dimensional Coulomb (logarithmic) interaction with each other and with a uniform neutralizing background with particle density $\rho_m = 1/2\pi l_0^2 m$. Therefore, in order to achieve charge neutrality, the plasma particles spread out uniformly in a disk with density ρ_m, corresponding to a filling factor $\nu =$

$1/m$, and m is an *odd* integer. The classical plasma provides strong support that the Laughlin state is indeed a translationally invariant liquid (Caillol et al. 1982).

The electrons however, in contrast to the classical plasma particles, interact via the *three-dimensional* Coulomb interaction, and the expectation value of the potential energy in a quantum state is given by

$$\frac{\langle V \rangle}{N_e} = \tfrac{1}{2} \int v(r)\, g(r)\, \mathrm{d}\mathbf{r}, \tag{48}$$

where $g(r)$ is the two-particle radial distribution function, which was calculated by Laughlin using the classical plasma approach. In this calculation, Laughlin used the hypernetted-chain (HNC) theory, which is a well-established technique for dealing with the classical plasma and other quantum systems (Chakraborty and Pietiläinen 1988a). The pair-correlation function in the Laughlin state shows a predominantly short-range behavior: for small r, $g(r)$ goes to zero as r^{2m} whereas for the crystal state it goes as r^2.

Immediately after Laughlin introduced his ground-state wave function (45), Halperin (1983) gave a very interesting interpretation of the significance of this wave function. Imagine freezing the positions of all but the first particles and let χ be the wave function for z_1, parametrized by the positions of the other particles:

$$\chi(z_1) = \prod_{j=1}^{L} (z_1 - Z_j) \exp(-\tfrac{1}{4}|z_1|^2). \tag{49}$$

From the theory of analytic functions one knows that the polynomial can be uniquely defined by the set of its zeros $\{Z_j; j = 1, ..., L\}$. In fact, the zeros of the wave function (45) look like two-dimensional Coulomb charges (in a plasma analogy) which repel the particle at z_1. These zeros must be uniformly distributed with density $1/2\pi$ in order to balance the effect of the Gaussian term which is attracting the particles into the origin. Halperin then observed that Laughlin's wave function places these zeros directly on the other particles. Each particle sees m zeros bound to the positions of the other particles. Let us define a *vortex* as a point where the wave function is zero, such that the phase changes by 2π as one circulates around the vortex in the negative direction. The special feature of the Laughlin wave function is that there are exactly m vortices at each electron position and no other *wasted* vortices elsewhere in the plane.

The ground-state energy for the $\nu = \frac{1}{3}$ state, obtained in the HNC scheme is $E(\nu = \frac{1}{3}) = -0.4056 e^2/\varepsilon l_0$. This result is quite close to the exact results shown for finite electron systems in the earlier section. There have also been very accurate Monte Carlo studies to confirm these results on quantitative grounds. Levesque et al. (1984) obtained the ground-state energy of the Laughlin wave function by evaluating the pair-correlation functions for about 256 particles, using the method described by Caillol et al. (1982). For the ground-state energy they obtained $E(\nu = \frac{1}{3}) = -0.4100 \pm 0.0001$ and $E(\nu = \frac{1}{5}) = -0.3277 \pm 0.0002$ (in units of $e^2/\varepsilon l_0$).

4.1.2. *Recent developments*

The spectroscopic method, based on radiative recombination of two-dimensional electrons with photoexcited holes, has generated considerable interest recently as a route to study the FQHE. Magneto-optical experiments in the FQHE regime are expected to reveal physics of the many-electron state that is not accessible in magneto-transport. While transport measurements study the electronic properties close to the Fermi energy, the method of radiative recombination provides information about the whole energy spectrum of the electrons.

Buhmann et al. (1990) studied the spectra of the radiative recombination of two-dimensional electrons in several GaAs–AlGaAs heterostructures with photoexcited holes localized in a δ-doped monolayer of acceptors (Be atoms) 25 nm away from the interface. They observed anomalies in the luminescence spectrum at $\nu = \frac{2}{3}, \frac{1}{3}, \frac{1}{5}, \frac{1}{7}$, and $\frac{1}{9}$. This was the first observation of the filling factor $\frac{1}{9}$.

Goldberg et al. (1990) also reported an interesting experiment which involved simultaneous measurement of the transport resistivity components and photoluminescence spectra from one-side-doped GaAs/AlGaAs single quantum wells. They observed a sudden shift of the luminescence peak and a minimum in peak intensity at $\nu = \frac{2}{3}$. Precisely at $\nu = \frac{2}{3}$ the peak position was found to shift abruptly to the blue by ~ 0.1 meV. In the high-field region about the FQHE states $\nu = \frac{2}{5}$ and $\frac{1}{3}$, Goldberg et al. observed a splitting of the luminescence. Anomalies in the energy of photoluminescence near $\nu = \frac{2}{3}$ were also observed by Turberfield et al. (1990). A popular account of these studies can be found in the article by Worlock (1990). Some theoretical work on optical recombination in the FQHE regime has appeared in the literature (Bychkov and Rashba 1989, Apal'kov and Rashba 1991, Chakraborty and Pietiläinen 1991).

At very low electron density, i.e., for very small ν, the ground state is expected to be a Wigner crystal. It should be noted that the two-dimensional one-component classical plasma has a crystallization transition which occurs at $m \simeq 70$ (Caillol et al. 1982). Therefore, it would be interesting to estimate the filling factor at which the liquid to solid transition takes place for the present quantum system. Most computations of the crystal energy have been within the HF approximation (Yoshioka and Lee 1983, Yoshioka et al. 1983). Comparing the accurate Monte Carlo results for the Laughlin liquid state with the HF crystal energies as a function of filling factor, the critical filling was found to occur at $\nu_c \sim \frac{1}{10}$ (Levesque et al. 1984), an estimate similar to that of Laughlin (1983a). With a variational wave function for a *correlated* Wigner crystal the crossover point was estimated by Lam and Girvin (1984) to be $\nu_c^{-1} = 6.5 \pm 0.5$.

A clear indication of the FQHE at $\nu = \frac{1}{5}$ has been observed in the experiments of Mallett et al. (1988), who determined the activation energy of this state [$\Delta \sim 50$ mK at 19 T] for the first time (see § 4.2.4 for a discussion on the energy gap). They also reported observation of weak ρ_{xx} minima at $\nu = \frac{2}{9}$, and $\frac{2}{11}$, thereby providing experimental confirmation of the $\frac{1}{5}$ hierarchy, which is different from the sequence of states obtained from $\frac{1}{3}$. Furthermore, the observation of the ρ_{xx} minimum at $\frac{2}{11}$ sets a lower limit to which the FQHE is shown to persist.

The difficulty of performing measurements in the low-ν, low-temperature range was greatly circumvented recently by Goldman et al. (1988). In recent magnetotransport measurements they observed a structure near the filling factor $\nu = \frac{1}{7}$. It was interpreted as evidence for a developing fractional quantum Hall state. The Wigner crystallization is, therefore, expected to occur for $\nu < \frac{1}{7}$. A structure in ρ_{xy}, indicating the FQHE state at $\nu = \frac{1}{7}$ in a magnetic field up to 18 T, was also observed by Wakabayashi et al. (1988). In a recent experiment, Jiang et al. (1990) also found that at filling factors below $\frac{1}{5}$ as well as in a narrow region above $\nu = \frac{1}{5}$, ρ_{xx} diverges exponentially as $T \to 0$. This is interpreted as indicative of a pinned solid-like phase in the underlying electronic system. This electronic phase is reentrant in a narrow region above $\nu \sim \frac{1}{5}$. The liquid state at $\nu = \frac{1}{5}$ is apparently embedded within a solid phase.

A two-dimensional electron solid has also been observed in very high-mobility Si-MOSFETs at large filling factors ($\nu > 1$) and low magnetic fields ($B < 5$ T) (D'Iorio et al. 1990, Kravchenko et al. 1992). Below a carrier concentration of $\approx 10^{11}$ cm^{-2}, the longitudinal resistance maxima near the filling factors 1.5 and 2.5 increases sharply by more than four orders of magnitude. The magnitude of the resistance maxima is thermally activated below 0.5 K and the activation energy at 8×10^{10} cm^{-2} is about 1 K. In the range of concentrations and filling factors where the electron solid is observed, the current–voltage characteristics show a sharp threshold at $\approx$100 mV/cm (D'Iorio et al. 1992).

Spectroscopic measurements in the FQHE regime have been recently employed to probe the Wigner crystallization in the strongly interacting electron system at low filling factors. A qualitative phase diagram $T_c(\nu)$ proposed by Buhmann et al. (1991) is shown in fig. 15 where the incompressible liquid state occurs at as low a density as $\nu = \frac{1}{9}$ but for $\nu < \frac{1}{5}$ all these states are surrounded by Wigner-crystal-like states (shaded regions).

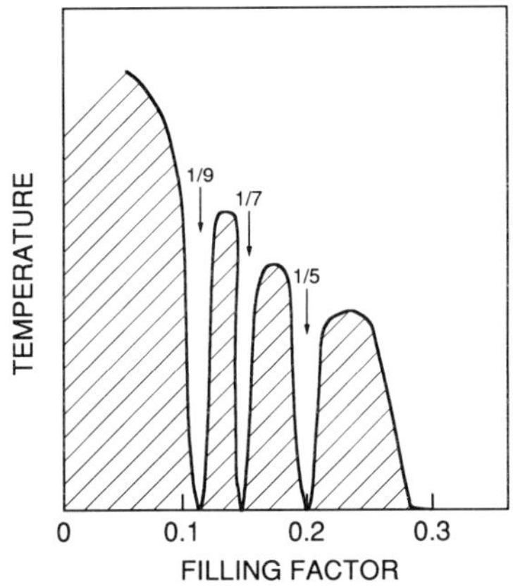

Fig. 15. Qualitative form of the phase diagram $T_c(\nu)$. Regions above the solid curve represent the incompressible liquid state. Shaded regions represent a solid phase.

4.2. Elementary excitations

One important result of Laughlin's theory was the observation that the elementary charged excitations in a stable state $\nu = 1/m$ are quasiparticles and quasiholes with *fractional* electron charge of $e^* = \pm e/m$. If one electron is added to the system, it amounts to adding m elementary excitations and, hence, the discontinuity in slope of the energy curve can be written as

$$\left.\frac{\partial E}{\partial N_e}\right|_{\nu_+} - \left.\frac{\partial E}{\partial N_e}\right|_{\nu_-} = m(\tilde{\varepsilon}_p + \tilde{\varepsilon}_h) = mE_g, \tag{50}$$

where E_g is the energy required to create one quasiparticle (with energy $\tilde{\varepsilon}_p$) and one quasihole (with energy $\tilde{\varepsilon}_h$) well separated from each other. The pinning of the density to $\nu = \frac{1}{3}$ suggests that there will be no low-frequency phonon-type excitations at long wavelengths. When the filling factor ν is slightly shifted from the stable state $1/m$, with m being an odd integer, the ground state of the system is expected to consist of a small density of quasiparticles or quasiholes, with charge $\pm e/m$ and Coulomb interactions. In the presence of impurities, these quasiparticles or quasiholes are expected, for low concentrations, to be trapped in potential fluctuations.

4.2.1. Quasiholes and quasiparticles

For densities slightly different from $\nu = 1/m$, we cannot construct a wave function with exactly m vortices tied to each electron. In order to have an electron density slightly *less* than $1/m$, either we add a few extra vortices not tied to electron positions, or have some electrons with more than m vortices (recall Halperin's interpretation of the Laughlin's wave function discussed above). Laughlin considered the first choice, and the wave function is written (in units where $l_0 = 1$) as

$$\psi_m^{(-)} = \exp\left(-\tfrac{1}{4}\sum_l |z_l|^2\right)\prod_j (z_j - z_0) \prod_{j<k} (z_j - z_k)^m, \tag{51}$$

where $z_0 = x_0 - \mathrm{i}y_0$. This wave function has a simple zero at $z_j = z_0$ for any j, as well as m-fold zeros at each point where $z_j = z_k$, for $k \neq j$. Writing

$$|\psi_m^{(-)}|^2 = \mathrm{e}^{-\mathscr{H}_m^{(-)}}, \tag{52}$$

we obtain

$$\mathscr{H}_m^{(-)} = \mathscr{H}_m + 2\sum_j \ln|z_j - z_0|, \tag{53}$$

which is just the Hamiltonian of a classical one-component plasma in the presence of an extra repulsive *phantom* point charge at point z_0, whose strength is less, by a factor $1/m$, than the charges in the plasma. The plasma will neutralize this phantom by a *deficit* of $1/m$ charge near z_0, while elsewhere in the interior of the system, the charge density will not be changed. However, the real three-dimensional electric charge is carried by the electrons and by the uniform positive background, and *not* by the phantom. As the electron charge density cancels the uniform background, a real net charge $-e/m$ is accumulated in the vicinity of z_0. The wave function (51), therefore, describes a *quasihole* at point z_0.

In the case where the electron density is slightly *higher* than the stable $1/m$ state, the choice of the wave function is not so clear. In this case, one clearly needs a state with one flux quantum (or equivalently, one zero of the wave function) missing. The wave function proposed by Laughlin for the quasiparticle state is written

$$\psi = \prod_{j=1}^{N_e} \left[e^{-|z_j|^2/4l_0^2} \left(2l_0^2 \frac{\partial}{\partial z_j} - z_0^* \right) \right] \prod_{l<k} (z_l - z_k)^m. \tag{54}$$

In this case, the square of the wave function is not directly interpretable as the distribution in a classical statistical-mechanics problem. However, Laughlin (1984a,b) has provided a means to calculate the charge density. Laughlin (1984a,b) (and later Chakraborty 1985) used a two-component HNC scheme to calculate the energy cost associated with the creation of the elementary excitations. The energies evaluated via the Monte Carlo scheme show a large discrepancy with the HNC results (Morf and Halperin 1986). For details, see Chakraborty and Pietiläinen (1988a).

The quasiparticle and quasihole size is the distance over which the one-component plasma screens. For a weakly coupled plasma this distance is the Debye length (Hansen and Levesque 1981), $\lambda_D = l_0/\sqrt{2}$. For the strongly coupled plasma relevant in the present case, a better estimate is the ion-disk radius associated with *charge* $1/m$, $R = \sqrt{2}\, l_0$. In this sense, the quasiparticles have the same *size* as electrons in the lowest Landau level.

4.2.2. Quasiparticle statistics

The fractional charge of the quasiparticles and quasiholes was derived by Laughlin from the plasma analogy discussed above. Later, Arovas et al. (1984) presented a direct method to determine the charge of excitations of the Laughlin state. Interestingly, that approach also revealed some surprising features about the *statistics* of the quasiparticles and quasiholes.

If the quasihole of charge e^* is carried around a loop enclosing the flux $\boldsymbol{\Phi}$ it will gain a change in phase of

$$\gamma = \frac{e^*}{\hbar c} \oint \boldsymbol{A} \cdot \mathrm{d}\boldsymbol{l} = 2\pi \frac{e^*}{e} \frac{\boldsymbol{\Phi}}{\boldsymbol{\Phi}_0}. \tag{55}$$

On the other hand, if we let the quasihole position z_0 move adiabatically around this loop, the rate of this change can be calculated from

$$\frac{\mathrm{d}\gamma(t)}{\mathrm{d}t} = \mathrm{i} \left\langle \psi_m^{(-)} \middle| \frac{\mathrm{d}\psi_m^{(-)}}{\mathrm{d}t} \right\rangle.$$

When we now substitute the time derivative

$$\frac{\mathrm{d}\psi_m^{(-)}}{\mathrm{d}t} = \sum_i \frac{\mathrm{d}}{\mathrm{d}t} \ln[z_i - z_0(t)] \psi_m^{(-)} \tag{56}$$

from eq. (51), we get

$$\frac{d\gamma}{dt} = i\left\langle \psi_m^{(-)} \left| \frac{d}{dt} \sum_i \ln(z_i - z_0) \right| \psi_m^{(-)} \right\rangle. \tag{57}$$

Introducing the one-particle density

$$\rho^{(-)}(z) = \left\langle \psi_m^{(-)} \left| \sum_i \delta(z_i - z) \right| \psi_m^{(-)} \right\rangle \tag{58}$$

of the quasihole state, this can be written as

$$\frac{d\gamma}{dt} = i \int dx\, dy\, \rho^{(-)}(z) \frac{d}{dt} \ln[z - z_0(t)], \tag{59}$$

where $z = x + iy$. We now expand $\rho^{(-)}(z)$ around the uniform density $\rho_0 = \nu\Phi/\Phi_0$ of the ground state, i.e., we write $\rho^{(-)}(z) = \rho_0 + \delta\rho^{(-)}(z)$. Let us first consider the ρ_0 term. When we integrate z_0 around a circle of radius R the total variation of $\ln[z - z_0(t)]$ will be $2\pi i$ if $|z| < R$ and zero if $|z| > R$. Thus, the corresponding change of phase is given by

$$\begin{aligned} \gamma_0 &= i \int_{|r|<R} dx\, dy\, \rho_0 2\pi i \\ &= -2\pi\langle n \rangle_R = -2\pi\nu \frac{\Phi}{\Phi_0}, \end{aligned} \tag{60}$$

where $\langle n \rangle_R$ is the mean number of electrons in a disk of radius R. The correction $\delta\rho^{(-)}(z)$ is localized in the vicinity of z_0, so its contribution to the change of phase is of the order of size of the quasihole, i.e., of the order of l_0^2. In comparison with the constant term contribution, which is of the order R^2, this can be clearly neglected. From eqs. (55) and (60) we can now deduce the charge of the quasihole to be $e^* = -\nu e$, in agreement with Laughlin's theory. Similar arguments can also be made for the quasiparticles.

Let us now consider two quasiholes located at z_a and z_b, a distance $|z_a - z_b| = R$ apart. When we carry the hole z_b adiabatically around the circle of radius R centered at the hole z_a the analysis presented above is still valid provided the mean electron number $\langle n \rangle_R$ in the disk is counted correctly. We find that the quasihole at z_a has a charge $-\nu e$ which we can interpret to mean that exactly ν electrons are removed from the disk, so we must substitute $\langle n \rangle_R - \nu$ for the mean electron number. Hence, an extra phase of amount $\Delta\gamma = 2\pi\nu$ is gained in the process. The interchange of two quasiparticles can be achieved by letting each of them make a turn of π around the other. From the treatment above it is obvious that the accumulation of this extra phase is continuous and, therefore, the total change of phase will be $\nu\pi$. For $\nu = 1$, the quasiholes behave like fermions. For fractional fillings, however, we end up with the striking conclusion that the quasiholes (and quasiparticles) obey neither Bose nor Fermi statistics, but *fractional* statistics. The concept of anomalous statistics in two dimensions was first introduced by Leinaas and Myrheim (1977). Indistinguishable quantum particles in two-dimensional space can, in general, have anomalous statistics. This is in contrast to three-dimensional space (and spaces of higher dimensionality), where there are only Bose and Fermi statistics. According to Wilczek (1982), particles obeying anomalous statistics are called *anyons* because the interchange of any two of them can result in *any* phase.

4.2.3. *Hierarchies of quasiparticles*

Thus far we have discussed the theoretical work on the FQHE aiming at an explanation of the behavior of the electron system at $\nu = 1/m$ (and its electron–hole symmetric fractions). But as shown in fig. 13, the FQHE appears at several higher order filling fractions, notably, $\frac{2}{5}, \frac{3}{7}, \frac{4}{9}$, etc. In order to explain the FQHE at these fractions, Laughlin's theory needs to be modified.

According to Haldane (1983), there is a hierarchical system in which p/q states with $1 < p < q$ are formed from new generation of elementary excitations in the same manner as the Laughlin state is formed by the electrons. Each new generation of elementary excitations appears against a *vacuum* formed by the preceding generation. In this picture, the elementary excitations are supposed to obey Bose statistics, unlike the electrons which obey Fermi statistics. The chain of equations are

$$n_s = mN_L + \alpha_1 N^{(1)}, \tag{61a}$$

$$N_L = p_1 N^{(1)} + \alpha_2 N^{(2)}, \tag{61b}$$

$$N^{(1)} = p_2 N^{(2)} + \alpha_3 N^{(3)}, \tag{61c}$$

$$\cdots \; \cdots \; \cdots \; \cdots$$

$$N^{(k-1)} = p_k N^{(k)}; \tag{61d}$$

where $m \geqslant 1$ is odd if all $p_j \neq 0$ are even and $\alpha_j = \pm 1$ and $n_s = 1/2\pi l_0^2$ is the degeneracy per unit area of the lowest Landau level. The above equations are understood as follows: in the first equation the imbalance between the electron density N_L and the density of the incompressible state n_s/m is compensated by the $N^{(1)}$ particles ($\alpha_1 = -1$) or holes ($\alpha_1 = +1$). The next equation is the same except that it describes the next generation with m replaced by an even quantity p_1. The solution of eqs. (61) gives the continued fraction for $\nu = N_L/n_s$:

$$\nu = \cfrac{1}{m + \cfrac{\alpha_1}{p_1 - \cfrac{\alpha_2}{p_2 - \cfrac{\alpha_3}{p_3 - \cdots}}}} \quad (\nu < 1). \tag{62}$$

A new liquid $[m, \alpha_1 p_1, \ldots, \alpha_j p_j]$ does not form before the appearance of the preceding liquid $[m, \alpha_1 p_1, \ldots, \alpha_{j-1} p_{j-1}]$.

The hierarchical scheme of Halperin (1984) is very much in the spirit of Laughlin's theory. In this scheme, the quasiparticles are required to obey *fractional statistics*. If ν_t is a stable filling factor obtained at level t of the hierarchy, the low-lying energy states for filling factors near to ν_t can be described by the addition of a small density of quasiparticle excitations to the ground state at ν_t. There are two types of elementary excitations: p-excitations (particle like) and h-excitations (hole like), with charges $q_t e$ and $-q_t e$, respectively. Halperin then constructed a *pseudo*-wave function (where the coordinates are for quasiparticles),

$$\psi(Z_1, \ldots, Z_{N_t}) = P[Z_j]\, Q[Z_j] \exp\left(-\frac{|q_t|}{4l_0^2} \sum_{j=1}^{N_t} |Z_j|^2 \right), \tag{63}$$

where Z_j is the complex coordinate of the quasiparticle j and N_t is the number of quasiparticles. The polynomial $P[Z_j]$ is chosen to be *symmetric*,

$$P[Z_j] = \prod_{i<j} (Z_i - Z_j)^{2p_{t+1}}, \tag{64}$$

where p_{t+1} is a positive integer, whose variational properties are known from Laughlin's theory. The function $Q[Z_j]$ determines the symmetry properties of ψ under the interchange of quasiparticles. As mentioned above, in contrast to three-dimensional space (and spaces of higher dimensionality), indistinguishable particles in the two-dimensional space can, in general, have anomalous statistics. The bound system of a spinless particle of charge q and an infinitesimally thin solenoid carrying a flux $\boldsymbol{\Phi}$ and cutting through the plane of the motion of the particle [the system was called an *anyon* by Wilczek (1982)] is such an example. The many-body wave function of N_t identical anyons is (Wu 1984a,b):

$$\psi^*(\boldsymbol{r}_1, \ldots, \boldsymbol{r}_{N_t}) = \prod_{i<j} \exp\left(\mathrm{i}\frac{\xi}{\pi} \varphi_{ij} \right) \psi(\boldsymbol{r}_1, \ldots, \boldsymbol{r}_{N_t}). \tag{65}$$

In eq. (65), φ_{ij} is the azimuthal angle of the relative vector $(\boldsymbol{r}_i - \boldsymbol{r}_j)$ and $\psi(\boldsymbol{r}_1, \ldots, \boldsymbol{r}_{N_t})$ is a single-valued wave function. According to eq. (65), a permutation of any two anyons $(\Delta\varphi_{ij} = \pi)$ multiplies ψ^* by the phase factor $e^{i\xi}$. Therefore, when $\eta = \xi/2\pi = q\boldsymbol{\Phi}/2\boldsymbol{\Phi}_0$ is an integer, the anyons are bosons and when η is a half-integer, they are fermions. Here $\boldsymbol{\Phi}$ is the usual magnetic flux.

In terms of the complex coordinates Z_j and Z_j^*, the multivalued wave function (65) takes the form

$$\psi^* = \prod_{i<j} (Z_i - Z_j)^{\xi/\pi} f(Z_i, Z_j^*), \tag{66}$$

where $f(Z_i, Z_j^*) = (r_{ij})^{-\xi/\pi}\, \psi(Z_j, Z_j^*)$ is a single-valued function of the coordinates.

Let us now consider the first step of the hierarchy when the anyon quasiparticles are derived from electrons. The phase change $\Delta\gamma$ of the wave function of a quasiparticle in a closed adiabatic path around another quasiparticle is $\Delta\gamma = 2\pi q$ and for $\nu = 1/m$, $|\Delta\gamma| = 2\pi/m$. Therefore, the change of phase that accompanies the interchange of quasiparticles is $\frac{1}{2}\Delta\gamma = \pi\nu = \xi$ (see § 4.2.2). According to the general structure of eq. (66), the factor $Q[Z_j]$ is now

$$Q[Z_j] = \prod_{i<j} (Z_i - Z_j)^{-\alpha/m_t}. \tag{67}$$

In eq. (67), $\alpha = \pm 1$, according to whether one is dealing with particle- or hole-type excitations, and m_t is a rational number $\geqslant 1$.

Assuming that at any stage of the hierarchy, the quasiparticles or quasiholes can be treated as point particles with pairwise Coulomb interactions, an estimate for the potential energy can be obtained. The resulting energy versus density curve was generated by Halperin for various filling fractions as shown in fig. 16. In this figure, the stable fractions appear with downward pointing cusps.

Finally, the hierarchical scheme described above has been employed by Zhang and Chakraborty (1986) to study the condensation of the spin-reversed quasiparticles.

4.2.4. Energy gap

We have seen in § 4.1.1 that the ground-state energy per electron in a finite-size calculation shows a cusp-like behavior at $\nu = \frac{1}{3}$. As has already been discussed above, the appearance of a cusp means a positive discontinuity in the chemical potential (Halperin 1986) [see eq. (50)]:

$$E_g = \tfrac{1}{3}(\mu_+ - \mu_-), \tag{68}$$

where the chemical potential is defined as

$$\mu = E_0(\nu) + \nu \frac{dE_0}{d\nu}, \tag{69}$$

and E_0 is the ground-state energy per particle. In eq. (68), $\mu_\pm$ corresponds to $\nu_\pm = N_e/(N_s \mp 1)$. The factor $\frac{1}{3}$ corresponds to the fractional charge of the quasiparticles discussed above.

The finite-size system result for the energy gap by Chakraborty et al. (1986) and Chakraborty and Pietiläinen 1986) is shown in fig. 17. Extrapolation of the results for spin-polarized three- to seven-electron systems (plotted as solid circles) leads to (approximately) $E_g \simeq 0.1e^2/\varepsilon l_0$. These studies also indicate that in the absence of the Zeeman energy, spin-reversed excitations cost even less energy (fig. 17) than the pure Coulomb excitations proposed by Laughlin (1983a), and will be discussed below.

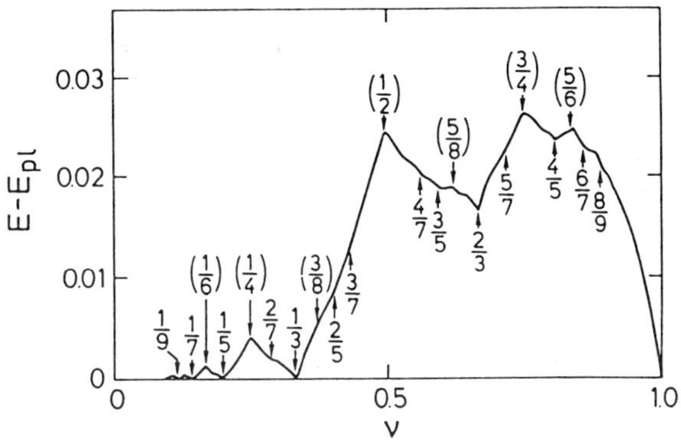

Fig. 16. Potential energy $E(\nu)$ per quantum of magnetic flux (in units of $e^2/\varepsilon l_0$) versus the filling factor ν of the lowest Landau level. The plasma energy E_{pl} has been subtracted at each filling fraction.

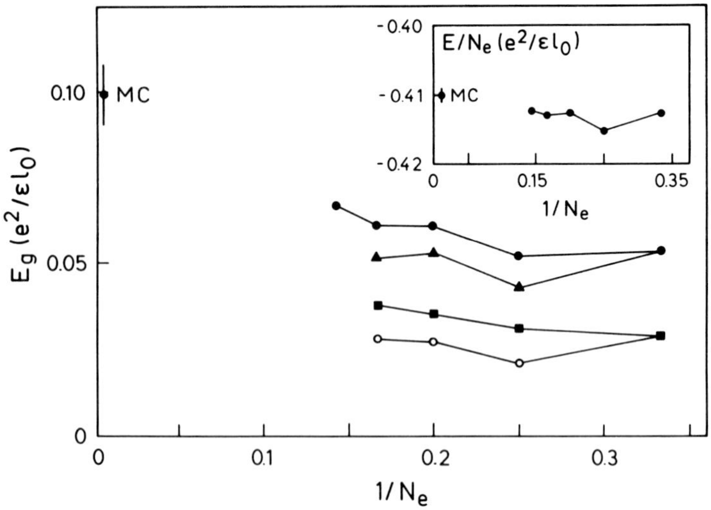

Fig. 17. The energy gap for different spin polarization of the quasiparticle (q.p.) and quasihole (q.h.) excitations for three- to seven-electron systems, ignoring the Zeeman energy. In the case of spin-aligned q.p. + q.h. gap (●), Monte Carlo (MC) results are from Morf and Halperin (1986). The other three cases are: spin-aligned q.p. + spin-reversed q.h. (▲); spin-reversed q.p. + spin-aligned q.h. (■); and spin-reversed q.p. + q.h. (○). The ground-state energies (Yoshioka et al. 1983) at $\nu = \frac{1}{3}$ are given in the inset for comparison.

In both the integer and fractional QHE, the vanishing of the diagonal resistivity implies a gap in the excitation spectrum. In the case of the IQHE, the gap is in the single-particle density of states, whereas in the FQHE, the gap lies in the excitation spectrum of the correlated many-electron ground state. The energy gap is usually obtained from the temperature dependence of the magnetoconductivity, σ_{xx} [or ρ_{xx} since near the ρ_{xx} minima, $\rho_{xx} \ll \rho_{xy}$, and $\sigma_{xx} = \rho_{xx}/(\rho_{xx}^2 + \rho_{xy}^2) \sim \rho_{xx}/\rho_{xy}^2$], as $\sigma_{xx} \propto \rho_{xx} \propto \exp(-W/k_B T)$, where $W = \frac{1}{2}E_g$ is the activation energy and k_B is the Boltzmann constant. In the case of the IQHE, Tausendfreund and von Klitzing (1984) found the energy gap obtained from activation energy measurements to be close to the cyclotron energy. For the FQHE, similar measurements for the energy gap have been undertaken by several experimental groups.

A systematic study of the energy gap for the filling fractions, $\nu = \frac{1}{3}, \frac{2}{3}, \frac{4}{3}$, and $\frac{5}{3}$ was reported by Boebinger et al. (1985, 1987). The activation energies for these filling factors are presented in fig. 18. The following features are noteworthy in these results:

(i) No apparent sample dependence was observed.

(ii) The data for $\nu = \frac{1}{3}$ (solid symbols with error bars) and $\frac{2}{3}$ (open symbols) overlap at $B \sim 20$ T. The data for $\nu = \frac{4}{3}$ and $\frac{5}{3}$ (■ at 5.9 and 7.4 T) are consistent with the data for $\nu = \frac{2}{3}$ at similar magnetic fields. Collectively, the experimental results therefore suggest a *single* activation energy (or *gap* E_g), for all the filling fractions mentioned above.

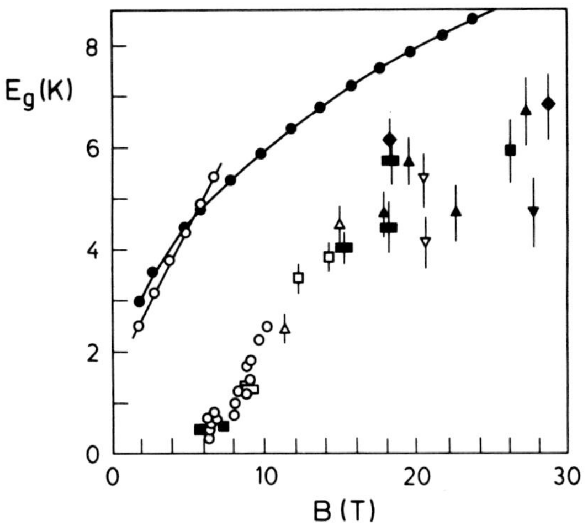

Fig. 18. Activation energies for the $p/3$ fractions (in units of K) as a function of magnetic field. Also shown are the theoretical results (solid lines) for the energy E_g (K) versus magnetic field B (Tesla) for a five-electron system. The open points on the theoretical curve (Chakraborty et al. 1986, Chakraborty and Pietiläinen 1986) are for spin-reversed quasiparticles and spin-polarized quasiholes, while the filled points are for the fully spin-polarized case.

(iii) E_g does not follow the expected $B^{1/2}$ dependence. For $B \lesssim 5.5$ T, E_g is vanishingly small, but at higher magnetic fields, there is a roughly *linear* increase in E_g up to $\sim$12 T.

In our discussions of the electron system thus far, we have ignored the finite spread of the electron wave function perpendicular to the two-dimensional plane. In the real systems, the wave function of an electron has a finite spread in the z-direction. It is well known that inclusion of the finite-thickness correction effectively softens the short-range divergence of the bare Coulomb interaction, when the interelectron spacing is comparable with the inversion layer width (MacDonald and Aers 1984, Zhang and Das Sarma 1986). This causes a reduction in the energy gap calculated for the ideal system. At low magnetic fields, mixing of the higher Landau levels also influences the energy gap somewhat (Yoshioka 1984b). The activation energy for $p/3$ states has also been measured by Willett et al. (1988) in an ultrahigh-mobility ($\mu = 5 \times 10^6\ \mathrm{cm^2\,V^{-1}\,s^{-1}}$) quasi-two-dimensional electron system in GaAs–(AlGa)As, for which the finite thickness of the carrier system is accurately known. The data are in remarkably good agreement with the theoretical results for the energy gap when Landau-level mixing and the finite-thickness effect are taken into account.

Finally, from the measurement of the activation energy, Clark et al. (1988) made a very important observation. They reported a systematic study of $\sigma_{xx}^{c} = \sigma_{xx}(1/T = 0)$:

$$\sigma_{xx}^{c} = \frac{\rho_{xx}^{c}}{(\rho_{xx}^{c})^2 + \rho_{xy}^2} = \frac{\rho_{xx}^{c}}{(\rho_{xx}^{c})^2 + \left[\left(\dfrac{q}{p}\right)\dfrac{h}{e^2}\right]^2}$$

which is valid at $\nu = p/q$ and ρ^c_{xx} is defined by $\rho_{xx} = \rho^c_{xx} e^{-E_g/kT}$. Clark et al. found that, within experimental errors, σ^c_{xx} is constant for p/q fractions of the same q. In fig. 19, the conductivity data σ_{xx} are plotted as a function of T^{-1} at both integer and fractional ν. Straight-line fits to the linear region, which corresponds to thermal activation of the quasiparticles and quasiholes across the energy gap, are shown and extrapolated to $T^{-1} = 0$. The extrapolated σ_{xx} intercept is close to e^2/h for the IQHE, whereas for $\nu = p/q$, $\sigma^c_{xx} \simeq (1/h)(e/q)^2 = e^{*2}/h$, which is consistent with Laughlin's prediction of the fractional charge of the quasiparticles and quasiholes discussed above.

4.2.5. Spin-reversed quasiparticles

In most of the theoretical work a major assumption was that, because of the strong magnetic fields involved, only one spin state is present. Halperin (1983) first pointed out that in GaAs, the electron *g*-factor is one-quarter of the free electron value. Therefore, the Zeeman energy is approximately sixty times smaller than the cyclotron energy. It might then be possible to have some electrons with *reversed* spins when the magnetic field is not too large. In the case when one half of the electrons have spins antiparallel to the field, Halperin constructed a simple Laughlin-type state of the form

$$\psi = \prod_{i<j} (z_i - z_j)^m \prod_{\alpha<\beta} (z_\alpha - z_\beta)^m \prod_{i,\alpha} (z_i - z_\alpha)^n \prod_i e^{-|z_i|^2/4l_0^2} \prod_\alpha e^{-|z_\alpha|^2/4l_0^2}, \tag{70}$$

where Roman and Greek indices correspond to spin-up and spin-down electrons,

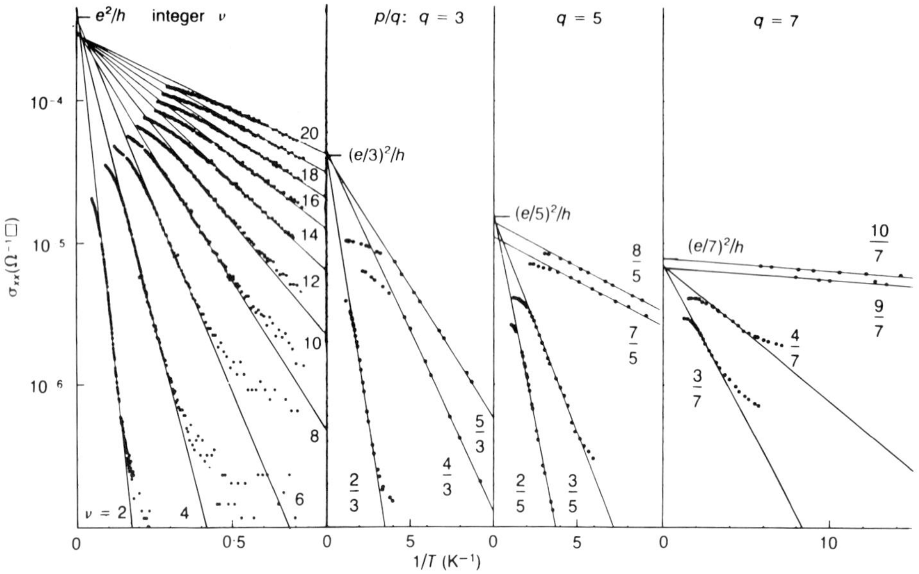

Fig. 19. Conductivity σ_{xx} versus T^{-1} results which measure the quasiparticle charge $e^* = e/q$ at the integer and fractional filling factors $\nu = p/q$.

respectively. Using the classical plasma approach, the filling factor is given by $\nu = 2/(m+n)$. Halperin suggested a study of the particular case of $m=3$, $n=2$ ($\nu = \frac{2}{5}$). The wave function then has the following desirable properties:

(a) all electrons are in the lowest Landau level, so the kinetic energy is an absolute minimum,

(b) electrons of the same spin are kept apart very well because the wave function vanishes as the cube of their separation; electrons of opposite spin are kept apart only slightly less well because the wave function vanishes as the square of the separation, and

(c) the wave function is antisymmetric under interchange of two electrons of the same spin, as required.

The wave function can also be shown to give a total spin $S=0$ state.

For $\nu = \frac{2}{5}$, the wave function given above was studied by Chakraborty and Zhang (1984) to examine the importance of unpolarized spin and it was concluded that, at a very low magnetic field, a spin-unpolarized state for $\nu = \frac{2}{5}$ would be energetically favored. In a later investigation, a systematic study of spin reversal in various filling fractions was attempted by these authors (Zhang and Chakraborty 1984). The finite-size calculations described in § 4.1.1, were employed for this purpose and were suitably generalized for various spin polarizations in a straightforward manner. The energy spectrum for the Hamiltonian is classified in terms of the total spin S and its z-component S_z. For a given S, the spectrum is identical for different values of S_z.

In table 1, we present the results for a finite electron system for the case of a polarized state ($S=2$), a partly polarized state ($S=1$) and the unpolarized state ($S=0$). As seen in table 1, except for $\nu = \frac{1}{3}$, the unpolarized states are energetically favored, as compared to the fully polarized state. For $\nu = \frac{1}{3}$, the spin state $S=2$ is found to be energetically favored compared with the other spin states, even in the absence of Zeeman energy. This result is quite supportive of Laughlin's state at $\nu = \frac{1}{3}$, which is fully antisymmetric. For other filling fractions, the possibility exists that the energy could be lowered by introducing the spin degree of freedom.

Table 1
Potential energy (per particle) for various filling fractions and spin polarization. The Zeeman energy is not included in the energy values which are in units of $e^2/\varepsilon l_0$.

ν	Potential energy				Ground state
	$S=0$	$S=1$	$S=2$	$S=3$	
$\frac{1}{3}$	−0.4135	−0.4120	−0.4152	–	Polarized
$\frac{2}{3}$	−0.5331	−0.5291	−0.5257	−0.5232	Unpolarized
$\frac{2}{5}$	−0.4464	−0.4410	−0.4403	–	Unpolarized
$\frac{2}{7}$	−0.3884	−0.3868	−0.3870	–	Unpolarized
$\frac{3}{5}$	−0.5074	−0.5096	−0.5044	−0.50104	Partially polarized
$\frac{4}{9}$	−0.4554	−0.4600	−0.4528	–	Partially polarized
$\frac{4}{11}$	−0.4241	−0.4278	−0.4219	–	Partially polarized
$\frac{4}{13}$	−0.3970	−0.3997	−0.3975	–	Partially polarized

In the preceding section, we discussed the results for the energy gap E_g from finite-size calculations in a periodic rectangular geometry and noticed that spin-reversal for the quasiparticles cost less energy (in the absence of the Zeeman energy) compared to fully spin-polarized quasiparticles and quasiholes. In the Laughlin approach, a spin-reversed quasiparticle is best visualized in the following wave function for the quasiparticles (Morf and Halperin 1986, Chakraborty 1986):

$$\psi = \prod_{j=2}^{N_e} (z_j - z_1)^{-1} \psi_m, \tag{71}$$

where ψ_m is the ground state given by eq. (45). Bringing in the plasma analogy one can show that $|\psi|^2$ is the distribution function for a two-dimensional plasma in which particle 1 has its charge *reduced* by a factor $1-(1/m)$ in its repulsive interaction with the other particles. Particle 1 has, however, the same interaction as the other particles in its attractive interaction with the background. Particle 1 will, therefore, be attracted to the center of the disk, while a two-dimensional *bubble* will be formed near the origin of size $1-(1/m)$. As a result, there will be an extra negative charge e/m near the origin. Furthermore, when eq. (71) is considered to be a function of the position of any electron other than the singled-out electron 1, there will be one less zero of the wave function than in the case of the ground-state ψ_m.

The magnetic field dependence of E_g for a five-electron system is shown in fig. 17, where we have presented only the lowest energy results. For low magnetic fields, the curve for the lowest energy excitations rises *linearly* as a result of the *spin-reversed quasiparticles*, which include the dominant contribution from the Zeeman energy, itself linear in magnetic field. As the magnetic field is increased, a crossover point is reached, beyond which the $B^{1/2}$ dependence (somewhat modified due to the finite-thickness correction) is then obtained due to the spin-polarized quasiparticles and quasiholes (Chakraborty et al. 1986, Chakraborty and Pietiläinen 1986). These theoretical predictions were made several years ago, the experimental evidence for the presence of spin-reversed quasiparticles has been reported only recently and will be discussed below.

The first clear-cut evidence for the existence of spin-unpolarized states in the ground state of several filling factors was from the experiments by the Oxford group (Clark et al. 1989, Clark and Maksym 1989) where the magnetic field was tilted from the direction perpendicular to the electron plane. It is known that the exchange and correlation contributions to the ground-state energy depend to a first approximation on $B_\perp$ alone. The Zeeman energy, on the other hand, depends on the total magnetic field B_{tot} (Haug et al. 1987, Chakraborty and Pietiläinen 1989, Halonen et al. 1990). They found that with increasing tilt angle, dramatic changes occur in the ρ_{xx} minima of various filling factors. In fig. 20 we present some of the experimental results of ρ_{xx} versus θ by Clark et al. Clearly, the $\frac{4}{3}$ state is first destroyed, followed by a reemergence as θ and hence the magnetic field is increased. The same effect was also observed for $\nu=\frac{2}{3}$. In contrast, the ρ_{xx} minima for $\frac{5}{3}$ and $\frac{1}{3}$ remain essentially unaltered with increasing tilt angle.

An important clue to understand these experimental results is the fact that, allowing for the spin degrees of freedom, the electron–hole symmetry (Zhang and Chakraborty

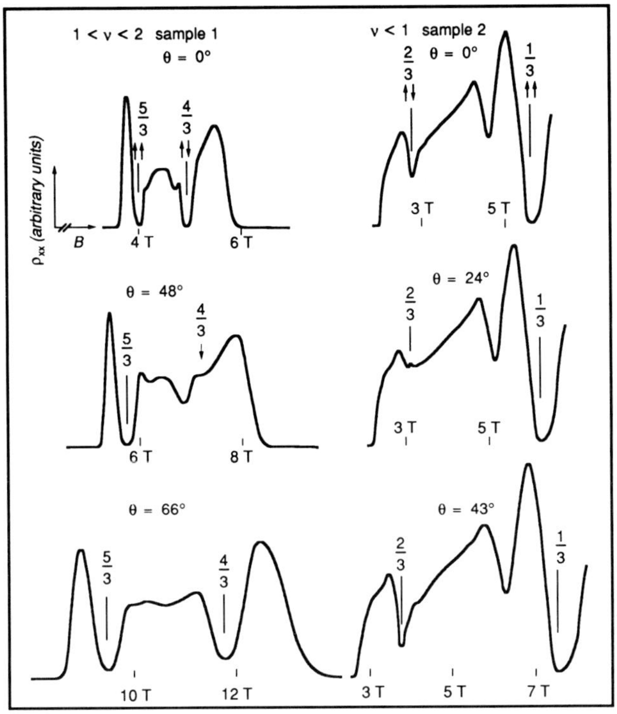

Fig. 20. Diagonal resistivity ρ_{xx} for various filling factors versus the tilted magnetic field.

1986) is not between ν and $1-\nu$, as described earlier, but between ν and $2-\nu$. Therefore, the $\frac{1}{3}$ and $\frac{5}{3}$ filling factors which are the spin-polarized states even at low magnetic fields, as predicted theoretically, remain unaffected by the tilted field. For the $\frac{2}{3}$ and $\frac{4}{3}$ states, the increasing magnetic field destroys the reversed-spin states (Maksym 1989) and, eventually, they reemerge as fully spin-polarized states.

Recently, the Cambridge group (Davies et al. 1991) reported very interesting FQHE experiments in high-quality p-type heterojunctions with tilted magnetic field. They found the same magnetic-field-dependent behavior of $\nu=\frac{4}{3}$ as observed in n-type heterojunctions by Clark et al. (1989). They also noted that for the two-dimensional hole system one requires a smaller magnetic field to destroy and return the $\frac{4}{3}$ state which suggests that the Zeeman splitting is larger for that system.

The other important tilted-field experiment indicative of spin-reversed states was by Eisenstein et al. (1989) who discovered a transition between two distinct FQHE states at the same filling factor $\nu=\frac{8}{5}$ (i.e., $2-\frac{2}{5}$). The transition is driven by tilting the magnetic field and as discussed below, the data quite consistently indicate a change from a spin-unpolarized fluid to a polarized fluid.

Eisenstein et al. then measured the activation energy W versus the tilt angle for $\nu=\frac{8}{5}$. Figure 21 depicts the angular dependence of the observed activation energy, which is the most interesting result as far as the spin-reversed quasiparticles are concerned. The results presented in fig. 21 are all obtained at a fixed filling factor

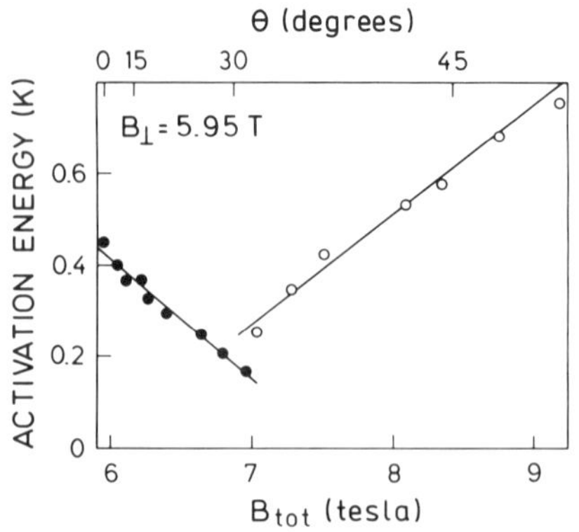

Fig. 21. Activation energy of the FQHE state at $\nu = \frac{8}{5}$ versus B_{tot}. Solid and open symbols refer to low-field and high-field components, respectively.

$\nu = \frac{8}{5}$ and hence a fixed perpendicular magnetic field $B_\perp \sim 5.95$ T. They are plotted against *total* magnetic field, $B_{\text{tot}} = B_\perp/\cos\theta$. As θ increases from zero, W *decreases* linearly. Beyond about 30°, W begins to rise again, eventually exceeding its value at $\theta = 0°$. For small angle (i.e., low magnetic field) the ground state is expected to be analogous to the two-spin $\frac{2}{5}$ state (see § 3.3). From the slope $\mathrm{d}W/\mathrm{d}B_{\text{tot}}$ at small angles in fig. 21, it was found that the g-factor is $g \sim 0.4$ in remarkable agreement with recent spin-resonance measurements on two-dimensional electrons in GaAs (Dobers et al. 1988).

At low magnetic fields, the ground state of $\nu = \frac{8}{5}$ is expected to be spin-unpolarized and presumably the low-lying excitations involve spin-reversed quasiparticles. This could explain the linear decrease of the activation energy as the tilt angle (or the magnetic field) is increased. As the Zeeman energy at $\nu = \frac{8}{5}$ is increased by tilting the magnetic field, the polarized state is energetically favored and eventually becomes the new ground state. The low-lying excitations might still involve spin-reversed quasiparticles. That could explain the linear increase of the activation energy (Chakraborty et al. 1986, Chakraborty and Pietiläinen 1986) discussed in the preceding section. Further increase of the magnetic field would lead to a fully spin-polarized ground state with Laughlin-type quasiparticle–quasihole excitations and we would expect the usual $B^{1/2}$ behavior of the activation energy (Coulomb gap) seen in fig. 18. There has been much theoretical and experimental work recently on the spin-reversed ground state and excitations at various filling fractions (Buckthought et al. 1991, Chakraborty and Pietiläinen 1990, Chakraborty 1990, Maksym 1989, Clark et al. 1990, Eisenstein et al. 1990, Furneaux et al. 1989, Sachrajda et al. 1990, Engel et al. 1992).

4.2.6. *Quantization condition*

The Kubo formula approach for the quantization condition of the IQHE has been discussed in § 3.2.3. There we discussed the topological approach by Niu et al. (1985). In an important paper, Tao and Haldane (1986) presented arguments on the integral and fractional quantization which are quite general and related to the above approaches. The arguments are, however, rather technical and cannot be reproduced here in their entirety. Therefore, we will mention briefly the essential results and refer the reader to the original article. For the electron system in a periodic rectangular geometry, Tao and Haldane (1986) and Haldane (1985) showed that the many-body Hamiltonian can be separated into two terms due to the center of mass motion and the internal relative motion. The center of mass moves as a charged particle in a periodic potential. If there is an energy gap at a filling factor, it is due to the relative motion. The Hall conductance is associated with the quantum motion of the center of mass and, therefore, the analysis of the quantization condition can be related to the topological approach discussed in § 3.2.3. Tao and Haldane (1986) also noticed that, in the presence of a weak concentration of impurities, the quantization condition remains unchanged. The Hall conductivity is still expressed in the topologically invariant form.

4.3. Other developments

4.3.1. Collective modes

In addition to the elementary charged excitations discussed in § 4.2.1, the FQHE system also supports collective excitations whose dispersion has been studied by various theoretical methods. In a spherical geometry, the energy-level spectrum for a finite-electron system at $\nu = \frac{1}{3}$ in the lowest Landau level was calculated by Haldane and Rezayi (1985) and Fano et al. (1986). In the case of a periodic rectangular geometry, the major contribution to the collective excitation spectrum was also calculated by Haldane (1985). He found that, at rational values of ν, the states could be characterized by a two-dimensional wave vector $\boldsymbol{k}$, and, hence, the collective excitation spectrum could be calculated in the periodic rectangular geometry.

The finite-size studies described in § 4.1.1 were carried out in an infinite lattice with a periodic rectangular geometry. The electrons in one cell of the lattice have identical mirror images in all other cells. This infinite repetition will introduce a symmetry which can be employed to classify the eigenstates of the Hamiltonian. This classification has the further advantage that the size of the matrix to be numerically diagonalized is reduced.

In the absence of a magnetic field, the symmetry analysis would be simple: we would have a translational group in the periodic lattice and the eigenstates could be labeled by the wave vectors in the inverse lattice. The physical interpretation of these quantum numbers would, of course, be the momentum. The presence of the magnetic field, however, slightly complicates the classification scheme. We could apply the well-known apparatus of group theory which would lead us to the so-called *ray representation* of the magnetic translation group (Brown 1968, Maksym 1985). The reason why we do not obtain an ordinary representation is that the symmetry operations of the group obey a noncommutative algebra where the product of operators is an operator of the same group only to within a phase factor. For every lattice vector $\boldsymbol{L}_{mn}$, there is a relative translation operator which commutes with the

Hamiltonian. The eigenvalues of this operator are $e^{2\pi i(ms+nt)/N}$, where N is the highest common divisor of N_e and N_s. The quantum numbers s and t ($s, t = 0, 1, \ldots, N-1$) are related to the physical momentum by

$$kl_0 = \left(\frac{2\pi}{N_s\lambda}\right)^{1/2} [s - s_0, \lambda(t - t_0)],$$

where the point (s_0, t_0), corresponding to the state $\boldsymbol{k} = 0$, is required to be the most symmetric point of the reciprocal lattice, and λ is the aspect ratio.

In fig. 22 we present the dispersion for the density wave mode obtained for four- to seven-electron systems in the lowest Landau level in a periodic rectangular geometry. Only the three lowest excitation energies are shown. The spectrum is, in fact, a function of the magnitude of the two-dimensional vector $\boldsymbol{k}$. The ground state is obtained at $\boldsymbol{k} = 0$, as expected. The lowest energy excitations are separated from the ground state by a large gap, which reflects the incompressible nature of the system and they clearly show a collective behavior with a minimum at finite kl_0. For small kl_0, the modes are not very well defined, as they are close to the continuum of the higher energy states. The numerical calculations were done by Yoshioka (1986a), but qualitatively similar results were also obtained earlier by Haldane (1985) for a six-electron system in a square geometry. Yoshioka (1986b) and Chakraborty and Pietiläinen (1988a) also calculated the excitation spectrum for spin-reversed systems.

The finite-size calculations discussed above provide quite accurate information for the collective excitations and the energy gap in the FQHE. However, not much physical insight is gained from these numerical calculations. The theoretical work by Girvin et al. (1985, 1986), drawing analogies from Feynman's well-known theory of liquid ^{4}He, is very helpful in this respect.

Given the exact ground-state ψ, the density-wave excited state at wave vector $\boldsymbol{k}$ is written as (Feynman 1972)

$$\varphi_{\boldsymbol{k}} = N^{-1/2} \rho_{\boldsymbol{k}} \psi, \tag{72}$$

with the density operator $\rho_{\boldsymbol{k}} \equiv \Sigma_{j=1}^{N} e^{-i\boldsymbol{k}\cdot\boldsymbol{r}_j}$, where N is the number of particles. Let

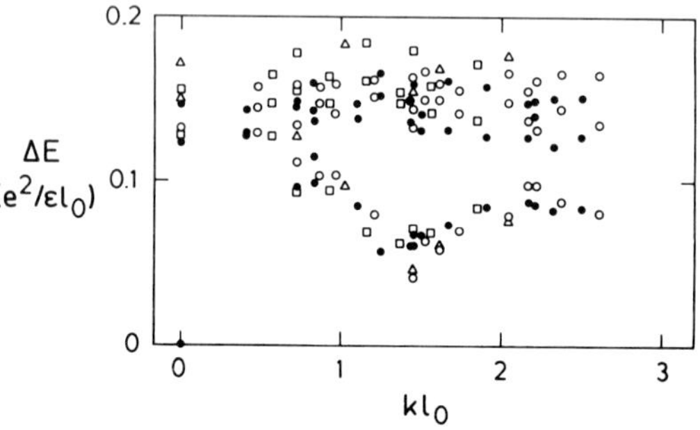

Fig. 22. Low-lying excitation energies for four- (△), five- (□), six- (○) and seven-electron (●) systems at $\nu = \frac{1}{3}$.

us define the oscillator strength as $f(k) \equiv N^{-1}\langle\psi|\rho_{\boldsymbol{k}}^{\dagger}(\mathscr{H}-E_0)\rho_{\boldsymbol{k}}|\psi\rangle$, where $\mathscr{H}$ is the Hamiltonian and E_0 is the ground-state energy. Let us also define the static structure function $s(k) \equiv N^{-1}\langle\psi|\rho_{\boldsymbol{k}}^{\dagger}\rho_{\boldsymbol{k}}|\psi\rangle$. The excitation energy is then given by the well-known formula

$$\Delta(k) = \frac{f(k)}{s(k)} = \frac{\hbar^2 k^2}{2m^* s(k)}. \tag{73}$$

The above result can be interpreted as saying that the collective mode energy is the single-particle energy $\hbar^2 k^2/2m^*$ renormalized by the static structure function representing correlations among the particles.

As is well known, for liquid ^{4}He there are no low-lying single-particle excitations and the only low-lying excitations are long-wavelength density oscillations – the *phonons*. The excitation energy versus wave vector curve vanishes linearly, its slope corresponding to the velocity of sound. Near $k = 2\ \text{Å}^{-1}$, the excitation energy shows a *roton minimum*, which arises due to the peak of thc static structure function.

In the case of the FQHE, if we insist that the excited state is entirely within the lowest Landau level, the density-wave excited state becomes

$$\varphi_{\boldsymbol{k}} = N^{-1/2}\bar{\rho}_{\boldsymbol{k}}\psi, \tag{74}$$

where $\bar{\rho}_{\boldsymbol{k}}$ is the *projection* of the density operator onto the subspace of the lowest Landau level,

$$\bar{\rho}_{\boldsymbol{k}} = \sum_{j=1}^{N} \exp\left[-\mathrm{i}k\frac{\partial}{\partial z_j}\right]\exp[-\tfrac{1}{2}\mathrm{i}k^* z_j], \tag{75}$$

where all derivatives operate to the left (Girvin et al. 1986). Also, the projected potential energy is

$$\bar{V} = \tfrac{1}{2}\int \frac{\mathrm{d}\boldsymbol{q}}{(2\pi)^2}\, v(q)(\bar{\rho}_{\boldsymbol{q}}^{\dagger}\bar{\rho}_{\boldsymbol{q}} - \rho\mathrm{e}^{-q^2 l_0^2/2}), \tag{76}$$

where $v(q)$ is the interaction potential [$v(q) = 2\pi/q$ in the present case]. The projected oscillator strength is

$$\bar{f}(k) = N^{-1}\langle 0|\bar{\rho}_{\boldsymbol{k}}^{\dagger}[\bar{\mathscr{H}}, \bar{\rho}_{\boldsymbol{k}}]|0\rangle, \tag{77}$$

where $|0\rangle$ is the ground state. Since the kinetic energy is constant, one can replace the Hamiltonian by the potential energy. Let us also define the projected static structure factor $\bar{s}(k)$ as

$$\bar{s}(q) = N^{-1}\langle 0|\bar{\rho}_{\boldsymbol{k}}^{\dagger}\bar{\rho}_{\boldsymbol{k}}|0\rangle. \tag{78}$$

In the single-mode approximation (so named because of the assumption that the density-wave alone saturates the full projected oscillator strength sum), the excitation energy is $\Delta(k) = \bar{f}(k)/\bar{s}(k)$. For small k, $\bar{f}(k)$ vanishes like $|k|^4$ and then for a gap to exist, $\bar{s}(k)$ must vanish as $\sim |k|^4$.

Using the Laughlin wave function for the ground state, the structure factor and the function $\bar{f}(k)$ can be computed numerically from the above relations. The resulting excitation energy obtained by Girvin et al. (1985, 1986) is presented in fig. 23 for

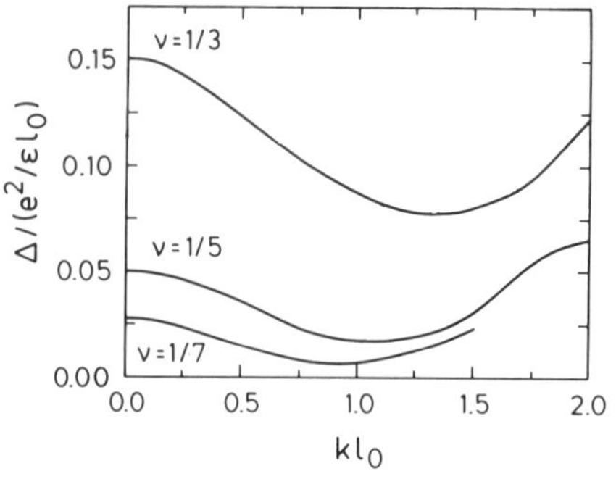

Fig. 23. Collective excitation curve in the single-mode approximation for $\nu = \frac{1}{3}$, $\frac{1}{5}$ and $\frac{1}{7}$ filling fractions.

$\nu = \frac{1}{3}$, $\frac{1}{5}$ and $\frac{1}{7}$. The low-lying excitation energy curve reveals several interesting features. The first thing to note is that, unlike ^{4}He, the collective mode has a finite gap at $k = 0$, i.e., the mode is *not* a massless Goldstone mode. It should be pointed out, however, that this gap is *not* due to the charged particles, since the Coulomb force is not sufficiently long-ranged to provide a finite plasma frequency in two dimensions (the plasma frequency goes to zero with a square root dependence on the wave vector). The finite gap originates from the incompressible nature of the electron system at some particular filling fractions. The collective mode also shows a minimum at a finite k. This minimum is due to the peak $\bar{s}(k)$ and is thus analogous to the *roton* minimum in liquid ^{4}He. For large wave vectors ($kl_0 \gg 1$) the density wave is no longer a suitable excitation and the single-mode approximation is no longer appropriate.

Unlike the elementary excitations discussed in § 4.2.1, no experimental results are available, as yet, for the collective mode described in this section. It should, however, be mentioned that roton-type structures expected in the inter-Landau-level excitations at integer filling factors (Kallin and Halperin 1984) have been observed by inelastic light-scattering experiments (Pinczuk et al. 1988, 1990). Similar excitations for a fractional filling of the lowest Landau level have also been studied theoretically (MacDonald et al. 1985, Pietiläinen and Chakraborty 1988, Pietiläinen 1988).

4.3.2. *Even-denominator filling fractions*

In our discussion of the theoretical work so far, we have only described the filling fractions with *odd* denominators. The fact that all investigations focused their attention mostly on these fractions is hardly surprising, because the experimental results clearly demonstrated that for the FQHE to occur, the odd denominators were apparently favored exclusively. As we recall, Laughlin's approach *explains* such a fact by the requirement of antisymmetry under interchange of particles and in the hierarchical scheme such filling factors are taken as the starting point in developing the higher-order filling factors with odd denominators. The possibility of observing the FQHE for *even* denominator filling factors is not excluded, however, in these theories.

The simplest filling factor with even denominator is $\nu = \frac{1}{2}$. In this case, the Laughlin-type wave function would describe a system of particles obeying *Bose* statistics. However, one can group the electrons into bound *pairs*, and the pairs can then transform as bosons under interchange of their positions (Halperin 1983) and a Laughlin-type wave function could still be used. For the finite-size system in a periodic rectangular geometry (§ 4.1.1), with particles obeying *Bose* statistics, a *cusp* at $\nu = \frac{1}{2}$ was, in fact, observed by Yoshioka (1984a). Finite-size studies in a spherical geometry were also performed by Fano et al. (1987). The pairing idea of Halperin has been recently revived in the theoretical work of Greiter et al. (1991, 1992) for this filling fraction.

The ground-state energy (per electron) at $\nu = \frac{1}{2}$, calculated for four- to ten-electron systems in a periodic rectangular geometry (Chakraborty and Pietiläinen 1988a) is shown in fig. 24. In contrast to the case of $\nu = \frac{1}{3}$, the results in this case show strong dependence on the electron number. The extrapolation of the results in the thermodynamic limit leads to the energy $\approx -0.472e^2/\varepsilon l_0$. The results are, of course, lower compared to the Wigner crystal (WC) energy in the HF limit: the energy difference is ≈ 0.028, while for $\nu = \frac{1}{3}$, the corresponding energy difference is ≈ 0.025. However, the energy difference is much smaller (~ 0.01) for the crystal energies obtained for the four-electron system by Yoshioka et al. (1983) (see fig. 14). Given such a small difference, it is not possible to entirely rule out the crystal state at $\nu = \frac{1}{2}$. The Laughlin state energy at $\nu = \frac{1}{2}$, which corresponds to the boson system, is also given in fig. 24.

The excitation spectrum at this filling fraction bears no similarity to that for an incompressible fluid. Here the ground state appears at a finite $\boldsymbol{k}$ and varies strongly with the particle number and geometry of the cell. The whole spectrum is, in fact, particle-number- and geometry-dependent. No general conclusion can be drawn from those numerical results (Chakraborty and Pietiläinen 1988a).

Experimentally, the possibility of observing the FQHE at even denominator filling factors was indicated by several groups. A minimum in ρ_{xx} at $\nu = \frac{3}{4}$ was first observed by Ebert et al. (1984). Clark et al. (1986) observed minima in the diagonal resistivity

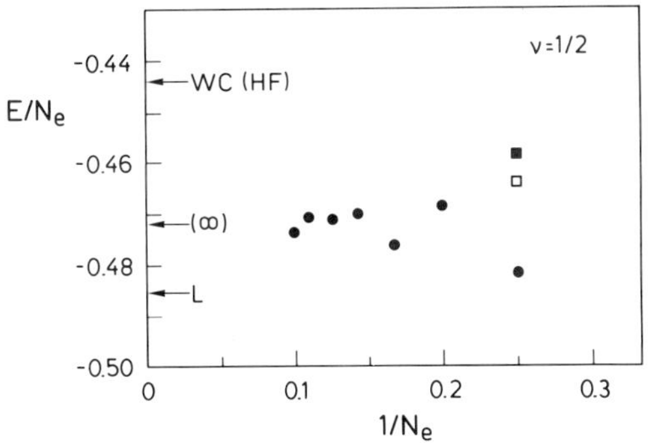

Fig. 24. Ground-state energy per electron at $\nu = \frac{1}{2}$ (in units of $e^2/\varepsilon l_0$) as a function of electron number in a periodic rectangular geometry. The closed and open squares correspond to the crystal energies of fig. 14, and the HF energy is given for comparison. The energy of the Laughlin state (depicted as L) and the extrapolation of the finite system results to the thermodynamic limit [depicted as (∞)] are also given.

in the second Landau level at $\nu = \frac{9}{4}, \frac{5}{2}$, and $\frac{11}{4}$. Correct quantization of ρ_{xy} to these fractional values was not achieved, however.

A thorough analysis of these filling factors has been performed by Willett et al. (1987). Their results for $\nu < 1$ (fig. 13) do not show any sign of the FQHE for even denominator fillings. While some features in ρ_{xx} were seen at $\nu = \frac{3}{4}$, two higher-order odd denominator fillings, $\nu = \frac{4}{5}$ and $\nu = \frac{5}{7}$, seem to converge toward this even denominator filling factor. For $\nu = \frac{1}{2}$, ρ_{xy} follows the classical straight line, while the broad minimum in ρ_{xx} is thought to be caused by, as yet, unresolved higher-order odd denominator filling factors.

For $3 > \nu > 2$ (the Landau level $n = 1$), however, the situation is entirely different. This region of filling factors in fig. 13 is presented in more detail in fig. 25. The ρ_{xy} curve shows a plateau at the magnetic field which corresponds to $\nu = \frac{5}{2}$, centered at $\rho_{xy} = (h/e^2)/\frac{5}{2}$ to within 0.5%. In the same region of magnetic field, a deep minimum is observed in ρ_{xx}.

There have been a few theoretical attempts to explain the experimental findings discussed above. Haldane and Rezayi (1988) proposed a spin-singlet wave function for an incompressible state which occurs at $\nu = \frac{1}{2}$. From small-system calculations (six electrons) with model pseudopotentials, they concluded that such a state might be responsible for the $\frac{5}{2}$ effect. Numerical calculations were performed for the Coulomb interaction by Chakraborty and Pietiläinen (1988b). The calculations were also based on finite-size systems (up to six electrons in a periodic rectangular geometry), and do not support the conclusions of Haldane and Rezayi. In table 2, we present the ground-state energies for four- and six-electron systems at $\nu = \frac{1}{2}$ in the second Landau level for different spin polarizations. The system has a fully spin-polarized ground state, even in the absence of the Zeeman energy. The energy difference between the various spin states is very small. However, for a spin-reversed system to be energetically favored, the energy of this state must be larger than that of the spin-

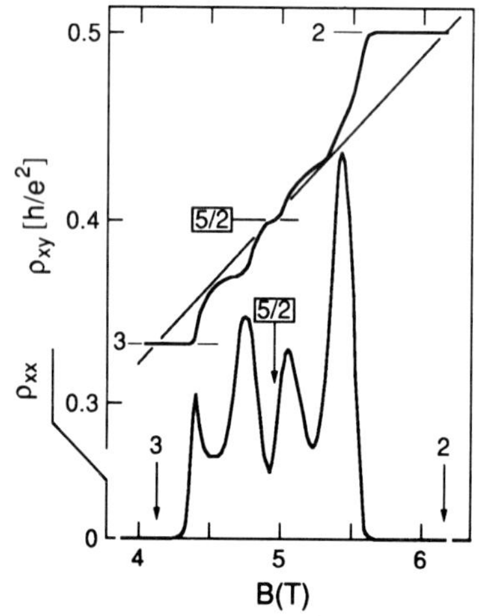

Fig. 25. The region of filling factors $3 > \nu > 2$ of fig. 13 shown in detail at $T = 25$ mK.

Table 2

Ground-state energies (in units of $e^2/\varepsilon l_0$) for four- and six-electron systems at $\nu = \frac{5}{2}$ for various values of the total spin S. The Zeeman energy is not included in the energy values.

N_e	$S=0$	$S=1$	$S=2$	$S=3$
4	−0.3644	−0.3655	−0.3849	–
6	−0.3782	−0.3783	−0.3785	−0.3797

polarized state by at least the Zeeman energy contribution. Such a situation might occur for larger systems than the ones considered here, as the even-denominator system results are known to be very system-size dependent.

In a tilted-field measurement on the $\frac{5}{2}$ state, Eisenstein et al. (1988) observed that the FQHE collapses rapidly at this filling factor. This indicates that the spin degree of freedom is perhaps playing a role. However, from the above discussions, it is fair to conclude that finite-size calculations performed so far are unable to provide a suitable explanation of the FQHE at this interesting filling factor.

Some recent experiments have unearthed several puzzling facts about the filling factors $\nu = \frac{1}{2}$ and $\frac{3}{2}$ whose origins are far from clear. Jiang et al. (1989) observed deep low-temperature minima in ρ_{xx} at these filling fractions. Although their strength exceeds the strength of neighboring FQHE states, there are no plateaus nor any discernible indication of plateau developments visible in ρ_{xy}. The states at half-filling are distinctly different from the FQHE states because of the unusual T-dependence in ρ_{xx} and the lack of plateaus in ρ_{xy}. Transport measurements in tilted magnetic fields seem not to alter the above features.

4.3.3. Multiple layer systems

In this section, we present the theoretical and experimental work on the FQHE in a *layered* electron system. Multilayer electron systems have been studied earlier quite extensively as an anisotropic model for an electron gas. Let us consider a model where two layers with equal density of electrons are embedded in an infinite dielectric. We consider the delta-function-localized electron density in each plane. The electrons move freely in each plane and the interaction of electrons in different planes is considered to be Coulombic. Tunneling of electrons between the two planes is not allowed. The electrons are also considered to be in their lowest subband. This model is often referred to in the literature as the Visscher–Falicov model (Visscher and Falicov 1971).

In fig. 26, we have presented the excitation spectrum for a two-layer system (layers separated by a distance C) with four electrons per layer in a periodic rectangular geometry (Chakraborty and Pietiläinen 1987, 1988c), with an aspect ratio $\lambda = 1.25$. The first important result is that the ground state is obtained uniquely at $\boldsymbol{k} = 0$. It remains so for different aspect ratios. The other interesting result is that a gap structure in the spectrum is obtained with a characteristic minimum at a finite kl_0, similar to that of the magnetoroton minimum, discussed in § 4.3.1.

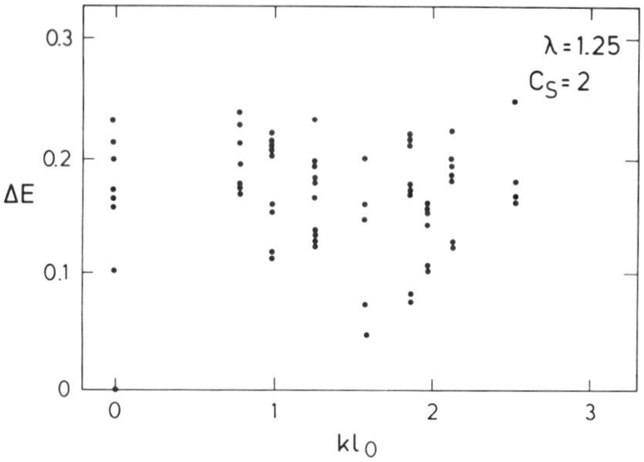

Fig. 26. Excitation energy (in units of $e^2/\varepsilon l_0$) of an eight-electron system in a two-layer geometry at $\nu = \frac{1}{2}$ for a dimensionless layer separation parameter $C_s = C/l_0 = 2.0$.

A comparison of the layered-system results with those of the single layer indicates that the introduction of an interacting electron layer has helped to reorganize the excitation energies of the system, particularly the $\boldsymbol{k} = 0$ state (Chakraborty and Pietiläinen 1988a). The observation of the roton-type minimum is also quite interesting. The results indicate the possibility of the occurrence of an incompressible fluid state at $\nu = \frac{1}{2}$ in a multiple layer system.

There have been a few other calculations on the two-layered systems. Yoshioka et al. (1989) obtained the ground-state wave functions for finite-size systems on a sphere and compared them with the Jastrow-type wave functions at filling factors $\nu = 1, \frac{1}{2}, \frac{2}{5}$, and $\frac{1}{3}$. Halperin's two-spin state wave function, eq. (70), was employed in that work. Fertig (1989) investigated the excitation spectrum of two- and three-layer systems with filling factors $\nu = \frac{1}{2}$ and $\nu = \frac{1}{3}$, respectively, in each layer. The excitation spectra were calculated following the scheme of Kallin and Halperin (1984) for magnetoplasmon work. He found that the excitation spectrum has soft modes when $C > 1.21 l_0$ for the two-layer system and $C_1 > 0.92 l_0$, $C_2 > 1.51 l_0$ for the three-layer system. This indicates that the system undergoes a phase transition as the layer spacing is increased through these critical separations.

Experimental attempts to explore the FQHE in two-layer systems have been reported recently. Lindelof et al. (1989) reported an interesting observation of $\nu = \frac{1}{2}$ in each of the two parallel layers of a selectively doped double heterostructure. Other experimental results with that sample indicate that the system indeed consists of two parallel and independent two-dimensional electron layers. Experiments performed by Eisenstein et al. (1992) also show results in agreement with the theoretical results discussed above.

4.3.4. Nature of long-range order in the Laughlin state

The analogy of the FQHE state with liquid ^{4}He has been quite apparent in the study of collective excitations discussed in § 4.3.1, in particular, the presence of magnetorotons in the excitation spectrum. It is, therefore, quite natural to look for some other interesting properties in the FQHE state which are known to exist in ^{4}He, notable among which is the so-called off-diagonal long-range order (ODLRO) in the density matrices (Sears 1985).

In the case of the QHE, the off-diagonal elements of the one-body density matrix in real space exhibit Gaussian decay (MacDonald and Girvin 1988)

$$\rho_1(\boldsymbol{r}, \boldsymbol{r}') = \frac{\nu}{2\pi} \mathrm{e}^{-|z-z'|^2/4} \mathrm{e}^{(z^*z' - zz'^*)/4}. \tag{79}$$

Therefore, no ODLRO exists in the one-body density matrix of the two-dimensional electron system.

To exhibit ODLRO, a fermionic system has to form at least pairs. This is the case in the BCS theory of traditional superconductivity where the two-body density matrix has extensive eigenvalues (Yang 1962). Some years ago, Thouless (1985) investigated, in the context of the FQHE, the corresponding density matrix

$$\rho_{m_1, m_2}(n) = \langle a^\dagger_{n+m_1} a^\dagger_{n-m_1} a_{n-m_2} a_{n+m_2} \rangle, \tag{80}$$

where, for fixed n, the eigenvalues were determined for finite-electron systems in a spherical geometry, as well as for the Laughlin wave function on a square with periodic boundary condition in one direction. The existence of Cooper pairs with *wave number* $2n$ is expected to be indicated by a large eigenvalue of $\rho(n)$. The largest eigenvalues in those calculations, however, were found to *decrease* rather than increase with the number of electrons. The results, therefore, indicate that there is no sign of any order in the two-body density matrix for the FQHE.

Although electrons in a strong magnetic field do not exhibit ODLRO, it seems likely that there might be some order hidden in the space of those quantum states. In the pursuit of such an order a singular gauge field $\mathscr{A}_j$ used in the study of *anyons* (see § 4.2.2), has been considered $\mathscr{A}_j(z_j) = (\lambda \Phi_0/2\pi)\Sigma_{i \neq j} \nabla_j \mathfrak{F} \ln(z_j - z_i)$ (Girvin and MacDonald 1987), where Φ_0 is the flux quantum and λ is a constant. It corresponds to a vector potential that would be included in the Hamiltonian if each particle had attached to itself a solenoid carrying $\frac{1}{2}\lambda$ flux quanta. It should be noted that adding this vector potential to the Hamiltonian we do not make a true gauge transformation since a flux tube is attached to each particle. However, if $\lambda = m$, an integer, the net effect is to change the phase of the wave function,

$$\psi_{\text{new}} = \exp\left[-\mathrm{i}m \sum_{i<j} \mathfrak{F} \ln(z_i - z_j)\right] \psi_{\text{old}}. \tag{81}$$

If ψ_{old} is the Laughlin wave function (45) we get the transformed state

$$\tilde{\psi}(z_1, \ldots, z_N) = \prod_{i<j} |z_i - z_j|^m \exp\left(-\tfrac{1}{4} \sum_k |z_k|^2\right), \tag{82}$$

which is purely real and is symmetric under particle exchange for both even and odd m. This is truly a remarkable result that both fermion and boson systems map onto bosons in this singular gauge.

Making use of the plasma analogy discussed in § 4.1, one obtains

$$\tilde{\rho}(z, z') = \left(\frac{\nu}{2\pi}\right) \exp[-\beta\, \Delta f(z, z')]\, |z - z'|^{-m/2}, \tag{83}$$

where $\beta \equiv 2/m$ and $\Delta f(z, z')$ is the difference in free energy between the two impurities of charge $\frac{1}{2}m$ (located at z and z') and a single impurity of charge m (with arbitrary location). The asymptotic value of Δf can be computed using thermodynamic integration of the screening charge density. For general values of m the screening charge distribution can be found using the ion-disk approximation or linear response based on the known static structure factor of the plasma (Caillol et al. 1982). As the plasma screens the impurities completely, the free energy difference $\Delta f(z, z')$ rapidly approaches a constant (fig. 27) as the separation $|z - z'| \rightarrow \infty$. The one-body density matrix, therefore, shows a *power-law* decay in the long-range. This implies that there is no ODLRO even in the transformed Laughlin state. It should be remembered that in the *ground state* the density matrix should attain a finite value for infinite separation if the ODLRO exists in that system *even in two dimensions* (Reatto and Chester 1967).

A physical interpretation of the power-law decay of the one-body density matrix for the transformed Laughlin state has been provided by Chakraborty and von der Linden (1990). In determining the one-body density matrix for the modified Laughlin wave function for N electrons, one can map the wave function onto a classical plasma with $N-1$ plasma particles and two additional *phantom* particles residing at sites z and z'. These particles have charges half as large as the original plasma particles and experience an interaction with the plasma particles but not with each other. The one-body density matrix can then be related to the difference in free energy $\Delta F(z, z')$ between this system and the N-particle classical plasma, with inverse temperature $\beta = 1/m$: $\rho(z, z') = \rho_0\, e^{\beta \Delta F(z, z')}$, and $\rho_0 = 1/2\pi m$ being the plasma density. In a two-component system, the one-body density matrix is, in fact, related to the pair-correlation function of the phantom particles in the zero-concentration limit,

$$\rho(z, z') = \lim_{\rho_\gamma \to 0} \{[g_{\gamma\gamma}(0, 0)]^{-1}\, g_{\gamma\gamma}(z, z')\}, \tag{84}$$

where γ stands for the phantom particles. For systems like ^{4}He, with interactions

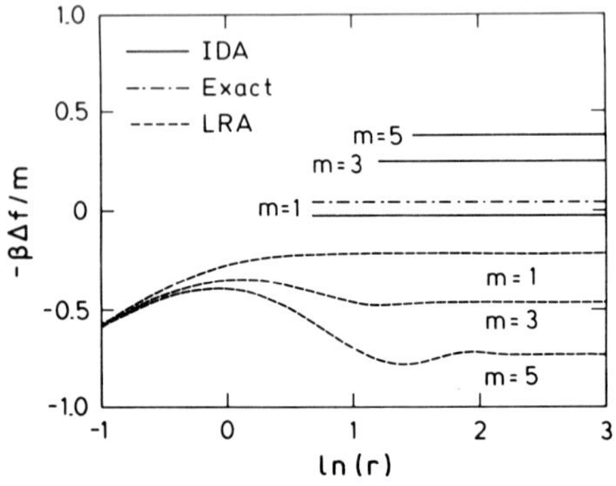

Fig. 27. Plot of $-\beta\, \Delta f(z, z')/m$ versus $r \equiv |z - z'|$ for filling factor $\nu = 1/m$. LRA is linear response approximation. IDA is ion-disk approximation.

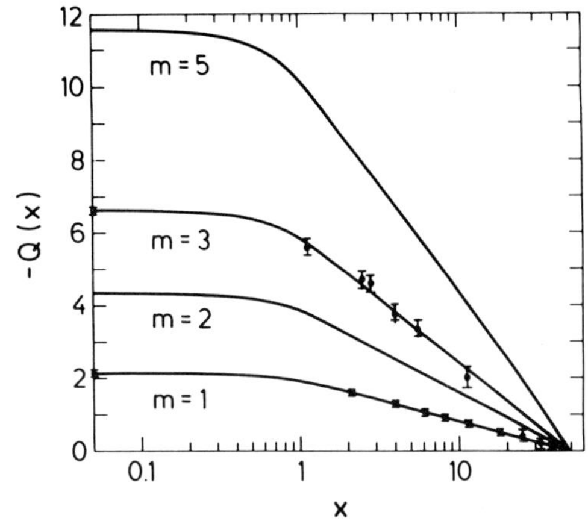

Fig. 28. Plot of $-Q(x)$ versus the interparticle separation x, measured in units of the mean electronic separation $r_0 = \sqrt{2m}\, l_0$. The HNC results (solid lines) are compared with the Monte Carlo data (solid circles).

vanishing at large distances, the asymptotic value of $g(z, z')$ is unity and, therefore, the asymptotic value of $\rho(z, z')$ is given by the inverse of the pair-correlation function at the origin. This is not true for systems with *increasing* interactions like the classical plasma in two dimensions.

Chakraborty and von der Linden (1990) performed both HNC calculations and Monte Carlo calculations for up to 1000 particles, in the case of logarithmically increasing interactions. The numerical results of the HNC and Monte Carlo studies are plotted in fig. 28, the conventional definition of the density matrix is used in the HNC scheme: $\rho(z, z') = n_c\, e^{-Q(|z-z'|)}$, where $n_c = g_{\gamma\gamma}(0, 0)^{-1}$ is the conventional condensate fraction and $Q(z, z')$ is related to the pair-correlation functions via the HNC equations (Puoskari and Kallio 1984, Lam and Ristig 1979, Fantoni 1978). The interparticle separation in these calculations is large enough that the asymptotic behavior of $\tilde{\rho}$ for various m can be extracted very accurately. From this plot we readily obtain the power $-\frac{1}{2}m$ in agreement with eq. (83). The values for $-\beta\Delta f_\infty/m$ are 0.02 for $m = 1$ and 0.24 for $m = 3$ and are in fairly good agreement with the earlier results (Girvin and MacDonald 1987). The screening length is of order $\sqrt{2m}\, l_0$. The values of n_c obtained in the HNC scheme are 0.121 for $m = 1$, 0.014 for $m = 2$, 0.0015 for $m = 3$, and ~ 0 for $m = 5$. The comparison of the density matrix results via HNC with the corresponding Monte Carlo data show once more that HNC is a highly accurate approximation scheme, even in the case of increasing interactions.

Acknowledgements

I wish to acknowledge long and fruitful collaboration with Dr. Pekka Pietiläinen, who co-authored most of our papers cited. Over the years, I have benefitted from discussions with numerous people, in particular, Bob Clark, Jim Eisenstein, Bert Halperin, Klaus von Klitzing, Allan MacDonald, Aron Pinczuk, and Horst Störmer. Excellent support from the colleagues at the Max-Planck-Institute, Stuttgart is gratefully acknowledged. I would like to thank Geof Aers for critically reading the manuscript and Marie D'Iorio and V.M. Pudalov for helpful discussions.

References

Abrahams, E., P.W. Anderson, D.C. Licciardello and T.V. Ramakrishnan, 1979, Phys. Rev. Lett. **42**, 673.
Aers, G.C., and A.H. MacDonald, 1984, J. Phys. C **17**, 5491.
Ando, T., 1982, Anderson Localization, eds Y. Nagaoka and H. Fukuyama (Springer, Berlin) p. 176.
Ando, T., 1983a, J. Phys. Soc. Jpn. **52**, 1740.
Ando, T., 1983b, Recent Topics in Semiconductor Physics, eds H. Kamimura and Y. Toyozawa (World Scientific, Singapore) p. 72.
Ando, T., 1984, J. Phys. Soc. Jpn. **53**, 3101, 3126.
Ando, T., 1986, Surf. Sci. **170**, 243.
Ando, T., and H. Aoki, 1985, J. Phys. Soc. Jpn. **54**, 2238.
Ando, T., and Y. Uemura, 1974, J. Phys. Soc. Jpn. **39**, 959.
Ando, T., Y. Matsumoto and Y. Uemura, 1975, J. Phys. Soc. Jpn. **39**, 279.
Ando, T., A.B. Fowler and F. Stern, 1982, Rev. Mod. Phys. **54**, 437.
Aoki, H., 1982, J. Phys. C **15**, L1227.
Aoki, H., 1987, Rep. Prog. Phys. **50**, 655.
Aoki, H., and T. Ando, 1981, Solid State Commun. **38**, 1079.
Aoki, H., and T. Ando, 1985, Phys. Rev. Lett. **54**, 831.
Aoki, H., and T. Ando, 1986a, Phys. Rev. Lett. **57**, 3093.
Aoki, H., and T. Ando, 1986b, Surf. Sci. **170**, 249.
Apal'kov, V.M., and E.I. Rashba, 1991, JETP Lett. **56**, 49.
Apenko, S.M., and Yu.E. Lozovik, 1985, Sov. Phys.-JETP **62**, 328.
Arovas, D., J.R. Schrieffer and F. Wilczek, 1984, Phys. Rev. Lett. **53**, 722.
Avron, J.E., and R. Seiler, 1985, Phys. Rev. Lett. **54**, 259.
Azbel, M.Ya., 1985, Solid State Commun. **53**, 147.
Baraff, G.A., and D.C. Tsui, 1981, Phys. Rev. B **24**, 2274.
Beenakker, C.W.J., and H. van Houten, 1991, Solid State Phys. **44**, 1.
Boebinger, G.S., A.M. Chang, H.L. Störmer and D.C. Tsui, 1985, Phys. Rev. Lett. **55**, 1606.
Boebinger, G.S., H.L. Störmer, D.C. Tsui, A.M. Chang, J.C.M. Hwang, A.Y. Cho, C.W. Tu and G. Weimann, 1987, Phys. Rev. B **36**, 7919.
Brézin, E., D.J. Gross and C. Itzykson, 1984, Nucl. Phys. B **235** [FS 11], 24.
Brown, E., 1968, Solid State Phys. **22**, 313.
Buckthought, A., R. Boulet, A. Sachrajda, Z. Wasilewski, P. Zawadzki and F. Guillon, 1991, Solid State Commun. **78**, 191.
Buhmann, H., W. Joss, K. von Klitzing, I.V. Kukushkin, G. Martinez, A.S. Plaut, K. Ploog and V.B. Timofeev, 1990, Phys. Rev. Lett. **65**, 1056.
Buhmann, H., W. Joss, K. von Klitzing, I.V. Kukushkin, A.S. Plaut, G. Martinez, K. Ploog and V.B. Timofeev, 1991, Phys. Rev. Lett. **66**, 926.
Büttiker, M., 1988, Phys. Rev. B **38**, 9375.
Bychkov, Yu.A., and E.I. Rashba, 1989, Sov. Phys.-JETP **69**, 430.
Bychkov, Yu.A., S.V. Iordanskii and G.M. Eliashberg, 1981, JETP Lett. **33**, 143.
Byers, N., and C.N. Yang, 1961, Phys. Rev. Lett. **7**, 46.
Cage, M.E., R.F. Dziuba, B.F. Field, E.R. Williams, S.M. Girvin, A.C. Gossard, D.C. Tsui and R.J. Wagner, 1983, Phys. Rev. Lett. **51**, 1374.
Caillol, J.M., D. Levesque, J.J. Weis and J.P. Hansen, 1982, J. Stat. Phys. **28**, 325.

Chakraborty, T., 1985, Phys. Rev. B **31**, 4026.
Chakraborty, T., 1986, Phys. Rev. B **34**, 2926.
Chakraborty, T., 1990, Surf. Sci. **229**, 16.
Chakraborty, T., and P. Pietiläinen, 1986, Phys. Scr. **T14**, 58.
Chakraborty, T., and P. Pietiläinen, 1987, Phys. Rev. Lett. **59**, 2784.
Chakraborty, T., and P. Pietiläinen, 1988a, The Fractional Quantum Hall Effect (Springer, New York, Berlin, Heidelberg).
Chakraborty, T., and P. Pietiläinen, 1988b, Phys. Rev. B **38**, 10 097.
Chakraborty, T., and P. Pietiläinen, 1988c, Recent Progress in Many-Body Theories, eds A. Kallio, E. Pajane and R.F. Bishop (Plenum, New York) p. 113.
Chakraborty, T., and P. Pietiläinen, 1989, Phys. Rev. B **39**, 7971.
Chakraborty, T., and P. Pietiläinen, 1990, Phys. Rev. B **41**, 10 862.
Chakraborty, T., and P. Pietiläinen, 1991, Phys. Rev. B **44**, 13 078.
Chakraborty, T., and W. von der Linden, 1990, Phys. Rev. B **41**, 7872.
Chakraborty, T., and F.C. Zhang, 1984, Phys. Rev. B **29**, 7032.
Chakraborty, T., P. Pietiläinen and F.C. Zhang, 1986, Phys. Rev. Lett. **57**, 130.
Chalker, J.T., and P.D. Coddington, 1988, J. Phys. C **21**, 2665.
Chang, A.M., P. Berglund, D.C. Tsui, H.L. Störmer and J.C.M. Hwang, 1984, Phys. Rev. Lett. **53**, 997.
Clark, R., and P. Maksym, 1989, Physics World **2**, 39.
Clark, R.G., R.J. Nicholas, A. Ulsher, C.T. Foxon and J.J. Harris, 1986, Surf. Sci. **170**, 141.
Clark, R.G., J.R. Mallett, S.R. Haynes, J.J. Harris and C.T. Foxon, 1988, Phys. Rev. Lett. **60**, 1747.
Clark, R.G., S.R. Haynes, A.M. Suckling, J.R. Mallett, P.A. Wright, J.J. Harris and C.T. Foxon, 1989, Phys. Rev. Lett. **62**, 1536.
Clark, R.G., S.R. Haynes, J.V. Branch, A.M. Suckling, P.A. Wright, P.M.W. Oswald, J.J. Harris and C.T. Foxon, 1990, Surf. Sci. **229**, 25.
Davies, A.G., R. Newbury, M. Pepper, J.E.F. Frost, D.A. Ritchie and G.A.C. Jones, 1991, Phys. Rev. B **44**, 13 128.
D'Iorio, M., V.M. Pudalov and S.G. Semenchinsky, 1990, Phys. Lett. **150**, 422.
D'Iorio, M., J.W. Campbell, V.M. Pudalov and S.G. Semenchinsky, 1992, Surf. Sci. **263**, 49.
Dobers, M., K. von Klitzing and G. Weimann, 1988, Phys. Rev. B **38**, 5453.
Ebert, G., K. von Klitzing, C. Probst and K. Ploog, 1982, Solid State Commun. **44**, 95.
Ebert, G., K. von Klitzing, K. Ploog and G. Weimann, 1983, J. Phys. C **16**, 5441.
Ebert, G., K. von Klitzing, J.C. Maan, G. Remenyi, C. Probst, G. Weimann and W. Schlapp, 1984, J. Phys. C **17**, L775.
Eisenstein, J.P., H.L. Störmer, V. Narayanamurti, A.Y. Cho, A.C. Gossard and C.W. Tu, 1985, Phys. Rev. Lett. **55**, 875.
Eisenstein, J.P., R. Willet, H.L. Störmer, D.C. Tsui, A.C. Gossard and J.H. English, 1988, Phys. Rev. Lett. **61**, 997.
Eisenstein, J.P., H.L. Störmer, L. Pfeiffer and K.W. West, 1989, Phys. Rev. Lett. **62**, 1540.
Eisenstein, J.P., H.L. Störmer, L.N. Pfeiffer and K.W. West, 1990, Phys. Rev. B **41**, 7910.
Eisenstein, J.P., G.S. Boebinger, L. Pfeiffer, K. West and Song He, 1992, Phys. Rev. Lett. **68**, 1383.
Engel, L.W., S.W. Hwang, T. Sajoto, D.C. Tsui and M. Shayegan, 1992, Phys. Rev. B **45**, 3418.
Englert, T., and K. von Klitzing, 1978, Surf. Sci. **73**, 70.
Fano, G., F. Ortolani and E. Colombo, 1986, Phys. Rev. B **34**, 2670.
Fano, G., F. Ortolani and E. Tosatti, 1987, Nuovo Cimento **9**, 1337.
Fantoni, S., 1978, Nuovo Cimento A**44**, 191.
Fertig, H., 1989, Phys. Rev. B **40**, 1087.
Feynman, R.P., 1972, Statistical Physics (Benjamin, Reading, MA) ch. 11.
Fock, V., 1928, Z. Phys. **47**, 446.
Fukuyama, H., P.M. Platzman and P.W. Anderson, 1979, Phys. Rev. B **19**, 5211.
Furneaux, J.E., D.A. Syphers and A.G. Swanson, 1989, Phys. Rev. Lett. **63**, 1098.
Girvin, S.M., and A.H. MacDonald, 1987, Phys. Rev. Lett. **58**, 1252.
Girvin, S.M., A.H. MacDonald and P.M. Platzman, 1985, Phys. Rev. Lett. **54**, 581.
Girvin, S.M., A.H. MacDonald and P.M. Platzman, 1986, Phys. Rev. B **33**, 2481.

Giuliani, G.F., J.J. Quinn and S.C. Ying, 1983, Phys. Rev. B **28**, 2969.
Goldberg, B.B., D. Heiman, A. Pinczuk, L. Pfeiffer and K. West, 1990, Phys. Rev. Lett. **65**, 641.
Goldman, V.J., M. Shayegan and D.C. Tsui, 1988, Phys. Rev. Lett. **61**, 881.
Gorkov, L.P., A.I. Larkin and D.E. Khmel'nitzkii, 1979, JETP Lett. **30**, 228.
Gornik, E., 1987, The Physics of The Two-Dimensional Electron Gas, eds J.T. Devreese and F.M. Peeters (Plenum, New York) p. 365.
Gornik, E., R. Lassnig, G. Strasser, H.L. Störmer, A.C. Gossard and W. Wiegmann, 1985, Phys. Rev. Lett. **54**, 1820.
Greiter, M., X.G. Wen and F. Wilczek, 1991, Phys. Rev. Lett. **66**, 3205.
Greiter, M., X.G. Wen and F. Wilczek, 1992, Nucl. Phys. B **374**, 567.
Grimes, C.G., 1978, Surf. Sci. **73**, 379.
Guldner, Y., J.P. Hirtz, J.P. Vieren, P. Voisin, M. Voos and M. Razeghi, 1982, J. Phys. Lett. **43**, L613.
Guldner, Y., J.P. Vieren, M. Voos, F. Delahaye, D. Dominguez, J.P. Hirtz and M. Razeghi, 1986, Phys. Rev. B **33**, 3990.
Gusev, G.M., Z.D. Kvon, I.G. Neizvestnyi, V.N. Ovsyuk and P.A. Cheremnykh, 1984, JETP Lett. **39**, 541.
Hajdu, J., and G. Landwehr, 1985, Strong and Ultrastrong Magnetic Fields and Their Applications, ed. F. Herlach (Springer, Berlin).
Haldane, F.D.M., 1983, Phys. Rev. Lett. **51**, 605.
Haldane, F.D.M., 1985, Phys. Rev. Lett. **55**, 2095.
Haldane, F.D.M., and E.H. Rezayi, 1985, Phys. Rev. Lett. **54**, 237.
Haldane, F.D.M., and E.H. Rezayi, 1988, Phys. Rev. Lett. **60**, 956.
Halonen, V., T. Chakraborty and P. Pietiläinen, 1990, Phys. Rev. B **41**, 10 202.
Halperin, B.I., 1982, Phys. Rev. B **25**, 2185.
Halperin, B.I., 1983, Helv. Phys. Acta **56**, 75.
Halperin, B.I., 1984, Phys. Rev. Lett. **52**, 1583, 2390(E).
Halperin, B.I., 1986, Surf. Sci. **170**, 115.
Hansen, J.P., and D. Levesque, 1981, J. Phys. C **14**, L603.
Haug, R.J., K. von Klitzing, R.J. Nicholas, J.C. Maan and G. Weimann, 1987, Phys. Rev. B **36**, 4528.
Haug, R.J., J. Kucera, P. Streda and K. von Klitzing, 1989, Phys. Rev. B **39**, 10892.
Hikami, S., 1984, Phys. Rev. B **29**, 3726.
Hikami, S., 1986, Prog. Theor. Phys. **76**, 1210.
Huckestein, B., 1990, Physica A**167**, 175.
Huckestein, B., and B. Kramer, 1990, Phys. Rev. Lett. **64**, 1437.
Huo, Y., and R.N. Bhatt, 1992, Phys. Rev. Lett. **68**, 1375.
Ioffe, L.B., and A.I. Larkin, 1981, Sov. Phys.-JETP **54**, 556.
Iordansky, S.V., 1982, Solid State Commun. **43**, 1.
Jiang, H.W., H.L. Störmer, D.C. Tsui, L.N. Pfeiffer and K.W. West, 1989, Phys. Rev. B **40**, 12 013.
Jiang, H.W., R.L. Willett, H.L. Störmer, D.C. Tsui, L.N. Pfeiffer and K.W. West, 1990, Phys. Rev. Lett. **65**, 633.
Joynt, R., and R.E. Prange, 1984, Phys. Rev. B **29**, 3303.
Kallin, C., and B.I. Halperin, 1984, Phys. Rev. B **30**, 5655.
Kazarinov, R.F., and S. Luryi, 1982, Phys. Rev. B **25**, 7626.
Khmel'nitzkii, D.E., 1983, JETP Lett. **38**, 552.
Khmel'nitzkii, D.E., 1984, Phys. Lett. A**106**, 182.
Kirk, W.P., P.S. Kobiela, R.A. Schiebel and M.A. Reed, 1986, J. Vac. Sci. & Technol. A **4**, 2132.
Komiyama, S., T. Takamasu, S. Hiyamizu and S. Sasa, 1985, Solid State Commun. **54**, 479.
Kravchenko, S.V., V.M. Pudalov, J. Campbell and M. D'Iorio, 1992, JETP Lett. **54**, 532.
Kubo, R., 1957, J. Phys. Soc. Jpn. **12**, 570.
Kubo, R., S.J. Miyake and N. Hashitsume, 1965, Solid State Phys. **17**, 269.
Lam, P.K., and S.M. Girvin, 1984, Phys. Rev. B **30**, 473.
Lam, P.M., and M.L. Ristig, 1979, Phys. Rev. B **20**, 1960.
Landau, L., 1930, Z. Phys. **64**, 629.
Landauer, R., 1970, Philos. Mag. **21**, 863.

Laughlin, R.B., 1981, Phys. Rev. B **23**, 5632.
Laughlin, R.B., 1983a, Phys. Rev. Lett. **50**, 1395.
Laughlin, R.B., 1983b, Phys. Rev. B **27**, 3383.
Laughlin, R.B., 1984a, Springer Series in Solid State Sciences, Vol. 53, eds G. Bauer, F. Kuchar and H. Heinrich (Springer, Heidelberg) p. 272.
Laughlin, R.B., 1984b, Surf. Sci. **142**, 163.
Laughlin, R.B., 1987, The Quantum Hall Effect, eds R.E. Prange and S.M. Girvin (Springer, Berlin) p. 233.
Laughlin, R.B., M.L. Cohen, M. Kosterlitz, H. Levine, S.B. Libby and A.M.M. Pruisken, 1985, Phys. Rev. B **32**, 1311.
Leinaas, J.M., and J. Myrheim, 1977, Nuovo Cimento **37**, 1.
Levesque, D., J.J. Weis and A.H. MacDonald, 1984, Phys. Rev. B **30**, 1056.
Levine, H., S.B. Libby and A.M.M. Pruisken, 1983, Phys. Rev. Lett. **51**, 1915.
Licciardello, D.C., and D.J. Thouless, 1978, J. Phys. C **11**, 925.
Lindelof, P.E., H. Bruus, R. Taboryski and C.B. Sørensen, 1989, Semicond. Sci. & Technol. **4**, 858.
Luryi, S., and R.F. Kazarinov, 1983, Phys. Rev. B **27**, 1386.
MacDonald, A.H., 1989, The Quantum Hall Effect: A Perspective (Jaca Books, Milano).
MacDonald, A.H., and G.C. Aers, 1984, Phys. Rev. B **29**, 5976.
MacDonald, A.H., and S.M. Girvin, 1988, Phys. Rev. B **38**, 6295.
MacDonald, A.H., and P. Streda, 1984, Phys. Rev. B **29**, 1616.
MacDonald, A.H., H.C.A. Oji and S.M. Girvin, 1985, Phys. Rev. Lett. **55**, 2208.
Maksym, P.A., 1985, J. Phys. C **18**, L433.
Maksym, P.A., 1989, J. Phys.: Condens. Matter **1**, L6299.
Mallett, J.R., R.G. Clark, R.J. Nicholas, R. Willett, J.J. Harris and C.T. Foxon, 1988, Phys. Rev. B **38**, 2200.
Mendez, E.E., L. Esaki and L.L. Chang, 1985, Phys. Rev. Lett. **55**, 2216.
Mil'nikov, G.V., and I.M. Sokolov, 1988, JETP Lett. **48**, 536.
Monarkha, Y.P., and V. Shikin, 1982, Sov. J. Low Temp. Phys. **8**, 279.
Morandi, G., 1988, Quantum Hall Effect (Bibliopolis, Napoli).
Morf, R., and B.I. Halperin, 1986, Phys. Rev. B **33**, 2221.
Müller, G., D. Weiss, S. Koch, K. von Klitzing, H. Nickel, W. Schlapp and R. Lösch, 1990, Phys. Rev. B **42**, 7633.
Nicholas, R.J., M.A. Brummell, J.C. Portal, M. Razeghi and M. Poission, 1982, Solid State Commun. **43**, 825.
Niu, Q., D.J. Thouless and Y.S. Wu, 1985, Phys. Rev. B **31**, 3372.
Ono, Y., 1982, J. Phys. Soc. Jpn. **51**, 237.
Paalanen, M.A., D.C. Tsui and A.C. Gossard, 1982, Phys. Rev. B **25**, 5566.
Pietiläinen, P., 1988, Phys. Rev. B **38**, 4279.
Pietiläinen, P., and T. Chakraborty, 1988, Europhys. Lett. **5**, 157.
Pinczuk, A., J.P. Valladares, D. Heiman, A.C. Gossard, J.H. English, C.W. Tu, L. Pfeiffer and K. West, 1988, Phys. Rev. Lett. **61**, 2701.
Pinczuk, A., J.P. Valladares, D. Heiman, A.C. Gossard, J.H. English, C.W. Tu, L. Pfeiffer and K. West, 1990, Surf. Sci. **229**, 384.
Prange, R.E., 1981, Phys. Rev. B **23**, 4802.
Prange, R.E., and S.M. Girvin, eds, 1987, The Quantum Hall Effect, (Springer, New York, Berlin, Heidelberg).
Pruisken, A.M.M., 1984, Nucl. Phys. **235** [FS11], 277.
Pruisken, A.M.M., 1985, Phys. Rev. B **32**, 2636.
Pudalov, V.M., and S.G. Semenchinsky, 1984a, Solid State Commun. **51**, 19.
Pudalov, V.M., and S.G. Semenchinsky, 1984b, JETP Lett. **39**, 170.
Pudalov, V.M., and S.G. Semenchinsky, 1985, Solid State Commun. **55**, 593.
Pudalov, V.M., S.G. Semenchinsky and V.S. Edelman, 1984, JETP Lett. **39**, 576.
Pudalov, V.M., S.G. Semenchinsky and V.S. Edelman, 1985, Sov. Phys.-JETP **62**, 1079.
Puoskari, M., and A. Kallio, 1984, Phys. Rev. B **30**, 152.

Quinn, T., 1989, Metrolgia **26**, 69.
Rashba, E.I., and V.B. Timofeev, 1986, Sov. Phys. Semicond. **20**, 617.
Reatto, L., and G.V. Chester, 1967, Phys. Rev. **155**, 88.
Sachrajda, A., R. Boulet, Z. Wasilewski, P. Coleridge and F. Guillon, 1990, Sol. State Commun. **74**, 1021.
Sajoto, T., Y.W. Suen, L.W. Engel, M.B. Santos and M. Shayegan, 1990, Phys. Rev. B **41**, 8449.
Sakaki, H., K. Hirakawa, J. Yoshino, S.P. Svensson, Y. Sekiguchi, T. Hotta and S. Nishii, 1984, Surf. Sci. **142**, 306.
Schweitzer, L., B. Kramer and A. MacKinnon, 1984, J. Phys. C **17**, 4111.
Sears, V.F., 1985, Can. J. Phys. **63**, 68.
Smith III, T.P., B.B. Goldberg, M. Heiblum and P.J. Stiles, 1986, Surf. Sci. **170**, 304.
Smrcka, L., 1984, J. Phys. C **17**, L63.
Stahl, E., D. Weiss, G. Weimann, K. von Klitzing and K. Ploog, 1985, J. Phys. C **18**, L783.
Störmer, H.L., 1984, Advances in Solid State Physics, Vol. 24, ed. P. Grosse (Vieweg, Braunschweig) p. 25.
Störmer, H.L., and D.C. Tsui, 1983, Science **220**, 1241.
Störmer, H.L., Z. Schlesinger, A.M. Chang, D.C. Tsui, A.C. Gossard and W. Wiegmann, 1983a, Phys. Rev. Lett. **51**, 126.
Störmer, H.L., A. Chang, D.C. Tsui, J.C.M. Hwang, A.C. Gossard and W. Wiegmann, 1983b, Phys. Rev. Lett. **50**, 1953.
Streda, P., 1982, J. Phys. C **15**, L717.
Streda, P., and K. von Klitzing, 1984, J. Phys. C **17**, L483.
Streda, P., J. Kucera and A.H. MacDonald, 1987, Phys. Rev. Lett. **59**, 1973.
Tao, R., and F.D.M. Haldane, 1986, Phys. Rev. B **33**, 3844.
Tausendfreund, B., and K. von Klitzing, 1984, Surf. Sci. **142**, 220.
Taylor, B.N., 1989, Physics Today, **42**, 23.
Thouless, D.J., 1981, J. Phys. C **14**, 3475.
Thouless, D.J., 1982, Anderson Localization, eds Y. Nagaoka and H. Fukuyama (Springer, Berlin) p. 191.
Thouless, D.J., 1985, Phys. Rev. B **31**, 8305.
Thouless, D.J., M. Kohmoto, M.P. Nightingale and M. den Nijs, 1982, Phys. Rev. Lett. **49**, 405.
Trugman, S.A., 1983, Phys. Rev. B **27**, 7539.
Tsui, D.C., and A.C. Gossard, 1981, Appl. Phys. Lett. **38**, 550.
Tsui, D.C., and H.L. Störmer, 1986, IEEE J. Quantum Electron. QE-**22**, 1711.
Tsui, D.C., H.L. Störmer and A.C. Gossard, 1982, Phys. Rev. Lett. **48**, 1559.
Turberfield, A.J., S.R. Haynes, P.A. Wright, R.A. Ford, R.G. Clark, J.F. Ryan, J.J. Harris and C.T. Foxon, 1990, Phys. Rev. Lett. **65**, 637.
Usov, N.A., and F.R. Ulinich, 1982, Sov. Phys.-JETP **56**, 877.
van Wees, B.J., E.M.M. Willems, C.J.P.M. Harmans, C.W.J. Beenakker, H. van Houten, J.G. Williamson, C.T. Foxon and J.J. Harris, 1989, Phys. Rev. Lett. **62**, 1181.
Visscher, P.B., and L.M. Falicov, 1971, Phys. Rev. B **3**, 2541.
von Klitzing, K., 1982, Surf. Sci. **113**, 1.
von Klitzing, K., 1986, Rev. Mod. Phys. **58**, 519.
von Klitzing, K., G. Dorda and M. Pepper, 1980, Phys. Rev. Lett. **45**, 494.
Wakabayashi, J., and S. Kawaji, 1980, Surf. Sci. **98**, 299.
Wakabayashi, J., A. Fukano, S. Kawaji, K. Hirakawa, H. Sakaki, Y. Koike and T. Fukase, 1988, J. Phys. Soc. Jpn. **57**, 3678.
Washburn, S., A.B. Fowler, H. Schmid and D. Kern, 1988, Phys. Rev. Lett. **61**, 2801.
Wegner, F., 1983, Z. Phys. B **51**, 279.
Wei, H.P., A.M. Chang, D.C. Tsui and M. Razeghi, 1985a, Phys. Rev. B **32**, 7016.
Wei, H.P., D.C. Tsui and A.M.M. Pruisken, 1985b, Phys. Rev. B **33**, 1488.
Wei, H.P., A.M. Chang, D.C. Tsui, A.M.M. Pruisken and M. Razeghi, 1986, Surf. Sci. **170**, 238.
Wilczek, F., 1982, Phys. Rev. Lett. **49**, 957.
Willett, R., J.P. Eisenstein, H.L. Störmer, D.C. Tsui, A.C. Gossard and J.H. English, 1987, Phys. Rev. Lett. **59**, 1776.
Willett, R., H.L. Störmer, D.C. Tsui, A.C. Gossard and J.H. English, 1988, Phys. Rev. B **37**, 8476.

Worlock, J.M., 1990, Physics World **3**, 26.
Wu, Y.S., 1984a, Phys. Rev. Lett. **52**, 2103.
Wu, Y.S., 1984b, Phys. Rev. Lett. **53**, 111.
Yang, C.N., 1962, Rev. Mod. Phys. **34**, 694.
Yoshioka, D., 1984a, Phys. Rev. B **29**, 6833.
Yoshioka, D., 1984b, J. Phys. Soc. Jpn. **53**, 3740.
Yoshioka, D., 1986a, J. Phys. Soc. Jpn. **55**, 885.
Yoshioka, D., 1986b, J. Phys. Soc. Jpn. **55**, 3960.
Yoshioka, D., and P.A. Lee, 1983, Phys. Rev. B **27**, 4986.
Yoshioka, D., B.I. Halperin and P.A. Lee, 1983, Phys. Rev. Lett. **50**, 1219.
Yoshioka, D., A.H. MacDonald and S.M. Girvin, 1989, Phys. Rev. B **39**, 1932.
Zallen, R., and H. Scher, 1971, Phys. Rev. B **4**, 4471.
Zhang, F.C., and T. Chakraborty, 1984, Phys. Rev. B **30**, 7320.
Zhang, F.C., and T. Chakraborty, 1986, Phys. Rev. B **34**, 7076.
Zhang, F.C., and S. Das Sarma, 1986, Phys. Rev. B **33**, 2903.

CHAPTER 18

Hot-Electron Transport Phenomena

DAVID K. FERRY

Department of Electrical Engineering
Arizona State University
Tempe, AZ 85287-5706, USA

Handbook on Semiconductors
Completely Revised Edition
Edited by T.S. Moss
Volume 1, edited by P.T. Landsberg

Contents

1. Introduction 1041
2. Physical observables 1043
   2.1. Velocity saturation 1044
   2.2. Transient transport 1046
   2.3. Impact ionization 1049
3. Analytical solutions to the Boltzmann equation 1051
4. Numerical ensemble Monte Carlo approaches 1059
   4.1. Path integral 1060
   4.2. Monte Carlo sampling techniques 1062
   4.3. Ensemble Monte Carlo 1064
   4.4. Molecular dynamics 1067
5. Femtosecond laser excitation of semiconductors 1070
Appendix A 1074
References 1077

1. Introduction

Essentially all theoretical treatments of electron and hole transport in semiconductors are based upon a one-electron transport equation, usually the Boltzmann transport equation. This has been discussed in a number of earlier chapters, where low-field transport and magnetotransport was discussed. This is especially true in the case of high-electric-field transport, which is the regime that is of interest in this chapter. In low-field transport, one utilizes the fact that the actual carrier distribution function is changed little from its basic Fermi–Dirac (or Maxwell–Boltzmann) form, and that averages can easily be taken using this distribution function. This is no longer true in the case of hot carriers, as the high field, or high energy of the carriers, causes a basic symmetry breaking which can significantly distort the actual distribution function. Thus, the overriding theoretical problem in such far-from-equilibrium transport is the process of obtaining the solution of the transport equation to yield the form of the distribution function in the presence of the field.

For transport purposes, the distribution function is not an end in itself, since integrals (over the distribution function) must be performed in order to evaluate the transport coefficients. It turns out, however, that in many cases, especially in numerical approaches, the appropriate averages can be computed more easily than the direct computation of the distribution function and subsequent integration for the transport average. This is especially true in the ensemble Monte Carlo technique introduced in the latter parts of this chapter, since the transport averages are computed from averages over an ensemble of semiclassical carriers, whose individual trajectories are followed in the numerical simulation.

For large values of the electric field, the average energy of the carriers is increased by the field, and the carriers are said to be "hot", since their equivalent "temperature" is greater than that of the lattice. Because the average energy of the carriers increases in the field, the net rate of phonon emission by the carriers must also increase, in order to transfer the energy gained from the field to the lattice if an overall energy balance situation is to be maintained.

The significance of the comments above can be illustrated by considering fig. 1, in which the various processes that contribute to the variation of the distribution function are shown. Consider a nondegenerate distribution characterized by an "electron" temperature T_e. Phonon emission and absorption processes connect states at energy ε with those at above $\varepsilon + \hbar\omega_0$. The rate of absorption of phonons out of the states at ε (and upward to the upper level) is given by

$$AN_{\boldsymbol{q}} \exp\left(-\frac{\varepsilon}{k_B T_e}\right), \tag{1}$$

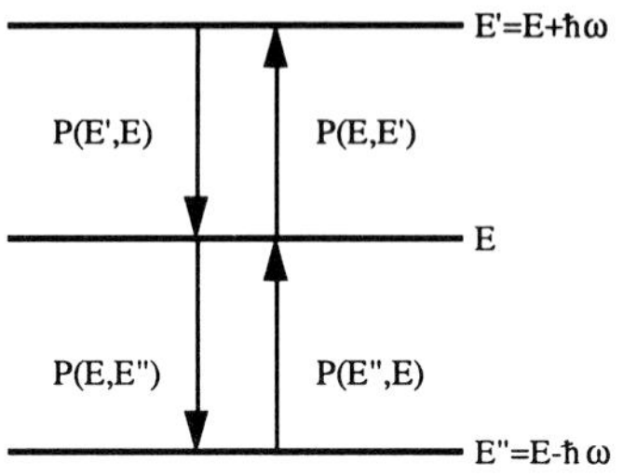

Fig. 1. The energy levels are depicted schematically for the detailed balance argument.

where $N_{\boldsymbol{q}}$ is the Bose–Einstein distribution function for the phonons and A is a proportionality constant. The rate of emission of phonons from the upper level is similarly given by

$$A(N_{\boldsymbol{q}}+1)\exp\left(-\frac{\varepsilon+\hbar\omega_0}{k_B T_e}\right). \tag{2}$$

It may be recalled that $(N_{\boldsymbol{q}}+1)=N_{\boldsymbol{q}}\exp(-\hbar\omega_0/k_B T)$, where T is the lattice temperature. Clearly, detailed balance is achieved in the steady state if $T_e = T$. However, when streaming terms due to the electric field are present, energy is gained from the field by the ensemble of carriers. In this simple tutorial example, the gain in energy is characterized by an increase in the electron temperature T_e, and the emission processes increase over the absorption processes by a factor

$$\exp\left[\frac{\hbar\omega_0}{k_B T}\left(1-\frac{T}{T_e}\right)\right], \tag{3}$$

provided that the distribution function remains a Maxwellian. The amount of the rise in electron temperature is governed by the energy gains and losses and a steady-state balance is achieved when the average rate of loss to the lattice equals the rate of gain of energy from the field, or

$$e\boldsymbol{v}_d\cdot\boldsymbol{E} = -\left.\frac{d\varepsilon}{dt}\right|_{\text{lattice}}. \tag{4}$$

Here, $\boldsymbol{E}$ is the electric field and $\boldsymbol{v}_d$ is the average, or drift, velocity of the ensemble of electrons. Equation (4) is presented here as a conceptual identity for this simple example. It turns out that it will be derived exactly in a later section, where transport is considered in the case where the distribution is taken to have a Maxwellian form. The validity of eq. (4) goes beyond this, however, since it is no more than a statement of the conservation of energy, which must also hold if the system reaches a stable steady state in the presence of the applied fields. The problem of evaluating eq. (4) in such a case usually reduces to one of determining the carrier distribution function.

In general, the transport of hot carriers is nonlinear in that the conductivity is itself a function of the applied electric field. This fact arises since the relationship between the velocity and field is expressed by a mobility, which depends on the average energy of the carriers, and the latter quantity is a function of the high electric field. In normal linear response theory, a linear conductivity is found by a small deviation from the equilibrium distribution function. This small deviation is linear in the electric field, and the equilibrium distribution function dominates the transport properties. Once the carriers begin gaining significant energy from the field, this is no longer the case. The dominant factor in the actual nonlinear transport does not arise directly from higher-order terms in the field, but rather from the implicit field dependence of the nonequilibrium distribution function, such as that of the electron temperature. Thus it is critical to ascertain this nonequilibrium distribution function correctly, because it is the spreading of this function (to higher average energy) in response to the field that dominates nonlinear response in semiconductors.

Equation (4) suggests that one can achieve a nonequilibrium steady state at high values of the electric field. However, under some conditions, such as intervalley transfer leading to negative differential conductance, this may not be a stable steady state. Indeed, in the case of negative differential conductance, space-charge oscillations occur with resulting traveling high-field domains. A solution to eq. (4) would exist under what could be called a "thermodynamic model" (i.e., one calculates the steady-state velocity-field curves and the distribution function without accounting for the time-dependent terms in the Boltzmann equation), but when the system is pushed further from equilibrium, this branch may become unstable, as suggested. In this case, the system evolves toward a new state involving coherent or ordered behavior – the domains and space-charge oscillations of the Gunn effect in intervalley transfer and negative differential conductivity. These coherent structures can be maintained only through a sufficient flow of energy and have been termed dissipative structures (Nicolis and Prigogine 1977). Still, the carrier distribution function must be carefully determined if these effects are to be properly treated.

In this chapter, a few approaches to solving the Boltzmann transport equation (and more appropriate modifications to it) are discussed. The general properties most often observed – velocity saturation and velocity overshoot – are discussed first.

2. Physical observables

Early studies of high-electric-field transport in solids focused mainly on the breakdown studies of dielectrics (Fröhlich and Seitz 1950). As a result, these studies were carried out at very high fields and generally resulted in the destructive breakdown of the sample. With the studies of Shockley (1951), though, emphasis shifted to the study of transport in semiconductors, especially as these materials were becoming useful at that time for new electron devices. The earliest studies of semiconductors focused on the dependence of the drift velocity on the applied electric field and the falloff of mobility at high fields, which led to the concept of velocity saturation. This remains one of the ways of evaluating high-field behavior of semiconductors since it directly evaluates the effectiveness of electron–phonon interaction. In the early 1960s attention was focused on the observation of negative differential conductivity (NDC) in several semiconductors (Gunn 1963), and this led to renewed interest in the field for the applicability of this effect to microwave devices. When individual semiconductor devices began to be fabricated with gate lengths in the 1 μm regime, velocity saturation was found to be important in the operation of these devices, as predicted earlier (Grosvalet et al. 1963, Trofimenkoff 1965). With the possibility of still shorter gate lengths, focus in recent years has shifted to the role of nonstationary transport and velocity overshoot. These effects provide a valuable overview of the general behavior that is associated with hot-carrier dynamics (Conwell 1967).

2.1. Velocity saturation

As the magnitude of the electric field applied to a semiconductor is raised, the carriers begin to gain energy from the field. To balance this energy gain, there is an increase in the energy loss to the lattice via a net emission of phonons. Due to the increased rate of phonon emission, the actual rate of scattering of the carriers by the phonons increases and the mobility is reduced. As a result, the velocity increases sublinearly with the electric field. At very high electric fields, the velocity almost saturates, continuing to increase only slightly with further increases in the electric field (Ryder 1953).

The saturated (or nearly saturated) velocity is an important parameter for electron device considerations. In many studies concerning the appropriate figures of merit for high-speed, high-frequency, or high-power devices, it is readily apparent that this parameter affects the frequency response (and hence the speed, through the transit time) and the power-handling capabilities (through the peak current). Thus it is an important parameter for the study of high-field effects, not only for the reasons stated above, but also because it is a direct mirror of the electron–phonon interaction. Indeed, the saturated velocity is a characterization of the far-from-equilibrium carrier system, which is dependent on the lattice interactions in governing the form of the nonequilibrium distribution function. The nonequilibrium distribution deforms to fit the nature of the electron–lattice interactions rather than maintaining the equilibrium Maxwellian form. A competing process is carrier–carrier scattering, which works to return the distribution function to a quasi-Maxwellian form – a Maxwellian form, but with an enhanced carrier temperature, and perhaps with other parameters as well.

It turns out that the saturation velocity is scalable from material to material, precisely because it is a property of the carrier–lattice interaction. The rate of energy loss can be estimated from eq. (4) to be approximately

$$ev_s E \sim \tfrac{3}{2}\hbar\omega_0(e^x - 1)N_{\boldsymbol{q}}, \tag{5}$$

where the last factors arise from the difference between emission and absorption processes, and $x = \hbar\omega_0/k_B T$. The details on the right-hand side are the difference between emission and absorption (rather than the sum used in calculating just a scattering rate), multiplied by the energy exchanged $\hbar\omega_0$. Most numerical factors have been ignored in eq. (5), as they will be divided out shortly. In a fashion similar to eq. (4), the momentum relaxation can be characterized by

$$eE \sim 2m^* v_s (e^x + 1) N_{\boldsymbol{q}}, \tag{6}$$

where the last factor arises from the sum of emission and absorption processes. The factor of 2 is a numerical factor correct only for nonpolar optical phonon scattering. Equations (5) and (6) can now be solved for the saturated velocity, by dividing the latter into the former (which causes the afore-mentioned constants to cancel each other), and then solving for the saturated velocity:

$$v_s = \sqrt{\frac{3\hbar\omega_0}{4m^*}} \tanh^{1/2}\left(\frac{\hbar\omega_0}{2k_B T}\right). \tag{7}$$

As remarked, eq. (7) is correct only for nonpolar scattering, and even for this scattering appears in slightly modified forms in a variety of other derivations. However, these various forms differ by only a small numerical factor. The result allows one to predict variations in materials, as it depends only on the effective mass, the phonon energy, and the lattice temperature. In fig. 2 the observed and/or calculated (by detailed calculations) saturation velocities are plotted against eq. (7) for a variety of semiconductors. Only the dominant phonon is used in this comparison.

In materials in which the polar-optical phonon is the dominant scattering mechanism, an interesting phenomenon has been predicted to occur – polar runaway. This effect is predicted to arise because the scattering rate (for the case of no screening) varies as (for phonon emission)

$$\frac{1}{\sqrt{\varepsilon}} \sinh^{-1}\left[\sqrt{\frac{\varepsilon}{\hbar\omega_0}} - 1\right], \tag{8}$$

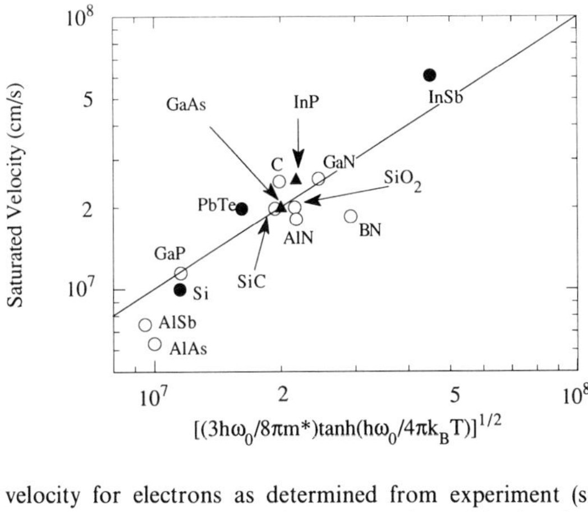

Fig. 2. Saturated drift velocity for electrons as determined from experiment (solid symbols) or from detailed calculations (open symbols), compared with eq. (7). The solid triangles are peak velocities in materials that exhibit negative differential conductivity, while the solid circles are actual saturated velocities. All values are at 300 K, except for InSb, which is at 77 K. (Note that Planck's constant is used here and not $\hbar$.)

so that when the energy of the carrier is sufficiently large compared to the phonon energy, the scattering rate actually decreases for further increases in the energy. This is an unstable situation for which no steady state exists, and the carriers run off to very high energies (Stratton 1958). Equations (5) and (6) assume that a stable steady state exists, although the existence of this steady state has not been fully investigated for all scattering processes, and certainly not for all semiconductors.

In the case of hot carriers, the considerable power given to the carriers by the field and relaxed to the lattice can drive the phonon distribution out of equilibrium, so that a hot-phonon distribution can occur. At room temperatures the phonon lifetimes are sufficiently short that such an effect is unlikely to occur in normal circumstances. At low temperatures, however, it can be expected that the situation is different. The long-wavelength acoustic phonons can achieve lifetimes as long as a few nanoseconds, so that this distribution could reasonably be expected to be driven out of equilibrium (Ferry 1974). This is the case in acoustoelectric effects. The optical phonons, though, have a shorter lifetime, so that it is unlikely that this distribution will be driven out of equilibrium except, e.g., under intense laser irradiation, where carriers are created high in the energy band and subsequently emit a shower of optical phonons (Kash et al. 1985). The optical-phonon lifetime, in fact, does not vary more than a factor of 2 to 4 with temperature. However, the difference between room temperature and low temperature (4.2 K) lies in the number of optical phonons that are thermally excited. If a number of phonons are emitted by the electrons (or holes), the importance of these excess phonons to the carrier–phonon scattering processes depends on how large the excess density is in comparison with the background. At room temperature, the background density of optical phonons is generally thought to be sufficient to wash out the hot-phonon effect, but this is not the case at low temperature. If the phonons are driven out of equilibrium, the saturation velocity can be expected to reflect this effect.

2.2. *Transient transport*

In high electric fields, there are at least two different relaxation times corresponding to the relaxation of both momentum and energy. The momentum relaxation time τ_m describes the decay of the velocity (and the velocity fluctuations about a local near-equilibrium state) and is the relaxation time dealt with predominantly in linear transport. However, the nonlinear transport in high electric fields arises primarily from the change in the distribution function in the presence of the high electric field, which leads to an increase in the average energy of the carriers. The response of the distribution function, which results in this increase in average energy, is characterized by its own relaxation time, which is referred to as the energy relaxation time τ_e, since the evolution of the distribution function represents the evolution of the average energy of the carrier ensemble.

If the energy relaxation process is slower than the momentum relaxation process, the velocity can overshoot its ultimate steady-state value (the saturation velocity) in high fields. This occurs because the distribution function first shifts in momentum space as the velocity rises to a value characterized mainly by its low-field mobility. As the distribution function then evolves to its nonequilibrium form, the mobility decreases to its ultimate high-field value, with a consequent decrease in the velocity.

It can readily be shown that this "overshoot" behavior requires a more complicated behavior than where the drift momentum is assumed to obey the standard Langevin equation

$$\frac{\mathrm{d}v}{\mathrm{d}t} = \frac{eE}{m^*} - \frac{v}{\tau_\mathrm{m}}. \tag{9}$$

When overshoot occurs, the left-hand side of eq. (9) must have at least two zeros – one at a time corresponding to the steady state and one at a time corresponding to the peak velocity. However, it is shown in fig. 3 that the relaxation rate is an increasing function of energy (or velocity), so that the right-hand side has only a single zero. Thus a second time scale must be involved, which is the characteristic time of the energy relaxation. In the latter case, the motion of the particles is governed by a retarded Langevin equation, written as (Zwanzig 1961)

$$\frac{\mathrm{d}v}{\mathrm{d}t} = -\int_0^t \gamma(t-u)v(u)\,\mathrm{d}u + \frac{R(t)}{m^*} + \frac{eE}{m^*}h(t), \tag{10}$$

where $R(t)$ is the random force symbolizing the nonregular part of the scattering processes, and it is assumed that the field is turned on at $t = 0$ [through the Heaviside function $h(t)$]. The function $\gamma(t)$ is a "memory function" for the non-equilibrium system, which will be related to the correlation function below.

Equation (10) is a non-Markovian form of the Langevin equation, since the rate of change of the velocity at time t depends not only on the present time but also on all past time. This equation can easily be solved by a Laplace transform technique,

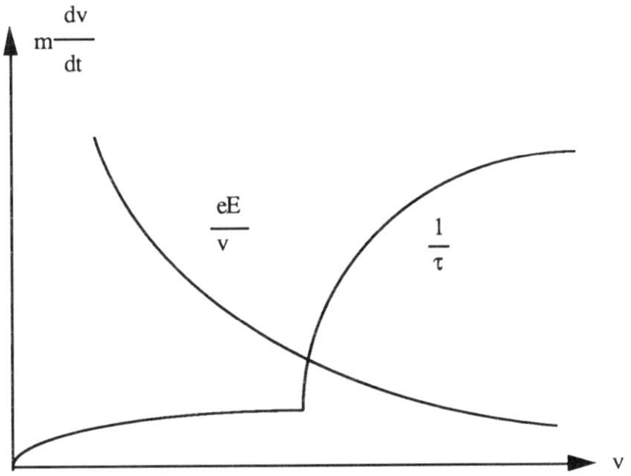

Fig. 3. The scattering relaxation rate is plotted as a function of the velocity, so that it is clear that eq. (9) has only a single zero, which must be the long-time steady state.

which generates the characteristic function (from the terms multiplying the transform of v)

$$X(s) \equiv \frac{1}{s + \hat{\gamma}(s)}. \tag{11}$$

The caret over γ implies the Laplace transform when capital letters are not used. This then leads to the solution

$$v(t) = v(0)x(t) + \frac{eE}{m^*}\int_0^t x(u)\,\mathrm{d}u + \frac{1}{m^*}\int_0^t R(t-u)x(u)\,\mathrm{d}u, \tag{12}$$

which is a general expression of the velocity of each carrier under the influence of the external field and the collisions. The function $x(t)$ may be found by averaging eq. (12) over the entire ensemble, under the assumption that the random force has a zero average and is uncorrelated with $x(t)$. This result leads to

$$v_{\mathrm{d}}(t) = \langle v(t)\rangle = \frac{eE}{m^*}\int_0^t x(u)\,\mathrm{d}u, \tag{13}$$

which tells us that $x(t)$ is the relaxation function discussed by Kubo (1957) (eq. (13) is a form of the Kubo formula). To understand this function, consider the two-time correlation function

$$\varphi_{\Delta v}(t', t) = \varphi_{\Delta v}(t, t') = \langle v(t')v(t)\rangle - v_{\mathrm{d}}(t)v_{\mathrm{d}}(t'). \tag{14}$$

The two-time behavior arises because the distribution function is nonstationary. If both sides of eq. (12) are multiplied by $v(0)$ and then the ensemble average is taken, under the assumptions that $\langle v(0)\rangle = 0$ and the initial velocity is uncorrelated with the random force, the result is

$$\varphi_{\Delta v}(0, t) = \langle v^2(0)\rangle x(t), \tag{15}$$

so that $x(t)$ is readily recognized as the normalized velocity autocorrelation function beginning at the initial time at which the field is applied.

The results given in eqs. (13) and (15) clearly indicate the need for a nonequilibrium system, characterized by at least two time constants, in order to show velocity overshoot. Clearly, eq. (13) indicates that the temporal behavior of the velocity is related to the integral of the autocorrelation function. If the velocity is to exhibit an overshoot behavior, the correlation function must be a nonmonotonic function, which is quite different from the behavior found in equilibrium systems. Indeed, if equilibrium behavior, with a single momentum relaxation time, is assumed so that $x(t) = \exp(-t/\tau_{\mathrm{m}})$, then eq. (11) clearly indicates that $\gamma(t) = \delta(t)/\tau_{\mathrm{m}}$, so that eq. (9) is recovered. Since it has already been shown that this cannot exhibit overshoot behavior, the observation of the latter effect must arise from a "far-from-equilibrium" system which exhibits at least two-time-constant behavior (and a negative excursion of the velocity autocorrelation function). If an energy relaxation time (which is the second time constant) is introduced through the definition

$$\gamma(t)=\frac{1}{\tau_m\tau_e}e^{-t/\tau_e}, \tag{16}$$

which has the limiting form of a delta function as τ_e goes to zero, it is found from eq. (11) that

$$X(s)=\frac{\tau_m(s\tau_e+1)}{s\tau_m(s\tau_e+1)+1}, \tag{17a}$$

for which the inverse transform is

$$\begin{aligned}x(t)=&\frac{1}{2}\left(1-\frac{1}{\sqrt{1-4\tau_e/\tau_m}}\right)\exp\left[-\frac{t}{2\tau_e}\left(1+\sqrt{1-4\tau_e/\tau_m}\right)\right]\\&+\frac{1}{2}\left[1+\frac{1}{\sqrt{1-4\tau_e/\tau_m}}\right]\exp\left[-\frac{t}{2\tau_e}\left(1-\sqrt{1-4\tau_e/\tau_m}\right)\right].\end{aligned} \tag{17b}$$

The latter equation exhibits the nonmonotonic behavior required to achieve velocity overshoot provided $\tau_e > \tau_m/4$.

These results illustrate a number of important aspects of hot-carrier behavior. First, the dynamics become retarded with a memory effect (the function $\gamma(t)$ is the memory function) because of the extra time behavior of the evolution of the distribution function. This, in turn, opens the door for velocity overshoot to occur. Moreover, this process, when coupled with the velocity saturation effect, clearly indicates the far-from-equilibrium nature of this nonlinear transport. In the field range where velocity saturation and velocity overshoot can occur, the distribution function is determined by carrier–lattice interactions and by boundary conditions in the form of, e.g., applied fields, and is not simply related to the equilibrium form.

2.3. *Impact ionization*

As electrons (or holes) drift through a semiconductor under the influence of the electric field, they gain energy from the field and lose it through collisions with the lattice. At relatively high electric fields, though, a few electrons with energies in the tail of the distribution will have gained sufficient energy for an additional type of energy dissipating collision – pair production through the *inverse Auger process*. An energetic electron (or hole) gives up its energy to a valence electron, raising the latter into the conduction band and leaving an additional hole in the valence band. This process is the reciprocal of Auger recombination, and this is the source of the name, but it is also called impact ionization, because the initial particle *knocks* the valence electron loose to create the electron–hole pair. The original particle has created two new particles (the new electron–hole pair). At low temperatures, collisions with neutral impurities can lead to the same effect, except that the hole is localized on the impurity so that only one new carrier is produced. The number of ionizing collisions produced will depend on the number of electrons (or holes) which have a sufficient energy for the pair production process and also will depend on the relative collision probabilities for other scattering processes.

The parameters that are desired for a discussion of the inverse Auger process are the ionization rate α (in units of cm^{-1}) and a related parameter, the generation rate $g(E)$ (in units of s^{-1}). The former can be expressed as

$$\alpha = \frac{1}{n_0 v_d} \sum_{\varepsilon > \varepsilon_i} n(\varepsilon)\Gamma_{\text{ion}}(\varepsilon) f(\varepsilon), \tag{18}$$

where $\Gamma_{\text{ion}}(\varepsilon) = v(\varepsilon)/l_{\text{ion}}$, $v(\varepsilon)$ is the velocity at energy ε, $n(\varepsilon)$ is the density of states, and l_{ion} is the mean free path for an ionizing collision. The latter quantity can roughly be calculated as $g = \alpha v$, although more properly, the multiplicative velocity term in g should be added to α prior to the summation over the energy, as

$$g(\varepsilon) = \frac{1}{n_0 v_d(\varepsilon)} \sum_{\varepsilon > \varepsilon_i} n(\varepsilon) v(\varepsilon)\Gamma_{\text{ion}}(\varepsilon) f(\varepsilon). \tag{19}$$

In both of these equations, ε_i is the threshold energy for the creation of an electron–hole pair.

Exact calculations for the ionization rate α, or generation rate g, must involve a knowledge of the distribution function of the carriers, which in itself is the major theoretical task involved. As observed above, solving for the distribution function can be a complicated task, since the added collision term for ionizing collisions must also be included. Most of the so-called "exact" theories for impact ionization which have appeared over the years are concerned with solving for the distribution function first. In this section, however, primary concern is with impact ionization itself, and a quite reasonable comprehension of the field dependence of the process can be obtained by considering two separate extreme cases, one for very high electric fields for which the distribution function is nearly spherically symmetric, and one for lower values of the electric field where concern is primarily with a few "lucky" electrons that are accelerated ballistically to a sufficient energy for ionization without having undergone any phonon collisions.

Wolff (1954), in one of the earliest theories applicable to semiconductors, assumed that a strong scattering interaction was always present in the semiconductor, and that this led to a quasi-Maxwellian distribution function, dominated by the spherically symmetric part. The field causes the distribution to become less and less spherically symmetric, and it is fair to ask if Wolff's assumptions could have any relevance. At very high electric fields, the diffusive effect (in energy space) dominates any anisotropy introduced by the scatterers. In fact, the electron temperature, if one can be defined, (or the average energy) rises very rapidly with electric field for fields of the order on the breakdown field. In turn, the scattering rates also rise very rapidly, which tends to increase the momentum randomization. Even polar scattering tends to be mainly backscattering at very high energies, which reduces the asymmetry. The major effect of all of this, however, is that the symmetric part of the distribution spreads so rapidly with the rapidly increasing electron temperature that it begins to provide the major fraction of the carriers undergoing impact ionization from electrons in the tail of this part of the distribution. The diffusion approximation leads to a quasi-Maxwellian for $f_0(\varepsilon)$ with an effective temperature related to the total mean free path L, or

$$k_B T_e \sim (eEL)^2/3\hbar\omega_0. \tag{20}$$

The form of eq. (20) is clearly that of a diffusion in "energy" since the numerator is the square of the energy gained over a mean free path and the denominator is the energy loss per collision. Since α (or g) follows from an integration over f_0, one then obtains (using the "temperature" of eq. (20) in a simple Maxwellian form, and then finding the fraction of the carriers with an energy greater than the threshold for ionization)

$$\alpha(E) \sim A \exp(-k/E^2). \tag{21}$$

Shockley (1961), on the other hand, assumed a very-weak interaction between the electron and the lattice, so that $f_0(\varepsilon)$ retained a strongly peaked form, with the peak in the direction of the electric field E. In his model, impact ionization is due to a few "lucky" electrons in the tail of the asymmetric part of the distribution. This first ballistic transport model assumed that these few electrons are accelerated to the threshold energy ε_i without collisions, so that the probability is proportional to (which is just the statistical probability that electrons reach the threshold energy if they only gain a fixed energy per mean free path)

$$\alpha(E) \sim B \exp(-\varepsilon_i/eEL). \tag{22}$$

In general, the fields are neither so high as to validate Wolff's model nor so low as to validate Shockley's approach.

It is clear that the exact distribution function is really required in evaluating any possible approach. Also, the threshold energy required to create the extra electron–hole pair is required. A more detailed treatment is beyond the present approach, as we concentrate here just on solving for the distribution function. A more detailed treatment has been given by Ferry (1991).

3. Analytical solutions to the Boltzmann equation

It is now apparent that it is necessary to find the distribution function in order to solve the hot carrier problem completely. While simple models give qualitative agreement with the expected results, full understanding requires the full and complete details of the carrier–lattice interactions. In this section, one analytical approach – the drifted Maxwellian technique – is discussed. This will be followed in the next section by a discussion of the numerical approaches that have become prevalent in recent years. The treatment is limited to a discussion of electrons.

The treatment of high-electric-field effects can generally be broken into two distinct regions. These two can be differentiated by the concept of the electron temperature introduced previously. The concept of such a temperature is valid only when the spherically symmetric part of the distribution function, denoted by $f_0(\varepsilon)$, remains in a Maxwellian form as $\exp(-\varepsilon/k_B T_e)$. If the distribution function differs from this form in high electric fields, the concept of the electron temperature becomes quite vague. However, Fröhlich and Paranjape (1956) showed many years ago that if the density of electrons is sufficiently high, energy and momentum exchanges are dominated primarily by inter-electronic collisions. These collisions provide the fast time scale that dominates the distribution function and forces it into a quasi-equilibrium *form* as described by Bogoliubov (1962). If this assumption is valid, the electron distribution function in momentum space will be in internal equilibrium at an electronic temperature T_e, which is a function of the electric field, although it may also be necessary to include other parameters which themselves are related to constants of the motion. In general, the electron temperature is greater than the lattice temperature. The distribution function can then be thought of as shifted in momentum space, as in the near-equilibrium case, but the spreading of the distribution is fully characterized by the electron temperature. The shift in momentum space is characterized by the drift velocity (or drift momentum, as complications due to nonparabolic bands will not be included).

In the electron–electron interaction, a simple argument illustrates the size of the carrier density needed to achieve the quasi-equilibrium distribution function. The energy loss due to energy exchange among carriers is characterized by the rate at which individual electrons are scattered. The energy loss by this process may be approximated by the product of the inter-electronic Coulomb energy and the plasma frequency, which characterizes the time scale of the interelectronic interactions. This gives

$$\left.\frac{d\varepsilon}{dt}\right|_{e-e} \sim -\frac{e^2\omega_p}{4\pi\varepsilon_0\langle r\rangle} \sim -\frac{e^2\omega_p^2}{4\pi\varepsilon_0 v} = -\frac{ne^4}{4\pi m^* v\varepsilon_0^2}, \tag{23}$$

where it has been assumed that the average interelectron distance may be expressed as $\langle r\rangle \sim v/\omega_p$. In fact, this result needs to be modified somewhat by the self-screening of the interaction. The more "correct" form is given by (Pines 1953)

$$\left.\frac{d\varepsilon}{dt}\right|_{e-e} = -\frac{ne^4}{4\pi m^* v\varepsilon_0^2}\ln\left(\frac{1}{bk_c}\right), \tag{24}$$

where b is an impact parameter ($\sim 1/n^{1/3}$) and k_c is a screening wave vector. However, the logarithmic term is generally *assumed to be* of the order of unity. In fact, neither of these forms are used for anything other than an illustrative argument over the physics of the interaction. It should be remarked that carriers which are moving faster than the average velocity generally lose energy by this process, whereas those that are traveling less than the average velocity gain energy by this process. In general, averaging eq. (23) or eq. (24) (the former is usually used) over a Maxwellian produces a zero result, since interelectronic collisions conserve total momentum, energy, and the number of carriers in the system. For this reason it is usually assumed that if a parameterized Maxwellian distribution is taken for the carriers, the details of the interelectronic interactions can be neglected under the premise that they are already included *de facto* within the distribution function.

In order for inter-electronic collisions to dominate the energy relaxation, it is necessary for the energy relaxation rate described above to dominate the lattice scattering rate, or

$$\left|\frac{d\varepsilon}{dt}\right|_{e-e} > \left|\frac{d\varepsilon}{dt}\right|_{e-L}. \tag{25}$$

If this inequality is satisfied, the assumption of an electron temperature is thought to be valid. If, on the other hand, eq. (25) is not satisfied, the distribution function can deviate markedly from a Maxwellian in form. For the latter situation, an electron temperature cannot be formally defined, and the form of the distribution function must be found from a full solution of the Boltzmann equation. For most semiconductors, the critical density for which eq. (25) becomes satisfied is of the order of 10^{15} to $10^{17}\ \mathrm{cm}^{-3}$.

The general approach that is followed, and which will be described in detail in the following, when the distribution function can be expressed as a generalized Maxwellian, is to multiply the Boltzmann equation with powers of the momentum $p = mv = \hbar k$, and then integrate over the momentum itself. This produces a series of equations that balance the various moments of the Boltzmann equation and can be used to evaluate the electron temperature and drift velocity, which have been included in the distribution function as parameters:

$$f(\boldsymbol{p}, \boldsymbol{v}_d, T_e) = C \exp(-(\boldsymbol{p} - \boldsymbol{p}_d)^2/2m^* k_B T_e), \tag{26}$$

where $\boldsymbol{p}_d = m^* \boldsymbol{v}_d = \hbar \boldsymbol{k}_d$ is the average drift momentum of the electron gas, C is a constant for normalization purposes, and T_e is the electron temperature. The first two terms of the generalized distribution function in eq. (26) may be found by expanding the exponential factor as

$$f_0(\varepsilon) = C \exp(-\varepsilon/k_B T_e) \tag{27}$$

for the spherically symmetric part, and

$$f_1(\varepsilon) = v\left(\frac{m^* v_d}{k_B T_e}\right) f_0(\varepsilon). \tag{28}$$

In arriving at eqs. (27) and (28), it has been assumed that $p_d^2/2m^* k_B T_e \ll 1$ and $\boldsymbol{p} \cdot \boldsymbol{p}_d/2m^* k_B T_e \ll 1$ are satisfied. The goal, of course, is to write the distribution function as a sum of a spherically symmetric term plus a perturbative term. The former is termed the energy distribution function and does not contribute to any average of odd powers of momentum, while the latter is the drift term.

The actual moments of the Boltzmann equation are complicated by the fact that the semiconductor is often describable by multiple valleys and hence is a coupled, and complicated, system. Indeed, there will be conditions in which even sets of valleys, which are equivalent in equilibrium, can become inequivalent in a high electric field. Thus full account for transfer between inequivalent valleys will be considered as the moment equations are developed. The starting point is, of course, the Boltzmann transport equation

$$\frac{\partial f}{\partial t} + \boldsymbol{v} \cdot \nabla f + e\boldsymbol{E} \cdot \frac{\partial f}{\partial \boldsymbol{p}} = \sum_{\boldsymbol{p}'} (W(\boldsymbol{p}, \boldsymbol{p}')f(\boldsymbol{p}') - W(\boldsymbol{p}', \boldsymbol{p})f(\boldsymbol{p})), \tag{29}$$

where $W(\boldsymbol{p}, \boldsymbol{p}')$ is the scattering rate from state $\boldsymbol{p}'$ to state $\boldsymbol{p}$. A function $\varphi(\boldsymbol{p})$ is assumed to be a simple power of the momentum $\boldsymbol{p}$ (e.g., p^n), and its average is described by

$$\langle \varphi(\boldsymbol{p}) \rangle = \frac{1}{n} \int \varphi(\boldsymbol{p}) f(\boldsymbol{p}) \, \mathrm{d}^3\boldsymbol{p}. \tag{30}$$

The Boltzmann equation is now multiplied by an arbitrary function $\varphi_i(\boldsymbol{p})$ (which will often just be written as φ_i, when there is no chance of confusion), in the ith valley, and then integrated (or summed as the case may be) over the momentum. This results in the form

$$\begin{aligned} \frac{\partial(n_i\langle \varphi_i \rangle)}{\partial t} &= n_i e\boldsymbol{E} \cdot \left\langle \frac{\partial \varphi_i}{\partial \boldsymbol{p}_i} \right\rangle - \frac{1}{m^*} \nabla \cdot (n_i \langle \boldsymbol{p}_i \varphi_i \rangle) \\ &+ \sum_{\boldsymbol{p}_i \boldsymbol{p}_i'} (W(\boldsymbol{p}_i, \boldsymbol{p}_i') f_i(\boldsymbol{p}_i') - W(\boldsymbol{p}_i', \boldsymbol{p}_i) f_i(\boldsymbol{p}_i)) \varphi(\boldsymbol{p}_i) \\ &+ \sum_{\boldsymbol{p}_i \boldsymbol{p}_j'} (W(\boldsymbol{p}_i, \boldsymbol{p}_j') f_j(\boldsymbol{p}_j') - W(\boldsymbol{p}_j', \boldsymbol{p}_i) f_i(\boldsymbol{p}_i)) \varphi(\boldsymbol{p}_i), \end{aligned} \tag{31}$$

where n_i is the density in the ith valley. Here the first term on the right-hand side has been integrated by parts to obtain the form shown here, and different sums have been provided for scattering within valley i and to/from valley j, which is assumed to be part of a different set of equivalent valleys. The second term on the right-hand side accounts for diffusive forces. When the initial and final states lie in the same valley (or an equivalent valley), they are describable by the same distribution function. However, when the initial and final states lie in nonequivalent valleys, the two distribution functions are, in general, not the same quasi-equilibrium distribution; e.g., a family of distributions may be required to describe all of the valleys that are involved in the far-from-equilibrium transport problem. With this in mind, the intravalley (or intra-equivalent valleys) terms on the right-hand side of eq. (31) (the summations over p_i') can be treated by a simple change of variables, since the summation is over both momenta, and this term becomes

$$n_i \langle \Gamma_\varphi(\boldsymbol{p}_i) \rangle, \tag{32}$$

where

$$\Gamma_\varphi(\boldsymbol{p}_i) = \sum_{\boldsymbol{p}_i'} [\varphi(\boldsymbol{p}_i') - \varphi(\boldsymbol{p}_i)] W(\boldsymbol{p}_i', \boldsymbol{p}_i). \tag{33}$$

If $\varphi = C$, this term vanishes and intravalley terms do not contribute to the lowest-order (density) balance equation. The remaining scattering term, for scattering between nonequivalent valleys, can be written as

$$n_i \frac{\mathrm{d}\langle\varphi\rangle}{\mathrm{d}t}\bigg|_{\mathrm{e-L}} = \sum_{\boldsymbol{p}_i} \varphi(\boldsymbol{p}_i) \sum_{\boldsymbol{p}_j} (W(\boldsymbol{p}_i, \boldsymbol{p}'_j) f_j(\boldsymbol{p}'_j) - W(\boldsymbol{p}'_j, \boldsymbol{p}_i) f_i(\boldsymbol{p}_i))$$
$$= \sum_{\boldsymbol{p}'_j} f_j(\boldsymbol{p}'_j)\varphi(\xi_j) \sum_{\boldsymbol{p}_i} [W(\boldsymbol{p}_i, \boldsymbol{p}'_j)] - \sum_{\boldsymbol{p}_i} f_i(\boldsymbol{p}_i)\varphi(\boldsymbol{p}_i) \sum_{\boldsymbol{p}'_j} [W(\boldsymbol{p}'_j, \boldsymbol{p}_i)], \tag{34}$$

where the energy-conserving delta function inherent in the scattering rates has been used to introduce ξ_j as a shifted momentum defined by

$$\varepsilon_j(\xi_j) = \varepsilon_i(\boldsymbol{p}_i) \pm \hbar\omega_0 + \Delta_{ji}. \tag{35}$$

The shift between the energy references has been included via Δ_{ij} to account for any energy separation between nonequivalent valley sets. Thus the momentum function $\varphi(\boldsymbol{p})$ has now been shifted to the other summation. Now, introducing the intervalley scattering rate as

$$\Gamma_{ij}(\boldsymbol{p}) = \sum_{\boldsymbol{p}'_j} [W(\boldsymbol{p}'_j, \boldsymbol{p}_i)], \tag{36}$$

eq. (34) now becomes

$$\left\langle \frac{\mathrm{d}(n\varphi)}{\mathrm{d}t} \right\rangle_{ij} = n_j \langle \varphi(\xi_j)\Gamma_{ji}(\boldsymbol{p}'_j)\rangle_j - n_i \langle \varphi(\boldsymbol{p}_i)\Gamma_{ij}(\boldsymbol{p}_i)\rangle_i, \tag{37}$$

and the subscript on the average refers to the particular distribution function for nonequivalent sets of valleys. By inserting the details of the various scattering processes, the individual moment equations can be readily set up for any situation. However, these equations can be quite complicated in the presence of multiple scattering processes. The first three moments become the density balance equation, where $\varphi(\boldsymbol{p}) = C$,

$$\frac{\mathrm{d}n_i}{\mathrm{d}t} + \nabla \cdot (n_i \boldsymbol{v}_{\mathrm{d}i}) = \sum_j (\langle \Gamma_{ji} \rangle_j - \langle \Gamma_{ij} \rangle_i), \tag{38}$$

where the first two terms (the left-hand side) lead to the normal continuity equation. The momentum balance equation, where $\varphi(\boldsymbol{p}) = \boldsymbol{p}$, is found to be

$$\frac{\partial(n_i m^* \boldsymbol{v}_{\mathrm{d}i})}{\partial t} + \nabla \cdot (n_i \langle m^* \boldsymbol{v}\boldsymbol{v}\rangle_i)$$
$$= e\boldsymbol{E}n_i - n_i\langle \Gamma_{\boldsymbol{p}}\rangle_i - n_i \sum_j \langle \boldsymbol{p}_i \Gamma_{ij}\rangle_i + \sum_j n_j \langle \sqrt{(2m_j(\varepsilon_i(\boldsymbol{p}i) \pm \hbar\omega_0 - \Delta_{ji})}\, \Gamma_{ij}\rangle_j, \tag{39}$$

and the energy balance equation, where $\varphi(\boldsymbol{p}) = \boldsymbol{p}^2/2m^*$, is similarly

$$\frac{\partial(n_i\langle\varepsilon\rangle_i)}{\partial t} + \nabla \cdot (n_i \langle \boldsymbol{v}\varepsilon\rangle_i)$$
$$= e\boldsymbol{E} \cdot \boldsymbol{v}_{\mathrm{d}i} n_i - n_i \langle \Gamma\varepsilon\rangle_i - n_i \sum_j \langle \varepsilon_i \Gamma_{ij}\rangle_i + \sum_j n_j \langle (\varepsilon_j - \Delta_{ji})\Gamma_{ij}\rangle_j. \tag{40}$$

In these equations, the quantity Δ represents the shift between the bottom of valley set j and valley set i (this is a positive quantity if the bottom of valley j lies above that of valley i, and conversely). In each case the last two terms on the right-hand side represent the transfer of particles (or momentum, or energy) in the intervalley scattering process between nonequivalent sets of valleys. The last term is that for "in" scattering, and the next-to-last term is for "out" scattering processes to the set of valleys i. In the case of a single set of equivalent valleys, these two terms cancel each other.

If the current density is defined by $\boldsymbol{J} = -ne\boldsymbol{v}_{\mathrm{d}}$ (for electrons the charge is negative, and the vector direction of the current is opposite that of the velocity), eq. (38) becomes just the normal continuity equation

$$\frac{\partial \rho_i}{\partial t} + \nabla \cdot \boldsymbol{J}_i = -e(g_i - r_i), \tag{38'}$$

where the "generation–recombination" terms represent, respectively, the transfer of particles from valley j to valley i (summed over the set of valleys j), and conversely. The other two equations are not so simply related to normal equations. For example, if $\langle \Gamma_{\boldsymbol{p}} \rangle$ is defined to be $m^* v_{\mathrm{d}}/\tau_{\mathrm{m}}$, eq. (39) can be written as (after multiplying by $-e$)

$$\boldsymbol{J}_i = \sigma_i \boldsymbol{E} - \tau_{mi}\rho_i \frac{\partial \boldsymbol{v}_{\mathrm{d}i}}{\partial t} + eD_{ei}\nabla n_i + e\tau_{mi}\left[\boldsymbol{v}_{\mathrm{d}i}(g_i - r_i) + \frac{n_i}{m^*}(\boldsymbol{G}_{pi} - \boldsymbol{R}_{pi})\right]. \tag{39'}$$

The first term on the right-hand side is obviously just the conductivity term normally found in the current equation. In the first term in brackets, the first term has been used with eq. (38′) to eliminate the partial derivative of the current with respect to time, the divergence of the velocity is taken to vanish, and the third term is used to *define* a diffusion constant. The last terms of the equation relate to the transfer of particles and momentum in the intervalley scattering process. It is important to note that it was necessary to define the diffusion term and that it does not really follow naturally unless the temperature is the lattice temperature. Also, if there is no intervalley process to consider, the sample is usually space-charge neutral, and the right-hand side consists of just the drift and diffusion terms. On the other hand, the time derivative of the drift velocity can be interpreted to include displacement currents as well.

As an example of the drifted Maxwellian approach, transport in a polar material such as GaAs will now be considered. In these materials, electronic motion in the central valley of the conduction band is dominated by the polar mode of the optical phonons. The example considered is to calculate the limiting velocity of the carriers (if such exists). In this single valley, the steady-state balance equations for the momentum and energy are

$$-e\boldsymbol{E} \cdot \boldsymbol{v}_{\mathrm{d}} = \frac{1}{n}\left\langle \frac{\mathrm{d}\varepsilon}{\mathrm{d}t} \right\rangle_{\mathrm{PO}}, \tag{41}$$

and

$$-e\boldsymbol{E}=\frac{1}{n}\left\langle\frac{\mathrm{d}\boldsymbol{p}}{\mathrm{d}t}\right\rangle_{\mathrm{PO}}. \tag{42}$$

The polar-mode scattering process for the optical phonons leads to the use of a scattering rate here for the unscreened interaction. In order to apply the balance equations, the average rates of energy and momentum loss due to the scattering process must be known. In the case of the optical phonons, the energy loss (or gain) with each scattering process is a constant, since the optical modes may be taken to be nondispersive. Thus the energy loss rate can be expressed simply as

$$-\left.\frac{\mathrm{d}\varepsilon}{\mathrm{d}t}\right|_{\mathrm{PO}}=\hbar\omega_0\left(\frac{1}{\tau_{\mathrm{em}}}-\frac{1}{\tau_{\mathrm{ab}}}\right)=\frac{m^*e^2\omega_0^2}{4\pi\hbar k}\left(\frac{1}{\varepsilon_\infty}-\frac{1}{\varepsilon_0}\right)$$
$$\times\left[(N_q+1)\ln\left(\frac{k+\sqrt{k^2-q_0^2}}{k-\sqrt{k^2-q_0^2}}\right)-N_q\ln\left(\frac{\sqrt{k^2+q_0^2}+k}{\sqrt{k^2+q_0^2}-k}\right)\right]. \tag{43}$$

The general forms (33)–(34) can now be used to compute the average energy loss rate by incorporating an average of the distribution function, as represented by eq. (27) for f_0, as

$$-\left\langle\frac{\mathrm{d}\varepsilon}{\mathrm{d}t}\right\rangle_{\mathrm{PO}}=C\frac{m^{*2}e^2\omega_0^2}{8\pi^3\hbar^3(\mathrm{e}^x-1)}\left(\frac{1}{\varepsilon_\infty}-\frac{1}{\varepsilon_0}\right)(\mathrm{e}^{x-x_\mathrm{e}}-1)k_\mathrm{B}T_\mathrm{e}\,\mathrm{e}^{x_\mathrm{e}/2}K_0(x_\mathrm{e}/2), \tag{44}$$

where only the spherically symmetric part of the distribution function has been used, as the energy-loss function is an even function of the momentum since it only depends upon the energy. The quantity $x_\mathrm{e}=\hbar\omega_0/k_\mathrm{B}T_\mathrm{e}$ (this differs from x by involving the electron temperature) and the conduction band has been approximated as a single spherically symmetric and parabolic band. Here K_0 is a modified Bessel function of the second kind. The details are given in Appendix A. The normalizing density may be evaluated from f_0 (the average of the first-order term vanishes, of course) as

$$n=\int_0^\infty n(\varepsilon)\,\mathrm{e}^{-x_\mathrm{e}}\,\mathrm{d}\varepsilon=\frac{1}{4}\left(\frac{2m^*k_\mathrm{B}T_\mathrm{e}}{\pi\hbar^2}\right)^{3/2}, \tag{45}$$

where an arbitrary normalization with the Fermi energy level has been omitted, as it will divide out of both numerator and denominator for these equations. This means, e.g., that now, the normalizing constant in eqs. (27) and (28) can be evaluated in terms of the density as $C=4n(\pi\hbar^2/2m^*k_\mathrm{B}T_\mathrm{e})^{3/2}$. This finally leads to the energy balance equation

$$-e\boldsymbol{E}\cdot\boldsymbol{v}_\mathrm{d}=\frac{m^{*2}e^2\omega_0^2(\mathrm{e}^{x-x_\mathrm{e}}-1)}{2\pi(2\pi m^*k_\mathrm{B}T_\mathrm{e})^{3/2}(\mathrm{e}^x-1)}\left(\frac{1}{\varepsilon_\infty}-\frac{1}{\varepsilon_0}\right)\mathrm{e}^{x_\mathrm{e}/2}K_0(x_\mathrm{e}/2). \tag{46}$$

Calculating the momentum loss rate is somewhat more difficult, as the momentum exchanged in a collision is not a constant that can be removed from the summation over final states. The quantity of interest is just

$$\left\langle \frac{\mathrm{d}\boldsymbol{p}}{\mathrm{d}t} \right\rangle_{\mathrm{PO}} = -\left\langle \frac{m^* \boldsymbol{v}}{\tau_{\mathrm{PO}}} \right\rangle, \tag{47}$$

according to the format above. However, the momentum exchange involves the phonon wave vector and must be included in the summation over the latter quantity. Indeed, the quantity that needs to be determined is just

$$\begin{aligned} \frac{\mathrm{d}p_E}{\mathrm{d}t} &= \frac{2\pi}{\hbar} \sum_{\boldsymbol{q}} [\hbar q_E |M(\boldsymbol{k}, \boldsymbol{k}')|^2 \delta(\varepsilon(\boldsymbol{k}+\boldsymbol{q}) - \varepsilon(\boldsymbol{k}) - \hbar\omega_0) \\ &\quad - \hbar q_E |M(\boldsymbol{k}', \boldsymbol{k})|^2 \delta(\varepsilon(\boldsymbol{k}-\boldsymbol{q}) - \varepsilon(\boldsymbol{k}) + \hbar\omega_0)] \\ &= -\frac{m^{*2} e^2 \omega_0 \cos\theta_E}{4\pi\hbar k(\mathrm{e}^x - 1)} \left(\frac{1}{\varepsilon_\infty} - \frac{1}{\varepsilon_0}\right) \left\{ k\sqrt{1 + \frac{\hbar\omega_0}{\varepsilon}} - \frac{2m^*\omega_0}{\hbar k} \sinh^{-1}\left(\frac{\varepsilon}{\hbar\omega_0}\right)^{1/2} \right. \\ &\quad \left. + \mathrm{e}^x u_0(\varepsilon - \hbar\omega_0) \left[k\sqrt{1 - \frac{\hbar\omega_0}{\varepsilon}} + \frac{2m^*\omega_0}{\hbar k} \sinh^{-1}\left(\frac{\varepsilon}{\hbar\omega_0} - 1\right)^{1/2} \right] \right\}. \end{aligned} \tag{48}$$

Here p_E and $\hbar q_E$ are the momentum along the field direction, and

$$\boldsymbol{q} \cdot \boldsymbol{E} = qE(\cos\theta \cos\theta_E + \sin\theta \sin\theta_E \cos(\varphi - \varphi_E)), \tag{49}$$

where q and q_E are the angles that $\boldsymbol{q}$ and $\boldsymbol{E}$ make with the direction of $\boldsymbol{k}$, and the latter is taken as the polar direction. This must still be averaged over the distribution function.

The averaging procedure outlined above can now be used to compute the average momentum loss rate. This gives (C has already been removed)

$$\begin{aligned} \frac{1}{n}\left\langle \frac{\mathrm{d}p_E}{\mathrm{d}t} \right\rangle_{\mathrm{PO}} &= \frac{m^* e^2 v_d}{6\pi\hbar(\mathrm{e}^x - 1)} \left(\frac{m^*\omega_0}{\pi\hbar}\right)^{1/2} \left(\frac{1}{\varepsilon_\infty} - \frac{1}{\varepsilon_0}\right) x_\mathrm{e}^{3/2} \mathrm{e}^{x_\mathrm{e}/2} \\ &\quad \times [(\mathrm{e}^{x - x_\mathrm{e}} + 1) K_1(x_\mathrm{e}/2) + (\mathrm{e}^{x - x_\mathrm{e}} - 1) K_0(x_\mathrm{e}/2)]. \end{aligned} \tag{50}$$

This can now be equated to the field term, which gives the momentum balance equation

$$\begin{aligned} e\boldsymbol{E} &= m^* \boldsymbol{v}_\mathrm{d} \frac{e^2 x_\mathrm{e}^{3/2} \mathrm{e}^{x_\mathrm{e}/2}}{6\pi\hbar(\mathrm{e}^x - 1)} \left(\frac{m^*\omega_0}{\pi\hbar}\right)^{1/2} \left(\frac{1}{\varepsilon_\infty} - \frac{1}{\varepsilon_0}\right) \\ &\quad \times [(\mathrm{e}^{x - x_\mathrm{e}} + 1) K_1(x_\mathrm{e}/2) + (\mathrm{e}^{x - x_\mathrm{e}} - 1) K_0(x_\mathrm{e}/2)]. \end{aligned} \tag{51}$$

Equations (46) and (51) can now be solved simultaneously to find the drift velocity and electron temperature at a given electric field $\boldsymbol{E}$. The velocity does not really saturate for polar-optical mode scattering, because if one solves the foregoing equations for the drift velocity, it is found that this quantity continues to depend on the rising electron temperature. This is in fact found for most polar materials; for example, the drift velocity is not found to exhibit the hard saturation found in silicon and germanium.

The nonsaturation can be demonstrated quite easily. The equations above can be combined to solve for the drift velocity as

$$
\begin{aligned}
v_{\mathrm{d}}^{2} &= \frac{3k_{\mathrm{B}}T_{\mathrm{e}}}{m^{*}}\left[1+\coth\left(\frac{x-x_{\mathrm{e}}}{2}\right)\frac{K_{1}(x_{\mathrm{e}}/2)}{K_{0}(x_{\mathrm{e}}/2)}\right] \\
&\approx \frac{3\hbar\omega_{0}}{m^{*}}\tanh\left(\frac{\hbar\omega_{0}}{2k_{\mathrm{B}}T_{\mathrm{e}}}\right)\ln\left(\frac{\hbar\omega_{0}}{2k_{\mathrm{B}}T_{\mathrm{e}}}\right),
\end{aligned}
\tag{52}
$$

where the latter form arises for the limiting case of $x_{\mathrm{e}} \ll 1, x$. Although the constants differ somewhat from those in eq. (7), it is mainly the logarithmic term that differs from that earlier form. Indeed, the temperature is found to continue to increase with increasing electric field, as it does for nearly all forms of scattering, but the presence of the logarithmic term here makes the argument of the latter term continue to decrease, and this in turn leads to a nonsaturating drift velocity. This is a weak increase, and can be characterized as a "soft" saturation. A further problem arises due to the fact that the temperature will begin to increase almost exponentially with the field at very high field values (typically on the order of a few hundred kilovolt per centimeter). In the latter case, the velocity will begin to increase rapidly with the field. This is the onset of "runaway", which was studied by Stratton (1958) in his early uses of the drifted Maxwellian approach. Although it is not clear that the method of solution has failed, it is also not clear that it applies very well to this situation, and this incongruity has never been resolved. Generally, approaches based on more exact dynamics, such as the ensemble Monte Carlo technique, have not indicated the presence of runaway except at much higher fields.

The accuracy of the balance equations obtained above ranges, in its applications to solving transport problems, from very good when the carrier density is relatively high, to exceedingly poor. The latter results arise in cases in which the assumptions about the form of the distribution function are just not valid, such as polar-mode scattering when the carrier density is low so that the distribution function is highly asymmetric with strong peaking along the field direction. It should be pointed out, however, that the validity of the balance equation approach goes beyond the assumption of the distribution function, and can be used to evaluate any parameterized distribution function if the integrals can be evaluated. It turns out that the drifted Maxwellian is the general form if only the density and two adiabatic invariants are introduced – the drift velocity and the electron temperature.

4. Numerical ensemble Monte Carlo approaches

It rapidly becomes difficult to solve the Boltzmann equation for the distribution function when many scattering processes are involved (and even for a single one in a complicated band structure). The scattering integrals become quite complicated and many approximations must be made. For this reason a number of computational methods have arisen over the past few decades, in which the inherent computational ability of digital computers is utilized. In these approaches, more effort is expended in characterizing the physics of the band structure and the scattering processes accurately, while this accuracy is traded against a greatly increased computational burden, although the latter is handled by the machine rather than the individual. Hence no attempt at an analytic form is made, and the results are obtained solely from the computations. These approaches have become extremely sophisticated, and we can study detailed effects such as the degree to which the anisotropy of the valence band affects electron cooling (through the electron–hole interaction). In this section these methods are introduced. The path integral approach is developed first, *primarily because the Monte Carlo technique is a numerical approach to solving the resultant Fredholm integral equation for the distribution function.* Although some authors have presented treatments in which the Monte Carlo approach is developed from statistics of free flights, these approaches are not developed from the underlying physics, and often overlook the fact that validity in the Monte Carlo approach relies solely on the validity of the underlying integral equation. As used in physics, the Monte Carlo approach is no more than the traditional numerical mathematics approach to statistical sampling of the kernal of the integral equation, except that the open-ended transport studies rely heavily on the use of the semiclassical trajectories of particles. Here the proper integral equation will be developed from the Boltzmann transport equation in the next section. This will then be used to develop the ensemble Monte Carlo approaches so prevalent today.

4.1. *Path integral*

The Boltzmann equation will first be written in terms of a path integral. In this, the streaming terms on the left-hand side will be written as partial derivatives of a general derivative of the time motion along a "path" in phase space; this is then used to develop a closed-form integral equation for the distribution function. This integral has itself been used to develop an iterative technique, but provides the basis of the connection between the Monte Carlo procedure and the Boltzmann equation. To begin, the Boltzmann equation is written as

$$\left\{\frac{\partial}{\partial t} + e\boldsymbol{E}\cdot\frac{\partial}{\partial \boldsymbol{p}} + \boldsymbol{v}\cdot\nabla\right\} f(\boldsymbol{p},\boldsymbol{r},t) = -\Gamma_0 f(\boldsymbol{p},\boldsymbol{r},t) + \int \mathrm{d}\boldsymbol{p}'\, W(\boldsymbol{p},\boldsymbol{p}') f(\boldsymbol{p},\boldsymbol{r},t), \tag{53}$$

where

$$\Gamma_0 = \int \mathrm{d}\boldsymbol{p}'\, W(\boldsymbol{p}',\boldsymbol{p}) \tag{54}$$

is the total out-scattering rate for all scattering processes (see, e.g., the form (29) where this simplification has not been made). The remaining scattering term provides the complementary scattering of particles into the state.

At this point it is convenient to transform to a variable that describes the motion of the distribution function along a trajectory in phase space. It is usually difficult to think of the motion of the distribution function, but perhaps easier to think of the motion of a typical particle that characterizes the distribution function. The coordinate along this trajectory is taken to be s, and the trajectory is rigorously defined by the semiclassical trajectory, which can be found by any of the techniques of classical mechanics (i.e., it corresponds to that path which is an extremum of the action). It is as easy to remember, however, that it follows Newton's laws, where the forces arise from all possible potentials – induced and self-consistent ones in device simulations. Each normal coordinate can be parameterized as a function of this variable as

$$\boldsymbol{r} \to \boldsymbol{x}^*(s), \quad \boldsymbol{p} = \hbar \boldsymbol{k} \to \boldsymbol{p}^*(s), \quad t \to s, \tag{55}$$

and for which the partial derivatives are constrained by the relationships

$$\frac{\mathrm{d}\boldsymbol{x}^*}{\mathrm{d}s} = \boldsymbol{v}, \quad \boldsymbol{x}^*(t) = \boldsymbol{r}, \tag{56a}$$

and

$$\frac{\mathrm{d}\boldsymbol{p}^*}{\mathrm{d}s} = e\boldsymbol{E}, \quad \boldsymbol{p}^*(t) = \boldsymbol{p}. \tag{56b}$$

With these changes, the Boltzmann equation (53) becomes simply

$$\frac{\mathrm{d}f}{\mathrm{d}s} + \Gamma_0 f = \int \mathrm{d}\boldsymbol{p}'\, W(\boldsymbol{p}, \boldsymbol{p}') f(\boldsymbol{p}'). \tag{57}$$

This is now a relatively simple equation to solve. It should be recalled at this point that $W(\boldsymbol{p}, \boldsymbol{p}')$ is the probability per unit time that a collision scatters a carrier from state $\boldsymbol{p}'$ to $\boldsymbol{p}$, and these variables will be retarded due to the phase-space variations described above. The form (57) immediately suggests the use of an integrating factor and eq. (57) can be rewritten as

$$\begin{aligned} f(\boldsymbol{p}, t) &= f(\boldsymbol{p}, 0) \exp\left(- \int_0^t \Gamma_0 \,\mathrm{d}s' \right) \\ &\quad + \int_0^t \mathrm{d}s \int \mathrm{d}\boldsymbol{p}'\, W(\boldsymbol{p}, \boldsymbol{p}') f(\boldsymbol{p}', s) \exp\left(- \int_s^t \Gamma_0 \,\mathrm{d}s' \right), \end{aligned} \tag{58}$$

where the momenta evolve in time as the energy increases in time along the path s due to the acceleration of the external fields. In fact, in the phase space defined by s, the energy does not increase, but as the "laboratory" coordinates are restored, this energy increase will appear. Indeed, the major time variation lies in the momenta themselves. With the change of variables $\boldsymbol{p}'(s) \to \boldsymbol{p}'(t) - e\boldsymbol{E}s$, this becomes

$$\begin{aligned} f(\boldsymbol{p}, t) &= f(\boldsymbol{p}, 0) \exp\left(- \int_0^t \Gamma_0 \,\mathrm{d}s \right) \\ &\quad + \int_0^t \mathrm{d}s \int \mathrm{d}\boldsymbol{p}'\, W(\boldsymbol{p}, \boldsymbol{p}' - e\boldsymbol{E}s) f(\boldsymbol{p}' - e\boldsymbol{E}s) \exp\left(- \int_0^s \Gamma_0 \,\mathrm{d}s' \right), \end{aligned} \tag{59}$$

This last form is often referred to as the Chambers–Rees path integral (Rees 1972). It is convenient to make some simplifications in the structure since Γ_0 in the exponential propagator is still a complicated energy-dependent quantity.

An important technical innovation introduced by Rees (1968) is the concept of self-scattering, a fictitious scattering process that does not alter either the energy or momentum (and hence not the physics) of a particle, but allows a technical simplification of the mathematical detail. To each side of eq. (57), a term is added of the form

$$\Gamma_s(\boldsymbol{p})\,\delta(\boldsymbol{p}-\boldsymbol{p}'). \tag{60}$$

Since this term does not alter the physics in any way (i.e., the dynamics of the particles are not disturbed by this term), the energy dependent coefficient can be chosen in an arbitrary fashion. The simplification arises if this function is chosen so that

$$\Gamma_s(\boldsymbol{p}) = \Gamma_T - \Gamma_0(\boldsymbol{p}), \tag{61}$$

where Γ_T is a constant sufficiently large that the left-hand side of eq. (61) is always positive semidefinite. Adequate results can be obtained for a weak violation of this rule, but this is not generally done. Then eq. (59) becomes (for a homogeneous system, although the result is easily extended to an inhomogeneous system by incorporating the additional propagation in position space in the distribution function within the integral)

$$f(\boldsymbol{p},t) = f(\boldsymbol{p},0)\,e^{-\Gamma_T t} + \int_0^t ds \int d\boldsymbol{p}'\, W^*(\boldsymbol{p},\boldsymbol{p}'-e\boldsymbol{E}s) f(\boldsymbol{p}'-e\boldsymbol{E}s) e^{-\Gamma_T s}, \tag{62}$$

where

$$W^*(\boldsymbol{p},\boldsymbol{p}') = W(\boldsymbol{p},\boldsymbol{p}') + \Gamma_s(\boldsymbol{p})\,\delta(\boldsymbol{p}-\boldsymbol{p}'). \tag{63}$$

The complexity of the propagator inside the collision integral has been greatly simplified by the introduction of the self-scattering, since it is no longer necessary to integrate the scattering function over the path "time." The role of this process is solely to achieve the desired simplification of this propagator. In so doing, the physics is not altered, but the mathematics has been simplified, as the role of the self-scattering serves only to ease the computations and does not alter the momentum and energy conservation within the state description of the electrons.

4.2. Monte Carlo sampling techniques

It can readily be observed that the path integral could be solved by a chain of integrations, which correspond alternately to the applications of a path integral and a scattering integral. The entire chain operates on an initial trial function, and the steady state evolves to a form that represents a balance between the accelerative effects of the field and the relaxation effects of the scatterers. Hence in the high-field, nonlinear transport case, the distribution function eventually is independent of the initial trial function. The chain of integrations suggests an alternative approach to the evaluation of eq. (62), and that is to evaluate the integrals with a Monte Carlo sampling technique (Binder 1979, Hockney and Eastwood 1981).

In the earliest forms of its application to semiconductor transport, the Monte Carlo technique was developed by following a single particle through a large number of two-step iterations. The first step is a path traversal, terminated at a time t selected on a random basis using the probability function $\exp(-\Gamma_T t)$ (which must still be properly normalized to be a probability density function). The second step involves scattering from the state resulting at the end of this traverse to a new state governed by the energy and momentum conservation rules of the particular scattering process, which itself is randomly selected from the group being considered, including self-scattering (Boardman et al. 1970).

This approach can now be expanded in some detail (a flowchart is presented in fig. 4). Consider that a typical electron is considered to be at time $t = 0$ (arbitrarily selected) in a state characterized by momentum $\boldsymbol{p}_0$, position $\boldsymbol{r}_0$, and energy ε. At this time the duration of the accelerated flight is determined from the probability of not being scattered given above with a random number R_1, which lies in the interval [0, 1], by

$$t_1 = -\frac{1}{\Gamma_T} \ln R_1. \tag{64}$$

At this time, the energy, momentum, and position are updated according to the energy gained from the field during the accelerative period to $(\boldsymbol{r}_1, \boldsymbol{p}_1, \varepsilon_1)$. Once these new dynamical variables are known, the various scattering rates can now be evaluated for this particle (in practice, these rates are usually stored as a table to enhance computational speed). A particular rate is selected as the germane scattering process according to a second random number R_2 and the rule

$$\frac{1}{\Gamma_T} \sum_{j=1}^{i-1} \Gamma_j(\varepsilon_1) < R_2 < \frac{1}{\Gamma_T} \sum_{j=1}^{i} \Gamma_j(\varepsilon_1). \tag{65}$$

Here the ith process is selected, and each scattering rate is normalized to the total scattering rate, including self-scattering. In essence, the various scattering processes are ordered (and this order must be maintained throughout), and the ith process is selected by the random number in eq. (65). Then the energy and momentum conservation relations are used to determine the postscattering momentum and energy $\boldsymbol{p}_2$ and ε_2. Additional random numbers are used to evaluate any individual parts of the momentum that are not well defined by the scattering process. For example, in polar scattering the polar angle is well defined by the $1/\boldsymbol{q}$ variation of the matrix element. On the other hand, the azimuthal angle change is not specified by the matrix element, so that it is randomly selected by a third random number. In isotropic scattering processes such as nonpolar optical and acoustic scattering, both angles are randomly selected. The final set of dynamical variables $(\boldsymbol{r}_2 = \boldsymbol{r}_1, \boldsymbol{p}_2, \varepsilon_2)$ are now used as the initial set $(\boldsymbol{r}_0, \boldsymbol{p}_0, \varepsilon)$ for the next iteration, and the process is continued for several hundred thousand cycles.

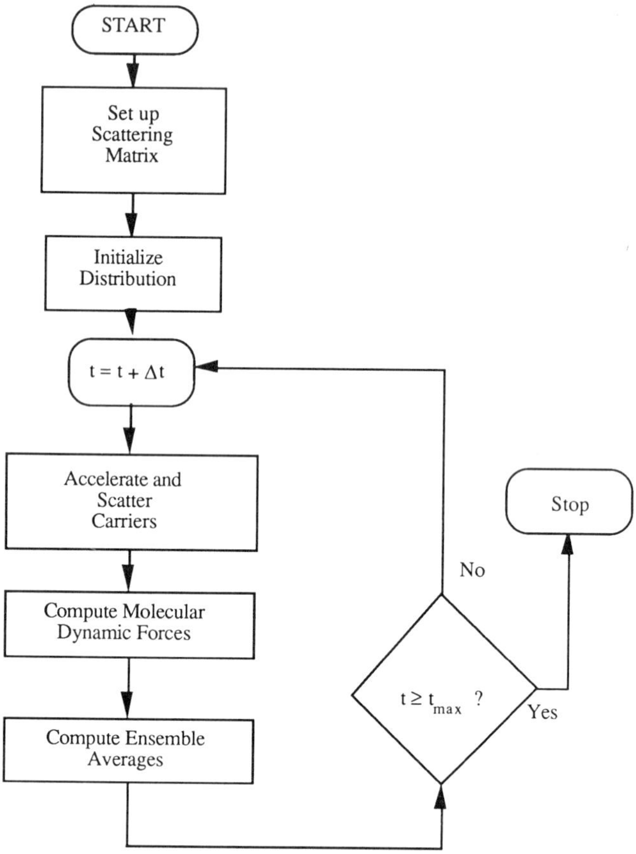

Fig. 4. Flowchart for a typical ensemble Monte Carlo program.

4.3. Ensemble Monte Carlo

The approach currently used by most practitioners of Monte Carlo techniques is the ensemble Monte Carlo procedure. Rather than following a single particle for several hundred thousand iterations, it is more efficient to follow several thousand particles at once (especially on vector computers). The single-particle approach above is used for each particle, but all dynamic averages are now computed as ensemble averages. For example, the drift velocity is now

$$v_{\mathrm{d}}(t) = \langle v(t) \rangle = \frac{1}{N} \sum_{i=1}^{N} v_i(t). \tag{66}$$

Here, N is the number of particles in the ensemble. This approach can now be used in the transient regime, since all averages are ensemble averages and do not depend on the existence of ergodicity. However, care must be exercised here. The averages are usually computed at regular time steps, but only a few of the particles are completing their accelerations exactly at these time steps. To require artificially terminating acceleration processes on the regular time steps introduces an artificial coarse-graining of time, which severely limits real-time calculations by requiring the time steps to be much smaller than the reciprocal of the highest frequency of interest. Most people today keep the two time scales separate.

A flowchart of a typical ensemble Monte Carlo program is shown in fig. 4. This particular form is one that is amenable to full vectorization and is relatively computationally efficient, in that it will utilize the hardware "scatter-gather" routines available on large vector computers. The program begins by creating the large scattering matrix in which all of the various scattering processes are stored as a function of the energy. This includes the self-scattering process. The energy is discretized, and the size of each elemental step in energy is set by the dictates of the physical situation that is being investigated. The initial distribution function is then established – the N electrons actually being simulated are given initial values of energy and momentum corresponding to the equilibrium ensemble, and they are given initial values of position and other possible variables corresponding to the physical structure being simulated. At this point, $t = 0$. If the intercarrier forces are being computed in real space by a molecular dynamics interaction (discussed below), the initial values of these forces, corresponding to the initial distributions in space are also computed. Part of the initialization process is also to assign to each of the N electrons a t_{1i}, which is its individual time to scattering. Next the time is incremented by the *observation* time step, and particles are accelerated and scattered (explained below) over this period. At the end of this time step, the ensemble averages are computed and the molecular dynamics forces are updated. If the time has reached its final simulation time, the process is ended. Otherwise, the process returns to the point where the observation time is incremented.

The process of acceleration and scattering is explained further in fig. 5. First, each electron is accelerated for a period of time determined by the minimum of either its t_1 or the time step of the observation time. The former group of electrons will undergo scattering within this time step, whereas the latter group does not undergo scattering in this period. A "mask" vector, whose value is either 0 or 1, is then set, in which a "flag" (value 1) is assigned to each vector whose t_1 lies in this time step and will therefore undergo scattering. This "mask" vector is used by the "gather" process to collect all the electrons (together with all their properties) undergoing scattering into a new vector. A scattering process is selected for each of the latter electrons by random numbers, and the details of the scattering are worked out with whatever new random numbers are required, just as in the preceding section. These electrons receive new t_1's and are then accelerated for these new t_1's or to the end of the observation time step, whichever is shorter. The electrons are then "scattered" back to the original matrix along with their new properties. Since some of the electrons may still have t_1 values lying in this observation time step, the loop is repeated until all electrons have moved out of the time step for their next scattering event. This process requires the interplay of the local time of each electron with the observation time of the entire ensemble.

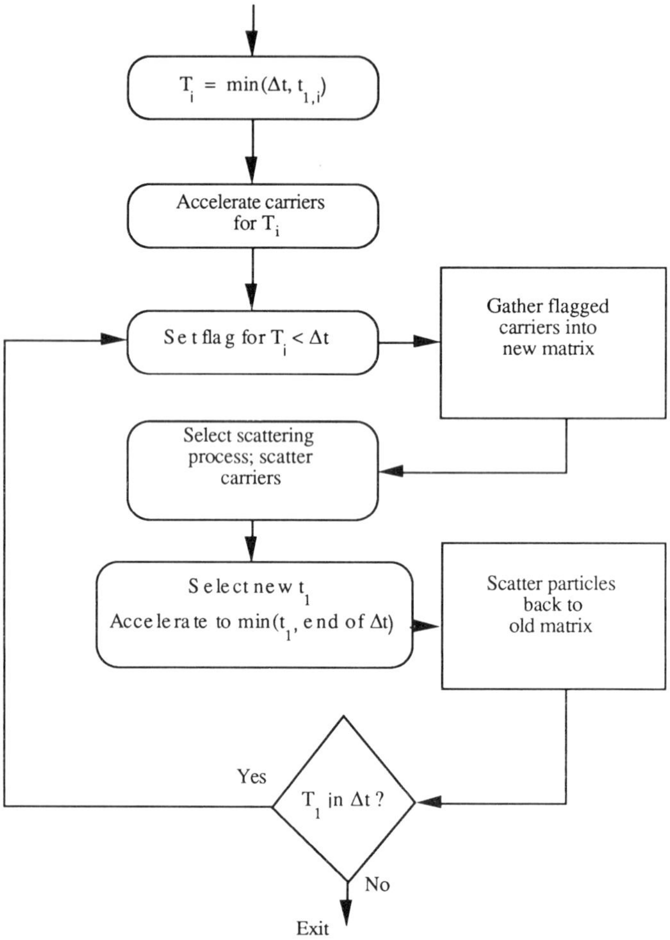

Fig. 5. Details of the acceleration–scattering loop for an ensemble Monte Carlo process.

The ensemble Monte Carlo process has become quite sophisticated in recent years, and a number of clever innovations have been introduced. Particle splitting, for example, can be used to improve the statistics in regions of phase space where rare events occur. In this process, a particle that moves into the desired part of phase space is split into M particles, each of which is assumed to be $1/M$ of a particle (Phillips and Price 1977). This provides an M-fold sampling of events in this part of phase space and allows a more effective incorporation of the rare processes such as impact ionization.

Degeneracy and Fermi–Dirac statistics have also been introduced through the concept of a secondary self-scattering process (Lugli and Ferry 1985, Bosi and Jacoboni 1976). Rather than recompute the scattering rates as the distribution function evolves in order to incorporate the degeneracy, all scattering rates are computed as if the final states were always empty. Then a grid in momentum space is maintained so that the number of particles in each state can be tracked (each cell of this grid has its population divided by the total number of states in the cell, which depends on the cell size, to provide the value of the distribution function in that cell). The scattering processes themselves are evaluated, but the acceptance of the process depends on a rejection technique. That is, an additional random number is used to accept the process if the final state is empty. Thus, as the state fills, most scattering events into that state are rejected and treated as a self-scattering process.

4.4. Molecular dynamics

One normally turns to a molecular dynamics treatment of electron–electron interaction in solids on the short time scale ($t < 100$ fs) since the dielectric function has always involved the need to approximate the full frequency-dependent and wave-vector-dependent distribution function in order to obtain results that could be interpreted. One way to avoid having to make these approximations is to treat a numerical simulation of the carriers and their transport, such as with an ensemble Monte Carlo technique. The molecular dynamics loop is complicated only by a need to shift the positions of the particles. Molecular dynamics differs from Monte Carlo in that it is a real-space simulation in which the actual inter-particle forces are utilized. The most common usage in Monte Carlo is to include the intercarrier Coulomb forces in real space rather than approximating them by a momentum space scattering rate. The fact that N electrons are used to simulate a carrier density n indicates that the simulation incorporates a volume $V = N/n$. Boundary conditions of this volume are usually taken to be periodic for a "homogeneous" system. However, when calculating the molecular dynamic force on an individual particle, a second volume is required. This second volume needs to be centered on the particle at which the force is being evaluated so that the other $N - 1$ electrons do not introduce any artificial directivity into the force. Thus the $N - 1$ electrons must be moved to positions in this centered box of volume V in which they interact with the selected particle through the shortest possible distance that exists in the periodically replicated original box. After the forces are computed, the particles are returned to the original simulation box. In fact, if the number of particles being simulated is not sufficiently large, errors can creep into the computation of the intercarrier Coulomb force due to the size of the simulation box and the periodicities inherent in the system. This leads to the need to include correction terms in the force computation, and these corrections have been discussed in the literature (Adams and Dubey 1987). Generally, if N is more than a few thousand, these correction terms are small, but should be evaluated for their importance in each molecular dynamics computation.

Consider an ensemble of electrons in which normal scattering and transport in a high electric field is treated through an ensemble Monte Carlo calculation. Inter-electronic Coulomb interaction is retained as a real-space potential, just as is the case with any self-consistent device potential and the applied electric fields, and the effect of interparticle potential on the motion of the electrons is computed through a molecular dynamics procedure in which the local force on each electron due to the electric field and the repulsion of all other electrons is calculated for each time step of the Monte Carlo process (Lugli and Ferry 1986). In essence, Coulomb interaction between electrons is now treated in real space, whereas phonon scattering processes are treated in momentum space. Such a procedure has the advantage that no approximations to the range of the scattering wave vector in the Coulomb interaction must be made. Moreover, there is no need to separate the interparticle force into direct Coulomb and plasmon terms. However, only a finite number of particles can be treated, so that a small cell of real space is considered and assumed to be replicated throughout the entire crystal. This is the simulation volume, which contains the number of particles being treated. The size of this volume is given by the ratio of the number of particles considered (N_0) to the simulated carrier density n, as $V = n/N_0$. Care must be taken that the simulated volume and the number of particles are sufficiently large that artifacts from periodic replication of this volume do not appear in the calculation results.

The Coulomb force is considered only through the shortest connecting vector between each pair of carriers, so that two primitive volume cells must be considered. The second volume is an equivalently sized cell which is centered on each carrier when the interparticle forces are being calculated at each time step. This volume ensures that each of the other simulated particles interacts with the particle of interest through the shortest distance between them (so that the latter may occur in one of the replications of the first volume). Since the simulated volume may not correspond to each electron interacting through the shortest distance with the carrier of interest, replicas of the electrons that lie in the second cell must be used to compute the force through the well-known

$$\boldsymbol{F} = \frac{-e^2}{4\pi\varepsilon_0} \sum_i \frac{1}{r_i^2} \boldsymbol{a}_{ri}, \tag{67}$$

where r_i is the distance to the interacting particle (from the reference particle) and $\boldsymbol{a}_{ri}$ is a unit vector pointing *to the interacting particle*, with both particles assumed to have the same charge.

The two boxes are explained best by considering that the replicated rectangular cells arise from the basic computational volume replicated over a so-called lattice. They form a "superlattice" defined by the so-called lattice vectors $\boldsymbol{L} = L_x\boldsymbol{a}_x + L_y\boldsymbol{a}_y + L_z\boldsymbol{a}_z$. Each of these cells contains the basic number of particles N_0 used in the ensemble Monte Carlo technique (typically, several thousand are used). The "on-site" particle (as opposed to the usual solid) is the one at which the force sum is being computed, and its local box is the one used in the force computation. A particle from the next "zone" is used for the calculation if it is closer to the on-site particle of interest than its image within the same zone (as the on-site particle). The reason for using this centered volume can be explained simply. The desire is to sum only over the actual N_0 electrons used in the summation. However, if this is done for the test site particle in an uncentered cell, the other $N_0 - 1$ are not uniformly distributed around this particle, and the force will have a net average value. This average has nothing to do with the true force but is an artifact of the test particle being off-center. Use of a centered cell avoids this problem. The appearance of the Ewald sums can also be illustrated. The total potential arising from the interparticle potential can be written as

$$\Phi = \frac{e^2}{4\pi\varepsilon_0}\frac{1}{2}\sum_{i\neq j}\frac{1}{r_{ij}}, \tag{68}$$

where the factor of 2 arises from double counting each pairwise force in the summation over both i and j. The summation in eq. (67) runs over all the particles in the solid, both those within the basic cell *and those that lie in all replicas of the basic cell.* Consider the interaction between the centered particle and one test particle and its replicas, which contribute the terms

$$\sum_{\text{cells}}\frac{1}{r} \to \frac{1}{r} + \sum_{L>0}\frac{1}{|\boldsymbol{L}+\boldsymbol{r}|} \approx \frac{1}{r} + \sum_{L>0}\frac{1}{L} - r\sum_{L>0}\frac{1}{L^2} + r^2\sum_{L>0}\frac{1}{L^3} + \cdots, \tag{69}$$

where the denominator has been expanded since, in general, $L > 2^{1/2}r$ when the centered cell is used for the direct force calculation. The summations over the lattice vectors are known as Ewald sums (Ewald 1921), since they were worked out (actually in reciprocal space) for X-ray scattering some years ago. The method of evaluating these interparticle forces and incorporating the Ewald sum terms to account for the longer-range replications of the principal cell have been studied for some time in connection with molecular dynamics calculations in other fields (Potter 1973, Brush et al. 1964). Note that when eq. (69) is multiplied by L, so that the expansion is in terms of the normalized distance L/r, the resulting sums are independent of r. Rather, they are sums over the defining vectors of the superlattice and become constants that are well known for each choice of superlattice for the boxes. With the expansion (69), the potential of eq. (68) and the force of eq. (67) now need only be calculated over the actual number of particles used in the basic cell, provided that the particle centered box is used in computing these sums. What these Ewald-sum terms contribute to the force (67) is correction terms that compensate for the fact that a finite simulation volume introduces certain Fourier periodicities that can upset the calculation accuracy of the force, as well as a limiting of the number of particles available for screening the interaction. For example, the leading correction to the potential is a term linear in r, rather than inversely proportional to r (Adams and Dubey 1987, Potter 1973, Brush et al. 1964). If the number of particles is small, say less than 100, these corrections are large. However, if the number of particles used is large, the corrections are small. This is translated to the casual observation that the primitive cell volume should have an edge that is significantly larger than the screening length, and this translates into a large number of particles if finite-size effects are to be avoided.

The basic coupled Monte Carlo and molecular dynamics techniques have been used to study the correlation functions of the electrons under high-field conditions (Lugli and Ferry 1986), and have been extended to the inclusion of the exchange interaction as well (Kriman et al. 1990). This latter is important as the direct Coulomb force overestimates the repulsion between particles at very close range, and the exchange interaction provides a correction (of course, acting only between particle of differing spin).

5. Femtosecond laser excitation of semiconductors

Under the pressure of technological competition, semiconductor devices inexorably shrink, typically by a factor of two in linear dimensions every 6.9 years. Both directly by decreasing the critical lengths, and indirectly by increasing the internal fields, this causes the characteristic times for the transit of an electron across the device also to decrease. As a result, current experimental devices require simulations to model times on the scale of a few tens of femtoseconds. These simulations are needed to model both the practical devices and the simplified systems which are used to study ultrafast electronic processes experimentally. On the other hand, laser excitation of semiconductors provides a classical example of a system in which the initial distribution function is decidedly different from what would exist in equilibrium. For example, laser excite–pump–probe and absorption bleaching experiments are conducted with laser pulses compressed to (as short as) 6 fs, and with photon energies of roughly 2–3 eV. During the first picosecond, electrons lose very roughly 10% of their energy every 100 fs by emitting a phonon. During this initial period of ultrafast relaxation, a finite density of electrons (holes) is usually modeled by an ensemble Monte Carlo (EMC). Because different carriers in the ensemble, with different momenta, have different rates for various scattering processes, the Monte Carlo time step would apparently depend on the element of the ensemble.

The development of laser pulse compression techniques has provided a technique to examine nonequilibrium electron distributions on time scales of a few tens of femtoseconds. In the standard pump-and-probe experiments used to study this regime, a short initial laser pulse creates a nonthermal electron–hole plasma; this "pump" pulse bleaches the absorption. By using a second, "probe" pulse to monitor the absorption, one observes the subsequent thermalization of the plasma. Recently, Becker et al. (1988) used femtosecond photon echo techniques to probe the electron–electron scattering relaxation process in dense electron–hole plasmas in semiconductors. Most recently, such fast laser excitation studies have focused on the high-density regime (Becker et al. 1988b). In this high-density regime, one cannot ignore the uncertainty principle. However, this is not the case in the short-time, low-density regime, which could be modeled effectively by using a single-particle picture of electrons which ignored the Fermi character of their statistics (Rosker et al. 1986). That is, one could choose a position uncertainty small compared to the interelectron length r_s, so the single-particle potential was well-defined, and simultaneously leave the momentum well-defined. These facts justified an ensemble Monte Carlo (EMC) approach based on classical electron motion interrupted by phonon interaction events. These EMC simulations could be coupled effectively with exact treatment of the direct Coulomb interaction between carriers through a molecular dynamics (MD) approach. Numerical values of the energy-dependent and density-dependent thermalization times for the various carrier–carrier and carrier–phonon processes have been worked out theoretically, and agree with the experimental data (Wise et al. 1987, Schoenlein et al. 1987, Lin et al. 1987).

Because of the high electron excitation energies occurring in these systems, it is difficult to find analytical treatments of screening which properly describe the nontrivial temporal and wavevector dependences of the screening. To overcome this difficulty, joint molecular dynamics and ensemble Monte Carlo simulations provide a real-space treatment of the Coulomb interaction and avoid such problems. In this approach, individual electrons interact through a Coulomb interaction which is screened only by the high-frequency (valence electron) dielectric constant. The remaining part of the screening, due to the conduction electrons, arises explicitly from the combined motion of the individual simulated electrons (Kann and Ferry 1989).

At densities near 10^{19} cm^{-3}, which can now be achieved routinely, there are strong theoretical and experimental (Osman and Ferry 1987, Baily et al. 1988) grounds to expect that many-electron effects will be important. In principle, these higher densities may be modeled by fully quantum mechanical descriptions which are well-understood. In practice, however, the far-from-equilibrium distributions are prohibitively expensive to study numerically, and accurate analytical approaches are not yet available even in the semi-classical limit. A desirable alternative, therefore, is to extend the existing Monte Carlo approaches into the slightly degenerate regime. This has the added advantage of retaining a picture which is accessible to (classical) intuition.

The present simulation incorporates only the electron–electron interaction. This is a justifiable approximation, however. Because the holes are much heavier than the electrons, and also much colder, the center-of-mass and laboratory frames for an electron–hole pair essentially coincide, and electron–hole scattering is approximately elastic. In other words, scattering from holes approximates ionized impurity scattering, so that essentially no energy exchange is involved and the event contributes primarily to momentum relaxation. This was confirmed numerically in earlier work by Osman and Ferry (1987) on the role of the electron–hole interaction on the picosecond timescale, which showed that the dominant part of the distribution relaxation occurs by electron–electron scattering. Electron–hole recombination is also negligible in the first picosecond after excitation, and that process is excluded from our simulations. However, it is not at all clear that the neglect of the *screening* role of the holes is justified (Bailey et al. 1988), and further work is necessary to assess this. Moreover, as is evident from the previous section, it would not be correct to ignore the holes if we were interested in the mobility or velocity of the electrons.

The neglect of electron–hole scattering is realized in simulations simply by not including within the MD computation the electrostatic forces exerted upon the electrons by holes. Since we are examining the electron statistics only, the MD does not need to time-evolve hole positions. As discussed earlier, within the MD/EMC approach the screening is determined in the same way as the carrier–carrier scattering: both arise in a unified manner from the explicit trajectories of the charged-carrier ensemble. Thus, by excluding the electron–hole scattering we also exclude the holes' contribution to the screening function (which, as mentioned above, may not be valid).

We justify this on the basis that the holes are heavier, so that their contribution to the high-frequency screening function is small. In general, of course, hole effects cannot be ignored. In particular, momentum relaxation effects are expected to be important in high fields, where they mediate energy exchange between carriers and field.

The time evolution of electrons in GaAs at 300 K, excited by a 20 fs pulse of 2 eV photons, is simulated. Masses, coupling constants and other physical parameters (Osman and Ferry 1987, Chamoun et al. 1989) that have been confirmed by several measurements using a variety of techniques are used. The pulse profile was rectangular, but for most analytical purposes this approximation does not have a large effect, and this process allows a more direct evaluation of scattering rates. For the photon energy chosen, the initial electron distribution is generated from all valence bands (light, heavy, and split-off). About 50% of the electrons are generated from the heavy-hole band. The electron population was followed in the central Γ and satellite L and X valleys. Excitation is direct only into the Γ valley, and electrons enter the satellite valleys only by subsequent phonon scattering.

Some qualitative results of the simulations provide a purely empirical argument supporting the importance of Coulomb interactions: while at one point, more than half of the electron population is in the satellite valleys (see fig. 6), the time for this is reached long after the excitation region is depleted to a few percent of its initial occupation. (The excitation region is the region of electron momentum space originally populated by photoexcitation.) The rate of polar-optical scattering events can be computed with greater confidence than intervalley scattering rates (if only because the electron–phonon matrix element is more accurately known), and this process also does not play an important role in the initial relaxation from the excitation region. In the pump–probe experiments, due to the large bandwidths of the ultrashort pump and probe pulses, optical phonon emission or absorption by either electrons or holes does not make a significant contribution to the initial relaxation. Ultimately, therefore, only electron–electron interaction can explain the initial relaxation.

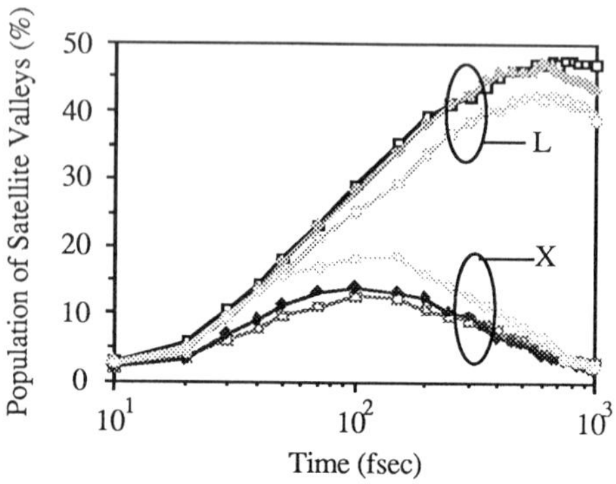

Fig. 6. Population of the satellite valleys for three densities. Upper three curves show L valley population, lower three show X valley. Symbol shapes correspond to densities according to the description in the inset of fig. 7.

One prediction of the model which is particularly sensitive to the electron–electron interaction, and therefore to the electron density, is the rate at which the electrons leave the Γ valley to enter the satellite valleys. From fig. 6, we find that the X valley population peaks around 160 fs which seems to agree with results inferred by Alfano's group (Katz and Alfano 1988, Wang et al. 1989), who reported evidence of the electron scattering rate to the X valley within 170 fs by using femtosecond pump and probe absorption measurements. The effects of a density dependence can be seen in the different satellite-valley occupancies occurring at different excitation densities (fig. 6).

Returning to the empirical argument, in principle electrons might relax from the excitation region by a mechanism that involved rapid transitions into and back out of the satellite valleys. This is a possibility because the short-time transition rates need not be such as to produce a distribution exponentially weighted toward lower energies. In contrast, at long times the approach to a thermal distribution makes transition rates back into the Γ valley much greater than those into the higher-energy satellite valleys, so the late occurrence of peaks in fig. 6 immediately rules out intervalley scattering as a significant contributor to the initial relaxation.

In the highly nonequilibrium femtosecond regime, occupancy data like those shown in fig. 6 are insufficient to determine transition rates. The simulations allow carriers to be exchanged between every pair of valleys. The population of the L valley, e.g., includes electrons that have arrived there after passing through the X valley as well as those that arrived directly from the Γ valley. Clearly, three valley occupancies (two independent, by particle conservation) cannot yield six intervalley transition rates. It is necessary to investigate directly in the code the rate of intervalley transitions.

The $\Gamma \rightarrow L$ rates are found to show no statistically significant density dependence. This is what one would expect in a simple picture, since the absorption of a phonon by deformation potential interaction is a single-electron process. In contrast, the $\Gamma \rightarrow X$ rates show a strong density dependence. Figure 7 shows the transition rate $R_{\Gamma \rightarrow X}$ for three simulations spanning almost two decades of density. The mechanism of this effect is the electron–electron scattering in the Γ valley: the threshold for transfer into the X valley lies at the upper edge of the excitation region. Even after

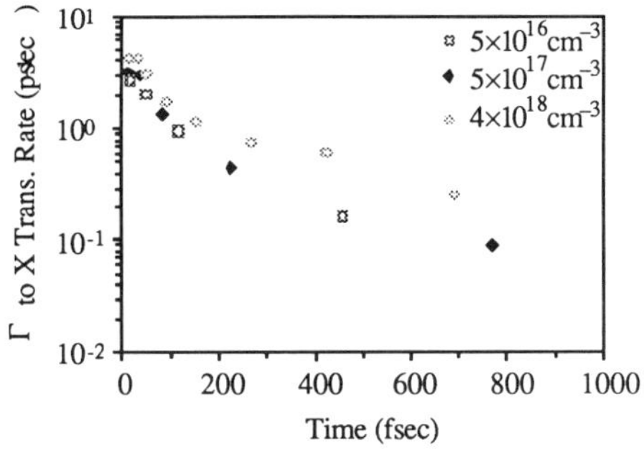

Fig. 7. Transition rates from the Γ valley to the X valley, for three electron densities.

accounting for thermal broadening, less than one tenth of the photoexcited electrons have sufficient energy to make the transition. At high density, electron–electron scattering is increased and the energy distribution is broadened. Because electron–electron scattering preserves the total energy within the electronic system, this broadening is approximately symmetrical towards greater and smaller energies. This increases the fraction of electrons above the threshold for $\Gamma \to X$ scattering. Transitions to the L valley do not exhibit a similar effect because, for the 2 eV laser energy used, essentially all of the photoexcited electrons are above threshold. Note that, although the $\Gamma \to L$ transition rates themselves are not appreciably dependent on the density, there is some density dependence in the L valley occupancies in fig. 6 due to the effect of $X \to L$ processes. At high densities, competition from $\Gamma \to X$ processes depletes the Γ valley, so virtually unchanged $\Gamma \to L$ rates yield a lower L valley population. Nevertheless, any density dependence in the $\Gamma \to L$ rates due to increased electron density in X is too small to detect at the current level of precision. The $X \to \Gamma$ and $L \to \Gamma$ rates are in the range of $(2–3) \times 10^{12}$ and $(6–10) \times 10^{11} s^{-1}$, respectively, for the three different densities. The scattering time constants are approximately 410 fs and 1.3 ps for $X \to \Gamma$ and $L \to \Gamma$ transitions, respectively.

At later times, the decay of the transition rate represents cooling of electrons in the Γ valley as a whole. The Γ valley loses energy in two ways: polar-optical phonon (POP) scattering and preferential transfer of the hottest electrons into satellite valleys. Since electrons just excited from the heavy-hole band typically lie $\sim 10–15\hbar\omega_0$ above the conduction band edge, and since POP scattering occurs at a rate slower than that of intervalley scattering, intervalley transfer predominates initially. It can be seen in fig. 6 that by about 500 fs, the population in the satellite valleys has reached a maximum; at the same time the transition rate $R_{\Gamma \to L}$ has leveled off. At these times, the $\Gamma \to L$ rate plays an important role in luminescence decay.

The general results obtained in these simulations may be summarized as follows: scattering out of the excitation volume is dominated in the initial tens of femtoseconds by electron–electron scattering, and the scattering rate increases with increasing density. *For low excitation densities*, this rate increase agrees both in its magnitude and in its density dependence with recent experimental measurements. The presence of electron–electron scattering modifies both the population transition rates and carrier densities in the satellite valleys, primarily by reshaping the energy distribution of carriers in the central valley. Intervalley scattering rates decrease by two orders of magnitude as the electronic system cools during the first picosecond. Thus, intervalley processes do play a role in the initial decay and the same processes play a modified role in the picosecond-scale luminescence decay. We find that rates, for particular mechanisms of carrier transfer to the satellite valleys, must be estimated carefully since the $\Gamma \to L$ population shift contains a significant fraction of electrons that reach the L valleys by way of the X valleys.

Appendix A

The general form (43) can now be used to compute the average energy loss rate by incorporating an average of the distribution function, as represented by f_0. We proceed by rewriting the expression in curly brackets in eq. (43) as

$$\{\cdot\} = 2\left[(N_q+1)\sinh^{-1}\left(\frac{E}{\hbar\omega_0}-1\right)^{1/2} u_0(E-\hbar\omega_0) - N_q \sinh^{-1}\left(\frac{E}{\hbar\omega_0}\right)^{1/2}\right]. \tag{A1}$$

Only the spherically symmetric part of the distribution function is used, as the energy-loss function is an even function of the momentum since it only depends upon the energy. The quantity $E-\hbar\omega_0$ is replaced by E' in the first term in braces, so that the total function can be written as (after including the density of states)

$$\{\cdot\} = C\frac{4m^*}{\hbar^3}\left(\frac{e^{x-x_e}-1}{e^x-1}\right)\int_0^\infty \sinh^{-1}\left(\frac{E}{\hbar\omega_0}\right)^{1/2} \exp\left(-\frac{E}{k_B T_e}\right) dE, \tag{A2}$$

where $x_e = \hbar\omega_0/k_B T_e$ (this differs from x by involving the electron temperature rather than the lattice temperature) and the conduction band has been approximated as a single spherically symmetric and parabolic band. The integral can be evaluated, with the change of variables $y = E/\hbar\omega_0$, and the subsequent change of variables $y = \sinh^2 u$, so that

$$\begin{aligned}\{\cdot\} &= C\frac{4m^*\omega_0}{\hbar^2}\left(\frac{e^{x-x_e}-1}{e^x-1}\right)\int_0^\infty \sinh^{-1}(\sqrt{y})^{1/2}\exp(-x_e y)\,dy \\ &= C\frac{4m^*\omega_0}{\hbar^2}\left(\frac{e^{x-x_e}-1}{e^x-1}\right)\int_0^\infty u\exp(-x_e\sinh^2 u)\,d(\sinh^2 u) \\ &= C\frac{4m^* k_B T_e}{\hbar^3}\left(\frac{e^{x-x_e}-1}{e^x-1}\right)\int_0^\infty \exp(-x_e\sinh^2 u)\,du.\end{aligned} \tag{A3}$$

The last expression has been obtained by integrating by parts. We can now use the double angle formula, and the well known expressions for the Bessel function integral forms to obtain

$$\{\cdot\} = C\frac{2m^* k_B T_e}{\hbar^3}\left(\frac{e^{x-x_e}-1}{e^x-1}\right)\exp(x_e/2)K_0(x_e/2). \tag{A4}$$

Here K_0 is a modified Bessel function of the second kind. This can now be inserted in eq. (43) to give eq. (44). Similarly, the density may be evaluated from f_0 (the average of the first-order term vanishes, and can be coupled together with eq. (A4) to yield eq. (46).

The averaging procedure outlined above can now be used to compute the average momentum loss rate. This is slightly more complicated, but we begin by rewriting eq. (48) as

$$\frac{1}{n}\left\langle\frac{\mathrm{d}p_E}{\mathrm{d}t}\right\rangle_{\mathrm{PO}} = \frac{Cm^* v_\mathrm{d}}{nk_\mathrm{B}T_\mathrm{e}}\int_0^\infty v(E)n(E)\exp\left(-\frac{E}{k_\mathrm{B}T_\mathrm{e}}\right)\frac{\mathrm{d}p_E}{\mathrm{d}t}\,\mathrm{d}E$$
$$= \Gamma_0\int_0^\infty \mathrm{d}E\, G(E)\exp\left(-\frac{E}{k_\mathrm{B}T_\mathrm{e}}\right), \tag{A5}$$

where

$$\Gamma_0 = \frac{2e^2 v_\mathrm{d}}{3\hbar k_\mathrm{B}T_\mathrm{e}(\mathrm{e}^x - 1)}\left(\frac{m^*}{\pi k_\mathrm{B}T_\mathrm{e}}\right)^{3/2}\left(\frac{1}{\varepsilon_\infty} - \frac{1}{\varepsilon_0}\right) \tag{A6}$$

and

$$G(E) = \sqrt{E(E+\hbar\omega_0)} - \hbar\omega_0 \sinh^{-1}\left(\frac{E}{\hbar\omega_0}\right)^{1/2}$$
$$+ \mathrm{e}^x\left[\sqrt{E(E-\hbar\omega_0)} + \hbar\omega_0\sinh^{-1}\left(\frac{E}{\hbar\omega_0}\right)^{1/2}\right]. \tag{A7}$$

The second term occurs only for $E > \hbar\omega_0$. The first set of integrals may be expressed simply as

$$I_1 = \int_0^\infty \sqrt{E(E+\hbar\omega_0)}\exp\left(-\frac{E}{k_\mathrm{B}T_\mathrm{e}}\right)\mathrm{d}E + \int_{\hbar\omega_0}^\infty \sqrt{E(E-\hbar\omega_0)}\exp\left(-\frac{E}{k_\mathrm{B}T_\mathrm{e}}\right)\mathrm{d}E$$
$$= (\hbar\omega_0)^2(\mathrm{e}^{x-x_\mathrm{e}}+1)\int_0^\infty \sqrt{y(y+1)}\,\mathrm{e}^{-x_\mathrm{e}y}\,\mathrm{d}y, \tag{A8}$$

where the change of variables $y = E/\hbar\omega_0$ has been used in the last form, and the origin has been shifted in the second term (in the first line) to illustrate the commonness in the two integrals. Letting $y = \sinh^2 u$, as above, this integral now becomes

$$I_1 = \frac{(\hbar\omega_0)^2}{x_\mathrm{e}}(\mathrm{e}^{x-x_\mathrm{e}}+1)\exp(x_\mathrm{e}/2)K_1(x_\mathrm{e}/2). \tag{A9}$$

Here K_1 is again a modified Bessel function of the second kind. The second set of integrals is just

$$I_2 = \hbar\omega_0(\mathrm{e}^x - 1)\int_0^\infty \sinh^{-1}\left(\frac{E}{\hbar\omega_0}\right)^{1/2}\exp\left(-\frac{E}{k_\mathrm{B}T_\mathrm{e}}\right)\mathrm{d}E$$
$$= \frac{(\hbar\omega_0)^2}{x_\mathrm{e}}(\mathrm{e}^{x-x_\mathrm{e}}-1)\exp(x_\mathrm{e}/2)K_0(x_\mathrm{e}/2). \tag{A10}$$

The two terms have been combined by shifting the origin of the integration for the emission term, and then integrated using the same steps as before. We can now combine eqs. (A5), (A6), (A9), and (A10) to yield eq. (50).

References

Adams, D.J., and G.S. Dubey, 1987, J. Comput. Phys. **72**, 156.
Bailey, D.W., C.J. Stanton, M.A. Artaki, K. Hess, F.W. Wise and C.L. Tang, 1988, Solid State Electronics **31**, 467.
Becker, P.C., H.L. Fragnito, C.H. Brito-Cruz, R.L. Fork, J.C. Cunningham, J.E. Henry and C.V. Shank, 1988a, Phys. Rev. Lett. **61**, 1647.
Becker, P.C., H.L. Fragnito, C.H. Brito Cruz, J. Shah, R.L. Fork, J.E. Cunningham, J.E. Henry and C.V. Shank, 1988b, Appl. Phys. Lett. **53**, 2089.
Binder, K., 1979, Monte Carlo Methods in Statistical Physics (Springer, Berlin).
Boardman, A.D., W. Fawcett and S. Swain, 1970, J. Phys. Chem. **31**, 1963.
Bogoliubov, N.N., 1962, in: Studies in Statistical Mechanics, Vol. 1, eds J. de Boer and G.E. Uhlenbeck (North-Holland, Amsterdam).
Bosi, S., and C. Jacoboni, 1976, J. Phys. C **9**, 315.
Brush, S.C., H.L. Salikin and E. Teller, 1964, J. Chem. Phys. **45**, 2101, and references therein.
Chamoun, S.N., R.P. Joshi, E.N. Arnold, R.O. Grondin, K.E. Meyer, M. Pessot and G.A. Mourou, 1989, J. Appl. Phys. **66**, 236.
Conwell, E.M., 1967, High Field Transport in Semiconductors (Academic Press, New York).
Ewald, P.P., 1921, Ann. Phys. **64**, 253.
Ferry, D.K., 1974, Phys. Rev. B **9**, 4279.
Ferry, D.K., 1991, Semiconductors (Macmillan, New York) ch. 11.
Fröhlich, H., and V.V. Paranjape, 1956, Proc. Phys. Soc. London B **69**, 21.
Fröhlich, H., and F. Seitz, 1950, Phys. Rev. **79**, 526.
Grosvalet, J., C. Motsch and R. Tribes, 1963, Solid State Electron. **6**, 65.
Gunn, J.B., 1963, Solid State Commun. **1**, 88.
Hockney, R.W., and J.W. Eastwood, 1981, Computer Simulation Using Particles (Wiley, New York).
Kann, M.J., and D.K. Ferry, 1989, in: Proc. of the Picosecond Electronics and Optoelectronics Conference, Vol. 4, eds T.C.L.G. Sollner and D.M. Bloom, Salt Lake City, March 1989 (Opt. Scc. Am., Washington, DC) pp. 153–157.
Kash, J.A., J.C. Tsang and J. Hvam, 1985, Phys. Rev. Lett. **54**, 2151.
Katz, A., and R.R. Alfano, 1988, Appl. Phys. Lett. **53**, 1065.
Kriman, A.M., M.J. Kann, D.K. Ferry and R. Joshi, 1990, Phys. Rev. Lett. **65**, 1619.
Kubo, R., 1957, J. Phys. Soc. Jpn. **12**, 570.
Lin, W.Z., J.G. Fujimoto, E.P. Ippen and R.A. Logan, 1987, Appl. Phys. Lett. **51**, 161.
Lugli, P., and D.K. Ferry, 1985, IEEE Trans. Electron Devices **ED-32**, 2431.
Lugli, P., and D.K. Ferry, 1986, Phys. Rev. Lett. **56**, 1295.
Nicolis, G., and I. Prigogine, 1977, Self-Organization in Nonequilibrium Systems (Wiley, New York).
Osman, M.A., and D.K. Ferry, 1987, Phys. Rev. B **36**, 6018.
Phillips, A., and P.J. Price, 1977, Appl. Phys. Lett. **30**, 528.
Pines, D., 1953, Phys. Rev. **92**, 626.
Potter, D., 1973, Computational Physics (Wiley, New York).
Rees, H.D., 1968, Solid State Commun. A **26**, 416.
Rees, H.D., 1972, J. Phys. C **5**, 64.
Rosker, M.J., F.W. Wise and C.L. Tang, 1986, Appl. Phys. Lett. **49**, 1726.
Ryder, E.J., 1953, Phys. Rev. **90**, 766.
Schoenlein, R.W., W.Z. Lin, E.P. Ippen and J.G. Fujimoto, 1987, Appl. Phys. Lett. **51**, 1442.
Shockley, W., 1951, Bell System Tech. J. **30**, 990.
Shockley, W., 1961, Solid State Electron. **2**, 36.
Stratton, R., 1958, Proc. R. Soc. London A **246**, 406.
Trofimenkoff, F.N., 1965, Proc. IEEE **53**, 1765.
Wang, W.B., N. Ockman, M. Yan and R.R. Alfano, 1989, Solid State Electron. **32**, 1337.
Wise, F.W., I.A. Walmsley and C.L. Tang, 1987, Appl. Phys. Lett. **52**, 605.
Wolff, P.A., 1954, Phys. Rev. **95**, 1415.
Zwanzig, R.W., 1961, in: Lectures in Theoretical Physics, eds W.E. Brittin, B.W. Downs and J. Downs (Interscience, New York).

CHAPTER 19

Fundamental Aspects of Quantum Transport Theory

J. R. BARKER

Nanoelectronics Research Centre
Department of Electronics and Electrical Engineering
University of Glasgow
Glasgow G12 8QQ, UK

Handbook on Semiconductors
Completely Revised Edition
Edited by T.S. Moss
Volume 1, edited by P.T. Landsberg

Contents

1. Introduction 1082
2. The transport problem 1083
   2.1. The structures 1083
      2.1.1. Bulk models 1083
      2.1.2. Low-dimensional structures and nanoelectronics 1084
      2.1.3. The confinement potential for lateral devices 1087
      2.1.4. The fluctuation potential 1087
      2.1.5. Recent structures 1088
   2.2. The transport problem 1088
   2.3. The Kubo formalism 1088
   2.4. The Landauer–Buttiker formalism 1089
   2.5. Kinetic equation and Green function methods 1089
3. Quantum phase-space distributions 1090
   3.1. The statistical density matrix 1090
   3.2. Quantum phase-space distributions 1091
   3.3. The Wigner phase-space distribution 1091
4. Quantum ballistic transport 1092
   4.1. Equation of motion of the Wigner function 1092
   4.2. Quasi-classical transport 1093
   4.3. Gauge invariant formulation 1094
   4.4. Technical difficulties with Wigner distributions 1094
   4.5. Restrictions on the solutions to the Wigner function equation of motion 1096
   4.6. Stationary Wigner functions and completeness 1097
   4.7. Extended phase-space picture 1098
5. Transport in electron waveguides 1099
   5.1. Introduction 1099
   5.2. Classical description 1099
   5.3. Quasi-classical description of the perfect channel 1100
   5.4. Multiple channels and conductance quantisation 1102
   5.5. Inclusion of potential scattering 1102
   5.6. The quantum picture and conductance quantisation 1103
   5.7. Coupled-mode theory in arbitrary electron waveguides 1104
      5.7.1. Introduction 1104
      5.7.2. Basic formulation for curved guides 1105
      5.7.3. Adiabatic limit 1106
      5.7.4. Coupled kinetic equations 1107

5.7.5. Effective fields 1107
5.7.6. Computational methods 1108
5.8. Interference phenomena and the Aharonov–Bohm effect 1109
5.8.1. Background 1109
5.8.2. The Aharonov–Bohm effect and transport in semiconductor rings 1111
5.9. Influence of the fluctuation potential 1113
6. Trajectory representations for single electrons 1115
6.1. The pilot-field picture 1115
6.2. Relation to the Wigner phase-space distributions 1118
7. High-field dissipative transport 1120
7.1. Introduction 1120
7.2. The Green function and Wigner function approaches 1122
7.2.1. Definitions 1122
7.2.2. Quantum kinetic equations 1122
7.2.3. The intra-collisional field effect 1123
7.2.4. Non-linear screening 1123
7.2.5. Inhomogeneous systems 1124
7.3. Monte Carlo evaluation of Feynman path integrals 1124
7.4. Other approaches 1124
8. Conclusions 1125
References 1125

1. Introduction

Since the first edition of this Handbook many important new phenomena have been discovered in semiconductors for which explicitly *quantum* transport theories, as opposed to the semi-classical Boltzmann–Bloch transport approach, are essential. Some of these new developments are sufficiently important that they merit their own specialist chapters in the Handbook. In some instances, the approaches to quantum transport theory described in the earlier edition have become overshadowed by different approaches which have necessarily evolved to meet the different circumstances which have occurred experimentally. Thus recent semiconductor materials science has perfected the means to achieve *ultra-high mobility* electron gases in a diverse range of *low-dimensional* and generally *spatially inhomogeneous* environments. Whereas the problems of quantum transport in the 1970s were concerned with *bulk*, generally *homogeneous* samples, the problems of the 1990s are concerned with structures in which the conducting channels may be significantly smaller than the inelastic mean free path and the full wave nature of electron transport becomes apparent. There has been a corresponding shift in interest towards *ballistic* transport problems. *Fluctuation phenomena* have become important as the extreme regimes of very low temperatures, very pure materials, very small number of carriers have been encountered.

Magnetotransport has come into its own in these new circumstances. Subtle effects of gauge fields, such as the Aharonov–Bohm and related effects, have been witnessed directly and indirectly in a wide range of magnetoresistance phenomena. Universal conductance fluctuations and the quantum Hall effect, originally studied in large samples, have maintained their fascination in nanostructures and nanodevices.

The possibility of novel semiconductor devices based on quantum transport effects have led to a resurgence of interest in *tunnelling phenomena* including resonant tunnelling and the correlated single-electron tunnelling controlled by *charging effects* now observable in very small *capacitative* devices. Tunnelling phenomena are dealt with separately in this volume (see chs. 12 and 16), although we shall allude to it where necessary in the present chapter (indeed tunnelling is an important component of many quantum transport problems). Electron diffraction and interference have all figured prominently in recent mesoscopic device proposals. Electron waveguide geometries have revealed interesting non-local transport effects and the breakdown of many classical concepts such as the way in which resistances combine or in the phenomenon of *conductance quantisation.*

In many of these cases the well-tried Kubo (1957) formalism of linear response theory has been overshadowed by the emergence of the Landauer–Buttiker formalism and the significant role of edge states in magnetotransport. The two approaches are related, and indeed both were conceived in their simplest forms in the mid-1950s, but as always the best theoretical framework is the one which is most closely tailored to the experimental situation. Indeed, it is the rise in importance of *ballistic* or *quasi-ballistic transport* which is the main distinction between quantum transport phenomena as described in the first edition and as presented in the present edition of the Handbook. With ballistic transport, we are concerned with the transmission of carriers between reservoirs, and the conduction through the channel may be conveniently lumped into a transmission matrix which links the input and output channels of the source and drain reservoirs much as one uses transmission matrices in lumped microwave systems. The very considerable problems of relaxive dissipative transport processes in bulk moderate-mobility sytems are replaced by quite different problems.

Despite the large shift in emphasis of quantum transport in the last ten years the basic concepts of quantum transport theory are still well established and relevant. In the present chapter we shall take a pedagogical approach to outline the foundations of the subject in a simple form for today's problems of interest. It will be shown, somewhat paradoxically, that there is a much closer relation between classical and quantum theories of electron transport than might be suspected.

We refer to the previous edition for the treatment of *homogeneous* quantum transport and many of the *linear* magnetotransport phenomena, such as the magnetophonon effect. Other chapters deal separately with mainline subjects, such as the quantum Hall effect, quantum interference phenomena, quantum point contact phenomena, hopping conduction and polaronic problems. The problems of high-field dissipative quantum transport remains acute and has become a playground for experts in Green function theory. The previous edition dealt extensively with that problem and we only attempt a brief update in the present edition for completeness. Similarly the general topics of magnetotransport in the first edition have been overtaken by a shift in interest towards the quantum Hall effect, effects of localisation and disorder on magnetotransport. Quantum transport theory is now a very large subject and in the present edition of the Handbook we shall attempt to underpin many of today's developments by a consistent approach based on the density matrix and Wigner function formalisms.

2. The transport problem

2.1. The structures

Let us first give a simplified picture of the transport scenarios encountered in modern semiconductor quantum transport.

2.1.1. Bulk models

The simplest bulk transport model for a semiconductor envisages a homogeneous semiconductor structure with macroscopically separated faces and distant contact regions. The transport of carriers through such a structure may be defined as bulk transport if there are negligible effects of the boundaries of the structure either through surface and interface scattering/charge trapping and when size quantisation may be ignored. *Bulk dissipative transport* occurs when the predominant energy loss due to the current flow occurs within the volume of the device: this requires the mean free path for inelastic scattering to be much less than the dimensions of the structures. Outside this regime there are several regimes of differing complexity which depend on the role of inter-carrier scattering for example in causing a diffusive motion of individual carriers.

A particular idealisation occurs in bulk ultra-high mobility semiconductors at temperatures such that the mean free path for dissipative scattering to the lattice and the mean free path for elastic scattering on defects/impurities are much larger than the structural dimensions. This regime is often called the *ballistic transport regime* for it is supposed that carriers evolve classically along parabolic trajectories in a uniform applied electric field. In fact, whenever the parameters for a structure are such that a carrier can explore the volume of the structure with only potential scattering being significant then boundary scattering and more significantly effects of size quantisation and interference phenomena between different possible paths will occur: true ballistic transport necessarily entails *quantum ballistic transport*.

Effects which depend critically on the phase of a carriers wavefunction require *coherent ballistic transport* for which the coherence length as determined by phase-breaking processes such as electron–electron scattering and phase-mixing potential scattering must be of the order of the dimensions of the structures. Coherent quantum ballistic transport is found experimentally to be available in mainly low-dimensional semiconductor structures at very low temperatures.

From a modelling viewpoint, the system Hamiltonian for a bulk transport model envisages a volume-independent picture in which the carrier kinetic energies are furnished by the appropriate band theory and the scattering by the appropriate bulk impurity distribution and phonon ensemble models plus inter-carrier Coulomb processes. The effects of charge carrier ordering will depend critically on the range of the screening of the Coulomb force relative to the dimensions of the structure.

Generically, the single-electron Hamiltonian for bulk models takes the form

$$H = T(\boldsymbol{p} - e\boldsymbol{A}) + V_{\text{field}}(\boldsymbol{r}) + H_{\text{scattering}}, \tag{1}$$

where T describes the energy band structure, V_{field} describes the local scalar potential of the applied field, the terms in $e\boldsymbol{A}$ describe the coupling to the vector potential of the applied field and the residual terms describe scattering Hamiltonians.

If the bulk material is disordered to the extent that localised states occur, then a bulk picture is still possible within the framework of hopping conduction (see the Chapter on Hopping conduction in the present Handbook edition by Gallagher and Butcher (ch. 14), and the section on the origin of hopping from a quantum transport viewpoint by Barker (ch. 11C) in the 1st edition of the Handbook).

2.1.2. Low-dimensional structures and nanoelectronics

The last decade has seen the beginning of thorough study of the science and potential device structures available with low-dimensional semiconductor heterostructure materials. The simplest such structures are vertical devices in which the current flow is perpendicular to the heterostructure layers. The vertical geometry is inhomogeneous due to the layers which may be only a few atomic layers in thickness (nm scales) and the lateral dimensions are large typically >100 μm × 100 μm. Such structures are dominated by the energy band profile as a function of the vertical depth.

Figure 1 illustrates the type of conduction-band profile that can be obtained in the vertical direction by band-gap engineering and modulation doping. This profile may be modelled by an effective potential $V_c(y)$. The large lateral dimensions lead to the adoption of one-dimensional models for the quantum mechanics of the vertical transport. Transport in such structures varies from hot-electron quasi-ballistic transport to resonant tunnelling quantum ballistic transport. The model Hamiltonian may be taken, without loss of generality, as

$$H = (\boldsymbol{p} - e\boldsymbol{A})^2/2m^* + V_{\text{field}}(\boldsymbol{r}) + V_c(y) + H_{\text{scattering}}. \tag{2}$$

The possibility of trapping a high-mobility quasi-two-dimensional electron gas in the confinement potential V_c at the interface between two heterostructure layers (such as an AlGaAs–GaAs interface) led to an interest in *lateral geometry* devices such as the HEMT and MODFET structures. The typical *effective confining potential* may be calculated self-consistently via full-scale band-structure calculations or more simply by solving the coupled Schrödinger and Poisson equations for charge transfer into the bare effective potential formed by the conduction-band profile. The simplest Hamiltonian models again have the form of eq. (2), but the quasi-two-dimensional electron gas occupies one or more sub-bands resulting from the deep confinement potential at the interface.

Towards the end of the eighties there appeared devices and structures which involved the re-configuration of the quasi-two-dimensional electron gas generated at the interfaces between layers in heterostructures into electron waveguide geometries. These *lateral geometry low-dimensional devices* led to a range of exciting physical phenomena: resistance quantization, quantum point contacts, non-classical addition of resistances, electron interference, diffraction and coherent focussing phenomena, non-local effects (recent reviews with an emphasis on transport include those of Reed and Kirk (1989) and Ferry et al. (1991)). The nanoelectronic structures here relied

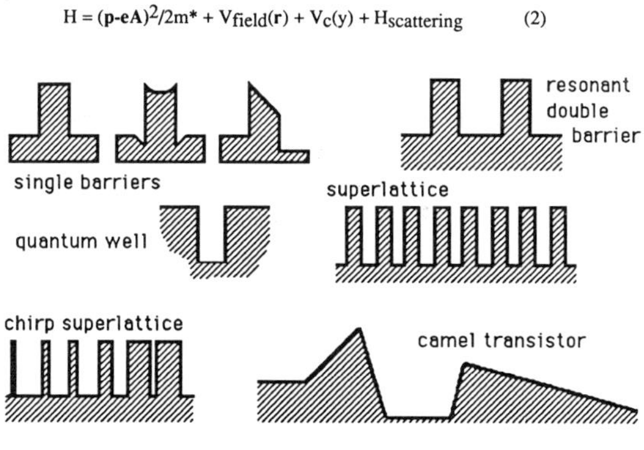

Fig. 1. One-dimensional device profiles obtainable in semiconductor heterostructures.

on both heterostructure engineering and the lateral patterning capabilities of electron beam and reactive ion etching to fully engineer small devices. Figures 2 and 3 illustrate the typical geometries of laterally patterned semiconductor heterostructures. Most of the heterostructure electron waveguides are based on a re-structuring of the two-dimensional electron gas using direct etching or a patterned split-gate to squeeze down the electron gas into a narrow channel or "quantum wire". This quasi-one-dimensional structure may be further restricted to form a quantum point contact or a quantum dot. The basic principles are illustrated in fig. 2.

Ideally, the resulting channel should have only one transverse confined state occupied corresponding to a monomode waveguide. The basic quantum wires can be combined into more complex structures: multi-port waveguides, tunnelling structures and interferometer devices such as Aharonov–Bohm rings and stub tuners.

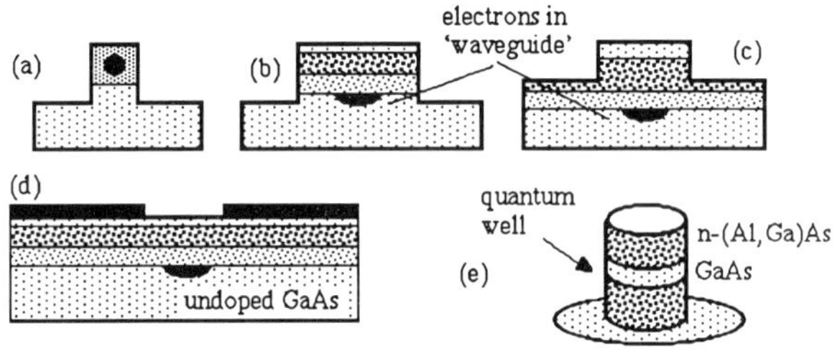

Fig. 2. The production of a quantum wire and quantum dot structures; (a) epitaxial n-GaAs wire; (b) modulation-doped deep-etched GaAs wire; (c) modulation-doped shallow-etched GaAs wire; (d) split-gate wire; and (e) quantum dot.

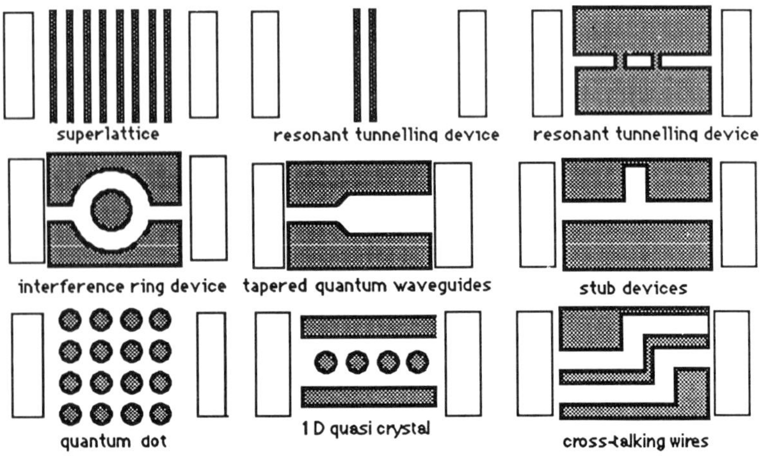

Fig. 3. Gate patterns for lateral nanostructures.

2.1.3. The confinement potential for lateral devices

For simplicity let us assume a parabolic effective-mass Hamiltonian to describe electrons confined to a quasi-two-dimensional layer (x–z plane) in a semiconductor heterostructure by a *confinement potential* $V_c(y)$,

$$H = (p_x^2 + p_y^2 + p_z^2)/2m^* + V_c(y) = (p_x^2 + p_z^2)/2m^* + (p_y^2/2m^* + V_c(y)). \tag{3}$$

In conventional analyses the confinement potential V_c is calculated from the Schrödinger equation and the Poisson equation by considering the fields arising from the band-edge discontinuities at the heterojunction(s), charge spill-over and the *mean field* of the remote donors and surface (including gate) charges. Consequently, V_c is a relatively smooth, idealised potential. Its most elementary form is a triangular quantum well.

Choosing the transverse direction z-axis as the axis of a possible electron waveguide we restrict the lateral (or transverse) degree of freedom in the x-direction by imposing further confinement forces. The corresponding model Hamiltonian is

$$\begin{aligned} H &= p_z^2/2m^* + (p_x^2 + p_y^2)/2m^* + V_c(\boldsymbol{R}, a_\mu(z)) + V(\boldsymbol{R}, z) \\ &= p_z^2/2m^* + H_c + V(\boldsymbol{R}, z), \end{aligned} \tag{4}$$

where $V_c(\boldsymbol{R}, a_\mu(z))$ is the full mean confinement potential for the quantum waveguide and $V(\boldsymbol{R}, z)$ is an external applied potential. The vector $\boldsymbol{R}$ lies in the x–y plane. We have allowed the confinement potential to depend on a finite number of parameters $a_\mu(z)$, $\mu = 1, \ldots, N$, which may vary with position z along the guide.

Typically, as described in detail elsewhere (Barker 1991a,b), V_c will have a roughly parabolic profile throughout, or at least at the boundaries of, the transverse region (see the definitive work of Laux and Stern (1986)). The addition of a uniform external electric field aligned along the z-direction results in the total potential $V_c + \mathrm{V}$. If this potential is a general quadratic function of the coordinates it may be termed a *gutter potential* V_g: a sloping channel with parabolic walls; it is parametrised by four constants, $a_1 = V_0$, $a_2 = E_e$, $a_3 = \omega_x$ and $a_4 = \omega_y$, and provides a good approximation for many real situations,

$$V_g = V_0 = eE_e z + \tfrac{1}{2}m^*[\omega_x^2 x^2 + \omega_y^2 y^2]. \tag{5}$$

2.1.4. The fluctuation potential

The confinement potential has so far been assumed to derive from the *mean field* of the donors and various image charges in the host heterostructure. It has been shown by Nixon et al. (1989), Davies and Nixon (1990) and Laughton et al. (1991) that it must be supplemented by a *fluctuation potential* V_f which arises because of the *discrete random* spatial distributions of donors. The effect is to distort the *full confining potential* $V_{cf} = V_c + V_f$ into a non-self-averaging *random* structure with peak rms values as large as a few milli-electron volts and an extent which may span several microns of material in the transverse direction. The fluctuation potential is a major source of problems in current devices (for an experimental review see Thornton et al. (1991)).

2.1.5. Recent structures

The starting point for heterostructure nanoelectronics is the production of the highest quality semiconductor layers. Molecular beam epitaxy has now progressed to the stage that a wide range of new growth techniques are about ready to enter an exploitation phase. These include atomic plane or delta-doping with silicon which holds good prospects for reducing the fluctuation potential in quasi-two-dimensional electron gases. Lower growth temperatures and techniques such as migration-enhanced epitaxy will be required to reduce the silicon impurity migration in such systems. Improvements in morphology and trap concentrations are expected to be important in the development of pseudo-morphic HEMT structures which are interesting as "conventional" devices and as the first stage in re-configuration into nanostructures.

The current approach to quantum dot and super-atom structures is inhibited by the restricted geometrical configuration of the underlying planar-layered superlattice structure. This limitation is being relaxed with the development of tilted or fractional layer superlattice on vicinal (001) GaAs surfaces as first proposed and demonstrated by Petroff et al. (1984).

2.2. The transport problem

The central problem of transport theory in semiconductors is to determine the currents induced through a device or structure by the applied fields. Transport systems are open systems, the transport parameters, such as conductance, are determined by connecting a semiconductor structure to two, or more, contacts to the macroscopic world where the contacts give the communication with the external thermal and electrical reservoirs and the device surfaces are in communication with the surrounding heat sources and sinks.

2.3. The Kubo formulation

In the early days of quantum transport theory considerable attention was focussed on homogeneous bulk systems in the linear-response regime. The fluctuation–dissipation theorem showed that the linear response to a generalised applied force could be related to an equilibrium correlation function. Thus, the most simple form of Kubo's formula relates conductivity to a velocity autocorrelation function. Attempts to evaluate such correlation functions led to the discovery of a wide variety of quantum kinetic equations and Green function hierarchies that required considerable ingenuity in developing suitable perturbation schemes. In many cases the procedures led to Boltzmann-like transport equations.

Quantum effects became important for galvanomagnetic phenomena in strong magnetic fields. The Landau quantisation of the energy bands, particularly in low effective mass materials, revealed a wealth of structure in the magnetoresistance and other transport coefficients. The effects derived from: (a) the quantisation of the energy bands; (b) from non-classical transport mechanisms such as cyclotron centre hopping in transverse magnetoresistance; (c) resonance phenomena such as the magnetophonon effect. Transport was discovered to be of three fundamental kinds: relaxive, diffusive or hopping. These effects are now well understood and are described in detail in the first edition of the Handbook (ch. 11C).

2.4. The Landauer–Buttiker formalism

An alternative view of transport, which is almost as old as the response theory formalism, was introduced by Landauer and refined by Buttiker and will be called the Landauer–Buttiker formalism. It remained dormant until its considerable successes in explaining aspects of universal conductance fluctuation phenomena, interference phenomena in multiply connected wires, conductance quantisation, the quantum Hall effect and the huge range of multi-terminal device experiments brought in with the development of nanoelectronics. The formalism is essentially a transmission approach which is at its best for quantum ballistic systems where the elastic scattering can be separated spatially from the inelastic scattering. In particular the ideal application would have elastic scattering confined to the volume of a device whereas the inelastic processes would occur solely in the thermal reservoirs each characterised by a well-defined electrochemical potential and forming the device contacts. The conductance may then be expressed in terms of a transmission coefficient describing the probability for carriers to cross the device in the direction of the applied force. The irreversibility of the transport is then associated with the dissipative effect of the contact reservoirs. In most cases the success of the transmission formalism has required low temperatures and quasi-ballistic devices for its simplest variants to be successful. We shall explore this formalism from a phenomenological point of view in later sections. The central problem is the determination of the transmission coefficient. In general this is actually a complicated matrix if a multi-terminal device is deployed.

2.5. Kinetic equation and Green function methods

The closest approach to traditional Boltzmann transport theory is found in the use of kinetic equations for carrier distribution functions in a generalised phase space. Since most of the useful kinetic equations involve dissipative collision integrals their derivation throws light on the origins of irreversibility. We shall use this approach as a unifying element in our discussion of quantum transport. Since position and momentum are conjugate dynamical variables it is not immediately obvious that useful kinetic equations can exist for phase-space distributions. Our choice of this approach is mainly determined by the current interest in inhomogeneous systems, especially small devices and structures often fabricated from heterostructures. The starting point is with the density matrix of quantum-statistical mechanics (see ter Haar 1961).

3. Quantum phase-space distributions

3.1. The statistical density matrix

Suppose a single-electron quantum system is in a general state described by the density matrix ρ. The most direct recipe for computing the quantum-statistical expectation value of an observable Λ for that system involves tracing the operator Λ over the density matrix:

$$\langle \Lambda \rangle = \mathrm{Tr}(\Lambda\rho) = \sum_{\boldsymbol{r}_1, \boldsymbol{r}_2} \langle \boldsymbol{r}_1 | \Lambda | \boldsymbol{r}_2 \rangle \langle \boldsymbol{r}_2 | \rho | \boldsymbol{r}_1 \rangle. \tag{6}$$

In the last term we choose the position representation to carry out the trace so that the matrix elements of Λ and ρ are functions of two positions $\boldsymbol{r}_1$ and $\boldsymbol{r}_2$. If the system is in a *pure state* described by a state vector $|\psi\rangle$, the density matrix has the simple bi-linear form

$$\rho = |\psi\rangle \langle \psi|, \tag{7}$$

and the generally *complex* matrix elements involve the wavefunctions in the position representation,

$$\langle \boldsymbol{r}_2 | \rho | \boldsymbol{r}_1 \rangle = \langle \boldsymbol{r}_2 | \psi \rangle \langle \psi | \boldsymbol{r}_1 \rangle = \psi(\boldsymbol{r}_2)\, \psi^*(\boldsymbol{r}_1). \tag{8}$$

The more general *mixed state* involves a probability distribution P_n over a set of states $\{|\psi_n\rangle\}$ and

$$\langle \boldsymbol{r}_2 | \rho | \boldsymbol{r}_1 \rangle = \sum_n P_n \psi_n(\boldsymbol{r}_2)\, \psi_n^*(\boldsymbol{r}_1). \tag{9}$$

It is a simple matter to use the Schrödinger equation $\mathrm{i}\hbar(\partial/\partial t)|\psi\rangle = H|\psi\rangle$ to derive the equation of motion of the density matrix as

$$\mathrm{i}\hbar(\partial/\partial t)\rho = [H, \rho], \tag{10}$$

which is known as the quantum Liouville equation.

For spatially homogeneous systems the density matrix becomes diagonal in momentum space and the positive-definite distribution $f(p) = \langle \boldsymbol{p} | \rho | \boldsymbol{p} \rangle$ was widely used in linear transport theories at one time. The methodology was extended to dissipative transport by eliminating the off-diagonal density matrix elements perturbatively in favour of the diagonal terms (reviewed by Dresden 1961, ter Haar 1961). There have been attempts to develop non-linear quantum transport theory directly via the spatial density matrix $\langle \boldsymbol{r}_2 | \rho | \boldsymbol{r}_1 \rangle$ (Brunnetti et al. 1989), or the corresponding momentum representation $\langle \boldsymbol{p}_2 | \rho | \boldsymbol{p}_1 \rangle$ (see, e.g., Frensley 1987), but these are difficult to interpret physically, especially if perturbative methods are used. In spatially inhomogeneous situations such as scattering beyond the point approximation, or for inhomogeneous driving fields or for non-uniform carrier densities, or for spatially varying confinement potentials, it becomes crucial to use the full off-diagonal density matrix. For this purpose a considerable simplification may be achieved, as outlined in the following section, by switching to a mixed position and momentum representation, one of which, the Wigner representation, has particularly valuable interpretative power.

3.2. *Quantum phase-space distributions*

From the two vectors $\boldsymbol{r}_1$ and $\boldsymbol{r}_2$, we may shift coordinates to a relative position vector $\boldsymbol{r}=\boldsymbol{r}_1-\boldsymbol{r}_2$ and a central position dependence on a vector $\boldsymbol{R}$ which lies on the line connecting $\boldsymbol{r}_2$ to $\boldsymbol{r}_1$:

$$\boldsymbol{r}=\boldsymbol{r}_1-\boldsymbol{r}_2, \qquad \boldsymbol{r}_2=\boldsymbol{R}-\sigma\boldsymbol{r}; \quad \boldsymbol{r}_1=\boldsymbol{R}+(1-\sigma)\boldsymbol{r},$$

where σ is a real parameter. The various matrix elements may then be denoted in terms of $\boldsymbol{R}, \boldsymbol{r}$ and σ by

$$\langle \boldsymbol{r}_1|\Lambda|\boldsymbol{r}_2\rangle=\Lambda(\boldsymbol{R},\boldsymbol{r},\sigma)=\langle \boldsymbol{R}+\sigma\boldsymbol{r}|\Lambda|\boldsymbol{R}-(1-\sigma)\boldsymbol{r}\rangle, \tag{11}$$

leading to

$$\langle\Lambda\rangle=\sum_{\boldsymbol{R},\boldsymbol{r}}\Lambda(\boldsymbol{R},\boldsymbol{r},\sigma)\,\rho(\boldsymbol{R},-\boldsymbol{r},\sigma). \tag{12}$$

The relative vector $\boldsymbol{r}$ is a natural candidate for a transformation to the momentum representation, so let us define the phase-space representation of an operator Λ by the pair of transformations:

$$\Lambda(\boldsymbol{R},\boldsymbol{P},\sigma)=\sum_{\boldsymbol{r}}\Lambda(\boldsymbol{R},\boldsymbol{r},\sigma)\exp(-\mathrm{i}\boldsymbol{P}\cdot\boldsymbol{r}/\hbar), \tag{13}$$

$$\Lambda(\boldsymbol{R},\boldsymbol{r},\sigma)=(2\pi\hbar)^{-N}\sum_{\boldsymbol{P}}\Lambda(\boldsymbol{R},\boldsymbol{P},\sigma)\exp(\mathrm{i}\boldsymbol{P}\cdot\boldsymbol{r}/\hbar), \tag{14}$$

where N is the number of dimensions. Then eq. (12) becomes:

$$\langle\Lambda\rangle=\sum_{\boldsymbol{R},\boldsymbol{P}}\Lambda(\boldsymbol{R},\boldsymbol{P},\sigma)\,f(\boldsymbol{R},\boldsymbol{P},\sigma), \tag{15}$$

$$f(\boldsymbol{R},\boldsymbol{P},\sigma)=\rho(\boldsymbol{R},\boldsymbol{P},\sigma)(2\pi\hbar)^{-N}. \tag{16}$$

Comparing eq. (15) with classical phase-space averages we see that our construction resembles a classical *phase-space average* over a distribution $f(\boldsymbol{R},\boldsymbol{P},\sigma)$. It includes (or can be used to construct) the majority of quantum distributions $f(\boldsymbol{R},\boldsymbol{P},\sigma)$ that have appeared in the literature (the cases $\sigma=0$ or 1 involve the product of momentum and direct space wavefunctions). The transformation also leads to a mixed position and momentum representation $\Lambda(\boldsymbol{R},\boldsymbol{P},\sigma)$ for the dynamical variables.

3.3. *The Wigner phase-space distribution*

The case $\sigma=\frac{1}{2}$ leads to the Wigner–Weyl transformation and corresponds to a "centre of mass"-like transformation (see fig. 4a). The function $f(\boldsymbol{R},\boldsymbol{P})=f(\boldsymbol{R},\boldsymbol{P},\sigma=\frac{1}{2})$ is known as the *Wigner distribution function* (Wigner 1932). Of the possible distributions obtainable by varying σ, only the Wigner distribution ($\sigma=\frac{1}{2}$) is *real*

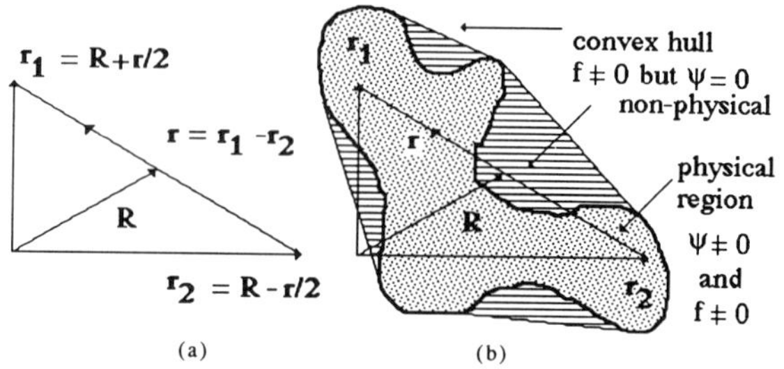

Fig. 4. (a) Geometry for construction of the Wigner function. (b) The Wigner function is not compact: it exists on the convex hull of the wavefunction, it therefore occupies spatial regions which are inaccessible to the wavefunction. The hatched region shows the non-physical area for a simple two-dimensional example.

(but not positive-definite), although real distributions could be built out of linear combinations of complex distribution functions with different σ parameters. A more compelling reason for choosing the Wigner distribution is its reduction to a classical *probability* distribution in the classical limit (the limit $h \to 0$ is approached non-uniformly). A comprehensive review of the Wigner distribution is given by Carruthers and Zachariasen (1983); an enlightening physical argument for its adoption is given by Wigner (1967).

Full second quantised many-body systems may also be described by Wigner functions which may be shown to have a precise correspondence with thermodynamic Green functions (Kadanoff and Baym 1962, Barker 1981, Jauho 1989).

4. Quantum ballistic transport

4.1. Equation of motion of the Wigner function

Let us first consider the ballistic transport problem, which we define as the situation when the effects of dissipative collisions are vanishingly small. For this case we adopt the simplified parabolic effective-mass approximation $H = p^2/2m + V$, where V is a potential energy which contains confinement potential, applied field and potential scattering contributions. Switching to the centre of mass coordinates (and set $\hbar = 1$ temporarily for readability) let us take the matrix elements $\langle \boldsymbol{R} - \frac{1}{2}\boldsymbol{r}|\ldots|\boldsymbol{R} + \frac{1}{2}\boldsymbol{r}\rangle$ of the quantum Liouville equation (10),

$$\mathrm{i}\partial\rho/\partial t = [p^2/2m^* + V(r), \rho], \tag{17}$$

$$[\mathrm{i}\partial/\partial t + (1/m^*)\frac{\partial}{\partial \boldsymbol{r}} \cdot \frac{\partial}{\partial \boldsymbol{R}} - V(\boldsymbol{R} + \tfrac{1}{2}\boldsymbol{r}, t) + V(\boldsymbol{R} - \tfrac{1}{2}\boldsymbol{r}, t)]\rho(\boldsymbol{R}, \boldsymbol{r}, t) = 0, \tag{18}$$

$$[\partial/\partial t + (\boldsymbol{P}/m) \cdot \partial/\partial \boldsymbol{R} + L]f(\boldsymbol{R}, \boldsymbol{P}, t) = 0, \tag{19}$$

where Lf is the "driving field term"

$$Lf(\boldsymbol{R}, \boldsymbol{P}, t) = -\int \mathrm{d}^3 r\, \mathrm{e}^{-\mathrm{i}\boldsymbol{P}\cdot\boldsymbol{r}} \{V(\boldsymbol{R} + \tfrac{1}{2}\boldsymbol{r}, t) - V(\boldsymbol{R} - \tfrac{1}{2}\boldsymbol{r}, t)\}\rho(\boldsymbol{R}, \boldsymbol{r}, t). \tag{20}$$

The general case may be put in a more elegant form, by defining an effective force $F_{\mathrm{eff}}(\boldsymbol{R}, \boldsymbol{r}, t)$ and it Wigner–Weyl transform $F_{\mathrm{eff}}(\boldsymbol{R}, \boldsymbol{P}, t)$,

$$F_{\mathrm{eff}}(\boldsymbol{R}, \boldsymbol{r}, t)\cdot\boldsymbol{r} = \{V(\boldsymbol{R} + \tfrac{1}{2}\boldsymbol{r}, t) - V(\boldsymbol{R} - \tfrac{1}{2}\boldsymbol{r}, t)\}, \tag{21}$$

$$F_{\mathrm{eff}}(\boldsymbol{R}, \boldsymbol{P}, t) = \int \mathrm{d}^3 r\, \mathrm{e}^{-\mathrm{i}\boldsymbol{P}\cdot\boldsymbol{r}/\hbar} F_{\mathrm{eff}}(\boldsymbol{R}, \boldsymbol{r}, t)/(2\pi\hbar)^3. \tag{22}$$

The resulting equation of motion is a *non-local* version of the classical Liouville equation or collisionless Boltzmann equation

$$[\partial/\partial t + (\boldsymbol{P}/m)\cdot\partial/\partial\boldsymbol{R}]f(\boldsymbol{R}, \boldsymbol{P}, t) + \int \mathrm{d}^3\boldsymbol{P}'\, F_{\mathrm{eff}}(\boldsymbol{R}, \boldsymbol{P}', t)\cdot\frac{\partial}{\partial\boldsymbol{P}} f(\boldsymbol{R}, \boldsymbol{P} - \boldsymbol{P}', t) = 0. \tag{23}$$

Planck's constant is re-exposed in the quantum force term, eq. (22).

The principal moments of the Wigner function yield the normalisation, carrier density and electrical current density,

$$1 = \int \mathrm{d}^3\boldsymbol{P} \int \mathrm{d}^3\boldsymbol{R}\, f(\boldsymbol{R}, \boldsymbol{P}, t), \tag{24}$$

$$n(\boldsymbol{R}, t) = \int \mathrm{d}^3\boldsymbol{P}\, f(\boldsymbol{R}, \boldsymbol{P}, t), \tag{25}$$

$$\boldsymbol{j}(\boldsymbol{R}, t) = 2e \int \mathrm{d}^3\boldsymbol{P}\, (\boldsymbol{P}/m^*) f(\boldsymbol{R}, \boldsymbol{P}, t). \tag{26}$$

4.2. *Quasi-classical transport*

The Wigner function has the extraordinary property that for a simple Hamiltonian $H = T(\boldsymbol{p}) + V(\boldsymbol{r})$ which is a *general quadratic* function of position and momentum the *non-local quantum* equation of motion reduces to the corresponding *local classical* Liouville equation. We can easily prove this by writing the potential as the general quadratic form or gutter potential (see § 5.7.5 for an example):

$$V(\boldsymbol{R}, t) = v_0 + \boldsymbol{v}_1\cdot\boldsymbol{R} + \boldsymbol{R}\cdot\boldsymbol{V}_2\cdot\boldsymbol{R}, \tag{27}$$

where v_0, $\boldsymbol{v}_1$, $\boldsymbol{V}_2$ are arbitrary constant scalars, vector and tensor quantities, respectively. The driving field term collapses to a familiar form,

$$\{V(\boldsymbol{R} + \tfrac{1}{2}\boldsymbol{r}, t) - V_{\mathrm{eff}}(\boldsymbol{R} - \tfrac{1}{2}\boldsymbol{r}, t)\} = (\boldsymbol{v}_1 + \boldsymbol{V}_2\cdot\boldsymbol{R} + \boldsymbol{R}\cdot\boldsymbol{V}_2)\cdot\boldsymbol{r} = -\boldsymbol{F}\cdot\boldsymbol{r}, \tag{28}$$

where $\boldsymbol{F}(\boldsymbol{R})$, the effective force, is identical to the classical force derivable from eq. (30).

In that case, eq. (29) may be Wigner transformed to the form

$$[\partial/\partial t + (\boldsymbol{P}/m)\cdot\partial/\partial \boldsymbol{R} + \boldsymbol{F}\cdot\partial/\partial \boldsymbol{P}]f(\boldsymbol{R},\boldsymbol{P},t) = 0, \tag{29}$$

which is just the classical Liouville equation.

Similarly, we recover the classical Liouville equation if $f(\boldsymbol{R}, \boldsymbol{P}, t)$ if independent of R, i.e., if the system is spatially homogeneous.

It is tempting to expand the non-local term in powers of the Planck constant but this leads to spurious results since the equation of motion is not analytic in $\hbar$.

4.3. *Gauge invariant formulation*

In the presence of an arbitrary electromagnetic field described by the vector and scalar potentials $A(\boldsymbol{R}, t)$ and $\varphi(\boldsymbol{R}, t)$, the electron velocity is determined by

$$m^*\boldsymbol{v} = \boldsymbol{p} - e\boldsymbol{A} \equiv \boldsymbol{\pi}, \tag{30}$$

and the Hamiltonian becomes

$$H = (\boldsymbol{p} - e\boldsymbol{A})^2/2m^* + V_c + V_f + V + e\varphi. \tag{31}$$

The theory is invariant under the gauge transformation

$$\boldsymbol{A} \to \boldsymbol{A} + \nabla\chi; \qquad \varphi \to \varphi - \partial\chi/\partial t, \tag{32}$$

provided the Wigner distribution is re-defined as

$$f(\boldsymbol{R}, \pi) = (2\pi\hbar)^{-3} \sum_{\boldsymbol{r}} \rho(\boldsymbol{R}, \boldsymbol{r}) \exp(-\mathrm{i}\boldsymbol{\pi}\cdot\boldsymbol{r}/\hbar) \exp\left(-\mathrm{i}e \int_0^1 d\sigma\, \boldsymbol{r}\cdot\boldsymbol{A}(\boldsymbol{R} - \tfrac{1}{2}\boldsymbol{r} + \sigma\boldsymbol{r}, t)\right). \tag{33}$$

For a uniform magnetic field and a general *linear* electric field we again obtain a classical transport equation for the Wigner distribution provided the total confining potential is at most quadratic,

$$[\partial/\partial t + (\boldsymbol{\pi}/m)\cdot\partial/\partial\boldsymbol{R} + (e\boldsymbol{E} + (e/m^*)\boldsymbol{\pi}\times\boldsymbol{B} + \boldsymbol{F})\cdot\partial/\partial\boldsymbol{\pi}]f(\boldsymbol{R}, \boldsymbol{\pi}, t) = 0. \tag{34}$$

This remarkable result shows once again that even with a uniform magnetic field and a harmonic potential the electron transport is exactly classical.

4.4. *Technical difficulties with Wigner distributions*

The general equation of motion given by eq. (23) may be formally solved by path-variable techniques along the lines of the Chambers–Rees method, but so far only the simplest of one-dimensional problems have been attempted. Convergence problems are very severe, a problem which may be traced to: (a) the breakdown of the area-preserving mapping property of classical phase space; and (b) the non-compactness of f.

There are other problems too, which may trap the unwary. First, the Wigner–Weyl transformation leads to Wigner functions which may assume *negative* as well as positive values. This rules out the interpretation of $f(\boldsymbol{R}, \boldsymbol{P})$ as a simple probability distribution, although it is a perfectly satisfactory *statistical* distribution function. As discussed by Wigner (1967) the requirement that f is generally not positive-definite is a consequence of the Heisenberg uncertainty relations. This point is particularly brought home if one computes the Wigner functions for stationary states, especially in strong confinement potentials.

One direct consequence of the "centre of mass" style construction is that $f(\boldsymbol{R}, \boldsymbol{P})$ can assume non-zero values at points $\boldsymbol{R}$ where the wavefunction is zero, i.e., in non-physical regions (as indeed can the "Λ"-density $\Lambda(\boldsymbol{R}, \boldsymbol{P})$); technically f has non-compact support. This possibility is sketched in fig. 4b. An extreme case is furnished by an electron confined to a long, curved, one-dimensional wire which is twisted in 3-space: the three-dimensional Wigner function displays strong fringes adjacent to concave sections of the wire. Similarly, the Wigner function may be non-zero in regions where the momentum wavefunction is zero.

Figure 5a illustrates the Wigner function for a wavepacket which was incident in the past on a resonant tunnelling barrier and which has subsequently divided into a reflected and a transmitted packet. Contributions to $f(\boldsymbol{R}, \boldsymbol{P})$ at the midpoint between the two exit wavepackets leads to a strong negative and positive going oscillation in f; in this "non-physical" region the particle density or momentum density obtained by either integrating f along the momentum or space axis is zero: all the probability density is concentrated where one expects – in the region occupied by the wavepackets. It is interesting to see the effect of removing this rapid oscillatory structure in a computer simulation: there is no effect on the outgoing wavepackets, but if the time evolution is reversed the system does not regress to a single wavepacket corresponding to the initial state but instead generates two "scattered" packets on either side of the barrier. This loss of time-reversibility occurs because of the loss of phase information which was contained in the oscillatory structure in the "non-physical" region. Therefore, it is not correct to say that the Wigner function is non-physical in regions where the wavefunction is zero: these regions may contain important phase information. Instead we are witnessing part of the inherent non-locality of quantum mechanics which appears whenever we attempt to force a classical particle-like picture.

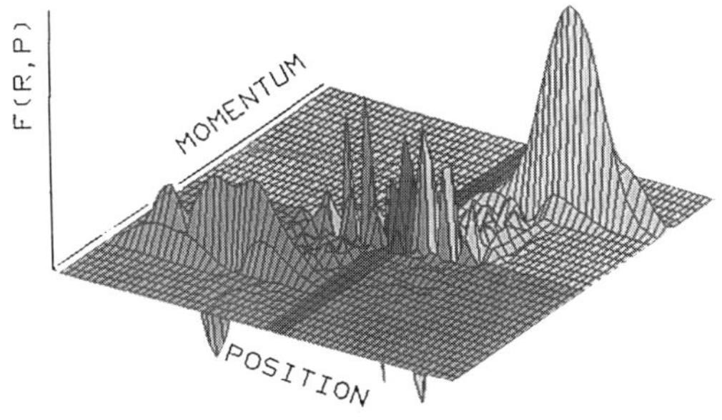

Fig. 5. The Wigner function for a Gaussian wavepacket scattering on a resonant tunnelling barrier (Barker 1986).

The source of the non-compactness may be traced to the "centre of mass" construction. The situation may usefully be compared with the movement of the centre of mass of two classical particles confined to opposite locations on a ring: the classical centre of mass coordinate remains in the physically inaccessible centre of the ring. In the classical case we can always switch back to the actual particle coordinates (corresponding to choosing $\sigma = 0$ or 1 in our construction) without losing the phase-space picture. In quantum theory this appears to be impossible: whereas we can obtain the probability density for locating the electron in space at a particular location $\boldsymbol{R}$ by integrating $f(\boldsymbol{R}, \boldsymbol{P})$ over momentum we cannot, at the same time, determine its momentum. Trajectories in phase space do not, therefore, generally exist in the Wigner picture.

We note that f is not a perfect differential except for quadratic Hamiltonians and hence it cannot generally be used to prove the general existence of phase-space trajectories. But in all the cases where the equation of motion of the distribution function does become local (i.e., reduces to classical form) it is possible to define an underlying ensemble of particle trajectories in phase space. However, this does not mean that we can prescribe a singular classical distribution of the form

$$f(\boldsymbol{R}, \boldsymbol{P}, t) = \delta(\boldsymbol{R} - \boldsymbol{R}(t))\, \delta(\boldsymbol{P} - \boldsymbol{P}(t)), \tag{35}$$

where $\boldsymbol{R}(t)$ and $\boldsymbol{P}(t)$ describe a classical point trajectory. Such a possibility would violate the Heisenberg uncertainty relations. The prevention of this embarrassment is possible because there are additional constraints on the form and values which f can assume other than those determined by the quantum Liouville equation. This subtle point is discussed in detail in the section on stationary Wigner functions.

These difficulties preclude using any simple classical technique such as Monte Carlo modelling, to evaluate the Wigner phase-space distributions. However, as we shall describe later there are rigorous ways of introducing phase-space trajectories in quantum transport, but at a price.

4.5. *Restrictions on the solutions to the Wigner function equation of motion*

The results of §§ 4.3 and 4.4 show that the equation of motion of the Wigner distribution function is exactly classical for a parabolic kinetic energy when the potential fields are quadratic and the magnetic field is uniform (of whatever strength). In these circumstances an initial (injected) distribution of carriers will propagate classically (i.e., via area-preserving transformations in phase space: therefore, admitting trajectory descriptions exactly). The underlying motion preserves the occupancy of the initial quantum states. It is not surprising then to find that many experimental observations of "quantum" ballistic transport admit a simple classical explanation because a phase-space of configuration-space *trajectory* description most certainly exists under the conditions we have just mentioned. Yet, *paradoxically*, we know that the eigenstates of an electron in a harmonic potential are quantised and, similarly, a magnetic field generates well-defined Landau states in the ballistic transport regime.

The "missing" quantum effects may be found hidden in the *boundary conditions* to the transport equations. The Wigner distribution cannot take on an arbitrary classical form; e.g., a delta-function in phase space would violate the uncertainty relations; indeed, the Wigner distribution, constructed directly from wavefunctions, has long been known to satisfy certain inequalities and sum rules which proscribe its behaviour (e.g., $|f| < 1/h^N$). Only initial distributions which satisfy stringent quantum conditions can be accepted as initial states which then evolve classically in time under quadratic potentials and uniform magnetic fields. We could construct those initial states directly by writing the initial density matrix as a bi-linear function of an appropriate complete set of wavefunctions (e.g., harmonic oscillator or Landau states or edge states) and then using the Wigner–Weyl transformation to determine the initial Wigner distribution. Alternatively, we can work entirely with Wigner distributions in extended formalism which provides the theory with a complete set of stationary Wigner functions.

It follows that the underlying quantised nature of the states will only be manifest if either: (a) transitions are induced between those states; in particular inelastic transitions most certainly reveal the discrete structure; or (b), the quantisation is revealed through the dependence of the current on the initial distribution, e.g., via a supply function (in the following sections we shall see that this is precisely why we can use semi-classical arguments to "derive" the conductance quantisation formulae discovered experimentally for quantum wires by van Wees et al. (1988) and Wharam et al. (1988)). Only if the Hamiltonian departs significantly from the conditions of uniform magnetic fields and quadratic potential fields do we pick up the effects of electron diffraction and interference phenomena.

4.6. *Stationary Wigner functions and completeness*

The initial state for an evolving quantum distribution must be a linear superposition of *stationary* Wigner functions made up from the eigenstates of the appropriate Hamiltonian. The Wigner equation for ballistic motion in a quadratic potential propagates an initial distribution classically. This shows that the Wigner equation of motion gives an incomplete picture of the physics – it cannot describe stationary states. A similar problem arises with the density matrix equation

$$i\hbar\, \partial\rho/\partial t = [H, \rho], \tag{36}$$

whose Wigner–Weyl transform generates the Wigner equation of motion directly; it is not completely equivalent to the Schrödinger equation, as can be seen by trying to set up an eigenvalue equation. In fact, the stationary states have to be obtained from an adjunct equation for the stationary density matrix (Dahl 1981, Carruthers and Zachariasen 1983, Barker 1989a,b)

$$\varepsilon\rho = \tfrac{1}{2}\{H, \rho\}, \tag{37}$$

which involves the anti-commutator $\{\,,\,\}$. The Wigner–Weyl transform of this equation yields

$$\tfrac{1}{2}\varepsilon_{mn}f_{mn}(\boldsymbol{R},\boldsymbol{P}) = [\boldsymbol{P}^2/2m - (\hbar^2/8m)\,\partial^2/\partial\boldsymbol{R}^2]f_{mn}(\boldsymbol{R},\boldsymbol{P}) + \int \mathrm{d}^3 r \int \mathrm{d}^3\boldsymbol{P}'\,\mathrm{e}^{-\mathrm{i}\boldsymbol{P}'\cdot\boldsymbol{r}/\hbar}\{V(\boldsymbol{R}+\tfrac{1}{2}\boldsymbol{r} + V(\boldsymbol{R}-\tfrac{1}{2}\boldsymbol{r})\} \times f_{mn}(\boldsymbol{R},\boldsymbol{P}-\boldsymbol{P}')/2(2\pi\hbar)^3. \tag{38}$$

where the Planck constant is firmly re-instated. Here the $\varepsilon_{mn} = \varepsilon_m + \varepsilon_n$, where the ε_n are the usual eigenvalues of H. For $m = n$ we obtain the usual stationary Wigner functions which could alternatively be constructed directly from the stationary wave-functions; they are *real* valued. The case $m \neq n$ gives *complex* functions which relate to the other eigenfunctions of the super-operator $\{H, \}$. The entire set of $f_{mn}(\boldsymbol{R},\boldsymbol{P})$ form a complete orthonormal set for all Wigner functions, stationary or otherwise. Any initial Wigner function should be projected onto this space if the correct boundary conditions for time-dependent transport are required. All the special sum rules of f, such as $|f| \leqslant 1/(h)^3$, are then taken care of.

4.7. *Extended phase-space picture*

The time-dependent Wigner function for a Hamiltonian $H = p^2/2m + V(r)$ may be written (following Barker and Murray 1983)

$$[\partial/\partial t + (\boldsymbol{P}/m)\cdot\partial/\partial\boldsymbol{R}]f(\boldsymbol{R},\boldsymbol{P},t) + (2\pi\hbar)^{-3} \int \mathrm{d}^3\boldsymbol{Q}' \int \mathrm{d}^3\boldsymbol{P}'\cos(\boldsymbol{Q}\cdot\boldsymbol{P}'/\hbar)\,F_{\mathrm{d}}(\boldsymbol{R},\boldsymbol{Q},t)\cdot\frac{\partial}{\partial\boldsymbol{P}}f(\boldsymbol{R},\boldsymbol{P}-\boldsymbol{P}',t) = 0, \tag{39}$$

where F_{d} is derived from the relation $F_{\mathrm{d}}(\boldsymbol{R},\boldsymbol{Q})\cdot\boldsymbol{Q} = V(\boldsymbol{R}-\tfrac{1}{2}\boldsymbol{Q}) - V(\boldsymbol{R}+\tfrac{1}{2}\boldsymbol{Q})$. In the limit $\hbar \to 0$, F_{d} goes over into the classical force and eq. (39) becomes the Liouville equation; the limit is approached non-uniformly.

This equation is non-local in momentum space (and in position space if the kinetic energy is non-quadratic). Ideally, we would use this equation in analogy with the collisionless Boltzmann equation (or Vlasov equation) to describe short-geometry structures. The effects of crystal structure can be taken into account through effective-mass theory and the band structure.

The quantum ballistic transport equation cannot be emulated by an ensemble of point particles in phase space except when the non-locality is removed. However, a classical-like picture does underly the representation. Barker and Murray (1983) have shown that f may be represented as the linear superposition of a set of distributions $f_{\mathrm{c}}(\boldsymbol{R},\boldsymbol{P},\boldsymbol{Q},\boldsymbol{P}',t)$ which obey classical equations,

$$f(\boldsymbol{R},\boldsymbol{P},t) = (2\pi\hbar)^{-3}\int \mathrm{d}^3\boldsymbol{Q}' \int \mathrm{d}^3\boldsymbol{P}'\cos(\boldsymbol{Q}'\cdot\boldsymbol{P}'/\hbar)\,f_{\mathrm{c}}(\boldsymbol{R},\boldsymbol{P},\boldsymbol{Q}',\boldsymbol{P}',t), \tag{40}$$

$$[\partial/\partial t + (\boldsymbol{P}/m)\cdot\partial/\partial\boldsymbol{R} + F_{\mathrm{d}}(\boldsymbol{R},\boldsymbol{Q}')\cdot\partial/\partial\boldsymbol{P}]f_{\mathrm{c}}(\boldsymbol{R},\boldsymbol{P},\boldsymbol{Q}',\boldsymbol{P}',t) = 0. \tag{41}$$

The causal boundary conditions are $f_{\mathrm{c}}(\boldsymbol{R},\boldsymbol{P},\boldsymbol{Q}',\boldsymbol{P}',0) = f(\boldsymbol{R},\boldsymbol{P}+\boldsymbol{P}',0)$.

It follows that quantum distributions can be emulated by ensembles of classical particles in a family of phase spaces and Monte Carlo techniques can be restored, but at extra computational cost. Phenomena like tunnelling can be readily pictured in this representation: for some values of the parameters P and Q the particle trajectories are at higher energies and the barrier potentials related to $F_d(\boldsymbol{R}, \boldsymbol{Q}')$ are smeared out in comparison to the classical limit $Q' = 0$, $P' = 0$; penetration to forbidden regions is thus allowed off the classical shell.

5. Transport in electron waveguides

5.1. Introduction

In the ballistic transport regime it is convenient to use transmission matrices for the description of the wave propagation of the electron in the confinement potential of the quantum wires. As with all quantum phenomena the details depend critically on the boundary conditions. Most recent studies have involved electron waveguide structures which merge into relatively large thermal reservoirs in the "contact" or "electrode" regions (see the generic picture in fig. 6). Such systems have proved to be well-described by a set of formulae for the effective conductance between the connected reservoirs, known as the Landauer or Landauer–Buttiker formulae (Landauer 1970, 1975, 1987, 1989, Buttiker 1986, 1988a,b, Buttiker et al. 1985, Stone and Szafer 1988), and which rely on the computation of appropriate transmission coefficients. This transmission approach has had spectacular successes including a detailed understanding of the quantum Hall effects (Buttiker 1988b). The simplest case arises for just two reservoirs connected by a single uniform quantum wire. It can be understood by a simple one-dimensional quasi-classical argument which we now outline.

5.2. Classical description

Classical ballistic transport may be described by phase-space techniques based on the Hamiltonian equations of motion. Since, classically, a particle can be described

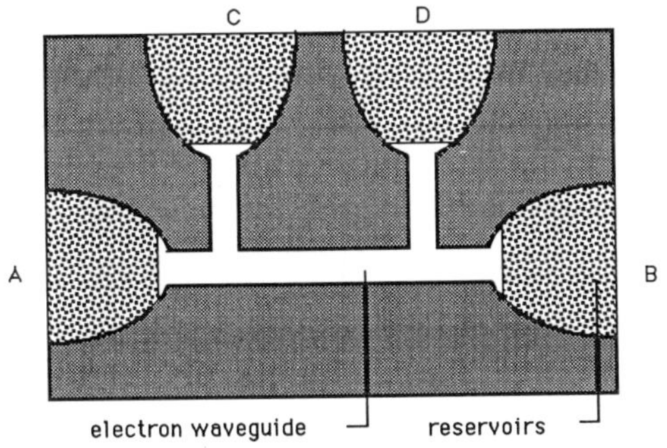

Fig. 6. A generic electron waveguide connected to four thermal reservoirs of electrons.

uniquely by its instantaneous position $\boldsymbol{R}$ and momentum $\boldsymbol{P}$ it is convenient to describe one or more non-interacting particles by a phase-space probability distribution $f(\boldsymbol{R}, \boldsymbol{P}, t)$ normalised to the number density of particles and which physically is conserved according to the Liouville theorem $\mathrm{d}f/\mathrm{d}t = 0$. Since f is a perfect differential we have, mathematically,

$$\left[\frac{\partial}{\partial t} + \frac{\mathrm{d}\boldsymbol{R}}{\mathrm{d}t}\cdot\frac{\partial}{\partial \boldsymbol{R}} + \frac{\mathrm{d}\boldsymbol{P}}{\mathrm{d}t}\cdot\frac{\partial}{\partial \boldsymbol{P}}\right] f(\boldsymbol{R}, \boldsymbol{P}, t) = 0, \tag{42}$$

or, physically, by substitution from the Hamiltonian equations,

$$\mathrm{d}\boldsymbol{R}/\mathrm{d}t = \partial H/\partial \boldsymbol{P}; \qquad \mathrm{d}\boldsymbol{P}/\mathrm{d}t = -\partial H/\partial \boldsymbol{R}, \tag{43}$$

$$\left[\frac{\partial}{\partial t} + \frac{\partial H}{\partial \boldsymbol{P}}\cdot\frac{\partial}{\partial \boldsymbol{R}} - \frac{\partial H}{\partial \boldsymbol{R}}\cdot\frac{\partial}{\partial \boldsymbol{P}}\right] f(\boldsymbol{R}, \boldsymbol{P}, t) = 0. \tag{44}$$

For a simple Hamiltonian $H = \boldsymbol{P}^2/2m^* + V(\boldsymbol{R})$ we set $\boldsymbol{F} = -\partial H/\partial \boldsymbol{R}$ to obtain the most simple form of the Liouville equation or the collisionless Boltzmann equation,

$$\left[\frac{\partial}{\partial t} + \frac{\boldsymbol{P}}{m}\cdot\frac{\partial}{\partial \boldsymbol{R}} + \boldsymbol{F}\cdot\frac{\partial}{\partial \boldsymbol{P}}\right] f(\boldsymbol{R}, \boldsymbol{P}, t) = 0. \tag{45}$$

If the solutions to eq. (44) or eq. (45) is known, one may compute the current density according to the statistical prescription

$$\boldsymbol{j}(\boldsymbol{R}, t) = e \int \mathrm{d}^3\boldsymbol{P}\,(\boldsymbol{P}/m) f(\boldsymbol{R}, \boldsymbol{P}, t) \tag{46}$$

(where we might also put in a factor of two for spin degeneracy).

Any point initial distribution in phase space, e.g., $f = \delta(\boldsymbol{R} - \boldsymbol{R}_0)\,\delta(\boldsymbol{P} - \boldsymbol{P}_0)$ will unfold as a unique phase-space trajectory which satisfies the Hamiltonian equations. Any area of phase space will be mapped by the Hamiltonian equations into an equal area of phase space at a later time. We may model the injection and extraction of particles by inserting suitable generation and recombination terms (source and sink descriptions) on the right-hand side of the Liouville equation which is a convenient way to manage open systems.

5.3. Quasi-classical description of the perfect channel

Let us apply the classical formalism to the electron waveguide problem, using, initially, a one-dimensional model for simplicity. Referring to fig. 6, suppose we ignore contacts C and D and suppose that the two reservoirs A and B are identical except that a potential difference V exists between them (fig. 7). The region between the reservoirs is conservative and carriers are injected or extracted at thermal sources G and perfectly absorbing sinks R located at the reservoir boundaries $x = 0$ and $x = L$. A phase-space portrait of this system is shown in fig. 7.

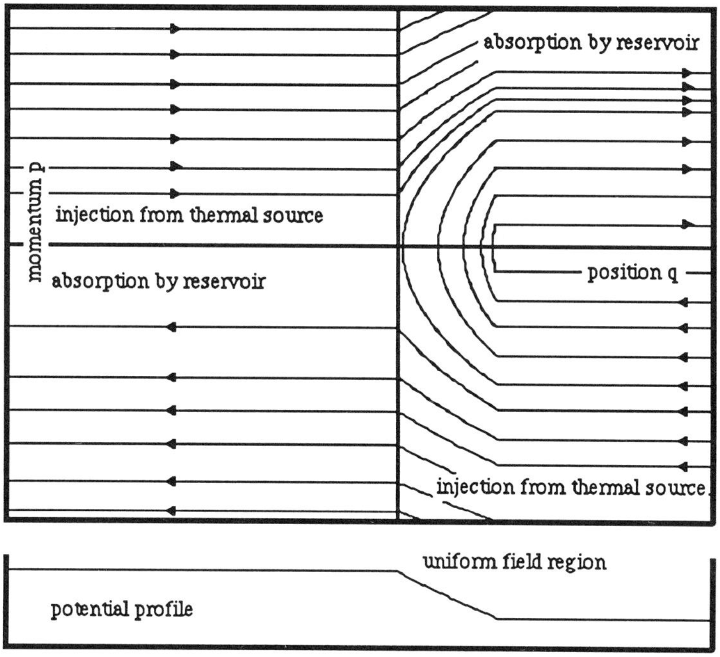

Fig. 7. Phase-space portrait of trajectories through a guide with a uniform field with thermal sources and sinks at outer extremities.

The carrier phase-space distribution $f(x, p, t)$ then satisfies the open-system Liouville equation

$$[\partial/\partial t + v(p)\partial/\partial x + F(x)\partial/\partial P]f(x, p, t) = G - R, \tag{47}$$

$$G = \{\theta(v)\, v\, \delta(x)\, f_0(ep) + \theta(-v)\, v\, \mathrm{d}(x-L)\, f_0(\varepsilon_p + eV)\}\rho_P, \tag{48}$$

$$R = \theta(v)\, v\, \delta(x-L)\, f(x, p, t) + \theta(-v)\, v\, \delta(x)\, f(x, p, t), \tag{49}$$

where θ is the unit step function, F is the local conservative force in the waveguide and f_0 is the Fermi–Dirac distribution for the reservoirs. ρ_P is the density of states in momentum space for the injected carriers; we take this as $1/h$. Here lies our only concession to quantum mechanics: we assume that the injecting reservoirs are maintained in thermal equilibrium with quantum distributions f_0, ρ_P. The current may be found from

$$J = 2e \int \mathrm{d}p\, v(p)\, f(x, p, t), \tag{50}$$

where the factor of 2 accounts for spin.

Equation (47) may be solved by path-variable techniques (just using an integrating factor in this trivial case) since we have conservative flow described by phase-space trajectories. Choosing $f(x, p, t=0) = 0$ within the wire and letting the low-temperature limit $f_0(\varepsilon_p) = \theta(\mu - \varepsilon)$ and solving for long times we find the current

$$j = (2e/h) \int d\varepsilon \, [\theta(\varepsilon - \mu - eVx/L) - \theta(\varepsilon - \mu - eVx/L + eV)]. \tag{51}$$

Finally,

$$j = (2e^2/h)V. \tag{52}$$

We might call this a "classical" Landauer formula. It should be stressed that the transport is classical but the supply of carriers by the reservoirs is quantised.

5.4. *Multiple channels and conductance quantisation*

For a two-dimensional channel with transverse states ε_n at the entry and exit to the guide we replace the supply function f_0 by

$$f_0 = \sum f_n; \quad f_n = \theta(\mu - \varepsilon_n - \varepsilon),$$

formula (52) trivially generalises to $j = (2Ne^2/h)V$, whence the conductance is found as

$$G = (2Ne^2/h), \tag{53}$$

where N is the number of occupied sub-bands. The *quantisation of conductance* is thus demonstrated. The experimental discovery of conductance quantisation occurred as recently as 1988 (van Wees et al. 1988, Wharam et al. 1988) and was explained by a quantum-mechanical analysis by Kirczenow (1988). The effect is only observable at very low temperatures. The inclusion of finite-temperature supply functions in our analysis may be shown to lead to progressive damping of the steps at high temperatures.

5.5. *Inclusion of potential scattering*

If we add a conservative "scattering potential" of maximum energy W into the guide with a classical transmission coefficient $T(\varepsilon)$, shown in figs. 8 and 9, we find the simple generalisation of eq. (52) to be

$$j = (2e^2/h)T(\mu)V. \tag{54}$$

The extension to a multi-channel system may be made by including a transmission coefficient T_{nn} for each channel and a coefficient T_{nm} for possible transmission from one channel to another,

$$j = (2e^2/h) \sum_n \sum_m T_{nm} V. \tag{55}$$

Technically, T_{nm} is the probability of flux input in channel n being transmitted out in channel m and the sums are over all input modes with energies below the Fermi energy.

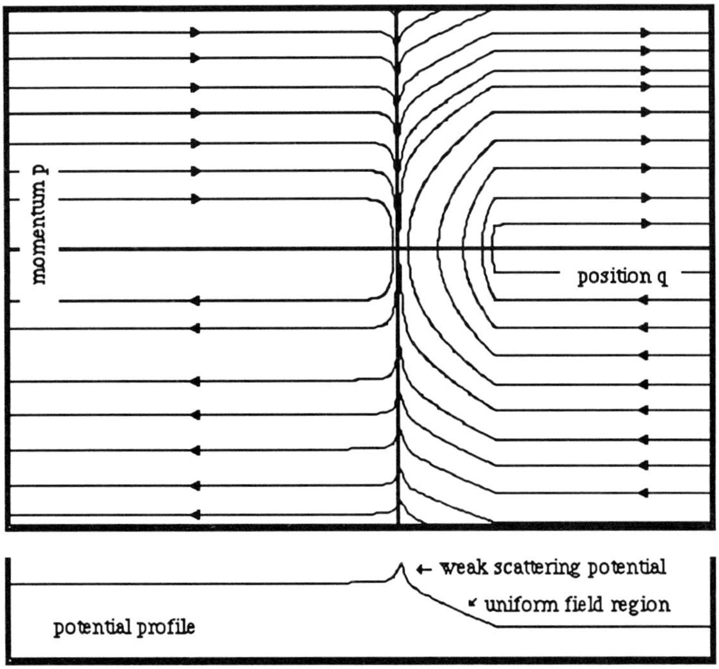

Fig. 8. Addition of a scattering potential to the guide.

The familiar finite-temperature Landauer formulae can be recovered similarly. The transition to the familiar mobility formulae obtained from the Boltzmann equation (the collisional version) can also be easily followed in this picture by varying the mean free path from infinity to less than the channel length.

5.6. *The quantum picture and conductance quantisation*

The precise quantum picture may be obtained phenomenologically by replacing the classical transmission coefficient T by the correct quantum-mechanical version computed for the guide. An elegant detailed analysis is given by Kirczenow (1988). Figure 10 shows the calculated conductance quantisation for a split-gate quantum

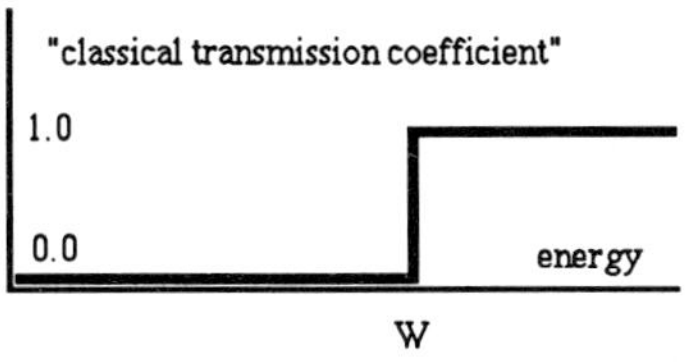

Fig. 9. Classical transmission coefficient for a barrier of height W.

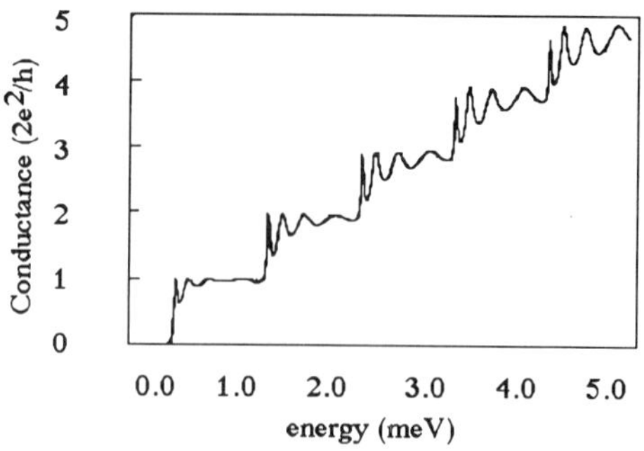

Fig. 10. Conductance quantisation in a perfect quantum point contact. The confining potential is parabolic and contact length is 0.65 μm at $T = 0$ K.

point contact based on a coupled-mode formalism (Laughton and Barker 1991, see also later sections).

The strong success of the semi-classical argument in obtaining the conductance quantisation for high-mobility quantum point contacts may be attributed to the confinement potentials being slowly-varying parabolic channels. As discussed in detail in § 4, the quantum analogue of the phase-space distribution function f is the Wigner distribution and this *evolves* classically in general quadratic potentials, although its initial state must satisfy quantum boundary conditions which, in our present analysis, are provided by the supply functions.

The Landauer–Buttiker formalism has been adapted to complicated geometries, such as rings and multi-port structures, and provided no strong inter-sub-band mixing takes place one may compute the appropriate transmission coefficients (or transfer matrices) as though we had a waveguide or transmission line problem and the conductance may be recovered with ease.

5.7. Coupled-mode theory in arbitrary electron waveguides

5.7.1. Introduction

The present interest in quantum ballistic electron transport in semiconductor structures has generated a requirement for accurate modelling of transport in single- and multiply-connected *spatially-extended* electron waveguide structures which may be *inhomogeneous*. Target devices are, e.g., the Aharonov–Bohm ring device (Ford et al. 1988) and the squeezed-channel or throttle device (van Houten et al. 1990, van Wees et al. 1988, Wharam et al. 1988). Until recently these have only been modelled as one-dimensional structures (we may quote the elegant work by Buttiker (1986) and Gefen et al. (1984) as examples) with a few numerical calculations based on the two-dimensional Schrödinger equation (Barker 1989a, Barker et al. 1989, Lent et al. 1989). Frohne and Datta (1988) have described an approximate numerical technique based on wavefunction matching to calculate the scattering matrix for electron transfer between two-dimensional channel regions with different confining potentials in the transverse direction. Kirczenow (1988) has described a wavefunction matching technique for the abrupt uniform channel case to model the quantised resistance. A general formalism has been developed by Barker and co-workers and has interesting analogies with electromagnetic tapered waveguide theory. In the following we develop the general wave-mechanical formalism for an arbitrary shaped electron waveguide.

5.7.2. Basic formulation for curved guides

Let the wavefunction $\Psi(\boldsymbol{r}, t) = \langle \boldsymbol{r} | \Psi(t) \rangle$ at a location $\boldsymbol{r} = (x, y, z) = (\boldsymbol{R}, z)$ be represented as a superposition of eigenstates $\varphi_m[a_\mu(z), \boldsymbol{R}]$ of H_c corresponding to the confinement potential in the two-dimensional region formed by the x–y plane passing through the point z. We use different transverse states at different cross-sectional slices in the guide.

Introducing an amplitude factor $\psi_m(z, t)$ we have

$$\Psi = \sum_m \psi_m(z, t)\, \varphi_m[a_\mu(z), \boldsymbol{R}]. \tag{56}$$

The set $\{\varphi_m[a_\mu(z), \boldsymbol{R}];\ \mu = 1, 2, \ldots, N\}$ is a complete orthonormal set parametrised by N parameters $a_\mu(z)$ which will generally depend on the z-spatial coordinate and which relate to the geometry of the confinement potential V_c (e.g., we might have a tapered three-dimensional channel, channel radius $a_1 = a(z)$ with perfectly reflecting walls). By substituting expression (56) into the time-dependent Schrödinger equation and projecting onto the z-domain by forming the partial scalar product on the complete set $\{\varphi_m[a_\mu(z), \boldsymbol{R}]\}$ we find that the amplitudes $\psi_m(z)$ obey coupled one-dimensional equations of the form:

$$i\hbar\, \partial\psi_m(z)/\partial t = [(-\hbar^2/2m^*)\partial^2/\partial z^2 + E_m(z)]\psi_m + \sum_n \{A_{mn}(z)\psi_n + B_{mn}(z)\, \partial\psi_n/\partial z\}. \tag{57}$$

Here, $E_m(z)$ is the eigenvalue of state $\varphi_m[a_\mu(z), \boldsymbol{R}]$. The coupling coefficients $A_{mn}(z)$ and $B_{mn}(z)$ are know functions of: the local guide eigenstates φ_m, the guide parameters a_μ and their derivatives and the partial matrix elements $V_{mn}(z)$ of the internal potential $V(\boldsymbol{R}, z)$,

$$A_{mn} = \int \mathrm{d}^2 R\, \varphi_m \left\{ V(\boldsymbol{R}, y)\varphi_n + (-\hbar^2/2m^*) \left\{ \sum_{\mu,\nu} \frac{\partial^2 \varphi_n}{\partial a_\mu\, \partial a_\nu} \cdot \left[\frac{\partial a_\mu}{\partial z}\right]\left[\frac{\partial a_\nu}{\partial z}\right] + \sum_\mu \frac{\partial \varphi_n}{\partial a_\mu} \cdot \frac{\partial^2 a_\mu}{\partial z^2} \right\}\right\} \tag{58}$$

$$= V_{mn}(z) + \sum_{\mu,\nu} W^{mn}_{\mu\nu}(z),$$

$$B_{mn} = \int \mathrm{d}^2 R\, \varphi_m \left\{ (-\hbar^2/2m^*)^2 \sum_\mu \frac{\partial \varphi_n}{\partial a_\mu}\left[\frac{\partial a_\mu}{\partial z}\right]\right\} = \sum_\mu U^{mn}_\mu(z), \tag{59}$$

$$V_{mn}(z) = \int \mathrm{d}^2 R\, \varphi_m \{V(R, z)\}\varphi_n, \qquad D_{mn} = A_{mn} - V_{mn}. \tag{60}$$

If the confinement parameters a_μ are time-dependent we should include an additional term

$$V^t_{mm} = \mathrm{i}\hbar \int \mathrm{d}^2 R\, \varphi_m \sum \frac{\partial \varphi_n}{\partial a_\mu}\left[\frac{\partial a_\mu}{\partial t}\right] \tag{61}$$

in the expression for A_{mn}. E_m, A_{mn} and B_{mn} then become functions of time.

Equation (57) is a set of coupled one-dimensional equations for the mode amplitudes. In solving the stationary-state version of these equations it is convenient to separate the contributions from forward propagating, backward propagating and evanescent states by the Ansatz

$$\psi_m(z) = \psi_m^{(+)}(z)\exp(\mathrm{i}k_m z) + \psi_m^{(-)}(z)\exp(-\mathrm{i}k_m z), \tag{62}$$

where $k_m^2 = 2m^*(E - E_m)/\hbar^2$, E is the total energy and the case $k_m^2 < 0$ describes evanescent states.

5.7.3. *Adiabatic limit*

In the case of the extreme quantum limit (lowest mode occupied) or the case of weak inter-mode scattering (slowly varying parameters) we obtain the *adiabatic approximation* in which a carrier will remain in the same mode n throughout the channel. Then expression (57) assumes an interesting structure if we group the coupling terms A and B into a manifestly Hermitian Hamiltonian form using the operator $p_z = -\mathrm{i}\hbar\,\partial/\partial z$ and preserving only diagonal terms $m = n$,

$$\begin{aligned} \mathrm{i}\hbar\,\partial\psi_m(z,t)/\partial t = \{&[p_z + \langle m|p_z|m\rangle]^2/2m^* - [\langle m|p_z|m\rangle]^2/2m^* \\ &- [\langle p_z m|p_z m\rangle]/2m^* + E_m(z) + V_{mm}(z)\}\psi_m(z,t), \end{aligned} \tag{63}$$

$$\langle m|p_z|m\rangle = \int \mathrm{d}^2 R\, \varphi_m \left\{(-\mathrm{i}\hbar)\sum \frac{\partial \varphi_m}{\partial a_\mu}\cdot\left[\frac{\partial a_\mu}{\partial z}\right]\right\}$$

$$= (-\mathrm{i}\hbar/2)\sum\left[\frac{\partial a_\mu}{\partial z}\right]\frac{\partial}{\partial a_\mu}\int \mathrm{d}^2 R\, \varphi_m\varphi_m = 0,$$

$$\langle p_z m|p_z m\rangle = \int \mathrm{d}^2 R\,(-\mathrm{i}\hbar)^2 \sum\sum \frac{\partial\varphi_m}{\partial a_\nu}\cdot\frac{\partial\varphi_m}{\partial a_\mu}\cdot\left[\frac{\partial a_\mu}{\partial z}\right]\left[\frac{\partial a_\nu}{\partial z}\right] = -T_{mm}.$$

We first observe that the diagonal momentum-dependent coupling terms $(\sim\partial\psi_m(z,t)/\partial z)$ vanish identically in this limit ($\langle m|p_z|n\rangle$ is generally non-zero); secondly, the term T_{mm}, which is positive, adds to the zero-point energy of the mode provided some of the gradients $[\partial a_\mu/\partial z]$ are non-zero,

$$\mathrm{i}\hbar\,\partial\psi_m(z,t)/\partial t = \{[p_z]^2/2m^* + T_{mm}(z) + E_m(z) + V_{mm}(z)\}\psi_m(z,t). \tag{64}$$

The terms T_{mm} and E_m play the role of one-dimensional quasi-potential fields whose gradients correspond to the effective force on a carrier due to the size quantization and the longitudinal interaction with the confinement potential.

A straightforward Wigner transform on the wavefunction $\psi_m(z,t)$ (following Barker 1989a) allows us to construct a relatively simple equation of motion for the Wigner distribution function $f_m(z, p_z, t)$: a non-local version of the collisionless Boltzmann equation:

$$[\partial/\partial t + (p_z/m^*)\cdot\partial/\partial z]f_m(z, p_z, t) + \int \mathrm{d}P'\, F_{\mathrm{eff}}(z, P', t)\cdot\partial f_m(z, p_z - P', t)/\partial P = 0,$$

$$F_m(z, r, t) = \{V_m(z + \tfrac{1}{2}r, t) - V_m(z - \tfrac{1}{2}r, t)\}/r; \qquad V_m = T_{mm}(z) + E_m(z) + V_{mm}(z), \qquad (65)$$

$$F_m(z, P, t) = \int \mathrm{d}r\, e^{-iP\cdot r/\hbar} F_m(z, r, t)/(2\pi\hbar).$$

In the case of slowly-varying applied and quasi-fields (i.e., very little quantum reflection, resonances or tunnelling) we obtain the local collisionless Boltzmann kinetic equation

$$[\partial/\partial t + (p_z/m^*)\cdot\partial/\partial z]f_m(z, p_z, t) + F_m(z, t)\cdot\partial f_m(z, p_z, t)/\partial p_z = 0;$$
$$F_m(z, t) \to -(\partial/\partial z)V_m(z). \qquad (66)$$

This equation is exact for the case that F_{eff} is a gutter potential.

5.7.4. Coupled kinetic equations

It is relatively easy to re-introduce the inter-mode coupling if the previous conditions for the local kinetic equations are met. For weak inter-mode coupling the standard projection super-operator calculus may be used in analogy with the case of longitudinal magnetotransport (Barker 1982) to obtain an inter-mode scattering integral within the Golden Rule approximation for the scattering rates $R(mn, z)$. The latter are straightforwardly related to matrix elements compounded from the coefficients A_{mn} and B_{mn} we find

$$[\partial/\partial t + (p_z/m^*)\cdot\partial/\partial z]f_m(z, p_z, t) + F_m(z, t)\cdot\partial f_m(z, p_z, t)/\partial p_z$$
$$= -\sum\{f_m(z, p_z, t)\, R(mn, z) - f_n(z, p_z, t)\, R(nm, z)\}. \qquad (67)$$

Interesting selection rules apply depending on the choice of confinement potential.

5.7.5. Effective fields

The quasi-fields $F_m(z)$ may be engineered by profiling the walls of the quantum wire. Thus a linearly-tapered section of a hard-walled wire gives rise to an effective constant force field (or linear potential); a succession of abruptly-connected uniform wires of different widths corresponds to a succession of step potentials. Figure 12 illustrates typical potentials that can ideally be achieved for different lateral patterning. It would appear that a number of architectures that have been difficult to achieve with vertical transport through heterostructures, such as staircase potentials, are realisable with laterally patterned nanostructures. There is, however, the question of how strong we can make the quasi-fields or, equivalently, how energetic can we make the quasi-potential barriers. A broad measure of scale is given by the zero-point energy, which for a 20 nm width hard-walled wire in GaAs gives $E_1 \sim 14$ meV; similarly for the parabolic potentials described by Kumar et al. (1989) we have $\hbar\omega \sim 2.6$ meV, corre-

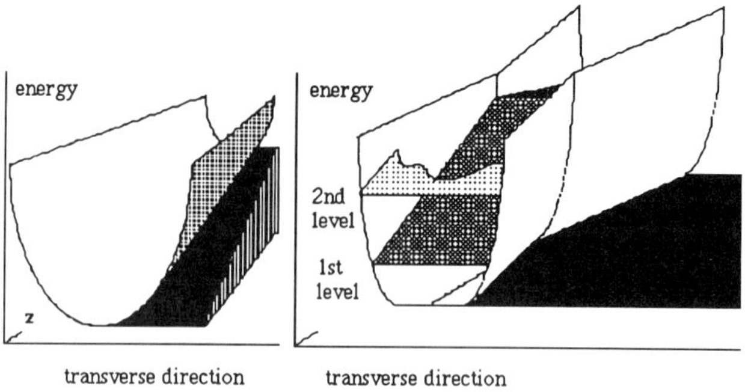

Fig. 11. Electron waveguides: (a) gutter potential (b) transverse sub-band energies as function of distance.

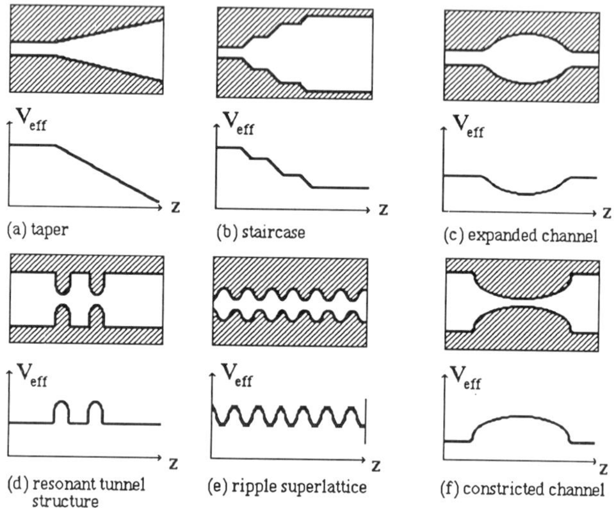

Fig. 12. Electron waveguide squeezed-gate geometries showing the effective potential for adiabatic motion in one of the transverse sub-bands ideal-confinement potential.

sponding to a squeezed-gate channel of width 0.4 μm and gate voltage −1.0 V. Figure 11b sketches the variation of the zero-point energy with distance along a pinched quantum wire. It is clear, therefore, that hot-electron effects induced by such fields will only be manifest at low temperatures. It is tantalising to note that with a suitable squeezed-gate geometry on the limiting scales of 10 nm defined channel widths it should be possible to achieve electron energies equivalent to 1000 K.

5.7.6. Computational methods

The most physical computational scheme (Barker 1989a,b, 1991a, Laughton et al. 1991) uses the coupled-mode theory of transport in electron waveguides which is suitable for describing the influence of realistic confinement potentials including the effects of fluctuations. Numerical modelling here is complicated by the need to include the evanescent and travelling wave modes, the number of which varies along the channels in extreme cases, such as a ripple superlattice or an Aharonov–Bohm ring device. The standard Thomson–Haskell propagator matrix methods are unstable here and a number of methods to remove the problem by decoupling algorithms have been discovered recently (Laughton and Barker 1991, Laughton et al. 1991, Laughton 1992). This method may be complemented by direct finite-difference solutions to the two-dimensional time-dependent effective-mass Schrödinger equation for the structure using a modified ADI algorithm (Barker 1989a, Finch 1989). Finite-element methods (Lent et al. 1989) have also been reported very recently for studying resistance quantisation in pinched channels. A further method utilises a network of one-dimensional lines to span the appropriate waveguide geometry and this method is a natural extension of traditional one-dimensional theory (Pepin 1990). Baranger and co-workers have used standard Green function propagator methods to numerically study electron waveguides.

5.8. Interference phenomena and the Aharonov–Bohm effect

Multiply-connected geometries, such as a waveguide ring, are ideal for observing quantum interference phenomena, particularly in the magnetotransport parameters. Good examples are given by the extensive observations of Aharonov–Bohm effects due to alteration of the phase difference between a split coherent electron flow by an electromagnetic flux threading the multiply-connected region (a detailed theory based on a one-dimensional analysis and the Landauer–Buttiker formalism was given by Gefen et al. (1984), with particular reference to metal rings; experimental observations on semiconductor ring devices have been reported by a number of groups, including Ford et al. (1988)).

5.8.1. Background

The original Aharonov–Bohm effect (Aharonov and Bohm 1959) considered the modification to the interference fringes of a classic two-slit electron interference experiment due to the presence of an electromagnetic flux. A gauge-invariant treatment of quantum mechanics reveals that the phase of an electron wavefunction at a space–time point $(\boldsymbol{r}, t)$ is advanced by a factor $\mathrm{e}\int[\varphi\,\mathrm{d}t - \boldsymbol{A}\cdot\mathrm{d}\boldsymbol{r}]/\hbar$ involving an integral of the scalar and vector electromagnetic potentials over a path in space–time up to the point in question. In the ideal two-slit experiment the amplitude of the wavefunction at an observation point is determined by the interference between waves travelling the two possible routes to the observation point. In the presence of an electromagnetic field the coupling to the potentials gives rise to an additional phase difference given by the space–time integral over a closed path along the routes,

$$I = e\oint_C [-\varphi\,\mathrm{d}t + \boldsymbol{A}\cdot\mathrm{d}\boldsymbol{r}]/\hbar. \tag{68}$$

This integral is generally non-zero. Stokes theorem generalised to four dimensions gives I as an integral over a surface S spanning the loop C,

$$I = e \int\int S\{\boldsymbol{E} \cdot \mathrm{d}\boldsymbol{r}\, \mathrm{d}t + \boldsymbol{B} \cdot \mathrm{d}\boldsymbol{S}\}/\hbar = e\Phi/\hbar. \tag{69}$$

The quantity Φ is the four-dimensional electromagnetic flux through the surface S. The surface S is arbitrary, except that it is bounded by C which itself lies outside the region where E and B are non-zero. *The quantum effects of electromagnetic fluxes are thus periodic functions of the amount of enclosed flux* Φ. The fundamental unit of flux is evidently

$$\Phi_0 = h/e = 4.13 \times 10^{-3}\ \mathrm{Tesla}/(\mu\mathrm{m})^2. \tag{70}$$

Some phenomena occur with half this period, i.e., $h/2e$, to be expected in superconducting systems, but less obvious in disordered systems and normal ring structures.

The most remarkable result of this analysis is that the phase-difference $e\Phi/\hbar$ may be non-zero, even if the electric and magnetic fields are totally confined within but not on the encircling path C. The phase difference does not depend on the precise arrangement of E and B fields but only on the net flux enclosed. The possible physical manifestation of the non-integrable phase factor controlled by $e\Phi/\hbar$ is attributed to Aharonov and Bohm (1959), although the effects were anticipated in a paper by Franz (1939) and, independently, by Ehrenberg and Siday (1949). The non-integrable phase factor has been shown to be responsible for a very wide range of transport-related phenomena including universal conductance fluctuations. These are described elsewhere in the Handbook.

The observation of the influence of the phase factor can only be expected in an interference experiment which allows the detection of phase differences between two waves. Aharonov and Bohm considered variants of the two-slit interference experiment, which, as Feynman has remarked, is the only truly significant experiment in quantum physics, and considered the electromagnetic flux to arise from either a pure magnetic field or a time-dependent electric field.

In the case of a two-slit experiment there are *two main effects of the non-integrable phase*: the first is a displacement of the envelope of the normal interference pattern and derives from the Lorentz force *on the physical paths*; secondly, a fringe shift within the envelope due to the *enclosed flux* is observed. Whereas the first effect disappears if the E and B fields are confined within the loop, the second effect depends on the enclosed flux and occurs always.

The Aharonov–Bohm effect is relatively easy to produce with an enclosed magnetic flux. It is less easy to consider with an electric flux. In the original theories an electric flux was produced in Gedanken experiments by imagining that a split beam of electrons would be passed between a pair of conducting cylinders, each at the same electrostatic potential and separated by a distance W.

Apart from interference experiments the non-integrable phase factor can lead to observable changes in the energies of bound quantum systems.

5.8.2. The Aharonov–Bohm effect and transport in semiconductor rings

The typical geometry of an experimental Aharonov–Bohm ring device is shown in fig. 13, typical ring diameters of $\sim$400 nm have been reported with channel widths of the order of 66 nm for split-gate GaAs structures (Ford et al. 1988).

An electron wavepacket incident from the left of the ring will divide onto the two arms of the ring and reconstruct at the right exit to the ring if there is no phase difference between the two paths. The details of scattering at the input and output to the ring determine the net transmission through the ring. If the channels are regarded as one-dimensional then if a net electromagnetic flux Φ threads the ring the phase difference between the two arms for a symmetrical ring will be $I = e\Phi/\hbar$. When $I = 2\pi N$ there is constructive interference and maximum transmission; when $I = 2\pi(N + \frac{1}{2})$ the interference is destructive and reflection occurs back down the ring. The low-temperature conductance may be determined by the Landauer formula as

$$G = 2e^2 T(\Phi)/h, \tag{71}$$

where T is the transmission coefficient for the ring. The one-dimensional theory is actually rather complicated as it depends critically on assumptions made about the scattering matrix for the two junction regions of the ring (Gefen et al. 1984), but nevertheless it predicts perfect modulation of the transmission coefficient and hence G as a function of the enclosed magnetic flux.

Experimentally, the best observed modulations of G have been <25–30%. A detailed study of wavefunction propagation and the magnetotransmission has been made by Barker and co-workers (Barker 1989a,b, Barker et al. 1989, Barker and Finch 1989, Pepin 1990, Laughton 1992) for a wide range of semiconductor ring structures based on two-dimensional models with the aim of discovering the influence of finite channel width and fluctuation potentials on the Aharonov–Bohm effect. It was observed that many experimental ring structures (see fig. 13) have a geometry such that electrons entering from the left must make a transition from a narrow to a wide waveguide region before settling into the arms of the ring; the first and second spatial derivatives of the guide radius $a(z)$ are, therefore, large at the ring entrance and exits. The coupled-mode theory then indicates that even if a wavepacket approaches from the left in a single (e.g., lowest) transverse mode it may be scattered into higher modes at the ring junction unless the mode energy separation is very large. This effect has been confirmed numerically for a range of realistic confinement

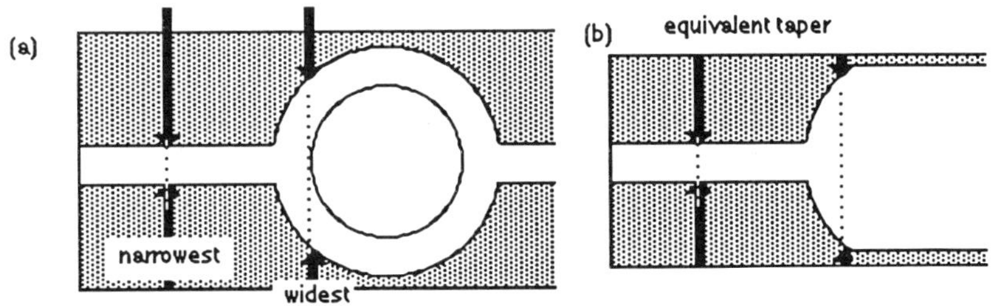

Fig. 13. Aharonov–Bohm ring geometries.

potentials in two-dimensional models evaluated independently by finite-difference methods, lattice theoretic methods and the coupled-mode formulation. Since the modes of the ring are spatially distributed, the path differences differ from mode to mode and if several modes are occupied it is not generally possible to obtain perfect modulation of the total transmission coefficient.

In a study of the experimental results of Ford et al. (1988), Barker (1989a,b) and Finch (1989) showed that the Aharonov–Bohm resonance is quite different from the pure one-dimensional result, only the central lobe of the exiting wavepacket exhibits destructive interference and the two side lobes are transmitted; the situation is closer to a two-slit interference experiment where the envelope of the interference pattern remains constant in shape and is shifted slightly by the Lorentz force, but the interference pattern develops a zero at the centre. At the most this can only lead to 30% current modulation. The typical computed magneto-transmission for a practical ring was found to agree qualitatively and quantitatively with the published experimental data.

An extension of this study (Barker and Finch 1989) is illustrated in fig. 14 where we display a selection of snapshots of a computer experiment to follow the propagation of a wavepacket through a typical ring device (flat-bottomed channels with strongly parabolic walls) but including a constriction in the output channel to filter out only the lowest mode in the exit channel. A very high modulation of the transmission coefficient is observed. The incident Gaussian wavepacket in the ground transverse state scatters predominantly into three transverse states in the ring and would maintain occupancy of those states if the exit channel was not constricted.

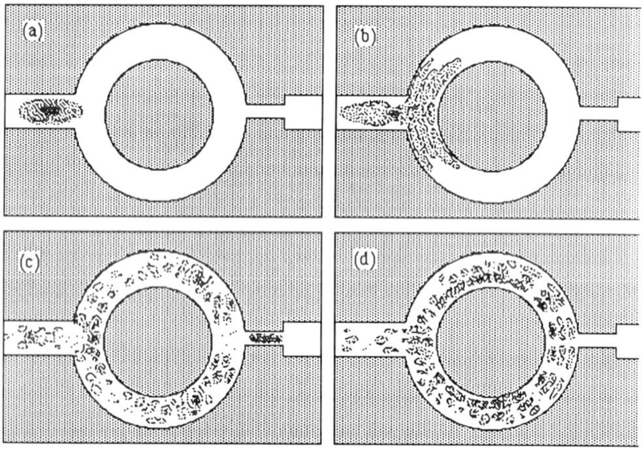

Fig. 14. The Aharonov–Bohm effect in semiconductor rings. Computed electron probability distributions for a Gaussian wavepacket incident in the ground state of a quantum ring GaAs waveguide of diameter 400 nm (Barker and Finch 1989). (a) Incident packet (10 meV) at $t = 0.0$ ps for both $B = 0$ and $B \neq 0$; (b) Wavepacket at $t = 1.0$ ps, case $B = 0$; (c) Wavepacket at $t = 9.0$ ps; case $B = 0$, partial transmission. (c) Wavepacket at $t = 7.0$ ps, case $B \neq 0$; on resonance: the packet has totally reflected. The notched exit guide is designed to filter out modes excited into higher energy states in the ring. This configuration exhibits almost perfect modulation of the current as the magnetic flux is varied.

A later study (Barker et al. 1989) showed that a weak modulation of conductance could be obtained in a circular domain if only a few modes were occupied due to the existence of well-defined closed paths in the domain arising from spatial quantisation of the models. The principle is illustrated in fig. 15.

5.9. Influence of the fluctuation potential

The Aharonov–Bohm effects, like most of the interference-based phenomena, have small amplitudes of modulation (typically 20–30%) and are observable only at very low temperatures. The amplitude dependence can be partly explained by the difficulty in achieving monomode conduction in the waveguide. The general theory of inhomogeneous electron waveguides shows that intermode scattering by the curved boundaries will occur easily except at very low energies.

But all the mentioned properties are features of electron waveguides describable by smooth confinement potentials (e.g., harmonic cross-section). However, it can be shown that the confinement potential which is assumed to derive ultimately from the mean field of the donors and various image charges must be supplemented by a fluctuation potential which arises because of the discrete spatial distributions of donors (Nixon et al. 1989, Davies and Nixon 1990, Laughton et al. 1991). This situation typically arises in a heterostructure device in which the quasi-two-dimensional electron gas originates in electron transfer from donors situated in a doping layer separated by a buffer layer from, for example, an AlGaAs–GaAs interface. Depending on the amount of screening, the rms fluctuation potential can be as large as 30 meV. The effect is to distort the electron gas into a random structure, which is illustrated in fig. 8b. Detailed calculations (Davies and Nixon 1990) for split-gate waveguides show that the pinch-off occurs randomly and generally occurs well before

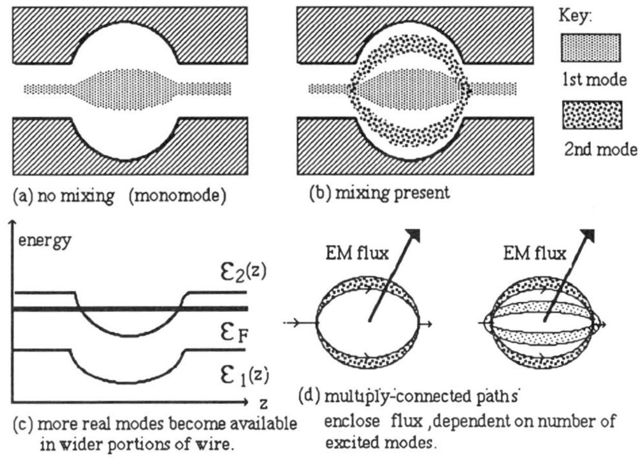

Fig. 15. Mode mixing can lead to multiply-connected "electron trajectories" which are manifest in phenomena such as the Aharonov–Bohm effect.

a scale at which monomode conduction would be established quantum mechanically – at typically 60–80 nm for 1 μm long wires with 0.4 μm split gates.

These semi-classical calculations neglect tunnelling and a quantum theory of the image force remains to be developed so they are probably pessimistic results; but they do indicate a serious problem for waveguide design (Nixon et al. 1990).

Very recently, the full coupled-mode theory has been applied to the problem of modal transport through a squeezed-gate quantum point contact using realistic potential profiles in the presence of the fluctuation potential (Laughton et al. 1991). Figure 16 illustrates some of the results based on experimental data from Timp et al. Different samples of the stochastic fluctuation potential lead to wide variation in conductance properties. The fluctuation potential was computed self-consistently from the semi-classical Thomas–Fermi approximation. The calculation includes the contribution from the donors, assumed to be fully ionised and distributed at random in a δ-doped layer (following the earlier study by Nixon et al. (1990) which used recursive Green function techniques). The confinement and fluctuation potentials are very device specific because of the random nature of the donor layer.

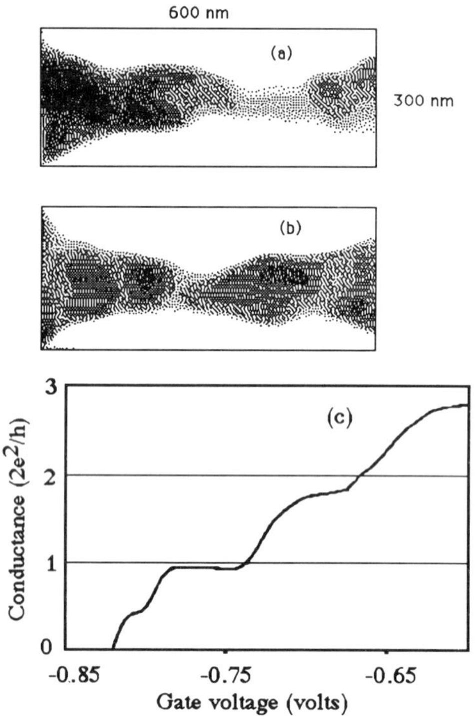

Fig. 16. Effect of the fluctuation potential on the effective channel confinement potential for a realistic quantum point contact based on the data of Timp et al. (1989), of dimensions 600 × 300 nm^2. (a) Electron density contour plot: this sample displays good quantisation of conductance. (b) Electron density contour plot: this sample displays poor quantisation of conductance; gate voltage = −0.724 V. (c) Conductance for sample (a). (After Laughton et al. 1991.)

Both the curvature of the confinement potential and the inhomogeneities introduced by the fluctuation potential lead to mode coupling. The detailed transport is quite complex and includes in-channel resonances and indirect processes which lead to strong back-scattering. The latter cannot be treated by Born approximation arguments for the mobility where the assumption of independent scatterers leads to contributions to back-scattering from direct processes only. The Born approximation thus seriously over-estimates the mobility. The *indirect* back-scattering is part of the breakdown of independent scattering which ultimately leads to exponential localisation in one-dimensional systems.

For realistic device potentials the transport is found to be *non-adiabatic* and yet good quantised conductance is still predicted. This surprising feature is attributed to "compensated scattering", an idea originally due to Payne but not thought to be significant in the region following the narrowest constriction of the device. Basically, flux scattering forward out of the lowest modes is compensated by scattering-in from the higher modes which are also occupied. The result is that the effective one-dimensionality of the transport is maintained and this, following the arguments of § 3, sustains the conductance quantisation effect: a one-dimensional channel has a maximum conductance of $2e^2/h$. The compensated scattering leads additionally to "mode scrambling" which reduces the coherence of the flux. This mechanism may be significant in reducing the amplitudes of interference processes in ring structures.

6. Trajectory representations for single electrons

6.1. The pilot-field picture

The controlling clocking of *single* electrons through an array of tunnel junctions in metal–insulator–metal nanostructures (Geerligs et al. 1990) and related single-electron charging effects in semiconductor quantum point contact nanostructures (Staring et al. 1992) raises the exciting prospect of a single-electronics technology in which one bit of information might be carried by one electron. Such structures rely on charging phenomena, such as the Coulomb blockade and related effects (especially correlated single-electron tunnelling, see e.g., Averin and Likhaerev (1986), Likhaerev (1988), and Bakhavalov et al. (1989)), to achieve stable-charge soliton transport in the devices. Perhaps even more interesting is the prospect for manipulating individual electrons and measuring their transport directly: a regime in which the interpretation of quantum mechanics becomes controversial (Schommers 1989, Barker 1991b). For these reasons it is appropriate to briefly review quantum transport theory for single electrons using the pilot-field interpretation of quantum mechanics to explore the possibility of a particle trajectory picture (Barker 1991b, Barker et al. 1992).

Quantum mechanics normally deals with predicting the statistical outcome of a series of measurements on identical systems or the outcome of a measurement on an ensemble of quantum systems. An example of the former might be the prediction of the statistical pattern of electron counts in a low-intensity two-slit experiment (which has a solid-state analogue in Aharonov–Bohm semiconductor rings (Ford et al. 1988, Barker 1989a,b, Barker et al. 1989). An example of the latter might be a conventional current transport measurement on a device containing very many carriers. It has been argued by some that quantum mechanics cannot describe single-quantum systems (see Ballentine's 1970 review). Although quantum mechanics provides a perfectly satisfactory framework for modelling the statistics of single-electron systems there is little in the orthodox *interpretation* to provide much intuition nor to suggest tests for the acknowledged problems with quantum mechanics, which become particularly apparent for single systems. Indeed, simple questions of the coherence of coupled Coulomb-blockade oscillator devices and related problems on the synchronisation of single-electron circuits by clock signals require an understanding of individual events as well as ensemble averages. At a deeper level there are difficult interplays between single systems and many-body dissipative systems when one considers phenomena ranging from the quantum image force arising from a single-electron tunnelling to the problem of the quantum measurement process itself.

In seeking insight into single-electronics there is considerable interpretive power and some very useful new computational advantages in adopting the *pilot-wave* or pilot-field alternative to orthodox Copenhagen theory, put forward originally by de Broglie and later revived by Bohm (1952). This approach predicts identical outcomes to conventional quantum mechanics but has the potential for significant deviation from the orthodox theory if "wavepacket collapse upon measurement" is rejected: an open question still in quantum measurement theory (Schommers 1989). Its practical significance, however, is that it: (a) retains a considerable element of the physical realism implicit in classical physics, notably the existence of well-defined *particle trajectories* in phase space; and (b) through the device of the *quantum potential* which describes the coupling of the pilot self-field to the electron, it provides a simple removal of very many of the paradoxes of quantum theory, such as the two-slit experiment dichotomy (see figs. 17 and 18), the delayed-choice problem and a simple picture of how interference and tunnelling phenomena occur.

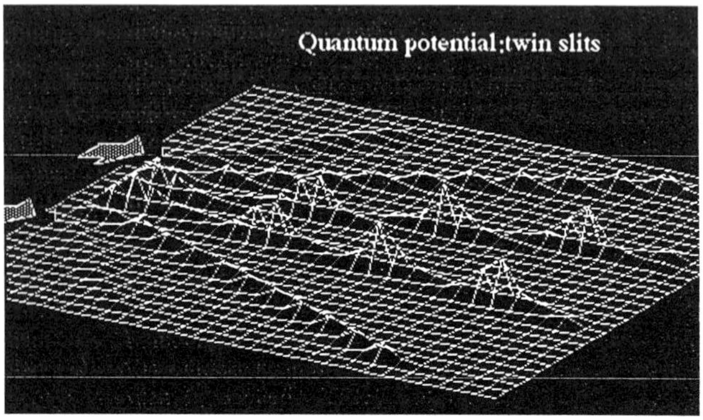

Fig. 17. The quantum potential for the steady-state flow of independent electrons from two parallel narrow electron waveguide channels into a very wide channel: a two-slit interference problem.

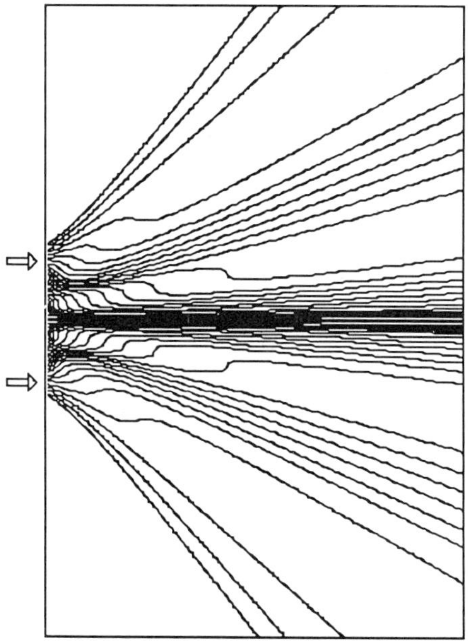

Fig. 18. Individual electron trajectories in the quantum potential of fig. 17 showing the bunching and anti-bunching into the well-known interference pattern.

The pilot-field picture provides a detailed picture of individual carrier histories which suggests a very useful framework for developing ideas in single-electronics. However, there have been a number of severe objections to the approach. It is shown elsewhere (Barker et al. 1992) that it is possible to overcome most of these difficulties and to reconcile the approach with conventional many-body quantum transport theory. Only a few points will be presented in the present article.

Let us first outline some aspects of the standard pilot-wave picture of quantum mechanics as a device for thinking about single-electron transport from a non-conventional viewpoint. There are three excuses for doing this. The first is that it is philosophically interesting; the second is that it offers an alternative computational and visualisation tool for some quantum transport problems in single-electronics; and the third is that is might provide an excellent stimulus for devising tests for weaknesses in quantum mechanics and its interpretative framework.

The pilot-wave model exploits the original view of both de Broglie and Einstein (and later Bohm) that objectively real quantum waves exist that propagate in space–time according to the relevant wave equation. Moreover, this picture is a dual picture in that an objectively real, spatially localised particle (or particles for a many-body wave field) exists in association with the extended quantum wave $\phi = c\psi$, where c is a constant and ψ is the normalised wavefunction satisfying the Schrödinger equation. The apparently stochastic behaviour of electrons undergoing phenomena such as diffraction or interference is then identified with deterministic particle trajectories resulting from the classical action of the classical forces on the electron plus the influence of a quantum force derived from a potential function V_Q determined by the wave field $\psi = R\exp(iS/\hbar)$ according to

$$V_Q = (-\hbar^2/2m)R^{-1}\nabla^2 R. \tag{72}$$

This result (which is shown here in the restricted form for the non-relativistic Schrödinger equation) is in accordance with the existence of a Hamilton–Jacobi-like equation for the action S of a particle which can be derived by separating the real and imaginary parts of the Schrödinger equation in terms of R and S

$$-\partial S/\partial t = (\nabla S)^2/2m + V + V_Q, \tag{73}$$

$$\partial R^2/\partial t + \nabla\cdot\{R^2[\nabla S]/m\} = 0. \tag{74}$$

The second equation is just the usual continuity equation. The possible real average dynamics of an electron may be described by classical trajectories on which the momentum is given by

$$\boldsymbol{p} = \nabla S. \tag{75}$$

Philippidis et al. (1979, 1982) and Vigier et al. (1987) have performed exact calculations for the particle trajectories present in two-slit interference and in two-slit interference in the presence of the Aharonov–Bohm effect. Figures 17 and 18 illustrate a similar calculation by us for the quantum potential and a sample of electron trajectories for a twin-source problem set up in a hypothetical squeezed 2DEG geometry. The intensity distribution depends on the occupation probability of each classical trajectory in the complicated quantum potential and, by construction, of course, agrees precisely with conventional quantum theory.

6.2. *Relation to the Wigner phase-space distributions*

There is an obvious conflict between conventional density matrix theory as represented by the *non-positive*-definite Wigner phase-space distributions $f_W(\boldsymbol{R}, \boldsymbol{P})$ of many-body quantum transport theory (or equivalent thermodynamic Green functions) and the *positive-definite* phase-space probability distributions $f_B(\boldsymbol{R}, \boldsymbol{P})$ implied by the existence of well-defined trajectories in the pilot-field picture. Indeed, explicit expressions for the two in the case of a single-electron pure state give

$$\begin{aligned} f_W(\boldsymbol{R}, \boldsymbol{P}) = (2\pi\hbar)^{-N} \sum_{\boldsymbol{r}} R(\boldsymbol{R}-\tfrac{1}{2}\boldsymbol{r})\, R(\boldsymbol{R}+\tfrac{1}{2}\boldsymbol{r}) \exp\{(i/\hbar)[S(\boldsymbol{R}+\tfrac{1}{2}\boldsymbol{r}) - S(\boldsymbol{R}-\tfrac{1}{2}\boldsymbol{r})]\} \\ \times \exp(-i\boldsymbol{P}\cdot\boldsymbol{r}/\hbar), \end{aligned} \tag{76}$$

where N is the number of dimensions and

$$\begin{aligned} f_B(\boldsymbol{R}, \boldsymbol{P}) &= R^2(\boldsymbol{R})\,\delta[\boldsymbol{P} - \nabla S(\boldsymbol{R})] \\ &= (2\pi\hbar)^{-N} \sum_{\boldsymbol{r}} R(\boldsymbol{R})\, R(\boldsymbol{R}) \exp\{(i/h)[\nabla S(\boldsymbol{R})]\} \exp(-i\boldsymbol{P}\cdot\boldsymbol{r}/\hbar). \end{aligned} \tag{77}$$

It is immediately obvious that, unlike the Wigner function, the function $f_B(\boldsymbol{R}, \boldsymbol{P})$ does have compact support: it exists precisely where the wavefunction exists. However, to obtain the correct statistical averages of dynamical variables great care is needed in taking account of the coupling of the electron to its pilot field (in fact, a similar problem occurs in the Wigner picture because a general classical variable $\Lambda(\boldsymbol{R}, \boldsymbol{P})$ will lead to complicated counterpart $\Lambda_W(\boldsymbol{R}, \boldsymbol{P})$ according to the Wigner–Weyl transformation). For example, any measurement of the electron kinetic energy must be considered as measuring a particle embedded in the pilot field and it is necessary to include the interaction energy V_Q with the pilot field as well as the kinetic term $p^2/2m$ in arriving at the observable kinetic energy. To illustrate let us use $f_W(x, p)$ and $f_B(x, p)$ to see how the Heisenberg uncertainty relations or statistical dispersion relations arise in the two approaches (see table 1).

In the table the averages $\langle\ldots\rangle_W$ and $\langle\ldots\rangle_B$ refer to the orthodox quantum-mechanical expectation value and the average value computed from the pilot-field picture statistical distribution f_B, respectively.

Here we obtain identical results in the two approaches by remembering to compute the interaction energy with the pilot field, i.e., the quantum potential, in determining the conventional kinetic energy. The interpretations are quite different, however. In the orthodox approach, the uncertainty relations describe the statistical scatter in complementary observables. In the pilot-wave picture the occupancy of a deterministic trajectory is determined randomly according to the initial position distribution.

Table 1
Comparison of orthodox approach and pilot-field approach.

Orthodox approach	Pilot-field approach
$\langle \Lambda\rangle_W = \int \Lambda f_W(x, p)\,\mathrm{d}x\,\mathrm{d}p$	
$\Delta_W x = \{\langle (x - \langle x\rangle_W)^2\rangle_W\}^{1/2}$	$\langle x\rangle_B = \int x f_B(x, p)\,\mathrm{d}x\,\mathrm{d}p = \langle x\rangle_W,$
	$\langle x^2\rangle_B = \int x^2 f_B(x, p)\,\mathrm{d}x\,\mathrm{d}p = \langle x^2\rangle_W,$
	$\Delta_B x = \{\langle (x - \langle x\rangle_B)^2\rangle_B\}^{1/2} = \Delta_W x$
$\Delta_W p = \{\langle (p - \langle p\rangle_W)^2\rangle_W\}^{1/2}$	$\langle p\rangle_B = \int p f_B(x, p)\,\mathrm{d}x\,\mathrm{d}p = \langle p\rangle_W$
	$\langle p^2/2m\rangle_B + \langle V_q\rangle_B = \int (p^2/2m) f_B(x, p)\,\mathrm{d}x\,\mathrm{d}p$
$\Delta_W x\,\Delta_W p > \frac{1}{2}\hbar$	$+ \int V_q(x) f_B(x, p)\,\mathrm{d}x\,\mathrm{d}p$
	$= \langle p^2/2m\rangle_W$
	let $\Delta_B p = \{\langle p^2 + 2mV_q\rangle_B - \langle p^2\rangle_B\} = \Delta_W p$
	$\therefore\ \Delta_B x\,\Delta_B p > \frac{1}{2}\hbar$

An exact non-linear relationship exists between the Wigner function and the pilot-field probability distribution in the case of pure states. We shall not develop that here but we note that

$$R^2(\boldsymbol{x}) = \int f_W(\boldsymbol{x}, p)\, d^3p \quad \text{and} \quad \nabla S(\boldsymbol{x}) = \int \boldsymbol{p} f_W(\boldsymbol{x}, \boldsymbol{p})\, d^3p, \tag{78}$$

so that the wavefunction and the pilot-field distributions are obtained from the first few moments of the Wigner function. Indeed, this provides a route to a more general use of phase-space trajectories when the Wigner function is derived from a mixed-state density matrix. R and S then appear as parameters in the quantum Hamilton–Jacobi equation and the continuity equation.

The pair of equations for R and S give a perfectly viable computational approach to computing quantum transport problems by solving the problem as a coupled-particle problem (to obtain $p = \nabla S(\boldsymbol{x})$) and a field problem to determine R.

7. High-field dissipative transport

7.1. Introduction

Classical and semi-classical high-field dissipative transport theory is the underpinning theory for most hot-electron phenomena and the starting point is the Boltzmann–Bloch transport equation (see the articles by Conwell and Mizes (ch. 11) and Ferry (ch. 18) in this volume) for a classical phase-space distribution $f(R, P, t)$ which is simply an extension of the Liouville equation to include phenomenological scattering rate terms. Since the carrier trajectories exist in a classical picture it is possible to circumvent the direct assault on the Boltzmann–Bloch equation by powerful Monte Carlo techniques (reviewed by Jacoboni and Reggiani 1983). By coupling the Boltzmann equation to the Poisson equation, it becomes possible to model space charge, transport and switching in semiconductor devices of interest in microelectronics technology. The need to go beyond the Boltzmann picture was spelled out by a number of workers in the late 1970s (Barker 1979, Barker and Ferry 1979, 1980) when it became apparent that the continual shrinking of device dimensions would lead to a serious breakdown of the conditions which appear to underpin the Boltzmann equation. For example, the onset of transient and ballistic transport effects when the device channel length is smaller than the bulk free path. It was soon realised that size quantisation effects become important under the latter condition and there emerged a surge of interest in low-dimensional structures produced by novel semiconductor heterostructure engineering. Typical device control voltages across very small (sub-micron) dimensions imply the existence of very high fields within the device, so the interest in high-field quantum transport is well founded. Some effects, such as conductance quantisation, can be partly understood by appeal to a semi-classical collisionless Boltzmann equation, as we have seen. However well the ballistic regime is now understood the problem of the dissipative high-field regime remains only partly understood and even that only for approximately homogeneous systems. To date, there have been some refinements of earlier known (reviewed in the first edition of the Handbook (ch. 11C)) high-field quantum transport effects such as the influence of the applied field on the collision processes (or intra-collisional field effect) and screening processes, and quasi-particle effects due to many-body dressing of the single-carrier states, but severe problems remain. In the following we shall connect the ballistic formalism described earlier with the dissipative regime and sketch very briefly the developments since our previous review. Excellent technical reviews of this field have been given recently by Reggiani (1985) and Jauho (1989, 1991).

The problem of dissipative quantum transport is a many-body problem. As such it requires either simple phenomenologies or the use of specialised tools such as the thermodynamic Green functions. The latter may be related to the many-body Wigner distributions but the formalism is much more complex than we have seen for the ballistic transport cases dealt with earlier. Thermodynamic Green function techniques reached prominence in linear transport theory in the early 1960s with the lucid treatment of Kadanoff and Baym (1962). Essentially, two different tactics were adopted; either the Green function equations were used to derive kinetic equations for a particle distribution function or they were used to compute transport coefficients directly. The latter tactic proved particularly useful for unravelling the Kubo–Greenwood formula of linear response theory, although the methods depended on computation of the two-electron Green function via a Bethe–Salpeter equation (the linear regime has been exhaustively explored recently by Mahan (1984, 1987)). Considerable successes were achieved in magneto-transport, particularly for linear transport in the disordered two-dimensional electron gas, but many important effects were elusive (such as the fractional quantum Hall effect).

By the early 1970s there were a number of attempts to tackle *non-linear* quantum transport theory (e.g., Barker 1973, Levinson 1970). The earliest studies focussed on clarifying the influence of a strong electric field on the relaxation processes as well as the processes of acceleration. The basic approach used a direct perturbative expansion of the many-body quantum Liouville equation for the electron density matrix using a variety of *projection and super-operator* techniques. Practically all the effort was concerned with macroscopically *uniform homogeneous systems.* Symmetry arguments for this case show immediately that the electron distribution function is obtained from the diagonal part of the density matrix. Considerable difficulties arose in attempting to "fix" the non-diagonal momentum matrix elements which proliferate in the maze of perturbation theory expansions. It was not always realised that the non-diagonal elements of the density matrix control the spatial variation of the quantum statistical distributions such as the Wigner function. Even for models which have macroscopic, or average uniformity, the basic dissipative processes are inhomogeneous on the microscopic scale; it is thus dangerous to exclude all the elements of the density matrix. Consequently, one must find a suitable book-keeping scheme to simplify and keep track of all the accounts of the full density matrix. Wigner function and Green function methods are particularly useful for this task but they still give rise to unsolved problems.

7.2. The Green function and Wigner function approaches

7.2.1. Definitions

In our earlier review (Barker 1982) we used the quantum kinetic equation based on super-operator perturbation theory applied to the quantum Liouville equation for the density matrix. That approach gives a fast method for deriving results in weak-coupling scattering theory but it is cumbersome and not well-suited to deal with multiple scattering, general strong-scattering problems or inhomogeneous systems. Most of the modern approaches use the Green function formalism, usually within the framework devised by Kadanoff and Baym (1962). The key quantities are the *non-equilibrium electron correlation function*

$$G^{<}(1, 1') = \mathrm{i}\langle \Psi^{+}(1')\,\Psi(1)\rangle, \tag{79}$$

and the retarded (r) and advanced (a) Green functions

$$G_{\mathrm{r}}(1, 1') = -\mathrm{i}\theta(t_1 - t_{1'})\langle\{\Psi^{+}(1'), \Psi(1)\}\rangle, \tag{80}$$

$$G_{\mathrm{a}}(1, 1') = +\mathrm{i}\theta(t_{1'} - t_1)\langle\{\Psi(1), \Psi^{+}(1')\}\rangle. \tag{81}$$

The braces { , } denote an anti-commutator, Ψ and Ψ^{+} are quantum field operators in the Heisenberg representation, corresponding to electron creation/annihilation at a point in space–time $(\boldsymbol{r}, t)$ where the shorthand notation $1 = (\boldsymbol{r}_1, t_1)$ is used. The angular brackets denote a many-body trace over the initial (usually equilibrium) density matrix. These formidable objects are related to each other and to the many-body Wigner distribution,

$$f(\boldsymbol{R}, \boldsymbol{P}, t) = -\mathrm{i}\int \frac{\mathrm{d}\omega}{2\pi}\, G(\boldsymbol{P}, \omega, \boldsymbol{R}, t), \tag{82}$$

where we have introduced the usual centre and relative coordinates $R = \frac{1}{2}(r_1 + r_{1'})$, $r = r_1 - r_{1'}$, and also $t = \frac{1}{2}(t_1 + t_{1'})$, $T = t_1 - t_{1'}$. A Fourier transform has been applied to each of the variables r ($\to P$) and T ($\to \omega$).

The quantity $G^{<}$ is determined exactly by the equation (in symbolic form):

$$[G_0^{-1} - U, G^{<}] - [\Sigma, G^{<}] - [\Sigma, G^{<}] = \tfrac{1}{2}[\Sigma^{<}, G^{>}] + \tfrac{1}{2}[\Sigma^{>}, G^{<}], \tag{83}$$

where $G = \frac{1}{2}(G_{\mathrm{a}} + G_{\mathrm{r}})$; $\Sigma = \frac{1}{2}(\Sigma_{\mathrm{a}} + \Sigma_{\mathrm{r}})$; U describes the self-consistent applied forces, including confining potentials and one-body potentials. The self-energies Σ ultimately describe the scattering processes. This generalised Kadanoff–Baym equation is exact, but to solve it one must obtain a closed set of equations by also deploying the Dyson equations for G_{a} and G_{r},

$$G_{\mathrm{r,a}} = G^{0}_{\mathrm{r,a}} + G^{0}_{\mathrm{r,a}} U G_{\mathrm{r,a}} + G^{0}_{\mathrm{r,a}} \Sigma_{\mathrm{r,a}} G_{\mathrm{r,a}}. \tag{84}$$

7.2.2. Quantum kinetic equations

The Kadanoff–Baym equation, eq. (83), gives rise under appropriate assumptions to a quantum kinetic equation for $f(r, P, t)$. Reading from left to right, the first term gives rise to the driving terms seen in the ballistic case, the second and third terms are renormalised terms; the right-hand side gives rise to a generalised collision integral. This reduction is non-trivial, however, and is still problematic. To obtain an equation for f we require a relation between f and $G^<$. In thermal equilibrium such a relation exists rigorously,

$$G^<(P,\omega)=\mathrm{i}A(P,\omega)\,f_0(\omega), \tag{85}$$

where $f_0(\omega)$ is the Fermi–Dirac function. Early attempts to generalise this result to make a non-equilibrium Ansatz have been fraught with difficulties (see the discussion in Jauho's review (1991)), including problems in recovering results already known from the kinetic equation approaches of Barker (1978) and Barker and Ferry (1979). The problem has been resolved for the case of uniform steady fields via a gauge-invariant formulation which systematically derives a non-equilibrium relation of the form

$$G^<(K,\tau)=\mathrm{i}A(K,\tau)\,f(K-\tfrac{1}{2}E\tau)), \tag{86}$$

where K is the kinematical momentum rather than the canonical momentum P and f is the Wigner function; note the relation is set up in τ-space not ω-space.

The corresponding equation of motion for the space-independent Wigner function was derived by Khan et al. (1987) within the self-consistent Born approximation for electron–phonon scattering and leads to a generalisation of the quantum kinetic equations found earlier by Barker and Ferry (1979). The equation of motion reads

$$\begin{aligned}\boldsymbol{F}\cdot\frac{\partial}{\partial \boldsymbol{K}}f(\boldsymbol{K})=\sum_{Q}\int \mathrm{d}\tau\,\{&P(\boldsymbol{K}-\boldsymbol{E}\tau,\boldsymbol{K}-\boldsymbol{Q}-\boldsymbol{E}\tau;\tau)\,F(\boldsymbol{K}-\boldsymbol{Q}-\boldsymbol{E}\tau)\\ &-P(\boldsymbol{K}+\boldsymbol{E}\tau,\boldsymbol{K}-\boldsymbol{Q}-\boldsymbol{E}\tau;\tau)\,F(\boldsymbol{K}-\boldsymbol{E}\tau)\},\end{aligned} \tag{87}$$

where

$$\begin{aligned}P(\boldsymbol{K}+\boldsymbol{Q},\boldsymbol{K};\tau)=2\pi|M_Q|^2\sum_{Q}\frac{1}{\pi}&(N_Q+\tfrac{1}{2}(1+\eta))\\ &\times\mathrm{Re}\{A(\boldsymbol{K}+\boldsymbol{Q}+\tfrac{1}{2}\boldsymbol{E}\tau)\,A(\boldsymbol{K}+\tfrac{1}{2}\boldsymbol{E}\tau,-t)\exp(-\mathrm{i}\omega Q\eta\tau)\}.\end{aligned} \tag{88}$$

Here the convolved A's are the convolved spectral density function describing generalised energy conservation. The limits $E\to 0$ and $A\to$ free spectral functions (i.e., delta-functions for energy conservation) give the conventional Boltzmann equation.

7.2.3. The intra-collisional field effect

Expression (88) illustrates the intra-collisional field effect discussed originally by Barker (1973, 1978, 1982) and reviewed by Jauho (1991). The scattering integral depends in a non-trivial way on the strength of the applied field: the collision takes place along an accelerated path due to the field.

7.2.4. Non-linear screening

The problem of screening in high fields has received little attention despite its importance for device modelling and in its control of phenomena such as the fluctuation potential. Lowe and Barker (1985) used the Green function/Wigner function approach to derive a non-equilibrium screened fully field-dependent scattering kernel involving a non-local field-dependent dielectric tensor; but the resulting equations were intractable without making certain model assumptions on the form of the phase-space distributions. An interesting development which does not make assumptions about the form of the phase-space distribution has been described by Hu and Wilkins (1989) and Hu et al. (1989) within a simplified model for the collisions.

7.2.5. Inhomogeneous systems

The most serious problem with high-field dissipative transport concerns the spatial homogeneity of the transport problem. As we have seen for the ballistic regimes, such as electron waveguides, the Wigner distribution functions are inherently inhomogeneous and may show the complex oscillatory behaviour associated with the non-compactness. Since it is the ballistic distribution functions which appear in deriving the perturbation approximations to the scattering kernels the task of producing manageable theory seems remote. Such a point is, of course, one where it is sensible to look for other approaches.

7.3. Monte Carlo evaluation of Feynman path integrals

In recent years there has been a resurgence of interest in using the Feynman path integral formulation of quantum mechanics (Thornber 1978) to evaluate the propagator for electrons coupled dissipatively to a phonon bath using Monte Carlo techniques (Fischetti 1984, Fischetti and Maria 1985, Mason and Hess 1989). Although the principle has been established it proved too massive a computational scheme to evaluate all but the simple one-dimensional transport problems. So far these methods appear fraught with difficulty of a fundamental kind. Calculations have been performed on superconducting systems using importance sampling to generate up to 28 million individual configuration-space trajectories for the evaluation of the path integrals. But very recently, a significant simplification has been devised (Pevzner et al. 1990) based on a conjecture by Schulman (1981) that for certain hard-walled electron waveguide problems it suffices to compute just the *classical* path contributions (ray tracing paths) to evaluate the quantum propagator. The method has been demonstrated for arbitrary two-dimensional structures with hard-wall boundary conditions. The resulting reduction in complexity is sufficient to bring this new method to the forefront as a scheme for computing quantum dissipative transport in semiconductor confined structures.

7.4. Other approaches

Among a number of promising new approaches to high-field quantum transport we finally refer to an interesting approach based on the quantum Langevin equation (a widely studied approach for *linear* problems in the 1950s and 60s) due to Hu and O'Connell (1987, 1988, 1989).

8. Conclusions

Quantum transport phenomena in semiconductors continue to be discovered and are likely to prove crucial in underpinning future nanoelectronic devices. The theoretical basis for quantum transport has expanded massively since the first edition of the Handbook, indeed only a fraction of the scope is covered in this article. Very many challenges remain: the understanding of many-body effects from the fractional quantum Hall effect to high-field quasi-particle transport; the advent of single-electronics and novel charge soliton states is likely to cause some surprises as ideas from metal–insulator systems are developed in the much richer semiconductor environments. Perhaps of most interest is the intimate relation of the now mature field of quantum transport with classical concepts.

References

Aharonov, A., and D. Bohm, 1959, Phys. Rev. **115**, 485.

Averin, A., and K.K. Likhaerev, 1986, J. Low Temp. Phys. **62**, 345.

Bakhvalov, N.S., G.S. Kazacha, K.K. Likhaerev and S.I. Serdyukova, 1989, Sov. Phys. JETP **68**, 581.

Ballentine, L., 1970, Rev. Mod. Phys. **42**, 358.

Barker, J.R., 1973, J. Phys. C **6**, 2663.

Barker, J.R., 1978, Solid State Electron. **21**, 267.

Barker, J.R., 1979, in: Physics of Non-linear Transport in Semiconductors, eds D. Ferry, J.R. Barker and C. Jacoboni (Plenum Press, New York) ch. 23.

Barker, J.R., 1981, J. Phys. Colloq. (France) C **7**, 245.

Barker, J.R., 1982, in: Handbook of Semiconductors, 1st Ed., Vol. 1, ed. W. Paul (North-Holland, Amsterdam) ch. 11C, p. 617.

Barker, J.R., 1986, in: Physics of Microstructures, eds M.J. Kelly and C. Weisbuch, Springer Proc. Phys., Vol. 13 (Springer, Berlin) p. 210.

Barker, J.R., 1989a, in: Semiconductor Device Modelling, ed. C.M. Snowden (Springer, Berlin) ch. 13, p. 207.

Barker, J.R., 1989b, in: Physics and Fabrication of Nanostructures, eds M.A. Reed and W.P. Kirk (Academic Press, New York) p. 253.

Barker, J.R., 1991a, in: Granular Nanoelectronics, eds D.K. Ferry, J.R. Barker and C. Jacoboni (Plenum Press, New York) ch. 2, p. 19.

Barker, J.R., 1991b, in: Granular Nanoelectronics, eds D.K. Ferry, J.R. Barker and C. Jacoboni (Plenum Press, New York) ch. 21, p. 327.

Barker, J.R., and D.K. Ferry, 1979, Phys. Rev. Lett. **42**, 1779.

Barker, J.R., and D.K. Ferry, 1980, Solid State Electron. **23**, 519, 531.

Barker, J.R., and M. Finch, 1989, unpublished.

Barker, J.R., and S.J. Murray, 1983, Phys. Lett. A **93**, 271.

Barker, J.R., J. Pepin, M. Finch and M.J. Laughton, 1989, Solid State Electron. **32**, 1155.

Barker, J.R., S. Roy and S. Babiker, 1992, in: Science and Technology of Mesoscopic Structures (Springer, Berlin) ch. 23, p. 213.

Bohm, D., 1952, Phys. Rev. **85**, 166, 180.

Brunetti, R., C. Jacoboni and F. Rossi, 1989, Phys. Rev. B **39**, 10781.

Büttiker, M., 1986, Phys. Rev. Lett. **57**, 1761.
Büttiker, M., 1988a, IBM J. Res. & Dev. **32**, 317.
Büttiker, M., 1988b, Phys. Rev. B **38**, 9375.
Büttiker, M., Y. Imry, R. Landauer and S. Pinhas, 1985, Phys. Rev. B **31**, 6207.
Carruthers, P., and F. Zachariasen, 1983, Rev. Mod. Phys. **55**, 245.
Dahl, J.P., 1981, Dynamical equations for Wigner functions, reprint (Technical University Denmark).
Davies, J.H., and J.A. Nixon, 1990, Phys. Rev. B **41**, 7929.
Dresden, M., 1961, Rev. Mod. Phys. **33**, 265.
Ehrenberg, W., and S.E. Siday, 1949, Proc. Phys. Soc. London B **62**, 8.
Ferry, D.K., and J.R. Barker, 1980, Solid State Electron. **23**, 545.
Ferry, D.K., J.R. Barker and C. Jacoboni, eds, 1991, Granular Nanoelectronics (Plenum, New York).
Finch, M., 1989, Ph.D. Thesis (University of Glasgow).
Fischetti, M.V., 1984, Phys. Rev. Lett. **53**, 1755.
Fischetti, M.V., and D.J. Maria, 1985, Phys. Rev. Lett. **55**, 2475.
Ford, C.J.B., T.J. Thornton, R. Newbury, M. Pepper, H. Ahmed, C.T. Foxon, J.J. Harris and C. Roberts, 1988, J. Phys. C **21**, L325.
Franz, W., 1939, Verh. Deutsch. Phys. Ges., Nr. 2, p. 65.
Frensley, W.R., 1987, Phys. Rev. B **36**, 1570.
Frohne, R., and S. Datta, 1988, J. Appl. Phys., **64**, 4086.
Geerligs, L.J., V.F. Anderegg, P.A. Holweg, J.E. Mooij, H. Pothier, C. Urbina, H. and M.H. Devoret, 1990, Phys. Rev. Lett. **64**, 2691.
Gefen, Y., Y. Imry and M.Ya. Azbel, 1984, Phys. Rev. Lett. **52**, 129.
Hu, B.Y.-K., and J.W. Wilkins, 1989, Phys. Rev. B **39**, 8464.
Hu, B.Y.-K., S.K. Sarker and J.W. Wilkins, 1989, Phys. Rev. B **39**, 8468.
Hu, G.Y., and R.F. O'Connell, 1987, Phys. Rev. B **36**, 5798.
Hu, G.Y., and R.F. O'Connell, 1988, Physica A **149**, 1.
Hu, G.Y., and R.F. O'Connell, 1989, Phys. Rev. B **39**, 12717.
Jacoboni, C., and L. Reggiani, 1983, Rev. Mod. Phys. **55**, 645.
Jauho, A.-P., 1989, Solid State Electron. **32**, 1265.
Jauho, A.-P., 1991, in: Granular Nanoelectronics, eds D.K. Ferry, J.R. Barker and C. (Plenum, New York) p. 133.
Kadanoff, L.P., and G. Baym, 1962, Quantum Statistical Mechanics (Benjamin, Reading, MA).
Khan, F.S., J.H. Davies and J.W. Wilkins, 1987, Phys. Rev. B **36**, 2578.
Kirczenow, G., 1988, Phys. Rev. B **38**, 10958.
Kubo, R., 1957, J. Phys. Soc. Jpn. **12**, 570.
Kumar, A., S.E. Laux and F. Stern, 1989, Appl. Phys. Lett. **54**, 1270.
Landauer, R., 1970, Philos. Mag. **21**, 863.
Landauer, R., 1975, Z. Phys. B **21**, 247.
Landauer, R., 1987, Z. Phys. B **68**, 217.
Landauer, R., 1989, Physiça D **38**, 226.
Laughton, M.J., 1992, Ph.D. Thesis (University of Glasgow).
Laughton, M.J., and J.R. Barker, 1991, unpublished results.
Laughton, M.J., J.R. Barker, J.A. Nixon and J.H. Davies, 1991, Phys. Rev. B **44**, 1150.
Laux, S.E., and F. Stern, 1986, Appl. Phys. Lett. **49**, 91.
Lent, C.S., S. Sivaprakasam and D.J. Kikner, 1989, in: Physics and Fabrication of Nanostructures, eds M.A. Reed and W.P. Kirk (Academic Press, New York) p. 279.
Levinson, I.B., 1970, Sov. Phys.-JETP **30**, 362.
Likhaerev, K.K., 1988, IBM J. Res. & Dev. **32**, 144.
Lowe, D., and J.R. Barker, 1985, J. Phys. C **18**, 2507.
Mahan, G.D., 1984, Phys. Rep. **110**, 321.
Mahan, G.D., 1987, Phys. Rep. **145**, 253.
Mason, B.A., and K. Hess, 1989, Phys. Rev. B **39**, 5051.
Nixon, J.A., J.H. Davies and J.R. Barker, 1989, in: Physics and Fabrication of Nanostructures, eds M.A. Reed and W.P. Kirk (Academic Press, New York) p. 123.

Nixon, J.A., J.H. Davies and H.U. Baranger, 1990, Phys. Rev. B **43**, 12638.
Pepin, J., 1990, Ph.D. Thesis (University of Glasgow).
Petroff, P.M., A.C. Gossard and W. Wiegmann, 1984, App. Phys. Lett. **45**, 620.
Pevzner, V., F. Sols and K. Hess, 1990, in: Granular Nanoelectronics, eds D.K. Ferry, J.R. Barker and C. Jacoboni (Plenum Press, New York) p. 223.
Philippidis, C., C. Dewdney and B.J. Hiley, 1979, Nuovo Cimento B **52**, 15.
Philippidis, C., D. Bohm and R.D. Kaye, 1982, Nuovo Cimento B **71**, 75.
Reed, M.A., and W.P. Kirk, eds, 1989, Physics and Fabrication of Nanostructures (Academic Press, New York).
Reggiani, L., 1985, Physica B **134**, 123.
Schommers, W., ed., 1989, Quantum Theory and Pictures of Reality (Springer, London) 279pp.
Staring, A.A.M., J.G. Williamson, H. van Houten, C.W.J. Beenakker, L.P. Kouwenhoven and C.T. Foxon, 1992, in: Proc. Int. Symp. on Analogies in Optics and Microelectronics, eds W. van Haeringen and D. Lenstra (North-Holland, Amsterdam) in press.
Stone, A.D., and A. Szafer, 1988, IBM J. Res. & Dev. **32**, 384.
Ter Haar, D., 1961, Rep. Prog. Phys. **24**, 304.
Thornber, K., 1978, Solid State Electron. **21**, 259.
Thornton, T.J., M.L. Roukes, A. Scherer and B.P. van der Gaag, 1991, in: Granular Nanoelectronics, eds D.K. Ferry, J.R. Barker and C. Jacoboni (Plenum, New York) p. 165.
Timp, G., S. Behringer, S. Sampere, J.E. Cunningham and R.E. Howard, 1989, in: Nanostructure Physics and Fabrication, eds M.A. Reed and W.P. Kirk (Academic Press, Boston) p. 331.
van Houten, H., C.J.W. Beenakker and B.J. van Wees, 1990, in: Semiconductors and Semimetals, ed. M. Reed (Academic Press, New York).
van Wees, B., H. van Houten, C.W.J. Beenakker, J.G. Williamson, L.P. Kouwenhoven, D. van der Marel and C.T. Foxon, 1988, Phys. Rev. Lett. **60**, 848.
Vigier, J.P., C. Dewdney, P.R. Holland and A. Kyprianidis, 1987, in: Quantum Implications, eds B.J. Hiley and F.D. Peat (Routledge/Kegan Paul, London/New York).
Wharam, D.A., T.J. Thornton, R. Newbury, M. Pepper, H. Ahmed, J.E.F. Frost, D.G. Hasko, D.C. Peacock, D.A. Ritchie and G.A.C. Jones, 1988, J. Phys. C **21**, L209.
Wigner, E.P., 1932, Phys. Rev. **40**, 749.
Wigner, E.P., 1967, in: Perspectives in Quantum Theory (Dover Publications, New York) p. 25.

Author Index

Aarts, J., *see* Hoeven, J.J. 380
Aaustuen, D.J.W., *see* Vos, M. 367
Abakumov, V.N. 239
Abarenkov, I., *see* Heine, V. 6, 8, 71
Abe, Y. 422
Abe, Y., *see* Hasagawa, J. 359
Abeles, B. 561
Abkowitz, M., *see* Epstein, A.J. 619
Aboie-Elfotouh, F.A. 305
Abragam, A. 183
Abraham, D. 387
Abrahams, E. 945, 947, 949, 950, 957, 989, 995
Abrahams, E., *see* Andereck, B. 945, 957
Abrahams, E., *see* Anderson, P.W. 932, 937, 947, 956
Abrahams, E., *see* Langreth, D.C. 932
Abrahams, M.S., *see* Landsberg, P.T. 265
Abram, R.A. 703
Abramowitz, M. 521, 634
Abrikosov, A.A. 5, 10, 19, 26, 867, 945, 957
Abstreiter, G. 385
Abstreiter, G., *see* Schuberth, G. 647, 655
Adams, D.J. 1067, 1069
Adams, E.N. 509
Adams, E.N., *see* Haering, R.R. 655
Adams, J.A., *see* Taylor, R.P. 877, 883, 888, 894, 904, 906, 919, 923, 924
Adams, M.J. 231, 232, 256
Adler, D. 211
Aers, G.C. 992
Aers, G.C., *see* MacDonald, A.H. 1015
Aggarwal, R.J., *see* Reed, M.A. 872, 922, 923, 925
Aggarwal, R.J., *see* Weiler, M.H. 521
Aggarwal, R.L., *see* Chang, L.L. 823, 836
Agranovich, V.M. 452
Agullo-Rueda, F. 850, 851, 857
Agullo-Rueda, F., *see* Mendez, E.E. 657, 850, 857
Aharonov, A. 1109, 1110
Ahlburn, B.T. 166
Ahmed, H., *see* Brown, R.J. 886, 899, 902, 904, 906, 907, 923, 924
Ahmed, H., *see* Ford, C.J.B. 867, 904, 906, 909, 911, 1104, 1109, 1111, 1112, 1115, 1116
Ahmed, H., *see* Greene, S.K. 916
Ahmed, H., *see* Kelly, M.J. 899, 902
Ahmed, H., *see* Smith, C.G. 659, 880, 889, 890, 923, 924, 930
Ahmed, H., *see* Thornton, T.J. 735, 869–871, 881
Ahmed, H., *see* van Houten, H. 902, 904, 906
Ahmed, H., *see* Wharam, D.A. 800, 801, 867, 870, 871, 881, 882, 888, 902–904, 913, 923, 924, 1097, 1102, 1104
Ahrenkiel, R.K., *see* Keyes, B.M. 238
Aina, L. 822
Aina, O., *see* Sarma, K. 380
Akagi, K. 591
Akera, H. 637, 646, 689, 895, 896, 906, 913, 915
Akera, H., *see* Ando, T. 674, 831, 895, 906
Akera, H., *see* Tsuchiya, T. 684, 685
Al-Dossary, O. 695, 696
Albers, R.C., *see* Steiner, M.M. 8, 10, 12
Albrektsen, O. 387
Alcock, C.B., *see* Kubaschewski, O. 332
Aldao, C.M., *see* Vitomirov, I.M. 367
Aldao, C.M., *see* Vos, M. 367
Aldao, C.M., *see* Waddill, G.D. 367
Aldao, C.M., *see* Xu, F. 326, 338
Alerhand, O.L. 304, 321
Alerhand, O.L., *see* Roukes, M.L. 910, 911
Aleshkin, Yu.A. 524
Alexander, H., *see* Hoffmann, D.M. 351
Alexandre, F., *see* Vuong, T.H.H. 730
Alfano, R.R., *see* Dorsinville, R. 615
Alfano, R.R., *see* Katz, A. 1073
Alfano, R.R., *see* Walser, A.D. 613, 614
Alfano, R.R., *see* Wang, W.B. 1073
Alibert, C., *see* Voisin, P. 850
Alkeev, N.V., *see* Kaminskii, A.S. 476
Allan, D.C. 106
Allan, D.C., *see* Shirley, E.L. 74
Allan, G. 819, 827
Allan, G., *see* Delerue, C. 118, 133, 135, 136, 140, 141
Allan, G., *see* Foulon, Y. 130, 145
Allan, G., *see* Krambrock, K. 156
Allan, G., *see* Petit, J. 137, 154, 155
Allee, D.R., *see* Chou, S.Y. 659
Allen, F.G. 312
Allen, J.W., *see* Baranowski, J.M. 170

Allen, P.B. 7, 65
Allen, P.B., *see* Li, X.P. 106
Allen, P.B., *see* Mackay, K.J. 380
Allen, R.E. 356
Allen, R.E., *see* Beres, R.P. 316
Allen, R.E., *see* Buisson, J.P. 118
Allen, R.E., *see* Dow, J.D. 313, 356
Allen Jr, S.J., *see* Roukes, M.L. 867, 870, 872, 908, 909, 911
Allen, W., *see* Wray, E.M. 170
Allgaier, R.S. 560
Allovon, M., *see* Bigan, E. 850, 852
Allovon, M., *see* Bleuse, J. 839, 840, 851
Alm, A., *see* Engström, O. 205, 211, 262
Almbladh, C.-O. 205
Almeier, J., *see* Merkt, U. 922
Alonso, M. 372
Alphenaar, B.W. 921
Alphenaar, B.W., *see* McEuen, P.L. 867, 892, 893, 898
Alphenaar, B.W., *see* McEwan, P.L. 796
Altarelli, M. 144, 670, 674, 675, 824, 827
Altarelli, M., *see* Fasolino, A. 824
Altshuler, B.L. 791, 933, 937, 946–948, 959
Altukhov, P.D. 455, 456, 464, 476, 478–482
Alvarado, S.F., *see* Abraham, D. 387
Alves, E.S. 930
Alves, E.S., *see* Beton, P.H. 745, 842
Alves, E.S., *see* Leadbeater, M.L. 655
Amand, T., *see* Barrau, J. 850, 857
Amer, N.M., *see* Olmstead, M.A. 319
Ammerlaan, C.A.J. 184
Ammerlaan, C.A.J., *see* Gregorkiewicz, T. 186
Ancilotto, F. 831, 841
Andereck, B. 945, 957
Anderegg, V.F., *see* Geerligs, L.J. 1115
Andersen, O.K., *see* Beeler, F. 132
Andersen, O.K., *see* Das, G.P. 356
Anderson, D.A. 703
Anderson, I.M., *see* Heiblum, M. 707
Anderson, J., *see* Smith, R.J. 312
Anderson, P.W. 6, 867, 932, 937, 947, 956
Anderson, P.W., *see* Abrahams, E. 945, 947, 949, 950, 989, 995
Anderson, P.W., *see* Fukuyama, H. 1001
Anderson, R.L. 384, 390, 645
Anderson, R.L., *see* Zeidenbergs, G. 645
Anderson, S. 303
Anderson, S.B., *see* Chambers, S.A. 326
Anderson, W.A., *see* Shi, Z.Q. 368
Andersson, G.I. 229
Andersson, G.I., *see* Bengtsson, S. 264
Andersson, M.O. 205, 245, 247, 262, 264
Andersson, M.O., *see* Bengtsson, S. 264
Andersson, T.G., *see* Svensson, S.P. 326, 344, 357
Ando, S., *see* Chang, S.S. 874
Ando, S., *see* Fukui, T. 875
Ando, T. 479, 481, 667, 674, 698, 708, 728, 732, 736, 831, 868, 869, 895, 906, 955, 979, 984, 985, 989, 994, 996
Ando, T., *see* Akera, H. 637, 646, 689, 895, 896, 906, 913, 915
Ando, T., *see* Aoki, H. 934, 984, 989, 990, 992, 994
Ando, T., *see* Bauer, G.E.W. 837
Ando, T., *see* Mori, N. 695
Ando, T., *see* Tsuchiya, T. 684, 685
André, J.M., *see* Brédas, J.-L. 595, 596, 603
André, J.P., *see* Lugagne-Delpon, E. 821, 822
André, J.P., *see* van Houten, H. 870
Andreani, L.C. 674
Andreatch Jr, P., *see* McSkimin, H.J. 83, 84, 94
Andrews, D., *see* Thornton, T.J. 735, 869–871, 881
Andrews, J.M. 352
Animalu, A.D.E. 6, 8
Ansems, J.P.M., *see* Henning, J.C.M. 144
Anthony, P.J. 273
Antoniadis, D.A., *see* Field, S.B. 924
Antoniadis, D.A., *see* Ismail, K. 659
Antoniadis, D.A., *see* Liu, C.T. 930
Antoniadis, D.A., *see* Scott-Thomas, J.H.F. 924
Antonini, A., *see* Sorba, L. 398, 399
Aoki, H. 930, 934, 943, 953, 955, 979, 984, 989, 990, 992, 994
Aoki, H., *see* Ando, T. 955, 984, 994
Aoki, K. 422, 432, 435, 437, 440
Aoki, K., *see* Schöll, E. 432
Aoki, M., *see* Suto, K. 184
Aono, M. 297
Aono, M., *see* Nishitani, R. 303
Aoyagi, Y., *see* Takagaki, Y. 867, 908, 911
Apal'kov, V.M. 1006
Apenko, S.M. 993, 997
Appel, J. 577
Applebaum, J.A. 294, 313, 347
Apsley, N., *see* Anderson, D.A. 703
Arakawa, Y. 874
Arakawa, Y., *see* Takahashi, T. 875
Arbuckle, G.A., *see* Tanner, D.B. 623
Arent, D.J., *see* Abraham, D. 387
Arent, D.J., *see* Albrektsen, O. 387
Argyres, P.N. 524, 640
Argyres, P.N., *see* Roth, L.M. 512, 517, 557
Arnold, E.N., *see* Chamoun, S.N. 1072
Aronov, A.G. 525, 937, 946, 948
Aronov, A.G., *see* Altshuler, B.L. 933, 937, 946–948
Arora, V.K., *see* Lee, J. 705
Arovas, D. 1009
Arps, M., *see* Niedernostheide, F.J. 443
Artacho, E. 314, 319
Artaki, M.A., *see* Bailey, D.W. 1071
Arthur, J.R. 298, 306
Arthur, J.R., *see* Cho, A.Y. 306

Asai, H., *see* Yamada, S. 875
Asano, S., *see* Fujitani, H. 356
Asaumi, K. 85, 94
Asbeck, P. 239
Asbel, M.Ya., *see* Lifshitz, I.M. 495
Ascarelli, G. 239
Asche, M. 422
Ashcroft, N.W. 6, 8, 71, 526
Ashen, D.J. 178, 180
Ashen, D.J., *see* White, A.M. 178
Asher, S.E., *see* Keyes, B.M. 238
Ashkinadze, B.M. 478
Asnin, V.M. 453, 457, 462, 463, 465–472
Aspnes, D.E. 307, 310, 320
Aspnes, D.E., *see* Frova, A. 656
Astemirov, T.H. 466
Astles, M.G., *see* Dean, P.J. 179
Aszodi, G. 139, 142
Aten, A.C. 179
Audenaert, M. 621
Aukerman, L.W. 350
Auret, F.D., *see* Snyman, L.W. 326
Austin, B.J. 6, 8, 62
Aven, M., *see* Halstead, R.E. 178, 351
Averin, A. 1115
Averin, D.V. 921, 924
Averkiev, N.S. 481
Avery, J., *see* Dahl, J.P. 116
Avishai, Y. 880, 885, 890, 897, 903, 914, 919
Avouris, Ph. 321, 322
Avouris, Ph., *see* Lyo, In-Whan 322
Avouris, Ph., *see* Wolkow, R. 319
Avron, J.E. 930, 992
Axmann, A., *see* Ennen, H. 139, 142
Azbel, M.Ya. 929, 997
Azbel, M.Ya., *see* Cota, E. 934
Azbel, M.Ya., *see* Gefen, Y. 1104, 1109
Azbel, M.Ya., *see* Ricco, B. 653

Baars, J. 332, 353
Babiker, M. 683, 695, 697, 699
Babiker, M., *see* Al-Dossary, O. 695, 696
Babiker, M., *see* Chamberlain, M.P. 694, 695
Babiker, M., *see* Ridley, B.K. 681, 682
Babiker, S., *see* Barker, J.R. 1115
Baccarani, G. 426
Bachelet, G.B. 72–74, 79, 80, 118, 130, 155
Bachrach, R.Z. 298, 299, 302
Bachrach, R.Z., *see* Bringans, R.D. 319, 320, 322, 324, 362, 381, 382
Bachrach, R.Z., *see* Hansson, G.V. 362
Bachrach, R.Z., *see* Northrup, J.E. 382
Badescu, V. 238
Badt, D. 322
Bagaev, V.S. 460, 464, 466
Bagaev, V.S., *see* Astemirov, T.H. 466
Bagwell, P.F. 893, 894
Bagwell, P.F., *see* Kumar, A. 893, 895
Bahder, T.B. 652
Baidyaroy, S. 349
Bailey, D.W. 1071
Bailey, D.W., *see* Stanton, C.J. 706
Bailey, S.M., *see* Wagman, D.D. 332
Bailyn, M. 750
Baj, M., *see* Dmowski, L. 177, 187, 188
Bak, P. 445
Baker, G.L., *see* Orenstein, J. 605, 606, 615
Baker, G.L., *see* Rothberg, L. 615, 616
Baker, G.L., *see* Vardeny, Z. 605, 606
Baker, G.L., *see* Weinberger, B.R. 605
Bakhvalov, N.S. 1115
Bakry, A.M., *see* Aboie-Elfotouh, F.A. 305
Bakshi, P., *see* Kempa, K. 922
Bakun, A.A., *see* Altukhov, P.D. 480–482
Baldereschi, A., *see* Gygi, F. 8, 10, 12
Baldereschi, A., *see* Marsi, M. 399
Baldereschi, A., *see* Peressi, M. 398
Baldluz Jr, J.L., *see* Perdew, J.P. 3, 4, 22, 29
Baldwin, K.W., *see* Pfeiffer, L.N. 874
Baldwin, K.W., *see* Spector, J. 872, 874, 878–880
Baldwin, K.W., *see* Störmer, H.L. 874
Bales, G.S. 304
Balk, P., *see* Reuters, P.J. 351
Balkan, N. 432, 700
Balkan, N., *see* Gupta, R. 716
Ballentine, L. 1116
Ballone, P. 106
Band, Y.B., *see* Avishai, Y. 880, 885, 890, 897, 903, 914, 919
Bandyopadhyay, S., *see* Subramaniam, S. 897
Bar-Joseph, I. 850
Bar-Yam, Y. 74, 117
Baraff, G.A. 117, 118, 152, 988
Baraff, G.A., *see* Applebaum, J.A. 294
Baraff, G.A., *see* Bachelet, G.B. 118, 130, 155
Baranger, H.U. 908, 910–912, 914–918
Baranger, H.U., *see* Behringer, R.E. 914
Baranger, H.U., *see* Chang, A.M. 916
Baranger, H.U., *see* Jalabert, R.A. 910, 911, 918, 919
Baranger, H.U., *see* Nixon, J.A. 895, 896, 906, 1114
Baranger, H.U., *see* Timp, G. 867, 904, 906, 908, 909, 911
Baranowski, J.M. 170, 171, 175
Baranowski, J.M., *see* Cieplak, M.Z. 184
Baranowski, J.M., *see* Kamińska, M. 173
Baranowski, J.M., *see* Kocot, K. 175
Baranowski, J.M., *see* Langer, J.M. 170
Baranowski, J.M., *see* Le Manh Hoang 181
Baranowski, J.M., *see* Uba, S. 170
Baranowski, J.M., *see* Vogl, P. 132
Baratoff, A., *see* Tekman, E. 878
Barber, H.D. 246
Barber, W.G. 473
Bardeen, J. 102, 343, 571, 610

Bardeen, J., *see* Brattain, W.H. 349
Bardeen, J., *see* Laidig, W.D. 380
Bardsley, J.N. 65
Bardsley, W., *see* White, A.M. 178
Barker Jr, A.S. 182
Barker, J.R. 1092, 1095, 1097, 1098, 1104, 1106, 1108, 1109, 1111–1113, 1115, 1116, 1120–1123
Barker, J.R., *see* Ferry, D.K. 1085
Barker, J.R., *see* Laughton, M.J. 1087, 1104, 1108, 1109, 1113, 1114
Barker, J.R., *see* Lowe, D. 1124
Barker, J.R., *see* Nixon, J.A. 1087, 1113
Barnard, R.D. 724
Barnscheidt, H.P., *see* Manzke, R. 317
Baroni, S., *see* Marsi, M. 399
Baroni, S., *see* McKinley, J.T. 398, 399
Baroni, S., *see* Peressi, M. 398
Barrau, J. 850, 857
Barrett, S.A., *see* Greene, J.E. 339
Bartels, F. 351
Bartelt, N.C., *see* Wang, X.S. 304
Bartelt, N.C., *see* Williams, E.D. 304
Bartoli, F.J., *see* Meyer, J.R. 703, 706
Bartynski, R.A., *see* Palmstrøm, C.J. 340
Basmaji, P., *see* von Bardeleben, H.J. 150
Basov, N.G. 232
Bassani, F. 61, 123, 144
Bassani, F., *see* Andreani, L.C. 674
Bassani, F., *see* Brust, D. 7, 61
Bassett, D., *see* Baars, J. 332, 353
Bastard, G. 168, 181, 670, 671, 819, 825, 827, 828, 831, 833, 837
Bastard, G., *see* Allan, G. 819, 827
Bastard, G., *see* Bleuse, J. 833, 844, 846, 849–851
Bastard, G., *see* Brun, J.A. 697
Bastard, G., *see* Chang, Y.C. 825
Bastard, G., *see* Chomette, A. 837, 838, 854–856
Bastard, G., *see* Deveaud, B. 838, 839
Bastard, G., *see* Ferreira, R. 839, 842, 843, 852, 854, 855
Bastard, G., *see* Soucail, B. 833, 845, 846
Bastard, G., *see* Voisin, P. 823, 832
Basu, P.K., *see* Kundu, S. 728, 734, 735
Batra, I.P. 314, 319
Batra, I.P., *see* Ciraci, S. 313
Batra, I.P., *see* Grant, P. 613
Batterman, B.W., *see* Durbin, S.M. 380
Bätz, P. 623
Bauer, E. 300, 329
Bauer, G., *see* Kuchar, F. 167
Bauer, G.E.W. 837
Bauer, R.S. 380, 390, 391
Bauer, R.S., *see* Bachrach, R.Z. 298, 299, 302
Bauer, R.S., *see* Chiaradia, P. 319, 394, 395
Bauer, R.S., *see* Hansson, G.V. 362
Bauer, R.S., *see* Katnani, A.D. 396, 399
Baughman, R.H. 592, 619
Baughman, R.H., *see* Murthy, N.S. 593
Bauhofer, W., *see* Ehinger, K. 621
Baumann, H., *see* Jäger, D. 440
Baumann, J.A., *see* Olego, D.J. 338
Bayerbach, M., *see* Rau, U. 443, 445
Baym, G., *see* Kadanoff, L.P. 1092, 1121, 1122
Bean, J.C., *see* Patel, J.R. 380
Beaumont, S.P. 867, 875
Beaumont, S.P., *see* Alves, E.S. 930
Beaumont, S.P., *see* Beton, P.H. 745, 842, 888, 889
Beaumont, S.P., *see* Gallagher, B.L. 787, 793–795
Beaumont, S.P., *see* Galloway, T. 791, 793
Beaumont, S.P., *see* Main, P.C. 888, 911
Bebb, H.B. 171, 229
Bechstedt, F. 681, 686, 687, 698
Bechstedt, F., *see* Gerecke, H. 682
Beck, W.A., *see* Davis, G.D. 332, 334
Becker, P.C. 1070
Becker, R.S. 296, 461
Becker, W.M., *see* Hoo, K. 187, 188
Becker, W.M., *see* Seiler, D.G. 502, 521
Bedrossian, P. 322
Beebe, E.D., *see* Roukes, M.L. 867, 870, 872, 908, 909, 911
Beebe, E.D., *see* Scherer, A. 870
Beeler, F. 132
Beenakker, C.W.J. 659, 867, 875, 877, 886, 888, 892, 905, 906, 910, 911, 915–917, 924, 998
Beenakker, C.W.J., *see* Kouwenhoven, L.P. 899, 900
Beenakker, C.W.J., *see* Molenkamp, L.W. 796, 804, 805, 872, 877, 878
Beenakker, C.W.J., *see* Montie, E.A. 886
Beenakker, C.W.J., *see* Staring, A.A.M. 888, 921, 1115
Beenakker, C.W.J., *see* van Houten, H. 806–808, 870–872, 876, 879, 882, 883, 886, 902, 904–906, 911, 924, 1104
Beenakker, C.W.J., *see* van Loosdrecht, P.H.M. 904
Beenakker, C.W.J., *see* van Wees, B.J. 800, 801, 867, 870, 871, 881, 882, 886, 902, 903, 913, 997, 1097, 1104
Beenakker, C.W.J., *see* Williamson, J.G. 880
Beer, A.C. 525, 526, 558, 559, 563
Begin, D., *see* Pouget, J.P. 622
Behringer, R.E. 914
Behringer, R.E., *see* Chang, A.M. 911
Behringer, R.E., *see* de Vegvar, P.G.N. 791
Behringer, R.E., *see* Prentiss, M. 875
Behringer, R.E., *see* Timp, G. 867, 872, 881, 883, 892, 894, 904, 906, 908, 909, 911, 1114
Bekman, M.H.P.Th., *see* Gregorkiewicz, T. 186
Bel'kov, V.V., *see* Ashkinadze, B.M. 478
Bel'kov, V.V., *see* Asnin, V.M. 469, 471
Bell, L.D. 365
Bell, L.D., *see* Kaiser, W.J. 363, 365
Belle, G. 841

Beltram, F. 855
Belyaev, A.D. 240
Belyantsev, A.M. 432
Bemski, G. 517
Ben-Daniel, D.J. 637, 826
Benedict, K.A. 868
Bengtsson, S. 264
Beni, G. 451, 456
Bennett, A.J. 656
Bennett, P.A. 296
Benoit à la Guillaume, C. 232, 462, 463, 472
Benoit à la Guillaume, C., *see* Scalbert, D. 825
Benoit à la Guillaume, C., *see* Voos, M. 451
Bensaid, B. 238
Benson, B.W., *see* Song, L.W. 250
Bensoussan, M., *see* Moisson, J.M. 303
Benton, P.H., *see* Alves, E.S. 930
Bentosela, F. 843
Beres, R.P. 316
Beresford, R. 647
Beresford, R., *see* Luo, L.F. 647, 651
Beresford, R., *see* Shanabrook, B.V. 825
Berggren, K.F. 871, 881, 896, 897, 902, 914
Berggren, K.F., *see* Ji, Z.-L. 890
Berggren, K.F., *see* Poole, D.A. 735, 736, 738
Berggren, K.F., *see* Rundquist, J. 896, 914
Berggren, K.F., *see* Sernelius, B.E. 736
Berggren, K.F., *see* Weisz, J.F. 896, 903, 914
Berglund, P., *see* Chang, A.M. 1000
Bergmann, G. 782, 784, 867, 933, 934, 936, 938, 940, 946, 948
Bergmann, G., *see* Kramer, B. 867, 868
Bergstresser, T.K., *see* Cohen, M.L. 7, 61
Berkemeier, J., *see* Purwins, H.G. 440
Berker, A.N., *see* Alerhand, O.L. 304, 321
Berman, L.E., *see* Durbin, S.M. 380
Bernard, M.G.A. 232
Bernard, W., *see* Roth, H. 655
Bernholc, J. 118
Bernstein, G., *see* Puechner, R.A. 930
Bernstein, G.H., *see* Kriman, A.M. 897
Berroir, J.M. 825, 828
Berry, J.G., *see* Laurich, B.K. 825
Bertoni, C.M. 313, 315, 362
Bertoni, C.M., *see* Manghi, F. 314
Bertoni, C.M., *see* Margaritondo, G. 322, 349
Beshers, D.N. 270
Bess, L. 230
Bethe, H., *see* von der Lage, F.C. 543
Bethea, G.G., *see* Levine, B.F. 697
Beton, P.H. 745, 842, 888, 889
Beton, P.H., *see* Gallagher, B.L. 787, 793–795
Beton, P.H., *see* Galloway, T. 791, 793
Beton, P.H., *see* Main, P.C. 888, 911
Beyer, W. 934, 935
Beyers, R.K. 336
Beyzavi, K., *see* Costa, J.C. 368
Bhat, R., *see* Sandroff, C.J. 339
Bhat, R., *see* Yablonovitch, E. 339
Bhatt, R.N., *see* Huo, Y. 995
Bhattacharya, P.K. 459
Bhattacharya, P.K., *see* Oh, J.E. 326
Bicknell, J., *see* Brotherton, S.R. 231
Biefeld, R.M., *see* Osbourn, G.C. 378
Biegelsen, D.K. 90, 294, 300, 302, 307
Bigan, E. 850, 852
Bigelow, N., *see* Prentiss, M. 875
Bigelow, R.W., *see* Epstein, A.J. 619–622
Bigot, J.Y. 22
Billaud, D., *see* Pouget, J.P. 622
Billingham, N.C., *see* Bott, D.C. 619
Bimberg, D. 697
Bimberg, D., *see* Christen, J. 874
Binder, K. 1063
Binh, V.T., *see* Sáenz, J.J. 878
Binnig, G. 296, 362
Bir, G.L. 144
Bir, G.L., *see* Pikus, G.E. 571, 674
Bir, G.L., *see* Vul, A.Ya. 188
Birch, F. 83
Birgeneau, R.J., *see* Chen, C.Y. 934
Birman, J.L., *see* Dorsinville, R. 615
Bischoff, J.-C., *see* Sandroff, C.J. 339
Bishop, A.R. 601, 616
Bishop, A.R., *see* Campbell, D.K. 603
Bishop, D.J. 809
Bishop, D.J., *see* Licini, J.C. 958
Bisi, O., *see* Calandra, C. 355
Biskupski, G. 934
Björklund, G. 177
Black, B.S., *see* Westervelt, R.M. 462
Blakely, J.M. 304
Blakely, J.M., *see* Durbin, S.M. 380
Blakemore, J.S. 187, 229
Blakemore, J.S., *see* Brown, W.J. 171
Blakemore, J.S., *see* Messenger, R.A. 172
Blakeslee, A.E., *see* Matthews, J.W. 378, 379
Blatt, F.J. 747
Bleaney, B., *see* Abragam, A. 183
Bleier, H., *see* Bradley, D.D.C. 617
Bleuse, J. 833, 839, 840, 844, 846, 849–851
Bleuse, J., *see* Voisin, P. 850
Blinowski, J., *see* Muller, F. 185, 186
Bloch, A.N. 96
Bloch, A.N., *see* Simons, G. 96
Bloch, A.N., *see* St. John, J. 96
Blöchl, P., *see* Das, G.P. 356
Block, S., *see* Piermarini, G.J. 85
Bloembergen, N., *see* Liu, J.M. 22
Bloembergen, N., *see* Lompre, L.A. 22
Blom, F.A.P., *see* Hendriks, P. 432
Blount, E.I. 494, 510
Blunier, S., *see* Maissen, C. 381
Boardman, A.D. 1063
Boccara, N., *see* Allan, G. 819, 827
Boebinger, G.S. 1014

Boebinger, G.S., *see* Eisenstein, J.P. 1028
Böer, K.W. 499, 567
Bogatina, N.I., *see* Sharvin, Yu.V. 883
Bogoliubov, N.N. 1051
Bohm, D. 634, 651, 1116
Bohm, D., *see* Aharonov, A. 1109, 1110
Bohm, D., *see* Philippidis, C. 1118
Böhm, W., *see* Weispfenning, M. 428
Bois, D. 170, 177
Bois, D., *see* Chantre, A. 151, 245, 250
Bois, D., *see* Mircea, A. 351
Bois, D., *see* Vincent, G. 151
Bokor, J., *see* Haight, R. 362
Boland, J.J. 324
Bolmont, D., *see* Chen, P. 380
Boltaks, B.I. 351
Bonch-Bruevich, V.L. 422, 433, 440, 867
Bonn, M., *see* Mark, P. 356
Bonnet, J.E., *see* Le Lay, G. 356
Bonnet, J.E., *see* Sauvage-Simkin, M. 302
Bonnot, A., *see* Benoit à la Guillaume, C. 462, 463
Boom, R., *see* Niessen, A.K. 332
Bordone, P., *see* Lugli, P. 698
Born, M. 575
Bose, S.S., *see* Skromme, B.J. 126
Bosi, S. 1066
Bott, D.C. 619
Bottka, N., *see* Seraphin, B.O. 656
Bottoms, W.R., *see* Baidyaroy, S. 349
Bouche, C., *see* Voisin, P. 850
Boudreaux, D.S., *see* Chance, R.R. 602
Boulet, R., *see* Buckthought, A. 1020
Boulet, R., *see* Sachrajda, A. 1020
Bourgoin, J.C. 144, 146, 147, 149
Bourgoin, J.C., *see* Delerue, C. 152
Bourgoin, J.C., *see* Lannoo, M. 128, 130
Bourgoin, J.C., *see* Mauger, A. 120
Bourgoin, J.C., *see* von Bardeleben, H.J. 150, 151, 153
Bowers, R. 530
Boyce, J.B., *see* Mikkelsen, J.C. 56
Boyle, B.L., *see* Picraux, S.T. 820
Boyn, R. 139, 142, 175
Bradley, D.D.C. 607, 617
Bradley, D.D.C., *see* Burroughes, J.H. 606
Bradshaw, J.T., *see* Tadayon, B. 326
Brafman, O., *see* Vardeny, Z. 607
Braicovitch, L. 336
Branch, J.V., *see* Clark, R.G. 1020
Brand, S., *see* Jaros, M. 118
Brandl, A. 422, 432, 437, 439, 440, 443
Brandsma, T.E.C., *see* Henning, J.C.M. 144
Brandt, R.C., *see* Stillman, G.E. 168
Bratina, G., *see* Sorba, L. 398, 399
Brattain, W.H. 349
Braud, C., *see* Bigan, E. 852
Braun, S. 190
Braun, W., *see* Alonso, M. 372
Bravman, J.C., *see* Farrow, R.F.C. 326
Brédas, J.-L. 595, 596, 603, 606
Brédas, J.-L., *see* Chance, R.R. 602
Bremond, G., *see* Delerue, C. 137, 138, 140
Brewer, L. 83, 84
Brews, J.R., *see* Nicollian, E.H. 261
Brey, L. 922
Brey, L., *see* Christensen, N.E. 397
Brey, L., *see* Dempsey, J. 922
Brézin, E. 984
Brézin, E., *see* Hikami, S. 953, 955
Briere, A., *see* Mircea, A. 151
Briggs, A., *see* Biskupski, G. 934
Brillson, L.J. 286, 312, 313, 315, 328, 330, 332–334, 337–339, 349–351, 353, 354, 358, 362, 366, 370–372
Brillson, L.J., *see* Brucker, C.F. 333, 366, 367
Brillson, L.J., *see* Chang, S. 340, 350, 372–374
Brillson, L.J., *see* Chiradia, P. 372
Brillson, L.J., *see* Kahn, A. 323
Brillson, L.J., *see* Margaritondo, G. 396, 398
Brillson, L.J., *see* Richter, H.W. 339, 340
Brillson, L.J., *see* Shapira, Y. 333, 337
Brillson, L.J., *see* Shaw, J.L. 333, 340, 351, 367, 370, 371
Brillson, L.J., *see* Viturro, R.E. 358, 362, 364, 370, 372
Bringans, R.D. 319, 320, 322, 324, 362, 381, 382
Bringans, R.D., *see* Biegelsen, D.K. 90, 294, 300, 302, 307
Bringans, R.D., *see* Northrup, J.E. 382
Brinkmann, W.F. 454, 459, 474
Brinkmann, W.F., *see* Rice, T.M. 454, 458
Brister, K.E., *see* Vohra, Y.K. 85, 87
Brito Cruz, C.H., *see* Becker, P.C. 1070
Brodde, A., *see* Badt, D. 322
Broekaart, M.E.I., *see* van Houten, H. 872, 876, 879, 911
Broekaart, M.E.I., *see* van Wees, B.J. 904, 906, 921, 924
Broekaart, M.E.I., *see* Williamson, J.G. 880
Broekaert, T.P.E., *see* Bagwell, P.F. 893
Broido, D.A. 837
Broido, D.A., *see* Kempa, K. 922
Broido, D.A., *see* Shanabrook, B.V. 825
Brooks, H. 211, 570
Broser, J. 181
Brotherton, S.R. 231
Broughton, J.Q., *see* Li, X.P. 106
Brousseau, M., *see* Barrau, J. 850, 857
Brovman, E.G. 27
Brown, A.R., *see* Burroughes, J.H. 606
Brown, C.S., *see* Bott, D.C. 619
Brown, E. 508, 510, 512, 513, 929, 988, 992, 1021
Brown, R.A. 239
Brown, R.D., *see* Koenig, S.H. 187
Brown, R.J. 886, 899, 902, 904, 906, 907, 923, 924

Brown, R.J., *see* Kelly, M.J. 899, 902
Brown, R.N., *see* Pidgeon, C.R. 520
Brown, W.J. 171
Browne, D.C., *see* Landsberg, P.T. 259
Brucker, C.F. 333, 366, 367
Brucker, C.F., *see* Brillson, L.J. 332, 333, 337, 338, 358, 366
Brueckner, K.A., *see* Ma, S.K. 30
Brum, J.A. 831, 841, 926
Brum, J.A., *see* Agullo-Rueda, F. 850, 851, 857
Brum, J.A., *see* Bastard, G. 819, 827, 831
Brum, J.A., *see* Marzin, J.Y. 824, 830
Brummell, M.A. 730
Brummell, M.A., *see* Nicholas, R.J. 986
Brummell, M.A., *see* Ruf, C. 741, 742
Brummell, M.A., *see* Vuong, T.H.H. 730
Brun, J.A. 697
Brun, J.A., *see* Bastard, G. 670
Brunel, L.C., *see* Gregorkiewicz, T. 186
Brunel, L.C., *see* Muller, F. 185, 186
Brunet, S., *see* van Cong, H. 256
Brunetti, R. 1090
Bruno, J.D., *see* Bahder, T.B. 652
Brush, S.C. 1069
Brust, D. 7, 61
Bruus, H., *see* Lindelof, P.E. 1028
Bruynseraede, Y., *see* Gijs, M. 937, 948
Bruynseraede, Y., *see* Kramer, B. 867, 868
Bruynseraede, Y., *see* van den Dries, L. 936
Bruynseraede, Y., *see* van Haesendonck, C. 948
Bryant, G.W. 659, 923, 925
Bube, R.H., *see* Iselev, G.W. 177, 188
Bube, R.H., *see* Lin, A.L. 188
Bube, R.H., *see* Wu, C.-h. 266
Buchanan, M., *see* Landheer, D. 674
Buchenauer, C.J. 94
Buckthought, A. 1020
Buhmann, H. 1006, 1007
Buhrman, R.A., *see* Farmer, K.R. 205, 247
Buisson, J.P. 118
Bujik, O.J.A., *see* Molenkamp, L.W. 806
Bumelienė, S.B. 432
Burgeat, J., *see* Quillec, M. 820
Burkey, B.C. 177, 188
Burns, M.J. 809, 810, 812
Burns, P.L., *see* Burroughes, J.H. 606
Burroughes, J.H. 606
Burrus, C.A., *see* Miller, D.A.B. 657
Burt, M.G. 670–672, 674
Busse, J.W. 181
Butcher, P.N. 440, 525, 568, 796, 800, 809
Butcher, P.N., *see* Cantrell, D.G. 736, 738–741, 750–752, 757, 759, 771, 783
Butcher, P.N., *see* Dharssi, I. 709
Butcher, P.N., *see* Fletcher, K. 544, 547, 548, 556
Butcher, P.N., *see* Fromhold, T.M. 762, 774–778
Butcher, P.N., *see* Gallagher, B.L. 732, 753, 754, 759
Butcher, P.N., *see* Karavolas, V.C. 728–730, 733
Butcher, P.N., *see* Kearney, M.J. 742, 743, 783, 784, 809
Butcher, P.N., *see* Kubakaddi, S.S. 734, 744, 771, 772
Butcher, P.N., *see* Milsom, P.K. 701
Butcher, P.N., *see* Oxley, J.P. 733, 754, 756, 758, 761, 768, 769, 774, 778
Butcher, P.N., *see* Qin, G. 777, 778
Butcher, P.N., *see* Smith, M.J. 747–749, 757, 758, 760
Butcher, P.N., *see* Timp, G. 809, 813
Butcher, P.N., *see* Zianni, X. 685
Butera, R.A. 331
Büttiker, M. 654, 796, 800, 877, 878, 880, 884, 885, 887, 894, 896, 902–904, 912–914, 932, 998, 1087, 1099, 1104
Büttiker, M., *see* Ford, C.J.B. 906, 909–911, 919
Button, K., *see* Zwerdling, D. 522
Bychkov, Yu.A. 1004, 1006
Byer, N.E., *see* Davis, G.D. 332, 334
Byers, N. 991
Bylander, D.M. 56, 78
Bylander, D.M., *see* Kleinman, L. 78

Cafolla, A.A., *see* Fowell, A.E. 365
Cage, M.E. 999
Caillol, J.M. 1005, 1006, 1030
Caine, E.J. 389, 822
Calandra, C. 355
Calandra, C., *see* Bertoni, C.M. 362
Calandra, C., *see* Manghi, F. 314
Calandra, C., *see* Margaritondo, G. 322, 349
Calecki, D. 855
Calecki, D., *see* Palmier, J.F. 855
Callaway, J. 494, 495, 505, 508, 522, 572, 575, 656
Calleja, J.M., *see* Tejedor, C. 389
Callen, H.B. 525, 576
Calverie, P., *see* Sauvage-Simkin, M. 302
Calvert, L.D., *see* Villars, P. 98
Calvert, P.D., *see* Bott, D.C. 619
Cammack, D., *see* Olego, D.J. 396
Campbell, D.K. 603
Campbell, D.K., *see* Bishop, A.R. 601, 616
Campbell, D.K., *see* Vogl, P. 596, 613
Campbell, J., *see* Kravchenko, S.V. 1007
Campbell, J.W., *see* D'Iorio, M. 1007
Camras, M.D., *see* Laidig, W.D. 380
Canagaratna, S.G., *see* Landsberg, P.T. 270
Canali, C. 335
Canter, K.F. 293
Cantrell, D.G. 736, 738–741, 750–752, 757, 759, 771, 783
Cantrell, D.G., *see* Gallagher, B.L. 753, 757, 811, 812
Cao, R.K. 349, 358
Cao, R.K., *see* Miyano, K. 372
Cao, R.K., *see* Spicer, W.E. 356, 357

Cao, Y. 591, 617
Capasso, C., *see* Perfetti, P. 389
Capasso, C., *see* Vitomirov, I.M. 367
Capasso, F. 387, 397, 399, 824
Capasso, F., *see* Beltram, F. 855
Capizzi, M., *see* Niles, D.W. 398
Capizzi, M., *see* Thomas, G.A. 474, 475
Car, R. 105, 106, 117
Car, R., *see* Galli, G. 106
Car, R., *see* Li, X.P. 106
Cardona, M. 498, 678
Cardona, M., *see* Buchenauer, C.J. 94
Cardona, M., *see* Christensen, N.E. 390
Carelli, J., *see* Duke, C.B. 322
Carelli, J., *see* Kahn, A. 323
Carenco, A., *see* Bigan, E. 852
Carey, G.P., *see* Friedman, D.J. 334
Caro, J. 938, 958
Carré, M., *see* Bigan, E. 850, 852
Carrol, J.E. 234
Carruthers, J.R. 306
Carruthers, P. 1092, 1097
Carstensen, H. 317
Carter, C.B., *see* Palmstrøm, C.J. 327, 340, 374, 375
Carter, M., *see* Gallagher, B.L. 787, 793–795
Carter, M., *see* Galloway, T. 791, 793
Casey Jr, H.C. 187, 233, 306, 377, 384
Castaño, E. 872, 884, 887, 889–892, 899, 900, 902, 917, 926, 928, 929
Castaño, E., *see* Kirczenow, G. 919–921, 928
Castaño, E., *see* Pendry, J.B. 946
Castaño, E., *see* Que, W. 922
Castaño, E., *see* Ulloa, S.E. 872, 926
Castellani, C. 785
Castner, T.G., *see* Seager, C.H. 265
Catalano, A., *see* Wyeth, N.C. 332, 353
Ceccone, G., *see* Sorba, L. 398, 399
Ceccone, G., *see* Yu, X. 386, 389
Celeste, A., *see* Leadbeater, M.L. 655
Celeste, A., *see* Sibille, A. 855
Celli, V., *see* Bassani, F. 61
Celotta, R.J., *see* Whitman, L.J. 305, 350
Celotti, G., *see* Canali, C. 335
Ceperley, D.M., *see* Bachelet, G.B. 80
Cerdeira, F., *see* Buchenauer, C.J. 94
Cernagora, J., *see* Scalbert, D. 825
Cerrina, F. 389
Cerullo, M., *see* Eaglesham, D.J. 378
Chabal, Y.J., *see* Higashi, G.S. 369
Chaban, E.E., *see* Ma, Y. 322
Chadi, D.J. 90, 91, 145, 147, 149, 152, 294, 296, 300, 304, 313, 314, 317, 318, 321, 390
Chadi, D.J., *see* Cohen, M.L. 377
Chadi, D.J., *see* Ihm, J. 90
Chadi, D.J., *see* Mailhiot, C. 314, 362
Chadi, D.J., *see* Nielsen, O.H. 319
Chadi, D.J., *see* Stone, A.D. 945
Chadi, D.J., *see* Zhang, S.B. 148
Chai, C.K., *see* Bott, D.C. 619
Chaiken, P.M., *see* Burns, M.J. 809, 810
Chakraborty, T. 922, 979, 1001, 1003, 1005, 1006, 1009, 1013, 1015, 1017, 1018, 1020, 1022, 1025–1028, 1030, 1031
Chakraborty, T., *see* Halonen, V. 1018
Chakraborty, T., *see* Maksym, P.A. 922
Chakraborty, T., *see* Pietiläinen, P. 1024
Chakraborty, T., *see* Zhang, F.C. 1013, 1017–1019
Chakravarty, S. 867
Chalker, J.T. 953, 955, 994
Chalker, J.T., *see* Benedict, K.A. 868
Chamberlain, J.M. 867, 875
Chamberlain, M.P. 694, 695
Chamberlain, M.P., *see* Babiker, M. 697
Chambers, R.G. 505, 553, 554
Chambers, S.A. 326, 374
Chambers, S.A., *see* del Guidice, M. 334
Chamoun, S.N. 1072
Chan, W.K., *see* Sands, T. 325, 327
Chan, W.W., *see* Sah, C.T. 189
Chance, R.R. 602
Chance, R.R., *see* Eckhardt, H. 605
Chandrasekar, V. 948
Chandrasekhar, S. 168
Chang, A.M. 911, 916, 1000
Chang, A.M., *see* Boebinger, G.S. 1014
Chang, A.M., *see* Störmer, H.L. 986, 999
Chang, A.M., *see* Timp, G. 872, 904, 906, 908, 911
Chang, A.M., *see* Wei, H.P. 984, 996
Chang, C.-A., *see* Chang, L.L. 824
Chang, C.-A., *see* Sai-Halasz, G.A. 822
Chang, C.-A., *see* Tejedor, C. 389
Chang, K.H., *see* Oh, J.E. 326
Chang, K.J. 85, 87, 96, 102, 103
Chang, K.J., *see* Chadi, D.J. 145, 147, 149, 152
Chang, K.J., *see* Cohen, M.L. 103, 104
Chang, K.J., *see* Dacorogna, M.M. 103
Chang, K.J., *see* Liu, A.Y. 85, 87
Chang, K.J., *see* Vohra, Y.K. 85, 87
Chang, K.J., *see* Wentzcovitch, R.M. 87
Chang, K.Y., *see* Brummell, M.A. 730
Chang, L.L. 371, 380, 646, 647, 651, 743, 823, 824, 836, 868, 869
Chang, L.L., *see* Bastard, G. 837
Chang, L.L., *see* Brum, J.A. 831, 841
Chang, L.L., *see* Esaki, L. 646, 819, 820
Chang, L.L., *see* Guldner, Y. 824
Chang, L.L., *see* Haug, R.J. 874, 926
Chang, L.L., *see* Ludeke, R. 325
Chang, L.L., *see* Mendez, E.E. 986
Chang, L.L., *see* Sai-Halasz, G.A. 822
Chang, L.L., *see* Voisin, P. 823
Chang, R. 314
Chang, S. 340, 350, 372–374
Chang, S., *see* Franciosi, A. 338

Chang, S.C. 302
Chang, S.C., *see* Lubinsky, A.R. 302
Chang, S.C., *see* Mark, P. 356
Chang, S.S. 874
Chang, T.Y., *see* Chang, A.M. 911, 916
Chang, T.Y., *see* Timp, G. 872, 904, 906, 908, 911
Chang, Y.-C. 673
Chang, Y.-C., *see* Shulman, J.N. 674
Chang, Y.A., *see* Lin, J.C. 337
Chang, Y.C. 825, 832, 833, 837, 838
Chang, Y.C., *see* Aspnes, D.E. 307, 320
Chang, Y.C., *see* Chu, H. 838
Chang, Y.C., *see* Deveaud, B. 838, 839
Chang, Y.C., *see* Jung, P.S. 840
Chang, Y.C., *see* Miller, R.C. 837
Chang, Y.C., *see* Sanders, G.D. 837
Chang, Y.C., *see* Schulman, J.N. 832
Chantre, A. 151, 245, 250
Chantre, A., *see* Vincent, G. 151
Chaplar, J., *see* Massies, J. 339, 340
Chaplik, A.V., *see* Wagner, M. 922
Chapman, R.A. 171
Chapman, R.A., *see* Bebb, H.B. 171
Chappell, T.I., *see* Offsey, S.D. 339, 368, 369
Charasse, M.N., *see* Marzin, J.Y. 824
Charbonneau, S., *see* Young, J.F. 825
Charfi, F.F., *see* Scalbert, D. 825
Charles, J.P., *see* Mialhe, P. 252
Chase, K.S. 959
Chassé, Th., *see* Alonso, M. 372
Chatterjea, A. 270
Chattopadhay, D. 703
Chaudhuri, S. 127
Chaussy, J., *see* Pannetier, B. 937, 948
Cheeks, T.L., *see* Palmstrøm, C.J. 327, 340, 374, 375
Cheeks, T.L., *see* Sands, T. 325–327, 336, 340, 365, 366, 374, 381
Chelikowsky, J.R. 61, 65, 68, 74, 75, 81, 96, 314
Chelikowsky, J.R., *see* Cohen, M.L. 7, 10–12, 15, 61, 63, 68, 71, 79, 89
Chelikowsky, J.R., *see* Louie, S.G. 118
Chelikowsky, J.R., *see* Schlüter, M. 313
Chelluri, B., *see* Chang, A.M. 911
Chel'tzov, V.F. 234
Chemla, D.S., *see* Bar-Joseph, I. 850
Chemla, D.S., *see* Miller, D.A.B. 657
Chen, C.T., *see* Ma, Y. 322
Chen, C.Y. 934
Chen, D., *see* Winokur, M.J. 592
Chen, D.M., *see* Bedrossian, P. 322
Chen, H.-W., *see* Chambers, S.A. 326
Chen, H.S., *see* Villars, P. 98
Chen, J., *see* Moraes, F. 602
Chen, J., *see* Moses, D. 619
Chen, J.F. 647, 651
Chen, J.F., *see* Yang, L. 647
Chen, L.J. 324
Chen, M.F., *see* Lucovsky, G. 714
Chen, P. 380
Chen, W.M., *see* Godlewski, M. 432
Cheng, H. 336, 357
Cheng, H.C., *see* Chen, L.J. 324
Cheremnykh, P.A., *see* Gusev, G.M. 986
Chester, G.V., *see* Reatto, L. 1030
Chester, M. 657
Chiang, C., *see* Hamann, D.R. 14, 15, 72, 73, 77
Chiaradia, P. 319, 394, 395
Chiaradia, P., *see* Bachrach, R.Z. 298, 299, 302
Chiaradia, P., *see* Brillson, L.J. 372
Chiaradia, P., *see* Hansson, G.V. 362
Chiaradia, P., *see* Katnani, A.D. 396, 399
Chiarotti, G., *see* Chiaradia, P. 319
Chien, C., *see* Farrow, R.F.C. 326
Chien, W.-Y., *see* Kroemer, H. 382
Chik, K.D., *see* Kriegler, R.J. 264
Chin, K.K., *see* Cao, R.K. 349, 358
Chin, K.K., *see* Newman, N. 343–345
Chin, M.A., *see* Narayanamurti, V. 678
Chinone, N., *see* Kishino, S. 178
Chiochetti, M.G.B., *see* Bachelet, G.B. 80
Chiradia, P. 372
Chisholm, M., *see* Woodall, J.M. 379, 382
Cho, A.Y. 302, 306, 326, 377
Cho, A.Y., *see* Beltram, F. 855
Cho, A.Y., *see* Boebinger, G.S. 1014
Cho, A.Y., *see* Brummell, M.A. 730
Cho, A.Y., *see* Caine, E.J. 389, 822
Cho, A.Y., *see* Capasso, F. 397, 399
Cho, A.Y., *see* Chen, J.F. 647, 651
Cho, A.Y., *see* Eisenstein, J.P. 984
Cho, A.Y., *see* Yang, L. 647
Cho, Y., *see* Katnani, A.D. 396, 399
Choi, H.-Y. 613, 614
Choi, K.K., *see* Levine, B.F. 697
Choi, K.K., *see* Wheeler, R.G. 869, 870, 958
Chomette, A. 837, 838, 854–856
Chomette, A., *see* Calecki, D. 855
Chomette, A., *see* Deveaud, B. 838, 839
Chomette, A., *see* Lambert, B. 855
Chomette, A., *see* Palmier, J.F. 855
Chomette, A., *see* Picoli, G. 132–134
Chou, M.Y., *see* Chelikowsky, J.R. 65, 74
Chou, S.Y. 659
Chouteau, G., *see* Chang, K.J. 87, 102, 103
Chow, D.H. 647
Christen, J. 874
Christen, J., *see* Bimberg, D. 697
Christensen, N.E. 390, 397
Christensen, N.E., *see* Das, G.P. 356
Christensen, O. 463
Christensen, O.B., *see* Christensen, N.E. 390
Christiansen, P.A. 65
Christman, S.B., *see* Rowe, J.E. 312
Chroboczek, J.A., *see* Porowski, S. 187, 188
Chroboczek, J.A., *see* Summerfield, S. 619

Chu, C.S. 1109, 1110
Chu, F.Y.F., *see* Scott, A.C. 601
Chu, H. 838
Chu, H., *see* Chang, Y.C. 838
Chu, H., *see* Deveaud, B. 838, 839
Chu, P., *see* Puechner, R.A. 930
Chu, S.-N.G., *see* Beltram, F. 855
Chu, S.-N.G., *see* Chen, J.F. 647, 651
Chu, X., *see* Berroir, J.M. 825
Chung, D.D.L., *see* Leung, S. 326
Chung, T.-C. 607, 619
Chung, T.-C., *see* Etemad, S. 614
Chung, T.-C., *see* Feldblum, A. 604
Chung, T.-C., *see* Moraes, F. 602
Chung, T.-C., *see* Moses, D. 619
Chung, T.-C., *see* Vardeny, Z. 607
Chung, Y.W., *see* Takatani, S. 389
Churilov, A.B., *see* Asnin, V.M. 457, 467, 470
Chye, P.W., *see* Lindau, I. 356
Chye, P.W., *see* Spicer, W.E. 349
Chynoweth, A.G. 638, 651, 656, 658
Ciampolini, P., *see* Baccarani, G. 426
Ciccotti, G. 105
Cieplak, M.Z. 184
Cimino, R., *see* Alonso, M. 372
Ciraci, S. 313
Ciraci, S., *see* Tekman, E. 878, 884–886, 893, 896
Claessen, R., *see* Carstensen, H. 317
Clark, R.G. 999, 1015, 1018–1020, 1025
Clark, R.G., *see* Mallett, J.R. 1006
Clark, R.G., *see* Turberfield, A.J. 1006
Clarke, R., *see* Oh, J.E. 326
Clarke, R.C., *see* White, A.M. 178
Clarke, T.C. 602, 607
Clausnitzer, M., *see* Dev, B.N. 380
Clauss, W., *see* Peinke, J. 433, 435, 437
Clauss, W., *see* Rau, U. 443, 445
Clemens, H.J., *see* Bartels, F. 351
Clerjaud, B. 132
Clerot, F., *see* Chomette, A. 837, 838
Clerot, F., *see* Deveaud, B. 838, 839
Clerot, F., *see* Guillemot, C. 684
Cobden, D.H. 245, 883, 892, 894
Cockayne, B., *see* Harris, I.R. 326
Coddington, P.D., *see* Chalker, J.T. 953, 955, 994
Cody, G.D. 258
Coghill, H.D., *see* Halstead, R.E. 351
Cohen, A. 959
Cohen, M.H. 6, 8, 62, 64, 512
Cohen, M.L. 7, 10–12, 14, 15, 61–63, 65, 67, 68, 71, 79, 81, 89, 98, 100, 102–104, 130, 377
Cohen, M.L., *see* Allen, P.B. 7, 65
Cohen, M.L., *see* Chadi, D.J. 317, 390
Cohen, M.L., *see* Chang, K.J. 85, 87, 96, 102, 103
Cohen, M.L., *see* Chelikowsky, J.R. 61, 314
Cohen, M.L., *see* Corkill, J.L. 85
Cohen, M.L., *see* Dacorogna, M.M. 103
Cohen, M.L., *see* Froyen, S. 87
Cohen, M.L., *see* Ihm, J. 82, 83, 89, 90, 92
Cohen, M.L., *see* Kerker, G.P. 65
Cohen, M.L., *see* Knittle, E. 100
Cohen, M.L., *see* Laughlin, R.B. 996
Cohen, M.L., *see* Liu, A.Y. 85, 87, 101
Cohen, M.L., *see* Louie, S.G. 14, 79, 81, 118
Cohen, M.L., *see* Martins, J.L. 70, 103
Cohen, M.L., *see* McMahan, A.K. 85
Cohen, M.L., *see* Northrup, J.E. 90, 319
Cohen, M.L., *see* Schlüter, M. 65, 313
Cohen, M.L., *see* Vanderbilt, D. 95
Cohen, M.L., *see* Vohra, Y.K. 85, 87
Cohen, M.L., *see* Wentzcovitch, R.M. 87, 89
Cohen, M.L., *see* Yin, M.T. 14, 67, 83–87, 90–95
Cohen, M.L., *see* Zhang, S.B. 87, 89, 90, 103, 105
Cohen, M.L., *see* Zunger, A. 14, 15, 72, 97, 98
Cohen, P.I. 307
Cohen, P.I., *see* van Hove, J.M. 307
Colas, E., *see* Christen, J. 874
Colbow, K.M. 389
Coldren, L.A., *see* Law, K.K. 852
Coldren, L.A., *see* Tsuchiya, M. 874
Coldren, L.A., *see* Yan, R.H. 850
Coleman, J.J. 380
Coleman, J.J., *see* Haase, M.A. 385
Coleman, J.J., *see* Kolodzey, J. 432
Coleman, J.J., *see* Laidig, W.D. 380
Coleridge, P., *see* Sachrajda, A. 1020
Coleridge, P., *see* Young, J.F. 825
Coleridge, P.T., *see* Taylor, R.P. 877, 883, 888, 894, 904, 906, 919, 923, 924
Colombo, E., *see* Fano, G. 1021
Colson, R. 621
Coluzza, C., *see* Marsi, M. 399
Colvard, C. 678
Colwell, P.J. 182
Comas, J., *see* Jarosik, N.C. 168
Combescot, M. 454, 459
Combescot, R. 658
Comes, R., *see* Pouget, J.P. 622
Condon, E.U., *see* Gurney, R.W. 629
Constantinou, N.C. 685
Constantinou, N.C., *see* Al-Dossary, O. 695, 696
Conwell, E.M. 550, 567, 570, 590, 597, 598, 604, 612, 617, 619–621, 623, 1044
Conwell, E.M., *see* Choi, H.-Y. 613, 614
Conwell, E.M., *see* Debye, P.P. 570
Conwell, E.M., *see* Jeyadev, S. 608–612, 616
Coon, D., *see* Chaudhuri, S. 127
Cooper, L.N., *see* Bardeen, J. 102
Copel, M., *see* LeGoues, F.K. 382
Corbett, J.W. 184
Corbett, J.W., *see* Lee, Y.H. 184
Corbett, J.W., *see* Watkins, G.D. 184
Corkill, J.L. 85
Cosman, E.C., *see* Montie, E.A. 886
Costa, J.C. 368
Costrini, G., *see* Coleman, J.J. 380

Cota, E. 934
Couder, Y., *see* Tuchendler, J. 167
Coulson, C.A. 119
Coutts, T.J. 252
Cox, G. 305
Craig, B.I. 314, 319
Craighead, H.G., *see* Roukes, M.L. 867, 870, 872, 908, 909, 911
Craighead, H.G., *see* Scherer, A. 870
Crawford, M.G. 177
Crawford, N.C., *see* Scifres, D.R. 179
Creighton, W.F., *see* Mark, P. 356
Cricenti, A. 91
Cricenti, A., *see* Chiaradia, P. 319
Cricenti, A., *see* Mårtensson, P. 297, 319
Crombeen, J.E., *see* van Bommel, A.J. 298, 299
Cronin, G.R., *see* Haisty, R.W. 171
Cuevas, M., *see* Falicov, L.M. 570
Culbertson, J.C., *see* Westervelt, R.M. 462
Cunningham, J.E., *see* Becker, P.C. 1070
Cunningham, J.E., *see* Behringer, R.E. 914
Cunningham, J.E., *see* Bigot, J.Y. 22
Cunningham, J.E., *see* Chang, A.M. 911
Cunningham, J.E., *see* de Vegvar, P.G.N. 791
Cunningham, J.E., *see* Goldman, V.J. 656
Cunningham, J.E., *see* Prentiss, M. 875
Cunningham, J.E., *see* Su, B. 659
Cunningham, J.E., *see* Timp, G. 867, 872, 881, 883, 892, 894, 904, 906, 908, 909, 911, 1114
Cusack, N.E. 867
Cuthbert, J.D., *see* Dean, P.J. 179
Czaja, W. 179
Czanderna, A.W. 286
Czycholl, G. 944

Dabiran, A., *see* Cohen, P.I. 307
Dabrowski, J. 149, 152, 155
Dacorogna, M.M. 103
Dacorogna, M.M., *see* Chang, K.J. 87, 102, 103
Dacorogna, M.M., *see* Cohen, M.L. 103, 104
Dagens, L. 14
Dahiya, R.P., *see* Mathur, V.K. 253
Dahl, J.P. 116, 1097
Dahlberg, E.D., *see* Davidson, J.S. 730, 767
Dahn, J.R., *see* Colbow, K.M. 389
Dakhovskii, I.V. 550, 551, 563
Dakhovskii, I.V., *see* Samoilovich, A.G. 493, 541, 550, 563
Dalby, R., *see* Sarma, K. 380
Damen, T.C., *see* Deveaud, B. 697, 854
Damen, T.C., *see* Mattos, J.V.C. 466
Damen, T.C., *see* Miller, D.A.B. 657
Damen, T.C., *see* Worlock, J.M. 464
Daniels, M.E. 701
Daniels, R.R. 323, 389
Daniels, R.R., *see* Brillson, L.J. 332, 333
Daniels, R.R., *see* Cerrina, F. 389
Daniels, R.R., *see* Davis, G.D. 334
Daniels, R.R., *see* Richter, H.W. 339
Dapkus, P.D., *see* Laidig, W.D. 380
Darwin, C.G. 922
Das, G.P. 356
Das Sarma, S. 118
Das Sarma, S., *see* Fertig, H.A. 657
Das Sarma, S., *see* He, S. 895
Das Sarma, S., *see* Jain, J.K. 699, 701, 820
Das Sarma, S., *see* Zhang, F.C. 1015
Datta, S. 897, 912
Datta, S., *see* Frohne, R. 1104
Davenport, J.W., *see* Qian, G.X. 106
Davidov, B. 342
Davidson, B.R., *see* Main, P.C. 911
Davidson, J.S. 730, 767
Davies, A.G. 1019
Davies, E.A., *see* Mott, N.F. 812
Davies, G. 152, 153
Davies, G.J., *see* Thornton, T.J. 735, 869–871, 881
Davies, J.F., *see* Westland, D.J. 697
Davies, J.H. 872, 1087, 1113
Davies, J.H., *see* Khan, F.S. 1123
Davies, J.H., *see* Laughton, M.J. 895, 917, 1087, 1108, 1109, 1113, 1114
Davies, J.H., *see* Nixon, J.A. 895, 896, 906, 1087, 1113, 1114
Davies, M., *see* Main, P.C. 888
Davies, M., *see* Taylor, R.P. 883, 888, 894, 904, 906, 919, 924
Davies, P., *see* Anderson, D.A. 703
Davies, R.A., *see* Uren, M.J. 937
Davis, E.A., *see* Mott, N.F. 621, 867, 934
Davis, G.D. 332, 334
Davis, G.D., *see* Patten, E.A. 389
Davis, L.C., *see* Bell, L.D. 365
Davis, R.H. 651
Davison, M., *see* Beton, P.H. 745, 842
Daw, M.S. 118, 356
Däweritz, L. 297, 298
Däweritz, L., *see* Notzel, R. 872
Dawson, L.R., *see* Casey Jr, H.C. 187
Dawson, L.R., *see* Osbourn, G.C. 378
Dawson, P., *see* Pulsford, N.J. 825
Day, B., *see* White, A.M. 178
de Aguiar, F.M. 926
de Blasi, C. 252
de Boer, F.R., *see* Miedema, A.R. 332, 347
de Boer, F.R., *see* Niessen, A.K. 332
de Châtel, P.F., *see* Miedema, A.R. 332, 347
de Châtel, P.F., *see* Niessen, A.K. 332
de Crémoux, B., *see* Soucail, B. 833
de Haas, W.J. 512
de Haas, W.J., *see* Shubnikov, L. 512
de Miguel, J.L., *see* Farrell, H.H. 382
de Miguel, J.L., *see* Meynadier, M.H. 825
de Raedt, H. 897
de Raedt, H., *see* García, N. 878

de Raedt, H., *see* Sáenz, J.J. 878
de Siqueira, M.L., *see* Fazzio, A. 118
de Vegvar, P.G.N. 791
de Vegvar, P.G.N., *see* Timp, G. 904, 906, 911
de Visschere, P. 259
de Wit, M. 184
Deal, B.E., *see* Grove, A.S. 262
Deal, M.D. 273
Dean, P. 940
Dean, P.J. 178–180, 182
Dean, P.J., *see* Ashen, D.J. 178, 180
Dean, P.J., *see* Manchon, D.D. 182
Dean, P.J., *see* White, A.M. 178, 189, 190
Debye, P.P. 570
Deckert, C.A., *see* Kern, W. 309, 310
Degtyarenko, N.N. 433
DeJoule, R.Y., *see* Skromme, B.J. 126
Dekker, C. 883, 894
del Alamo, J.A. 886
del Alamo, J.A., *see* Eugster, C.C. 886
del Guidice, M. 334
del Guidice, M., *see* Butera, R.A. 331
del Guidice, M., *see* Grioni, M. 330, 331
del Sole, R., *see* Selloni, A. 297, 319
Delahaye, F., *see* Guldner, Y. 986
Delalande, C., *see* Bastard, G. 819, 827, 833
Deleo, G.G. 132
Delerue, C. 118, 126, 133, 135–138, 140–142, 152
Delerue, C., *see* Krambrock, K. 156
Delerue, C., *see* Langer, J.M. 135, 136
Delerue, C., *see* Liro, Z. 136, 137
Delerue, C., *see* von Bardeleben, H.J. 150
Delescluse, P., *see* Massies, J. 325, 326
Dellepiane, G., *see* Dorsinville, R. 615
Dellow, M., *see* Beton, P.H. 745, 842
Delves, R.T. 493, 544, 548
Delyon, F. 934
Demel, T., *see* Grambow, P. 870, 872
Demel, T., *see* Heitmann, D. 922
Dempsey, J. 922
Demuth, J.E. 290
Demuth, J.E., *see* Alerhand, O.L. 304, 321
Demuth, J.E., *see* Hamers, R.J. 297, 319, 321
Demuth, J.E., *see* Kohler, V.K. 321, 322, 324
Demuth, J.E., *see* Tromp, R.M. 297, 319
den Nijs, M., *see* Thouless, D.J. 930, 993
Dench, W.A., *see* Seah, M.P. 286, 288
Denenstein, A., *see* Moses, D. 609, 619
Dernier, P.D., *see* Cho, A.Y. 326
Derrida, B. 943
Desgreniers, S., *see* Vohra, Y.K. 85, 87
Deuling, H. 264
Deutscher, G., *see* Burns, M.J. 810
Deutscher, G., *see* van den Dries, L. 936
Dev, B.N. 380
Deveaud, B. 697, 838, 839, 854
Deveaud, B., *see* Chomette, A. 837, 838, 854–856
Deveaud, B., *see* Lambert, B. 855
deVegvar, P., *see* Timp, G. 867, 908, 909, 911
Devenyi, T.F., *see* Kriegler, R.J. 264
DeVillard, P., *see* Makram-Ebeid, S. 351, 371
Devoret, M.H., *see* Geerligs, L.J. 1115
Devreese, J.T., *see* Huant, S. 169
Devreux, F., *see* Nechtschein, M. 608, 612
Dewdney, C., *see* Philippidis, C. 1118
Dewdney, C., *see* Vigier, J.P. 1118
Dexter, D.L. 209, 243, 245
Dexter, R.L. 171
Dexter, R.N. 499, 504, 506
Dexter, R.N., *see* Zeiger, H.J. 506
Dezaki, K., *see* Nishizawa, J. 357
Dezaly, F., *see* Massies, J. 298, 306, 339
Dhariwal, S.R. 239, 241–243, 259
Dhariwal, S.R., *see* Landsberg, P.T. 239, 240
Dharssi, I. 709
Dharssi, I., *see* Zianni, X. 685
Dhere, R.G., *see* Aboie-Elfotouh, F.A. 305
DiCastro, C., *see* Castellani, C. 785
Dietz, R.E., *see* Pappalardo, R. 170
Dignam, M.M. 838, 851, 857
Dimmock, J.O., *see* Hemstreet, L.A. 132
Dimmock, J.O., *see* Stillman, G.E. 187, 548, 556
Dinan, J.H., *see* Zahn, D.T.R. 380
Dinardo, N.J., *see* Wong, T.M. 322
Ding, J. 326
Dingle, R. 819
Dingle, R.B. 569
D'Iorio, M. 761, 770, 780, 1007
D'Iorio, M., *see* Fletcher, R. 738, 740, 771, 772
D'Iorio, M., *see* Kravchenko, S.V. 1007
Dischler, B., *see* Schneider, J. 184
DiVincenzo, D.P., *see* Baranger, H.U. 910, 911, 915–918
Djurado, D. 590
Dmowski, L. 177, 187, 188
Dobers, M. 1020
Dobson, P.J., *see* Joyce, B.A. 306–309
Dobson, P.J., *see* Larsen, P.K. 302, 319
Dobson, P.J., *see* Neave, J.H. 307, 308
Döhler, G.H. 820, 855
Döhler, G.H., *see* Tsu, R. 855
Dohmen, R., *see* Niedernostheide, F.J. 443
Dolan, G.J. 936
Doll, G.L., *see* Tanner, D.B. 623
Dolling, G. 93, 95
Dominguez, D., *see* Guldner, Y. 986
Donnay, J.D.H. 321
Donohue, J. 83, 84
Dorda, G., *see* von Klitzing, K. 953, 979, 985, 986
Döring, W., *see* Becker, R.S. 461
Dorn, R., *see* Ibach, H. 350
Dorsinville, R. 615
Dorsinville, R., *see* Walser, A.D. 613, 614
Dos Passos, W., *see* Oh, J.E. 326
Dose, V. 312
Döttling, R. 435

Dow, J.D. 313, 356
Dow, J.D., *see* Allen, R.E. 356
Dow, J.D., *see* Beres, R.P. 316
Dow, J.D., *see* Buisson, J.P. 118
Dow, J.D., *see* Hjalmarson, H.P. 117, 118
Dow, J.D., *see* Lederman, F.L. 837, 843
Do, Xuan Than, *see* Barrau, J. 850, 857
Dragoset, R.A., *see* Whitman, L.J. 305, 350
Drasdo, D., *see* Schöll, E. 432, 440, 443, 444
Drathen, P. 298, 302
Dreicer, H. 706
Dresden, M. 1090
Dresselhaus, G. 500–502, 504, 505, 643
Drube, W. 317
Druy, M.A., *see* Epstein, A.J. 622
Dubenskii, K.K. 181
Dubey, G.S., *see* Adams, D.J. 1067, 1069
Dubois, J.G.A., *see* Hendriks, P. 432
Dugdale 725
Duggan, G. 385
Duggan, G., *see* Nicholas, R.J. 832
Duggan, G., *see* Pulsford, N.J. 825
Duke, C.B. 289, 293, 294, 302, 313, 314, 316, 322, 344, 345, 359, 360, 373, 643, 645, 650, 661, 662
Duke, C.B., *see* Ben-Daniel, D.J. 637, 826
Duke, C.B., *see* Bennett, A.J. 656
Duke, C.B., *see* Canter, K.F. 293
Duke, C.B., *see* Ford, W.K. 323
Duke, C.B., *see* Holland, B.W. 297
Duke, C.B., *see* Kahn, A. 323
Duke, C.B., *see* Lubinsky, A.R. 302
Duke, C.B., *see* Mailhiot, C. 314, 347, 362, 384
Duke, C.B., *see* Scifres, D.R. 179
Dumke, W.P. 238
Dunlavy, D.J., *see* Keyes, B.M. 238
Dupuis, N., *see* Soucail, B. 833, 838, 840, 850, 851, 857
Dupuis, R.D., *see* Scifres, D.R. 179
Duraffourg, B., *see* Bernard, M.G.A. 232
Duran, J.C., *see* Flores, F. 347
Durbin, S.M. 380
Dyakin, V.V., *see* Karyagin, V.V. 764, 778
Dynes, R.C., *see* Bishop, D.J. 809
Dziesiaty, J., *see* Boyn, R. 175
Dziuba, R.F., *see* Cage, M.E. 999
Dzyaloshinski, I.E., *see* Abrikosov, A.A. 5, 10, 19, 26

Eagle, G., *see* Orenstein, J. 615
Eagles, D.M. 171
Eagles, P.M. 541, 562
Eaglesham, D.J. 378
Eastman, D.E. 311, 312
Eastman, D.E., *see* Grobman, W.D. 311
Eastman, D.E., *see* Gudat, W. 315
Eastman, D.E., *see* Himpsel, F.J. 319
Eastman, D.E., *see* Knapp, J.A. 316
Eastman, L.F., *see* Lester, L.F. 679
Eastman, L.F., *see* Okamoto, K. 326
Eastman, L.F., *see* Tadayon, B. 326
Eastman, L.F., *see* Welch, D.F. 673
Eastwood, J.W., *see* Hockney, R.W. 1063
Eaves, L., *see* Alves, E.S. 930
Eaves, L., *see* Beton, P.H. 745, 842, 888, 889
Eaves, L., *see* Chamberlain, J.M. 867, 875
Eaves, L., *see* Leadbeater, M.L. 655
Eaves, L., *see* Main, P.C. 888, 911
Eaves, L., *see* Sheard, F.W. 871, 881
Ebeling, W. 421
Eberhardt, W., *see* Colbow, K.M. 389
Ebert, G. 955, 986, 998, 999, 1025
Echenique, P.M. 658
Echenique, P.M., *see* Anderson, S. 303
Eckhardt, H. 605
Economou, E.N. 884, 904, 912, 932
Economou, E.N., *see* Papaconstantopoulos, D.A. 118
Edelman, V.S., *see* Pudalov, V.M. 984, 989
Edwall, D.D., *see* Kroemer, H. 382
Edwards, D.M., *see* Eagles, P.M. 541, 562
Edwards, J.T. 943
Efetov, K.B. 867, 903, 951, 959
Efros, A.L. 896, 935
Efros, A.L., *see* Shklovskii, B.I. 254, 867
Ehinger, K. 621
Ehrenberg, W. 1110
Ehrenfreund, E., *see* Flood, J.D. 606
Ehrenfreund, E., *see* Vardeny, Z. 607
Ehrenfreund, E., *see* Weinberger, B.R. 602
Ehrenreich, H. 544, 576
Einstein, T.L., *see* Wang, X.S. 304
Eisenstein, J.P. 984, 1019, 1020, 1027, 1028
Eisenstein, J.P., *see* Pfeiffer, L.N. 874
Eisenstein, J.P., *see* Willett, R. 999, 1000, 1026
Eizenberg, M., *see* Heiblum, M. 385
Ejder, E. 177, 178
Ekenberg, U. 673, 674
Ekenberg, U., *see* Altarelli, M. 675
Ekenberg, U., *see* Wharam, D.A. 903
Eklund, P.C., *see* Tanner, D.B. 623
Elcess, K., *see* Laurich, B.K. 825
Elconin, M.H., *see* van Hove, M.A. 303
Elesin, V.F., *see* Degtyarenko, N.N. 433
Eliashberg, G.M., *see* Bychkov, Yu.A. 1004
Elkin, E.L. 184
Elliott, M., *see* Shen, T.-H. 398
Elliott, R.J. 867, 940
Elsenbaumer, R.L., *see* Baughman, R.H. 592, 619
El'tsov, K.N., *see* Altukhov, P.D. 455, 476, 478, 479
Emanuel, M.A., *see* Kolodzey, J. 432
Emery, M., *see* Daniels, M.E. 701
Emery, M., *see* Gupta, R. 716
Emin, D. 619
Emunds, A., *see* Thoma, R. 426

Emura, S. 714
Enderlein, R. 690
Enderlein, R., *see* Bechstedt, F. 698
Enderlein, R., *see* Bonch-Bruevich, V.L. 867
Engel, G.E. 89
Engel, L.W. 956, 1020
Engel, L.W., *see* Sajoto, T. 1000
Engl, W.L., *see* Thoma, R. 426
Englert, T. 985
English, J.H., *see* Eisenstein, J.P. 1027
English, J.H., *see* Pinczuk, A. 1024
English, J.H., *see* Willett, R. 999, 1000, 1015, 1026
Engström, O. 205, 211, 231, 245, 248, 262, 264
Engström, O., *see* Andersson, G.I. 229
Engström, O., *see* Andersson, M.O. 205, 245, 247, 262, 264
Engström, O., *see* Bengtsson, S. 264
Engström, O., *see* Ricksand, A. 245, 263, 264
Ennen, H. 139, 142
Ennen, H., *see* Aszodi, G. 139, 142
Ennen, H., *see* Pomrenke, G. 139, 142
Ennen, H., *see* Wagner, J. 139, 142
Ennen, H., *see* Weber, E.R. 151, 357
Ensslin, K. 930
Eppenga, R., *see* Dekker, C. 883, 894
Eppenga, R., *see* Molenkamp, L.W. 796, 804, 805, 872, 877, 878
Epstein, A.J. 619–622
Epstein, A.J., *see* Gibson, H.W. 611
Epstein, A.J., *see* Pouget, J.P. 622
Epworth, R.W., *see* Ralls, K.S. 205, 247
Erdös, P. 867
Erginsoy, C. 553, 570
Ermanis, F., *see* Casey Jr, H.C. 187
Erskine, D., *see* Zhang, S.B. 105
Esaki, L. 638, 646, 656, 819, 820, 869
Esaki, L., *see* Bastard, G. 837
Esaki, L., *see* Brum, J.A. 831, 841
Esaki, L., *see* Chang, L.L. 646, 647, 651, 743, 823, 824, 836
Esaki, L., *see* Döhler, G.H. 855
Esaki, L., *see* Guldner, Y. 824
Esaki, L., *see* Ludeke, R. 317, 325
Esaki, L., *see* Mendez, E.E. 655, 986
Esaki, L., *see* Sai-Halasz, G.A. 822
Esaki, L., *see* Segmuller, A. 820
Esaki, L., *see* Tejedor, C. 389
Esaki, L., *see* Tsu, R. 651
Esaki, L., *see* Voisin, P. 823
Esaki, L., *see* Yajima, T. 658
Escapa, L. 884, 886
Escapa, L., *see* García, N. 884
Esipov, S.E. 698
Esposito, F.P. 792
Esser, B., *see* Bonch-Bruevich, V.L. 867
Esteve, H., *see* Geerligs, L.J. 1115
Estle, T.L., *see* de Wit, M. 184
Estle, T.L., *see* Holton, W.C. 184
Estle, T.L., *see* Schneider, J. 184
Estrup, P.J. 324
Etemad, S. 614
Etemad, S., *see* Feldblum, A. 604
Etemad, S., *see* Orenstein, J. 615
Etemad, S., *see* Rothberg, L. 615, 616
Etemad, S., *see* Weinberger, B.R. 605
Etienne, B. 462, 463
Etienne, B., *see* Barrau, J. 850, 857
Etienne, B., *see* Huant, S. 169
Etienne, B., *see* Stepniewski, R. 181
Etienne, P., *see* Massies, J. 298, 306, 325, 326
Eugster, C.C. 886
Eugster, C.C., *see* del Alamo, J.A. 886
Evans, D.A. 226
Evans, K.R., *see* Look, D.C. 368
Evans, W.H., *see* Wagman, D.D. 332
Ewald, P.P. 1069
Ewing, R.D., *see* Fair, H.D. 184

Fabricius, N., *see* von der Linde, D. 22
Fadden, C.M., *see* Sernelius, B.E. 736
Fahy, S. 68, 84
Fair, H.D. 184
Fairhurst, K.M., *see* White, A.M. 178
Faist, J. 894
Falicov, L.M. 140, 570
Falicov, L.M., *see* Cohen, M.H. 512
Falicov, L.M., *see* Visscher, P.B. 1027
Fal'ko, V. 945
Falks, R.T., *see* Brillson, L.J. 339
Fallahi, M., *see* Taylor, R.P. 883, 888, 894
Fan, H.Y., *see* Laff, R.A. 562
Fan, H.Y., *see* Taylor, W.E. 264
Fan, H.Y., *see* Thomas, S.R. 181
Fan, L.-Y., *see* Troost, D. 339
Fang, H. 930
Fano, G. 1021, 1025
Fano, V., *see* Cerrina, F. 389
Fantner, E., *see* Kuchar, F. 167
Fantoni, S. 1031
Farmer, K.R. 205, 247
Farnsworth, H.E., *see* Schlier, R.E. 296
Farrell, H.H. 382
Farrell, H.H., *see* Aspnes, D.E. 307, 320
Farrow, R.F.C. 326
Fasolino, A. 678, 822, 824
Fasolino, A., *see* Altarelli, M. 675
Fasolino, A., *see* Ancilotto, F. 831, 841
Fasolino, A., *see* Molinari, E. 678, 680
Fathy, D., *see* Goodnick, S.M. 708
Faulkner, R.A. 124, 131, 146, 165, 179
Faulkner, R.A., *see* Dean, P.J. 178
Faurie, J.P., *see* Berroir, J.M. 825
Faurie, J.P., *see* Chow, D.H. 647
Faurie, J.P., *see* Guldner, Y. 828
Fauster, Th., *see* Himpsel, F.J. 317, 319

Fawcett, W., *see* Boardman, A.D. 1063
Fazzio, A. 118, 132
Feast, W.J., *see* Bott, D.C. 619
Fedorus, G.A., *see* Gorbik, P.P. 389
Feenstra, R.M. 290, 305, 313, 322, 362, 363
Feenstra, R.M., *see* Mårtensson, P. 322, 323
Feenstra, R.M., *see* Stroscio, J.A. 305, 322
Feher, G. 499
Feher, G., *see* Wilson, D.K. 499
Feidenhans'l, R., *see* Grey, F. 324
Feigenbaum, M.J. 436
Fein, A.P., *see* Stroscio, J.A. 305, 322
Feistel, R., *see* Ebeling, W. 421
Feldblum, A. 604
Feldblum, A., *see* Epstein, A.J. 621, 622
Feldblum, A., *see* Pouget, J.P. 622
Feldman, B.J., *see* Lo, T.K. 461
Feldman, L.C., *see* Bennett, P.A. 296
Feldman, W.L., *see* Chynoweth, A.G. 658
Feng, S. 912
Feofilov, P.P., *see* Dubenskii, K.K. 181
Fermi, E. 61, 62
Fernandez, R., *see* Lewis, B.F. 309
Fernando, G.W., *see* Qian, G.X. 106
Ferreira, R. 839, 842, 843, 852, 854, 855
Ferreira, R., *see* Bastard, G. 819, 827
Ferreira, R., *see* Soucail, B. 833, 838, 840, 845, 846, 850, 851, 857
Ferry, D.K. 706, 872, 1046, 1051, 1085
Ferry, D.K., *see* Barker, J.R. 1120, 1123
Ferry, D.K., *see* Goodnick, S.M. 708
Ferry, D.K., *see* Kann, M.J. 1071
Ferry, D.K., *see* Kluksdahl, N.C. 653
Ferry, D.K., *see* Kriman, A.M. 893, 897, 1069
Ferry, D.K., *see* Lugli, P. 1066, 1068, 1069
Ferry, D.K., *see* Maracas, G.N. 432
Ferry, D.K., *see* Osman, M.A. 1071, 1072
Ferry, D.K., *see* Puechner, R.A. 930
Ferry, D.K., *see* Takagaki, Y. 918
Ferry, D.K., *see* Tsen, K.T. 712
Fertig, H. 1028
Fertig, H.A. 657
Fetter, L.A., *see* Ralls, K.S. 205, 247
Fetter, L.A., *see* Skocpol, W.J. 869, 870
Feucht, D.L., *see* Milnes, A.G. 390
Fewster, P.F. 709
Feynman, R.P. 1022
Field, B.F., *see* Cage, M.E. 999
Field, S.B. 924
Field, S.B., *see* Scott-Thomas, J.H.F. 924
Fimland, B.-O., *see* Palmstrøm, C.J. 340
Finch, M. 1109, 1112
Finch, M., *see* Barker, J.R. 1104, 1111–1113, 1116
Finlayson, D.M. 868
Finstad, T.G., *see* Palmstrøm, C.J. 327
Firment, L.E. 303
Fischer, J.E., *see* Djurado, D. 590
Fischer, R., *see* Colvard, C. 678
Fischetti, M.V. 1124
Fisher, D.S. 884, 904, 912, 932
Fisher, D.S., *see* Anderson, P.W. 932, 956
Fisher, J.R., *see* Burkey, B.C. 177, 188
Fisher, P., *see* Jones, R.L. 166, 167
Fisher, P., *see* Onton, A. 166
Fisher, P., *see* Reuszer, J.K. 164
Fishman, I.M., *see* Asnin, V.M. 469, 471
Flepp, L., *see* Ospelt, M. 355
Fletcher, K. 544, 547, 548, 556
Fletcher, R. 738, 740, 754, 756, 757, 760, 761, 770–773, 775, 776, 778–781
Fletcher, R., *see* D'Iorio, M. 761, 770, 780
Fletcher, R.J. 519
Flodström, S.A., *see* Magnusson, K.O. 317
Flodstrom, S.A., *see* Uhrberg, R.I.G. 319
Flood, J.D. 606
Flood, J.D., *see* Chung, T.-C. 619
Flores, F. 347, 390
Flores, F., *see* Echenique, P.M. 658
Flores, F., *see* Tejedor, C. 326, 346, 347
Florez, L.T., *see* Aspnes, D.E. 307, 320
Florez, L.T., *see* Kash, K. 875
Florez, L.T., *see* Marshall, E.D. 337, 380
Florez, L.T., *see* Palmstrøm, C.J. 327
Florez, L.T., *see* Sands, T. 325, 327
Florez, L.T., *see* Yablonovitch, E. 339
Fock, V. 922, 982
Fockele, M. 150
Fofana, B., *see* Bensaid, B. 238
Foiles, C.L., *see* Blatt, F.J. 747
Föll, H., *see* Schmid, P.E. 335, 358
Fonstad, C.G., *see* Bagwell, P.F. 893
Fonstad, C.G., *see* Laurich, B.K. 825
Foot, P.J.S., *see* Bott, D.C. 619
Forbes, L., *see* Sah, C.T. 177, 189, 220, 229, 241, 247
Ford, C.J.B. 867, 883, 894, 904, 906, 909–911, 919–921, 928, 1104, 1109, 1111, 1112, 1115, 1116
Ford, R.A., *see* Turberfield, A.J. 1006
Ford, W.K. 323
Ford, W.K., *see* Duke, C.B. 322
Foreman, B.A., *see* Lester, L.F. 679
Fork, R.L., *see* Becker, P.C. 1070
Fornari, R. 357
Forrest, S.R. 385, 389
Fortin, S., *see* Taylor, R.P. 883, 888, 894
Foulon, Y. 130, 145
Fowell, A.E. 365
Fowler, A.B. 735, 809, 938, 939, 957
Fowler, A.B., *see* Ando, T. 479, 481, 667, 698, 708, 728, 732, 736, 869, 979
Fowler, A.B., *see* Timp, G. 809, 813
Fowler, A.B., *see* Washburn, S. 998
Fowler, R.H. 629
Fowler, W.B., *see* Deleo, G.G. 132
Foxman, E.B., *see* McEuen, P.L. 924

Foxon, C.T. 710
Foxon, C.T., *see* Clark, R.G. 999, 1015, 1018–1020, 1025
Foxon, C.T., *see* Dekker, C. 883, 894
Foxon, C.T., *see* Fletcher, R. 738, 740, 772
Foxon, C.T., *see* Ford, C.J.B. 904, 906, 1104, 1109, 1111, 1112, 1115, 1116
Foxon, C.T., *see* Kouwenhoven, L.P. 744, 872, 888, 889, 899, 900, 921, 923, 926, 927
Foxon, C.T., *see* Mallett, J.R. 1006
Foxon, C.T., *see* Molenkamp, L.W. 796, 804–806, 872, 877, 878
Foxon, C.T., *see* Nicholas, R.J. 832
Foxon, C.T., *see* Pulsford, N.J. 825
Foxon, C.T., *see* Staring, A.A.M. 888, 921, 1115
Foxon, C.T., *see* Turberfield, A.J. 1006
Foxon, C.T., *see* van Houten, H. 806–808, 872, 876, 879, 902, 904, 906, 911
Foxon, C.T., *see* van Loosdrecht, P.H.M. 904
Foxon, C.T., *see* van Wees, B.J. 800, 801, 867, 870, 871, 881, 882, 886, 902–904, 906, 913, 919, 921, 924, 997, 1097, 1104
Foxon, C.T., *see* Williamson, J.G. 880
Foxton, C.T., *see* Fletcher, R. 740, 778, 780, 781
Foy, P.W., *see* Capasso, F. 397, 399
Foyt, A.G. 188
Fragnito, H.L., *see* Becker, P.C. 1070
Franciosi, A. 338
Franciosi, A., *see* Sorba, L. 398, 399
Franciosi, A., *see* Yu, X. 386, 389
Frank, D.J., *see* Laux, S.E. 872
Frank, W. 272
Franz, W. 524, 639, 640, 656, 1110
Franz, W., *see* Homelius, J. 642
Frazier, G.F., *see* Randall, J.N. 874, 875
Fredkin, D.R., *see* Wannier, G.H. 510
Freeouf, J.L. 358, 359, 390
Freeouf, J.L., *see* Eastman, D.E. 312
Freeouf, J.L., *see* Grobman, W.D. 311
Freeouf, J.L., *see* Woodall, J.M. 336, 358, 381
Frenkel, D., *see* Ciccotti, G. 105
Frenkel, J. 629
Frenken, J.W.M., *see* van Loenen, E.J. 331
Frensley, W.R. 390, 653, 1090
Frensley, W.R., *see* Randall, J.N. 874, 875
Frensley, W.R., *see* Reed, M.A. 925
Friday, W., *see* Pashley, M.D. 299, 302
Friedel, J. 119
Friedman, D.J. 334
Friedman, L.R. 867
Friedmann, L. 744
Friedmann, L., *see* Tao, Z. 743, 744
Friend, R.H., *see* Bott, D.C. 619
Friend, R.H., *see* Bradley, D.D.C. 607
Friend, R.H., *see* Burroughes, J.H. 606
Fritsche, L., *see* Chester, M. 657
Fritz, J., *see* Osbourn, G.C. 378
Fritzsche, H., *see* Tiemann, J.J. 644
Fröhlich, H. 1043, 1051
Frohne, R. 1104
Froitzheim, H. 328
Fromhold, T.M. 762, 774–778
Fromhold, T.M., *see* Karavolas, V.C. 728, 733
Fromhold, T.M., *see* Oxley, J.P. 733, 754, 756, 758, 761, 768, 769, 774, 778
Fromhold, T.M., *see* Qin, G. 777, 778
Fronc, K., *see* Godlewski, M. 432
Frost, J.E.F., *see* Brown, R.J. 886, 899, 902, 904, 906, 907, 923, 924
Frost, J.E.F., *see* Cobden, D.H. 883, 892, 894
Frost, J.E.F., *see* Davies, A.G. 1019
Frost, J.E.F., *see* Ford, C.J.B. 867, 904, 906, 909, 911
Frost, J.E.F., *see* Greene, S.K. 916
Frost, J.E.F., *see* Kelly, M.J. 899, 902
Frost, J.E.F., *see* Patel, N.K. 899–901
Frost, J.E.F., *see* Smith, C.G. 659, 880, 889, 890, 923, 924, 930
Frost, J.E.F., *see* Tewordt, M. 922, 923
Frost, J.E.F., *see* Wharam, D.A. 800, 801, 867, 870, 871, 881, 882, 888, 902–904, 913, 923, 924, 1097, 1102, 1104
Frova, A. 232, 656
Froyen, S. 87
Froyen, S., *see* Louie, S.G. 14, 79
Fu, H.S., *see* Sah, C.T. 189
Fuchs, R. 688
Fuji, K. 432
Fujimoto, J.G., *see* Lin, W.Z. 1071
Fujimoto, J.G., *see* Schoenlein, R.W. 1071
Fujitani, H. 356
Fukai, Y.K., *see* Yamada, S. 875
Fukano, A., *see* Wakabayashi, J. 1007
Fukase, T., *see* Wakabayashi, J. 1007
Fukui, T. 875
Fukui, T., *see* Chang, S.S. 874
Fukui, T., *see* Yamada, S. 875
Fukuyama, H. 937, 946, 1001
Fukuyama, H., *see* Ando, T. 868
Fukuyama, H., *see* Lee, P.A. 790, 791, 934
Fukuyama, H., *see* Nagaoka, Y. 867
Fuller, C.S., *see* Reiss, H. 271
Funshor, R.L., *see* Qiu, J. 380, 396
Furmanov, V.A., *see* Degtyarenko, N.N. 433
Furneaux, J.E. 1020

Gabbe, D.R., *see* Chen, C.Y. 934
Gaillard, S., *see* Voisin, P. 850
Gaines, J.M., *see* Tsuchiya, M. 874
Gaines, J.W., *see* Pashley, M.D. 300, 301, 374
Gajewska, M., *see* Godlewski, M. 432
Gallagher, B.L. 732, 753, 754, 757, 759, 787, 793–795, 811, 812
Gallagher, B.L., *see* Fromhold, T.M. 762, 774–778

Gallagher, B.L., *see* Galloway, T. 791, 793
Gallagher, B.L., *see* Hartland, A. 882
Gallagher, B.L., *see* Howson, M.A. 750, 783, 785
Gallagher, B.L., *see* Karavolas, V.C. 728, 733
Gallagher, B.L., *see* Oxley, J.P. 733, 754, 756, 758, 761, 768, 769, 774, 778
Gallagher, B.L., *see* Qin, G. 777, 778
Galli, G. 106
Galli, G., *see* Ballone, P. 106
Galloway, T. 791, 793
Galloway, T., *see* Gallagher, B.L. 732, 753, 754, 759, 787, 793–795
Galloway, T., *see* Hartland, A. 882
Galloway, T., *see* Oxley, J.P. 768, 769, 774, 778
Galtier, P., *see* Gołdys, E. 165
Gammon, D. 825
Gammon, D., *see* Shanabrook, B.V. 825
Gamo, K., *see* Takagaki, Y. 867, 906, 908, 911
Gamow, G. 629, 634, 652
Gandit, P., *see* Pannetier, B. 937, 948
Gantmakher, V.F. 568, 571, 573, 575, 577
Gao, J.A., *see* Caro, J. 938, 958
Gao, Y., *see* Colbow, K.M. 389
Garcia, A., *see* Corkill, J.L. 85
García, N. 878, 884
García, N., *see* de Raedt, H. 897
García, N., *see* Escapa, L. 884, 886
García, N., *see* Sáenz, J.J. 878
Garner, C.M., *see* Lindau, I. 356
Garner, C.M., *see* Spicer, W.E. 349
Garrison, K.C., *see* Palmstrøm, C.J. 340
Gatos, H.C. 309, 310, 312, 362
Gatos, H.C., *see* Kamińska, M. 151, 170
Gauneau, M., *see* Lambert, B. 139, 142
Gautard, D., *see* Makram-Ebeid, S. 351, 371
Geballe, T.H., *see* Puri, S.M. 566
Geerligs, L.J. 1115
Gefen, Y. 1104, 1109
Geisel, T., *see* Brandl, A. 432
Geldart, D.J.W. 25, 26, 30
Geldart, D.J.W., *see* Rasolt, M. 27, 30
Gell, M.A. 832
Gell, M.A., *see* Jaros, M. 674, 832
Gell, M.A., *see* Ninno, D. 832
Genoud, F., *see* Nechtschein, M. 608, 612
Gerber, Ch., *see* Binnig, G. 296, 362
Gere, E.A., *see* Feher, G. 499
Gerecke, H. 682
Gerecke, H., *see* Bechstedt, F. 681, 686, 687
Gerhardts, R.R. 768, 922, 930
Gerl, M., *see* Kauffer, E. 118
Germano, F.A., *see* Frova, A. 656
Gershenzon, E.M. 174, 175
Gershoni, D., *see* Pfeiffer, L.N. 874
Gershonzon, M., *see* Thomas, D.G. 180
Geurts, J., *see* Zahn, D.T.R. 380
Ghaisas, S.V., *see* Madhukar, A. 307
Ghosal, A., *see* Babiker, M. 697, 699
Gibala, R., *see* Oh, J.E. 326
Gibart, P., *see* Fockele, M. 150
Gibart, P., *see* von Bardeleben, H.J. 150
Gibb, K., *see* Soucail, B. 833, 838, 840, 850, 851, 857
Gibb, R.P. 239
Gibbings, C.J., *see* Gallagher, B.L. 753, 757, 811, 812
Gibson, H.W. 611
Gibson, H.W., *see* Epstein, A.J. 619–622
Gibson, H.W., *see* Pouget, J.P. 622
Gibson, J.M., *see* Tung, R.T. 324, 325, 355
Giessen, B.C., *see* Chang, L.L. 868, 869
Gijs, M. 937, 948
Gilchrist, H.L., *see* Palmstrøm, C.J. 327, 340, 374, 375
Gilchrist, H.L., *see* Sands, T. 325, 327
Giles, P.L., *see* Anderson, D.A. 703
Gillin, W.P., *see* Tang, Y.S. 139, 142
Girvin, S.M. 764, 1022, 1023, 1029, 1031
Girvin, S.M., *see* Cage, M.E. 999
Girvin, S.M., *see* Jonson, M. 762, 764–766
Girvin, S.M., *see* Lam, P.K. 1006
Girvin, S.M., *see* MacDonald, A.H. 1024, 1029
Girvin, S.M., *see* Prange, R.E. 869, 881, 904, 913, 979
Girvin, S.M., *see* Yoshioka, D. 1028
Giuliani, G.F. 992
Glaser, E. 150
Glaser, E., *see* Kennedy, A. 150
Glazman, L.I. 658, 872, 886–888, 899, 900, 902, 903, 906, 917, 924
Glicksman, M. 422, 473
Glyde, H.R., *see* Sa-Yakanit, V. 256–258
Glyde, H.R., *see* Sritrakool, W. 257
Gmelin, E., *see* Ruf, C. 730, 731, 741, 742, 760, 762, 763, 767
Gmitter, T.J., *see* Yablonovitch, E. 339, 368, 370
Gobeli, G.W., *see* Allen, F.G. 312
Godby, R.W. 3, 4, 19, 22, 116
Goddard III, W.A., *see* Chang, R. 314
Goddard III, W.A., *see* Kahn, L.R. 65
Goddard III, W.A., *see* Melius, C.F. 65
Goddard III, W.A., *see* Swarts, C.A. 314
Godlewski, M. 139, 142, 432
Godlewski, M., *see* Cieplak, M.Z. 184
Godlewski, M., *see* Kamińska, M. 173
Godlewski, M., *see* Przybylinska, H. 139, 142
Godz, A.S., *see* Kash, K. 875
Goel, A., *see* Wheeler, R.G. 869, 870, 958
Goetzberger, A., *see* Deuling, H. 264
Gogolin, A.A. 867, 935
Goguenheim, D. 264
Gold, L. 561
Goldberg, B.B. 1006
Goldberg, B.B., *see* Smith III, T.P. 984
Goldberg, C. 561, 562
Goldberg, J.L., *see* Wang, X.S. 304

Golding, D.T., *see* Zahn, D.T.R. 380
Goldman, V.J. 656, 1007
Goldman, V.J., *see* Su, B. 659
Goldstein, L., *see* Quillec, M. 820
Gołdys, E. 165
Golovchenko, J.A. 295
Golovchenko, J.A., *see* Becker, R.S. 296
Golovchenko, J.A., *see* Bedrossian, P. 322
Golovchenko, J.A., *see* Patel, J.R. 380
Goltsman, G.N., *see* Gershenzon, E.M. 174, 175
Golubev, L.V., *see* Vul, A.Ya. 188
Gombia, E., *see* Fornari, R. 357
Gomma, N., *see* Kamieniecki, E. 264
Gonda, S., *see* Emura, S. 714
Gonsalves, J.M., *see* Li, D. 380
Gonzalves da Silva, C.E.T., *see* Voisin, P. 823
Gonze, X. 78
Goodman, B., *see* Esposito, F.P. 792
Goodman, B., *see* Serota, R.A. 793
Goodman, R.R. 519
Goodnick, S.M. 706, 708
Goodnick, S.M., *see* Lugli, P. 698
Goodnick, S.M., *see* Weisshaar, A. 896–899, 926
Goodnick, S.M., *see* Wu, J.C. 898
Goodwin, M.W., *see* Reed, M.A. 647
Goodwin, T.A. 349
Goossen, K.W., *see* Bar-Joseph, I. 850
Göpel, W., *see* Bätz, P. 623
Goradia, C. 252
Gorbik, P.P. 389
Gorbovitskii, B.M. 524
Gorczyca, I., *see* Christensen, N.E. 390
Gordon, J.P., *see* Worlock, J.M. 464
Gorkov, L.P. 945–947, 995
Gorkov, L.P., *see* Abrikosov, A.A. 5, 10, 19, 26
Gornik, E. 181, 984
Gornik, E., *see* Schuberth, G. 647, 655
Gossard, A.C., *see* Cage, M.E. 999
Gossard, A.C., *see* Colvard, C. 678
Gossard, A.C., *see* Dingle, R. 819
Gossard, A.C., *see* Eisenstein, J.P. 984, 1027
Gossard, A.C., *see* Gornik, E. 984
Gossard, A.C., *see* Miller, D.A.B. 657
Gossard, A.C., *see* Miller, R.C. 181, 837
Gossard, A.C., *see* Narayanamurti, V. 678
Gossard, A.C., *see* Paalanen, M.A. 986
Gossard, A.C., *see* Petroff, P.M. 382, 1088
Gossard, A.C., *see* Pinczuk, A. 1024
Gossard, A.C., *see* Störmer, H.L. 986, 999
Gossard, A.C., *see* Tsui, D.C. 979, 986, 999
Gossard, A.C., *see* Willett, R. 999, 1000, 1015, 1026
Gossard, A.C., *see* Yan, R.H. 850
Gossman, H.-H., *see* Eaglesham, D.J. 378
Goto, N., *see* Yano, H. 432
Gourley II, P.L., *see* Osbourn, G.C. 378
Gozzo, F., *see* Marsi, M. 399
Grabbe, P., *see* Kash, K. 875
Graf, K.H., *see* Cox, G. 305
Graham, S.C., *see* Kaiser, A.B. 623
Grahn, H.T. 694
Grahn, H.T., *see* Schneider, H. 848, 850, 854, 857
Grambow, P. 870, 872
Grambow, P., *see* Heitmann, D. 922
Grambow, P., *see* Kern, K. 930
Grambow, P., *see* Weiss, D. 922, 930
Grandpierre, G., *see* Lambert, B. 139, 142
Grant, J.T. 303
Grant, P. 613
Grant, R.W. 339, 368, 369, 385–387, 393, 394
Grant, R.W., *see* Harrison, W.A. 381, 382, 393
Grant, R.W., *see* Kowalczyk, S.P. 389
Grant, R.W., *see* Kraut, E.A. 386
Grant, R.W., *see* Waldrop, J.R. 326, 368, 369, 389, 393
Grant, T.A., *see* Colvard, C. 678
Grard, J.M., *see* Marzin, J.Y. 824, 830
Gravesteijn, D.J., *see* Zalm, P.C. 380
Gravier, Th., *see* Molenkamp, L.W. 806
Grebe, G. 181
Grecchi, V., *see* Bentosela, F. 843
Gredescul, S.A., *see* Lifshitz, I.M. 867
Green, P.D., *see* Ashen, D.J. 178, 180
Green, P.D., *see* White, A.M. 178
Greene, J.E. 339
Greene, S.K. 916
Greene, W.J., *see* Kubby, J.A. 322
Greenside, H.S. 73
Gregorkiewicz, T. 186
Gregory, P.E., *see* Spicer, W.E. 349
Greig, D., *see* Blatt, F.J. 747
Greiter, M. 1025
Grey, F. 324
Grey, F., *see* Dev, B.N. 380
Gribnikov, Z.S. 432
Gribnikov, Z.S., *see* Asche, M. 422
Griffith, J.E., *see* Wieranga, P.E. 305
Griffith, J.S. 170
Griffiths, C.H., *see* Brillson, L.J. 351
Grilli, M., *see* Castellani, C. 785
Grimes, C.G. 980
Grimmeiss, H.G. 131, 132, 175, 176, 178, 189, 190
Grimmeiss, H.G., *see* Björklund, G. 177
Grimmeiss, H.G., *see* Braun, S. 190
Grimmeiss, H.G., *see* Ejder, E. 177, 178
Grimmeiss, H.G., *see* Engström, O. 231, 245, 264
Grimmeiss, H.G., *see* Kleverman, M. 229, 230, 245
Grimmeiss, H.G., *see* Lang, D.V. 231
Grimmeiss, H.G., *see* Rees, G.J. 239
Grinberg, A.A. 233
Grioni, M. 330–332, 334
Grioni, M., *see* del Guidice, M. 334
Grobman, W.D. 311
Grondin, R.O., *see* Chamoun, S.N. 1072

Gronsky, R., *see* Lilienthal-Weber, Z. 336, 358
Groshev, A. 924
Gross, D.J., *see* Brézin, E. 984
Gross, E.K.U. 16
Gross, E.K.U., *see* Runge, E. 16, 20
Grossman, B. 465
Grosvalet, J. 1044
Grove, A.S. 262
Groves, S.H. 529
Groves, S.H., *see* Pidgeon, C.R. 521
Groves, S.H., *see* Roth, L.M. 521
Groves, W.D., *see* Scifres, D.R. 179
Grubin, H.L., *see* Shaw, M.P. 422, 440
Grundmann, M., *see* Christen, J. 874
Grunthaner, F.J., *see* Grunthaner, P.J. 336
Grunthaner, F.J., *see* Lewis, B.F. 309
Grunthaner, P.J. 336
Grynberg, M., *see* Baranowski, J.M. 171
Grynberg, M., *see* Karpierz, K. 168
Grynberg, M., *see* Muller, F. 185, 186
Grynberg, M., *see* Tuchendler, J. 167
Grynberg, M., *see* Wittlin, A. 185
Gu, B.Y., *see* Gu, L. 654
Gu, L. 654
Gu, S.-W., *see* Sun, H. 659
Guckenheimer, J. 421
Gudat, W. 312, 315
Guénault, A.M. 747
Guéret, P. 659
Guerrieri, R., *see* Baccarani, G. 426
Guglielmi, M., *see* Nechtschein, M. 608, 612
Guichar, G.M. 302, 303
Guillemot, C. 684
Guillon, F., *see* Buckthought, A. 1020
Guillon, F., *see* Sachrajda, A. 1020
Guillot, G., *see* Delerue, C. 137, 138, 140
Guinea, F. 601
Guinea, F., *see* Menendez, C. 620
Guldner, Y. 824, 828, 986
Guldner, Y., *see* Bastard, G. 819, 827, 833
Guldner, Y., *see* Berroir, J.M. 825
Guldner, Y., *see* Chang, Y.C. 825
Gummlich, H.E., *see* Busse, J.W. 181
Gundlach, A., *see* Syme, R.T. 754, 755, 785
Gundlach, K.H. 634, 650
Gunn, J.B. 500, 577, 1043
Gunnarson, O. 3, 4
Gunnarson, O., *see* Das, G.P. 356
Gunnarson, O., *see* Schönhammer, K. 12
Günnarsson, O. 84
Gunshor, R.L., *see* Li, D. 380
Guo, T., *see* Ford, W.K. 323
Gupta, R. 709, 711, 715, 716
Gupta, R., *see* Ridley, B.K. 711
Guret, P., *see* Faist, J. 894
Gurnee, M.N., *see* Glicksman, M. 473
Gurney, R.W. 629
Gusev, G.M. 795, 986
Gutman, F. 251
Gygi, F. 8, 10, 12

Haanappel, E.G. 885, 893
Haanappel, E.G., *see* van der Marel, D. 885, 893
Haarer, D., *see* Winters, H.F. 310
Haas, T.W., *see* Grant, J.T. 303
Haase, M.A. 385
Haberern, K.W. 302
Haberern, K.W., *see* Pashley, M.D. 299–302, 374
Hadley, G.R., *see* Olbright, G.R. 852
Haering, R.R. 655
Haga, T., *see* Hasagawa, J. 359
Hagan, L., *see* Martin, W.C. 139
Haight, R. 362
Haisty, R.W. 171
Hajdu, J. 979
Haken, H. 421
Haldane, F.D.M. 1003, 1011, 1021, 1022, 1026
Haldane, F.D.M., *see* Tao, R. 993, 1020, 1021
Hale, E.B. 183
Haller, E.E., *see* Held, G.A. 432
Haller, E.E., *see* Nolte, D.D. 759
Haller, E.E., *see* Teitsworth, S.W. 422, 432
Haller, E.E., *see* Westervelt, R.M. 461
Halonen, V. 1018
Halonen, V., *see* Chakraborty, T. 922
Halow, I., *see* Wagman, D.D. 332
Halperin, B.I. 257, 764, 904, 979, 992, 997, 1003, 1005, 1011, 1013, 1016, 1025
Halperin, B.I., *see* Brey, L. 922
Halperin, B.I., *see* Dempsey, J. 922
Halperin, B.I., *see* Kallin, C. 347, 1024, 1028
Halperin, B.I., *see* Morf, R. 1009, 1014, 1018
Halperin, B.I., *see* Rammal, R. 930
Halperin, B.I., *see* Rasolt, M. 3, 5, 43
Halperin, B.I., *see* Tešanović, Z. 3, 43
Halperin, B.I., *see* Yoshioka, D. 1001, 1002, 1006, 1013, 1025
Halstead, R.E. 178, 351
Halstead, R.E., *see* Foyt, A.G. 188
Ham, F.S. 170
Hamaguchi, C., *see* Yamada, K. 432
Hamann, D.R. 14, 15, 72, 73, 77, 79, 116, 239
Hamann, D.R., *see* Applebaum, J.A. 294, 313, 347
Hamann, D.R., *see* Bachelet, G.B. 72–74
Hamers, R.J. 297, 319, 321
Hamers, R.J., *see* Alerhand, O.L. 304, 321
Hamers, R.J., *see* Demuth, J.E. 290
Hamers, R.J., *see* Kohler, V.K. 321, 322, 324
Hamers, R.J., *see* Tromp, R.M. 297, 319
Hamilton, B., *see* Gibb, R.P. 239
Hamilton, E.M. 621
Hanamura, E., *see* Inoue, M. 454, 459
Hanbücken, M., *see* Le Lay, G. 324
Handler, P., *see* Frova, A. 656
Haneman, D. 303

Hangleiter, A. 229
Hangleiter, A., *see* Kröber, W. 139, 142
Hanke, W. 8, 10, 12
Hansen, J.P. 1009
Hansen, J.P., *see* Caillol, J.M. 1005, 1006, 1030
Hansen, M. 335
Hansson, G.V. 362
Hansson, G.V., *see* Bachrach, R.Z. 298, 299, 302
Hansson, G.V., *see* Mårtensson, P. 297, 319
Hansson, G.V., *see* Ni, W.-X. 369, 394
Hansson, G.V., *see* Uhrberg, R.I.G. 314, 319
Hara, S. 352
Harbison, J.P., *see* Aspnes, D.E. 307, 320
Harbison, J.P., *see* Kash, K. 875
Harbison, J.P., *see* Marshall, E.D. 337, 380
Harbison, J.P., *see* Palmstrøm, C.J. 327
Harbison, J.P., *see* Roukes, M.L. 867, 870, 872, 908, 909, 911
Harbison, J.P., *see* Sands, T. 325–327, 336, 340, 365, 366, 374, 381
Harbison, J.P., *see* Scherer, A. 870
Harbison, J.P., *see* Yablonovitch, E. 339
Harman, T.C. 567
Harmans, C.J.P.M., *see* Kouwenhoven, L.P. 744, 872, 888, 889, 899, 900, 921, 923, 924, 926, 927
Harmans, C.J.P.M., *see* Molenkamp, L.W. 872, 877, 878
Harmans, C.J.P.M., *see* van Wees, B.J. 882, 886, 903, 904, 906, 919, 921, 924, 997
Harper, P.G. 513
Harris, G.L., *see* Tadayon, B. 326
Harris, I.R. 326
Harris, J. 32
Harris, J.J. 710, 867
Harris, J.J., *see* Clark, R.G. 999, 1015, 1018–1020, 1025
Harris, J.J., *see* Fletcher, R. 738, 740, 772, 778, 780, 781
Harris, J.J., *see* Ford, C.J.B. 904, 906, 1104, 1109, 1111, 1112, 1115, 1116
Harris, J.J., *see* Foxon, C.T. 710
Harris, J.J., *see* Kouwenhoven, L.P. 899, 900
Harris, J.J., *see* Mallett, J.R. 1006
Harris, J.J., *see* Turberfield, A.J. 1006
Harris, J.J., *see* van Houten, H. 872, 876, 879, 902, 904, 906, 911
Harris, J.J., *see* van Loosdrecht, P.H.M. 904
Harris, J.J., *see* van Wees, B.J. 902–904, 906, 919, 921, 924, 997
Harris Jr, J.S., *see* Chou, S.Y. 659
Harris, J.S., *see* Kroemer, H. 382
Harrison, W.A. 62, 117, 134, 137, 294, 296, 347, 348, 381, 382, 390–393, 575
Harrison, W.A., *see* Grant, R.W. 393
Harrison, W.A., *see* Tersoff, J. 135
Harshman, P., *see* Landsberg, P.T. 202
Harstein, A., *see* Fowler, A.B. 735
Harstein, A., *see* Timp, G. 809, 813
Hartke, J.L. 242
Hartland, A. 882
Hartman, R.L., *see* Dean, P.J. 182
Hartstein, A., *see* Fowler, A.B. 938, 939, 957
Harzstein, C., *see* Sivan, U. 925
Hasagawa, J. 359
Hasagawa, Y. 322
Hasegawa, H. 359
Hasegawa, Y., *see* Hashizume, T. 322
Hashitsume, N., *see* Kubo, R. 982, 984, 989
Hashizume, T. 322
Hashizume, T., *see* Hasagawa, Y. 322
Hasko, D.G., *see* Brown, R.J. 886, 899, 902, 904, 906, 907, 923, 924
Hasko, D.G., *see* Greene, S.K. 916
Hasko, D.G., *see* Kelly, M.J. 899, 902
Hasko, D.G., *see* Smith, C.G. 659, 880, 889, 890, 923, 924, 930
Hasko, D.G., *see* Wharam, D.A. 800, 801, 867, 870, 871, 881, 882, 888, 902–904, 913, 923, 924, 1097, 1102, 1104
Hass, K.C., *see* Liu, A.Y. 101
Hauenstein, R.J. 355
Haug, H. 422
Haug, R.J. 872–874, 926, 927, 998, 1018
Haugstad, G., *see* Yu, X. 386, 389
Haukness, B.S., *see* Kriman, A.M. 893, 897
Haupt, R., *see* Wendler, L. 689
Hauser, J.R., *see* Chatterjea, A. 270
Havlova, H. 769, 770
Haydl, W., *see* Pomrenke, G. 139, 142
Hayes, J.R., *see* Levi, A.F.J. 660, 707
Haynes, J.R. 454, 469
Haynes, S.R., *see* Clark, R.G. 1015, 1018–1020
Haynes, S.R., *see* Turberfield, A.J. 1006
Haywood, S.K., *see* Warburton, R.J. 824
He, L., *see* Hasagawa, J. 359
He, S. 895
He, S., *see* Fertig, H.A. 657
He, Song, *see* Eisenstein, J.P. 1028
Headrick, R.L., *see* Ma, Y. 322
Heasell, E.L., *see* Sharan, R. 170
Heasman, K.C., *see* Tang, Y.S. 139, 142
Hecht, M.H. 312, 372
Hecht, M.H., *see* Bell, L.D. 365
Hedin, L. 17–19
Hedin, L., *see* von Barth, U. 84
Heeger, A.J. 586, 595, 598, 605, 607
Heeger, A.J., *see* Cao, Y. 591, 617
Heeger, A.J., *see* Chung, T.-C. 607, 619
Heeger, A.J., *see* Epstein, A.J. 622
Heeger, A.J., *see* Etemad, S. 614
Heeger, A.J., *see* Feldblum, A. 604
Heeger, A.J., *see* Flood, J.D. 606
Heeger, A.J., *see* Mo, Z. 607
Heeger, A.J., *see* Moraes, F. 602, 607
Heeger, A.J., *see* Moses, D. 609, 619
Heeger, A.J., *see* Sinclair, M. 612, 615, 616

Heeger, A.J., *see* Su, W.-P. 599, 605, 609
Heeger, A.J., *see* Vardeny, Z. 607
Heeger, A.J., *see* Weinberger, B.R. 602
Heiblum, M. 385, 660, 707, 872, 880
Heiblum, M., *see* Meirav, U. 875, 924
Heiblum, M., *see* Sivan, U. 659, 872, 879, 880
Heiblum, M., *see* Smith III, T.P. 984
Heijman, M.G.J., *see* van Houten, H. 870
Heilan, G., *see* Kohl, D. 302
Heiman, D., *see* Goldberg, B.B. 1006
Heiman, D., *see* Pinczuk, A. 1024
Heimann, R.B. 309, 310
Heine, V. 6, 8, 71, 91, 346
Heine, V., *see* Animalu, A.D.E. 6, 8
Heine, V., *see* Austin, B.J. 6, 8, 62
Heine, V., *see* Cohen, M.H. 6, 8, 62, 64
Heine, V., *see* Cohen, M.L. 61, 65, 71, 130
Heine, V., *see* Littlewood, P.B. 57
Heinke, W. 178
Heinrich, H., *see* Langer, J.M. 135, 136, 390
Heinrich, V.E. 303
Heinz, G., *see* Richter, R. 438
Heitmann, D. 922
Heitmann, D., *see* Grambow, P. 870, 872
Heitmann, D., *see* Kern, K. 930
Hekking, F.W.J., *see* Kouwenhoven, L.P. 744, 872, 926, 927
Held, G.A. 432
Heller, G.S., *see* Rauch, F.J. 499
Heller, G.S., *see* Stickler, T.J. 519
Helmholtz, L., *see* Wolfsberg, M. 141
Hemenger, P.M., *see* Peterson, T.L. 553, 554, 563
Hemstreet, L.A. 132, 143
Hendriks, P. 432
Henini, M., *see* Alves, E.S. 930
Henini, M., *see* Beton, P.H. 888, 889
Henini, M., *see* Leadbeater, M.L. 655
Henini, M., *see* Main, P.C. 888
Henneberger, K., *see* Wünsche, H.-J. 475
Henning, J.C.M. 144
Henry, C.H. 156, 182, 189, 209, 245, 248
Henry, C.H., *see* Dingle, R. 819
Henry, C.H., *see* Kukimoto, H. 189
Henry, J.E., *see* Becker, P.C. 1070
Hensel, J.C. 451, 499, 519
Henz, J., *see* Ospelt, M. 355
Henzler, M. 293, 300
Henzler, M., *see* Kohl, D. 302
Herbert, D.C. 944
Herbert, D.C., *see* Rorison, J.M. 707
Herbst, J.F. 139, 140, 142, 143
Heremans, J.J., *see* Li, Y.P. 883, 894
Herman, F. 140, 141
Herman III, J.M. 190
Herndon, R.C., *see* Erdös, P. 867
Herrenden-Harker, W., *see* Mackay, K.J. 380
Herring, C. 64, 493, 503, 537, 539–541, 566, 570, 576, 749
Herring, C., *see* Pearson, G.L. 562, 563
Herzog, A.H., *see* Scifres, D.R. 179
Heskett, D., *see* Wong, T.M. 322
Heslinga, D.R. 324, 356
Hesnault, J.C., *see* Sibille, A. 855
Hess, K. 432
Hess, K., *see* Bailey, D.W. 1071
Hess, K., *see* Kolodzey, J. 432
Hess, K., *see* Laidig, W.D. 380
Hess, K., *see* Mason, B.A. 1124
Hess, K., *see* Pevzner, V. 1123
Hess, K., *see* Sols, F. 897, 898
Hess, K., *see* Stanton, C.J. 706
Hewitt, J., *see* Foxon, C.T. 710
Hey, R., *see* Däweritz, L. 297, 298
Heymann, G., *see* Ludwig, M. 347
Hibma, T., *see* Heslinga, D.R. 324, 356
Hickmott, T.W. 655
Hiesinger, P. 351
Higashi, G.S. 369
Higashi, G.S., *see* Becker, R.S. 296
Higman, T.K., *see* Kolodzey, J. 432
Hikami, S. 933, 947, 948, 953–955, 994, 996
Hildebrand, O. 273
Hiley, B.J., *see* Philippidis, C. 1118
Hill, D.M., *see* Trafas, B.M. 323
Hill, G., *see* Leadbeater, M.L. 655
Hill, G., *see* Poole, D.A. 735, 736, 738
Hill, G., *see* Smith, C.G. 889, 890, 930
Hilton, D., *see* Foxon, C.T. 710
Himpsel, F.J. 312, 315–317, 319, 362
Himpsel, F.J., *see* Drube, W. 317
Himpsel, F.J., *see* Kaxiras, E. 317
Himpsel, F.J., *see* Magnusson, K.O. 317
Himpsel, F.J., *see* Straub, D. 316
Hirai, H., *see* Komiyama, S. 919
Hirakawa, K. 656, 701
Hirakawa, K., *see* Sakaki, H. 708, 709, 998
Hirakawa, K., *see* Wakabayashi, J. 1007
Hiraki, A. 336
Hirayama, Y. 870, 879, 886, 888–890, 911
Hirayama, Y., *see* Nakata, S. 870
Hirayama, Y., *see* Tarucha, S. 922, 923
Hirlimann, C., *see* Shank, C.V. 22
Hirose, K. 366
Hirtz, J.P., *see* Guldner, Y. 986
Hiyamizu, S., *see* Komiyama, S. 999
Hjalmarson, H.P. 117, 118
Ho, K.M., *see* Louie, S.G. 81
Ho, K.M., *see* Schlüter, M. 65
Ho, P.S., *see* Liehr, M. 355
Ho, P.S., *see* Schmid, P.E. 335, 358
Hockney, R.W. 1063
Hoenstra, J.J., *see* Aten, A.C. 179
Hoeser, I., *see* Weispfenning, M. 428
Hoeven, J.J. 380
Hoffman, D.M., *see* Epstein, A.J. 619, 620
Hoffmann, D.M. 351

Hoffmann, H.J. 211, 212
Hofmann, D.M., *see* Meyer, B.K. 152, 153, 155
Hofmann, D.M., *see* Spaeth, J.-M. 154, 156
Hofstadter, D. 513, 929
Hohenberg, P. 4, 12, 13, 19, 72, 115
Hohl, D., *see* Jones, R.O. 106
Hökelek, E. 353
Holczer, K., *see* Nechtschein, M. 608, 612
Hollan, L., *see* Mircea, A. 151
Holland, B.W. 297
Holland, M.G. 188
Holland, P.R., *see* Vigier, J.P. 1118
Holmes, A.B., *see* Burroughes, J.H. 606
Holmes, P., *see* Guckenheimer, J. 421
Holonyak Jr, N., *see* Crawford, M.G. 177
Holonyak Jr, N., *see* Laidig, W.D. 380
Holonyak Jr, N., *see* Scifres, D.R. 179
Holstein, T. 613, 614
Holt, D.B., *see* Yacobi, B.G. 312
Holton, W.C. 184
Holton, W.C., *see* de Wit, M. 184
Holton, W.C., *see* Schneider, J. 184
Holtz, P.O., *see* Tsuchiya, M. 874
Holweg, P.A., *see* Geerligs, L.J. 1115
Holzapfel, W.B., *see* Olijnyk, H. 87
Homelius, J. 642
Hommel, D., *see* Godlewski, M. 139, 142
Hong, J.M., *see* Agullo-Rueda, F. 850, 851, 857
Hong, J.M., *see* Ford, C.J.B. 883, 894, 906, 909–911, 919–921, 928
Hong, J.M., *see* Haug, R.J. 872–874, 926, 927
Hong, J.M., *see* Mendez, E.E. 850, 857
Honig, J.M., *see* Harman, T.C. 567
Hoo, K. 187, 188
Hopfield, J.J. 179
Hopfield, J.J., *see* Henry, C.H. 182
Hopfield, J.J., *see* Thomas, D.G. 179
Hörhold, H.H. 617
Horikoshi, Y., *see* Hirayama, Y. 879, 886, 911
Horikoshi, Y., *see* Nakata, S. 870
Horn, K., *see* Alonso, M. 372
Horn, K., *see* Ibach, H. 350
Horn, K., *see* Wilke, W.G. 389
Horng, S.F., *see* Kahn, A. 349, 350
Horovitz, B. 605
Horovitz, B., *see* Bishop, A.R. 601, 616
Horsch, P., *see* von der Linden, W. 8, 10, 12, 19, 22
Hosack, H.H., *see* Davis, R.H. 651
Hott, R. 8, 10, 12, 19–22
Hotta, T., *see* Sakaki, H. 998
Hou, Y., *see* Aono, M. 297
Houghton, A., *see* Ting, C.S. 785
Houle, F.A. 310
Howard, R.E. 270
Howard, R.E., *see* Chang, A.M. 911
Howard, R.E., *see* de Vegvar, P.G.N. 791
Howard, R.E., *see* Ralls, K.S. 205, 247
Howard, R.E., *see* Skocpol, W.J. 735, 869, 870, 958
Howard, R.E., *see* Timp, G. 867, 872, 881, 904, 906, 908, 909, 911, 1114
Howard, W.E., *see* Goldberg, C. 561, 562
Howson, M.A. 750, 783, 785
Hricovini, K., *see* Le Lay, G. 356
Hsieh, J., *see* Hsieh, S. 389
Hsieh, S. 389
Hsieh, S.J., *see* Patten, E.A. 389
Hu, B.Y.-K. 1124
Hu, E.L., *see* Skocpol, W.J. 869, 870
Hu, G.W.Y., *see* Xu, F. 302
Hu, G.Y. 1125
Hu, J.Z. 87
Huam, J.M., *see* Kash, J.A. 712
Huang, H., *see* Tong, S.Y. 296
Huang, K. 681, 683
Huang, K., *see* Born, M. 575
Huant, S. 169
Huant, S., *see* Stepniewski, R. 181
Huber, A., *see* von Bardeleben, H.J. 151, 153
Huckestein, B. 953, 955, 956, 995
Huebener, R.P., *see* Mayer, K.M. 440
Huebener, R.P., *see* Peinke, J. 432
Huebener, R.P., *see* Rau, U. 433, 443, 445
Huebener, R.P., *see* Richter, R. 438
Huebener, R.P., *see* Röhricht, B. 428
Hughes, A.E., *see* Vallin, J.T. 170, 173
Hughes, G.J. 332, 353
Hughes, O.H., *see* Alves, E.S. 930
Hughes, O.H., *see* Beton, P.H. 888, 889
Hughes, O.H., *see* Leadbeater, M.L. 655
Hughes, O.H., *see* Main, P.C. 888
Huijser, A. 305, 315, 316, 350
Hunt, B.D., *see* Hauenstein, R.J. 355
Huo, Y. 995
Hüpper, G. 432, 435, 437
Hüpper, G., *see* Schöll, E. 437, 438
Hurle, D.T.J., *see* Ashen, D.J. 178, 180
Huser, J., *see* Merkt, U. 922
Hutchinson, A., *see* Beltram, F. 855
Hutchinson, W.G., *see* Chapman, R.A. 171
Huzinaga, S., *see* Sakai, Y. 65
Hvam, J., *see* Kash, J.A. 1046
Hvam, J.V., *see* Christensen, O. 463
Hwang, D.M., *see* Christen, J. 874
Hwang, D.M., *see* Yablonovitch, E. 339
Hwang, J.C.M., *see* Boebinger, G.S. 1014
Hwang, J.C.M., *see* Chang, A.M. 1000
Hwang, J.C.M., *see* Skromme, B.J. 126
Hwang, J.C.M., *see* Störmer, H.L. 999
Hwang, S.W., *see* Engel, L.W. 1020
Hwang, S.W., *see* Simmons, J.A. 890
Hwu, Y., *see* Le Lay, G. 307
Hwu, Y., *see* Marsi, M. 399
Hwu, Y., *see* McKinley, J.T. 398, 399
Hybertsen, M.S. 8, 10, 12, 19–22, 30, 81, 116

Hybertsen, M.S., *see* Northrup, J.E. 90, 91, 314
Hyodo, S., *see* Hashizume, T. 322

Iadonisi, G., *see* Altarelli, M. 144
Iadonisi, G., *see* Bassani, F. 123, 144
Ibach, H. 350
Ibuki, S., *see* Suto, K. 184
Ide, T., *see* Hashizume, T. 322
Idzerda, Y.U., *see* Lind, D.M. 326
Ignatiev, A. 303
Ignatov, A.A., *see* Belyantsev, A.M. 432
Ihm, J. 82, 83, 89, 90, 92, 323
Ihm, J., *see* Northrup, J.E. 90
Iijima, T., *see* Kawamura, T. 307
Ikoma, T., *see* Hirakawa, K. 656
Ilegems, M., *see* Dean, P.J. 178
Iller, A., *see* Dmowski, L. 177, 187, 188
Imada, H., *see* Yamada, K. 432
Imry, Y. 884, 886, 887, 912, 959
Imry, Y., *see* Büttiker, M. 932, 1099
Imry, Y., *see* Gefen, Y. 1104, 1109
Imry, Y., *see* Sivan, U. 796, 809, 911, 925
Imry, Y., *see* Yacoby, A. 887
Inkson, J.C. 349, 658
Inoue, M. 454, 459
Ioffe, A.F. 942
Ioffe, L.B. 984
Iordanskii, S.V. 993, 994
Iordanskii, S.V. 953
Iordanskii, S.V., *see* Bychkov, Yu.A. 1004
Ipotova, I.P., *see* Maradudin, A.A. 684
Ippen, E.P., *see* Lin, W.Z. 1071
Ippen, E.P., *see* Schoenlein, R.W. 1071
Iselev, G.W. 177, 188
Ishibashi, K., *see* Takagaki, Y. 867, 908, 911
Ishida, S., *see* Takagaki, Y. 867, 908, 911
Ishii, K. 867, 943
Ishizawa, Y., *see* Aono, M. 297
Iskra, V.D., *see* Samoilovich, A.G. 493, 541, 550, 563
Ismail, A. 344
Ismail, K. 659, 930
Ismail, K., *see* Liu, C.T. 930
Ito, R., *see* Kishino, S. 178
Itzykson, C., *see* Brézin, E. 984
Ivanov, A.V., *see* Altukhov, P.D. 479–481
Ivanov, T., *see* Groshev, A. 924
Iwanomoto, M. 178
Iwasaki, H., *see* Nishitani, R. 303
Izabelle, A., *see* Schubert, E.F. 703

Jackel, L.D., *see* Ralls, K.S. 205, 247
Jackel, L.D., *see* Skocpol, W.J. 735, 869, 870, 958
Jackel, M.T., *see* Rammal, R. 930
Jackson, T.N., *see* Woodall, J.M. 351, 378, 381
Jacob, J.M., *see* Jung, P.S. 840
Jacobi, K. 302, 320
Jacobi, K., *see* Drathen, P. 298, 302
Jacobi, K., *see* Ranke, W. 306, 317
Jacoboni, C. 538, 653, 1120
Jacoboni, C., *see* Bosi, S. 1066
Jacoboni, C., *see* Brunetti, R. 1090
Jacoboni, C., *see* Ferry, D.K. 1085
Jacobsen, K.W., *see* Wingreen, N.S. 656
Jäger, D. 440
Jäger, D., *see* Symanczyk, R. 440
Jain, J.K. 699, 701, 820, 904, 913
Jain, J.K., *see* McEwan, P.L. 796
Jain, J.K., *see* Timp, G. 904, 906
Jalabert, R.A. 910, 911, 918, 919
Jalabert, R.A., *see* Baranger, H.U. 910, 911, 915–918
Jalabert, R.A., *see* Jain, J.K. 699, 701
Jamieson, J.C. 85
Janowitz, C., *see* Manzke, R. 317
Janssen, J.T.M.B., *see* Patel, N.K. 899–901
Janzén, E., *see* Kleverman, M. 229, 230, 245
Janzén, E., *see* Rees, G.J. 239
Jaros, M. 118, 172, 390, 674, 832
Jaros, M., *see* Gell, M.A. 832
Jaros, M., *see* Lang, D.V. 231, 245, 249
Jaros, M., *see* Ninno, D. 832
Jaros, M., *see* Wolford, D.J. 385
Jarosik, N.C. 168
Jastrzebski, L. 277
Jauho, A.-P. 1092, 1121, 1123
Jayaraman, A., *see* Kosicki, B.B. 188
Jeanloz, R., *see* Knittle, E. 100
Jedju, T.M., *see* Rothberg, L. 615, 616
Jedrecy, N., *see* Sauvage-Simkin, M. 302
Jeffries, C.D. 451
Jeffries, C.D., *see* Held, G.A. 432
Jeffries, C.D., *see* Lo, T.K. 461
Jeffries, C.D., *see* Westervelt, R.M. 459
Jeng, S.J., *see* Coleman, J.J. 380
Jenssen, H.P., *see* Chen, C.Y. 934
Jeyadev, S. 608–613, 616
Jeyadev, S., *see* Conwell, E.M. 612, 617, 620, 623
Jezequel, G., *see* Ludeke, R. 357
Jezewski, M., *see* Knap, W. 432
Ji, Z.-L. 890
Ji, Z.-L., *see* Berggren, K.F. 896, 897, 914
Ji, Z.-L., *see* Rundquist, J. 896, 914
Jiang, H.W. 1007, 1027
Joannopoulos, J.D., *see* Alerhand, O.L. 304, 321
Joannopoulos, J.D., *see* Bar-Yam, Y. 74, 117
Joannopoulos, J.D., *see* Ihm, J. 323
Joannopoulos, J.D., *see* Mele, G. 349
Joannopoulos, J.D., *see* Payne, M.C. 314, 319
Joannopoulos, J.D., *see* Rappe, A.M. 76
Joannopoulos, J.D., *see* Shirley, E.L. 74
Joannopoulos, J.D., *see* Starkloff, T. 14, 15, 73
Joannopoulos, J.D., *see* Stone, A.D. 945, 957
Jochler, J. 466
Joe, Y.S. 893, 894, 898
Joffe, A., *see* Frenkel, J. 629

Johnson, A.T., *see* Kouwenhoven, L.P. 921
Johnson, B.L. 922, 931
Johnson, D.A., *see* Maracas, G.N. 432
Johnson, D.L. 20
Johnson, M.H. 517
Johnson, N.F. 922
Johnson, N.F., *see* Brey, L. 922
Johnson, N.F., *see* Dempsey, J. 922
Johnson, N.M. 263
Johnson, N.M., *see* Goguenheim, D. 264
Johnson, N.M., *see* Schulz, M. 264
Johnson, R.A. 270
Johnson, R.L., *see* Dev, B.N. 380
Johnson, R.L., *see* Grey, F. 324
Jona, F., *see* Yang, W.S. 297
Jones, G.A.C., *see* Brown, R.J. 886, 899, 902, 904, 906, 907, 923, 924
Jones, G.A.C., *see* Cobden, D.H. 883, 892, 894
Jones, G.A.C., *see* Davies, A.G. 1019
Jones, G.A.C., *see* Ford, C.J.B. 867, 904, 906, 909, 911
Jones, G.A.C., *see* Greene, S.K. 916
Jones, G.A.C., *see* Kelly, M.J. 899, 902
Jones, G.A.C., *see* Patel, N.K. 899–901
Jones, G.A.C., *see* Smith, C.G. 659, 880, 889, 890, 923, 924, 930
Jones, G.A.C., *see* Tewordt, M. 922, 923
Jones, G.A.C., *see* Wharam, D.A. 800, 801, 867, 870, 871, 881, 882, 888, 902–904, 913, 923, 924, 1097, 1102, 1104
Jones, K., *see* Hartland, A. 882
Jones, R., *see* Herbert, D.C. 944
Jones, R.L. 166, 167
Jones, R.O. 106, 643
Jones, R.O., *see* Harris, J. 32
Jonker, B.T., *see* Krebs, J.J. 326
Jonker, B.T., *see* Lind, D.M. 326
Jonson, M. 762, 764–766
Jonson, M., *see* Girvin, S.M. 764
Jonson, M., *see* Glazman, L.I. 887, 888, 906, 917
Jonson, M., *see* Rudberg, B.G.R. 658
José, J.V., *see* Cota, E. 934
Joshi, R.P., *see* Chamoun, Ŝ.N. 1072
Joshi, R.P., *see* Ferry, D.K. 706
Joshi, R.P., *see* Kriman, A.M. 1069
Joshi, R.P., *see* Tsen, K.T. 712
Joss, W., *see* Buhmann, H. 1006, 1007
Jou, C., *see* Lilienthal-Weber, Z. 326
Joyce, B.A. 306–309
Joyce, B.A., *see* Larsen, P.K. 302, 319
Joyce, B.A., *see* Neave, J.H. 299, 307, 308
Joyce, J.J., *see* del Guidice, M. 334
Joyce, J.J., *see* Grioni, M. 330–332, 334
Joyce, J.J., *see* Ruckman, M.W. 326, 334
Joynt, R. 993
Jung, H. 273
Jung, P.S. 840
Junge, B., *see* Ruf, C. 730, 731, 760, 762, 763, 767
Jusserand, B. 822
Justice, R.J., *see* Seiler, D.G. 432

Kachaturyan, K. 147
Käckell, P., *see* Gonze, X. 78
Kadanoff, L.P. 1092, 1121, 1122
Kafalas, J.A., *see* Iselev, G.W. 177, 188
Kagan, Y., *see* Brovman, E.G. 27
Kaganov, M.I., *see* Lifshitz, I.M. 495
Kahn, A. 292, 293, 300, 323, 349, 350
Kahn, A., *see* Duke, C.B. 322
Kahn, A., *see* Le Lay, G. 307
Kahn, A., *see* Stiles, K. 349
Kahn, L.R. 65
Kaiser, A.B. 623
Kaiser, W.J. 363, 365
Kaiser, W.J., *see* Bell, L.D. 365
Kajiyama, K. 349
Kalia, R.K., *see* Serota, R.A. 959
Kallin, C. 347, 1024, 1028
Kallio, A., *see* Puoskari, M. 1031
Kalmeyer, V. 658
Kalos, M.H. 68, 80
Kamieniecki, E. 264
Kamińska, M. 151–153, 170, 173
Kamińska, M., *see* Kachaturyan, K. 147
Kaminskii, A.S. 476
Kaminskii, A.S., *see* Pokrovskii, Ya.E. 459
Kaminura, H., *see* Sugano, S. 132
Kamiya, I., *see* Hasagawa, Y. 322
Kamiya, I., *see* Hashizume, T. 322
Kamiya, T., *see* Kuriyama, T. 238
Kanatzidis, M.G. 586, 590
Kane, E.O. 8, 10, 18, 19, 124, 256, 311, 456, 497, 501, 502, 523, 639–643, 650, 827
Kann, M.J. 1071
Kann, M.J., *see* Ferry, D.K. 706
Kann, M.J., *see* Kriman, A.M. 1069
Kanski, J., *see* Svensson, S.P. 357
Kantor, Y. 957
Kao, Y.C., *see* Randall, J.N. 874, 875
Kao, Y.C., *see* Yang, C.H. 656
Kapitulnik, A., *see* Kantor, Y. 957
Kaplan, H. 144
Kaplan, M.L., *see* Forrest, S.R. 389
Kaplan, R. 167
Kaplan, R., *see* Mrstik, B.J. 303
Kaplan, S.B. 791
Kaplan, S.B., *see* Gibson, H.W. 611
Kapon, E., *see* Christen, J. 874
Karasz, F.E., *see* Winokur, M.J. 592
Karavolas, V.C. 728–730, 733
Karavolas, V.C., *see* Oxley, J.P. 768, 769, 774, 778
Kariotis, R. 304
Kariotis, R., *see* Swartzentruber, B.S. 304
Kariotis, Y., *see* Mo, Y.-M. 321
Karlik, I.Ya., *see* Mirlin, D.N. 153
Karlsson, U.O., *see* Magnusson, K.O. 317

Karpierz, K. 168
Karris, Ya.E., *see* Dubenskii, K.K. 181
Karyagin, V.V. 764, 778
Kasani, A., *see* Iwanomoto, M. 178
Kasatkin, V.A. 139, 142
Kash, J.A. 712, 1046
Kash, J.A., *see* Tsang, J.C. 712
Kash, K. 875
Kastner, M.A., *see* Chen, C.Y. 934
Kastner, M.A., *see* Field, S.B. 924
Kastner, M.A., *see* Licini, J.C. 958
Kastner, M.A., *see* McEuen, P.L. 924
Kastner, M.A., *see* Meirav, U. 921, 924
Kastner, M.A., *see* Scott-Thomas, J.H.F. 924
Katayama-Yoshida, H. 132
Katilius, R., *see* Mirlin, D.N. 153
Katnani, A.D. 389, 391, 392, 396, 399
Katnani, A.D., *see* Brillson, L.J. 332, 333, 338, 339, 358, 366
Katnani, A.D., *see* Chiaradia, P. 319, 394, 395
Katnani, A.D., *see* Daniels, R.R. 323
Katnani, A.D., *see* Kahn, A. 323
Katnani, A.D., *see* Margaritondo, G. 396, 398
Katnani, A.D., *see* Shapira, Y. 337
Katnani, A.D., *see* Yeh, J.L. 323
Katsumoto, S. 934
Katz, A. 1073
Katz, W., *see* Sarma, K. 380
Katzer, D.C., *see* Gammon, D. 825
Kauffer, E. 118
Kaufman, E. 150
Kaufman, J.H., *see* Chung, T.-C. 607
Kaufman, J.H., *see* Feldblum, A. 604
Kaufmann, U. 154, 155, 170, 185
Kaufmann, U., *see* Aszodi, G. 139, 142
Kaufmann, U., *see* Koidl, P. 170
Kaufmann, U., *see* Weber, E.R. 151, 357
Kavanaugh, K.L., *see* Woodall, J.M. 351, 378
Kaveh, M. 934
Kaveh, M., *see* Avishai, Y. 880, 890
Kaveh, M., *see* Moses, D. 619
Kaveh, M., *see* Yosefin, M. 887
Kawabata, A. 885, 946, 948
Kawai, N.J. 307
Kawai, N.J., *see* Chang, L.L. 824
Kawai, S., *see* Nishitani, R. 303
Kawaji, S., *see* Wakabayashi, J. 985, 1007
Kawamura, T. 307
Kawasaki, K., *see* Tsukamoto, J. 617, 618
Kawase, Y., *see* Aoki, K. 437
Kaxiras, E. 317
Kaxiras, E., *see* Lyo, In-Whan 322
Kaxiras, E., *see* Rappe, A.M. 76
Kaye, R.D., *see* Philippidis, C. 1118
Kazacha, G.S., *see* Bakhvalov, N.S. 1115
Kazarinov, R.F. 855, 953, 993, 994
Kazarinov, R.F., *see* Luryi, S. 953, 993, 994
Kazmerski, L.L. 264
Kazmerski, L.L., *see* Aboie-Elfotouh, F.A. 305
Kearney, M.J. 742, 743, 783–785, 787–789, 793, 809
Kearsley, M.J., *see* Coulson, C.A. 119
Keiper, R., *see* Bonch-Bruevich, V.L. 867
Keldysh, L.V. 451–453, 456, 465, 466, 471, 472, 524, 639, 641, 642, 644, 656
Keldysh, L.V., *see* Bagaev, V.S. 460, 464, 466
Keldysh, L.V., *see* Jeffries, C.D. 451
Kelly, M.J. 678, 899, 902
Kelly, M.J., *see* Brillson, L.J. 339, 349, 372
Kelly, M.J., *see* Brown, R.J. 886, 899, 902, 904, 906, 907, 923, 924
Kelly, M.J., *see* Chiradia, P. 372
Kelly, M.J., *see* Syme, R.T. 754
Kelly, M.J., *see* Tewordt, M. 922, 923
Kelly, M.K., *see* Aspnes, D.E. 307
Kelly, M.K., *see* Davis, G.D. 332
Kelly, M.K., *see* Richter, H.W. 339
Kelly, M.K., *see* Shaw, J.L. 333, 340, 367
Kelly, M.K., *see* Turowski, M. 389
Kelly, P.J., *see* Car, R. 117
Kemble, E.C. 631
Kempa, K. 922
Kendelewicz, T. 334, 337
Kendelewicz, T., *see* Cao, R.K. 349, 358
Kendelewicz, T., *see* Miyano, K. 372
Kendelewicz, T., *see* Newman, N. 343–345
Kendelewicz, T., *see* Spicer, W.E. 356, 357
Kennedy, A. 150
Kennedy, A., *see* Glaser, E. 150
Keramidas, V.G., *see* Ding, J. 326
Keramidas, V.G., *see* Palmstrøm, C.J. 327
Keramidas, V.G., *see* Sands, T. 325–327, 336, 340, 365, 366, 374, 381
Kerker, G.P. 14, 15, 65, 72, 77
Kerker, G.P., *see* Schlüter, M. 65
Kern, D., *see* Washburn, S. 998
Kern, K. 930
Kern, K., *see* Grambow, P. 872
Kern, K., *see* Heitmann, D. 922
Kern, W. 309, 310
Kerner, B.S. 422
Kerr, T., *see* Vuong, T.H.H. 730
Kesamanly, F.P., *see* Kasatkin, V.A. 139, 142
Keyes, B.M. 238
Khaetskii, A.V., *see* Glazman, L.I. 899, 900, 902, 903
Khan, F.S. 1123
Khartsiev, V.E., *see* Altukhov, P.D. 456, 482
Khartsiev, V.E., *see* Wünsche, H.-J. 475
Khilko, G.I., *see* Dubenskii, K.K. 181
Khirouni, K., *see* Barrau, J. 850, 857
Khmel'nitskii, D.E. 995
Khmel'nitskii, D.E., *see* Altshuler, B.L. 791, 959
Khmel'nitskii, D.E., *see* Glazman, L.I. 886, 887
Khmel'nitskii, D.E., *see* Gorkov, L.P. 945–947, 995

Khmel'nitskii, D.E., *see* Lesovik, G.B. 793, 795
Khosla, R.P., *see* Burkey, B.C. 177, 188
Khoury, A., *see* Mialhe, P. 252
Khurana, A. 881
Kikner, D.J., *see* Lent, C.S. 1104, 1109
Kilday, D.G., *see* Brillson, L.J. 372
Kilday, D.G., *see* Chiradia, P. 372
Kilday, D.G., *see* Davis, G.D. 332
Kilday, D.G., *see* Kahn, A. 349, 350
Kilday, D.G., *see* Shaw, J.L. 333, 340, 367
Kilday, D.G., *see* Stiles, K. 349
Kim, B., *see* Beyers, R.K. 336
Kim, H.J., *see* Murakami, M. 337, 365, 366
Kim, O.K. 714
Kime, Y.J., *see* Chang, S. 340, 350, 372
Kimerling, L.C. 191
Kimerling, L.C., *see* Chantre, A. 250
Kimerling, L.C., *see* Miller, G.L. 191
Kimura, S., *see* Dean, P.J. 178
Kimura, T., *see* Tarucha, S. 922, 923
Kip, A.F., *see* Dresselhaus, G. 504, 505, 643
Kirchner, P.D., *see* Brillson, L.J. 339, 349, 370, 371
Kirchner, P.D., *see* Chang, S. 340, 350, 372–374
Kirchner, P.D., *see* Farrow, R.F.C. 326
Kirchner, P.D., *see* Offsey, S.D. 339, 368, 369
Kirchner, P.D., *see* Pashley, M.D. 299, 302
Kirchner, P.D., *see* Viturro, R.E. 358, 372
Kirchner, P.D., *see* Woodall, J.M. 379, 381, 382
Kirczenow, G. 476, 478, 884–886, 896, 905, 906, 908, 913–915, 917–921, 928, 930, 931, 1102–1104
Kirczenow, G., *see* Castaño, E. 872, 884, 887, 889–892, 899, 900, 902, 917, 926, 928, 929
Kirczenow, G., *see* Johnson, B.L. 922, 931
Kirczenow, G., *see* Que, W. 922
Kirczenow, G., *see* Shi, H. 912, 928
Kirczenow, G., *see* Sun, Y. 872
Kirczenow, G., *see* Ulloa, S.E. 872, 926
Kirk, W.P. 986
Kirk, W.P., *see* Reed, M.A. 1085
Kirkman, P.D. 945, 957
Kirkman, P.D., *see* Pendry, J.B. 946
Kirkner, D.J., *see* Lent, C.S. 896, 899, 900, 902
Kirkpatrick, S., *see* Thouless, D.J. 944
Kirtley, J.P. 870
Kirton, M.J. 205, 245, 247
Kirton, M.J., *see* Cobden, D.H. 245
Kishino, S. 178
Kisielowski-Kemmerich, C., *see* Hoffmann, D.M. 351
Kittel, A., *see* Rau, U. 443, 445
Kittel, A., *see* Richter, R. 438
Kittel, C. 610, 636
Kittel, C., *see* Dresselhaus, G. 504, 505, 643
Kittel, C., *see* Wand, J.S. 456
Kivelson, S. 618, 619
Kivelson, S., *see* Heeger, A.J. 586, 595, 598, 605, 607
Kivelson, S.A., *see* Jain, J.K. 904, 913
Klapwijk, T.M., *see* Heslinga, D.R. 324, 356
Klausmann, E., *see* Deuling, H. 264
Klaver, T., *see* van Hall, P.J. 703
Kleiman, G.G., *see* Scifres, D.R. 179
Klein, M.V. 678
Klein, M.V., *see* Colvard, C. 678
Klein, M.V., *see* Colwell, P.J. 182
Klein, P.B. 139, 142
Kleinman, L. 3, 4, 22, 28, 64, 78, 79
Kleinman, L., *see* Bylander, D.M. 56, 78
Kleinman, L., *see* Phillips, J.C. 6, 8, 62, 64
Klem, J., *see* Olbright, G.R. 852
Klemens, P.G. 712
Klempt, G., *see* Purwins, H.G. 440
Kleverman, M. 229, 230, 245
Kliewer, K.L., *see* Fuchs, R. 688
Klingenberger, H., *see* von der Linde, D. 712
Kloc, A., *see* Kamieniecki, E. 264
Kluksdahl, N.C. 653
Knap, W. 432
Knap, W., *see* Gregorkiewicz, T. 186
Knap, W., *see* Huant, S. 169
Knap, W., *see* Wittlin, A. 185
Knapp, J.A. 312, 316
Knittle, E. 100
Knoedler, C.M., *see* Ford, C.J.B. 883, 894, 906, 909–911, 919–921, 928
Knoedler, C.M., *see* Heiblum, M. 707
Kobayashi, K., *see* Osaka, J. 151
Kobayashi, M. 165
Kobayashi, M., *see* Li, D. 380
Kobayashi, M., *see* Mo, Z. 607
Kobayashi, M., *see* Moraes, F. 607
Kobayashi, M., *see* Qiu, J. 380, 396
Kobayashi, S. 936
Kobayashi, T., *see* Aoki, K. 422, 432
Kobiela, P.S., *see* Kirk, W.P. 986
Kocevar, P. 698
Kocevar, P., *see* Lugli, P. 698
Koch, S., *see* Müller, G. 997
Kochanski, G.P., *see* Yurke, B. 894
Kocka, J. 251
Kocot, K. 175
Koelling, D.D., *see* Norman, M.R. 3, 4
Koendens, L., *see* Troost, D. 339
Koenig, S.H. 187
Koestner, R.J., *see* Reed, M.A. 647
Kogan, Sh. M., *see* Volkov, A.F. 422
Kohjima, T., *see* Kawai, N.J. 307
Kohl, D. 302
Kohl, M., *see* Grambow, P. 870, 872
Kohler, M. 538
Kohler, V.K. 321, 322, 324
Kohler, V.K., *see* Hamers, R.J. 321
Kohmoto, M. 930
Kohmoto, M., *see* Thouless, D.J. 930, 993
Kohn, W. 4, 13, 72, 116, 123, 126, 127, 872

Kohn, W., *see* Gross, E.K.U. 16
Kohn, W., *see* Hohenberg, P. 4, 12, 13, 19, 72, 115
Kohn, W., *see* Luttinger, J.M. 123, 126, 127, 492, 508, 513, 669
Kohn, W., *see* Sham, L.J. 4, 8, 12, 14, 19
Koidl, P. 170
Koidl, P., *see* Kaufmann, U. 170
Koike, Y., *see* Wakabayashi, J. 1007
Koke, P., *see* Mönch, W. 296
Kolodzey, J. 432
Kolodziejczak, J. 548
Koltenbah, B.E.C., *see* McKinley, J.T. 398, 399
Koma, A., *see* Chang, L.L. 380
Koma, A., *see* Ludeke, R. 317
Komashchenko, V.N., *see* Gorbik, P.P. 389
Kometer, K. 423
Komiyama, S. 919, 999
Komori, F., *see* Kobayashi, S. 936
Kończykowski, M., *see* Porowski, S. 187, 188
Kondrat'ev, A.S., *see* Zelenin, S.P. 763
Kontsevoi, Yu.A., *see* Altukhov, P.D. 480–482
Kool, W., *see* Kouwenhoven, L.P. 888, 889, 923
Kopaev, Yu.V., *see* Keldysh, L.V. 451, 452, 456, 471, 472
Kopf, R.F., *see* Bar-Joseph, I. 850
Kopley, T.E. 893
Kopylov, A.A. 172, 173
Korbutyak, D.V., *see* Litovchenko, V.G. 483
Korenblit, I.Ya., *see* Samoilovich, A.G. 493, 541, 550, 563
Korn, D.M., *see* Stillman, G.E. 174
Kosevich, A.M., *see* Lifshitz, I.M. 512
Kosicki, B.B. 188
Koss, R.W. 523
Kossut, J., *see* Wittlin, A. 185
Koster, G.F. 660, 661
Kosterlitz, M., *see* Laughlin, R.B. 996
Kotthaus, J.P., *see* Lorke, A. 922, 930
Kousik, G.S., *see* Landsberg, P.T. 276
Kouwenhoven, L.P. 744, 872, 888, 889, 899, 900, 921, 923, 924, 926, 927
Kouwenhoven, L.P., *see* Staring, A.A.M. 888, 921, 1115
Kouwenhoven, L.P., *see* van Wees, B.J. 800, 801, 867, 870, 871, 881, 882, 886, 902–904, 906, 913, 919, 921, 924, 1097, 1104
Kowalczyk, S.P. 389
Kowalczyk, S.P., *see* Grant, R.W. 339, 385–387, 393, 394
Kowalczyk, S.P., *see* Kraut, E.A. 386
Kowalczyk, S.P., *see* Waldrop, J.R. 368, 389, 393
Kozlov, A.N. 453
Kozlov, A.N., *see* Keldysh, L.V. 453
Kozyrev, S.V. 687
Krambrock, K. 156
Krambrock, K., *see* Spaeth, J.-M. 154, 156
Kramer, B. 867, 868, 940, 943, 954, 955
Kramer, B., *see* Czycholl, G. 944
Kramer, B., *see* Huckestein, B. 953, 955, 956, 995
Kramer, B., *see* MacKinnon, A. 933, 952, 955
Kramer, B., *see* Mašek, J. 894, 895, 946, 959
Kramer, B., *see* Ono, Y. 954
Kramer, B., *see* Sak, J. 945, 957
Kramer, B., *see* Schreiber, M. 959
Kramer, B., *see* Schweitzer, L. 955, 994
Kraut, E.A. 386
Kraut, E.A., *see* Grant, R.W. 339, 385–387, 393, 394
Kraut, E.A., *see* Harrison, W.A. 381, 382, 393
Kraut, E.A., *see* Kowalczyk, S.P. 389
Kraut, E.A., *see* Waldrop, J.R. 368, 389, 393
Kravchenko, S.V. 1007
Kravtsov, V.E. 959
Kravtsov, V.E., *see* Altshuler, B.L. 959
Krebs, J.J. 185, 326
Krebs, J.J., *see* Lind, D.M. 326
Krebs, J.J., *see* Prinz, G.A. 325, 326
Kree, R. 945, 957
Kressel, H. 256
Krieger, J.B. 3, 4, 118, 639, 642, 643, 651
Kriegler, R.J. 264
Kriman, A.M. 893, 897, 1069
Kriman, A.M., *see* Kluksdahl, N.C. 653
Kriman, A.M., *see* Puechner, R.A. 930
Krimchansky, S., *see* Dorsinville, R. 615
Krisna, P., *see* Segmuller, A. 820
Krispin, P. 277
Krivanek, *see* Goodnick, S.M. 708
Kröber, W. 139, 142
Kroemer, H. 376, 382, 385, 390, 391, 393
Kroemer, H., *see* Caine, E.J. 389, 822
Kroemer, H., *see* Frensley, W.R. 390
Krokhin, O.N., *see* Basov, N.G. 232
Kröninger, W., *see* Brandl, A. 437, 439
Krueger, S., *see* Mönch, W. 296
Krüger, J., *see* Hoffmann, D.M. 351
Krüger, P., *see* Pollmann, J. 313, 314
Krumhansl, J.A., *see* Elliott, R.J. 867, 940
Kubakaddi, S.S. 734, 744, 771, 772
Kubaschewski, O. 332
Kubby, J.A. 322
Kubby, J.A., *see* Wieranga, P.E. 305
Kubo, R. 653, 982, 984, 989, 1048, 1082
Kubota, M., *see* Hashizume, T. 322
Kucera, J., *see* Haug, R.J. 998
Kucera, J., *see* Streda, P. 904, 913, 998
Kuchar, F. 167, 955, 956
Kuchma, A.E., *see* Zelenin, S.P. 763
Kuech, T.F., *see* Kaufman, E. 150
Kuech, T.F., *see* Wolford, D.J. 385
Kühl, H. 948
Kuhle, J., *see* von der Linde, D. 712
Kuhn, M. 262
Kuhn, T. 697
Kuk, Y., *see* Bennett, P.A. 296
Kukimoto, H. 189

Kukushkin, I.V., *see* Buhmann, H. 1006, 1007
Kumar, A. 872, 893, 895, 922, 1107
Kume, K., *see* Mizoguchi, K. 608, 612
Kunc, K. 96
Kunc, K., *see* Fasolino, A. 678
Kunc, K., *see* Molinari, E. 678, 680
Kunc, K., *see* Nielsen, O.H. 319
Kundu, S. 728, 734, 735
Kunz, A.B., *see* Scifres, D.R. 179
Kunz, C., *see* Gudat, W. 312
Kunz, H. 944, 955
Kunz, R. 440
Kuo, J.M., *see* Bar-Joseph, I. 850
Kuriyama, T. 238
Kurtin, S. 345, 346, 354
Kuryla, M.S., *see* Keyes, B.M. 238
Kurz, H., *see* Liu, J.M. 22
Kurz, H., *see* Lompre, L.A. 22
Kurz, H., *see* Rasolt, M. 22
Kusumi, Y., *see* Takagaki, Y. 911
Kuszko, W., *see* Kamińska, M. 152, 153, 170
Kuszko, W., *see* Knap, W. 432
Kuznetsov, Yu.A., *see* Altukhov, P.D. 480–482
Kvon, Z.D., *see* Gusev, G.M. 795, 986
Kyotani, H., *see* Akagi, K. 591
Kyprianidis, A., *see* Vigier, J.P. 1118

Lacelle, C., *see* Soucail, B. 833, 838, 840, 850, 851, 857
LaFemina, J. 313, 315
Laff, R.A. 562
Lagally, M.G. 300
Lagally, M.G., *see* Kariotis, R. 304
Lagally, M.G., *see* Mo, Y.-M. 321
Lagally, M.G., *see* Savage, D.E. 326
Lagally, M.G., *see* Swartzentruber, B.S. 304
Lagowski, J., *see* Gatos, H.C. 312, 362
Lagowski, J., *see* Kamińska, M. 151, 170
LaGraffe, D., *see* Shaw, J.L. 333, 351, 367, 370, 371
Laibowitz, R.B., *see* Umbach, C.P. 958
Laibowitz, R.B., *see* Webb, R.A. 948
Laidig, W.D. 380
Lal, P. 239
Lam, P.K. 1006
Lam, P.K., *see* Wentzcovitch, R.M. 87, 89
Lam, P.M. 1031
Lambert, B. 139, 142, 855
Lambert, B., *see* Chomette, A. 837, 838
Lambert, B., *see* Deveaud, B. 854
Lambert, L.M., *see* Koss, R.W. 523
Lampert, M.A. 452
Landau, L. 516, 982
Landau, L.D. 521, 631, 678
Landauer, R. 796, 884, 897, 912, 932, 951, 998, 1099
Landauer, R., *see* Büttiker, M. 932, 1099
Lander, J.J. 295
Landgren, G., *see* Svensson, S.P. 326
Landheer, D. 674
Landsberg, P.T. 123, 187, 191, 200, 202, 205, 209–211, 215, 220, 222, 226, 229–234, 238–240, 251, 252, 254, 256, 259, 265, 268, 270, 274–276, 422, 423, 428, 495, 674
Landsberg, P.T., *see* Adams, M.J. 231, 232
Landsberg, P.T., *see* Dhariwal, S.R. 239, 241–243
Landsberg, P.T., *see* Evans, D.A. 226
Landsberg, P.T., *see* Lal, P. 239
Landsberg, P.T., *see* Robbins, D.J. 422
Landsberg, P.T., *see* Schöll, E. 234, 236, 422, 428, 440
Landsberg, P.T., *see* Smith, E.F. 239
Landwehr, G., *see* Hajdu, J. 979
Lang, D.V. 177, 188–191, 229, 231, 241, 245, 247–250, 263, 385
Lang, D.V., *see* Henry, C.H. 189, 209, 245, 248
Lang, D.V., *see* Miller, G.L. 191
Langer, J.M. 135, 136, 170, 390
Langer, J.M., *see* Baranowski, J.M. 175
Langer, J.M., *see* Delerue, C. 135, 136
Langer, J.S. 946, 947
Langerak, C.J.G.M., *see* Huant, S. 169
Langreth, D.C. 932
Lannoo, F., *see* Mauger, A. 120
Lannoo, M. 3, 4, 22, 116, 118, 119, 128, 130, 132, 152
Lannoo, M., *see* Allan, G. 819, 827
Lannoo, M., *see* Baraff, G.A. 152
Lannoo, M., *see* Delerue, C. 118, 126, 133, 135–138, 140–142, 152
Lannoo, M., *see* Foulon, Y. 130, 145
Lannoo, M., *see* Friedel, J. 119
Lannoo, M., *see* Krambrock, K. 156
Lannoo, M., *see* Langer, J.M. 135, 136
Lannoo, M., *see* Liro, Z. 136, 137
Lannoo, M., *see* Makram-Ebeid, S. 151, 658
Lannoo, M., *see* Mauger, A. 120
Lannoo, M., *see* Pêcheur, P. 117, 118
Lannoo, M., *see* Petit, J. 137, 154, 155
Lannoo, M., *see* Picoli, G. 132–134
Lannoo, M., *see* von Bardeleben, H.J. 150
Lanza, C., *see* Solomon, P.M. 646
Lanza, C., *see* Woodall, J.M. 351, 378
Lapeyre, G.J., *see* Knapp, J.A. 312, 316
Lapeyre, G.J., *see* Smith, R.J. 312
Lapujoulade, J. 305
Larkin, A.I., *see* Glazman, L.I. 872
Larkin, A.I., *see* Gorkov, L.P. 945–947, 995
Larkin, A.I., *see* Hikami, S. 933, 947, 948
Larkin, A.I., *see* Ioffe, L.B. 984
LaRosa, S., *see* Marsi, M. 399
Larsen, D.M., *see* Stillman, G.E. 168
Larsen, P.K. 302, 319
Larsen, P.K., *see* Hoeven, J.J. 380
Larsen, P.K., *see* Joyce, B.A. 306–309

Larsson, K., *see* Grimmeiss, H.G. 131, 132
Laruelle, F., *see* Barrau, J. 850, 857
Lary, J., *see* Weisshaar, A. 896–899, 926
Laskar, J., *see* Kolodzey, J. 432
Lassabetere, L., *see* Ismail, A. 344
Lassnig, R. 693, 703
Lassnig, R., *see* Gornik, E. 984
Lassnig, R., *see* Zawadzki, W. 763, 765, 767
Lau, S.S., *see* Marshall, E.D. 337, 380
Lau, S.S., *see* Mayer, J.W. 328, 332, 378
Laufer, P.M., *see* Krieger, J.B. 118
Laughlin, R.B. 991, 992, 996, 1003, 1004, 1006, 1009, 1013
Laughlin, R.B., *see* Kalmeyer, V. 658
Laughton, M.J. 895, 917, 1087, 1104, 1108, 1109, 1111, 1113, 1114
Laughton, M.J., *see* Barker, J.R. 1104, 1111, 1113, 1116
Laurant, J.M., *see* Benoit à la Guillaume, C. 462, 463
Laurich, B.K. 825
Laux, S.E. 872, 1087
Laux, S.E., *see* Kumar, A. 872, 922, 1107
Lavine, M.C., *see* Gatos, H.C. 309, 310
Laviron, M., *see* Massies, J. 339, 340
Law, K.K. 852
Law, V.J., *see* Tewordt, M. 922, 923
Lawaetz, P. 493, 499, 519, 541, 543, 552, 563
Lax, B. 529, 550, 552, 563
Lax, B., *see* Dexter, R.N. 499, 504, 506
Lax, B., *see* Mavroides, J.G. 562, 563
Lax, B., *see* Reine, M. 521, 525
Lax, B., *see* Roth, L.M. 517, 520
Lax, B., *see* Weiler, M.H. 521, 525
Lax, B., *see* Zawadzki, W. 507, 517, 525
Lax, B., *see* Zeiger, H.J. 506
Lax, B., *see* Zwerdling, D. 522
Lax, M. 239, 242
Lax, M., *see* Halperin, B.I. 257
Lazzaroni, R., *see* Löglund, M. 623
Le Corre, A., *see* Lambert, B. 139, 142
Le Lay, G. 307, 324, 356
Le Manh Hoang 181
Le Person, H., *see* Palmier, J.F. 855
Le Roux, G., *see* Quillec, M. 820
Le Vu Ky, *see* Bonch-Bruevich, V.L. 433
Leadbeater, M.L. 655
Leath, P.L., *see* Elliott, R.J. 867, 940
Leavens, C.R., *see* Taylor, R.P. 888, 923, 924
Leavitt, R.P., *see* Shanabrook, B.V. 825
Ledebo, L.Å. 231
Ledebo, L.Å., *see* Grimmeiss, H.G. 175, 176
Ledentsov, N.N., *see* Notzel, R. 872
Lederman, F.L. 837, 843
Lee, B., *see* Skromme, B.J. 126
Lee, B.W., *see* Ignatiev, A. 303
Lee, B.W., *see* Lubinsky, A.R. 302
Lee, B.W., *see* Mark, P. 356
Lee, B.W., *see* Xu, F. 302
Lee, J. 705
Lee, K.-B., *see* Mo, Z. 607
Lee, K.Y., *see* Haug, R.J. 872–874, 926, 927
Lee, P.A. 782, 783, 790, 791, 867, 934, 940, 946, 951, 959
Lee, P.A., *see* Altshuler, B.L. 933, 937, 948
Lee, P.A., *see* Fisher, D.S. 884, 904, 912, 932
Lee, P.A., *see* Meir, Y. 924
Lee, P.A., *see* Serota, R.A. 959
Lee, P.A., *see* Yoshioka, D. 1001, 1002, 1006, 1013, 1025
Lee, T.C., *see* Lewis, B.F. 309
Lee, Y.H. 184
Lee, Y.S., *see* Christiansen, P.A. 65
Legendy, C., *see* Bowers, R. 530
LeGoues, F.K. 382
LeGoues, F.K., *see* Liehr, M. 355
Lehr, M., *see* Rau, U. 443, 445
Leinaas, J.M. 1010
Leising, G. 617
Leite, J.R., *see* Fazzio, A. 118, 132
Leman, G., *see* Friedel, J. 119
Leng, M., *see* Lent, C.S. 928
Lenglart, P., *see* Lannoo, M. 119
Lent, C.S. 896, 897, 899, 900, 902, 928, 1104, 1109
Lent, C.S., *see* van Hove, J.M. 307
Leo, J. 657
Leo, J.L. 934
Lepkowsky, T.R., *see* Skromme, B.J. 126
Lerner, I.V. 942, 959
Lerner, I.V., *see* Altshuler, B.L. 959
Lerner, I.V., *see* Kravtsov, V.E. 959
Leroux, M., *see* Bensaid, B. 238
Lesee, D.L., *see* Burkey, B.C. 177, 188
Lesovik, G.B. 793, 795, 894
Lesovik, G.B., *see* Glazman, L.I. 886, 887
Lessor, D.L., *see* Ford, W.K. 323
Lester, L.F. 679
LeToullec, R., *see* Tuchendler, J. 167
Leung, S. 326
Levesque, D. 1005
Levesque, D., *see* Caillol, J.M. 1005, 1006, 1030
Levesque, D., *see* Hansen, J.P. 1009
Levi, A.F.J. 660, 707
Levine, B.F. 697
Levine, H. 955, 995
Levine, H., *see* Laughlin, R.B. 996
Levine, J.D. 296
Levine, Z.H. 19
Levinson, I.B. 885, 1121
Levinson, Y.B. 893
Levinson, Y.B., *see* Esipov, S.E. 698
Levinson, Y.B., *see* Gantmakher, V.F. 568, 571, 573, 575, 577
Levinson, Y.B., *see* Sukhorukov, E.V. 908
Levy, F.J., *see* Daniels, R.R. 389

Levy, F.J., *see* Tung, R.T. 355
Levy, M., *see* Perdew, J.P. 3, 4, 22, 29, 116
Lewis, B.F. 309
Lewis, N., *see* Sarma, K. 380
Leysen, R., *see* van Hove, H. 302
Lhomer, C., *see* Lambert, B. 139, 142
Li, D. 380
Li, D., *see* Qiu, J. 380, 396
Li, Q. 913
Li, X.P. 106
Li, Y., *see* Krieger, J.B. 3, 4
Li, Y.P. 883, 894
Libby, S.B., *see* Laughlin, R.B. 996
Libby, S.B., *see* Levine, H. 955, 995
Licciardello, D.C. 994
Licciardello, D.C., *see* Abrahams, E. 945, 947, 949, 950, 989, 995
Lichtman, N.P., *see* Lifshitz, T.M. 174
Licini, J.C. 958
Lidiard, A.B., *see* Howard, R.E. 270
Liefrink, F., *see* Dekker, C. 883, 894
Liehr, M. 355
Lieski, N.P. 313
Lifshitz, E.M., *see* Landau, L.D. 631, 678
Lifshitz, I.M. 495, 512, 867, 940
Lifshitz, I.M., *see* Landau, L.D. 521
Lifshitz, T.M. 174
Li'in 118
Likhaerev, K.K. 1115
Likhaerev, K.K., *see* Averin, A. 1115
Likhaerev, K.K., *see* Bakhvalov, N.S. 1115
Likharev, K.K., *see* Averin, D.V. 921, 924
Lilienthal, Z., *see* Goodnick, S.M. 708
Lilienthal-Weber, Z. 326, 336, 358
Lilienthal-Weber, Z., *see* Newman, N. 349
Lilienthal-Weber, Z., *see* Spicer, W.E. 356, 357
Lin, A.L. 188
Lin, J.C. 337
Lin, P.S.D., *see* Kash, K. 875
Lin, W.T., *see* Chen, L.J. 324
Lin, W.Z. 1071
Lin, W.Z., *see* Schoenlein, R.W. 1071
Lin, Z., *see* Weaver, J.H. 331
Lin, Zhangda 331
Lin-Liu, Y.R., *see* Takayama, M. 599–601
Lince, J.R., *see* Williams, R.S. 336
Lind, D.M. 326
Lindau, I. 356
Lindau, I., *see* Cao, R.K. 349, 358
Lindau, I., *see* Friedman, D.J. 334
Lindau, I., *see* Kendelewicz, T. 334, 337
Lindau, I., *see* Miyano, K. 372
Lindau, I., *see* Skeath, P. 323
Lindau, I., *see* Spicer, W.E. 349, 356, 357, 370, 371
Lindefelt, U., *see* Singh, V.A. 118
Lindefelt, U., *see* Zunger, A. 118, 132
Lindelof, P.E. 1028
Linh, N.T., *see* Massies, J. 298, 306, 325, 326, 339, 340
Lipari, N.O., *see* Bernholc, J. 118
Lipavski, P., *see* Mašek, J. 894, 895
Lippmann, B.A., *see* Johnson, M.H. 517
Liro, Z. 136, 137
List, R.S. 368
List, R.S., *see* Kendelewicz, T. 334
Litovchenko, V.G. 483
Little, W.J., *see* Neuringer, L.J. 563
Littler, C.L., *see* Seiler, D.G. 432
Littlewood, P.B. 57
Litwin, A., *see* Kleverman, M. 229, 230, 245
Liu, A.Y. 85, 87, 101
Liu, C.T. 930
Liu, D.D., *see* Keyes, B.M. 238
Liu, H.C., *see* Landheer, D. 674
Liu, J.M. 22
Liu, J.M., *see* Lompre, L.A. 22
Liu, S.H. 518
Liu, W.P., *see* Puechner, R.A. 930
Lloyd, P. 205
Lo, T.K. 461
Lo, T.K., *see* Westervelt, R.M. 459
Loebs, V.A., *see* Chambers, S.A. 374
Logan, R.A., *see* Chynoweth, A.G. 656, 658
Logan, R.A., *see* Lang, D.V. 177, 188, 189, 245, 249, 250
Logan, R.A., *see* Lin, W.Z. 1071
Löglund, M. 623
Loloee, M.R., *see* Song, X.N. 437
Lomasov, Yu.N., *see* Altukhov, P.D. 479–481
Lomdahl, P.S., *see* Bishop, A.R. 601, 616
Lompre, L.A. 22
Long, D. 551, 562, 563
Longenbach, K.F. 647
Look, D.C. 188, 368
Lorenz, M.R. 177
Lorke, A. 922, 930
Lösch, R., *see* Müller, G. 997
Loudon, R., *see* Barker Jr, A.S. 182
Louie, S.G. 14, 79, 81, 118
Louie, S.G., *see* Chelikowsky, J.R. 75, 81, 314
Louie, S.G., *see* Fahy, S. 68, 84
Louie, S.G., *see* Hybertsen, M.S. 8, 10, 12, 19–22, 30, 81, 116
Louie, S.G., *see* Levine, Z.H. 19
Louie, S.G., *see* Northrup, J.E. 90, 91, 314
Louie, S.G., *see* Schlüter, M. 313
Louie, S.G., *see* Tejedor, C. 326, 346, 347
Louie, S.G., *see* Vanderbilt, D. 95
Louie, S.G., *see* Zhang, S.B. 90
Louie, S.G., *see* Zhu, X. 81
Low, T.S., *see* Skromme, B.J. 126
Lowe, D. 1124
Lowy, D.N., *see* Herbst, J.F. 139, 140, 142, 143
Lozovik, Yu.E., *see* Apenko, S.M. 993, 997
Lubin, M.I., *see* Levinson, Y.B. 893

Lubinsky, A.R. 302
Lubinsky, A.R., *see* Duke, C.B. 302
Lucovsky, G. 171, 714
Ludeke, R. 317, 325, 326, 334, 357, 372
Ludeke, R., *see* Drube, W. 317
Ludwig, M. 347
Lugagne-Delpon, E. 821, 822
Lugli, P. 698, 1066, 1068, 1069
Lugli, P., *see* Goodnick, S.M. 706, 708
Lugli, P., *see* Rücker, H. 695
Lundqvist, B.I., *see* Günnarsson, O. 84
Lundqvist, B.I., *see* Hedin, L. 17–19
Lundqvist, S., *see* Hedin, L. 17–19
Luo, L.F. 647, 651
Luo, L.F., *see* Beresford, R. 647
Luo, L.F., *see* Longenbach, K.F. 647
Luryi, S. 953, 993, 994
Luryi, S., *see* Kazarinov, R.F. 953, 993, 994
Lusakowski, J., *see* Knap, W. 432
Luscombe, J.H., *see* Randall, J.N. 874, 875
Luscombe, J.H., *see* Reed, M.A. 872, 874, 875, 925
Lüth, H., *see* Ibach, H. 350
Luther, L.C., *see* Casey Jr, H.C. 187
Luther, L.C., *see* Henry, C.H. 182
Luttinger, J.M. 33, 123, 126, 127, 492, 501, 508, 509, 513, 518, 519, 669, 675
Luy, J.F., *see* Schuberth, G. 647, 655
Lyapilin, I.I., *see* Karyagin, V.V. 764, 778
Lynch, R.T., *see* Dean, P.J. 179
Lynch, R.T., *see* Hopfield, J.J. 179
Lyo, In-Whan 322
Lyo, In-Whan, *see* Avouris, Ph. 321, 322
Lyo, S.K. 759, 760, 764, 772, 773
Lyon, S.A. 698
Lyons, L.E., *see* Gutman, F. 251

Ma, J., *see* Djurado, D. 590
Ma, J., *see* Puechner, R.A. 930
Ma, M., *see* Esposito, F.P. 792
Ma, M., *see* Serota, R.A. 793
Ma, S.K. 30
Ma, Y. 322
Maan, J.C. 513, 841, 842
Maan, J.C., *see* Ancilotto, F. 831, 841
Maan, J.C., *see* Belle, G. 841
Maan, J.C., *see* Deveaud, B. 838, 839
Maan, J.C., *see* Ebert, G. 1025
Maan, J.C., *see* Fletcher, R. 754, 756, 757, 760, 761, 771–773, 775, 776, 778, 779
Maan, J.C., *see* Haug, R.J. 1018
Maaref, M., *see* Scalbert, D. 825
Maas, D.J., *see* Kouwenhoven, L.P. 921
Mabesoone, M.A.A., *see* Molenkamp, L.W. 806
MacColl, L.A. 634
MacDiarmid, A.G., *see* Epstein, A.J. 622
MacDiarmid, A.G., *see* Etemad, S. 614
MacDiarmid, A.G., *see* Feldblum, A. 604
MacDiarmid, A.G., *see* Flood, J.D. 606
MacDiarmid, A.G., *see* Moses, D. 609, 619
MacDiarmid, A.G., *see* Tanner, D.B. 623
MacDiarmid, A.G., *see* Weinberger, B.R. 602
MacDonald, A.H. 8, 10, 12, 930, 979, 997, 1015, 1024, 1029
MacDonald, A.H., *see* Aers, G.C. 992
MacDonald, A.H., *see* Girvin, S.M. 1022, 1023, 1029, 1031
MacDonald, A.H., *see* Leo, J. 657
MacDonald, A.H., *see* Levesque, D. 1005
MacDonald, A.H., *see* Rasolt, M. 3, 5, 43
MacDonald, A.H., *see* Streda, P. 904, 913, 998
MacDonald, A.H., *see* Yoshioka, D. 1028
MacEwan, W.R., *see* Harris, I.R. 326
Mackay, K., *see* Burroughes, J.H. 606
Mackay, K.J. 380
Mackay, K.J., *see* Zahn, D.T.R. 380
MacKinnon, A. 867, 933, 944, 952, 955
MacKinnon, A., *see* Chase, K.S. 959
MacKinnon, A., *see* Czycholl, G. 944
MacKinnon, A., *see* Kramer, B. 955
MacKinnon, A., *see* Oakeshott, R.B.S. 918
MacKinnon, A., *see* Schweitzer, L. 955, 994
Macksey, H.M., *see* Scifres, D.R. 179
MacLaren, J.M. 293
MacRae, A.U. 302, 303
Macucci, M., *see* Sols, F. 897, 898
Madhukar, A. 307
Madhukar, A., *see* Das Sarma, S. 118
Madhukar, A., *see* Lewis, B.F. 309
Magerramov, E.M., *see* Baranowski, J.M. 171
Magnusson, K.O. 317
Magnusson, K.O., *see* Cricenti, A. 91
Mahan, G.D. 1121
Mahler, G., *see* Kuhn, T. 697
Mahoney, D.D., *see* Kash, K. 875
Mahowald, P., *see* Katnani, A.D. 396, 399
Mahowald, P., *see* Spicer, W.E. 356, 357
Mahowald, P.H., *see* List, R.S. 368
Mai, G., *see* Huant, S. 169
Maier, M., *see* Broser, J. 181
Maier-Hosch, D., *see* Busse, J.W. 181
Maierhofer, Ch., *see* Alonso, M. 372
Mailhiot, C. 314, 347, 362, 378, 384, 825
Mailhiot, C., *see* Brillson, L.J. 370, 371
Mailhiot, C., *see* Duke, C.B. 344, 345, 359, 360, 373
Mailhiot, C., *see* Laurich, B.K. 825
Mailhiot, C., *see* Smith, D.L. 378, 385, 819, 825, 827
Mailhiot, C., *see* Viturro, R.E. 372
Mailly, D. 938, 958
Main, P.C. 888, 911
Main, P.C., *see* Alves, E.S. 930
Main, P.C., *see* Beton, P.H. 745, 842, 888, 889
Maissen, C. 381
Majri, G., *see* Canali, C. 335

Maki, K., *see* Takayama, M. 599–601
Makram-Ebeid, S. 151, 351, 371, 658
Makram-Ebeid, S., *see* Martin, G.M. 150
Maksimov, L.A., *see* Kozlov, A.N. 453
Maksym, P.A. 922, 1019–1021
Maksym, P.A., *see* Clark, R.G. 1018
Maksym, P.A., *see* Kawamura, T. 307
Malik, R.J., *see* Levine, B.F. 697
Mallett, J.R. 1006
Mallett, J.R., *see* Clark, R.G. 1015, 1018, 1019
Malvezzi, A.M., *see* Rasolt, M. 22
Manass, J., *see* Guldner, Y. 828
Manchon, D.D. 182
Manghi, F. 314
Manghi, F., *see* Bertoni, C.M. 362
Manghi, G., *see* Margaritondo, G. 322, 349
Mankiewich, P.M., *see* Chang, A.M. 911
Mankiewich, P.M., *see* de Vegvar, P.G.N. 791
Mankiewich, P.M., *see* Skocpol, W.J. 735, 958
Mankiewich, P.M., *see* Timp, G. 867, 872, 904, 906, 908, 909, 911
Manzke, R. 317
Manzke, R., *see* Carstensen, H. 317
Mao, D., *see* Kahn, A. 349, 350
Mao, D., *see* Le Lay, G. 307
Maracas, G.N. 432
Maracas, G.N., *see* Puechner, R.A. 930
Maradudin, A.A. 684
Marcus, P.M., *see* Yang, W.S. 297
Margaritondo, G. 311, 322, 349, 386–389, 392, 393, 396, 398
Margaritondo, G., *see* Brillson, L.J. 328, 332–334, 337–339, 349, 354, 358, 366, 370–372
Margaritondo, G., *see* Capasso, F. 387, 824
Margaritondo, G., *see* Chiradia, P. 372
Margaritondo, G., *see* Daniels, R.R. 323, 389
Margaritondo, G., *see* Davis, G.D. 332, 334
Margaritondo, G., *see* Kahn, A. 349, 350
Margaritondo, G., *see* Katnani, A.D. 389
Margaritondo, G., *see* Le Lay, G. 307
Margaritondo, G., *see* Marsi, M. 399
Margaritondo, G., *see* McKinley, J.T. 398, 399
Margaritondo, G., *see* Niles, D.W. 398, 399
Margaritondo, G., *see* Perfetti, P. 389
Margaritondo, G., *see* Richter, H.W. 339
Margaritondo, G., *see* Rowe, J.E. 312
Margaritondo, G., *see* Shapira, Y. 337
Margaritondo, G., *see* Shaw, J.L. 333, 340, 367
Margaritondo, G., *see* Stiles, K. 349
Margaritondo, G., *see* Turowski, M. 389
Margaritondo, G., *see* Viturro, R.E. 372
Margaritondo, G., *see* Yeh, J.L. 323
Maria, D.J., *see* Fischetti, M.V. 1124
Marinace, J.C. 645
Mark, P. 356
Mark, P., *see* Baidyaroy, S. 349
Mark, P., *see* Chang, S.C. 302
Mark, P., *see* Goodwin, T.A. 349
Mark, P., *see* Lubinsky, A.R. 302
Markham, J.J. 243
Marks, R.N., *see* Burroughes, J.H. 606
Marks, R.N., *see* Farrow, R.F.C. 326
Marshall, E.D. 337, 380
Marshall, P., *see* Jung, H. 273
Marshall, P., *see* Taylor, R.P. 877, 888
Marsi, M. 399
Martelli, F. 479
Mårtensson, P. 297, 319, 322, 323
Mårtensson, P., *see* Feenstra, R.M. 362, 363
Martin, G.M. 150, 170, 191
Martin, G.M., *see* Makram-Ebeid, S. 351, 371
Martin, G.M., *see* Mitonneau, A. 191
Martin, R.M., *see* Chadi, D.J. 91
Martin, R.M., *see* Galli, G. 106
Martin, R.M., *see* Kunc, K. 96
Martin, R.M., *see* Needs, R.J. 85
Martin, R.M., *see* Nielsen, O.H. 319
Martin, R.M., *see* Shirley, E.L. 74
Martin, R.M., *see* van de Walle, C.G. 56
Martin, R.M., *see* Wendel, H. 91
Martin, W.C. 139
Martín-Moreno, L. 886, 924
Martín-Moreno, L., *see* Patel, N.K. 899–901
Martín-Moreno, L., *see* Pendry, J.B. 898
Martín-Moreno, L., *see* Prêtre, A.B. 903
Martinez, G. 138, 377
Martinez, G., *see* Buhmann, H. 1006, 1007
Martinez, G., *see* Chang, K.J. 87, 102, 103
Martinez, G., *see* Gołdys, E. 165
Martinez, G., *see* Gregorkiewicz, T. 186
Martinez, G., *see* Huant, S. 169
Martinez, G., *see* Muller, F. 185, 186
Martinez, G., *see* Stepniewski, R. 181
Martins, J.L. 70, 103
Martins, J.L., *see* Troullier, N. 76, 77
Marzin, J.Y. 674, 824, 830
Mašek, J. 894, 895, 946, 959
Mašek, J., *see* Maissen, C. 381
Maserjian, J. 644
Mashkevich, V.S. 232
Mason, B.A. 1124
Mason, N.J., *see* Warburton, R.J. 824
Masselink, W.T., *see* Ismail, K. 930
Massey, H.S.W. 570
Massida, V., *see* Bhattacharya, P.K. 459
Massies, J. 298, 306, 325, 326, 339, 340
Massies, J., *see* Sauvage-Simkin, M. 302
Masson, J.M., *see* Vuong, T.H.H. 730
Masterov, V.F., *see* Li'in 118
Materlik, G., *see* Dev, B.N. 380
Mathur, V.K. 253
Matsumoto, Y., *see* Ando, T. 985
Matsusue, T., *see* Sakaki, H. 708, 709
Mattens, W.C.M., *see* Niessen, A.K. 332
Matthews, J.W. 378, 379
Mattingly, M., *see* Aina, L. 822

Mattos, J.V.C. 466
Matulis, A. 885
Matyi, R.J., *see* Randall, J.N. 874, 875
Matyi, R.J., *see* Reed, M.A. 872, 922, 923, 925
Mauger, A. 120
Mauger, A., *see* Bourgoin, J.C. 146, 149
Mavroides, J.G. 512, 520, 562, 563
Mavroides, J.G., *see* Lax, B. 529, 550, 552, 563
Mayer, J.W. 328, 332, 335, 378
Mayer, J.W., *see* Grunthaner, P.J. 336
Mayer, J.W., *see* Ottaviani, G. 352
Mayer, J.W., *see* Poate, J.M. 328, 335, 339
Mayer, J.W., *see* Woodall, J.M. 351, 378
Mayer, K.M. 440
Mazur, A., *see* Larsen, P.K. 302, 319
Mazur, A., *see* Pollmann, J. 313, 314
McCaldin, J.O., *see* Yu, E.T. 389
McCants, C., *see* Spicer, W.E. 356, 357
McCants, C.E., *see* Newman, N. 343–345
McClure, J.W. 554, 563
McCombe, B.D., *see* Jarosik, N.C. 168
McDonald, I.R., *see* Ciccotti, G. 105
McEuen, P.L. 867, 892, 893, 898, 924
McEuen, P.L., *see* Alphenaar, B.W. 921
McEuen, P.L., *see* Kopley, T.E. 893
McEwan, P.L. 796
McGill, T.C., *see* Chow, D.H. 647
McGill, T.C., *see* Hauenstein, R.J. 355
McGill, T.C., *see* Kurtin, S. 345, 346, 354
McGill, T.C., *see* Schulman, J.N. 825
McGill, T.C., *see* Swarts, C.A. 314
McGill, T.C., *see* Yu, E.T. 389
McGilp, J.F. 332
McGilp, J.F., *see* McLean, A.B. 358
McGinnis, W.C., *see* Burns, M.J. 810
McGovern, I.T., *see* Hughes, G.J. 332, 353
McGovern, I.T., *see* McGilp, J.F. 332
McInnes, J.A., *see* Butcher, P.N. 809
McKay, K.G., *see* Chynoweth, A.G. 638, 651
McKinley, J., *see* Kahn, A. 349, 350
McKinley, J.T. 398, 399
McKinley, J.T., *see* Brillson, L.J. 370, 371
McKinley, J.T., *see* Hughes, G.J. 332, 353
McKinley, J.T., *see* Niles, D.W. 399
McKinley, J.T., *see* Viturro, R.E. 372
McLaughlin, D.W., *see* Scott, A.C. 601
McLean, A.B. 357, 358
McLean, A.B., *see* Ludeke, R. 372
McLean, A.B., *see* McGilp, J.F. 332
McMahan, A.K. 85
McMillan, H.P., *see* Iselev, G.W. 177, 188
McMillan, W.G. 102
McMillan, W.L., *see* Steele, A.G. 478
McRae, A.V. 300
McRae, E.G., *see* Bennett, P.A. 296
McSkimin, H.J. 83, 84, 94
McWorther, A.L., *see* Hamann, D.R. 239
Mead, C.A. 344, 346, 353
Mead, C.A., *see* Kurtin, S. 345, 346, 354
Meade, R.D., *see* Alerhand, O.L. 304
Meade, R.D., *see* Bedrossian, P. 322
Mearns, D. 12, 14
Mecheri, K., *see* Derrida, B. 943
Mehrotra, D.R., *see* Dhariwal, S.R. 259
Mehta, M.L. 943
Mei, W.N., *see* Tong, S.Y. 302
Meiboom, S., *see* Abeles, B. 561
Meier, H.P., *see* Abraham, D. 387
Meier, H.P., *see* Albrektsen, O. 387
Meigs, G.M., *see* Ma, Y. 322
Meijer, E., *see* Lang, D.V. 231
Meinerzhagen, B., *see* Thoma, R. 426
Meir, Y. 924
Meir, Y., *see* McEuen, P.L. 924
Meirav, U. 875, 921, 924
Meirav, U., *see* Field, S.B. 924
Meirav, U., *see* McEuen, P.L. 924
Meissner, B., *see* Busse, J.W. 181
Mele, G. 349
Melius, C.F. 65
Mello, P.A. 959
Melngailis, J., *see* Licini, J.C. 958
Mel'nikov, V.I. 945, 956, 957
Mendez, E.E. 655, 657, 850, 857, 986
Mendez, E.E., *see* Agullo-Rueda, F. 850, 851, 857
Mendez, E.E., *see* Bastard, G. 837
Mendez, E.E., *see* Chang, L.L. 824
Mendez, E.E., *see* Esaki, L. 646, 820
Mendez, E.E., *see* Ohno, H. 655
Mendez, E.E., *see* Tejedor, C. 389
Mendez, E.E., *see* Wang, W.I. 385
Menendez, C. 620
Menendez, J. 678
Menke, D.R., *see* Qiu, J. 380, 396
Menschig, A., *see* Weiss, D. 922, 930
Merkt, U. 922
Merkt, U., *see* Wagner, M. 922
Merlin, R., *see* Oh, J.E. 326
Mermin, D. 12
Mermin, N.D., *see* Ashcroft, N.W. 526
Merrit, F.R., *see* Fletcher, R.J. 519
Merrit, F.R., *see* Kukimoto, H. 189
Merz, J.L., *see* Caine, E.J. 389, 822
Merz, J.L., *see* Law, K.K. 852
Merzbacher, E. 631, 634, 653
Meseguer, F., *see* Tejedor, C. 389
Messenger, R.A. 172
Messmer, R.P. 119
Messmer, R.P., *see* Watkins, G.D. 119
Mestres, N., *see* Oh, J.E. 326
Metzger, R.M., *see* Baughman, R.H. 592, 619
Meyer, B.K. 152, 153, 155, 156
Meyer, B.K., *see* Hoffmann, D.M. 351
Meyer, J.R. 703, 706
Meyer, J.R., *see* Glicksman, M. 473
Meyer, K.E., *see* Chamoun, S.N. 1072

Meyer, R.J., *see* Kahn, A. 323
Meynadier, M.H. 825
Mezenner, R., *see* Puechner, R.A. 930
Mialhe, P. 252
Micocci, G. 252
Micocci, G., *see* de Blasi, C. 252
Micovic, M., *see* Sorba, L. 398, 399
Midelhoek, J., *see* Caro, J. 938, 958
Miedema, A.R. 332, 347
Miedema, A.R., *see* Niessen, A.K. 332
Mieher, R.L., *see* Hale, E.B. 183
Mignot, J.M., *see* Chang, K.J. 87, 102, 103
Mikhai, E.F., *see* Dakhovskii, I.V. 563
Miki, K., *see* Tokumoto, H.S. 305
Mikkelsen, J.C. 56
Miller, B., *see* Brown, R.J. 886
Miller, D.A.B. 657
Miller, D.A.B., *see* Bar-Joseph, I. 850
Miller, G.G., *see* Baughman, R.H. 592, 619
Miller, G.G., *see* Murthy, N.S. 593
Miller, G.L. 191
Miller, R.C. 181, 837
Miller, T.J., *see* Costa, J.C. 368
Milliken, F.P., *see* Kirtley, J.P. 870
Mills Jr, A.P., *see* Canter, K.F. 293
Mills, K.C. 332
Milnes, A.G. 187, 351, 390
Milnes, A.G., *see* Cheng, H. 336, 357
Milnes, A.G., *see* Leung, S. 326
Mil'nikov, G.V. 955, 994
Milonni, P.W., *see* Seiler, D.G. 432
Milsom, P.K. 701
Minnhagen, P. 17
Minomura, S., *see* Asaumi, K. 85, 94
Minot, C., *see* Palmier, J.F. 855
Mintmire, J.W. 30
Mircea, A. 151, 351
Mircea, A., *see* Martin, G.M. 191
Mircea, A., *see* Mitonneau, A. 151, 191
Miret-Goutier, A., *see* Lilienthal-Weber, Z. 326
Mirlin, D.N. 153
Mironov, A.G. 523
Mironov, A.G., *see* Bonch-Bruevich, V.L. 422, 440, 867
Missous, M. 326, 344, 349
Mitani, T., *see* Etemad, S. 614
Mitchell, A.H., *see* Kittel, C. 636
Mitin, V.V., *see* Asche, M. 422
Mitin, V.V., *see* Reggiani, L. 422, 427
Mitin, V.V., *see* Shaw, M.P. 422, 440
Mitonneau, A. 151, 191
Mitonneau, A., *see* Martin, G.M. 191
Mitonneau, A., *see* Mircea, A. 151
Miura, N., *see* Yamada, K. 432
Miyahara, Y., *see* Esaki, L. 656
Miyake, S.J., *see* Kubo, R. 982, 984, 989
Miyano, K. 372
Miyano, K., *see* Cao, R.K. 349, 358
Miyano, K., *see* Spicer, W.E. 356, 357
Mizes, H.A., *see* Conwell, E.M. 590, 604, 623
Mizoguchi, K. 608, 612
Mizushima, Y., *see* Kajiyama, K. 349
Mizutani, T., *see* Hirose, K. 366
Mo, Y.-M. 321
Mo, Y.-M., *see* Davis, G.D. 332
Mo, Y.-M., *see* Swartzentruber, B.S. 304
Mo, Z. 607
Mock, G.B., *see* Thomas, G.A. 474, 475
Modesti, S., *see* Ma, Y. 322
Mogab, C.J. 309
Mohammed, K., *see* Capasso, F. 397, 399
Moiseiwitsch, B.L., *see* Massey, H.S.W. 570
Moisson, J.M. 303
Molenkamp, L.W. 796, 804–806, 872, 877, 878
Molenkamp, L.W., *see* Staring, A.A.M. 888
Molenkamp, L.W., *see* van Houten, H. 806–808
Molinari, E. 678, 680
Molinari, E., *see* Bertoni, C.M. 362
Molinari, E., *see* Fasolino, A. 678, 822
Molinari, E., *see* Manghi, F. 314
Molinari, E., *see* Rücker, H. 695
Moll, J.L., *see* Shockley, W. 273
Mollot, F., *see* Sibille, A. 855
Monakhov, A.M. 456, 482, 483
Monakhov, A.M., *see* Altukhov, P.D. 456, 482
Monarkha, Y.P. 980
Mönch, W. 296–299, 339, 347–349, 352, 358
Mönch, W., *see* Bartels, F. 351
Mönch, W., *see* Troost, D. 339
Monemar, B. 178
Monemar, B., *see* Godlewski, M. 432
Monemar, B., *see* Grimmeiss, H.G. 178, 190
Montelius, L., *see* Grimmeiss, H.G. 131, 132
Montgomery, V. 367
Montgomery, V., *see* Srivastava, G.P. 314
Montgomery, V., *see* Williams, R.H. 316, 332, 338, 353, 356
Montie, E.A. 886
Montie, E.A., *see* Henning, J.C.M. 144
Montroll, E.W., *see* Maradudin, A.A. 684
Montroll, E.W., *see* Scher, H. 614
Moody, P.L., *see* Gatos, H.C. 310
Mooij, J.E., *see* Geerligs, L.J. 1115
Mooij, J.E., *see* van Houten, H. 872, 876, 879, 911
Mooij, J.E., *see* van Loosdrecht, P.H.M. 904
Mooij, J.E., *see* van Wees, B.J. 882, 886, 902, 903
Moon, Y.B., *see* Mo, Z. 607
Mooney, P.M. 144
Mooney, P.M., *see* Kaufman, E. 150
Mooradian, A., *see* Wright, G.B. 177, 182
Moore, K.J., *see* Nicholas, R.J. 832
Moore, K.J., *see* Pulsford, N.J. 825
Moore, T.M., *see* Randall, J.N. 874, 875
Moore, T.M., *see* Reed, M.A. 872, 922, 923, 925
Moore, W.T., *see* Fletcher, R. 771, 772

Moraes, F. 602, 607
Moraes, F., *see* Chung, T.-C. 619
Morandi, G. 979, 992, 993
Morf, R. 1009, 1014, 1018
Morgan, D.V., *see* Palmstrøm, C.J. 325, 343, 365
Morgan, T. 146, 147
Morgan, T.N. 180, 209, 243, 248
Morgan, T.N., *see* Grimmeiss, H.G. 176
Mori, N. 695
Morigaki, K. 184
Morin, F.J., *see* Reiss, H. 271
Morkoç, H., *see* Colvard, C. 678
Morkoç, H., *see* Costa, J.C. 368
Morkoç, H., *see* Hess, K. 432
Morkoç, H., *see* Tsen, K.T. 712
Morozova, V.A., *see* Mironov, A.G. 523
Morris, D., *see* Soucail, B. 833, 838, 840, 850, 851, 857
Morris, R.J., *see* Patel, J.R. 380
Morrison, C.A., *see* Bahder, T.B. 652
Morrison, J., *see* Estrup, P.J. 324
Morrison, J., *see* Lander, J.J. 295
Morrison, S.R. 310
Mortensen, K., *see* Bedrossian, P. 322
Mosca, R., *see* Fornari, R. 357
Moses, D. 609, 619
Moses, D., *see* Sinclair, M. 612, 615, 616
Moses, D., *see* Vardeny, Z. 607
Mosher, R.A., *see* Gibson, H.W. 611
Moskalenko, S.A. 452
Moss, T.S. 238
Motsch, C., *see* Grosvalet, J. 1044
Mott, N.F. 342, 620, 621, 810, 812, 867, 932, 934, 941
Mott, N.F., *see* Gibb, R.P. 239
Mott, N.F., *see* Kaveh, M. 934
Mounier, S., *see* Palmstrøm, C.J. 327
Mourou, G.A., *see* Chamoun, S.N. 1072
Moutinho, H.R., *see* Aboie-Elfotouh, F.A. 305
Movaghar, B., *see* Leo, J.L. 934
Mrstik, B.J. 303
Mugibayashi, N., *see* Aoki, K. 432, 437
Mühlbach, A., *see* Peinke, J. 432
Mui, D.S.L., *see* Costa, J.C. 368
Mukai, T. 256
Mulhern Jr, J.E., *see* Roth, H. 655
Mulimani, B.G., *see* Fromhold, T.M. 762, 774–778
Mulimani, B.G., *see* Karavolas, V.C. 728, 733
Mulimani, B.G., *see* Kubakaddi, S.S. 734, 744, 771, 772
Mulimani, B.G., *see* Qin, G. 777, 778
Muller, F. 185, 186
Müller, G. 997
Müller, H.D., *see* Ennen, H. 139, 142
Müller, H.D., *see* Wagner, J. 139, 142
Mullin, J.B., *see* Ashen, D.J. 178, 180
Mullin, J.B., *see* White, A.M. 178
Munekata, H., *see* Haug, R.J. 874, 926
Munteanu, O., *see* Miller, R.C. 181
Murakami, M. 337, 365, 366
Murarka, S.P. 335
Murase, K., *see* Takagaki, Y. 867, 906, 908, 911
Murata, Y., *see* Hashizume, T. 322
Murnaghan, F.D. 83
Muro, K., *see* Nishida, Y. 175
Muroz, A.H., *see* Flores, F. 347
Murray, S.J., *see* Barker, J.R. 1098
Murthy, N.S. 593
Murthy, N.S., *see* Baughman, R.H. 592, 619
Mushwitz, C.B., *see* Jacobi, K. 302, 320
Muttalib, K.A. 943, 959
Myers, J., *see* Long, D. 551, 562, 563
Myrheim, J., *see* Leinaas, J.M. 1010
Myron, H.W., *see* Poole, D.A. 735, 736, 738

Naber, H. 433, 439
Naber, H., *see* Schöll, E. 433
Nag, B.R. 538, 544, 546, 569, 570
Nagaoka, Y. 867
Nagaoka, Y., *see* Hikami, S. 933, 947, 948
Nagels, P., *see* Colson, R. 621
Nagle, J., *see* Soucail, B. 833
Nahory, R.E., *see* Meynadier, M.H. 825
Nahory, R.E., *see* Palmstrøm, C.J. 327, 340, 374, 375
Najda, S.P. 168
Najda, S.P., *see* Huant, S. 169
Nakada, M., *see* Suto, K. 184
Nakagawa, T., *see* Emura, S. 714
Nakagawa, T., *see* Kawai, N.J. 307
Nakajima, M., *see* Yahata, A. 351
Nakamura, A. 872
Nakamura, K., *see* Liu, C.T. 930
Nakamura, S., *see* Nishitani, R. 303
Nakashima, H., *see* Kishino, S. 178
Nakashima, H., *see* Yokotsuka, T. 357
Nakata, S. 870
Namba, S., *see* Takagaki, Y. 867, 906, 908, 911
Narayanamurti, V. 678
Narayanamurti, V., *see* Eisenstein, J.P. 984
Narusawa, J., *see* Yokotsuka, T. 357
Nassau, K., *see* Henry, C.H. 182
Nathan, M.I., *see* Costa, J.C. 368
Nathan, M.I., *see* Eastman, D.E. 311
Nathan, M.I., *see* Heiblum, M. 385, 707
Neal, T., *see* Langer, J.S. 946, 947
Neave, J.H. 299, 307, 308
Neave, J.H., *see* Joyce, B.A. 306
Neave, J.H., *see* Larsen, P.K. 302, 319
Nechtschein, M. 608, 612
Neddermeyer, H., *see* Badt, D. 322
Neddermeyer, H., *see* Tosch, St. 322
Nedozerov, S.S. 676
Needels, M., *see* Payne, M.C. 314, 319
Needs, R.J. 85

Needs, R.J., *see* Engel, G.E. 89
Needs, R.J., *see* Payne, M.C. 314, 319
Neizvestnyi, I.G., *see* Gusev, G.M. 986
Nelin, G., *see* Nilsson, G. 93, 95
Nelson, A.J., *see* Aboie-Elfotouh, F.A. 305
Nenciu, G. 510, 523, 843
Neuman, E., *see* Busse, J.W. 181
Neumark, G.F. 230
Neuringer, L.J. 563
Neves, A.J.M., *see* Beton, P.H. 888, 889
Neves, A.J.M., *see* Main, P.C. 888, 911
Newbury, R., *see* Brown, R.J. 886, 904, 906, 907, 923, 924
Newbury, R., *see* Davies, A.G. 1019
Newbury, R., *see* Ford, C.J.B. 867, 883, 894, 904, 906, 909, 911, 920, 921, 928, 1104, 1109, 1111, 1112, 1115, 1116
Newbury, R., *see* Kelly, M.J. 899, 902
Newbury, R., *see* Patel, N.K. 899–901
Newbury, R., *see* Smith, C.G. 659, 880, 889, 890, 930
Newbury, R., *see* Tewordt, M. 922, 923
Newbury, R., *see* Wharam, D.A. 800, 801, 867, 870, 871, 881, 882, 902, 904, 913, 923, 1097, 1102, 1104
Newman, N. 343–345, 349
Newman, N., *see* Kendelewicz, T. 334
Newman, N., *see* Lilienthal-Weber, Z. 326, 336, 358
Newman, N., *see* Spicer, W.E. 356, 357
Newman, P., *see* Puechner, R.A. 930
Newns, D.M., *see* Stroscio, J.A. 305, 322
Newson, D.J., *see* Berggren, K.F. 871, 881, 902
Ng, K.K., *see* Tung, R.T. 355
Ng, T.K. 3, 4
Ngai, K.L., *see* Emin, D. 619
Ngoc, T.C., *see* Poppendieck, T.D. 296
Ni, W.-X. 369, 394
Nicholas, R.J. 749, 832, 986
Nicholas, R.J., *see* Brummell, M.A. 730
Nicholas, R.J., *see* Clark, R.G. 999, 1025
Nicholas, R.J., *see* Haug, R.J. 1018
Nicholas, R.J., *see* Mallett, J.R. 1006
Nicholas, R.J., *see* Pulsford, N.J. 825
Nicholas, R.J., *see* Vuong, T.H.H. 730
Nicholas, R.J., *see* Warburton, R.J. 824
Nicholls, J.M. 317, 321
Nicholls, J.M., *see* Perfetti, P. 91, 317
Nicholls, J.M., *see* Uhrberg, R.I.G. 319
Nickel, H., *see* Müller, G. 997
Nickerson, S.B., *see* Rasolt, M. 14
Nicolis, G. 421, 1043
Nicollian, E.H. 261
Niedernostheide, F.J. 443
Nieh, C.W., *see* Chow, D.H. 647
Nielsen, M., *see* Grey, F. 324
Nielsen, O.H. 319
Niessen, A.K. 332
Nightingale, M.P., *see* Thouless, D.J. 930, 993
Niklas, J.R., *see* Meyer, B.K. 152, 153, 155
Niles, D.W. 398, 399
Nilsson, G. 93, 95
Nilsson, P.O., *see* Svensson, S.P. 357
Ninno, D. 832
Ninno, D., *see* Gell, M.A. 832
Nishida, Y. 175
Nishii, S., *see* Sakaki, H. 998
Nishioka, M., *see* Takahashi, T. 875
Nishitani, R. 303
Nishizawa, J. 357
Nitecki, R., *see* Kamieniecki, E. 264
Niu, Q. 930, 992, 993, 1020
Nixon, J.A. 895, 896, 906, 1087, 1113, 1114
Nixon, J.A., *see* Davies, J.H. 1087, 1113
Nixon, J.A., *see* Laughton, M.J. 895, 917, 1087, 1108, 1109, 1113, 1114
Noda, T., *see* Sakaki, H. 708, 709
Nogami, J. 321, 322
Nolte, D.D. 759
Nordheim, L.W. 629, 659
Nordheim, L.W., *see* Fowler, R.H. 629
Norita, S., *see* Kobayashi, M. 165
Norman, M.R. 3, 4
Norrman, S., *see* Andersson, M.O. 205, 247
Northrup, J.E. 90, 91, 296, 314, 319, 323, 381, 382
Northrup, J.E., *see* Biegelsen, D.K. 90, 294, 300, 302, 307
Northrup, J.E., *see* Nicholls, J.M. 321
Northrup, J.E., *see* Zhang, S.B. 90
Norton, N., *see* Neave, J.H. 307
Nottenburg, R.N., *see* Sandroff, C.J. 339
Notzel, R. 872
Nouailhat, A., *see* Delerue, C. 137, 138, 140
Nowak, M., *see* Vardeny, Z. 607
Nozières, P. 10, 16
Nozières, P., *see* Combescot, M. 454, 459
Nygren, B. 173

Oakeshott, R.B.S. 918
Obermair, G., *see* Brandl, A. 437, 439
Obermair, G.M., *see* Rauh, A. 929
Obermair, G.M., *see* Schellnhuber, H.J. 929
Obermair, G.M., *see* Wannier, G.H. 929
Obloh, H. 730, 767, 768
Obloh, H., *see* Ruf, C. 730, 731, 760, 762, 763, 767
Obraztsov, Y.N. 764
Ockman, N., *see* Wang, W.B. 1073
O'Connell, R.F., *see* Hu, G.Y. 1125
O'Connor, A.J. 956
Odell, N.H., *see* Taylor, W.E. 264
O'Dwyer, J.J., *see* Lloyd, P. 205
Offenberg, M., *see* Reuters, P.J. 351
Offsey, S.D. 339, 368, 369
Offsey, S.D., *see* Lester, L.F. 679
Ogletree, D.F., *see* van Hove, M.A. 286, 293

Oh, J.E. 326
Ohdomari, I., *see* Hara, S. 352
Ohno, H. 655
Ohno, H., *see* Hasagawa, J. 359
Ohno, H., *see* Hasegawa, H. 359
Ohno, H., *see* Mendez, E.E. 655
Ohno, T., *see* Yamaguchi, E. 130, 145, 146, 149, 150
Ohno, T.R. 305
Ohno, Y., *see* Yano, H. 432
Ohta, K., *see* Kawai, N.J. 307
Ohtsuki, T. 955
Ohtsuki, T., *see* Kramer, B. 943, 954
Ohtsuki, T., *see* Ono, Y. 954
Ohyama, T., *see* Fuji, K. 432
Oji, H. 762–765
Oji, H.C.A., *see* MacDonald, A.H. 1024
Okamoto, H., *see* Hirayama, Y. 870
Okamoto, H., *see* Osaka, J. 151
Okamoto, K. 326
Okayama, S., *see* Tokumoto, H.S. 305
Okiji, A., *see* Nakamura, A. 872
Okuyama, Y. 759, 760
Olbright, G.R. 852
Olego, D.J. 338, 396
Olijnyk, H. 87
Olmstead, M.A. 319
Olmstead, M.A., *see* Northrup, J.E. 382
Olofsson, G. 190
O'Neill, D.G., *see* Franciosi, A. 338
Ono, Y. 896, 914, 953, 954, 988, 994
Ono, Y., *see* Kramer, B. 943, 954
Ono, Y., *see* Ohtsuki, T. 955
Onsager, L. 510, 511
Onton, A. 166, 180
Ootuka, Y., *see* Kobayashi, S. 936
Opfermann, J., *see* Hörhold, H.H. 617
Oppenheimer, J.R. 629
O'Reilly, E.P. 674
Orenstein, J. 605, 606, 615
Orenstein, J., *see* Vardeny, Z. 605, 606
Orenstein, J., *see* Weinberger, B.R. 605
Orlando, T.P., *see* Bagwell, P.F. 893
Ortolani, F., *see* Fano, G. 1021, 1025
Osaka, J. 151
Osbourn, G.C. 378, 820
Osheroff, G.D., *see* Dolan, G.J. 936
Oshima, C., *see* Aono, M. 297
Oshima, C., *see* Nishitani, R. 303
Oshiyama, A., *see* Car, R. 117
Osipov, V.V., *see* Kerner, B.S. 422
Osman, M.A. 1071, 1072
Ospelt, M. 355
Oswald, P.M.W., *see* Clark, R.G. 1020
Otsuka, E., *see* Fuji, K. 432
Otsuka, N., *see* Li, D. 380
Otsuka, N., *see* Qiu, J. 380, 396
Ottaviani, G. 335, 352
Ottaviani, G., *see* Calandra, C. 355
Overhof, H., *see* Fockele, M. 150
Ovren, C., *see* Grimmeiss, H.G. 176
Ovsyuk, V.N., *see* Gusev, G.M. 986
Owers-Bradley, J.R., *see* Beton, P.H. 888, 889
Owers-Bradley, J.R., *see* Main, P.C. 888, 911
Oxley, J.P. 733, 754, 756, 758, 761, 768, 769, 774, 778
Oxley, J.P., *see* Fromhold, T.M. 762, 774–778
Oxley, J.P., *see* Gallagher, B.L. 732, 753, 754, 759, 787, 793–795
Oxley, J.P., *see* Galloway, T. 791, 793
Oxley, J.P., *see* Karavolas, V.C. 728, 733
Oxley, J.P., *see* Qin, G. 777, 778
Oyama, Y., *see* Nishizawa, J. 357
Ozaki, M., *see* Etemad, S. 614

Paalanen, M.A. 986
Paalanen, M.A., *see* Wei, H.P. 955
Paasch, G. 391
Paduchih, L.J., *see* Astemirov, T.H. 466
Palau, J.M., *see* Ismail, A. 344
Palevski, A., *see* Sivan, U. 659
Palik, E.D. 529, 530
Palmateer, L.F., *see* Kirtley, J.P. 870
Palmier, J.F. 855
Palmier, J.F., *see* Calecki, D. 855
Palmier, J.F., *see* Sibille, A. 855
Palmstrøm, C.J. 325, 327, 340, 343, 365, 374, 375
Palmstrøm, C.J., *see* Marshall, E.D. 337, 380
Palmstrøm, C.J., *see* Sands, T. 325–327, 336, 340, 365, 366, 374, 381
Pals, J.A., *see* Harris, J.J. 710, 867
Pandey, K.C. 90, 296, 297, 313, 314, 319
Pandey, K.C., *see* Kaxiras, E. 317
Pandey, K.C., *see* Knapp, J.A. 316
Panish, J.B. 306
Panish, M.B. 326
Panish, M.B., *see* Casey Jr, H.C. 306, 377, 384
Pankove, J.I. 254, 258
Pannetier, B. 937, 948
Pantelides, S.T. 123
Pantelides, S.T., *see* Bar-Yam, Y. 74
Pantelides, S.T., *see* Bernholc, J. 118
Pantelides, S.T., *see* Car, R. 117
Papaconstantopoulos, D.A. 118
Pappalardo, R. 170
Paquet, D., *see* Jusserand, B. 822
Paranjape, V.V. 896, 913
Paranjape, V.V., *see* Fröhlich, H. 1051
Parisi, J., *see* Mayer, K.M. 440
Parisi, J., *see* Peinke, J. 422, 432, 433, 435, 437
Parisi, J., *see* Rau, U. 433, 443, 445
Parisi, J., *see* Richter, R. 438
Parisi, J., *see* Röhricht, B. 428
Parisi, J., *see* Schöll, E. 433
Park, D.H., *see* Takiguchi, T. 617
Park, R.L., *see* Duke, C.B. 289

Parker, V.B., *see* Wagman, D.D. 332
Parkin, S.S.P., *see* Farrow, R.F.C. 326
Parks, C.C., *see* Murakami, M. 337, 365, 366
Park, Sang-il, *see* Nogami, J. 321, 322
Parr, R.G., *see* Perdew, J.P. 3, 4, 22, 29
Parrinello, M., *see* Car, R. 105, 106
Parrinello, M., *see* Galli, G. 106
Parrinello, M., *see* Li, X.P. 106
Parsey, J.M., *see* Kamińska, M. 151, 170
Partain, L.D., *see* Keyes, B.M. 238
Pashley, M.D. 299–302, 374
Pashley, M.D., *see* Haberern, K.W. 302
Pasquarello, A., *see* Andreani, L.C. 674
Pastawski, H.M. 912
Pastur, L.A., *see* Lifshitz, I.M. 867
Pate, M.A., *see* Leadbeater, M.L. 655
Patel, J.R. 380
Patel, N.K. 899–901
Patel, N.K., *see* Cobden, D.H. 883, 892, 894
Patella, F., *see* Knapp, J.A. 316
Patella, F., *see* Perfetti, P. 389
Patillon, J.N., *see* Lugagne-Delpon, E. 821, 822
Paton, A., *see* Duke, C.B. 322
Paton, A., *see* Holland, B.W. 297
Paton, A., *see* Kahn, A. 323
Patten, E.A. 389
Patten, E.A., *see* Hsieh, S. 389
Patterson, M.H., *see* Williams, R.H. 366
Paul, W. 165, 187, 188
Paul, W., *see* Foyt, A.G. 188
Paul, W., *see* Groves, S.H. 529
Paul, W., *see* Holland, M.G. 188
Paul, W., *see* Kosicki, B.B. 188
Pauling, L.N. 347
Payne, M.C. 314, 319, 887
Payne, M.C., *see* Johnson, N.F. 922
Payne, R.T. 656
Peacock, D.C., *see* Brown, R.J. 886, 899, 902, 904, 906, 907, 923, 924
Peacock, D.C., *see* Ford, C.J.B. 867, 904, 906, 909, 911
Peacock, D.C., *see* Greene, S.K. 916
Peacock, D.C., *see* Kelly, M.J. 899, 902
Peacock, D.C., *see* Smith, C.G. 659, 880, 889, 890, 923, 924, 930
Peacock, D.C., *see* Tewordt, M. 922, 923
Peacock, D.C., *see* Wharam, D.A. 800, 801, 867, 870, 871, 881, 882, 888, 902–904, 913, 923, 924, 1097, 1102, 1104
Pearsal, T.P. 820
Pearsall, N.M., *see* Coutts, T.J. 252
Pearson, G.L. 560, 562, 563
Pearson, G.L., *see* Baranowski, J.M. 170
Pease, R.F.W., *see* Chou, S.Y. 659
Pêcheur, P. 117, 118, 132
Pêcheur, P., *see* Kauffer, E. 118
Pêcheur, P., *see* van der Rest, J. 118
Pechstedt, R., *see* Wendler, L. 699
Peeters, E.M., *see* Huant, S. 169
Peeters, F.M. 745, 746, 896, 905, 913, 914
Peierls, R.E. 508
Peifer, H.J., *see* Thoma, R. 426
Peinke, J. 422, 432, 433, 435, 437, 440
Peinke, J., *see* Rau, U. 433, 443, 445
Peinke, J., *see* Richter, R. 438
Peinke, J., *see* Schöll, E. 433
Pendry, J.B. 292, 293, 898, 946
Pendry, J.B., *see* Kirkman, P.D. 945, 957
Pendry, J.B., *see* MacLaren, J.M. 293
Pendry, J.B., *see* Prêtre, A.B. 903
Pendry, T.B., *see* Anderson, S. 303
Penzkofer, A. 233
Pepin, J. 1109, 1111
Pepin, J., *see* Barker, J.R. 1104, 1111, 1113, 1116
Pepper, M., *see* Berggren, K.F. 871, 881, 902
Pepper, M., *see* Brown, R.J. 886, 899, 902, 904, 906, 907, 923, 924
Pepper, M., *see* Cobden, D.H. 883, 892, 894
Pepper, M., *see* Davies, A.G. 1019
Pepper, M., *see* Ford, C.J.B. 867, 904, 906, 909, 911, 1104, 1109, 1111, 1112, 1115, 1116
Pepper, M., *see* Gallagher, B.L. 753, 757, 811, 812
Pepper, M., *see* Greene, S.K. 916
Pepper, M., *see* Kearney, M.J. 785, 787–789
Pepper, M., *see* Kelly, M.J. 899, 902
Pepper, M., *see* Patel, N.K. 899–901
Pepper, M., *see* Poole, D.A. 735, 736, 738
Pepper, M., *see* Smith, C.G. 659, 880, 889, 890, 923, 924, 930
Pepper, M., *see* Syme, R.T. 754, 755, 760, 761, 783–788
Pepper, M., *see* Tewordt, M. 922, 923
Pepper, M., *see* Thornton, T.J. 735, 869–871, 881
Pepper, M., *see* Uren, M.J. 937
Pepper, M., *see* van Houten, H. 902, 904, 906
Pepper, M., *see* von Klitzing, K. 953, 979, 985, 986
Pepper, M., *see* Wharam, D.A. 800, 801, 867, 870, 871, 881, 882, 888, 902–904, 913, 923, 924, 1097, 1102, 1104
Perdew, J.P. 3, 4, 22, 29, 116
Perel', V.I., *see* Abakumov, V.N. 239
Perenboom, J.A.A.J., *see* Patel, N.K. 899–901
Peressi, M. 398
Peretti, J., *see* Le Lay, G. 324
Perfetti, P. 91, 317, 389
Perfetti, P., *see* Daniels, R.R. 389
Perfetti, P., *see* Margaritondo, G. 386–389, 392, 393
Perfetti, P., *see* Niles, D.W. 398
Perrot, F., *see* Rasolt, M. 3, 5, 22, 43
Pessot, M., *see* Chamoun, S.N. 1072
Peterson, T.L. 553, 554, 563
Petit, J. 137, 154, 155
Petrich, G.S., *see* Cohen, P.I. 307

Petro, W.G., *see* Kendelewicz, T. 334, 337
Petro, W.G., *see* Newman, N. 343–345
Petroff, P.M. 296, 382, 1088
Petroff, P.M., *see* Ensslin, K. 930
Petroff, P.M., *see* Tsuchiya, M. 874
Pettit, G.D., *see* Brillson, L.J. 370, 371
Pettit, G.D., *see* Chang, S. 340, 350, 372–374
Pettit, G.D., *see* Offsey, S.D. 339, 368, 369
Pettit, G.D., *see* Viturro, R.E. 358, 372
Pettit, G.D., *see* Woodall, J.M. 351, 378, 381
Pevzner, V. 1123
Pfannkuche, D., *see* Gerhardts, R.R. 930
Pfeiffer, L.N. 874
Pfeiffer, L.N., *see* Eisenstein, J.P. 1019, 1020, 1028
Pfeiffer, L.N., *see* Goldberg, B.B. 1006
Pfeiffer, L.N., *see* Jiang, H.W. 1007, 1027
Pfeiffer, L.N., *see* Pinczuk, A. 1024
Pfeiffer, L.N., *see* Schubert, E.F. 703
Pfeiffer, L.N., *see* Spector, J. 872, 874, 878–880
Pfeiffer, L.N., *see* Störmer, H.L. 874
Pfeiffer, R.S., *see* Rose, J.H. 475, 476
Pfeiffer, R.S., *see* Shore, H.B. 454, 475
Philip, P., *see* Franciosi, A. 338
Philippidis, C. 1118
Phillips, A. 1066
Phillips, J.C. 6, 8, 50, 55, 56, 62, 64, 98, 100, 137, 291
Phillips, J.C., *see* Andrews, J.M. 352
Phillips, J.C., *see* Brust, D. 7, 61
Phillips, J.C., *see* Chelikowsky, J.R. 96
Phillips, J.C., *see* Kleinman, L. 64
Phillips, J.C., *see* Pandey, K.C. 313
Phillips, J.C., *see* Villars, P. 98
Phillips, M.C., *see* Yu, E.T. 389
Phillips, T.G., *see* Hensel, J.C. 451
Phillpot, S.R., *see* Bishop, A.R. 601, 616
Pianetta, P., *see* Katnani, A.D. 396, 399
Pianetta, P., *see* Lindau, I. 356
Pianetta, P., *see* Spicer, W.E. 349
Pichard, J.-L. 938, 943, 959
Pichard, J.-L., *see* Derrida, B. 943
Pichard, J.-L., *see* Muttalib, K.A. 943, 959
Pichard, J.-L., *see* Zanon, N. 959
Pickett, W.E. 8, 10, 12, 22
Pickett, W.E., *see* Wang, C.S. 8, 10, 12, 19, 22
Picoli, G. 132–134
Picone, P.J., *see* Chen, C.Y. 934
Picraux, S.T. 820
Pidgeon, C.R. 520, 521
Piermarini, G.J. 85
Piermarini, G.J., *see* Weinstein, B.A. 85, 94
Pietiläinen, P. 1024
Pietiläinen, P., *see* Chakraborty, T. 922, 979, 1001, 1003, 1005, 1006, 1009, 1013, 1015, 1018, 1020, 1022, 1025–1028
Pietiläinen, P., *see* Halonen, V. 1018
Pikus, F.G., *see* Efros, A.L. 896
Pikus, G.E. 571, 674
Pikus, G.E., *see* Altukhov, P.D. 455, 476, 478, 479
Pikus, G.E., *see* Aronov, A.G. 525
Pikus, G.E., *see* Averkiev, N.S. 481
Piller, H. 530
Pimpale, A., *see* Landsberg, P.T. 422, 428
Pinchaux, R., *see* Sauvage-Simkin, M. 302
Pinczuk, A. 384, 1024
Pinczuk, A., *see* Goldberg, B.B. 1006
Pines, D. 706, 1052
Pinhas, S., *see* Büttiker, M. 932, 1099
Piskarev, V.I., *see* Belyantsev, A.M. 432
Pithkin, A.N., *see* Kopylov, A.A. 172, 173
Pitsina, N.G., *see* Gershenzon, E.M. 174, 175
Pitt, G.D. 188
Pitzer, K.S., *see* Christiansen, P.A. 65
Planel, R., *see* Scalbert, D. 825
Platzman, P.M., *see* Fukuyama, H. 1001
Platzman, P.M., *see* Girvin, S.M. 1022, 1023
Platzman, P.M., *see* Levi, A.F.J. 660, 707
Plaut, A.S., *see* Buhmann, H. 1006, 1007
Ploog, K., *see* Buhmann, H. 1006, 1007
Ploog, K., *see* Ebert, G. 955, 986, 998, 999
Ploog, K., *see* Fletcher, R. 754, 756, 757, 760, 775, 776, 778, 779
Ploog, K., *see* Grahn, H.T. 694
Ploog, K., *see* Grambow, P. 870, 872
Ploog, K., *see* Heitmann, D. 922
Ploog, K., *see* Kern, K. 930
Ploog, K., *see* Lorke, A. 922, 930
Ploog, K., *see* Notzel, R. 872
Ploog, K., *see* Obloh, H. 730, 767, 768
Ploog, K., *see* Ruf, C. 730, 731, 741, 742, 760, 762, 763, 767
Ploog, K., *see* Schneider, H. 848, 850, 857
Ploog, K., *see* Stahl, E. 984
Ploog, K., *see* Weiss, D. 744, 930
Ploog, K., *see* Wieck, A.D. 870
Plummer, E.W., *see* Wong, T.M. 322
Poate, J.M. 328, 335, 339
Poate, J.M., *see* Tung, R.T. 324, 325
Pogosov, A.G., *see* Gusev, G.M. 795
Poiarkov, A.G., *see* Astemirov, T.H. 466
Poission, M., *see* Nicholas, R.J. 986
Pokrovskii, Ya.E. 459, 463, 467–469, 476
Pokrovskii, Ya.E., *see* Kaminskii, A.S. 476
Polasko, J., *see* Kroemer, H. 382, 385
Pollmann, J. 313, 314
Pollmann, J., *see* Larsen, P.K. 302, 319
Pomrenke, G. 139, 142
Pomrenke, G., *see* Ennen, H. 139, 142
Poole, D.A. 735, 736, 738
Popov, Y.M. 254
Popov, Y.M., *see* Basov, N.G. 232
Poppe, U., *see* Cox, G. 305
Poppendieck, T.D. 296
Porod, W., *see* Maracas, G.N. 432
Porod, W., *see* Subramaniam, S. 897

Porowski, S. 187, 188
Porowski, S., *see* Dmowski, L. 177, 187, 188
Portal, J.C., *see* Brummell, M.A. 730
Portal, J.C., *see* Chamberlain, J.M. 867, 875
Portal, J.C., *see* Leadbeater, M.L. 655
Portal, J.C., *see* Nicholas, R.J. 986
Portal, J.C., *see* Sibille, A. 855
Portal, J.C., *see* Vuong, T.H.H. 730
Portella, M.T., *see* Bigot, J.Y. 22
Porteous, P., *see* White, A.M. 189, 190
Pothier, H., *see* Geerligs, L.J. 1115
Potter, D. 1069
Pouget, J.P. 590, 592, 593, 622
Pouxviel, J.C., *see* Pouget, J.P. 622
Požela, J. 422
Požela, J., *see* Bumelienė, S.B. 432
Požela, J., *see* Pyragas, K.A. 432
Praddaude, H.C. 525
Prange, R.E. 869, 881, 904, 913, 979, 990
Prange, R.E., *see* Joynt, R. 993
Pratt, G.W., *see* Zeiger, H.J. 495, 511, 512
Prentiss, M. 875
Prest Jr, W.M., *see* Gibson, H.W. 611
Prêtre, A.B. 903, 912
Prêtre, A.B., *see* Pendry, J.B. 898
Prettl, W., *see* Brandl, A. 422, 432, 437, 439, 440, 443
Prettl, W., *see* Weispfenning, M. 428
Preyer, N.W., *see* Chen, C.Y. 934
Preziosi, B., *see* Bassani, F. 123, 144
Price, P.J. 641, 646, 652, 661, 701, 702, 710, 897
Price, P.J., *see* Jacoboni, C. 653
Price, P.J., *see* Phillips, A. 1066
Price, W.H., *see* Murakami, M. 337, 365, 366
Priester, C. 824
Prigogine, I., *see* Nicolis, G. 421, 1043
Primot, J., *see* Quillec, M. 820
Prinz, G.A. 325, 326
Prinz, G.A., *see* Farrow, R.F.C. 326
Prinz, G.A., *see* Krebs, J.J. 326
Prinz, G.A., *see* Lind, D.M. 326
Prober, D.E., *see* Chandrasekar, V. 948
Prober, D.E., *see* Wheeler, R.G. 869, 870, 958
Probst, C., *see* Ebert, G. 955, 986, 1025
Proc. Int. Conf. on Electronic Properties of Two-Dimensional Systems IV to IX 867, 869, 875
Proc. Int. Conf. on Modulated Semiconductor Structures 867, 875
Pron, A., *see* Moses, D. 609
Pron, A., *see* Weinberger, B.R. 602
Pruisken, A.M.M. 953, 955, 995
Pruisken, A.M.M., *see* Laughlin, R.B. 996
Pruisken, A.M.M., *see* Levine, H. 955, 995
Pruisken, A.M.M., *see* Wei, H.P. 955, 996
Przybylinska, H. 139, 142
Pu-Lin, L., *see* Aszodi, G. 139, 142
Pudalov, V.M. 984, 989, 998, 999
Pudalov, V.M., *see* D'Iorio, M. 1007
Pudalov, V.M., *see* Kravchenko, S.V. 1007
Puechner, R.A. 930
Puga, M.W., *see* Xu, F. 302
Pugh, J.H. 326
Pugh, J.H., *see* Williams, R.S. 336
Pukite, P.R., *see* Cohen, P.I. 307
Pukite, P.R., *see* van Hove, J.M. 307
Pulsford, N.J. 825
Pulsford, N.J., *see* Nicholas, R.J. 832
Puoskari, M. 1031
Puri, S.M. 566
Purwins, H.G. 440
Purwins, H.G., *see* Niedernostheide, F.J. 443
Pyragas, K.A. 432
Pyragas, K.A., *see* Bumelienė, S.B. 432
Pytte, E. 422

Qian, G.X. 106
Qian, Q-D., *see* Qiu, J. 380, 396
Qin, G. 777, 778
Qin, G., *see* Fromhold, T.M. 762, 774–778
Qin, G., *see* Oxley, J.P. 733, 754, 756, 758, 761
Qiu, J. 380, 396
Qiu, J., *see* Li, D. 380
Quade, W. 426
Quade, W., *see* Schöll, E. 426
Quaresima, C., *see* Daniels, R.R. 389
Quaresima, C., *see* Niles, D.W. 398
Quaresima, C., *see* Perfetti, P. 389
Quate, C.F., *see* Nogami, J. 321, 322
Que, W. 922
Queisser, H.J. 171, 172, 178
Queisser, H.J., *see* Chattopadhay, D. 703
Quillec, M. 820
Quillec, M., *see* Bleuse, J. 839, 840, 851
Quinn, J.J., *see* Giuliani, G.F. 992
Quinn, T. 986

Rabe, K.M., *see* Rappe, A.M. 76
Radcliffe, J.M., *see* Price, P.J. 641, 661
Radelaar, S., *see* Caro, J. 938, 958
Radliński, A.P. 181
Raether, H. 312
Raghavachari, K., *see* Higashi, G.S. 369
Raisanen, A., *see* Chang, S. 372
Raisanen, A., *see* Franciosi, A. 338
Raisanen, A., *see* Yu, X. 386, 389
Ralls, K.S. 205, 247
Ralph, H.I. 553, 571
Ralston, J., *see* Jarosik, N.C. 168
Ramakrishnan, T.V., *see* Abrahams, E. 945, 947, 949, 950, 989, 995
Ramakrishnan, T.V., *see* Anderson, P.W. 937, 947
Ramakrishnan, T.V., *see* Lee, P.A. 782, 783, 867, 940, 946, 951
Ramdas, A.K. 163
Ramdas, A.K., *see* Ahlburn, B.T. 166

Ramdas, A.K., *see* Onton, A. 166
Ramesh, R., *see* Palmstrøm, C.J. 327
Ramesh, R., *see* Sands, T. 325–327, 336, 340, 365, 366, 374, 381
Ramirez, J.V., *see* Miller, G.L. 191
Rammal, R. 930
Rammal, R., *see* Pannetier, B. 937, 948
Randall, J.N. 874, 875
Randall, J.N., *see* Reed, M.A. 872, 874, 875, 922, 923, 925
Ranke, W. 306, 317
Ranke, W., *see* Drathen, P. 298, 302
Ranke, W., *see* Jacobi, K. 302, 320
Rao, A.M., *see* Tanner, D.B. 623
Rappe, A.M. 76
Rashba, E.I. 979
Rashba, E.I., *see* Apal'kov, V.M. 1006
Rashba, E.I., *see* Bychkov, Yu.A. 1006
Rasolt, M. 3–5, 8, 10, 12, 14, 19, 22, 27, 30, 31, 33, 36, 42, 43
Rasolt, M., *see* Dagens, L. 14
Rasolt, M., *see* Geldart, D.J.W. 25, 26, 30
Rasolt, M., *see* Vignale, G. 14
Rasolt, M., *see* Wang, J.S.Y. 8, 10, 12, 14
Rathbun, L., *see* Okamoto, K. 326
Rathbun, L., *see* Tadayon, B. 326
Ratner, M.A., *see* Topiol, S. 72
Rau, U. 433, 443, 445
Rau, U., *see* Peinke, J. 433, 435, 437
Rau, U., *see* Richter, R. 438
Räuber, A., *see* Schneider, J. 184
Rauch, F.J. 499
Rauh, A. 929
Ravaioli, U., *see* Sols, F. 897, 898
Ravenhall, D.G. 905, 913–915
Ravenhall, D.G., *see* Schult, R.L. 896, 905, 908, 913–915
Ray, B.K. 512
Ray, R., *see* Wannier, G.H. 929
Rayleigh, Lord 678
Raymond, F., *see* Bensaid, B. 238
Razeghi, M., *see* Brummell, M.A. 730
Razeghi, M., *see* Guldner, Y. 986
Razeghi, M., *see* Nicholas, R.J. 986
Razeghi, M., *see* Wei, H.P. 984, 996
Read, W.T., *see* Shockley, W. 222, 223
Reader, J., *see* Martin, W.C. 139
Reatto, L. 1030
Reed, M.A. 647, 659, 867, 872, 874, 875, 922, 923, 925, 1085
Reed, M.A., *see* Kirk, W.P. 986
Reed, M.A., *see* Randall, J.N. 874, 875
Rees, G.J. 239, 259
Rees, G.J., *see* Abram, R.A. 703
Rees, G.J., *see* Almbladh, C.-O. 205
Rees, G.J., *see* Gibb, R.P. 239
Rees, G.J., *see* Warburton, R.J. 824
Rees, H.D. 1062
Regel, A.R., *see* Ioffe, A.F. 942
Reggiani, L. 422, 427, 1121
Reggiani, L., *see* Hüpper, G. 432, 435
Reggiani, L., *see* Jacoboni, C. 538, 1120
Reggiani, L., *see* Lugli, P. 698
Regreny, A., *see* Chomette, A. 837, 838, 854–856
Regreny, A., *see* Deveaud, B. 838, 839, 854
Regreny, A., *see* Lambert, B. 855
Regreny, A., *see* Palmier, J.F. 855
Regreny, A., *see* Voisin, P. 850
Reihl, B., *see* Cricenti, A. 91
Reihl, B., *see* Nicholls, J.M. 317, 321
Reihl, B., *see* Perfetti, P. 91, 317
Rein, A., *see* Hüpper, G. 437
Rein, A., *see* Schöll, E. 437, 438
Reine, M. 521, 525
Reinecke, T.L., *see* Mrstik, B.J. 303
Reinecke, T.L., *see* Rudlin, S. 695
Reiss, H. 127, 271
Remenyi, G., *see* Ebert, G. 1025
Resca, L. 127, 128
Resta, R., *see* Marsi, M. 399
Resta, R., *see* McKinley, J.T. 398, 399
Resta, R., *see* Peressi, M. 398
Resta, R., *see* Resca, L. 127, 128
Reuszer, J.K. 164
Reuters, P.J. 351
Rezayi, E.H., *see* Haldane, F.D.M. 1003, 1021, 1026
Rhines, F.N. 332
Rhoderick, E.H. 340, 343, 344
Rhoderick, E.H., *see* Missous, M. 326, 344, 349
Ribot, H., *see* Yan, R.H. 850
Ricco, B. 653
Rice, T.M. 451, 454, 458
Rice, T.M., *see* Beni, G. 451, 456
Rice, T.M., *see* Brinkmann, W.F. 454, 459, 474
Richardson, B.E., *see* Fowell, A.E. 365
Richter, C.A., *see* McEwan, P.L. 796
Richter, H.J., *see* Solzbach, U. 303
Richter, H.W. 339, 340
Richter, H.W., *see* Brillson, L.J. 339
Richter, R. 438
Richter, R., *see* Peinke, J. 433, 435, 437
Richter, W., *see* Zahn, D.T.R. 380
Ricksand, A. 245, 263, 264
Riddoch, F.A. 692
Ridley, B.K. 676, 677, 681, 682, 684, 685, 693, 694, 696, 698, 699, 702, 703, 705, 706, 710, 711
Ridley, B.K., *see* Babiker, M. 697, 699
Ridley, B.K., *see* Balkan, N. 432, 700
Ridley, B.K., *see* Chamberlain, M.P. 695
Ridley, B.K., *see* Constantinou, N.C. 685
Ridley, B.K., *see* Daniels, M.E. 701
Ridley, B.K., *see* Gupta, R. 709, 711, 712, 715, 716
Ridley, B.K., *see* Lester, L.F. 679
Ridley, B.K., *see* Riddoch, F.A. 692
Riedel, R.A., *see* Davis, G.D. 332

Rieger, M. 423
Rieger, M., *see* Lugli, P. 698
Riffar, J.R., *see* Westland, D.J. 697
Ringhofer, C., *see* Kluksdahl, N.C. 653
Rioux, D.S., *see* Chang, S. 340, 350, 372–374
Ristig, M.L., *see* Lam, P.M. 1031
Ritchie, D.A., *see* Brown, R.J. 886, 899, 902, 904, 906, 907, 923, 924
Ritchie, D.A., *see* Cobden, D.H. 883, 892, 894
Ritchie, D.A., *see* Davies, A.G. 1019
Ritchie, D.A., *see* Ford, C.J.B. 867, 904, 906, 909, 911
Ritchie, D.A., *see* Greene, S.K. 916
Ritchie, D.A., *see* Kelly, M.J. 899, 902
Ritchie, D.A., *see* Patel, N.K. 899–901
Ritchie, D.A., *see* Smith, C.G. 659, 880, 889, 890, 923, 924, 930
Ritchie, D.A., *see* Tewordt, M. 922, 923
Ritchie, D.A., *see* Wharam, D.A. 800, 801, 867, 870, 871, 881, 882, 888, 902–904, 913, 923, 924, 1097, 1102, 1104
Ritchie, R.H., *see* Echenique, P.M. 658
Ritchie, R.H., *see* Zheng, X.-Y. 658
Rizzo, A., *see* de Blasi, C. 252
Rizzo, A., *see* Micocci, G. 252
Robbins, D.J. 422
Robbins, D.J., *see* Landsberg, P.T. 422
Roberts, C., *see* Ford, C.J.B. 904, 906, 1104, 1109, 1111, 1112, 1115, 1116
Roberts, C., *see* Foxon, C.T. 710
Roberts, C., *see* Nicholas, R.J. 832
Roberts, J.S., *see* Balkan, N. 700
Roberts, J.S., *see* Vickers, A.J. 432
Roberts, N., *see* Payne, M.C. 314, 319
Roberts, R.B. 724
Roberts, S., *see* Vallin, J.T. 170, 173
Robin, P., *see* Pouget, J.P. 622
Robinson, D.A.H., *see* Miller, G.L. 191
Robinson, G.Y., *see* Davidson, J.S. 730, 767
Robinson, G.Y., *see* Hökelek, E. 353
Robinson, H.G., *see* Deal, M.D. 273
Robinson, I.L., *see* Sauvage-Simkin, M. 302
Rockett, A., *see* Greene, J.E. 339
Rode, D.L. 493, 544, 551, 562, 574, 576
Rodriguez, S., *see* Ascarelli, G. 239
Rodriguez, S., *see* Brown, R.A. 239
Rodriguez, S., *see* Ramdas, A.K. 163
Roessler, O.E., *see* Peinke, J. 422, 435
Rogachev, A.A. 453, 454, 464, 465, 472
Rogachev, A.A., *see* Altukhov, P.D. 455, 456, 464, 476, 478–482
Rogachev, A.A., *see* Asnin, V.M. 453, 457, 462, 463, 465–472
Rogachev, A.A., *see* Grinberg, A.A. 233
Rogachev, A.A., *see* Monakhov, A.M. 456, 482, 483
Rogers, C.T., *see* Farmer, K.R. 205, 247
Rogers, D.L., *see* Woodall, J.M. 379, 382
Rohrer, H., *see* Binnig, G. 296, 362
Röhricht, B. 428
Röhricht, B., *see* Schöll, E. 433
Romanov, Yu.A., *see* Aleshkin, Yu.A. 524
Romanova, T.L., *see* Altukhov, P.D. 480–482
Romestain, R., *see* Deveaud, B. 838, 839
Rommelmann, H., *see* Epstein, A.J. 619–622
Rondi, D., *see* Soucail, B. 833
Rooks, M.J., *see* Chandrasekar, V. 948
Rorison, J.M. 707
Rose, F., *see* Bowers, R. 530
Rose, J.H. 475, 476
Rose, K., *see* Sarma, K. 380
Röseler, J., *see* Enderlein, R. 690
Rosenberg, J.J., *see* Woodall, J.M. 379, 382
Rosier, L.L. 190, 242
Rosier, L.L., *see* Sah, C.T. 177, 189, 220, 229, 241, 247
Rosker, M.J. 1070
Ross, S.F., *see* Jaros, M. 172
Rossi, F., *see* Brunetti, R. 1090
Rossi, J.A., *see* Crawford, M.G. 177
Roth, A.P., *see* Soucail, B. 833, 838, 840, 850, 851, 857
Roth, H. 655
Roth, L.M. 495, 509, 510, 512, 517, 518, 520, 521, 557
Roth, L.M., *see* Gold, L. 561
Roth, L.M., *see* Zwerdling, D. 522
Roth, S., *see* Bradley, D.D.C. 617
Roth, S., *see* Ehinger, K. 621
Roth, Y., *see* Cohen, A. 959
Rothberg, L. 615, 616
Rothuizen, H., *see* Faist, J. 894
Roukes, M.L. 867, 870, 872, 888, 892, 906, 908–912, 915, 916, 919
Roukes, M.L., *see* Scherer, A. 870
Roukes, M.L., *see* Shepard, K.L. 913, 914
Roukes, M.L., *see* Thornton, T.J. 877, 892, 906, 1087
Roukes, M.L., *see* Weiss, D. 922, 930
Rous, P.J., *see* MacLaren, J.M. 293
Rous, P.J., *see* Pendry, J.B. 898
Roussos, G., *see* Grebe, G. 181
Rowe, J.E. 312
Rowe, J.E., *see* Bennett, P.A. 296
Rowe, J.E., *see* Ma, Y. 322
Rowe, J.E., *see* Margaritondo, G. 322, 349
Roxlo, C.B., *see* Weinberger, B.R. 605
Roy, S., *see* Barker, J.R. 1115
Rubloff, G.W. 352
Rubloff, G.W., *see* Schmid, P.E. 335, 358
Rubtsov, G.P., *see* Altukhov, P.D. 480–482
Rücker, H. 695
Ruckman, M.W. 326, 334
Ruckman, M.W., *see* del Guidice, M. 334
Rudan, M., *see* Baccarani, G. 426
Rudan, M., *see* Quade, W. 426

Rudberg, B.G.R. 658
Rudlin, S. 695
Ruf, C. 730, 731, 741, 742, 760, 762, 763, 767
Rühle, W.W., *see* Grahn, H.T. 694
Rundquist, J. 896, 914
Runge, E. 16, 20
Ruoff, A.L., *see* Vohra, Y.K. 85, 87
Rupert, A., *see* Lambert, B. 139, 142
Ruppel, W. 232
Ruthen, R.M., *see* Roukes, M.L. 867, 870, 872, 908, 909, 911
Ruthen, R.M., *see* Scherer, A. 870
Ruthvan, A., *see* Syme, R.T. 754, 755, 785
Ryan, J.C., *see* Gammon, D. 825
Ryan, J.F., *see* Turberfield, A.J. 1006
Ryan, J.F., *see* Westland, D.J. 697
Ryder, E.J. 1044
Ryskin, A., *see* Dubenskii, K.K. 181
Rytov, S.M. 678
Ryvkin, S.M., *see* Asnin, V.M. 453
Ryvkin, S.M., *see* Grinberg, A.A. 233
Ryzkhin, I.A., *see* Abrikosov, A.A. 867, 945

Sa-Yakanit, V. 256–258
Sa-Yakanit, V., *see* Sritrakool, W. 257
Sablina, N.I., *see* Asnin, V.M. 462, 463, 465, 466
Sachdev, S. 3, 43
Sachrajda, A. 1020
Sachrajda, A., *see* Buckthought, A. 1020
Sachrajda, A.S., *see* Fletcher, R. 772
Sachrajda, A.S., *see* Taylor, R.P. 877, 883, 888, 894, 904, 906, 919, 923, 924
Sacks, R.N., *see* McEuen, P.L. 867, 892, 893, 898
Sacks, R.N., *see* McEwan, P.L. 796
Sadowski, M.L., *see* Karpierz, K. 168
Sáenz, J.J. 878
Sáenz, J.J., *see* de Raedt, H. 897
Sáenz, J.J., *see* García, N. 878
Sah, C.T. 177, 189, 220, 229, 241, 247
Sah, C.T., *see* Grove, A.S. 262
Sah, C.T., *see* Herman III, J.M. 190
Sah, C.T., *see* Rosier, L.L. 190, 242
Sai-Halasz, G.A. 822
Sai-Halasz, G.A., *see* Chang, L.L. 823, 836
Sajoto, T. 1000
Sajoto, T., *see* Engel, L.W. 1020
Sak, J. 945, 957
Sakai, Y. 65
Sakaki, H. 708, 709, 998
Sakaki, H., *see* Arakawa, Y. 874
Sakaki, H., *see* Hirakawa, K. 656, 701
Sakaki, H., *see* Wakabayashi, J. 1007
Sakamoto, T., *see* Kawai, N.J. 307
Sakamoto, T., *see* Takagaki, Y. 908
Sakata, S., *see* Kajiyama, K. 349
Saku, T., *see* Hirayama, Y. 870, 879, 886, 888–890, 911
Saku, T., *see* Tarucha, S. 922, 923
Sakurai, T., *see* Hasagawa, Y. 322
Sakurai, T., *see* Hashizume, T. 322
Salaneck, W.R., *see* Löglund, M. 623
Saldin, D.K., *see* MacLaren, J.M. 293
Salem, S. 593, 595
Salemink, H.W.M., *see* Albrektsen, O. 387
Salikin, H.L., *see* Brush, S.C. 1069
Salvan, F., *see* Benoit à la Guillaume, C. 462, 463, 472
Samoilovich, A.G. 493, 541, 550, 563
Samorukov, B.E., *see* Kasatkin, V.A. 139, 142
Sampere, S., *see* Timp, G. 881, 906, 1114
Samsonidze, G.G., *see* Efros, A.L. 896
Samuelson, B., *see* Monemar, B. 178
Sanders, G.D. 837
Sanders, G.D., *see* Miller, R.C. 837
Sanders, L.M., *see* Rose, J.H. 475, 476
Sandroff, C.J. 339
Sandroff, C.J., *see* Yablonovitch, E. 339
Sands, T. 325–327, 335, 336, 340, 365, 366, 374, 381
Sands, T., *see* Ding, J. 326
Sands, T., *see* Marshall, E.D. 337, 380
Sands, T., *see* Palmstrøm, C.J. 327, 340
Sang Jr, H.W., *see* Bauer, R.S. 380, 390, 391
Sang Jr, H.W., *see* Chiaradia, P. 319, 394, 395
Sankey, O., *see* Dow, J.D. 313
Sanquer, M., *see* Mailly, D. 938, 958
Sanquer, M., *see* Pichard, J.-L. 938, 959
Santos, M.B., *see* Sajoto, T. 1000
Sapega, V.F., *see* Mirlin, D.N. 153
Sarbey, O.G., *see* Asche, M. 422
Sarkar, C.K., *see* Kundu, S. 728, 734, 735
Sarker, S.K., *see* Hu, B.Y.-K. 1124
Sarma, K. 380
Sasa, S., *see* Komiyama, S. 999
Sasaki, W., *see* Kobayashi, S. 936
Sato, F., *see* Kawai, N.J. 307
Sauer, R. 476, 477
Sauvage-Simkin, M. 302
Savage, D.E. 326
Savoia, A., *see* Perfetti, P. 389
Sawada, T., *see* Hasagawa, J. 359
Scalapino, D.J., *see* Steiner, M.M. 8, 10, 12
Scalbert, D. 825
Schachter, R., *see* Olego, D.J. 338
Schaffer, H., *see* Moraes, F. 607
Schaffer, H., *see* Vardeny, Z. 607
Schaffer, W.J., *see* Kowalczyk, S.P. 389
Schaffer, W.J., *see* Tadayon, B. 326
Schäffler, F., *see* Schuberth, G. 647, 655
Schaft, W.J., *see* Lester, L.F. 679
Schauer, F., *see* Kocka, J. 251
Scheer, J.J., *see* van Laar, J. 349
Scheffler, M., *see* Beeler, F. 132
Scheffler, M., *see* Dabrowski, J. 149, 152, 155
Scheffler, M., *see* Gonze, X. 78
Scheffler, M., *see* Meyer, B.K. 155, 156

Scheider, H., *see* Grahn, H.T. 694
Schellnhuber, H.J. 929
Scher, H. 614
Scher, H., *see* Zallen, R. 994
Scherer, A. 870
Scherer, A., *see* Kash, K. 875
Scherer, A., *see* Roukes, M.L. 867, 870, 872, 892, 906, 908, 909, 911, 915, 916, 919
Scherer, A., *see* Thornton, T.J. 877, 892, 906, 1087
Schiavone, L.M., *see* Christen, J. 874
Schiebel, R.A., *see* Kirk, W.P. 986
Schillinger, W., *see* Koenig, S.H. 187
Schimansky-Geier, L. 443
Schirmer, O.F., *see* Koidl, P. 170
Schlapp, W., *see* Bimberg, D. 697
Schlapp, W., *see* Ebert, G. 1025
Schlapp, W., *see* Müller, G. 997
Schlesinger, T.E., *see* Hauenstein, R.J. 355
Schlesinger, Z., *see* Kirtley, J.P. 870
Schlesinger, Z., *see* Störmer, H.L. 986
Schlier, R.E. 296
Schlögl, F. 428, 433
Schlüter, M. 3, 4, 65, 116, 117, 296, 297, 313, 314, 319, 346, 348, 354
Schlüter, M., *see* Bachelet, G.B. 72–74, 79, 118, 130, 155
Schlüter, M., *see* Baraff, G.A. 117, 118, 152
Schlüter, M., *see* Godby, R.W. 3, 4, 19, 22, 116
Schlüter, M., *see* Greenside, H.S. 73
Schlüter, M., *see* Hamann, D.R. 14, 15, 72, 73, 77
Schlüter, M., *see* Kerker, G.P. 65
Schlüter, M., *see* Lannoo, M. 3, 4, 22, 116
Schlüter, M., *see* Louie, S.G. 118
Schlüter, M., *see* Sham, L.J. 3, 4, 22, 29, 33, 116
Schmeisser, D., *see* Bätz, P. 623
Schmid, A., *see* Chakravarty, S. 867
Schmid, A., *see* Kree, R. 945, 957
Schmid, H., *see* Washburn, S. 998
Schmid, P.E. 335, 347, 348, 352, 358
Schmid, P.E., *see* Liehr, M. 355
Schmid, U., *see* Christensen, N.E. 390
Schmidt, M., *see* Stocker, H.J. 175
Schmidt, P.-H., *see* Forrest, S.R. 389
Schnatterly, S.E., *see* Henry, C.H. 156
Schneider, H. 848, 850, 854, 857
Schneider, J. 184
Schneider, J., *see* Aszodi, G. 139, 142
Schneider, J., *see* de Wit, M. 184
Schneider, J., *see* Ennen, H. 139, 142
Schneider, J., *see* Holton, W.C. 184
Schneider, J., *see* Kaufmann, U. 185
Schneider, J., *see* Weber, E.R. 151, 357
Schneider, T., *see* Bishop, A.R. 601
Schoenlein, R.W. 1071
Schoenlein, R.W., *see* Bigot, J.Y. 22
Schöll, E. 234, 236, 421, 422, 426, 428–440, 442–444
Schöll, E., *see* Döttling, R. 435
Schöll, E., *see* Hüpper, G. 432, 435, 437
Schöll, E., *see* Kunz, R. 440
Schöll, E., *see* Landsberg, P.T. 422, 423
Schöll, E., *see* Naber, H. 433, 439
Schöll, E., *see* Quade, W. 426
Schöll, E., *see* Rau, U. 433
Schöll, E., *see* Robbins, D.J. 422
Schöll, E., *see* Schimansky-Geier, L. 443
Schöll, E., *see* Schlögl, F. 433
Schöll, E., *see* Shaw, M.P. 422, 440
Schöll, E., *see* Symanczyk, R. 440
Schöll, E., *see* Wacker, A. 432, 440
Schöll, E., *see* Weispfenning, M. 428
Scholten, A.J., *see* Dekker, C. 883, 894
Schommers, W. 1116
Schön, G., *see* Kramer, B. 868
Schönenberg, Ch., *see* Abraham, D. 387
Schönhammer, K. 12
Schönhammer, K., *see* Gunnarson, O. 3, 4
Schottky, W. 341, 342, 629
Schowalter, L.J., *see* Hauenstein, R.J. 355
Schreiber, M. 959
Schrey, F., *see* Tung, R.T. 305
Schrieffer, J.R., *see* Arovas, D. 1009
Schrieffer, J.R., *see* Bardeen, J. 102
Schrieffer, J.R., *see* Heeger, A.J. 586, 595, 598, 605, 607
Schrieffer, J.R., *see* Jeyadev, S. 613
Schrieffer, J.R., *see* Su, W.-P. 599, 605, 609, 615
Schroeder, D.K. 229, 242, 286
Schroeder, P.A., *see* Blatt, F.J. 747
Schubert, E.F. 703
Schuberth, G. 647, 655
Schüle, R., *see* Grambow, P. 870, 872
Schulman, J.N. 825, 832
Schulman, J.N., *see* Chang, Y.C. 825, 832, 833, 837
Schulman, J.N., *see* Miller, R.C. 837
Schult, R.L. 896, 905, 908, 913–915
Schult, R.L., *see* Ravenhall, D.G. 905, 913–915
Schulz, H.J., *see* Busse, J.W. 181
Schulz, H.J., *see* Grebe, G. 181
Schulz, J.H., *see* Broser, J. 181
Schulz, M. 184, 264
Schulz, M., *see* Baars, J. 332, 353
Schumm, R.H., *see* Wagman, D.D. 332
Schuster, H.G. 421, 424, 435, 437–439
Schwartz, C.L., *see* Marshall, E.D. 337, 380
Schwarz, S.A., *see* Marshall, E.D. 337, 380
Schweitzer, L. 955, 994
Schweitzer, L., *see* MacKinnon, A. 955
Scifres, D.R. 179
Scott, A.C. 601
Scott, J.C., *see* Brédas, J.-L. 606
Scott, J.C., *see* Clarke, T.C. 602, 607
Scott, M.D., *see* Westland, D.J. 697
Scott-Thomas, J.H.F. 924

Scott-Thomas, J.H.F., *see* Field, S.B. 924
Seabaugh, A.C., *see* Randall, J.N. 874, 875
Seager, C.H. 265
Seah, M.P. 286, 288
Sealy, B.J., *see* Tang, Y.S. 139, 142
Sears, V.F. 1029
Seas, A., *see* Walser, A.D. 613, 614
Sébenne, C.A., *see* Chen, P. 380
Sébenne, C.A., *see* Guichar, G.M. 302, 303
Seedorf, R., *see* Wilke, W.G. 389
Seeger, K. 525, 526, 562, 563, 566, 567
Segall, B., *see* Lorenz, M.R. 177
Segmuller, A. 820
Segzda, D., *see* Matulis, A. 885
Seiler, D.G. 432, 502, 521
Seiler, D.G., *see* Song, X.N. 437
Seiler, R., *see* Avron, J.E. 930, 992
Seitz, F. 560
Seitz, F., *see* Fröhlich, H. 1043
Seiwatz, R. 296
Sekiguchi, Y., *see* Sakaki, H. 998
Selci, S., *see* Chiaradia, P. 319
Selci, S., *see* Cricenti, A. 91
Selloni, A. 297, 319
Semenchinsky, S.G., *see* D'Iorio, M. 1007
Semenchinsky, S.G., *see* Pudalov, V.M. 984, 989, 998, 999
Senna, J.R., *see* Ting, C.S. 785
Seraphin, B.O. 656
Serdyukova, S.I., *see* Bakhvalov, N.S. 1115
Sermage, B., *see* Marzin, J.Y. 824
Sernelius, B.E. 22, 736
Serota, R.A. 793, 959
Sette, F., *see* Ma, Y. 322
Sette, F., *see* Perfetti, P. 389
Shaban, E.H., *see* Landsberg, P.T. 191
Shacklee, K.L., *see* Grossman, B. 465
Shacklee, K.L., *see* Mattos, J.V.C. 466
Shacklee, K.L., *see* Voos, M. 464
Shacklee, K.L., *see* Worlock, J.M. 464
Shacklette, L.W., *see* Baughman, R.H. 592, 619
Shah, J. 698
Shah, J., *see* Becker, P.C. 1070
Shah, J., *see* Deveaud, B. 697, 854
Sham, L.J. 3, 4, 8, 12, 14, 19, 22, 29–33, 116
Sham, L.J., *see* Austin, B.J. 6, 8, 62
Sham, L.J., *see* Broido, D.A. 837
Sham, L.J., *see* Godby, R.W. 3, 4, 19, 22, 116
Sham, L.J., *see* Hanke, W. 8, 10, 12
Sham, L.J., *see* Kohn, W. 4, 13, 72, 116, 872
Sham, L.J., *see* Lannoo, M. 3, 4, 22, 116
Sham, L.J., *see* Schlüter, M. 3, 4, 116
Sham, L.J., *see* Steiner, M.M. 8, 10, 12
Sham, L.J., *see* White, S. 827
Shanabrook, B.V. 825
Shanabrook, B.V., *see* Gammon, D. 825
Shanabrook, B.V., *see* Jarosik, N.C. 168
Shank, C.V. 22
Shank, C.V., *see* Becker, P.C. 1070
Shank, C.V., *see* Bigot, J.Y. 22
Shapira, Y. 333, 337
Shapiro, B. 959
Shapiro, B., *see* Cohen, A. 959
Shappir, J., *see* Kriegler, R.J. 264
Sharan, R. 170
Sharvin, Yu. 937, 948
Sharvin, Yu.V. 883
Sharvin, Yu.V., *see* Aronov, A.G. 937, 946, 948
Sharvin, Yu.V., *see* Sharvin, Yu. 937, 948
Shashkin, V.I., *see* Belyantsev, A.M. 432
Shatz, S., *see* Avishai, Y. 890
Shaw, D. 351
Shaw, J.L. 333, 340, 351, 367, 370, 371
Shaw, J.L., *see* Brillson, L.J. 370, 371
Shaw, J.L., *see* Chang, S. 372
Shaw, J.L., *see* Viturro, R.E. 358, 372
Shaw, M.P. 341, 361, 422, 440
Shayegan, M., *see* Engel, L.W. 956, 1020
Shayegan, M., *see* Goldman, V.J. 1007
Shayegan, M., *see* Sajoto, T. 1000
Shayegan, M., *see* Simmons, J.A. 890
Sheard, F.W. 871, 881
Sheard, F.W., *see* Leadbeater, M.L. 655
Sheinkman, M.K. 230
Shekhter, R.I., *see* Glazman, L.I. 658, 886, 887, 924
Shen, T.-H. 398
Shen, T.-H., *see* Fowell, A.E. 365
Shen, Y.Q., *see* Bradley, D.D.C. 617
Shepard, K.L. 913, 914
Shi, H. 912, 928
Shi, Z.Q. 368
Shibuya, M. 561
Shichijo, H., *see* Hess, K. 432
Shih, C.K., *see* Friedman, D.J. 334
Shih, Y.-C., *see* Murakami, M. 337, 365, 366
Shik, Y.A., *see* Kozyrev, S.V. 687
Shikin, V., *see* Monarkha, Y.P. 980
Shikin, V.B. 886
Shimizu, S., *see* Emura, S. 714
Shimomura, M., *see* Akagi, K. 591
Shiraishi, K., *see* Yamaguchi, E. 130, 145, 146, 149, 150
Shirakawa, H., *see* Akagi, K. 591
Shirakawa, H., *see* Mizoguchi, K. 608, 612
Shirley, E.L. 74
Shklovskii, B.I. 254, 867
Shklovskii, B.I., *see* Efros, A.L. 935
Shmartsev, Yu.V., *see* Vul, A.Ya. 188
Shockley, W. 222, 223, 273, 505, 1043, 1051
Shockley, W., *see* Bardeen, J. 571, 610
Shockley, W., *see* van Roosbroeck, W. 229
Shore, H.B. 454, 475
Shore, H.B., *see* Rose, J.H. 475, 476
Shronova, L.V., *see* Vul, A.Ya. 188
Shtrikman, H., *see* Sivan, U. 872, 879

Shubnikov, L. 512
Shuey, R.T. 641
Shukhorukov, E.V., *see* Levinson, Y.B. 893
Shukla, P., *see* Landsberg, P.T. 423
Shulman, J.N. 674
Sibeldin, N.N., *see* Bagaev, V.S. 460, 464, 466
Sibille, A. 855
Siday, S.E., *see* Ehrenberg, W. 1110
Sidorow, V.J., *see* Lifshitz, T.M. 174
Siefert, R.L., *see* Trafas, B.M. 323
Sikka, S.K., *see* Olijnyk, H. 87
Sikorski, Ch., *see* Merkt, U. 922
Silberberg, Y., *see* Sands, T. 325–327, 336, 340, 365, 366, 374, 381
Silbey, R., *see* Chance, R.R. 602
Silin, A.P., *see* Keldysh, L.V. 456
Sillmon, R.S., *see* Glaser, E. 150
Silver, R.N. 460
Silverstein, S.D., *see* Bennett, A.J. 656
Simes, R.J., *see* Tsuchiya, M. 874
Simes, R.J., *see* Yan, R.H. 850
Simmons, J.A. 890, 908
Simmons, J.A., *see* Li, Y.P. 883, 894
Simmons, J.A., *see* Roukes, M.L. 870, 911
Simon, B., *see* Avron, J.E. 930
Simon, B., *see* Delyon, F. 934
Simon, R.W., *see* Burns, M.J. 810
Simons, G. 96
Simons, G., *see* Bloch, A.N. 96
Sinclair, M. 612, 615, 616
Sinclair, R., *see* Beyers, R.K. 336
Singelton, J., *see* Huant, S. 169
Singer, K.E., *see* Missous, M. 326, 344, 349
Singh, I., *see* Srivastava, G.P. 314
Singh, J., *see* Oh, J.E. 326
Singh, V.A. 118
Singleton, J., *see* Patel, N.K. 899–901
Singwi, K.S. 459
Singwi, K.S., *see* Bhattacharya, P.K. 459
Sinha, S.K. 95
Sinitsyn, M.A., *see* Belyantsev, A.M. 432
Sipe, J.E., *see* Dignam, M.M. 838, 851, 857
Sivan, U. 659, 796, 809, 872, 879, 880, 911, 925
Sivan, U., *see* Heiblum, M. 872, 880
Sivaprakasam, S., *see* Lent, C.S. 896, 897, 899, 900, 902, 1104, 1109
Sivco, D.L., *see* Beltram, F. 855
Skarstam, B., *see* Rees, G.J. 239
Skeath, P. 323
Skeath, P., *see* Spicer, W.E. 356, 357, 370, 371
Skibowski, M., *see* Carstensen, H. 317
Skibowski, M., *see* Manzke, R. 317
Skibowski, M., *see* Straub, D. 316
Skillman, S., *see* Herman, F. 140, 141
Skocpol, W.J. 735, 869, 870, 958
Skocpol, W.J., *see* Ralls, K.S. 205, 247
Skolnick, M.S., *see* Whittaker, D.M. 851, 852
Skowroński, M., *see* Kamińska, M. 151–153, 170
Skromme, B.J. 126
Slack, G.A., *see* Ham, F.S. 170
Slack, G.A., *see* Nygren, B. 173
Slack, G.A., *see* Vallin, J.T. 170, 173
Slade, M.L., *see* Brillson, L.J. 339, 349, 372
Slade, M.L., *see* Chiradia, P. 372
Slade, M.L., *see* Viturro, R.E. 362, 364, 370
Slater, J.C. 66, 71, 155
Slater, J.C., *see* Koster, G.F. 661
Slichter, C.B., *see* Henry, C.H. 156
Slowik, J., *see* Brillson, L.J. 333, 337, 366
Smith, A.C. 526
Smith, C.G. 659, 880, 889, 890, 923, 924, 930
Smith, C.G., *see* Brown, R.J. 904, 906, 907, 923, 924
Smith, C.G., *see* Kelly, M.J. 899, 902
Smith, C.G., *see* Martín-Moreno, L. 886, 924
Smith, D.L. 378, 385, 819, 825, 827
Smith, D.L., *see* Daw, M.S. 118, 356
Smith, D.L., *see* Laurich, B.K. 825
Smith, D.L., *see* Mailhiot, C. 378, 825
Smith, E.F. 239
Smith, G.W., *see* Whittaker, D.M. 851, 852
Smith, H.I., *see* Field, S.B. 924
Smith, H.I., *see* Ismail, K. 659, 930
Smith, H.I., *see* Liu, C.T. 930
Smith, H.I., *see* Scott-Thomas, J.H.F. 924
Smith, M.J. 747–749, 757, 758, 760
Smith, M.J., *see* Gallagher, B.L. 732, 753, 754, 759
Smith, M.J., *see* Karavolas, V.C. 728, 733
Smith, N.A., *see* Harris, I.R. 326
Smith, P., *see* Cao, Y. 591, 617
Smith, P.V., *see* Craig, B.I. 314, 319
Smith, R.J. 312
Smith, R.S., *see* Ennen, H. 139, 142
Smith, S.C., *see* Haase, M.A. 385
Smith III, T.P. 922, 984
Smith III, T.P., *see* Haug, R.J. 872, 873, 926, 927
Smith III, T.P., *see* Ismail, K. 930
Smrcka, L. 764, 997
Smrcka, L., *see* Havlova, H. 769, 770
Snell, B.R., *see* Beton, P.H. 888, 889
Snell, B.R., *see* Main, P.C. 888
Snow, E.H., *see* Grove, A.S. 262
Snyman, L.W. 326
So, E., *see* Mark, P. 356
Söderström, J.R., *see* Chow, D.H. 647
Sokolov, I.M., *see* Mil'nikov, G.V. 955, 994
Solomon, P.M. 646
Solomon, P.R., *see* Shaw, M.P. 440
Sols, F. 884, 897, 898, 912
Sols, F., *see* Pevzner, V. 1123
Solzbach, U. 303
Somorjai, G.A. 293, 324
Somorjai, G.A., *see* MacLaren, J.M. 293
Somorjai, G.A., *see* van Hove, M.A. 286, 293
Song, J.J., *see* Jung, P.S. 840

Song, L.W. 250
Song, X.N. 437
Sorba, L. 398, 399
Sorbello, R.S., *see* Chu, C.S. 1109, 1110
Sørensen, C.B., *see* Lindelof, P.E. 1028
Sotomayor-Torres, C.M., *see* Beaumont, S.P. 867, 875
Sou, I.K., *see* Chow, D.H. 647
Soucail, B. 833, 838, 840, 842, 845, 846, 850, 851, 857
Soucail, B., *see* Ferreira, R. 839, 842, 843, 852, 854
Souillard, B., *see* Delyon, F. 934
Souillard, B., *see* Kunz, H. 944, 955
Soukassian, P., *see* Franciosi, A. 338
Soukoulis, C.M., *see* Economou, E.N. 884, 904, 912, 932
Spaeth, J.-M. 154, 156
Spaeth, J.-M., *see* Fockele, M. 150
Spaeth, J.-M., *see* Hoffmann, D.M. 351
Spaeth, J.-M., *see* Krambrock, K. 156
Spaeth, J.-M., *see* Meyer, B.K. 152, 153, 155, 156
Spain, I.L., *see* Hu, J.Z. 87
Spector, H.N., *see* Lee, J. 705
Spector, J. 872, 874, 878–880
Spector, J., *see* Pfeiffer, L.N. 874
Spector, J., *see* Störmer, H.L. 874
Spencer, M.G., *see* Glaser, E. 150
Spencer, M.G., *see* Tadayon, B. 326
Spendeler, L.I.A., *see* Williamson, J.G. 880
Speriosu, V.S. 820
Speriosu, V.S., *see* Farrow, R.F.C. 326
Spicer, W.E. 349, 356, 357, 370, 371
Spicer, W.E., *see* Cao, R.K. 349, 358
Spicer, W.E., *see* Friedman, D.J. 334
Spicer, W.E., *see* Kendelewicz, T. 334, 337
Spicer, W.E., *see* Lilienthal-Weber, Z. 326, 336, 358
Spicer, W.E., *see* Lindau, I. 356
Spicer, W.E., *see* List, R.S. 368
Spicer, W.E., *see* Miyano, K. 372
Spicer, W.E., *see* Newman, N. 343–345, 349
Spicer, W.E., *see* Skeath, P. 323
Spinnewyn, J. 432, 437
Spitzer, W.G., *see* Kim, O.K. 714
Spitzer, W.G., *see* Mead, C.A. 344
Spivak, B.Z., *see* Altshuler, B.L. 937, 947
Sritrakool, W. 257
Srivasta, G.P. 118
Srivastava, G.P. 314, 356
Srivastava, G.P., *see* Montgomery, V. 367
St. John, J. 96
Staehli, J.L., *see* Westervelt, R.M. 459, 461
Stafström, S., *see* Löglund, M. 623
Stahl, E. 984
Stanton, C.J. 706
Stanton, C.J., *see* Bailey, D.W. 1071
Stapor, A., *see* Przybylinska, H. 139, 142
Staring, A.A.M. 888, 921, 1115
Staring, A.A.M., *see* Kouwenhoven, L.P. 888, 889, 921, 923
Staring, A.A.M., *see* Molenkamp, L.W. 872, 877, 878
Stark, J.P. 271
Starkloff, T. 14, 15, 73
Stauss, G.H., *see* Krebs, J.J. 185
Steckenborn, A., *see* Bimberg, D. 697
Stecker, L., *see* Aina, L. 822
Steele, A.G. 478
Stefanowa, S., *see* Baranowski, J.M. 175
Stegun, I.A., *see* Abramowitz, M. 521, 634
Steiner, M.M. 8, 10, 12
Stepanov, V.I., *see* Asnin, V.M. 457, 466, 467, 469–471
Stephen, M., *see* Abrahams, E. 945, 957
Stepniewski, R. 181
Stern, F. 659, 703
Stern, F., *see* Ando, T. 479, 481, 667, 698, 708, 728, 732, 736, 869, 979
Stern, F., *see* Casey Jr, H.C. 233
Stern, F., *see* Kumar, A. 872, 922, 1107
Stern, F., *see* Laux, S.E. 872, 1087
Stern, F., *see* Meirav, U. 875
Stern, F., *see* Wang, W.I. 385
Stewart, H.B., *see* Thompson, J.M.T. 421, 435
Stickler, T.J. 519
Stickler, T.J., *see* Rauch, F.J. 499
Stievenard, D., *see* Delerue, C. 152
Stievenard, D., *see* von Bardeleben, H.J. 151, 153
Stiles, K. 349
Stiles, K., *see* Kahn, A. 349, 350
Stiles, P.J., *see* Fang, H. 930
Stiles, P.J., *see* Smith III, T.P. 984
Stillman, G.E. 168, 174, 187, 548, 556
Stillman, G.E., *see* Crawford, M.G. 177
Stillman, G.E., *see* Haase, M.A. 385
Stillman, G.E., *see* Skromme, B.J. 126
Stirn, R.J. 499, 552, 561, 563
Stocker, H.J. 175
Stoffel, N.G., *see* Brillson, L.J. 332, 333, 337, 338, 354, 358, 366
Stoffel, N.G., *see* Margaritondo, G. 396, 398
Stolwijk, N.A., *see* Frank, W. 272
Stone, A.D. 789, 790, 801, 884, 912, 945, 957, 959, 1099
Stone, A.D., *see* Baranger, H.U. 910–912, 914–918
Stone, A.D., *see* Jalabert, R.A. 910, 911, 918, 919
Stone, A.D., *see* Lee, P.A. 790, 791, 934, 959
Stone, A.D., *see* McEwan, P.L. 796
Stone, A.D., *see* Muttalib, K.A. 943, 959
Stone, A.D., *see* Skocpol, W.J. 735, 958
Stone, A.D., *see* Szafer, A. 884, 886, 912
Stoneham, A.M. 239, 243
Stoneham, A.M., *see* Tasker, P.W. 178
Stoner, R., *see* D'Iorio, M. 761, 770, 780

Stoner, R., *see* Fletcher, R. 771, 772
Stoner, R., *see* Landheer, D. 674
Stoop, R., *see* Peinke, J. 422, 435
Stopa, M.P. 872
Störmer, H.L. 874, 979, 981, 986, 987, 999
Störmer, H.L., *see* Boebinger, G.S. 1014
Störmer, H.L., *see* Chang, A.M. 1000
Störmer, H.L., *see* Eisenstein, J.P. 984, 1019, 1020, 1027
Störmer, H.L., *see* Gornik, E. 984
Störmer, H.L., *see* Jiang, H.W. 1007, 1027
Störmer, H.L., *see* Narayanamurti, V. 678
Störmer, H.L., *see* Pfeiffer, L.N. 874
Störmer, H.L., *see* Spector, J. 872, 874, 878–880
Störmer, H.L., *see* Tsui, D.C. 979, 999
Störmer, H.L., *see* Willett, R. 999, 1000, 1015, 1026
Stradling, R.A. 174, 701
Strait, J., *see* Vardeny, Z. 607
Strasser, G., *see* Gornik, E. 984
Stratton, R. 649, 1046, 1059
Straub, D. 316
Straub, D., *see* Drube, W. 317
Straub, D., *see* Magnusson, K.O. 317
Straub, W.D., *see* Roth, H. 655
Strauss, A.J., *see* Iselev, G.W. 177, 188
Strauven, H., *see* Spinnewyn, J. 432, 437
Straw, A., *see* Vickers, A.J. 432
Streda, P. 764, 765, 796, 803, 904, 913, 990, 998, 999
Streda, P., *see* Haug, R.J. 998
Streda, P., *see* MacDonald, A.H. 997
Streda, P., *see* Oji, H. 765
Streda, P., *see* Smrcka, L. 764
Street, G.B., *see* Brédas, J.-L. 606
Street, R.A. 251
Streetman, B.G., *see* Hess, K. 432
Strehlov, R., *see* Dabrowski, J. 149
Strinati, G.S., *see* Castellani, C. 785
Strite, S., *see* Costa, J.C. 368
Stroscio, J.A. 305, 322
Stroscio, J.A., *see* Feenstra, R.M. 305
Stroscio, J.A., *see* Whitman, L.J. 305, 350
Strutz, T. 186
Studna, A.A., *see* Aspnes, D.E. 307, 310, 320
Sturge, M.D., *see* Kash, K. 875
Sturge, M.D., *see* Meynadier, M.H. 825
Stutz, C.E., *see* Look, D.C. 368
Su, B. 659
Su, C.Y., *see* Lindau, I. 356
Su, C.Y., *see* Skeath, P. 323
Su, C.Y., *see* Spicer, W.E. 356, 357, 370, 371
Su, W.-P. 599, 605, 609, 615
Su, W.-P., *see* Heeger, A.J. 586, 595, 598, 605, 607
Su, Z. 613
Subbanna, S., *see* Caine, E.J. 389, 822
Subramaniam, S. 897
Suckling, A.M., *see* Clark, R.G. 1018–1020
Suen, Y.W., *see* Sajoto, T. 1000
Suezaki, M., *see* Akagi, K. 591
Sugano, S. 132
Sugano, T., *see* Yamasaki, K. 263
Sugar, J., *see* Martin, W.C. 139
Suhl, H., *see* Pearson, G.L. 560, 562
Suisky, D., *see* Enderlein, R. 690
Sukhorukov, E.V. 908
Sumita, I., *see* Hashizume, T. 322
Summerfield, S. 619
Summerfield, S., *see* Ehinger, K. 621
Sun, H. 659
Sun, Y. 872
Sundaram, G.M., *see* Warburton, R.J. 824
Sundgren, J.-E., *see* Greene, J.E. 339
Suris, R.A., *see* Kazarinov, R.F. 855
Suto, K. 184
Suzuki, K., *see* Hensel, J.C. 499, 519
Suzuki, Y., *see* Hirayama, Y. 870
Svensson, S.P. 326, 344, 357
Svensson, S.P., *see* Sakaki, H. 998
Svistunova, K.I., *see* Pokrovskii, Ya.E. 459, 463, 467–469
Swain, S., *see* Boardman, A.D. 1063
Swanson, A.G., *see* Furneaux, J.E. 1020
Swarts, C.A. 314
Swartz, L.-E., *see* Biegelsen, D.K. 90, 294, 300, 302, 307
Swartzentruber, B.S. 304
Swartzentruber, B.S., *see* Becker, R.S. 296
Swartzentruber, B.S., *see* Mo, Y.-M. 321
Symanczyk, R. 440
Symanczyk, R., *see* Jäger, D. 440
Syme, R.T. 754, 755, 760, 761, 783–788
Syme, R.T., *see* Kearney, M.J. 785, 787–789
Syphers, D.A., *see* Furneaux, J.E. 1020
Szafer, A. 884, 886, 912
Szafer, A., *see* McEwan, P.L. 796
Szafer, A., *see* Stone, A.D. 801, 884, 912, 1099
Sze, S.M. 341, 343, 345
Szmulowicz, F. 502, 543, 552, 553, 563
Szmulowicz, F., *see* Peterson, T.L. 553, 554, 563
Szynka, *see* Cox, G. 305

't Hooft, G.W., *see* Montie, E.A. 886
Tabatabaie, N., *see* Sands, T. 325–327, 336, 340, 365, 366, 374, 381
Taboryski, R., *see* Lindelof, P.E. 1028
Tache, N., *see* Brillson, L.J. 339, 349, 370, 371
Tache, N., *see* Chiradia, P. 372
Tache, N., *see* Davis, G.D. 332
Tache, N., *see* Viturro, R.E. 372
Tadayon, B. 326
Tadayon, S., *see* Tadayon, B. 326
Takagaki, Y. 867, 906, 908, 911, 918
Takahashi, A., *see* Tsukamoto, J. 617, 618
Takahashi, H., *see* Hasagawa, J. 359
Takahashi, M., *see* Takayanagi, K. 90, 295, 296

Takahashi, S., *see* Takayanagi, K. 90, 295, 296
Takahashi, T. 875
Takamasu, T., *see* Komiyama, S. 999
Takano, F., *see* Tankei, K. 957
Takaoka, S., *see* Takagaki, Y. 867, 906, 908, 911
Takara, N., *see* Yamada, K. 432
Takatani, S. 389
Takayama, M. 599–601
Takayanagi, K. 90, 295, 296
Takiguchi, T. 617
Takimoto, N. 703
Taleb-Ibrahimi, A., *see* Ludeke, R. 357, 372
Talwar, D.N. 118, 154
Tamargo, M.C., *see* Farrell, H.H. 382
Tamargo, M.C., *see* Meynadier, M.H. 825
Tamaševičius, A., *see* Pyragas, K.A. 432
Tamaševičius, A.V., *see* Bumelienė, S.B. 432
Tamor, M.A., *see* Liu, A.Y. 101
Tanabe, Y., *see* Akagi, K. 591
Tanabe, Y., *see* Sugano, S. 132
Tanaka, M., *see* Sakaki, H. 708, 709
Tanaka, T., *see* Nishitani, R. 303
Tang, C.L., *see* Bailey, D.W. 1071
Tang, C.L., *see* Bak, P. 445
Tang, C.L., *see* Rosker, M.J. 1070
Tang, C.L., *see* Wise, F.W. 1071
Tang, M., *see* Niles, D.W. 399
Tang, Y.S. 139, 142
Tanishiro, Y., *see* Takayanagi, K. 90, 295, 296
Tankei, K. 957
Tanner, D.B. 623
Tanner, D.B., *see* Epstein, A.J. 619, 620
Tao, R. 993, 1020, 1021
Tao, R., *see* Widom, A. 885
Tao, Z. 743, 744
Tarucha, S. 922, 923
Tarucha, S., *see* Hirayama, Y. 870, 879, 911
Tarucha, S., *see* Nakata, S. 870
Tasch Jr, A.F., *see* Sah, C.T. 177, 189, 220, 229, 241, 247
Tasker, P.W. 178
Tasker, P.W., *see* Tadayon, B. 326
Tauc, J. 726
Tausendfreund, B. 1014
Taylor, B.N. 882, 986
Taylor, L.L., *see* White, A.M. 178
Taylor, R., *see* Dagens, L. 14
Taylor, R., *see* Geldart, D.J.W. 26
Taylor, R., *see* Rasolt, M. 8, 10, 14
Taylor, R.C., *see* Onton, A. 180
Taylor, R.P. 877, 883, 888, 894, 904, 906, 919, 923, 924
Taylor, R.P., *see* Beton, P.H. 745, 842
Taylor, W.E. 264
Teitsworth, S.W. 422, 432
Tejedor, C. 326, 346, 347, 389
Tejedor, C., *see* Flores, F. 390
Tekman, E. 878, 884–886, 893, 896
Teller, E., *see* Brush, S.C. 1069
Tennant, D.M., *see* Ralls, K.S. 205, 247
Tennant, D.M., *see* Skocpol, W.J. 735, 958
Tepore, A., *see* de Blasi, C. 252
Tepore, A., *see* Micocci, G. 252
Ter Haar, D. 1089, 1090
Tersoff, J. 135, 346, 347, 390, 392
Tešanović, Z. 3, 43
Teter, M., *see* Allan, D.C. 106
Tewordt, M. 922, 923
Tharmalingham, K. 524
Theis, T.N., *see* Kirtley, J.P. 870
Thémans, B., *see* Brédas, J.-L. 595, 596, 603
Theophilou, N., *see* Djurado, D. 590
Thewalt, M.L.W. 477
Thewalt, M.L.W., *see* Steele, A.G. 478
Thio, T., *see* Chen, C.Y. 934
Thoma, R. 426
Thomas, B.W., *see* Gibb, R.P. 239
Thomas, D.C., *see* Heiblum, M. 707
Thomas, D.E., *see* Chynoweth, A.G. 656
Thomas, D.G. 179, 180
Thomas, D.G., *see* Hopfield, J.J. 179
Thomas, G.A. 451, 474, 475, 934
Thomas, G.A., *see* Hensel, J.C. 451
Thomas, H. 422
Thomas, H., *see* Pytte, E. 422
Thomas, S.R. 181
Thome, H., *see* Tuchendler, J. 167
Thomeer, R.A.J., *see* Huant, S. 169
Thompson, G.H.B. 233
Thompson, J.M.T. 421, 435
Thompson, M.J., *see* Street, R.A. 251
Thoms, S., *see* Gallagher, B.L. 787, 793–795
Thoms, S., *see* Galloway, T. 791, 793
Thornber, K. 1124
Thornton, T.J. 735, 869–871, 877, 881, 892, 906, 1087
Thornton, T.J., *see* Berggren, K.F. 871, 881, 902
Thornton, T.J., *see* Ford, C.J.B. 867, 904, 906, 909, 911, 1104, 1109, 1111, 1112, 1115, 1116
Thornton, T.J., *see* Roukes, M.L. 870, 892, 906, 908, 911
Thornton, T.J., *see* van Houten, H. 902, 904, 906
Thornton, T.J., *see* Wharam, D.A. 800, 801, 867, 870, 871, 881, 882, 902, 904, 913, 1097, 1102, 1104
Thouless, D.J. 867, 930, 932, 937, 944, 947, 948, 990, 993, 997, 1029
Thouless, D.J., *see* Anderson, P.W. 932, 956
Thouless, D.J., *see* Edwards, J.T. 943
Thouless, D.J., *see* Li, Q. 913
Thouless, D.J., *see* Licciardello, D.C. 994
Thouless, D.J., *see* Niu, Q. 930, 992, 993, 1020
Thuault, C.D., *see* Guichar, G.M. 302, 303
Thurmond, C.D. 249
Thurmond, C.D., *see* van Vechten, J.A. 211, 244
Tiedje, T., *see* Colbow, K.M. 389

Tiemann, J.J. 644
Tikhodeev, S.G., *see* Keldysh, L.V. 465
Timmering, C.E., *see* Kouwenhoven, L.P. 744, 872, 924, 926, 927
Timmering, C.E., *see* Molenkamp, L.W. 872, 877, 878
Timmering, C.E., *see* van Wees, B.J. 904, 906, 921, 924
Timofeev, V.B., *see* Buhmann, H. 1006, 1007
Timofeev, V.B., *see* Rashba, E.I. 979
Timp, G. 809, 813, 867, 872, 875, 881, 883, 892, 894, 904, 906, 908, 909, 911, 1114
Timp, G., *see* Behringer, R.E. 914
Timp, G., *see* Chang, A.M. 911
Timp, G., *see* de Vegvar, P.G.N. 791
Timp, G., *see* Prentiss, M. 875
Ting, C.S. 785
Ting, C.S., *see* Talwar, D.N. 118, 154
Ting, D.Z.-Y., *see* Chow, D.H. 647
Title, R.S. 139, 142
Tochihara, H., *see* Hashizume, T. 322
Tokuda, N., *see* Okuyama, Y. 759, 760
Tokumoto, H.S. 305
Tokura, Y., *see* Fukui, T. 875
Tomak, M., *see* Sernelius, B.E. 736
Tomlinson, R.D., *see* Turowski, M. 389
Tong, S.Y. 296, 302
Tong, S.Y., *see* Mrstik, B.J. 303
Tong, S.Y., *see* van Hove, M.A. 293, 303
Tong, S.Y., *see* Xu, F. 302
Toombs, G.A., *see* Alves, E.S. 930
Topiol, S. 72
Toriyama, T., *see* Fukui, T. 875
Tosatti, E., *see* Fano, G. 1025
Tosch, St. 322
Tosch, St., *see* Badt, D. 322
Toshich, B.S., *see* Agranovich, V.M. 452
Tosi, M.P., *see* Singwi, K.S. 459
Toudic, Y., *see* Lambert, B. 139, 142
Toulouse, G., *see* Rammal, R. 930
Toussaint, G., *see* Pêcheur, P. 117, 118, 132
Toyozawa, Y. 128, 129
Trafas, B.M. 323
Tribes, R., *see* Grosvalet, J. 1044
Tric, C., *see* Benoit à la Guillaume, C. 232
Tripathi, V.K., *see* Weisshaar, A. 896–899, 926
Trofimenkoff, F.N. 1044
Tromp, R.M. 297, 319
Tromp, R.M., *see* Demuth, J.E. 290
Tromp, R.M., *see* Hamers, R.J. 297, 319
Tromp, R.M., *see* Kaxiras, E. 317
Tromp, R.M., *see* LeGoues, F.K. 382
Troost, D. 339
Troullier, N. 76, 77
Troullier, N., *see* Franciosi, A. 338
Troullier, N., *see* Yu, X. 386, 389
Trucks, G.W., *see* Higashi, G.S. 369
Trugman, S.A. 953, 954, 993, 994
Trumbore, F.A., *see* Thomas, D.G. 180
Tsai, T.C., *see* Williams, R.S. 336
Tsang, J.C. 712
Tsang, J.C., *see* Kash, J.A. 712, 1046
Tsang, W.T., *see* Deveaud, B. 697
Tsang, W.T., *see* Miller, R.C. 181
Tsao, J.Y., *see* Picraux, S.T. 820
Tsen, K.T. 712
Tsen, S.Y., *see* Tsen, K.T. 712
Tsu, R. 651, 855
Tsu, R., *see* Chang, L.L. 647, 651
Tsu, R., *see* Döhler, G.H. 855
Tsu, R., *see* Esaki, L. 819
Tsu, R., *see* Sai-Halasz, G.A. 822
Tsuchiya, M. 874
Tsuchiya, T. 684, 685
Tsuda, H., *see* Hirose, K. 366
Tsui, D.C. 979, 986, 999
Tsui, D.C., *see* Baraff, G.A. 988
Tsui, D.C., *see* Bishop, D.J. 809
Tsui, D.C., *see* Boebinger, G.S. 1014
Tsui, D.C., *see* Cage, M.E. 999
Tsui, D.C., *see* Chang, A.M. 1000
Tsui, D.C., *see* Eisenstein, J.P. 1027
Tsui, D.C., *see* Engel, L.W. 956, 1020
Tsui, D.C., *see* Goldman, V.J. 656, 1007
Tsui, D.C., *see* Jiang, H.W. 1007, 1027
Tsui, D.C., *see* Li, Y.P. 883, 894
Tsui, D.C., *see* Liu, C.T. 930
Tsui, D.C., *see* Paalanen, M.A. 986
Tsui, D.C., *see* Simmons, J.A. 890, 908
Tsui, D.C., *see* Störmer, H.L. 986, 987, 999
Tsui, D.C., *see* Wei, H.P. 955, 984, 996
Tsui, D.C., *see* Willett, R. 999, 1000, 1015, 1026
Tsukada, M. 953, 954
Tsukamoto, J. 617, 618
Tsvetkov, V.A., *see* Bagaev, V.S. 460, 464, 466
Tu, C.W., *see* Boebinger, G.S. 1014
Tu, C.W., *see* Eisenstein, J.P. 984
Tu, C.W., *see* Jung, P.S. 840
Tu, C.W., *see* Pinczuk, A. 1024
Tu, K.N., *see* Mayer, J.W. 335
Tu, K.N., *see* Ottaviani, G. 352
Tu, K.N., *see* Poate, J.M. 328, 335
Tubino, R., *see* Dorsinville, R. 615
Tubino, R., *see* Walser, A.D. 613, 614
Tuchendler, J. 167
Tung, R.T. 305, 324, 325, 355, 374, 381
Tunstall, D.P., *see* Friedman, L.R. 867
Turberfield, A.J. 1006
Turowski, M. 389
Twose, W.D., *see* Mott, N.F. 941

Uba, S. 170
Uba, S., *see* Schöll, E. 433
Uchida, Y., *see* Yokotsuka, T. 357
Uemura, Y., *see* Ando, T. 984, 985
Ueno, H., *see* Takiguchi, T. 617

Uhrberg, R.I.G. 314, 319
Uhrberg, R.I.G., *see* Northrup, J.E. 382
Ulbikas, J., *see* Pyragas, K.A. 432
Ulinich, F.R., *see* Usov, N.A. 989
Ulloa, S.E. 872, 926
Ulloa, S.E., *see* Castaño, E. 872, 899, 926, 928, 929
Ulloa, S.E., *see* Joe, Y.S. 893, 894, 898
Ulsher, A., *see* Clark, R.G. 999, 1025
Umbach, C.P. 958
Umbach, C.P., *see* Blakely, J.M. 304
Umbach, C.P., *see* Sivan, U. 659, 872, 879, 880
Umbach, C.P., *see* Webb, R.A. 791, 948
Upadyaya, U.N., *see* Wannier, G.H. 512
Urban, K., *see* Cox, G. 305
Urbina, C., *see* Geerligs, L.J. 1115
Ure Jr, R.W. 565, 566
Uren, M.J. 937
Uren, M.J., *see* Cobden, D.H. 245
Uren, M.J., *see* Kirton, M.J. 205, 245, 247
Usov, N.A. 989

Vainus, B. 252
Valladares, J.P., *see* Pinczuk, A. 1024
Vallin, J.T. 170, 173, 185
Vallin, J.T., *see* Nygren, B. 173
Valois, A.J., *see* Davidson, J.S. 730, 767
Valtchinov, V., *see* Groshev, A. 924
van Alphen, P.M., *see* de Haas, W.J. 512
van Bommel, A.J. 298, 299
van Cong, H. 256
van de Walle, C.G. 56
van den Dries, L. 936
van der Enden, B., *see* Kouwenhoven, L.P. 924
van der Gaag, B.P., *see* Kash, K. 875
van der Gaag, B.P., *see* Roukes, M.L. 870, 892, 906, 908, 911, 915, 916, 919
van der Gaag, B.P., *see* Shepard, K.L. 913, 914
van der Gaag, B.P., *see* Thornton, T.J. 877, 892, 906, 1087
van der Marel, D. 885, 893
van der Marel, D., *see* Haanappel, E.G. 885, 893
van der Marel, D., *see* van Wees, B.J. 800, 801, 867, 870, 871, 881, 882, 913, 1097, 1104
van der Mark, M.B., *see* Montie, E.A. 886
van der Rest, J. 118
van der Vaart, N.C., *see* Kouwenhoven, L.P. 921
van der Veen, J.F. 297
van der Veen, J.F., *see* Larsen, P.K. 302, 319
van der Veen, J.F., *see* van Loenen, E.J. 331
van derWalle, G.F.A., *see* Zalm, P.C. 380
van Dirschot, T.G.J., *see* van Bommel, A.J. 298, 299
van Gorkum, A.A., *see* Zalm, P.C. 380
van Haesendonck, C. 948
van Haesendonck, C., *see* Gijs, M. 937, 948
van Haesendonck, C., *see* van den Dries, L. 936
van Hall, P.J. 703
van Houten, H. 806–808, 870–872, 876, 879, 882, 883, 886, 902, 904–906, 911, 924, 1104
van Houten, H., *see* Beenakker, C.W.J. 659, 867, 875, 877, 886, 888, 892, 905, 906, 910, 911, 915–917, 924, 998
van Houten, H., *see* Dekker, C. 883, 894
van Houten, H., *see* Kouwenhoven, L.P. 899, 900
van Houten, H., *see* Molenkamp, L.W. 796, 804–806
van Houten, H., *see* Staring, A.A.M. 921, 1115
van Houten, H., *see* van Loosdrecht, P.H.M. 904
van Houten, H., *see* van Wees, B.J. 800, 801, 867, 870, 871, 881, 882, 886, 902, 903, 913, 997, 1097, 1104
van Houten, H., *see* Williamson, J.G. 880
van Hove, H. 302
van Hove, J.M. 307
van Hove, M.A. 286, 293, 303
van Hove, M.A., *see* Ignatiev, A. 303
van Hove, M.A., *see* MacLaren, J.M. 293
van Hove, M.A., *see* Mrstik, B.J. 303
van Hove, M.A., *see* Somorjai, G.A. 293, 324
van Kampen, N.G. 427
van Kanel, H., *see* Ospelt, M. 355
van Laar, J. 349
van Laar, J., *see* Huijser, A. 305, 315, 316, 350
van Loenen, E.J. 331
van Loosdrecht, P.H.M. 904
van Loosdrecht, P.H.M., *see* van Houten, H. 872, 876, 879, 902, 904, 906, 911
van Roosbroeck, W. 229
van Rooy, T.L., *see* Huijser, A. 316
van Vechten, J.A. 55, 211, 244, 356, 380, 390
van Wees, B.J. 800, 801, 867, 870, 871, 881, 882, 884, 886, 902–906, 912, 913, 919, 921, 924, 997, 1097, 1104
van Wees, B.J., *see* Beenakker, C.W.J. 905, 906
van Wees, B.J., *see* Kouwenhoven, L.P. 744, 872, 888, 889, 899, 900, 923, 924, 926, 927
van Wees, B.J., *see* van Houten, H. 870–872, 876, 879, 882, 883, 902, 905, 911, 1104
van Wees, B.J., *see* van Loosdrecht, P.H.M. 904
van Wees, B.J., *see* Williamson, J.G. 880
Vander Plas, H., *see* Brillson, L.J. 339
vander Werf, D.P., *see* Heslinga, D.R. 324, 356
Vanderbilt, D. 14, 15, 75–78, 95
Vanderbilt, D., *see* Alerhand, O.L. 304, 321
Vanderbilt, D., *see* Bedrossian, P. 322
Vanderbilt, D., *see* Rasolt, M. 3, 5, 43
Vanececk, M., *see* Kocka, J. 251
Vardeny, Z. 605–607
Vardeny, Z., *see* Orenstein, J. 606, 615
Varma, R.R., *see* Williams, R.H. 316, 332, 338, 353, 356
Vashishta, P., *see* Bhattacharya, P.K. 459
Vasiliadou, E., *see* Grambow, P. 872
Vasilopoulos, P., *see* Peeters, F.M. 745, 746
Vavilov, V.S., *see* Mironov, A.G. 523

Veider, L.A., *see* Abraham, D. 387
Verbeke, O.B., *see* Spinnewyn, J. 432, 437
Verbruggen, A.H., *see* Caro, J. 938, 958
Vèrié, C., *see* Bensaid, B. 238
Vermaak, J.S., *see* Snyman, L.W. 326
Vickers, A.J. 432, 702
Vieren, J.P., *see* Berroir, J.M. 825
Vieren, J.P., *see* Guldner, Y. 824, 828, 986
Vieren, J.P., *see* Lugagne-Delpon, E. 821, 822
Vigier, J.P. 1118
Vignale, G. 14
Vignale, G., *see* Rasolt, M. 14
Vihlein, Ch., *see* Aszodi, G. 139, 142
Villars, P. 98
Vincent, G. 151
Vincent, G., *see* Bois, D. 170, 177
Vincent, G., *see* Chantre, A. 151
Vincent, G., *see* Goguenheim, D. 264
Vinetskii, V.L. 270
Vinetskii, V.L., *see* Mashkevich, V.S. 232
Vinter, B., *see* Weil, T. 655
Visscher, P.B. 1027
Vitomirov, I.M. 367
Vitomirov, I.M., *see* Chambers, S.A. 326
Vitomirov, I.M., *see* Chang, S. 340, 350, 372–374
Vitomirov, I.M., *see* Waddill, G.D. 367
Vitomirov, I.M., *see* Xu, F. 326, 338
Viturro, R.E. 358, 362, 364, 370, 372
Viturro, R.E., *see* Brillson, L.J. 312, 334, 339, 349, 370–372
Viturro, R.E., *see* Chang, S. 340, 350, 372
Viturro, R.E., *see* Chiradia, P. 372
Viturro, R.E., *see* Shaw, J.L. 333, 340, 351, 367, 370, 371
Vloeberghs, H., *see* van Haesendonck, C. 948
Vogl, P. 132, 596, 613
Vogl, P., *see* Hjalmarson, H.P. 117, 118
Vogl, P., *see* Kometer, K. 423
Vogl, P., *see* Rieger, M. 423
Vogt, E., *see* Herring, C. 493, 503, 537, 539–541
Vohra, Y.K. 85, 87
Voisin, P. 823–825, 832, 833, 849–852
Voisin, P., *see* Bastard, G. 819, 827, 833
Voisin, P., *see* Bigan, E. 850, 852
Voisin, P., *see* Bleuse, J. 833, 839, 840, 844, 846, 849–851
Voisin, P., *see* Brum, J.A. 831, 841
Voisin, P., *see* Ferreira, R. 839, 842, 843, 852, 854
Voisin, P., *see* Guldner, Y. 824, 986
Voisin, P., *see* Lugagne-Delpon, E. 821, 822
Voisin, P., *see* Marzin, J.Y. 824, 830
Voisin, P., *see* Soucail, B. 833, 838, 840, 845, 846, 850, 851, 857
Völcker, M., *see* Brandl, A. 440, 443
Volkov, A.F. 422
Vollhardt, D. 867, 940, 947
von Bardeleben, H.J. 150–153
von Bardeleben, H.J., *see* Delerue, C. 152
von Bardeleben, H.J., *see* Mauger, A. 120
von Barth, U. 84
von der Lage, F.C. 543
von der Linde, D. 22, 712
von der Linden, W. 8, 10, 12, 19, 22
von der Linden, W., *see* Chakraborty, T. 1030, 1031
von Faber, E., *see* Paasch, G. 391
von Klitzing, K. 953, 979, 985, 986
von Klitzing, K., *see* Buhmann, H. 1006, 1007
von Klitzing, K., *see* Dobers, M. 1020
von Klitzing, K., *see* Ebert, G. 955, 986, 998, 999, 1025
von Klitzing, K., *see* Englert, T. 985
von Klitzing, K., *see* Grahn, H.T. 694
von Klitzing, K., *see* Haug, R.J. 998, 1018
von Klitzing, K., *see* Müller, G. 997
von Klitzing, K., *see* Obloh, H. 730, 767, 768
von Klitzing, K., *see* Schneider, H. 848, 850, 854, 857
von Klitzing, K., *see* Stahl, E. 984
von Klitzing, K., *see* Streda, P. 999
von Klitzing, K., *see* Tausendfreund, B. 1014
von Klitzing, K., *see* Weiss, D. 744, 922, 930
von Roos, O. 239
Voos, M. 451, 464
Voos, M., *see* Allan, G. 819, 827
Voos, M., *see* Benoit à la Guillaume, C. 462, 463, 472
Voos, M., *see* Berroir, J.M. 825
Voos, M., *see* Brum, J.A. 831, 841
Voos, M., *see* Chang, Y.C. 825
Voos, M., *see* Etienne, B. 462, 463
Voos, M., *see* Grossman, B. 465
Voos, M., *see* Guldner, Y. 824, 828, 986
Voos, M., *see* Lugagne-Delpon, E. 821, 822
Voos, M., *see* Mattos, J.V.C. 466
Voos, M., *see* Soucail, B. 833
Voos, M., *see* Voisin, P. 823, 832
Vos, M. 367
Vosko, S.H., *see* Geldart, D.J.W. 26
Vosko, S.H., *see* Rasolt, M. 8, 10, 12, 14, 19, 31
Vrehen, Q.H.F. 520, 525
Vrehen, Q.H.F., *see* Reine, M. 521, 525
Vuillaume, D., *see* Goguenheim, D. 264
Vul, A.Ya. 188
Vuong, T.H.H. 730
Vuong, T.H.H., *see* Brummell, M.A. 730
Vvedensky, D.P., *see* MacLaren, J.M. 293

Wacker, A. 432, 440
Waddill, G.D. 367
Waddill, G.D., *see* Vitomirov, I.M. 367
Waddill, G.D., *see* Weaver, J.H. 328, 339, 356
Wagman, D.D. 332
Wagner, J. 139, 142
Wagner, J., *see* Ennen, H. 139, 142
Wagner, M. 922

Wagner, M., *see* Merkt, U. 922
Wagner, R.J. 185
Wagner, R.J., *see* Cage, M.E. 999
Wahi, A.K., *see* Miyano, K. 372
Wainer, J.J., *see* Fowler, A.B. 809, 938
Wakabayashi, J. 985, 1007
Wakahara, S., *see* Akera, H. 637, 646
Wakahara, S., *see* Ando, T. 674
Wakaya, F., *see* Takagaki, Y. 911
Wakiyama, *see* Tokumoto, H.S. 305
Waldrop, J.R. 326, 338, 344, 347, 368, 369, 389, 393
Waldrop, J.R., *see* Grant, R.W. 339, 368, 369, 385–387, 393, 394
Waldrop, J.R., *see* Harrison, W.A. 381, 382, 393
Waldrop, J.R., *see* Kraut, E.A. 386
Walker, J.F., *see* Levine, B.F. 697
Walker, J.F., *see* Sorba, L. 398, 399
Walker, J.W., *see* Sah, C.T. 189
Walker, N.S., *see* Bott, D.C. 619
Walker, P.J., *see* Warburton, R.J. 824
Wall, A., *see* Franciosi, A. 338
Wallace, R.L., *see* Shi, Z.Q. 368
Walmsley, I.A., *see* Wise, F.W. 1071
Walser, A.D. 613, 614
Walukiewicz, W. 350, 357
Walukiewicz, W., *see* Nolte, D.D. 759
Wand, J.S. 456
Wang, C.S. 8, 10, 12, 19, 22
Wang, C.S., *see* Pickett, W.E. 8, 10, 12, 22
Wang, H., *see* Sibille, A. 855
Wang, J.S.Y. 8, 10, 12, 14
Wang, K.L. 263
Wang, S.-W., *see* van Hove, M.A. 286, 293
Wang, S.R., *see* Xu, F. 302
Wang, W.B. 1073
Wang, W.I. 385
Wang, W.I., *see* Beresford, R. 647
Wang, W.I., *see* Longenbach, K.F. 647
Wang, W.I., *see* Luo, L.F. 647, 651
Wang, W.I., *see* Ohno, H. 655
Wang, W.I., *see* Shanabrook, B.V. 825
Wang, X.S. 304
Wang, X.W., *see* Fahy, S. 68, 84
Wang, Y.-R., *see* Duke, C.B. 294
Wang, Y.R., *see* Kubby, J.A. 322
Wang, Z.G., *see* Ledebo, L.Å. 231
Wannier, G.H. 492, 508, 510, 512, 522, 843, 929
Wannier, G.H., *see* Rauh, A. 929
Warburton, R.J. 824
Ward, J. 26
Ward, J.C., *see* Luttinger, J.M. 33
Warren, A.C., *see* Offsey, S.D. 339, 368, 369
Washburn, J., *see* Ding, J. 326
Washburn, J., *see* Lilienthal-Weber, Z. 326
Washburn, J., *see* Newman, N. 349
Washburn, S. 922, 930, 934, 937, 938, 948, 958, 998
Washburn, S., *see* Ford, C.J.B. 883, 894, 906, 909–911, 919–921, 928
Washburn, S., *see* Umbach, C.P. 958
Washburn, S., *see* Webb, R.A. 791, 948
Washburn, T., *see* Lilienthal-Weber, Z. 336, 358
Wasilewski, Z., *see* Buckthought, A. 1020
Wasilewski, Z., *see* Sachrajda, A. 1020
Wasinki, T., *see* Weber, E.R. 357
Watkins, G.D. 119, 120, 183, 184
Watkins, G.D., *see* Ammerlaan, C.A.J. 184
Watkins, G.D., *see* Corbett, J.W. 184
Watkins, G.D., *see* Deleo, G.G. 132
Watkins, G.D., *see* Elkin, E.L. 184
Watkins, G.D., *see* Messmer, R.P. 119
Watkins, G.D., *see* Song, L.W. 250
Watkins, G.D., *see* Vallin, J.T. 173, 185
Watson, R.E., *see* Herbst, J.F. 139, 140, 142, 143
Wattenbach, M., *see* Hoffmann, D.M. 351
Watts, R.K. 184
Wayman, C.M., *see* Coleman, J.J. 380
Weagley, R.J., *see* Epstein, A.J. 621, 622
Weagley, R.J., *see* Gibson, H.W. 611
Weaire, D., *see* Cohen, M.L. 130
Weaire, D., *see* Heine, V. 91
Weaire, D., *see* Kramer, B. 867, 940
Weakliem, D.H. 170
Weaver, J.H. 328–332, 339, 356
Weaver, J.H., *see* Butera, R.A. 331
Weaver, J.H., *see* Chambers, S.A. 326
Weaver, J.H., *see* del Guidice, M. 334
Weaver, J.H., *see* Franciosi, A. 338
Weaver, J.H., *see* Grioni, M. 330–332, 334
Weaver, J.H., *see* Lin, Zhangda 331
Weaver, J.H., *see* Ruckman, M.W. 326, 334
Weaver, J.H., *see* Trafas, B.M. 323
Weaver, J.H., *see* Vitomirov, I.M. 367
Weaver, J.H., *see* Vos, M. 367
Weaver, J.H., *see* Waddill, G.D. 367
Weaver, J.H., *see* Xu, F. 326, 338
Webb, M., *see* White, A.M. 178
Webb, M.B., *see* Kariotis, R. 304
Webb, M.B., *see* Mo, Y.-M. 321
Webb, M.B., *see* Poppendieck, T.D. 296
Webb, M.B., *see* Swartzentruber, B.S. 304
Webb, R.A. 791, 948, 958
Webb, R.A., *see* Fowler, A.B. 735, 809, 938, 939, 957
Webb, R.A., *see* Umbach, C.P. 958
Webb, R.A., *see* Washburn, S. 934, 937, 938, 948, 958
Weber, E.R. 151, 357
Weber, E.R., *see* Kachaturyan, K. 147
Weber, E.R., *see* Lilienthal-Weber, Z. 326, 336, 358
Weber, E.R., *see* Newman, N. 349
Weber, E.R., *see* Spicer, W.E. 356, 357
Weber, J., *see* Aszodi, G. 139, 142
Weber, J., *see* Sauer, R. 477

Wegner, F.J. 940, 942, 950, 951, 984
Wei, C.M., *see* Tong, S.Y. 296
Wei, H.P. 955, 984, 996
Wei, H.P., *see* Engel, L.W. 956
Wei, S.-H. 56
Weibel, E., *see* Binnig, G. 296, 362
Weidenmüller, H.A. 953
Weil, T. 655
Weiler, M.H. 521, 525
Weimann, G., *see* Belle, G. 841
Weimann, G., *see* Boebinger, G.S. 1014
Weimann, G., *see* Dobers, M. 1020
Weimann, G., *see* Ebert, G. 998, 999, 1025
Weimann, G., *see* Fletcher, R. 754, 756, 757, 760, 761, 771–773, 775, 776, 778, 779
Weimann, G., *see* Haug, R.J. 1018
Weimann, G., *see* Ruf, C. 730, 731, 760, 762, 763, 767
Weimann, G., *see* Simmons, J.A. 908
Weimann, G., *see* Stahl, E. 984
Weimann, G., *see* Weiss, D. 744, 922, 930
Weimann, G.W., *see* Li, Y.P. 883, 894
Weinberger, B.R. 602, 605
Weinert, M., *see* Qian, G.X. 106
Weinstein, B.A. 85, 94
Weis, J.J., *see* Caillol, J.M. 1005, 1006, 1030
Weis, J.J., *see* Levesque, D. 1005
Weise, M., *see* Röhricht, B. 428
Weispfenning, M. 428
Weiss, D. 744, 922, 930
Weiss, D., *see* Gerhardts, R.R. 922, 930
Weiss, D., *see* Müller, G. 997
Weiss, D., *see* Stahl, E. 984
Weiss, G.H., *see* Maradudin, A.A. 684
Weisshaar, A. 896–899, 926
Weisshaar, A., *see* Wu, J.C. 898
Weisskopf, V.F., *see* Conwell, E.M. 570
Weisz, J.F. 896, 903, 914
Weitering, H.H., *see* Heslinga, D.R. 324, 356
Weizer, V.G., *see* Goradia, C. 252
Welch, D.F. 673
Weller, W. 868
Welmann, G., *see* Bimberg, D. 697
Welter, J.M., *see* Sai-Halasz, G.A. 822
Wen, X.G., *see* Greiter, M. 1025
Wendel, H. 91
Wendler, L. 683, 689, 693, 699
Wentzcovitch, R.M. 87, 89
Wentzcovitch, R.M., *see* Knittle, E. 100
Wepfer, G.G., *see* Schulz, M. 184
West, K.W., *see* Eisenstein, J.P. 1019, 1020, 1028
West, K.W., *see* Goldberg, B.B. 1006
West, K.W., *see* Jiang, H.W. 1007, 1027
West, K.W., *see* Pfeiffer, L.N. 874
West, K.W., *see* Pinczuk, A. 1024
West, K.W., *see* Schubert, E.F. 703
West, K.W., *see* Spector, J. 872, 874, 878–880
West, K.W., *see* Störmer, H.L. 874
Westervelt, R.M. 459, 461, 462, 464, 465, 470, 471
Westervelt, R.M., *see* Teitsworth, S.W. 422, 432
Westland, D.J. 697
Westwood, D.I., *see* Fowell, A.E. 365
Westwood, D.I., *see* Shen, T.-H. 398
Wetsel, A.E., *see* Reed, M.A. 872, 922, 923, 925
Wharam, D.A. 800, 801, 867, 870, 871, 881, 882, 888, 902–904, 913, 923, 924, 1097, 1102, 1104
Wharam, D.A., *see* de Aguiar, F.M. 926
Wharam, D.A., *see* Greene, S.K. 916
Wharam, D.A., *see* Kelly, M.J. 899, 902
Wheeler, R.G. 869, 870, 958
Wheeler, R.G., *see* Alphenaar, B.W. 921
Wheeler, R.G., *see* Kopley, T.E. 893
Wheeler, R.G., *see* McEuen, P.L. 867, 892, 893, 898
Wheeler, R.G., *see* McEwan, P.L. 796
White, A.M. 178, 189, 190
White, A.M., *see* Ashen, D.J. 178, 180
White, A.M., *see* Dean, P.J. 179
White, A.M., *see* Wagner, R.J. 185
White, C.T., *see* Mintmire, J.W. 30
White, C.W., *see* Zehner, D.M. 296
White, S. 827
Whitehouse, C.R., *see* Mackay, K.J. 380
Whitehouse, C.R., *see* Whittaker, D.M. 851, 852
Whitlock, P.A., *see* Kalos, M.H. 68, 80
Whitman, L.J. 305, 350
Whitsett, C.R. 512, 513, 521
Whittaker, D.M. 838, 851, 852
Wicks, G., *see* Jarosik, N.C. 168
Wicks, G.W., *see* Welch, D.F. 673
Widom, A. 885
Wieck, A.D. 870
Wieder, H.H. 356
Wieder, H.H., *see* Puechner, R.A. 930
Wiegmann, W., *see* Dingle, R. 819
Wiegmann, W., *see* Gornik, E. 984
Wiegmann, W., *see* Levi, A.F.J. 660, 707
Wiegmann, W., *see* Miller, D.A.B. 657
Wiegmann, W., *see* Narayanamurti, V. 678
Wiegmann, W., *see* Petroff, P.M. 382, 1088
Wiegmann, W., *see* Störmer, H.L. 986, 999
Wieranga, P.E. 305
Wiesenfeld, K., *see* Bak, P. 445
Wight, D.R., *see* Gibb, R.P. 239
Wigner, E.P. 1091, 1092, 1095, 1102
Wilamowski, Z., *see* Wittlin, A. 185
Wilczek, F. 1011, 1012
Wilczek, F., *see* Arovas, D. 1009
Wilczek, F., *see* Greiter, M. 1025
Wiley, J.D. 501, 519, 550, 552, 569, 575
Wiley, W.P., *see* Reed, M.A. 867, 875
Wilke, W.G. 389
Wilkening, W., *see* Kaufman, E. 150
Wilkins, J.W., *see* Günnarsson, O. 84
Wilkins, J.W., *see* Herbst, J.F. 139, 140, 143

Wilkins, J.W., *see* Hu, B.Y.-K. 1124
Wilkins, J.W., *see* Khan, F.S. 1123
Wilkins, J.W., *see* Wingreen, N.S. 656
Wilkinson, C.D.W., *see* Alves, E.S. 930
Wilkinson, C.D.W., *see* Beton, P.H. 745, 842, 888, 889
Wilkinson, C.D.W., *see* Gallagher, B.L. 787, 793–795
Wilkinson, C.D.W., *see* Galloway, T. 791, 793
Wilkinson, C.D.W., *see* Main, P.C. 888, 911
Willardson, R.K., *see* Beer, A.C. 563
Willebrand, H., *see* Niedernostheide, F.J. 443
Willems, E.M.M., *see* van Wees, B.J. 882, 886, 903, 919, 997
Willet, R., *see* Eisenstein, J.P. 1027
Willett, R. 999, 1000, 1015, 1026
Willett, R., *see* Mallett, J.R. 1006
Willett, R.L., *see* Jiang, H.W. 1007
Williams, E.D. 304
Williams, E.D., *see* Ohno, T.R. 305
Williams, E.D., *see* Wang, X.S. 304
Williams, E.D., *see* Yang, Y-N. 305
Williams, E.M., *see* Dean, P.J. 179
Williams, E.R., *see* Cage, M.E. 999
Williams, E.W., *see* White, A.M. 178
Williams, F.E., *see* Fair, H.D. 184
Williams, G.M., *see* Mackay, K.J. 380
Williams, J.M., *see* Hartland, A. 882
Williams, M.D., *see* Kendelewicz, T. 334, 337
Williams, M.D., *see* Newman, N. 343–345
Williams, R.H. 316, 332, 338, 353, 356, 366
Williams, R.H., *see* Fowell, A.E. 365
Williams, R.H., *see* Hughes, G.J. 332, 353
Williams, R.H., *see* Mackay, K.J. 380
Williams, R.H., *see* McLean, A.B. 357, 358
Williams, R.H., *see* Montgomery, V. 367
Williams, R.H., *see* Rhoderick, E.H. 340, 343, 344
Williams, R.H., *see* Shen, T.-H. 398
Williams, R.H., *see* Srivastava, G.P. 314
Williams, R.H., *see* Zahn, D.T.R. 380
Williams, R.S. 336
Williams, R.S., *see* Pugh, J.H. 326
Williamson, F., *see* Costa, J.C. 368
Williamson, J.G. 880, 919, 921
Williamson, J.G., *see* Kouwenhoven, L.P. 899, 900, 921
Williamson, J.G., *see* Molenkamp, L.W. 872, 877, 878
Williamson, J.G., *see* Staring, A.A.M. 921, 1115
Williamson, J.G., *see* van Houten, H. 872, 876, 879, 911
Williamson, J.G., *see* van Loosdrecht, P.H.M. 904
Williamson, J.G., *see* van Wees, B.J. 800, 801, 867, 870, 871, 881, 882, 886, 903, 904, 906, 913, 919, 921, 924, 997, 1097, 1104
Wilmsen, C.W., *see* Goodnick, S.M. 708
Wilson, A.H. 629
Wilson, B.A. 385
Wilson, B.L.H., *see* Abram, R.A. 703
Wilson, B.L.H., *see* Gibb, R.P. 239
Wilson, D.K. 499
Wilson, D.K., *see* Feher, G. 499
Wilson, R.B., *see* Forrest, S.R. 389
Wilson, R.J., *see* Petroff, P.M. 296
Wilson, T.A., *see* Friedman, D.J. 334
Wind, S.J., *see* Chandrasekar, V. 948
Wind, S.J., *see* Field, S.B. 924
Wind, S.J., *see* McEuen, P.L. 924
Wind, S.J., *see* Meirav, U. 921, 924
Windscheif, J., *see* Aszodi, G. 139, 142
Windscheif, J., *see* Kaufmann, U. 154, 155
Windscheif, J., *see* Weber, E.R. 151, 357
Wingreen, N.S. 656
Wingreen, N.S., *see* McEuen, P.L. 924
Wingreen, N.S., *see* Meir, Y. 924
Winokur, M.J. 592
Winters, H.F. 310
Wintle, H.J. 252
Wise, F.W. 1071
Wise, F.W., *see* Bailey, D.W. 1071
Wise, F.W., *see* Rosker, M.J. 1070
Wisnieff, R., *see* Wheeler, R.G. 869, 870, 958
Wissel, J.M., *see* Aboie-Elfotouh, F.A. 305
Witchlow, G.P., *see* O'Reilly, E.P. 674
Witowski, A.M., *see* Strutz, T. 186
Witt, A.F., *see* Carruthers, J.R. 306
Witt, T.J., *see* Taylor, B.N. 882
Wittlin, A. 185
Woicik, J., *see* List, R.S. 368
Wolf, E.L. 658, 662
Wolfe, C.M., *see* Hsieh, S. 389
Wolfe, C.M., *see* Patten, E.A. 389
Wolfe, C.M., *see* Stillman, G.E. 168, 174, 187, 548, 556
Wolff, P.A. 1050
Wolfgarten, G., *see* Pollmann, J. 313, 314
Wölfle, P., *see* Vollhardt, D. 867, 940, 947
Wolford, D.J. 385
Wolfsberg, M. 141
Wolkow, R. 319
Wolter, J.H., *see* Hendriks, P. 432
Wolter, J.H., *see* van Hall, P.J. 703
Woltjer, R., *see* Harris, J.J. 867
Woltzer, R., *see* Harris, J.J. 710
Wong, K.B., *see* Gell, M.A. 832
Wong, K.B., *see* Jaros, M. 674, 832
Wong, K.B., *see* Ninno, D. 832
Wong, T.M. 322
Wood, C.E.C., *see* Okamoto, K. 326
Wood, R.A., *see* Stradling, R.A. 701
Wood, T.H., *see* Miller, D.A.B. 657
Woodall, J.M. 336, 351, 358, 378, 379, 381, 382
Woodall, J.M., *see* Brillson, L.J. 339, 349, 370, 371
Woodall, J.M., *see* Chang, S. 340, 350, 372–374

Woodall, J.M., *see* Freeouf, J.L. 358, 359, 390
Woodall, J.M., *see* Offsey, S.D. 339, 368, 369
Woodall, J.M., *see* Pashley, M.D. 299, 302
Woodall, J.M., *see* Viturro, R.E. 358, 372
Woodbridge, K., *see* Nicholas, R.J. 832
Woodbury, H.H., *see* Lorenz, M.R. 177
Woolf, D.A., *see* Fowell, A.E. 365
Worlock, J.M. 464, 1006
Worlock, J.M., *see* Etienne, B. 462, 463
Worlock, J.M., *see* Jochler, J. 466
Worlock, J.M., *see* Kash, K. 875
Worlock, J.M., *see* Mattos, J.V.C. 466
Worlock, J.M., *see* Meynadier, M.H. 825
Worlock, J.M., *see* Voos, M. 464
Wosinski, T., *see* Weber, E.R. 151
Wray, E.M. 170
Wright, G.B. 177, 182
Wright, G.B., *see* Palik, E.D. 529, 530
Wright, P.A., *see* Clark, R.G. 1018–1020
Wright, P.A., *see* Turberfield, A.J. 1006
Wright, S.J., *see* Kroemer, H. 382, 385
Wright, S.J., *see* Viturro, R.E. 372
Wright, S.L., *see* Brillson, L.J. 339, 349, 370, 371
Wright, S.L., *see* Kirtley, J.P. 870
Wright, S.L., *see* Solomon, P.M. 646
Wright, S.L., *see* Viturro, R.E. 358, 372
Wruck, D., *see* Boyn, R. 175
Wu, C.-h. 266
Wu, J.C. 898
Wu, M.C., *see* Chen, J.F. 647, 651
Wu, Y.S. 1012
Wu, Y.S., *see* Niu, Q. 930, 992, 993, 1020
Wudl, F., *see* Chung, T.-C. 607
Wudl, F., *see* Mo, Z. 607
Wudl, F., *see* Moraes, F. 607
Wudl, F., *see* Vardeny, Z. 607
Wulf, U., *see* Gerhardts, R.R. 922, 930
Wünsche, H.-J. 475
Würfel, P. 234
Würfel, P., *see* Ruppel, W. 232
Wyatt, P.W., *see* Roth, L.M. 521
Wybourne, B.G. 139
Wybourne, M.N., *see* Wu, J.C. 898
Wyder, P., *see* Strutz, T. 186
Wyeth, N.C. 332, 353
Wyld, H.W., *see* Ravenhall, D.G. 905, 913–915
Wyld, H.W., *see* Schult, R.L. 896, 905, 908, 913–915

Xiao, Z., *see* Andersson, M.O. 205, 247
Xin, S., *see* Longenbach, K.F. 647
Xu, F. 302, 326, 338
Xu, F., *see* Chambers, S.A. 326
Xu, F., *see* Lin, Zhangda 331
Xu, F., *see* Weaver, J.H. 331
Xu, G., *see* Tong, S.Y. 302

Yablonovitch, E. 339, 368, 370
Yacobi, B.G. 312
Yacoby, A. 887
Yafet, Y. 516, 518
Yager, W.A., *see* Fletcher, R.J. 519
Yahata, A. 351
Yajima, T. 658
Yakovlev, M.L., *see* Belyantsev, A.M. 432
Yakushi, K., *see* Brédas, J.-L. 606
Yamada, K. 432
Yamada, S. 875, 888
Yamaguchi, E. 130, 145, 146, 149, 150
Yamamoto, K., *see* Aoki, K. 422, 432, 437, 440
Yamamoto, M., *see* Yamada, S. 888
Yamamoto, Y., *see* Mukai, T. 256
Yamasaki, K. 263
Yan, M., *see* Wang, W.B. 1073
Yan, R.H. 850
Yan, R.H., *see* Law, K.K. 852
Yan, R.H., *see* Tsuchiya, M. 874
Yanai, H., *see* Kuriyama, T. 238
Yang, C.H. 656
Yang, C.N. 1029
Yang, C.N., *see* Byers, N. 991
Yang, L. 647
Yang, L., *see* Chen, J.F. 647, 651
Yang, M.J., *see* Yang, C.H. 656
Yang, W.S. 297
Yang, Y-N. 305
Yano, H. 432
Yassievich, I.N., *see* Abakumov, V.N. 239
Yavich, B.S., *see* Belyantsev, A.M. 432
Yeh, J.-J. 325
Yeh, J.L. 323
Yeh, J.L., *see* Kahn, A. 323
Yeh, J.L., *see* Xu, F. 302
Yen, R., *see* Shank, C.V. 22
Yin, M.T. 14, 67, 83–87, 90–95
Yin, M.T., *see* Ihm, J. 92
Yin, M.T., *see* McMahan, A.K. 85
Yindeepol, W., *see* Wu, J.C. 898
Ying, S.C., *see* Giuliani, G.F. 992
Yndurain, F., *see* Artacho, E. 314, 319
Yoffa, E.J., *see* Adler, D. 211
Yokotsuka, T. 357
Yosefin, M. 887
Yoshida, M. 270
Yoshida, M., *see* Yamasaki, K. 263
Yoshino, J., *see* Sakaki, H. 998
Yoshino, K., *see* Takiguchi, T. 617
Yoshioka, D. 1001, 1002, 1006, 1013, 1015, 1022, 1025, 1028
Young, J.F. 825
Young, K., *see* Kahn, A. 349, 350
Yu, E.T. 389
Yu, E.T., *see* Chow, D.H. 647
Yu, L., *see* Su, Z. 613
Yu, P.Y., *see* Zhang, S.B. 105
Yu, X. 386, 389
Yurke, B. 894

Zachariasen, F., *see* Carruthers, P. 1092, 1097
Zack, G.W., *see* Scifres, D.R. 179
Zahn, D.T.R. 380
Zak, J. 510, 523, 843
Zallen, R. 377, 867, 994
Zalm, P.C. 380
Zamani, N., *see* Maserjian, J. 644
Zamkovets, N.V., *see* Bagaev, V.S. 460, 464
Zandler, G., *see* Kometer, K. 423
Zangwill, A. 286, 290, 291, 315
Zangwill, A., *see* Bales, G.S. 304
Zanon, N. 959
Zanoni, R., *see* Niles, D.W. 399
Zanzucci, P., *see* Jastrzebski, L. 277
Zaremba, E., *see* Fletcher, R. 738, 740
Zavaritskii, N.V. 721, 731, 753, 755, 760
Zawadzki, P., *see* Buckthought, A. 1020
Zawadzki, P., *see* Taylor, R.P. 877, 883, 888, 894, 904, 906, 919, 923, 924
Zawadzki, W. 498, 502, 507, 517, 525, 546, 548, 549, 562, 568, 571–573, 575, 577, 763, 765, 767
Zawadzki, W., *see* Weiler, M.H. 525
Zehner, D.M. 296
Zeidenbergs, G. 645
Zeiger, H.J. 495, 506, 511, 512
Zeiger, H.J., *see* Dexter, R.N. 499, 504, 506
Zeiger, H.J., *see* Rauch, F.J. 499
Zeiger, H.J., *see* Stickler, T.J. 519
Zelenin, S.P. 763
Zener, C. 638
Zesch, J., *see* Street, R.A. 251
Zhan, X.D., *see* Song, L.W. 250
Zhang, F.C. 1013, 1015, 1017–1019
Zhang, F.C., *see* Chakraborty, T. 1013, 1017, 1020
Zhang, J., *see* Neave, J.H. 308
Zhang, S.B. 87, 89, 90, 103, 105, 148
Zhang, X-J., *see* Cheng, H. 336, 357
Zhang, Y.H., *see* Heitmann, D. 922
Zhang, Y.H., *see* Kern, K. 930
Zhao, T.-X., *see* Daniels, R.R. 323
Zheng, X.-Y. 658
Zhu, B., *see* Huang, K. 681, 683
Zhu, J.G., *see* Palmstrøm, C.J. 327, 340, 374, 375
Zhu, X. 81
Zianni, X. 685
Ziesche, P., *see* Weller, W. 868
Zilberman, G.E. 929
Ziman, J. 494, 525, 533, 538, 539, 564, 567
Ziman, J.M. 23, 867
Zipperian, T.E., *see* Olbright, G.R. 852
Zipperian, T.E., *see* Osbourn, G.C. 378
Zirnbauer, M.R. 955
Zironi, F., *see* Bentosela, F. 843
Zogg, H., *see* Maissen, C. 381
Zohta, Y. 189
Zonker, B.T., *see* Farrow, R.F.C. 326
Zook, J. 539, 574
Zülicke, Ch., *see* Schimansky-Geier, L. 443
Zunger, A. 14, 15, 72, 97, 98, 118, 132, 314, 323, 330, 356, 390
Zunger, A., *see* Daniels, R.R. 323
Zunger, A., *see* Ihm, J. 82
Zunger, A., *see* Katayama-Yoshida, H. 132
Zunger, A., *see* Kerker, G.P. 65
Zunger, A., *see* Schlüter, M. 65
Zunger, A., *see* Singh, V.A. 118
Zunger, A., *see* Topiol, S. 72
Zunger, A., *see* Wei, S.-H. 56
Zürcher, P., *see* Bauer, R.S. 390, 391
Zvyagin, I.P., *see* Belyaev, A.D. 240
Zvyagin, I.P., *see* Bonch-Bruevich, V.L. 422, 440, 867
Zwaal, E.A.E., *see* Hendriks, P. 432
Zwanzig, R.W. 1047
Zwerdling, D. 522
Zwerdling, S., *see* Roth, L.M. 517, 520

List of Main Abbreviations

AES: Auger electron spectroscopy 286
ARPES: angle-resolved photoemission spectroscopy 311
ATM: atomic force microscopy 287

BCS: Bardeen–Cooper–Schrieffer theory of superconductivity 102
BEEM: ballistic electron energy microscopy 287
BEP: beam equivalent pressure 297
BMEC: bound multi-exciton complexes 476

c-PA: *cis*-polyacetylene 587, 606
– definition 587
CFS: constant-final-state spectroscopy 312
CIS: constant-initial-state spectroscopy 312
CLS: cathodoluminescence spectroscopy 312

DC: dielectric continuum 680, 685
DLTS: deep-level transient spectroscopy 264

EDC: energy distribution curve 311
EPM: empirical pseudopotential method 5, 61
EPR: electron paramagnetic resonance 287
ETB: empirical tight-binding method 117

FET: field effect transistor 261
FIM: field ion microscopy 287

GW: a first-order expansion of the electron self-energy 116

HD: hydrodynamic model 683
HNC: hypernetted chain 1005
HRTEM: high-resolution transmission electron microscopy 287

IP: interface polariton 689
IPS: inverse photoemission spectroscopy 312
IQHE: integer quantum Hall effect 998
IRAV: infrared active vibrations 605
IRS: infrared absorption spectroscopy 287

LAPS: laser excited photoemission spectroscopy 287
LDA: local density approximation 116, 144, 596
LEED: low-energy electron diffraction 287
LEELS (or LELS): low-energy electron-loss spectroscopy 312
LEEM: low-energy electron microscopy 287
LELS (or LEELS): low-energy electron-loss spectroscopy 286
LEPD: low-energy positron diffraction 291
LLR: large-lattice relaxation model 148
LMTO: linear muffin tin orbital
LO: longitudinal optical 694

MBE: molecular beam epitaxy 297, 868
MISFET: metal–insulator–silicon field effect transistor 868
MNDO: modified neglect of differential overlap 596
MOS: metal–oxide semiconductor 261
MOSFET: metal–oxide–silicon field effect transistor 261, 645

NDGS: nondegenerate ground state 589

ODLRO: one-dimensional long-range order 1029
OMCVD: organometallic vapor deposition 380, 868

PA: polyacetylene 587
PPP: poly(paraphenylene) 587, 592
– definition 587
PPV: polyphenylene vinylene 587, 606
– definition 587
PT: polythiopene 587, 606
– definition 587

RBS: Rutherford backscattering spectroscopy 287
RE: rare earth 138
RHEED: reflection high-energy electron diffraction 289
rms: root mean square 957

SB: symmetry breaking 41
SEXAFS: surface extended X-ray absorption fine structure 287

SIMS: secondary ion mass spectroscopy 287
SL: superlattice 820
SNDC: S-shaped negative differential conductivity 430
SPS: surface photovoltage spectroscopy 287
SRS: surface reflectance spectroscopy 287
SRV: surface recombination velocity 368
SSH: Su–Schrieffer–Heeger model 599
STM: scanning tunneling microscopy 290
STS: scanning tunneling microscopy 287
SXPS: soft X-ray photoemission spectroscopy 312

t-PA: *trans*-polyacethylene 587
TBA: tight-binding approximation 144
TCR: truncated cascade recombination 239
TD: thermal donors 250
TEXRD: total external X-ray diffraction 287
TM: transition metal 132
TM: tunneling microscopy 132, 135
2DEG: two-dimensional electron gas 868

UCF: universal conductance fluctuations 939
UHV: ultra-high vacuum 291
UPS: UV photoelectron spectroscopy 286
UV: ultraviolet 286

VEH: valence effective Hamiltonian 596
VRH: variable range hopping 620

WKB: Wentzel–Kramers–Brillouin approximation 631

XPS: X-ray photoelectron spectroscopy 286

Subject Index

Chemical elements and compounds are only indexed when they are particularly relevant.

ab initio pseudopotential method 12, 67
ab initio quantum chemical calculations 314
absorption bleaching experiment 1070
absorption coefficient 172, 257
absorption of light 163
absorption spectroscopies 312
acoustic deformation potential scattering 533, 546
acoustic phonons 547, 550, 573, 608, 609
- energy 572
- interaction 700
- scattering 550, 562, 570
acoustic piezoelectric scattering 533, 546
acousto-electric effects 701
activation barriers 356
activation energy 614, 1006, 1014, 1015, 1020
- of sublimation 307
adatom–adatom bonding 328
adatom–substrate bonding 328
adiabatic collimation 915
adiabatic invariants 1059
adiabatic limit 1106
adiabatic transport 886
adjunct equation 1097
adsorbate 320
adsorbate chemisorption 356
adsorption 306
agglomeration of atoms at surface 322, 336
Aharonov–Bohm effect 1109–1113
Aharonov–Bohm oscillations 904
Airy function 521, 634, 843
AlGaAs(11 Å)–GaAs(39 Å) 841
alignment of TM levels 135
Al(In)As–InP 822
all-electronic potential 64
alloy fluctuations and neutral impurities 713
alloy scattering 709, 728
alloying 332
$Al_xGa_{1-x}As$ 869
$Al_xIn_{1-x}As$(100 Å)–InP(40 Å) 821
amorphous materials 254
amorphous Si 258
Anderson localization 956
Anderson model 940
angle-resolved photoemission spectroscopy (ARPES) 287, 311
angular harmonics 542
anisotropy 563
anisotropy and interband coupling 552
anisotropy parameter K 550, 561, 563
anomalous statistics 1012
antidots 930
antisite As_{Ga} 151, 155
antisite defects 356
antisoliton 588
anyon 1012
$\boldsymbol{A}\cdot\boldsymbol{p}$ interaction 691, 695
Arrhenius plot 247, 251, 269
arsenic antisite As_{Ga} 151–155
As_{Ga}–As_i pair 156
asymmetric dimer 297
atomic force microscopy (AFM) 287
atomic orbitals 49
atomic-scale control 365, 397
atomic-scale tight-binding methods 832
attempt frequency 640, 653
attractor 424, 949
Auger electron spectroscopy (AES) 286, 287
Auger processes 215, 216, 220, 221
autocatalytic process 428
average of relaxation time 545

ballistic electron energy microscopy (BEEM) 287, 363
ballistic transport 875–880, 1051
ballistic transport problem 1092
band bending 312, 341
band edge alignment 822
band-edge crossing 914
band engineering 869
band gap discontinuities 4, 22–33
band gaps, some numerical 21
band mixing 674, 831, 837, 852
"band motion" 613, 614
band offset 56, 136, 376, 383, 822, 824
band structure
- Ge 498
- polymers 596

band tails, microscopic 256
band theory
– chemical models of Chapter 2
– topics in Chapter 1
band transitions 219
band-trap transitions 219
banding of magnetic levels 513
barrier height in CdS (n-type) as a function of illumination 267
barrier heights as a function of illumination 266
barrier heights at the boundary of polysilicon grains 266
barrier in polymers 611, 612, 614, 620
BCS theory 102, 452
beam collimation 878
beam equivalent pressures (BEP) 297
"bends" 897
– bend resistance of a junction 906
biexcitons 456
– different from exciton condensate 468
bifurcation 424
billiard-ball trajectories 910
bipolaron 589, 603–607
Birch equation of state 83
bistabilities 654
bit-number variance 433
Bloch functions 494
Bloch oscillations 928
Bloch representation 508, 522, 637
Bloch state 6
Bloch symmetry 37, 38
Bloch–Grüneisen regime 700
Bohr–Sommerfeld quantization condition 511
Boltzmann equation 531–544, 564
Boltzmann transport equation, moment expansion of 425
bond charges 49
bond hybridization 352
Born approximation, self-consistent, for thermopower 768, 771, 772
Bose condensation 451, 453
Bose–Einstein distribution 202, 1042
bound exciton 178
boundary conditions 668, 671–685
Bravais lattice 292
breather (a lattice excitation) 616
breathing filaments 443
Breit–Wigner relation 652
Brillouin zone 33
– pockets of electrons in 33, 34
broken gap in band line-up 376
broken symmetry 42
Brooks–Herring theory 553, 570
buckling of surface chain 293
buffer layers to bridge lattice constant difference 381
built-in potential 384
bulk dissipative transport 1084
bulk moduli 98
bunching of defects 270
Burgers vector 375
buried III–V in epitaxy 327
Burstein–Moss shift 468
Büttiker's equations 912

camel-back structure 499
canonical partition function 205
capacitance transient spectroscopy 188–191
capacitance–voltage characteristic 341
capture and emission coefficients for r-electron trap 220
capture cross section 151, 239, 242
carrier injection by tunneling 659
carrier–carrier scattering 577, 706–708
cathodoluminescence spectroscopy (CLS) 312
"cavity" related to quantum dot 896
cellular automata 423
center coordinates of the cyclotron motion 982
central and directional forces compared 56
central-cell correction 125, 127, 146
Chambers' solution in transport theory 554
Chambers–Rees method 1094
chaos 437
chaotic attractor 424
chaotic oscillations 432, 436
Chapman–Kolmogoroff equation 423
charge conjugation symmetry 606
charge-density wave 42, 597
charge exchange across contacts 341
charge neutrality level 346
charge of excitations in quantum Hall effect 1009
charge of quasihole 1010
charged-impurity scattering 702, 712
charging phenomena in single-electronics 1115–1118
chemical bond charge transfer 352
chemical contamination of interfaces 349
chemical indices based on pseudopotentials 96
chemical parameters 49
chemical potential 1013
chemical shift 126, 165
chemical theories of semiconductors 49–57
chemical trapping at surface 332, 333
chemisorption 321, 328
classical approximation in electron statistics 203, 531, 532, 545
classical chaos 911
classical equation of motion 495
classical Hall conductivity 987, 988
classical one-component plasma 1008
classical percolation 953
classical plasma 1005
classical transport 525
classical turning point 631
closed orbit 505
cluster deposition 367

cluster formation in interface growth 323, 331, 358
coherence length 590
coherent ballistic transport 1084
cohesive energy 309
collective modes 922, 1021, 1022
collision integrals 426
collision term 532, 535
collisionless Boltzmann equation 1093
combined electric and magnetic fields 524
commensurate structure 324
commutation relations 510
complex dielectric constant 528
computational methods 1108
condensation of paths in limit cycle 424
conductance quantization 1102–1104
conduction and displacement currents 427
conduction bands, ellipsoidal 504
conduction in polymers 618, 620
conductivity 525, 527, 528, 555, 617–620, 623, 986
conductivity anisotropies 591
conductivity in warped hole bands 550–553
conductivity tensor 440, 525, 527, 528, 550, 555, 558, 559
conductivity/thermopower cross correlation function 795
configurational coordinate 243
confinement 871
confinement mass in non-parabolic band 673
confinement of optical phonons 678–690
confinement potential 1087
conjugated-chain model 296
conjugated polymers 593
constant initial (CIS) and final (CFS) state spectroscopy 287, 312
constant-pressure ensemble 205
constrictions
- in parallel 871, 889–892
- in series 871, 888, 889, 897
- nonlinear transport in 899–902
contact potential difference 341
contact rectification 343–350
contamination imperfections 345
contamination of facets 305
continuous symmetries 3, 34, 35, 42
continuous trap distributions 251
continuum model for polarons 599, 603
control parameters 423
Cooper pairs 452
Copenhagen theory 1116
Corbino disk 525
core–valence interactions 79
correlation energy of exciton 455
correlation energy of negative-U center 211
correlation function for conductance fluctuations 792
correlation function for energy 791, 795
correlation magnetic field 791
Coulomb attraction for solitons 618
Coulomb blockade 659, 921, 924, 1115
Coulomb effects in polymers 602
Coulomb enhancement of oscillator strength 837
Coulomb forces, *see* electron–electron interactions
Coulomb gap 810
Coulomb modes 686
coupled hole bands 569
coupled kinetic equations 1107
coupled-mode theory for electron waveguides 1104–1109
coupled valence bands 543
coupling constant integral 32
covalent and ionic structures compared 54
covalent bond 164
creation operators 41
critical disorder 933
critical epitaxical layer thickness 378
critical exponent 955, 994, 995
critical nucleus 444
critical point 431
critical slowing down 431
crossed electric and magnetic fields 507, 524
crystal growth 306–310
- kinetic models of 307
crystal momentum 6, 495
crystal structure of polymers 590
crystalline potential 495
crystallization transition 1006
cubic harmonic 543, 552
cumulant approximation for thermopower 768
current-carrying edge states 997
current density functional theory 14
current filaments 439
current–current correlation function 15
current–current response function 9, 20, 30
current–voltage characteristic 341
curved guides 1105
cusp-type behavior 1001, 1002, 1013, 1025
cycloidal motion in Hall effect 507
cyclotron energy 1001
cyclotron period 505, 982
cyclotron resonance 503–507, 512, 513, 529

dangling bonds 296
de Haas–van Alphen effect 65, 512
Debye temperature 102
decay laws for band–band recombination 223
decay laws for band–trap recombination 225
decay rates of excess carriers 224
decay times for recombination at traps 224
deep donors 130
deep-level transient spectroscopy (DLTS) 190, 191, 264, 385
deep states 130
defect density as a function of doping 275
defect molecule models 119

defect scattering 623
defects, bunching of 270
defects at surface 300–305
definition of recombination coefficients 216
deformability of polymers 586
deformation potential 570–575, 609, 610, 690, 759
deformation-potential optical-phonon scattering 550
degeneracy 208, 517, 518, 1066
– of Landau levels 983, 991
– of quantum orbits 511
degenerate approximation in electron statistics 203
delocalization fixed points 995
delta doping 397
density balance equation 1055
density fluctuation correlation function 27
density functional theory 12, 13
density matrix 1029–1031, 1089
density of states 203, 521, 522, 620, 621, 735, 984
density operator 1023
density–density response function 20
depletion layer 343, 868
desorption 306
detailed balance 216, 429, 532, 535
– relation to measurements 228–231
diagonalization, exact 1002
dielectric constant, real or complex 528
dielectric-continuum (DC) model 680, 685
dielectric model of semiconductor 50, 51
dielectric relaxation time 435
dielectric theory and chemical indices 98
difference equation for transport theory 544
differential conductivity 434
diffraction streaks due to surface disorder 306
diffraction techniques in surface studies 289
diffusion 307, 608, 615
diffusion barrier 323, 333
diffusion coefficient 308, 534, 610, 1056
diffusion magnetothermopower 762
– quantized 767
diffusion of defects 272
diffusion rate in polymers 607, 611
diffusion thermopower 725–742
diffusive forces in Boltzmann equation 1054
dimensionless conductance 943
dimer-adatom-stacking fault 295
dimer model 296, 297, 315
dimerization 318, 594, 598–600
diode conductance 653
direct transitions 385
directional and central forces compared 56
discontinuity in chemical potential 1013
dislocation
– densities 378
– loops 351
– scattering 577
dislocations 296, 310, 351
dislocations and strain 375
dispersion relation 680
dispersive transport 614
displacement and conduction currents 1056
displacement currents 427
dissipative dynamic systems 424
dissociation 306
distribution functions 425, 532
domains at surfaces 300–305
donor at surface 356
donor-doped semiconductors, band structure of 33–43
donor–acceptor pairs 180
doping
– effect on residual defect density 275
– for polymers 590–592, 604–606, 617–621
– heavy 253
– nonuniform 622
double donor 132, 231
double hybridization 685, 687
double layer 341
downward pointing cusps 1002, 1013
drift-diffusion equation 427
drift instability 428
drift mobility 534
drift velocity 537, 555, 556, 987, 1001
Drude model 526
dry etching 309
DX center 144–150
dynamic Hall effect 437
dynamic response 16
dynamic systems, semiconductors as 421–427
dynamical response 20
Dyson equation 9, 15, 24, 31

edge currents 997
edge states 764, 796, 903, 904, 992
effective charge 575, 576
effective confining potential 1085
effective emission rate for TCR 240
effective fields 1107
effective Hamiltonian 36–39, 509–514
effective local external periodic potential 5
effective-mass approximation 120–125, 130, 203, 496–502, 513–521, 549, 558, 559, 636, 669–671
effective mass calculations 145–147
effective-mass Hamiltonian 34–38, 41, 42
Einstein relation 427, 534
EL2 center 150–156
elastic-continuum models 683
elastic scattering 536, 540, 541
– length 790
– time 783
electric current, expression for 564
electric field
– electron motion in 503–507

electric field (*cont'd*)
– in transport 495, 526–564, 680, 689, 843
see also high-field transport
electric scattering 539
electrical conductivity 545
electro-optical properties in superlattices 843–852
electrochemical potential 200, 534, 564
electroluminescence 606
electrolytic etching 309
electromagnetic boundary conditions 680–682, 685
electron affinity rule 384
electron beam lithography 870
electron beam refraction 879
electron lifetime 222
electron localization 993, 994
electron motion in electric and magnetic fields 503–507
electron-nuclear double resonance 183
electron paramagnetic resonance (EPR) 287
electron pockets 34
electron scattering length 286
electron solid 1007
electron spectroscopy techniques 286
electron spin resonance (ESR) 183, 605
electron temperature 425, 1041
electron transmission probabilities 796
electron waveguides, transport in 1099
electron–electron interaction 215, 576, 784, 785, 809, 1000, 1052, 1067
electron–electron scattering, *see* carrier–carrier scattering
electron–hole channel 9
electron–hole correlation function 459
electron–hole drops 451
– condensation of excitons into them 459–471
– electrical conductivity 471–473
electron–hole liquid Chapter 9
– binding energy 457–459
electron–hole scattering, *see* carrier–carrier scattering
electron–hole symmetry 1000, 1002, 1018
electron–phonon interaction 102, 128, 597, 598, 619, 750
electronegativity 347
electronic resonances 697
electronic structure of surfaces 340
elementary excitations 1008, 1012
ellipsoidal energy surfaces 498, 503, 539, 543, 549, 559
ellipsometric measurement 307, 312
emission rate 245–249
emission spectroscopies 312
emission times 190
empirical laws for impurity levels 135–137
empirical pseudopotential method (EPM) 5, 6, 61, 65, 68
empirical tight-binding method 117, 313, 314
energy balance equation 1055
energy band parameters, numerical 499
energy correlation function 795
energy distribution function 311, 1053
energy gap in quantum Hall effect 1013, 1014
energy processing techniques for surfaces 330, 339
energy relaxation 707
energy relaxation time 1046
ensemble Monte Carlo procedure 1041, 1064
entropy 205, 209, 725, 764
envelope function 123, 514, 521, 635, 636, 668, 671, 674, 827
epitaxial interface 636, 645
epitaxy 321–327, 355, 376
equal areas rule 440–443
equations of motion in electric and magnetic fields 503, 504, 526, 531
equilibrium phase diagrams 335
equivalent minima 123, 127
Esaki diode 638, 645, 651
etching, discussion of 309, 1086
Ettingshausen effect 567
Ettingshausen–Nernst effect 567
eutectic temperature 352
evanescent states 631, 884
evaporation effects at surfaces 307
even-denominator filling fractions 1024
– experiments 1026
evidence for existence of spin-unpolarized states 1018
Ewald sphere 293
Ewald summation 82, 1069
excess charge carriers 223
excess current in tunneling 645
exchange and correlation energy 13, 26
exchange energy of exciton 455
exchange interaction 1069
excitation dependence in radiative processes 232
excitation energy 1023
excitation spectrum 164, 1028
excitation spectrum for spin-reversed systems 1022
excite–pump–probe experiment 1070
excited state models 313
excited states 169, 175, 239–243
exciton 10, 229–231, 606
exciton binding energy 454
exciton condensate, different from biexcitons 468
exciton condensation into electron–hole drops 454, 459–471
– drop radius 461–466
– kinetics 460, 461
exciton drop, photon drag 461–466
exciton liquid 451, 452, 456

exciton liquid as different from ordinary liquids 460
exciton number in a stable electron–hole drop 476
exciton solid 451, 455
exciton work function in an electron–hole drop 469, 470
excitons in recombination processes 230
excitons in superlattices 837–839
exponential band tails 253, 254
extended phase-space picture 1098
extended states 941, 984–992
extremum principle 539
extrinsic surface states 314

4f level in a solid 141
facetting, droplet formation 297
Faraday rotation 530
Farey tree 433, 439
Fermi degeneracy 203
Fermi distribution function 201, 531
Fermi energy 565, 620–623
Fermi golden rule 568, 1107
Fermi level 202–209, 343, 531, 534
Fermi liquid 10, 43
Feynman graphs 37
Feynman path integrals 1022, 1124
fibrils in growth of polymers 590, 622
field domains 439
field ion microscopy (FIM) 287
field operators 6, 9
filling factor 983, 1024, 1027
finite-frequency effects 528
finite-size scaling 995
first-order perturbation 28
first return map 436
fixed points 423, 949
fluctuation potential 1087
fluctuations 956
flux quantum 983, 1029
focused ion beams 870
focusing 872
force balance argument 747
force constant models 90
form factor 63
four-band model 643
Fourier transform spectroscopy 164
fractal dimension 424, 943
fractional bond ionicity 54, 55
fractional charge 1009–1013
– experimental observation of 1016
fractional filling factors 1000
fractional quantization 1021
fractional quantum Hall effect 979
fractional statistics 1010, 1011
Franck–Condon shift 245, 613
Franz–Keldysh effect 656, 843
Fredholm integral equation 1060
free mode 917
freezing-in temperature T_f 275
Frenkel defect 268
Fröhlich interaction 690, 693, 697
frozen 4f shell approximation 140
frozen-phonon approach 91, 103

g-tensor 183, 499, 510–518
GaAs 499, 869
GaAs–AlGaAs 835, 838
GaAs–$Al_xGa_{1-x}As$ 822
GaInAs–AlGaInAs 842
galvanomagnetic effects 567
Γ–*X* barrier 674
gap discontinuities 25
gap in excitation spectrum 1014
gate voltage 985
gauge invariance 991, 992, 1094
gauge transformation 1029
Gaussian-like trap distribution 252, 253
Ge 498, 499, 504, 505
generalized eigenvalue problem 78
generalized specific heat 433
generation–recombination
– instability 428
– rate 429, Chapter 6
generic quenching 915
geometric effects at interfaces 355
geometrical resonances 892, 896
ghost states 78
Gibbs free energies 205
Gibbs free energy differences 55
glancing angle X-ray diffraction 291
global adiabatic transport 886
global bifurcations 435
globally stable 424
Goldstone modes 42
graded composition 365
grain boundary recombination theory, assumptions 264, 265
grand canonical ensemble 200
grand canonical partition function 200, 204
Green function 15, 154, 1122
grey tin 502
ground-state energy 1002–1005
ground-state models 313
group-IV semiconductors 87
group velocity 494
groups of carriers 558
Grüneisen parameters 94
Gunn instability 440
GW approximation 17, 21, 116

half-plateaus in transport in constrictions 899
Hall conductance 992
Hall conductivity 979, 988–992, 1021
– plateaus 913, 979, 986
Hall constant 526, 527, 555, 558

Hall current 988, 990
Hall effect 567
- dynamic 437
- quantum Chapter 17
Hall factor 556–564
Hall mobility 556
Hall resistance 906, 985, 987
Hamiltonian equations of motion in ballistic transport 1099
hard-core pseudopotentials 72
hard-mode instability 424
harmonic oscillator, in *w* dimensions 209
harmonic oscillator, recombination model 243
Harrison rule 134–137
Hartree energy 13
Hartree–Fock approximation 17, 25, 80
- to ground state energy of electron–hole liquid 457–459
He beam scattering 287
heat current 533
heat of formation 332
heat of semiconductor compound formation 332
heat of solution ΔH_{sol} 332
heavy- and light-hole bands 569
heavy doping 253
heavy holes 501, 505, 506, 558, 675
Heisenberg uncertainty relations 1119
helicon wave 530
Hellmann–Feynman forces 90
Helmholtz free energy 206
Herring–Vogt transformation 540
heterogeneous growth 331
heterostructure engineering 1086
heterostructure–hot-electron-diode model 432
heterostructures 432, 728, 868
HgTe 502
HgTe–CdTe 825
hierarchies of quasiparticles 1011–1013
high-energy electron diffraction (RHEED) 289
high-field transport 899, 1120
high-injection lifetime 227
high-mobility 2DEG 999
high-pressure structural phases 85
high-resolution transmission electron microscopy (HRTEM) 287
higher-order filling fractions 1011
higher-order response functions 27
Hofstadter butterfly 929
hole lifetime 222
hole–electron pair 605
homoclinic orbit 424
homojunction 398
homopolar gap E_{h} 100
Hopf bifurcation 424
hopping transport 613–621, 934
hot-carrier dynamics 1044
hot-electron thermopower 793, 794, 804
hot phonons 697, 1046
hot polaron 616
hot soliton 616
Hückel model of hydrocarbons 50, 51
hybrid orbital energy 359
hybridization 49, 51
hybridon 695
hydrodynamic balance equations 425
hydrodynamic (HD) model 683–685
hydrogenic states 120
hypernetted chain (HNC) theory 1005
hysteresis of exciton condensation 461

impact ionization 216, 428, 1049
imperfections 314, 377
impurities 309
- experimental Chapter 5
- theory Chapter 4
impurity breakdown 429, 435, 440
impurity gradients 385
impurity levels 357
impurity scattering 547–550
impurity–lattice coupling 187
$In_{0.15}Ga_{0.85}As$–GaAs 838
in-plane dispersion relations 831
in-plane mass 673, 676
InAs–GaSb 822, 836
incommensurate structure 324
incompressibility 1024
incompressible electron system 1001
index of interface behavior 354
indirect transitions 385
inelastic mean-free path 286
inelastic scattering 543
infrared absorption 312
infrared absorption spectroscopy (CLS) 287
infrared active vibrations 605–607, 622
InGaAs-based SLs 838
InGaAs–AlGaInAs 850
injection laser 254
InP–AlInAs 836
InP–AlInAs SLs 852
InSb 499, 502
insulator–metal transition for polymers 590, 622
insulator–semiconductor interfaces 261
integer quantum Hall effect (IQHE) 979
- breakdown 998
integral equation for transport theory 544
integral quantization 986, 988
interband Faraday rotation 512
interband magneto-absorption 520
interband transitions 833
interband tunnel diodes 645
interdiffusion 314, 346
interelectronic collisions, *see* electron–electron interactions
interface dipoles 390
interface heat of reaction ΔH_{R} 332
interface LO modes 693

interface matrix 831
interface modes 684, 695, 697
interface polaritons 685, 688–690, 695
interface-roughness scattering 707, 714
interface states, density of states 262
interfacial stress 351
interference phenomena in quantum transport 1109–1113
interlayer passivation 337
interlayers 368
intermittency 438
intermixing 380
internal photoemission spectroscopy 341, 385
internal potential part of work function 343
internal symmetries 35
interstitials 268, 350
intersubband scattering 698
intervalley coupling 124
intervalley phonon scattering 533, 546, 550, 576
intra-collisional field effect 1123
intracenter optical absorption band 151
intrasubband processes 693, 694
intrasubband scattering 701
intrinsic semiconductor 214, 726
intrinsic surface states 314
inverse Auger process (impact ionization) 1049
inverse participation number 942
inverse photoemission spectroscopy (IPS) 287, 312
inverse Seitz coefficients 560–562
inversion asymmetry splittings 502
inversion electric field 521
inversion layer 868, 980, 981, 985
inversion symmetry 34
inverted splitting 131
$In_xGa_{1-x}As$–GaAs 824
$In_xGa_{1-x}Sb$–GaSb 824
iodine-doped polyacetylene 620
Ioffe–Regel criterion 942
ion-beam damage 870
ion beam techniques 291
ion bombardment etching 309, 330, 339
ion-core pseudopotential 66
ionic and covalent structures compared 54
ionicity spans 353
ionization rate 1050
ionized-impurity scattering 533, 546, 550, 552, 553, 562, 569
irreducible representation of the symmetry group 542, 543
isoelectronic impurities 178, 179
isolated arsenic defect 156
isospin 5
isospin properties 37–43
isotope scattering 713
isotropic scattering 536
iterated maps 423
iterative method for transport theory 544

Jahn–Teller distortion 183–185
Jahn–Teller effect
– dynamic 152
Jastrow-type wave function 1004
jellium 346
Jones zone 50, 51

k-randomizing scattering 536, 539–541
k-space orbit 505
Kadanoff–Baym equation 1122
Kane two-band model 650
Kelvin probe 362
Kelvin relations 564–567
kinetic models of crystal growth 307
kinetic momentum 496
kink site 301
Kirczenow shell model for bound multi-exciton complexes (BMEC) 476–478
Kohn and Sham equations 13
Kohn anomaly 752, 782
$\boldsymbol{k}\cdot\boldsymbol{p}$ perturbation method 497, 514
Kramers degeneracy 516
Kramers theorem 500
Kronig–Penney problem 826
Kubo formula 988, 989, 1048, 1088

Landau gauge 983
Landau-level degeneracy 983, 1002, 1004
Landau-level filling factor 983
Landau levels 517, 841, 982
Landau quantization 852
Landauer formulas 884
Landauer–Büttiker formalism 796, 1089
Langevin equation 1047
– quantum 1124
– retarded 1047
laser annealing 330
laser-excited photoemission spectroscopy (LAPS) 287
"last plateaus" in quantum Hall effect 913
lateral geometry low-dimensional devices 1085
lateral patterning 1086
lattice defects 267–273, 345
– approach to theoretical identification 269
lattice dynamics 685
lattice matching conditions 377
lattice mismatch 374, 377
lattice parameters 327
lattice relaxation or disortion 147
Laughlin wave function 1004–1009, 1023–1029
– significance 1005
– transformed 1029
Laughlin's theory of quantum Hall effect 1003
Lawaetz method for band calculations 552
Lax–Mavroides relaxation time 550
layered electron system 1027
LEED spot intensities 293
Lever rule 332

lifetime-broadening effects 738
lifetime of charge carriers 221, 222, 226–228
lifetime of optical phonon 711
light holes 501, 505, 506, 558, 675
light scattering 385
limit cycles 424
limiting soliton velocity 601, 609
Lindhard dielectric function 39
linear *k* terms 502
linear muffin tin orbital (LMTO) 85
linear terms in effective-mass Hamiltonian 521
linearized Boltzmann equation 534, 535, 542, 545
LO interface modes 695
local density approximation (LDA) 15, 67, 71, 148, 596
local density pseudopotentials 71
local density theory 115
local field effects 18, 20
local impurity scattering 703, 704
local pseudopotential 77
localization 843, 845, 867, 932
localization fixed points 995
localization in a magnetic field 953
localization length 942, 994, 995
localized description 130, 144
localized electron 991
localized states 314, 984, 989, 990, 992
locally asymptotically stable 424
locally stable 423
longitudinal magnetoresistance 526, 527, 557, 561
longitudinal modes 685
longitudinal Nernst effect 567
longitudinal resonances 885
low-dimensional structures 1084
low-energy electron diffraction 289
low-energy electron (LEED) and positron (LEPD) diffraction 287
low-energy electron-loss spectroscopy (LELS) 286, 287, 312
low-energy electron microscopy (LEEM) 287, 300
low-energy positron diffraction (LEPD) 291
low-field Hall effect 560
low-lying excitations 1020
"lucky" electrons 1050
luminescence 177–182, 606
luminescence peak
– shift of 1006
luminescence spectroscopy 312
Luttinger–Kohn representation 514

macrofield versus microfield 573
magnetic backscattering reduction 902
magnetic breakdown 512
magnetic circular dichroism 154
magnetic depopulation 902, 903
magnetic diffraction effect 919
magnetic energy levels near band edges 515
magnetic field
– electron motion in 503–507
– in transport 508–538
– tilted 1018
magnetic focusing 876
magnetic orbits 504
magnetic resonance 149, 607
magnetic susceptibility 602, 604
magnetic translation 508, 513
magneto-exciton 852
magneto-optical effects 512, 520, 1006
magnetoabsorption 512, 520, 525
magnetoplasma effect 529
magnetoresistance 437, 526, 527, 555–567, 937, 947
– of metals 554
magnetothermopower 762–778
– quantized diffusion 767
majority carrier 343
majority carrier inversion 334
many-electron Hamiltonian 1003
matrix element 568, 572
matrix element of the coordinate 495
matrix methods for transport theory 544
maximally crossed diagrams 946
Maxwell–Boltzmann distribution 202, 531, 608
McClure solution in transport theory 554
MCDA spectrum 155
McMillan equation in superconductivity 102
mean free path 1051
mean occupation number for fermions 201
mechanical boundary conditions 683, 685
mechanical modes 686
memory function 1047
mesa etching 872
mesh vectors of overlayers 295
metal-induced gap states 345
metal–insulator transition 809, 933
metal–semiconductor junction 342
metallic limit of localization problem 867
metallic state of polymers 622
metastability of defect states 249
metastable behavior 129, 152
Meyer–Neldel rule 251
microfield versus macrofield 573
microstructures, transport in 796–809
miniband structure 826
minimum metallic conductivity 950
MISFET 868
misfit dislocations 351, 378
misorientation angle 340
missing row model 296
MNDO, a quantum mechanical calculation method 596
mobility 556, 607, 611, 614, 616, 617, 620
mobility edge 932, 941, 985

mobility effective mass 550
mobility for Meyer–Neldel rule 252
mobility gap 985, 991, 992
mode-locking 433
model of a grain boundary 265
modified Bloch representation 509
modulated-beam mass spectrometry 307
molecular-beam epitaxy (MBE) 297, 306, 645, 868
molecular dynamical simulations 105
molecular dynamics 1064
molecular model of transition-metal impurities 133
moment equations in transport 425
momentum balance equation 1055
momentum mass 673
momentum matrix element 494
momentum relaxation time 1046
Monte Carlo simulation 433, 1005
– of Feynman path integrals 1124
MOS structure 262, 263, 644
MOSFET 261, 770–778, 980, 981
Mott criterion 454, 474
Mott formula for thermopower 727, 762, 784, 785, 791
multi-electron trap 205
multi-phonon processes 243–248
multiband scattering 542
multiexciton complexes 454–456, 475–479
multiple channels, transport in 1102
multiple constrictions 888
multiple layer systems in quantum Hall effect 1027, 1028
multiple scattering in crystal growth 307
multivalley semiconductors 3, 42, 43
Murnaghan equation of state 83

nanoelectronics 1084
nanostructures 867–875
narrow-gap semiconductors 502
narrowing band gap 349
native defects 350
NDGS polymer 589, 603–606, 614, 623
negative absorption 232
negative differential conductance 1043
negative differential conductivity (SNDC) 430
negative Hall resistance 910
negative Hall voltage 910
negative quantum Hall plateau 931
negative-*U* center 147, 211, 212, 347
Nernst–Ettingshausen coefficient 762
neutral-impurity scattering 533, 546, 553, 570
non-integrable phase factor 1110
non-linear screening 1123
non-linear transport in constrictions 899–902
non-local behavior 778
non-parabolicity 517, 547, 548, 557, 564, 636, 673
non-polar optical-phonon interaction 699
non-radiative processes 239
nonequilibrium electron correlation function 1122
nonequilibrium phase transition 428
nonequilibrium radiation 234
nonequilibrium transport 425
nonlocal self-energy 9
nonuniform doping 622
nonuniform interacting many Fermion system 12
norm conserving wavefunctions 72
nucleation at surfaces 328, 440

"oblique" radiative transition 849
occupation probabilities of localized states 204–213
occupation probability compared with mean number of particles in that state 218
"octet" compounds 97
off-diagonal long-range order in quantum Hall effect 1029
one-body density matrix 1029
1D electron systems 734
one-electron Hamiltonian 63, 66
one-parameter scaling theory 933, 949
Onsager relations 525, 564
open orbits 505, 507, 525
optical absorption 9, 152
see also radiative transitions
optical absorption by polymers 604–607
– in quantum well structures 385
optical cross sections 137
optical deformation potential scattering 533, 546
optical lithography 870
optical modes 549
optical-phonon energy 619
optical-phonon scattering 575
optical phonons 547, 608, 621
optical polar scattering 533, 546
optical transition, *see* radiative transitions
orbit, open or periodic 525
orbit in real space 505
orbital radii 96
order parameter 390, 599–603
organ pipe resonances 885
organometallic vapor phase deposition (OMCVD) 380, 868
orthogonalized plane wave 64
oscillations in the susceptibility 512
oscillations of magnetoresistance 948
oscillations of surface electric field 482
oscillator, partition function 210
oscillator entropy 210
oscillator strength 1022
oscillatory instability 434
oscillatory magnetoresistance 512
Ostwald ripening 442
overheating instabililty 428
overlap integral 568

overscreening 456
overshoot 349
oxidation state of rare-earth impurities 139–142

p–n junction
– as tunnel diode 645
– gigantic current fluctuations 462
p–n junction techniques for study of impurities 177
pair potentials 14
pair production 1049
paramagnetic state 149
parity 834
partial Gibbs free energies 207
participation number 954
particle kinetic model 526, 537
particle splitting in Monte Carlo simulation 1066
partition function of defect, relation to enthalpies and entropies of defect 274
partition functions Q or Z 207
passivation of surfaces 368
path integral 1062
Pauli principle 66
Pauli susceptibility 621
Peierls gap 597, 598
Peierls phase factor 509
Peltier and Thomson effects 724
Peltier effect 564, 807
Peltier tensor 762
Penn model 50, 51
percolation theory 954
period-doubling 435, 437
periodic orbit 525
periodically modulated 2D conductors 744
persistent photoconductivity 177
phantom point charge 1008
phase breaking 918
phase breaking time 783
phase coherence length 783, 790
phase diagram of electron–hole liquid 470, 473–475
phase portraits for electrons and photons in two-level systems 235
phenyl ring 595
Phillips–Kleinman cancellation theorem 65
phonon-assisted hopping 934
phonon drag 466, 566
phonon-drag magnetothermopower 770
phonon-drag thermopower 727, 742–761
phonon scattering 611–614, 710
phonon self-energies 95
phonon wind 466
phonon wings 179
phonon–phonon processes 710
photoconductivity
– in the study of impurities 173–177
– in the study of polymers 606, 612–617
photoemission of soft X-rays 312
photoionization 170–176
photoluminescence spectroscopy 287
photon and electrons in two-level systems 234
photon-assisted tunneling 524, 525
photon number, non-equilibrium 237
photon recycling 238
photospin resonance 357
photostimulated capacitance 312
photovoltage spectroscopy, *see* surface photovoltage spectroscopy (SPS)
photovoltaic effects 386
physisorption at a surface 328
π and π^* bonds and bands 593–595
π-bonded chain model 90, 297
piezoabsorption 170
piezoelectric fields 377, 378, 573, 759, 825
pilot-field picture 1115
pinned Fermi level 343
pinned mode of soliton 605, 606
pinning frequency 619
planar Hall effect 526
plasma etching 309
plasma frequency 40, 528, 707
plasmon LO coupling 698
plateaus of Hall conductivity 913, 979, 986
– quantized 999
point contacts 869
point defects 305
point-symmetry operations 34
Poisson ratio 375
Poisson's equation 342
polar excitons 606
polar-optical mode scattering 1058
polar-optical-phonon scattering 547, 574
polar runaway 1045
polaron 587, 588, 603–619
polaron velocity
– maximum 612
poly(3-hexylthiophene) 590
polyacetylene (PA) 586–588, 613–615
polymers Chapter 11
polyparaphenylene (PPP) 590, 592–596
poly(phenylene vinylene) (PPV) 591–596, 606, 607, 617
polypyrrole (PPy) 590, 595, 606, 623
polythiopene 593–596, 607, 623
polytype heterostructures 647
Poole–Frenkel effect 242
population inversion 232
– generalized condition for 233
positive-U center 212
potential fluctuations 894, 895
potential screening 347
power-law decay 226, 1030
power-law localization in disordered systems 934
pressure effects 188
probability distributions 200, 1063
projected oscillator strength 1023

projected potential energy 1023
projected static structure factor 1023
projection and super-operator techniques in transport theory 1121
pseudodensity 14
pseudomorphic two-dimensional growth 379
pseudopotential 6, 14, 63, 130, 314
pseudowavefunction 14

quantization condition in quantum Hall effect 986, 992, 1020, 1021
quantization in an electric field 522
quantization of conductance in narrow constrictions 881
quantized conductance 881
quantized diffusion magnetothermopower 765
quantum ballistic transport 1092
quantum boxes, field-induced 852–854
quantum chaos 918
quantum collimation and scrambling 915
quantum dots 919–921
quantum Hall effect 803, Chapter 17
quantum interference 782, 946
quantum kinetic equations 1122
quantum Langevin equation 1124
quantum levels
- bending of 513
- degeneracy of 517
quantum limit in heterojunction 778
quantum Liouville equation 1090
quantum-modulated transistor 897
quantum Monte Carlo method 68, 84
quantum phase space distributions 1091
quantum point contacts 796, 800, 871, 1115
quantum potential 1116
quantum states in the presence of fields 524
quantum wells 385, 397, 652, 826
quantum wire junctions 906
quantum wires 659, 1086
- intersection of narrow 896
quasi-bound states 893, 923
quasi-classical transport 1093
quasi-Fermi level 214, 218
- assumption implied by its use 213
quasi-fields $F_m(z)$ 1107
quasi-level 652
quasi-one-dimensional crystal 926
quasi-periodic route 438
quasi-periodicity 433
quasi-potential fields 1106
quasiballistic regime 892
quasicrystals 98
quasihole size 1009
quasihole wave function 1008, 1009
quasiholes 1008, 1009, 1013
quasiparticle 81, 1008–1013
quasiparticle amplitude 16
quasiparticle statistics 1009–1011
quenching behavior 151
quenching of the Hall resistance 908

radiative transition 231, 833, 1006
Raman scattering 182, 312
Raman scattering spectroscopy 287
random Landau model 955
random matrices 943
random phase approximation (RPA) 17, 35, 459
random potential 256, 989
random telegraph signals 894
rare-earth (RE) impurities 138
ray representation 513, 1021
Rayleigh surface waves 684, 685
reactive ion etching 309
real-space potential 1067
real-space transfer 432
reciprocal lattice vector 38
recombination by exciton processes 229
recombination coefficients 220, 221
recombination in exciton drops 460–464
recombination instability 457
recombination rates Chapter 6
recombination statistics 215–231
reconstruction of surface 291–300
rectification 343–350
rectification fluctuations 791
reflectance 312
relativistic pseudopotentials 79
relaxation of surface 291–300
relaxation semiconductors 432
relaxation time approximation 553
relaxation time tensor 539, 541, 559
relaxation times 536–554
remote impurity scattering 703, 704
renormalization group equations 995
resident lifetime in crystal growth 306
residual defect density 275
residual defect in Si 276
resistivity 525, 527, 565, 986
resonant hydrogenic states 124
resonant space charge 654
resonant states 165–167, 914
resonant tunneling 634, 651–655, 826
retarded Langevin equation 1047
reversed-spin states 1016–1019
RHEED oscillations 307–309
Righi–Leduc effect 568
ring diagrams for RPA 36
rms fluctuations 957
roton minimum 1023–1028
roton-type structures 1024
routes to chaos 435
Rutherford backscattering spectroscopy (RBS) 287, 335

saddle-point exciton 838, 843
saturation solubilities of impurities 273

scalar potential 495, 514, 680, 682, 689
scanning tunneling microscopy (STM) 287, 290
scanning tunneling spectrosocpy (STS) 287
scattering 546, 608–610, 612
scattering integral 1062
scattering mechanisms 568, 728
scattering rate 690
scattering time 610
Schottky barrier formation 327, 341
Schottky barrier technique for study of impurities 177
Schottky defect 268, 269
Schrödinger equation, numerical solution by transfer matrix 833
scrambled classical trajectories 911, 916
screened exchange 18
screened potentials 895
screening 40, 701, 703, 732, 758, 775, 1052
– nonlinear 1123
screening factor 574
second-order nonequilibrium phase transition 429
secondary-ion mass spectroscopy (SIMS) 287, 380
Seebeck effect 723
Seitz coefficients 560
self-averaging 942
self-consistent device potential 1067
self-consistent Green function 16
self-consistent potential 66
self-consistent RPA 16
self-energy 15, 16
self-organized criticality 445
self-scattering 1062
self-similar structure 955
semiclassical quantization 510, 512
semiclassical quantization of magnetic orbits 510
semiconductor electron affinity 343
semiconductor heterostructures 645, 869, 980, 981
semiconductor outdiffusion 332
semiconductor passivation 368
serpentine superlattices 874
shallow acceptors 165
shallow and deep impurities 163
shallow donors 163
shallow donors in quantum wells 168
shallow states 120
shallow–deep instability 125–131
shallow–deep transition 127
Shockley and Read statistics, generalized 222
Shockley tube integral for cyclotron period 505
Shockley–Read–Hall statistics 221
Shubnikov–de Haas effect 512, 700
Si 499, 504, 505
Si MOSFET 729, 980, 985
σ bands and bonds 593
silicon-on-sapphire MOSFETs 760, 783
simple bands, transport in 526–529, 545–549
simulation box in Monte Carlo technique 1067
single donor 131
single-electronics 1115–1118
single-mode approximation 1023
single parameter scaling 949
single-particle Green function 9
single-particle propagators 36
singular gauge 1029
skeleton diagram 33
skeleton expansion 15
SL exciton 850
soft-mode instability 424
soft pseudopotentials 74
soft X-ray photoemission spectroscopy (SXPS) 287
solid-phase regrowth 337, 380
soliton 588–623
soliton band 620
soliton decay 615, 616
soliton defects 602
soliton mass 608–610
soliton velocity 609
– maximum 611
solubility 267
solubility of defects 274
Sommerfeld factor 837
sp^2 orbitals in polymers 593
spectroscopic method 1006
spectrum of traps, effect of, on surface recombination 259
specular reflections 877
spherical bands 555
spin degree of freedom 1017
spin-reversed excitations 1013
spin-reversed ground state 1020
spin-reversed quasiparticles 1013–1020
– experimental evidence 1018
spin splitting 512, 517, 903
spin susceptibility 619, 620
spin-unpolarized states 1017
– evidence for existence 1018
spin waves 42
spin–orbit interaction 500, 502
spin–orbit splitting 79
spinless conductivity in polymers 602, 619, 622
split-gate 871
split-off band 166, 501
spontaneous emission 232
sputtering etching 309
squeezed gate 1108
SSH Hamiltonian 599, 601, 605
stacking fault 296
staggered band line-up 376
Stark ladder 522, 640, 844, 849, 928
states in semiconductors 163
static dielectric function 20
static properties of Si and Ge 84

static structure function 1023
stationary-phase method 641
stationary Wigner functions 1096
statistical density matrix 1090
statistical screening 703, 704
steady state of the electron–photon system 236
steering of electrons 872, 876
steps at surfaces 300–305
sticking probability 242
stimulated emission 232
Stirling's approximation 268
stoichiometry 306
straddling in band line-up 376
strain fields at surface 304
strained-layer superlattices 378, 674, 824, 825
strange attractor 424
stress splitting of energy levels 154
stretching in polymers 591, 615–618
strong localization 808
structure factor 63
subbands 734, 778, 883
substrate atom solubility 331
superconductivity 102
superlattice, types I, II, III 822, 825, 851
superlattice, with electric field 843–848
– serpentine 874
superlattice band structure 826–833
superlattice optical properties 833–837
superlattice structural properties 820
superlattices 378, 647, 743, 826, 1068
– in strong magnetic fields 840–843
– in strong electric fields 843–848
– – electro-optical absorption in 848–852
supersymmetric formulation 867
surface boundaries 264
surface core–hole relaxation 315
surface diffusion 307
surface dipole part of work function 343
surface electron–hole drops 479–483
surface energy minimization 313
surface extended X-ray absorption fine structure (SEXAFS) 287
surface mobility 328
surface photoconductivity 312
surface photoconductivity spectroscopy 287
surface photovoltage effects 372
surface photovoltage spectroscopy (SPS) 287, 312, 362
surface recombination 258, 259, 264
surface recombination velocity (SRV) 368
surface reflectance spectroscopy (SRS) 287
surface roughness scattering 729
surface segregation 331
surface space charge region 341
surface state dispersion 316
surface structures 89
surface topology 296
symmetric gauge vector potential 1003
symmetric superlattices 834
symmetry-breaking (SB) corrections 41
symmetry properties 834
synchrotron radiation 311
Szmulowisz calculation for warped bands 552

tapered waveguide 1105
temperature dependence of vacancy distribution 271
temperature gradients in Boltzmann equation 531
template structures 285, 381
tensor relaxation time 549
ternary and quaternary alloy systems 377
terraces 304
tetrahedrally coordinated semiconductors 50
– numerical parameters 53
theory of pairing of defects 271
thermal annealing 330
thermal capture and emission for multiphonon process 245
thermal conductivity 564–568, 802, 806
thermal current, expression for 564
thermal donors (TD) 250
thermal etching 309
thermal length 791
thermionic work function 342
thermodynamic compensation law 252
thermodynamic Green function techniques 1121
thermoelectric effects 564–567, 619
thermomagnetic effects 567
thermopower 723
– sign of 727
Thomas–Fermi approximation 1114
Thomas–Fermi screening 68, 774
Thouless number 943, 994
three-band model 502, 517
III–V compounds 502
III–V zincblendes 87
tight-binding approximation 117
tight-binding Hamiltonian 944
tight-binding model for polymers 594
tight binding of the envelope functions 832
tight-binding techniques 90
tilted-field experiment 1019
tilted-field measurement on the $\frac{5}{2}$ state 1027
tilted superlattices 382
time-reversal symmetry 34
time-reversibility 1095
topological invariance 992, 993, 1021
total energy calculations 81–83
total external X-ray diffraction (TEXRD) 287
trajectory representations 1115
trans-polyacetylene (*t*-PA) 587–622
transfer Hamiltonian 643, 646
transfer integral 594, 832
transfer matrix 831–833
transferability in pseudopotential calculations 74

transient behavior of centers with one excited state 239
transit time 1044
transition metal (TM) impurities 132, 169
transition probability 216, 567–572
translation group 513, 1021
transmission coefficient 802
transmission resonance 892
transport in electric and magnetic fields 553–563
transverse electronic bandwidth 613
transverse magnetoresistance 526, 527, 557, 561
transverse modes 685, 883
transverse Nernst effect 567
transverse polaron bandwidth 613
trapping of electron in a junction 917
trapping of solitons 607–616
trimers 315
triple hybrid 686, 689
triple hybridization 685
triplet, quadruplet, etc. correlation functions 27
truncated cascade recombination 239
tunnel diode 638
tunneling 523, 524, 843, Chapter 12
– amorphous material 644
– Anderson diode 645
– diode current 648, 654
– double-barrier structure 651, 652
– dwell time 654
– Esaki diode 650
– evanescent wave 631
– excess current 645, 651, 658
– four-band model 651
– Fowler–Nordheim 633, 646
– image effects 658
– indirect tunneling 644, 657
– interband 637
– interband tunneling 638, 640
– localized state 658
– magnetic field effects 655
– momentum conservation 644
– phonon-assisted tunneling 644
– phonon-induced tunneling 656
– phonon replica 656
– prefactor 631, 648
– resonant tunneling 634, 651
– reverse bias 651
– Schottky diode 644
– sequential tunneling 655
– simple tunneling barrier 646
– transit time 634, 653
– tunneling time 659
– zero-bias anomaly 656
tunneling probability 629–633
turning point 639
two-band model 517, 639, 643
two-body density matrix 1029
two-dimensional arrays 928
two-dimensional classical plasma 1004
two-dimensional electron systems 979, 991
two-dimensional interacting electrons 1001
two-layer system 1027, 1028
– experiments 1028
two-parameter scaling theory 995, 996
two-particle Green function 9

U-processes 744
UHV high-resolution transmission electron diffraction (HRTEM) 291
ultrahigh-vacuum technology 286
ultralow compressibilities 102
unit cell of the semiconductor surface 293
unitary matrix 35
universal conductance fluctuations (UCF) 782, 789, 790, 938
universal thermopower fluctuations 791
universality classes 952
unscreened interaction 1057
Urbach tails 257
UV irradiation as surface treatment 330
UV photoelectron spectroscopy 286
UV photoemission spectrosocpy 287

vacancies 271, 334, 350, 356
vacancies, various types 184
valence band degeneracy 643
valence band Landau levels 842
valence effective Hamiltonian (VEH) 596
valley degeneracy 754
valley electrons 5, 36–41
– wavevector of 39
valley number as quantum number 451, 455
"valley waves" in band theory 42
valleys 35, 37, 39, 42
van Hove singularities 940
variable range hopping 620–622, 812, 988
variational principle 538
vector potential 495, 514, 689, 983
velocity autocorrelation function 1048, 1088
velocity saturation 1043
Verdet coefficient 530
vertex corrections 17
vertical quantum dots 922
vertical radiative transition 849
vertical transport in superlattices 854, 855
vibrational properties 90
vibronic entropy 208
vibronic states 209, 244, 246
Visscher–Falicov model 1027
von Klitzing constant 986
"vortex" of wave function 1005

Wannier functions 508
Wannier levels 522, 523
Wannier quantization 843, 852
Ward–Pitayevski identity 26
warped energy surfaces 541

warped hole bands 551–553
warping of bands 501
wave function 516, 517, 526
  see also Laughlin wave function
wave-function tailing 314
wave packet 494, 634, 653
wavevector conservation 751
weak localization 782, 787, 867, 934, 937
well capture 697
well-width resonance 697
wet etching 309
Wigner crystal energy 1025
Wigner crystallization 1007
Wigner distribution function 653, 1091–1098, 1118–1120
Wigner–Weyl transformation 1091
WKB approximation 631–634
work function 341

X-ray diffraction 287
X-ray double diffraction 820
X-ray photoelectron spectroscopy 286
X-ray photoemission spectrosocpy (XPS) 287

Zeeman energy 1016
Zeeman interaction 510, 511
Zeeman splitting 174, 186
Zener tunneling 524
zero-gap semiconductors 167
zero of wave function 1018
zero-phonon line 152, 170, 179
zero-point oscillation 452
zig-zag chains at surfaces 322
zone folding 678